DICTIONARY OF CHEMISTRY AND CHEMICAL TECHNOLOGY

English-German

DICTIONARY OF CHEMISTRY AND CHEMICAL TECHNOLOGY

English
German

Editor in Chief
HELMUT GROSS
Department of Applied Linguistics, Technische Universität Dresden,
German Democratic Republic

Compiled by
Wolfgang Borsdorf
Rolf Cramer
Helmut Gross
Helmut Hildebrand
Gunter Neubert
Fridrun Pfeifer

ELSEVIER
Amsterdam–Oxford–New York–Tokyo
1984

PUBLISHED IN COEDITION WITH VEB VERLAG TECHNIK, BERLIN 1984

This book is exclusively distributed in all non-socialist countries with the exception of the Federal Republic of Germany, West-Berlin, Austria and Switzerland by Elsevier Science Publishers B.V.
Molenwerf 1
P.O. Box 211, 1000 AE Amsterdam, The Netherlands

Distributors for the United States and Canada
Elsevier Science Publishing Company, Inc.
52 Vanderbilt Avenue
New York, NY 10017

Library of Congress Cataloging in Publication Data
Main entry under title:

Dictionary of chemistry and chemical technology, English-
    German.

    Includes index.
    1. Chemistry — Dictionaries. 2. Chemistry, Technical —
Dictionaries. 3. English language — Dictionaries — German.
I. Gross, Helmut, Dipl.-Sprachlehrer. II. Borsdorf,
Wolfgang.
QD5.D53   1984   540'.3'21   84-1587
ISBN 0-444-99618-4 (U.S.)

5-6-85

PREFACE

This dictionary has been compiled by a team of scientific co-workers of the Department of Applied Linguistics at the Technical University of Dresden. It comprises the special terms of an important branch of the natural sciences. It is intended for all those working in the fields of chemistry and chemical technology as an aid in their study of the literature pertinent to these disciplines.

About 55,000 terms have been gathered from all branches of chemistry and chemical technology as well as from related fields of science.

The authors have provided additional information for many of the entries by indicating the scientific branch concerned and adding short definitions or examples. This information is intended to help the user select the most appropriate German equivalent.

Numerous scientists of our University have assisted us in defining difficult terms. Their help is gratefully acknowledged.

All suggestions for improving this dictionary are welcome, and these should be directed to the publisher.

Helmut Gross

DIRECTIONS FOR USE · HINWEISE FÜR DIE BENUTZUNG

1. Examples of the alphabetical arrangement
Beispiele für die alphabetische Ordnung

cesium	fire / to
cetane	f. to a white colour
c. index	fire assay
cetyl alcohol	f.-dried
CFR	f.-finish / to
CFR engine	f. point
C/H ratio	f.-polish / to
chabasite	f. polish
china	f. polishing
c. clay	f.-proof
c. stone	f.-tube boiler
China bark	firebox
C. oil	fireclay
C. silver	fired body
C. wood oil	f. glaze
chinaware	f. property
Chinese anise	firedamp

Italicized symbols denoting position or configuration like *asym-*, *sym-*, *prim-*, *sec-*, *tert-*, *cis-*, *trans-*, *m-*, *o-*, *p-*, *N-*, *d-*, *l-*, α-, β-, γ- are disregarded in alphabetization.

Kursiv gesetzte Stellungs- und Konfigurationssymbole wie *asym-*, *sym-*, *prim-*, *sec-*, *tert-*, *cis-*, *trans-*, *m-*, *o-*, *p-*, *N-*, *d-*, *l-*, α-, β-, γ- bleiben bei der alphabetischen Einordnung unberücksichtigt.

2. Signs used in the dictionary
Bedeutung der Zeichen

/   chlorinate / to = to chlorinate

( )   chromium tannage (tanning) = chromium tannage *or* chromium tanning
Zerreiblichkeit des Kokses (der Kokssubstanz) = Zerreiblichkeit des Kokses *oder* Zerreiblichkeit der Kokssubstanz
coordinate covalence (covalency, link) = coordinate covalence *or* coordinate covalency *or* coordinate link
kompressive (erzwungene) Schrumpfung (Krumpfung) = kompressive Schrumpfung *oder* erzwungene Schrumpfung *oder* kompressive Krumpfung *oder* erzwungene Krumpfung

[ ]   contributing [mesomeric] form = contributing mesomeric form *or* contributing form
[oberflächlich] zerstören = oberflächlich zerstören *oder* zerstören
cotton dry[er] felt = cotton dryer felt *or* cotton dry felt
Wellrohr[dehnungs]ausgleicher = Wellrohrdehnungsausgleicher *oder* Wellrohrausgleicher

( )   These brackets contain explanations
Diese Klammern enthalten Erklärungen

ABBREVIATIONS · ABKÜRZUNGEN

Abk.	Abkürzung / abbreviation
Am	amerikanisches Englisch / American English
Bakt	Bakteriologie / bacteriology
Bau	Baustoffchemie / chemistry of building materials
Bgb	Bergbau / mining
Bio	Biologie / biology
Bioch	Biochemie / biochemistry
Bot	Botanik / botany
bzw.	beziehungsweise / respectively
chem	chemisch / chemical
Chem	Chemie / chemistry
Ech	Elektrochemie / electrochemistry
Erdöl	Erdölchemie und -technologie / petroleum chemistry and technology
ET	Elektrotechnik / electrical engineering
f	Femininum / feminine noun
Farb	Farbenchemie / dye chemistry
Flot	Flotation / floatation
Forst	Forstchemie / forest chemistry
Foto	fotografische Chemie / photographic chemistry
fpl	Femininum Pluralis / feminine plural
Gär	Gärungschemie und -gewerbe / fermentation chemistry and industry
Geoch	Geochemie / geochemistry
Geol	Geologie / geology
Gerb	Gerbereichemie und -technologie / chemistry and technology of leather manufacture
Glas	Glasindustrie / glass industry
Gum	Gummi- und Kautschukindustrie / rubber industry
Holz	Holzchemie / wood chemistry
i.e.S.	im engeren Sinne / in a narrower sense, specifically
i.w.S.	im weiteren Sinne / in a broader sense
Kch	Kernchemie / nuclear chemistry
Ker	Keramik / ceramics
Koll	Kolloidchemie / colloid chemistry
Kosmet	Kosmetik und Parfümerie / cosmetics and perfumery
Kph	Kernphysik / nuclear physics
Krist	Kristallografie / crystallography
Kt	Kerntechnik / nuclear engineering
Landw	Landwirtschaftschemie / agricultural chemistry
Lebm	Lebensmittelchemie und -technologie / food chemistry and technology
m	Maskulinum / masculine noun
Math	Mathematik / mathematics
Med	Medizin / medicine

Met	Metallurgie und Metallkunde / metallurgy and science of metals
Min	Mineralogie / mineralogy
mpl	Maskulinum Pluralis / masculine plural
n	Neutrum / neuter noun
npl	Neutrum Pluralis / neuter plural
org Ch	organische Chemie / organic chemistry
Pap	Zellstoff- und Papierindustrie / pulp and paper industry
Pharm	pharmazeutische Chemie / pharmaceutical chemistry
Phys	Physik / physics
phys Ch	physikalische Chemie / physical chemistry
pl	Plural / plural
Plast	Kunststoffchemie / plastics chemistry
s.	siehe / see
s. a.	siehe auch / see also
Tech	Technik / technology
Text	Textilchemie / textile chemistry
z. B.	zum Beispiel / for example
Zucker	Zuckerindustrie / sugar industry

A

a s. 1. activity 2. ; 2. ampere
a. s. acid
A s. 1. absolute temperature; 2. ampere; 3. angstrom [unit]; 4. atomic weight
Å s. angstrom [unit]
A acid A-Säure f, 6-Amino-1-naphthol-5-sulfonsäure f
a-bond a-Bindung f, axiale Bindung f
A-horizon (Bodenkunde) A-Horizont m, Eluvialhorizont m, Auswaschungshorizont m, Oberboden m
A stage (Plast) A-Zustand m
A-stage resin Harz n im A-Zustand, A-Harz n, (bei Phenolharzen auch) Resol n
AA s. 1. adenylic acid; 2. allyl alcohol
aanerodite, aan[n]erödite s. ånnerödite
AATCC (Abk. für) American Association of Textile Chemists and Colorists
abaca 1. Abaka m, Faserbanane f, Musa textilis Née; 2. Abaka m, Manilahanf m (Blattfasern von 1.)
Abbe number Abbesche Zahl f, ν-Wert m, reziproke relative Dispersion f
Abbe refractometer Abbe-Refraktometer n, Zeiß-Abbe-Refraktometer n
Abbe spectrometer Abbe-Spektrometer n
Abbe theory Abbesche Theorie f
Abbe value s. Abbe number
Abbot-Cox process Abbot-Cox-Verfahren n, Abbot-Cox-Färbeverfahren n
abbreviated chemical name chemischer Kurzname m, chemische Kurzbezeichnung f
Abegg rule Abeggscher Satz m
Abel apparatus, Abel closed apparatus (tester), Abel flash-point apparatus s. Abel tester
Abel-Pensky apparatus (tester) Abel-Pensky-Flammpunkt[s]prüfer m, Abel-Penskyscher Flammpunkt[s]prüfer m, Abel-Pensky-Apparat m
Abel test Abel-Test m (1. zur Flammpunktbestimmung; 2. zur Bestimmung der Stabilität eines Explosivstoffes)
Abel tester Abel-Flammpunkt[s]prüfer m, Abel-Gerät n, Abel-Apparat m
Abel's reagent Abels (Abelsches) Reagens n
abernathyite (Min) Abernathyit m
aberration Aberration f, Abweichung f, Abbildungsfehler m
abhesion Abhäsion f
abhesive abhäsiv, adhäsionsfeindlich
abhesiveness abhäsive Eigenschaft f, Trennvermögen n
abichite (Min) Abichit m, (veraltet für) Klinoklas m (Kupfer(II)-trihydroxidorthoarsenat)
abietate Abietat n
abietene Abieten n
abietic acid Abietinsäure f (Harzsäure, Diterpenderivat)

abietin Koniferin n, Abietin n, Larizin n, Koniferosid n (β-Glukosid des Koniferylalkohols)
abietinic acid s. abietic acid
ability Fähigkeit f, Vermögen n; Leistungsfähigkeit f
a. of foam formation Schaum[bildungs]fähigkeit f, Schaumvermögen n, Schaumkraft f, Schaumigkeit f, Schäumigkeit f
a. of wetting Benetzungsfähigkeit f, Netzfähigkeit f, Benetzungsvermögen n, Netzvermögen n, Netzkraft f; Benetzbarkeit f
a. to flow Fließfähigkeit f, Fließvermögen n
a. to storage Lagerfähigkeit f
a. to swell Quellfähigkeit f, Quellvermögen n, Quellbarkeit f
abnormal abnorm, anomal, regelwidrig
a. transference number (phys Ch) anomale Überführungszahl f
abomasum, abomasus Labmagen m
ABR s. acrylate-butadiene rubber
abradability Zerreiblichkeit f (des Kokses, der Kohle)
abradant s. abrasive
abrade / to abreiben; abschleifen; verschleißen
abrader s. abrasion-testing machine
abrasion Abreiben n, (Text meist) Scheuern n; Abschleifen n; Abrieb m, Abnutzung f [durch Abrieb], Abriebsabnutzung f, [reibender] Verschleiß m, Reibungsverschleiß m, (Text meist) Scheuerverschleiß m; (Geol) Abnutzung f, Abschleifung f, Abrasion f; (Geol) Korrasion f
a. index Abriebzahl f, Abrasion-Index m
a. loss Abriebverlust m
a. machine s. a.-testing machine
a. resistance Abriebwiderstand m, Abriebbeständigkeit f, Abriebfestigkeit f, Widerstandsfähigkeit f gegen Abrieb, Abnutzungswiderstand m, Abnutzungsbeständigkeit f, Verschleißwiderstand m, Verschleißfestigkeit f, (Text meist) Scheuerbeständigkeit f, Scheuerfestigkeit f
a.-resistant abriebbeständig, abriebfest, abnutzungsbeständig, verschleißfest, (Text meist) scheuerbeständig, scheuerfest
a. test Abriebprüfung f, Abriebversuch m, Abnutzungsprüfung f, (Text meist) Scheuerprüfung f
a. tester s. a.-testing machine
a.-testing machine Abrieb[prüf]maschine f, Abrieb[prüf]gerät n, Abnutzungs[prüf]maschine f, Abnutzungs[prüf]apparat m, Abnutzungsprüfer m
abrasive [ab]reibend, schleißend; abschleifend, Schleif...
abrasive Schleifmittel n, (Text meist) Scheuermittel n
a. body paper Schleifrohpapier n
a. paper Schleifpapier n
a. powder Schleifpulver n
a. stone Schleif[er]stein m
a. wear Abnutzung f [durch Abrieb], Ab-

riebsabnutzung f, [reibender] Verschleiß m, Reibungsverschleiß m, Abrieb m, (Text meist) Scheuerverschleiß m

abraum salt Abraumsalz n, (veraltet für) Kalisalz n

abrine Abrin n, N-Methyltryptophan n

ABS s. acrylonitrile butadiene styrene

abscisin Abszisin n

absinth Wermut m, Absinth m, Artemisia absinthium L.; Absinth[branntwein] m, Absinthlikör m

a. **oil** Wermutöl n

absinthe s. absinth

absinthiin Absinthiin n (ein Bitterstoffglykosid von Artemisia absinthium L.)

absinthin Absinthin n (ein Bitterstoff von Artemisia absinthium L.)

absinthine absinthartig

absolute absolut, Absolut...; absolut, rein, unvermischt, (oft auch) wasserfrei, abs.

absolute (Kosmet) absolutes Öl n, Absolu [de concret] n, Essence absolue [de concrète] f

a. **alcohol** absoluter (wasserfreier) Alkohol m, wasserfreier Äthylalkohol m, wasserfreies Äthanol n

a. **atomic weight** absolute Atommasse f, (früher) absolutes Atomgewicht n

a. **chassis** (Kosmet) absolutes Châssis (Blütenöl „Châssis") n, Essence absolue châssis f

a. **configuration** absolute Konfiguration f

a. **essence** (Kosmet) absolutes Öl n, Essence absolue [de concrète] f, Absolu [de concret] n

a. **ether** absoluter (wasserfreier) Äther m

a. **flower oil** (Kosmet) absolutes Blütenöl n

a. **from concrete** s. a. essence

a. **humidity** absolute Feuchte (Feuchtigkeit) f

a. **method** Absolutmethode f

a. **of enfleurage** (Kosmet) Enfleurageöl n, Absolu d'enfleurage n, Essence absolue d'enfleurage f

a. **temperature** absolute Temperatur f, Kelvin-Temperatur f

a. **temperature scale** absolute (thermodynamische) Temperaturskale f

a. **value** absoluter Wert (Betrag) m, Absolutwert m, Absolutbetrag m, Betrag m

a. **viscosity** absolute Viskosität f

a. **zero** absoluter Nullpunkt m

absorb / **to** (Gase) absorbieren, aufsaugen, einsaugen, aufnehmen, in sich einziehen; (Strahlen) absorbieren, [ver]schlucken; (Farbstoffe) aufziehen

absorbability Absorbierbarkeit f, Aufsaugbarkeit f

absorbable absorbierbar, aufsaugbar, aufnehmbar

absorbance (Kolorimetrie) Extinktion f

absorbancy s. absorbency

absorbate Absorptiv n, Absorbat n, absorbierter (aufgenommener) Stoff m

absorbed material (substance) s. absorbate

absorbefacient absorbierend, aufsaugend, einsaugend, aufnehmend

absorbency 1. Absorptionsvermögen n, Absorptionsfähigkeit f, Saugvermögen n, Aufsaugfähigkeit f, Saugfähigkeit f, Aufnahmevermögen n, Aufnahmefähigkeit f; 2. (Kolorimetrie) Extinktion f

a. **index** (Kolorimetrie) [spezieller] Extinktionskoeffizient m

absorbent absorbierend, aufsaugend, einsaugend, aufnehmend, absorptionsfähig, absorptiv, [auf]saugfähig, aufnahmefähig

absorbent Absorbens n, Absorptionsmittel n, absorbierender (aufnehmender, aufsaugender) Stoff m, Aufnehmer m

a. **board** Saugpappe f; Löschkarton m

a. **oil** Absorptionsöl n, Waschöl n

a. **paper** Saugpapier n, Fließpapier n

absorber Absorber m, Absorptionsapparat m, Aufsauger m; Absorbens n, Absorptionsmittel n, absorbierender (aufnehmender, aufsaugender) Stoff m

absorbing absorbierend, aufsaugend, einsaugend, aufnehmend, Absorptions..., Aufnahme...

a. **agent** s. absorbent

a. **apparatus** Absorptionsapparat m, Absorber m, Aufsauger m

a. **capacity** 1. s. absorbency; 2. (Farb) Aufziehvermögen n, Ziehvermögen n

a. **column** Absorptionskolonne f, Absorbierkolonne f, Absorptionssäule f

a. **liquid** Absorptionsflüssigkeit f

a. **material (substance)** s. absorbent

a. **tower** Absorptionsturm m

absorptiometer Absorptiometer n

absorptiometric absorptiometrisch

absorptiometry Absorptiometrie f, Absorptionsanalyse f

absorption Absorption f, Absorbieren n, Absorbierung f, Aufsaugen n, Einsaugen n, Aufnehmen n, Aufnahme f (von Gasen); Absorption f, Schluckung f, Verschluckung f (von Strahlen); Aufziehen n (von Farbstoffen)

a. **apparatus** Absorptionsapparat m, Absorber m, Aufsauger m

a. **band** Absorptionsbande f

a. **base** (Pharm) Absorptionsgrundlage f, Absorptionsbasis f

a. **bottle** Waschflasche f, Absorptionsflasche f

a. **capacity** 1. s. absorbency; 2. (Farb) Aufziehvermögen n, Ziehvermögen n

a. **cell** Absorptionsküvette f

a. **coefficient** Absorptionskoeffizient m

a. **column** Absorptionskolonne f, Absorbierkolonne f, Absorptionssäule f

a. **curve** Absorptionskurve f

a. **flask** Absorptionskolben m

a. **gas pipet** Absorptionspipette f

a. **gasoline** durch Absorption gewonnenes Naturbenzin n

a. **line** Absorptionslinie f

a. **liquid** Absorptionsflüssigkeit f

a. machine Absorptionsmaschine *f*, Rotationsabsorber *m*
a. maximum Absorptionsmaximum *n*
a. of light Lichtabsorption *f*
a. oil Absorptionsöl *n*, Waschöl *n*
a. pipet Absorptionspipette *f*
a. plant Absorptionsanlage *f*
a. process Absorptionsverfahren *n*, Absorptionsprozeß *m*; Absorptionsvorgang *m*
a. rate Absorptionsgeschwindigkeit *f*
a. refrigerating (refrigeration) machine Absorptionskältemaschine *f*
a. refrigeration system Absorptionskälteanlage *f*
a. region Absorptionsbereich *m*
a. spectrometer Absorptionsspektrometer *n*
a. spectroscopy Absorptionsspektroskopie *f*
a. spectrum Absorptionsspektrum *n*
a. stack Absorptionssäule *f*, Absorptionskolonne *f*, Absorbierkolonne *f*
a. system Absorptionssystem *n*, Absorbiersystem *n*; Absorptionsanlage *f*
a. tower Absorptionsturm *m*; *(Pap)* Säureturm *m* *(beim Sulfitverfahren)*; *s. a.* column
a. train Absorptionsbatterie *f* *(für Gase)*
a. tube Absorptionsrohr *n*, Absorptionsröhre *f*, Absorptionsröhrchen *n*
a. unit Absorptionsanlage *f*; Absorptionsapparat *m*
a. velocity Absorptionsgeschwindigkeit *f*
a. vessel Absorptionsgefäß *f*
absorptive absorptiv, absorptionsfähig, [auf]saugfähig, aufnahmefähig, absorbierend, aufsaugend, einsaugend, aufnehmend, Absorptions..., Aufsaug..., Saug..., Aufnahme...
a. capacity 1. *s.* absorbency; 2. *(Farb)* Aufziehvermögen *n*, Ziehvermögen *n*
a. power 1. *s.* absorbency; 2. *(Farb)* Aufziehvermögen *n*, Ziehvermögen *n*; 3. Absorptionsvermögen *n*, Absorptionszahl *f*, Absorptionsgrad *n*
absorptiveness *s.* absorptivity
absorptivity 1. *s.* absorbency; 2. Absorptionsvermögen *n*, Absorptionszahl *f*, Absorptionsgrad *m*; 3. *(Kolorimetrie)* [spezieller] Extinktionskoeffizient *m*
abstract / to 1. absondern, aussondern, [ab]trennen, [ab]scheiden, ausscheiden, separieren, loslösen; extrahieren, [her]ausziehen, entziehen, entfernen; destillieren; 2. *(die wesentlichsten Ergebnisse aus einer Veröffentlichung)* [her]ausziehen, *(referatähnlich)* zusammenfassen, referieren
abstract Auszug *m* *(aus einer Veröffentlichung)*, *(referatähnliche)* Zusammenfassung *f*, Referat *n*
a. journal Referatezeitschrift *f*, referierende Zeitschrift *f*, Referatorgan *n*, Referateblatt *n*
abstracting Herausziehen *n* *(der wesentlichsten Ergebnisse aus einer Veröffentlichung)*, Zusammenfassen *n*, Referieren *n*

a. journal *s.* abstract journal
abstraction Absonderung *f*; Entziehung *f*, Entzug *m*
a. of water Wasserentziehung *f*, Wasserentzug *m*
abukumalite *(Min)* Abukumalit *m*
abyssal rock abyssisches Gestein *n*
Abyssinian myrrh *(Pharm)* Commiphora abyssinica (Berg.)Engl.
ac. s. 1. acid; 2. acetic acid
a.c., AC *s.* alternating current
a.c. arc *s.* alternating-current arc
acacetin Akazetin *n*, 5,7-Dihydroxy-4'-methoxyflavon *n*
acacia gum Akaziengummi *n*
acanthite *(Min)* Akanthit *m* *(Silbersulfid)*
acaricide Akarizid *n*, Milbengift *n*
acaroid gum Grasbaumharz *n*, Akaroidharz *n* *(von Xanthorrhoea-Arten)*
accelerate / to beschleunigen
accelerated *(Gum)* beschleunigt, beschleunigerhaltig
a. ageing beschleunigte (künstliche) Alterung *f*, Kurzalterung *f*
a. ageing test beschleunigte (künstliche) Alterungsprüfung *f*
a. tannage beschleunigte Gerbung *f*, Schnellgerbung *f*
a. test Kurzversuch *m*, Kurzprüfung *f*
accelerating ability Beschleunigungsvermögen *n*
a. activity *(Gum)* Beschleunigerwirkung *f*
a. chamber Beschleunigungskammer *f*
a. effect Beschleunigungseffekt *m*
a. potential Beschleunigungsspannung *f.*
a. system *(Gum)* Beschleunigersystem *n*
acceleration Beschleunigung *f*
a. due to gravity Gravitationsbeschleunigung *f*, Schwerebeschleunigung *f*, Fallbeschleunigung *f*, Erdbeschleunigung *f*
a. head Beschleunigungsdruckhöhe *f*, Beschleunigungshöhe *f*
a. of gravity *s. a.* due to gravity
a. zone Beschleunigungszone *f*
accelerator Beschleuniger *m*; *(Ker)* Mineralisator *m*, Kristallisator *m*
a. activator *(Gum)* Beschleunigeraktivator *m*, Aktivator *m* für den Beschleuniger
a. level *(Gum)* Beschleunigerdosierung *f*
a. masterbatch *(Gum)* Beschleunigerbatch *m*, Beschleunigervormischung *f*
a. of vulcanization Vulkanisationsbeschleuniger *m*
accept / to *(z.B. Elektronen)* aufnehmen
acceptable daily intake *(Toxikologie)* duldbare (zulässige) Tagesdosis *f*
acceptance Aufnahme *f*, Annahme *f*
a. of electrons Elektronenaufnahme *f*
accepted stock *(Pap)* Feinstoff *m*, Gutstoff *m*, büttenfertiger Stoff *m*
acceptor Akzeptor *m*
a. ability Akzeptoreigenschaften *fpl*

a. behaviour Akzeptorverhalten n
a. impurity Akzeptorstörstelle f, Akzeptorverunreinigung f
a. level Akzeptorniveau n, Akzeptorterm m
a. power s. a. ability
accepts s. accepted stock
access of air Luftzutritt m
accessibility Akzessibilität f, Zugänglichkeit f
accessory components (constituents) s. a. minerals
a. minerals akzessorische (beigemengte) Minerale npl, akzessorische Gemengteile mpl, Akzessorien npl, Nebengemengteile mpl, Übergemengteile mpl
a. tube Zuleitungsrohr n
accidental degeneracy (phys Ch) zufällige Entartung f
a. error zufälliger Fehler m, Zufallsfehler m
a. moisture freies Wasser n
accidentally degenerate (phys Ch) zufällig entartet
accommodate / to akkommodieren, angleichen, anpassen
accommodation Akkommodation f, Akkommodierung f, Angleichung f, Anpassung f
a. coefficient Akkommodationskoeffizient m
account book paper Bücher[schreib]papier n, Geschäftsbücherpapier n
Accra copal (Farb) Ogia-Kopal m (von Daniella ogea Rolfe)
accroides gum Grasbaumharz n, Akaroidharz n (von Xanthorrhoea-Arten)
accumulate / to [an]sammeln, anhäufen, aufhäufen, [auf]speichern, anreichern; sich ansammeln (sammeln, anhäufen, aufhäufen, aufspeichern, speichern, anreichern)
accumulate layer (Bodenkunde) Anreicherungshorizont m
accumulation Ansammlung f, Anhäufung f, Aufhäufung f, Aufspeicherung f, Speicherung f, Anreicherung f; (Geol) Akkumulation f, Anhäufung f
a. of humus (Landw) Humusanreicherung f
a. of slime (Pap) Schleimansammlung f, Schleimbatzen m
accumulator Akku[mulator] m, Sammler m, Batterie f; Hydraulikspeicher m, Druckspeicher m, Speicher m; Sammelflasche f; (Pap) Drucksäurebehälter m, Druckbehälter m, Kochsäuredruckspeicher m, Rezipient m
a. acid (Pap) (Kochsäure während des Aufgasens im Drucksäurebehälter)
accuracy in (of) measurement Meßungenauigkeit f
acet. s. acetone
acet. a. s. acetic acid
acetal R·CH(OR')2 Azetal n, (i.e.S.) CH3CH(OC2H5)2 Azetal n, 1,1-Diäthoxyäthan n
a. resin Azetalharz n
acetaldehyde CH3CHO Azetaldehyd m, Essigsäurealdehyd m, Äthanal n

a. cyanohydrin CH3CH(OH)CN Azetaldehydzyanhydrin n, Laktonitril n, Milchsäurenitril n, 2-Hydroxypropannitril n
acetaldol CH3CH(OH)CH2CHO Azetaldol n, Aldol n (i.e.S.), 3-Hydroxybutanal n, 3-Hydroxybutyraldehyd m, Butanol-(3)-al-(1) n
acetamide CH3CONH2 Azetamid n, Äthanamid n, Essigsäureamid n
acetamido group —NHCOCH3 Azetamidogruppe f, Äthanoylaminogruppe f, Azetylaminogruppe f
p-acetamidobenzenesulphonyl chloride p-Azetamidobenzolsulfochlorid n, p-Azetylaminobenzolsulfochlorid n, p-Azetylaminobenzolsulfonsäurechlorid n
acetanilide C6H5NHCOCH3 Azetanilid n, N-Phenylazetamid n
acetaniline s. acetanilide
acetarsol Azetarsol n, Acetarsol n, 4-Hydroxy-3-azetylaminophenylarsinsäure f
acetate 1. Azetat n; 2. Azetatfaserstoff m, AZ
a. buffer Azetatpuffer m
a. complex Azetatokomplex m
a. fibre Azetatfaser f; Azetatfaserstoff m, AZ
a. film Azetatfilm m
a. ion Azetat-Ion n
a. multifilament yarn polyfile Azetatseide f
a. rayon Azetatseide f, Azetatrayon m (n)
a. staple fibre Azetat[stapel]faser f
acetic acid CH3COOH Essigsäure f, Azetsäure f, Äthansäure f
a.-acid bacteria Essigsäurebakterien npl, Essigbakterien npl
a.-acid fermentation Essigsäuregärung f
a. aldehyde CH3CHO Azetaldehyd m, Äthanal n
a. anhydride (CH3CO)2O Essigsäureanhydrid n, Äthansäureanhydrid n, Azetanhydrid n
a. oxide s. a. anhydride
acetification Essigsäurebildung f, Essigsäuregärung f; Essigherstellung f
acetimeter Azetimeter n, Azetometer n, Essigmesser m, Essigprüfer m
acetimetry Essiguntersuchung f
acetoacetic acid CH3COCH2COOH Azetessigsäure f, Azetylessigsäure f, 3-Ketobuttersäure f
a. carboxylase Azetessigsäurekarboxylase f
a. decarboxylase Azetessigsäuredekarboxylase f
a. ester CH3COCH2COOC2H5 Azetessigsäureäthylester m, Azetessigester m
acetoarsenite Arsenitazetat n
acetoin CH3CH(OH)COCH3 Azetoin n, 3-Hydroxybutanon-(2) n, Dimethylketol n
acetolysis Azetolyse f
acetone CH3COCH3 Azeton n, Aceton n, Dimethylketon n, 2-Propanon n, Essiggeist m
a. body Azetonkörper m
a. chloroform (CH3)2C(OH)CCl3 Azetonchloroform n, Chloreton n, Trichlorbutylalkohol m
a.-extracted mit Azeton extrahiert
a. resin Azetonharz n
a.-soluble azetonlöslich

a.-soluble matter azetonlöslicher Bestandteil *m* *(bei der Bestimmung des Aushärtungsgrades von Phenoplasten)*

acetonedicarboxylic acid $O=C(CH_2 \cdot COOH)_2$ Azetondikarbonsäure *f*, 3-Ketoglutarsäure *f*, Pentanon-(3)-disäure *f*

acetophenone $C_6H_5COCH_3$ Azetophenon *n*, Methylphenylketon *n*, Azetylbenzol *n*

acetotartrate Azetotartrat *n*

acetotoluide *s.* acetotoluidide

acetotoluidide $CH_3CONHC_6H_4CH_3$ Azet[o]toluidid *n*, Azetotoluid *n*, N-Azetyltoluidin *n*

acetous fermentation Essigsäuregärung *f*
  a. odour Essiggeruch *m*

acettoluidide *s.* acetotoluidide

acetyl —$COCH_3$ Azetyl *n*, Äthanoyl *n*
  a. chloride CH_3COCl Azetylchlorid *n*, Äthanoylchlorid *n*, Essigsäurechlorid *n*
  a. coenzyme A Azetyl-Koenzym A *n*, Azetyl-Koferment A *n*, aktivierte Essigsäure *f*
  a. esterase Azetylesterase *f*
  a. group —$COCH_3$ Azetylgruppe *f*, Azetylrest *m*
  a. number Azetylzahl *f*, AZ
  a. oxide $(CH_3CO)_2O$ Essigsäureanhydrid *n*, Azetanhydrid *n*, Äthansäureanhydrid *n*
  a. radical —$COCH_3$ Azetylgruppe *f*, Azetylrest *m*, Azetylradikal *n*; [freies] Azetylradikal *n*
  a. value *s.* a. number

acetylable azetylierbar

acetylacetic acid CH_3COCH_2COOH Azetylessigsäure *f*, Azetessigsäure *f*, 3-Ketobuttersäure *f*

acetylacetonate Azetylazetonat *n*

acetylacetone $CH_3COCH_2COCH_3$ Azetylazeton *n*, 2,4-Pentandion *n*

acetylaminoanthraquinone Azetylaminoanthrachinon *n*

p-acetylaminobenzenesulphonyl chloride $C_8H_8O_3NSCl$ p-Azetylaminobenzolsulfochlorid *n*, p-Azetylaminobenzolsulfonsäurechlorid *n*, p-Azetamidobenzolsulfochlorid *n*

acetylatable azetylierbar

acetylate / to azetylieren

acetylated cellulose $[C_6H_7O_5(COCH_3)_3]_x$ Azetylzellulose *f*, Zelluloseazetat *n*
  a. paper Azetylpapier *n*, azetyliertes Papier *n* *(Chromatografie)*

acetylating agent Azetylierungsmittel *n*

acetylation Azetylieren *n*, Azetylierung *f*, Azetylation *f*
  a. agent Azetylierungsmittel *n*
  a. catalyst Azetylierungskatalysator *m*

acetylbenzene $C_6H_5COCH_3$ Azetylbenzol *n*, Methylphenylketon *n*, Azetophenon *n*

acetylene Azetylenkohlenwasserstoff *m*, Azetylen *n*, Alkin *n*, Äthinkohlenwasserstoff *m*, *(i.e.S.)* $HC≡CH$ Azetylen *n*, Äthin *n*
  a. black Azetylenschwarz *n*, Azetylenruß *m*
  a. hydrocarbon Azetylenkohlenwasserstoff *m*, Azetylen *n* *(i.w.S.)*, Alkin *n*, Äthinkohlenwasserstoff *m*

a. lime Karbidkalk *m*

a. tetrachloride $CHCl_2CHCl_2$ Azetylentetrachlorid *n*, 1,1,2,2-Tetrachloräthan *n*

acetylenedicarboxylic acid $HOOC \cdot C≡C \cdot COOH$ Azetylendikarbonsäure *f*, Äthindikarbonsäure *f*, Butindisäure *f*

acetylenic Azetylen..., Äthin...
  a. acid Azetylenkarbonsäure *f (Karbonsäure mit Dreifachbindung)*
  a. compound *s.* a. hydrocarbon
  a. hydrocarbon Azetylenkohlenwasserstoff *m*, Azetylen *n* *(i.w.S.)*, Alkin *n*, Äthinkohlenwasserstoff *m*

acetylenyl —$C≡CH$ Azetylenyl *n*, Äthinyl *n*

acetylenylbenzene $C_6H_5C≡CH$ Azetylenylbenzol *n*, Äthinylbenzol *n*, Phenylazetylen *n*

acetylenylcarbinol $HC≡CCH_2OH$ Azetylenylkarbinol *n*, Äthinylkarbinol *n*, 2-Propin1-ol *n*, Propargylalkohol *m*

acetylide $M^I{}_2C_2$ Azetyl[en]id *n*

acetylization *s.* acetylation

acetylmethylcarbinol $CH_3CH(OH)COCH_3$ Azetylmethylkarbinol *n*, Azetoin *n*, 3-Hydroxy-2-butanon *n*, Dimethylketol *n*

acetylphenylglycine $C_6H_5CH_2CONHCH_2COOH$ Azetylphenylglyzin *n*, Phenazetaminoessigsäure *f*, Phenazetylglykokoll *n*, Phen[yl]azetursäure *f*

acetylsalicylate Azetylsalizylat *n*

acetylsalicylic acid $CH_3COOC_6H_4COOH$ Azetylsalizylsäure *f*, Azetoxybenzoesäure *f*

N-acetyltoluidine $CH_3CONHC_6H_4CH_3$ N-Azetyltoluidin *n*, N-Äthanoyltoluidin *n*

acetylurea $CH_3CONHCONH_2$ Azetylharnstoff *n*, Azetylureid *n*, Essigsäureureid *n*, Äthanoylharnstoff *m*

achirite *(Min)* Achirit *m*, *(veraltet für)* Dioptas *m* *(Kupfer(II)-trisilikat)*

achondrite *(Min)* Achondrit *m*

aci form aci-Form *f*, Säureform *f (von Nitroverbindungen)*

acicular nadelförmig, nadelartig, nadelig

acid 1. sauer; 2. *s.* a.-lined
  to be a. to sauer reagieren gegen

acid Säure *f*, S, Sre
  γ-acid $C_{10}H_5NH_2OHSO_3H$ Gammasäure *f*, γ-Säure *f*, 7-Amino-1-naphthol-3-sulfonsäure *f*, 2-Amino-8-naphthol-6-sulfonsäure *f*
  δ-acid $NH_2C_{10}H_5(SO_3H)_2$ δ-Säure *f*, Naphthylamindisulfonsäure S *f*, Amino-S-Säure *f*, 1-Naphthylamin-4,8-disulfonsäure *f*
  ε-acid $C_{10}H_5(NH_2)(SO_3H)_2$ ε-Säure *f*, Amino-ε-Säure *f*, 1-Naphthylamin-3,8-disulfonsäure *f*
  a.-absorption tower Säureabsorptionsturm *m*
  a. ageing *(Text)* Säuredämpfen *n*
  a. ager *(Text)* Säuredämpfer *m*
  a. amide $RCONH_2$ Säureamid *n*
  a. ammonium tartrate $NH_4HC_4H_4O_6$ Ammoniumhydrogentartrat *n*
  a. ammonium valerate $C_4H_9COONH_4 \cdot 2C_4H_9COOH$ Ammoniumdihydrogenvalerat *n*

a. anhydride Säureanhydrid *n*
a. anthracene brown Säureanthrazenbraun *n*
a. apron Säureschutzschürze *f*
a. arsenate $M^IH_2AsO_4$ Dihydrogenarsenat(V) *n*, primäres Arsenat(V) *n*; $M^I_2HAsO_4$ Hydrogenarsenat(V) *n*, sekundäres Arsenat(V) *n*
a. azide $R \cdot CO \cdot N{=}N{=}N$ Säureazid *n*
a.-base balance *s.* a.-base equilibrium
a.-base catalysis Säure-Basen-Katalyse *f*
a.-base catalyzed säure-basen-katalysiert
a.-base dissociation curve Säure-Basen-Dissoziationskurve *f*
a.-base equilibrium Säure-Basen-Gleichgewicht *n*
a.-base indicator Säure-Basen-Indikator *m*
a.-base pair Säure-Basen-Paar *n*
a.-base reaction Säure-Basen-Reaktion *f*
a.-base theory Säure-Basen-Theorie *f*
a.-base titration Säure-Basen-Titration *f*, Neutralisationstitration *f*
a. bath Säurebad *n*
a. behaviour Säureverhalten *n*
a. Bessemer converter Bessemer-Konverter *m*, Bessemer-Birne *f*
a. Bessemer process Bessemer-Verfahren *n*, Bessemer-Konverterverfahren *n*, saures Windfrischverfahren (Bessemer-Verfahren) *n*
a. Bessemer steel Bessemer-Stahl *m*, Bessemer-Konverterstahl *m*
a.-binding säurebindend
a. black Säureschwarz *n*
a. blue Säureblau *n*
a. bromide Säurebromid *n*
a. brown Säurebraun *n*
a. caesium sulphate $CsHSO_4$ Zäsiumhydrogensulfat *n*
a. calcium phosphate $Ca(H_2PO_4)_2$ Monokalziumdihydrogenphosphat *n*, Kalziumdihydrogen[ortho]phosphat *n*; $CaHPO_4$ Dikalziumdihydrogenphosphat *n*, Kalziumhydrogen[ortho]phosphat *n*
a. carbonate M^IHCO_3 Hydrogenkarbonat *n*
a. carboy Säureglasballon *m*, Säureballon *m*
a. catalysis Säurekatalyse *f*, Reaktionsbeschleunigung *f* durch Säuren
a.-catalyzed säurekatalysiert, sauer (durch Säure) katalysiert
a.-catalyzed cure *(Farb)* Säurehärtung *f*
a. chloride Säurechlorid *n*
a. chloride reduction Rosenmund-Saizews Säurechlorid-Reduktion *f*, Rosenmund-Saizew-Reduktion *f*
a. clay säureaktivierter Ton *m*, säureaktivierte Tonerde (Bleicherde) *f (zum Raffinieren)*
a. cleavage Säurespaltung *f*
a. colour *s.* a. dye
a. component saure Komponente *f*, Säurekomponente *f*, saurer Anteil *m*
a.-containing säurehaltig
a. content Säuregehalt *m*
a. converter process *s.* a. Bessemer process

a.-cured *(Farb)* säurehärtend
a. dissociation constant Säuredissoziationskonstante *f*
a. dye saurer Farbstoff *m*, Säurefarbstoff *m*
a. egg Säuredruckvorlage *f*; Druckfaß *n*, Druckbirne *f*, Montejus *n*; Pulsometer *m*
a. endurance Säurebeständigkeit *f*, Säurefestigkeit *f*, Säurewiderstandsfähigkeit *f*, Säureresistenz *f*
a. equivalent Säureäquivalent *n*
a. etching Säureätzung *f*, Säuremattierung *f*
a. ethylsulphate $C_2H_5HSO_4$ Äthylhydrogensulfat *n*, Äthylschwefelsäure *f*, Schwefelsäuremonoäthylester *m*
a. exchange Säureaustausch *m*
a.-extracted mit Säure extrahiert
a.-fast säurefest, säurebeständig, beständig gegen Säuren; *(Farb)* säureecht
a. fastness Säurefestigkeit *f*, Säurebeständigkeit *f*, Säurewiderstandsfähigkeit *f*; *(Farb)* Säureechtheit *f*
a. fixer *(Foto)* saures Fixiersalz *n*; saures Fixierbad *n*
a. fixing bath *(Foto)* saures Fixierbad *n*
a. formers *s.* a.-forming bacteria
a.-forming säurebildend
a.-forming bacteria säurebildende Bakterien *npl*, Säuerungsbakterien *npl*, Säurebildner *mpl*
a.-free säurefrei
a. fuchsine Säurefuchsin *n*, Fuchsin *n*
a. fulling *(Text)* saure Walke *f*
a. fume Säuredampf *m*
a. gold tribromide *s.* bromoauric acid
a. halide Säurehalogenid *n*
a. hardening Säurehärtung *f*
a. hydrolysis saure Hydrolyse *f*, Säurehydrolyse *f*
a. indicator saurer Indikator *m*, Indikatorsäure *f*
a.-insoluble säureunlöslich
a.-inverted sugar durch Hydrolyse mit Säuren gewonnener Invertzucker *m*
a. iodate $M^IH[JO_3]_2$ Hydrogenbisjodat *n*; $M^IH_2[JO_3]_3$ Dihydrogentrisjodat *n*
a. layer Säureschicht *f*
a. lead sulphate $Pb(HSO_4)_2$ Bleidihydrogenbisulfat *n*
a.-lined sauer zugestellt (ausgekleidet), sauer *(metallurgischer Ofen)*
a. lining saure Zustellung (Auskleidung) *f*, saures Futter *n (eines metallurgischen Ofens)*
a. magnesium carbonate $Mg(HCO_3)_2$ Magnesiumhydrogenkarbonat *n*
a. magnesium phosphate $Mg(H_2PO_4)_2$ Magnesiumdihydrogen[ortho]phosphat *n*, Monomagnesiumdihydrogenphosphat *n*
a. making *s.* a. preparation
a.-making tower *s.* a. tower
a. milling *(Text)* saure Walke *f*
a. mist Säurenebel *m*
a. nitrile *(veraltet für)* nitrile

a. **number** Säurezahl f, SZ, Neutralisationszahl f
a. **open-hearth process** saures Herdfrischverfahren (Siemens-Martin-Verfahren, SM-Verfahren) n
a. **open-hearth steel** saurer Siemens-Martin-Stahl m
a. **orange** Säureorange n
a. **oxalate** $M^I HC_2O_4$ Hydrogenoxalat n, saures Oxalat n
a. **phosphatase** saure Phosphatase f
a. **phosphate** $M^I H_2PO_4$ Dihydrogen[ortho]phosphat n, primäres Phosphat n; $M^I_2 HPO_4$ Hydrogen[ortho]phosphat n, sekundäres Phosphat n
a. **phosphite** $M^I_2 HPO_3$ Monohydrogenphosphit n, sekundäres Phosphit n
a. **plant** (Pap) Säurestation f, Kochsäureanlage f
a. **polishing** Säurepolieren n, Säurepolitur f
a. **potassium fluoride** KHF_2 Kaliumhydrogenfluorid n
a. **potassium oxalate** KHC_2O_4 Kaliumhydrogenoxalat n, Monokaliumoxalat n
a. **potassium sulphate** $KHSO_4$ Kaliumhydrogensulfat n
a.**-precipitated** säuregefällt
a. **precipitation** Säurefällung f
a.**-preferring** (Bot) azidophil, säureliebend
a. **preparation** (Pap) Sulfitkochsäureherstellung f, Sulfitsäureherstellung f, Kochsäureherstellung f, Säurebereitung f
a. **preparation plant** s. a. plant
a. **process** saures Verfahren n; (Met) saures Windfrischverfahren (Bessemer-Verfahren) n, Bessemer-Verfahren n, Bessemer-Konverterverfahren n; (Gum) Säureverfahren n
a.**-producers** s. a.-producing bacteria
a.**-producing bacteria** säurebildende Bakterien npl, Säurungsbakterien npl, Säurebildner m [pl]
a.**-proof** säurefest, säurebeständig, beständig gegen Säuren
a.**-proof stoneware** säurefestes Steinzeug n
a. **radical** Säurerest m, Säureradikal n
a. **ratio** s. a.-to-wood ratio
a. **reclaim** (Gum) Säureregenerat n
a. **reclaiming process** (Gum) Säureverfahren n
a. **refractory [material]** saures feuerfestes (ff.) Material n
a. **residue** Säurerest m, Säureradikal n
a. **resistance** Säurebeständigkeit f, Säurefestigkeit f, Säurewiderstandsfähigkeit f, Säureresistenz f
a.**-resistant, a.-resisting** säurebeständig, beständig gegen Säuren , säurefest
a. **rock** saures Gestein n
a. **rubber** Karboxyl[at]kautschuk m
a. **salt** saures Salz n, Hydrogensalz n
a. **scarlet** Säurescharlach m
a. **separator** Säureabscheider m
a. **siphon** Säureheber m

a. **size** (Pap) Freiharzleim m
a. **slag** saure Schlacke f; Bessemer-Schlacke f
a. **sludge** Säureschlamm m, Säureteer m
a. **smut** (Luftverunreinigung in Form schwefeltrioxidhaltiger, durch Luftfeuchtigkeit zusammengeballter schwebender Kohleteilchen)
a. **sodium tartrate** $NaHC_4H_4O_6$ Natriumhydrogentartrat n
a. **solution** saure Lösung f; Säurelösung f
a. **souring** s. a. stage
a. **stability** Säurebeständigkeit f, Säurefestigkeit f, Säurewiderstandsfähigkeit f, Säureresistenz f
a.**-stable** säurebeständig, beständig gegen Säuren, säurefest, säurestabil
a. **stage** (Pap) Absäuern n, Absäuerung f (des Stoffs mit SO_2-Lösung; eine Bleichstufe)
a. **steel** saurer Stahl m; Bessemer-Stahl m
a. **steeping** Säurebehandlung f
a. **stop bath** (Foto) saures Unterbrecherbad n, [saures] Stoppbad n
a. **storage tank** Säurevorratsbehälter m
a. **strength** Säurestärke f, Säurekonzentration f; Azidität f, Säuregrad m, Säuregehalt m (Maß für die Säurestärke)
a. **sulphide** $M^I HS$ Hydrogensulfid n, saures (primäres) Sulfid n
a. **sulphite** $M^I HSO_3$ Hydrogensulfit n, saures (primäres) Sulfit n
a. **tank** Säurebehälter m, Säuretank m
a. **tar** Säureteer m, Säuregoudron m
a. **tartrate** $COOH \cdot (CHOH)_2 \cdot COOM^I$ Hydrogentartrat n, saures Tartrat n
a. **test** Säureprüfung f, Säureprobe f
a.**-to-wood ratio** (Pap) (Verhältnis der Menge der sauren Aufschlußchemikalie, i.e.S. des Gesamt-SO_2, zur Holzmenge)
a. **tower** Säureturm m
a.**-treated** mit Säure behandelt
a. **treatment** Säurebehandlung f, (bei Ölen auch) Säureraffination f
a. **urate** $M^I HC_5H_2N_4O_3$ Hydrogenurat n, saures Urat n
a. **value** Säurezahl f, SZ, Neutralisationszahl f
a. **vapour** Säuredampf m
a. **washing** Säurewaschen n, Säurewäsche f, Waschen (Auswaschen) n mit Säure
acidic sauer, säurehaltig; Säure...
to make a. [an]säuern
a. **accelerator** (Gum) saurer Beschleuniger m
a. **auxochromic group** negatives Auxochrom n
a. **catalysis** Säurekatalyse f, Reaktionsbeschleunigung f durch Säuren
a. **catalyst** Säurekatalysator m
a. **character** Säurecharakter m
a. **chlorination** (Pap) Chlorierung f in saurem Medium
a. **dissociation** Säuredissoziation f
a. **dissociation constant** Säuredissoziationskonstante f
a. **dye** saurer Farbstoff m, Säurefarbstoff m

a. fraction Säurefraktion *f*
a. function Säurefunktion *f*
a. hydrogen Säurewasserstoff *m*
a. ion Säureion *n*
a. oxide saures Oxid *n*
a. strength Säurestärke *f*
a.-type accelerator *s.* a. accelerator
acidiferous säurehaltig
acidification Ansäuern *n*, Ansäuerung *f*, Säuern *n*, Säuerung *f*, Azidifizierung *f*, Azidifikation *f*; *(Text)* Absäuern *n*, Absäuerung *f*; *(Landw)* Versauerung *f (des Bodens)*
acidified milk *(mit Säureweckern oder natürlich)* gesäuerte Milch *f*
acidify / to [an]säuern, sauer machen, azidifizieren; *(Text)* absäuern; *(den Boden)* versauern *(z.B. von Düngemitteln)*
acidimeter Azidimeter *n*
acidimetric azidimetrisch
acidimetry Azidimetrie *f*
acidity Azidität *f*, Säuregrad *m*, Säuregehalt *m (Maß für die Säurestärke)*
a. constant Aziditätskonstante *f*, Säurekonstante *f*
a. test Säuregradbestimmung *f*
acido complex Azidokomplex *m*
acidoid Azidoid *n (mit H⁺-Ionen gesättigte kolloidale Substanz)*
acidometer *s.* acidimeter
acidophilic, acidophilous *(Bot)* azidophil, säureliebend; azidophil *(durch saure Farbstoffe leicht färbbar)*
acidophilus milk Azidophilusmilch *f*
acidosis Azidose *f (Zustand einer abnormen Anhäufung saurer Körper im Blut)*
acidulate / to [an]säuern, sauer machen; mit Säure aufschließen
acidulation Ansäuern *n*, Ansäuerung *f*, Säuern *n*, Säuerung *f*; Aufschluß *m* mit Säure, Säureaufschluß *m*
acmite *(Min)* Akmit *m (Natriumeisen(III)-disilikat)*
aconine Akonin *n*
aconitic acid $C_3H_3(COOH)_3$ Akonitsäure *f*, β-Karboxyglutakonsäure *f*, α-Propentrikarbonsäure(1,2,3) *f*
aconitine Akonitin *n*, Azetylbenzoylakonin *n (Alkaloid)*
acorn gall *(Gerb)* Knopper *f (Eichelgalle)*
acquire / to annehmen, erreichen; aufnehmen
acquisition Annahme *f*, Erreichung *f*; Aufnahme *f*
acraldehyde *s.* acrolein
acrid scharf, beißend, herb
acridan[e] Akridan *n*, 9,10-Dihydroakridin *n*
acridine Akridin *n*
a. dye Akridinfarbstoff *m*
acridinyl *s.* acridyl
acridity Schärfe *f*, Herbheit *f*, Herbe *f*, Herbigkeit *f*
acridonation Akridonringschluß *m*
acridone Akridon *n*
a. dye Akridonfarbstoff *m*

acridyl Akridyl *n*
acriflavine Akriflavin *n*, Acriflavin *n*, 3,6-Diamino-10-methylakridiniumhydroxid *n*
acrolein $CH_2{=}CHCHO$ Akrolein *n*, Akrylaldehyd *m*, Allylaldehyd *m*, Propenal *n*
acrylaldehyde *s.* acrolein
acrylamide $CH_2{=}CHCONH_2$ Akrylamid *n*, Akrylsäureamid *n*, Propenamid *n*
acrylate Akrylat *n*
a.-butadiene rubber Akryl-Butadienkautschuk *m*, Akrylatkautschuk *m (Butadien-Akrylsäureester-Mischpolymerisat)*
a. elastomer Akrylelastomer[es] *n*, Polyakrylatelastomer[es] *n*
a. resin *s.* acrylic resin
a. rubber *s.* a.-butadiene rubber
acrylic Akryl...
a. acid $CH_2{=}CHCOOH$ Akrylsäure *f*, Propensäure *f*, Vinylkarbonsäure *f*, Äthenkarbonsäure *f*
a.-acid amide *s.* acrylamide
a.-acid ester *s.* a. ester
a.-acid resin *s.* a. resin
a. aldehyde *s.* acrolein
a. amide *s.* acrylamide
a. elastomer *s.* acrylate elastomer
a. ester Akrylsäureester *m*
a. fibre Akrylfaser *f*, Polyakrylnitrilfaser *f*; Akrylfaserstoff *m*, Polyakrylnitrilfaserstoff *m*, PVY
a. lacquer Akryllack *m*
a. resin Akrylharz *n*, Polyakrylat *n*
a. rubber *s.* acrylate-butadiene rubber
acrylonitrile $CH_2{=}CHCN$ Akrylonitril *n*, Akryl[säure]nitril *n*, Propennitril *n*, Vinylzyanid *n*
a.-butadiene rubber Akrylnitril-Butadienkautschuk *m*, Nitrilkautschuk *m (Butadien-Akrylnitril-Mischpolymerisat)*
a.-butadiene styrene Akrylnitril-Butadien-Styrol *n*, ABS *n*
a.-butadiene-styrene copolymer Akrylnitril-Butadien-Styrol-Kopolymer[es] *n*, ABS-Kopolymer[es] *n*, Akrylnitril-Butadien-Styrol-Mischpolymer[es] *n*, ABS-Mischpolymer[es] *n*
a.-butadiene-styrene plastic Akrylnitril-Butadien-Styrol-Kunststoff *m*, ABS-Kunststoff *m*, Akrylnitril-Butadien-Styrol-Plast *m*, ABS-Plast *m*
a.-butadiene-styrene polymer Akrylnitril-Butadien-Styrol-Polymer[es] *n*, ABS-Polymer[es] *n*
a.-butadiene-styrene resin Akrylnitril-Butadien-Styrol-Harz *n*, ABS-Harz *n*
acryl[o]yl —COCH=CH₂ Akrylyl *n*, Propenoyl *n*
a. chloride $CH_2{=}CHCOCl$ Akrylylchlorid *n*, Propenoylchlorid *n*
ACS *(Abk. für)* American Chemical Society
act / to [ein]wirken
act. *s.* acetone
ACTH *s.* adrenocorticotrop[h]ic hormone
actinic aktinisch, fotochemisch wirksam
actinide *s.* actinoid
actinium Ac Aktinium *n*

actinium A AcA Aktinium A n (Poloniumisotop 215)

actinium B AcB Aktinium B n (Bleiisotop 211)

actinium C AcC Aktinium C n (Wismutisotop 211)

actinium D AcD Aktinium D n, Aktiniumblei n (Bleiisotop 207)

actinium K AcK Aktinium K n (Franziumisotop 223)

actinium X AcX Aktinium X n (Radiumisotop 223)

a. decay series Aktiniumzerfallsreihe f, Aktiniumreihe f, Uran-Aktinium-Zerfallsreihe f, Uran-Aktinium-Reihe f

a. emanation AcEm, An Aktiniumemanation f, Aktinon n (Radonisotop 219)

a. fluoride AcF_3 Aktiniumfluorid n

a. hydroxide $Ac(OH)_3$ Aktiniumhydroxid n

a. oxalate $Ac_2(C_2O_4)_3$ Aktiniumoxalat n

a. phosphate $AcPO_4$ Aktiniumphosphat n

a. series s. a. decay series

a. sulphide Ac_2S_3 Aktiniumsulfid n

actinoid Aktinoid n, Aktinoidenelement n, (bisher) Aktinid n, Aktinidenelement n, Element n der Aktiniumreihe

a. contraction Aktinoidenkontraktion f

a. element s. actinoid

a. group (series) Aktinoidenreihe f, Aktiniumreihe f, Aktinoidengruppe f

actinolite (Min) Aktinolith m, Strahlenstein m, Strahlstein m

actinometer Aktinometer n

actinometry Aktinometrie f

actinomycetin Aktinomyzetin n (antibiotischer Wirkstoff aus Aktinomyzeten)

actinomycin Aktinomyzin n (antibiotischer Farbstoff aus Aktinomyzeten)

actinon An, AcEm Aktinon n, Aktiniumemanation f (Radonisotop 219)

actinorhodine Aktinorhodin n (Chinonfarbstoff aus Aktinomyzeten)

actinospectacin Aktinospektazin n (Antibiotikum aus Streptomyces spectabilis)

actinote s. actinolite

actinouranium AcU Aktinouran n

action Einwirken n, Einwirkung f, Wirkung f

a. constant Plancksches Wirkungsquantum n, Plancksche Konstante f

a. of enzymes Enzymwirkung f, Fermentwirkung f

a. value Wirkungswert m (z.B. von Düngemitteln)

actithlazic acid Aktithiazinsäure f, 4-Thiazolidon-2-kapronsäure f

activate / to aktivieren, anregen, erregen

activated adsorption aktivierte (aktive) Adsorption f

a. alumina [künstlich] aktivierte Tonerde f, künstliche [aktive] Tonerde f, Aktivtonerde f

a. atom angeregtes Atom n

a. carbon aktivierter (aktiver) Kohlenstoff m, aktivierte (aktive) Kohle f, Aktivkohle f

a. charcoal aktivierte (aktive) Kohle f, Aktivkohle f; aktivierte (aktive) Holzkohle f

a. clay aktivierter Ton m, (bei der Erdölraffination auch) [künstlich] aktivierte Tonerde (Bleicherde) f, künstliche [aktive] Tonerde (Bleicherde) f

a. complex aktivierter Komplex m

a. earth aktivierte Erde f, Aktiverde f, Edelerde f

a. formulation Wirkstoffzubereitung f mit aktivierendem Zusatz

a. molecule aktiviertes (aktives, energiereiches) Molekül n

a. sewage sludge s. a. sludge

a. sintering aktiviertes Sintern n

a. sludge belebter (aktivierter, biologischer) Schlamm m, Belebtschlamm m

a.-sludge chamber s. a.-sludge tank

a.-sludge plant Belebungsanlage f, Belebtschlammanlage f

a.-sludge process Belebungsverfahren n, Schlammbelebungsverfahren n, Belebtschlammverfahren n

a.-sludge tank Belebungsbecken n, Belebtschlammbecken n

activating Beleben n, Aktivieren n, Aktivierung f (bei der Flotation)

a. accelerator (Gum) aktivierender Zusatzbeschleuniger (Sekundärbeschleuniger, Zweitbeschleuniger) m

a. agent Aktivierungsmittel n, Aktivator m

a. group aktivierende Gruppe f, Substituent m mit aktivierender Wirkung

a. substance Aktivator m, Aktivierungsmittel n; Aktivator m, Promotor m, [synergetischer] Verstärker m

activation Aktivieren n, Aktivierung f, Aktivation f; Aktivatoreffekt m (eine Form der Verbundwirkung von Wirkstoffen)

a. analysis Aktivierungsanalyse f, Aktivitätsanalyse f

a. cross section Aktivierungsquerschnitt m

a. energy Aktivierungsenergie f

a. entropy Aktivierungsentropie f

a. method Aktivierungsmethode f

a. process Aktivierungsprozeß m

activator Aktivator m, Aktivierungsmittel n; Aktivator m, Promotor m, [synergetischer] Verstärker m; (Flot) belebendes (aktivierendes) Mittel n, belebender (aktivierender) Zusatz m, Beleber m; (Krist) Aktivator m, Phosphorogen n

a. level s. a. term

a. of cure (Gum) Beschleunigeraktivator m, Aktivator m für den Beschleuniger

a. of polymerization Polymerisationsaktivator m

a. of vulcanization s. a. of cure

a. term Aktivatorterm m

active aktiv, wirksam

a. acidity aktuelle Azidität f

a. agent s. a. substance

a. alkali (Pap) aktives (wirksames) Alkali n (NaOH + Na_2S, ausgedrückt als Na_2O)

a. alkalinity aktuelle Basizität f

a. black aktiver Ruß m, Aktivruß m

a. **carbon** s. activated carbon
a. **centre** aktives Zentrum n (z.B. einer Katalysatoroberfläche)
a. **charcoal** s. activated charcoal
a. **chemical** (Pap) aktives (wirksames) Alkali n (NaOH + Na₂S, ausgedrückt als Na₂O)
a. **chlorine** aktives (wirksames) Chlor n, Aktivchlor n; Aktivchlorgehalt m
a. **earth** Aktiverde f, aktivierte Erde f, Edelerde f
a. **filler** (Gum) aktiver (verstärkender) Füllstoff m, Verstärker[füllstoff] m
a. **group** aktive (prosthetische) Gruppe f, Wirkungsgruppe f, Wirkgruppe f, Koferment n, Koenzym n, Agon n
a. **hydrogen** aktiver Wasserstoff m
a. **ingredient** s. a. substance
a. **mass** aktive Masse f
a. **material** s. a. substance
a. **molecule** aktives (aktiviertes, energiereicheres) Molekül n
a. **nitrogen** aktiver Stickstoff m
a. **principle** s. a. substance
a. **screen area** nützliche Siebfläche f, Siebnutzfläche f
a. **solvent** aktives (echtes) Lösungsmittel n, aktiver (echter) Löser m
a. **substance** aktiver (wirksamer) Stoff (Bestandteil) m, aktives (wirksames) Agens (Prinzip) n, aktive (wirksame) Substanz f, Aktivsubstanz f, Aktivstoff m, Wirkstoff m, Wirksubstanz f
a. **transport** (Bioch) aktiver Transport m
activity 1. Aktivität f, Wirksamkeit f, Tätigkeit f; 2. [chemische] Aktivität f
a. **coefficient** Aktivitätskoeffizient m
a. **of bacteria** Bakterientätigkeit f
a. **of enzymes** Enzymaktivität f, Fermentaktivität f
actual tatsächlich, wirklich, real, praktisch (z.B. Versuch), praktisch erzielbar (z.B. Temperatur)
a. **gas** reales (wirkliches) Gas n
a. **plate** (Destillation) praktischer Boden m
a. **plate number** (Destillation) praktische (wirkliche, tatsächliche) Bodenzahl f
actuating signal (Regelungstechnik) Signal n der Regelabweichung
actuator Betätigungsorgan n; (Regelungstechnik) Stellantrieb m, Stellmotor m
acutance (Foto) Konturenschärfe f, Kantenschärfe f
acute toxicity akute Toxizität (Giftigkeit) f
acyclic azyklisch, nichtzyklisch
acyl Azyl n
a. **group** Azylgruppe f, Azylrest m
a. **halide** Azylhalogenid n (Säurehalogenid von Karbonsäuren)
a. **migration** Azylwanderung f, Azylverschiebung f
a. **radical** Azylgruppe f, Azylrest m, Azylradikal n; [freies] Azylradikal n
acylable azylierbar

acylate / to azylieren
acylation Azylierung f
acylium ion Azylium-Ion n
acyloin Azyloin n
a. **condensation** Azyloinkondensation f, Azyloinsynthese f
a. **method** Azyloinmethode f
a. **synthesis** s. a. condensation
A.D. s. air-dried
adamantine Diamant..., diamanten, diamantartig; diamanthart, sehr hart
a. **lustre** (Min) Diamantglanz m
a. **spar** (Min) Diamantspat m
adamine, adamite (Min) Adamin m (Zinkhydroxidorthoarsenat)
adamsite 1. (Min) Adamsit m; 2. NH(C₆H₄)₂AsCl Adamsit n (Phenarsazinchlorid, Arsenkampfstoff)
adapter Anpaßstück n, Paßstück n, Verlängerungsstück n, Zwischenstück n; Reduzierstück n; Vorstoß m (einer Destilliervorrichtung)
a. **heater** (Plast) Heizeinsatzstück n
adaptor s. adapter
adatom adsorbiertes Atom n
add / to [hin]zufügen, [hin]zugeben, zusetzen, beimengen, beimischen; anlagern, addieren; sich anlagern (addieren)
a. **a correction** berichtigen
a. **to form a lower layer** unterschichten
added sulphur (Gum) zugesetzter Schwefel m
addend Addend m, Ligand m
addition Zufügen n, Zufügung f, Zugeben n, Zugabe f, Zusatz m, Beimengung f, Beimischung f; Addition f, Anlagerung f, Anlagern n; Zufuhr f (z.B. von Luft)
1,2-addition 1,2-Addition f
1,4-addition 1,4-Addition f
a. **agent** Zusatzmittel n, Zusatz[stoff] m, (manchmal auch) Hilfsmittel n, Hilfsstoff m
a.-**capable** additionsfähig
a. **complex** Anlagerungskomplex m, Normalkomplex m
a. **compound** Additionsverbindung f, Anlagerungsverbindung f, additive Verbindung f
a. **of sulphur** (Gum) Aufschwefelung f
a. **polymer** Additionspolymer[es] n, Polyaddukt n
a. **polymerization** Additionspolymerisation f, Polyaddition f
a. **product** Additionsprodukt n, Anlagerungsprodukt n
a. **reaction** Additionsreaktion f, Anlagerungsreaktion f
a. **reactivity** Additionsfähigkeit f, Additionsvermögen n, Additivität f
a. **solid solution** Additionsmischkristall m
additional zusätzlich
a. **finish** (Text) Nachappretur f
additive additiv
additive Zusatzmittel n, Zusatz[stoff] m, (manchmal

aucn) Hilfsmittel n, Hilfsstoff m; Additiv[e] n (Zusatzstoff in Mineralölprodukten, der in Mengen von 1 bis 10% benutzt wird; s. a. dope)
a. capacity Additionsvermögen n, Additionsfähigkeit f, Additivität f
a. compound additive Verbindung f, Additionsverbindung f, Anlagerungsverbindung f
a. effect Additionseffekt m
a. for adhesive Leimzusatzstoff m
a. name additiver Name m, Additivname m, Additionsname m
a. property additive Eigenschaft f
a. substance Zusatz[stoff] m, Zusatzmittel n
additivity Additivität f, Additionsvermögen n, Additionsfähigkeit f
adduct Adduct n
a. rubber Adduktkautschuk m
adenase Adenase f
adenine Adenin n, 6-Aminopurin n
adenosine diphosphate Adenosindiphosphat n, Adenosindiphosphorsäure f, ADP, Adenosinpyrophosphat n, Adenosinpyrophosphorsäure f
a. monophosphate Adenosinmonophosphat n, Adenosinmonophosphorsäure f, AMP
a. phosphate Adenosinphosphat n, Adenosinphosphorsäure f
a. triphosphatase Adenosintriphosphatase f, ATPase f
a. triphosphate Adenosintriphosphat n, Adenosintriphosphorsäure f, ATP, Adenylpyrophosphat n, Adenylpyrophosphorsäure f
adenosinediphosphoric acid s. adenosine diphosphate
adenosinephosphoric acid s. adenosine phosphate
adenosinetriphosphoric acid s. adenosine triphosphate
adenyl Adenyl n
a. pyrophosphatase s. adenosine triphosphatase
a. pyrophosphate s. adenosine triphosphate
adenylic acid s. adenosine phosphate
adenylpyrophosphoric acid s. adenosine triphosphate
adhere / to [an]haften, aneinanderhaften, adhärieren; [an]kleben
adherence Adhäsion f, Haften n, Haftung f, Aneinanderhaften n, Anhaften n; Kleben n, Ankleben n; Adhäsionsfähigkeit f, Adhäsionsvermögen n, Adhäsionskraft f, Haftfähigkeit f, Haftvermögen n; Kleb[e]fähigkeit f, Kleb[e]kraft f
adherend Adhärend m, Packstoff m, Schichtträger m (zu verklebender Stoff)
adherent [an]haftend, aneinanderhaftend, adhärierend, adhäsionsfähig, haftfähig, adhäsiv; [an]klebend, klebfähig
adhesion Adhäsion f, Haften n, Haftung f, Aneinanderhaften n, Anhaften n; Kleben n, Ankleben n
a. agent s. adhesive agent

a. energy Adhäsionsenergie f
a. force s. adhesive force
a. power s. adhesiveness
a. promoter Adhäsionsbeschleuniger m
a. strength s. adhesive strength
a. tension s. adhesive tension
a. term Haftterm m
a. tester Adhäsionsprüfer m
adhesional work Adhäsionsarbeit f, Haftarbeit f
adhesive adhäsiv, adhäsionsfähig, haftfähig, [an]haftend, aneinanderhaftend, adhärierend, Adhäsions..., Haft...; klebfähig, [an]klebend, Kleb[e]...
adhesive s. a. agent
a. agent Haftmittel n, Haftstoff m, Haftvermittler m, Haftlösung f; Klebemittel n, Kleb[e]stoff m, Kleber m, Adhärens n; Bindemittel n
a. bandage Schnellverband m, Schnellverbandpflaster n
a. bond Klebverbindung f
a. capacity s. adhesiveness
a. film Klebfilm m, Klebfolie f
a. force Adhäsionskraft f
a. layer Klebstoffschicht f
a. paper gummiertes Papier n, Klebpapier n
a. plaster Heftpflaster n, Zinkkautschukpflaster n
a. power s. adhesiveness
a. resin Klebharz n
a. solution Leimlösung f
a. strength Adhäsionsfestigkeit f, Haftfestigkeit f
a. stress s. a. tension
a. substance s. a. agent
a. tape Klebeband n, Klebstreifen m
a. tension Adhäsionsspannung f, Haftspannung f
adhesiveness Adhäsionsfähigkeit f, Adhäsionsvermögen n, Adhäsionskraft f, Haftfähigkeit f, Haftvermögen n; Klebvermögen n, Kleb[e]fähigkeit f, Kleb[e]kraft f
adhesivity s. adhesiveness
adiabat Adiabate f
adiabatic adiabatisch
adiabatic Adiabate f
a. calorimeter adiabatisches Kalorimeter n
a. change adiabatische Zustandsänderung f
a. curve s. a. line
a. demagnetization adiabatische Entmagnetisierung f
a. expansion adiabatische Expansion f
a. invariant adiabatische Invariante f
a. line Adiabate f
adion Adion n (adsorbiertes Ion)
adipate Adipat n
adipic acid HOOC(CH$_2$)$_4$COOH Adipinsäure f, 1,4-Butandikarbonsäure f, Hexandisäure f
adipose fettig, fetthaltig, talgig, Fett...
a. cell Fettzelle f
a. tissue (Gerb) Fettgewebe n, Unterhautbindegewebe n

adip[o]yl Adipyl n, Hexandioyl n
adjacency effects *(Foto)* Nachbareffekte mpl
adjacent anliegend, anstoßend, angrenzend, benachbart, nachbarständig, vizinal, Neben..., Rand..., Nachbar...
a. atom Nachbaratom n
a. molecule Nachbarmolekül n
a. position Nachbarstellung f, 1,2,3-Stellung f
a. rock Nebengestein n, Nachbargestein n
adjective dye adjektiver (beizenfärbender) Farbstoff m, Beizenfarbstoff m
adjoining rock s. adjacent rock
adjust / to einstellen (z.B. pH-Wert)
adjustable-speed drive stufenlos stellbarer Antrieb m
a.-zero thermometer Einstellthermometer n
adjustment Einstellung f
adjuvant Hilfsmittel n, Hilfsstoff m, Zusatzmittel n, Zusatz[stoff] m; *(Pharm)* Adjuvans n *(unterstützendes Mittel)*
Adkins' catalyst Adkins-Katalysator m
administer / to *(Arzneimittel)* applizieren, verabreichen
administration Applikation f, Verabreichung f *(von Arzneimitteln)*
Admiralty steam coal Admiralitätskohle f
admix / to beimischen, beimengen, zusetzen
admixture Beimischen n, Beimengen n; Beimischung f, Beimengung f, Begleitstoff m, Fremdstoff m, Fremdbestandteil m, Zusatz m, Verunreinigung f
adobe 1. Adobe m, Luftziegel m; 2. s. a. clay
a. clay Ton m für Adoben
ADP s. adenosine diphosphate
adrenal cortex Nebennierenrinde f, NNR
a. cortex (cortical) hormone Nebennierenrindenhormon n
a. corticotrop[h]ic hormone s. adrenocorticotrop[h]ic hormone
adrenaline Adrenalin n, Epinephrin n, 1-(3′,4′-Dihydroxyphenyl)-2-methylaminoäthanol n
adrenalone Adrenalon n, 4-(Methylaminoazeto)-brenzkatechin n
adrenochrome $C_9H_9NO_3$ Adrenochrom n
adrenocorticotrop[h]ic hormone adrenokortikotropes Hormon n, ACTH n, Adrenokortikotropin n, Kortikotrop[h]in n
A.D.S. s. air-dried sheet
adsorb / to adsorbieren
adsorbability Adsorbierbarkeit f
adsorbable adsorbierbar
adsorbate Adsorptiv n, Adsorbat n, adsorbierter (aufgenommener) Stoff m; Adsorbat n *(System aus Adsorptiv plus Adsorbens)*
adsorbed film Adsorptionsfilm m
a. layer Adsorptionsschicht f
a. material (substance) s. adsorbate
adsorbent adsorbierend, adsorptionsfähig, adsorptiv
adsorbent Adsorbens n, Adsorptionsmittel n,

Adsorber m, adsorbierender (aufnehmender) Stoff m
a.-adsorbate complex (system) s. adsorption complex
a. bed Adsorbensschicht f, Adsorptionsmittelschicht f
a. column s. adsorbing column
a. material s. adsorbent
adsorber Adsorber m, Adsorptionsapparat m
adsorbing adsorbierend, aufnehmend, Adsorptions...
a. agent s. adsorbent
a. column Adsorptionskolonne f, Adsorptionssäule f
a. material (substance) s. adsorbent
adsorption Adsorption f, Adsorbieren n, Adsorbierung f
a. analysis Adsorptionsanalyse f
a. balance Adsorptionswaage f
a. chromatography Adsorptionschromatografie f, chromatografische Adsorptionsanalyse f
a. column Adsorptionskolonne f, Adsorptionssäule f
a. complex Adsorbat n *(System aus Adsorptiv plus Adsorbens)*
a. compound Adsorptionsverbindung f
a. curve Adsorptionskurve f
a. displacement Adsorptionsverdrängung f
a. energy Adsorptionsenergie f
a. equation Adsorptionsgleichung f
a. equilibrium Adsorptionsgleichgewicht n
a. force Adsorptionskraft f
a. heat Adsorptionswärme f
a. indicator Adsorptionsindikator m
a. isostere Adsorptionsisostere f
a. isotherm Adsorptionsisotherme f
a. medium Adsorptionsmedium n
a. of ions Ionenadsorption f
a. phenomenon Adsorptionserscheinung f
a. potential Adsorptionspotential n
a. process Adsorptionsverfahren n, Adsorptionsprozeß m; Adsorptionsvorgang m
a. site Adsorptionszentrum n
a. swelling Adsorptionsquellung f
a. tower Adsorptionsturm m
a. zone Adsorptionszone f
adsorptive adsorptiv, adsorptionsfähig, adsorbierend, Adsorptions...
adsorptive Adsorptiv n, Adsorbat n, adsorbierter (aufgenommener) Stoff m
a. capacity (power) Adsorptionsvermögen n, Adsorptionsfähigkeit f, Adsorptionskraft f
adsorptiveness s. adsorptive power
adularia *(Min)* Adular m *(Kaliumalumotrisilikat)*
adulterate / to verfälschen; verschneiden
adulteration Verfälschen n, Verfälschung f; Verschneiden n, Verschnitt m
advance / to fortschreiten, steigen, wachsen
advance of wood *(Pap)* Holzvorschub m *(beim Schleifen)*

advancing contact angle vorwärtsschreitender Randwinkel *m*
advection Advektion *f*
a. fog Advektionsnebel *m*
aegirine *(Min)* Ägirin *m (Natriumeisen(III)-disilikat)*
aeolotropic anisotrop
aeolotropism Anisotropie *f*
aerate / to mit Luft versetzen (vermischen), belüften, durchlüften; *(Getränke)* mit CO_2 imprägnieren (sättigen), karbonisieren
aerated concrete Schaumbeton *m*
aeration Lüftung *f*, Belüftung *f*
a. flask Belüftungskolben *m*
a. line Belüftungsleitung *f*
aerator Belüftungsanlage *f*
aerial application Applikation *f* aus der Luft *(z.B. von Pflanzenschutzmitteln)*
a. filter Luftfilter *n*
a. fog Luftschleier *m*
a. oxidation Oxydation *f* durch Luftsauerstoff
a. spraying Besprühen *n* aus der Luft
aero-mist sprayer Nebelblaser *m (Pflanzenschutzgerät)*
aerobe *(Bio)* Aerobier *m*, Aerobiont *m*
aerobic aerob
under a. conditions unter aeroben Bedingungen (Verhältnissen), im aeroben Medium, unter Luftzutritt, aerob
a. bacteria aerobe Bakterien *npl*
a. fermentation aerobe (oxydative, unter Mitwirkung von Sauerstoff verlaufende) Gärung *f*
aeroclay windgesichteter Ton *m*
aerodynamics Aerodynamik *f*
Aerofall mill Aerofall-Mühle *f (Mühle für autogene Mahlung)*
aerogel Aerogel *n (Gel mit gasförmigem Dispersionsmittel)*
aerolite, aerolith *(Geol)* Aerolith *m*, Steinmeteorit *m*, Meteorstein *m*
aeroplane application Applikation *f* (Ausbringen *n*) vom Flugzeug aus, Flugzeugausbringung *f (z.B. von Pflanzenschutzmitteln)*
a. fertilization Düngung *f* durch Flugzeuge, Flugzeug-Düngerstreuen *n*, aviotechnisches Düngerstreuen *n*
a. spraying Besprühen *n* vom Flugzeug aus *(Pflanzenschutz)*
aerosol Aerosol *n*, Wirkstoffnebel *m (Sol mit gasförmigem Dispersionsmittel)*
a. bomb 1. Aerosolsprühgerät *n*, Aerosolbombe *f*; 2. Druckzerstäuberdose *f*, Aerosolzerstäuber *m*, Sprühdose *f*, Aerosolsprühdose *f*, Aerosoldose *f*
a. paint Aerosolfarbe *f*
a. projector *s.* a. bomb 1.
a. shave Rasierschaumaerosol *n*
a. spray Aerosolspray *m*
aerosolize / to aerosolieren, aerolisieren, vernebeln
aerostatics Aerostatik *f*
aerugo Grünspan *m*

AES *(Abk. für)* American Electrochemical Society
aeschynite Aeschynit *m (Zer, Titan und Thorium enthaltendes Mineral)*
aesculin Äskulin *n*, Bikolorin *n*, 6,7-Dihydroxykumarin-6-(β-D-glukopyranosid) *n*
AET *s.* aminoethylisothiuronium bromide hydrobromide
aethacridine Äthakridin *n*, 2-Äthoxy-6,9-diaminoakridin *n*
affect / to angreifen, beeinflussen, [ein]wirken auf; *(Foto) (die lichtempfindliche Schicht)* schwärzen
affinated sugar affinierter Zucker *m*, Affinade *f (im Affinierverfahren gewaschener Rohzucker)*
affination Affination *f (Zucker-Reinigungsverfahren)*
a. sugar *s.* affinated sugar
affinity [chemische] Affinität *f*, Triebkraft *f (maximale Nutzarbeit einer Reaktion, Reaktionsarbeit)*
a. constant Affinitätskonstante *f*, Dissoziationskonstante *f*
a. curve Affinitätskurve *f*
a. for cotton Baumwollaffinität *f*
a. for dyestuffs Farbstoffaffinität *f*
a. to the fibre Affinität *f* zur Faser, Faseraffinität *f*
affix / to fixieren, festheften
afflux Zufluß *m*
African ammoniac afrikanisches Ammoniakgummi *n (von Ferula brevifolia)*
A. elemi *(Harz von afrikanischen Burseraceen)*
A. kino Pterocarpus erinaceus Lam. *(kinoharzliefernder Baum)*
A. locust Parkia africana R.Br. *(gerbstoffliefernder Baum)*
A. tragacanth afrikanischer Tragant *m (Gummiharz von Sterculia tragacantha Lindl.)*
after-shave lotion nachbehandelndes Rasierwasser *n*, After-shave-Lotion *f*
afterbake / to *(Plast)* nachhärten
afterbake *(Plast)* Nachhärten *n*, Nachhärtung *f*
afterblow / to nachblasen *(beim Windfrischen)*
afterblow Nachblasen *n (beim Windfrischen)*
afterchlorinate / to nachchlorieren
afterchrome / to nachchromieren
afterchrome nachchromierbar, Nachchromier[ungs]...
a. dye Nachchromierungsfarbstoff *m*, Nachchromierfarbstoff *m*, Chromentwicklungsfarbstoff *m*
a. [dyeing] method Nachchromierungsverfahren *n*, Nachchromierverfahren *n*
a. process *s.* a. [dyeing] method
afterchromed dye *s.* afterchrome dye
afterchroming Nachchromieren *n*, Nachchromierung *f*
aftercontraction *(Ker)* Nachschwinden *n*, Nachschwindung *f*
aftercopper / to nachkupfern
aftercoppering Nachkupfern *n*, Nachkupferung *f*
aftercrystallization Nachkristallisation *f*

aftercure s. aftervulcanization
afterdamp Nachschwaden m, Stickwetter pl (Kohlendioxidanreicherung nach Grubenexplosionen)
afterdryer (Pap) Nachtrockenzylinder m, Nachtrockner m
aftereffect Nachwirkung f
afterexpansion (Ker) Nachwachsen n
afterfermentation Nachgärung f, Lagerung f
afterfiltration Nachfiltration f
afterflaming (Text) Aufflammen n
afterglow Nachleuchten n; Nachglühen n, Nachglimmen n
afterhardening Nachhärtung f
afterhydrolysis Nachhydrolyse f
aftermassecuite (Zucker) Nachproduktfüllmasse f, C-Füllmasse f
afterpolymerization Nachpolymerisation f
afterproduct (Zucker) Nachprodukt n
afterpurification Nachreinigung f
aftershaping Nachformung f
aftershrinkage Nachschwinden n, Nachschwindung f, Nachverformung f
afterstretching (Text) Nachverstrecken n, Nachverstreckung f
aftertaste Nachgeschmack m
aftertreat / to nachbehandeln
aftertreated dye Nachbehandlungsfarbstoff m
aftertreating, aftertreatment Nachbehandeln n, Nachbehandlung f
aftervulcanization Nachvulkanisation f, Nachheizung f
afterwort (Lebm) Nachwürze f
agalmatolite (Min) Agalmatolith m, Bildstein m (Aluminiumdihydrogentetrasilikat)
agar Agar m(n), Agar-Agar m(n), Gelose f
  a.-agar s. agar
  a. gel s. agar
  a. medium Agarnährboden m, Nähragar m(n)
  a.-tube method Agarröhrchenmethode f
agaric acid $C_{19}H_{36}OH(COOH)_3$ Agarizinsäure f, Zetylzitronensäure f
  a. mineral (Min) Bergmilch f (Kalziumkarbonat)
agaricic acid s. agaric acid
agaricin Agarizin n
agarobilli (Gerb) Algarobilla m (Früchte von Caesalpinia brevifolia (Baill.)Clos.)
agarslant tube (Bioch) Kulturröhrchen n
agarweed Gelidium cartilagineum (L.)Gaill. (Agar liefernde Rotalge)
agate (Min) Achat m
  a. burnisher Achatpolierstein m
  a. jet s. a. tube
  a. mortar Achatmörser m
  a. plate Achatpfanne f, (bei Feinwaagen) Achatplanlager n
  a. tube Achatausflußrohr n, Achatablaufdüse f (des Redwood-Viskosimeters)
agathene dicarboxylic acid $HOOC \cdot C_{10}H_{12}(CH_3)_2$ $(=CH_2) \cdot CH_2 \cdot C(CH_3) \cdot CH \cdot COOH$ Agathendisäure f, Agathsäure f

agave Agave f, Agave L.; Agavefaser f, Sisalfaser f
age / to altern, (Lebm auch) ablagern, (Ker auch) mauken, faulen, lagern, (Pap auch) vergilben, (Text auch) dämpfen (Färbungen und Drucke fixieren); (Gerb) in Borke [ab]lagern
age determination Altersbestimmung f
  a.-harden / to aushärten
  a. hardening Aushärten n, Aushärtung f
  a.-resister s. a.-resistor
  a.-resisting alterungsbeständig
  a.-resistor (Gum) Alterungsschutzmittel n
aged wine ausgebauter (geschulter) Wein m
ageing Altern n, Alterung f, (Lebm auch) Ablagerung f, (Ker auch) Mauken n, Faulen n, Lagern n, (Pap auch) Vergilben n, (Text auch) Dämpfen n (Fixieren von Färbungen und Drucken)
  a. characteristics (performance, properties) Alterungseigenschaften fpl, Alterungsverhalten n
  a. resistance Alterungsbeständigkeit f, Alterungswiderstand m
  a. test Alterungsprüfung f, Alterungstest m, Alterungsversuch m
  a. vat (Lebm) Rahmreifer m, Säuerungswanne f, Säuerungsgefäß n
agency treibende Kraft f, Triebkraft f, Faktor m
agent Agens n, [wirksames] Mittel n
ager (Text) Dämpfer m
agglomerate / to agglomerieren, zusammenbacken, zusammenballen, stückigmachen, (mittels Flüssigkeit) granulieren, krümeln, (mittels Hitze) sintern, fritten; agglomerieren, sich zusammenballen, zusammenbacken, klumpen, (bei Hitzeeinwirkung) sintern, [zusammen]fritten
agglomerate Agglomerat n, Anhäufung f; Granulatkorn n, Granalie f; Sinterstoff m, Sintererzeugnis n, (Ker) Fritte f; (Geol) Agglomerat n
  a. breaker Ballenzerteiler m (z.B. in Mischern)
agglomeration Agglomerieren n, Agglomeration f, Agglomerierung f, Zusammenbacken n, Zusammenballen n, Stückigmachen n, (mittels Flüssigkeit) Granulieren n, Granulation f, Granulierung f, Granulatformen n, Krümeln n, Krümelung f, (mittels Hitze) Sintern n, Sinterung f, Fritten n, Zusammenfritten n
  a. tabling Granulieren (Stückigmachen) n auf einem Rütteltisch
agglutinant agglutinierend, klebend, leimend
agglutinant Bindemittel n, Klebemittel n; (Geol) Bindemittel n, Zement m
agglutinate / to agglutinieren, [ver]kleben, verkitten, binden, [zusammen]leimen, zusammenballen; agglutinieren, verkleben, klumpen, sich zusammenballen
agglutinating value Agglutinating-Index m (Maß für die Bindefähigkeit der Kohle)
agglutination Agglutination f, Verklebung f, Verkittung f, Zusammenballung f, Verklumpung f

aggrading *(Geol)* Aufschüttung *f*, Schüttung *f*
aggregate / to anhäufen; sich ansammeln; *(Landw)* *(Bodenpartikel zu Krümeln)* verkitten
aggregate 1. Aggregat *n*; Zuschlag[stoff] *m*; *(Bodenkunde)* Krümel *n*; 2. *(Nomenklatur)* polyzyklische Verbindung *f*
aggregating agent *(Bodenkunde)* Krümelbildner *m*
aggregation Aggregation *f*, Aggregierung *f*, Zusammenballung *f*
aging *s.* ageing
agitate' / **to** bewegen, [um]rühren, aufrühren, durchrühren, schütteln, auflockern
agitated crystallizer Rührkristallisator *m*, Rührkristaller *m*
  a. extractor Rührextraktor *m*, Rührextrakteur *m*
  a.-film evaporator Dünnschichtverdampfer *m* mit rotierenden Wischern
  a. line mixer Rohrleitungsrührwerk *n*
  a. pan dryer Rührtrockner *m*
  a. vessel Rührgefäß *n*
agitation Bewegung *f*, Rühren *n*, Umrühren *n*, Durchrühren *n*, Aufrühren *n*, Schütteln *n*
agitator Rührwerk *n*, Rührapparat *m*, Rührmaschine *f*, Rühreinrichtung *f*, Rührer *m*, Mischwerk *n*; *s.* a. blade
  a. blade Rührerschaufel *f*, Rührschaufel *f*, Rührerstab *m*, Rührstab *m*, Rührerflügel *m*, Rührflügel *m*, Knetschaufel *f*, Mischflügel *m*
  a. kettle Rührkessel *m*
  a. power Rührleistung *f*
  a. shaft Rührerwelle *f*, Rührwelle *f*
aglucone Aglukon *n* *(zuckerfreier Bestandteil eines Glukosids)*
aglycone Aglykon *n*, Genin *n* *(zuckerfreier Bestandteil eines Glykosids)*
agmatine $NH=C(NH_2)NH(CH_2)_4NH_2$ Agmatin *n*, 1-Amino-4-guanidinobutan *n*
agon Agon *n*, Wirkungsgruppe *f*, Wirkgruppe *f*, prosthetische (aktive) Gruppe *f*, Koferment *n*
agricolïte *(Min)* Agricolit *m* *(Wismut(III)-orthosilikat)*
agricultural chemical Agrochemikalie *f*
  a. chemistry Agrikulturchemie *f*, landwirtschaftliche Chemie *f*, Landwirtschaftschemie *f*
  a. control chemical Pflanzenschutzmittel *n*
  a. hydrate *(Landw)* Löschkalk *m*, gelöschter Kalk *m*
  a. limestone Kalkdünger *m* *(aus gemahlenem Kalkstein)*
  a. triad *(Sammelname für die Elemente N, P und K als wichtigste Stoffe bei Düngungen)*
agro-chemical *s.* agricultural chemical
agrochemistry Agrikulturchemie *f*, landwirtschaftliche Chemie *f*, Landwirtschaftschemie *f*
agstone *s.* agricultural limestone
aguilarite *(Min)* Aguilarit *m* *(Silbersulfid)*
Ahrens process Ahrens-Verfahren *n*, Gaserzeugungsverfahren *n* von Ahrens
AIC *(Abk. für)* 1. American Institute of Chemists; 2. Associate of the Institute of Chemistry

AICE *(Abk. für)* 1. American Institute of Chemical Engineers; 2. American Institute of Consulting Engineers
aid Hilfsstoff *m*, Hilfsmittel *n*, Hilfe *f*
aikenite, aikinite *(Min)* Aikenit *m*, Aikinit *m*, Patrinit *m*
aipi *(Lebm)* Süßer Maniok *m*, Manihot dulcis (J.F. Gmel.)Pax
air Luft *f*
  a.-agitated mixer Gebläsemischer *m*
  a.-atomizing burner Freiluftbrenner *m*
  a.-atomizing nozzle Druckluftdüse *f*
  a.-avid aerophil
  a. bag *(Gum)* Heizschlauch *m*
  a.-bag mould *(Gum)* Heizschlauchform *f*
  a.-bag stock *(Gum)* Heizschlauchmischung *f*
  a. bath Luftbad *n*
  a. bell Luftblase *f* *(Fabrikationsfehler)*
  a. blade coater Luftrakelauftragmaschine *f*
  a. blast Wind *m*, Gebläsewind *m*, Gebläseluft *f*
  a.-blast freezer Luftgefrierapparat *m*
  a.-blast freezing Gefrieren *n* in bewegter Luft
  a.-blast main Windleitung *f*, Luftleitung *f*
  a.-blast sprayer *(Landw)* Sprühblaser *m*, Atomiseur *m*
  a.-blast stove Winderhitzer *m*, Hochofenwinderhitzer·*m*
  a.-blast system Luftstrahlgebläse *n*, Gebläse *n*
  a.-blast temperature Windtemperatur *f*
  a.-blast tuyere Winddüse *f*, Blasform *f*, Windform *f*
  a. blister Lufteinschluß *f*
  a.-blow / to mit Druckluft [trocken]blasen
  a. blower Luftgebläse *n*, Windgebläse *n*, Gebläse *n*, Luftventilator *m*
  a. blowing Blasen *n* mit Luft; Ausblasen *n* mit Druckluft
  a.-blowing process *(Glas)* Luftdüsenblasverfahren *n*, Düsenblasverfahren *n*
  a.-blown asphalt Blasbitumen *n*, geblasenes Bitumen *n*
  a. bomb ageing (test) *(Gum)* Luftbombenalterung *f*, Luft-Druckalterung *f*
  a. brush *(Pap)* Luftbürste *f*
  a. brush coater *(Pap)* Luftbürstenstreichmaschine *f*
  a. bubble Luftblase *f*, Luftbläschen *n*
  a.-bulked *(Text)* lufttexturiert, düsentexturiert
  a.-carbon reaction Reaktion *f* des Kohlenstoffs mit Luft, Generatorgasreaktion *f*, Boudouard-Reaktion *f*, Boudouardsche Reaktion *f*
  a. chamber Luftkammer *f*; Windkessel *m*
  a.-circulating oven Heißluftschrank *m*
  a. classification *s.* a. classifying
  a. classifier Windsichter *m*
  a. classifying Windsichten *n*, Windsichtung *f*
  a. cleaner Luftreiniger *m*, Luftfilter *n*
  a. cock Lufthahn *m*
  a.-condition / to klimatisieren, konditionieren
  a.-conditioned klimatisiert, konditioniert; *(Gerb)* luftgetrocknet, lufttrocken

a. **conditioner** Klimagerät n
a. **conditioning** Klimatisierung f, Klimaregelung f, Luftkonditionierung f
a. **conditioning plant** Klima[tisierungs]anlage f, Luftkonditionieranlage f
a. **contaminant** Schmutzstoff m (in der Luft)
a. **converting** (Met) Windfrischen n
a. **conveying** pneumatisches Fördern n
a. **conveyor** pneumatischer Förderer m, Luftförderer m
a.**-cooled** luftgekühlt
a. **cooler** Luftkühler m, Luftkühlgerät n
a. **cooling** Luftkühlung f
a. **cure** Heißluftvulkanisation f, Vulkanisation f in Heißluft, Heißluftheizung f
a. **curing** 1. Lufttrocknung f, Trocknen n an der Luft; 2. (Gum) s. a. cure
a. **damping device** Luftdämpfungseinrichtung f (z.B. an Feinwaagen)
a. **dew point** Lufttaupunkt m
a. **dome** Brüdenhaube f, Dunsthaube f *
a.**-dried** luftgetrocknet, lufttrocken, lutro
a.**-dried moisture** Feuchte f der lufttrockenen Probe, hygroskopische Feuchte f
a.**-dried sheet** (Gum) luftgetrockneter Sheet m
a.**-driven** luftangetrieben, luftgetrieben, mit Luft betrieben
a.**-driven vibrator** Druckluftrüttler m, Preßluftrüttler m, Druckluftvibrator m
a.**-dry / to** an der Luft trocknen
a.**-dry** lufttrocken, luftgetrocknet, lutro
a. **dryer** Lufttrockner m, atmosphärischer Trockner m
a.**-drying** lufttrocknend
a. **drying** Lufttrocknen n, Lufttrocknung f, Freilufttrocknen n, Freilufttrocknung f, natürliches Trocknen n, atmosphärische Trocknung f, Trocknen n (Trocknung f) an der Luft
a. **elutriation** Windsichten n, Windsichtung f
a.**-entrained concrete** belüfteter Beton m, Luftporenbeton m, LP-Beton m, Bläschenbeton m
a.**-entraining additive (admixture, agent)** Luftporenbildner m, luftporenbildender Zusatzstoff m, LP-Bildner m, Belüftungsmittel n, Betonbelüfter m
a.**-entraining cement** Luftporenzement m
a.**-entraining compound** s. a.-entraining additive
a.**-entraining concrete** s. a.-entrained concrete
a. **entrainment** Luftporenbildung f, Porenbildung f
a.**-filled** luftgefüllt
a. **filter** Luftfilter n
a.**-floated** windgesichtet
a.**-floated clay** (Pap) trockenaufbereitetes Kaolin n
a. **floating** Windsichten n, Windsichtung f
a. **flow** Luftstrom m
a.**-flow classification** Luftstromsichtung f, Windsichtung f
a. **furnace** Flammofen m

a. **gap** Luftspalt m
a. **gas** Luftgas n
a.**-harden / to** lufthärten, in Luft härten
a. **hardening** Lufthärten n, Lufthärtung f, Härten n in Luft
a.**-hardening steel** lufthärtender Stahl m, Lufthärtestahl m, Lufthärter m
a. **heater** Lufterhitzer m; Luftvorwärmer m; Winderhitzer m
a. **impermeability** Luftdichtigkeit f, Luftundurchlässigkeit f
a.**-impermeable** luftdicht, luftundurchlässig
a. **improver** Luftverbesserungsmittel n
a. **inlet** Luftzuführung f
a. **intake** Lufteingang m (bei der Untertagevergasung)
a. **jacket** Luftmantel m, Luftbad n
a. **jet crimping** (Text) Luftstromtexturieren n
a. **knife** (Pap) Luftmesser n
a. **knife coater** (Pap) Luftmesserstreichmaschine f; (Plast) Luftbürstenauftragmaschine f, Schlitzdüsenauftragmaschine f
a.**-knife coating** (Pap) Luftmesserstreichverfahren n
a.**-lay dryer** Düsentrockner m
a.**-leak tube** Sied ♪apillare f
a. **lift** Druckluftheber m, Luftheber m, Airlift m, Mammutpumpe f; Airlift m (Kontaktheber beim Airliftkrackverfahren)
a.**-lift method** s. a.-lift process
a.**-lift mixer** Mammutrührwerk n
a.**-lift process** Druckluftförderverfahren n, Airliftverfahren n, Mammutpumpenprinzip n; Airliftkrackverfahren n, Airliftverfahren n
a.**-lift pump** s. a. lift
a.**-lift system** Drucklufthebersystem n, Airliftsystem n
a. **lifting** Druckluftförderung f, Liften n mit Druckluft, Airliftförderung f, Förderung f mittels Druckluft (Airlift)
a. **line** Luftleitung f; Druckluftleitung f, Preßluftleitung f
a. **liquefaction** Luftverflüssigung f
a.**-mail paper** Luftpostpapier n
a. **moistening** Luftbefeuchtung f
a.**-operated** pneumatisch betätigt, druckluftbetätigt, luftbewegt; luftgesteuert
a.**-operated die-casting machine** Druckluftgießmaschine f
a.**-operated jig** luftgesteuerte (luftbewegte, Baumsche) Setzmaschine f
a. **out / to** entlüften, durch Belüftung entfernen (z.B. Entwesungsmittel)
a. **oven ageing** (Gum) Ofenalterung f, Alterung f im Wärmeschrank, Heißluftalterung f [im Geer-Ofen]
a. **oxidation** Oxydation f durch Luftsauerstoff
a. **permeability** Luftdurchlässigkeit f
a.**-permeable** luftdurchlässig
a. **pocket** Lufteinschluß m

a. pollutant Schmutzstoff *m (in der Luft)*
a. pollution Luftverunreinigung *f*, Luftverschmutzung *f*
a. preheater Luftvorwärmer *m*, Luvo *m*
a. pressure Luftdruck *m*, Atmosphärendruck *m*
a. pressure heat test *(Gum)* Luftbombenalterung *f*, Luft-Druckalterung *f*
a. pressure regulator Luftdruckregler *m*
a. pressure test *s. a.* pressure heat test
a.-proof luftdicht, luftundurchlässig
a. purifier Luftreiniger *m*, Luftfilter *n*
a. quenching *(Glas)* Luftabschreckung *f*
a. rate Luftdurchsatz *m*, Luftmassestrom *m*; Luftvolumenstrom *m*
a. receiver Windkessel *m*
a. recirculation Luftrückführung (Luftführung) *f* im Kreislauf, Luftführung *f* im Umluftbetrieb; Luftzirkulation *f*
a. refining *(Met)* Windfrischen *n*
a. refining method *(Met)* Windfrischverfahren *n*
a. regenerator chamber Luftregenerativkammer *f*, Luftkammer *f*
a. register Luftklappe *f*, Luftschieber *m*
a. removal *(Pap)* Entlüftung *f (des Stoffs)*
a. requirements Luftbedarf *m*
a.-sensitive luftempfindlich
a. separating *s. a.* separation 1.
a. separation 1. Windsichten *n*, Windsichtung *f*; 2. Luftzerlegung *f*, Lufttrennung *f*
a. separator Windsichter *m*, Separator *m*
a.-shrinkage *(Ker)* Trockenschwindung *f*
a.-slaked lime durch feuchte Luft gelöschter Kalk *m*
a. slaking Löschen *n* von Branntkalk durch feuchte Luft
a.-slip forming *(Plast)* Streckformen *n* mit pneumatischer Vorstreckung, Air-slip-Verfahren *n*
a. slug Luftkolben *m (einer Mammutpumpe)*
a. stone Filterstein *m*, Gasverteilungsfritte *f*
a. stream Luftstrom *m*
a. supply Luftzufuhr *f*, Luftzuführung *f*, Windzufuhr *f*
a. sweeping Windsichten *n*, Windsichtung *f*
a.-swept classifier Luftstromsichter *m*, Windsichter *m*
a.-swept mill Stromsichtermühle *f*, Windsichtermühle *f*
a. table Lufthed *m (in der Aufbereitungstechnik)*
a.-through circulation dryer Trockner *m* mit Durchbelüftung
a.-tight luftdicht, luftundurchlässig
a. tightness Luftdichtigkeit *f*, Luftundurchlässigkeit *f*
a.-to-ground application *(Landw)* Applikation *f* (Ausbringen *n*) aus der Luft
a. tube *(Gum)* Luftschlauch *m*
a. valve Pneumatikventil *n*, Druckluftventil *n*, Preßluftventil *n*
a. velocity Luftgeschwindigkeit *f*

a. vessel Windkessel *m*
a. void Lufteinschluß *m*, Luftpore *f*, Luftblase *f*
a.-water steep Weichschema *n (abwechselndes Bewässern, Belüften und Trockenstehen des Weichguts beim Mälzen)*
airheater *s.* air heater
airless spraying luftloses (druckluftloses, druckluftfreies) Spritzen *n*, [druckluftloses] Höchstdruckspritzen *n (in der Farbspritztechnik)*
airproof luftdicht, luftundurchlässig
airswept mill Luftstrommühle *f*
airtight luftdicht, luftundurchlässig
Ajax-Northrup [coreless induction] furnace kernloser Induktionsofen *m* nach Ajax-Northrup, Ajax-Northrup-Mittelfrequenzinduktionsofen *m*, Ajax-Northrup-Ofen *m*
Ajax-Northrup high-frequency induction furnace *s.* Ajax-Northrup coreless induction furnace
Ajax-Wyatt furnace Ajax-Wyatt-Ofen *m*, Ajax-Wyatt-Induktionsofen *m*, Ajax-Wyatt-Niederfrequenzinduktionsofen *m*
ajmalicine Ajmalizin *n*, Raubasin *n*, δ-Yohimbintetrahydroserpentin *n*
ajmaline Ajmalin *n*, Neoajmalin *n*, Rauwolfin *n*
ajowan oil Ajowanöl *n (ätherisches Öl von Carum copticum (L.) Benth. et Hook.)*
akonit *(Min)* Akonit *m*, *(veraltet für)* Glaukodot *m*
alabamine, alabamium Alabamin *n*, *(veraltet für)* At Astat *n*
alabandine, alabandite *(Min)* Alabandin *m*, Manganblende *f (Mangan(II)-sulfid)*
alabaster *(Min)* Alabaster *m (Kalziumsulfat-2-Wasser)*
a. board (cardboard) Alabasterkarton *m*
a. glass Alabasterglas *n*, Reisglas *n*
alalite *(Min)* Alalith *m*, *(veraltet für)* Diopsid *m (Kalziummagnesiumdisilikat)*
alanate Alanat *n (Metall-Mischhydrid der allgemeinen Formel M(AlH$_4$)$_n$)*
alane (AlH$_3$)$_x$ Alan *n*, Aluminiumwasserstoff *m*
alanine CH$_3$CH(NH$_2$)·COOH Alanin *n*, 2-Aminopropionsäure *f*
alant camphor Helenin *n*, Alantkampfer *m*, Inulakampfer *m*
a. starch *s.* alantin
alantin Inulin *n (Kohlenhydrat)*
alantolactone Helenin *n*, Alantkampfer *m*
alarm Meldeeinrichtung *f*, Melder *m*, Warnapparat *m*, Warnvorrichtung *f*, Alarmapparat *m*, Alarmvorrichtung *f*
Albany clay Albany-Ton *m (stark verunreinigter Glazialton aus der Nähe von Albany/New York)*
A. slip *(Ker)* Albany-Schlicker *m (aus Albany-Ton hergestellt)*
Albert reversal Albert-Effekt *m*
albertite Albertit *m (natürlich vorkommende Bitumensorte)*
albite *(Min)* Albit *m*, Natronfeldspat *m (Natriumalumotrisilikat)*
albumen 1. Eiweiß *n*, Eiklar *n*, Weißei *n*; *(seltener)* 2. *s.* albumin

a. **finish** *(Gerb)* Eiweißappretur f
albumin Albumin n *(Sammelname für eine Gruppe von Eiweißstoffen)*
a. **foam** Eiweißschaum m
a. **process** *(Foto)* Albuminverfahren n
a. **sol** Albuminsol n
albuminoid Albuminoid n, Proteinoid n, Gerüsteiweiß n
albuminous Eiweiß..., eiweißartig; albuminartig
a. **cloudiness** Eiweißtrübung f
a. **matter** Eiweißstoff m, Eiweiß n
a. **nitrogen** Eiweißstickstoff m
albumose Albumose f *(Eiweißspaltprodukt)*
alc. s. alcohol
alcalosis *(Med)* Alkalose f, Baseose f
alchlor process *(Reinigung von Schmierölen mittels Aluminiumchlorid)*
alcogel *(Koll)* *(Gel mit Alkohol als Dispersionsmittel)*
alcohol Alkohol m, *(i.e.S.)* CH_3CH_2OH Äthylalkohol m, Äthanol n
a. **burner** Spiritusbrenner m
a. **C-12** $CH_3(CH_2)_{11}OH$ Alkohol C 12 m, n-Dodezylalkohol m, 1-Dodekanol n, Laurylalkohol m
a. **coagulation test** Alkoholprobe f *(zur Milchuntersuchung)*
a. **concentration (content)** Alkoholgehalt m, *(i.e.S.)* Äthanolgehalt m
a. **dehydrogenase** Alkoholdehydrogenase f
a. **deterrent** Antialkoholikum n, Alkoholentwöhnungsmittel n
a.-**free** alkoholfrei
a.-**free beverage** alkoholfreies Getränk n
a. **fuel** Alkoholkraftstoff m
a. **gauge** Alkoholmesser m, Alkohol[o]meter n
a. **lamp** Spirituslampe f
a.-**like** alkoholartig
a. **plant** Spiritusbrennerei f, Branntweinbrennerei f, Brennereibetrieb m, Brennerei f
a.-**soluble** alkohollöslich
a. **strength** s. a. content
a. **test** Alkoholprobe f, Alkoholtest m
a. **vapour** Alkoholdampf m
alcoholate Alkoholat n
alcoholdehydrogenase Alkoholdehydrogenase f
alcoholic alkoholisch, spirituos, spirituös, geistig, Alkohol..., Spiritus..., Sprit...
a. **content** s. alcohol content
a. **extract** alkoholischer Auszug (Extrakt) m, Alkoholauszug m, Alkoholextrakt m
a. **fermentation** alkoholische Gärung f
a. **liquor** alkoholisches (geistiges) Getränk n
a. **potash** alkoholische Kalilauge f
a. **solution** alkoholische Lösung f
alcoholizable alkoholisierbar
alcoholization Alkoholisierung f
alcoholize / **to** alkoholisieren
alcoholometer Alkohol[o]meter n, Alkoholmesser m
alcoholometric alkoholometrisch

alcoholometry Alkoholometrie f
alcoholysis Alkoholyse f, *(meist)* Umesterung f
alcohometry s. alcoholometry
alcosol Alkosol n *(kolloidales System mit Alkohol als Dispersionsmittel)*
aldanite *(Min)* Aldanit m
aldehyde Aldehyd m
a. **acid** Aldehydkarbonsäure f, Aldehyd[o]säure f
a.-**amine accelerator** *(Gum)* Aldehydaminbeschleuniger m
a. **C-12 [lauric]** $CH_3(CH_2)_{10}CHO$ Aldehyd C 12 [l] m, Lauraldehyd m, Laurinaldehyd m, Dodezylaldehyd m, Dodekanal n
a. **condensation** Aldehydkondensation f
a. **dehydrogenase** Aldehyddehydrogenase f
a.-**free** aldehydfrei
a. **group** —CHO Aldehydgruppe f, Formylgruppe f
a. **resin** Aldehydharz n
a. **tannage** Aldehydgerbung f
aldehydic acid s. aldehyde acid
a. **group** s. aldehyde group
aldehydine $(CH_3)(C_2H_5)C_5H_3N$ 2-Methyl-5-äthylpyridin n
alder buckthorn (dogwood) bark Faulbaumrinde f *(von Rhamnus frangula L.)*
aldime Aldimin n
aldohexose Aldohexose f *(Hexose mit einer Aldehydgruppe)*
aldoketene RCH=C=O Aldoketen n
aldol Aldol n *(Sammelname für 3-Hydroxyaldehyde)*; *(i.e.S.)* $CH_3CH(OH)CH_2CHO$ Aldol n, Azetaldol n, 3-Hydroxybutanal n, 3-Hydroxybutyraldehyd m, Butanol-(3)-al-(1) n
a. **addition** s. a. condensation
a. **condensation** Aldolkondensation f, Aldoladdition f, Aldolisation f, Aldolisierung f, Aldehyddimerisation f
aldolase Aldolase f, Zymohexase f *(Ferment)*
aldolization s. aldol condensation
aldonic acid Aldonsäure f, Onsäure f *(Sammelname für Aldose-Zuckerderivate der allgemeinen Formel* $HOCH_2(CHOH)_nCOOH$)
aldose Aldose f *(Sammelname für Monosaccharide mit der Gruppe -CHO)*
aldosterone Aldosteron n *(Hormon)*
aldoxime $R \cdot CH=NOH$ Aldoxim n
ale Ale n *(starkes, hopfenreiches Bier)*
Aleppo galls *(Gerb)* Aleppogallen fpl, Türkische (Smyrnaer) Gallen fpl, *(von Quercus infectoria Oliv.)*
A. **pine** Aleppokiefer f, Seekiefer f, Pinus halepensis Mill.
aleuritic acid $CH_2OH(CH_2)_5CHOHCHOH(CH_2)_7COOH$ Aleuritinsäure f, 9,10,16-Trihydroxyhexadekansäure f, 9,10,16-Trihydroxypalmitinsäure f
aleurometer *(Lebm)* *(Gerät zur Messung der Quellfähigkeit des Klebers)*
aleurone grains Aleuronkörner npl, Proteinkörner npl

a. layer Aleuronschicht f
alexandrite *(Min)* Alexandrit m *(Berylliumaluminat)*
Alfa butter Alfa-Butter f *(nach dem Alfa-Verfahren hergestellte Butter)*
A. grass Espartogras n, Esparto m, Halfagras n, Stipa tenacissima L.
A. process Alfa-Butterung f, Alfa-Verfahren n
Alfin catalyst *(Gum)* Alfin-Katalysator m
Alfin catalyzed polymer s. Alfin polymer
Alfin catalyzed polymerization s. Alfin polymerization
Alfin polymer Alfin-Kautschuk m
Alfin polymerization *(Gum)* Alfin-Polymerisation f
algal Algen...
algarob[ill]a *(Gerb)* Algarobilla m *(Früchte von Caesalpinia brevifolia (Baill.)Clos.)*
algaroth [powder] *(Pharm)* Algarotpulver n *(Antimon(III)-pentoxydichlorid)*
algic (algin) acid s. alginic acid
alginate Alginat n *(Salz oder Ester der Alginsäure)*
a. cream *(Pharm)* Alginatkrem f
a. fibre *(Text)* Alginatfaser f; Alginatfaserstoff m, AL
a. thread *(Text)* Alginatfaden m
a. yarn Alginatseide f, ALS
alginic acid Alginsäure f, Algensäure f
a. ester Alginsäureester m
alginite Alginit m *(Kohleart)*
alicyclic alizyklisch, zykloaliphatisch
a. compounds Alizyklen pl, alizyklische (zykloaliphatische) Verbindungen fpl
align / **to** ausrichten; sich ausrichten, sich *(regelmäßig)* anordnen
alignment Ausrichtung f, Orientierung f, *(regelmäßige, gerichtete)* Anordnung f, Ordnung f *(von Kettenmolekülen)*
alimentary canal (tract) Verdauungstrakt m, Magen-Darm-Trakt m, Verdauungskanal m
aliphatic aliphatisch
a. aldehyde aliphatischer Aldehyd m, Alkanal n
a. bridge aliphatische Brücke f
a. class Fettreihe f
a. compounds aliphatische (azyklische) Verbindungen fpl, Verbindungen fpl der Fettreihe, Aliphaten pl
a.-cyclic zykloaliphatisch, alizyklisch
a. hydrocarbon aliphatischer (azyklischer, kettenförmiger, offenkettiger) Kohlenwasserstoff m
aliphatics Aliphaten pl, aliphatische (azyklische) Verbindungen fpl, Verbindungen fpl der Fettreihe
aliquot aliquot
a. part aliquoter Teil m
alite Alit m *(Kristallart im Portlandzementklinker)*
alitizing Alitieren n *(Metallveredelungsverfahren)*
alizarin Alizarin n, 1,2-Dihydroxyanthrachinon n
a.-alcohol test Alizarolprobe f *(zur Milchuntersuchung)*
a. blue Alizarinblau n
a. bordeaux Alizarinbordeaux n, Chinalizarin n, 1,2,5,8-Tetrahydroxyanthrachinon n

a. cyanine Alizarinzyanin n
a. cyanine green Alizarinzyaningrün n, Alizarinbrillantgrün n
a. dye Alizarinfarbstoff m
a. fluorine blue Alizarinfluorblau n, 1,2-Dihydroxyanthrachinon-3-ylmethylamin-N,N-diessigsäure f
a. indigo blue Alizarinindig[o]blau n
a. orange Alizarinorange n, 3-Nitroalizarin n, 1,2-Dihydroxy-3-nitroanthrachinon n
a. red Alizarinrot n
a. saphirol B Alizarinsaphirol B n, 4,8-Diamino-1,5-dihydroxy-9,10-anthrachinon-2,6-dinatriumsulfonat n
a. test Alizarinprobe f *(zur Milchuntersuchung)*
a. yellow Alizaringelb n
a. yellow A Alizaringelb A n, 2,3,4-Trihydroxybenzophenon n
a. yellow R Alizaringelb R n *(Natriumsalz der p-Nitrobenzolazosalizylsäure)*
alizarine s. alizarin
alizarol test Alizarolprobe f *(zur Milchuntersuchung)*
alk. s. alkali
alkalescent schwach alkalisch
alkali alkalisch
alkali Alkali n, Laugensalz n
a. alkoxide Alkalialkoholat n
a.-binding agent Alkalibindemittel n
a. blue Alkaliblau n, Nicholson-Blau n
a. builder Enthärtungsmittel n *(Seifenzusatz)*
a. carbonate Alkali[metall]karbonat n
a. cellulose Alkalizellulose f, *(meist)* $C_6H_{10}O_5 \cdot NaOH$ Natronzellulose f
a. chelate Alkalichelat n
a. cleavage Alkalispaltung f
a. content Alkaligehalt m, Laugengehalt m
a. cyanide Alkalizyanid n
a. extraction stage s. a. stage
a.-fast alkalifest, alkalibeständig, beständig gegen Alkalien; *(Farb)* alkaliecht
a. feldspar Alkalifeldspat m
a. fusion Alkalischmelze f, alkalische Verschmelzung f, Schmelzen n mit Alkali
a. glass Alkaliglas n, alkalihaltiges Glas n
a. granite Alkaligranit m
a. halide Alkalimetallhalogenid n, Alkalihalogenid n
a. hydroxide Alkalimetallhydroxid n, Alkalihydroxid n
a.-insoluble alkaliunlöslich
a. lignin *(Pap)* Alkalilignin n
a.-like alkaliähnlich
a. make-up *(Pap)* zur Deckung der Alkaliverluste zugesetzte Chemikalie f, Zusatzchemikalie f, Alkalizusatz m
a. melt Alkalischmelze f
a. metal Alkalimetall n
a. metal acetylide Alkaliazetylid n, Alkaliazetylenid n, Alkalikarbid n

a.-metal catalyzed polymerization s. a. metal polymerization
a. metal polymer (Gum) Alkalimetallpolymerisat n, Alkalimetallpolymer[es] n
a. metal polymerization (Gum) Alkalimetallpolymerisation f
a. metal salt Alkalimetallsalz n, Alkalisalz n
a. milling (Text) alkalisches Walken n, alkalische Walke f
a. oxide Alkalimetalloxid n
a. process [of regeneration] (Gum) Alkaliverfahren n
a. proof alkalibeständig, alkalifest
a. pulp alkalisch gekochter (erkochter, aufgeschlossener) Zellstoff m
a. pulp mill Zellstoffabrik f für alkalischen Aufschluß, (i.e.S.) Sulfatzellstoffabrik f
a. pulping process s. alkaline process
a. ratio s. a.-to-wood ratio
a. reclaim (Gum) Alkaliregenerat n
a. reclaiming process s. a. process [of regeneration]
a. recovery (Pap) Alkalirückgewinnung f
a. refining (Pap) Veredelung f, Alkalisierung f
a. refining liquor (Pap) alkalische Zellstoffveredelungslauge f, Alkali-Veredelungslauge f
a. reserve (Med) Alkalireserve f
a. resistance Alkalibeständigkeit f, Laugenbeständigkeit f
a.-resistant alkalibeständig, alkalifest, alkaliunempfindlich, laugenbeständig
a. rock Alkaligestein n, Gestein n der Alkalireihe
a.-sensitive alkaliempfindlich
a. sensitivity Alkaliempfindlichkeit f
a. silicate Alkalisilikat n
a. soil Alkaliboden m
a.-soluble alkalilöslich
a. solution alkalische Lösung f
a. stage (Pap) Alkalibehandlung f, Alkaliextraktion f, alkalische Wäsche f, Alkaliwäsche f (Bleichstufe)
a.-to-wood ratio (Pap) Alkaliverhältnis n
a.-unstable alkaliunbeständig, nicht alkalibeständig (alkalifest)
a. wash Alkaliwäsche f, Lauge[n]wäsche f, Laugenbehandlung f, Laugen n, Laugung f
alkalic alkalisch, Alkali..., basisch
alkalies Alkalien npl
alkalifiable alkalisierbar
alkalify / to alkalisieren, alkalisch machen; alkalisch werden
alkaligenous alkalibildend
alkalimeter Alkalimeter n (zur Bestimmung des Laugengehalts einer Flüssigkeit); Kaliapparat m (zur Kohlensäurebestimmung)
alkalimetric alkalimetrisch
alkalimetry Alkalimetrie f
alkaline alkalisch, basisch
  to be a. to alkalisch reagieren gegen
  to make a. alkalisch machen, alkalisieren, (Farb auch) alkalisch stellen

a. bath alkalisches Bad n
a. cook (Pap) Alkalikochung f, alkalische Kochung f
a.-cooked (Pap) alkalisch gekocht (erkocht, aufgeschlossen)
a.-cooked pulp alkalisch gekochter (erkochter, aufgeschlossener) Zellstoff m
a. cooking liquor (Pap) alkalische Kochflüssigkeit f, Kochlauge f
a. detergent alkalisches Waschmittel n
a. earth Erdalkali n
a.-earth chelate Erdalkalichelat n
a.-earth metal Erdalkalimetall n, Metall n der Erdalkaligruppe
a.-earth phosphate Erdalkaliphosphat n
a.-earth phosphor Erdalkaliphosphor m
a. earths Erdalkalien npl
a. extraction tower (Pap) Alkalisierungsturm m, Alkaliturm m
a. fusion Alkalischmelze f, alkalische Verschmelzung f, Schmelzen n mit Alkali
a. glaze (Ker) Alkaliglasur f
a. milling (Text) alkalisches Walken n, alkalische Walke f
a. process (Pap) alkalisches Aufschlußverfahren n
a. pulping (Pap) alkalischer Aufschluß m
a. purification (Pap) Alkalisierung f, Veredelung f
a. reaction alkalische (basische) Reaktion f
a. rubber (Butadien-Vinylpyridin-Kautschuk)
a. saponification alkalische Verseifung f
a. solution alkalische Lösung f
a. type of reclaim (Gum) Alkaliregenerat n
a.-washing stage (Pap) Alkalibehandlung f, Alkaliextraktion f, alkalische Wäsche f, Alkaliwäsche f (Bleichstufe)
alkalinity Alkalität f, Basizität f, (für gelöste Hydrogenkarbonate auch) Alkalinität f, (Bodenchemie auch) Basengehalt m
alkalinization Alkalisierung f
alkalinize / to alkalisieren, alkalisch machen, (Farb auch) alkalisch stellen
alkalis s. alkalies
alkalizable alkalisierbar
alkalization s. alkalinization
alkalize / to s. to alkalinize
alkaloid Alkaloid n
a. base Alkaloidbase f
a.-like alkaloidartig
a. poisoning Alkaloidvergiftung f
alkaloidal alkaloidisch, Alkaloid...
a. solution Alkaloidlösung f
alkamine Alkamin n, Alkanolamin n, Aminoalkohol m
alkanal Alkanal n, aliphatischer Aldehyd m
alkanation Alkylation f, Alkylieren n, Alkylierung f
alkane Alkan n, Paraffinkohlenwasserstoff m, paraffinischer (gesättigter) Kohlenwasserstoff m, Paraffin n, Grenzkohlenwasserstoff m
a. family Alkanreihe f, Paraffinreihe f, Grenzkohlenwasserstoffreihe f

alkanet Alkanna f (Farbstoffdroge aus Alkanna tinctoria (L.)Tausch.)

alkanization (Petrolchemie) (Überführung von Isobuten in Trimethylbuten und Isooktan)

alkanna s. alkanet

alkannin Alkannin n, Alkannarot n, Alkannafarbstoff m

alkanolamine Alkanolamin n, Alkamin n, Aminoalkohol m

alkanone Alkanon n, aliphatisches Keton n

alkazid plant Alkazidanlage f, Sulfosolvananlage f (zur Reinigung von Gasen)

alkene Alken n, Alkylen n, Äthylenkohlenwasserstoff m, Äthylen n (i.w.S.), Olefin n, Monoolefin n

  a. family Alkenreihe f, Olefinreihe f

alkenoic acid Alkensäure f (Sammelname für einfach ungesättigte Fettsäuren)

alkenylbenzene Alkenylbenzol n

alkimine chelate Alkiminchelat n

alkine s. alkyne

alkiodide Alkyljodid n

alkoxide Alkoholat n

alkoxy group s. alkoxyl

alkoxyl [radical] $C_nH_{2n+1}O-$ Alkoxyl n, Alkoxy[l]gruppe f

alkyd [resin] Alkyd[harz] n

  a. varnish Alkydharzlack m

alkyl Alkyl n

  a. bridge Alkylbrücke f

  a. compound Alkylverbindung f

  a. cyanide Alkylzyanid n

  a. derivative Alkylderivat n

  a. dihalide Alkyldihalogenid n, Dihalogenalkan n

  a. group Alkylgruppe f, Alkylrest m

  a. halide Alkylhalogenid n, Alkanhalogenid n, Halogenalkan n

  a. monohalide Alkylmonohalogenid n, Monohalogenalkan n

  a. peroxide Alkylperoxid n

  a. phenol novolak Alkylphenolnovolak m

  a. radical Alkylgruppe f, Alkylrest m, Alkylradikal n; [freies] Alkylradikal n

  a. shift 1,2-Alkylkarbonium-Umlagerung f

  a. silicone Alkylsilikon n

  a. sulphide Alkylsulfid n, Thioäther m

  a. sulphonate Alkylsulfonat n

  a. sulphonic acid Alkylsulfonsäure f

alkylamine Alkylamin n

alkylaryl sulphonate Alkylarylsulfonat n, Aralkylsulfonat n (Waschrohstoff)

alkylate / to alkylieren

alkylate Alkylat n

alkylating agent Alkylierungsmittel n, alkylierendes Mittel n, alkylierende Verbindung f, Alkylans n

alkylation Alkylation f, Alkylieren n, Alkylierung f

  a. process Alkylierungsverfahren n

alkylbenzene Alkylbenzol n

alkylene Alken n, Alkylen n, Olefin n, Monoolefin n, Äthylenkohlenwasserstoff m, Äthylen n (i.w.S.)

alkyleneimine Alkylenimin n

alkylolamine Alkanolamin n, Alkamin n, Aminoalkohol m

alkylphenol resin Alkylphenolharz n

alkylsulphuric acid Alkylschwefelsäure f

alkyne Alkin n, Azetylen n (i.w.S.), Azetylenkohlenwasserstoff m, Äthinkohlenwasserstoff m

alkynoic acid Alkinsäure f

alkynylbenzene Alkinylbenzol n

all-cis all-cis-..., ganz-cis-..., nur-cis-...

all-cis isomer all-cis-Isomer[es] n

all-electric furnace (Glas) vollelektrisch beheizter Schmelzofen m

all-purpose adhesive Alleskleber m, Vielzweckkleber m

all-purpose cleaner Allesreiniger m, Mehrzweckreinigungsmittel n

all-purpose washing agent Universalwaschmittel n

all-rag furnish (Pap) Hadernhalbstoff m, Hadernstoff m, Lumpenhalbstoff m, Hadernhalbzeug n

all-rag paper Hadernhalbstoffpapier n, Hadernpapier n, Lumpenpapier n

all-round efficiency umfassende Wirkungsbreite f (z.B. von Pflanzenschutzmitteln)

all-round fastness Gesamtechtheit f, Universalechtheit f

all-sliming process All-sliming-Verfahren n (Goldgewinnung)

all-trans all-trans-..., ganz-trans-..., nur-trans-...

all-trans isomer all-trans-Isomer[es] n

allanite (Min) Allanit m, Orthit m

allantoic acid Allantoinsäure f

allantoin Allantoin n, 5-Ureido-hydantoin n

allelopathic allelopathisch

allelopathic Allelopathikum n (auf benachbarte Pflanzen wirkender Inhaltsstoff höherer Pflanzen)

allelopathy Allelopathie f (gegenseitige Beeinflussung von Pflanzen durch Inhaltsstoffe)

allelotropic allelotrop

allelotropism Allelotropie f

allemontite Allemontit m (Antimonmineral)

Allen cone classifier Allen-Konus m, Allen-Kegel m (Klassierapparat)

Allen-Moore cell Allen-Moore-Zelle f

allene Allen n, (i.e.S.) $CH_2=C=CH_2$ Allen n, Propadien n

  a. isomerism Allen-Isomerie f

allergen (Med) Allergen n

alliaceous lauchartig

allied verwandt (chemische Stoffe)

  a. compound verwandte Verbindung f

Allihn condenser Allihn-Kühler m, Allihnscher Kühler m, Kugelkühler m nach Allihn

allocatalysis Allokatalyse f (Gegenteil von Autokatalyse)

allochthonous (Geol) allochthon, bodenfremd, ortsfremd

allocrotonic acid $CH_3CH=CHCOOH$ Allokroton-

säure f, β-Krotonsäure f, Isokrotonsäure f, cis-Buten-(2)-säure-(1) f, cis-3-Methylakrylsäure f
allogonite (Min) Allogonit m, (veraltet für) Herderit m
alloisoleucine CH₃CH₂CH(CH₃)CH(NH₂)COOH Alloisoleuzin n
allopalladium (Min) Allopalladium n
allophane (Min) Allophan m
allose Allose f (eine Aldohexose)
allothigenic mineral allothigenes (nicht am Fundort entstandenes) Mineral n
allotriomorphic (Krist) allotriomorph, xenomorph, fremdgestaltig
allotrope allotrope Modifikation f
allotropic allotrop
  a. form s. allotrope
allotropism, allotropy Allotropie f (Auftreten eines chemischen Elements in mehreren Modifikationen)
allowed band erlaubtes Band n, erlaubter (zugelassener) Energiebereich m
  a. transition (phys Ch) erlaubter Übergang m
alloy /to legieren, zusammenschmelzen
alloy Legierung f
  a. case-hardening steel legierter Einsatzstahl m
  a. constituent (element) s. alloying element
  a. steel legierter Stahl m
alloying Legieren n
  a. agent (constituent) s. a. element
  a. element Legierungsbestandteil m, Legierungskomponente f, Legierungselement n, Legierungszusatz m
  a. ingredient s. a. element
allspice Piment n, Nelkenpfeffer m, Jamaikapfeffer m, Neugewürz n, Allgewürz n (von Pimenta dioica (L.)Merr.)
allyl allylständig
allyl Allyl n
  a. alcohol CH₂=CHCH₂OH Allylalkohol m. Propenol-(3) n, 2-Propen-1-ol n
  a. bromide CH₂=CHCH₂Br Allylbromid n, 3-Brompropylen n, 3-Brompropen-(1) n
  a. chloride CH₂=CHCH₂Cl Allylchlorid n, 3-Chlorpropylen n, 3-Chlorpropen-(1) n
  a. diglycol carbonate Allyldiglykolkarbonat n, Polydiallylglykolkarbonat n
  a. free radical [freies] Allylradikal n
  a. group Allylgruppe f, Allylrest m
  a. migration Allylwanderung f
  a. mustard oil Allylsenföl n
  a. plastic Allylharz-Plast m, Allylharz-Kunststoff m, Allylester-Kunststoff m
  a. radical Allylgruppe f, Allylrest m, Allylradikal n; [freies] Allylradikal n
  a. rearrangement Allylumlagerung f
  a. resin Allylharz n
allylation Allylierung f
allylene CH₃C≡CH Allylen n, Methyläthin n, Methylazetylen n, Propin n

allylic Allyl...
  a. alcohol s. allyl alcohol
  a. bromination Allylbromierung f
  a. resin Allylharz n
almandine (Min) Almandin m, Eisentongranat m, gemeiner Granat m (Aluminiumeisen(II)-orthosilikat)
  a. spinel (Min) Almandinspinell m
almandite s. almandine
Almen-Nylander test Almén-Nylander-Probe f (Zuckernachweis)
almendro Katappenbaum m, Katappa-Terminalie f, Indischer Mandelbaum m, Terminalia catappa L.
almond Mandelbaum m, Prunus amygdalus Batsch; Mandel f
  a. milk Mandelmilch f
  a. oil Mandelöl m
aloin Aloin n
  a. test Aloinprobe f (Blutnachweis)
alpha brass α-Messing n
  a. cellulose Alpha-Zellulose f, α-Zellulose f
  a. cellulose content α-Zellulosegehalt m, α-Gehalt m
  a. decay (disintegration) Alpha-Zerfall m
  a. disintegration energy Alpha-Zerfallsenergie f
  a. fraction α-Fraktion f, α-Bestandteile mpl (bei der Pyridinextraktion der Steinkohle)
  a.-halogenated α-halogeniert, in α-Stellung halogeniert
  a.-halogenation α-Halogenierung f
  a.-hydroxy acid α-Hydroxysäure f
  a. particle Alphateilchen n, α-Teilchen n
  a.-particle spectrum Alpha-Spektrum n, α-Spektrum n
  a. pulp Edelzellstoff m
  a. radiation Alphastrahlung f
  a. ray Alphastrahl m, α-Strahl m
alphina particle (Kch) Alphina-Teilchen n
alphyl Alphyl n, (veraltet für) Alkyl n
Alpine Kolloplex mill Alpine-Kolloplex-Mühle f (siebllose Stiftmühle)
alstonite (Min) Alstonit m
altaite (Min) Altait m (Bleitellurid)
alterant Alterans n (den Stoffwechsel umstimmendes Medikament)
alteration Änderung f, Wechsel m
  a. in the colour Farbänderung f, Farbumschlag m, Farbwechsel m
  a. of shade Farbtonänderung f, Farbtonumschlag m
alterative s. alterant
alternate abwechselnd
  a. immersion test Wechseltauchversuch m, Wechseltauchtest m, Wechseltauchprüfung f (Korrosionsprüfung)
alternating bending test (Plast) Wechselbiegeprüfung f
  a. copolymer alternierendes Kopolymeres n
  a. copolymerization alternierende Kopolymerisation (Mischpolymerisation) f

a. **current** Wechselstrom *m*
a.**-current arc** Wechselstromlichtbogen *m*, Wechselstrombogen *m*
a.**-current polarography** Wechselstrompolarografie *f*
a. **stress** Wechselbeanspruchung *f*
alternative parameter Alternativparameter *m*
altrose Altrose *f (eine Aldohexose)*
alum 1. $M^IM^{III}(SO_4)_2 \cdot 12H_2O$ Alaun *m*; 2. [konzentrierter] Alaun *m*, technisches (handelsübliches) Aluminiumsulfat *n*
a. **cake** *(Pap) (durch Einwirkung von Schwefelsäure auf Ton oder Bauxit erzeugte schwefelsaure Tonerde)*
a.**-dressed** *(Gerb)* alaungar, weißgar
a. **flour** s. a. **meal**
a. **leather** Alaunleder *n*
a. **liquor** *(Pap)* Alaunlösung *f (Aluminiumsulfatlösung)*
a. **meal** Alaunmehl *n (feinkristalliner Kalialaun)*
a. **schist (shale)** Alaunschiefer *m*, Vitriolschiefer *m*
a. **slate** s. a. **shale**
a. **stone** *(Min)* Alaunstein *m*, Alunit *m (Kaliumaluminiumhexahydroxiddisulfat)*
a. **tannage** Alaungerbung *f*, Weißgerbung *f*
a.**-tanned** *(Gerb)* alaungar, weißgar
alumed s. **alum-tanned**
alumetize / to [ver]aluminieren
alumina Al_2O_3 Tonerde *f*, Aluminiumoxid *n*
a. **boat** Tonerdeboot *n*, Tonerdeschiffchen *n*, Substanzschiffchen *n* aus Tonerde *(beim Zonenschmelzen)*
a. **gel** Aluminogel *n*, Tonerdegel *n (Aluminiumoxidhydrat)*
a. **porcelain** Tonerdeporzellan *n*
a. **trihydrate** s. **aluminium trihydroxide**
a. **whiteware** *(Ker)* Tonerde-Weißware *f*
aluminate Aluminat *n*
a. **liquor (solution)** Aluminatlauge *f*, [Natronlauge-]Aluminatlösung *f (Bayer-Verfahren)*
aluminic Aluminium..., Alu...
aluminide Aluminid *n*
aluminiferous 1. aluminiumhaltig; 2. tonerdehaltig, aluminiumoxidhaltig; 3. alaunhaltig
aluminite *(Min)* Aluminit *m (Aluminiumtetrahydroxidsulfat-7-Wasser)*
aluminium Al Aluminium *n*
a. **acetate** $Al(CH_3COO)_3$ Aluminiumazetat *n*
a. **acetylacetonate** $Al[CH(CO \cdot CH_3)_2]_3$ Aluminiumazetylazetonat *n*
a. **alkyl** AlR_3 Aluminiumalkyl *n*, Aluminiumtrialkyl *n*
a. **alloy** Aluminiumlegierung *f*
a.**-ammonium chloride** $NH_4Cl \cdot AlCl_3$ Ammoniumaluminiumchlorid *n*
a.**-ammonium sulphate** $NH_4Al(SO_4)_2$ Ammoniumaluminiumsulfat *n*
a. **arsenate** $AlAsO_4$ Aluminiumarsenat (V) *n*
a.**-base** auf Aluminiumbasis (Aluminiumgrundlage)

a. **bromide** $AlBr_3$ Aluminiumbromid *n*
a. **bronze** Aluminiumbronze *f*
a. **bronze powder** Aluminiumbronzepulver *n*
a.**-caesium sulphate** $CsAl(SO_4)_2$ Zäsiumaluminiumsulfat *n*
a. **carbide** Al_4C_3 Aluminiumkarbid *n*
a. **chlorate** $Al(ClO_3)_3$ Aluminiumchlorat *n*
a. **chloride** $AlCl_3$ Aluminiumchlorid *n*
a. **diformate** $Al(OH)(HCOO)_2$ Aluminiumhydroxiddiformiat *n*
a. **fluoride** AlF_3 Aluminiumfluorid *n*
a. **fluoride hydrate** Aluminiumfluorid-Hydrat *n*
a. **fluorosilicate** $Al_2[SiF_6]_3$ Aluminiumhexafluorosilikat *n*
a. **foil** Aluminiumfolie *f*
a. **foil backing paper** Aluminiumfolien-Kaschierpapier *n*
a. **hydrate (hydroxide)** s. a. **trihydroxide**
a. **hydroxyacetate** $Al(OH)(CH_3COO)_2$ Aluminiumhydroxidazetat *n*
a. **iodide** AlJ_3 Aluminiumjodid *n*
a. **lake** Aluminiumlack *m*
a.**-lined** mit Aluminium umkleidet
a. **monohalide** Aluminiummonohalogenid *n*, Aluminium(I)-halogenid *n*
a. **naphthenate** Aluminiumnaphthenat *n*
a. **nitrate** $Al(NO_3)_3$ Aluminiumnitrat *n*
a. **nitride** AlN Aluminiumnitrid *n*
a. **ore** Aluminiumerz *n*
a. **oxide** Al_2O_3 Aluminiumoxid *n*
a. **paper** Aluminiumpapier *n*, Silberpapier *n*
a.**-potassium silicate** $K[AlSi_3O_8]$ Kaliumalumotrisilikat *n*, Kaliumalumosilikat *n*
a.**-potassium sulphate** $KAl(SO_4)_2$ Kaliumaluminiumsulfat *n*
a. **powder** Aluminiumpulver *n*, Aluminiumstaub *m*
a. **silicofluoride** s. a. **fluorosilicate**
a. **soap** Aluminiumseife *f*
a. **sodium chloride** $AlCl_3 \cdot NaCl$ Natriumaluminiumchlorid *n*
a. **sodium fluoride** $AlF_3 \cdot 3NaF$ Natriumaluminiumfluorid *n*, Natriumhexafluoroaluminat *n*
a.**-sodium silicate** $Na[AlSi_3O_8]$ Natriumalumotrisilikat *n*, Natriumalumosilikat *n*
a.**-sodium sulphate** $NaAl(SO_4)_2$ Natriumaluminiumsulfat *n*
a. **stearate** $Al(C_{17}H_{35}COO)_3$ Aluminiumstearat *n*
a. **sulphate** $Al_2(SO_4)_3$ Aluminiumsulfat *n*
a. **sulphide** Al_2S_3 Aluminiumsulfid *n*
a. **sulphocyanate (thiocyanate)** $Al(SCN)_3$ Aluminiumthiozyanat *n*, Aluminiumrhodanid *n*
a. **trialkyl** AlR_3 Aluminiumtrialkyl *n*, Aluminiumalkyl *n*
a. **triethyl** $Al(C_2H_5)_3$ Aluminiumtriäthyl *n*, Triäthylaluminium *n*, Äthylaluminium *n*
a. **triformate** $Al(HCOO)_3$ Aluminiumtriformiat *n*
a. **trihalide** Aluminium(III)-halogenid *n*
a. **trihydrate** s. a. **trihydroxide**
a. **trihydroxide** $Al(OH)_3$ Aluminiumtrihydroxid *n*,

Aluminiumorthohydroxid *n*, Aluminiumhydroxid *n*
a. trimethyl $Al(CH_3)_3$ Aluminiumtrimethyl *n*, Trimethylaluminium *n*
a. tristearate *s*. a. stearate
a. works Aluminiumhütte *f*
aluminize / to [ver]aluminieren
aluminizing Aluminieren *n*, Veraluminieren *n*, Veraluminierung *f*
aluminosilicate Alumosilikat *n*
a. glass Tonerdesilikatglas *n*, Aluminiumsilikatglas *n*
aluminothermic aluminothermisch
a. process aluminothermisches Verfahren *n*, Thermitverfahren *n*, Thermitprozeß *m*
aluminothermics, aluminothermy Aluminothermie *f*, Thermitverfahren *n*, Goldschmidt-Verfahren *n*
aluminous aluminiumhaltig, Aluminium...; alaunhaltig, Alaun...
a. cake *s*. alum cake
a. cement Tonerdeschmelzzement *m*, Tonerdezement *m*
aluminum *(Am) s*. aluminium
alumogel *(Min)* Alumogel *n(n)*, Kliachit *m*, Sporogelit *m (Gel des Aluminiumhydroxids)*
alumstone, alunite *(Min)* Alunit *m*, Alaunstein *m (Kaliumaluminiumhexahydroxiddisulfat)*
alunitization Alunitisierung *f*, Alunitisation *f*
alunogene *(Min)* Alunogen *m (Aluminiumsulfat-18-Wasser)*
am., Am *s*. amyl
Amagat law Amagatsches Gesetz *n*
Amagat unit Amagat-Einheit *f*, Amagat *n (Molvolumen von Gasen bei 0°C und 1 at)*
amalgam / to *s*. to amalgamate
amalgam Amalgam *n (Quecksilberlegierung); (Min)* Amalgam *m*
a. electrode Amalgamelektrode *f*
amalgamate / to amalgamieren
amalgamation Amalgamation *f*, Amalgamieren *n*, Amalgamierung *f*
a. pan Amalgamierpfanne *f*, Amalgamationspfanne *f*
a. process Amalgamationsverfahren *n*, Amalgamierverfahren *n*
amanitines Amanitine *npl (Sammelname für Giftstoffe des grünen Knollenblätterpilzes)*
amaranth Amarant[h] *n (dunkelroter Farbstoff)*
amarogentin Amarogentin *n (Bitterstoff aus Enzian und Swertia)*
amazonite, amazonstone *(Min)* Amazonit *m*, Amazonenstein *m*, Mikroklin *m*
amber bernsteingelb
amber Bernstein *m*, Sukzinit *m*; [natürliche, graue] Ambra *f*, Naturambra *f*
a. acid $HOOCCH_2CH_2COOH$ Bernsteinsäure *f*, Sukzinsäure *f*, Äthandikarbonsäure-(1,2) *f*, Butandisäure *f*
a. glass kohlegelbes (goldgelbes, braunes) Glas *n*, Kohlegelbglas *n*, Braunglas *n*
a. mica Amberglimmer *m*

a. oil Bernsteinöl *n*
a. varnish Bernsteinlack *m*
ambergris [natürliche, graue] Ambra *f*, Naturambra *f*
ambient umgebend, einschließend, ambient, Umgebungs...
a. humidity Umgebungsfeuchtigkeit *f*
a. pressure Umgebungsdruck *m*
a. temperature Umgebungstemperatur *f*
amblygonite Amblygonit *m (Lithiummineral)*
ambrette seed oil Moschuskörneröl *n*
amebicidal amöbizid
amebicide Amöbizid *n*
ameliorate / to meliorieren
amelioration Melioration *f*, Bodenverbesserung *f*
a. injector Meliorationslanze *f (für flüssige Düngemittel)*
amendment Melioration *f*, Bodenverbesserung *f*; Bodenverbesserungsmittel *n*
American ashes nordamerikanische Pottasche *f (Kaliumkarbonat)*
A. chestnut 1. Amerikanische Kastanie *f*, Castanea dentata *(Marsh.)Borkh*.; 2. Frucht von 1.
A. elm Amerikanische Ulme *f*, Weißulme *f*, Ulmus americana L.
A. mastic amerikanischer Mastix *m*, Molleharz *n (Pflanzengummi von Schinus molle L.)*
A. sweet gum Amerikanischer Amberbaum *m*, Liquidambar styraciflua L.
A. vermilion (vermillion) *(Farb)* Chromrot *n (basisches Bleichromat); (Farb) (Bleimolybdat als Pigmentfarbe)*
americium Am Amerizium *n*
amethopterin Amethopterin *n*, 4-Amino-N^{10}-methylpteroylglutaminsäure *f*
amethyst amethyst[farben]
amethyst *(Min)* Amethyst *m (Silizium(IV)-oxid)*
amianthus *(Min)* Amiant *m*
amicron Amikron *n (mikroskopisch unsichtbares Schwebeteilchen)*
amidase Amidase *f (C-N-Bindung spaltendes Ferment)*
amidate / to amidieren
amidation Amidierung *f*
amide Amid *n*; $M^I NH_2$ Metallamid *n*, Amid *n*
a. linkage Amidbindung *f*
a. nitrogen Amidstickstoff *m*
a. powder *s*. ammonium powder
amidine Amidin *n*
a. group Amidingruppe *f*, Guanylgruppe *f*
amido group $—CONH_2$ Amidogruppe *f*
a. mercuric chloride $[Hg(NH_2)]_n Cl_n$ Amidoquecksilber(II)-chlorid *n*, Quecksilber(II)-amidochlorid *n*, unschmelzbares weißes Präzipitat *n*
amidogen *s*. amino group
amidol $C_6H_3OH(NH_2)_2$ Amidol *n*, 2,4-Diaminophenol *n*; $C_6H_3(OH)(NH_2)_2 \cdot 2HCl$ Amidol *n*, 2,4-Diaminophenoldihydrochlorid *n*
a. developer *(Foto)* Amidol-Entwickler *m*
amidopyrine Amidopyridin *n*, 1-Phenyl-2,3-dimethyl-4-dimethylamino-5-pyrazolon *n*

amidosulphonic acid s. p-aminobenzenesulphonic acid

aminase Aminase f

amination Aminierung f, Aminolyse f

amine Amin n

a.-**like** aminartig

a. **oxide** Aminoxid n, N-Oxid n

a. **salt** Aminsalz n

a. **value** Aminzahl f

amino accelerator (Gum) Aminbeschleuniger m

a. **acid** Amino[karbon]säure f

a.-**acid anion** Aminosäureanion n

a.-**acid chelate** Aminosäurechelat n

a.-**acid oxidase** Aminosäureoxydase f

a.-**acid residue** Aminosäurerest m

a.-**acid sequence** Aminosäuresequenz f, Aminosäure[reihen]folge f, Reihenfolge f [der Verknüpfung] der Aminosäuren

a.-**acid unit** Aminosäureneinheit f

a. **alcohol** Aminoalkohol m, Alkanolamin n, Alkamin n

a. **compound** Aminoverbindung f

a. **dicarboxylic acid** Aminodikarbonsäure f

a. **G acid** $C_{10}H_5(NH_2)(SO_3H)_2$ Amino-G-Säure f, 2-Naphthylamin-6,8-disulfonsäure f

a. **G salt** Amino-G-Salz n

a. **group** —NH_2 Aminogruppe f

a. **methanamidine** $NH=C(NH_2)_2$ Aminomethanamidin n, Iminoharnstoff m, Guanidin n

a. **moiety** Amino[an]teil m, Aminokomponente f

a. **oxidase** Aminooxydase f

a. **polycarboxylic acid** Aminopolykarbonsäure f

a. **R acid** $C_{10}H_5(NH_2)(SO_3H)_2$ Amino-R-Säure f, 2-Naphthylamin-3,6-disulfonsäure f

a. **R salt** Amino-R-Salz n

a. **resin** Aminoplastharz n

a. **succinic acid** $HOOCCH(NH_2)CH_2COOH$ Aminobernsteinsäure f, Asparaginsäure f, Aminobutandisäure f

a. **sugar** Aminozucker m

a.-**sugar browning** (Lebm) Amin-Zucker-Bräunung f, Bräunung f vom Maillard-Typus, nichtenzymatische Bräunungsreaktion f, Maillard-Reaktion f

aminoacetal $NH_2CH_2CH(OC_2H_5)_2$ Aminoazetaldehyddiäthylazetal n, Aminoazetal n

aminoacetanilide $CH_3CONHC_6H_4NH_2$ Aminoazetanilid n

aminoacetic acid H_2NCH_2COOH Aminoessigsäure f, Glykokoll n, Glyzin n, Leimsüß n

aminoalcohol Aminoalkohol m, Alkanolamin n, Alkamin n

aminoanthraquinone $C_6H_4(CO)_2C_6H_3\cdot NH_2$ Aminoanthrachinon n

aminoazo compound Aminoazoverbindung f

p-**aminoazobenzene** $C_6H_5N=NC_6H_4NH_2$ p-Aminoazobenzol n, 4-Aminoazobenzol n, Anilingelb n, Spritgelb n

aminoazobenzene sulphonic acid $NH_2C_6H_4N_2C_6H_4SO_3H$ Aminoazobenzolsulfonsäure f

aminobenzene $C_6H_5NH_2$ Aminobenzol n, Phenylamin n, Anilin n

p-**aminobenzenesulphonamide** $H_2NC_6H_4SO_2NH_2$ p-Aminobenzolsulfonamid n, p-Aminophenylsulfonamid n, Sulfanilamid n

p-**aminobenzenesulphonic acid** $NH_2C_6H_4SO_3H$ p-Aminobenzolsulfonsäure f, Anilin-p-sulfonsäure f, Sulfanilsäure f

m-**aminobenzoic acid** $C_6H_4(NH_2)COOH$ m-Aminobenzoesäure f, Benzaminsäure f

o-**aminobenzoic acid** $C_6H_4(NH_2)COOH$ o-Aminobenzoesäure f, Anthranilsäure f

p-**aminobenzoic acid** $NH_2C_6H_4COOH$ p-Aminobenzoesäure f (Vitamin H')

aminobenzoyl J acid Aminobenzoyl-J-Säure f, Aminobenzoyl-I-Säure f

α-**aminobenzylpenicillin** α-Aminobenzylpenizillin n

aminocarbonyl J acid $(NHCONH_2)(OH)(SO_3H)C_{10}H_5$ I-Säureharnstoff m, Karbamid-I-Säure f

aminodimethylaniline $(CH_3)_2NC_6H_4NH_2$ Aminodimethylanilin n, Aminodimethylaminobenzol n

aminodimethylbenzene $(CH_3)_2C_6H_3NH_2$ Aminodimethylbenzol n, Dimethylanilin n, Aminoxylol n, Xylidin n

2-amino-4,6-dinitrophenol $NH_2C_6H_2(NO_2)_2OH$ 2-Amino-4,6-dinitrophenol n, 4,6-Dinitro-2-aminophenol n, Pikraminsäure f

1-aminododecane $CH_3(CH_2)_{11}NH_2$ 1-Aminododekan n, Dodezylamin n, Laurylamin n

aminoethane $C_2H_5NH_2$ Aminoäthan n, Monoäthylamin n, Äthylamin n

aminoethanoic acid H_2NCH_2COOH Aminoessigsäure f, Glykokoll n, Glyzin n, Gly, Leimsüß n

1-aminoethanol $CH_3CH(NH_2)OH$ 1-Aminoäthanol n, 1-Aminoäthylalkohol m

2-aminoethanol $NH_2CH_2CH_2OH$ 2-Aminoäthanol-(1) n, 2-Aminoäthylalkohol m, Kolamin n, Äthanolamin n

α-**aminoethyl alcohol** s. 1-aminoethanol

β-**aminoethyl alcohol** s. 2-aminoethanol

aminoethylbenzene Aminoäthylbenzol n, Phen[yl]äthylamin n

aminoethylisothiuronium bromide hydrobromide 2-Aminoäthylisothio-uroniumbromid-hydrobromid n, AET

aminoformate H_2NCOOM^I Aminoformiat n, Karbaminat n, Karbamat n

aminohydroxybenzoic acid $HOC_6H_3(NH_2)COOH$ Aminohydroxybenzoesäure f

aminohydroxynaphthalenesulphonic acid $H_2N\cdot C_{10}H_5(OH)(SO_3H)$ Aminohydroxynaphthalinsulfo[n]säure f, Aminonaphtholsulfo[n]säure f

aminomethane CH_3NH_2 Aminomethan n, Methylamin n, Monomethylamin n

aminonaphthalenesulphonic acid $NH_2C_{10}H_6SO_3H$ Aminonaphthalinsulfo[n]säure f, Naphthylaminsulfo[n]säure f

aminonaphtholdisulphonic acid $H_2N \cdot C_{10}H_4(OH)(SO_3H)_2$ Aminonaphtholdisulfo[n]säure f

aminonaphtholsulphonic acid $H_2N \cdot C_{10}H_5(OH)(SO_3H)$ Aminonaphtholsulfo[n]säure f

6-aminopenicillanic acid 6-Aminopenizillansäure f, 6-APS

aminophenetole Aminophenetol n, Aminophenoläthyläther m, Phenetidin n, Äthoxyanilin n

p-aminophenol $NH_2C_6H_4OH$ p-Aminophenol n, 4-Amino-1-hydroxybenzol n

aminophenol ethyl ether s. aminophenetole

aminophenylacetic acid Aminophenylessigsäure f

p-aminophenylsulphonamide $H_2NC_6H_4SO_2NH_2$ p-Aminophenylsulfonamid n, p-Aminobenzolsulfonamid n, Sulfanilamid n

aminophylline Aminophyllin n, Theophyllin-Äthylendiamin n (Additionsverbindung von Theophyllin mit Äthylendiamin)

aminoplastic Aminoplast m

aminopterin Aminopterin n, 4-Aminopteroylglutaminsäure f, 4-Aminofolsäure f

aminopyridine Aminopyridin n, Pyridylamin n

aminoquinoline Aminochinolin n, Chinolylamin n

aminosalicylic acid $HOC_6H_3(NH_2)COOH$ Aminosalizylsäure f, Aminohydroxybenzoesäure f

aminosulphate H_2NSO_2OM Amidosulfat n, Sulfaminat n, Sulfamat n

aminothiazole Aminothiazol n, Thiazolylamin n

aminothiophenol Aminothiophenol n

aminothiourea $H_2NCSNHNH_2$ Thiosemikarbazid n, Hydrazinthiokarbonsäureamid n

aminotoluene $CH_3C_6H_4NH_2$ Aminotoluol n, Toluidin n

ammeter Amperemeter n, Ammeter n, Strommesser m

ammine Amminverbindung f, Amminsalz n, Ammoniakat n

a. complex Amminkomplex m

ammonate s. ammine

ammonia NH_3 Ammoniak n

a.-air mixture Ammoniak-Luft-Gemisch n

a. alum $NH_4Al(SO_4)_2 \cdot 12H_2O$ Ammoniakalaun m, Ammoniumalaun m, Ammonalaun m (Ammoniumaluminiumsulfat-12-Wasser)

a.-base liquor s. a.-base sulphite acid

a.-base sulphite acid (liquor) (Pap) Ammoniumbisulfitkochsäure f

a. condenser Ammoniakverflüssiger m, Ammoniakkondensator m

a. converter Ammoniaksyntheseofen m

a.-distillation apparatus Ammoniakdestillationsapparat m

a. drum Ammoniakflasche f

a. evaporation Ammoniakverdampfung f

a. fume Ammoniakdampf m

a. fumigation (Landw) Ammoniakbegasung f

a. gas Ammoniak n (in Gasform), Ammoniakgas n

a. hydrate Ammoniakhydrat n

a. in aqueous solution s. a. solution

a. liquor Ammoniakwasser n, NH_3-Wasser n, Gaswasser n; s. a. solution

a. nitrogen Ammoniumstickstoff m, Ammoniakstickstoff m

a. oxidation Ammoniakverbrennung f

a.-preserved (Gum) mit Ammoniak konserviert

a. refrigerating machine Ammoniakkältemaschine f

a. scrubber Ammoniakwäscher m

a. scrubbing Ammoniakwäsche f

a. soda process Ammoniak-Soda-Verfahren n, Solvay-Verfahren n, Solvay-Prozeß m

a. solution (spirit) [wäßrige] Ammoniaklösung f, Ammoniak n in wäßriger Lösung, wäßriges Ammoniak n, Ammoniakwasser n, Ammoniakflüssigkeit f, Salmiakgeist m

a. synthesis Ammoniaksynthese f

a. triacetic acid $HOOCCH_2N(CH_2COOH)_2$ Nitrilotriessigsäure f, NTE f, Trimethylamin-α,α',α''-trikarbonsäure f

a. vaporizer Ammoniakverdampfer m, NH_3-Verdampfer m

a. vapour Ammoniakdampf m

a. vat (Text) ammoniakalische Küpe f, Ammonsalzküpe f

a. washer Ammoniakwäscher m

a. waste Ammoniakabwasser n, Kolonnenablauf m (eines Gaswerks)

a. water Ammoniakwasser n, NH_3-Wasser n, Gaswasser n

ammoniac s. ammoniacal

ammoniac Ammoniak-Gummiharz n (von mehreren Doldenblütlerarten) (s. z.B. African ammoniac, Persian ammoniac)

a. of Cyrenaica (Gummiharz von Ferula marmarica Asch. et Taub.)

ammoniacal ammoniakalisch, ammoniakhaltig, Ammoniak enthaltend

a. copper hydroxide $[Cu(NH_3)_4](OH)_2$ ammoniakalisches Kupferhydroxid n, Kupferoxidammoniak n, Kupfer(II)-tetramminhydroxid n, Kuoxam n, Kuoxammon n

a. liquor Ammoniakwasser n, NH_3-Wasser n, Gaswasser n

a. nitrogen Ammoniakstickstoff m

ammoniacate Ammoniakat n, Amminverbindung f, Amminsalz n

ammoniate / to (Düngemittel) ammonisieren

ammoniate s. ammoniacate

ammoniated copper sulphate $[Cu(NH_3)_4]SO_4$ Tetramminkupfer(II)-sulfat n

a. nickel nitrate $[Ni(NH_3)_4](NO_3)_2$ Tetramminnickel(II)-nitrat n

ammoniation Ammonisierung f (von Düngemitteln)

a. drum (Düngemittelindustrie) Ammonisiertrommel f

ammoniator (Düngemittelindustrie) Ammonisator m, Ammonisierapparat m

a.-granulator (Düngemittelindustrie) Ammonisiergranulator m

a.-granulator drum Ammonisier-Granuliertrommel f

ammonification 1. *(Landw)* Ammonifizierung f, Ammonifikation f *(Ammoniakbildung durch bakterielle Zersetzung organischer Stoffe im Boden)*; 2. Ammonisierung f *(von Düngemitteln)*

ammonifyers, ammonifying bacteria Ammoniakbildner *mpl*

ammonio-cupric sulphate $[Cu(NH_3)_4]SO_4$ Tetramminkupfer(II)-sulfat n

a.-ferric sulphate $Fe(NH_4)(SO_4)_2$ Eisen(III)-ammoniumsulfat n

a. group $-NH_3$ Ammonio-Gruppe f

ammonium Ammonium n

a. acetate CH_3COONH_4 Ammoniumazetat n

a. acid carbonate s. a. hydrogen carbonate

a. acid fluoride s. a. hydrogen fluoride

a. acid phosphate s. a. dihydrogen phosphate

a. acid phosphite s. a. dihydrogen orthophosphite

a. acid sulphate s. a. hydrogen sulphate

a. acid sulphide s. a. hydrogen sulphide

a. acid sulphite s. a. hydrogen sulphite

a. acid tartrate s. a. hydrogen tartrate

a. alum $NH_4Al(SO_4)_2 \cdot 12H_2O$ Ammoniakalaun m, Ammoniumalaun m *(Ammoniumaluminiumsulfat-12-Wasser)*

a.-aluminium chloride $NH_4Cl \cdot AlCl_3$ Ammoniumaluminiumchlorid n

a.-aluminium sulphate $NH_4Al(SO_4)_2$ Ammoniumaluminiumsulfat n

a. amalgam Ammoniumamalgam n

a. aminoformate s. a. carbamate

a. azide $[NH_4]N_3$ Ammoniumazid n, stickstoffwasserstoffsaures Ammonium n

a. benzoate $C_6H_5COONH_4$ Ammoniumbenzoat n

a. biborate s. a. tetraborate

a. bicarbonate s. a. hydrogen carbonate

a. bichromate s. a. dichromate

a. bifluoride s. a. hydrogen fluoride

a. biphosphate s. a. dihydrogen phosphate

a. bisulphate s. a. hydrogen sulphate

a. bisulphide s. a. hydrogen sulphide

a. bisulphite s. a. hydrogen sulphite

a. bitartrate s. a. hydrogen tartrate

a. borofluoride s. a. fluoroborate

a. bromide NH_4Br Ammoniumbromid n

a. carbamate $[NH_4](H_2NCOO)$ Ammoniumkarbaminat n, Ammoniumkarbamat n, Ammoniumamidokarbonat n

a. carbazotate s. a. picrate

a. carbonate $(NH_4)_2CO_3$ Ammoniumkarbonat n; [gewöhnliches] Ammoniumkarbonat n, Hirschhornsalz n *(Gemisch aus Ammoniumhydrogenkarbonat und Ammoniumkarbamat)*

a. chlorate NH_4ClO_3 Ammoniumchlorat n

a. chloride NH_4Cl Ammoniumchlorid n, Salmiak m

a. chlorostannate $(NH_4)_2[SnCl_6]$ Ammoniumhexachlorostannat(IV) n, Pinksalz n

a.-chromium sulphate $NH_4Cr(SO_4)_2 \cdot 12H_2O$ Ammoniumchrom(III)-sulfat-12-Wasser n, Ammoniumchromalaun m

a.-copper chloride $CuCl_2 \cdot 2NH_4Cl$ Ammoniumkupfer(II)-chlorid n

a.-cupric chloride s. a.-copper chloride

a. cyanate NH_4OCN Ammoniumzyanat n

a. cyanide NH_4CN Ammoniumzyanid n

a. decaborate s. a. pentaborate

a. dichromate $(NH_4)_2Cr_2O_7$ Ammoniumdichromat(VI) n

a. dihydrogen orthophosphate s. a. dihydrogen phosphate

a. dihydrogen orthophosphite $NH_4H_2PO_3$ Ammoniumdihydrogen[ortho]phosphit n

a. dihydrogen phosphate $NH_4H_2PO_4$ Ammoniumdihydrogen[ortho]phosphat n

a. dithionate $(NH_4)_2S_2O_6$ Ammoniumdithionat n

a. ferrocyanide $(NH_4)_4[Fe(CN)_6]$ Ammoniumhexazyanoferrat(II) n, Ammoniumeisen(II)-zyanid n

a. fluoborate s. a. fluoroborate

a. fluoride NH_4F Ammoniumfluorid n

a. fluoroborate $[NH_4][BF_4]$ Ammoniumfluoroborat n, Ammoniumborfluorid n

a. fluosilicate $(NH_4)_2[SiF_6]$ Ammoniumhexafluorosilikat n

a. formate $HCOONH_4$ Ammoniumformiat n

a. group Ammoniumgruppe f, Ammoniumrest m

a. heptamolybdate $(NH_4)_6[Mo_7O_{24}] \cdot 4H_2O$ Ammoniumheptamolybdat-4-Wasser n, Hexammoniumheptamolybdat-Tetrahydrat n

a. hydrate s. a. hydroxide

a. hydrogen carbonate NH_4HCO_3 Ammoniumhydrogenkarbonat n

a. hydrogen fluoride $[NH_4]HF_2$ Ammoniumhydrogenfluorid n, *(Tech auch)* Ammoniumbifluorid n

a. hydrogen sulphate $(NH_4)HSO_4$ Ammoniumhydrogensulfat n

a. hydrogen sulphide NH_4HS Ammoniumhydrogensulfid n

a. hydrogen sulphite NH_4HSO_3 Ammoniumhydrogensulfit n, *(Tech auch)* Ammoniumbisulfit n

a. hydrogen tartrate $NH_4OOC(CHOH)_2COOH$ Ammoniumhydrogentartrat n

a. hydrosulphide s. a. hydrogen sulphide

a. hydroxide NH_4OH Ammoniumhydroxid n

a. hypophosphite $NH_4H_2PO_2$ Ammoniumhypophosphit n

a. hyposulphite s. a. thiosulphate

a. iodate NH_4JO_3 Ammoniumjodat n

a. iodide NH_4J Ammoniumjodid n

a. lactate $CH_3CHOHCOONH_4$ Ammoniumlaktat n

a. metaantimonate $NH_4SbO_3 \cdot 2H_2O$ Ammoniummetaantimonat(V) n, Ammoniumtrioxoantimonat(V) n

a. metaarsenite NH_4AsO_2 Ammoniummetaarsenat(III) n

a. **metavanadate** NH_4VO_3, $(NH_4)_4[V_4O_{12}]$ Ammoniummetavanadat n, Ammoniumtetravanadat n

a. **molybdate** $(NH_4)_2MoO_4$ [normales] Ammoniummolybdat(VI) n

a. **monosulphide** $(NH_4)_2S$ Ammonium[mono]sulfid n

a. **muriate** s. a. chloride

a.**-nickel sulphate** $(NH_4)_2SO_4 \cdot NiSO_4 \cdot 6H_2O$ Ammoniumnickel(II)-sulfat-6-Wasser n

a. **nitrate** NH_4NO_3 Ammoniumnitrat n

a. **nitrate explosive** Ammonsalpetersprengstoff m

a. **nitrite** NH_4NO_2 Ammoniumnitrit n

a. **oxalate** $(COONH_4)_2$ Ammoniumoxalat n

a. **paramolybdate** Ammoniumparamolybdat n, Ammoniummolybdat n des Handels

a. **pentaborate** $(NH_4)_2B_{10}O_{16}$ Ammoniumpentaborat n

a. **pentasulphide** $(NH_4)_2S_5$ Ammoniumpentasulfid n

a. **perchlorate** NH_4ClO_4 Ammoniumperchlorat n

a. **perchromate** $(NH_4)_3CrO_8$ Ammoniumperoxochromat n

a. **periodate** NH_4JO_4 Ammoniummetaperjodat n, Ammoniumtetroxojodat n

a. **permanganate** NH_4MnO_4 Ammoniumpermanganat n, Ammoniummanganat(VII) n

a. **peroxodisulphate (peroxydisulphate, persulphate)** $(NH_4)_2S_2O_8$ Ammoniumperoxodisulfat n, Ammoniumpersulfat n

a. **picrate** $NH_4OC_6H_2(NO_2)_3$ Ammoniumpikrat n

a. **picronitrate** s. a. picrate

a. **powder** Ammonpulver n

a. **radical** Ammoniumgruppe f, Ammoniumrest m, Ammoniumradikal n

a. **reineckate** s. a. tetrathiocyanate diaminochromite

a. **rhodanate** s. a. thiocyanate

a. **rhodanide** s. a. thiocyanate

a. **salicylate** $C_6H_4OHCOONH_4$ Ammoniumsalizylat n

a. **salt** Ammoniumsalz n

a. **selenate** $(NH_4)_2SeO_4$ Ammoniumselenat n

a. **sesquicarbonate** $(NH_4)_2CO_3 \cdot 2NH_4CHO_3$ Ammoniumdihydrogenkarbonat n

a. **silicofluoride** s. a. fluosilicate

a. **soap** Ammoniumseife f, Ammoniakseife f

a. **stearate** $C_{17}H_{35}COONH_4$ Ammoniumstearat n

a. **sulphamate** $NH_4OSO_2NH_2$ Ammoniumsulfam[in]at n

a. **sulphate** $(NH_4)_2SO_4$ Ammoniumsulfat n, Ammonsulfat n

a. **sulphide** $(NH_4)_2S$ Ammoniummonosulfid n, Ammoniumsulfid n

a. **sulphite** $(NH_4)_2SO_3$ Ammoniumsulfit n

a. **sulphocyanate** s. a. thiocyanate

a. **sulphocyanide** s. a. thiocyanate

a. **tartrate** $(CHOHCOONH_4)_2$ Ammoniumtartrat n

a. **tellurate** $(NH_4)_2TeO_4$ Ammoniumtellurat n

a. **tetraborate** $NH_4HB_4O_7$ Ammoniumhydrogenborat n

a. **tetrathiocyanate diaminochromite** $NH_4[Cr(SCN)_4(NH_3)_2] \cdot H_2O$ Ammoniumtetrathiozyanatodiamminchromat(III) n, Ammoniumtetrarhodanodiamminchromat(III) n, Reinecke-Salz n

a. **tetrathiocyanodiammonochromate** s. a. tetrathiocyanate diaminochromite

a. **thiocarbonate** $(NH_4)_2CS_3$ Ammoniumtrithiokarbonat n

a. **thiocyanate** NH_4SCN Ammoniumthiozyanat n, Ammoniumrhodanid n

a. **thioglycolate** $HSCH_2COONH_4$ Ammoniumthioglykolat n

a. **thiosulphate** $(NH_4)_2S_2O_3$ Ammoniumthiosulfat n

a. **tungstate** $(NH_4)_2WO_4$ Ammoniumwolframat n

a. **valerate (valerianate)** $C_4H_9COONH_4$ Ammoniumvalerat n

a. **vanadate** s. a. metavanadate

a. **wolframate** s. a. tungstate

ammonobase $M'NH_2$ Metallamid n, Amid n

ammonolysis Ammonolyse f

ammonolytic ammonolytisch

amor. s. amorphous

amorphous amorph, formlos, gestaltlos, nicht kristallin[isch], ohne Kristallform

a. **material** amorpher Stoff m

a. **phase** amorphe Phase f, amorpher Zustand m

a. **phosphorus** amorpher (roter) Phosphor m (des Handels)

a. **polymer** amorphes Polymer[es] n

a. **sulphur** amorpher Schwefel m, μ-Schwefel m

a. **wax** amorphes Wachs (Paraffin) n, (veraltet für) mikrokristallines Wachs (Paraffin, Paraffinwachs) n, Mikrowachs n, Mikroparaffin n

amount Menge f, Gehalt m; Grad m, Betrag m, Summe f; Anteil m; Wert m; Größe f

a. **of alkali required** (Pap) Alkaliaufwand m, Alkalibedarf m

a. **of application** Aufwandmenge f

a. **of charge** Ladungsmenge f

a. **of energy** Energiebetrag m

a. **of heat** Wärmemenge f

a. **of heat required** Wärmebedarf m

a. **of liquor** (Text) Flottenmenge f

a. **of moisture** Feuchtigkeitsgehalt m, Feuchtegehalt m, Feuchteanteil m

a. **of precipitate** Bodenkörpermenge f, Niederschlagsmenge f

a. **of rotation** Drehwert m, [optischer] Drehungswinkel m, [optischer] Drehwinkel m

a. **of steam used** verbrauchte Dampfmenge f

a. **of substance** Stoffmenge f, Substanzmenge f

a. **of swelling** Quellungsgrad m, Quellungsbetrag m

a. **to the total conductance** Leitfähigkeitsanteil m

amp s. ampere

AMP s. adenosine monophosphate

ampere Ampere n, A
a.-hour Amperestunde f, Ah
a.-second Amperesekunde f, As, Coulomb n, C
amperometer Amperemeter n, Strommesser m
amperometric amperometrisch
a. titration amperometrische Titration f
a. titrimetry amperometrische Titrimetrie f
amperometry Amperometrie f
amphetamine Amphetamin n, β-Phenylisopropylamin n, 1-Phenyl-2-aminopropan n, α-Methyl-β-phenyläthylamin n
a. sulphate Amphetaminsulfat n
amphi-position amphi-Stellung f
amphibole (Min) Amphibol m, Hornblende f; s. a. asbestos
a. asbestos (Min) Amphibolasbest m, Hornblendeasbest m
amphibolite Amphibolit m, Hornblendegestein n
amphigene (Min) Amphigen m, (veraltet für) Leuzit m (Kaliumalumodisilikat)
amphion Ampho-Ion n, Zwitterion n
amphiprotic amphoter
a. substance s. ampholyte
ampholyte Ampholyt m, amphoterer Stoff (Elektrolyt) m
a. ion Zwitterion n, Ampho-Ion n
ampholytoid Ampholytoid n (amphoteres Bodenkolloid)
amphoteric amphoter
a. electrolyte amphoterer Elektrolyt (Stoff) m, Ampholyt m
a. ion Zwitterion n, Ampho-Ion n
amphotericin B Amphotericin B n, Fungizon n (Antibiotikum)
amplified (amplifying) distillation (Destillation unter Zusatz von Hilfsstoffen gleicher Siedebereiche)
amplitude Amplitude f, Schwing[ungs]weite f
a. of vibration Schwingungsamplitude f, Schwingamplitude f; Schwing[ungs]weite f, Schwingweg m; Hublänge f, Hub m
ampoule Ampulle f
ampul[e], ampulla s. ampoule
Amritsar gum (Pflanzengummi von Acacia modesta Wall.)
AMS s. ammonium sulphamate
amu, a.m.u. s. atomic mass unit
amygdala amara oil Bittermandelöl n
amygdale (Geol) Mandel f, Geode f
amygdalic acid $C_6H_5CH(OH)COOH$ Mandelsäure f, α-Hydroxyphenylessigsäure f, Phenylglykolsäure f
amygdalin Amygdalin n (Glykosid aus bitteren Mandeln)
amygdaloid (Geol) Amygdaloid m, Mandelstein m
amygdule s. amygdale
amyl Amyl n
a. acetate $CH_3COOC_5H_{11}$ Essigsäureamylester m, Amylazetat n
n-a. acetylene $HC{\equiv}C(CH_2)_4CH_3$ n-Amyläthin n, n-Amylazetylen n, 1-Heptin n, Oenanthyliden n

prim-a. alcohol $CH_3(CH_2)_3CH_2OH$ prim-n-Amylalkohol m, 1-Pentanol n
prim-act-a. alcohol $CH_3CH_2CH(CH_3)CH_2OH$ optisch aktiver prim-Amylalkohol m, 2-Methylbutanol-(1) n
sec-act-a. alcohol $CH_3CH_2CH_2CHOHCH_3$ optisch aktiver sek-Amylalkohol m, 2-Pentanol n
tert-a. alcohol $CH_3CH_2C(CH_3)(OH)CH_3$ tert-Amylalkohol m, 2-Methyl-2-butanol n, Amylenhydrat n, Dimethyläthylkarbinol n
n-a. aldehyde $CH_3(CH_2)_3CHO$ n-Amylaldehyd m, n-Valeraldehyd m, n-Pentanal n
a. carbinol $CH_3(CH_2)_4CH_2OH$ Amylkarbinol n, 1-Hexanol n, n-Hexylalkohol m, n-Kapronalkohol m
n-a. chloride $CH_3(CH_2)_3CH_2Cl$ n-Amylchlorid n, 1-Chlorpentan n
a. group Amylgruppe f, Amylrest m
a. halide Amylhalogenid n
a. nitrite $C_5H_{11}O_2N$ Salpetrigsäureamylester m, Amylnitrit n
a. radical Amylgruppe f, Amylrest m, Amylradikal n; [freies] Amylradikal n
amylaceous stärkehaltig
amylase Amylase f, α-1,4-Glukanase f (stärkespaltendes Ferment)
α-n-amylene $CH_3CH_2CH_2CH{=}CH_2$ n-Amylen-(1) n, Penten-(1) n, n-Propyläth[yl]en n
β-n-amylene $CH_3CH_2CH{=}CHCH_3$ n-Amylen-(2) n, Penten-(2) n, 1-Methyl-2-äthyläth[yl]en n
amylene hydrate $CH_3CH_2C(CH_3)(OH)CH_3$ Amylenhydrat n, tert-Amylalkohol m, Dimethyläthylkarbinol n, 2-Methyl-2-butanol n
amyliferous stärkehaltig
amylo fermentation process Amyloverfahren n, Pilzmaischverfahren n (Stärkeverzuckerung und Vergärung ohne diastatische Fermente)
amyloclastic s. amylolytic
amylodextrin Amylodextrin n
amyloid amyloid, stärkeähnlich, stärkeartig
amyloid (Pap) Amyloid n (mit konzentrierter H_2SO_4 behandelte Zellulose); (Med) Amyloid n (außerhalb der Blutbahn im Gewebe ausgefälltes Antikörperglobulin)
amyloidal s. amyloid
amylolytic stärkespaltend
a. enzyme stärkespaltendes Ferment n, Amylase f
amylopectin Amylopektin n, Stärkegranulose f
amylose Amylose f
an. s. anhydrous
anabasine Anabasin n, 3(2-Piperidyl)-pyridin n (Alkaloid)
anabolism Anabolismus m, Assimilation f
anaerobe Anaerobier m, Anaerobiont m
anaerobic anaerob[isch]
under a. conditions unter anaeroben Bedingungen (Verhältnissen), im anaeroben Medium, unter Luftabschluß, anaerob
a. bacteria anaerobe Bakterien npl, Anaerobier mpl, Anaerobionten mpl

a. fermentation anaerobe *(in Abwesenheit von Luftsauerstoff verlaufende)* Gärung f
a. respiration anaerobe (luftfreie, sauerstofffreie) Atmung f
anaerobiosis Anaerobiose f, Anoxybiose f *(Leben bei Abwesenheit von Sauerstoff)*
anaesthesia Anästhesie f
anaesthetic Anästhetikum n, Anästhesie[rungs]mittel n
a. action anästhetisierende (anästhesierende, betäubende) Wirkung f
a. chloroform Narkosechloroform n
a. ether Narkoseäther m, Diäthyläther m
anaesthetize / to anästhetisieren, anästhesieren
analcime, analcite *(Min)* Analzim m *(Natriumalumodisilikat)*
analgesic analgetisch, schmerzstillend
analgesic Analgetikum n, analgetisches (schmerzstillendes, schmerzlinderndes) Mittel n
analyse / to s. to analyze
analyser s. analyzer
analysis Analyse f, Zergliederung f, Zerlegung f, Auflösung f, Untersuchung f; Gehalt m *(an wirksamer Substanz)*
a. by titration Titrieranalyse f, Titrimetrie f, Maßanalyse f, Volumetrie f
a. in a dissolved state Naßanalyse f
a. in dry state (way) Trockenanalyse f
a. in wet way s. a. in a dissolved state
a. of crystal structure Kristallstrukturanalyse f
analyst Analytiker m
analytic s. analytical
analytical analytisch, Analysen...
a. balance Analysenwaage f, analytische Waage f
a. chemist s. analyst
a. chemistry analytische Chemie f
a. error Analysenfehler m
a. factor Faktor m *(Masseanalyse)*
a. result Analysenbefund m
a. sample Analysenprobe f
a. value Analysenwert m
analyzable analysierbar
analyze / to analysieren, zergliedern, zerlegen, auflösen, untersuchen; *(an wirksamer Substanz)* enthalten
analyzer Analysator m, Analysiergerät n; Analytiker m
analyzing range Analysenbereich m
a. speed Analysengeschwindigkeit f
anamorphism *(Geol)* Anamorphose f; Weiterbildung f *(Fossilien)*
anaphoresis Anaphorese f
anatase *(Min)* Anatas m *(Titandioxid)*
anatoxin Anatoxin n *(Formoltoxoid)*
anchimonomineralic anchimonomineralisch
anchor agitator (mixer) Ankerrührer m, Ankermischer m
anchoring agent *(Plast)* Haftmittel n
ancillary equipment Hilfseinrichtung f, Hilfsausrüstung f

a. operation Nebenvorgang m *(z.B. bei der Aufbereitung)*
ancylite *(Min)* Ankylit m
andalusite, andaluzite *(Min)* Andalusit m *(Aluminiumoxidorthosilikat)*
andersonite *(Min)* Andersonit m
andesine *(Min)* Andesin m
andesite Andesit m *(Ergußgestein)*
andradite *(Min)* Andradit m, Kalkeisengranat m *(Kalziumeisen(III)-triorthosilikat)*
andreolite *(Min)* Andreolith m, *(veraltet für)* Harmotom m
andromedotoxin *(Pharm)* Andromedotoxin n, Asebotoxin n
anelectrode s. anode
anellate / to anellieren, kondensieren
anellated-ring hydrocarbon anellierter (kondensierter) Ringkohlenwasserstoff m
anellation Anellierung f, Kondensation f, Ringkondensation f
anemometer Anemometer n, Windmesser m; Gasdurchflußmesser m
anerobic s. anaerobic
aneroid barometer Aneroid[barometer] n, Dosenbarometer n, Metallbarometer n, Federbarometer n
anesthesia ether s. anaesthetic ether
anesthetic s. anaesthetic
anethol[e] $H_3COC_6H_4CH=CHCH_3$ Anethol n, Aniskampfer m, Allylphenylmethyläther m, p-Methoxypropenylbenzol n
angelic acid $CH_3CH=C(CH_3)COOH$ Angelikasäure f, 2-Methylisokrotonsäure f, cis-2-Methyl-2-butensäure f
Anger mill Anger-Mühle f, Anger-Prallmühle f
angiospermous wood Laubholz n
angle Winkel m
a. between the faces *(Krist)* Flächenwinkel m, Kantenwinkel m
a. connector Winkelstück n
a.-cut paper Diagonalschnittpapier n, Schrägschnittpapier n
a. cutter s. a. cutting machine
a. cutting *(Pap)* Diagonalschnitt m
a. cutting machine *(Pap)* Diagonal-Querschneider m, Diagonalschneidemaschine f
a. head Schrägspritzkopf m; Winkelkopf m *(einer Winkelzentrifuge)*
a. method *(Gum)* Angle-Methode f, Winkelprobe f *(zur Bestimmung der Weiterreißfestigkeit)*
a. of contact *(phys Ch)* Kontaktwinkel m, Randwinkel m
a. of crease recovery *(Text)* Knittererholungswinkel m, Knitterwinkel m
a. of deflection Auslenkungswinkel m, Auslenkwinkel m
a. of diffraction Beugungswinkel m
a. of incidence Einfallswinkel m, Einfallwinkel m
a. of inclination (incline) Neigungswinkel m
a. of nip Einzugswinkel m, Fassungswinkel m *(am Walzenbrecher)*

a. of polarization Polarisationswinkel *m*
a. of recovery *(Text)* Knittererholungswinkel *m*, Knitterwinkel *m*
a. of reflection Reflexionswinkel *m*
a. of refraction Brechungswinkel *m*
a. of repose Schüttwinkel *m*, Gleitwinkel *m*, Rutschwinkel *m*
a. of rotation Drehwert *m*, [optischer] Drehungswinkel *m*, [optischer] Drehwinkel *m*
a. of scattering Streuungswinkel *m*, Streuwinkel *m*
a. of slide Scherwinkel *m*, Schiebungswinkel *m*, Gleitwinkel *m*
a. of spray Sprühkegelwinkel *m*, Sprühwinkel *m*, Spritzwinkel *m*
a. of valency Valenzwinkel *m*
a. of wrap Umschlingungswinkel *m*
a. paper *s.* a.-cut paper
a. press Winkelpresse *f*
a. strain Ringspannung *f (bei Deformation des Bindungswinkels in einer Ringverbindung)*
a. test piece *(Gum)* Winkelprobe *f*, Angle-Probe *f (zur Bestimmung der Weiterreißfestigkeit)*
a. thermometer Winkelthermometer *n*
a. valve Eckventil *n*
anglesite *(Min)* Anglesit *m (Blei(II)-sulfat)*
Angola weed tangartige Färberflechte *f*, Roccella fuciformis (L.)Lam.
angstrom (ångstrom, Ångström) [unit] Ångström *n*, Å, Ångström-Einheit *f*
angular angular, winkelig, winklig, gewinkelt, winkelförmig, eckig, kantig, spitz, Winkel...
a. acceleration Winkelbeschleunigung *f*, Radialbeschleunigung *f*
a. cathode Wellenkatode *f (in der Billiter-Zelle)*
a. distribution Winkelverteilung *f*
a. extruder head Schrägspritzkopf *m*
a. mixer Rührmaschine *f* mit schräger Rührwelle
a. momentum Orbitaldrehimpuls *m*, Drehimpuls *m*, Impulsmoment *n*, Drall *m*
a. momentum expansion Drehimpulsentwicklung *f*
a. paper Diagonalschnittpapier *n*, Schrägschnittpapier *n*
a. spectrum Winkelspektrum *n*
a. velocity Winkelgeschwindigkeit *f*
angustifoline Angustifolin *n (Alkaloid)*
anh. *s.* anhydrous
anhalonidine Anhalonidin *n*, 7,8-Dimethoxy-6-hydroxy-1-methyl-1,2,3,4-tetrahydroisochinolin *n*
anharmonic unharmonisch, anharmonisch
a. binding force anharmonische Bindungskraft *f*
a. constant *s.* anharmonicity constant
anharmonicity Anharmonizität *f*
a. constant Anharmonizitätskonstante *f*
anhedral xenomorph, allotriomorph, fremdgestaltig
anhedron xenomorpher Kristall *m*
anhydride Anhydrid *n*
anhydrite *(Min)* Anhydrit *m (wasserfreies Kalziumsulfat)*

a. binder Anhydritbinder *m*
anhydrobase Anhydrobase *f*
anhydroglucose Anhydroglukose *f*
anhydrosugar Anhydrozucker *m*
anhydrous nicht wäßrig, nichtwäßrig, nichtwässerig, wasserfrei, entwässert, kristallwasserfrei
a. alcohol absoluter Alkohol *m*, wasserfreies (100%iges) Äthanol *n*
a. aluminium fluoride AlF_3 wasserfreies Aluminiumfluorid *n*
a. aluminium hydroxide $Al(OH)_3$ Aluminiumhydroxid *n*, Aluminiumtrihydroxid *n*, Aluminiumorthohydroxid *n*
a. ammonia NH_3 wasserfreies Ammoniak *n*
a. borax $Na_2B_4O_7$ wasserfreies Natriumtetraborat *n*, Pyroborax *m*
a. hydrazine $H_2N \cdot NH_2$ wasserfreies Hydrazin *n*
a. hydrogen chloride wasserfreier (trockener) Chlorwasserstoff *m*
a. lime CaO gebrannter Kalk *m*, Branntkalk *m*
a. milk fat wasserfreies Milchfett *n*
a. sodium carbonate Na_2CO_3 wasserfreies Natriumkarbonat *n*, kalzinierte (kristallwasserfreie) Soda *f*
a. wolframic acid WO_3 Wolframsäureanhydrid *n*, Wolfram(VI)-oxid *n*, Wolframtrioxid *n*
anilide C_6H_5NHCOR Anilid *n (Säureamidderivat des Anilins)*
anilin *s.* aniline
aniline $C_6H_5NH_2$ Anilin *n*, Aminobenzol *n*, Phenylamin *n*
a. black Anilinschwarz *n*
a. blue [spritlösliches] Anilinblau *n*, Spritblau *n (Cl-Salz des Triphenylrosanilins)*
a. dye Anilinfarbstoff *m*, Anilinfarbe *f*
a.-formaldehyde resin Anilinformaldehydharz *n*, Anilinharz *n*
a. green Anilingrün *n*
a. hydrochloride $C_6H_5NH_2 \cdot HCl$ Anilinhydrochlorid *n*, Anilinchlorhydrat *n*, salzsaures Anilin *n*, Anilinsalz *n*
a. ink Anilingummidruckfarbe *f*, Farbe *f* für Flexodruck (Flexografie, Anilingummidruck)
a. oil Anilinöl *n*, technisches Anilin *n*
a. point Anilinpunkt *m*, A.P., Trübungspunkt *m*
a. printing Anilingummidruck *m*, Anilindruck *m*, Flexodruck *m*, flexografischer Druck *m*, Flexografie *f*
a. purple Anilinpurpur *m*, Perkinviolett *n*, Mauvein *n*
a. resin Anilinharz *n*; *s.* a.-formaldehyde resin
a. salt *s.* a. hydrochloride
a.-ω-sulphonate Anilin-ω-sulfonat *n*
a. water Anilinwasser *n*
a. yellow $C_6H_5N=NC_6H_4NH_2$ Anilingelb *n*, p-Aminoazobenzol *n*
m-anilinesulphonic acid $H_2NC_6H_4SO_3H$ Anilin-m-sulfonsäure *f*, m-Aminobenzolsulfonsäure *f*, Metanilsäure *f*

p-**anilinesulphonic acid** $H_2NC_6H_4SO_3H$ Anilin-*p*-sulfonsäure *f*, *p*-Aminobenzolsulfonsäure *f*, Sulfanilsäure *f*
anilino benzene $(C_6H_5)_2NH$ Anilinobenzol *n*, Diphenylamin *n*, *N*-Phenylanilin *n*
  a. derivative Anilinoderivat *n*
animal tierisch, animalisch, Tier...
  a. black Knochenschwarz *n*, Beinschwarz *n*, Kölner Schwarz *n*
  a. char[coal] Tierkohle *f*, *(i.e.S.)* Knochenkohle *f*
  a. dip Bademittel *n* *(zur Ungezieferbekämpfung)*
  a. drug tierische Droge *f*
  a. fat tierisches Fett *n*, Tierfett *n*
  a. fat margarine aus Tierfetten bestehende Margarine *f*
  a. fibre tierische Faser *f*, Tierfaser *f*
  a. gelatine (glue) tierischer Leim *m*, Tierleim *m*
  a. hair Tierhaar *n*
  a. oil Tieröl *n*
  a. parchment tierisches (animalisches) Pergament *n*, Hautpergament *n*, Pergament *n*
  a. size *(Pap)* tierischer Leim *m*, Tierleim *m*
  a.-sized *(Pap)* mit Tierleim (Gelatine) geleimt, gelatinegeleimt
  a. starch tierische Stärke *f*, Tierstärke *f*, Leberstärke *f*, Glykogen *n*
  a. sterol tierisches Sterin *n*, Zoosterin *n*
  a. tankage Fleischdüngemehl *n*, Fleischmehl *n*, Tierkörpermehl *n*, Kadavermehl *n*
  a. test Tierversuch *m*
  a. tub-sized *s.* a.-sized
  a. tub-sizing *(Pap)* Oberflächenleimung (Leimung *f* mit Tierleim (Gelatine), Gelatineleimung *f*
  a. wax tierisches Wachs *n*, Tierwachs *n*
animalize / to *(Text)* animalisieren
animalizing *(Text)* Animalisieren *n*
Anime copal Anime-Kopal *m* *(von Hymenaea courbaril L. oder Trachylobium hornemannianum Hayne)*
anion Anion *n*, negatives (negativ geladenes) Ion *n*
  a.-active anionenaktiv, anionaktiv, anionisch, Anion[en]...
  a. base Anionbase *f*
  a. complex Anionenkomplex *m*
  a. electrode Anionelektrode *f*
  a. exchange Anionenaustausch *m*
  a. exchange resin Anionenaustauscher *m* auf Kunstharzbasis, Anion[en]austausch[er]harz *n*
  a.-exchange substance Anionenaustauscher *m*
  a. exchangeability Anionenaustauschfähigkeit *f*
  a. exchanger Anionenaustauscher *m*
  a. vacancy Anionenleerstelle *f*, Anionenlücke *f*, Anion[en]fehlstelle *f*
anionic anionisch, anionenaktiv, anionaktiv, Anion[en]...
  a. current Anionenstrom *m*
  a. nature anionischer Charakter *m*, Anioncharakter *m*
  a. polymerization anionische Polymerisation *f*
  a. tenside anionisches Tensid *n*, Aniontensid *n*

anionite Anionenaustauscher *m*
anionoid anionoid
anise Anis *m*, Pimpinella anisum L.; *s.* aniseed
  a. camphor $H_3COC_6H_4CH=CHCH_3$ Aniskampfer *m*, Anisöl *n*, Anethol *n*, Allylphenylmethyläther *m*, *p*-Methoxypropenylbenzol *n*
aniseed Anissaat *f*, Anis *m*
  a. oil Anisöl *n* *(von Pimpinella anisum L. und Illicium verum L.)*
anisic acid $CH_3OC_6H_4COOH$ Anissäure *f*, *p*-Methoxybenzoesäure *f*
anisidine $CH_3OC_6H_4NH_2$ Anisidin *n*, Aminophenylmethyläther *m*, Aminoanisol *n*
anisodesmic *(Krist)* anisodesmisch
  a. structure *(Krist)* anisodesmische Struktur *f*
anisodimensional anisodimensional
anisole $CH_3OC_6H_5$ Anisol *n*, Methylphenyläther *m*, Methoxybenzol *n*
anisotopic elements anisotope Elemente *npl* *(nur aus einer einzigen natürlichen Atomart bestehende Elemente)*
anisotropic anisotrop
anisotropy Anisotropie *f*
  a. factor Anisotropiefaktor *m*
  a. term Anisotropieglied *n*
ankerite *(Min)* Ankerit *m* *(Kalziumeisen(II)-karbonat)*
annabergite *(Min)* Annabergit *m*, Nickelblüte *f* *(Nickelorthoarsenat)*
annatto Annatto *n(m)*, Anatto *n(m)*, Orlean *m*, Annattofarbstoff *m* *(von Bixa orellana L.)*
anneal / to glühen; *(Plast)* tempern, spannungsfreimachen; *(Glas)* kühlen
anneal *s.* annealing
annealing Glühen *n*, Glühung *f*; *(Plast)* Tempern *n*, Spannungsfreimachen *n*; *(Glas)* Kühlen *n*, Kühlung *f*
  a. box Glühkasten *m*, Glühkiste *f*
  a. can *s.* a. pot
  a. cup *(ein unglasierter, hoher)* Schmelztiegel *m*
  a. colour Glühfarbe *f*
  a. furnace Glühofen *m*
  a. lehr (oven) *(Glas)* Kühlofen *m*
  a. point *(Glas)* oberer Kühlpunkt *m*, obere Kühltemperatur *f*
  a. pot Glühtopf *m*
  a. range *(Glas)* Kühlbereich *m*
  a. temperature Glühtemperatur *f*; *(Glas)* obere Kühltemperatur *f*, oberer Kühlpunkt *m*
ånnerödite *(Min)* Ånneröit *m* *(ein Samarskit)*
annihilate / to vernichten
annihilation *(phys CH)* Annihilation *f*, Vernichtung *f*
  a. radiation Vernichtungsstrahlung *f*, Zerstrahlung *f*, Paarvernichtung *f*, Annihilation *f*
  a. rate Vernichtungsrate *f*
  a. spectrum Zerstrahlungsspektrum *n*, Vernichtungsspektrum *n*, Annihilationsspektrum *n*
annotta *s.* annatto
annual Annuelle *f*, Einjahrespflanze *f*, Einjahrpflanze *f*, Einjährige *f*, Therophyt *m*

a. ring *(Forst)* Jahresring *m*
annular ringförmig, Ring...
a. chamber ringförmige Kammer *f*, Ringkammer *f*, Ringraum *m*
a. chamber kiln Kammerringofen *m*
a. electrode ringförmige Elektrode *f*, Ringelektrode *f*
a. kiln Ringofen *m*
a. solids-discharge disk centrifuge Tellerzentrifuge *f* mit Schlitzaustrag, Ringspalt-Tellerzentrifuge *f*
a. space [freier] Ringraum *m* *(z.B. bei Kolbenpumpen)*
a. structure ringförmige Struktur *f*, Ringstruktur *f*
annulization *s.* anellation
annulize / to *s.* to anellate
annulus Ring *m*; Ringrohr *n*, Mantelrohr *n*
anode Anode *f*
a. coating *s.* anodic oxidation
a. compartment Anodenraum *m*
a. effect Anodeneffekt *m*
a. life Anodenlaufzeit *f*
a. mud *s. a.* sludge
a. potential Anodenpotential *n*
a. process Anodenverfahren *n*, Anodenprozeß *m*
a. ray Anodenstrahl *m*
a. slime (sludge) Anodenschlamm *m*
anodic anodisch, Anoden...
a. coating *s. a.* oxidation
a. inhibitor anodischer Inhibitor *m*
a. oxidation anodische (elektrochemische, elektrolytische) Oxydation *f*, *(Met auch)* Anodisieren *n*, Anodisierung *f*, anodische Behandlung *f*
a. passivity anodische (elektrochemische) Passivität *f*
a. treatment *s.* anodizing
anodization *s.* anodizing
anodize / to *(Met)* anodisieren, anodisch behandeln, anodisch (elektrochemisch, elektrolytisch) oxydieren
anodizing *(Met)* Anodisieren *n*, Anodisierung *f*, anodische Behandlung *f*, anodische (elektrochemische, elektrolytische) Oxydation *f*
anogenic *(Geol)* anogen
anoint / to einölen
anolyte Anolyt *m* *(Elektrolyt im Anodenraum)*
anomalous anomal, abnorm, regelwidrig; *(Tech)* normwidrig
a. rotatory dispersion anomale Rotationsdispersion *f*
anomaly Anomalie *f*, Regelwidrigkeit *f*, Unregelmäßigkeit *f*
anomite *(Min)* Anomit *m*
anorogenic *(Geol)* anorogen
anorthic *(Krist)* triklin
anorthite *(Min)* Anorthit *m*, Kalziumfeldspat *m*
anorthoclase *(Min)* Anorthoklas *m*
anorthosite *(Geol)* Anorthosit *m*
ansa-compound Ansa-Verbindung *f*

antacid säurewidrig
antacid Antazidum *n* *(Mittel gegen Hyperazidität)*
antagonism Antagonismus *m*
antagonist Antagonist *m*, antagonistisch wirkender Stoff *m*
antagonistic action antagonistische Wirkung *f*
antagonize / to entgegenwirken
anthanthrone *(Farb)* Anthanthron *n*
a. dye Anthanthronfarbstoff *m*
anthelmintic anthelmint[h]isch, wurmvertreibend, wurmwidrig, gegen Würmer wirksam
anthelmintic Anthelmint[h]ikum *n*, Wurmmittel *n*, wurmwidriges Mittel *n*
anthocyan *s.* anthocyanin
anthocyanidin Anthozyanidin *n* *(Farbstoff)*
anthocyanin Anthozyan[in] *n* *(Farbstoff)*
anthocyanogen Anthozyanogen *n*, Leukoanthozyan *n*
anthophyllite *(Min)* Anthophyllit *m*
a. asbestos Anthophyllitasbest *m*
anthracene $C_6H_4=(CH_2)=C_6H_4$ Anthrazen *n*
a. blue Anthrazenblau *n*
a. oil Anthrazenöl *n*
a. series Anthrazenreihe *f*
anthracite Anthrazit *m*
a. group Anthrazitgruppe *f*
anthracitic coal anthrazitische Kohle *f*
anthraconite *(Min)* Anthrakonit *m* *(Kalziumkarbonat)*
anthranilate Anthranilat *n*, Anthranilsäureester *m*
anthranilic acid $C_6H_4(NH_2)COOH$ Anthranilsäure *f*, o-Aminobenzoesäure *f*
anthranol Anthranol *n*
anthrapurpurine Anthrapurpurin *n*, 1,2,7-Trihydroxy-9,10-anthrachinon *n*
α-anthraquinoline α-Anthrachinolin *n*, Naphtho[2,3-f]chinolin *n*
anthraquinolinequinone Anthrachinolinchinon *n*, Anthrachinonchinolin *n*
anthraquinone $C_6H_4(CO)_2C_6H_4$ Anthrachinon *n*, 9,10-Dioxo-dihydroanthrazen *n*
a. class Anthrachinonklasse *f*
a. dye Anthrachinonfarbstoff *m*, anthrachinoider Farbstoff *m*
a. series Anthrachinonreihe *f*
a. vat dye Anthrachinonküpenfarbstoff *m*, Küpenfarbstoff *m* der Anthrachinonreihe
anthraquinoneacridone Anthrachinonakridon *n*
anthraquinonecarboxylic acid $C_6H_4(CO)_2C_6H_3COOH$ Anthrachinonkarbonsäure *f*
anthraquinonedisulphonic acid Anthrachinondisulfo[n]säure *f*
anthraquinonesulphonic acid $C_6H_4(CO)_2C_6H_3 \cdot SO_3H$ Anthrachinonsulfo[n]säure *f*
anthraquinonoid anthrachinoid, Anthrachinon...
a. dye *s.* anthraquinone dye
anthraxylon *(Kohlechemie)* Anthraxylon *n*
anthrimid *(Farb)* Anthrimid *n*

anthrone Anthron n, 9,10-Dihydro-9-oxoanthrazen n

anti anti-..., in anti-Stellung befindlich
a.-ager Alterungsschutzmittel n
a.-coincidence (Kph) Anti-Koinzidenz f
a. conformation gestaffelte ($\pm$ anti-periplanare) Konformation (Konstellation) f, anti-Konformation f, anti-Konstellation f, trans-Konformation f, trans-Konstellation f
a.-dimmer s. antidimming agent
a.-dust staubverhindernder Zusatz m
a.-fatiguing agent Ermüdungsschutzmittel n
a.-flex-cracking antioxidant (Gum) Ermüdungsschutzmittel n, Alterungsschutzmittel n gegen Oxydation und Biegerisse
a.-gad-fly oil Bremsenöl n (gegen Bremsen und andere Insekten)
a.-game protective agent Wildverbißschutzmittel n
a.-gas-gangrene serum (Pharm) Gasbrandserum n, Gasgangränserum n
a.-halo backing (layer) s. antihalation coating
a.-icing additive (agent) Anti-Icing-Additiv n, Anti-Icing-Zusatz m, Vereisungsverhinderer m
a.-Markovnikov addition Anti-Markownikow-Addition f, Addition f im Anti-Markownikow-Sinn, durch Peroxide initiierte (ausgelöste) Addition f
a.-position anti-Position f
a.-proteolytic anti-proteolytisch
a.-Stokes line Antistokessche (Anti-Stokessche) Linie f
a.-syn isomerism anti-syn-Isomerie f, syn-anti-Isomerie f
antiabrasion layer Schutzschicht f gegen Abrieb, Verschleißschutzschicht f
antiacid paper säurefreies Papier n
antiageing agent Alterungsschutzmittel n
antiaphrodisiac Antiaphrodisiakum n, An[t]aphrodisiakum n
antiarin $C_{29}H_{42}O_{11}$ Antiarin n (Glykosid aus dem Milchsaft von Antiaris toxicarea)
antiarthritic Antiarthritikum n, Mittel n gegen Gicht
antibacterial antibakteriell, antibakteriell wirkend (wirksam)
antibase Antibase f
antibiosis Antibiose f
antibiotic antibiotisch, antibiotisch wirkend (wirksam)
antibiotic Antibiotikum n, antibiotischer Stoff (Wirkstoff) m, antibiotisches Heilmittel n
a. agent (substance) s. antibiotic
antiblocking das Zusammenbacken verhindernd (bei Folien und Papier)
a. agent Antihaftmittel n (für Folien und Papier)
antibody Antikörper m, Immunkörper m
a. protein Antikörperprotein n, Immunprotein n
antibond s. antibonding
antibonding Lockerung f
a. orbital antibindendes (lockerndes) Orbital n
anticaking agent (Düngemittelindustrie) Antibackmittel n

anticatalysis Antikatalyse f, negative Katalyse f, (i.w.S.) Inhibition f, Hemmung f, Verzögerung f
anticatalyst Antikatalysator m, negativer Katalysator m, (i.w.S.) Inhibitor m, Hemmstoff m, Verzögerer m, Passivator m
anticatalytic antikatalytisch
anticathode Antikatode f
anticavitation agent (Met) Lunkerverhütungsmittel n
anticephalalgic Migränemittel n
antichlor Antichlor n (Mittel zur Chlorentfernung nach Chlorbleiche)
anticlinal trap Antiklinalfalle f, antiklinale Falle f
anticoagulant Antikoaguliermittel n, Antikoagulans n
anticoincidence Antikoinzidenz f
a. counter Antikoinzidenzzähler m
anticonvulsive Antikonvulsivum n, krampflinderndes Mittel n
anticorrosive agent Korrosionsschutzmittel n, Korrosionsinhibitor m, Korrosionshemmer m, Korrosionshemmstoff m, Korrosionsverzögerer m
a. paper Korrosionsschutzpapier n, korrosionsschützendes Papier n, Rostschutzpapier n
anticreaming agent (Farb) Hautverhütungsmittel n, Hautverhinderungsmittel n, Hautverhinderer m, Antihautbildungsmittel n, Antihautmittel n
anticrease finish Knitterarmausrüstung f, Knitterfestausrüstung f, Knitterechtausrüstung f, Knitterarmappretur f, Knitterechtappretur f, knitterbeständige Ausrüstung f
anticreased knitterfrei (knitterarm, knitterfest, knitterecht, knitterbeständig) ausgerüstet
antidepressant Antidepressivum n, Thymoleptikum n, Mittel n gegen depressive Verstimmung
antidiabetic Antidiabetikum n, Mittel n gegen Zuckerkrankheit
antidiarrhoeal Antidiarrhoikum n, Mittel n gegen Durchfall
antidiazo compound $R \cdot N = NOM^I$ Antidiazotat n
antidimming agent Mittel n gegen Beschlagen (z.B. von Schaufensterscheiben)
antidiphtheria serum Diphtherieserum n
antidiuretic antidiuretisch
antidote Gegengift n, Antidot n
antiedrite (Min) Antiedrit m, (veraltet für) Edingtonit m
antielectron Positron n, positives Elektron n
antiemetic Antiemetikum n, Mittel n gegen Erbrechen
antienzyme s. antiferment
antiepileptic Antiepileptikum n, Mittel n gegen Epilepsie
antifatigue agent Ermüdungsschutzmittel n
antifeedant (Landw) fraßverhinderndes Mittel n (gegen Schadinsekten)
antifeeding compound s. antifeedant
antifelting treatment Antifilzausrüstung f, Filzfreiausrüstung f

antiferment Antiferment *n*, Antienzym *n*
antifermentative gärungshemmend, gärungsverhindernd
antiferromagnetism Antiferromagnetismus *m*
antifertility agent *(Pharm)* Antifertilitätspräparat *n*, Ovulationshemmer *m*
antiflare coating Vergütung *f*, reflexmindernde Schicht *f*, reflexmindernder Belag *m*, Antireflexschicht *f*, Antireflexbelag *m*
antifly preparation Fliegenbekämpfungsmittel *n*
antifoam [agent, compound] *s.* antifoaming agent
antifoamer *s.* antifoaming agent
antifoaming agent Antischaummittel *n*, Mittel *n* gegen Schaumbildung, Schaumverhinderungsmittel *n*, Schaumverhütungsmittel *n*, Schaumzerstörungsmittel *n*, Entschäumungsmittel *n*, Entschäumer *m*, Schaumdämpfungsmittel *n*, Schaumdämpfer *m*
antifog agent, antifoggant *s.* antifogging agent
antifogging agent *(Foto)* Antischleiermittel *n*, schleierdämpfendes (schleierverhütendes, schleierwidriges) Mittel *n*; *(Tech)* Klarsichtmittel *n*
antifouling paint Antifoulingfarbe *f*, Antifouling[anstrichmittel] *n* *(anwuchsverhindernde Unterwasseranstrichfarbe)*
antifreeze *s.* a. agent
  a. agent Frostschutzmittel *n*, Gefrierschutzmittel *n*
  a. cardboard Frostschutzpappe *f*
  a. compound (fluid) *s.* a. agent
  a. paper Frostschutzpapier *n*
antifreezing agent (dope) *s.* antifreeze agent
antifriction bearing grease Wälzlagerfett *n*
antifroth agent *s.* antifoaming agent
  a. oil Schaumverhütungsöl *n*, Schaumöl *n*, Schaumbekämpfungsöl *n*
antifrothing agent *s.* antifoaming agent
antifungal antifungal, fungizid [wirksam], pilzwirksam, pilztötend, *(Pharm auch)* antimykotisch
  to have a. activity antifungal (fungizid, gegen Pilze) wirken (wirksam sein), pilzwirksam sein, *(Pharm auch)* antimykotisch wirken
  a. drug Fungizid *n*, *(Pharm auch)* Antimykotikum *n*
antigen Antigen *n*
antigenicity *(Med)* Antigenwirkung *f*
antigorite *(Min)* Antigorit *m*, Blätterserpentin *m*
antihaemorrhagic vitamin *s.* antihemorrhagic vitamin
antihalation backing (coating, layer) *(Foto)* Lichthofschutzschicht *f*
antihemorrhagic vitamin antihämorrhagisches Vitamin *n*, Koagulationsvitamin *n*, Vitamin K$_1$ *n*, Phyllochinon *n*
antihidrotic *s.* antiperspirant
antihistamine, antihistaminic drug Antihistaminikum *n*, Antihistaminpräparat *n*
antihormone Antihormon *n*
antiknock Antiklopfmittel *n*, Klopfbremse *f*

  a. additive (agent) *s.* antiknock
  a. characteristic Klopffestigkeit *f*
  a. compound (dope) *s.* antiknock
  a. effect klopfhindernder (klopfhemmender) Effekt *m*, klopfhindernde (klopfhemmende) Wirkung *f*, Antiklopfwirkung *f*, Klopfbremswirkung *f*
  a. fuel klopffester Kraftstoff *m*
  a. gasoline klopffestes Benzin *n*
  a. material *s.* antiknock
  a. properties Antiklopfeigenschaften *fpl*
  a. quality Klopffestigkeit *f*
  a. rating Klopfwertbestimmung *f*, Klopf[wert]prüfung *f*, Ermittlung *f* der Klopffestigkeit; Klopfwert *m*, Klopffestigkeit *f*
  a. reagent (substance) *s.* antiknock
  a. value Klopfwert *m*, Klopffestigkeit *f*
antileucocytic antileukozytär
antilivering agent Eindick[ungs]verhinderungsmittel *n* *(Lackindustrie)*
antimagnetic antimagnetisch
antimalarial Antimalaria..., gegen Malaria wirksam
antimalarial Malariabekämpfungsmittel *n*, Malariamittel *n*, Malariaheilmittel *n*, Malariapräparat *n*, Antimalariamittel *n*, Antimalariapräparat *n*, Mittel *n* gegen Malariainfektion
  a. activity Antimalariawirksamkeit *f*, Malariawirksamkeit *f*
  a. drug *s.* antimalarial
antimatter Antimaterie *f*
antimer *s.* optical isomer
antimetabolite Antimetabolit *m*
antimicrobial antimikrobiell, antimikrobisch
antimildew agent *(Text)* Antiseptikum *n* gegen Schimmel, verrottungshemmendes Ausrüstungsmittel *n*, Verrottungsschutzmittel *n*
antimonate Antimonat *n*
antimonial cinnabar *s.* antimony cinnabar
  a. lead Hartblei *n*, Blei-Antimon-Legierung *f*
  a. nickel *(Min)* Antimonnickel *n*, *(veraltet für)* Breithauptit *m* *(Nickelantimonid)*
  a. saffron Sb$_2$S$_5$ Antimon(V)-sulfid *n*, Antimonpentasulfid *n*, Goldschwefel *m*
  a. silver *(Min)* Antimonsilber *n*, *(veraltet für)* Dyskrasit *m*, Diskrasit *m*
  a. speiss *(Met)* Antimonspeise *f*
antimoniate *s.* antimonate
antimonic Antimon..., *(i.e.S.)* Antimon(V)-...
  a. acid HSb(OH)$_6$ Antimon(V)-säure *f*
  a. chloride SbCl$_5$ Antimon(V)-chlorid *n*, Antimonpentachlorid *n*
  a. oxide Sb$_2$O$_5$ Antimon(V)-oxid *n*, Diantimonpentoxid *n*, Antimonpentoxid *n*
  a. oxychloride SbOCl$_3$ Antimon(V)-oxidchlorid *n*
  a. sulphide Sb$_2$S$_5$ Antimon(V)-sulfid *n*, Antimonpentasulfid *n*, Goldschwefel *m*
antimonide Antimonid *n*
antimonious *s.* antimonous
antimonite M^I_n(SbO$_2$)$_n$ Antimonat(III) *n*; *(Min)* Antimonit *m*, Stibnit *m*, Antimonglanz *m*, Grauspießglanz *m* *(Antimon(III)-sulfid)*

antimonous Antimon..., *(i.e.S.)* Antimon(III)-...
  a. chloride $SbCl_3$ Antimon(III)-chlorid *n*, Antimontrichlorid *n*, Antimonbutter *f*
  a. nickel *s.* antimonial nickel
  a. oxide Sb_2O_3 Antimon(III)-oxid *n*, Diantimontrioxid *n*, Antimontrioxid *n*
  a. oxychloride $SbOCl$ Antimon(III)-oxidchlorid *n*
  a. sulphate $Sb_2(SO_4)_3$ Antimon(III)-sulfat *n*, Antimontrisulfat *n*
  a. sulphide Sb_2S_3 Antimon(III)-sulfid *n*, Antimontrisulfid *n*
antimony Sb Antimon *n*; *(Glas)* Antimonoxid *n*
  a. blende *(Min)* Antimonblende *f*, *(veraltet für)* Kermesit *m* *(Antimon(III)-oxidsulfid)*
  a. bloom *(Min)* Antimonblüte *f*, Valentinit *m* *(Antimon(III)-oxid)*
  a. butter $SbCl_3$ Antimon(III)-chlorid *n*, Antimontrichlorid *n*, Antimonbutter *f*
  a. cinnabar *(Farb)* Antimonkarmin *n*, Antimonzinnober *m* *(Antimon(III)-sulfid mit Beimengungen von Antimon(III)-oxid und Schwefel)*
  a. electrode Antimonelektrode *f*
  a. glance *(Min)* Antimonglanz *m*, Grauspießglanz *m*, Antimonit *m*, Stibnit *m* *(Antimon(III)-sulfid)*
  a. halide Antimonhalogenid *n*
  a. hydride Antimonhydrid *n*, *(i.e.S.)* SbH_3 Antimonwasserstoff *m*, Stibin *n*, Antimon(III)-hydrid *n*
  a. ochre *(Min)* Antimonocker *m*
  a. pentachloride $SbCl_5$ Antimonpentachlorid *n*, Antimon(V)-chlorid *n*
  a. pentafluoride SbF_5 Antimonpentafluorid *n*, Antimon(V)-fluorid *n*
  a. pentaiodide SbJ_5 Antimonpentajodid *n*, Antimon(V)-jodid *n*
  a. pentasulphide Sb_2S_5 Antimonpentasulfid *n*, Antimon(V)-sulfid *n*, Goldschwefel *m*
  a. pentoxide Sb_2O_5 Antimonpentoxid *n*, Diantimonpentoxid *n*, Antimon(V)-oxid *n*
  a. perchloride *s.* a. pentachloride
  a. persulphide *s.* a. pentasulphide
  a. potassium tartrate $K[C_4H_2O_6Sb(OH_2)] \cdot 0,5H_2O$ Kaliumantimon(III)-tartrat-0,5-Wasser *n*
  a. red Sb_2S_5 Antimonpentasulfid *n*, Antimon(V)-sulfid *n*, Goldschwefel *m*
  a. regulus Antimonregulus *m*
  a. tetroxide Sb_2O_4 Antimontetroxid *n*, Antimon(III,V)-oxid *n*
  a. tribromide $SbBr_3$ Antimontribromid *n*, Antimon(III)-bromid *n*
  a. trichloride $SbCl_3$ Antimontrichlorid *n*, Antimon(III)-chlorid *n*, Antimonbutter *f*
  a. trifluoride SbF_3 Antimontrifluorid *n*, Antimon(III)-fluorid *n*
  a. triiodide SbJ_3 Antimontrijodid *n*, Antimon(III)-jodid *n*
  a. trioxide Sb_2O_3 Antimontrioxid *n*, Diantimontrioxid *n*, Antimon(III)-oxid *n*
  a. triselenide Sb_2Se_3 Antimontriselenid *n*, Antimon(III)-selenid *n*

  a. trisulphate $Sb_2(SO_4)_3$ Antimontrisulfat *n*, Antimon(III)-sulfat *n*
  a. trisulphide Sb_2S_3 Antimontrisulfid *n*, Antimon(III)-sulfid *n*
  a. tritelluride Sb_2Te_3 Antimontritellurid *n*, Antimon(III)-tellurid *n*
  a. vermilion *s.* a. cinnabar
  a. white Sb_2O_3 *(oder)* Sb_4O_6 Antimonweiß *n*, Antimon(III)-oxid *n*
  a. yellow Antimongelb *n*, Neapelgelb *n*, Bleiantimonat(V) *n*
antimonyl SbO— Antimonoxid...
  a. chloride $SbOCl$ Antimon(III)-oxidchlorid *n*
  a. potassium tartrate $K[C_4H_2O_6Sb(OH_2)] \cdot 0,5H_2O$ Kaliumantimon(III)-tartrat-0,5-Wasser *n*
antimutagen Antimutagen *n*
antimycin Antimyzin *n* *(Antibiotikum)*
  a. A $C_{28}H_{40}N_2O_9$ Antimyzin A *n*
  a. lactone Antimyzinlakton *n*
antimycotic Antimykotikum *n*
antineuralgic Antineuralgikum *n*, Mittel *n* gegen Kopf-und Nervenschmerzen
antineuritic antineuritisch
antineutrino Antineutrino *n*
antineutron Antineutron *n*
antinucleon Antinukleon *n*
antioxidant [agent] Antioxydationsmittel *n*, Antioxydans *n*, Antioxygen *n*, Oxydationsverhinderer *m*
antioxidizer *s.* antioxidant
antioxidizing agent *s.* antioxidant
antioxygen *s.* antioxidant
antiozidant, antiozonant Ozonschutzmittel *n*, Antiozonant *m*, Antiozonisator *m*
antiparallax card Ableseblatt *n* *(für Büretten)*
antiparticle Antiteilchen *n*, Antipartikel *f*
antiperiplanar conformation *s.* anti conformation
antiperspirant transpirationshemmend, schweißhemmend, schweißlindernd, antiperspirierend
antiperspirant Antitranspirationsmittel *n*, Antischweißmittel *n*, Schweißhemmungsmittel *n*, Schweißmittel *n*, Antiperspirant *n*, Antihidrotikum *n*, schweißhemmendes (schweißlinderndes, antiperspirierendes) Mittel *n*
  a. lotion Antitranspirationslotion *f*, transpirationsverringernde Lotion *f*, Antischweißlotion *f*
antiperthite *(Geol)* Antiperthit *m*
antiphlogistic antiphlogistisch, entzündungshemmend
antiphlogistic Antiphlogistikum *n*, entzündungshemmendes Mittel *n*, Mittel *n* gegen örtliche Entzündungen
antiplasmodial antiplasmodisch
antipode [optischer] Antipode *m*, enantiomorphe Form *f*, Enantiomer[es] *n*, Spiegelbildisomer[es] *n*
antiprothrombin Antiprothrombin *n*
antiproton Antiproton *n*
antipyretic antipyretisch, fiebersenkend

antipyretic Antipyretikum *n*, fiebersenkendes Mittel *n*, Antifiebermittel *n*, Fiebermittel *n*, Antifebrilium *n*

antique glass Antikglas *n*

antirachitic antirachitisch

a. vitamin antirachitisches Vitamin *n*, Vitamin D *n*

antiredeposition agent *(Text)* Antivergrauungsmittel *n*, Mittel *n* gegen Rückvergrauung (Wiederaufziehen des Schmutzes)

antireflection coating Vergütung *f*, reflexmindernde Schicht *f*, reflexmindernder Belag *m*, Antireflexschicht *f*, Antireflexbelag *m*

antiresistant *(Bioch)* Widerstandsabschwächer *m*

antirheumatic Antirheumatikum *n*, Mittel *n* gegen Rheuma

antirot fäulnisverhindernd, fäulnishindernd, fäulnishemmend, fäulniswidrig

a. substance Antiseptikum *n*

antirust agent Rostverhütungsmittel *n*, Rostschutzmittel *n*

antiscorbutic vitamin antiskorbutisches Vitamin *n*, Vitamin C *n*

antiscorcher, antiscorching agent Antiscorcher *m*, Vulkanisationsverzögerer *m*, Verzögerer *m*

antiseize agent Mittel *n* gegen Festfressen *(von Plasten z.B. in Preßwerkzeugen)*

antiseptic antiseptisch

antiseptic [agent] Antiseptikum *n*, antiseptischer Stoff *m*, antiseptisches Mittel *n*

antisettling agent Absetzverhütungsmittel *n*, Absetzverhinderungsmittel *n*, Antiabsetzmittel *n*, Schwebemittel *n*, absetzverhinderndes Mittel *n*

antiskimming agent *(Farb)* Hautverhütungsmittel *n*, Hautverhinderungsmittel *n*, Hautverhinderer *m*, Antihautbildungsmittel *n*, Antihautmittel *n*

antislip effect Gleitschutzwirkung *f*

a. finish Schiebefestappretur *f*, Schiebefestausrüstung *f*

antisoiling schmutzabweisend, schmutzabstoßend

antispasmodic Antispasmodikum *n*, Spasmolytikum *n*, krampflösendes Mittel *n*

antispatterer *(Lebm)* Antispritzmittel *n*, Mittel *n* zur Verhinderung des [unerwünschten] Spritzens

antistatic antistatisch

antistatic [additive, agent] Antistatikum *n*, Antistatikmittel *n*, antistatisches Mittel *n*

a. finish[ing] antistatische Ausrüstung (Präparation, Appretur) *f*

antisterility vitamin Antisterilitätsvitamin *n*, Fertilitätsvitamin *n*, Vitamin E *n*

antistick abweisend gegen klebrige (klebende) Stoffe, Antikleb..., Antihaft...

a. properties Antiklebvermögen *n*, Antihaftvermögen *n*

antisticking agent Antiklebemittel *n*, Antikleber *m*

antistress mineral *(Geol)* Antistreßmineral *n*

antiswelling finish Quellfestausrüstung *f*

antisymmetric antisymmetrisch

antitack agent Trennmittel *n*

antitarnish paper Korrosionsschutzpapier *n*, korrosionsschützendes Papier *n*, Rostschutzpapier *n*

antitetanus serum Tetanusserum *n*

antithrombin Antithrombin *n*

antitoxic antitoxisch

antitoxin Antitoxin *n*

antitrypsin Antitrypsin *n*

antitussive Antitussivum *n*, Hustenmittel *n*

antiurease Antiurease *f*

antiviral, antivirus virozid

antivitamin Antivitamin *n*, Vitaminantagonist *m*

antlerite *(Min)* Antlerit *m* *(Kupfer(II)-tetrahydroxidsulfat)*

Antonoff rule Antonowsche Regel *f*

Antu, ANTU ANT *n*, Antu *n*, α-Naphthylthioharnstoff *m* *(Rodentizid)*

anvil jaw feste Backe *f* *(des Backenbrechers)*

AO *s.* atomic orbital

AP *(Abk. für)* American Patent

A.P. *s.* annealing point

6-APA *s.* 6-aminopenicillanic acid

apargylic acid HOOC(CH$_2$)$_7$COOH Lepargylsäure *f*, Azelainsäure *f*, Nonandisäure *f*, 1,7-Heptandikarbonsäure *f*

apatite *(Min)* Apatit *m*

aperture Apertur *f*, Öffnung *f*; Öffnungswinkel *m*

a. of a lens Linsenapertur *f*, Linsenöffnung *f*

apex Apex *m* *(eines Hydrozyklons)*

a. opening Apexdüse *f*

aphanesite *(Min)* Aphanesit *m*, *(veraltet für)* Klinoklas *m* *(Kupfer(II)-trihydroxidorthoarsenat)*

aphanin Aphanin *n* *(Algenfarbstoff)*

aphicide Aphizid *n*, Blattlausbekämpfungsmittel *n*

aphrodine Aphrodin *n*, Yohimbin *n*, Quebrachin *n* *(Alkaloid)*

aphrodisiac Aphrodisiakum *n*

aphthitalite *(Min)* Aphthitalit *m*, Glaserit *m* *(Kaliumnatriumsulfat)*

aphtonite *(Min)* Aphtonit *m*, *(veraltet für)* Freibergit *m* *(silberreicher Tetraedrit)*

API *(Abk. für)* American Petroleum Institute

API degree API-Grad *m*, Grad API *m*, °API

API gravity API-Dichte *f*, Dichte *f* in API-Graden

API scale API-Skale *f*

APIC *(Abk. für)* American Petroleum Industries Committee

apigenidin Apigenidin *n*, 5,7,4′-Trihydroxyflavylium *n* *(Anthozyanidin)*

apigenin Apigenin *n*, 5,7,4′-Trihydroxyflavon *n*

aplite *(Geol)* Aplit *m*

aplome *(Min)* Aplom *m*, Andradit *m* *(ein Kalkeisengranat)*

APO *s.* trisaziridinyl phosphine oxide

apo rearrangement Apo-Umlagerung *f*

apoatropine Apoatropin *n* *(Tropinester der Atropasäure; Alkaloid)*

apocrenic acid *(Bodenchemie)* Apokrensäure *f* *(eine Fulvosäure)*

apoenzyme, apoferment Apoenzym n, Apoferment n

apolar apolar, nichtpolar, unpolar

  a. adsorption apolare Adsorption f

apomorphine Apomorphin n (Alkaloid)

  a. hydrochloride Apomorphinhydrochlorid n, 3,4-Dihydroxyapomorphinhydrochlorid n

apophyllite (Min) Apophyllit m

apophysis (Geol) Apophyse f

aporphine Aporphin n

  a. alkaloid Aporphinalkaloid n

apparatus Apparat m, Gerät n, Vorrichtung f, Einrichtung f

  a. clamp Stativklemme f, Apparateklemme f

  a. constant Apparatekonstante f, Apparatkonstante f, Gerätekonstante f

  a. for determining carbon residue Gerät n für den Verkokungstest, Verkokungsapparat m

  a. rack Ablaufbrett n, Abtropfbrett n

apparent scheinbar

  a. density scheinbare Dichte f, Schüttdichte f, Korndichte f (z. B. des Kokses)

  a. porosity (Ker) scheinbare Porosität f, Scheinporosität f

  a. solubility scheinbare Löslichkeit f, Scheinlöslichkeit f

  a. viscosity scheinbare Viskosität f, Scheinviskosität f

appearance potential Erscheinungspotential n, Appearance-Potential n

apple acid HOOCCH(OH)CH₂COOH Äpfelsäure f, Apfelsäure f, Hydroxybernsteinsäure f, Hydroxybutandisäure f

  a. essence (CH₃)₂CH·CH₂COO·C₅H₁₁ Apfelessenz f, Apfeläther m, Isoamylvalerianat n, Valeriansäureisoamylester m

  a. juice Apfelsaft m, Apfelmost m

  a. oil s. a. essence

applicability Anwendbarkeit f, Verwendbarkeit f, Einsetzbarkeit f, Anwendungsmöglichkeit f, Verwendungsmöglichkeit f

application Anwendung f, Einsatz m; (Landw) Ausbringen n (z.B. von Düngemitteln); Auftragen n, Aufbringen n (z.B. von Farben); (Med) Applikation f, Verabreichung f (von Medikamenten)

  a. apparatus s. applicator

  a. of heat Wärmeanwendung f

  a. roll Auftragewalze f, Auftragwalze f, Auftragswalze f

applicator (Med) Applikator m (Gerät zum Verabreichen von Medikamenten); (Landw) Ausbringungsgerät n (für Chemikalien), (bei Pflanzenschutzmitteln auch) Pflanzenschutzgerät n, Applikationsgerät n

  a. roll s. application roll

apply / to anwenden, verwenden, einsetzen; (z.B. Düngemittel) ausbringen; (z.B. Farben) auftragen, aufbringen

  a. a skim-coat skimmen, belegen (bereits friktionierte Gewebe beschichten)

  a. broadcast (Landw) breitwürfig (flächenhaft) ausbringen

appraisal well Erweiterungsbohrung f

approach (Restdifferenz der Temperatur des gekühlten Mediums und des feuchten Thermometers)

approval (staatliche) Anerkennung f (z.B. von Pflanzenschutzmitteln)

approved name offizieller Name m (einer organischen Verbindung)

approximate / to be angenähert gelten

approximate formula angenäherte Formel f, Näherungsformel f

  a. procedure Näherungsverfahren n

  a. value Näherungswert m

approximation method Näherungsmethode f, Näherungsverfahren n

  a. of linear combinations of atomic orbitals LCAO-Näherung f

aprobarbital Aprobarbital n, 5-Allyl-5-isopropylbarbitursäure f

aprofene Aprofen n, 2-Diäthylaminoäthyl-2,2-diphenylpropionat-HCl n

apron Band n, Förderband n; (Pap) Siebleder n, Auflaufleder n

  a. conveyor Plattenbandförderer m, Plattenförderer m, Gliederbandförderer m

  a. dryer Bandtrockner m, Förderbandtrockner m

APS (Abk. für) American Physical Society

apurinic acid Apurinsäure f, APS (Ribonukleinsäurederivat)

aq. s. aqueous

aq. reg. s. aqua regia

aqua ammonia s. aqueous ammonia

  a. fortis HNO₃ Aqua fortis, Scheidewasser n (Salpetersäure)

  a.-igneous (Geol) hydratopyrogen

  a. regia Aqua regia, Königswasser n, Kö., Kw. (Salzsäure-Salpetersäure-Gemisch)

aquamarine (Min) Aquamarin n

aquametry Aquametrie f

aquarel[le] paper Aquarellpapier n

aquasol Hydrosol n

aquated hydratisiert

aquation, aquatization Hydratation f

aqueous, wäßrig, wässerig, wssr., wss., Wasser..., wasserhaltig; wasserartig; (Geol) aquatisch

  a. ammonia [wäßrige] Ammoniaklösung f, Ammoniak n in wäßriger Lösung, wäßriges Ammoniak n, Ammoniakwasser n, Ammoniakflüssigkeit f, Salmiakgeist m

  a. deposit (Geol) aquatischer Absatz m

  a. emulsion wäßrige Emulsion f

  a. iodine solution (Pharm) wäßrige Jodlösung f, Lugolsche Lösung f

  a. phase wäßrige Phase f, Wasserphase f

  a. pulp of ground ore wäßrige Erztrübe f

  a. solubility Wasserlöslichkeit f, Löslichkeit f in Wasser

  a. solution wäßrige Lösung f

a. solution of iodine s. a. iodine solution

a. stuff *(Pap)* Stoffsuspension *f*, Stoffwasser *n*

aquo compound Aquoverbindung *f*

a.-ion Aquóion *n*, hydratisiertes Ion *n*

araban ($C_5H_{10}O_5$)$_n$ Araban *n (zu den Pentosanen gehörendes Polysaccharid)*

Arabian gum arabisches Gummi *n*, Gummiarabikum *n*, Sudangummi *n*, Gummi arabicum

arabinose Arabinose *f*, Pektinose *f*, Gummizucker *m*

arabinosic Arabinose...

arabitol $CH_2OH(CHOH)_3CH_2OH$ Arabit *m*

arachic (arachidic) acid $CH_3(CH_2)_{18}COOH$ Arachinsäure *f*, Erdnußsäure *f*, *n*-Eikosansäure *f*

arachidonic acid $C_{19}H_{31}COOH$ Arachidonsäure *f*, 5,8,11,14-Eikosantetraensäure *f*

arachis Erdnuß *f*, Arachis hypogaea L.; Erdnuß *f*

a. oil Arachisöl *n*, Erdnußöl *n*

arachnolysine Arachnolysin *n (ein Spinnengift)*

aragonite, aragonspath *(Min)* Aragonit *m (Kalziumkarbonat)*

aralkyl Aralkyl *n*

a. sulphonate Alkylarylsulfonat *n*, Aralkylsulfonat *n (Waschrohstoff)*

araroba *(Pharm)* Goapulver *n*, Ararobapulver *n (von Andira araroba Aguiar)*

arbitration[al] analysis Schiedsanalyse *f*, Schiedsprobe *f*, Schiedsuntersuchung *f*

arbutin $HOC_6H_4 \cdot O \cdot C_6H_{11}O_5$ Arbutin *n (Glukosid des Hydrochinons)*

arc Lichtbogen *m*

a. carbon Lichtbogenkohle *f*, Bogenlampenkohle *f*

a. discharge Lichtbogenentladung *f*, Bogenentladung *f*

a.[-heated] furnace Lichtbogenofen *m*

a. heating Lichtbogenbeheizung *f*, Lichtbogenheizung *f*

a. lamp Lichtbogenlampe *f*, Bogenlampe *f*

a. process Lichtbogenverfahren *n (zur Bindung des Luftstickstoffs)*

a.-proof lichtbogenfest, lichtbogenbeständig

a. resistance Lichtbogenfestigkeit *f*, Lichtbogenbeständigkeit *f*

a. resistance furnace Lichtbogenwiderstandsofen *m*

a.-resistant lichtbogenfest, lichtbogenbeständig

a. spectrum Bogenspektrum *n*

a. tracking Kriechwegbildung *f* durch Lichtbogen

arcanite *(Min)* Arkanit *m (Kaliumsulfat)*

archil Archil *n*, Orseille *f*, Cudbear *n*, Persio *n*, Pourpre Française *f (ein Flechtenfarbstoff)*

arching *(Pap)* Brückenbildung *f (der Hackschnitzel im Silo)*

architectural terra-cotta *(Ker)* Bauterrakotta *f*

ardent spirits [hochprozentige] alkoholische Getränke *npl*

aera flowmeter (Durchflußmengenmesser (Mengenstrommesser),bei dem die Strömungskraft auf einen umströmten Körper gemessen wird)

a. of contact Berührungsfläche *f*, Kontaktfläche *f*

a. of cooling surface Kühlfläche *f*

a. of spread Größe *f* der bedeckten Fläche *(Netzmittelprüfung)*

areca nut Betelnuß *f*, Arekanuß *f*

arecai[di]ne Arekai[di]n *n*, N-Methyl-tetrahydronikotinsäure *f*

arecolidine Arekolidin *n (Alkaloid der Betelnuß)*

arecoline Arekolin *n*, Arekaidinmethylester *m*

arenaceous sandig, sandhaltig; sandartig

a. rocks *(Geol)* Sandgestein *n*

arenarious s. arenaceous

arendalite *(Min)* Arendalit *m*, *(veraltet für)* Epidot *m*

arenite Psammit *m (ein mittelklastisches Gestein)*

areometer Aräometer *n*, Senkwaage *f*, Spindel *f*

areometric aräometrisch

areometry Aräometrie *f*

arfvedsonite *(Min)* Arfvedsonit *m (eine Alkalihornblende)*

argal s. argol

Argan gum *(Pflanzengummi von Argania sideroxylon Roem., dem Marokkanischen Eisenholzbaum)*

Arge synthesis Arge-Synthese *f*

argentic Silber..., *(i.e.S.)* Silber(II)-...

a. fluoride AgF_2 Silberdifluorid *n*, Silber(II)-fluorid *n*

argentiferous silberhaltig, silberführend

argentine *(Min)* Argentin *m (Kalziumkarbonat)*

argentite *(Min)* Argentit *m (Silbersulfid)*

argentocyanide Zyanoargentat(I) *n*

argentometric argentometrisch

argentometry Argentometrie *f*

argentopyrite *(Min)* Argentopyrit *m*

argentous Silber..., *(i.e.S.)* Silber(I)-...

a. oxide Ag_2O Silber(I)-oxid *n*

argil Tonerde *f*, Ton *m*

argilla Ton *m*; Kaolin *n*, Porzellanerde *f*

argillaceous lehmig, lehmhaltig, Lehm...; tonig, tonhaltig, Ton...

a. hematite Toneisenstein *m (Hämatit mit Ton vermengt)*

a. limestone Tonkalk *m*

argilliferous lehmhaltig; tonhaltig

argillite Argillit *m*, Tonschiefer *m*

arginase Arginase *f (manganhaltiges Ferment)*

arginine Arginin *n*, Arg, α-Amino-δ-guanidovaleriansäure *f*

argipressin Argipressin *n*, 8-Argininvasopressin *n*

argol [roher] Weinstein *m*, Rohweinstein *m (Kaliumhydrogentartrat)*

argon Ar Argon *n*

a. atmosphere Argonatmosphäre *f*

a. clathrate Argonklathrat *n*

argyrodite *(Min)* Argyrodit *m (Ge-haltiges Silbererz)*

aricite *(Min)* Arizit *m*, *(veraltet für)* Gismondin *m (Kalziumdialumodisilikat)*

Aridye process Aridye-Verfahren *n (Pigmentdruck-verfahren)*

aristolochic acid, aristolochin Aristolochiasäure *f*, Aristolochiagelb *n*, Aristolochin *n*, 1-Methoxy-5,6-methylendioxy-9-nitro-8-phenanthroinsäure *f*

arizonite *(Min)* Arizonit *m (Eisen(III)-metatitanat)*

arkansite *(Min)* Arkansit *m (Titandioxid)*

arkose *(Geol)* Arkose *f*

arm Schaufel *f*, Arm *m*, Flügel *m*, Paddel *n*; Abzweigung *f*
  a. mixer Schaufelmischer *m*, Flügelmischer *m*, Schaufelrührer *m*, Schaufelrührwerk *n*, Paddelrührer *m*

Armco iron Armcoeisen *n (technisch reines Eisen)*

armoured glass Panzerglas *n*

Armstrong method Armstrong-Verfahren *n (zur Herstellung von 2-Naphthol-1-sulfonsäure)*

arnatto Annatto *n(m)*, Anatto *n(m)*, Orlean *m*, Annattofarbstoff *m (von Bixa orellana L.)*

Arndt-Eistert synthesis Arndt-Eistert-Synthese *f*, Arndt-Eistertscher Karbonsäureaufbau *m*

Arndt's alloy Arndtsche Legierung *f (Cu-Mg-Legierung; Reduktionsmittel)*

arnica flowers oil Arnika[blüten]öl *n*

Arnold-Mentzel formaldehyde test Prüfung *f* auf Formaldehyd nach Arnold und Mentzel, Reaktion *f* von Arnold und Mentzel

Arnold's test Arnoldsche Farbreaktion *f*, Arnold-Probe *f (Azetessigsäure-Nachweis)*

arnotta *s.* arnatto

Arnu-Audibert's dilatometer test Dilatometertest *m* nach Arnu-Audibert

aroma Aroma *n*, Geruch *m*, Wohlgeruch *m*, Duft *m*; Würze *f*, Blume *f (des Weins)*
  a. bacteria aromabildende Bakterien *npl*, Aromabakterien *npl*, Aromabildner *mpl*
  a. body *s.* a. substance
  a. compound Aromakomposition *f*, Aromaträger *m*
  a. development Aromabildung *f*
  a.-forming bacteria *s.* a. bacteria
  a. ingredient *s.* a. substance
  a. loss Aromaverlust *m*
  a. organisms (producers) aromabildende Mikroben *fpl*, Aromabildner *mpl*
  a. substance Aromastoff *m*, Aroma *n*

aromadendrin Aromadendrin *n*, 2,3-Dihydro-kämpferol *n (Flavanol)*

aromatic 1. aromatisch *(den Benzolring enthaltend)*; 2. aromatisch, wohlriechend, duftig, würzig
  a.-alkane compound Alkylbenzol *n*
  a.-alkene compound Alkenylbenzol *n*
  a.-alkyne compound Alkinylbenzol *n*
  a. compounds *s.* a. hydrocarbons
  a. ether aromatischer Äther *m*, Aryläther *m*
  a. halogenation Halogenierung *f* aromatischer Kohlenwasserstoffe
  a. hydrocarbons aromatische Kohlenwasser-

stoffe *mpl*, Benzolkohlenwasserstoffe *mpl*, Aromaten *pl*
  a. nucleus *s.* a. ring
  a. ring Benzolring *m*, Benzolkern *m*, aromatischer Kern *m*
  a. substance Aromastoff *m*, Aroma *n*
  a. waters *(Pharm)* aromatische Wässer *npl*

aromatics *s.* aromatic hydrocarbons

aromatization 1. Aromatisieren *n*, Aromatisierung *f (Überführung z.B. von Alkanen und Alkenen in aromatische Kohlenwasserstoffe)*; 2. Aromatisieren *n*, Aromatisierung *f (Versehen mit Aroma)*

aromatize / to 1. aromatisieren *(in Kohlenwasserstoffe überführen)*; 2. aromatisieren *(mit Aroma versehen)*

aromatizing *s.* aromatization
  a. product Aromastoff *m*, Aroma *n*

aronium ion Aroniumion *n*

aroyl Aroyl *n*

arrack Arrak *m*

arrangement Ordnung *f*, Anordnung *f*, Einrichtung *f*, Einteilung *f*; Aufstellung *f*, Zusammenstellung *f*; System *n*; [räumliche] Anordnung *f*, Konfiguration *f*
  a. of orbital electrons Elektronenanordnung *f*, Elektronenkonfiguration *f*

arrest / to arretieren, feststellen, festhalten; *(eine Reaktion)* zum Stehen bringen, abbrechen

arrest device Arretierung *f*, Arretiervorrichtung *f*, Sperrvorrichtung *f*, Sperre *f*

arrester Staubabscheider *m*

arrhenite *(Min)* Arrhenit *m*

Arrhenius equation Arrheniussche Gleichung (Formel) *f*

arrojadite *(Min)* Arrojadit *m*

arrow poison Pfeilgift *n*

arrowroot Arrow-root *n*, Pfeilwurzelmehl *n*, Marantastärke *f (von Maranta arundinacea L. u. a.)*

arsanilic acid $C_6H_4NH_2 \cdot AsO(OH)_2$ Arsanilsäure *f*, p-Aminophenylarsinsäure *f*

arsenate Arsenat(V) *n*

arsenian arsenhaltig

arseniate *s.* arsenate

arsenic Arsen..., *(i.e.S.)* Arsen(V)-...; Arsenik...

arsenic As Arsen *n*; As_2O_3 Arsenik *n*, Diarsentrioxid *n*, Arsentrioxid *n*, Arsen(III)-oxid *n*
  α-arsenic gelbes Arsen *n*
  β-arsenic schwarzes Arsen *n*
  a. acid H_3AsO_4 Arsen(V)-säure *f*, Arsensäure *f*
  a. anhydride As_2O_5 Arsensäureanhydrid *n*, Arsenpentoxid *n*
  a. bisulphide *s.* a. disulphide
  a. bloom *s.* arsenolite
  a. disulphide Diarsendisulfid *n*, *(veraltet für)* As_4S_4 Tetrarsentetrasulfid *n*
  a. iron *(Min)* Arsenopyrit *m*, Arsenkies *m*, Mißpickel *m*, Giftkies *m (Eisenarsensulfid)*; *(Min)* Arsenikalkies *m*, Arseneisen *n*, *(veraltet für)* Löllingit *m (Eisenarsenid)*
  a. limit *(Toxikologie)* Grenzwert *m* für Arsen

(ausgedrückt in As$_2$O$_3$)
a. mirror Arsenspiegel *m*
a. ore Arsenerz *n*
a. paint *(Gerb)* Arsenikschwöde *f*
a. pentachloride AsCl$_5$ Arsenpentachlorid *n*, Arsen(V)-chlorid *n*
a. pentafluoride AsF$_5$ Arsenpentafluorid *n*, Arsen(V)-fluorid *n*
a. pentaiodide AsJ$_5$ Arsenpentajodid *n*, Arsen(V)-jodid *n*
a. pentasulphide As$_2$S$_5$ Arsenpentasulfid *n*, Diarsenpentasulfid *n*, Arsen(V)-sulfid *n*
a. pentoxide As$_2$O$_5$ Arsenpentoxid *n*, Diarsenpentoxid *n*, Arsen(V)-oxid *n*
a. poisoning Arsenvergiftung *f*; Arsenikvergiftung *f*
a. sulphide *(Gerb)* Arsensulfid *n* *(Gemisch mit Hauptanteilen Arsendisulfid und Arsentrisulfid)*
a. tersulphide *s. a.* trisulphide
a. test Arsenprobe *f*
a. tribromide AsBr$_3$ Arsentribromid *n*, Arsen(III)-bromid *n*
a. trichloride AsCl$_3$ Arsentrichlorid *n*, Arsen(III)-chlorid *n*
a. trihydride AsH$_3$ Arsen(III)-hydrid *n*, Arsen(III)-wasserstoff *m*, Arsin *n*, Monoarsin *n*
a. triiodide AsJ$_3$ Arsentrijodid *n*, Arsen(III)-jodid *n*
a. trioxide As$_2$O$_3$ Arsentrioxid *n*, Diarsentrioxid *n*, Arsen(III)-oxid *n*, Arsenik *n*
a. trisulphide As$_2$S$_3$ Arsentrisulfid *n*, Diarsentrisulfid *n*, Arsen(III)-sulfid *n*
a. vapour Arsendampf *m*
a. yellow gelbes Arsenik (Schwefelarsen) *n*, Auripigment *n*, Operment *n*, Königsgelb *n*, Rauschgelb *n*, Chinagelb *n*, Gelbglas *n*
arsenical arsenhaltig, Arsen...
arsenical Arsenikalie *f*, arsenhaltiger Stoff *m*, arsenhaltige Substanz *f*; Arsenverbindung *f*
a. cobalt *(Min)* Arsenkobalt *m* *(Kobaltarsenid)*
a. copper *(Min)* Arsenkupfer *n*, *(veraltet für)* Domeykit *m* *(Kupferarsenid)*
a. insecticide Arseninsektizid *n*
a. iron *s.* arsenic iron
a. nickel *(Min)* Rotnickelkies *m*, Rotnickel *n*, Nickelin *n*, Niccolit *m* *(Nickelarsenid)*
a. pyrite *s.* arsenopyrite
a. speiss *(Met)* Arsenspeise *f*
arsenicate / to mit Arsen behandeln
arsenide Arsenid *n*
arsenious Arsen..., *(i.e.S.)* Arsen(III)-...
a. acid H$_3$AsO$_3$ Arsen(III)-säure *f*, arsenige Säure *f*; *s.* arsenic trioxide
a. anhydride *s.* a. oxide
a. bromide AsBr$_3$ Arsen(III)-bromid *n*, Arsentribromid *n*
a. chloride AsCl$_3$ Arsen(III)-chlorid *n*, Arsentrichlorid *n*
a. fluoride AsF$_3$ Arsen(III)-fluorid *n*, Arsentrifluorid *n*

a. hydride AsH$_3$ Arsen(III)-hydrid *n*, Arsen(III)-wasserstoff *m*, Arsin *n*, Monoarsin *n*
a. iodide AsJ$_3$ Arsen(III)-jodid *n*, Arsentrijodid *n*
a. oxide As$_2$O$_3$ Arsen(III)-oxid *n*, Diarsentrioxid *n*, Arsentrioxid *n*, Arsenik *n*
a. oxychloride AsOCl Arsen(III)-oxidchlorid *n*
a. phosphide AsP Arsen(III)-phosphid *n*
a. selenide As$_2$Se$_3$ Arsen(III)-selenid *n*, Diarsentriselenid *n*, Arsentriselenid *n*
a. sulphide As$_2$O$_3$ Arsen(III)-sulfid *n*, Diarsentrisulfid *n*, Arsentrisulfid *n*
a. sulphide sol Arsentrisulfidsol *n*, As$_2$S$_3$-Sol *n*
arsenite Arsenat(III) *n*
arseniuretted hydrogen *s.* arsenious hydride
arsenobenzene C$_6$H$_5$As=AsC$_6$H$_5$ Arsenobenzol *n*
arsenoferrite *(Min)* Arsenoferrit *m* *(Eisenarsenid)*
arsenolite *(Min)* Arsenolith *m*, Arsenikblüte *f*, Arsenblüte *f* *(Arsen(III)-oxid)*
arsenopyrite *(Min)* Arsenopyrit *m*, Arsenkies *m*, Mißpickel *m*, Giftkies *m* *(Eisenarsensulfid)*
arsenosulphide M^{II}AsS Arsenosulfid *n*
arsenous *s.* arsenious
arsenphyllite *(Min)* Arsenphyllit *m*, *(veraltet für)* Claudetit *m* *(Arsen(III)-oxid)*
arsine AsH$_3$ Arsin *n*, Monoarsin *n*, Arsen(III)-wasserstoff *m*, Arsen(III)-hydrid *n*
arsino-acetic acid (HO)$_2$AsCH$_2$COOH Arsinoessigsäure *f*
arsonium group Arsoniumgruppe *f*, Arsoniumrest *m*
arsphenamine C$_{12}$H$_{12}$As$_2$N$_2$O$_2$·2HCl·2H$_2$O Arsphenamin *n*, 3,3'-Diamino-4,4'-dihydroxidarsenobenzol-Dihydrochlorid *n*
art base paper Kunstdruckrohpapier *n*
a. cardboard Kunstdruckkarton *m*
a. paper Kunstdruckpapier *n*
arterite, arteritic migmatite Arterit *m* *(Adergneis)*
artifact Artefakt *n*
artificial künstlich, Kunst..., *(auch)* synthetisch, Synthese...
a. ageing künstliche (beschleunigte) Alterung *f*, Kurzalterung *f*
a. ageing test künstliche (beschleunigte) Alterungsprüfung *f*
a. ant oil synthetisches Ameisenöl *n*, Furfural *n*, Fural *n*, 2-Furfuraldehyd *m*, 2-Furaldehyd *m*, 2-Furankarbonal *n*
a. butter Kunstbutter *f*, Margarine *f*
a. camphor künstlicher Kampfer *m*, Bornylchlorid *n*, 2-Chlor-1,7,7-trimethylbizyklo-[1,2,2]-heptan *n*
a. coal künstliche (synthetische) Kohle *f*, Kunstkohle *f*
a. cotton Zellstoffwatte *f*
a. cream Kunstrahm *m*
a. drying künstliches Trocknen *n*
a. essential oil of almond C$_6$H$_5$CHO künstliches Bittermandelöl *n*, Bittermandelessenz *f*, Benzaldehyd *m*
a. fat Synthesefett *n*, Kunstfett *n*

a. fertilizer Handelsdünger *m*, Kunstdünger *m*
a. fibre *(veraltet für)* man-made fibre
a. gold [unechtes] Muschelgold *n*, Musivgold *n*, Zinn(IV)-sulfid *n*
a. gum ($C_6H_{10}O_5)_x$ Dampfgummi *n*, Kristallgummi *n*, Stärkegummi *n*, Dextrin *n*
a. hair Kunsthaar *n*
a. honey Kunsthonig *m*
a. horn Kunsthorn *n*
a. ice Kunsteis *n*
a. isotope künstlich erzeugtes Isotop *n*
a. leather Kunstleder *n*
a. light künstliches Licht *n*, Kunstlicht *n*
a. manure Kunstmist *m*
a. musk künstlicher (synthetischer) Moschus *m*, 1-(Dimethyläthyl)-3-methyl-2,4,6-trinitrobenzol *n*
a. oil of ants *s. a.* ant oil
a. oil of wintergreen $HOC_6H_4COOCH_3$ künstliches Wintergrünöl *n*, Methylsalizylat *n*
a. parchment Pergamentersatzpapier *n*, Pergamentersatz *m*
a. radioactivity künstliche (induzierte) Radioaktivität *f*
a. resin Kunstharz *n*
a. rubber künstlicher (synthetischer) Kautschuk *m*, Kunstkautschuk *m*, Synthesekautschuk *m*, SK
a. silk Regeneratseide *f*, Kunstseide *f*, Chemieseide *f*
a. soil *(Text)* künstlicher Schmutz *m*
a. transmutation [künstliche] Kernumwandlung *f*
artificially soiled *(Text)* künstlich angeschmutzt
artists' cardboard Zeichenkarton *m*, Malerpappe *f*, Malpappe *f*, Malkarton *m*, Ölkarton *m*
a. colour Künstlerfarbe *f*
Arundo donax cane *(Pap)* Riesenschilf *n*, Pfahlrohr *n*, Arundo donax L.
aryl Aryl *n*
a. alkyl silicone Arylalkylsilikon *n*, Aralkylsilikon *n*, Alkylarylsilikon *n*
a. coupling Arylverknüpfung *f*
a. ether Aryläther *m*, aromatischer Äther *m*
a. group Arylgruppe *f*, Arylrest *m*
a. halide Arylhalogenid *n*
a.-phenol rearrangement Arylphenolumlagerung *f*
a. radical Arylgruppe *f*, Arylrest *m*, Arylradikal *n*; [freies] Arylradikal *n*
a. residue Arylrest *m*, Arylgruppe *f*
a. silicone Arylsilikon *n*
arylazo group Ar·N=N— Arylazogruppe *f*, Arylazorest *m*
arylene Arylen *n*
arylesterase Arylesterase *f* *(ein Ferment des menschlichen Blutserums)*
arylsulphonic acid Arylsulfonsäure *f*
as-mined ore Fördererz *n*, [aus dem Bergwerk kommendes] Roherz *n*
as-received im (bezogen auf den) Anlieferungszustand

as-sintered im (bezogen auf den) gesinterten Zustand
ASA *(Abk. für)* American Standards Association
asafetida Asa foetida *f*, Asant *m*, Stinkasant *m*, Teufelsdreck *m* *(Gummiharz von Ferula-Arten)*
asbestine *(Pap)* Asbestine *f*
asbestos Asbest *m*
a. board Asbestpappe *f*; Asbestplatte *f*
a. card Asbestpappe *f*
a. cement Asbestzement *m*
a. cloth Asbesttuch *n*, Asbestgewebe *n*, Asbeststoff *m*
a. clothing Asbestkleidung *f*
a. cord Asbestschnur *f*
a. diaphragm Asbestdiaphragma *n*, Asbestmembran[e] *f*
a. disk Asbestscheibe *f*
a. felt Asbestfilz *m*; *(Pap)* Baumwollfilz *m* mit Asbestzusatz
a. fibre Asbestfaser *f*
a. filter Asbestfilter *n*
a. finger cot Asbestfingerling *m*
a. gasket Asbestdichtung *f*
a. gauze Asbestdrahtnetz *n*
a. glove Asbesthandschuh *m*
a. joint Asbestdichtung *f*
a. mat Asbestplatte *f*, Asbestunterlage *f*
a. milk Asbestaufschlämmung *f*
a. mitt[en] Asbestfausthandschuh *m*
a. packing Asbestdichtung *f*, Asbestpackung *f*
a. paper Asbestpapier *n*
a. pipe Asbestrohr *n*, Asbeströhre *f*
a. plate Asbestplatte *f*
a.-pulp disk Asbestmasseplatte *f*
a.-rubber material It-Stoff *m* *(Dichtungsstoff)*
a. slate Asbestschiefer *m*
a. suspension *s. a.* milk
a. tape Asbestband *n*
a. wool Asbestwolle *f*
a. yarn Asbestgarn *n*
asbestosis *(Med)* Asbestose *f*
asbestus *s.* asbestos
asbolane, asbolite *(Min)* Asbolan *m*, Erdkobalt *m* *(Kobaltmanganerz)*
ascaridole Askaridol *n*, 1,4-Peroxido-β-menthen-(2) *n*
ascending chromatography aufsteigende Papierchromatografie *f*
a. chromatography tank Chromatografiegefäß *n* (Entwicklungsgerät *n*, Entwicklungskammer *f*) für die aufsteigende Methode, Steiggefäß *n*
a. current aufwärtsgerichtete Strömung *f*, Aufstrom *m*, aufsteigender Strom *m*
ascension pipe Steig[e]rohr *n*
aschistic *(Geol)* aschist, ungespalten
aschistite *(Geol)* aschistes (ungespaltenes) Ganggestein *n*
ascorbic acid Askorbinsäure *f*
a. [acid] oxidase Askorbinsäureoxydase *f*
ascorbigen Askorbigen *n*

asebotoxin Asebotoxin *n*, Andromedotoxin *n*
asepsis Asepsis *f*, Keimfreiheit *f*, Sterilität *f*
aseptic aseptisch, keimfrei, steril
ash / to veraschen
ash 1. Asche *f*; 2. Na_2CO_3 kristallwasserfreie (wasserfreie, kalzinierte) Soda *f*, wasserfreies Natriumkarbonat *n*; 3. Esche *f*, Fraxinus L.; Eschenholz *n*
   a. content Asche[n]gehalt *m*
   a. curve Asche[n]gehaltskurve *f*
   a.-discharging vessel Ascheaustragsschleuse *f*, Asche[n]schleuse *f*
   a. fluid temperature Asche[n]erweichungspunkt *m*
   a.-free asche[n]frei, af
   a.-free filter asche[n]freies Filter *n*
   a. hopper Aschentrichter *m*, Schlackentrichter *m*
   a. pan Asche[n]kasten *m*, Asche[n]schüssel *f*
   a. pit Asche[n]raum *m*, Asche[n]fall *m*, Asche[n]keller *m*, Asche[n]grube *f*
   a. removal Asche[n]entfernung *f*, Asche[n]austrag *m*, Asche[n]austragung *f*, Entaschung *f*
   a. scale Aschenwaage *f*
   a. tuff *(Geol)* Aschentuff *m*
   a. zone Aschenzone *f*
ashing Veraschen *n*, Veraschung *f*
ashlar lime Quaderkalk *m*
ashless asche[n]frei, af; asche[n]arm, nahezu asche[n]frei
   a. filter *s.* ash-free filter
asiderite *(Geol)* Asiderit *m*
ASLE *(Abk. für)* American Society of Lubrication Engineers
asparagic acid *s.* aspartic acid
asparaginase Asparaginase *f* *(zu den Azylamidasen gehörendes Ferment)*
asparagine $HOOCCHNH_2CH_2CONH_2$ Asparagin *n*, Asp-NH₂
asparaginic acid *s.* aspartic acid
asparagus stone *(Min)* Asparagusstein *m* *(Apatitvarietät)*
aspartic acid $HOOCCHNH_2CH_2COOH$ Asparaginsäure *f*, Asp, Aminobernsteinsäure *f*, Aminobutandisäure *f*
aspen Zitterpappel *f*, Aspe *f*, Espe *f*, Populus tremula L.; *(Am)* Amerikanische Zitterpappel *f*, Populus tremuloides Michx.
aspergillic acid Aspergillsäure *f*
asphalt Asphalt *m*, Erdpech *n*, Erdharz *n* *(Naturasphalt)*; Erdölasphalt *m*, Pertroleumasphalt *m*, Petrolasphalt *m* *(künstlicher Asphalt)*
   a. base Asphaltbasis *f*, asphaltische Basis *f*, asphaltischer Charakter *m* *(eines Roherdöls)*
   a.-base crude *s.* a.-base petroleum
   a.-base petroleum asphaltbasisches (asphaltisches, asphaltöses) Erdöl (Rohöl) *n*, Asphaltbasisöl *n*, Asphaltöl *n*, Erdöl *n* auf Asphaltbasis
   a. concrete Asphaltbeton *m*
   a. lake *(Geol)* Asphaltsee *m*, Pechsee *m*
   a. macadam Asphaltmakadam *m(n)*

   a. mastic Asphaltmastix *m*
   a. paint Asphaltfarbe *f*
   a. paper Asphaltpapier *n*, Bitumenpapier *n*, bituminiertes Papier *n*, Teerpapier *n*; Doppelpechpapier *n*
   a. process Asphaltverfahren *n*, Asphaltfotografie *f*, Heliografie *f*
   a. rock Asphaltgestein *n*
asphalte *s.* asphalt
asphaltene Asphalten *n*
asphaltic Asphalt..., asphaltisch, asphalthaltig
   a.-base crude oil *s.* asphalt-base petroleum
   a. binder Asphaltbinder *m*
   a. bitumen Asphaltbitumen *n*, Bitumen *n* *(i.e.S.)*
   a. cement Asphaltzement *m*
   a. concrete Asphaltbeton *m*
   a. felt Bitumendachpappe *f*, Teerdachpappe *f*
   a. petroleum *s.* asphalt-base petroleum
   a. residue Erdölasphalt *m*, Petroleumasphalt *m* *(künstlicher Asphalt)*
asphaltite Asphaltit *m* *(Naturasphalt)*
asphaltogenic acid Asphaltogensäure *f*
asphaltum *s.* asphalt
aspic jelly Aspik *m(n)*
   a. oil *(Pharm)* Spiköl *n* *(von Lavandula latifolia)*
aspidocarpine Aspidokarpin *n*
aspidolimine Aspidolimin *n*
aspidospermine Aspidospermin *n* *(Alkaloid)*
aspirate / to ansaugen
aspiration Ansaugen *n*
   a. psychrometer Aspirationspsychrometer *n*
aspirator Sauggebläse *n*, Saugventilator *m*, Absauger *m*, Aspirator *f*
   a. bottle Saugflasche *f*
Asplund process *(Pap)* Asplund-Defibrator-Verfahren *n*
Assam rubber Assamkautschuk *m* *(von Ficus elastica Roxb.)*
assay / to prüfen, untersuchen, analysieren; *(Met)* Erzprobe durchführen
assay Probe *f*, Versuch *m*, Analyse *f*, Bestimmung *f*; *(Met)* Erzprobe *f*
   a. at source vom Hersteller vorgenommene Bestimmung *f*
   a. balance chemische Waage *f*, *(i.e.S.)* analytische Waage *f*, Analysenwaage *f*
assemblage *(Geol)* Vergesellschaftung *f*
assemble / to *(Gum)* konfektionieren, zusammenbauen
assembling department *(Gum)* Konfektionsabteilung *f*, Konfektionierraum *m*
   a. solution *(Gum)* Konfektionierlösung *f*, Konfektionslösung *f*
   a. table *(Gum)* Konfektioniertisch *m*
assembly 1. (Gum) Konfektion[ierung] *f*, Zusammenbau *m*; 2. Stativharfe *f*, Harfe *f*, Universalstativ *n*
   a. department *s.* assembling department
assimilate / to assimilieren
assimilation Assimilation *f*; *(Geol)* Assimilation *f*, Einschmelzung *f*, Aufzehrung *f*

associate / to 1. assoziieren; sich assoziieren; 2. *(Min)* vergesellschaften, assoziieren; sich vergesellschaften (assoziieren)
associated liquid assoziierte Flüssigkeit f, Assoziationsflüssigkeit f
association Assoziation f; *(Min)* Vergesellschaftung f, Assoziation f
a. constant Assoziationskonstante f ·
assortment of dyes Farbstoffsortiment n
astatic astatisch
a. galvanometer astatisches Galvanometer n
astatine At Astat n
asterism *(Krist)* Asterismus m
asteroite *(Min)* Asteroit m, (veraltet für) Hedenbergit m (Kalziumeisen(II)-disilikat)
asthenosphere *(Geol)* Asthenosphäre f
asthma weed *(Pharm)* Lobelienkraut n (von Lobelia inflata L.)
ASTM *(Abk. für)* American Society for Testing and Materials
ASTM distillation ASTM-Destillation f, Destillation f nach ASTM
ASTM method ASTM-Methode f, ASTM-Verfahren n
ASTM specification ASTM-Spezifikation f
Aston dark space Astonscher Dunkelraum m
Aston mass spectrograph Astonscher Massenspektrograf m, Aston-Massenspektrograf m
Aston whole number rule Astonsche Regel f
Astra pressure cooler *(Lebm)* Astra-Druckschneckenkühler m
astrachanite s. astrakanite
astragalin Astragalin n, Kämpferol-3-glukosid n
astrakanite *(Min)* Astrakanit m, Blödit m *(Magnesiumnatriumsulfat)*
astringency Adstringenz f
astringent adstringierend, zusammenziehend
astringent Adstringens n, adstringierendes (zusammenziehendes) Mittel n, adstringierende Substanz f
astro dyeing Astro-Dyeverfahren n *(örtliche Färbung von Kreuzspulen)*
Astrom chain barker *(Pap)* Kettenentrinder m nach Astrom, Astrom-Entrindungsmaschine f
astrophyllite *(Min)* Astrophyllit m
asymmetric asymmetrisch, asym., unsymmetrisch, uns.
a. centre Asymmetriezentrum m
a. synthesis asymmetrische Synthese f ·
asymmetry Asymmetrie f, Unsymmetrie f
a. effect Relaxationseffekt m
a. potential Asymmetriepotential n
at s. atmosphere
ATA s. ammonia triacetic acid
atacamite *(Min)* Atakamit m *(basisches Kupferchlorid)*
atactic ataktisch
a. polymer ataktisches Polymeres n
ataxite Ataxit m *(ein Eisenmeteorit)*
ATE. s. aluminium triethyl

atesine s. atisine
atherm[an]ous adiatherman, atherman, keine Wärmestrahlen durchlassend, wärmeundurchlässig
aticine, atisine *(Pharm)* Atisin n
atlantic suite *(Geol)* atlantische Sippe (Reihe) f
atm s. atmosphere
ATM. s. aluminium trimethyl
atmidometer s. atmometer
atmolysis Atmolyse f
atmometer Atmometer n, Verdunstungsmesser m
atmophile atmophil, zur Lufthülle gehörend
a. element atmophiles Element n *(Element der Lufthülle)*
atmophilic s. atmophile
atmosphere 1. Atmosphäre f, Lufthülle f; *(Phys)* [physikalische] Atmosphäre f, atm (1 atm = 101325 N/m²); *(Tech)* [technische, metrische, neue] Atmosphäre f, at (1 at = 10⁴ kp/m²); 2. Atmosphäre f, Medium n, Mittel n
atmospheric atmosphärisch
a. air atmosphärische Luft f
a. attack Witterungseinfluß m
a. column atmosphärische Kolonne f
a. condenser Berieselungsverflüssiger m, Berieselungskondensator m
a. cooling tower Kühlturm m mit natürlichem Zug, selbstbelüfteter Kühlturm m, Kaminkühlturm m
a. corrosion Korrosion f durch die Atmosphäre (den Luftsauerstoff), Luftkorrosion f
a. dryer Normaldrucktrockner m, atmosphärischer Trockner m, Lufttrockner m
a. drying atmosphärische Trocknung f, Trocknen n an der Luft, Lufttrocknung f
a. electricity atmosphärische Elektrizität f, Luftelektrizität f
a. humidity Luftfeuchtigkeit f, Luftfeuchte f
a. nitrogen Luftstickstoff m, Stickstoff m der Luft, atmosphärischer Stickstoff m
a. oxygen Luftsauerstoff m, Sauerstoff m der Luft, atmosphärischer Sauerstoff m
a. pressure atmosphärischer Druck m, Atmospärendruck m, Luftdruck m
atom Atom n
a. % Atomprozent n, Atom-%
a. arrangement Atomanordnung f, Atomkonfiguration f, Konfiguration f
a. bomb Atombombe f, A-Bombe f
a. disintegration Atomzerfall m, Kernzerfall m
a. form factor *(Krist)* Atomformfaktor m, Atomfaktor m
a. lattice *(Krist)* Atomgitter n
a. model Atommodell n
a. polarization Atompolarisation f
a. sheath Atomhülle f, Elektronenhülle f des Atoms
a. splitting Atom[auf]spaltung f, Atomzertrümmerung f
atomic atomar, einatomig, aus einem Atom bestehend; atomar *(in der Größenordnung eines Atoms)*; atomistisch

a. absorption coefficient atomarer Absorptionskoeffizient *m*
a. arrangement Atomanordnung *f*, Atomkonfiguration *f*, Konfiguration *f*
a. beam Atomstrahl *m*
a.-beam apparatus Atomstrahlapparatur *f*, Atomstrahlapparat *m*
a.-beam method Atomstrahlmethode *f*
a.-beam spectroscopy Atomstrahlspektroskopie *f*
a. bomb Atombombe *f*, A-Bombe *f*
a. bond Atombindung *f (als Zustand)*, Elektronenpaarbindung *f*, kovalente (unpolare, einpolare, homöopolare, unitarische) Bindung *f*
a. bonding Atombindung *f (als Vorgang)*
a. charge Atomladung *f*
a. configuration *s.* a. arrangement
a. core Atomrumpf *m*, Atomrest *m*
a. decay Atomzerfall *m*, Kernzerfall *m*
a. diameter Atomdurchmesser *m*
a. disintegration *s.* a. decay
a. displacement Atomverschiebung *f*, Atomumlagerung *f*
a. distance Atomabstand *m*, Kernabstand *m*
a. electron Atomelektron *n*
a. electron shell Elektronenschale *f* des Atoms, Atomschale *f*; Elektronenhülle *f* des Atoms, Atomhülle *f*
a. energy 1. Atom[kern]energie *f*, Kernenergie *f*; 2. Bindungsenergie *f*
a. energy level Energieniveau *n* des Atoms, Atomniveau *n*
a. envelope *s.* a. electron shell
a. excitation Anregung *f* von Atomen
a. field atomares Feld *n*, Kräftefeld *n* des Atoms
a. form factor *(Krist)* Atomformfaktor *m*, Atomfaktor *m*
a. framework Atomgerüst *n*
a. frequency Atomfrequenz *f*
a. heat Atomwärme *f*
a. heat of formation atomare Bildungswärme *f*
a. hydrogen atomarer (einatomiger) Wasserstoff *m*, Monowasserstoff *m*
a. hydrogen [arc] welding atomares Lichtbogenschweißen *n*, Arcatom-Schweißverfahren *n*, Arcatom-Verfahren *n*, Arcatom-Schweißung *f*
a. ion Atomion *n*, ionisiertes Atom *n*
a. kernel *s.* a. core
a. lattice Atomgitter *n*
a. level *s.* a. energy level
a. link (linkage) *s.* a. bond
a. linking Atomverkettung *f*; *s.* a. bonding
a. mass Massenwert *m*, Isotopengewicht *n*, Atommasse *f*, atomare Masse *f*
a. mass number Massenzahl *f*
a. mass unit atomare Masseeinheit *f*, amu
a. model Atommodell *n*
a. molecule Atommolekül *n*
a. nuclear energy Atom[kern]energie *f*, Kernenergie *f*

a. nucleus Atomkern *m*
a. number Atomnummer *f*, Ordnungszahl *f*, Kernladungszahl *f*, OZ, Z
a. orbit Elektronenbahn *f* um den Atomkern
a. orbital Atomorbital *n*, atomares Orbital *n*, AO
a. parachor Atomparachor *m*
a. parameters *(Krist)* Parameter *mpl* (Koordinaten *fpl*) der Atomschwerpunkte im Gitter
a. percentage Atomprozent *n*, Atom-%
a. plane *(Krist)* Netzebene *f*, Gitterebene *f*
a. polarization Atompolarisation *f*
a.-powered mit Atomkraft betrieben (angetrieben), atombetrieben, atomgetrieben, mit Atomantrieb, mit Kernenergie betrieben (angetrieben)
a. proportion Atomverhältnis *n*, Atomverhältniszahl *f*
a. radiation Atomstrahlung *f*
a. radius Atomradius *m*
a. ratio [value] *s.* a. proportion
a. ray *s.* a. beam
a. refraction Atomrefraktion *f*
a. research Atomforschung *f*
a. residue Atomrest *m*, Atomrumpf *m*
a. rotation Atomrotation *f*, Atomdrehung *f*
a. scattering factor *(Krist)* Atomformfaktor *m*, Atomfaktor *m*
a. scattering power Beugungsvermögen *n* eines Atoms *(für Röntgen- oder Elektronenstrahlung)*
a. shift *s.* a. displacement
a. size Atomgröße *f*
a. spacing *s.* a. distance
a. species Atomart *f*
a. spectroscopy Atomspektroskopie *f*
a. spectrum Atomspektrum *n*
a. spectrum of hydrogen Wasserstoffatomspektrum *n*, Wasserstoffspektrum *n*
a. state atomarer Zustand *m*, Atomzustand *m*, Energiezustand *m* des Atoms
a. structure Atomstruktur *f*, Atomaufbau *m*, Atombau *m*
a. structure factor *s.* a. form factor
a. susceptibility Atomsuszeptibilität *f*
a. term symbol Termsymbol *n* eines Atoms
a. theory Atomtheorie *f*, Atomlehre *f*
a. torso *s.* a. trunk
a. transmutation Atomumwandlung *f*
a. trunk Atomrumpf *m*, Atomrest *m*
a. union Atomverband *m*, Atomverbindung *f*
a. units [Hartree] Hartree-Einheiten *fpl (Untersuchung der Elektronenstruktur von Atomen und Molekülen)*
a. vibration Atomschwingung *f*
a. volume Atomvolumen *n*
a. weight relative Atommasse *f*, *(bisher)* Atomgewicht *n*
a. weight determination Bestimmung *f* der relativen Atommasse, *(bisher)* Atomgewichtsbestimmung *f*
atomical *s.* atomic

atomicity 1. Atomigkeit *f*, Atomizität *f*, *(veraltet für)* Wertigkeit *f*, Valenz *f*; 2. Basizität *f*
atomistic atomistisch
atomization Zerstäuben *n*, Zerstäubung *f*, Atomisieren *n*; Versprühen *n*, Versprühung *f*, Zersprühen *n*; Verspritzen *n*, Verspritzung *f*; Verdüsen *n*, Verdüsung *f*
atomize / **to** zerstäuben, atomisieren; versprühen, zersprühen; verspritzen; verdüsen
atomizer Zerstäuber *m*; Sprüher *m*, Versprüher *m*; Düse *f*
atomizing *s.* atomization
  a. burner Zerstäubungsbrenner *m*
  a. chamber Spritzkabine *f*
  a. dryer *(Ker)* Zerstäubungstrockner *m*
  a. nozzle Sprühdüse *f*, Zerstäuberdüse *f*
atoxylic acid $C_6H_4NH_2 \cdot AsO(OH)_2$ Arsanilsäure *f*, *p*-Aminophenylarsinsäure *f*
ATP *s.* adenosine triphosphate
atrolactic acid $CH_3C(C_6H_5)(OH)COOH$ Atrolaktinsäure *f*, 2-Phenylmilchsäure *f*, 2-Oxyhydratropasäure *f*
atromentin Atromentin *n*, 2,5-[Di-*p*-oxydiphenyl]-3,6-dioxybenzochinon *n*
atropia *s.* atropine
atropic acid $CH_2=C(C_6H_5)COOH$ Atropasäure *f*, 2-Phenylakrylsäure *f*
atropina, atropine Atropin *n*, DL-Hyoszyamin *n*
atropo-isomer Atropisomer[es] *n*
A.T.S. *s.* animal tub-sizing
attach / **to** befestigen, anbringen, anlagern, anheften; *(Farbstoffe)* aufziehen
attached to / **to be** gebunden sein an, hängen (sitzen) an
attachment Befestigung *f*, Anbringung *f*; Zusatzgerät *n*; Ansatz *m*
attack / **to** angreifen
attack Angriff *m*
  a. by sulphates Sulfatangriff *m*
attacking agent angreifendes Agens *n*
attapulgite Attapulgit *m*, Polygorskit *m* *(Tonmineral)*
attar of roses Rosenöl *n*
attemperator Kühlrohrleitung (Kühlschlange) *f* in Gärbottichen
attenuate / **to** verdünnen; dämpfen, [ab]schwächen
attenuation Verdünnen *n*, Verdünnung *f*; Dämpfung *f*, Schwächung *f*, Abschwächung *f*; Extraktnahme *f (der Bierwürze durch Vergärung)*
  a. degree Vergärungsgrad *m* *(Maß für die Extraktnahme der Bierwürze)*
attract / **to** anziehen; *(Elektronen)* aufnehmen
attractant Lockmittel *n*, Lockstoff *m*, Anlockmittel *n*, Anlockstoff *m*, Attraktivstoff *m*
  a. action Lockwirkung *f*
attraction Anziehung *f*, Attraktion *f*
attractive action *s.* attractant action
  a. force (power) Anziehungskraft *f*, Attraktionskraft *f*

a. properties Lockstoffeigenschaften *fpl*
attrition Abreiben *n*, Abrieb *m*, Abtragen *n*, Abtragung *f*; Abnutzung *f* durch Reibung; *(Geol)* Abtragung *f*, Abrieb *m*, Abschleifung *f*, Abrasion *f*
  a. mill Mahlscheibenmühle *f*, Mahlgang *m*, Scheibenmühle *f*; Reibmühle *f*
Attritor [mill, pulverizer] Attritor *m* *(eine Reibmühle)*
Å.U., A.U. *s.* angstrom [unit]
Audibert-Arnu dilatometer Audibert-Arnu-Dilatometer *n*
Audibert-Arnu dilatometer test Dilatometertest *m* nach Audibert-Arnu
Audibert-Arnu method Audibert-Arnu-Dilatometerverfahren *n*
auerbachite *(Min)* Auerbachit *m* *(Zirkoniumorthosilikat)*
aufbau principle Aufbauprinzip *n*
  a. process Aufbauverfahren *n*
augen structure *(Geol)* Augentextur *f*
auger Strangpresse *f*; *(Ker)* Schneckenpresse *f* *(eine Art Strangpresse)*; *(großer)* Bohrer *m* *(z.B. für Bodenproben)*
Auger effect Auger-Effekt *m*, innere Absorption *f*
Auger electron Auger-Elektron *n*
Auger yield Auger-Elektronen-Ausbeute *f*, Auger-Ausbeute *f*
augite *(Min)* Augit *m*
augment / **to** *(z.B. eine Farbe)* verstärken
auramine $[(CH_3)_2NC_6H_4]_2C=NH \cdot HCl \cdot H_2O$ Auramin *n*, Bis-(4-dimethylaminophenyl)methylenimin *n*
aurate Aurat *n*
aureole *(Geol)* Kontaktaureole *f*, Kontakthof *m*
auribromide $M^I[AuBr_4]$ Bromoaurat(III) *n*
auric Gold..., *(i.e.S.)* Gold(III)-...
  a. acid $Au(OH)_3$ Goldsäure *f*, Gold(III)-hydroxid *n*
  a. bromide $AuBr_3$ Gold(III)-bromid *n*, Goldtribromid *n*
  a. chloride $AuCl_3$ Gold(III)-chlorid *n*, Goldtrichlorid *n*
  a. cyanide $Au(CN)_3$ Gold(III)-zyanid *n*, Goldtrizyanid *n*
  a. hydroxide $Au(OH)_3$ Gold(III)-hydroxid *n*, Goldsäure *f*
  a. iodide AuJ_3 Gold(III)-jodid *n*, Goldtrijodid *n*
  a. oxide Au_2O_3 Gold(III)-oxid *n*, Goldtrioxid *n*
  a. sulphide Au_2S_3 Gold(III)-sulfid *n*
  a. trioxide Au_2O_3 Goldtrioxid *n*, Gold(III)-oxid *n*
aurichalcite *(Min)* Aurichalzit *m*, Messingblüte *f*
aurichloride $M^I[AuCl_4]$ Chloroaurat(III) *n*
auricyanide $M^I[Au(CN)_4]$ Zyanoaurat(III) *n*, Tetrazyanoaurat(III) *n*
auride Aurid *n*
auriferous goldhaltig, goldführend
aurin[e] $(C_6H_4OH)_2CC_6H_4O$ Aurin *n*, *p*-Rosolsäure *f*
auripigment *(Min)* Auripigment *n* *(Arsen(III)-sulfid)*
aurobromide $M^I[AuBr_2]$ Bromoaurat(I) *n*

aurochloride $M^I[AuCl_2]$ Chloroaurat(I) n
aurocyanide $M^I[Au(CN)_2]$ Zyanoaurat(I) n, Dizyanoaurat(I) n
aurone Auron n *(Vertreter einer Stoffgruppe innerhalb der Flavonoide)*
aurora yellow CdS Kadmiumgelb n, Kadmiumsulfid n
auroso-auric bromide Au_2Br_4 Gold(I,III)-bromid n
  a.-auric chloride Au_2Cl_4 Gold(I,III)-chlorid n
  a.-auric sulphide Au_2S_2 Gold(I,III)-sulfid n
aurothiosulphate $M^I_3[Au(S_2O_3)_2]$ Thiosulfatoaurat(I) n
aurous Gold..., *(i.e.S.)* Gold(I)-...
  a. bromide AuBr Gold(I)-bromid n, Goldmonobromid n, Goldbromid n
  a. chloride AuCl Gold(I)-chlorid n, Goldmonochlorid n, Goldchlorid n
  a. cyanide AuCN Gold(I)-zyanid n
  a. hydroxide AuOH Gold(I)-hydroxid n
  a. iodide AuJ Gold(I)-jodid n
  a. oxide Au_2O Gold(I)-oxid n, Goldmonoxid n, Goldoxid n
  a. sulphide Au_2S Gold(I)-sulfid n
austenite *(Met)* Austenit m
austenitic austenitisch, Austenit...
austenitize / to austenitisieren
austenitizing Austenitisierung f
  a. temperature Austenitisierungstemperatur f
Australian gum *(Pflanzengummi von australischen Acaciaarten)*
  A. kino Rotgummi n *(gummihaltiges Kinoharz von Eucalyptus camaldulensis Dehnh.)*
australite Australit m *(ein Tektit)*
Austrian cinnabar $Pb(OH)_2 \cdot PbCrO_4$ Chromrot n, Chromzinnober m, Persischrot n, Derbyrot n, Wienerrot n
authigene *(Geol)* authigen
  a. mineral authigenes Mineral n
authigenesis *(Geol)* Authigenese f
authigen[et]ic, authigenous *s.* authigene
autocatalysis Autokatalyse f
autocatalytic autokatalytisch
  a. reaction autokatalytische Reaktion f
autochthonous *(Geol)* autochthon, bodeneigen, bodenständig, ortseigen, am Bildungsort (Ort der Entstehung) liegend, an Ort und Stelle entstanden, in natürlicher Lage, in situ
autoclast, autoclastic rock autoklastisches Gestein n
autoclave / to autoklavieren, *(besser)* im Autoklaven behandeln (kochen), *(Med)* sterilisieren
autoclave Autoklav m, Druckgefäß n, Druckkessel m, *(Med auch)* Dampfdrucksterilisator m; *(Gum)* Vulkanisierkessel m, Vulkanisationskessel m; *(Gum)* s. a. press
  a.moulding *(Plast)* abgewandeltes Gummisackverfahren n *(Niederdruck-Preßverfahren)*
  a. press *(Gum)* Autoklav[en]presse f, Autoklavheizpresse f, Autoklav m
  a. sterilizer Drucksterilisator m, Drucksterilisierapparat m

autoclaving Autoklavieren n, *(besser)* Autoklavenbehandlung f
autocollimating spectrograph Autokollimationsspektrograf m
autodisintegration *(Landw)* Bodendispergierung f *(bei Überwiegen von Natriumionen)*
autofermentation Spontangärung f
autofining Autofining n *(katalytische Druckwasserstoffraffination, bei welcher der für die Hydrierung benötigte Wasserstoff im gleichen Prozeß erhalten wird)*
  a. process Autofining-Verfahren n
Autoform Autoform-Heizer m *(Vulkanisierpresse für Autoreifen)*
autogenous autogen
  a. cutting autogenes Schneiden n, Autogenschneiden n
  a. grinding autogenes Mahlen n, autogene Mahlung f
  a. smelting autogenes Schmelzen (Verschmelzen, Schwebeschmelzen) n, Autogenschmelzen n
  a. tumbling mill autogen arbeitende Sturzmühle f, Sturzmühle f für das autogene Mahlen, Trommelfallmühle f
  a. welding autogenes Schweißen n, Autogenschweißen n, Autogenschweißung f, Gas[schmelz]schweißen n
autohydration *(Geol)* Autohydratation f
autoionization Autoionisation f, Selbstionisation f, Eigenionisation f, Präionsation f
autolith *(Geol)* Autolith m, endogener (homöogener) Einschluß m
autolysate Autolysat n
autolysis *(Bio)* Autolyse f, Selbstauflösung f
autolytic autolytisch, selbstauflösend
  a. decomposition *s.* autolysis
autolyzed yeast medium Hefeautolysat n, Hefeextrakt m
automate / to automatisieren
automatic automatisch, selbsttätig
  a. air filter *(ein Metallfilter mit selbsttätigem Austrag)*
  a. buret automatische Bürette f
  a. control selbsttätige (automatische) Steuerung f; selbsttätige (automatische) Regelung f, Regelung f
  a. cycle control automatische Programmsteuerung f
  a. ejection *(Plast)* automatisches Ausdrücken n, automatischer Ausstoß m
  a. feed automatische (selbsttätige) Beschickung f
  a. feeder Beschickungsautomat m, automatische (selbsttätige) Beschickung f
  a. moulding machine Preßautomat m
  a. rotary compression press *(Am)* Karussellpressenautomat m, automatische Karussellpresse f, Rundläufer-Preßautomat m
  a. screen printing machine automatische Filmdruckmaschine f, Filmdruckautomat m

a.-start-and-stop control Stillsetzregelung *f*, Ein-Aus-Regelung *f*
a. titration apparatus *s*. a. titrator
a. titrator automatisch arbeitendes Titriergerät *n*, Titrierautomat *m*
a. tyre curing press of the Bag-O-Matic type Bag-o-matic-Heizer *m* *(Vulkanisierpresse für Autoreifen)*
a. zero burette Bürette *f* mit automatischer Nullpunktseinstellung
autometamorphism *(Geol)* Autometamorphose *f*
autometasomatism *(Geol)* Autometasomatose *f*
automorphic *(Geol)* automorph, idiomorph
a.-granular *(Geol)* panidiomorph[körnig]
automotive diesel fuel Fahrdieselkraftstoff *m*, Kraftstoff *m* für Fahrdieselmotoren
autooxidation *s*. autoxidation
autopneumatolysis *(Geol)* Autopneumatolyse *f*
autoprotolysis Autoprotolyse *f*
autopurification Selbstreinigung *f*
a. power Selbstreinigungskraft *f*, Selbstreinigungsvermögen *n*
autoracemization Autorazemisierung *f*, Autorazemisation *f*
autoradiochromatography Autoradiochromatografie *f*
autoradiograph Autoradiogramm *n*, autoradiografische Aufnahme *f*
autoradiography Autoradiografie *f*, Radioautografie *f*
autotrophic *(Bioch)* autotroph
autotype paper Autotypiedruckpapier *n*, Autotypiepapier *n*
autoxidation Autoxydation *f*
autoxidator Autoxydator *m*
autrometer Autrometer *n* *(Analysenautomat)*
autumn fertilization Herbstdüngung *f*
autunite *(Min)* Autunit *m* *(Kalziumbisuranylorthophosphat)*
Auwers-Skita rule Auwers-Skitasche Regel *f*
auxiliary zusätzlich, Zusatz..., Hilfs..., Neben...
a. coal hopper Kohlezwischenbunker *m*, Zwischenbehälter *m*, Zwischenbunker *m* *(eines Gaswerksofens)*
a. condition Nebenbedingung *f*
a. electrode Hilfselektrode *f*
a.-fluid atomizer Zerstäuber *m* mit Hilfsstoff, Zweistoffzerstäuber *m*, Druckluftzerstäuber *m*; Versprüher *m* mit Hilfsstoff, Zweistoffversprüher *m*, Druckluftversprüher *m*
a. group Nebengruppe *f* *(im Periodensystem)*
a. material Hilfsstoff *m*
a. material for cheese preparing Käsereihilfsstoff *m*
a. ram Hilfskolben *m*
a. reference electrode Hilfsbezugselektrode *f*
a. series Nebenreihe *f* *(im Periodensystem)*
a. solution Hilfslösung *f*
a. spray material Spritzhilfsmittel *n*
a. strainer *(Pap)* Hilfsknotenfänger *m*

a. syntan synthetischer Hilfsgerbstoff *m*
a. thermometer Hilfsthermometer *n*
a. valency Nebenvalenz *f*
auxin Auxin *n* *(pflanzlicher Wuchsstoff)*
auxochrome Auxochrom *n*, auxochrome (farbvermehrende, farbverstärkende) Gruppe *f*
auxochromic auxochrom, farbverstärkend, farbvermehrend
a. group *s*. auxochrome
auxotrophic *(Bioch)* auxotroph
av. *s*. average
A.V. *s*. acid value
availability Verfügbarkeit *f*; *(Landw)* Pflanzenverfügbarkeit *f*, Aufnehmbarkeit *f* *(von Nährstoffen)*
available verfügbar; *(Landw)* pflanzenverfügbar, zugänglich, aufnehmbar *(Nährstoffe)*
a. chlorine *(Pap)* wirksames (aktives) Chlor (Bleichchlor) *n*, Aktivchlor *n*
a. chlorine content *(Pap)* Gehalt *m* an wirksamem (aktivem) Chlor (Bleichchlor), Aktivchlorgehalt *m*
a. chlorine demand *(Pap)* Aktivchlorbedarf *m*
a. oxygen aktiver Sauerstoff *m*, Aktivsauerstoff *m*
a. sulphur dioxide *(Pap)* freies (loses) SO_2 *n*
avalanche *(phys Ch)* Lawine *f*
a. of electrons Elektronenlawine *f*
a. of ions Ionenlawine *f*
avalent nullwertig
Avaram bark *(Gerb)* Avaram-Rinde *f* *(von Cassia auriculata L.)*
avenized paper Frischhaltepapier *n*
aventurine feldspar *(Min)* Aventurinfeldspat *m*
a. glass Aventuringlas *n*
a. glaze *(Ker)* Aventuringlasur *f*
a. quartz *(Min)* Aventurinquarz *m* *(Silizium(IV)-oxid)*
average durchschnittlich, gemittelt, Durchschnitts..., Mittel...
average Durchschnitt *m*, Mittelwert *m*
a. amount Durchschnittsbetrag *m*
a. fat content Durchschnittsfettgehalt *m*
a. life [period] *(phys Ch)* [mittlere] Lebensdauer *f*
a. molecular weight mittlere relative Molekülmasse *f*
a. particle size mittlere Korngröße *f*, Mittelfeinheit *f*, Kornmittel *n*
a. retention time mittlere Verweilzeit *f*
a. sample Durchschnittsprobe *f*
a. yield Durchschnittsertrag *m*, Durchschnittsausbeute *f*
aviation gasoline Flugmotorenbenzin *n*, Flugbenzin *n*
a. spirit *s*. a. gasoline
Avignon berry (grain) französische Gelbbeere *f* *(von Rhamnus infectorius L.)*
avitaminosis Avitaminose *f*
avocado oil Avocadoöl *n*, Avocadofett *n*
Avogadro constant *s*. Avogadro number
Avogadro hypothesis (law) Avogadrosche Hy-

pothese (Regel) *f*, Avogadrosches Gesetz *n*, Satz *m* des Avogadro
Avogadro number Loschmidt-Zahl *f*, Loschmidtsche Zahl *f* *(Anzahl der Moleküle in 1 Mol einer Substanz bei 0°C und 1 atm; hingegen: Avogadrosche Zahl = Loschmidt number)*
A.W. *s.* atomic weight
axial axial
  a. angle *(Krist)* Achsenwinkel *m*
  a. bond axiale Bindung *f*, a-Bindung *f*
  a. compressor Axialverdichter *m*, Axialkompressor *m*
  a. conformation axiale Konformation *f*
  a. extruder head Längsspritzkopf *m*
  a.-flow compressor *s. a.* compressor
  a.-flow fan Flügelradlüfter *m*, Propellerventilator *m*, Axialventilator *m*, Axiallüfter *m*
  a.-flow pump Axialpumpe *f*, Propellerpumpe *f*
  a.-piston pump Axialkolbenpumpe *f*
  a. ratio *(Krist)* Achsenverhältnis *n*
  a. substituent axialer Substituent *m*
  a. symmetry Achsensymmetrie *f*
axially symmetric[al] achsensymmetrisch, axialsymmetrisch
axinite *(Min)* Axinit *m*
axis Achse *f*
  a. of a crystal Kristallachse *f*, kristallografische Achse *f*
  a. of fourfold symmetry *(Krist)* vierzählige Achse (Drehachse) *f*
  a. of rotation Rotationsachse *f*, Drehachse *f*
  a. of sixfold symmetry *(Krist)* sechszählige Achse (Drehachse) *f*
  a. of symmetry *(Krist)* Symmetrieachse *f*, Drehachse *f*
  a. of threefold symmetry *(Krist)* dreizählige Achse (Drehachse) *f*
  a. of twofold symmetry *(Krist)* zweizählige Achse (Drehachse) *f*
axisymmetric achsensymmetrisch, axialsymmetrisch
axonometry *(Krist)* Axonometrie *f*
azelaic acid $HOOC(CH_2)_7COOH$ Azelainsäure *f*, Lepargylsäure *f*, Nonandisäure *f*, 1,7-Heptandikarbonsäure *f*
azeotrope Azeotrop *n*, azeotropisches (azeotropes) Gemisch *n*, azeotropische (azeotrope) Mischung *f*
  a.-former Zusatzstoff *m* bei der Azeotropdestillation, Schleppmittel *n*, Mitnehmer *m*
azeotropic azeotrop[isch]
  a. column Kolonne *f* zur Azeotropdestillation
  a. distillation azeotropische (azeotrope) Destillation *f*, Azeotropdestillation *f*
  a. mixture azeotropische (azeotrope) Mischung *f*, azeotropisches (azeotropes) Gemisch *n*, Azeotrop *n*
  a. point azeotropischer (azeotroper) Punkt *m*
azeotropy Azeotropie *f*
azide $M^I N_3$ Azid *n*; $R \cdot CO \cdot N = N = N$ Säureazid *n*

azidivinylphosphine oxide Azidivinylphosphinoxid *n*
azimethylene CH_2N_2 Azimethylen *n*, Diazomethan *n*
aziminobenzene $C_6H_5N_3$ Aziminobenzol *n*, Benzolazimid *n*, 1,2,3-Benzotriazol *n*
azimuthal quantum number Azimutalquantenzahl *f*, azimutale (sekundäre) Quantenzahl *f*, Orbitaldrehimpulsquantenzahl *f*
azine Azin *n* *(Sammelname für organische Verbindungen mit mehreren N-Atomen)*
azo colour *s. a.* dye
  a. component Azokomponente *f*
  a. compound $R \cdot N = N \cdot R'$ Azoverbindung *f*, Azokörper *m*
  a. dye Azofarbstoff *m*
  a.-dye unit Azoapparatur *f*
  a. group —N=N— Azogruppe *f*
  a. link[age] Azobindung *f*, Azobrücke *f*
  a.-pyrazolone dye Pyrazolon-Azofarbstoff *m*, Pyrazolonfarbstoff *m*
  a. series Azoreihe *f*
  a. shed Azofarbenbetrieb *m*
  a.-unit *s.* a.-dye unit
  a. yellow Azogelb *n*
azobenzene $C_6H_5N_2C_6H_5$ Azobenzol *n*, Diphenyldiimid *n*
azobenzide *s.* azobenzene
azodicarbonamide Azodikarbonamid *n*
azofication *(Landw)* nichtsymbiotische Stickstoffbindung *f* (Fixierung *f* von Luftstickstoff), Stickstoffbindung *f* durch freilebende Bakterien
azoic unlöslicher (auf der Faser erzeugter) Azofarbstoff *m*
  a. base Base *f* zur Herstellung der unlöslichen Azofarbstoffe
  a. dye *s.* azoic
azoimide HN_3 Azoimid *n*, Stickstoffwasserstoffsäure *f*
azole Azol *n*, *(i.e.S.)* Pyrrol *n*, Imidol *n*
azomethane $CH_3 \cdot N = N \cdot CH_3$ Azomethan *n*
azomethine dye Azomethinfarbstoff *m*
azophenylene Azophenylen *n*, Phenazin *n*, Dibenzopyrazin *n*
azotic acid HNO_3 Salpetersäure *f*
azotometer Azotometer *n*, Nitrometer *n*
azoxy compound Azoxyverbindung *f*
  a. group —N(O)N— Azoxygruppe *f*
azoxybenzene $C_6H_5(NON)C_6H_5$ Azoxybenzol *n*
azoxybenzide *s.* azoxybenzene
azurite *(Min)* Azurit *m*, Kupferlasur *m* *(Kupfer(II)-dihydroxiddikarbonat)*

B

B-horizon *(Geol)* B-Horizont *m*, Illuvialhorizont *m*
B stage *(Plast)* B-Zustand *m*

B-stage resin Harz *n* im B-Zustand, B-Harz *n*, *(bei Phenolharzen auch)* Resitol *n*
babassu oil Babassuöl *n*, Babassufett *n* *(Palmöl von Orbignya speciosa Berk.)*
Babbit metal Babbit-Metall *n*, Weißmetall *n*
Babcock mill *s.* Babcock & Wilcox ball-and-ring mill
Babcock test Babcock-Methode *f* *(zur Milchuntersuchung)*
Babcock & Wilcox ball-and-ring mill, Babcock & Wilcox pulverizer Babcock-Kugelringmühle *f*
Babinet absorption rule *(Krist)* Babinetsche Regel *f*, Babinetsches Prinzip (Theorem) *n*
babingtonite *(Min)* Babingtonit *m*
bablah *(Gerb)* *(Hülsen und Rinde von Acacia nilotica (L.)Del.)*
Babo boiling plate Babo-Blech *n* *(des Babo-Trichters)*
Babo funnel Babo-Trichter *m*
Babul bark *(Gerb)* Babulrinde *f* *(von Acacia nilotica (L.)Del.)*
babussu oil *s.* babassu oil
baby Bessemer converter Bessemer-Kleinkonverter *m*, Bessemer-Kleinbirne *f*, Klein-Bessemer-Konverter *m*, Klein-Bessemer-Birne *f*
b. converter Kleinkonverter *m*
b. dryer *(Pap)* Vortrockenzylinder *m*, Vortrockner *m*
b. powder Kinderpuder *m*
b. press *(Pap)* Babypresse *f*, Vorpresse *f*
b. press roll *(Pap)* Vorpreßwalze *f*
BAC *(Abk. für)* British Association of Chemists
bacitracin Bazitrazin *n* *(Antibiotikum)*
back / to kaschieren
back Rückstand *m*
b. cloth *(Text)* Mitläufer *m*, Stützgewebe *n*
b. coating *(Text)* Rückenbeschichtung *f*
b.-diffusion Rückdiffusion *f*
b. electromotive force gegenelektromotorische Kraft *f*, Gegen-EMK *f*, Gegenurspannung *f*
b. e.m.f. *s.* b. electromotive force
b. end Nachlauf *m* *(bei der Destillation)*
b. flow Rückfluß *m*, Rücklauf *m*, Rückstrom *m*
b.-flow condenser Rückflußkühler *m*
b.-flow valve Rückschlagventil *n*, Rückstromventil *n*, Rückstromsperre *f*
b.-flushing Rückspülen *n*, Rückspülung *f*
b. grey [cloth] *(Text)* Mitläufer *m*, Stützgewebe *n*
b. knotter *(Pap)* Hilfsknotenfänger *m*
b. pressure Gegendruck *m*, Rückdruck *m*
b. reflection powder diagram (pattern) Rückstrahlpulverdiagramm *n*, Rückstrahldiagramm *n*, Pulver-Diagramm *n*, Debye-Scherrer-Diagramm *n*
b. run Backrun *m*, Rückwärtsgasen *n*, Rückwärtsgasung *f*, Abwärtsgasen *n* (Abwärtsgasung *f*) mit Dampfüberhitzung
b.-sizing *(Text)* Rückenappretur *f*
b. stamp *(Fabrikmarke oder Warenzeichen auf der Unterseite eines keramischen Gegenstandes)*

b. stand *(Pap)* Abrollgestell *n*
b.-tanning Tannin-Nachbehandlung *f*
b. titration Rücktitration *f*, Rücktitrieren *n*, Zurücktitrieren *n*
b. wall Einlegewand *f* *(im Glasschmelzofen)*
b.-washing *s.* backwashing
b. with rollers *(Text)* Rollenkufe *f*
backbone chain Hauptkette *f* *(eines verzweigten Moleküls)*
backfall *(Pap)* Kropf *m* *(beim Holländer)*
b. crest (crown) *(Pap)* Kropfkrone *f*, Sattel *m* *(beim Holländer)*
background *(Kph)* Untergrund *m*, Hintergrund *m*, Nulleffekt *m*, Nullrate *f*, Fond *m*, Leerwert *m*
b. radiation Untergrundstrahlung *f*
backing 1. Rückenappretur *f*; 2. Kaschieren *n* *(Verstärken einer Plastfolie durch Gewebeunterlage)*; 3. *(Foto)* Rückschicht *f*, Lichthofschutzschicht *f*, Farbstoffschutzschicht *f*
b. layer *s.* backing 3.
b. plate *(Plast)* Zwischenplatte *f*
b. pump Hilfspumpe *f*
b.-up Hintermauerung *f* *(z.B. eines Ofens)*
backrinding *(Gum)* Ausfressungen *fpl* *(an Trennfugen der Formen)*
backscattering *(Kph)* Rückstreuung *f*
backtitrate / to zurücktitrieren
backup plate Stützplatte *f*, Stützsieb *n* *(am Filter)*
backward-curved-blade fan Lüfter (Ventilator) *m* mit rückwärtsgekrümmten Schaufeln
b.-feed operation Gegenstromführung *f* *(z.B. von Mehrkörperverdampfern)*
b. reaction Rückreaktion *f*
backwash / to rückspülen, *(durch Rückspülung)* waschen
backwash Rückspülen *n*, Rückspülung *f*, Rückwäsche *f*; Rückwasch *m* *(zum Sauberhalten des Extrakts in die Extraktionsbatterie zurückgepumpter Teil der Extraktlösung)*
backwashing Rückspülen *n*, Rückspülung *f*, Rückwäsche *f*; *(Text)* Lissieren *n*
b. machine *(Text)* Lisseuse *f*
backwater *(Pap)* Siebabwasser *n*, Siebwasser *n*
b. box *s.* b. tank
b. pump *(Pap)* Siebwasserpumpe *f*, Abwasserpumpe *f*
b. tank *(Pap)* Siebwasserbehälter *m*
Bacon high-pressure hydrogen cell Bacon-Hochdruckzelle *f*, Hochdruckzelle *f* von Bacon, Bacon-Zelle *f*
bacteria preparation Bakterienpräparat *n*, Impfpräparat *n*
b. staining Bakterienfärbung *f*
b.-tight bakteriendicht
bacterial bakteriell, Bakterien...
b. activity Bakterientätigkeit *f*
b. attack bakterieller Angriff *m*, Bakterieneinwirkung *f*
b. cell Bakterienzelle *f*
b. culture Bakterienkultur *f*

b. fermentation Bakteriengärung f
b. manure Bakteriendünger m
b. pigment Bakterienfarbstoff m
bactericidal bakterizid, bakterientötend, bakterienvernichtend, keimtötend
bactericide Bakterizid n, bakterizide (bakterientötende) Substanz f
bacteriocidal s. bactericidal
bacteriological bakteriologisch
b. filter bakteriendichtes Filter n, Bakterienfilter n
bacteriology Bakteriologie f
bacteriolysin Bakteriolysin n
bacteriolysis Bakteriolyse f
bacteriophage Bakteriophag[e] m
bacteriostasis Bakteriostase f
bacteriostatic bakteriostatisch, das Bakterienwachstum hemmend, bakterienwachstumshemmend
bacteriostatic Bakteriostatikum n, bakteriostatisches Mittel n
bacterium Bakterie f, Bakterium n
bad flow schlechter Verlauf m (Anstrichmangel); Apfelsinenschaleneffekt m, Orangenschaleneffekt m, Apfelsinenschalenhaut f, Spritznarben fpl (Oberflächenstörung des Anstrichfilms bei Spritzlackierungen)
badan (Gerb) Badan m (Wurzeln von Bergenia-Arten, vor allem B. crassifolia (L.)Fritsch)
baddeleyite (Min) Baddeleyit m (Zirkoniumdioxid)
badian Badian m, Sternanis m, Illicium verum Hooker fil.
Baekeland process Baekeland-Verfahren n (Kondensation von Phenolen und Formaldehyd zu Bakelit)
Baeyer [angle] strain theory Baeyersche Spannungstheorie f
Baeyer system Baeyersches System n (der Benennung heterozyklischer Brückenverbindungen)
Baeyer tension theory s. Baeyer strain theory
Baeyer-Villiger oxidation Baeyer-Villiger-Reaktion f (Oxydation von Ketonen zu Estern)
Baeyer's angle strain Baeyer-Spannung f
Baeyer's indophenine reaction Baeyers Indopheninreaktion f
baffle / to mit Prallflächen ausstatten
baffle 1. Prallblech n, Prallfläche f, Prallwand f, Prallplatte f, Leitblech n, Umlenkblech n, Umlenkplatte f, Zwischenwand f; 2. (Glas) Vorformboden m
b. board (Pap) Spritzbrett n, Schutzbrett n
b. mark (Glas) Vorformbodennaht f, Vorformbodennarbe f
b. plate s. baffle 1.
b. separator Staubkammer f mit Prallflächen
b. wall Schattenwand f (im Glasschmelzofen)
baffled unterteilt, mit Einbauten (Prallflächen, Leitblechen, Umlenkplatten) versehen
bag Beutel m; Sack m; Schlauch m (z.B. im Schlauchfilter); Heizschlauch m

b. collector Schlauchabscheider m, Schlauchfilter n (zur Gasreinigung)
b. filter Schlauchabscheider m, Schlauchfilter n, Beutelfilter n, Seihfilter n, Sackfilter n
b. moulding Pressen n mit Gummisack, Gummisack-Preßverfahren n, Gummisack-Formverfahren n, Gummisackverfahren n
b. paper Tütenpapier n, Beutelpapier n; Sackpapier n
b. sealing machine Beutelschweißmaschine f, Beutelschließmaschine f
bagasse Bagasse f, Begasse f, Zuckerrohrbagasse f (Zuckerrohrabfälle)
b. furnace Bagassefeuerung f
bagging (Gum) Einführen n des Heizschlauchs, Heizschlaucheinziehen n
Bagomatic Bag-o-matic-Heizer m (Vulkanisierpresse für Autoreifen)
bagrationite (Min) Bagrationit m
baikalite (Min) Baikalit m (eine Salit-Abart)
bainite [structure] (Met) Bainit m, Zwischenstufengefüge n, bainitisches Gefüge n, Gefüge n der Zwischenstufe
bainitic structure s. bainite
bait 1. Köder m, Ködermittel n; 2. (Glas) Fangstück n
baiting agent s. bait 1.
bakability Backfähigkeit f, backtechnische Eigenschaft f, Backeigenschaft f, Backqualität f
bakable backfähig
bake / to 1. wärmebehandeln, thermisch (mit Hitze) behandeln; (mittels Wärme) trocknen; 2. (Amine zwecks Sulfonierung) [ver]backen; 3. (keramische Erzeugnisse) brennen; glühen, schrühen; 4. (Aufglasurfarben, Lacke usw.) einbrennen; 5. (Plast) härten; 6. (Lebm) backen; 7. [zusammen]backen (Kohle)
bake-out losses Backverluste mpl
b.-out period Härtezeit f
bakelite paper Hartpapier n
baker's yeast s. bakery yeast
bakery margarine Bäckermargarine f, Backmargarine f
b. yeast Bäckerhefe f, Backhefe f
baking 1. Wärmebehandlung f, thermische Behandlung f, Hitzebehandlung f, Trockenhitzebehandlung f; Trocknen n, Trocknung f (mittels Wärme); 2. Backen n, Verbacken n (von Aminen zwecks Sulfonierung); 3. (Ker) Brennen n; Glühen n, Schrühen n; 4. Einbrennen n, Einbrand m (von Aufglasurfarben, Lacken usw.); 5. (Plast) Härten n, Härtung f; 6. (Lebm) Backen n; 7. Backen n, Zusammenbacken n (z.B. der Kohle)
b. characteristics s. b. quality
b. coal Backkohle f
b. finish s. b. varnish
b. losses Backverluste mpl
b. oven Backofen m
b. powder Backpulver n
b. process Backverfahren n, Backprozeß m (zur Sulfonierung von Aminen)

b. property (quality, value) Backeigenschaft f, Backqualität f, Backfähigkeit f, backtechnische Eigenschaft f

b. varnish (Am) Einbrennlack m, ofentrocknender Lack m

balance / to austarieren, ausgleichen

balance Waage f; Gleichgewicht n; Bilanz f

b. beam Waagebalken m, Waagbalken m

b. column Waagensäule f

b. desiccator Trockenpatrone f, Lufttrockner m (für Feinwaagen)

b. dish Wägeschiffchen n

b. method Abgleichmethode f, Kompensationsmethode f

b. pan Waagschale f

b. rider Reiter m (der Analysenwaage)

b. room Wägezimmer n, Wägeraum m, Wägungszimmer n

b. table Wägetisch m

balanced-jaw crusher Schwingbackenbrecher m, Schwingbrecher m

b. reaction reversible (umkehrbare) Reaktion f; Gleichgewichtsreaktion f

b. relief valve entlastetes Sicherheitsventil n

b. seal druckentlastete (druckausgeglichene) Dichtung f

b. tray thickener Mehrkammer-Eindicker m mit parallelgeschalteten Kammern

b.-weigh belt Dosierbandwaage f

balas ruby (Min) Balasrubin m

balata Balata f (kautschukähnlicher Rohstoff, von Mimusops balata Crueger)

Balbuch process Balbach-Elektrolyse f

bale backwashing unit (Text) Ballenlisseuse f

b. breaker (Text) Ballenöffner m

b. cutter Ballenspaltmaschine f, Ballenspalter m, Kautschukspalter m

b. of rubber Kautschukballen m

b. splitting machine s. b. cutter

ball / to [zusammen]ballen; sich [zusammen]ballen; (Met) Luppen bilden, zu Luppen zusammenbacken

b. up s. to ball

ball Schliffkugel f, Schliffball m (einer Schliffverbindung); Stahlkugel f (Mahlkörper); (Met) Luppe f

b.-and-race mill s. b.-and-ring mill

b.-and-ring coal pulverizer Kugelringmühle f für Kohlenstaubmahlung

b.-and-ring mill Kugelringmühle f

b.-and-socket joint clamp Halteklemme (Halteschelle, Ligatur) f für Kugelschliffe ·

b. bearing Kugellager n

b. charge Kugelfüllung f

b. check s. b. check valve

b.-check nozzle Kugelverschlußdüse f

b.-check valve Kugelrückschlagventil n

b. clay Ball Clay m (sehr plastischer feuerfester Ton)

b. condenser Kugelkühler m

b. crusher Kugelmühle f

b. float trap Schwimmerkondenstopf m, Schwimmerkondens[at]wasserableiter m, Kondenstopf m

b. impact test Schlaghärteprüfung f

b.-mill / to kugelmahlen, in der Kugelmühle vermahlen, mit der Kugelmühle mahlen

b. mill Kugelmühle f

b.-mill method Kugelmühlenverfahren n, Kugelmühlenmethode f

b. milling Kugelmühlenaufbereitung f

b. moulding press (Gum) Ballformpresse f, Ballpresse f

b. powder Kugelpulver n (Treibladungspulver in Kugelform)

b.-puncture resistance Kugeldruckhärte f

b.-race mill s. b.-and-ring mill

b.-ring mill s. b.-and-ring mill

b. roller mill s. b.-and-ring mill

b. test Kugeldruckprüfung f, Kugeldrucktest m

b. valve Kugelventil n

b. viscosimeter Kugelfallviskosimeter n

ballast tray Ballastventilboden m

balling Zusammenballen n, Zusammenballung f, Kugelbildung f; Pellet[is]ieren n, Pellet[is]ierung f; (Met) Luppenbildung f; (Zucker) s. Balling

b. cone Pellet[is]ierkonus m

b. device Pelletformeinrichtung f

b. disk Pellet[is]ierteller m

b. drum Pellet[is]iertrommel f

Balling Balling-Grad m

ballistic ballistisch

b. galvanometer ballistisches Galvanometer n, Stromstoßgalvanometer n

b. pendulum test Prüfung f mit dem ballistischen Pendel (Messung der relativen Sprengkraft im ballistischen Mörser)

balloon flask Kurzhals-Rundkolben m

Bally-Scholl reaction Bally-Scholl-Reaktion f

balm 1. Balsam m; 2. Melisse f, Melissa L., (i.e.S.) Zitronenmelisse f, Melissa officinalis L.; 3. Propolis f, Bienenharz n, Bienenwachs n, Kittharz n, Vorwachs n, Stopfwachs n, Halbwachs n

b.-mint Zitronenmelisse f, Melissa officinalis L.

b. of Gilead Mekkabalsam m (von Commiphora opobalsamum (L.)Engl.)

b. oil Melissenöl n (aus Melissa officinalis L.)

Balmer formula Balmer-Formel f, Balmersche Formel f

Balmer series Balmer-Serie f

Balmer terms Balmer-Terme mpl

balsam Balsam m

b. fir Balsamtanne f, Abies balsamea (L.)Mill.

b. of fir Kanadabalsam m (von Abies balsamea (L.)Mill.)

b. of Peru Perubalsam m, Peruanischer (Indischer, schwarzer) Balsam m (von Myroxylon balsamum (L.) Harms var. pereira)

b. of Tolu Tolubalsam m (von Myroxylon balsamum (L.)Harms var. balsamum)

Baly cell (tube) Baly-Gefäß n (zur Absorptionsmessung)
bamboo Bambus m, Bambusa Schreb.
Bambuk butter Schibutter f, Galambutter f, Karitebutter f (von Butyrospermum parkii (Don)Kotschy)
banana oil $CH_3COOC_5H_{11}$ Birnenöl n, Birnenäther m, Essigsäureisoamylester m, Amylazetat n
Banbury [mixer] Banbury-Mischer m, Banbury-Innenmischer m, Banbury-Kneter m
band 1. Band n, Streifen m; 2. Streifen m, Bande f, Band n (Spektrum); 3. Band n, Energieband n, Energiebereich m, Zone f; 4. (Gum) Fell n; 5. (Geol) Band n, [sehr dünne] Schicht f
b. aid Schnellverbandpflaster n, Schnellverband m
b. conveyor Gurt[band]förderer m, Bandförderer m
b. dryer Bandtrockner m, Förderbandtrockner m
b. edge Bandenkante f, Bandenkopf m (Spektrum)
b. filter Bandfilter n, Umlauffilter n
b. group Bandengruppe f, Bandensystem n (Spektrum)
b. head s. b. edge
b. heater Bandheizkörper m, Heizband n
b. knife Bandmesser n
b. knife splitting machine (Gerb) Bandmesserspaltmaschine f
b. placement [of fertilizer] Streifendüngung f
b. sequence Bandenfolge f, Bandenreihe f (Spektrum)
b. spectrum Bandenspektrum n, (i.e.S.) Molekülspektrum n
b. system s. b. group
b. theory of solids Bandtheorie (Bändertheorie) f der Festkörper
b. tire (Am) Vollgummireifen m, Massivreifen m
b. treatment (Landw) Streifenbehandlung f, Streifenbegiftung f (oder) Streifendüngung f
banded gestreift, streifig, streifenartig [ausgebildet], mit Streifen, Streifen...; (Geol) gebändert
b. coal Streifenkohle f, streifige Kohle f
b. component (constituent) Streifenart f, Mikrolithotype f
b. constituent of coal Kohlenstreifenart f
b. spectrum s. band spectrum
b. structure Streifenstruktur f, Streifung f, streifiger Aufbau m; (Geol) Lagentextur f, gebänderte Textur f
banding (Geol) Bänderung f; (Glas) Bandschliff m (Dekorationstechnik); (Ker) Rändern n (Dekorationstechnik)
b. grease Raupenleim m
bank (Tech) Reihe f, Gruppe f, Serie f, Satz m, Anordnung f; (Gum) Wulst m(f) (im Walzenspalt)
b. of cans Zylindertrockenaggregat n
b. paper (post) Bankpostpapier n, Hartpostpapier n, Feinpostpapier n
banknote paper Banknotenpapier n

baptifoline Baptifolin n (Alkaloid)
baptitoxine Baptitoxin n, Zytisin n, Sophorin n, Ulexin n
bar 1. Bar n, bar (Einheit des Drucks); 2. Stange f, Stab m; Barren m; Strang m
b. grizzly Stangenrost m, Stabrost m
b. mill Desintegrator m, Schleudermühle f, Schlagkorbmühle f
b. mould Backenwerkzeug n
b. of margarine Margarinestrang m
b. of soap Seifenstrang m, Seifenstange f, endloser Seifenriegel m
b. of solvent Flüssigkeitsbarren m (beim Zonenschmelzen)
b. screen Stangenrostsieb n, Stangen[sieb]rost m, Stab[sieb]rost m, Rostsieb n, Siebrost m
b. sealer (Plast) Hochfrequenzschweißgerät n, Hochfrequenzschweißanlage f
baratte Baratte f, Sulfidiertrommel f, Xanthatkneter m
Barbados aloe (Pharm) Barbados-Aloe f, Curaçao-Aloe f (von Aloe vera L.)
barbaloin Barbaloin n (Aloeglykosid)
barber-dried (Pap) luftgetrocknet, lutro
Barbier-Wieland degradation Barbier-Wielandscher Abbau (Karbonsäure-Abbau) m, Barbier-Wieland-Reaktion f
barbierite (Min) Barbierit m (Mikroklin-Mikroperthit)
barbital Barbital n, 5,5-Diäthylbarbitursäure f
b. sodium Barbitalnatrium n
barbiturate Barbiturat n
barbituric acid Barbitursäure f, N,N'-Malonylharnstoff m, 2,4,6-Trioxo-hexahydropyrimidin n
barbotage Barbotage f, Druckluftmischung f, pneumatisches Mischen (Rühren) n
barcenite (Min) Barcenit m
Barcol hardness Barcol-Härte f
bare tube glattes Rohr n, Glattrohr n
Barfoed's reagent Barfoeds Reagens n (zur Bestimmung von Monosacchariden)
Barfoed's test Barfoedsche Bestimmung f (von Monosacchariden)
Bari-Sol process s. Barisol process
barion s. baryon
Barisol process Barisol-Verfahren n, Bari-Sol-Prozeß m, Dichloräthan-Benzol-Verfahren n (Entparaffinierungsverfahren)
barite (Min) Baryt m, Schwerspat m (Bariumsulfat)
b. paper Barytpapier n
barium Ba Barium n
b. acetate $Ba(CH_3COO)_2$ Bariumazetat n
b. acid arsenate s. b. hydrogen arsenate
b. arsenate $Ba_3(AsO_4)_2$ Bariumarsenat n
b. arsenide Ba_3As_2 Bariumarsenid n
b. azide $Ba(N_3)_2$ Bariumazid n
b. bichromate s. b. dichromate
b. bi[n]oxide s. b. peroxide
b. boride BaB_6 Bariumborid n
b. bromate $Ba(BrO_3)_2$ Bariumbromat n

b. bromide BaBr$_2$ Bariumbromid n
b. carbide BaC$_2$ Bariumkarbid n, Bariumazetylid n
b. carbonate BaCO$_3$ Bariumkarbonat n
b. chlorate Ba(ClO$_3$)$_2$ Bariumchlorat n
b. chloride BaCl$_2$ Bariumchlorid n
b. chloride fluoride BaF$_2$ · BaCl$_2$ Bariumchloridfluorid n
b. chloroplatinate Ba[PtCl$_6$] Bariumhexachloroplatinat(IV) n
b. chloroplatinite Ba[PtCl$_4$] Bariumtetrachloroplatinat(II) n
b. chromate BaCrO$_4$ Bariumchromat n
b. crown [glass] Baritkronglas n, Bariumkronglas n
b. cyanide Ba(CN)$_2$ Bariumzyanid n
b. cyanoplatinite Ba[Pt(CN)$_4$] Bariumtetrazyanoplatinat(II) n
b. dichromate BaCr$_2$O$_7$ Bariumdichromat(VI) n
b. dioxide s. b. peroxide
b. dithionate BaS$_2$O$_6$ Bariumdithionat n
b. ferrocyanide Ba$_2$[Fe(CN)$_6$] Bariumhexazyanoferrat(II) n
b. flint [glass] Baritflintglas n, Bariumflintglas n
b. fluorosilicate Ba[SiF$_6$] Bariumhexafluorosilikat n
b. formate Ba(HCOO)$_2$ Bariumformiat n
b. getter Bariumgetter m
b. hexanitride s. b. azide
b. hydrate s. b. hydroxide
b. hydride BaH$_2$ Bariumhydrid n
b. hydrogen arsenate BaHAsO$_4$ Bariumhydrogenarsenat n
b. hydrogen phosphate BaHPO$_4$ Bariumhydrogenphosphat n
b. hydrosulphide Ba(HS)$_2$ Bariumhydrogensulfid n
b. hydroxide Ba(OH)$_2$ Bariumhydroxid n
b. hydroxide pentahydrate Ba(OH)$_2$·5H$_2$O Bariumhydroxid-5-Wasser n
b. hypochlorite Ba(ClO)$_2$ Bariumhypochlorit n
b. hypophosphite Ba(H$_2$PO$_2$)$_2$ Bariumhypophosphit n
b. hyposulphite s. b. thiosulphate
b. iodate Ba(JO$_3$)$_2$ Bariumjodat n
b. iodide BaJ$_2$ Bariumjodid n
b. lake Bariumlack m
b. manganate BaMnO$_4$ Bariummanganat(VI) n
b. meal (Med) Barium sulfuricum purissimum, Bariumsulfat, reinst n (Röntgenkontrastmittel)
b. metasilicate BaSiO$_3$ Bariumtrioxosilikat n, Bariummetasilikat n
b. metatungstate BaW$_4$O$_{13}$·9H$_2$O Bariummetawolframat(VI)-9-Wasser n
b. molybdate BaMoO$_4$ Bariummolybdat(VI) n
b. monosulphide BaS Bariummonosulfid n, Bariumsulfid n
b. monoxide s. b. oxide
b. nitrate Ba(NO$_3$)$_2$ Bariumnitrat n
b. nitride Ba$_3$N$_2$ Bariumnitrid n, Bariumdinitrid n

b. nitrite Ba(NO$_2$)$_2$ Bariumnitrit n
b. orthosilicate Ba$_2$SiO$_4$ Bariumtetroxosilikat n, Bariumorthosilikat n
b. oxide BaO Bariumoxid n
b. pentahydrate s. b. hydroxide pentahydrate
b. perchlorate Ba(ClO$_4$)$_2$ Bariumperchlorat n
b. permanganate Ba(MnO$_4$)$_2$ Bariumpermanganat n, Bariummanganat(VII) n
b. peroxide BaO$_2$ Bariumperoxid n
b. peroxodisulphate BaS$_2$O$_8$ Bariumperoxodisulfat n, Bariumoxodisulfat n
b. persulphate s. b. peroxodisulphate
b. platinochloride s. b. chloroplatinite
b. platinocyanide s. b. cyanoplatinite
b. protoxide s. b. oxide
b. pyrophosphate Ba$_2$P$_2$O$_7$ Bariumdiphosphat n, Bariumpyrophosphat n
b. pyrovanadate Ba$_2$V$_2$O$_7$ Bariumpyrovanadat(V) n, Bariumheptoxodivanadat(V) n, Bariumdivanadat(V) n
b. rhodanate (rhodanide) s. b. thiocyanate
b. saccharate process (Zucker) Barytverfahren n
b. selenate BaSeO$_4$ Bariumselenat n
b. silicofluoride s. b. fluorosilicate
b. sulphate BaSO$_4$ Bariumsulfat n
b. sulphide BaS Bariummonosulfid n, Bariumsulfid n
b. sulphite BaSO$_3$ Bariumsulfit n
b. sulphocyanate (sulphocyanide) s. b. thiocyanate
b. sulphydrate s. b. hydrosulphide
b. superoxide s. b. peroxide
b. tellurate BaTeO$_4$ Bariumtellurat n
b. tetrasulphide BaS$_4$ Bariumtetrasulfid n
b. thiocyanate Ba(SCN)$_2$ Bariumthiozyanat n, Bariumrhodanid n
b. thiocyanide s. b. thiocyanate
b. thiosulphate BaS$_2$O$_3$ Bariumthiosulfat n
b. trisulphide BaS$_3$ Bariumtrisulfid n
b. tungstate BaWO$_4$ Bariumwolframat n
b. wolframate s. b. tungstate
bariumcalcite (Min) Bariumkalzit m
bark / to (Pap) entrinden, schälen
bark 1. Rinde f; (Gerb) Lohe f; 2. Wanne f, Kufe f
b. cork Naturkork m, [echter] Kork m
b.-free entrindet
b. leather pflanzlich gegerbtes Leder n
b. mill (Gerb) Lohmühle f
b. of tan (Gerb) Lohe f
b. removal s. barking
b. speck (spot) Rindenfleck m (Papierfehler)
b. tannage (tanning) pflanzliche Gerbung f, (i.e.S.) Lohgerbung f
barker (Pap) Entrindungsmaschine f, Entrinder m, Schälmaschine f, Schäler m
Barker tower (Pap) Barker-Turm m (Kalkmilchverfahren)
barking (Pap) Entrinden n, Entrindung f, Schälen n, Schälung f

b. drum *(Pap)* Entrindungstrommel *f*
b. machine *(Pap)* Entrindungsmaschine *f*, Entrinder *m*, Schälmaschine *f*, Schäler *m*
b.-plant *(Pap)* Entrindungsanlage *f*
barkometer, barktrometer *(Gerb)* Brühenmesser *m*, Gerbsäuremesser *m*, Barkometer *n* *(Aräometertyp)*
barley Gerste *f*, Hordeum L.
b. flour Gerstenmehl *n*
b. groats Gerstengraupen *fpl*, Perlgraupen *fpl*; Gerstengrütze *f*
b. malt Gerstenmalz *n*
b. meal Gerstenmehl *n*
b. starch Gerstenstärke *f*
b. sugar Gerstenzucker *m*
barm Bärme *f*, Hefe *f*, Bierhefe *f*, Bäckerhefe *f*
barn Barn *n*, b *(Einheit für den Wirkungsquerschnitt des Atomkerns, 10^{-24} cm^2)*
Barnard reagent Barnardsches Reagens *n* *(Hydrazin-Säurefuchsinlösung zum Aldehydnachweis)*
barograph Barograf *m*, Luftdruckschreiber *m*
barometer Barometer *n*, Luftdruckmesser *m*
b. formula Barometerformel *f*
b. tube Barometerrohr *n*, barometrisches Fallrohr *n*
barometric barometrisch
b. condenser barometrischer Kondensator *m*, Fallrohrkondensator *m*
b. pressure Barometerdruck *m*, barometrischer Druck *m*
barometry Barometrie *f*, Luftdruckmessung *f*
Baros camphor Baroskampfer *m*, Borneokampfer *m*, Sumatrakampfer *m* *(von Dryobalanops aromatica Gaertn. f.)*
barotropic barotrop
Barral solution *(Phenolsulfonsäurelösung zum Nachweis von Eiweiß und Gallenfarbstoffen in Harn)*
barré effect *(Text)* Barré-Effekt *m*, Streifigfärben *n* *(Fehler)*
barrel 1. Faß *n*, Tonne *f*; 2. Brennerrohr *n*, Brennrohr *n* *(z.B. am Bunsenbrenner)*; 3. Hahnhülse *f* *(des Glashahns)*; Spritzgehäuse *n* *(einer Spritzmaschine)*
b. amalgamation process Amalgamation *f* *(Edelmetall-Aufbereitungsverfahren)*
b. blender (mixer) Trommelmischer *m*, Mischtrommel *f*
b. polishing *(Plast)* Trommelpolieren *n*, Polieren *n* in Trommeln, Trommeln *n*
barren *(Bio, Bodenkunde)* unfruchtbar, steril; *(Geol)* taub
b. solution Armlauge *f* *(Zyanidlaugerei)*
Barret's [wash] test Waschprobe *f* nach Barret
barrier Schwelle *f*, Schranke *f*, Barriere *f*; Potentialbarriere *f*, Potentialschwelle *f*, Potentialberg *m*, Potentialwall *m*
b. cream Gewerbeschutzsalbe *f*
b. layer Sperrschicht *f*

b.-layer cell Sperrschicht[foto]zelle *f*, Sperrschicht[foto]element *n*, Fotoelement *n*
b.-layer photoelectric (photovoltaic) cell *s.* b.-layer cell
b.-layer rectifier Halbleitergleichrichter *m*, Sperrschichtgleichrichter *m*
bars *(Pap)* Holländermesser *npl*, Messer *npl*
b. in the plug *s.* b. on the rotor
b. in the shell *s.* b. on the casing
b. of the beater roll *s.* b. on the roll
b. on the bedplate *(Pap)* Grundwerksmesser *npl* *(beim Holländer)*
b. on the casing *(Pap)* Gehäusemesser *npl*, Mantelmesser *npl* *(bei der Kegelstoffmühle)*
b. on the roll *(Pap)* Walzenmesser *npl* *(beim Holländer)*
b. on the rotor *(Pap)* Kegelmesser *npl*, Rotormesser *npl* *(bei der Kegelstoffmühle)*
Bart reaction Bartsche Reaktion *f*, Bart-Reaktion *f* *(Arsenitarylierung)*
Bartlett force *(Kph)* Bartlett-Kraft *f*
Barvoys process Barvoys-Verfahren *n*, Sophia-Jacoba-Verfahren *n*, Verfahren *n* der Grube Sophia-Jacoba, Schwerflüssigkeitsaufbereitungsverfahren *n* nach Sophia-Jacoba
barwood afrikanisches Rotholz *n* *(Sammelname für rote Farbhölzer westafrikanischer Pterocarpus- und Baphia-Arten)*
barylite *(Min)* Barylith *m* *(Bariumberylliumpyrosilikat)*
baryon *(Kch)* Baryon *n*
barysphere *(Geol)* Barysphäre *f*, Erdkern *m*
baryta BaO Baryterde *f*, *(veraltet für)* Bariumoxid *n*
b. coated paper Barytpapier *n*
b. coating (layer) Barytschicht *f*, Barytageschicht *f*
b. paper *s.* b. coated paper
b. process *(Zucker)* Barytverfahren *n*
b. water Barytwasser *n*
b. yellow BaCrO$_4$ Barytgelb *n*, Ultramaringelb *n*, gelbes Ultramarin *n*, Steinbühler Gelb *n*, Bariumchromat *n*
baryte *(Min)* Baryt *m*, Schwerspat *m* *(Bariumsulfat)*
b. concrete Barytbeton *m*
barytes *s.* baryte
barytic bariumhaltig; Barium...
barytocalcite *(Min)* Barytokalzit *m* *(Bariumkalziumkarbonat)*
basal dressing Grunddüngung *f*
b. fertilizer Grunddünger *m*
basalt Basalt *m*
b. grit Basaltgrus *m*
b. ware *(Ker)* Basaltware *f*
b. wool Basaltwolle *f*
basaltic basaltisch, Basalt...; basalthaltig
basanite *(Min)* Basanit *m*, *(veraltet für)* Lydit *m*
base geringwertig; unecht, unedel *(z.B. Metall)*
base Basis *f*, Grundlage *f*; Grundbestandteil *m*, Hauptbestandteil *m*; *(Chem)* Base *f*; *(Math)* Basis

f, Grundlinie f; Basisfläche f, Grundfläche f;
(Math) Basis f, Grundzahl f; (Min) Endfläche f
(eines Kristalls); (Foto) Schichtträger m; (Erdöl)
Basis f, Charakter m (eines Rohöls); (Glas) Boden
m (eines Gefäßes)
b.-bearing centrifuge von unten angetriebene
Zentrifuge f, stehende Zentrifuge f, Stehzentrifuge f
b.-binding capacity (power) Basenbindungs-
vermögen n
b. catalysis Basenkatalyse f, Reaktionsbeschleunigung f durch Basen
b.-catalyzed basenkatalysiert
b. compound s. b. stock
b. exchange Basenaustausch m
b.-exchange capacity Basenaustauschfähigkeit
f, Basenaustauschkapazität f
b.-exchange complex (Landw) adsorbierender
Bodenkomplex m, Austauschmaterial n
b.-exchange method Kationenaustauschverfah-
ren n
b. exchanger Basenaustauscher m
b. metal unedles Metall n, Unedelmetall n,
Nichtedelmetall n; Grundmetall n (einer Legierung)
b.-metal catalyst Nichtedelmetallkatalysator m
b.-mineral index Basenmineralindex m
b. mix s. b. stock
b. molecule Grundmolekül n, Struktureinheit f,
Staudinger-Einheit f, Grundeinheit f
b. of crude petroleum Erdölbasis f, Rohölbasis
f, Basis f (Charakter m) des Roh[erd]öls
b. of the bushing Düsen[loch]boden m (Glas-
faserstoffherstellung)
b. paper Rohpapier n
b. paper for vulcanized fibre Vulkanfiberroh-
papier n, Vulkanfiberrohstoff m
b. plate Grundplatte f, Sohlplatte f, Fundament-
platte f
b. region (phys Ch) Basiszone f, Basisgebiet n,
Basisraum m, Basis[schicht] f
b.-saturated (Bodenkunde) basengesättigt
b. saturation (Bodenkunde) Basensättigung f
b.-saturation percentage (Bodenkunde) Ba-
sensättigungsgrad m
b. stock (Gum) Grundmischung f
b. stone (Pap) Bodenstein m (des Kollergangs)
b. unit s. b. molecule
b.-unsaturated (Bodenkunde) basenungesättigt
b.-weakening die Basizität schwächend
basic basisch, alkalisch; (Met) basisch zugestellt
(ausgekleidet)
b. accelerator (Gum) basischer Beschleuniger m
b. anhydride Basenanhydrid n
b. auxochromic group basische auxochrome
Gruppe f, positives Auxochrom n
b. Bessemer converter Thomas-Konverter m,
Thomas-Birne f
b. Bessemer process basisches Windfrischverfahren (Bessemer-Verfahren) n, Thomas-
Verfahren n, Thomas-Konverterverfahren n

b. Bessemer steel Thomas-Stahl m, Thomas-
Konverterstahl m
b. capacity Basizität f (von Säuren)
b. catalysis Basenkatalyse f, Reaktionsbeschleunigung f durch Basen
b. compound (Gum) Grundmischung f
b. converter process s. b. Bessemer process
b. converter steel s. b. Bessemer steel
b. dye basischer Farbstoff m
b. fibre (Glas) Rohfaser f
b. fuchsin Fuchsin n, Rosanilin n, Magenta n
b. indicator basischer Indikator m, Indikatorbase
f
b.-lined basisch zugestellt (ausgekleidet),
basisch (metallurgischer Ofen)
b. lining basische Zustellung (Auskleidung) f,
basisches Futter n (eines metallurgischen
Ofens)
b. magenta s. b. fuchsin
b. model Rohmodell n
b. open-hearth process basisches Herdfrischverfahren (Siemens-Martin-Verfahren, SM-
Verfahren) n
b. open-hearth steel basischer Siemens-
Martin-Stahl m
b. oxide basisches Oxid n
b. oxygen converter Sauerstoffaufblas[e]kon-
verter m, O_2-Aufblas[e]konverter m, Sauerstoff-
[blas]konverter m, Aufblas[e]konverter m, Blasstahlkonverter m, Oberwindkonverter m
b. oxygen converter process Sauerstoff[auf]-
blasverfahren n, Sauerstoff-Frischverfahren n,
O_2-Aufblasverfahren n, Aufblas[e]verfahren n,
Blasstahlverfahren n, Oberwindfrischverfahren n
b. oxygen furnace s. b. oxygen converter
b. oxygen furnace process s. b. oxygen
converter process
b. oxygen furnace steel s. b. oxygen steel
b. oxygen process s. b. oxygen converter
process
b. oxygen steel Sauerstoffblasstahl m, Blasstahl
m, sauerstoffgefrischter Stahl (Konverterstahl) m
b. oxygen steel plant Sauerstoffblaswerk n,
Blasstahlwerk n
b. oxygen steel process s. b. oxygen converter
process
b. process basisches Verfahren n; basisches
Windfrischverfahren (Bessemer-Verfahren) n,
Thomas-Verfahren n, Thomas-Konverterverfahren n
b. reaction Hauptreaktion f
b. refractory [material] basisches Futter (Ofenfutter) n, (i.w.S.) basisches feuerfestes Material
n
b. rock basisches Gestein (Erstarrungsgestein) n,
Basit m
b. salt basisches Salz n, (inkorrekt für) Hydroxidsalz n (oder) Oxidsalz n
b. slag basische Schlacke f, (i.e.S.) Thomas-
Schlacke f

b. steel basischer Stahl *m*, *(i.e.S.)* Thomas-Stahl *m*

b. strength Basenstärke *f*, Basizität *f*, Alkalität *f*

b. structure Grundstruktur *f*

b. term Grundbegriff *m*; *(phys Ch)* Grundterm *m*, Grundniveau *n*, Grundzustand *m*, Normalzustand *m*

b. test orientierende Prüfung *f*

b. topping *(Gerb)* Übersetzen *n* mit basischen Farbstoffen *(nach saurer Färbung)*

b.-type accelerator s. b. accelerator

b. weight s. basis weight

basicity Basizität *f*, Alkalität *f*, Basengehalt *m* *(einer Lösung)*

basification Basischstellen *n*

basify / to basisch stellen

basil oil Basilikumöl *n* *(von Ocimum basilicum L.)*

basin Becken *n*, Schale *f*, Schüssel *f*, Behälter *m*, Bassin *n*

basis Basis *f*, Grundlage *f*; Grundbestandteil *m*, Hauptbestandteil *m*; *(Math)* Basis *f*, Grundlinie *f*; Basisfläche *f*, Grundfläche *f*

b. solution Grundlösung *f*

b. weight *(Pap)* Masse *f* je Flächeneinheit, *(bisher)* Flächengewicht *n*, Quadratmetergewicht *n*, Basisgewicht *n*; Masse *f* je Bogen, *(bisher)* Bogengewicht *n*

basite Basit *m*, basisches Gestein (Erstarrungsgestein) *n*

basket Trommel *f*; Siebtrommel *f*

b. band extractor Becherwerkextrakteur *m*, Becherwerkextraktor *m*

b. bottle Korbflasche *f*

b. centrifugal Trommelzentrifuge *f*

b. evaporator Verdampfer *m* mit eingehängtem Rohrkorb, Rohrkorbverdampfer *m*

b. extractor s. b. band extractor

b. press Korbpresse *f*

basnasite s. bastnaesite

basoid Basoid *n* *(mit OH⁻-Ionen gesättigte kolloidale Substanz)*

basophil *(Med)* Basophiler *m*, basophile Zelle *f*, *(meist)* basophiler Leukozyt *m*

basophile s. 1. basophil; 2. basophilic

basophilia Basophilie *f*

basophilic basophil *(durch basische Farbstoffe leicht färbbar)*

basophilous *(Bot)* basiphil, basophil *(auf Böden alkalischer Reaktion gedeihend)*; s. basophilic

bassora gum Bassoragummi *n* *(unscharf abgegrenzter Sammelname für geringwertige Tragantsorten verschiedener Herkunft)*

bassorin Bassorin *n* *(pektinähnlicher Pflanzeninhaltsstoff)*

basswood Linde *f*, Tilia L.; Lindenholz *n*

bast fibre Bastfaser *f*

b. fibre bundle Bastfaserbündel *n*, Faserzellbündel *n*

bastard cedar Indische Bastardzedrele *f*, Soymida febrifuga Juss.

b. Siamese cardamom Bastard-Kardamom *m(n)* *(von Amomum xanthioides Wall.)*

b. teak Plossobaum *m*, Butea superba Roxb.

bastite *(Min)* Bastit *m*, Schillerspat *m*

bastnaesite *(Min)* Bastnäsit *m* *(Hauptanteil Zerfluoridkarbonat)*

bat 1. Brocken *m*, Klumpen *m*; 2. *(Ker)* Platte *f*, Brennunterlage *f*, Brennplatte *f*; Blatt *n* *(zur Herstellung von Flachware)*

b.-making machine *(Ker)* Blattscheibe *f*, Blattformermaschine *f*

batate Batate *f*, Süßkartoffel *f*, ipomoea batatas (L.)Poir.

Batavia cinnamon *(Zimtsorte, von Cinnamomum burmani Blume)*

batch diskontinuierlich (periodisch, unstetig, [ab]satzweise, chargenweise arbeitend *(Gerät)*, diskontinuierlich, periodisch, unstetig, [ab]satzweise, chargenweise *(Betrieb)*

batch Charge *f*, Partie *f*, Posten *m*, Füllung *f*, Ladung *f*, Einsatz *m*, Eintrag *m*, Beschickung *f*, Satz *m*; Ansatz *m*; *(Glas)* Gemenge *n*, Glassatz *m*; *(Ker)* Versatz *m*, Masseversatz *m*; *(Gum)* Vormischung *f*, Batch *m*; *(Text)* Docke *f*

b. blending *(Erdöl)* Mischen (Vermischen) *n* im Fertigtank

b. book Ansatzbuch *n*

b. box *(Text)* Dockenkasten *m*

b. carbonation *(Zucker)* periodische Saturation *f*, Einzelpfannensaturation *f*

b. charger *(Glas)* Gemengespeiser *m*, Einlegevorrichtung *f*, Einlegemaschine *f*

b. composition *(Ker)* Versatz *m*

b. distillation diskontinuierliche Destillation *f*, Chargendestillation *f*

b. dryer Chargentrockner *m*

b. feeder s. b. charger

b. formula *(Ker)* Versatzformel *f*; *(Glas)* Gemengesatz *m*

b. house *(Glas)* Gemengehaus *n*

b. mixer Chargenmischer *m*, Periodenmischer *m*, Stoßmischer *m*; *(Glas)* Gemengemischer *m*

b. nitration Chargennitrierung *f*, Nitrierung *f* in Ansätzen, diskontinuierliche Nitrierung *f*

b. of liquid Flüssigkeitseinsatz *m*

b.-off equipment *(Gum)* Batch-Off-Vorrichtung *f*

b. operation s. 1. b. process; 2. b. processing

b. pasteurizer Trommelerhitzer *m*, Kreiselerhitzer *m*

b. process diskontinuierliches (periodisches, mit Unterbrechung arbeitendes) Verfahren *n*, Chargenverfahren *n*

b. processing diskontinuierliche (periodische) Arbeitsweise (Fahrweise) *f*, diskontinuierlicher Betrieb *m*, Chargenbetrieb *m*, Satzbetrieb *m*

b. rectification diskontinuierliche (unstetige) Rektifikation (Gegenstromdestillation) *f*

b. scale Chargenwaage *f*

b. size Stückgröße *f*

b. stone Gemengestein *m* *(Glasfehler)*

b. weight Stückmasse f, Chargenmasse f
batching equipment (Text) Aufdockvorrichtung f
  b. oil Batschingöl n, Batschöl n (zum Tränken der Jutefasern)
batchwise diskontinuierlich, periodisch, unstetig, [ab]satzweise, chargenweise
bate / to (Blößen vor dem Gerben) beizen
bate (Gerb) Beize f
bath Bad n, (Text meist) Flotte f
  b.-carburize / to badaufkohlen, badzementieren, badeinsetzen, im Salzbad (in flüssigen Mitteln) aufkohlen (zementieren, einsetzen)
  b. carburizing Badaufkohlen n, Salzbadaufkohlen n, flüssige Aufkohlung f, Badzementieren n, Salzbadzementieren n, Badeinsetzen n, Aufkohlen (Zementieren, Einsetzen) n im Salzbad (in flüssigen Mitteln)
  b. metal Bathmetall n (Kupfer-Zink-Legierung)
  b. of sulphuric acid Schwefelsäurebad n
  b. oil Badeöl n
  b. powder Badepulver n
  b. preparation Badepräparat n, Badezusatz m
  b. ratio Flottenverhältnis n
  b. salt Badesalz n
bathmometry Bathmometrie f (Messung der Wasserstoffionenkonzentration)
bathochromic bathochrom, farbvertiefend
  b. displacement bathochrome Verschiebung (Farbtonverschiebung, Farbverschiebung) f
  b. effect bathochromer Effekt m, Bathochromie f, Farbvertiefung f, positive Farbänderung f
  b. group bathochrome Gruppe f
batholith (Geol) Batholith m (ausgedehnter Tiefengesteinskörper)
bathyal (Geol) bathyal
bathylith s. batholith
batik dyeing Batikfärberei f
  b. paper Aderpapier n, Batikpapier n, Javakunstpapier n
  b. printing Batikdruck m
bating (Gerb) Beizen n
bats-wing burner Schnittbrenner m, Fischschwanzbrenner m, Schlitzbrenner m
batter gehämmerte Oberfläche f (Fehler bei gezogenem Glas)
battery Batterie f, Sammler m, Akku[mulator] m; Gruppe f, Serie f, Satz m, Batterie f
  b. acid Akkumulatorensäure f, Akku-Säure f
  b. carbon Batteriekohle f
  b. of diffusion cells Diffusionsbatterie f
  b.-operated batteriegespeist
batteuse (Kosmet) Batteuse f (Rührgefäß aus verzinntem Kupfer zur Extraktion mit Alkohol)
batting Watte f in Lagen
  b.-out machine (Ker) Blattscheibe f, Blattformermaschine f
batyl alcohol Batylalkohol m, α-Oktadezylglyzeryläther m
Baudouin reaction (test) Baudouinsche Probe f (Nachweis von Sesamöl)

Bauer double-disk refiner Doppelscheibenmühle f System Bauer, Bauer-Mühle f
Baum jig [washer] Baumsche (luftgesteuerte, luftbewegte) Setzmaschine f (zur Kohleaufbereitung)
Baum wash box Baumscher Setzkasten m
Baumé scale Baumé-Skale f
Baur musk Baur-Moschus m, Moschus m Baur (künstlicher Moschus)
bauxite Bauxit m
bauxitic bauxitisch
bave Kokondoppelfaden m, Kokonfaden m
bay fat s. b. oil 2.
  b. leaf Lorbeerblatt n, Lorbeerblätter npl
  b. oil 1. Bayöl n (aus Beeren und Blättern von Pimenta racemosa (Mill.)I. W. Moore); 2. (meist sweet bay oil) Lorbeeröl n, Lorbeerfett n, Lorbeerbutter f (von Laurus nobilis L.)
  b. salt Baisalz n, aus Meersalzsalinen (Meersalinen) gewonnenes Salz n
bayberry oil Bayöl n (aus den Beeren von Pimenta racemosa (Mill.)I. W. Moore)
  b. tallow (wax) Bayberrytalg m, Kapbeerenwachs n, Myrikawachs n, Myrikatalg m, Myrtenwachs n
Bayer process Bayer-Verfahren n (Aufschluß von Bauxit mit Natronlauge)
bayleaf oil Bayöl n (aus den Blättern von Pimenta racemosa (Mill.) I. W. Moore)
bayonet coupling (joint) Bajonettkupplung f
  b.-tube heat exchanger Nadelwärmeaustauscher m, Nadelwärmeübertrager m
B.D. s. 1. bone-dry; 2. bulk density
BDWB s. bone-dry-weight basis
beach section Trockenabschnitt m (der Schneckenzentrifuge)
bead / to [um]bördeln
bead 1. Perle f, Kügelchen n; (Ionenausschlußverfahren) Harzkorn n; 2. Wulst m(f), Rand m, Bördelrand m, Umbördelung f
  b. catalyst perlförmiger Katalysator (Kontakt) m, Perlkontakt m
  b. compound (Gum) Wulstmischung f
  b. cutter (Gum) Entwulster m, Wulstschneidemaschine f
  b. polymerisate Perlpolymerisat n
  b. polymerization Perlpolymerisation f, Suspensionspolymerisation f, Kornpolymerisation f
  b. removal (Gum) Entwulsten n
  b. test Perlenprobe f, Perlenreaktion f
beaded perlenförmig
  b. black geperlter Ruß m, Perlruß m
  b. material Perlkontaktmasse f (beim katalytischen Kracken)
beaker Becherglas n; Becher m
  b. brush Gläserbürste f
Beale reagent (Med) Bealesches Reagens n (Karminlösung in ammoniakalischem Glyzerin-Alkohol-Gemisch)
beam 1. Strahl m, Lichtstrahl m; Strahlenbündel n; 2. Waag[e]balken m; 3. Gerberbaum m (zur mechanischen Bearbeitung der Blößen)

b. autoclave *(Text)* Baumfärbeautoklav *m*
b. dyeing *(Text)* Baumfärben *n*, Baumfärberei *f*
b. dyeing machine *(Text)* Baumfärbeapparat *m*
b. of electrons Elektronenstrahl *m*
b. of light Lichtstrahl *m*, Lichtbündel *n*, Strahlenbündel *n*
b. of X-rays Röntgenstrahlbüschel *n*, Röntgenstrahlbündel *n*
beamhouse *(Gerb)* Wasserwerkstatt *f*
beaming *(Gerb)* Strecken *n* auf dem Baum
bean Bohne *f*
 b. cake Sojabohnenkuchen *m*
bearing Lager *n*, Lagerung *f*, Lagerstelle *f*; Halterung *f*
 b. capacity Tragfähigkeit *f*, Belastbarkeit *f (eines Lagers)*
 b. metal Lagermetall *n*
 b. plate Tragplatte *f*
 b. strength *s.* b. capacity
beat / to *(Pap)* mahlen
 b. dead *(Pap)* totmahlen
beat frequency Schwebungsfrequenz *f*
 b.-frequency method Schwebungsmethode *f*, Schwebungsverfahren *n*
beaten stuff tester *(Pap)* Mahlgradprüfer *m*
beater Schläger *m*, Stampfer *m*, Hammer *m (der Hammermühle)*, Flügel *m (des Rührwerks)*; *(Text)* Walkhammer *m*, Walkmaschine *f*; *(Pap)* Mahlmaschine *f*, Stoffmühle *f*, Mahlgeschirr *n*, *(i.e.S.)* Holländer *m*, Mahlholländer *m*, Messerholländer *m*, Ganzzeugholländer *m*
 b. additives *(Pap)* Mahlhilfsmittel *npl*
 b. charge *(Pap)* *(zur Mahlung im Holländer)* eingetragener Faserhalbstoff *m*, Mahlgut *n*, Charge *f*
 b. colour *(Pap)* Farbstoff *m* für Massefärbung (Holländerfärbung)
 b. colouring (dyeing) *(Pap)* Färben *n* in der Masse, Massefärbung *f*, Färben *n* im Stoff (Holländer), Holländerfärbung *f*
 b. hood *(Pap)* Haube *f (des Holländers)*
 b. house *s.* b. room
 b. mill Schlägermühle *f*, Schlagmühle *f*
 b. plate *(Pap)* Grundwerk *n*, Messerwerk *n*, Messerblock *m (eines Holländers)*
 b. roll *(Pap)* Holländerwalze *f*, Messerwalze *f*, Mahlwalze *f*
 b. roll bars *(Pap)* Walzenmesser *npl (beim Holländer)*
 b. roll pressure *(Pap)* Mahlwalzendruck *m*, Walzendruck *m*
 b. room *(Pap)* Holländersaal *m*
 b.-sized *(Pap)* massegeleimt, stoffgeleimt
 b. sizing *(Pap)* Leimung *f* im Stoff, Leimung *f* in der Masse, Masseleimung *f*
 b. tank *s.* b. tub
 b.-treated stock *(Pap)* im Holländer gemahlener Stoff *m*
 b. tub (vat) *(Pap)* Holländertrog *m*, Stoffwanne *f*
beaterman *(Pap)* Holländermüller *m*
beating *(Pap)* Mahlen *n*, Mahlung *f*, Stoffmahlung

b. action *(Pap)* Mahlwirkung *f*
b. engine *(Pap)* Mahlmaschine *f*, Stoffmühle *f*, Mahlgeschirr *n*, *(i.e.S.)* Holländer *m*, Mahlholländer *m*, Messerholländer *m*, Ganzzeugholländer *m*
b. machine *(Gum)* Schaumschlagmaschine *f*, Schaummaschine *f*
b. material *(Pap)* Mahlgut *n*, Mahlstoff *m*; Mahlzeug *n*
b. of stock *(Pap)* Ganzzeugmahlung *f*
b. pressure *(Pap)* Mahldruck *m*
b. rate *(Pap)* Mahlgeschwindigkeit *f*
b. roll *(Pap)* Holländerwalze *f*, Messerwalze *f*, Mahlwalze *f*
b. time *(Pap)* Mahldauer *f*
Beattie and Bridgeman equation Beattie- und Bridgemansche Gleichung (Zustandsgleichung) *f*, Beattie-Bridgeman-Gleichung *f*
beauxite *s.* bauxite
bebeerine Bebirin *n*, Bebeerin *n*, Chondrodendrin *n*, Kurin *n (Alkaloid)*
beccarite *(Min)* Beccarit *m*, *(veraltet für)* olivgrüner Zirkon *m*
Béchamp method (reduction) Béchampsches Verfahren *n (Nitroarylreduktion)*
Bechi solution alkoholische Silbernitratlösung *f* nach Bechi *(zum Nachweis von Baumwollsamenöl)*
beck Kufe *f*, Färbekufe *f*, Bottich *m*, Wanne *f*, Trog *m*
Becker combination underjet coke oven Koppers-Becker-Verbundkoksofen *m* mit Unterbrennern
Becker oven [Koppers-]Becker-Ofen *m*, Koppers-Becker-Verbund[koks]ofen *m*
beckite *s.* beekite
Beckmann apparatus Beckmannsche Apparatur *f*
Beckmann method Beckmannsche Methode *f (der Molekülmassenbestimmung)*
Beckmann molecular transformation *s.* Beckmann rearrangement
Beckmann rearrangement Beckmannsche Umlagerung *f (eines Ketoxims in ein Amid)*
Beckmann thermometer Beckmann-Thermometer *n*, Beckmannsches Thermometer *n*
beclamide Beclamid *n*, N-Benzyl-β-chlorpropionamid *n*
bed 1. Bett *n*; Grundplatte *f*; 2. Schüttung *f*, Schüttschicht *f*; Haufwerk *n*; 3. *(Geol)* Bank *f*, Bett *n*, Lager *n*, Flöz *n*, Schicht *f*
b. depth Betthöhe *f*, Schichtdicke *f*
b. filter Filter *n* mit loser Schicht, loses Filter *n*, Haufwerkfilter *n*
b. knife Statormesser *n (einer Rotorschneidemaschine)*; *(Pap)* Untermesser *n*, Anschlagmesser *n (des Querschneiders)*
b. material Schüttgut *n*
b. of coal being washed Setzbett *n (einer Setzmaschine für Kohlenaufbereitung)*
b. of crystals Kristallbett *n*, Kristallpolster *n*, Kristallschüttung *f*, Kristall[schütt]schicht *f*

b. of granular solids Körnerhaufwerk *n*, körnige Schüttschicht *f*, Körnerschüttung *f*
b. plate Grundplatte *f*, Unterlagplatte *f*, Auflageplatte *f*
b. stone Bodenstein *m*, Unterstein *m*, unterer Mühlstein *m*
b.-washer *(Pap)* Sandfang *m*, Sandfänger *m*
bedder *(Pap)* Bodenstein *m (des Kollergangs)*
bedding plane *(Geol)* Schichtungsfläche *f*, Schichtfläche *f*, Schichtebene *f*
bedplate *(Pap)* Grundwerk *n*, Messerwerk *n*, Messerblock *m (eines Holländers)*
b. bars *(Pap)* Grundwerksmesser *npl (beim Holländer)*
b. box *(Pap)* Grundwerkfassung *f*, Grundwerkkasten *m*
b. knives *s.* b. bars
bedrock Grundgestein *n*, Muttergestein *n*, Untergrund *m*, anstehendes Gestein *n*, Anstehendes *n*
bedstone *(Pap)* Bodenstein *m (des Kollergangs)*
bee glue Bienenvorwachs *n*, Bienenharz *n*, Kittharz *n*, Vorwachs *n*, Stopfwachs *n*, Halbwachs *n*, Propolis *f*
b. plant honigende Pflanze *f*, Honigpflanze *f*, Trachtpflanze *f*, Nektarspender *m*, Nektarlieferant *m*
b. poison Bienengift *n*
beech Buche *f*, Fagus L.
beechnut oil Bucheckernöl *n*, Buchöl *n*
beechwood Buchenholz *n*
b. creosote Buchenholzteerkreosot *n*
beef fat (tallow) Rindertalg *m*, Rindstalg *m*
beehive *(Destillation)* Spirale *f (als Füllkörperrost)*
b. coke *s.* b.-oven coke
b. coke (coking) oven *s.* b. oven
b. kiln *(Ker)* Rundofen *m*
b. oven Bienenkorbofen *m*
b.-oven coke Bienenkorbofenkoks *m*, Bienenkorbkoks *m*
beekite *(Min)* Beekit *m*
beer column Maischekolonne *f*, Maischesäule *f*
b. inspection glass Bierbeschauglas *n*
b. scale Bierstein *m (in Bierleitungen)*
b. vinegar Bieressig *m*
b. wort Bierwürze *f*, Würze *f*
b. yeast Bierhefe *f*
Beer law Beersches Gesetz *n (der Lichtabsorption)*
beery bierartig, Bier...
beestings Biestmilch *f*, Erstlingsmilch *f*, Vormilch *f*, Kolostralmilch *f*, Kolostrum *n*
beeswax Bienenwachs *n*
beeswing 1. Kahmhaut *f*, Kahm *m (auf nährstoffhaltigen Flüssigkeiten)*; 2. Futtermehl *n*, Kleiemehl *n*
beet Rübe *f*, Beete *f*, Bete *f*, Beta L.; Zuckerrübe *f*
b. cleaning Rübenreinigung *f*, Rübenwäsche *f*
b. cossettes Rübenschnitzel *npl*, Zuckerrübenschnitzel *npl*
b. cutter Rübenschneidmaschine *f*, Rübenschnitzelmaschine *f*

b. diffuser *(Zucker)* Diffuseur *m*
b. drying Rübentrocknung *f*
b. growing Zuckerrübenanbau *m*, Rübenanbau *m*
b. juice Rübensaft *m*
b. leaf catcher Rübenkrautfänger *m*
b. molasses Rübenmelasse *f*, Zuckerrübenmelasse *f*
b. pectin Rübenpektin *n*
b. pulp [ausgelaugte] Zuckerrübenschnitzel (Rübenschnitzel) *npl*
b. pulp catcher *(Zucker)* Pülpefänger *m*
b. pulp dryer *(Zucker)* Schnitzeltrockner *m*, Schnitzeltrocknungsanlage *f*
b. pulp drying *(Zucker)* Schnitzeltrocknung *f*, Schnitzeltrocknungsverfahren *n*
b. pulp press *(Zucker)* Schnitzelpresse *f*
b. pulp press water *(Zucker)* Schnitzelpreßwasser *n*
b. pulp pump *(Zucker)* Schnitzelpumpe *f*
b. pulp water *(Zucker)* Diffusionsabwasser *n*
b. raw sugar Rübenrohzucker *m*
b. slicer *s.* b. slicing machine
b. slices Rübenschnitzel *npl*, Zuckerrübenschnitzel *npl*
b. slicing knife *(Zucker)* Schnitzelmesser *n*
b. slicing machine Rübenschnitzelmaschine *f*, Rübenschneidmaschine *f*, Schnitzelmaschine *f*
b. slicing season Rübenkampagne *f*
b. sugar $C_{12}H_{22}O_{11}$ Rübenzucker *m*, Sa[c]charose *f*
b. sugar factory Rübenzuckerfabrik *f*
b. sugar house Rübenzuckerfabrik *f*, Zuckerfabrik *f*, *(i.e.S.)* Zuckerhaus *n*
b. sugar industry Rübenzuckerindustrie *f*
b. sugar manufacture Rübenzuckerfabrikation *f*
b. tail catcher Rübenschwanzfänger *m*
b. tops Rübenköpfe *mpl*
b. washer (washing machine) Rübenwäsche *f*, Rübenwaschmaschine *f*
beetle / to *(Text)* beeteln *(Mangeleffekt erzeugen)*
beetler *(Text)* Beetle-Maschine *f*, Stampfkalander *m*
beetling *(Text)* Beeteln *n (Erzeugen eines Mangeleffekts)*
b. machine *s.* beetler
beetroot sugar *s.* beet sugar
begass[e] *s.* bagasse
beginning of cure Vulkanisationseinsatz *m*
b. of the bleaching period Bleichbeginn *m*
behave / to sich verhalten, arbeiten, funktionieren
behavior *s.* behaviour
behaviour Verhalten *n*, Reaktion *f (eines chemischen Stoffes)*
b. of a real gas Realgasverhalten *n*
behaviouristic resistance *(Bio)* Verhaltensresistenz *f (z.B. gegenüber Schädlingsbekämpfungsmitteln)*
behenic acid $CH_3(CH_2)_{20}COOH$ Behensäure *f*, Dokosansäure *f*
behenolic acid $CH_3(CH_2)_7C{\equiv}C(CH_2)_{11}COOH$ Be-

henolsäure f, Dokosin-(13)-säure f
behenyl alcohol $CH_3(CH_2)_{20}CH_2OH$ Behenylalkohol m, prim-n-Dokosylalkohol m, 1-Dokosanol n
behn oil s. ben oil
Beilstein's test Beilsteinprobe f
Belgian kiln (Ker) belgischer Ofen m
  B. slag Thomas-Schlacke f
Belgium roll machine Walzenbrikett[ier]presse f, Brikettwalzenpresse f
belite Belit m (Kristallart im Portlandzementklinker)
bell 1. Glocke f, Verschlußglocke f, Gichtglocke f (eines Hochofens); 2. Blase f (Papierfehler)
  b. and hopper Glocken- und Trichter[gicht]verschluß m, Parry-Gichtverschluß m, Parry-Verschluß m (eines Hochofens)
  b. beam Glockenstange f (einer Gichtglocke)
  b. brass Glockenmessing n
  b. bronze Glockenbronze f
  b. cap Rundglocke f
  b. cap tray Glockenboden m
  b.-jar Glasglocke f
  b. metal Glockenmetall n
  b.-metal ore (Min) Zinnkies m, Stannin m
Bell process (Met) Bell-Verfahren n (Entfernung von P und Si mittels Eisenoxid)
belladonna Tollkirsche f, Atropa belladonna L.
  b. alkaloid Belladonnaalkaloid n
  b. herb (Pharm) Tollkirschenkraut n (von Atropa belladonna L.)
  b. root (Pharm) Tollkirschenwurzel f, Belladonnawurzel f (von Atropa belladonna L.)
Belleek china Belleek-Porzellan n
Bellmer [bleacher] (Pap) Bleichholländer m
bellows-type expansion joint Wellrohrkompensator m, Wellrohr[dehnungs]ausgleicher m
belonite Belonit m (nadelförmiger Mikrolith)
belt Band n, Gurt m; Riemen m, Treibriemen m
  b. and bucket elevator s. b. elevator
  b. building machine (Gum) Transportband-Konfektioniermaschine f
  b. cleaner Gurtreiniger m, Gurtreinigungseinrichtung f (am Gurtförderer)
  b. conveyor Gurt[band]förderer m, Bandförderer m, Förderband n
  b. dryer Bandtrockner m, Förderbandtrockner m
  b. elevator Gurtbecherwerk n
  b. filter Bandfilter n, Umlauffilter n
  b. sag Banddurchhängen n, Banddurchhang m, Gurtdurchhängen n, Gurtdurchhang m
  b. tension Bandspannung f, Gurtspannung f
  b. tunnel dryer s. b. dryer
  b. weigher Bandwaage f
belting Riemenwerkstoff m, Gurtwerkstoff m, Bandwerkstoff m; Bänder npl und Riemen mpl
bemegride (Pharm) Bemegrid n, 4-Äthyl-4-methyl-2,6-dioxopiperidin n
ben oil Behenöl n (von Moringa-Arten)
benactyzine [hydrochloride]
  $(C_6H_5)_2COHCOO \cdot CH_2CH_2N(C_2H_5)_2 \cdot HCl$ Benaktysin n, Benzilsäurediäthylaminoäthylester-HCl n

Bence-Jones protein Bence-Jones-Eiweißkörper m
bench Batterie f (z.B. Verkokungsbatterie); Bühne f, Bedienungsbühne f; Rampe f; (Glas) Bank f, Hafenbank f, Gesäß n, Ofensohle f (eines Hafenofens)
  b. blowpipe Gebläselampe f, Gebläsebrenner m
  b. photometer Fotometerbank f (visuelles Fotometer)
  b. sink Ausgußbecken n, Spülbecken n, Abflußbecken n
bend (Gerb) Kernstückhälfte f
  b. roller (Gerb) Lederwalze f
  b. test Biegeprüfung f, Biegeprobe f
Benda solution (Bio) Bendasche Flüssigkeit (Fixierungsflüssigkeit) f (aus Osmiumsäure, Chromsäure und Eisessig)
bendazol Bendazol n, 2-Benzylbenzimidazol n
Bender lead sulphide process s. Bender process
Bender process Bender-Prozeß m, Bleisulfidverfahren n (Süßungsverfahren)
Bender sweetening Bleisulfidsüßen n
bending endurance Dauerbiegefestigkeit f
  b. furnace (Glas) Biegeofen m
  b. roll (Glas) Biegewalze f (beim Colburn-Verfahren)
  b. strength Biegefestigkeit f
  b. test s. bend test
  b. vibration Deformationsschwingung f, Biegeschwingung f, Biegungsschwingung f
bene s. benne
  b. oil s. benne oil
Benedict metal Benedict-Metall n (Kupfer-Nickel-Legierung)
Benedict nickel Benedict-Nickel n (Legierung)
Benedict solution Benedictsches Reagens n, Benedictsche Lösung f (zum Nachweis reduzierender Zucker)
beneficiate / to (Erz) aufbereiten, anreichern
beneficiation Aufbereitung f, Anreicherung f (von Erz)
Bengal cardamom Bengalischer Kardamom m(n), Nepalkardamom m(n) (von Amomum aromaticum Roxb. und A. subulatum Roxb.)
  B. kino Bengalisches Kinoharz n, Bengalisches Kino n, Bengalkino n, Buteakino n (von Butea superba Roxb.)
Bengough-Stuart process Bengough-Stuart-Verfahren n, Chromsäureverfahren n (mit Chromsäure arbeitendes Verfahren zur anodischen Oxydation)
beni oil s. benne oil
benihinol Myrtenol n, 6,6-Dimethyl-2-hydroxymethylbizyklo[1,1,3]-2-hepten n (Pinenderivat)
benihiol Dihydromyrtenol n, 6,6-Dimethyl-2-hydroxymethylbizyklo-[1,1,3]-heptan n
benitoite (Min) Benitoit m (ein Zyklosilikat)
Benjamin gum Benzoeharz n, Benzoe f (von Styrax-Arten)
benne Indischer Sesam m, Sesamum indicum L.

b. oil Sesamöl n *(von Sesamum indicum L.)*
benni oil s. benne oil
benoxinate Benoxinat n, 3-Butoxy-4-aminoben-
zoesäure-β-diäthylaminoäthylester m
bent bond gebogene Bindung f, Bananenbindung
f *(C=C-Doppelbindung)*
b. glass gebogenes Glas n
bentonite [clay] Bentonit[ton] m
benzal acetone $C_6H_5CH=CHCOCH_3$ Benzalazeton
n
b. chloride $C_6H_5CHCl_2$ Benzalchlorid n, Benzylidenchlorid n, ω,ω-Dichlortoluol n
b. green Malachitgrün n, Benzalgrün n, Benzoylgrün n, Bittermandelölgrün n
benzaldehyde C_6H_5CHO Benzaldehyd m, For-
mylbenzol n, α-Oxotoluol n
b. green s. benzal green
benzamide $C_6H_5CONH_2$ Benzamid n
benzamido acetic acid $C_6H_5CO \cdot NHCH_2COOH$
Benzoylaminoessigsäure f, Benzoylglykokoll n,
Benzoylglyzin n, Hippursäure f
benzamine blue Trypanblau n
benzanthrone Benzanthron n
b. dye Benzanthronfarbstoff m
b. series Benzanthronreihe f
benzanthronequinoline Benzanthronchinolin n
1-benzazine 2,3-Benzopyridin n, Chinolin n
2-benzazine 3,4-Benzopyridin n, Isochinolin n
benzazoline Tolazolin n, 2-Benzyl-4,5-imidazolin n
benzene C_6H_6 Benzol n
b. azimide $C_6H_4NHN_2$ Benzolazimid n,
Aziminobenzol n, 1,2,3-Benzotriazol n
b. carbonal C_6H_5CHO Benzaldehyd m, For-
mylbenzol n, α-Oxotoluol n
b. carbonitrile C_6H_5CN Benzonitril n, Zyanbenzol n
b. carboxylic acid C_6H_5COOH Benzolkarbonsäure f, Benzoesäure f
b. chloride C_6H_5Cl Chlorbenzol n, Monochlorbenzol n
b. derivative Benzolderivat n, Benzolabkömm-
ling m
1,4-b. diamine p-Phenylendiamin n, p-Diamino-
benzol n
b.-diazoanilide $C_6H_5N=NNHC_6H_5$ Diazoaminobenzol n, Diazoamidobenzol n, 1,3-Diphenyltri-
azen n, Benzoldiazoanilid n
1,2-b. dicarboxylic acid $C_6H_4(COOH)_2$ 1,2-Ben-
zoldikarbonsäure f, Benzol-o-dikarbonsäure f,
o-Phthalsäure f, Phthalsäure f
1,3-b. dicarboxylic acid $C_6H_4(COOH)_2$ 1,3-Ben-
zoldikarbonsäure f, Benzol-m-dikarbonsäure f,
Isophthalsäure f, m-Phthalsäure f
1,4-b. dicarboxylic acid $C_6H_4(COOH)_2$ 1,4-Ben-
zoldikarbonsäure f, Benzol-p-dikarbonsäure f,
Terephthalsäure f, p-Phthalsäure f
b. hexacarboxylic acid $C_6(COOH)_6$ Benzolhexakarbonsäure f, Mellit[h]säure f, Honigsteinsäure f

b. hexachloride $C_6H_6Cl_6$ Hexachlorzyklohexan
n, HCH, Benzolhexachlorid n
b. hexol $C_6(OH)_6$ Hexahydroxybenzol n
b. hydrocarbon Benzolkohlenwasserstoff m,
aromatischer Kohlenwasserstoff m
b.-like benzolähnlich
b. nucleus s. b. ring
b.-pressure extraction Benzoldruckextraktion f,
Druckextraktion f mit Benzol, Benzolextraktion f
unter Druck
b. propionic acid $C_6H_5CH_2CH_2COOH$ Hydrozimtsäure f, β-Phenylpropionsäure f
b. ring Benzolring m, Benzolkern m, aromatischer Kern m
b. series Benzolreihe f
b. sulphinic acid $C_6H_5SO_2H$ Benzolsulfinsäure f
b. sulphonamide $C_6H_5SO_2NH_2$ Benzolsulfon-
säureamid n, Benzolsulfonamid n
1,2,4,5-b. tetracarboxylic acid $C_6H_2(COOH)_4$
1,2,4,5-Benzoltetrakarbonsäure f, Pyromellit[h]-
säure f
1,2,3-b. tricarboxylic acid $C_6H_3(COOH)_3$
1,2,3-Benzoltrikarbonsäure f, Hemimellit[h]säure
f
b. vapour Benzoldampf m
benzeneboronic acid $C_6H_5B(OH)_2$ Benzolboronsäure f, Phenylboronsäure f
1,2-benzenediol $C_6H_4(OH)_2$ o-Dihydroxybenzol n,
Brenzkatechin n
1,3-benzenediol $C_6H_4(OH)_2$ m-Dihydroxybenzol n,
Resorzin n
1,4-benzenediol HOC_6H_4OH p-Dihydroxybenzol n,
Hydrochinon n
benzenephosphinic acid $C_6H_5H_2PO_2$ Benzolphosphinsäure f, Phenylphosphinsäure f
benzenephosphonic acid $C_6H_5H_2PO_3$ Benzolphosphonsäure f, Phenylphosphonsäure f
benzenesulphonic acid $C_6H_5SO_3H$ Benzolsulfonsäure f, *(i.e.S.)* Benzolmonosulfonsäure f
1,2,3-benzenetriol $C_6H_3(OH)_3$ 1,2,3-Trihydroxyben-
zol n, Pyrogallol n, Pyrogallussäure f
benzenoid benzoid
b. ring benzoider Ring m
benzfuran s. benzofuran
benzhydrol $(C_6H_5)_2CHOH$ Benzhydrol n
benzidine $NH_2 \cdot C_6H_4 \cdot C_6H_4 \cdot NH_2$ Benzidin n, 4,4-Di-
aminodiphenyl n
b. base Benzidinbase f
b. conversion (rearrangement) Benzidin-
Umlagerung f
b. test Benzidinprobe f
b. transformation s. b. rearrangement
benzidinesulphonedisulphonic acid Benzidinsul-
fondisulfonsäure f
benzil $C_6H_5COCOC_6H_5$ Benzil n, Diphenylglyoxal
n, Diphenyldiketon n
benzilic acid $HO \cdot C(C_6H_5)_2 \cdot COOH$ Benzilsäure f,
Diphenylglykolsäure f, α-Hydroxydiphenylessig-
säure f
b. acid rearrangement Benzilsäureumlagerung
f

bis-benzimidazole brightener *Bis*-Benzimidazol-aufheller *m*

benzin *s.* benzine

benzine Benzin *n*; Leichtbenzin *n*, leichtes Benzin *n (Siedebereich 20 bis 135°C)*; Petroleumäther *m*, Petroläther *m (Siedebereich 40 bis 70°C)*

benzinoform CCl_4 Kohlenstofftetrachlorid *n*, Tetra[chlorkohlenstoff] *m*, Tetrachlormethan *n*

benzoaric acid Ellagsäure *f*, 4,5,6,4',5',6'-Hexa-hydroxydiphensäuredilakton *n*

benzoate Benzoat *n*

benzoated tincture of opium *(Pharm)* benzoesäurehaltige Opiumtinktur *f*

benzoazurin *s.* benzoazurine

benzoazurine Benzoazurin *n*

1,2-benzodiazine 1,2-Benzodiazin *n*, Zinnolin *n*

1,3-benzodiazine 1,3-Benzodiazin *n*, Chinazolin *n*, Benzopyrimidin *n*

1,4-benzodiazine 1,4-Benzodiazin *n*, Benzopyrazin *n*, 1,4-Diazanaphthalin *n*, Chinoxalin *n*

benzofuran Benzofuran *n*, Kumaron *n*

benzohydrol *s.* benzhydrol

benzoic acid C_6H_5COOH Benzoesäure *f*, Benzolkarbonsäure *f*

  b. aldehyde *s.* benzaldehyde

  b. amide *s.* benzamide

  b. anhydride $(C_6H_5CO)_2O$ Benzoesäureanhydrid *n*

  b. ether $C_6H_5COOC_2H_5$ Benzoesäureäthylester *m*, Äthylbenzoat *n*

benzoin $C_6H_5 \cdot CHOH \cdot CO \cdot C_6H_5$ 1. Benzoin *n*, Benzoylphenylkarbinol *n*, α-Hydroxy-α-phenyl-azetophenon *n*, Bittermandelölkampfer *m*; 2. *s.* b. gum

  b. condensation Benzoinkondensation *f*

  b. gum (resin) Benzoeharz *n*, Benzoe *f (von Styrax-Arten)*

  b. Siam Siambenzoe *f (von Styrax-Arten)*

benzol Benzol *n (als Handelsprodukt)*

  b.-acetone process Azeton-Benzol-Verfahren *n (Entparaffinierungsverfahren)*

  b. condenser Benzolkondensator *m*

  b. forerunnings Benzolvorlauf *m*

  b.-ketone process Keton-Benzol-Verfahren *n (Entparaffinierungsverfahren)*

  b. mixture (motor spirit) Motorenbenzol *n*

  b. plant Benzolanlage *f*

  b. pump Benzolpumpe *f*

  b. recovery Benzolgewinnung *f*

  b. rectification still Rektifizierapparat *m* für Benzol

  b. scrubber *s.* b. washer

  b. separator Benzolabscheider *m*

  b. still Benzoldestillationsanlage *f*, Benzoldestillieranlage *f*, Benzolabtreiber *m*

  b. thermometer Benzolthermometer *n*

  b. washer Benzolwäscher *m*, Benzolwascher *m*

90's gum 90er Benzol (Handelsbenzol) *n*

benzole *s.* benzol

benzolize / to mit Benzol beladen (anreichern, sättigen)

benzolized mit Benzol beladen (angereichert), benzolgesättigt

benzonitrile C_6H_5CN Benzonitril *n*, Zyanbenzol *n*

benzoorange R Benzoorange R *n*

benzophenol C_6H_5OH Phenol *n*, Hydroxybenzol *n*

benzophenone $C_6H_5COC_6H_5$ Benzophenon *n*, Benzoylbenzol *n*, Diphenylketon *n*

benzopyrene Benzpyren *n*

benzo[*b*]pyridine 2,3-Benzopyridin *n*, Chinolin *n*

benzo[*c*]pyridine Isochinolin *n*

1,2-benzopyrone 1,2-Benzopyron *n*, 5,6-Benzokumalin *n*, 2-Oxo-1,2-chromen *n*, 1,2-Chromenon *n*, Kumarsäurelakton *n*, Kumarin *n*

1,4-benzopyrone 1,4-Benzopyron *n*, Chromon *n*

benzo[*f*]quinoline 5,6-Benzochinolin *n*, β-Naphthochinolin *n*

p-benzoquinone $C_6H_4O_2$ *p*-Benzochinon *n*, *p*-Chinon *n*, Chinon *n*, 1,4-Zyklohexadiendion *n*

1,2,3-benzotriazole 1,2,3-Benzotriazol *n*, Aziminobenzol *n*, Benzolazimid *n*

benzotrichloride $C_6H_5CCl_3$ Benzotrichlorid *n*, α-Trichlortoluol *n*, α-Trichlormethylbenzol *n*, Phenylchloroform *n*

benzoyl Benzoyl *n*

  b. benzene $C_6H_5COC_6H_5$ Benzoylbenzol *n*, Benzophenon *n*, Diphenylketon *n*

  n-b. glycine $C_6H_5CONHCH_2COOH$ Benzoylglyzin *n*, Benzoylglykokoll *n*, Benzoylaminoessigsäure *f*, Hippursäure *f*

  b. group C_6H_5CO- Benzoylgruppe *f*, Benzoylrest *m*

  b. hydroperoxide $C_6H_5COO_2H$ Benzoylhydroperoxid *n*, Perbenzoesäure *f*, Benzopersäure *f*

  b. J acid Benzoyl-J-Säure *f*, Benzoyl-I-Säure *f*

  b. peroxide $C_6H_5 \cdot OCO \cdot O \cdot CO \cdot C_6H_5$ Benzoylperoxid *n*, Dibenzoylperoxid *n*

  b.-phenyl carbinol $C_6H_5 \cdot CHOH \cdot CO \cdot C_6H_5$ Benzoylphenylkarbinol *n*, α-Hydroxy-α-phenylazetophenon *n*, Benzoin *n*, Bittermandelölkampfer *m*

benzoylate / to benzoylieren

benzoylation Benzoylierung *f*

benzoylformic acid $C_6H_5COCOOH$ Benzoylameisensäure *f*, Phenylglyoxylsäure *f*

benzoylmethylecgonine Benzoylmethylekgonin *n*, Benzoylekgoninmethylester *m*, Methylbenzoylekgonin *n*, Kokain *n*

benzoylnaphthalene $C_{10}H_7COC_6H_5$ Benzoylnaphthalin *n*

1,2-benzphenanthrene 1,2-Benz[o]phenanthren *n*, Chrysen *n*

benzpyrene Benzpyren *n*

benzyl Benzyl *n*

  b. acetate $CH_3COOCH_2C_6H_5$ Essigsäurebenzylester *m*, Benzylazetat *n*

  b. acetic acid $C_6H_5CH_2CH_2COOH$ β-Phenylpropionsäure *f*, Hydrozimtsäure *f*

  b. alcohol $C_6H_5CH_2OH$ Benzylalkohol *m*, Phenylkarbinol *n*

  b. benzoate $C_6H_5COO \cdot CH_2C_6H_5$ Benzoesäurebenzylester *m*, Benzylbenzoat *n*

b. butyrate $CH_3(CH_2)_2COO \cdot CH_2C_6H_5$ Buttersäurebenzylester *m*, Benzylbutyrat *n*
b. carbinol $C_6H_5CH_2CH_2OH$ 2-Phenyläthanol *n*, Phenyläthylalkohol *m*
b. cellulose Benzylzellulose *f*
b. chloride $C_6H_5CH_2Cl$ Benzylchlorid *n*, α-Chlortoluol *n*
b. cinnamate $C_6H_5CH=CHCOO \cdot CH_2C_6H_5$ Benzylzinnamat *n*, Zimtsäurebenzylester *m*, Zinnamein *n*
b. cyanide $C_6H_5CH_2CN$ Benzylzyanid *n*, Phenylazetonitril *n*, Phenylessigsäurenitril *n*
b. dichloride $C_6H_5CHCl_2$ Benzylidenchlorid *n*, Benzalchlorid *n*, α,α-Dichlortoluol *n*
b. ether $(C_6H_5CH_2)_2O$ Dibenzyläther *m*
b. group Benzylgruppe *f*, Benzylrest *m*
b. mandelate $C_6H_5CHOHCOO \cdot CH_2C_6H_5$ Mandelsäurebenzylester *m*, Phenylglykolsäurebenzylester *m*
b. nicotinate Nikotinsäurebenzylester *m*
p-**b. phenol** $C_6H_5CH_2 \cdot C_6H_4OH$ *p*-Benzylphenol *n*
b. propionate $C_2H_5COO \cdot CH_2C_6H_5$ Propionsäurebenzylester *m*, Benzylpropionat *n*
b. radical Benzylgruppe *f*, Benzylrest *m*, Benzylradikal *n*; [freies] Benzylradikal *n*
b. salicylate $HOC_6H_4COO \cdot CH_2C_6H_5$ Salizylsäurebenzylester *m*, Benzylsalizylat *n*
b. succinate $(CH_2COO \cdot CH_2C_6H_5)_2$ Bernsteinsäuredibenzylester *m*, Dibenzylsukzinat *n*
benzylbenzene $CH_2(C_6H_5)_2$ Benzylbenzol *n*, Diphenylmethan *n*
benzylidene acetone $C_6H_5CH=CHCOCH_3$ Benzalazeton *n*
b. chloride $C_6H_5CHCl_2$ Benzylidenchlorid *n*, Benzalchlorid *n*, α,α-Dichlortoluol *n*
benzylisoquinoline Benzylisochinolin *n*
benzylpenicillin Benzylpenizillin *n*, Penizillin G *n*, Penizillin II *n*
benzyne C_6H_4 Dehydrobenzol *n*, Benz-in *n*, Benzyn *n*, Arin *n* (*i.e.S.*), Aryn *n* (*i.e.S.*)
berberine Berberin *n* (*Isochinolinalkaloid*)
berberonic acid $C_5H_2N(COOH)_3$ 2,4,5-Pyridintrikarbonsäure *f*
bergamot oil Bergamottöl *n* (*von Citrus aurantium L. ssp. bergamia*)
berginization *s.* Bergius process
Berglus process Bergius-Hydrierverfahren *n*, Bergius-Verfahren *n*, Bergius-Hochdruckverfahren *n*, [katalytisches] Hochdruckhydrier[ungs]verfahren *n* nach Bergius, Bergius-Pier-Verfahren *n*, Berginverfahren *n*, Leuna-Verfahren *n*
Bergmann series (*phys Ch*) Bergmann-Serie *f*
Berkefeld filter Berkefeld-Filter *n*
berkelium Bk Berkelium *n*
Berl saddle Berl-Sattel[körper] *m* (*Füllkörper*)
Berlin blue Berliner Blau *n*, Pariser Blau *n*, Preußisch Blau *n*
Berthelot bomb Berthelot-Bombe *f*, kalorimetrische Bombe *f* nach Berthelot

Berthelot calorimeter Berthelot-Kalorimeter *n*
Berthelot equation Berthelot-Gleichung *f*, Berthelotsche Gleichung (Zustandsgleichung) *f*
Berthelot-Mahler bomb calorimeter Berthelot-Mahler-Bombenkalorimeter *n*
berthollide compounds *s.* berthollid[e]s
berthollid[e]s Berthollide *npl*, Berthollidverbindungen *fpl*, berthollide (nichtdaltonide, nichtdaltonische) Verbindungen *fpl*, Verbindungen *fpl* von nichtkonstanter Zusammensetzung
bertrandite (*Min*) Bertrandit *m*
beryl (*Min*) Beryll *m* (*Berylliumaluminiumsilikat*)
beryllate Beryllat *n*
beryllia BeO Beryllerde *f*, Süßerde *f*, Gluzinerde *f*, Glyzinerde *f*, (*veraltet für*) Berylliumoxid *n*
beryllide Beryllid *n*
beryllium Be Beryllium *n*
b. bromide $BeBr_2$ Berylliumbromid *n*
b. carbide Be_2C Berylliumkarbid *n*
b. carbonate $BeCO_3$ Berylliumkarbonat *n*
b. chloride $BeCl_2$ Berylliumchlorid *n*
b. fluoride BeF_2 Berylliumfluorid *n*
b. halide Berylliumhalogenid *n*
b. hydrate *s.* b. hydroxide
b. hydride BeH_2 Berylliumhydrid *n*
b. hydroxide $Be(OH)_2$ Berylliumhydroxid *n*
b. iodide BeJ_2 Berylliumjodid *n*
b. nitrate $Be(NO_3)_2$ Berylliumnitrat *n*
b. orthosilicate $Be_2[SiCO_4]$ Berylliumorthosilikat *n*, Berylliumtetroxosilikat *n*
b. oxide BeO Berylliumoxid *n*
b.-potassium fluoride $K_2[BeF_4]$ Kaliumfluoroberyllat *n*
b. sodium fluoride $Na_2[BeF_4]$ Natriumfluoroberyllat *n*
b. sulphate $BeSO_4$ Berylliumsulfat *n*
b. sulphide BeS Berylliumsulfid *n*
b. target Berylliumtarget *n*
beryllonite (*Min*) Beryllonit *m*
berzelianite (*Min*) Berzelianit *m*, Selenkupfer *n* (*Kupfer(I)-selenid*)
B.E.S.A. (*Abk. für*) British Engineering Standards Association
Bessemer converter Windfrischkonverter *m*; Bessemer-Konverter *m*, Bessemer-Birne *f*
Bessemer converter process *s.* Bessemer process
Bessemer iron *s.* Bessemer pig
Bessemer ore phosphorarmes Eisenerz *n* (*P-Gehalt kleiner oder gleich 0,09 %*)
Bessemer pig Bessemer-Roheisen *n*
Bessemer process Windfrischverfahren *n*, Konverter[frisch]verfahren *n*, Bessemer-Verfahren *n* (*i.w.S.*); Bessemer-Verfahren *n*, Bessemer-Konverterverfahren *n*, saures Windfrischverfahren (Bessemer-Verfahren) *n*
Bessemer slag Bessemer-Schlacke *f*
Bessemer steel Windfrischstahl *m*, windgefrischter Stahl *m*, Konverterstahl *m*; Bessemer-Stahl *m*, Bessemer-Konverterstahl *m*
Bessemer-type converter *s.* Bessemer converter

bessemerize / to windfrischen, bessemern, im Konverter verblasen; [Kupferstein] im Konverter verblasen, [kupfer]bessemern

bessemerizing Windfrischen *n*, Bessemern *n*, Verblasen *n* im Konverter; Verblasen *n* [von Kupferstein] im Konverter, Kupferbessemern *n*, Bessemern *n*

best gold *(Ker)* Poliergold *n*

BET equation BET-Gleichung *f*, Gleichung *f* von Brunauer, Emmett und Teller

BET method BET-Methode *f (Bestimmung der Oberfläche von Absorbentien mit Hilfe der Gleichung von Brunauer, Emmett und Teller)*

beta-absorption gauge Betastrahlen-Dickenmesser *m*, Beta-Dickenmesser *m*

b. acid β-Anthrachinonsulfonsäure *f*

b. antimony schwarzes Antimon *n*

b. arsenic schwarzes Arsen *n*

b. brass β-Messing *n (45,5 bis 50 % Zinkgehalt)*

b. cellulose Beta-Zellulose *f*, β-Zellulose *f*

b. decay (disintegration) Beta-Zerfall *m*, β-Zerfall *m*, Beta-Umwandlung *f*

b. disintegration energy Beta-Zerfallsenergie *f*

b.-disintegrator *(Kch)* Beta-Strahler *m*, β-Strahler *m*

b.-emitter *s.* b.-disintegrator

b. fraction β-Fraktion *f*, β-Bestandteile *mpl (bei der Pyridinextraktion der Steinkohle)*

b. gauge *s.* b.-absorption gauge

b.-hydroxy acid β-Hydroxysäure *f*

b.-methylumbelliferone β-Methylumbelliferon *n*, 4-Methylumbelliferon *n*, 7-Hydroxy-4-methylumarin *n*, 7-Hydroxy-4-methyl-1,2-chromenon *n*

b. particle Betateilchen *n*, β-Teilchen *n*, Elektron *n*

b. phase β-Phase *f*

b.-ray spectrum Beta-Spektrum *n*, β-Spektrum *n*

b.-rays Betastrahlen *mpl*, β-Strahlen *mpl*

b. tin β-Zinn *n (metallisches Zinn)*

betafite *(Min)* Betafit *m*

betaine Betain *n*, Oxyneurin *n*, Lyzin *n*, Trimethylglykokoll *n (Alkaloid)*

b. compound Betain *n (i.w.S.)*

betatopic betatop

betatron Betatron *n*, Elektronenschleuder *f*, Wirbelbeschleuniger *m*, Rheotron *n*

betel nut Betelnuß *f*, Arekanuß *f (von Areca cathechu L.)*

betelnut palm Betelnußpalme *f*, Betelpalme *f*, Areca cathechu L.

bethabar[r]a wood Bethabarraholz *n*

betol Betol *n*, Salizyl-β-naphtholester *m*, Naphthalol *n*

Bettendorf's test for arsenic Bettendorfsche Arsenprobe (Probe) *f*

Betts process Betts-Verfahren *n (zur Bleiraffination)*

betula camphor Betulin *n*, Birkenkampfer *m (Triterpenalkohol)*

betulinol *s.* betula camphor

bevel / to schräg abschneiden

bevel Schrägfläche *f*

b. gear Kegelradgetriebe *n*

bevelling Abschrägung *f*

beverage Getränk *n*

b. alcohol Trinkbranntwein *m*

Bewoid process *(Pap)* Bewoid-Verfahren *n*

Bewoid size *(Pap)* Bewoid-Leim *m*

Bey reagent Reagens *n* nach Bey *(ätherische Resorzinlösung zum Kadmium- und Zinn-Nachweis)*

beyrichite *(Min)* Beyrichit *m*, *(veraltet für)* Violarit *m*

Bezssonov reagent Reagens *n* nach Bezssonow *(zum Nachweis von Polyphenolen)*

BHC *s.* benzene hexachloride

BHT *s.* butylated hydroxytoluene

BI *s.* buffer index

biacetyl $CH_3COCOCH_3$ Diazetyl *n*, Butandion-(2,3) *n*, Dimethylglyoxal *n*

biacetylene $HC{\equiv}CC{\equiv}CH$ Diazetylen *n*, Diäthin *n*, Butadiin *n*

biacid zweisäurig *(von Basen)*

Bial reagent Bials Reagens *n*, Bialsche Lösung *f (zum Nachweis von Pentosen)*

biallyl $CH_2{=}CHCH_2CH_2CH{=}CH_2$ Biallyl *n*, Diallyl *n*, 1,5-Hexadien *n*

p,p'-bàniline $NH_2C_6H_4 \cdot C_6H_4NH_2$ Benzidin *n*, 4,4-Diaminodiphenyl *n*

bias control Kompensator *m (z.B. eines Polarografen)*

biatomic zweiatomig

biaxial *(Krist)* [optisch] zweiachsig, biaxial

b. orientation biaxiale Orientierung *f*

b. stretching zweiachsiges Strecken (Recken) *n*

o,o'-benzoic acid $HOOC \cdot C_6H_4 \cdot C_6H_4 \cdot COOH$ Diphensäure *f*, 2,2'-Diphenyldikarbonsäure *f*

bibenzoyl $C_6H_5COCOC_6H_5$ Bibenzoyl *n*, Dibenzoyl *n*, Benzil *n*, Diphenylglyoxal *n*, Diphenyldiketon *n*

sym-bibenzyl $C_6H_5CH_2CH_2C_6H_5$ *sym*-Diphenyläthan *n*, 1,2-Diphenyläthan *n*, Dibenzyl *n*, Bibenzyl *n*

biberite *s.* bieberite

bible paper Bibeldruckpapier *n*, Dünndruckpapier *n*

bibliometer *(Pap)* Saughöhenprüfgerät *n*

biborate $M^I_2B_4O_7$ Tetraborat *n*

bibromide $M^{II}Br_2$ Dibromid *n*

bibulous *(Pap)* saugfähig

bibulousness *(Pap)* Saugfähigkeit *f*

bicalcic phosphate $CaHPO_4$ Kalziumhydrogenphosphat *n*, sekundäres Kalziumphosphat *n*

bicarbonate M^IHCO_3 Hydrogenkarbonat *n*

b. of soda $NaHCO_3$ Natriumhydrogenkarbonat *n*

bicarburetted hydrogen $H_2C{=}CH_2$ Äthen *n*, Äthylen *n*

bicentric molecular orbital Zweizentrenmolekülorbital *n*

Bicheroux process Bicheroux-Verfahren *n (Flachglasherstellung)*

Bicheroux semi-continuous casting s. Bicheroux process

bichloride Dichlorid n

bichromate $M^I_2Cr_2O_7$ Dichromat n

b. of soda $Na_2Cr_2O_7$ Natriumdichromat n

bicolorin Bikolorin n, Äskulin n, 6,7-Dihydroxykumarin-6-(β-d-glukopyranosid) n

bicolour effect 1. Zweifarbeneffekt m; 2. Bikoloreffekt m (unterschiedlich gefärbte Faserstoffmischung)

bicomponent fibre Bikomponentenfaser f, Bifaser f

bicyclic bizyklisch

bidentate (Komplexchemie) zweizahnig, zweizähnig, zweizählig (Koordinationswert des Liganden)

Bidery (Bidri) metal (Cu- und Pb-haltige Zinklegierung)

bieberite (Min) Bieberit m, Kobaltvitriol m (Kobalt(II)-sulfat-7-Wasser)

Biebrich [scarlet] red Biebricher Scharlach m, Doppelscharlach m, Neurot n

biebrite s. bieberite

Bierer-Davis oxygen bomb Bierer-Davis-Bombe f, Sauerstoffbombe f nach Bierer und Davis

biethylene $CH_2=CHCH=CH_2$ Diäthylen n, Diäthen n, Vinyläthylen n, Vinyläthen n, Divinyl n, Butadien-(1,3) n

bifluoride Hydrogenfluorid n

biformyl $OHC \cdot CHO$ Glyoxal n, Äthandial n, Oxalaldehyd m

bifunctional difunktionell, D-, bifunktionell

b. unit bifunktionelles Grundmolekül n, difunktionelle Einheit f, D-Einheit f (Struktureinheit)

big series große (lange) Periode f, Langperiode f

biguanide $H_2NC(=NH)NHC(=NH)NH_2$ Biguanid n

bihexyl $CH_3(CH_2)_{10}CH_3$ Bihexyl n, Dihexyl n, n-Dodekan n

bilberry Blaubeere f, Heidelbeere f, Vaccinium myrtillus L.

bile Galle f

b. acid Gallensäure f

b. pigment Gallenfarbstoff m

bilifulvin s. bilirubin

bilirubin Bilirubin n

biliverdin Biliverdin n, Dehydrobilirubin n, Uteroverdin n

bill paper Wechselformularpapier n, Wechselpapier n

billet (Plast) Puppe f, Walzpuppe f; (Met) Knüppel m, Barren m, Strang m, Walzblock m

Billiter cell Billiter-Zelle f, Horizontalzelle f (Elektrolyse)

billitonite Billitonit m (ein Tektit)

bimetal Bimetall n

bimetallic bimetallisch, Bimetall...

b.thermometer Bimetallthermometer n

bimolecular bimolekular, dimolekular, 2molekular

b. reaction bimolekulare (dimolekulare, 2molekulare) Reaktion f

bin Silo m (n), Bunker m, Behälter m, Zellenspeicher m

b. curing Anvulkanisation f, Anvulkanisieren n, Anspringen n, Anbrennen n (während der Lagerung)

b. dryer Schachttrockner m

binary binär, zweigliedrig, aus zwei Einheiten bestehend, Zweistoff...

b. alloy binäre Legierung f, Zweistofflegierung f

b. collision (phys Ch) Zweierstoß m

b. column Kolonne f für Zweistoffgemische

b. compound binäre Verbindung f

b. distillation Destillation f eines Zweistoffgemischs (Zweikomponentensystems)

b. feed (Destillation) Zulauf m eines Zweistoffgemisches (binären Gemisches)

b. mixture binäres Gemisch n, Zweistoffgemisch n

b. system binäres System n, Zweistoffsystem n, Zweikomponentensystem n

bind / to binden

binder Bindemittel n, Binder m, (Farb auch) Bindekörper m, (Plast auch) Harzträger m, (Geol auch) Zement m, (Glas auch) Schmälze f

b. spray (Glas) Schmälznebel m

binderless bindemittellos, bindemittelfrei, ohne Bindemittel[zusatz]

binder's board Buchbinderpappe f, Buchbinderdeckel m, Bücherpappe f

binding Bindung f

b. ability s. b. power

b. agent Bindemittel n, Binder m

b. capacity s. b. power

b. electron Bindungselektron n, Bindeelektron n

b. energy Bindungsenergie f

b. force bindende Kraft f, Bindungskraft f, Bindekraft f

b. link Bindeglied n, Brückenglied n

b. material Bindemittel n, Binder m

b. power Bindevermögen n, Bindefähigkeit f, Bindekraft f; (Bindungstheorie) Bindungsvermögen n, Bindungsfähigkeit f, Bindungskraft f, Bindekraft f, bindende Kraft f

b. strength Bindefestigkeit f; (Bindungstheorie) Bindungsstärke f, Bindungsfestigkeit f, Bindefestigkeit f

binegative zweifach negativ

Bingham body (Koll) Bingham-Körper m

Bingham flow Binghamsches Fließen n, Strukturviskosität f

biniodide Dijodid n

binnite (Min) Binnit m

binodal curve binodale Kurve f

binoxalate $M^IHC_2O_4$ Hydrogenoxalat n

binoxide Dioxid n, (manchmal auch) Peroxid n

binuclear, binucleate zweikernig

bioaeration Schlammbelebung f

bioassay s. biological assay

biocatalyst Biokatalysator m, Ergin n

biochemical biochemisch

b. catalyst s. biocatalyst
b. oxygen demand biochemischer Sauerstoffbedarf m, BSB m
biochemist Biochemiker m
biochemistry Biochemie f, biologische Chemie f
biochemy s. biochemistry
biochrome Naturfarbstoff m, natürlicher Farbstoff m
biocide Biozid n (Chemikalie zur Bekämpfung schädlicher Mikroben)
bioclast (Geol) bioklastisches Gestein n
bioclastic (Geol) bioklastisch
biocrystal Biokristall m
biodegradability biologische Abbaubarkeit f
biodegradable biologisch abbaubar
biofilter Biofilter n
bioflavonoids Bioflavonoide npl (Vitamin-P-Faktoren)
biogenesis Biogenese f, Biosynthese f, biologische Synthese f
biogenetic[al] s. biogenic
biogenic biogen
  b. amines biogene Amine npl, Protoalkaloide npl
  b. rock biogenes (organogenes) Gestein (Sediment) n, Biolith m
biogenous biogen
biolith s. biogenic rock
biological biologisch
  b. assay Tierversuch m, Tierexperiment n; biologische Wertbestimmung f
  b. chemistry biologische Chemie f, Biochemie f
  b. control biologische Bekämpfung (Schädlingsbekämpfung) f
  b. elimination s. b. purification of sewage
  b. filter Biofilter n
  b. haze biologische Trübung f (z.B. des Bieres)
  b. purification of sewage biologische Abwasserreinigung f
  b. retting (Text) biologische Röste f
  b. sludge biologischer (belebter, aktivierter) Schlamm m, Belebtschlamm m
bioluminescence Biolumineszenz f
biophile element biophiles Element n
biose Biose f (Monosaccharid mit zwei Sauerstoffatomen)
bioside Biosid n
biosynthesis Biosynthese f, biologische Synthese f, Biogenese f
biotic biotisch
biotin Biotin n, Bios II n, Vitamin H n
biotite (Min) Biotit m, Kali-Magnesiaeisenglimmer m, Magnesiaeisenglimmer m
bioxalate s. binoxalate
biphenyl Diphenyl n, Biphenyl n
biphosphate $M^IH_2PO_4$ Dihydrogenorthophosphat n, Dihydrogenphosphat n
biquartz Biquarz m, Doppelquarz m, Soleil-Doppelplatte f
biradical Biradikal n, Diradikal n
birch Birke f, Betula L.

b. bark oil Birkenrindenöl n, (Gerb auch) Juchtenöl n
b. camphor Birkenkampfer m, Betulin n (Triterpenalkohol)
b. hair lotion Birkenhaarwasser n
b. oil Birkenöl n; Birkenteeröl n
b. tar Birkenteer m
b. tar oil Birkenteeröl n
b. water Birkenwasser n, Birkenwein m
Birch reduction Birch-Hückel-Reaktion f (Hydrierung aromatischer Verbindungen mit Natrium und Ammoniak)
bird repellent Vogelfraß-Abwehrmittel n, Saatschutzmittel n (gegen Vögel)
b. swing s. birdcage
birdcage (Glas) Affenschaukel f (freihängender Glasfaden in einer Flasche)
birefringence Doppelbrechung f
birefringent doppelbrechend
Birkeland-Eyde process Birkeland-Eyde-Verfahren n (Herstellung von Stickoxiden durch Luftoxydation)
Bisabol myrrh (Balsamharz von Commiphora erythraea Engl.)
bisabolene Bisabolen n (monozyklisches Sesquiterpen)
bisazo dye Disazofarbstoff m
bisbeeite (Min) Bisbeeit m (Aluminiumkupfer(II)-dodekahydroxidsulfat)
Bischler-Napieralski reaction Bischler-Napieralski-Reaktion f (Isochinolinsynthese)
bischofite (Min) Bischofit m (Magnesiumchlorid)
biscuit 1. (Ker) Schrühware f, geschrühte Ware f; Biskuitbrandware f, Biskuitware f, Biskuitporzellan n; 2. (Plast) Tablette f, Preßkuchen m, Rohling m (für Schallplattenpressen)
  b. earthenware (Ker) Schrühware f, geschrühte Ware f
  b.-fire / to (Ker) verschrühen, verglühen
  b.-fired ware s. biscuit 1.
  b. firing (Ker) Rohbrand m, Verglühbrand m, Glühbrand m, Schrühbrand m, Rauhbrand m, (beim Porzellan) Biskuitbrand m
  b. kiln (oven) (Ker) Schrühbrandofen m, Schrühofen m
  b. placing s. b. setting
  b. porcelain (Ker) Biskuitporzellan n
  b. setting (Ker) Setzen n von Schrühware beim Schrühbrand
  b. ware s. biscuit 1.
biscuitted body (Ker) geschrühter (verschrühter) Scherben m, Schrühbrandscherben m
  b. ware s. biscuit 1.
biscuitting s. biscuit firing
bisethyleneurea Bisäthylenharnstoff m, N,N-Diäthylenharnstoff m
bishydrazone Bishydrazon n, Dihydrazon n
bisimidazole Bisimidazol n, Diimidazol n
Bismarck brown Bismarckbraun n
bismite (Min) Bismit m, Wismutocker m (Wismut(III)-oxid-3-Wasser)

bismuth Bi Wismut n
  b. blende *(Min)* Kieselwismut m, Eulytin m *(Wismut(III)-orthosilikat)*
  b. bronze Wismutbronze f
  b. glance *(Min)* Wismutglanz m, Bismuthinit m, Bismut[h]in m *(Wismut(III)-sulfid)*
  b. gold *(Min)* Wismutgold n, Maldonit m
  b. iodate $Bi(JO_3)_3$ Wismut(III)-jodat n
  b. monobromide BiBr Wismutmonobromid n
  b. monosulphide BiS Wismutmonosulfid n, Wismut(II)-sulfid n
  b. nickel *(Min)* Wismutnickelkobaltkies m, Wismutnickelkies m, Grünauit m
  b. ochre *(Min)* Wismutocker m *(Wismut(III)-oxid)*
  b. ore Wismuterz n
  b. orthophosphate s. b. phosphate
  b. oxalate $Bi_2(C_2O_4)_3$ Wismut(III)-oxalat n
  b. oxybromide BiOBr Wismutoxidbromid n
  b. oxycarbonate $(BiO)_2CO_3$ Wismutoxidkarbonat n
  b. oxychloride BiOCl Wismutoxidchlorid n
  b. oxyfluoride BiOF Wismutoxidfluorid n
  b. oxyiodide BiOJ Wismutoxidjodid n
  b. oxynitrate $BiO(NO_3)$ Wismutoxidnitrat n
  b. pentoxide Bi_2O_5 Wismutpentoxid n, Wismut(V)-oxid n
  b. phosphate $BiPO_4$ Wismut(III)-phosphat n
  b. putty Wismut-Glaskitt m *(Legierung aus Bi, Pb und Sn)*
  b. selenide Bi_2Se_3 Wismut(III)-selenid n, Diwismuttriselenid n
  b. silver *(Min)* Wismutsilber n, Chilenit m
  b. spar s. bismutite
  b. sulphate $Bi_2(SO_4)_3$ Wismut(III)-sulfat n
  b. tetroxide Bi_2O_4 Wismuttetroxid n, Wismutdioxid n
  b. tribromide $BiBr_3$ Wismuttribromid n, Wismut(III)-bromid n
  b. trichloride $BiCl_3$ Wismuttrichlorid n, Wismut(III)-chlorid n
  b. trifluoride BiF_3 Wismuttrifluorid n, Wismut(III)-fluorid n
  b. trihydrate s. b. trihydroxide
  b. trihydroxide $Bi(OH)_3$ Wismuttrihydroxid n, Wismut(III)-hydroxid n
  b. triiodide BiJ_3 Wismuttrijodid n, Wismut(III)-jodid n
  b. trinitrate $Bi(NO_3)_3$ Wismut(III)-nitrat n
  b. trioxide Bi_2O_3 Wismuttrioxid n, Wismut(III)-oxid n
  b. trisulphide Bi_2S_3 Wismuttrisulfid n, Wismut(III)-sulfid n
  b. white Wismutweiß n, Perlweiß n, Schminkweiß n, Spanischweiß n *(Wismutoxidnitrat)*
bismuthal Wismut...
bismuthate M^IBiO_3 Bismutat(V) n
bismuthic Wismut..., *(i.e.S)* Wismut(V)-...
  b. acid $HBiO_3$ Wismutsäure f
  b. oxide Bi_2O_5 Wismut(V)-oxid n, Wismutpentoxid n

bismuthiferous wismuthaltig
bismuthine 1. s. bismuthinite; 2. BiH_3 Wismutwasserstoff m
bismuthinite *(Min)* Bismuthinit m, Bismut[h]in m, Wismutglanz m *(Wismut(III)-sulfid)*
bismuthiol Wismuthiol n, 2,5-Dimerkapto-1,3,4-thiodiazol n
bismuthite s. 1. bismuth trinitrate; 2. bismuth oxynitrate
bismuthous Wismut..., *(i.e.S.)* Wismut(III)-...
bismuthyl BiO— Wismutoxid...
bismutite *(Min)* Bismutit m, Wismutspat m *(Wismutoxidkarbonat)*
bisphenol A $HOC_6H_4 \cdot C(CH_3)_2 \cdot C_6H_4OH$ Bisphenol A n, 2,2-Bis(p-Hydroxyphenyl-)propan n, Dian n
bisque s. biscuit
bissy nut Kolanuß f *(von Cola-Arten)*
bisulphate M^IHSO_4 Hydrogensulfat n
bisulphide M^IHS Hydrogensulfid n; $M^I_4S_2$ Disulfid n
bisulphite M^IHSO_3 Hydrogensulfit n, *(Tech auch)* Bisulfit n
  b. [cooking] liquor *(Pap)* Sulfitkochsäure f, Sulfitsäure f, Kochsäure f
  b. liquor base *(Pap)* Base f für Sulfitkochsäure
  b. of lime *(Pap)* Kalziumbisulfit n
bitartrate $HOOC \cdot (CHOH)_2 \cdot COOM^I$ Hydrogentartrat n
bite Walzenspalt m
  b.-type fitting joint Schneidringverbindung f, Schneidringverschraubung f
α-bitter acid α-Bittersäure f, α-Hopfenbittersäure f, Humulon n
β-bitter acid β-Bittersäure f, β-Hopfenbittersäure f, β-Lupulinsäure f, Lupulon n
bitter almond bittere Mandel f
  b. almond oil Bittermandelöl n
  b. almond oil camphor $C_6H_5CHOHCOC_6H_5$ Bittermandelölkampfer m, Benzoylphenylkarbinol n, α-Hydroxy-α-phenylazetophenon n, Benzoin n
  b. apple Koloquinte f *(Frucht von Citrullus colocynthis(L.)Schrad.)*
  b. ash Quassiaholz n, Bitterholz n, Fliegenholz n
  b. cucumber (gourd) s. b. apple
  b. orange-peel oil [bitteres] Pomeranzenschalenöl n *(von Citrus aurantium L. ssp. aurantium)*
  b. principle *(Pharm)* Bitterstoff m
  b. salt *(Min)* Bittersalz n, Epsomsalz n, Epsomit m *(Magnesiumsulfat-7-Wasser)*
  b. spar *(Min)* Bitterspat m
  b. spring Bitterquelle f
  b. substance Bitterstoff m
  b. substance in hops Hopfenbitterstoff m
  b. water Bitterwasser n
  b. wood s. bitterwood
bittern Bitterstoff m; *(bei der Speisesalzgewinnung aus Meerwasser verbleibende)* restliche Mutterlauge f
bitterness of hops Hopfenbittere f

b. value Bitterstoffwert *m*, Bitterwert *m (des Hopfens)*
bitterwood Quassiaholz *n*, Bitterholz *n*, Fliegenholz *n*
bitumen Bitumen *n*
b. binder bituminöses Bindemittel *n*
b. board Bitumenpappe *f*, Asphalthartpappe *f*
b. cutback Verschnittbitumen *n*, Cutback-bitumen *n*
b. process Asphaltverfahren *n*, Asphaltfotografie *f*, Heliografie *f*
bitumenite *(Min)* Bituminit *m*, Bogheadkohle *f*
bituminous bituminös, bitumisch, bitumenhaltig
b. coal bituminöse Kohle *f*
b.-coal tar Steinkohlenteer *m*
b. coating Bitumenanstrich *m*
b. emulsion Bitumenemulsion *f*
b. lignite bituminöse (steinkohlenähnliche) Braunkohle *f*, Glanz[braun]kohle *f*, subbituminöse Kohle *f*
b. material (matter) s. b. substance
b. moulding composition s. b. plastic
b. plastic Bitumen-Preßmasse *f*
b. sand bituminöser (bitumenhaltiger) Sand *m*, Bitumensand *m*
b. shale bituminöser Schiefer *m*, Bitumenschiefer *m*
b. substance bituminöser Stoff *m*
b. varnish Bitumenlack *m*
biurate M^{I}HC$_5$H$_2$N$_4$O$_3$ Hydrogenurat *n*
biuret H$_2$N·CO·NH·CO·NH$_2$ Biuret *n*, Allophansäureamid *n*
b. reaction Biuretreaktion *f*
b. test Biuretprobe *f*
bivalency Zweiwertigkeit *f*, Bivalenz *f*
bivalent zweiwertig, bivalent
bivariant bivariant, divariant, zweifachfrei
bivinyl CH$_2$=CHCH=CH$_2$ Bivinyl *n*, Divinyl *n*, Vinyläthylen *n*, Diäthylen *n*, 1,3-Butadien *n*
bixin Bixin *n (Farbstoff des Orleans)*
bk s. black
Bk figure *(die Dichte einer Lösung anzeigender)* Barkometerwert *m*
black 1. Schwarz *n (Farbempfindung)*; schwarzfärbender (schwarzer) Farbstoff *m*, Schwarz *n*; Schwärze *f*; 2. Ruß *m*
b. alder Schwarzerle *f*, Roterle *f*, Alnus glutinosa (L.)Gaertn.
b.-and-white schwarzweiß
b.-and-white print Schwarzweißkopie *f*
b. antimony Sb$_2$S$_3$ schwarzes Schwefelantimon *n*, Antimon(III)-sulfid *n*, Antimontrisulfid *n*
b. arsenic schwarzes Arsen *n*
b. ash *(Pap)* Sodaschmelze *f*, Schmelzsoda *f*, Schwarzschmelze *f*, Schwarzsoda *f*; Rohsoda *f*; BaS Bariummonosulfid *n*
b. balsam Perubalsam *m*, Peruanischer (indischer, schwarzer) Balsam *m (von Myroxylon balsamum (L.)Harms var. pereirae)*
b. batch s. b. masterbatch

b. beer dunkles Bier *n*, Dunkles *n*
b. berry Brombeere *f*, Rubus fruticosus L.
b. body schwarzer Körper *m*
b. body radiation schwarze Strahlung *f*, Hohlraumstrahlung *f*
b. body radiator schwarzer Strahler *m*, Hohlraumstrahler *m*
b.-boy gum Grasbaumharz *n*, Akaroidharz *n (von Xanthorrhoea-Arten)*
b. catechu Braunes Katechu *n*, Pegukatechu *n*, Bombaykatechu *n (Extrakt aus Acacia catechu Willd.)*
b. chalk Schieferschwarz *n*, Erdschwarz *n*, Ölschwarz *n*, Mineralschwarz *n*
b. coal Schwarzkohle *f*, *(veraltet für)* Steinkohle *f*
b. cobalt *(Min)* Kobaltschwärze *f*, schwarzer Erdkobalt *m*, Asbolan *m*
b. cohosh Schwarze Schlangenwurzel *f (Wurzelstock von Cimicifuga racemosa (L.) Nutt.)*
b. compound *(Gum)* rußgefüllte (rußhaltige, mit Ruß gefüllte, Ruß enthaltende) Mischung *f*
b. cook *(Pap)* Schwarzkochung *f*
b. copper 1. *(Min)* Kupferschwärze *f*, Tenorit *m*, *(veraltet)* Schwarzkupfererz *n (Kupfer(II)-oxid)*; 2. *(Met)* Schwarzkupfer *n*, Schwarz-Cu *n*
b. copper oxide CuO Kupfer(II)-oxid *n*
b. core *(Ker)* schwarzer Kern *m*
b. crottle *(Farb)* Parmelia omphalodes (L.)Ach. *(Schildflechten-Art)*
b. cummin Schwarzkümmel *m*, Nigella sativa L.
b. currant Schwarze Johannisbeere *f*, Ribes nigrum L.
b. cutch s. b. catechu
b. dammar resin Schwarzes Dammarharz *n (von Canarium-Arten)*
b.-damp *(Kohlendioxidanreicherung nach Grubenexplosionen)*
b. dispersion *(Gum)* Rußdispergierung *f*, Rußverteilung *f*
b. dogwood bark Faulbaumrinde *f (von Rhamnus frangula L.)*
b. earth *(Landw)* Schwarzerde *f*, Tschernosjom *m(n)*, Tschernosem *m(n)*
b.-filled s. b.-loaded
b. gambier *(Gerb)* Gambir *n (von Uncaria gambir Roxb.)*
b. glass Schwarzglas *n*
b. gum Tupelobaum *m*, Wasser-Tupelo *m*, Nyssa sylvatica Marsh.
b. heart s. b. core
b. jack *(Am)* *(Min)* Sphalerit *m*, Zinkblende *f*, Blende *f (Zinksulfid)*
b. lead *(Min)* Reißblei *n*, *(veraltet für)* Graphit *m*
b. lignite schwarze Braunkohle *f*, Glanz[braun]kohle *f*, subbituminöse Kohle *f*
b. liquor 1. *(Pap)* Schwarzlauge *f*; 2. *(Farb)* Schwarzbeize *f*, Eisenbeize *f (Eisenazetatlösung)*
b. liquor solids *(Pap)* Trockensubstanz *f (Festgehalt m)* der Schwarzlauge

b. liquor storage [tank] *(Pap)* Schwarzlaugenbehälter *m*, Schwarzlaugenvorratstank *m*
b.-loaded *(Gum)* rußhaltig, rußgefüllt, mit Ruß gefüllt, Ruß enthaltend
b. loading *(Gum)* Rußzusatz *m*, Rußzuschlag *m*, Füllung *f* mit Ruß, Rußdosierung *f*
b. lye *s.* b. liquor 1.
b. malt Farbmalz *n*
b. masterbatch *(Gum)* Polymer-Ruß-Batch *m*, Vormischung *f* aus Polymer und Ruß, Rußbatch *m*, Rußvormischung *f*
b. mercury oxide Hg + HgO Quecksilber-Quecksilber(II)-oxid-Gemisch *n*
b. mica *(Min)* Magnesiaeisenglimmer *m*, Kali-Magnesiaeisenglimmer *m*, Biotit *m*
b. mordant *(Farb)* Schwarzbeize *f*, Eisenbeize *f* *(Eisenazetatlösung)*
b. mustard Schwarzer (Brauner, Roter) Senf *m*, Brassica nigra Koch
b. oak *(Gerb)* Färbereiche *f*, Quercus velutina Lam.
b. ocher *(Min)* Manganschaum *m*, Wad *n* *(m)* *(Mangandioxid)*
b. oil Dunkelöl *n*, dunkles Schmieröl *n*
b. peat *(Landw)* Schwarztorf *m*
b. pepper Schwarzer Pfeffer *m*; [Echter] Pfeffer *m*, Schwarzer Pfeffer[strauch] *m*, Piper nigrum L.
b. pepper oil Pfefferöl *n*
b. phosphorus schwarzer Phosphor *m*
b. pigmented *s.* b.-loaded
b. plant Rußfabrik *f*
b. plate Schwarzblech *n*
b. poplar Schwarzpappel *f*, Populus nigra L.
b. powder Schwarzpulver *n*, Sprengpulver *n*
b.-reinforced *s.* b.-loaded
b. rot Schwarzfäule *f*
b. rouge *(Farb)* Eisenoxidschwarz *n*, Eisenschwarz *n* *(Eisen(II,III)-oxid)*
b. silver *(Min)* Schwarzsilberglanz *m*, Schwarzgültigerz *n*, Sprödglaserz *n*, *(veraltet für)* Stephanit *m* *(Antimon(III)-silbersulfid)*
b. snake root Schwarze Schlangenwurzel *f* *(Wurzelstock von Cimicifuga racemosa (L.)Nutt.)*
b. spruce Schwarzfichte *f*, Picea mariana (Mill.)B.S.P.
b. stock *s.* b. compound
b. tellurium *(Min)* Blättertellur *m*, Nagyagit *m* *(ein Goldtellurid)*
b. wattle Schwarzakazie *f*, Acacia decurrens Willd. var. mollis Lindl.
b. willow Salix nigra Marsh. *(Weiden-Art)*
blackberry *s.* black berry
blacken / to schwärzen; schwarz färben; sich schwärzen
blackening Schwärzen *n*, Schwärzung *f*; Schwarzfärben *n*, Schwarzfärbung *f*
blackish schwärzlich
b.-brown schwärzlichbraun
blackjack [oak] Quercus marilandica Muenchh. *(amerikanische Eichen-Art)*

blackout *(Foto)* Verdunklung *f*
bladder *(Gum)* Balg *m*, Heizbalg *m*, Bladder *m*
blade 1. Flügel *m*, Schaufel *f* *(z.B. einer Mischmaschine)*; Knetarm *m*, Arm *m*; 2. Messer *n*; *(Pap)* Holländermesser *n*; 3. *(Krist)* flacher Stengel *m*
b. mixer Flügelmischer *m*, Schaufelmischer *m*, Schaufelrührer *m*, Schaufelrührwerk *n*, Paddelrührer *m*
b. tip Schaufelende *n*
bladed *(Krist)* plattstengelig
Blake jaw crusher Kniehebelbackenbrecher *m*
blakeite *(Min)* Bla[c]keit *m*, *(veraltet für)* Coquimbit *m* *(Eisen(III)-sulfat-9-Wasser)*
blanc de perle, b. d'Espagne Spanischweiß *n*, Schminkweiß *n*, Perlweiß *n*, Wismutweiß *n* *(Wismutoxidnitrat)*
b. fixe Blanc fixe *n*, Blankfix *n*, Permanentweiß *n*, Barytweiß *n* *(gefälltes Bariumsulfat)*
Blanc chloromethylation reaction Blancsche Reaktion *f* *(Chlormethylierung)*
Blanc rule Blancsche Regel *f*
blanch / to bleichen
bland mild
blank 1. Rohling *m*, Rohteil *n*, Rohstück *n*, Formling *m*, Formteil *n*; 2. *(Glas)* Külbel *m*; Vorform *f*; 3. *(Ker)* Hubel *m*, Batzen *m*; 4. *(Tech)* Blindflansch *m*; 5. *s.* b. experiment
b. boiling *(Zucker)* Blankkochen *n*
b. determination Blindbestimmung *f*
b. experiment Blindversuch *m*, Blindprobe *f*, Nullversuch *m*
b. mix *s.* b. stock
b. mould *(Glas)* Vorform *f*
b. stock *(Gum)* Grundmischung *f*
b. test *s.* b. experiment
b. titration Blindtitration *f*
b. trial *s.* b. experiment
b. vat blinde Küpe *f*
blanket / to abdecken, zudecken, abschirmen; *(Feuer)* ersticken
blanket *(Text)* Druckdecke *f*, Drucktuch *n*
b. application *(Landw)* Flächenbehandlung *f* *(z.B. mit Herbiziden)*
b. table *(Aufbereitungstechnik)* Tuchherd *m*
b. washer Drucktuchwäscher *m*, Druckdeckenwäscher *m*, Deckenwaschmaschine *f*
blanketing *(Plast)* *(Abschirmen des Monomeren gegen Sauerstoffzutritt durch Inertgas)*
blanking Ausschneiden *n*, Ausstanzen *n*, Stanzen *n* *(Rohling, Fassonteil)*
blast 1. Wind *m*, Gebläsewind *m*, Gebläseluft *f*; 2. Explosion *f*; Sprengung *f*; Druckwelle *f*
b. burner Gebläsebrenner *m*, Gebläselampe *f*
b. freezer Luftgefrierapparat *m*
b. freezing Gefrieren *n* in bewegter Luft
b. furnace Gebläseschachtofen *m*, Gebläseofen *m*, Blashochofen *m*, *(i.e.S.)* Hochofen *m* *(zur Gewinnung von Eisen)*
b.-furnace coke Hochofenkoks *m*, Hüttenkoks *m*
b.-furnace cube Hochofenwürfel *m*

b.-furnace dust Gichtstaub *m*
b.-furnace gas Gichtgas *n*, Hochofengichtgas *n*, Hochofengas *n*
b.-furnace plant Hochofenanlage *f*, Hochofenwerk *n*
b.-furnace process Hochofenverfahren *n*
b.-furnace slag Hochofenschlacke *f*
b.-furnace stove Hochofenwinderhitzer *m*, Winderhitzer *m*
b. inlet Lufteingang *m* (z.B. bei der Untertagevergasung)
b. lamp Gebläselampe *f*
b. main Windleitung *f*
b. of air Luftstrom *m*
b. process (refining) (Met) Windfrischen *n*
b. roasting Verblaserösten *n*, Verblaseröstung *f*
b. temperature Windtemperatur *f*, Gebläselufttemperatur *f*, Gebläsegastemperatur *f*
blasting Sprengen *n*, Sprengung *f*; (Brg) Schießen *n*, Schießarbeit *f*; (Gerb) Kalkschatten *mpl*, Schatten *mpl*, Kalkflecken *mpl*
b. cap Sprengkapsel *f*
b. charge Sprengladung *f*
b. fuse Zündschnur *f*
b. gelatin Sprenggelatine *f*
b. oil Sprengöl *n*
b. powder Sprengpulver *n*, Schwarzpulver *n*
blastogranitic (Geol) blastogranitisch
blastophitic (Geol) blastophitisch
blastoporphyritic (Geol) blastoporphyrisch, reliktporphyrisch
Blau gas Blau-Gas *n*
Blaw-Knox mill Blaw-Knox-Mühle *f* (eine Strahlmühle)
bleach / to bleichen, entfärben, aufhellen
b. out ausbleichen, sich entfärben
bleach Bleichen *n*, Bleichung *f*, Bleiche *f*; Bleichmittel *n*; Bleichflüssigkeit *f*, Bleichlauge *f*, Bleichlösung *f*, Bleichflotte *f*
b. bath Bleichbad *n*, Bleichflotte *f*
b. boiler (Pap) Hadernkocher *m*
b. consumption (Pap) Bleichmittelverbrauch *m*
b. demand s. b. requirement
b.-fastness Bleichechtheit *f*
b. liquor Bleichflüssigkeit *f*, Bleichlauge *f*, Bleichlösung *f*, Bleichflotte *f*
b. liquor tank Bleichlaugenbehälter *m*
b.-out process Bleichprozeß *m*, Bleichverfahren *n*, Bleichung *f*
b. plant Bleicherei[anlage] *f*, Bleichanlage *f*
b. ratio (Pap) Bleichverhältnis *n*
b. requirement (Pap) Bleichmittelbedarf *m*, Bleichbedarf *m*, Bleichmittelaufwand *m*
b. sludge (Pap) Bleichschlamm *m*
b. solution s. b. liquor
bleachability Bleichfähigkeit *f*, Bleichbarkeit *f*
bleachable bleichfähig, bleichbar
bleached beeswax weißes Wachs *n*, gebleichtes Bienenwachs *n*
b. fat gebleichtes Fett *n*

b. lac s. b. shellac
b. sand Bleichsand *m*
b. shellac gebleichter (weißgebleichter, weißer) Schellack *m*
bleacher Bleicher *m*, Bleichapparat *m*; Bleichholländer *m*; Bleichturm *m*; Bleicher *m* (Beruf)
bleachery Bleichanlage *f*, Bleicherei[anlage] *f*
bleaching Bleichen *n*, Bleichung *f*, Bleiche *f*, Entfärben *n*, Entfärbung *f*
b. action Bleichwirkung *f*
b. agent Bleichmittel *n*
b. apparatus Bleichapparat *m*, Bleicher *m*
b. assistant Bleichhilfsmittel *n*
b. at high consistency (Pap) Dickstoffbleiche *f*
b. bath Bleichbad *n*, Bleichflotte *f*
b. chamber Bleichkammer *f*
b. chest (Pap) Bleichbottich *m*
b. clay Bleichton *m*, (bei der Mineralölraffination auch) Bleicherde *f*, Entfärbungserde *f*
b. earth Bleicherde *f*, Entfärbungserde *f*
b. efficiency Bleichwirkung *f*, Bleicheffekt *m*
b. engine (Pap) Bleichholländer *m*
b. liquid (liquor, lye) Bleichflüssigkeit *f*, Bleichlauge *f*, Bleichlösung *f*, Bleichflotte *f*
b. material Bleichmittel *n*
b. of groundwood Holzschliffbleiche *f*
b. of rag pulp (stock) Hadernhalbstoffbleiche *f*, Hadernbleiche *f*
b. plant Bleichanlage *f*, Bleicherei[anlage] *f*
b. potcher (Pap) Bleichholländer *m*
b. powder Bleichpulver *n*, (i.e.S.) Bleichkalk *m*, Chlorkalk *m*
b. procedure (process) Bleichprozeß *m*, Bleichverfahren *n*, Bleichung *f*
b.-pulp [leicht] bleichbarer Zellstoff *m*
b. soda Bleichsoda *f* (Einweichmittel)
b. solution Bleichflüssigkeit *f*, Bleichlauge *f*, Bleichlösung *f*, Bleichflotte *f*
b. stage Bleichstufe *f*
b. temperature Bleichtemperatur *f*
b. time Bleichzeit *f*, Bleichdauer *f*
b. tower (Pap) Bleichturm *m*, Reaktionsturm *m*
b. vat Bleichbottich *m*, Bleichkessel *m*, Bleichkufe *f*
bleed / to [aus]bluten, durchbluten, anbluten, [farbig] durchschlagen, abschmutzen (z.B. von Farb- oder Gerbstoffen); (Tech) Luft ablassen, entlüften
b. off (through) s. to bleed
bleed hole (Tech) Entlüftungsbohrung *f*, Entlüftungsloch *n*
bleeding Bluten *n*, Ausbluten *n*, Durchbluten *n*, Anbluten *n*, Durchschlagen *n*, Abschmutzen *n* (z.B. von Farb- oder Gerbstoffen); Schwitzen *n* (z. B. des Zements); (Filtration) Trüblauf *m*, Durchlaufen *n* der Feststoffpartikel; (Gaschromatografie) Abdampfen *n*
blend / to [ver]mischen, zusammenmischen, [ver]mengen, (Gum auch) verschneiden, (Lebm auch) verschneiden, kupieren, (Basisprodukte in der Erdölindustrie) blenden

blend Mischung *f*, *(Gum, Lebm auch)* Verschnitt *m*
b. component Mischkomponente *f*, Mischungsanteil *m*
b. for margarine Fettmischung *f* (Fettkomposition *f*, Fettansatz *m)* zur Margarineherstellung, Margarinerezeptur *f*
b. tank *s.* blending tank
b. tempering tank *(Lebm)* Temperiergefäß *n*, Temperierkessel *m*
blende *(Min)* Blende *f*, *(i.e.S.)* Blende *f*, Zinkblende *f*, Sphalerit *m (Zinksulfid)*
blended brandy Weinbrandverschnitt *m*
blender Mischer *m*, Mischmaschine *f*
blending Mischen *n*, Vermischen *n*, Mischung *f*, Mengen *n*, Vermengen *n*, *(Gum auch)* Verschneiden *n*, *(Lebm auch)* Verschneiden *n*, Verschnitt *m*, Kupieren *n*, Coupage *f*; Blenden *n*, Blending *n (von Basisprodukten in der Erdölindustrie)*
b. agent Verschnittmittel *n*, Zusatzmittel *n*
b. drum Mischtrommel *f*, Trommelmischer *m*
b. medium Mischkomponente *f*
b. octane number Mischoktanzahl *f*, Misch-OZ *f*
b. plant Mischanlage *f*
b. ratio Mischungsverhältnis *n*
b. tank Mischtank *m*, Mischbehälter *m*, Mischgefäß *n*, Mischer *m*
blendor Mazerator *m*
bleu de Lyon Bleu de Lyon *n (Cl-Salz des Triphenylrosanilins)*
blind / to (z.B. *Filtertuch)* zusetzen, verstopfen
blind *(Tech)* Blindscheibe *f*
b. flange Blindflansch *m*
blinding Zusetzen *n*, Verstopfen *n* (z.B. *von Filtertuch)*; *(unerwünschtes)* Mattwerden *n*, Glanzloswerden *n*
b. agent *(Flot)* Schlammblockierungsmittel *n*
blister / to Blasen bilden
blister Blase *f*, Gasblase *f*, Gaseinschluß *m (oder)* Luftblase *f*, Lufteinschluß *m (Materialfehler)*, *(Glas auch)* Blatter *f*, *(Met auch)* Gußblase *f*
b. copper Blasenkupfer *n*, Blisterkupfer *n*
b. formation Blasenbildung *f*, Blasenziehen *n*
b. gas blasenziehender Gaskampfstoff *m*
blistering Blasenbildung *f*, Blasenziehen *n*; *(Gerb)* Selbstspalten *n (von Häuten)*
b. plaster *(Med)* Zugpflaster *n*
blistery blasig
bloating *(Ker)* Aufblähen *n*, Blähen *n*
Bloch wall *(Krist)* Bloch-Wand *f*, Blochsche Wand *f*
block / to blockieren, abriegeln, verriegeln, sperren; *(Chem)* neutralisieren, inaktivieren, desaktivieren, unwirksam machen
block Block *m*; *(Pap) (veraltet für)* bedplate
b. briquette Stückbrikett *n*
b. copolymer Blockkopolymer[es] *n*, Blockkopolymerisat *n*, Blockmischpolymerisat *n*
b. copolymerization Blockkopolymerisation *f*
b. diagram Blockschaltbild *n*, Blockbild *n*

b. ice Blockeis *n*, Formeneis *n*
b. mould *(Glas)* einteilige Form *f*
b. polymerization 1. Block[misch]polymerisation *f*, Blockkopolymerisation *f*; 2. *s.* mass polymerization
b. press Blockpresse *f*
b. printing Blockdruck *m*, Handdruck *m*, Modeldruck *m*
b. valve Absperrventil *n*
blocking Blockierung *f*, Blockung *f*, Hemmung *f*; *(Plast)* Blocking *n*, Blocken *n (unerwünschtes Haften der Oberflächen zweier lose aufeinanderliegender Folien oder Schichten)*; *(Glas)* Bülwern *n*, Blasenlassen *n*
blockpolymerizate Blockpolymerisat *n*
blocky blockig, großstückig
blödite *(Min)* Blödit *m*, Astrakanit *m (Magnesiumnatriumsulfat)*
blondizing agent Blondiermittel *n*, Blondierpräparat *n*
blood Blut *n*
b. albumin Blutwasseralbumin *n*, Blutserumalbumin *n*, Serumalbumin *n*
b. char[coal] Blutkohle *f*
b. clotting *s.* b. coagulation
b. coagulation Blutgerinnung *f*
b. dilution Blutverdünnung *f*
b. glue Blutalbuminleim *m*, Albuminleim *m*
b. level of penicillin Blutpenizillinspiegel *m*
b. meal Blutmehl *n*
b. pigment Blutfarbstoff *m*, Hämoglobin *n*, Hb
b. plasma Blutplasma *n*
b.-red blutrot
b. scent Blutduftstoff *m*
b. serum Blutserum *n*, Serum *n*
b. serum albumin Blutserumalbumin *n*, Blutwasseralbumin *n*, Serumalbumin *n*
b. stain Blutfleck *m*
b.-stone *(Min)* Blutstein *m*, Roteisenstein *m*, Roteisenerz *n (Eisen(III)-oxid)*; *(Min)* Blutjaspis *m*, Heliotrop *m (Silizium(IV)-oxid)*
b. sugar Blutzucker *m*
bloom / to 1. ausblühen; 2. [blau] anlaufen, Hauchbildung zeigen *(besonders von Öllacken)*
b. out *s.* to bloom 1.
bloom 1. Ausblühung *f*; Beschlag *m*; 2. *(Lebm)* Flaum *m*, Hauch *m*, Belag *m*, Reif *m*, feine Schicht *f* (z.B. *auf Früchten oder Kakaoerzeugnissen)*; Gärungsschaum *m*, Kräusen *pl*; 3. *(Gerb)* Blume *f*, Lederausschlag *m*, Mud *m*
blooming Ausblühen *n*; Beschlagen *n*; Anlaufen *n*, Hauchbildung *f*, Nebligwerden *n*, Schleierbildung *f*, Blauanlaufen *n (besonders von Öllacken)*
b. of sulphur Ausblühen *n* von Schwefel, Ausschwefeln *n*
blot / to *(mit Fließpapier)* abtupfen
blotch *(Text)* Fleck *m (in einem gedruckten Muster)*
b. printing *(Text)* Deckerdruck *m*, Deckdruck *m*, Gründeldruck *m*
blotter press Fließpapier-Filterpresse *f*

blotting s. b. paper
b. board Löschkarton m
b. paper Löschpapier n
b. paper tester Löschpapierprüfgerät n
blow / to blasen; (z.B. eine Flüssigkeit in einen
Behälter) drücken; [heiß]blasen, warmblasen (bei
der Gaserzeugung); (z.B. Zellstoffkocher) ausblasen, leerblasen, abblasen
b. back (Glas) vorblasen, gegenblasen
b. down (Glas) festblasen, niederblasen
b. in einblasen; eindrücken
b. off (z.B. Zellstoffkocher) ausblasen, leerblasen,
abblasen
b. out ausblasen; [her]ausdrücken, hinausdrücken; eruptieren (z.B. Erdgas)
b. up explodieren, zerplatzen, aufplatzen
blow Schlag m, Stoß m; (Gaserzeugung) Blasen n,
Heißblasen n, Warmblasen n; (Pap) (durch
Ausblasen bewirkte) Kocherleerung f, Kocherausblasen n, Ausblasen (Leerblasen, Abblasen)
n des Zellstoffkochers; (Pap) ausgeblasener
Kocherinhalt m
b. air Blasluft f
b.-and-blow process (Glas) Blas-Blas-Verfahren
n
b. case s. blowcase
b. discharge (Filtration) Abblasen n
b. gas Blasegas m
b. gun Druckluftpistole f
b. head Blaskopf m; (Plast) Folienblaskopf m
b. hole Blase f, Gasblase f, Gaseinschluß m (im
Guß), Gußblase f
b. iron s. blowpipe 2.
b. line Ausblasleitung f, Ausdrückleitung f
b. mould (Glas) Fertigform f; (Plast) Blas-
werkzeug n
b. mould bottom plate (Glas) Fertigformboden m
b. moulder Blasmaschine f, Blasanlage f
b. moulding (Plast) Blasformen n, Hohlkörperblasen n; Blasformteil m
b. moulding machine s. b. moulder
b.-off (Pap) 1. Abblasprodukt n, (i.e.S.) beim
Ausblasen des Zellstoffkochers freiwerdendes
Gas n, aus der Stoffgrube abgezogenes Gas n,
gasförmiges Abblasprodukt n, Ausblasgas n
(oder) flüssiges Abblasprodukt n, Ablauge f; 2.
Kocherabgas n, Abgas n, Übertriebgas n;
Übertriebsäure f, Übertrieb m, Rücklauge f; 3.
Ausblasschieber m, Ausblasventil n; 4. (Landw)
Abdrift f, Abtrift f
b.-off valve s. b. valve
b.-out (Erdöl) Ölausbruch m, Ausbruch m,
Eruption f
b.-out line s. b. line
b.-out preventer (Erdöl) Ausbruchpreventer m,
Preventer m, Bohrlochabsperrungsvorrichtung f
b. period Blaseperiode f, Heißblaseperiode f,
Warmblaseperiode f
b. pipe s. blowpipe
b. pit (Pap) Stoffgrube f, Stoffkasten m,

Kochergrube f, Kocherbütte f, Blastank m,
Ausblasbehälter m
b.-pit dilution (Pap) Verdünnung f des Stoffs in
der Stoffgrube
b.-pit gas (Pap) beim Ausblasen des Zellstoffkochers freiwerdendes Gas n, aus der
Stoffgrube abgezogenes Gas n, gasförmiges
Abblasprodukt n, Ausblasgas n (zum Abblasen)
b. port Drucklufteintritt m (zum Abblasen)
b. tank s. b. pit
b.-tank dilution s. b.-pit dilution
b.-through valve (Pap) Abgasventil n
b.-up ratio (Plast) Aufblasverhältnis n
b. valve Ausblasschieber m, Ausblasventil n
b. vat s. b. pit
blowback Rückstoß m, Rückschlag m
blowcase Druckfaß n, Druckbirne f, Montejus n;
Pulsometer m
blower 1. Lüfter m, Ventilator m; Gebläse n; 2.
Glasbläser m, Glasmacher m
blowhead s. blow head
blowing Blasen n; (Gaserzeugung) Blasen n,
Heißblasen n, Warmblasen n; Ausblasen n,
Leerblasen n, Abblasen n (z.B. eines Zellstoffkochers)
b. agent (Gum, Plast) Treibmittel n, Blähmittel n
b. iron s. blowpipe 2.
b. mould (Glas) Fertigform f; (Plast) Blas-
werkzeug n
b. of the digester (Pap) (durch Ausblasen
bewirkte) Kocherleerung f, Kocherausblasen n,
Ausblasen (Leerblasen, Abblasen) n des Zellstoffkochers
b. of tubular parisons (Plast) Blasen n von
schlauchförmigen Vorformlingen (Hohlkörper-
blasverfahren)
b. pressure Blasdruck m
b. process Blasverfahren n, Blasprozeß m
b. torch Lötrohr n
blowlamp Gebläsebrenner m, Gebläselampe f;
Lötlampe f
blown asphalt geblasenes Bitumen n, Blasbitumen
n
b. film (Plast) Blasfolie f, Schlauchfolie f
b. glass geblasenes Glas n
b. oil geblasenes (oxydiertes) Öl n, Blasöl n
b. tubing (Plast) Blasfolie f, Schlauchfolie f
blowpipe / to mit dem Lötrohr arbeiten
blowpipe 1. Lötrohr n; 2. (Glas) Glasmacherpfeife
f, Glasbläserpfeife f, Pfeife f; 3. (Tech) Entlüftungsrohr n; Ausblasrohr n, Blasrohr n, Abblas-
leitung f; Schweißbrenner m
b. analysis (assay) Lötrohranalyse f, Lötrohr-
probierkunde f, Lötrohrprobe f
b. charcoal Lötrohrkohle f
b. flame Lötrohrflamme f
b. mouthpiece Lötrohrmundstück n
b. proof s. b. analysis
blowtorch Gebläselampe f, Gebläsebrenner m;
Lötlampe f

blubber Blubber *m*, Walspeck *m*
b. oil Waltran *m*
blue / to [an]bläuen, blau färben; blauen, blau werden
blue Blau *n (Farbempfindung)*; blaufärbender (blauer) Farbstoff *m*, Blau *n*
b. asbestos *(Min)* Krokydolith *m (Natriumeisensilikat)*
b. berry Blaubeere *f*, Heidelbeere *f*, Vaccinium myrtillus L.
b. brick Eisenschmelzklinker *m*, Eisenklinker *m*
b. carbonate of copper s. b. copper 2.
b. cast *(Text)* Blaustich *m*
b. chalcocite *(Min)* blauer isotroper Kupferglanz *m*, α-Chalkosin *m*, α-Kupferglanz *m (Kupfer(I)-sulfid)*
b. cheese Blauschimmelkäse *m*, Blaukäse *m*
b. chrome leather frisch gegerbtes Chromleder *n*
b. copper 1. *(Min)* Kupferindig[o] *m*, Kovellin *m (Kupfer(II)-sulfid)*; 2. *(Min)* Kupferlasur *m*, Azurit *m (Kupfer(II)-dihydroxiddikarbonat)*
b. copper carbonate s. b. copper 2.
b. dust *(Met)* Krätze *f (durch ZnO verunreinigtes Zinkpulver)*
b. fining *(Gär)* Blauschönung *f*
b. gas Blauwassergas *n*, Blaugas *n*, blaues Wassergas *n*, *(veraltet für)* Kokswassergas *n*
b. glass Kobaltglas *n*, Blauglas *n*
b. iron earth (ore) *(Min)* Blaueisenerde *f*, Blaueisenerz *n*, *(veraltet für)* Vivianit *m (Eisen-(II)-orthophosphat)*
b. ironstone s. b. iron earth
b. malachite *(Min)* Kupferlasur *m*, Azurit *m (Kupfer(II)-dihydroxiddikarbonat)*
b. mallet *(Gerb)* Eucalyptus gardneri Maiden
b. mud Blauschlick *m*, blauer Schlick *m*, Blauschlamm *m*
b. oil abgepreßtes Öl *n*, Preßöl *n (bei der Entparaffinierung vor der Aufarbeitung zu Neutralöl erhaltenes Öl)*
b. powder s. b. dust
b. print paper Blaupauspapier *n*
b.-printing Blaudruck *m*, Blaupause *f*, Zyanotypie *f*
b.-sensitive, b.-sensitized blausensibilisiert, blauempfindlich
b. shade Blauton *m*, blauer Farbton *m*, blaue Farbnuance *f*
b. spar *(Min)* Blauspat *m*, *(veraltet für)* Lazulith *m (Magnesium-Aluminium-Eisen-Hydroxidphosphat)*
b. stone $CuSO_4 \cdot 5H_2O$ Kupfervitriol *m(n)*, Kupfer(II)-sulfat-5-Wasser *n*
b. value Blauwert *m*
b.-veined cheese s. b. cheese
b. verdigris Grünspan *m (blaue Varietät; Gemisch basischer Kupfer(II)-azetate)*
b. verditer Bremerblau *n*, Braunschweiger Blau *n*, Kalkblau *n*, Neuwieder Blau *n (Kupfer(II)-hydroxid)*
b. vitriol $CuSO_4 \cdot 5H_2O$ Kupfervitriol *m(n)*, Kupfer(II)-sulfat-5-Wasser *n*; *(Min)* Kupfervitriol *m*, Chalkanthit *m (Kupfer(II)-sulfat-5-Wasser)*
b. water gas Blauwassergas *n*, Blaugas *n*, blaues Wassergas *n*, *(veraltet für)* Kokswassergas *n*
b. wattle *(Gerb)* Acacia decurrens (J.C. Wendl.)Willd. var. dealbata
b. wet *(Gerb)* chromfeucht
blued sugar geblauter Zucker *m*
blueing Bläuen *n*, Bläuung *f*, Anbläuen *n*; Blauanlaufen *n (z.B. eines Werkzeugs)*; Bläuungsmittel *n*, blauer Farbstoff *m (zum Verbessern des Weißgrades)*
blueness Bläue *f*, blaue Färbung *f*
bluestone s. blue stone
bluish bläulich; blaustichig
b.-green bläulichgrün
blunger *(Ker)* Quirl *m*, Massequirl *m*
blunging *(Ker)* Quirlen *n*, Naßrühren *n*
blush / to [weiß] anlaufen, weiß werden *(besonders von Nitrolacken)*
blushing Anlaufen *n*, Weißanlaufen *n*, Weißwerden *n (besonders von Nitrolacken)*
board / to *(Gerb)* krispeln
board 1. Platte *f*; 2. Pappe *f*; Karton *m*
b. [making] machine Pappenmaschine *f*, Kartonmaschine *f*, *(i.e.S.)* Pappenrundsiebmaschine *f*, Rundsiebkartonmaschine *f*
b. of chemical wood pulp *(Pap)* Zellstoff *m* in Bogenform (Pappenform), Zellstoffpappe *f*, Zellstoffblätter *npl*
b. of mechanical wood pulp *(Pap)* Holzschliff *m* in Bogenform (Pappenform), Holzschliffpappe *f*, Holzschliffblätter *npl*
boarding machine *(Gerb)* Krispelmaschine *f*
boart s. boort
boat 1. Schiffchen *n*, Substanzschiffchen *n*, Glühschiffchen *n*, Boot *n*, Kahn *m (Laborgerät)*; 2. s. b. form
b. conformation (form) *(Stereochemie)* Bootform *f*, Wannenform *f*, Wanne *f*, flexible Form *f*
bob Filzpolierscheibe *f*, Filzscheibe *f*
bobbin spinning Spulenspinnverfahren *n*
bobby Gummiwischer *m (für Rührstäbe)*
bocking s. bowking
B.O.D. s. biochemical oxygen demand
bodied eingedickt
body / to *(Farb)* Körper geben, füllend wirken; *(Farb)* verdicken, eindicken
body 1. Stoff *m*, Substanz *f*, Körper *m*; 2. *(Farb)* Körper *m*, Konsistenz *f*, Festigkeit *f (der Farbmasse)*; 3. *(Ker)* Masse *f*; Scherben *m*; 4. *(Tech)* Konvertermittelstück *n*; Brüdenraum *m (eines Verdampfers)*; Verdampfereinheit *f*, Verdampferkörper *m*; Körper *m*, Gehäuse *n (z.B. eines Ventils)*
b. board Rohpappe *f*
b.-centred *(Krist)* raumzentriert, innenzentriert
b.-centred cubic lattice (pattern) *(Krist)* kubisch-raumzentriertes (kubisch-innenzentriertes,

raumzentriertes kubisches) Gitter *n*, raumzentriertes Würfelgitter *n*
b.-centred grating (lattice) raumzentriertes (innenzentriertes) Gitter *n*
b. colour Deckfarbe *f*, deckende Farbe *f*
b. composition *(Ker)* Massezusammensetzung *f*
b. compound *s.* b. stock
b. fat Körperfett *n*
b. fluid Körperflüssigkeit *f*
b. label Bauchetikett *n*
b. paper Rohpapier *n*
b. powder Körperpuder *m*
b. preparation *(Ker)* Masseaufbereitung *f*
b. scrap *(Ker)* Masseabfall *m*
b. slip *(Ker)* Masseschlicker *m*
b. spar *(Ker) (als Massebestandteil verwendeter)* Feldspat *m*
b. stock *(Gum)* Karkaßmischung *f*
b. surface *(Ker)* Scherbenoberfläche *f*
b. temperature Körpertemperatur *f*, Körperwärme *f*
bodying *(Farb)* körpergebend
boehmite Böhmit *m (kristallines Aluminiummetahydroxid)*
bog iron [ore] *(Min)* Raseneisenerz *n*, Raseneisenstein *m*
b. lime *(Bodenkunde)* Wiesenkalk *m*
b. manganese *(Min)* Manganschaum *m*, Wad *n(m) (Mangandioxid)*
b. ore *s.* 1. b. iron [ore]; 2. b. manganese
b. peat Moortorf *m*
b. spruce Schwarzfichte *f*, Piceà mariana (Mill.)B.S.P.
boghead *s.* b. coal
b. cannel [coal] Boghead-Kannel-Kohle *f*
b. coal Bogheadkohle *f*, Boghead *m*
bogie kiln *(Ker)* Herdwagenofen *m*
Bogomolets serum Bogomolez-Serum *n (gegen Altern des Bindegewebes)*
Bohn's synthesis Bohnsche Synthese *f*
Bohr atom [model] Bohrsches (Bohr-Rutherfordsches) Atommodell *n*
Bohr atom model for hydrogen Bohrsches Wasserstoffatommodell *n*, Bohrsches Modell *n* des Wasserstoffatoms
Bohr frequency condition Bohrsche Frequenzbedingung *f*
Bohr magneton Bohrsches Magneton *n*
Bohr orbit Bohrsche Bahn *f*
Bohr radius Bohrscher Wasserstoffradius *m*
Bohr-Rutherford atom [model] *s.* Bohr atom [model]
Bohr-Sommerfeld atom [model] Bohr-Sommerfeldsches Atommodell *n*
boil / to 1. *(im allgemeinen in Wasser, jedoch auch in Gegenwart von Chemikalien)* kochen [lassen], zum Kochen bringen, abkochen; auskochen; einkochen [lassen], verkochen; *(Seife)* sieden; 2. kochen, sieden, [auf]wallen
b. away (down) eindampfen, verdampfen, ein-

kochen [lassen], verkochen, eindicken, kondensieren
b. off (out) auskochen, abkochen; *(Text)* entbasten, degummieren, entschälen
b. over überkochen, überlaufen, überschäumen
b. to grain *(Zucker)* auf Korn [ver]kochen, Korn bilden
b. up aufkochen [lassen]
boil Kochung *f*
b. house Kocherei *f*, Kochstation *f*
b.-off *(Text)* Abkochen *n*, Entbasten *n*, Degummieren *n*
b. test Kochprobe *f*
boildown Eindampfen *n*, Verdampfen *n*, Einkochen *n*, Verkochen *n*, Eindicken *n*, Kondensieren *n*
boiled linseed oil Leinölfirnis *m*
b. oil Ölfirnis *m*, Firnis *m*
boiler Destillationsblase *f*, Destillierblase *f*, Blase *f*, Destillationsgefäß *n*, Destillationsapparat *m*; Dampferzeuger *m*, Dampfkessel *m*, Kessel *m*; *(Pap)* Kocher *m*
b. ash Kesselschlacke *f*
b. feed preheater Speisewasservorwärmer *m*, Wasservorwärmer *m*, Ekonomiser *m*, Eko *m*
b. feed pump Kesselspeisepumpe *f*
b. feed[ing] water Kesselspeisewasser *n*
b. make-up evaporator Speisewasserverdampfer *m*
b. plant Kesselanlage *f*, Dampferzeugungsanlage *f*
b. plate Kesselblech *n*
b. return trap Kondensat-Rückleiter *m*
b. scale Kesselstein *m*
b. scale removal Kesselsteinentfernung *f*
b. steam Kesseldampf *m*
boilerman *(Pap)* Kochermeister *m*
boilfast kochecht, kochfest, kochbeständig
boiling 1. Kochen *n*, Kochung *f*, Abkochen *n*; Verkochen *n*; Sieden *n*, Sud *m*; 2. Kochen *n*, Sieden *n*, Wallen *n*, Aufwallen *n*
b. agent *(Pap)* Aufschlußmittel *n*, Aufschlußchemikalie *f*, Aufschließungsreagens *n*
b.-bed technique Fließbett-Technik *f*, Wirbelschichttechnik *f*, Staubfließtechnik *f*, Fluid-Technik *f*
b. chip Siedesteinchen *n*
b.-down *s.* boildown
b.-down pan Verdampfungspfanne *f*, Eindampfpfanne *f*
b. fermentation kochende Gärung *f*
b. flask Kochkolben *m*, Kochflasche *f*
b. heat Siedehitze *f*
b. house Kocherei *f*, Kochstation *f*
b. limit Siedegrenze *f*
b. lye Kochlauge *f*
b.-off Auskochen *n*, Abkochen *n*; *(Text)* Entbasten *n*, Degummieren *n*, Entschälen *n*
b.-off bath *(Text)* Entbastungsflotte *f*
b.-off process Siedeverfahren *n (ein Butterungsverfahren)*

b. point Kochpunkt *m*, Kp., K, Siedepunkt *m*, Siedetemperatur *f*

b. point constant ebullioskopische Konstante *f*, molale Siedepunktserhöhung *f*

b. point curve Siedepunktskurve *f*, Siedekurve *f*

b. point diagram Siedediagramm *n*, Siedeschaubild *n*

b. point elevation Siedepunktserhöhung *f*

b. point method Ebullioskopie *f*

b. point of sulphur Siedepunkt *m* des Schwefels, Schwefelpunkt *m*

b. point of water Siedepunkt *m* des Wassers, Dampfpunkt *m*

b. point range Siedepunktsbereich *m*, Siedebereich *m*, Siedeintervall *n*

b. point rise *s.* b. point elevation

b. reactor Siedewasserreaktor *m*, Siedereaktor *m*, Verdampferreaktor *m*, Kochendwasserreaktor *m*

b. resistance Kochfestigkeit *f*, Kochbeständigkeit *f*

b. room Kochstation *f*, Kocherei *f*

b. stone Siedesteinchen *n*, Siedestein *m*

b. temperature Siedetemperatur *f*, Siedepunkt *m*, Kochpunkt *m*, Kp., K

b. test Kochprobe *f*

b. to grain (*Zucker*) Kochen *n* auf Korn, Kornkochen *n*

b. tube Siederohr *n*

b. under pressure Kochen *n* unter Druck, Druckkochen *n*

b.-up Aufkochen *n*; Verkochen *n*

b. vessel Kochkessel *m*, Kocher *m*

b. water reactor *s.* b. reactor

b. zone Siedezone *f*

boilover Überkochen *n*, Überlaufen *n*, Überschäumen *n*

boilproof kochecht, kochfest, kochbeständig

Bokalahy Marsdenia verrucosa Decne (*Kautschukpflanze*)

Bokhara galls (*Gerb*) (*Gallen von Pistacia vera L.*)

bole (*Min*) Bol[us] *m*

boletic acid HOOCCH=HCCOOH Boletsäure *f*, *trans*-Butendisäure *f*

Bollmann extractor Bollmann-Extrakteur *m*, Extraktionsapparat *m* nach Bollmann

bolometer Bolometer *n*

bolometric bolometrisch

bolopherite (*Min*) Bolopherit *m*, (*veraltet für*) Hedenbergit *m* (*Kalziumeisen(II)-disilikat*)

bolt / to sieben, [durch]beuteln

bolting Sieben *n*, Beuteln *n*, Durchbeuteln *n*

b. cloth Siebgewebe *n*, Siebtuch *n*, Seihtuch *n*

boltonite (*Min*) Boltonit *m*, (*veraltet für*) Forsterit *m* (*Magnesiumorthosilikat*)

Boltzmann constant Boltzmannsche Konstante *f*, Boltzmann-Konstante *f*, k

Boltzmann distribution law Boltzmannsches Energieverteilungsgesetz (Verteilungsgesetz) *n*, Boltzmann-Theorem *n*

Boltzmann factor Boltzmannscher Faktor *m*, Boltzmann-Faktor *m*

Boltzmann statistics Boltzmannsche (klassische) Statistik *f*, Boltzmann-Statistik *f*

bolus alba (*Min*) Kaolin *m*, Porzellanerde *f*, weißer (reiner) Ton *m*, China Clay *m(n)*, Bolus alba

bomb Bombe *f*, Druckbombe *f*; (*Geol*) [vulkanische] Bombe *f*, Lavabombe *f*

b. calorimeter Bombenkalorimeter *n*, Kalorimeterbombe *f*, kalorimetrische Bombe *f*, Verbrennungsbombe *f*

b. method Bombenmethode *f* (*zur Schwefelbestimmung*)

bombard / to (*Kch*) beschießen, bestrahlen, bombardieren

bombardment (*Kch*) Beschuß *m*, Beschießen *n*, Bombardement *n*, Bombardierung *f*

Bombay mace Bombay-Macis *m* (*von Myristica malabarica Lam.*)

B. mastic (*Pflanzengummi von Pistacia mutica Fisch. et Mey*)

bond / to binden; verbinden; [ver]kleben; die Bindung eingehen, sich verbinden

b. together verbinden, verknüpfen

bond 1. (*Bindungstheorie*) Bindung *f* (*als Zustand*); 2. Verbindung *f*; Klebung *f*, Verklebung *f*; 3. (*Pap*) Bankpostpapier *n*, Hartpostpapier *n*, Feinpostpapier *n*

π bond π-Bindung *f*, Pi-Bindung *f*, π-Elektronenpaarbindung *f*

σ bond σ-Bindung *f*, Sigma-Bindung *f*, σ-Elektronenpaarbindung *f*

b. angle Bindungswinkel *m*

b. axis Bindungsachse *f*

b. breaking Auftrennen (Lösen) *n* der Bindung, Bindungsbruch *m*

b. character Bindungscharakter *m*

b. clay Bindeton *m*

b. dipole Bindungsdipol *m*

b. dipole moment Bindungsdipolmoment *n*

b. direction Bindungsrichtung *f*

b.-dissociation energy Bindungsdissoziationsenergie *f*, Trennungsenergie *f* [der Bindung]

b. distance *s.* b. length

b. energy Bindungsenergie *f*

b. fission Bindungsspaltung *f*, Aufspaltung (Sprengung) *f* der Bindung

b. force Bindungskraft *f*, Bindekraft *f*, bindende Kraft *f*

b. formation Bindungsbildung *f*, Ausbildung *f* von Bindungen; Knüpfen *n* von Bindungen

b. hindrance Behinderung *f* der freien Drehbarkeit einer Bindung (*Atropisomerie*)

b. ink Wertpapierdruckfarbe *f*

b. length Bindungslänge *f*, Atomabstand *m*

b. migration *s.* b. shift[ing]

b. moment Bindungsmoment *n*

b. number Bindungszahl *f*, Atombindungszahl *f*, Bindungswertigkeit *f*, Bindigkeit *f*, Atombindigkeit *f*, Atomwertigkeit *f*, kovalente Wertigkeit *f*, Kovalenz *f*; Bindungsgradzahl *f*, Bindungsgrad *m*

b. orbital Bindungsorbital *n*, für Bindungen benötigtes Orbital *n*; Valenzorbital *n*
b. order Bindungsordnung *f*, Bindungsgrad *m*
b. orientation *s.* b. direction
b. paper Bankpostpapier *n*, Hartpostpapier *n*, Feinpostpapier *n*
b. printing Wertpapierdruck *m*
b. refraction Bindungsrefraktion *f*
b. rupture *s.* b. fission
b. shift[ing] Bindungsverschiebung *f*
b. strength Bindefestigkeit *f*, Haftfestigkeit *f*; *(Bindungstheorie)* Bindungsstärke *f*, Bindungsfestigkeit *f*, Bindefestigkeit *f*
b. system Bindungssystem *n*
b. type Bindungstyp *m*, Bindungsart *f*
b. valence (valency) Bindungswertigkeit *f*, Bindigkeit *f*, Atombindigkeit *f*, Atomwertigkeit *f*, kovalente Wertigkeit *f*, Kovalenz *f*, Atombindungszahl *f*, Bindungszahl *f*
bondability *s.* bonding power
bondable bindungsfähig
bondage *s.* bond valence
bonded fabric Textilverbundstoff *m*, nicht gewebte Textilie *f*
bonding 1. *(Bindungstheorie)* Bindung *f (als Vorgang)*; 2. Binden *n*; Verbinden *n*, Verbindung *f*; Kleben *n*, Klebung *f*, Verkleben *n*, Verklebung *f*; 3. Bindemittel *n*, bindender Zusatz *m*
b. ability *s.* b. power
b. agent Bindemittel *n*; *(Gum)* Haftmittel *n*, Bindemittel *n*
b. capacity *s.* b. power
b. cement Klebkitt *m*
b. clay Bindeton *m*
b. dash Bindungsstrich *m*, Bindestrich *m*
b. electron Bindungselektron *n*, Bindeelektron *n*
b. electron pair bindendes Elektronenpaar *n*, Bindungselektronenpaar *n*, Bindeelektronenpaar *n*, Bindungsdublett *n*
b. energy Bindungsenergie *f*
b. force bindende Kraft *f*, Bindungskraft *f*, Bindekraft *f*
b. material Bindemittel *n*
b. of metals Metallverklebung *f*, Metallklebung *f*
b. of rubber to metals Gummi-Metall-Verbindung *f*
b. of rubber to textiles Gummi-Textil-Verbindung *f*
b. of the fibres *(Pap)* Zwischenfaserbindung *f*, Faserverkettung *f*
b. orbital bindendes Orbital *n*
b. pair of electrons *s.* b. electron pair
b. power Bindevermögen *n*, Bindefähigkeit *f*, Bindekraft *f*; *(Bindungstheorie)* Bindungsvermögen *n*, Bindungsfähigkeit *f*, Bindungskraft *f*, Bindekraft *f*, bindende Kraft *f*
b. state Bindungszustand *m*
b. strength Bindefestigkeit *f*, Haftfestigkeit *f*; *(Bindungstheorie)* Bindungsstärke *f*, Bindungsfestigkeit *f*, Bindefestigkeit *f*

b. to metals Metallverbindung *f*
b. type Bindungstyp *m*, Bindungsart *f*
Bond's law Zerkleinerungsgesetz *n* nach Bond und Wang
bone ash Knochenasche *f*
b. black Knochenschwarz *n*, Beinschwarz *n*, Kölner Schwarz *n*
b. char[coal] Knochenkohle *f*, Spodium *n*
b. china Knochenporzellan *n*
b.-dry knochentrocken, absolut trocken, atro
b.-dry weight Trockensubstanzmasse *f*, Trockenmasse *f*, Trockenstoffmasse *f*, *(bisher)* Trockengewicht *n*
b.-dry-weight basis Bezugsbasis *f* Trockensubstanzmasse (Trockenmasse, Trockenstoffmasse)
b. dust *s.* b. meal
b. fat Knochenfett *n*
b. gelatin Knochengelatine *f*
b. glue Knochenleim *m*
b. grease *s.* b. fat
b. meal Knochenmehl *n*
b. oil Knochenöl *n*
b. tar Knochenteer *m*
book end paper Vorsatzpapier *n*, Vorstoßpapier *n*
b.-lining paper *s.* b. end paper
b. paper Werkdruckpapier *n*, Buchdruckpapier *n*, Bücher[schreib]papier *n*, Geschäftsbücherpapier *n*
b.-printing paper *s.* b. paper
bookbinder's board Buchbinderpappe *f*, Buchbinderdeckel *m*, Bücherpappe *f*
boom sprayer Feldspritzrohr *n (Pflanzenschutz)*
boort Bort *m*, Poort *m (nur für Industriezwecke verwertbare Diamanten oder Diamantenabfälle)*
boost / to *(Gum) (einen Beschleuniger)* aktivieren
booster Verstärker *m*, Aktivator *m*; *(Gum)* aktivierender Zusatzbeschleuniger (Sekundärbeschleuniger, Zweitbeschleuniger) *m*
boot *(Glas)* Stiefel *m (Durchlaß zwischen Schmelz- und Arbeitswanne)*
boracic acid *s.* boric acid
boracite *(Min)* Borazit *m*
borane B_nH_{n+4} *oder* B_nH_{n+6} Boran *n*, Borwasserstoff *m*, Borhydrid *n*
borate Borat *n*
b. buffer Boratpuffer *m*
b. ester Borsäureester *m*
b. glass Boratglas *n*
b. phosphor Boratphosphor *m*
borax $Na_2B_4O_7 \cdot 10H_2O$ Borax *m*, Natriumtetraborat-Dekahydrat *n*; *(Min)* Borax *m*, Tinkal *m (Natriumtetraborat-Dekahydrat)*
b. bead Boraxperle *f*
b. glass Boraxglas *n*
b. lake Boraxsee *m*
b. pentahydrate $Na_2B_4O_7 \cdot 5H_2O$ Natriumtetraborat-5-Wasser *n*, oktaedrischer Borax *m*, Juwelierborax *m*
b. usta $Na_2B_4O_7$ Borax usta, kalzinierter (gebrannter) Borax *m*

borazene, borazine s. borazole
borazole Borazol n, Triborintriamin n, anorganisches Benzol n, Pseudobenzol n
Borazon Borazon n (Warenzeichen für kubisches Bornitrid)
Bordeaux mixture Bordeauxbrühe f, Bordelaiser Brühe f, Kupferkalkbrühe f
B. turpentine Bordeaux-Terpentin n(m) (von Pinus pinaster Ait.)
border effect Randeffekt m
bore / to [durch]bohren
bored stopper durchbohrter Gummistopfen m
borehole Bohrloch n
  b. producer method Bohrlochverfahren n (bei der Untertagevergasung)
  b. wall Bohrlochwand[ung] f
borer Bohrer m, Bohrvorrichtung f
borethane B_2H_6 Boräthan n, (veraltet für) Diboran n
borethyl $B(CH_2CH_3)_3$ Boräthyl n, Bortriäthyl n, Triäthylborin n
boric Bor...
  b. acid $B(OH)_3$ [normale] Borsäure f, Orthoborsäure f, Monoborsäure f, Trioxoborsäure f
  b.-acid anhydride B_2O_3 Borsäureanhydrid n, Bortrioxid n, Bor(III)-oxid n, Boroxid n
  b.-acid ester Borsäureester m
  b. anhydride s. b. acid anhydride
  b. oxide B_2O_3 Boroxid n, Bor(III)-oxid n, Bortrioxid n, Borsäureanhydrid n
boride Borid n
borine B_nH_{n+2} Borin n, (i.e.S.) BH_3 Borin n, Monoborin n
  b. carbonyl BH_3CO Borinkarbonyl n, Borkarbonylhydrid n
boring Bohrung f, Durchbohrung f
borings Bohrspäne mpl
Born approximation Bornsche Näherung f
Born equation Bornsche Gleichung f
Born exponent Bornscher Exponent m
Born-Haber [thermochemical] cycle Born-Haberscher Kreisprozeß m, Born-Haber-Kreisprozeß m
bornane Bornan n, 1,7,7-Trimethylbizyklo-[1,2,2]-heptan n
Borneo camphor Borneokampfer m, Sumatrakampfer m, Baroskampfer m (von Dryobalanops aromatica Gaertn. f.)
borneol $C_{10}H_{17}OH$ Borneol n, Bornylalkohol m, 2-Hydroxy-1,7,7-trimethylbizyklo[2,2,1]heptan n
  b. acetate $C_{10}H_{17}OOCCH_3$ Borneolazetat n, Bornylazetat n, Essigsäurebornylester m
bornesitol $C_6H_{11}O_5 \cdot OCH_3$ Bornesit m (Monomethyläther des i-Inosits)
bornite (Min) Bornit m, Buntkupfererz n, Buntkupferkies m (Eisen(II)-kupfer(II)-sulfid)
bornyl Bornyl n
  b. acetate s. borneol acetate
  b. alcohol s. borneol
bornylamine Bornylamin n, 2-Aminobornan n,

2-Aminokamphan n, 2-Amino-1,7,7,-trimethylbizyklo-[1,2,2]-heptan n
bornylane Bornylan n, (veraltet für) Bornan n, 1,7,7-Trimethylbizyklo-[1,2,2]-heptan n
bornylene Bornylen n, l-Bornylen n, 1,7,7-Trimethylbizyklo-[1,2,2]-2-hepten n
borobutane B_4H_{10} Borbutan n, Borobutan n, (veraltet für) Tetraboran n
boroethane B_2H_6 Boräthan n, (veraltet für) Diboran n
borofluoric acid $H[BF_4]$ Borfluorwasserstoffsäure f, Fluorborsäure f, Fluo[ro]borsäure f
borofluoride Fluo[ro]borat n
borohydride $M^I_n[BH_4]_n$ Boranat n, Metallborwasserstoff m, Metallborhydrid n, Tetrahydridoborat n, Hydridoborat n
boron B Bor n
  b. acetate $(CH_3COO)_2BOB(OOCCH_3)_2$ Borazetat n
  b. carbide B_4C Borkarbid n
  b. deficiency Bormangel m
  b. fertilizer Bordüngemittel n
  b. hydride B_nH_{n+4} oder B_nH_{n+6} Borwasserstoff m, Borhydrid n, Boran n
  b. mononitride BN Bormononitrid n, Bornitrid n
  b. phosphate BPO_4 Borphosphat n
  b. phosphide BP Borphosphid n
  b. sesquioxide B_2O_3 Borsesquioxid n, (veraltet für) Bortrioxid n, Bor(III)-oxid n, Boroxid n, Borsäureanhydrid n
  b. steel Borstahl m
  b. tribromide BBr_3 Bortribromid n, Bor(III)-bromid n, Borbromid n
  b. trichloride BCl_3 Bortrichlorid n, Bor(III)-chlorid n, Borchlorid n
  b. trifluoride BF_3 Bortrifluorid n, Bor(III)-fluorid n, Borfluorid n
  b. triiodide BJ_3 Bortrijodid n, Bor(III)-jodid n, Borjodid n
  b. trioxide B_2O_3 Bortrioxid n, Bor(III)-oxid n, Boroxid n, Borsäureanhydrid n
boronate / to (Düngemittel) mit Borverbindungen versetzen
boronatrocalcite (Min) Boronatrokalzit m, Natroborokalzit m, (veraltet für) Ulexit m (Kalziumnatriumpentaborat)
boronotungstic acid s. borotungstic acid
borosilicate Borosilikat n
  b. crown [glass] Borosilikatkronglas n
  b. glass Borosilikatglas n
borotartrate $M^IC_4H_4BO_7$ Borotartrat n
borotungstate Wolframatoborat n
borotungstic acid Wolframatoborsäure f
borowolframic acid s. borotungstic acid
bort s. boort
Bose-Einstein distribution Bose-Einsteinsche Verteilung f
Bose-Einstein gas Bose-Einstein-Gas n
Bose-Einstein statistics Bose-Einstein-Statistik f, Bose-Statistik f

bosh Rast f *(eines Hochofens)*
boson Boson n, Bose-Teilchen n
boss Wulst m, Vorsprung m; Muffe f; *(Geol)* Stock m, Eruptivstock m
bosshead Doppelmuffe f *(am Stativ)*
boswell[in]ic acid Boswellinsäure f
botanical pflanzliches Insektizid n, insektizider Naturstoff m
Botany Bay gum *(Farb)* Gelbes Akaroidharz n, Gelbharz n, Botanybayharz n, Erdschellack m *(von Xanthorrhoea hastilis R.Br.)*
botesite *(Min)* Botesit m, *(veraltet für)* Hessit m *(Silbertellurid)*
botryoidal traubenförmig, traubig
botryolite *(Min)* Botryolith m, *(veraltet für)* Datolith m *(Kalziumhydroxidborosilikat)*
bottle / to auf (in) Flaschen [ab]füllen, auf Flaschen [ab]ziehen
  b. in (up) s. to bottle
bottle Flasche f
  b. beer Flaschenbier n
  b. bobbin *(Text)* Flaschenspule f
  b. brush Flaschenbürste f
  b. cap Flaschenkappe f, Flaschenverschluß m
  b.-cap board Milchflaschen-Verschlußkarton m
  b. carrier Transportbehälter (Tragkasten) m für Flaschen
  b. champagnization Flaschengärverfahren n, Flaschengärung f, Methode champenoise f *(zur Schaumweinherstellung)*
  b. cleaner, b.-cleaning machine s. b.-washing machine
  b. delivery *(Glas)* Flaschenabgabe f *(bei der Flaschenblasmaschine)*
  b.-fermented in der Flasche vergoren
  b.-fermented champagne im Flaschengärverfahren (nach der Methode champenoise) hergestellter Schaumwein m
  b. filler, b.-filling machine Flaschen[ab]füllmaschine f, Flaschenfüller m; Flaschenbierabfüllmaschine f
  b. gas Flaschengas n
  b. glass Flaschenglas n
  b.-green flaschengrün
  b. green Flaschengrün n
  b. house Flaschen[ab]füllerei f, Abfüllabteilung f, Füllstation f
  b. kiln *(Ker)* Flaschenofen m
  b. label Flaschenetikett n
  b. labeller Flaschenetikettiermaschine f
  b. of resistance glass Hartglasgefäß n
  b. oven s. b. kiln
  b. rinser s. b.-washing machine
  b.-ripe abfüllreif
  b. stopper Flaschenstöpsel m, Flaschenstopfen m
  b. tissue Flaschenseidenpapier n
  b. tray Untersatzschale f für Säureflaschen
  b. washer, b.-washing machine Flaschenwaschmaschine f, Flaschenreinigungsmaschine f, Flaschenspülmaschine f

  b. wine Flaschenwein m
  b. wrapping s. b. tissue
bottled beer Flaschenbier n
  b. gas Flaschengas n
  b. milk Flaschenmilch f
bottler s. bottling machine
bottlery s. bottling room
bottling Abfüllen n auf Flaschen, Flaschenfüllung f, Flaschenabzug m
  b. department s. b. room
  b. machine Flaschen[ab]füllmaschine f, Flaschenfüller m; Flaschenbierabfüllmaschine f
  b. plant (room) Flaschen[ab]füllerei f, Abfüllabteilung f, Füllstation f
bottom / to *(Gerb, Text)* grundieren
bottom Boden m, unterster Teil m; Sohle f, Boden m *(beim Kopf- und Bodenschmelzen)*; *(Glas)* Form f
  b. blowing Bodenblasen n, Blasen n mit Bodenwind, Durchblasen n
  b.-blown converter bodenblasender (normal blasender) Konverter m, Unterwindfrischkonverter m
  b.-blown-converter process bodenblasendes Verfahren n, Unterwindfrischverfahren n
  b. brickwork Bodensteine mpl
  b. calender roll *(Pap)* untere Glättwerkswalze f
  b. couch-press roll *(Pap)* untere Gautschwalze f, Kopfwalze f
  b. discharge Untenaustrag m, Bodenentleerung f
  b. discharge centrifuge unten entleerende Zentrifuge (Schleuder) f, Zentrifuge f mit Untenaustrag
  b.-driven centrifuge von unten angetriebene Zentrifuge (Schleuder) f, stehende Zentrifuge f, Stehzentrifuge f
  b. ejection *(Plast)* Ausdrücken n von unten *(aus dem Gesenk)*
  b.-feeding filter Filter n mit Untenaufgabe
  b. felt *(Pap)* Unterfilz m, unterer Abnahmefilz m *(bei Rundsieb-Kartonmaschinen)*
  b. fermentation Untergärung f
  b.-fermentation yeast untergärige Hefe f, Unterhefe f
  b. fraction Bodenfraktion f
  b. lacquer Grundlack m
  b. layer untere Schicht f, Bodenschicht f
  b. leach *(Gerb)* *(letztes Gefäß einer Extraktionsbatterie)*
  b. millstone Bodenstein m, Unterstein m
  b. plate Bodenplatte f, Grundplatte f; *(Glas)* Bodenplatte f, Formboden m, *(i.e.S.)* Fertigformboden m
  b. plug *(Plast)* Unterteil n, Matrizeneinsatz m
  b. product s. bottoms
  b. punch *(Ker)* Unterstempel m
  b. ram *(Plast)* Unterstempel m
  b. ram press *(Plast)* Unterkolbenpresse f, Unterdruckpresse f

b. roll Unterwalze f
b. roll of the couch press s. b. couch-press roll
b. roll of the size press s. b. size-press roll
b. roller s. b. roll
b. sediment[s] s. b. settlings
b. settlings Bodenkörper m, Niederschlag m, Bodensatz m, Satz m, Sediment n, Ausscheidung f, Ausscheidungsprodukt n, Ablagerung f
b. size-press roll (Pap) Unterwalze f der Leimpresse
b. valve Bodenventil n
b. vat unterster Bottich m
b. yeast untergärige Hefe f, Unterhefe f
bottoming (Text) Vorwaschen n, Vorwäsche f (vor dem Bleichen); Grundieren n, Grundierung f (vor dem Überfärben)
b. dyestuff Grundierfarbe f
bottoms 1. s. bottom settlings; 2. (Lebm) Faß-geläger n; 3. (Destillation) Bodenprodukt n, Sumpfprodukt n, Bodenkörper m
Bouguer law Bouguer-Lambertsches Gesetz (Adsorptionsgesetz) n, Lambertsches Gesetz (Adsorptionsgesetz) n
boulangerite (Min) Boulangerit m, Schwefelantimonblei n (Blei(II)-antimon(III)-sulfid)
boulder clay Blocklehm m, Geschiebelehm m, Geschiebemergel m
bouldery blockig
bounce-back Zurückprallen n (des Farbstaubs in der Farbanstrichtechnik)
bouncing putty Springkitt m, hüpfender Kitt m, Rückprallkitt m, Silikonspringkitt m
bound moisture gebundene Feuchte (Feuchtigkeit) f, gebundenes Wasser n
b. rubber Bound Rubber m (im Rußgel fest an den Füllstoff gebundener Kautschuk)
b. sulphur gebundener Schwefel m
b. water gebundenes Wasser n
boundary Grenze f
b. condition Randbedingung f
b. layer Grenzschicht f
b. stone (Bodenkunde) Ortstein m
b. strap (Pap) Deckelriemen m
b. surface Grenzfläche f, Trennungsfläche f, Phasengrenzfläche f; (Bindungstheorie) Bindungssphäre f
bounding surface s. boundary surface
bouquet Bukett n, Blume f (z.B. des Weins)
bourbonal $C_6H_3(OH)(OC_2H_5)CHO$ Bourbonal n, Vanillal n, Protokatechualdehyd-3-äthyläther m
Bourdon [pressure] gauge Bourdon-Röhre f, Bourdonsche Röhre f, Röhrenfedermanometer n
bournonite (Min) Bournonit m (ein Bleikupferspießglanz)
Bouveault-Blanc method s. Bouveault-Blanc reduction
Bouveault-Blanc reaction Bouveault-Blanc-Reaktion f, Bouveault-Blancsche Reaktion f
Bouveault-Blanc reduction Bouveault-Blanc-Reduktion f, Bouveault-Blanc-Methode f, Bou-

veault-Blancsche Reduktion f (von Karbonsäure-estern zu primären Alkoholen)
B.O.V. s. brown oil of vitriol
bowenite (Min) Bowenit m, (veraltet für) Antigorit m, Blätterserpentin m
bowing (Text) Bogenverzug m
bowk / to beuchen, abkochen, brühen
bowking Beuchen n, Abkochen n, Brühen n
bowl Schale f, Schüssel f; Schüssel f, Mahlschüssel f (einer Schüsselmühle); Kessel m (einer Zentrifuge); Speiserbecken n, Speiserkopf m; Walze f, (oft) Kalanderwalze f; Trommel f
b. classifier Schüsselklassierer m
b. desiltor Schüsselklassierer m mit seitlichem Austrag
b. mill Schüsselmühle f
b.-mill coal pulverizer Schüsselmühle f für Kohlenstaubmahlung
b. paper Kalanderwalzenpapier n
b. ring-roller mill s. b. mill
b.-type roller mill s. b. mill
b. wall Trommelwand[ung] f, Trommelmantel m
bowsprit (Stereochemie) Bugspriet m
box Kasten m, Kammer f; Buchse f, Büchse f, Packung f; Kasten m, Kiste f (zur Wärmebehandlung)
b.-anneal / to kastenglühen, kistenglühen
b. annealing Kastenglühen n, Kastenglühung f, Kistenglühen n, Kistenglühung f
b.-annealing furnace Kastenglühofen m, Kistenglühofen m
b.-carburize / to kastenaufkohlen, kastenzementieren, kasteneinsetzen, in Zementationskästen einsetzen
b. carburizing Kastenaufkohlen n, Kastenzementieren n, Kasteneinsetzen n, Einsetzen n in Zementationskästen
b. feeder Kastenbeschicker m
b. kiln Kammerofen m
b. malting Kastenmälzerei f
b. press Etagenpresse f, Schachtelpresse f
boxboard Kartonagenpappe f, Faltschachtelkarton m, Schachtelkarton m
Boyle curve Boyle-Kurve f
Boyle law Boyle-Mariottesches Gesetz n
Boyle point (temperature) Boyle-Temperatur f, B.T., BT, Boyle-Punkt m
bp, b.p. s. boiling point
BP (Abk. für) 1. British Patent; 2. British Pharmacopoeia
BPh (Abk. für) British Pharmacopoeia
BR s. 1. butyl rubber; 2. butadiene rubber
Brace-Lemon spectrophotometer Bracesches Spektralfotometer n
Brackelsberg furnace Brackelsberg-Ofen m (kohlenstaubbeheizter Trommelofen besonders für Temperguß)
Brackett series Brackett-Serie f
brackish brackig, etwas (leicht) salzig, Brack...
b. water Brackwasser n

Bradley mill Bradley-Mühle f *(eine Pendelmühle)*
Bradley three-roll[er] mill Bradley-Mühle f mit drei Pendelrollen (Pendelwalzen, Pendeln)
bradykinin Bradykinin n *(ein Nonapeptid)*
Bragg angle Braggscher Winkel m, Glanzwinkel m
Bragg equation Braggsche Gleichung (Formel, Reflexionsgleichung) f
Bragg law *(Krist)* Braggsches Reflexionsgesetz (Gesetz) n
Bragg method [of crystal analysis] Bragg-Methode f, Bragg-Verfahren n, Drehkristallmethode f (Drehkristallverfahren n) von Bragg
Bragg rule Braggsche Regel f
Bragg spectrometer Bragg-Spektrometer n, Braggsches Spektrometer n
Bragg treatment s. Bragg method
braided hose *(Gum)* geklöppelter Schlauch m
brake kit *(elektrische)* Bremsvorrichtung f *(z.B. für Zentrifugen)*
bran Kleie f
    b. culture method Kleie[kultur]verfahren n, Kleiekulturmethode f, Kleieprozeß m, Bran-culture-Verfahren n
    b. drench[ing] *(Gerb)* Kleienbeize f, Schrotbeize f
    b. shorts Nachmehl n, Futtermehl n
branch / to abzweigen, sich verzweigen
branch Zweig m, Abzweig m, Abzweigung f, Verzweigung f
    b. point Verzweigungsstelle f
branched verzweigt
    b.-chain verzweigtkettig, mit verzweigter Kette
    b. chain verzweigte Kette f
    b.-chain hydrocarbon verzweigtkettiger Kohlenwasserstoff m, Kohlenwasserstoff m mit verzweigter Kette
    b.-chain mechanism Kettenverzweigungsmechanismus m
    b.-chain structure verzweigte Struktur f *(eines Kohlenwasserstoffs)*
    b. disintegration Dualzerfall m, dualer (verzweigter) Zerfall m, [radioaktive] Verzweigung f, Mehrfachzerfall m
    b. molecule verzweigtes Molekül n
    b. polyethylene verzweigtes Polyäthylen n, Hochdruckpolyäthylen n, Polyäthylen n niedriger Dichte
    b. polymer verzweigtes Polymeres n
branching Abzweigung f, Verzweigung f; *(Kph)* Dualzerfall m, dualer (verzweigter) Zerfall m, Verzweigung f, Mehrfachzerfall m
    b. fraction Verzweigungsanteil m
    b. of chains Kettenverzweigung f
    b. ratio Verzweigungsverhältnis n
brand / to *(Leder)* mit Brandzeichen markieren
brand Marke f
    b. name Markenname m, Markenbezeichnung f
brandy Weinbrand m; Brandy m *(durch Destillation von vergorenem Fruchtsaft erhaltenes Getränk)*
brasilic acid s. brazilic acid

brasilin s. brazilin
brass Messing n, Ms
    b. gasket Messingdichtung f
    b. weight Messingwägestück n, *(bisher)* Messinggewicht n
brass[id]ic acid $C_8H_{17}CH=CHC_{11}H_{22}COOH$ Brassidinsäure f, Brassinsäure f, trans-13-Dokosensäure f
brassylic acid Brassylsäure f, Tridekandisäure f
brattice dryer Siebbandtrockner m
Braun degradation Braunscher (von-Braunscher) Abbau m *(tertiärer Amine mit Bromzyanid)*
braunite *(Min)* Braunit m
Bravais lattice *(Krist)* Bravais-Gitter n, Bravaissches Translationsgitter n
Bravais-Miller indices *(Krist)* Bravaissche Symbole npl, Bravaissche Indizes mpl
bravaisite *(Min)* Bravaisit m, Hydromuskovit m *(Aluminiumdihydrogentetrasilikat)*
brazan Brasan n
braze / to hartlöten
Brazil copal Amerikanischer Kopal m, Anime-Kopal m *(i.e.S.) (von Hymenaea courbaril L.)*
    B. nut Paranuß f, Brasilnuß f *(Samen von Bertholletia excelsa Humb. et Bonpl.)*
    B. redwood *(Farb)* Brasilholz n *(von Caesalpinia-Arten)*
    B. wax Karnaubawachs n *(von Palme Copernicia prunifera (Muell.)H.E. Moore)*
brazilein Brasilein n, Brazilein n
Brazilian emerald *(Min)* Turmalin m
    B. ipecacuanha brasilianische Brechwurz f, Rio-Brechwurz f *(Cephaëlis ipecacuanha (Brot.)A.Rich.)*
    B. redwood s. Brazil redwood
    B. sapphire *(Min)* Turmalin m
brazilic acid Brasilsäure f, 3-Oxy-7-methoxychromanonessigsäure-(3) f
brazilin Brasilin n, Brazilin n
brazilite *(Min)* Brazilit m
brazilwood *(Farb)* Brasilholz n, Pernambuko n, Pernambukholz n *(von südamerikanischen Caesalpinia-Arten, i.e.S. von Caesalpinia echinata Lam.); (Farbholz von Haematoxylum brasiletto Karst.)*
brazing Hartlöten n, Hartlötung f
    b. alloy Lotlegierung f, Lot n
    b. metal (solder) Hartlot n
bread dough Brotteig m
    b. flour Brotmehl n
breadfruit tree Brotfruchtbaum m, Artocarpus communis J. R. et G. Forst.
breadth of a spectral line *(phys Ch)* Linienbreite f
break / to 1. [zer]brechen; zerreißen; *(Licht)* brechen; brechen *(bei der Leinölherstellung)*; *(eine Kette)* abbrechen, beenden; *(eine chemische Bindung)* lösen, [auf]spalten, sprengen; *(Emulsionen)* brechen, spalten, entmischen; *(Hadern)* aufbereiten, *(Altpapier)* auflösen;

(Gerb) strecken *(nach der Bleiche)*; 2.[zer]brechen; zerreißen; brechen *(von Emulsionen)*; [ab]reißen *(von einer Papierbahn)*
b. down 1. [auf]spalten, zertrümmern, abbauen, *(ein Molekül)* aufbrechen; *(Gum)* abbauen, plastizieren; *(Pap)* aufschlagen; 2. sich spalten; zerfallen; zusammenbrechen
b. in *(Hadern)* aufbereiten, *(Altpapier)* auflösen
b. irregularly *(Min)* einen unebenen Bruch haben
b. up 1. zerkleinern, zerschlagen, brechen; *(Hadern)* aufbereiten, *(Altpapier)* auflösen; 2. sich aufspalten *(von chemischen Verbindungen)*
break Umschlag *m* *(z.B. bei der Titration)*; Knickpunkt *m* *(z.B. einer Kurve)*; *(Pap)* Abreißen *n*, Abriß *m* *(der Papierbahn)*
b. in the web *(Pap)* Bahn[ab]riß *m*, Papierbahn[ab]riß *m*, Abreißen *n* der Papierbahn
b.-off reaction Abbruchreaktion *f*, Kettenabbruchreaktion *f*
b. point Durchbruch *m* *(bei der Adsorption)*
b. seal [tube] Schlenk-Rohr *n* mit Zerschlagventil
b.-through curve *(Kurve des Adsorptivgehalts des austretenden Gases über der Zeit)*
breakage Brechen *n* *(z.B. des Kokses)*
b. by impact Bruch *m* durch Stoß
breakdown Spaltung *f*, Aufspaltung *f*, Abbau *m*; Zerfall *m*; *(Gum)* Abbau *m*, Plastizierung *f*; Durchschlag *m* *(eines Dielektrikums)*
b. mill Brecherwalzwerk *n*
b. of size Zerfall *m*
b. product Spaltprodukt *n*, Abbauprodukt *n*
b. strength Durchschlag[s]festigkeit *f*, [di]elektrische Festigkeit *f*
b. voltage Durchschlag[s]spannung *f*
breaker 1. Zerteiler *m*; 2. *(Gum)* Breaker *m*; Brecherwalzwerk *n*; Formenöffner *m*, Formenbrecher *m*, Brecheisen *n* *(zum Öffnen von Vulkanisierformen)*; 3. *(Pap)* Halbzeugholländer *m*, Halbstoffholländer *m*
b. beater *(Pap)* Auflöseholländer *m*
b. compound s. b. stock
b. drum s. b. roll
b. engine *(Pap)* Halbzeugholländer *m*, Halbstoffholländer *m*
b. plate Brechplatte *f*; *(Plast)* Lochscheibe *f*, Stauscheibe *f*
b. roll *(Pap)* Halbzeugholländerwalze *f*, Halbstoffholländerwalze *f*
b. stock *(Gum)* Breakermischung *f*
breakfast cocoa Trinkkakao *m*
breaking Brechen *n*, Zerbrechen *n*; Zerreißen *n*; *(Text)* Brechen *n*, Knicken *n*; Abbrechen *n*, Abbruch *m* *(einer Kette)*; Aufspalten *n*, Aufspaltung *f* *(einer chemischen Bindung)*; Aufbereitung *f* *(von Hadern)*, Auflösung *f* *(von Altpapier)*; *(Gerb)* Strecken *n* *(nach der Bleiche)*; Abreißen *n*, Abriß *m* *(der Papierbahn)*
b. down *(Gum)* Abbau *m*, Plastizierung *f*
b. elongation Bruchdehnung *f*, Zerreißdehnung *f*, Reißdehnung *f*

b. engine *(Pap)* Halbzeugholländer *m*, Halbstoffholländer *m*
b. extension s. b. elongation
b.-in engine s. b. engine
b. length Reißlänge *f*
b. mill Brecherwalzwerk *n*
b. mill for scrap rubber Altgummibrecher *m*
b. of emulsions Brechen (Entmischen, Spalten) *n* von Emulsionen, Emulsionsentmischung *f*, Emulsionsspaltung *f*, Entemulsionieren *n*, Demulgieren *n*, Demulgierung *f*, Dismulgieren *n*
b. of foams Schaumzerstörung *f*, Schaumverhütung *f*, Entschäumen *n*
b. of reaction chains Kettenabbruch *m*
b.-off Abbrechen *n*, Abbruch *m*
b. point Brechpunkt *m*, Bruchpunkt *m*, Zerreißpunkt *m*
b. strain *(Pap)* Bruchlast *f* *(bei der Zugfestigkeitsprüfung)*
b. strength Bruchfestigkeit *f*; Reißfestigkeit *f*, *(Pap auch)* Zugfestigkeit *f*
b. tenacity *(Text)* Reißfestigkeit *f*
breast box *(Pap)* Stoffauflaufkasten *m*, Auflaufkasten *m*, Stoffauflauf *m*
b. milk Frauenmilch *f*, Muttermilch *f*
b. roll *(Pap)* Brustwalze *f*
b. roll doctor *(Pap)* Schaber *m* an der Brustwalze
b. wall Oberbauseitenwand *f* *(im Glasschmelzofen)*
breathability Atmungsfähigkeit *f*, Atmung *f* *(eines Überzugs)*
breathe / to [ein]atmen; arbeiten *(sich ausdehnen oder zusammenziehen)*
breathing Atmen *n*, Atmung *f*; Arbeiten *n* *(Ausdehnen oder Zusammenziehen)*; *(Plast)* Lüften *n*, Entlüften *n*, Entlüftung *f*, Entgasen *n*, Entgasung *f* *(des Werkzeugs, der Form)*
breccia Brekzie *f*, Breccie *f* *(Trümmergestein)*
brecciated brekzienartig, brekziös, brecciös
brecciation Brekzienbildung *f*; brekzienartige Beschaffenheit *f*
bredigite *(Min)* Bredigit *m* *(Kalziumorthosilikat)*
Bredt rule Bredtsche Regel *f*
breeding reactor *(Kch)* Brutreaktor *m*
breeze Grus *m*; Asche *f*, Lösche *f*
b. concrete Koksaschenbeton *m*, Kokslöschebeton *m*, Leichtbeton *m* mit Koksaschenzusatz
breithauptine, breithauptite *(Min)* Breithauptit *m*, Antimonnickel *n* *(Nickelantimonid)*
Bremen blue Bremer Blau *n*, Braunschweiger Blau *n*, Kalkblau *n*, Neuwieder Blau *n* *(Kupfer(II)-hydroxid)*
bremsstrahlung Bremsstrahlung *f*, kontinuierliche Röntgenstrahlung *f*
breunnerite *(Min)* Breunnerit *m*, *(veraltet für)* Mesitinspat *m*
brevifolin Brevifolin *n* *(Ellagengerbstoff)*
brevifolincarboxylic acid Brevifolinkarbonsäure *f* *(Gerbstoffbaustein)*
brevilagin Brevilagin *n* *(Ellagengerbstoff)*

brew / to brauen
brew Sud *m*, Gebräu *n*; Brühe *f*
  b. kettle Braukessel *m*, Braupfanne *f*, Würzekochkessel *m*, Würzekessel *m*, Würzepfanne *f*
  b. water Brauereiwasser *n*, Brauwasser *n*
brewer Brauer *m*, Bierbrauer *m*
brewer's barley Braugerste *f*, BG
  b. grains Biertreber *pl*, Treber *pl*
  b. malt Brauereimalz *n*, Braumalz *n*
  b. rice Braureis *m*, Reis *m* zur Bierherstellung
  b. sugar Brauzucker *m*
  b. yeast Bierhefe *f*
brewery Brauerei *f*, Bierbrauerei *f*
  b. wastes Brauereiabwasser *n*
  b. yeast Bierhefe *f*
brewhouse Brauhaus *n*, Brauerei *f*, *(i.e.S.)* Sudhaus *n*
brewing Brauen *n*, Bierbrauen *n*; Sud *m*
  b. barley Braugerste *f*, BG
  b. chemistry Brauereichemie *f*
  b. industry Brauereiindustrie *f*, Brauindustrie *f*
  b. liquor *s.* b. water
  b. quality Brauqualität *f*, Brauwert *m* *(z.B. der Gerste)*
  b. sugar Brauzucker *m*
  b. value *s.* b. quality
  b. water Brauereiwasser *n*, Brauwasser *n*
  b. yeast Bierhefe *f*
Brewster angle Brewsterscher Winkel *m*
Brewster law Brewstersches Gesetz *n*
brewsterite *(Min)* Brewsterit *m* *(ein Tektosilikat)*
brick Ziegel[stein] *m*; Stein *m* *(aus geformten und gebrannten keramischen Rohstoffen)*
  b. cheese Ziegelkäse *m*, Backsteinkäse *m*
  b. clay Ziegelton *m*
  b. dust Ziegelmehl *n*
  b. earth Ziegelerde *f*
  b. kiln Ziegelofen *m*
  b.-line / to ausmauern, aussteinen; mit Ziegeln ausmauern
  b. lining Ausmauern *n*, Ausmauerung *f*, Aussteinen *n*; Ziegelauskleidung *f*, Ziegelausmauerung *f*, Ziegelausfütterung *f*
  b. machine *(Ker)* Ziegelformmaschine *f*, Ziegelpresse *f*
  b. maker Ziegler *m*, Ziegelbrenner *m*
  b.-making clay *s.* b. clay
  b.-red ziegelrot
  b. red Ziegelrot *n*
  b. retort [aus Steinen] gemauerte Retorte *f*, steinerne (keramische) Retorte *f*, Steinretorte *f*
  b. sugar Würfelzucker *m*
  b. tea Ziegeltee *m*, Backsteintee *m*
brickware Ziegeleierzeugnis *n*
brickwork Mauerwerk *n*
brickworks Ziegelei *f*, Ziegelbrennerei *f*
bridge / to überbrücken, untereinander verbinden
bridge Brücke *f*; *(Glas)* Brücke *f*, Brückenwand *f*
  b. atom Brückenatom *n*
  b. bond Brückenbindung *f*

b. connection Brückenschaltung *f*, Meßbrücke *f*
b. method Brückenmethode *f*, Brückenverfahren *n*
b. ring Brückenring *m*
b.-ring compound Brückenringverbindung *f*
b.-ring system Brückenringsystem *n*
b.-type bond Brückenverbindung *f*
b.-type cross-link Brückenglied *n*, Vernetzungsbrücke *f*
b.-type furnace *(Glas)* Brückenwannenofen *m*
b. wall *(Glas)* Brückenwand *f*, Brücke *f*
bridged überbrückt, Brücken...
  b. hydrocarbon Brücken-Kohlenwasserstoff *m*, Brückenringsystem *n*
  b. ion Brückenion *n*
  b. ring Brückenring *m*
bridgehead Brückenkopf *m* *(bei Brückenringsystemen)*
  b. atom Brückenkopfatom *n*
bridging Brückenbildung *f*; *(Plast)* Zusammenbacken *n* des Materials beim Einfüllen
  b. bond Brückenbindung *f*
  b. group brückenbildende Gruppe *f*
  b. oxygen Brückensauerstoff *m*
bright glänzend, Glanz...; leuchtend, lebhaft; hell; klar; blank
  b.-anneal / to blankglühen, in Schutzgas glühen
  b. anneal[ing] Blankglühen *n*, Schutzgasglühen *n*, Glühen *n* (Glühung *f*) in Schutzgas
  b.-annealing furnace Blankglühofen *m*
  b. beer helles Bier *n*, Helles *n*
  b. coal Glanzkohle *f*, glänzende Kohle *f*
  b. enamel paper Hochglanzpapier *n*
  b. glaze *(Ker)* Glanzglasur *f*
  b. gold Glanzgold *n*
  b. palladium Glanzpalladium *n*
  b. plating *(Galvanotechnik)* elektrolytisches Glänzen *n*
  b. platinum Glanzplatin *n*
  b.-red leuchtendrot, granatrot
  b. silk Cuite *f* *(voll entbastete Seide)*
  b. silver Glanzsilber *n*
  b. stock Brightstock *m*, Brightstock-Öl *n* *(hochviskoser Schmieröldestillationsrückstand)*
  b.-yellow hellgelb
brightener [optischer] Aufheller *m*, [optisches] Aufhellungsmittel *n*, optisches Bleichmittel *n*, Weißtöner *m*; *(Galvanotechnik)* Glanzbildner *m*
brightening Aufhellen *n*; *(Pap)* Bleichen *n*, Bleichung *f*, Entfärben *n*, Entfärbung *f*, Schaffung (Herstellung) *f* eines hohen Weißgehalts, Erhöhung *f* des Weißgehalts
  b. agent *s.* brightener
  b. effect Aufhell[ungs]effekt *m*, Aufhellung *f*
brightness Glanz *m*; Leuchtkraft *f*, Lebhaftigkeit *f*; Helligkeit *f*; Klarheit *f*; Leuchtdichte *f*; *(Pap)* *s.* b. level
  b. increase *(Pap)* Weißgehaltserhöhung *f*, Weißgehaltssteigerung *f*
  b. level *(Pap)* Weißgehalt *m*, Weißgrad *m*, Weiße *f*

b. range *(Foto)* Helligkeitsumfang *m*, Kontrastumfang *m*

brilliance, brilliancy Brillanz *f*, Farbbrillanz *f*, Leuchtkraft *f*, Farb[en]schönheit *f*

brilliant brillant, leuchtend, farbstark

b. alizarin blue Brillantalizarinblau *n*

b. cut Glasschleifen *n*, Glasschliff *m*

b. dye Brillantfarbstoff *m*

b. green Brillantgrün *n* *(Sulfat des 4,4'-Bis(diäthylamino)triphenylmethans)*

b. lustre Diamantglanz *m*

b. yellow Brillantgelb *n*

brilliantine Brillantine *f*

Brillouin polyhedron Brillouin-Polyeder *n*

Brillouin zone Brillouin-Zone *f*

brim Rand *m*

brimstone Schwefel *m*

brine / **to** mit Salzlake (Salzlösung, Lake) behandeln; naßpökeln; naßsalzen

brine Sole *f*, Lake *f*, Salzsole *f*, Salzlake *f*, Salzlösung *f*, Salzlauge *f*; Sole *f*, Kühlsole *f*, Kältesole *f*

b. cooler Solekühler *m*

b. cooling Solekühlung *f*

b. cure *s.* b. curing

b.-cured *(Gerb)* [salz]lakenkonserviert; *(Lebm)* naßgepökelt

b. curing *(Gerb)* Lakenkonservierung *f*, Salzlakenkonservierung *f*, Konservierung *f* durch Salzlakenbehandlung; *(Lebm)* Naßpökeln *n*, Naßpökelung *f*

b. refrigeration *s.* b. cooling

b. salting *(Lebm)* Naßpökeln *n*, Naßpökelung *f*

b. solution Salzlösung *f*, Salzlake *f*, Salzsole *f*, Salzlauge *f*, Lake *f*, Sole *f*

b. spring Solquelle *f*, *(i.e.S.)* Kochsalzquelle *f*, muriatische Quelle *f*

Brinell hardness Brinellhärte *f*, HB

bring about / **to** hervorrufen, *(eine Reaktion)* auslösen

b. down *(Gerb)* verfallen machen *(z.B. in der Beize)*

b. into solution in Lösung bringen, lösen

b. out entwickeln

b. to einstellen auf *(z.B. die gewünschte Stärke)*

b. to standard strength *(einen Farbstoff)* auf Typ bringen

b. to the boil zum Kochen bringen

brining Behandlung *f* mit Salzlake (Salzlösung, Lake); *(Gerb)* Lakenbehandlung *f*, Salzlakenbehandlung *f*; *(Lebm)* Naßpökeln *n*, Naßpökelung *f*; Naßsalzung *f*

b. of seed Saatbeize *f*, Saatbeizung *f*

b. process *(Lebm)* Naßpökelverfahren *n*

Brin's [oxygen] process Brinsches Verfahren *n (zur Sauerstoffgewinnung)*

briny salzig, salzhaltig, solehaltig, Salz...

briquet *s.* briquette

briquettability Brikettierbarkeit *f*, Verpreßbarkeit *f*, Brikettierfähigkeit *f*

briquettable brikettierbar, verpreßbar, brikettierfähig

briquette / **to** brikettieren, verpressen, *(auch)* zu Briketts pressen

b. warm warmverpressen

briquette Preßling *m*, Preßstück *n*, Preßkörper *m*; Brikett *n*

b. coal *s.* briquetting coal

b. coke Brikettkoks *m*

b. machine *s.* briquetting machine

b. plant *s.* briquetting plant

b. press *s.* briquetting press

b. strength Brikettfestigkeit *f*

briquetted coal Preßkohle *f*, brikettierte Kohle *f*

briquetting Brikettieren *n*, Brikettierung *f*, Verpressen *n*, Verpressung *f*

b. coal Brikettierkohle *f*, Brikettkohle *f*; Brikettiersteinkohle *f*

b. installation Brikettieranlage *f*

b. machine Brikettierpresse *f*, Brikettpresse *f*, Brikettiermaschine *f*

b. method Brikettier[ungs]verfahren *n*, Brikettier[ungs]methode *f*

b. operation Brikettier[ungs]vorgang *m*, Brikettierungsprozeß *m*

b. pitch Brikettierpech *n*, Brikettpech *n*

b. plant Brikettfabrik *f*, Brikettierwerk *n*, Brikettwerk *n*; Brikettieranlage *f*

b. press Brikettierpresse *f*, Brikettpresse *f*, Brikettiermaschine *f*; Pastillenpresse *f (einer Kalorimetereinrichtung)*

b. pressure Brikettierpreßdruck *m*, Brikettierdruck *m*, Preßdruck *m (beim Brikettieren)*

b. process Brikettier[ungs]verfahren *n*, Brikettier[ungs]methode *f*; Brikettierungsprozeß *m*, Brikettier[ungs]vorgang *m*

b. properties (qualities) Brikettiereigenschaften *fpl*, Brikettierbarkeit *f*, Brikettierfähigkeit *f*, Verpreßbarkeit *f*

b. rolls Brikettierwalzen *fpl*, Preßwalzen *fpl (einer Walzenbrikettierpresse)*

b. technique *s.* b. method

brisance Brisanz *f*, Sprengkraft *f*

b. value Brisanzwert *m*

brisk lebhaft; schäumend, perlend, prickelnd

briskly boiling lebhaft kochend

bristle Borste *f*

Bristol [board] Bristolkarton *m*

Brit. Pat. *(Abk. für)* British Patent

Brit. Pharm. *(Abk. für)* British Pharmacopoeia

British thermal unit *(kalorische Arbeits-, Energie- oder Wärmemengeneinheit in Großbritannien)*

brittle spröde, zerbrechlich; brüchig

b. failure Sprödbruch *m*

b. mica Sprödglimmer *m*

b. point Sprödigkeitspunkt *m*, Kältesprödigkeitspunkt *m*, Kältefestigkeit *f*

b.-point temperature *s.* b. temperature

b. silver ore *(Min)* Sprödglaserz *n*, Schwarzgültigerz *n*, Schwarzsilberglanz *m*, *(veraltet für)* Stephanit *m (Antimon(III)-silbersulfid)*

b. temperature Sprödbruchtemperatur *f*, Versprödungstemperatur *f*, Kältebruchtemperatur *f*
brittleness Spröde *f*, Sprödigkeit *f*; Brüchigkeit *f*
b. temperature *s.* brittle temperature
brnsh. *s.* brownish
broad cloth *(Text)* Stückware *f*
b.-spectrum Breitspektrum..., Breitband..., mit breitem Wirkungsspektrum, mit großer Wirkungsbreite
b.-spectrum antibiotic Breitspektrumantibiotikum *n*, Breitbandantibiotikum *n*
broadcast treatment *(Landw)* Flächenbehandlung *f* (z.B. mit Herbiziden)
broadening of spectral lines *(phys Ch)* Linienverbreiterung *f*
broadleaved lavender Großer Speik *m*, Lavandula latifolia (L.f.)Medik.
brocchite *(Min)* Brocchit *m*, *(veraltet für)* Chondrodit *m*
brochantite *(Min)* Brochantit *m* *(Kupfer(II)-hexahydroxidsulfat)*
bröggerite *(Min)* Bröggerit *m*
broke Ausschußpapier *n*, Ausschuß *m*, Papierausschuß *m*, Kollerstoff *m*
b. beater *(Pap)* Auflöseholländer *m*
broken *(Pap)* *s.* broke
b. brick Ziegelsplitt *m*
b. coke Brechkoks *m*
b. degumming liquor gebrochenes Entbastungsbad *n*
b. material (paper) *s.* broke
b. rice Bruchreis *m*
brokes *s.* broke
bromacetol $CH_3CBr_2CH_3$ Bromazetol *n*, 2,2-Dibrompropan *n*
bromacetone *s.* bromoacetone
bromal CBr_3CHO Bromal *n*, Tribromazetaldehyd *m*, Tribromäthanal *n*
bromalide Bromalid *n*
bromallylene *s.* bromoallylene
bromamine acid Bromaminsäure *f*, 1-Amino-4-bromanthrachinon-2-sulfonsäure *f*
bromanil $O=C_6Br_4=O$ Bromanil *n*, Tetrabrom-1,4-zyklohexadiendion *n*
bromanilic acid Bromanilsäure *f*, 2,5-Dibrom-3,6-dihydroxy-1,4-zyklohexadiendion *n*
bromargyrite *(Min)* Bromargyrit *m*, Bromyrit *m* *(Silberbromid)*
bromate / **to** mit Bromat behandeln; bromieren
bromate M^IBrO_3 Bromat *n*
bromated camphor *s.* bromocamphor
bromatometric bromatometrisch
bromatometry Bromatometrie *f*
bromaurate(III) *s.* bromoaurate(III)
bromauric acid *s.* bromoauric acid
bromcamphor *s.* bromocamphor
bromchlorargyrite *(Min)* Bromchlorargyrit *m*, Embolit *m*
bromcresol *s.* bromocresol
bromelin Bromelin *n* *(ein aus Ananas gewonnenes Ferment)*

bromellite *(Min)* Bromellit *m* *(Berylliumoxid)*
brometone $(CH_3)_2C(OH)CBr_3$ Brometon *n*, 2-(Tribrommethyl)-2-propanol *n*
bromic Brom..., *(i.e.S.)* Brom(V)-...
b. acid $HBrO_3$ Bromsäure *f*
b. silver AgBr Silberbromid *n*; *s.* bromargyrite
b.-silver paper *s.* bromide paper
bromide M^IBr Bromid *n*
b. paper Bromsilberpapier *n*, Bromidpapier *n*
b. print Bromsilberdruck *m*
brominate / **to** bromieren
brominated butyl rubber *s.* bromobutyl rubber
b. camphor *s.* bromocamphor
brominating agent Bromierungsmittel *n*
bromination Bromieren *n*, Bromierung *f*
b. reaction Bromierungsreaktion *f*
bromine Br Brom *n*
b.-containing bromhaltig
b. hydrate $Br_2 \cdot 10H_2O$ Bromdekahydrat *n*
b. iodide BrJ Jod(I)-bromid *n*, Jodmonobromid *n*
b. monofluoride BrF Brommonofluorid *n*
b. pentafluoride BrF_5 Brompentafluorid *n*, Brom(V)-fluorid *n*
b. salt Bromsalzgemisch *n* *(Mischung aus NaBr und NaBrO_3)*
b. trifluoride BrF_3 Bromtrifluorid *n*, Brom(III)-fluorid *n*
b. vapour Bromdampf *m*
b. water Bromwasser *n*
bromism Bromismus *m* *(Symptomenkomplex der Bromvergiftung)*
bromite *(Min)* Bromit *m*, *(veraltet für)* Bromargyrit *m* *(Silberbromid)*
bromlite *(Min)* Bromlit *m*, *(veraltet für)* Alstonit *m*
bromo acid Eosinfarbstoffsäure *f*, Eosinsäure *f*, Bromoacid *f*
bromoacetaldehyde $BrCH_2CHO$ Bromazetaldehyd *m*, Bromäthanal *n*
bromoacetic acid $CH_2BrCOOH$ Bromessigsäure *f*, Bromäthansäure *f*
bromoacetone CH_3COCH_2Br Bromazeton *n*, 1-Brom-2-propanon *n*
bromoalkane Bromalkan *n*
bromoallyl alcohol Bromallylalkohol *n*
bromoallylene $CH_2=CHCH_2Br$ Bromallyl *n*, Allylbromid *n*, 3-Brompropen *n*
bromoantimonate Bromoantimonat *n*
bromoaurate(III) $M^I[AuBr_4]$ Bromoaurat(III) *n*
bromoauric acid $H[AuBr_4]$ Bromgoldsäure *f*, Goldbromsäure *f*, Goldbromwasserstoff *m*, Gold(III)-hydrogenbromid *n*
bromobenzene Brombenzol *n*, *(i.e.S.)* C_6H_5Br Brombenzol *n*, Monobrombenzol *n*, Phenylbromid *n*
bromobenzoic acid BrC_6H_4COOH Brombenzoesäure *f*
1-bromobutane $CH_3CH_2CH_2CH_2Br$ 1-Brombutan *n*, *n*-Butylbromid *n*, *prim-n*-Butylbromid *n*
2-bromobutane $CH_3CH_2CHBrCH_3$ 2-Brombutan *n*, *sek*-Butylbromid *n*

bromobutyl rubber Brombutylkautschuk *m*
bromobutyric acid Brombuttersäure *f*, Brombutansäure *f*
bromocamphor Bromkampfer *m*, Monobromkampfer *m*
bromochloroiodate(I) M'[JBrCl] Bromchlorjodat(I) *n*
bromocresol Bromkresol *n*
  b. green Bromkresolgrün *n*, 3',3'',5',5''-Tetrabrom-*m*-kresolsulfonphthalein *n*
  b. purple Bromkresolpurpur *m*, 5',5''-Dibrom-*o*-kresolsulfonphthalein *n*
bromocuprate Bromokuprat *n*
bromodesulphonation Bromdesulfonierung *f*
bromoethane CH_3CH_2Br Bromäthan *n*, Äthylbromid *n*
2-bromoethanol $BrCH_2CH_2OH$ 2-Bromäthanol *n*, 2-Bromäthylalkohol *m*, Äthylenbromhydrin *n*
bromoethene $CH_2{=}CHBr$ Bromäthen *n*, Bromäthylen *n*, Vinylbromid *n*
2-bromoethyl alcohol *s.* 2-bromoethanol
bromoethylene *s.* bromoethene
bromoform $CHBr_3$ Bromoform *n*, Tribrommethan *n*
bromohydrin Bromhydrin *n*
bromohypoiodite M'[JBr₂] Bromojodit(I) *n*
bromoiodate(I) M'[JBr₂] Bromojodat(I) *n*
bromoiodide M'[JBr₂] Bromojodat(I) *n*
bromoiridate(III) M'₃[IrBr₆] Bromoiridat(III) *n*
bromoiridate(IV) M'₂[IrBr₆] Bromoiridat(IV) *n*
bromoiridite M'₃[IrBr₆] Bromoiridat(III) *n*
bromoisobutyric acid $C_4H_7O_2Br$ Bromisobuttersäure *f*, Brommethylpropansäure *f*
bromomethane CH_3Br Brommethan *n*, Monobrommethan *n*, Methylbromid *n*
bromometric bromometrisch
bromometry Bromometrie *f*
bromonium Bromonium *n*
  b. ion Bromoniumion *n*
bromoosmate(III) M'₃[OsBr₆] Bromoosmat(III) *n*
bromoosmate(IV) M'₂[OsBr₆] Bromoosmat(IV) *n*
bromopalladate(II) M'₂[PdBr₄] Bromopalladat(II) *n*
bromopalladate(IV) M'₂[PdBr₆] Bromopalladat(IV) *n*
bromophenol BrC_6H_4OH Bromphenol *n*
  b. blue Bromphenolblau *n*, 3',3'',5',5''-Tetrabromphenolsulfonphthalein *n*
bromophosgene $COBr_2$ Bromphosgen *n*, Kohlenoxidbromid *n*, Karbonylbromid *n*
bromoplatinate(II) M'₂[PtBr₄] Bromoplatinat(II) *n*, Tetrabromoplatinat(II) *n*
bromoplatinate(IV) M'₂[PtBr₆] Bromoplatinat(IV) *n*, Hexabromoplatinat(IV) *n*
bromoplatinic acid $H_2[PtBr_6]$ Bromoplatin(IV)-säure *f*, Hexabromoplatin(IV)-säure *f*, Platin(IV)-bromwasserstoffsäure *f*
bromoplatinite M'₂[PtBr₄] Bromoplatinat(II) *n*, Tetrabromoplatinat(II) *n*
bromoplatinous acid $H_2[PtBr_4]$ Bromoplatin(II)-säure *f*, Tetrabromoplatin(II)-säure *f*, Platin(II)-bromwasserstoffsäure *f*

bromopropanoic acid *s.* bromopropionic acid
bromopropionic acid Brompropionsäure *f*, Brompropansäure *f*
bromorhenate(IV) M'₂[ReBr₆] Bromorhenat(IV) *n*
bromorhodate(III) M'₃[RhBr₆] Bromorhodat(III) *n*
bromoselenate(IV) M'₂[SeBr₆] Bromoselenat(IV) *n*
bromoselenite M'₂[SeBr₆] Bromoselenat(IV) *n*
bromosilane, bromosilicane SiH_3Br Bromsilan *n*
bromostannate Bromostannat *n*
N-bromosuccinimide N-Bromsukzinimid *n*, N-Brombernsteinsäureimid, Sukzinbromimid *n*, NBS
bromotellurate M'₂[TeBr₆] Bromotellurat *n*
bromothymol Bromthymol *n*, 4-Brom-2-isopropyl-5-methylphenol *n*
  b. blue Bromthymolblau 3',3''-Dibromthymolsulfonphthalein *n*
bromotitanate M'₂[TiBr₆] Bromotitanat *n*
bromotoluene Bromtoluol *n*
bromotrichlorosil[ic]ane $SiBrCl_3$ Bromtrichlorsilan *n*
bromotungstate(V) M'₂[WOBr₅] Bromowolframat(V) *n*
bromotungstate(VI) M'₂[WO₂Br₄] Bromowolframat(VI) *n*
bromous Brom..., *(i.e.S.)* Brom(III)-...
  b. acid $HBrO_2$ bromige Säure *f*
bromozincate Bromozinkat *n*
bromphenol *s.* bromophenol
bromthymol *s.* bromothymol
bromyrite *(Min)* Bromyrit *m*, Bromargyrit *m* *(Silberbromid)*
Brönner's acid Brönner-Säure *f (2-Naphthylamin-6-sulfonsäure)*
bronze / *to* bronzieren
bronze Bronze *f*, Zinnbronze *f (Kupfer-Zinn-Legierung)*, *(i.w.S.)* Bronze *f (kupferreiche Legierung)*
  b. paper Bronzepapier *n*
  b. powder Bronzepulver *n*
  b. speck Bronzefleck *m (Papierfehler)*
bronzing Bronzieren *n*
  b. fluid *s.* b. liquid
  b. lacquer Bronzelack *m*
  b. liquid Bronzetinktur *f*
bronzite *(Min)* Bronzit *m*
brookite *(Min)* Brookit *m (Titan(IV)-oxid)*
broth Fleischbrühe *f*, Brühe *f*, Bouillon *f*; Nährbouillon *f*, Bouillon *f*, Nährflüssigkeit *f*, Nährlösung *f*, Nährboden *m*, Kulturflüssigkeit *f*, Kulturlösung *f*
  b. culture Bouillonkultur *f*
  b. dilution method Verdünnungsmethode *f*, Bouillonverdünnungsmethode *f*
brown / *to* (*Lebm*) bräunen; *(Metalloberflächen)* brünieren; braun werden
brown Braun *n (Farbempfindung)*; braunfärbender (brauner) Farbstoff *m*, Braun *n*; braunes Packpapier *n*
  b.-black braunschwarz

b. coal Braunkohle *f*; nichtverfestigte (un-verfestigte, nicht erhärtete, lockere) Braunkohle *f*
b.-coal briquette Braunkohlenbrikett *n*
b.-coal coke Braunkohlenkoks *m*
b.-coal generator Braunkohlengaserzeuger *m*
b.-coal tar Braunkohlenteer *m*
b. coloration Braunfärbung *f*
b. earth *(Bodenkunde)* Braunerde *f*
b. forest soil *(Bodenkunde)* Brauner Waldboden *m*
b. glaze *(Ker)* Braunglasur *f*
b. hematite *s.* b. iron ore
b. humic acid *(Bodenkunde)* Braunhuminsäure *f*
b. iron ore (stone) *(Min)* Brauneisenerz *n*, Brauneisenstein *m*, Brauneisen *n*, brauner Glaskopf *m*, Limonit *m*
b. lead ore *(Min)* Braunbleierz *n*, Pyromorphit *m*
b. lead oxide PbO_2 Blei(IV)-oxid *n*, Bleidioxid *n*
b. lignite braune Braunkohle *f* *(im Gegensatz zur schwarzen Glanzbraunkohle)*
b. loam Braunlehm *m*
b. mallee (mallet) *(Gerb)* Eucalyptus astringens Maiden
b. mechanical pulp *(Pap)* Braunschliff *m*, brauner Holzschliff *m*
b. mechanical pulp board Braunschliffpappe *f*
b. mixed pulp board braune Mischpappe *f*
b. mustard Brauner (Roter, Schwarzer) Senf *m*, Brassica nigra (L.)W.D.J. Koch
b. oil of vitriol Glover-Säure *f*, Glover-Turmsäure *f* *(konzentrierte Schwefelsäure mit 77 bis 80% H_2SO_4)*
b. pulp *s.* b. stock
b.-red braunrot
b.-ring test Ringprobe *f* *(zum NO_3-Nachweis)*
b. shade Braunton *m*, brauner Farbton *m*, braune Farbnuance *f*
b. spar *(Min)* Braunspat *m*
b. stock ungewaschener Sulfatzellstoff *m*
b.-stock washing Sulfatzellstoffwäsche *f*
b. substitute *(Gum)* brauner Faktis *m*
b. sugar brauner Rohzucker (Zucker) *m*; Fa-rin[zucker] *m*
b.-sugar medium Nährboden *m* mit braunem Zucker
b. ware *(Ker)* braunes Steinzeug *n*
Brownian motion (movement) Brownsche Be-wegung (Molekularbewegung) *f*
browning *(Lebm)* Bräunen *n*, Bräunung *f*; Brünie-ren *n* *(von Metalloberflächen)*; Braunwerden *n*
b. aid (ingredient, material) *(Lebm)* Bräunungs-zusatz *m*, Bräunungsmittel *n*
b. reaction *(Lebm)* Bräunungsreaktion *f*
brownish bräunlich
b.-red bräunlichrot
brownmillerite *(Min)* Brownmillerit *m*
brucine Bruzin *n*, 2,3-Dimethoxystrychnin *n* *(Al-kaloid)*

b. paper Bruzinpapier *n* *(HNO$_2$-Nachweis)*
brucite *(Min)* Brucit *m* *(Magnesiumhydroxid)*
Brücke reagent for proteins Brückesches Reagens *n* *(Milchuntersuchung)*
Brühl receiver Brühlsche Glocke *f*, Brühlscher Apparat *m*
Brunauer-Emmett-Teller relationship Gleichung *f* von Brunauer, Emmett und Teller, BET-Gleichung *f*
brush / to bürsten; [an]streichen
b. on aufstreichen, [be]streichen
b. out *(Pap)* aufschlagen; *(Pap)* mahlen; *(Pap)* feinmahlen, fertigmahlen
brush 1. Bürste *f*; Pinsel *m*; 2. *(Et)* Bürste *f*, Stromabnehmerbürste *f*; 3. Gebüsch *n*, Gehölz *n*
b. bloodwood Balognia lucida Endl. *(Farb-stoffpflanze)*
b. coater Bürstenstreichmaschine *f*
b. coating Bürstenstrich *m*
b. damper *(Pap)* Bürstenfeuchter *m*
b. doctor *(Pap)* [rotierender] Bürstenschaber *m*
b. dyeing Bürstfärberei *f*, Bürstfärbung *f*
b. enamel paper gebürstetes Glacépapier *n*
b. machine Bürstmaschine *f*
b. marks Pinselstriche *mpl* *(Anstrichfehler)*; senkrechte Falten *fpl* *(Glasfehler)*
b. polishing *(Pap)* Bürstenglättung *f*, Glätten *n* auf der Bürstenmaschine
b. sifter Bürstenentstauber *m*
b. spreader (spreading machine) Bürsten-streichmaschine *f*
brushing Bürsten *n*; Streichen *n*, Anstreichen *n*
b. machine Bürstmaschine *f*; Streichmaschine *f*
b. paint Streichfarbe *f*
brushkiller Gehölzvernichtungsmittel *n*, Arborizid *n*
brushless hydro-soaker machine bürstenlose Reinigungsmaschine *f*
BS *(Abk. für)* 1. British Standard; 2. Biochemical Society
B.S. *s.* bottom settlings
BSDS *s.* benzidinesulphonedisulphonic acid
BSI *(Abk. für)* British Standards Institution
B.Th.U., BTU *(veraltet für)* Btu
Btu *s.* British thermal unit
Bu *s.* butyl 1.
bubble / to *(Gas in Flüssigkeiten)* in Blasen aufsteigen lassen, hindurchperlen lassen; Blasen bilden, in Blasen aufsteigen, sprudeln, perlen
bubble Blase *f*, Bläschen *n*; *(Glas)* Luftsack *m*, Luftblase *f* *(im Külbel)*; Luftblase *f* *(Papierfehler)*; *(Met, Plast)* Lunker *m*
b. bath sprudelndes (brausendes) Bad (Badepa-rat) *n*
b. cap Glocke[nkappe] *f*, Austauschglocke *f* *(einer Glockenbodenkolonne)*
b.-cap [distillation] column Destillierkolonne *f* mit Glockenböden, Glockenbodenkolonne *f*, Glockenkolonne *f*
b.-cap plate Glockenboden *m*

b.-cap plate column s. b.-cap distillation column
b.-cap tray s. b.-cap plate
b. chamber Blasenkammer f
b. counter Blasenzähler m
b. fermentation Blasengärung f
b. glass Glas n mit Luftblasendekor
b. method Blasendruckmethode f, Blasenmethode f
b. plate s. b.-cap plate
b.-plate column s. b.-cap distillation column
b.-point line Siedelinie f
b.-point plane Siedefläche f
b. pressure Blasendruck m
b. tray s. b.-cap plate
b.-tray column s. b.-cap distillation column
bubbler Barboteur m, Druckmischer m, pneumatisches Rührwerk n; Waschflasche f, Gasspüler m; Glocke[nkappe] f (einer Glockenbodenkolonne)
bubbling Blasenbildung f, Bläschenbildung f
b. cap column Glockenbodenkolonne f, Glockenkolonne f, Destillierkolonne f mit Glockenböden
buccocamphor s. buchucamphor
Bucherer reaction Bucherer-Reaktion f, Bucherersche Reaktion f, Bucherer-Lepetit-Reaktion f (Naphthol-Naphthylamin-Umwandlung)
Büchner filter (funnel) Büchner-Trichter m, Büchner-Nutsche f, Filternutsche f nach Büchner
bucholzite (Min) Bucholzit m, (veraltet für) Sillimanit m (Aluminiumalumoorthosilikat)
buchucamphor Bukkokampfer m, Diosphenol n, 2-Hydroxypiperiton m
buck / to (Text) laugen, laugieren, beuchen, in Lauge kochen
buck (Text) Lauge f, Beuche f (Bleichlauge)
bucket Becher m, Eimer m, Kübel m, Behälter m; Löffel m, Kelle f
b. elevator Becherwerk n, Becheraufzug m, Becherelevator m
bucking Beuchen n, Beuche f (Kochen in Lauge)
buckling 1. (Tech) Knickung f, Knicken n; Stauchen n, Stauchung f; Verziehen n; 2. (Lebm) Bückling m, Pökling m
buckstay Ankersäule f (eines Schmelzofens)
buckthorn bark Faulbaumrinde f (von Rhamnus frangula L.)
b. berry Gelbbeere f
buckwheat Buchweizen m, Heidekorn n, Fagopyrum sagittatum Gilib.
b. flour Buchweizenmehl n
b. groats Buchweizengrütze f
b. honey Buchweizenhonig m
Budde effect Budde-Effekt m (Fotochemie)
budding Sprossung f
b. cell (Bio) Sproßzelle f, durch Sprossung entstandene Zelle f
b. fungi Sproßpilze mpl
buff / to polieren, schwabbeln; (Gerb) abbuffen, den Narben abschleifen (abziehen), buffieren

buff ledergelb, lederfarbig, gelbbraun, braungelb
buff Schwabbel[scheibe] f
b.-coloured s. buff
buffer / to puffern
buffer Puffer m, Pufferlösung f
b. acid Puffersäure f
b. action Pufferwirkung f
b. anion Pufferanion n
b. capacity Puffer[ungs]kapazität f, Pufferungsvermögen n
b. index Pufferwert m
b. kit Puffersubstanzen fpl
b. mixture Puffermischung f, Puffergemisch n
b. reagent Puffersubstanz f
b. region Pufferbereich m, Pufferzone f
b. salt Puffersubstanz f
b. solution Pufferlösung f, Puffer m
b. system Puffersystem n
b. value s. b. index
buffered solution gepufferte Lösung f, Pufferlösung f
buffering Abpuffern n, Puffern n, Pufferung f
b. action Pufferwirkung f
b. agent Puffersubstanz f
b. effect Pufferwirkung f, Pufferung f
b. capacity (power) Puffer[ungs]kapazität f, Pufferungsvermögen n
b. substance s. b. agent
buffing Polieren n, Schwabbeln n; (Gerb) Schleifen n, Bimsen n, Dollieren n
b. composition (compound) Poliermittel n
b. machine Schleifmaschine f, Bimsmaschine f, Poliermaschine f, Schwabbelmaschine f
b. wheel Polierscheibe f, Schwabbelscheibe f
bufotenine Bufotenin n, 5-Oxoindoläthyldimethylamin n
bugbane, bugroot, bugwort Schwarze Schlangenwurzel f (von Cimicifuga racemosa (L.)Nutt.)
buhr s. burstone
buhrmill s. burstone mill
buhrstone s. burstone
build Bau m, Aufbau m, Bauart f
b.-up Aufbau m (z.B. von Farbstoffen); Bildung f (z.B. von statischer Elektrizität)
b.-up of pressure Druckaufbau m
builder Aufbaustoff m, Gerüstsubstanz f, Gerüststoff m, Builder m (zum Aufbau synthetischer Waschmittel)
building (Gum) Zusammenbau m, Konfektion[ierung] f
b. block Baustein m, Baueinheit f (z.B. einer chemischen Verbindung)
b. board Baupappe f; Bauplatte f, Leichtbauplatte f
b. brick Bauziegel m, Mauerziegel m
b. cement Bauzement m
b. ceramics Baukeramik f
b. drum (Gum) Wickeltrommel f, Reifenwickeltrommel f
b. lime Baukalk m

b. principle Aufbauprinzip *n*, Bauprinzip *n*

b. stone Baustein *m* (*z.B. einer chemischen Verbindung*)

b. tack (*Gum*) Konfektionsklebrigkeit *f*

b.-up 1. Aufbau *m*, Aufbauen *n* (*z.B. einer chemischen Verbindung*); 2. *s.* building

b.-up principle *s.* b. principle

built-in eingebaut, Einbau...

b.-in producer Einbaugenerator *m*, eingebauter Generator *m*

b. soap aufgebaute Seife *f*

bulb 1. Kolben *m*, Ballon *m*, [kugelförmiges] Gefäß *n*, Kugel *f*; 2. Knolle *f*; Zwiebel *f*; Zwiebelgewächs *n*; 3. Thermometergefäß *n*; 4. (*Tech*) Zwiebel *f*, Spinnzwiebel *f*, Ziehzwiebel *f*

b. condenser Kugelkühler *m*, Allihn-Kühler *m*, Allihnscher Kühler *m*

b. edge Borte *f* (*beim Fensterglas*)

b. tube Kugelrohr *n*, Kugelröhre *f*

b. tube mill Kugelrohrmühle *f*

bulk / to (*Pap*) griffig sein

b. high (*Pap*) dickgriffig (bauschig, auftragend) sein

b. low (*Pap*) dünngriffig (wenig griffig) sein

bulk 1. Volumen *n*, Umfang *m*, Masse *f*, Menge *f*; Hauptmasse *f*, Hauptmenge *f*, Großteil *m*; 2. (*Pap*) Rohdichte *f*, (*bisher*) Raumgewicht *n*; Griff *m*, Griffigkeit *f*

b. cement loser (ungesackter) Zement *m*

b. champagnization Großraumgärverfahren *n*, Tankgärverfahren *n*, Tankgärung *f*

b. density Raummasse *f*, (*bisher*) Raumgewicht *n*; (*Holz*) Rohdichte *f*; (*Chem, Bgb*) Schüttdichte *f*, (*bisher*) Schüttgewicht *n*

b. factor Verdichtungsgrad *m*, Füllfaktor *m*

b. floatation kollektive Flotation *f*, Kollektivflotation *f*

b. handling Tankbehandlung *f* (*z.B. der Milch*)

b. liquid Hauptteil *m* der flüssigen Phase

b. material Schüttgut *n*

b. milk Sammelmilch *f*

b.-milk collection Sammlung *f* der Milch in Milchtanks

b.-milk handling *s.* b. handling

b. mixing plant Mischdüngerfabrik *f*

b. moulding machine Formmaschine (Strangpresse) *f* zur Herstellung von Großpackungen

b. package Großpackung *f*, Blockpackung *f*

b. polymerization Polymerisation *f* in Masse (Substanz), Massepolymerisation *f*

b. polymerization with the alkali metals Alkalimetallpolymerisation *f*

b. process Großraumgärverfahren *n*, Tankgärverfahren *n*, Tankgärung *f*

b. process champagne im Großraumgärverfahren (Tankgärverfahren) hergestellter Schaumwein *m*

b. starter (*Lebm*) Betriebssäurewecker *m*

b. volume Schüttvolum[en] *n*; (*Ker*) Rohvolumen *n*, Volumen *n*, Rauminhalt *m*

b. wine Faßwein *m*

bulked yarn texturierte Seide *f*, Texturseide *f*

bulkiness Feinheitsgrad *m*; (*Text*) Bauschigkeit *f*

bulking agent Füllstoff *m*, Füllmittel *n*, Füllmaterial *n*

b. paper (*Am*) Dickdruckpapier *n*, Federleichtpapier *n*, Daunendruckpapier *n*

b. sludge Blähschlamm *m*

bulky groß, dick, umfangreich, voluminös; sperrig

b. filter medium Filtermasse *f*

b. paper auftragendes Papier *n*

b. yarn 1. Hohlseide *f*, Profilseide *f* (*Chemiefaserstoff*); 2. Hochbauschgarn *n*

bull ring Reibbarren *m* (*im Reibwerk*); Mahlring *m* (*in der Wälzmühle*)

b. screen (*Pap*) Splitterfänger *m*

Bullers ring Bullers-Ring *m* (*Brennring*)

bullet Druckgasflasche *f*, Gasflasche *f*, Flasche *f*, Bombe *f*

b.-proof glass kugelsicheres (kugelfestes, schußfestes) Glas *n*, Panzerglas *n*

b.-resistant glass, b.-resisting glass *s.* b.-proof glass

bullhead Gewölbeschlußstein *m*

bump / to stoßen (*von siedenden Flüssigkeiten*)

bumping Stoßen *n* (*beim Siedeverzug*)

buna 1. *s.* b. rubber; 2. Morgenländische Platane *f*, Platanus orientalis L.

b. latex Bunalatex *m*

b. rubber Bunakautschuk *m*, Buna *m(n)*

bundle of fibres (*Pap*) Faserbündel *n*

b. of molecules Molekülschwarm *m*, Schwarm *m*

bung / to [ver]spunden, spünden, mit Spund versehen

bung Pfropfen *m*, Spund *m*, Stöpsel *m*, Zapfen *m*, Stopfen *m*; Spundloch *n*, Zapfloch *n*; (*Ker*) Stapel *m* (*z.B. von Tellern, Ziegeln, Kapseln*)

bunghole Spundloch *n*, Zapfloch *n*

bunker Bunker *m*; Kohle[n]bunker *m*

b. C fuel [oil], b. C oil Bunker-C-Öl *n*, Bunkeröl C *n*

b. coal Bunkerkohle *f*

b. fuel [oil], b. oil Bunkeröl *n*

Bunsen absorption coefficient *s.* Bunsen coefficient

Bunsen burner Bunsenbrenner *m*

Bunsen coefficient Bunsenscher Absorptionskoeffizient *m*

Bunsen flame Bunsenflamme *f*

Bunsen funnel Bunsentrichter *m*, Trichter *m* mit langem Stiel

Bunsen ice calorimeter Eiskalorimeter *n* nach Bunsen

Bunsen-Roscoe [reciprocity] law Bunsen-Roscoesches Gesetz *n*, Reziprozitätsgesetz *n*

Bunsen valve Bunsenventil *n*

bunsenine (*Min*) Bunsenin *m*, (*veraltet für*) Krennerit *m* (*Gold(III)-tellurid*)

Bunte gas burette Buntesche Bürette *f*, Bunte-Bürette *f*

Bunte salt Buntesches Salz *n*
buoy up / to schwimmend erhalten, emporheben, an die Oberfläche treiben
buoyancy [statischer] Auftrieb *m (in einer Flüssigkeit oder einem Gas)*
  b. force Auftriebskraft *f*, Auftrieb *m*
  b. method [for gas density] Auftriebsmethode *f (zur Gasdichtemessung)*
  b. of the air Luftauftrieb *m*
  b. process Schwimmaufbereitungsverfahren *n*, Schwimmverfahren *n*, Flotationsverfahren *n*
buoyant force Auftriebskraft *f*, Auftrieb *m*
burden / to *(Met)* beschicken, begichten, möllern
burden Beschickung *f*, Gicht *f*, Möller *m*, Hochofenmöller *m*
bureau mixture *(Schädlingsbekämpfungsmittel mit Hexaäthyltetraphosphat als Wirkstoff)*
buret *s.* burette
burette Bürette *f*
  b. brush Bürettenbürste *f*
  b. clamp Bürettenklemme *f*, Bürettenhalter *m*, *(meist)* Dreifingerklemme *f*
  b. filler (funnel) Bürettentrichter *m*
  b. holder *s.* b. clamp
  b. jet Bürettenspitze *f (für Quetschhahnanschluß)*
  b. meniscus magnifier Ableselupe *f (für Büretten)*
  b. meniscus reader *s.* b. reader
  b. outlet tube Bürettenspitze *f (Ersatzteil)*
  b. pinchcock Bürettenquetschhahn *m*
  b. reader Ableseblatt *n (für Büretten)*
  b. stopcock Bürettensperrhahn *m*, Bürettenhahn *m*
  b. tip *s.* b. jet
  b. valve *s.* b. stopcock
Burger-Dorgelo-Ornstein sum rule Burger-Dorgelo-Ornsteinsche Summenregel *f*
Burgundy mixture Burgunder Brühe *f*, Kupfersodabrühe *f (Pflanzenschutzmittel)*
  B. pitch Burgunderharz *n*, Burgunderpech *n*, *(fälschlich oft)* Fichtenharz *n (vorwiegend von der Strandkiefer Pinus pinaster Ait.)*
buried unterirdisch, erdverlegt, Erd...
  b. duct unterirdische (erdverlegte) Leitung *f*, Erdleitung *f*
burl *(Text)* Noppe *f*, Knötchen *n*
  b. dyeing *(Text)* Noppenfärben *n*
Burmese lacquer Firnis *m* von Martaban *(von Melanorrhoea usitata Wall.)*
  B. varnish tree Hinterindischer Firnisbaum *m*, Melanorrhoea usitata Wall.
burn / to [ver]brennen; brennen, kalzinieren; *(Gum)* anbrennen, anvulkanisieren, anspringen
  b. in *(z.B. Farben)* einbrennen
  b. off wegbrennen, abbrennen, ausbrennen
  b. on anbrennen *(z.B. am Gußstück)*
  b. out ausbrennen
  b. the cook *(Pap)* schwarzkochen
  b. to ashes veraschen

burn Verbrennungen *fpl (z.B. durch Pflanzenschutzmittel)*
  b.-off Abbrand *m*; *(Glas)* Absprengen *n*
  b.-out Ausbrand *m*
  b.-up *(Kch)* Abbrand *m*
burnable [ver]brennbar
burned cook *(Pap)* Schwarzkochung *f*
  b. lime Branntkalk *m*, gebrannter Kalk *m*, Ätzkalk *m (Kalziumoxid)*
  b. spot *(Plast)* Brandstelle *f*, Verbrennungsmarkierung *f (Spritzfehler)*
burner Brenner *m*; *(Tech)* Ofen *m*, Verbrennungsofen *m*
  b. channel Brennerrinne *f*
  b. chimney Schornsteinaufsatz *m*, Schornstein *m (für Brenner)*
  b. flame Brennerflamme *f*
  b. flame spreader Breitbrenneraufsatz *m*, Schnittbrenneraufsatz *m*, Schlitzaufsatz *m*
  b. gas Verbrennungsabgas *n*, Verbrennungsgas *n*, Ofengas *n*
  b. guard Tonesse *f (für Bunsenbrenner)*
  b. house Brennerhaus *n (Rußherstellung)*
  b. lighter Gasanzünder *m*
  b. throat Brennermund *m*
  b. tile Brennerstein *m*
  b. tip Brennermundstück *n*
  b. tube Brennerrohr *n*, Brennrohr *n (z.B. am Bunsenbrenner)*
  b. wick Docht *m* für Spirituslampen
  b. wingtop *s.* b. flame spreader
burning Brennen *n*, Verbrennen *n*; Brennen *n*, Kalzinieren *n*, Kalzinierung *f*; *(Tech)* Rösten *n*, Röstung *f*, Abrösten *n*, Abröstung *f*; *(Gum)* Anbrennen *n*, Anvulkanisieren *n*, Anvulkanisation *f*, Anspringen *n*; *(Pap)* Schwarzkochung *f*
  b. of pyrites Kiesabröstung *f*
  b.-off Wegbrennen *n*, Abbrennen *n*, Abbrennung *f*, Ausbrennen *n*, Ausbrennung *f*
  b. oil Brennöl *n*
  b.-on Anbrennen *n (z.B. von Formsand am Gußstück)*
  b.-out Ausbrennen *n*
  b. point Brennpunkt *m*
  b. rate Verbrennungsgeschwindigkeit *f*, Brenngeschwindigkeit *f*
  b. test Brennprobe *f*, Brennversuch *m*
  b. time Brennzeit *f*, Brenndauer *f*
burnish[ed] gold *(Ker)* Poliergold *n*
burnt alum gebrannter Alaun *m*
  b. borax $Na_2B_4O_7$ gebrannter (kalzinierter) Borax *m*, wasserfreies Natriumtetraborat *n*
  b. centre *(Pap)* dunkler Kern *m (im Hackschnitzel)*
  b. cook *(Pap)* Schwarzkochung *f*
  b. lime gebrannter Kalk *m*, Branntkalk *m*, Ätzkalk *m (Kalziumoxid)*
  b.-out fabric Ausbrennartikel *m*, Ausbrenner *m*
burr *(Tech)* Grat *m*, Bart *m*
burrstone *s.* burstone

burry wool klettenhaltige Wolle f
burst into flame / to entflammen
burst strength s. bursting strength
bursting disk Berstscheibe f, Reißscheibe f, Platzscheibe f
b. strength Berst[druck]festigkeit f, Bruchfestigkeit f, Zerreißfestigkeit f, Zerreißgrenze f; (Gerb) Knickfestigkeit f, Knickbeständigkeit f
burstone Mühlstein m, Mahlstein m
b. mill Steinmahlgang m
Burt filter Burt-Filter n (Blattfilter mit rotierender Filtertrommel)
Burton-Clark process Burton-Clark-Verfahren n, Burton-Clark-Spaltverfahren n
Burton [cracking] process Burton-Verfahren n, Spaltverfahren n nach Burton
burtonize / to (Brauwasser) burtonisieren, gipsen
burtonizing Burtonisieren, Gipsen n (des Brauwassers)
bustle pipe (Met) Heißwindringleitung f, Heißwindring m
busulphan (Pharm) Busulfan n, Butandiol-(1,4)-bis-methylsulfat n
butacaine (Pharm) Butakain n, 3-Dibutylaminopropyl-p-aminobenzoat n
b. sulphate Butakainsulfat n (Oberflächenanästhetikum)
butadiene $CH_2=CHCH=CH_2$ Butadien n, Divinyl n, Vinyläthylen n, Vinyläthen n, Diäthylen n, Diäthen n
b.-acrylonitrile copolymer Butadien-Akrylnitril-Mischpolymerisat n
b. copolymer Butadienmischpolymerisat n
b. plant Butadienanlage f
b. polymer Butadienpolymerisat n, Butadienpolymer[es] n, Polybutadien n
b. rubber Butadienkautschuk m
b.-sodium rubber Natrium-Butadienkautschuk m (Butadien-Na-Polymerisat)
b.-styrene copolymer Butadien-Styrol-Mischpolymerisat n
b.-styrene-latex Butadien-Styrol-Latex m, SBR-Latex m
b.-styrene rubber Butadien-Styrolkautschuk m, Styrol-Butadienkautschuk m, Styrolkautschuk m, SBK (Butadien-Styrol-Mischpolymerisat)
b.-vinylpyridine-copolymer Butadien-Vinylpyridin-Mischpolymerisat n
butadiyne $HC{\equiv}CC{\equiv}CH$ Diäthin n, Butadiin n, Diazetylen n
n-butanal $CH_3CH_2CH_2CHO$ n-Butanal n, n-Butyraldehyd m
butane Butan n, (i.e.S.) s. n-butane
n-butane $CH_3CH_2CH_2CH_3$ n-Butan n, Normalbutan n, Butan n
b.-containing butanhaltig
b.-free butanfrei
1,4-butanedicarboxylic acid $COOH(CH_2)_4COOH$ 1,4-Butandikarbonsäure f, Hexandisäure f, Adipinsäure f

butanedioic acid $HOOCCH_2CH_2COOH$ Butandisäure f, Äthandikarbonsäure-(1,2) f, Bernsteinsäure f, Sukzinsäure f
2,3-butanedione $CH_3COCOCH_3$ Diazetyl n, Butandion-(2,3) n, Dimethylglyoxal n
anti-1,2,3,4-butanetetrol $(HOCH_2CHOH)_2$ Butantetrol n, Erythrit m, Threit m
butanoic acid C_3H_7COOH Butansäure f, n-Buttersäure f, Normalbuttersäure f, Buttersäure f
1-butanol $CH_3(CH_2)_2CH_2OH$ 1-Butanol n, n-Butylalkohol m
2-butanol $CH_3CH_2CHOHCH_3$ 2-Butanol n, sek-Butanol n, sek-Butylalkohol m
2-butanone $CH_3COC_2H_5$ 2-Butanon n, Methyläthylketon n
butea gum Buteakino n, Bengalkino n, Bengalisches Kino n (von Butea superba Roxb.)
butein $(HO)_2C_6H_3{\cdot}CO{\cdot}CH=CH{\cdot}C_6H_3(OH)_2$ Butein n (Chalkonderivat)
2-butenal $CH_3CH=CHCHO$ Buten-(2)-al-(1) n, Krotonaldehyd m, 3-Methylakrolein n
1-butene $CH_2=CHCH_2CH_3$ 1-Buten n, 1-Butylen n, Äthyläthylen n
2-butene $CH_3CH=CHCH_3$ 2-Buten n, 2-Butylen n, sym-Dimethyläthylen n, sym-Dimethyläthen n
cis-butenedioic acid $HOOCCH=CHCOOH$ cis-Butendisäure f, cis-1,2-Äthylendikarbonsäure f, Maleinsäure f
trans-butenedioic acid $HOOCCH=HCCOOH$ trans-Butendisäure f, trans-1,2-Äthylendikarbonsäure f, Fumarsäure f
cis-butenedioic anhydride cis-Butendisäureanhydrid n, Maleinsäureanhydrid n, 2,5-Furandion n
cis-2-butenoic acid $CH_3CH=CHCOOH$ cis-Buten-(2)-säure-(1) f, Isokrotonsäure f
trans-2-butenoic acid $CH_3CH=CHCOOH$ trans-Buten-(2)-säure-(1) f, Krotonsäure f
butin Butin n, 7,3′,4′-Trihydroxyflavonon n
butine s. butyne
butt / to (Gerb) crouponieren, kruponieren
butt (Gerb) Croupon m, Krupon m, Kern m, Kernstück n
b. fusion (Plast) Stumpfschweißen n, Stumpfschweißung f
b. roller (Gerb) Lederwalze f
butter aroma Butteraroma n, butterähnliches Aroma n
b.-blending machine Buttermischmaschine f
b. churn Butterfaß n, Butterfertiger m, Butterherstellungsmaschine f, Butterungsanlage f
b. colouring Butterfärbung f (mit Farbstoffen); Butterfarbe f (Farbstoff)
b. cutter Butterschneider m
b. factory Butterei f, (i.w.S.) Molkerei f
b.-fat content Fettgehalt m der Butter
b. grain (granule) Butterkorn n
b.-like butterähnlich
b. making Buttern n, Butterung f, Verbuttern n, Verbutterung f, Butterherstellung f, Buttererzeugung f, Butterbereitung f

b.-making plant Butterei f, (i.w.S.) Molkerei f
b.-moulding machine Butterformmaschine f
b. of antimony SbCl₃ Antimonbutter f, Antimon(III)-chlorid n, Antimontrichlorid n
b. of arsenic AsCl₃ Arsenbutter f, Arsen(III)-chlorid n, Arsentrichlorid n
b. of tin SnCl₄·5H₂O Zinnbutter f, Zinntetrachlorid-5-Wasser n
b. of zinc ZnCl₂ Zinkbutter f, Zinkchlorid n
b. [parchment] paper Butterpapier n
b.-pat machine s. b.-moulding machine
b. salt Buttersalz n
b. serum Butterserum n
b. tree Westafrikanischer Talgbaum m, Pentadesma butyraceum Sabine
b. yellow C₆H₅N=NC₆H₄N(CH₃)₂ Buttergelb n, Dimethylgelb n, p-Dimethylaminoazobenzol n
butterfat Butterfett n, Butterschmalz n, Schmelzbutter f
butterfly valve Drosselklappe f, Drossel[klappen]ventil n
buttermaking process Butterungsverfahren n
buttermilk Buttermilch f
b. powder Buttermilchpulver n, getrocknete Buttermilch f
buttery butterartig, butterähnlich, butt[e]rig, Butter...
b. flavour Butteraroma n, butterähnliches Aroma n
button catcher s. b. trap
b. lac Knopfschellack m, Knopflack m, Blutlack m
b. test (Ker) Knopfprobe f, Knopfprüfung f (der Fließbarkeit einer Fritte)
b. trap (Pap) Nagelfang m
butyl 1. Butyl n; 2. s. b. rubber
n-b. acetate CH₃COO(CH₂)₃CH₃ n-Butylazetat n, Essigsäure-n-butylester m, Butyl-Äthanoat n
n-b. alcohol CH₃(CH₂)₂CH₂OH n-Butylalkohol m, 1-Butanol n
prim-b. alcohol s. n-b. alcohol
sec-b. alcohol CH₃CH₂CHOHCH₃ sek-Butylalkohol m, sek-Butanol n, 2-Butanol n
tert-b. alcohol (CH₃)₃COHCH₃ tert-Butylalkohol m, tert-Butanol n, 2-Methylpropanol-(2) n
b. aminobenzoate Butylaminobenzoat n, 4-Aminobenzoesäurebutylester m
n-b. bromide CH₃CH₂CH₂CH₂Br n-Butylbromid n, prim-n-Butylbromid n, 1-Brombutan n
sec-b. bromide CH₃CH₂CHBRCH₃ sek-Butylbromid n, 2-Brombutan n
tert-b. bromide (CH₃)₃CBr tert-Butylbromid n, Trimethylbrommethan n, 2-Brom-2-methylpropan n
n-b. carbinol CH₃(CH₂)₃CH₂OH Butylkarbinol n, 1-Pentanol n, n-Amylalkohol m
b. cement (Gum) Butyllösung f
n-b. chloride CH₃(CH₂)₂CH₂Cl prim-Butylchlorid n, 1-Chlorbutan n
b. compound s. b. rubber compound

b. group Butylgruppe f, Butylrest m
b. halide Butylhalogenid n
b. inner tube (Gum) Butylschlauch m
b. latex Butyllatex m
tert-b. methyl ketone CH₃COC(CH₃)₃ Methyl-tert-butylketon n, 2,2-Dimethylbutanon-(3) n, Trimethylazeton n, Pinakolon n
b. perbenzoate Butylperbenzoat n
b. radical Butylgruppe f, Butylrest m, Butylradikal n; [freies] Butylradikal n
b. reclaim Butylregenerat n
b. rubber Butylkautschuk m
b. rubber compound Butylkautschukmischung f, Butylmischung f
b. stearate C₁₇H₃₅·COO·C₄H₉ Butylstearat n
b. tube s. b. inner tube
b. tyre Butylreifen m
b. vulcanizate Butylkautschukvulkanisat n
n-butylacetylene HC≡C(CH₂)₃CH₃ Butylazetylen n, Hexin-(1) n
butylated hydroxytoluene 2,6-Di-tert-butyl-p-kresol n, DBPC
2-sec-butyl-4,6-dinitrophenol CH₃(C₂H₅)CHC₆H₂(NO₂)₂OH 2-sek-Butyl-4,6-dinitrophenol n, 2,4-Dinitro-o-sek-butylphenol n, Dinitro-sek-butylphenol n, Dinoseb n, DNBP
butylene:
α-butylene CH₂=CHCH₂CH₃ 1-Buten n, 1-Butylen n, Äthyläthylen n
β-butylene CH₃CH=CHCH₃ 2-Buten n, 2-Butylen n, sym-Dimethyläth[yl]en n
b. polymer Butylenpolymer[es] n, Butenpolymer[es] n
butylic fermentation Butanol-Azeton-Gärung f
1-butyne HC≡CCH₂CH₃ 1-Butin n, Äthylazetylen n
2-butyne CH₃C≡CCH₃ 2-Butin n, Dimethylazetylen n, Krotonylen n
bytynedioic acid HOOCC≡CCOOH Butindisäure f, Äthindikarbonsäure f, Azetylendikarbonsäure f
n-butyraldehyde CH₃CH₂CH₂CHO n-Butyraldehyd m, n-Butanal n
butyrate Butyrat n
butyric:
n-b. acid C₃H₇COOH n-Buttersäure f, Normalbuttersäure f, Buttersäure f, Butansäure f
b. acid bacteria Buttersäurebakterien npl
n-b. aldehyde s. n-butyraldehyde
b. fermentation Buttersäuregärung f
butyrometer Butyrometer n
butyrone CH₃CH₂CH₂COCH₂CH₂CH₃ Butyron n, Dipropylketon n, 4-Heptanon n
bypass / to umgehen; vorbeiführen, herumführen
bypass valve Umführungsventil n
byproduct Nebenprodukt n, Nebenerzeugnis n
b. coke oven Koksofen m mit Nebenproduktengewinnung
b. lime Abfallkalk m
b. oven Nebenproduktenofen m, Nebengewinnungsofen m

b. plant Nebenproduktenanlage *f*, Wertstoffanlage *f*

b. recovery Nebenproduktengewinnung *f*, Wertstoffgewinnung *f*

bytownite *(Min)* Bytownit *m*

C

c *s.* 1. cold; 2. concentrate; 3. concentrated; 4. concentration

C *s.* 1. concentrate; 2. concentrated; 3. concentration

C acid C-Säure *f*, 2-Naphthylamin-4,8-disulfonsäure *f*

C-alkaloid C-Alkaloid *n* *(Kalebassenalkaloid)*

C-frame press einhüftige Presse *f*, Maulpresse *f*

C-horizon *(Bodenkunde)* C-Horizont *m*, Untergrund *m*, Muttergestein *n*

C^{14}-method *s.* radiocarbon dating

C process *s.* Croning process

C stage *(Plast)* C-Zustand *m*

C-stage resin Harz *n* im C-Zustand, C-Harz *n*, *(bei Phenolharzen auch)* Resit *n*

C-terminal C-terminal, C-endständig *(von Aminosäuren in Proteinen)*

C-terminal group (residue) C-terminaler Rest *m*, C-terminale Endgruppe *f*, Karboxylende *n* *(in Proteinen)*

C.A. *(Abk. für)* Chemical Abstracts

CAB *s.* cellulose acetate butyrate

C.A.B. *s.* critical air blast

cabal glass Cabalglas *n*

cabinet Kammer *f*, Raum *m*

c. dryer Kammertrockner *m* *(mit einer Kammer)*

c.-type air dryer *s. c.* dryer

c.-type smoke house Rauchschrank *m*

cable covering *(Gum)* Umspritzen *n* von Kabeln

c. oil Kabelisolieröl *n*, Kabelöl *n*

c. paper Kabelisolierpapier *n*, Kabelpapier *n*

c. tool Seilbohrwerkzeug *n*

c. tool drilling Kabelbohren *n*, [pennsylvanisches] Seilbohren *n*

c. tool installation Kabelbohranlage *f*, Seilbohranlage *f*, Kabelbohrgerät *n*, Seilbohrgerät *n*

c. tool method Kabelbohrverfahren *n*, Kabelbohrmethode *f*, Seilschlag[bohr]verfahren *n*, pennsylvanisches Bohrsystem *n*

c. tool rig *s. c.* tool installation

c. tool system *s. c.* tool method

c. tool well Seilbohrloch *n*, Seilschlagbohrung *f*, Seilbohrung *f*

cacao Kakaobaum *m*, Theobroma cacao L.; Kakaobohne *f*; Kakaopulver *n*; Kakao *m* *(Getränk)*

c. bean Kakaobohne *f*

c. butter (oil) Kakaobutter *f*, Kakaoöl *n*, Kakaofett *n*

c. shells Kakaoschalen *fpl*

c. tree Kakaobaum *m*, Theobroma cacao L.

cachou de Laval Cachou de Laval *n* *(ein Schwefelfarbstoff)*

cacodyl (CH$_3$)$_2$AsAs(CH$_3$)$_2$ Kakodyl *n*, Tetramethyl-[di]arsin *n*, bis-Dimethylarsyl *n*

c. chloride (CH$_3$)$_2$AsCl Kakodylchlorid *n*, Chlordimethylarsin *n*, Dimethylchlorarsin *n*, Dimethylarsinchlorid *n*

c. oxide [As(CH$_3$)$_2$]$_2$O Kakodyloxid *n*, Alkarsin *n*

cacodylic acid (CH$_3$)$_2$AsOOH Kakodylsäure *f*, Dimethylarsinsäure *f*

cacoxene, cacoxenite *(Min)* Kakoxen *m* *(Eisen(III)-trihydroxidorthophosphat)*

cadaverine H$_2$N(CH$_2$)$_5$NH$_2$ Kadaverin *n*, Pentamethylendiamin *n*

cadinene Kadinen *n* *(bizyklisches Sesquiterpen)*

cadinol Kadinol *n* *(bizyklischer Sesquiterpenalkohol)*

cadmic Kadmium...

cadmium Cd Kadmium *n*

c. acetate Cd(CH$_3$COO)$_2$ Kadmiumazetat *n*

c. arsenide Cd$_3$As$_2$ Kadmiumarsenid *n*

c.-base auf Kadmiumbasis, auf Kadmiumgrundlage

c.-bearing kadmiumhaltig

c. blende *(Min)* Kadmiumblende *f*, Kadmiumokker *m*, *(veraltet für)* Greenockit *m* *(Kadmiumsulfid)*

c. borotungstate Cd$_5$(BW$_{12}$O$_{40}$)$_2$ Kadmiumborowolframat *n*

c. bromate Cd(BrO$_3$)$_2$ Kadmiumbromat *n*

c. bromide CdBr$_2$ Kadmiumbromid *n*

c. carbonate CdCO$_3$ Kadmiumkarbonat *n*

c. chlorate Cd(ClO$_3$)$_2$ Kadmiumchlorat *n*

c. chloride CdCl$_2$ Kadmiumchlorid *n*

c. cyanide Cd(CN)$_2$ Kadmiumzyanid *n*

c. dithionate CdS$_2$O$_6$ Kadmiumdithionat *n*

c. electrode Kadmiumelektrode *f*

c. ferrocyanide Cd$_2$[Fe(CN)$_6$] Kadmiumhexazyanoferrat(II) *n*

c. fluoride CdF$_2$ Kadmiumfluorid *n*

c. hydrate *s. c.* hydroxide

c. hydroxide Cd(OH)$_2$ Kadmiumhydroxid *n*

c. iodate Cd(JO$_3$)$_2$ Kadmiumjodat *n*

c. iodide CdJ$_2$ Kadmiumjodid *n*

c. nitrate Cd(NO$_3$)$_2$ Kadmiumnitrat *n*

c. ochre *(Min)* Kadmiumocker *m*, Kadmiumblende *f*, *(veraltet für)* Greenockit *m* *(Kadmiumsulfid)*

c. orthophosphate Cd$_3$(PO$_4$)$_2$ Kadmiumorthophosphat *n*

c. oxide CdO Kadmium(II)-oxid *n*

c. permanganate Cd(MnO$_4$)$_2$ Kadmiumpermanganat *n*, Kadmiummanganat(VII) *n*

c. pyrophosphate Cd$_2$P$_2$O$_7$ Kadmiumdiphosphat *n*, Kadmiumpyrophosphat *n*

c. red Kadmiumrot *n*

c. selenate CdSeO$_4$ Kadmiumselenat *n*

c. selenide CdSe Kadmiumselenid *n*

c. silicate CdSiO$_3$ Kadmiummetasilikat *n*, Kadmiumsilikat *n*

c. **stearate** $Cd(C_{16}H_{35}COO)_2$ Kadmiumstearat n
c. **suboxide** Cd_2O Kadmiumsuboxid n
c. **sulphate** $CdSO_4$ Kadmiumsulfat n
c. **sulphide** CdS Kadmiumsulfid n
c. **sulphite** $CdSO_3$ Kadmiumsulfit n
c. **telluride** CdTe Kadmiumtellurid n
c. **tungstate** $CdWO_4$ Kadmiumwolframat n
c. **wolframate** s. c. tungstate
c. **yellow** CdS Kadmiumgelb n, Kadmiumsulfid n
caesium Cs Zäsium n
c. **acetate** CH_3COOCs Zäsiumazetat n
c. **acid carbonate** s. c. hydrogen carbonate
c. **acid sulphate** s. c. hydrogen sulphate
c. **acid tartrate** s. c. hydrogen tartrate
c. **alum** s. c.-aluminium sulphate
c.-**aluminium sulphate** $CsAl(SO_4)_2 \cdot 12H_2O$ Zäsiumaluminiumsulfat-12-Wasser n, Zäsiumalaun m
c. **bicarbonate** s. c. hydrogen carbonate
c. **bichromate** s. c. dichromate
c. **bisulphate** s. c. hydrogen sulphate
c. **bitartrate** s. c. hydrogen tartrate
c. **bromate** $CsBrO_3$ Zäsiumbromat n
c. **bromide** CsBr Zäsiumbromid n
c. **carbonate** Cs_2CO_3 Zäsiumkarbonat n
c. **chlorate** $CsClO_3$ Zäsiumchlorat n
c. **chloride** CsCl Zäsiumchlorid n
c. **chromate** Cs_2CrO_4 Zäsiumchromat n
c. **cyanide** CsCN Zäsiumzyanid n
c. **dichromate** $Cs_2Cl_2O_7$ Zäsiumdichromat(VI) n
c. **dihydronitrate** $CsNO_3 \cdot 2HNO_3$ Zäsiumdihydrogennitrat n
c. **dioxide** Cs_2O_2 Zäsiumperoxid n
c. **disulphate** s. c. hydrogen sulphate
c. **disulphide** Cs_2S_2 Zäsiumdisulfid n
c. **fluoride** CsF Zäsiumfluorid n
c. **fluorosilicate** $Cs_2[SiF_6]$ Zäsiumhexafluorosilikat n
c. **hexasulphide** Cs_2S_6 Zäsiumhexasulfid n
c. **hydrate** s. c. hydroxide
c. **hydride** CsH Zäsiumhydrid n
c. **hydrogen carbonate** $CsHCO_3$ Zäsiumhydrogenkarbonat n
c. **hydrogen sulphate** $CsHSO_4$ Zäsiumhydrogensulfat n
c. **hydrogen tartrate** Zäsiumhydrogentartrat n
c. **hydronitrate** $CsNO_3 \cdot HNO_3$ Zäsiumhydrogennitrat n
c. **hydroxide** CsOH Zäsiumhydroxid n
c. **iodide** CsJ Zäsiumjodid n
c. **mercuric bromide** $CsBr \cdot 2HgBr_2$ Zäsiumquecksilber(II)-bromid n
c. **mercuric chloride** $CsCl \cdot HgCl_2$ Zäsiumquecksilber(II)-chlorid n
c. **monoxide** Cs_2O Zäsiummonoxid n, Zäsiumoxid n
c. **nitrate** $CsNO_3$ Zäsiumnitrat n
c. **nitrite** $CsNO_2$ Zäsiumnitrit n
c. **pentaiodide** CsJ_5 Zäsiumpentajodid n
c. **pentasulphide** Cs_2S_5 Zäsiumpentasulfid n

c. **perchlorate** $CsClO_4$ Zäsiumperchlorat n
c. **periodate** $CsJO_4$ Zäsiummetaperjodat n, Zäsiumtetroxoperjodat n
c. **permanganate** $CsMnO_4$ Zäsiumpermanganat n, Zäsiummanganat(VII) n
c. **peroxide** s. c. tetroxide
c. **silicofluoride** s. c. fluorosilicate
c. **sulphate** Cs_2SO_4 Zäsiumsulfat n
c. **sulphide** Cs_2S Zäsiumsulfid n
c. **tetrasulphide** Cs_2S_4 Zäsiumtetrasulfid n
c. **tetroxide** Cs_2O_4 Zäsiumtetroxid n
c. **tribromide** $CsBr_3$ Zäsiumtribromid n
c. **triiodide** CsJ_3 Zäsiumtrijodid n
c. **trioxide** Cs_2O_3 Zäsiumtrioxid n
c. **trisulphide** Cs_2S_3 Zäsiumtrisulfid n
caffeic acid $(HO)_2C_6H_3CH=CHCOOH$ Kaffeesäure f, Kaffeinsäure f, 3,4-Dihydroxyzimtsäure f, 3-(3,4-Dihydroxyphenyl)propensäure f
caffeine Koffein n, Kaffein n, 1,3,7-Trimethylxanthin n
c. **and sodium benzoate** Koffein-Natriumbenzoat n (Gemisch aus Koffein und Natriumbenzoat)
c. **and sodium salicylate** Koffein-Natriumsalizylat n (Gemisch aus Koffein und Natriumsalizylat)
c.-**free** koffeinfrei
caffeinic koffeinhaltig
caffeoylquinic acid Kaffeesäureester m der Chinasäure
cage (Krist) Käfig m; Seiher m (an der Seiherpresse)
c. **compound** Käfigeinschlußverbindung f, Käfigverbindung f, Klathratverbindung f, Klathrat n, Clathrat n
c. **disintegrator (mill)** Schlagkorbmühle f, Desintegrator f (i.e.S.), Desintegratormühle f (i.e.S.), Schleudermühle f
c. **press** Seiherpresse f
c. **wall** (Krist) Käfigwand f
c. **zone refining** Cage-zone-refining n (Abart des tiegelfreien Zonenreinigens)
caged ring system endozyklisches System n
cairngorm [stone] (Min) Cairngormstone m, Rauchquarz m (Silizium(IV)-oxid)
cajeput oil (Pharm) Kajeputöl n (von Melaleuca leucadendron L.)
cajeputole Eukalyptol n, 1,8-Zineol n, Zineol n
cake / to [zusammen]backen
cake Kuchen m; Filterkuchen m; Preßkuchen m; Spinnkuchen m, Zentrifugenwickel m; Ölkuchen m; Stück n, Seifenstück n
c. **capacity** Filterkuchenleistung f
c. **compressor** Kuchenquetschapparat m
c. **discharge** Kuchenabnahme f, Kuchenaustrag m
c. **filter** Rückstandsfilter n, Schlammfilter n
c.-**filter operation** s. c. filtration
c. **filtration** Schlammfiltration f, Scheidefiltration f
c. **ice** Formeneis n, Blockeis n
c. **margarine** Backmargarine f, Bäckermargarine f

c. mill Ölkuchenbrecher *m*
c. moisture Kuchenfeuchte *f*, Kuchenfeuchtigkeit *f*, Feuchtigkeitsgehalt *m* des Kuchens
c. operation s. c. filtration
c. resistance Kuchenwiderstand *m*
c. thickness Kuchendicke *f*, Kuchenstärke *f*, Kuchenhöhe *f*
c. thickness detector (sensing device) Kuchendickefühler *m*; Kuchendickenmeßeinrichtung *f*
caking backend *(z.B. Kohle)*
caking Backen *n (z.B. der Kohle)*, Anbacken *n*, Festbacken *n*, Zusammenbacken *n*, Sintern *n*; *(Text)* Balligwerden *n*, Zusammenbacken *n*, Zusammenballen *n*
c. coal backende Kohle *f*; Backkohle *f*
c. index Backfähigkeitszahl *f*, Backzahl *f*, BZ
c. power Backvermögen *n*, Backfähigkeit *f*
c. properties Backeigenschaften *fpl*, backende Eigenschaften *fpl*, Backverhalten *n*
calabar bean Kalabarbohne *f*, Gottesurteil[s]bohne *f*, Gottesgerichtsbohne *f*, Physostigma venenosum Balf.
calabash curare Kalebassenkurare *n*
calaite *(Min)* Kallait *m*, Türkis *m (Aluminiumkupfer(II)-oktahydroxidtetraorthophosphat)*
calamine *(Min)* Kalamin *m*, *(veraltet für)* Hemimorphit *m (Zinkdihydroxidpyrosilikat)*; *(Min)* Kalamin *m*, *(veraltet für)* Smithsonit *m (Zinkkarbonat)*; *(Min)* Kalamin *m*, *(veraltet für)* Hydrozinkit *m (Zinktrihydroxidkarbonat)*; *(Pharm)* Galmei[stein] *m*
calandria Heizkammer *f (des Vertikalrohrverdampfers)*
c. evaporator Vertikalrohrverdampfer *m* mit Innenheizkammer, Robert-Verdampfer *m*
calaverite *(Min)* Calaverit *m (Gold(III)-tellurid)*
calc-alkali[c] rock Alkalikalkgestein *n*, Kalkalkaligestein *n*, Gestein *n* der Alkalikalkreihe (Kalkalkalireihe)
c.-spar *(Min)* Kalkspat *m*, Doppelspat *m*, Kalzit *m (Kalziumkarbonat)*
calcareous kalkartig, kalkig, kalkhaltig, kalkreich, Kalk...
c. crust *(Geol)* Kalkkruste *f*
c. mud *(Geol)* Seekreide *f*, Wiesenkalk *m*, Wiesenmergel *m*; Kalksilt *m*, Kalkmudde *f*
c. rock *(Geol)* Kalkgestein *n*
c. sinter *(Geol)* Kalksinter *m*
c. soil Kalkboden *m*
c. tufa Kalktuff *m*
calciferol Kalziferol *n*, Calciferol *n (Vitamin D_2)*
calcimeter Kalzimeter *n*
calcimine Leimfarbe *f*
calcination Kalzinieren *n*, Kalzinierung *f*, Kalzination *f*, Brennen *n*; Rösten *n*, Röstung *f*; Glühaufschluß *m*
calcine / to kalzinieren, brennen; rösten; glühen
calcine Kalzinationsprodukt *n*; Röstgut *n*, abgeröstetes Gut (Material) *n*; Abbrand *m*

c. collecting tank Abbrandsammelbehälter *m*
c. cooler Abbrandkühler *m*
c. discharge Röstgutaustrag *m*
c.-discharge outlet Abbrandauslauf *m*
calcined baryta BaO Bariumoxid *n*
c. borax $Na_2B_4O_7$ kalzinierter (gebrannter) Borax *m*, wasserfreies Natriumtetraborat *n*
c. magnesium oxide MgO kalzinierte (gebrannte) Magnesia *f*, Magnesiumoxid *n*
c. phosphate Glühphosphat *n (Düngemittel)*
c. soda Na_2CO_3 kalzinierte (kristallwasserfreie) Soda *f*, wasserfreies Natriumkarbonat *n*
calciner Kalzinierofen *m*; Röstofen *m*
calcining Kalzinieren *n*, Kalzinierung *f*, Kalzination *f*, Brennen *n*; Rösten *n*, Röstung *f*; Glühaufschluß *m*
c. compartment s. c. zone
c. kiln Kalzinierofen *m*; Röstofen *m*
c. zone Kalzinierzone *f*, Brennzone *f*
calciphilous kalkliebend, kalziphil *(Pflanzen)*
calciphobous kalkfliehend *(Pflanzen)*
calcite *(Min)* Kalzit *m*, Kalkspat *m*, Doppelspat *m (Kalziumkarbonat)*
calcitic limestone *(Kalkstein mit nur geringen Beimengungen von Magnesiumkarbonat)*
calcium Ca Kalzium *n*
c. acetate $Ca(CH_3COO)_2$ Kalziumazetat *n*
c. acid sulphite s. c. hydrogen sulphite
c. alginate fibre Kalziumalginatfaser *f*; Kalziumalginatfaserstoff *m*
c. aluminate $Ca(AlO_2)_2$ Kalziumaluminat *n*
c.-aluminium silicate $CaO \cdot Al_2O_3 \cdot 2SiO_2$ Kalziumdialumodisilikat *n*
c.-ammonium arsenate NH_4CaAsO_4 Ammoniumkalziumarsenat *n*
c.-ammonium phosphate NH_4CaPO_4 Ammoniumkalziumphosphat *n*
c. arsenate s. c. orthoarsenate
c. arsenide Ca_3As_2 Kalziumarsenid *n*
c.-base acid (liquor) *(Pap)* Kalziumbisulfitkochsäure *f*
c. bicarbonate s. c. hydrogen carbonate
c. bichromate s. c. dichromate
c. biphosphate $Ca(H_2PO_4)_2$ Kalziumdihydrogenorthophosphat *n*, Kalziumdihydrogenphosphat *n*, Monokalziumdihydrogenphosphat *n*, Monokalziumphosphat *n*
c. bisulphide s. c. hydrogen sulphide
c. bisulphite $Ca(HSO_3)_2$ Kalziumhydrogensulfit *n*, *(Tech noch)* Kalziumbisulfit *n*
c. bisulphite cooking liquor *(Pap)* Kalziumbisulfitkochsäure *f*
c. boride CaB_6 Kalziumborid *n*
c. bromate $Ca(BrO_3)_2$ Kalziumbromat *n*
c. bromide $CaBr_2$ Kalziumbromid *n*
c. carbide CaC_2 Kalziumkarbid *n*, Kalziumazetylid *n*, Karbid *n*
c. carbonate $CaCO_3$ Kalziumkarbonat *n*
c. chelate Kalziumchelat *n*
c. chlorate $Ca(ClO_3)_2$ Kalziumchlorat *n*

c. chloride $CaCl_2$ Kalziumchlorid *n*
c. chloride tube Kalziumchloridrohr *n*, Chlorkalziumröhrchen *n* *(Absorptionsgefäß)*
c. chromate $CaCrO_4$ Kalziumchromat *n*
c. cyanamide $CaCN_2$ Kalziumzyanamid *n*
c. cyanide $Ca(CN)_2$ Kalziumzyanid *n*
c. dichromate $CaCr_2O_7$ Kalziumdichromat(VI) *n*
c. dioxide *s.* c. peroxide
c. dithionate CaS_2O_6 Kalziumdithionat *n*
c. ferricyanide $Ca_3[Fe(CN)_6]_2$ Kalziumhexazyanoferrat(III) *n*
c. ferrocyanide $Ca_2[Fe(CN)_6]$ Kalziumhexazyanoferrat(II) *n*
c. fluoride CaF_2 Kalziumfluorid *n*
c. fluorosilicate $Ca[SiF_6]$ Kalziumhexafluorosilikat *n*
c. gluconate Kalziumglukonat *n*
c.-hard kalziumhart *(Wasser)*
c. hardness Kalkhärte *f (des Wassers)*
c. hydrate *s.* c. hydroxide
c. hydride CaH_2 Kalziumhydrid *n*
c. hydrogen carbonate $Ca(HCO_3)_2$ Kalziumhydrogenkarbonat *n*
c. hydrogen phosphate $CaHPO_4$ Kalziumhydrogenphosphat *n*
c. hydrogen sulphide $Ca(SH)_2$ Kalziumhydrogensulfid *n*, *(Gerb noch)* Kalziumsulfhydrat *n*
c. hydrogen sulphite $Ca(HSO_3)_2$ Kalziumhydrogensulfit *n*
c. hydrophosphate *s.* c. hydrogen phosphate
c. hydrosulphide *s.* c.hydrogen sulphide
c. hydroxide $Ca(OH)_2$ Kalziumhydroxid *n*
c. hypochlorite $Ca(ClO)_2$ Kalziumhypochlorit *n*
c. hypochlorite bleach liquor *(Pap)* Kalziumhypochlorit-Bleichlauge *f*
c. hypophosphate $Ca_2P_2O_6$ Kalziumhypophosphat *n*
c. hypophosphite $Ca(H_2PO_2)_2$ Kalziumhypophosphit *n*
c. hyposulphite *s.* c. thiosulphate
c. iodate $Ca(JO_3)_2$ Kalziumjodat *n*
c. iodide CaJ_2 Kalziumjodid *n*
c. lake Kalziumlack *m*
c. level Kalziumspiegel *m (des Blutes)*
c. lignosulphonate Kalziumligninsulfonat *n*, Kalziumlignosulfonat *n*
c. metaborate $Ca(BO_2)_2$ Kalziummetaborat *n*
c. metaplumbate $CaPbO_3$ Kalziumtrioxoplumbat(IV) *n*, Kalziummetaplumbat(IV) *n*
c. metasilicate $CaSiO_3$ Kalziumtrioxosilikat *n*, Kalziummetasilikat *n*
c. molybdate $CaMoO_4$ Kalziummolybdat(VI) *n*
c. nitrate $Ca(NO_3)_2$ Kalziumnitrat *n*
c.-nitrate-urea $Ca(NO_3)_2 \cdot 4CO(NH_2)_2$ Harnstoff-Kalziumnitrat *n*, Harnstoffkalksalpeter *m*
c. nitride Ca_3N_2 Kalziumnitrid *n*
c. nitrite $Ca(NO_2)_2$ Kalziumnitrit *n*
c. orthoarsenate $Ca_3(AsO_4)_2$ Kalziumorthoarsenat(V) *n*, Kalziumarsenat(V) *n*, Trikalziumorthoarsenat(V) *n*, Trikalziumarsenat(V) *n*

c. orthophosphite $CaHPO_3$ Dikalziumorthophosphit *n*
c. orthoplumbate Ca_2PbO_4 Kalziumtetroxoplumbat(IV) *n*, Kalziumorthoplumbat *n*
c. orthosilicate Ca_2SiO_4 Kalziumtetroxosilikat *n*, Kalziumorthosilikat *n*
c. orthotungstate *s.* c. tungstate
c. oxide CaO Kalziumoxid *n*, Ätzkalk *m*, gebrannter Kalk *m*
c. oxychloride *s.* c. hypochlorite
c. pectate Kalziumpektat *n*
c. perchlorate $Ca(ClO_4)_2$ Kalziumperchlorat *n*
c. permanganate $Ca(MnO_4)_2$ Kalziumpermanganat *n*, Kalziummanganat(VII) *n*
c. peroxide CaO_2 Kalziumperoxid *n*
c. phosphide Ca_3P_2 Kalziumphosphid *n*
c. plumbite $CaPbO_2$ Kalziumplumbit *n*
c.-potassium sulphate $K_2Ca(SO_4)_2$ Kaliumkalziumsulfat *n*
c. pyroborate *s.* c. tetraborate
c. pyrophosphate $Ca_2P_2O_7$ Kalziumdiphosphat *n*, Dikalziumdiphosphat *n*, Kalziumpyrophosphat *n*
c. rhodanate (rhodanide) *s.* c. thiocyanate
c. selenate $CaSeO_4$ Kalziumselenat *n*
c. silicofluoride *s.* c. fluorosilicate
c. soap Kalziumseife *f*
c. stearate Kalziumstearat *n*
c. sulphate $CaSO_4$ Kalziumsulfat *n*
c. sulphide CaS Kalziumsulfid *n*
c. sulphite $CaSO_3$ Kalziumsulfit *n*
c. sulphocyanate *s.* c. thiocyanate
c. sulphydrate *s.* c. hydrogen sulphide
c. superoxide *s.* c. peroxide
c. tetraborate CaB_4O_7 Kalziumtetraborat *n*
c. thiocyanate $Ca(SCN)_2$ Kalziumthiozyanat *n*, Kalziumrhodanid *n*
c. thiosulphate CaS_2O_3 Kalziumthiosulfat *n*
c. tungstate $CaWO_4$ Kalziumwolframat *n*
c. wolframate *s.* c. tungstate
calciumsodium borate *(Min)* Boronatrokalzit *m*, Natroborokalzit *m*, *(veraltet für)* Ulexit *m (Kalziumnatriumpentaborat)*
caledonite *(Min)* Kaledonit *m*
calender / to kalandern, kalandrieren, *(Pap auch)* glätten, satinieren, *(Plast auch)* auswalzen
calender Kalander *m*, Glätt[walzen]werk *n*, Walzenglättwerk *n*, Glättmaschine *f*, Maschinenglättwerk *n*, Maschinenkalander *m*, *(Pap auch)* Papierkalander *m*, Trockenglättwerk *n*, Satinierkalander *m*, Satinier[walz]werk *n*
c. bowl Kalanderwalze *f*, Glättwerkswalze *f*
c. colour *(Pap)* Farbstoff *m* für Kalanderfärbung
c. colouring *(Pap)* Kalanderfärbung *f*, Oberflächenfärbung *f* im Kalander
c. cuts Satinierfalten *fpl*, Quetschfalten *fpl* *(Papierfehler)*
c. department Kalandersaal *m*
c. frame *(Pap)* Kalanderständer *m*
c. grain *(Gum)* Kalandereffekt *m*

c. machine *s.* calender
c. operator Kalanderführer *m*
c. roll Kalanderwalze *f*, Glättwerkswalze *f*
c. roll paper Kalanderwalzenpapier *n*
c. section *s.* calender
c. shrinkage *(Gum)* Kalanderschrumpfung *f*
c. sizing *(Pap)* Kalanderleimung *f*
c. solution *(Pap)* Farblösung *f* für Kalanderfärbung
c. spots Satinierflecken *mpl (Papierfehler)*
c. stack *(Pap)* Kalander[walzen]satz *m, (i.w.S.) s.* calender
c. staining *s. c.* colouring
calendered paper kalandriertes (satiniertes) Papier *n*
c. sheet kalandrierte Folie *f, (Gum)* Kalanderplatte *f*
calendering Kalandrieren *n*, Kalandern *n, (Pap auch)* Glätten *n*, Glättung *f*, Satinieren *n*, Satinierung *f*, Satinage *f, (Plast auch)* Auswalzen *n*
calf paper Kalbslederpapier *n*
calibrate / to *(Meßgeräte)* eichen, kalibrieren, graduieren, mit genauer Einteilung versehen
calibrated to contain auf Einguß geeicht (graduiert)
c. to deliver auf Ablauf geeicht (graduiert)
c. to jet bis zur Spitze geteilt *(Meßpipetten)*
calibrating machine Kalibriermaschine *f*
c. plot Eichkurve *f*
c. solution Eichlösung *f*
calibration Eichen *n*, Eichung *f*, Kalibrieren *n*, Kalibrierung *f*
c. curve Eichkurve *f*
caliche Caliche *m, (roher)* Chilesalpeter *m; (Geol)* Kalkkruste *f*
California redwood Mammutbaum *m*, Sequoia sempervirens (D. Don)Endl.
californium Cf Kalifornium *n*
caliper Taster *m*, Mikrometerschraube *f; (Pap)* Dicke *f*
calk / to [ab]dichten
calking compound Dichtungsmasse *f*
c. lead *(Weichblei mit 99,7 % Pb)*
callochrome *(Min)* Kallochrom *m, (veraltet für)* Krokoit *m (Blei(II)-chromat)*
calomel Hg_2Cl_2 Kalomel *n*, Quecksilber(I)-chlorid *n*
c. electrode Kalomelelektrode *f*
caloric kalorisch, Kalorien...
c. radiation Wärmeausstrahlung *f*, Wärmestrahlung *f*, thermische Strahlung *f*
calorie Kalorie *f*, cal
c. content Kaloriengehalt *m*
calorific wärmeerzeugend, Wärme...
of high c. value heizwertreich, heizkräftig, mit hohem Heizwert
of low c. value heizwertarm, mit niedrigem Heizwert
c. capacity Wärmekapazität *f*
c. effect Heizeffekt *m*

c. power Heizkraft *f; s. c.* value
c. radiation Wärmeausstrahlung *f*, Wärmestrahlung *f*, thermische Strahlung *f*
c. value Heizwert *m; (Lebm)* kalorischer Wert *m*, Kalorie[n]wert *m*
calorigenic wärmeerzeugend
calorimeter Kalorimeter *n*
c. bomb *s.* calorimetric bomb
c. fluid Kalorimeterflüssigkeit *f*
c. vessel Kalorimetergefäß *n*
calorimetric kalorimetrisch
c. bomb Kalorimeterbombe *f*, kalorimetrische Bombe *f*, Bombenkalorimeter *n*, Verbrennungsbombe *f*
c. liquid Kalorimeterflüssigkeit *f*
c. vessel *s.* calorimeter vessel
calorimetrical kalorimetrisch
calorimetry Kalorimetrie *f*
calorize / to *(Met)* kalorisieren; *(z.B. Leinöl)* durch Hitze eindicken
calorized linseed oil Leinöl-Standöl *n*
calorizing *(Met)* Kalorisieren *n*, Kalorisierung *f (Veredelungsverfahren)*
calotype process *(Foto)* Kalotypie *f*, Talbotypie *f*
calx *s.* calcium oxide
calycanine Kalykanin *n (Alkaloid)*
calycanthidine Kalykanthidin *n (Alkaloid)*
calycanthine Kalykanthin *n (Alkaloid)*
camber / to wölben, biegen, krümmen; *(Walzen, Profile)* bombieren, ballig bearbeiten
camber Wölbung *f*, Krümmung *f*, Abrundung *f*; Bombage *f (bei Walzen und Profilen)*
cambering Bombieren *n (von Walzen und Profilen)*
cambium *(Bot)* Kambium *n*
cambogia Gummigutt *n (von Garcinia-Arten, vorwiegend G. hanburgi Hook. f.)*
camel hair Kamelhaar *n*, Kamelwolle *f*
camelback *(Gum)* roher Laufstreifen *m*, Rohlaufstreifen *m (zur Runderneuerung)*
c. compound *(Gum)* Rohlaufstreifenmischung *f (für Runderneuerungen)*
Camelia metal *(Legierung aus Cu, Pb, Zn, Sn und Fe mit Kupfer als Grundmetall)*
cameline oil Leindotteröl *n*, Dotteröl *n (aus den Samen von Camelina sativa Crantz)*
Cameroon cardamom Kamerun-Kardamom *m(n) (von Aframomum hanburgi Schum.)*
C. copal *(Kopalharz von Copaifera demeusii Harms)*
camomile 1. Kamille *f*, Matricaria L.; Kamillenblüten *fpl;* 2. Hundskamille *f*, Anthemis L.
Camota rubber *(Kautschuk von Sapium taburu Ule)*
camouflage of a trace element Tarnung *f* eines Spurenelements
Campeachy (Campechy) wood Kampescheholz *n*, Campecheholz *n*, Blauholz *n*, Blutholz *n (von Haematoxylum campechianum L.)*
camphane Bornan *n*, Kamphan *n*, 1,7,7-Trimethyl-bizyklo-[1,2,2]-heptan *n*
camphanol *(veraltet für)* borneol

camphene Kamphen *n*, 2,2-Dimethyl-3-methylen-norbornan *n* *(bizyklischer Terpenkohlenwasserstoff)*

camphol *(veraltet für)* borneol

campholic acid $C_5H_5(CH_3)_4COOH$ Kampholsäure *f*, 1,2,2,3-Tetramethylzyklopentankarbonsäure *f*

camphonanic acid Kamphonansäure *f*, 1,2,2-Trimethylzyklopentankarbonsäure-(1) *f*

camphor Kampfer *m*, Kamphanon-(2) *n*, dl-1,7,7-Trimethylbizyklo-[2.2.1]-heptanon-(2) *n*; *s.* natural camphor

  c. liniment *s.* camphorated oil

  c. method Kampfermethode *f* [nach Rast] *(zur Bestimmung der relativen Molekülmasse)*

  c. oil Kampferöl *n* *(von Cinnamomum camphora (L.)Sieb.)*

camphoraceous kampferartig

α-camphoram[id]ic acid α-Kamphoramsäure *f*, 3-Karbamyl-1,2,2-trimethylzyklopentankarbonsäure *f*, Kampfersäure-3-monoamid *n*

camphorated oil Kampferliniment *n*

camphoric acid Kampfersäure *f*, *cis*-1,2,2-Trimethyl-1,3-zyklopentandikarbonsäure *f*

  c. anhydride Kampfersäureanhydrid *n*

camphoronic acid $(CH_3)_2C(COOH)C(CH_3)(COOH)CH_2COOH$ Kamphoronsäure *f*, 2,3-Dimethyl-1,2,3-butantrikarbonsäure *f*

Campredon index Backfähigkeitszahl (Backzahl) *f* nach Campredon

Camps reaction Camps-Reaktion *f (Hydroxychinolin-Ringschluß)*

campylite *(Min)* Kampylit *m*

camwood Camholz *n*, Gabanholz *n*, Cabanholz *n*, Camba[l]holz *n.* afrikanisches Rotholz *n (von Baphia nitida Afzelius)*, *(i.w.S.) s.* barwood

can / **to** eindosen, in Büchsen konservieren (einmachen)

can 1. Dose *f*, Büchse *f*, Kanne *f*, Kanister *m*; Konservendose *f*, Konservenbüchse *f*; 2. Walze *f*, Zylinder *m (des Walzenbandtrockners)*; 3. *(Text)* Kanne *f*, Spinnkanne *f*, Topf *m*

  c. buckling Bombage (Aufbauchung, Auftreibung) *f* von Konservendosen

  c. dryer *(Pap, Text)* Trockenwalze *f*, Trockenzylinder *m*, Zylindertrockner *m*, ZylinderTrockenpartie *f*, Walzenbandtrockner *m*, Trommeltrockner *m*

  c. for starter culture *(Lebm)* Muttersäurekulturgefäß *n*

  c. ice Formeneis *n*, Blockeis *n*

Canada balsam Kanadabalsam *m*, Kanadaterpentin *n(m)*, kanadisches (kanadischer) Terpentin *n(m) (von Abies balsamea (L.)Mill.)*

  C. pitch kanadisches Pech *n (meist von Tsuga canadensis (L.)Carr.)*

  C. turpentine Kanadaterpentin *n(m)*, kanadisches (kanadischer) Terpentin *n(m)*, Kanadabalsam *m (von Abies balsamea (L.)Mill.)*

Canadian asbestos *(Min)* Chrysotil *m*, Faserserpentin *m*

canadol Kanadol *n (leichte Benzinfraktion)*

canaigre *(Gerb)* Canaigre *n (Wurzeln von Rumex hymenosepalus Torr.)*

canal Kanal *m*

  c. dryer Kanaltrockner *m*, Tunneltrockner *m*, Tunneltrockenanlage *f*

  c. ray Kanalstrahl *m*, positiver Strahl *m*

  c.-ray analysis Kanalstrahlanalyse *f*

cananga oil Kanangaöl *n (von Cananga odorata (Lam.)Hook. f. et Thoms.)*

cancer-producing krebserzeugend, krebserregend, krebsauslösend, karzinogen, kanzerogen

cancrinite *(Min)* Cancrinit *m*

candela Candela *f*, cd *(Einheit der Lichtstärke)*

candelilla wax Kandelillawachs *n (von Pedilanthus pavonis (Klotzsch et Gcke.)Boiss.)*

candelite *s.* cannel coal

candle coal *s.* cannel coal

  c. filter Filterkerze *f*

  c. paper Kerzenpapier *n*

candlenut oil *(Samenöl von Aleurites moluccana (L.)Willd.)*

candy *(Am)* Zuckerwerk *n*, Süßwaren *fpl*, Süßigkeiten *fpl*; *s.* c. sugar

  c. depositing machine Konfektgießmaschine *f*, Gießmaschine *f* für Zuckerwaren

  c. sugar Kandis[zucker] *m*, Zuckerkant *m*, Zuckerkand[is] *m*, Kandelzucker *m*

cane 1. Rohr *n*, Zuckerrohr *n*; 2. Glasstab *m*

  c. juice *(Zucker)* Rohrsaft *m*

  c. molasses Rohrmelasse *f*, Zuckerrohrmelasse *f*

  c. sugar $C_{12}H_{22}O_{11}$ Rohrzucker *m*, Sa[c]charose *f*

  c. sugar factory Rohrzuckerfabrik *f*

  c. ware *(Ker)* Gelbware *f*

  c. wax Zuckerrohrwachs *n*

canella oil *(ätherisches Öl der Kaneelrinde von Canella alba Murr.)*

canescine Kaneszin *n*, Deserpidin *n*, 11-Desmethoxyreserpin *n*

cannabic oil Hanfsamenöl *n*, Hanföl *n*

canned eingedost, in Büchsen konserviert (eingemacht), Dosen..., Büchsen...

  c. beer Dosenbier *n*, Büchsenbier *n*

  c. butter Dosenbutter *f*

  c. cheese Dosenkäse *m*

  c. food Dosenkonserven *fpl*

  c. foods industry Konservenindustrie *f*

  c. meat Büchsenfleisch *n*

  c. milk Dosenmilch *f*, Büchsenmilch *f*

  c.-motor pump Spaltrohrpumpe *f*

  c. preserved fish Fischkonserve *f*

  c. vegetables Dosengemüse *n*, Gemüsekonserve *f*

cannel coal Cannelkohle *f*, Kannelkohle *f*, Kännelkohle *f*, Kennelkohle *f*, Candelkohle *f*, Kandelkohle *f*, Kerzenkohle *f*, Blätterkohle *f*, Gasschiefer *m*, Kandelit *m*

canneloid coal Canneloidkohle *f*

cannery Konservenfabrik *f*
canning Konservieren *n*, Konservierung *f*, Haltbarmachung *f*, Eindosen *n*, Konservenherstellung *f*, Konservenfabrikation *f*
  c. industry Konservenindustrie *f*
  c. plant Konservenfabrik *f*
Cannizzaro reaction Cannizzaro-Reaktion *f*, Cannizzarosche Reaktion *f* *(Aldehyd-Dismutation)*
canopy hood Rauchfangdach *n*
canoxinite *(Min) (fälschlich für)* cancrinite
caoutchouc Kautschuk *m(n)*
cap / to überdecken
cap 1. Deckel *m*, Haube *f*, Kappe *f*; Glocke *f (in Destillieranlagen)*; 2. *(Geol)* Deckschicht *f*, Hut *m*
  c.-and-riser assembly *(Destillation)* Glocke *f* mit Dampfkamin
capability of reaction Reaktionsfähigkeit *f*, Reaktionsvermögen *n*, Reaktivität *f*
  c. of swelling Quellbarkeit *f*
capable of being vaporized verdampfungsfähig
  c. of existence existenzfähig
  c. of reacting (reaction) reaktionsfähig, reaktiv
capacitance Kapazität *f*, kapazitiver Widerstand *m*
capacitor Kondensator *m*
capacity Fassungsvermögen *n*, Kapazität *f*; Aufnahmevermögen *n*, Aufnahmefähigkeit *f*; Inhalt *m*, Volumen *n*; Leistungsfähigkeit *f*, Leistungsvermögen *n*, Produktionsvermögen *n*; Durchsatzleistung *f*
  c. for addition Additionsvermögen *n*, Additionsfähigkeit *f*, Additivität *f*
  c. of water evaporation Wasserleistung *f (des Trockners)*
cape aloe *(Pharm)* Kap-Aloe *f (von Aloe ferox Mill.)*
  c. asbestos (blue) Krokydolith *m (Abart der Hornblende)*
  c. gum Kapgummi *n (von Acacia giraffae Willd. und Acacia horrida Willd.)*
capillarity Kapillarität *f*, Kapillarattraktion *f*, Kapillarwirkung *f*, Haarröhrchenwirkung *f*
capillary kapillar, Kapillar..., sehr eng, haarfein
capillary Kapillare *f*, Kapillarrohr *n*, Kapillarröhrchen *n*, Haarröhrchen *n*
  c. action s. capillarity
  c. active kapillaraktiv
  c. activity Kapillaraktivität *f*
  c. analysis Kapillaranalyse *f*
  c. attraction s. capillarity
  c. breaking *(Text)* Kapillarbruch *m*
  c. column Kapillarsäule *f*
  c. condensation Kapillarkondensation *f*, kapillare Kondensation *f*
  c. constant Kapillaritätskonstante *f*, Kapillarkonstante *f*
  c. cuprite *(Min)* Kupferblüte *f*, Chalkotrichit *m (Kupfer(I)-oxid)*
  c. depression Kapillardepression *f*
  c. electrometer Kapillarelektrometer *n*
  c. inactive kapillarinaktiv
  c. pipette Kapillarpipette *f*

  c. pressure Kapillardruck *m*
  c. pyrite *(Min)* Haarkies *m*, Millerit *m (Nikkel(II)-sulfid)*
  c. red oxide of copper s. c. cuprite
  c. rise Kapillaraszension *f*
  c. rise method Kapillarmethode *f*
  c. sorption Kapillarkondensation *f*, kapillare Kondensation *f*
  c. stopcock Kapillar[sperr]hahn *m*
  c. suction Saugwirkung *f* der Kapillaren
  c. tap s. c. stopcock
  c. theory of separation Kapillaritätstheorie *f* der Gastrennung
  c. tube Kapillarrohr *n*, Kapillarröhrchen *n*, Haarröhrchen *n*, Kapillare *f*
  c. tubing Kapillarrohre *npl*, Kapillarröhren *fpl*; *(manchmal auch für)* c. tube
  c. viscosimeter Kapillarviskosimeter *n*
  c. water *(Bodenkunde)* Kapillarwasser *n*
capper, capping machine Kappenverschließmaschine *f*, Verschließmaschine *f*, Verschlußmaschine *f*
capraldehyde $CH_3(CH_2)_8CHO$ Kaprinaldehyd *m*, Dekanal *n*
capramide $CH_3(CH_2)_8CONH_2$ Kaprinamid *n*, Dekanamid *n*
caprate Kaprinat *n (Salz oder Ester der n-Kaprinsäure)*
capric acid $CH_3(CH_2)_8COOH$ n-Kaprinsäure *f*, n-Dezylsäure *f*, n-Dekansäure *f*
  c. aldehyde s. capraldehyde
  c. amide s. capramide
  c. anhydride $(C_9H_{19}CO)_2O$ Kaprinsäureanhydrid *n*, Dekansäureanhydrid *n*
caprilic acid s. caprylic acid
caproaldehyde $CH_3(CH_2)_4CHO$ n-Kapronaldehyd *m*, n-Hexanal *n*, n-Hexylaldehyd *m*
caproate Kapronat *n (Salz oder Ester der n-Kapronsäure)*
caproic acid $CH_3(CH_2)_4COOH$ n-Kapronsäure *f*, n-Hexansäure *f*, Hexylsäure *f*
  c. aldehyde s. caproaldehyde
  c. anhydride $(C_5H_{11}CO)_2O$ Kapronsäureanhydrid *n*
  c. nitrile s. capronitrile
caprolactam Kaprolaktam *n*
capronate s. caproate
capronic acid s. caproic acid
capronitrile $CH_3(CH_2)_4CN$ Kaprononitril *n*
capryl alcohol s. caprylic alcohol
  c. aldehyde s. caprylic aldehyde
caprylate Kaprylat *n (Salz oder Ester der n-Kaprylsäure)*
caprylene C_8H_{16} Okten-(1) *n*
caprylic acid $CH_3(CH_2)_6COOH$ n-Kaprylsäure *f*, n-Oktansäure *f*, Heptankarbonsäure-(1) *f*, Hexylessigsäure *f*
  c. alcohol $CH_3(CH_2)_6CHOH$ n-Kaprylalkohol *m*, n-Oktylalkohol *m*, Oktanol-(1) *n*
  c. aldehyde $CH_3(CH_2)_6CHO$ n-Kaprylaldehyd *m*, n-Oktanal *n*, n-Oktylaldehyd *m*

caprylidene HC≡C(CH₂)₅CH₃ Kapryliden n, n-Hexylazetylen n, 1-Oktin n
caprylonitrile CH₃(CH₂)₆CN Kaprylonitril n
capsule Kapsel f; Abdampfschale f, Abdampftiegel m
capture / to einfangen; abfangen
capture (Kph) Einfang m, Einfangen n (z.B. von Elektronen); (Krist) Abfangen n (z.B. eines Spurenelements)
  c. cross section (Kph) Einfangquerschnitt m, Absorptionsquerschnitt m
car-bottom furnace (kiln) (Ker) Herdwagenofen m
caracurine Karakurin n (Alkaloid)
  caracurine VII Karakurin VII n, Wieland-Gumlich-Aldehyd m
caramel Karamel m, Karamelle f, gebrannter Zucker m, (als Lösung) Zuckercouleur f, Kulör f
  c. flavour Karamelgeschmack m
  c. malt Karamelmalz n
caramelization Karamelisierung f
  c. browning Karamelisationsbräunung f
caramelize/to karamelisieren
caramelized sugar karamelisierter Zucker m, Karamel m, Karamelle f
carane Karan n, 2,2,5-Trimethylbizykloheptan n (ein Terpen)
caraway oil Kümmelöl n (von Carum carvi L.)
carbachol NH₂COOCH₂CH₂N(CH₃)₃Cl Karbachol n, N-[2-Karbamoyloxyäthyl]-trimethylammoniumchlorid n
carbamate H₂NCOOM^I Karbamat n
carbamic acid H₂NCOOH Karbamidsäure f, Amidokohlensäure f, Kohlensäuremonamid n
carbamide NH₂CONH₂ Karbamid n, Kohlensäurediamid n, Harnstoff m
carbamidine HN=C=(NH₂)₂ Guanidin n, Imidoharnstoff m, Iminoharnstoff m, Aminomethanamidin n
carbamonitrile CNNH₂ Zyanamid n
carbamyl urea H₂N·CO·NHCO·NH₂ Biuret n, Allophanamid n
carbanilic acid C₆H₅NHCOOH Karbanilsäure f, Phenylkarbaminsäure f, Phenylkarbamidsäure f
carbanion Karbanion n, Karbeniation n, Karbeniat-Anion n
carbarsone NH₂CONH·C₆H₄·AsO(OH)₂ Karbarson n, 4-Karbamidophenylarsinsäure f
carbazide RNH·NH·CO·NH·NHR' Karbonohydrazid n, Karbazid n, (i.e.S.) H₂N·NH·CO·NH·NH₂ Karbonohydrazid n
carbazole Karbazol n, Dibenzopyrrol n, Diphenylenimin n, Diphenylenimid n, Diphenylimid n
  c. dye Karbazolfarbstoff m
  c. indophenol Karbazolindophenol n, Indophenolkarbazol n
  c. ring Karbazolring m
carbazone R·N=N·CO·NH·NHR' Karbazon n, (i.e.S.) H₂N·NH·CO·N=NH Karbazon n
carbazotate s. picrate
carbazotic acid C₆H₂(NO₂)₃OH Pikrinsäure f, 2,4,6-Trinitrophenol n

carbazylic acid Amidin n
carbene Karben n, Methylen n, (i.e.S.) CH₂ Karben n, Methylen n
carbeniate formula Karbeniatform[el] f
  c. ion Karbeniation n, Karbeniat-Anion n, Karbanion n
  c. limiting formula Karbeniatgrenzform[el] f
  c. structure Karbeniatstruktur f
carbenium structure Karbeniumstruktur f
carbide Karbid n, (i.e.S.) CaC₂ Kalziumkarbid n, Kalziumazetylid n, Karbid n
  c. furnace Karbidofen m
  c. plant Karbidfabrik f
  c. precipitation Karbidausscheidung f
  c. skeleton Karbidskelett n, Karbidgerüst n
carbinol Karbinol n (verzweigtkettiges Derivat des Methanols), (i.e.S.) CH₃OH Karbinol n, (veraltet für) Methylalkohol m
carbocerine (Min) Karbozerit m, (veraltet für) Lanthanit f
carbocyanine (Farb) Karbozyanin n
carbocyclic karbozyklisch, isozyklisch
  c. compound karbozyklische (isozyklische) Verbindung f, ringförmige Kohlenstoffverbindung f
carbodinicotinic acid C₅H₂N(COOH)₃ Pyridin-2,3-5-trikarbonsäure f
carbohydrate Kohlenhydrat n, Kohlehydrat n, Sa[c]charid n
  c. chemistry Kohlenhydratchemie f, Kohlehydratchemie f
  c. metabolism Kohlenhydratstoffwechsel m, Kohlehydratstoffwechsel m
carbolic acid C₆H₅OH Karbolsäure f, Phenol n, Hydroxybenzol n
  c. oil Karbolöl n
  c. paper karbolisiertes Papier n
carboline Karbolin n (Alkaloid), (i.w.S.) Karbolinbase f, Karbolinalkaloid n
carbomycin Karbomyzin n (Antibiotikum)
carbon 1. C Kohlenstoff m; 2. Kohle[elektrode] f
  c. arc Kohlelichtbogen m, Kohlebogen m
  c. balance (Landw) Kohlenstoffbilanz f
  c. base paper Karbonrohpapier n, Karbonrohseide f
  c. bisulphide CS₂ Kohlendisulfid n, Kohlenstoffdisulfid n, Schwefelkohlenstoff m
  c. black Ruß m, Kohleschwarz n, Rußschwarz n, Carbon-Black n
  c. black batch s. c. black masterbatch
  c. black compound s. c. black stock
  c. black dispersion Rußdispergierung f, Rußverteilung f
  c. black-filled, c. black-loaded (Gum) rußhaltig, rußgefüllt, mit Ruß gefüllt, Ruß enthaltend
  c. black loading (Gum) Rußzusatz m, Rußzuschlag m, Füllung f mit Ruß, Rußdosierung f
  c. black masterbatch (Gum) Polymer-Ruß-Batch m, Vormischung f aus Polymer und Ruß, Rußbatch m, Rußvormischung f

c. black-pigmented s. c. black-loaded
c. black plant Rußfabrik f
c. black-reinforced s. c. black-loaded
c. black stock (Gum) rußgefüllte (rußhaltige, mit Ruß gefüllte, Ruß enthaltende) Mischung f
c. block (Met) Kohlenstoffblock m
c. boat Kohleschiffchen n
c. brick Kohlenstoffstein m
c.-carbon bond Kohlenstoff-Kohlenstoff-Bindung f, C-C-Bindung f
c.-carbon cross-link C-C-Vernetzungsstelle f, C-C-Verknüpfungsstelle f
c.-carbon cross-linking C-C-Vernetzung f, Verkettung f über C-C-Bindungen, Vernetzung f über C-C-Verknüpfung
c.-carbon double bond Kohlenstoff-Doppelbindung f, doppelte Kohlenstoffbindung f, C=C-Doppelbindung f, C=C-Bindung f
c.-carbon single bond Kohlenstoff-Einfachbindung f, einfache Kohlenstoffbindung f, C-C-Einfachbindung f, [einfache] C-C-Bindung f
c.-carbon triple bond Kohlenstoff-Dreifachbindung f, dreifache Kohlenstoffbindung f, C≡C-Dreifachbindung f, C≡C-Bindung f
c.-carburizing steel, c. casehardening steel Kohlenstoffeinsatzstahl m
c. chain Kohlenstoffkette f
c. compound Kohlenstoffverbindung f
c. copy paper Durchschlagpapier n
c. crucible Kohletiegel m
c. deposit Kohlenstoffablagerung f
c. dioxide CO_2 Kohlendioxid n, (ungenau oft) Kohlensäure f
c. dioxide assimilation Kohlendioxidassimilation f
c. dioxide carrier Kohlendioxidüberträger m
c. dioxide fire extinguisher Kohlendioxidlöscher m, Kohlensäurelöscher m, CO_2-Löscher m
c. dioxide ice (snow) Kohlendioxidschnee m, Kohlensäureschnee m, Trockeneis n
c. disulphide CS_2 Kohlendisulfid n, Kohlenstoffdisulfid n, Schwefelkohlenstoff m
c. double bond s. c.-carbon double bond
c. electrode Kohleelektrode f
c. filament lamp Kohle[n]fadenlampe f
c. framework Kohlenstoffgerüst n
c.-functional karbofunktionell, organofunktionell
c.-functional silicone karbofunktionelles Silikon n; karbofunktionelle Organosiliziumverbindung f
c. gel (Gum) Rußgel n
c.-halogen bond Kohlenstoff-Halogen-Bindung f
c. hexachloride CCl_3CCl_3 Hexachloräthan n, Perchloräthan n
c.-hydrogen bond Kohlenstoff-Wasserstoff-Bindung f, C-H-Bindung f
c. lining Kohlenstoffzustellung f, C-Futter n (eines Hochofens)
c. monosulphide CS Kohlenstoffmonosulfid n, Kohlenmonosulfid n

c. monoxide CO Kohlenmonoxid n, Kohlenoxid n
c.-monoxide free kohlenmonoxidfrei
c. monoxy hemoglobin Kohlenoxidhämoglobin n
c.-nitrogen skeleton Kohlenstoff-Stickstoff-Gerüst n
c. oxybromide $COBr_2$ Kohlenoxidbromid n, Karbonylbromid n, Bromphosgen n
c. oxychloride $COCl_2$ Kohlenoxidchlorid n, Karbonylchlorid n, Phosgen n
c. oxysulphide COS Kohlenoxidsulfid n, Karbonylsulfid n
c. paper Kohle[n]papier n, Karbonpapier n; Durchschreibepapier n
c. residue Kohlenstoffrückstand m, Verkokungsrückstand m, Koksrückstand m (beim Verkokungstest nach Conradson)
c.-residue test Verkokungstest m, Carbon-Test m
c.-residue value Verkokungswert m, Verkokungszahl f, Carbon-Wert m
c. ring Kohlenstoffring m
c. selenosulphide CSeS Thiokarbonylselenid n, Kohlenstoffselenidsulfid n, Kohlenselenidsulfid n
c. silicide CSi_2 Karbonsilizid n
c.-silicon bond Kohlenstoff-Silizium-Bindung f, Silizium-Kohlenstoff-Bindung f
c. single bond s. c.-carbon single bond
c. skeleton Kohlenstoffgerüst n
c. spot Kohlefleck m (Papierfehler)
c. steel Kohlenstoffstahl m, C-Stahl m
c. suboxide C_3O_2 Kohlensuboxid n, Karbodikarbonyl n
c. subsulphide C_3S_2 Kohlensubsulfid n
3-c.-tautomerism Drei-Kohlenstoff-Tautomerie f
c. tellurosulphide CSTe Thiokarbonyltellurid n, Kohlenstofftelluridsulfid n, Kohlenstoffsulfidtellurid n
c. tet s. c. tetrachloride
c. tetrabromide CBr_4 Kohlenstofftetrabromid n, Tetrabromkohlenstoff m, Tetrabrommethan n
c. tetrachloride CCl_4 Kohlenstofftetrachlorid n, Tetra[chlorkohlenstoff] m, Tetrachlormethan n
c. thionyl chloride $CSCl_2$ Thiokarbonyldichlorid n, Thiophosgen n
c. thionyl perchloride CCl_3SCl Thiokarbonyltetrachlorid n, Perchlormethylmerkaptan n
c.-to-carbon cross-link s. c.-carbon cross-link
c. to carbon double bond s. c.-carbon double bond
c. to carbon single bond s. c.-carbon single bond
c. treatment (Lebm) Kohleschönung f
carbonaceous kohlenstoffhaltig, kohlenstoffreich; kohlenhaltig, kohlig
c. coal kohlenstoffreiche Kohle f
c. sediment Kohlengestein n
c. shale Brandschiefer m
carbonado Karbonado m, schwarzer Diamant m
carbonate / to 1. karbonisieren, (Getränke) mit CO_2

imprägnieren (sättigen); 2. *(Zuckerrübensaft)*
saturieren; 3. *s.* to carboxylate; 4. in ein Karbonat
umwandeln
carbonate Karbonat *n*
  c. hardness Karbonathärte *f*, temporäre
(schwindende, vorübergehende) Härte *f*
  c. of lime *s.* calcium carbonate
  c. of potash *s.* potassium carbonate
  c. of soda *s.* sodium carbonate
  c. rock Karbonatgestein *n*
  c. sediment Karbonatsediment *n*
  c. sludge *(Pap)* Kalkschlamm *m*
carbonated beverage karbonisiertes (mit CO_2
imprägniertes *oder* gesättigtes) Getränk *n*
  c. water kohlensäurehaltiges Wasser *n*, Sodawasser *n, (i.e.S.)* Selterswasser *n*
  c. wine Perlwein *m*
carbonating tank *(Zucker)* Saturationsgefäß *n*
(Saturationspfanne *f*) für CO_2
  c. tower Karbonisierungskolonne *f*, Karbonisator
m, Fällkolonne *f*, Fällturm *m*
carbonation 1. *(Lebm)* Karbonisieren *n*, Imprägnieren *n (von Getränken)* mit CO_2; 2.
Saturation *f (Entkalkung des Zuckerrübensaftes
mittels CO_2)*; 3. Überführung (Umwandlung) *f* in
Karbonat; 4. *s.* carboxylation
  c. juice *(Zucker)* Schlammsaft *m*; Saturationssaft
m
  c. mud *(Zucker)* Saturationsschlamm *m*
  c. pan *(Zucker)* Saturationspfanne *f*
  c. process *(Zucker)* Kalk-Kohlensäure-Verfahren
n, Scheidesaturation *f*
  c. scum *(Zucker)* Saturationsschlamm *m*
  c. tank *(Zucker)* Saturationsgefäß *n* (Saturationspfanne *f*) für CO_2
carbonatization Umwandlung *f* in ein Karbonat;
(Geoch) Karbonatisierung *f*, Karbonatisation *f*
carbonato chelate Karbonatochelat *n*
  c. complex Karbonatokomplex *m*
carbonic Kohlenstoff...; Kohlensäure...; Kohlendioxid...
  c. acid H_2CO_3 Kohlensäure *f*
  c. acid gas CO_2 Kohlendioxid[gas] *n*, Kohlen-
stoffdioxid *n*
  c. anhydrase Kohlensäureanhydra[ta]se *f*,
Karboanhydra[ta]se *f*, Karbonatanhydratase *f*
  c. paper *s.* carbon paper
carboniferous coal Karbonkohle *f*, karbonische
Kohle *f*
carbonification *(Geol)* Inkohlung *f*, Kohlenreifung
f, Karbonifikation *f*
  c. series Inkohlungsreihe *f*
carbonitride / to karbonitrieren, gaszyanieren,
trockenzyanieren
carbonitriding Karbonitrieren *n*, Karbonitrierung *f*,
Gaszyanieren *n*, Trockenzyanieren *n*
carbonium ion Karboniumion *n*, Karbeniumion *n*,
Karbeniumkation *n*
  c. salt Karboniumsalz *n*, Karbeniumsalz *n*
carbonization 1. Umwandlung *f* in Kohlenstoff;

Kohlenstoffanreicherung *f*, C-Anreicherung *f*; 2.
(Geoch) *s.* carbonification; 3. Verkokung *f*
(i.w.S.), Verkoken *n (i.w.S)*, Entgasung *f*, Entgasen *n*; Verkokung *f (i.e.S.)*, Verkoken *n (i.e.S.)*,
Hochtemperaturverkokung *f*, HT.-Verkokung *f*,
Normalverkokung *f*, Vollverkokung *f*, Hochtemperaturentgasung *f*, HT.-Entgasung *f*; Schwelung *f*, Schwelen *n*, Verschwelung *f*, Verschwelen
n, Tieftemperaturverkokung *f*, Tieftemperaturentgasung *f*; 4. Verkohlung *f*, Verkohlen *n*,
Entgasung *f*, Entgasen *n (z. B. von Holz)*; 5.
Karbonisieren *n*, Karbonisation *f*, Auskohlen *n*,
Entkohlen *n (von Wolle)*
  c. chamber Entgasungsschacht *m*, Schwelschacht *m (eines Schwelgenerators)*
  c. gas Verkokungsgas *n*, Entgasungsgas *n*;
Schwelgas *n*
  c. plant *s.* carbonizing plant
  c. process Entgasungsverfahren *n*, Verkokungsverfahren *n*, Kokungsprozeß *m*, Koksprozeß *m*;
Schwelverfahren *n*
  c. zone Schwelzone *f (eines Schwelgenerators)*
carbonize / to 1. in Kohlenstoff umwandeln; mit
Kohlenstoff anreichern; 2. inkohlen *(Pflanzen-
reste in Kohle umwandeln)*; 3. verkoken, entgasen; [ver]schwelen; 4. verkohlen *(z.B. Holz)*;
(Wolle) karbonisieren, auskohlen, entkletten
carbonized briquette Schwelbrikett *n*
  c. brown-coal briquette Braunkohlenschwelbri-
kett *n*
  c. paper *s.* carbon paper
carbonizer Schweler *m*, Schwelofen *m*
carbonizing *s.* carbonization
  c. assistant *(Text)* Karbonisierhilfsmittel *n*,
Karbonisiernetzmittel *n*
  c. bench Verkokungsbatterie *f*, Koksofenbatterie
f
  c. industry Kokereiindustrie *f*; Schwelindustrie *f*
  c. plant Koksofenanlage *f*, Kokerei[anlage] *f*,
Verkokungsanlage *f*; Schwelanlage *f*, Schwelwerk *n*, Schwelerei *f*
carbonyl Karbonyl *n*, Metallkarbonyl *n*; *s.* c. group
  c. bromide $COBr_2$ Karbonylbromid *n*, Kohlenoxidbromid *n*, Bromphosgen *n*
  c. carbon Kohlenstoff *m* (C-Atom *n*) der
Karbonylgruppe (C=O-Gruppe *f*)
  c. chloride $COCl_2$ Karbonylchlorid *n*, Kohlenoxidchlorid *n*, Phosgen *n*
  c. compound Karbonylverbindung *f*
  c. diamide NH_2CONH_2 Kohlensäurediamid *n*,
Karbamid *n*, Harnstoff *m*
  c. double bond C=O-Doppelbindung *f*
  c. group =C=O-Karbonylgruppe *f*, C=O-Gruppe
f, CO-Gruppe *f*, Keto[n]gruppe *f (i.w.S.)*
  c. iron Karbonyleisen *n*
  c. method (process) Karbonylverfahren *n*, Kar-
bonylprozeß *m*
  c. radical *s.* c. group
  c. sulphide COS Karbonylsulfid *n*, Kohlenoxidsulfid *n*

carbonyttrine *(Min)* Karbonyttrin *m*, *(veraltet für)* Tengerit *m*
carboxazylic acid Karbonsäureamid *n*
carboxy-modified nitrile rubber *s.* carboxynitrile rubber
carboxybenzene C_6H_5COOH Benzoesäure *f*, Benzolkarbonsäure *f*
carboxyl [group] Karboxylgruppe *f*, Karboxyl *n*, COOH-Gruppe *f*
carboxylase Karboxylase *f*, Pyruvatdekarboxylase *f*
carboxylate / to karboxylieren *(in organische Verbindungen die Karboxylgruppe -COOH einführen)*
carboxylate Karboxylat *n* *(Salz oder Ester einer Karbonsäure)*
  c. complex Karboxylatkomplex *m*
  c. ion Karboxylat-Ion *n*
carboxylated rubber *s.* carboxylic rubber
carboxylation Karboxylierung *f*
carboxylic acid Karbonsäure *f*
  c. acid amide Karbonsäureamid *n*
  c. rubber Karboxylatkautschuk *m*, Karboxylkautschuk *m*
carboxymethylcellulose Karboxymethylzellulose *f*, Zelluloseglykolsäure *f*
carboxynitrile rubber karboxylierter (karboxylgruppenhaltiger) Nitrilkautschuk *m*
carboxypeptidase Karboxypeptidase *f*
carboy Korbflasche *f*, Glasballon *m*, Ballon *m* *(besonders für Säuren)*
  c. emptier Ballonentleerer *m*
  c. pourer Ballonausgießer *m*
  c. tipper Ballonkipper *m*
carburant Vergaserkraftstoff *m*, Vergasertreibstoff *m*, VK
carburet / to *(Gase)* karburieren
carburetor *s.* carburettor
carburetted water gas karburiertes Wassergas *n*
carburetter *s.* carburettor
carburetting fuel Vergaserkraftstoff *m*, Vergasertreibstoff *m*, VK
carburettor Karburator *m* *(zum Karburieren von Gasen)*; Vergaser *m* *(eines Motors)*
  c. icing Vergaservereisung *f*
carburization *(Met)* Aufkohlen *n*, Aufkohlung *f*, Kohlen *n*, Kohlung *f*, Zementieren *n*, Zementierung *f*, Zementation *f*, Einsetzen *n*
  c. depth *s.* carburizing depth
carburize / to *(Met)* [auf]kohlen, zementieren, einsetzen
carburized case *(Met)* aufgekohlte (gekohlte, zementierte, eingesetzte) Randschicht (Randzone, Oberflächenschicht, Oberflächenzone, Schicht) *f*, aufgekohlter Rand *m*, Aufkohlungsschicht *f*, Aufkohlungszone *f*, Zementationsschicht *f*, Einsatzschicht *f*, Einsatzzone *f*, Einsatz *m*
  c. steel aufgekohlter (zementierter) Stahl *m*
carburizer *(Met)* Aufkohlungsmittel *n*, Aufkohl-

mittel *n*, Kohlungsmittel *n*, Zementationsmittel *n*, Einsatzmittel *n*, [auf]kohlendes Mittel *n*
carburizing *s.* carburization
  c. action *(Met)* Aufkohlungswirkung *f*, Kohlungswirkung *f*, Zementationswirkung *f*, zementierende Wirkung *f*
  c. agent *s.* c. medium
  c. bath *(Met)* Aufkohlungsbad *n*, Kohlungsbad *n*, aufkohlendes Salzbad *n*, Zementierbad *n*, Zementationsbad *n*, Einsatzbad *n*
  c. box *(Met)* Zementationskasten *m*, Zementierkasten *m*, Einsatzkasten *m*
  c. compound *s.* c. medium
  c. depth *(Met)* Aufkohlungstiefe *f*, Kohlungstiefe *f*, Zementationstiefe *f*, Einsatztiefe *f*, Tiefe *f* der Einsatzschicht
  c. furnace *(Met)* Aufkohlungsofen *m*, Kohlungsofen *m*, Zementier[ungs]ofen *m*
  c. gas *(Met)* aufkohlendes Gas *n*, Aufkohlungsgas *n*, Kohlungsgas *n*, Zementationsgas *n*
  c. heat *(Met)* Aufkohlungshitze *f*, Einsatzhitze *f*
  c. level *(Met)* Kohlungspegel *m*, Kohlenstoffpotential *n*
  c. material (medium) *(Met)* Aufkohlungsmittel *n*, Aufkohlmittel *n*, Zementationsmittel *n*, Einsatzmittel *n*, [auf]kohlendes Mittel *n*
  c. mixture *(Met)* Aufkohlungsgemisch *n*, Einsatzgemisch *n*, Zementationsgemisch *n*
  c. oven *s.* c. furnace
  c. pot *(Met)* Einsatztopf *m*
  c. powder *(Met)* Aufkohlungspulver *n*, Kohlungspulver *n*, Zementationspulver *n*, Einsatzpulver *n*
  c. process *(Met)* Aufkohlungsverfahren *n*, Einsatzverfahren *n*; Aufkohlungsvorgang *m*
  c. rate *(Met)* Aufkohlungsgeschwindigkeit *f*
  c. salt *(Met)* Aufkohlungssalz *n*, Kohlungssalz *n*
  c. steel Stahl *m* für Einsatzhärtung, Einsatzstahl *m*
  c. temperature *(Met)* Aufkohlungstemperatur *f*, Kohlungstemperatur *f*, Zementationstemperatur *f*
  c. time *(Met)* Aufkohlungszeit *f*, Kohlungszeit *f*, Zementationszeit *f*
carbyl oxime $C\equiv N \cdot OH$ Knallsäure *f*, Fulminsäure *f*
carbylamine $R \cdot N \equiv C$ Isonitril *n*, Isozyanid *n*, Karbylamin *n*
  c. test Karbylaminreaktion *f*, Isonitrilreaktion *f* *(zum Nachweis primärer Aminogruppen)*
carbylic acids *(Sammelname für Karbonsäuren, Karbonsäureamide und Amidine)*
carcase *s.* carcass
carcass *(Gum)* 1. Karkasse *f*, Unterbau *m* *(eines Reifens)*; 2. Kern *m*
  c. compound *s.* c. stock
  c. ply *(Gum)* Karkaßlage *f*, Karkasseneinlage *f*, Kordlage *f*, Gewebelage *f*
  c. rubber Karkassengummi *m*
  c. stock *(Gum)* Karkaßmischung *f*
carcinogen Karzinogen *n*, karzinogener (kanzerogener) Stoff *m*, karzinogene (kanzerogene) Substanz *f*

carcinogenic karzinogen, kanzerogen, krebserzeugend, krebserregend, krebsauslösend

card paper *(veraltet für)* cardboard

cardamom, cardamon Kardamom *m(n)* *(von Elettaria- und Amomum-Arten)*

c. oil Kardamomöl *n* *(von Elettaria cardamomum (L.)White et Maton)*

cardboard Kartonpapier *n*, Halbkarton *m*; Karton *m*; Pappe *f*

c. box Faltschachtel *f*, Pappschachtel *f*, Pappkarton *m*

c. culture disk Nährkartonscheibe *f*

cardiac glykoside Herzglykosid *n*, herzaktives Glykosid *n*

c. tonic *s.* cardiotonic

cardioactive *s.* cardiotonic

cardioid condenser Kardioidkondensor *m*

cardiotonic herzwirksam, herzaktiv

cardiotonic Kardiotonikum *n* *(die Herzmuskulatur kräftigendes Mittel)*

carene Δ^4-Karen *n*, Pinonen *n*, 3,7,7-Trimethyl-bizyklo-[0,1,4]-2-hepten *n*; Δ^3- Karen *n*, Isodipren *n*, 3,7,7-Trimethylbizyklo-[0,1,4]-3-hepten *n*

Carius furnace Schießofen *m*, Bombenofen *m*

Carius method Carius-Methode *f*, Carius-Aufschluß *m*

Carius tube Bombenrohr (Einschmelzrohr, Schießrohr) *n* nach Carius, Carius-Rohr *n*

carminative *(Pharm)* Karminativum *n*, blähungstreibendes Mittel *n*

carminazarin *(Farb)* Karminazarin *n*

carmine-red karmesinrot, karminrot, karmoisinrot, feurigrot

carminic acid Karminsäure *f*

carnallite *(Min)* Karnallit *m*

carnauba wax Karnaubawachs *n* *(von der Palme Copernicia prunifera (Muell.)H.E. Moore)*

carnelian *(Min)* Karneol *m*

Carnot cycle Carnotscher Kreisprozeß *m*, Carnot-Prozeß *m*

Carnot theorem Carnotsches Prinzip *n*

carob galls *(Farb)* Judäakaroben *fpl*, Pistaziengallen *fpl*, Terpentingallen *fpl* *(von Pistacia terebinthus L.)*

carob[-seed] gum, caroban Johannisbrotgummi *n* *(von Ceratonia siliqua L.)*

Caro's acid H_2SO_5 Carosche Säure *f*, Peroxomonoschwefelsäure *f*

carotene Karotin *n*

c. extract Karotinextrakt *m*

carotin *s.* carotene

carotinoid Karotinoid *n*

carpaine Karpain *n* *(Alkaloid von Carica papaya L.)*

carpamic acid
$HNC_4H_7 \cdot C(CH_3)(OH) \cdot (CH_2)_7 \cdot COOH$ Karpamsäure *f*

Carpathian turpentine Karpatenbalsam *m* *(von Pinus cembra L.)*

carpet felt Abdeckpapier *n*, Unterlagspapier *n*; Abdeckpappe *f*, Unterlagspappe *f*

c. grade (quality) Teppichtyp *m* *(Ausführungsart von Chemiefaserstoffen)*

carpholite *(Min)* Karpholit *m*, Strohstein *m*

carrageen, carragheen Karrag[h]een *n*, Carrageen *n*, Irländisches Moos *n* *(Meeresalgen Chondrus crispus und Gigartina mamillosa)*

carrier 1. Träger *m*, Überträger *m*, Trägerstoff *m*, Trägersubstanz *f*, Trägermittel *n*, Carrier *m*, *(Kch auch)* Trägerelement *n*; Trägergas *n*; 2. *(Text)* Färbebeschleuniger *m*

c. bed Speichergestein *n*, Trägergestein *n*

c. distillation Trägerstoffdestillation *f*

c. electrode Trägerelektrode *f*

c.-free trägerfrei

c. gas Trägergas *n*, Schleppgas *n* *(Gaschromatografie)*

c. gas stream Trägergasstrom *m*

c. liquid Trägerflüssigkeit *f*

c. of heat Wärmeüberträger *m*, Wärmeträger *m*, Wärmeübertragungsmittel *n*, Wärmemittel *n*

c. storage Trägerspeicherung *f*

c. sublimation Trägergas-Sublimation *f*

c. thread *(Text)* Trägerfaden *m*, Stützfaden *m*

carry out / to *(ein Experiment)* ausführen, durchführen

c. over *(eine Substanz in die andere)* verschleppen; *(Dämpfe)* überleiten, übertreiben

carrying roll *(Pap)* Tragwalze *f*

Cartagena ipecac[uanha] Kartagena-Brechwurzel *f*, Nikaragua-Brechwurzel *f*, Panama-Brechwurzel *f* *(von Cephaelis acuminata Karsten)*

carthamus oil Safloröl *n*, Carthamusöl *n* *(aus den Samen von Carthamus tinctorius L.)*

carton Faltschachtel *f*, Pappschachtel *f*, Pappkarton *m*; Faltschachtelkarton *m*, Schachtelkarton *m*, Kartonagenpappe *f*

catridge Patrone *f*

c. filter Patronenfilter *n*

c. heater Einsatzheizkörper *m*, Einschubheizkörper *m*, Patronenheizkörper *m*, Heizpatrone *f*

carvacrol $CH_3(OH)C_6H_3CH(CH_3)_2$ Karvakrol *n*, 2-Hydroxy-4-isopropyl-1-methylbenzol *n*, 2-p-Zymenol *n*, Zymophenol *n*

carvene d-Limonen *n*, d-1-Methyl-4-isopropyl-zyklohexadien-(1,8) *n*, Zitren *n*, Karven *n*, Hesperiden *n*

carvol *s.* d-carvone

d-carvone d-Karvon *n* *(monozyklisches Terpenketon)*

caryophil oil Nelkenöl *n*, Gewürznelkenöl *n* *(von Syzygium aromaticum (L.)Merr. et L.M. Perry)*

caryophyllic acid $CH_2=CHCH_2 \cdot C_6H_3(OCH_3)OH$ Eugenol *n*, 4-Hydroxy-3-methoxy-1-allylbenzol *n*

Casale process Casale-Verfahren *n* *(Ammoniaksynthese)*

cascade burner Kaskadenofen *m*

c. cooler Rieselkühler *m*

c. dryer Kaskadentrockner *m*; Brüdenschlottrockner *m*

c. evaporator Kaskadenverdampfer *m*, Mehr-

stufenverdampfer *m*, Mehrkörperverdampfer *m*
c. grate Treppenrost *m*
c. method Kaskadenmethode *f (Gasverflüssigung)*
c. mill Kaskadenmühle *f*
c. system Kaskadenschaltung *f*
Cascade fir Purpurtanne *f*, Abies amabilis (Loud.)Farbes.
cascara sagrada [bark] *(Pharm)* Cascarasagrada-Rinde *f*, Kaskararinde *f*, Sagrada[rinde] *f*, Amerikanische Faulbaumrinde *f (von Rhamnus purshiana DC.)*
cascarilla bark Kaskarillarinde *f (von Croton eluteria Benn.)*
case 1. Kasten *m*, Kiste *f*; Gehäuse *f*; 2. *(Met)* Randschicht *f*, Randzone *f*, Oberflächenschicht *f*, Oberflächenzone *f*, Rand *m*; aufgekohlte (gekohlte, zementierte, eingesetzte) Randschicht (Schicht) *f*, aufgekohlter Rand *m*, Aufkohlungsschicht *f*, Aufkohlungszone *f*, Zementationsschicht *f*, Einsatzschicht *f*, Einsatzzone *f*, Einsatz *m*; gehärtete Randschicht (Einsatzschicht) *f*, Härteschicht *f*; 3. *(Gum)* Unterbau *m*, Karkasse *f (eines Reifens)*; 4. *(Ker)* Boms *m (Einlagekörper, Brennhilfsmittel)*
c. carburization *s.* c. carburizing
c.-carburize / to *(Met)* rand[auf]kohlen, in der Randschicht (Randzone) aufkohlen (zementieren)
c. carburizing *(Met)* Randaufkohlung *f*, Aufkohlen *n* in der Randschicht (Randzone)
c. depth *(Met)* Einsatztiefe *f*, Tiefe *f* der Einsatzschicht, Zementationstiefe *f*, Aufkohlungstiefe *f*, Aufkohlungstiefe *f*; Einsatzhärtungstiefe *f*, Einsatzhärtetiefe *f*
c. hardenability *(Met)* Einsatzhärtbarkeit *f*
c. hardness *(Met)* Einsatzhärte *f*, Härte *f* der [gehärteten] Randschicht
c. thickness *(Met)* Einsatzschichtdicke *f*
caseate / to verkäsen, käsig werden
caseharden / to 1. *(Met)* einsatzhärten, im Einsatz[verfahren] härten; 2. *(Gerb)* totgerben
casehardened glass vorgespanntes Glas *n*, *(i.e.S.)* Einscheibensicherheitsglas *n*, Einschichtensicherheitsglas *n*
c. steel einsatzgehärteter Stahl *m*, Einsatzstahl *m*
casehardening 1. *(Met)* Einsatzhärten *n*, Einsatzhärtung *f*, Zementationshärten *n*, Zementationshärtung *f*; Oberflächenhärten *n (i.w.S.)*, Oberflächenhärtung *f (i.w.S.)*; 2. *(Gerb)* Totgerben *n*, Totgerbung *f*
c. bath *(Met)* Einsatzhärtebad *n*
c. box *(Met)* Einsatzhärtekasten *m*, Einsatzkasten *m*, Härtekasten *m*
c. carburizer *(Met)* Aufkohlungsmittel *n*, Aufkohlmittel *n*, Kohlungsmittel *n*, Zementationsmittel *n*, Einsatzmittel *n*, [auf]kohlendes Mittel *n*
c. compound *s.* c. material
c. furnace *(Met)* Einsatzhärteofen *m*

c. material *(Met)* Einsatzmittel *n*
c. pot *(Met)* Einsatztopf *m*
c. powder *(Met)* Härtepulver *n*
c. process *(Met)* Einsatzhärtungsverfahren *n*, Einsatzhärteverfahren *n*; Einsatzhärtungsvorgang *m*
c. steel Stahl *m* für Einsatzhärtung, Einsatzstahl *m*
casein Kasein *n*
c. coating colour *(Gerb)* Kaseindeckfarbe *f*
c. dope *(Text)* Kaseinspinnlösung *f*
c. fibre Kaseinfaser *f*; Kaseinfaserstoff *m*, KA
c. glue Kaseinleim *m*
c. paint Kaseinfarbe *f*
c. plastic Kunsthorn *n*
c. sodium *(Pharm)* Kaseinnatrium *n*
c. staple *(Text)* Kaseinfaser *f*
caseinate Kaseinat *n*
caseinic acid Kaseinsäure *f*, Diaminotrihydroxydodekansäure *f*
caseinogen Kaseinogen *n*
Casella benzol thermometer Benzolthermometer *n* nach Casella
casemaking drum *(Gum)* Wickeltrommel *f*, Reifenwickeltrommel *f*
casement wall Oberbauseitenwand *f (im Glasschmelzofen)*
caseous käseartig, käsig
cashawa gum *s.* cashew gum
cashew gum Acajugummi *n*, Acajou *n*, Anacardiumgummi *n (von Anacardium occidentale L.)*
c. nut Kaschunuß *f (von Anacardium occidentale L.)*
casing 1. Gehäuse *n*; Behälter *m*, Behältnis *n*; 2. *(Erdöl)* Verrohren *n*, Verrohrung *f*, Futterrohreinbau *m*; Verrohrung *f*, Futterrohre *npl*, Casing *n*; 3. *(Gum)* Unterbau *m*, Karkasse *f (eines Reifens)*
c. compound *s.* c. stock
c. effect Rampenbildung *f*, Wellenbildung *f (Glasfehler)*
c. head Rohrkopf *m*
c.-head gas Naturgas *n*, Bohrlochkopfgas *n*, Casinghead-Gas *n*
c.-head gasoline (spirit) Rohrkopfbenzin *n*, Natur[gas]benzin *n*, Rohrkopfgasolin *n*, Casinghead-Benzin *n*
c. ply *(Gum)* Karkaßlage *f*, Karkasseneinlage *f*, Kordlage *f*, Gewebelage *f*
c. rubber Karkassengummi *m*
c. stock *(Gum)* Karkaßmischung *f*
c. wall Oberbauseitenwand *f (im Glasschmelzofen)*
cask Faß *n*, Tonne *f*, Bottich *m*
c. beer Faßbier *n*
c. racker, c.-racking machine Faßabfüllmaschine *f*, Faßabfüller *m*
c. washer, c.-washing machine Faßwaschmaschine *f*, Faßreinigungsmaschine *f*
c. wine Faßwein *m*
cassava Maniok *m*, Kassave *f*, Kassavestrauch *m*, Manihot esculenta Crantz

casse

casse *(Wein)* Bruch *m*, Trübung *f*
Cassel green Kasseler (Rosenstiehls, Böttgers) Grün *n*, Mangangrün *n* *(Bariummanganat)*
Cassella's acid Cassellasche Säure *f* *(2-Naphthol-7-sulfonsäure oder 2-Naphtholamin-4,8-disulfonsäure)*
casserole tongs Kasserollenzange *f*
cassia oil chinesisches Zimtöl *n*, Kassiaöl *n* *(von Cinnamomum aromaticum Nees)*
cassiopeium Cp Cassiopeium *n*, *(veraltet für)* Lu Lutetium *n*
cassiterite *(Min)* Kassiterit *m*, Zinnstein *m* *(Zinn(IV)-oxid)*
Cassius purple Cassiusscher (Kassiusscher) Goldpurpur *m* *(purpurfarbene kolloidale Goldlösung)*
Cassumunar ginger Blockzitwer *m*, Gelber Zitwer *m*, Zingiber cassumunar Roxb.
cast / to gießen; *(Text)* nuancieren, ausmustern
cast *(Text)* Nuance *f*, Farbschattierung *f*, Farbton *m*
 c. alloy Gußlegierung *f*
 c. coating *(Pap)* Streichgießverfahren *n*, Kontaktverfahren *n*
 c. film *(Plast)* gegossene Folie *f*, Gießfolie *f*, Gießfilm *m*
 c.-in-place concrete,c.-in-situ concrete Ortbeton *m*
 c.-iron gußeisern
 c. iron Gußeisen *n*
 c.-iron armouring Gußeisenarmierung *f*
 c.-iron borings Gußeisenspäne *mpl*, Gußspäne *mpl*
 c.-iron pipe Graugußrohr *n*
 c.-iron retort Gußeisenretorte *f*, gußeiserne Retorte *f*
 c. moulding Gießen *n*; gegossenes Formteil *n*
 c. pipe gegossenes Rohr *n*, Gußrohr *n*
 c. resin Gießharz *n*, Vergußharz *n*, Schmelzharz *n*
 c. sheet gegossene Folie *f*, Gießfolie *f*, Gießfilm *m*
 c. steel Gußstahl *m*, Stahlformguß *m*, Stahlguß *m*
 c.-steel filter press Gußstahlfilterpresse *f*
 c.-steel high-pressure autoclave Gußstahl-Hochdruckautoklav *m*
castability Gießbarkeit *f*, Gießfähigkeit *f*
caster Gießer *m*
castile [soap] kastilianische Seife *f*, *(i.w.S.)* spanische Seife *f*
casting Gießen *n*, Guß *m*, Vergießen *n*; Gußstück *n*, Gußteil *n*, Formgußstück *n*, Gießling *m*
 c. alloy Gußlegierung *f*
 c. area Gießfläche *f*
 c. copper Raffinatkupfer *n*, Garkupfer *n*
 c. hole Gießloch *n*
 c. line *s.* **c. seam**
 c. machine Gießmaschine *f*; *(Glas)* Walzmaschine *f*
 c. mould Gießform *f*

 c. process Gießverfahren *n*
 c. resin Gießharz *n*, Vergußharz *n*, Schmelzharz *n*
 c. scrap *(Ker)* Abfallschlicker *m*
 c. seam Gießnaht *f*, Gußnaht *f*, Formfugennaht *f*
 c. skin *(Ker)* Gießhaut *f*
 c. slip *(Ker)* Gießschlicker *m*, Gießmasse *f*
 c. spot (stain) *(Ker)* Gießfleck *m* *(Fehler)*
 c. table *(Glas)* Gußtisch *m*, Gießtisch *m*, Aufnahmetisch *m*
 c. temperature Gießtemperatur *f*
Castner cell Castner-Zelle *f* *(Elektrolyse)*
castor 1. *(Pharm)* Bibergeil *n*; 2. *(Min)* Kastor[it] *m*, *(veraltet für)* Petalit *m*
 c. cake (meal) *s.* **c. pomace**
 c. oil Rizinusöl *n*, Kastoröl *n*
 c. oil acid $CH_3(CH_2)_5CH(OH)CH_2CH=CH \cdot (CH_2)_7COOH$ Rizinoleinsäure *f*, Rizinolsäure *f*, Rizinus[öl]säure *f*, Hydroxyölsäure *f*
 c. pomace Rizinussaatkuchen *m*, Rizinuspreßkuchen *m*, Rizinusrückstände *mpl*
 c. sugar Kastor[zucker] *m*, Puderzucker *m*, Staubzucker *m*
castorite *(Min)* Kastor[it] *m*, *(veraltet für)* Petalit *m*
cat-crack / to katalytisch kracken (spalten)
cat-cracked gasoline katalytisches Krackbenzin *n*, Cat-Benzin *n*
 c. cracker *s.* **catalyst cracker**
 c. cracking *s.* **catalytic cracking**
 c. eye Blase *f* *(Glasfehler)*
 c.-forming *s.* catforming
 c. gold *(Min)* Katzengold *n* *(angewitterter Biotit)*
 c.-reformed katalytisch reformiert
 c. scratch Kratzer *m* *(Glasfehler)*
 c. silver *(Min)* Katzensilber *n*, Muskovit *m*, Kaliglimmer *m*
catabolism *s.* katabolism
cataclasis *(Geol)* Kataklase *f*
cataclasite *(Geol)* Kataklasit *m*
cataclastic *(Geol)* kataklastisch
catalase Katalase *f* *(Ferment)*
 c. action Katalasewirkung *f*
 c. activity Katalaseaktivität *f*
 c. complex Katalasekomplex *m*
 c. complex with hydrogen peroxide Katalase-H_2O_2-Komplex *m*
 c.-like katalaseähnlich
 c. test Katalaseprobe *f*
catalyse / to *s.* **to catalyze**
catalysis Katalyse *f*
catalyst Katalysator *m*, *(bei heterogener Katalyse auch)* Kontakt[stoff] *m*
 c. bed Katalysatorbett *n*, Kontaktbett *n*, Katalysatorschicht *f*, Kontaktschicht *f*
 c. carrier Katalysatorträger *m*, Kontaktträger *m*
 c. chamber Katalysatorkammer *f*, Kontaktkammer *f*
 c. circulation Katalysatorumlauf *m*, Katalysatorkreislauf *m*, Kontaktumlauf *m*

c. cracker katalytische Krackapparatur (Krackanlage) f, Catcracker m
c. cracking s. catalytic cracking
c. cycle s. c. circulation
c. dust Katalysatorstaub m, Katalysatorpulver n, Kontaktstaub m
c./oil ratio Katalysator-Öl-Verhältnis n, Kontakt-Öl-Verhältnis n, Verhältnis n von Katalysator zu Öleinsatz, Verhältnis n von Kontakt zu Öl
c. poison Katalysatorgift n, Kontaktgift n
c. recycle s. c. circulation
c. removal column Katalysatorstripper m
c. scrubber column Katalysatorauswaschkolonne f
c. slurry Katalysatorschlamm m
c. support Katalysatorträger m, Kontaktträger m
c. surface Katalysatoroberfläche f
c. surge hopper Auffangbehälter m für Katalysator
c.-to-oil ratio s. c./oil ratio
catalytic katalytisch
c. action (activity) katalytische Wirkung (Wirksamkeit) f, Katalysatorwirkung f
c. cracking katalytisches (ionisches) Kracken (Spalten) n, katalytische (ionische) Krackung (Spaltung) f, Katkracken n, Catcracking n, Kracken n am Katalysator, Kracken n über Ionen
c.-cracking process katalytisches Krackverfahren (Spaltverfahren) n, Catcrack-Verfahren n
c. distillation katalytische Destillation f
c. efficiency s. c. action
c. hydrogenation katalytische Hydrierung f
c. oxidation katalytische Oxydation f
c. poison s. catalyst poison
c. polymerization katalytische Polymerisation f
c. reactor Katalyseofen m
c. reforming katalytisches Reformieren (Reformen) n, katalytische Reformierung f, katalytisches Reforming n
c. surface Katalysatoroberfläche f
catalyze / to katalysieren
catalyzed by acid[s] säurekatalysiert
c. by acids and bases säure-basenkatalysiert
c. by base[s] basenkatalysiert
catalyzer s. catalyst
cataphoresis Kataphorese f
cataphoretic kataphoretisch
catapleiite (Min) Katapleit m (Natriumzirkonium(IV)-trisilikat)
catarole process Catarol-Prozeß m (zur Pyrolyse von Erdölfraktionen)
catch / to abfangen, auffangen
catch pot Auffanggefäß n; Schlammfänger m; Filtereinsatz m, Siebeinsatz m
catcher Auffänger m, Fänger m, Fang m, Abscheider m
catechin Katechin n, (i.e.S.) s. catechinic acid
catechinic acid Katechin n, 3,5,7,3',4'-Flavanpentol n, 3,5,7,3',4'-Pentahydroxyflavan n
catechol s. 1. catechin; 2. pyrocatechol

c. monoethyl ether $C_2H_5OC_6H_4OH$ Brenzkatechinmonoäthyläther m, o-Äthoxyphenol n, Guäthol n
c. tan Katechingerbstoff m
catechu [gum] [braunes] Katechu n (Gerbstoffextrakt aus Acacia catechu Willd.); Gambir n, gelbes Katechu n (von Kucaria gambir Roxb.); (i.w.S.) (Sammelname für Gerbstoffextrakte zahlreicher Pflanzenarten)
catechuic acid s. catechinic acid
catelectrode Katode f
catenation compound Catenan n, Catena-Verbindung f
cater-cornered paper Diagonalschnittpapier n, Schrägschnittpapier n
caterpillar-powered scraper Schürfkübelraupe f
catforming Catformen n, Catforming n, katalytisches Reformieren (Reformen) n, katalytische Reformierung f
c. process Catformingverfahren n
cathartic Kathartikum n, Abführmittel n
cathedral glass Kathedralglas n
cathode Katode f
c. compartment Katodenraum m
c. copper Katodenkupfer n
c. dark space Katodendunkelraum m, innerer (Crookesscher, Hittorffscher) Dunkelraum m
c. drop Katodenfall m (Abfall der EMK in Katodennähe)
c. luminescence Katodenlumineszenz f, Katodolumineszenz f
c. ray Katodenstrahl m
c. sputtering Katodenzerstäubung f
cathodic katodisch, Katoden...
c. deposition of metals katodische Metallabscheidung f
c. inhibitor katodischer Inhibitor m
c. overvoltage katodische Überspannung f
c. protection katodischer Schutz m
c. sputtering s. cathode sputtering
cathodoluminescence s. cathode luminescence
catholyte Katolyt m (Elektrolyt des Katodenraumes)
cation Kation n, positives (positiv geladenes) Ion n
c. acid (phys Ch) Kationsäure f
c.-active kation[en]aktiv, kationkapillaraktiv, kationisch, Kationen..., Kation...
c. exchange (Bodenkunde) Kationenaustausch m, Kationenumtausch m
c.-exchange adsorption Kationenaustauschadsorption f
c.-exchange bed Kationenaustauschbett n
c.-exchange capacity Kationenaustauschkapazität f, Kationenumtauschkapazität f, KUK
c.-exchange column Kationenaustausch[er]säule f
c.-exchange resin Kationenaustauscher m auf Kunstharzbasis, Kationenaustausch[er]harz n
c.-exchangeability Kationenaustauschfähigkeit f
c. exchanger Kationenaustauscher m

c. hole s. c. vacancy
c. swarm *(Koll)* Kationenschwarm *m*
c. transference number Kation[en]überführungszahl *f*
c. vacancy Kationenleerstelle *f*, Kationenfehlstelle *f*
cationic kationisch, Kationen..., Kation..., kation[en]aktiv
  c. collector kationischer (kationaktiver) Sammler *m*
  c. exchanger s. cation exchanger
  c. nature kationischer Charakter *m*, Kationcharakter *m*
  c. polymerization kationische Polymerisation *f*
  c. surfactant kationisches (kationaktives) Netzmittel *n*
  c. tenside kationisches Tensid *n*, Kationtensid *n*
cationoid kationoid
  c. dye kationischer Farbstoff *m*
cat's eye *(Min)* Katzenauge *n (Abart des Quarzes und des Chrysoberylls)*
cattle lick (salt) Viehsalz *n*
  c. spraying Viehbesprühung *f*, Spraybehandlung *f* des Viehs
caucho *(Kautschuk von Castilloa ulei Warb. und C. elastica Cerv.); (s.a.)* c. blanco, c. verde
  c. blanco *(Kautschuksorte von Sapium-Arten)*
  c. verde *(Kautschuksorte von Sapium stylare Muell. Arg.)*
  c. virgin rubber Jungfernkautschuk *m (von Sapium thomsoni God.)*
cauliflower appearance blumenkohlähnliches Aussehen *n*, Blumenkohlstruktur *f (des Kokskuchens)*
  c. end Blumenkohlende *n*, Blumenkohlkopf *m (Wandseitenende eines Koksstücks)*
caulk / to [ab]dichten
caulking compound Dichtungsmasse *f*
caustic ätzend, brennend, kaustisch
caustic Alkali *n*, *(meist)* NaOH Natriumhydroxid *n*, Ätznatron *n*; *(Med)* Ätzmittel *n*
  c. alcohol Natriumalkoholat *n*, *(i.e.S.)* C_2H_5ONa Natriumäthylat *n*, Natriumäthoxid *n*
  c. antimony $SbCl_3$ Antimon(III)-chlorid *n*, Antimontrichlorid *n*
  c. arsenic chloride $AsCl_3$ Arsen(III)-chlorid *n*, Arsentrichlorid *n*
  c. baryta $Ba(OH)_2$ Bariumhydroxid *n*, Ätzbaryt *m*
  c. cracking (embrittlement) *(Met)* Laugensprödigkeit *f*, Laugenbrüchigkeit *f (interkristalline Spannungsrißkorrosion)*
  c. extraction [stage] *(Pap)* Alkalibehandlung *f*, Alkaliextraktion *f*, alkalische Wäsche *f*, Alkaliwäsche *f (Bleichstufe)*
  c. hydrosulphite solution Dithionit-Natronlauge *f*
  c. lime Ätzkalk *m*, gebrannter Kalk *m*, Branntkalk *m*, Kalziumoxid *n*; *(manchmal auch für)* hydrated lime
  c. lye of soda Natronlauge *f*

c. potash KOH Ätzkali *n*, Kaliumhydroxid *n*
c. potash solution Kalilauge *f*, Kaliumhydroxidlösung *f*
c. silver Höllenstein *m*, Silbernitrat *n*
c. soda Ätznatron *n*, Natriumhydroxid *n*, *(als Produkt des Kalk-Soda-Verfahrens auch)* kaustische Soda *f*, Sodastein *m*
c.-soda fusion Ätznatronschmelze *f*
c.-soda solution s. c. lye of soda
c.-soda wash *(Erdöl)* Lauge[n]wäsche *f*, Laugenbehandlung *f*, Alkaliwäsche *f*, Laugen *n*, Laugung *f*
c. solution alkalische Lauge *f*
c. stage s. c. extraction [stage]
c. tower *(Pap)* Alkalisierungsturm *m*, Alkaliturm *m*
c. wash s. c.-soda wash
causticize / to kaustifizieren, *(Pap auch)* aussüßen *(Soda oder Pottasche in NaOH bzw. KOH überführen)*; *(Text)* merzerisieren
causticizer s. causticizing tank
causticizing department (plant, room) *(Pap)* Kaustifizieranlage *f*
  c. tank *(Pap)* Kaustifizierbehälter *m*, Kaustifizierbottich *m*
causticproof laugenbeständig, laugenfest
causto[bio]lith *(Geol)* Kaustobiolith *m (brennbare biogene Ablagerung)*
caution label Warnetikett *n*
cava s. kava
cavitation Kavitation *f*, Hohlsog *m*, Hohlraumbildung *f*
cavity Hohlraum *m*, Aushöhlung *f*, *(Plast auch)* Werkzeughohlraum *m*, Werkzeughöhlung *f*, Formhöhlung *f*, Formnest *n*
  c. block *(Plast)* Gesenkblock *m*, Gesenkplatte *f*, Matritze *f*
  c. plate Tüpfelplatte *f*
  c.-preventing agent *(Met)* Lunkerverhütungsmittel *n*
  c. retainer s. c. block
cay-cay fat Cay-Cay-Butter *f (von Irvingia oliveri Pierre)*
Cayenne pepper Cayenne-Pfeffer *m*, Spanischer Pfeffer *m*, Gewürzpaprika *m*
cb. s. cubic
C-C bond C-C-Bindung *f*, Kohlenstoff-Kohlenstoff-Bindung *f*
C-C bond length C-C-Bindungslänge *f*, C-C-Abstand *m*, C-C-Kernabstand *m*
C-C bond rupture Bruch *m* (Sprengung *f*) der C-C-Bindung
C-C cross-link C-C-Vernetzungsstelle *f*, C-C-Verknüpfung *f*
C-C rupture s. C-C bond rupture
CCH s. calcium hypochlorite
C.D. s. current density
CDAA $ClCH_2 \cdot CO \cdot N(CH_2CH=CH_2)_2$ CDAA *n*, 2-Chlor-N,N-diallylacetamid *n (Herbizid)*
CDEC CDEC *n*, 2-Chlor-allyl-N,N-diäthyldithiokarbamat *n (Herbizid)*

CDW *s.* commercial dry basis
CE *(Abk. für)* Chemical Engineer
Ceara rubber Ceara-Kautschuk *m*, Manicoba-Kautschuk *m (von Manihot glaziovi Muell. Arg.)*
CEC, C.E.C. *s.* cation-exchange capacity
C.E.D. *s.* cohesive energy density
cedar camphor *s.* cedrol
  c.-leaf oil Zedernblätteröl *n*, Thujaöl *n (meist von Thuja occidentalis L.)*
  c. lichen *(Farb)* Cetraria juniperina (L.)Ach. *(Flechten-Art)*
  c. nut Zirbelnuß *f (von Pinus cembra L.)*
cedarwood camphor *s.* cedrol
  c. oil Zedernholzöl *n (meist von Cedrus atlantica Manetti)*; Amerikanisches Zedernholzöl *n (von Juniperus-Arten)*
cedrene Zedren *n (trizyklisches Sesquiterpen)*
cedrol Zedrol *n*, Zedernkampfer *m*, Zypressenkampfer *m (Sesquiterpenalkohol)*
Celdecor-Pomilio process *(Pap)* Chloraufschluß *m* nach Pomilio-Celdecor, Celdecor-Pomilio-Verfahren *n*
celery salt Selleriesalz *n*
  c. [seed] oil Sellerieöl *n (aus Samen von Apium graveolens L.)*
celestial blue Himmelblau *n*, Reinblau *n*
celestine, celestite *(Min)* Zölestin *m (Strontiumsulfat)*
cell 1. Zelle *f*, Kammer *f*, Raum *m*; 2. Küvette *f*; 3. *(Ech)* Elektrolyse[n]zelle *f*, Elektrolysierzelle *f*, elektrolytische Zelle *f*, Elektrolyseraum *m*; 4. *(phys Ch)* [galvanische, elektrochemische] Zelle *f*, [galvanisches, elektrochemisches] Element *n*; 5. *(Bio)* Zelle *f*; 6. *(Gum)* Pore *f*, Zelle *f*
  c. constant Widerstandskapazität *f (einer Leitfähigkeitszelle)*
  c. contents *(Bio)* Zellinhalt *m*
  c. control agent *(Plast)* Zellkontrollmittel *n*, Zellsteuermittel *n*
  c. for measurement of conductance Leitfähigkeitsgefäß *n*, Leitfähigkeitszelle *f*
  c.-free fermentation zellfreie Gärung *f*
  c. hormone Zellhormon *n*, Zytohormon *n*
  c. liquor *(Ech)* Zellenflüssigkeit *f*
  c. membrane Zellmembran[e] *f*
  c. reaction Zellenreaktion *f*
  c. reaction charge number Zellenreaktionsladungszahl *f*
  c. sap *(Bio)* Zellsaft *m*, Zellflüssigkeit *f*
  c. structure of wood Holzzell[en]struktur *f*, Holzzell[en]gefüge *n*
  c.-type filter Zellenfilter *n*
  c.-type oven Zellenofen *m*
  c. voltage Zellenspannung *f*
  c. wall *(Bot)* Zellwand *f*
cellar treatment Kellerbehandlung *f*
cellobiose $C_{12}H_{22}O_{11}$ Zellobiose *f (ein Disaccharid)*
cellophane Zellglas *n*
cellotetrose $C_{24}H_{42}O_{21}$ Zellotetraose *f*

cellotriose $C_{18}H_{32}O_{16}$ Zellotriose *f*
cellucotton Zellstoffwatte *f*
cellular zellig, zellenartig, zellenähnlich, zellulär, Zell[en]..., *(manchmal auch)* blasig, porig, Schaum...
  c. board Wellpappe *f*
  c. concrete Zellenbeton *m*, Porenbeton *m*, *(i.e.S.)* Schaumbeton *m*
  c. dolomite Zellendolomit *m*, Zellenkalk *m*, Rauchwacke *f*, Rauhwacke *f*
  c. ebonite zelliger Hartgummi *m*, Zellhartgummi *m*
  c.-expanded concrete *s.* c. concrete
  c. filter Zellenfilter *n*
  c. fluid Zellsaft *m*, Zellflüssigkeit *f*
  c. glass Schaumglas *n*
  c. lime *(Geol)* Zellenkalk *m*
  c. plastic Schaum[kunst]stoff *m*
  c. respiration Zellatmung *f*
  c. rubber poröser Gummi *m*, Zellgummi *m*
  c. structure Zell[en]struktur *f*, Zell[en]gefüge *n*, zell[art]ige Struktur *f*
  c. tissue *(Text)* Zellgewebe *n*
celluloid Zelluloid *n*, Zellhorn *n*
cellulose $(C_6H_{10}O_5)_x$ Zellulose *f*; *(Tech)* Zellstoff *m*
  α-cellulose α-Zellulose *f*
  β-cellulose β-Zellulose *f*
  γ-cellulose γ-Zellulose *f*
  c. acetate Zelluloseazetat *n*, CA, Azetylzellulose *f*
  c. acetate butyrate Zelluloseazetatbutyrat *n*, CAB, Azetylbutyrylzellulose *f*
  c. acetate dope Zelluloseazetatspinnlösung *f*
  c. acetate fibre Zelluloseazetatfaser *f*; Zelluloseazetatfaserstoff *m*, AZ
  c. acetate propionate Zelluloseazetopropionat *n*, CAP
  c. acetate rayon Zelluloseazetatseide *f*, Azetatseide *f*
  c. coal Zellulosekohle *f*
  c. decomposition *(Landw)* Zelluloseabbau *m*
  c. derivative Zelluloseabkömmling *m*, Zellulosederivat *n*
  c. diacetate Zellulosediazetat *n*, Sekundärzelluloseazetat *n*, Hydrozelluloseazetat *n*
  c. ester Zelluloseester *m*
  c. ether Zelluloseäther *m*
  c. fibre Zellulosefaser *f*; Zellulosefaserstoff *m*, CZ
  c. fibril Zellulosefibrille *f*
  c. film Zellglas *n*, Zellhaut *f*, Glashaut *f*
  c. formate Zelluloseformiat *n*
  c. gasket Zellulosedichtung *f*
  c. gum Natrium-Zelluloseglykolat *n*, Natrium-Karboxymethylzellulose *f*
  c. hydrate Zellulosehydrat *n*, Hydratzellulose *f*, regenerierte Zellulose *f*
  c. hydration Hydrolyse *f* der Zellulose
  c. lacquer Zelluloselack *m*
  c. nitrate Zellulosenitrat *n*, Nitratzellulose *f*, Zellulosesalpetersäureester *m*

c. nitrate lacquer Nitro[zellulose]lack *m*
c. propionate Zellulosepropionat *n*, CP
c. triacetate $[C_6H_7O_2(OCOCH_3)_3]_x$ Zellulosetri-azetat *n*, Triazetat *n*
c. triacetate fibre Triazetatfaser *f*; Triazetat-faserstoff *m*, TA
c. trinitrate $[C_6H_7O_5(NO_2)_3]_x$ Zellulosetrinitrat *n*
c. wadding *(Am)* Zellstoffwatte *f*
c. xanthate $[C_6H_9O_4 \cdot O \cdot CSSNa]_x$ Zellulose-xanth[ogen]at *n*
c. xanthic acid Zellulosexanthogensäure *f*
c. xanthogenate *s.* c. xanthate
cellulosic fibre Zellulosefaser *f*; Zellulosefaserstoff *m*, CZ
cellulosics Zelluloseerzeugnisse *npl*, Zellulosede-rivate *npl*
celtium Ct Celtium *n*, *(veraltet für)* Hf Hafnium *n*
cement / to 1. [ver]kitten, [ver]kleben; 2. *(Bau)* zementieren; 3. *(Met)* zementieren, *(durch Dif-fusion)* oberflächenhärten; 4. [aus]zementieren, *(Metall mittels unedleren Metalls aus Lösungen)* ausfällen; 5. *(mit Gummilösung)* einstreichen
c. together zusammenkitten
cement Kleb[e]stoff *m*, Kleber *m*, *(i.e.S.)* Kitt *m*; *(Gum)* Klebelösung *f*, Lösung *f*, Konfektionier-lösung *f*, Zement *m*, Klebezement *m*; *(Bau)* Zement *m*; *(Med)* Zahnzement *m*, Zement *m*; *(Geol)* Bindemittel *n*, Zement *m*
c.-bonded sand Zement[form]sand *m*
c. burning Zementbrennen *n*
c. clinker Zementklinker *m*
c. copper Zementkupfer *n*, Fällkupfer *n*
c. kiln Zement[schacht]ofen *m*
c. mortar Zementmörtel *m*
c. paste Zementleim *m*
c. pat Zementkuchen *m*
cementation 1. Kitten *n*, Verkitten *n*, Kleben *n*, Verkleben *n*; 2. *(Bau)* Zementieren *n*; 3. *(Met)* Zementation *f*, Zementieren *n*, Oberflächen-härten *n* (Oberflächenhärtung *f*) durch Diffusion; 4. *(Met, Geol)* Zementation *f* *(Ausfällen oder Ausfallen eines Metalls aus seiner Lösung durch Einbringen oder Anwesenheit eines unedleren)*; 5. *(Geol)* Verkittung *f*
cemented carbide Sinterkarbid *n*, Sinterhartmetall *n*, Sinterkarbidmetall *n*, Karbidhartmetall *n*, Hartmetall *n*; Zementit *m* *(metallografische Bezeichnung für Eisenkarbid, Fe_3C)*
c. carbide material Hartmetallwerkstoff *m*, Werkstoff *m* auf Hartmetallbasis
c. hard carbide (metal) *s.* c. carbide
c. tube *(Plast)* Rohr *n* mit Klebnaht
c. tungsten carbide gesintertes Wolframkarbid *n*, Wolframsinterkarbid *n*, Wolframkarbidhart-metall *n*
cementing *s.* cementation
c. accelerator *(Bau)* Abbindebeschleuniger *m*
c. agent Bindemittel *n*, *(Geol auch)* Zement *m*
c. box Zementationskasten *m*, Zementierkasten *m*

c. material Bindemittel *n*, Klebstoff *m*
c. medium *(Met)* Zementationsmittel *n*, Einsatz-mittel *n*
c. technique *(Plast)* Klebtechnik *f*
cementite *(Met)* Fe_3C Zementit *m*, Eisenkarbid *n*
cementitious zementartig; verkittend
cementuin Zahnzement *m*, Zement *m*
center / to zentrieren, [ein]mitten
center *(Am)* *s.* centre
centigrade heat unit *(kalorisches Energiemaß, entspricht 453,59 cal)*
c. scale Celsius-Thermometerskale *f*, Celsius-Skale *f*
c. thermal unit *s.* c. heat unit
centigram[me] method Zentigramm-Methode *f* *(Halbmikroanalyse)*
centipoise Zentipoise *n*, cP *(Maßeinheit der dynamischen Viskosität)*
centistoke Zentistokes *n*, cSt *(Maßeinheit der kinematischen Viskosität)*
central atom Zentralatom *n*
c. carbon atom Zentralkohlenstoffatom *n*
c. ion Zentralion *n*
c. knife edge Mittelschneide *f*, Stützschneide *f* *(an Waagen)*
c. metal Zentralmetall *n* *(einer Komplex-verbindung)*
c. pipe Zentralrohr *n*
c. shaft Zentralwelle *f*, zentrale Welle (Rührwelle) *f*, Hauptwelle *f*, Königswelle *f*
c. tube Zentralrohr *n*
centrally located mittelständig
centre Zentrum *n*, Mitte *f*, Mittelpunkt *m*; *(Krist)* Zentrum *n*, Kern *m*, Keim *m*; *(Pap)* Hülse *f*, Wickelhülse *f*; *(Pap)* Einlage *f* *(beim Triplex-karton)*
c. board Trennwand *f* *(einer Kolbensetz-maschine)*
c.-column[-supported] thickener Eindicker *m* mit Mittelsäule
c. division *(Pap)* Mittelwand *f*, Zwischenwand *f*, Scheidewand *f* *(des Holländers)*
c. gate *(Plast)* Zentralanguß *m*
c. gated mould *(Plast)* Spritzgießwerkzeug *n* mit Zentralanguß
c. heat zone *(Plast)* mittlere Heizzone *f*
c. of asymmetry Asymmetriezentrum *n*
c. of crystallization Kristallisationszentrum *n*, Kristall[isations]keim *m*, Kristall[isations]kern *m*
c. of symmetry *(Krist)* Symmetriezentrum *n*, Inversionszentrum *n*
c. of the band Bandenmitte *f*, Nullstelle *f* *(eines Bandenspektrums)*
c. rewind method *(Pap)* axiale Papieraufwicklung *f*, Kernwicklung *f*
c. rewinder (winder) *(Pap)* Rollenschneid-maschine *f* mit Kernwicklung (axialer Papierauf-wicklung)
centricleaner *(Pap)* Centricleaner *m*, Zentrireiniger *m* *(ein Zentrifugalreiniger)*

centrifiner *(Am)* Fliehkraftreiniger *m*, Wirbelreiniger *m*, Wirbelsichter *m*, Wirbelschleuder *f*, Rohrschleuder *f*
centrifugal zentrifugal, Fliehkraft...
centrifugal Zentrifuge *f*, Trennschleuder *f*, Schleuder *f*; Siebzentrifuge *f*, Filterzentrifuge *f*
c. acceleration Zentrifugalbeschleunigung *f*
c. atomization Fliehkraftzerstäubung *f*, Fliehkraftversprühung *f*
c. atomizer Fliehkraftzerstäuber *m*, Fliehkraftversprüher *m*, Zentrifugalzerstäuber *m*, Zentrifugalversprüher *m*
c. attrition mill Fremdkraftwälzmühle *f*, Fremdkraftrollmühle *f*, Fremdkraftringmühle *f*, Ringrollenmühle *f (i.w.S.)*, Ringwalzenmühle *f (i.w.S.)*
c. blower Kreiselgebläse *n*
c. casting Schleudergußteil *n*; *s.* c.-casting process
c.-casting process Schleuder[gieß]verfahren *n*, Schleudergußverfahren *n*, Schleuderguß *m*, Zentrifugalguß *m*, Zentrifugalgießen *n*, *(Glas auch)* Rotationsguß *m*
c. chromatography Zentrifugalchromatografie *f*
c. clarifier Klärzentrifuge *f*
c. classifier Zentrifugal[kraft]sichter *m*, Fliehkraftsichter *m*, Zentrifugal[kraft]klassierer *m*, Fliehkraftklassierer *m*
c. cleaner Zentrifugalreiniger *m*, Fliehkraftreiniger *m*, Wirbelreiniger *m*, Wirbelsichter *m*, Wirbelschleuder *f*, Rohrschleuder *f*
c. collector Fliehkraft[ab]scheider *m*, Zentrifugal[ab]scheider *m, (meist)* Zyklon[ab]scheider *m*, Wirbel[ab]scheider *m*, Abscheidezyklon *m*, Zyklon *m*; Spiral[ab]scheider *m (mit umlaufender Abscheidefläche)*
c. compressor Kreisel[rad]verdichter *m*, Kreisel[rad]kompressor *m*
c. concrete Schleuderbeton *m*
c. cyclone separator Zyklon[ab]scheider *m*, Wirbel[ab]scheider *m*, Abscheidezyklon *m*, Zyklon *m*
c.-discharge bucket elevator Zentrifugalbecherwerk *n*
c. dryer Trockenzentrifuge *f*
c. extractor Zentrifugalextraktor *m*
c. fan Ventilator *m*, Kreiselradlüfter *m*; Radialventilator *m*
c. fertilizer distributor *(Landw)* Schleuderstreuer *m*
c. field Zentrifugalfeld *n*
c. filter Zentrifugalfilter *n*
c. force Zentrifugalkraft *f*, Fliehkraft *f*
c. grinder *s.* c. attrition mill
c. honey Schleuderhonig *m*
c. mill Schleudermühle *f*
c. moulding *(Plast)* Schleuderguß *m*, Rotationsguß *m*
c. pot spinning Topf-Zentrifugenspinnverfahren *n*, Topfspinnverfahren *n*, Zentrifugen[spinn]verfahren *n*, Zentrifugenspinnen *n*, Spinntopfverfahren *n*

c. process Schleuderverfahren *n (Glasfaserherstellung)*
c. pump Kreiselpumpe *f*, Zentrifugalpumpe *f*
c. pump mixer Kreiselpumpenmischapparat *m*
c. separator *s. c.* collector
c. still Rotationskolonne *f*
c. strainer *(Pap)* Zentrifugalsortierer *m*, Zentrifugalreiniger *m*, Schleudersortierer *m (Stoffreinigung)*
c. syrup *(Zucker)* Ablaufsirup *m*
c. type screen *(Pap)* Zentrifugalsortierer *m*, Zentrifugalsichter *m (Holzschliffreinigung)*
centrifugally cast concrete Schleuderbeton *m*
c. cast pipe Schleudergußrohr *n*
centrifugate / *to* zentrifugieren, schleudern
centrifugate Zentrifugat *n*
centrifugation Zentrifugieren *n*, Zentrifugierung *f*, Schleudern *n*, Abschleudern *n*
centrifuge / *to* zentrifugieren, schleudern
c. off abzentrifugieren, abschleudern
centrifuge Zentrifuge *f*, Trennschleuder *f*, Schleuder *f*
c. basket Siebtrommel *f* einer Zentrifuge, Zentrifugentrommel *f*
c. brake kit *(elektrische)* Bremsvorrichtung *f* der Zentrifuge
c. cup Schleuderbecher *m*
c. efficiency Ausnutzungsgrad *m* der Zentrifuge
c. field Zentrifugalfeld *n*
c. head Zentrifugenrotor *m (bei Laborzentrifugen)*
c. rotor Zentrifugenrotor *m, (i.e.S.)* Zentrifugentrommel *f*
c. tube Zentrifugenglas *n*
centrifuged latex zentrifugierter Latex *m*
centrifuging Zentrifugieren *n*, Schleudern *n*, Abschleudern *n*
centring Zentrierung *f*, Zentrieren *n*, Einmitten *n*, Mitten *n*
centripetal zentripetal
c. force Zentripetalkraft *f*
centron Atomkern *m*
centrosphere *(Geol)* Zentrosphäre *f*, Erdkern *m*
centrosymmetric zentrosymmetrisch, zentralsymmetrisch
centurium Ct Centurium *n, (veraltet für)* Fm Fermium *n*
cephaelin[e] Zephaelin *n*, Cephaelin *n (Alkaloid)*
cephalin Kephalin *n (ein Phosphatid)*
cephalosporanic acid Zephalosporansäure *f*
cephalosporin N Zephalosporin N *n*
ceramal, ceramel, ceramet *s.* cermet
cerametallic metallkeramisch
ceramic keramisch, Keramik...
c. art Kunstkeramik *f*
c. article keramisches Erzeugnis *n*
c. body keramischer Körper *m*; Scherben *m*; keramische Masse *f*
c. bonded keramisch gebunden
c. building material keramischer Baustoff *m*

c. capacitor Keramikkondensator *m*, keramischer Kondensator *m*
c. colour keramische Farbe *f*
c. dielectric keramisches Dielektrikum *n*, keramischer Isolierstoff *m*
c. fibre keramische Faser *f*, Keramikfaser *f*; keramischer Faserstoff *m*
c. filter Keramikfilter *n*
c. industry keramische Industrie *f*
c. material keramischer Werkstoff *m*
c.-metal adhesive metallkeramisches Haftmittel *n*
c. mix keramische Masse *f*
c. packing keramische Schüttung *f*, Keramikschüttung *f*
c. paste keramische Masse *f*
c. product s. c. article
c. staple keramische Faser *f*, Keramikfaser *f*
ceramics Keramik *f*
ceramist Keramiker *m*
cerargyrite *(Min)* Chlorargyrit *m*, Kerargyrit *m*, Hornsilber *n*, Silberhornerz *n (Silberchlorid)*
cereal seed oil Getreidekeimöl *n*
c. starch Getreidestärke *f*
c. straw Getreidestroh *n*
cerebroside Zerebrosid *n (ein Lipoid)*
cererite s. cerite
ceresin [wax] Zeresin *n*, Hartparaffin *n*, gereinigtes Erdwachs *n*
ceresine s. ceresin [wax]
ceria CeO_2 Zererde *f*, *(veraltet für)* Zer(IV)-oxid *n*
ceric Zer..., *(i.e.S.)* Zer(IV)-...
c. fluoride CeF_4 Zer(IV)-fluorid *n*, Zertetrafluorid *n*
c. hydroxide $Ce(OH)_4$ Zer(IV)-hydroxid *n*
c. hydroxynitrate $Ce(OH)(NO_3)_3$ Zer(IV)-hydroxidnitrat *n*
c. nitrate $Ce(NO_3)_4$ Zer(IV)-nitrat *n*
c. oxide CeO_2 Zer(IV)-oxid *n*, Zerdioxid *n*
c. silicide Zersilizid *n*
c. sulphate $Ce(SO_4)_2$ Zer(IV)-sulfat *n*
cerimetry Zerimetrie *f*
cerin s. ceresin [wax]
cerine s. cerite
Cerini dialyzer Cerini-Dialysator *m*
cerinic acid s. cerotic acid
cerinstein s. cerite
ceriometry s. cerimetry
cerite *(Min)* Zerit *m (ein Zersilikat)*
c. earth Zeriterde *f*
cerium Ce Zer *n*
c. carbide CeC_2 Zerkarbid *n*
c. dioxide CeO_2 Zerdioxid *n*, Zer(IV)-oxid *n*
c. earth Zererde *f*
c. tribromide $CeBr_3$ Zertribromid *n*, Zer(III)-bromid *n*
cermet Cermet *n*, Kerametall *n*, mischkeramischer (metallkeramischer, keramometallischer) Werkstoff *m*, Mischkeramik *f*
cerotic acid $C_{25}H_{51}COOH$ Zerotinsäure *f*, *n*-Hexakosansäure *f*

cerous Zer..., *(i.e.S.)* Zer(III)-...
c. bromide $CeBr_3$ Zer(III)-bromid *n*, Zertribromid *n*
c. carbonate $Ce_2(CO_3)_3$ Zer(III)-karbonat *n*
c. chloride $CeCl_3$ Zer(III)-chlorid *n*, Zertrichlorid *n*
c. fluoride CeF_3 Zer(III)-fluorid *n*, Zertrifluorid *n*
c. hydroxide $Ce(OH)_3$ Zer(III)-hydroxid *n*
c. nitrate $Ce(NO_3)_3$ Zer(III)-nitrat *n*
c. orthophosphate $CePO_4$ Zer(III)-orthophosphat *n*
c. oxide Ce_2O_3 Zer(III)-oxid *n*
c. oxychloride $CeOCl$ Zer(III)-oxidchlorid *n*
c. sulphate $Ce_2(SO_4)_3$ Zer(III)-sulfat *n*
c. sulphide Ce_2S_3 Zer(III)-sulfid *n*
certificated hop Qualitätshopfen *m*
certified milk Vorzugsmilch *f*
cerulein s. coerulein
ceruse $2PbCO_3 \cdot Pb(OH)_2$ Bleiweiß *n*, basisches Bleikarbonat *n*
cerusite s. cerussite
cerussa s. ceruse
cerussite *(Min)* Zerussit *m*, Weißbleierz *n (Blei(II)-karbonat)*
cervantite *(Min)* Cervantit *m (Antimon(III,V)-oxid)*
ceryl alcohol $CH_3(CH_2)_{24}CH_2OH$ Zerylalkohol *m*, 1-Hexakosanol *n*
cesium s. caesium
cetane $C_{16}H_{34}$ Zetan *n*, Hexadekan *n*
c. index CFR-Zetanindex *m*
c. number Zetanzahl *f*, Zetanziffer *f*, CaZ
c. rating s. c. number
cetin $C_{15}H_{31}COO \cdot C_{16}H_{33}$ Palmitinsäurezetylester *m*, Zetylpalmitat *n*
cetoleic acid $C_{21}H_{41}COOH$ 11-Zetoleinsäure *f*, 11-Dokosensäure *f*
cetyl alcohol $CH_3(CH_2)_{15}OH$ Zetylalkohol *m*, Zetanol *n*, 1-Hexadekanol *n*, *n*-Hexadezylalkohol *m*
cetylacetic acid $CH_3(CH_2)_{16}COOH$ Zetylessigsäure *f*, Oktadekansäure *f*, Stearinsäure *f*
cetylic acid $CH_3(CH_2)_{14}COOH$ Zetylsäure *f*, Hexadekansäure *f*, Palmitinsäure *f*
cevadilla Sabadille *f*, Schoenocaulon officinale (Schl. et. Ch.) A. Gray; s. c. seeds
c. seeds Sabadillsamen *mpl (von Schoenocaulon officinale (Schl. et. Ch.) A. Gray, Insektizid)*
cevitamic acid s. ascorbic acid
ceylanite *(Min)* Ceylanit *m*, Ceylonit *m*, Pleonast *m*
Ceylon cardamom Ceylon-Kardamom *m(n) (von Elettaria major Sm.)*
C. cinnamon Ceylonzimt *m*, Echter Zimt (Kaneel) *m (von Cinnamomum zeylanicum Bl.)*
ceylonite s. ceylanite
CFR *(Abk. für)* Cooperative Fuel Research
CFR engine CFR-Motor *m*, CFR-Prüfmotor *m*, Cooperative-Fuel-Research-Motor *m (Klopfprüfmotor)*
CFR motor method CFR-Motormethode *f (zur Bestimmung der Oktanzahl)*

CFR test engine *s.* CFR engine
C/H ratio C/H-Verhältnis *n*
chabasite, chabazite *(Min)* Chabasit *m (Tektosilikat)*
chadacryst Chedakristall *m*, Gastkristall *m*
chafer [strip] *(Gum)* Wulstschutzstreifen *m*, Friktionsstreifen *m*
chafing corrosion Reibkorrosion *f*
chain Kette *f;* Atomkette *f;* Reaktionskette *f;* [radioaktive] Zerfallsreihe (Reihe) *f*
  c. and bucket elevator Kettenbecherwerk *n*
  c. axis Kettenachse *f*
  c. barker *(Pap)* Kettenentrinder *m*
  c. branching Kettenverzweigung *f*
  c. breakage (breaking) Kettenabbruch *m*
  c.-breaking reaction Kettenabbruchreaktion *f*, Abbruchreaktion *f*
  c. carrier Kettenträger *m*, kettentragendes Radikal *n*
  c. clamp Kettenklemme *f (Stativzubehör)*
  c. compound Kettenverbindung *f*
  c. conveyor Kettenförderer *m*
  c. elevator *s.* c. and bucket elevator
  c. end Kettenende *n*
  c. folding Kettenfaltung *f*
  c.-formed kettenförmig
  c. grate Kettenrost *m*, Wanderrost *m*
  c.-grate stoker Kettenrostfeuerung *f*, Wanderrostfeuerung *f*
  c. grinder *(Pap)* Kettenschleifer *m*
  c. growth Kettenwachstum *n*, Wachstumsperiode *f (Polymerisation)*
  c. initiation Kettenstart *m*, Startreaktion *f*
  c. initiator *s.* c. carrier
  c. lattice *(Krist)* Fadengitter *n*
  c. length Kettenlänge *f*
  c. lengthening Kettenverlängerung *f*
  c.-like kettenartig
  c. mark Bandeindruck *m (Glasfehler)*
  c. mercerizer *(Text)* Kettenmerzerisiermaschine *f*
  c. mobility Kettenbeweglichkeit *f*
  c. molecule Kettenmolekül *n*
  c. of atoms Atomkette *f*
  c. of carbon atoms Kohlenstoffkette *f*, C-Kette *f*
  c. of reactions Reaktionskette *f*
  c. polymerization Kettenpolymerisation *f*
  c.-propagating kettentragend, eine Kette fortpflanzend
  c.-propagating reaction Kettenfortpflanzungsreaktion *f*
  c. propagation Kettenwachstum *n*, Wachstumsperiode *f (Polymerisation)*
  c. propagator *s.* c. carrier
  c. reaction Kettenreaktion *f*
  c. reduction Kettenreduktion *f*
  c. scission Kettenspaltung *f*, Kettensprengung *f*
  c. segment Kettensegment *n*
  c. splitting *s.* c. scission
  c. stopper Kettenabbrecher *m*, Kettenabbruchmittel *n*, Stopper *m*

  c.-stopping reaction *s.* c.-terminating reaction
  c. structure Kettenstruktur *f*
  c.-terminating reaction Kettenabbruchreaktion *f*, Abbruchreaktion *f*
  c. termination Kettenabbruch *m*
  c. transfer Kettenübertragung *f*
  c.-transfer agent Übertragungsregler *m*, Kettenüberträger *m*
  c.-type grinder *s.* c. grinder
  c. unzipping reaction *(Plast)* Reißverschlußreaktion *f*
  c. without branches unverzweigte (geradlinige, normale) Kette *f*
chainlength Kettenlänge *f*
chainless kettenlos
  c. mercerizing machine *(Text)* kettenlose Merzerisiermaschine *f*
chair [conformation] *s.* c. form
  c. form *(Stereochemie)* Sesselform *f*, Sessel *m*, starre Form *f*
chalcanthite *(Min)* Chalkanthit *m*, Kupfervitriol *m (Kupfer(II)-sulfat-5-Wasser)*
chalcedony *(Min)* Chalzedon *m (Abart des Quarzes)*
chalcocite Chalkozit *m*, *(veraltet für)* Chalkosin *m*, Kupferglanz *m (Mineralgruppe)*
  α-**chalcocite** *(Min)* α-Chalkosin *m*, α-Kupferglanz *m (Kupfer(I)-sulfid)*
chalcogen Chalkogen *n*
chalcogenide Chalkogenid *n (binäre Verbindung eines Chalkogens)*
chalcomenite *(Min)* Chalkomenit *m (Kupfer(II)-selenit)*
chalcone Chalkon *n*, *(i.e.S.)* $C_6H_5COCH=CHC_6H_5$ Chalkon *n*, Benzalazetophenon *n*
chalcophane, chalcophanite *(Min)* Chalkophanit *m (Zinkmanganerz)*
chalcophile element chalkophiles Element *n*
chalcophyllite *(Min)* Chalkophyllit *m*
chalcopyrite *(Min)* Chalkopyrit *m*, Kupferkies *m*, Kupferpyrit *m (Eisen(II)-kupfer(II)-sulfid)*
chalcosine Chalkosin *m*, Kupferglanz *m (Mineralgruppe)*
chalcosphere *(Geol)* Chalkosphäre *f*, Zwischenschicht *f*
chalcostaktite *(Min)* Chalkostaktit *m*, *(veraltet für)* Chrysokoll *m*
chalcotrichite *(Min)* Chalkotrichit *m*, Kupferblüte *f (Abart von Kuprit, Rotkupfererz)*
chalk / to [ab]kreiden, auskreiden *(von Anstrichen)*; *(Gum)* pudern
chalk *(Min)* Kreide *f*, Seekreide *f*, Schreibkreide *f (Kalziumkarbonat)*; *(Gum)* Pudermittel *n*
  c. overlay paper Kreide[relief]papier *n*, Kreidezurichtepapier *n*
  c. paper Kreidepapier *n*
chalkiness Kreidigkeit *f*
chalking Kreiden *n*, Abkreiden *n*, Auskreiden *n (Abfärben von Anstrichen)*
chalkone *s.* chalcone

chalkophyllite s. chalcophyllite
chalkopyrite s. chalcopyrite
chalky kreidig, kreideartig; kreidehaltig
  to become c. [ab]kreiden, auskreiden *(von Anstrichen)*
chalybeate spring Eisenquelle f
  c. water eisenhaltiges Wasser n, Eisenwasser n
chalybite *(Min)* Chalybit m, *(veraltet für)* Eisenspat m, Spateisenstein m *(Eisen(II)-karbonat)*
chamber Kammer f, Zelle f, Raum m
  c. acid Kammersäure f *(im Bleikammerverfahren erzeugte, etwa 60%ige Schwefelsäure)*
  c. crystals Bleikammerkristalle mpl *(Nitrosylhydrogensulfat)*
  c. dryer Kammertrockner m
  c. filter press Kammer[filter]presse f
  c. furnace *(Met)* Kammerofen m
  c. kiln *(Ker)* Kammerofen m
  c. plate Einsatzplatte f *(eines chromatografischen Entwicklungsgeräts)*
  c. press s. c. filter press
  c. process Bleikammerverfahren n, Kammerverfahren n
  c. reaction Kammerreaktion f *(Bleikammerverfahren)*
  c. space Kammerraum m *(Bleikammerverfahren)*
  c.-type furnace s. c. furnace
chamecin Chamaezin n *(ein Hydroxy-β-thujaplizin-Tropolonderivat)*
chameleon mineral *(Min)* mineralisches Chamäleon n *(Kaliummanganat)*
chamenol Chamaenol n, 1-Methoxy-2-hydroxy-4-isopropylbenzol n
chamic acid Chamsäure f *(Monoterpenderivat)*
chaminic acid Chaminsäure f *(Monoterpenderivat)*
chamois sämischgar, fettgar
  c. leather Sämischleder n
  c. process s. c. tannage
  c. tannage Sämischgerbung f, Fettgerbung f
chamoising process s. chamois tannage
chamoisite *(veraltet für)* chamosite
chamomile s. camomile
chamosite *(Min)* Chamosit m *(ein Phyllosilikat)*
Champaca oil *(Kosmet)* Champacablütenöl n *(von Michelia longifolia Blume und M. champaca L.)*
  C. wood oil *(Kosmet)* *(Holzöl von Bulnesia sarmienti Lorentz)*
champagne Champagner m, Schaumwein m, Sekt m
champagnization Schaumweinherstellung f, Sektherstellung f
Chance cone Chance-Kegel m *(Tieftrogscheider nach dem Chance-Verfahren)*
Chance process s. Chance sand-floatation process
Chance sand-floatation process Chance-Sandflotationsverfahren n, Chance-Sandschwimmverfahren n, Chance-Verfahren n
chandoo Tschandu n, Rauchopium n
change / to 1. *(Farbe, Gestalt, Ladung)* [ver]ändern, wechseln; *(Struktur)* verwandeln, umwandeln,

umformen; 2. sich [ver]ändern, wechseln *(Farbe)*; sich verwandeln (umwandeln, umformen, umlagern) *(Struktur)*; übergehen; umschlagen *(Farbe, Reaktion)*
change Änderung f, Veränderung f, Wechsel m, Übergang m, Umschlag m; Umwandlung f, Umformung f, Umlagerung f
  c.-can mixer Mischer m mit Wechselbehälter
  c. in colour Farbänderung f, Farbumschlag m, Farbwechsel m
  c. in state Zustandsänderung f, *(i.e.S.)* Aggregatzustandsänderung f
  c. in temperature Temperaturänderung f
  c. of colour s. c. in colour
  c. of direction of flow Strömungsumkehr f, Wechsel m (Umkehr f) der Strömungsrichtung
  c. of entropy Entropieänderung f
  c. of pH pH-Änderung f
  c. of place Platzwechsel m
  c. of state s. c. in state
changing of felts *(Pap)* Filzwechsel m
  c. of wires *(Pap)* Siebwechsel m
channel Kanal m, Rinne f
  c. black Kanalruß m, Kanalschwarz n, Channel-Black n *(im Channel-Verfahren hergestellter Gasruß)*
  c. black plant Channel-[Black-]Anlage f
  c. inclusion compound Kanaleinschlußverbindung f
  c. iron U-Eisen n *(zur Herstellung von Kanalruß)*
  c. method (process) Channel[-Black]-Verfahren n, Channel-Prozeß m, Kanalprozeß m, Kanalrußverfahren n
  c.-roller pulverizer Rollenwälzmühle f, Ringmühle f
channelling Kanalbildung f *(z.B. in Staubfließsystemen)*, Bachbildung f *(z.B. in Füllkörperschüttungen)*
chapmanizing *(Met)* *(Nitrierhärtung mit Ammoniak in Zyanidschmelze)*
chappe silk Schappe[seide] f
char / to [ver]kohlen, ankohlen, verschwelen
char künstliche Kohle f *(z.B. Holz-, Blut-, Knochen- oder Tierkohle)*; Halbkoks m
  c. filter Kohlefilter m
  c. kiln Holzverkohlungsofen m
characteristic Kennzeichen n; Kennziffer f
  c. curve Kennlinie f, Charakteristik f; *(Foto)* Schwärzungskurve f, fotografische (charakteristische) Schwärzungskurve f, charakteristische (sensitometrische) Kurve f, Dichtekurve f, *(fälschlich)* Gradationskurve f
  c. frequency charakteristische Frequenz f, Eigenfrequenz f
  c. number s. eigenvalue
  c. rotational momentum Eigendrehimpuls m
  c. state *(phys Ch)* Eigenzustand m
  c. temperature charakteristische Temperatur f, Debye-Temperatur f
characteristics Verhalten n, Charakteristik f

charcoal künstliche Kohle f (z.B. Holz-, Blut-, Knochen- oder Tierkohle), (i.e.S.) Holzkohle f
  c. burning Holzverkohlung f
  c. drawing paper Kohlezeichenpapier n
  c.-fired furnace Holzkohlenofen m, holzkohlengefeuerter Ofen m; Holzkohlenhochofen m
  c. pig iron Holzkohlen[roh]eisen n
  c. plant (Erdöl) Aktivkohleanlage f
chardonnet silk Chardonnet-Seide f
charge / to 1. (z.B. Ofen) beschicken, besetzen, speisen, füllen, ' chargieren, (Hochofen) begichten, (Beschickungsmaterial) eintragen, einsetzen, einlegen, einspeisen, einfüllen, einschütten, aufgeben; chargenweise (portionsweise) zugeben, zuteilen, dosieren; 2. (phys Ch) [auf]laden
  c. by pumping (Gefäße) vollpumpen
charge 1. Charge f, Beschickungsmaterial n, Beschickung f, Füllung f, Füllmasse f, Schüttung f, Einsatz m, Einsatzgut n, Einsatzstoff m, Einsatzmaterial n, Einsatzprodukt n, Eintrag m, Ansatz m, Satz m, Aufgabegut n, Speisung f, Einspeisung f, Gicht f (des Hochofens); 2. (phys Ch) Ladung f, Aufladung f
  c. carrier (phys Ch) Ladungsträger m
  c. cloud (phys Ch) Ladungswolke f
  c. density (phys Ch) Ladungsdichte f
  c. difference (phys Ch) Ladungsunterschied m, Ladungsdifferenz f
  c.-dipole interaction Dipol-Ladungs-Wechselwirkung f
  c. distribution (phys Ch) Ladungs[dichte]verteilung f
  c. dosage (phys Ch) Ladungsdosis f
  c. floor (Met) Gichtbühne f
  c. heater Einsatzofen m
  c. number (phys Ch) Ladungszahl f
  c. oil Einsatzöl n, Ausgangsöl n
  c. reversal (phys Ch) Umladung f
  c. separation (phys Ch) Ladungsabtrennung f
  c. stock s. charge 1.
  c. transfer complex Donator-Akzeptor-Komplex m
charging 1. Beschicken n, Beschickung f, Besetzen n, Speisen n, Füllen n, Chargieren n (z.B. eines Ofens), Begichten n, Begichtung f (des Hochofens), Eintragen n, Eintrag m, Einsetzen n, Einsatz m, Einlegen n, Einlage f, Einspeisen n, Einfüllen n, Einschütten n, Aufgeben n, Aufgabe f (des Beschickungsmaterials); chargenweises (portionsweises) Zugeben n, Zuteilen n, Dosieren n; 2. (phys Ch, Et) Aufladen n, Laden n
  c. bin Füllbunker m
  c. box Einsatzmulde f, Beschickungsmulde f
  c. door Chargiertür f, Beschickungstür f, Füllklappe f; s. c. hole
  c. hole Füllöffnung f, Fülloch n, Beladeöffnung f, Eintragöffnung f, Beschickungsöffnung f, Einwurföffnung f, Aufgabeöffnung f
  c. hopper Fülltrichter m, Aufgabetrichter m

  c. pipe Zuflußrohr n, Zulaufrohr n, Füllrohr n; Zuflußstutzen m, Zulaufstutzen m, Füllstutzen m
  c. platform Beschickungsbühne f
  c. point Guteintrag m, Gutaufgabe f
  c. stock s. charge 1.
  c. tray (Plast) Fülltablett n, Füllvorrichtung f (für Preßwerkzeuge)
charmat process Großraumgärverfahren n, Tankgärverfahren n, Tankgärung f
Charpy test Charpy-Prüfung f, Schlagzähigkeitsprüfung f nach Charpy
chart Diagramm n, Schaubild n
  c. drive Bandantrieb m, Diagrammpapierantrieb m (des Bandschreibers)
  c. paper Landkartenpapier n, Kartenpapier n; Seekartenpapier n
Chatelier-Braun principle s. Le Chatelier-Braun principle
chaulmoogra oil (Pharm) Chaulmoograöl n, Chaulmugraöl n (von Hydnocarpus-Arten)
d-chaulmoogric acid $C_5H_7(CH_2)_{12}COOH$ d-Chaulmoograsäure f, d-13-(2-Zyklopentenyl-)tridekansäure f
chavicol methyl ether $CH_2=CHCH_2C_6H_4OCH_3$ Chavikolmethyläther m, Estragol n, 4-Methoxy-1-allylbenzol n, p-Allylanisol n
Ch D (Abk. für) Doctor of Chemistry
Ch E (Abk. für) Chemical Engineer
cheapener (Gum) Streckmittel n; inaktiver Füllstoff m
chebulagic acid Chebulagsäure f (Ellagengerbstoff)
chebulic acid Chebulsäure f (den Gallotanninen zugehöriger Gerbstoff)
chebulinic acid Chebulinsäure f (den Gallotanninen zugehöriger Gerbstoff)
check / to prüfen, kontrollieren
  c. for purity auf Reinheit prüfen
check 1. Rückschlagventil n; 2. Riß m, Spalt m, Spalte f, (Glas auch) Oberflächenriß m
  c. paper (Am) s. cheque paper
  c. plot Kontrollfläche f, Nullfläche f, Nullparzelle f (z.B. bei Düngungsversuchen)
  c. valve Rückschlagventil n
  c.-work s. checkerwork
checker Gitter n, Gitterraum m (z.B. im Winderhitzer)
  c. brick Gitter[werk]stein m
  c. chamber s. checker
checkerbrick s. checker brick
checkerwork Steingitterwerk n, Gitter[mauer]werk n

Cheddar cheese Cheddarkäse m
cheese 1. Käse m; 2. (Text) Kreuzspule f
  c. colouring Käsefärbung f (mit Farbstoffen); Käsefarbe f (Farbstoff)
  c. curd Käsebruch m
  c. dyeing (Text) Färben n von Kreuzspulen
  c. dyeing machine (Text) Kreuzspulfärbeapparat m

c. factory Käsefabrik f, Käserei f, Käsereibetrieb m
c. harp Käseharfe f
c. knife Käsesäbel m
c. making Käseherstellung f, Käseerzeugung f, Käsefabrikation f, Käserei f
c.-making plant s. c. factory
c. milk Käsereimilch f
c. powder Käsepulver n
c. ripening Käsereifung f
c. ripening room Käsereifungsraum m
c. wax Käsewachs n
c. whey Käse[rei]molke f
cheesecloth Mull m
cheesemaking s. cheese making
chelate / to Chelate bilden, Chelatbindung eingehen
chelate Chelat n, Chelatverbindung f, Scherenverbindung f, Chelatkomplex m
c. arm Scherenarm m
c. catalysis Chelatkatalyse f
c. complex (compound) s. chelate
c. donor group Chelatdonatorgruppe f
c. formation s. chelation
c. formation constant Chelatbildungskonstante f
c.-forming chelatbildend
c. ion-exchange resin s. c. resin
c.-like chelatartig
c. linkage Chelatbindung f
c. pigment Chelatfarbstoff m
c. resin Chelatharz n, chelatbildendes Austauscherharz n, Chelataustauscher m
c. ring Chelatring m
c. stability Chelatstabilität f
c. stability constant Chelat-Stabilitätskonstante f
c. structure Chelatstruktur f
chelated chelatgebunden
chelating chelatbildend
c. agent Chelatbildner m, Chelator m, chelatbildende Substanz f, chelatbildendes Reagens n
c. group Chelatgruppe f
c. resin s. chelate resin
chelation Chelatbildung f, Chelation f, Scherenbildung f
chelatometric chelatometrisch
c. titration s. chelatometry
chelatometry Chelatometrie f, chelatometrische Titration f
chelidonic acid Chelidonsäure f, Jervasäure f, Pyron-(4)-dikarbonsäure-(2,6) f
chelometric s. chelatometric
chem s. 1. chemical; 2. chemist; 3. chemistry
chemical chemisch, Chemie...
chemical Chemikalie f
c. adsorption chemische Adsorption f, Chemisorption f, Chemosorption f
c. affinity [chemische] Affinität f, Triebkraft f (maximale Nutzarbeit einer Reaktion, Reaktionsarbeit)

c. agent chemisches (chemisch wirksames) Mittel (Agens) n, chemischer Wirkstoff m
c. analysis chemische Analyse f
c. balance chemische Waage f, (i.e.S.) Analysenwaage f, analytische Waage f
c. binding s. c. bonding
c. bond chemische Bindung f (als Zustand)
c. bonding chemische Bindung f (als Vorgang)
c. cell (Ech) [galvanische, elektrochemische] Zelle f, [galvanisches, elektrochemisches] Element n
c. ceramics chemische Keramik f, Keramik f für chemische Zwecke
c. charge (Pap) zugegebene (zugeführte) Menge f an Aufschlußchemikalien
c. composition chemische Zusammensetzung f
c. consumption Chemikalienverbrauch m
c. conversion chemische Umsetzung f
c. dating radioaktive (absolute, physikalische, physikalisch-chemische) Altersbestimmung f, Altersbestimmung f durch Radionuklide
c. debarking (Pap) chemische Entrindung f
c. decolorization (Glas) chemische Entfärbung f
c. denudation (Geol) Auslaugung f
c. development (Foto) Entwickeln n, Entwicklung f
c. dipper Schöpfgefäß n (Schöpfer m) für Chemikalien
c. displacement chemische Verschiebung f
c. engineer Chemieingenieur m, (i.e.S.) Verfahrensingenieur m, Verfahrenstechniker m, Verfahrenschemiker m
c. engineering Chemieingenieurtechnik f, Chemieingenieurwesen n, chemische Ingenieurtechnik f, chemisches Ingenieurwesen n, (i.e.S.) chemische Verfahrenstechnik f
c.-engineering unit operation Grundoperation f der chemischen Verfahrenstechnik
c. entity chemisch einheitlicher Stoff m, chemisch einheitliche Substanz f, einheitliche chemische Substanz f
c. equation chemische Gleichung (Reaktionsgleichung, Umsatzgleichung) f, Reaktionsgleichung f
c. equipment Chemieausrüstungen fpl
c. fibre Chemiefaser f; Chemiefaserstoff m, CN
c. fog (Foto) Entwicklungsschleier m
c. funnel Einfülltrichter m
c. group Gruppe f, Rest m
c. industry chemische Industrie f, Chemieindustrie f
c. inertness chemische Trägheit (Reaktionsträgheit, Inertie, Indifferenz) f
c. intensification chemische Verstärkung f
c. kinetics chemische Kinetik (Reaktionskinetik) f, Reaktionskinetik f
c. lead Reinblei n, Weichblei n (Pb-Gehalt größer als 99,9%)
c. machinery Chemiemaschinen fpl
c. nomenclature chemische Nomenklatur f, Nomenklatur f der chemischen Stoffe

c. **notation** chemische Notation f, Notation f chemischer Strukturen

c. **passivity** chemische Passivität f

c. **plant** chemische Fabrik f, chemisches Werk n, Chemiewerk n, chemischer Betrieb m, Chemiebetrieb m; chemische Anlage f, Chemieanlage f

c. **plasticization** chemische Plastizierung (Plastifikation, Weichmachung) f

c. **plasticizer** chemisches Plastizierungsmittel (Plastiziermittel, Abbaumittel) n, Plastifikator m, (Gum auch) Peptisier[ungs]mittel n

c. **plasticizing agent** s. c. plasticizer

c. **porcelain** chemisches Porzellan n, Laboratoriumsporzellan n

c. **potential** chemisches Potential n

c. **precipitation** chemisches Fällen n, chemische Fällung f

c. **processing industry** chemische Industrie f, Chemieindustrie f

c. **proofing** (Text) chemische Ausrüstung f

c. **pulp** (chemisch vollkommen aufgeschlossener) Zellstoff m, Vollzellstoff m, klassischer Zellstoff m

c. **pulp factory wastes** Zellstoffabrikabwasser n, Zellstoffabwasser n

c. **pulping** Zellstoffaufschluß m, chemischer Aufschluß m

c. **pump** Chemiepumpe f

c. **purifier** (nach dem Zonenschmelzverfahren arbeitende Apparatur zur Reinigung von Chemikalien)

c. **ratio** s. c.-to-wood ratio

c. **reacting column (tower)** Kolonne f, Reaktionsturm m

c. **reaction** chemische Reaktion f, chemischer Vorgang m

c. **reaction engineering** chemische Reaktionstechnik f

c.-**reaction paint** chemisch (durch chemische Reaktion) trocknende Anstrichfarbe f

c. **recovery** Chemikalienrückführung f, Chemikalienrückgewinnung f

c.-**recovery coke oven** Koksofen m mit Nebenproduktgewinnung

c.-**recovery oven** Nebenproduktenofen m, Nebengewinnungsofen m

c. **red** Eisenoxidrot n

c. **reduction** chemische Reduktion f

c. **resistance** chemische Widerstandsfähigkeit (Festigkeit, Stabilität, Beständigkeit, Resistenz, Unangreifbarkeit) f, Beständigkeit f gegen chemische Einwirkungen; Chemikalienbeständigkeit f, Chemikalienfestigkeit f, Widerstandsfähigkeit (Resistenz) f gegen Chemikalien

c. **retting** (Text) chemische Röste f

c. **rubber** synthetischer (künstlicher) Kautschuk m, Synthesekautschuk m, Kunstkautschuk m, SK

c. **sensitization** (Foto) chemische Sensibilisierung f

c. **sensitizer** (Foto) chemischer Sensibilisator m

c. **sensitizing** s. c. sensitization

c. **shaker** Schüttelmaschine f

c. **shorthand** chemische Zeichensprache f

c. **sign** chemisches Zeichen n, Elementsymbol n, Symbol n

c. **space-geometry** Stereochemie f, Raumchemie f

c. **stability** s. c. resistance

c. **stoneware** chemisches Steinzeug n, Steinzeug n für die chemische Industrie

c. **storage tank** (Pap) Vorratsbehälter m für Aufschlußchemikalien

c. **symbol** s. c. sign

c. **taxonomy** s. chemotaxonomy

c. **technology** chemische Technik f, Chemietechnik f; chemische Technologie f

c. **tracer** chemischer Tracer m

c. **wood[-pulp]** Holzzellstoff m, Zellstoff m

c. **works** chemisches Werk n, Chemiewerk n, chemischer Betrieb m, Chemiebetrieb m, chemische Fabrik f

Chemical Abstracts system Chemical-Abstracts-System n, System n der Chemical Abstracts (zur Benennung und Registrierung chemischer Verbindungen)

chemicalization Chemisierung f (verstärkter Einsatz chemischer Mittel z.B. in der Landwirtschaft)

chemically bound chemisch gebunden

c. **foamed plastic** nach chemischem Treibverfahren hergestellter Schaumstoff m

c. **held** s. c. bound

c. **indifferent** s. c. inert

c. **inert** reaktionsträge, chemisch träge (inert, indifferent)

c. **pure** chemisch rein

c. **resistant glass** chemisch widerstandsfähiges (haltbares) Glas n, Laborglas n

c. **treated tree** (Pap) begifteter (mit Gift behandelter) Baum m (chemische Entrindung)

chemick / to (Text) bleichen, chloren

chemick (Text) Chlorbleiche f

chemicking (Text) Bleichen n, Chloren n

chemigraph Ätzung f (durch Ätzen hergestellter Druckstock)

chemigraphy Chemigrafie f (Ätzverfahren zur Herstellung von Druckstöcken für Hochdruck)

chemigroundwood (Pap) chemischer Holzschliff (Schliff) m, Chemieschliff m

c. **process** (Pap) chemisches Schleifen n, Erzeugung f von chemischem Holzschliff

chemiluminescence Chemilumineszenz f, Chemolumineszenz f

chemipulper (Pap) Chemipulper m

chemism Chemismus m

chemisorb / to chemisch adsorbieren, chemosorbieren

chemisorption Chemisorption f, Chemosorption f, chemische (aktivierte) Adsorption f

chemist Chemiker *m*
  c. in charge verantwortlicher Chemiker *m*
chemistry Chemie *f*
  c. of fats Fettchemie *f*
  c. of fibres Faserstoffchemie *f*
  c. of photography Chemie *f* der Fotografie (fotografischen Prozesse), fotografische Chemie *f*
  c. of surfaces Oberflächenchemie *f*
chemitype 1. Chemotypie *f (veraltetes drucktechnisches Ätzverfahren ohne Anwendung der Fotografie), (i.w.S.)* s. chemigraphy; 2. *s.* chemigraph
chemosorption Chemosorption *f*, Chemisorption *f*
chemosterilant Chemosteril[is]ans *n*, Chemosterilisierungsmittel *n*
chemosterility *(Landw) (durch Chemikalien (Chemosterilantien) herbeigeführte Sterilität)*
chemotaxis *(Bio)* Chemotaxis *f (durch Chemikalien hervorgerufene Ortsbewegung von Lebewesen)*
chemotaxonomy *(Bio)* Chemotaxonomie *f (Einordnen von Lebewesengruppen in das Pflanzenbzw. Tiersystem anhand der Inhaltsstoffe)*
chemotherapeutic chemotherapeutisch
  c. agent Chemotherapeutikum *n*, chemotherapeutisches Mittel *n*
chemotherapy Chemotherapie *f*
chemotropic chemotropisch
  c. trap chemotropische Falle *f (zur Schädlingsbekämpfung)*
chemotropism Chemotropismus *m (durch Chemikalien hervorgerufene gerichtete Wachstumsbewegung von Pflanzen)*
chemurgy Chemurgie *f (Gewinnung chemischer Produkte aus pflanzlichen oder tierischen Rohstoffen)*
chenevixite *(Min)* Chenevixit *m*
chenocoprolite *(Min)* Chenokoprolith *m*, *(veraltet für)* Ganomatit *m*
chenopodium oil *(Pharm)* Wurmsamenöl *n*, Chenopodiumöl *n (von Chenopodium ambrosioides L. var. anthelminthicum Gray)*
cheque paper Scheckpapier *n*, Sicherheitspapier *n*
chequer brick Gitter[werks]stein *m*
  c. brickwork *s.* chequerwork
  c. chamber Gitter *n*, Gitterraum *m (eines Winderhitzers)*
chequerwork Steingitterwerk *n*, Gitter[mauer]werk *n*
Cherenkov radiation Tscherenkow-Strahlung *f*
chernozem Tschernosjom *m(n)*, Tschernosem *m(n)*, Schwarzerde *f*, Steppenboden *m*
cherry birch Zuckerbirke *f*, Betula lenta L.
  c. gum Kirschgummi *n*
  c.-red kirschrot
  c.-red heat Kirschrotglut *f*
chert *(Min)* Hornstein *m (Abart des Opals)*
  c. nodule *(Geol)* Silexknolle *f*, Kieselknolle *f*
chessy copper *s.* chessylite

chessylite *(Min)* Chessylit *m*, *(veraltet für)* Kupferlasur *m*, Azurit *m* *(Kupfer(II)-dihydroxiddikarbonat)*
chest-type hand duster Bauchstäuber *m (für Pflanzenschutzmittel)*
chestnut 1. Echte Kastanie *f*, Edelkastanie *f*, Castanea sativa Miller; 2. Marone *f (Frucht von 1.)*; 3. *s.* American chestnut; 4. *s.* horse chestnut
  c. oak Kastanieneiche *f*, Quercus montana Willd.; *s.* yellow chestnut oak
chevilling Chevillieren *n (Glanzsteigerung durch mechanisches Glätten der Garnoberfläche)*
chewing gum Kaugummi *m*
  c. tobacco Kautabak *m*
chiastoline, chiastolite, chiastolith Chiastolith *m (Aluminiumsilikatmineral, Abart von Andalusit)*
Chicago acid $C_{10}H_4(NH_2)(OH)(SO_3H)_2$ Chicagosäure *f*, 1-Amino-8-naphthol-2,4-disulfonsäure *f*, 1-Naphthylamin-8-hydroxy-2,4-disulfonsäure *f*
  C. blue Chicagoblau *n*
Chichibabin reaction Tschitschibabin-Reaktion *f*
chicle gum Chiclegummi *n*, Chicle *m (von Achras sapota L.); (i.w.S. Sammelname für Pflanzengummi von mehreren südamerikanischen Baumarten)*
chief component (constituent) Hauptbestandteil *m*, Hauptkomponente *f*, Grundkomponente *f*
  c. product Hauptprodukt *n*
Chile mill *s.* Chilean mill
  C. niter (nitrate, nitre) *s.* C. saltpetre
  C. saltpeter *s.* C. saltpetre
  C. saltpetre Chilesalpeter *m*, Natronsalpeter *m (Natriumnitrat)*
Chilean mill chilenische Mühle *f*, Kollergang *m*, Kollermühle *f*
  C. nitrate *s.* Chile saltpetre
chilenite *(Min)* Chilenit *m*, Wismutsilber *n*
Chili saltpeter *s.* Chile saltpetre
Chilian nitrate *s.* Chile saltpetre
chill / to [ab]kühlen, erkalten lassen; abschrecken; abgeschreckt (hart) werden
chill Abschrecken *n*, Abschreckung *f*; Schreckplatte *f (beim Schalenhartguß)*; Schreckschicht *f (am Schalengußstück)*
  c.-cast iron Schalen[hart]guß *m*, Hartguß *m (als Produkt)*
  c. casting Schalen[hart]guß *m (als Verfahren)*
  c. core Schalen[hart]gußkern *m*
  c. haze Kältetrübung *f (z.B. des Bieres)*
  c. mark *(Glas)* Runzel *f*, Kühlfalte *f (Oberflächenfehler)*
  c. roll Kühlwalze *f*, Kühltrommel *f*
  c.-roll extrusion *(Plast)* Extrudieren *n* mit Kühlwalzen
  c.-roll method Kühltrommelverfahren *n*, Abkühlverfahren *n* mit Kühltrommeln, Trockenverfahren *n*, Trommelverfahren *n (Margarineherstellung)*
chilled cast iron *s.* chill-cast iron
  c. cast iron roll *s.* c.-iron roll

c. casting s. chill-cast iron
c. core s. chill core
c. fruit Gefrierobst n
c.-iron roll Hartgußwalze f
c. meat Gefrierfleisch n
c. portion Schreckschicht f *(beim Schalenhart-guß)*
c. roll s. c.-iron roll
chiller Chiller m, Kühlapparat m *(für Temperaturen, die mit Wasser nicht mehr erreichbar sind)*
chilli Spanischer Pfeffer m, Cayennepfeffer m, Gewürzpaprika m
chilling Kühlen n, Kühlung f, Abkühlen n, Abkühlung f; Abschrecken n, Abschreckung f; Erkalten n, Kaltwerden n, Erstarren n
c. cylinder s. c. tube
c. tower Kühlturm m
c. tube Kühlrohr n
chimney Schornstein m, Esse f, Schlot m, Kamin m; *(Destillation)* Kamin m, Dampfkamin m, Steigrohr n *(einer Glocke)*
c. pull Schornsteinzug m
chimonanthine Chimonanthin n *(Alkaloid)*
chimyl alcohol $C_{16}H_{33}OCH_2CH(OH)CH_2OH$ Chimylalkohol m, α-Hexadezylglyzeryläther m, Testriol n
china Kunstporzellan n, [nichttechnisches] Porzellan n
c. clay *(Min)* Kaolin m, Porzellanerde f, weißer (reiner) Ton m, China Clay m(n), *(i.e.S.)* Kaolin m aus Cornwall; *(Tech)* Kaolin n, Schlämmkaolin n, geschlämmte Porzellanerde f
c. process Porzellanverfahren n *(zur Herstellung von Steingut und Knochenporzellan)*
c. stone Feldspatpegmatit m
China bark Quillajarinde f, Seifenrinde f *(von Quillaja saponaria Mol.)*
C. grass Ramie f, Chinagras n, Indische Nessel f, Boehmeria nivea (L.)Gaudich
C. oil Perubalsam m, Indischer Balsam m *(von Myroxylon balsamum (L.)Harms var. pereirae)*
C. paper Chinapapier n, Chinesisches Papier n
C. silver Chinasilber n *(Legierung aus Cu, Sn, Ni und Ag)*
C.wood oil Chinaholzöl n, Chinesisches Holzöl n *(meist Samenöl von Aleurites fordii Hemsl.)*
chinaware s. china
Chinawood oil s. China wood oil
Chinese anise Stern-Anis m, Illicium verum Hook. fil.
C. aniseed *(Lebm)* Stern-Anis m *(Sammelfrüchte von Illicium verum Hook. fil.)*
C. bean Sojabohne f, Glycine max (L.)Merr.
C. bean oil Soja[bohnen]öl n
C. bronze *(Legierung aus Cu, Ag, Pb und Au)*
C. cinnamon Chinesischer Zimt m, Kassiazimt m, Zimtkassia f *(von Cinnamomum aromaticum Nees)*
C. galls Chinesische (Japanische) Gallen fpl *(von Rhus chinensis Mill.)*

C. gelatin s. agar
C. grass s. China grass
C. green Chinesisches Grün n, Chinagrün n, Lo kao *(Naturfarbstoff aus Rhamnus-Arten)*
C. lacquer Japanischer Firnis m, Japanlack m *(i.e.S.) (von Rhus verniciflua Stokes)*
C. oil 1. Chinesisches Zimtöl n, Kassiaöl n *(von Cinnamomum aromaticum Nees)*; 2. s. Peru balsam
C. paper s. China paper
C. potato Chinesische Kartoffel f, Bataten-Yams m, Dioscorea batatas Decne
C. silver s. China silver
C. [tree] wax Chinesisches Wachs n, Chinawachs n, Insektenwachs n, Cera chinensis *(durch Schildläuse abgeschiedenes Wachs von Fraxinus chinensis Roxb.)*
C. white Chinesischweiß n, Chinaweiß n *(Zinkoxid)*
C. wood oil s. China wood oil
C. yam s. C. potato
chiniofon Chiniofon n *(7-Jod-8-hydroxychinolin-5-sulfonsäure in Mischung mit $NaHCO_3$)*
chinoline Chinolin n, 2,3-Benzopyridin n
chinone Chinon n, *(i.e.S.)* $C_6H_4O_2$ p-Benzochinon n, 1,4-Zyklohexadiendion n, p-Chinon n, Chinon n
Chios mastic Mastix m *(Harz von Pistacia lentiscus L.)*
chip / to *(Pap)* [durch]hacken, zerhacken
chip Splitter m, Span m; *(Pap)* Hackspan m, Hackschnitzel n, Holzschnitzel n, Kochschnitzel n, Schnitzel n
c. bin s. c. silo
c. board s. chipboard
c. capacity *(Pap)* Fülldichte f, Holzfülldichte f, Packungsdichte f
c. crusher *(Pap)* Desintegrator m, Desintegratormühle f, Schlagmühle f, Schlägermühle f, Schleudermühle f
c. distributor *(Pap)* Kocherfüllapparat m
c.-filled *(Pap)* mit Hackschnitzeln gefüllt (beschickt)
c. filling *(Pap)* Holzfüllung f, Hackschnitzelfüllung f, Einbringen n der Hackschnitzel
c. length *(Pap)* Hackspanlänge f
c. loft *(Pap)* Hackschnitzelspeicher m, Spänehaus n
c. packer s. c. distributor
c. packing s. c. filling
c. screen *(Pap)* Hackschnitzelsortiermaschine f
c. screening *(Pap)* Sortierung (Sichtung) f der Hackschnitzel
c. silo *(Pap)* Hackschnitzelsilo m, Hackschnitzelbehälter m
c. storage *(Pap)* Hackschnitzellagerung f
c. storage bin s. c. silo
chipboard 1. Maschinengraupappe f; 2. Span[holz]platte f
chipbreaker s. chip crusher

chipped ice Splittereis *n*
chipper *(Pap)* Hackmaschine *f*, Hacke *f*, Hacker *m*
  c. knife *(Pap)* Hackmesser *n*
chipping *(Pap)* Hacken *n*, Durchhacken *n*, Zerhacken *n*; *(Ker)* Absplittern *n*, Abschuppen *n (z.B. der Glasur)*
  c. machine *s.* chipper
chippings *s.* chips
chips *(Pap)* Hackspäne *mpl*, Hackschnitzel *npl*, Holzschnitzel *npl*, Kochschnitzel *npl*, Schnitzel *npl*
Chir pine *(Gerb)* Pinus voxburghii Sarg. *(asiatische Kiefernart)*
chirality *(Stereochemie)* Chiralität *f*
chit malt Kurzmalz *n*, Spitzmalz *n*
chitinous chitinös, chitinig, Chitin...
chittam (chittem, chittim) bark Cascara-sagrada-Rinde *f*, Kaskararinde *f*, Sagrada[rinde] *f*, Amerikanische Faulbaumrinde *f (von Rhamnus purshiana DC.)*
  chl. *s.* chloroform
chladnite *(Min)* Chladnit *m (Magnesiumdimetasilikat)*
chloanthite *(Min)* Chloanthit *m*, Weißnickelkies *m (Nickelarsenid)*
chloracetic acid *s.* chloroacetic acid
chloracetophenone *s.* chloroacetophenone
chloral $CCl_3 \cdot CHO$ Chloral *n*, Trichloräthanal *n*, Trichlorazetaldehyd *m*
  c. hydrate $CCl_3 \cdot CH \cdot (OH)_2$ Chloralhydrat *n*, 1,1,1-Trichlor-2,2-dihydroxyäthan *n*, Trichlorazetaldehydhydrat *n*
chlorallylene $CH_2=CHCH_2Cl$ Allylchlorid *n*, 3-Chlorpropylen *n*, 3-Chlorpropen-(1) *n*
chloralose Chloralose *f (ein Narkotikum)*
chlorambucil $(ClC_2H_4)_2NC_6H_4(CH_2)_3COOH$ Chlorambuzil *n*, 4-[*p*-Di-(2-chloräthyl)aminophenyl]-buttersäure *f*
chloramine Chloramin *n*, *(i.e.S.)* NH_2Cl Chloramin *n*; *s.* chloramine-T
chloramine-T $CH_3C_6H_4SO_2NNaCl \cdot 3H_2O$ Monochloramin T *n*, Toluolsulfonsäurechloramidnatrium *n*
chloramphenicol Chloramphenikol *n*, *d*-Threo-1-*p*-nitrophenyl-2-dichlorazetylaminopropandiol-1,3 *n*
chloranil $C_6Cl_4O_2$ Chloranil *n*, Tetrachlor-*p*-benzochinon *n*
chloraniline Chloranilin *n*
chlorapatite *(Min)* Chlorapatit *m*
chlorargyrite *(Min)* Chlorargyrit *m*, Kerargyrit *m*, Chlorsilber *n*, Hornsilber *n*, Silberhornerz *n (Silberchlorid)*
chlorate $M^ICl O_3$ Chlorat *n*
  c. explosive Chloratsprengstoff *m*
chloratite Chloratit *n (ein Chloratsprengstoff)*
chloraurate *s.* chloroaurate
chlorauric acid *s.* chloroauric acid
chlorazotic acid *s.* chloroazotic acid
chlorbenzene, chlorbenzol *s.* chlorobenzene

chlore / to chloren
chlorellagic acid Chlorellagsäure *f*
chlorendic acid *cis*-Hexachlor-endo-methylen-tetrahydrophthalsäure *f*
chlorhydrin *s.* chlorohydrin
chloric acid $HClO_3$ Chlorsäure *f*
chloride M^ICl Chlorid *n*
  c. brine Chloridsole *f*
  c. melt Chloridschmelze *f*
  c. of lime Chlorkalk *m*, Bleichkalk *m*, Bleichpulver *n*
  c. of potash *(Landw)* 50er Kalidüngesalz *n (mit etwa 50% K_2O-Gehalt)*
  c. paper *(Foto)* Chloridpapier *n*, Chlorsilberpapier *n*
chloridize / to mit Chlor *(oder* Chlorid*)* behandeln, chlorieren
chloridizing roasting chlorierendes Rösten *n*, chlorierende Röstung *f*
chlorinate / to chlorieren; chloren
chlorinated butyl rubber Chlorbutylkautschuk *m*
  c. hydrocarbon Chlorkohlenwasserstoff *m*
  c. lignin *(Pap)* Chlorlignin *n*, chloriertes Lignin *n*
  c. lime Chlorkalk *m*, Bleichkalk *m*, Bleichpulver *n*
  c. paraffin Chlorparaffin *n*
  c. rubber Chlorkautschuk *m*
  c.-rubber lacquer Chlorkautschuklack *m*
  c.-rubber paint Chlorkautschuk[anstrich]farbe *f*
chlorinating Chlorieren *n*, Chlorierung *f*; Chloren *n*, Chlorung *f*
  c. agent Chlorierungsmittel *n*; Chlorungsmittel *n*
  c. vessel Chlorierungsgefäß *n*, Chlorierungsbehälter *m*, Chlorierungskessel *m*, Chlorierungsapparat *m*, Chlorierer *m*, Chlorator *m*
chlorination Chlorieren *n*, Chlorierung *f*; Chloren *n*, Chlorung *f*
  c. stage Chlorierungsstufe *f*
  c. tower Chlor[ierungs]turm *m*
chlorinator Chlorierungsgefäß *n*, Chlorierungsbehälter *m*, Chlorierungskessel *m*, Chlorierungsapparat *m*, Chlorierer *m*, Chlorator *m*; Chlor[ierungs]turm *m*
chlorine Cl Chlor *n*
  c. bleaching Chlorbleiche *f*
  c. cell Chlorzelle *f*
  c. consumption Chlorverbrauch *m*
  c. cyanide ClCN Chlorzyan *n*, Zyanchlorid *n*
  c. detonating gas Chlorknallgas *n*
  c. dioxide ClO_2 Chlordioxid *n*, Chlor(IV)-oxid *n*
  c. dioxide bleaching *(Pap)* Chlordioxidbleiche *f*
  c. dioxide bleaching liquor *(Pap)* Chlordioxid-Bleichlauge *f*
  c. dioxide stage *(Pap)* Chlordioxidstufe *f (Bleichstufe)*
  c. drying Chlortrocknung *f*
  c. electrode Chlorelektrode *f*
  c. fastness *(Text)* Chlorechtheit *f*
  c. gas Chlorgas *n*
  c. heptoxide Cl_2O_7 Chlorheptoxid *n*, Dichlorheptoxid *n*, Chlor(VII)-oxid *n*

c. hydrate $Cl_2 \cdot 8H_2O$ Chlor-8-Wasser n, Chloroktahydrat n
c. liquefaction Chlorverflüssigung f
c. monoxide Cl_2O Chlor[mon]oxid n, Dichloroxid n, Chlor(I)-oxid n
c. number (Pap) Chlor[verbrauchs]zahl f
c. retention Chlorzurückhaltung f, Chlorrückhaltevermögen n, Chlorretention f
c. stage Chlorierungsstufe f
c. tetroxide ClO_4 Chlortetroxid n
c. water Chlorwasser n
c.-water solution Chlorwasser n
chloriridate s. chloroiridate
chlorite $M^I ClO_2$ Chlorit n; (Min) Chlorit m
c. bleaching (Text) Chloritbleiche f, Natriumchloritbleiche f
c. bleaching fastness (Text) Chloritbleichechtheit f
c. ferrugineuse (Min) Delessit m
c. schist (Geol) Chlor[it]schiefer m
chloritization (Geol) Chloritisierung f, Chloritisation f
chloritoid (Min) Chloritoid m
chlormagnesite (Min) Chlor[o]magnesit m (Magnesiumchlorid)
chlormethine $C_5H_{11}Cl_2N$ Chlormethin n, Di-(2-chloräthyl)methylamin n, Methylbis-(2-chloräthyl)amin n, Mechloräthamin n
chloro acetaldehyde s. 2-chloroethanal
chloro-acetaldehyde-diethyl-acetal s. chloroacetal
chloroacetal $ClCH_2 \cdot CH(OC_2H_5)_2$ Chlorazetal n, Chlorazetaldehyddiäthyl-azetal n
chloroacetic acid $CH_2ClCOOH$ Chloressigsäure f, Chloräthansäure f
chloroacetone CH_3COCH_2Cl Chlorazeton n, Monochlorazeton n, Azetonylchlorid n, Chlorpropanon n
chloroacetophenone Chlorazetophenon n, (i.e.S.) $C_6H_5COCH_2Cl$ α-Chlorazetophenon n, ω-Chlorazetophenon n, Phenazylchlorid n
2-chloroallyldiethyldithiocarbamate 2-Chlor-allyl-N,N-diäthyldithiokarbamat n, CDEC n (Herbizid)
chloroaniline Chloranilin n
chloroanthraquinone Chloranthrachinon n
chloroantimonate Chloroantimonat n
chloroargentate Chloroargentat n
chloroaurate Chloroaurat n
chloroauric acid $HAuCl_4 \cdot 4H_2O$ Tetrachlorogold-(III)-säure f, Goldchlorwasserstoffsäure f, Chlorogoldsäure f
chloroazotic acid Königswasser n
chlorobenzene C_6H_5Cl Chlorbenzol n, Monochlorbenzol n, Phenylchlorid n
2-chloro-1,4-benzenediol $C_6H_3(OH)_2Cl$ Chlorbenzoldiol-(1,4) n, Chlor-1,4-dihydroxybenzol n, Chlorhydrochinon n
chlorobenzilate Chlorbenzilat , 4,4'-Dichlorbenzilsäureäthylester m
chlorobenzoic acid ClC_6H_4COOH Chlorbenzoesäure f, Chlorbenzolkarbonsäure f

chlorobenzol s. chlorobenzene
chlorobromate Chlorobromat n
chlorobromide s. chlorobromate
c. paper (Foto) Chlorbromsilberpapier n
1-chlorobutane $CH_3(CH_2)_2CH_2Cl$ 1-Chlorbutan n, Butylchlorid n
chlorobutyl rubber Chlorbutylkautschuk m
chlorocadmate Chlorokadmat n
chlorocarbonic acid $ClCOOH$ Chlorameisensäure f, Chlorkohlensäure f
chlorochromate $M^I[CrO_3Cl]$ Chlorochromat n
chlorochromic anhydride CrO_2Cl_2 Chrom(VI)-oxidchlorid n, Chromylchlorid n
chlorocomplex Chlorokomplex m
chlorocresol Chlorkresol n
chlorocruorin Chlorokruorin n (Pigment im Blut von Borstenwürmern)
chlorocuprate Chlorokuprat n
2-chloro-1,4-dihydroxybenzene $C_6H_3(OH)_2Cl$ Chlor-1,4-dihydroxybenzol n, Chlorbenzoldiol-(1,4) n, Chlorhydrochinon n
chloro-dimethyl arsine $(CH_3)_2AsCl$ Chlordimethylarsin n, Dimethylchlorarsin n, Dimethylarsinchlorid n, Kakodylchlorid n
1-chloro-2,4-dinitrobenzene $C_6H_3(NO_2)_2Cl$ 1-Chlor-2,4-dinitrobenzol n, 2,4-Dinitrochlorbenzol n, Dinitrochlorbenzol n
2-chloroethanal $CH_2Cl \cdot CHO$ Chlorazetaldehyd m
chloroethane CH_3CH_2Cl Chloräthan n, Monochloräthan n, Äthylchlorid n, Chloräthyl n
chloroethanoic acid $CH_2ClCOOH$ Chloräthansäure f, Chloressigsäure f
2-chloroethanol $ClCH_2CH_2OH$ 2-Chloräthanol n, 2-Chloräthylalkohol m, Äthylenchlorhydrin n, Glykolchlorhydrin n
chloroethene s. chloroethylene
β-chloroethyl alcohol s. 2-chloroethanol
chloroethyl group $ClCH_2CH_2-$ Chloräthylgruppe f, Chloräthylrest m
bis-β-chloroethyl sulphide $(ClCH_2CH_2)_2S$ Dichlordiäthylsulfid n, Senfgas n, Schwefel-Yperit n
chloroethylene $CH_2=CHCl$ Chloräthen n, Chloräthylen n, Vinylchlorid n
chloroform $CHCl_3$ Chloroform n, Trichlormethan n, Chlf.
chloroformic acid ester Chlorameisensäureester m, Chlorkohlensäureester m
chloroformyl chloride $COCl_2$ Kohlenoxidchlorid n, Karbonylchlorid n, Phosgen n
chlorogenic acid $(HO)_2C_6H_3 \cdot CH=CH \cdot COO \cdot C_6H_2(OH)_3COOH$ Chlorogensäure f, 3,4-Dihydroxyzinnamoylchinasäure f, 3-Kaffoyl-1,4,5-trihydroxyzyklohexankarbonsäure f (Depsid aus China- und Kaffeesäure)
chloroguanide Chloroguanid n, Proguanil n, N^1-p-Chlorphenyl-N^5-isopropylbiguanid n
1-chlorohexane $CH_3(CH_2)_5Cl$ 1-Chlorhexan n, n-Hexylchlorid n
chlorohydrin Chlorhydrin n, Chlorwasserstoffsäureglyzerinester m, (i.e.S.) s. α-chlorohydrin

chlorohydrin

α-**chlorohydrin** HOCH₂CH(OH)CH₂Cl α-Monochlorhydrin *n*, Glyzerin-α-chlorhydrin *n*, 3-Chlor-1,2-propandiol *n*
chlorohydroquinone C₆H₃(OH)₂Cl Chlorhydrochinon *n*, Chlor-1,4-dihydroxybenzol *n*, Chlorbenzoldiol-(1,4) *n*
chloroiodate Chlorojodat *n*
chloroiodide s. chloroiodate
chloroiodite Mᴵ[JCl₄] Chlorojodat(III) *n*, Tetrachlorojodat(III) *n*
chloroiridate Chloroiridat *n*
chloroiridite Mᴵ₃[IrCl₆] Chloroiridat(III) *n*, Hexachloroiridat(III) *n*
chlorolignin *(Pap)* Chlorlignin *n*, chloriertes Lignin *n*
chloromagnesite *(Min)* Chlor[o]magnesit *m* (Magnesiumchlorid)
chloromanganate Chloromanganat *n*
chloromelane *(Min)* Chlormelan *m*, *(veraltet für)* Cronstedtit *m*
chloromercurate Chloromerkurat *n*
chlorometallic acid Chlorosäure *f*
chloromethane CH₃Cl Chlormethan *n*, Methylchlorid *n*
chloromethylation Chlormethylierung *f*
chloromolybdate Chloromolybdat *n*
chloronaphthalene Chlornaphthalin *n*
chloroniobate Chloroniobat *n*
chloro-nitrobenzene C₆H₄Cl(NO₂) Chlornitrobenzol *n*, Nitrochlorbenzol *n*
chloro-nitroparaffin Chlornitroparaffin *n*
chloronitrous acid Königswasser *n*
chloroosmate Chloroosmat *n*
chloroosmite Mᴵ₃[OsCl₆] Chloroosmat(III) *n*, Hexachloroosmat(III) *n*
chloropalladate Chloropalladat *n*
chloropalladite Mᴵ₂[PdCl₄] Chloropalladat(II) *n*, Tetrachloropalladat(II) *n*
chloropentammine-platinic chloride [Pt(NH₃)₅Cl]Cl₃ Chloropentamminplatin(IV)-chlorid *n*
1-chloropentane CH₃(CH₂)₃CH₂Cl 1-Chlorpentan *n*, *n*-Amylchlorid *n*
chlorophanerite *(Min)* Chlorophanerit *m*, *(veraltet für)* Glaukonit *m*
chlorophenol Chlorphenol *n*, Chloroxybenzol *n*
chlorophenylenediamine ClC₆H₃(NH₂)₂ Chlorphenylendiamin *n*
Nᴵ-p-chlorophenyl-N⁵-isopropylbiguanide C₁₁H₁₆ClN₅ Nᴵ-*p*-Chlorphenyl-N⁵-isopropylbiguanid *n*, Chlorguanid *n*, Proguanil *n*
chlorophosphate Chlorophosphat *n*
chlorophyll[I] Chlorophyll *n*, Blattgrün *n*
chlorophyllase Chlorophyllase *f*
chlorophyllide Chlorophyllid *n*
chlorophyllin Chlorophyllin *n*
chlorophyllite *(Min)* Chlorophyllit *m*
chloropicrin CCl₃NO₂ Chlorpikrin *n*, Trichlornitromethan *n*, Nitrochloroform *n*
chloroplast Chloroplast *m*, Chlorophyllkorn *n*

chloroplatinate Chloroplatinat *n*
chloroplatinic acid H₂[PtCl₆]·6H₂O Chloroplatin(IV)-säure *f*, Hexachloroplatin(IV)-säure *f*, Platin(IV)-chlorwasserstoffsäure *f*
chloroplatinite Mᴵ₂[PtCl₄] Chloroplatinat(II) *n*, Tetrachloroplatinat(II) *n*
chloroplumbate Chloroplumbat *n*
chloroprene CH₂=CClCH=CH₂ Chloropren *n*, 2-Chlor-1,3-butadien *n*
c. rubber Chloroprenkautschuk *m*
1-chloropropane CH₃CH₂CH₂Cl 1-Chlorpropan *n*, Propylchlorid *n*, Chlorpropyl *n*
2-chloropropane CH₃CHClCH₃ 2-Chlorpropan *n*, Isopropylchlorid *n*
3-chloro-1,2-propanediol HOCH₂CH(OH)CH₂Cl 3-Chlor-1,2-propandiol *n*, α-Monochlorhydrin *n*, Glyzerin-α-chlorhydrin *n*
3-chloropropene CH₂=CHCH₂Cl 3-Chlorpropen-(1) *n*, 3-Chlorpropylen *n*, Allylchlorid *n*
α-**chloropropionic acid** CH₃CHClCOOH 2-Chlorpropionsäure *f*
β-**chloropropionic acid** CH₂ClCH₂COOH 3-Chlorpropionsäure *f*
chloropropylene oxide γ-Chlorpropylenoxid *n*, Epichlorhydrin *n*, 1-Chlor-2,3-epoxypropan *n*
6-chloropurine 6-Chlorpurin *n*
chloroquine Chloroquin *n*, 7-Chlor-4-(4'-diäthylamino-1'-methylbutylamino)-chinolin *n*
chloroquinol C₆H₃(OH)₂Cl Chlorhydrochinon *n*, Chlor-1,4-dihydroxybenzol *n*, Chlorbenzoldiol-(1,4) *n*
chlororhenate Chlororhenat *n*
chlororhodate Chlororhodat *n*
chlororhodite Mᴵ₃[RhCl₆] Chlororhodat(III) *n*, Hexachlororhodat(III) *n*
chlororuthenate Chlororuthenat *n*
chloros NaOCl Natriumhypochlorit *n*
chloroselenite Chloroselenat *n*
chlorosilicane SiH₃Cl Chlorsilan *n*, Monochlorsilan *n*
chlorosis *(Bot)* Chlorose *f*, Bleichsucht *f*, Gelbsucht *f*
chlorostannate Chlorostannat *n*
chlorostannic acid H₂[SnCl₆] Hexachlorozinn(IV)-säure *f*
chlorosuccinic acid Chlorbernsteinsäure *f*
chlorosulphonate Chlorsulfat *n*
chlorosulphonated chlorsulfoniert, sulfochloriert
c. polyethylene chlorsulfoniertes (sulfochloriertes) Polyäthylen *n*
chlorosulphonation Chlorsulfonierung *f*, Sulfochlorierung *f*
chlorosulphonic acid SO₂(OH)Cl Chlorsulfonsäure *f*, *(veraltet für)* Chloroschwefelsäure *f*
c. anhydride S₂O₅Cl₂ Chloroschwefelsäureanhydrid *n*, Disulfurylchlorid *n*
chlorotellurate(IV) Mᴵ₂[TeCl₆] Chlorotellurat(IV) *n*, Hexachlorotellurat(IV) *n*
chlorotellurite Mᴵ₂[TeCl₆] Chlorotellurat(IV) *n*, Hexachlorotellurat(IV) *n*

chlorotetracycline Chlortetrazyklin n *(Antibiotikum)*
chlorothallate Chlorothallat n
chlorothionite *(Min)* Chlorothionit m
chlorotitanate(IV) $M^I_2[TiCl_6]$ Chlorotitanat(IV) n, Hexachlorotitanat(IV) n
o-chlorotoluene $CH_3C_6H_4Cl$ o-Chlortoluol n, o-Chlormethylbenzol n
α-chlorotoluene $C_6H_5CH_2Cl$ α-Chlortoluol n, Benzylchlorid n
chlorotrifluoroethylene $ClFC=CF_2$ Chlortrifluoräthylen n, Chlortrifluoräthen n
chlorotungstate Chlorowolframat n
chlorous acid $HClO_2$ chlorige Säure f
chlorozincate Chlorozinkat n
chlorozirconate $M^I_2[ZrCl_6]$ Chlorozirkonat(IV) n, Hexachlorozirkonat(IV) n
chlorpromazine Chlorpromazin n, 10-(γ-Dimethylaminopropyl)-3-chlorphenothiazin n
chlortetracycline *s.* chlorotetracycline
chocolate schokoladen, Schokolade[n]...; mit Schokolade[n]geschmack; mit Schokolade[n]überzug; schokolade[n]farben, schokolade[n]farbig, schokolade[n]braun
chocolate Schokolade f; Schokolade[n]braun n
  c. bloom Fettreif m *(dünner Belag aus Kakaobutterkristallen auf Schokolade)*
  c.-brown schokolade[n]braun, schokolade[n]farben, schokolade[n]farbig
  c. industry Schokoladenindustrie f
choice lard Liesenschmalz n, Flomenschmalz n
choke / to 1. verstopfen, versperren, verschmutzen, verschmieren; drosseln; *(Feuer)* hemmen, ersticken; 2. sich verstopfen
  c. up verstopfen, versperren, verschmutzen, verschmieren
  c. with mud verschlammen
choke crushing Verreiben n, Verreibung f
  c. mechanism *(Tech)* Drosseleinrichtung f
chokedamp Nachschwaden m *(Kohlendioxidanreicherungen nach Grubenexplosionen)*
cholesteric cholesterisch
  c. phase cholesterische Phase f
cholesterin Cholesterin n
cholesterol *s.* cholesterin
cholic acid Cholsäure f, 3α,7α,12α-Trihydroxy-5β-cholansäure f
chondrite Chondrit m *(ein Steinmeteorit)*
chondrodite *(Min)* Chondrodit m
chondrules *(Geol)* Chondren npl
chop / to [zer]hacken, zerkleinern, schneiden, schnitzeln, *(Rinden)* brechen, *(Stroh)* häckseln, *(Eis)* stoßen
chop length *(Pap)* Schnittlänge f
chopped ice zerkleinertes (feinzerschlagenes) gestoßenes) Eis n
  c. strand geschnittener Glasspinnfaden m
  c. strand mat Glasseidenmatte f, Glasfaservlies n
  c. strands Stapelglasseide f, gehackte Glasseidenstränge mpl

  c. straw gehäckseltes Stroh n, Strohhäcksel m
chopper Hackmaschine f, Hacke f, Hacker m; Häckselmaschine f *(Strohaufschluß)*
  c. strand mat *s.* chopped strand mat
chopping *(Pap)* Häckseln n, Häckselei f *(von Stroh)*
  c. machine *s.* chopper
chorismic acid Chorisminsäure f, Chorismasäure f *(3-Enolbrenztraubensäureäther der trans-3,4-Dihydroxyzyklohexadien-(1,5)-karbonsäure)*
chorismite *(Geol)* Chorismit m, chorismatisches Gestein n
Christmas tree *(Erdöl)* Eruptionskreuz n, Eruptivkreuz n, Eruptionskopf m, Christbaum n, Weihnachtsbaum m
chromable chromierbar
chromate $M^I_2CrO_4$ Chromat n, Monochromat n
  c. [dyeing] method *s.* c. process
  c. process *(Text)* Chromatverfahren n, Metachromverfahren n, Beizenverfahren n
  c. red $Pb(OH)_2 \cdot PbCrO_4$ Chromrot n, Chromzinnober m, Persischrot n
chromatic aberration chromatische Aberration f, Farbenabweichung f, Farb[en]fehler m
  c. gelatin Chrom[at]gelatine f
chromatite *(Min)* Chromatit m *(Kalziumchromat)*
chromatize / to chromat[is]ieren
chromatogram Chromatogramm n
chromatograph / to chromatografieren
chromatographic chromatografisch
  c. adsorption [analysis] chromatografische Adsorptionsanalyse f, Adsorptionschromatografie f
  c. analysis Chromatografie f
  c. cabinet Chromatografiekammer f
  c. column Chromatografiesäule f, chromatografische Säule f
  c. method *s.* c. analysis
  c. packing Füllstoff m, Adsorptionsmittel n *(für die Adsorptionschromatografie)*
  c. paper chromatografisches Papier n, Chromatografiepapier n
  c. tube Chromatografierohr n
chromatography Chromatografie f
  c. apparatus Geräte npl zur Chromatografie; chromatografisches Entwicklungsgerät n
  c. chamber *(chromatografische)* Entwicklungskammer f, Chromatografiekammer f
chromatometric method Chromatometrie f
chromatometry Chromatometrie f
chromatopack method Chromatopackverfahren n, Filterpaketchromatografie f
chromatophore *(Biol)* Chromatophor n, Farbstoffträger m
chromatopile method Chromatopileverfahren n, Rundfilterchromatografie f
chromatosprayer Sprüher m *(bei der Papier- und Dünnschichtchromatografie)*
chrome / to 1. *(Farb)* chromieren; 2. mit Chromsalzen gerben, chromgerben; 3. *(Metalle)* verchromen
chrome *s.* chromium

c. alum $M^ICr(SO_4)_2 \cdot 12H_2O$ Chromalaun *m*, *(i.e.S.)* $KCr(SO_4)_2 \cdot 12H_2O$ Kaliumchromalaun *m*, Kaliumchrom(III)-sulfat-12-Wasser *n*
c. alum ammonium *s.* c. ammonium alum
c.-ammine Chromiak *n (Ammin-chrom(III)-Kom-plex)*
c. ammonium alum $NH_4Cr(SO_4)_2 \cdot 12H_2O$ Ammoniumchromalaun *m*, Ammoniumchrom(III)-sulfat-12-Wasser *n*
c. aventurine Chromaventuringlas *n*
c. black Chromschwarz *n*
c. brown Chrombraun *n*
c. colour *s.* c. dye
c.-developed dye Chromentwicklungsfarbstoff *m*, Nachchromier[ungs]farbstoff *m*
c. dye Chromfarbstoff *m*, Chromier[ungs]farb-stoff *m*
c. gelatin Chrom[at]gelatine *f*
c. glue Chromleim *m*
c. green Chromgrün *n*, Deckgrün *n*, Druckgrün *n*, Ölgrün *n (grüne Mischfarbe aus Chromgelb und Berlinerblau)*; $Cr_2O_3 \cdot xH_2O$ Chromoxid-hydratgrün *n*, Smaragdgrün *n*, Brillantgrün *n*, Guignetgrün *n*; Cr_2O_3 Chromoxidgrün *n*, Laub-grün *n*, grüner Zinnober *m (Chrom(III)-oxid)*
c. iron ore *(Min)* Chromeisenerz *n*, Chromit *m (Eisen(II)-chrom(III)-oxid)*
c. leather Chromleder *n*
c. liquor *(Gerb)* Chrom[gerb]brühe *f*
c.-mica *(Min)* Chromglimmer *m*
c. molybdenum steel Chrommolybdänstahl *m*
c. nickel steel Chromnickelstahl *m*
c. ochre *(Min)* Chromocker *m*
c. orange $PbO \cdot PbCrO_4$ Chromorange *n*
c. oxide green Cr_2O_3 Chromoxidgrün *n*, Laub-grün *n*, grüner Zinnober *m (Chrom(III)-oxid)*
c. pit tannage Chromgrubengerbung *f*
c. potash alum $KCr(SO_4)_2 \cdot 12H_2O$ Kalium-chromalaun *m*, Chromalaun *m*, Kalium-chrom(III)-sulfat-12-Wasser *n*
c. re-tannage *s.* c. retan[ning]
c. re-tanned *(mit Chromsalzen gegerbt und mit pflanzlichen oder synthetischen Gerbstoffen nachgegerbt)*
c. red Chromrot *n*, Chromzinnober *m*, Persisch-rot *n (basisches Bleichromat schwankender Zusammensetzung)*
c. retan[ning] *(Kombinationsgerbung mit Chrom-salzen und pflanzlichen oder synthetischen Gerbstoffen)*
c.-spinel *(Min)* Chromspinell *m*, Pikotit *m*
c. steel Chromstahl *m*
c. tannage *s.* c. tanning
c. tanned chromgar, chromgegerbt
c. tanning Chromgerbung *f*
c. topped nachchromiert
c. violet Chromviolett *n*
c. yellow $PbCrO_4$ Chromgelb *n*
chromiate *(Gerb)* Chromverbindung *f*
chromic Chrom..., *(i.e.S.)* Chrom(III)-... *(s. dagegen* c. acid)

c. acetate $Cr(CH_3COO)_3$ Chrom(III)-azetat *n*, Chromtriazetat *n*
c. acid H_2CrO_4 Chromsäure *f*, Monochromsäure *f*
c.-acid oxidation Chromsäureoxydation *f*
c.-acid process Chromsäureverfahren *n*, Bengough-Stuart-Verfahren *n (mit Chromsäure arbeitendes Verfahren zur anodischen Oxyda-tion)*
c. ammonium sulphate $NH_4Cr(SO_4)_2 \cdot 12H_2O$ Ammoniumchrom(III)-sulfat-12-Wasser *n*, Am-moniumchromalaun *m*
c. anhydride CrO_3 Chrom(VI)-oxid *n*, Chrom-trioxid *n*
c. arsenide CrAs Monochromarsenid *n*
c. boride CrB Monochromborid *n*
c. bromide $CrBr_3$ Chrom(III)-bromid *n*, Chrom-tribromid *n*
c. carbide Cr_3C_2 Trichromdikarbid *n*
c. chloride $CrCl_3$ Chrom(III)-chlorid *n*, Chrom-trichlorid *n*
c. fluoride CrF_3 Chrom(III)-fluorid *n*, Chrom-trifluorid *n*
c. formate $Cr(HCOO)_3$ Chromformiat *n*
c. hydrate *s.* c. hydroxide
c. hydroxide $Cr(OH)_3$ Chrom(III)-hydroxid *n*
c. iron *(Min)* Chromeisenerz *n*, Chromit *m (Eisen(II)-chrom(III)-oxid)*
c. nitrate $Cr(NO_3)_3$ Chrom(III)-nitrat *n*
c. nitride CrN Chromnitrid *n*
c. orthophosphate $CrPO_4$ Chrom(III)-ortho-phosphat *n*
c. oxide Cr_2O_3 Chrom(III)-oxid *n*, Dichromtrioxid *n*
c. phosphide CrP Chromphosphid *n*
c. potassium sulphate $KCr(SO_4)_2 \cdot 12H_2O$ Ka-liumchrom(III)-sulfat-12-Wasser *n*, Kaliumchrom-alaun *m*, Chromalaun *m*
c. sulphate $Cr_2(SO_4)_3$ Chrom(III)-sulfat *n*
c. sulphide Cr_2S_3 Chrom(III)-sulfid *n*
chromicyanide $M^I_3[Cr(CN)_6]$ Zyanochromat(III) *n*, Hexazyanochromat(III) *n*
chroming 1. *(Farb)* Chromierung *f*; 2. Chrom-gerbung *f*; 3. Verchromen *n*, Verchromung *f (von Metallen)*
chromioxalate $M^I_3[Cr(C_2O_4)_3]$ Oxalatochromat(III) *n*, Trioxalatochromat(III) *n*
chromite Chromat(III) *n*; *(Min)* Chromit *m*, Chrom-eisenerz *n (Eisen(II)-chrom(III)-oxid)*
chromium Cr Chrom *n*
c.-ammonium sulphate $NH_4Cr(SO_4)_2 \cdot 12H_2O$ Ammoniumchrom(III)-sulfat-12-Wasser *n*, Am-moniumchromalaun *m*
c. carbonyl $Cr(CO)_6$ Chromkarbonyl *n*
c. complex Chromkomplex *m*, komplexe Chrom-verbindung *f*
c. dioxide CrO_2 Chromdioxid *n*, Chrom(IV)-oxid *n*
c. intensifier *(Foto)* Chromverstärker *m*
c. mordant Chrombeize *f*

c.-nickel steel Chrom-Nickel-Stahl *m*
c. ore Chromerz *n*
c. oxychloride CrO_2Cl_2 Chrom(VI)-oxidchlorid *n*, Chromylchlorid *n*
c.-plated verchromt
c. plating Verchromen *n*, Verchromung *f*
c.-potassium sulphate $KCr(SO_4)_2 \cdot 12H_2O$ Kaliumchrom(III)-sulfat-12-Wasser *n*, Kaliumchromalaun *m*, Chromalaun *m*
c.-rubidium sulphate $RbCr(SO_4)_2$ Rubidiumchrom(III)-sulfat *n*
c. sesquioxide *s.* chromic oxide
c. steel Chromstahl *m*
c. tannage (tanning) Chromgerbung *f*
c. tetrasulphide Cr_3S_4 Chromtetrasulfid *n*, Trichromtetrasulfid *n*, Chrom(II,III)-sulfid *n*
c. trioxide CrO_3 Chromtrioxid *n*, Chrom(VI)-oxid *n*
chromize / to inchromieren, inkromieren, chromdiffundieren
chromized steel Inchromierstahl *m*, Inkromierstahl *m*, inchromierter (inkromierter) Stahl *m*
chromizing Inchromieren *n*, Inchromierung *f*, Inkromieren *n*, Inkromierung *f*, Chromdiffundieren *n*, Diffusionsverchromung *f*, Einsatzverchromung *f*
chromo base paper Chromorohpapier *n*
c. board Chromokarton *m*
c. body paper *s.* c. base paper
c. paper Chromopapier *n*
chromocher *s.* chrome ochre
chromocyanide $M^I_4[Cr(CN)_6]$ Zyanochromat(II) *n*, Hexazyanochromat(II) *n*
chromoferrite *(Min)* Chromoferrit *m*, *(veraltet für)* Chromit *m*
chromogen Chromogen *n* *(Verbindung mit chromophorer Gruppe)*
chromoisomer Chromoisomer[es] *n*
chromoisomerism Chromoisomerie *f*
chromone Chromon *n*, 1,4-Benzopyron *n*
chromophore Chromophor *m*, Farbträger *m*, chromophore (farbtragende, farbgebende) Gruppe *f*
chromophoric chromophor, farbtragend, farbgebend
c. group *s.* chromophore
chromoproteid Chromoproteid *n*
chromotrope acid Chromotropsäure *f*, 1,8-Dihydroxynaphthalin-3,6-disulfonsäure *f*
chromous Chrom..., *(i.e.S.)* Chrom(II)-...
c. acetate $Cr(CH_3COO)_2$ Chrom(II)-azetat *n*
c. bromide $CrBr_2$ Chrom(II)-bromid *n*, Chromdibromid *n*
c. carbonate $CrCO_3$ Chrom(II)-karbonat *n*
c. chloride $CrCl_2$ Chrom(II)-chlorid *n*, Chromdichlorid *n*
c. fluoride CrF_2 Chrom(II)-fluorid *n*, Chromdifluorid *n*
c. hydroxide $Cr(OH)_2$ Chrom(II)-hydroxid *n*
c. iodide CrJ_2 Chrom(II)-jodid *n*, Chromdijodid *n*

c. oxalate CrC_2O_4 Chrom(II)-oxalat *n*
c. oxide CrO Chrom(II)-oxid *n*, Chrommonoxid *n*
c. sulphate $CrSO_4$ Chrom(II)-sulfat *n*
c. sulphide CrS Chrom(II)-sulfid *n*, Chrom[mono]sulfid *n*
chromyl chloride CrO_2Cl_2 Chromylchlorid *n*, Chrom(VI)-oxidchlorid *n*
chronic toxicity chronische Toxizität (Giftigkeit) *f*
chronopotentiometry Chronopotentiometrie *f*
chrysamine Chrysamin *n*
chrysaniline Chrysanilin *n*, 3-Amino-9-(4-aminophenyl)-akridin *n*
chrysanthemine Chrysanthemin *n*, Zyanidin-3-monoglukosid *n*
chrysanthemumic acid Chrysanthemumsäure *f*
chrysarobin *(Pharm)* Chrysarobin *n* *(aus Andira araroba Aguiar)*
chrysatropic acid Chrysatropasäure *f*, Gelseminsäure *f*, Skopoletin *n*, β-Methyläskuletin *n*, 6-Methoxy-7-hydroxykumarin *n*
chryselectrum *s.* chrysoberyl
chrysene Chrysen *n*, 1,2-Benzphenanthren *n*, Benzophenanthren *n*
chrysergonic acid Chrysergonsäure *f*, Sekalonsäure *f*
chrysin Chrysin *n*, 5,7-Dihydroxyflavon *n*
chrysoberyl *(Min)* Chrysoberyll *m* *(Berylliumaluminat)*
chrysocolla *(Min)* Chrysokoll *m* *(Kupfer(II)-metasilikat)*
chrysogen Tetrazen *n*, Naphthazen *n*, 2,3-Benzanthrazen *n*
chrysoidine $(NH_2)_2C_6H_3N=NC_6H_5$ Chrysoidin *n*, 2,4-Diaminoazobenzol *n*
chrysolite *(Min)* Chrysolith *m*
chrysophanic acid Chrysophansäure *f*, Chrysophanol *n*, 1,8-Dihydroxy-3-methylanthrachinon *n*
chrysophanol *s.* chrysophanic acid
chrysophenine Chrysophenin [G] *n*, Aurophenin O *n*
c. G *s.* chrysophenine
chrysoprase *(Min)* Chrysopras *m* *(Quarzabart)*
chrysotile *(Min)* Chrysotil *m*, Faserserpentin *m*
Chu, CHU *s.* centigrade heat unit
Chugaev reaction Tschugajew-Reaktion *f*, Tschugajewsches Xanthogenatverfahren *n*
churn / to [ver]buttern; *(Margarine)* [ver]kirnen
c. to butter *s.* [ver]buttern
churn Butterfaß *n*; *(Margarineherstellung)* Kirne *f*, Kirnapparat *m*, Kirnmaschine *f*; *(Text)* Baratte *f*, Sulfidiertrommel *f*, Xanthatkneter *m*
churnability Butterungsfähigkeit *f*
churning Buttern *n*, Butterung *f*, Verbuttern *n*, Verbutterung *f*, Butterbereitung *f*, Butterherstellung *f*, Buttererzeugung *f*; *(Margarineherstellung)* Kirnen *n*, Verkirnen *n*, Kirnung *f*; *(Text)* Sulfidieren *n*, Sulfidierung *f*, Xanthogenieren *n*, Xanthogenierung *f*
c. period *s.* c. time
c. process Butterungsverfahren *n*; *(Margarineherstellung)* Kirnprozeß *m*

c. temperature Butterungstemperatur *f*; *(Margarineherstellung)* Kirntemperatur *f*
c. time Butterungszeit *f*, Butterungsdauer *f*; *(Margarineherstellung)* Kirndauer *f*
chute Rutsche *f*, Schurre *f*, Rinne *f*, Schütte *f*
CI *(Abk. für)* 1. Colour Index; 2. Chlorine Institute
CI No. *s.* Colour Index number
cider Apfelmost *m*, *(i.w.S.)* Süßmost *m*; Apfelwein *m*, Zider *m*, *(i.w.S.)* Obstwein *m*
cigar wrapping paper Zigarrendeckblattpapier *n*, Zigarrenumblattpapier *n*, Kunstumblattpapier *n*, Umblattpapier *n*
cigarette paper Zigarettenpapier *n*
c. tissue *(Am) s.* c. paper
cinchona Chinarindenbaum *m*, Cinchona L.; Chinarinde *f*, Fieberrinde *f*
c. alkaloid Cinchonaalkaloid *n*, China[rinden]alkaloid *n*
c. bark Chinarinde *f*, Fieberrinde *f*
cinchonidine Zinchonidin *n*, α-Chinidin *n*
cinchonine Zinchonin *n* *(Alkaloid)*
cinchoratine *s.* cinchonidine
cinder Schlacke *f*; Zunder *m*
c. notch Schlacken[ab]stichloch *n*, Schlacken[ab]stich *m*, *(meist)* Schlackenform *f*
cineole 1,8-Zineol *n*, Zineol *n*, Eukalyptol *n*, 1,8-Oxido-*p*-menthan *n*
cinnabar[ite] *(Min)* Zinnabarit *m*, Zinnober *m* *(Quecksilber(II)-sulfid)*
cinnamal *s.* cinnamaldehyde
cinnamaldehyde $C_6H_5CH=CHCHO$ Zinnam[yl]aldehyd *m*, Zimtaldehyd *m*, Zinnamal *n*, 3-Phenylpropenal *n*
cinnamate Zinnamat *n* *(Salz oder Ester der Zimtsäure)*
cinnamein $C_6H_5CH=CHCOOCH_2C_6H_5$ Zinnamein *n*, Benzylzinnamat *n*, Zimtsäurebenzylester *m*
cinnamene $C_6H_5 \cdot CH=CH_2$ Styrol *n*, Phenyläthylen *n*, Vinylbenzol *n*
cinnamic acid $C_6H_5CH=CHCOOH$ [gewöhnliche] Zimtsäure *f*, *trans*-Zimtsäure *f*, *trans*-3-Phenylakrylsäure *f*, *trans*-3-Phenylpropensäure *f*
***allo*-c. acid** $C_6H_5CH=CHCOOH$ Allozimtsäure *f*, *cis*-3-Phenylpropensäure *f*, *cis*-3-Phenylakrylsäure *f*
***cis*-c. acid** *s. allo*-c. acid
***trans*-c. acid** *s.* c. acid
c. alcohol $C_6H_5CH=CHCH_2O$ Zimtalkohol *m*, Zinnamylalkohol *m*, Styron *n*, Styryl-3-phenylpropen-2-ol *n*
c. aldehyde *s.* cinnamaldehyde
cinnamol *s.* cinnamene
cinnamon Zimtstrauch *m*, Zimtbaum *m*, Cinnamomum Schaeffer; Zimtrinde *f*, Zimt *m*; [gemahlener] Zimt *m*
c. bark Zimtrinde *f*, Zimt *m*
c. leaf oil Zimtblätteröl *n*
c. oil Zimtöl *n*, *(i.e.S.)* Zimtrindenöl *n*,
cinnamyl acid *s.* cinnamic acid
c. alcohol *s.* cinnamic alcohol

c. aldehyde *s.* cinnamaldehyde
cinnamylic acid *s.* cinnamic acid
CIPC $ClC_6H_4NHCOOCH(CH_3)_2$ CIPC *n*, Chlor-IPC *n*, Chlorpropham *n*, Isopropyl-N-(3-chlorphenyl)karbamat *n* *(Herbizid)*
cipher / to chiffrieren *(z.B. chemische Strukturen)*
cipher Chiffre *f*
c. notation *s.* ciphering
ciphering Chiffrierung *f*
c. system Chiffrier[ungs]system *n*, Chiffresystem *n*
circle-throw screen Kreisschwingsieb *n*; Steilwurfsieb *n*
circular cap Rundglocke *f*
c.-chart recorder Kreisblattschreiber *m*
c. chromatography Zirkularchromatografie *f*, Ringchromatografie *f*, Rundfilterchromatografie *f*
c. component Zirkularkomponente *f*
c. couche *(Lebm)* Rundreibe[maschine] *f*
c.-deck table Rundherd *m*
c. die Ringdüse *f* *(Strangpreßwerkzeug)*
c. grate Kreisrost *m*
c. knife Rundmesser *n*, Tellermesser *n*, Kreismesser *n*
c. orbit Kreisbahn *f*, kreisförmige Umlaufbahn *f*
c. paper chromatography *s.* c. chromatography
c. polarization zirkulare Polarisation *f*, Zirkularpolarisation *f*
c. slitting knife *(Pap)* Tellermesser *n*, Kreismesser *n*
circularly polarized zirkularpolarisiert, zirkular polarisiert
circulate / to im Kreis[lauf] führen, zirkulieren [lassen], umlaufen [lassen], umwälzen
circulating-liquor dyeing machine Zirkulationsfärbeapparat *m*
c. load Umlaufbeladung *f*, Umlaufgut *n*, Umlaufanteil *m*
c. pump Umwälzpumpe *f*, Umlaufpumpe *f*, Kreislaufpumpe *f*, Zirkulationspumpe *f*
c. water Umlaufwasser *n*, Rückwasser *n*
circulation Kreislauf *m*, Zirkulation *f*, Umlauf *m*, *(manchmal auch)* Umwälzung *f*
c. heater Umwälzheizeinrichtung *f*
c. of liquor *(Pap)* Kochsäurezirkulation *f*, Kochsäureumlauf *m*, Kochsäureumwälzung *f*, *(beim alkalischen Verfahren)* Laugenzirkulation *f*, Laugenumlauf *m*, Laugenumwälzung *f*
c. of oil Ölumlauf *m*
c. rate Umwälz[förder]strom *m*; Umwälzgeschwindigkeit *f*
c. system Kreislaufsystem *n*, Zirkulationssystem *n*, Umlaufsystem *n*, *(bei Rotary-Bohranlagen auch)* Spülungskreislauf *m*
circulator *s.* circulating pump
circumneutral reaction im Neutralbereich liegende Reaktion *f*
cire-perdue process Wachsausschmelzverfahren *n*, Lost-Wax-Prozeß *m*

cis cis-..., cis-ständig, in cis-Stellung befindlich
to be c. cis (cis-ständig, in cis-Stellung) stehen, cis-ständig [angeordnet] sein
c. addition cis-Addition f
c.-cis isomer cis-cis-Isomer[es] n
c. form cis-Form f, Cisform f
c. isomer cis-Isomer[es] n
c.-oriented cis-orientiert
c. position cis-Stellung f, cis-Lage f
c.-trans isomer cis-trans-Isomer[es] n
c.-trans isomerism cis-trans-Isomerie f, geometrische Isomerie f
cisoid cisoid
citraconic acid $CH_3C(COOH)=CHCOOH$ Zitrakonsäure f, Methylmaleinsäure f, Methylbutendisäure f
citral a, α-citral $(CH_3)_2C=CHCH_2CH_2C(CH_3)=CHCHO$ Zitral A n, trans-Zitral n, Geranial n
c. b, β-citral $(CH_3)_2C=CHCH_2CH_2C(CH_3)=CHCHO$ Zitral B n, cis-Zitral n, Neral n
citrate Zitrat n
c.-solubility Zitratlöslichkeit f
c.-soluble zitratlöslich
citrazinic acid Zitrazinsäure f, 2,6-Dihydroxyisonikotinsäure f
citrene Zitrin n, α-Limonen n, Karven n, Hesperiden n, d-1-Methyl-4-isopropylzyklohexadien-(1,8) n
citric acid $HOC(CH_2COOH)_2COOH$ Zitronensäure f, 3-Hydroxytrikarballylsäure f, 2-Hydroxypropantrikarbonsäure-(1,2,3) f
c.-acid cycle Zitronensäurezyklus m, Zitratzyklus m, Trikarbonsäurezyklus m, Krebs-Zyklus m, Szent-Györgyi-Krebs-Zyklus m
c. acid fermentation Zitronensäuregärung f
citridic acid Akonitsäure f, α-Propen-trikarbonsäure-(1,2,3) f
citrine (Min) Zitrin m (Quarzvarietät)
citrinin Zitrinin n (Antibiotikum)
citron-yellow zitronengelb
citronella oil Zitronellöl n, Bartgrasöl n (vom Gras Cymbopogon nardus (L.)Rendle)
citronellal
$CH_3 \cdot C(CH_3)=CH(CH_2)_2 \cdot C(CH_3)H \cdot CH_2 \cdot CHO$ Zitronellal n, Zitronellaldehyd m
citronin A Citronin A n, Naphtholgelb S n (Na_2- oder K_2-Salz der 2,4-Dinitro-1-naphthol-7-sulfonsäure)
citrovorum factor Zitrovorumfaktor m, Leukovorin n, Folinsäure f, Formyltetrahydrofolsäure f
citrus Zitrusfrucht f; Pflanze f der Gattung Citrus L.
c. flavonoid compounds Bioflavonoide pl (Vitamin-P-Faktoren)
c. fruit Zitrusfrucht f
c. oil Zitrusöl n, Zitronenöl n
city garbage Stadtmüll m
c. gas Stadtgas n
civet (Kosmet) Zibet m
cladding Plattierung f; Verkleidung f, Umhüllung f, Überzug m; Auskleidung f

c. material Plattierungswerkstoff m; Verkleidungswerkstoff m, Umhüllungswerkstoff m, Überzugwerkstoff m; Auskleidungswerkstoff m
cladinose Cladinose f (ein Zucker)
clairce (Zucker) Kochkläre f, Raffinadekochkläre f
Claisen condensation Claisen-Kondensation f, Claisensche Kondensation (Esterkondensation) f
Claisen [distilling] flask Claisen-Kolben m
Claisen rearrangement Claisen-Umlagerung f (O-Allyl-C-Allyl-Orthoumlagerung)
Claisen-Schmidt condensation Claisen-Schmidt-Kondensation f (Chalkon-Kondensation)
Claisen stillhead Destillationsaufsatz m nach Claisen
clamp / to festhalten, festklemmen, [ein]klemmen, [ein]spannen
clamp Haltevorrichtung f, Halter m, Klemmvorrichtung f, Klemme f, Zwinge f, Klammer f, Schelle f, Feststellvorrichtung f; Quetschhahn m; (Ker) Feld[brenn]ofen m
c. holder Doppelmuffe f (für Stativklemmen)
clamping frame (Plast) Einspannrahmen m, Spannrahmen m
c. mechanism (Plast) Werkzeugschließsystem n
c. plate (Plast) Aufspannplatte f
c. pressure (Plast) Schließdruck m, Spanndruck m (beim Spritzgießen, Spritzpressen)
Clapeyron[-Clausius] equation Clapeyron-Clausiussche Gleichung (Formel) f
ciarain Klarit m, Clarain m
claret weinrot
c. brown s. c. red
c. red Weinrot n (Farbempfindung); weinroter Farbstoff m, Weinrot n
clarificant Klär[hilfs]mittel n
clarificate / to s. to clarify
clarification Klären n, Klärung f, Abklärung f, Klarifikation f; Abschlämmen n; Läutern n, Läuterung f; Schönen n, Schönung f; Klärfiltration f; Reinigen n, Reinigung f, Raffinieren n, Raffination f
c. plant Kläranlage f, Klärwerk n, Abwasserreinigungsanlage f, Abwasserbeseitigungsanlage f
c. zone Klärzone f
clarified liquid (liquor) geklärte Flüssigkeit f, Klarflüssigkeit f
c. water geklärtes Wasser n, Klärwasser n, Klarwasser n
clarifier Klär[hilfs]mittel n; Klärvorrichtung f, Klarifikator m, Klärseparator m, Klärapparat m; Klärgefäß n
clarify / to 1. [ab]klären; abschlämmen; läutern; schönen; reinigen; raffinieren; 2. sich klären (abklären), klar werden; sich läutern; sich reinigen
clarifying agent Klär[hilfs]mittel n
c. capacity Klärleistung f
c. filter Klärfilter n

c. filter beds Kläranlage f
c. tank Klärtank m
clarity Klarheit f, Durchsichtigkeit f, Lichtdurchlässigkeit f; Reinheit f
Clarke numbers Clarke-Zahlen fpl, Clarkes npl (geochemische Zusammensetzung der Erdkruste)
clarkeite (Min) Clarkeit m
clarkes s. Clarke numbers
clarodurite Klarodurit m (Streifenart der Steinkohle)
class / to s. to classify
class name Klassenname m, Klassenbenennung f
  c. number Klassennummer f; Klassenziffer f, Kodenummer f der Klasse, Erste Kodeziffer f (des internationalen Kohlenklassifikationssystems)
  c. of crystal symmetry Kristallklasse f, Symmetrieklasse f
  c. of dyestuffs Farbstoffklasse f
  c. parameter Klassenparameter m, Klassenkennzeichen n
classical chromatography Säulenadsorptionschromatografie f
  c. distribution (phys Ch) klassische Verteilung f, (i.e.S.) Boltzmann-Verteilung f
  c. distribution law klassisches (Maxwell-Boltzmannsches) Verteilungsgesetz (Impulsverteilungsgesetz) n
classification Klassifizierung f, Klassifikation f, Einteilung f, Unterteilung f, Eingruppierung f, Einstufung f, Einordnung f; (Tech) Klassieren n, Klassierung f, Sortierung f, Sichtung f, Siebung f, Separierung f, Separation f
  c. of hard coals Steinkohlenklassifizierung f
classificator s. classifier
classifier Klassierer m, Klassierapparat m, Klassiergerät n, Klassifikator m, Klassiersieb n, Sichter m, Scheider m
  c. mill Sicht[er]mühle f
  c. trough Klassiertrog m
classify / to klassifizieren, einteilen, unterteilen, eingruppieren, einstufen, einordnen; (Tech) klassieren, sortieren, sichten, scheiden, sieben, separieren
classifying chamber Sichtkammer f
  c. crystallizer klassifizierender Kristallisator m
  c. pool Klassierbecken n
clastic (Geol) klastisch
  c. sediments (Geol) klastische Sedimente npl
clastics s. clastic sediments
clathrate [compound] Klathrat n, Klathratverbindung f, Käfig[einschluß]verbindung f
  c. formation Klathratbildung f, Bildung f von Einschlußverbindungen
  c. inclusion compound s. clathrate
Claude process Claude-Verfahren n (Gewinnung von NH_3 aus der Luft)
claudetite (Min) Claudetit m (Arsen(III)-oxid)
clausius Clausius m, Cl (Maßeinheit der Entropieänderung, 1 Cl = 1 cal/°K)

Clausius-Clapeyron equation s. Clapeyron-Clausius equation
Clausius equation of state Clausiussche Zustandsgleichung f
Clausius-Mosotti equation Clausius-Mosotti-Gleichung f, Clausius-Mosotti-Formel f
clausthalite (Min) Clausthalit m, Selenblei n (Bleiselenid)
clay / to mit Ton (oder Lehm) behandeln; durch Ton filtrieren
clay Ton m; Lehm m; Tonerde f, Bleicherde f (zum Raffinieren); (Pap) Kaolin n, Schlämmkaolin n, geschlämmte Porzellanerde f
  c. bank Tonschicht f
  c. bin Bleicherdevorratsbehälter m (beim Bleicherde-Kontaktverfahren)
  c. body (Ker) Tonmasse f
  c.-bonded (Ker) tongebunden
  c. [building] brick Tonziegel[stein] m
  c.-burning kiln Regenerierofen (Regenerator) m für verbrauchte Bleicherde
  c. catalyst Tonerdekatalysator m, Tonerdekontakt m, Bleicherdekontakt m
  c.-coated (Pap) kaolingestrichen
  c. column (Ker) Massestrang m
  c. combustion boat Glühschiffchen (Verbrennungsschiffchen, Schiffchen) n aus Ton
  c. contact s. c. catalyst
  c. contacting Bleicherdebehandlung f nach dem Kontaktverfahren
  c. cover tonige Deckschicht f
  c. crucible Tontiegel m
  c. cutter (Ker) Tonhobel m
  c. deposit Tonlagerstätte f, Tonlager n, Tonvorkommen n
  c. dryer Trockner m für gewaschene Bleicherde
  c.-filled (Pap) kaolingefüllt
  c. filler (Pap) Kaolinfüllstoff m
  c. gall (Geol) Tongalle f
  c.-humus complex (Bodenkunde) Ton-Humus-Komplex m, organomineralischer Komplex m
  c. ironstone (Min) Toneisenstein m
  c. marl (Geol) Tonmergel m
  c. milk (Pap) Kaolindispersion f, Kaolinmilch f, Kaolintrübe f
  c. mill (Ker) Tonschneider m, Tonknetmaschine f, Tonkneter m
  c. mineral Tonmineral n
  c. mining Tonabbau m, Tongewinnung f
  c. pipe Tonrohr n, Tonröhre f
  c. pit Tongrube f
  c. preparation Tonaufbereitung f
  c. preparation plant Tonaufbereitungsanlage f
  c. shredder (Ker) Tonraspler m
  c. silo (Ker) Tonsilo m (n)
  c. slate (Geol) Tonschiefer m
  c. slurry Tonschlicker m, Tonschlempe f; (Pap) Kaolinbrei m
  c. soil Lehmboden m; Tonboden m
  c. substance Tonsubstanz f

c.-tile pipe Tonrohr n, Tonröhre f

c. treating (treatment) Bleicherdebehandlung f, Bleicherderaffination f, Behandlung f mit Bleicherde

c. ware s. clayware

c.-water slurry s. c. slurry

Clayden effect (Foto) Clayden-Effekt m

clayey tonhaltig, tonig; tonartig; lehmhaltig, lehmig; lehmartig

clayish s. clayey

claystone (Geol) Tongestein n; (Min) erdiger Feldspat m

Clayton gas Clayton-Gas n (SO_2-N_2-Gemisch)

Clayton yellow Thiazolgelb n, Titangelb n

clayware Tonwaren fpl

clean / to reinigen, säubern

clean circulation sauberer Umlauf m (beim thermischen Kracken)

c. coal Reinkohle f, reine Kohle f (aschenärmste Kohle)

c. culture Reinkultur f, Reinzucht f

c. gas Reingas n

c.-grained (Gerb) narbenrein

c. print scharfer (reiner) Druck m

c. screen machine (Trennapparat, bei dem sich der Kuchen nicht auf dem Filtermedium ablagert)

c.-up s. cleaning

c.-up of the stock (Pap) Stoffreinigung f, Ganzstoffreinigung f

c.-up procedure Reinigungsprozeß m

c.-up technique Reinigungsverfahren n

cleaned gas Reingas n

cleaner Reiniger m, Reinigungseinrichtung f, Reinigungsanlage f; Reiniger m, Reinigungsmittel n, (i.e.S.) Entfettungsmittel n

cleaner's naphtha (solvent) Reinigungsbenzin n, Waschbenzin n, Fleckenbenzin n

cleaning Reinigen n, Reinigung f, Säubern n, Säuberung f

c. effect (Text) Reinigungseffekt m, Reinigungswirkung f, Wascheffekt m, Waschwirkung f

c. efficiency (Text) Reinigungsvermögen n, Waschvermögen n, Reinigungskraft f, Waschkraft f, Reinigungsleistung f, Waschleistung f

c. intensifyer (Text) Reinigungsverstärker m

c. plant Reinigungsanlage f, Waschanlage f, Wäsche f

c. promoter (Text) Reinigungsverstärker m

cleanliness Reinheit f, Sauberkeit f; Reinheitsgrad m

c. of cut Schärfe f der Absiebung

cleanse / to s. to clean

cleanser Reiniger m, Reinigungsmittel n

cleansing Reinigen n, Reinigung f, Säubern n, Säuberung f

c. agent Reinigungsmittel n, Reiniger m

c. cream Reinigungscreme f, Hautreinigungscreme f

c. lotion Reinigungslotion f

c. tissue Abschminkpapier n, Lippentupfpapier n

clear / to 1. klären, filtern; reinigen; schönen; 2. aufhellen, bleichen; 3. (Pap) aufschlagen; (Pap) mahlen; (Pap) feinmahlen, fertigmahlen

clear glaze (Ker) Klarglasur f

c. ice Klareis n, Kristalleis n

c. lacquer Klarlack m

c. opening lichte Weite f, Maschenweite f (eines Siebes)

c. overflow Überlauf m der geklärten Flüssigkeit, Wasserüberlauf m (z.B. eines Eindickers)

c. point Klarpunkt m (Fällungsanalyse)

c.-solution zone Klärzone f (z.B. im Eindicker)

c. varnish Klarlack m

c. water Klarwasser n

c.-yellow klar gelb

clearance (Tech) freier Raum m, Spielraum m, Spiel n, Luft f; relativer Schadraum m (an der Hubkolbenmaschine)

c. between the rolls Walzenspalt m

c. control unloading Zuschaltraumregulierung f (eines Verdichters)

c. volume schädlicher Raum m, Schadraum m

clearing 1. Klären n, Filtern n, Reinigen n, Reinigung f; Schönen n, Schönung f; 2. Aufhellen n, Bleichen n; 3. (Pap) Aufschlagen n; (Pap) Mahlen n; (Pap) Feinmahlen n, Fertigmahlen n

c. agent (Mikroskopiertechnik) Aufhellungsmittel n

c. bath Klärbad n, Klarwaschbad n, Nachspülbad n, Reinigungsbad n

c. liquor (Zucker) Kochkläre f, Raffinadekochkläre f

c. time Klärzeit f

cleavability (Min) Spaltbarkeit f

cleavable (Min) spaltbar

cleavage 1. Spalten n, Spaltung f, Aufspalten n, Aufspaltung f; Abspalten n, Abspaltung f; Aufblättern n (von Schichtstoffen); 2. (Min) Bruch m; Spaltbarkeit f (blätteriger Bruch); Richtung f der Spaltebene; 3. (Geol) Schieferung f

c. by acids Säurespaltung f

c. by ozone Ozonspaltung f, Ozonidspaltung f, Ozonolyse f

c. face Spaltfläche f, Spaltebene f

c. plane Spaltfläche f, Spaltebene f; (Geol) Schiefer[ungs]fläche f, Schieferungsebene f

c. product Spalt[ungs]produkt n

c. reaction Spalt[ungs]reaktion f

c. surface Spaltfläche f, Spaltebene f

cleave / to 1. [auf]spalten; abspalten; 2. sich [auf]spalten; sich abspalten; aufblättern (von Schichtstoffen)

cleavelandite (Min) Cleavelandit m (Natriumalumotrisilikat)

Clemmensen method s. Clemmensen reduction

Clemmensen reduction Clemmensen-Reduktion f, Clemmensen-Methode f (Karbonyl-Methylen-Reduktion)

Clerici's solution Clerici-Lösung f (Gemisch aus Thalliumformiat und Thalliummalonat)

Cleve's acid Cleve-Säure f, Clevesche Säure f *(Benennung für 1-Naphthol-5-sulfonsäure und mehrere 1-Naphthylaminsulfonsäuren)*
Cleve's 1,6 acid Cleve-Säure-1,6 f, Cleve-Säure-6 f, 1-Naphthylamin-6-sulfonsäure f
Cleve's 1,7 acid Cleve-Säure-1,7 f, Cleve-Säure-7 f, 1-Naphthylamin-7-sulfonsäure f
cleveite *(Min)* Cleveit m
Cleveland apparatus (open cup) s. Cleveland open tester
Cleveland open tester Cleveland-Flammpunkt[s]-prüfer m, Cleveland-Prüfer m, Cleveland-Gerät n, Cleveland-Tester m, offener Tiegel m nach Cleveland
clevelandite s. cleavelandite
climatic chamber Klimakammer f
clingmanite *(Min)* Clingmanit m, *(veraltet für)* Margarit m *(Kalziumaluminiumdihydroxiddialumodisilikat)*
clinker Klinker m, *(i.e.S.)* Zementklinker m, Portlandzementklinker m; Schlacke f
  c. brick Klinker m
  c. concrete Schlackenbeton m
  c. discharge Schlackenziehen n, Entschlackung f, Schlackenentfernung f
  c. formation Schlackenbildung f, Verschlackung f
  c. grinder Schlackenbrecher m, Schlackenmühle f
  c. process Klinkerverfahren n *(der Davison Chemical Company zur Phosphorsäureherstellung)*
clinkering Schlackenbildung f, Verschlackung f
clinkstone *(Min)* Klingstein m, Phonolith m
clinochlore, clinochlorite *(Min)* Klinochlor m
clinoclase, clinoclasite *(Min)* Klinoklas m *(Kupfer-(II)-trihydroxidorthoarsenat)*
clinoenstatite *(Min)* Klinoenstatit m *(Magnesiummetasilikat)*
clinozoisite *(Min)* Klinozoisit m
clintonite *(Min)* Clintonit m
clip Klemmvorrichtung f, Klemme f, Klammer f, Haltevorrichtung f, Halter m, Zwinge f, Feststellvorrichtung f
  c. chain *(Text)* Kluppenkette f
  c. frame *(Text)* Kluppenspannrahmen m
  c. wool Schurwolle f
clock glass Uhrglas n, Uhrglasschale f
clockwise im Uhrzeigersinn, rechtsdrehend
  c. numbering Bezifferung (Numerierung) f im Uhrzeigersinn
clog / to 1. verstopfen, zusetzen, verlegen; 2. sich verstopfen (zusetzen); [ver]klumpen, klumpig werden, sich zusammenballen, Klumpen bilden
  c. up s. to clog 1.
clogging Verstopfen n
cloromagnesite s. chloromagnesite
clorotionite s. chlorothionite
close-coupled pump Maulwurfpumpe f
  c.-grained feinkörnig, Feinkorn..., dicht *(Gefüge)*

c. grinding Flachmüllerei f
c.-meshed engmaschig, feinmaschig, kleinmaschig
c. package s. c. packing
c.-packed *(Krist)* dichtgepackt
c.-packed spheres s. c. packing
c. packing *(Krist)* dichteste Kugelpackung (Packung) f, d.P.
c. sand wenig gasdurchlässiger Sand m, Sand m geringer Gasdurchlässigkeit
c.-up effect *(phys Ch)* Nah[e]wirkung f
closed assembly time geschlossene Wartezeit f
  c.-belt conveyor Schlauchbandförderer m
  c.-cell geschlossenzellig *(Schaumstoff)*
  c. chain geschlossene (ringförmige, zyklische) Kette f
  c. chain of carbon atoms geschlossene Kohlenstoffkette f
  c. circuit geschlossener Kreislauf m
  c.-circuit grinding Mahlen n im geschlossenen Kreislauf
  c. cup geschlossener Tiegel m, g.T. *(eines Flammpunktprüfers)*
  c.[-cup] flash point Flammpunkt m im geschlossenen Tiegel, Fl. g. T.
  c.[-cup] flash tester geschlossener Flammpunkt[s]prüfer (Flammpunkt[s]apparat) m
  c. impeller geschlossenes Laufrad n
  c.-loop control system Regelung f, Regelsystem n
  c. mixer geschlossener Mischer (Kneter) m, Innenmischer m
  c. mould *(Plast)* geschlossenes Werkzeug n
  c. pore *(Ker)* geschlossene Pore f
  c. pot *(Ker)* geschlossener (verdeckter, gedeckter) Hafen m
  c. shell abgeschlossene Schale f
  c.-suction control Saug-Absperr-Regulierung f *(eines Verdichters)*
  c. system *(phys Ch)* [ab]geschlossenes System n
  c.-type headbox *(Pap)* geschlossener Stoffauflauf[kasten] m, Stoffauflauf m geschlossener Bauart
  c.-vessel furnace Gefäßofen m
closely graded engklassiert
closest package s. c. packing
  c. packing *(Krist)* dichteste Kugelpackung (Packung) f, d.P.
  c. sphere packing s. c. packing
closing head Verschließkopf m, Verschließstempel m
  c. of a ring Ringschluß m, Zyklisierung f
  c. travel Schließweg m *(einer Presse)*
closure Schließen n, Verschließen n, Schluß m, Verschluß m; Verschlußvorrichtung f, Verschluß m
  c. of the carbazole ring Karbazolringschluß m
clot / to [ver]klumpen, klumpig werden, Klumpen bilden, sich zusammenballen, gerinnen, koagulieren, [aus]flocken

clot Klumpen *m*, Klümpchen *n*; Koagulat *n*, Koagulum *n*, Gerinnsel *n*; *(Ker)* Batzen *m*, Hubel *m*
  c.-on-boiling test Kochprobe *f*
cloth Gewebe *n*, Stoff *m*, Tuch *n*, Zeug *n*
  c. adhesive Tuchkleber *m*
  c.-centred paper *s.* c.-faced paper
  c.-covered tuchbespannt, stoffbespannt
  c.-faced paper Gewebepapier *n*, Leinenpapier *n*, Papyrolin *n*, Packgewebe *n* *(Papierbahn mit ein- oder zweiseitiger Gewebebahn)*
  c. filter Tuchfilter *n*, Stoffilter *n*, Gewebefilter *n*
  c. finish *(Pap) (Am)* Leinenprägung *f*
  c.-lined paper, c.-mounted paper *s.* c.-faced paper
  c. opener *(Text)* Strangöffner *m*
  c. opening *(Text)* Strangöffnen *n*
  c. passage *(Text)* Warendurchlauf *m*
  c. tree Papiermaulbeerbaum *m*, Broussonetia papyrifera (L.)L'Hérit.
clotting Klumpen *n*, Verklumpen *n*, Klumpenbildung *f*, Gerinnen *n*, Gerinnung *f*, Koagulieren *n*, Koagulation *f*, Koagulierung *f*, Flockung *f*, Ausflockung *f*, Flockenbildung *f*
  c. of blood Blutgerinnung *f*
cloud / to sich trüben, trüb[e] werden; matt (glanzlos) werden; *(Text)* moirieren
cloud Wolke *f*; Nebel *m*; Trübe *f*, Trübung *f*, Trub *m*, Satz *m*, Bodensatz *m*; Geläger *n*, Gärniederschlag *m*, Drusen *fpl*
  π **cloud** Ladungswolke *f* einer π-Bindung, π-Wolke *f*
  c. chamber [Wilsonsche] Nebelkammer *f*, Wilson-Kammer *f*
  c.-chamber photograph Nebelkammeraufnahme *f*, Nebelkammerbild *n*, Wilson-Aufnahme *f*
  c. effect *(Pap)* Wolkigkeit *f*
  c. point Trübungspunkt *m*
  c. track *(phys Ch)* Nebelspur *f*
  c.-track apparatus [Wilsonsche] Nebelkammer *f*, Wilson-Kammer *f*
cloudiness Trübung *f*, Trübheit *f*, Trübsein *n*, Trübe *f*; Mattheit *f*, Glanzlosigkeit *f*
cloudpoint Trübungspunkt *m*
cloudy trüb[e], getrübt, unrein, verunreinigt, undurchsichtig; matt, glanzlos; *(Pap)* wolkig; *(Text)* moiriert
  to become c. sich trüben
clove Gewürznelke *f* *(von Syzygium aromaticum (L.)Merr. et L.M. Perry)*
  c. bark Nelkenrinde *f*, Nelkenzimt *m* *(von Dicypellum caryophyllatum Nees)*
  c. oil Nelkenöl *n*, Gewürznelkenöl *n* *(von Syzygium aromaticum (L.)Merr. et L.M. Perry)*
clover honey Kleehonig *m*
clupanodonic acid Klupanodonsäure *f*, 4,8,12,15,19-Dokosapentaensäure *f*, *(i.w.S. Bezeichnung für hochungesättigte Fettsäuren mit 20 bis 22 C-Atomen)*
clupeine Klupein *n* *(ein Protamin)*

Clusius column Clusius-Trennrohr *n* *(zur Trennung gasförmiger Isotope)*
cluster cardamon *s.* cardamon
  c. formation *s.* clustering
  c. pine Strandkiefer *f*, Pinus pinaster Ait.
clustering Traubenbildung *f* *(z.B. von Fettkügelchen beim Aufrahmen)*
clusters Trauben *fpl* *(beim Aufrahmen entstehende Fettkügelchen)*
clytocybine Clitocybin *n* *(Antibiotikum)*
C.M.C. *s.* carboxymethylcellulose
C/N ratio C/N-Verhältnis *n*, Kohlenstoff-Stickstoff-Verhältnis *n* *(des Bodens)*
cnicin Knizin *n* *(Bitterstoff aus Cnicus benedictus L.)*
CO$_2$ process *(Met)* CO$_2$-Verfahren *n*, CO$_2$-Erstarrungsverfahren *n*, Kohlensäureerstarrungsverfahren *n* *(Formherstellungsverfahren)*
CO$_2$ recorder Kohlensäureschreiber *m*
coacervate Koazervat *n*
coacervation Koazervation *f*, Koazervierung *f*
coagel Koagel *n*
coagulability Koagulierbarkeit *f*, Gerinnbarkeit *f*
coagulable koagulierbar, gerinnbar
coagulant Koagulans *n*, Koagulator *m*, koagulierendes Mittel *n*, Koagulationsmittel *n*, Ausflockungsmittel *n*
  c. dipping process Koagulationstauchverfahren *n*, Koagulans-Verfahren *n*, Tauchen *n* mit Koagulationsmittel
coagulase Koagulase *f* *(ein Bakterienferment)*
coagulate / to 1. koagulieren, gerinnen lassen, zur Ausflockung bringen; *(Text)* fällen; 2. koagulieren, [aus]flocken, gerinnen
coagulate Koagulat *n*
coagulating agent Koagulationsmittel *n*, koagulierendes Mittel *n*, Ausflockungsmittel *n*, Koagulans *n*, Koagulator *m*
  c. bath *(Text)* Koagulationsbad *n*, Fällbad *n*
  c. power Koagulationsvermögen *n*, Flockungsvermögen *n*, Flockungskraft *f*, Flockungsfähigkeit *f*
coagulation 1. Koagulieren *n*, Koagulation *f*, Ausflocken *n*, Ausflockung *f*, Flockung *f*, Gerinnen *n*, Gerinnung *f*; *(Text)* Fällung *f*, Fällen *n*; 2. Koagulat *n*
  c. bath *s.* coagulating bath
  c. period *s.* c. time
  c. point Gerinnungspunkt *m*
  c. rate Koagulationsgeschwindigkeit *f*
  c. time Koagulationsdauer *f*, Koagulationszeit *f*
  c. value Flockungswert *m*
  c. vitamin Koagulationsvitamin *n*, Vitamin K$_1$ *n*, antihämorrhagisches Vitamin *n*, Phyllochinon *n*
coagulator Koagulator *m*, Koagulans *n*, Koagulationsmittel *n*, koagulierendes Mittel *n*, Ausflockungsmittel *n*
coagulatory Koagulator..., Koagulations..., Ausflockungs...
coagulum Koagulat *n*

coal Kohle f; Steinkohle f
  c. ball Dolomitknolle f, Torfdolomit m
  c. band Kohlenband n, Kohlenstreifen m
  c.-bearing kohle[n]führend
  c. blend Kohle[n]gemisch n, Kohle[n]mischung f
  c. breeze Kohlengrus m
  c. briquette Kohlenbrikett n, Kohlenpreßling m
  c. carbonization Kohlenverkokung f, Kohlenentgasung f; Kohlen[ver]schwelung f; Steinkohlenverkokung f, Steinkohlenentgasung f; Steinkohlen[ver]schwelung f
  c. charge Kohlencharge f, Kohle[n]füllung f, Kohlenladung f, Kohle[n]schüttung f, Kohle[n]füllmasse f
  c.-charging car Kohlefüllwagen m, Koksofen-Füllwagen m
  c.-charging machine Kohlenbeschick[ungs]maschine f
  c.-charging vessel Kohlenaufgabeschleuse f, Kohle[n]schleuse f (z.B. eines Lurgi-Druckgasgenerators)
  c. chart Kohle[n]klassifizierungsschaubild n
  c. chemical Kohle[n]wertstoff m
  c. chemicals production Kohle[n]wertstoffgewinnung f
  c. class Kohle[n]klasse f
  c. classification Kohle[n]klassifikation f, Kohle[n]klassifizierung f, Kohle[n]einteilung f
  c. cleaning Kohlenwäsche f; Kohlenaufbereitung f; Steinkohlenwäsche f; Steinkohlenaufbereitung f
  c. cleaning plant Kohlenwäsche[anlage] f
  c. coke Steinkohlenkoks m
  c. component Kohle[n]komponente f, Kohle[n]bestandteil m
  c. deposit Kohlenlager n, Kohlenvorkommen n
  c.-derived chemical product s. c. chemical
  c. dust Kohlenstaub m
  c.-dust mill Kohlenstaubmühle f, Kohle[n]mühle f; Steinkohlenmühle f
  c. extraction Kohleextraktion f
  c. feed Kohlebeschickung f, Kohlezufuhr f, Kohle[n]aufgabe f
  c. fines Feinkohle f
  c.-fired kohlegeheizt, kohlebeheizt, mit Kohle geheizt (beheizt, gefeuert), mit Kohleheizung (Kohlefeuerung)
  c. formation Kohle[n]bildung f, Kohlenentstehung f, Kohlenwerdung f, Kohlegenesis f; Steinkohle[n]bildung f, Steinkohlenentstehung f, Steinkohlenwerdung f, Steinkohlegenesis f
  c.-forming kohle[n]bildend
  c.-forming process Kohle[n]bildungsprozeß m, Kohle[n]bildungsvorgang m
  c. gas Kohlengas n; Steinkohlengas n
  c. gasification Kohle[n]vergasung f
  c. genesis s. c. formation
  c. grinding s. c. milling
  c.-grinding mill s. c. mill

  c. handling Kohlentransport m
  c. hopper Kohlentrichter m, Kohlenbunker m
  c. hydrogenation Kohle[n]hydrierung f; Steinkohle[n]hydrierung f
  c. industry Kohle[n]industrie f
  c. inlet valve Kohlenfüllhahn m, Trommelschieber m (eines Gaswerksofens)
  c. jig Kohlensetzmaschine f
  c. leveller bar Planierstange f, Einebnungsstange f (zum Einebnen der in die Kokskammer eingefüllten Kohle)
  c.-like kohle[n]ähnlich, kohleartig
  c. liquefaction Kohle[n]verflüssigung f
  c.-load instrument Kohlenlademaschine f
  c. maceral Kohlenmazeral n, Kohle[n]gefügebestandteil m; Steinkohlenmazeral n, Steinkohlengefügebestandteil m
  c. material Kohlenmaterial n, Kohle[n]substanz f
  c. measure Kohlenlager n, Kohlenvorkommen n
  c. measures produktives (flözführendes) Kohlengebirge (Karbon) n, produktive Steinkohlenformation f, Hauptsteinkohlenformation f
  c. microscopy Kohlenmikroskopie f
  c. mill Kohle[n]mühle f, Kohlenstaubmühle f; Steinkohlenmühle f
  c. milling Kohle[nver]mahlung f, Kohlenstaubmahlung f
  c. mixture Kohle[n]mischung f, Kohle[n]gemisch n
  c. oil Kohlenteeröl n; Steinkohlen[teer]öl n
  c.-ore briquette Kohle-Erz-Brikett n
  c.-ore mixture Kohle-Erz-Gemisch n
  c. particle Kohlenteilchen n, Kohlenpartikel n
  c. paste Kohle[n]paste f, Kohle[n]brei m
  c. petrography Kohlenpetrografie f
  c. petrology Kohlenpetrologie f, Kohlengesteinskunde f
  c. preparation Kohlenaufbereitung f; Steinkohlenaufbereitung f
  c.-preparation plant Kohlenaufbereitungsanlage f
  c. product Kohlefolgeprodukt n
  c. pulverization Kohlenstaubmahlung f, Kohle[nver]mahlung f
  c. pulverizer Kohlenstaubmühle f, Kohle[n]muhle f; Steinkohlenmühle f
  c. pulverizing s. c. pulverization
  c.-pulverizing mill s. c. pulverizer
  c. rank Kohlenrang m, Kohlenrangstufe f, Inkohlungsgrad m der Kohle
  c. sample Kohle[n]probe f, Kohlenprobekörper m
  c. slurry Kohle[n]schlamm m, Kohle[n]brei m
  c. stall Kohlekammer f (bei der Kohlehydrierung)
  c. storage hopper Kohlen[vorrats]bunker m
  c. store Kohlenlager n
  c. substance s. c. material
  c. tar Kohlenteer m; Steinkohlenteer m
  c.-tar colour s. c.-tar dye
  c.-tar creosote Steinkohlen[teer]kreosot n, Kreosotöl n

c.-tar distillate Kohlenteerdestillat *n*; Steinkohlenteerdestillat *n*
c.-tar dye Teerfarbstoff *m*, Teerfarbe *f*
c.-tar naphtha Kohlenteer-Solventnaphtha *n(f)*
c.-tar oil Kohlenteeröl *n*; Steinkohlen[teer]öl *n*
c.-tar pitch Kohlenteerpech *n*; Steinkohlen[teer]pech *n*
c. utilization Kohlenausnutzung *f*
c. washing Kohlenwäsche *f*; Steinkohlenwäsche *f*
c./water slurry Kohlenstaub/Wasser-Trübe *f (bei der Druckvergasung von Kohlenstaub mit O₂)*
coalesce / to sich vereinigen (verbinden), koal[is]ieren, koaleszieren; verschmelzen; zusammenlaufen, zusammenfließen; zusammenwachsen
coalesced copper stranggepreßtes Kupfer *n*
coalescence Koaleszenz *f*, Vereinigung *f*, Verbindung *f*; Verschmelzung *f*; Zusammenfließen *n*; Zusammenwachsen *n*
coalification *(Geol)* Inkohlung *f*, Karbonifikation *f*, Kohlenreifung *f*
c. band Inkohlungsband *n*, Inkohlungsstreifen *m*
c. process Inkohlungsprozeß *m*, Inkohlungsvorgang *m*
c. series Inkohlungsreihe *f*
coalify / to inkohlen
Coalite Coalite *m (Warenzeichen für einen rauchlosen Brennstoff)*
Coalite [carbonization] process Coalite-Schwelverfahren *n*, Coalite-Verfahren *n*
Coalite retort Coalite-Schwelretorte *f*
coarse grob, Grob...
c. aggregate grober Zuschlag *m*, Grobzuschlag *m*
c.-clastic *(Geol)* grobklastisch
c. clay Grobton *m*, Schluff *m*
c. copper Schwarzkupfer *n*, Schwarz-Cu *n*
c. crusher Grobbrecher *m*
c. crushing Grobbrechen *n*
c.-crystalline grobkristallin
c.-disperse grobdispers
c. dispersion grobe Dispersion *f*
c. flour Schrotmehl *n*
c. grain *(Foto)* grobes Korn *n*
c.-grain emulsion *(Foto)* grobkörnige Emulsion *f*
c.-grained grobkörnig; grobkristallin; grobfas[e]rig; *(Gerb)* grobnarbig; *(Holz)* weitlumig
c.-grained clastics *(Geol)* grobklastische (grobkörnige klastische, makroklastische) Sedimente *npl*
c.-grained emulsion s. c.-grain emulsion
c. grinding Grobmahlen *n*, Grob[aus]mahlung *f*, grobe Mahlung (Ausmahlung) *f*
c. material Grobgut *n*; *(Pap)* Grobstoff *m*, Spuckstoff *m*, „Sauerkraut" *n*
c. material screening Grobsortierung *f*
c. meal Schrotmehl *n*
c. sand Grobsand *m*

c. screen Grobsortierer *m*
c. sizing Grobklassieren *n*, Grobklassierung *f*
c. sludge Grobtrub *m*, Heißtrub *m*, Kochtrub *m* *(Bierherstellung)*
c. soybean meal Soja[bohnen]schrot *n(m)*
c. straw pulp [gelber] Strohstoff *m*, Gelbstrohstoff *m*
c. suspension grobe Suspension *f*
coarsely disperse grobdispers
coarseness Grobkörnigkeit *f*
coarsening crystallization Sammelkristallisation *f*
coaster board Bierfilzpappe *f*, Bierdeckelpappe *f*, Saugpappe *f*
coat / to belegen, beschichten, überziehen, kaschieren; gummieren; anstreichen; umhüllen, einhüllen, ummanteln; auftragen; *(Düngemittel)* pudern; *(Pap)* streichen, beschichten; *(Gerb)* abdecken *(mit Deckfarbe)*; *(Ker)* aufgießen
c. on both sides *(Pap)* zweiseitig streichen
c. on one side *(Pap)* einseitig streichen
coat Belag *m*, Schicht *f*, Beschichtung *f*, Schutzschicht *f*, Überzug *m*, Auftrag *m*; Anstrich *m*; *(Pap)* Aufstrich *m*, Strich *m*, Beschichtung *f*
coated *(Pap)* gestrichen, Streich...
to become c. sich überziehen
c. base paper Streichrohpapier *n*
c. board gestrichener Karton *m*, Streichkarton *m*
c. both sides s. c. on both sides
c. cardboard s. c. board
c. fabric gestrichenes (beschichtetes, kaschiertes) Gewebe *n*; gummiertes Gewebe *n*, gummierter Stoff *m*; Gewebekunstleder *n*
c. on both sides *(Pap)* zweiseitig gestrichen, mit beidseitigem Strich
c. on one side *(Pap)* einseitig gestrichen, mit einseitigem Strich
c. paper gestrichenes Papier *n*, Streichpapier *n*
coater s. coating machine
c. unit *(Pap)* Streichvorrichtung *f*
coating 1. Belegen *n*, Beschichten *n*, Überziehen *n*, Kaschieren *n*; Gummieren *n*; Anstreichen *n*; Umhüllen *n*, Einhüllen *n*, Ummanteln *n*; Auftragen *n*; Pudern *n*, Puderung *f (von Düngemitteln)*; *(Gerb)* Abdecken *n (mit Deckfarben)*; *(Ker)* Aufgießen *n*; 2. Belag *m*, Schicht *f*, Beschichtung *f*, Schutzschicht *f*, Überzug *m*, Auftrag *m*; Anstrich *m*; Gummierung *f*; 3. *(Pap)* Streichen *n*, Strich *m*; Streichmasse *f*; Aufstrich *m*, Strich *m*, Beschichtung *f*
c. application *(Pap)* Strichauftrag *m*
c. base paper Streichrohpapier *n*
c. calender Kalander *m* zum Belegen von Geweben
c. colour Deckfarbe *f*
c. finish Deckappretur *f*, Deckauftrag *m*
c. machine Streichmaschine *f*, Auftragmaschine *f*; Lackiermaschine *f*
c. mill s. c. plant
c. mixture *(Pap)* Streichmasse *f*, Streichfarbe *f*
c. of rubber Gummibelag *m*, Gummierung *f*

c. pan *(Pap)* Streich[massen]trog *m*; Dragierkessel *m*; Lackiertrommel *f*
c. paper Streichrohpapier *n*
c. plant *(Pap)* Streichanlage *f*, Streicherei *f*
c. press *(Pap)* Streichpresse *f*
c. resin Überzugharz *n*, Lackharz *n*
c. roll *(Pap)* Streichwalze *f*
c. slip *(Pap)* Streichmasse *f*, Streichfarbe *f*; *(Ker)* Anstrichmasse *f*
c. slurry *s. c.* mixture
c. substance *(Pap)* Streichmasse *f*, Streichfarbe *f*; Pudermittel *n*, Puderstoff *m* *(zum Konditionieren von Düngemitteln)*
c. system *(Text)* Auftragswerk *n*
c. unit *(Pap)* Streichvorrichtung *f*
c. varnish Überzug[s]lack *m*
c. vehicle Anstrichbindemittel *n*
cobalt Co Kobalt *n*
c. arsenosulphide CoAsS Arsenkobaltsulfid *n*
c.-base auf Kobaltbasis (Kobaltgrundlage)
c. bloom *(Min)* Kobaltblüte *f*, Erythrin *m* *(Kobalt(II)-orthoarsenat)*
c. blue Kobaltblau *n*, Kobaltultramarin · *n*, Thénards Blau *n* *(Kobaltaluminat)*
c. catalyst Kobaltkatalysator *m*, Kobaltkontakt *m*, Co-Kontakt *m*
c. chelate Kobaltchelat *n*
c. chloride test paper Kobaltchloridpapier *n*
c. disulphide CoS$_2$ Kobaltdisulfid *n*
c. drier *(Farb)* Kobalttrockner *m*, Kobalttrockenstoff *m*, Kobaltsikkativ *n*
c. glance *(Min)* Kobaltglanz *m*, Kobaltin *m*
c. glass Kobaltglas *n*
c. green Kobaltgrün *n*, Rinmanns Grün *n*
c.-histidine chelate Histidinkobaltchelat *n*
c. monoboride CoB Kobalt[mono]borid *n*
c. monoxide CoO Kobalt[mon]oxid *n*, Kobalt(II)-oxid *n*
c. orthotitanate Co$_2$TiO$_4$ Kobalt(II)-orthotitanat *n*
c. phthalocyanine Kobaltphthalozyanin *n*
c. powder Kobaltpulver *n*
c. pyrites *(Min)* Kobaltkies *m*, Linneit *m* *(Kobalt(II,III)-sulfid)*
c. sesquioxide Co$_2$O$_3$ Kobalt(III)-oxid *n*, Dikobalttrioxid *n*, Kobalttrioxid *n*
c. soap Kobaltseife *f*
c. tetracarbonyl Co(CO)$_4$, [Co(CO)$_4$]$_2$ Kobalttetrakarbonyl *n*
c. tricarbonyl Co(CO)$_3$, [Co(CO)$_3$]$_4$ Kobalttrikarbonyl *n*
c. trifluoride CoF$_3$ Kobalttrifluorid *n*, Kobalt(III)-fluorid *n*
c. ultramarine *s. c.* blue
c. vitriol *(Min)* Kobaltvitriol *m*, Bieberit *m* *(Kobalt(II)-sulfat-7-Wasser)*
cobaltammines Kobaltiake *npl*, Kobaltammine *npl*
cobaltate Kobaltat *n*
cobaltic Kobalt..., *(i.e.S.)* Kobalt(III)-...
c. acetate Co(CH$_3$COO)$_3$ Kobalt(III)-azetat *n*
c. boride CoB Kobalt[mono]borid *n*

c. chelate Kobalt(III)-chelat *n*
c. fluoride CoF$_3$ Kobalt(III)-fluorid *n*, Kobalttrifluorid *n*
c. hydroxide Co(OH)$_3$ Kobalt(III)-hydroxid *n*
c. oxide Co$_2$O$_3$ Kobalt(III)-oxid *n*, Dikobalttrioxid *n*, Kobalttrioxid *n*
c. potassium nitrite K$_3$[Co(NO$_2$)$_6$] Kaliumhexanitrokobaltat(III) *n*
c. sulphate Co$_2$(SO$_4$)$_3$ Kobalt(III)-sulfat *n*
c. sulphide Co$_2$S$_3$ Kobalt(III)-sulfid *n*, Dikobalttrisulfid *n*
cobaltichloride aquapentammine [Co(NH$_3$)$_5$H$_2$O]Cl$_3$ Aquopentamminkobalt(III)-chlorid *n*, Roseokobaltchlorid *n*
c. chloropentammine [Co(NH$_3$)$_5$Cl]Cl$_2$ Chloropentamminkobalt(III)-chlorid *n*
c. hexammine [Co(NH$_3$)$_6$]Cl$_3$ Hexamminkobalt(III)-chlorid *n*, *(veraltet)* Luteokobaltchlorid *n*
cobalticyanide M$^I_3$[Co(CN)$_6$] Zyanokobaltat(III) *n*, Hexazyanokobaltat(III) *n*
cobaltine *(Min)* Kobaltin *m*, Kobaltglanz *m*
cobaltinitrite M$^I_3$[Co(NO$_2$)$_6$] Nitrokobaltat(III) *n*, Hexanitrokobaltat(III) *n*
cobaltioxalate M$^I_3$[Co(C$_2$O$_4$)$_3$] Oxalatokobaltat(III) *n*, Trioxalatokobaltat(III) *n*
cobaltite *s.* cobaltine
cobalto-cobaltic oxide Co$_3$O$_4$ Kobalt(II,III)-oxid *n*, Trikobalttetroxid *n*, Kobalttetroxid *n*
c.-histidine chelate Histidinkobalt(II)-chelat *n*
cobaltocyanide Zyanokobaltat(II) *n*
cobaltonitrite M$^I_4$[Co(NO$_2$)$_6$] Nitrokobaltat(II) *n*, Hexanitrokobaltat(II) *n*
cobaltothiocyanate M$^I_2$[Co(SCN)$_4$] Thiozyanatokobaltat(II) *n*, Rhodanokobaltat(II) *n*, Tetrarhodanokobaltat(II) *n*
cobaltous Kobalt..., *(i.e.S.)* Kobalt(II)-...
c. acetate Co(CH$_3$COO)$_2$ Kobalt(II)-azetat *n*
c. acid arsenite Co$_3$(AsO$_3$)$_2$·H$_3$AsO$_3$ Kobalt(II)-hydrogenorthoarsenit(III) *n*
c. acid fluoride CoF$_2$·5HF Kobalt(II)-hydrogenfluorid *n*
c. arsenate Co$_3$(AsO$_4$)$_2$ Kobalt(II)-[ortho]arsenat(V) *n*
c. arsenite *s. c.* acid arsenite
c. bromate Co(BrO$_3$)$_2$ Kobalt(II)-bromat *n*
c. bromide CoBr$_2$ Kobalt(II)-bromid *n*, Kobaltdibromid *n*
c. carbonate CoCO$_3$ Kobalt(II)-karbonat *n*
c. chelate Kobalt(II)-chelat *n*
c. chlorate Co(ClO$_3$)$_2$ Kobalt(II)-chlorat *n*
c. chloride CoCl$_2$ Kobalt(II)-chlorid *n*, Kobaltdichlorid *n*
c. chromate CoCrO$_4$ Kobalt(II)-chromat *n*
c. cyanide Co(CN)$_2$ Kobalt(II)-zyanid *n*
c. ferricyanide Co$_3$[Fe(CN)$_6$]$_2$ Kobalt(II)-hexazyanoferrat(III) *n*
c. ferrocyanide Co$_2$[Fe(CN)$_6$] Kobalt(II)-hexazyanoferrat(II) *n*
c. fluoride CoF$_2$ Kobalt(II)-fluorid *n*, Kobaltdifluorid *n*

c. fluorosilicate $Co[SiF_6]$ Kobalt(II)-hexafluoro-silikat n

c. hydroxide $Co(OH)_2$ Kobalt(II)-hydroxid n

c. iodate $Co(JO_3)_2$ Kobalt(II)-jodat n

c. iodide CoJ_2 Kobalt(II)-jodid n, Kobaltdijodid n

c. nitrate $Co(NO_3)_2$ Kobalt(II)-nitrat n

c. orthophosphate $Co_3(PO_4)_2$ Kobalt(II)-[ortho]-phosphat n

c. orthophosphite $CoHPO_3$ Kobalt(II)-[ortho]-phosphit n

c. orthosilicate Co_2SiO_4 Kobalt(II)-orthosilikat n, Kobalt(II)-tetroxosilikat n

c. oxide CoO Kobalt(II)-oxid n, Kobalt[mon]oxid n

c. perchlorate $Co(ClO_4)_2$ Kobalt(II)-perchlorat n

c. perrhenate $Co(ReO_4)_2$ Kobalt(II)-perrhenat n, Kobalt(II)-tetroxorhenat(VII) n

c. potassium sulphate $K_2SO_4 \cdot CoSO_4$ Kalium-kobalt(II)-sulfat n

c. rhodanide s. c. thiocyanate

c. salt Kobalt(II)-salz n, Salz n des zweiwertigen Kobalts

c. selenate $CoSeO_4$ Kobalt(II)-selenat n

c. selenide $CoSe$ Kobalt(II)-selenid n, Kobalt[mono]selenid n

c. silicofluoride s. c. fluorosilicate

c. sulphate $CoSO_4$ Kobalt(II)-sulfat n

c. sulphide CoS Kobalt(II)-sulfid n, Kobalt[mono]sulfid n

c. sulphite $CoSO_3$ Kobalt(II)-sulfit n

c. sulphocyanate s. c. thiocyanate

c. thiocyanate $Co(SCN)_2$ Kobalt(II)-thiozyanat n, Kobalt(II)-rhodanid n

c. tungstate $CoWO_4$ Kobalt(II)-wolframat n

cobwebbing Fadenziehen n, Fadenbildung f *(beim Auftragen von Anstrichstoffen)*

coca Kokastrauch m, Koka f, Erythroxylum coca Lam.; Kokablätter npl

c. alkaloid Kokaalkaloid n

c. leaves Kokablätter npl *(von Erythroxylum coca Lam.)*

cocaine Kokain n, Benzoylmethylekgonin n, Benzoylekgoninmethylester m

c. hydrochloride Kokainhydrochlorid n

c. nitrate Kokainnitrat n

cocatalysis Kokatalyse f

cocatalyst Kokatalysator m

coccolite *(Min)* Kokkolith m

cochenillic acid Koschenillesäure f, m-Kresol-4,5,6-trikarbonsäure f, 6-Hydroxy-4-methyl-1,2,3-benzoltrikarbonsäure f

cochineal Koschenille f, Cochenille f

c. insect Koschenille[schild]laus f, Echte Koschenille[schild]laus f, Nopalschildlaus f, Dactylopius coccus

c. red A Cochenillerot A n

Cochrane test Cochrane-Trommelprüfung f, Trommelprüfung f (Abriebprobe f in der Trommel) nach Cochrane

cock Hahn m, Hahnventil n

c. plug Hahnkegel m, Hahnküken n

c.-spur *(Ker)* Hahnenfuß m *(Brennhilfsmittel)*

cockle / to *(Pap)* wellig werden

cockling *(Pap)* Welligwerden n; *(Text)* Kräuseln n, Kräuselung f

cockur Unechte Lackmusflechte f, Ochrolechia tartarea (L.)Mass.

coco palm s. coconut palm

cocoa s. cacao

cocoanut s. coconut

cocondensation Mischkondensation f, Kokondensation f

coconut Kokosnuß f

c. butter s. c. oil

c. cake Kokoskuchen m, Kokos[nuß]preßkuchen m

c. fibre Kokosfaser f

c. meal Kokoskuchenmehl n

c. milk Kokosmilch f

c. oil Kokos[nuß]öl n, Kokos[nuß]fett n, Kokosbutter f

c. palm Kokospalme f, Cocos nucifera L.

c. shell Kokos[nuß]schale f

c. shell flour Kokosnußschalenmehl n

c. tree s. c. palm

c. water s. c. milk

cocoon Kokon m, Seidenkokon m

co-current im Gleichstrom (Parallelstrom), nach dem Gleichstromprinzip (Parallelstromprinzip), im Gleichstrom (Parallelstrom) geführt

co-current flow Gleichstrom m, Parallelstrom m

cod-liver oil Lebertran m, *(i.e.S.)* Dorschlebertran m, Dorschleberöl n

c.-liver oil emulsion Lebertranemulsion f

c. oil *(Gerb)* (geringwertige Sorte von) Lebertran m

codeine Kodein n, Codein n, Morphiummethyläther m, Methylmorphin n

c. phosphate Kodeinphosphat n, 6-Hydroxy-3-methoxy-N-methyl-4,5-epoxymorphinen-(7)-dihydrogenphosphat n

codeinone Kodeinon n

codistillant *(Destillation)* Zusatzstoff m, Zusatzkomponente f

codistillation Destillation f mit Zusatzstoff[en], Kodestillation f

coefficient Koeffizient m, Kennzahl f, Kennziffer f; Faktor m, Beizahl f, Vorzahl f

c. of absorption Absorptionskoeffizient m

c. of amplification Verstärkungsfaktor m, Verstärkungsgrad m

c. of compressibility Kompressibilitätskoeffizient m

c. of contraction Kontraktionskoeffizient m

c. of cubic[al] expansion kubischer (räumlicher) Ausdehnungskoeffizient m

c. of diffusion Diffusionskoeffizient m, Diffusionskonstante f

c. of digestibility Verdauungskoeffizient m, Verdauungsquotient m, Verdauungswert m

c. of discharge Durchflußzahl f, Durchfluß-koeffizient m, Durchflußbeiwert m
c. of electrolyte (electrolytic) dissociation Dissoziationsgrad m
c. of expansion Ausdehnungskoeffizient m, Ausdehnungszahl f
c. of friction Reib[ungs]koeffizient m, Reib[ungs]faktor m, Reibwert m
c. of hardness Härtezahl f
c. of heat transfer Wärmeübergangszahl f, Wärmetransmissionskoeffizient m
c. of linear expansion linearer Ausdehnungs-koeffizient m
c. of mass absorption Massenabsorptions-koeffizient m
c. of mass-transfer Stoffübergangszahl f
c. of performance Leistungszahl f
c. of purity Reinheitsquotient m (des Zuk-kerrübensaftes)
c. of recombination Rekombinationskoeffizient m
c. of reflection Reflexionskoeffizient m, Re-flexionszahl f
c. of roughness Rauhigkeitszahl f, Rauhigkeits-grad m
c. of solubility Löslichkeitskoeffizient m, Löslich-keitskonstante f
c. of thermal conduction (conductivity) Wär-meleitzahl f, [spezifisches] Wärmeleit[ungs]ver-mögen n, Wärmeleitfähigkeit f
c. of thermal expansion thermischer Aus-dehnungskoeffizient m, Wärmeausdehnungs-koeffizient m, Wärmeausdehnungszahl f
c. of variation (Statistik) Variabilitätskoeffizient m, Variationskoeffizient m
c. of viscosity Viskositätskoeffizient m, Vis-kositätskonstante f, Konstante f der inneren Reibung
c. of vulcanization Vulkanisationskoeffizient m, Vulkanisationsgrad m, VK
coelestine s. celestine
coenzyme Koenzym n, Koferment n, Agon n, Wirk[ungs]gruppe f, prosthetische (aktive) Gruppe f
coenzyme A Koenzym A n, Koferment A n, CoA
coerulein (Farb) Coerulein n
coexist / to nebeneinander bestehen
coextract / to gemeinsam extrahieren
coffee bean Kaffeebohne f
c. berry (cherry) Kaffeekirsche f
c. cream Kaffeesahne f
c. nib s. c. bean
coffeine Koffein n, Kaffein n, 1,3,7-Trimethyl-xanthin n
cognate inclusion (Geol) endogener (homöogener) Einschluß m, Autolith m
cohere / to zusammenhängen, kohärieren
coherence, coherency (Phys) Kohärenz f; Zusam-menhalt m (z.B. des Kokses)
coherent zusammenhängend, kohärent

cohesion Kohäsion f
c. force s. cohesive force
c. pressure Kohäsionsdruck m, Binnendruck m, innerer Druck m
cohesional force s. cohesive force
cohesive kohäsiv, zusammenhaltend, Kohäsions...
c. energy density kohäsive Energiedichte f, CED
c. force Kohäsionskraft f
c. strength Kohäsionsfestigkeit f
cohoba Niopo, Parica (Samenmehl von Piptadenia peregrina (L.)Benth., Genußmittel)
cohobate / to kohobieren
cohobation Kohobation f
cohune oil Cohuneöl n (Samenöl von Orbignya cohune (Mart.)Dahlgr.)
coil Rohrschlange f, Schlangenrohr n, Schlange f, Rohrspirale f, Spirale f; Knäuel n (z.B. Molekülknäuel)
c. condenser s. coiled-tube condenser
c. heat exchanger Schlangenwärmeaustauscher m, Schlangenwärmeübertrager m
coiled-coil Superhelix f (Feinstruktur von Po-lypeptidketten)
c. condenser s. c.-tube condenser
c. molecule geknäueltes Molekül n, Knäuel-molekül n, Molekülknäuel n
c.-tube condenser Schlangenkühler m
c.-tube evaporator Verdampfapparat m mit Heizschlange
coiling Knäuelung f (z.B. einer Molekülkette)
coin gold Münz[en]gold n
c. technique Coin-Technik f (Kaschieren zweier Textilien unter streifen- oder rasterförmigem Klebemittelauftrag)
coinage alloy Münzlegierung f
c. bronze Münzbronze f
c. metal Münz[en]metall n
coincide / to übereinstimmen, zusammenfallen, sich (einander) decken, koinzidieren
to make c. zur Deckung bringen
coincidence Koinzidenz f, [zeitliche] Übereinstim-mung f, Gleichzeitigkeit f; Deckung f
c. arrangement Koinzidenzanordnung f
c. circuit Koinzidenzschaltung f
c. counter Koinzidenzzähler m
c. counting Koinzidenzzählung f
c. method Koinzidenzmethode f
c. system s. c. circuit
coincident koinzident, zeitlich zusammenfallend (übereinstimmend); einander deckend
coir Kokosfaser f
coke / to verkoken, entgasen; [ver]koken, zu Koks werden, entgasen
coke Koks m, (i.e.S. auch) Hochtemperaturkoks m
c. belt conveyor Koks[förder]band n, Koks-transportband n, Rampenabzugsband n zur Kokssieberei
c. black Koksschwarz n
c. breaker Koksbrecher m
c. breeze Koksgrus m

c. **briquette** Koksbrikett *n*
c. **bunker** Koksbunker *m*
c. **button** Kokskuchen *m*
c. **car** Koks[lösch]wagen *m*, Löschwagen *m*
c. **chamber** *(Kokerei)* Koks[ofen]kammer *f*, Verkokungskammer *f*; *(Erdöl)* Kokskammer *f*, Verkokungskammer *f (beim Kracken auf Koks)*
c.-**contaminated** koksbeladen, mit Koks beladen *(Katalysator)*
c. **discharge** Koksaustrag *m*; Koksdrücken *n*, Koksausstoß *m*
c.-**discharge side** Koks[ausstoß]seite *f*
c. **discharger** Koksaustrag[e]vorrichtung *f*; Koks[aus]drückmaschine *f*, Koksausstoßmaschine *f*
c.-**discharging machine** Koks[aus]drückmaschine *f*, Koksausstoßmaschine *f*
c. **dust** Koksstaub *m*
c. **extractor** Koksschleuse *f*, Koksaustragungswalze *f*, Koksextraktor *m*
c.-**fired** koksgefeuert
c. **fires** Koksfeuerung *f*
c. **formation** Koksbildung *f*
c. **friability** Zerreiblichkeit *f* des Kokses (der Kokssubstanz)
c. **guide** Koksführungsschild *m*; Koks[kuchen]führungswagen *m*, Kuchenführungswagen *m*
c. **guide and door machine** Koks[kuchen]führungswagen *m* mit Türabhebemaschine
c. **industry** Kokereiindustrie *f*, Koksindustrie *f*
c. **manufacture** Koksherstellung *f*, Brzeugung *f*, Koksproduktion *f*, Koksfabrikation *f*
c. **number** Verkokungszahl *f*, Verkokungswert *m*
c. **oven** Kokereiofen *m*, Verkokungsofen *m*, Kok[ung]sofen *m*
c.-**oven battery** Koks[ofen]batterie *f*, Verkokungsbatterie *f*
c.-**oven benzol** Kokereibenzol *n*
c.-**oven gas** Koks[ofen]gas *n*, Kokereigas *n*
c.-**oven industry** s. c. industry
c.-**oven plant** Koksofenanlage *f*, Kokereianlage *f*, Verkokungsanlage *f*, Kokerei *f*
c.-**oven tar** Koksofenteer *m*, Kokereiteer *m*
c.-**oven works** s. c.-oven plant
c. **plant** s. c.-oven plant
c. **producer** Koksgenerator *m*
c. **properties** Kokseigenschaften *fpl*, Koksbeschaffenheit *f*
c. **pusher** Koks[aus]drückmaschine *f*, Koksausstoßmaschine *f*
c. **pusher ram** Koksausdrückstange *f*, Koksdruckstange *f*, Druckstange *f*, Stoßstange *f (zum Ausstoßen des Kokses)*
c. **quenching** Kokslöschen *n*, Kokslöschung *f*
c. **quenching car** Koks[lösche]wagen *m*, Löschwagen *m*
c. **residue** Verkokungsrückstand *m*, Koksrückstand *m*, Kohlenstoffrückstand *m (beim Verkokungstest nach Conradson)*

c. **side** Koks[ausstoß]seite *f*
c.-**side bench** koksseitige Bedienungsbühne *f*
c. **type** Kokstyp[us] *m*
c. **wharf** Koks[kühl]rampe *f*, Koksabwerframpe *f*, Abwurframpe *f*, Schrägrampe *f*, schräge Rampe (Abwurframpe) *f*
coking Verkokung *f*, Verkoken *n*, Entgasung *f*, Entgasen *n*, *(i.e.S. auch)* Hochtemperaturverkokung *f*, HT.-Verkokung *f*, Normalverkokung *f*, Vollverkokung *f*, Hochtemperaturentgasung *f*, HT.-Entgasung *f*; Verkokung *f*, Kracken *n* auf Koks *(beim thermischen Kracken)*
c. **chamber** *(Kokerei)* Verkokungskammer *f*, Koks[ofen]kammer *f*; *(Erdöl)* Verkokungskammer *f*, Kokskammer *f (beim Kracken auf Koks)*
c. **coal** Kokskohle *f*, Verkokungskohle *f*
c. **heat** Verkokungswärme *f*
c. **oven** s. coke oven
c. **plant** Verkokungsanlage *f*, Koksofenanlage *f*, Kokerei[anlage] *f*
c. **power** Koksbildungsvermögen *n*, Kokungsvermögen *n*, Verkokungsvermögen *n*, Verkokungsfähigkeit *f*
c. **process** Verkokungsverfahren *n*, Verkokungsprozeß *m*, Kok[ung]sprozeß *m*; Verkokungsvorgang *m*
c. **properties** Verkokungseigenschaften *fpl*, Verkokungsverhalten *n*, verkokungstechnische Eigenschaften *fpl*, verkokungstechnisches Verhalten *n*
c. **steam coal** kokende Kesselkohle *f*
c. **still** Kokungsdestillationsanlage *f*, Verkokungsblase *f*
c. **test** Verkokungstest *m*, Verkokungsprobe *f*
c. **time** Verkokungszeit *f*, Garungszeit *f*
col. s. colourless
cola [nut] Kolanuß *f*
c. **tree** Kolabaum *m*, *(Gattung)* Cola Schott et Endl.
colamine Kolamin *n*, 2-Aminoäthylalkohol *m*, 2-Aminoäthanol-(1) *n*, Äthanolamin *n*
colander / to durchseihen, abseihen, kolieren
colander Seiher *m*, Durchschlag *m*, Koliertuch *n*
colation Durchseihen *n*, Abseihen *n*, Kolieren *n*
colature Seihflüssigkeit *f*, Kolatur *f*
Colburn [sheet] process Colburn-Verfahren *n*, Libbey-Owens-Verfahren *n (Tafelglasherstellung)*
colchicine Kolchizin *n (Alkaloid)*
colchicum seed *(Pharm)* Zeitlosensamen *m*, Herbstzeitlosensamen *m*
cold kalt, kühl, Kalt...; kalt, nicht [radio]aktiv, aktivitätsfrei
cold Kälte *f*
c. **acid process** Kaltsäureverfahren *n (zur katalytischen Polymerisation)*
c. **adhesive** Kaltkleber *m*
c. **air** kalte Luft *f*, Kaltluft *f*
c.-**air blast** kalter Wind (Gebläsewind) *m*, Kaltwind *m*

c. alkali refining *(Pap)* Kaltveredelung *f*, Kaltalkalisierung *f*
c. asphalt Kaltasphalt *m*
c. bath Kältebad *n (eines Gefriertrocknungsapparats)*
c.-bend test Kältebiegeprüfung *f*, Biegeprüfung *f* in der Kälte
c. blast *s.* c.-air blast
c.-blast pig iron kalterblasenes Roheisen *n*
c. bleach[ing] *(Pap)* Kaltbleiche *f*, kalte Bleiche *f*
c. brine Kühlsole *f*
c. caustic semichemical process *s.* c. caustic soda process
c. caustic soda process *(Pap)* Kalt-Soda-Verfahren *n*, Kalt-Natron-Verfahren *n*, CCSC-Verfahren *n*
c.-chamber die casting Kaltkammerdruckgießen *n*, Kaltkammerdruckguß *m*, *(veraltet)* Preßguß *m*
c.-chamber [die-casting] machine Kaltkammer-Druckgießmaschine *f*, Kaltkammer-Druckgußmaschine *f*, Druckgießmaschine *f* für Kaltkammerverfahren, Kaltkammermaschine *f*, *(veraltet)* Preßgußmaschine *f*
c.-chamber pressure casting *s.* c.-chamber die casting
c.-chamber process Kaltkammerverfahren *n (des Druckgießens)*
c. cream Coldcreme *f*, Kühlcreme *f*
c. cure Kaltvulkanisation *f*, Kaltvulkanisieren *n*
c.-cure method Kaltvulkanisierverfahren *n*
c.-cured kaltvulkanisiert
c.-cured vulcanizate Kaltvulkanisat *n*
c.-curing kalthärtend, kaltabbindend
c. curing Kaltvulkanisation *f*; Kalthärtung *f*
c.-cut varnish Kalt[ansatz]lack *m*
c. defecation *(Zucker)* kalte Scheidung *f*
c.-draw / to *(Met)* kaltziehen; *(z.B. Öle)* kaltpressen, kaltschlagen
c. drawing *(Met)* Kaltziehen *n*; *(Plast)* Kalt[ver]strecken *n*, Kaltverstreckung *f*, Kaltrecken *n*; Kaltpressen *n*, Kaltschlagen *n (von Ölen)*
c.-drawn oil kaltgepreßtes (kaltgeschlagenes) Öl *n*
c.-drawn steel kaltgezogener Stahl *m*
c.-dyeing kaltfärbend
c. dyeing colour Kaltfärber *m*
c. finger [condenser] Kühlfinger *m*, kalter Finger *m*, Einhängekühler *m*
c. flex Biegsamkeit *f* bei niedriger Temperatur
c. flow kalter Fluß *m*, kaltes Fließen *n (bei thermoplastischen Klebstoffen)*
c. forming Kalt[ver]formen *n*, Kalt[ver]formung *f*, Kaltformgebung *f*, Kaltumformung *f*
c. galvanizing galvanisches Verzinken *n*
c. gas Kaltgas *n*, kaltes Gas *n*
c. glue Kaltleim *m*
c. grinding *(Pap)* Kaltschliffverfahren *n*, Kaltschleifen *n*

c.-ground pulp *(Pap)* Kaltschliff *m*
c. hardening Kalthärten *n*
c.-hardening varnish kalthärtender Lack *m*
c. impulse rendering [method] Kaltschmelzprozeß *m (Fettgewinnung)*
c. light Kaltlicht *n*
c. liming *s.* c. defecation
c. main defecation *(Zucker)* kalte Hauptscheidung *f*
c. mastication kalte Mastikation *f*, Kaltmastikation *f*, kaltes Mastizieren *n*, Kaltmastizierung *f*
c. milk separator Kaltmilchzentrifuge *f*, Kaltmilchseparator *m*
c. milling Kaltmahlen *n*
c. moulded *(Plast)* kaltgepreßt, kaltverpreßt
c. moulding *(Plast)* Kaltpressen *n*; Kaltformteil *n*
c.-moulding compound (material) Kaltpreßmasse *f*
c. neutron kaltes (unterthermisches) Neutron *n*
c. orientation *(Plast)* Kalt[ver]strecken *n*, Kaltverstreckung *f*, Kaltrecken *n*
c.-pack method *(Lebm)* Gefrierkonservierung *f*
c. painting *(Glas)* Kaltmalerei *f*
c. pass Kalt[walz]stich *m (beim Pulverwalzen)*
c. polymer Tieftemperaturpolymerisat *n*, Tieftemperaturpolymeres *n*
c. polymerization Tieftemperaturpolymerisation *f*, Kaltpolymerisation *f*
c. polymerized rubber *s.* c. rubber
c. predefecation (preliming) *(Zucker)* kalte Vorscheidung *f*
c.-press / to kaltpressen, kalt pressen (verpressen); *(z.B. Öle)* kaltpressen, kaltschlagen
c.-press method *s.* c.-pressing method
c.-press welding Kaltpreßschweißen *n*, Kaltpreßschweißung *f*
c.-pressed oil *s.* c.-drawn oil
c. pressing Kaltpressen *n*
c.-pressing method (technique) Kaltpreßmethode *f*, Kaltpreßverfahren *n*
c. refining *s.* c. alkali refining
c. resistance Kältebeständigkeit *f*, Tieftemperaturbeständigkeit *f*, Kältefestigkeit *f*
c.-resistant kältebeständig, tieftemperaturbeständig, kältefest
c. rolling Kaltwalzen *n*
c. rubber Kaltkautschuk *m*, Tieftemperaturkautschuk *m*, kalt polymerisierter Kautschuk *m*, Coldrubber *m*
c. rubber latex Kalt-Latex *m*
c.-setting kalthärtend, kaltabbindend *(z.B. Leim)*
c.-settling process Tank-Absetz-Verfahren *n*
c. sludge Kühltrub *m*, Feintrub *m*, Kältetrub *m* *(Bierherstellung)*
c. slug *(Plast)* kalter Stopfen (Pfropfen) *m*
c.-smoke / to kalträuchern
c. smoking Kalträuchern *n*, Kalträucherung *f*, Kalträucherei *f*
c. soda process *s.* c. caustic soda process
c. soda pulp Kaltsodastoff *m*, Kaltnatron-Halbzellstoff *m*

c.-soldering flux Kaltlötmittel n
c. sterilization Kaltsterilisation f
c. storage Kühl[raum]lagerung f, Kaltlagerung f, Kühlhausaufbewahrung f, Kühlraumaufbewahrung f; s. c. store
c.-storage egg Kühlhausei n
c.-storage house s. c. store
c.-storage meat Gefrierfleisch n
c.-storage room Kühlraum m
c.-store / to kühl (kalt) lagern
c. store Kühlhaus n, Kühlhalle f
c. stretching s. c. orientation
c. test Kälteprüfung f, Kälteprobe f
c.-tested neatsfoot oil geklärtes (gereinigtes) Rinderklauenöl n
c.-tolerant yeast Kaltgärhefe f
c. vulcanizate Kaltvulkanisat n
c. vulcanization Kaltvulkanisation f, Kaltvulkanisieren n
c.-vulcanize / to kaltvulkanisieren
c. vulcanizing s. c. vulcanization
c.-vulcanizing machine Kaltvulkanisiermaschine f
c. water extract Kaltwasserextrakt m
c. waving Kaltwellbehandlung f
c. welding s. c.-press welding
c. welding agent Kaltnetzer m
c.-work / to kaltverarbeiten
c.-work harden / to kalthärten
c. working Kaltverarbeitung f; s. c. forming
c.-working process Kaltverarbeitungsverfahren n
colemanite (Min) Kolemanit m (Kalziumhexaborat)
collagen Kollagen n (Eiweißstoff)
  c. fibre Keratinfaser f; (Med) kollagene Faser f (Bindegewebsfaser)
collagenase (Gerb) Kollagenase f (Ferment)
collapse (Ker) Erweichen n, Zusammensinken n (bei der Druckfeuerbeständigkeitsprüfung)
collar Ringrohr n
collect / to sammeln, (Gase) auffangen, (Elektronen) einfangen; sich sammeln (ansammeln, anhäufen)
collecting agent (Flot) Sammler m, Kollektor m
  c. cylinder (Text) Aufspulvorrichtung f
  c. electrode Niederschlagselektrode f, Sammelelektrode f; s. collector 2.
  c. main Sammelleitung f
  c. plate Niederschlagsplatte f, Sammelplatte f
  c. system Abscheidungssystem n
  c. tank Sammelbehälter m
  c. trough Sammeltrog m
collection Sammeln n, Sammlung f; Ansammlung f (z.B. von Erdöl)
  c. cup (Gum) Sammelbecher m, Latexbecher m
  c. efficiency Abscheidegrad m, Ausscheidungsgrad m, Entstaubungsgrad m
collective floatation kollektive Flotation f, Kollektivflotation f
  c. index Sammelregister n, Generalregister n, Gesamtregister n

collector 1. (Flot) Sammler m, Kollektor m; Abscheider m, Entstauber m; 2. Kollektor m, Kollektorelektrode f, Sammelelektrode f, Auffangelektrode f, Fangelektrode f
  c. electrode s. collector 2.
  c. junction Kollektorübergang m, Kollektorschicht f
collide / to zusammenstoßen, zusammentreffen, kollidieren
α-collidine α-Kollidin n, 4-Äthyl-2-methylpyridin n
β-collidine β-Kollidin n, 4-Äthyl-3-methylpyridin n
γ-collidine γ-Kollidin n, sym-Kollidin n, 2,4,6-Trimethylpyridin n
Collin oven Collin-Ofen m, Vollstromwechselofen m nach Collin
collinite Kollinit m (Komponente der Steinkohle)
collision Zusammenstoß m, Stoß m, Zusammentreffen n, Zusammenprall m, Kollision f
  c. complex Stoßkomplex m
  c. cross section Stoßquerschnitt m
  c. density Stoß[zahl]dichte f
  c. diameter Stoßdurchmesser m
  c. excitation Stoßanregung f
  c. factor Stoßfaktor m
  c. frequency Stoßhäufigkeit f
  c. integral Stoßintegral n
  c. ionization Stoßionisation f, Stoßionisierung f
  c. number Stoßzahl f [je Zeiteinheit]
  c. of the first kind Stoß m erster Art, anregender (aktivierender) Stoß m
  c. of the second kind Stoß m zweiter Art, desaktivierender Stoß m
  c. parameter Stoßparameter m
  c. theory Stoßtheorie f
collochemistry s. colloid chemistry
collodion Kollodium n
  c. cotton Kollodiumwolle f, Kolloxylin n, Nitrozellulose f
  c. membrane Kollodiummembran f
  c. ultrafilter Kollodium-Ultrafilter n
  c. wool s. c. cotton
colloid Kolloid n
  c. carrier kolloidaler Träger m, Trägersubstanz f, Apoferment n, Apoenzym n, Pheron n
  c. chemical kolloidchemisch
  c. chemist Kolloidchemiker m
  c. chemistry Kolloidchemie f, Kolloidik f, Kolloidlehre f
  c.-disperse kolloiddispers
  c. mill Kolloidmühle f
colloidal kolloid[al], fein verteilt, kolloiddispers, Kolloid...
  c. chemistry s. colloid chemistry
  c. complex (Bodenkunde) Ton-Humus-Komplex m, organomineralischer Komplex m
  c. electrolyte Kolloidelektrolyt m, kolloider Elektrolyt m
  c. kaolin Kolloidkaolin n, kolloidales Kaolin n
  c. particle Kolloidteilchen n, Kolloidpartikel f
  c. solution kolloid[al]e Lösung f, Kolloidlösung f

c. state Kolloidzustand *m*
c. sulphur kolloid[al]er Schwefel *m*, Kolloidschwefel *m*
c. system kolloid[dispers]es System *n*, Kolloidsystem *n*
collotype paper Lichtdruckpapier *n*
colocynth Koloquinte *f (Frucht von Citrullus colocynthis (L.)Schrad.)*
cologne [water] Kölnischwasser *n*, Eau de Cologne *n*
Colombia copal Amerikanischer Kopal *m*, Anime-Kopal *m (i.e.S., von Hymenaea courbaril L.)*
colophonite *(Min)* Kolophonit *n*
colophony Kolophonium *n*, Geigenharz *n (von Pinus-Arten)*
color / to *(Am) s.* to colour
color *(Am) s.* colour
coloration *(Am) s.* colouration
colorimeter Kolorimeter *n*, Tintometer *n*
c. tube Kolorimeterrohr *n*
colorimetric kolorimetrisch
c. analysis kolorimetrische Analyse (Bestimmung, Methode) *f*, Kolorimetrie *f*
c. coulometer kolorimetrisches Coulometer *n*
c. determination (method) *s.* c. analysis
colorimetry Kolorimetrie *f*, fotometrische Analyse (Bestimmung, Methode) *f*
coloring *(Am) s.* colouring
colorless *(Am) s.* colourless
colostrum Kolostrum *n*, Kolostralmilch *f*, Biestmilch *f*, Erstlingsmilch *f*, Vormilch *f*
colour / to 1. färben; *(Foto)* kolorieren; 2. sich [ver]färben, Farbe annehmen; 3. *(Gerb)* angerben, anfärben, abfärben *(schwach gerben)*
colour 1. Farbe *f (als Qualität der Lichtempfindung)*; Färbung *f (Zustand)*; 2. *s.* colouring matter
c. additive Farbstoffzusatz *m*
c. atlas *s.* c. chart
c. base Farb[stoff]base *f*, Färbebase *f*
c. change Farbänderung *f*, Farb[en]umschlag *m*, Farbwechsel *m*
c. chart Farbenatlas *m*
c.-code pipette Pipette *f* mit Farbmarkierung
c. comparator Farbkomparator *m*, Farbvergleicher *m*
c.-comparator tube Farbvergleichszylinder *m*
c. defect Farbfehler *m (z.B. im Glas)*
c. developer *(Foto)* Farb[en]entwickler *m*
c. development *(Foto)* Farb[en]entwicklung *f*
c. fastness Farbechtheit *f*
c. filter Farbfilter *n*
c. former Farbbildner *m*
c. index *(Min)* Farbzahl *f*
c. indicator Farbindikator *m*
c. intensity Farbintensität *f*
c. lake Farblack *m*
c. photography Farb[en]fotografie *f*
c. pits *(Gerb)* Farbengang *m*
c. printing Farbendruck *m*, Mehrfarbendruck *m*, Chromodruck *m*

c. range Farb[en]skale *f*
c. retention Farb[ton]beständigkeit *f*, Farbenbeständigkeit *f*, Farbtonechtheit *f*
c. salt Farbsalz *n*
c. sensitive *(Foto)* farbempfindlich
c. sensitivity *(Foto)* Farbempfindlichkeit *f*
c. separation Entmischung *f* der Körperfarben
c. spot Farbfleck *m (Papierfehler)*
c. stability *s.* c. retention
c. temperature Farbtemperatur *f*
c. two-sidedness *(Pap)* Zweiseitigkeit *f* in der Färbung
c. vision Farbensehen *n*, Chromatopsie *f*
c. vision chemistry Chemie *f* des Farbensehens
Colour Index Colour Index *m*, CI *(von der Society of Dyers and Colourists und der American Association of Textile Chemists and Colorists herausgegebenes Nachschlagewerk für Handelsfarbstoffe)*
Colour Index number Colour-Index-Nummer *f*, Nummer *f* im Colour Index, CI-Nummer *f*
colouration Färben *n*, Färbung *f*
c. of paper *s.* colouring of paper
coloured coated paper farbig gestrichenes Papier *n*, Buntpapier *n*
c. discharge Buntätzen *n*
c. glass farbiges Glas *n*, Farbglas *n*, Buntglas *n*
c. lake Farblack *m*
c. malt Farbmalz *n*
c. paper Buntpapier *n*, farbiges (gefärbtes) Papier *n*, *(i.e.S.)* farbig gestrichenes Papier *n (oder)* in Stoff gefärbtes Papier *n*
c. solution Farblösung *f*
colouring 1. Färben *n*, Färbung *f*; 2. *s.* c. matter
c. agent *s.* c. matter
c. machine *(veraltet für)* coating machine
c. material *s.* c. matter
c. matter färbender (farbgebender) Stoff *m*, färbende Substanz *f*, Farbmittel *n*, *(i.e.S.)* Farbstoff *m*, Farbe *f (oder)* Pigment *n*
c. matter annatto Anattofarbstoff *m*, An[n]atto *n*, Orlean *m (aus Bixa orellana)*
c. of cheese Färben (Anfärben) *n* von Käse, Käsefärbung *f (mit Farbstoffen)*
c. of paper Papierfärbung *f*, Papierfärben *n*, Färben *n* von Papier, Papierfärberei *f*
c. principle färbendes (färberisches) Prinzip *n*, Farbprinzip *n*, färbender Bestandteil *m*, Farbbestandteil *m*
c. strength Farbintensität *f*
c. substance *s.* c. matter
colourist *(Text)* Kolorist *m*
colourless farblos, farbfrei, nichtfarbig
colourlessness Farblosigkeit *f*
colourwoven buntgewebt
columbate Niobat *n*
columbite *(Min)* Kolumbit *m*
columbium Cb Kolumbium *n*, *(veraltet für)* Nb Niob *n (Verbindungen s. unter niobium)*
column Kolonne *f*, Säule *f*; *(Krist)* Stengel *m*; *(Geol)* Schichtfolge *f*

c. chromatography Säulenchromatografie *f*
c. diameter Kolonnendurchmesser *m*
c. distillation Kolonnendestillation *f*
c. efficiency Kolonnenwirkungsgrad *m*
c. height Kolonnenhöhe *f*, Säulenhöhe *f*
c. hold-up *(Destillation)* Flüssigkeitsinhalt *m*, Betriebsinhalt *m*, in der Säule enthaltene Flüssigkeitsmenge *f*
c. of coal Kohlensäule *f*, Kolonne *f* gepulverter Kohle
c. of liquid Flüssigkeitssäule *f*
c. of smoke Rauchsäule *f*
c. operation Säulenbetrieb *m* *(im Ionenverzögerungsverfahren)*
c. packing Säulenfüllung *f*
c. separation Säulentrennung *f*, säulenchromatografische Trennung *f*
c. wall Säulenwand *f*
columnar *(Krist)* steng[e]lig, säulenförmig, säulenartig
c. crystallization Stengelkristallisation *f*
colza oil Kolzaöl *n*, Kohlsaatöl *n*, *(i.e.S.)* Rüb[sen]öl *n (aus Brassica rapa L.em. Metzg.) (oder)* Rapsöl *n (aus Brassica rapus L.em. Metzg.)*
comagmatic region *(Geol)* komagmatische Region *f*
comb honey Wabenhonig *m*, Scheibenhonig *m*
Combee resin *(Harz von Gardenia gummifera L.f.)*
Combes quinoline synthesis Combessche Chinolinsynthese *f*
combination Verbindung *f*, Vereinigung *f*, Kombination *f*
c. drier *(Farb)* Kombinationstrockner *m*
c. lacquer Kombinationslack *m*
c. nitrocellulose lacquer Nitrokombinationslack *m*
c. oven Verbund[koks]ofen *m*
c. principle *(phys Ch)* [Ritzsches] Kombinationsprinzip *n*
c. pyroxylin lacquer *s.* c. nitrocellulose lacquer
c. syntan Kombination *f* synthetischer Gerbstoffe
c. tannage Kombinationsgerbung *f*
c. underjet coke oven Verbund-Unterbrennerofen *m*, Verbundkoksofen *m* mit Unterbrennern
c. vibration Kombinationsschwingung *f*
combine / to verbinden, vereinigen, zusammensetzen, kombinieren; sich verbinden (vereinigen), eine Verbindung eingehen
combined:
in c. form gebunden, in gebundener Form
c. fabric kaschiertes Gewebe *n*
c. material *(Plast)* Verbundstoff *m*
c. plastic Verbundpreßstoff *m*
c. sewer system *s.* c. system of sewerage
c. sulphur gebundener Schwefel *m*
c. sulphur dioxide *(Pap)* gebundenes SO_2 *n*
c. system of sewerage Mischsystem *n*, Mischverfahren *n (Stadtentwässerung)*
c. water chemisch gebundenes Wasser *n*
combing *(Text)* Kämmen *n*

combining ability (capacity) *s.* c. power
c. force Verbindungskraft *f*, Bindungskraft *f*, Bindekraft *f*, bindende Kraft *f*
c. power Verbindungsfähigkeit *f*, Verbindungskraft *f*, Bindungsvermögen *n*, Bindungsfähigkeit *f*, Bindungskraft *f*, Bindekraft *f*, bindende Kraft *f*
c. weight Äquivalentmasse *f*, *(bisher)* Verbindungsgewicht *n*, Äquivalentgewicht *n*
combustibility Verbrennlichkeit *f*, Verbrennbarkeit *f*, Brennbarkeit *f*
combustible verbrennlich, [ver]brennbar
combustible Brennbares *n*; Brennmaterial *n*
c. gas brennbares Gas *n*, Brenngas *n*
c. powder Räucherpulver *n*
c. principle *(historisch)* brennbares Prinzip *n*, Feuerstoff *m*, Feuermaterie *f*, Phlogiston *n*
combustion Verbrennen *n*, Verbrennung *f*
c. air Verbrennungsluft *f*
c. analysis quantitative Elementaranalyse *f (zur Bestimmung organischer Substanzen)*, Verbrennungsanalyse *f*
c. barge rechteckige Glühschale (Verbrennungsschale) *f*
c. boat Glühschiffchen *n*, Verbrennungsschiffchen *n*, Schiffchen *n*
c. chamber Verbrennungskammer *f*, Verbrennungsraum *m*, Brennkammer *f*, Brennraum *m*, Feuerraum *m*; Brennschacht *m*, Brennkanal *m (eines Winderhitzers)*; *(Pap)* Nachverbrennungskammer *f*
c. fuel Brennstoff *m*
c. furnace Verbrennungsofen *m*
c. gas Verbrennungsgas *n*, Rauchgas *n*, Abgas *n*
c. heat Verbrennungswärme *f*
c. method *s.* c. analysis
c. process Verbrennungsprozeß *m*, Verbrennungsvorgang *m*
c. product Verbrennungsprodukt *n*
c. residue Verbrennungsrückstand *m*
c. space Verbrennungskammer *f*, Verbrennungsraum *m*, Brennkammer *f*, Brennraum *m*, Feuerraum *m*
c. spoon Verbrennungslöffel *m*
c. temperature Verbrennungstemperatur *f*
c. tube Verbrennungsrohr *n*, Verbrennungsröhre *f*
c. zone Verbrennungszone *f*, Brennzone *f*, Oxydationszone *f (eines Hochofens)*
come / to:
c. down *(als Niederschlag)* ausfallen
c. over übergehen, überdestillieren
comestible eßbar, genießbar
comestibles Eßwaren *pl*, Nahrungsmittel *pl*, Lebensmittel *pl*
coming-up time Anwärmzeit *f*, Anheizzeit *f*, Aufheizzeit *f*
Commanducci formic-acid test Nachweis *m* von Ameisensäure nach Commanducci
commercial handelsüblich, Handels...; technisch, industriell

to bring to the c. stage zur Betriebsreife entwickeln, fabrikreif gestalten

c. ammonium carbonate [NH₄]HCO₃·[NH₄][CO₂NH₂] Ammoniumkarbonat *n* des Handels, käufliches Ammoniumkarbonat *n*, Hirschhornsalz *n (Doppelverbindung von Ammoniumhydrogenkarbonat mit Ammoniumamidokarbonat)*

c. ammonium molybdate Ammoniummolybdat *n* des Handels, Ammoniumparamolybdat *n*

c. analysis technische (praktische) Analyse *f*

c. annealing Zwischenglühen *n*, Zwischenglühung *f*

c. benzol Handelsbenzol *n*, technisches Benzol *n*

c. carbide Handelskarbid *n*

c. chemist Handelschemiker *m*

c. coal Handelskohle *f*, Verkaufskohle *f*, handelsübliche Kohle *f*

c. dry basis Bezugsbasis *f* Masse der handelsüblich trockenen Substanz

c. fertilizer Handelsdünger *m*, Mineraldünger *m*, mineralischer (anorganischer Dünger) *m*

c. grade handelsübliche Qualität *f*, Handelsqualität *f*

c. grease technisches Fett *n*

c. name Handelsname *m*, Handelsbezeichnung *f*, kommerzieller Name *m*

c. process technisches Verfahren *n*

c. product Handelsprodukt *n*, Handelsware *f*

c. spirit technischer Alkohol *m*, Industriealkohol *m*

c. synthesis technische Synthese *f*, technisch brauchbare (verwertbare, erfolgreiche) Synthese *f*

c. tissue Packseiden[papier] *n*, Einschlagseidenpapier *n*, Einwickelseidenpapier *n*

commercially available im Handel erhältlich

c. pure technisch rein

comminute / to zerkleinern, *(besonders)* feinzerkleinern

comminuting machine Zerkleinerungsmaschine *f*, Zerkleinerungsapparat *m*

comminution Zerkleinern *n*, Zerkleinerung *f*, *(besonders)* Feinzerkleinern *n*, Feinzerkleinerung *f*

c. plant Zerkleinerungsanlage *f*

commission on nomenclature Kommission *f* für Nomenklatur, Nomenklaturkommission *f*

common brick [gewöhnlicher] Mauerziegel *m*

c. hop [Gemeiner] Hopfen *m*, Humulus lupulus L.

c. hornblende *(Min)* gemeine Hornblende *f (i.e.S.)*

c. ion effect Löslichkeitsverminderung *f (durch Anwesenheit eines Elektrolyten mit gleicher Ionenart)*

c. malic acid HOOCCH(OH)CH₂COOH natürliche Äpfelsäure *f*, *l*-Äpfelsäure *f*, *l*-Hydroxybutandisäure *f*

c. name Trivialname *m*

c. ring gewöhnlicher (normaler) Ring *m (5- bis 7gliedrig)*

c. rosin Balsamkolophonium *n*, Balsamharz *n*, Harz *n* aus Rohterpentin

c. salt Kochsalz *n*, Siedesalz *n*, Salz *n*

c. wheat Brotweizen *m*, Saatweizen *m*, Weichweizen *m*, Triticum aestivum L.

compact / to kompaktieren, preßverdichten; verdichten, komprimieren, [zusammen]pressen; verfestigen

compact kompakt, preßdicht; dicht, *(Geol auch)* massiv

compact Preßling *m*, Preßkörper *m*, Preßstück *n*, Pulverpreßling *m*, Pulverpreßkörper *m*, Metallpulverpreßling *m (in der Pulvermetallurgie)*

c. powder Kompakt[puder] *m*, kompakter Puder *m*

compacted strip Preßband *n*, dichtgepreßtes Pulverband *n (beim Pulverwalzen)*

compacting Kompaktieren *n*, Kompaktierung *f*, Preßverdichten *n*, Preßverdichtung *f*; Verdichten *n*, Verdichtung *f*, Komprimieren *n*, Pressen *n*, Zusammenpressen *n*; Verfestigen *n*, Verfestigung *f*

c. mill s. compactor [mill]

c. pressure Preßdruck *m (beim Kompaktieren)*, Verdichtungsdruck *m*

compaction s. compacting

c. of concrete Betonverdichtung *f*

compactor [mill] Kompaktiermaschine *f*

companion Begleiter *m*, Begleitstoff *m*, Begleitsubstanz *f*

c. alkaloid Begleitalkaloid *n*, Nebenalkaloid *n*

c. substance s. companion

comparator *(Kolorimetrie)* Komparator *m*

c. tube Vergleichszylinder *m*

comparison Vergleich *m*

c. compound Vergleichsverbindung *f*

c. electrode Vergleichselektrode *f*, Bezugselektrode *f*

c. microscope Vergleichsmikroskop *n*

c. solution Vergleichslösung *f*

c. spectrum Vergleichsspektrum *n*

compartment / to unterteilen

compartment Kammer *f*; Zelle *f*

c. dryer Kammertrockner *m (mit mehreren Kammern)*

c. mill Verbund[rohr]mühle *f*, Mehrkammer[rohr]mühle *f*

compatibility Verträglichkeit *f*, Kompatibilität *f*

compatible verträglich, kompatibel

c. dyes kombinierbare Farbstoffe *mpl*

compensated semiconductor gemischter Halbleiter *m*, Kompensationshalbleiter *m*

compensating developer *(Foto)* Ausgleichsentwickler *m*

c. ion Gegenion *n*

compensation Ausgleich *m*, Kompensation *f*

c. fertilization Ausgleichsdüngung *f*, Ersatzdüngung *f*

c. of errors Fehlerkompensation *f*, Fehlerausgleich *m*

c. pH-meter Kompensations-pH-Meter *n*

compensator Kompensator *m*

competing reaction *s*. competitive reaction

competition *s*. competitive reaction

competitive inhibition kompetitive Hemmung *f* (*reversible Hemmung einer Fermentreaktion*)

c. reaction Konkurrenzreaktion *f*, Parallelreaktion *f*

competitor ion (*Bodenkunde*) eintauschendes Ion *n*

complementarity Komplementarität *f*

complementary komplementär

c. colour Komplementärfarbe *f*

complete / to vervollständigen, ergänzen; beenden

complete vollständig, vollkommen; geschlossen (*Kreislauf*)

c. analysis Vollanalyse *f*

c. combustion vollständige Verbrennung *f*

c. cure Ausvulkanisation *f*, Ausheizung *f*

c. fermentation Durchgärung *f*

c. fertilization Volldüngung *f*

c. fertilizer Volldünger *m*

c. gasification vollständige (vollkommene, restlose, rückstandslose) Vergasung *f*

c. partition function vollständige Verteilungsfunktion *f*

c. reaction Gesamtreaktion *f*

c. reflection Totalreflexion *f*

c. softening Vollentsalzung *f*, völlige Entsalzung *f* (*des Wassers*)

c. wetting vollkommene Benetzung *f*

completely saponified vollverseift

completeness Vollständigkeit *f*, Vollkommenheit *f*

completing step letzte Stufe *f*

completion Vervollständigung *f*, Ergänzung *f*

complex / to einen Komplex bilden, komplex (im Komplex) binden; in einen Komplex überführen, komplexieren

complex komplex, Komplex...

complex Komplex *m*, komplexe Gruppe *f*

π-complex π-Komplex *m*

1 : 1 complex 1 : 1-Komplex *m*

c. anion komplexes Anion *n*, Komplexanion *n*

c. bond komplexe Bindung *f*, Komplexbindung *f*

c. cation komplexes Kation *n*, Komplexkation *n*

c. coacervation Komplexkoazervation *f*

c. compound komplexe Verbindung *f*, Komplexverbindung *f*

c. formation Komplexbildung *f*

c.-formation constant Komplexbildungskonstante *f*

c.-forming komplexbildend

c. hydride komplexes Hydrid *n*, Hydridkomplex *m*

c. ion komplexes Ion *n*, Komplex-Ion *n*, Molekelion *n*

c. isomer Komplexisomeres *n*

c. isomerism Komplexisomerie *f*

c. of bivalent platinum Platin(II)-komplex *m*

c. ore Komplexerz *n*, komplexes (polymetallisches, gemengtes, zusammengesetztes) Erz *n*, Mischerz *n*

c. out verbrauchter Komplex *m*, inaktiver Aluminiumchlorid-Kohlenwasserstoff-Komplex (AlCl$_3$-KW-Komplex) *m*, Katalysatorabfall *m* (*bei der Flüssigphaseisomerisierung*)

c. reaction komplexe (zusammengesetzte) Reaktion *f*

c. salt Komplexsalz *n*, komplexes Salz *n*

complexation Komplexbildung *f*

c. titration Komplexbildungstitration *f*, Titration *f* mit Komplexbildnern

compleximetric *s*. complexometric

complexing komplexbildend

complexing Komplexbildung *f*

c. agent Komplexbildner *m*, komplexbildender Stoff *m*

complexometric komplexometrisch

c. titration *s*. complexometry

complexometry Komplexometrie *f*, komplexometrische Titration *f*

complexone Komplexon *n*

compliance (*Text*) Nachgiebigkeit *f*

c. ratio Nachgiebigkeitsverhältnis *n*

comply with / to entsprechen

component am Aufbau beteiligt

component Komponente *f*, Bestandteil *m*, Anteil *m*; (*phys Ch*) Komponente *f* (*z.B. die Molekülart als Teil eines Systems*)

c. glyceride gemischtes (gemischtsäuriges) Glyzerid *n*

componental movement (*Geol*) Teilbewegung *f*

composite (*Plast*) Verbund[werk]stoff *m*

c. batholith (*Geol*) zusammengesetzter Batholith *m*

c. film Verbundfolie *f*

c. mould (*Plast*) zusammengesetztes Preßwerkzeug (Werkzeug) *n*, Mehrfachwerkzeug *n*

c. plastic (*Plast*) Verbund[werk]stoff *m*

composition Zusammensetzung *f*; (*Gum*) Mischung *f*

c. [weighing] tank Wiegebehälter *m*, Wägegefäß *n*, Tankwaage *f* (*Margarineherstellung*)

compost (*Landw*) Kompost *m*

compound / to mischen, eine Mischung herstellen; ein Mischungsrezept aufbauen (aufstellen); (*Öle*) compoundieren, fetten

compound Verbindung *f*, Verb.; Mischung *f*, Gemisch *n*; Kompound *n*, Kompoundmasse *f*; (*Gum*) Mischung *f*, (*auch*) Mischungsrezept *n*

c. designer (*Gum*) Mischungsentwickler *m*, Mischungsfachmann *m*

c. fat Mischfett *n* (*Gemisch fester Tierfette mit flüssigen Tierfetten oder pflanzlichen Ölen*)

c. fertilizer Mehrnährstoffdünger *m*, Kombinationsdünger *m*

c. formation Verbindungsbildung *f*

c. formula *s*. compounding formula

c. formulation Mischungsrezeptur *f*
c. lard *s.* c. fat
c. lime fertilizer Düngemischkalk *m*, Mischkalk *m*
c.   mill Verbund[rohr]mühle   *f*,   Mehrkammer[rohr]mühle *f*
c. semiconductor zusammengesetzter Halbleiter *m*, Verbindungshalbleiter *m*
c. with a congruent melting point kongruent schmelzende Verbindung *f*
c. with an incongruent melting point inkongruent schmelzende Verbindung *f*
compounded oil Compoundöl *n*, compoundiertes (gefettetes) Öl *n*
compounder *(Gum)* Mischungsentwickler *m*, Mischungsfachmann *m*
compounding Mischen *n*, Mischungsherstellung *f*; Rezeptaufstellunng *f*, Aufstellung *f* eines Mischungsrezepts; Compoundieren *n (von Ölen)*
c. formula Mischungsrezept *n*
c. ingredient Mischungsbestandteil *m*
c. recipe *s.* c. formula
c. room Mischkammer *f*
compreg Compreg *n*, Preßschichtholz *n (harzgetränktes Preßholz)*
compress / to komprimieren, [zusammen]pressen, zusammendrücken, verdichten
compressed air Druckluft *f*, Preßluft *f*
c.-air ejector Luftstrahlpumpe *f*, Luftstrahlgebläse *n*, Luftstrahlverdichter *m*
c.-air line Druckluftleitung *f*
c.-air spraying Druckluftspritzen *n (in der Anstrichtechnik)*
c. cartridge *(Sprengtechnik)* Preßkörper *m*
c. powder [fest]gepreßter Puder *m*
c. resin-impregnated wood *s.* compreg
c. wood Preß[voll]holz *n*, Druckholz *n*
c. yeast Preßhefe *f*
compressibility Kompressibilität *f*, Zusammendrückbarkeit *f*, Komprimierbarkeit *f*, Verdichtbarkeit *f*, Verdichtungsfähigkeit *f*, Preßbarkeit *f*
c. coefficient (factor) Kompressibilitätskoeffizient *m*
compressible kompressibel, zusammendrückbar, komprimierbar, verdichtbar, preßbar
compressing *s.* compression
compression Kompression *f*, Komprimieren *n*, Zusammenpressen *n*, Zusammendrücken *n*, Pressen *n*, Verdichten *n*, Verdichtung *f*
c. air Druckluft *f*, Preßluft *f*
c. blanket Drucktuch *n*
c.-fitting joint Klemmringverbindung *f*
c. heat Kompressionswärme *f*
c. mould *(Plast)* Preßwerkzeug *n*, Preßform *f*
c. moulding *(Plast)* Formpressen *n*, Pressen *n*; *(Gum)* Druckverformung *f*, Kompressions-Verdrängungsverfahren *n (zur Herstellung von Gummi-Formteilen)*
c.-moulding material Formpreßstoff *m*, Preßmasse *f*
c.-moulding pressure Preßdruck *m (beim Formpressen)*

c.-moulding resin Preßharz *n*
c. press Formpresse *f*
c. process Kompressionsverfahren *n*
c. ratio Kompressionsverhältnis *n*, Verdichtungsverhältnis *n*, Druckverhältnis *n*, Verdichtungsgrad *m*
c. refrigerating machine Kompressions[kälte]-maschine *f*, Kompressionskälteanlage *f*
c. roll Quetschwalze *f*, Preßwalze *f*, Druckwalze *f*
c. set Druckverformungsrest *m*, Zusammendrückungsrest *m*, [bleibende] Druckverformung *f*, bleibende Verformung *f* nach Druckbelastung (Druckeinwirkung), Formänderungsrest *m* bei Druckbeanspruchung
c. shrinkage *s.* compressive shrinkage
c. strength *s.* compressive strength
c.-type fitting Klemmverschraubung *f*, Klemmverbindung *f*
c. wood Preß[voll]holz *n*, Druckholz *n*
c. zone Kompressionszone *f*, Verdichtungszone *f*
compressive modulus Kompressionsmodul *m*
c.   shrinkage kompressive   (erzwungene) Schrumpfung (Krumpfung) *f*, Kompressionsschrumpf *m*
c. strength Druckfestigkeit *f*
c. stress Druckspannung *f*
compressor Verdichter *m*, Kompressor *m*; Quetschkolben *m (z.B. des Membranventils)*
c. plant Kompressionsanlage *f*
Compton effect Compton-Effekt *m*
Compton electron Compton-Elektron *n*, Rückstoßelektron *n*
Compton recoil particle Compton-Rückstoßteilchen *n*
Compton scattering Compton-Streuung *f*
Compton shift Compton-Verschiebung *f*
Compton wavelength Compton-Wellenlänge *f (des Elektrons)*
Comstock   and   Wescott   process Comstock-Wescott-Verfahren *n (zum chlorierenden Aufschluß von Pyrit)*
conalbumin Konalbumin *n (eisenbindendes Glykoprotein aus dem Eiweiß von Vogeleiern)*
conc, conc. *s.* 1. concentration; 2. concentrated
concave grating Konkavgitter *n*
c.-grating spectrograph Konkavgitterspektrograf *m*
concd. *s.* concentrated
Concentra[-type] mill Concentra-Mühle *f (eine Rohrmühle)*
concentrate / to konzentrieren, *(eine Lösung)* einengen, *(eine Flüssigkeit)* verdichten, eindicken, *(Erz)* anreichern; *(Strahlen)* bündeln
c. by evaporation eindampfen
concentrate Konzentrat *n*, *(in der Aufbereitungstechnik auch)* Aufbereitungskonzentrat *n*
c. discharge Konzentrataustrag *m*
c.-laden froth Konzentratschaum *m*, Schaumkonzentrat *n*

c. table *s.* concentrating table
concentrated konzentriert, k, konz.

  c. ammoniacal liquor verdichtetes (konzentriertes) Ammoniakwasser (NH_3-Wasser) *n*

  c. black liquor *(Pap)* Dick[ab]lauge *f*, eingedickte Schwarzlauge *f*

  c. fertilizer *(Am) (Düngemittel mit mehr als 30 % Nährstoffgehalt)*

  c. latex konzentrierter Latex *m*

  c. milk Kondensmilch *f*, kondensierte (eingedickte) Milch *f*

  c. oil of vitriol konzentrierte Schwefelsäure *f* *(92- bis 99%ig)*

  c. superphosphate Doppelsuperphosphat *n* *(Düngemittel aus Phosphorsäureaufschluß von Rohphosphaten)*

concentrating machine Anreicherungsmaschine *f*, Anreicherungsapparat *m*, Anreicher[ungs]gerät *n*, Anreicherungsanlage *f*

  c. table Herd *m*, Anreicher[ungs]herd *m*, Aufbereitungsherd *m*

concentration 1. Konzentrieren *n*, Konzentrierung *f*, Einengen *n (einer Lösung)*, Verdichten *n*, Eindicken *n (einer Flüssigkeit)*, Anreichern *n*, Anreicherung *f (von Erz)*; Bündelung *f (von Strahlen)*; 2. Konzentration *f*; Gehalt *m*, Beladung *f*

  c. cell Konzentrationselement *n*, Konzentrationszelle *f*, Konzentrationskette *f*

  c. cell with transference Konzentrationselement *n* mit Überführung

  c. cell without transference Konzentrationselement *n* ohne Überführung

  c. change Konzentrations[ver]änderung *f*

  c. gradient Konzentrationsgefälle *n*, Konzentrationsgradient *m*

  c. method *s.* c. process

  c. of active constituents Wirkstoffkonzentration *f*

  c. of ores Erzanreicherung *f*, Erzkonzentrierung *f*

  c. of ozone Ozonkonzentration *f*

  c. of solute impurity Konzentration *f* des gelösten Stoffes, Konzentration *f* der Beimengung (Verunreinigung), Fremdstoffkonzentration *f (in der Zonenschmelztheorie)*

  c. of the peptizing agent Peptisatorkonzentration *f*

  c. plant Konzentrierungsanlage *f*, Konzentrationsanlage *f*; Anreicherungsanlage *f (in der Aufbereitungstechnik)*

  c. polarization Konzentrationspolarisation *f*

  c. process Anreicher[ungs]verfahren *n*, Anreicher[ungs]prozeß *m*, Anreicher[ungs]methode *f*

  c. profile Konzentrationsprofil *n*

concentrator Eindicker *m*, Eindickapparat *m*, Eindickzylinder *m*, Eindickmaschine *f*, Entwässerungsmaschine *f*, Entwässerungsanlage *f*; Anreicher[ungs]gerät *n*, Anreicherungsapparat

m, Anreicherungsmaschine *f*, Anreicherungsanlage *f*

  c. table Herd *m*, Anreicher[ungs]herd *m*, Aufbereitungsherd *m*

concept Begriff *m*, Vorstellung *f*

  c. of pH pH-Begriff *m*

  c. of resonance Resonanzbegriff *m*, Resonanzvorstellung *f*, Mesomeriebegriff *m*, Mesomerievorstellung *f*

conche / to *(Lebm)* konchieren

conche *s.* conching machine

conching *(Lebm)* Konchieren *n*, Konchierung *f*

  c. machine *(Lebm)* Konche *f*, Längsreibe[maschine] *f*

conchoidal *(Min)* musch[e]lig *(z.B. Bruch)*

concn. *s.* concentration

concrete *(Kosmet)* konkret *(z.B. Öl)*

concrete Beton *m*; *(Kosmet)* konkretes Öl *n*, Konkret *n*, Concret *n*, Essence concrète *f*

  c. acid tower *(Pap)* Säureturm *m* aus Beton, Betonturm *m*

  c. aggregate Betonzuschlag[stoff] *m*

  c. block Betonblock *m*

  c. brick Beton[mauer]stein *m*

  c. for [atomic] radiation shielding Strahlenschutzbeton *m*, Abschirmbeton *m*

  c. mass Betonmasse *f*

  c. oil *(Kosmet)* konkretes Öl *n*, Essence concrète *f*, Konkret *n*, Concret *n*

  c. pump Betonpumpe *f*

  c. strength Betonfestigkeit *f*

concreting Betonieren *n*

concretion *(Geol)* Konkretion *f*

concurchine Konkurchin *n*, Kurchenin *n (ein Alkaloid der Kurchirinde)*

concurrent flow Parallelstrom *m*, Gleichstrom *m*

  c. reaction Konkurrenzreaktion *f*, Parallelreaktion *f*; Nebenreaktion *f*, Begleitreaktion *f*

condensability Kondensierbarkeit *f*, *(Phys auch)* Niederschlagbarkeit *f*, Verdichtbarkeit *f*

condensable kondensierbar, *(Phys auch)* niederschlagbar, verdichtbar

condensate Kondensat *n*, Kondensationsprodukt *n*, *(Phys auch)* Niederschlag *m*; Kondenswasser *n*, Niederschlagswasser *n*, Schwitzwasser *n*

  c. removal *(Pap)* Kondensatableitung *f*, Kondensatabführung *f*

  c. reservoir *(Erdöl)* Kondensatlagerstätte *f*

  c. returns Kondensatrücklauf *m*

  c. water Kondenswasser *n*, Niederschlagswasser *n*, Schwitzwasser *n*

condensation Kondensation *f*, Kondensieren *n*, Kondensierung *f*, *(Phys auch)* Niederschlagen *n*, Niederschlagung *f*, Verdichten *n*, Verdichtung *f*

  c. bulb Kondensationskolben *m*

  c. centre *s.* c. nucleus

  c. method Kondensationsmethode *f*, Kondensationsverfahren *n*

  c. nucleus Kondensationskeim *m*, Kondensationskern *m*, Kondensationszentrum *n*

c. polymer Kondensationspolymeres *n*, Polykondensat *n*
c. polymerization Kondensationspolymerisation *f*, Polykondensation *f*
c. pressure Kondensationsdruck *m*; Verflüssigungsdruck *m*
c. product Kondensationsprodukt *n*, Kondensat *n*, *(Phys auch)* Niederschlag *m*
c. reaction Kondensationsreaktion *f*
c. resin Kondensationsharz *n*
c. water Kondenswasser *n*, Niederschlagswasser *n*, Schwitzwasser *n*
condense / to kondensieren, *(Phys auch)* niederschlagen, verdichten; *(Lebm)* kondensieren, eindicken; [sich] kondensieren, *(Phys auch)* sich niederschlagen (verdichten)
c. out [sich] auskondensieren
condensed film kondensierter Film (Oberflächenfilm) *m*
c. juice Dicksaft *m*, eingedickter Saft *m*
c. milk Kondensmilch *f*, kondensierte (eingedickte) Milch *f*
c. nucleus kondensierter Kern *m*
c.-ring system kondensiertes Ringsystem *n*
c. skim milk Kondensmagermilch *f*, kondensierte (eingedickte) Magermilch *f*
c. system *(phys Ch)* kondensiertes System *n*
c. tannin kondensierter Gerbstoff *m*
c. water *s.* condensation water
c. whole milk Kondensvollmilch *f*, kondensierte (eingedickte) Vollmilch *f*
condenser Kondensator[kühler] *m*, Kühler *m*; Verflüssiger *m*, Kondensator *m* *(einer Kältemaschine)*; Sublimiervorlage *f*, Sublimationsvorlage *f*, *(fälschlich)* Kondensator *m*; *(ET)* Kondensator *m*; Kondensor *m* *(sammelndes optisches System)*
c. adapter Vorstoß *m* *(der Destillationsapparatur)*
c. clamp Kühlerklemme *f*, Stativklemme *f* mit halbrunden Spannbacken, große Klemme *f*
c. jacket Kühlermantel *m*, Schweinchen *n*, Kühlerschweinchen *n*
c. lens Kondensorlinse *f*
c. paper Kondensator[seiden]papier *n*
c. pressure Verflüssigerdruck *m*
c. tissue [paper] *s.* c. paper
c. water Kondensatorkühlwasser *n*
condensing Kondensieren *n*, Kondensierung *f*, Kondensation *f*, *(Phys auch)* Niederschlagen *n*, Niederschlagung *f*, Verdichten *n*, Verdichtung *f*
c. agent Kondensationsmittel *n*, Kondensationsagens *n*
c. chamber Kondensationskammer *f*
c. pressure Kondensationsdruck *m*; Verflüssigungsdruck *m*
c. surface Kondensationsfläche *f*
c. system Kondensationsanlage *f*
c. tower Kondensationsturm *m*
c. tube Kondensationsrohr *n*, Kühlrohr *n*
condiment Würzmittel *n* *(Würze oder Gewürz)*

condition / to konditionieren; den Feuchtegrad bestimmen; klimatisieren
condition Bedingung *f*; Zustand *m*
c. of crystallization Kristallisationsbedingung *f*
c. of growth Wachstumsbedingung *f*
c. of preparation Darstellungsbedingung *f*
conditional acceptable daily intake *(Toxikologie)* bedingt duldbare Tagesdosis *f*
conditioner 1. Zusatzstoff *m* *(zur besseren Handhabung einer Chemikalie)*, *(i.e.S.)* Umhüllungsmittel *n* *(oder)* Füllstoff *m* *(oder)* Trägerstoff *m* *(oder)* Lockerungsmittel *n*; 2. Konditionierapparat *m*; Klimagerät *n*; 3. Vorgranulator *m* *(z.B. für Düngemittel)*
conditioning Konditionieren *n*, Konditionierung *f*; Klimatisierung *f*, Klimaregelung *f*
c. apparatus Konditionierapparat *m*
c. plant Konditionieranlage *f*
c. section *(Glas)* Aufbereitungsteil *m*, Vorbereitungsabschnitt *m* *(des Speiserkanals)*
c. unit *s.* c. plant
c. zone *s.* c. section
condorvine *(Pharm)* Marsdenia reichenbachi Triana
condrodite *s.* chondrodite
conduct / to *(Elektrizität, Wärme)* leiten
conductance Leitfähigkeit *f*, Leit[ungs]vermögen *n*; Wirkleitwert *m*, Konduktanz *f*
c. band Leitungsband *n*, Leitfähigkeitsband *n*, L-Band *n*
c. bridge Leitfähigkeitsmeßbrücke *f*
c. cell Leitfähigkeitsgefäß *n*, Leitfähigkeitszelle *f*
c. curve Leitfähigkeitskurve *f*
c. measurement Leitfähigkeitsmessung *f*
c. meter Leitfähigkeitsmesser *m*, Leitfähigkeitsmeßgerät *n*
c. method Leitfähigkeitsmethode *f*, konduktometrische Methode *f*, konduktometrisches Verfahren *n*
c. ratio Leitfähigkeitskoeffizient *m*
c. titration Leitfähigkeitstitration *f*, konduktometrische Titration *f*
c. water Leitfähigkeitswasser *n*
conductimeter *s.* conductometer
conductimetric *s.* conductometric
conductimetry *s.* conductometry
conducting channel black Conducting-Channel-Ruß *m*, CC-Ruß *m* *(leitfähiger Kanalruß)*
c. power Leitfähigkeit *f*, Leit[ungs]vermögen *n*
conduction Leitung *f* *(von Elektrizität oder Wärme)*
c. band Leitungsband *n*, Leitfähigkeitsband *n*, L-Band *n*
c. current Leitfähigkeitsstrom *m*
c. drying Kontakttrocknen *n*, Kontakttrocknung *f*, Berührungstrocknen *n*, Berührungstrocknung *f*
c. electron Leitungselektron *n*, Leitfähigkeitselektron *n*
conductive leitfähig, leitend
c. channel black *s.* conducting channel black

conductivity [spezifische] Leitfähigkeit *f*, [spezifisches] Leit[ungs]vermögen *n*
  c. band Leitfähigkeitsband *n*, Leitungsband *n*, L-Band *n*
  c. cell Leitfähigkeitsgefäß *n*, Leitfähigkeitszelle *f*
  c.-concentration curve Leitfähigkeits-Konzentrations-Kurve *f*
  c. measurement Leitfähigkeitsmessung *f*, Konduktometrie *f*
  c. of heat Wärmeleitfähigkeit *f*, Wärmeleitvermögen *n*, Wärmeleitzahl *f*
  c. ratio Leitfähigkeitskoeffizient *m*
  c. water Leitfähigkeitswasser *n*
conductometer Leitfähigkeitsmesser *m*, Leitfähigkeitsmeßgerät *n*
conductometric konduktometrisch
  c. analysis konduktometrische Maßanalyse *f*
  c. measurement Leitfähigkeitsmessung *f*
  c. method konduktometrische Methode *f*, konduktometrisches Verfahren *n*, Leitfähigkeitsmethode *f*
  c. titration konduktometrische Titration *f*, Leitfähigkeitstitration *f*
conductometry Konduktometrie *f*
conductor Leiter *m*, *(i.e.S.)* [elektrischer] Leiter *m*, Elektrizitätsleiter *m*, Stromleiter *m*
  c. of electricity [elektrischer] Leiter *m*, Elektrizitätsleiter *m*, Stromleiter *m*
  c. of heat Wärmeleiter *m*
condurangin Kondurangin *n* *(Bitterstoffglykosid aus Kondurango)*
Condy's fluid Condysche Desinfektionsflüssigkeit *f*
cone Kegel *m*, Konus *m*; Kegel *m*, Eisenkegel *m*, Glocke *f*, Parry-Kegel *m*, Parry-Glocke *f* *(eines Hochofens)*; Kern *m*, Kegel *m*, Konus *m*, Kegelrotor *m*, Rotor *m* *(der Kegelstoffmühle)*; *(Text)* keglige Spule (Hülse) *f*, konische Kreuzspule *f*
  c. classifier Klassierkegel *m*, [konische] Klärspitze *f*, Spitztrichter *m*
  c. crusher Kegelbrecher *m*
  c. dyeing apparatus Konusfärbeapparat *m*
  c. mill Glockenmühle *f*, Kegelmühle *f*; Konusmühle *f*, konische (zylindrisch-konische) Kugelmühle *f*, Doppelkegelmühle *f*
  c. pelletizer Pelletisierkonus *m*
  c. save-all *(Pap)* Trichterstoffänger *m*
  c.-shaped kegelförmig, keg[e]lig, konusartig, konisch
conessine Conessin *n*, 3β-Dimethylamino-18,20-methylimino-5-pregnen *n*
confectionary margarine Bäckermargarine *f*, Backmargarine *f*
configuration Konfiguration *f*, Atomkonfiguration *f*, Atomanordnung *f*
  D configuration D-Konfiguration *f*
  L configuration L-Konfiguration *f*
configurational Konfigurations...
  c. symbol Konfigurationssymbol *n*

confirmatory test Nachweis *m*
conform to a pattern / to typkonform sein
conformation Konformation *f*, Konstellation *f*
conformational Konformations...
  c. analysis Konformationsanalyse *f*
  c. formula Konformationsformel *f*, Konstellationsformel *f*
conformer Konformeres *n*
congeal / to erstarren, fest werden, gerinnen, gefrieren; erstarren (gerinnen, gefrieren) lassen
congealable gerinnbar, gefrierbar
congealing Erstarren *n*, Festwerden *n*, Gerinnen *n*, Gefrieren *n*
  c. point (temperature) Erstarrungspunkt *m*
congelation *s.* congealing
conglomerate Konglomerat *n*
Congo brown Kongobraun *n*
  C. copal Kongokopal *m* *(halbfossiles Harz von Copaifera-Arten)*
  C. corinth Kongokorinth *n*
  C. dye Kongofarbstoff *m*
  C. gum *s.* C. copal
  C. paper Kongopapier *n*
  C. red Kongorot *n*
  C. rubin[e] Kongorubin *n*
  C. rubine dye solution Kongorubinlösung *f*
  C. rubine number Rubinzahl *f* *(zur Messung der Schutzkolloidwirkung)*
  C. rubine sol Kongorubinsol *n*
congress method Kongreßverfahren *n* *(zur Bestimmung der Extraktverhältnisse bei der Malzanalyse)*
  c. wort Kongreßwürze *f*
conhydrine Konhydrin *n*, α-Hydroxykoniin *n*, 2-(1-Hydroxypropyl)-piperidin *n*
conical keg[e]lig, konisch, kegelförmig, konusartig
  c. ball mill *s.* c. mill
  c. crusher Kegelbrecher *m*
  c. flask Erlenmeyer-Kolben *m*
  c. grinder Glockenmühle *f*, Kegelmühle *f*
  c. mill Konusmühle *f*, konische (zylindrisch-konische) Kugelmühle *f*, Doppelkegelmühle *f*
  c. refiner *(Pap)* Kegelrefiner *m*
  c.-screen centrifuge Siebzentrifuge *f* mit Konussiebtrommel (konischer Trommel, konischer Siebtrommel)
α-coniceine α-Konizein *n*, 2-Methylkonidin *n*
β-coniceine β-Konizein *n*, 2-(1-Propenyl)-piperidin *n*
γ-coniceine γ-Konizein *n*, 6-Propyl-1,2,3,4-tetrahydropyridin *n*
δ-coniceine δ-Konizein *n*, Piperolidin *n*, 1,2-Trimethylenpiperidin *n*
ε-coniceine ε-Konizein *n*, Iso-2-methylkonidin *n*
conicine *s.* coniine
conidendrin Konidendrin *n* *(ein Lignan)*
conidine Konidin *n*, 1,2-Äthylenpiperidin *n*
coniferin Koniferin *n*, Abietin *n*, Larizin *n*, Koniferosid *n* *(β-Glukosid des Koniferylalkohols)*
coniferous wood Nadelholz *n*

coniferyl alcohol $HOC_6H_3(OCH_3)CH=CHCH_2OH$ Koniferylalkohol *m*, 3-(4-Hydroxy-3-methoxy-phenyl-)2-Propen-1-ol *n*, 4-Hydroxy-3-methoxy-zimtalkohol *m*, γ-Oxyeugenol *n*

coni[i]ne Koniin *n*, 2-Propylpiperidin *n*

conjugate / to konjugieren, in Wechselwirkung treten

conjugate konjugiert, korrespondierend
 c. acid korrespondierende Säure *f*
 c. base korrespondierende Base *f*
 c. fibre Bi[komponenten]faser *f*
 c. solution konjugierte Mischung *f*

conjugated konjugiert
 c. dienoid system System *n* mit konjugierten Doppelbindungen
 c. double bond (linkage) konjugierte Doppelbindung *f*
 c. protein zusammengesetzter (konjugierter) Eiweißstoff *m*, Proteid *n*
 c. system konjugiertes System *n*

conjugation Konjugation *f*
 c. energy Konjugationsenergie *f*

conjunctive name konjunktiver (zusammengesetzter) Name *m*, Verbundname *m*

connect to / to anschließen an

connecting tube Zwischenstück *n*

connection Verbindung *f*, Anschluß *m*; Verbindungsstück *n*

connector [adapter] Zwischenstück *n*, Verbindungsstück *n*

connellite *(Min)* Konnelit *m*, Connellit *m*, Tallingit *m*

connexion *s.* connection

Conrad-Limpach synthesis Conrad-Limpachsche Synthese *f (von 4-Hydroxychinolin)*

Conradson carbon residue *s.* Conradson coke residue

Conradson carbon residue method *s.* Conradson method

Conradson carbon test *s.* Conradson test

Conradson coke number Conradson-Verkokungszahl *f*, Conradson-Zahl *f*, Conradson-Verkokungswert *m*, Conradson-[Carbon-]Wert *m*

Conradson coke residue Verkokungsrückstand (Koksrückstand, Kohlenstoffrückstand) *m* nach Conradson, Conradson-Carbon *n*

Conradson coke value *s.* Conradson coke number

Conradson [coking] method Conradson-Methode *f (zur Bestimmung der Verkokungsneigung)*

Conradson test Conradson-Test *m*, Conradson-Carbon-Test *m*, CCT, Conradsonsche Verkokungsprobe *f*, Verkokungstest *m* nach Conradson

Conradson value *s.* Conradson coke number

consecutive aufeinanderfolgend
 c. reaction Folgereaktion *f*

consequent reaction *s.* consecutive reaction

conservation Erhaltung *f*, Aufrechterhaltung *f*; Konservierung *f*, Haltbarmachung *f*
 c. law Erhaltungssatz *m*

c. of angular momentum Erhaltung *f* des Drehimpulses

c. of chemicals Chemikalienrückführung *f*, Chemikalienrückgewinnung *f*

c. of energy Energieerhaltung *f*, Erhaltung *f* der Energie

c. of mass (matter) Massenerhaltung *f*, Erhaltung *f* der Masse

conserve / to [aufrecht]erhalten; konservieren, haltbar machen

consistence *s.* consistency

consistency Konsistenz *f*
 c. of margarine Margarinekonsistenz *f*
 c. regulator *(Pap)* Konsistenzregler *m*, Stoffdichteregler *m*

consistent konsistent

consistometer Konsistometer *n*

consolidated verfestigt *(Braunkohle)*; erstarrt *(Lava)*

consolute temperature kritische Lösungstemperatur (Mischungstemperatur) *f*, kritischer Lösungspunkt (Mischungspunkt) *m*

constant konstant
 with c. shaking unter ständigem Schütteln

constant Konstante *f*, konstante Größe *f*
 c.-boiling konstant siedend, konstantsiedend
 c.-displacement pump Konstantpumpe *f (Pumpe mit nichtstellbarem Förderstrom)*
 c.-level device Wasserstandsregler *m (z.B. für Wasserbäder)*
 c.-level tank Niveaukonstanthalter *m (beim Entwachsungsprozeß)*
 c. of gravitation Gravitationskonstante *f*
 c. of proportionality Proportionalitätsfaktor *m*
 c.-pressure filtration Filtration *f* unter konstantem Filtrationsdruck
 c.-pressure gas thermometer Gasthermometer *n* konstanten Drucks
 c.-pressure steam Dampf *m* konstanten Drucks
 c. quantity konstante Größe *f*, Konstante *f*
 c.-rate drying period Abschnitt *m* konstanter Trocknungsgeschwindigkeit
 c.-rate filtration Filtration *f* mit konstanter Filtriergeschwindigkeit
 c.-temperature transformation Umwandlung *f* bei gleichbleibender Temperatur (Unterkühlungstemperatur), isotherm[isch]e Umwandlung *f*
 c. term fester (konstanter) Term *m*
 c.-volume mit konstantem (gleichbleibendem) Volumen
 c.-volume gas thermometer Gasthermometer *n* konstanten Volumens (konstanter Dichte)
 c. weight Massekonstanz *f*, *(bisher)* Gewichtskonstanz *f*

constantan Konstantan *n*
 c. wire Konstantandraht *m*

constituent am Aufbau beteiligt

constituent Bestandteil *m*, Teil *m*, Anteil *m*; Gefügebestandteil *m*, Mazeral *n*

c. element Anteil *m*, Komponente *f*, Elementarbestandteil *m*
c. part Bestandteil *m*, Teil *m*, Anteil *m*
c. particle Bauelement *n*, Baustein *m*
c. particle of the nucleus Kernbaustein *m*
constitution Konstitution *f*, Struktur *f*, Anordnung *f*, Bau *m*, Beschaffenheit *f*
c. diagram Zustandsdiagramm *n*
c. water s. constitutional water
constitutional formula Konstitutionsformel *f*, Strukturformel *f*, Bauformel *f*
c. water konstitutiv gebundenes Wasser *n*, Konstitutionswasser *n*
constitutive konstitutiv
c. property konstitutive Eigenschaft *f*
constricted end verjüngtes Ende *n* (*einer Röhre*)
constriction Einschnürung *f*, Drosselstelle *f*, Drossel[ung] *f*
constringence Abbesche Zahl *f*, *ν*-Wert *m*, reziproke relative Dispersion *f*
construction material Baustoff *m*
consumption Verbrauch *m*
c. material Verbrauchsstoff *m*
c. sugar Verbrauchszucker *m*
contact / **to** 1. berühren, mit etwas in Kontakt kommen; in Kontakt bringen, [an]drücken, anlegen; 2. sich berühren, miteinander in Kontakt kommen
contact Kontakt *m*
out of c. with air unter Luftabschluß
c. action Kontaktwirkung *f*
c. adhesive Kontaktklebstoff *m*, Kontaktkleber *m*, Haftkleber *m*
c. angle Kontaktwinkel *m*, Randwinkel *m*, Berührungswinkel *m* (*bei der Prüfung grenzflächenaktiver Stoffe*)
c. aureole (*Geol*) Kontaktaureole *f*, Kontakthof *m*
c. boundary Berührungsgrenze *f*
c. catalysis Kontaktkatalyse *f*, Oberflächenkatalyse *f*, heterogene Katalyse *f*
c. catalyst Kontaktkatalysator *m*, heterogener Katalysator *m*, Kontakt *m*
c.-catalytic kontaktkatalytisch
c. cement s. c. adhesive
c. corrosion Kontaktkorrosion *f*
c. drying Kontakttrocknen *n*, Kontakttrocknung *f*, Berührungstrocknen *n*, Berührungstrockung *f*
c. effect (*phys Ch*) Nah[e]wirkung *f*
c. emulsion (*Landw*) Kontaktmittelemulsion *f*, Ätzmittelemulsion *f* (*zur Unkrautbekämpfung*)
c. fertilization Kontaktdüngung *f*
c. filtration Kontaktfiltration *f*
c. freezing Kontaktgefrieren *n*
c. gettering (*Vakuumtechnik*) Kontaktgetterung *f*
c. goniometer (*Krist*) Anlegegoniometer *n*
c. herbicide Kontaktherbizid *n*
c. insecticide Kontaktinsektizid *n*, Berührungsgift *n* (*für Insekten*)
c. metamorphism (**metamorphosis**) (*Geoch*)

Kontaktmetamorphose *f*, Kontaktmetamorphismus *m*, Berührungsmetamorphose *f*
c. method s. c. process
c. mineral Kontaktmineral *n*
c. moulding (*Plast*) Kontaktpressen *n*, Kontaktpreßverfahren *n*, Niederdruckpreßverfahren *n*, Handauflegeverfahren *n*
c. paper Kontaktpapier *n*
c. poison Kontaktgift *n*, Berührungsgift *n*
c. pressure resin (*Plast*) Harz *n* für Kontaktpressen (Kontaktpreßverfahren, Niederdruckpreßverfahren, Handauflegeverfahren)
c. print (*Foto*) Kontaktabzug *m*, Kontaktkopie *f*, Abzug *m*
c. printer (*Foto*) Kopiergerät *n*, Kontaktkopiergerät *n*, Kontaktkopierapparat *m*, Kontaktkopiermaschine *f*
c. printing (*Foto*) Kontaktkopieren *n*, Kopiertechnik *f*, Kontaktverfahren *n*
c.-printing paper s. c. paper
c. process Kontaktverfahren *n*, Kontaktprozeß *m*; Kontaktschwefelsäureverfahren *n*, Schwefelsäurekontaktverfahren *n*
c. reactor Kontaktofen *m*
c. space Kontaktraum *m*
c. toxicant s. c. poison
c. treatment Kontaktraffination *f*
c. zone (*Geol*) Kontakthof *m*, Kontaktaureole *f*
contactor Kontaktor *m*, Extraktor *m*, Extrakteur *m*, Extraktionskolonne *f*, Extraktionsapparat *m*, Extraktionsmaschine *f* (*zur Solventextraktion*); Kontaktor *m*, Kontaktautoklav *m*
container Behälter *m*, Container *m*, Transportbehälter *m*; Substanzbehälter *m* (*beim Zonenschmelzen*)
c. board Kistenpappe *f*, Behälterpappe *f*, Containerpappe *f*
c. glass Hohlglas *n*
contaminant Verunreinigungsstoff *m*, Verunreinigungssubstanz *f*, Schmutzstoff *m*, Verunreinigung *f*
contaminate / **to** verunreinigen, verschmutzen, vergiften, verseuchen, kontaminieren
contaminated rock (*Geol*) Intrusionsgestein *n*
contamination Verunreinigung *f*, Verschmutzung *f*, Vergiftung *f*, Verseuchung *f*, Kontamination *f*
content Gehalt *m*, (*auch*) Beladung *f*, Anteil *m*
c. of active component Wirkstoffgehalt *m*
c. of free rosin (*Pap*) Freiharzgehalt *m*
contents of the digester (*Pap*) Kocherinhalt *m*, Kochereintrag *m*, Kocherfüllung *f*, Kochgut *n*, Charge *f*
continuous kontinuierlich (stetig, ununterbrochen) arbeitend (betrieben) (*Gerät*); kontinuierlich, stetig, ununterbrochen, fortschreitend, durchlaufend (*Betrieb*); anhaltend, ständig (z. B. *Schütteln*)
by a c. process in laufender Betriebsanordnung, im Fließbetrieb (Dauerbetrieb), (*Farb auch*) kontinuemäßig ablaufend, im Kontinuebetrieb

c. annealing *(Met)* Durchlaufglühen *n*, Durchlaufglühung *f*; *(Glas)* Durchlaufkühlen *n*, Durchlaufkühlung *f*

c.-annealing furnace *(Met)* Durchlaufglühofen *m*

c.-annealing lehr *(Glas)* Durchlaufkühlofen *m*, kontinuierlich arbeitender Kühlofen *m*

c. bake oven Kettenbackofen *m*, Durchgangsofen *m*

c. binary distillation stetige (kontinuierliche) Destillation *f* von Zweistoffgemischen

c. bleaching Kontinuebleiche *f*

c.-bucket elevator Becherwerk *n* mit Becherstrang

c. buttermaking machine kontinuierlich arbeitende Butterungsmaschine *f*, Fritz-Buttermaschine *f*

c. carbonizing plant Koksofen *m* mit wandernder Ladung, Koksofen *m* mit bewegter Beschickung

c. casting kontinuierliches (ununterbrochenes) Gießen *n*, kontinuierlicher (ununterbrochener) Guß *m*

c.-casting process kontinuierliches Walzverfahren *n (Gußglasherstellung)*

c. churn *s.* c. buttermaking machine

c. column distillation stetige Rektifikation *f*

c. compounder *(Plast)* kontinuierlich arbeitender Mischer *m*

c. contact coking Continuous-Contact-Coking *n*, kontinuierliches katalytisches Verkoken *n*

c. conveyor *s.* c.-flow conveyor

c.-cooling transformation Umwandlung *f* bei kontinuierlicher (stetiger, fortschreitender) Abkühlung

c.-cooling transformation diagram Zeit-Temperatur-Umwandlungsdiagramm (Zeit-Temperatur-Umwandlungsschaubild, ZTU-Diagramm, ZTU-Schaubild, Umwandlungsschaubild) *n* für kontinuierliche Abkühlung

c. cure (curing) kontinuierliche Vulkanisation *f*

c. dialysis Fließdialyse *f*

c. distillation kontinuierliche Destillation *f*

c.-distillation unit kontinuierlich arbeitende Destillationsanlage *f*

c. dryer kontinuierlich arbeitender Trockner *m*

c. dyeing *(Text)* Kontinuefärben *n*, kontinuierliches Färben *n*

c.-dyeing machine *(Text)* Kontinuefärbemaschine *f*

c. electrode Dauerelektrode *f*

c. equilibrium vaporization [kontinuierliche] Entspannungsdestillation *f*

c.-extraction apparatus *s.* c. extractor

c. extractor kontinuierlich arbeitender Extraktor (Extrakteur) *m*, Apparat *m* für kontinuierliche Extraktion

c. fibre (filament) Elementarfaden *m*, Endlosfaser *f*, Filament *n*

c.-filament glass yarn *s.* c. glass filament

c.-flow calorimeter Strömungskalorimeter *n*

c.-flow chromatography Durchlaufchromatografie *f*

c.-flow conveyor Trogkettenförderer *m*

c.-flow method Strömungsmethode *f*

c. frother kontinuierlich arbeitende Schaum[herstellungs]maschine *f*

c. furnace Durchlaufofen *m*, Ofen *m* für durchlaufenden (kontinuierlichen) Betrieb, kontinuierlicher (kontinuierlich arbeitender) Ofen *m*

c.-furnace method *(Gum)* Furnace-Verfahren *n*, Furnace-Black-Prozeß *m*, Ofenverfahren *n*, Ofenprozeß *m (zur Rußerzeugung)*

c. glass filament Glasseide *f*

c. glass strand mat Glasseidenmatte *f*

c. grinder *(Pap)* stetiger Schleifer *m*, Stetigschleifer *m*

c. heat resistance Dauerwärmebeständigkeit *f*

c. hot-air vulcanization kontinuierliche Heißluftvulkanisation (Heißluftheizung) *f*

c. laminating *(Plast)* kontinuierliche Schichtstoffherstellung *f*

c.-line recorder Linienschreiber *m*

c. mixer kontinuierlich arbeitender Mischer *m*, Stetigmischer *m*

c. neutralization kontinuierliche Entsäuerung *f (Margarineherstellung)*

c. nitration kontinuierliche Nitrierung *f*

c. open-width bleaching machine *(Text)* Kontinue-Breitbleichanlage *f*

c. operation kontinuierlicher (stetiger, ununterbrochener) Betrieb *m*, Fließbetrieb *m*, Dauerbetrieb *m*, *(Farb auch)* Kontinuebetrieb *m*

c. paper Rollenpapier *n*

c. pasteurization *(Lebm)* kontinuierliche Pasteurisation *f*, Durchflußpasteurisation *f*, Durchflußerhitzung *f (zur Pasteurisation)*

c. percolation kontinuierliche Perkolation *f*

c.-percolation process (system) kontinuierliches Perkolationsverfahren *n*

c. phase zusammenhängende (geschlossene) Phase *f*, Dispersionsphase *f*

c. process kontinuierliches (ununterbrochenes) Verfahren *n*, *(Farb auch)* Kontinueverfahren *n*

c. rectification stetige (kontinuierliche) Gegenstromdestillation (Rektifikation) *f*

c.-recuperative system Rekuperativsystem *n*

c. rotary cure Rotationsvulkanisationsverfahren *n*

c. screw-feed process of polymerization kontinuierliche Polymerisation *f* in einer Druckschnecke

c. shell still kontinuierliche Blasendestillationsanlage (Batteriedestillationsanlage) *f*, Kaskadendestillationsanlage *f*

c. spectrum kontinuierliches Spektrum *n*

c. spinning Kontinue-Spinnverfahren *n*

c. steam vulcanization kontinuierliche Vulkanisation *f* im Dampfrohr

c. steamer *(Text)* Kontinuedämpfer *m*

c. tank *(Glas)* kontinuierliche Wanne *f*, Dauerwanne *f*

c.-type furnace Durchlaufofen *m*, Ofen *m* für durchlaufenden (kontinuierlichen) Betrieb, kontinuierlicher (kontinuierlich arbeitender) Ofen *m*

c. vertical retort kontinuierliche (stetige, stetig betriebene) Vertikalretorte *f*, Vertikalretorte *f* mit kontinuierlichem Betrieb (mit kontinuierlicher Beschickung, mit wandernder Ladung)

c. vulcanization kontinuierliche Vulkanisation *f*

c. vulcanizer (vulcanizing machine) Vulkanisiermaschine *f*

c. working *s.* c. operation

continuum Kontinuum *n*

contraceptive antikonzeptionelles (empfängnisverhütendes, schwangerschaftsverhütendes) Mittel *n*, Antikonzipiens *n*, Antikonzeptionsmittel *n*

contract / **to** zusammenziehen, kontrahieren; sich zusammenziehen, [ein]schrumpfen, schwinden, sich verkleinern, kleiner werden; *(Text)* einlaufen, eingehen

contraction Kontraktion *f*, Zusammenziehung *f*, Schrumpfung *f*, Einschrumpfung *f*, Schwindung *f*, Schwund *m*; *(Text)* Einlaufen *n*, Eingehen *n*; *(Ker)* Schwindung *f*

c. cavity Schwindungshohlraum *m*

contractor Lieferfirma *f*

contraries Fremdstoffe *mpl*

contrarotating gegenläufig, gegeneinanderlaufend

c. rolls gegenläufige Walzen *fpl*

contrast *(Foto)* Kontrast *m*

c. range *(Foto)* Kontrastumfang *m*

contrasty *(Foto)* kontrastreich

contribute / **to** beisteuern, liefern; beitragen, mitwirken, beteiligt sein, teilhaben

contributing [mesomeric] form mesomere Grenzstruktur *f*

control / **to** lenken, steuern, kontrollieren, leiten, zügeln *(z.B. die Oxydation)*; steuern; regeln

control agent Schädlingsbekämpfungsmittel *n*; Pflanzenschutzmittel *n*

c. computer Steuerungsrechner *m*; Regelungsrechner *m*

c. measure *(Landw)* Bekämpfungsmaßnahme *f*

c. of slime *(Pap)* Schleimkontrolle *f*

c. rod *(Kch)* Kontrollstab *m*, Regelstab *m*

c. test Kontrollprüfung *f*

c. valve Steuerventil *n*, Stellventil *n*; Regelventil *n*; Dosierventil *n*

controlled atmosphere kontrollierte Atmosphäre *f* *(z.B. eines Ofens)*; [kontrollierte] Schutzgasatmosphäre *f*, Schutzatmosphäre *f*

c.-potential electrodeposition Elektroanalyse *f* bei kontrolliertem Katodenpotential

c. pyrolysis gelenkte Pyrolyse *f*

c.-volume pump Dosier[ungs]pumpe *f*, Zumeßpumpe *f*

controller Steuergerät *n*, Steuereinrichtung *f*, Steuerung *f*; Regeleinrichtung *f*, Regelanlage *f*, Regler *m*; Stellglied *n*

convallamarin Konvallamarin *n* *(ein Glykosid)*

convallarin Konvallarin *n* *(ein Glykosid)*

convallatoxin Konvallatoxin *n* *(ein Glykosid)*

convected heat *s.* convection heat

convection Konvektion *f*

c. current Konvektionsstrom *m*

c. dryer Konvektionstrockner *m*; *(Ker)* Umwälztrockner *m*

c. drying Konvektionstrocknen *n*, Konvektionstrocknung *f*

c. heat Konvektionswärme *f*

c. heating Konvektionsheizung *f*

convective konvektiv

c. mixing Konvektions[ver]mischen *n*

conventional cloud chamber Expansionsnebelkammer *f*

c. twisting *(Text)* Echtdrahtverfahren *n*

convergence Konvergenz *f*

c. frequency Seriengrenzfrequenz *f*, Frequenz *f* der Seriengrenze *(einer Spektralserie)*

c. limit Seriengrenze *f* *(einer Spektralserie)*

convergency *s.* convergence

conversion Umwandeln *n*, Umwandlung *f*, Überführen *n*, Überführung *f*, Konvertierung *f*, Konversion *f*; Umsetzung *f*; Umrechnung *f*

c. coating *(Pap)* Streichen *n* außerhalb der Maschine (Papiermaschine), Separatstreichen *n*, Separatstrich *m*

c. factor Umrechnungsfaktor *m*

c. method *s.* c. process

c. of starch Stärkeverzuckerung *f* *(durch Diastase)*

c. operation *(Arbeitsvorgang innerhalb eines technologischen Prozesses, der außerhalb der Hauptmaschine durchgeführt wird)*

c. point Umwandlungspunkt *m*

c. process Umwandlungsverfahren *n*, Konvertierungsverfahren *n*

c. product Umwandlungsprodukt *n*

c. rate Umwandlungsgeschwindigkeit *f*; Umsetzungsgeschwindigkeit *f*

c. reactor *(Kch)* Konverter *m*, Konversionsreaktor *m*, Konverterreaktor *m*

c. saltpetre Konversionssalpeter *m*

c. scale Umwandlungsskale *f*

c. table Umrechnungstabelle *f*

convert / **to** 1. umwandeln, überführen, konvertieren; umsetzen; umrechnen; *(Met)* windfrischen, im Konverter verblasen, bessemern; 2. sich umwandeln

c. back zurückverwandeln

c. into ash veraschen

c. into paper zu (auf) Papier verarbeiten

c. into vapour verdampfen

converted:

to be c. into übergehen in

c. saltpetre Konversionssalpeter *m*

converter Reaktor *m*, Reaktionsofen *m*, Kontaktofen *m*, *(bei der Hydrierung auch)* Hydrier[ungs]ofen *m*; *(Met)* Konverter *m*; *(Kch)*

Konverter *m*, Konversionsreaktor *m*, Konverterreaktor *m*; *(Text)* Konverter *m*, Spinnbandschneidemaschine *f*
c. **lining** *(Met)* Konverterzustellung *f*, Konverterauskleidung *f*, Konverterfutter *n*
c. **process** Windfrischverfahren *n*, Windfrischprozeß *m*, Konverter[frisch]verfahren *n*, Konverter[frisch]prozeß *m*, Bessemer-Verfahren *n* *(i.w.S.)*, Bessemer-Prozeß *m (i.w.S.)*
c. **reactor** *(Kch)* Konverter *m*, Konversionsreaktor *m*, Konverterreaktor *m*
c. **steel** Konverterstahl *m*, Windfrischstahl *m*, windgefrischter Stahl *m*
convertible chemisch (durch chemische Reaktion) trocknend *(z.B. Anstrichmittel)*
converting Windfrischen *n*, Verblasen *n* im Konverter, Bessemern *n*
c. **operation** *s.* conversion operation
c. **process** *s.* converter process
Convertol process Convertolverfahren *n (Sortierverfahren für allerfeinstes Korn)*
conveyed length Förderstrecke *f*, Förderweg *m*
conveyer *s.* conveyor
conveying Fördern *n*
c. **air** Förderluft *f*
c. **belt** *s.* conveyor belt
c. **characteristics** Fördereigenschaften *fpl*
c. **screw** *s.* conveyor screw
conveyor Fördergerät *n*, Förderanlage *f*, Fördereinrichtung *f*, Förderer *m*, *(i.e.S.)* Stetigförderer *m*
c. **belt** Förderband *n*, Fördergurt *m*, Transportband *n*
c. **belting** Förderbänder *npl*, Fördergurte *mpl*, Transportbänder *npl*
c. **dryer** Bandtrockner *m*, Förderbandtrockner *m*
c. **felt** *(Pap)* Transportfilz *m*, Überführfilz *m*
c. **length** *(nutzbare)* Förderlänge *f*
c. **scale** Bandwaage *f*
c. **screw (worm)** Förderschnecke *f*, Transportschnecke *f*, Schneckenförderer *m*
convolvulin Konvolvulin *n (ein glykosidhaltiges Harz)*
convolvulinic acid Konvolvulinsäure *f*
convolvulinolic acid Konvolvulinolsäure *f (Aglukon der Konvolvulinsäure)*
convulsant Konvulsionen (Schüttelkrämpfe) erregend (auslösend)
convulsant Konvulsionen (Schüttelkrämpfe) erregendes (auslösendes) Mittel *n*
cook / **to** kochen; *(Pap) (in Gegenwart von Chemikalien)* kochen, aufschließen; *(Zellstoff)* erkochen
c. **raw** *(Zellstoff)* hartkochen, wenig (unvollständig) aufschließen
c. **soft** *(Zellstoff)* weichkochen, weit herunterkochen, intensiv aufschließen
c. **thoroughly** durchkochen
cook *(Pap)* Kochung *f*; *(Pap)* Kocherinhalt *m*, Kochereintrag *m*, Kocherfüllung *f*, Kochgut *n*, Endkochgut *n*, Kochstoff *m*; *(Pap)* Kochermeister *m*

c. **cheese** Kochkäse *m*, Topfkäse *m*
cooked cheese *s.* cook cheese
c. **flavour** Kochgeschmack *m*
c. **mash** Kochmaische *f*
c. **taste** *s. c.* flavour
cooker Kocher *m*; *(Pap)* Kochermeister *m*, Kocherführer *m*
cooking Kochen *n*; *(Pap)* Kochen *n*, Kochung *f*, Aufschließen *n*, Aufschluß *m*; Erkochung *f (von Zellstoff)*
c. **acid** *(Pap)* Kochsäure *f*, Sulfit[koch]säure *f*
c.-**acid circulation** *(Pap)* Kochsäurezirkulation *f*, Kochsäureumlauf *m*, Kochsäureumwälzung *f*
c.-**acid composition** *(Pap)* Kochsäurezusammensetzung *f*
c. **agent (chemical)** *(Pap)* Aufschlußmittel *n*, Aufschlußchemikalie *f*, Aufschließungsreagens *n*, Kochchemikalie *f*
c. **condition** *(Pap)* Koch[ungs]bedingung *f*
c. **cycle** *(Pap)* Kochzyklus *m*
c. **fat** Kochfett *n*, Speisefett *n*
c. **kettle** Kochkessel *m*
c. **liquor** *(Pap)* Kochflüssigkeit *f*, Kochlösung *f*, Aufschlußlösung *f*, *(beim Sulfitverfahren)* Kochsäure *f*, Sulfit[koch]säure *f*, *(beim alkalischen Verfahren)* Kochlauge *f*, Aufschlußlauge *f*
c.-**liquor composition** *(Pap)* Kochsäurezusammensetzung *f*; *(Pap)* Kochlaugenzusammensetzung *f*
c.-**liquor manufacture** *(Pap)* Sulfit[koch]säureherstellung *f*, Kochsäureherstellung *f*, Säureherstellung *f*; *(Pap)* Kochlaugenherstellung *f*
c. **of rags** Hadernkochung *f*
c. **of straw** Strohkochung *f*
c. **of wood** Holzkochung *f*
c. **operation** *s. c.* process
c. **process** *(Pap)* Kochprozeß *m*, Kochvorgang *m*, Kochung *f*
c. **reagent** *s. c.* agent
c. **salt** Kochsalz *n*, Siedesalz *n*, Salz *n*
c. **temperature** Kochtemperatur *f*
c. **time** Kochzeit *f*, Kochdauer *f*
c. **vat** *s. c.* kettle
cool / **to** [ab]kühlen; sich abkühlen, kaltwerden, erkalten
to allow to c. abkühlen lassen
c. **down** herunterkühlen, abkühlen; sich abkühlen, kaltwerden, erkalten
c. **rapidly** abschrecken
cool brine *(Kaliindustrie)* Kaltlauge *f*
c. **plate** Kühlplatte *f*
coolant *s.* cooling agent
cooler Kühler *m*, Kühlvorrichtung *f*; *(Gär)* Kühlschiff *n*
c. **crystallizer** Kühlungskristallisator *m*
c. **tun** Kühlbottich *m*
coolhouse Kühlhaus *n*, Kühlhalle *f*
cooling Kühlen *n*, Kühlung *f*, Abkühlen *n*, Abkühlung *f*, Kaltwerden *n*, Erkalten *n*; Kälteerzeugung *f*, Kühlung *f*

c. agent Kühlmittel *n*, Kältemittel *n*, Kühlmedium *n*
c. air Kühlluft *f*
c. bath Kühlbad *n*
c. brine Kühlsole *f*
c. by evaporation Verdampfungskühlung *f*, Verdunstungskühlung *f*
c. chamber Kühlkammer *f*
c. channel Kühlkanal *m*
c. coil Kühlschlange *f*
c. compartment Kühlzone *f*, Abkühlzone *f*
c. crack (*Ker*) Kühlriß *m*
c) crystallizer Kühlungskristallisator *m*
c. curve Abkühlungskurve *f*
c. cylinder Kühlzylinder *m*
c.-down time Rückkühlzeit *f*
c. drum Kühltrommel *f*
c. effect Kühlwirkung *f*
c. fixture (*Plast*) Abkühlungsvorrichtung *f*, Schrumpfvorrichtung *f* (*für Preßteile*)
c. jacket Kühlmantel *m*
c. jig *s.* c. fixture
c. liquid Kühlflüssigkeit *f*
c. loss Abkühlverlust *m*
c. medium *s.* c. agent
c. mixture Kältemischung *f*
c. plant Kühlvorrichtung *f*, Kühlanlage *f*, Kälte[erzeugungs]anlage *f*, Kältemaschinenanlage *f*
c. pond Kühlteich *m*
c. rate Kühlgeschwindigkeit *f*
c. roll Kühlzylinder *m*; Kühlwalze *f*
c. section Kühlzone *f*, Abkühlzone *f*
c. surface Kühl[ober]fläche *f*
c. system Kühlsystem *n*
c. tank Kühltank *m*
c. tower Kühlturm *m*, Turmkühler *m*
c.-tower casing Kühlturmkamin *m*
c. tube Kühlrohr *n*
c. vat Kühlwanne *f*
c. water Kühlwasser *n*
c.-water outlet Kühlwasseraustritt *m*
c. wind Kühlluft *f*
c. zone Kühlzone *f*, Abkühlzone *f*, (*Text auch*) Kühlfeld *n*
coolship (*Gär*) Kühlschiff *n*
Cooper reagent Coopersches Reagens *n* (*2,4-Dihydroxyazetophenon zum Nachweis von Fe*$^{+++}$)
coordinate / to (*z.B. Moleküle*) koordinativ anlagern (binden); sich [koordinativ] anlagern, koordinativ gebunden werden
coordinate Koordinate *f*
c. bond koordinative Bindung *f*, semipolare (halbpolare) Doppelbindung *f*, dative Bindung *f*, Donator-Akzeptor-Bindung *f*
c. covalence (covalency, link) *s.* c. bond
coordination bond *s.* coordinate bond
c. chemistry *s.* c. theory
c. complex Koordinationskomplex *m*
c. compound Koordinationsverbindung *f*

c. entity Komplex *m*, komplexe Gruppe *f*
c. formula Koordinationsformel *f*
c. lattice (*Krist*) Koordinationsgitter *n*
c. number Koordinationszahl *f*, Zähligkeit *f*, koordinative Wertigkeit *f*
c. polymerization Koordinationspolymerisation *f*, koordinative Polymerisation *f*
c. position (site) Koordinationsstelle *f*
c. theory Koordinationslehre *f*, Koordinationstheorie *f*, Wernersche Theorie *f*
coordinative koordinativ, Koordinations...
c. bond *s.* coordinate bond
cop (*Text*) Kops *m*, (*manchmal auch*) Kötzer *m*, Bobine *f*
c. dyeing machine (*Text*) Kopsfärbeapparat *m*
copaiba balsam Kopaivabalsam *m*, (*ungenau für*) Kopaivaterpentin *n(m)* (*von Copaifera-Arten*)
c. oil Kopaivaöl *n* (*von Copaifera-Arten*)
c. resin *s.* c. balsam
copal [resin] Kopal *m* (*Sammelname für hochschmelzende pflanzliche Harze*)
copaline, copalite (*Min*) Kopalin *m* (*bernsteinähnliches fossiles Harz*)
copiapite (*Min*) Kopiapit *m* (*Magnesium(III)-dihydroxidhexasulfat*)
copigment (*Bioch*) Kopigment *n*
copolyaddition (*Plast*) Kopolyaddition *f*
copolycondensation (*Plast*) Kopolykondensation *f*
copolymer Kopolymer[es] *n*, Kopolymerisat *n*, Mischpolymer[es] *n*, Mischpolymerisat *n*
c. fibre Kopolymerisatfaser *f*, Mischpolymerisatfaser *f*; Kopolymerisatfaserstoff *m*, Mischpolymerisatfaserstoff *m*
copolymerization Kopolymerisation *f*, Mischpolymerisation *f*
c. with cross-linking Kopolymerisation *f* mit Vernetzung
copolymerize / to kopolymerisieren, mischpolymerisieren
Coppée oven Coppée-Ofen *m*, Coppée-Flammofen *m*, Verbundkoksofen *m* von Evence-Coppée
copper / to verkupfern
copper 1. Cu Kupfer *n*; 2. Braukessel *m*, Braupfanne *f*, Würze[koch]kessel *m*, Würzepfanne *f*
c. acetoarsenite $Cu(CH_3COO)_2 \cdot 3Cu(AsO_2)_2$ Kupfer(II)-arsenitazetat *n*, Kupfer(II)-azetatarsenit *n*
c. after-treatment Kupfernachbehandlung *f*, Nachkupfern *n*, Nachkupferung *f*
c. alloy Kupferlegierung *f*
c. amalgam electrode Kupferamalgamelektrode *f*
c. aminosulphate $[Cu(NH_3)_4]SO_4$ Tetramminkupfer(II)-sulfat *n*
c.-ammonium chloride $CuCl_2 \cdot 2NH_4Cl$ Ammoniumkupfer(II)-chlorid *n*
c.-base auf Kupferbasis (Kupfergrundlage)
c. bath kupfernes Wasserbad *n*; Verkupferungsbad *n*, Kupferbad *n*

c.-bearing kupferhaltig
c. bichromate s. c. dichromate
c. blast furnace Kupferschachtofen m, Kupferschmelzofen m
c. blue Bremer (Braunschweiger) Blau n, Kalkblau n, Neuwieder Blau n (Kupfer(II)-hydroxid)
c. boride Cu_3B_2 Kupfer(II)-borid n
c. bottom (Met) reicher Kupferstein m
c. carbide Cu_2C_2 Kupferkarbid n, Kupfer(I)-azetylid n
c. casse Kupferbruch m, Kupfertrübung f (Weinfehler)
c. chloride process (Erdöl) Kupferchloridverfahren n, $CuCl_2$-Verfahren n (Süßungsverfahren)
c. chloride sweetening (Erdöl) Kupferchloridsüßen n, Kupferchloridsüßung f
c. chloride sweetening process s. c. chloride process
c. complex Kupferkomplex m
c. concentration cell Kupferkonzentrationszelle f
c. converter Kupfer[stein]konverter m
c. converting Kupfersteinverblasen n, Verblasen n von Kupferstein im Konverter, Kupferbessemern n
c. coulometer Kupfercoulometer n
c. deficiency Kupfermangel m
c. dichloride $CuCl_2$ Kupferdichlorid n, Kupfer(II)-chlorid n
c. dichromate $CuCr_2O_7$ Kupfer(II)-dichromat(VI) n
c. dust Kupferstaub m (Fungizid)
c.-dye chelate Kupfer-Farbstoffchelat n
c. electrode Kupferelektrode f
c. electroplating galvanisches Verkupfern n
c. engraving Kupferstich m
c. enzyme Kupferenzym n
c. ferrocyanide membrane Ferrozyankupfermembran f
c. filings Kupferfeilspäne mpl
c. foil Kupferfolie f
c. froth (Min) Kupferschaum m, Tirolit m
c. fungicide Kupferfungizid n
c. gasket Kupferdichtung f
c. glance Kupferglanz m, Chalkosin m (Mineralgruppe)
c. glycine s. c. glycocoll
c.-glycine chelate Kupferglyzinchelat n
c. glycocoll Glykokollkupfer n
c. graphite Kupfergraphit m
c. gray (Am) s. c. grey
c. grey (Min) Antimonfahlerz n, Tetraedrit m
c. halide Kupferhalogenid n
c. head Kupferkopf m (Emailfehler)
c. hemioxide Cu_2O Kupfer(I)-oxid n
c. hydride Cu_2H_2 Kupfer(I)-hydrid n, Kupferwasserstoff m
c. hydroxide $Cu(OH)_2$ Kupfer(II)-hydroxid n

c. index s. c. number
c. matte (Met) armer Kupferstein m
c. metaborate $Cu(BO_2)_2$ Kupfer(II)-metaborat n
c. mica (Min) Kupferglimmer m, (veraltet für) Chalkophyllit m
c. mineral Kupfermineral n
c. monoxide CuO Kupfer[mon]oxid n, Kupfer(II)-oxid n
c. naphthenate Kupfernaphthenat n
c. nickel (Min) Kupfernickel m, (veraltet für) Rotnickelkies m (Nickelarsenid)
c. nitride Cu_3N Kupfernitrid n
c. number (Pap, Text) Kupferzahl f, CuZ
c. ore Kupfererz n
c. oxidase Kupferoxydase f, kupferhaltige Oxydase f
c.-oxide rectifier Kupfer(I)-oxid-Gleichrichter m, Kuproxgleichrichter m
c. peroxide CuO_2 Kupferperoxid n
c. phthalocyanine Kupferphthalozyanin n
c.-plated verkupfert
c. plating Verkupfern n, Verkupferung f
c. powder Kupferpulver n
c. protoxide s. c. hemioxide
c. pyrites (Min) Kupferkies m, Kupferpyrit m, Chalkopyrit m (Eisen(II)-kupfer(II)-sulfid)
c. rayon Kupfer[kunst]seide f, Kupfer-Reyon n, Glanzstoff m
c.-red kupferrot
c. red (Min) Kupferrot n, (veraltet für) Kuprit m, Rotkupfererz n (Kupfer(I)-oxid)
c. reducing power (Zucker) Reduktionsvermögen n gegenüber Fehlingscher Lösung, Kupferzahl f
c. ruby glass Kupferrubinglas n
c. salt Kupfersalz n
c. soap Kupferseife f
c. spiral Kupferspirale f
c. spray (Landw) Kupferbrühe f
c. stain (Glas) Kupferbeize f, Kupferätze f, Rotbeize f, Rotätze f
c. suboxide s. c. hemioxide
c. sweetening (Erdöl) Kupfersüßen n, Kupfersüßung f
c. sweetening process (Erdöl) Kupferverfahren n, Kupferprozeß m (Süßungsverfahren)
c.-tin alloy Kupfer-Zinn-Legierung f
c. turnings Kupferdrehspäne mpl
c. value s. c. number
c. vitriol Kupfersulfat n, (i.e.S.) Kupfervitriol n, (veraltet für) $CuSO_4 \cdot 5H_2O$ Kupfer(II)-sulfat-5-Wasser n; (Min) Kupfervitriol m, Chalkanthit m
c.-yield (Gär) Sudhausausbeute f
copperas $FeSO_4 \cdot 7H_2O$ Eisenvitriol n, Eisen(II)-sulfat-7-Wasser n; (Min) Eisenvitriol m, Melanterit m
coppering Verkupfern n, Verkupferung f
copperization (Farb) Kupferung f
copperize / **to** (Farb, Text) kupfern
copperoid (galvanisch verkupfertes Zinkblech)
copperplate engraving Kupferstich m

c. engraving ink Druckfarbe *f* für Kupferstich
c. ink Kupferdruckfarbe *f*
c. printing Kupferdruck *m*
copperplating bath Verkupferungsbad *n*, Kupferbad *n*
copra Kopra *f (getrocknetes Nährgewebe der Kokosnuß)*
  c. cake Kokoskuchen *m*, Kokos[nuß]preßkuchen *m*
  c. meal Kokoskuchenmehl *n*
  c. oil Kokos[nuß]öl *n*, Kokos[nuß]fett *n*, Kokosbutter *f*
coprecipitate / to mitfällen
coprecipitation [induzierte] Mitfällung *f*
coproduct Nebenprodukt *n*, Beiprodukt *n*
coprogen Koprogen *n (Bakterienwuchsstoff)*
coprolite *(Geol)* Koprolith *m (fossiler Kotballen)*
copy paper Durchschlagpapier *n*
copying ink Kopiertinte *f*, Kopierdruckfarbe *f*
  c. paper *s.* copy paper
  c. paste Hektografenmasse *f*
  c. post *(veraltet für)* copy paper
  c. tissue paper Kopierseiden[papier] *n*
copyings *s.* copy paper
coquimbite *(Min)* Coquimbit *m (Eisen(III)-sulfat-9-Wasser)*
coquina *(Geol)* Muschelkalk *m*
coquinoid limestone *(Geol)* Muschelkalkstein *m*
coral lime Korallenkalk *m*
  c. mud Korallenschlick *m*
coralline Korallin *n*
cord 1. *(Gum)* Kord[faden] *m*; 2. *(Glas)* Schliere *f (Fehler)*
  c. fabric *s.* cordage
  c.-rubberizing compound Mischung *f* für Kordgummierung
cordage *(Gum)* Kordgewebe *n*
cordierite *(Min)* Kordierit *m (Magnesiumaluminiumalumopentasilikat)*
  c. ceramics Kordieritkeramik *f*
  c. porcelain Kordieritporzellan *n*
  c. whiteware Kordieritweißware *f*
cordiness *(Glas)* Schlierigkeit *f (Fehler)*
cordite Cordit *m (Sprengstoff)*
corduroy blanket table *(Erzaufbereitung)* Kordherd *m*, Plachentisch *m*
cordy glass schlieriges Glas *n*
core 1. Kern *m*, *(Ker, Met auch)* Gießkern *m*, *(beim Einsatzhärten auch)* Kernzone *f*, *(Geol auch)* Bohrkern *m*, *(Bot auch)* Kernholz *n*; 2. *(Pap)* Kern *m*, Kegel *m*, Konus *m*, Rotor *m*, Kegelrotor *m (der Kegelstoffmühle)*; 3. *(Pap)* Hülse *f*, Wickelhülse *f*; Einlage *f (beim Triplexkarton)*; 4. Seele *f*, Kern *m (eines Kabels oder Seiles)*, Pulverseele *f (der Zündschnur)*; 5. Dorn *m (bei Strangpreßdüsen)*; 6. *(phys Ch)* Atomrumpf *m*, Rumpf *m*; *(Kph)* Spaltzone *f*
  c. bars *(Pap)* Kegelmesser *npl*, Rotormesser *npl (bei der Kegelstoffmühle)*
  c. binder Kern[sand]bindemittel *n*, Kern[sand]binder *m (Gießereihilfsmittel)*

c. binding agent *s. c.* binder
c. drill Kernbohrer *m*
c. iron Kerneisen *n (Verstärkung aus Gußeisen oder Stahl für schwere oder bruchanfällige Gießkerne)*
c. making Kernherstellung *f*, Herstellung *f* von Gießkernen
c. of the earth *(Geol)* Erdkern *m*, Zentrosphäre *f*
c. oil *(Met)* Kernöl *n*
c. sand *(Met)* Kernsand *m*
c. spun yarn umsponnenes Garn *n*, Kerngarn *n*, Kernmantelfaden *m*
c. strength *(Met)* Kernfestigkeit *f*
c. thread *(Text)* Kernfaden *m*
cored mould *(Plast)* Form *f* (Werkzeug *n*) mit Heizkanälen
coreless induction furnace kernloser Induktionsofen *m*, Mittelfrequenzinduktionsofen *m*, MF-Induktionsofen *m*
coriander Koriander *m*, Coriandrum sativum L.; *s.* c. seed
c. seed Koriander *m (Früchte von Coriandrum sativum L.)*
corilagin Korilagin *n (Ellagengerbstoff)*
coring *(Geol)* Kernen *n*, Kernbohren *n*, Ziehen *n* von Bohrkernen
corium *(Gerb)* Korium *n*, Lederhaut *f*
cork / to zukorken, verkorken, *(i.w.S.)* zupfropfen, zustöpseln, verstöpseln
cork 1. Kork *m*, Korkrinde *f (von Quercus suber L.)*; Kork *m*, Korkgewebe *n*, Phellem *n*; 2. Kork[en] *m*, Korkstopfen *m*, *(i.w.S.)* Stopfen *m*, Pfropf[en] *m*, Stöpsel *m*
c. borer Korkbohrer *m*
c. borer sharpener Korkbohrerschärfer *m*
c. gasket Korkdichtung *f*
c. oak Korkeiche *f*, Quercus suber L.
c. paper Kork[mehl]papier *n*
c. powder Korkmehl *n*
c. press Korkpresse *f*
c. ring Korkring *m*
c. roller Walzenpresse *f (für Korkstopfen)*
c. softener (squeezer) *s. c.* press
c. wax Korkwachs *n*
corky korken, Kork...
corn Samenkorn *n*, *(i.e.S.)* Getreidekorn *n*; Körnerfrüchte *fpl*, Getreide *n*, *(in England)* Weizen *m*, Triticum aestivum L., *(in Schottland und Irland)* Hafer *m*, Avena sativa L., *(Am)* Mais *m*, Zea mays L.
c. crusher Schrotmühle *f*
c. cuckle Kornrade *f*, Agrostemma githago L.
c. flakes *(geröstete)* Maisflocken *fpl*
c. meal Getreidemehl *n*, *(in England)* Weizenmehl *n*, *(in Schottland und Irland)* Hafermehl *n*, *(Am)* Maismehl *n*
c. oil Mais[keim]öl *n*
c. remedy Hühneraugenmittel *n*
c. starch Maisstärke *f*

c.-steep liquor *(Pharm)* Maisquellwasser *n*, Maiseinweichwasser *n*
c. sugar Maiszucker *m*
c. syrup Stärkezuckersirup *m* aus Mais
cornelian *s.* carnelian
corneous lead *(Min)* Hornblei *n*, Bleihornerz *n*, Phosgenit *m*
corning ribbon machine *(Glas)* Corning-Band-Maschine *f* *(Kolbenblasmaschine)*
Cornish stone Cornishstone *m* *(Feldspatpegmatit aus Cornwall)*
Cornu prism Cornu-Prisma *n* *(zur Spektralanalyse)*
cornwallite *(Min)* Cornwallit *m* *(Kupfer(II)-tetrahydroxiddiorthoarsenat)*
corona [discharge] Korona *f*, Koronaentladung *f*
c. resistance Koronabeständigkeit *f*, Beständigkeit *f* gegen den Koronaeffekt
coronizing Koronisieren *n* *(Art der Knitterarmappretur von Glasseidengewebe)*
corozo-nut oil *s.* cohune oil
corpuscular radiation Korpuskularstrahlung *f*, Teilchenstrahlung *f*
corrected grain *(Gerb)* geschliffener Narben *m*
correction:
to add a c. to berichtigen
c. factor Korrekturfaktor *m*, Korrekturkoeffizient *m*
c. term Korrektionsgröße *f*, Korrekturglied *n*
corrective Hilfsstoff *m*, Beistoff *m* *(bei der Konfektionierung von Wirkstoffgemischen)*; *(Landw)* Bodenverbesserungsmittel *n*; *(Pharm)* Korrigens *n*, geschmacksverbessernder Zusatz *m*
correlation Korrelation *f*, Wechselbeziehung *f*, gegenseitige Beziehung *f*
correspondence paper *(Am)* Briefpapier *n*, Ausstattungspapier *n*, Postpapier *n*
corresponding state *(phys Ch)* übereinstimmender Zustand *m*
corridor *(Ker)* Flur *m*, Kammer *f* *(eines Kammertrockners)*
c. dryer *(Ker)* Kanaltrockner *m*, Gangtrockner *m*, *(manchmal auch)* Kammertrockner *m*
corrigent Korrigens *n*, geschmacksverbessernder Zusatz *m*
corrode / **to** korrodieren, zerfressen, anfressen, angreifen, [oberflächlich] zerstören, ätzen; korrodieren, *(vom Eisen)* rosten
corrodibility Korrodierbarkeit *f*, Korrosionsempfindlichkeit *f*, Angreifbarkeit *f*, Ätzbarkeit *f*
corrodible korrodierbar, korrosionsempfindlich, angreifbar, ätzbar
corroding medium *s.* corrosive agent
corrosion Korrosion *f*, Korrodieren *n*, Fressen *n*, Zerfressen *n*, Fraß *m*, oberflächliche Zerstörung *f*; Rosten *n*, Rostung *f*; *(Geol)* Korrosion *f*
c. cell Korrosionselement *n*
c. cracking Rißbildung *f* (Reißen *n*) infolge Korrosion
c. fatigue Korrosionsermüdung *f*

c. figure *(Krist)* Ätzfigur *f*, Lösungsfigur *f*
c. inhibitor Korrosionsschutzmittel *n*, Korrosionsinhibitor *m*, Korrosionshemmer *m*, Korrosionshemmstoff *m*, Korrosionsverzögerer *m*
c. meter Korrosionsprüfgerät *n*
c. product Korrosionsprodukt *n*
c. rate Korrosionsgeschwindigkeit *f*
c. resistance Korrosionsbeständigkeit *f*, Korrosionsfestigkeit *f*, Korrosionswiderstand *m*
c.-resistant, c.-resisting korrosionsbeständig, korrosionsfest
c. test Korrosionsversuch *m*, Korrosionsprüfung *f*, Korrosionstest *m*
corrosive korrosiv, korrodierend [wirkend], zerfressend, angreifend, ätzend, Korrosions..., Ätz...
c. agent Korrosionsmittel *n*, Korrosionsmedium *n*, korrodierendes Mittel *n*
c. environment korrodierend wirkende Umgebung *f*
c. mercuric (mercury) chloride $HgCl_2$ Ätzsublimat *n*, Sublimat *n*, Quecksilber(II)-chlorid *n*
c. sublimate *s.* c. mercuric chloride
corrosiveness korrodierende Wirkung *f*, Korrosivität *f*
corrugated asbestos Wellasbest *m*
c. bellow Faltenbalg *m*; Wellrohr *n*
c. board Wellpappe *f*
c. brown *s.* c. paper
c. cardboard Wellkarton *m*
c. expansion joint Wellrohrkompensator *m*, Wellrohr[dehnungs]ausgleicher *m*
c. paper Well[pappen]papier *n*
c. roll geriffelte Walze *f*, Riffelwalze *f*
c.-roll crusher Walzenbrecher *m* mit geriffelten Walzen
c. sheet *(Plast)* Wellplatte *f*
corrugating paper Well[pappen]papier *n*
cortex *(Pharm, Med)* Rinde *f*; *(Bot)* primäre Rinde *f*, Cortex *m*; *(Pharm)* Fruchtschale *f*
cortical cell *(Bot)* primäre Rindenzelle *f*
c. hormone *s.* corticoid hormone
corticoid hormone Nebennierenrindenhormon *n*, Kortikoid *n*, Kortikosteroid *n*
corticosterone Kortikosteron *n* *(Hormon der Nebennierenrinde)*
corticotrop[h]in Kortikotrop[h]in *n*, Adrenokortikotropin *n*, adrenokortikotropes Hormon *n*, ACTH *n*
cortisol Kortisol *n*, Hydrokortison *n*, 17α-Hydroxykortikosteron *n*
corundellite *(Min)* Korundellit *m*, *(veraltet für)* Margarit *m* *(Kalziumaluminiumdihydroxiddialumodisilikat)*
corundum *(Min)* Korund *m* *(α-Aluminiumoxid)*
corydine Korydin *n* *(Alkaloid)*
corynine Corynin *n*, Yohimbin *n*, Quebrachin *n* *(Alkaloid)*
coslettize / **to** *(Metalle)* phosphatieren
coslettizing Phosphatieren *n*, Phosphatierung *f* *(Korrosionsschutzbehandlung)*

cosmetic kosmetisches Präparat n, Kosmetikum n, Körperpflegemittel n, *(i.e.S.)* Schönheitsmittel n
c. bismuth Schminkweiß n, Perlweiß n, Spanischweiß n, Wismutweiß n *(Wismutoxidnitrat)*
c. chemist Kosmetikchemiker m
c. preparation kosmetisches Präparat n, kosmetische Zubereitung f, Kosmetikpräparat n
cosmetology Kosmetologie f
cosmic chemistry Kosmochemie f
c. radiation kosmische Strahlung f, Höhenstrahlung f, Weltraumstrahlung f
c. ray inducted durch kosmische Strahlung induziert
c. rays kosmische Strahlen mpl, Höhenstrahlen mpl
cosmochemistry Kosmochemie f
cosolvent Verschnittmittel n *(bei Lösungen)*
cossettes Rübenschnitzel npl, Zuckerrübenschnitzel npl
cossyrite *(Min)* Cossyrit m, *(veraltet für)* Aenigmatit m *(titanhaltiges Silikat)*
cost of maintenance (upkeep) Unterhaltungskosten pl
cotoin Kotoin n, *(i.e.S.)* $C_6H_2(OH)_2(OCH_3)COC_6H_5$ Kotoin n, 2,6-Dihydroxy-4-methoxybenzophenon n, Phlorbenzophenon-4-methyläther m
cotton Baumwolle f
c. blue Baumwollblau n
c. boll Baumwollkapsel f
c. bowl *(Pap)* Baumwoll[kalander]walze f
c. cake Baumwollsaatkuchen m
c. cloth Baumwollgewebe n
c. colour s. c. dye
c. cord *(Gum)* Baumwollkord m
c. count Baumwollfeinheit f
c. dry[er] felt *(Pap)* Baumwolltrockenfilz m
c. dye Baumwollfarbstoff m
c. fabric Baumwollgewebe n, baumwollenes Gewebe n, Baumwollstoff m, baumwollener Stoff m
c. felt *(Pap)* Baumwollfilz m
c. fibre Baumwollfaser f
c. fibre gasket Baumwollfaserdichtung f
c. fuzz Baumwollkurzhaar n
c. gin Baumwollegreniermaschine f, Egreniermaschine f, Baumwollentkörnungsmaschine f
c. gum Wasser-Tupelo m, Tupelobaum m, Nyssa aquatica L.
c. lint egrenierte Baumwolle f, Lintbaumwolle f
c. linters Baumwollinters pl
c. oil s. cottonseed oil
c. packing Baumwolldichtung f, Baumwollpackung f
c. paper Baumwollpapier n
c. [rag] pulp *(Pap)* Baumwollhalbstoff m
c. rags Baumwollhadern mpl, Baumwollumpen mpl
c. roll *(Pap)* Baumwollwalze f *(eines Kalanders)*
c. seed oil s. cottonseed oil
c. thread Baumwollfaden m, *(in Kalorimeterbomben auch)* Zündfaden m aus Baumwolle

c. wax Baumwollwachs n
c. wool Watte f
Cotton effect *(phys Ch)* Cotton-Effekt m *(anomale Rotationsdispersion im Bereich von Absorptionsbanden)*
cottonin *(Text)* Flockenbast m
cottonize / to kotonisieren *(Flachs und Hanf zu Flockenbast aufschließen)*
cottonized bast fibre Flockenbast m
cottonseed lecithin Baumwollsaatlezithin n
c. oil Baumwollsaatöl n, Baumwollsamenöl n, Baumwollkernöl n, Cottonöl n
c.-oil green *(Komplexverbindung aus Kupferarsenit und verseiftem Baumwollsaatöl) (Insektizid)*
Cottrell electric precipitation process Cottrell-Entstaubungsverfahren n
Cottrell precipitator Cottrell-Elektroabscheider m, Cottrell-Abscheider m, Cottrell-Staubfilter n, Cottrell-Filter n
Cottrell process s. Cottrell electric precipitation process
cotunnite *(Min)* Kotunnit m *(Blei(II)-chlorid)*
couch / to *(Pap)* [ab]gautschen
c. together *(Pap)* zusammengautschen
couch *(Pap)* Gautschbrett n; s. c. roll; s. c. press
c. box (pit) *(Pap)* Gautschbruchbütte f
c. press *(Pap)* Gautschpresse f, Gautsche f
c.-press roll s. c. roll
c. roll *(Pap)* Gautschwalze f
couched board gedeckte (gegautschte) Pappe f
coucher *(Pap)* Gautscher m
couching *(Pap)* Gautschen n
c. roll s. couch roll
couchman *(Pap)* Gautscher m
cough-provoking zum Husten reizend
coulomb Coulomb n, C, Amperesekunde f, As
Coulomb field Coulomb-Feld n
Coulomb force [of attraction] Coulombsche Kraft (Anziehungskraft) f *(elektrostatische Kraft)*
Coulomb integral s. Coulomb term
Coulomb law Coulomb-Gesetz n, Coulombsches Gesetz n
Coulomb potential Coulomb-Potential n
Coulomb term Coulomb-Glied n, Coulomb-Integral n
coulombic force s. Coulomb force
coulombmeter s. coulometer
coulometer Coulometer n, Voltameter n
coulometric coulometrisch
c. analysis coulometrische Analyse f
c. titration coulometrische Titration f
coulometry Coulometrie f
coumalic acid Kumalinsäure f, α-Pyron-(5)-karbonsäure f
coumaric acid $HOC_6H_4CH=CHCOOH$ Kumarsäure f, trans-o-Hydroxyzimtsäure f
coumarilic acid Kumarilsäure f, Kumaron-2-karbonsäure f
coumarin Kumarin n, Kumarinsäurelakton n,

2-Oxo-1,2-chromen *n*, 5,6-Benzokumalin *n*, Tonkabohnenkampfer *m*
coumarinic acid Kumarinsäure *f*, *cis-o*-Hydroxyzimtsäure *f*
c. lactone *s.* coumarin
coumarone Kumaron *n*, Benzofuran *n*
c.[-indene] resin Kumaron-Indenharz *n*, Kumaronharz *n*
coumaroyl quinic acid Kumarylchinasäure *f* (*Depsid*)
count (*Text*) Feinheit *f*, Nummer *f*, Titer *m*
counter Zählgerät *n*, Zählwerk *n*, Zählvorrichtung *f*, Zählapparat *m*, Zählmaschine *f*, Zähler *m*; (*Kch*) Zählrohr *n*, Zähler *m*
c. blow (*Glas*) Gegenblasen *n*, Vorblasen *n*
c. current *s.* countercurrent
c.-electromotive force gegenelektromotorische Kraft *f*, Gegen-EMK *f*, Gegenurspannung *f*
c. flow *s.* counterflow
c. heat exchanger *s.* counterflow heat exchanger
c.-rotating gegenläufig
c. tube (*Kch*) Zählrohr *n*, Zähler *m*
counteract / to entgegenwirken, rückwirken
counterbalance / to ausgleichen, kompensieren; ausbalancieren, auswuchten, austarieren
counterbalancing weights (*Pap*) Kompensationsgewichte *npl*, Entlastungsgewichte *npl*
counterclockwise entgegen dem Uhrzeigersinn, gegen den Uhrzeigersinn, im Gegenzeigersinn, linksdrehend
countercurrent *s.* countercurrently
countercurrent Gegenstrom *m*, Gegenfluß *m*
in **c.** *s.* countercurrently
c. condenser Gegenstromkühler *m*; Gegenstromkondensator *m*, Gegenstromverdichter *m*
c. contactor Gegenstromextraktionsapparat *m*
c. cooler Gegenstromkühler *m*
c. decantation Gegenstromdekantation *f*
c. distillation Gegenstromdestillation *f*
c. distribution Gegenstromverteilung *f*
c. extraction Gegenstromextraktion *f*
c. flow *s.* countercurrent
c. flow condenser *s.* c. condenser
c. hydrolysis Gegenstromhydrolyse *f*
c. partial condenser Teilkondensator *m*, Dephlegmator *m*
c. principle Gegenstromprinzip *n*
c. system Gegenstromsystem *n*
c. washing Gegenstromwäsche *f*, Gegenstrom[aus]waschung *f*
countercurrently im Gegenstrom, nach dem Gegenstromprinzip
counterdraft (*Plast*) Hinterschneidung *f*
counterelectrode Gegenelektrode *f*
counterflow Gegenfluß *m*, Gegenstrom *m*
c. condenser *s.* countercurrent condenser
c. cooler Gegenstromkühler *m*
c. cooling tower Gegenstromkühlturm *m*
c. dryer Gegenstromtrockner *m*
c. heat exchanger Gegenstromwärmeaustauscher *m*, Gegenströmer *m*

counterion Gegenion *n*
counterpressure filler (racker) Gegendruckfüller *m*, Gegendruckfüllmaschine *f*, Gegendruckabfüllapparat *m*
counterweight Gegengewicht *n*
counting apparatus (device) Zählapparat *m*, Zählgerät *n*, Zählwerk *n*, Zählvorrichtung *f*, Zählmaschine *f*, Zähler *m*
c. roll (*Pap*) Zählwalze *f*
c. tube (*Kch*) Zählrohr *n*, Zähler *m*
couple / to verknüpfen; (*Farb*) [an]kuppeln, kombinieren; (*Farb*) kuppeln, sich kombinieren
c. in acid solution sauer (in saurer Lösung, in saurem Medium) kuppeln
c. in alkaline solution alkalisch (in alkalischer Lösung, in alkalischem Medium) kuppeln
coupled reaction gekoppelte Reaktion *f*
coupling Verknüpfung *f*; (*Farb*) Kuppeln *n*, Kupp[e]lung *f*, Kopp[e]lung *f*, Kombinieren *n*; (*Tech*) (nicht schaltbare) Kupplung *f*
c. agent (*Text, Plast*) Haftmittel *n*
c. component (*Farb*) Kupplungskomponente *f*, passive Komponente *f*, Entwickler *m*
c. rate Kupplungsgeschwindigkeit *f*
c. vat (*Farb*) Kupplungsbütte *f*, Kupplungsbottich *m*, Kupplungskufe *f*, Kupplungsgefäß *n*
course Verlauf *m*, Lauf *m*, Ablauf *m*, Gang *m*; (*Ker*) Steinschicht *f*, Schicht *f*
c. of reaction Reaktionsverlauf *m*, Reaktionsablauf *m*
c. of vulcanization Vulkanisationsverlauf *m*
C.O.V. *s.* concentrated oil of vitriol
covalence, covalency kovalente Wertigkeit *f*, Kovalenz *f*, Bind[ungswert]igkeit *f*, Atombindigkeit *f*, Atomwertigkeit *f*, Atombindungszahl *f*, Bindungszahl *f*; (manchmal auch für) covalent bond
c. number Koordinationszahl *f*, Zähligkeit *f*, koordinative Wertigkeit *f*
covalent kovalent, homöopolar, unpolar, einpolar, unitarisch
c. bond kovalente (unpolare, homöopolare, einpolare, unitarische) Bindung *f*, Elektronenpaarbindung *f*, Atombindung *f* (als Zustand)
c. bond angle Kovalenzbindungswinkel *m*
c. bonding kovalente (unpolare, homöopolare, einpolare, unitarische) Bindung *f*, Elektronenpaarbindung *f*, Atombindung *f* (als Vorgang)
c. crystal kovalenter (homöopolarer) Kristall *m*, Atomkristall *m*
c. link (linkage) *s.* c. bond
c. radius [of atoms] kovalenter Atomradius *m*
covalently bonded kovalent (homöopolar, unpolar, einpolar) gebunden
covelline; covellite (*Min*) Kovellin *m*, Kupferindig[o] *m* (*Kupfer(II)-sulfid*)
cover / to [be]decken, abdecken, belegen, überziehen, umhüllen, ummanteln; (*Fläche*) bedecken, einnehmen, umfassen, sich erstrecken auf

cover Decke *f*, Belag *m*, Schicht *f*, Überzug *m*, Bezug *m*, Hülle, *f*, Umhüllung *f*, Mantel *m*, Ummantelung *f*; Deckel *m*, Deckplatte *f*, Abdeckplatte *f*, Verschluß *m*; *(Gum)* Decke *f*, Mantel *m*
c. coat Deckmasse *f*; Überzug *m*
c. glass Deckglas *n*, Deckgläschen *n*
c. paper Umschlagpapier *n*
c. slide Deckplättchen *n*
coverage 1. Deckvermögen *n*, Deckfähigkeit *f*, Deckkraft *f*; 2. *(Text)* Decken *n*, Egalfärben *n*; 3. *(mit einer gegebenen Chemikalienmenge)* behandelte (oder behandelbare) Fläche *f*; 4. deckender Belag *m* *(von Insektiziden)*
c. of barré *(Text)* Decken *n* von streifigfärbendem Material
covered pot (Glas) verdeckter (gedeckter, geschlossener) Hafen *m*
covering *s.* cover
c. agent Deckmittel *n*
c. lye *(Düngemittelindustrie)* Decklauge *f*
c. power Deckvermögen *n*, Deckfähigkeit *f*, Deckkraft *f*
covolume *(phys Ch)* Kovolumen *n*
cow berry Preiselbeere *f*, Vaccinium vitis-idaea L.
c. milk Kuhmilch *f*
c. milk fat Kuhbutterfett *n*
c. skim milk Kuhmagermilch *f*
cowrie *s.* kauri
coxoba snuff Niopo, Parica *(Samenmehl von Piptadenia peregrina (L.)Benth., Genußmittel)*
cp, c.p., CP, C.P. *s.* 1. chemically pure; 2. continuous percolation
cr. *s.* 1. crystal; 2. crystalline
CR *s.* chloroprene rubber
crab / to *(Text)* krabben, krappen, brennen
crab *(Text)* Krabbmaschine *f*
c. orchard salt *(Abführmittel mit Magnesiumsulfat als Hauptbestandteil)*
crabbing *(Text)* Krabben *n*, Krappen *n*, Brennen *n*
crabeye lichen *(Farb)* Parelleflechte *f*, Erdorseille *f*, Lecanora parella Mass.
crab's claw complex Chelatkomplex *m*, Chelat *n*, Chelatverbindung *f*, Scherenverbindung *f*
crack / to 1. *(eine chemische Bindung)* sprengen, [auf]spalten, lösen; *(in einfachere Verbindungen)* spalten; *(Erdöl)* kracken, spalten; *(Emulsionen)* brechen, spalten, entmischen; 2. sich [auf]spalten *(von Verbindungen)*; brechen *(von Emulsionen)*; [zer]platzen, [zer]springen, bersten, aufspringen, [auf]reißen, rissig werden, zerreißen
c. up aufspalten
crack Sprung *m*, Riß *m*, Spalt *m*, Bruch *m*
c. growth Rißwachstum *n*
c. growth resistance Widerstand *m* gegen Rißwachstum
c. initiation Rißbildung *f*
c. of the grain *(Gerb)* Sprung *m* *(Elastizitätseigenschaft des Leders)*
cracked distillate Krackdestillat *n*
c. fraction Krackfraktion *f*

c. gasoline (naphtha) Krackbenzin *n*, Spaltbenzin *n*
cracker Krackapparatur *f*, Krackanlage *f*
c. gas Krackgas *n*
c. mill *(Gum)* Brecherwalzwerk *n*
cracking Sprengen *n* *(einer Bindung)*, Spalten *n* *(in einfachere Verbindungen)*; *(Erdöl)* Kracken *n*, Krackung *f*, Spalten *n*, Spaltung *f*; Brechen *n*, Spalten *n*, Entmischen *n* *(von Emulsionen)*, Entemulsionieren *n*, Emulsionsspaltung *f*, Emulsionsentmischung *f*, Demulgieren *n*, Demulgierung *f*, Dismulgieren *n*; Platzen *n*, Zerplatzen *n*, Springen *n*, Zerspringen *n*, Bersten *n*, Aufspringen *n*, Reißen *n*, Aufreißen *n*, Rissigwerden *n*, Rißbildung *f*, Rissebildung *f*, Zerreißen *n*
to give a mild c. treatment mild (gelinde) kracken
c. chamber Krackraum *m*, Krackreaktor *m*
c. conditions Krackbedingungen *fpl*
c. feed[stock] Krackgut *n*, Krackeinsatz *m*, Krackrohstoff *m*, Spaltgut *n*, Einsatzgut (Einsatzmaterial, Ausgangsmaterial) *n* für Krackverfahren (Krackprozesse)
c. furnace Krackofen *m*
c. gas Krackgas *n*
c. mill *s.* cracker mill
c. process Krackverfahren *n*, Spaltverfahren *n*; Krackvorgang *m*, Spaltvorgang *m*
c. reaction Krackreaktion *f*
c. stock *s.* c. feed[stock]
c. treatment Krackbehandlung *f*
c. zone Krackzone *f*, Spaltzone *f*
crackle *(Pap)* *(Am)* Klang *m*
c. glaze *(Ker)* Krackglasur *f*, Craqueléeglasur *f*, Krakeleeglasur *f*, Haarrißglasur *f*
crackled glass Craqueléeglas *n*, Krakeleeglas *n*
crackling salt Knistersalz *n*
cracklings Grieben *fpl*, Fettgrieben *fpl*; Griebenrückstände *mpl*
craftsman in rubber Gummifachmann *m*
Cramer solution Cramersche Lösung *f* *(aus KJ und HgO, zum Nachweis reduzierender Zucker)*
crank *(Ker)* Sparkapsel *f* *(Brennhilfsmittel)*
c. angle Kurbelwinkel *m*
c. press Kurbelpresse *f*
crankshaft Kurbelwelle *f*
crater *(Plast)* Krater *m*
crawling *(Ker)* Kriechen *n*, Abrollen *n*, Aufrollen *n* *(Glasurfehler)*
crayon Pastellkreide *f*, Zeichenkreide *f*; Pastellstift *m*, Zeichenstift *m*, Buntstift *m*
craze / to *(Ker)* reißen *(Glasur)*
crazing *(Ker)* Haarrißbildung *f*, Rissebildung *f*, Haarrissigwerden *n*; Haarrisse *mpl* *(Glasurfehler)*; *(Gum)* Crazing-Effekt *m*, Elefantenhautbildung *f*
cream / to aufrahmen, Sahne ansetzen lassen; abrahmen, entrahmen, Rahm abschöpfen; *(z.B. Eiweiß)* zu Schaum schlagen; *(Gum, Plast)* aufrahmen

c. off abrahmen, entrahmen, Rahm abschöpfen
cream s. c.-coloured
cream Rahm m, Sahne f; Creme f, Krem f; dicker Brei m
  c. ageing vat s. c. ripener
  c. cheese Rahmkäse m
  c. chilling vat Rahmkühlwanne f
  c.-coloured cremefarben, cremefarbig, zart gelblich
  c. cooling vat s. c. chilling vat
  c. holding vat s. c. ripener
  c. ice Kremeis n, Rahmeis n, Sahneeis n
  c. layer Rahmschicht f
  c. make-up Make-up-Creme f, Make-up-Krem f
  c. margarine Sahnemargarine f
  c. nut Paranuß f, Brasilnuß f (Samen von Bertholletia excelsa Humb. et Bonpl.)
  c. of lime (Gerb) Kalkmilch f
  c. of tartar KOOC(CHOH)$_2$COOH [gereinigter] Weinstein m, Weinsteinrahm m, Kaliumhydrogentartrat n
  c. paper (Foto) Chamoispapier n
  c. plasma Rahmplasma n
  c. ripener Rahmreifer m, Säuerungsgefäß n, Säuerungswanne f
  c. ripening Rahmreifung f
  c. ripening tank (vat) s. c. ripener
  c. separator Entrahmungszentrifuge f, Entrahmungsschleuder f, Rahmseparator m, Milchzentrifuge f, Milchschleuder f, Milchseparator m
creamability Aufrahmungsfähigkeit f, Aufrahmungsvermögen n, Aufrahmungspotential n
creamed latex (Gum) aufgerahmter Latex m
creamery Molkerei f, Meierei f
  c. butter Molkereibutter f
creaming (Lebm, Gum, Plast) Aufrahmen n, Aufrahmung f, Rahmen n (von Emulsionen), Emulsionsverdichtung f
  c. ability s. creamability
  c. agent Aufrahmungsmittel n
  c. potential (power) s. creamability
  c. process Aufrahmungsvorgang m, Aufrahmungsprozeß m
creamy cremig, cremeförmig, cremeartig
crease / to knittern
crease Knitter m, Kniff m, Falte f
  c. angle Knitter[erholungs]winkel m
  c.-proofed knitterfest ausgerüstet
  c. recovery resistance Knittererholungsvermögen n
  c. resistance Knitterfestigkeit f, Knitterechtheit f, Knitterbeständigkeit f, Knitterwiderstand m, Knitterfreiheit f
  c.-resistant knitterarm, knitterfest, knitterecht, knitterbeständig
  c.-resistant finish Knitterarmausrüstung f, Knitterfestausrüstung f, Knitterechtausrüstung f, Knitterarmappretur f, Knitterfestappretur f, Knitterechtappretur f, knitterbeständige Ausrüstung f

c.-resisting finish (treatment) s. c.-resistant finish
creased paper Aderpapier n, Knitterpapier n
creasing resistance s. crease resistance
creasote s. creosote
creation of nucleation centres Kristall[isations]keimbildung f
creep / to kriechen
creep Kriechen n, Kriechvorgang m
  c. elongation Kriechdehnung f
  c. point Kriechpunkt m
  c. resistance Kriechfestigkeit f, Kriechwiderstand m, Dauerstandfestigkeit f
  c. test (Gum) Creep-Test m
crenic acid (Bodenchemie) Krensäure f (eine Fulvosäure)
creosote / to mit Kreosotöl durchtränken (imprägnieren), kreosotieren
creosote Kreosot n, Holzteerkreosot n; Steinkohlen[teer]kreosot n, Kreosotöl n, Kreosot n (i.w.S.)
  c. carbonate Kreosotkarbonat n, Kreosotal n
  c. oil Kreosotöl n (Holzdestillationsprodukt, das auf Reinkreosot weiter verarbeitet wird); Kreosotöl n, Kreosot n (i.w.S.), Steinkohlen[teer]kreosot n
crepe / to (Am) s. to crêpe
crepe / to (Gum) zu Crepe aufarbeiten (aufbereiten); (Pap) kreppen; (Text) kreppen, kräuseln, kreponieren
crepe (Am) s. crêpe
crêpe (Gum) Crepe f, Crêpe m, Krepp m, Crepekautschuk m, Kreppkautschuk m; (Text) Krepp[stoff] m, gekrepptes Gewebe n
  c. paper Kreppapier n, gekrepptes Papier n
  c. rubber Crepekautschuk m, Kreppkautschuk m, Crepe f, Crêpe m, Krepp m
  c. tissue paper Kreppseidenpapier n, Seidenkreppapier n, gekrepptes Seidenpapier n
  c. wrapping paper Krepp-Pack[papier] n, Packkrepp m
crêpeing (Gum) Aufbereitung (Aufarbeitung) f zu Crepe; (Pap, Text) Kreppen n
  c. bath (liquor) (Text) Kreponierbad n, Kreppbad n
crepenynic acid CH$_3$(CH$_2$)$_4$C≡CCH$_2$CH=CH$_2$(CH$_2$)$_7$COOH Krepeninsäure f, cis-9-Oktadezen-12-in-säure f
creping s. crêpeing
crescent method s. c. tear test
  c. tear test (Gum) Crescent-Methode f (zur Bestimmung der Weiterreißfestigkeit)
  c. test piece (Gum) Crescent-Probe f, bogenförmige Probe f, bogenförmiger Probekörper m
cresol CH$_3$C$_6$H$_4$OH Kresol n, Methylphenol n, Hydroxytoluol n
  c. purple m-Kresolpurpur m, m-Kresolsulfophthalein n
  c. red Kresolrot n, o-Kresolsulfophthalein n, 3',3''-Dimethylphenolsulfophthalein n

c. resin Kresolharz n
o-cresolphthalein o-Kresolphthalein n, 3,3'-Dimethylphenolphthalein n
m-cresolsulphonephthalein m-Kresolsulfophthalein n, m-Kresolpurpur m
o-cresolsulphonephthalein o-Kresolsulfophthalein n, 3',3''-Dimethylphenolsulfophthalein n, Kresolrot n
cresotic (cresotinic) acid C6H3(CH3)(OH)COOH Kresotinsäure f, (i.e.S.) o-Kresotinsäure f, 2-Kresol-3-karbonsäure f
cresyl alcohol s. cresol
cresylic acid rohe Karbolsäure f (Gemisch aus Phenolen, Kresolen und Xylenolen), (i.e.S.) Teerkresol n, Kresylsäure f (Gemisch aus o-, m- und p-Kresol)
c. resin Kresolharz n
crevice Spalt m, Riß m
c. corrosion Spaltkorrosion f
Criegee reaction Criegeesche Glykolspaltung f, Criegee-Oxydation f
crimp / to kräuseln
crimp Kräuseln n, Kräuselung f
c. rigidity (stability) Kräuselungsbeständigkeit f, Kräuselfestigkeit f
crimped yarn Kräuselgarn n, Kräuselfaden m, Krinkelgarn n
crimping Kräuseln n, Kräuselung f
c. force Kräuselungskraft f
crimson karm[es]inrot, karmoisinrot, feurigrot
crinkled yarn s. crimped yarn
crisp knusp[e]rig, bröck[e]lig, mürbe (z.B. Gebäck); frisch, fest, saftig (z.B. Gemüse); (Text) hart (im Griff)
c. fat Fettgrieben fpl, Grieben fpl
crisped mint Krauseminze f (krausblättrige Varietäten mehrerer Mentha-Arten)
cristobalite (Min) Cristobalit m (Siliziumdioxid-Modifikationen)
critical air blast (Kohlechemie) kritische Luftmenge f, CAB, C.A.B.
c. air blast method (Kohlechemie) Methode f der kritischen Luftmenge
c. air blast test (Kohlechemie) CAB-Test m, C.A.B.-Test m
c. air blast value (Kohlechemie) CAB-Wert m, C.A.B.-Wert m
c. concentration for micelle formation (Text) kritische Mizellbildungskonzentration f
c. constants (data) kritische Konstanten fpl (Daten pl, Größen fpl)
c. diameter Trennkorngröße f, Kornscheide f
c. minerals (Geol) kritische Minerale npl
c. moisture content Knickpunktfeuchte f, Knickpunkt-Feuchtebeladung f, kritische Feuchte[beladung] f
c. moisture point Knickpunkt m, kritischer Punkt m (bei der Trocknung)
c. point kritischer Punkt m
c. pressure kritischer Druck m
c. solution temperature kritische Lösungstemperatur (Mischungstemperatur) f, kritischer Lösungspunkt (Mischungspunkt) m
c. speed kritische Drehzahl f; kritische Geschwindigkeit f
c. temperature kritische Temperatur f
c. volume kritisches Molvolumen (Volumen) n
crizzle (Glas) Haarrisse mpl
croceic (crocein) acid s. croceine acid
croceine acid Krozeinsäure f, 2-Naphthol-8-sulfonsäure f
c. scarlet Krozeinscharlach m
crocidolite Krokydolith m (Abart der Hornblende)
crock fastness (Text) (Am) Reibechtheit f
c.-meter (Text) (Am) Reibechtheitsprüfer m, Crockmeter n
crocoise, crocoisite, crocoite (Min) Krokoit m, Rotbleierz n (Blei(II)-chromat)
cromfordite (Min) Cromfordit m, (veraltet für) Phosgenit m, Bleihornerz n, Hornblei n
Croning process Croning-Verfahren n, Croning-Formmaskenverfahren n, Formmaskenverfahren n [nach Croning], Maskenformverfahren n, Shell-Moulding-Verfahren n
cronstedtite (Min) Cronstedtit m (ein Phyllosilikat)
Crookes dark space Crookesscher (Hittorffscher) Dunkelraum m, Katodendunkelraum m, innerer Dunkelraum m
Crookes glass Crookes-Glas n (UV-absorbierendes Glas)
crookesite (Min) Crookesit m
crop nutrition Pflanzenernährung f
c. of crystals Kristallisationsprodukt n, Kristallisat n
c. out / to (Geol) ausstreichen, ausgehen, zutage treten
c.-producing power Ertragsfähigkeit f (des Bodens)
c. protection Pflanzenschutz m
c. rotation (Landw) Fruchtwechsel m, Fruchtfolge f
c. sprayer (Landw) Sprühgerät n
cross Kreuzstück n
c. air circulation Überlüftung f (beim Trocknen)
c.-axis mounting Walzenschränkung f
c.-belt [magnetic] separator Kreuzband[magnet]scheider m
c.-bend[ing] test Biegeprüfung f, Prüfung f auf Biegefestigkeit
c.-breaking strength Knickfestigkeit f
c. circulation s. c. air circulation
c.-cut / to (Pap) querschneiden
c.-cutter (Pap) Querschneider m
c. direction (Pap) Querrichtung f, Querlauf m
c. dyeing Überfärben n, Überfärbung f, Nachfärben m, Nachfärbung f (einer Komponente in Faserstoffmischungen)
c. extruder head Querspritzkopf m
c. flow Kreuzstrom m
c.-flow tray Kreuzstromboden m
c.-gallery Querschlag m (bei der Untertagevergasung)

c.-grinder *(Pap)* Querschleifer *m*
c.-grinding *(Pap)* Querschleifen *n*
c.-laminate Kreuzschichtstoff *m*
c.-link / to [quer]vernetzen, querverbinden
c.-link Vernetzungsstelle *f*, Brücke *f*, Querverbindung *f*
c.-link density Vernetzungsdichte *f*
c.-linkage 1. Vernetzung *f*, Quervernetzung *f*, Querverbindung *f* *(als Vorgang)*; 2. *s.* c.-link
c.-linking *s.* c.-linkage 1.
c.-linking agent Vernetzungsmittel *n*, Vernetzer *m*
c.-linking reaction Vernetzungsreaktion *f*
c.-linking system Vernetzungssystem *n*, Vernetzersystem *n*, Vernetzerkombination *f*
c.-printing *(Text)* Überdrucken *n*, Überdruck *m*
c. resistance *(Bio)* Kreuzresistenz *f (durch Einwirkung eines Giftes indirekt erworbene Widerstandskraft gegen andersartige Gifte)*
c. section Querschnitt *m*
c. section at break *(Gum)* Querschnitt *m* im Augenblick des Bruches, wirklicher Querschnitt *m*
c.-section target area *(Kph)* Wirkungsquerschnitt *m*
c.-sectional area Querschnittsfläche *f*, Querschnitt *m*
c.-wall Querwand *f*
c.-way *s.* c. direction
Cross cellulose Cross-Zellulose *f*
Cross process *(Erdöl)* Cross-Verfahren *n*, Krackverfahren *n* nach Cross
crossarm paddle mixer Kreuzbalkenrührer *m*
crossed aldol condensation gekreuzte Aldolkondensation *f*
c.-axes machine *(Gum)* Kalander *m* mit Walzenschränkung
c. double bond gekreuzte Doppelbindung *f*
crossflow cooling tower Kreuzstrom-Kühlturm *m*, Querstrom-Kühlturm *m*
crosshead Querspritzkopf *m*
crosslink / to *s.* to cross-link
crosslinkable vernetzbar
crosslinked molecule vernetztes Molekül *n*
crosslinking *s.* cross-linkage 1.
c. agent *s.* cross-linking agent
crossover flue Überführungskanal *m*, Überströmkanal *m*, Überführung *f (eines Koppers-Becker-Verbundkoksofens)*
crotal *s.* crottle
croton oil Krotonöl *n (von Croton tiglium L.)*
crotonaldehyde $CH_3CH=CHCHO$ Krotonaldehyd *m*, 3-Methylakrolein *n*, Buten-(2)-al-(1) *n*
crotonic acid $CH_3CH=CHCOOH$ Krotonsäure *f*, α-Krotonsäure *f*, *trans*-Buten-(2)-säure-(1) *f*, *trans*-3-Methylakrylsäure *f*
α-**c. acid** *s.* crotonic acid
β-**c. acid** *s.* isocrotonic acid
c. aldehyde *s.* crotonaldehyde
crotonolic acid $CH_3CH=C(CH_3)COOH$ Tiglinsäure *f*, *trans*-2-Methyl-2-butensäure *f*

crotonylene $CH_3C\equiv CCH_3$ Krotonylen *n*, Dimethylazetylen *n*, 2-Butin *n*
crottal, crottle *(Sammelname für farbstoffliefernde Flechten)*; *s.* dark crottle
Crow receiver Meßzylinder *m* nach Crow *(mit konisch verjüngter Basis)*
crown / to *(Walzen, Profile)* bombieren, ballig bearbeiten
crown 1. Bombage *f*, Bombierung *f*, Balligkeit *f (von Walzen, Profilen)*; 2. Gewölbe *n*, Haube *f*, Glocke *f (von Schmelzöfen)*, Kuppel *f (des Glasofens)*; 3. *s.* c. glass; 4. *s.* c. cap
c. block Kronenblock *m*, Turmrollenblock *m (einer Rotary-Bohranlage)*
c. cap (closure, cork) Kronenverschluß *m*
c. corking machine Kronenverschließmaschine *f*
c. flint glass Kronflintglas *n*
c. glass 1. Kronglas *n*; 2. Mondglas *n*
c. optical glass Kronglas *n*
crowner Kronenverschließmaschine *f*
c. head Verschließkopf *m*, Verschließstempel *m (einer Kronenverschließmaschine)*
crowning *s.* crown 1.
c. head *s.* crowner head
crucible Tiegel *m*, Schmelztiegel *m*; Tiegel *m (Unterteil eines Schachtschmelzofens für Nichteisenmetalle)*; Gestell *n (eines Hochofens)*
c. adapter Tulpe *f*, Einsatztulpe *f*, Vorstoß *m* für Frittentiegel; Gummimanschette *f* für Frittentiegel
c. cast steel *s.* c. steel
c. coke Tiegelkoks *m*
c. furnace Tiegelofen *m*
c. holder Gummimanschette *f* für Frittentiegel
c. lid Tiegeldeckel *m*
c. melting furnace *s.* c. furnace
c. melting process *s.* c. process
c. method Methode *f* der Tiegelverkokung
c. process Tiegel[ofen]verfahren *n*, Tiegelstahlverfahren *n*, Tiegelschmelzverfahren *n*
c. ring Tiegelring *m*
c. stand Tiegeluntersatz *m*
c. steel Tiegelstahl *m*
c. swelling test Tiegel-Blähprobe *f*
c. tongs Tiegelzange *f*
crude roh, Roh...
crude Rohgut *n*, Rohprodukt *n*; Rohöl *n*, rohes Öl *n*; Roh[erd]öl *n*, rohes Erdöl *n*, Crude[oil] *n*
c. alcohol Rohspiritus *m*, Rohsprit *m*; Rohbranntwein *m*
c. ammonia liquor Ammoniakwasser *n*, Gaswasser *n*, NH_3-Wasser *n*
c. asbestos Crudeasbest *m*, Crude *n (zusammenhängende Asbestfasern)*
c. beet juice Rübenrohsaft *m*
c. benzol Rohbenzol *n*
c.-benzol plant Rohbenzolanlage *f*
c.-benzol still Rohbenzoldestillationsanlage *f*, Rohbenzoldestillieranlage *f*, Rohbenzolabtreiber *m*

c. **calcium acetate** rohes (unreines) Kalziumazetat *n, (technisches)* Kalziumazetat *n*, Graukalk *m*
c. **clay** Rohton *m*
c. **concentrate** Rohkonzentrat *n*
c. **creosote** Rohkreosot *n*
c. **cymene** Rohzymol *n*
c. **distillate** Rohdestillat *n*
c. **drug** Droge *f*
c. **extract** Rohextrakt *m*
c. **fat** Rohfett *n*, Rohtalg *m*; Rohfett *n (Gesamtheit ätherlöslicher Stoffe in Nahrungsmitteln)*
c. **fibre** Rohfaser *f*
c. **fibre material** *(Pap)* Faserrohstoff *m*
c. **gas** Rohgas *n*
c. **juice** Rohsaft *m*
c. **kaolin** Rohkaolin *m*
c. **lead** Rohblei *n*, Werkblei *n*
c. **oil** Rohöl *n*, rohes Öl *n*; Roh[erd]öl *n*, rohes Erdöl *n*, Crude[oil] *n*
c.**-oil cooler** Rohölkühler *m*
c.**-oil pipeline** Rohölpipeline *f*, Rohölleitung *f*
c.**-oil tank** Rohöllagergefäß *n*, Rohöltank *m*
c. **ore** Roherz *n*, Fördererz *n*
c. **petroleum** rohes Erdöl *n*, Roh[erd]öl *n*
c. **product** Rohprodukt *n*
c. **protein** Rohprotein *n*
c. **resin** Rohharz *n*
c. **rubber** Rohkautschuk *m*
c. **sewage** rohes Abwasser *n*, Rohabwasser *n*
c. **shale oil** Rohschieferöl *n*, Schieferrohöl *n*
c. **sugar** Rohzucker *m*
c. **sulphur** Rohschwefel *m*
c. **test** Vorprobe *f*, Vorprüfung *f*
c. **waste** *s. c.* sewage
crumb structure *(Landw)* Krümelgefüge *n*, Krümelstruktur *f*
crumble / **to** zerkrümeln, zerbröckeln, zermalmen; zerbröckeln, zerfallen
c. **away** zerbröckeln, zerfallen
c. **to pieces** in Stücke zerfallen
crumbly krüm[e]lig, bröck[e]lig, leicht bröckelnd
crumbs 1. Krümel *mpl(npl)*; 2. zerfaserte Alkalizellulose *f*
c. **of buna synthetic rubber** Bunakrümel *mpl(npl)*, Krümelbuna *m*
crush / **to** zerquetschen; brechen, grobzerkleinern; *(Text)* zerdrücken, knittern; zerstoßen, *(Eis)* stoßen, *(Malz)* schroten
crush *s.* crushing
c. **proof** *(Am)* knitterarm, knitterbeständig, knitterecht, knitterfest
c. **proofing** *(Am)* Knitterarmausrüstung *f*, knitterbeständige Ausrüstung *f*, Knitterechtausrüstung *f*, Knitterfestausrüstung *f*, Knitterarmappretur *f*, Knitterechtappretur *f*
c. **resistant** *s. c.* proof
crushed cocoa Kakaobruch *m*
c. **ice** gestoßenes (feinzerschlagenes, zerkleinertes) Eis *n*
c. **malt** geschrotetes Malz *n*, Malzschrot *m*

c. **resin** Stückenharz *n*
c. **rocks** *(Landw)* Gesteinsmehl *n*
crusher Brecher *m*
c. **roll** *s.* crushing roll
c. **rolls** *s.* crushing rolls
crushing Zerquetschen *n*; Brechen *n*, Grobzerkleinerung *f*; *(Text)* Zerdrücken *n*, Knittern *n*; Zerstoßen *n*, Stoßen *n (von Eis)*, Schroten *n (von Malz)*
c. **and grinding** Zerkleinern *n*, Zerkleinerung *f*
c. **cone (head)** Brechkegel *m*
c. **mortar** Diamantmörser *m*
c. **plate** Brechplatte *f*
c. **roll** Brechwalze *f*
c. **rolls** Brechwalzen *fpl*; Brechwalzwerk *n*, Walzenbrecher *m*
c. **strength** Bruchfestigkeit *f*
crust Kruste *f*, Ansatz *m*; Rinde *f*; Schale *f*
c. **condition** *(Gerb)* Borkezustand *m (von ausgegerbtem und getrocknetem, aber noch nicht zugerichtetem Leder)*
c. **of the earth** Erdkruste *f*, Erdrinde *f*
crustless cheese Käse *m* ohne Rinde, rindenloser Käse *m*
cryogenic engineering Tieftemperaturtechnik *f*
c. **process** Verfahren *n* der Tieftemperaturtechnik, Tieftemperaturverfahren *n*
cryogenics Kryogenik *f*
cryohydrate Kryohydrat *n*
cryohydric kryohydratisch
c. **point** kryohydratischer Punkt *m*
cryolite Kryolith *m (Natriumfluoroaluminat)*
cryoscope Kryoskop *n*, Gefrierpunktmesser *m*
cryoscopic kryoskopisch
c. **constant** kryoskopische Konstante *f*, molare Gefrierpunktserniedrigung *f*
c. **method** *s.* cryoscopy
cryoscopy Kryoskopie *f*
crypto-ionic kryptoionisch
c.**-ionic reaction** Kryptoionenreaktion *f*
cryptoclastic sediment *(Geol)* kryptoklastisches Sediment *n*, feinkörniges klastisches Sediment *n*
cryptocrystalline *(Min)* kryptokristallin
cryptolepine Kryptolepin *n*
α-**cryptomerene** Kryptomeren *n (Diterpenderivat)*
cryptopleurine Kryptopleurin *n*
crysaniline *s.* chrysaniline
cryst. *s.* 1. crystal; 2. crystalline
crystal Kristall *m*
c. **analysis** Kristallstrukturanalyse *f*, Feinstrukturanalyse *f*, Röntgenstrukturanalyse *f*
c. **angle** Kristallwinkel *m*
c. **axis** Kristallachse *f*, kristallografische Achse *f*
c. **bed** Kristallbett *n*, Kristallpolster *n*, Kristallschüttung *f*, Kristall[schütt]schicht *f*
c. **boiling** *(Zucker)* Kochen *n* auf Korn, Kornkochen *n*
c. **chemistry** Kristallchemie *f*
c. **defect** Kristall[bau]fehler *m*; Fehlstelle *f*
c. **discharge** Kristallaustrag *m*

c. **dislocation** Kristallversetzung f, Versetzung f, Dislokation f
c. **druse** Kristalldruse f
c. **face** Kristallfläche f, Kristallebene f
c. **field theory** Kristallfeldtheorie f
c. **form** Kristallform f
c. **glass** Kristallglas n
c. **glaze** (Ker) Kristallglasur f
c. **grating** Kristallgitter n
c. **grating spectrograph** Kristall[gitter]spektrograf m
c. **growth** Kristallwachstum n; Kristallzüchtung f
c. **habit** Kristallhabitus m
c. **ice** Kristalleis n, Klareis n
c. **imperfection** Kristallstörung f, Kristallfehler m (i.w.S.)
c. **lattice** Kristallgitter n
c. **malt** Karamelmalz n
c. **nucleus** Kristall[isations]kern m, Kristall[isations]keim m, Kristallisationszentrum n
c. **of fat** Fettkristall m
c. **of ice** Eiskristall m
c. **of sugar** Zuckerkristall m
c. **phosphor** Kristallphosphor m
c. **phosphorescence** Kristallphosphoreszenz f
c. **property** Kristalleigenschaft f
c. **structure** Kristallstruktur f, Kristallbau m
c.**-structure analysis** Kristallstrukturanalyse f
c.**-structure determination** Kristallstrukturbestimmung f
c. **surface** Kristalloberfläche f
c. **system** Kristallsystem n, kristallografisches System n
c. **tuff** (Geol) Kristalltuff m
c. **violet** [(CH₃)₂NC₆H₄]₂C=C₆H₄=N(CH₃)₂Cl·9H₂O $[(CH_3)_2NC_6H_4]_2C=C_6H_4=N(CH_3)_2Cl \cdot 9H_2O$ Kristallviolett n, Hexamethylpararosanilinchlorhydrat n
c. **water** Kristallwasser n
crystalline kristallin[isch], kristallen, kristallisch, Kristall-, kristallartig, kristallförmig
c. **aggregate** kristallinisches Aggregat n, Kristallaggregat n, Kristallhaufwerk n
c. **area** kristalliner Bezirk (Bereich) m
c. **crop** Kristallisationsprodukt n, Kristallisat n
c. **form** Kristallform f
c. **fraction** kristalliner Anteil m
c. **glaze** (Ker) Kristallglasur f
c. **liquid** kristalline (anisotrope) Flüssigkeit f, flüssiger Kristall m, Fastkristall m
c. **phase** kristalline Phase f, Kristallphase f
c. **powder** Kristallpulver n
c. **region** s. c. area
c. **schist** (Geol) kristalliner Schiefer m
c. **solid** kristalliner Feststoff m
crystallinity Kristallinität f
crystallite (Krist) Kristallit m; (Krist) Mosaikblöckchen n, Mosaikblock m; (Bioch) Mizelle f, Mizell n
c. **theory** Kristallittheorie f (Theorie vom kristallitischen Aufbau des Glases)

crystallizability Kristallisationsfähigkeit f, Kristallisationsvermögen n, Kristallisierbarkeit f
crystallizable kristallisationsfähig, kristallisierbar
crystallizate Kristallisat n
crystallization Kristallisation f, Kristallisierung f, Kristallisieren n
c. **centre** Kristallisationszentrum n, Kristall[isations]kern m, Kristall[isations]keim m
c. **differentiation** Kristallisationsdifferentiation f
c. **dish** Kristallisierschale f
c. **in motion** (Zucker) Kristallisation f in Bewegung
c. **rate** Kristallisationsgeschwindigkeit f
c. **resistance** Kristallisationswiderstand m, Widerstand m gegen Kristallisation
c. **truck [for wet cooling]** Kristallisationswagen m (Margarineherstellung)
c. **wag[g]on** s. c. truck [for wet cooling]
c. **water** Kristallwasser n
crystallize [out] / **to** [aus]kristallisieren, Kristalle bilden; (etwas) auskristallisieren, zur Kristallisation bringen
crystallized sugar Kristallzucker m
crystallizer Kristallisator m, Kristallisierapparat m, Kristaller m, Kristallierer m
crystallizing chamber Kristallisationsraum m, Kristallisierraum m, Kristallabscheideraum m, Kristallabscheider m
c. **dish** Kristallisierschale f
c. **evaporator** Verdampfungskristallisator m
c. **pond** Kristallisierbecken n (z.B. in Salzgärten)
c. **process** Kristallisationsverfahren n
c. **solution** Kristallisationslösung f, Kristallisierlösung f
c. **tank** Kristallisierbehälter m, Kristallisierkasten m, Kristallisiermulde f, Kristallisierpfanne f
c. **vessel** Kristallisiergefäß n, Kristallisationsgefäß n
crystalloblastic (Geol) kristalloblastisch
c. **series** (Geol) kristalloblastische (idioblastische) Reihen fpl
c. **texture** (Geol) kristalloblastisches Gefüge n
crystallochemical kristallchemisch
crystallographic kristallografisch
c. **axial ratio** (Krist) Achsenverhältnis n
c. **axis** kristallografische Achse f, Kristallachse f
c. **class** Kristallklasse f, Symmetrieklasse f
c. **plane** Kristallebene f
c. **system** Kristallsystem n, kristallografisches System n
crystallography Kristallografie f
crystalloid Kristalloid n
crystalloluminescence Kristallolumineszenz f
crystn. s. crystallization
CS (Abk. für) Chemical Society
CTC s. chlorotetracycline
CTFE s. chlorotrifluoroethylene
CTU s. centigrade thermal unit
cub. s. cubic
Cuba wood Kuba[gelb]holz n, kubanisches Gelbholz n

cube Würfel *m*, Kubus *m*, *(Krist)* Elementarwürfel *m*
  c. gambier *(Gerb)* Gambir *n (von Uncaria gambir Roxb.)*
  c. mixer Würfelmischer *m*
  c. strength Würfelfestigkeit *f (Betonprüfung)*
  c. sugar Würfelzucker *m*
cubeba *(Pharm)* Kubebenpfeffer *m*, Piper cubeba L.
cubebs Kuben *fpl*, Fructus cubebae *(von Piper cubeba L.f.)*
cubic kubisch, würfelförmig, Würfel...
  c. lattice kubisches Gitter *n*, Würfelgitter *n*
  c. nitre NaNO$_3$ Natronsalpeter *m*, Natriumnitrat *n*, *(früher)* kubischer Salpeter *m*
  c. shrinkage *(Ker)* kubische (räumliche) Schwindung *f*, Volumenschwindung *f*
  c. system *(Krist)* kubisches System *n*, kubisches Gitter *n*, Würfelgitter *n*
cubical *s.* cubic
  c. blender Würfelmischer *m*
  c. expansion kubische (räumliche) Ausdehnung *f*, Volumenausdehnung *f*
cudbear Cudbear *m*, Persio *m (Flechtenfarbstoff)*; *(Bot) (Sammelname für Cudbear liefernde Flechten)*, *(i.e.S.)* Unechte Lackmusflechte *f*, Ochrolechia tartarea (L.)Mass.
cuit Cuite *f (voll entbastete Seide)*
cull / to aussuchen, wählen, aussortieren
cull *(Plast)* Abfall *m*, Ausschuß *m*
cullender / to *s.* to colander
cullender *s.* colander
cullet *(Glas)* Scherben *fpl*, Bruch *m*, Bruchglas *n*
cultivate / to kultivieren, züchten *(z.B. Bakterien)*
cultivated hop Kulturhopfen *m*
  c. yeast Kulturhefe *f*, Zuchthefe *f*
cultivation Kultivieren *n*, Züchten *n*, Züchtung *f (z.B. von Bakterien)*
culture Kultur *f*, Aufzucht *f*, Zucht *f*
  c. bottle *(Bioch)* Kulturkolben *m*
  c. dish Petrischale *f*
  c. filtrate Kulturfiltrat *n*
  c. flask Kulturkolben *m*
  c. medium Nährmedium *n*, Nährboden *m*, Nährsubstrat *n*, Kulturmedium *n*
  c. plate *(Bakt)* Kulturplatte *f*
  c. tube Kulturröhrchen *n*
  c. yeast Kulturhefe *f*, Zuchthefe *f*
cultured butter Sauerrahmbutter *f*
  c. cream saure Sahne *f*, saurer Rahm *m*
  c. dairy product gesäuertes Milchprodukt *n*, Sauermilchprodukt *n*, Sauermilcherzeugnis *n*
  c. milk mit Säureweckern gesäuerte (versetzte) Milch *f*, saure (dickgelegte) Milch *f*, Sauermilch *f*, Dickmilch *f*
  c. milk product *s.* c. dairy product
cumaric acid *s.* coumaric acid
cumarin *s.* coumarin
cumarinic acid *s.* coumarinic acid
cumene C$_6$H$_5$CH(CH$_3$)$_2$ Kumol *n*, Isopropylbenzol *n*, 2-Phenylpropan *n*

cum[m]in Kreuzkümmel *m*, Römerkümmel *m*, Cuminum cyminum L.
cumol *s.* cumene
cumulated double bond kumulierte (angehäufte, gehäufte) Doppelbindung *f*
cumulative kumulativ, kumulierend
  c. action (effect) kumulative Wirkung *f (z.B. von Insektiziden)*
  c. index Generalregister *n*, Gesamtregister *n*, Sammelregister *n*
  c. ionization lawinenartige Ionisation (Ionisierung) *f*
  c. sample Sammelprobe *f*
cundurango bark Kondurangorinde *f*, Kondorrinde *f*, Geierrinde *f*, Cortex condurango *(von Marsdenia reichenbachi Triana)*
cup 1. Tiegel *m*, Petroleumgefäß *n (eines Flammpunktprüfers)*; 2. Formmulde *f (einer Brikettpresse)*; 3. Trichter *m*, Zuführungstrichter *m*, Fülltrichter *m*, Schütttrichter *m (eines Hochofens)*; 4. Topfmanschette *f*, Napfmanschette *f*; Stulpe *f*, Stulpdichtung *f*
  c. and cone Glocken- und Trichter[gicht]verschluß *m*, Parry-Gichtverschluß *m*, Parry-Verschluß *m (eines Hochofens)*
  c. flow figure *(Plast)* Becher[schließ]zeit *f*, Becherfließzahl *f*, Prüfzeit *f*
  c. flow test *(Plast)* Schließzeitbestimmung *f (mittels Prüfbecher)*
  c.-handling machine *(Ker)* Tassengarniermaschine *f*
  c. jolley *(Ker)* Tassendrehmaschine *f*
  c. method Zylinder[platten]methode *f*, Oxford-Methode *f (zur Penizillinwertbestimmung)*
  c. ring Topfmanschette *f*, Napfmanschette *f*, Stulpe *f*, Stulpdichtung *f*
  c.-type atomizer Sprühkorb *m*, Zentrifugalkorb *m*, Düsenkorb *m*
cuparene Kuparen *n*, *p*-(1′,2′,2′-Trimethylzyklopentyl-)toluol *n*
cuparenic acid Kuparensäure *f*, *p*-(1′,2′,2′-Trimethylzyklopentyl-)benzoesäure *f*
cupel *(Met)* Kapelle *f*
  c. test *(Met)* Kapellenprobe *f*
cupellation Kupellation *f*, Kupellieren *n*, Treiben *n*, Abtreiben *n*, Treibarbeit *f*; *(Met)* Kapellenprobe *f*
cupferron [C$_6$H$_5$N(NO)O]NH$_4$ Kupferron *n*, N-Nitroso-phenylhydroxylamin-Ammonium *n*
cupid's darts *(Min)* Crinis veneris, Venushaar *n (faseriger Rutil; Titandioxid)*
cupola Kupolofen *m*, Kuppelofen *m*
  c. brick Kupolofenstein *m*, Kuppelofenstein *m*
  c. furnace *s.* cupola
  c. linings Kupolofenfutter *n*, Kupolofenausmauerung *f*, Kuppelofenfutter *n*, Kuppelofenausmauerung *f*
cuprammonia *s.* cuprammonium hydroxide solution
cuprammonium cellulose Kupferoxidammoniakzellulose *f*, Kuoxamzellulose *f*, Blaumasse *f*

c. hydroxide solution [Cu(NH₃)₄](OH)₂ Kupfertetramminhydroxidlösung f, Kuoxamlösung f, ammoniakalische Kupferhydroxidlösung f, Schweizers Reagens n
c. rayon Kuoxamfaserstoff m, KU
c. solution s. c. hydroxide solution
c. sulphate [Cu(NH₃)₄]SO₄ Tetramminkupfer(II)-sulfat n
cuprate Kuprat n
cupreous sulphuret of silver (Min) Kupfersilberglanz m, Silberkupferglanz m, (veraltet für) Stromeyerit m (Kupfer(I)-silbersulfid)
cupric Kupfer..., (i.e.S.) Kupfer(II)-...
c. acetate Cu(CH₃COO)₂ Kupfer(II)-azetat n
c. acetoarsenite Cu(CH₃COO)₂·3Cu(AsO₂)₂ Kupfer(II)-arsenitazetat n, Kupfer(II)-azetatarsenit n
c. acid arsenate s. c. acid orthoarsenate
c. acid orthoarsenate Cu₅H₂(AsO₄)₄ Kupfer(II)-hydrogen[ortho]arsenat(V) n
c. ammine (ammonia, ammonio) sulphate s. c. tetrammine sulphate
c. ammonium chloride CuCl₂·2NH₄Cl Ammoniumkupfer(II)-chlorid n
c. ammonium sulphate (NH₄)₂SO₄·CuSO₄ Ammoniumkupfer(II)-sulfat n
c. arsenate s. c. orthoarsenate
c. arsenite Cu₃(AsO₃)₂ Kupfer(II)-[ortho]arsenat(III) n
c. benzoate Cu(C₆H₅COO)₂ Kupfer(II)-benzoat n
c. bromate Cu(BrO₃)₂ Kupfer(II)-bromat n
c. bromide CuBr₂ Kupfer(II)-bromid n
c. butyrate Cu(C₃H₇COO)₂ Kupfer(II)-butyrat n
c. carbonate 1. CuCO₃·Cu(OH)₂ Kupfer(II)-dihydroxidkarbonat n; 2. 2CuCO₃·Cu(OH)₂ Kupfer(II)-dihydroxiddikarbonat n
c. chelate Kupfer(II)-chelat n
c. chloride CuCl₂ Kupfer(II)-chlorid n
c. citrate Cu₂C₆H₄O₇ Kupfer(II)-zitrat n
c. cyanide Cu(CN)₂ Kupfer(II)-zyanid n, Kupferdizyanid n
c. dichromate CuCr₂O₇ Kupfer(II)-dichromat(VI) n
c. ferricyanide Cu₃[Fe(CN)₆]₂ Kupfer(II)-hexazyanoferrat(III) n
c. ferrocyanide Cu₂[Fe(CN)₆] Kupfer(II)-hexazyanoferrat(II) n
c. fluoride CuF₂ Kupfer(II)-fluorid n, Kupferdifluorid n
c. fluorosilicate Cu[SiF₆] Kupfer(II)-hexafluorosilikat n
c. formate Cu(HCOO)₂ Kupfer(II)-formiat n
c. hydrogen arsenite CuHAsO₃ Kupfer(II)-hydrogen[ortho]arsenat(III) n
c. hydroxide Cu(OH)₂ Kupfer(II)-hydroxid n, Kupferdihydroxid n
c. iodate Cu(JO₃)₂ Kupfer(II)-jodat n
c. ion Kupfer(II)-ion n
c. lactate Cu(CH₃CH(OH)COO)₂ Kupfer(II)-laktat n

c. laurate Cu(C₁₂H₂₃O₂)₂ Kupfer(II)-laurat n
c. nitrate Cu(NO₃)₂ Kupfer(II)-nitrat n
c. nitroferricyanide s. c. nitroprusside
c. nitroprussiate s. c. nitroprusside
c. nitroprusside Cu[Fe(CN)₅NO] Kupfer(II)-nitroprussid n, Kupfer(II)-nitroprussiat n, Kupfer(II)-pentazyanonitrosylferrat n
c. oleate Cu(C₁₇H₃₃COO)₂ Kupfer(II)-oleat n
c. orthoarsenate Cu₃(AsO₄)₂ Kupfer(II)-[ortho]arsenat(V) n
c. orthophosphate Cu₃(PO₄)₂ Kupfer(II)-[ortho]phosphat n
c. orthophosphite CuHPO₃ Kupfer(II)-[ortho]phosphit n
c. oxalate Cu(COO)₂ Kupfer(II)-oxalat n
c. oxide CuO Kupfer(II)-oxid n
c. oxychloride Kupfer(II)-oxidchlorid n
c. periodate Cu₂HJO₆ Kupfer(II)-hydrogenorthoperjodat n
c. phosphate s. c. orthophosphate
c. phosphide Cu₃P₂· Kupfer(II)-phosphid n, Trikupferdiphosphid n
c. potassium chloride K₂CuCl₄ (oder) KCl·CuCl₂ Kaliumkupfer(II)-chlorid n
c. salicylate Cu(HOC₆H₄COO)₂ Kupfer(II)-salizylat n
c. salt complex Kupfersalzkomplex m
c. selenate CuSeO₄ Kupfer(II)-selenat n
c. silicofluoride s. c. fluorosilicate
c. stearate Cu(C₁₇H₃₅COO)₂ Kupfer(II)-stearat n
c. sulphate CuSO₄ Kupfer(II)-sulfat n
c. sulphide CuS Kupfer(II)-sulfid n
c. tartrate Cu(OOC·CH(OH)·CH(OH)COO) Kupfer(II)-tartrat n
c. tetrammine sulphate [Cu(NH₃)₄]SO₄ Tetramminkupfer(II)-sulfat n
c. tetramminosulphate s. c. tetrammine sulphate
c. thiocyanate Cu(SCN)₂ Kupfer(II)-thiozyanat n, Kupfer(II)-rhodanid n
c. tungstate CuWO₄ Kupfer(II)-wolframat n
c. wolframate s. c. tungstate
cupriethylenediamine (Text) Kupferäthylendiamin n, Cuen n
cupriferous kupferhaltig, Cu-haltig
c. pyrite (Min) Chalkopyrit m, Kupferkies m, Kupferpyrit m (Eisen(II)-kupfer(II)-sulfid)
cuprite (Min) Kuprit m, Rotkupfererz n, Kupferrot n (Kupfer(I)-oxid)
cupro Kupferseide f, KUS
c. filament Kupferseidenelementarfaden m
c.-nickel Kupfer-Nickel-Legierung f, Kupfernickel n
c. staple Kupferfaser f, KUF
c.-uranite (Min) Kuprouranit m, Torbernit m (Kupfer(II)-bisuranylorthophosphat)
cuproscheelite (Min) Kuproscheelit m
cuprous Kupfer..., (i.e.S.) Kupfer(I)-...
c. acetylide Cu₂C₂ Kupfer(I)-azetylid n, Kupferkarbid n, Azetylenkupfer n, Kupferazetylen n
c. antimonide Cu₃Sb Kupfer(I)-antimonid n

c. arsenide Cu_3As Kupfer(I)-arsenid n
c. catalyst Kupfer(I)-katalysator m
c. chloride $CuCl$ Kupfer(I)-chlorid n
c. cyanide $CuCN$ *(oder)* $Cu_2(CN)_2$ Kupfer(I)-zyanid n
c. ferricyanide $Cu_3[Fe(CN)_6]$ Kupfer(I)-hexa-zyanoferrat(III) n
c. ferrocyanide $Cu_4[Fe(CN)_6]$ Kupfer(I)-hexa-zyanoferrat(II) n
c. fluorosilicate $Cu_2[SiF_6]$ Kupfer(I)-hexafluoro-silikat n
c. ion method Kupfer(I)-ionen-Verfahren n, Kuproionenmethode f
c. oxide Cu_2O Kupfer(I)-oxid n
c. oxide rectifier Kupfer(I)-oxid-Gleichrichter m, Kuproxgleichrichter m
c. silicide Cu_4Si Kupfer(I)-silizid n
c. sulphate Cu_2SO_4 Kupfer(I)-sulfat n
c.-sulphide cloud Kupferbruch m, Kupfertrübung f *(ein Weinfehler)*
c. sulphite Cu_2SO_3 Kupfer(I)-sulfit n
c. thiocyanate $Cu(SCN)$ Kupfer(I)-thiozyanat n, Kupfer(I)-rhodanid n
cuprovanadite *(Min)* Cuprovanadit m, *(veraltet für)* Descloizit m
cuprox *s.* cuprous oxide rectifier
c. cell Kupfer(I)-oxid-Zelle f, Kupfer(I)-oxid-Element n
curable *(Gum)* vulkanisierbar, vernetzungsfähig, vernetzbar
Curacao aloe *(Pharm)* Curaçao-Aloe f, Barbados-Aloe f *(von Aloë vera L.)*
curare Kurare n, Curare n *(Pfeilgift von südameri-kanischen Menispermaceen und Loganiaceen)*
c. alkaloid Kurarealkaloid n
curares Kurararten fpl
curative action heilende Wirkung f, Heilwirkung f
c. drug Heilmittel n
c. system *(Gum)* Vulkanisationssystem n, Ver-netzungssystem n, Vernetzersystem n, Ver-netzerkombination f
curatives *(Gum)* 1. *s.* curative system; 2. Vulkani-sationsmittel npl, Vulkanisationsagenzien npl, Vulkanisiermittel npl
curb Seiher[körper] m; Preßkorb m, *(i.w.S.)* Preßmantel m
c. press Korbpresse f; Seiherpresse f
curcas oil *(Öl von Jatropha curcas L.)*
curcuma Kurkuma f, Gelbwurz[el] f, Gelber Ingwer m, Curcuma longa L.
curd / to gerinnen, dick werden, koagulieren, [aus]flocken, [ver]klumpen; gerinnen lassen, koagulieren, zur Ausflockung bringen
curd Gerinnsel n, Klumpen m; Käsebruch m, Bruch m; geronnene Milch f, Quark m
c. characteristics Gerinnungseigenschaften fpl *(z.B. der Milch)*
c. cheese Quarkkäse m
c. firmness *(Lebm)* Bruchfestigkeit f
c. formation *(Lebm)* Bruchbildung f

c. grain *(Lebm)* Bruchkorn n
c. knife *(Lebm)* Bruchmesser n
c. ladle Käsebruchschöpfkelle f
c. particle *s.* c. grain
c. strength (tension) *s.* c. firmness
c.-tension test *(Lebm)* Bruchfestigkeitsbestim-mung f
c. treatment *(Lebm)* Bruchbearbeitung f
curdle / to *s.* to curd
curdling Gerinnen n, Gerinnung f, Koagulieren n, Koagulierung f, Koagulation f, Flockung f, Ausflockung f, Flockenbildung f, Klumpen n, Verklumpen n
c. point Gerinnungspunkt m
curdly leicht gerinnbar; geronnen, käsig
curds geronnene Milch f, Quark m; käsiger Niederschlag m
curdy geronnen, dick[lich]; klumpig; flockig; quarkig, käsig, Quark...
c. cheese Quarkkäse m
c. precipitate käsiger Niederschlag m
cure / to *(Gum)* vulkanisieren, heizen, vernetzen; *(Farb, Plast)* [aus]härten; *(Landw)* nachreifen *(z.B. Superphosphat)*; konservieren, haltbar machen *(durch Trocknen, Salzen, Pökeln, Räu-chern oder Säuern)*, *(i.e.S.)* [ein]pökeln; nach-behandeln, *(i.e.S.)* warmbehandeln
cure *s.* 1. curing; 2. curing salt
c. accelerator *(Gum)* Vulkanisationsbeschleuni-ger m
c.-all Allheilmittel n
c. by high-energy radiation *(Gum)* Vulkanisation (Vernetzung) f durch energiereiche Strahlung
c. curve *(Gum)* Vulkanisationskurve f
c. plateau *(Gum)* Vulkanisationsplateau n, Plateau n, Plateaueffekt m
c. rate *(Gum)* Anvulkanisationsgeschwindigkeit f; Vulkanisationsgeschwindigkeit f, Heizge-schwindigkeit f
c. system *(Gum)* Vulkanisationssystem n, Ver-netzungssystem n, Vernetzersystem n, Ver-netzerkombination f
c. temperature *(Gum)* Vulkanisationstemperatur f, Heiztemperatur f; *(Farb,Plast)* Härtetemperatur f
c. time *(Gum)* Vulkanisationszeit f, Heizzeit f, Gesamtheizzeit f; *(Farb, Plast)* Härtezeit f, Härtungszeit f, Aushärtungszeit f
cured malt Darrmalz n
c. meat Pökelfleisch n, [ein]gepökeltes Fleisch n
c. moulding ausgehärtetes Preßteil n
c. resin gehärtetes Harz n
c. rubber vulkanisierter Kautschuk m, Kautschuk-vulkanisat n, Gummi m
c. weight *(Gerb)* Salzmasse f *(der Häute)*
Curie point (temperature) Curie-Punkt m, Curie-Temperatur f
curine Kurin n, Bebeerin n, Bebirin n, Chondro-dendrin n *(Alkaloid)*
curing *(Farb, Plast)* Härten n, Härtung f, Aushärten

n, Aushärtung f; (Gum) Vulkanisation f, Vulkanisierung f, Vernetzung f, Heizung f; (Landw) Nachreifen n (z.B. von Superphosphat); Konservieren n, Konservierung f, Haltbarmachen n, Haltbarmachung f (durch Trocknen, Salzen, Pökeln, Räuchern oder Säuern), (i.e.S.) Pökeln n, Einpökeln n, Fleischpökeln n; Nachbehandlung f, (i.e.S.) Warmbehandlung f

c. adhesive (Holz) härtender Klebstoff m

c. agent (Gum) Vulkanisationsmittel n, Vulkanisationsagens n, Vulkanisiermittel n, Vernetzungsmittel n, Vernetzer m; (Farb, Plast) Härtungsmittel n, Härtemittel n, Härter m; (Lebm) Konservierungsmittel n, (i.e.S.) Pökelsalz n

c. bag (Gum) Heizschlauch m

c. catalyst (Plast) Härtekatalysator m

c. characteristics (Gum) Vulkanisationsverhalten n, Vulkanisationseigenschaften fpl

c. curve (Gum) Vulkanisationskurve f

c. cycle (Plast) Härtungsperiode f, Härtezyklus m, Preßdauer f

c. drum (Gum) Vulkanisiertrommel f

c. ingredients (Gum) Vulkanisationssystem n, Vernetzungssystem n, Vernetzersystem n, Vernetzerkombination f; Vulkanisationsmittel npl, Vulkanisationsagenzien npl, Vulkanisiermittel npl

c. mould (Gum) Vulkanisierform f

c. oven (Gum) Vulkanisierofen m

c. pan (Gum) Vulkanisierkessel m, Vulkanisationskessel m

c. press (Gum) Vulkanisierpresse f

c. reaction (Gum) Vulkanisationsreaktion f; (Plast) Härtungsreaktion f

c. salt Konservierungssalz n

c. solution (Lebm) Pökelsalzlösung f, Pökellake f, Pökelflüssigkeit f

c. system (Gum) Vulkanisationssystem n, Vernetzungssystem n, Vernetzersystem n, Vernetzerkombination f

c. temperature (Gum) Vulkanisationstemperatur f, Heiztemperatur f; (Farb, Plast) Härtetemperatur f

c. time (Gum) Vulkanisationszeit f, Heizzeit f, Gesamtheizzeit f; (Farb, Plast) Härtezeit f, Härtungszeit f, Aushärtungszeit f

c. tube s. c. bag

c.-up (Gum) Anvulkanisation f, Anvulkanisieren n, Anbrennen n, Anspringen n

curium Cm Curium n

curlate / to (Pap) curlatieren, kräuseln

curlation (Pap) Curlatieren n, Kräuseln n

curlator (Pap) Curlator m (Maschine zum Kräuseln des Faserstoffs)

curled mint Krauseminze f

curling (Ker) Kräuseln n (Fehler)

curometer Curometer n (Apparat zur Bestimmung der Vulkanisationskurve)

currency paper (Am) Banknotenpapier n

current 1. [elektrischer] Strom m; Stromstärke f; 2. Strom m, Strömung f, Fluß m

c.-carrying stromführend, stromdurchflossen, unter Strom

c. density Stromdichte f

c. efficiency Stromausbeute f

c. flow Stromfluß m

c. intensity s. c. strength

c. meter Flügelradzähler m, Turbinenzähler m (Mengenstrommessung)

c. of air Luftstrom m

c. of heated air Heißluftstrom m

c. of oxygen Sauerstoffstrom m

c. of water Wasserstrom m

c. passage Stromdurchgang m

c.-scan voltammetry stromgeregelte Voltametrie f

c. source Stromquelle f

c. strength Stromstärke f

c. term Laufterm m, variabler Term m

c.-voltage curve Strom-Spannungs-Kurve f, Strom-Spannungs-Kennlinie, f, U-I-Kennlinie f

currier (Gerb) Lederzurichter m

curry / to (Gerb) zurichten

currying (Gerb) Zurichtung f, Zurichten n

curtain Vorhang m, Gardine f, Läufer m (fehlerhafter Anstrich); Lackvorhang m (bei der Gießlackierung)

c. coating (Farb) Lackgießen n, Gießlackierung f, Gießen n

curtaining Vorhangbildung f, Gardinenbildung f, Läuferbildung f, Ablaufen n, Laufen n (von Anstrichmitteln)

Curtius method (reaction) Curtiussche Methode f, Curtiusscher Abbau m (von Säureaziden zu primären Aminen)

curvature Krümmung f, Biegung f; Kurvenform f

cuscohygrine $C_{13}H_{24}NO_2$ Kuskhygrin n, Cuscohygrin n, Cuschygrin n, Bellaradin n

cushion board (Am) Wellpappe f

c. stock (Gum) Polstermischung f, Mischung f für Polsterplatten

custard ice Eierkremeis n

cut / to (z.B. Glas, Papier, Hadern) schneiden; (Glas) schleifen; (Stroh) häckseln; (Kautschukmischung) einschneiden; (Kautschukballen) spalten, schneiden

c. across (Pap) querschneiden

c. back verschneiden, fluxen

c. into sheets (Pap) in Bogen (Format) schneiden

c. lengthways (Pap) längsschneiden, längstrennen

c. off (Kautschukfell) von der Walze schneiden

c. single sheet (Pap) (Bogen) einlagig schneiden

cut Schnitt m; Schnitt m, Trennschnitt m, Fraktion f (bei der Destillation); (Glas) Schliff m

to make a c. trennen, scheiden (z.B. beim Klassieren)

c. glass geschliffenes Glas n

c. growth (Gum) Schnittwachstum n, Rißwachstum n

c.-growth resistance (Gum) Widerstand m gegen Schnittwachstum

c.-off Absperrventil *n*; *(Plast)* Abquetschfläche *f*, Abquetschrand *m (einer Preßform)*
c. point Trennkorngröße *f*, Kornscheide *f (z.B. beim Klassieren)*
c. rubber thread Gummischnittfaden *m*
c. size *s.* c. point
c. sugar Stückzucker *m*
c. surface Schnittfläche *f*
c. wine Verschnittwein *m*
cutback [bitumen] Verschnittbitumen *n*, Cutback-bitumen *n*
c. product Verschnittprodukt *n*
cutch [braunes] Katechu *n (Gerbstoffextrakt aus Acacia catechu Willd.)*, *(manchmal auch)* Gambir *n*, gelbes Katechu *n (von Uncaria gambir Roxb.)*, *(i.w.S. Sammelname für Gerbstoffextrakte zahlreicher Pflanzenarten)*
cuticle Kutikula *f*, Blatt[ober]haut *f*
c. remover Nagelhautentferner *m*
c. residues kutikuläre Rückstände *mpl (von Pflanzenschutzmitteln)*
cutinite Kutinit *m (Einzelmazeral der Kohle)*
cutlery-marking *(Ker)* Silberstrichbildung *f (Fehler)*
cutoff *(Glas)* Abschnitt *m (des Postens)*
cutter Schneid[e]apparat *m*, Schneid[e]maschine *f*, Schneider *m*; Häckselmaschine *f (Strohaufschluß)*; *(Pap)* Querschneider *m*; *(Ker)* Abschneider *m*, Abschneidetisch *m*; *(Glas)* Glasschleifer *m*, Glaskugler *m*
c. knife *(Pap)* Messer *n* des Querschneiders
c. trim *(Pap)* Randstreifen *m*
cutters *(veraltet für)* bars
cutting Schneiden *n (z.B. von Glas, Papier, Hadern)*; Schleifen *n*, Schliff *m (von Glas)*; Häckseln *n (z.B. von Stroh)*; Einschneiden *n (einer Kautschukmischung)*; Spalten *n*, Schneiden *n (von Kautschukballen)*; *(Kosmet)* Aussalzen *n (der Seife)*
c. action *(Pap)* Schnittwirkung *f*
c. ceramics Schneidkeramik *f*
c. device *s.* c. machine
c. diamond Glasschneiderdiamant *m*, Glaserdiamant *m*
c. fluid Schneidflüssigkeit *f*
c. length *(Pap)* Schnittlänge *f*
c. machine Schneid[e]maschine *f*, Schneid[e]vorrichtung *f*, Schneider *m*
c. mill Schneidmühle *f*
c.-off table *(Ker)* Abschneidetisch *m*, Abschneider *m*
c. oil Schneidöl *n*
c. reducer *(Foto)* subtraktiver (subtraktiv wirkender) Abschwächer *m*
cuttings Späne *mpl*, Schnitzel *npl(mpl)*, Abfälle *mpl*; *(Min)* Gesteinsabrieb *m*, Gesteinsstückchen *npl*
cuvet *s.* cuvette
cuvette Küvette *f*
c. well adapter Küvettenwechselvorrichtung *f (eines Fotometers)*

c.v. *s.* calorific value
CV *s.* continuous vulcanization
CV unit CV-Anlage *f (zur Kabelherstellung)*
CWP, C.W.P. *s.* chemical wood-pulp
cyanamide H_2NCN Zyanamid *n*
c. process Kalkstickstoffverfahren *n*, Zyanamidverfahren *n (zur Ammoniakgewinnung)*
cyanate $M^I OCN$ Zyanat *n*
cyanhydrin $R' \cdot C(OH)(CN) \cdot R''$ Zyanhydrin *n*, α-Hydroxynitril *n*
cyanic acid HOCN Zyansäure *f*
cyanidation *(Met)* Zyanidlaugung *f*
c. vat *(Met)* Pachuca-Tank *m (bei der Zyanidlaugung zur Gold- und Silbergewinnung)*
cyanide / to zyanieren, im Zyan[salz]bad härten
cyanide $M^I CN$ Zyanid *n*
c. bath Zyanid[salz]bad *n*, Zyan[salz]bad *n*, Zyansalzschmelze *f*, zyan[id]haltiges Salzbad *n*
c. [case] hardening Zyanbadhärten *n*, Zyanbadhärtung *f*, Zyanieren *n*
c. salt bath *s.* c. bath
cyaniding *s.* 1. cyanide hardening; 2. cyanidation
cyanin Zyanin *n*, Zyanidin-3,5-diglykosid *n*
cyanine 1. Zyanin *n*, Zyaninfarbstoff *m*; 2. Zyaninblau *n*, Chinolinblau *n*, 1,1'-Diäthyl-4,4'-zyaninjodid *n*
c. blue *s.* cyanine 2.
c. dye Zyaninfarbstoff *m*, Zyanin *n*
cyanit[e] *(Min)* Zyanit *m*, Kyanit *m*, Disthen *m (Aluminiumoxidorthosilikat)*
cyano group CN— Zyangruppe *f*, Zyanrest *m*
c. paper Blaupauspapier *n*
c. radical CN— Zyangruppe *f*, Zyanrest *m*, Zyanradikal *n*; [freies] Zyanradikal *n*
c. silicone rubber Nitrilsilikongummi *m*
cyanoaurate(I) $M^I[Au(CN)_2]$ Zyanoaurat(I) *n*, Dizyanoaurat(I) *n*
cyanoaurate(III) $M^I[Au(CN)_4]$ Zyanoaurat(III) *n*, Tetrazyanoaurat(III) *n*
cyanoaurite $M^I[Au(CN)_2]$ Zyanoaurat(I) *n*, Dizyanoaurat(I) *n*
cyanocadmate Zyanokadmat *n*
cyanochromate Zyanochromat *n*
cyanocobaltate Zyanokobaltat *n*
cyanocuprate Zyanokuprat *n*
cyanoethylated cotton zyanäthylierte Baumwolle *f*
cyanoethylation Zyanäthylierung *f*
cyanoferrate(II) $M^I_4[Fe(CN)_6]$ Zyanoferrat(II) *n*, Hexazyanoferrat(II) *n*
cyanoferrate(III) $M^I_3[Fe(CN)_6]$ Zyanoferrat(III) *n*, Hexazyanoferrat(III) *n*
cyanoferrite $M^I_4[Fe(CN)_6]$ Zyanoferrat(II) *n*, Hexazyanoferrat(II) *n*
cyanogen 1. CN— Zyanradikal *n*; 2. $(CN)_2$ Dizyan *n*, Zyan *n*
c. chloride ClCN Zyanchlorid *n*, Chlorzyan *n*, Chlorinzyanid *n*, Zyanogenchlorid *n*
c. gas *s.* cyanogen 2.
c. iodide JCN Zyanjodid *n*, Jodzyan *n*, Jodzyanid *n*, Zyanogenjodid *n*

cyanogenetic glycoside Blausäureglykosid n
cyanohydrin R' · C(OH)(CN) · R″ Zyanhydrin n, α-Hydroxynitril n
cyanoiridate Zyanoiridat n
cyanomanganate Zyanomanganat n
cyanomercurate Zyanomerkurat n
cyanomolybdate Zyanomolybdat n
cyanonickelate Zyanoniccolat n
cyanoosmate Zyanoosmat n
cyanoosmite $M^I_4[Os(CN)_6]$ Zyanoosmat(II) n, Hexazyanoosmat(II) n
cyanopalladate Zyanopalladat n
cyanophoric glycoside Blausäureglykosid n
cyanoplatinate(II) $M^I_2[Pt(CN)_4]$ Zyanoplatinat(II) n, Tetrazyanoplatinat(II) n
cyanoplatinate(IV) $M^I_4[Pt(CN)_6]$ Zyanoplatinat(IV) n, Hexazyanoplatinat(IV) n
cyanoplatinite $M^I_2[Pt(CN)_4]$ Zyanoplatinat(II) n, Tetrazyanoplatinat(II) n
cyanorhodate Zyanorhodat n
cyanoruthenate Zyanoruthenat n
cyanosis Zyanose f (Blaufärbung der Haut durch CO_2-Überladung des Blutes)
cyanosite (Min) Cyanosit m, (veraltet für) Chalkanthit m (Kupfer(II)-sulfat-5-Wasser)
cyanotungstate(IV) $M^I_4[W(CN)_8]$ Zyanowolframat(IV) n, Oktazyanowolframat(IV) n
cyanotype Zyanotypie f, Blaudruck m, Blaupause f
cyanovanadate Zyanovanadat n
cyanozincate Zyanozinkat n
cyanuric acid Zyanursäure f, 2,4,6-Trioxohexahydro-1,3,5-triazin n
  c. chloride $C_3Cl_3N_3$ Zyanurchlorid n, Zyanursäurechlorid n, Trichlor-sym-triazin n
cybotactic (phys Ch) zybotaktisch
  c. group zybotaktische Gruppe f (Gebiet geordneter Molekularstruktur)
cybotaxis zybotaktische Struktur f, zybotaktischer Zustand m
cyclane Zykloalkan n, Zykloparaffin n, Naphthen n
cycle Kreis m; Kreislauf m, Zyklus m, Kreisprozeß m; (Erdöl) zyklische Verbindungen fpl (Kohlenwasserstoffe mpl)
  c. oil Kreislauföl n, Umlauföl n
cyclic zyklisch, ringförmig, Zyklo..., Ring...
  c. aliphatic hydrocarbon zykloaliphatischer (alizyklischer) Kohlenwasserstoff m
  c. alkane Zykloalkan n, Zykloparaffin n, Naphthen n
  c. alkene Zykloalken n, Zykloolefin n
  c. compound zyklische (ringförmige) Verbindung f, Ringverbindung f, Ring m
  c. ester zyklischer Ester m
  c. ether zyklischer Äther m
  c. hydrocarbon zyklischer (ringförmiger) Kohlenwasserstoff m, Zyklokohlenwasserstoff m, Ringkohlenwasserstoff m
  c. hydrocarbon with side chains Ring-Ketten-Kombination f

c. process Kreisprozeß m, Kreislauf m
  c. siloxane zyklisches (ringförmiges) Siloxan n, Zyklosiloxan n
cycling Kreislaufführung f
cyclitol Zyklit m (isozyklischer Polyalkohol)
cyclization Zyklisierung f, Ringschließung f, Ringschluß m, Ringbildung f; (Gum) Zyklisierung f, Molekülverkettung f, Vernetzung f
  c. reaction Zyklisierungsreaktion f, Ringschlußreaktion f; (Gum) Vernetzungsreaktion f, Molekülverkettungsreaktion f
cyclize /to zyklisieren, einem Ringschluß unterwerfen, „ringschließen"; sich zum Ring schließen
cyclized latex Zyklokautschuklatex m
  c. rubber zyklisierter Kautschuk m, Zyklokautschuk m
cycloalkane Zykloalkan n, Zykloparaffin n, Naphthen n
cycloalkyne Zykloalkin n
cyclobutane Zyklobutan n, Tetramethylen n
cyclodecane Zyklodekan n
cyclodecapeptide zyklisches Dekapeptid n, Zyklodekapeptid n
cyclodehydration Zyklodehydratisierung f
cyclodepsipeptide Zyklodepsipeptid n
cycloheptane Zykloheptan n, Heptamethylen n, Suberan n
1,3-cyclohexadiene 1,3-Zyklohexadien n, 1,2-Dihydrobenzol n
1,4-cyclohexadiene 1,4-Zyklohexadien n, 1,4-Dihydrobenzol n
1,4-cyclohexadienedione $C_6H_4O_2$ 1,4-Zyklohexadiendion n, p-Chinon n, Chinon n, p-Benzochinon n
cyclohexadienylidene $C_6H_6=$ Zyklohexadienyliden n, Phenyliden n
cyclohexane Zyklohexan n, Hexahydrobenzol n, Hexamethylen n
  c. ring Zyklohexanring m
cyclohexane-1,2-dicarboxylic acid 1,2-Zyklohexandikarbonsäure f, Hexahydrophthalsäure f
cyclohexanol Zyklohexanol n, Hexahydrophenol n
cyclohexanone peroxide Zyklohexanonperoxid n
cyclohexene Zyklohexen n, 1,2,3,4-Tetrahydrobenzol n
cycloheximide Zykloheximid n
cyclohexylmethane $(C_6H_{11})CH_3$ Zyklohexylmethan n, Methylzyklohexan n, Hexahydrotoluol n, Heptanaphthen n
cycloidal blower Roots-Gebläse n, Wälzkolbengebläse n
cyclone s. c. collector
  c. air separator Aerozyklon m
  c. classifier Klassierzyklon m
  c. collector Zyklon[ab]scheider m, Wirbel[ab]scheider m, Abscheidezyklon m, Zyklon m
  c. dust collector Staub[abscheide]zyklon m, Zyklonentstauber m
  c. evaporator Zyklon-Oberflächenverdampfer m
  c. firing Wirbelfeuerung f; Zyklonfeuerung f

c. separation Zyklonieren *n*, Fliehkraftabscheiden *n*
c. separator *s.* c. collector
c. washer Waschzyklon *m*, Zyklonwäscher *m*
cyclonic scrubber *s.* cyclone washer
c. separator *s.* cyclone collector
cyclononane Zyklononan *n*
cyclooctane Zyklooktan *n*
cycloparaffin Zykloparaffin *n*, Zykloalkan *n*, Naphthen *n*
cyclopentane Zyklopentan *n*, Pentamethylen *n*
cyclopropane Zyklopropan *n*, Trimethylen *n*
c. ring Zyklopropanring *m*, Trimethylenring *m*
cyclopropanedicarboxylic acid Zyklopropandikarbonsäure *f*
cyclorubber Zyklokautschuk *m*, zyklisierter Kautschuk *m*
cyclosiloxane Zyklosiloxan *n*, zyklisches (ringförmiges) Siloxan *n*
cyclotron Zyklotron *n*
cycloversion process Cycloversion-Verfahren *n* *(katalytisches Reformierungsverfahren)*
cylinder Zylinder *m*, Walze *f*, Trommel *f*; Druckgasflasche *f*, Bombe *f*; *(Pap)* Holländerwalze *f*, Messerwalze *f*, Mahlwalze *f*; *(Plast)* Spritzgehäuse *n* *(einer Spritzmaschine)*; *(Gum)* Vulkanisiertrommel *f* *(einer Vulkanisiermaschine)*
c. board Handpappe *f*
c. board machine Pappenrundsiebmaschine *f*, Rundsiebkartonmaschine *f*
c. cover *(Pap)* Siebmantel *m*, Siebüberzug *m* *(des Rundsiebzylinders)*
c.-dried *(Pap)* maschinengetrocknet, auf der Maschine getrocknet, maschinentrocken
c. dryer Zylindertrockner *m*, Trockenzylinder *m*, Walzentrockner *m*, Trockenwalze *f*, Trommeltrockner *m*, Trockentrommel *f*
c. drying machine *s.* c. dryer
c. lubricating oil Zylinderöl *n*
c. machine *s.* c. paper machine
c. machine-made paper *s.* c.-made deckle-edge paper
c.-made deckle-edge paper Maschinenbütten[papier] *n*, Rundsieb-Büttenpapier *n*
c. mould *(Pap)* Siebzylinder *m*, Rundsiebzylinder *m*
c. oil Zylinderöl *n*
c. paper machine Rundsieb[papier]maschine *f*
c. part Siebpartie *f* der Rundsiebpapiermaschine
c. process Zylinderverfahren *n* *(Fensterglas- und Spezialglasherstellung)*
c. sizing machine *(Text)* Zylinderschlichtmaschine *f*, Trommelschlichtmaschine *f*
c. stock Zylinder[öl]stock *m*
c. vat *(Pap)* Siebtrog *m*, Stofftrog *m*, Rundsiebbütte *f* *(der Rundsiebpapiermaschine)*
c. vat machine *s.* c. paper machine
c. washer *(Am)* Waschtrommel *f*
cylindrical zylindrisch, zylinderförmig, Zylinder...
c. batch pasteurizer *(Lebm)* Trommelerhitzer *m*, Kreiselerhitzer *m*

c.-screen centrifuge Siebzentrifuge *f* mit zylindrischer Siebtrommel
cylpebs Cylpeps *pl* *(zylindrische Mahlkörper in Kugelmühlen)*
cymene Zymol *n*
cysteine Zystein *n*, Thioserin *n*, α-Amino-β-merkaptopropionsäure *f*
cysteinic acid Zysteinsäure *f*
cystine Zystin *n*
c. link Zystin-Bindeglied *n*, Zystin-Brücke *f*
cytase Zytase *f*, Hemizellulase *f* *(Hemizellulose spaltendes Ferment)*
cytisine Zytisin *n*, Babtitoxin *n*, Sophorin *n*, Ulexin *n*
cytochemistry Zytochemie *f*
cytochrome Zytochrom *n*
c. oxidase Zytochromoxydase *f*, Warburgsches Atmungsferment *n*, Zytochrom a *n*, Eisenoxygenase *f*
cytoplasm Zytoplasma *n*
cytotoxic zytotoxisch, zellschädigend
cytotoxin Zytotoxin *n*, Zellgift *n*
Czapek-Dox medium Czapek-Dox-Medium *n*, Czapek-Dox-Nährboden *m*, Czapek-Dox-Nährlösung *f*, Czapek-Dox-Lösung *f*

D

D *s.* 1. difunctional; 2. Debye; 3. density
2,4-D 2,4-D *n*, 2,4-Dichlorphenoxyessigsäure *f* *(Herbizid)*
d-form d-Form *f*, (+)-Form *f*, rechtsdrehende Form *f*, Rechtsform *f*
D-line of sodium Natrium-D-Linie *f*, Na-D-Linie *f*
D unit *s.* difunctional unit
dacite *(Geol)* Dazit *m* *(ein Ergußgestein)*
d.a.f. *s.* dry, ash-free
Dahomey rubber *(Kautschuk von Ficus vogeli Miq. Dob.)*
daily intake [dose] *(Toxikologie)* Tagesdosis *f*
dairy milchwirtschaftlich
dairy Molkerei *f*, Molkereibetrieb *m*
d. butter Molkereibutter *f*
d. chemistry Milchchemie *f*, Chemie *f* der Milch [und Milchprodukte], milchwirtschaftliche Chemie *f*
d. foods *s.* d. products
d. industry Milchindustrie *f*, Milchwirtschaft *f*
d. products Milchprodukte *npl*, Molkereiprodukte *npl*, Molkereierzeugnisse *npl*
dairying Milchwirtschaft *f*, Molkereiwesen *n*
Dakin reaction Dakin-Reaktion *f* *(Phenolaldehydoxydation zu Polyphenolen)*
Dakin-West reaction Dakin-West-Reaktion *f* *(α-Azylaminoketon-Synthese)*
dalapon Dalapon *n* *(Herbizid; 2,2-Dichlorpropionsäure)*
Dalton law Daltonsches (Daltons) Gesetz *n*

daltonide compounds s. daltonides
daltonides Daltonide npl, daltonide (stöchiometrisch zusammengesetzte) Verbindungen fpl
Dalton's law of partial pressures Daltonsches (Daltons) Gesetz n der Partialdrücke (Teildrücke), Daltonsches Partialdruckgesetz n
dam (Pap) Staubrett n (Holzschliffherstellung)
   d. retting Wasserröste f in stehenden Gewässern (Flachsverarbeitung)
damage factor Schädigungsfaktor m
   d. to fibres Faserschaden m, Faserschädigung f
damar s. dammar
Damask finish (Pap) (Am) Leinenprägung f
dammar Dammar[harz] n
   d. batu (Dammarharz von Hopea maranti Miq.)
   d. hiroe (Dammarharz von Vatica papuana Dyer)
   d. kedemut (Dammarharz von Hopea fagifolia Miq.)
   d. kumus (Dammarharz von Shorea glauca King)
   d. penak (Dammarharz von Balanocarpus heimi King)
   d. resin s. dammar
   d. sengal (Dammarharz von Canarium hirsutum Willd.)
   d. temak (Dammarharz von Shorea hypochra Hance)
   d. tenang (Dammarharz von Shorea koordersii Brandis)
damp / to anfeuchten, befeuchten, feucht (naß) machen, benetzen; dämpfen
damp feucht, naß, dumpfig
damp Feuchtigkeit f, Dunst m; Schwaden m
   d. product Feuchtgut n
   d. warm feuchtwarm
damped balance Dämpfungswaage f, Waage (Analysenwaage) f mit Luftdämpfung (Dämpfungseinrichtung)
dampen / to s. to damp
dampening s. damping
damper (Pap) Feuchtapparat m, Feuchtwalze f, Feuchter m; (Ker) Schieber m (im Brennofen)
damping Anfeuchten n, Befeuchten n, Feuchtung f, Benetzen n, Netzen n; Dämpfen n
   d. device Dämpfungseinrichtung f (z.B. an Feinwaagen)
   d. factor Dämpfungsfaktor m
   d. fluid Dämpfungsflüssigkeit f
   d. machine Befeuchtungsmaschine f, Einsprengmaschine f
   d. medium Dämpfungsmittel n, Dämpfungsmedium n, Dämpfungsmaterial n
   d. properties Dämpfungsverhalten n
   d. roll (Pap) Feuchtwalze f
dampness Feuchte f, Feuchtigkeit f, Nässe f
danaite (Min) Danait m
danalite (Min) Danalith m
danburite (Min) Danburit m (Kalziumdiborodisilikat)
dandy s. d. roll
   d. mark durch den Egoutteur hervorgerufener Papierfehler m

   d. roll (Pap) Vordruckwalze f, Vorpreßwalze f, Wasserzeichenwalze f, Egoutteur m
   d. roll watermark (Pap) Egoutteur[wasser]zeichen n
dangerous to health gesundheitsgefährdend, gesundheitsschädlich, gesundheitsschädigend
Daniell cell Daniell-Element n, Daniell-Kette f
Danner process Danner-Verfahren n (Glasröhrenziehverfahren)
dark dunkel
   d. antimony (Min) Antimonit m, Stibnit m, Antimonglanz m, Grauspießglanz m (Antimon(III)-sulfid)
   d. beer dunkles Bier n, Dunkles n
   d. catechu Braunes Katechu n, Pegukatechu n, Bombaykatechu n (Extrakt aus Acacia catechu Willd.)
   c.-coloured dunkel gefärbt
   d. crottle (Farb) Rindenflechte f, Schlüsselflechte f, Parmelia physodes (L.)Ach.
   d. crystal Rohzucker m
   d. current Dunkelstrom m
   d. factice (Gum) brauner Faktis m
   d.-field illumination s. d.-ground illumination
   d.-field microscope Dunkelfeldmikroskop n
   d.-ground illumination Dunkelfeldbeleuchtung f
   d. malt dunkles Malz n, Dunkelmalz n
   d. mica (Min) Biotit m
   d. reaction Dunkelreaktion f
   d. red silver [ore] (Min) dunkles Rotgültigerz n, Pyrargyrit m (Antimon(III)-silbersulfid)
   d. space Dunkelraum m
   d. speck (Plast) dunkler Punkt m (z.B. in Härtgewebe)
darken / to dunkel färben; sich dunkel (dunkler) färben, dunkel (dunkler) werden, nachdunkeln, (Foto auch) sich schwärzen
darkening Dunkelfärbung f, Dunkelwerden n, Nachdunkeln n, (Foto auch) Schwärzung f
   d. of wine Braunwerden n, Rahnwerden n (ein Weinfehler)
darkroom Dunkelkammer f
   d. illumination Dunkelkammerbeleuchtung f
D'Arsonval galvanometer Drehspulgalvanometer n von D'Arsonval
Darzens [glycidic ester] condensation Darzens-Erlenmeyer-Claisen-Kondensation f, Darzens-Reaktion f (Glyzidesterkondensation)
dash Bindungsstrich m
   d. pot Stoßdämpfer m, Dämpfungszylinder m
   d. valence formula Valenzstrichformel f
dasymeter Dasymeter n, Gaswaage f
date of manufacture Herstellungsdatum n
dative bond s. d. covalence
   d. covalence (covalency) 1. koordinative Wertigkeit f, Koordinationszahl f, Zähligkeit f; 2. dative (koordinative) Bindung f, semipolare (halbpolare) Doppelbindung f, Donator-Akzeptor-Bindung f
datolite (Min) Datolith m (Kalziumhydroxidborosilikat)

dawsonit *(Min)* Dawsonit *m (Natriumaluminium-dihydroxidkarbonat)*

28-day strength 28-Tage-Druckfestigkeit *f (eines Betons)*

d. **tank** *(Glas)* Tageswanne *f*, periodische Wanne *f*

Day pinchcock Quetschhahn *m* nach Day

daylight [lichte] Einbauhöhe *f*; *(Gum)* Etagenhöhe *f (einer Vulkanisierpresse)*

d. **curing press** *(Gum)* Etagen[vulkanisier]presse *f*

d. **opening** [lichte] Einbauhöhe *f*

d. **paper** *(Foto)* Tageslichtpapier *n*

d. **press** Etagenpresse *f*, *(Gum auch)* Etagen-vulkanisierpresse *f*

2,4-DB 2,4-DB *n*, 2,4-Dichlorphenoxybuttersäure *f (Herbizid)*

DBP *s.* dibutyl phthalate

d.c., D.C. *s.* direct current

d.c. arc *s.* direct-current arc

d.c. polarograph *s.* direct-current polarograph

DCA *s.* deoxycorticosterone acetate

DD *(Gemisch aus Dichlorpropen und Dichlorpropan) (Schädlingsbekämpfungsmittel)*

D.D.N.P. *s.* diazodinitrophenol

DDP *s.* didecyl phthalate

DDT, D.D.T. $(ClC_6H_4)_2CHCCl_3$ DDT *n*, Dichlordiphenyltrichloräthan *n (Kontaktinsektizid)*

De Broglie equation De-Broglie-Gleichung *f*

De Broglie relationship De-Broglie-Beziehung *f*

De Broglie waves De-Broglie-Wellen *fpl*

De Mattia [flexing] machine De-Mattia-Biegeprüfmaschine *f*, De-Mattia-Knickermüdungsprüfer *m*

De Nora cell De-Nora-Zelle *f (eine Quecksilberzelle)*

deacetylate / to entazetylieren

deacetylation Entazetylieren *n*, Entazetylierung *f*

deacidification Entsäuern *n*, Entsäuerung *f*; Neutralisation *f* einer Säure

deacidify / to entsäuern; eine Säure neutralisieren

deacidifying *s.* deacidification

Deacon process Deacon-Prozeß *m*, Deacon-Verfahren *n*

deactivate / to desaktivieren, entaktivieren, inaktivieren

deactivating group Substituent *m* mit desaktivierender Wirkung

deactivation Desaktivierung *f*, Entaktivierung *f*, Inaktivierung *f*

d. **period** Desaktivierungsperiode *f*, Desaktivierungszeit *f*

d. **process** Desaktivierungsprozeß *m*, Entaktivierungsprozeß *m*

dead abgestanden, schal, fad[e] *(Getränke)*; matt, glanzlos, stumpf *(Farben)*; *(Et)* stromlos; spannungslos; *(Min)* taub

d. **air** *(Bgb)* matte Wetter *pl*

d.-**beaten** *(Pap)* totgemahlen

d.-**burn / to** totbrennen, totrösten

d.-**burned dolomite** Sinterdolomit *m*

d. **cotton** tote (unreife) Baumwolle *f*

d. **dyeing** tote (leere, stumpfe, matte) Färbung *f*

d. **finish (gloss)** Mattglanz *m*

d. **knife** *(Pap)* feststehendes Messer *n*

d. **lime** *(Gerb)* Stinkäscher *m*, Fauläscher *m*, fauler (toter) Äscher *m*

d. **lustre** *s.* d. finish

d.-**milled** *(Gum)* totgewalzt, übermastiziert, totmastiziert

d.-**milled rubber** *s.* d. rubber

d. **milling** *(Gum)* Totwalzen *n*, Übermastizieren *n*, Totmastizieren *n*

d. **plate** *(Pap)* Grundwerk *n*, Messerwerk *n*, Messerblock *m (eines Holländers)*; *(Glas)* Absetzplatte *f*

d.-**plate knife** *(Pap)* Grundwerksmesser *n (beim Holländer)*

d.[-**rolled] rubber** totgewalzter (übermastizierter, totmastizierter) Kautschuk *m*

d. **space** *s.* d. volume

d. **spot** Totraum *m (im Extruder)*

d.-**stop method** Dead-stop-Methode *f (Amperometrie)*

d.-**stop titration** Dead-stop-Titration *f*, Stillstandstitration *f*

d.-**stop titration curve** Dead-stop-Titrationskurve *f*

d. **volume** *(phys Ch)* Totvolumen *n*, Totraum *m*, toter (schädlicher) Raum *m*

deaden / to abstumpfen, mattieren

deadly nightshade Tollkirsche *f*, Atropa belladonna L.

deaerate / to entlüften; entgasen

deaerating *s.* deaeration

deaeration Entlüften *n*, Entlüftung *f*; Entgasen *n*, Entgasung *f*

d. **of paper stock** *(Pap)* Stoffentlüftung *f*

deaerator Entlüfter *m*, Entlüftungsapparat *m*, Entlüftungseinrichtung *f*, Entlüftungsanlage *f*; Entgaser *m*, Entgasungsanlage *f*

de-air / to *(Ker)* entlüften

de-aired *(Ker)* entlüftet, luftfrei

de-airing *(Ker)* Entlüften *n*, Entlüftung *f*

d. **auger [machine]** *s.* d. pug mill

d. **chamber** *(Ker)* Entlüftungskammer *f (einer Vakuumstrangpresse)*

d. **machine** *s.* d. pug mill

d. **mixer** *(Ker)* Vakuummischer *m*

d. **pug mill** *(Ker)* Vakuumstrangpresse *f*, Vakuumtonschneider *m*, Vakuumpresse *f* zur Masseentlüftung

dealginate / to Alginatseide aus Grundgewebe herauslösen

dealgination Herauslösen *n* der Alginatseide aus Grundgewebe

dealkalization Entbasung *f*

dealkalize / to entbasen

dealkylate / to dealkylieren, entalkylieren

dealkylation Dealkylieren *n*, Dealkylierung *f*, Entalkylierung *f*

deamidate / **to** desamidieren, entamidieren
deamidation Desamidierung f, Entamidierung f
deaminase Desaminase f (hydrolytische Desaminierung bewirkendes Ferment)
deaminate / **to** desaminieren
deamination Desaminierung f
deash / **to** entaschen
deasphalt / **to** entasphaltieren, asphaltartige Anteile ausfällen
deasphalting Entasphaltieren n, Entasphaltierung f, Ausfällen n asphaltartiger Anteile
de-bagging [operation] (Gum) Herausnehmen (Ausziehen) n des Heizschlauchs
debark / **to** (Pap) entrinden, schälen
debarking (Pap) Entrinden n, Entrindung f, Schälen n, Schälung f
debeader (Gum) Entwulster m, Wulstschneidemaschine f
debeading (Gum) Entwulsten n
d. machine s. debeader
debenzolization s. debenzolizing
debenzolize / **to** entbenzol[ier]en, Benzol abscheiden
debenzolizing Entbenzol[ier]en n, Entbenzolierung f
debiteuse (Glas) Ziehdüse f, Schwimmdüse f (beim Fourcault-Verfahren)
debitter / **to** entbittern
debittering Entbittern n
debitterize / **to** s. to debitter
debris Trümmer pl, Bruchstücke npl, Reste mpl, Relikte npl
debrominate / **to** entbromen, Brom abspalten
debromination Entbromen n, Bromabspaltung f
deburr / **to** (Plast) entgraten, abgraten
deburring (Plast) Entgraten n, Abgraten n
debutanization Entbutanisieren n, Entbutanisierung f, Debutanisierung f, Butanabtrennung f
debutanize / **to** entbutanisieren, debutanisieren
debutanized gasoline butanfreies Benzin n
debutanizer Entbutaner m, Entbutanisierkolonne f, Butantrennkolonne f, Debutanisator m, Debutanisierungskolonne f
Debye Debye-Einheit f, Debye n, D (Maßeinheit des Dipolmoments)
Debye equation Debyesche Gleichung (Formel) f
Debye-Falkenhagen effect Debye-Falkenhagen-Effekt m, Dispersionseffekt m der Leitfähigkeit
Debye function Debye-Funktion f, Debyesche Funktion f
Debye-Hückel-Onsager equation Debye-Hückel-Onsagersche Gleichung f
Debye-Hückel theory Debye-Hückel-Theorie f
Debye length Debye-Länge f
Debye relation Debyesche Beziehung f
Debye-Scherrer diagram Debye-Scherrer-Diagramm n
Debye-Scherrer-Hull method Debye-Scherrer-Verfahren n, Debye-Scherrer-Methode f, Pulvermethode f, Pulververfahren n

Debye-Scherrer-X-ray method s. Debye-Scherrer-Hull method
Debye T^3 law Debyesches T^3-Gesetz n
Debye theory of specific heat Debyesche Theorie f der spezifischen Wärme
Debye unit s. Debye
decaborane $B_{10}H_{14}$ Dekaboran n
decaffeinate / **to** koffeinfrei (i.w.S. koffeinarm) machen, Koffein entziehen
decaffeinated koffeinfrei, (i.w.S.) koffeinarm
decaffeinize / **to** s. to decaffeinate
decahydrate Dekahydrat n, 10-Hydrat n
decahydrated ...dekahydrat n, ...10-Wasser n
decal s. decalcomania
decalcification Entkalken n, Entkalkung f; (Med) Kalziumentzug m
decalcify / **to** entkalken; (Med) Kalzium entziehen
decalcomania (Ker) Dekorieren n (Dekoration f) mit Abziehbildern; Abziehbild n
d. paper Abziehbilderpapier n
decane $C_{10}H_{22}$ Dekan n
decane-1,10-dicarboxylic acid $HOCO(CH_2)_{10}COOH$ Dekan-1,10-dikarbonsäure f, Dodekandisäure f
decanedioic acid $HOOC(CH_2)_8COOH$ Dekandisäure f, Oktandikarbonsäure-(1,8) f, Sebazinsäure f, Sebazylsäure f
decanoic acid $CH_3(CH_2)_8COOH$ n-Dekansäure f, Dekansäure-(1) f, n-Dezylsäure f, n-Kaprinsäure f
decant / **to** dekantieren, [vorsichtig] abgießen; abfüllen, umfüllen
decantate dekantierte Flüssigkeit f
decantation Dekantieren n, Dekantierung f, Dekantation f, [vorsichtiges] Abgießen (Abfließenlassen) n; Abfüllen n, Umfüllen n
decanter Dekantiergefäß n, Dekantiertopf m, Dekantierapparat m, Dekanter m, Dekanteur m
decanting s. decantation
d. jar s. decanter
decapeptide Dekapeptid n (Peptid mit 10 Aminosäureeinheiten)
decarbonate / **to** CO_2 oder H_2CO_3 austreiben (entfernen)
decarbonation Austreiben (Entfernen) n von CO_2 oder H_2CO_3
decarbonization Entkohlen n, Entkohlung f, Dekarbonisieren n, Dekarbonisierung f; Entkarbonisieren n (Entfernen der Karbonathärte des Wassers)
d. plant Entkarbonisierungsanlage f
decarbonize / **to** entkohlen, dekarbonisieren; (Wasser) entkarbonisieren
decarbonizing s. decarbonization
decarboxylase Dekarboxylase f, Karbolyase f (zu den Lyasen gehörendes Ferment)
decarboxylate / **to** dekarboxylieren
decarboxylation Dekarboxylieren n, Dekarboxylierung f
decarburization (Met) Entkohlen n, Entkohlung f

decarburize / to *(Met)* entkohlen
decate / to *s.* to decatize
decating *s.* decatizing
decatize / to *(Text)* dekatieren
  d. with dry steam trocken dekatieren
decatizing *(Text)* Dekatieren *n*, Dekatur *f*
decay / to abnehmen, geringer werden, schwinden; abklingen, ausschwingen; absterben, verfallen, zugrunde gehen, zerstört werden; verwesen, sich zersetzen; verderben *(z.B. Lebensmittel)*; *(Kch)* zerfallen; *(Geol)* verwittern
decay Abnahme *f*; Abklingen *n*; Verfall *m*; Verwesung *f*, Zersetzung *f*; Verderben *n* *(der Lebensmittel)*; *(Kch)* Zerfallen *n*, Zerfall *m*; *(Geol)* Verwittern *n*, Verwitterung *f*
  d. coefficient (constant) *(Kch)* Zerfallskonstante *f*
  d. curve *(Kch)* Zerfallskurve *f*
  d. electron Zerfallselektron *n*
  d. energy *(Kch)* Zerfallsenergie *f*
  d. factor *s.* d. coefficient
  d. law *(Kch)* Zerfallsgesetz *n*
  d. luminescence *(Kch)* Zerfallsleuchten *n*
  d. particle *(Kch)* Zerfallsteilchen *n*
  d. period *(Kch)* Zerfallszeit *f*
  d. probability *(Kch)* Zerfallswahrscheinlichkeit *f*
  d. product *(Kch)* Zerfallsprodukt *n*
  d. rate *(Kch)* Zerfallsgeschwindigkeit *f*
  d. resistance Fäulniswidrigkeit *f*, Fäulnisbeständigkeit *f*
  d. series *(Kch)* Zerfallsreihe *f*
  d. time *s.* d. period
decaying *(Kch)* zerfallend, instabil, radioaktiv
decelerate / to verzögern, verlangsamen, bremsen; sich verzögern (verlangsamen)
deceleration Verzögerung *f*, Verlangsamung *f*, Abbremsung *f*, negative Beschleunigung *f*
1-decene $CH_2{=}CH(CH_2)_7CH_3$ 1-Dezen *n*, n-Dezylen *n*
1-decene-1,10-dicarboxylic acid $C_{10}H_{18}(COOH)_2$ 1-Dezen-1,10-dikarbonsäure *f*, 2-Dodezendisäure *f*, Traumatinsäure *f*
dechlorinate / to entchloren
dechlorinating *s.* dechlorination
  d. agent Entchlorungsmittel *n*
dechlorination Entchloren *n*, Entchlorung *f*
deciduous wood Laubholz *n*
decimal balance Dezimalwaage *f*
1-decine *s.* 1-decyne
5-decine *s.* 5-decyne
decinormal solution zehntelnormale Lösung *f*, Zehntelnormallösung *f*, 1/10-Normallösung *f*, n/10-Lösung *f*
deck Deck *m*, Herdplatte *f*, Herdtafel *f* *(eines Sortierherdes)*; Boden *m* *(z.B. eines Siebes)*; Rost *m* *(eines Siebes)*
deckel *s.* deckle
decker / to *(Pap)* eindicken, verdicken, entwässern
decker *(Pap)* Eindicker *m*, Eindickapparat *m*, Eindickzylinder *m*, Eindickmaschine *f*, Entwässerungsmaschine *f*

deckering *(Pap)* Eindicken *n*, Eindickung *f*, Verdicken *n*, Entwässern *n*
deckle *(Pap)* 1. Schöpfrahmen *m*, Auflaufrahmen *m*; 2. Deckelriemen *m*; 3. Deckelrahmen *m*; 4. [echter] Büttenrand *m*, Schöpfrand *m*, *(i.e.S.)* Handbüttenrand *m*; künstlicher (imitierter) Büttenrand *m*; 5. Maschinenbreite *f*, Arbeitsbreite *f*
  d. board *(Pap)* *(unbewegliche Begrenzungslineale zur Formatbegrenzung)*
  d. edge *s.* deckle 4.
  d. frame *(Pap)* Deckelrahmen *m*
  d. pulley *(Pap)* Deckelriemenrolle *f*, Deckelriemenführungsrad *n*, Deckelleitrad *n*
  d. strap *(Pap)* Deckelriemen *m*
deckled *(Pap)* mit Büttenrand (Schöpfrand)
decoct / to abkochen, absieden, auskochen; digerieren, *(mit heißem Lösungsmittel)* ausziehen
decoction Abkochen *n*, Absieden *n*, Auskochen *n*; Abkochung *f*, Dekokt *n*, Absud *m*; Aufguß *m*
  d. mash Kochmaische *f*
  d. mashing (method) *s.* d. process
  d. process Dekoktionsverfahren *n*, Kochverfahren *n*, Verfahren *n* mit Maischekochung
decoctum Dekokt *n*, Abkochung *f*, Absud *m*
decoic acid *s.* decanoic acid
decoke / to entkohlen, dekarbonisieren
decoking Entkohlen *n*, Entkohlung *f*, Dekarbonisieren *n*, Dekarbonisierung *f*
decolor / to *s.* to decolourize
decolorant *s.* decolourant
decoloration *s.* decolourization
decolorise / to *s.* to decolourize
decolorization *s.* decolourization
decolorize / to *s.* to decolourize
decolorizer *s.* decolourizer
decolour / to *s.* to decolourize
decolourant entfärbend, bleichend
decolourant Entfärber *m*, Entfärbungsmittel *n*, Bleichmittel *n*
decolouration *s.* decolourization
decolourization Entfärben *n*, Entfärbung *f*, Bleichen *n*
decolourize / to entfärben, bleichen; Farbe abziehen (ablösen); sich entfärben
decolourizer Entfärber *m*, Entfärbungsmittel *n*, Bleichmittel *n*; *(Farb)* Abzieh[hilfs]mittel *n*
decolourizing Entfärben *n*, Entfärbung *f*, Bleichen *n*; Abziehen (Ablösen) *n* der Farbe
  d. agent *s.* decolourizer
  d. carbon (charcoal) Entfärbungskohle *f*
  d. earth Entfärbungserde *f*, Bleicherde *f*
decomposable zersetzlich, zersetzbar
decompose / to zersetzen, abbauen; zerlegen, [auf]spalten; sich zersetzen (auflösen), zerfallen; verwesen, verfaulen
decomposer Zersetzerzelle *f*, Zersetzungszelle *f*, Zersetzer *m*, Pile *f*
decomposition Zersetzung *f*, Abbau *m*; Zerlegung *f*, Spaltung *f*, Aufspaltung *f*; Zerfall *m*, Auflösung *f*; Verwesung *f*

d. catalyst Zersetzungskatalysator *m*
d. of nitramide Nitramidzerfall *m*
d. phenomenon Zersetzungserscheinung *f*
d. point Zersetzungspunkt *m*
d. potential Zersetzungspotential *n*
d. product Abbauprodukt *n*, Zersetzungsprodukt *n*; Zerfallsprodukt *n*
d. rate Abbaugeschwindigkeit *f*, Zersetzungsgeschwindigkeit *f*; Zerfallsgeschwindigkeit *f*
d. reaction Abbaureaktion *f*, Zersetzungsreaktion *f*; Zerfallsreaktion *f*
d. voltage Zersetzungsspannung *f*
decontaminate / to *(Kch)* dekontaminieren, entaktivieren, entgiften, *(i.e.S.)* entgasen *(oder)* entseuchen *(oder)* entstrahlen; reinigen, säubern
decontaminating *s.* decontamination
d. agent (chemical, substance) *(Kch)* Entseuchungsmittel *n*, Dekontaminierungsmittel *n*
decontamination *(Kch)* Dekontaminierung *f*, Entaktivierung *f*, Entgiftung *f*, *(i.e.S.)* Entgasung *f* *(oder)* Entseuchung *f* *(oder)* Entstrahlung *f*; Reinigung *f*, Säuberung *f*
d. factor *(Kch)* Entseuchungsfaktor *m*, Entseuchungsgrad *m*; logarithmischer Entstaubungsgrad *m*
d. index *(Kch)* Entseuchungsindex *m*, Dekontaminationsindex *m*
deconvolution count Entwindungszahl *f* *(beim Merzerisieren von Baumwolle)*
decopperate / to *s.* to decopperize
decopperize / to entkupfern
decopperizing Entkupfern *n*, Entkupferung *f*
decorated paper *(Am)* Fantasiepapier *n*; Ausstattungspapier *n*
decorating firing *(Ker)* Dekorbrand *m*
d. kiln *(Ker)* Dekor[brand]ofen *m*
d. paper Dekorationspapier *n*
d. sheeting Dekor[ations]folie *f*
d. tissue paper Dekorationsseidenpapier *n*
decoration firing *(Ker)* Dekorbrand *m*
d. tissue Dekorationsseidenpapier *n*
decorative foil Dekor[ations]folie *f*
d. laminate Dekorationsplatte *f*, Dekorationsschichtstoff *m*
d. sheet *s.* d. foil
d. tile Zierfliese *f*
decorators' size Abziehlack *m*, Klebelack *m*
decorticate / to entrinden; *(z.B. Getreide)* enthülsen, schälen, putzen; *(Text)* entbasten, entholzen
decorticate entrindet; enthülst, geschält, geputzt *(z.B. Getreide)*; *(Text)* entbastet, entholzt
decortication Entrinden *n*, Entrindung *f*; Enthülsen *n*, Schälen *n*, Putzen *n* *(z.B. von Getreide)*; *(Text)* Entbasten *n*, Entholzen *n*
decorticator Schälmaschine *f*, Schälmühle *f*
decrease / to 1. abnehmen, sich vermindern (verringern), kleiner (kürzer, schwächer, geringer) werden; 2. [ab]fallen, [ab]sinken; 3. vermindern, verringern, verkleinern, verkürzen, herabsetzen, reduzieren

decrease Abnahme *f*, Verminderung *f*, Verringerung *f*, Verkleinerung *f*, Erniedrigung *f*, Verkürzung *f*; Fallen *n*, Abfallen *n*, Abfall *m*, Sinken *n*, Absinken *n*; Herabsetzung *f*, Reduzierung *f*
d. in contrast Kontrastverminderung *f*
d. in solubility Löslichkeitserniedrigung *f*
d. in volume Volum[en]abnahme *f*
decrepitate / to *(Krist)* dekrepitieren
decrepitation *(Krist)* Dekrepitieren *n*, Dekrepitation *f*
deculator *(Pap)* Dekulator *m*, Stoffentlüfter *m*
n-decyl alcohol $CH_3(CH_2)_8CH_2OH$ n-Dezylalkohol *m*, 1-Dekanol *n*
n-decylene $CH_2=CH(CH_2)_7CH_3$ 1-Dezen *n*, n-Dezylen *n*
n-decylic acid $CH_3(CH_2)_8COOH$ n-Dezylsäure *f*, Dekansäure-(1) *f*, n-Kaprinsäure *f*
1-decyne $HC\equiv C(CH_2)_7CH_3$ 1-Dezin *n*, n-Oktylazetylen *n*
5-decyne $CH_3(CH_2)_3C\equiv C(CH_2)_3CH_3$ 5-Dezin *n*, Dibutylazetylen *n*
dedolomitization *(Geol)* Dedolomitisation *f*, Dedolomitisierung *f*
dedust / to entstauben
dedusting Entstauben *n*, Entstaubung *f*, Staubabscheidung *f*
deed paper *(Am)*, Aktenpapier *n*, Dokumentenpapier *n*, Kanzleipapier *n*
de-emulsify / to *s.* to demulsify
de-enamelling Entemaillieren *n*
deep cooling Tiefkühlung *f*
d.-cooling plant Tiefkühlanlage *f*, Tieftemperaturanlage *f*
d.-cut screw tiefgeschnittene Schnecke *f*
d. drawing Tiefziehen *n*; Tiefziehteil *n*
d. dyeing tieffärbend, stark anfärbend, mit erhöhter Farbstoffaffinität
d. etching *(Glas)* Tiefätzen *n*, Tiefätzung *f*
d.-freeze / to tiefgefrieren, frosten, durch Kälte konservieren
d. freezing Tiefgefrieren *n*, Frosten *n*, Konservieren *n* durch Kälte
d.-frozen milk Tiefgefriermilch *f*
d. mining Tiefbauförderung *f*, Tiefbaubetrieb *m*, Förderung *f* im Tiefbau (Untertagebau)
d.-sea deposit *(Geol)* pelagische Ablagerung *f*, Tiefseeablagerung *f*
d.-seated rock *(Geol)* Tiefengestein *n*
d. trap *(phys Ch)* tiefe (tiefliegende) Haftstelle *f*, Rekombinationszentrum *n*
dees *(Kph)* Duanten *mpl*, Dees *pl*, D's *(D-förmige Elektroden eines Zyklotrons)*
de-ethanization Entäthanisierung *f*, Deäthanisierung *f*, Äthanabtrennung *f*
de-ethanize / to entäthanisieren, deäthanisieren, Äthan abtrennen
de-ethanizer Entäthaner *m*, Deäthanisator *m*
defat / to entfetten
defatting Entfetten *n*, Entfettung *f*
defecate / to reinigen, klären, läutern; *(Zucker)* scheiden

defecated juice *(Zucker)* geschiedener Saft *m*, Scheidesaft *m*

defecation *(Zucker)* Defäkation *f*, Scheiden *n*, Scheidung *f*, Kalkung *f*

  d. mud *(Zucker)* Scheideschlamm *m*

  d. pan s. d. tank

  d. scum s. d. mud

  d. tank *(Zucker)* Scheidepfanne *f*

  d. with dry lime *(Zucker)* trockene Scheidung *f*, Trockenscheidung *f*

  d. with milk of lime *(Zucker)* nasse Scheidung *f*, Naßscheidung *f*, Kalkmilchscheidung *f*

defecator *(Zucker)* Scheidepfanne *f*

defecocarbonation s. defecosaturation

defecosaturation *(Zucker)* Scheidesaturation *f*, Kalk-Kohlensäure-Verfahren *n*

defect *(Krist)* Fehler *m*, Baufehler *m*, Defekt *m*; Fehlstelle *f*

  d. concentration Fehlstellenkonzentration *f*

  d. conductor *(phys Ch)* Defektleiter *m*, Mangelleiter *m*

  d. electron Defektelektron *n*, Loch *n*

  d. in paint films Anstrichschaden *m*

  d. in [the] paper Papierfehler *m*

  d. lattice Defektgitter *n*, Fehl[stellen]gitter *n*

  d. motion Fehlstellenwanderung *f*

  d. of paper s. d. in [the] paper

defective fehlerhaft

defensive enzyme Abwehrferment *n*

deferred-curing finish *(Text)* Kunstharzausrüstung *f* mit verzögerter Formfixierung

deferrization Enteisenung *f*, Enteisenen *n* *(z.B. des Wassers)*

deferrize / to *(z.B. Wasser)* enteisenen

defiber / to s. to defibre

defibering, defiberization, defiberizing s. defibring

defibre / to *(Pap)* defibrieren, zerfasern, in Einzelfasern zerlegen

defibrinate / to *(Blut)* defibrinieren

defibrination Defibrinieren *n* *(des Blutes)*

defibring *(Pap)* Defibrierung *f*, Zerfaserung *f*

deficiency Mangel *m*, Fehlen *n*; Fehler *m*

  d. of nutrients Nährstoffmangel *m*

  d. symptom Mangelerscheinung *f*, Mangelsymptom *n*, Anzeichen *n* von Mangelkrankheiten

defined ausgeprägt, erkennbar, definiert

definite bestimmt, eindeutig, präzis, genau, definitiv

deflagrate / to deflagrieren, *(explosionsfrei)* abbrennen, niederbrennen, verpuffen

deflagrating agent *(Landw)* Abbrennmittel *n*

  d. powder Pulversprengstoff *m*, Schießstoff *m*, Schießmittel *n*, Treibstoff *m*, Treibmittel *n*

  d. spoon s. deflagration spoon

deflagration Deflagration *f*, *(explosionsfreies)* Abbrennen *n*, Niederbrennen *n*, Verpuffung *f*

  d. spoon Verbrennungslöffel *m*

  d. temperature Verpuffungstemperatur *f*

deflashing *(Plast)* Entgraten *n*, Abgraten *n*

  d. machine *(Plast)* Entgratmaschine *f*, Abgratmaschine *f*

deflect / to ablenken, auslenken, ableiten; abweichen; ausschlagen

deflection Ablenken *n*, Ablenkung *f*, Auslenken *n*, Auslenkung *f*; Abweichung *f*; Ausschlag *m*, Ausschlagen *n* *(eines Zeigers)*; Durchbiegung *f*

  d. instrument Ausschlag[meß]instrument *n*

  d. method Ausschlagmethode *f*

deflector 1. Ablenkplatte *f*, Ablenkwand *f*, Ablenkblech *n*, Prallplatte *f*, Prallwand *f*, Prallblech *n*, Leitblech *n*, Verteilerplatte *f*, Verteiler *m*; 2. *(Glas)* Ablenk[ungs]rinne *f*, Umlenkrinne *f*

  d. nozzle Pralldüse *f*

  d. plate s. deflector 1.

deflexion s. deflection

deflocculant Entflocker *m*, Dispergens *n*, Dispersionsmittel *n*, Dispergiermittel *n*, Peptisator *m*, Peptisationsmittel *n*; *(Ker)* Verflüssigungsmittel *n*

deflocculate / to *(geflockte Kolloide)* entflocken, zerteilen, dispergieren, peptisieren; *(Ker)* *(die Konsistenz eines Glasurschlickers)* verflüssigen, verdünnen; sich entflocken

deflocculating agent s. deflocculant

deflocculation Entflocken *n*, Entflockung *f*, Zerteilung *f*, Dispergierung *f*, Peptisation *f* *(von geflockten Kolloiden)*; *(Ker)* Verflüssigen *n*, Verdünnung *f* *(der Konsistenz eines Glasurschlickers)*

deflocculent s. deflocculant

deFlorez process deFlorez-Krackprozeß *m*, Spaltverfahren *n* nach deFlorez

defluorinate / to entfluorieren, Fluor austreiben

defluorination Entfluorieren *n*, Fluoraustreibung *f*

Defo elasticity *(Gum)* Defo[meter]elastizität *f*, DE

Defo hardness *(Gum)* Defo[meter]härte *f*, DH

Defo number *(Gum)* Defo[meter]zahl *f*

Defo plastometer *(Gum)* Defomeßgerät *n*, Defoprüfgerät *n*, Defoapparat *m*

Defo value *(Gum)* Defo[meter]wert *m*

defoam / to entschäumen

defoamer s. defoaming agent

defoaming Entschäumen *n*

  d. agent Antischaummittel *n*, Mittel *n* gegen Schaumbildung, Schaumverhinderungsmittel *n*, Schaumverhütungsmittel *n*, Schaumdämpfungsmittel *n*, Schaumdämpfer *m*, Schaumzerstörungsmittel *n*, Entschäumungsmittel *n*, Entschäumer *m*

defoliant Entlaubungsmittel *n*, Entblätterungsmittel *n*, Defoliationsmittel *n*

deformation Deformieren *n*, Verformen *n*; Deformation *f*, Verformung *f*, Form[ver]änderung *f*, Gestalts[ver]änderung *f*

defrost / to entfrosten, auftauen, abtauen, enteisen

defroster Enteisungsanlage *f*

defrosting Entfrosten *n*, Auftauen *n*, Abtauen *n*, Enteisen *n*

degas / to entgasen

degasification Entgasen *n*, Entgasung *f*

degasify / to s. to degas

degasifying s. degasification
degasser Entgasungsgerät n, Entgaser m
degassing Entgasen n, Entgasung f, (Plast auch)
Entlüften n, Entlüftung f, Lüften n (z.B. der Form,
des Werkzeugs)
degate / to (Plast) Anguß entfernen
degating (Plast) Entfernen n des Angusses
degeneracy (phys Ch) Entartung f
  d. temperature (phys Ch) Entartungstemperatur
  f
degenerate / to (phys Ch) entarten
degenerate (phys Ch) entartet
degeneration (phys Ch) Entartung f
degerm / to entkeimen, keimfrei machen, sterilisieren; entkeimen, Keime (z.B. Wurzelkeime
beim Malz) entfernen
degermation Entkeimen n, Sterilisieren n, Sterilisation f; keimfreier Zustand m, Sterilität f
degerminate / to entkeimen, Keime (z.B. Wurzelkeime beim Malz) entfernen
degerminating machine s. degerminator
degerminator (Lebm) Entkeimungsmaschine f
degeroïte (Min) Degeroït m, (veraltet für) Hisingerit
m
degradable abbaubar
degradation Abbau m, Zerlegung f; (Landw)
Degradation f, Degradierung f (des Bodens)
  d. of amides Säureamidabbau m
  d. of energy Degradation f der Energie,
  Energiedegradation f, Entwertung (Zerstreuung)
  f von Energie
  d. product Abbauprodukt n
degradative reaction Abbaureaktion f
degrade / to abbauen, zerlegen; sich abbauen
(zersetzen); (Landw) degradieren, den Wert des
Bodens mindern; (Phys) degradieren, Energie
zerstreuen
degrain / to (Gerb) Narben abstoßen (abziehen)
degras Degras m(n), Moellon n (Lederfettungsmittel)
degrease / to entfetten
degreaser s. degreasing agent
degreasing Entfetten n; (Text) Entschweißen n (der
Wolle)
  d. agent Entfettungsmittel n; (Text) Entschweißungsmittel n
degree Grad m
  d. API (Erdöl) Grad API m, API-Grad m, °API
  d. Balling Balling-Grad m
  d. Baumé Grad Baumé m, Baumé-Grad m, °Bé
  d. Celsius Grad Celsius m, °C
  d. centigrade s. d. Celsius
  d. Fahrenheit Grad Fahrenheit m, °F
  d. Kelvin Grad Kelvin m, °K
  d. of accuracy Genauigkeitsgrad m
  d. of acidity Säuregrad m, Azititätsgrad m,
  Säurehaltigkeit f
  d. of admission Füllungsgrad m
  d. of agitation Mischungsgrad m
  d. of association (phys Ch) Assoziationsgrad m

d. of attenuation Vergärungsgrad m (Maß für die
Extraktabnahme der Bierwürze)
d. of base saturation (Bodenkunde) Basensätti-
gungsgrad m
d. of beating (Pap) Mahlgrad m
d. of bleaching Bleichgrad m
d. of branching Verzweigungsgrad m
d. of browning (Lebm) Bräunungsgrad m
d. of coalification Inkohlungsgrad m
d. of condulation s. d. of crimp
d. of conversion Umsetz[ungs]grad m, Um-
satzgrad m, Umwandlungsgrad m
d. of convolution s. d. of crimp
d. of cooking (Pap) Aufschlußgrad m
d. of coverage (phys Ch) Bedeckungsgrad m
d. of crimp (Text) Einkräuselungsgrad m
d. of crosslinking Vernetzungsgrad m
d. of crystallinity Kristallinitätsgrad m
d. of cure s. d. of vulcanization
d. of degeneracy (phys Ch) Entartungsgrad m
d. of degradation Abbaugrad m
d. of depolarization Depolarisationsgrad m,
Entpolarisierungsgrad m
d. of dilution Verdünnungsgrad m
d. of dispersion Dispersionsgrad m, Dispersitätsgrad m, Zerteilungsgrad m
d. of dissociation Dissoziationsgrad m
d. of drying Austrocknungsgrad m
d. of enrichment Anreicherungsgrad m
d. of esterification Veresterungsgrad m
d. of extraction Extraktionsgrad m
d. of fermentation Vergärungsgrad m
d. of filling Füllungsgrad m; (phys Ch) Be-
setzungsgrad m (von Energieniveaus)
d. of fineness Feinheitsgrad m, Feinheit f;
Mahl[feinheits]grad m, Ausmahlungsgrad m,
Mahlfeinheit f
d. of freedom (phys Ch) Freiheitsgrad m, Freiheit
f, (i.e.S.) Anzahl f der Freiheitsgrade (Freiheiten)
d. of frost Kältegrad m
d. of hardening Aushärtungsgrad m
d. of hardness Härtegrad m, Härtestufe f
d. of hydration Hydratationsgrad m
d. of hydrolysis Hydrolyse[n]grad m
d. of ionization Ionisationsgrad m, Ionisierungs-
grad m
d. of metamorphosis Metamorphosegrad m,
Umwandlungsgrad m
d. of mixing Mischungsgrad m
d. of moisture Feuchtigkeitsgrad m, Feuchtegrad
m
d. of ondulation s. d. of crimp
d. of order (orientation) Orientierungsgrad m,
Ordnungsgrad m
d. of oxidation Oxydationsgrad m
d. of polarization Polarisationsgrad m
d. of pollution Verschmutzungsgrad m, Ver-
unreinigungsgrad m
d. of polymerization Polymerisationsgrad m
d. of purification Reinigungsgrad m

d. of purity Reinheitsgrad *m*
d. of reduction Zerkleinerungsgrad *m*
d. of retentivity Absperrgröße *f (eines Filters)*
d. of ripeness *(Text)* Reifegrad *m*, Reifezahl *f*
d. of ripening *(Lebm)* Reifungsgrad *m*
d. of rotational freedom *(phys Ch)* Rotationsfreiheitsgrad *m*
d. of saturation Sättigungsgrad *m*
d. of separation Trenn[ungs]grad *m*, Trennschärfe *f*
d. of shrinkage *(Pap)* Schrumpfungsgrad *m*
d. of sizing *(Pap)* Leimungsgrad *m*
d. of softness Weichheitsgrad *m*
d. of soiling Verschmutzungsgrad *m*
d. of solvation Solvationsgrad *m*
d. of steeping Weichgrad *m*, Quellreife *f (des Malzes)*
d. of stretching Reckgrad *m*, Verstreckungsgrad *m*
d. of substitution Substitutionsgrad *m*
d. of sulphonation Sulfonierungsgrad *m*
d. of supersaturation Übersättigungsgrad *m*
d. of swelling Quell[ungs]grad *m*
d. of symmetry Symmetriegrad *m*, Symmetrieordnung *f*
d. of temperature Temperaturgrad *m*
d. of thermal dissociation thermischer Dissoziationsgrad *m*
d. of thickening Eindickungsgrad *m*
d. of translational freedom *(phys Ch)* Translationsfreiheitsgrad *m*
d. of turbidity Trübungsgrad *m*
d. of unsaturation Ungesättigtheitsgrad *m*
d. of vulcanization Vulkanisationsgrad *m*, Vulkanisationskoeffizient *m*, VK
d. of white[ness] Weißgrad *m*, Weißgehalt *m*, Weiße *f*
d. S/H *s.* d. Soxhlet-Henkel
d. Soxhlet-Henkel Säuregrad *m* nach Soxhlet-Henkel (SH)
d. sugar solution Grad Sugar *m*, °S *(zur Angabe des Zuckergehalts von Lösungen)*
d. Twaddell Twaddell-Grad *m*, Grad Twaddell *m*, °Tw, *(fälschlich)* Twaddle-Grad *m*
degum / to entbasten, degummieren, entschälen; *(Öl)* raffinieren, entschleimen
degumming Entbasten *n*, Degummieren *n*, Entschälen *n*
d. of oil Ölraffination *f*, Entschleimung *f* (Entschleimen *n*) von Öl
dehair / to *(Gerb)* enthaaren
dehairing *(Gerb)* Enthaaren *n*
dehalogenation Dehalogenierung *f*, Halogenentzug *m*
dehumidification Entfeuchtung *f*, Trocknung *f (von Gasen)*
dehumidifier Trockenmittel *n*; Entfeuchtungsapparat *m*
dehumidify / to *(Gase)* entfeuchten, trocknen
dehydrase *s.* dehydrogenase

dehydrate / to 1. dehydratisieren, entwässern, entfeuchten, Wasser entziehen, trocknen; 2. dehydrieren, *(einer Verbindung)* Wasserstoff entziehen, Wasserstoff abspalten
dehydrated alcohol absoluter (reiner) Alkohol *m*, [nahezu] 100%iges (wasserfreies) Äthanol *n*, Alcoholus (Alcohol, Spiritus) absolutus, abs. A.
d. borax $Na_2B_4O_7$ wasserfreies Natriumtetraborat *n*, Pyroborax *m*
d. eggs Eipulver *n*, Trockenei *n*
dehydrater *s.* dehydrator
dehydrating *s.* dehydration
d. agent wasserentziehendes (wasserabspaltendes) Mittel *n*, Entwässerungsmittel *n*, Dehydratisierungsmittel *n*, Dehydratationsmittel *n*, Trockenmittel *n*
dehydration Dehydratisierung *f*, Dehydratation *f*, Entwässern *n*, Entwässerung *f*, Wasserentzug *m*, Wasserabspaltung *f*, Trocknen *n*, Trocknung *f*
d. plant Entwässerungsanlage *f*
dehydrator 1. *s.* dehydrating agent; 2. Entwässerer *m*, Entwässerungsgerät *n*; Trockner *m*, Trockenapparat *m*
dehydro base Dehydrobase *f*
dehydroacetic acid Dehydr[o]azetsäure *f*, Dehydroessigsäure *f*, 3-Äthanoyl-6-methyl-2,4-pyrandion *n*
dehydrobromination Bromwasserstoffabspaltung *f*
dehydrocyclization Dehydrozyklisierung *f*
dehydrofreeze / to *(Lebm)* durch Gefriertrocknung haltbar machen
dehydrofreezing *(Lebm)* Gefriertrocknung *f*
dehydrogenase Dehydr[ogen]ase *f*
dehydrogenate / to dehydrieren, *(einer Verbindung)* Wasserstoff entziehen, Wasserstoff abspalten
dehydrogenating catalyst *s.* dehydrogenation catalyst
dehydrogenation Dehydrieren *n*, Dehydrierung *f*, Dehydration *f*, Wasserstoffabspaltung *f*, Wasserstoffentzug *m*
d. catalyst Dehydrier[ungs]katalysator *m*
dehydrogenative dehydrierend
dehydrogenization *s.* dehydrogenation
dehydrogenize / to *s.* to dehydrogenate
dehydrogeranic acid $(CH_3)_2C=CHCH=CHC(CH_3)=CHCOOH$ Dehydrogeraniumsäure *f*
dehydrohalogenate / to dehydrohalogenieren, Halogenwasserstoff abspalten (entziehen)
dehydrohalogenation Dehydrohalogenierung *f*, Halogenwasserstoffabspaltung *f*, Halogenwasserstoffentzug *m*
dehydroisomerization Dehydroisomerisierung *f*
dehydroquinic acid Dehydrochinasäure *f*
dehydrotosylation *(indirekte Wasserabspaltung über den Tosylester eines Alkohols zwecks Einführung von Doppelbindungen)*
deice / to enteisen, Eisansatz entfernen
deicer Enteisungsanlage *f*

deink / to Druckfarbe *(aus Altpapier)* entfernen, entfärben, deinken

deinker *(Pap)* Chemikalie *f* für den Deinking-Prozeß

deinking Druckfarbenentfernung *f (aus Altpapier)*, Deinking *n*

  d. plant *(Pap)* Deinking-Anlage *f*

  d. solution *(Pap)* Chemikalienlösung *f* für den Deinking-Prozeß

deionization Deionisation *f*, Deionisierung *f*, Entionisation *f*, Entionisierung *f*

  d. time Deionisationszeit *f*, Entionisierungszeit *f*

  d. potential Deionisationspotential *n*, Entionisierungspotential *n*

deionize / to deionisieren, entionisieren

deionizer Deionisierungsmittel *n*, Entionisierungsmittel *n*

deisobutanization Entisobutanisierung *f*, Deisobutanisierung *f*, Isobutanentfernung *f*

deisobutanize / to entisobutanisieren, deisobutanisieren

deisobutanizer Entisobutaner *m*, Deisobutanisator *m*, Entisobutanisierkolonne *f*

dekameter Dekameter *n*, DK-Meter *n*, Dielektrizitätskonstante-Messer *m*

delaminate / to aufspalten, aufblättern, abschichten, in Schichten zerfallen; [auf]spalten, in Schichten zerlegen (trennen, spalten)

delamination Aufspaltung *f*, Aufblättern *n*, Abschichtung *f (von Schichtstoffen)*; Schichtentrennung *f*, Schicht[en]spaltung *f*

delay / to verzögern, verlangsamen

delay Verzögerung *f*, Verzug *m*

  d. in boiling Siedeverzug *m*, Siedeverzögerung *f*

delayed action *(Gum)* verzögernde Wirkung *f*, Einsatzverzögerung *f*

  d.-action accelerator Beschleuniger *m* mit verzögertem Vulkanisationseinsatz, Sicherheitsbeschleuniger *m*

  d. boiling *s.* delay in boiling

  d. coking verzögertes Verkoken *n*, Delayed Coking *n*

  d. coking process Delayed-Coking-Verfahren *n*

  d. crazing *(Ker)* Haarrißbildung *f* (Haarrissigwerden *n*) nach längerer Zeit

  d. disintegration verzögerter Zerfall *m*

  d. elasticity elastische Nachwirkung *f*

  d. ignition Zündverzögerung *f*, Zündverzug *m*

  d. neutron verzögertes Neutron *n*

  d. neutron emission verzögerte Neutronenemission *f*

delessite *(Min)* Delessit *m*

deleterious impurity schädliche (gesundheitsschädliche) Beimengung *f*

deleteriousness Schädlichkeit *f*, Giftigkeit *f*

delf[t] *s.* delftware

Delft test piece *(Gum)* Delfter Probe *f*

delftware Delftware *f*, Delfter Ware *f* (Steinzeug *n*, Fayence *f*)

delicate control Feinregulierung *f*

delignification *(Pap)* Delignifizierung *f*, Ligninentfernung *f*, Lignin[her]auslösung *f*

delignify / to *(Pap)* delignifizieren, Lignin entfernen (herauslösen)

delignifying agent *(Pap)* Delignifizierungsmittel *n*

delime / to entkalken; *(Gerb)* entkälken

deliming Entkalken *n*, Entkalkung *f*; *(Gerb)* Entkälken *n*

  d. agent Entkälkungsmittel *n*

delineate / to *(organische Strukturen)* in linearer Form wiedergeben

deliquefy / to Flüssigkeit entfernen

deliquefying Entfernen *n* von Flüssigkeit

deliquesce / to zerfließen, zergehen; wegschmelzen, zerschmelzen

deliquescence Zerfließen *n*, Zergehen *n*; Zerschmelzen *n*; Zerfließlichkeit *f*; Schmelzflüssigkeit *f*, Schmelzprodukt *n*

deliquescent zerfließlich; zerfließend, zergehend; zerschmelzend

deliver / to fördern

delivery Fördermenge *f*, Förderleistung *f*; Förderstrom *m*, Lieferstrom *m*

  d. flask Meßkolben *m*, Meßflasche *f*

  d. line Druckleitung *f*, Steigleitung *f*, Förderleitung *f*

  d. of the stock *(Pap)* Stofführung *f*, Stoffzuführung *f*

  d. pipe Druckrohr *n*, Steigrohr *n*, Förderrohr *n*; Fallrohr *n (einer Destillierkolonne)*

  d. pressure Förderdruck *m*, Lieferdruck *m*

  d. rate Fördergeschwindigkeit *f (z.B. einer Pumpe)*; *(Text)* Liefergeschwindigkeit *f*; Förderstrom *m*, Lieferstrom *m*

  d. side Druckseite *f*, Förderseite *f*, Lieferseite *f*

  d. speed *(Text)* Liefergeschwindigkeit *f*

  d. stroke Druckhub *m*, Förderhub *m*

  d. table *(Ker)* Abnahmetisch *m*

  d. tape *(Pap)* Transportband *n (des Querschneiders)*

  d. tube Einleitungsrohr *n*, Zuleitungsrohr *n*; Ableitungsrohr *n*; Vorstoß *m (der Destillationsapparatur)*

  d. valve Druckventil *n (am Kompressor oder an der Pumpe)*

delocalization *(phys Ch)* Delokalisierung *f*, Nichtlokalisierung *f*

  d. effect Delokalisierungseffekt *m*

  d. energy Delokalisierungsenergie *f*, Mesomerieenergie *f*, Resonanzenergie *f*, Konjugationsenergie *f*

delocalize / to delokalisieren

delocalized bond delokalisierte (nichtlokalisierte) Bindung *f*

delph *s.* delftware

delphinic acid $(CH_3)_2CHCH_2COOH$ Delphinsäure *f*, Isovaleriansäure *f*, 3-Methylbutansäure *f*

delphinine Delphinin *n*

delphonine Delphonin *n*

delphware *s.* delftware

delq. *s.* deliquescent

delta electron (ray) Delta-Elektron *n*, Delta-Strahl *m*

dense

Delthirna process *(Pap)* Delthirna-Verfahren *n*, Delthirna-Kaltverfahren *n*
Delthirna size *(Pap)* Delthirna-Leim *m*
delustrant *(Text)* Mattierungsmittel *n*
delustre / to *(Text)* mattieren
delustring *(Text)* Mattieren *n*
  d. agent *s.* delustrant
demagnetization Entmagnetisieren *n*, Entmagnetisierung *f*; Entmagnetisierung *f (als Zustand)*
demagnetize / to entmagnetisieren
demagnetizing factor Entmagnetisierungsfaktor *m*
  d. field entmagnetisierendes Feld *n*
demand for paper Papierbedarf *m*
demanganization Entmanganung *f*
demanganize / to entmanganen
demargarinate / to *(Öle)* demargarinieren, entstearin[is]ieren, ausfrieren, wintern
demargarination Demargarinieren, *n*, Demargarinisation *f*, Entstearin[is]ierung *f*, Ausfrieren *n*, Winterung *f*, Winterisation *f*
demarginate / to *s.* to demargarinate
demethanization Entmethanisieren *n*, Entmethanisierung *f*, Demethanisierung *f*
demethanize / to entmethanisieren, demethanisieren
demethanizer Entmethaner *m*, Demethanisator *m*
demethylate / to entmethylieren, demethylieren
demethylation Entmethylieren *n*, Entmethylierung *f*, Demethylierung *f*, Demethylation *f*
demethylchlortetracycline Demethylchlortetrazyklin *n*, 7-Chlor-6-demethyltetrazyklin *n*
demijohn Demijohn *m*, große Korbflasche *f*
demineralization Demineralisation *f*, Entmineralisieren *n*, Entmineralisierung *f*, Entfernen *n* mineralischer Substanzen
  d. of water Wasserentsalzung *f*
demineralize / to entmineralisieren, demineralisieren, mineralische Substanzen entfernen, *(Wasser)* entsalzen
demineralizer Wasserentsalzungsapparat *m*
demineralizing *s.* demineralization
Demjanov rearrangement Demjanow-Umlagerung *f*, Demjanowsche Methode *f*
demonstration lecture Experimentalvorlesung *f*
demoulding Herausnehmen *n (der Vulkanisate aus der Form)*
demulsification Demulgieren *n*, Demulgierung *f*, Dismulgieren *n*, Brechen (Spalten, Entmischen) *n* einer Emulsion, Emulsionsspaltung *f*, Entemulsionieren *n*, Emulsionsentmischung *f*
demulsifier Demulgator *m*, Emulsionsentmischer *m*, Emulsionsspalter *m*, Emulsionsbrecher *m*
demulsify / to demulgieren, entemulsionieren, eine Emulsion brechen (spalten, entmischen)
denaphthalization Entnaphthal[is]ierung *f*
denaphthalize / to entnaphthal[is]ieren
denaturant Denaturierungsmittel *n*, Vergällungsmittel *n*
denaturate / to *s.* to denature

denaturation Denaturieren *n*, Denaturierung *f*, Denaturation *f*, Vergällen *n*, Vergällung *f (von Äthanol, Kochsalz usw.)*; Denaturieren *n*, Denaturierung *f*, Denaturation *f (Strukturänderung nativer Proteine)*
denature / to *(Äthanol, Kochsalz usw.)* denaturieren, vergällen; *(bei nativen Proteinen)* denaturieren
denatured alcohol denaturierter (vergällter) Alkohol (Spiritus, Branntwein) *m*, Spiritus denaturatus *m*
denaturing Denaturieren *n*, Denaturierung *f*, Denaturation *f*, Vergällen *n*, Vergällung *f*
  d. agent *s.* denaturant
  d. of salt Salzdenaturierung *f*, Salzvergällung *f*
denaturize / to *s.* to denature
dendrite Dendrit *m*, Baumkristall *m*
dendritic[al] *(Krist)* dendritisch, verzweigt, verästelt
Denier, denier Denier *n*, den *(Masse von 9000 m Faden in g)*
Denigès formic-acid test Nachweis *m* von Ameisensäure nach Denigès
denitrate / to denitrieren
denitration Denitrieren *n*, Denitrierung *f*, Denitration *f*, Entstickung *f*
  d. tower Denitrier[ungs]turm *m*, Denitrator *m*
denitrator Denitrier[ungs]apparat *m*, *(i.e.S.)* Denitrier[ungs]turm *m*, Denitrator *m*
denitrification Denitrifikation *f (Reduktion von Nitraten durch Bodenbakterien)*
denitrifiers *s.* denitrifying bacteria
denitrify / to denitrifizieren
denitrifying bacteria denitrifizierende Bakterien *npl*, Denitrifikationsbakterien *npl*, Denitrifikanten *mpl*, Denitrifikatoren *mpl*
denitrogenate / to Stickstoff entziehen
denitrogenation Stickstoffentzug *m*
denitrogenize / to *s.* to denitrogenate
dense arsenic Stück[en]arsenik *n*, Arsenikbrocken *mpl*
  d. bed *(Erdöl)* Festbett *n*
  d.-bed column Festbettkolonne *f*
  d.-burning *(Ker)* dichtbrennend
  d. glass optisches Glas *n* mit hohem Brechungsindex
  d.-media *s. unter* d.-medium
  d. medium schwere Flüssigkeit *f*, Schwerflüssigkeit *f*, Trennflüssigkeit *f*, Trennmedium *n*, *(i.e.S.)* schwere Trübe *f*, Schwertrübe *f (unechte Schwerflüssigkeit)*
  d.-medium cleaning *s.* d.-medium separation
  d.-medium cyclone Schwertrübe-Waschzyklon *m*
  d.-medium plant *s.* d.-medium separation plant
  d.-medium process *s.* d.-medium separation
  d.-medium separation Schwerflüssigkeitsaufbereitung *f*, Schwerflüssigkeitssortieren *n*, Schwimm-und-Sink-Aufbereitung *f*, Sinkscheideverfahren *n*, *(i.e.S.)* Trennung *f* in

Schweretrüben, Schwertrübeaufbereitung *f*, Schwertrübescheidung *f*

d.-medium separation plant Schwerflüssigkeits[aufbereitungs]anlage *f*, Sinkscheideranlage *f*

d.-medium separator Schwerflüssigkeitsscheider *m*, Sinkscheider *m*, *(i.e.S.)* Schwertrübescheider *m*

d.-medium separatory vessel *s.* d.-medium separator

d.-medium unit *s.* d.-medium separation plant

d.-medium vessel (washer) *s.* d.-medium separator

d.-medium washing *s.* d.-medium separation

densification Verdichtung *f*

densified laminated wood Preßschichtholz *n*

densifier Eindicker *m* zur Regeneration der Schwerflüssigkeit

densify / to verdichten

densimeter Densimeter *n*, Dichtemesser *m*, Aräometer *n*, Senkwaage *f*, Senkspindel *f*; Luftdurchlässigkeitsprüfer *m*

densimetry Densimetrie *f*, Dichtemessung *f*, Dichtebestimmung *f*

densitometer 1. *(Foto)* Dens[it]ometer *n*, Densograf *m*, Schwärzungsmesser *m*; 2. *s.* densimeter

densitometry Densitometrie *f*, Schwärzungsmessung *f*

density Dichte *f*, D *(oder)* d, Raumdichte *f (Masse je Volumeneinheit)*; Wichte *f*, Artgewicht *n*, spezifisches Gewicht *n (Gewicht je Volumeneinheit)*; Schwärzungsdichte *f*, [optische] Dichte *f*, Schwärzung *f*, Extinktion *f*

d. balance Dichtewaage *f*

d. bottle Wägefläschchen *n*, Pyknometer *n*

d. current Dichteströmung *f*

d. cut *s.* d. separation

d. in raw state Rohdichte *f*

d. of a gas Gasdichte *f*

d. of a liquid Flüssigkeitsdichte *f*

d. of air Luftdichte *f*

d. of electrons Elektronendichte *f*, Elektronenbelegung *f*

d. of loading Ladungsdichte *f*, Ladungskonzentration *f (von Sprengstoffen)*

d. of setting *(Ker)* Besatzdichte *f*

d. of vapour Dampfdichte *f*

d. of water Dichte *f* des Wassers

d. range *(Foto)* Schwärzungsbereich *m*

d. separation Trennung *f* nach der Dichte, Dichtesortierung *f*

densometer Luftdurchlässigkeitsprüfer *m*

dental cement Zahnzement *m*

d. enamel Zahnschmelz *m*

d. porcelain Zahnporzellan *n*, Dentalporzellan *n*

dentifrice Zahnputzmittel *n*, Zahnreinigungsmittel *n*, Zahnpflegemittel *n*

dentin[e] Dentin *n*, Zahnbein *n*

deodorant desodor[is]ierend, geruchsbeseitigend, geruchszerstörend

d. lotion desodor[is]ierende Lotion *f*, Desodorant-Lotion *f*

d. powder desodor[is]ierender Puder *m*

d. stick desodor[is]ierender Stift *m*

deodorization Desodor[is]ieren *n*, Desodor[is]ierung *f*, Desodor[is]ation *f*, Geruchlosmachen *n*, Geruchfreimachen *n*, Geruchsbeseitigung *f*, Geruchsentfernung *f*, Geruchszerstörung *f*

deodorize / to desodor[is]ieren, geruchlos (geruchfrei) machen, den Geruch beseitigen (entfernen, zerstören)

deodorizer 1. Desodor[is]ierungsmittel *n*, Desodorans *n*, Deodorant *n*, desodor[is]ierendes (geruchsbeseitigendes, geruchszerstörendes) Mittel *n*, Geruchsverbesserer *m*; 2. Desodorierer *m*, Desodoreur *m*, Dämpfer *m (Gerät zur Desodorierung von Fetten und Ölen)*

deoil / to entölen

deoiling Entölen *n*, Entölung *f*

de-olation *(Gerb)* Entolung *f*

deoxidant Desoxydationsmittel *n*, Reduktionsmittel *n*

deoxidate / to *s.* to deoxidize

deoxid[iz]ation Desoxydieren *n*, Desoxydation *f*, Sauerstoffentzug *m*, Reduktion *f*

deoxidize / to desoxydieren, reduzieren, Sauerstoff entfernen (abspalten); *(Stahl)* beruhigen

deoxidizer, deoxidizing agent *s.* deoxidant

deoxycorticosterone Desoxykortikosteron *n*, Desoxykorton *n*, Δ^4-Pregnen-21-ol-3,20-dion *n*

d. acetate Desoxykortikosteronazetat *n*, Desoxykortonazetat *n*, 21-Azetoxy-Δ^4-pregnen-3,20-dion *n*

deoxycortone *s.* deoxycorticosterone

deoxygenate / to desoxydieren, reduzieren, Sauerstoff entfernen (abspalten)

deoxygenated blood venöses Blut *n*

d. hemoglobin reduziertes [dunkelrotes] Hämoglobin *n*

deoxygenation Desoxydation *f*, Sauerstoffabspaltung *f*, Sauerstoffentzug *m*, Reduktion *f*

deoxypentose Desoxypentose *f*

d. nucleic acid Desoxypentosenukleinsäure *f*

deoxyribonuclease Desoxyribonuklease *f (zu den Phosphatasen gehörendes Ferment)*

deoxyribonucleic acid Desoxyribonukleinsäure *f*, DNS *f*

deoxyribonucleoprotein Desoxyribonukleoprotein *n*

deoxyribose Desoxyribose *f*, 2-Desoxy-D-ribose *f*, 2-Ribodesose *f*, Thyminose *f (ein Desoxyzucker)*

deozonize / to desozoni[si]eren, entozonisieren

D.E.P. *s.* diethyl phthalate

2,4-DEP *tris*-(2,4-Dichlorphenoxyäthyl)phosphit *n (Herbizid)*

deparaffin / to *s.* to deparaffinize

deparaffinization Entparaffinieren *n*, Entparaffinierung *f*

deparaffinize / to entparaffinieren

depart / to abweichen

departure Abweichen *n*, Abweichung *f*; Abweichung *f (als Zustand)*

dependence Abhängigkeit *f*

dependent joint action *(Toxikologie)* Abhängigkeitsverbundwirkung *f (Wirkung zweier physiologisch verschieden angreifender Gifte, wobei ein Gift den Effekt des anderen verstärkt)*

depentanization Entpentanisieren *n*, Entpentanisierung *f*, Depentanisierung *f*

depentanize / to entpentanisieren, depentanisieren

depentanizer Entpentaner *m*, Depentanisator *m*

dephenol[iz]ation Entphenol[ier]ung *f*

dephenolize / to entphenol[ier]en

dephenolizer Phenolabscheider *m*

dephenolizing Entphenol[ier]ung *f*

   d. plant Entphenol[ier]ungsanlage *f*

dephlegmate / to dephlegmieren, mit dem Dephlegmator behandeln, rektifizieren

dephlegmating cooler Entwässerungskühler *m*

dephlegmation Dephlegmierung *f*, Dephlegmation *f*, Teilkondensation *f*, teilweise Kondensation (Verflüssigung) *f*, Aufstärkung *f*

dephlegmator *(Destillation)* Dephlegmator *m (Rücklaufkondensator mit nur teilweiser Kondensation)*

dephlogisticated marine acid s. d. muriatic acid

   d. muriatic acid *(historisch)* dephlogisti[si]erte (von Phlogiston befreite) Muriatsäure *f*

dephosphorization Entphosphoren *n*, Entphosphorung *f*

dephosphorize / to entphosphoren

dephosphorylate / to dephosphorylieren

dephosphorylation Dephosphorylieren *n*, Dephosphorylierung *f*

depickle / to *(Gerb)* entpickeln

depickling *(Gerb)* Entpickeln *n*

depilate / to enthaaren, das Haar entfernen, depilieren

depilation Enthaaren *n*, Enthaarung *f*, Haarentfernung *f*, Haarlockerung *f*, Depilieren *n*, Depilierung *f*, Depilation *f*

depilator Enthaarungsmittel *n*, Haarentfernungsmittel *n*, Haarlockerungsmittel *n*, depilierendes Mittel *n*, Depilatorium *n*

depilatory haarentfernend, enthaarend, haarlockernd, depilierend

depilatory s. depilator

   d. agent s. depilator

   d. cream Haarentfernungscreme *f*, Enthaarungscreme *f*, enthaarende Creme *f*, Depiliercreme *f*

   d. method Enthaarungsverfahren *n*, *(i.e.S.)* Entwollungsverfahren *n*

   d. powder Enthaarungspulver *n*, pulverförmiges Enthaarungsmittel *n*

depilitant s. depilator

deplete / to *(Gerb)* entquellen, verfallen machen

depletion *(Gerb)* Verfallen *n (der Blöße)*

   d. layer *(phys Ch)* [träger]verarmte Schicht *f*

depolarization Depolarisation *f*

depolarize / to depolarisieren

depolarizer Depolarisator *m*

depolymerization Entpolymerisieren *n*, Depolymerisieren *n*, Depolymerisierung *f*, Depolymerisation *f*

depolymerize / to entpolymerisieren, depolymerisieren

deposit / to abscheiden, ausscheiden, ablagern, sedimentieren, niederschlagen; sich abscheiden (ausscheiden, absetzen, setzen, ablagern, niederschlagen), sedimentieren, einen Niederschlag (Bodensatz) bilden, zur Ausscheidung gelangen

deposit Abscheidung *f*, Niederschlag *m*, Ablagerung *f*; *(Geol)* Lager *n*, Lagerstätte *f*, Vorkommen *n*, Depot *n*; *(z.B. auf Blättern)* aufgelagertes Pflanzenschutzmittel *n*

   d. builder Haftmittel *n*

   d. of sulphur Schwefelablagerung *f*

   d. tracking Kriechwegbildung *f* durch leitende Ablagerungen

deposited metal aufgetragenes Metall *n*, Metallauftrag *m*

deposition Abscheidung *f*, Niederschlag *m*, Ablagerung *f*

   d. potential Abscheidungspotential *n*

depot fat Depotfett *n*

depress / to erniedrigen, herabsetzen, senken

depressant *(Flot)* Drücker *m*, drückendes Schwimmittel , drückend wirkendes Flotationsmittel *n*, drückender Sammler *m*, drückendes (passivierendes) Mittel (Reagens) *n*, drückender (passivierender) Zusatz *m*

depressing *(Flot)* Drücken *n*, Passivieren *n*, Passivierung *f*

depression Depression *f*, Erniedrigung *f*, Senkung *f*; Vertiefung *f*

   d. of the freezing point Gefrierpunktserniedrigung *f*

   d. of the melting point Schmelzpunktserniedrigung *f*, Schmelzpunkt[s]depression *f*

deprive of / to entziehen

depropanization Entpropanisieren *n*, Entpropanisierung *f*, Depropanisierung *f*, Propanabtrennung *f*

depropanize / to entpropanisieren, depropanisieren

depropanizer Entpropaner *m*, Entpropanisier[ungs]kolonne *f*, Depropanisator *m*, Depropanisierungskolonne *f*

deproteinization Deproteinisieren *n*, Enteiweißen *n*, Eiweißabtrennung *f*

deproteinize / to deproteinisieren, enteiweißen, Eiweiß abtrennen

deproteinized rubber enteiweißter (eiweißarmer) Kautschuk *m*

depside Depsid *n (Ester einer Phenolkarbonsäure)*

depsidone Depsidon *n (ein aus zwei Polyoxybenzolkarbonsäuren mit ätherartiger Bindung zusammengesetzter Stoff)*

depth Tiefe *f*

d. developer *(Foto)* Tiefenentwickler *m*
d. development *(Foto)* Tiefenentwicklung *f*
d. of application Einbringungstiefe *f (z.B. für Düngemittel)*
d. of bed Schichthöhe *f (z.B. des Brennstoffs)*
d. of case Einsatztiefe *f*, Tiefe *f* der Einsatzschicht, Zementationstiefe *f*, Aufkohlungstiefe *f*, Kohlungstiefe *f*; Einsatzhärtungstiefe *f*, Einsatzhärtetiefe *f*
d. of colour Farbtiefe *f*
d. of hardening Härtungstiefe *f*, Härtetiefe *f*
d. of liquid Füllhöhe *f*
d. of packing Schüttungshöhe *f*
d. of penetration Eindringtiefe *f*
d. of submergence Eintauchtiefe *f*
deragger *(Pap)* Zopfwinde *f*
derbylite *(Min)* Derbylit[h] *m*
derbyshire spar *(Min)* Flußspat *m*, Fluorit *m (Kalziumfluorid)*
derivation Ableitung *f*, Herleitung *f*
derivative Derivat *n*, Abkömmling *m*
d. action Vorhaltwirkung *f*, differenzierende Wirkung *f (Regelungstechnik)*
derrick Bohrturm *m*
desactivate / to *s.* to deactivate
desactivation *s.* deactivation
desaerate / to *s.* to deaerate
desaerating, desaeration *s.* deaeration
desalination *s.* desalinization
desalinization Entsalzen *n*, Entsalzung *f*
d. of sea water Meerwasserentsalzung *f*
d. of water Wasserentsalzung *f*
desalt / to entsalzen
desalter Wasserentsalzungsapparat *m*
desalting Entsalzen *n*, Entsalzung *f*
desamidase Desamidase *f (zu den Amidasen gehörendes Ferment)*
desamidate / to *s.* to deamidate
desamidation *s.* deamidation
desaminase *s.* deaminase
desamination *s.* deamination
descale / to Kesselstein entfernen
descaling agent Mittel *n* gegen Kesselstein, Kesselsteingegenmittel *n*, Kesselsteinverhütungsmittel *n*
descent plate *(Pap)* Kropf *m (beim Holländer)*
descloizite *(Min)* Descloizit *m*
desemulsification *s.* demulsification
desemulsify / to *s.* to demulsify
desensitization *(Foto)* Desensibilisieren *n*, Desensibilisierung *f*, Hellicht-Entwicklung *f*; Phlegmatisierung *f (Herabsetzung der Empfindlichkeit eines Explosivstoffes)*
desensitize / to *(Foto)* desensibilisieren, unempfindlich machen; einen Explosivstoff phlegmatisieren *(gegen mechanische Einwirkung unempfindlich machen)*
desensitizer *(Foto)* Desensibilisator *m*
deserpidine Deserpidin *n*, Kaneszin *n*, 11-Desmethoxyreserpin *n*

desert varnish *(Geol)* Wüstenlack *m*, Wüstenkruste *f*, Wüstenrinde *f*
desiccant Sikkativ *n*, Trockenstoff *m*, Trockenmittel *n*, Trocknungsmittel *n*, Trockenmedium *n*, Trockner *m*; *(Landw)* Abbrandmittel *n (zur Verätzung von Blattflächen)*
d. chamber Trockenmittelkolben *m (einer Trockenpistole)*
desiccate / to trocknen, entfeuchten, entwässern, dehydratisieren, Wasser entziehen; *(Obst)* dörren, *(Malz)* darren, rösten
desiccated eggs Eipulver *n*, Trockenei *n*
d. milk Trockenmilch *f*, Milchpulver *n*
desiccation Trocknung *f*, Austrocknung *f*
desiccative *s.* desiccant
desiccator Exsikkator *m (Gerät)*; *s.* desiccant
d. cage Drahtkorb *m* (Schutzhaube *f*, Schutzkorb *m*) für Exsikkatoren
d. disk *s.* d. plate
d. guard *s.* d. cage
d. lid Exsikkatordeckel *m*
d. plate Exsikkatorplatte *f*
design feature Konstruktionsmerkmal *n*
d. of compound *(Gum)* Rezeptaufstellung *f*, Aufstellung *f* eines Rezepts
d. pressure Entwurfsdruck *m*
d. temperature Entwurfstemperatur *f*
desilicate / to *(Geol)* desilifizieren
desilication *(Geol)* Desilifizierung *f*
desilicification Entkieselung *f*
desilicify / to entkieseln
desilver / to *s.* to desilverize
desilvering *s.* desilverization
desilverization Entsilbern *n*, Entsilberung *f*
desilverize / to entsilbern
desize / to *(Text)* entschlichten
desizing *(Text)* Entschlichten *n*, Entschlichtung *f*
d. agent *(Text)* Entschlichtungsmittel *n*
d. bath *(Text)* Entschlichtungsbad *n*
d. compound *s.* d. agent
deslime / to entschleimen; entschlämmen
desliming Entschleimen *n*, Entschleimung *f*; Entschlämmen *n*, Entschlämmung *f*
desmine *(Min)* Desmin *m*
desmo-enzyme Desmoenzym *n*, Desmoferment *n*, extrazelluläres Ferment *n*
desmolase Desmolase *f*
desmolyses *s.* desmolysis
desmolysis Desmolyse *f*
desmotropic desmotrop
desmotropism *s.* desmotropy
desmotropy Desmotropie *f*, Tautomerie *f*
desodorant *s.* deodorizer 1.
desolventize / to Lösungsmittel entfernen (austreiben)
desooting Entrußen *n*
desorb / to desorbieren
desorbable desorbierbar
desorbing Desorbieren *n*
desorption Desorption *f (Entweichen oder Entfernen sorbierter Gase vom Sorptionsmittel)*

d. curve Desorptionskurve *f*
desosamine Desosamin *n*
desoxycholic acid Desoxycholsäure *f*
desoxycorticosterone, desoxycortone *s.* deoxycorticosterone
desoxyribonucleic acid *s.* deoxyribonucleic acid
desoxyribose *s.* deoxyribose
desoxystreptamine Desoxystreptamin *n*
destabilize / to instabil machen, entstabilisieren
destaticization Entfernen *n* (Entfernung *f*) elektrostatischer Ladungen
destearinate / to *(Öle)* entstearin[is]ieren, demargarinieren, ausfrieren, wintern
destearinization Entstearin[is]ierung *f*, Demargarinisation *f*, Demargarinieren *n*, Winterung *f*, Winterisation *f*, Ausfrieren *n*
destearinize / to *s.* to destearinate
destruction Zerstören *n*, Zerstörung *f*, Vernichtung *f*, Abbau *m*, Zerfall *m*, Destruktion *f*
d. of positrons Positronenzerstrahlung *f*
destructive destruktiv, zersetzend, abbauend
d. distillation destruktive (zersetzende, abbauende, trockene) Destillation *f*, Zersetzungsdestillation *f*
d. hydrogenation destruktive (spaltende, abbauende) Hydrierung *f*
destructor Müllverbrennungsofen *m*
desublimation Solidensieren *n*
desugar / to entzuckern
desugarization Entzuckern *n*, Entzuckerung *f*
desugarize / to *s.* to desugar
desugarizing *s.* desugarization
d. of molasses Melasseentzuckerung *f*
desuinting Entschweißen *n* *(von Wolle)*
desulfurization *s.* desulphurization
desulfurize (desulphur) / to *s.* to desulphurize
desulphur[iz]ation Desulfurieren *n*, Desulfurierung *f*, Entschwefeln *n*, Entschwefelung *f*
desulphurize / to desulfurieren, entschwefeln
desulphurizing *s.* desulphurization
desyl chloride $C_6H_5COCHClC_6H_5$ Desylchlorid *n*
detach / to [ab]lösen, abnehmen; abbauen
detachment Ablösung *f*, Lösung *f*
d. method *(phys Ch)* Abreißmethode *f*, Lamellenmethode *f*, Bügelmethode *f*
d. of electrons Elektronenablösung *f*
detail paper Detailzeichenpapier *n*
de-tan / to entgerben
detar / to entteeren
detarrer Teer[ab]scheider *m*
detarring Entteeren *n*, Entteerung *f*, Teerabscheidung *f*, Teerentfernung *f*
detearing Tropfenabziehen *n* *(beim Tauchlackieren)*
detect / to nachweisen, feststellen, auffinden
detectability Nachweisbarkeit *f*, Erkennbarkeit *f*; Erfassungsgrenze *f* *(Analytik)*
detectable nachweisbar, feststellbar, wahrnehmbar
detectible *s.* detectable

detection Nachweis *m*, Feststellung *f*
d. method Nachweismethode *f*
d. of carbon Kohlenstoffnachweis *m*
d. of hydrogen Wasserstoffnachweis *m*
d. of nitrogen Stickstoffnachweis *m*
d. of radiation Strahlennachweis *m*
d. sensitivity Nachweisempfindlichkeit *f*
d. unit Nachweisgerät *n*
detector Meßfühler *m*, Fühler *m*; Meßeinrichtung *f*; Detektor *m* *(Bauelement eines Gaschromatografen)*
d. substance Erkennungsmittel *n*
detention period (time) Verweilzeit *f*, Aufenthaltszeit *f*, Haltezeit *f*, Stehzeit *f*
deter / to *(tierische Schädlinge)* abschrecken
detergency reinigende Eigenschaften *fpl*, Reinigungsvermögen *n*, Reinigungskraft *f*; Wascheigenschaften *fpl*, Waschvermögen *n*, Waschkraft *f*
d. builder Aufbaustoff *m*, Builder *m*, Gerüstsubstanz *f*, Gerüststoff *m* *(zum Aufbau synthetischer Waschmittel)*
d. process Reinigungsprozeß *m*; Waschprozeß *m*
detergent reinigend, Reinigungs...; Wasch...
detergent Reinigungsmittel *n*; Waschmittel *n*; Detergens *n*, Detergent *n*, Syndet *n*, synthetisches Reinigungsmittel *(oder)* Waschmittel *n*; Detergent *n*, Schlamminhibitor *m*, absetzverhinderndes Mittel *n*
d. action *s.* d. effect
d. additive Detergent-Additiv[e] *n*, Detergentzusatz *m*, Reinigungszusatz *m*, reinigender Zusatz *m*
d. effect Reinigungseffekt *m*, Reinigungswirkung *f*, reinigende Wirkung *f*; Wascheffekt *m*, Waschwirkung *f*; Detergenteffekt *m*, Detergentwirkung *f* *(eines Schlamminhibitors)*
d. for the coloured wash Buntwaschmittel *n*, Waschmittel *n* für Buntwäsche
d. industry Waschmittelindustrie *f*
d. manufacture Waschmittelherstellung *f*
d. performance Reinigungsleistung *f*; Waschleistung *f*
d. power (properties) *s.* detergency
d. solution Waschflotte *f*, Waschlauge *f*
d. surfactant waschaktive Substanz *f*, waschaktiver Stoff *m*, WAS
deteriorate / to 1. schlechter (unbrauchbar) werden, sich verschlechtern, an Wert verlieren, *(i.e.S.)* verderben; entarten; sich zersetzen; sich entmischen *(von Emulsionen)*; 2. verschlechtern, im Wert herabsetzen (mindern)
deterioration Verschlechterung *f*, Herabminderung *f*, Verminderung *f*, Minderung *f*, Verringerung *f*; Wertminderung *f*; Verderben *n*, Verderb *m*, Schlechtwerden *n*; Entartung *f*; Zersetzung *f*; Entmischung *f* *(von Emulsionen)*
d. of quality Qualitätsminderung *f*
determination Bestimmung *f*, Feststellung *f*, Ermittlung *f*

d. of acidity Aziditätsbestimmung *f*
d. of ashes Aschebestimmung *f*
d. of atomic weights Bestimmung *f* der relativen Atommassen, *(bisher)* Atomgewichtsbestimmung *f*
d. of constitution Konstitutionsbestimmung *f*, Konstitutionsaufklärung *f*
d. of formulae Formelbestimmung *f*
d. of grain size Korngrößenbestimmung *f*
d. of hardness Härtebestimmung *f*
d. of hydroxyl value Hydroxylzahlbestimmung *f*
d. of molecular weights Bestimmung *f* der relativen Molekülmassen, *(bisher)* Molekulargewichtsbestimmung *f*
d. of pH pH-[Wert-]Bestimmung *f*, pH-[Wert-]Ermittlung *f*
d. of position Ortsbestimmung *f*, Lokalisieren *n*, Lokalisation *f*
d. of the limit Grenzwertbestimmung *f*
d. of vapour density Dampfdichtebestimmung *f*, Dampfdichtemessung *f*
determine / to bestimmen, feststellen, ermitteln
deterrent Abschreck[ungs]stoff *m*, Abschreck[ungs]mittel *n* *(gegen tierische Schädlinge)*
d. action Abschreckwirkung *f* *(auf tierische Schädlinge)*
detoluate / to 1. enttoluolen; 2. mit Mononitrotoluol ausrühren *(Trinitrotoluol-Herstellung)*
detoluation 1. Enttoluolen; 2. Ausrühren *n* mit Mononitrotoluol *(Trinitrotoluol-Herstellung)*
detonable zur Detonation fähig
detonate / to detonieren; *(i.w.S.)* explodieren; detonieren lassen, zur Detonation bringen
detonating cap Sprengkapsel *f*
d. fuse detonierende Zündschnur *f*, Knallzündschnur *f*, Sprengschnur *f*
d. gas Knallgas *n*
detonation Detonation *f*, *(i.w.S.)* heftige Explosion *f*; Sprengung *f*; Klopfen *n* *(des Motors)*
d. rate Detonationsgeschwindigkeit *f*
d. wave Stoßwelle *f*
detonator Zündstoff *m*, Initialsprengstoff *m*; Sprengkapsel *f*
detoxicate / to entgiften
detoxication, detoxification Entgiften *n*, Entgiftung *f*
detoxify / to s. to detoxicate
detrimental schädlich
detrital material s. detritus 1.
detritus 1. *(Geol)* Detritus *m*, Verwitterungsschutt *m*, Gesteinsschutt *m*; 2. *(Wasserwirtschaft)* Detritus *m*, Tripton *n*
Detroit rocking [arc] furnace Detroit-Lichtbogenschaukelofen *m*, Detroit-Schaukelofen *m*, Detroit-Ofen *m*
deuterate / to deuterieren, Deuterium einbauen
deuteration Deuterierung *f*, Einbau *m* von Deuterium (schwerem Wasserstoff)
deuteric *(Geol)* deuterisch

deuterium D, $^2_1$D, $^2_1$H Deuterium *n*, schwerer Wasserstoff *m*
d. oxide D_2O Deuteriumoxid *n*, schweres Wasser *n*
deuterize / to s. to deuterate
deut[er]on *(Kch)* Deuteron *n*
Devarda's alloy Devardasche Legierung *f* *(Reduktionsmittel)*
deveilite *(fälschlich für)* deweylite
develop / to entwickeln; *(Min)* aufschließen, erschließen
developed dye Entwicklungsfarbstoff *m*
developer *(Farb, Foto)* Entwickler *m*
d. formula Entwicklerformel *f*, Entwicklervorschrift *f*
d. improver *(Foto)* Entwicklerzusatz *m*
d. solution *(Foto)* Entwicklerlösung *f*, Entwicklungslösung *f*
d. stains *(Foto)* Entwicklerflecken *mpl*
developing *(Foto)* Entwickeln *n*, Entwicklung *f*
d. agent *(Foto)* Entwicklersubstanz *f*, Entwicklungssubstanz *f*, Entwickler *m*
d. bath *(Foto)* Entwicklerbad *n*, Entwicklungsbad *n*
d. chamber Entwicklungskammer *f*, Steiggefäß *f* *(Chromatografie)*
d. dish Entwicklerschale *f*, Fotoschale *f*
d. paper s. development paper
d. solution s. developer solution
d. tank *(Foto)* Entwicklertank *m*; Entwicklungsdose *f*
development Entwicklung *f*; *(Foto)* Entwickeln *n*, Entwicklung *f*; *(Min)* Aufschließung *f*, Erschließung *f*
d. by inspection *(Foto)* Entwicklung *f* nach Sicht, Sichtentwicklung *f*
d. by time *(Foto)* Entwicklung *f* nach Zeit, Zeitentwicklung *f*
d. fog *(Foto)* Entwicklungsschleier *m*
d. laboratory Entwicklungslabor[atorium] *n*
d. paper *(Foto)* Entwicklungspapier *n*
d. rate *(Foto)* Entwicklungsgeschwindigkeit *f*
d. stage Entwicklungsstadium *n*
d. technique *(Foto)* Entwicklungsverfahren *n*
d. time *(Foto)* Entwicklungszeit *f*, Entwicklungsdauer *f*
d. well *(Erdöl)* Produktionsbohrung *f*, Förderbohrung *f*
deviate / to ablenken; abweichen
deviation Ablenkung *f*; Abweichung *f*, Deviation *f*
device Apparat *m*, Einrichtung *f*, Vorrichtung *f*, Gerät *n*
devil *(Pap)* Haderndrescher *m*
devil's dung Teufelsdreck *m*, Asa foetida *f*, Asant *m*, Stinkasant *m* *(Gummiharz von Ferula-Arten)*
devitrification Entglasen *n*, Entglasung *f*
devitrified glass Glaskeramik *f*, Keramik *f* aus Glas, glaskeramischer Stoff *m*, Vitrokeram *n*
devitrify / to entglasen
devitrite Devitrit *m* *(Entglasungsprodukt)*

devoid of structure strukturlos, gefügelos, unstrukturiert
devolatilization Abnahme *f* der flüchtigen Bestandteile; Entfernen *n* der flüchtigen Bestandteile
devolatilize / to flüchtige Bestandteile entfernen, von flüchtigen Bestandteilen befreien
devolution Umwandlung *f (eines Elements durch radioaktiven Zerfall)*
devulcanization Devulkanisation *f*
devulcanize / to devulkanisieren
dew-point Taupunkt *m*
  d.-point curve *(Destillation)* Taulinie *f*, Kondensationslinie *f*
  d.-point method Taupunktsmethode *f*, Taupunktsverfahren *n*
  d.-ret[ting] *(Text)* Tauröste *f*, Taurotte *f*, Rasenröste *f*
Dewar *s.* 1. Dewar flask; 2. Dewar vessel
Dewar flask (jar) Dewar-Gefäß *n*, Weinhold-Gefäß *n*, Weinhold-Dewarsches Gefäß *n*
Dewar vessel Dewar-Gefäß *n*, Metall-Dewar-Gefäß *n (zum Transport flüssiger Gase)*
dewater / to entwässern, Wasser entziehen, entfeuchten, *(manchmal auch)* eindicken, verdicken
dewatering Entwässern *n*, Entwässerung *f*
dewax / to entwachsen, entparaffinieren, Wachs (Paraffin) abtrennen (entfernen)
dewaxed-oil tank Tank *m* für entwachstes (entparaffiniertes) Öl
dewaxing Entwachsen *n*, Entparaffinieren *n*, Entparaffinierung *f*, Abtrennen (Entfernen) *n* von Wachs (Paraffin)
  d. plant Entparaffinierungsanlage *f*
  d. process Entwachsungsprozeß *m*, Entparaffinierungsprozeß *m*
deweylite *(Min)* Deweylith *m (ein Serpentin)*
dewing Anfeuchten *n*, Befeuchten *n*, Netzen *n*, Benetzen *n*, Berieseln *n*, Besprühen *n*, Einsprengen *n*
dexanthation Dexanthogenierung *f*, Abspalten *n* von Xanthatgruppen
dextran Dextran *n (ein Polysaccharid)*
dextranase Dextranase *f (zu den Polyasen gehörendes Ferment)*
dextrane *s.* dextran
dextrin ($C_6H_{10}O_5$)x Dextrin *n*, Stärkegummi *n*, Dampfgummi *n*
  d. adhesive (glue) Dextrinleim *m*
  d. kettle Rührwerkpfanne *f (Etagenofen zur Dextringewinnung)*
dextrinate / to zu Dextrin abbauen, in Dextrin umwandeln (überführen)
dextrine *s.* dextrin
dextrinization Dextrinherstellung *f*, Dextringewinnung *f*, Überführung *f* in Dextrin; Dextrinbildung *f*
dextrinize / to *s.* to dextrinate
dextrinizing *s.* dextrinization
dextrinogenic dextrinogen

  d. amylase dextrinogene Amylase *f*, Dextrinogenamylase *f*
dextro *s.* dextrorotatory
  d. acid *s.* dextrorotatory acid
  d. carbonato compound *d*-Karbonatverbindung *f*
  d. form *s.* dextrorotatory form
dextrogyrate, dextrogyratory, dextrogyre, dextrogyrous *s.* dextrorotatory
dextrolactic acid $CH_2CHOHCOOH$ Rechtsmilchsäure *f*, *d*-Milchsäure *f*, Fleischmilchsäure *f*, Paramilchsäure *f*
dextronic acid $CH_2OH(CHOH)_4COOH$ Dextronsäure *f*, *d*-Glukonsäure *f*
dextropimaric acid Dextropimarsäure *f*, *d*-Pimarsäure *f*, α-Pimarsäure *f*
dextrorotary, dextrorotating *s.* dextrorotatory
dextrorotation Rechtsdrehung *f*
dextrorotatory rechtsdrehend, *d*-drehend, *d*-
  d. acid rechtsdrehende Säure *f*, Rechtssäure *f*, *d*-Säure *f*
  d. form rechtsdrehende Form *f*, Rechtsform *f*, *d*-Form *f*, (+)-Form *f*
  d. lactic acid *s.* dextrolactic acid
  d. tartaric acid *s.* dextrotartaric acid
dextrose $C_6H_{12}O_6$ Dextrose *f*, *d*-Glukose *f*, Glukose *f*, Stärkezucker *m*, Traubenzucker *m*, Blutzucker *m*
dextrotartaric acid $HOOC(CHOH)_2COOH$ Rechtsweinsäure *f*, rechtsdrehende Weinsäure *f*, *d*-Weinsäure *f*, Wein[stein]säure *f*, *d*-2,3-Dihydroxybutandisäure *f*
dezincification Entzinken *n*, Entzinkung *f*
DHS *s.* dihydrostreptomycin
Dhupa fat Vateriafett *n*, Pineytalg *m*, Malabartalg *m (Samenfett von Vateria indica L.)*
diabase Dolerit *m (grobkörnige Abart des Basalts)*; *(Am)* Diabas *m*, Grünstein *m*
diablastic *(Krist)* diablastisch
diacetic acid *(fälschlich für)* acetoacetic acid
  d. ether *s.* ethyl acetoacetate
diacetin $CH_2O(OCCH_3)CHOHCH_2O(OCCH_3)$ Glyzerindiazetat *n*
diacetone alcohol $CH_3COCH_2C(CH_3)_2OH$ Diazetonalkohol *m*, 4-Hydroxy-4-methyl-2-pentanon *n*
diacetyl $CH_3COCOCH_3$ Diazetyl *n*, Butandion-(2,3) *n*, Dimethylglyoxal *n*
diacid zweisäurig *(Base)*; zweifachsauer, primär, Dihydrogen...
diacid zweibasige (zweibasische, zweiwertige) Säure *f*
  d. phosphate $M^IH_2PO_4$ Dihydrogen[ortho]phosphat *n*, Dihydrogenmonophosphat *n*, primäres Phosphat *n*
  d. salt zweifachsaures (primäres) Salz *n*, Dihydrogensalz *n (einer dreibasigen Säure)*
diacidic *s.* diacid
diacolation Perkolation *f* unter Druck
diactinic *(durchlässig für kurzwellige, chemisch aktive Strahlung)*

diactinism *(Durchlässigkeit für kurzwellige, chemisch aktive Strahlung)*
diacyl peroxide Diazylperoxid *n*
diad *(Krist)* zweizählig; zweiwertig
diad zweiwertiges Element *n*; zweiwertige Atomgruppe *f*
  d. axis *(Krist)* zweizählige Achse (Drehachse) *f*
diadic *s.* diad
diagenesis *(Geol)* Diagenese *f*
diagenetic *(Geol)* diagenetisch
diagonal relationship Diagonalbeziehung *f*, Schrägbeziehung *f (im Periodensystem)*
diagonite *(Min)* Diagonit *m*, *(veraltet für)* Brewsterit *m (ein Tektosilikat)*
diagram Diagramm *n*, Schaubild *n*
dialdehyde Dialdehyd *m*
dialkenes Dien *n*, Diolefin *n*
dialkyl ether Dialkyläther *m*
dialkylbenzenes Dialkylbenzol *n*
dialkylborane Dialkylboran *n*
dialkylmalonic ester Dialkylmalon[säure]ester *m*
diallage *(Min)* Diallag *m*
diallogite *s.* dialogite
diallyl $CH_2=CHCH_2CH_2CH=CH_2$ Diallyl *n*, 1,5-Hexadien *n*
  d. isophthalate Diallylisophthalat *n*
  d. orthophthalate Diallylorthophthalat *n*
  d. phthalate $C_6H_4(COOCH_2CH=CH_2)_2$ Diallylphthalat *n*; *s.* d. phthalate resin
  d. phthalate resin Diallylphthalatharz *n*, DAP
dialogite *(Min)* Dialogit *m*, Manganspat *m*, Rhodochrosit *m (Mangan(II)-karbonat)*
dialuramide Uramil *n*, 5-Aminobarbitursäure *f*
dialuric acid Dialursäure *f*, Tartronylharnstoff *m*
dialyse / *to s.* to dialyze
dialysis Dialyse *f*
  d. tubing Dialysierhülse *f*
dialytic cell *s.* dialyzer
dialyzable dialysierbar
dialyzate Dialysat *n*; *(manchmal auch)* Dialysiergut *n*, zu dialysierende Flüssigkeit *f*
dialyze / *to* dialysieren
dialyzer Dialysator *m*, Dialysierzelle *f*
dialyzing area Dialysierfläche *f*
  d. membrane Dialysiermembran *f*
diamagnetic diamagnetisch
  d. material (substance) diamagnetischer Stoff (Körper) *m*, diamagnetische Substanz *f*, Diamagnetikum *n*
diamagnetism Diamagnetismus *m*
diamide $H_2N \cdot NH_2$ Diamid *n*, Hydrazin *n*
  d. hydrate $H_2N \cdot NH_2 \cdot H_2O$ Hydrazinhydrat *n*
diamidogen sulphate $[N_2H_5]HSO_4$ Hydraziniumsulfat *n*
diamine Diamin *n*
  d. chelate Diaminchelat *n*
  d.-cross-linked *(Gum)* diaminvernetzt
  d. hydrate $H_2N \cdot NH_2 \cdot H_2O$ Hydrazinhydrat *n*
  d. hydrochloride $[N_2H_6]Cl_2$ Hydraziniumdichlorid *n*

  d. sulphate $[N_2H_5]HSO_4$ Hydraziniumsulfat *n*
diaminoanthraquinone Diaminoanthrachinon *n*
2,4-diaminoazobenzene 2,4-Diaminoazobenzol *n*, Chrysoidin *n*
1,4-diaminobenzene $C_6H_4(NH_2)_2$ 1,4-Diaminobenzol *n*, p-Phenylendiamin *n*
4,4'-diaminobiphenyl $NH_2C_6H_4C_6H_4NH_2$ 4,4'-Diaminodiphenyl *n*, Benzidin *n*
2,4-diaminophenol $C_6H_3(OH)(NH_2)_2$ 2,4-Diaminophenol *n*
2,4-diaminophenol dihydrochloride $C_6H_3(OH)(NH_2)_2 \cdot 2HCl$ 2,4-Diaminophenoldihydrochlorid *n*
diaminopimelic acid $HOOC \cdot CH(NH_2) \cdot CH_2CH_2CH_2CH(NH_2) \cdot COOH$ Diaminopimelinsäure *f*, 2,6-Diaminoheptandisäure *f*
diaminostilbene $NH_2C_6H_4CH=CHC_6H_4NH_2$ Diaminostilben *n*, Bis(aminophenyl)-äthen *n*
diaminotoluene $CH_3C_6H_3(NH_2)_2$ Diaminotoluol *n*, Toluylendiamin *n*
diammine mercuric chloride $[Hg(NH_3)_2]Cl_2$ Diamminquecksilber(II)-chlorid *n*, schmelzbares weißes Präzipitat *n*
diammonium hydrogen phosphate $(NH_4)_2HPO_4$ Diammonium[hydrogen]phosphat *n*, Ammoniumhydrogen[ortho]phosphat *n*
diamond Diamant *m*
  d. black Diamantschwarz *n*
  d. cutter Diamantschneider *m*
  d. dye Diamantfarbstoff *m*
  d. ink Diamanttinte *f (zum Ätzen von Glas)*
  d. lattice *(Krist)* Diamantgitter *n*
  d.-like diamantartig
  d.-like structure *s.* d. structure
  d. mortar Diamantmörser *m*, Stahlmörser *m*, Mineralmörser *m*
  d. packing *(Krist)* Diamantpackung *f*
  d. pyramid hardness Vickers-Härte *f*, HV
  d. structure *(Krist)* Diamantstruktur *f*
diamyl phthalate $C_6H_4(COOC_5H_{11})_2$ Diamylphthalat *n*
dianisidine Dianisidin *n*, Dimethoxybenzidin *n*
  d. blue Dianisidinblau *n*
dianthranilide Dianthranilid *n*
dianthraquinoneindigo Dianthrachinonindigo *m*
dianthrimide Dianthrimid *n*, Dianthrachinonylamin *n*
diaphoretic Diaphoretikum *n*, schweißtreibendes Mittel *n*
  d. antimony Kaliumantimonat(V) *n*
diaphorite *(Min)* Diaphorit *m*
diaphragm Scheidewand *f*, Trenn[ungs]wand *f*, Zwischenwand *f*, Membran *f*; Zwischenboden *m*; *(Elektrolyse, Dialyse)* Diaphragma *n*, Membran *f*, Scheidewand *f*; *(Gum)* Balg *m*, Heizbalg *m*
  d.-actuated jig Membrankolbensetzmaschine *f*
  d. cell *(Elektrolyse)* Diaphragmazelle *f*, Diaphragmenzelle *f*, Diaphragmaelement *n*

d. cell process (*Elektrolyse*) Diaphragmaverfahren *n*, Diaphragmenverfahren *n*
d. compressor Membranverdichter *m*, Membrankompressor *m*
d. motor valve membranbetätigtes Ventil *n*
d. process *s.* d. cell process
d. pump Membranpumpe *f*
d. screen Membransortierer *m*
d. valve Saunders-Ventil *n*, Membranventil *n*
diaphthorite (*Geol*) Diaphthorit *m*
diarsenic tetramethyl $(CH_3)_2AsAs(CH_3)_2$ Tetramethyl[di]arsin *n*, Tetramethylbiarsyl *n*, *bis*-Dimethylarsyl *n*, Kakodyl *n*
diarsenobenzene Diarsenobenzol *n*
diaschistite (*Geol*) diaschistes (gespaltenes) Ganggestein *n*, Spaltungsgestein *n*, Schizolith *m*
diaspore (*Min*) Diaspor *m* (α-Aluminiumoxidhydroxid)
d. clay Diasporton *m*
diastase Diastase *f*, (*veraltet für*) . Amylase *f*, α-1,4-Glukanase *f* (*stärkespaltendes Ferment*)
diastatic Diastase..., (*veraltet für*) Amylase...
d. action Amylasewirkung *f*
d. activity Amylaseaktivität *f*
d. malt Diastasemalz *n*
diastereo[iso]mer Diastereomer[es] *n*, diastereomere Verbindung *f*
diastereo[iso]meric diastereomer
diatherma[n]cy Diathermansie *f*, Wärmedurchlässigkeit *f*
diathermanous, diathermic diatherm[an], wärmedurchlässig, durchlässig für Wärmestrahlen
diatom Diatomee *f*, Kieselalge *f*
d. earth *s.* diatomaceous earth
d. ooze Diatomeenschlamm *m*
diatomaceous earth Diatomeenerde *f*, Infusorienerde *f*, Kieselgur *f*
diatomic zweiatomig
diatomite Diatomit *m* (*aus Kieselalgen entstandene Opalsubstanz*)
diaxial diaxial
diazacyanine (*Farb*) Diazazyanin *n*
diazine Diazin *n*
1,3-diazine 1,3-Diazin *n*, Miazin *n*, Pyrimidin *n*
diazo anhydride Diazoanhydrid *n*
d. component Diazo[tierungs]komponente *f*, aktive Komponente *f*, Erstkomponente *f*
d. compound Diazoverbindung *f*
d. coupling Diazokupplung *f*
d. dye Diazofarbstoff *m*
d. oxide Diazoxid *n*
d. reaction Diazoreaktion *f*
d. solution Diazolösung *f*
diazoamino compound $RN = NNHR$ Diazoaminoverbindung *f*
diazoaminobenzene $C_6H_5N=NNHC_6H_5$ Diazoaminobenzol *n*, 1,3-Diphenyltriazen-(1) *n*
diazoate $RN = NOM$ Diazotat *n*
diazobenzene chloride $C_6H_5N_2Cl$ Benzoldiazoniumchlorid *n*

diazodinitrophenol Diazodinitrophenol *n*, 4,6-Dinitrobenzol-2-diazo-1-oxid *n*
diazomethane $N≡N=CH_2$ Diazomethan *n*
d. reaction Diazomethanreaktion *f*
diazonium component *s.* diazo component
d. compound *s.* d. salt
d. group $[ArN≡N]^+$ Diazoniumgruppe *f*
d. ion Diazonium-Ion *n*
d. nitrogen Stickstoff *m* der Diazoniumgruppe
d. salt Diazoniumsalz *n*, Diazoniumverbindung *f*
d. solution Diazonium[salz]lösung *f*
diazotate *s.* diazoate
diazotizable diazotierbar
diazotization Diazotieren *n*, Diazotierung *f*
diazotize / **to** diazotieren
diazotizing salt (*zur Diazotierung verwendetes*) Natriumnitrit *n*
diazotype paper Diazo[typie]papier *n*
diaphanic paper Buntglaspapier *n*, Diaphaniepapier *n*, Fensterpapier *n*
dibasic zweibasig, zweibasisch
d. acid zweibasige (zweibasische, zweiwertige) Säure *f*
d. ammonium citrate $(NH_4)_2HC_6H_5O_7$ Ammoniumhydrogenzitrat *n*, Diammoniumzitrat *n*, Ammoniumzitratsäure *f*
d. ammonium phosphate $(NH_4)_2HPO_4$ Ammoniumhydrogen[ortho]phosphat *n*, Diammoniumhydrogenphosphat *n*
d. barium phosphate $BaHPO_4$ Bariumhydrogen[ortho]phosphat *n*
d. calcium phosphate $CaHPO_4$ Kalziumhydrogen[ortho]phosphat *n*
d. lead arsenate $PbHAsO_4$ Blei(II)-hydrogenarsenat(V) *n*
d. magnesium phosphate $MgHPO_4$ Magnesiumhydrogen[ortho]phosphat *n*
d. manganous phosphate $MnHPO_4$ Mangan(II)-hydrogen[ortho]phosphat *n*
d. phosphate Me'_2HPO_4 sekundäres Phosphat *n*, Hydrogen[ortho]phosphat *n*
d. potassium orthoarsenate K_2HAsO_4 Kaliumdihydrogen[ortho]arsenat(V) *n*
d. potassium orthophosphate K_2HPO_4 Kaliumhydrogen[ortho]phosphat *n*, Dikaliumhydrogenphosphat *n*
d. sodium orthoarsenate Na_2HAsO_4 Dinatriumhydrogen[ortho]arsenat(V) *n*
d. sodium orthoarsenite Na_2HAsO_3 Dinatriumhydrogen[ortho]arsenat(III) *n*
d. sodium orthophosphate Na_2HPO_4 Natrium[mono]hydrogenphosphat *n*, Dinatriumhydrogenphosphat *n*
dibenzanthracene Dibenzanthrazen *n*
dibenzanthraquinone Dibenzanthrachinon *n*
dibenzopyran Dibenzopyran *n*, Xanthen *n*
dibenzopyrone Dibenzo-γ-pyron *n*, 9-Oxoxanthen *n*, Xanthon *n*
dibenzopyrrole Dibenzopyrrol *n*, Karbazol *n*, Diphenylimid *n*

dibenzoyl $C_6H_5COCOC_6H_5$ Dibenzoyl n, Benzil n, Diphenylglyoxal n, Diphenyldiketon n
  d. peroxide $C_6H_5OCOOCOC_6H_5$ Dibenzoylperoxid n
dibenzoylation Dibenzoylieren n, Dibenzoylierung f, zweifache Benzoylierung f
dibenzpyrenequinone Dibenzpyrenchinon n
dibenzyl $C_6H_5CH_2CH_2C_6H_5$ Dibenzyl n, Bibenzyl n, 1,2-Diphenyläthan n
  d. ether $(C_6H_5CH_2)_2O$ Dibenzyläther m
diborane $(BH_3)_2$ oder B_2H_6 Diboran n
diboranide Diboranid n
diboride Diborid n
diboron hexahydride s. diborane
  d. tetrachloride B_2Cl_4 Dibortetrachlorid n
dibromide Dibromid n
dibrominate / to dibromieren, zweifach bromieren
dibromination Dibromieren n, Dibromierung f, zweifache Bromierung f
dibromo compound Dibromverbindung f
dibromoanthraquinone Dibromanthrachinon n
dibromobenzene $C_6H_4Br_2$ Dibrombenzol n
1,2-dibromoethane $BrCH_2CH_2Br$ 1,2-Dibromäthan n, Äthylen[di]bromid n
dibromofluorescein Dibromfluoreszein n
dibromoindigo Dibromindigo m
dibromomethane CH_2Br_2 Dibrommethan n, Methylenbromid n
1,2-dibromopropane $CH_3CHBrCH_2Br$ 1,2-Dibrompropan n, Propylen[di]bromid n
dibromothymolsulphonephthalein Dibromthymolsulfophthalein n, Bromthymolblau n (ein pH-Indikator)
dibutyl phthalate $C_6H_4(COOC_4H_9)_2$ Dibutylphthalat n
  d. sebacate $C_4H_9OCO(CH_2)_8OCOC_4H_9$ Dibutylsebazat n
dibutylacetylene $CH_3(CH_2)_3C\equiv C(CH_2)_3CH_3$ Dibutylazetylen n, 5-Dezin n
dicalcium phosphate $CaHPO_4$ Kalziumhydrogen[ortho]phosphat n
dicarbide Dikarbid n
dicarboxylic acid Dikarbonsäure f
dicatechin Dikatechin n (Hydroxyflavan-Derivat)
dice / to in Würfel schneiden
dice (Glas) würfelförmiger Bruch m, Würfelbruch m (bei gehärtetem Glas)
  d. block (Glas) Durchlaß-Seitenstein m
dicer s. dicing cutter
dichlone Dichlone n, 2,3-Dichlor-1,4-naphthochinon n (Herbizid)
dichloride Dichlorid n
dichloro derivative Dichlorderivat n
α,α-**dichloroacetamide** $CHCl_2CONH_2$ 2,2-Dichlorazetamid n, 2,2-Dichloräthansäureamid n
dichloroacetic acid $Cl_2CHCOOH$ Dichloressigsäure f, Dichloräthansäure f
dichloroaniline $Cl_2C_6H_3NH_2$ Dichloranilin n
dichloroanthraquinone Dichloranthrachinon n
dichlorobenzene $C_6H_4Cl_2$ Dichlorbenzol n

dichlorodiethyl ether $ClCH_2CH_2OCH_2CH_2Cl$ Dichlordiäthyläther m, Bis-[2-chlor-äthyl]-äther m
  d. sulphide $(ClCH_2CH_2)_2S$ Dichlordiäthylsulfid n, Senfgas n
dichlorodifluoromethane Cl_2CF_2 Dichlordifluormethan n
dichlorodiphenyltrichloroethane Dichlordiphenyltrichloräthan n, 1,1,1-Trichlor-2,2-bis(p-chlorphenyl)äthan n, DDT n
2,2-dichloroethanamide s. α,α-dichloroacetamide
1,1-dichloroethane CH_3CHCl_2 1,1-Dichloräthan n, Äthylidenchlorid n
1,2-dichloroethane $ClCH_2CH_2Cl$ 1,2-Dichloräthan n, Äthylen[di]chlorid n
dichloroethanoic acid s. dichloroacetic acid
dichloroether s. 1,2-dichloro-1-ethoxy ethane
1,2-dichloro-1-ethoxy ethane $ClCH_2CH(Cl)OC_2H_5$ 1-Äthoxy-1,2-dichloräthan n, 1,2-Dichlordiäthyläther m, Dichloräther m
α,β-**dichloroethyl ether** s. 1,2-dichloro-1-ethoxy ethane
sym-dichloroethyl ether $(ClCH_2CH_2)_2O$ sym-Dichloräthyläther m, Bis-β-chloräthyläther m, β,β'-Dichlordiäthyläther m, 1-Chlor-2-(2-chloräthoxy)-äthan n
dichloroethyl sulphide s. dichlorodiethyl sulphide
dichloromethane CH_2Cl_2 Dichlormethan n, Methylenchlorid n
dichloromonofluoromethane Cl_2CHF Dichlorfluormethan n
dichloronaphthalene $C_{10}H_6Cl_2$ Dichlornaphthalin n
2,4-dichlorophenoxyacetic acid $Cl_2C_6H_3OCH_2COOH$ 2,4-Dichlorphenoxyessigsäure f, 2,4-D n (Herbizid)
4-(2,4-dichlorophenoxy)butyric acid $Cl_2C_6H_3O(CH_2)_3COOH$ 4(2',4'-Dichlorphenoxy)-buttersäure f, 2,4-DB n (Herbizid)
dichloropropane $C_3H_6Cl_2$ Dichlorpropan n
dichloropyrimidine Dichlorpyrimidin n
dichlorotetrammineplatinic chloride $[Pt(NH_3)_4Cl_2]Cl_2$ Dichlorotetramminplatin(IV)-chlorid n
dichlorotriazine Dichlortriazin n
dichlorprop $Cl_2C_6H_3OCH(CH_3)COOH$ Dichlorprop n, 2,4-DP n, 2-(2',4'-Dichlorphenoxy)propionsäure f (Herbizid)
dichroic (Krist, Koll) dichroitisch, doppelfarbig; s. dichromatic
  d. fog (Foto) dichroitischer Schleier m (Fehler)
dichroism (Krist, Koll) Dichroismus m, Doppelfarbigkeit f
dichroite (Min) Dichroit m, (veraltet für) Cordierit m (Magnesiumaluminiumalumopentasilikat)
dichromate $M^I_2Cr_2O_7$ Dichromat n
  d. titration Chromatometrie f
dichromatic zweifarbig, dichromatisch
dichromatism Zweifarbigkeit f
dicing Würfeln n (Schneiden in Würfel)
  d. cutter (machine) Schnitzelmaschine f; (Plast)

Würfelschneider *m*; *(Halbleitertechnik)* Plätt-chenschneidemaschine *f*
dickinsonite *(Min)* Dickinsonit *m* *(Natriumman-gan(II)-hydrogenorthophosphat)*
dickite *(Min)* Dickit *m* *(Aluminiumhydroxidsilikat)*
dicoumarol Dikumarol *n*, 3,3'-Methylen-bis-4-hy-droxykumarin *n*
dicrotaline Dikrotalin *n* *(Alkaloid)*
dicrotalic acid HOOC·$CH_2C(CH_3)(OH)CH_2$·COOH Dikrotolsäure *f*, Dikrotalinsäure *f*
dicumyl peroxide
$C_6H_5C(CH_3)_2$·O·O·$C(CH_3)_2C_6H_5$ Dikumylper-oxid *n*
dicy *s.* dicyandiamide
dicyandiamide H_2N·C(=NH)·NH·CN Dizyandiamid *n*
dicyanogen $(CN)_2$ Dizyan *n*, Zyan *n*
didecyl phthalate $C_6H_4(COOC_{10}H_{21})_2$ Didezyl-phthalat *n*
di-derivative Diderivat *n*
Didier-Bubiag process Didier-Bubiag-Verfahren *n*, Bubiag-Didier-Gaserzeugungsverfahren *n*, Gleichstromverfahren *n* nach Didier-Bubiag
DIDP *s.* diisodecyl phthalate
didymia Di_2O_3 Didymerde *f*, *(veraltet für)* Didym-oxid *n*
didymium Didym *n* *(Nd-Pr-Gemisch)*; Didym[me-tall] *n* *(Nd-Pr-Legierung)*
die 1. Matrize *f*, *(meist)* Unterwerkzeug *n*, Form *f*, Preßform *f* *(oder)* Stanzform *f*, Schnittplatte *f*, Schneidplatte *f*, Stanzwerkzeug *n*, Schnittwerk-zeug *n*; 2. Hohlform *f*, Form *f*, *(Metallverarbeitung auch)* Gesenk *n* *(oder)* [metallische] Dauergieß-form (Dauerform) *f*, Metallform *f*, Kokille *f*; 3. *(Ker)* Mundstück *n* *(einer Strangpresse)*; *(Gum)* Spritz-mundstück *n*; *(Plast)* Düse *f*
   d. adapter *(Plast)* Düsenpaßstück *n*, Düsenhalter *m*
   d. base *(Am)* *s.* d. body
   d. body *(Plast)* Düsenkörper *m*
   d.-box *(Gum)* Spritzkopf *m*
   d. for tubing *(Gum)* Schlauchspritzmundstück *n*
   d. head *(Plast)* Extruderkopf *m*, Strangpressen-kopf *m*, Spritzkopf *m*
   d. lips *(Plast)* Austrittsspalt *m* des Extruderkopfes
   d.-pressed *(Ker)* trockengepreßt
   d. pressure Druck *m* in der Düse
   d. swell *(Gum)* Spritzquellung *f*
diecast / **to** in Kokille gießen, in der Kokille vergießen *(i.w.S.)*; druckgießen, unter Druck gießen, durch Druck (auf Druckgießmaschinen) vergießen, im Druckguß herstellen
diecasting Kokillengießen *n*, Kokillenguß *m* *(i.w.S.)*; Druckgießen *n*, Druckguß *m*; Druckguß-stück *n*, Druckgußteil *n*, Dauerformgußstück *n*
   d. alloy Kokillengußlegierung *f*; Druckgußlegie-rung *f*
   d. die Druckgießform *f*, Druckgußform *f*
   d. machine Kokillengießmaschine *f*; Druckgieß-maschine *f*, Druckgußmaschine *f*

   d. process Kokillengußverfahren *n* *(i.w.S.)*; Druckgießverfahren *n*, Druckgußverfahren *n*
Dieckmann condensation (reaction) Dieckmann-Reaktion *f*, Dieckmannsche [intermolekulare] Esterkondensation *f*
dielectric dielektrisch, Dielektrizitäts...
dielectric Dielektrikum *n*, Nichtleiter *m*
   d. constant Dielektrizitätskonstante *f*, DK
   d. dryer Hochfrequenztrockner *m*
   d. drying dielektrische Trocknung *f*, Hochfre-quenztrocknung *f*
   d. effect dielektrischer Effekt *m*
   d. fluid dielektrische Flüssigkeit *f*
   d. heating dielektrische Beheizung *f*, Dielektro-[be]heizung *f*, Hochfrequenz[be]heizung *f*
   d. loss dielektrischer Verlust *m*
   d. loss factor [dielektrischer] Verlustwinkel *m*
   d. material Dielektrikum *n*, Nichtleiter *m*
   d. polarization dielektrische Polarisation *f*
   d. preheating dielektrische Vorwärmung *f*, Hochfrequenzvorwärmung *f*
   d. sealing Hochfrequenzsiegeln *n* *(von Folien)*
   d. strength Durchschlag[s]festigkeit *f*, dielektri-sche Festigkeit *f*
dielectrometer DK-Meter *n*, Dekameter *n*, Dielektri-zitätskonstante-Messer *m*
dielectrometry DK-Metrie *f*, Dekametrie *f*, Dielek-trometrie *f*, Bestimmung (Messung) *f* der Dielektrizitätskonstante
Diels-Alder reaction Diels-Alder-Reaktion *f*, Diels-Alder-Synthese *f*, Diels-Aldersche Diensynthese *f*, Dien-1,4-Addition *f*
diene Dien *n*, Diolefin *n*
   d.[-based] polymer Dienpolymerisat *n*, Dienpoly-mer[es] *n*
   d. rubber Dienkautschuk *m*
   d. synthesis Diensynthese *f*
dienophile Dienophil *n*, Philodien *n*, dienophile Komponente *f*, philodiener Partner *m*
dienophilic dienophil, philodien
diesel engine oil Dieselschmieröl *n*
   d. fraction *s.* d. oil fraction
   d. fuel Dieselkraftstoff *m*, DK, Dieseltreibstoff *m*
   d. index Dieselindex *m*, D.I.
   d. knock Nageln *n*, Dieselklopfen *n* *(Klopfen im Dieselmotor)*
   d. oil Dieselkraftstoff *m*, DK, Dieseltreibstoff *m*, *(veraltet)* Dieselöl *n*
   d. oil for road vehicles Fahrdieselkraftstoff *m*, Kraftstoff *m* für Fahrdieselmotoren, *(veraltet)* Fahrdieselöl *n*
   d. oil fraction Dieselölfraktion *f*
diester Diester *m* •
Diesulforming process Diesulforming-Verfahren *n* *(katalytischer Entschwefelungsprozeß)*
diet Nahrung *f*, Speise *f*, Kost *f*, Ernährung *f*; Diät *f*, Schonkost *f*, Krankenkost *f*
dietary diätetisch, Diät...
   d. fat Nahrungsfett *n*
   d. food diätetische Lebensmittel *npl*

dietetic[al] s. dietary
dietherate Diätherat n
diethoxymethane $CH_2(OC_2H_5)_2$ Diäthoxymethan n, Diäthylformal n, Formaldehyddiäthylazetal n, Äthylal n
diethyl acetylene $CH_3CH_2C≡CCH_2CH_3$ Diäthylazetylen n, Hexin-(3) n
  d. disulphide $(C_2H_5)_2S_2$ Diäthyldisulfid n, Äthyldithioäthan n
  d. ether $C_2H_5OC_2H_5$ Diäthyläther m, Äthyläther m, Äther m (i.e.S.), Äthoxyäthan n
  d. hexyl phthalate Diäthylhexylphthalat n
  d. hydroxy butenedioate s. d. oxal[o]acetate
  d. malonate $CH_2(COOC_2H_5)_2$ Malonsäurediäthylester m, Äthylmalonat n
  d. oxal[o]acetate $C_2H_5OOC·C(OH)=CH·COOC_2H_5$ Oxalessigsäurediäthylester m, Oxalessigester m
  d. phthalate $C_6H_4(COOCH_2CH_3)_2$ Phthalsäurediäthylester m, Äthylphthalat n
  d. propanedioate s. d. malonate
  d. sebacate $C_2H_5COO(CH_2)_8COOC_2H_5$ Sebazinsäurediäthylester m, Diäthylsebazat n
  d. succinate $C_2H_5O·CO(CH_2)_2COOC_2H_5$ Bernsteinsäurediäthylester m, Diäthylsukzinat n
  d. sulphate $(C_2H_5)_2SO_4$ Schwefelsäurediäthylester m, Diäthylsulfat n, Äthylsulfat n
  d. sulphide $(C_2H_5)_2S$ Diäthylsulfid n, Äthylsulfid n, Äthylthioäthan n
  d. zinc $Zn(C_2H_5)_2$ Diäthylzink n
diethylamine $(C_2H_5)_2NH$ Diäthylamin n
diethylcarbocyanine iodide Diäthylkarbozyaninjodid n
diethyldithiocarbamic acid $(C_2H_5)_2·NCS·SH$ Diäthyldithiokarbaminsäure f
diethylene glycol $HOCH_2CH_2OCH_2CH_2OH$ Diäthylenglykol n, 2-(2-Hydroxyäthoxy)äthanol n
diethylenetriamine chelate Diäthylentriaminchelat n
diethylformal $CH_2(OC_2H_5)_2$ Diäthylformal n, Formaldehyddiäthylazetal n, Diäthoxymethan n, Äthylal n
diethylmagnesium $Mg(C_2H_5)_2$ Diäthylmagnesium n, Magnesiumdiäthyl n
diethylrhodamine Diäthylrhodamin n, Rhodamin 6 G n, Brillantrosa n, Rosamin n
diethylstilboestrol Diäthylstilböstrol n, trans-p,p'-Dihydroxydiäthylstilben n
difference Unterschied m, Verschiedenheit f, Abweichung f, Differenz f
  d. in charge Ladungsunterschied m, Ladungsdifferenz f
  d. in colour Farbunterschied m
  d. in solubility Löslichkeitsunterschied m
  d. in specific gravity Wichteunterschied m
  d. of phase Phasendifferenz f, Phasenunterschied m
  d. of potential Potentialdifferenz f, Potentialunterschied m
  d. of potential on direct contact Kontaktpotentialdifferenz f

  d. of velocity Geschwindigkeitsdifferenz f, Geschwindigkeitsgefälle n
differential calorimeter Differentialkalorimeter n, Zwillingskalorimeter n
  d. distillation Differentialdestillation f, offene (differentielle) Destillation f
  d. dyeing (Text) Kontrastfärbung f
  d. floatation differentielle (selektive, sortenweise) Flotation f, Differentialflotation f, Selektivflotation f
  d. head Differenzdruckhöhe f, Differenzdruck m, Wirkdruck m
  d. heat differentielle (differentiale, intermediäre) Wärme f
  d. heat of adsorption differentielle (differentiale) Adsorptionswärme f
  d. heat of dilution differentielle (differentiale) Verdünnungswärme (Verdünnungsenthalpie) f
  d. heat of solution differentielle (differentiale) Lösungswärme (Lösungsenthalpie) f
  d. Joule-Thomson-effect differentieller Joule-Thomson-Koeffizient (Joule-Thomson-Effekt) m
  d. manometer Differentialmanometer n, Differenz[druck]manometer n
  d. photometry Differentialfotometrie f
  d. pressure Differenzdruck m, Wirkdruck m
  d.-pressure meter Differenzdruckmeßgerät n, Differenzdruckmesser m, Wirkdruckmesser m
  d. shrinkage (Text) differentielle Schrumpfung f
  d.-speed ungleich schnell laufend, mit Friktion (Walzen)
  d. thermal analysis Differentialthermoanalyse f, DTA
  d. thermogravimetry derivate Thermogravimetrie f, Differential-Thermogravimetrie f
  d. titration Differentialtitration f
difficulty fusible schwerschmelzbar
  d. soluble schwerlöslich, schwer (wenig) löslich, slö, wl.
diffraction Beugung f, Diffraktion f
  d. angle Beugungswinkel m
  d. grating Beugungsgitter n
  d. maximum Interferenzmaximum n
  d. pattern Beugungsbild n
  d. phenomenon Beugungserscheinung f, Diffraktionserscheinung f
  d. ring Beugungsring m
  d. spectrum Beugungsspektrum n, Gitterspektrum n, Normalspektrum n
diffusate Diffusat n
diffuse / to ausbreiten, (Flüssigkeiten) ausgießen, diffundieren; eindringen lassen; diffundieren, (in feiner Verteilung) eindringen
diffuse double layer (phys Ch) diffuse Doppelschicht f
  d. series (Spektroskopie) diffuse (erste) Nebenserie f
diffused junction diffundierter Übergang m, Diffusionsübergang m, Diffusionsschicht f
diffuser (Pap) Diffuseur m

d. pump Turbinenpumpe f, Pumpe f mit Leitrad \
d. stone Filterstein m, Fritte f *(zum Lösen von Gasen in Flüssigkeiten)* \
diffusible diffusionsfähig \
diffusing tank s. diffuser \
diffusion Diffusion f; *(mit poröser Trennwand)* Transfusion f \
d. battery Diffusionsbatterie f \
d. cell Diffuseur m, Diffusionsapparat m \
d. cloud chamber Diffusionsnebelkammer f \
d. coefficient Diffusionskoeffizient m, Diffusionskonstante f \
d. cossettes *(Zucker)* Diffusionsschnitzel npl \
d. current Diffusionsstrom m, Grenzstrom m \
d.-current constant Diffusionsstromkonstante f \
d. juice *(Zucker)* Diffusionssaft m, Rohsaft m \
d. layer Diffusionsschicht f \
d. length Diffusionslänge f \
d. loss *(Zucker)* Diffusionsverlust m \
d. method Diffusionsmethode f, Diffusionsverfahren n \
d. potential Diffusionspotential n, Flüssigkeitspotential n \
d. process Diffusionsvorgang m, Austauschvorgang m, Ausgleichvorgang m; *(Met)* Diffusionsverfahren n, Zementierverfahren n, Zementationsverfahren n \
d. pulp water s. d. waste water \
d. pump Diffusionspumpe f \
d.-pump fluid Betriebsflüssigkeit f für Diffusionspumpen \
d.-pump oil Diffusionspumpenöl n \
d. rate Diffusionsgeschwindigkeit f; Transfusionsgeschwindigkeit f \
d. ring Leitring m, Leitrad n *(z.B. bei Pumpen)* \
d. through a porous diaphragm (membrane) Transfusion f \
d. velocity s. d. rate \
d. [waste] water *(Zucker)* Diffusionsabwasser n \
diffusional diffusorisch, Diffusions... \
diffusive mixing Diffusions[ver]mischen n \
diffusivity Diffusionsfähigkeit f, Diffusionsvermögen n \
diffusor s. diffuser \
difluoride Difluorid n \
difluorodichloromethane CF_2Cl_2 Difluordichlormethan n \
diformate Diformiat n \
difunctional difunktionell, bifunktionell \
d. unit difunktionelle Einheit f, D-Einheit f *(beim molekularen Aufbau)* \
dig out / **to** *(z.B. Filterkuchen)* abstechen \
m-digallic acid m-Digallussäure f, 3-Digallussäure f, Gallussäure-3-monogallat n \
digenite *(Min)* Digenit m *(ein Kupfersulfid)* \
digermanate $M_2^IGe_2O_5$ Digermanat n \
digermane Ge_2H_6 Digerman n, Germaniumhexahydrid n \
digest / **to** *(durch Hitze und/oder Lösungsmittel)*

aufschließen, digerieren; ausziehen, auslaugen, digerieren; *(Bio)* verdauen \
digested sludge ausgefaulter Schlamm m, Faulschlamm m \
digester *(Pap)* Zellstoffkocher m, Kocher m; *(Text)* Eiweißverdauer m, Eiweißspalter m *(für Vordetachur)*; *(Pharm)* Digestivum n, verdauungsförderndes Mittel n; *(Wasseraufbereitung)* Faulbehälter m, Schlammfaulbehälter m; *(Kosmet)* Digestor m, Digerierkolben m, Digestionskolben m \
d. blow *(Pap)* *(durch Ausblasen bewirkte)* Kocherleerung f, Kocherausblasen n, Ausblasen (Leerblasen, Abblasen) n des Zellstoffkochers \
d. capacity *(Pap)* Kochervolumen n, Koch[er]raum m \
d. charge (contents) *(Pap)* Kocherinhalt m, Kochereintrag m, Kocherfüllung f, Kochgut n, Charge f \
d. cover *(Pap)* Kocherdeckel m \
d. cycle *(Pap)* Kocherturnus m, Kocherumtrieb m \
d. gas Faulgas n \
d. house *(Pap)* Kocherei f, Kochstation f \
d. liquor *(Pap)* *(im Kocher befindliche)* Kochsäure f *(oder)* Kochlauge f \
d. pressure *(Pap)* Kocherdruck m \
d. relief gas *(Pap)* Kocherabgas n, Abgas n, Übertriebgas n \
d. space *(Pap)* Koch[er]raum m, Kochervolumen n \
d. temperature *(Pap)* Temperatur f im Kocher, Kochtemperatur f \
d. test *(Pap)* Kontrolle f der Kochsäurezusammensetzung \
d. top relief *(Pap)* Abgasen n am Kocherdeckel \
d. wall *(Pap)* Kocherwand[ung] f \
digestibility Verdaulichkeit f, Bekömmlichkeit f \
d. coefficient Verdauungskoeffizient m, Verdauungsquotient m, Verdauungswert m \
digestible aufschließbar; ausziehbar, auslaugbar; *(Bio)* verdaulich, verdaubar, bekömmlich \
digesting s. digestion \
d. agent Aufschlußmittel n, Aufschlußchemikalie f, Aufschließungsreagens n \
d. compartment Faulraum m, Schlammfaulraum m \
digestion Aufschließen n, Aufschluß m *(durch Hitze und/oder Lösungsmittel)*, Digerieren n, *(Pap auch)* Kochen n, Kochung f; Ausziehen n, Auslaugen n, Digerieren n; *(Bio)* Verdauung f, Digestion f; *(Foto)* Reifen n, Reifung f *(der Emulsion)* \
d. at high temperatures *(Pap)* Hochtemperaturkochung f \
d. chamber Faulraum m, Schlammfaulraum m \
d. coefficient s. digestibility coefficient \
d. flask *(Kosmet)* Digerierkolben m, Digestionskolben m, Digestor m \
d. liquor *(Pap)* Kochflüssigkeit f, Kochlösung f, Aufschlußlösung f, *(beim Sulfitverfahren)* Koch-

säure f, Sulfit[koch]säure f, *(beim alkalischen Verfahren)* Kochlauge f, Aufschlußlauge f
d. mix Aufschlußmittelgemisch n, Reaktionsgemisch n, *(beim Kjeldahl-Aufschluß auch)* Katalysatorgemisch n
d. period s. **d. time**
d. stand Aufschließgestell n *(der Kjeldahl-Apparatur)*
d. time *(Wasseraufbereitung)* Faulzeit f
digestive ferment Verdauungsferment n
d. juice Verdauungssaft m
digestor s. **digester**
digitonin precipitation Digitoninfällung f *(zum Nachweis von Pflanzenfetten)*
digitoxin Digitoxin n, 3,14-Dihydroxykardenolidtridigitoxosid n *(Alkaloid)*
diglyceride Diglyzerid n
diglycidyl ether Diglyzidyläther m
diglycolic acid $O(CH_2COOH)_2$ Diglykolsäure f
digonal hybrid digonales Hybrid n
d. hybridization digonale Hybridisierung f
digoxin Digoxin n *(Alkaloid)*
dihalide Dihalogenid n
diheptadecyl ketone Diheptadezylketon n, Stearon n, 18-Pentatriakontanon n
dihexyl $CH_3(CH_2)_{10}CH_3$ Dihexyl n, Bihexyl n, n-Dodekan n
d. phthalate $C_6H_4(COOC_6H_{13})_2$ Dihexylphthalat n, DHP
dihydrate Dihydrat n
dihydric zweiwertig, mit zwei Hydroxylgruppen (OH-Gruppen); zweifachsauer, primär, Dihydrogen...
d. alcohol $C_nH_{2n}(OH)_2$ zweiwertiger Alkohol m, Diol n, Glykol n *(i.w.S.)*
d. phenol zweiwertiges Phenol n
d. phosphate $M^IH_2PO_4$ Dihydrogen[ortho]phosphat n, primäres Phosphat n
dihydrite *(Min)* Dihydrit m *(Kupfer(II)-tetrahydroxiddiorthophosphat)*
dihydro stage Dihydrostufe f
N,N-**dihydro-1,2,1',2'-anthraquinoneazine** *N,N*-Dihydro-1,2,1',2'-anthrachinonazin n, Indanthron n
1,2-dihydrobenzene 1,2-Dihydrobenzol n, Zyklohexadien-(1,3) n
1,4-dihydrobenzene 1,4-Dihydrobenzol n, Zyklohexadien-(1,4) n
dihydrocorynantheine Dihydrokorynanthein n *(Alkaloid)*
dihydrogen Dihydrogen...
dihydrogen H_2 Diwasserstoff m
d. phosphate $M^IH_2PO_4$ Dihydrogen[ortho]phosphat n, Dihydrogenmonophosphat n, primäres Phosphat n
d. potassium orthophosphite KH_2PO_3 Kaliumdihydrogen[ortho]phosphit n
dihydropinosylvin Dihydropinosylvin n *(Stilbenderivat)*
dihydroquinine Dihydrochinin n, Hydrochinin n

dihydrostreptomycin Dihydrostreptomyzin n *(Antibiotikum)*
dihydroxide Dihydroxid n
dihydroxy Dihydroxy...
d. compound Dihydroxyverbindung f
dihydroxyanthraquinone $C_{14}H_6O_2(OH)_2$ Dihydroxyanthrachinon n
o,o'-**dihydroxyazo compound** *o,o'*-Dihydroxyazoverbindung f
m-**dihydroxybenzene** $C_6H_4(OH)_2$ *m*-Dihydroxybenzol n, 1,3-Dihydroxybenzol n, Resorzin n
o-**dihydroxybenzene** $C_6H_4(OH)_2$ *o*-Dihydroxybenzol n, 1,2-Dihydroxybenzol n, Brenzkatechin n
p-**dihydroxybenzene** $C_6H_4(OH)_2$ *p*-Dihydroxybenzol n, 1,4-Dihydroxybenzol n, Hydrochinon n
2,5-dihydroxybenzoic acid $(HO)_2C_6H_3COOH$ 2,5-Dihydroxybenzoesäure f, Gentisinsäure f
3,4-dihydroxybenzoic acid $(HO)_2C_6H_3COOH$ 3,4-Dihydroxybenzoesäure f, Protokatechusäure f
2,3-dihydroxybutanedioic acid $HOOC(CHOH)_2COOH$ 2,3-Dihydroxybutandisäure f, Dihydroxybernsteinsäure f, Wein[stein]säure f
3,3'-dihydroxy-α-carotene 3,3'-Dihydroxy-α-karotin n, Xanthophyll n, Lutein n *(gelber Naturfarbstoff)*
3,4-dihydroxycinnamic acid $(OH)_2C_6H_3CH=CHCOOH$ 3,4-Dihydroxyzimtsäure f, 3-(3,4-Dihydroxyphenyl)propensäure f, Kaffeesäure f, Kaffeinsäure f
1,8-dihydroxy-3-methyl anthraquinone 1,8-Dihydroxy-3-methylanthrachinon n, Chrysophansäure f, Chrysophanol n
dihydroxynaphthalene $C_{10}H_6(OH)_2$ Dihydroxynaphthalin n
dihydroxyphenylalanine $(HO)_2C_6H_3CH_2CH(NH_2)COOH$ Dihydroxyphenylalanin n, Dopa n, 2-Amino-3-(3,4-dihydroxyphenyl)propansäure f
2,3-dihydroxypropanal $CH_2OHCHOHCHO$ 2,3-Dihydroxypropanal n, Propandiol-(2,3)-al-(1) n, Glyzer[in]aldehyd m
dihydroxypyridine $C_5H_3N(OH)_2$ Dihydroxypyridin n, Pyridindiol n
d. carboxylic acid 2,6-Dihydroxyisonikotinsäure f, Zitrazinsäure f
dihydroxysuccinic acid $HOOC(CHOH)_2COOH$ Dihydroxybernsteinsäure f, 2,3-Dihydroxybutandisäure f, Wein[stein]säure f
diiodide Dijodid n
diiodomethane CH_2J_2 Dijodmethan n, Methylenjodid n
diisocyanate Diisocyanat n *(Verbindung mit zwei* $-N=C=O$-*Gruppen)*
diisodecyl phthalate Diisodezylphthalat n
diisooctyl phthalate Diisooktylphthalat n
diisopropyl ether $(CH_3)_2CHOCH(CH_3)_2$ Diisopropyläther m, Isopropyläther m, 2-Isopropoxypropan n

diisopropylmethane $(CH_3)_2CHCH_2CH(CH_3)_2$
Diisopropylmethan n, 2,4-Dimethylpentan n
Dika butter Dikabutter f, Dikafett n *(von Irvingia gabonensis Baill.)*
dike Gesteinsgang m, Gang m
   d. rock Ganggestein n
diketo ester Diketoester m
diketone Diketon n
Dikka butter s. Dika butter
Dikkamaly resin *(Harz von Gardenia gummifera L.f.)*
dil. s. dilute
dilatancy *(Koll)* Dilatanz f *(Gegensatz zur Thixotropie)*
dilatant *(Koll)* dilatant
dilatation Dilatation f, Ausdehnung f, Expansion f
dilatometer Dilatometer n
   d. test Dilatometertest m, Dilatometerversuch m, Dilatometerprobe f
dilatometric dilatometrisch
   d. method s. dilatometry
dilatometry Dilatometrie f, Dehnungsmessung f, dilatometrische Untersuchung f
dilituric acid Dilitursäure f, 5-Nitrobarbitursäure f
dill Dill m, Anethum graveolens L.
   d. seed oil Dillöl n
diluent Verdünnungsmittel n, Verschnittmittel n, Streckmittel n, Streckstoff m; Ballaststoff m, Ballast m *(z.B. des Brennstoffs)*
dilute / **to** verdünnen, strecken; *(Landw) (Nährstoffe aus dem Oberboden)* auswaschen
dilute verdünnt, verd., *(von Säuren, Laugen auch)* schwach
   d. black liquor *(Pap)* Dünn[ab]lauge f
diluting agent Verdünnungsmittel n; Verschnittmittel n, Streckmittel n, Streckstoff m
dilution Verdünnen n, Verdünnung f; Strecken n, Streckung f; *(verdünnte)* Lösung f; *(Landw)* Auswaschung f *(von Nährstoffen)*
   d. factor Verdünnungsfaktor m
   d. law Ostwaldsches Verdünnungsgesetz n
   d. ratio Verdünnungsverhältnis n, Verdünnungsgrad m, Verdünnungsfaktor m; Verschnittfähigkeit f
dimensional stability Dimensionsstabilität f, Maßbeständigkeit f, Maßhaltigkeit f; *(Plast, Text)* Formbeständigkeit f
   d. stabilization Dimensionsstabilisierung f
dimer Dimer[es] n
dimeric dimer
dimerization Dimerisation f, Dimerisierung f
dimesoperiodate $M^I_4J_2O_9$ Enneaoxodiperjodat n, Diperjodat n
dimethoxymethane $CH_2(OCH_3)_2$ Dimethoxymethan n, Dimethylformal n, Formaldehyddimethylazetal n, Formal n, Methylal n
dimethoxystrychnine 2,3-Dimethoxystrychnin n, Bruzin n *(Alkaloid)*
dimethyl acetic acid $(CH_3)_2CHCOOH$ Dimethylessigsäure f, Isobuttersäure f, Buttersäure f,

2-Methylpropionsäure f, 2-Methylpropansäure f
   d. arsenic monochloride $(CH_3)_2AsCl$ Dimethylarsinchlorid n, Dimethylchlorarsin n, Chlordimethylarsin n, Kakodylchlorid n
   d. arsinic acid $(CH_3)_2AsOOH$ Dimethylarsinsäure f, Kakodylsäure f
   d. carbinol $CH_3CHOHCH_3$ Dimethylkarbinol n, Isopropylalkohol m, sek-Propylalkohol m, 2-Propanol n
   d. ether CH_3OCH_3 Dimethyläther m, Methyläther m, Methoxymethan n
   d. ethyl carbinol $CH_3CH_2C(CH_3)(OH)CH_3$ Dimethyläthylkarbinol n, 2-Methyl-2-butanol n, tert-Amylalkohol m
   d. formamide $H\cdot CO\cdot N(CH_3)_2$ Dimethylformamid n
   d. glyoxal $CH_3COCOCH_3$ Dimethylglyoxal n, Diazetyl n, Butandion-(2,3) n
   d. phthalate $C_6H_4(COOCH_3)_2$ Phthalsäuredimethylester m, Dimethylphthalat n
   2,2-d. propanoic (propionic) acid $(CH_3)_3CCOOH$ 2,2-Dimethylpropansäure f, 2,2-Dimethylpropionsäure f, Trimethylessigsäure f, Pivalinsäure f
   d. pyridine Dimethylpyridin n, Lutidin n
   d. sulphate $(CH_3)_2SO_4$ Schwefelsäuredimethylester m, Dimethylsulfat n, Methylsulfat n
   d. sulphoxide $(CH_3)_2SO$ Dimethylsulfoxid n, Methylsulfoxid n, Methylsulfinylmethan n
   d. terephthalate Dimethylterephthalat n
   d. zinc $Zn(CH_3)_2$ Dimethylzink n, Zink[di]methyl n, Zinkmethid n
dimethylacetylene $CH_3C\equiv CCH_3$ Dimethylazetylen n, 2-Butin n, Krotonylen n
dimethylamine $(CH_3)_2NH$ Dimethylamin n
p-dimethylaminoazobenzene
   $C_6H_5N=NC_6H_4N(CH_3)_2$ p-Dimethylaminoazobenzol n, Buttergelb n, Dimethylgelb n
p-dimethylaminoazobenzene-o-carboxylic acid $C_6H_4(COOH)N=NC_6H_4N(CH_3)_2$ p-Dimethylaminoazobenzol-o-karbonsäure f, Anthranilsäureazodimethylanilin n, Methylrot n
dimethylaniline $(CH_3)_2C_6H_3NH_2$ Dimethylanilin n, Aminodimethylbenzol n, Aminoxylol n, Xylidin n
dimethylbenzene $C_6H_4(CH_3)_2$ Dimethylbenzol n, Xylol n
2,2-dimethyl-3-butanone $CH_3COC(CH_3)_3$ 2,2-Dimethylbutanon-(3) n, Trimethylazeton n, Methyl-tert-butylketon n, Pinakolon n
dimethyldichlorosilane $(CH_3)_2SiCl_2$ Dimethyldichlorsilan n
asym-dimethylethylene $CH_2=C(CH_3)_2$ asym-Dimethyläth[yl]en n, Isobut[yl]en n, 2-Methylpropen n
sym-dimethylethylene $CH_3CH=CHCH_3$ sym-Dimethyläth[yl]en n, 2-But[yl]en n
dimethylethylmethane $CH_3CH(CH_3)CH_2CH_3$ Äthyldimethylmethan n, 2-Methylbutan n, Isopentan n
dimethylformal s. dimethoxymethane
dimethylformamide $HCON(CH_3)_2$ Dimethylformamid n, DMF

dimethylketone CH_3COCH_3 Dimethylketon n, Azeton n, Aceton n, 2-Propanon n

2,3-dimethylnaphthalene $C_{10}H_6(CH_3)_2$ 2,3-Dimethylnaphthalin n, Guajen n

dimethylnitrobenzene $NO_2C_6H_3(CH_3)_2$ Dimethylnitrobenzol n, Nitroxylol n

3,7-dimethyl-2,6-octadiene-1-ol $(CH_3)_2C=CHCH_2CH_2C(CH_3)=CHCH_2OH$ 3,7-Dimethyl-2,6-oktadien-1-ol n, Geraniol n, Nerol n

dimethylol urea $HOCH_2 \cdot NH \cdot CO \cdot NH \cdot CH_2OH$ Dimethylolharnstoff m

2,4-dimethylpentane $(CH_3)_2CHCH_2CH(CH_3)_2$ 2,4-Dimethylpentan n, Diisopropylmethan n

dimethylphenol $(CH_3)_2C_6H_3OH$ Dimethylphenol n, Dimethylhydroxybenzol n, Hydroxyxylol n, Xylenol n

2,2-dimethylpropane $C(CH_3)_4$ 2,2-Dimethylpropan n, Tetramethylmethan n, Neopentan n

diminution Verringerung f

dimolybdate $M^I_2Mo_2O_7$ Dimolybdat n

dimolybdenum trioxide Mo_2O_3 Dimolybdäntrioxid n, Molybdäntrioxid n, Molybdän(III)-oxid n

dimorphic, dimorphous dimorph

Dimroth condenser Dimroth-Kühler m

dineric interface Flüssigkeit-Flüssigkeit-Grenzfläche f, Grenzfläche f flüssig-flüssig

dinitraniline $s.$ dinitroaniline

dinitrate Dinitrat n

dinitration Dinitrieren n, Dinitrierung f

dinitride Dinitrid n

dinitrile Dinitril n

  d. fibre Dinitrilfaserstoff m; Dinitrilfaser f

dinitro Dinitroverbindung f (als Herbizid verwendetes Phenolderivat)

  d. body Dinitrokörper m

  d. compound Dinitroverbindung f

  d. mixed acid Dinitromischsäure f

dinitroaminophenol $NH_2C_6H_2(NO_2)_2OH$ Dinitroaminophenol n, 4,6-Dinitro-2-aminophenol n, 2-Amino-4,6-dinitrophenol n, Pikraminsäure f

dinitroamylphenol Dinitroamylphenol n, 4,6-Dinitro-2-sek-amylphenol n, Dinosam n

dinitroaniline $(NO_2)_2C_6H_3NH_2$ Dinitr[o]anilin n

dinitroanthraquinone $C_{14}H_6O_2(NO_2)_2$ Dinitroanthrachinon n

dinitrobenzoic acid $C_7H_8O_6N_2$ Dinitrobenzoesäure f, Dinitrobenzolkarbonsäure f

dinitrobiphenyl $C_{12}H_8O_4N_2$ Dinitrobiphenyl n, Dinitrodiphenyl n

  d.-dicarboxylic acid $C_{14}H_8O_8N_2$ Dinitrobiphenyldikarbonsäure f, Dinitrodiphensäure f

2,4-dinitro-ortho-sec-butylphenol $CH_3(C_2H_5)CHC_6H_2(NO_2)_2OH$ 2,4-Dinitro-o-sek-butylphenol n, 2-sek-Butyl-4,6-dinitrophenol n, Dinoseb n, DNBP n

dinitrochlorobenzene $C_6H_3(NO_2)_2Cl$ Dinitrochlorbenzol n, 2,4-Dinitrochlorbenzol n, 1-Chlor-2,4-dinitrobenzol n

dinitrocresol $C_7H_6O_5N_2$ Dinitrokresol n

dinitro-ortho-cresol $CH_3C_6H_2(NO_2)_2OH$ Dinitro-

orthokresol n, 4,6-Dinitro-o-kresol n, DNOK n, DNOC n, DNC n

dinitro-ortho-cyclohexylphenol $C_6H_{11}C_6H_2(NO_2)_2OH$ Dinitroorthozyklohexylphenol n, 2,4-Dinitro-6-zyklohexylphenol n, DNOCHP n

2,4-dinitro-6-cyclohexylphenol $s.$ dinitro-orthocyclohexylphenol

dinitrodimethylaniline Dinitrodimethylanilin n

dinitrodiphenic acid $s.$ dinitrobiphenyl-dicarboxylic acid

dinitrodiphenyl $s.$ dinitrobiphenyl

  d. disulphide $s.$ dinitrophenyl disulphide

dinitrofluorobenzene Dinitrofluorbenzol n

dinitrogen monoxide N_2O Distickstoff[mon]oxid n, Stickstoff(I)-oxid n

  d. tetroxide N_2O_4 Distickstofftetroxid n

dinitromonomethylaniline Dinitromonomethylanilin n

dinitrophenate Dinitrophenolat n

dinitrophenol $C_6H_4O_5N_2$ Dinitrophenol n, Hydroxydinitrobenzol n

dinitrophenyl Dinitrophenyl n, DNP n

  d. derivative Dinitrophenylderivat n, DNP-Derivat n

  d. disulphide $C_{12}H_8O_4N_2S_2$ Dinitrophenyldisulfid n, Dinitrodiphenyldisulfid n

dinitrophenylhydrazine $C_6H_6O_4N_4$ Dinitrophenylhydrazin n

dinitroresorcinol $C_6H_4O_6N_2$ Dinitroresorzin n, Dihydroxydinitrobenzol n

dinitrotoluene $C_7H_6O_4N_2$ Dinitrotoluol n

dinitrotoluic acid $C_8H_6O_6N_2$ Dinitrotoluylsäure f

dinoseb $s.$ 2,4-dinitro-ortho-sec-butylphenol

dioctadecylamine $CH_3(CH_2)_{17}NH(CH_2)_{17}CH_3$ Dioktadezylamin n

dioctyl ketone $(C_8H_{17})_2CO$ Dioktylketon n, Pelargon n, Nonylon n, 9-Heptadekanon n

  d. phthalate $C_6H_4(COOC_8H_{17})_2$ Dioktylphthalat n, Phthalsäure-di-n-oktylester m

  d. sodium sulphosuccinate $C_{20}H_{37}NaO_7S$ Dioktylnatriumsulfosukzinat n, Natriumdioktylsulfosukzinat n

diol $C_nH_{2n}(OH)_2$ Diol n, Glykol n (i.w.S.), zweiwertiger Alkohol m

diolefin[e] Diolefin n

diolefinic diolefinisch, Diolefin...

DIOP $s.$ diisooctyl phthalate

diopside (Min) Diopsid m (Kalziummagnesiumdisilikat)

dioptase, dioptasite (Min) Dioptas m (Kupfer(II)-trisilikat)

diorite (Min) Diorit m

diorsellinic acid Diorsellinsäure f, Lekanorsäure f, Glabratsäure f

dioscin Dioszin n (ein Steroidsaponin)

dioscorine Dioskorin n (Alkaloid)

diosgenin Diosgenin n (ein Sapogenin)

diosmetin Diosmetin n, 5=7=3'-Trihydroxy-4'-methoxyflavon n

diosmin Diosmin n *(ein Glykosid)*
diosphenol Diosphenol n, Bukkokampfer m, 2-Hydroxypiperiton n
dioxan[e] Dioxan n *(Diäther des Glykols)*
dioxide Dioxid n
dioxindole Dioxindol n, 2=3-Dihydro-3-hydroxy-2-oxindol n
dioxygen O_2 Disauerstoff m
dip / **to** [ein]tauchen, abtauchen, eintunken, einbringen
  d. out *(Pap)* schöpfen
dip Tauchen n, Eintauchen n, Abtauchen n, Eintunken n, Einbringen n; Peilen n *(Messung des Ölstandes in einem Behälter)*; *(Landw)* Bademittel n; Tauchbad n
  d.-aluminize / **to** tauch[ver]aluminieren
  d. aluminizing Tauch[ver]aluminieren n
  d. bleaching *(Gerb)* Tauchbleiche f
  d.-braze / **to** tauchlöten, im Lötbad hartlöten
  d. brazing Tauchlöten n, Tauchlötung f, Hartlöten n im Lötbad
  d.-coat / **to** durch Tauchen beschichten, *(Überzüge)* heißtauchen
  d. coat getauchter Überzug m
  d. coater *(Plast)* Tauchbeschichtungseinrichtung f, Tauchüberzugseinrichtung f; *(Pap)* Tauchstreichmaschine f
  d. coating Beschichten n durch Tauchen, Heißtauchen n *(von Überzügen)*
  d. coating in powder *(Plast)* Wirbelsintern n
  d.-dye / **to** tauchfärben, im Tauchverfahren färben
  d. dyeing Tauchfärben n, Tauchfärbung f, Färben n im Tauchverfahren, *(Gerb auch)* Tunkfärbung f
  d.-dyeing machine Tauchfärb[e]maschine f
  d. enamelling Tauchemaillieren n
  d. feed Tauchauftrag m
  d.-feed drum dryer Tauchwalzentrockner m
  d. glazing Tauchglasieren n
  d.-glazing machine Tauchglasiermaschine f
  d.-harden / **to** tauchhärten
  d. hardening Tauchhärten n, Tauchhärtung f
  d. mix Tauchmischung f
  d. moulding *(Plast)* Heißtauchen n
  d.-patent / **to** tauchpatentieren
  d. patenting Tauchpatentieren n
  d. [reclaiming] process *(Gum)* Dip-Verfahren n *(Regenerierverfahren)*
  d.-solder / **to** tauchlöten
  d. soldering Tauchlöten n, Tauchlötung f
  d. tank Tauchbehälter m, Tauchtank m, Tauchgefäß n, Tauchwanne f, Tauchbottich m, *(Gum auch)* Lösungsbehälter m, Lösungstrog m, Lösungskasten m
dipentene Dipenten n, dl-Limonen n, dl-1,8(9)-p-Menthadien n
dipeptidase Dipeptidase f
dipeptide Dipeptid n *(Peptid mit zwei Aminosäureeinheiten)*

diphanite *(Min)* Diphanit m, *(veraltet für)* Margarit m *(Kalziumaluminiumdihydroxydialumodisilikat)*
diphenic acid $HOOC \cdot C_6H_4 \cdot C_6H_4 \cdot COOH$ Diphensäure f, 2,2'-Diphenyldikarbonsäure f
diphenyl $C_6H_5C_6H_5$ Diphenyl n, Biphenyl n
  d. diketone $C_6H_5COCOC_6H_5$ Diphenyldiketon n, Diphenylglyoxal n, Benzil n, Bibenzoyl n, Dibenzoyl n
  d. disulphide $C_6H_5SSC_6H_5$ Diphenyldisulfid n, Phenyldisulfid n, Phenyldithiobenzol n
  d. dye Diphenylfarbstoff m
  d. ether $C_6H_5OC_6H_5$ Diphenyläther m, Phenyläther m, Diphenyloxid n
  d. isomerism Diphenylisomerie f
  d. ketone $C_6H_5COC_6H_5$ Diphenylketon n, Benzophenon n, Benzoylbenzol n
  d. oxide s. d. ether
diphenylacetic acid $(C_6H_5)_2CHCOOH$ Diphenylessigsäure f, Diphenyläthansäure f
diphenylacetylene s. diphenylethyne
diphenylamine $C_6H_5NHC_6H_5$ Diphenylamin n
  d. blue $(C_6H_5NHC_6H_4)_3COH$ Diphenylaminblau n, Tris-(4-anilinophenyl)methanol n
  d. orange $C_{18}H_{14}O_3N_3NaS$ Diphenylaminorange n, Orange IV n, Orange N n, Tropäolin OO n
diphenylaminechlorarsine Diphenylaminchlorarsin n, Phenarsazinchlorid n, 10-Chlor-5,10-dihydrophenarsazin n
diphenylbenzene $(C_6H_5)_2C_6H_4$ Diphenylbenzol n, Diphenylphenylen n, Terphenyl n
diphenylcarbinol $C_6H_5CH(OH)C_6H_5$ Diphenylkarbinol n, Benzhydrol n
diphenyldiimide $C_6H_5N=NC_6H_5$ Diphenyldiimid n, Azobenzol n
diphenyleneimine $C_6H_4NHC_6H_4$ Diphenylenimin n, Diphenylenimid n, Karbazol n, Dibenzopyrrol n
1,2-diphenylethane $C_6H_5CH_2CH_2C_6H_5$ 1,2-Diphenyläthan n, Dibenzyl n, Bibenzyl n
diphenylethyne $C_6H_5C{\equiv}CC_6H_5$ Diphenyläthin n, Diphenylazetylen n, Tolan n
diphenylglyoxal s. diphenyl diketone
diphenylhydrazine Diphenylhydrazin n
diphenyline $H_2NC_6H_5C_6H_4NH_2$ Diphenylin n, 2,4'-Diaminobiphenyl n
diphenylmethane $CH_2(C_6H_5)_2$ Diphenylmethan n, Benzylbenzol n
  d. dye Diphenylmethanfarbstoff m
  d. oxide Diphenylmethanoxid n, Dibenzo-1,4-pyran n, o,o'-Methylendiphenyläther m, Xanthen n
diphenylphenylene s. diphenylbenzene
N,N'-diphenyl-p-phenylenediamine $(C_6H_5NH)_2C_6H_4$ N,N'-Diphenyl-p-phenylendiamin n, DPPD
diphenylthiocarbazone Diphenylthiokarbazon n, Dithizon n
diphenyltriazine Diphenyltriazin n
diphosgene $ClCOOCCl_3$ Diphosgen n, Trichlormethylchlormethanoat n
diphosphate Diphosphat n, Pyrophosphat n

diphosphoglyceric acid *(Bioch)* Diphosphoglyze-
rinsäure *f*
diphosphopyridine nucleotide Diphosphopyridin-
nukleotid *n*, DPN⁺, Nikotinamid-adenin-
dinukleotid *n*, NAD
diphosphoric acid $H_4P_2O_7$ Diphosphorsäure *f*,
Pyrophosphorsäure *f*
diphosphorous acid $H_4P_2O_5$ diphosphorige (pyrophosphorige) Säure *f*
diphthalyl Diphthalyl *n*, Biphthaliden *n*
diphthalylic acid Diphthalylsäure *f*, Biphthaliden-
säure *f*, 2,2'-Benzildikarbonsäure *f*
diphtheria toxin Diphtherietoxin *n*
dipicolinic acid Dipikolinsäure *f*, 2,6-Pyridindikar-
bonsäure *f*
dipolar dipolar, Dipol...
   d. ion Zwitterion *n*, Ampho-Ion *n*, Dipolion *n*
   d. molecule Dipolmolekül *n*, Dipolmolekel *f*
   d. structure dipolare Struktur *f*
dipole Dipol *m*; Dipolmolekül *n*, Dipolmolekel *f*
   d. association Dipolassoziation *f*
   d. axis Dipolachse *f*
   d.-dipole interaction Dipol-Dipol-Wechselwir-
kung *f*
   d. force Dipolkraft *f*
   d. measurement Dipolmessung *f*, Messung *f* des
Dipolmoments
   d. molecule *s.* dipolar molecule
   d. moment Dipolmoment *n*
   d. moment of linkage Bindungsdipolmoment *n*
   d. radiation Dipolstrahlung *f*
   d. term Dipolglied *n*
dipotassium disulphate $K_2S_2O_7$ Dikaliumdisulfat
n, Kaliumdisulfat *n*, Kaliumpyrosulfat *n*
   d. hydrogen phosphate K_2HPO_4 Dikaliumhydrogenphosphat *n*, Kaliumhydrogen[ortho]-
phosphat *n*
   d. orthophosphate *s.* d. hydrogen phosphate
dipped articles *s.* d. goods
   d. electrode Tauchelektrode *f*
   d. goods Tauch[gummi]waren *fpl*, Tauchartikel
mpl
Dippel's oil Dippels Öl (Tieröl) *n*, Dippelsches Öl *n*
dipper 1.Schöpfgefäß *n*, Schöpfer *m*; 2.Eintaucher
*m (Arbeiter, der das Tauchen eines Gegen-
standes ausführt); (Pap)* Schöpfer *m*, Büttgeselle
m
   d. sample Schöpfprobe *f*
dipping Tauchen *n*, Eintauchen *n*, Abtauchen *n*,
Eintunken *n*, Einbringen *n*; Tauchen *n*, Tauchlackieren *n*, Tauchlackierung *f (in der Anstrichtechnik); (Pap)* Oberflächenfärbung *f*
   d. apparatus *s.* d. machine
   d. bath Tauchbad *n; (Gerb)* Tunkbad *n*
   d. colorimeter Eintauchkolorimeter *n*
   d. compound Tauchmischung *f*
   d. form Tauchform *f*
   d. machine Tauchapparat *m*
   d. method Tauchverfahren *n*
   d. plant Tauchanlage *f*

   d. process Tauchverfahren *n*, Tauchprozeß *m*;
Tauchvorgang *m*
   d. pyrometer Eintauchpyrometer *n*
   d. rack Tauchgestell *n; (Gum)* Formenrahmen *m*
   d. refractometer Eintauchrefraktometer *n*
   d. roll *(Pap)* Tauchwalze *f*, Leitwalze *f (Oberflächenleimung)*
   d. solution Tauchlösung *f*
   d. tank Tauchbehälter *m*, Tauchtank *m*, Tauchgefäß *n*, Tauchwanne *f*, Tauchbottich *m, (Gum
auch)* Lösungsbehälter *m*, Lösungstrog *m*,
Lösungskasten *m*
   d. time Tauchzeit *f*, Tauchdauer *f*
   d. varnish Tauchlack *m*
   d. vat *(Pap)* Schöpfbütte *f*, Tauchbütte *f*
dipropyl ketone $CH_3CH_2CH_2COCH_2CH_2CH_3$ Dipropylketon *n*, Butyron *n*, 4-Heptanon *n*
dipropylacetylene $CH_3(CH_2)_2C{\equiv}C(CH_2)_2CH_3$ Dipropylazetylen *n*, 4-Oktin *n*
dipyrazolanthrone Dipyrazolanthron *n*, Dianhy-
dro-1,5-dihydrazinoanthrachinon *n*
dipyrazolanthronyl Dipyrazolanthronyl *n*
dipyre *(Min)* Dipyr *m (ein Skapolith)*
Dirac electron theory Diracsche Theorie *f* [des
Elektrons]
diradical Diradikal *n*
direct / **to** bestimmen, vorschreiben; *(einen
Substituenten nach der m- bzw. o-, p-Stellung)*
dirigieren, lenken
direct direkt, unmittelbar, Direkt...; *(Farb)* direktziehend, substantiv, Direkt...
   to be d. to cotton direkt auf Baumwolle
[auf]ziehen, Baumwolle direkt [an]färben
   d. arc[-heated] furnace direkter Lichtbogenofen
m, Lichtbogenofen *m* mit direkter Beheizung
(Lichtbogenheizung)
   d. burner Brenner *m* mit Gas-Luft-Mischung am
Brennermund
   d. combustion direkte (unmittelbare) Verbrennung *f*
   d. condenser Mischkondensator *m*
   d.-contact heat interchanger direkter Wär-
meaustauscher *m*
   d. contact poison Kontaktgift *n*, Berührungsgift
n
   d. cook process *(Pap)* direkte Kochung *f*
   d. current Gleichstrom *m*
   d.-current arc Gleichstrom[licht]bogen *m*
   d.-current polarogram Gleichstrompolarogramm
n
   d.-current polarograph Gleichstrompolarograf *m*
   d. drying Konvektionstrocknen *n*, Konvektionstrocknung *f*
   d. dye[stuff] Direktfarbstoff *m*, direktziehender
(substantiver) Farbstoff *m*, Substantivfarbstoff *m*
   d.-expansion chiller Direktexpansionskühler *m*
   d.-fire / **to** direkt befeuern (beheizen)
   d.-fired reboiler Destillierblase *f* mit direkter
Beheizung
   d. fungicide direktes *(durch Kontakt mit dem*

disazo

Pilzmyzel wirkendes) Fungizid *n*, Fungizid *n* mit kurativer Wirkung
d. gate *(Plast)* direkter Anguß *m*
d. heating direktes Heizen (Beheizen) *n*
d.-indirect rotary dryer Verbundtrommeltrockner *m*
d.-light electron microscope Auflichtelektronenmikroskop *n*
d.-light microscopy Auflichtmikroskopie *f*
d. method *s.* d. process 2.
d.-moulded footwear Preßschuhwerk *n*, formgepreßtes Schuhwerk *n*
d. moulding [process] Preßmethode *f (Herstellung von formgepreßtem Schuhwerk)*
d. nitration *(Farb)* direkte Nitrierung *f*
d. printing Direktdruck *m*
d. process 1. direktes Verfahren (Ammoniakgewinnungsverfahren, NH_3-Gewinnungsverfahren) *n*; 2. direktes Verfahren *n*, Direktverfahren *n*, direkte Methode *f*, Rochow-Verfahren *n*, Müller-Rochow-Verfahren *n (zur Herstellung von Chlorsilanen)*; 3. *(Met)* Verfahren *n* zur direkten Eisengewinnung; 4. *s.* d. reaction
d.-quench / to direkt abschrecken
d. quenching direktes Abschrecken *n*, direkte Abschreckung *f*, Direktabschreckung *f*
d. reaction Hinreaktion *f*
d. reduction direkte Reduktion *f*
d.-reduction process *(Met)* Verfahren *n* zur direkten Eisengewinnung
d. resistance-heated furnace direkter Widerstandsofen *m*
d. rolling Walzen *n (von Metallpulvern)*, Pulverwalzen *n*
d. rolling method Walzverfahren *n*, Pulverwalzverfahren *n (in der Pulvermetallurgie)*
d. rolling of powders *s.* d. rolling
d. rotary dryer Drehtrommeltrockner *m*, Trommeltrockner *m*
d. smelting direktes (unmittelbares) Schmelzen (Verschmelzen) *n*
d. spinning system Direktspinnverfahren *n*
d. steam direkter Dampf *m*, Direktdampf *m*, Frischdampf *m*
d. synthesis direkte Synthese *f* [nach Müller-Rochow], Direktsynthese *f*, Rochow-Synthese *f*, Müller-Rochow-Synthese *f (zur Herstellung von Chlorsilanen)*
d. system *s.* d. process 1.
d.-vision spectroscope geradsichtiges Spektroskop *n*, Geradsichtspektroskop *n*
d. vulcanization process *s.* d. moulding [process]
directed application *(Landw)* Applikation *f* (Ausbringen *n*) unter Abschirmung
d. interesterification gelenkte Umesterung *f*
d. spray Sprühstrahl *m*, Nebelstrahl *m*
d. transesterification *s.* d. interesterification
d. valence gerichtete Valenz *f*
direction Richtung *f*; Steuerung *f*, Führung *f*, Leitung *f*, Lenkung *f*

d. focusing Richtungsfokussierung *f*
d. in space Raumrichtung *f*
d. of addition Anlagerungsrichtung *f*
d. of dehydration Dehydratisierungsrichtung *f*
d. of flow Strömungsrichtung *f*, Stromrichtung *f*, Fließrichtung *f*
d. of motion Bewegungsrichtung *f*
d. of propagation Ausbreitungsrichtung *f*, Fortpflanzungsrichtung *f*
d. of rotation Drehrichtung *f*
d. of travel *(Pap)* Laufrichtung *f*
d. of vibration Schwingungsrichtung *f*
d. of zone travel Wanderungsrichtung *f* der Schmelzzone, Schmelzzonenrichtung *f*, Zonenrichtung *f*, Richtung *f* des Zonendurchgangs *(beim Zonenschmelzen)*
d. of zoning *s.* d. of zone travel
directional Richtungs..., Richt..., gerichtet
d. distribution [räumliche] Ausrichtung *f*, Orientierung *f*
d. drilling Richtbohren *n*, Richtbohrung *f*, gerichtete Bohrung *f*
d. frictional effect richtungsabhängiger Reibungseffekt *m*, differentieller Friktionseffekt *m*
d. preference Vorzugsrichtung *f*
d. quantization Richtungsquantelung *f*, Raumquantelung *f*, räumliche Quantelung (Quantisierung) *f*
d.-throw conveyor Wurfförderrinne *f*, Wurfförderer *m*
directive influence dirigierender Einfluß *m*, dirigierende Wirkung *f*
directly fed bucket elevator Aufgabebecherwerk *n*
dirt Schmutz *m*
d. content Schmutzgehalt *m*
d.-holding capacity Staubaufnahmevermögen *n*
d.-holding space Schlammraum *m*, Schmutzraum *m*
d. particle Schmutzteilchen *n*
d. remover *(Pap)* Schmutzfang *m*, Schmutzfänger *m*, Schmutzschleuse *f*
d.-repellent schmutzabweisend, schmutzabstoßend
d.-repellent treatment *(Text)* schmutzabweisende (schmutzabstoßende) Behandlung (Ausrüstung) *f*
d. speck Schmutzfleck *m (Papierfehler)*
d.-suspending power Schmutztragevermögen *n*
dirtiness Verschmutzung *f*, verschmutzter Zustand *m*
dirty schmutzig, verschmutzt
d. paper Papier *n* mit Unreinheiten
d. water Schmutzwasser *n*
disaccharide $C_{12}H_{22}O_{11}$ Disa[c]charid *n*
disaccharose *s.* disaccharide
disagreeable unangenehm, schlecht, widerlich, übel *(Geruch)*
disappearing-filament pyrometer Glühfadenpyrometer *n*
disazo dye Disazofarbstoff *m*

disc *s.* disk

discard / **to** *(als unbrauchbar)* verwerfen; *(Berge bei der Aufbereitung)* ausscheiden, abscheiden, aussortieren, aussondern, austragen; abstoßen

discard Ausscheiden *n*, Abscheiden *n*, Aussortieren *n*, Aussondern *n*, Austrag *m (der Berge bei der Aufbereitung)*

discarded verbraucht, gebraucht *(z.B. Lösung)*

discernible wahrnehmbar, sichtbar, merklich

discharge / **to** 1. austragen, ausbringen, entnehmen; ablassen, ausströmen lassen; abwerfen *(bei Fördererh)*; entladen, entleeren; *(Koks beim Koksofenbetrieb)* austragen, [aus]drücken, [her]ausstoßen; ablaufen, auslaufen, abfließen, ausströmen; 2. fördern *(von Pumpen, Kompressoren)*; 3. *(Phys)* entladen; 4. *(die Färbung)* beseitigen, entfernen, zerstören; *(Text)* ätzen; 5. *(Raupenseide)* entbasten

discharge 1. *(als Vorgang)* s. discharging; 2. *(als Stelle oder Einrichtung)* Austrag *m*, Gutaustrag *m*, Auslaß *m*, Ausfall *m*, Gutausfall *m*, Abzug *m*, Entnahme *f*, Gutentnahme *f*, Austritt *m*, Gutaustritt *m*; Abwurf *m (eines Förderers)*; Ablauf *m*, Auslauf *m*, Abfluß *m*, Abgang *m*; 3. Spalt *m (z.B. am Backenbrecher)*; 4. Fördermenge *f (von Pumpen, Kompressoren)*

d. angle Austrittswinkel *m (z.B. an den Schaufeln von Strömungsmaschinen)*

d. aperture s. d. opening

d. chute Austragrutsche *f*, Austragschurre *f*, Austragrinne *f*; Auslaufrutsche *f*, Auslaufschurre *f*, Ablaufschurre *f*

d. coefficient Durchflußzahl *f*, Durchflußkoeffizient *m*, Durchflußbeiwert *m*

d. cone Austragkonus *m*, Ausfallkonus *m*

d. device s. discharging device

d. door Austragklappe *f*, Entladeklappe *f*, Entleerungsklappe *f*, Ausfallklappe *f*; Austragöffnung *f*, Ausfallöffnung *f*, Entnahmeöffnung *f*; Abwurföffnung *f (eines Förderers)*

d. electrode Sprühelektrode *f*, Entladungselektrode *f*

d. end Austrag[s]ende *n*, Austrag[s]seite *f*, Ausfallseite *f*, Abzugsende *n*, Abgangsseite *f*, Entnahmeende *n*; Ablaufende *n*, Auslaufende *n*, Austrittsende *n*; Abwurfende *n (eines Förderers)*

d. friction head Druckreibungshöhe *f*

d. gate Austragschieber *m*, Austragschlitzschieber *m*, Entleerungsschieber *m*

d. grate (grating) Austragrost *m*

d. head Förderhöhe *f (z.B. bei Pumpen)*

d. manifold *(Filtration)* Filtratsammelleitung *f*, Sammelablaßleitung *f*, Sammelkanal *m*

d. nozzle Austragdüse *f*

d. onto the wire *(Pap)* Stoffauflauf *m*

d. opening (outlet) Austragöffnung *f*, Ausfallöffnung *f*, Entnahmeöffnung *f*; Ablaßöffnung *f*, Abflußöffnung *f*, Ausflußöffnung *f*, Austrittsöffnung *f*, Ausströmöffnung *f*; Abwurföffnung *f (eines Förderers)*

d. pipe Austrag[s]rohr *n*, Abflußrohr *n*, Ausflußrohr *n*, Abzugsrohr *n*, Auslaufrohr *n*, Ablaufrohr *n*, Austrittsrohr *n*, Abflußstutzen *m*

d. plasma Gasentladungsplasma *n*

d. point Austragstelle *f*, Entnahmestelle *f*; Austrittsstelle *f*; Abwurfstelle *f (eines Förderers)*

d. port s. d. opening

d. potential Entladungspotential *n*

d. pressure Förderdruck *m*, Lieferdruck *m*

d. print[ing] Ätzdruck *m*

d.-printing paste Ätz[druck]paste *f*

d. resist *(Text)* Ätzreserve *f*

d. roll Austrag[s]walze *f*, Abnahmewalze *f*

d. scroll Austrag[s]schnecke *f*

d. stroke Druckhub *m*, Förderhub *m*

d. tube 1. s. d. pipe; 2. Entladungsröhre *f*, Gasentladungsröhre *f*

d.-tube phenomenon Entladungserscheinung *f*

d. valve Austragventil *n*, Auslaßventil *n*, Abflußventil *n*, Ablaßventil *n*, Leerventil *n*

d. vane Nachleitschaufel *f*

d. velocity Austrittsgeschwindigkeit *f*, Ausströmungsgeschwindigkeit *f*, Auslaufgeschwindigkeit *f*

discharged / **to become** verschwinden *(von einer Färbung)*

discharger s. discharging apparatus

discharging 1. Austragen *n*, Austrag *m*, Ausbringen *n*, Ausfallen *n*, Entnahme *f*; Ablassen *n*; Abwerfen *n*, Abwurf *m (bei Förderern)*; Entladen *n*, Entleeren *n*; Austragen *n*, Austrag *m*, Ausdrücken *n*, Ausstoßen *n*, Ausstoß *m*, Drücken *n* *(des Kokses beim Koksofenbetrieb)*; Ablaufen *n*, Auslaufen *n*, Abfließen *n*, Ausströmen *n*; 2. Förderung *f (von Pumpen, Kompressoren)*; 3. *(Phys)* Entladung *f*; 4. Beseitigung *f*, Entfernung *f*, Zerstörung *f (der Färbung)*; *(Text)* Ätzen *n*; 5. Entbasten *n (von Raupenseide)*

d. agent *(Text)* Ätzmittel *n*

d. apparatus (device) Austrag[s]vorrichtung *f*, Austrag[s]einrichtung *f*, Austrag[s]apparat *m*; Ablaßvorrichtung *f*; Entladevorrichtung *f*, Entleerungsvorrichtung *f*

d. door s. discharge door

d. point s. discharge point

Disco process Disco-Verfahren *n*, Disco-Schwelverfahren *n*

discolor / **to** s. to discolour

discoloration s. discolouration

discolour / **to** verfärben; entfärben; sich verfärben; *(Pap)* vergilben; sich entfärben, die Farbe verlieren, verbleichen, verblassen, bleichen, verschießen

discolouration Verfärben *n*, Verfärbung *f*, Farb[ver]änderung *f*; *(Glas)* Mißfärbung *f*; *(Pap)* Vergilben *n*, Vergilbung *f*, Gelbwerden *n*; Entfärben *n*, Entfärbung *f*, Farbverlust *m*, Verbleichen *n*, Verblassen *n*, Bleichung *f*, Verschießen *n*

discolouring clay Bleichton *m*, *(bei der Mineralöl-raffination auch)* Entfärbungserde *f*, Bleicherde *f*

d. earth Entfärbungserde *f*, Bleicherde *f*

disconnect / to trennen

discontinuous diskontinuierlich, aussetzend, unterbrochen, unstetig

d. phase disperse (innere, offene) Phase *f*, Dispersum *n (einer Emulsion)*

discovery well Fundbohrung *f*, fündige Bohrung *f*, Erfolgsbohrung *f*, Entdeckungsbohrung *f*

discrasite *(Min)* Dyskrasit *m*, Diskrasit *m (Silberantimonid)*

discrete spectrum diskretes Spektrum *n*; Linienspektrum *n*

disecondary disekundär

disgorge / to *(zur Entfernung des Trubs)* degorgieren, entkorken

disgorgment Degorgement *n*, Degorgieren *n*, Entkorken *n (zur Entfernung des Trubs)*

disgusting unangenehm, schlecht, widerlich, übel *(Geruch)*

dish Schale *f*

to be d.-ended gewölbte Böden haben, Klöpperböden haben *(von einer Destillierkolonne)*

d. butter Tafelbutter *f*

d. development *(Foto)* Schalenentwicklung *f*

d. tongue Abdampfschalenhalter *m*

dished [flach] gewölbt *(z.B. Boden)*; *(Plast)* konkav *(Verformung)*

disilane Si_2H_6 Disilan *n*, *(veraltet)* Disilikan *n*, Disilikoäthan *n*

disilicane *s.* disilane

disilicate Disilikat *n*

disilicide Disilizid *n*

disilicoethane *s.* disilane

disiloxane $H_3Si \cdot O \cdot SiH_3$ Disiloxan *n*

disilver fluoride Ag_2F Disilberfluorid *n*

disinfect / to desinfizieren, entkeimen, entseuchen

disinfectant Desinfektionsmittel *n*, Desinfiziens *n*, desinfizierendes (keimtötendes) Mittel *n*

disinfection Desinfizierung *f*, Desinfektion *f*, Entkeimung *f*, Entseuchung *f*

disinfest / to entwesen

disinfestation Entwesung *f*

disintegrate / to *(weiche bis mittelharte Stoffe)* [ver]mahlen; zersetzen, abbauen; *(Kch)* zerfallen; sich zersetzen, zerfallen, *(Bodenkunde auch)* verwittern

d. by a kollergang kollern

disintegration Mahlen *n*, Vermahlen *n (weicher bis mittelharter Stoffe)*; *(Kch)* Zerfall *m*; Zersetzung *f*, Zerfall *m*, Abbau *m*, *(Bodenkunde auch)* Verwitterung *f*; Entmischung *f (von Düngemitteln)*, Zerrieselung *f (von granulierten Düngemitteln)*

d. by solution *(Bodenkunde)* Lösungsverwitterung *f*

d. constant Zerfallskonstante *f*

d. electron Zerfallselektron *n*

d. energy Zerfallsenergie *f*

d. of atom Atomzerfall *m*, Kernzerfall *m*

d. particle Zerfallsteilchen *n*

d. product Zerfallsprodukt *n*

d. rate Zerfallsgeschwindigkeit *f*

d. series Zerfallsreihe *f*

d. velocity *s.* d. rate

disintegrator 1. Desintegrator *m*, Desintegratormühle *f (für weiche bis mittelharte Stoffe)*; Desintegrator *m (i.e.S.)*, Desintegratormühle *f (i.e.S.)*, Schlagkorbmühle *f*, Schleudermühle *f*; 2. *s.* d. washer

d. washer Desintegrator[gas]wäscher *m*, Desintegrator *m*, Zentrifugal[gas]wäscher *m*

disk Scheibe *f*; Platte *f*; Schieber *m (eines Ventils)*; Teller *m (eines Sitzventils)*; Klappe *f (eines Klappenventils)*

d.-and-doughnut tray *(Destillation)* *(Kombination aus einem Siebboden mit zentralem Rücklauf und einem ähnlichen mit peripherem ringförmigen Rücklauf)*

d. atomizer Sprühscheibe *f*, Zerstäuberscheibe *f*, Zerstäubungsscheibe *f*, Zentrifugalteller *m*

d. attrition mill *s.* d. mill

d. barker *(Pap)* Messer[scheiben]entrinder *m*

d. [bowl] centrifuge Tellerzentrifuge *f*, Tellerseparator *m*, Tellerschleuder *f*, Trommelzentrifuge *f* mit Einsatztellern

d. crusher Tellerbrecher *m*, Scheibenbrecher *m*

d. dryer Tellertrockner *m*

d. evaporator Scheibenverdampfer *m*

d. feeder Tellerbeschicker *m*, Tellerspeiser *m*, Telleraufgeber *m*, Telleraufgabegerät *n*

d. filter Scheibenfilter *n*

d. gate *(Plast)* Regenschirmanguß *m*

d. grinder *s.* d. mill

d. knife *(Pap)* Tellermesser *n*, Kreismesser *n*

d. meter Scheibenzähler *m*, Taumelscheibenzähler *m*

d. mill Scheibenmühle *f*, Mahlscheibenmühle *f*, Tellermühle *f*; Scheibenkolloidmühle *f*

d. refiner *(Pap)* Scheibenrefiner *m*, Scheibenaufschläger *m*

d. separator Tellerseparator *m*, Tellerzentrifuge *f*, Tellerschleuder *f*, Trommelzentrifuge *f* mit Einsatztellern; Scheibentrieur *m*, Scheibenseparator *m*

d.-shaped scheibenförmig

d. stack Tellersatz *m*

d.-type agitator Scheibenrührer *m*

d.-type atomizer *s.* d. atomizer

d.-type fan Schaufelradventilator *m*

dislocation *(Krist)* Versetzung *f*, Dislokation *f*

d. metamorphism *(Geol)* Dislokationsmetamorphose *f*, Dynamometamorphose *f*, kinetische (mechanische) Metamorphose *f*, Kinetometamorphose *f*

dismutate / to dismutieren

dismutation Dismutation *f*, Dismutierung *f*

d. reaction Dismutierungsreaktion *f*
dismutative dismutativ
disodium dihydropyrophosphate $Na_2H_2P_2O_7$ Dinatriumdihydrogendiphosphat *n*, Natriumdihydrogenpyrophosphat *n*
d. methyl arsonate $CH_3As(=O)(ONa)_2$ Dinatriummethylarsonat *n*
d. orthophosphate Na_2HPO_4 Dinatriumhydrogenphosphat *n*, Natriumhydrogen[ortho]phosphat *n*
d. salt Dinatriumsalz *n*
disorder *(Krist)* Fehlordnung *f*, Unordnung *f*
disordered *(Krist)* fehlgeordnet, ungeordnet
d. fashion *(phys Ch)* ungeordnete Lage *f*, Zufallslage *f*
disorient / to entorientieren
disoxidation, disoxygenation *s.* deoxidation
dispel / to *(flüchtige Stoffe)* austreiben, verjagen, vertreiben
dispensation *(Pharm)* Dispensation *f*
dispensatory *(Pharm)* Dispensatorium *n*
dispense / to *(Pharm)* dispensieren
dispensing buret Vorratsbürette *f*
d. spigot Ablaßhahn *m*
dispergate / to *s.* to disperse
dispergator *s.* dispersing agent
dispersal *s.* dispersion
d. gettering *(Vakuumtechnik)* Volumengetterung *f*
dispersant *s.* dispersing agent
d. action *s.* dispersing effect
d. additive *s.* dispersing additive
disperse / to dispergieren, fein verteilen (zerteilen); *(Bindemittel für Druckfarbe von Papierfasern)* weglösen
disperse dispers, fein verteilt (zerteilt)
d. azo dye Dispersionsazofarbstoff *m*, Azodispersionsfarbstoff *m*
d. dye Dispersionsfarbstoff *m*, dispergierter Farbstoff *m*
d. medium *s.* dispersion medium
d. particles *s.* d. phase
d. phase disperse (dispergierte, zerteilte) Phase *f*, disperser Bestandteil (Anteil) *m*, Dispersum *n*, disperse (dispergierte) Teilchen *npl*
d.-phase rule *(Koll)* Bodenkörperregel *f*
d. system disperses System *n*
dispersed dye *s.* disperse dye
d. particles *s.* disperse phase
d. phase *s.* disperse phase
d. reclaimed rubber Regeneratdispersion *f*
dispersibility Dispergierbarkeit *f*
dispersible dispergierbar
dispersing Dispergieren *n*, Dispergierung *f*, Feinzerteilen *n*, Feinverteilen *n*
d. action *s.* d. effect
d. additive Dispersionszusatz *m*, Dispersantadditiv *n*, Dispersantzusatz *m*
d. agent Dispergier[üngs]mittel *n*, Dispergierhilfsmittel *n*, Dispersionsmittel *n*, Dispergens

n, Dispersant *n*, dispergierendes Mittel *n*, Dispergator *m*
d. effect Dispergierwirkung *f*, dispergierende Wirkung *f*, Dispersantwirkung *f*, Dispersanteffekt *m*
d. element Dispersionsmittel *n (z.B. ein Prisma, Gitter)*
d. mill Dispersionsmühle *f*
d. power (property) Dispergier[ungs]vermögen *n*, Dispersionsvermögen *n*, Dispersionskraft *f*
dispersion Dispersion *f*, Feinverteilung *f*, feine Verteilung (Zerteilung) *f*; Dispergieren *n*, Dispergierung *f*, Feinzerteilen *n*, Feinverteilen *n* *(Herstellen einer Dispersion)*; *(Phys)* Dispersion *f (von Wellen)*
d. agent *s.* dispersing agent
d. analysis Dispersionsanalyse *f*
d. colloid Dispersionskolloid *n*
d. curve Dispersionskurve *f*
d. degree Dispersionsgrad *m*, Dispersitätsgrad *m*
d. effect Dispersionseffekt *m*
d. force Dispersionskraft *f*
d. formula Dispersionsformel *f*
d. medium Dispergier[ungs]mittel *n*, Dispersionsmittel *n*, Dispersionsmedium *n*, Dispergens *n*, dispergierender Bestandteil *m (in einem dispersen System)*
d. method Dispersionsmethode *f*, Dispersionsverfahren *n*, Dispergierungsmethode *f*, Verteilungsverfahren *n*
d. of colloids Kolloiddispersion *f*
d. of conductance *(Ech)* Dispersionseffekt *m* der Leitfähigkeit, Debye-Falkenhagen-Effekt *m*
d. of filler Füllstoffdispergierung *f*, Füllstoffverteilung *f*
d. of gas Gasdispersion *f*
d. of light Dispersion *f* des Lichtes (der Lichtwellen)
d. of rotation Rotationsdispersion *f*, R.D.
d. resin dispergiertes Harz *n*
d. theory Dispersionstheorie *f*
dispersity Dispersität *f*, Dispersionszustand *m*, disperser Zustand *m*; Dispersionsgrad *m*, Dispersitätsgrad *m*
dispersive dispergierend, Dispersions...
d. constant Dispersionskonstante *f*
d. medium *s.* dispersion medium
d. power *s.* dispersing power
d. system dispergierendes System *n*
dispersivity *(Phys)* Dispersion *f*
dispersoid Dispersoid *n*
d. analysis Dispersoidanalyse *f*
dispersoidology Dispersoidologie *f*, *(ungebräuchlich für)* Kolloidlehre *f*, Kolloidkunde *f*, Kolloidik *f*
displace / to verlagern, verschieben; verdrängen; ersetzen
displaceable verschiebbar; verdrängbar; ersetzbar
displacement 1. Verlagerung *f*, Verschiebung *f*;

distillate

Verdrängung f; Ersetzung f; 2. Fördermenge f je Umdrehung, Fördervolumen n
d. analysis Verdrängungsanalyse f
d. chamber Verdrängerraum m
d.-chromatographic analysis s. d. analysis
d. chromatography Verdrängungschromatografie f
d. current Verschiebungsstrom m
d. development Verdrängungsentwicklung f
d. law Verschiebungsgesetz n, Verschiebungssatz m, Verschiebungsregel f
d. meter Verdrängungszähler m, Verdrängungsvolumenzähler m
d. method Verdrängungsmethode f
d. reaction Verdrängungsreaktion f, Substitutionsreaktion f, Reaktion f mit wechselseitiger Substitution, Austauschreaktion f
d. technique Verdrängungstechnik f
d. wash[ing] Verdrängungswaschung f
displacer Verdränger m, Verdrängungsmittel n (Chromatografie); Verdrängungskörper m (Füllstandmessung)
disposal crock (irdener) Abfallkübel m (im Labor)
disproportionate / to disproportionieren; sich disproportionieren
disproportionation Disproportionierung f
d. reaction Disproportionierungsreaktion f
dissd. (Abk. für) dissolved
dissecting blowpipe Schneidbrenner m
d. microscope Präpariermikroskop n
disseminated (Geol) eingesprengt
dissemination (Geol) Einsprengung f
dissimilar ungleich[artig], andersartig, verschieden[artig]
d. joint action (Toxikologie) (Verbundwirkung zweier physiologisch an verschiedenen Stellen angreifender Gifte)
dissimilation Dissimilation f
dissipate / to (Wärme) ableiten, abführen
dissipation Dissipation f; Ableitung f (von Wärme)
d. factor [dielektrischer] Verlustfaktor m
d. of energy Energiedissipation f
d. of heat Wärmeableitung f
dissipative dissipativ
dissociate / to dissoziieren, [auf]spalten, auflösen, trennen; dissoziieren, sich aufspalten (spalten, auflösen), zerfallen
dissociation Dissoziation f, Zersetzung f, Zerfall m, Spaltung f, Auflösung f, Trennung f
d. constant Dissoziationskonstante f, Affinitätskonstante f
d. continuum Dissoziationskontinuum n
d. energy Dissoziationsenergie f
d. equilibrium Dissoziationsgleichgewicht n
d. limit Dissoziationsgrenze f
d. pressure Dissoziationsdruck m
d. to radicals Radikaldissoziation f
dissoluble löslich, [auf]lösbar, l, l., lö
dissolute [auf]gelöst

dissolution Lösen n, Lösung f, Auflösen n, Auflösung f
d. equilibrium Lösungsgleichgewicht n
d. of carbohydrates (Pap) Kohlehydratauslösung f
d. of lignin (Pap) Lignin[her]auslösung f, Ligninentfernung f
d. rate Auflösungsgeschwindigkeit f
dissolvability Löslichkeit f, Lösbarkeit f
dissolvable löslich, [auf]lösbar
dissolve / to [auf]lösen; sich [auf]lösen, in Lösung gehen
d. away (out) herauslösen
dissolved-air floatation machine Druckluft-Flotationsapparat m, Druckluftzelle f, pneumatische Flotationszelle f
d. bones (mit Schwefelsäure behandeltes Knochenmehl) (Düngemittel)
d.-gas drive Gasentlösungsdruck m, Expansionsdruck m des im Öl gelösten Gases
d. substance [auf]gelöster Stoff m, Gelöstes n
dissolvent Lösungsmittel n, Lösemittel n, Lsgm., Lm, Löser m, Solvens n
dissolver s. 1. dissolvent; 2. dissolving tank
dissolving chest s. d. tank
d. process Lösungsvorgang m, Auflösungsprozeß m, Löseprozeß m
d. pulp Textilzellstoff m, Kunstfaserzellstoff m, Chemiezellstoff m, Zellstoff m für die Chemiefaserindustrie, Reyonzellstoff m
d. tank (Pap) Schmelzlöser m, Lösetank m, Lösebehälter m
dissymmetric asymmetrisch, asym., unsymmetrisch, uns.
dissymmetry Asymmetrie f, Unsymmetrie f
d. factor Anisotropiefaktor m
dist. s. distillable
distance Abstand m
d. thermometer Fernthermometer n
distasteful unschmackhaft, von unangenehmem Geschmack, widerlich schmeckend
distd. (Abk. für) distilled
distg. (Abk. für) distilling
disthene (Min) Disthen m, Zyanit m, Kyanit m (Aluminiumoxidorthosilikat)
distil / to destillieren, umsieden; (bei der Alkoholdestillation) destillieren, brennen
d. along with gemeinsam destillieren mit
d. off abdestillieren, abtreiben, abziehen, wegdestillieren; (bei der Alkoholdestillation) abdestillieren, abbrennen
d. out [her]ausdestillieren
d. over überdestillieren, übertreiben; überdestillieren, übergehen
distill / to s. to distil
distillability Destillierbarkeit f
distillable destillierbar
distilland Destillationsgut n, Destilliergut n, Destillationsmaterial n, Destillans n
distillate Destillat n

d. collection gutter s. d. gutter
d. composition Destillatzusammensetzung f
d. cooler Destillatkühler m
d. drain Destillatabzug m
d. fraction Destillat[ions]fraktion f
d. fuel oil Destillatheizöl n, destilliertes Heizöl n
d. gasoline Destillat[ions]benzin n, direkt herausdestilliertes Benzin n, Straightrun-Benzin n
d. gutter Destillatsammelrinne f, Destillatfangrinne f
d. lubricating oil Destillatschmieröl n
d. oil Destillatöl n
d. receiver Destillat[ions]vorlage f, Destilliervorlage f, Destillatsammler m, Vorlage f
d. stock Destillatstock m (Rückstandsöl bei der Erdöldestillation)
distillation Destillieren n, Destillation f, Umsieden n; (bei der Alkoholdestillation) Destillieren n, Destillation f, Brennen n
by [means of] d. durch Destillation, destillativ
d. analysis Siedeanalyse f
d. apparatus Destilliergerät n, Destillationsgerät n, Destillierapparat m, Destillationsapparat m, (bei der Alkoholdestillation auch) Brenngerät n, Brennapparat m
d. assembly Destillierapparatur f, Destillationsapparatur f
d. at constant pressure isobare Destillation f
d. boiler Destillierblase f, Destillationsblase f, Blase f
d. by ascent aufsteigende Destillation f
d. by descent absteigende Destillation f
d. column Destillierkolonne f, Destillationskolonne f, Destilliersäule f
d. connecting tube Destillieraufsatz m
d. curve Destillationskurve f, Destillationslinie f
d. equipment Destilliereinrichtungen fpl, Destillationseinrichtungen fpl
d. flask Destillierkolben m, Destillationskolben m
d. furnace Destillierofen m, Destillationsofen m
d. gas Destillationsgas n; Schwelgas n (bei der Untertagevergasung)
d. gas main Schwelgasleitung f, Abgangsleitung f des Schwelgases
d. head Destillationskopf m, Destillierkopf m, Destillationsaufsatz m, Destillieraufsatz m, Destillierhelm m, Destillationsdom m, Helm m, Dom m
d. in steam Dampfdestillieren n, Dampfdestillation f, Wasserdampfdestillation f
d. loss Destillationsverlust m
d. method Destillationsmethode f, Destillationsverfahren n
d. of coal Kohlendestillation f, Trockendestillation f der Kohle; Steinkohlendestillation f, Trockendestillation f der Steinkohle
d. of plants Pflanzendestillation f
d. of wood Holzdestillation f, Trockendestillation f des Holzes
d. operation Destillierbetrieb m, Destillierarbeit

f; Brennereibetrieb m (bei der Alkoholdestillation)
d. over a short path Kurzwegdestillation f
d. plant Destillieranlage f, Destillationsanlage f, Destillierbetrieb m, Destillationsbetrieb m
d. pressure Destillationsdruck m
d. process Destillationsverfahren n, (Met auch) Retortenverfahren n; Destillationsvorgang m; Brennverfahren n (zur Alkoholdestillation)
d. product Destillationsprodukt n, Destillationserzeugnis n
d. range Destillationsbereich m, Siedebereich m, Siedegrenzen fpl
d. rate Destillationsgeschwindigkeit f
d. receiver s. distillate receiver
d. residue Destillationsrückstand m
d. retort Destillationsretorte f
d. stage Destillationsstufe f, Trennstufe f
d. stillhead s. d. head
d. temperature Destillationstemperatur f, Siedetemperatur f
d. test Destillationsprobe f
d. tower Destillationsturm m
d. tube Destillationsrohr n, Destillierrohr n
d. under reduced pressure s. d. under vacuum
d. under vacuum Vakuumdestillation f, Unterdruckdestillation f, Destillation f im Vakuum (unter vermindertem Druck)
d. unit Destillationseinheit f
d. water Destillationswasser n
d. zone Entgasungszone f, Schwelzone f (im Gaserzeuger)
distilled in vacuo vakuumdestilliert, im (unter) Vakuum destilliert
d. oil destilliertes (gebranntes) Öl n
d. tar destillierter Teer m
d. water destilliertes Wasser n
distiller Destillateur m, Destillierer m, (bei der Alkoholdestillation auch) Brenner m (Beruf); Destilliergerät n, Destillationsgerät n, Destillierapparat m, Destillationsapparat m
distillers' [spent] grains Getreideschlempe f
d. wash s. distillery slop
d. yeast s. distillery yeast
distillery Brennerei f, Spiritusbrennerei f, Branntweinbrennerei f; Destillieranlage f, Destillationsanlage f, Destillierbetrieb m, Destillationsbetrieb m
d. malt Brennmalz n
d. mash Brennereimaische f
d. operation Brennereibetrieb m
d. slop (vinasse) Brennereischlempe f, Schlempe f
d. yeast Brennereihefe f, Branntweinhefe f
distilling column s. distillation column
d. flask s. distillation flask
d. furnace s. distillation furnace
d. industry Brennereiindustrie f, Spiritusindustrie f
d. operation s. distillation operation

d. plant s. distillation plant
d. retort s. distillation retort
d. temperature s. distillation temperature
d. unit s. distillation unit
d. vessel Destilliergefäß n, Destillationsgefäß n
distn. s. distillation
distort / to verzerren; verdrehen; verziehen
distorted grain (Gerb) gezogener Narben m
distortion Verzerrung f; Verdrehung f; Verziehen n; (Ker) Verziehung f (Formfehler)
d. polarization Verschiebungspolarisation f
distribute / to verteilen
distribution Verteilung f
d. coefficient Verteilungskoeffizient m, Verteilungskonstante f; Verteilungskoeffizient m, Segregationskonstante f, Abscheidungskonstante f (in der Zonenschmelztheorie)
d. curve Verteilungskurve f
d. gas main Gasverteilleitung f, Verteilungsgasleitung f
d. law Verteilungsgesetz n; Nernstsches Verteilungsgesetz n, Nernstscher Verteilungssatz m
d. method Ausbringungsweise f (z.B. von Düngemitteln)
d. of electrons Elektronenverteilung f
d. of frequencies Frequenzverteilung f; Häufigkeitsverteilung f
d. of particle size Korn[größen]verteilung f, Teilchen[größen]verteilung f
d. of velocities Geschwindigkeitsverteilung f
distributor Verteiler m (z.B. für Flüssigkeiten); Verteiler m, Gichtverteiler m (eines Hochofens); (Landw) Streugerät n
d. plate Verteil[er]platte f
disturbance Störung f; Störgröße f
d. variable Störgröße f
disubstituted disubstituiert, zweifach substituiert
disulf... s. disulph...
disulphate $M^I_2S_2O_7$ Disulfat n
disulphide MS_2 Disulfid n
d. bond (bridge) Disulfidbrücke f, Disulfidbindung f, -S-S-Brücke f, -S-S-Bindung f
d. crosslink (link) s. d. bond
disulphiram Disulfiram n, Bis(diäthylthiokarbamyl)-disulfid n, Tetraäthylthiuramdisulfid n
disulphonate Disulfonat n
disulphonation Disulfonierung f
disulphonic acid Disulfo[n]säure f
disulphur dichloride S_2Cl_2 Dischwefel[di]chlorid n, Schwefelmonochlorid n
disulphuric acid $H_2S_2O_7$ Dischwefelsäure f, Pyroschwefelsäure f
disulphurous acid $H_2S_2O_5$ dischweflige (pyroschweflige) Säure f
disulphuryl chloride $S_2O_5Cl_2$ Disulfurylchlorid n, Pyrosulfurylchlorid n, Dischwefelpentoxiddichlorid n
diterpene Diterpen n
ditertiary ditertiär
dithian[e] Dithian n

dithiazine Dithiazin n, 3,3'-Diäthylthiadikarbozyanin n
dithiazolanthraquinone dye Dithiazolanthrachinonfarbstoff m
dithio acid $C_nH_{2n+1}CSSH$ Dithiosäure f
d. compound Dithioverbindung f
dithiocarbamate NH_2CSSM^I Dithiokarbamat n
d. accelerator (Gum) Dithiokarbamatbeschleuniger m
dithiocarbamic acid NH_2CSSH Dithiokarbaminsäure f, Dithiokarbamidsäure f
dithiocarbonic acid CH_2OS_2 Dithiokohlensäure f
dithiolcarbonic acid $HSCOSH$ Dithiolkohlensäure f
dithionate $M^I_2S_2O_6$ Dithionat n
dithionic acid $H_2S_2O_6$ Dithionsäure f
dithionite $M^I_2S_2O_4$ Dithionit n; (Farb) $Na_2S_2O_4$ Dithionit n, Natriumdithionit n
dithionous acid $H_2S_2O_4$ dithionige Säure f
dithiooxalic acid $HSCOCOSH$ oder $HOCSCSOH$ Dithiooxalsäure f
dithiooxamide $H_2NSCCSNH_2$ Dithiooxamid n, Dithiooxalsäurediamid n, Rubeanwasserstoffsäure f, Rubeanwasserstoff m
dithiophosphate $M^I_3PS_2O_2$ Dithiophosphat n
dithiophosphoric acid $H_3PS_2O_2$ Dithiophosphorsäure f
dithiosalicylic acid $(C_6H_4COOH)_2S_2$ Dithiosalizylsäure f, 2,2'-Dithiodibenzoesäure f, Diphenyldisulfid-2,2'-dikarbonsäure f
dithizone Dithizon n, Diphenylthiokarbazon n
ditrigonal-scalenohedral (Krist) ditrigonal-skalenoedrisch
ditungstate Diwolframat n
diunsaturated doppelt (zweifach) ungesättigt
diuranate $M^I_2U_2O_7$ Diuranat n
diurea $C_2H_4O_2N_4$ Diharnstoff m, 4-Urazin n
diuretic diuretisch, harntreibend
diuretic Diuretikum n, harntreibendes (diuretisches) Mittel n
d. salt Kaliumazetat n (Diuretikum)
divacancy Doppelleerstelle f
divalence, divalency Zweiwertigkeit f, Bivalenz f
divalent zweiwertig, bivalent
divariant divariant, bivariant, zweifachfrei
d. system divariantes System n
divaric acid Divarsäure f, 2,4-Dihydroxy-6-propylbenzolkarbonsäure f
divaricatic acid Divarikatsäure f
divaricatinic acid Divarikatinsäure f
divergent (Krist) radialstrahlig, radialstengelig
Divers liquid Diverssche Flüssigkeit f
divi-divi (Gerb) Dividivi pl, Dividihülsen fpl, Libidibi pl (von Caesalpinia coriaria (Jacq.)Willd.)
divided circle Teilkreis m (z.B. eines Refraktometers)
divider (Pap) Holzzwischenleiste f (für Bemesserung von Mahlmaschinen)
divinyl $CH_2=CHCH=CH_2$ Divinyl n, Bivinyl n, 1,3-Butadien n, Vinyläth[yl]en n, Diäth[yl]en n

d. ether $CH_2=CHOCH=CH_2$ Divinyläther m, Vinyläther m, Äthenyloxyäthen n

d. rubber Butadienkautschuk m

divisible teilbar; spaltbar

division Teilung f, Einteilung f, Gradeinteilung f, Strichteilung f (eines Meßinstruments); Skalenteil m, Teilstrichabstand m; Teilstrich m

d. line (mark) Teilstrich m

d. scale Strichskale f

djenkolic acid $CH_2(SCH_2CHNH_2COOH)_2$ Djenkolsäure f, β,β'-Methylendithiodialanin n

DK-meter DK-Meter n, Dekameter n, Dielektrizitätskonstante-Messer m

DMA s. disodium methyl arsonate

DME s. dropping-mercury electrode

DMF s. dimethyl formamide

D.M.F., d.m.m.f. s. dry, mineral-matter-free

D.M.P. s. dimethyl phthalate

DMSO s. dimethyl sulphoxide

DMT s. dimethyl terephthalate

DNA s. deoxyribonucleic acid

DNAP s. dinitroamylphenol

DNBP s. 2,4-dinitro-*ortho-sec*-butylphenol

DNC s. dinitro-*ortho*-cresol

DNFB s. dinitrofluorobenzene

DNOC s. dinitro-*ortho*-cresol

DNOCHP s. dinitro-*ortho*-cyclohexylphenol

DNP s. dinitrophenyl

dobbin Karusselltrockner m (Trockenofen für Feinkeramik)

Döbereiner's lamp Döbereinersches (Döbereiners) Feuerzeug n, Döbereinersche Zündmaschine f

Döbereiner's law of triads Döbereinersche (Döbereiners) Triadenregel f

Döbereiner's triad Döbereinersche Triade f

Döbner's violet Döbners (Döbnersches) Violett n

DOCA, D.O.C.A. s. deoxycorticosterone acetate

docosane Dokosan n, (i.e.S.) $CH_3(CH_2)_{20}CH_3$ Dokosan n, n-Dokosan n

docosanoic acid $CH_3(CH_2)_{20}COOH$ Dokosansäure f

1-docosanol $CH_3(CH_2)_{20}CH_2OH$ 1-Dokosanol n, prim-n-Dokosylalkohol m

docosenoic acid Dokosensäure f

13-docosenol $CH_3(CH_2)_7CH=CH(CH_2)_{12}OH$ 13-Dokosenol n, Eruzylalkohol m

docosenyl alcohol s. 13-docosenol

13-docosinoic acid s. 13-docosynoic acid

docosoic acid s. docosanoic acid

docosyl alcohol s. 1-docosanol

13-docosynoic acid $CH_3(CH_2)_7C\equiv C(CH_2)_{11}COOH$ Dokosin-(13)-säure f, Behenolsäure f

doctor / to (eine Walzenoberfläche) mit einem Schaber reinigen

d. off abschaben, mittels Schaber abnehmen

doctor Schaber m, Abnahmeschaber m (zum Abnehmen des Gutes von einer Walzenoberfläche)

d. blade Schabermesser n, Schab[e]messer n, Schaberklinge f; (Plast, Gum) Streichmesser n,

Rakelmesser n; Rakel f (beim Textildruck und Rakeltiefdruck)

d. coater Rakelstreichmaschine f, Rakelauftragmaschine f

d. finish (Text) Rakelappretur f, Streichappretur f

d. gum (Pharm) Schweinsgummi n (von Symphonia globulifera L.)

d. kiss coater s. d. coater

d. knife s. d. blade

d. process (Erdöl) Doktor[süßungs]verfahren n, Doktor[süßungs]prozeß m

d. roll Schaberwalze f; (Plast,Gum) Streichwalze f

d. solution (Erdöl) Doktorlösung f, Doktorlauge f

d. streak Rakelstreifen m

d.-sweet (Erdöl) doktornegativ, gedoktert

d. sweetening (Erdöl) Doktorsüßen n, Doktorbehandlung f, Behandlung (Nachbehandlung) f mit Doktorlösung (Doktorlauge)

d. test (Erdöl) Doktortest m (zur qualitativen Prüfung von Erdöldestillaten auf Merkaptane)

d. treating s. d. sweetening

d.-treating unit (Erdöl) Anlage f zur Doktorbehandlung

d. treatment s. d. sweetening

document paper Dokumentenpapier n

dodder oil Dotteröl n, Leindotteröl n (aus dem Samen von Camelina sativa Crantz)

dodecahedron (Krist) Dodekaeder n, Zwölfflächner m, Zwölfflach n

dodecahydrate Dodekahydrat n

dodecanal $CH_3(CH_2)_{10}CHO$ Dodekanal n, Dodezylaldehyd m, Lauraldehyd m, Laurinaldehyd m, Aldehyd C12[L] m

dodecane Dodekan n, (i.e.S.) $CH_3(CH_2)_{10}CH_3$ Dodekan n, n-Dodekan n, Dihexyl n, Bihexyl n

dodecane-1-carboxylic acid $CH_3(CH_2)_{11}COOH$ Dodekan-1-karbonsäure f, Tridekansäure f

dodecanedioic acid $HOCO(CH_2)_{10}COOH$ Dodekandisäure f, Dekan-1,10-dikarbonsäure f

1-dodecanesulphonic acid $H_3C(CH_2)_{10}CH_2SO_3H$ Dodekan-1-sulfonsäure f

1-dodecanethiol $CH_3(CH_2)_{11}SH$ 1-Dodekanthiol n, n-Dodezylmerkaptan n, Laurylmerkaptan n

dodecanoic acid $CH_3(CH_2)_{10}COOH$ Dodekansäure f, Laurinsäure f

dodecanol Dodekanol n

dodecanoyl peroxide Dodekanoylperoxid n, Lauroylperoxid n

1-dodecene $CH_3(CH_2)_9CH=CH_2$ 1-Dodezen n, α-Dodezylen n

2-dodecenedioic acid $C_{10}H_{18}(COOH)_2$ 2-Dodezendisäure f, 1-Dezen-1,10-dikarbonsäure f, Traumatinsäure f

dodecenoic acid $C_{11}H_{21}COOH$ Dodezensäure f

dodecine s. dodecyne

dodecoic acid s. dodecanoic acid

dodecyl $-CH_2(CH_2)_{10}CH_3$ Dodezyl *n*
  n-d. alcohol $CH_3(CH_2)_{11}OH$ *n*-Dodezylalkohol *m*, 1-Dodekanol *n*, Laurylalkohol *m*, Alkohol C 12 *m*
  d. mercaptan Dodezylmerkaptan *n*, *(i.e.S.)* $CH_3(CH_2)_{11}SH$ *n*-Dodezylmerkaptan *n*, Laurylmerkaptan *n*, 1-Dodekanthiol *n*
dodecylaldehyde *s.* dodecanal
dodecylamine $CH_3(CH_2)_{11}NH_2$ Dodezylamin *n*, 1-Aminododekan *n*, Laurylamin *n*
α-dodecylene *s.* 1-dodecene
dodecyne Dodezin *n*
Dodge [jaw] crusher Dodge-Brecher *m*, Dodge-Backenbrecher *m*
dodge reaction Ausweichreaktion *f*
Doebner-Miller reaction (synthesis) Doebner-[v.]Miller-Reaktion *f*, Doebner-[v.]Miller-Synthese *f (zur Darstellung von Chinolin und dessen Derivaten)*
Doebner synthesis Doebner-Synthese *f*, Doebner-Reaktion *f*, Doebnersche Synthese (Reaktion) *f*
dog-puered *(Gerb)* mit Hundekot gebeizt
doghouse *(Glas)* Einlegevorbau *m*
dohexacontane Dohexakontan *n*
β-dolabrin[e] β-Dolabrin *n (ein Tropolon)*
dolerofano, dolerophane *s.* dolerophanite
dolerophanite *(Min)* Dolerophanit *m (Kupfer(II)-oxidsulfat)*
dolly [washer] Strangwaschmaschine *f*
dolomite *(Min)* Dolomit *m (Kalziummagnesiumkarbonat)*; *(Geol)* Dolomit *m*, Dolomitgestein *n*
  d. brick Dolomitstein *m*
  d. lime *s.* dolomitic lime
  d. rock Dolomitgestein *n*, Dolomit *m*
dolomitic dolomitisch, Dolomit...
  d. lime Dolomitkalk *m*, Magnesiumbranntkalk *m*, Graukalk *m (Kalzium-Magnesium-Oxid)*
  d. limestone dolomitischer Kalkstein *m*, Dolomitkalkstein *m*
dolomitization *(Geol)* Dolomitisierung *f*
dolomitize / to *(Geol)* dolomitisieren
dolphin oil Delphintran *m*
domain Bereich *m*, Bezirk *m*; *(Krist)* Weiß-Bezirk *m*, Weißscher Bezirk (Bereich) *m*; *(Krist)* Mosaikblöckchen *n*
dome *(Tech)* Haube *f*; *(Destillation)* Glocke[n-kappe] *f*, Austauschglocke *f*
domed gewölbt; *(Plast)* konvex *(Verformung)*
domestic coke Hausbrandkoks *m*
  d. cullet *(Glas)* [fabrik]eigene Scherben *fpl*
  d. fuel Haus[brand]brennstoff *m*, Haushaltsbrennstoff *m*, Hausbrandmaterial *n*, Hausbrand *m*
  d. porcelain Haushaltporzellan *n*
  d. refrigerator Haushaltkühlschrank *m*
  d. sewage häusliches Abwasser *n*
  d. softener Haushalt-Enthärtungsapparat *m*, Hausenthärter *m (zur Enthärtung des Wassers)*
domeykite *(Min)* Domeykit *m*, α-Domeykit *m (Kupferarsenid)*

donate / to abgeben, liefern, spenden
donation Abgabe *f*
  d. of electrons Elektronenabgabe *f*
donator *s.* donor
Donnan effect Donnan-Effekt *m*
Donnan [membrane] equilibrium Donnansches Membrangleichgewicht (Gleichgewicht) *n*, Donnan-Gleichgewicht *n*
Donnan potential Donnan-Potential *n*
Donnan swelling *(Gerb) (Säurequellung aufgrund des Donnanschen Membrangleichgewichts)*
Donnelly process Donnelly-Verfahren *n*, Donnelly-Krackverfahren *n*
donor Don[at]or *m*
  d.-acceptor bond Donator-Akzeptor-Bindung *f*, koordinative (dative, semipolare, halbpolare) Bindung *f*
  d. atom Donatoratom *n*
  d. group Donatorgruppe *f*
  d. impurity Donatorstörstelle *f*, Donatorverunreinigung *f*
  d. level Donatorniveau *n*, Donatorterm *m*
  d. molecule Donatormolekül *n*, Donatormolekel *f*
  d. power Donatorstärke *f*
  d. reagent Donatorsubstanz *f*
  d. solvent Donatorsolvens *n*
door Tür *f*, Klappe *f*; Öffnung *f*
  d. extractor Tür[ab]heber *m*, Türabhebevorrichtung *f*
dopa, DOPA *s.* dihydroxyphenylalanine
dope / to dopen *(Mineralölprodukte mit einem Dope-Stoff versetzen)*
dope Dope *m(n)*, Dope-Stoff *m*, Dope-Mittel *n (Zusatzstoff in Mineralölprodukten, der in Mengen von unter 1 % benutzt wird; s.a. additive)*; *(Gum)* Imprägnierlösung *f*; *(Text)* Spinnlösung *f*, Erspinnlösung *f*; *(Glas)* Formenschmiere *f*, Formenschmiermittel *n*
  d.-dyed [er]spinngefärbt, düsengefärbt
  d. dyeing Spinnfärbung *f*, Düsenfärbung *f*
  d. viscosity Spinnviskosität *f*
Doppler broadening Doppler-Verbreiterung *f*
Doppler effect Doppler-Effekt *m*
Doppler half-width of a spectral line Doppler-Breite *f*
Doppler shift Doppler-Verschiebung *f*
dopplerite *(Min)* Dopplerit *m*
dormant oil [spray] öliges Winterspritzmittel *n*
  d. spray 1. Winterspritzmittel *n*; 2. *s.* d. spraying
  d. spraying Winterspritzung *f (mit Pflanzenschutzmitteln)*
  d. treatment Behandlung *f (mit Pflanzenschutzmitteln) während der Winterruhe, (meist)* Winterspritzung *f*
Dorr bowl classifier Dorr-Klassierer *m* mit Schüssel
Dorr [rake] classifier Dorr-Rechenklassierer *m*, Dorr-Klassierer *m*
Dorr thickener Dorr-Eindicker *m*
Dorrco FluoSolids reactor Dorrco-FluoSolids-

Reaktor *m*, Dorrco-FluoSolids-Wirbelschichtofen *m*, FluoSolids-Reaktor *m*, FluoSolids-Ofen
m

Dorrco FluoSolids system Dorrco-FluoSolids-
System *n*, FluoSolids-System *n* *(Wirbelschichtofensystem der Dorrco)*

dosage Dosieren *n*, Dosierung *f*, Abmessen *n*,
Zuteilen *n*; Dosis *f*; Einsatzmenge *f*; Verabreichung *f*, Applikation *f*

d. **constant** Dosiskonstante *f*

d. **measurement** s. dosimetry

d. **meter** s. dosimeter

d. **of filler** Füllstoffdosierung *f*

d. **of sulphur** Schwefeldosierung *f*

d. **rate** s. dose rate

d. **response** s. dose response

dose / to dosieren, abmessen, zuteilen

dose Dosis *f*

d.-**effect curve** s. d.-response curve

d. **of radiation** Strahlungsdosis *f*, Strahlendosis
f

d. **protraction** Dosisprotrahierung *f*

d. **rate** Dosisrate *f*, Dosisleistung *f*

d. **rate meter** Dosisleistungsmesser *m*

d. **response** Dosiswirkung *f*, Dosiseffekt *m*

d.-**response curve** Dosis-Wirkungs-Kurve *f*, Do-
sis-Effekt-Kurve *f*

dosemeter s. dosimeter

dosimeter Dosimeter *n*, Dosismesser *m*, Dosismeß-
gerät *n*, Strahlungsdosimeter *n*

dosimetry Dosimetrie *f*, Dosismessung *f*

dosing Dosieren *n*, Dosierung *f*, Abmessen *n*,
Zuteilen *n*

d. **apparatus** Dosiergerät *n*, Dosierapparat *m*,
Zuteiler *m*

d. **equipment** Dosiereinrichtungen *fpl*, Zu-
teileinrichtungen *fpl*

d. **machine** Dosiermaschine *f*

d. **plant** Dosieranlage *f*, Zuteilanlage *f*

d. **pump** Dosier[ungs]pumpe *f*

dot *(Ker)* Brennstütze *f* *(Brennhilfsmittel)*

d. **formula** Elektronenformel *f*

dottling *(Ker)* *(Brennstützenanordnung, bei der
zwischen die Flachware Fingerhüte oder Drei-
füße gesetzt werden, die meist Abdruckpunkte
hinterlassen)*

double-acting doppeltwirkend

d.-**action compacting** doppelseitige (beidseitige,
zweiseitige) Verdichtung *f*

d. **agitator** Doppelrührwerk *n*

d.-**arm mixer** Schaufelkneter *m*, Doppelarmkneter *m*, Doppelpaddelmischer *m*

d.-**base powder** zweibasiges Pulver (Treibladungspulver) *n*

d.-**beam burette holder** zweiarmiger Bürettenhalter *m*, zweiarmige Bürettenklemme *f*

d.-**beam instrument** Doppelstrahlinstrument *n*

d.-**beam spectrometer** Zweistrahlspektrometer *n*

d.-**beam spectrophotometer** Zweistrahlspektral-
fotometer *n*

d.-**bell distributor** drehbarer Gichtverteiler *m*
nach McKee

d. **bond** Doppelbindung *f*, doppelte Bindung *f*,
Zweifachbindung *f*

d.-**bond character** Doppelbindungscharakter *m*

d.-**bond electron** Doppelbindungselektron *n*

d.-**bond system** Doppelbindungssystem *n*

d.-**bore stopcock** Zweiweghahn *m*

d. **brewhouse** Sudhaus *n* mit doppeltem Sudwerk
(Sudzeug), Sudhaus *n* mit Mehrgerätesudwerk
(Mehrgerätesudzeug)

d. **bulb blower** Doppelgebläse *n*

d.-**coat / to** zweiseitig (beidseitig) belegen; *(Pap)*
zweiseitig (beidseitig) streichen

d.-**coated** *(Pap)* zweiseitig gestrichen, mit beidseitigem Strich

d.-**coated film** Zweischicht[en]film *m*, Dop-
pelschichtfilm *m*

d. **coater** *(Pap)* Streichmaschine *f* für doppelseitigen Strich, Streichmaschine *f* für zweiseitige Beschichtung

d. **coating** *(Pap)* doppelseitiges Streichen *n*,
doppelseitiger Strich *m*, zweiseitige Beschichtung *f*

d. **column** Doppelsäule *f*, Doppelkolonne *f*

d.-**cone blender (mixer)** Doppelkonusmischer *m*,
Doppelkegel-Trommelmischer *m*

d.-**cream cheese** Doppelrahmkäse *m*

d. **crêpe paper** Doppelkreppapier *m*

d. **deckle** *(Pap)* zweiseitiger Büttenrand *m*

d.-**deckle kiln** *(Ker)* Zweietagenofen *m*

d.-**deckled** *(Pap)* mit zweiseitigem Büttenrand

d. **decomposition** doppelte Umsetzung *f*, Wechselzersetzung *f*, Metathese *f*

d.-**decomposition reaction** doppelte Umsetzung
f, Wechselzersetzung *f*, Metathese *f*; *(Met)*
Röstreaktion *f*

d.-**disk gate valve** Zweiplattenschieber *m*

d.-**disk refiner** *(Pap)* Doppelscheibenrefiner *m*,
Doppelscheibenmühle *f*

d. **displacement** s. d. decomposition

d.-**drum dryer** Zweiwalzentrockner *m*

d.-**drum system** Kühlsystem *n* mit Walzenpaar
(nach Schon)

d.-**effect evaporator** Zweikörperverdampfer *m*,
Zweistufenverdampfer *m*

d.-**face fabric** s. d.-faced fabric

d.-**face printing** *(Text)* beidseitiges Bedrucken *n*,
Duplexdruck *m*

d.-**faced fabric** zweiseitiger (doppelseitiger, auf
beiden Seiten tragbarer) Stoff *m*, Zweiseitenstoff
m, Double-face *n*

d. **faucet** *(Am)* Doppel-Auslaufventil *n*

d.-**floor kiln** Zweihordendarre *f*

d. **focusing** Doppelfokussierung *f*

d.-**focusing mass spectrograph** doppel[t]fokus-
sierender Massenspektrograf *m*

d. **folds** *(Pap)* Doppelfalzungen *fpl*

d.-**headed arrow** Doppelpfeil *m*, *(in der theoretischen organischen Chemie auch)* Mesomerie-
pfeil *m*, Resonanzpfeil *m*

d.-lactate method *(Bodenkunde)* Doppellaktatmethode *f*, Doppellaktatverfahren *n*, Laktatmethode *f*
d. layer Doppelschicht *f*
d.-layer potential elektrokinetisches Potential *n*, Zeta-Potential *n*, ζ-Potential *n*
d.-lined *(Pap)* zweiseitig beklebt; zweiseitig gedeckt
d. link *s.* d. bond
d. molecule Doppelmolekül *n*
d.-motion agitator zweiachsiger Rührer *m*, gegenläufiges Rührwerk *n*, Planetenmischer *m*
d. nickel salt $(NH_4)_2SO_4 \cdot NiSO_4 \cdot 6H_2O$ Ammoniumnickel(II)-sulfat-6-Wasser *n*
d. oil of vitriol [hoch]konzentrierte Schwefelsäure *f (enthält 93 bis 98 % H_2SO_4)*
d.-pipe condenser Doppelrohrverflüssiger *m*, Doppelrohrkondensator *m*
d.-pipe crystallizer Doppelrohrkristallisator *m*, Doppelrohrkristaller *m*
d.-pipe heat exchanger Doppelrohr[wärme]austauscher *m*, Doppelrohrwärmeübertrager *m*
d.-ram press Doppelkolbenpresse *f*
d.-refracting doppelbrechend
d. refraction Doppelbrechung *f*
d. refraction of flow Strömungsdoppelbrechung *f*
d. replacement [reaction] doppelte Umsetzung *f*, Wechselzersetzung *f*, Metathese *f*
d. ring closure doppelter Ringschluß *m*
d.-roll press Doppelwalzenpresse *f*
d. salt Doppelsalz *n*, Dps.
d.-seat[ed] valve doppelsitziges Ventil *n*, Doppelsitzventil *n*
d.-shaft mixer Doppelwellenmischer *m*
d.-shaft pulp (steam) mixer *(Pap)* Doppelwellendampfmischer *m*
d.-shaft mixer *s.* d.-shaft mixer
d.-shot moulding *(Plast)* Zweifarbenspritzgießen *n*, Zweistufenspritzgießen *n*
d.-sided coating *s.* d. coating
d.-sized *(Pap)* doppelt geleimt
d. slit Doppelspalt *m*
d.-stage zweistufig, Zweistufen...
d.-strength *(Chem)* doppelt konzentriert, von doppelter Konzentration
d.-strength glass Fensterglas *n* doppelter Dicke
d.-suction pump Pumpe *f* mit zweiseitigem (doppelseitigem) Flüssigkeitseintritt, zweiseitig (doppelseitig) saugende Pumpe *f*
d. sulphate sulfatisches Doppelsalz *n*, Doppelsulfat *n*
d. thickness Doppeldicke *f*, DD *(des Fensterglases)*
d. tyre press *(Gum)* Doppel-Reifeneinzelheizer *m*
d.-walled doppelwandig
doubled molecule Doppelmolekül *n*
doublet Dublett *n*, Elektronendublett *n*, gemeinsames Elektronenpaar *n* zweier Atome; Dublett *n*, Spektrallinendublett *n*, Doppellinie *f*; Dublett *n (Benennung für die Duplizität)*

d. spectrum Dublettspektrum *n*
d. splitting Dublettaufspaltung *f*
d. state Dublettzustand *m*
d. structure Dublettstruktur *f*
d. system Dublettsystem *n*
d. term Dublett-Term *m*
doubly bonded doppelt (zweifach) gebunden, doppelt verbunden
d. charged zweifach geladen
d. ionized zweifach ionisiert
d. linked *s.* d. bonded
d. refracting (refractive) doppelbrechend
d. unsaturated doppelt ungesättigt
dough / to zu Teig kneten (machen), einteigen; [ein]maischen; teigig (zu Teig) werden
d. in zu Teig kneten (machen), einteigen; [ein]maischen
dough teigartige Masse *f*, Paste *f*, Teig *m*; Teig *m (Backmasse)*; *(Gum)* Streichlösung *f*, Streichmischung *f*, Streichteig *m*
to make into a d. anteigen
d. dispensing machine *s.* d. dividing machine
d. divider (dividing machine) *(Lebm)* Teigteilmaschine *f*
d. forming machine *(Lebm)* Teigformmaschine *f*, Aufwirkmaschine *f*, Wirkmaschine *f*
d. kneader (kneading machine) Knetmaschine *f*, Knetwerk *n*, Knetapparat *m*, Kneter *m*, *(oft auch)* Mischmaschine *f*, Mischer *m*; *(Lebm)* Teigknetmaschine *f*, Teigkneter *m*, *(oft auch)* Teigmischmaschine *f*, Teigmischer *m*
d.-like teigartig, teigförmig, teigig
d. mixer (mixing machine) Mischmaschine *f*, Mischer *m*, *(oft auch)* Knetmaschine *f*, Knetwerk *n*, Knetapparat *m*, Kneter *m*, *(i.e.S.)* Schaufelkneter *m*, Doppelarmkneter *m*, Doppelpaddelmischer *m*; *(Lebm)* Teigmischmaschine *f*, Teigmischer *m*, *(oft auch)* Teigknetmaschine *f*, Teigkneter *m*
d. moulding compound *(Plast)* kittförmige Formmasse *f*
d. moulding machine *s.* d. forming machine
d. raising power Triebkraft *f (z.B. der Backhefe)*
doughiness teigige (pastöse) Konsistenz *f*; *(Lebm)* Teigkonsistenz *f*
doughy teigartig, teigförmig, teigig
Douglas fir Douglasie *f*, Pseudotsuga menziesi (Mirbel) Franco
D.O.V. *s.* double oil of vitriol
doverite *(Min)* Doverit *m (Yttriumfluorkarbonat)*
down / to be *(Ker)* gefallen sein *(pyrometrischer Kegel)*
down-draught kiln *(Ker)* Ofen *m* mit überschlagender Flamme
d.-flow high density tower *(Pap)* Dickstoff-Abwärts[bleich]turm *m*, Abwärts-Dickstoffturm *m*
d. period *s.* downtime
d.-run[ning] Abwärtsgasen *n*, Abwärtsgasung *f*, Gasen *n* (Gasung *f*) in absteigender Richtung,

Gasen *n* von oben *(bei der Wassergasherstellung)*

d.-steaming s. d.-run

d. stroke press Oberkolbenpresse *f*

downcomer *(Destillation)* Ablaufrohr *n*, Rücklaufrohr *n*, Fallrohr *n*, Rückflußrohr *n*; fallendes Gasabzug[s]rohr *n*, Schrägrohr *n* zum Staubsack *(eines Hochofens)*

downdraft s. downdraught

downdraught Unterwind *m*, Saugzug *m*

d. firing überschlagende Flammenführung *f*, Feuerung *f* mit umgekehrter Flamme

d. sintering Saugzugsinterung *f*

d. sintering machine Saugzugsinterapparat *m*, Saugzugsintermaschine *f*, Saugzugsinteranlage *f*

downflow Rück[lauf]strom *m*

d. evaporator Fallfilmverdampfer *m*, Fallstromverdampfer *m*

downpipe, downspout, downtake *(Destillation)* Ablaufrohr *n*, Rücklaufrohr *n*, Fallrohr *n*, Rückflußrohr *n*

downtime Betriebsunterbrechung *f*; Maschinenausfallzeit *f*, Stillstandszeit *f*; Nebenzeit *f*; Wartezeit *f*

downward current abwärtsgerichtete Strömung *f*, abwärtsgerichteter Strom *m*

2,4-DP $Cl_2C_6H_3OCH(CH_3)COOH$ 2,4-DP *n*, 2-(2,4-Dichlorphenoxy)-propionsäure *f*, Dichlorprop *n* *(Herbizid)*

DPA s. diphenylamine

DPH s. diamond pyramid hardness

DPN s. diphosphopyridine nucleotide

DPPD s. N,N'-diphenyl-p-phenylenediamine

dracorhodin Drakorhodin *n*

dracorubin Drakorubin[harz] *n*

draft *(Am)* s. draught

drag-chain conveyor Stegkettenförderer *m* *(i.e.S.)* *(Kette auf der Bahn gleitend)*

d. classifier Kratzband *n*, Bandklassierer *m*

d. conveyor s. d.-chain conveyor

d. effect Bremswirkung *f* *(Debye-Hückelsche Theorie starker Elektrolyte)*

d.-flight conveyor Kratzerförderer *m* *(i.e.S.)* *(Kette oberhalb des Fördergutes laufend)*

d. flow *(Plast)* Hauptfluß *m*, Schleppströmung *f* *(am Extruder)*

d. force hydrodynamische Widerstandskraft *f*; aerodynamische Widerstandskraft *f*

d.-in Eintrag *m*, eingeschleppte Lösung *f*

d.-ladle / to *(Glas)* fritten *(Scherben)*

d.-link conveyor s. d.-chain conveyor

d.-out Austrag *m*, herausgeschleppte Lösung *f*

dragade / to s. to drag-ladle

dragée Dragée *n*

Dragendorff alkaloid reagent s. Dragendorff's solution

Dragendorff's solution Dragendorffs Reagens *n* *(Alkaloidnachweis)*

dragline [excavator] Schürfkübelbagger *m*,

Schleppschaufelbagger *m*, Schlepplöffelbagger *m*, Schürfkübelwagen *m*

dragon's blood [resin] Drachenblutharz *n*, Resina draconis, Sanguis draconis *(von Daemonorops draco Blume oder von Dracaena-Arten)*

d. blood resin paper Drakorubinpapier *n*

drain / to 1. dränieren, entwässern; entleeren, ablassen; abtropfen lassen; 2. ablaufen; abtropfen; leerlaufen

d. away (off) ableiten, abführen, abziehen, abgießen; ablaufen lassen

d. through durchfließen, abfließen; durchfließen (abfließen) lassen

drain Abfluß *m*, Ablauf *m*, Ablaß *m*, Auslaß *m*, Ausfluß *m*, Abflußöffnung *f*; Ausguß *m*; Dränrohr *n*, Drän *m*

d. board Abtropfbrett *n*

d. casting *(Ker)* Hohlguß *m*

d. line Abflußleitung *f*, Ablaßleitung *f*, Abzugsleitung *f*; Abwasserleitung *f*; Dränageleitung *f*, Entwässerungsleitung *f*

d. of nutrients *(Bodenchemie)* Nährstoffentzug *m*

d. screen Entwässerungssieb *n*, Entbrühungssieb *n*

d. valve Ablaßventil *n*

drainage Dränage *f*, Entwässerung *f*; Entleerung *f*, Ablassen *n*; Abtropfenlassen *n*; Ablaufen *n*; Abtropfen *n*; Leerlaufen *n*

d. channel Ablaufkanal *m*; Entwässerungskanal *m*

d. of oil Ölabfluß *m*

d. of water Entwässerung *f*

d. period *(Pap)* Entwässerungszeit *f*, Entwässerungsperiode *f*

d. pipe Ablaufrohr *n*, Ablaufleitung *f*, Abflußrohr *n*, Abflußleitung *f*; Dränrohr *n*, Drän *m*

d. resistance *(Pap)* Entwässerungswiderstand *m* *(des Stoffs)*

d. tube s. d. pipe

d. water Sickerwasser *n*

drainer *(Pap)* Absetzbütte *f*, Eindickbütte *f*

draining board Abtropfbrett *n*, Ablaufbrett *n*, Trockenbrett *n*

d. chest s. d. tank

d. pan Tropfwanne *f*, Abtropfkanal *m*, Dräntunnel *m*, Dränkammer *f* *(beim Flow-Coating-Verfahren)*

d. tank *(Pap)* Absetzbütte *f*, Eindickbütte *f*

drains tank Tropfbenzoltank *m* *(einer Benzolanlage)*

drape forming *(Plast)* Streckformen *n*, Streckziehen *n*

d. mould *(Plast)* Streckformwerkzeug *n*

Draper law Drapersches Gesetz *n*, Drapersche Regel *f*

draping *(Plast)* Streckformen *n* *(einer Folie über einem Streckformwerkzeug)*

draught Zug *m*, Schornsteinzug *m*, Kaminzug *m*

d. air eingeblasene Luft *f*

d.-free zugfrei

d. gauge Zugmesser *m* *(z.B. für Schornsteine)*

d. of air Luftstrom *m*
d. paper Konzeptpapier *n*
d. tube Leitrohr *n (Rührer)*
dravite *(Min)* Dravit *m (Turmalinart)*
draw / to 1. ziehen *(z.B. Glas, Kunststoff)*; 2. [an]saugen, anziehen *(z.B. eine Flüssigkeit)*
d. in [an]saugen, anziehen *(z.B. eine Flüssigkeit)*
d. off *(Flüssigkeiten)* abziehen, *(flüssige Metalle)* abstechen
draw bar *(Glas)* Ziehbalken *m*
drawing Ziehen *n (z.B. von Glas, Kunststoff)*; Saugen *n*, Ansaugen *n*, Anziehen *n (z.B. einer Flüssigkeit)*
d. [card]board Zeichenkarton *m*
d. chamber *(Glas)* Ziehkammer *f*
d.-in roll *(Pap)* Walze *f* der Zugpresse (Transportpresse, Greiferpresse, Vorzugspresse)
d. machine *(Glas)* Ziehmaschine *f*
d. mark Ziehstreifen *m (Glasfehler)*
d. of the kiln *(Ker)* Ausfahren *n* des Ofens
d. paper Zeichenpapier *n*
d. pot *(Glas)* Ziehherd *m*
d. process *(Glas)* Ziehverfahren *n*
d. properties Aufziehvermögen *n (von Farbstoffen)*
d. shaft *(Glas)* Ziehschacht *m*
d. speed Ziehgeschwindigkeit *f (bei der Tafelglasherstellung)*
drawn glass gezogenes Glas *n*, Ziehglas *n*
d. grain *(Gerb)* gezogener Narben *m (Lederfehler)*
d.-rod method *(Glas)* Stabziehverfahren *n*
d. yarn gereckte Seide *f*
drawworks Hebewerk *n*, Getriebehebewerk *n*
DRC *s.* dry rubber content
dregs Bodensatz *m*, Satz *m*, Sediment *n*
d. washer *(Pap)* Bodensatzwäscher *m (bei der Kaustizierung)*
Dreiding nitrogen model Dreidingsches Stickstoffmodell *n*
drench / to *(Gerb)* in Schrotbeize (Kleienbeize) behandeln; *(Gerb)* mit Säure behandeln
drench shower Notbrause *f*, Löschbrause *f*, Feuerlöschbrause *f*
drenching *(Gerb)* Schrotbeize *f*, Kleienbeize *f*; *(Gerb)* Beizbehandlung *f* mit Säure
dress / to *(Erz)* aufbereiten; *(Landw)* düngen; *(Gerb)* appretieren, *(i.w.S.)* zurichten; *(Sämischleder)* weichmachen; *(Text)* appretieren, schlichten
dressing *(Erz)* Aufbereitung *f*; *(Landw)* Düngung *f*; *(Gerb)* Appretur *f*, *(i.w.S.)* Zurichtung *f*; *(Text)* Appretur *f*, Appretieren *n*, Schlichten *n*; *(Text)* Appreturmittel *n*, Appret *n*, Schlichte *f*
d. agent *(Gerb)* Appreturmittel *n*, Zurichtmittel *n*
d. by spray irrigation Beregnungsdüngung *f*
d. grease *(Gerb)* Walkfett *n*
d. leather Oberleder *n*
Dressler kiln *(Ker)* Dressler-Ofen *m*
Dressler muffle *(Ker)* Dressler-Muffel *f*

Drewboy separator Drewboy[-Scheider] *m*
dried alum gebrannter Alaun *m*
d. beet Trockenrübe *f*
d. beet pulp *s.* d. pulp
d. blood Blutmehl *n (Düngemittel)*
d. buttermilk Buttermilchpulver *n*
d. cream Trockenrahm *m*, Trockensahne *f*, Sahnepulver *n*, Rahmpulver *n*
d. egg yolk Trockeneigelb *n*
d. eggs Eipulver *n*, Trockenei *n*
d. fat-free milk Magermilchpulver *n*, Trockenmagermilch *f*
d. fruit Trockenobst *n*, Dörrobst *n*
d. malt Darrmalz *n*
d. milk Trockenmilch *f*, Trockenmilchpulver *n*, Milchpulver *n*
d. molassed beet pulp Melasseschnitzel *npl*
d. pulp Trockenschnitzel *npl*
d. rennet Labpulver *n*
d. silica gel Silikagel *n*, Silika-Gel *n*, Kiesel[säure]gel *n*
d. vegetables Trockengemüse *n*, Dörrgemüse *n*
d. whey Trockenmolke *f*, Molkenpulver *n*
drier 1. Trockenstoff *m*, Trockenmittel *n*, Trockenmedium *n*, Trocknungsmittel *n*, Trockner *m*, Sikkativ *n*; 2. *s.* dryer
drift Drift *f*, Wanderung *f (z.B. des Nullpunktes)*; Drift *f*, Verschwemmung *f (bei der Kohleentstehung)*
drill / to 1. bohren; 2. *(Landw)* drillen, streuen *(z.B. Düngemittel)*
drill collar Schwerstange *f (einer Rotary-Bohranlage)*
d. cuttings Bohrklein *n*, Bohrschmand *m*
d. pipe Bohrrohr *n*
d. pipe elevator Gestängeanheber *m*, Elevator *m (einer Rotary-Bohranlage)*
drillability 1. Bohrbarkeit *f*; 2. *(Landw)* Drillfähigkeit *f*, Streubarkeit *f*, Rieselfähigkeit *f (z.B. von Düngemitteln)*
drillable 1. bohrbar; 2. *(Landw)* drillfähig, streubar *(z.B. Düngemittel)*
drilling 1. Bohren *n*; 2. *(Landw)* Drillen *n*, Streuen *n (von Düngemitteln)*
d. bit Bohrmeißel *m (besonders zum Erdölbohren)*
d. cable Bohrseil *n*
d. derrick Bohrturm *m*
d. floor Arbeitsbühne *f (einer Rotary-Bohranlage)*
d. fluid Bohrflüssigkeit *f*, Bohrspülung *f*
d. fluid circulating system Spülungsumlaufsystem *n*, Spülungskreislauf *m (einer Rotary-Bohranlage)*
d. mast Bohrmast *m (einer Rotary-Bohranlage)*
d. mud Bohrspülung *f*, [schlammartige] Bohrflüssigkeit *f*, Bohrschlamm *m*, [schlammartige] Spülflüssigkeit *f*, Spülschlamm *m*, Spülung *f*
d. platform Bohrplattform *f*
d. rig Bohranlage *f*, Bohrgerät *n*
d. string Bohrstrang *m*, Bohrgarnitur *f*

d. tool Bohrwerkzeug n
drillo (Gerb) Trillo m (Schuppen der Fruchtbecher mehrerer orientalischer Eichenarten)
drillometer Drillometer n, Bohrdruckmesser m
drinking water Trinkwasser n
d.-water supply Trinkwasserversorgung f
drip / to tropfen, tröpfeln
drip Tropfen n, Tröpfeln n; Tropfen m, Lacktropfen m (beim Tauchlackieren)
d.-dry / to trocknen unter Abtropfen, tropfnaß zum Trocknen aufhängen
d. pipe Tropfrohr n
d.-proof tropfwassergeschützt
drive / to antreiben, betreiben
d. off (out) [her]austreiben, vertreiben, verjagen
drive Antrieb m
d. shaft Antriebswelle f
driver Antriebsmaschine f
driving force (potential) Triebkraft f
d. pulley Antriebstrommel f, Antriebsrolle f, Treibtrommel f, Treibrolle f
drop / to 1. tropfen; 2. sinken, abfallen (Temperatur, Spannung usw.); 3. [hinab]stürzen, frei fallen lassen (z.B. Koks)
d. in zutropfen
drop 1. Tropfen m; 2. Sinken n, Abfallen n, Fallen n, Abfall m, Fall m, Sturz m (der Temperatur, Spannung usw.); Gefälle n
d. analysis Tüpfelanalyse f, Tüpfelprobe f
d. by drop tropfenweise
d. chalk (in konische Form gepreßte) Schlämmkreide f
d. culture (Bioch) Hängetropfenkultur f, Deckglaskultur f
d. diameter Tropfendurchmesser m
d. electrode Tropfelektrode f
d. formation Tropfenbildung f
d. funnel Tropftrichter m
d. hole Falloch n, Durchfalloch n, Durchfallöffnung f
d. ignition temperature Tropfzündpunkt m
d. in head Druckabfall m
d. in voltage Spannungsabfall m, Spannungsabnahme f, Spannungsrückgang m, Spannungsminderung f, Spannungs[ab]senkung f, Spannungsverlust m
d. of potential Potentialabfall m, Potentialrückgang m; Potentialgefälle n
d. of temperature Temperaturabfall m; Temperaturgefälle n
d.-out box Abscheidekammer f, Staubkammer f; Kammerabscheider m
d. period (Gum) Abstehzeit f
d. reaction Tüpfelreaktion f
d.-shaped tropfenförmig
d. shatter test Sturzprobe f, Sturzversuch m, Sturzprüfung f, Fallprobe f, Fallprüfung f
d. size Tropfengröße f
d.-size distribution Tropfenverteilung f
d. throat (Glas) versenkter (tiefliegender, tiefer) Durchlaß m

d. time Tropfzeit f (z.B. bei einer Titration)
d. weight 1. Tropfengewicht n; 2. Fallkugel f
d.-weight device Kugelfallwerk n
d.-weight method (phys Ch) Tropfengewichtsmethode f
droplet Tröpfchen n, kleiner Tropfen m
d. formation Tropfenbildung f
d. size Tröpfchengröße f
dropper Tropfer m, (i.w.S.) Tropfenzähler m
d. teat Pipettenhütchen n
dropping-ball viscometer Kugelfallviskosimeter n
d. bottle Tropffläschchen n, Tropfflasche f, Tropfglas n, (i.w.S.) Tropfenzähler m
d. cathode Tropfkatode f
d. electrode Tropfelektrode f
d. funnel Tropftrichter m
d.-mercury electrode Quecksilbertropfelektrode f
d. pipette Tropfpipette f, Stechpipette f, (i.w.S.) Tropfenzähler m
d. point Tropfpunkt m
dropwise tropfenweise
d. addition tropfenweises Zugeben (Zusetzen) n, Zutropfen n
d. condensation Tropfenkondensation f
droserone Droseron n, 3,5-Dihydroxy-2-methyl-1,4-naphthochinon n
dross Schaum m, Metallschaum m
drug Droge f; Arzneimittel n; Narkotikum n, (i.e.S.) Rauschgift n
d. research Arzneimittelforschung f
drum / to (Gerb) (Häute) in der Lattentrommel (oder) im Faß durcharbeiten (behandeln)
drum Trommel f, Walze f; (Gerb) (rotierendes) Faß n
d. barker (Pap) Entrindungstrommel f
d. barking (Pap) Trommelentrindung f
d. beet slicer (Zucker) Trommelschneidmaschine f
d.-built tyre (Gum) Trommelreifen m
d.-chart recorder Trommelschreiber m
d. cooling Trommelkühlung f
d. dryer Trommeltrockner m, Trockentrommel f, Walzentrockner m, Trockenwalze f, Zylindertrockner m, Trockenzylinder m
d.-dye / to (Text) in der Trommel färben; (Gerb) im Faß färben
d.-dyeing (Text) Trommelfärbung f; (Gerb) Faßfärbung f
d.-dyeing machine (Text) Trommelfärbemaschine f
d. film dryer Walzen[dünnschicht]trockner m
d. filter Trommelfilter n
d. layer filter Trommelschichtenfilter n
d. liming (Gerb) Faßäscher m, Faßäscherung f
d. malting Trommelmälzerei f
d. mill Trommelmühle f
d. mixer Trommelmischer m, Mischtrommel f
d. pelletizer Pelletisiertrommel f
d. recorder s. d.-chart recorder

d. screen Trommelsieb n
d. separator Trommelscheider m
d. setting machine (Gerb) Trommelausstoß-maschine f
d. slope Trommelneigung f
d. strainer (Pap) Drehknotenfänger m, rotierender Knotenfänger m
d. stuffing (Gerb) Fetten n im Faß, Faßfettung f, Faßschmiere f
d. tannage Faßgerbung f
d. washer Waschtrommel f
drumming (Gerb) Durcharbeiten n (Behandlung von Häuten) in der Lattentrommel (oder) im Faß
Drummond's limelight Drummonds (Drummond-sches) Kalklicht n
druse (Geol) Druse f, Kristalldruse f; miarolitischer Hohlraum (Drusenraum) m
drusy kleindrusig, miarolitisch; löch[e]rig, blasig, kavernös
dry / to 1. trocknen, entfeuchten, entwässern, dehydratisieren, Wasser entziehen; (eine Lösung) eindampfen, einengen; (Obst) dörren, (Malz) darren, rösten; 2. austrocknen, abtrocknen, eintrocknen
d. hard (Pap) stark trocknen, übertrocknen
d. off [ab]trocknen
d. out austrocknen
d. soft (Pap) schwach trocknen, nicht völlig trocknen
d. to constant weight bis zur Massekonstanz trocknen, (bisher) bis zur Gewichtskonstanz trocknen
d. to leather hard (Ker) lederhart trocknen
dry trocken
d. adhesive Trockenkleber m
d. air cure Heißluftvulkanisation f, Vulkanisation f in Heißluft, Heißluftheizung f
d. analysis Trockenanalyse f
d., ash-free wasser- und aschefrei, waf
d.-ashed trocken verascht
d. assay (Met) trockene Probe f, Trockenprobe f
d. bait Trockenköder m
d. basis s. d. weight basis
d. battery Trockenbatterie f, (manchmal auch für) d. cell
d. blend Trockenmischung f
d.-blend extrusion (Plast) Strangpressen (Extrudieren) n von Trockenmischung, Dry-blend-Strangpressen n
d. blender Trockenmischer m
d. blending Mischen (Vermischen, Vermengen) n von Feststoffkomponenten
d. blotting paper Trockenfließpapier n
d.-bone ore (Min) Smithsonit m, Zinkspat m (Zinkkarbonat)
d.-bottom furnace Feuerung f mit trockenem Schlackenabzug
d. break (Pap) Bahnriß m in der Trockenpartie
d. broke (Pap) Trockenausschuß m

d.-bulb temperature Temperatur f des trockenen Thermometers, Trockenkugeltemperatur f
d. buttermilk Buttermilchpulver n
d. carbonizing (Text) trockenes Karbonisieren n, Trockenkarbonisieren n, trockene Karbonisation f, Trockenkarbonisation f
d. cell Trockenelement n
d. chemical fire extinguisher Trockenlöscher m
d.-clean / to (z.B. Gas) trocken reinigen; (Met) trocken (pneumatisch) aufbereiten; (Kleidung) chemisch (trocken) reinigen
d. cleaner Trockenreiniger m; (Met) Luftsetzmaschine f, Luftsetzapparat m; Chemischreiniger m, Trockenreiniger m
d. cleaning trockene Reinigung f (z.B. von Gas); (Met) Luftwäsche f, Luftaufbereitung f, Trockenaufbereitung f, trockene (pneumatische) Aufbereitung f, Aufbereitung f auf trockenem Wege; Chemischreinigen n, Chemischreinigung f, Trockenreinigen n, Trockenreinigung f (von Kleidung)
d.-cleaning solvent Lösungsmittel n für Chemischreinigung (Trockenreinigung)
d. colour Pigment n
d. colouring (Plast) Trockeneinfärben n (von Formmassen)
d. concentrate Trockenkonzentrat n, Staubkonzentrat n (Wirkstoffzubereitung)
d. content Gehalt m an Trockensubstanz
d. cooling method (process) s. d. drum-cooling method
d. cream Trockenrahm m, Trockensahne f, Sahnepulver n, Rahmpulver n
d.-cure / to trockensalzen, trocken einsalzen, (Fleisch auch) trockenpökeln; dörren
d. cure s. d. curing
d. curing Trockensalzen n, (bei Fleisch auch) Trockenpökeln n, Trockenpökelung f; Dörren n
d. curing method (process) Trockensalzverfahren n, (bei Fleisch auch) Trockenpökelverfahren n; Dörrverfahren n
d. curve Trocknungskurve f
d.-cyanide / to trockenzyanieren, gaszyanieren, karbonitrieren
d. cyaniding Trockenzyanieren n, Gaszyanieren n, Karbonitrieren n, Karbonitrierung f
d. decatizing (Text) Trockendekatieren n, Trockendekatur f, Trockendämpfen n
d. decomposition process trockener Aufschluß m
d. desulphuration Trockenentschwefelung f (von Gasen)
d. distillation trockene Destillation f, Trockendestillation f, Entgasung f
d. drum-cooling method Kühltrommelverfahren n, Abkühlverfahren n mit Kühltrommeln, Trockenverfahren n, Trommelverfahren n (Margarineherstellung)
d. edging (Ker) (Bildung von trockenen, rauhen Kanten infolge unzureichenden Deckens der Glasur)

d. **emulsion** *(Foto)* Trockenemulsion *f*

d. **end of the dryer section** *(Pap)* Ende *n* der Trockenpartie

d. **end of the paper-making machine** *s.* d. part

d. **extract** Trockenextrakt *m*, trockener Extrakt *m*

d. **extrusion** *(Plast)* Strangpressen *n* ohne Anwendung von Lösungsmitteln

d. **fat melting** Trockenschmelze *f*, Trockenschmelzverfahren *n (Fettgewinnungsverfahren)*

d. **felt** *(Pap)* Trockenfilz *m*

d. **felt roll** *(Pap)* Trockenfilzleitwalze *f*

d. **filter** Trockenfilter *n*

d. **fruit** Trockenobst *n*, Dörrobst *n*

d. **fumigant** Räuchermittel *n (zur Schädlingsbekämpfung)*

d. **furfural** Trockenfurfural *n*

d.-**gage / to** *s.* to drag-ladle

d. **gas** Trockengas *n*, trockenes Gas (Erdgas, Naturgas) *n*

d. **gas cleaning** Trockengasreinigung *f*, Trockenreinigung *f* eines Gases

d. **gas meter** trockener Gasmesser (Gaszähler) *m*

d. **gel** Trockengel *n*

d. **grinding** Trocken[ver]mahlen *n*, Trocken[ver]mahlung *f*

d. **heat cure (vulcanization)** *s.* d. air cure

d. **hole** *(Erdöl)* trockene Bohrung *f*, Fehlbohrung *f*, erfolglose Aufschlußbohrung *f*

d. **ice** Trockeneis *n*, Kohlensäureschnee *m*

d. **liming** *(Zucker)* trockene Scheidung *f*, Trockenscheidung *f*

d. **matter** Trockensubstanz *f*; Abdampfrückstand *m*, Rückstand *m*

d. **milk** Trockenmilch *f*, Trockenmilchpulver *n*, Milchpulver *n*

d. **milling** Trocken[ver]mahlen *n*, Trocken[ver]mahlung *f*

d., **mineral-matter-free** trocken (lufttrocken) und mineral[stoff]frei

d. **mix** *s.* d. mixing

d. **mixer** Trockenmischer *m*

d. **mixing** Mischen (Vermischen, Vermengen) *n* von Feststoffkomponenten; *(Ker)* Trockenaufbereitung *f*

d. **moulding** Trockensandformen *n*, Trockengußformen *n*, Formen *n* in Trocken[guß]sand

d. **natural gas** trockenes Erdgas (Naturgas) *n*, Trockengas *n*

d. **offset ink** Druckfarbe *f* für Hochoffset (Trockenoffset)

d. **offset printing** indirekter Hochdruck *m*, Hochoffsetdruck *m*, Trockenoffsetdruck *m*

d. **pan** Trockenkollergang *m*

d. **part** *(Pap)* Trockenpartie *f*

d. **peat** Trockentorf *m*

d. **peat substance** Torftrockensubstanz *f*, Torftrockenmasse *f*

d. **petrolatum** Hartpetrolat[um] *n*

d. **plate** *(Foto)* Trockenplatte *f*

d.-**plate pressure drop** trockener Druckverlust *m* eines Kolonnenbodens

d. **precipitator** Trockenelektroabscheider *m*, Trockenelektrofilter *n*

d. **preparation** Trockenaufbereitung *f*; Trockenpräparat *n*

d. **press body (mix)** *(Ker)* Trockenpreßmasse *f*

d. **press process** *(Ker)* Trockenpreßverfahren *n*

d.-**pressed** *(Ker)* trocken gepreßt

d. **pressing** *(Ker)* Trockenpressen *n*, Trockenpressung *f*

d. **pressing mix** *s.* d. press body

d. **pressure drop** trockener Druckverlust *m*

d. **process** Trockenverfahren *n*, trockenes Verfahren *n*, Trockenaufbereitung *f*

d. **processing** Trockenverarbeitung *f*

d. **product** getrocknetes (trockenes) Gut *n*, Trockengut *n*; *(Lebm)* Trockenware *f*

d. **pulp** *(Pap)* Halbstoff *m* in trockenen Bogen, trockener Halbstoff *m*

d. **purification** trockene Reinigung *f*

d. **purifier** Trockenreiniger *m*

d. **quenching** trockene Kokslöschung (Kokskühlung) *f*

d.-**rendered lard** im Trockenschmelzverfahren ausgeschmolzenes Fett *n*

d. **rendering** Trockenschmelzen *n*, Trockenschmelze *f*

d. **rendering method** Trockenschmelzverfahren *n*

d. **residue** Trockensubstanz *f*; Abdampfrückstand *m*, Rückstand *m*

d. **rosin size** *(Pap)* Harzleimpulver *n*

d. **rot** Trockenfäule *f*

d. **rubber** Festkautschuk *m*

d. **rubber content** Trockenkautschukgehalt *m*, Kautschuktrockengehalt *m*

d. **rubber manufacture** „trockene" Kautschukverarbeitung *f*

d.-**salt / to** trockensalzen, trocken einsalzen, *(Fleisch auch)* trockenpökeln

d.-**salt cure** *s.* d. salting

d. **salting** Trockensalzen *n*, *(bei Fleisch auch)* Trockenpökeln *n*, Trockenpökelung *f*

d. **sand** Trockensand *m*, Trockenguß[form]sand *m*

d.-**sand casting** Trockenguß *m*, Gießen *n* in getrocknete Sandformen

d.-**sand mould** Trockensandform *f*, trockene (getrocknete) Sandform (Form) *f*, Trockengußform *f*

d.-**sand moulding** Trockensandformen *n*, Trockengußformen *n*, Formen *n* in Trocken[guß]sand

d. **saturated vapour** trockengesättigter Dampf *m*, Trockendampf *m*

d. **screening** Trockensieben *n*

d. **seal** Trockenverschluß *m*, trockener Abschluß *m*

d. **seal gasholder** Scheibengasbehälter *m*

d. **seed treatment** *(Landw)* Trockenbeize *f*

d. **shampoo** Trockenschampun *n*, Trockenshampoo[n] *n*

drying

d. **skim milk** Trockenmagermilch f, Magermilchpulver n
d. **spent grains** Trockentreber pl
d. **spinning** Trockenspinnen n, Trockenspinn-
verfahren n
d. **spray** Trockenspritzen n (Farbspritztechnik)
d. **steam coal** Magerkohle f, Halbanthrazit m
d. **steaming** (Text) Trockendämpfen n, Trok-
kendekatieren n, Trockendekatur f
d. **strength** Trockenfestigkeit f
d. **substance** Trockensubstanz f; Abdampfrückstand m, Rückstand m
d. **table** Luftherd m (Aufbereitungstechnik)
d. **tannage** Trockengerbung f
d. **tensile strength** Reißfestigkeit f in trockenem
Zustand
d. **treatment** s. d. seed treatment
d.**-type precipitator** Trockenelektroabscheider
m, Trockenelektrofilter n
d. **vinegar** Trockenessig m
d. **volume** Schüttvolum[en] n
d. **weight** Trocken[substanz]masse f, Trok-
kenstoffmasse f, (bisher) Trockengewicht n
d. **weight basis** Bezugsbasis f Trocken[sub-
stanz]masse (Trockenstoffmasse)
d. **well** (Erdöl) trockene Bohrung f, Fehlbohrung
f, erfolglose Aufschlußbohrung f
d. **whey** Trockenmolke f, Molkenpulver n
d. **whole milk** Trockenvollmilch f, Vollmilchpulver
n, Milchpulver n
d. **wine** trockener Wein m
d. **wood weight** (Pap) Trockenmasse f des Holzes,
Darrmasse f, (bisher) Darrgewicht n
d. **yeast** Trockenhefe f
dryer 1. Trockner m, Trockenapparat m, Trockenmaschine f, (i.e.S.) s. d. cylinder; 2. s. drier
d. **arrangement** (Pap) Trockenzylinderanord-
nung f
d. **corridor** (Ker) Trocknerkammer f
d. **cylinder** Trockenzylinder m, Trocknerzylinder
m, Trockentrommel f, Trockenwalze f
d. **doctor** (Pap) Schaber m am Trockenzylinder
d. **drum** Trockentrommel f, Trocknertrommel f,
Trockenzylinder m, Trockenwalze f
d. **felt** (Pap) Trockenfilz m
d.**-group** (Pap) Trocken[zylinder]gruppe f, Trocknungsgruppe f, Zylindergruppe f, Heizgruppe f
d. **journal** (Pap) Zylinderzapfen m (des Trockenzylinders)
d. **part** (Pap) Trockenpartie f
d. **roll** Trockenwalze f, Trocknerwalze f, Trockenzylinder m, Trockentrommel f
d. **section** (Pap) Trockenpartie f; (Pap) Trok-
ken[zylinder]gruppe f, Trocknungsgruppe f,
Zylindergruppe f, Heizgruppe f
d. **shell** (Pap) Zylindermantel m (des Trockenzylinders)
d. **surface** (Pap) Trockenzylinderoberfläche f
d. **surface temperature** (Pap) Temperatur f der
Trockenzylinderoberfläche, Heizoberflächen-
temperatur f

drying Trocknen n, Trocknung f, Entfeuchtung f,
Feuchtigkeitsentfernung f, Wasserauftrocknung
f; Dörren n, Darren n (von Malz); Eindampfen n,
Einengen n; Austrocknen n, Abtrocknung f
d. **agent** Trockenmittel n, Trockenmedium n,
Trocknungsmittel n, Trockenstoff m, Trockner m,
Sikkativ n
d. **air** Trocknungsluft f
d. **apparatus** Trockenapparat m
d. **area** Trockenfläche f
d. **box** s. d. chamber
d. **by contact** (Pap) Kontakttrocknung f
d. **by evaporation** Verdampfungstrocknung f;
Verdunstungstrocknung f
d. **cabinet** s. d. chamber
d. **chamber** Trockenkammer f, Trocknungs-
kammer f, Trockenraum m
d. **contraction** (Ker) Trockenschwindung f
d. **crack** (Ker) Trockenriß m
d. **cupboard** Trockenschrank m
d. **curve** Trocknungskurve f
d. **cycle time** Gutverweilzeit f, Verweilzeit f (bei
der Trocknung)
d. **cylinder (drum)** Trockenzylinder m, Zylindertrockner m, Trockentrommel f, Trommeltrockner m, Trockenwalze f, Walzentrockner
m
d. **duct** Stromrohr n (eines Trockners)
d. **equipment** Trocknungsanlage f, Trockenanlage f
d. **gas** Trocknungsgas n, Trockengas n
d. **ground** Trocknungsfeld n, Trockenfeld n,
Trockenplatz m
d. **jar** Trockenturm m, Chlorkalziumzylinder m
d. **kiln** Trockenofen m; Darrofen m, Darre f
d. **loft** Trockenboden m, Trockenspeicher m,
Trockenkammer f
d. **loss** (Ker) Trockenverlust m
d. **machine** Trockenmaschine f, Trockenapparat
m, Trockner m
d. **marks** (Foto) Trockenflecken mpl, Was-
serflecken mpl; Kalkflecken mpl
d. **medium** s. d. agent
d. **of pulp** Zellstofftrocknung f
d. **oil** trocknendes Öl n
d.**-out time** s. d. time
d. **oven** Trockenofen m; (Gerb) Austrockner m
d. **part** (Pap) Trockenpartie f
d. **period** s. d. time
d. **pistol** Trockenpistole f
d. **plant** Trockenanlage f, Trocknungsanlage f
d. **potential** Trocknungspotential n, Trocknungs-
triebkraft f
d. **process** Trocknungsprozeß m, Trocknungs-
verfahren n; Trocknungsvorgang m
d. **rack** Abtropfgestell n, Abtropfbrett n; (Ker)
Trockengestell n
d. **rate** Trockengeschwindigkeit f, Trocknungsgeschwindigkeit f, spezifische Austauschfeuchtemenge f

d. reel *(Text)* Trockenhaspel *f*
d. roll *s.* d. cylinder
d. room Trockenraum *m*, Trocknungsraum *m*, Trockenkammer *f*
d. section *(Pap)* Trockenpartie *f*
d. shed Trockenschuppen *m*
d. shrinkage *(Ker)* Trockenschwindung *f*
d. stove Trockenofen *m*
d. surface Trockenfläche *f*
d. temperature Trockentemperatur *f*, Trocknungstemperatur *f*
d. time Trockenzeit *f*, Trocknungszeit *f*, Trocknungsdauer *f*
d. to constant weight Trocknen *n* bis zur Massekonstanz, *(bisher)* Trocknen *n* bis zur Gewichtskonstanz
d. tower Trockenturm *m*
d. tray Trockenschale *f*; Trockenhorde *f*
d. tube Trockenrohr *n*, Trockenröhre *f*
d. tunnel Trockentunnel *m*, Kanaltrockner *m*
dryness Trockenheit *f*; Trockne *f*
d.s. *s.* difficultly soluble
D.S. *s.* degree of substitution
DTA *s.* differential thermal analysis
dual collision *(phys Ch)* Zweierstoß *m*
d. ion Zwitterion *n*, Ampho-Ion *n*, Dipolion *n*
d. press Doppelpresse *f*
d. temperature process *(Kch)* Zweitemperaturverfahren *n*, Heiß-Kalt-Verfahren *n*
Duane and Hunt law Duane-Huntsches Gesetz *n*
dub / to *(Gerb)* [ein]schmieren
dubbin *(Gerb)* Tafelschmiere *f*, Fettschmiere *f*, Aasschmiere *f* *(Gemisch aus Lebertran und Talg)*
dubbing *(Gerb)* Kaltfetten *n*, Kaltschmieren *n*, Schmieren *n* auf der Tafel; *s.* dubbin
Dubbs process Dubbs-Verfahren *n*, Dubbs-Krackprozeß *m*
duct Leitung *f*, Leitungsrohr *n*, Kanal *m*, Schacht *m*
d. fan Ringlüfter *m*
ductile duktil, dehnbar, streckbar, ziehbar, [ver]formbar, plastisch, biegsam
ductility Duktilität *f*, Dehnbarkeit *f*, Streckbarkeit *f*, Ziehbarkeit *f*, Verformbarkeit *f*, Formbarkeit *f*, Plastizität *f*, Biegsamkeit *f*
due to / to be hervorgerufen werden durch, zurückzuführen sein auf
duff *(Bodenkunde)* Moder *m* *(Humusform)*
Duff reaction Duff-Reaktion *f* *(Phenolformylierung)*
Dufour effect *(phys Ch)* Dufour-Effekt *m*
dufrenoysite *(Min)* Dufrenoysit *m* *(Blei(II)-arsen(III)-sulfid)*
Duhem-Margules equation *(phys Ch)* Duhem-Margulessche Gleichung *f*, Gibbs-Duhem-Margulessche Gleichung *f*
Dühring's rule *(phys Ch)* Dühringsche Regel *f*
dulcification Absüßen *n*, Versüßen *n*, Versüßung *f*
dulcify / to absüßen, versüßen
dulcin $C_2H_5OC_6H_4NHCONH_2$ Dulzin *n*, Sucrol *n*, 4-Äthoxyphenylharnstoff *m*

dulcite $CH_2OH(CHOH)_4CH_2OH$ Dulzit *m*, Melampyrit *m*, Melampyrin *m* *(ein Hexit)*
dulcitol *s.* dulcite
dull / to 1. mattieren, den Glanz verringern; abstumpfen; trüben; 2. matt werden; stumpf werden; trüb werden
dull matt; stumpf; trüb[e]
d.-black mattschwarz
d. coal Mattkohle *f*, matte Kohle *f*
d.-green schmutziggrün
d. print Mattdruck *m*
d. red heat dunkle Rotglut *f*, Dunkelrotglut *f*
d. redness *s.* d. red heat
d. spun *(Text)* spinnmattiert
d.-white mattweiß
dulling 1. Mattieren *n*; Abstumpfen *n*; Trübung *f*, Abtrübung *f*; 2. Mattwerden *n*; Stumpfwerden *n*; Trübwerden *n*
d. agent Mattierungsmittel *n*
dullness Mattheit *f*, mattes Aussehen *n*, Glanzlosigkeit *f*; Stumpfheit *f*; Trübheit *f*
Dulong and Petit's law *(phys Ch)* Dulong-Petit-Regel *f*, Dulong-Petitsche Regel *f*
Dumas' method for nitrogen Stickstoffbestimmung *f* nach Dumas
Dumas' method of determining molecular weights Bestimmung *f* der relativen Molekülmasse nach Dumas
dumb bell *s.* d.-bell test piece
d.-bell model Hantelmodell *n* *(eines Moleküls)*
d.-bell strip *s.* d.-bell test piece
d.-bell test piece *(Gum)* hantelförmiger Schulterstab *m*, stabförmiger Probekörper *m*, Stäbchenprobe *f*, Stabprobe *f*, Hantelprüfkörper *m*
dummy tube blindes Rohr *n*, Blindrohr *n*
dumontite *(Min)* Dumontit *m*
dumortierite *(Min)* Dumortierit *m*
dump / to ausschütten, auskippen, ausstoßen, auswerfen, abwerfen, entleeren
dump temperature Ausstoßtemperatur *f*
dumped packings geschüttete Füllkörper *mpl*
dumping Ausschütten *n*, Auskippen *n*, Ausstoßen *n*, Auswerfen *n*, Abwerfen *n*, Entleeren *n*
d. floor Kipphorde *f*
d. grate Klapprost *m*
Dundas blackbutt *(Gerb)* Eucalyptus dundasi Maiden
dung Dung *m*, Mist *m*
dunite Dunit *m* *(ein Tiefengestein)*
Dunlop pendulum Dunlop-Pendel *n* *(Gerät zur Bestimmung der Stoßelastizität)*
Dunlop process Dunlop-Verfahren *n* *(Schaumgummiherstellungsverfahren)*
Dunlop tripsometer Dunlop-Tripsometer *n* *(Gerät zur Bestimmung der Stoßelastizität)*
Dunnachie kiln *(Ker)* Dunnachie-Ofen *m* *(gasbeheizter Kammerofen nach J. Dunnachie)*
dunt *(Ker)* Kühlriß *m*
dunting *(Ker)* Kühlrißbildung *f*
Duo-sol Duosol *n*, Duo-Sol *n*, Lösungsmittelpaar *n*, Lösemittelpaar *n*

D.-sol extraction Duosolextraktion f
D.-sol process Duosolverfahren n, Duo-Sol-Verfahren n, Zweilösungsmittelverfahren n
D.-sol solvent extraction plant Duosolanlage f
D.-sol solvent extraction process s. D.-sol process
duopulper (Pap) Duopulper m, Pulper (Stofflöser) m mit zwei Auflösescheiben
duplet Dublett n, Elektronendublett n, gemeinsames Elektronenpaar n zweier Atome
d. shell Zweierschale f
duplex blender Duplexmischer m
d. cardboard Duplexkarton m, gegautschter Karton m
d. coater (Pap) Streichmaschine f für doppelseitigen Strich, Streichmaschine f für zweiseitige Beschichtung
d. paper Duplexpapier n, gegautschtes Papier n
d. process Duplex[schmelz]verfahren n, Zweistufenverfahren n (Zusammenarbeiten von zwei Schmelzöfen)
d. pump Duplexpumpe f, Zwillingspumpe f
d. transfer paper Duplexpapier n (Sonderqualität des Abziehbilderpapiers)
duplicating paper Vervielfältigungspapier n; Abzugpapier n, Saugpostpapier n
duplicator paper s. duplicating paper
Du Pont-Grasselli-Williams machine (Gum) Du-Pont-Grasselli-Abriebmaschine f, Grasselli-Prüfer m, Williams-Prüfer m, Du-Pont-Prüfer m
Du Pont machine (Gum) Du-Pont-Maschine f, Du-Pont-Biegemaschine f, Du-Pont-Ermüdungsmaschine f, Du-Pont-Kettenermüdungsmaschine f, Du-Pont-Biegeprüfmaschine f
Du Pont process Du-Pont-Verfahren n (Ammoniakverbrennung)
Dupré equation (phys Ch) Duprésche Beziehung f, Young-Dupré-Beziehung f
durability Haltbarkeit f, Dauerhaftigkeit f, Lebensdauer f; Standfestigkeit f (von Werkzeugen)
d. test Haltbarkeitsprüfung f, Dauerhaftigkeitsprüfung f
durable haltbar, dauerhaft, beständig
d. press (Text) Permanent-Appretur f, Permanent-Ausrüstung f
durain Durit m, Durain m
durangite (Min) Durangit m
duration Dauer f, Zeit[dauer] f
d. of collision Stoßdauer f
d. of discharge Entladungszeit f, Entladezeit f
d. of exposure Belichtungszeit f
d. of test Prüfdauer f, Versuchsdauer f
durene $C_6H_2(CH_3)_4$ Duren n, 1,2,4,5-Tetramethylbenzol n
Durham tube Gärröhrchen n nach Durham
durmast oak (Gerb) Traubeneiche f, Wintereiche f, Quercus petraea (Matt.)Liebl.
durmolize / to (z. B. Leinöl) durch Hitze eindicken
duroclarite Duroklarit m
durol s. durene

duromer Duromer[es] n
durometer Härteprüfer m, Härtemesser m, Durometer n
d. hardness Durometerhärte f (Shore-Härte)
durum wheat Hartweizen m, Glasweizen m, Triticum durum Desf.
Durville casting process Durville-Gießverfahren n
durylic acid Durylsäure f, 2,4,5-Trimethylbenzolkarbonsäure f
dusky trüb
dust / to 1. stauben; verstauben; (Pap) stäuben (von hochgefüllten Papieren); 2. bestäuben, einstäuben, [ein]pudern; (mit Stäubemitteln) stäuben, (Stäubemittel) verstäuben; 3. (Ker) zerrieseln (von Materialien mit hohem Kalziumorthosilikat-Gehalt); 4. (Ker) (Ware vor dem Brennen) abblasen, abstauben; 5. (Pap) (Hadern, Altpapier) entstäuben
dust 1. Staub m; 2. (Landw) Stäubemittel n (z.B. Insektizid); 3. (Met) (sich zunächst im kälteren Teil eines Bleischachtofens anreichernder und von Gichtgasen mitgerissener) Flugstaub m; 4. (Ker) Stanzmasse f, Preßmasse f
d.-absorption system Entstaubungsanlage f
d. apparatus s. d. impinger
d. catcher Staubsammler m, Staubfänger m, Staubabscheider m, Entstauber m, Staubsack m
d. chamber Staubkammer f, Staubtasche f, Flugstaubkammer f
d. collecting fan Spiral[ab]scheider m, Gebläseentstauber m
d. collecting liquid Staubbindeflüssigkeit f
d. collector s. d. catcher
d. concentration Staubgehalt m, Staubbeladung f
d. deposition Staubniederschlag m
d. elimination Entstauben n, Entstaubung f, Staubabscheidung f
d. explosion Staubexplosion f
d. extraction s. d. elimination
d. filter Staubfilter n
d. filtration s. d. elimination
d. formation Staubbildung f; Staubanfall m
d. formulation (Landw) (mit Streckmitteln versetztes) kombiniertes Stäubemittel n, Feststofformulierung f
d.-free staubfrei; nichtstaubend
d. hazard Gefahr f von Staubschäden
d. hopper Staub[sammel]bunker m
d. impinger Staubprobensammler m
d.-laden staubhaltig, staubig, staubbeladen
d.-laden gas staubhaltiges Gas n, Staubgas n, Rohgas n (bei der Staubabscheidung)
d.-laying oil staubbindendes Öl n, Staubbindeöl n
d. loading Staubgehalt m, Staubbeladung f
d. loss Verstaubungsverlust m
d. monitor Luftüberwachungsanlage f
d. outlet Staubaustrag m, Staubaustritt m, Staubaustragöffnung f

d. particle Staubteilchen n, Staubpartikel f, Staubkorn n

d.-pressed tile (Ker) trockengepreßte Fliese (Kachel) f

d. pressing (Ker) Trockenpressen n, Trockenpressung f

d. proofing Besprühung (Berieselung) f zur Staubbindung (z.B. von Kohle)

d. removal s. d. elimination

d. residue Staubrückstand m, Staubbelag m, Staubniederschlag m

d. room s. d. chamber

d. separator (settler) s. d. catcher

d. shield Staubschutz m

d. trap Ablenkkammer f (ein Staubabscheider)

d. treatment Einstäubung f, Trockenbeize f (von Saatgut)

d. tuff (Geol) Staubtuff m

dustability Verstäubbarkeit f

duster s. dusting machine; (Gerb) Versteckfarbe f, Abtränkbrühe f

dusting 1. Stauben n; Verstauben n, Verstaubung f; (Pap) Stäuben n (hochgefüllter Papiere); 2. Bestäuben n, Einstäuben n, Einpudern n, Pudern n; Stäuben n (mit Stäubemitteln), Verstäuben n, Verstäubung f (von Stäubemitteln); 3. (ker) Zerrieseln n (von Materialien mit hohem Kalziumorthosilikat-Gehalt); 4. (Ker) Abblasen n, Abstauben n (der Ware vor dem Brennen); 5. (Pap) Entstäuben n (von Hadern, Altpapier)

d. agent s. d. powder

d. appliance (Landw) Verstäubungsgerät n

d. machine 1. (Pap) Entstäubungsapparat m, Stäuber m (Hadernaufbereitung); 2. (Ker) Abstaubmaschine f; 3. (Landw) Verstäubungsgerät n, Stäubegerät n, Pulverstäuber m

d. powder Pudermittel n; Stäubemittel n; Streupuder m

d. system Stäubegerät n

dustless staubfrei; nichtstaubend

dustproof staubdicht, staubundurchlässig

Dutch liquid Dutch-Flüssigkeit f, Äthylendichlorid n, 1,2-Dichloräthan n, Öl n der holländischen Chemiker

D. metal Tombak m

D. State Mines process Staatsmijnen-Verfahren n, Staatsmijnen-Lößverfahren n, S.M.-Verfahren n (der holländischen Staatsmijnen)

dwell (Plast) Entlüftungspause f (beim Pressen)

Dwight-Lloyd sintering machine Dwight-Lloyd-Sintermaschine f, Dwight-Lloyd-Apparat m

dy (Geol) Dy m (ein organischer Schlamm)

dyad zweiwertig

dyad zweiwertiges Element n; zweiwertige Atomgruppe f

dyadic s. dyad

dye / **to** [an]färben

d. directly direkt (unmittelbar) färben (anfärben)

dye Farbstoff m, (unexakt) Farbe f

d.-affinitive farbstoffaffin

d. affinity Farbstoffaffinität f

d. back Färberkufe f, Färbebottich m, Haspelkufe f

d. bath Färbebad n, Färbeflotte f, Farbbad n, Farbflotte f, Flotte f

d. chelate Farbstoffchelat n

d. chemistry Farbstoffchemie f

d. coat (Gerb) Farbüberzug m

d. fastness Farbstoffechtheit f

d. fixative s. d.-fixing agent

d.-fixing agent Farbstoffixiermittel n

d. industry Farbstoffindustrie f

d. intermediate Farbstoffzwischenprodukt n, Farbenzwischenstoff m

d. layer Farbschicht f, Farbstoff[schutz]schicht f

d. liquor s. d. bath

d. penetration Durchfärbung f

d. receptivity Farb[stoff]aufnahmevermögen n, Anfärbbarkeit f

d. retardant Retarder m, Färbungsbremsmittel n, Egalisiermittel n

d. sensitization (sensitizing) Farbensensibilisierung f, optische Sensibilisierung f

d. uptake Farbstoffaufnahme f

d. vessel Färbeflottenbehälter m, Färbekessel m, Färbekufe f

dyeability s. dye receptivity

dyeable [an]färbbar

dyed in the beater (stuff) (Pap) in der Masse (im Stoff, im Holländer) gefärbt

dyehouse Färberei f (Betrieb)

dyeing Färben n, Färberei f

d. accelerant Färbebeschleuniger m

d. affinity Farbstoffaffinität f

d. apparatus Färbeapparat m

d. assistant Färbe[rei]hilfsmittel n

d. cone (Text) Färbehülse f

d. cycle Färbezeit f

d. drum Färbefaß n

d. machine Färbemaschine f

d. power Farbkraft f, Färbekraft f, Färbefähigkeit f

d. properties färberische Eigenschaften fpl, färberisches Verhalten n, färberische Natur f, Färbeeigenschaften fpl, Färbeverhalten n, Farbstoffnatur f, Farbstoffcharakter m

d. time Färbezeit f

d. vessel Färbekessel m, Färbekufe f, Färbeflottenbehälter m

dyer Färber m

dyer's chamomille Färberkamille f, Anthemis tinctoria L.

d. greenwood Färberginster m, Genista tinctoria L.

d. Indian mulberry Morinda tinctoria Roxb. (Rubiacee mit farbstoffhaltigen Wurzeln)

d. mulberry Färbermaulbeerbaum m, Gelbholz n, Gelbes Brasilholz n, Alter (Echter) Fustik m, Fustikholz n, Chlorophora tinctoria Gaudich

d. oak Färbereiche f, Quercus velutina Lam.

d. woodruff Färber-Meister *m*, Färber-Meier *m*, Asperula tinctoria L.
dyestuff Farbstoff *m*, *(unexakt)* Farbe *f*
  d. analysis Farbstoffanalyse *f*
  d. industry Farbstoffindustrie *f*
dyewood Farbholz *n*
dyke Gesteinsgang *m*, Gang *m*
  d. rock Ganggestein *n*
dykite *s.* dyke rock
dynamic equilibrium dynamisches Gleichgewicht *n*
  d. isomer Tautomer[es] *n*, tautomere Form *f*
  d. isomerism Tautomerie *f*, Desmotropie *f*
  d. metamorphism *s.* dynamometamorphism
  d. method dynamische Methode *f* *(zur Messung von Dampfdrücken)*
  d. pressure Staudruck *m*
  d. properties dynamisches Verhalten *n*
  d. seal Dichtung *f* *(zur Abdichtung gegeneinander bewegter Teile)*
  d. stress dynamische Beanspruchung *f*
  d. test dynamische Prüfung *f*
  d. viscosity dynamische Viskosität (Zähigkeit) *f*, absolute Viskosität (Zähigkeit) *f*
dynamics Dynamik *f*
dynamite Dynamit *n*
dynamometamorphism *(Geol)* Dynamometamorphose *f*, Dislokationsmetamorphose *f*, kinetische (mechanische) Metamorphose *f*, Kinetometamorphose *f*
dypnone $C_6H_5 \cdot C(CH_3) = CH \cdot CO \cdot C_6H_5$ Dypnon *n*, β-Methylchalkon *n*
dysanalite, dysanalyte *(Min)* Dysanalyt *m*
dyscrasite *(Min)* Dyskrasit *m*, Diskrasit *m* *(Silberantimonid)*
Dyson notation Dyson-Notation *f*
Dyson [notation] system Dyson-Notationssystem *n*, Dyson-System *n*
dysprosia Dy_2O_3 Dysprosium(III)-oxid *n*
dysprosium Dy Dysprosium *n*
  d. acetate $Dy(CH_3COO)_3$ Dysprosiumazetat *n*
  d. bromate $Dy(BrO_3)_3$ Dysprosiumbromat *n*
  d. carbonate $Dy_2(CO_3)_3$ Dysprosiumkarbonat *n*
  d. chloride $DyCl_3$ Dysprosiumchlorid *n*
  d. chromate $Dy_2(CrO_4)_3$ Dysprosiumchromat *n*
  d. nitrate $Dy(NO_3)_3$ Dysprosiumnitrat *n*
  d. orthophosphate $DyPO_4$ Dysprosium[ortho]phosphat *n*
  d. oxide Dy_2O_3 Dysprosium(III)-oxid *n*
  d. selenate $Dy_2(SeO_4)_3$ Dysprosiumselenat *n*
  d. sulphate $Dy_2(SO_4)_3$ Dysprosiumsulfat *n*
dystectic mixture dystektisches Gemisch *n* *(Stoffgemisch mit höchstem Schmelzpunkt)*

E

e-bond e-Bindung *f*, äquitoriale Bindung *f*
E-glass E-Glas *n* *(alkaliarmes Borosilikatglas)*

Eagle mill Eagle-Mühle *f* *(eine Ringleitungsstrahlmühle)*
E.A.N. *s.* effective atomic number
ear Ähre *f*, Getreideähre *f*
earth Erde *f*, Erdboden *m*, Boden *m*; Erde *f*, Erdkugel *f*, Erdball *m*
  e. almond Erdmandel *f*, Cyperus esculentus L.
  e. colour Erdfarbe *f*
  e. filtration Erdbehandlung *f*, Filtration *f* mit reinigenden (adsorptiv wirksamen) Erden
  e. globe Erdkugel *f*, Erdball *m*, Erde *f*
  e. interior Erdinneres *n*
  e.-moist erdfeucht
  e.-moist concrete erdfeuchter (steifer) Beton *m*
  e. oil Erdöl *n*
  e. pitch Asphalt *m*, Erdpech *n*, Erdharz *n*
  e. wax Erdwachs *n*, Bergwachs *n*, Ozokerit *m*
earthen basin Erdbecken *n*
earthenware Irdengut *n*, Irdenware *f*, Irdengeschirr *n*, Tongut *n*, Tonware *f*; Töpferware *f*; Steingut *n*
  e. cock Steinguthahn *m*, Tonhahn *m*
  e. filter Steingutfilter *n*, Tonfilter *n*
  e. pipe Tonrohr *n*
  e. pot (vessel) Tontopf *m*
earthnut oil Erdnußöl *n*, Arachisöl *n*, Oleum Arachidis
earth's core *(Geol)* Erdkern *m*, Nife-Kern *m*, Nickeleisenkern *m*, Siderosphäre *f*, Barysphäre *f*
  e. crust Erdrinde *f*, Erdkruste *f*
  e. nucleus *s.* e. core
earthy erdig, Erd..., erd[e]haltig; erdartig
  e. brown lignite erdige Braunkohle *f*, Erdbraunkohle *f*
  e. cobalt *(Min)* Erdkobalt *m*, Asbolan *m* *(Kobaltmanganerz)*
ease of care *(Text)* leichte Pflegbarkeit *f*, Pflegeleichtigkeit *f*
  e. of dehydration Dehydratisierungsgeschwindigkeit *f*
  e. of ignition niedriger Zündpunkt *m*
easily decomposable leicht zersetzlich
  e. wetted leicht benetzbar
East India kino indisches Kino *n*, Malabarkino *n*, Amboinokino *n* *(Kinoharz von Pterocarpus marsupium Roxb.)*
  E. Indian copal *(Kopal von Canarium-Arten)*
  E. Indian tanning *(pflanzliche Vorgerbung mit Myrobalanen und Akazienrinde)*
eastern hemlock Kanadische Hemlockstanne (Schierlingstanne) *f*, Tsuga canadensis (L.)Carr.
easy-bleaching pulp leicht bleichbarer Zellstoff *m*
  e. care *s.* ease of care
  e.-care finish *(Text)* Pflegeleicht-Ausrüstung *f*, Easy-care-Ausrüstung *f*
  e. processing channel black Easy-Processing-Channel-Ruß *m*, EPC-Ruß *m* *(gut verarbeitbarer Kanalruß)*
  e. to ignite leicht entzündbar (zündend)

eat / **to** zerfressen, anfressen, angreifen, [oberflächlich] zerstören, ätzen, korrodieren; sich hineinfressen
  e. away zerfressen
  e. through durchfressen
eatable eßbar, genießbar
eatables Eßwaren *pl*, Lebensmittel *pl*, Nahrungsmittel *pl*
eating maturity Genußreife *f*
  e.- ripe genußreif
eau de Javel[le] Eau de Javelle *n(f)*, Javellesche Lauge *f (wäßrige Kaliumhypochloritlösung oder wäßrige Natriumhypochloritlösung)*
  e. de Labarraque Eau de Labarraque *n(f)*, Natronbleichlauge *f (wäßrige Natriumhypochloritlösung)*
Ebelmen producer Abstichgenerator (Abstichgaserzeuger) *m* von Ebelmen
Eberhard effect *(Foto)* Eberhard-Effekt *m*, Durchmessereffekt *m*
ebonite Ebonit *n*, Hartgummi *m*, Hartkautschuk *m*
  d. bonding Hartgummiverfahren *n*, Hartkautschukverfahren *n*
  e. dust (powder) Hartgummistaub *m*, Hartkautschukstaub *m*
  e. sheet Hartgummiplatte *f*, Hartkautschukplatte *f*
ebullience, ebulliency Kochen *n*, Aufkochen *n*, Sieden *n*, Aufwallen *n*
ebullient [auf]kochend, siedend, aufwallend
ebulliometer, ebullioscope Ebullioskop *n*
ebullioscopic ebullioskopisch
  e. constant ebullioskopische Konstante *f*, molale Siedepunktserhöhung *f*
  e. method ebullioskopisches Verfahren *n*, Ebullioskopie *f*
ebullioscopy Ebullioskopie *f*
ebullition Kochen *n* Aufkochen *n*, Sieden *n*, Aufwallen *n*
eccentric drive Exzenterantrieb *m*
  e. press Exzenterpresse *f*
eccentrically driven vibrating screen Exzenterschwingsiebmaschine *f*
ecgonine Ekgonin *n*
echelette grating Echelettegitter *n*
echelon grating Echelongitter *n*, Stufengitter *n*
echimidine Echimidin *n*
echimidinic acid Echimidinsäure *f (C₇-Trihydroxysäure)*
echinatine Echinatin *n (Alkaloid von Rindera echinata und Eupatorium maculatum L.)*
echinochrome A Echinochrom A *n*, 2-Äthyl-3,5,6,7,8-pentahydroxy-1,4-naphthochinon *n*
eclipsed *(Stereochemie)* ekliptisch, verdeckt, ±syn-periplanar
  e. configuration (conformation) ekliptische (±syn-periplanare) Konformation *f*
eclogite Eklogit *m (ein kristalliner Schiefer)*
  e.- shell Eklogithülle *f*, Eklogitschale *f*

economic poison Pflanzenschutzmittel *n (i.w.S.)*
economizer Ekonomiser *m*, Speisewasservorwärmer *m*, Abgasvorwärmer *m*, Rauchgasvorwärmer *m*
economy in fuel Brennstoffeinsparung *f*
ecru silk Ecruseide *f*, Hartseide *f (wenig entbastete Naturseide)*
ectohormone Ektohormon *n*, Pheromon *n*
ectotoxin Ektotoxin *n*
ectylurea Ektylharnstoff *m*, 2-Äthyl-*cis*-krotonylharnstoff *m*
ecuelle method *(Kosmet)* Nadelverfahren *n (zum Auspressen von Ölen mit der Hand)*
EDB *s.* ethylene dibromide
eddy Wirbel *m*, Strudel *m*
  e. current heating Wirbelstromheizung *f*
  e. flow Wirbelströmung *f*
Edeleanu extract Edeleanu-Extrakt *m (durch Extraktion von Ölen mit flüssigem Schwefeldioxid erhaltener Extrakt)*
Edeleanu process Edeleanu-Verfahren *n (Solventextraktion)*
edge Kante *f*, Rand *m*
  e.-crimped *(Text)* kantentexturiert
  e. crimping *(Text)* Kantentexturieren *n*
  e. dislocation *(Krist)* Kantenversetzung *f*, Stufenversetzung *f*, Taylor-Orowan-Versetzung *f*
  e. effect *(Foto)* Kanteneffekt *m*
  e. filter Oberflächenfilter *n*; Spaltfilter *n*
  e. filtration Oberflächenfiltration *f*
  e. mill *s.* e.-runner mill
  e. of the wire *(Pap)* Siebrand *m*, Siebkante *f*
  e. rolls Randwalzen *fpl*, seitliche Halterollen *fpl (bei der düsenlosen Tafelglasherstellung)*
  e.-runner [mill] Kollergang *m*, Kollermühle *f*; *(Met)* chilenische Mühle *f*
  e. water Randwasser *n*, Ölfeldwasser *n*
  e. watermark *(Pap)* Randwasserzeichen *n*
edgemill, edgerunner *s.* edge-runner mill
edging *(Glas)* Kantenschliff *m*
edibility Eßbarkeit *f*, Genießbarkeit *f*
edible eßbar, genießbar
  e. fat Speisefett *n*
  e. gelatin Speisegelatine *f*
  e. mushroom Speisepilz *m*
  e. oil Speiseöl *n*
  e. purposes Speisezwecke *mpl*
  e. syrup Speisesirup *m*
  e. tallow Speisetalg *m*
edibleness *s.* edibility
edibles Eßwaren *pl*, Lebensmittel *pl*, Nahrungsmittel *pl*
edingtonite *(Min)* Edingtonit *m (Bariumalumotrisilikat)*
Edison accumulator (cell) Edison-Akkumulator *m*, NiFe-Akkumulator *m*, Stahlakkumulator *m*
EDTA *s.* ethylenediamine tetraacetic acid
eductor Ejektor *m*, Saugstrahlpumpe *f*
edulcorate / **to** absüßen, versüßen
edulcoration Absüßen *n*, Versüßen *n*, Versüßung *f*

Edwards roaster Edwards-Ofen *m (dreiherdiger Röstofen)*
edwardsite *(Min)* Edwardsit *m, (veraltet für)* Monazit *m*
edwarsit *(fälschlich für)* edwardsite
EEA *s.* ethylene ethyl acrylate copolymer
effect Effekt *m*, Wirkung *f*; Stufe *f*, Körper *m (eines Mehrstufenverdampfers)*
  e. law of Mitscherlich Wirkungsgesetz *n* der Wachstumsfaktoren, Wirkungsgesetz *n* von Mitscherlich
  e. of beating *(Pap)* Mahleffekt *m*
  e. of chelation Chelateffekt *m*
  e. of impact Stoßwirkung *f*; Schlagwirkung *f*
  e. of light Licht[ein]wirkung *f*
  e. of solvent Lösungsmitteleffekt *m*
  e. on yield Ertragswirkung *f (z.B. von Düngemitteln)*
  e. value Wirkungswert *m (z.B. von Düngemitteln)*
  e. varnish Effektlack *m*
  e. yarn Effektgarn *n*
effective wirksam, effektiv, nutzbar, Wirk..., Nutz...
  e. alkali *(Pap)* effektives Alkali *n (NaOH + 1/2 Na₂S, ausgedrückt als Na₂O)*
  e. area wirksame Oberfläche *f*
  e. atomic number effektive Atomnummer (Ordnungszahl) *f*
  e. charge effektive Kernladung (Kernladungszahl) *f*
  e. chemical *s.* e. alkali
  e. cross section Wirk[ungs]querschnitt *m*, effektiver Querschnitt *m*
  e. cross section of carriers trapping effektiver Einfangquerschnitt *m*
  e. distribution coefficient effektiver Verteilungskoeffizient *m (in der Zonenschmelztheorie)*
  e. partition (segregation) coefficient *s.* e. distribution coefficient
  e. surface wirksame Oberfläche *f*
effectiveness Aktivität *f*, Wirksamkeit *f*
effervesce / **to** [auf]brausen, [auf]schäumen, sprudeln, moussieren, perlen
effervescence, effervescency Aufbrausen *n*, Brausen *n*, Aufschäumen *n*, Sprudeln *n*, Moussieren *n*, Perlen *n*
effervescent [auf]brausend, [auf]schäumend, sprudelnd, moussierend, perlend
  e. lemonade *s.* e. soft drink
  e. powder (salt) Brausepulver *n*
  e. soft drink Brauselimonade *f*
  e. wine Schaumwein *m*, Sekt *m*
effervescing *s.* effervescent
efficiency Nutzleistung *f*; Wirkungsgrad *m*; Gütegrad *m*, Güte *f*; Ausnutzungsgrad *m*; Ausbeute *f*; Wirksamkeit *f*
  e. factor Wirkungsgrad *m*
  e. of cut Trenngüte *f*, Trenngütegrad *m*
  e. of screening Siebgütegrad *m*, Siebwirkungsgrad *m*, Sieberfolg *m*

e. of separation Abscheidegrad *m*, Ausscheidungsgrad *m*; Entstaubungsgrad *m*
efficient brauchbar, wirksam, wirkungsvoll, leistungsfähig; mit hohem Wirkungsgrad
effloresce / **to** *(Min)* effloreszieren, ausblühen; *(Krist)* unter Kristallwasserverlust verwittern
efflorescence *(Min)* Effloreszenz *f*, Ausblühung *f*; *(Krist)* Verwitterung *f* unter Kristallwasserverlust
effluence Ausfließen *n*, Ausfluß *m*, Abfließen *n*, Abfluß *m*, Ablaufen *n*, Ablauf *m*, Ausströmen *n*; Ausstrahlung *f*
effluent ausfließend, ausströmend; ausstrahlend
effluent Ausfließen *n*, Ausfluß *m*, Abfließen *n*, Abfluß *m*, Ablaufen *n*, Ablauf *m*, Ausströmen *n*; ausfließendes Medium *n*, Ausfluß *m*, abgehender Strom *m*; Abwasser *n*
  e. disposal Abwasserbeseitigung *f*
  e. pipe Ausflußrohr *n*, Ablaufrohr *n*, Austrittsrohr *n*
  e. water Abwasser *n*
efflux Ausfluß *m*, Abfluß *m*
  e. time Ausflußzeit *f*, Ausflußdauer *f*
  e. viscosimeter Ausflußviskosimeter *n*
effuse / **to** ausströmen, ausfließen
effusion Effusion *f*, Ausströmen *n*, Ausströmung *f*, Ausfließen *n*, Ausfluß *m*, Erguß *m*
effusive *(Geol)* effusiv, Effusiv..., Erguß...
  e. rock Effusivgestein *n*, Ergußgestein *n*, Oberflächengestein *n*
egg 1. Ei *n*; 2. *s.* acid egg
  e. albumen 1. Eiklar *n*, Weißei *n*; 2. *s.* e. albumin 1.
  e. albumin 1. Ei[er]albumin *n*, Ovalbumin *n*; 2. *s.* e. albumen 1.
  e. fat *s.* e.-yolk oil
  e. globulin Eiglobulin *n*
  e. powder Eipulver *n*, Trockenei *n*
  e. vitellin Vitellin *n*, Ovovitellin *n (Phosphoproteid aus Eidotter)*
  e. yolk Eigelb *n*, Eidotter *n*, Dotter *n*
  e.-yolk oil Eieröl *n*, Eigelböl *n*
egging *(Gerb)* Eigelbnachgare *f*
eggonite *(Min)* Eggonit *m*, Sterrettit *m (Skandiumphosphat)*
eggshell finish *(Ker)* eierschaliger Fertigbrand *m*
  e. porcelain *(Ker)* Eierschalenporzellan *n (sehr dünnes, durchsichtiges Porzellan)*
eggshelling *(Ker)* Eierschaligkeit *f (der Glasur)*
Egyptian privet Henna *f*, Hennastrauch *m*, Lawsonia inermis L.; Henna *f (gepulverte Blätter von Lawsonia inermis L.)*
ehlite *(Min)* Ehlit *m*, Pseudomalachit *m (Kupfer(II)-trihydroxyorthophosphat)*
E.I.-tanning *s.* East Indian tanning
eicosane $C_{20}H_{42}$ Eikosan *n*
eicosanoic acid $CH_3(CH_2)_{18}COOH$ Eikosansäure *f*, Arachinsäure *f*, Erdnußsäure *f*
5,8,11,14-eicosatetraenoic acid 5,8,11,14-Eikosantetraensäure *f*, Arachidonsäure *f*

eicos-9-enoic acid $C_{19}H_{37}COOH$ 9-Eikosensäure f, Gadoleinsäure f
n-**eicosoic acid** *s.* eicosanoic acid
eigenfunction Eigenfunktion f
eigenstate Eigenzustand m
eigenvalue Eigenwert m
eight-electron configuration Elektronenoktett-Anordnung f, Edelgaskonfiguration f
e.-**membered ring** achtgliedriger Ring (Kohlenstoffring) m, Acht[er]ring m, 8-Ring m
Einstein equation Einstein-Gleichung f
Einstein transition probability Einsteinsche Übergangswahrscheinlichkeit f
einsteinium Es Einsteinium n
Einthoven galvanometer Saitengalvanometer n nach Einthoven
Eirich mixer Eirich-Mischer m (ein Naßkollergang)
eject /to ausdrücken, auswerfen, ausstoßen; hinausdrücken, verdrängen; (Teilchen) emittieren, ausstoßen
ejection Ausdrücken n, Auswerfen n, Ausstoßen n
e. **connecting bar** (Plast) Ausdrücktraverse f
e. **pad** (Plast) Ausdrückstempel m, Auswerfstempel m
e. **plate** (Plast) Ausdrückplatte f
e. **ram** (Plast) Ausdrückkolben m
ejector 1. Ausdrücker m, Ausstoßer m, Auswerfer m, Auswurfvorrichtung f; 2. Ejektor m, Saugstrahlpumpe f, Saugstrahlgebläse n
e. **frame** (Plast) Ausdrückrahmen m
e. **pin** (Plast) Ausdrückstift m, Auswerferstift m
e. **plate** (Plast) Ausdrückplatte f
elaborate /to bilden, erzeugen, aufbauen
elaeolite (Min) Eläolith m (getrübter Nephelin, ein Tektosilikat)
elaeostearic acid Eläostearinsäure f, 9,11,13-Oktadekatriensäure f
elaidic acid $C_8H_{17}CH=CH(CH_2)_7COOH$ Elaidinsäure f, trans-9-Oktadezensäure f
elaidin test Elaidinprobe f
elaidinization Elaidini[si]erung f
elastase Elastase f (ein Ferment des Pankreassaftes)
elastic aftereffect elastische Nachwirkung f
e. **axis** (Krist) Elastizitätsachse f
e. **bitumen** (Min) Erdpech n, Elaterit m
e. **compliance** elastische Nachgiebigkeit f
e. **constant** Elastizitätskonstante f
e. **deformation** elastische Verformung (Deformation) f
e. **gel** elastisches Gel n
e. **limit** Elastizitätsgrenze f
e. **modulus** Elastizitätsmodul m, E-Modul m, (früher auch) Youngscher Modul m
e. **recovery** [elastische] Erholung f; Rückverformung f, Rückfederung f, Zurückspringen n
e. **scattering** elastische Streuung f
e. **strain energy** elastische Formänderungsarbeit f
elasticator Elastikator m (ein Kautschukweichmacher)

elasticity Elastizität f; Biegsamkeit f, Geschmeidigkeit f, Dehnfähigkeit f, Nachgiebigkeit f, Spannkraft f, Federkraft f, Schnellkraft f
e. **test** Elastizitätsprüfung f
elastin Elastin n (ein Protein des Bindegewebes)
elastomer elastomer
elastomer Elastomer[es] n, Elast m
elastomeric elastomer
e. **fibre** Elastomerfaser f
e. **yarn** Elastomerfaden m
elaterite (Min) Elaterit m, Erdpech n
elayl $H_2C=CH_2$ Äthen n, Äthylen n
elbow [fitting] Krümmer m, Rohrkrümmer m, Knie[stück] n, Winkelstück n
e. **meter** Krümmer-Durchfluß[mengen]messer m
Elbs persulphate oxidation Elbssche Persulfatoxydation f (von Phenolen)
Elbs reaction Elbs-Reaktion f (Anthrazen-Ringschluß)
elecampane Altwurzel f, Donnerwurzel f (getrocknete Wurzelstöcke von Inula helenium L.)
e. **camphor** Helenin n, Alantkampfer m, Inulakampfer m
electric elektrisch (s.a. unter electrical)
e. **arc** [elektrischer] Lichtbogen m
e.-**arc furnace** Elektrolichtbogenofen m, Lichtbogen[elektro]ofen m, elektrischer Lichtbogenofen m
e.-**arc heating** Lichtbogen[be]heizung f
e. **blanket heating** (Plast) Heizung f mit elektrischer Heizdecke
e. **boosting** elektrische Zusatzbeheizung f
e. **burner** Elektrobrenner m
e. **current** [elektrischer] Strom m
e. **decantation** Elektrodekantierung f, Elektrodekantation f
e. **direct arc furnace** direkter Lichtbogenofen m, Lichtbogenofen m mit direkter Beheizung
e. **discharge** elektrische Entladung f
e. **drying oven** elektrischer Trockenschrank m
e. **furnace** Elektroofen m, elektrischer (elektrisch beheizter, elektrothermischer) Ofen m, Ofen m mit elektrischer Beheizung
e.-**furnace process** Elektro[stahl]verfahren n, Elektroschmelzverfahren n (Stahlherstellung)
e.-**furnace steel** Elektrostahl m
e. **heating** Elektro[be]heizung f, elektrische Heizung (Beheizung) f
e. **indirect arc furnace** indirekter Lichtbogenofen m, Lichtbogenofen m mit indirekter Beheizung (mit reiner Strahlungsbeheizung), Lichtbogenstrahlungsofen m
e. **log** elektrisches Kerndiagramm n (bei der elektrischen Bohrlochvermessung)
e. **logging** elektrische Bohrlochvermessung f, elektrisches Kernen n
e. **process** s. e.-furnace process
e.-**resistance furnace** [elektrischer] Widerstandsofen m, widerstandsbeheizter Ofen m, Ofen m mit Widerstandsheizung

e. steel s. e.-furnace steel
e. storage ability Aufladefähigkeit f
e. strength elektrische (dielektrische) Festigkeit (Durchschlag[s]festigkeit) f, Durchschlag[s]festigkeit f
e. tunnel kiln Elektrotunnelofen m
e. vibrator Elektrorüttler m, Elektrovibrator m
e. wind elektrischer Wind m, Ionenwind m
electrical elektrisch (s.a. unter electric)
e. attraction elektrische Anziehung f
e. circuit Stromkreis m
e. conductance elektrischer Leitwert m
e. conducting paper leitendes (elektrisch leitendes, stromleitendes) Papier n
e. conductivity elektrische Leitfähigkeit f, elektrisches Leitvermögen n
e. contact material [elektrischer] Kontaktwerkstoff m, Kontaktbaustoff m
e. coring elektrisches Kernen n, elektrische Bohrlochvermessung f
e. discharge elektrische Entladung f
e.-discharge heating Beheizung f durch Bogenentladung
e. dispersion Elektrodispersion f, elektrische Zerstäubung f
e. double layer elektrische (elektrochemische) Doppelschicht f
e. double refraction elektrische Doppelbrechung f
e. insulating oil Elektroisolieröl n, Isolieröl n
e. insulating paper Isolierpapier n
e. insulator [elektrischer] Isolator m, Nichtleiter m
e. porcelain Elektroporzellan n
e. precipitation elektrisches (elektrostatisches) Abscheiden (Reinigen, Entstauben) n, elektrische (elektrostatische) Abscheidung (Reinigung, Entstaubung) f, Elektroabscheiden n, Elektroreinigen n
e. precipitator Elektroabscheider m, Elektrofilter n
e. pressboard Preßspan m für Elektrotechnik
e. resistance [elektrischer] Widerstand m
electrically driven vibrator Elektrorüttler m, Elektrovibrator m
e. neutral elektroneutral
electricity Elektrizität f
electrification Elektrisieren n, Elektrisierung f; Elektrifizieren n, Elektrifizierung f
electrify / to elektrisieren; elektrifizieren
electroaffinity Elektro[nen]affinität f, E.A.
electroanalysis Elektroanalyse f, elektrochemische Analyse f, elektroanalytische Bestimmung f
electroanalytical elektroanalytisch
electrocapillarity Elektrokapillarität f
electrocapillary curve Elektrokapillarkurve f
electroceramics Elektrokeramik f
electrochemical elektrochemisch

e. initiation of polymerization elektrochemische Startreaktion f des Polymerisationsvorgangs
e. series elektrochemische Spannungsreihe f
e. valency positive Elektrovalenz f, [positive] elektrochemische Wertigkeit f, Oxydationsstufe f, Oxydationszahl f
electrochemist Elektrochemiker m
electrochemistry Elektrochemie f
electrochromatography Elektrochromatografie f
electrocoating s. electroplating
electrocorrosion Korrosion f durch Fremdstrom, Fremdstromkorrosion f, elektrochemische Korrosion f durch Streuströme, Streustromkorrosion f
electrocortin Elektrokortin n, (veraltet für) Aldosteron n (ein Nebennierenrindenhormon)
electrocremage s. electrodecantation
electrode Elektrode f
e. carbon Elektrodenkohle f
e. chamber (compartment) Elektrodenkammer f (des Elektrodialysators)
e. material Elektrodenmaterial n, Elektrodenwerkstoff m; Elektroden[stampf]masse f
e. potential Elektrodenpotential n, Nullpotential n
e. reaction Elektrodenreaktion f
e. surface Elektroden[ober]fläche f
e. voltage Elektrodenspannung f
electrodecantation Elektrodekantierung f, Elektrodekantation f
electrodecanted elektrodekantiert
electrodeless elektrodenlos
electrodeposit / to galvanisieren, galvanische Überzüge (Niederschläge) erzeugen, (metallische oder nichtmetallische Schichten) elektrolytisch (elektrochemisch) abscheiden (niederschlagen, fällen) (zum Zwecke eines Oberflächenschutzes usw.)
electrodeposit galvanisch (elektrolytisch, elektrochemisch) aufgebrachte (metallische oder nichtmetallische) Schicht (Abscheidung) f, galvanischer Überzug (Niederschlag) m
electrodeposition galvanisches Auftragen n, elektrolytisches (elektrochemisches) Abscheiden n (metallischer oder nichtmetallischer Schichten)
e. of metals s. electroplating
electrodialysis Elektrodialyse f
electrodialyzer, electrodializing device Elektrodialysator m
electrodispersing, electrodispersion Elektrodispersion f; elektrische Zerstäubung f
electroendosmosis Elektro[end]osmose f
electroengraving elektrolytisches Ätzen n, Elektroätzen n
electrofiltration Elektrofiltration f, Elektrofilterung f
electroformed galvanoplastisch hergestellt
electroforming Galvanoplastik f
electrofusion Elektroschmelze f

electrogalvanize / to galvanisch verzinken
electrogalvanizing galvanisches Verzinken *n*
electrographite Elektrographit *m*
electrogravimetric analysis Elektrogravimetrie *f*
electrokinetic elektrokinetisch
  e. potential elektrokinetisches Potential *n*, Zeta-Potential *n*, ζ-Potential *n*
electrolinkage Electrolinking *n (Verfahren, der Untertagevergasung)*
electroluminescence Elektrolumineszenz *f*
electrolysis Elektrolyse *f*
  e. cell Elektrolyse[n]zelle *f*, elektrolytische Zelle *f*, Elektrolysierzelle *f*
  e. of a fused salt Schmelz[fluß]elektrolyse *f*
  e. of alkali-metal chlorides Chloralkalielektrolyse *f*
  e. of water Wasserelektrolyse *f*
electrolyte Elektrolyt *m*
  e. solution Elektrolytlösung *f*
electrolytic elektrolytisch, Elektrolyt..., Elektrolyse...
  e. analysis elektrolytische Analyse *f*, Elektroanalyse *f*
  e. bleach *(Pap)* Elektrolytbleiche *f*, elektrolytisches Bleichen *n*
  e. capacitor elektrolytischer Kondensator *m*, Elektrolytkondensator *m*, Elko *m*
  e. cell Elektrolyse[n]zelle *f*, elektrolytische Zelle *f*, Elektrolysierzelle *f*
  e. conductance *s.* e. conductivity
  e. conduction elektrolytische Leitung *f*
  e. conductivity elektrolytische Leitfähigkeit *f*, elektrolytisches Leitvermögen *n*
  e. conductor elektrolytischer Leiter *m*, Elektrolyt *m*, Leiter *m* zweiter Ordnung
  e. copper Elektrolytkupfer *n*, E-Kupfer *n*, E-Cu
  e. degreasing elektrolytische Entfettung *f*
  e. deposition analysis Elektrogravimetrie *f*
  e. dissociation elektrolytische Dissoziation (Zersetzung) *f*
  e. etch[ing] elektrolytisches Ätzen *n*, Elektroätzen *n*
  e. initiation of polymerization Kettenstart *m* durch Elektrolyse
  e. iron Elektrolyteisen *n*, E-Eisen *n*
  e. method elektrolytisches Verfahren *n*, Elektrolytverfahren *n*
  e. nickel Elektrolytnickel *n*, E-Nickel *n*, Katodennickel *n*
  e. oxidation elektrolytische (elektrochemische, anodische) Oxydation *f*
  e. polarization elektrolytische (galvanische) Polarisation *f*
  e. polishing elektrolytisches Polieren *n*
  e. powder elektrolytisches (elektrolytisch erzeugtes, katodisches) Pulver *n*, Elektrolytpulver *n*
  e. process *s.* e. method
  e. rectifier elektrolytischer Gleichrichter *m*, Elektrolytgleichrichter *m*

  e. reduction elektrolytische Reduktion *f*
  e. refining elektrolytische Raffination *f*, Elektroraffination *f*
  e. separation elektrolytische Trennung *f*
  e. silver Elektrolytsilber *n*, E-Silber *n*
  e. solution Elektrolytlösung *f*
  e. solution pressure elektrolytischer Lösungsdruck *m*
  e. tank elektrolytischer Trog *m*
  e. technique *s.* e. method
  e. zinc Elektrolytzink *n*, E-Zink *n*
electrolytical *s.* electrolytic
electrolyze / to elektrolysieren, elektrolytisch zersetzen (zerlegen)
electrolyzer Chlorzelle *f*
electromagnetic elektromagnetisch
  e. pulley Elektromagnetrolle *f (in Förderbändern)*
  e. pump elektromagnetische Pumpe *f*
  e. vibrator Magnetrüttler *m*, Magnetvibrator *m*
electromagnetism Elektromagnetismus *m*
electromeric elektromer
  e. effect Mesomerie-Effekt *m*, M-Effekt *m*, mesomerer Substituenteneffekt *m*, elektromerer Effekt *m*
electrometer Elektrometer *n*
electrometric titration elektrometrische (potentiometrische) Titration (Maßanalyse) *f*
electromotive force elektromotorische Kraft *f*, EMK *f*, [elektrische] Urspannung *f*
electron Elektron *n*, Negatron *n*
  π **electron** π-Elektron *n*
  σ **electron** σ-Elektron *n*
  e. acceleration Elektronenbeschleunigung *f*
  e.-accepting elektronenaufnehmend, Elektronen aufnehmend
  e. acceptor Elektronenakzeptor *m*, Akzeptor *m*, Auffänger *m*
  e. acceptor strength Elektronenakzeptorstärke *f*
  e.-affinitive elektronenaffin
  e. affinity Elektro[nen]affinität *f*, E.A.
  e. arrangement Elektronenanordnung *f*, Elektronenkonfiguration *f*
  e.-attracting elektronenanziehend, elektronensuchend, elektronenfreundlich, elektrophil, kationoid
  e. avalanche Elektronenlawine *f*
  e. beam Elektronenstrahl *m*
  e.-beam melting Elektronenstrahlschmelzen *n*
  e. bombardment Elektronenbeschuß *m*, Elektronenbeschießung *f*, Elektronenbombardement *n*
  e.-bombardment heating Beheizung *f* durch Elektronenbeschuß
  e. bond Elektronenbindung *f*
  e. capture Elektroneneinfang *m*
  e. carrier Elektronen[über]träger *m*
  e. charge Elektronenladung *f*
  e. cloud Elektronenwolke *f*
  e. collision Elektronenzusammenstoß *m*

e. conduction elektronische Leitfähigkeit *f*, Elektronenleitfähigkeit *f*, Elektronenleitung *f*, *(i.e.S.)* Überschußleitung *f*, *n*-Leitung *f*, *n*-Halbleitung *f*

e. configuration Elektronenkonfiguration *f*, Elektronenanordnung *f*

e. correlation Elektronenkorrelation *f*

e. deficiency Elektronenmangel *m*

e.-deficient elektronenarm, Elektronenmangel...

e.-deficient compound Elektronenmangelverbindung *f*

e.-deficient hydride Elektronenmangelhydrid *n*

e. density Elektronendichte *f*, Elektronenbelegung *f*, Elektronenkonzentration *f*

e. diffraction Elektronenbeugung *f*

e. diffraction analysis Elektronenbeugungsanalyse *f*

e. diffraction experiment Elektronenbeugungsversuch *m*

e. diffraction pattern Elektronenbeugungsbild *n*, Elektronenbeugungsdiagramm *n*

e. displacement Elektronenverschiebung *f*

e.-donating Elektronen abgebend, elektronenliefernd, elektronenspendend

e. donor Elektronendon[at]or *m*, Elektronenspender *m*

e. energy Elektronenenergie *f*

e. excess Elektronenüberschuß *m*

e. exchange Elektronenaustausch *m*

e.-exchange resin Elektronenaustauscherharz *n*

e. exchanger Elektronenaustauscher *m*

e. excitation Elektronenanregung *f*

e. formula Elektronenformel *f*

e. gap Elektronenlücke *f*

e. gas Elektronengas *n*

e. group Elektronengruppe *f*

e. hole Defektelektron *n*, Elektronenloch *n*, Loch *n*, Mangelelektron *n*

e.-hole pair Elektronen-Defektelektronen-Paar *n*, Loch-Elektronen-Paar *n*

e. impact Elektronenstoß *m*

e. impact method Elektronenstoßmethode *f*

e. jump Elektronensprung *m*

e. jump spectrum Elektronensprungspektrum *n*

e. lattice Elektronengitter *n*

e. mass Elektronenmasse *f*

e. micrograph elektronenmikroskopische Abbildung *f*

e. microscope Elektronenmikroskop *n*, Übermikroskop *n*

e. microscopy Elektronenmikroskopie *f*

e. migration Elektronenwanderung *f*

e. octet Elektronenoktett *n*

e. orbit Elektronen[kreis]bahn *f*

e. orbital Elektronenorbital *n*, elektronisches Orbital *n*

e. pair Elektronenpaar *n*

e.-pair acceptor Akzeptor *m* des Elektronenpaares

e.-pair bond Elektronenpaarbindung *f*, Atom-

bindung *f*, Kovalenz *f*, homöopolare (kovalente, einpolare, unpolare, unitarische) Bindung *f*

e.-pair donor Donator *m* des Elektronenpaares

e.-pair linkage *s.* e.-pair bond

e.-pair method Elektronenpaarmethode *f*, Valenzbindungsmethode *f*, Valenzstrukturmethode *f*, VB-Methode *f*, Heitler-London-Slater-Pauling-Methode *f*, HLSP-Methode *f*, Spinmethode *f*

e.-pair theory Theorie (Bindungstheorie) *f* der Elektronenpaarbindungen (Valenzstrukturen), Valenzbindungstheorie *f*, Valenzstrukturtheorie *f*, VB-Theorie *f*, Heitler-London-Slater-Pauling-Theorie *f*, HLSP-Theorie *f*

e.-paramagnetic resonance paramagnetische Elektronenresonanz *f*

e. polarization Elektronenpolarisation *f*

e.-positron pair Elektron-Positron-Paar *n*

e.-positron-pair production Elektron-Positron-Paarbildung *f*

e. ray Elektronenstrahl *m*

e. release Elektronenauslösung *f*, Elektronenloslösung *f*, Elektronenablösung *f*

e.-releasing elektronenabstoßend, Elektronen auslösend (ablösend, loslösend)

e.-releasing group nukleophile Gruppe *f*

e.-releasing potency Elektronenabgabevermögen *n*

e. removal Elektronenentzug *m*

e.-repelling elektronenabstoßend

e. rest mass Elektronenruh[e]masse *f*

e. ring Elektronenschale *f*

e. sextet Elektronensextett *n*

e. sheath Elektronenhülle *f*

e. shell Elektronenschale *f*

e. shift Elektronenverschiebung *f*

e. source Elektronenquelle *f*

e. spectrum Elektronenspektrum *n*

e. spin Elektronenspin *m*, Elektronendrall *m*, Spin *m*

e. spin resonance Elektronenspinresonanz *f*, ESR

e. spin resonance spectroscopy Elektronenspinresonanzspektroskopie *f*, paramagnetische Resonanzspektroskopie *f*

e. transfer[ence] Elektronenübertragung *f*, Elektronenübergang *m*

e. trap Elektronenfalle *f*, Elektronenfänger *m*, Elektronenhaftstelle *f*

e. trapping Elektroneneinfang *m*

e.-volt Elektronenvolt *n*, eV

e.-withdrawing *s.* e.-attracting

electronegative elektronegativ

electronegativity Elektronegativität *f*

e. difference Elektronegativitätsunterschied *m*

electroneutrality Elektroneutralität *f*

electronic elektronisch, Elektronen...

e. arrangement *s.* e. configuration

e. band spectrum *s.* e. spectrum

e. charge Elektronenladung *f*

e. conductivity Elektronenleitfähigkeit *f*, elek-

tronische Leitfähigkeit f, Elektronenleitung f, (i.e.S.) Überschußleitung f, n-Leitung f, n-Halbleitung f
e. conductor Elektronenleiter m, Leiter m erster Ordnung
e. configuration Elektronenkonfiguration f, Elektronenanordnung f
e. configuration of an inert gas Edelgaskonfiguration f
e. emission Elektronenemission f, Elektronenaustritt m
e. energy Elektronenenergie f
e. formula Elektronenformel f
e. gluing Hochfrequenzverleimung f
e. heating Hochfrequenz[be]heizung f, Hochfrequenzerhitzung f, Hochfrequenzerwärmung f, dielektrische Beheizung (Heizung, Erhitzung, Erwärmung) f
e. level Elektronenniveau n
e. orbit Elektronen[kreis]bahn f
e. sealing Hochfrequenzsiegeln n (von Folien)
e. semiconductor [elektronischer] Halbleiter m, Elektronenhalbleiter m
e. spectrum Elektronenspektrum n
e. state Elektronenzustand m
e. structure elektronische Struktur f, Elektronenstruktur f, Elektronenaufbau m
e. theory Elektronentheorie f
e. theory of valency Elektronentheorie f der Valenz
e. transition Elektronenübergang m
e. work function Austrittsarbeit f, Ablösearbeit f, Abtrennungsarbeit f
electrooptical Kerr effect elektrooptischer Kerr-Effekt m
electroosmosis Elektro[end]osmose f
electropherogram Elektropherogramm n
electrophile Elektrophil n, elektrophiles Reagens (Agens) n
electrophilic elektrophil, elektronensuchend, elektronenfreundlich, elektronenanziehend, kationoid
e. addition elektrophile (kationoide) Addition f
e. attack elektrophile Attacke f, elektrophiles Angreifen n
e. reaction elektrophile (kationoide) Reaktion f
e. substitution elektrophile (kationoide) Substitution f
electrophoresing Elektrophoretisieren n
electrophoresis Elektrophorese f
e. apparatus Elektrophoresegerät n
e. cabinet Elektrophoresekammer f
e. on paper Papierelektrophorese f
e. tank Elektrophoresetrog m
electrophoretic elektrophoretisch
e. apparatus Elektrophoresegerät n
e. effect elektrophoretischer (kataphoretischer, longitudinaler) Effekt m
e. mobility elektrophoretische Beweglichkeit f
electroplate / to elektroplattieren

electroplating Galvanostegie f, Elektroplattierung f; (als Tätigkeit) Galvanisieren n, Elektroplattieren n
e. and electroforming technology Galvanotechnik f
electropolishing elektrolytisches Polieren n
electropositive elektropositiv
electroreclamation (Landw) (Urbarmachung von Alkaliböden durch Elektrodialyse)
electrorefining elektrolytische Raffination f, Elektroraffination f
electroscope Elektroskop n
electrostatic elektrostatisch
e. bond elektrostatische (elektrovalente, polare, heteropolare, ionogene) Bindung f, Ionenbeziehung f, Ionenbindung f, Elektrovalenz f
e. charge elektrostatische Ladung (Aufladung) f
e. coater (Plast) Vorrichtung f für elektrostatisches Beschichten
e. detearing elektrostatisches Tropfenabziehen n (beim Tauchlackieren)
e. field elektrostatisches Feld n
e. filter Elektrofilter n
e. interactions between the ions interionische elektrostatische Wechselwirkungskräfte fpl, [elektrische] Ionenkräfte fpl, interionische Wechselwirkung f (Kräfte fpl), Ionenwechselwirkung f
e. precipitation elektrisches (elektrostatisches) Abscheiden (Reinigen, Entstauben) n, elektrische (elektrostatische) Abscheidung (Reinigung, Entstaubung) f, Elektroabscheidung f, Elektroreinigung f
e. precipitator Elektroabscheider m, Elektrofilter n
e. repulsion elektrostatische Abstoßung f
e. separation Elektrosortieren n, Elektroscheiden n, elektrostatisches Scheiden n, Elektroscheidung f
e. spray coating (Plast) elektrostatisches Sprühbeschichten n
e. spray gun elektrostatische Spritzpistole (Farbspritzpistole) f
e. spraying elektrostatisches Besprühen (Sprühen) n; elektrostatisches Spritzen (Farbspritzen) n, (i.e.S.) elektrostatisches Spritzlackieren n, elektrostatische Spritzlackierung f
e. tar filter elektrostatischer Teer[ab]scheider m, Elektro-Teerfilter n, Elektro-Teerscheider m, Teer-Elektrofilter n
e. unit elektrostatische Einheit f, e.s.E.
electrostatics Elektrostatik f
electrosteel Elektrostahl m
electrostriction Elektrostriktion f
electrothermal, electrothermic elektrothermisch
electrothermics Elektrothermie f
electrothermodynamic potential elektrothermodynamisches (thermodynamisches) Potential n
electrotin / to galvanisch verzinnen
electrotinning galvanische Verzinnung f
electrotyping Galvanoplastik f

electroultrafiltration Elektro-Ultrafiltration *f*
electrovalence, electrovalency 1. elektrochemische Wertigkeit (Valenz) *f*, Elektrovalenz *f*; 2. Ionenbeziehung *f*, Ionenbindung *f*, elektrovalente (elektrostatische, heteropolare, ionogene, polare) Bindung *f*, Elektrovalenz *f*
electrovalent elektrovalent
e. bond *s.* electrovalence 2.
electroviscous effect elektroviskoser Effekt *m*
electrum *(Min)* Elektrum *m* *(natürliche Gold-Silber-Legierung)*
electuary Latwerge *f*, Elektuarium *n* *(teigförmige Arzneizubereitung)*
element Element *n*, Grundbestandteil *m*, wesentlicher Bestandteil *m*; [chemisches] Element *n*, [chemischer] Grundstoff *m*; Bauelement *n*, Bauteil *n*; Schalt[ungs]element *n*, Schaltorgan *n*
e. of symmetry *(Krist)* Symmetrieelement *n*
elemental elementar[isch], Elementar..., natürlich, rein, *(Min auch)* gediegen
e. copper elementares (gediegenes) Kupfer *n*
e. sulphur cure Schwefelvulkanisation *f* *(mit elementarem Schwefel)*, Vernetzung *f* mit *(elementarem)* Schwefel
elementary elementar[isch], grundlegend, Elementar..., Grund...
e. analysis Elementaranalyse *f*
e. charge Elementarladung *f*, elektrisches Elementarquantum *n*, e
e. particle Elementarteilchen *n*
e. quantum *s.* e. charge
elemi Elemi[harz] *n* *(von Burserazeen, Rutazeen und Humiriazeen)*
e. frankincense *(Weihrauch von Boswellia frereana Birdw.)*
e. of Guiana Guayana-Elemi *n* *(Harz von Icica viridiflora Lam.)*
e. of Mexico (Yucatan) *(Elemiharz von Amyris plumieri DC.)*
eleolite *s.* elaeolite
eleonorite *(Min)* Eleonorit *m* *(Eisen(III)-trihydroxiddiorthophosphat)*
eleostearic acid *s.* elaeostearic acid
eleuthera bark *s.* eluteria bark
eleutherinol Eleutherinol *n*
elevate / to [in die Höhe] heben, hochfördern, nach oben tragen; erhöhen, steigern
elevated tank Hochbehälter *m*
e. temperature erhöhte Temperatur *f*
e.-temperature dye machine *(Text)* Hochtemperaturfärbemaschine *f*, HT-Färbemaschine *f*
elevation Anhebung *f*; Erhöhung *f*, Steigerung *f*
e. of boiling point Siedepunktserhöhung *f*
elevator Steilförderer *m*, Senkrechtförderer *m*, Elevator *m*; Hebevorrichtung *f*, Elevator *m*, Gestängeanheber *m* *(einer Rotary-Bohranlage)*
elfwort Altwurzel *f*, Donnerwurzel *f* *(getrockneter Wurzelstock von Inula helenium L.)*
eliminate / to eliminieren, beseitigen, entfernen; abscheiden, ausscheiden, absondern; abspalten

elimination Eliminierung *f*, Beseitigen *n*, Beseitigung *f*, Entfernen *n*, Entfernung *f*, Abscheidung *f*, Ausscheidung *f*, Absonderung *f*; Abspaltung *f*
e. of mist Entnebeln *n*, Entnebelung *f*
e. of water Wasserabspaltung *f*, Wasserentzug *m*, Entwässern *n*, Entwässerung *f*, Dehydratisierung *f*, Dehydratation *f*
e. reaction Eliminierung[sreaktion] *f*, Abspaltungsreaktion *f*, Verdrängungsreaktion *f*
elixir Elixier *n*
ell *s.* elbow
ellagic acid Ellagsäure *f*, 4,5,6,4',5',6'-Hexahydroxydiphensäuredilakton *n*
ellagitannin Ellagengerbstoff *m*
elliptical orbit Ellipsenbahn *f*
e. polarization elliptische Polarisation *f*
elliptically polarized elliptisch polarisiert
Elmendorf tester *(Pap)* Elmendorf-Prüfgerät *n*, Durchreißprüfer *m* nach Elmendorf
elongation Verlängerung *f*, Dehnung *f*, Ausdehnung *f*, Streckung *f*
e. at break (failure, rupture) Bruchdehnung *f*, Zerreißdehnung *f*, Reißdehnung *f*
e. set Dehnungsrest *m*, Zugverformungsrest *m*, Formänderungsrest *m* bei Dehnungsbeanspruchung
elpasolite *(Min)* Elpasolith *m* *(Kaliumnatriumhexafluoroaluminat)*
elpidite *(Min)* Elpidit *m* *(Natriumzirkoniumhexahydrogenditrisilikat)*
Elsholtzia oil *(Kosmet)* Elsholtziaöl *n* *(aus Elsholtzia ciliata (Thunb.) Hyl.)*
eluant Elu[a]tionsmittel *n*, Eluent *m*, Eluant *m*
eluate / to eluieren, *(adsorbierte Stoffe aus festen Adsorptionsmitteln)* herausspülen, herauslösen
eluate Eluat *n* *(durch Herauslösen adsorbierter Stoffe gewonnene Flüssigkeit)*
elucidate / to aufklären, ermitteln, erforschen, erschließen
eluent, elutant *s.* eluant
elute / to *s.* to eluate
eluteria bark Kaskarillarinde *f* *(von Croton eluteria Benn.)*
eluting agent (solvent) *s.* eluant
elution Elution *f*, Eluieren *n*; Elutionsanalyse *f*; Elutionschromatografie *f*
e. analysis Elutionsanalyse *f*
e. chromatography Elutionschromatografie *f*
e. time Elutionszeit *f*
e. volume Elutionsvolumen *n*
elutriate / to [ab]schlämmen, aufschlämmen, ausschlämmen, abschwemmen, auswaschen, auslaugen
elutriation Schlämmen *n*, Abschlämmen *n*, Aufschlämmen *n*, Ausschlämmen *n*, Abschwemmen *n*, Auswaschen *n*, Auslaugen *n*
e. analysis Schlämmanalyse *f*
elutriator 1. Schlämmapparat *m*; 2. Entstauber *m*, Elutriator *m* *(zur Abtrennung der feinsten Anteile*

des Katalysators beim Kracken mit bewegtem Katalysatorbett)

eluvial horizon *(Bodenkunde)* Eluvialhorizont *m*, A-Horizont *m*, Auswaschungshorizont *m*, Oberboden *m*

eluviation *(Geol)* Auslaugung *f (des Bodens)*

emanate / to emanieren, ausfließen, ausströmen, ausstrahlen, *(z.B. radioaktive Strahlen)* aussenden

emanation Emanation *f*, Ausfließen *n*, Ausströmen *n*, Ausstrahlen *n (z.B. von radioaktiven Gasen)*; Em Emanation *f*, Niton *n*, *(veraltet für)* Rn Radon *n*

emanon Emanon *n*, *(ungebräuchlich für)* Rn Radon *n*

embed / to einbetten

embedding medium Einbettungsmittel *n*, Einschlußmittel *n*

embitter / to *(z.B. Bier)* einen bitteren Geschmack verleihen, bitter machen

embolite *(Min)* Embolit *m*, Bromchlorargyrit *m*

embonic acid Embonsäure *f*, Pamoasäure *f*, Methylen-bis(2-hydroxynaphthoesäure-3) *f*

emboss / to *(Pap, Text, Gerb)* prägen, gaufrieren, einpressen, aufpressen; *(Ker)* bossieren

embossed paper gaufriertes (geprägtes) Papier *n*

embosser Gaufriermaschine *f*

embossing *(Pap, Text)* Prägen *n*, Prägung *f*, Gaufrieren *n*, Gaufrage *f*, Einpressen *n*, Aufpressen *n*, *(Gerb auch)* Narbenpressen *n*; *(Ker)* Bossieren *n (Dekortechnik)*

  e. calender *(Pap)* Gaufrierkalander *m*, Prägekalander *m*

  e. paper Prägepapier *n*

embrittle / to 1. spröd[e] machen; brüchig machen; 2. spröd[e] werden, verspröden; brüchig werden

embrittlement Versprödung *f*, Sprödwerden *n*

Emde degradation Emde-Abbau *m (quartärer Ammoniumsalze)*

emerald *(Min)* Smaragd *m*

  e. copper *(Min)* Dioptas *m (Kupfer(II)-trisilikat)*

  e.-green smaragdgrün

  e. green $Cr_2O_3 \cdot xH_2O$ Chromoxidhydratgrün *n*, Guignetgrün *n*, Mittlers Grün *n*, Smaragdgrün *n*, Viridian *n*; Emeraldgrün *n (Mischung von Schweinfurter Grün mit Teerfarbstoffen)*; Brillantgrün *n*, Diamantgrün *n (basischer Triphenylmethanfarbstoff)*

  e. malachite *(Min)* Dioptas *m (Kupfer(II)-trisilikat)*

  e. nickel *(Min)* Nickelsmaragd *m*, *(veraltet für)* Zaratit *m (Nickeltetrahydroxidkarbonat)*

emergency shower Notbrause *f*, Löschbrause *f*, Feuerlöschbrause *f*

emery Schmirgel *m*

  e. paper Schmirgelpapier *n*

emerylite *(Min)* Emerylith *m*, *(veraltet für)* Margarit *m (Kalziumaluminiumdihydroxiddialumodisilikat)*

emetic Emetikum *n*, Brechmittel *n*

emetine Emetin *n (ein Benzylisochinolinalkaloid)*

emf, e.m.f., EMF, E.M.F. *s.* electromotive force

emission Emittieren *n*, Emission *f*, Aussendung *f*, Ausstrahlung *f*, Abstrahlung *f*; Austritt *m*; Ausströmung *f*

  e. capability *s.* emissive power

  e. coefficient Emissionskoeffizient *m*

  e. line Emissionslinie *f*

  e. of radiation Strahlungsemission *f*, Ausstrahlung *f*, Abstrahlung *f*

  e. spectrometer Emissionsspektrometer *n*

  e. spectroscopy Emissionsspektroskopie *f*

  e. spectrum Emissionsspektrum *n*

emissive emittierend, Emissions..., ausstrahlend, aussendend

  e. power Emissionsvermögen *n (ausgestrahlte Energie je Flächen- und Zeiteinheit)*

emissivity Emissionsvermögen *n (Verhältnis der Strahlung einer Oberfläche zur Strahlung eines schwarzen Körpers bei der gleichen Temperatur)*

emit / to emittieren, aussenden, ausstrahlen, abstrahlen; ausströmen; ausstoßen, auswerfen

emittance spezifische Ausstrahlung *f*; *s.* emissivity

emitter Emitter *m*, Strahler *m*, Strahlungsquelle *f*; Emitterelektrode *f*, Emissionselektrode *f*

  α-emitter *(Kch)* α-Strahler *m*, Alpha-Strahler *m*

  e. electrode Emitterelektrode *f*, Emissionselektrode *f*

  e. junction Emitterübergang *m*, Emitterübergangsschicht *f*

  e. region Emitterzone *f*, Emitterbereich *m*

emmonsite *(Min)* Emmonsit *m*

emollient Erweichungsmittel *n*, erweichendes Mittel *n*

emperor green $Cu(CH_3COO)_2 \cdot 3Cu(AsO_2)_2$ Kaisergrün *n*, Schweinfurter Grün *n*, Kupferarsenitazetat *n*

empirical empirisch [abgeleitet], erfahrungsgemäß, Erfahrungs...

  e. equation empirische Gleichung (Beziehung) *f*

  e. formula 1. empirische Formel *f*, Summenformel *f*, Bruttoformel *f*; 2. *(Math, Phys)* empirisch abgeleitete Formel *f*

  e. molecular formula empirische Molekularformel *f*, wahre Summenformel (Bruttoformel) *f*, Summenformel *f (i.e.S.)*, Bruttoformel *f (i.e.S.)*

emplectite *(Min)* Emplektit *m (Wismut(III)-kupfer(I)-sulfid)*

empty / to [ent]leeren

empty leer

  e. band leeres (unbesetztes) Energieband (Leitungsband, Leitfähigkeitsband) *n*, leerer (unbesetzter, nicht besetzter) Energiebereich *m*

  e. grain *(Gerb)* rinnender Narben *m*, *(manchmal auch)* loser Narben *m*

  e. leather leeres Leder *n (Gerbfehler)*

  e. place *(Krist)* Leerstelle *f*

emptying cycle Entleerungszyklus *m*

empyreumatic empyreumatisch, brenzlig, brenzlich *(Geruch)*

emulgator 1. Emulgator m, Emulgier[ungs]mittel n, Emulsionsbildner m, Emulsionsvermittler m, Emulgens n; 2. Emulgator m, Emulsor m, Emulgiermaschine f
emulsible s. emulsifiable
emulsifiability Emulgierbarkeit f
emulsifiable emulgierbar
  e. concentrate emulgierbares Konzentrat n, mischbares Öl n (als Pflanzenschutzmittel)
  e. oil emulgierbares („wasserlösliches") Öl n
emulsification Emulgieren n, Emulgierung f, Emulsionieren n, Emulsionsbildung f
  e. machine s. emulgator 2.
emulsifier s. emulgator
  e. churn s. emulsion churn
emulsify / to emulgieren, emulsionieren
emulsifying agent s. emulgator 1.
  e. mill Emulgiermühle f
  e. power Emulgiervermögen n, Emulgierfähigkeit f
emulsin Emulsin n (Glykosidase der Mandel)
emulsion Emulsion f; fotografische (lichtempfindliche) Emulsion f, Fotoemulsion f; fotografische (lichtempfindliche) Schicht f, Emulsion f
  e. breakdown Brechen n einer Emulsion
  e. churn Kirne f, Kirnapparat m, Kirnmaschine f
  e. copolymerization Emulsionsmischpolymerisation f
  e. fog (Foto) Emulsionsschleier m
  e. layer (Foto) Emulsionsschicht f
  e. of fat in water Fett-in-Wasser-Emulsion f
  e. oil Emulsionsöl n
  e. polymer Emulsionspolymerisat n, Emulsionspolymer[es] n
  e. polymerization Emulsionspolymerisation f, Polymerisation f in Emulsion
  e. polyvinylchloride Emulsions-Polyvinylchlorid n, E-PVC n
  e. scouring Emulsionswäsche f
  e. side (Foto) Schichtseite f
  e. speed (Foto) Schichtempfindlichkeit f, Allgemeinempfindlichkeit f
  e. spinning (Text) Emulsionsspinnverfahren n
  e. stability Emulsionsstabilität f, Emulsionsbeständigkeit f
  e. stabilizer Emulsionsstabilisator m
  e. support (Foto) Emulsionsunterlage f, Emulsionsträger m, Schichtträger m
  e. surface (Foto) Schichtseite f, Schichtoberfläche f
  e. thickening Emulsionsverdickung f
  e. type Emulsionstyp m
  e.-type mud (Erdöl) Emulsionsspülung f
emulsoid [colloid] Emulsoid n, Emulsionskolloid n, lyophiles Kolloid n
enamel / to emaillieren, (i.w.S.) mit einem email[le]ähnlichen Überzug versehen, lackieren; (Pap) satinieren, glätten; (Gerb) glanzstoßen; (Kosmet) (z.B. Nägel) lackieren
enamel Email n, Emaille f, (i.w.S.) email[le]ähn-

licher Überzug m, Lacküberzug m; emaillierter Gegenstand m; Emaillelack m, (Farb auch) Emaille f; (Bio) Zahnschmelz m, Schmelz m
  e. colour Emailfarbe f, Aufglasurfarbe f, Schmelzfarbe f
  e. paper Glacépapier n
  e. varnish Email[le]lack m
  e. ware Email[le]waren fpl
enamelled leather Lackleder n
  e. paper s. enamel paper
enamelling kiln Emaillierofen m
enamine Enamin n
enanthaldehyde $CH_3(CH_2)_5CHO$ Oenanthaldehyd m, Heptaldehyd m, Heptanal n
enanthate Oenanthat n, Heptylat n (Salz oder Ester der Oenanthsäure)
enanthic acid $CH_3(CH_2)_5COOH$ Oenanthsäure f, n-Heptylsäure f
  e. alcohol $CH_3(CH_2)_5CH_2OH$ Oenanthalkohol m, Heptylalkohol m, n-Heptanol n
enanthole s. enanthaldehyde
enanthylic acid s. enanthic acid
enanthylidene $HC\equiv C(CH_2)_4CH_3$ Oenanthyliden n, n-Amyläthin n, n-Amylazetylen n, 1-Heptin n
enantiomer s. enantiomorph
enantiomeric s. enantiomorphous
enantiomorph enantiomorphe Form f, Enantiomer[es] n, optisches Isomer[es] n, Spiegelbildisomer[es] n
enantiomorphic s. enantiomorphous
enantiomorphism Enantiomorphie f, optische Isomerie f, Spiegelbildisomerie f
enantiomorphous enantiomorph, enantiomer, optisch isomer, spiegelbildisomer
  e. form (isomer) s. enantiomorph
enantiotropic enantiotrop, wechselseitig (ineinander) umwandelbar (von Modifikationen)
enantiotropy Enantiotropie f, wechselseitige Umwandelbarkeit f
enargite (Min) Enargit m (Arsen(III)-kupfer(I,II)-sulfid)
encapsulate / to (Am) (Plast) einbetten
encapsulating (Am) (Plast) Einbetten n
enclosable einschließbar
enclose / to einschließen
enclosed headbox (Pap) geschlossener Stoffauflauf[kasten] m, Stoffauflauf m geschlossener Bauart
  e. impeller geschlossenes Laufrad n
enclosing rock Nebengestein n, umgebendes Gestein n
enclosure Einschluß m
  e. compound Einschlußverbindung f
encrust / to inkrustieren, eine Kruste bilden, verkrusten; inkrustieren, mit einer Kruste überziehen (bedecken), überkrusten
encrustants Inkrusten pl, inkrustierende Substanzen fpl, Inkrustsubstanzen fpl
encrustation Inkrustierung f, Inkrustation f, Krustenbildung f, (i.e.S.) Kesselsteinbildung f;

Inkrustation *f*, Kruste *f*, Belag *m*, *(i.e.S.)* Kesselstein *m*

encrusting constituents (materials, substances) *s.* encrustants

end atom Endatom *n*, endständiges Atom *n*

e. **component** Endkomponente *f*

e. **condition** Endzustand *m*

e. **fermentation** Endvergärung *f*

e.**-fired furnace** *(Glas)* U-Flammen-Wanne *f*, Ofen *m* mit U-Flammenführung, Glaswanne *f* mit Längsfeuerung

e. **gas** Endgas *n*

e. **group** Endgruppe *f*

e. **group analysis** Endgruppenanalyse *f*

e. **group assay** Endgruppenbestimmung *f*

e. **of bleaching** *s.* e. of the bleaching period

e. **of the bleaching period** Bleichende *n*

e. **paper** Vorsatzpapier *n*, Vorstoßpapier *n*

e. **phase** Endphase *f*

e. **point** Endpunkt *m*, *(Titration auch)* Umschlagspunkt *m*; Endsiedepunkt *m*, Siedeendpunkt *m*, Siedeende *n*, SE, Endkochpunkt *m*

e.**-point detection** Endpunktserkennung *f*

e.**-point determination** Endpunktsbestimmung *f*

e.**-port furnace** *s.* e.-fired furnace

e. **product** Endprodukt *n*

e. **temperature** Endtemperatur *f*

e. **wall** Einlegewand *f (im Glasschmelzofen)*

endeiolite *(Min)* Endeiolith *m*, *(veraltet für)* Pyrochlor *m*

endellione, endellionite *(Min)* Endellionit *m*, *(veraltet für)* Bournonit *m (ein Bleikupferspießglanz)*

ending *(Text)* ungleichmäßiges (ungleiches) Färben *n (Endenungleichheit)*

endless felt *(Pap)* endloser Filz *m*, endloses Filztuch *n*, Filzschleife *f*

e. **wire** *(Pap)* Endlossieb *n*, endloses Sieb *n*

endo product endo-Produkt *n*

endocellular intrazellular, intrazellulär

e. **enzyme** intrazelluläres Ferment *n*, Endoenzym *n*

endocrine endokrin, innersekretorisch

e. **gland** endokrine (inkretorische) Drüse *f*

endocyclic endozyklisch

endoenzyme Endoenzym *n*, intrazelluläres Ferment *n*

endogenetic rock endogenes Gestein *n*

endogenous endogen, *(Geol auch)* innenbürtig, im Erdinnern entstanden

endolytic insecticide endolytisches [systemisches] Insektizid *n*

endometatoxic insecticide endometatoxisches [systemisches] Insektizid *n*

endomorphism *(Geol)* Endomorphose *f*, endomorphe Kontaktwirkung *f*

endopeptidase Endopeptidase *f (proteolytisches Ferment)*

endosmosis Endosmose *f*

endosmotic endosmotisch

endosperm *(Bot)* Endosperm *n*, Nährgewebe *n (im Samen)*

endothermic endotherm, wärmeaufnehmend, wärmeverbrauchend, wärmebindend

endotoxin Endotoxin *n*

endurance limit *(Plast)* Dauer[schwing]festigkeit *f*, Haltbarkeitsgrenze *f*

energizer *(Met)* Aktivierungsmittel *n*, aktivierendes Mittel *n*, Aktivierungszusatz *m*, Verstärker *m*, Aktivator *m (Zusatz zum Kohlungsmittel)*

energy Energie *f*, Arbeitsvermögen *n (eines physikalischen Systems)*

e. **balance** Energiebilanz *f*, Energiehaushalt *m*; Energiegleichgewicht *n*; *(Bio)* Energieumsatz *m*

e. **band** Energieband *n*, Energiebereich *m*

e. **barrier** Energieschranke *f*, Energieberg *m*, Energieschwelle *f*

e. **change** Energieumsatz *m*, Energieumformung *f*, Energiewandlung *f*

e. **content** Energieinhalt *m*

e. **contribution** Energiebeitrag *m*

e. **density** Energiedichte *f*

e. **diagram** Energie[niveau]diagramm *n*

e. **difference** Energieunterschied *m*, Energiedifferenz *f*

e. **distribution** Energieverteilung *f*

e. **gain** Energiegewinn *m*

e. **gap** Energielücke *f*, verbotenes Band *n*, verbotene Zone *f*, verbotener (nicht zugelassener) Energiebereich *m*

e. **input** zugeführte Energie *f*

e. **level** Energieniveau *n*, Energiezustand *m*, Energiestufe *f*, Energieterm *m*, Term *m*, Elektronenniveau *n*

e. **loss** Energieverlust *m*

e. **of activation** Aktivierungsenergie *f*

e. **of adsorption** Adsorptionsenergie *f*, Adsorptionswärme *f*

e. **of binding** Bindungsenergie *f*

e. **of dislocation** Versetzungsenergie *f*

e. **of dissociation** Dissoziationsenergie *f*

e. **of excitation** Anregungsenergie *f*, Erregungsenergie *f*

e. **of ionization** Ionisationsenergie *f*, Ionisierungsenergie *f*, Ionisierungsarbeit *f*, Ionisierungsarbeit *f*

e. **of light** Lichtenergie *f*

e. **of radiation** Strahlungsenergie *f*, Strahlenenergie *f*

e. **output** Energieabgabe *f*

e. **principle** Energieprinzip *n*, Energieerhaltungssatz *m*, Erhaltungssatz *m* der Energie, Gesetz *n* (Prinzip *n*, Satz *m*) [von] der Erhaltung der Energie

e. **quantum** Energiequant *n*

e. **range** Energiebereich *m*

e. **reduction** Energieabnahme *f*, Energiesenkung *f*

e. **source** Energiequelle *f*

e. **state** Energiezustand *m*

e. transfer Energieübertragung *f*, Energieüberführung *f*, Energieübergang *m*
e. unit Energieeinheit *f*
enflame / to *s.* to inflame
enfleurage *(Kosmet)* Enfleurage *f (Gewinnung von Duftstoffen mittels Adsorption an Fette)*
e. absolute *(Kosmet)* Enfleurageöl *n*, Absolu d'enfleurage *n*, Essence absolue d'enfleurage *f*
engaging chamber Zwischenbehälter *m (beim Pebble-Heater-Verfahren)*
Engel process *(Plast)* Engel-Verfahren *n (Pulver-Sinterverfahren)*
engine oil Schmieröl *n (für Verbrennungsmotoren und Dampfmaschinen)*
e.-sized *(Pap)* massegeleimt, stoffgeleimt
e. sizing *(Pap)* Leimung *f* im Stoff, Leimung *f* in der Masse, Masseleimung *f*
e. sludge Ölschlamm *m*
engineering brick Klinker[stein] *m*, Klinkerziegel *m*
e. chemistry Chemieingenieurtechnik *f*, Chemieingenieurwesen *n*, chemische Ingenieurtechnik *f*, chemisches Ingenieurwesen *n*, *(i.e.S.)* chemische Verfahrenstechnik *f*
Engler flask Engler-Kolben *m*
engobe / to *(Ker)* engobieren, *(mit einer Deckschicht)* überziehen
engobe *(Ker)* Engobe *f*, Beguß *m*, Begußmasse *f*, Angußmasse *f*, Angußfarbe *f*
engobing *(Ker)* Engobieren *n*
engrave / to [ein]gravieren
engraver's acid Ätzflüssigkeit *f*, Ätzmittel *n (meist Salpetersäure)*
engraving Gravieren *n*, Eingravieren *n*, Gravur *f*
enhydritic, enhydrous *(Min)* wasserhaltig
enlarging adapter Übergangsstück *n (mit großer Hülse auf kleinem Kern)*
e. paper *(Foto)* Vergrößerungspapier *n*
enol Enol *n*
e. form (structure) Enolform *f*
enolase Enolase *f (ein Ferment)*
enolate Enolat *n*
e. resonance Enolatmesomerie *f*, Enolatresonanz *f*
enolic enolisch, Enol...
e. form (structure) Enolform *f*
enolizable enolisierbar
enolization Enolisierung *f*
e. tendency Enolisierungstendenz *f*
enolize / to enolisieren
enometry Enometrie *f (Bestimmung der Doppelbindungen in Fetten durch Halogenanlagerung)*
enrich / to anreichern, *(Brennstoffe)* im Heizwert steigern
e. by vitamins vitamin[is]ieren, mit Vitaminen anreichern
enriched layer *(phys Ch)* angereicherte Schicht *f*, Anreicherungsschicht *f*
e. water gas karburiertes Wassergas *n*
enriching section *(Destillation)* Rektifizierteil *m*,

Rektifikationsteil *m*, Rektifizierzone *f*, Rektifikationszone *f*, Verstärkungsteil *m*
enrichment Anreicherung *f*
e. factor Anreicherungsgrad *m*, Anreicherungsfaktor *m*
e. horizon *(Bodenkunde)* Illuvialhorizont *m*, Anreicherungshorizont *m*
e. line Verstärkungsgerade *f (im Destillationsdiagramm)*
e. of ores Erzanreicherung *f*
e. section *s.* enriching section
enstatite *(Min)* Enstatit *m (Magnesiummetasilikat)*
entanglement Verhakung *f (von Polymerenketten)*
enter / to eintreten
e. into chemical combination eine chemische Verbindung eingehen
e. onto the wire *(Pap)* auf das Sieb auflaufen *(Stoff)*
enterokinase Enterokinase *f (ein Ferment des Darmsaftes)*
enthalpy Enthalpie *f*, Wärmeinhalt *m*, Gibbssche Wärmefunktion *f*, H
e. change Enthalpieänderung *f*
e.-concentration chart Wärmeinhalts-Konzentrations-Diagramm *n*
e.-entropy chart i-s-Diagramm *n*, Enthalpie-Entropie-Diagramm *n*, Mollier-Diagramm *n*
e. of formation Bildungsenthalpie *f*
e. of reaction Reaktionsenthalpie *f*
entity Stoff *m*, Substanz *f*, Körper *m*, Gebilde *n*; einheitlicher (einheitlich zusammengesetzter) Stoff (Körper) *m*, einheitliche Substanz (Verbindung) *f*, Ganzheit *f*
entrain / to mitreißen, mitführen; aufströmen *(Kohlenstaub bei der Vergasung)*
entrained catalyst system Flugstaubverfahren *n*
e. gasification process Staubvergasungsverfahren *n*
e. liquor mitgerissene Flüssigkeit *f*
entrainer Zusatzstoff *m* bei der Azeotropdestillation, Schleppmittel *n*, Mitnehmer *m*
e. gas Trägergas *n*
e. sublimation Trägergas-Sublimation *f*
entrainment Mitreißen *n*, Mitführen *n*; Aufströmen *n (von Kohlenstaub bei der Vergasung)*
e. loss Verlust *m* durch Mitreißen
e. separator Abscheider *m* für mitgerissene Flüssigkeit
entrance loss Eintrittsverlust *m*, Einlaufverlust *m*
e. slit Eintrittsspalt *m*
entrap / to *(z.B. Kristallwasser)* einschließen
entrapped air Lufteinschluß *m*
entropy Entropie *f*, S
e. change Entropieänderung *f*
e. chart T-s-Diagramm *n*, Temperatur-Entropie-Diagramm *n*; i-s-Diagramm *n*, Enthalpie-Entropie-Diagramm *n*, Mollier-Diagramm *n*
e. contribution Entropieanteil *m*
e. decrease Entropieabnahme *f*
e. diagram *s.* e. chart

e. effect Entropieeffekt *m*
e. increase Entropiezunahme *f*
e. of absolute zero Nullpunktsentropie *f*, S_0
e. of activation Aktivierungsentropie *f*
e. of evaporation Verdampfungsentropie *f*
e. of fusion Schmelzentropie *f*
e. of mixing Mischungsentropie *f*
e. of reaction Reaktionsentropie *f*
e. of vaporization Verdampfungsentropie *f*
e. production Entropieerzeugung *f*
e. unit Entropieeinheit *f*
e. value Entropiewert *m*
entry lock Eintragzelle *f*, Beschickungsschleuse *f*
e. loss Eintrittsverlust *m*, Einlaufverlust *m*
enumerate / to zählen (*z.B. die Atome eines Ringsystems*)
enumeration Zählung *f* (*z.B. der Atome eines Ringsystems*)
e. system Zählweise *f*
envelope Hülle *f*, Umhüllung *f*, Mantel *m*
e. filter Rahmenfilter *n*, Flächenabscheider *m*
e. gasket eingefaßte Dichtung *f*; umhüllte Dichtung *f*
e. lining [tissue] Futterseidenpapier *n*, Briefumschlag[futter]seide *f*, Briefhüllenfutterseide *f*
e. paper Briefumschlagpapier *n*, Briefhüllenpapier *n*, Kuvertpapier *n*
environment Umgebung *f*, Milieu *n*
e.-sensitive umgebungsempfindlich
environmental factors Umweltfaktoren *mpl*, Umweltbedingungen *fpl*, Umwelteinflüsse *mpl*
e. stress cracking (*Plast*) Spannungsrißbildung *f* unter dem Einfluß des umgebenden Mediums, Spannungsrißkorrosion *f*
e. temperature Umgebungstemperatur *f*
enzymatic enzymatisch, fermentativ, Enzym..., Ferment...
e. decomposition enzymatische (fermentative) Spaltung *f*
e. filter aid Filtrationsenzym *n*
e. hydrolysis enzymatische Hydrolyse *f*
e. reaction enzymatische (enzymkatalysierte) Reaktion *f*, Enzymreaktion *f*
enzyme Enzym *n*, Ferment *n*
e. action Enzymwirkung *f*, Fermentwirkung *f*
e. activity Enzymaktivität *f*, Enzymtätigkeit *f*
e.-catalyzed enzymkatalysiert, fermentkatalysiert
e. chemistry Enzymchemie *f*, Fermentchemie *f*
e.-converted starch (*Pap*) enzymatisch (fermentativ) abgebaute Stärke *f*
e. desizing Entschlichten *n* durch Enzyme (Fermente), enzymatisches (fermentatives) Entschlichten *n*
e. detergent enzymatisches Waschmittel *n*
e. donor Enzymdonator *m*, Fermentdonator *m*
e. reaction Enzymreaktion *f*, Fermentreaktion *f*
e.-resistant enzymresistent, fermentresistent
e. system Enzymsystem *n*, Fermentsystem *n*

e. treatment enzymatische (fermentative) Behandlung *f*, Enzymbehandlung *f*
e. unhairing (*Gerb*) Enzymäscher *m*, Fermentäscher *m*, enzymatische (fermentative) Haarlockerung *f*
enzymic *s.* enzymatic
e. action *s.* enzyme action
e. activity *s.* enzyme activity
enzymize / to enzym[is]ieren, mit Enzymen (Fermenten) anreichern
enzymology Enzymologie *f*
eolotropic anisotrop
eolotropism Anisotropie *f*
eosin[e] Eosin *n*, 2,4,5,7-Tetrabromfluoreszein *n*
eosinophil[e], eosinophilic (*Bio*) eosinophil (*durch Eosin leicht färbbar*)
eosphorite (*Min*) Eosphorit *m*
Eötvös rule Eötvössche Regel *f*
EP lubricant *s.* extreme-pressure lubricant
EP-rubber *s.* ethylene-propylene rubber
EPC black *s.* easy processing channel black
ephedrine $C_6H_5CH(OH)CH(NHCH_3)CH_3$ Ephedrin *n*, 1-Phenyl-2-methylaminopropan-1-ol *n*
e. hydrochloride Ephedrinhydrochlorid *n*, L-*erythro*-1-Phenyl-2-methylaminopropanol-hydrochlorid *n*
epi *s.* epichlorohydrin
epibaptifoline Epibaptifolin *n*
epichlorohydrin Epichlorhydrin *n*, 1-Chlor-2,3-epoxypropan *n*, Methyloxiran *n*
epidote, epidotite (*Min*) Epidot *m* (*ein Kalziumaluminiumsilikat*)
epilate / to enthaaren, epilieren
epilating wax Enthaarungswachs *n*, Epilierwachs *n*
epilation Enthaarung *f*, Epilation *f*
epilator Enthaarungsmittel *n*, Epiliermittel *n*, Epilatorium *n*
epilupinine Epilupinin *n* (*Alkaloid*)
epimer Epimer[es] *n*, Diastereomer[es] *n*
epimeric epimer
epimeride *s.* epimer
epimerism (*Stereochemie*) Epimerie *f*
epimerization Epimerisierung *f*
epimerize / to epimerisieren
epinephrine Adrenalin *n*, Epinephrin *n*
epistilbite (*Min*) Epistilbit *m* (*ein Tektosilikat*)
epitaxy Epitaxie *f* (*orientierte Kristallabscheidung auf Fremdkristallen*)
epithermal reaction epithermische Reaktion *f*
epoxidation Epoxydation *f*, Epoxydieren *n*, Epoxydierung *f*
epoxide Epoxid *n*
e. group Epoxidgruppe *f*, Epoxygruppe *f*, Epoxidring *m*
e. plasticizer Epoxidweichmacher *m*
e. resin Epoxidharz *n*, Epoxyharz *n*, EP
e. resin adhesive Epoxidkleber *m*
e. ring *s.* e. group
epoxidize / to epoxydieren

epoxidizing Epoxydieren *n*, Epoxydierung *f*
epoxies Epoxidharze *npl*, Epoxyharze *npl*
epoxy cure Vulkanisation *f* mit Epoxidharzen, Epoxidharzvernetzung *f*
 e. **group** *s*. epoxide group
 e. **resin** *s*. epoxide resin
 e. **ring** *s*. epoxide group
EPR *s*. ethylene-propylene rubber
epsilon acid ε-Säure *f*, Amino-ε-Säure *f*, 1-Naphthylamin-3,8-disulfonsäure *f*
Epsom salt Bittersalz *n*, *(Min auch)* Epsomit *m* *(Magnesiumsulfat-7-Wasser)*
EPT *s*. ethylene-propylene terpolymer
EPTC $(C_3H_7)_2N \cdot CO \cdot SCH_2CH_3$ EPTC *n*, Eptam *n*, Äthyldipropylthiokarbamat *n* *(Herbizid)*
equal to type typkonform
equation Gleichung *f*
 e. **of state** Zustandsgleichung *f*
equatorial / to be äquatorial stehen
 e. **bond** äquatoriale Bindung *f*, e-Bindung *f*
 e. **conformation** äquatoriale Konformation *f*, e
 e. **substituent** äquatorialer Substituent *m*
equigranular gleichmäßig körnig
equilibrate / to ins Gleichgewicht bringen (setzen), äquilibrieren, ausbalancieren, ausgleichen; auswuchten; im Gleichgewicht halten
equilibration Herstellung *f* des Gleichgewichts, Äquilibrierung *f*, Ausbalancieren *n*; Auswuchten *n*; Aufrechterhaltung *f* des Gleichgewichts
 e. **process** Äquilibrierungsverfahren *n*, Äquilibrierungsmethode *f*
equilibrium Gleichgewicht *n*; Ausgleich *m*; Ruhelage *f*
 e. **box** *(phys Ch)* Gleichgewichtskasten *m* *(Modell für Gasreaktionen)*
 e. **concentration** Gleichgewichtskonzentration *f*
 e. **condition[s]** Gleichgewichtszustand *m*, Gleichgewichtsbedingungen *fpl*
 e. **constant** Gleichgewichtskonstante *f*, Massenwirkungskonstante *f*
 e. **diagram** Zustandsdiagramm *n*, Phasendiagramm *n*
 e. **distillation** Gleichgewichtsdestillation *f*, geschlossene (integrale) Destillation *f*
 e. **distribution coefficient** Gleichgewichtsverteilungskoeffizient *m*, idealer Verteilungskoeffizient *m* *(in der Zonenschmelztheorie)*
 e. **flash vaporization** *s*. e. vaporization
 e. **method** Gleichgewichtsmethode *f*
 e. **moisture content** Gleichgewichtsfeuchte[beladung] *f*
 e. **partial condensation** geschlossene (integrale) Teilkondensation *f*
 e. **partition coefficient** *s*. e. distribution coefficient
 e. **position** Gleichgewichtslage *f*
 e. **pressure** Gleichgewichtsdruck *m*
 e. **ratio** Gleichgewichtsverhältnis *n*
 e. **segregation coefficient** *s*. e. distribution coefficient

 e. **still** *(Destillation)* Gleichgewichtsapparatur *f*
 e. **state** Gleichgewichtszustand *m*
 e. **swelling** *(Gum)* Gleichgewichtsquellung *f*
 e. **temperature** Gleichgewichtstemperatur *f*
 e. **value** Gleichgewichtswert *m*, [idealer] Ruhewert *m* *(in der Zonenschmelztheorie)*
 e. **vaporization** Gleichgewichtsverdampfung *f*, geschlossene Verdampfung *f*
 e. **vapour pressure** Gleichgewichtsdampfdruck *m*
equimolar äquimolar
 e. **solution** äquimolare Lösung *f*
equimolecular *s*. equimolar
equipartition Gleichverteilung *f*
 e. **principle** Gleichverteilungssatz *m*, Äquipartitionstheorem *n*, Äquipartitionsprinzip *n*
equipment Ausrüstung *f*, Ausrüstungsgegenstände *mpl*, Ausstattung *f*, Einrichtung *f*, Gerätschaften *pl*, Gerät *n*, Apparatur *f*
 e. **of chemical laboratories** Labor[atoriums]einrichtung *f*, Labor[atoriums]ausstattung *f*
equipotential surface Äquipotentialfläche *f*, Potentialfläche *f*, Niveaufläche *f*
equipped with bars *(Pap)* bemessert, mit Messern besetzt
equisetic acid *s*. aconitic acid
equivalence, equivalency Äquivalenz *f*, Gleichwertigkeit *f*
 e. **point** Äquivalenzpunkt *m*
equivalent äquivalent, gleichwertig, Äquivalent...
equivalent Äquivalent *n*, äquivalente (äquimolekulare) Menge *f*
 e. **charge** Äquivalentladung *f*
 e. **conductance** Äquivalentleitfähigkeit *f*, Äquivalenzleitfähigkeit *f*
 e. **conductance at infinite dilution** Äquivalentleitfähigkeit *f* bei unendlicher Verdünnung, Grenzleitfähigkeit *f*
 e. **ion[ic] conductance** Ionenäquivalentleitfähigkeit *f*, Äquivalentleitfähigkeit *f* der Ionen
 e. **point** Äquivalenzpunkt *m*, Umschlagspunkt *m* *(bei der Titration)*
 e. **roentgen (röntgen)** Röntgenäquivalent *n*
 e. **weight** Äquivalentmasse *f*, *(bisher)* Äquivalentgewicht *n*, Verbindungsgewicht *n*
eradicant fungicide direktes Fungizid *n*
erasability *(Pap)* Radierbarkeit *f*, Radierfestigkeit *f*, Radierfähigkeit *f*
erasing qualities *s*. erasability
erbia Er_2O_3 Erbinerde *f*, Erbia *f*, *(veraltet für)* Erbiumoxid *n*
erbium Er Erbium *n*
 e. **chloride** $ErCl_3$ Erbiumchlorid *n*
 e. **nitrate** $Er(NO_3)_3$ Erbiumnitrat *n*
 e. **oxide** Er_2O_3 Erbiumoxid *n*
 e. **sulphate** $Er_2(SO_4)_3$ Erbiumsulfat *n*
ercinite *(Min)* Erzinit *m*, *(veraltet für)* Harmotom *m* *(ein Tektosilikat)*
eremite *s*. monazite
erepsin Erepsin *n* *(eiweißspaltendes Fermentsystem)*

erfc *s.* error function
erg Erg *n*, erg
ergobasine, ergometrine Ergometrin *n*, Ergonovin *n*, Ergobasin *n*, Ergostetrin *n*, Ergotozin *n* (Alkaloid)
ergone Biokatalysator *m*, Wirkstoff *m*, Ergin *n*; Wirkungsgruppe *f (eines Ferments)*
ergonovine *s.* ergometrine
ergosterin, ergosterol Ergosterin *n*, Ergosterol *n*
ergostetrine *s.* ergometrine
ergot Mutterkorn *n*; Mutterkornpilz *m*, Claviceps purpurea (Fr.) Tul.
e. alkaloid Mutterkornalkaloid *n*
ergotamine Ergotamin *n (Alkaloid)*
ergotism Ergotismus *m*, Mutterkornvergiftung *f*
ergotocine *s.* ergometrine
erinite *(Min)* Erinit *m, (veraltet für)* Cornwallit *m (ein Kupferarsenat)*
eriochalcite *(Min)* Eriochalzit *m (Kupfer(II)-chlorid-2-Wasser)*
erkensator *(Pap)* Erkensator *m*, Zentrifugalsortierer *m*, Schleudersortierer *m*
Erlenmeyer flask Erlenmeyer-Kolben *m*
Erlenmeyer-Plöchl azlactone synthesis Erlenmeyer-Plöchlsche Azlakton-Kondensation *f*, Erlenmeyersche Synthese *f (von α-Aminosäuren)*
erosion corrosion *(durch Erosion begünstigte Korrosion)*
error function Fehlerfunktion *f*, Fehlerintegral *n*
e. of method methodischer Fehler *m*, Verfahrensfehler *m*
e. of observation Beobachtungsfehler *m*
erubescite *(Min)* Erubeszit *m, (veraltet für)* Bornit *m (Eisen(II)-kupfer(II)-sulfid)*
erucic acid $CH_3(CH_2)_7CH=CH(CH_2)_{11}COOH$ Erukasäure *f*, 13-Dokosensäure *f*
erucyl alcohol $CH_3(CH_2)_7CH=CH(CH_2)_{11}CH_2OH$ Eruzylalkohol *m*, 13-Dokosenol *n*
eruptive rock Eruptivgestein *n*
erysodine Erysodin *n (Alkaloid)*
erysotrine Erysotrin *n (Alkaloid)*
erysovine Erysovin *n (Alkaloid)*
erythorbic acid Isoaskorbinsäure *f*
erythraline Erythralin *n (Alkaloid)*
erythrene $CH_2=CHCH=CH_2$ Erythren *n*, Butadien-(1,3) *n*, Butadien *n*
erythrina alkaloid Erythrinaalkaloid *n*
erythrine, erythrite *(Min)* Erythrin *m*, Kobaltblüte *f (Kobalt(II)-orthoarsenat)*
i-erythritol $(HOCH_2CHOH)_2$ Erythrit *m*, Butantetrol *n*
erythro form *(Stereochemie)* Erythroform *f*
erythroaphin Erythroaphin *n*
erythrocalcite *s.* eriochalcite
erythrodextrin[e] Erythrodextrin *n*
erythrogenic acid $CH_2=CH(CH_2)_4C\equiv CC\equiv C(CH_2)_7COOH$ Erythrogensäure *f*, Oktadezen-(17)-diin-(9,11)-säure *f*
erythroglucin *s.* i-erythritol
erythroidine Erythroidin *n (Alkaloid)*

erythrol *s.* i-erythritol
erythromycin Erythromyzin *n (Antibiotikum)*
erythrose $CH_2OHCHOHCHOHCHO$ Erythrose *f.* 2,3,4-Trihydroxybutanal *n*
erythrosine Erythrosin *n (Dinatriumsalz des Tetrajodfluoreszeins)*
ES *(Abk. für)* Electrochemical Society
E.S. *s.* 1. engine sizing; 2. engine-sized
e.s. unit elektrostatische Einheit *f*; elektrostatische Ladungseinheit *f*
escape / to entweichen, austreten, ausströmen, abströmen, abfließen
Eschka method Eschka-Verfahren *n*, Eschka-Methode *f (Aufschlußverfahren zur Gesamtschwefelbestimmung)*
esculent eßbar, genießbar
esculin Äskulin *n*, Bikolorin *n*, 6,7-Dihydroxykumarin-(β-d-glukopyranosid)-(6) *n*
esculinic acid *s.* esculin
eserine Eserin *n*, Physostigmin *n (Alkaloid)*
eseroline Eserolin *n (Alkaloid)*
esparto [grass] Espartogras *n*, Esparto *m*, Halfagras *n*, Stipa tenacissima L.
e. mill Espartozellstoffabrik *f*
e. paper Espartopapier *n*, Alfapapier *n*
e. pulp Espartozellstoff *m*
e. wax *(Pap)* Espartowachs *n*
esr, ESR *s.* electron spin resonance
essence Essenz *f*, Auszug *m*, Extrakt *m*; Parfüm *n*, Wohlgeruch *m*
e. of mirbane (myrbane) *(Kosmet)* $C_6H_5NO_2$ Mirbanessenz *f*, Mirbanöl *n*, Nitrobenzol *n*
e. of roses Rosenöl *n*
essential wesentlich, essentiell, *(Bio auch)* unentbehrlich, lebensnotwendig; ätherisch, flüchtig
e. element *(Landw)* unentbehrlicher Grundstoff (Nährstoff) *m*
e. fatty acid essentielle Fettsäure *f*
e. oil ätherisches Öl *n*
essentiality *(Landw)* Unentbehrlichkeit *f (z.B. von Nährstoffen)*
essonite *s.* hessonite
establish / to herstellen *(z.B. Gleichgewicht, Kontakt)*
e. the structure die Konstitution feststellen (aufklären, ermitteln, sichern)
establishment Herstellung *f*, Einstellung *f (z.B. des Gleichgewichts, Kontakts)*
estate rubber Plantagenkautschuk *m*, Pflanzungskautschuk *m*
ester Ester *m*
e.-based auf Esterbasis
e. bond Esterbindung *f*
e. enolate Esterenolat *n*
e. exchange Esteraustausch *m*
e. formation Esterbildung *f*
e. gum Esterharz *n*
e. interchange Umesterung *f*
e. linkage *s.* e. bond

e. number Esterzahl *f*, EZ
e. of a sulphonic acid Sulfonsäureester *m*
e. pool *(Bioch)* Ester-pool *m (Sammelname für eine Gruppe von Phosphorsäure-Zuckerestern)*
e. rubber Polyesterkautschuk *m*
e. saponification Esterverseifung *f*, Verseifung *f (i.e.S.)*, Esterspaltung *f*
e. value *s. e.* number
esterase Esterase *f (esterspaltendes Ferment)*
esterification Verestern *n*, Veresterung *f*
e. catalyst Veresterungskatalysator *m*
esterify / to verestern
esterifying Verestern *n*, Veresterung *f*, Esterbildung *f*
estimate / to bestimmen
estimation Bestimmung *f*
e. of moisture Feuchtigkeitsbestimmung *f*, Feuchtebestimmung *f*
e. of sulphur Schwefelbestimmung *f*
estradiol *(Am)* Östradiol *n (ein Sexualhormon)*
estragon Estragon *m*, Artemisia dracunculus L.
estrane *(Am)* Östran *n*
et. *s.* ethyl ether
Étard reaction Étardsche Reaktion *f (Überführung aromatisch oder heterozyklisch gebundener Methylgruppen in Aldehyd- oder Ketogruppen)*
etch / to ätzen
etch figure *(Krist)* Ätzfigur *f*, Lösungsfigur *f*
etchant Ätzmittel *n*
etching Ätzen *n*, Ätzung *f*, Anätzen *n*; Beizen *n*
e. acid (fluid) Ätzflüssigkeit *f*
e. paper Kupferdruckpapier *n*
e. reagent Ätzmittel *n*
e. solution Ätzlösung *f*
e. time Ätzzeit *f*
ethal $C_{16}H_{33}OH$ Zetylalkohol *m*, Äthal *n*, Hexadezylalkohol *m*, Hexadekanol-(1) *n*
ethanal CH_3CHO Äthanal *n*, Azetaldehyd *m*, Essigsäurealdehyd *m*
ethanamide CH_3CONH_2 Äthanamid *n*, Azetamid *n*, Essigsäureamid *n*
ethanamine bridge Äthanaminbrücke *f*
ethane $CH_3 \cdot CH_3$ Äthan *n*, Dimethyl *n*, Bimethyl *n*, Methylmethan *n*
1,2-e. diamine $H_2NCH_2CH_2NH_2$ 1,2-Äthandiamin *n*, Äthylendiamin *n*
e. dinitrile $(CN)_2$ Dizyan *n*, Zyan *n*, Oxalsäurenitril *n*
ethanedial CHOCHO Äthandial *n*, Oxalaldehyd *m*, Oxalsäuredialdehyd *m*, Glyoxal *n*
ethanedioic acid HOOCCOOH Äthandisäure *f*, Oxalsäure *f*, Kleesäure *f*
1,2-ethanediol Äthandiol-(1,2) *n*, Glykol *n (i.e.S.)*, Äthylenglykol *n*
ethanethiol C_2H_5SH Äthanthiol *n*, Äthylmerkaptan *n*, Thioäthanol *n*, Äthylhydrosulfid *n*, Äthylthioalkohol *m*
ethanethiolic acid CH_3COSH Thioessigsäure *f*
ethanoic acid CH_3COOH Äthansäure *f*, Azetsäure *f*, Essigsäure *f*

e. anhydride $(CH_3CO)_2O$ Äthansäureanhydrid *n*, Azetanhydrid *n*, Essigsäureanhydrid *n*
ethanol CH_3CH_2OH Äthanol *n*, Äthylalkohol *m*
e.-insoluble äthanolunlöslich
e.-soluble äthanollöslich
ethanolamine Äthanolamin *n, (i.e.S.)* $CH_2OHCH_2NH_2$ Monoäthanolamin *n*, Aminoäthylalkohol *m*, 2-Aminoäthanol-(1) *n*, Kolamin *n*
ethanoldimethylamine Äthanoldimethylamin *n*, Dimethylaminoäthylalkohol *m*
ethanolic äthanolisch
ethanolysis Äthanolyse *f*
ethanoyl chloride CH_3COCl Äthanoylchlorid *n*, Azetylchlorid *n*, Essigsäurechlorid *n*
ethene $H_2C=CH_2$ Äthen *n*, Äthylen *n*
ethenoid plastics *(Sammelname für Akrylate, Vinyl- und Styrolplaste)*
ether $R \cdot O \cdot R'$ Äther *m*, Ä., *(i.e.S.)* $C_2H_5 \cdot O \cdot C_2H_5$ Äther *m*, Äthyläther *m*, Diäthyläther *m*, Äthoxyäthan *n; (fälschlich für)* ester
e. bond Ätherbindung *f*
e. cleavage Ätherspaltung *f*
e. extraction Ätherextraktion *f*
e. formation Ätherbildung *f*, Äthersynthese *f*
e.-insoluble ätherunlöslich
e. link Ätherbindung *f (als Zustand)*
e. linkage Ätherbindung *f (als Vorgang)*
e.-soluble ätherlöslich
etherate Ätherat *n*
ethereal ätherisch, Äther...
e. blue Himmelblau *n*, Reinblau *n*
etherial *s.* ethereal
etherification Veräthern *n*, Verätherung *f*
etherin $H_2C=CH_2$ Äthen *n*, Äthylen *n*
ethical preparation *(Am)* rezeptpflichtiges Arzneimittel *n*
ethine CH≡CH Äthin *n*, Azetylen *n*
ethinyl group —C≡CH Äthinylgruppe *f*, Äthinylrest *m*
ethiodide CH_3CH_2J Jodäthanverbindung *f*, Äthyljodidverbindung *f*
ethisterone Äthisteron *n*, 17-Äthinyl-17-hydroxy-3-ketoandrosten-(4) *n*, Anhydroxyprogesteron *n*
ethol *s.* ethal
ethoxide $C_2H_5OM^I$ Äthylat *n*
ethoxyaniline Äthoxyanilin *n*, Phenetidin *n*, Aminophenetol *n*, Aminophenoläthyläther *m*
ethoxybenzene $C_6H_5OC_2H_5$ Äthoxybenzol *n*, Äthylphenyläther *m*, Phenetol *n*
ethoxybenzoic acid $C_2H_5OC_6H_4COOH$ Äthoxybenzoesäure *f*
ethoxyethane $C_2H_5OC_2H_5$ Äthoxyäthan *n*, Äthyläther *m*, Diäthyläther *m*, Äther *m (i.e.S.)*
2-ethoxyethanol $C_2H_5O \cdot CH_2CH_2OH$ Äthylglykol *n*, Glykolmonoäthyläther *m*
ethoxylene resin Epoxidharz *n*, Epoxyharz *n*, EP, Äthoxylinharz *n*
m-ethoxyphenol $C_2H_5OC_6H_4OH$ *m*-Äthoxyphenol *n*, Resorzinmonoäthyläther *m*
o-ethoxyphenol $C_2H_5OC_6H_4OH$ *o*-Äthoxyphenol *n*, Guäthol *n*, Brenzkatechinmonoäthyläther *m*

p-ethoxyphenol p-Äthoxyphenol n, Hydrochinonmonoäthyläther m
ethyl Äthyl n
  e. acetate $CH_3COOC_2H_5$ Essigsäureäthylester
m, Essigester m, Äthylazetat n
  e. acetic acid s. n-butyric acid
  e. acetoacetate $CH_3COCH_2COO \cdot C_2H_5$ Azetessigsäureäthylester m, Azetessigester m
  e. acetone $CH_3COC_3H_7$ Methylpropylketon n,
Pentanon-(2) n
  e. acetylene $HC \equiv CCH_2CH_3$ Äthylazetylen n,
1-Butin n
  e. acrylate $CH_2 = CH \cdot COOC_2H_5$ Akrylsäureäthyl-
ester m, Äthylakrylat n
  e. alcohol CH_3CH_2OH Äthylalkohol m, Äthanol n,
Alkohol m (i.e.S.), Weingeist m
  e. aminobenzoate $NH_2C_6H_4COO \cdot C_2H_5$
Aminobenzoesäureäthylester m, Äthylaminoben-
zoat n
  e. bromide CH_3CH_2Br Äthylbromid n, Bromäthan n
  e. butanoate s. e. butyrate
  e. butyrate $CH_3CH_2CH_2COO \cdot C_2H_5$ Buttersäure-
äthylester m, Äthylbutyrat n
  e. caprate $CH_3(CH_2)_8COOC_2H_5$ Kaprinsäure-
äthylester m, Dekansäureäthylester m, Äthyl-
kaprinat n
  e. caproate $CH_3(CH_2)_4COOC_2H_5$ Kapronsäure-
äthylester m, Hexansäureäthylester m, Äthyl-
kapronat n, Äthylhexylat n
  e. capronate s. e. caproate
  e. carbamate $NH_2COOC_2H_5$ Karbamidsäureäthylester m, Karbamat n, Äthylurethan n
  e. cellulose s. ethylcellulose
  e. chloride CH_3CH_2Cl Äthylchlorid n, Chloräthan n
  e. cinnamate $C_6H_5CH=CHCOOC_2H_5$ Zimtsäu-
reäthylester m, Äthylzinnamat n
  e. citrate s. triethyl citrate
  e. crotonate $CH_3 \cdot CH=CHCOOC_2H_5$ Kroton-
säureäthylester m, Äthylkrotonat n
  e. cyanide CH_3CH_2CN Äthylzyanid n, Propannitril n, Propionitril n, Propionsäurenitril n
  e. disulphide $(C_2H_5)_2S_2$ Äthyldisulfid n,
Diäthyldisulfid n, Äthyldithioäthan n
  e. dithioethane s. e. disulphide
  e. dodecanoate s. e. laurate
  e. ester Äthylester m
  e. ethanoate s. e. acetate
  e. ether $C_2H_5OC_2H_5$ Äthyläther m, Diäthyläther
m, Äther m (i.e.S.), Äthoxyäthan n
  e. fluid Äthylfluid n, TEL-Fluid n (Antiklopfmittel,
Hauptanteil Bleitetraäthyl)
  e. formate $HCOOC_2H_5$ Ameisensäureäthylester
m, Äthylformiat n
  e. gasoline Äthylbenzin n, gebleites (verbleites,
mit Bleitetraäthyl versetztes) Benzin n, Bleibenz-
zin n
  e. group C_2H_5- Äthylgruppe f, Äthylrest m
  e. halide Äthylhalogenid n, Halogenäthan n

  e. hexanoate s. e. caproate
  e. hydrogen sulphate $C_2H_5HSO_4$ Äthylhydrogensulfat n, Äthylschwefelsäure f, Schwefelsäuremonoäthylester m
  e. hydrosulphide C_2H_5SH Äthylhydrosulfid n,
Äthylmerkaptan n, Äthylthioalkohol m, Äthanthiol n, Thioäthanol n
  e. hydroxide s. e. alcohol
  e. iodide CH_3CH_2J Äthyljodid n, Jodäthan n
  e. isopropyl carbinol $(CH_3)_2CHCHOHCH_2CH_3$
Äthylisopropylkarbinol n, 2-Methylpentanol-(3) n
  e. lactate $CH_3CHOHCOOC_2H_5$ Milchsäureäthyl-
ester m, Äthyllaktat n
  e. laurate $CH_3(CH_2)_{10}COOC_2H_5$ Laurinsäure-
äthylester m, Dodekansäureäthylester m, Äthyl-
laurat n
  e. malonate $CH_2(COOC_2H_5)_2$ Malonsäure[di-
äthyl]ester m, Äthylmalonat n
  e. mercaptan s. e. hydrosulphide
  e. methyl acetylene $CH_3C \equiv CCH_2CH_3$ Äthylmethylazetylen n, Pentin-(2) n
  e. methyl ketone $CH_3COC_2H_5$ Methyläthylketon
n, Butanon-(2) n
  e. nitrate $C_2H_5NO_3$ Salpetersäureäthylester m,
Äthylnitrat n
  e. nitrite C_2H_5ONO Salpetrigsäureäthylester m,
Äthylnitrit n
  e. orthoformate $HC(OC_2H_5)_3$ Orthoameisen-
säureäthylester m
  e. oxal[o]acetate
$C_2H_5OOC \cdot C(OH)=CH \cdot COOC_2H_5$ Oxalessigsäurediäthylester m, Oxalessigester m
  e. oxide s. e. ether
  e. phenyl acrylate s. e. cinnamate
  e. phenyl ether $C_2H_5OC_6H_5$ Äthylphenyläther m,
Äthoxybenzol n, Phenetol n
  e. phenylacetate $C_6H_5CH_2COO \cdot C_2H_5$ Phenyles-
sigsäureäthylester m, Äthylphenylazetat n
  e. phosphate $(C_2H_5)_3PO_4$ Phosphorsäuretri-
äthylester m, Triäthylphosphat n
  e. propionate $CH_3CH_2COOC_2H_5$ Propionsäure-
äthylester m, Äthylpropionat n
  e. radical Äthylgruppe f, Äthylrest m, Äthylradikal
n; [freies] Äthylradikal n
  e. red Äthylrot n
  e. sebacate $C_2H_5OOC(CH_2)_8COOC_2H_5$ Sebazinsäurediäthylester m, Diäthylsebazat n
  e. silicate casting process (Ker) Äthylsilikat-
Gießverfahren n
  e. silicone Äthylsilikon n
  e. stearate $CH_3(CH_2)_{16}COOC_2H_5$ Stearinsäure-
äthylester m, Äthylstearat n
  e. succinate $C_2H_5O \cdot CO(CH_2)_2COOC_2H_5$ Bernsteinsäurediäthylester m, Diäthylsukzinat n
  e. sulphate $(C_2H_5)_2SO_4$ Schwefelsäure[di]äthyl-
ester m, Diäthylsulfat n, Äthylsulfat n
  e. sulphide s. ethylthioethane
  e. thioalcohol s. e. hydrosulphide
  e. n-valerate $CH_3(CH_2)_3COOC_2H_5$ Valerian-
säureäthylester m, Baldriansäureäthylester m,
Äthylvaler[ian]at n

ethylacetic acid $CH_3CH_2CH_2COOH$ Äthylessigsäure f, Butansäure f, n-Buttersäure f, Buttersäure f

ethylacetoacetate $CH_3COCH_2COOC_2H_5$ Äthylazetoazetat n, Azetoessigsäureäthylester m, Azetessigester m

ethylal $CH_2(OC_2H_5)_2$ Äthylal n, Diäthylformal n, Formaldehyddiäthylazetal n, Diäthoxymethan n

ethylamine $C_2H_5NH_2$ Äthylamin n, Monoäthylamin n, Aminoäthan n

ethylate / to äthylieren

ethylate Äthylat n

ethylating reagent Äthylierungsmittel n

ethylation Äthylierung f

ethylbenzene, ethylbenzol $C_2H_5C_6H_5$ Äthylbenzol n, Phenyläthan n

ethylcellulose Äthylzellulose f, EC, Triäthylzellulose f, Zelluloseäthyläther m

ethylcyclohexane $C_2H_5C_6H_{11}$ Äthylzyklohexan n

ethylene $H_2C=CH_2$ Äthen n, Äthylen n;
—CH_2CH_2— Äthylen n (Radikal)
- **e. alcohol** s. e. glycol
- **e. bridge** Äthylenbrücke f
- **e. bromide** s. e. dibromide
- **e. bromohydrin** $BrCH_2CH_2OH$ Äthylenbromhydrin n, 2-Bromäthylalkohol m, 2-Bromäthanol n
- **e. carboxylic acid** $CH_2=CHCOOH$ Äthenkarbonsäure f, Vinylkarbonsäure f, Propensäure f, Akrylsäure f
- **e. chloride** s. e. dichloride
- **e. chlorohydrin** $ClCH_2CH_2OH$ Äthylenchlorhydrin n, Glykolchlorhydrin n, 2-Chloräthylalkohol m, 2-Chloräthanol n
- **e. dibromide** $BrCH_2CH_2Br$ Äthylen[di]bromid n, 1,2-Dibromäthan n
- **e. dichloride** $ClCH_2CH_2Cl$ Äthylen[di]chlorid n, 1,2-Dichloräthan n
- **e. ethyl acrylate copolymer** Äthylen-Äthylakrylat-Kopolymerisat n
- **e. glycol** $HO \cdot CH_2CH_2 \cdot OH$ Äthylenglykol n, 1,2-Glykol n, Glykol n (i.e.S.), Äthandiol-(1,2) n
- **e. glycol terephthalate** Äthylenglykolterephthalat n
- **e. imine** s. ethyleneimine
- **e. isomerism** Äthylenisomerie f
- **e. oxide** $(CH_2)_2O$ Äthylenoxid n, Oxiran n, 1,2-Epoxyäthan n
- **e.-oxidic** äthylenoxidisch
- **e.-propylene rubber** Äthylen-Propylen-Kautschuk m, AP-Kautschuk m
- **e.-propylene terpolymer** Äthylen-Propylen-Terpolymer[es] n, APT-Kautschuk m
- **e. series** Äthylenreihe f
- **e. urea** Äthylenharnstoff m, Imidazolidon-(2) n
- **e.-vinylacetate copolymer** Äthylen-Vinylazetat-Kopolymerisat n

ethylenediamine $H_2NCH_2CH_2NH_2$ Äthylendiamin n, 1,2-Äthandiamin n
- **e. chelate** Äthylendiaminchelat n

- **e. tetraacetate** Äthylendiaminotetraazetat n
- **e. tetraacetic acid** $(HOOC \cdot CH_2)_2NCH_2CH_2N(CH_2COOH)_2$ Äthylendiamintetraessigsäure f, AeDTE, ÄDTE

ethylenedicarboxylic acid $HOOCCH_2CH_2COOH$ Äthandikarbonsäure-(1,2) f, Butandisäure f, Bernsteinsäure f, Sukzinsäure f

2,2'-ethylenedioxydiethanol $(CH_2OCH_2CH_2OH)_2$ 2,2'-Äthylendioxydiäthanol n, Triäthylenglykol n, Triglykol n

ethyleneimine C_2H_5N Äthylenimin n, Dimethylenimin n

ethylenelactic acid $HOCH_2CH_2COOH$ Äthylenmilchsäure f, 3-Hydroxypropionsäure f, Hydrakrylsäure f

ethylenic hydrocarbon Äthylenkohlenwasserstoff m, Äthylen n (i.w.S.), Alk[yl]en n, Olefin n, Monoolefin n

ethylethylene $CH_2=CHCH_2CH_3$ Äthyläthen n, Äthyläthylen n, 1-Buten n, 1-Butylen n

2-ethylhexyl alcohol $CH_3(CH_2)_3CHC_2H_5CH_2OH$ 2-Äthyl-1-hexanol n

ethylhydrogensulphate $C_2H_5OSO_3H$ Äthylhydrogensulfat n, Äthylschwefelsäure f

ethylide cyanohydrin $CH_3CH(OH)CN$ Azetaldehydzyanhydrin n, Laktonitril n, Milchsäurenitril n, 2-Hydroxypropannitril n

ethylidene $=CHCH_3$ Äthyliden...
- **e. chloride** CH_3CHCl_2 Äthylidenchlorid n, 1,1-Dichloräthan n

ethylidenelactic acid $CH_3CH(OH)COOH$ Äthylidenmilchsäure f, Milchsäure f, 2-Hydroxypropionsäure f

ethylized fuel gebleiter (verbleiter, mit Bleitetraäthyl versetzter) Kraftstoff m

ethylmethylcarbinol $CH_3CH_2CHOHCH_3$ Äthylmethylkarbinol n, sek-Butylalkohol m, sek-Butanol n, 2-Butanol n

ethylsulphonic acid $C_2H_5SO_3H$ Äthansulfonsäure f

ethylsulphuric acid $C_2H_5HSO_4$ Äthylschwefelsäure f, Äthylsulfursäure f, Äthylhydrogensulfat n, Schwefelsäuremonoäthylester m

ethylthioethane $(C_2H_5)_2S$ Äthylthioäthan n, Diäthylsulfid n, Äthylsulfid n

ethylvanillin $C_6H_3(O \cdot C_2H_5)(OCH_3)CHO$ Äthylvanillin n, 3-Methyl-4-äthylprotokatechualdehyd m, Protokatechualdehyd-4-äthyl-3-methyläther m; (fälschlich für) vanillal

ethyne $HC \equiv CH$ Äthin n, Azetylen n

ethynyl carbinol $CH \equiv CCH_2OH$ Äthinylkarbinol n, Azetylenylkarbinol n, 2-Propin-1-ol n, Propargylalkohol n
- **e. group** $—C \equiv CH$ Äthinylgruppe f, Äthinylrest m

ethynylbenzene $C_6H_5 \cdot C \equiv CH$ Äthinylbenzol n, Azetylenylbenzol n, Phenylazetylen n

etruria marl (Ker) Etruria-Mergel m

ettringite (Min) Ettringit m (Kalziumaluminiumhydroxidsulfat)

eucairite (Min) Eukairit m (Kupfersilberselenid)

eucalyptole Eukalyptol n, 1,8-Zineol n, Zineol n, 1,8-Epoxy-p-menthan n

eucalyptus gum Rotgummi *n (von Eucalyptus camaldulensis Dehnhardt)*

euchroite *(Min)* Euchroit *m (Kupfer(II)-hydroxidor-thoarsenat)*

euclase, euclasite *(Min)* Euklas *m (Aluminium-berylliumhydroxidorthosilikat)*

eucolloid Eukolloid *n*

eucrasite *(Min)* Eukrasit *m (Thoriumsilikat)*

eucryptite *(Min)* Eukryptit *m (Lithiumaluminium-silikat)*

eudalene $C_3H_7 \cdot C_{10}H_6CH_3$ Eudalin *n (Sesquiter-penderivat)*

eudesmol Eudesmol *n (bizyklisches Sesquiterpen)*

eudialite, eudialyte *(Min)* Eudialyt *m (zirkonhalti-ges Zyklosilikat)*

eudiometer Eudiometer *n*

eugenic acid *s.* eugenol

eugenol $C_6H_3(OH)(OCH_3)CH_2CH=CH_2$ Eugenol *n*, 4-Hydroxy-3-methoxy-1-allylbenzol *n*

euhedral *(Min)* idiomorph, automorph *(in der typischen Kristallform vorliegend)*

eulytine, eulytite *(Min)* Eulytin *m*, Kieselwismut *m (Wismut(III)-orthosilikat)*

euphyllite *(Min)* Euphyllit *m (Mischkristalle aus Paragonit und Muskovit)*

European alder Schwarzerle *f*, Roterle *f*, Alnus glutinosa (L.) Gaertn.

E. aspen *(Am)* Zitterpappel *f*, Aspe *f*, Espe *f*, Populus tremula L.

E. beech Rotbuche *f*, Fagus sylvatica L.

E. larch Gemeine Lärche *f*, Larix decidua Mill.

europia Eu_2O_3 Europiumoxid *n*

europium Eu Europium *n*

e. dichloride $EuCl_2$ Europiumdichlorid *n*, Euro-pium(II)-chlorid *n*

e. oxide Eu_2O_3 Europiumoxid *n*

e. sulphate $Eu_2(SO_4)_3$ Europiumsulfat *n*

e. trichloride $EuCl_3$ Europiumtrichlorid *n*, Euro-pium(III)-chlorid *n*

eusynchite *(Min)* Eusynchit *m (krustenförmiger Deskloizit)*

eutectic eutektisch

eutectic *s.* 1. e. mixture; 2. e. point

e. alloy eutektische Legierung *f*

e. brine eutektische Sole *f*

e. halt eutektischer Haltepunkt *m*

e. mixture eutektische Mischung *f*, eutektisches Gemisch *n*, Eutektikum *n*

e. point eutektischer Punkt *m*

e. temperature eutektische Temperatur *f*

eutectoid eutektoid[isch]

eutectoid Eutektoid *n*

euxenite *(Min)* Euxenit *m (Seltenerdmineral)*

eV, EV., E.V. *s.* electron-volt

EVA *s.* ethylene-vinylacetate copolymer

evacuate / **to** evakuieren, auspumpen, abpumpen, entleeren, *(i.e.S.)* luftleer machen

evacuated leer, *(i.e.S.)* luftleer

e. flask *(doppelwandiger Kolben mit evakuiertem Zwischenraum)*

evacuation Evakuieren *n*, Evakuierung *f*

evaporability Verdampfbarkeit *f*, Verdampfungs-fähigkeit *f*; Verdunstbarkeit *f*

evaporable verdampfbar, verdampfungsfähig; ver-dunstbar

evaporate / **to** 1. eindampfen, abdampfen, ein-engen, evaporieren, verdampfen lassen; ver-dunsten lassen; 2. verdampfen; *(unterhalb des normalen Siedepunktes)* verdunsten

e. to dryness zur Trockne eindampfen

evaporated black liquor *(Pap)* Dick[ab]lauge *f*, eingedickte Schwarzlauge *f*

e. latex eingedampfter (eingedickter) Latex *m*

e. milk evaporierte (eingedampfte, eingedickte, kondensierte) Milch *f*, Kondensmilch *f*

e. salt Siedesalz *n*

evaporating area *s.* evaporative area

e. basin (dish) Abdampfschale *f*

e. efficiency Verdampfungsleistung *f*

e. pan Verdampfungspfanne *f*, Eindampfpfanne *f*

e. station *(Zucker)* Verdampfstation *f*, Verdampf-anlage *f*

e. surface *s.* evaporative area

e. temperature *s.* evaporation temperature

evaporation Eindampfen *n*, Abdampfen *n*, Ein-engen *n*, Evaporieren *n*, Evaporisation *f*, Ver-dampfen *n*, Verdampfung *f*; Verdunsten *n*, Verdunstung *f*

e. coil *s.* evaporator coil

e. cooling *s.* evaporative cooling

e. enthalpy Verdampfungsenthalpie *f*

e. loss Verdampfungsverlust *m*, Verdunstungs-verlust *m*

e. number Verdunstungszahl *f*

e. rate Verdampfungsgeschwindigkeit *f*, Ver-dunstungsgeschwindigkeit *f*

e. surface *s.* evaporative area

e. temperature Verdampfungstemperatur *f*, Ver-dunstungstemperatur *f*

evaporative Verdampfungs..., Verdunstungs...

e. area Verdampfungsfläche *f*, Verdunstungs-fläche *f*

e. cooling Verdampfungskühlung *f*, Verdun-stungskühlung *f*

e. crystallization Verdampfungskristallisation *f*, Kochen *n* auf Korn

e. crystallizer Verdampf[ungs]kristallisator *m*

e. loss *s.* evaporation loss

evaporator Verdampfungsapparat *m*, Verdampf[er]apparat *m*, Eindampfapparat *m*, Verdampfer *m*, Eindampfer *m*

e. area Verdampferfläche *f*

e. coil Verdampferschlange *f*

e. crystallizer *s.* evaporative crystallizer

e. pressure Verdampferdruck *m*

e. station *s.* evaporating station

evaporimeter, evaporometer Evaporimeter *n*, Evaporometer *n*, Verdunstungsmeßgerät *n*, Ver-dunstungsmesser *m*

evasion reaction Ausweichreaktion *f*
Evelyn tube Evelyn-Röhrchen *n (ein Prüfröhrchen in der Turbidimetrie)*
even gleichmäßig, konstant
  e.-speed gleichschnell laufend, mit Gleichlauf *(Walzen)*
evidence of structure Strukturbeweis *m*, Konstitutionsbeweis *m*
evolution of gas Gasentwicklung *f*, Gasabscheidung *f*
  e. of heat Wärmeentwicklung *f*, Wärmeabgabe *f*
  e. of hydrogen Wasserstoffentwicklung *f*, Wasserstoffabscheidung *f*
evolve / to entwickeln, erzeugen, hervorrufen, abgeben, abscheiden, ausscheiden; sich entwickeln, entstehen
evolved / to be sich entwickeln
exacting *(Landw)* anspruchsvoll *(hinsichtlich der Nährstoffversorgung)*
exaltation [optische] Exaltation *f (der Molrefraktion)*
examination Prüfung *f*, Überprüfung *f*, Untersuchung *f*, Beobachtung *f*
exceed / to übersteigen
exceedingly fast *(Farb)* äußerst echt
excess überschüssig, Überschuß..., übermäßig, Über...
excess Überschuß *m*, Übermaß *n*
  in slight e. of mit einem geringen Überschuß an
  e. air überschüssige Luft *f*, Überschußluft *f*, Luftüberschuß *m*
  e. alum *(Pap)* Überschuß *m* an essigsaurer Tonerde
  e. carrier *(phys Ch)* Überschuß[ladungs]träger *m*
  e. carrier resorption *(phys Ch)* Verringerung (Verminderung) *f* der Überschußladungsträger
  e. carrier storage *(phys Ch)* Überschußträgerspeicherung *f*
  e. chlorination Über[schuß]chlorung *f*, Hochchloren *n*, Hochchlor[ier]ung *f*
  e. electron Überschußelektron *n*
  e. energy Überschußenergie *f*, Energieüberschuß *m*
  e. gas Abgas *n*
  e. hydrogen Überschußwasserstoff *m*
  e. lime process Kalküberschußverfahren *n*
  e. moisture Wasserüberschuß *m*, überschüssige Wassermenge *f*
  e. of acid Säureüberschuß *m*
  e. of air Luftüberschuß *m*
  e. of electrons Elektronenüberschuß *m*
  e. potential [elektrochemische, elektrolytische] Überspannung *f*
  e. pressure Überdruck *m*
  e. retting *(Text)* Überröste *f*
  e. rubber s. e. stock
  e. size *(Pap)* Leimüberschuß *m*, überschüssige Leimlösung *f*
  e. stock *(Gum)* Austrieb *m*
  e. water Wasserüberschuß *m*; *(Pap)* Zusatzwasser *n*

excessive überschüssig, Überschuß...; übermäßig, Über...
  e. drying zu langes Trocknen *n*, Übertrocknen *n*
  e. liming *(Gerb)* Überäscherung *f*
  e. pressure Überdruck *m*
  e. stuffing *(Gerb)* Überfettung *f*
exchange / to austauschen
exchange Austausch *m*
  e. acidity *(Bodenkunde)* Austauschazidität *f*
  e.-adsorbed *(Bodenkunde)* eingetauscht
  e. adsorption Austauschadsorption *f*
  e. capacity Austauschkapazität *f*, Austauschfähigkeit *f*, Austauschvermögen *n*, Umtauschkapazität *f*
  e. chromatography Austauschchromatografie *f*, *(i.e.S.)* Ionenaustauschchromatografie *f*
  e. column Austauschersäule *f*
  e. complex *(Bodenkunde)* Austauschkomplex *m*, Sorptionskomplex *m*
  e. degeneracy *(phys Ch)* Austauschentartung *f*
  e. fertilization Austauschdüngung *f*
  e. force Austauschkraft *f*
  e. integral *(Quantenchemie)* Austauschintegral *n*, Austauschglied *n*
  e. material *(Bodenkunde)* Austauschmaterial *n*, adsorbierender Bodenkomplex *m*
  e. of bases Basenaustausch *m*
  e. of heat Wärmeaustausch *m*, Wärmeübertragung *f*, Wärmeübergang *m*
  e. of place Platzwechsel *m*
  e. plate Austauschboden *m*, Rektifizierboden *m*
  e. possibility Austauschmöglichkeit *f*
  e. property s. e. capacity
  e. rate Austauschgeschwindigkeit *f*
  e. reaction Austauschreaktion *f*, *(i.e.S.)* Ionenaustauschreaktion *f*
  e. separation Austauschtrennung *f*
  e. substitution rate s. e. rate
  e. syntan Austauschgerbstoff *m*
  e. term s. e. integral
exchangeability Austauschbarkeit *f*, Auswechselbarkeit *f*
exchangeable austauschbar, auswechselbar
  e. calcium *(Bodenkunde)* Austauschkalk *m*
exchanger Austauscher *m*
excitability Anregbarkeit *f*; Erregbarkeit *f*
excitable anregbar; erregbar
excitation Anregung *f*; Erregung *f*
  e. energy Anregungsenergie *f*; Erregungsenergie *f*
  e. function Anregungsfunktion *f*
  e. level Anregungsniveau *n*, angeregtes Niveau *n*
  e. potential Anregungspotential *n*
  e. probability Anregungswahrscheinlichkeit *f*
  e. voltage Anregungsspannung *f*
excite / to anregen; erregen
excited atom angeregtes Atom *n*
  e. level angeregtes Niveau *n*, Anregungsniveau *n*

e. **state** angeregter Zustand *m*, Anregungs- zustand *m*

exciton *(phys Ch)* Exciton *n*, Exziton *n*

exclusion Ausschluß *m*, Abschluß *m*

e. **of air** Luftabschluß *m*

e. **principle** Ausschließungsprinzip *n*, Paulisches Ausschließungsprinzip (Prinzip) *n*, Pauli-Prinzip *n*

excorticate / to entrinden, abschälen; *(Getreide usw.)* enthülsen, schälen, putzen

excrete / to ausscheiden

excretion Exkretion *f*, Ausscheidung *f*

e. **product** Ausscheidungsprodukt *n*, Ausschei- dung *f*

excretory ausscheidend, exkretorisch, Ausschei- dungs...

exert / to (z.B. einen Druck) ausüben

exfoliate / to abblättern, aufblättern, sich abschie- fern, abbröckeln, abplatzen, abspringen

exfoliation Abblättern *n*, Aufblätterung *f*, Abschie- ferung *f*, Abbröckeln *n*, Abplatzen *n*, Abspringen *n*

exhaust / to 1. absaugen, ablassen; exhaustieren, auspumpen, luftleer machen, evakuieren; 2. erschöpfend extrahieren; *(z.B. Färbeflotte)* aus- ziehen, erschöpfen, verbrauchen

exhaust gas Abgas *n*; Auspuffgas *n*

e. **steam** Abdampf *m*; indirekter Dampf *m*

exhausted beet pulp ausgelaugte Zuckerrüben- schnitzel (Rübenschnitzel) *npl*

e. **cossettes** *s.* e. beet pulp

e. **developer** *(Foto)* erschöpfter (verbrauchter) Entwickler *m*

exhauster Saugzuglüfter *m*, Exhaustor *m*

exhausting section Abtreibeteil *m*, Abtriebsäule *f*, Abtriebteil *m*, Abstreiferzone *f*, Abstrippzone *f*

exhaustion 1. Auspumpen *n*, Evakuieren *n*, Exhaustieren *n*, Exhaustierung *f*; 2. Erschöpfung *f*, Ausnutzung *f*

exhaustive erschöpfend, durchgreifend

e. **methylation** erschöpfende Methylierung *f*

exhibit / to zeigen, aufweisen

exine *(Bakt)* Exine *f*

exinite Exinit *m (Kohlenmazeral)*

exinitic exinitisch, Exinit...

existence / of fleeting vorübergehend existenz- fähig

existent gum vorhandener (aktueller, vorgebilde- ter) Gum *m*, vorhandenes (aktuelles, vorgebilde- tes) Harz *n*, aktuelle Verharzungsprodukte *npl*, Existent Gum *m*

exit Ableitung *f*, Austritt *m*, Auslaß *m*, Auslauf *m*

e **air** Abluft *f*

e. **beam** Austrittsstrahl *m*

e. **gas** Abgas *n*

e. **gas temperature** Gasaustrittstemperatur *f*, Austrittstemperatur *f* des Gases

e. **lock** Austragschleuse *f*

e. **loss** Austrittsverlust *m*, Auslaufverlust *m*, Ausströmverlust *m*

e. **slit** Austrittsspalt *m*, Ausgangsspalt *m*

e. **threshold** Austrittsschwelle *f*

e. **tube** Saugansatz *m (einer Saugflasche)*

exocrine exokrin

e. **gland** Drüse *f* mit äußerer Sekretion

exocyclic exozyklisch

exoelectron Exoelektron *n*

exoenzyme Exoenzym *n (vom Molekülende her abbauendes Ferment)*

exoergic exoergonisch, energieabgebend, ener- giefreigebend; *s.* exothermic

exogenetic rock exogenes Gestein *n*

exogenous enclosure *(Geol)* exogener (fremder, enallogener) Einschluß *m*, Xenolith *m*

exomorphism *(Geol)* Exomorphose *f*, exomorphe Kontaktwirkung *f*

exopeptidase Exopeptidase *f (proteolytisches Ferment)*

exosmosis Exosmose *f*

exothermic exotherm[isch], wärmeabgebend, wärmeliefernd

e. **reaction** exotherme Reaktion *f*

exotoxin Exotoxin *n*

expand / to ausdehnen, expandieren, erweitern, ausweiten, entspannen; sich ausdehnen (erwei- tern, ausweiten, vergrößern), zunehmen, an- wachsen; *(Plast)* [auf]schäumen

expandability *s.* expansibility

expandable *(Plast)* schaumfähig, aufschäumbar, verschäumbar

e. **polystyrene** schaumfähiges Polystyrol *n*

e. **polystyrene bead** aufschäumbare Polystyrol- perle *f*, aufschäumbares Polystyrolkügelchen *n*

expanded clay Blähton *m*, Porensinter *m*

e. **joint** Aufweitverbindung *f (Rohrverbindung)*

e. **metal** Streckmetall *n*

e.-**metal tray** *(Destillation)* Streckmetallboden *m*

e. **plastic** Schaum[kunst]stoff *m*

e. **polystyrene** Schaumpolystyrol *n*, geschäum- tes Polystyrol *n*

e. **polyvinyl chloride** Schaumpolyvinylchlorid *n*, Schaum-PVC *n*

e. **rubber** Zellgummi *m*, poröser Gummi *m*

e. **shale** Blähschiefer *m*, Porensinter *m*

e. **sheet** geschäumte Folie *f*, Schaumstoffolie *f*

expander 1. Entspannungsmaschine *f*, Expan- sionsmaschine *f*; 2. *(Kältetechnik)* Refrigerator *m*, Verdampfer *m*; 3. *(Text)* Breithalter *m*

e. **machine** Entspannungsmaschine *f*, Expan- sionsmaschine *f*

expandible *s.* expandable

expanding adapter Übergangsstück *n (mit großer Hülse auf kleinem Kern)*

e. **agent** Treibmittel *n*, Aufblähungsmittel *n*

e. **band** Spannring *m*

e. **cement** treibender Zement *m*, Quellzement *m*, Schwellzement *m*, Expansivzement *m*

expansibility Ausdehnbarkeit *f*, Dehnbarkeit *f*, Ausdehnungsvermögen *n*, Ausdehnungsfähig- keit *f*, Dehn[ungs]fähigkeit *f*, Expansionsfähig- keit *f*

expansible [aus]dehnbar, [aus]dehnungsfähig, expansibel

expansion Ausdehnen n, Ausdehnung f, Entspannen n, Entspannung f, Expansion f, Expandieren n; Zunahme f, Anwachsen n; Treiben n, Treibneigung f; Schwellen n, Schwellvorgang m, Schwellwirkung f
  e. cloud chamber Expansionsnebelkammer f
  e. coefficient Ausdehnungskoeffizient m, Ausdehnungszahl f
  e. coil (Kältetechnik) Verdampferschlange f
  e. engine Expansionsmaschine f, Entspannungsmaschine f
  e. joint Kompensator m, Dehnungsausgleicher m, Ausgleicher m; Dehn[ungs]fuge f
  e. percentage prozentuale Ausdehnung f
  e. thermometer Ausdehnungsthermometer n
  e. trap Ausdehnungskondens[at]wasserableiter m, Dampfstauer m, Stauer m
  e. valve Expansionsventil n, Entspannungsventil n, Entspanner m

expansive cement s. expanding cement

expectancy (expectation, expected) value Erwartungswert m

expectorant Expektorans n, Expektorantium n, auswurfförderndes (schleimlösendes) Mittel n

expel / **to** (Gase) [her]austreiben, vertreiben, entfernen, verdrängen, verjagen; (Öl) abpressen, herauspressen; (radioaktive Teilchen) ausstoßen, emittieren, aussenden, ausstrahlen

expenditure of energy Energieaufwand m
  e. of work Arbeitsaufwand m

expenses of fertilizers Düngeraufwand m

experiment / **to** experimentieren, Versuche anstellen

experiment Experiment n, Versuch m, Probe f
  e. boil Versuchskochung f
  e. station Versuchsstation f, Versuchsstelle f

experimental experimentell, experimental, Experimental..., Versuchs...

experimental Experimentelles n, Methodik f (in Abhandlungen)
  e. animal Versuchstier n
  e. chemistry Experimentalchemie f
  e. error Versuchsfehler m
  e. material s. e. substance
  e. plant Versuchsanlage f
  e. reactor Versuchsreaktor m; Forschungsreaktor m
  e. research Experimentalforschung f
  e. solution Versuchslösung f, Probelösung f
  e. station Versuchsstation f, Versuchsstelle f
  e. substance Versuchssubstanz f, Probesubstanz f, Untersuchungssubstanz f
  e. value experimenteller Wert m, Versuchswert m

experimentalist s. experimenter

experimentalize / **to** experimentieren, Versuche anstellen

experimentally determined experimentell bestimmt

  e. true experimentell bewiesen

experimentation Experimentieren n

experimentator, experimenter Experimentator m

expert committee Fachausschuß m

expl. (Abk. für) explodes

explode / **to** explodieren, (i.w.S. auch) in die Luft fliegen, [zer]platzen; explodieren lassen, zur Explosion bringen

exploitation Exploitation f, Ausbeutung f, Produktion f, Förderung f
  e. well Produktionsbohrung f, Förderbohrung f

exploration Exploration f, Aufschluß m; (Bgb) Schürfung f
  e. drilling Explorationsbohrung f, Aufschlußbohrung f, Untersuchungsbohrung f, Mutungsbohrung f, Bohrung f auf Neuland (unerforschtem Boden), Wildcat[-Bohrung] f
  e. well Explorationsbohrloch n

exploratory well s. exploration well

explosibility Explosibilität f, Explosionsfähigkeit f, Explodierbarkeit f, Explosivität f, Sprengfähigkeit f

explosible explosibel, leicht explodierend, explosionsgefährlich, explosiv, explodierbar

explosion Explosion f
  e. bomb Verbrennungsbombe f, Kalorimeterbombe f, kalorimetrische Bombe f, Bombenkalorimeter n
  e. chain Explosionskette f
  e. front Explosionsfront f
  e. limit Explosionsgrenze f
  e. method Explosionsmethode f (zur Molwärmebestimmung)
  e. process (Pap) Explosionsverfahren n, Masonite-Verfahren n
  e. product Explosionsprodukt n
  e.-proof explosionsgeschützt, explosionssicher, nicht explosionsgefährdet
  e. temperature Explosionstemperatur f
  e. velocity Explosionsgeschwindigkeit f

explosive explosiv, Explosions..., Spreng...; explosibel, leicht explodierend, explosionsgefährlich, explosiv, explodierbar; explosionsartig

explosive Explosivstoff m, Sprengstoff m
  e. action Explosionswirkung f, Sprengwirkung f
  e. charge Spreng[stoff]ladung f
  e. effect s. e. action
  e. force s. e. power
  e. gelatin Sprenggelatine f
  e. limits s. e. range
  e. mixture Sprengstoffgemisch n
  e. oil Sprengöl n (Glyzerintrinitrat, Äthylenglykoldinitrat oder Gemisch beider)
  e. power Sprengkraft f
  e. properties explosive Eigenschaften fpl, Sprengstoffeigenschaften fpl
  e. range Explosionsgrenzen fpl
  e. saltpetre Sprengsalpeter m
  e. wave Stoßwelle f

explosively explosionsartig, unter Explosion

explosiveness, explosivity s. explosibility
expose / to 1. (einer Strahlung) aussetzen; 2. (Foto) belichten, exponieren
 e. to light s. to expose 2.
 e. to X-rays einer Röntgenstrahlung aussetzen
exposed thread correction (phys Ch) Korrektion f wegen des heraushängenden Fadens
exposure Aussetzen n; Ausgesetztsein n; Bestrahlen n, Bestrahlung f; (Foto) Belichten n, Belichtung f, Exposition f
 on e. to air an der Luft, unter Lufteinwirkung, bei Luftzutritt; beim Aushängen (Verhängen) an der Luft (von Geweben zur Entwicklung des Indigofarbstoffs)
 on e. to light unter Lichteinwirkung (Lichteinfluß, Einstrahlung von Licht), am (im) Licht, bei Belichtung
 e. latitude (range) (Foto) Belichtungsbreite f, Belichtungsumfang m, Belichtungsspielraum m
 e. scale (Foto) Belichtungsskale f
 e. time Einwirkungszeit f; Bestrahlungszeit f; (Foto) Belichtungszeit f, Belichtungsdauer f
 e. to gas Begasung f
 e. to light (Foto) Belichten n, Belichtung f, Exposition f
express / to auspressen, abpressen, ausdrücken
expressed fat Preßfett n
 e. oil Preßöl n
expression Auspressen n, Abpressen n, Ausdrücken n
expulsion Ausstoßen n
exsiccate / to [aus]trocknen
exsiccated alum gebrannter Alaun m
exsiccation Trocknen n, Austrocknen n, Trocknung f, Austrocknung f
exsiccator s. desiccator
extend / to [aus]dehnen; (Chemikalien durch Zusatzstoffe) strecken
extender Streck[ungs]mittel n, Verschnittmittel n, Füllmittel m, Füllstoff m, Beschwerungsmittel n, Extender m
 e. pigment Verschnittpigment n
extending filler s. extender; (Gum) inaktiver Füllstoff m
 e. oil (Gum) Extenderweichmacheröl n
extensibility Dehnbarkeit f, Ausdehnbarkeit f, Streckbarkeit f
extension Dehnung f, Ausdehnung f, Längenzunahme f
 e. at break Bruchdehnung f, Zerreißdehnung f, Reißdehnung f, Längenänderung f bei der Reißkraft (Bruchkraft)
 e. limb Ansatz m, Ansatzstück n
extensive property (quantity) extensive Eigenschaft (Größe) f, Quantitätseigenschaft f
extensometer Ausdehnungsmesser m, Dehnungsmesser m, Dehnungsmeßgerät n, Tensometer n
extent of penetration (Pap) Imprägnierungsgrad m, Durchtränkungsgrad m (der Hackschnitzel)
exterior durability Außenbeständigkeit f, Wit-

terungsbeständigkeit f, Wetterbeständigkeit f, Wetterfestigkeit f
 e. paint Außenanstrichfarbe f
 e. varnish Außenlack m, Lack m für Außenanstriche (außen)
external außenseitig, äußerer, außen befindlich, Außen...; äußerlich
 e. air classifier äußerer (angebauter) Windsichter m
 e. bactericide äußerlich anwendbares Bakterizid n
 e. chromatogram äußeres (fließendes, flüssiges) Chromatogramm n
 e. coil Außenrohrschlange f, Mantelrohrschlange f, Rohrschlangenmantel m
 e. electron Außenelektron n, äußeres Elektron n
 e.-gear pump Außenzahnradpumpe f
 e. grinding (Ker) Außenschleifen n
 e. heating Außenbeheizung f, indirekte Beheizung f, außenliegende Feuerung f
 e. ignition Fremdzündung f
 e. indicator externer (außerhalb verwendeter) Indikator m
 e. lubricant (Plast) äußeres Trennmittel n, Gleitmittel n
 e. pressure Außendruck m
 e. screen classifier äußere (angebaute) Siebvorrichtung f
 e. shell äußere Schale f, Außenschale f
 e. symmetry (Krist) äußere Symmetrie f
 e. vibrator Außenrüttler m
 e. work äußere Arbeit[sleistung] f
externally heated außenbeheizt, von außen beheizt
 e. heated oven außenbeheizter Ofen m, Heizflächenofen m, Ofen m mit Außenbeheizung (indirekter Beheizung, äußerer Wärmezufuhr, mittelbarer Wärmezufuhr, außenliegender Feuerung)
extinction Extinktion f, logarithmische Opazität f, [optische] Dichte f, (Foto auch) Schwärzung f, Deckung f
 e. coefficient Extinktionskoeffizient m, Extinktionskonstante f
 e. curve Extinktionskurve f
extinguish / to [aus]löschen, zum Erlöschen bringen
extra-cellular extrazellulär
 e.-cellular enzyme extrazelluläres Ferment n, Desmoferment n, Desmoenzym n
 e. long-oil varnish überfetter Öllack (Lack) m
 e. white hochweiß, extra weiß
extract / to extrahieren, ausziehen, (i.e.S.) auslaugen; (Bgb) abbauen, fördern, gewinnen
 e. with ether ausäthern, mit Äther ausschütteln
extract Extrakt m(n), Auszug m, Extrakt[iv]stoff m
 e. end Extraktende n, Extraktseite n
 e. evaporator Extraktverdampfer m
 e. factory Extraktfabrik f
 e. liquor Extraktbrühe f
 e. manufacture Extraktherstellung f, Extraktion f

e. **phase** Extraktphase f
e. **solvent evaporator** Verdampfer m für Extraktlösung
e. **stripper** Extraktstripper m
extractability Extrahierbarkeit f
extractable extrahierbar, extrahierfähig, ausziehbar, auslaugbar
e. **sulphur** *(Gum)* extrahierbarer Schwefel m
extractant Extraktionsmittel n
extracted coarse soybean meal Sojabohnenextraktionsschrot n(m), extrahiertes Sojaschrot n
e. **honey** Schleuderhonig m
extractibility s. extractability
extractible s. extractable
extracting liquid Extraktionsmittel n
e. **vessel** Extraktionsgefäß n, Extraktor m
extraction Extrahieren n, Extraktion f, Ausziehen n, *(i.e.S.)* Auslaugen n, Auslaugung f *(Extraktion eines Stoffes aus einer festen Substanz mit H_2O)*
e. **analysis** Extraktionsanalyse f
e. **apparatus** Extraktionsapparat m, Extrakteur m, Extraktor m
e. **battery** Extraktionsbatterie f
e. **by shaking with solvent** Ausschütteln n
e. **column** Extraktionssäule f, Extraktionskolonne f
e. **cycle** Extraktionszeit f je Charge
e. **flask** Extraktionskolben m
e. **kettle** s. e. vessel
e. **material** Extraktionsgut n
e. **naphtha** Extraktionsbenzin n
e. **plant** Extraktionsanlage f
e. **process** Extraktionsverfahren n, Extraktionsprozeß m
e. **rate** Extraktionsgeschwindigkeit f
e. **system** Extraktionssystem n
e. **temperature** Extraktionstemperatur f
e. **thimble** Extraktionshülse f
e. **tower** Extraktionsturm m
e. **vessel** Extraktionsgefäß n, Extraktor m
e. **with ether** Ausäthern n, Ausschütteln n mit Äther
extractive extraktiv, durch Extraktion erfolgend, extrahierend
extractive Extrakt[iv]stoff m, Extrakt m(n), Auszug m
e. **crystallization** extraktive Kristallisation f
e. **distillation** extraktive (extrahierende) Destillation f, Extraktivdestillation f, Distex-Prozeß m
e. **material (matter, substance)** Extrakt[iv]stoff m, Extrakt m(n), Auszug m
extractor Extraktor m, Extrakteur m, Extraktionsapparat m; *(Plast)* Hilfsvorrichtung f zum Ausdrücken, Ausdrück-Hilfsvorrichtung f; Extraktor m, Schleuder f, Zentrifuge f
extraneous fremd
e. **ash** äußere Asche f, Fremdasche f
e. **material (substance)** Fremdstoff m, fremder Stoff m, fremde Substanz f

extranuclear extranuklear, außerhalb des Kerns liegend, Hüllen...
e. **electron** Hüllenelektron n
extraordinary ray außerordentlicher (extraordinärer) Strahl m
e. **wave** *(Krist)* außerordentliche (extraordinäre) Welle f
extrasurface residues Oberflächenrückstände mpl *(von Pflanzenschutzmitteln)*
extreme [point] s. e. value
e.**-pressure lubricant** Höchstdruckschmiermittel n, Höchstdruckschmierstoff m, Extreme-Pressure-Schmiermittel n, EP-Schmiermittel n
e. **value** Extremwert m, Spitzenwert m, Extremum n
extremely insoluble sehr schwer (wenig) löslich, swl., swl., ssl.
extremum s. extreme value
extrinsic factor Extrinsic-Faktor m, Hämogen n, Vitamin B_{12} n
e. **semiconductor** Störstellen[halb]leiter m, Störhalbleiter m, Fremd[stoffhalb]leiter m, Fehlstellenhalbleiter m
extrudability *(Gum, Plast)* Spritzbarkeit f
extrudate Strangpreßerzeugnis n, Extrudat n
extrude / to extrudieren, strangpressen, spritzen, auspressen
extruded articles s. e. goods
e. **bead sealing** Siegeln n mit stranggepreßtem (gespritztem) Zusatzdraht, Extrudersiegeln n
e. **goods** Spritzartikel mpl
e. **parison** extrudierter Vorformling m *(beim Hohlkörperblasen)*, Külbel n
e. **section** stranggepreßtes (gespritztes) Profil n
e. **sheet** stranggepreßte (gespritzte) Folie (Platte) f
e. **tube** stranggepreßtes (extrudiertes) Rohr n
extruder Strangpresse f, Schnecken[strang]presse f, Spritzmaschine f, Extruder m
e. **barrel** Strangpressenzylinder m, Extruderzylinder m
e. **core** Dorn m *(einer Spritzmaschine)*, Spritzdorn m
e. **die** Extrudermundstück n, Strangpressenmundstück n, Spritzmundstück n
e. **head** Extruderkopf m, Strangpressenkopf m, Strangpreßkopf m, Spritzkopf m
extruding Extrudieren n, Auspressen n, Strangpressen n, Spritzen n
e. **machine** s. extruder
extrusion Extrudieren n, Auspressen n, Strangpressen n, Spritzen n; Spinnen (Erspinnen) n aus der Schmelze; *(Geol)* Extrusion f
e. **auger** s. extruder
e. **blowing** Extrusionsblasformen n, Extrusionsblasen n
e. **coating** Beschichten n mittels Extruders, Beschichten n über Schneckenpresse, Kaschieren n mit stranggepreßter Folie
e. **compound** Strangpreßmischung f

e. die Extrudermundstück *n*, Strangpressenmundstück *n*, Spritzmundstück *n*
e. head Extruderkopf *m*, Strangpressenkopf *m*,
Strangpreßkopf *m*, Spritzkopf *m*
e. moulding Strangpressen *n*, Spritzen *n*
e. press *s.* extruder
e. properties Spritzeigenschaften *fpl*, Spritzbarkeit *f*
e. ratio Reckverhältnis *n* beim Extrudieren
e. speed Spritzgeschwindigkeit *f*
e. tube Preßkopf *m*
extrusive rock Extrusivgestein *n*, Ergußgestein *n*
exudate / to *s.* to exude
exudate Exsudat *n*, Ausscheidungsprodukt *n*,
Ausscheidung *f*, Ausschwitzungsprodukt *n*
exudation Ausscheiden *n*, Absondern *n*, Ausschwitzen *n*, Austreten *n*, Exsudation *f*; Exsudat
n, Ausscheidungsprodukt *n*, Ausscheidung *f*,
Ausschwitzungsprodukt *n*; *(Gerb)* Blume *f*,
Lederausschlag *m*, Mud *m (vorwiegend aus
Ellagsäure bestehende Abscheidungen der
Ellagengerbstoffe)*
exude / to ausscheiden, absondern, *(Harze)*
ausschwitzen; sich ausscheiden (absondern),
austreten
exuding Ausscheiden *n*, Absondern *n*, Ausschwitzen *n*, Austreten *n*, Exsudation *f*
eye drops Augentropfen *pl*
e.-make-up pencil *s.* eyebrow pencil
e. ointment Augensalbe *f*
eyebrow pencil Augenbrauenstift *m*
eyepiece Okular *n (des Mikroskops)*
eyeshadow Augenschatten *m*, Lidschatten *m*,
Augenschattenschminke *f*
eyewash Augenwasser *n*
eyrite *(Min)* Eyrit *m*

F

f. s. face-centred
F *s.* 1. Fahrenheit; 2. faraday; 3. magnetomotive
force
F acid $C_{10}H_6(OH)(SO_3H)$ β-Naphtholsulfonsäure *f*,
F-Säure *f*, 2-Naphthol-7-sulfonsäure *f*
F-layer *(Bodenkunde)* Vermoderungshorizont *m*
F1 method F-1-Methode *f*, Research-Methode *f*,
Research-Verfahren *n (zur Oktanzahlbestim-
mung)*
F2 method F-2-Methode *f*, Motormethode *f*,
Motorverfahren *n (zur Oktanzahlbestimmung)*
F3 method F-3-Methode *f*, Fliegermethode *f (zur
Oktanzahlbestimmung)*
F4 method F-4-Methode *f*, Überlademethode *f (zur
Oktanzahlbestimmung)*
fabric [cloth] Textilgewebe *n*, Fasergewebe *n*,
[textiler] Stoff *m*, Gewebe *n*
f.-filled moulding compound (material) Ge-
webeschnitzelpreßmasse *f*, Gewebeschnitzel-
preßstoff *m*

f. filter Tuchfilter *n*, Stoffilter *n*, Gewebefilter *n*,
Gewebeabscheider *m*
f. filter medium Filtergewebe *n*, Filtertuch *n*,
Filtermedium *n* aus faserigen Stoffen
f.-free gewebefrei
f. press *(Pap)* Fabric-Presse *f*, Siebtuchpresse *f*
fabrication stress Bearbeitungsspannung *f*
face Fläche *f*, Oberfläche *f*; Kristallfläche *f*;
(Strukturlehre) Seite *f (eines Ringes)*
f.-centred *(Krist)* flächenzentriert, fl.z.
f.-centred cube *(Krist)* flächenzentrierter Elementarwürfel *m*
f.-centred cubic *(Krist)* kubisch-flächenzentriert
f.-centred cubic lattice (structure, system)
kubisch-flächenzentriertes Gitter *n*, flächenzentriertes kubisches Gitter *n*, flächenzentriertes
Würfelgitter *n*
f.-centred grating (lattice) flächenzentriertes
Gitter *n*
f. centreing *(Krist)* Flächenzentrierung *f*
f. mask Gesichtsmaske *f (Atemschutzgerät)*
f. of the dryer *(Pap)* Trockenzylinderoberfläche
f
f. of the roll Walzenoberfläche *f*
f. pack *(Kosmet)* Gesichtspackung *f*
f. powder *(Kosmet)* Gesichtspuder *m*
f. protector Gesichtsschutzschale *f*, Gesichts-
schutz *m*
f. tonic *(Kosmet)* Gesichtswasser *n*
facial tissue [paper] Abschminkpapier *n*, Lippentupfpapier *n*
facies *(Geol)* Fazies *f*
facing brick Verblender *m*, Verblendziegel *m*,
Verblendstein *m*
f. paper Bezugspapier *n*
factice, factis *(Gum)* Faktis *m*
factitious künstlich, nachgemacht, unecht
factor Faktor *m*, Koeffizient *m*, Beiwert *m*
factory cullet *(Glas)* [fabrik]eigene Scherben *fpl*
facultative anaerobes (anaerobic bacteria) fa-
kultative Anaerobier (Anaerobionten) *mpl*
FAD *s.* flavin adenine dinucleotide
fade / to verschießen, ausbleichen, verbleichen,
verblassen, sich verfärben, *(in der Farbe)*
schwächer werden, verwaschen
f. away verschwinden; abklingen
fade covering Abdecken *n* von verschossenen
Farbpartien
fadeless lichtecht, farbecht, nicht verschießend
(verbleichend, ausbleichend, verblassend)
fademeter *(Text)* Fadeometer *n*, Fade-Ometer *n*,
Farbechtheitsprüfer *m*, Farbechtheitsmesser *m*,
Lichtechtheitsmesser *m*
fading Verschießen *n*, Ausbleichen *n*, Verbleichen
n, Verblassen *n*, Verfärben *n*, Verfärbung *f*
fadometer *s.* fadeometer
faecal matter, faeces Fäkalien *pl*, Fäzes *pl*, Kot *m*
Fagergren floatation machine Fagergren-Zelle *f*
(ein Flotationsapparat)
fagopyrine Fagopyrin *n (roter Farbstoff in Buch-
weizenblüten)*

fast

Fah s. Fahrenheit
fahlerz, fahlore (Min) Fahlerz n, (i.e.S.) Antimonfahlerz n, Tetraedrit m
fahlunite (Min) Falunit m
Fahr s. Fahrenheit
Fahrenheit 1. Fahrenheitsgrad m, Grad Fahrenheit m, °F; 2. Fahrenheitskale f
Fahrenheit degree s. Fahrenheit 1.
Fahrenheit scale s. Fahrenheit 2.
faience Fayence f
  f. tile Fayencefliese f, Fayenceplatte f
  f. ware Fayenceware f
failure of the reciprocity law Abweichung f vom Reziprozitätsgesetz, Schwarzschild-Effekt m
faint schwach
  f.-blue schwachblau
faintly acid schwach sauer
  f. defined schwach ausgeprägt, schwer erkennbar
  f. luminous schwach leuchtend
Fair process (Pap) Fair-Verfahren n (Zusatz der Füllstoffsuspension zum Faserstoff unmittelbar auf dem Sieb der Papiermaschine)
fair yield brauchbare Ausbeute f
fairfieldite (Min) Fairfieldit m (Kalziummangan(II)-orthophosphat)
fall / to [ab]fallen, [ab]sinken, abnehmen, sich vermindern (verringern), zurückgehen; (Gerb) verfallen (Blöße); (eine Blöße) verfallen machen
fall Fallen n, Abfallen n, Abfall m, Sinken n, Absinken n, Abnehmen n, Abnahme f, Rückgang m, Minderung f, Verminderung f
  f. of potential Potentialabfall m
  f.-out chamber Abscheidekammer f, Staubkammer f; Kammerabscheider m
falling (Gerb) Verfallen n (der Blöße)
  f.-ball impact test Kugelfallprüfung f
  f.-ball method Kugelfallmethode f
  f.-ball test s. f.-ball impact test
  f.-ball viscosimeter Kugelfallviskosimeter n
  f.-drop method s. f.-ball method
  f. film fallender Film m (Molekulardestillation)
  f.-film concentrator Fallfilmschwefelsäurekonzentrierer m
  f.-film distillation Destillation f mit fallendem Film
  f.-film evaporator Fallfilmverdampfer m, Fallstromverdampfer m
  f.-film still Fallfilmkolonne f; Destillierapparat m mit fallendem Film
  f.-on (Text) Überfall m, Überdeckung f, Überdruck m
  f.-rate drying period Abschnitt m abnehmender Trocknungsgeschwindigkeit
  f.-sphere method s. f.-ball method
  f.-sphere viscosimeter s. f.-ball viscosimeter
fallout Fallout n (das Absetzen fester Sedimente, z.B. radioaktiver Asche, aus Aerosolen); Fallout m, (atmosphärischer) radioaktiver Niederschlag m (als festes Sediment)
false body (Plast) scheinbarer Körpergehalt m

  f. bottom Zwischenboden m, Blindboden m; Läuterboden m
  f. grain (Zucker) Feinkorn n
  f. perforated bottom Siebboden m
  f. topaz (Min) Zitrin m
  f. twist (Text) falscher Draht m
  f.-twist crimping (Text) Falschdraht-Texturierverfahren n
  f.-twist process s. f. twisting
  f. twisting (Text) Falschdrahtverfahren n
famatinite (Min) Famatinit m (Antimon(III)-kupfer-(I,II)-sulfid)
family Familie f, Gruppe f, Verwandschaftsgruppe f (im Periodensystem); Verbindungsklasse f, Stoffklasse f, Reihe f
  f. mould (Plast) Mehrfachwerkzeug n
  f. of hydrocarbons Kohlenwasserstoffreihe f
fan [down] / to fächern (Papiersortierung)
fan Ventilator m, Lüfter m, Gebläse n; (Plast) Austrieb m, Preßgrat m, Formgrat m, Grat m
  f. gate (Plast) fächerförmiger Anschnitt m
  f. impeller type collector Spiral[ab]scheider m, Gebläseentstauber m
  f. nozzle Fächerdüse f, Pralldüse f, Zungendüse f
fancy board Ausstattungskarton m
  f. paper Fantasiepapier n; Ausstattungspapier n
  f. stained paper Effektpapier n
fango (Med) Fango m (ein Heilschlamm)
fantail firing U-Feuerung f (eine Kohlenstaubfeuerung)
far-infrared radiation element Infrarot-Dunkelstrahler m
faraday Faraday n, F
Faraday effect Faraday-Effekt m, Magnetorotation f, magnetische Drehung f [der Polarisationsebene], magnetisches Dreh[ungs]vermögen n
Faraday law Faradaysches Gesetz n
Faraday-Tyndall phenomenon Faraday-Tyndall-Effekt m, Tyndall-Effekt m, Tyndall-Phänomen n
farina feines Mehl n; Kartoffelstärke f, Kartoffelmehl n, Stärkemehl n
farinaceous mehlig, mehlartig; mehlhaltig, mehlreich; stärkehaltig, stärkereich
farinha Kassavastärke f
farinose s. farinaceous
farm manure Mist m, Dung m, Dünger m, Wirtschaftsdünger m
Farmer's reducer (Foto) Farmerscher Abschwächer m
fast accelerator (Gum) schnellwirkender (schneller, starker) Beschleuniger m
  f. base s. f. colour base
  f. breeder reactor (Kch) schneller Brutreaktor m
  f. coagulation rasche Koagulation f
  f. colour s. f. dye
  f. colour base Echt[färbe]base f, Echtfarbbase f
  f. colour salt Echt[färbe]salz n
  f.-curing (Gum) rasch (schnell) vulkanisierend, schnellvulkanisierend, rasch (schnell) heizend

f. curing accelerator s. f. accelerator
f. drying Schnelltrocknen n, Schnelltrocknung f
f. dye Echtfarbstoff m, echter Farbstoff m
f.-dyed echtfarbig, echtgefärbt
f. dyeing Echtfärben n, Echtfärberei f
f. effect (Kch) Schnellspaltungseffekt m
f. emulsion (Foto) hochempfindliche Emulsion f
f. evaporating schnellflüchtig, leichtflüchtig
f. extrusion furnace black (Gum) Fast-Extrusion-Furnace-Ruß m, FEF-Ruß m (Ölruß, der die Kautschukmischung gut verformbar macht)
f. filter Schnellfilter n
f. freezing schnelles Gefrieren (Einfrieren) n, Schnellgefrierverfahren n, Schnellgefrieren n
f. neutron (Kch) schnelles Neutron n
f. neutron reactor s. f. reactor
f. nitro dye Nitroechtfarbstoff m
f. orange Echtorange n
f. orange base Echtorangebase f
f. orange R base Echtorange-R-Base f
f. pink Echtrosa n
f. pulping (Pap) Schnellaufschluß m
f. reactor (Kch) schneller Reaktor m
f. red Echtrot n
f. red A Echtrot A n, Rokzellin n
f. red B Echtrot B n, Bordeaux B n
f. red base Echtrotbase f
f. red GL base Echtrot-GL-Base f
f. running machine Schnelläufer m
f. salt s. f. colour salt
f. scarlet Echtscharlach m
f. scarlet base Echtscharlachbase f
f. scarlet G base Echtscharlach-G-Base f
f. setting schnellbindend (z. B. Zement); schnellhärtend (z. B. Plast)
f. solvent schnellflüchtiges (leichtflüchtiges) Lösungsmittel n
f. stock (stuff) (Pap) röscher (wasserlässiger) Stoff m
f. to acid säurefest, säurebeständig, beständig gegen Säuren; (Farb) säureecht
f. to alkali alkalifest, alkalibeständig, beständig gegen Alkalien; (Farb) alkaliecht
f. to bleaching bleichecht
f. to boiling kochecht, kochfest, kochbeständig
f. to chlorine chlorecht
f. to cross-dyeing überfärbeecht
f. to hydrogen peroxide hydrogenperoxidecht
f. to ironing bügelecht
f. to light lichtecht, lichtbeständig
f. to lime kalkecht
f. to oil ölfest, ölbeständig
f. to perspiration schweißecht
f. to soaping seifecht
f. to solvents lösungsmittelbeständig
f. to topping s. f. to cross-dyeing
f. to washing waschecht, waschbeständig
f. to water wasserecht
fastness Festigkeit f, Haltbarkeit f, Beständigkeit f,

Widerstandsfähigkeit f; Echtheit f (z.B. bei Farben)
f. of colour Farbechtheit f
f. test Echtheitsprüfung f
f. to bleaching Bleichechtheit f
f. to boiling Kochechtheit f, Kochfestigkeit f, Kochbeständigkeit f
f. to burnt gas fumes Abgasechtheit f, Rauchgasechtheit f, Gasechtheit f, Abgasbeständigkeit f
f. to chlorinated water Beständigkeit f gegen gechlortes Wasser
f. to chlorine Chlorechtheit f
f. to cold washing Kaltwaschechtheit f
f. to cross-dyeing Überfärbeechtheit f, Säurekochechtheit f
f. to decatizing Dekatierechtheit f, Dekaturechtheit f
f. to dry cleaning Trockenreinigungsechtheit f
f. to dry rubbing Trockenreibechtheit f
f. to gas fading s. f. to burnt gas fumes
f. to hypochlorite bleaching Hypochloritbleichechtheit f
f. to ironing Bügelechtheit f
f. to kier-boiling Beuchechtheit f
f. to light Lichtechtheit f, Lichtbeständigkeit f
f. to lime Kalkechtheit f (von Farben und Farbstoffen)
f. to mercerizing Merzerisierechtheit f
f. to milling Walkechtheit f
f. to perspiration Schweißechtheit f
f. to processing Fabrikationsechtheit f
f. to rain Regenechtheit f
f. to refinishing Überlackierechtheit f
f. to rubbing Reibechtheit f (von Farben)
f. to soaping Seifechtheit f
f. to topping s. f. to cross-dyeing
f. to washing Waschechtheit f, Waschbeständigkeit f
f. to water Wasserechtheit f
f. to wearing Gebrauchsechtheit f, Tragechtheit f
f. to wet rubbing Naßreibechtheit f
f. to wetting Naßechtheit f
fat fett[ig], fetthaltig, ölhaltig, Fett..., Öl...; fett, fruchtbar (Boden)
fat Fett n
f. acid höhere Fettsäure f (mit 12 bis 24 C-Atomen im Molekül)
f. blend Fettmischung f, Fettansatz m, Fettkomposition f (z.B. zur Margarineherstellung), (i.e.S.) Margarinerezeptur f
f. bloom Fettreif m (dünner Belag aus Kakaobutterkristallen auf Schokolade)
f. cell Fettzelle f
f. cheese Fettkäse m
f. chemistry Fettchemie f
f. coal Fettkohle f, fette Kohle f
f. coal, properly so called eigentliche fette Kohle f, Schmiedekohle f

257 feculent

f. content Fettgehalt *m*
f. crystal Fettkristall *m*
f. deposition Fettablagerung *f*
f.-dissolving soap Fettlöserseife *f*
f.-dissolving washing agent Fettlöserwasch-mittel *n*
f. emulsion Fettemulsion *f*
f.-free fettfrei
f.-free milk entrahmte Milch *f*, Magermilch *f*
f. gland Talgdrüse *f*
f. globule Fettkügelchen *n*, Fetttröpfchen *n*
f. globule membrane Fettkügelchenmembran *f*, Fettkügelchenhülle *f*
f. globule protein Fettkügelchenprotein *n*
f. hardening Fetthärtung *f*, Fetthydrierung *f*
f. hasher Rohfettzerkleinerungsmaschine *f*
f. hydrolysis Fetthydrolyse *f*, Hydrolyse *f* der Fette
f.-in-water emulsion Fett-in-Wasser-Emulsion *f*
f. interceptor *s.* f. trap
f.-liking fettliebend, lipophil, sich mit Fett mischend, in Fett löslich
f. lime Fettkalk *m*, Weißkalk *m*
f.-liquor / to (Gerb) fettlickern
f. liquor (Gerb) Fettlicker *m*, Fettbrühe *f*, Licker *m*
f.-liquoring (Gerb) Fettlickern *n*
f., long-flame coal langflammige Fettkohle *f*, fette Kohle (Steinkohle) *f* mit langer Flamme
f. loss Fettverlust *m*
f.-loving *s.* f.-liking
f. melter (melting kettle) Fettkocher *m*
f. metabolism Fettstoffwechsel *m*
f. of bones Knochenfett *n*
f. of wool Wollfett *n*
f. oil fettes Öl *n*, Fettöl *n*
f. particle Fetteilchen *n*
f. resistance Fettdichtigkeit *f*, Fettundurch-lässigkeit *f*
f. sample Fettprobe *f*
f., short-flame coal kurzflammige Fettkohle *f*, fette Kohle (Steinkohle) *f* mit kurzer Flamme
f.-soluble fettlöslich
f. solvent Fettlösungsmittel *n*, Fettlöser *m*
f.-splitting fettspaltend, lipolytisch
f. splitting Fettspaltung *f*, Lipolyse *f*
f. spoilage Fettverderb *m*
f. spue (Gerb) Fettflecke[n] *mpl* (Lederfehler)
f. synthesis Fettsynthese *f*
f. trap Fettfang *m*, Fettfänger *m*, Fettabscheider *m*
fatal dose letale (tödliche) Dosis *f*, Tödlichkeits-dosis *f*, dosis letalis, LD
fatigue / to (Tech) ermüden
fatigue (Tech) Ermüdung[serscheinung] *f*
f. endurance Ermüdungs-Dauerfestigkeit *f*
f. failure Ausfall *m* (Versagen *n*) durch Ermüdung
f. limit Ermüdungsgrenze *f*; Dauer[schwing]-festigkeit *f*
f.-preventing agent Ermüdungsschutzmittel *n*
f. resistance Ermüdungsbeständigkeit *f*, Er-müdungswiderstand *m*

f. resisting ermüdungsbeständig
f. strength Ermüdungsfestigkeit *f*
fatless fettlos, ohne Fett, mager
fatlike fettartig, fettähnlich
fatliquor / to *s.* to fat-liquor
fatliquor *s.* fat liquor
fatliquoring *s.* fat-liquoring
fatty fettig, fett[haltig]; fettartig, fettähnlich; Fett...
f. acid Fettsäure *f*
f. acid composition Fettsäurezusammensetzung *f*
f. acid ester Fettsäureester *m*
f. acid ester of sugar Zucker-Fettsäureester *m*
f. acid pitch Fettpech *n*
f. acid radical Fettsäurerest *m*
f. acid salt fettsaures Salz *n*
f. acid spew *s.* f. spew
f. alcohol Fettalkohol *m*
f. alkyl sulphate Fettalkoholsulfat *n*, Fett-alkylsulfat *n*
f. alkylolamide Fettsäurealkylolamid *n*
f. amide Fettsäureamid *n*
f. amine Fettsäureamin *n*
f. gas Fettgas *n*, Ölgas *n*
f. matter Fettstoff *m*, Fettsubstanz *f*
f. oil fettes Öl *n*, Fettöl *n*
f. phase Fettphase *f*
f. series Fettreihe *f*
f. spew (Gerb) Fettausschlag *m*
Faugeron kiln Faugeron-Ofen *m* (kohlenbeheizter Tunnelofen nach Faugeron)
faujasite (Min) Faujasit *m*
fault Mangel *m*, Fehler *m*, Defekt *m*; (Geol) Verwerfung *f*
f. trap (Geol) Verwerfungsfalle *f*
faulty fehlerhaft, defekt
f. dyeing Fehlfärbung *f*
f. fermentation Fehlgärung *f*
Fauser [ammonia] process Fausersche Am-moniaksynthese *f*, Fauser-Verfahren *n*
Faworskii rearrangement Faworski-Umlagerung *f* (von α-Halogenketonen in Karbonsäuren)
fayalite (Min) Fayalit *m* (Eisen(II)-orthosilikat)
F.B.P. *s.* final boiling point
FC *s.* fixed carbon
F.D.A. (Abk. für) Food and Drug Administration
feasible [technisch] durchführbar
feather ore (Min) Federerz *n*, Bleiantimonspieß-glanz *m*, Jamesonit *m* (Blei(II)-antimon(III)-sulfid)
feathering (Ker) Federbildung *f* (Glasurfehler)
f. of writing ink (Pap) Auslaufen *n* der Tinte
featherweight paper Dickdruckpapier *n*, Fe-derleichtpapier *n*, Daunendruckpapier *n*
feathery fed[e]rig, schuppenförmig
febrifuge Fiebermittel *n*, Antipyretikum *n*
febrifugine Febrifugin *n* (wirksames Alkaloid gegen Malaria)
feculence Trübung *f*, Trübe *f*; Bodensatz *m*, Satz *m*
feculent trüb[e], getrübt, unrein, undurchsichtig, verunreinigt, schlammig

17 Chemie, E-D

Feder solution for aldehydes Feders Aldehydreagens n
feebly acid[ic] schwach sauer
f. basic schwach basisch
f. caking coal schwach backende Kohle f
f. coloured schwach [an]gefärbt, farbschwach, schwachfarbig
feed / to *(Material)* zuführen, zuspeisen, einspeisen, aufgeben, eintragen, einfüllen; *(z.B.* einen Ofen) speisen, beschicken, füllen; *(Gerb)* füllen, nachbessern
f. in einspeisen, eintragen, einfüllen; *(Gerb)* füllen, nachbessern
f. on zuführen
feed 1. Zuführen n, Zufuhr f, Zuführung f, Zuspeisen n, Einspeisen n, Aufgeben n, Aufgabe f, Eingeben n, Eingabe f, Beschicken n, Beschickung f, Eintragen n, Eintrag m, Einfüllen n, Einsetzen n, Einsatz m, Zulauf m; 2. Charge f, Beschickungsmaterial n, Beschickung f, Füllung f, Füllmasse f, Schüttung f, Einsatz m, Einsatzgut n, Einsatzstoff m, Einsatzmaterial n, Einsatzprodukt n, Eintrag m, Ansatz m, Satz m, Aufgabegut n, Speisung f, Einspeisung f, Gicht f *(des Hochofens)*; 3. Futter[mittel] n
f. barley Futtergerste f
f.-belt weigher Bandwaage f
f. box Eintragkasten m, Aufgabekasten m; Eingußkasten m; Zulaufgefäß n
f. bush *(Plast)* Angußbuchse f
f. chute Aufgabeschurre f, Beschickungsschurre f, Aufgaberinne f, Beschickungsrinne f, Aufgaberutsche f, Beschickungsrutsche f
f. coal Aufgabekohle f, Einsatzkohle f, aufgegebene (eingesetzte) Kohle f
f. composition *(Destillation)* Zulaufzusammensetzung f
f. end Aufgabeende n, Aufgabeseite f, Eintragseite f, Zulaufseite f, Beschickungsseite f
f. funnel s. f. hopper
f. gate Aufgabeschieber m, Zulaufschieber m
f. heater Vorwärmer m
f. hole s. f. inlet
f. hopper Aufgabetrichter m, Fülltrichter m, Füllrumpf m, Einfülltrichter m, Schütttrichter m, Beschickungstrichter m, Beschickungsbunker m
f. inlet Aufgabeöffnung f, Einwurföffnung f, Eintragöffnung f, Beschickungsöffnung f, Beladeöffnung f, Füllöffnung f, Fülloch n; *(Filtration)* Trübezulauf m, Suspensionszulauf m
f. launder Aufgaberinne f; Trübeaufgaberinne f *(bei der Aufbereitung)*
f. lime Futterkalk m
f. liquor *(Destillation)* Ausgangsflüssigkeit f
f. mechanism Dosierapparat m, Aufgabeapparat m, Zubringervorrichtung f
f. of the stock *(Pap)* Stoffzuführung f
f. opening Brechmaul n, Brechmaulöffnung f
f. orifice *(Plast)* Anschnitt m
f. pipe Zulaufrohr n, Eintragrohr n, Beschickungsrohr n

f. plate s. f. tray
f. preservation Futterkonservierung f
f. pulp s. f. slurry
f. rate Eintragmenge f in der Zeiteinheit; Durchsatz m
f. roll Aufgabewalze f, Auftrag[e]walze f, Auftragswalze f; Transportwalze f
f. section s. f. zone
f. slurry Aufgabegut n, [zulaufende] Trübe f *(z.B. beim Zentrifugieren)*
f.-splitter box Eintragverteilerkasten m, Einlaufverteilerkasten m
f. steamer Futterdämpfer m
f. stock Aufgabegut n
f.-stock coal s. f. coal
f. stream *(Destillation)* Zulauf m, Zufluß m
f. supplement Futterzusatz m
f. tank Aufgabebehälter m, Zulaufbehälter m, Beschickungsbehälter m, Speisebehälter m, Aufgabetrog m
f. throat Fülltrichter-Auslaufstutzen m
f. tray *(Destillation)* Zulaufboden m, Speiseboden m, Zuflußboden m, Aufgabeboden m
f. tube s. f. pipe
f. value Futterwert m
f. vapours dampfförmiger Öleinsatz m *(beim Thermofor-Catalytic-Cracking)*
f. water Speisewasser n
f.-water heater Speisewasservorwärmer m
f. well Eintragzylinder m *(in Form eines Tauchrohrs)*
f. wheat Futterweizen m
f. yeast Futterhefe f
f. zone *(Plast)* Beschickungszone f, Speisezone f
feedback 1. Rückführung f; 2. s. f. inhibition
f. control Regeln n, Regelung f
f. control system Regel[ungs]system n, Regelung f
f. inhibition *(Bioch)* Endprodukthemmung f, Rückkoppelungshemmung f
feeder Zuteileinrichtung f, Zuteiler m, Aufgabevorrichtung f, Aufgeber m, Eintragvorrichtung f, Beschickungsvorrichtung f, Einspeisevorrichtung f, Speiser m, Feeder m
f. bowl *(Glas)* Speiserbecken n
f. channel *(Glas)* Speiserkanal m, Speiserrinne f
f. gate *(Glas)* Absperrstein m des Speisers
f. nose s. f. bowl
f. plug s. f. gate
f. process *(Glas)* Speiserverfahren n, Feederverfahren n
f. sleeve *(Glas)* Speiserrohr n
f. spout s. f. bowl
f. tube s. f. sleeve
feeding s. feed 1.
f. funnel (hopper) s. feed hopper
f. roll s. feed roll
f. stuff s. feed 3.
f. sugar (syrup) Futterzucker m

f. trough Beschickertrog *m*
feedingstuff *s.* feed 3.
feedstock *s.* feed 2.
feedstuff *s.* feed 3.
feedwater *s.* feed water
feel *(Pap, Text)* Griff *m*, Griffigkeit *f*
FEF black *s.* fast extrusion furnace black
Fehling solution, Fehling's reagent Fehlingsche Lösung *f*
Feigl micro-reaction Feiglsche Reaktion *f*
Feist acid Feistsche Säure *f*
feldspar *(Min)* Feldspat *m*
feldspathic *(Min)* feldspathaltig, Feldspat...
feldspathoid *(Min)* Feldspatoid *m*, Feldspatvertreter *m*
felite Felit *m* *(Kristallart im Portlandzementklinker)*
fellmongered wool Hautwolle *f*
fellmongering [treatment] Entwollungsverfahren *n*
felsite *(Min)* Felsit *m*, *(veraltet für)* Orthoklas *m* *(Kaliumalumotrisilikat)*
felspar *s.* feldspar
felt / to filzen, zu Filz machen; filzen, sich verfilzen; mit Filz überziehen (auskleiden)
  f. together *(Pap)* filzen, sich verfilzen
felt Filz *m*; *(Pap)* *s.* f. blanket
  f. blanket *(Pap)* Filztuch *n*, Filz *m*
  f. brown Abdeckpappe *f*
  f. calender Filzkalander *m*
  f.-carrying roll *(Pap)* Filzleitwalze *f*
  f. changing *(Pap)* Filzwechsel *m*
  f. cleaner *(Pap)* Filzwäscher *m*, Filzwäsche *f*, Filzwascheinrichtung *f*
  f. cleaning *(Pap)* Filzreinigung *f*, Filzwäsche *f*
  f. conditioner *(Pap)* Filzinstandhalter *m*
  f. conditioning *(Pap)* Filzinstandhaltung *f*
  f.-covered couch roll *(Pap)* Filzwalze *f*
  f. dryer (drying cylinder) *(Pap)* Filztrockenzylinder *m*, Filztrockner *m*
  f.-fabric filter Faser[vlies]filter *n*
  f. filter *s.* f.-fabric filter
  f. gasket Filzdichtung *f*
  f. guide (leading) roll *s.* f.-carrying roll
  f. life *(Pap)* Laufzeit *f* des Filzes, Filzlaufzeit *f*, Filzlaufdauer *f*
  f.-like filzartig, filzig
  f. mark Filzmarkierung *f*, Filzmarke *f* *(Papierfehler)*
  f. packing Filzdichtung *f*, Filzpackung *f*
  f. roll *s.* f.-carrying roll
  f. run *(Pap)* Filztrum *m(n)*, Filzstrang *m*, *(i.e.S.)* vorlaufender Filztrum *m*
  f. seal *s.* f. packing
  f. side Filzseite *f*, Oberseite *f*, Schönseite *f* *(des Papiers)*
  f. sleeve *(Gerb)* Filzärmel *m* *(der Abwelkpresse)*
  f. stretching roll *(Pap)* Filzspannwalze *f*, Spannwalze *f*
  f. suction box *(Pap)* Filzsauger *m*
  f. tightener roll *s.* f. stretching roll
  f. travel *(Pap)* Filzlauf *m*

  f. washer *s.* f. cleaner
felting Filzen *n*, Filzbildung *f*, Verfilzen *n*, Verfilzung *f*
  f. of the fibres *(Pap)* Faserverfilzung *f*
  f. power Verfilzbarkeit *f*, Verfilzungsvermögen *n*, Filzvermögen *n*, Filzfähigkeit *f*, Filzneigung *f*
  f. property Filzeigenschaft *f*
  f. shrinkage Filzschrumpfung *f*
feltless *(Pap)* filzlos
female form (mould) *(Plast)* Gesenk *n*, Matrize *f*
fen peat Nieder[ungs]moortorf *m*
fenchol *s.* fenchyl alcohol
fenchyl alcohol 2-Fenchylalkohol *m*, Fenchol *n*, 1,3,3-Trimethylbizyklo-[1,2,2]-2-heptanol *n*
fennel Fenchel *m*, Foeniculum vulgare Mill.
  f.[-seed] oil Fenchelöl *n*
fenuron $C_6H_5NHCON(CH_3)_2$ Fenuron *n*, 1,1-Dimethyl-3-phenylharnstoff *m* *(Herbizid)*
FEP *s.* fluorinated ethylene-propylene resin
ferberite *(Min)* Ferberit *m* *(Eisen(II)-wolframat)*
fergusonite *(Min)* Fergusonit *m*
Fermat's principle Fermatsches Prinzip (Gesetz) *n*
ferment / to [ver]gären; fermentieren
ferment Ferment *n*, Enzym *n*
  f. of rennet Lab[ferment] *n*, Chymosin *n*, Chymase *f*, Rennin *n*
  f. system Fermentsystem *n*, Enzymsystem *n*
fermentability Vergärbarkeit *f*, Gär[ungs]fähigkeit *f*
fermentable vergärbar, gär[ungs]fähig
fermentation Gären *n*, Gärung *f*, Vergären *n*, Vergärung *f*; Fermentieren *n*, Fermentierung *f*, Fermentation *f*
  f. alcohol Gärungsalkohol *m*, [gewöhnlicher] Alkohol *m*, Äthanol *n*
  f. amyl alcohol Gärungsamylalkohol *m*
  f. broth Gärungsflüssigkeit *f*; *(bei der Antibiotika-Herstellung)* Fermentationsbrühe *f*
  f. bung Gärspund *m*
  f. cask Gärbottich *m*
  f. cellar Gärkeller *m*, Gärraum *m*
  f. cellar output Gärkellerausbeute *f*
  f. chemistry Gärungschemie *f*
  f. enzyme Gärungsenzym *n*, Gärungsferment *n*
  f. flask Gärkolben *m*
  f. from below Untergärung *f*
  f. from top Obergärung *f*
  f. gas Biogas *n*, Mistgas *n*, Faulgas *n*
  f. industry Gärungsindustrie *f*, Gärungsgewerbe *n*
  f. lactic acid Gärungsmilchsäure *f*
  f. layer *(Forst)* Vermoderungshorizont *m*
  f. losses Gärverluste *mpl*
  f. method Gärführung *f*, Gärverfahren *n*
  f. plant Gärkelleranlage *f*
  f. process Gär[ungs]prozeß *m*
  f. reductase test Gärreduktaseprobe *f*
  f. room *s.* f. cellar
  f. salt Gärsalz *n*
  f. tank Gärtank *m*, Gärbottich *m*

f. test Gärprobe *f*
f. time Gärdauer *f*
f. tube Gär[ungs]röhrchen *n*, Gärröhre *f*
f. vat *s.* f. tank
f. vinegar Gärungsessig *m*
fermentative fermentativ, enzymatisch, Ferment...,
Enzym...; gärend, Gärung erregend, gärungs-
erregend, Gär[ungs]...
f. power Gärvermögen *n*, Gärkraft *f*, Gärwirkung
f
fermented beet pulp eingesäuerte Zuckerrübenschnitzel *npl*
f. from below untergärig
f. from top obergärig
f. manure *(Landw)* Edelmist *m*
f. milk mit Säureweckern gesäuerte (versetzte)
Milch *f*, saure (dickgelegte) Milch *f*, Sauermilch
f, Dickmilch *f*
f. tea fermentierter Tee *m*
fermenter Gärbottich *m*, Gärtank *m*; Fermenter *m*,
Fermentator *m* (*z.B. bei der Antibiotika-
Herstellung*)
f. set temperature Anstelltemperatur *f*, *(i.w.S.)*
Gärtemperatur *f*
fermenting cellar *s.* fermentation cellar
f. power *s.* fermentative power
f. room *s.* fermentation cellar
f. tank (tub, vessel) *s.* fermentation tank
fermentor *s.* fermenter
Fermi characteristic energy level *s.* Fermi level
Fermi constant Fermi-Konstante *f*
Fermi-Dirac gas Fermi-Gas *n*
Fermi-Dirac statistics Fermi[-Dirac]-Statistik *f*
Fermi distribution Fermi-Verteilung *f*
Fermi energy Fermi-Energie *f*
Fermi level Fermi-Niveau *n*, Fermi-Kante *f*, Fer-
mische Grenzenergie *f*
Fermi statistics *s.* Fermi-Dirac statistics
Fermi surface Fermi-Fläche *f*
Fermi temperature Fermi-Temperatur *f*
Fermi theory [of beta decay] Fermi-Theorie *f* [des
β-Zerfalls]
fermion Fermion *n*
fermium Fm Fermium *n*
ferrate Ferrat *n*
ferric Eisen-..., *(i.e.S.)* Eisen(III)-...
f. ammonium alum *s.* f. ammonium sulphate
f. ammonium oxalate $(NH_4)_3[Fe(C_2O_4)_3]$ Am-
moniumeisen(III)-oxalat *n*, Eisen(III)-ammonium-
oxalat *n*
f. ammonium sulphate $(NH_4)Fe(SO_4)_2 \cdot 12H_2O$
Eisen(III)-ammoniumsulfat-12-Wasser *n*, Ammoniumeisenalaun *m*, Eisenammoniakalaun *m*
f. arsenate *s.* f. orthoarsenate
f. bichromate *s.* f. dichromate
f. bromide $FeBr_3$ Eisen(III)-bromid *n*, Eisentribromid *n*
f. chelate Eisen(III)-chelat *n*
f. chloride $FeCl_3$ Eisen(III)-chlorid *n*, Eisentrichlorid *n*

f. chromate $Fe_2(CrO_4)_3$ Eisen(III)-chromat *n*
f. compound Eisen(III)-Verbindung *f*
f. cyanide $Fe(CN)_3$ Eisen(III)-zyanid *n*,
Eisentrizyanid *n*
f. dichromate $Fe_2(Cr_2O_7)_3$ Eisen(III)-dichro-
mat(VI) *n*
f. ferrocyanide $Fe_4[Fe(CN)_6]_3$ Eisen(III)-hexa-
zyanoferrat(II) *n*
f. fluoride FeF_3 Eisen(III)-fluorid *n*, Eisentrifluorid
n
f. formate $Fe(HCOO)_3$ Eisen(III)-formiat *n*
f. hydrate *s.* f. hydroxide
f. hydroxide $Fe(OH)_3$ Eisen(III)-hydroxid *n*,
(besser) $Fe_2O_3 \cdot nH_2O$ Eisen(III)-oxidhydrat *n*
f. hydroxide sol Eisen(III)-hydroxidsol *n*
f. hypophosphite $Fe(H_2PO_2)_3$ Eisen(III)-hypo-
phosphit *n*
f. ion Ferri-Ion *n*, *(veraltet für)* Eisen(III)-ion *n*
f. iron dreiwertiges Eisen *n*
f. nitrate $Fe(NO_3)_3$ Eisen(III)-nitrat *n*, Eisentri-
nitrat *n*
f. orthoarsenate $FeAsO_4$ Eisen(III)-[ortho]ar-
senat(V) *n*
f. orthophosphate $FePO_4$ Eisen(III)-[ortho]phos-
phat *n*
f. oxalate $Fe_2(C_2O_4)_3$ Eisen(III)-oxalat *n*
f. oxalate chelate Eisen(III)-oxalat-Chelat *n*
f. oxide Fe_2O_3 Eisen(III)-oxid *n*, Eisentrioxid *n*,
Dieisentrioxid *n*
f. oxide gel Eisen(III)-oxidgel *n*
f. perchloride *s.* f. chloride
f. phosphate *s.* f. orthophosphate
f. phosphate haze Eisenphosphattrübung *f (des
Bieres)*
f. potassium chloride $2KCl \cdot FeCl_3$ Kaliumei-
sen(III)-chlorid *n*
f. potassium oxalate $KFe(C_2O_4)_2$ Kaliumei-
sen(III)-oxalat *n*
f. potassium sulphate $KFe(SO_4)_2 \cdot 12H_2O$ Ka-
liumeisen(III)-sulfat-12-Wasser *n*, Kaliumeisenen-
alaun *m*
f. pyrophosphate $Fe_4(P_2O_7)_3$ Eisen(III)-diphos-
phat *n*, Eisen(III)-pyrophosphat *n*
f. resinate Eisen(III)-resinat *n*
f. salt Eisen(III)-salz *n*
f. sesquibromide *s.* f. bromide
f. sesquichloride *s.* f. chloride
f. sesquisulphate *s.* f. sulphate
f. sodium oxalate $Na_3[Fe(C_2O_4)_3]$ Natrium-
eisen(III)-oxalat *n*
f. sulphate $Fe_2(SO_4)_3$ Eisen(III)-sulfat *n*,
Eisentrisulfat *n*, Dieisentrisulfat *n*
f. sulphide Fe_2S_3 Eisen(III)-sulfid *n*, Dieisentrisul-
fid *n*
f. sulphocyanate *s.* f. thiocyanate
f. tartarate $Fe_2(C_4H_4O_6)_3$ Eisen(III)-tartrat *n*
f. thiocyanate $Fe(SCN)_3$ Eisen(III)-thiozyanat *n*,
Eisen(III)-rhodanid *n*
f. tribromide *s.* f. bromide
f. trichloride *s.* f. chloride

f. trioxide s. f. oxide
f. trisulphate s. f. sulphate
ferrichrome Ferrichrom n
ferricoproporphyrin hydroxide Eisen(III)-koproporphyrinhydroxid n
ferricyanid acid $H_3[Fe(CN)_6]$ Hexazyanoeisen(III)-säure f
ferricyanide $M^I_3[Fe(CN)_6]$ Zyanoferrat(III) n, Hexazyanoferrat(III) n
f. complex Zyanoferrat(III)-komplex m, Hexazyanoferrat(III)-komplex m
ferriheme Eisen(III)-häm n
ferrihemochrome s. ferrihemochromogen
ferrihemochromogen Eisen(III)-hämochromogen n
ferrihemoglobin Eisen(III)-hämoglobin n
ferrimolybdite (Min) Ferrimolybdit m, Molybdänocker m
ferrimycin Ferrimyzin n
ferrioxamine Ferrioxamin n
ferrite 1. Ferrit m (Eisen-Kohlenstoff-Mischkristall im α- und δ-Bereich); 2. $M^I FeO_2$ Ferrat(III) n; 3. Ferrit m (keramischer Magnetwerkstoff)
f. formation Ferritbildung f
f. yellow Eisengelb n, Eisenoxidgelb n
ferritic ferritisch
ferroalloy Ferrolegierung f
ferroboron Ferrobor n
ferrocarbon s. pig iron
ferrocene $Fe(C_5H_5)_2$ Ferrozen n, Eisen-bis-Zyklopentadienyl n
f. polymer Ferrozenpolymer[es] n
ferrochromium Ferrochrom n
ferrocolumbium Ferroniob n
ferrocyanic acid $H_4[Fe(CN)_6]$ Hexazyanoeisen(II)-säure f
ferrocyanide $M^I_4[Fe(CN)_6]$ Zyanoferrat(II) n, Hexazyanoferrat(II) n
ferroelectric ferroelektrisch
ferroelectric s. f. material
f. hysteresis ferroelektrische Hysterese f
f. material (substance) ferroelektrisches Material n, ferroelektrischer Stoff m, Ferroelektrikum n
ferroheme Eisen(II)-häm n
ferrohemoglobin Eisen(II)-hämoglobin n
ferroin Ferroin n
ferromagnetic ferromagnetisch
ferromagnetic s. f. material
f. material (substance) ferromagnetisches Material n, ferromagnetischer Stoff m, Ferromagnetikum n
ferromagnetism Ferromagnetismus m
ferromanganese Ferromangan n
ferromolybdenum Ferromolybdän n
ferronickel Ferronickel n
ferroniobium Ferroniob n
ferrophosphorus Ferrophosphor m
ferrosilicon Ferrosilizium n
ferroso-ferric chloride $FeCl_2 \cdot 2FeCl_3$ Eisen(II,III)-chlorid n
f.-ferric ferricyanide $Fe^{III}_4 Fe^{II}_3[Fe(CN)_6]_6$ Eisen(II,III)-hexazyanoferrat(III) n

f.-ferric oxide $FeO \cdot Fe_2O_3$ oder Fe_3O_4 Eisen(II,III)-oxid n
f.-ferric sulphide Fe_3S_4 Eisen(II,III)-sulfid n
ferrotantalum Ferrotantal n
ferrotellurite (Min) Ferrotellurit m
ferrotitanium Ferrotitan n
ferrotungsten Ferrowolfram n
ferrous Eisen-..., (i.e.S.) Eisen(II)-...
f. acetate $Fe(CH_3COO)_2$ Eisen(II)-azetat n
f. ammonium sulphate $(NH_4)_2Fe(SO_4)_2 \cdot 6H_2O$ Ammoniumeisen(II)-sulfat-6-Wasser n, Mohrsches Salz n
f. arsenate s. f. orthoarsenate
f. bromide $FeBr_2$ Eisen(II)-bromid n, Eisendibromid n
f. carbonate $FeCO_3$ Eisen(II)-karbonat n
f. chelate Eisen(II)-chelat n
f. chloride $FeCl_2$ Eisen(II)-chlorid n, Eisendichlorid n
f. chloroplatinate $Fe[PtCl_6]$ Eisen(II)-hexachloroplatinat(IV) n
f. ferricyanide $Fe_3[Fe(CN)_6]_2$ Eisen(II)-hexazyanoferrat(III) n
f. ferrocyanide $Fe_2[Fe(CN)_6]$ Eisen(II)-hexazyanoferrat(II) n
f. fluoride FeF_2 Eisen(II)-fluorid n, Eisendifluorid n
f. formate $Fe(HCOO)_2$ Eisen(II)-formiat n
f. gluconate Eisen(II)-glukonat n
f. glutamate Eisen(II)-glutam[in]at n
f. hydroxide $Fe(OH)_2$ Eisen(II)-hydroxid n, Eisendihydroxid n
f. iodide FeJ_2 Eisen(II)-jodid n, Eisendijodid n
f. iron zweiwertiges Eisen n
f. lactate Eisen(II)-laktat n
f. metal Eisenmetall n, Eisenwerkstoff m
f. nitrate $Fe(NO_3)_2$ Eisen(II)-nitrat n, Eisendinitrat n
f. orthoarsenate $Fe_3(AsO_4)_2$ Eisen(II)-[ortho]arsenat(V) n
f. orthophosphate $Fe_3(PO_4)_2$ Eisen(II)-[ortho]phosphat n
f. oxalate $Fe(C_2O_4)$ Eisen(II)-oxalat n
f. oxalate chelate Eisen(II)-oxalat-Chelat n
f. oxide FeO Eisen(II)-oxid n, Eisen[mon]oxid n
f. perchlorate $Fe(ClO_4)_2$ Eisen(II)-perchlorat n
f. phosphate s. f. orthophosphate
f. phthalocyanine Eisen(II)-phthalozyanin n
f. pyroarsenite $Fe_2As_2O_5$ Eisen(II)-diarsenat(III) n, Eisen(II)-pyroarsenat(III) n
f. salt Eisen(II)-salz n
f. silicate $FeSiO_3$ Eisen(II)-silikat n
f. sulphate $FeSO_4$ Eisen(II)-sulfat n
f. sulphide FeS Eisen(II)-sulfid n, Eisen[mono]sulfid n
f. sulphite $FeSO_3$ Eisen(II)-sulfit n
f. sulphocyanate s. f. thiocyanate
f. tartrate $Fe(C_4H_4O_6)$ Eisen(II)-tartrat n
f. thiocyanate $Fe(SCN)_2$ Eisen(II)-thiozyanat n, Eisen(II)-rhodanid n

f. thiosulphate FeS$_2$O$_3$ Eisen(II)-thiosulfat *n*
ferrovanadium Ferrovanadin *n*, Ferrovanadium *n*
ferroxyl indicator Ferroxylindikator *m*
ferrozirconium Ferrozirkon[ium] *n*
ferrugin[e]ous 1. eisenhaltig, eisenführend; 2. rostfarbig
f. water eisenhaltiges Wasser *n*, Eisenwasser *n*
fertile fruchtbar, fertil; *(Kph)* brutfähig, brütbar, in Spaltstoff umwandelbar
f. material *(Kph)* Brutmaterial *n*, Brutstoff *m*
fertility Fruchtbarkeit *f*, Fertilität *f*; *(Kph)* Brutfähigkeit *f*, Brütbarkeit *f*, Vermehrungsfähigkeit *f*
fertilization *(Landw)* Düngung *f*; *(Kph)* Brüten *n*, Brutvorgang *m*, Vermehrung *f*
f. by soil injection Lanzendüngung *f*
f. in stages Stadiendüngung *f*
f. of subsoil Untergrunddüngung *f*
f. recommendation Düngungsempfehlung *f*
fertilize / to *(Landw)* düngen
fertilizer *(Landw)* Düngemittel *n*, Dünger *m*
f. dressing Düngung *f*
f. for plant forcing Treibdünger
f. formula Mischungsverhältnis *n* von Mischdüngern
f. industry Düngemittelindustrie *f*
f. mill Düngermühle *f*
f. needs (requirement) Dünge[mittel]bedarf *m*, Düngerbedarf *m*, Dünge[r]bedürfnis *n*
f. roller Düngerwalze *f*
fertilizing machine Düngemaschine *f*
f. plough Düngungspflug *m*
ferulic acid CH$_3$OC$_6$H$_3$(OH)CH = CHCOOH Ferulasäure *f*, Kaffeesäure-3-methyläther *m*, 3-(4-Hydroxy-3-methoxyphenyl)propensäure *f*
feruloylquinic acid Feruloylchinasäure *f* *(Depsid)*
festoon ager *(Text)* Festoondämpfer *m*, Hängedämpfer *m*
f. dryer Hänge[band]trockner *m*, Laufbandtrockner *m*, *(Text auch)* Trockenhänge *f*
fettle / to [ver]putzen, entgraten
fettler Putzer *m*
fettling Verputzen *n*, Putzen *n*, Entgraten *n*
f. machine Putzmaschine *f* *(z.B. in der keramischen Industrie)*
Feulgen reaction (staining method, test) Feulgensche Reaktion (Nuklealreaktion) *f* *(Methode zur Bestimmung von Desoxyribonukleinsäure)*
fever tree Blaugummibaum *m*, Eucalyptus globulus Lab.
FF black *s.* fine furnace black
F.F.A. *s.* free fatty acid
fiber *(Am)* *s.* fibre
fiberboard *(Am)* *s.* fibreboard
fiberization Defibrierung *f*, Zerfaserung *f*, Zerlegung *f* in Einzelfasern
fiberize / to defibrieren, zerfasern, in Einzelfasern zerlegen
fibre *(Bio)* Faser *f*, Fiber *f*; *(Text)* Faserstoff *m*; *(Text)* Faser *f*
f.-bearing liquid (water) *(Pap)* Faser[stoff]suspension *f*, Stoffsuspension *f*

f. blend Fasermischung *f*
f. bundle Faserbündel *n*
f. diagram Faserdiagramm *n*
f. disintegration Faserabbau *m*
f. drawing process Fadenziehprozeß *m* *(Herstellung von Chemiefaserstoffen)*
f.-filled moulding material *(Plast)* Formmasse *f* mit Faserstoffverstärkung (Faserstoffüllung)
f. flax Faserlein *m*, Linum usitatissimum L.
f. formation Faserstoffbildung *f*; Faserbildung *f*
f.-forming faserstoffbildend; faserbildend
f. glass *s.* fibreglass
f. incrustation Faserinkrustierung *f*
f. length Faserlänge *f*
f. loss *(Pap)* Faserverlust *m*
f. of glass Glasfaserstoff *m*, GL; Glasfaser *f*
f. of wood Holzfaser *f*
f. photograph Faserdiagramm *n*
f. preparation *(Pap)* Ganzstoffaufbereitung *f*, Ganzzeugbereitung *f*, Stoffaufbereitung *f*
f.-protective agent Faserstoffschutzmittel *n*; Faserschutzmittel *n*
f. recovery *(Pap)* Faserrückgewinnung *f*, Faserwiedergewinnung *f*, Stoffrückgewinnung *f*, Stoffwiedergewinnung *f*
f. staple 1. Faser *f*, Spinnfaser *f*, *(bei Chemiefaserstoffen)* Stapelfaser *f*; 2. Faserlänge *f*, Stapellänge *f*, Stapel *m*
f. structure *(Text)* Faserstoffstruktur *f*; Faserstruktur *f*
f. suspension *(Pap)* Fasersuspension *f*
fibreboard Faserplatte *f*, *(meist)* Holzfaserplatte *f*
fibreglass Glasfaserstoff *m*, *(manchmal auch)* Faserglas *n*; Glasfaser *f*
f. product Glasfaserstofferzeugnis *n*; Glasfasererzeugnis *n*, Glasfaserprodukt *n*
fibril[la] Fibrille *f*, Fäserchen *n*, Teilfäserchen *n*
fibrillate / to fibrillieren, in Teilfäserchen (Fibrillen) aufspalten, defibrillieren
fibrillated polypropylene Polypropylen *n* mit vorgebildeter Faserstruktur
fibrillation Fibrillieren *n*, Fibrillierung *f*, Fibrillenbildung *f*, Defibrillierung *f*
fibrin Fibrin *n*
fibrinogen Fibrinogen *n*
fibroin Fibroin *n* *(Eiweißkörper der Naturseide)*; Seidenstoff *m*
fibrolite *(Min)* Fibrolith *m* *(filziger Sillimanit)*
fibrous fas[e]rig, faserartig, Faser..., fibrös
f. brown coal faserige Braunkohle *f*, Faserbraunkohle *f*
f. coal Faserkohle *f*
f. colloid Linearkolloid *n*
f. furnish *(Pap)* eingetragener Faserhalbstoff *m*, Mahlgut *n*
f. glass *s.* fibreglass
f. mass *(Pap)* Fasermasse *f*, Faserbrei *m*
f. material Fasermaterial *n*, Faserstoff *m*
f. polymer [elementar]fadenbildendes Polymer[es] *n*

film

f. protein Faserprotein n, fibrilläres (faserartiges) Protein n

f. pulp (Pap) Faser[stoff]brei m, Stoffbrei m, Fasermasse f, Fasersuspension f, Stoff m

f. raw material (Pap) Faserrohstoff m

f. structure Faserstruktur f, Fasergefüge n

fichtelite (Min) Fichtelit m

ficin Fizin n (proteolytisches Ferment aus dem Milchsaft südamerikanischer Ficus-Arten)

Fick's law Ficksches Gesetz n (der Diffusion)

fidelity Maßgenauigkeit f

f. of reproduction (Foto) Wiedergabetreue f

fiducial limit Vertrauensgrenze f

field direction Feldrichtung f

f. efficiency s. f. performance

f. electron microscope s. f. emission microscope

f. emission Feld[elektronen]emission f, kalte Emission f, Kalt[katoden]emission f

f. emission microscope Feldemissionsmikroskop n, Feldelektronenmikroskop n

f. experiment Feldversuch m, Freilandversuch m, Feldtest m, Naturversuch m

f. fertilization test Felddüngungsversuch m

f.-free region feldfreier Raum m (z.B. im Spektrografen)

f. latex frisch gezapfter Latex m

f. of force Kraftfeld n

f. performance Feldleistung f, Feldwirksamkeit f (von Pflanzenschutzmitteln)

f. strength Feldstärke f

f. test (trial) s. f. experiment

f. valency compound Feldvalenzverbindung f

fiery principle (historisch) brennbares Prinzip n, Feuerstoff m, Feuermaterie f, Phlogiston n

figured glass Ornamentglas n

Fikentscher K-value K-Wert m nach Fikentscher (zur Kennzeichnung der Molekularmasse von Hochpolymeren)

filament 1. (Text) Elementarfaden m, Filament n, Endlosfaser f; 2. Glühfaden m, Leuchtfaden m (einer Glühlampe); Heizfaden m (einer Elektronenröhre)

f. forming (Text) Elementarfadenbildung f, Endlosfaserbildung f

f. yarn Seide f, Endlosgarn n

filicic acid s. 1. filixic acid; 2. filicinic acid

filicin s. filicinic acid

filicinic acid Filizinsäure f, 1,1-Dimethyl-2,4,6-zyklohexantrion n

filigic acid s. filicinic acid

filings Feilspäne mpl

filixic acid Filixsäure f

fill / **to** füllen, (Tech auch) beschicken; füllen, Füllstoffe zusetzen; (Gerb) füllen, nachgerben

f. up nachfüllen; (Text) nachdecken, überfärben

fill Füllung f; Einbauten mpl, Einbau m (z.B. in einem Kühlturm)

f. height Füllhöhe f, Füllstand m

f. member Einbauteil n (z.B. des Kühlturms)

f. orifice (Plast) Anschnitt m

filled desiccator Exsikkator m mit Einsatz

f. roll (Pap) [elastische] Papierwalze f, elastische Walze f (des Superkalanders)

filler 1. Füllstoff m, Füllmittel n, Füllmaterial n, Füller m, Füllkörper m, Füllmasse f, (Plast auch) Streckmittel n, Harzträger m, (Pap auch) Papierfüllstoff m; 2. (Pap) Einlage f (beim Triplexkarton); 3. (Anstrichtechnik) Spachtelmasse f, Spachtel m(f); 4. (Bau) Zuschlag[stoff] m

f. board (Pap) Einlage f (beim Triplexkarton)

f. clay (Pap) Füllstoffkaolin n

f. dispersion Füllstoffdispergierung f, Füllstoffverteilung f

f.-free füllstofffrei, ungefüllt, ohne Füllstoffe

f. gas Füllgas n

f. loading Zusatz (Zuschlag) m von Füllstoffen, Füllung f, Füllstoffdosierung f

f. material s. filler 1.

f.-reinforced (Gum) mit aktiven Füllstoffen gefüllt

f. rod Schweißstab m

f.-rubber stock Kautschuk-Füllstoff-Mischung f

f. specks Füllstoffnester npl, Stippen fpl

fillet 1. (Pap) Holzzwischenleiste f (für Bemesserung von Mahlmaschinen); 2. (Plast) Abrundung f von Übergängen

filling 1. Füllen n, Füllung f, (Tech auch) Beschicken n; Füllen n, Zusetzen n von Füllstoffen; (Gerb) Füllen n, Nachgerben n; 2. s. filler 1.; 3. s. fillet 1.; 4. (Pap) Bemesserung f, Messergarnierung f

f. agent s. filler 1.

f. degree Füllungsgrad m

f. door Füllklappe f, Beschickungstür f, Chargiertür f; s. f. hole

f. funnel Pulvertrichter m

f. hole Füllöffnung f, Fülloch n, Beladeöffnung f, Eintragöffnung f, Beschickungsöffnung f, Einwurföffnung f, Aufgabeöffnung f

f. level Füllhöhe f, Füllstand m

f. machine Füllmaschine f, Abfüllmaschine f, Abfüllautomat m, Füller m

f. material s. filler 1.

f. of the chips with liquor (Pap) Hackschnitzeldurchtränkung f, Holzdurchtränkung f

f. pigment (Gum) inaktiver Füllstoff m

f. room Abfüllerei f, Abfüllabteilung f

f. with chips (Pap) Holzfüllung f, Hackschnitzelfüllung f, Einbringen n der Hackschnitzel

fillmass (Zucker) Füllmasse f

film Film m, Häutchen n, Belag m, [dünne] Schicht f, [dünner] Überzug m; (Plast) Folie f, (i.e.S.) Feinfolie f (Dicke unter 0.01 inch = 0,25 mm); (Foto) Film m

f. base (Foto) Film[schicht]träger m, Schichtträger m, Filmgrundlage f, Filmunterlage f

f. blowing machine Folienblasmaschine f

f. casting machine Feinfoliengießmaschine f, Filmgießmaschine f

f. cooler Rieselkühler m

f. coverage Filmdeckung f (von Netzmitteln)

f. dryer Dünnschichttrockner *m*
f. extrusion Strangpressen *n* von Folien
f. fault *(Farb)* Anstrichfehler *m*, Anstrichmangel *m*
f. fill Rieseleinbauten *mpl*, Rieseleinbau *m*
f. formation Filmbildung *f*
f. former Filmbildner *m*, filmbildender Stoff *m*, filmbildende Substanz *f*, filmbildendes Material *n*
f.-forming filmbildend
f. forming *s.* f. formation
f.-forming component (material, substance) *s.* f. former
f. of lubricant Schmier[mittel]schicht *f*, *(i.e.S.)* Schmier[öl]film *m*
f. of moisture Feuchtigkeitsfilm *m*
f. of oxide Oxidfilm *m*, Oxidhaut *f*, Oxidbelag *m*, dünne Oxidschicht *f*
f. pack *s.* f. fill
f. precipitator Naß-Elektroabscheider *m*, Naß-Elektrofilter *n* *(zur elektrischen Gasreinigung)*
f. processing *(Foto)* Filmbearbeitung *f*
f. still Dünnschicht-Destilliergerät *n*
f. strip *(Foto)* Filmstreifen *m*, Filmband *n*, Bildband *n*
f. thickness Filmdicke *f*, Schichtdicke *f*
f.-type evaporator Dünnschichtverdampfer *m*
filter / to filtrieren, filtern
f. by suction *(durch Filter)* absaugen, [ab]nutschen
f. off abfiltrieren, abfiltern
f. off by suction *s.* to f. by suction
f. under suction *s.* to f. by suction
f. until bright klar filtrieren (filtern)
f. with the help of slight suction durch einen schwachen Unterdruck absaugen
filter Filter *n(m)*
f. adapter Gummimanschette *f* für Frittentiegel
f. aid poröses Filtermittel (Filtermedium) *n*; Filterhilfsstoff *m*; Filterhilfe *f*, Filter[ungs]hilfsmittel *n*
f. area Filterfläche *f*
f. assembly Filtriervorrichtung *f*
f. bag Filtersack *m*, Filterbeutel *m*, Filterschlauch *m*
f. bed Filterbett *n*, Filter[schütt]schicht *f*, Filterschüttung *f*
f. cake Filterkuchen *m*; Zentrifugenkuchen *m*
f. candle Filterkerze *f*
f. carriage Filterwagen *m*; Filterrahmen *m*
f. chamber Filterkammer *f*
f. cloth Filtergewebe *n*, Filtertuch *n*, Filterstoff *m*
f. cone Trichtereinlage *f* zum Filtrieren
f. crucible Filtertiegel *m*, Filtriertiegel *m*
f. disk Filterscheibe *f*, Filterplatte *f*
f. drum Filtertrommel *f*
f. element Filterelement *n*, Filtereinsatz *m*; Filterkerze *f*; Filterstein *m*
f. fabric *s.* f. cloth
f. felt *(Pap)* Filterfilz *m* *(zur Faserrückgewinnung)*

f. flakes Filterflocken *fpl* *(Filterhilfe für Kolloide)*
f. flask Saugflasche *f*
f. funnel Filtertrichter *m*
f. gravel Filterkies *m*
f. leaf Filterblatt *n* *(des Blattfilters)*
f. mass Filtermasse *f*, Filtriermasse *f*, Filtrierstoff *m*
f. mat Filtermatte *f*, Preßtuchmatte *f*
f. material (medium) Filtermaterial *n*, Filtriermaterial *n*, Filtermedium *n*, Filtermittel *n*
f. medium filtration Tiefenfiltration *f*
f. pad Filtermassekuchen *m*, Filterpreßmasse *f*, Filtermasseplatte *f*
f. paper Filterpapier *n*, Filtrierpapier *n*
f. paper chromatography Papierchromatografie *f*
f. paper disk Filterpapierscheibe *f*, Rundfilter *n*
f. photometer Filterfotometer *n*
f. plate Filterplatte *f*, Filterboden *m*
f.-press / to in Filterpressen abpressen (filtrieren, abfiltrieren)
f. press Filterpresse *f*
f. press paper Filterpressenpapier *n*
f. press-type dialyzer Dialyse[n]presse *f*
f. pressing Filtration *f* in Filterpressen
f. pulp *s.* f. mass
f. pump Wasserstrahlpumpe *f*
f. save-all *(Pap)* Filterstoffänger *m*
f. screen Filtersieb *n*
f. stick Fiilterstäbchen *n*, Filterstab *m*
f. surface *s.* f. area
f. tank Filterkasten *m*, Filterwanne *f*, Filtertrog *m*, Filterbottich *m*, Läuterbottich *m*
f. thickener Eindickfilter *n*, Filter-Eindicker *m*
f. tube Filterkerze *f*; Filterröhrchen *n*, Filtrierröhrchen *n*
f. valve Steuerkopf *m* eines Filters
f. vat *s.* f. tank
filterability Filtrierbarkeit *f*, Filtrationsfähigkeit *f*, Filtrierfähigkeit *f*
filterable filtrierbar, filtrierfähig, filterfähig, filtrationsfähig, filtrierend
filterableness *s.* filterability
filteraid *s.* filter aid
filtercake Filterkuchen *m*; Zentrifugenkuchen *m*
f. washing Filterkuchenwäsche *f*
filtered stock filtriertes Rückstandsöl *n*
f. water *(Pap)* geklärtes Abwasser *n*, Klarwasser *n*, Filterwasser *n*
filtering Filtrieren *n*, Filtration *f*, Filtern *n*, Filterung *f* *(Zusammensetzungen s. a. unter filter)*
f. centrifugal Filterzentrifuge *f*, Filtrierzentrifuge *f*
f. element Filterelement *n*
f. jar Filtrierstutzen *m*
f. pressure Filtrationsdruck *m*
f. rate Filtergeschwindigkeit *f*, Filtriergeschwindigkeit *f*
f. time Filtrierzeit *f*
filtermasse Filtermassekuchen *m* aus Baumwollfasern

filtrable *s.* filterable
filtrate / to filtrieren, filtern
filtrate Filtrat *n*; Fugat *n* *(beim Zentrifugieren)*
  f. outlet Filtratauslauf *m*, Filtratablauf *m*, Filtrataustritt *m*
filtration Filtration *f*, Filterung *f*, Filtrieren *n*, Filtern *n*
  f. accelerator Filterhilfe *f*, Filterhilfsstoff *m*, Filter[ungs]hilfsmittel *n*
  f. by gravity Filtration *f* unter Wirkung (Ausnutzung) der Schwerkraft
  f. constant Filtrationskonstante *f*
  f. curve Filtrationskurve *f*, Durchflußkurve *f*
  f. cycle Filtrationszyklus *m*
  f. equation Filtriergleichung *f*, Filtergleichung *f*
  f. extractor Extraktionspresse *f*
  f. fabric *s.* filter cloth
  f. of water Wasserfiltration *f*, Wasserreinigung *f*
  f. pressure *s.* filtering pressure
  f. rate *s.* filtering rate
  f. residue Filterrückstand *m*
  f. sterilization Sterilfiltration *f*
  f. time *s.* filtering time
  f. under reduced pressure Unterdruckfiltration *f*, Filtration *f* unter vermindertem Druck
filty Grubengas *n*; schlagende Wetter *pl*, Schlagwetter *pl*
fin 1. *(Tech)* Rippe *f*; Kühlrippe *f*; 2. *(Plast)* Preßgrat *m*, Formgrat *m*, Grat *m*, Austrieb *m*
  f. drum dryer Rillenwalzentrockner *m*
  f. tube Rippenrohr *n*
final bleaching Endbleiche *f*, Fertigbleiche *f*, Nachbleiche *f*
  f. bleaching at high consistency *(Pap)* Dickstoffendbleiche *f*
  f. blow[ing] *(Glas)* Fertigblasen *n*
  f. boiling point Siedeendpunkt *m*, Siedeende *n*, SE, Endsiedepunkt *m*, Endkochpunkt *m*
  f. brightness *(Pap)* Endweiße *f*
  f. concentration Endkonzentration *f*
  f. container Aufbewahrungsbehälter *m*
  f. coupling *(Farb)* Schlußkupplung *f*
  f. drying Endtrocknung *f*
  f. firing *(Ker)* Garbrand *m*
  f. hypochlorite stage *(Pap)* Hypochlorit-Endbleiche *f*, Endhypochloritbleiche *f*
  f. lye Endlauge *f* *(z.B. in der Düngemittelindustrie)*
  f. magma Kristallisationsprodukt *n*, Kristallisat *n*
  f. moisture content Endfeuchte[beladung] *f*
  f. molasses *(Zucker)* Endmelasse *f*
  f. period Nachperiode *f*, Nachversuch *m* *(einer kalorimetrischen Messung)*
  f. point Umschlagspunkt *m* *(von Indikatoren)*; Endpunkt *m* *(einer Titration)*
  f. pressure Enddruck *m*
  f. product Endprodukt *n*, Finalprodukt *n*, Enderzeugnis *n*; Fertigprodukt *n*, Fertigerzeugnis *n*; *(phys Ch)* Endprodukt *n* *(z.B. einer radioaktiven Umwandlung)*

f. saturation *(Zucker)* Nachsaturation *f*
f. sintering Fertigsintern *n*, Fertigsinterung *f*, Hochsintern *n*, Hochsinterung *f*, Formieren *n*
f. solubility Endlöslichkeit *f*
f. spreading coefficient *(Koll)* Endspreitungskoeffizient *m*
f. state Endzustand *m*
f. strength Endfestigkeit *f*; Endkonzentration *f*
f. temperature Endtemperatur *f*
f. tower bleaching *(Pap)* Turmfertigbleiche *f*
f. vat Schlußbütte *f*
f. washing Endwäsche *f*
finally shaped fertiggeformt
fine fein, *(Schüttgut auch)* feinkörnig, *(Edelmetalle auch)* rein, *(Folien auch)* dünn, *(Einstellung von Meßgeräten auch)* genau, präzis
  in a f. state in fein verteiltem Zustand
  in f. powder feingepulvert, feinpulverisiert, feinpulverig
  f. aggregate feinkörniger Zuschlag *m*, Feinzuschlag *m*
  f. annealing *(Glas)* Feinkühlen *n*, Feinkühlung *f*
  f.-banded feinstreifig, mit feinen Streifen
  f. board Feinpappe *f*
  f. ceramic industry feinkeramische Industrie *f*
  f. ceramic ware Feinkeramik *f*, feinkeramische Erzeugnisse *npl*
  f. ceramics *s.* f. ceramic ware
  f. coal Feinkohle *f*
  f. crusher Feinbrecher *m*
  f. crushing Feinbrechen *n*
  f. filter Feinfilter *n*
  f. furnace black *(Gum)* Fine-Furnace-Ruß *m*, FF-Ruß *m* *(feinteiliger Furnace-Ruß)*
  f.-grain feinkörnig, Feinkorn..., dicht *(Gefüge)*
  f.-grain developer *(Foto)* Feinkornentwickler *m*
  f.-grain development *(Foto)* Feinkornentwicklung *f*
  f.-grain emulsion *(Foto)* feinkörnige Emulsion *f*, Feinkornemulsion *f*
  f.-grain image *(Foto)* Feinkornbild *n*
  f.-grain structure feinkörnige Struktur *f*, feinkörniges (dichtes) Gefüge *n*
  f.-grained *s.* f.-grain
  f.-granular *s.* f.-grain
  f. grinder *s.* f.-grinding mill
  f. grinding Fein[ver]mahlen *n*, Fein[aus]mahlung *f*, feine Mahlung (Ausmahlung) *f*; Feinschleifen *n*, Feinschliff *m*
  f.-grinding mill Feinmühle *f*
  f. material Feingut *n*
  f.-meshed feinmaschig, engmaschig, kleinmaschig
  f. mill Feinpapierfabrik *f*
  f. optical annealing Feinkühlung *f* von optischem Glas
  f. ore Feinerz *n*, feines Erz *n*
  f. paper Feinpapier *n*
  f.-pored feinporig, mit feinen Poren, feinporös, engporig, kleinlückig

f. reduction jaw crusher Einschwingenbacken-
brecher *m*
f. sand Feinsand *m (Korngröße 0,2 bis 0,02 mm)*
f. screen Feinsortierer *m*, Nachsortierer *m*
f. sieve Feinsieb *n*
f.-size fractionating (separation) Feinklassieren
n, Feinklassierung *f*
f. sizes fein[st]e Kornfraktionen *fpl*, Fein[st]gut *n*,
Fein[st]korn *n*, Feines *n*, Feinanteile *mpl*, feinerer
Teil *m*
f. soil Feinboden *m*, Feinerde *f (Bodenfraktion
mit Teilchendurchmesser kleiner als 2 mm)*
f. stoneware *(Ker)* Feinsteinzeug *n*
f. straw pulp [vollaufgeschlossener] Stroh[zell]-
stoff *m*
f. structure *(phys Ch)* Feinstruktur *f*
f.-structure constant *(phys Ch)* Feinstrukturkonstante *f*
f. thermal black *(Gum)* Fine-Thermal-Ruß *m*,
FT-Ruß *m (Gasruß von kleiner Teilchengröße)*
finely banded *s.* fine-banded
f. dispersed feindispers, feindispergiert
f. divided feinzerteilt, feinzerkleinert; feinverteilt,
fein verteilt, in feiner Verteilung
f. ground feinvermahlen, fein[aus]gemahlen
f. pored (porous) feinporig, mit feinen Poren,
feinporös, engporig, kleinlückig
f. powdered feingepulvert, feinpulverisiert, feinpulverig
fineness Feinheit *f*, *(i.e.S.)* Mahlfeinheit *f*
f. degree Feinheitsgrad *m*, Feinheit *f*, *(i.e.S.)*
Mahl[feinheits]grad *m*, Ausmahlungsgrad *m*,
Mahlfeinheit *f*
f. module *(Am) s.* f. modulus
f. modulus Feinheitsmodul *m*, Körnungsmodul *m*
f. of grinding Mahlfeinheit *f*
fines fein[st]e Kornfraktionen *fpl*, Fein[st]gut *n*,
Fein[st]korn *n*; Feines *n*, Feinanteile *mpl*, feinerer
Teil *m*, Unterkorn *n*; *(Bgb)* Feinkohle *f*; *(Aufbereitung)* Feingut *n*, Schlamm *m*; *(Pap)* Staub
m, feines Holzmehl *n*; *(Plast)* Feinstoffe *mpl*,
pulvriges Material *n*, Feinpulveriges *n (z. B. einer
Formmasse)*
finger agitator Fingerrührer *m*
Fingotra rubber *(Kautschuk von Landolphia
crassipes Radl.)*
fining 1. *(Gär)* Schönen *n*, Schönung *f*, Klären *n*,
Klärung *f*; *(Glas)* Läuterung *f*; 2. *(Gär)* Klärmittel
n, Schönungsmittel *n*, Schöne *f*; *(Glas)* Läuter[ungs]mittel *n*
f. agent *s.* fining 2.
finish / to 1. beend[ig]en, vollenden, zu Ende
führen; 2. enden, aufhören; 3. *(Tech)* fertigbear-
beiten, fertigstellen, fertigmachen, nachbe-
arbeiten, veredeln, die Oberflächenbeschaffenheit (Oberflächengüte) verbessern, die Oberfläche behandeln; *(Gerb)* zurichten; *(Gerb)*
appretieren; *(Text)* appretieren, ausrüsten,
veredeln; *(Farb)* mit einem Schlußanstrich (Deckanstrich) versehen; *(Pap)* fertigstellen, ausrüsten

finish 1. Oberflächenzustand *m*, Oberflächenbeschaffenheit *f*, Oberflächengüte *f*, Finish *n*;
(Gerb) Appretur *f*, Finish *n*; *(Text)* Appretur *f*,
Ausrüstung *f*; *(Farb)* Schlußanstrich *m*, Deckanstrich *m*; *(Pap)* *(Oberbegriff für den Bearbei-
tungszustand der Oberfläche eines Papiers)*; 2.
(Gerb, Text) Appret *n*, Appreturmittel *n*; 3. *s.*
finishing
f.-breaking machine *(Text)* Appret[ur]brech-
maschine *f*
f. mould *(Glas)* Kopfform *f*
f. of transformation Umwandlungsende *n*, Ende
n (Vollendung *f*) der Umwandlung
finished product Endprodukt *n*, Finalprodukt *n*,
Enderzeugnis *n*; Fertigprodukt *n*, Fertigerzeugnis *n*
f. stock Fertigprodukt *n*, Fertigerzeugnis *n*
f. stuff *(Pap)* [fertiger] Papierstoff *m*, Ganzstoff *m*,
Ganzzeug *n*, Stoff *m*
finisher *(Text)* Appreteur *m*, Ausrüster *m*, Veredler
m
finishing 1. Beend[ig]en *n*, Vollenden *n*, Vollendung *f*; 2. Ende *n*, Schluß *m*; 3. *(Tech)*
Fertigbearbeitung *f*, Fertigstellung *f*, Fertigmachen *n*, Nachbearbeitung *f*, Veredelung *f*,
Verbesserung *f* der Oberflächenbeschaffenheit
(Oberflächengüte), Oberflächenbehandlung *f*;
(Gerb) Zurichten *n*; *(Gerb)* Appretieren *n*,
Appretur *f*; *(Text)* Appretieren *n*, Appretur *f*,
Ausrüsten *n*, Ausrüstung *f*, Veredelung *f*; *(Farb)*
Schlußanstrich *m*, Deckanstrich *m* (als Vorgang);
(Pap) *(Verfahren zur Fertigstellung bzw. Ausrüstung des Papiers nach der Trockenpartie bis
zum Verlassen der Papierfabrik)*
f. auxiliary *(Gerb)* Zurichthilfsmittel *n*
f. coat *(Farb)* Schlußanstrich *m*, Deckanstrich *m*
f. department *(Pap)* Ausrüstungsabteilung *f*
f. padder *(Text)* Appreturfoulard *m*
f. plant *s.* f. department
f. room *(Pap)* Ausrüstungssaal *m*
f. temperature Endtemperatur *f*
finned heater Rippenheizkörper *m*, Rippen-
heizelement *n*
f. tube Rippenrohr *n*
fiorite *(Min)* Fiorit *m*
fir Tanne *f*, *(Gattung)* Abies Mill.
fire / to anzünden, entzünden; *(Kessel)* feuern,
[be]heizen; *(Sprengstoff)* zünden; *(Ker)* brennen
f. to a white colour *(Ker)* weiß brennen
fire assay *(Met)* Brandprobe *f*
f. blanket Feuerlöschdecke *f*
f. brick *s.* firebrick
f. bridge Feuerbrücke *f (eines Flammofens)*
f. clay *s.* fireclay
f. crack *(Glas, Ker)* Brandriß *m*, Brennriß *m*
f. damp *s.* firedamp
f.-dried beet pulp feuergetrocknete Zuckerrübenschnitzel *npl*
f. extinguisher Feuerlöschmittel *n*; Feuerlöscher
m, Feuerlöschapparat *m*, Feuerlöschgerät *n*

f.-fighting foam Feuerlösch-Schaummittel n
f.-finish / to s. to f.-polish
f. finish s. f. polish
f. gilding Feuervergolden n, Feuervergoldung f
f. glazing s. f. polishing
f. hole Schürloch n (z.B. eines Brennofens)
f. mouth Feueröffnung f (z.B. eines Brennofens)
f. opal (Min) Feueropal m
f. point Brennpunkt m, BP
f.-polish / to (Glas) feuerpolieren; (Labortechnik) glattschmelzen (z.B. Glasröhren)
f. polish (Glas) Feuerpolitur f
f. polishing (Glas) Feuerpolieren n, Feuerpolitur f; (Labortechnik) Glattschmelzen n (z.B. von Glasröhren)
f.-proof s. fireproof
f. protection Brandschutz m
f. pump Feuerlöschpumpe f
f.-refine / to feuerraffinieren, im Schmelzfluß raffinieren
f. refining Feuerraffination f, Raffination f im Schmelzfluß
f. resistance Feuerbeständigkeit f, Feuerwiderstand m
f.-resistant feuerbeständig, feuerwiderstandsfähig
f.-resistant (f.-resisting) paper s. fireproof paper
f.-retardant feuerhemmend
f. retardant feuerhemmendes Mittel n
f.-retardant agent s. f. retardant
f.-retarding s. f.-retardant
f. shrinkage (Ker) Brennschwindung f
f. silvering Feuerversilbern n, Feuerversilberung f
f. tinning Feuerverzinnen n, Feuerverzinnung f
f. travel (Ker) Feuerfortschritt m (im Brennofen)
f. tube Rauchrohr n, Heizrohr n; Flammrohr n
f.-tube boiler Rauchrohrkessel m, Heizrohrkessel m, Flammrohrkessel m
f. zone Feuerzone f, Feuerstrecke f (z.B. bei der Untertagevergasung)
firebox Feuerkammer f, Feuerraum m, Feuerung f, Brennkammer f (eines Industrieofens)
firebrick Schamottestein m, Schamotteziegel m, feuerfester Stein m
f.-lined feuerfest ausgemauert
f. lining Auskleidung f aus feuerfesten Steinen
fireclay Schamotteton m, Feuerton m, feuerfester Ton m
f. baffle Prallplatte f aus feuerfestem Ton
f. box (Ker) Schamottekapsel f, Brennkapsel f
f. brick s. firebrick
f. plate Schamotteplatte f
fired body (Ker) gebrannte Masse f
f. brick gebrannter Ziegel[stein] m
f. colour (Ker) Brennfarbe f
f. density (Ker) Dichte f nach dem Brand
f. glaze (Ker) gebrannte Glasur f
f. property (Ker) Brenneigenschaft f
firedamp Grubengas n; schlagende Wetter pl, Schlagwetter pl

fireplace tile Ofenkachel f
fireproof / to feuerfest (feuersicher, feuerbeständig, brandsicher, unbrennbar) machen
fireproof feuerfest, feuersicher, feuerbeständig, brandsicher, unbrennbar
f. finish (impregnation) (Text) Flamm[en]schutzimprägnierung f, Flamm[en]schutzausrüstung f, Flammfestimprägnierung f, Flammfestausrüstung f
f. paper feuerfestes (feuersicheres, unentflammbares, nicht entflammbares) Papier n, (i.e.S.) flammsicheres (flammensicher imprägniertes, schwer entflammbares) Papier n
fireproofing Feuerfestmachen n, Feuersichermachen n, Feuerbeständigmachen n, Brandsichermachen n, Unbrennbarmachen n
f. agent Flamm[en]schutzmittel n, Feuerschutzmittel n
firestone (Min) Feuerstein m, Flint m
Firestone-Dillon plastometer s. Firestone plastometer
Firestone flexometer (Gum) Firestone-Flexometer n
Firestone plastometer (Gum) Firestone-Plastometer n, Firestone-Extrusionsplastometer n, Spritzplastometer n nach Dillon-Firestone
fireworks Feuerwerkerei f, Pyrotechnik f; Feuerwerkskörper mpl
firing Anzünden n, Entzünden n; Feuern n, Heizen n, Beheizen n; Zünden n, Zündung f (von Sprengstoffen); (Ker) Brennen n, Brand m, Flammenführung f (eines Industrieofens)
f. behaviour (Ker) Brennverhalten n
f. channel (Ker) Brennkanal m
f. contraction (Ker) Brennschwindung f
f. crack (Glas, Ker) Brandriß m, Brennriß m
f. curve (Ker) Brennkurve f
f. defect (fault) (Ker) Brennfehler m
f. hole Schürloch n (z.B. eines Brennofens)
f. range (Ker) Brennintervall n, Brennbereich m
f. shrinkage s. f. contraction
f. temperature (Ker) Brenntemperatur f
f. terminal Zündpol m (z.B. einer Kalorimeterbombe)
f. time (Ker) Brennzeit f
f. up (Gum) Anbrennen n, Anvulkanisieren n, Anvulkanisation f, Anspringen n
f. wire Zünd[ungs]draht m
firmness (Gerb) Stand m (Elastizitätseigenschaft des Leders)
f. of the head Schaumhaltigkeit f, Schaumstabilität f, Schaumdauer f, Stabilität (Haltbarkeit, Beständigkeit) f des Schaumes (beim Bier)
first break (Text) Anreißen n (erste wahrnehmbare Änderung der Farbe einer Probe)
f. carbonation juice (Zucker) erster Schlammsaft (Saturationssaft) f
f. coat Grund[ier]anstrich m, Untergrundanstrich m, Grundierung f
f. exposure (Foto) Erstbelichtung f

f. fillmass *(Zucker)* A-Füllmasse *f*, Weißzukkerfüllmasse *f*, Erstproduktfüllmasse *f*

f. ionization potential erstes Ionisierungspotential *n*, erste Ionisierungsspannung *f*, erste Ionisierungsarbeit *f*, *(in eV ausgedrückt)* erste Ionsierungsenergie *f*

f. law of thermodynamics erster Hauptsatz *m* der Thermodynamik

f. milk Erstlingsmilch *f*, Vormilch *f*, Biestmilch *f*, Kolostralmilch *f*, Kolostrum *n*

f. molasses *(Zucker)* Grünsirup *m*, Grünablauf *m*

f. order reaction Reaktion *f* erster Ordnung

f. order transition Phasenumwandlung *f* (Phasenübergang *m*) erster Ordnung

f. order transition temperature Temperatur *f* der Phasenumwandlung erster Ordnung, Umwandlungstemperatur *f* erster Ordnung

f. perceptible step *(Am)* s. f. break

f. product sugar s. f. raw sugar

f. quantum number Hauptquantenzahl *f*

f. raw sugar Rohzuckererstprodukt *n*, Rohzucker I *m*, Erstproduktzucker *m*

f. run[nings] Vorlauf *m* *(bei der Destillation)*

f. stuff *(veraltet für)* half-stuff

f. substituent Erstsubstituent *m*

f. wort Vorderwürze *f*, Hauptwürze *f*

Fischer base Fischersche Base *f*, Fischer-Base *f*

Fischer-Hepp rearrangement Fischer-Hepp-Umlagerung *f* *(aromatischer Nitrosamine zu p-Nitrosaminen in Gegenwart von Säuren)*

Fischer indol synthesis Fischersche Indolsynthese *f* *(aus Phenylhydrazonen)*

Fischer projection formula Fischersche Projektionsformel *f*

Fischer-Tropsch plant Fischer-Tropsch-Anlage *f*

Fischer-Tropsch process (synthesis) Fischer-Tropsch-Verfahren *n*, Fischer-Tropsch-Synthese *f*, Kogasinverfahren *n*

fischerite *(Min)* Fischerit *m* *(Aluminiumtrihydroxidorthophosphat)*

fisetin Fisetin *n*, 3,7,3′,4′-Tetrahydroxyflavon *n*

fish 1. *(Lebm)* Fisch *m*; 2. *(Erdöl)* Fisch *m* *(im Bohrloch verlorengegangenes Gerät)*

f. curing Fischkonservierung *f* *(durch Trocknen, Salzen, Räuchern und Säuern)*

f. eye Fischauge *n* *(Materialfehler bei durchsichtigem oder durchscheinendem Plast)*

f. fat s. f. oil

f. flour Fischmehl *n*

f. gelatine (glue, isinglass) Fischleim *m*; Hausenblasenleim *m*, Hausenblase *f*

f. liver oil Fischleberöl *n*, Fischlebertran *m*

f. oil Fischöl *n*, Fischtran *m*

f.-oil spew (spue) *(Gerb)* Tranausharzung *f*

f.-oil stuffing *(Gerb)* Tranfüllung *f*, Tranfettung *f*

f. reduction plant Fischverwertungsanlage *f*

f. scale 1. Fischschuppe *f*; 2. *(Ker)* Fischschuppen *fpl* *(Emaillierfehler)*

f. scale essence *(Kosmet)* Fischschuppenessenz *f*

f.-tail bit *(Erdöl)* Fischschwanzmeißel *m*

f.-tail burner Fischschwanzbrenner *m*, Schlitzbrenner *m*, Schnittbrenner *m*

f. tankage Fischmehl *n* *(als Düngemittel)*

fishiness *(Lebm)* fischiger (fischartiger) Geschmack *m*, Fischgeschmack *m*; Fischigwerden *n* *(von Milch und Milchprodukten)*

fishing *(Erdöl)* Fischen *n*, Fisch-Arbeit *f*, Fangen *n*, Fangarbeit *f*, Fangarbeiten *n*

f. magnet *(Erdöl)* Fangmagnet *m*

f. tap Fangdorn *m* *(beim Bohren auf Erdöl)*

fishy fischig, fischartig *(Geruch)*

fissile s. fissionable

fissility s. fissionability

fission Spalten *n*, Spaltung *f*, Aufspalten *n*, Aufspaltung *f*

f. capture *(Kch)* Spalt[ungs]einfang *m*

f. chain *(Kch)* Spaltkette *f*, Spaltproduktreihe *f*

f. chamber *(Kch)* Spaltungskammer *f*

f. cross section *(Kch)* Spaltungsquerschnitt *m*

f. decay chain s. f. chain

f. energy Spalt[ungs]energie *f*, Kernspaltungsenergie *f*

f. fragment *(Kch)* Spaltstück *n*, Bruchstück *n*

f. poison Reaktorgift *n*, Spaltgift *n*

f. product Spalt[ungs]produkt *n*

f. reaction Abspaltungsreaktion *f*, Spaltungsreaktion *f*, Spaltprozeß *m*

f. yield Spalt[produkt]ausbeute *f*

f. yield curve Spalt[produkt]ausbeutekurve *f*

fissionability Spaltbarkeit *f*

fissionable spaltbar, spaltfähig

f. material spaltbares Material *n*, Spaltmaterial *n*, Spaltstoff *m*

fitted with bars (knives) bemessert, mit Messern besetzt

Fittig reaction (synthesis) Fittigsche Synthese *f* *(von Di- und Polyarylen)*

fitting Formstück *n*, Fitting *m(n)*, Verbindungsstück *n*, Verbindung *f*, Verbinder *m*, Verschraubung *f* *(für Rohr und Schläuche)*

f. board Dichtungspappe *f*

f. cardboard Dichtungskarton *m*

five-carbon chain C_5-Kette *f*

f. days biochemical oxygen demand fünftägiger biochemischer Sauerstoffbedarf *m*, BSB_5[-Wert] *m*

f.-membered fünfgliedrig

f.-membered carbon ring Kohlenstoff-Fünfring *m*

f.-membered ring fünfgliedriger Ring *m*, Fünf[er]ring *m*

f.-roll mill *(Farb)* Fünfwalzenmühle *f*, Fünfwalzenstuhl *m*

f.-stage bleaching *(Pap)* Fünfstufenbleiche *f*

fix / to binden, fixieren; *(Foto)* fixieren; *(Landw)* *(Nährstoffe im Boden)* festlegen

fixation Bindung *f*, Fixierung *f*; *(Foto)* Fixieren *n*, Fixierung *f*, Fixage *f*; *(Landw)* Festlegung *f* *(von Nährstoffen im Boden)*

                                             flameproof

f. of nitrogen Stickstoffbindung *f*, Stickstoff-Fixierung *f*

fixative *(Farb)* Fixativ *n*; *(biologische Konservierungstechnik)* Fixierungsmittel *n*; *(Kosmet)* Fixateur *m (Fixiermittel für Parfüme)*

fixed-angle head Winkelkopf *m (einer Laborzentrifuge)*

f.-angle rotor Winkelrotor *m (einer Laborzentrifuge)*

f. bed Festbett *n*, ruhendes (statisches) Bett (Feststoffbett) *n*, ruhende (statische) Schüttung *f*

f.-bed adsorber Festbettadsorber *m*

f.-bed catalyst Festbettkatalysator *m*, Festbettkontakt *m*, festliegender (fester, ruhender, fest angeordneter) Katalysator (Kontakt) *m*

f.-bed column Festbettkolonne *f*

f.-bed process Festbettverfahren *n*

f. carbon fixer (fester) Kohlenstoff *m*

f. catalyst bed fest[liegend]es (ruhendes, stationäres) Katalysatorbett *n*

f. derrick stationärer Bohrturm *m*

f. dryer Trockner *m* mit unbewegtem Trockengut

f. fuel bed festes (ruhendes) Brennstoffbett *n*, ruhende Brennstoffschicht (Brennstoffschüttung) *f*

f. jaw feststehende Brechbacke *f*, Stirnwand-Brechbacke *f (eines Backenbrechers)*

f. millstone Bodenstein *m*, Unterstein *m*

f. nitrogen gebundener (fixierter) Stickstoff *m*

f. oil fettes Öl *n*, Fettöl *n*

f. plate *(Plast)* Stammplatte *f*, Stammgesenk *n*

f. point Fixpunkt *m*

f. seat Brecherrahmen *m (eines Backenbrechers)*

f.-sieve jig Setzmaschine *f* mit festem Sieb

f. tan gebundener Gerbstoff *m*

f. white Permanentweiß *n*, Barytweiß *n (Bariumsulfat)*

fixer *(Foto)* Fixiermittel *n*, Fixiersalz *n*; Fixierbad *n*

fixing Fixieren *n*, Fixierung *f*, *(Foto auch)* Fixage *f*

f. agent *(Foto, Text)* Fixiermittel *n*; *(Text, Farb)* Nachbehandlungsmittel *n (zur Echtheitserhöhung von Färbungen)*; *s.* fixative

f. bath Fixierbad *n*

f. process Fixiervorgang *m*

f. salt Fixiersalz *n*

f. solution Fixierlösung *f*

f. speed Fixiergeschwindigkeit *f*

f. time Fixierdauer *f*

flag root Kalmus *m*, Acorus calamus L.

flake / to 1. in Schichten zerlegen, Schichten abspalten; 2. *s.* to f. off

f. off abblättern, abschuppen, abplatzen, abspringen

flake Flocke *f*, Schuppe *f*

f. ice Scherbeneis *n*, Splittereis *n*

f.-like flockig, flockenartig, flockenförmig

f. sulphide geschupptes Sulfid *n*

f. white Schieferweiß *n*, *(feines)* Bleiweiß *n*

flaked oats Haferflocken *fpl*

flaker *(Lebm)* Flockenstuhl *m*

flakice *s.* flake ice

flaking Flocken *n*, Flockung *f*; Abblättern *n*, Abschuppen *n*, Schuppenbildung *f*, Abplatzen *n*, Abspringen *n*; Klumpen *n*; *(Pap) (Am)* Rupfen *n*

flaky ashes Flugasche *f*

f. pastry margarine Ziehmargarine *f*

flamability *s.* flammability

flame Flamme *f*

f.-blowing process *(Glas)* Flammenblasverfahren *n*

f. coal Flammkohle *f*

f. colouration Flammenfärbung *f*

f. cutoff *(Glas)* Absprengen *n*

f. descaling Flammstrahlen *n (zur Reinigung von Metalloberflächen)*

f. failure ungewolltes Erlöschen (Verlöschen) *n* einer Flamme

f. failure safeguard Flammenwächter *m*

f. front Flamm[en]front *f*

f.-harden / to flamm[en]härten, brennhärten

f. hardening Flamm[en]härten *n*, Flamm[en]härtung *f*, Brennhärten *n*, Brennhärtung *f*

f. impingement Flammenaufprall *m*

f. ionization detector Flammenionisationsdetektor *m*

f. lamination Flammkaschierung *f*, Schmelzbeschichtung *f*

f. photometer Flammenfotometer *n*

f.-photometric flammenfotometrisch

f. photometry Flamm[en]fotometrie *f*, Flammenspektrometrie *f*

f.-proof *s.* flameproof

f. propagation Flammenfortpflanzung *f*

f. reactor Reaktionsofen *m* für Flammenreaktionen

f. resistance Flammbeständigkeit *f*, Flammwidrigkeit *f*

f.-resistant flammbeständig, flammwidrig

f.-resistant finish *s.* flameproof finish

f. resistivity *s.* f. resistance

f. retardancy Flammenverzögerungsvermögen *n*

f.-retardant flammenhemmend

f. retardant (retarder) Flammenverzögerungsmittel *n*

f. spectroscopy Flammenspektroskopie *f*

f. spectrum Flammenspektrum *n*

f. spraying Flammspritzen *n*, *(Plast auch)* Wärmespritzen *n*

f.-spraying gun Flammspritzpistole *f*

f. spreader Breitbrenneraufsatz *m*, Schnittbrenneraufsatz *m*, Schlitzaufsatz *m*

f. temperature Flammentemperatur *f*

f. thrower Flammenwerfer *m*

f. velocity Flammengeschwindigkeit *f*

flameproof flammsicher; flammfest, nicht *(oder)* schwer entflammbar, unentflammbar

f. finish (impregnation) *(Text)* Flammfestausrüstung *f*, Flammfestimprägnierung *f*, Flamm[en]schutzausrüstung *f*, Flamm[en]schutzimprägnierung *f*

f. paper flammsicheres (flammsicher imprägniertes, schwer entflammbares) Papier *n*
flameproofed flammsicher (flammfest) ausgerüstet (imprägniert)
flameproofing Flammsichermachen *n*, Flammfestmachen *n*, Unentflammbarmachen *n*
f. agent Flamm[en]schutzmittel *n*, Feuerschutzmittel *n*
f. paint Feuerschutzfarbe *f*
flameproofness Flammsicherheit *f*, Flammfestigkeit *f*, Nichtentflammbarkeit *f*, Unentflammbarkeit *f*
flaming arc Flammenbogen *m*
flammability Entflammbarkeit *f*, Entzündbarkeit *f*, Entzündlichkeit *f*; Brennbarkeit *f*
flammable entflammbar, entzündbar, entzündlich; brennbar
flange Flansch *m*
f. bolt Flanschschraube *f*, Flanschbolzen *m*
f. face Flansch[stirn]fläche *f*
f. gasket Flanschdichtung *f*
f. seal Hutmanschette *f*; Flanschdichtung *f*
f.-type heater Heizflansch *m*
flanged angeflanscht, Flansch...
f.-end pipe Flanschenrohr *n*
f. fitting Flansch[form]stück *n*, Flanschfitting *m(n)*
f. joint Flanschverbindung *f*
f. pipe Flanschenrohr *n*
flap *(Gum)* Felgenband *n*
f. mould (vulcanizer) *(Gum)* Felgenbandheizer *m*
flare *(Foto)* Reflexionsfleck *m*, Reflex *m*, Spiegelfleck *m (auf dem Negativ)*
flareback Zurückschlagen *n (einer Flamme)*, Flammenrückschlag *m*
flared-fitting joint Bördelverbindung *f (von Rohren)*
flash / to entflammen; aufflammen, aufblitzen; sehr rasch verdampfen *(z.B. durch Entspannung)*; *(Glas)* überfangen
f. back zurückschlagen *(Flammen in Brennern)*
flash 1. Aufflammen *n*, Aufblitzen *n*; 2. Grat *m*, *(Plast auch)* Austrieb *m*
f. ageing *(Text)* Blitzdämpfen *n*, Schnelldämpfen *n*
f. ager *(Text)* Blitzdämpfer *m*, Schnelldämpfer *m*
f. boiler Zwangdurchlauf-Dampfkessel *m*
f. chamber Entspann[ungs]kammer *f*, Verdampfungskammer *f*, Brüdenraum *m (des Entspannungsverdampfers)*, Flashkammer *f*, Flashraum *m*
f. cooler Entspannungskühler *m*
f. cup Flammpunktstiegel *m*
f. curve Gleichgewichtssiedekurve *f*, Siedekurve *f* bei geschlossener Verdampfung, Flashkurve *f*
f. distillation [kontinuierliche] Entspannungsdestillation *f*
f. dryer Zerstäubungstrockner *m*, Sprühtrockner *m*
f. drying Zerstäubungstrocknung *f*, Sprühtrocknung *f*

f. evaporation Entspannungsverdampfung *f*
f. evaporator Entspannungsverdampfer *m*, Schnellverdampfer *m*
f. line Gratkante *f*, Gratlinie *f*
f. mould *(Plast)* Abquetschform *f*, Abquetschwerkzeug *n*
f. pasteurization *(Lebm)* Momentanpasteurisation *f*, Hocherhitzung *f*, Momenterhitzung *f (zur Pasteurisation)*
f. pasteurizer (Lebm) Hocherhitzer *m (zur Pasteurisation)*, Blitzpasteurisierapparat *m*
f. photolysis Blitz[licht]fotolyse *f (Analysenverfahren)*
f. plate *(Galvanotechnik)* Schnellüberzug *m*
f. point Flammpunkt *m*, FP.
f.-point apparatus (tester) Flammpunktprüfer *m*, Flammpunkt[prüf]gerät *n*, Flammpunktapparat *m*
f. pulverizer Strahlprallmühle *f*
f. removal *(Plast)* Abgraten *n*, Entgraten *n*
f. roaster *(Met)* Blitzröstofen *m*, Flashröster *m*
f. roasting *(Met)* Blitzröstung *f*, *(i.w.S.)* Schweberöstung *f*
f. smelting Schweberöstschmelzen *n*
f. spectrum Flashspektrum *n (der Sonnenkorona)*; Funkenspektrum *n*
f. tester *s.* f.-point apparatus
flashback Zurückschlagen *n (einer Flamme)*, Flammenrückschlag *m*
flashed glass Überfangglas *n*
flashing *(Lebm)* s. flash pasteurization; *(Glas)* Überfangen *n*; *(Ker)* Gießfleck *m (Fehler)*
f. powder *(Foto)* Blitzlichtpulver *n*
flashlight *(Foto)* Blitzlicht *n*; Blitzlichtpulver *n*
flask Kolben *m*; Flasche *f*
f. clamp Stativklemme *f* mit halbrunden Spannbacken
f. combustion *(Analytik)* Kolbenverbrennung *f*
f. holder (support) Kolbenträger *m*
flat / to *(Farb)* mattieren
flat flach, eben; fahl, matt, stumpf, trübe *(Farbton)*; schal, fad[e], abgestanden *(Geschmack)*
f.-bed printing screen Flachdruckschablone *f*
f. belt Flachband *n*, Flachgurt *m*; Flachriemen *m*
f.-bottom[ed] flask Stehkolben *m*
f.-bottom[ed] flask with long neck Langhalsstehkolben *m*
f.-bottom[ed] flask with short neck Kurzhalsstehkolben *m*
f. burner head Breitbrenneraufsatz *m*
f.-cone crusher Flachkegelbrecher *m*
f.-curing *(Gum)* mit breitem Vulkanisationsplateau (Plateau) *(mit großem Optimalbereich der Vulkanisation)*
f.-curing effect *(Gum)* Plateaueffekt *m*
f. die Düse *f* zum Strangpressen von Folien
f.-frame burner Breitbrenner *m*, Schnittbrenner *m*
f. glass Flachglas *n*
f.-glass furnace Flachglasofen *m*

f.-plate evaporator Verdampfer *m* mit Plattenheizkörpern
f. screen Flachsortierer *m*, Plansortierer *m*; *(Pap)* Planknotenfänger *m*
f. setting *(Text)* Flächenfixieren *n*, Flächenfixierung *f*
f. sheet drawing process Tafelglasziehverfahren *n*
f. sieve Flachsieb *n*, Plansieb *n*
f. spray nozzle Flachstrahldüse *f*, Pralldüse *f*, Breitstrahldüse *f*
f. strainer *(Pap)* Planknotenfänger *m*
f. varnish Mattlack *m*
f. ware *(Glas, Ker)* Flachware *f*
flatness *(Foto)* Kontrastlosigkeit *f*
flatten / to *(Farb)* abstumpfen; *(Glas)* glätten, bügeln; *(Glas)* strecken
flattening *(Farb)* Abstumpfen *n*; *(Glas)* Glätten *n*, Bügeln *n*; *(Glas)* Strecken *n*
f. kiln (oven) *(Glas)* Streckofen *m*
flatting agent *(Farb)* Mattierungsmittel *n*
f. varnish Spachtellack *m*
flavan Flavan *n*
flavanoid Flavanoid *n*, Flavan-Derivat *n*
flavanone Flavanon *n*
flavanthrone Flavanthron *n*
flavianic acid Flaviansäure *f*, 2,4-Dinitro-1-naphthol-7-sulfonsäure *f*
flavin 1. Flavin *n*, Isoalloxazin *n*; 2. Querzetin *n*, 5,7,3′,4′-Tetrahydroxyflavonol *n*
f. adenine dinucleotide Flavin-adenin-dinukleotid *n*, FAD, Alloxazin-adenin-dinukleotid *n*
f. enzyme *s.* flavoprotein
f. mononucleotide Flavinmononukleotid *n*, FMN
flavine 1. *s.* flavin; 2. *(Pharm)* Akriflavin *n*
flavoenzyme *s.* flavoprotein
flavone Flavon *n*, 2-Phenylchromon *n*, 2-Phenyl-1,4-chromenon *n*
f. pigment Flavonfarbstoff *m*
flavoprotein Flavoprotein *n*, Flavinferment *n*, Gelbes Ferment *n*
flavor / to *s.* to flavour
flavor *s.* flavour
flavour / to schmackhaft machen, aromatisieren, würzen; parfümieren *(z.B. Tabak)*
flavour Aroma *n*, Geschmack *m*, Wohlgeschmack *m*; Duft *m*, Geruch *m*, Wohlgeruch *m*; Würze *f*, Blume *f* *(des Weins)*
f. defect Geschmacksfehler *m*
f. deterioration Geschmacksverschlechterung *f*
f. reversion Geschmacksabwertung *f*
flavouring material (matter, substance) Geruchsstoff *m*, Geschmacksstoff *m*, Aromastoff *m*
flavourless geschmacklos; geruchlos
flavylogen Flavylogen *n*, Anthozyanogen *n*, Leukoanthozyan *n*
flaw Materialfehler *m*, Fehler *m*, Defekt *m*; *(Krist)* Lockerstelle *f*; *(Ker)* Riß *m*, Sprung *m*
flax Lein *m*, Linum L., *(i.e.S.)* Flachs *m*, [Gemeiner] Lein *m*, Linum usitatissimum L.

f. fibre Flachsfaser *f*
f. oil *s.* f.-seed oil
f. packing Flachsdichtung *f*, Flachspackung *f*
f. rolling Brechen *n*, Knicken *n* *(von Flachs mit Walzen)*
f.-seed oil Leinöl *n*
f. straw Flachsstroh *n*
flèche d'amour *(Min)* Crinis veneris, Venushaar *n* *(faseriger Rutil; Titandioxid)*
fleece wool Vlieswolle *f*
Fleig reagent for blood Fleigsches Reagens *n* zum Blutnachweis im Harn
Fleming method Fleming-Methode *f* *(zur Penizillinbestimmung)*
Flemming solution *(Bio)* Flemmingsche Lösung *f*, Flemmings Chromosmiumessigsäurelösung *f* *(Fixierungsmittel)*
flesh / to *(Gerb)* entfleischen, ausstoßen
flesh-coloured fleischfarben, fleischfarbig
f. finish *(Gerb)* Aasappretur *f*
f. layer *(Gerb)* Aasseite *f*
f. split *(Gerb)* Fleischspalt *m*
fleshing knife *(Gerb)* Scherdegen *m*
Fletcher bleacher *(Pap)* Fletcher-Bleichturm *m*
fletton Fletton-Ziegel *m*
Fletton brick clay Fletton-Ziegelton *m*
flex cracking Biegerißbildung *f*, Biegeermüdung *f*
f.-cracking resistance Biegerißwiderstand *m*, Biegerißfestigkeit *f*
f.-cracking test Prüfung *f* der Biegerißfestigkeit, Dauerknickversuch *m*, Ermüdungsprüfung *f*
f. life Dauerbiegefestigkeit *f*, Ermüdungsbeständigkeit *f*
f.-life time *(Gum)* Bruchzeit *f*
f.-resistance *s.* f.-cracking resistance
flexibility Biegsamkeit *f*, Flexibilität *f*
flexibilize / to *(Plast)* weichmachen, flexibilisieren
flexibilizer Weichmacher *m* *(für Epoxidharze)*
flexible biegsam, flexibel, elastisch, federnd; nicht starr *(z.B. Kupplung)*
f. foam *(Plast)* weicher (weich-elastischer) Schaumstoff *m*
f. metal hose Metallschlauch *m*
f. plastic flexibler (weicher, weichgestellter) Plast *m*
f. PVC Weich-PVC *n*, PVC weich *n*
flexing life *s.* flex life
flexographic ink Farbe *f* für Flexodruck (Flexografie, Anilingummidruck), Anilingummidruckfarbe *f*
f. printing Flexodruck *m*, flexografischer Druck *m*, Flexografie *f*, Anilin[gummi]druck *m*
flexography *s.* flexographic printing
flexometer Flexometer *n* *(Gummiprüfgerät)*
flexural cracking *s.* flex cracking
f. impact strength test Schlagbiegeversuch *m*
f. modulus Biegemodul *m* *(aus dem Biegeversuch ermittelter E-Modul)*
f. property Biegeeigenschaft *f*
f. rigidity Biegesteifigkeit *f*

f. strength Biegefestigkeit f
Flick solution Flicksche Lösung f *(aus HCl und H₂F₂ zum Ätzen von Aluminium)*
flight Hebeleiste f, Mitnehmerblech n, Schaufel f *(z.B. in einer Trockentrommel)*; Gang m *(z.B. eines Schraubenrades)*; Mitnehmer m *(eines Kratzerförderers)*
f. conveyor Kratzer[förderer] m, Kratz[en]band n, Kratzerkette f
flighted conveyor s. flight conveyor
f. wheel Schöpfrad n
flighty element *(Element mit mehreren, leicht ineinander überführbaren Wertigkeitsstufen)*
flinger Schleuderfinger m, Schleuder f
flint *(Min)* Flint[stein] m, Feuerstein m
f. glass Flintglas n
f.-glazed paper Glanzpapier n
f. glazer s. f.-glazing machine
f. glazing *(Pap)* Glätten n mit Achatstein
f.-glazing machine *(Pap)* Achatsteinglätteinrichtung f, Achatsteinglätte f, Steinglätte f
f. paper s. f.-glazed paper
f. stone s. flint
flinting s. flint glazing
flipping-over Umklappen n *(von Bindungen bei der Waldenschen Umkehrung)*
float / to flotieren *(durch Aufschwemmen aufbereiten)*; aufschwimmen
f. off ausschwimmen, fortschwemmen, abschwemmen, forttragen
float Schwimmer[körper] m
f.-and-sink method Schwimm-und Sinkverfahren n, Schwimmsinkverfahren n, *(meist)* Schwertrübeverfahren n
f. dryer Düsentrockner m
f. gauge Schwimmermesser m
f. material (product) *(Flot)* Schwimmgut n, aufschwimmendes Gut n
f. stone *(Min)* Schwimmkiesel m
f.-type manometer Schwimmerdruckmesser m, Schwimmermanometer n
f. valve Schwimmerventil n
floatability Schwimmfähigkeit f; Flotierbarkeit f
floatable schwimmfähig; flotierbar
floatation Flotieren n, Flotation f, Schwimmaufbereitung f
f. agent s. f. reagent
f. apparatus s. f. machine
f. assay Flotationsprobe f
f. benefication flotative Aufbereitung f, Aufbereitung f durch Flotation
f. cell Flotationszelle f
f. machine Flotationsmaschine f, Flotationsapparat m, Flotationsgerät n, Schwimmaschine f, Schwimmgerät n
f. method s. f. process
f. oil Flotationsöl n
f. plant Flotationsanlage f
f. process Flotationsverfahren n, Flotationsprozeß m, Schwimm[aufbereitungs]verfahren n

f. reagent Flotationsmittel n, Flotationsreagens n
f. recovery Gewinnung f durch Flotation
f. save-all *(Pap)* Flotationsstoffänger m
f. separation flotative Trennung f
f. sulphur Flotationsschwefel m
f. tailings Flotationsberge pl, Flotationsabgänge mpl, Abgänge mpl
f. unit Einzelzelle f *(einer Flotationsanlage)*, Flotationszelle f
floater Schwimmer m, Pegel m
floating Aufschwimmen n [in horizontaler Richtung] *(Entmischen der Pigmente in Anstrichstoffen)*
f. fraction Schwimmanteil m, Schwimmgut n, aufschwimmendes Gut n
f. head schwimmender Kopf m, Schwimmkopf m, beweglicher Rohrboden m, frei bewegliche Umkehrkammer f *(eines Wärmeaustauschers)*
f. lime *(Gerb)* Schwimmäscher m
f. material s. float material
f. plate *(Plast)* bewegliche Platte (Zwischenplatte) f
f. punch *(Plast)* beweglicher Stempel m
f. soap Schwimmseife f
f. zone [frei]schwebende Zone f, Schwebezone f *(beim tiegelfreien Zonenschmelzen)*
f.-zone apparatus Schwebezonenapparatur f, Apparatur f zum tiegelfreien (tiegellosen) Zonenschmelzen
f.-zone melting Zonenschmelzen n nach dem Schwebezonenverfahren, tiegelfreies (tiegelloses) Zonenschmelzen n, Zonenschmelzen n ohne Substanzbehälter
f.-zone method Schwebezonenverfahren n, Methode f der [frei]schwebenden Zone, tiegelfreies (tiegelloses) Zonenschmelzverfahren n
f.-zone refining Zonenreinigung f nach dem Schwebezonenverfahren, tiegelfreie (tiegellose) Zonenreinigung f
floats Schwimmgut n, aufschwimmendes Gut n
f. curve Schwimmkurve f
floc Flocke f *(besonders in Suspensionen)*
flocculant Flockungsmittel n, Ausflockungsmittel n, Flocker m, *(Koll auch)* Koagulationsmittel n, Koagulator m
flocculate / to 1. [aus]flocken, zusammenballen, *(Koll auch)* koagulieren; 2. [aus]flocken, sich zusammenflocken (zusammenballen), *(Koll auch)* koagulieren
flocculate Flocke f *(eines Niederschlags)*; flockiger Niederschlag m, *(Koll)* Koagulat n
flocculating agent s. flocculant
f. chamber Flockungsraum m
. **f. power** s. flocculation power
flocculation 1. Ausflocken n, Ausflockung f, Flockung f, Flokkulation f, Zusammenballen n, Zusammenballung f, *(Koll auch)* Koagulation f, Pektisation f, *(Bodenkunde auch)* Krümelbildung f, Krümelung f; 2. s. flocculate

f. by electrolytes *(Koll)* Elektrolytkoagulation *f*
f. power flockende Kraft *f*, Flockungskraft *f*, Flockungsfähigkeit *f*
f. process Flockungsablauf *m*, Flockungsverlauf *m*, *(Koll auch)* Koagulationsablauf *m*, Koagulationsverlauf *m*
f. rate Flockungsgeschwindigkeit *f*, *(Koll auch)* Koagulationsgeschwindigkeit *f*
f. value Flockungswert *m*, *(Koll auch)* Koagulationswert *m*
f. zone Flockungszone *f*, *(Koll auch)* Koagulationszone *f*
floccule Flocke *f*, Flöckchen *n*
flocculence Ausflockung *f*, *(Koll auch)* Koagulation *f*
flocculent flockig, flockenartig, flockenförmig
flocculent *s.* flocculant
flock Flocke *f*; *(Text)* Kurzfaser *f*, Flocke *f*
f. formation Flockenbildung *f*
f. paper Velourpapier *n*, Plüschpapier *n*, Tuchpapier *n*, Samtpapier *n*
f. print *(Text)* Flockdruck *m*
f. silk Flockseide *f* *(von Kokonabfällen)*
f. spraying *(Plast)* Beflocken *n*
flood */ to* [über]fluten; *(Gerb)* anschwöden; *(Farb)* [in vertikaler Richtung] ausschwimmen *(sich entmischen)*
flooding Überfluten *n*, Fluten *n*; *(Gerb)* Anschwöden *n*; *(Farb)* Ausschwimmen *n* [in vertikaler Richtung] *(Entmischen der Pigmente)*
f. point *(Destillation)* Überflutungsgrenze *f*, Spuckgrenze *f*
floor Boden *m*; Fußboden *m*; Horde *f*, Rösthorde *f*, Darrhorde *f*; Malztenne *f*
f. covering Fußbodenbelag *m*, Bodenbelag *m*
f. malting Tennenmälzerei *f*, Tennenvermälzung *f*
f. of the pan (tub) *(Pap)* Trogsohle *f*, Trogboden *m* *(beim Holländer)*
f. plough Tennenwender *m*, Wender *m*, Wendevorrichtung *f* *(Mälzerei)*
f. polish Bohnerwachs *n*, *(i.w.S.)* Fußbodenpflegemittel *n*
f. tile Bodenfliese *f*, Bodenplatte *f*
f. wax Bohnerwachs *n*
flooring *s.* 1. floor covering; 2. floor malting
floral concrete *(Kosmet)* konkretes Blütenöl *n*
Florence reagent Reagens *n* nach Florence *(zum Nachweis von Blut und Gallenfarbstoffen in Urin)*
Florentine receiver Florentiner Flasche *f* *(eine Vorlage)*
floret[te] silk Florettseide *f*, Kokonseide *f*
Florey unit Florey-Einheit *f*, Oxford-Einheit *f*, OE *(veraltete biologische Maßeinheit des Penizillins)*
floridean starch Florideenstärke *f*
florspar *s.* fluorspar
floss silk Kokonseide *f*; Stickseide *f*
flotation *s.* floatation
flour Mehl *n*, Staub *m*, feines Pulver *n*; *(Pap)* Mehlstoff *m*; *(Lebm)* [feines] Mehl (Weizenmehl) *n*

f. dust Dunstmehl *n*, Dunst *m*; Mehlstaub *m*
f. improvement Mehlverbesserung *f* *(Verbesserung der backtechnischen Eigenschaften)*; Mehlbleichung *f*
f. improver Mehlverbesserungsmittel *n* *(zur Verbesserung der backtechnischen Eigenschaften)*; Mehlbleich[ungs]mittel *n*
f. mixer Mehlmischer *m*
f. sand Staubsand *m*, feinster Sand *m*
f. sifting and dressing machine Beutelmaschine *f*, Beutelwerk *n*, Sichtmaschine *f*, Sichter *m*
f. sulphur Schwefelmehl *n*, Ventilatoschwefel *m*
floury mehlig, mehlartig; mehlhaltig, stärkehaltig, mehlreich, stärkereich
flow */ to* fließen, strömen, rinnen; laufen, flüssig werden; *(Farb)* verlaufen, fließen *(von Anstrichen)*
f. back zurückfließen
f. round umspülen
flow Fließen *n*, Fluß *m*, Strömen *n*, Strömung *f*; Strom *m*; Verlauf *m* *(z.B. eines Anstrichmittels)*
f. area Durchlaßquerschnitt *m*, Durchströmquerschnitt *m*, Strömungsquerschnitt *m*
f. behaviour Fließverhalten *n*
f. box *(Pap)* Stoffauflauf[kasten] *m*, Auflaufkasten *m*
f. button test *(Ker)* Knopfprobe *f*, Knopfprüfung *f* *(der Fließbarkeit einer Fritte)*
f. casting *(Gum, Plast)* Gießverfahren *n*, Gußverfahren *n*
f. coating Flutlackieren *n*, Flutlackierung *f*, Fluten *n*; Flow-Coating *n* *(moderne Weiterentwicklung des Flutlackierens)*
f.-coating section Flutzone *f*, Fluttunnel *m*, Flutkammer *f* *(beim Flow-Coating-Verfahren)*
f. coefficient Durchflußzahl *f*, Durchflußkoeffizient *m*, Durchflußbeiwert *m*
f.-control agent *(Farb)* Verlaufmittel *n*
f. controller Stellglied *n* für Massenstrom (Mengenstrom), Durchflußstrom-Stellglied *n*
f. counter Durchflußzählrohr *n*
f. cup *(Plast)* Prüfbecher *m*
f. cup test *(Plast)* Schließzeitbestimmung *f* *(mittels Prüfbechers)*
f. curve Fließkurve *f*
f. diagram Fließbild *n*, Fließdiagramm *n*, Strömungsbild *n*
f. distributor *(Pap)* Stoffverteiler *m*
f. feeder *(Glas)* Fließspeiser *m*
f. hole *(Glas)* Durchlaß *m*, Durchfluß *m*
f. inducer Schlauch[quetsch]pumpe *f*
f. line *(Glas)* Runzel *f*, Preßrunzel *f* *(Oberflächenfehler)*
f. measurement Durchfluß[mengen]messung *f*, Mengenstrommessung *f*
f. meter *s.* flowmeter
f. mixer kontinuierlicher Flüssigkeitsmischer *m*
f. moulding *(Plast)* Preßspritzen *n*, Spritzpressen *n*
f. nozzle Meßdüse *f* *(zur Mengenstrommessung)*

f. of pulp (stock) *(Pap)* Stofffluß *m*, Stoffstrom *m*, Stofflauf *m*
f. onto the wire *(Pap)* Stoffauflauf *m*
f. path Fließweg *m (besonders bei Spritzgußwerkzeugen)*
f. pattern s. f. diagram
f. point *(Met)* Fließpunkt *m*, Fließgrenze *f*, Fließtemperatur *f*
f. property s. flowability
f. rate Durchsatz[strom] *m*, Durchfluß[strom] *m*, Förderstrom *m*, Strom *m*; Fließgeschwindigkeit *f*, Strömungsgeschwindigkeit *f*, Durchflußgeschwindigkeit *f*, Durchlaufgeschwindigkeit *f*
f. sheet s. f. diagram
f. temperature Fließtemperatur *f*
f. texture *(Geol)* Fluidaltextur *f*, Fließtextur *f*, Fließgefüge *n*
flowability Fließvermögen *n*, Fließfähigkeit *f*, Fließbarkeit *f*, Fließverhalten *n*, Fließeigenschaft *f*
flower absolute *(Kosmet)* absolutes Blütenöl *n*
f. fertilizer Blumendünger *m*
f. oil *(Kosmet)* Blütenöl *n*
f. tissue Blumen[seiden]papier *n*
flowers of antimony *(Min)* Antimonblüte *f*, Valentinit *m (Antimon(III)-oxid)*
f. of camphor Kampferblüte *f*, Kampferblume *f*
f. of sulphur Schwefelblüte *f*, Schwefelblume *f*
f. of tin SnO₂ Zinnasche *f*, Zinn(IV)-oxid *n*
f. of zinc Zinkblumen *fpl (weißes, wollartiges Zinkoxid)*
flowing *(Ker)* Ablaufen *n (der Glasur)*
f. property s. flowability
f. well *(Erdöl)* fließender Brunnen *m*, freifließende Bohrung *f*
flowmeter Durchfluß[mengen]meßgerät *n*, Durchfluß[mengen]messer *m*, Flußmesser *m*, Mengenstrommesser *m*
flue Zug *m*, Heizzug *m*, Feuerzug *m*, Ofenzug *m*
f. ash Flugasche *f*
f. dust monitor Rauchmeldeanlage *f*
f. gas Rauchgas *n*, Abgas *n*, Verbrennungs[ab]gas *n*, Ofengas *n*
f. temperature Heizzugtemperatur *f*
fluellite *(Min)* Fluellit *m (Aluminiumfluorid-1-Wasser)*
fluff / **to** flockig machen, auflockern; flockig werden; *(Gerb)* auf der Fleischseite abschleifen
fluffy flockig
f. black *(Gum)* loser Ruß *m*
fluid flüssig
fluid gestaltloses Medium *n*; Flüssigkeit *f*; Gas *n*
f. bed Fließbett *n*, Fließschicht *f*, Wirbelbett *n*, Wirbelschicht *f*
f.-bed catalyst s. f. catalyst
f.-bed dryer Fließbetttrockner *m*, Wirbelbetttrockner *m*, Wirbelschichttrockner *m*
f.-bed roaster (roasting furnace) Wirbelschichtröstofen *m*, Wirbelschichtröster *m*
f.-bed vulcanization Fließbettvulkanisation *f*

f. catalyst Fließbettkatalysator *m*, Wirbelbettkatalysator *m*, Staubfließkatalysator *m*, Fließstaubkontakt *m*
f. catalytic cracking katalytisches Wirbelschichtkracken *n*, katalytisches Kracken *n* in [der] Wirbelschicht, Kracken *n* in Wirbelschicht mit Fließbettkatalysator
f. catalytic cracking plant katalytische Krackanlage *f* nach dem Wirbelbettverfahren, FCC-Anlage *f*
f. catalytic [cracking] process fluidkatalytisches Krackverfahren *n*, Fluidkrackverfahren *n*, FCC-Verfahren *n*
f. coal Staubkohle *f*
f. coking *(Erdöl)* Fluid-Coking *n (ein Krackverfahren)*
f. culture Flüssigkeitskultur *f*
f.-energy mill Strahlmühle *f*
f. flow 1. Flüssigkeitsströmung *f*; Gasströmung *f*; 2. Flüssigkeitsstrom *m*; Gasstrom *m*
f. friction hydrodynamischer Widerstand *m*
f. hydroforming Fluid-Hydroformen *n*, Fluid-Hydroforming *n (Reformieren nach dem Wirbelschichtverfahren)*
f. milk Trinkmilch *f*
f. process Fluidverfahren *n*, Wirbelschichtverfahren *n*, Staubfließverfahren *n*, Fließ[bett]verfahren *n*, *(Kohlevergasung auch)* Winkler-Verfahren *n*
f. pulp Klassiertrübe *f*, Trübe *f*
f.-solid transition Übergang *m* vom flüssigen in den festen Zustand
f. system Fluidsystem *n*, Staubfließsystem *n*
fluidal texture s. flow texture
fluidextract *(Pharm)* Fluidextrakt *m*
fluidify / **to** verflüssigen
fluidity Fließvermögen *n*, Fließfähigkeit *f*, Fließbarkeit *f*, Fließverhalten *n*, Fließeigenschaft *f*, *(phys Ch)* Fluidität *f*, Flüssigkeitscharakter *m (Kehrwert der Viskosität)*
fluidization 1. Verflüssigung *f*; 2. Fluidisation *f (Herbeiführen des Fließbettzustands)*
fluidize / **to** 1. fluidisieren, in den Fließbettzustand überführen; 2. s. to fluidify
fluidized adsorption Fließbettadsorption *f*, Wirbelschichtadsorption *f*
f. bed s. fluid bed
f.-bed coating *(Plast)* Wirbelsintern *n*
f.-bed gasification Fließbettvergasung *f*, Winkler-Gaserzeugung *f*
f.-bed technique Fließbett-Technik *f*, Wirbelschichttechnik *f*, Staubfließtechnik *f*, Fluid-Technik *f*
f. carbonization Schwelung *f* im Wirbelbett
f. catalyst s. fluid catalyst
f. fuel bed wirbelndes (fluidisiertes, kochendes) Brennstoffbett *n*
f. gas gasifier Wirbelschichtgenerator *m*, Winkler-Generator *m*
f. mass s. fluid bed

f. operation s. fluid process
f. oxidation of coal Oxydation f von Kohle im Wirbelbett
f. process s. fluid process
f. system s. fluid system
f. technique s. f.-bed technique
fluidizing velocity Wirbelgeschwindigkeit f *(im Fließbett)*
flume Venturikanal m, Meßgerinne n *(Mengenstrommessung)*
f. water Schwemmwasser n
fluming water s. flume water
fluo... *(in chem. Benennungen)* s. fluoro...
fluocerite *(Min)* Fluozerit m, *(veraltet für)* Tysonit m
fluochlore *(Min)* Fluochlor m, *(veraltet für)* Pyrochlor m
fluophotometer s. fluorometer
fluor crown glass Fluorkronglas n
f. spar s. fluorspar
fluoracetic acid FCH_2COOH Fluoressigsäure f
fluorapatite *(Min)* Fluorapatit m
fluorcarbon s. fluorocarbon
fluorcrown s. fluor crown glass
fluoresce / to fluoreszieren
fluorescein Fluoreszein n, Resorzinphthalein n
fluorescence Fluoreszenz f; Fluoreszenzstrahlung f; *(Kch)* Resonanzstreuung f
f. analysis Fluoreszenzanalyse f
f. microscopy Fluoreszenzmikroskopie f
f. radiation Fluoreszenzstrahlung f
f. spectrum Fluoreszenzspektrum n
f. titration Fluoreszenztitration f
fluorescent fluoreszierend, Fluoreszenz..., Leucht...
f. agent Fluoreszenzstoff m, Leuchtstoff m
f. brightener (brightening agent) optischer Aufheller m, optisches Aufhellungsmittel (Bleichmittel) n, Weißtöner m, aufhellender Leuchtstoff m
f. dye Fluoreszenzfarbstoff m
f. indicator Fluoreszenzindikator m
f. light Fluoreszenzlicht n
f. paint Fluoreszenzfarbe f, fluoreszierende Farbe (Leuchtfarbe) f
f. standard Fluoreszenzstandard m
f. substance fluoreszierende Substanz f, Fluoreszenzstoff m, Leuchtstoff m
f. white dye, f. whitening agent s. f. brightener
fluorescin s. fluorescein
fluoridate / to *(Trinkwasser)* fluoridieren, fluorisieren
fluoridation Fluoridierung f, Fluorisierung f *(des Trinkwassers)*
fluoride M^IF Fluorid n
f. glass Fluoridglas n
fluoridize / to s. to fluoridate
fluorimeter s. fluorometer
fluorimetric s. fluorometric
fluorimetry s. fluorometry

fluorinate / to fluorieren
fluorinated ethylene-propylene resin fluoriertes Äthylen-Propylen-Kopolymerisat n
f. rubber Fluorkautschuk m, Fluorelastomer[es] n
fluorinating agent Fluorierungsmittel n, Fluorüberträger m
fluorination Fluorieren n, Fluorierung f
fluorine F Fluor n
f.-containing fluorhaltig
f. oxide OF_2 Fluoroxid n, *(fälschlich für)* Sauerstoff[di]fluorid n
f. rubber s. fluorinated rubber
fluorite *(Min)* Fluorit m, Flußspat m *(Kalziumfluorid)*
f. structure *(Krist)* Fluoritstruktur f
fluoroaluminate Fluoroaluminat n
fluoroantimonate Fluoroantimonat n
fluoroarsenate Fluoroarsenat n
fluorobenzene C_6H_5F Fluorbenzol n
fluoroberyllate Fluoroberyllat n
fluoroborate $M^I[BF_4]$ Fluoroborat n
fluoroboric acid $H[BF_4]$ Fluoroborsäure f, Borfluorwasserstoffsäure f
fluorocarbon Fluorkohlenwasserstoff m, fluorierter Kohlenwasserstoff m
f. fibre Fluorkarbonfaserstoff m; Fluorkarbonfaser f
f. polymer (rubber) Fluorkautschuk m, Fluorelastomer[es] n, *(i.e.S.)* Fluorkarbonkautschuk m
fluorochromate Fluorochromat n
fluoroelastomer s. fluorocarbon polymer
fluoroferrate Fluoroferrat n
fluoroform CHF_3 Trifluormethan n, Fluoroform n
fluorogen s. fluorophore
fluorogermanate $M^I_2[GeF_6]$ Fluorogermanat n
fluorohafnate Fluorohafnat n
fluorohydric fluorwasserstoffsauer
f. acid s. hydrofluoric acid
fluoroiodate $M^I[JO_2F_2]$ Fluorojodat(V) n
fluoromanganite $M^I_2[MnF_6]$ Fluoromanganat(IV) n, Hexafluoromanganat(IV) n
fluorometer Fluorometer n, Fluorimeter n, Fluorofotometer n *(Gerät für Fluoreszenzmessungen)*
fluorometric fluorometrisch
fluorometry Fluorometrie f, Fluorimetrie f, Fluoreszenzmessung f
fluoromolybdate Fluoromolybdat n
fluoroniobate Fluoroniobat n
fluoro-olefin Fluorolefin n, Fluoralken n, fluoriertes Alken n, fluorierter ungesättigter Kohlenwasserstoff m
fluorophore Fluorophor m, fluoreszierendes Prinzip n *(Fluoreszenz verursachender Molekülanteil)*
fluorophosphate Fluorophosphat n
fluoroplastic Fluorkarbonplast m
fluororhenate Fluororhenat n
fluororhodate Fluororhodat n

fluoroscope Röntgenbildschirm *m*
fluorosilic acid *s.* fluorosilicic acid
fluorosilicate $M^I_2[SiF_6]$ Fluorosilikat *n*, Hexafluorosilikat *n*, Silikofluorid *n*
fluorosilicic acid $H_2[SiF_6]$ Fluorokieselsäure *f*, Hexafluorokieselsäure *f*, Kieselfluorwasserstoffsäure *f*, Kieselflußsäure *f*, Siliziumfluorwasserstoffsäure *f*, Silikofluorwasserstoffsäure *f*
fluorosilicone rubber Fluorsilikonkautschuk *m*
fluorostannate $M^I_2[SnF_6]$ Fluorostannat(IV) *n*, Hexafluorostannat(IV) *n*
fluorotantalate Fluorotantalat *n*
fluorotellurate Fluorotellurat(VI) *n*
fluorotellurite Fluorotellurat(IV) *n*
fluorothoriate Fluorothorat *n*
fluorotitanate Fluorotitanat *n*
fluorotungstate Fluorowolframat *n*
5-fluorouracil 5-Fluor[o]urazil *n*
fluorouranate Fluorouranat *n*
fluorovanadate Fluorovanadat *n*
fluorozirconate Fluorozirkonat *n*
fluorspar *(Min)* Flußspat *m*, Fluorit *m* *(Kalziumfluorid)*
FluoSolids lime kiln FluoSolids-Kalkbrennofen *m*
FluoSolids method (process) FluoSolids-Verfahren *n* *(Wirbelschichtverfahren zur Röstung in FluoSolids-Öfen)*
FluoSolids reactor FluoSolids-Reaktor *m*, FluoSolids-Ofen *m*, Dorrco-FluoSolids-Reaktor *m*, Dorrco-FluoSolids-Wirbelschichtofen *m*
FluoSolids roasting FluoSolids-Röstung *f*, Wirbelschichtröstung *f* in FluoSolids-Öfen
FluoSolids system FluoSolids-System *n*, Dorrco-FluoSolids-System *n* *(Wirbelschichtofensystem der Dorrco)*
flush / to 1. [aus]spülen, auswaschen; strömen lassen; *(Farb)* flushen *(Pigmente durch Kneten mit hydrophoben Bindemitteln entwässern und in Pastenform überführen)*; 2. strömen, sich ergießen
f. away fortspülen
f. out [aus]spülen, auswaschen
f. over überschäumen
flush:
to be f. glatt abschließen
f.-plate [filter] press Rahmenfilterpresse *f*
f. water *(Pap)* Spritzwasser *n* *(Holzschliffherstellung)*
flushed colour geflushtes Pigment *n*, Flushpaste *f*
flusher Flushkneter *m* *(Knetmaschine zum Bereiten von Pigmentpasten)*
flushing *(Farb)* Flushing *n* *(Überführung wäßriger Pigmente in Pastenform durch Kneten mit hydrophoben Bindemitteln)*
f. process Flush[ing]verfahren *n*, Direktverfahren *n* *(zum Flushen von Farbpigmenten)*
fluted filter Faltenfilter *n*
f. funnel Rippentrichter *m*
f. glass Rippenglas *n*, geripptes (geriffeltes) Glas *n*

f. roll Riffelwalze *f*, geriffelte Transportwalze *f*
flux / to schmelzen, flüssig machen; *(durch Schmelzen)* aufschließen; *(Plast)* plastifizieren, weichmachen, erweichen; *(Erdöl)* verschneiden, fluxen
f. on aufschmelzen
flux 1. Fließen *n*, Fluß *m*; 2. Flußmittel *n*, Fluß *m*, Schmelzmittel *n*, *(Met auch)* Zuschlag *m*; 3. Lötmittel *n*; 4. *s. f.* oil
f.-free flußmittelfrei
f. oil *(Erdöl)* Fluxöl *n*, Verschnittöl *n*
fluxing agent Flußmittel *n*, Fluß *m*, Schmelzmittel *n*
fly ash Flugasche *f*
f.-ash collection Flugaschenabscheidung *f*
f.-ash precipitator Flugaschenabscheider *m*, Rauchgasentstauber *m*
f. bars *(Pap)* Walzenmesser *npl* *(beim Holländer)*
f. knife Rotormesser *n* *(einer Rotorschneidmaschine)*; *(Pap)* Obermesser *n* *(des Querschneiders)*
f. paper Fliegen[fänger]papier *n*
f. roll Papierleitwalze *f*
f. spray Sprühmittel *n* gegen Fliegen
flyer *(Text)* Flyer *m*, Vorspinnmaschine *f*
flywheel Schwungrad *n*
FMC *s.* final moisture content
foam / to [auf]schäumen, sich mit Schaum bedecken; zum Schäumen bringen, schaumig (schäumend) machen; *(Plast)* [ver]schäumen
f. over überschäumen
foam Schaum *m*; *(Plast) s.* foamed plastic
f.-backed mit Schaumstoff laminiert (beschichtet, kaschiert), schaumstoffbeschichtet, schaumstoffkaschiert
f. beater *s.* foaming machine
f. breakage Zusammenbruch *m* (Zusammensinken *n*) des Schaumes
f. breaker (destroyer) *s. f.* inhibitor
f. draining Schaumentwässerung *f*
f. extinguisher Schaumlöscher *m*
f. fermentation Schaumgärung *f*
f. filament *(Text)* Schaumfaden *m*, Schaumstoff[schnitt]faden *m*
f. film *(Text)* Schaumlamelle *f*
f. formation Schaumbildung *f*, Schaumentwicklung *f*, Schäumen *n*, Verschäumen *n*
f. glass Schaumglas *n*
f. glass block Schaumglasblock *m*
f.-holding capacity *s. f.* stability
f. improver Schaumverbesserer *m*
f. inhibitor (killer) Mittel *n* gegen Schaumbildung, Schaumverhinderungsmittel *n*, Schaumverhütungsmittel *n*, Antischaummittel *n*, Schaumdämpfungsmittel *n*, Schaumdämpfer *m*, Schaumzerstörungsmittel *n*, Entschäumungsmittel *n*, Entschäumer *m*
f. latex *s. f.* rubber
f.-like schaumartig, schaumähnlich
f. line *(Glas)* Schaumlinie *f*

f. marks Luftblasen *fpl (Papierfehler)*
f. meter Schaummeßgerät *n*
f. moulding Schaumstoff-Formen *n*
f. number Schaumzahl *f*
f. plastic *s.* foamed plastic
f.-producing schaumbildend
f.-promoting schaumfördernd
f. resin Harzschaum *m*
f. rubber Schaumgummi *m*, Latexschaum[-gummi] *m*
f. stability Schaumbeständigkeit *f*
f. stabilizer (stabilizing compound) Schaumstabilisator *m*
f. volume Schaumvolumen *n*
foambacks schaumstoffkaschierte Textilien *fpl*
foamclay *s.* foamed clay
foamed clay Schaumton *m*, Blähton *m*
f. concrete Schaumbeton *m*
f. glass Schaumglas *n*
f. latex rubber *s.* foam rubber
f. plastic Schaum[kunst]stoff *m*
f. polystyrene Schaumpolystyrol *n*, geschäumtes Polystyrol *n*
f. slag Schaumschlacke *f, (i.w.S.)* Hüttenbims *m*, Thermosit *m*
foamer *s.* foaming agent
foaminess Schaumigkeit *f*, Schäumigkeit *f*; Schaum[bildungs]fähigkeit *f*, Schaum[bildungs]vermögen *n*
foaming Schäumen *n*, Aufschäumen *n*, Schaumbildung *f; (Plast)* Schäumen *n*, Verschäumung *f*
f. ability *s.* f. power
f. agent Schaum[erzeugungs]mittel *n*, schaumerzeugendes Mittel *n*, Schaumerzeuger *m*, Schaumbildungsmittel *n*, Schaumbildner *m*, Schäummittel *n*, Schäumer *m; (Plast)* Treibmittel *n*
f. characteristic Schäumeigenschaft *f*
f. in place (situ) Formverschäumung *f*, Fertigteilverschäumung *f*
f. loss Verlust *m* durch Schäumen, Schäumverlust *m*
f. machine Schaum[schlag]maschine *f*
f. power Schaum[bildungs]vermögen *n*, Schaum[bildungs]fähigkeit *f*, Schaumkraft *f*
foamless foamback „schaumloser Schaumstoff" *m (durch Schmelzen dünner Laminate erhaltene Bindeschicht)*
foamy schaum[art]ig, schäumig
fodder beet Futterrübe *f*, Runkelrübe *f*, Dickrübe *f*, Beta vulgaris L., var. crassa
f. salt Viehsalz *n*
f. yeast Futterhefe *f*
fog / to *(Foto)* [ver]schleiern
fog Nebel *m; (Foto)* Schleier *m*, Schleierschwärzung *f*, Schwärzung *f* im unbelichteten Teil *(der Schwärzungskurve)*
f. appliance *s.* f. generator
f. application *(Landw)* Nebeln *n*
f. density *(Foto)* Schleierdichte *f*

f. droplet Nebeltröpfchen *n*
f.-free *(Foto)* schleierfrei, schleierlos
f. generator *(Landw)* Nebelgerät *n*, Nebelblaser *m*, Atomiseur *m*
f. inhibition *s.* f. prevention
f. level *(Foto)* Stärke *f* des Schleiers, Schleier *m*
f. prevention *(Foto)* Schleierverhinderung *f*, Schleierverhütung *f*
f. track *(phys Ch)* Nebelspur *f*
fogging *(Foto)* Schleierbildung *f*
f. machine Nebelgerät *n*
f. machinery Nebelgeräte *npl*
foil Folie *f;* Blech *n*
f. paper Folienpapier *n, (i.e.S.)* Metallpapier *n (oder)* kunststoffbeschichtetes Papier *n*
fold / to falten; *(Pap)* falzen
f. back *(Gum) (eine Kautschukmischung)* aufrollen
fold Falte *f; (Glas)* Falte *f*, Verfaltung *f (Fehler)*
f. formation Faltenbildung *f*
f.-testing machine *(Pap)* Falzwiderstandsprüfgerät *n*
folded filter Faltenfilter *n*
folding Falten *n; (Pap)* Falzen *n; (Gum)* Aufrollen *n (einer Kautschukmischung)*
f. boxboard Faltschachtelkarton *m*, Schachtelkarton *m*, Kartonagenpappe *f*
f. endurance Falzwiderstand *m*, Falzfestigkeit *f*
f.-endurance test *(Pap)* Falzwiderstandsprüfung *f*
f. resistance (strength) *s.* f. endurance
f. tester *s.* fold-testing machine
folds *(Pap)* Falzungen *fpl*
foliage burn Blattverbrennungen *fpl (z.B. durch Pflanzenschutzmittel)*
foliated blätt[e]rig, geblättert, schieferig, schuppig
f. gypsum Schiefergips *m*
f. tellurium *(Min)* Blättertellur *m*, Nagyagit *m*
folic acid Folsäure *f*, Pteroylglutaminsäure *f*
folinic acid Folinsäure *f*, Leukovorin *n*, Formyltetrahydrofolsäure *f*, Zitrovorumfaktor *m*
folio board Aktendeckelkarton *m*
folium *(Krist)* Blatt *n*
follicle-stimulating hormone follikelstimulierendes Hormon *n*, Follikelstimulierungshormon *n*, Follikelreifungshormon *n*, FSH
follicular hormone Follikelhormon *n*
fony oil *(Samenöl des Affenbrotbaumes Adansonia digitata L.)*
food additive Lebensmittelzusatz[stoff] *m*, Fremdstoff *m* in Lebensmitteln
f. alcohol Trinkbranntwein *m*
f.-canning factory Konservenfabrik *f*
f. chemistry Lebensmittelchemie *f*, Nahrungsmittelchemie *f*
f. colour *s.* f. dye
f. control Lebensmittelüberwachung *f*
f. dye Lebensmittelfarbstoff *m*, Lebensmittelfarbe *f*
f. industry Lebensmittelindustrie *f*, Nahrungsmittelindustrie *f*

f. intoxication s. f. poisoning
f. microbiology industrielle (technische) Mikrobiologie f
f. of the gods Asa foetida f, Asant m, Stinkasant m, Teufelsdreck m (Gummiharz von Ferula-Arten)
f. package Lebensmittelpackung f
f. poisoning Nahrungsmittelvergiftung f
f. potato Speisekartoffel f
f. preservation Lebensmittelkonservierung f, Haltbarmachung f von Lebensmitteln
f. preservative Konservierungsmittel n (Konservierungsstoff m) für Lebensmittel
f. processing Lebensmittelverarbeitung f
f.-processing industry s. f. industry
f. protein Nahrungsmitteleiweiß n
f. technology Lebensmitteltechnologie f
f. value Nährwert m
f. yeast Nährhefe f
foodstuff industry s. food industry
foot (Foto) Durchhang m, Kurvendurchhang m, durchhängender Teil m (der Schwärzungskurve), Gebiet n der Unterexposition (Unterbelichtung)
f. bellows Blasebalg m mit Fußbetrieb, Tretgebläse n
f. valve Fußventil n
foots Bodenkörper m, Niederschlag m, Bodensatz m, Satz m, Sediment n, Ausscheidung[1] f, Ausscheidungsprodukt n, Ablagerung f; (Destillation) Nachlauf m
f. oil Schwitzöl n
forbidden band (gap) verbotenes Band n, verbotene Zone f, verbotener (nicht zugelassener) Energiebereich m, Energielücke f
f. line verbotene Linie f (im Spektrum)
f. transition (phys Ch) verbotener Übergang m
force / to:
f. into hineinpressen
f. out auspressen, abpressen
f. through hindurchpressen, [hin]durchdrücken
force 1. Kraft f; 2. (Plast) Formunterteil n, Gesenk n
f. constant Kraftkonstante f
f. effect Kraftwirkung f
f.-feed / to zwangsläufig beschicken; zwangsläufig eintragen
f. field Kraftfeld n
f. of adhesion Adhäsionskraft f
f. of attraction Anziehungskraft f, Attraktionskraft f
f. of cohesion Kohäsionskraft f
f. of crystallization Kristallisationskraft f
f. of gravitation (gravity) Gravitationskraft f, Schwerkraft f, Erdanziehungskraft f
f. of interaction Wechselwirkungskraft f
f. of repulsion Abstoßungskraft f
f. plate (Plast) Stempelplatte f
f. plug (Plast) Stempel m, Stempelprofil n, Preßstempel m, Patritze f
f. retainer s. f. plate

forced circulation Zwang[s]umlauf m, Zwanglauf m, Zwangszirkulation f
f.-circulation evaporator Zwang[s]umlaufverdampfer m
f.-circulation reboiler (Destillation) Umlaufverdampfer m, Röhrenbündelverdampfer m
f. configuration erzwungene Konfiguration f
f. draught künstlicher (künstlich erzeugter) Zug m, Saugzug m (z.B. eines Ofens)
f.-draught cooling tower Ventilatorkühlturm m
f.-draught fan Luftgebläse n, Unterwindgebläse n
f. penetration (Pap) Druckdurchtränkung f (der Hackschnitzel)
f.-vibration method [of dynamic measurement] Dämpfungsprüfung f mittels erzwungener Schwingungen
forceps Zange f; Pinzette f
forcing machine Spritzmaschine f, Schneckenpresse f
forebath (Foto) Vorbad n
forehearth (Met, Glas) Vorherd m
foreign atom Fremdatom n, Stör[stellen]atom n, Fremdstörstelle f, [chemische] Störstelle f
f. cullet (Glas) fremde Scherben fpl
f. flavour Beigeschmack m, Fremdgeschmack m
f. gas Fremdgas n
f. material (matter) s. f. substance
f. molecule Fremdmolekül n
f. note paper Überseepostpapier n
f. odour Fremdgeruch m, fremder Geruch m
f. substance Fremdstoff m, Fremdsubstanz f, fremder Stoff m, fremde Substanz f; Verunreinigung f
f. taste s. f. flavour
forensic chemistry forensische (gerichtliche) Chemie f, Gerichtschemie f
forepump Vorvakuumpumpe f
forerun[s], forerunning[s] Vorlauf m
foreshot[s] Vorlauf m (bei der Alkoholdestillation)
forest peat Waldtorf m
foresterite (fälschlich für) forsterite
forevacuum Vorvakuum n
forewarming Vorwärmung f; Vorerhitzung f
forge pig iron Puddelroheisen n
forged fitting Schmiedeformstück n, Schmiedefitting m(n)
forging Schmieden n
f. furnace Schmiedeofen m
fork truck Gabelstapler m
forked chain verzweigte Kette f
forklift truck s. fork truck
form / to bilden, darstellen, formen; sich bilden (ausbilden, entwickeln), entstehen
form Form f, Gestalt f; (Plast) Werkzeug n, Form f
formability Formbarkeit f
formal Formal n, Formaldehydazetal n, (i.e.S.) $CH_2(OCH_3)_2$ Formal n, Dimethylformal n, Formaldehyddimethylazetal n, Dimethoxymethan n, Methylal n

f. charge formale Ladung *f*, Formalladung *f*
formaldehyde HCHO Formaldehyd *m*, Methanal *n*
f. acetal Formaldehydazetal *n*, Formal *n*
f.-in-milk test Prüfung *f* der Milch auf Formaldehyd, Nachweis *m* von Formaldehyd in Milch
f. oxime *s.* formaldoxime
f. sodium sulphoxylate $NaHSO_2 \cdot HCHO \cdot 2H_2O$ Formaldehydnatriumsulfoxylat *n*, Natriumformaldehydsulfoxylat *n*, Natriumsulfoxylatformaldehyd *m*
f. solution *(Pharm)* Formaldehydlösung *f*
f. tanning Formaldehydgerbung *f*
f. test Prüfung *f* auf Formaldehyd, Formaldehydnachweis *m*
formaldehydesulphoxylate Formaldehydsulfoxylat *n*
formaldehydesulphoxylic acid $HOCH_2SO_2H$ Formaldehydsulfoxylsäure *f*
formaldoxime HCH=NOH Formaldoxim *n*, Formaldehydoxim *n*
Formalin Formalin *n*, Formol *n* *(Warenzeichen für eine 35- bis 40%ige wäßrige Formaldehydlösung)*
F. reaction *s.* formolite reaction
formalinization Formalinbehandlung *f*
formalinize / *to* mit Formalin behandeln
formamide $HCONH_2$ Formamid *n*, Methanamid *n*
formamidine $HN=CHNH_2$ Formamidin *n*, Methanamidin *n*
formamido —NHCH(=O) Formamido...
formanilide $HCONHC_6H_5$ Formanilid *n*, *N*-Phenylformamid *n*, *N*-Phenylmethanamid *n*
formate HCOOM^I Formiat *n*
f. masking *(Gerb)* Maskierung *f* *(von Chrombrühen)* mit Formiaten
formation Bildung *f*, Ausbildung *f*, Entwicklung *f*, Entstehung *f*
f. constant Bildungskonstante *f*
f. function Bildungsfunktion *f*
f. of bonds Bindungsbildung *f*, Ausbildung *f* von Bindungen; Knüpfen *n* von Bindungen
f. of bubbles Blasenbildung *f*
f. of chelate rings Chelatringbildung *f*
f. of droplets Tropfenbildung *f*, Tröpfchenbildung *f*
f. of dye Farbstoffbildung *f*
f. of gas hydrates Gashydratbildung *f*, Bildung *f* von Gashydraten
f. of mix-crystals Mischkristallbildung *f*
f. of nuclei Kristallkeimbildung *f*, Keimbildung *f*
f. of the Grignard reagent Grignardierung *f*
f. reaction Bildungsreaktion *f*
formazyl $-C(=NNHC_6H_5)(N=NC_6H_5)$ Formazyl *n*
formed coke Formkoks *m*
former *(Gum)* Form *f*, Tauchform *f*
f.-moulding process *(Gum)* Gießverfahren *n*
formhydroxamic acid HCONHOH Formhydroxamsäure *f*
formhydroximic acid HOCH=NOH Formhydroximsäure *f*
formiate *s.* formate

formic acid HCOOH Ameisensäure *f*, Methansäure *f*
f.-acid green *(dem Schweinfurter Grün analoge Komplexverbindung aus Kupferarsenit und Kupferformiat) (Insektizid)*
f.-acid test Prüfung *f* auf Ameisensäure, Ameisensäure-Nachweis *m*
f. aldehyde HCHO Formaldehyd *m*, Methanal *n*
forming Formen *n*, Formung *f*, Formgebung *f*; *(Plast)* Umformen *n*, spanloses Formen *n*
f. of a blank *(Plast)* Formen *n* eines Rohlings
f. pressure *(Plast)* Formungsdruck *m* *(beim Niederdruckverfahren)*
f. roll Gußwalze *f*, Ziehwalze *f* *(bei der Tafelglasherstellung)*
f. table *(Pap)* Siebtisch *m*
f. temperature *(Plast)* Formungstemperatur *f*
Formol Formalin *n*, Formol *n* *(Warenzeichen für eine 35- bis 40%ige wäßrige Formaldehydlösung)*
F. test (titration) Formoltitration *f*
F. titration by Sörensen Formoltitration *f* nach Sörensen, Sörensen-Titration *f*
formolite Formolit *n* *(Kondensationsprodukt bei der Behandlung von Rohöl mit Formaldehyd und konzentrierter Schwefelsäure)*
f. number Formolitzahl *f*
f. reaction Formolitreaktion *f*
formolization Formolisierung *f*, Formalinbehandlung *f*
formolize / *to* formolisieren, mit Formalin behandeln
formonitrile HCN Formonitril *n*, *(veraltet für)* Zyanwasserstoff *m*, Wasserstoffzyanid *n*, Blausäure *f*
formose Formose *f* *(Gemisch zuckerähnlicher Stoffe bei Einwirkung von schwachen Alkalien auf Formaldehyd)*
formula Formel *f*; Rezept *n*, Vorschrift *f*; *(Landw)* Mischungsverhältnis *n* *(z.B. bei Mischdüngern)*
f. conversion Formelumsatz *m*
f. index Formelregister *n*
f. language Formelsprache *f*
f. weight Formelmasse *f*, *(bisher)* Formelgewicht *n*
formulate / *to* formulieren, durch eine Formel charakterisieren (zum Ausdruck bringen, symbolisieren), die Formel aufstellen; formulieren, rezeptmäßig aufbauen, sachgemäß zubereiten, konfektionieren, finalisieren; ein Rezept aufstellen (aufbauen)
formulation 1. Formulierung *f*, Charakterisierung (Symbolisierung) *f* durch eine Formel, Aufstellung *f* der Formel; Formulierung *f*, rezeptmäßiger Aufbau *m*, sachgemäße Zubereitung *f*, Konfektionierung *f*, Finalisierung *f*; Aufstellung *f* *(Aufbau m)* eines Rezepts, Rezeptur *f*; 2. Formulierung *f*, *(nach festgelegter Rezeptur gefertigte)* Zubereitung *f*; *(Plast)* Ansatz *m* *(Ausgangsgemisch)*
formulator *(Hersteller von Wirkstoffzubereitungen, z.B. kombinierten Pflanzenschutzmitteln)*

formyl —C(=O)H Formyl *n*
f. group Formylgruppe *f*, Formylrest *m*
f. radical Formylgruppe *f*, Formylrest *m*, Formylradikal *n*; [freies] Formylradikal *n*
f. tribromide $CHBr_3$ Formyltribromid *n*, Tribrommethan *n*, Bromoform *n*
f. trichloride $CHCl_3$ Formyltrichlorid *n*, Formylterchlorid *n*, Trichlormethan *n*, Chloroform *n*
f. triiodide CHJ_3 Formyltrijodid *n*, Trijodmethan *n*, Jodoform *n*
formylacetic acid $OHCCH_2COOH$ Formylessigsäure *f*
formylacetone CH_3COCH_2CHO Formylazeton *n*
formylaniline $HCONHC_6H_5$ Formanilid *n*, *N*-Phenylformamid *n*, *N*-Phenylmethanamid *n*
formylate / to formylieren
formylation Formylierung *f*
2-formylfuran $C_4H_3O \cdot CHO$ Fur[fur]al *n*, 2-Fur[fur]aldehyd *m*, 2-Furfurylaldehyd *m*, 2-Furankarbonal *n*
formyltetrahydrofolic acid Formyltetrahydrofolsäure *f*, Folinsäure *f*, Leukovorin *n*, Zitrovorumfaktor *m*
forsterite *(Min)* Forsterit *m* *(Magnesiumorthosilikat)*
f. porcelain Forsteritporzellan *n*
f. refractory [feuerfestes] Forsteriterzeugnis *n*
f. refractory brick [feuerfester] Forsteritstein *m*
f. whiteware Forsterit-Weißware *f*
fortification Verstärkung *f*, Stärkung *f*; Anreicherung *f*; Wirkungssteigerung *f* *(z.B. eines Insektizids durch Kombination mit anderem Mittel)*; *(Pap)* Verstärken *n*, Verstärkung *f*, Anreicherung *f*, Aufkonzentrieren *n*, Aufgasen *n* *(der Kochsäure)*
fortified by vitamins vitamin[is]iert, mit Vitaminen angereichert, vitaminreich
f. wine alkoholisierter (gespriteter) Wein *m*
fortify / to [ver]stärken, aufstärken; anreichern; die Wirkung *(z.B. eines Insektizids durch Kombination mit anderem Mittel)* steigern; *(Pap)* *(Kochsäure)* verstärken, anreichern, aufkonzentrieren, aufgasen
forward-curved blade fan Lüfter (Ventilator) *m* mit vorwärtsgekrümmten Schaufeln
f.-feed operation Parallelstromführung *f* *(z.B. von Mehrkörperverdampfern)*
f. reaction Hinreaktion *f*
fossil *(Geol)* fossil, versteinert
fossil *(Geol)* Fossil *n*, Versteinerung *f*
f.-bearing *(Geol)* fossilienhaltig, fossilführend
f. charcoal fossile (mineralische) Holzkohle *f*
f. coal fossile (mineralische, natürliche) Kohle *f*, Mineralkohle *f*, Naturkohle *f*
f. fuel fossiler (mineralischer) Brennstoff *m*, Fossilbrennstoff *m*
f. oil Steinöl *n*, Bergöl *n*, *(veraltet für)* Erdöl *n*
f. resin fossiles Harz *n*
f. wax mineralisches Wachs *n*
fossiliferous s. fossil-bearing

fossilize / to fossilieren
foul sludge Faulschlamm *m*
f.-smelling übelriechend, unangenehm (schlecht, widerlich) riechend
f. water Abwasser *n*
fouled with carbon mit Kohlenstoff beladen *(Katalysator)*
fouling Verschmutzung *f*; Ablagerung *f*, Bewuchs *m*, Niederschlag *m*
found in nature natürlich vorkommend
foundation cream Grundlagencreme *f*, Fondcreme *f*
f. lotion flüssige Grundlagencreme *f*, Foundation-Lotion *f*
founder *(Glas)* Schmelzer *m*
founding *(Glas)* Läuterung *f*
foundry Gießerei *f*
f. coke Gießerei[schmelz]koks *m*
f. cupola Gießereikupolofen *m*
f. [pig] iron Gießerei[roh]eisen *n*
f. resin Gießerei[kern]harz *n*
fountain roll *(Pap)* Tauchwalze *f*
four-bowl calender Vierwalzenkalander *m*
f.-bowl stack type of calender Vierwalzenkalander *m* mit übereinanderliegenden Walzen, I-Kalandèr *m*
f.-colour[-process] printing Vierfarbendruck *m*
f.-component system Vierkomponentensystem *n*, Vierstoffsystem *n*, quaternäres System *n*
f.-membered viergliedrig
f.-membered ring viergliedriger Ring (Kohlenstoffring) *m*, Vier[er]ring *m*, 4-Ring *m*
f.-roll calender s. f.-bowl calender
f.-stage pump vierstufige Pumpe *f*
f.-step process Vierstufenverfahren *n*
f.-step reaction Vierstufenreaktion *f*
f.-way juncture Vierwegkreuzung *f*
f.-way valve Vierwegeventil *n*, Vierweg[e]hahn *m*
Foucault [sheet-drawing] process Foucault-Verfahren *n*, Fourcaultsches Tafelglasziehverfahren *n*
Fourdrinier [paper machine] Langsieb[papier]maschine *f*
Fourdrinier part (section) *(Pap)* Langsiebpartie *f*, Siebpartie *f*
Fourdrinier wire Sieb (Langsieb) *n* der Papiermaschine, Maschinensieb *n*
Fourier series Fourier-Reihe *f*
Fourier synthesis Fourier-Synthese *f*
fousel oil s. fusel oil
Fowler solution Fowlersche Lösung *f*
fowlerine, fowlerite *(Min)* Fowlerit *m*
F.p. s. fine paper
Fraass breaking point Brechpunkt *m* nach Fraass
fraction / to s. to fractionate
fraction Fraktion *f* *(z.B. eines Destillats)*; Fraktion *f*, Kornklasse *f*, Klasse *f*; Bruchteil *m*
f. collector Fraktionssammler *m*
f. of dissociation Dissoziationsgrad *m*
fractional fraktioniert

f. bond Teilbindung *f*
f. column *s.* fractionating column
f. crystallization fraktionierte Kristallisation *f*, fraktioniertes Kristallisieren *n*; *(Geol)* Kristallisationsdifferentiation *f*, fraktionierte Kristallisation *f*
f. distillation fraktionierte (fraktionierende) Destillation *f*, Fraktionieren *n*, Fraktionierung *f*
f. efficiency *s.* f.-weight collection efficiency
f. order *(Reaktionskinetik)* gebrochene Ordnung *f*
f. precipitation fraktionierte (stufenweise) Fällung *f*
f. washing Stufenwäsche *f*
f.-weight collection efficiency Fraktionsabscheidegrad *m*, Fraktionsentstaubungsgrad *m*, Teilentstaubungsgrad *m*
fractionate / to fraktionieren, *(i.e.S. auch)* fraktioniert (stufenweise) destillieren; klassieren *(in der Aufbereitungstechnik)*
fractionating apparatus Fraktioniergerät *n*, Fraktionierapparat *m*
f. column Fraktionierkolonne *f*, Fraktioniersäule *f*, Trennkolonne *f*, Fraktionator *m*
f. distillation *s.* fractional distillation
f. flask Fraktionierkolben *m*, Fraktionskolben *m*
f. tower Fraktionierturm *m*
fractionation Fraktionieren *n*, Fraktionierung *f*, *(i.e.S. auch)* fraktionierte (fraktionierende) Destillation *f*
f. column *s.* fractionating column
f. tower *s.* fractionating tower
f. with salts fraktionierte Aussalzung *f*
fractionator *s.* fractionating column
fractioning *s.* unter fractionating
fracture Bruch *m*
f. surface Bruchfläche *f*
fragment Bruchstück *n*, Spaltstück *n*, Rest *m*
fragmental rock Trümmergestein *n*
fragmented zerbrochen; zerklüftet *(Oberfläche)*
fragments of fibres *(Pap)* Faserbruchstücke *npl*
fragrant wohlriechend, duftend, angenehm (aromatisch) riechend
frame Rahmen *m*
f. filter Rahmenfilter *n*
f. filter press Rahmen[filter]presse *f*
framework Gerüst *n*; Rahmen *m*
framycetin Framyzetin *n* *(Antibiotikum aus Streptomyces decaris)*
francium Fr Frankium *n*, Franzium *n*
Franck-Condon principle Franck-Condon-Prinzip *n*, Franck-Condonsches Prinzip *n*
frangula Faulbaum *m*, Rhamnus frangula L.; Faulbaumrinde *f*
f. emodin Frangulaemodin *n*, 2-Methyl-4,5,7-trihydroxy-9,10-anthrachinon *n*
frangulic acid *s.* frangula emodin
frangulin, franguloside Frangulin *n*, Frangulosid *n* *(ein Glykosid)*
Frankfort black Frankfurter Schwarz *n*

frankincense Weihrauch *m* *(Gummiharz von Boswellia-Arten)*
f. oil *(Pharm)* Weihrauchöl *n*, Olibanumöl *n*
franklinite *(Min)* Franklinit *m*
Frasch process Frasch-Verfahren *n* *(zur Schwefelgewinnung)*
Fraunhofer lines Fraunhofer-Linien *fpl*, Fraunhofersche Linien *fpl*
free / to befreien
free frei, unverbunden
f.-beaten stock (stuff) *s.* f. stock
f. beating *(Pap)* rösche Mahlung *f*, Schneidmahlung *f*
f.-blown glass freihandgeblasenes Glas *n*
f.-burning freibrennend, ungestört, frei [brennend]
f. carbon freier Kohlenstoff *m*, Unlösliches *n* *(im Teer)*
f. electron freies (frei bewegliches) Elektron *n*
f. electron pair freies (einsames) Elektronenpaar *n*
f.-electron theory Theorie *f* der frei beweglichen Elektronen
f.-electron theory of metals Elektronentheorie *f* der Metalle
f. energy freie Energie *f*, F
f. fatty acid freie Fettsäure *f*
f.-fatty-acid content Gehalt *m* an freien Fettsäuren
f.-filtering leicht filtrierend (filtrierbar)
f. flame offene Flamme *f*
f. flowing Rieselvermögen *n* *(von Schüttgütern)*
f. from acidity säurefrei
f. from angle strain spannungsfrei
f. from aromatic hydrocarbons aromatenfrei
f. from ash aschefrei
f. from bitterness ohne bitteren Beigeschmack
f. from blow-holes blasenfrei *(Stahl)*
f. from carbon kohlenstofffrei
f. from chloride chloridfrei
f. from grittiness nicht körnig
f. from iron eisenfrei
f. from isomers isomerenfrei
f. from odour geruchlos, geruchfrei
f. from oil ölfrei
f. from oxygen sauerstofffrei
f. from phenol phenolfrei
f. from suspended matter schwebestofffrei
f. from voids hohlraumfrei, blasenfrei *(z.B. Folie)*
f. from water wasserfrei
f. gold Freigold *n*
f.-hand throwing *(Ker)* Freidrehen *n*
f.-machining steel leicht spanbarer Stahl *m*, Automatenstahl *m*
f. moisture freies Wasser *n*
f. of ... *s.* unter f. from ...
f. path freie Weglänge *f*
f. pulp *s.* f. stock
f. radical freies Radikal *n*, Radikal *n* *(i.e.S.)*
f.-radical chain reaction über freie Radikale

verlaufende Kettenreaktion f, Radikalkettenreaktion f, radikalische Kettenreaktion f

f.-radical polymerization durch freie Radikale ausgelöste Polymerisation f, Radikal[ketten]polymerisation f, radikalische Polymerisation f

f.-radical reaction über freie Radikale verlaufende Reaktion f, radikalische Reaktion f

f. resin s. f. rosin

f. rosin (Pap) Freiharz n, freies Harz n

f.-rosin size (Pap) Freiharzleim m

f. rotation freie Rotation (Drehbarkeit) f

f.-running stock (stuff) s. f. stock

f. settling freies Absetzen (Sedimentieren) n (bei der Aufbereitung)

f.-settling classification Klassierung f nach dem Prinzip des freien Absetzens

f.-settling zone Zone f freien Absetzens, Klärzone f

f. state freier Zustand m

f. stock (stuff) (Pap) röscher (wasserlässiger) Stoff m

f. sulphur freier Schwefel m, Freischwefel m

f. sulphur dioxide (Pap) freies (loses) SO_2 n

f. surface energy freie Oberflächenenergie f, Oberflächenarbeit f

f.-swelling index freier Blähungsgrad m, Blähungsgrad m ohne Belastung der Kohle

f.-venting sand gut gasdurchlässiger Sand (Formsand) m, Sand m guter (hoher) Gasdurchlässigkeit

f.-vibration method [of dynamic measurement] Dämpfungsprüfung f mittels freier Schwingungen

f. water freies Wasser n, grob disperses (verteiltes) Wasser n (bei der Erdölentwässerung); (Pap) Sieb[ab]wasser n

f. working stock (stuff) s. f. stock

freedom from ash Aschefreiheit f

f. from oxygen Sauerstofffreiheit f

freely filt[e]rable leicht filtrierend (filtrierbar)

f. miscible leicht mischbar

f. soluble leicht löslich, leichtlöslich, ll

freeness (Pap) Mahlgrad m, Entwässerungsgrad m

f. test (Pap) Mahlgradprüfung f, Mahlgradbestimmung f, Entwässerungsgradprüfung f, Entwässerungsgradbestimmung f

f. tester (Pap) Mahlgradprüfer m, Entwässerungsgradprüfer m

f. value s. freeness

freeze / to 1. gefrieren; erstarren (von einer Schmelze); [zusammen]frieren (von Kohle); 2. (Lebm) gefrieren, einfrieren

f. out ausfrieren

freeze brand (Gerb) Gefrier-Brandzeichen n

f.-dry / to gefriertrocknen, lyophil trocknen, lyophilisieren

f. dryer Gefriertrockner m, Gefriertrocknungsanlage f

f. drying Gefriertrocknen n, Gefriertrocknung f, Sublimationstrocknen n, Sublimationstrocknung f, Lyophilisierung f, Lyophilisation f

f. drying apparatus s. f. dryer

freezer s. freezing apparatus

freezing Gefrieren n; Erstarren n (einer Schmelze); Frieren n, Zusammenfrieren n (von Kohle); (Lebm) Gefrieren n, Einfrieren n

f. apparatus Gefrierapparat m

f. curve s. f.-point curve

f. machine Kältemaschine f

f. microtome Gefriermikrotom n

f. mixture Kältemischung f

f. point Gefrierpunkt m, Gefriertemperatur f; Erstarrungspunkt m, Erstarrungstemperatur f (einer Schmelze)

f.-point curve Gefrierkurve f; Erstarrungskurve f (einer Schmelze)

f.-point depression (lowering) Gefrierpunktserniedrigung f, Gefrierpunktsdepression f

f.-point method Methode f der Gefrierpunktserniedrigung

f. preservation Gefrierkonservierung f, Frosten n

f. process Gefrierverfahren n

f. rate Gefriergeschwindigkeit f; Erstarrungsgeschwindigkeit f (einer Schmelze)

f. temperature s. f. point

f. velocity s. f. rate

Freiberg value for basicity (Gerb) Freiberger Basizität f (von Chrombrühen)

freibergite (Min) Freibergit m (silberreicher Tetraedrit)

freieslebenite (Min) Freieslebenit m (ein Bleisilberspießglanz)

Frémy's salt $(KO_3S)_2NO$ Frémysches Salz n; KHF_2 Kaliumhydrogenfluorid n

French berry Französische Gelbbeere f

F. chalk (Min) Talk m, Talkum n, Talkstein m (Magnesiumdihydrogentetrasilikat)

F. chestnut Echte Kastanie f, Edelkastanie f, Marone f, Castanea sativa Miller

F. cleaning Chemischreinigen n, Chemischreinigung f, Trockenreinigen n, Trockenreinigung f (von Kleidung)

F. lavender Schopflavendel m, Lavandula stoechas L.

F. oil turpentine Bordeaux-Terpentin n(m), Terebinthina gallica (von Pinus pinaster Ait.)

Frenkel defect Frenkel-Defekt m; Frenkel-Fehlstelle f, Frenkelsche Fehlstelle f

Frenkel disorder Frenkel-Fehlordnung f, Frenkelsche Fehlordnung f

Frenkel pair Frenkelsches Fehlstellenpaar n

frenzelite (Min) Frenzelit m, (veraltet für) Guanajuatit m

frequency Frequenz f, Schwingungszahl f, Schwing[ungs]frequenz f; Frequenz f, Häufigkeit f

f. band Frequenzband n

f. distribution Frequenzverteilung f; Häufigkeitsverteilung f

f.-distribution curve Häufigkeitsverteilungskurve f

f. factor Frequenzfaktor *m*, Entropiefaktor *m*; Häufigkeitsfaktor *m*
f. function Frequenzgang *m*
f.-modulated cyclotron frequenzmoduliertes Zyklotron *n*, FM-Zyklotron *n*, Synchrozyklotron *n*
f. of the incident light Primärfrequenz *f*, Frequenz *f* des eingestrahlten Lichtes
fresh frisch, Frisch...; *(Gerb)* ungesalzen
f. acid Frischsäure *f*, frische Säure *f*
f. catalyst Frischkatalysator *m*, frischer Katalysator *m*
f. cheese Frischkäse *m*
f. concrete Frischbeton *m*
f. cooking liquor *(Pap)* Frischlauge *f*, Weißlauge *f*
f. egg Frischei *n*
f. egg yolk Frischeigelb *n*
f. lime *(Gerb)* frischer Äscher *m*, Nachäscher *m*, Weißkalkäscher *m*, Schwelläscher *m*
f. meat Frischfleisch *n*
f. milk Frischmilch *f*
f. vegetables Frischgemüse *n*
f. water Frischwasser *n*; Süßwasser *n*; *(Foto)* reines Wasser *n*
freshly boiled frisch [aus]gekocht
f. drawn milk *s.* fresh milk
f. prepared frischbereitet, frisch bereitet (zubereitet, hergestellt)
fretting corrosion Reibkorrosion *f*
Freundlich adsorption isotherm Freundlichsche Adsorptionsisotherme *f*
Freundlich equation Freundlichsche Gleichung *f*
Freundlich isotherm *s.* Freundlich adsorption isotherm
Freund's acid Freundsche Säure *f*, 1-Naphthylamin-3,6-disulfonsäure *f*
freyalite *(Min)* Freyalith *m*
friability Bröcklichkeit *f*; Zerreiblichkeit *f* *(z.B. des Kokses, der Kohle)*
friable bröck[e]lig, krüm[e]lig, mürbe; zerreiblich, zerreibbar *(z.B. Koks, Kohle)*
f. humus Nährhumus *m*
friction / to *(Gum)* friktionieren
friction Reibung *f*, Friktion *f*
f. calender Friktionskalander *m*, Friktionierkalander *m*, Reibungskalander *m*
f. compound *(Gum)* Friktionsmischung *f*
f. drag Reibungswiderstand *m* *(gegenüber einem flüssigen oder gasförmigen Medium)*
f. drop Druckabfall *m* infolge Reibung
f. finish *(Gerb)* Stoßappretur *f*
f. force Reib[ungs]kraft *f*
f. glazing calender *s.* f. calender
f. head Reibungshöhe *f*
f. heat Reibungswärme *f*
f. loss Reibungsverlust *m*
f. ratio Reibungsverhältnis *n*; Friktionsverhältnis *n* *(von gegeneinanderlaufenden Walzen)*
f. resistance Reibungswiderstand *m*
f. soldering Reib[ungs]löten *n*

f. welding Reib[ungs]schweißen *n*
frictional Reibungs..., Friktions... *(Zusammensetzungen s. unter friction)*
frictioning *(Gum)* Friktionieren *n*
f. calender *s.* friction calender
frictionless reibungslos, reibungsfrei
Friedel-Crafts acylation Friedel-Crafts-Azylierung *f*
Friedel-Crafts agent *s.* Friedel-Crafts catalyst
Friedel-Crafts alkylation Friedel-Crafts-Alkylierung *f*
Friedel-Crafts catalyst Friedel-Crafts-Katalysator *m*, Katalysator *m* der Friedel-Crafts-Synthese
Friedel-Crafts condensation Friedel-Crafts-Kondensation *f*
Friedel-Crafts ketone synthesis Friedel-Craftssche Ketonsynthese *f*
Friedel-Crafts reaction (synthesis) Friedel-Crafts-Reaktion *f*, Friedel-Crafts-Synthese *f*, Friedel-Craftssche Reaktion (Synthese) *f*
friedelite *(Min)* Friedelit *m*
Friedländer synthesis Friedländer-Synthese *f*, Friedländersche Chinolinsynthese *f*
Fries reaction (rearrangement) Friessche Umlagerung (Verschiebung) *f*, Fries-Reaktion *f*
Fries rule Friessche Regel *f*
frigorific kälteerzeugend
f. mixture Kältemischung *f*
Frigorifico curing (method) *(Gerb)* *(Kombination von Salzlakenbehandlung und Einstreuen von festem Salz zur Häutekonservierung)*
frill / to sich kräuseln
frilling Kräuseln *n*, Ablösen *n*
fringe effect *(Foto)* Saumeffekt *m*
fringed fibril[la] Fansenfibrille *f*
f. micelle Fransenmizelle *f*
frit / to *(Ker)* fritten
frit *(Ker)* Fritte *f*; *(Landw)* Fritte *f* *(Mikronährstoffdüngemittel mit gefritteter keramischer Masse als Trägersubstanz)*
f. kiln *(Ker)* Fritteofen *m*
fritted disk funnel Filtertrichter *m*
f. glaze *(Ker)* Fritteglasur *f*, gefrittete Glasur *f*
f. porcelain Frittenporzellan *n*
fritting *(Ker)* Fritten *n*
Fritz method Fritz-Verfahren *n*, Fritz-Methode *f* *(ein Butterungsverfahren)*
Fröhde reagent Fröhdes Reagens *n* *(zum Alkaloidnachweis)*
front end Vorlauf *m* *(bei der Destillation)*
f. heat zone *(Plast)* vordere Heizzone *f*
f. reflection pattern *(Krist)* Durchstrahlungsaufnahme *f*
frontal analysis Frontalanalyse *f*
frost / to *(Glas)* mattätzen, mattieren
frost glass *(Glas)* Flitter *m*
f. resistance Frostbeständigkeit *f*
frosted glass Eis[blumen]glas *n*
frosting *(Glas)* Mattätzen *n*, Mattätze *f*, Mattätzung *f*, Mattieren *n*, Mattierung *f*; *(Plast)* Eisblumenbildung *f* *(Fehler)*

f. agent *(Glas)* Mattsalz *n*
f. effect Frosting-Effekt *m*, Grauschleier *m*, Zweifarbigkeit *f (nach Scheuerung von bicolorgefärbten Mischgeweben)*
frostproof frostbeständig
froth / to zum Schäumen bringen, schaumig (schäumend) machen; [auf]schäumen, sich mit Schaum bedecken
f. over überschäumen
f. up aufschäumen
froth Schaum *m*
f. copper *(Min)* Kupferschaum *m*, Tirolit *m*
f. fermentation Schaumgärung *f*
f. floatation Schaumflotieren *n*, Schaumflotation *f*, Schaumschwimmaufbereitung *f*
f. floatation method Schaumschwimmverfahren *n*
f.-forming agent *s.* frothing agent
f. mark *s.* f. spot
f. oil *(Pap)* Schaum[verhütungs]öl *n*, Schaumbekämpfungsöl *n*
f.-preventing agent Mittel *n* gegen Schaumbildung, Schaumverhinderungsmittel *n*, Schaumzerstörungsmittel *n*, Antischaummittel *n*, Schaumdämpfungsmittel *n*, Schaumdämpfer *m*, Schaumverhütungsmittel *n*, Entschäumungsmittel *n*, Entschäumer *m*
f. product Schaumprodukt *n*, Schaumkonzentrat *n (als Flotationsaustrag)*
f. skimmer Schaumabstreifer *m*
f. spot Schaumfleck *m (Papierfehler)*
f. suppression Schaumzerstörung *f*, Entschäumen *n*
f. suppressor *s.* f.-preventing agent
frother *s.* 1. frothing agent; 2. frothing machine
frothiness Schaumigkeit *f*, Schäumigkeit *f*; Schaum[bildungs]fähigkeit *f*, Schaumvermögen *n*, Schaumkraft *f*
frothing Schäumen *n*, Aufschäumen *n*; Schaumbildung *f*
f. agent Schaum[erzeugungs]mittel *n*, schaumerzeugendes Mittel *n*, Schaumerzeuger *m*, Schäummittel *n*, Schaumbildner *m*, Schäumer *m*
f. machine Schaum[schlag]maschine *f*
frothy schaum[art]ig, schäumig, schäumend
frozen fruit Gefrierobst *n*
f.-in *(Plast)* eingefroren
f.-in stress *(Plast)* eingefrorene Spannung *f*
f. lava erstarrte Lava *f*
f. meat Gefrierfleisch *n*
f. rubber gefrorener Kautschuk *m*
f. storage Gefrierlagerung *f*
f. vegetables Gefriergemüse *n*
f. water Eiswasser *n*
FRS *(Abk. für)* Fuel Research Station
fructosan Fruktosan *n (ein Polysaccharid)*
fructose Fruktose *f*, Fruchtzucker *m*, Lävulose *f*, *(i.e.S.)* D-Fruktose *f*
fructosidase Fruktosidase *f*
fructoside Fruktosid *n (ein Glykosid)*

frue vanner Frue-Vanner *m (Aufbereitungsherd mit umlaufender Plane und Schüttelbewegung)*
fruit-coat fat Fruchtfleischfett *n*
f. essence Fruchtessenz *f*, Fruchtäther *m*, Fruchtarom[a] *n*
f. fibre Fruchtfaser *f*
f. flesh Fruchtfleisch *n*
f. juice Fruchtsaft *m*, Obstsaft *m*
f. kiln Obsttrockner *m*, Obsttrocknungsapparat *m*
f. paper Fruchtseidenpapier *n*, Obsteinwickelpapier *n*
f. pickles Essigfrüchte *fpl*
f. pulp Fruchtfleisch *n*, Fruchtmark *n*, Obstpülpe *f*, Pülpe *f*, Pulpe *f*, Pulp *m*
f. sugar Fruchtzucker *m*, Fruktose *f*, Lävulose *f*
f. tissue, f. tissue wrapper (wrapping) *s.* f. paper
f. vinegar Obstessig *m*, Fruchtessig *m*
fruity frucht[art]ig, obstartig, Frucht...
Fry reagent Fry-Reagens *n*, Fry-Ätzmittel *n*, Frysches Reagens (Ätzmittel) *n*
FS *(Abk. für)* Federal Specifications
FSB *(Abk. für)* Federal Specifications Board
FSH *s.* follicle-stimulating hormone
FT black *s.* fine thermal black
fuchsin[e] Fuchsin *n*, Magenta *n*, Rosanilinchlorhydrat *n*, Rosanilin *n*
fuchsinesulphurous acid fuchsinschweflige Säure *f*
fuchsite *(Min)* Fuchsit *m*, Chrom-Muskovit *m*
fuchsonimine *(Farb)* Fuchsonimin *n*
fuel Brennstoff *m*, Brennmaterial *n*, Heiz[brenn]stoff *m*, Feuerungsmaterial *n*; Kraftstoff *m*, Treibstoff *m*
f. additive Kraftstoffadditiv[es] *n*
f. bed Brennstoffbett *n*, Brennstoffschicht *f*, Brennstoffschüttung *f*
f.-bed firing Rostfeuerung *f*
f. cell *(Ech)* Brennstoffelement *n*, Brennstoffkette *f*, Brennstoffzelle *f*
f. chemistry Brennstoffchemie *f*
f. element *(Kch)* Brenn[stoff]element *n*, Spaltstoffelement *n*; *s.* f. cell
f. for jet planes Düsenkraftstoff *m*, Düsentreibstoff *m*
f. gas Heizgas *n*, Beheizungsgas *n*
f.-heated furnace brennstoffbeheizter Ofen *m*, Brennstoffofen *m*
f. hole Schürloch *n*
f. knock Kraftstoffklopfen *n*
f. oil Heizöl *n*
f. oil heating Ölheizung *f*
f. research Brennstofforschung *f*
f. tank Kraftstoffbehälter *m*
fugacity Fugazität *f*, Flüchtigkeit *f*
fugitive flüchtig; *(Farb)* unecht
f. constituent flüchtiger (gasförmiger) Bestandteil *m*
f. dye unechter Farbstoff *m*
full annealing vollständiges Ausglühen *n*, Vollständigglühen *n (Umwandlungsglühung mit verzögerter Abkühlung)*

f.-bodied stark, schwer *(Wein)*
f. chemical pulping Zellstoffaufschluß *m*, chemischer Aufschluß *m*
f.-cream cheese s. f.-fat cheese
f.-cream milk s. f. milk
f. cure *(Gum)* Ausvulkanisation *f*, Ausheizung *f*
f. drum revolving strainer *(Pap)* Drehknotenfänger *m*, rotierender Knotenfänger *m*
f.-fat cheese Vollfettkäse *m*
f. fire *(Ker)* Vollfeuer *n*, Hochfeuer *n*
f. line ausgezogene Linie *f*
f. maturity Vollreife *f*
f. milk Vollmilch *f*
f.-scale chromatography Chromatografie *f* mit normaler Substanzmenge
f.-scale factory production großtechnische Herstellung *f*
f.-scale model Vollrohrmodell *n*
f. sintering Fertigsintern *n*, Fertigsinterung *f*, Hochsintern *n*, Hochsinterung *f*, Formieren *n*
f.-strength cooking liquor *(Pap)* hochkonzentrierte (hochprozentige) Kochsäure *f*
f. vulcanization s. f. cure
fuller board Isolierpappe *f*
Fuller-Lehigh mill Fuller-Lehigh-Mühle *f*
Fuller mill Fuller-Mühle *f* *(eine Kugelringmühle)*
Fuller-Peters mill Fuller-Peters-Mühle *f*
fuller's earth Fullererde *f*, Walk[er]erde *f*
fulling Walken *n*
f. machine Walkmaschine *f*, Walke *f*
f. stocks *(Text)* Hammerwalke *f*; *(Gerb)* Walke *f*
fullness *(Lebm)* Fülle *f*, Aromafülle *f*; Vollmundigkeit *f* *(z.B. beim Bier)*
fullonite *(Min)* Fullonit *m*, *(veraltet für)* Goethit *m* *(Eisenoxidhydroxid)*
fully cured *(Gum)* ausvulkanisiert, ausgeheizt
f. reinforcing *(Gum)* hochverstärkend, hochaktiv
fulminate CNOM^I Fulminat *n*
f. of mercury s. fulminating mercury
fulminating mercury Hg(ONC)$_2$ Knallquecksilber *n*, Quecksilber(II)-fulminat *n*
f. silver AgONC Silberfulminat *n*, knallsaures Silber *n*
fulminic acid C=NOH Knallsäure *f*, Fulminsäure *f*
fulminuric acid Isozyanursäure *f*
fulness s. fullness
fulvene Fulven *n*, 5-Methylen-1,3-zyklopentadien *n*; Fulven *n*, Fulvenkohlenwasserstoff *m*
fulvic acid Fulvosäure *f*, Gelbstoff *m* *(Sammelbezeichnung für eine Gruppe von Huminstoffen)*
fumagillin Fumagillin *n* *(Antibiotikum aus Aspergillus fumigatus)*
fumarase Fumarase *f* *(Ferment, das die Einstellung des Gleichgewichts zwischen Fumarsäure und L-Apfelsäure bewirkt)*
fumarate M$^I_2$C$_4$H$_2$O$_4$ Fumarat *n*
fumaric acid HOOCCH = CHCOOH Fumarsäure *f*, *trans*-Butendisäure *f*, *trans*-1,2-Äth[yl]endikarbonsäure *f*

f.-acid fermentation Fumarsäuregärung *f*
fumarine Fumarin *n*, Protopin *n*
fumarole *(Geol)* Fumarole *f*
fume / **to** rauchen; räuchern
fume Dunst *m*, Rauch *m*, Dampf *m*, Abdampf *m*; *(Met)* *(direkt von den Abgasen eines Bleischachtofens mitgeführter)* Flugstaub *m*
f. chamber s. 1. f. exhaust manifold; 2. f. hood 1.
f. closet s. f. hood 1.
f. duct s. f. exhaust manifold
f. exhaust manifold Abzug[s]rohr *n*, Absaugrohr *n*, Absaugvorrichtung *f* *(z.B. am Aufschließgestell der Kjeldahlapparatur)*
f. hood 1. Abzug[s]schrank *m*, Abzug *m*, Digestorium *n*, Kapelle *f*; 2. Rauchfangdach *n*; 3. s. f. exhaust manifold
f. pipe Abzug[s]rohr *n*; Gasabzug[s]rohr *n*
f. scrubber Absorptionsanlage *f* für Rauchgase
fumigacin Fumigazin *n*, Helvolinsäure *f*
fumigant Räuchermittel *n*; Vergasungsmittel *n*, Begasungsmittel *n*, Durchgasungsmittel *n*
fumigate / **to** [aus]räuchern, beräuchern, durchräuchern, dem Rauch aussetzen; begasen, durchgasen
fumigating paper Räucherpapier *n* *(zur Vertreibung von Insekten)*
fumigation Räuchern *n*, Räucherung *f*, Ausräuchern *n*; Begasen *n*, Begasung *f*, Durchgasung *f*
fumigator 1. Räucherapparat *m*; 2. s. fumigant
fuming rauchend, rauch-
f. nitric acid rauchende (konzentrierte) Salpetersäure *f*
f. sulphuric acid rauchende Schwefelsäure *f*, Oleum *n*
function / **to** wirken *(z.B. als Promotor)*
functional group funktionelle (charakteristische) Gruppe *f*
functionality Funktionalität *f*
fundamental band Grundschwingungsbande *f*
f. frequency Grundfrequenz *f*
f. particle Elementarteilchen *n*
f. research Grundlagenforschung *f*
f. series Fundamentalserie *f*, *(i.e.S.)* Bergmann-Serie *f*
fungal pilzartig, Pilz...
f. diastase Pilzdiastase *f*
f. pigment Pilzfarbstoff *m*
fungichromin Fungichromin *n* *(ein Polyenantibiotikum)*
fungicidal fungizid [wirksam], antifungal, pilztötend, pilzwirksam, *(Pharm auch)* antimykotisch
f. activity (efficiency) fungizide Wirksamkeit *f*
fungicide Fungizid *n*, pilztötendes Mittel *n*
f.-resistant fungizidresistent
f. spray fungizides Spritzmittel *n*
fungistat Fungistatikum *n*

fungistatic fungistatisch, das Pilzwachstum hemmend
fungitoxic fungitoxisch
fungus Pilz *m*
  f. pigment Pilzfarbstoff *m*
  f.-proof pilzfest
funnel Trichter *m*; Fülltrichter *m* *(eines Flammpunktprüfers)*
  f. heater Heißwassertrichter *m*, Warmwassertrichter *m*
  f. holder Trichterhalter *m*
  f. rack (stand) Filtrierstativ *n*
  f. tube Trichterrohr *n*
fur / to Kesselstein ansetzen
fur Pelz *m*; Kesselstein *m*
  f. dye Pelzfarbstoff *m*
furacrylic acid *s.* furylacrylic acid
fural *s.* furfural
furaldehyde *s.* furfuraldehyde
furan *s.* furfuran
  2-f. carbonal *s.* furfural
  f. resin Furanharz *n*
furancarbox[y]aldehyde *s.* furfuraldehyde
furancarboxylic acid Furankarbonsäure *f*, *(i.e.S.)* 2-Furankarbonsäure *f*, Brenzschleimsäure *f*, Pyroschleimsäure *f*
  f. aldehyde *s.* furfuraldehyde
2,5-furandione 2,5-Furandion *n*, *cis*-Butendisäureanhydrid *n*, Maleinsäureanhydrid *n*
furane *s.* furfuran
furanose Furanose *f*
furanoside Furanosid *n*
furazan Furazan *n*, 1,2,5-Oxadiazol *n*
  f. ring Furazanring *m*
furfural Fur[fur]al *n*, 2-Fur[fur]aldehyd *m*, 2-Fur[fur]ylaldehyd *m*, 2-Furankarbonal *n*, *(fälschlich)* Fur[fur]ol *n*; Furfuryliden *n*, Fur[fur]al *n*
  f. extraction Furfuralextraktion *f*, Extraktion *f* mit Furfural
  f. extraction plant Furfural[extraktions]anlage *f*
  f. extraction process *s.* f. process
  f.-free furfuralfrei
  f. process Furfuralverfahren *n*, Furfuralprozeß *m*
  f. refining Furfuralraffination *f*
  f. refining process *s.* f. process
  f. resin Furfuralharz *n*
  f. solvent extraction plant *s.* f. extraction plant
  f. solvent process *s.* f. process
  f. stripper Furfuralstripper *m*
  f. treating tower Furfuralwaschturm *m*
  f.-water stripper Stripper *m* zur Trennung des Furfurals vom Wasser
furfuralacetic acid *s.* furylacrylic acid
furfuralcohol *s.* furfuryl alcohol
furfuraldehyde Fur[fur]aldehyd *m*, Furfurylaldehyd *m*, Furanaldehyd *m*, *(i.e.S.)* 2-Fur[fur]aldehyd *m*, 2-Fur[fur]ylaldehyd *m*, Fur[fur]al *n*, 2-Furankarbonal *n*
furfuran Fur[fur]an *n*

furfurol[e] *s.* furfural
furfuryl Furfuryl *n*
  f. alcohol Furfurylalkohol *m*, Furfuralkohol *m*, 2-Furylkarbinol *n*, 2-Furankarbinol *n*, 2-Hydroxymethylfuran *n*
furfurylidene Furfuryliden *n*, Fur[fur]al *n* *(Atomgruppe)*
furil Furil *n*, Difuryl(2)-äthandial *n*
furnace Ofen *m*; *s.* f. chamber
  f. atmosphere Ofenatmosphäre *f*
  f. black Ofenruß *m*, Ofenschwarz *n*, Furnace-Ruß *m* *(im Furnace-Verfahren hergestellter Gasruß)*
  f. blast Hochofenwind *m*
  f. chamber Feuerung *f*, Feuerraum *m*, Feuerstätte *f*, Brennraum *m*, Ofenraum *m*, Verbrennungsraum *m*, Verbrennungskammer *f*
  f. charge Ofeneinsatz *m*, Ofenbeschickung *f*, Ofenladung *f*, Ofencharge *f*
  f. clinker Ofenschlacke *f*
  f. coke Hochofenkoks *m*
  f. combustion black *s.* f. black
  f. combustion process *s.* f. process
  f. gas Verbrennungs[ab]gas *n*, Ofengas *n*; Gichtgas *n*
  f. hearth Ofengestell *n*; Hochofengestell *n*
  f. lining Ofenzustellung *f*, Ofenauskleidung *f*, Ofenfutter *n*, Ofenausmauerung *f*; Hochofenfutter *n*
  f. process Furnace-Verfahren *n*, Furnace-Black-Prozeß *m*, Ofenverfahren *n*, Ofenprozeß *m* *(zur Rußerzeugung)*
  f. shaft Ofenschacht *m*
  f. shell Ofenmantel *m*
  f. thermal black [thermischer] Spaltruß *m*, Thermalruß *m*
  f. thermal process Thermalspaltprozeß *m*
  f. transformer Ofentransformator *m*
furnish / to *(z. B. einen Kocher)* beschicken, füllen; *(z. B. Kochgut)* eintragen
furnish 1. Papierrohstoffe *mpl*, Rohstoffe *mpl* für die Papiererzeugung; *(zur Mahlung im Holländer)* eingetragener Faserhalbstoff *m*, Mahlgut *n*, Charge *f*; Kochereintrag *m*, Kocherinhalt *m*, Kocherfüllung *f*, Kochgut *n*; 2. *s.* furnishing
furnishing *(Pap)* Beschicken *n* des Holländers, Holländerfüllung *f*, Holländereintrag *m*, Stoffeintrag *m*
furoate Furoat *n*
furoic acid *s.* furancarboxylic acid
  f. aldehyde *s.* furfuraldehyde
furoin Furoin *n*, 1,2-Difuryl(2)-2-oxoäthanol *n*
furol[e] *s.* furfural
furoyl Furoyl *n*
  f. chloride Furoylchlorid *n*, 2-Furankarbonylchlorid *n*
furrow fertilization Furchendüngung *f*
further nitration Weiternitrieren *n*, Weiternitrierung *f*, weitere Nitrierung *f*
  f. tearing *(Pap)* Weiterreißen *n*, Durchreißen *n*
furyl Furyl *n*

furylacrylic acid Fur[yl]akrylsäure *f*, Furalessigsäure *f*
2-furylaldehyde *s.* furfural
2-furylcarbinol *s.* furfuryl alcohol
furylidene Furyliden *n*
fusain Fusit *m*, Fusain *m* (*Streifenart der Kohle*)
fusaric acid Fusar[in]säure *f*, 5-Butylpyridin-2-karbonsäure *f*
fuse / to schmelzen; (*schwerlösliche Substanzen durch Schmelzen*) aufschließen; (*Ringe*) kondensieren, anellieren
  f. to (*Ringe*) ankondensieren
  f. together verschmelzen, zusammenschmelzen; miteinander kondensieren (*von Ringen*)
fused electrolyte Schmelze *f*
  f. nucleus kondensierter (anellierter) Kern *m*
  f.-on colour Schmelzfarbe *f*
  f. phosphate Glühphosphat *n* (*Düngemittel*)
  f. quartz Quarzglas *n*, durchsichtiges (klares) Kieselglas *n*
  f. ring kondensierter (anellierter) Ring *m*
  f.-ring aromatic system kondensiertes aromatisches Ringsystem *n*
  f.-ring hydrocarbon kondensierter (anellierter) Ringkohlenwasserstoff *m*
  f.-ring system kondensiertes (anelliertes) Ringsystem *n*
  f. rock (*Geol*) Schmelzfluß *m*, Schmelze *f*
  f. salt Salzschmelze *f*
  f. silica Quarzgut *n*, durchscheinendes (undurchsichtiges) Kieselglas *n*
fusel oil Fuselöl *n*
  f.-oil test Prüfung *f* auf Fuselöl
fusibility Schmelzbarkeit *f*
fusible schmelzbar
  f. alloy leichtschmelzende (leichtschmelzbare, niedrigschmelzende) Legierung *f*
  f. alloy pattern Ausschmelzmodell *n*, ausschmelzbares (verlorenes) Modell *n* (*beim Investmentguß*)
  f. clay leichtschmelzender (unter 1380 °C erweichender) Ton *m*
  f. cone Schmelzkegel *m*, Brennkegel *m*, Segerkegel *m*
  f. metal *s.* f. alloy
  f. white precipitate [$H_3N \cdot Hg \cdot NH_3$]Cl_2 schmelzbares weißes Präzipitat *n*, Diamminquecksilber(II)-chlorid *n*
fusibleness *s.* fusibility
fusidic acid Fusidinsäure *f* (*Antibiotikum*)
fusinite Fusinit *m* (*Einzelmazeral der Kohle*)
fusion Schmelzen *n*, Verschmelzen *n*, Verschmelzung *f*, Schmelze *f*; Schmelze *f* (*als Zustand*); Schmelzaufschluß *m* (*schwerlöslicher Substanzen*); (*Kch*) Fusion *f*, Kernfusion *f*, Kernverschmelzung *f*, Kernsynthese *f*; Kondensation *f*, Ringkondensation *f*, Anellierung *f*
  f.-cast refractory schmelzgeformtes Erzeugnis *n*
  f. casting Schmelzgießen *n*, Schmelzformen *n*

f. crucible Schmelztiegel *m*
f. cup Kalorimeterschälchen *n*, Verbrennungstiegel *m*
f.-flow test (*Ker*) Knopfprobe *f*, Knopfprüfung *f* (*auf Fließbarkeit einer Fritte*)
f. name Fusionsname *m*, Verschmelzungsname *m*, Anellierungsname *m*
f. pan Schmelzpfanne *f*, Schmelzkessel *m*, Schmelzgefäß *n*
f. prefix Anellierungspräfix *n*
f. with alcoholic potassium hydroxide alkoholische Kali[umhydroxid]schmelze *f*, Schmelzen (Verschmelzen) *n* mit alkoholischem Kaliumhydroxid (Kali)
fustet Ungarisches Gelbholz *n*, Junger Fustik *m*, Jungfustik *m*, Fiset[te]holz *n*, Fustet *m*
fustic Gelbholz *n*, Fustikholz *n*, Fustik *m* (*von Chlorophora tinctoria Gaud.*)
  f. extract Gelbholzextrakt *m*
fustin Fustin *n*, Dihydrofisetin *n*

G

G acid $HOC_{10}H_5(SO_3H)_2$ G-Säure *f*, 2-Naphthol-6,8-disulfonsäure *f*
g. equiv. *s.* gram equivalent
g-m counter, G-M counter *s.* Geiger-Müller counter
G salt G-Salz *n*, Dikaliumsalz *n* der 2-Naphthol-6,8-disulfonsäure
G-strophantin g-Strophantin *n*, Strophantin G *n*, Gratus-Strophantin *n*, Quabain *n*
gabbro Gabbro *m* (*ein Tiefengestein*)
gable wall Einlegewand *f* (*im Glasschmelzofen*)
Gabriel [phthalimide] synthesis [of primary amines] Gabriel-Synthese *f*, Gabrielsche Aminsynthese *f* (*durch Phthalimidspaltung*)
gadoleic acid $C_{19}H_{37}COOH$ Gadoleinsäure *f*, 9-Eikosensäure *f*
gadolinia Gd_2O_3 Gadolinerde *f*, Gadolinia *f*, (*veraltet für*) Gadoliniumoxid *n*
gadolinite (*Min*) Gadolinit *m* (*Yttrium-Eisen-Beryllosilikat*)
gadolinium Gd Gadolinium *n*
  g. acetate $Gd(CH_3COO)_3$ Gadoliniumazetat *n*
  g. bromide $GdBr_3$ Gadoliniumbromid *n*
  g. chloride $GdCl_3$ Gadoliniumchlorid *n*
  g. fluoride GdF_3 Gadoliniumfluorid *n*
  g. nitrate $Gd(NO_3)_3$ Gadoliniumnitrat *n*
  g. oxide Gd_2O_3 Gadoliniumoxid *n*
  g. sulphate $Gd_2(SO_4)_3$ Gadoliniumsulfat *n*
  g. sulphide Gd_2S_3 Gadoliniumsulfid *n*
gagate (*Min*) Gagat *m*, Jet[t] *m*(*n*) (*Varietät von Braunkohle*)
gage *s.* gauge
gahnite (*Min*) Gahnit *m*, Zinkspinell *m* (*Zinkaluminat*)
Gaillard disperser Gaillard-Turbozerstäuber *m*
Gaillard-Parrish chamber Turmkammer *f* von Gaillard

gain / **to** anlagern, aufnehmen
 g. moisture Feuchtigkeit anziehen
 g. weight an Masse zunehmen
gain in weight Massezunahme f
galactagogue Galaktagogum n *(die Milchsekretion förderndes Mittel)*
galactan Galaktan n *(ein Polysaccharid)*
galactaric acid s. galactosaccharic acid
galactomannan Galaktomannan n, Mannogalaktan n
galactometer Galaktometer n, Laktometer n
galactonic acid HOCH$_2$(CHOH)$_4$COOH Galaktonsäure f
galactosaccharic acid HOOC(CHOH)$_4$COOH Galaktozuckersäure f, Galaktarsäure f, Schleimsäure f, Muzinsäure f, Tetrahydroxyadipinsäure f
galactose Galaktose f *(ein Monosaccharid)*
galactosidase Galaktosidase f, Laktase f *(zu den Glykosidasen gehörendes Ferment)*
galactowaldenase Galaktowaldenase f, Waldenase f *(ein Galaktose vergärendes Ferment)*
d-**galacturonic acid** HOC(CHOH)$_4$·COOH D-Galakturonsäure f
Galam butter Schibutter f, Galambutter f, Karitebutter f *(von Butyrospermum parkii (Don) Kotschy)*
galanga, galangal Galgant m, Galgantwurzel f *(von Alpinia officinarum Hance und A. galanga (L.) Willd.)*
galbanum Galbanum n *(Gummiharz von Ferula-Arten)*
galena *(Min)* Galenit m, Bleiglanz m *(Blei(II)-sulfid)*
galenical *(Pharm)* galenisches Mittel (Präparat) n, galenische Zubereitung f, Galenikum f
galenite s. galena
gall *(Gerb, Bot)* Galle f, Pflanzengalle f; *(Glas)* Galle f, Glasgalle f; *(Med)* Galle f, Gallenflüssigkeit f
 g. nut s. gallnut
gallamic acid s. gallamide
gallamide (HO)$_3$C$_6$H$_2$CONH$_2$ Gallussäureamid n
gallamidic acid s. gallamide
gallanilid, gallanol C$_6$H$_5$NHCOC$_6$H$_2$(OH)$_3$ Gallanilid n, Gallussäureanilid n, 3,4,5-Trihydroxybenzanilid n
gallein Gallein n, Alizarinviolett n, 4,5-Dihydroxyfluoreszein n *(ein Chinonfarbstoff)*
galley proof paper Abziehpapier n, Abzug[s]papier n, Probeabzug[s]papier n, Andruckpapier n
gallic Gallium(III)-...; gallussauer, Gallussäure...
 g. acid C$_6$H$_2$(OH)$_3$COOH Gallussäure f, 3,4,5-Trihydroxybenzoesäure f
 g. acid 3-monogallate Gallussäure-3-monogallat n, 3-Digallussäure f
galling Reiben n, Scheuern n; Fressen n
gallium Ga Gallium n
 g.-ammonium sulphate Ga$_2$(SO$_4$)$_3$ · (NH$_4$)$_2$SO$_4$ Ammoniumgallium(III)-sulfat n
 g. dichloride GaCl$_2$ Galliumdichlorid n, Gallium(II)-chlorid n

 g. ferrocyanide Ga$_4$[Fe(CN)$_6$]$_3$ Galliumhexazyanoferrat(II) n
 g. hexammine chloride [Ga(NH$_3$)$_6$]Cl$_3$ Hexammingallium(III)-chlorid n
 g. hydroxide Ga(OH)$_3$ Gallium(III)-hydroxid n
 g. monoammine chloride [Ga(NH$_3$)]Cl$_3$ Monammingallium(III)-chlorid n
 g. monoselenide GaSe Gallium[mono]selenid n, Gallium(II)-selenid n
 g. monosulphide GaS Gallium[mono]sulfid n, Gallium(II)-sulfid n
 g. monoxide GaO Gallium[mon]oxid n, Gallium(II)-oxid n
 g. nitrate Ga(NO$_3$)$_3$ Gallium(III)-nitrat n
 g. oxychloride GaOCl Galliumoxidchlorid n
 g.-potassium sulphate KGa(SO$_4$)$_2$ Kaliumgallium(III)-sulfat n
 g. sesquioxide Ga$_2$O$_3$ Digalliumtrioxid n, Gallium(III)-oxid n
 g. sesquiselenide Ga$_2$Se$_3$ Gallium(III)-selenid n
 g. sesquisulphide Ga$_2$S$_3$ Gallium(III)-sulfid n
 g. sesquitelluride Ga$_2$Te$_3$ Gallium(III)-tellurid n
 g. suboxide Ga$_2$O Galliumsuboxid n, Gallium(I)-oxid n
 g. subselenide Ga$_2$Se Galliumsubselenid n, Gallium(I)-selenid n
 g. subsulphide Ga$_2$S Galliumsubsulfid n, Gallium(I)-sulfid n
 g. sulphate Ga$_2$(SO$_4$)$_3$ Gallium(III)-sulfat n
 g. tribromide GaBr$_3$ Galliumtribromid n, Gallium(III)-bromid n
 g. trichloride GaCl$_3$ Galliumtrichlorid n, Gallium(III)-chlorid n
 g. trifluoride GaF$_3$ Galliumtrifluorid n, Gallium(III)-fluorid n
 g. triiodide GaJ$_3$ Galliumtrijodid n, Gallium(III)-jodid n
gallnut Gallapfel m
gallogen s. ellagic acid
gallotannic acid, gallotannin Gallotannin n, Tanningerbstoff m, Gallusgerbsäure f, Tannin n *(i.e.S.) (Estergemisch aus Glukose und Gallussäure und (oder) deren Depsiden)*
gallous Gallium(II)-...
galmei, galmey *(Min)* Galmei m, Edler Galmei m, Zinkspat m, Smithsonit m *(Zinkkarbonat); (Min)* Galmei m, Gemeiner Galmei m, Kieselgalmei m, Kieselzinkerz n, Hemimorphit m *(Zinkdihydroxypyrosilikat)*
galvanic galvanisch
 g. cell s. g. element
 g. corrosion galvanische Korrosion f
 g. element galvanisches (elektrochemisches) Element n, galvanische (elektrochemische) Zelle (Kette) f
galvanism *(Ech)* Galvanismus m; *(Med)* Galvanismus m, Galvanotherapie f
galvanization Verzinken n, Verzinkung f, *(meist i.e.S.)* Feuerverzinken n; *(veraltet für)* electroplating; *(Med)* Galvanisation f

galvanize / to verzinken, *(meist i.e.S.)* feuerverzinken; *(veraltet für)* to electroplate; *(Med)* galvanisieren, mit galvanischem Strom behandeln
galvanizing s. galvanization
galvanoluminescence Galvanolumineszenz f
galvanometer Galvanometer n
  g. recorder schreibendes Galvanometer n, Schreibgalvanometer n, Galvanometerschreiber m
galvanoplastic galvanoplastisch
galvanoplastics, galvanoplasty Galvanoplastik f, *(i.w.S.)* Elektroplattierung f
galvanoscope Galvanoskop n
galvanostegy Galvanostegie f
galvanotechnics Galvanotechnik f
gambier Gambir m, gelbes Katechu n *(Extrakt aus Uncaria gambir Roxb.)*
gamboge Gummigutt n *(von Garcinia-Arten, vorwiegend G. hanburyi Hook. f.)*
gamma acid $C_{10}H_5NH_2OHSO_3H$ Gammasäure f, γ-Säure f, 7-Amino-1-naphthol-3-sulfonsäure f, 2-Amino-8-naphthol-6-sulfonsäure f
  g.-aminopropyl triethoxysilane γ-Aminopropyltriäthoxysilan n, γ-APT
  g. antimony graues (metallisches) Antimon n
  g. arsenic graues (metallisches) Arsen n
  g. fraction γ-Fraktion f, γ-Bestandteile mpl *(bei der Pyridinextraktion der Steinkohle)*
  g. helix γ-Helix f *(Strukturvariante von Polypeptidketten)*
  g. infinity *(Foto)* Gammagrenzwert m, Gamma unendlich n
  g. iron γ-Eisen n, Austenit m
  g. radiation Gammastrahlung f
  g. rays Gammastrahlen mpl, γ-Strahlen mpl
  g. value *(Foto)* Gammawert m, Gamma n, Kontrastfaktor m, Entwicklungsfaktor m, Gradation f
  g. vulcanizate *(Gum)* γ-Vulkanisat n
gangue [mineral] Gangart f, Gangmineral n, taubes Gestein n, Nichterz n
gangway Verbindungsgang m
gan[n]ister *(Geol)* Ganister m *(kieselsäurereicher Ton)*
ganomatite *(Min)* Ganomatit m, Gänsekötigerz n *(silber- und kobalthaltiges Eisenarsenat)*
gap Hohlraum m, Lücke f, Spalt m, Spalte f
  g. type press Maulpresse f, einhüftige Presse f
garbage Abfall m, Müll m, Abfälle mpl, Abfallstoffe mpl, Rückstände mpl
  g. disposal plant Müllaufbereitungsanlage f
garbyite *(Min)* Garbyit m, *(veraltet für)* Enargit m *(Arsen(III)-kupfer(I,II)-sulfid)*
garden balm Zitronenmelisse f, Milissa officinalis L.
gardjan (gargan) balsam oil s. gurjun balsam oil
gargarism, gargle *(Pharm)* Gargarisma n, Gurgelmittel n
garlic Knoblauch m, Allium sativum L.

  g. oil Knoblauchöl n
garment leather Bekleidungsleder n
garnet *(Min)* Granat m
  g. lac Granat[schel]lack m, Rubin[schel]lack m
  g. lime chrome *(Min)* Kalkchromgranat m *(Kalziumchrom(III)-orthosilikat)*
  g.-red granatrot, leuchtendrot
garnierite *(Min)* Garnierit m, Nickel-Antigorit m *(Nickelsilikat)*
gas / to gasen, Gas abgeben; *(Met) (eine Schmelze)* vergasen, zur Gasaufnahme veranlassen; *(Landw, Lebm)* begasen; mit Gas vergiften; *(Text)* sengen, gasieren
gas. s. gaseous
gas Gas n; *(Am)* Benzin n, Kraftstoff m
  g. absorption Gasaufnahme f
  g. adsorption chromatography Gasadsorptionschromatografie f
  g. agitation Gebläsemischen n, pneumatisches Rühren n
  g. analysis Gasanalyse f
  g.-analytical gasanalytisch
  g.-atomizing nozzle Zweistoffdüse f, Injektionszerstäuber m; Druckluftdüse f, Druckluftzerstäuber m
  g. balance Gas[dichte]waage f
  g.-bearing gasführend
  g. black Gasruß m
  g. bleaching *(Pap)* Gasbleiche f
  g. blowtorch Gebläsebrenner m
  g. bubble Gasblase f, Gasbläschen n
  g. buret[te] Gasbürette f
  g. burner Gasbrenner m
  g. calorimeter Gaskalorimeter n, Kalorimeter n für Gase
  g. cap *(Erdöl)* Gaskappe f, Gaskopf m
  g. cap drive *(Erdöl)* Gaskappendruck m, Druck m der expandierenden Gaskappe
  g. cap-drive field (reservoir) Gaskappenlagerstätte f, Lagerstätte f *(Erdölfeld n)* unter Gaskappendruck (mit Gaskappe)
  g. carbon Retortenkohle f, Retortengraphit m
  g.-carburize / to gasaufkohlen, gaszementieren, gaseinsetzen, in gasförmigen Mitteln aufkohlen (zementieren, einsetzen)
  g. carburizing Gasaufkohlen n, Gaskohlung f, Gaszementieren n, Gaseinsetzen n, Aufkohlen (Zementieren, Einsetzen) n in gasförmigen Mitteln
  g.-carburizing furnace Gaskohlungsofen m
  g.-carburizing process Gasaufkohlungsverfahren n, Gaszementationsverfahren n
  g. cavity Gasblase f, Blase f *(im Guß)*
  g. cell *(Spektroskopie)* Gaskammer f; *(Ech)* [elektrisches] Gaselement n, [elektrische] Gaszelle f, Gaskette f
  g. centrifugation *(Kch)* Gaszentrifugenverfahren n, Schleuderverfahren n *(zur Isotopentrennung)*
  g. centrifuge Gaszentrifuge f
  g. centrifuge method s. g. centrifugation

g. chamber kiln Gaskammerofen *m*
g. checking *(Farb)* Eisblumenbildung *f*, Holzöl-
Eisblumenbildung *f*, Holzölerscheinung *f*, Holz-
ölkrankheit *f*
g. chromatograph Gaschromatograf *m*
g. chromatography Gaschromatografie *f*
g. chromizing Inchromieren (Inkromieren, Chromieren) *n* aus der Gasphase (gasförmigen Phase)
g. clarification *s.* **g. cleaning**
g. clathrate compound Gashydrat *n*, Eishydrat *n*
(Käfigeinschlußverbindung)
g. cleaning Gasreinigung *f*
g. coal Gaskohle *f*
g. cock Gashahn *m*
g. coke Gas[werks]koks *m*
g. collecting main Gassammelleitung *f*, Gas-
vorlage *f*, Sammelvorlage *f*, Ofenvorlage *f (einer
Koksofenbatterie)*
g. collection tube Gassammelröhre *f*
g. column Gassäule *f*
g. compressor Gasverdichter *m*, Gaskompressor
m
g. concentration Gaskonzentration *f*
g. concrete Gasbeton *m*
g. connecting tube Verbindungsrohrleitung *f*
(Verbindungsrohr *n*) für Gase
g. constant Gaskonstante *f*
g. container Gasbehälter *m*
g. content Gasgehalt *m*
g. cooler Gaskühler *m*
g. cooling Gaskühlung *f*
g. coulometer Knallgascoulometer *n*, Gas-
coulometer *n*
g.-cyanide / to *(Met)* gaszyanieren, trocken
zyanieren, karbonitrieren
g. cyaniding (cyanization) *(Met)* Gaszyanieren *n*,
Trockenzyanieren *n*, Karbonitrieren *n*
g.-cycle refrigeration system Kaltgasanlage *f*
g. cylinder Gasflasche *f (Druckflasche)*
g.-cylinder support Rohrschelle *f* für Gas-
flaschen
g. degeneracy *(phys Ch)* Gasentartung *f (nahe
dem absoluten Nullpunkt)*
g. density Gasdichte *f*
g. density balance Gas[dichte]waage *f*
g. detector Gasspürgerät *n*
g. diffuser stone Filterstein *m*
g. discharge tube Gasentladungsröhre *f*, Entladungsröhre *f*, Entladungsrohr *n*, Entladerohr *n*
g. dispersion Gasdispersion *f*
g. dispersion (distribution) tube Begasungsfilter
n, Begasungsröhrchen *n*, *(mit zylindrischem
Frittenteil auch)* Begasungsfilterkerze *f*
g. drive Gastrieb *m*, Gasdruck *m*
g.-drive field (reservoir) Gastrieblagerstätte *f*,
Lagerstätte *f* (Erdölfeld *n*) mit Gastrieb, unter
Gastrieb (Gasdruck) stehende Lagerstätte *f*
g. drying jar Trockenturm *m*, Chlorcalcium-
zylinder *m*
g. electrode Gaselektrode *f*
g. enrichment Gasanreicherung *f*

g. equation Gasgleichung *f*
g. equilibrium Gasgleichgewicht *n*
g. exchange quotient Gaswechselquotient *m*
g. fading *(Text)* Gasschwund *m*, Abgasempfind-
lichkeit *f (Ausbleichen von Färbungen)*
g. fan Gasventilator *m*
g. filter (filtering tube) Gasfilter *n*
g. filtration Filtrieren *n* von Gas, Gasfiltration *f*
g.-fired gasbeheizt, mit Gas geheizt (beheizt,
betrieben)
g. flame Gasflamme *f*
g. flame coal Gasflammkohle *f*
g. flow Gasströmung *f*
g. flowmeter Strömungsmesser *m* für Gase
g.-forming gasbildend
g.-fuel firing Gas[be]heizung *f*, Gasfeuerung *f*
g.-fume fastness Abgasechtheit *f*, Gasechtheit *f*,
Abgasbeständigkeit *f*, Rauchgasechtheit *f*
g. furnace process *(Gum)* Gas-Furnace-
Verfahren *n*
g.-gangrene toxin Gasgangräntoxin *n*
g. generator Gasentwicklungsapparat *m*, Gasentwickler *m*, Gaserzeuger *m*, Gasgenerator *m*
g.-heated *s.* **g.-fired**
g. heating *s.* **g.-fuel firing**
g. hydrate Gashydrat *n*, Eishydrat *n (Käfigeinschlußverbindung)*
g. in solution gelöstes Gas *n*
g. industry Gasindustrie *f*
g. injection Gasinjektion *f*, Gaseinpressen *n*,
Einpressen *n* von Gas
g. inlet Gaseintritt *m*, Gaszuleitung *f*
g. inlet pipe Gaszuführungsrohr *n*
g. jar Standzylinder *m*
g. law Gasgesetz *n*
g. levelling bulb Niveaukugel *f*
g. lift Gaslift *m*, Gasheber *m*
g.-lift method (process) Gasliftverfahren *n*
g.-lift system Gasliftsystem *n*
g.-lift valve Gasliftventil *n*
g. lifting Gasliften *n*, Gas[lift]förderung *f*,
Förderung *f* mittels Gaslifts (Druckgases)
g.-light paper *s.* gaslight paper
g. liquefaction Gasverflüssigung *f*
g.-liquid chromatogram Gas-Flüssig[keits]-Chro-
matogramm *n*
g.-liquid chromatography Gas-Flüssig[keits]-
Chromatografie *f*, Gas-Liquidus-Chromatografie
f, GLC
g.-liquid partition chromatography Gasverteï-
lungschromatografie *f*
g. liquor Gaswasser *n*, Ammoniakwasser *n*
g. main Gasleitung *f*
g. making Gaserzeugung *f*, Gasherstellung *f*
g.-making coal Gaskohle *f*
g.-making plant Gaserzeugungsanlage *f*, Gas-
werk *n*, Gasanstalt *f*
g. mantle Gasglühkörper *m*, Gasglühlichtstrumpf
m, Auer-Glühstrumpf *m*, Auer-Strumpf *m*
g. mask Gasmaske *f*
g. meter Gasmesser *m*, Gaszähler *m*

g. mixture Gasgemisch n, Gasmischung f
g. muffle kiln Gas-Muffelofen m
g. offtake Gasabgang m, Gasabzug m, Gasaustritt m, Gasabführung f, Gasentnahme f
g. offtake pipe Gasabgangsrohr n, Gasabzugsrohr n, Gasableitungsrohr n; Liegerohr n (eines Coalite-Schwelofens)
g. oil Gasöl n, Dieselöl n, Treiböl n
g.-oil distillate Gasöldestillat n
g./oil ratio Gas/Öl-Verhältnis n, GÖV, G.Ö.V.
g. outlet Gasaustritt m, Gasaustrittsöffnung f, Gasausgang m, Gasabgang m, Gasabzug m
g. output Gasausbringen n, Gasausbringung f; Gasausbeute f
g. passage Gasdurchgang m
g. permeability Gasdurchlässigkeit f
g. phase gasförmige Phase f, Gasphase f
g. pipette Gaspipette f
g. plant s. g.-making plant
g. plating Vakuummetallisierung f, Hochvakuumbedampfung f, Herstellung f metallischer Überzüge nach dem Aufdampfverfahren (Vakuumaufdampfverfahren, Gasverfahren)
g. pliers Rohrzange f
g. poisoning Gasvergiftung f
g. polymerization s. gaseous polymerization
g. pressure Gasdruck m
g. pressure gauge Manometer n
g. producer Gaserzeuger m, Gasentwickler m, Gasgenerator m
g. production Gaserzeugung f, Gasherstellung f
g. pump Pumpe f für Gase (Verdichter oder Vakuumpumpe)
g. purification Gasreinigung f
g. purifier Gasreiniger m
g. reaction Gasreaktion f
g. reservoir Gaslagerstätte f
g. retort Gas[werks]retorte f
g.-retort coke Gasretortenkoks m
g.-retort gas Retortengas n
g.-retort tar Gasretortenteer m
g. sampling pipette (tube) Gassammelröhre f
g.-saturation method Saturationsmethode f (dynamische Methode zur Dampfdruckmessung)
g. scrubbing s. g. washing
g. seal Gasdichtung f
g. separation Gas[ab]trennung f, (Erdöl auch) Gas-Öl-Trennung f, Erdgasabtrennung f, Entgasung f (des geförderten Roherdöls)
g. separator Gasseparator m, Gas[ab]scheider m, Gastrennanlage f, (Erdöl auch) Gas-Öl-Separator m, Gas-Öl-Trennvorrichtung f
g. singeing (Text) Gassengen n, Sengen n, Gasieren n
g.-singeing machine (Text) Gassengmaschine f
g.-solid chromatography Gas-Fest[stoff]-Chromatografie f, Gas-Solidus-Chromatografie f, GSC
g. solubility Gaslöslichkeit f
g. space Gasraum m
g. spirit Gasbenzin n

g. stopcock Gashahn m
g. stream Gasstrom m
g. strength Gaskonzentration f
g. sulphur Flotationsschwefel m
g. tank Gasbehälter m
g. tar Gas[werks]teer m
g. thermometer Gasthermometer n
g. thermometer of the constant pressure type Gasthermometer n konstanten Drucks
g. thermometer of the constant volume type Gasthermometer n konstanten Volumens (konstanter Dichte)
g.-tight gasdicht, gasundurchlässig
g. to be compressed Fördergas n (im Verdichter)
g. uptake steigendes Gasabzug[s]rohr n, Gasabzug m
g. valve Gasventil n, Gasschieber m
g. volumeter Gasvolumeter n
g. volumetry Gasvolumetrie f
g. wash bottle s. g. washing bottle
g. washer Gaswäscher m, Wäscher m, Wascher m, Skrubber m, Turmwäscher m, Waschturm m
g. washing Gaswäsche f
g. washing bottle Gaswaschflasche f
g.-washing system Gaswaschsystem n, Gasreinigungssystem n, Gaswäsche f (als Anlage)
g.-washing tower Gaswaschturm m
g. welding Gas[schmelz]schweißen n, autogenes Schweißen n, Autogenschweißen n, Autogenschweißung f
g. well Gasbohrung f, Gasbrunnen m; Gasquelle f, Erdgasquelle f
gaseous gasförmig, gasartig
g. ammonia gasförmiges Ammoniak n, Ammoniakgas n
g. component Gaskomponente f
g. condition Gaszustand m
g. density Gasdichte f
g. diffusion method (Kch) Gasdiffusionsverfahren n (zur Isotopentrennung)
g. film gasanaloger Oberflächenfilm m
g. fuel gasförmiger Brennstoff m
g. ion Gasion n
g. mixture Gasgemisch n, Gasmischung f
g. polymerization Polymerisation f in der Gasphase
gasholder Gasbehälter m
gasifiable motor fuel Vergaserkraftstoff m, Vergasertreibstoff m, VK
gasification Vergasen n, Vergasung f
g. chamber Vergasungskammer f, Vergasungsraum m
g. medium Vergasungsmedium n, Vergasungsmittel n
g. plant Vergasungsanlage f
g. rate Vergasungsgeschwindigkeit f
g. zone Vergasungszone f
gasifier Vergaser m, Vergasungsapparat m
gasify / to vergasen
gasifying agent (material) Schaum[bildungs]mittel

n, Schaumbildner m (z.B. zur Herstellung von Schaumglas)

gasket Dichtungsmanschette f, Dichtung f (für nicht gegeneinander bewegte Teile)
- **g. material** s. gasketing material

gasketing area Dichtfläche f (einer ruhenden Dichtung)
- **g. material** Dicht[ungs]werkstoff m, Dicht[ungs]stoff m, Dicht[ungs]material n

gaslifting Gasliften n, Gas[lift]förderung f, Förderung f mittels Gaslifts (Druckgases)

gaslight paper Gaslichtpapier n, (veraltet für) Kontaktpapier n

gasoline Benzin n (besonders als Vergaserkraftstoff); Leichtbenzin n, leichtes Benzin n (Siedebereich 30 bis 100 °C bzw. 14 bis 90 °C)
- **g. additive** Benzinadditiv[e] n
- **g. fraction** Benzinfraktion f
- **g. plant** Benzingewinnungsanlage f
- **g. range** Benzin[siede]bereich m
- **g. refining** Benzinraffination f
- **g.-soluble** benzinlöslich

gassing Gasen n, Gasentwicklung f; (Met) Vergasen n, Vergasung f (einer Schmelze), Schmelzvergasung f; (Landw, Lebm) Begasung f; Gasvergiftung f; (Text) Gassengen n, Sengen n, Gasieren n

gassy concrete s. gas concrete

gastight gasdicht, gasundurchlässig

gastric acid Magensäure f
- **g. antacid** Antazidum n (Mittel gegen Hyperazidität)
- **g. juice** Magensaft m

gasworks s. gas-making plant
- **g. tar** Gas[werks]teer m

gate Schieber m; (Pap) Austrittsspalt m, Ausflußspalt m, Ausflußschlitz m (des Stoffauflaufs)
- **g. agitator** Gitterrührer m, Gatterrührer m
- **g. control** Schiebersteuerung f
- **g. discharge** Schieberaustrag m
- **g. mixer** s. g. agitator
- **g. valve** Schieber m, Absperrschieber m, Schieberventil n

gather [glass] / to Glas (aus der Schmelze) aufnehmen

gather [of glass] Glasposten m, Posten m

gatherer (Glas) Anfänger m

gathering hole (Glas) Arbeitsloch n, Arbeitsöffnung f, Entnahmeloch n, Entnahmestelle f
- **g. iron** (Glas) Anfangeisen n, Aufnahmeeisen n
- **g. opening** s. g. hole
- **g. station** Sammelstation f

Gattermann aldehyde synthesis Gattermannsche Aldehydsynthese f

Gattermann-Koch reaction Gattermann-Koch-Synthese f

Gattermann-Kohl reaction Gattermann-Kohl-Synthese f, Gattermann-Kohlsche Synthese f (von Phenolaldehyden)

Gattermann reaction Gattermann-Reaktion f (Diazoniumaustausch)

gauche form (Stereochemie) schiefe (windschiefe, syn-clinale) Form f, syn-Form f

gauge 1. Meßgerät n, Meßapparat m, Meßinstrument n, Messer m; 2. (Text) Warendichte f
- **g. glass** Flüssigkeitsstand[s]anzeiger m, Wasserstand[s]anzeiger m, Schauglas n
- **g. roller** Führungswalze f
- **g. stick** Meßlatte f, Peilstab m

Gaussian curve (Statistik) Gaußsche Verteilungskurve f

gauze Netz n, Siebgewebe n, Siebtuch n, Siebbespannung f; Mull m, Gaze f
- **g. cathode** Netzkatode f
- **g. disk** Drahtnetz n (eines Flammpunktprüfers)
- **g. electrode** Netzelektrode f
- **g. plug** Netzspirale f, Drahtnetzspirale f, (meist) Kupferdrahtnetzspirale f (im Verbrennungsrohr)

Gay-Lussac law Gay-Lussacsches Gesetz n, Gay-Lussac-Gesetz n

Gay-Lussac method Endpunkt[s]bestimmung (Bestimmung) f nach Gay-Lussac

Gay-Lussac tower Gay-Lussac-Turm m

gay-lussite, gaylussite (Min) Gaylussit m (Natriumkalziumkarbonat)

Gaz Integral predistillation [gas] producer Schwelgenerator (Doppelgasgenerator, Doppelgaserzeuger) m der Gaz Integral, G.I.-Schwelgenerator m, G.I.-Generator m

gear oil Getriebeöl n
- **g. pump** Zahnradpumpe f

Geer oven (Gum) Geer-Ofen m

Geer oven ageing (Gum) Geer-Ofenalterung f, Geer-Alterung f, Alterung (Heißluftalterung) f im Geer-Ofen

Geer oven test (Gum) Alterungsprüfung f nach Geer, Geer-Alterungsprüfung f

gegenion Gegenion n

gehlenite (Min) Gehlenit m (ein Sorosilikat)

Gehman torsion test (Gum) Gehman-Test m

Geiger[-Müller] counter (tube) Geiger-Müller-Zählrohr n, Geiger-Zähler m

Geiger-Nuttall rule Geiger-Nuttallsche Regel f, Geiger-Nuttall-Beziehung f

Geissler-Mohr absorption (potash) bulb Geißlersches (Mohrsches) Gefäß n, Kaliapparat m nach Geißler (Mohr)

Geissler tube Geißler-Röhre f, Geißlersche Röhre f

gel / to erstarren, festwerden, steifwerden, (i.e.S.) gel[atin]ieren, zu Gelee erstarren, (Koll auch) in den Gelzustand übergehen, ausflocken, koagulieren

gel. s. gelatinous

gel (Koll) Gel n
- **g. chromatography** s. g. permeation chromatography
- **g. coat** gelartiger Überzug m, Gelschicht f
- **g. column** Gelsäule f
- **g. content** (Plast) Gelteilchengehalt m
- **g. count** (Plast) Gelteilchenzählung f

g. filtration Gel-Filtration *f (Chromatografie)*
g. formation Gelbildung *f*
g. of the silicic acid type Kiesel[säure]gel *n*
g. paint thixotrop[isch]e Anstrichfarbe *f*
g. particle Gelteilchen *n*
g. permeation chromatography Gel-Permeations-Chromatographie *f*, Gelchromatographie *f*
g. point Gel[atin]ierungstemperatur *f*
g.-sol-gel transformation Gel-Sol-Gel-Umwandlung *f*
g. time s. gelling time
gelate / to s. to gel
gelatification s. gelatinization
g. oven *(Text)* Gelierkanal *m*
gelatin Gelatine *f*; Gallerte *f*, Gallert *n*
g. bromide plate *(Foto)* Bromgelatineplatte *f*
g. culture Gelatinekultur *f*
g. dynamite Sprenggelatine *f*
g. effect *(Foto)* Gelatineeffekt *m*
g. emulsion *(Foto)* Gelatine-Emulsion *f*
g. layer *(Foto)* Gelatineschicht *f*
g. protective layer *(Foto)* Gelatineschutzschicht *f*
g.-sized *(Pap)* mit Gelatine (Tierleim) geleimt, gelatinegeleimt
g. sizing *(Pap)* Gelatineleimung *f*, Oberflächenleimung (Leimung) *f* mit Gelatine (Tierleim)
g. sol Gelatinesol *n*
gelatinate / to s. to gelatinize
gelatination s. gelatinization
gelatine s. gelatin
gelatiniform gelatineartig, gelatinös, gallertartig, gallertähnlich,
gelatinization Gelatinieren *n*, Gelatinierung *f*
gelatinize / to gelatinieren, in Gelee überführen; gel[atin]ieren, zu Gelee erstarren
gelatinizer s. gelatinizing agent
gelatinizing agent Gelatinierungsmittel *n*, Geliermittel *n*, Gelierstoff *m*
gelatinous gelatineartig, gelatinös, gallertartig, gallertähnlich, gallertig; gelatinehaltig
g. aluminium hydroxide Aluminogel *n*, Tonerdegel *n* *(Aluminiumoxidhydrat)*
g. mass Gallerte *f*, Gallert *n*
g. silica Kiesel[säure]gel *n*
g. substance s. g. mass
gelation Erstarren *n*, Festwerden *n*, Steifwerden *n*; Gel[atin]ieren *n*, Gel[atin]ierung *f*, Gallertbildung *f*, *(Koll auch)* Gelbildung *f*
g. temperature s. gel point
gelbin Gelbin *n*, Kalkchromgelb *n* *(Kalziumchromat)*; Barytgelb *n*, Ultramaringelb *n*, gelbes Ultramarin *n*, Steinbühler Gelb *n* *(Bariumchromat)*
geling s. gelling
gelling Gel[atin]ieren *n*, Gel[atin]ierung *f*, Gallertbildung *f*, *(Koll auch)* Gelbildung *f*
g. agent s. gelatinizing agent
g. time Gelzeit *f*
gelseminic acid Gelseminsäure *f*, Skopoletin *n*,

Äskuletin-6-monomethyläther   *m*,   6-Methoxy-7-hydroxykumarin *n*
gem[stone] Edelstein *m*
general acid catalysis allgemeine Säurekatalyse *f*
g. base catalysis allgemeine Basenkatalyse *f*
g. corrosion allgemeine Korrosion *f*
g. formula allgemeine Formel *f*, Allgemeinformel *f*
g.-purpose allgemein (universell) verwendbar, Allzweck..., Universal...
g.-purpose cream Creme *f* für alle Zwecke, Tag- und Nachtcreme *f*
g.-purpose furnace black *(Gum)* General-Purpose-Furnace-Ruß *m*, GPF-Ruß *m* *(vielseitig verwendbarer Öl-Gas-Ruß)*
g.-purpose pH meter Universal-pH-Meter *n*
g.-purpose rubber universeller Kautschuk *m*, Allzweckkautschuk *m*
g.-purpose synthetic rubber universeller synthetischer Kautschuk *m*, Allzwecksynthesekautschuk *m*
generate / to erzeugen, generieren
generation Erzeugung *f*, Generation *f*; Entwicklung *f*; Entstehung *f*
g. of gases Gasentwicklung *f*
generator Generator *m (Schachtofen zur Gaserzeugung)*, *(i.e.S.)* Generatorschacht *m*; Dampferzeuger *m*; Austreiber *m (einer Absorptionskälteanlage)*
g. gas Generatorgas *n*
g. method of acetification [englisches] Generator-Verfahren *n*, Rundpumpverfahren *n (Essigherstellung)*
generic name (term) Sammelname *m*, Sammelbezeichnung *f*, allgemeiner (generischer) Name *m*, Gattungsname *m*; nichtgeschützte (freie) Bezeichnung *f*, nicht wortgeschützter Name. *m*, freier Warenname *m*, Freiname *m*
genesis Genesis *f*, Genese *f*, Entstehung *f*, Bildung *f* (*z. B. von Kohle, Erdöl*)
Geneva name Genfer Name (Systemname) *m*, nach der Genfer Nomenklatur gebildete Benennung *f*
G. nomenclature Genfer Nomenklatur *f*
G. system of naming (nomenclature) Genfer Nomenklatursystem *n*
genthite *(Min)* Genthit *m*
gentian Enzian *m*, Gentiana L.; Enzianwurzel *f*, Bitterwurzel *f*, Fieberwurzel *f*, Bergfieberwurzel *f*
g. violet Gentianaviolett *n*, Enzianviolett *n*
gentianic acid s. gentisin
gentisic acid $(HO)_2C_6H_3COOH$ Gentisinsäure *f*, 2,5-Dihydroxybenzoesäure *f*
gentisin Gentisin *n*, 1,7-Dihydroxy-3-methoxyxanthon *n*
genuine handmade paper handgeschöpftes Papier *n*, Büttenpapier *n*, Handpapier *n*, Schöpfpapier *n*, Handbütten *n*, Echt-Bütten *n*
g. watermark *(Pap)* echtes Wasserzeichen *n*
geochemical geochemisch

g. stage geochemische (geologische, zweite) Phase f [der Inkohlung], zweite Inkohlungsstufe f, zweiter Inkohlungsabschnitt m, geochemische Inkohlung f

geochemistry Geochemie f

geocronite (Min) Geokronit m (ein Blei-Arsen-Antimon-Sulfid)

geode Geode f, Mandel f; Druse f, Kristalldruse f

geography paper (Am) Landkartenpapier n, Kartenpapier n; Seekartenpapier n

geokronite s. geocronite

geologize / to geologisch untersuchen

geology Geologie f

geometric isomer geometrisch isomere Verbindung (Modifikation) f, geometrisches Isomer[es] n

g. isomerism geometrische Isomerie f, cis-trans-Isomerie f

geometrical isomer s. geometric isomer

geosphere Geosphäre f

geranial $(CH_3)_2C=CHCH_2CH_2C(CH_3)=CHCHO$ Geranial n, Zitral A n, trans-Zitral n

geranialdehyde s. geranial

geranic acid $(CH_3)_2C=CHCH_2CH_2C(CH_3)=CHCOOH$ Geraniumsäure f, 3,7-Dimethyl-2,6-oktadiensäure f, 2,6-Dimethyl-heptadien-(1,5)-karbonsäure-(1) f

geraniol $(CH_3)_2C=CHCH_2CH_2C(CH_3)=CHCH_2OH$ Geraniol n, 3,7-Dimethyl-2,6-oktadien-1-ol n

gerhardtite (Min) Gerhardtit m (Kupfer(II)-trihydroxidnitrat)

germ oil Keimöl n

g.-proofing filter s. g.-tight filter

g.-separating machine (Lebm) Entkeimungsmaschine f

g. separator (Lebm) Keimseparator m

g.-tight filter bakteriendichtes Filter n, Bakterienfilter n

German method of acetification Schnellessigverfahren n

germanate Germanat(IV) n

germane GeH_4 Monogerman n, Germaniumtetrahydrid n

germanic Germanium..., (i.e.S.) Germanium(IV)-...

g. acid s. germanium dioxide

g. oxide GeO_2 Germanium(IV)-oxid n, Germaniumdioxid n

germanite Germanat(II) n

germanium Ge Germanium n

g. bromoform $GeHBr_3$ Germaniumbromoform n, Tribromgerman n, Germaniumhydrogentribromid n

g. chloroform Germaniumchloroform n, Trichlorgerman n, Germaniumhydrogentrichlorid n

g. dibromide $GeBr_2$ Germaniumdibromid n, Germanium(II)-bromid n

g. dichloride $GeCl_2$ Germaniumdichlorid n, Germanium(II)-chlorid n

g. difluoride GeF_2 Germaniumdifluorid n, Germanium(II)-fluorid n

g. dihydro dichloride GeH_2Cl_2 Germaniumdihydrogendichlorid n, Dichlorgerman n

g. diiodide GeJ_2 Germaniumdijodid n, Germanium(II)-jodid n

g. dinitride Ge_3N_2 Trigermaniumdinitrid n, Germanium(II)-nitrid n

g. dioxide Germaniumdioxid n, Germanium(IV)-oxid n

g. disulphide GeS_2 Germaniumdisulfid n, Germanium(IV)-sulfid n

g. hexahydride Ge_2H_6 Germaniumhexahydrid n, Digerman n

g. imide $Ge(NH)_2$ Germaniumdiimid n

g. monosulphide GeS Germaniummonosulfid n, Germanium(II)-sulfid n

g. monoxide GeO Germaniummonoxid n, Germanium(II)-oxid n

g. octahydride Ge_3H_8 Germaniumoktahydrid n, Trigerman n

g. oxychloride $GeOCl_2$ Germaniumoxidchlorid n

g. tetrabromide $GeBr_4$ Germaniumtetrabromid n, Germanium(IV)-bromid n

g. tetrachloride $GeCl_4$ Germaniumtetrachlorid n, Germanium(IV)-chlorid n

g. tetrafluoride GeF_4 Germaniumtetrafluorid n, Germanium(IV)-fluorid n

g. tetrahydride GeH_4 Germaniumtetrahydrid n, Monogerman n

g. tetraiodide GeJ_4 Germaniumtetrajodid n, Germanium(IV)-jodid n

g. tetranitride Ge_3N_4 Germaniumtetranitrid n, Trigermaniumtetranitrid n, Germanium(IV)-nitrid n

g. trihydro monochloride GeH_3Cl Germaniumtrihydrogenchlorid n

germanous Germanium..., (i.e.S.) Germanium(II)-...

g. hydroxide $GeO \cdot xH_2O$ Germanium(II)-oxidhydrat n

g. oxide GeO Germanium(II)-oxid n, Germaniummonoxid n

germicidal keimtötend

germicide Germizid n, keimtötendes (keimfreimachendes) Mittel n

germinal brush (Lebm) Entkeimungsmaschine f

germinating apparatus (Lebm) Keimapparat m

g. drum (Lebm) Keimtrommel f

g. power (Lebm) Keimfähigkeit f, Keimkraft f

g. test (Lebm) Keimprobe f, Keimprüfung f

germination Keimen n, Keimung f

g. injury Keimschädigung f (durch Saatgutbeize)

g. test s. germinating test

germinative activity (capacity) s. germinating power

g. energy (Lebm) Keimenergie f

g. power s. germinating power

gersdorffite (Min) Gersdorffit m (Nickelarsensulfid)

Gerstenberg complector plant Gerstenberg-Komplektor m (Margarineherstellung)

get out of hand / to außer Kontrolle geraten

getter (Vakuumtechnik) Getter m(n), Getterstoff m, Fangstoff m, Gettermetall n

gettering *(Vakuumtechnik)* Gettern *n*, Getterung *f*
geyserite *(Min)* Geyserit *m*, Kieselsinter *m*, Opalsinter *m* *(als Absatz von Geysiren)*
gg. *s.* gauge
ghatti gum *(Pflanzengummi von Anogeissus latifolia Wall.)*
ghost image *(Foto)* Doppelbild *n*, Geisterbild *n*, Reflexbild *n*, Phantombild *n*, Nebenbild *n*
ghosts *(Spektroskopie)* Geister *mpl (vorgetäuschte Spektrallinien)*
G.I. predistillation [gas] producer *s.* Gaz Integral predistillation [gas] producer
giant fir Riesentanne *f*, Abies grandis Lindl.
g. molecule Makromolekül *n*, Riesenmolekül *n*
gibberellic acid *(Bioch)* Gibberellinsäure *f*
Gibbs adsorption equation Gibbssche Adsorptionsgleichung *f*
Gibbs cell Gibbs-Zelle *f (zur Elektrolyse)*
Gibbs-Duhem equation Gibbs-Duhemsche Gleichung *f*
Gibbs-Helmholtz equation Gibbs-Helmholtz-Gleichung *f*, Gibbs-Helmholtzsche Gleichung *f*
Gibbs reagent Gibbs' Reagens *n (2,6-Dibromchinon-N-chlorimid, Hydroxyflavon-Nachweis)*
gibbsite *(Min)* Gibbsit *m*, Hydrargillit *m (Aluminiumorthohydroxid)*
Giemsa reagent Giemsasches Reagens *n (aus $HgCl_2$ und KJ, Chininnachweis)*
Gieseler plastometer Gieseler-Plastometer *n*, Gieselersches Plastometer *n*
gild / to vergolden
gilding Vergolden *n*, Vergoldung *f*
gin 1. *(Text)* Egreniermaschine *f*, Entkernungsmaschine *f*; 2. Gin *m (Wacholderbranntwein)*
gingelly oil Sesamöl *n (von Sesamum indicum L.)*
ginger Ingwer *m*, Ingwerpflanze *f*, Zingiber officinale Rosc.; Ingwerwurzel *f*
g. ale (beer) Ingwerbier *n*
g. oil Ingweröl *n*
gingerol Gingerol *n*
ging[i]li (gingily) oil *s.* gingelly oil
ginning *(Text)* Egrenieren *n*, Entkernen *n*
ginseng Ginseng *m*, Kraftwurz *f (Panax quinquefolius L. und P. schin-seng Nees)*
Girbotol process Girbotol-Verfahren ·n *(zum Entfernen von Schwefelwasserstoff aus Gasen)*
gismondine, gismondite *(Min)* Gismondin *m (Kalziumdialumodisilikat)*
give off (up) / to abgeben
glabratic acid Glabratsäure *f*, Diorsellinsäure *f*, Lekanorsäure *f*
glacial acetic acid Eisessig *m*
g. phosphoric acid $(HPO_3)_n$ Metaphosphorsäure *f*
glair[e] Eiweiß *n*, Eiklar *n*; Klebstoff *m* auf Eiweißgrundlage, Eiweißleim *m*; schleimig-klebrige Masse *f*
glaireous, glairy eiweißartig, Eiweiß...; zähflüssig, schleimig, klebrig
glancing angle *(Krist)* Glanzwinkel *m*

gland 1. Drüse *f*, Glandula *f*; 2. Stopfbuchse *f*, Stopfbuchspackung *f*; Stopfbuchsenbrille *f*
g. of external secretion Drüse *f* mit äußerer Sekretion
glandless stopfbuchslos
glarimeter *(Pap)* Glanzmesser *m*
glaserite *(Min)* Glaserit *m*, Aphthitalit *m (Kaliumnatriumsulfat)*
glass Glas *n*; *(Tech)* Schauglas *n*, Glas *n*, Fenster *n*; *(Geol)* Glas *n*, Gesteinsglas *n*
glass A A-Glas *n*, Alkaliglas *n*, alkalihaltiges Glas *n*
glass E E-Glas *n (alkaliarmes Borsilikatglas)*
g. adhesive Glasklebstoff *m*
g.-air interface *(Foto)* Luft-Glas-Fläche *f*, Fläche *f* Glas-Luft
g. ampoule Glasampulle *f*
g. basin Glasschälchen *n*
g. batch Glasgemenge *n*
g. bead Glasperle *f*
g. beaker Glasbecher *m*
g. bell Glasglocke *f*
g. block Glasbaustein *m*, Glasziegel *m*
g. blower Glasbläser *m*
g. blower's lamp Glasbläserlampe *f*, Gebläselampe *f*
g. blower's pipe Glasbläserpfeife *f*
g. blowing Glasblasen *n*
g. brick *s.* g. block
g. brush Glasbürste *f*
g. bulb Glaskolben *m*
g. cap Glaskappe *f*
g. cement Glaszement *m*
g. ceramic Glaskeramik *f*, Keramik *f* aus Glas, glaskeramischer Stoff *m*, Vitrokeram *n*
g.-clear glasklar
g. cloth *s.* g. fabric
g. colour Glasfarbe *f*
g. composition Glaszusammensetzung *f*
g. connection Glasverbindung *f*
g. crucible *s.* g. filter crucible
g. cutter Glasschneider *m*
g. cutting Glasschneiden *n*
g. cuvette Glasküvette *f*
g. defect Glasfehler *m*
g. depth Glasbadtiefe *f*
g. dish Glasschale *f*
g. drill Glasbohrer *m*
g. drop *s.* g. tear
g. electrode Glaselektrode *f*
g. enamel Glasemaille *f (Glasüberzug von emailleartiger Zusammensetzung)*
g. etching *s.* glassware etching
g. fabric Glas[faser]gewebe *n*
g. fabrication *s.* g. making
g. factory Glasfabrik *f*, Glaswerk *n*, Glashütte *f*
g. fibre Glasfaserstoff *m*, GL; Glasfaser *f*
g.-fibre fabric *s.* g. fabric
g.-fibre filler Glasfaserfüllstoff *m*
g.-fibre laminate Glasfaserschichtstoff *m*, GFS, Glasfaserlaminat *n*

g.-fibre moulding compound s. g.-fibre reinforced moulding compound
g.-fibre paper Glasfaserpapier n
g.-fibre product Glasfaserprodukt n, Glasfasererzeugnis n
g.-fibre reinforced glasfaserverstärkt, glasarmiert
g.-fibre reinforced moulding compound Glasfaserpreßmasse f
g.-fibre reinforced plastic glasfaserverstärkter Plast (Kunststoff) m, GFP, GFK
g.-fibre reinforced unsaturated polyester resin Faserglasharz n
g.-fibre reinforcement Glasfaserverstärkung f
g.-fibre roving Glasseidenroving m, Roving m, Glasseidenstrang m
g.-fibre strand Glasseidenspinnfaden m
g.-fibre tape Glasfaserband n
g.-fibre veil Glasfaservlies n
g.-fibre yarn Glas[stapel]fasergarn n
g. filament Glasfaden m
g. filler rings Glasringe mpl, Absprengperlen fpl, Raschig-Ringe mpl
g. filter Glasfilter n
g. filter crucible Glasfiltertiegel m
g. filter disk Glasfilterplatte f
g. filter[ing] funnel Glasfiltertrichter m
g. fitting Glasformstück n
g. flask Glaskolben m
g. flow Glasfluß m
g. former s. g.-forming element
g.-forming glasbildend
g. forming Glasformgebung f, Glasformung f
g.-forming element Glasformer m, Glasbildner m
g.-forming machine Glasverarbeitungsmaschine f
g.-forming substance s. g.-forming element
g. frost (Glas) Flitter m
g. funnel Glastrichter m
g. furnace Glas[schmelz]ofen m
g. gall Glasgalle f
g. gob Glasposten m, Posten m, Glastropfen m, Speisertropfen m
g.-hard glashart
g. hardness Glashärte f
g. house Glashütte f, Glasfabrik f, Glaswerk n
g. industry Glasindustrie f
g. ink s. g.-marking ink
g. knife Glasmesser n
g.-like glasähnlich, glasartig, glasig
g.-like state glasiger (glasartiger) Zustand m, Glaszustand m
g.-lined mit Glasauskleidung
g. maker Glasmacher m, Glashersteller m
g. making Glasherstellung f, Glasfabrikation f
g. marble Glaskugel f (Glasfaserherstellung)
g.-marking ink Glastinte f
g. mass Glasmasse f
g. measure Meßglas n
g. melt Glasschmelze f

g.-melting furnace s. g. furnace
g.-melting pot Glashafen m
g.-melting tank Glasschmelzwanne f
g.-metal seal s. g.-to-metal seal
g. packing Füllkörper mpl aus Glas
g. paper Glaspapier n (ein Schleifpapier); Nur-Glas-Papier n, Glaspapier n (Isolierungsmaterial)
g. pipe Glasrohr n, Glasröhre f
g. pipeline Glasrohrleitung f
g. piping Glasrohr[material] n, Röhrenglas n; Glasrohrsystem n, Glasrohranlage f
g. pot s. g.-melting pot
g. pot furnace Glashafenofen m
g. powder Glaspulver n
g.-reinforced s. g.-fibre reinforced
g.-reinforced plastic s. g.-fibre reinforced plastic
g. reinforcement s. g.-fibre reinforcement
g. rod Glasstab m
g. silk Glasseide f
g. silk coating Glasseidenbeschichtung f
g. silk fabric Glasseidengewebe n
g. silk tape Glasseidenband n
g. soap Glas[macher]seife f
g. spar (zur Glasherstellung verwendeter) Feldspat m
g. spiral Glasschlange f
g.-spring manometer Glasfedermanometer n
g. stirrer Glasrührer m
g. stone Stein m, Steinchen n (Glasfehler)
g. stopcock Glashahn m
g. stopper Glasstöpsel m, Glasstopfen m
g.-stoppered mit Glasstöpsel (Glasstopfen), Glasstöpsel..., Glasstopfen...
g.-stoppered bottle Glasstöpselflasche f, Glasstopfenflasche f
g. strand Glasseidenspinnfaden m
g. structure Glasstruktur f
g. suction filter Glas[filter]nutsche f
g. tank furnace Glaswannenofen m, Glas[schmelz]wanne f
g. tap s. g. stopcock
g. tear Glasträne f, Glastropfen m
g. tissue Glasseidenpapier n
g.-to-metal seal Glas-Metall-Verschmelzung f, Glas-Metall-Verschweißung f
g. transition Glasumwandlung f, Glastransformation f, Glasübergang m, Umwandlung f 2. Ordnung, Gamma-Umwandlung f, α-Anomalie f
g. transition region Glasübergangsbereich m
g. transition temperature Glasumwandlungstemperatur f, Glasumwandlungspunkt m, Transformationstemperatur f, Transformationspunkt m, Einfriertemperatur f, Umwandlungspunkt m 2. Ordnung
g. tube Glasrohr n, Glasröhre f
g. tubing 1. s. g. piping; 2. Füllringe mpl aus Glas
g. vessel Glasgefäß n
g. wadding Glaswatte f
g. wool Glaswolle f

g. works s. glassworks
glassblower Glasbläser m
glassblowing Glasblasen n
glassed glasemailliert, emailliert
glasshouse s. glassworks
glassine [paper] Pergamin n, Pergamyn n, Pergaminpapier n, hochsatiniertes Pergamentersatzpapier n
glassmaker Glasmacher m, Glashersteller m
glassmakers' soap Glas[macher]seife f
glassmaking Glasherstellung f, Glasfabrikation f
glassman s. glassmaker
glassware Glasware f, Glaswaren fpl, Glasartikel mpl, Glas n; Glasgeräte npl
g. etching Glasätzen n, Glasätzung f
glassworks Glaswerk n, Glasfabrik f, Glashütte f
glassy glasartig, glasähnlich, glasig
g. arsenic (Glas) Stück[en]arsenik m(n), Arsenikbrocken mpl
g. feldspar (Min) Eisspat m, (veraltet für) Sanidin m
g. lustre Glasglanz m
g. phase Glasphase f, glasige Phase f
g. state glasiger (glasartiger) Zustand m, Glaszustand m
Glauber salt $Na_2SO_4 \cdot 10H_2O$ Glaubersalz n, Natriumsulfat-10-Wasser n
glauberite (Min) Glauberit m (Kalziumnatriumsulfat)
glaucarubin Glaukarubin n
glaucine Glauzin n (Alkaloid)
glaucodot[e], glaucodotite (Min) Glaukodot m
glauconite (Min) Glaukonit m
glaucophane (Min) Glaukophan m
glaucosiderite (Min) Glaukosiderit m, (veraltet für) Vivianit m (Eisen(II)-orthophosphat)
glaze / to (Lebm) glasieren, mit Zuckerglasur überziehen; glasieren (z.B. Fisch gefrieren lassen); (Ker) glasieren, mit Glasur (glasartigem Überzug) versehen; (Pap) glätten, satinieren; (Gerb) glanzstoßen, glänzen, blankstoßen
glaze Glasur f; (Pap) Glätte f; Glanz m, Oberflächenglanz m
g. batch (Ker) Glasurschmelze f
g. coating (Ker) Glasurüberzug m
g. composition (Ker) Glasurzusammensetzung f
g. fault (Ker) Glasurfehler m
g. fibre (Text) Glanzfaser f
g. fit (Ker) Glasursitz m
g. penetration (Ker) Eindringtiefe f der Glasur
g. slip (Ker) Glasurschlicker m
g. spar (Ker) Glasurspat m
glazed board Glanzpappe f
g. brick glasierter Ziegel m
g. finish (Text) Glanzappretur f, Glanzausrüstung f; (Gerb) Glanzstoßzurichtung f
g. on both sides (Pap) beidseitigglatt
g. on one side (Pap) einseitigglatt, egl
g. paper satiniertes (sat) Papier n
g. pot (Glas) glasierter (eingeglaster) Hafen m

glazier's putty Glaserkitt m, Fensterkitt m
glazing Glasieren n; (Pap) Glätten n, Glättung f, Satinieren n, Satinierung f, Satinage f; (Gerb) Glanzstoßen n, Glänzen n, Blankstoßen n
g. felt (Pap) Glättfilz m
g. machine (Ker) Glasiermaschine f; (Gerb) Glanzstoßmaschine f
G.L.C. s. gas-liquid chromatography
glei s. gley
gleissite Glessit m (Bernsteinsorte)
gley (Geol) Gleiboden m
gliadin Gliadin n (Prolamin des Weizens und Roggens)
glide plane (Krist) Gleitebene f, Translationsebene f
gliotoxin Gliotoxin n (Antibiotikum)
glistening glänzend
globe boiler s. g. digester
g. digester (Pap) Kugelkocher m
g. mill Kugelmühle f
g. rotary digester s. g. digester
g. valve Sitzventil n, Hubventil n, Ventil n mit kugeligem Gehäuse
globigerina ooze (Geol) Globigerinenschlamm m
globin Globin n (Eiweißkomponente des Hämoglobins)
globular globulär, globular, kugelförmig, kugelig, rund
g. colloid Sphärokolloid n, globulares Kolloid n
g. form Kugelform f
g. protein globuläres (kugelförmiges) Protein n, Sphäroprotein n, Kugelprotein n
globule Kügelchen n
globulin Globulin n (zu den Sphäroproteinen gehörender Eiweißstoff)
globulite (Geol) Globulit m
gloriosine Gloriosin n (Alkaloid aus Gloriosa superba)
glory hole (Glas) Anwärm[e]loch n, Aufwärm[e]loch n
gloss Glanz m, glänzendes Aussehen n, Oberflächenglanz m
g. ink Glanzfarbe f, glänzende Farbe f
g. meter (Pap) Glanzmesser m
g. point (Ker) Gloss point m (Temperatur, bei der die Glasur so weit geschmolzen ist, daß die Oberfläche glänzt)
g. printing ink s. g. ink
g. retention Glanz[er]haltung f, Glanzbeständigkeit f
g. varnish Glanzlack m; Glanzfirnis m (in der grafischen Technik)
glossimeter s. glossmeter
glossiness Glanz m, Glätte f
glossing (Text) Glanzpressen n, Glanzausrüstung f
glossmeter (Pap) Glanzmesser m
glossy glänzend, Glanz...
g. finish Glanzappretur f
g. print (Foto) Glanzabzug m

glost-fire / to *(Ker)* glattbrennen
glost firing *(Ker)* Glattbrand *m*, Glasurbrand *m*
  g. kiln *(Ker)* Glatt[brand]ofen *m*, Glasur[brand]-
    ofen *m*
  g. pitchers *(Ker)* Glattscherben *fpl*
glove box Schutzkammer *f* mit [eingebauten]
  Handschuhen, Handschuhbox *f*, Manipulations-
  kammer *f*, Isotopenkasten *m*
Glover and West coking retort *s.* Glover-West
  continuous vertical retort
Glover tower Glover[turm] *m*
Glover-West continuous vertical retort Glover-
  West-Retorte *f*, Vertikalretorte *f* mit kontinuierlicher Beschickung von Glover-West
Glover-West system Glover-West-Ofensystem *n*
glow / to glimmen, glühen
glow Glimmen *n*, Glühen *n*; Glut *f*
  g. bar test Glutbeständigkeitsprobe *f*, Glutfestig-
    keitsprobe *f*
  g. discharge Glimmentladung *f*
glowing hot-body test *s.* glow bar test
glucagon Glukagon *n* *(Bauchspeicheldrüsen-*
  hormon)
glucaric acid COOH(CHOH)$_4$COOH *D*-Zuckersäure
  f, *D*-Glukozuckersäure *f* *(eine Form der 2,3,4,5-*
  Tetrahydrohexandisäure)
glucina Gluzinerde *f*, Glyzinerde *f*, Süßerde *f*,
  Beryllerde *f*, *(veraltet für)* Berylliumoxid *n*
glucine Gluzin *n* *(ein Süßstoff)*
glucinite *(Min)* Gluzinit *m*, *(veraltet für)* Herderit *m*
glucin[i]um Gl Gluzinium *n*, Glyz[in]ium *n*, *(veraltet*
  für) Beryllium *n* *(Verbindungen s. unter* beryllium)
glucobrassicin Glukobrassizin *n* *(Indolylsenföl-*
  glukosid)
glucocorticoid Glukokortikoid *n* *(Nebennieren-*
  rindenhormon)
glucofrangulin Glukofrangulin *n*
glucogallin Glukogallin *n*, 1-Galloylglukose *f*
glucomannan Glukomannan *n*
gluconeogenesis Glukoneogenese *f* *(Aufbau von*
  Glykogen aus Eiweißen)
d-**gluconic acid** CH$_2$OH(CHOH)$_4$COOH *D*-Glukon-
  säure *f*
glucono delta lactone Glukonsäure-δ-lakton *n*,
  Glukonsäure-delta-lakton *n*
glucoprotein Glukoproteid *n*, Glykoproteid *n*,
  Eiweißzucker *m*
glucopyranose Glukopyranose *f*
glucosamine Glukosamin *n*, Chitosamin *n*,
  Aminoglukose *f*
glucosazone Glukosazon *n*
glucose C$_6$H$_{12}$O$_6$ Glukose *f*, *D*-Glukose *f*, Glykose
  f, Traubenzucker *m*, Dextrose *f*
  g. oxydase Glukose-Oxydase *f* *(ein Flavin-*
    ferment)
  g. phosphate Glukosephosphat *n*
  g. residue Glukoserest *m*
  g. syrup Glukosesirup *m*, Stärkesirup *m*
glucosic Glukose...

glucosid *s.* glucoside
glucosidal *s.* glucosidic
glucosidase Glukosidase *f* *(glukosidspaltendes*
  Ferment, Untergruppe der Glykosidasen)
glucoside Glukosid *n* *(ein Glykosid, in dem als*
  Zuckerkomponente Glukose auftritt)
glucosidic glukosidisch, Glukosid...
glucovanillin Glukovanillin *n*
glucuronic acid OHC(CHOH)$_4$COOH Glukuron-
  säure *f*
glucuronidase Glukuronidase *f* *(zu den Glykosida-*
  sen gehörendes Ferment)
glucuronide Glukuronid *n*
glucuronolactone Glukuronsäurelakton *n*
glucuronoside *s.* glucuronide
glue / to [ver]leimen, [ver]kleben; mit Leim
  bestreichen
glue Leim *m*
  g. line Klebfuge *f*
  g.-sized *(Pap)* gelatinegeleimt, mit Gelatine
    (Tierleim) geleimt
glutamate Glutamat *n* *(Salz der Glutaminsäure)*
glutamic acid HOOCCH(NH$_2$)CH$_2$CH$_2$COOH
  Glutaminsäure *f*, Glu, 2-Aminoglutarsäure *f*,
  2-Aminopentandisäure *f*
  g. acid dehydrogenase Glutamin[säure]de-
    hydrogenase *f*, Glutaminsäuredehydrase *f*
glutaminase Glutaminase *f* *(zu den Azylamidasen*
  gehörendes Ferment)
glutamine H$_2$NCOCH$_2$CH$_2$CH(NH$_2$)COOH
  Glutamin *n*, Glu-NH$_2$
glutaminic acid *s.* glutamic acid
glutar[di]aldehyde OHC·CH$_2$·CH$_2$·CH$_2$·CHO
  Glutaraldehyd *m*, Glutar[säure]dialdehyd *m*
glutaric acid HOOC(CH$_2$)$_3$COOH Glutarsäure *f*,
  Pentandisäure *f*
glutathione C$_{10}$H$_{17}$N$_3$SO$_6$ Glutathion *n*, Glutamin-
  zysteinglykokoll *n*, Glutaminylzysteinylglyzin *n*
glutelin Glutelin *n*, Glutenin *n* *(Eiweißkörper des*
  Getreidemehls)
gluten Gluten *n*, Kleber *m*, Klebereiweiß *n*
glutenin *s.* glutelin
glutoform, glutol Glutoform *n*, Glutol *n* *(Trok-*
  kenantiseptikum)
glutose Glutose *f*
gly. *s.* 1. glycerin; 2. glycerol
glycan Glykan *n*, Polysaccharid *n*
glycaric acid *(Sammelname für Hexose-Zuk-*
  kerdikarbonsäuren der allgemeinen Formel
  HOOC(CHOH)$_4$COOH)
glyceraldehyde CH$_2$OHCHOHCHO Glyzer[in]al-
  dehyd *m*, Propandiol-(2,3)-al-(1) *n*, 2,3-Di-
  hydroxypropanal *n*
glycerate Glyzerat *n* *(Verbindung des Glyzerins mit*
  Alkalimetallen)
glyceric acid CH$_2$OH·CHOH·COOH Glyzerin-
  säure *f*, 2,3-Dihydroxypropansäure *f*
  g. aldehyde *s.* glyceraldehyde
glyceride Glyzerid *n*
  g. molecule Glyzeridmolekül *n*

g. oil Glyzeridöl *n*
glycerin[e] Glyzerin *n (als Handelsprodukt)*
g. bath Glyzerinbad *n*
glycerol $CH_2OHCHOHCH_2OH$ Glyzerin *n*, Trihydroxypropan *n*, 1,2,3-Propantriol *n*
   g. α-chlorohydrin $HOCH_2CH(OH)CH_2Cl$ Glyzerin-α-chlorhydrin *n*, α-Monochlorhydrin *n*, 3-Chlor-1,2-propandiol *n*
   g. fermentation Glyzeringärung *f*
   g. monostearate Glyzerinmonostearat *n*, Monostearin *n*
   g. phthalic resin Glyzerin-Phthalsäureharz *n*, Glyptal[harz] *n*
   g. trimyristate Glyzerintrimyristat *n*, Glyzeroltrimyristat *n*, Trimyristin *n*, Glyzeryltetradekanoat *n*
   g. trinitrate $C_3H_5(ONO_2)_3$ Glyzerintrinitrat *n*
glycerophosphate Glyzerophosphat *n*
glycerophosphoric acid Glyzerinphosphorsäure *f*
glycerophthalate Glyzerophthalat *n*
glyceryl α-chlorohydrin s. glycerol α-chlorohydrin
   g. monostearate s. glycerol monostearate
glycidyl ether resin Glyzidylätherharz *n*
glycin *(Foto)* Glyzin *n (Entwicklersubstanz)*
   g. developer *(Foto)* Glyzin-Entwickler *m*
glycine s. 1. glycocoll; 2. glycin
glycinin Glyzinin *n (Globulin der Sojabohne)*
glycocholic acid Glykocholsäure *f (eine Gallensäure)*
glycocoll H_2NCH_2COOH Glykokoll *n*, Glyzin *n*, Gly, Leimsüß *n*, Leimzucker *m*, Aminoessigsäure *f*, Aminoäthansäure *f*
   g. copper Glykokollkupfer *n*
glycocyamine Glykozyamin *n*, Guanidinessigsäure *f*
glycogen Glykogen *n*, Leberstärke *f*
glycogenase Glykogenase *f*
glycogenesis Glykogenese *f*
glycogenic acid *(veraltet für)* d-gluconic acid
glycogenolysis Glykogenolyse *f*, Glykogenabbau *m*
glycohol alcohol s. glycol *(i.e.S.)*
glycol $C_nH_{2n}(OH)_2$ Glykol *n*, Diol *n*, zweiwertiger Alkohol *m*, *(i.e.S.)* $HOCH_2CH_2OH$ Glykol *n*, 1,2-Glykol *n*, Äthylenglykol *n*, Äthandiol-(1,2) *n*
   g. alcohol s. glycol *(i.e.S.)*
   g. chlorohydrin $ClCH_2CH_2OH$ Glykolchlorhydrin *n*, Äthylenchlorhydrin *n*, 2-Chloräthylalkohol *m*, 2-Chloräthanol *n*
   g. dibromide $BrCH_2CH_2Br$ 1,2-Dibromäthan *n*, Äthylen[di]bromid *n*, Glykoldibromid *n*
   g. monostearate Glykolmonostearat *n*, 2-Hydroxyäthyloktadekanoat *n*
   g.-splitting glykolspaltend
glycolaldehyde CH_2OHCHO Glykolaldehyd *m*, Hydroxyazetaldehyd *m*, Äthanolal *n*
glycolic acid $CH_2OHCOOH$ Glykolsäure *f*, Hydroxyessigsäure *f*, Hydroxyäthansäure *f*, Äthanolsäure *f*
   g. aldehyde s. glycolaldehyde

glycolipid[e] Glykolipid *n*
glycollic acid s. glycolic acid
glycolysis Glykolyse *f*
glyconeogenesis s. gluconeogenesis
glyconic acid Aldohexonsäure *f (Sammelname für Aldose-Zuckerderivate der allgemeinen Formel $HOCH_2(CHOH)_4COOH$)*
   d-glyconic acid s. d-gluconic acid
glycopeptide, glycoproteid s. glycoprotein
glycoprotein Glykoproteid *n*, Glukoproteid *n*, Eiweißzucker *m*
glycose s. glucose
glycosidase Glykosidase *f (zu den Hydratasen gehörendes Ferment)*
glycoside Glykosid *n*
glycosidic glykosidisch
   g. bond (linkage) glykosidische Bindung (Verknüpfung) *f*, Glykosidbindung *f*
glycosuria *(Med)* Glykosurie *f*
glycuronic acid Uronsäure *f (Sammelname für Aldehydkarbonsäuren der Zuckerreihe)*
glyoxal $OHC \cdot CHO$ Glyoxal *n*, Oxalaldehyd *m*, Äthandial *n*
glyoxalase Glyoxalase *f*, Aldoketomutase *f*
glyoxalic (glyoxylic) acid $CHOCOOH$ Glyoxylsäure *f*, Glyoxalsäure *f*, Äthanalsäure *f*, Oxoäthansäure *f*, Oxalaldehydsäure *f*, Oxoessigsäure *f*
glyptal [resin] Glyptal[harz] *n*, Glyzerin-Phthalsäure-Harz *n*
gmelinite *(Min)* Gmelinit *m*
GMS s. glycerol monostearate
gneiss *(Geol)* Gneis *m*
gneissic (gneissose) structure *(Geol)* Gneistextur *f*
go-devil *(Erdöl)* Reinigungsmolch *m*, Molch *m*, Go-devil *m*
go yellow / to gelb werden, [ver]gilben
Goa butter Kokumbutter *f (von Garcinia indica Choisy)*
   G. powder *(Pharm)* Goapulver *n*, Ararobapulver *n (von Andira araroba Aguiar)*
goat fat Ziegenfett *n*
   g. hair Ziegenhaar *n*, Ziegenwolle *f*
   g. milk Ziegenmilch *f*
goat's milk cheese Ziegenkäse *m*
goatskin leather Ziegenleder *n*
gob Glasposten *m*, Posten *m*, Glastropfen *m*, Speisertropfen *m*
   g. delivery *(Glas)* Tropfenabgabe *f (in die Vorform)*
   g.-fed machine *(Glas)* Speisermaschine *f*, Feedermaschine *f*
   g. feeder *(Glas)* Tropfenspeiser *m*, Postenspeiser *m*
   g. feeding *(Glas)* Tropfspeisung *f*, Postenspeisung *f*
   g. process *(Glas)* Speiserverfahren *n*, Feederverfahren *n*
   g. shape *(Glas)* Tropfenform *f*, Postenform *f*
   g. weight *(Glas)* Tropfengewicht *n*, Postengewicht *n*

goethite *(Min)* Goethit *m*, Nadeleisenerz *n* *(Eisenoxidhydroxid)*

goffer / to gaufrieren, prägen, einpressen

goffering Gaufrieren *n*, Gaufrage *f*, Prägen *n*, Prägung *f*, Einpressen *n*

  g. calender Gaufrierkalander *m*, Prägekalander *m*

goggles Schutzbrille *f*

gold Au Gold *n*

  g. amalgam *(Min)* Goldamalgam *m*

  g. aventurine Kupferaventurin *m*, Goldaventurin *m*

  g. beryl *(Min)* Chrysoberyll *m* *(Berylliumaluminat)*

  g. bronze Goldbronze *f*

  g. bronze powder Goldbronzepulver *n*

  g. content Goldgehalt *m*

  g. decoration *(Ker)* Golddekor *n*

  g. dust Goldstaub *m*

  g. foil Goldfolie *f*; Blattgold *n*

  g. hydrosol Goldhydrosol *n*

  g. leaf Blattgold *n*

  g. mine Goldbergwerk *n*, Goldmine *f*, Goldgrube *f*

  g. monobromide AuBr Gold[mono]bromid *n*, Gold(I)-bromid *n*

  g. monochloride AuCl Gold[mono]chlorid *n*, Gold(I)-chlorid *n*

  g. nugget Goldklumpen *m*, Nugget *n*

  g. number *(Koll)* Goldzahl *f*

  g. orange Goldorange *n*, Methylorange *n*, Orange III *n*, Helianthin *n*

  g.-placer *(Geol)* Goldseife *f*

  g.-plate / to vergolden, *(i.e.S.)* galvanisch vergolden

  g. plating Vergolden *n*, Vergoldung *f*, *(i.e.S.)* galvanisches Vergolden *n*

  g.-potassium cyanide K[Au(CN)$_2$] Kaliumgold(I)-zyanid *n*, Kaliumdizyanoaurat(I) *n*

  g. preparation Goldpräparat *n*

  g. ruby [glass] Goldrubinglas *n*

  g. sesquioxide s. g. trioxide

  g. size Anlegeöl *n* *(zum Befestigen von Blattgold)*

  g.-sodium chloride Na[AuCl$_4$] Natriumgold(III)-chlorid *n*, Natriumtetrachloroaurat(III) *n*

  g.-sodium sulphide NaAuS Natriumgold(I)-sulfid *n*

  g.-sodium thiosulphate Na$_3$[Au(S$_2$O$_3$)$_2$] Natriumgoldthiosulfat *n*, Natriumdithiosulfatoaurat(I) *n*

  g. sol Goldsol *n*, kolloidales Gold *n*

  g. sol test *(Med)* Goldsolreaktion *f*

  g. tellur *(Min)* Goldtellur *m*, *(veraltet für)* Sylvanit *m* *(Silbergoldtellurid)*

  g. toning *(Foto)* Goldtonung *f*, Goldtönung *f*

  g. tribromide AuBr$_3$ Goldtribromid *n*, Gold(III)-bromid *n*

  g. trichloride AuCl$_3$ Goldtrichlorid *n*, Gold(III)-chlorid *n*

  g. trichloride acid H[AuCl$_4$] Tetrachlorogold(III)-säure *f* Chlorogold(III)-säure *f*, Gold(III)-hydrogenchlorid *n*

  g. trioxide Au$_2$O$_3$ Goldtrioxid *n*, Gold(III)-oxid *n*

  g. washing Goldwäsche[rei] *f*

  g. wattle Acacia pycnantha Benth.; Acacia penninervis Sieber *(Gerberakazien)*

golden antimony sulphide Sb$_2$S$_5$ Goldschwefel *m*, Antimon(V)-sulfid *n*, Antimonpentasulfid *n*

  g.-brown goldbraun

  g. chamomille Färberkamille *f*, Anthemis tinctoria L.

  g. sulphide of antimony s. g. antimony sulphide

  g. wattle s. gold wattle

  g.-yellow goldgelb

Goldschmidt radius Goldschmidt-Radius *m*

Goldschmidt's process Goldschmidt-Verfahren *n*, Thermitverfahren *n*, Aluminothermie *f*

Gomberg-Bachmann-Hey reaction Gomberg-Reaktion *f* *(Diarylsynthese)*

goniometer *(Krist)* Goniometer *n*, Winkelmesser *m*

goniometric *(Krist)* goniometrisch, Goniometer...

goober Erdnuß *f*, Arachis hypogaea L.

  g. cake Erdnußkuchen *m*

Gooch crucible (filter) Gooch-Tiegel *m*, Filtriertiegel *m* nach Gooch

good concrete fetter Beton *m*

  g. tilth *(Landw)* Gare *f*

Goodrich flexometer *(Gum)* Goodrich-Flexometer *n*

Goodyear angle machine *(Gum)* Goodyear-Winkelmaschine *f*

goora nut Kolanuß *f* *(von Cola vera u.a.)*

goose dung ore *(Min)* Gänsekötigerz *n*, Ganomatit *m* *(silber- und kobalthaltiges Eisenarsenat)*

GOR s. gas/oil ratio

Gordon-plasticator *(Gum)* Gordon-Plastikator *m*

goslarite *(Min)* Goslarit *m*, Zinkvitriol *m* *(Zinksulfat-7-Wasser)*

gossan *(Geol)* Eiserner Hut *m*

göthite s. goethite

gotthardite *(Min)* Gotthardit *m*, *(veraltet für)* Dufrenoysit *m* *(Blei(II)-arsen(III)-sulfid)*

Gottignies kiln *(Ker)* Gottignies-Ofen *m* *(elektrischer Mehrkanaldurchschubofen)*

Goudsmit and Uhlenbeck assumption Goudsmit-Uhlenbeckscke Annahme (Hypothese) *f* *(vom rotierenden Elektron)*

Goulard's extract Goulards Extrakt *m*, Goulardsches Wasser *n*, Bleiextrakt *m*, Bleiessig *m*

Gould-Jacobs reaction Gould-Jacobs-Reaktion *f* *(Hydroxychinolin-Ringschluß)*

gourd curare Kalebassenkurare *n*

governor diaphragm Reglermembran *f*

gozzan s. gossan

GPC s. gel permeation chromatography

GPF black s. general-purpose furnace black

gradation *(Foto)* Gradation *f*, Abstufung *f*, Abtönung *f*, Schwärzungsabstufung *f*

grade / to sortieren; klassieren; trennen

grade Grad *m*, Stufe *f*, Sorte *f*, Qualität *f*, Güteklasse *f*, Klasse *f*, Gütegrad *m*; Korngröße *f*, Körnung *f*, Siebfeinheit *f*

g. of paper Papiersorte f
g. of rubber Kautschukqualität f
graded junction *(phys Ch)* stetiger (allmählicher, abgestufter) Übergang m
grader Sortiermaschine f
gradient Gradient m, Neigung f, Gefälle n, Steilheit f
g. elution Gradient[en]elution f, Stufeneluierung f
g. elution analysis Gradient-Elutionschromatografie f, Gradientenchromatografie f
g. layer Gradientschicht f
g. layer chromatography s. g. thin-layer chromatography
g. layer technique *(phys Ch)* Gradientschichttechnik f
g. thin-layer chromatography Gradient-Dünnschichtchromatografie f, Gradientschichtchromatografie f
grading Sortieren n, Sortierung f; Klassieren n, Klassierung f; Trennen n, Trennung f
g. curve Siebkurve f, Sieblinie f
g. into size Klassieren n, Klassierung f, Trennen n (Trennung f) nach Größenklassen (Korngrößen)
g. limit s. g. curve
gradual gradweise, stufenweise, graduell
graduate / to 1. graduieren, mit genauer Einteilung versehen, in Grade [ein]teilen, unterteilen; 2. *(Lösungen durch Abdunsten)* gradieren, verstärken, konzentrieren
graduate Meßgefäß n, Mensur f
graduated cylinder Meßzylinder m, Maßzylinder m, Mensur f
g. flask Meßkolben m, Maßkolben m, Meßflasche f
g. pipette Meßpipette f, Teilpipette f
g. to deliver auf Ablauf (Auslauf) geteilt
graduation 1. Graduierung f, Gradeinteilung f; Teilstrich m, *(i.w.S.)* Teilungsbild n *(z.B. an Meßgefäßen)*; 2. Gradieren n, Verstärken n, Konzentrieren n *(von Lösungen durch Abdunsten)*
g. houses s. g. works
g. mark Teilstrich m
g. works Gradierwerk n
Graebe-Ullmann synthesis Graebe-Ullmannsche Karbazolsynthese f
graft / to [auf]pfropfen, aufpolymerisieren, anpolymerisieren
graft s. g. polymer
g. copolymer Pfropfkopolymer[es] n, Pfropfkopolymerisat n
g. copolymerization Pfropfkopolymerisation f
g. elastomer Pfropfelastomer[es] n
g. polymer Propfpolymer[es] n, Graftpolymer[es] n, Pfropfpolymerisat n
g. polymerization Pfropfpolymerisation f
grafting Pfropfen n, Pfropfung f, Aufpfropfen n, Aufpolymerisation f, Anpolymerisation f
g. degree Pfropfgrad m
g. reaction Pfropfreaktion f

g. wax Baumwachs n
Graham law of diffusion Grahamsches Gesetz n der Transfusionsgeschwindigkeiten
Graham's salt Grahamsches Salz n *(ein Natriumpolyphosphat)*
grain / to granulieren, körnen, körnigmachen; aussalzen; *(Pap)* grainieren; *(Gerb)* pantoffeln; *(Gerb)* krispeln
grain Korn n, Körnchen n; Korn n, Körnung f, körnige Beschaffenheit f; *(Foto)* Korn n; *(Foto)* Körnigkeit f; *(Gerb)* Narben m; *(Gerb)* Narbenspalt m; *(Zucker)* Korn n, Kristallkörnchen n; *(Krist)* Korn n; *(Geol)* Feinkorn n, Gefüge n; *(Landw)* Samenkorn n, *(i.e.S.)* Getreidekorn n; Getreide n, Korn n
g. alcohol Kornbranntwein m, Getreidebranntwein m
g.-alcohol plant Kornbrennerei f
g. boundary Korngrenze f
g. brush Bürstmaschine f *(zum Polieren des Malzes)*
g. buffing *(Gerb)* Abbuffen n *(Abschleifen der Narbenseite)*
g. crack resistance Narbenfestigkeit f *(des Leders)*
g. density Korndichte f
g. direction *(Pap)* Längsrichtung f, Laufrichtung f, Maschinenrichtung f, Arbeitsrichtung f
g. dryer Getreidetrockner m, Getreidetrocknungsanlage f
g. effect *(Gum)* Orientierungserscheinung f
g.-free kornlos
g. growing Kristallzüchtung f
g. layer *(Gerb)* Narbenseite f
g. oil Fuselöl n
g. pipeyness *(Gerb)* Losnarbigkeit f
g. size Korngröße f
g. size check Korngrößenkontrolle f
g. size determination Korngrößenbestimmung f, Korngrößenmessung f
g. size distribution Korngrößenverteilung f
g. size range Körnungsbereich m
g. split *(Gerb)* Narbenspalt m
g. structure Gefüge n
g. surface Kornoberfläche f
g. syrup Glukosesirup m, Stärkesirup m
g. washing machine Getreidewaschmaschine f
grained gekörnt, körnig, granuliert, granulös; gemasert; genarbt; *(Pap)* grainiert
g. board gemaserte Pappe f, Maserpappe f
g. lac Körnerlack m, Samenlack m
g. paper grainiertes Papier n, durch Prägung mit einer Körnung versehenes Papier n
graininess Körnigkeit f
graining Granulieren n, Granulierung f, Granulation f, Körnen n, Körnigmachen n; Aussalzen n, Aussalzung f; *(Pap)* Grainieren n; *(Gerb)* Pantoffeln n; *(Gerb)* Krispeln n; *(Zucker)* Kornbildung f
g. kettle Granuliertrommel f

grains *(Gär)* Treber *pl*
  g. dryer Trebertrocknungsanlage *f*, Trebertrockner *m*
  g. of kermes Kermeskörner *npl*, Scharlachkörner *npl*, Kermes *m*, Alkermes *m (getrocknete weibliche Kermesschildläuse)*
  g. of paradise Paradieskörner *npl*, Guineakörner *npl*, Malagettapfeffer *m (von Aframomum melegueta Schum.)*
  g. settling Treberschicht *f*
grainy körnig, gekörnt, granulös, granuliert
gram atom Grammatom *n*, Atomgramm *n*
  g.-atomic parachor Atomparachor *m*
  g.-atomic weight *s.* g. atom
  g. calorie Grammkalorie *f, (veraltet für)* Kalorie *f*, cal
  g. equivalent Grammäquivalent *n*, Val *n*
  g. ion Grammion *n*
  g. mole *s.* g. molecule
  g.-molecular volume Mol[norm]volumen *n*, Molvolum *n*
  g.-molecular weight *s.* g. molecule
  g. molecule Grammolekül *n*, Mol *n*, Grammol *n*
Gram-negative *(Bakt)* gramnegativ
Gram-positive *(Bakt)* grampositiv
Gram stain *(Bakt)* Gram-Farbstoff *m*
Gram staining *(Bakt)* Gram-Färbung *f*, Gramsche Färbung *f (zur Differenzierung der Bakterien)*
gramicidin Gramizidin *n (Bestandteil des Tyrothrizins)*
  gramicidin-S Garmizidin S *n (Antibiotikum)*
graminin Graminin *n (aus Fruktose zusammengesetztes pflanzliches Polysaccharid)*
grammatite *(Min)* Grammatit[strahlstein] *m*, Tremolit *m*
Gram's method *s.* Gram staining
gran *s.* granular
grand fir Riesentanne *f*, Abies grandis Lindl.
granite Granit *m*
  g. paper meliertes (gefasertes) Papier *n*
  g. roll *(Pap)* Granitwalze *f*
granitic granitisch, graniten, Granit...
granitization *(Geol)* Granitisierung *f*, Granitisation *f*
granitize / to *(Geol)* granitisieren
granoblastic *(Geol)* granoblastisch
granodiorite *(Min)* Granodiorit *m*
granophyre *(Geol)* Granophyr *m*
granophyric *(Geol)* granophyrisch
granular körnig, gekörnt, granulös, granuliert
  g. aggregate *(Geol)* körniges Aggregat *n*
  g. bed körnige Schüttschicht *f*, Körnerhaufwerk *n*, Körnerschüttung *f*
  g.-bed separator Schüttschicht-Staubabscheider *m*, Hohlraumfilter *n*
  g. coke Grudekoks *m*
  g. formulation granuliertes Pflanzenschutzpräparat *n*
  g. material Granulat *n*
  g. variation *(Geol)* Körnungsschwankung *f*

granularity Körnigkeit *f*, körnige Struktur *f*, Körnung *f*
granulate / to granulieren, körnen, körnigmachen
granulate *s.* granulated
granulate Granulat *n*
granulated körnig, gekörnt, granuliert, granulös
  g. sugar Kristallzucker *m, (i.e.S.)* mittelkörniger Kristallzucker *m*
granulater *s.* granulator
granulating *s.* granulation
  g. machine *s.* granulator
granulation Granulieren *n*, Granulierung *f*, Granulation *f*, Körnigmachen *n*, Körnen *n*, Körnchenbildung *f, (i.e.S.)* Abbaugranulieren *n*, Schneiden *n (oder)* Aufbaugranulieren *n*, Krümeln *n*, Krümelung *f*
  g. plant Granulierungsanlage *f*
granulator Granulierapparat *m*, Granuliermaschine *f*, Granuliereinrichtung *f*, Granulator *m*, Granulatformer *m; (Zucker)* Granulator *m*, Trockentrommel *f*, Trommeltrockner *m*
granule Granalie *f*, Körnchen *n*, Korn *n*
  g. density Korndichte *f (von Schüttgütern)*
  g. method *(Ker)* Granulierverfahren *n*
granulite *(Geol)* Granulit *m*
granulitic *(Geol)* granulitisch
granulometric granulometrisch
  g. analysis Korngrößenanalyse *f*
granulose *s.* granoblastic
grape Weintraube *f, (i.e.S.)* Weinbeere *f*; Rebe *f*, Vitis L., *(i.e.S.)* Weinrebe *f*, Weinstock *m*, Vitis vinifera L.
  g. juice Traubensaft *m*, Traubenmost *m*
  g. pomace Traubentrester *pl*, Trester *pl*, Traubenmaische *f*
  g. press Traubenpresse *f*, Traubenquetsche *f*, Kelter *f*
  g.-seed oil Weintraubenkernöl *n*, Traubenkernöl *n*
  g.-stone oil *s.* g.-seed oil
  g. sugar $C_6H_{12}O_6$ Traubenzucker *m*, Glukose *f*, D-Glukose *f*, Glykose *f*, Dextrose *f*
  g. vinegar Weinessig *m*
  g. wine Traubenwein *m*
grapevine Weinrebe *f*, Weinstock *m*, Vitis vinifera L.
graph Diagramm *n*, grafische Darstellung *f*, Schaubild *n; (Math)* Kurve *f*
  g. paper Diagrammpapier *n*
graphic formula Formelbild *n*, Strukturformel *f*, Konstitutionsformel *f*
  g. granite Schriftgranit *m*
  g. tellurium *(Min)* Schrifttellur *m*, Schrifterz *n, (veraltet für)* Sylvanit *m (Silbergoldtellurid)*
graphite / to graphitieren, mit Graphit einstäuben *(oder)* mit Graphit auskleiden (überziehen)
graphite Graphit *m*
  g. bisulphate Graphithydrogensulfat *n*
  g. boat Graphitboot *n*, Graphitschiffchen *n*,

Substanzschiffchen *n* aus Graphit *(beim Zonen-schmelzen)*
g. burner Graphitbrenner *m*
g. crucible Graphittiegel *m*
g. deposition Graphitablagerung *f*
g. electrode Graphitelektrode *f*
g.-faced graphitüberzogen, graphitausgekleidet, graphitiert
g. heat exchanger Graphit-Wärmeaustauscher *m*, Graphit-Wärmeübertrager *m*
g. lattice Graphitgitter *n*
g. layer Graphitschicht *f*
g.-like graphitartig
g.-moderated *(Kch)* graphitmoderiert, graphit-gebremst
g.-moderated reactor graphitmoderierter Reaktor *m*, Graphitreaktor *m*
g. paper Graphitpapier *n*
g. rod Graphitstab *m*
g. salt Graphitsalz *n*
graphitic graphitisch, Graphit...; graphithaltig
g. acid Graphitsäure *f*
g. compound Graphitverbindung *f*
g. salt *s.* graphite salt
graphitization Graphit[is]ieren *n*, Graphit[is]ierung *f*, Graphitbildung *f*
graphitize / to graphit[is]ieren, Graphit bilden; graphitieren, mit Graphit einstäuben *(oder)* mit Graphit auskleiden (überziehen)
graphitizing Graphitglühen *n*, graphitisierendes Glühen *n*
grappier cement Grappierzement *m*, Krebszement *m*
grass bleaching Rasenbleiche *f*, Naturbleiche *f*, natürliche Bleiche *f*
g. butter Grasbutter *f*
g. killer Herbizid *n* gegen Gräser
g. tree gum Akaroidharz *n*, Akaroidgummi *n*, Grasbaumharz *n*, Resina Acaroidis *(von Xanthorrhoea-Arten)*
grassing *s.* grass bleaching
grate / to [zer]reiben, [zer]mahlen
grate Rost *m*; Gitter *n*
g. bar Roststange *f*
g. drive Rostantrieb *m*
g. firing Rostfeuerung *f*
g. heat release Rostwärmebelastung *f*
grating Rost *m*; Gitter *n*; *(phys Ch)* Beugungsgitter *n*, Gitter *n*
g. discharge Rostaustrag *m*
g. spectrograph Gitterspektrograf *m*
g. spectrophotometer Gitterspektrofotometer *n*
g. spectroscope Gitterspektroskop *n*
g. spectrum Gitterspektrum *n*
gratiolin Gratiolin *n* *(Glykosid aus Gratiola officinalis)*
gravel Kies *m*, *(i.e.S.)* Geröll *n* *(oder)* Schotter *m*
g. [packed] filter Kiesfilter *n*
gravelly kiesig
gravimetric gravimetrisch

g. analysis Gewichtsanalyse *f*, gravimetrische Analyse *f*, Gravimetrie *f*
g. factor Faktor *m* *(Gewichtsanalyse)*
g. method gravimetrisches Verfahren *n*, gravimetrische Methode *f* *(bei der Erdölsuche)*
gravimetrical *s.* gravimetric
gravimetry *s.* gravimetric analysis
gravitation Gravitation *f*, Schwerkraft *f*, Massenanziehung *f*
gravitational Gravitations..., Schwere...
g. acceleration Erdbeschleunigung *f*
g. constant Gravitationskonstante *f*
g. field Gravitationsfeld *n*, Schwerefeld *n*
g. force (pull) Gravitationskraft *f*, Schwerkraft *f*, Erdanziehungskraft *f*
g. sedimentation Sedimentieren (Absetzen) *n* im Schwerefeld
g. separator Schwerkraftabscheider *m*
gravitative differentiation *(Geol)* gravitative Differentiation *f*, Gravitationsdifferentiation *f*, Differentiation (Sonderung) *f* nach der Schwere
gravity Gravitation *f*, Schwerkraft *f*, Massenanziehung *f*; Schwere *f*
g. chamber Fallraum *m*
g. closing *(Plast)* Schließen *n* durch Schwerkraft (Eigenmasse des Stempels)
g. collector Schwerkraftabscheider *m*
g. concentrating apparatus Schwerkraftaufbereitungsanlage *f*, *(i.e.S.)* Schwerflüssigkeits[aufbereitungs]anlage *f*, Sinkscheideranlage *f*
g. concentration Schwerkraftaufbereitung *f*, Dichtesortierung *f*, Dichteanreicherung *f*
g. concentration apparatus *s.* g. concentrating apparatus
g. convection incubator Brutschrank *m* mit natürlicher Luftumwälzung
g. die metallische Dauer[gieß]form *f*, Metallform *f*, Kokille *f*
g. die-cast / to in [metallischen] Dauerformen [ver]gießen, in Kokille gießen, in der Kokille vergießen
g. die casting Dauerformguß *m*, [normales] Kokillengießen *n*, [normaler] Kokillenguß *m*; Dauerformgußstück *n*, Kokillengußstück *n*, Kokillengußteil *n*
g. die-casting process Dauerformgießverfahren *n*, Dauerformgußverfahren *n*, [normales] Kokillengußverfahren *n*
g. drainage reservoir *(Erdöl)* unter Schwerkraft entölende Lagerstätte *f*
g. dust catcher Staubfänger *m*, Staubsammler *m*, Staubsack *m* *(eines Hochofens)*
g.-fed schwerkraftzugeführt
g.-fed spray *(Landw)* Gießmittel *n*
g. feed Schwerkraftzuführung *f*, Gefällezuführung *f*
g. filling *(Pap)* Holzfüllung *f* ohne Füllapparat
g. filter Schwerkraftfilter *n*
g. filtration Filtration *f* unter Wirkung der Schwerkraft

g.-flow burette Ausflußbürette f
g. fluidizing conveyor pneumatische Rinne f
g.-operated gewichtsbetätigt, lastbetätigt, Gewichts..., Last...
g.-return system Naturumlaufsystem n
g.-roller conveyor (Fördertechnik) Rollenbahn f
g. save-all (Pap) Sedimentationsstoffänger m, Absetzstoffänger m, nach dem Sedimentationsprinzip (Absetzprinzip) arbeitender Stoffänger m
g. sedimentation s. gravitational sedimentation
g. separation Schwerkraftabscheidung f, Schwerkrafttrennung f, Schweretrennung f
g. separator s. gravitational separator
g. settling chamber Kammerabscheider m; Abscheidekammer f, Abscheideraum m
g. settling tank Schwerkraftabsetzer m, Schwerkraftabsetzbehälter m
g.-type headbox (Pap) offener Stoffauflauf[kasten] m, Stoffauflauf m offener Bauart
Gray-King assay Gray-King-Test m, Gray-King-Versuch m, Gray-King-Verkokungstest m
Gray-King assay coke type s. Gray-King coke type
Gray-King assay test s. Gray-King assay
Gray-King coke type Gray-King-Kokstyp[us] m, Kokstyp m nach Gray-King
Gray-King index Gray-King-Index m
Gray process Gray-Prozeß m (zur Behandlung von Benzin mit Bleicherde in der Dampfphase)
graying s. greying
grease / to [ein]fetten; [ein]schmieren
grease ausgelassenes (ausgeschmolzenes, zerlassenes) tierisches Fett n; Schmierfett n, Schmiere f
g. interceptor s. g. trap
g. oil Fettöl n, Speeköl n, Lardöl n, Schmalzöl n
g. repellency Fettabweisungsvermögen n, fettabweisendes Verhalten n
g. resistance Fettdichtigkeit f, Fettundurchlässigkeit f
g. resistant fettdicht, fettundurchlässig
g. resistant paper fettdichtes Papier n, Butterbrotpapier n
g. solvent Fettlösungsmittel n, Fettlöser m
g. spot Fettfleck m
g.-spot photometer Fettfleckfotometer n
g. trap Fettfang m, Fettfänger m, Fettabscheider m
g. wool Schmutzwolle f, Schweißwolle f, Rohwolle f
greaseband (Landw) Leimgürtel m, Leimring m
greaseless cream fettfreie (nichtfettende) Creme f
greaseproof fettdicht, fettundurchlässig
g. paper fettdichtes Papier n, Butterbrotpapier n
greasiness Fettigkeit f, Schmierigkeit f
greasy fettig, schmierig
g. lustre (Min) Fettglanz m
g. pulp (stock, stuff) (Pap) schmieriger Stoff m
great centrifugal breaker Großkreiselbrecher m

Great Falls converter Great-Falls-Konverter m (stehender Konverter vom Typ Great Falls)
Great Northern grinder (Pap) Great-Northern-Schleifer m
greaves Grieben fpl, Fettgrieben fpl
Greaves-Etchells furnace Greaves-Etchells-Ofen m (ein Lichtbogenofen)
Greek turpentine (Terpentin von Pinus halepensis Mill.)
green 1. grün; 2. grün, frisch (z. B. Holz); grün, roh; (Ker) ungebrannt; ungesintert
green Grün n (Farbempfindung); grün[färbend]er Farbstoff m, Grün n
g. aventurine Chromaventurin m, Chromaventuringlas n
g. cheese grüner (ungereifter) Käse m; Kräuterkäse m
g. cinnabar Cr_2O_3 Chrom[oxid]grün n, Ölgrün n, grüner Zinnober m, Chrom(III)-oxid n
g. coffee Rohkaffee m
g.-coloured grüngefärbt
g. compact grünes Teil n, Grünling m (ungesinterter Preßling in der Pulvermetallurgie)
g. compound (Gum) Rohmischung f, unvulkanisierte Mischung f
g. concrete grüner (junger) Beton m, Frischbeton m
g. copper (Min) Kupfergrün n, (veraltet für) Chrysokoll m (Kupfer(II)-metasilikat)
g. copperas $FeSO_4 \cdot 7H_2O$ Eisenvitriol n, Eisen(II)-sulfat-7-Wasser n
g. earth (Min) Grünerde f, Seladonit m
g.-edged grün gesäumt
g. flax Grünflachs m (verspinnbare Fasern aus nichtgeröstetem, mechanisch zerteiltem Strohflachs)
g. glass Grünglas n
g. hide (Gerb) Rohhaut f
g. hop Junghopfen m
g. house (Ker) Trocken- und Lagerhaus n für Rohware
g. lead ore (Min) Grünbleierz n, Pyromorphit m
g. liquor (Pap) Grünlauge f
g. liquor clarification (Pap) Grünlaugenklärung f
g. liquor clarifier (Pap) Grünlaugenklärtank m, Grünlaugenklärer m
g. liquor storage (Pap) Grünlaugenbehälter m, Grünlaugenvorratstank m
g. malachite (Min) Kupferspat m, Malachit m (Kupfer(II)-dihydroxidkarbonat)
g. malt Grünmalz n
g.-manuring Gründüngung f
g. moulding sand Grün[form]sand m, Grüngußsand m, grüner (nasser) Sand (Formsand) m, Naßguß[form]sand m
g. mud Grünschlick m
g. oil Grünöl n
g. paste grüne (rohe) Elektroden[stampf]masse f, grüne (rohe) Söderberg-Masse f

g. pellet Grünpellet n
g.-saltet (Gerb) (durch Einstreuen von Salz konserviert)
g. salting (Gerb) Stapelsalzung f (Konservierung frischer Häute mit festem Salz)
g. sand s. g. moulding sand
g.-sand casting Naßguß m, Gießen n in grüne (ungetrocknete) Sandformen
g.-sand mould Grün[sand]form f, grüne (nasse, ungetrocknete) Form f, Naßgußform f
g.-sand moulding Grünsandformen n, Naßgußformen n, Formen n in Grünsand (grünem Sand)
g.-shave / **to** (Gerb) aus dem Kalk falzen, kalkfalzen, im Kalkzustand egalisieren (falzen)
g. state (Ker) grüner (roher) Zustand m
g. strength Grünfestigkeit f (z.B. eines Formsandes); (Ker) Grünfestigkeit f, Trockenfestigkeit f, Rohbruchfestigkeit f
g. sulphur Grünschwefel m (aus Gasreinigungsmasse)
g. tyre (Gum) Reifenrohling m
g. vitriol s. g. copperas
g. wattle Acacia decurrens (I. C. Wendl.)Willd. (eine Gerberakazie)
g. weight (Gerb) Grünmasse f (der Häute)
greenalite (Min) Greenalith m (kalifreies Eisensilikat)
Greenawalt sintering machine Greenawalt-Sinterpfanne f, Greenawalt-Pfanne f
greenfly repellent Abschreckmittel n gegen Blattläuse
greening Grünen n, Grünung f (von Gemüsekonserven)
greenish grünlich
g.-blue grünlichblau
Greenland spar (Min) Kryolith m (Natriumfluoroaluminat)
greenockite (Min) Greenockit m (Kadmiumsulfid)
greenoughite, greenovite (Min) Greenovit m (Titanit-Varietät)
greensand Grünsand m
greenstone (Geol) Grünstein m; (Min) s. nephrite
grege, greige Grège[seide] f, Rohseide f, Bastseide f, Haspelseide f
greisen (Geol) Greisen m
greisening, greisenization (Geol) Greisenbildung f
grey grau
grey Grau n (Farbempfindung); grau[färbend]er Farbstoff m, Grau n
g. acetate s. g. lime
g. antimony (Min) Grauspießglanz m, Antimonglanz m, Antimonit m, Stibnit m (Antimon(III)-sulfid)
g. board Graupappe f
g. body (phys Ch) grauer Körper (Strahler) m
g. cardboard Graukarton m
g. cast iron s. g. iron
g. copper [ore] (Min) Antimonfahlerz n, Tetraedrit m

g. cotton Rohbaumwolle f
g. cutting (Glas) Rauhschliff m
g. finish Vorappretur f, Vorausrüstung f
g. floatation sulphur grauer Flotationsschwefel m
g. fog (Foto) Grauschleier m
g. goods (Text) Rohware f
g. humic acid (Bodenkunde) Grauhuminsäure f
g. iron Grauguß m
g. lime Graukalk m, rohes (unreines) Kalziumazetat n
g. mould Grauschimmel m, Botrytis cinerea
g. paper (veraltet für) blotting paper
g. pig iron graues Roheisen n
g. souring (Text) Säurebehandlung f
g. state (Text) Rohzustand m
greying (Text) Vergrauen n
greyish gräulich
greywacke (Geol) Grauwacke f
grid Gitter[werk] n, Netz n, Rost m
g. tray Gitter[rost]boden m, Turbogridboden m
Griffin mill Griffin-Mühle f (eine Pendelrollenmühle)
Griffin ring-roll[er] mill s. Griffin mill
Griffith's white Griffith-Weiß n, (veraltet für) Lithopone f, Lithopon n (Weißpigment aus ZnS und BaSO$_4$)
Grignard compound Grignard-Verbindung f
Grignard method (process) Grignard-Methode f, Grignard-Verfahren n
Grignard reaction s. Grignard synthesis
Grignard reagent Grignards (Grignardsches) Reagens n, Grignard-Reagens n
Grignard synthesis Grignard-Synthese f, Grignard-Reaktion f
grignardization Grignardisierung f
grime Schmutz m
grind / **to** zerkleinern; [zer]mahlen, vermahlen; [zer]schleifen, verschleifen; [zer]reiben; verreiben; (Farb) anreiben; (Lebm) schroten (z.B. Malz, Getreide)
g. across the grain (Pap) querschleifen
g. fine feinmahlen
g. level spiegelschleifen
g. smooth glattschleifen
grindability Mahlbarkeit f, Vermahlbarkeit f; Schleifbarkeit f
g. index Mahlbarkeitszahl f
g. test Mahlbarkeitstest m, Mahlbarkeitsprüfung f
g. value s. g. index
grindable mahlbar; schleifbar
grinder Mühle f; Schleifmaschine f; (Pap) Schleifer m, Holzschleifer m, Holzschleifmaschine f
g. bar Reibbarren m
g. house (Pap) Schleiferei f, Holzschleiferei f, Schleifereibetrieb m
g. pit (Pap) Schleifertrog m
g. pit temperature (Pap) Trogtemperatur f
g. room s. g. house

g. wheel Schleifscheibe *f*

grinding Zerkleinern *n*, Zerkleinerung *f*; Mahlen *n*, Mahlung *f*, Zermahlen *n*, Vermahlen *n*; Schleifen *n*, Zerschleifen *n*, Verschleifen *n*, Schliff *m*; Reiben *n*, Zerreiben *n*, Verreiben *n*; *(Farb)* Anreiben *n*, Anreibung *f*; *(Lebm)* Schroten *n* (z.B. von Malz, Getreide)

g. aid Mahlhilfsstoff *m*, Mahlhilfe *f*

g. area *(Pap)* Schleifzone *f*, Schleiffläche *f*

g. ball Mahlkugel *f*

g. chamber Mahlkammer *f*, Mahlraum *m*

g. element Mahlorgan *n*

g. equipment Zerkleinerungsanlage *f*

g. machine Mahlmaschine *f*; Schleifmaschine *f*

g. medium Mahlkörper *m*

g. mill Mühle *f*

g. pan Mahlteller *m*, Mahlschüssel *f*

g. pressure *(Pap)* Schleifdruck *m*

g. resin dispergiertes Harz *n*

g. resistance Mahlwiderstand *m*

g. ring Mahlring *m*

g. roll Mahlwalze *f*

g. sand Schleifsand *m*

g. surface Mahlfläche *f*, Mahlbahn *f*; *(Pap)* Schleiffläche *f*, Schleifzone *f*

g. table Schleiftisch *m*

g. wheel Schleifscheibe *f*

grindstone Mahlstein *m*, Mühlstein *m*; Schleifstein *m*, *(Pap auch)* Schleiferstein *m*

grip roll *(Gerb)* Druckwalze *f*, Führungswalze *f*

gripping jaw Backe *f* *(einer Stativklemme)*

grisein Grisein *n* *(Antibiotikum)*

grist Mahlgut *n*; Malzschrot *n(m)*, geschrotetes Malz *n*

grit Grieß *m*; Schrot *n* *(m)*; Grobsand *m*; Grus *m*; Splitt *m*

grittiness grießartige (grießförmige, körnige) Beschaffenheit *f*, Sandigkeit *f*

gritty grießig, grießartig, grießförmig, körnig; sandig, sandartig

grizzly [screen] Stangenrostsieb *n*, Stangen[sieb]rost *m*, Stab[sieb]rost *m*; Rostsiebmaschine *f*

groats Grütze *f*

grog zerkleinerte Schamotte *f*

grogging *(Ker)* Magern *n*

groove spew *(Plast)* Austriebnut *f*

grooved genutet; gerieffelt, gerillt

g. roll gerieffelte Walze *f*, Riffelwalze *f*

gross calorific value Verbrennungswärme *f*, *(veraltet)* oberer Heizwert *m*, H_o

g. coking heat obere Verkokungswärme *f*

g. composition Makrozusammensetzung *f*

g. heating value *s.* g. calorific value

grossular[ite] *(Min)* Grossular *m* *(Kalziumaluminiumorthosilikat)*

Grotthus[s]-Draper law Grotthus[s]-Drapersches Gesetz *n*

ground / to *(Farb)* grundieren

ground Grund *m*, Boden *m*; *(Bgb)* Gestein *n*, Gebirge *n*; *(Farb)* Untergrund *m*

g. band *(phys Ch)* Grundband *n*, G-Band *n*

g. coat Untergrundanstrich *m*, Grund[ier]anstrich *m*, Grundierung *f*

g.-glass equipment Schliffgeräte *npl*, Schliffapparaturen *fpl*

g.-glass joint Schliffverbindung *f*

g.-glass lid Glasdeckel *m* mit Schliff, Wägeglaskappe *f*

g.-glass stopper Schliffstopfen *m*, eingeschliffener Glasstopfen *m*

g.-in aufgeschliffen, eingeschliffen

g. level *s.* g. state

g. malt geschrotetes Malz *n*, Malzschrot *n* *(m)*

g.-mass *(Ker)* Grundmasse *f*, Einbettungsmasse *f*, Matrix *f*

g. oats Hafergrütze *f*

g. rice Bruchreis *m*

g. state *(phys Ch)* Grundzustand *m*, Normalzustand *m*, Grundniveau *n*, Grundterm *m*

g. stopper *s.* g.-glass stopper

g. term *s.* g. state

g. water Grundwasser *n*

groundmass *(Geol)* Grundmasse *f*

groundnut Erdnuß *f*, Arachis hypogaea L.; *(manchmal auch)* Erdbirne *f*, Apios americana Medik.

g. cake Erdnußkuchen *m*

g. fibre Erdnußfaser *f*; Erdnußfaserstoff *m*

g. lecithin Erdnußlezithin *n*

g. oil Erdnußöl *n*, Arachisöl *n*

g. protein Erdnußprotein *n*, Erdnußeiweiß *n*

g. protein staple Erdnußproteinfaser *f*, Erdnußeiweißfaser *f*

groundwood *(Pap)* [mechanischer] Holzschliff *m*, [mechanischer] Holzstoff *m*, Holzmasse *f*, Schleifmasse *f*

g. bleaching *(Pap)* Holzschliffbleiche *f*

g. fibre *(Pap)* Holzschliffaser *f*

g. mill *(Pap)* Holzschleiferei *f*, Schleiferei *f*, Schleifereibetrieb *m*

g. paper holzhaltiges Papier *n*, Holzschliffpapier *n*

g. process *(Pap)* Holzschliff-Verfahren *n*

g. pulp *s.* groundwood

g. pulping *(Pap)* mechanischer Aufschluß *m* des Holzes, Holzschlifferzeugung *f*, Holzschliffgewinnung *f*

g. rejects *(Pap)* Grobstoff *m*, Spuckstoff *m*, „Sauerkraut" *m*

group / to in eine Gruppe zusammenfassen, in Gruppen anordnen, gruppieren

group Gruppe *f* *(z.B. des Periodensystems)*; Gruppe *f*, Rest *m* *(eines Moleküls)*

g. analysis Gruppenanalyse *f*

g.-cut / to *(Pap)* *(Bogen)* mehrlagig schneiden

g. displacement law [radioaktiver] Verschiebungssatz *m*, Verschiebungsgesetz *n*

g. number Gruppennummer *f*, Gruppenziffer *f*, Kodenummer *f* der Gruppe, Zweite Kodeziffer *f* *(des internationalen Kohlenklassifikationssystems)*

g. **of atoms** Atomgruppe *f*
g. **parameter** Gruppenparameter *m (des internationalen Kohlenklassifikationssystems)*
g. **reagent** Gruppenreagens *n*
g. **velocity [of waves]** Gruppengeschwindigkeit *f*
grouping Gruppierung *f*
grout Schrotmehl *n*, grobes Mehl *n*
grouted concrete vorgepackter Beton *m*, Prepaktbeton *m*, Schlämmbeton *m*
grow / to [auf]wachsen, zunehmen, ansteigen, sich vergrößern; *(Kristalle)* züchten
growth Wachstum *n*, Wachsen *n*, Wuchs *m*; Zunahme *f*, Anstieg *m*, Vergrößerung *f*
g. **factor** Wachstumsfaktor *m*
g. **hormone** Wachstumshormon *n*, Wuchshormon *n*
g. **inhibitor** Hemmstoff *m*
g. **medium** Nährboden *m*, Nährmedium *n*, Nährsubstrat *n*
g. **modifier** s. g. regulator
g.**-promoting** wachstumsfördernd
g.**-promoting factor** s. g. factor
g. **rate** Wachstumsgeschwindigkeit *f*
g.**-regulating substance** s. g. regulator
g. **regulator** Wachstumsregulator *m*, Wuchsstoffmittel *n*, *(mit hemmender Wirkung auch)* Wuchsstoffherbizid *n*
g. **stage** *(Plast)* Wachstumsstadium *n*, Fortpflanzungsstadium *n*
g.**-stimulating factor** s. g. factor
g. **substance** Wuchsstoff *m*, wachstumsfördernder Stoff *m*
g.r.p., G.R.P. s. glass-fibre reinforced plastic
grünauite *(Min)* Grünauit *m*, Wismutnickel[kobalt]kies *m*
Grundmann synthesis Grundmann-Reaktion *f*, Grundmann-Synthese *f (von Aldehyden)*
G.S.C. s. gas-solid chromatography
guaethol $C_2H_5OC_6H_4OH$ Guäthol *n*, Äthoxyphenol *n*, Brenzkatechinmonoäthyläther *m*
guaiac [rosin] Guajakharz *n (von Guaiacum officinale L. und G. sanctum L.)*
guaiacol $CH_3OC_6H_4OH$ Guajakol *n*, o-Oxyanisol *n*, 2-Methoxyphenol *n*, Brenzkatechinmonomethyläther *m*, Pyrokatecholmonomethyläther *m*
guaiene $C_{10}H_6(CH_3)_2$ Guajen *n*, 2,3-Dimethylnaphthalin *n*
guaiol Guajol *n (Sesquiterpen-Alkohol)*
guajen s. guaiene
guanajuatite *(Min)* Guanajuatit *m*, Selenwismutglanz *m*
guanidine $NH=C(NH_2)_2$ Guanidin *n*, Iminoharnstoff *m*, Imidoharnstoff *m*, Aminomethanamidin *n*
g. **accelerator** Guanidinbeschleuniger *m*
guanidoacetic acid Guanidinessigsäure *f*, Glykozyamin *n*
guanine Guanin *n*, 2-Amino-6-hydroxypurin *n*, 2-Aminohypoxanthin *n*
guano Guano *m (Düngemittel aus Vogelkotablagerungen)*

guanosine Guanosin *n (Nukleosid)*
guarana Guarana *n*, Guaranapaste *f (Genußmittel aus den Samen von Paullinia cupana Kunth)*
guaruma wax Guaruma-Wachs *n (aus den Blättern von Calathea lutea G.F.W. Mey.)*
guayacanite *(Min)* Guayacanit *m*, *(veraltet für)* Enargit *m (Arsen(III)-kupfer(I,II)-sulfid)*
guayule rubber Guayule-Kautschuk *m*, Guayule *f (von Parthenium argentatum A. Gray)*
g. **shrub** Guayulestrauch *m*, Guayule *f*, Huayule *f*, Parthenium argentatum A. Gray
guest Gast *m (bei Einschlußverbindungen)*; Gastelement *n*
g. **component** Gastkomponente *f*, Gastsubstanz *f*
g. **element** Gastelement *n*
Guggenheim process Guggenheim-Verfahren *n (Gewinnung von Natriumnitrat aus Chilesalpeter)*
guhr dynamite Gur-Dynamit *n*
guide Führung *f*; *(Pap)* Sieblaufregler *m*
g. **funnel** *(Glas)* Führungstrichter *m*
g. **roll** Führungswalze *f*, Führungsrolle *f*, Leitwalze *f*, Leitrolle *f*; *(Pap)* Sieblaufregulierwalze *f*; *(Pap)* Siebleitwalze *f*
g. **value** Richtwert *m*
g. **vane** Leitschaufel *f (z.B. einer Turbine)*
Guignet green $Cr_2O_3 \cdot xH_2O$ Guignetgrün *n*, Chromoxidhydratgrün *n*, Chromoxidgrün feurig *n*, Mittlers Grün *n*, Viridian *n*, Smaragdgrün *n*
guillotine cutter (cutting machine, press) *(Pap)* Guillotineschneider *m*; *(Pap)* Planschneider *m*, Formatschneider *m*, Riesbeschneidemaschine *f*
g. **trimmed** *(Pap)* mit Planschneider (Hubmesser) geschnitten
g. **trimmer** s. g. cutter
g. **trimming** *(Pap)* Planschneiden *n*, Schneiden *n* mit Hubmesser
g.**-type rag cutter** *(Pap)* Guillotinehadernschneider *m*
guitermanite *(Min)* Guitermanit *m (unreiner Jordanit)*
Gulf HDS process Gulf-HDS-Verfahren *n (zur hydrierenden Entschwefelung)*
gum Pflanzengummi *m*, Gummi *n*; *(ungenau für)* Kautschuk *m*; *(ungenau für)* Harz *n*; *(in Zusammensetzungen auch gummi- oder harzliefernde Baumart)*; *(Gum) (ungefüllte)* Mischung *f*; *(Erdöl)* Gum *m*, Harz *n*
g. **accroides** Grasbaumharz *n*, Akaroidharz *n (von Xanthorrhoea-Arten)*
g. **ammoniac** Ammoniak-Gummiharz *n (von Dorema ammoniacum Don und Verwandten)*
g. **arabic** Gummiarabikum *n*, arabisches Gummi *n*, Sudangummi *n*, Akaziengummi *n*, Mimosengummi *n*
g. **bassora** Bassoragummi *n (Sammelname für geringwertige Tragantsorten verschiedener Herkunft)*
g. **benjamin** s. g. benzoin
g. **benzoin** Benzoe *f*, Benzoeharz *n (von Styrax-Arten)*

g. butea Buteakino *n*, Bengalkino *n*, Bengalisches Kino *n* (*von Butea superba Roxb.*)

g. compound *s.* g. stock

g. content (*Erdöl*) Gumgehalt *m*, Harzgehalt *m*

g. copal Kopal *m* (*Sammelname für hochschmelzende pflanzliche Harze*)

g. damar *s.* dammar

g. formation (*Erdöl*) Gumbildung *f*, Harzbildung *f*, Verharzung *f*

g.-forming (*Erdöl*) gumbildend, harzbildend

g. gamboge Gummigutt *n* (*von Garcinia-Arten, vorwiegend G. hanburyi Hook. f.*)

g. ghatti Ghatti Gum *m* (*von Anogeissus latifolia Wall.*)

g. guaiac Guajakharz *n* (*von Guiacum officinale L. und G. sanctum L.*)

g. juniper Wacholderteer *m* (*von Juniperus-Arten*); Sandarak *m*, Sandarakharz *n* (*von Tetraclinis articulata (Vahl.) Mast.*)

g. karaya Karayagummi *n*, Sterkuliagummi *n*, Indischer (Ostindischer) Tragant *m* (*meist von Sterculiaurens Roxb.*)

g. kateera (kateira) *s.* g. kuteera

g. kuteera Kuteragummi *n* (*von Sterculia-, Cochlospermum- oder Astragalus-Arten*)

g. lac (lacquer) Gummilack *m*, Rohschellack *m*

g. mastic Mastix *m*, Mastixharz *n* (*von Pistacia lentiscus L.*)

g. myrrh Myrrhe *f* (*Gummiharz von Commiphora-Arten*)

g. pontianak (*halbfossiler*) Manila-Kopal *m* (*von Agathis alba (Lam.)Foxw.*); Pontianak *m*, (*veraltet für*) Djelutung *m*, Djelutungharz *n* (*Kopal von Dyera-Arten und von Parthenium argentatum A. Gray*)

g. resin Gummiharz *n*

g. rosin Balsamkolophonium *n*, Balsamharz *n*, Harz *n* aus Rohterpentin

g. sandarac Sandarak *m*, Sandarakharz *n* (*von Tetraclinis articulata (Vahl.)Mast.*)

g. shiraz (*aus Schiras exportiertes Pflanzengummi*)

g. silk Rohseide *f*, Grège[seide] *f*, Bastseide *f*, Haspelseide *f*

g. stock (*Gum*) ungefüllte Mischung *f*

g. test (*Erdöl*) Gum-Test *m*, Gumbildungstest *m*, Harzbildnertest *m*, HBT

g. tragacanth Tragantgummi *n*, Tragant *m* (*von Astralagus-Arten*)

g. tragasol (*Gerb*) Tragasol *n* (*gallertiges Lederappretiermittel aus Samenschalen von Ceratonia siliqua*)

g. vulcanizate (*Gum*) Reinvulkanisat *n*

g. zanzibar Sansibar-Kopal *m* (*von Trachylobium verrucosum Oliv. und Tr. hornemannianum Heyne*)

gummed paper gummiertes Papier *n*, Klebpapier *n*

gumming *s.* gum formation

gummite (*Min*) Gummit *m*, rotes Pechuran *n*

gun Spritzpistole *f*, Farbspritzpistole *f* (*in der Farbspritztechnik*); Pistole *f*, Brenner *m* (*in der Schweißtechnik*)

g. cotton *s.* guncotton

g. perforating (*Erdöl*) Schußperforierung *f*

guncotton Schieß[baum]wolle *f*

gunite / **to** torkretieren, Spritzbeton im Torkretverfahren (Mörtelspritzverfahren) anbringen

gunite Torkretbeton *m*, Spritzbeton *m*

gunk (*Am*) Polyester-Glasfasermasse *f*

gunned concrete *s.* gunite

gunpowder Schwarzpulver *n*, Sprengpulver *n*, Schießpulver *n*

gurjun balsam Gurjunbalsam *m*, Gardschanbalsam *m* (*von Dipterocarpus alatus Roxb.*)

g. balsam oil Gurjunbalsamöl *n*

gut fat Eingeweidefett *n*, Darmfett *n*

gutta-percha Guttapercha *f(n)*, Guttapertja *f(n)*

gutter Rinne *f*, Sammelrinne *f*; (*Met*) Gießrinne *f*

Gutzeit test Gutzeitsche Probe *f* (*Arsen-Nachweis*)

gyle Sud *m* (*die auf einmal gebraute Biermenge*); Gärbottich *m*

gymnite (*Min*) Gymnit *m*, (*veraltet für*) Deweylith *m*

gymnospermous wood Nadelholz *n*

gynocardia oil (*Pharm*) Chaulmoograöl *n* (*von Hydnocarpus anthelminthica Pierre und H. kurzii (King) Warb.*)

gynocardic acid Gynokardsäure *f* (*Säuregemisch aus Chaulmoograöl*); (*manchmal auch für*) d-chaulmoogric acid

gypseous gipshaltig, gipsreich (*z.B. Brauwasser*); gipsähnlich; aus Gips [bestehend], Gips...

gypsiferous (*Geol*) gipshaltig, gipsführend

gypsum / **to** (*Brauwasser*) gipsen, burtonisieren; (*Landw*) gipsen

gypsum (*Min*) Gips *m*, Selenit *m* (*Kalziumsulfat-2-Wasser*)

g. cement *s.* g. plaster

g. mortar Gipsmörtel *m*

g. plaster Gipsputz *m*

gyratory crusher Kreiselbrecher *m*, Rundbrecher *m*, Kegelbrecher *m*

g. screen (sifter) Plansichter *m*, Rätter *m*, Siebrätter *m*

Gyro [vapour-phase] process Gyro-Dampfphase[krack]verfahren *n*, Gyro-Spaltverfahren *n*, Gyro-Verfahren *n*

gyromagnetic ratio (*Kch*) gyromagnetisches Verhältnis *n*

gyttja Gyttja *f* (*Halbfaulschlamm*)

H

H acid $C_{10}H_4NH_2OH(SO_3H)_2$ H-Säure *f*, 1-Amino-8-naphthol-3,6-disulfonsäure *f*, 1-Naphthylamin-8-hydroxy-3,6-disulfonsäure *f*

H and D curve (*Foto*) Schwärzungskurve *f*, Gradationskurve *f*

H-layer (*Bodenkunde*) Humusstoffhorizont *m* (*Symbol: H*)

Haber[-Bosch] process Haber-Bosch-Verfahren *n*

habit Habitus *m*, Kristallhabitus *m*

h. formation *(Pharm)* Suchtbildung *f*
h.-forming *(Pharm)* suchterzeugend
hack drying *(Ker)* Freilufttrocknung *f*; Haldentrocknung *f (von Ziegeln)*
hackle / to *(Text)* hecheln
hackling *(Text)* Hecheln *n*
hackly fracture *(Min)* hakiger Bruch *m*
hackmatack Amerikanische Lärche *f*, Larix laricina (Duroi) K. Koch
Hadsel mill Hadsel-Mühle *f*, Hadsel-Prallmühle *f*
haem Häm *n*
h. protein Hämprotein *n*
haemin Hämin *n*, *(i.e.S.)* Hämin[chlorid] *n*, Chlorhämin *n*, Protohämin *n*
haemoglobin Hämoglobin *n*, Hb, roter Blutfarbstoff *m*
haemolymph Hämolymphe *f*
haemolysin Hämolysin *n*
haemolysis Hämolyse *f*
haemolyze / to hämolysieren
HAF black *s.* high abrasion furnace black
hafnate $M^I_2[HfO_3]$ Hafnat *n*
hafnia HfO_2 Hafnia *f*, *(veraltet für)* Hafnium(IV)-oxid *n*
hafnium Hf Hafnium *n*
h. carbide HfC Hafniumkarbid *n*
h. oxide HfO_2 Hafniumdioxid *n*, Hafnium(IV)-oxid *n*
h. oxychloride $HfOCl_2$ Hafniumoxidchlorid *n*
h. sulphate $Hf(SO_4)_2$ Hafniumsulfat *n*
h. tetrachloride $HfCl_4$ Hafniumtetrachlorid *n*
Hägglund process Hägglund-Verfahren *n (Holzverzuckerung)*
Hägglund's solid lignosulphonic acid nach Hägglund bestimmte feste Ligninsulfonsäure *f*
haidingerite *(Min)* Haidingerit *m (Kalziumhydrogenorthoarsenat)*
Haine reagent Hainesche Lösung *f (Traubenzuckernachweis)*
hair bleach (bleaching agent) Haarbleichmittel *n*, Blondiermittel *n*, Blondierpräparat *n*
h. cosmetic Haarkosmetikum *n*, Haarpflegemittel *n*
h. dye Haarfärbemittel *n*, Haarfärbepräparat *n*, Haarfarbe *f*
h. fibre Wolle *f (außer Schafwolle)*, Haar *n*
h. lacquer Haarlack *m*
h. loosening by amines *(Gerb)* Aminäscher *m*
h. loosening by ammonia *(Gerb)* Ammoniakäscher *m*
h. loosening by enzymes *(Gerb)* Enzymäscher *m*, Fermentäscher *m*
h. lotion Haarlotion *f*
h. preparation Haarpflegemittel *n*, Haarbehandlungsmittel *n*
h. remover Haarentfernungsmittel *n*, Enthaarungsmittel *n*, depilierendes Mittel *n*, Depilatorium *n*
h. salt *(Min)* Bittersalz *n*, Epsomsalz *n*, Epsomit *m (Magnesiumsulfat-7-Wasser)*; Alunogen *m (Aluminiumsulfat-18-Wasser)*

h. tint Haartöner *m*
h. wash Haarwaschmittel *n*, Haarreinigungsmittel *n*, Kopfwaschmittel *n*, Haarwäsche *f*
hairiness Haarigkeit *f*
hairpin coil U-Rohr *n*, Haarnadelrohr *n*; Haarnadelrohrbündel *n*
h. tube U-Rohr *n*, Haarnadelrohr *n*
half-baked halb gebacken
h. band width *(phys Ch)* Halbwert[s]breite *f*
h.-bleached *(Pap)* halbgebleicht
h.-cell *s.* h.-element
h.-chair form Halbsesselform *f*
h. change value *(Kch)* Halbwert[s]zeit *f*
h.-condensation product Kondensationszwischenprodukt *n*
h.-crystal Halbkristallglas *n*
h.-element Halbelement *n*
h. ester Halbester *m*
h.-hard rubber halbharter Gummi *m*, Halbhartgummi *m*, Semiebonit *n*
h.-hour synthetic Halbstundenlack *m (schnelltrocknender Nitrozelluloselack)*
h.-intensity width *s.* h. band width
h.-life [period] *s.* h. change value
h.-mat[t] *(Foto)* halbmatt
h.-neat soap Halbkernseife *f*
h. neutralization point Halbneutralisationspunkt *m*
h.-polar halbpolar, semipolar
h.-polar bond halbpolare (semipolare) Doppelbindung *f*, koordinative (dative) Bindung *f*, Donator-Akzeptor-Bindung *f*
h.-sandwich compound Halbsandwichverbindung *f*
h.-shade Nicol *(phys Ch)* Halbschattennicol *m*
h.-shadow polarimeter Halbschattenpolarimeter *n*, Halbschattenpolarisationsgerät *n*, Halbschattenpolarisator *m*
h.-sized *(Pap)* halbgeleimt, $^1/_2$geleimt, mit mittlerer Leimung
h.-stock *s.* h.-stuff
h.-stuff *(Pap)* Faserhalbstoff *m*, Halbstoff *m*, Halbzeug *n*, Stoff *m*
h.-stuff beater *(Pap)* Halbzeugholländer *m*, Halbstoffholländer *m*
h.-stuff board *(Pap)* Halbstoff *m* in Bogenform (trockenen Bogen)
h.-stuff machine *(Pap)* Holzschliffentwässerungsmaschine *f*; Zellstoffentwässerungsmaschine *f*
h.-thickness *(phys Ch)* Halbwerts[schicht]dicke *f*
h.-tone paper Illustrationsdruckpapier *n*, Autotypie[druck]papier *n*
h. wave Halbwelle *f*
h.-wave potential Halbstufenpotential *n*, Halbwellenpotential *n*
h.-white oil Halbweißöl *n*
h.-width [value] *s.* h. band width
halibut liver oil Heilbuttlebertran *m*, Heilbuttleberöl *n*

halide Halogenid *n*
  h. leak detector Gasspürgerät *n* für Halogenide
  h. of phosphorus Phosphorhalogenid *n*
  h. phosphor Halogenidphosphor *m* (*Phosphoreszenz zeigendes Halogenid*)
halite (*Min*) Halit *m*, Steinsalz *n* (*Natriumchlorid*)
haliver oil *s.* halibut liver oil
Hall effect Hall-Effekt *m*
Hall process Hall-Verfahren *n* (*zur Al-Gewinnung*); Hall-Verfahren *n* (*zur Ölvergasung*)
halloysite Halloysit *m* (*Tonmineral*)
hallucinogen Halluzinogen *n*, halluzinogenisierender Stoff *m*
hallucinogenic halluzinogen[isierend]
halmyrolysis Halmyrolyse *f* (*submarine Verwitterung*)
halo Halogen...
halo Halo *m*, Hof *m*, Lichthof *m*; (*Foto*) Schwärzungsring *m*, Schwärzungshof *m*
  h. compound Halogenverbindung *f*
α-**haloacid** α-Halogenkarbonsäure *f*
haloalkane Halogenalkan *n*, Alkanhalogenid *n*, Alkylhalogenid *n*
haloanthraquinone Halogenanthrachinon *n*
halobenzene Halogenbenzol *n*
halocarbon compound Halogenkohlenstoffverbindung *f*
halochromic effect Halochromieerscheinung *f*
halochromism, halochromy Halochromie[erscheinung] *f*
haloform Haloform *n* (*Trihalogenmethanderivat*)
halogen Halogen *n*
  h. addition Halogenaddition *f*
  h. bridge Halogenbrücke *f*
  h. carrier Halogenüberträger *m*
  h. compound Halogenverbindung *f*
  h.-containing halogenhaltig
  h. derivative Halogenderivat *n*, Halogenabkömmling *m*
  h. derivative of the hydrocarbons Halogenkohlenwasserstoff *m*
  h. derivative of the paraffin hydrocarbons Halogenalkan *n*, Alkanhalogenid *n*, Alkylhalogenid *n*
  h. electrode Halogenelektrode *f*
  h. hydracid Halogenwasserstoffsäure *f*
  h.-metal exchange Halogen-Metall-Austausch *m*
  h.-nitrogen compound Halogenstickstoffverbindung *f*
  h.-substituted halogensubstituiert
halogenate / to halogenieren
α-**halogenated** α-halogeniert, in α-Stellung halogeniert
  α-**h. acid** *s.* α-haloacid
  h. hydrocarbon Halogenkohlenwasserstoff *m*
halogenation Halogenation *f*, Halogen[is]ierung *f*, Halogen[is]ieren *n*
  h. at the alpha position α-Halogenierung *f*
halohydrin Halogenhydrin *n* (*Glyzerinderivat*)
haloketone Halogenketon *n*

halophyte Halophyt *m*, Salzpflanze *f*
halotrichine, halotrichite (*Min*) Halotrichit *m* (*Aluminiumeisen(II)-sulfat-22-Wasser*)
halt [eutektischer] Haltepunkt *m*
hamameli tannin (*Gerb*) Hamamelitannin *n*
hamamelis Hamamelis *f*, Zaubernuß *f*, Hamamelis L.; Hamamelisblätter *npl*
hamamelose Hamamelose *f*, 2-C-Hydroxymethylribose *f*
hamartite (*Min*) Hamartit *m*, (*veraltet für*) Bastnäsit *m*
hammer crusher Hammerbrecher *m*
  h. disintegrator Hammermühle *f* (*für weiche bis mittelharte Stoffe*)
  h.-finished paper gehämmertes Papier *n*
  h. grinder Hammerbrecher *m*; Hammermühle *f*
  h. mill Hammermühle *f*; (*Pap*) Stampfwerk *n*, „Deutsches Geschirr" *n*
  h.-type crusher *s.* h. crusher
  h.-type mill *s.* h. mill
hand (*Text*) Griff *m*, Griffigkeit *f*
  h. balance Apothekerwaage *f*
  h. block (*Text*) Model *m*, Druckstock *m*
  h. centrifuge Handzentrifuge *f*
  h. charging Handbeschickung *f*
  h. cleaning *s.* h. picking
  h.-cleaning paste Handwaschpaste *f*
  h. compaction Verdichten *n* von Hand
  h. cream Handcreme *f*
  h.-cut peat handgestochener Torf *m*, Handstichtorf *m*
  h.-cutting Handstich *m*, Torfstechen *n* von Hand
  h. cylinder method (*Glas*) Walzenblasverfahren *n*
  h. dipping (*Ker*) Tauchen *n* von Hand
  h.-driven centrifuge *s.* h. centrifuge
  h. dust gun (*Landw*) (*pneumatisch betriebenes*) Handstäubegerät *n*
  h. duster (*Landw*) Handstäubegerät *n*, Hand[ver]stäuber *m*, Kleinverstäuber *m*
  h. ejection (*Plast*) Ausdrücken *n* von Hand
  h. fettling (*Ker*) Putzen *n* von Hand
  h. gun *s.* 1. h. spray gun; 2. h. dust gun
  h. hole Handloch *n*
  h. lance (*Landw*) Stabspritze *f*
  h. lay-up [technique] (*Plast*) Handauflegeverfahren *n*, Kontaktpreßverfahren *n*, Kontaktpressen *n*, Niederdruckpreßverfahren *n*
  h. lens Lupe *f*
  h. lotion Handlotion *f*
  h.-made brick Handstrichziegel *m*, Handformziegel *m*, Handstrichstein *m*, Handformstein *m*
  h.-made glass *s.* handblown glass
  h.-made grey board Handgraupappe *f*
  h.-made leather board Handlederpappe *f*
  h.-made mill *s.* h.-made paper mill
  h.-made paper handgeschöpftes Papier *n*, Büttenpapier *n*, Handpapier *n*, Handbütten *n*, Echt-Bütten *n*, Schöpfpapier *n*
  h.-made-paper making Echt-Büttenpapierherstellung *f*

h.-made-paper mill Büttenpapierfabrik *f*; Papiermühle *f*
h.-made wood board Handholzpappe *f*
h. modelling *(Ker)* Handformen *n*, Handformgebung *f*, Modellieren (Einformen) *n* von Hand
h. mould *(Plast)* Handwerkzeug *n*, Handform *f*; *(Pap)* Form *f*, Schöpfform *f*, Schöpfrahmen *m*
h.-moulded handgeformt, von (mit der) Hand geformt
h. moulding s. h. modelling
h.-operated handbetrieben, Hand...
h.-operated duster s. h. duster
h. operation Handbetrieb *m*
h. painting *(Ker)* Handmalerei *f*
h.-pick / to [von Hand] ausklauben (klauben, auslesen) *(bei der Kohlenaufbereitung)*
h. picking Handklaubung *f*, Handscheidung *f*, Ausklauben (Klauben, Auslesen) *n* [von Hand] *(bei der Kohlenaufbereitung)*
h. power air blower Handgebläse *n*
h. preparation Handpflegemittel *n*
h. press Handpresse *f*
h. pressing *(Ker)* Handpressen *n*
h.-propelled sprayer *(Landw)* Karrenspritzgerät *n*, Karrenspritze *f*; *(Landw)* Karrensprühgerät *n*, Karrenzerstäuber *m*
h. scooping Handaustrag *m*, Ausschaufeln *n* [von Hand]
h. sheet Prüfbogen *m*, Handmuster *n*, Schöpfpapiermuster *n*, Laborpapierblatt *n*
h. sieving Handsieben *n*, Handsiebung *f*
h. sizing *(Pap)* Handleimung *f*
h.-sorted handsortiert
h. sorting Handsortierung *f*, Sortierung *f* von Hand; s. h. picking
h. spray gun Handspritzpistole *f*
h. sprayer *(Landw)* Handspritzgerät *n*, Handspritze *f*; *(Landw)* Handsprühgerät *n*, Handzerstäuber *m*
h. stoking Schüren *n* von Hand
h. stuffing *(Gerb)* Kaltfetten *n*, Kaltschmieren *n*
h. throwing *(Ker)* Freidrehen *n*
h. washing Handwäsche *f*
h. wedging *(Ker)* Masseschlagen *n* von Hand
h. welding Handschweißen *n*
h. welding unit Handschweißgerät *n*
handblown *(Glas)* mundgeblasen
h. glass mundgeblasenes Glas *n*, Mundblaseglas *n*
h. process *(Glas)* Mundblasverfahren *n*
handle / to 1. *(Werkzeug)* handhaben, gebrauchen, bedienen; sich handhaben lassen, funktionieren; 2. *(Stoffe)* [be]fördern, transportieren, weiterleiten; *(Abgase)* [über]leiten, überführen; 3. *(eine Stoffmenge)* durchsetzen; 4. *(Gerb)* aufschlagen, aufziehen *(Häute aus der Brühe ziehen)*; 5. *(Ker)* [an]henkeln, garnieren
handle 1. Handgriff *m*, Griff *m*; 2. *(Text, Gerb, Pap)* Griff *m*, Griffigkeit *f*
h. sticking machine *(Ker)* Garniermaschine *f*

handler *(Gerb)* Lohgrube *f*; *(Ker)* Garnierer *m*
handling 1. Handhabung *f*, Gebrauch *m*, Bedienung *f (eines Werkzeugs)*; 2. Beförderung *f*, Förderung *f*, Transport *m*, Weiterleitung *f (von Stoffen)*; Leitung *f*, Überleitung *f*, Überführung *f (von Abgasen)*; 3. *(Ker)* Henkeln *n*, Anhenkeln *n*, Garnieren *n*
h. vat *(Gerb)* Aufschlagfarbe *f (Gefäß zum Einhängen von Hautblößen)*
handwheel Handrad *n*
hang up / to sich ansetzen, hängenbleiben
hanging s. h. paper
h. base paper Tapetenrohpapier *n*
h.-drop culture Hängetropfenkultur *f*, Deckglaskultur *f*
h. paper Tapetenpapier *n*
hank *(Text)* Strang *m*, Strähn *m*
h.-drying machine *(Text)* Strang[garn]trockenmaschine *f*
h.-dye / to *(Text)* strangfärben
h. dyeing *(Text)* Strangfärben *n*, Färben *n* im Strang (in Strangform)
h.-dyeing machine Strang[garn]färbemaschine *f*
h.-mercerizing machine Strang[garn]merzerisiermaschine *f*
h. reel *(Text)* Stranghaspel *f*
h. reeling *(Text)* Haspeln *n* zum Strang
h. sizing *(Text)* Schlichten *n* im Strang (in Strangform)
h.-sizing machine *(Text)* Strang[garn]schlichtmaschine *f*
h. washer, h.-washing machine *(Text)* Strang-[garn]waschmaschine *f*
hanksite *(Min)* Hanksit *m*
Hansa yellow Hansagelb *n (Azofarbstoffgruppe)*
Hantzsch pyridine synthesis Hantzschsche Pyridinsynthese *f*
haplite *(Geol)* Aplit *m*
hapten Hapten *n*, Halbantigen *n*
haptoglobin Haptoglobin *n (Gruppe von Glykoproteiden des Blutplasmas)*
hard hart *(z.B. Metalle, Wasser)*; hart, durchdringend *(Strahlung)*; sauer, herb, streng *(im Geschmack)*; *(Am)* hochprozentig, stark *(Getränke)*
h. anodizing *(Met)* Hartoxydation *f (Variante der anodischen Oxydation)*
h. asphalt Hartasphalt *m*
h.-bleaching pulp schwer bleichbarer Zellstoff *m*
h. board gehärtete Pappe *f*, Hartpappe *f*; Hartfaserplatte *f*
h.-burned *(Ker)* scharfgebrannt, hochgebrannt
h.-burned brick *(Ker)* Hartbrandstein *m*, scharfgebrannter Stein *m*
h. carbide hartes Karbid (Metallkarbid) *n*, Hartkarbid *n*
h. cheese Hartkäse *m*
h. chrome-plating Hartverchromen *n*, Hartverchromung *f*
h. coal Steinkohle *f*, „Hartkohle" *f*; Anthrazit *m*

h.-coal briquette Steinkohlenbrikett n
h.-coal constituent Steinkohlengefügebestandteil m
h. coke Hartkoks m
h. facing Auftragschweißen n *(einer harten Oberflächenschicht)*; Panzern n
h. fat Hartfett n, gehärtetes (hydriertes) Fett n
h. ferrite *(Ker)* Hartferrit m
h. fibre Hartfaser f
h. fire *(Ker)* hartes (hohes) Feuer n, Scharffeuer n
h.-fired *(Ker)* scharfgebrannt, hochgebrannt
h.-fired brick s. h.-burned brick
h.-formation bit *(Erdöl)* Meißel m für harte Formationen, Meißel m für hartes Gestein (Gebirge)
h. glass Hartglas n
h. lead Hartblei n
h. material Hartstoff m, harter Stoff m
h. metal Hartmetall n, *(i.e.S.)* Sinterhartmetall n, Karbidhartmetall n, Sinterkarbid[metall] n, gesintertes Hartmetall (Karbidhartmetall, Karbid) n
h.-metal alloy Hartmetallegierung f
h. metal carbide hartes Metallkarbid (Karbid) n, Hartkarbid n
h. paper Hartpapier n
h. paraffin Hartparaffin n, hartes Paraffin n
h. paste *(Ker)* Hartmasse f
h.-paste porcelain Hartporzellan n
h. pitch Hartpech n, hartes Pech n
h. plaster Hartgips m
h. porcelain s. h.-paste porcelain
h. processing channel black Hard-Processing-Channel-Ruß m, HPC-Ruß m *(schwer verarbeitbarer Kanalruß)*
h. pulp harter (hartgekochter, wenig aufgeschlossener) Zellstoff m
h. resin hartes Naturharz n, Hartharz n
h. rubber Hartgummi m, Hartkautschuk m, Ebonit n
h. rubber clay harter Kaolin m *(für kautschuktechnische Zwecke)*
h. rubber-covered roll Hartgummiwalze f
h.-rubber mix Hartgummimischung f
h. salt Hartsalz n *(Rohkalisalz mit $MgSO_4$)*
h. settling hartes Absetzen n, Bildung f eines harten (festen) Bodensatzes
h. silk Hartseide f, unentbastete Rohseide f
h.-sized *(Pap)* starkgeleimt, vollgeleimt, mit starker Leimung
h. soap harte (feste) Seife f, Hartseife f
h. solder Hartlot n, Schlaglot n, Strenglot n
h. spirit Hartspiritus m
h.-surfacing Oberflächenhärten n; Auftragschweißen n *(einer harten Oberflächenschicht)*
h. water hartes (kalkhaltiges) Wasser n
h. wax Hartwachs n
h. wheat Hartweizen m, Glasweizen m, Triticum durum Desf.
h. wood Hartholz n

hardboard gehärtete Pappe f, Hartpappe f; Hartfaserplatte f
harden / to härten, hart (härter) machen; hart (fest) werden, erhärten, erstarren, *(manchmal auch)* abbinden
Harden and Young ester Harden-Young-Ester m, Fruktose-1,6-diphosphorsäureester m
Harden-Young fermentation equation Harden-Youngsche Gärungsgleichung (Gleichung) f
hardenability Härtbarkeit f
hardenable härtbar
hardened case gehärtete Randschicht (Randzone, Einsatzschicht) f, Härteschicht f
h. emulsion *(Foto)* gehärtete Emulsion f
h. fat gehärtetes (hydriertes) Fett n, Hartfett n
h. lead Hartblei n
h. oil gehärtetes (hydriertes) Öl (Fett) n, Hartfett n
h. peanut oil gehärtetes Erdnußöl n, Erdnußhartfett n
h. vulcanite Hartgummi m
hardener Härter m, Härtungsmittel n, Härtemittel n, Härtezusatz m; *(manchmal auch)* Härtebad n
h. bath Härtebad n
hardening Härten n, Härtung f, Erhärten n, Erhärtung f, Hartwerden n, Festwerden n, Erstarrung f, Verfestigen n, Verfestigung f; *(manchmal auch)* Abbinden n *(z.B. von Bindemitteln)*; *(Met)* Härten n, Härtung f, Abschrecken n; *(Lebm)* Härten n, Härtung f, Aushärtung f *(z.B. von Fetten)*
h. agent Härter m, Härtungsmittel n, Härtemittel n, Härtezusatz m
h. and tempering Vergüten n, Vergütung f
h. bath Härtebad n
h. constituent Härtebildner m
h. fixer (fixing bath) *(Foto)* härtendes Fixierbad n, Härtefixierbad n
h. flavour Härtungsgeruch m; Härtungsgeschmack m
h. furnace *(Ker)* Härteofen m
h. of fat Fetthärtung f, Fetthydrierung f
h. oil Härteöl n
h.-on *(Ker)* Verglühen n, Verschrühen n
h.-on kiln *(Ker)* Glühofen m
h. process Härtungsverfahren n, *(bei der Fetthärtung auch)* Hydrierverfahren n
h. solution *(Foto)* Härtelösung f
h. vessel Härtungsautoklav m, Hydrierautoklav m *(Fetthärtung)*
hardface / to *(eine harte Oberflächenschicht)* anschweißen; panzern
Hardgrove grindability index Mahlbarkeitsindex m nach Hardgrove
Hardgrove machine Hardgrove-Maschine f, Hardgrove-Mühle f *(zur Bestimmung der Mahlbarkeit)*
Hardgrove method Hardgrove-Verfahren n *(zur Bestimmung der Mahlbarkeit mit Hilfe einer Standard-Kugelringmühle)*

Hardgrove mill s. Hardgrove machine
Hardinge cascade mill Hardinge-Kaskadenmühle f
Hardinge conical [ball] mill s. Hardinge mill
Hardinge mill Hardinge-Mühle f (eine Konusmühle)
hardness Härte f (z.B. der Metalle, des Wassers); (Phys) Härte f, Stärke f (z.B. von Röntgenstrahlen); Herbe f, Herbheit f, Herbigkeit f (z.B. von Getränken)
 h. constituent Härtebildner m
 h. degree Härtegrad m, Härtestufe f
 h. element s. h. constituent
 h. factor Härtefaktor m
 h. meter s. h. tester
 h. of water Wasserhärte f
 h. removal Enthärtung f
 h. tester Härteprüfer m, Härtemesser m, Durometer n, Weichheitsprüfer m, Weichheitsprüfgerät n
 h. testing Härteprüfung f, Härtebestimmung f
hardpan (Bodenchemie) Ortstein m
hardrubber article Hartgummiartikel m
hardwood Laubholz n
 h. groundwood Laubholzschliff m
 h. pulp Laubholzzellstoff m
hare hair Hasenhaar n
Hargreaves-Bird cell Hargreaves-Bird-Zelle f (Chloralkalielektrolyse)
Hargreaves process Hargreaves-Verfahren n (Darstellung von Na_2SO_4 bzw. K_2SO_4 und HCl)
harmal[a] Harmalraute f, Harmelraute f, [Syrische] Steppenraute f, Harmelstaude f, Peganum harmala L.
 h. alkaloid Harmalaalkaloid n
harmaline Harmalin n, Dihydroharmin n
harmel s. harmala
harmine Harmin n, Harmolmethyläther m, Banisterin n, Telepathin n, Yagein n, 1-Methyl-7-methoxy-β-karbolin n
harmless unschädlich
harmlessness Unschädlichkeit f
harmotome, harmotomite (Min) Harmotom m
harobol myrrh (Myrrhenharz von Commiphora myrrha (Nees) Engl.)
harpagoside Harpagosid n (Glykosid aus den Wurzeln von Harpagophytum procumbens DC.)
Harris process Harris-Verfahren n, Harris-Raffinationsverfahren n
Harrison red Helioechtrot n (besonders lichtechter Teerfarbstoff aus 3-Nitro-4-toluidin)
Harristrip process Harristrip-Verfahren n (Wolleentfärbung)
Harrop kiln Harrop-Ofen m (ein Tunnelofen)
harshness Herbe f, Herbheit f, Herbigkeit f
hartshorn Hirschhornsalz n
harvest maturity Pflückreife f, Baumreife f
Harz [fixed-sieve] jig (Aufbereitungstechnik) Harzer Setzmaschine (Kolbensetzmaschine) f
haschisch, hasheesh, hashish Haschisch m(n), Marihuana n (von Cannabis indica Lam.)

hatchet[t]ine, hatchettite (Min) Hatchettin m
hatchettolite (Min) Hatchettolith m (Uran-Pyrochlor)
hatching factor Schlüpffaktor m (das Schlüpfen der Nematoden veranlassende Wurzelausscheidung)
hauchecornite (Min) Hauchecornit m
hauerite (Min) Hauerit m (Mangan(IV)-sulfid)
haul [out] / to (Gerb) aufschlagen, aufziehen (Häute aus der Brühe ziehen)
haul-off rate (Plast) Abzugsgeschwindigkeit f, Abziehgeschwindigkeit f, Abnahmegeschwindigkeit f
 h.-off roll Abzugswalze f
hausmannite (Min) Hausmannit m (Mangan(II,III)-oxid)
Haüy law (Krist) Haüysches Gesetz n, Rationalitätsgesetz n, Gesetz n der rationalen Parameterverhältnisse
hauyne, hauynite (Min) Haüyn m
HAV s. hot-air vulcanizing
hawthorn sap Weißdornsaft m (von Crataegus oxyacantha L.)
hay butter Winterbutter f
hazard of explosion Explosionsgefahr f
hazardous gefährlich
haze [feiner] Nebel m; (Foto) Schleier m; Trübung f (z.B. des Biers)
hazelnut Haselnuß f
 h. butter (oil) Haselnußbutter f, Haselnußöl n
hazy trüb[e], getrübt, matt, glanzlos
Hb s. hemoglobin
HCR s. high-consistency refining
HD oil s. heavy-duty oil
H.D. polythene s. high-density polyethylene
HDS s. hydrodesulphurization
HDS process s. hydrodesulphurization process
H.E. s. high explosive
head 1. Druckhöhe f; 2. (Tech) Haube f, Helm m; 3. Spritzkopf m, Kopf m (einer Spritzmaschine); 4. (Lebm) Schaum m, Schaumkrone f (beim Bier)
 h. fat s. h. oil
 h. flowmeter Wirkdruck-Durchfluß[mengen]messer m, Wirkdruck-Mengenstrommesser m
 h. formation Schaumbildung f, Schaumentwicklung f (beim Bier)
 h. leach (Gerb) (erstes Gefäß einer Extraktionsbatterie)
 h. lime (Gerb) frischer Äscher m
 h. of stock [in the headbox] (Pap) Stauhöhe f, Stau m
 h. of the band (phys Ch) Bandenkopf m, Bandenkante f
 h. oil Kopföl n
 h. pulley Kopfrolle f, Kopftrommel f (beim Bandförderer)
 h. retention Schaumhaltigkeit f, Schaumstabilität f, Schaumdauer f, Stabilität (Haltbarkeit, Beständigkeit) f des Schaumes (beim Bier)
 h.-to-head polymerization Kopf-Kopf-Polymerisation f
 h.-to-head structure Kopf-Kopf-Struktur f, Kopf-Kopf-Verkettung f

h.-to-tail polymerization Kopf-Schwanz-Polymerisation f

h.-to-tail structure Kopf-Schwanz-Struktur f, Kopf-Schwanz-Verkettung f

headbox (Pap) Stoffauflauf[kasten] m, Auflaufkasten m

header Sammelrohr n

heading (Pharm) Hauptname m

heap Haufen m; Meiler[haufen] m

h. fermentation Fermentation f (Rotten n, Rottung f) im Haufen (z.B. von Kakaobohnen)

h. leaching (Met) Haufenlaugung f

heart fat Herzfett n

hearth Herd m, Arbeitsherd m (eines Flammofens); Gestell n (eines Hochofens)

h. furnace Herdofen m

heartwood Kernholz n, Kern m

heat / to 1. erhitzen, erwärmen, heiß machen; [be]heizen; [aus]glühen; 2. sich erhitzen, heiß werden

h. to constant weight bis zur Massekonstanz glühen

h. to redness auf (bis zur) Rotglut erhitzen, rotglühen

h. up aufheizen, hochheizen

heat 1. (Phys) Wärme f (als Energieform); (Phys) (manchmal unkorrekt für) temperature; 2. Hitze f, Wärme f, Glut f (als Empfindung eines Zustandes); 3. Erhitzen n, Erhitzung f, Erwärmen n, Erwärmung f; Heizung f, Beheizung f; 4. Schärfe f (z.B. von Gewürzen)

in intense h. bei hoher Temperatur

h.-absorbing glass Wärmeschutzglas n, wärmeabsorbierendes (Wärmestrahlen absorbierendes, Ultrarot absorbierendes) Glas n

h. ageing thermische Alterung f, Wärmealterung f, Hitzealterung f

h.-altered coal pyrogene Kohle f (Kohle, die durch Einwirkung heißen Gesteins ihr Backvermögen verloren hat)

h. balance Wärmebilanz f, Wärmehaushalt m

h. barrier Hitzebarriere f, Hitzemauer f, Hitzegrenze f

h.-bodied oil durch Hitze eingedicktes Öl n

h.-body / to (Farb) durch Erhitzen (Hitzebehandlung, Hitze) eindicken, thermisch eindicken

h. build-up (Gum) Temperaturentwicklung f, Wärmeentwicklung f (bei dynamischer Beanspruchung)

h. capacity Wärmekapazität f

h. change Wärmeänderung f, (i.e.S.) Wärmetönung f

h. change in chemical reactions Wärmetönung f

h. cleaning thermisches Entschlichten n (von Glasfaserstoffen)

h. coagulation Hitzekoagulation f

h. conduction Wärmeleitung f

h. conductivity Wärmeleitfähigkeit f, Wärmeleitvermögen n, Wärmeleitzahl f

h. conductor Wärmeleiter m

h. content Wärmeinhalt m, Enthalpie f, Gibbssche Wärmefunktion f

h.-content—temperature diagram Wärmeinhalt-Temperatur-Diagramm n, It-Diagramm n, I-t-Diagramm n

h.-curable (Gum) wärmevulkanisierbar, heißvulkanisierbar

h.-cured (Gum) wärmevulkanisiert, heißvulkanisiert

h. damage Hitzeschädigung f

h. deflection temperature (Plast) Formbeständigkeit f in der Wärme

h. denaturation Hitzedenaturierung f

h. displacement Wärmeverschiebung f

h. distortion point (temperature) s. h. deflection temperature

h. distribution Wärmeverteilung f

h. effect Wärmeeffekt m

h. energy Wärmeenergie f

h. exchange Wärmeaustausch m, Wärmeübertragung f, Wärmeübergang m

h. exchange medium s. h.-exchanging medium

h. exchange surface s. h.-exchanging surface

h. exchanger Wärme[aus]tauscher m, Wärmeaustauschapparat m, Wärmeübertrager m; Gegenströmer m (Tieftemperaturtechnik)

h.-exchanging medium Wärme[übertragungs]mittel n, Wärmeübertragungsmedium n, Wärme[über]träger m

h.-exchanging surface Wärmeaustauschfläche f, Wärmeübertragungsfläche f

h. flocculation Hitzeflockung f

h. flow Wärmefluß m

h. function at constant pressure s. h. content

h.-generating wärmeerzeugend

h.-giving wärmeliefernd

h. in the relief (Pap) Kocherabgaswärme f

h. inactivation Hitzeinaktivierung f

h. input Wärmezufuhr f

h.-insulated wärmeisoliert

h.-insulating wärmedämmend

h.-insulating material Wärmeisolierstoff m, Wärmeisoliermittel n, Wärmedämmstoff m, Wärmeschutzstoff m, Wärmeschutzmittel n, Wärmeisolator m

h. insulation Wärmeisolierung f, Wärmeisolation f, Wärmedämmung f, Wärmeschutz m

h. insulator s. h.-insulating material

h. interchange s. h. exchange

h. interchanger s. h. exchanger

h. lamp Wärmestrahler m

h. load Wärmebelastung f

h. loss Wärmeverlust m

h. of activation Aktivierungswärme f

h. of adsorption Adsorptionswärme f

h. of association Assoziationswärme f

h. of combination Verbindungswärme f

h. of combustion Verbrennungswärme f

h. of combustion at constant pressure Verbrennungsenthalpie f

h. of combustion per mole molare Verbrennungswärme *f*
h. of compression Kompressionswärme *f*
h. of condensation Kondensationswärme *f*
h. of crystallization Kristallisationswärme *f*, *(manchmal auch)* Erstarrungswärme *f*
h. of decomposition Zersetzungswärme *f*
h. of dilution Verdünnungswärme *f*
h. of dissociation Dissoziationswärme *f*
h. of evaporation Verdunstungswärme *f*, Verdampfungswärme *f*
h. of explosion Explosionswärme *f*
h. of foods Kalorie[n]wert *m* der Nährstoffe
h. of formation Bildungswärme *f*, *(i.e.S.)* molare Bildungswärme *f*
h. of formation at constant pressure Bildungsenthalpie *f*
h. of formation per mole molare Bildungswärme *f*
h. of fusion Schmelzwärme *f*
h. of hydration Hydratationswärme *f*
h. of hydrogenation Hydrier[ungs]wärme *f*
h. of ionization Ionisationswärme *f*
h. of linkage Verbindungswärme *f*
h. of melting *s.* h. of fusion
h. of mixing (mixture) Mischungswärme *f*
h. of neutralization Neutralisationswärme *f*
h. of neutralization at constant pressure Neutralisationsenthalpie *f*
h. of oxygenation Sauerstoffabsorptionswärme *f*
h. of radioactivity Zerfallswärme *f*
h. of reaction Reaktionswärme *f*
h. of reaction at constant pressure Reaktionsenthalpie *f*
h. of reaction at constant volume Reaktionsenergie *f*
h. of recombination Rekombinationswärme *f*
h. of solidification Erstarrungswärme *f*
h. of solution Lösungswärme *f*, Lösewärme *f*
h. of sublimation Sublimationswärme *f*
h. of swelling *(Koll)* Quellungswärme *f*
h. of transformation (transition) Umwandlungswärme *f*
h. of vaporization Verdampfungswärme *f*, Verdunstungswärme *f*
h. of welding Schweißwärme *f*
h. of wetting Benetzungswärme *f*, Netzwärme *f*
h.-plasticized thermisch · (durch thermischen Abbau) plastiziert
h. polymer Hochtemperaturpolymerisat *n*, Wärmepolymerisat *n*, Hochtemperaturpolymer[es] *n*, Wärmepolymer[es] *n*
h. polymerization Hochtemperaturpolymerisation *f*, Wärmepolymerisation *f*
h. processing *(Chem)* Wärmebearbeitung *f*, Wärmebehandlung *f*, Erhitzungsbehandlung *f*
h.-proof *s.* h.-stable
h. pump Wärmepumpe *f*
h. radiation Wärme[aus]strahlung *f*, thermische Strahlung *f*

h. regenerator Regenerator *m*, regenerativer Vorwärmer *m*
h. release Wärmefreisetzung *f*; Wärmebelastung *f (bei Feuerungen)*
h. requirement[s] Wärmebedarf *m*
h. reservoir Wärmespeicher *m*, Wärmebehälter *m*, Wärmereservoir *n*
h. resistance *s.* h. stability
h.-resistant, h.-resisting *s.* h.-stable
h.-resisting glass hitzebeständiges (feuerfestes) Glas *n*
h. rise Wärmeanstieg *m*, Wärmeentwicklung *f*
h. sealability *(Plast)* Heißsiegelfähigkeit *f*
h. sealable *(Plast)* heißsiegelfähig
h. sealing *(Plast)* Heißsiegeln *n*, Heißkleben *n*, Heißverschweißen *n*
h.-sensitive hitzeempfindlich, wärmeempfindlich
h. sensitivity Hitzeempfindlichkeit *f*, Wärmeempfindlichkeit *f*
h. sensitization Wärmesensibilisierung *f*
h.-sensitized wärmesensibel eingestellt
h. sensitizer Wärmesensibilisierungsmittel *n*
h.-set ink Heißtrockenfarbe *f*
h. setting Thermofixieren *n*, Thermofixierung *f*, Heißfixieren *n*, Heißfixierung *f*
h.-soften / to thermisch erweichen (plastizieren)
h. softening thermische Erweichung (Plastizierung) *f*
h. source Wärmequelle *f*
h. stability Hitzebeständigkeit *f*, Erhitzungsbeständigkeit *f*, Hitzefestigkeit *f*, Hitzeresistenz *f*, Wärmebeständigkeit *f*, Wärmefestigkeit *f*, thermische Beständigkeit (Stabilität, Festigkeit, Widerstandsfähigkeit) *f*, Thermostabilität *f*
h. stabilization Wärmestabilisierung *f*
h. stabilizer Wärmestabilisierungsmittel *n*, Wärmestabilisator *m*
h.-stable hitzebeständig, hitzefest, hitzestabil, hitzeresistent, wärmebeständig, wärmefest, wärmestabil, thermostabil, thermoresistent, thermisch beständig (stabil, widerstandsfähig)
h. sterilization Hitzesterilisation *f*, Hitzesterilisierung *f*
h.-sterilized hitzesterilisiert
h. storage Wärmespeicherung *f*
h. storer Wärmespeicher *m*
h.-strengthened glass *s.* h.-treated glass
h.-thicken / to durch Erhitzen (Hitzebehandlung, Hitze) eindicken, thermisch eindicken
h. tonality Wärmetönung *f*
h.-toughened glass *s.* h.-treated glass
h. transfer Wärmeübertragung *f*, Wärmeaustausch *m*, Wärmeübergang *m*
h.-transfer cement Wärmeübertragungsmasse *f*
h.-transfer coefficient Wärmeübergangszahl *f*
h.-transfer fluid (medium) Wärme[übertragungs]mittel *n*, Wärmeübertragungsmedium *n*, Wärme[über]träger *m*
h.-transfer oil Wärmeübertragungsöl *n*

h.-transfer salt Wärmeübertragungssalz n
h. transmission s. h. transfer
h.-treat / to wärmebehandeln, warmbehandeln, hitzebehandeln
h.-treated wärmebehandelt, warmbehandelt, hitzebehandelt
h.-treated glass vorgespanntes Glas n, *(i.e.S.)* Einscheibensicherheitsglas n, Einschichtensicherheitsglas n
h. treatment Wärmebehandlung f, Warmbehandlung f, Hitzebehandlung f, thermische Behandlung f
h.-treatment furnace Wärmebehandlungsofen m, Warmbehandlungsofen m
h. unit Wärmeeinheit f
h.-up time s. heating-up time
h. value Heizwert m
h. volatilization Verflüchtigung f durch Hitze
h. welding Verschweißen n durch Wärme, Heißverschweißen n
heatable [be]heizbar
heated air Heißluft f; *(Met)* Heißwind m
h.-air welding gun Heißluftschweißpistole f, Heißluftschweißbrenner m
h. bar *(Plast)* Schweißlineal n
h.-bar sealing *(Plast)* Siegeln n mit Schweißlineal (Heizstab)
h. chamber Wärmekammer f
h. cylinder Heizzylinder m
h. tool welding Heizelementschweißen n, HE-Schweißen n, Heizkeilschweißen n
h. wedge *(Plast)* Heizkeil m
h. with gas gasbeheizt, mit Gas geheizt (beheizt, betrieben)
h. with oil ölgefeuert, ölbeheizt, mit Öl geheizt (beheizt, betrieben)
h. zone heiße Zone f
heater Heizelement n, Heizkörper m; Heizeinrichtung f, Heizvorrichtung f, Heizapparat m, Beheizungsapparat m, Heizgerät n, Heizer m, Erhitzer m; *(Gum)* Heizer m, Heizapparat m, Vulkanisierapparat m, Vulkanisationsapparat m, Vulkanisiergerät n, Vulkanisator m
h. band Heizband n, Bandheizkörper m
h. process *(Gum)* Heater-Prozeß m, Heater-Verfahren n *(Regenerierverfahren)*
heather honey Heide[kraut]honig m
h. peat Heidetorf m
heating Erhitzen n, Erhitzung f, Erwärmen n, Erwärmung f; Aufheizen n, Aufheizung f, Anwärmen n, Anwärmung f; Heizen n, Heizung f, Beheizen n, Beheizung f
h. band s. h. tape
h. bath Heizbad n, Hitzebad n, Wärmebad n
h. blanket Heizgeflecht n, Heizmantel m
h. chamber Heizraum m, Heizkammer f; Trocknungsrohr n *(einer Trockenpistole)*
h. coil Heizschlange f; Heizspule f; Heizspirale f
h. cord Heizschnur f
h. cycle Heizzyklus m, Erhitzungszyklus m

h. cylinder Heizzylinder m
h. element Heizkörper m, Heizelement n; Heizkammer f *(am Verdampfer)*
h. flue Heizzug m
h. funnel Heißwassertrichter m, Warmwassertrichter m
h. furnace Glühofen m; Warmhalteofen m, Wärmeofen m
h. gas Heizgas n, Beheizungsgas n
h. jacket Heizmantel m
h. load Heizgut n
h. mantle Heizmantel m, Flächenheizkörper m
h. medium Heizmittel n; s. heat-exchanging medium
h. oil Heizöl n
h. plate[n] Heizplatte f
h. rate Aufheiz[ungs]geschwindigkeit f, Erwärmungsgeschwindigkeit f, Erhitzungsgeschwindigkeit f
h. resistor Heizleiter m, Heizwiderstand m
h. section Aufheizsektion f, Anwärmsektion f, Erhitzungsabteilung f
h. steam Heizdampf m
h. steam inlet Heizdampfeintritt m
h. surface Heizfläche f
h. tape Heizband n
h. tube Heizrohr n
h. unit Heizelement n, Heizkörper m
h.-up period (time) Anheizzeit f, Aufheizzeit f, Anwärmzeit f
h. value Heizwert m
h. wall Heizwand f
h. zone Heizzone f, Wärmzone f
Heatley unit Heatley-Einheit f, Oxford-Einheit f *(veraltete biologische Maßeinheit des Penizillins)*
heatproof s. heat-stable
heavier-than-heavy component am schwersten siedender Stoff m *(eines Mehrstoffgemisches)*
heaviest crown glass Schwerstkron[glas] n
heavily filled *(Pap)* hochgefüllt
h. insulated stark isoliert
h. loaded stark belastet, hochbelastet; *(Gum)* stark gefüllt, hochgefüllt
heavy 1. schwer; 2. schwerflüssig, zäh[flüssig]; steif *(z.B. Paste)*, hart, fest *(z.B. Teig)*; 3. hochsiedend, höhersiedend, schwer[er] flüchtig
h. alloy Schwerlegierung f, Schwermetallegierung f
h. benzine s. h. gasoline
h. benzol Schwerbenzol n
h. ceramics Grobkeramik f
h. chemicals Schwerchemikalien fpl *(in großem Umfang hergestellte Säuren, Salze und Alkalien)*
h.-clay industry grobkeramische Industrie f, Grobkeramikindustrie f
h.-clay product (ware) grobkeramisches Erzeugnis n, Grobkeramik f
h. concrete Schwerbeton m
h. crown glass Schwerkron[glas] n
h.-duty detergent Grobwaschmittel n, Vollwaschmittel n; Starkreiniger m

h.-duty extruder Extruder *m* für schwere Beanspruchung, Hochleistungsextruder *m*
h.-duty oil Heavy-Duty-Öl *n*, HD-Öl *n*, Hochleistungsöl *n* *(Motorenöl mit Zusatzstoffen, die schwere Betriebsbelastungen zulassen)*
h. fraction schwere (hochsiedende, schwerer flüchtige) Fraktion *f*; *s. h.* material
h. gasoline Schwerbenzin *n*, schweres Benzin *n*
h. hydrogen $_1^2$H *oder* D schwerer Wasserstoff *m*, Deuterium *n*
h. isotope schweres Isotop *n*
h. key höhersiedende Schlüsselkomponente *f* *(bei der Destillation)*
h. liquor filter *(Zucker)* Dicksaftfilter *n*
h. mash Dickmaische *f*
h. material Schwergut *n*, Sinkgut *n*, schweres Gut (Material) *n*
h.-media ... *s.* h.-medium ...
h. medium schwere Flüssigkeit *f*, Schwerflüssigkeit *f*, Trennflüssigkeit *f*, Trennmedium *n*, *(i.e.S.)* schwere Trübe *f*, Schwertrübe *f* *(unechte Schwerflüssigkeit)*
h.-medium cleaning *s.* h.-medium separation
h.-medium cyclone Schwertrübe-Waschzyklon *m*
h.-medium separation Schwerflüssigkeitsaufbereitung *f*, Schwerflüssigkeitssortieren *n*, Schwimm[-und]-Sink-Aufbereitung *f*, Sinkscheideverfahren *n*, *(i.e.S.)* Trennung *f* in Schwereträüben, Schwertrübeaufbereitung *f*, Schwertrübescheidung *f*
h.-medium separation plant Schwerflüssigkeits-[aufbereitungs]anlage *f*, Sinkscheideranlage *f*
h. metal Schwermetall *n*
h. metal acetylide Schwermetallazetylid *n*
h. metal alloy Schwermetallegierung *f*, Schwerlegierung *f*
h. metal salt Schwermetallsalz *n*
h. mineral schweres Mineral *n*, Schwermineral *n*
h. naphtha Schwerbenzin *n*, schweres Benzin *n*, Naphtha *n(f)*
h. oil Schweröl *n*, schweres Öl *n*
h. paste harter (fester, zäher) Teig *m*
h. phase schwere[re] Komponente *f*
h. spar *(Min)* Schwerspat *m*, Baryt *m* *(Bariumsulfat)*
h.-wall dickwandig
h. water D$_2$O schweres Wasser *n*, Deuteriumoxid *n*
hebronite *(Min)* Hebronit *m* *(Amblygonit von Hebron)*
hectographic ink Hektografentinte *f*
hectorite *(Min)* Hectorit *m*
hedenbergite *(Min)* Hedenbergit *m* *(Kalziumeisen-(II)-disilikat)*
hederin Hederin *n* *(ein Saponin)*
hedyphane, hedyphanite *(Min)* Hedyphan *m*, Kalziumbarium-Mimetesit *m*
hefting Triebkraft *f* *(z.B. der Hefe)*
Hehner formaldehyde in milk test Nachweis *m* von

Formaldehyd in Milch nach Hehner, Reaktion *f* von Hehner
height Höhe *f*
h. equivalent to a theoretical plate äquivalente Füllkörperhöhe *f*, Trennstufenhöhe *f*, HETP-Wert *m*
h. of one transfer unit *(Destillation)* Höhe *f* einer Übertragungseinheit, HTU
h. of packing Füllkörperschichthöhe *f*
Heisenberg indeterminacy principle Heisenbergsche Unbestimmtheitsrelation (Ungenauigkeitsrelation, Unschärferelation, Ungenauigkeitsbeziehung, Unschärfebeziehung) *f*
Heisenberg representation Heisenberg-Darstellung *f*
Heisenberg uncertainty principle *s.* Heisenberg indeterminacy principle
Heitler-London method *(Quantenchemie)* Heitler-London-Methode *f*, Heitler-London-Verfahren *n*
Heitler-London-Slater-Pauling method Heitler-London-Slater-Pauling-Methode *f*, HLSP-Methode *f*, Valenzbindungsmethode *f*, Valenzstrukturmethode *f*, VB-Methode *f*, Spinmethode *f*, Elektronenpaarmethode *f*
Heitler-London-Slater-Pauling theory Heitler-London-Slater-Pauling-Theorie *f*, HLSP-Theorie *f*, Valenzbindungstheorie *f*, Valenzstrukturtheorie *f*, VB-Theorie *f*, Theorie (Bindungstheorie) *f* der Valenzstrukturen (Elektronenpaarbindungen)
Heitler-London theory Heitler-London-Theorie *f*, Heitler-Londonsche Theorie *f*
helenin Helenin *n*, Alantkampfer *m*, Inulakampfer *m*
helianthin[e] 1. $(CH_3)_2$ $NC_6H_4N=NC_6H_4SO_3H$ Helianthin *n*, 4-Dimethylaminoazobenzol-4'-sulfonsäure *f*; 2. Methylorange *n*, Helianthin *n*, Orange III *n* *(Natriumsalz von 1.)*
helical coil Zylinder[rohr]schlange *f*
h.-conveyor centrifuge Schnecken[austrag]-zentrifuge *f*, Schneckenschleuder *f*
h. structure Helixstruktur *f* *(schraubenlinienartiger Aufbau von Eiweißmolekülen)*
helices Wendeln *fpl* *(Füllkörper)*
helicoidal structure Helixstruktur *f* *(der Eiweißkörper)*
heliotrope *(Min)* Heliotrop *m*, Blutjaspis *m* *(Silizium(IV)-oxid)*
heliotropin $CH_2(O_2)C_6H_3CHO$ Heliotropin *n*, Piperonal *n*
helium He Helium *n*
helium I He I Helium I *n* *(flüssige Modifikation des He)*
helium II He II Helium II *n* *(flüssige Modifikation des He)*
h. liquefaction Heliumverflüssigung *f*
h. nucleus Heliumkern *m*, α-Teilchen *n*
helix Wendel *f* *(Füllkörper)*
h. agitator Mischschnecke *f*, Schneckenrührer *m*; Schneckenkneter *m*, Schraubenkneter *m*
h. angle Gangsteigungswinkel *m*, Steigungswinkel *m* *(z.B. der Extruderschnecke)*

h. conveyor Förderschnecke f, Schneckenförderer m, Transportschnecke f
h. structure s. helicoidal structure
helixin 1. Endomyzin n, Helixin n *(eine Gruppe von Antibiotika)*; 2. Helixin n *(veraltet für ein Glykosidgemisch mit Hederagenin als Hauptbestandteil)*
Hell-Volhard-Zelinsky reaction Hell-Volhard-Zelinsky-Reaktion f *(α-Halogenierung aliphatischer Karbonsäuren)*
Heller layer test Hellersche Schichtprobe f *(Eiweißnachweis im Harn)*
Heller process Wassergas[erzeugungs]verfahren n von Heller
helminthicide Helminthizid n, wurmtötendes Mittel n, Wurmmittel n
helvin[e], helvite *(Min)* Helvin m
helvolic acid Helvolinsäure f, Fumigazin n
hem s. heme
hematic acid Hämatinsäure f
hematite *(Min)* Hämatit m, Haematit m *(Eisen(III)-oxid)*
h. [pig] iron Hämatit[roh]eisen n
heme Häm n
h. protein Hämprotein n
hemellitene $C_6H_3(CH_3)_3$ Hemellitol n, Hemimellitol n, 1,2,3-Trimethylbenzol n
hemellitic acid $(CH_3)_2C_6H_3COOH$ Hemellitsäure f, 2,3-Xylylsäure f, 2,3-Dimethylbenzoesäure f
hemiacetal Halbazetal n, Hemiazetal n
hemialdol Hemialdol n
hemicellulose Hemizellulose f
hemichalcite *(Min)* Hemichalzit m, *(veraltet für)* Emplektit m
hemicolloid Hemikolloid n
hemicrystalline halbkristallin[isch]
hemiformal $HOCH_2OR$ Halbformal n *(Halbazetal des Formaldehyds)*
hemihedral *(Krist)* hemiedrisch, halbflächig
hemihydrate Halbhydrat n
hemimellitene s. hemellitene
hemimellitic acid $C_6H_3(COOH)_3$ Hemimellit[h]säure f, 1,2,3-Benzoltrikarbonsäure f
hemimorphite *(Min)* Hemimorphit m, Kieselzinkerz n, Kalamin m *(Zinkdihydroxypyrosilikat)*
hemin Hämin n, *(i.e.S.)* Hämin[chlorid] n, Chlorhämin n, Protohämin n
hemip[in]ic acid $(CH_3O)_2C_6H_2(COOH)_2$ Hemipinsäure f, 3,4-Dimethoxyphthalsäure f, 3,4-Dimethoxy-1,2-benzoldikarbonsäure f
hemiterpene C_5H_8 Hemiterpen n, *(i.e.S.)* Isopren n
hemitoxiferine Hemitoxiferin n
hemlock 1. Schierling m, Conium L.; s. poison hemlock; 2. Hemlockstanne f, Schierlingstanne f, *(Gattung)* Tsuga Carr., *(i.e.S.)* Tsuga canadensis (L.)Carr.
h. extract *(Gerb)* Extrakt m(n) von Hemlockstannenrinde
h. pitch Kanadisches Pech n *(meist von Tsuga canadensis (L.)Carr.)*

hemochrome, hemochromogen Hämochromogen n
hemocyanin Hämozyanin n
hemoglobin Hämoglobin n, Hb, roter Blutfarbstoff m
hemolymph Hämolymphe f
hemolysin Hämolysin n
hemolysis Hämolyse f
hemolyze / to hämolysieren
hemp Hanf m, Cannabis sativa L.
h. fibre Hanffaser f
h.-mallow fibre Kenaffaser f, Hanfeibischfaser f *(Faser von Hibiscus cannabinus L.)*
h. oil Hanf[samen]öl n
h. paper Hanfpapier n
Hempel gas burette Hempelsche Gasbürette f, Hempel-Bürette f
Hempel gas pipette Hempelsche Gaspipette f, Hempel-Pipette f
hempseed oil s. hemp oil
henbane Bilsenkraut n, Hyoscyamus niger L.
hendecanal $CH_3(CH_2)_9CHO$ Hendekanal n, Undekanal n, Undezylaldehyd m
9-hendecenoic acid $CH_3CH=CH(CH_2)_7COOH$ 9-Hendezensäure f, 9-Undezensäure f, 9-Undezylensäure f
Henderson stove Schwitzapparat (Zylinderzellenapparat) m von Henderson *(zur Reinigung von Rohparaffin)*
heneicosane $CH_3(CH_2)_{19}CH_3$ Heneikosan n
henna Henna f, Hennastrauch m, Lawsonia inermis L.; Henna f *(gepulverte Blätter von Lawsonia inermis L.)*
Henry's law Henrysches Gesetz (Absorptionsgesetz) n
hentriacontane $CH_3(CH_2)_{29}CH_3$ Hentriakontan n
hepar sulphuris Hepar sulfuris n, Schwefelleber f *(technisches Kaliumsulfid)*
h. test Heparprobe f *(zum Schwefelnachweis)*
heparin Heparin n
hepatic starch Leberstärke f, Glykogen n, tierische Stärke f, Tierstärke f
hepatocupreine Hepatokuprein n *(Kupferproteid)*
heptachlor Heptachlor n, Heptachlorendomethylentetrahydroinden n *(Insektizid)*
n-heptadecanoic acid $CH_3(CH_2)_{15}COOH$ n-Heptadekansäure f, Margarinsäure f
heptafluoride Heptafluorid n
heptamethylene C_7H_{14} Heptamethylen n, Zykloheptan n, Suberan n
heptamolybdate $M^I_6Mo_7O_{24}$ Heptamolybdat n
heptane Heptan n, *(i.e.S.)* s. n-heptane
n-heptane C_7H_{16} n-Heptan n, Normalheptan n, Heptan n
1,7-h. dicarboxylic acid $HOOC(CH_2)_7COOH$ 1,7-Heptandikarbonsäure f, Nonandisäure f, Azelainsäure f, Lepargylsäure f
heptanedioic acid $HOOC(CH_2)_5COOH$ Heptandisäure f, 1,5-Pentandikarbonsäure f, Pimelinsäure f

heptanoic acid $CH_3(CH_2)_5COOH$ n-Heptylsäure f, Önanthsäure f

1-heptanol s. n-heptyl alcohol

heptasulphide Heptasulfid n

heptavalent siebenwertig, heptavalent

1-heptine s. 1-heptyne

heptoic acid s. heptanoic acid

heptose Heptose f (Monosaccharid mit sieben Sauerstoffatomen)

heptoxide Heptoxid n

n-heptyl acetylene $CH=C(CH_2)_6CH_3$ n-Heptylazetylen n, 1-Nonin n

n-h. alcohol n-Heptylalkohol m, n-Heptanol n, Önanthalkohol m

h. carbinol $CH_3(CH_2)_7OH$ Heptylkarbinol n, n-Oktylalkohol m, 1-Oktanol n, n-Kaprylalkohol m

h. phenol Heptylphenol n

heptylic acid s. heptanoic acid

n-heptylpenicillin n-Heptylpenizillin n, Penizillin K n, Penizillin IV n

1-heptyne $HC = C(CH_2)_4CH_3$ 1-Heptin n, n-Amyläthin n, n-Amylazetylen n, Önanthyliden n

herbarium paper Herbariumpapier n, Pflanzenpreßpapier n

herbicidal herbizid, Herbizid...

h. activity herbizide Wirkung f, Herbizidwirkung f

h. dust herbizides Stäubemittel n, staubförmiges Herbizid n

herbicide Herbizid n, Unkrautbekämpfungsmittel n

hercinite (fälschlich für) hercynite

hercynite (Min) Hercynit m, Ferrospinell m (Eisen(II)-aluminat)

herd [bulk] milk Herdenmilch f

herderite (Min) Herderit m

hermetic[al] hermetisch [dicht], hermetisch (gas- und wasserdicht, luft- und feuchtigkeitsdicht) abgeschlossen (verschlossen)

heroin Heroin n, Diazetylmorphin n, Diamorphin n (Rauschgift)

Héroult furnace Héroult-Ofen m, Héroult-Lichtbogenofen m

Herreshoff furnace Herreshoff-Ofen m, Mehretagenröstofen m nach Herreshoff

herring oil Heringsöl n, Heringstran m

herringbone coil grätenförmige Rohrschlange f

Herschel effect (Foto) Herschel-Effekt m

Herz compound Herz-Verbindung f, Herzsche Verbindung (Zwischenverbindung) f

Herz reaction Herzsche Reaktion (Chlorschwefelreaktion) f

herzenbergite (Min) Herzenbergit m

hesperetic $HO(CH_3O)C_6H_3CH=CHCOOH$ Hesperetinsäure f, Kaffeesäure-4-methyläther m, 3-(3-Hydroxy-4-methoxyphenyl)propensäure f

hesperetin Hesperetin n (Flavonfarbstoff)

hesperidene Hesperiden n, d-Limonen n, Karven n, Zitren n, d-1-Methyl-4-isopropylzyklohexadien-(1,8) n

hesperidin Hesperidin n

hesperitin s. hesperetin

hessite (Min) Hessit m (Silbertellurid)

hessonite (Min) Hessonit m

Hess's law of heat summation Heßscher Satz m, Heßsches Gesetz n, Gesetz n der konstanten Wärmesummen

hetaerolite (Min) Hetaerolith m

heteroatom Heteroatom n

heteroauxin (Bioch) Heteroauxin n, β-Indolylessigsäure f

heteroblastic (Geol) heteroblastisch

heterocycles s. heterocyclic compounds

heterocyclic heterozyklisch

h. atom s. heteroatom

h. chemistry Chemie f der Heterozyklen (heterozyklischen Verbindungen)

h. compounds Heterozyklen pl, heterozyklische Verbindungen fpl

h. system heterozyklisches System n

heterodyne beat method Überlagerungsmethode f (Verfahren in der Dekametrie)

heterofermentative heterofermentativ

heterogeneous heterogen, inhomogen, verschiedenartig, ungleichartig zusammengesetzt, innerlich uneinheitlich

h. catalysis heterogene Katalyse f, Kontaktkatalyse f, Oberflächenkatalyse f

h. polymer s. heteropolymer

h. reaction heterogene Reaktion f

h. reactor (Kph) heterogener Reaktor m

h. system heterogenes (inhomogenes) System n

heterolite s. hetaerolite

heterolysis Heterolyse f (Dissoziation eines Moleküls in entgegengesetzt geladene Ionen)

heterolytic heterolytisch (in entgegengesetzt geladene Ionen dissoziierend)

h. reaction heterolytische Reaktion f

heterometry Heterometrie f (nephelometrische Titration)

heteromorphic rocks heteromorphe Gesteine npl

heteromorphism (Geol) Heteromorphismus m, Heteromorphie f

heteropolar heteropolar

h. bond heteropolare (polare, elektrostatische, elektrovalente, ionogene) Bindung f, Elektrovalenz f, Ionenbeziehung f, Ionenbindung f

h. colloid heteropolares Kolloid n

h. compound [hetero]polare Verbindung f, Ionenverbindung f

h. linkage s. h. bond

heteropoly acid Heteropolysäure f

heteropolymer Heteropolymer[es] n

heteropolymeric heteropolymer

heteropolymerization Heteropolymerisation f

heterotrophic (Bio) heterotroph (sich von organischen Stoffen ernährend)

hetisine Hetisin n

HETP s. 1. height equivalent to a theoretical plate; 2. hexaethyltetraphosphate

H.E.T.P. s. height equivalent to a theoretical plate

heulandite *(Min)* Heulandit *m (Kalziumdialumo-hexasilikat)*

Heumann method Heumannsches Verfahren *n*, Heumannsche Methode *f*

Hevea latex Hevea-Latex *m*

H. rubber Hevea-Kautschuk *m*

H. tree Heveabaum *m*, Hevea *f*, Hevea brasiliensis (H.B.K.)Muell. Arg.

Heveacrumb Heveacrumb *m (Krümelkautschuk)*

hewettite *(Min)* Hewettit *m*

hex. *s.* hexagonal

hexa *s.* hexamethylenetetramine

hexaborane B_6H_{10} oder B_6H_{12} Hexaboran *n*

hexaboride Hexaborid *n*

hexabromide Hexabromid *n*

hexabromodisilicoethane Si_2Br_6 Siliziumhexabromid *n*, Disiliziumhexabromid *n*, Hexabromdisilan *n*

hexacarbonyl Hexakarbonyl *n*

hexachloride Hexachlorid *n*

hexachlorobenzene C_6Cl_6 Hexachlorbenzol *n*, Perchlorbenzol *n*

hexachlorobutadiene Hexachlorbutadien *n*

hexachlorodisilicoethane Si_2Cl_6 Siliziumhexachlorid *n*, Disiliziumhexachlorid *n*, Hexachlordisilan *n*

hexachloroethane C_2Cl_6 Hexachloräthan *n*, Perchloräthan *n*

hexachloroosmate $M^I_3[OsCl_6]$ Hexachloroosmat(III) *n*; $M^I_2[OsCl_6]$ Hexachloroosmat(IV) *n*

hexachloropalladate $M^I_2[PdCl_6]$ Hexachloropalladat(IV) *n*

hexachlorophene Hexachlorophen *n*, 2,2'-Dihydroxy-3,5,6,3',5',6'-hexachlorodiphenylmethan *n*

hexachloroplatinate $M^I_2[PtCl_6]$ Hexachloroplatinat(IV) *n*

hexachlororuthenate $M^I_2[RuCl_6]$ Hexachlororuthenat(IV) *n*

hexacoordinated sechsfach koordiniert

n-hexacosanoic acid $C_{25}H_{51}COOH$ *n*-Hexakosansäure *f*, Zerotinsäure *f*

hexacovalent koordinativ sechswertig

hexacyanoferrate $M^I_4[Fe(CN)_6]$ Hexazyanoferrat(II) *n*; $M^I_3[Fe(CN)_6]$ Hexazyanoferrat(III) *n*

hexad *(Krist)* sechszählig, 6zählig

hexadecane $CH_3(CH_2)_{14}CH_3$ Hexadekan *n*, Zetan *n*

hexadecanoic acid *s.* *n*-hexadecylic acid

hexadeca-6,10,14-trienoic acid 6,10,14-Hexadekatriensäure *f*

n-hexadecyl mercaptan *n*-Hexadezylmerkaptan *n*, 1-Hexadekanthiol *n*, Zetylmerkaptan *n*

n-hexadecylic acid $CH_3(CH_2)_{14}COOH$ *n*-Hexadezylsäure *f*, Hexadekansäure *f*, Zetylsäure *f*, Palmitinsäure *f*

1,5-hexadiene $CH_2=CHCH_2CH_2CH=CH_2$ 1,5-Hexadien *n*, Diallyl *n*, Biallyl *n*

2,4-hexadienoic acid $CH_3CH=CHCH=CHCOOH$ 2,4-Hexadiensäure *f*, Sorbinsäure *f*

hexaethyltetraphosphate $(C_2H_5)_6P_4O_{13}$ Hexaäthyltetraphosphat *n*

hexafluoride Hexafluorid *n*

hexafluoroferrate(III) $M^I_3[FeF_6]$ Hexafluoroferrat(III) *n*

hexafluorophosphate $M^I[PF_6]$ Hexafluorophosphat *n*

hexafluorophosphoric acid $H[PF_6]$ Hexafluorophosphorsäure *f*

hexagonal *(Krist)* hexagonal, hex., sechseckig; *(bei Symmetrieachsen)* sechszählig, 6zählig

h. close (closest) packing *(Krist)* hexagonal[e] dichteste Kugelpackung (Packung) *f*

h. lattice *(Krist)* hexagonales Gitter *n*

hexahedrite Hexaedrit *m (Ni-haltiges Meteoreisen)*

hexahydrate Hexahydrat *n*

hexahydride Hexahydrid *n*

hexahydrobenzene C_6H_{12} Hexahydrobenzol *n*, Hexamethylen *n*, Zyklohexan *n*

hexahydromesitylene $C_6H_9(CH_3)_3$ Hexahydromesitylen *n*, 1,3,5-Trimethylzyklohexan *n*

hexahydrophenol $C_6H_{11}OH$ Hexahydrophenol *n*, Zyklohexanol *n*

hexahydrophthalic acid $C_6H_{10}(COOH)_2$ Hexahydrophthalsäure *f*, 1,2-Zyklohexandikarbonsäure *f*

hexahydropyridine Hexahydropyridin *n*, Pentamethylenimin *n*, Piperidin *n*

hexahydrotoluene $C_6H_{11}CH_3$ Hexahydrotoluol *n*, Methylzyklohexan *n*

hexahydroxoantimonate $M^I[Sb(OH)_6]$ Hexahydroxoantimonat(V) *n*

hexahydroxostannate $M^I_2[Sn(OH)_6]$ Hexahydroxostannat(IV) *n*

hexahydroxyanthraquinone Hexahydroxyanthrachinon *n*

hexahydroxycyclohexane $C_6H_6(OH)_6$ Hexahydroxyzyklohexan *n*, Inositol *n*, Inosit *m*

hexaiodide Hexajodid *n*

hexamethylene C_6H_{12} Hexamethylen *n*, Hexahydrobenzol *n*, Zyklohexan *n*

h. diamine $H_2N(CH_2)_6NH_2$ Hexamethylendiamin *n*, 1,6-Diaminohexan *n*

h. diammonium adipate adipinsaures Hexamethylendiamin *n*, AH-Salz *n*

h. di-isocyanate Hexamethylendiisozyanat *n*

hexamethylenetetramine $(CH_2)_6N_4$ Hexamethylentetramin *n*, Hexamin *n*, Methenamin *n*

hexamine *s.* 1. hexamethylenetetramine; 2. hexanitrodiphenylamine

hexammine cobalt chloride $[Co(NH_3)_6]Cl_3$ Hexamminkobalt(III)-chlorid *n*, *(veraltet)* Luteokobaltchlorid *n*

h. platinic chloride $[Pt(NH_3)_6]Cl_4$ Hexamminplatin(IV)-chlorid *n*

hexane Hexan *n*, *(i.e.S.) s.* *n*-hexane

n-hexane *n*-Hexan *n*, Normalhexan *n*, Hexan *n*

hexanedioic acid $HOOC(CH_2)_4COOH$ Hexandisäure *f*, Butandikarbonsäure *f*, Adipinsäure *f*

hexanetriol ester of fatty acids Fettsäurehexantriolester *m*

hexanitride Hexanitrid *n*

hexanitrocobaltate $M^I_4[Co(NO_2)_6]$ Hexanitrokobaltat(II) n; $M^I_3[Co(NO_2)_6]$ Hexanitrokobaltat(III) n
hexanitrodiphenylamine 2,4,6,2',4',6'-Hexanitrodiphenylamin n, Hexanitrodiphenylamin n, Hexamin n, Dipikrylamin n
hexanoic acid $CH_3(CH_2)_4COOH$ n-Hexansäure f, Hexylsäure f, n-Kapronsäure f
1-hexanol $C_6H_{13}OH$ 1-Hexanol n, n-Hexanol n, n-Hexylalkohol m, n-Kapronalkohol m, Amylkarbinol n
hexasilane Si_6H_{14} Hexasilan n
hexasilicate $M^I_{12}Si_6O_{18}$ Hexasilikat n
hexasulphide Hexasulfid n
hexatantalate $M^I_8Ta_6O_{19}$ Hexatantalat n
hexathionate $M^I_2[S_6O_6]$ Hexathionat n
hexatomic sechsatomig
hexatungstate Hexawolframat n
hexavalent sechswertig, hexavalent
2-hexene C_6H_{12} Hexen-(2) n, Hexylen-(2) n
1-hexine s. 1-hexyne
2-hexine s. 2-hexyne
3-hexine s. 3-hexyne
hexite s. hexanitrodiphenylamine
hexitol Hexit m, Hexanhexol n (sechswertiger Alkohol)
n-hexoic acid s. hexanoic acid
hexokinase Hexokinase f (Ferment)
hexon[e] base Hexonbase f
hexosan $(C_6H_{10}O_5)_n$ Hexosan n
hexose Hexose f (Monosaccharid mit 6 Sauerstoffatomen)
hexyl acetic acid $CH_3(CH_2)_6COOH$ Hexylessigsäure f, n-Oktansäure f, n-Kaprylsäure f, Heptankarbonsäure-(1) f
n-h. acetylene $HC≡C(CH_2)_5CH_3$ n-Hexylazetylen n, Kapryliden n, 1-Oktin n
n-h. alcohol $C_6H_{13}OH$ n-Hexylalkohol m, Hexanol-(1) n, n-Hexanol n, n-Kapronalkohol m
n-h. chloride $CH_3(CH_2)_5Cl$ n-Hexylchlorid n, 1-Chlorhexan n
hexylic acid s. hexanoic acid
1-hexyne $HC≡C(CH_2)_3CH_3$ Hexin-(1) n, Butylazetylen n
2-hexyne $CH_3C≡C(CH_2)_2CH_3$ Hexin-(2) n, Methylpropylazetylen n
3-hexyne $CH_3CH_2C≡CCH_2CH_3$ Hexin-(3) n, Diäthylazetylen n
h. f., HF s. high frequency
HF alkylation HF-Alkylierung f, Fluorwasserstoffalkylierung f, Flußsäurealkylierung f, Alkylierung f mit Fluorwasserstoff (Fluorwasserstoffsäure, Flußsäure)
HF process s. hydrofluoric-acid process
hi-bulk yarn s. high-bulk yarn
HI press s. high-intensity press
hibiscus fibre Kenaffaser f, Hanfeibischfaser f (von Hibiscus cannabinus L.)
hidden maximum (Krist) verdecktes Maximum n
h. maximum system (Krist) Verbindung f mit verdecktem Maximum

hiddenite (Min) Hiddenit m (Lithiumaluminiumdisilikat)
hide glue Hautleim m, Lederleim m
h. powder (Gerb) Hautpulver n
h. scrapings (shavings) (Gerb) Leimleder n
hiding power Deckvermögen n, Deckfähigkeit f, Deckkraft f
hielmite s. hjelmite
high abrasion furnace black (Gum) High-Abrasion-Furnace-Ruß m, HAF-Ruß m (hochabriebfester Ölruß)
h.-acid food (good) (Lebm) (Konserve mit einem pH-Wert kleiner als 4,5)
h.-activity waste (Kph) hochaktiver (heißer) Abfall m
h.-alloy steel hochlegierter Stahl m
h. alpha pulp Edelzellstoff m
h.-alumina body (Ker) Masse f mit hohem Aluminiumoxidgehalt (Tonerdegehalt)
h.-alumina cement Tonerde[schmelz]zement m
h.-alumina clay tonerdereicher (aluminiumoxidreicher, hoch aluminiumhaltiger) Ton m
h.-alumina porcelain Porzellan n mit hohem Aluminiumoxidgehalt (Tonerdegehalt)
h.-alumina refractory hochtonerdehaltiges feuerfestes Erzeugnis n
h.-analysis hochprozentig
h.-analysis fertilizer (Am) (Düngemittel mit 20 bis 30% Nährstoffgehalt)
h.-ash hochaschehaltig, stark aschehaltig, asche[n]reich, mit hohem Aschegehalt
h.-beryllia body (Ker) Masse f mit hohem Gehalt an Berylliumoxid
h.-boiled pulp (Pap) weicher (weichgekochter, weit heruntergekochter) Zellstoff m
h. boiler Hochsieder m, hochsiedendes Lösungsmittel n
h.-boiling hochsiedend, höhersiedend, schwer[er]siedend
h.-boiling component hochsiedende (höhersiedende, schwer[er]siedende) Komponente f, Schwer[er]siedendes n
h.-brightness pulp Zellstoff m mit hohem Weißgehalt
h.-bulk yarn Hochbauschgarn n, hochvoluminöses Garn n
h.-caking coal starkbackende Kohle f
h.-capacity centrifuge Großraumzentrifuge f, Großleistungszentrifuge f
h.-carbon hochgekohlt, hochkohlenstoffhaltig, kohlenstoffreich, mit hohem Kohlenstoffgehalt (C-Gehalt) (z.B. Stahl)
h.-chalcocite (Min) Hoch-Chalkosin m, Hochkupferglanz m, Chalkosin(-H) m (Kupfer(I)-sulfid)
h.-chlorinating method Hochchlor[ier]ungsverfahren n
h. chlorination Hochchlor[ier]en n, Hochchlor[ier]ung f, Über[schuß]chlor[ier]ung f
h.-class table margarine Delikateßmargarine f
h.-concentration hochkonzentriert

h.-consistency refining *(Pap)* Dickstoffmahlung *f*, HCR-Mahlung *f*

h.-contrast *(Foto)* kontrastreich

h.-contrast developer *(Foto)* Kontrastentwickler *m*, konstrastreich (hart) arbeitender Entwickler *m*

h.-contrast emulsion *(Foto)* kontrastreich (hart) arbeitende Emulsion *f*

h.-density bleacher *(Pap)* Dickstoffbleichturm *m*

h.-density bleaching *(Pap)* Dickstoffbleiche *f*

h.-density hypochlorite stage *(Pap)* Hochchloritbehandlung (Hypochloritbleiche) *f* bei hoher Stoffdichte

h.-density peroxide bleaching *(Pap)* Peroxidbleiche *f* bei hoher Stoffdichte

h.-density polyethylene Polyäthylen *n* mit hoher Dichte, Niederdruckpolyäthylen *n*

h.-density pulp *(Pap)* Dickstoff *m*

h.-density second caustic extraction *(Pap)* zweite Alkalibehandlung (Alkaliextraktion, Alkaliwäsche) *f* bei hoher Stoffdichte

h.-density stage *(Pap)* Dickstoffbleichstufe *f*

h.-dielectric-strength mit hoher dielektrischer (elektrischer) Festigkeit, mit hoher Durchschlag[s]festigkeit

h.-duty dryer Hochleistungtrockner *m*

h.-efficiency Hochleistungs...

h. elasticity Hochelastizität *f*

h.-energy alpha particle energiereiches α-Teilchen *n*, α-Teilchen *n* hoher Energie

h.-energy fission schnelle Spaltung *f*, Schnellspaltung *f*, Spaltung *f* durch schnelle Neutronen

h.-energy irradiation Bestrahlen *n* mit energiereicher (harter) Strahlung

h.-energy neutron energiereiches Neutron *n*

h.-energy radiation energiereiche (harte) Strahlung *f*

h.-energy source starke Energiequelle *f*

h.-ester pectin hochverestertes Pektin *n*, Pektin H *n*

h. explosive brisanter Explosivstoff (Sprengstoff) *m*

h.-fired *(Ker)* hochgebrannt

h.-firing porcelain glace Porzellanscharffeuerglasur *f*

h. free protected size *(Pap)* stabilisierter Freiharzleim *m*, stabilisierte Harzemulsion *f*

h. free rosin size *(Pap)* hochfreiharzhaltiger (hochfreiharzreicher, freiharzreicher) Leim *m*

h. frequency Hochfrequenz *f*, HF

h.-frequency coagulation Hochfrequenzkoagulation *f*

h.-frequency dryer Hochfrequenztrockner *m*

h.-frequency drying Hochfrequenztrocknung *f*

h.-frequency induction furnace Hochfrequenzinduktionsofen *m*, *(besser)* Mittelfrequenzinduktionsofen *m*, kernloser Induktionsofen *m*

h.-frequency moulding *(Plast)* Formpressen *n* mit Hochfrequenzvorwärmung (Hochfrequenzheizung)

h.-frequency preheating Hochfrequenzvorwärmung *f*

h.-frequency sealing *(Plast)* Hochfrequenzsiegeln *n* *(von Folien)*

h.-frequency sealing machine *(Plast)* Hochfrequenzsiegelgerät *n*

h.-frequency titration Hochfrequenztitration *f*

h.-frequency titrator Hochfrequenztitrator *m*

h.-frequency welding Hochfrequenzschweißen *n*, Hochfrequenzschweißung *f*

h.-fusion *s.* h.-melting

h.-gloss hochglänzend, Hochglanz...

h. gloss *(Pap)* Hochglanz *m*

h.-grade hochwertig

h.-grade massecuite A-Füllmasse *f*, Weißzuckerfüllmasse *f*, Erstproduktfüllmasse *f*

h.-grade ore hochwertiges (reichhaltiges, reiches) Erz *n*

h.-grade sugar Rohzuckererstprodukt *n*, Rohzucker I *m*, Erstproduktzucker *m*

h.-gravity bath (medium) schwere Trübe *f*, Schwertrübe *f* *(unechte Schwerflüssigkeit)*

h.-gravity solid Schwerstoff *m*, Beschwerungsstoff *m*

h. green syrup *(Zucker)* Grünsirup *m*, Grünablauf *m*, Vorlauf *m*, Ablauf I *m*

h. grinding Hochmahlverfahren *n*, Hochmüllerei *f*, Grießmüllerei *f*

h.-head pump Hochdruckpumpe *f*, Preßpumpe *f*

h.-impact polystyrene hochschlagfestes Polystyrol *n*

h.-impact polyvinyl chloride hochschlagfestes Polyvinylchlorid (PVC) *n*

h.-intensity magnetic separation Starkfeldscheidung *f*

h.-intensity press *(Pap)* HI-Presse *f* *(eine Hochleistungspresse)*

h.-iron hocheisenhaltig, eisenreich, mit hohem Eisengehalt

h.-leaded hochbleihaltig, mit hohem Bleigehalt

h.-lustrous hochglänzend, Hochglanz...

h.-magnesia body *(Ker)* Masse *f* mit hohem Magnesiumoxidgehalt

h.-magnesium [lime]stone dolomitischer Kalkstein *m*

h.-melting hochschmelzend, schwerschmelzbar

h.-melting-point *s.* h.-melting

h. milling Hochmahlverfahren *n*, Hochmüllerei *f*, Grießmüllerei *f*

h.-modulus furnace black *(Gum)* High-Modulus-Furnace-Ruß *m*, HMF-Ruß *m* *(Gasruß, der Vulkanisat hohen Spannungswert verleiht)*

h.-modulus rubber Kautschuk *m* mit (von) hohem Spannungswert

h.-molecular hochmolekular

h.-octane hochoktan[zahl]ig, mit hoher Oktanzahl, hochklopffest, Hochoktan...

h.-octane fuel Hochoktankraftstoff *m*, hochoktan[zahl]iger (hochklopffester) Kraftstoff *m*

h.-octane gasoline (petrol) Hochoktanbenzin *n*, hochoktan[zahl]iges (hochklopffestes) Benzin *n*

h.-oxygen hochsauerstoffhaltig, sauerstoffreich, mit hohem Sauerstoffgehalt
h.-phosphorus hochphosphorhaltig, phosphorreich, mit hohem Phosphorgehalt
h. polymer Hochpolymer[es] *n*
h.-polymeric chemistry Chemie *f* der Hochpolymeren
h.-potassium kalireich
h. pressure Hochdruck *m*
h.-pressure boiling kier *(Text)* Hochdruckkochkessel *m*
h.-pressure compressor Hochdruckverdichter *m*, Hochdruckkompressor *m*
h.-pressure headbox *(Pap)* Hochdruckstoffauflauf *m*
h.-pressure hydrogen cell Hochdruck-Wasserstoff-Sauerstoff-Brennstoffelement *n*
h.-pressure hydrogenation Hochdruckhydrierung *f*
h.-pressure jig *(Text)* Hochdruckjigger *m*
h.-pressure laminate Hochdruckschicht[preß]stoff *m*
h.-pressure laminating Hochdrucklaminieren *n*, Herstellen *n* von Hochdruckschicht[preß]stoffen
h.-pressure line Hochdruckleitung *f*
h.-pressure liquid-phase hydrogenation Hochdruckhydrierung *f* in flüssiger Phase
h.-pressure moulding *(Plast)* Hochdruckpressen *n*, Hochdruck-Preßverfahren *n*
h.-pressure packing Hochdruckdichtung *f*, Hochdruckpackung *f*
h.-pressure process Hochdruckverfahren *n*; *(Gum)* Hochdruckdampfverfahren *n*, Palmer-Verfahren *n* *(Regenerierverfahren)*
h.-pressure-process polyethylene Hochdruckpolyäthylen *n*, verzweigtes Polyäthylen *n*, Polyäthylen *n* niedriger Dichte
h.-pressure pump Hochdruckpumpe *f*, Preßpumpe *f*
h.-pressure soil injector *(Landw)* Hochdruckdüngelanze *f*
h.-pressure sprayer Hochdrucksprühgerät *n*
h.-pressure steam Hochdruckdampf *m*
h.-pressure steam curing *(Bau)* Hochdruck-Dampferhärtung *f*
h.-pressure water jet Hochdruckwasserstrahl *m*, scharfer Wasserstrahl *m*
h.-purity hochrein
h.-rank hochinkohlt, hoch (stark) inkohlt, von hohem Inkohlungsgrad
h.-ratio fat *(Backfett mit hohem Gehalt an Mono- und Diglyzeriden)*
h.-shrinking *(Text)* hochschrumpfend
h.-silica clay kieselsäurereicher Ton *m*
h.-speed schnell; schnellaufend; schnellumlaufend, hochtourig
h.-speed centrifugation Ultrazentrifugieren *n*
h.-speed centrifuge Ultrazentrifuge *f*, Hochleistungszentrifuge *f*
h.-speed developer *(Foto)* Schnellentwickler *m*

h.-speed drying Schnelltrocknung *f*, Schnelltrocknen *n*
h.-speed electron schnelles Elektron *n*, Elektron *n* hoher Geschwindigkeit
h.-speed emulsion *(Foto)* hochempfindliche Emulsion *f*
h.-speed fixing bath *(Foto)* Schnellfixierbad *n*
h.-speed ion schnelles Ion *n*
h.-speed neutron schnelles (unverzögertes) Neutron *n*
h.-speed news[print] paper machine schnellaufende Zeitungsdruckpapiermaschine *f*
h.-speed paper[making] machine schnellaufende Papiermaschine *f*, Schnelläufer *m*
h.-speed recorder Schnellschreiber *m*
h.-strength acid hochkonzentrierte (hochprozentige) Säure *f*
h.-strength cooking acid *(Pap)* hochkonzentrierte (hochprozentige) Kochsäure *f*
h.-strength pulp unaufgeschlossener (unvollständig aufgeschlossener, sehr roher, sehr harter) Zellstoff *m*
h.-structure [carbon] black *(Gum)* High-structure-Ruß *m*, Hochstruktur-Ruß *m*
h.-styrene hochstyrolhaltig, mit hohem Styrolgehalt
h.-styrene copolymer (polymer, resin) hochstyrolhaltiges Polymerisat (Polymeres) *n*, Mischpolymerisat *n* mit hohem Styrolgehalt
h.-sugar beet zuckerreiche Rübe (Zuckerrübe) *f*, Z-Rübe *f*, ZZ-Rübe *f*
h.-sulphur hochgeschwefelt, mit hohem Schwefelgehalt
h.-temperature Hochtemperatur...
h.-temperature alloy Hochtemperaturlegierung *f*
h.-temperature carbonization Hochtemperaturverkokung *f*, HT-Verkokung *f*, Normalverkokung *f*, Verkokung *f* (i.e.S.), Hochtemperaturentgasung *f*, HT-Entgasung *f*
h.-temperature chemistry Hochtemperaturchemie *f*
h.-temperature coke Hochtemperaturkoks *m*
h.-temperature coking s. h.-temperature carbonization
h.-temperature corrosion Hochtemperaturkorrosion *f*
h.-temperature cure s. h.-temperature vulcanization
h.-temperature distillation Hochtemperaturdestillation *f*
h.-temperature durability s. h.-temperature stability
h.-temperature dyeing *(Text)* Hochtemperaturfärben *n*, HT-Färben *n*, HT-Färbeverfahren *n*, Färben *n* unter HT-Bedingungen, Heißfärben *n*
h.-temperature dyeing machine *(Text)* Hochtemperaturfärbemaschine *f*, HT-Färbemaschine *f*
h.-temperature furnace Hochtemperaturofen *m*
h.-temperature gas-cooled reactor s. h.-temperature reactor

h.-temperature grease Hochtemperaturschmierfett *n*

h.-temperature heat treatment Hochtemperatur-Hitzebehandlung *f*

h.-temperature kiln *s.* h.-temperature furnace

h.-temperature material warmfester (hochtemperaturbeständiger) Werkstoff *m*

h.-temperature polymer Hochtemperaturpolymerisat *n*, Wärmepolymerisat *n*, Hochtemperaturpolymer[es] *n*, Wärmepolymer[es] *n*

h.-temperature polymerization Hochtemperaturpolymerisation *f*, Wärmepolymerisation *f*

h.-temperature porcelain Porzellan *n* für hohe Temperaturen

h.-temperature pressure dyeing *(Text)* Hochtemperatur-Überdruckfärben *n*, HT-Druckfärben *n*, HT-Druckfärbung *f*

h.-temperature properties Wärmeverhalten *n*, Verhalten *n* bei hohen Temperaturen

h.-temperature reactor *(Kph)* Hochtemperaturreaktor *m*

h.-temperature resistance *s.* h.-temperature stability

h.-temperature short-time heat treatment *(Lebm)* Kurzzeiterhitzung *f*, Hochkurzerhitzung *f*, HTST-Erhitzung *f*

h.-temperature short-time method *(Lebm)* Kurzzeiterhitzungsverfahren *n*, Hochkurzerhitzungsverfahren *n*, HTST-Verfahren *n*

h.-temperature short-time pasteurization Kurzzeitpasteurisation *f*, Hochkurzpasteurisation *f*, HTST-Pasteurisation *f*, Kurzzeiterhitzung *f*, Hochkurzerhitzung *f* *(zur Pasteurisation)*

h.-temperature short-time pasteurizer *(Lebm)* Kurzzeitpasteurisierapparat *m*, Kurzzeiterhitzer *m*

h.-temperature short-time sterilization *(Lebm)* Hochkurzsterilisation *f*, Hochkurzsterilisierung *f*

h.-temperature stability Hochtemperaturbeständigkeit *f*, Beständigkeit (Widerstandsfähigkeit) *f* gegen hohe Temperaturen

h.-temperature steamer Hochtemperaturdämpfer *m*, HT-Dämpfer *m*

h.-temperature steaming Hochtemperaturdämpfen *n*, HT-Dämpfen *n*

h.-temperature tar Hochtemperaturteer *m*

h.-temperature tunnel kiln *(Ker)* Hochtemperatur-Tunnelofen *m*

h.-temperature vulcanization Hochtemperaturvulkanisation *f*

h.-tension electrical porcelain Hochspannungs-Elektroporzellan *n*

h.-tension separation Elektroscheiden *n* mit Ionisation

h.-titania body *(Ker)* hoch titanoxidhaltige Masse *f*

h. vacuum Hochvakuum *n*

h.-vacuum distillation Hochvakuumdestillation *f*

h.-vacuum work Arbeiten *n* im Hochvakuum

h.-viscosity hochviskos, hochviskös, hochzähflüssig, von hoher Viskosität

h.-volatile leichtflüchtig, hochflüchtig

h.-volatile [bituminous] coal hochflüchtige (hochgashaltige, gasreiche) bituminöse Kohle *f*

h. volatility Leichtflüchtigkeit *f*

h.-voltage electrophoresis Hochspannungselektrophorese *f*

h.-voltage paper electrophoresis Hochspannungspapierelektrophorese *f*

h.-volume spray *(mit reichlich Wasser verdünntes)* Spritzmittel *n*

h.-volume spraying Spritzen *n*, Spritzung *f* *(Ausbringen von Pflanzenschutzmitteln mit viel Wasser als Träger- und Dispersionsmittel)*

h.-wash syrup *(Zucker)* Weißsirup *m*, Weißablauf *m*, Deckablauf I *m*

h. wet modulus fibre Hochnaßmodulfaser *f*, HWM-Faser *f*; Hochnaßmodulfaserstoff *m*, HWM-Faserstoff *m*

h.-yield mit hoher Ausbeute, Hochausbeute...

h.-yield pulp Hochausbeute[zell]stoff *m*, High-Yield-Stoff *m*

h.-yield sulphite pulp Hochausbeute-Sulfitzellstoff *m*

h.-zinc brass Messing *n* mit hohem Zinkgehalt

h.-zirconia body *(Ker)* Masse *f* mit hohem Zirkoniumoxidgehalt

higher alcohol höherer Alkohol *m*

h. alkyl silicone höheres Alkylsilikon *n*

h.-boiling hochsiedend, höhersiedend, schwer[er]siedend

h.-boiling fraction hochsiedende (höhersiedende, schwer[er]siedende) Fraktion *f*

h. heating value Verbrennungswärme *f*, oberer Heizwert *m*

h.-melting höherschmelzend, schwererschmelzbar

h.-membered höhergliedrig

h.-order reaction Reaktion *f* höherer Ordnung

highest useful compression ratio höchstes nutzbares Kompressionsverhältnis *n*

highgate resin *(Min)* Kopalin *m* (bernsteinähnliches fossiles Harz)

highly active 1. hochaktiv, hochwirksam, sehr wirksam; 2. *s.* h. radioactive

h. aromatic hocharomatisch

h. branched stark verzweigt

h. colloidal hochkolloidal

h. coloured farbstark

h. compressed hochkomprimiert

h. disperse hochdispers

h. loaded stark belastet, hochbelastet; *(Gum)* hochgefüllt, stark gefüllt

h. luminous stark leuchtend

h. lustrous hochglänzend, Hochglanz...

h. oriented hochorientiert

h. plastic hochplastisch

h. polarizable hochpolarisierbar

h. polymerized hochpolymer

h. porous hochporös

h. purified hochgereinigt, hochrein

h. radioactive stark radioaktiv, hochaktiv, heiß
h. reactive hochreaktiv, hochreaktionsfähig
h. refined hoch[aus]raffiniert, hoch (stark) [aus]raffiniert
h. refractive stark lichtbrechend
h. refractory 1. *s.* h. refractive; 2. hochfeuerfest
h. refractory brick hochfeuerfester Stein (Ziegel) *m*
h. selective hochselektiv
h. sensitive hochempfindlich
h. strained hochgespannt *(z.B. Ringsysteme)*
h. stressed hochbeansprucht
h. viscous hochviskos, hochviskös, hochzähflüssig, von hoher Viskosität
h. volatile leichtflüchtig, hochflüchtig
Hildebrand rule *(phys Ch)* Hildebrandsche Regel *f*
Hildebrandt extractor Hildebrandt-Extraktor *m*, Hildebrandt-Extrakteur *m (ein kontinuierlich arbeitender Feststoffextraktor)*
Hill reaction *(Bioch)* Hillsche Reaktion *f*, Hill-Reaktion *f*
Hill's system Hillsches System *n (Reihungssystem für die Summenformel von chemischen Verbindungen)*
Hilt's law (rule) Hiltsche Regel *f*, Gesetz *n* von Hilt
hindered settling behindertes (gestörtes) Absetzen *n*, gestörtes Sedimentieren *n*
h.-settling classification Klassierung *f* nach dem Prinzip des behinderten (gestörten) Absetzens
hindrance [sterische] Hinderung *f*
hinokiflavone Hinokiflavon *n (Biflavonyl)*
hinokiic acid Hinokisäure *f (Sesquiterpensäure)*
hinokinin Hinokinin *n (Lignan in Nadelhölzern)*
hinokiol Hinokiol *n (Diterpen in Nadelhölzern)*
Hinsberg amine test, Hinsberg method [of amine separation] *s.* Hinsberg test
Hinsberg test Hinsberg-Probe *f*, Hinsbergsche Reaktion *f (zur Trennung primärer, sekundärer und tertiärer Amine)*
hiortdahlite *(Min)* Hjortdahlit *m*, Guarinit *m*
hippuric acid $C_6H_5CO \cdot NHCH_2COOH$ Hippursäure *f*, Benzoylglykokoll *n*, Benzoylaminoessigsäure *f*, Benzoylglyzin *n*
hiragonic acid 6,10,14-Hexadekatriensäure *f*
Hirsch funnel Trichter *m* nach Hirsch
hisingerite *(Min)* Hisingerit *m*
histamine Histamin *n*, 4-(2'-Aminoäthyl)imidazol *n*
h. phosphate Histaminphosphat *n*
histidine Histidin *n*, His, α-Amino-β-imidazolylpropionsäure *f*, β-Imidazolylalanin *n*
histone Histon *n (Protein)*
histozyme Histozym *n*, Hippurikase *f (Ferment)*
hitch roll *(Pap)* Wendelwalze *f*; *(Pap)* Filzspannwalze *f*, Spannwalze *f*
Hittorf number Hittorfsche Überführungszahl *f*
hjelmite *(Min)* Hjelmit *m*
HL method *s.* Heitler-London method
HL theory *s.* Heitler-London theory
HLB *s.* hydrophilic-lipophilic balance
HLSP method *s.* Heitler-London-Slater-Pauling method

HLSP theory *s.* Heitler-London-Slater-Pauling theory
HMF black *s.* high-modulus furnace black
HMO *s.* Hückel molecular orbital
Hoagland solution *(Bio) (Spurenelemente enthaltende zusätzliche)* Nährlösung *f* nach Hoagland, A-Z-Lösung *f*
hoaling power *(Ker)* Heilvermögen *n (einer Glasur)*
hob *(Plast)* Prägestempel *m*, Pfaff *m*
hobbing *(Plast)* Prägen *n*, Prägung *f*
h. press *(Plast)* Prägepresse *f*
Hoechst [continuous] coking process Hoechst-Koker-Prozeß *m*, Hoechster Koker-Prozeß *m (zur kontinuierlichen Pyrolyse von Rohölen und Erdölrückständen)*
Hoechst high-temperature pyrolysis process Hoechster Hochtemperatur-Pyrolyseverfahren *n*
Hoechst process Hoechst-Prozeß *m*
Hoffmann clamp Quetschhahn *m* (Schlauchklemme *f*) nach Hoffmann, Schraubenquetschhahn *m*, Schraubklemme *f*
Hoffmann kiln *(Ker)* Hoffmann-Ofen *m*, Hoffmannscher Ringofen *m*
Hofmann degradation of amides Hofmannscher Säureamidabbau *m*
Hofmann degradation of quaternary ammonium hydroxides *s.* Hofmann elimination
Hofmann elimination Hofmannscher Abbau *m (quartärer Ammoniumhydroxide)*
Hofmann's violet Hofmanns Violett *n*
Hofmeister series Hofmeistersche Reihe *f*, lyotrope Reihe *f (der Ionen im Verhalten gegenüber Eiweiß)*
hog fat Schweineschmalz *n*, Schweinefett *n*
h. gum Schweinsgummi *m (von Clusia flava L.)*
h. pit *(Pap)* Siebwasserbehälter *m*, Siebwassersammelbecken *n*
HOH angle HOH-Winkel *m*
hoist bridge Gichtbrücke *f*, Schrägbrücke *f*
h. machine Aufzugswinde *f*
hoisting gear Hebezeug *n*, Flaschenzugsystem *n (z.B. einer Rotary-Bohranlage)*
h. rope Hubseil *n*, Förderseil *n*
hold / to binden; enthalten
h. together zusammenhalten
hold-back carrier *(Kch)* Rückhalteträger *m*
h. tank Vorratsbehälter *m*, Speicherbehälter *m*
h.-up Haftinhalt *m*, Ruheinhalt *m (z.B. einer Kolonne)*
h.-up time Verweilzeit *f*, Aufenthaltszeit *f*, Retentionszeit *f*, Rückhaltezeit *f*, Haltezeit *f*, Stehzeit *f*, Standzeit *f*, Durchgangszeit *f*, Durchlaufzeit *f*; *(Milchpasteurisation)* Heißhaltezeit *f*
Holdcroft bars *(Ker)* Holdcroft-Stäbe *mpl*
holder Halter *m*; Behälter *m*
h. pasteurization (pasteurizing treatment) *(Lebm)* Dauerpasteurisation *f*, Dauererhitzung *f (zur Pasteurisation)*
h. tube Heißhalteplatte *f (im Plattenerhitzer für Milchpasteurisierung)*
holding furnace Warmhalteofen *m*

h. pasteurization s. holder pasteurization
h. period s. hold-up time
h. temperature Aufbewahrungstemperatur f, Lager[ungs]temperatur f
h. time s. hold-up time
holdup s. hold-up
hole 1. Loch n, Öffnung f; Bohrung f; 2. (Krist) Loch n, Leerstelle f, Leerplatz m, Lücke f, Gitterleerstelle f, Gitterloch n, Gitterlücke f; (phys Ch) Loch n, Defektelektron n, Elektronenloch n, Mangelelektron n
h. conduction Lochleitung f, Löcherleitung f, Störstellenleitung f, Mangelleitung f, Defekt[elektronen]leitung f
h. dressing Lochdüngung f
h.-electron pair Loch-Elektronen-Paar n, Elektronen-Defektelektronen-Paar n
h. for removal of slag Schlacken[ab]stichloch n, Schlacken[ab]stich m, (meist) Schlackenform f
h. position Leerstelle f, Leerplatz m, Loch n, Lücke f, Gitterleerstelle f, Gitterloch n, Gitterlücke f
h. theory of liquids Löchertheorie f der Flüssigkeiten
holey roll Lochwalze f
holiday freigelassene Stelle f (beim Auftragen von Farbanstrichen)
Holland beater s. Hollander
H. beater tub (Pap) Holländertrog m, Stoffwanne f
Hollander [beater, beating engine] (Pap) Holländer m, Mahlholländer m, Messerholländer m, Ganzzeugholländer m
H. roll (Pap) Holländerwalze f, Messerwalze f, Mahlwalze f
H. washer (Pap) Waschholländer m; Halbzeugholländer m
hollow article (Plast) Hohlkörper m
h. block Hohlblock m
h. body s. h. article
h. brick Hohlziegel m, Hohlstein m, Lochziegel m
h. casting Hohlguß m
h.-cone nozzle Hohl[sprüh]kegeldüse f, Kegelstrahldüse f, Dralldüse f
h. electrode Fangtaschenelektrode f
h. fibre Hohlfaser f, Hohlseide f
h. glassware Hohlglas n
h. journal Hohlzapfen m
h. part s. h. article
h. profile Hohlprofil n
h. shaft Hohlwelle f
h. space Hohlraum m
h. spray cone Hohlsprühkegel m
h. tile s. h. brick
h. ware (Ker) Hohlware f; Hohlglas n
Holmes-Manley process Holmes-Manley-Verfahren n, Krackverfahren (Spaltverfahren) n nach Holmes und Manley
holmia Ho_2O_3 Holmiumerde f, (veraltet für) Holmiumoxid n

holmium Ho Holmium n
h. chloride $HoCl_3$ Holmium(III)-chlorid n
h. hydroxide $Ho(OH)_3$ Holmium(III)-hydroxid n
h. oxalate $Ho_2(C_2O_4)_3$ Holmiumoxalat n
h. oxide Ho_2O_3 Holmiumoxid n
h. sulphate $Ho_2(SO_4)_3$ Holmiumsulfat n
holocellulose Holozellulose f
holocrystalline vollkristallin[isch], holokristallin
h. rock holokristallines Gestein n
holohedral (Krist) holoedrisch, vollflächig
h. crystal Vollflächnerkristall m
holohyaline (Geol) [holo]hyalin
holosiderite Siderit m, Meteoreisen n, Eisenmeteorit m, Holosiderit m
Holst process (Pap) Holst-Verfahren n (Chlordioxidbleichlaugenherstellung)
homatropine Homatropin n, Mandelsäuretropylester m, Tropinmandelsäureester m
h. hydrobromide Homatropinhydrobromid n, Mandelsäuretropylesterhydrobromid n
homeoblastic (Geol) homöoblastisch
homeomilite (Min) Homilit n
homocyclic homozyklisch, isozyklisch, karbozyklisch
h. compounds Isozyklen pl, Karbozyklen pl, isozyklische (karbozyklische, homozyklische) Verbindungen fpl
homofermentative homofermentativ
homogeneity Homogenität f, Einheitlichkeit f, Gleichartigkeit f, Gleichverteilung f
homogeneous homogen, einheitlich, gleichartig, gleichmäßig zusammengesetzt
h. batholith (Geol) homogener Batholith m
h. catalysis homogene Katalyse f
h. reaction homogene Reaktion f, Homogenreaktion f
h. system homogenes (einphasiges) System n, Einphasensystem n
homogeneously in homogener Phase (Reaktion)
homogenization 1. Homogenisieren n, Homogenisierung f (z.B. von Emulsionen), gleichmäßiges (gutes) Durcharbeiten (Verteilen, Vermischen) n; 2. (Met) Homogenisierungsglühen n, homogenisierendes Glühen n, Homogenisieren n, Homogenisierung f, Diffusionsglühen n, Diffusionsglühung f
h. pressure Homogenisationsdruck m, Homogenisierungsdruck m
homogenize / to 1. homogenisieren, homogen machen, gleichmäßig (gut) durcharbeiten (verteilen, vermischen); 2. (Met) homogenisierend glühen, homogenisieren, diffusionsglühen
homogenizer Homogenisator m, Homogenisiermaschine f, Homogenisiervorrichtung f, Homogenisierungsapparat m
homogenizing s. homogenization
h. head Homogenisierkopf m (einer Homogenisiermaschine)
h. roll Mischwalze f
h. valve s. h. head

homolog s. homologue
homologous homolog, gleichliegend, gleichnamig, gleichlautend, gleiche Beziehung habend, übereinstimmend, gleichen Ursprungs
   h. series homologe Reihe f
homologue Homolog[es] n, homologe Verbindung f
homolysis Homolyse f (Aufspaltung einer kovalenten Bindung unter Bildung gleichsinnig geladener Bruchstücke)
homolytic homolytisch
   h. reaction homolytische Reaktion f
homopolar homöopolar, kovalent, unpolar, einpolar, unitarisch
   h. bond (linkage) homöopolare (unpolare, einpolare, unitarische, kovalente) Bindung f, Atombindung f, Elektronenpaarbindung f
   h. molecule homöopolares (homöopolar gebundenes) Molekül n
homopolymer Homopolymer[es] n, Homopolymerisat n
homopolymeric homopolymer
homopolymerization Homopolymerisation f, Eigenpolymerisation f
homopolymerize / to homopolymerisieren
4-homosulphanilamide 4-Homosulfanilamid n, p-Aminomethylbenzolsulfonamid n
honey Honig m
   h. vinegar Honigessig m
honeycomb / to mit einem Gitterwerk auskleiden
honeycomb Wabe f
   h. core Wabenkern m (Sandwichkonstruktion)
   h. graphitic structure (Krist) wabenförmige Graphitstruktur f
   h. sandwich Wabenmittellage f (für Verbundkonstruktionen)
   h. sandwich material Verbundwerkstoff m mit Wabenkern
   h. structure wabenförmige (wabenartige) Struktur f, Wabenstruktur f
honeycombed wabenförmig (wabenartig) strukturiert, mit Wabenstruktur; wabenförmig (wabenartig) gemustert
Hongay oil Kurrunjeöl n (von Pongamia pinnata (L.)Merr.)
hood Haube f, Kappe f, Deckel m, Glocke f; Abzug m, Abzug[s]schrank m, Digestorium n; Rauchfangdach n; Dunsthaube f, Schwadenhaube f, Verdunstungshaube f; (Glas) Stiefel m (Durchlaß zwischen Schmelz- und Arbeitswanne)
   h.-type annealing furnace Haubenglühofen m
hooded pot (Glas) verdeckter (gedeckter, geschlossener) Hafen m
Hooker cell Hooker-Zelle f (eine Diaphragmenzelle)
Hoopes [electrolytic-refining] process Hoopes-Verfahren n, Aluminium-Raffinationselektrolyse f nach Hoopes
hop / to (Würze) hopfen, Hopfen zusetzen
hop [Gemeiner] Hopfen m, Humulus lupulus L.
   h. aroma Hopfenaroma n

   h. back Hopfenseiher m
   h. bine Hopfenranke f, Hopfenrebe f
   h. bitter acid Hopfenbittersäure f (Sammelname)
   h. bitter substance Hopfenbitterstoff m
   h. bitters Hopfenbitterstoffe mpl
   h. cone Hopfendolde f, Hopfenzäpfchen n
   h. dryer Hopfendarre f
   h. drying Trocknen n (Trocknung f, Darren n) des Hopfens
   h. extract[ive] Hopfenextrakt m
   h. flour Hopfenmehl n, Lupulin n
   h. for brewing Brauhopfen m, Hopfen m als Brauware
   h. garden Hopfengarten m
   h. growing Hopfen[an]bau m, Hopfenzüchtung f
   h.-growing district Hopfenanbaugebiet n
   h. jack s. h. back
   h. kiln s. h. dryer
   h. mould Hopfenmehltau m
   h. oil Hopfenöl n
   h. pectin Hopfenpektin n
   h. plant Hopfenpflanze f
   h. rate Hopfengabe f
   h. resin Hopfenharz n
   h. strainer s. h. back
   h. vine s. h. bine
   h. yard s. h. garden
hopeite (Min) Hopeit m (Zinkorthophosphat-4-Wasser)
hopped wort gehopfte Würze f
hopper 1. Trichter m, Fülltrichter m, Einfülltrichter m, Zuführungstrichter m, Schütttrichter m, Beschickungstrichter m; 2. Silo m(n), Bunker m, Behälter m; Hopper m (Zwischenbehälter zum Margarinetransport); 3. Hopfenkufe f
   h. dryer Trichtertrockner m
hopping Hopfen n, Zusetzen n des Hopfens; Hopfenernte f
Höppler [falling-ball] viscosimeter Höppler-Viskosimeter n, Höppler-Kugelfallviskosimeter n
hoppy hopfenreich, Hopfen...
hordein Hordein n (Eiweißkörper der Gerste)
horizon (Bodenkunde) Horizont m
horizontal horizontal, waagerecht
   h. chamber Horizontalkammer f
   h. clarifyer Horizontalklärer m
   h. digester (Pap) liegender Kocher m
   h.-draught firing horizontale (waagerechte) Flammenführung f (in einem Industrieofen)
   h.-draught kiln Ofen m mit horizontaler (waagerechter) Flammenführung
   h. evaporator Horizontalrohrverdampfer m
   h. extractor liegender Extraktor (Extrakteur) m
   h. flue Horizontalzug m, horizontaler (waagerechter) Zug m; Horizontalkanal m
   h. head (Plast) Längsspritzkopf m
   h. kiln (Lebm) Horizontaldarre f
   h. lehr (Glas) [horizontaler] Kühlkanal m
   h. plate filter Etagennutsche f
   h. relationship Horizontalbeziehung f (im Periodensystem)

h. retort Horizontalretorte *f*, horizontale (waagerechte, liegende) Retorte *f*
h. sheet drawing process Horizontalziehverfahren (Waagerechtziehverfahren) *n* für Tafelglas
h. still Horizontalkolonne *f*
h.-tank sluicing filter Vallez-Filter *n*
h.-tube evaporator s. h. evaporator
h. vulcanizer liegender Vulkanisierkessel *m*, Vulkanisierkessel *m* in liegender Ausführung
horizontally perforated brick Langlochziegel[stein] *m*
hormone Hormon *n*
h. weedkiller Wuchsstoffherbizid *n*, Unkrautbekämpfungsmittel *n* auf Wuchsstoffbasis, Wuchsstoffpräparat *n*, Wuchsstoffmittel *n*
horn lead *(Min)* Hornblei *n*, Bleihornerz *n*, Phosgenit *m*
h.-like hornähnlich
h. meal *(Landw)* Hornmehl *n*
h. mercury *(Min)* Hornquecksilber *n*, Quecksilberhornerz *n*, Kalomel *n* *(Quecksilber(I)-chlorid)*
h. silver *(Min)* Hornsilber *n*, Silberhornerz *n*, Chlorargyrit *m*, Kerargyrit *m*, Chlorsilber *n* *(Silberchlorid)*
h. spatula Hornspatel *m*
hornblende *(Min)* [gemeine] Hornblende *f*, *(i.e.S.)* Hornblende *f*, Amphibol *m*
hornfels *(Geol)* Hornfels *m*
hornstone *(Min)* Hornstein *m*
horse *(Gerb)* Bock *m*, Gerberbock *m*
h. chestnut Roßkastanie *f*, Aesculus L., *(i.e.S.)* [Gemeine] Roßkastanie *f*, Aesculus hippocastanum L.; Roßkastaniensamen *m*, Kastanie *f*
h. fat Pferdefett *n*
h.-flesh ore *(Min)* Buntkupfererz *n*, Buntkupferkies *m*, Bornit *m* *(Eisen(II)-kupfer(II)-sulfid)*
h. hair Roßhaar *n*
h. radish Meerrettich *m*, Armoracia rusticana Ph. Gaertn.; *(Lebm)* Meerrettich *m*, Kren *m*
h. radish peroxidase Meerrettichperoxydase *f*
h. up / to *(Gerb)* Leder zum Abtropfen überschüssiger Gerbbrühe auf einen Bock legen)
horsemint oil Monardaöl *n* (ätherisches Öl von Monarda-Arten)
horseshoe mixer Ankerrührer *m*
horticultural spray Sprühmittel *n* (Spray *m*) für den Gartenbau
hose Schlauch *m*
h. clamp (clip) Schlauchschelle *f*
h. cock Schlauchklemme *f*, Quetschhahn *m*
h. connection (connector) Schlauchtülle *f*
host [component] Wirtskomponente *f*, Wirtssubstanz *f*, Wirt *m*
h. element Wirtselement *n*
h. lattice *(Krist)* Wirtsgitter *n*
h. plant Wirtspflanze *f*, Wirt *m*
h. rock Wirtsgestein *n*
hot heiß, warm; heiß, hochgradig (stark) radioaktiv
h.-acid process Heißsäureverfahren *n* *(zur katalytischen Polymerisation)*

h. air Heißluft *f*, Warmluft *f*
h.-air ageing Heißluftalterung *f*
h.-air blast Heißwind *m*, heißer Wind (Gebläsewind) *m*
h.-air-blast main Heißwindleitung *f*
h.-air chamber Heißluftraum *m*, Heißluft[trocken]kammer *f*
h.-air cure (curing) s. h.-air vulcanizing
h.-air inlet Warmlufteintritt *m*
h.-air sterilizer [Trockenschrank-]Heißluftsterilisator *m*, Heißluftsterilisierschrank *m*
h.-air stuffing *(Gerb)* Warmfetten *n*
h.-air vulcanizing Heißluftvulkanisation *f*, Vulkanisation *f* in Heißluft, Heißluftheizung *f*
h. alkali refining *(Pap)* Heißveredelung *f*, Heißalkalisierung *f*
h. atom heißes (hochenergiereiches, hoch angeregtes) Atom *n*
h.-atom chemistry heiße Chemie *f*
h. Banbury process *(Gum)* Banbury-Lancaster-Verfahren *n* *(Regenerierverfahren)*
h. blast s. h.-air blast
h.-blast line (main) s. h.-air-blast main
h.-blast pig iron heißerblasenes Roheisen *n*
h.-blast stove Winderhitzer *m*, Hochofenwinderhitzer *m*
h. brine *(Kaliindustrie)* Heißlauge *f*
h. cathode Glühkatode *f*, Heizkatode *f*
h.-chamber die casting Warmkammerdruckgießen *n*, Warmkammerdruckguß *m*, *(veraltet)* Spritzguß *m*
h.-chamber [die-casting] machine Warmkammer-Druckgießmaschine *f*, Warmkammer-Druckgußmaschine *f*, Druckgießmaschine *f* für Warmkammerverfahren, Warmkammermaschine *f*, *(veraltet)* Spritzgußmaschine *f*
h.-chamber pressure casting s. h.-chamber die casting
h.-charge pump Pumpe *f* für heiße Medien
h. chromatography Heißchromatografie *f*
h. cure s. h. vulcanization
h.-cured s. h.-vulcanized
h. curing s. h. vulcanization
h. defecation *(Zucker)* heiße (warme) Scheidung *f*
h.-dip aluminizing Feuer[ver]aluminierung *f*
h.-dip galvanized steel feuerverzinkter Stahl *m*
h.-dip galvanizing Feuerverzinken *n*, Feuerverzinkung *f*
h.-dip tinning Feuerverzinnen *n*, Feuerverzinnung *f*
h. dissolution Heißverlösen *n* (z.B. von chloridischen Düngesalzen)
h. drawing *(Text)* Heißverstreckung *f*
h. fermentation of manure *(Landw)* Heißmistverfahren *n*
h.-filament pyrometer Glühfadenpyrometer *n*
h. floor *(Ker)* Trockenboden *m*
h. flue *(Text)* Heißlufttrockenmaschine *f*, Hotflue *f*

h. foil stamping Heißprägen n von Folien
h. former dipping (Gum) Tauchen n mit heißen Formen
h. forming Warm[ver]formen n, Warm[ver]formung f, Warmformgebung f, Warmumformung f
h.-forming technique Warm[ver]formungsverfahren n, Warmformgebungsverfahren n, Warmumformungsverfahren n
h. galvanizing s. h.-dip galvanizing
h. gas Heißgas n, heißes Gas n
h.-gas inlet Heißgaseintritt m
h.-gas producer Heißgaserzeuger m
h.-gas sealing Heißgassiegeln n (von Folien)
h.-gas welding Heißgasschweißen n
h. glue s. h.-setting adhesive
h. grinding (Pap) Heißschliffverfahren n, Heißschleifen n
h.-ground pulp (Pap) Heißschliff m
h. hardness Warmhärte f
h. house Brennerhaus n (Rußherstellung)
h. laboratory (Kph) heißes Labor[atorium] n
h. liming s. h. defecation
h. mastication heiße Mastikation f, Heißmastikation f, heißes Mastizieren n, Heißmastizierung f
h.-melt adhesive s. h.-setting adhesive
h.-melt coating heißschmelzender Überzug m
h.-metal ladle Roheisenpfanne f (eines Hochofens)
h. mixing Heißmischen n
h.-oil dyeing Heißölfärben n, Heißölfärbeverfahren n
h. pepper Gewürzpaprika m, Spanischer Pfeffer m, Cayennepfeffer m
h.-pitted (Gerb) im Hot-pit-Verfahren (bei erhöhter Temperatur mit Extrakten) gegerbt
h. pitting (Gerb) Hot-pit-Verfahren n, Hot-pit-Gerbung f (Grubengerbung bei erhöhter Temperatur)
h. plate Heizplatte f
h. polymer Hochtemperaturpolymerisat n, Wärmepolymerisat n, Hochtemperaturpolymer[es] n, Wärmepolymer[es] n
h. polymerization Wärmepolymerisation f, Hochtemperaturpolymerisation f
h. predefecation (preliming) (Zucker) heiße (warme) Vorscheidung f
h. preparation (Ker) Heißaufbereitung f, Dampfaufbereitung f
h.-press / to heißpressen; warmpressen; (Metallpulver) heißpressen, heiß pressen (verpressen)
h. press Heizpresse f, beheizbare Presse f; (Pap) Heißpresse f
h.-press method s. h.-pressing method
h.-pressed naphthalene (durch Abpressen) gereinigtes Naphthalin n
h. pressing Heißpressen n; Warmpressen n; Heißpressen n, Drucksintern n, Preßsintern n, Heißsintern n (von Metallpulver)
h.-pressing method (technique) Heißpreß-

methode f, Heißpreßverfahren n (in der Pulvermetallurgie)
h. refining s. h. alkali refining
h. runner (Plast) beheizter Angußverteiler m
h.-runner mould (Plast) Heißkanal-Spritzgießwerkzeug n
h.-runner moulding (Plast) Heißkanal-Spritzgießen n, Heißkanal-Spritzguß m
h.-sealing adhesive Heißsiegelkleber m
h.-set resin heißhärtendes Harz n
h.-setting adhesive Warmkleber m, Heißkleber m, Schmelzkleber m, heißhärtender (heißabbindender) Kleber m
h.-smoke / to warmräuchern (bei 25 °C); heißräuchern (bei 80 bis 100 °C)
h. smoking Warmräuchern n, Warmräucherei f (bei 25 °C); Heißräuchern n (bei 80 bis 100 °C)
h. spot (Glas) Quellpunkt m
h.-spray / to heißspritzen (in der Farbspritztechnik)
h. spray s. h. spraying
h.-spray lacquer Heißspritzlack m
h. spraying Heißspritzen n (in der Farbspritztechnik)
h. spraying process Heißspritzverfahren n
h.-steam cylinder oil Heißdampfzylinderöl n
h. strength Warmfestigkeit f
h. stretched heißverstreckt, heißgereckt
h. stuffing (Gerb) Warmfetten n
h. tinning s. h.-dip tinning
h. vulcanization Heißvulkanisation f
h.-vulcanized heißvulkanisiert
h.-vulcanizing heißvulkanisierend
h. waste (Kph) hochaktiver (heißer) Abfall m
h. water Heißwasser n
h.-water accumulator Heißwasserbehälter m, Sammelbehälter m für heißes Wasser
h.-water bath Wasserbad n
h.-water funnel Heißwassertrichter m, Warmwassertrichter m
h.-water pipe Heißwasserrohr n
h.-water pump Heißwasserpumpe f
h. wetting agent Heißnetzer m, Heißnetzmittel n
h.-wire anemometer Hitzdrahtanemometer n
h.-wire katharometer Katharometer n (Wärmeleitfähigkeitszelle f) mit Heizdrähten
h.-wire reference and detector cell Wärmeleitfähigkeitszelle f, Katharometer n
h.-work / to warmverarbeiten, warm verarbeiten
h. working 1. Warmverarbeitung f; 2. s. h. forming
h.-working process 1. Warmverarbeitungsverfahren n; 2. s. h.-forming technique
hotness Hitze f, Wärme f
Hottenroth number Hottenroth-Zahl f (Maß der Spinnreife einer Viskoselösung)
Houdresid process Houdresid-Verfahren n (katalytisches Krack- oder Reformierverfahren)
Houdriflow [catalytic cracking] process Houdriflow-Verfahren n (katalytisches Krackverfahren)
houdriformer Houdriformer m, Houdriforming-Anlage f

houdriforming Houdriformen *n*, Houdriformierung *f (Abart des katalytischen Reformierens)*
h. process Houdriforming-Verfahren *n*
Houdry catalytic cracking process katalytisches Houdry-Krackverfahren *n*
Houdry fixed-bed process Houdry-Festbettverfahren *n*
Houdry process Houdry-Verfahren *n*
house coal *s.* household coal
household ammonia Salmiakgeist *m*, Ammoniakwasser *n*
h. china Haushaltporzellan *n*
h. coal Hausbrandkohle *f*, Kohle *f* für Hausbrandzwecke
h. fire extinguisher Handfeuerlöscher *m*
h. margarine Haushaltmargarine *f*
h. refrigerator Haushaltkühlschrank *m*
h. stove coal *s.* h. coal
h. waste water häusliches Abwasser *n*
housing Gehäuse *n (z.B. von Pumpen)*; Ständer *m*, Rahmen *m*
Howard crystallizer Howard-Kristallisator *m (ein Säulenkristallisator)*
HP steam *s.* high-pressure steam
HPC black *s.* hard processing channel black
H.S. *s.* hard-sized
HT *s.* high-temperature
HTS *s.* heat-transfer salt
HTST method *s.* high-temperature short-time method
HTST pasteurization *s.* high-temperature short-time pasteurization
HTU *s.* height of one transfer unit
huantajayite *(Min)* Huantajayit *m*
Huanuca coca Huanaco-Koka *f*, Bolivianische Koka *f (von Erythroxylum coca Lam.)*
hübnerite *s.* huebnerite
Hückel molecular orbital Molekülorbital (Molekelorbital) *n* nach Hückel, Hückelsches Molekülorbital (Molekelorbital) *n*
H.U.C.R. *s.* highest useful compression ratio
Hudson lactone rule Hudsonsche Laktonregel *f*, Laktonregel *f* [der optischen Drehung]
hue Ton *m*, Farbton *m*, Farbtönung *f*, Farbschattierung *f*, Nuance *f*, Stich *m*
Huebl number *s.* iodine number
huebnerite *(Min)* Hübnerit *m (Mangan(II)-wolframat)*
hull / to schälen, enthülsen, entspelzen
hull 1. Schale *f*, Hülse *f*, Hülle *f*, Spelze *f*; 2. *(Plast)* dunkler Punkt *m (z.B. im Hartgewebe)*
hulled rice geschälter (entspelzter) Reis *m*
huller Schälmaschine *f*; Schälgang *m*
hulless barley Nacktgerste *f*
hulling mill *s.* huller
human milk Frauenmilch *f*, Muttermilch *f*
humate *(Bodenkunde)* Humat *n*
humboldtilite *(Min)* Humboldtilith *m*
humboldtine *(Min)* Humboldtin *m (Eisenoxalat)*
Hume-Rothery phase Hume-Rothery-Phase *f*, Hume-Rotherysche Phase *f*

Hume-Rothery rule Hume-Rotherysche Regel *f*
humectant Feuchthaltemittel *n*; Netzmittel *n*, Netzer *m*
humic Humus...
h. acid Huminsäure *f*, Humussäure *f*
h. coal Humuskohle *f*, Humit *m*, Huminkohle *f*, humitische Kohle *f*
h. gley soil Gleyboden *m*
h. material *s.* h. substance
h. ortstein *(Bodenkunde)* Humusortstein *m*
h. substance Huminstoff *m*, Huminsubstanz *f*
humid feucht, humid
humidification Befeuchtung *f*, Anfeuchtung *f*; Durchfeuchtung *f*
humidifier Befeuchter *m*, Befeuchtungsanlage *f*, Befeuchtungsapparat *m*
humidify / to anfeuchten, befeuchten, feucht machen, benetzen; durchfeuchten
humidifying nozzle Befeuchtungsdüse *f*
humidity Feuchtigkeit *f*, Feuchte *f*, Feuchtigkeitsgehalt *m*, Feuchtegehalt *m*, *(Tech auch)* Feuchtebeladung *f*; Wassergehalt *m*
h. chamber feuchte Kammer *f*
h. chart Feuchtigkeitsdiagramm *n*, Feuchtigkeitstafel *f*
h. content Feuchtigkeitsgehalt *m*, Feuchtegehalt *m*, *(Tech auch)* Feuchtebeladung *f*
h. dryer *(Ker)* Feucht[luft]trockner *m*
h. drying *(Ker)* Feucht[luft]trocknung *f*
humification Humifizierung *f*, Humifikation *f*
humify / to humifizieren
humite *(Min)* Humit *m*
humulon[e] Humulon *n*, α-Hopfenbittersäure *f*, α-Bittersäure *f*, α-Lupulinsäure *f*
humus Humus *m*
h. formation Humusbildung *f*
h. soil Humusboden *m*
h. substance Humusstoff *m*, Humussubstanz *f*
Hund maximum-multiplicity principle (rule) [Hundsches] Prinzip *n* der größten Multiplizität, [Hundsche] Regel *f* der größten Multiplizität, Hundsches Multiplizitätsprinzip (Prinzip) *n*, [erste] Hundsche Regel *f*
Hund-Mulliken-Lennard-Jones-Hückel method Hund-Mulliken-Lennard-Jones-Hückel-Methode *f*, Molekülorbitalmethode *f*, MO-Methode *f*, Methode *f* der Molekülorbitale
Hund-Mulliken-Lennard-Jones-Hückel theory Hund-Mulliken-Lennard-Jones-Hückel-Theorie *f*, Molekülorbitaltheorie *f*, MO-Theorie *f*, Theorie *f* der Molekülorbitale
Hund principle (rule) of maximum multiplicity, Hund's first rule *s.* Hund maximum-multiplicity principle
Hungarian turpentine Karpatenbalsam *m (von Pinus cembra L.)*
hunger sign *(Landw)* Nährstoffmangelerscheinung *f*
Hunsdiecker reaction Hunsdiecker-Reaktion *f (Silbersalz-Dekarboxylierung)*

hunteriamine Hunteriamin *n (Alkaloid)*
huntilite *(Min)* Huntilith *m (Silberarsenid)*
Huntington-Heberlein process *(Met)* Windröst-
verfahren (Verblaseröstverfahren) *n* [nach Huntington und Heberlein]
Huntington mill Huntington-Mühle *f*, Huntington-
Pendel[rollen]mühle *f*
Huntington ring-roll mill, Huntington roller mill *s.*
Huntington mill
hureaulite *(Min)* Huréaulith *m*
Hurter and Driffield curve *(Foto)* Schwärzungskurve *f*, Gradationskurve *f*
husk / to schälen, enthülsen, entspelzen
husk Schale *f*, Hülse *f*, Hülle *f*, Spelze *f*
husked rice geschälter (entspelzter) Reis *m*
husky barley bespelzte Gerste *f*
hutch Mulde *f*, Faß *n*; Setzfaß *n*, Setzraum *m*,
Setzkasten *m*
h. product Faßgut *n (beim Setzen)*
HWM *s.* high wet modulus fibre
hyacinth *(Min)* Hyazinth *m (Zirkoniumorthosilikat)*
hyalite *(Min)* Hyalit *m (Silizium(IV)-oxid)*
hyalophane *(Min)* Hyalophan *m*
hyalopilitic *(Geol)* hyalopilitisch
hyaluronic acid Hyaluronsäure *f (Polysaccharid)*
hyaluronidase Hyaluronidase *f*
hybrid hybrid
hybrid Hybrid *n*
 sp-hybrid sp-Hybrid *n*
 h. atomic state hybridisierter Atomzustand *m*
 h. bond Hybridzustand *m*
 h. bond orbital *s.* h. orbital
 h. ion Zwitterion *n*, Ampho-Ion *n*
 h. orbital Hybridorbital *n*, Bastardorbital *n*
 h. sp orbital sp-Hybrid-Orbital *n*, sp-hybridisier-
tes Orbital *n*
 h. rock hybrides Gestein *n*, Mischgestein *n*
 h. state Hybridzustand *m*
 h. structure Hybridstruktur *f*, Bastardstruktur *f*
hybridization Hybridisierung *f*, Bastardisierung *f*,
Verwitterung *f*, Mischung *f* von Valenzzu-
ständen
 sp²-hybridization sp²-Hybridisierung *f*, sp²-Ba-
stardisierung *f*, trigonale Hybridisierung (Ba-
~rdisierung) *f*
 ~ize/to hybridisieren, bastardisieren
 ~toin Hydantoin *n*, Glykolharnstoff *m*
hydatogenous *(Geol)* hydatogen
hydnocarpic acid $C_5H_7(CH_2)_{10}COOH$ Hydnokarp-
säure *f*, Hydnocarpussäure *f*
hydnocarpus oil *(Pharm)* Chaulmoograöl *n (von
Hydnocarpus anthelminthica Pierre und H. kurzii
(King)Warb.)*
hydracid Wasserstoffsäure *f*
hydracrylic acid $CH_2OH \cdot CH_2 \cdot COOH$ Hydrakrylsäure *f*, 3-Hydroxypropionsäure *f*, 3-Propanol-
säure *f*, Äthylenmilchsäure *f*
hydrangeic acid Hydrangeasäure *f*, 3,4'-Di-
hydroxy-stilben-karbonsäure *f*
hydrapulper *(Pap)* Stoff[auf]löser *m*, Pulper *m*

hydrargillite *(Min)* Hydrargillit *m*, Gibbsit *m*
(Aluminiumtrihydroxid)
hydrargyrism Quecksilbervergiftung *f*
hydrate/to hydratisieren
hydrate Hydrat *n*; Hydrat *n*, Gashydrat *n (bei der
Erdgasaufbereitung)*
 h. cellulose *s.* hydrated cellulose
 h. water Hydratwasser *n*
hydrated hydratisiert, Hydrat...; gelöscht *(Kalk)*
 h. aluminium-ammonium sulphate
$NH_4Al(SO_4)_2 \cdot 12H_2O$ Ammoniumaluminiumsul-
fat-12-Wasser *n*, Ammoniakalaun *m*, Ammoniumalaun *m*
 h. auric chloride $AuCl_3 \cdot 2H_2O$ Gold(III)-chlorid-
2-Wasser *n*
 h. basic copper carbonate *(Min)* Kupferlasur *m*,
Azurit *m* *(Kupfer(II)-dihydroxiddikarbonat)*
 h. cellulose Hydratzellulose *f*, regenerierte
Zellulose *f*
 h. copper oxide $Cu(OH)_2$ Kupfer(II)-hydroxid *n*
 h. ferric oxide $Fe_2O_3 \cdot nH_2O$ Eisen(III)-oxidhydrat
n
 h. ion hydratisiertes Ion *n*, Aquoion *n*
 h. iron sesquioxide *s.* h. ferric oxide
 h. lime gelöschter Kalk *m*, Löschkalk *m*
 h. magnesium lime *(Landw)* Magnesiumlösch-
kalk *m*
 h. oxide Oxidhydrat *n*
 h. tungstic acid $H_2WO_4 \cdot H_2O$ [weiße] Wolframsäure *f*
hydration Hydra[ta]tion *f*, Hydratisierung *f*, Wasseranlagerung *f*
 h. energy Hydra[ta]tionsenergie *f*, Hydratisie-
rungsenergie *f*
 h. number [of ions] Hydra[ta]tionszahl *f*
 h. of ions Hydra[ta]tion *f* der Ionen, Ionen-
hydra[ta]tion *f*
 h. water Hydratwasser *n*
 h. water of exchangeable ions *(Bodenkunde)*
Schwarmwasser *n*
hydrator Hydrator *m*, Löscher *m*, Löschmaschine
f
hydraulic barker *(Pap)* hydraulischer Entrinder *m*,
Wasserstrahlentrinder *m*, Streambarker *m*
 h. barking *(Pap)* hydraulische Entrindung *f*,
Wasserstrahlentrindung *f*
 h. binder hydraulischer Binder *m*
 h. cement hydraulischer (wasserbindender) Zement *m*, Wasserzement *m*
 h. clamp *(Plast)* hydraulische Werkzeugzu-
haltung *f*
 h. classification Gegenstromklassieren *n*, Auf-
stromklassieren *n*, Stromklassieren *n*
 h. classifier Gegenstromklassierer *m*, Auf-
stromklassierer *m*, Stromklassierer *m*, Strom-
apparat *m*
 h. cure Wasservulkanisation *f*, Vulkanisation *f* in
Wasser
 h. cyclone separator Hydrozyklon *m*, Naßzyklon
m

h. cylinder Hydraulikzylinder *m*
h. efficiency hydraulischer Wirkungsgrad *m*
h. fluid Hydraulikflüssigkeit *f*
h. jig hydraulischer Setzapparat *m*, hydraulische Setzmaschine *f*
h. lime hydraulischer Kalk *m*, Wasserkalk *m*, Zementkalk *m*
h. magazine grinder *(Pap)* hydraulischer Magazinschleifer *m*
h. medium *s.* h. fluid
h. mortar hydraulischer Mörtel *m*, Wassermörtel *m*
h.-operated hydraulisch betätigt (bewegt, betrieben, angetrieben)
h. ram Hydraulikkolben *m*; hydraulische Stempelpresse *f*
h. vulcanizing press hydraulische Vulkanisierpresse *f*
hydrazide Hydrazid *n*
hydrazine NH_2NH_2 Hydrazin *n*, Diamid *n*
h. azide (azoimide) $[N_2H_5][N_3]$ *oder* N_5H_5 Hydrazin[ium]azid *n*
h. base *s.* hydrazine
h. di[hydro]chloride $[N_2H_6]Cl_2$ Hydrazin[ium]dichlorid *n*
h. dinitrate $N_2H_4 \cdot 2HNO_3$ Hydrazin[ium]dinitrat *n*
h. hydrate $H_2N \cdot NH_2 \cdot H_2O$ Hydrazin[ium]hydrat *n*
h. hypophosphate $N_2H_4 \cdot H_4P_2O_6$ Hydrazin[ium]hypophosphat *n*
h. monochloride $N_2H_4 \cdot HCl$ *oder* $[N_2H_5]Cl$ Hydrazin[ium]monochlorid *n*, Hydrazin[ium]chlorid *n*, Hydrazin[ium]hydrogenchlorid *n*
h. mononitrate $N_2H_4 \cdot HNO_3$ Hydrazin[ium]mononitrat *n*, Hydrazin[ium]hydrogennitrat *n*
h. sulphate $[N_2H_6]SO_4$ Hydrazin[ium]sulfat *n*
h. yellow Hydrazingelb O *n*, Tartrazin *n*, Echtwollgelb *n*, Säuregelb *n*, Echtlichtgelb *n*, Flavazin T *n*
hydrazinobenzene $C_6H_5 \cdot NH \cdot NH_2$ Phenylhydrazin *n*
hydrazo compound Hydrazoverbindung *f*
hydrazoate M^IN_3 Azid *n*
hydrazobenzene $C_6H_5NHNHC_6H_5$ Hydrazobenzol *n*, 1,2-Diphenylhydrazin *n*
hydrazoic acid HN_3 Stickstoffwasserstoffsäure *f*, Azoimid *n*
hydrazoxime Hydrazoxim *n* *(Hydrazid der Hydroxamsäure)*
hydride Hydrid *n*
h. displacement law Hydridverschiebungssatz *m*
h. ion H^- Hydridion *n*
h. phase Hydridphase *f*
h. process Hydridverfahren *n*
h. shift Hydridverschiebung *f*
hydrindene Hydrinden *n*, Indan *n*
hydrindone Hydrindon *n*, Indanon *n*, Ketoindan *n*
hydriodic acid HJ Jodwasserstoffsäure *f*
hydrion Wasserstoffion *n*, H-Ion *n*
hydro bottle washing machine Spritzmaschine *f*
h.-gutta-percha Hydroguttapercha *f(n)*

h. peat Hydrotorf *m*, Schwemmtorf *m*, Spritztorf *m*
h.-peat process Hydrotorfverfahren *n*, Spritz[torf]verfahren *n*, hydraulisches Verfahren *n*
h. rubber Hydrokautschuk *m*
hydroaromatic hydroaromatisch
h. compounds Hydroaromaten *pl*, hydroaromatische Verbindungen *fpl*
hydroboracite *(Min)* Hydroborazit *m*
hydroborate $M^I[BH_4]$ Boranat *n*, Metallborwasserstoff *n*, Metallborhydrid *n*, Tetrahydridoborat *n*, Hydridoborat *n*
hydroboration Hydroborierung *f* *(Addition von Diboran an Olefine)*
hydroboron Borwasserstoff *m*, Boran *n*, Borhydrid *n*
hydrobromic acid HBr Bromwasserstoffsäure *f*
hydrobromide Hydrobromid *n*
hydrocarbon Kohlenwasserstoff *m*, KW-Stoff *m*
h. chain Kohlenwasserstoffkette *f*
h. gas Kohlenwasserstoffgas *n*, kohlenwasserstoffhaltiges Gas *n*
h. mixture Kohlenwasserstoffgemisch *n*
h. oil Kohlenwasserstofföl *n*
h. polymer polymerer Kohlenwasserstoff *m*
h. radical Kohlenwasserstoffgruppe *f*, Kohlenwasserstoffrest *m*, Kohlenwasserstoffradikal *n*; [freies] Kohlenwasserstoffradikal *n*
h. resin Kohlenwasserstoffharz *n*
h.-soluble kohlenwasserstofflöslich
h. tail Kohlenwasserstoffrest *m*
h. wax Kohlenwasserstoffwachs *n*
hydrocarbonate M^IHCO_3 Hydrogenkarbonat *n*, primäres (saures) Karbonat *n*
hydrocellulose Hydrozellulose *f*
hydrocerussite *(Min)* Hydrozerussit *m* (basisches Bleikarbonat)
hydrochinone $C_6H_4(OH)_2$ Hydrochinon *n*, p-Dihydroxybenzol *n*, 1,4-Dihydroxybenzol *n*
hydrochlore *(Min)* Hydrochlor *m*, (veraltet für) Pyrochlor *n*
hydrochloric acid HCl Chlorwasserstoffsäure *f*, Salzsäure *f*,
h.-acid extract *(Bodenkunde)* Salzsäureauszug *m*
h.-acid furnace Salzsäureofen *m*
h.-acid gas Chlorwasserstoffgas *n*, Salzsäuregas *n*, HCl-Gas *n*
hydrochloride Hydrochlorid *n*
hydrochlorination Behandlung *f* mit Chlorwasserstoff; Chlorwasserstoff-Anlagerung *f*, Hydrochlorierung *f*
hydrocinnamic acid $C_6H_5CH_2CH_2COOH$ Hydrozimtsäure *f*, Benzylessigsäure *f*, 3-Phenylpropansäure *f*
h. aldehyde $C_6H_5CH_2CH_2CHO$ Hydrozimtaldehyd *m*, β-Phenylpropylaldehyd *m*
hydrocinnamyl alcohol $C_6H_5CH_2CH_2CH_2OH$ Hydrozimtalkohol *m*, Phenylpropylalkohol *m*

hydroclastic rock hydroklastisches Gestein *n*

Hydrocol process Hydrocol-Verfahren *n (Wirbelschichtverfahren zur Fischer-Tropsch-Synthese)*

Hydrocone crusher Hydrocone-Brecher *m (ein Feinbrecher)*

hydrocortisone Hydrokortison *n*, Kortisol *n*, 17α-Hydroxykortikosteron *n*

o-hydrocoumaric acid $HOC_6H_4CH_2CH_2COOH$ o-Hydrokumarsäure *f*, Melilotsäure *f*

hydrocracking Hydrokracken *n*, Hydrokrackung *f*, hydrierendes Kracken (Spalten) *n*, hydrierende Krackung (Spaltung) *f*

hydrocyanic acid HCN Zyanwasserstoffsäure *f*, Blausäure *f*

hydrocyanite *(Min)* Hydrozyanit *m (Kupfer(II)-sulfat)*

hydrodesulphurization Hydrodesulfurierung *f*, hydrierende Entschwefelung *f*, Entschwefelung *f* durch Hydrierung, Wasserstoffentschwefelung *f*, HDS

  h. plant Anlage *f* zur Hydrodesulfurierung (hydrierenden Entschwefelung)

  h. process Hydrodesulfurierungsverfahren *n*, HDS-Verfahren *n (zur hydrierenden Entschwefelung)*

hydrodisodium phosphate Na_2HPO_4 Dinatriumhydrogenphosphat *n*, Natriumhydrogen[ortho]phosphat *n*

hydroextract / to zentrifugieren, [ab]schleudern

hydroextraction Zentrifugieren *n*, Abschleudern *n*, Schleudern *n*

hydroextractor Trockenzentrifuge *f*, Trockenschleuder *f*, Schleudertrockner *m*

hydroferricyanic acid $H_3[Fe(CN)_6]$ Hexazyanoeisen(III)-säure *f*

hydroferrocyanic acid $H_4[Fe(CN)_6]$ Hexazyanoeisen(II)-säure *f*

hydrofiner Hydrofiner *m*, Hydrofining-Anlage *f*

hydrofining Hydrofining *n (katalytische Entschwefelung und Produktverbesserung von Erdölfraktionen bei Anwesenheit von Wasserstoff)*

hydrofluoboric acid $H[BF_4]$ Fluoroborsäure *f*, Tetrafluoroborsäure *f*, Borfluorwasserstoffsäure *f*, Borhydrogenfluorid *n*

hydrofluoric acid HF Fluorwasserstoffsäure *f*, Flußsäure *f*

  h.-acid alkylation Alkylierung *f* mit Fluorwasserstoffsäure (Flußsäure, Fluorwasserstoff), Flußsäurealkylierung *f*, Fluorwasserstoffalkylierung *f*, HF-Alkylierung *f*

  h.-acid process Fluorwasserstoff[säure]verfahren *n*, Flußsäureprozeß *m*

hydrofluoride Hydrofluorid *n*

hydrofluorosilicic acid $H_2[SiF_6]$ Fluorokieselsäure *f*, Hexafluorokieselsäure *f*, Kieselfluorwasserstoffsäure *f*, Kieselflußsäure *f*, Siliziumfluorwasserstoffsäure *f*, Silikofluorwasserstoffsäure *f*

hydroform / to hydroformieren

hydroformate Hydroformat *n*, Hydroforming- Produkt *n*

hydroformer Hydroformer *m*, Hydroforming-Anlage *f*

hydroforming Hydroform[ier]en *n*, Hydroforming *n (ein katalytisches Reformieren mit Wasserstoffumlauf)*

hydroformylation Hydroformylierung *f*, Oxosynthese *f*, Roelen-Reaktion *f*

hydrofranklinite *(Min)* Hydrofranklinit *m*, *(veraltet für)* Chalkophanit *m*

hydrogel Hydrogel *n*

hydrogen H Wasserstoff *m*

  h. acceptor Wasserstoffakzeptor *m*

  h. acid Wasserstoffsäure *f*

  h. arsenate $M^I_2HAsO_4$ Hydrogenarsenat *n*, sekundäres Arsenat *n*

  h. arsenide AsH_3 Arsin *n*, Monoarsin *n*, Arsen(III)-wasserstoff *m*, Arsen(III)-hydrid *n*

  h. atmosphere Wasserstoffatmosphäre *f*

  h. atom Wasserstoffatom *n*

  h. azide HN_3 Stickstoffwasserstoffsäure *f*, Azoimid *n*

  h. blistering Abblättern *n (von Stahl)* infolge Wasserstoffeinflusses

  h. bond Wasserstoff[brücken]bindung *f*, Wasserstoffbrücke *f*, H-Bindung *f*

  h.-bonded über Wasserstoff verbunden

  h. bonding s. h. bond

  h. bridge Wasserstoffbrücke *f*

  h. bridge bond (linkage) s. h. bond

  h. bromide HBr Bromwasserstoff *m*, Hydrogenbromid *n*

  h. carbonate M^IHCO_3 Hydrogenkarbonat *n*, primäres Karbonat *n*

  h. carboxylic acid HCOOH Hydrokarbonsäure *f*, Methansäure *f*, Ameisensäure *f*

  h. chloride HCl Chlorwasserstoff *m*, Hydrogenchlorid *n*

  h. compound Wasserstoffverbindung *f*

  h. cracking *(Korrosion)* Rißbildung *f* infolge Wasserstoffbrüchigkeit

  h. cyanide HCN Zyanwasserstoffsäure *f*, Blausäure *f*

  h.-deficient condition Wasserstoffunterschuß *m*

  h. dioxide s. h. peroxide

  h. disulphide H_2S_2 Wasserstoffdisulfid *n*, Dischwefelwasserstoff *m*, Disulfan *n*

  h. donor Wasserstoffdon[at]or *m*

  h. electrode Wasserstoffelektrode *f*

  h. evolution Wasserstoffentwicklung *f*, Wasserstoffabscheidung *f*

  h. ferrocyanide $H_4[Fe(CN)_6]$ Hexazyanoeisen(II)-säure *f*

  h. fluoride HF Fluorwasserstoff *m*, Hydrogenfluorid *n*

  h. gas Wasserstoffgas *n*

  h. gas electrode Wasserstoffelektrode *f*

  h. half-cell Wasserstoffhalbelement *n*

  h. halide Halogenwasserstoff *m*

  h.-hydrochloric acid cell Chlorknallgaskette *f*

  h. iodide HJ Jodwasserstoff *m*, Hydrogenjodid *n*

h.-iodide equilibrium Jodwasserstoffgleichgewicht *n*
h. ion Wasserstoffion *n*, H-Ion *n*
h.-ion activity Wasserstoffionenaktivität *f*
h.-ion buffer Wasserstoffionenpuffer *m*
h.-ion concentration Wasserstoffionenkonzentration *f*
h.-ion dissociation constant Wasserstoffionen-Dissoziationskonstante *f*
h.-ion measurement Wasserstoffionenmessung *f*
h.-like wasserstoffähnlich
h. nitrate HNO_3 Salpetersäure *f*
h. nucleus Wasserstoffkern *m*, Proton *n*
h. overvoltage *(Ech)* Wasserstoffüberspannung *f*
h. oxalate $M^IHC_2O_4$ Hydrogenoxalat *n*
h. oxide H_2O Wasserstoffoxid *n*, Wasser *n*
h.-oxygen fuel cell Knallgaselement *n*
h.-oxygen reaction Knallgasreaktion *f*
h. pentasulphide H_2S_5 Wasserstoffpentasulfid *n*, Pentaschwefelwasserstoff *m*, Pentasulfan *n*
h. peroxide H_2O_2 Wasserstoffperoxid *n*, Hydrogenperoxid *n*
h. phosphate Hydrogen[ortho]phosphat *n*, Phosphat *n*
h. phosphide Phosphin *n*, Phosphor(III)-hydrid *n*, Phosphorwasserstoff *m*, *(i.e.S.)* PH_3 Monophosphin *n*, gasförmiger Phosphorwasserstoff *m*
h.-poor wasserstoffarm
h. pressure Wasserstoffdruck *m*
h. reduction Wasserstoffreduktion *f*, Reduktion *f* mit Wasserstoff
h.-rich wasserstoffreich
h. salt Hydrogensalz *n*, saures Salz *n*
h. scale for electrode potentials Wasserstoffskale *f* für Elektrodenpotentiale
h. selenide H_2Se Selenwasserstoff *m*, Selenwasserstoffsäure *f*, Hydrogenselenid *n*, Wasserstoffselenid *n*
h. spectrum Wasserstoff[atom]spektrum *n*
h. sulphate M^IHSO_4 Hydrogensulfat *n*, primäres Sulfat *n*
h. sulphide M^IHS Hydrogensulfid *n*, primäres Sulfid *n*; H_2S Wasserstoffsulfid *n*, Schwefelwasserstoff *m*, Monoschwefelwasserstoff *m*, Sulfan *n*, Monosulfan *n*
h. sulphite M^IHSO_3 Hydrogensulfit *n*, primäres Sulfit *n*
h. telluride H_2Te Tellurwasserstoff *m*
h. tetrasulphide H_2S_4 Wasserstofftetrasulfid *n*, Tetrasulfan *n*
h. thiocyanate HSCN Rhodanwasserstoffsäure *f*, Thiozyansäure *f*
h. trisulphide H_2S_3 Wasserstofftrisulfid *n*, Trischwefelwasserstoff *m*, Trisulfan *n*
h. urate $M^IHC_5H_2N_4O_3$ Hydrogenurat *n*
hydrogenability Hydrierbarkeit *f*
hydrogenable hydrierbar
hydrogenase Hydrogenase *f*

hydrogenate / to hydrieren, *(Öle und Fette auch)* härten
hydrogenated coconut oil Kokoshartfett *n*
h. cottonseed oil Baumwollsaathartfett *n*, Cottonhartfett *n*
h. fat hydriertes (gehärtetes) Fett *n*, Hartfett *n*
h. groundnut oil gehärtetes Erdnußöl *n*, Erdnußhartfett *n*
h. oil hydriertes (gehärtetes) Öl (Fett) *n*, Hartfett *n*
h. palm kernel oil Palmkernhartfett *n*
h. peanut oil *s.* **h. groundnut oil**
h. rapeseed oil Rapshartfett *n*
h. rubber Hydrokautschuk *m*
h. soy[a]bean oil Sojahartfett *n*
h. sunflower oil Sonnenblumenhartfett *n*
h. whale oil Waltranhartfett *n*
hydrogenating catalyst Hydrier[ungs]katalysator *m*
h. gas Hydriergas *n*
h. reaction Hydrierreaktion *f*
hydrogenation Hydrieren *n*, Hydrierung *f*, Hydrogen[is]ieren *n*, Hydrogenisation *f*, *(bei Ölen und Fetten auch)* Härten *n*, Härtung *f*
h. catalyst Hydrier[ungs]katalysator *m*
h. cracking hydrierendes Kracken (Spalten) *n*, hydrierende Krackung (Spaltung) *f*, Hydrokracken *n*, Hydrokrackung *f*
h. flavour Härtungsgeruch *m*; Härtungsgeschmack *m*
h. gasoline Hydrierbenzin *n*
h. of coal Kohlehydrierung *f*
h. of fat Fetthydrierung *f*, Fetthärtung *f*
h. of oil Ölhärtung *f*
h. process *(Lebm)* Hydrierverfahren *n*, Härtungsverfahren *n*
h. reaction Hydrierungsreaktion *f*
h. spirit *s.* **h. gasoline**
h. under pressure Druckhydrierung *f*
hydrogenic wasserstoffähnlich
hydrogenide Hydrid *n*
hydrogenize / to hydrieren, *(Öle und Fette auch)* härten
hydrogenolysis Hydrogenolyse *f*, hydrogenolytische Spaltung *f*
hydrogenous wasserstoffhaltig
hydrohalide Halogenwasserstoff *m*
hydrohalite *(Min)* Hydrohalit *m* (Natriumchlorid)
hydrohematite *(Min)* Hydrohämatit *m* (Eisen(III)-oxid)
hydroiodic acid HJ Jodwasserstoffsäure *f*
hydrojuglone $C_{10}H_5(OH)_3$ Hydrojuglon *n*, Trihydroxynaphthalin *n*
hydrolase Hydrolase *f* (hydrolytische Spaltungen katalysierendes Ferment)
hydrolysable *s.* **hydrolyzable**
hydrolysate *s.* **hydrolyzate**
hydrolyse / to *s.* **to hydrolyze**
hydrolysis Hydrolyse *f*; Aquotisierung *f* (Einbau von Aquoliganden in Komplexionen)

h. constant Hydrolysenkonstante f
h. precipitating Hydrolysenfällung f
h. rate Hydrolysengeschwindigkeit f
hydrolyte Hydrolyt m
hydrolytic hydrolytisch
h. acidity (Bodenkunde) hydrolytische Azidität f
h. decomposition (degradation) hydrolytische Spaltung f
h. rancidity hydrolytische Ranzigkeit f, Fettsäureranzigkeit f
h. splitting s. h. decomposition
h. stability Hydrolysebeständigkeit f, Hydrolysestabilität f, Hydrolyseresistenz f
hydrolyting enzyme s. hydrolase
hydrolyzability Hydrolysierbarkeit f
hydrolyzable hydrolysierbar
hydrolyzate Hydrolysat n
hydrolyze / to hydrolysieren
hydromagnesite (Min) Hydromagnesit m (Magnesiumhydroxidkarbonat)
hydrometallurgical hydrometallurgisch, naßmetallurgisch
hydrometallurgy Hydrometallurgie f, Naßmetallurgie f
hydrometer Aräometer n, Senkwaage f, Senkspindel f, Spindel f, Densimeter n
h. scale Aräometerskale f
hydrometric[al] aräometrisch, Aräometer...
hydrometry Dichtebestimmung f von Flüssigkeiten [mit Aräometern]
hydromica (Min) Hydroglimmer m
hydronalium Hydronalium n (Al-Mg-Legierung)
hydronaphthalene Hydronaphthalin n
hydronepheline, hydronephelite (Min) Hydronephelit m
hydronium ion H_3O^+ Hydro[xo]niumion n, Oxoniumion n
hydroperoxidate Peroxohydrat n
hydroperoxide Hydroperoxid n (Sammelname für organische Verbindungen mit der Gruppe $-O\cdot OH$)
h. rearrangement Hydroperoxidumlagerung f
hydrophane (Min) Hydrophan m, Milchopal m (Silizium(IV)-oxid)
hydrophile s. hydrophilic
hydrophilic hydrophil, wasserliebend, wasserfreundlich, wasseraufnehmend, wasseranziehend; benetzbar
h.-lipophilic balance (Text) HLB-Wert m
hydrophiling Hydrophilieren n, Hydrophilierung f
hydrophilite (Min) Hydrophilit m (Kalziumchlorid)
hydrophobe s. hydrophobic
hydrophobic hydrophob, wasserabstoßend, wasserfeindlich, wasserabweisend; nicht in Wasser löslich; nicht benetzbar
h. bonding hydrophobe Bindung f
h. cement wasserabweisender (wasserdichter) Zement m
hydrophobing Hydrophobieren n, Hydrophobierung f

h. agent Hydrophobier[ungs]mittel n, hydrophobierendes Mittel n
hydropigenous hydropigen, wasserfixierend
hydroponic culture, hydroponics (Landw) Hydroponik f, Wasserkultur f
hydroquinine Hydrochinin n, Dihydrochinin n, Chinäthylin n
hydroquinol s. hydroquinone
hydroquinone $C_6H_4(OH)_2$ Hydrochinon n, p-Dihydroxybenzol n, 1,4-Dihydroxybenzol n
h. developer (Foto) Hydrochinonentwickler m
h. monoethyl ether $C_2H_5OC_6H_4OH$ Hydrochinonmonoäthyläther m, p-Äthoxyphenol n
hydrorefining hydrierende Raffination f, Druckwasserstoffraffination f
hydrorubber Hydrokautschuk m
hydroseparator Hydroseparator m (in Überlastung arbeitender Rundeindicker)
hydrosetting (Text) Hydrofixierung f, Heißwasserfixierung f
hydrosiderite (Min) Hydrosiderit m, (veraltet für) Limonit m
hydrosilicofluoric acid s. hydrofluorosilicic acid
hydrosol Hydrosol n
hydrosoluble wasserlöslich
hydrostatic hydrostatisch
h. head filter Schwerkraftfilter n, druckloses Filter n
hydrosulphate $M^I HSO_4$ Hydrogensulfat n, primäres (saures) Sulfat n; $M^I_2S_2O_6$ Dithionat n, (bisher) Hydrosulfat n
hydrosulphide $M^I HS$ Hydrogensulfid n, primäres (saures) Sulfid n
hydrosulphite $M^I HSO_3$ Hydrogensulfit n, primäres (saures) Sulfit n; Hydrosulfit n, (veraltet für) $M^I_2S_2O_4$ Dithionit n; (Farb) $Na_2S_2O_4$ Dithionit n, Natriumdithionit n
hydrosulphuric acid H_2S Wasserstoffsulfid n, Schwefelwasserstoff m, Monoschwefelwasserstoff m, Sulfan n, Monosulfan n
hydrosulphurous acid $H_2S_2O_4$ dithionige Säure f, (bisher) hyposchweflige Säure f
hydrothermal hydrothermal
h. stage (Min) hydrothermales Stadium n
h. synthesis Hydrothermalsynthese f
hydrotropic solution hydrotrope Lösung f
hydrotropy Hydrotropie f
hydrous wasserhaltig; wäßrig, wässerig
h. aluminium oxide Aluminiumoxidhydrat n
h. mica s. hydromica
hydrox cell Knallgaselement n
hydroxamic acid $C_nH_{2n+1}C(=O)NHOH$ Hydroxamsäure f
hydroxide Hydroxid n
h. ion Hydroxidion n
h. ion activity Hydroxidionenaktivität f
hydroxo complex Hydroxokomplex m
h. salt Hydroxosalz n
hydroxoaluminate Hydroxoaluminat n
hydroxoantimonate $M^I[Sb(OH)_6]$ Hydroxoantimonat(V) n, Hexahydroxoantimonat(V) n

hydroxonium ion H_3O^+ Hydro[xo]niumion *n*, Oxoniumion *n*
hydroxostannate Hydroxostannat(IV) *n*
hydroxostannite Hydroxostannat(II) *n*
hydroxy hydroxylhaltig, Hydroxy...
  h. acid Hydroxy[karbon]säure *f*
  β-**h. ester** *β*-Hydroxy[karbon]säureester *m*
  h.-fatty acid Hydroxyfettsäure *f*
  h. group Hydroxy[l]gruppe *f*, Hydroxy[l]rest *m*, OH-Gruppe *f*
  h. ketone Hydroxyketon *n*, Ketonalkohol *m*
  h.-substituted hydroxysubstituiert
hydroxyacetic acid $CH_2OHCOOH$ Hydroxyessigsäure *f*, Hydroxyäthansäure *f*, Äthanolsäure *f*, Glykolsäure *f*
hydroxyalanine $C_3H_7NO_3$ Hydroxyalanin *n*
hydroxyaldehyde Hydroxyaldehyd *m*
  β-**hydroxyaldehyde** *β*-Hydroxyaldehyd *m*, 3-Hydroxyaldehyd *m*, Aldol *n*
p-hydroxyaniline $NH_2C_6H_4OH$ 4-Amino-1-hydroxybenzol *n*, 4-Aminophenol *n*, *p*-Aminophenol *n*
hydroxyanthraquinone Hydroxyanthrachinon *n*
hydroxybenzaldehyde HOC_6H_4CHO Hydroxybenzaldehyd *m*
hydroxybenzene C_6H_5OH Hydroxybenzol *n*, Phenol *n*
o-**hydroxybenzoic acid** $C_6H_4(OH)COOH$ *o*-Hydroxybenzoesäure *f*, Salizylsäure *f*, Spir[oyl]säure *f*
p-**hydroxybenzylpenicillin** *p*-Hydroxybenzylpenizillin *n*, Penizillin X *n*, Penizillin III *n*
β-**hydroxybutanal** *s.* *β*-hydroxybutyraldehyde
hydroxybutanedioic acid $HOOCCH(OH)CH_2COOH$ Hydroxybutandisäure *f*, Hydroxybernsteinsäure *f*, Äpfelsäure *f*, Apfelsäure *f*
β-**hydroxybutyraldehyde** $CH_3CH(OH)CH_2CHO$ 3-Hydroxybutyraldehyd *m*, Butanol-(3)-al-(1) *n*, Azetaldol *n*, Aldol *n* *(i.e.S.)*
hydroxycarboxylic acid Hydroxy[karbon]säure *f*
hydroxycinnamic acid $HOC_6H_4CH=CHCOOH$ Hydroxyzimtsäure *f*, Kumarsäure *f*, Hydroxyphenylpropensäure *f*
17*α*-**hydroxycorticosterone** 17*α*-Hydroxykortikosteron *n*, Kortisol *n*, Hydrokortison *n*
hydroxydroserone Hydroxydroseron *n*
2-**hydroxyethane sulphonic acid** $CH_2(OH)CH_2SO_2OH$ 2-Hydroxyäthansulfonsäure *f*, Äthanol-(1)-sulfonsäure-(2) *f*, Isäthionsäure *f*
hydroxyethanoic acid *s.* hydroxyacetic acid
hydroxyethylcellulose Hydroxyäthylzellulose *f*
hydroxyfluoroborate $M^I[B(OH)F_3]$ Hydroxofluoroborat *n*
hydroxyketone Hydroxyketon *n*, Ketonalkohol *m*
hydroxyl group Hydroxy[l]gruppe *f*, Hydroxy[l]rest *m*, OH-Gruppe *f*
  h. number *s.* h. value
  h. radical Hydroxy[l]gruppe *f*, OH-Gruppe *f*, Hydroxy[l]rest *m*, Hydroxy[l]radikal *n*; [freies] Hydroxy[l]radikal *n*
  h. value Hydroxylzahl *f*, OHZ *(Kennzahl der Fette und fetten Öle)*

hydroxylamine H_2NOH Hydroxylamin *n*
  h. hydrochloride *s.* hydroxylammonium chloride
  h. hydrosulphate *s.* hydroxylammonium sulphate
  h. nitrate *s.* hydroxylammonium nitrate
  h. sulphate *s.* hydroxylammonium sulphate
hydroxylammonium chloride $[H_3NOH]Cl$ Hydroxylammoniumchlorid *n*
  h. nitrate $[H_3NOH]NO_3$ Hydroxylammoniumnitrat *n*
  h. sulphate $[H_3NOH]_2SO_4$ Hydroxylammoniumsulfat *n*
hydroxylate */ to* hydroxylieren
hydroxylation Hydroxylieren *n*, Hydroxylierung *f*
hydroxylic oxygen Hydroxylsauerstoff *m*
2-**hydroxymethylfuran** 2-Hydroxymethylfuran *n*, Furfur[yl]alkohol *m*
2-**hydroxy-3-methyl-1,4-naphthoquinone** 2-Hydroxy-3-methyl-1,4-naphthochinon *n*, Phthiokol *n*
1-**hydroxynaphthalene** 1-Hydroxynaphthalin *n*, *α*-Naphthol *n*
hydroxynaphthoic acid $C_{10}H_6OHCOOH$ Hydroxynaphthoesäure *f*, Hydroxynaphthalinkarbonsäure *f*
hydroxynaphthoquinone $(C_{10}H_5O_2)OH$ Hydroxynaphthochinon *n*
hydroxyphenanthrene Hydroxyphenanthren *n*, Phenanthrol *n*
α-**hydroxy-*α*-phenyl acetophenone** $C_6H_5CHOHCOC_6H_5$ 1-Hydroxy-1-phenylazetophenon *n*, Benzoylphenylkarbinol *n*, Benzoin *n*, Bittermandelölkampfer *m*
hydroxyproline $C_4H_7N(OH)COOH$ Hydroxyprolin *n*, Hydroxy-2-pyrrolidinkarbonsäure *(Eiweißbaustein)*
2-**hydroxypropane nitrile** $CH_3CH(OH)CN$ 2-Hydroxypropannitril *n*, Laktonitril *n*, Milchsäurenitril *n*, Azetaldehydzyanhydrin *n*
2-**hydroxy-1,2,3-propane tricarboxylic acid** $(COOH)CH_2C(OH)(COOH)CH_2COOH$ 2-Hydroxypropantrikarbonsäure-(1,2,3) *f*, 3-Hydroxytrikarballylsäure *f*, Zitronensäure *f*
2-**hydroxyproplonic acid** $CH_3CHOHCOOH$ 2-Hydroxypropionsäure *f*, 2-Hydroxypropansäure *f*, Äthylidenmilchsäure *f*, Laktinsäure *f*, Milchsäure *f* *(i.e.S.)*
d-2-**hydroxypropionic acid** $CH_3CH(OH)COOH$ *d*-2-Hydroxypropionsäure *f*, *d*-Milchsäure *f*, Fleischmilchsäure *f*, Paramilchsäure *f*
3-**hydroxypropionic acid** CH_2OHCH_2COOH 3-Hydroxypropionsäure *f*, Äthylenmilchsäure *f*, Hydrakrylsäure *f*, 3-Propanolsäure *f*
6-**hydroxypurine** 6-Hydroxypurin *n*, 6(1)-Purinon *n*, Hypoxanthin *n*, Sarkin *n*
3-**hydroxy-4-pyrone-2,6-dicarboxylic acid** 3-Hydroxy-4-oxo-1,4-pyran-2,6-dikarbonsäure *f*, Mekonsäure *f*. Mohnsäure *f*, Opiumsäure *f*
hydroxyquinoline Hydroxychinolin *n*
8-**hydroxyquinoline** 8-Hydroxychinolin *n*, Oxin *n*
hydroxysalt Hydroxidsalz *n*

hydroxysuccinic acid $HOOCCH(OH)CH_2COOH$ Hydroxybernsteinsäure *f*, Hydroxybutandisäure *f*, Äpfelsäure *f*, Apfelsäure *f*
hydroxytoluene $CH_3C_6H_4OH$ Hydroxytoluol *n*, Methylphenol *n*, Kresol *n*
β-hydroxytricarballylic acid $(COOH)CH_2C(OH)(COOH)CH_2COOH$ 3-Hydroxytrikarballylsäure *f*, 2-Hydroxypropantrikarbonsäure-(1,2,3) *f*, Zitronensäure *f*
5-hydroxytryptamine 5-Hydroxytryptamin *n*, Serotonin *n*, Enteramin *n*
hydroxyvaline Hydroxyvalin *n*
hydroxyzincate Hydroxozinkat *n*
hydrozincite *(Min)* Hydrozinkit *m*, Zinkblüte *f* *(Zinktrihydroxidkarbonat)*
hyg. s. hygroscopic
hygrinic acid Hygrinsäure *f*, 1-Methyl-2-pyrrolidinkarbonsäure *f*
hygromycin Hygromyzin *n (Antibiotikum)*
hygroscopic[al] hygroskopisch, hygr., wasseranziehend
h. moisture (water) hygroskopisches Wasser *n*
hygroscopicity Hygroskopizität *f*
hymatomelanic acid *(Bodenkunde)* Hymatomelansäure *f*
hyoscine Hyoszin *n*, Atroszin *n*, Skopolamin *n (Alkaloid)*
h. hydrobromide Skopolaminhydrobromid *n*, Tropasäure-6,7-epoxytropylesterhydrobromid *n*
dl-**hyoscyamine** DL-Hyoszyamin *n*, Atropin *n*
hyoscyamus herb *(Pharm)* Bilsenkraut *n (von Hyoscyamus niger L.)*
hypargyrite *(Min)* Hypargyrit *m*, *(veraltet für)* Miargyrit *m (Antimon(III)-silbersulfid)*
hyperconjugation Hyperkonjugation *f*, Baker-Nathan-Effekt *m*
hypereutectic, hypereutectoid übereutektoid[isch]
hyperfine spectrum Hyperfeinspektrum *n*
h. structure *(phys Ch)* Hyperfeinstruktur *f*, HFS
hyperforming Hyperform[ier]en *n*, Hyperformierung *f (Reformieren bei schonendem Katalysatortransport durch ,,Hyperflow lift line")*
hypergol s. hypergolic fuel
hypergolic fuel (rocket propellant) hypergoler (selbstzündender) Treibstoff (Raketentreibstoff) *m*, Hypergol *m*
hyperligated complex stark gebundener Komplex *m*
hypermature überreif
hypernucleus Hypernukleus *m*
hyperon Hyperon *n (überschweres Elementarteilchen)*
hyperosmic acid OsO_4 Osmium(VIII)-oxid *n*, Osmiumtetroxid *n*, *(in der Mikroskopie oft noch)* Osmiumsäure *f*
hypersensitivity Hypersensibilität *f*, Überempfindlichkeit *f*
hypersensitization Hypersensibilisierung *f*, Hypersensibilisation *f*, Übersensibilisierung *f*
hypersonic Hyperschall..., Überschall...

hypersorption Hypersorption *f*
hypersthene *(Min)* Hypersthen *m*
hypertonic solution hypertonische Lösung *f*
hypervitaminosis Hypervitaminose *f*
hypidiomorphic hypidiomorph
hypnotic Hypnotikum *n*, Schlafmittel *n*
hypo $Na_2S_2O_3$, Hypo *n (Kurzbezeichnung für)* Natriumthiosulfat *n*, *(Foto auch)* Fixiernatron *n*
h. bath *(Foto)* Fixierbad *n*
h. elimination *(Foto)* Fixiernatronzerstörung *f*, Thiosulfatentfernung *f*
h. eliminator *(Foto)* Fixiernatronzerstörer *m*
hypobromite M^IOBr Hypobromit *n*
hypobromous acid $HOBr$ hypobromige (unterbromige) Säure *f*
hypochlorite M^IOCl Hypochlorit *n*
h. bleach [liquor] Hypochloritbleichlauge *f*
h. bleaching Hypochloritbleiche *f*
h. of lime $Ca(OCl)_2$ Kalziumhypochlorit *n*
h. process *(Erdöl)* Hypochloritverfahren *n*
h. stage *(Pap)* Hypochloritbehandlung *f*, Hypochloritbleiche *f (Bleichstufe)*
h. sweetening *(Erdöl)* Hypochloritsüßen *n*, Hypochloritbehandlung *f*
h. treating (treatment) s. 1. h. sweetening; 2. h. stage
hypochlorous acid $HOCl$ hypochlorige (unterchlorige) Säure *f*
hypocrystalline hypokristallin, halbkristallinisch
hypodiphosphoric acid s. hypophosphoric acid
hypoeutectic, hypoeutectoid untereutektoid[isch]
hypogene *(Geol)* hypogen, in der Tiefe gebildet
h. rock Tiefengestein *n*, plutonisches Gestein *n*, Plutonit *m*
hypoid lubricant Hypoidöl *n*
hypoiodite Hypojodit *n*
hypoiodous acid HOJ hypojodige (unterjodige) Säure *f*
hypoligated complex schwach gebundener Komplex *m*
hypomagma Tiefenmagma *n*, Hypomagma *n*
hyponitrite Hyponitrit *n*
hyponitrous acid $H_2N_2O_2$ hyposalpetrige (untersalpetrige) Säure *f*
hypophosphate $M^I_4P_2O_6$ Hypophosphat *n*
hypophosphite $M^IH_2PO_2$ Hypophosphit *n*
hypophosphoric acid $H_4P_2O_6$ Hypophosphorsäure *f*, Unterdiphosphorsäure *f*
hypophosphorous acid H_3PO_2 hypophosphorige (unterphosphorige) Säure *f*
hyposiderite *(Min)* Hyposiderit *m*, *(veraltet für)* Limonit *m*
hyposulphate $M^I_2S_2O_6$ Dithionat *n*, *(bisher)* Hyposulfat *n*
hyposulphite $M^I_2S_2O_4$ Dithionit *n*, *(bisher)* Hyposulfit *n*; $M^I_2S_2O_3$ Thiosulfat *n*; *(Foto)* $Na_2S_2O_3$ Fixiersalz *n*, Natriumthiosulfat *n*
hyposulphurous acid $H_2S_2O_4$ dithionige Säure *f*, *(bisher)* hyposchweflige Säure *f*
hypotensive *(Pharm)* blutdrucksenkend

hypothetical hypothetisch, angenommen, auf einer Annahme (Vermutung) beruhend
hypotonic solution hypotonische Lösung f
hypovanadic oxide VO_2 Vanadin(IV)-oxid n, Vanadindioxid n
hypovanadous oxide VO Vanadin(II)-oxid n, Vanadin[mon]oxid n
h. sulphate VSO_4 Vanadin(II)-sulfat n
hypovitaminosis Hypovitaminose f
hypoxanthine Hypoxanthin n, 6-Hydroxypurin n, 6(1)-Purinon n, Sarkin n
hypoxanthosine Hypoxanthosin n, Inosin n, Ino
hypsochromic hypsochrom, farberhöhend
h. effect hypsochromer Effekt m, Hypsochromie f, Farberhöhung f, Farbaufhellung f, negative Farbänderung f
h. group hypsochrome (farberhöhende) Gruppe f
hyssop oil Ysopöl n *(aus Hyssopus officinalis L.)*
hysteresis Hysterese f, Hysteresis f; *(Gum)* Hysteresis f, Arbeitsverlust m, Dämpfung f
h. loop Hystereseschleife f, Hysteresisschleife f
h. loss Hystereseverlust m, Hysteresisverlust m
Hytor pump Flüssigkeitsringverdichter m; Flüssigkeitsringpumpe f; Flüssigkeitsringgebläse n; Elmo-Pumpe f

I

i. s. insoluble
iboga alkaloid Ibogaalkaloid n
ibogaine Ibogain n *(Alkaloid)*
ibogamine Ibogamin n *(Alkaloid)*
ibogine s. ibogaine
I.B.P. s. initial boiling point
IC *(Abk. für)* Institute of Chemistry
ice 1. Eis n, Natureis n; Eisschicht f, Eisdecke f; 2. s. i. cream
i. bath Eisbad n
i. bordeaux Eisbordeaux n, α-Naphthylaminbordeaux n
i. cake Eisblock m
i. calorimeter Eiskalorimeter n
i.-cold eiskalt, eisgekühlt
i. colour Eisfarbe f
i.-colour base Eisfarbbase f
i.-cooled s. i.-cold
i. cooling Eiskühlung f
i. cream Speiseeis n, Gefror[e]nes n, Eis n, *(i.e.S.)* Eiskrem f
i.-cream freezer (freezing machine) Eismaschine f
i.-cream making Speiseeisbereitung f
i. crusher Eismühle f
i. crystal Eiskristall m
i. dye Eisfarbstoff m, Eisfarbe f
i. freezing machine s. i.-cream freezer
i.-house Eiskeller m, Eislager n

i.[-making] machine Eiserzeugungsanlage f
i. paper Eispapier n, Kristallisationspapier n
i. plant Eiserzeugungsanlage f, Eisfabrik f
i. refrigeration Eiskühlung f
i. spar *(Min)* Eisspat m, *(veraltet für)* Sanidin m
i. stone *(Min)* Kryolith m *(Natriumfluoroaluminat)*
i. water Eiswasser n, eiskaltes Wasser n; Schmelzwasser n
i.-water method Naßverfahren n, Duschverfahren n *(bei der Margarineherstellung)*
ICE *(Abk. für)* Institution of Chemical Engineers
icebox Eisschrank m
iced water Eiswasser n, eiskaltes Wasser n
Iceland moss Isländisches Moos n, Cetraria islandica (L.) Acharius
I. spar *(Min)* Islandspat m, isländischer Doppelspat m *(Kalziumkarbonat)*
ichthyocol[l] Fischleim m; Hausenblasenleim m, Hausenblase f
ichthyotoxin Ichthyotoxin n, Fischgift n
icing 1. Vereisen n, Vereisung f; 2. Zuckerglasur f, Zuckerguß m
i. sugar Puderzucker m, Staubzucker m
ICSH s. interstitial-cell-stimulating hormone
i.d., I.D. s. inside diameter
idaein s. idein
ideal crystal Idealkristall m
i. gas ideales (vollkommenes) Gas n
i. gas equation [thermische] Zustandsgleichung f idealer Gase, ideale Zustandsgleichung f für Gase
i. plate *(Destillation)* theoretischer (idealer) Boden m, *(i.w.S.)* theoretische Trennstufe f
i. plate number *(Destillation)* theoretische Bodenzahl f
i. solution ideale Lösung f
ideality Idealzustand m
idein Idaein n *(ein Pflanzenfarbstoff)*
identification Identifizierung f, Identitätsprüfung f; Nachweis m
i. test Identitätsprüfung f
identify / to identifizieren; nachweisen; bestimmen, feststellen
identifying reaction Identitätsreaktion f, Identifizierungsreaktion f
idiochromatic idiochromatisch
idiomorphic, idiomorphous idiomorph, automorph
idle außer Betrieb befindlich, stillstehend
idler [roller] Tragrolle f, Stützrolle f *(z.B. eines Förderbandes)*
idocrase *(Min)* Idokras m, *(veraltet für)* Vesuvian m
idose Idose f *(Monosaccharid)*
idrialine, idrialite *(Min)* Idrialin n
igneous feurig; glühend; *(Geol)* vulkanisch, magmatisch, eruptiv
i. electrolysis Schmelz[fluß]elektrolyse f
i. intrusion Einwirkung f heißen Gesteins, vulkanischer Einfluß m
i. magma [feuerflüssiges] Magma n
i. rock magmatisches Gestein n, Magmagestein n, Magmatit m, Eruptivgestein n

ignitability Entzündbarkeit f
ignitable entzündbar
ignite / to entzünden, anzünden; sich entzünden; zünden; glühen
 i. to constant weight bis zur Massekonstanz glühen
igniter Zündeinrichtung f, Zündvorrichtung f, Zünder m; Zündmittel n; Zündofen m, Zündhaube f (einer Sintermaschine)
ignitibility s. ignitability
ignitible s. ignitable
igniting s. ignition
ignition Entzünden n, Anzünden n; Zünden n, Zündung f; Glühen n
 i. cap Zündhütchen n
 i. crucible Schmelztiegel m
 i. delay Zündverzug m, Zündverzögerung f
 i. delay period Zündverzugszeit f, Zündverzug m
 i. furnace Zündofen m
 i. loss (Ker) Glühverlust m, Glv., GV
 i. pellet Zündkirsche f
 i. performance (quality) Zündeigenschaften fpl, Zündverhalten n, Zündwilligkeit f
 i. system Zündsystem n, Zündanlage f
 i. temperature Zünd[ungs]temperatur f, Entzündungstemperatur f; Verpuffungstemperatur f
 i. test tube Glühröhrchen n
 i. time Zündzeit f
 i. tube test Glührohrprobe f
IIR s. isobutylene-isoprene rubber
ilesite (Min) Ilesit m (Mangansulfat-7-Wasser)
ill-defined von unklarer Zusammensetzung
 i.-smelling übelriechend, unangenehm riechend
Illingworth process Illingworth-Verfahren n (Schwelverfahren)
illinium ll Illinium n, (veraltet für) Pm Promethium n
illite series Illit-Gruppe f (Sammelname für glimmerartige Tonminerale)
Illorin gum Ilhurinbalsam m (von Daniella thurifera Bennett)
illuminance Beleuchtungsstärke f
illuminant Beleuchtungsmittel n, Leuchtmittel n, Leuchtstoff m
illuminating gas Leuchtgas n, (veraltet für) Stadtgas n
 i. lens Beleuchtungslinse f, Beleuchtungsobjektiv n
 i. oil Leuchtöl n, Leuchtpetroleum n, Lampenpetroleum n
illumination / under unter Lichteinwirkung (Lichteinfluß, Einstrahlung von Licht), am (im) Licht, bei Belichtung
illuvial horizon (Geol) Illuvialhorizont m, Einschwemmungshorizont m, Anreicherungshorizont m, B-Horizont m
illuviation (Bodenkunde) Einwaschung f
ilmenite (Min) Ilmenit m (Eisen(II)-metatitanat)
ilvaite (Min) Ilvait m, Lievrit m (ein Sorosilikat)
image contrast (Foto) Bildkontrast m

imbibe / to (Flüssigkeiten) aufnehmen, aufsaugen, einsaugen, absorbieren, imbibieren; (mit einer Flüssigkeit) [durch]tränken, imprägnieren, imbibieren
imbibition Flüssigkeitsaufnahme f, Imbibition f; Tränkung f, Durchtränkung f, Imprägnierung f, Imbibition f, Durchdringung f (mit Flüssigkeiten)
imbitter / to (z. B. Bier durch Hopfenzugabe) einen bitteren Geschmack verleihen, bitter machen
IMC s. initial moisture content
imenes Imene npl, Nitrene npl (Molekülfragmente mit einem Elektronensextett am Stickstoff)
Imhoff tank Emscherbrunnen m, Imhoff-Brunnen m
imidazole Imidazol n, Glyoxalin n
imide Imid n
 i. chloride Imidchlorid n
imidic acid Imidsäure f
imido ester s. imino ether
 i. residue s. imino group
imine Imin n, Iminoverbindung f
imino acid Iminosäure f
 i. ether Iminoäther m
 i. group Iminogruppe f, Imidogruppe f, =NH-Gruppe f
 i. urea $NH=C(NH_2)_2$ Iminoharnstoff m, Aminomethanamidin n, Guanidin n
imitation art paper Naturkunstdruckpapier n
 i. chromo board Chromoersatzkarton m
 i. handmade paper imitiertes Büttenpapier n, Büttenersatzpapier n, Büttenpapier n mit imitiertem Büttenrand
 i. kraft [paper] imitiertes Kraftpapier n
 i. leather Kunstleder n
 i. leather board imitierte Lederpappe f
 i. parchment Pergamentersatzpapier n, Pergamentersatz m
 i. pressboard Preßspanersatz m
Immergan process Immergangerbung f (mit Paraffinsulfochloriden)
immerge / to s. to immerse
immerse / to [ein]tauchen, untertauchen, versenken
immersion Tauchen n, Eintauchen n, Untertauchen n, Versenken n, Immersion f
 i. colorimeter Eintauchkolorimeter n, Tauchkolorimeter n
 i. electrode Eintauchelektrode f, Tauchelektrode f
 i. filter tube Eintauchnutsche f
 i. freezing (Lebm) Eintauch[gefrier]verfahren n
 i. heater Tauchsieder m
 i. jigger (Text) Unterflottenjigger m, Unterwasserjigger m
 i. lens Immersionslinse f
 i. liquid Immersion[sflüssigkeit] f
 i. method Tauchverfahren n
 i. objective Immersionsobjektiv n
 i. oil Immersionsöl n

i. refractometer Eintauchrefraktometer *n*
i. roll Eintauchwalze *f*, Tauchwalze *f*
i. ultramicroscope Immersionsultramikroskop *n*
i. vibrator Tauchrüttler *m*, Tiefenrüttler *m*, Innenrüttler *m*, Innenvibrator *m*
immiscibility Un[ver]mischbarkeit *f*, Nichtmischbarkeit *f*
immiscible un[ver]mischbar, nicht mischbar
immix / to sich vermischen
immixture Vermischen *n*, Vermischung *f*
immobile unbeweglich, stationär
i. phase unbewegliche (stationäre) Phase *f* *(Chromatografie)*
immobilization *(Landw)* Immobilisierung *f*, biologische Festlegung *f (von Nährstoffen in Bodenmikroorganismen)*
immobilize / to *(Landw) (Nährstoffe)* immobilisieren, biologisch festlegen
immune immun, *(gegen Krankheitserreger)* unempfindlich
i. body Immunkörper *m*, Antikörper *m*
i. globulin Immunglobulin *n*
i. milk Immunmilch *f*, Antikörpermilch *f*
i. protein Immunprotein *n*, Antikörperprotein *n*
i. serum Immunserum *n*
i. serum globulin s. i. globulin
i. substance Immunstoff *m*
immunity Immunität *f*
immunization Immunisierung *f*
immunize / to immunisieren, immun machen
immunobiology Immunbiologie *f*
immunochemical immunochemisch
immunochemistry Immunochemie *f*
immunogenic immunogenetisch
immunogenicity Immunogenetik *f*
immunology Immunologie *f*
impact Stoß *m*, Anstoß *m*, Zusammenstoß *m*, Aufprall *m*, Anprall *m*, Zusammenprall *m*; Schlag *m*, Aufschlag *m*
i. bending strength Schlagbiegefestigkeit *f*, Schlagbiegezähigkeit *f*
i. cross section Stoßquerschnitt *m*
i. crusher Prallbrecher *m*
i. crushing Prallzerkleinerung *f*
i. effect Stoßwirkung *f*; Schlagwirkung *f*
i. electron Stoßelektron *n*, Anstoßelektron *n*
i. excitation Stoßanregung *f*
i. fatigue testing machine Dauerschlagwerk *n*
i. figures *(Krist)* Schlagfiguren *fpl*
i. fluorescence Stoßfluoreszenz *f*
i. hardness Schlaghärte *f*
i. idler gepolsterte Tragrolle *f*
i. ionization Stoßionisation *f*, Stoßionisierung *f*
i. mill Prallmühle *f*
i. moulding Schlagpressen *n*
i. nozzle Pralldüse *f*
i. parameter Stoßparameter *m*
i. pendulum machine Pendelschlagwerk *n*
i. polystyrene schlagfestes Polystyrol *n*
i. pressure Stoßdruck *m*

i. radiation Stoßstrahlung *f*
i. resilience Rückprallelastizität *f*, Stoßelastizität *f*, elastischer Wirkungsgrad *m* bei Stoß-Druck-Beanspruchung
i. resistance Schlagfestigkeit *f*, Schlagzähigkeit *f*, Stoßfestigkeit *f*
i.-resistant schlagfest, stoßfest
i. spectrum Stoßspektrum *n*
i. strength s. i. resistance
i. test Schlagprüfung *f*, Schlagprobe *f*
impalpable nicht tastbar, sehr fein
impeller Laufrad *n*; Schnellrührer *m*
i. agitator mechanischer Rührer (Rührapparat) *m*
i. blade Laufschaufel *f*
i. Reynolds number Reynoldszahl *f* des Rührvorgangs
i. vane s. i. blade
impenetrability s. impermeability
impenetrable s. impermeable
impenetrate / to ganz (vollständig) durchdringen
imperatorin Imperatorin *n (aus Peucedanum ostruthium L.)*
imperfect cleavage *(Krist)* unvollkommene Spaltbarkeit *f*
i. crystal gestörter (realer) Kristall *m*, Realkristall *m*
imperfection *(Krist)* Störung *f*, Fehler *m (i.w.S.)*
imperial green $Cu(CH_3COO)_2 \cdot 3Cu(AsO_2)_2$ Kaisergrün *n*, *(veraltet für)* Schweinfurter Grün *n*, Kupferarsenitazetat *n*
impermeability Impermeabilität *f*, Undurchlässigkeit *f*, Undurchdringlichkeit *f*
impermeable impermeabel, undurchlässig, undurchdringlich, undurchgängig, dicht
i. to water wasserdicht, wasserundurchlässig
impervious s. impermeable
imperviousness s. impermeability
impinge / to aufprallen, auftreffen, anstoßen, [zusammen]stoßen; schlagen *(von einer Flamme)*
impingement Stoß *m*, Anstoß *m*, Zusammenstoß *m*, Aufprall *m*, Anprall *m*, Zusammenprall *m*
i. black Channel-Ruß *m*, Channel Black *n*, Kanalruß *m*
i. collector Prall[ab]scheider *m*
i. corrosion Korrosion *f* infolge Oberflächenangriffs durch turbulent strömende Medien
i. resistance Stoßfestigkeit *f*
i. separation Prallabscheiden *n*
i. separator Prall[ab]scheider *m*; Stoßabscheidekammer *f*, Stoßabscheider *m*
impoverish / to *(Bodenkunde)* verarmen
impoverishing *(Bodenkunde)* Verarmung *f*
impregnant s. impregnating agent
impregnate / to imprägnieren, [durch]tränken
i. beverages Getränke mit CO_2 imprägnieren (sättigen), karbonisieren
impregnating s. impregnation
i. agent Imprägnier[ungs]mittel *n*
i. liquor *(Pap)* Tränklauge *f*
i. material Imprägnierungsmasse *f*, Tränkmasse *f*

i. resin Imprägnierharz *n*, Tränkharz *n*
i. solution Imprägnierlösung *f*
i. stage *(Pap)* Imprägnierungsperiode *f*, Durchtränkungsperiode *f*
impregnation Imprägnieren *n*, Imprägnierung *f*, Tränken *n*, Durchtränken *n*, Tränkung *f*, Durchtränkung *f*
i. material Tränkstoff *m*, Imprägnierungsmittel *n*
i. period *(Pap)* Imprägnierungsperiode *f*, Durchtränkungszeit *f*; *(Pap)* Ankochperiode *f*
impressed e.m.f. method Fremdstromverfahren *n* *(katodischer Korrosionsschutz)*
i. watermark *(Pap)* Molettewasserzeichen *n*
impression moulding *(Plast)* Kontaktpreßverfahren *n*, Kontaktpressen *n*, Niederdruckpreßverfahren *n*, Handauflegeverfahren *n*
improve / to verbessern; verfeinern, veredeln
improvement 1. Verbesserung *f*; Verfeinerung *f*, Vered[e]lung *f*; 2. *s.* flour improvement
improver *(Lebm)* Verbesserungsmittel *n*, *(i.e.S.) s.* flour improver
impulse rendering Impulsschmelzen *n* *(ein Fettextraktionsverfahren)*
i. sealing Impulssiegeln *n* *(z.B. von Folien)*
impure unrein, nicht rein, verunreinigt, verschmutzt, mit Beimischungen, verfälscht
impurity Verunreinigung *f*, Verschmutzung *f*, Fremdbestandteil *m*, Fremdstoff *m*, Beimengung *f*, Beimischung *f*
i. atom Fremdatom *n*, Stör[stellen]atom *n*, Fremdstörstelle *f*, [chemische] Störstelle *f*
i. band Störstellenband *n*, Verunreinigungszone *f*
i. electric conductivity Stör[stellen]leitung *f*, Extrinsic-Leitfähigkeit *f*, Extrinsic-Leitung *f*
i. interstitial Einlagerungsfremdatom *n*
i. level Stör[stellen]niveau *n*, Stör[stellen]term *m*, Verunreinigungsniveau *n*
i. semiconductor Stör[stellen]halbleiter *m*, Störleiter *m*, Fremd[halb]leiter *m*, Fremdstoffhalbleiter *m*
in-between alkali stage *(Pap)* alkalische Zwischenbehandlung *f* *(Bleichstufe)*
in-flow of pulp *(Pap)* Stoffeintritt *m*
in-glaze decoration *(Ker)* Zwischenglasurdekor *n*, Inglasurdekor *n*
in-glaze painting *(Ker)* Inglasurmalerei *f*
in-line agitator Rohrleitungsrührwerk *n*
in-line blending Mischen (Vermischen) *n* in der Pumpleitung, „in-line"-Mischung *f*
in place (situ) in situ, in natürlicher Lage, an Ort und Stelle entstanden, am Ort des Entstehens liegend, am Bildungsort liegend, bodenständig, bodeneigen, ortseigen, autochthon
in situ concrete Ortbeton *m*
in situ polymerization Polymerisation *f* an Ort und Stelle, In-situ-Polymerisation *f*
inactivate / to inaktivieren, desaktivieren, unwirksam machen, neutralisieren
inactivation Inaktivierung *f*, Desaktivierung *f*

inactive inaktiv, reaktionslos, nichtreaktionsfähig, reaktionsträge, inert; optisch inaktiv, razemisch; nichtradioaktiv
i. black inaktiver Ruß *m*, Inaktivruß *m*
i. compound inaktive Verbindung *f*, *(i.e.S.)* razemisches (optisch-inaktives) Gemisch *n*, Razemat *n*; nichtradioaktive (stabile) Verbindung *f*
i. filler inaktiver (inerter, passiver) Füllstoff *m*
i. gas inertes Gas *n*, Inertgas *n*, Edelgas *n*
i. tartaric acid $HOOC(CHOH)_2COOH$ razemische Weinsäure (Säure) *f*, *dl*-Weinsäure *f*, Traubensäure *f*, para-Weinsäure *f*
inactivity Inaktivität *f*, Trägheit *f*
INAH *s.* isonicotinic acid hydrazide
incandesce / to weißglühen, auf (bis zur) Weißglut erhitzen
incandescence, incandescency Weißglühen *n*; Weißglut *f*
incandescent weißglühend; glühend, Glüh...
i. filament Glühfaden *m*; Glühdraht *m*, Leuchtdraht *m*
i. light lamp Glühlampe *f*
i. mantle Glühkörper *m*, Glühstrumpf *m*
i. zone Glühzone *f*, glühende Zone *f*
incendiary gel Brandgel *n*
incense Weihrauch *m*, Olibanum *n*
i. paper Räucherpapier *n*; Weihrauchpapier *n*
incidence Einfall *m* *(z.B. eines Strahls)*
i. angle Einfall[s]winkel *m*
incident einfallend, auffallend, auftreffend, Einfalls...
i. light einfallendes (auffallendes) Licht *n*; Primärlicht *n*
incinerate / to veraschen
incinerating *s.* incineration
incineration Veraschen *n*, Veraschung *f*
i. dish Veraschungsschälchen *n*
i. procedure Veraschungsverfahren *n*
incinerator Müllverbrennungsanlage *f*, Müllverbrennungsofen *m*; *(Pap)* Verascher *m*
inclined bucket elevator Schrägbecherwerk *n*
i. evaporator Schrägrohrverdampfer *m*
i. hoist Schrägaufzug *m*
i.-seat valve Schrägsitzventil *n*, Freiflußventil *n*, Vollflußventil *n*
inclosing rock Nebengestein *n*, umgebendes Gestein *n*
inclusion Einschluß *m*, Einlagerung *f*, Inklusion *f*
i. compound Einschlußverbindung *f*, Inklusionsverbindung *f*
i. of air Lufteinschluß *m*
incombustible un[ver]brennbar, nicht brennbar
incombustible Unverbrennbares *n*, Unverbrennliches *n*
i. paper feuerfestes (feuersicheres, unentflammbares, nicht entflammbares) Papier *n*, *(i.e.S.)* flammensicheres (flammensicher imprägniertes, schwer entflammbares) Papier *n*
incompatibility Unverträglichkeit *f*, Inkompatibilität *f*

incompatible unverträglich, inkompatibel
incomplete unvollständig, unvollkommen
i. miscibility begrenzte (teilweise) Mischbarkeit f
incompletely miscible begrenzt (teilweise) mischbar
incongruent melting *(phys Ch)* inkongruentes Schmelzen n
i. melting point *(phys Ch)* peritektischer Punkt m
incorporate / to einmischen
incorporation Einmischen n, Einmischung f; Einbau m *(z. B. von Nährstoffen in organische Substanz)*
increase / to vergrößern, vermehren, verstärken, erhöhen, steigern; sich vergrößern (vermehren, verstärken, erhöhen, steigern), [an]wachsen, zunehmen, [an]steigen
increase Vergrößerung f, Vermehrung f, Verstärkung f, Erhöhung f, Zunehmen n, Zunahme f, Wachsen n, Anwachsen n, Zuwachs m, Wachstum n, Steigen n, Ansteigen n, Anstieg m, Steigerung f
i. in brightness *(Pap)* Weißgehaltserhöhung f, Weißgehaltssteigerung f
i. in conductance Leitfähigkeitserhöhung f
i. in pressure Druckerhöhung f, Drucksteigerung f, Druckanstieg m
i. in temperature Temperaturerhöhung f, Temperatursteigerung f, Temperaturanstieg m
i. in volume Volumenvergrößerung f
i. of efficiency Leistungssteigerung f
i. of surface Oberflächenzuwachs m, Oberflächenvergrößerung f
incrust / to inkrustieren, eine Kruste bilden, verkrusten; inkrustieren, mit einer Kruste überziehen (bedecken), überkrusten
incrustants Inkrusten pl, inkrustierende Substanzen fpl, Inkrustsubstanzen fpl
incrustation Inkrustierung f, Inkrustation f, Krustenbildung f, (i.e.S.) Kesselsteinbildung f; Inkrustation f, Kruste f, (i.e.S.) Kesselstein m
incrusting material (matter, substance) s. incrustants
incubate / to bebrüten, in der Brutkammer aufbewahren *(z.B. Pilzkulturen)*
incubation Bebrütung f, Aufbewahrung f in der Brutkammer
i. period Inkubationszeit f
i. temperature Inkubationstemperatur f
i. test Inkubationsprobe f, Inkubationsversuch m
i. time s. i. period
incubator Brutschrank m
indan Indan n, Hydrinden n
indanthrene s. indanthrone
indanthrone Indanthron n, N-Dihydro-1,2,1′,2′-anthrachinonazin n
i. vat dye Indanthron-Küpenfarbstoff m
indent / to riffeln, rippen
indentation Eindruck m, Einprägung f; Rille f
i. hardness Eindruckhärte f, Kugeldruckhärte f, Eindringhärte f, Druckhärte f

independent joint action *(Toxikologie)* Unabhängigkeitsverbundwirkung f *(ohne gegenseitige Beeinflussung der Komponenten)*
i. migration (mobility) of ions unabhängige Ionenwanderung f
i. producer Zentralgenerator m *(zur Erzeugung von Generatorgas)*
indeterminacy principle Unbestimmtheitsrelation f, Unbestimmtheitsbeziehung f, Unschärferelation f, Ungenauigkeitsrelation f, Ungenauigkeitsbeziehung f, Unsicherheitsbeziehung f
index / to indizieren, beziffern; registrieren
index Index m, Unterscheidungszeichen n; Kennzahl f, Beiwert m, Faktor m; Zeiger m *(eines Geräts)*; Zunge f *(einer Waage)*; Register n, Verzeichnis n, Index m
i. board Karteikarton m, Kartothekkarton m
i. name Registriername m
i. of caking power Back[fähigkeits]zahl f, BZ
i. of refraction Brechungsindex m, Brech[ungs]zahl f, Brechungsquotient m, Brechungskoeffizient m
indexing Indizierung f, Bezifferung f, Registrierung f
i. system Registrierungssystem n
India gum *(Pflanzengummi von Anogeissus latifolia Wall.)*
I. ink [chinesische] Tusche f, Ausziehtusche f
I. paper Bibeldruckpapier n, Dünndruckpapier n; s. I. proof paper
I. proof paper Chinapapier n, Chinesisches Papier n
Indian almond Katappenbaum m, Katappa-Terminalie f, Indischer Mandelbaum m, Terminalia catappa L.
I. aloe *(Aloe-Sorte von Aloe vera L.)*
I. balsam Indischer Balsam m, Perubalsam m *(von Myroxylon balsamum (L.)Harms var. pereirae)*
I. bdellium Indisches Bdellium n, Gugul n *(Balsamharz von Commiphora mukul Engl.)*
I. copal Pineyharz n *(von Vateria indica L.)*; s. East Indian copal
I. corn Mais m, Zea mays L.; Kolbenmais m; Körnermais m
I. geranium (grass) oil Palmarosaöl n, Ostindisches Geranium[gras]öl n *(von Cymbopogon martini (Roxb.)Stapf)*
I. hemp 1. *(Bot)* Indischer Hanf m, Cannabis indica Lam.; 2. *(Pharm)* indischer Hanf m *(getrocknete Triebspitzen von 1.)*; Haschischpräparat n *(i.w.S.)*
I. lake Lack-Lack m, Lac-Lac m, [roter] Färbelack m
I. meal Maismehl n
I. mustard Indischer Senf m, Rai m, Brassica juncea (L.)Coss.
I. paper Chinapapier n, Chinesisches Papier n
I. rubber Assamkautschuk m *(von Ficus elastica Roxb.)*

343 **indium**

l. saffron Kurkuma *f*, Gelber Ingwer *m*, Gelbwurz[el] *f*, Curcuma longa L.
l. sumac *(Gerb)* *(Blätter und Zweige von Anogeissus latifolia Wall.)*
l. tragacanth Indischer (Ostindischer) Tragant *m*, Karayagummi *n*, Sterkuliagummi *n* *(meist von Sterculia urens Roxb.)*
l. turpentine *(Terpentin von Pinus roxburghii Sarg.)*
l. wood oil Gurjunbalsamöl *n*
l. yellow [echtes] Indischgelb *n*, Piuri *n* *(von Mangifera indica L.)*; Aureolin *n*, Kobaltgelb *n*, Indischgelb *n* *(Kaliumhexanitrokobaltat)*
indican 1. $C_{14}H_{17}NO_6$ Pflanzenindikum *n*, Indikan *n* *(Glukosid des Indoxyls)*; 2. $C_8H_7NSO_4$ Harnindikan *n*, Indikan *n* *(Schwefelsäureester des Indoxyls)*
i. reaction Indikanreaktion *f* *(Nachweis von Glukosidase mittels Pflanzenindikans)*
indicate / **to** anzeigen; angeben, kennzeichnen; indizieren, beziffern
indicated hydrogen indizierter Wasserstoff *m*, Wasserstoff *m* mit Zahlenangabe
indicating instrument Anzeigeinstrument *n*, Anzeigegerät *n*, anzeigendes Meßgerät *n*
indication Anzeige *f* *(von Meßgeräten)*; Angabe *f*, Kennzeichnung *f*; Indizierung *f*, Bezifferung *f*
i. of valence Wertigkeitsbezeichnung *f*, Valenzbezeichnung *f*
indicator *(Chem)* Indikator *m*; Anzeigevorrichtung *f*, Anzeiger *m*, Zeiger *m*
i. change Indikatorumschlag *m*
i. constant Indikatorkonstante *f*
i. diagram Indikatordiagramm *n*
i. dye Indikatorfarbstoff *m*
i. electrode Indikatorelektrode *f*
i. for change of hydrogen ion concentration pH-Indikator *m*; pH-Anzeigegerät *n*
i. ingredient *(Lebm)* Erkennungsmittel *n*, Indikator *m*
i. method Indikatormethode *f*, Tracer-Methode *f*
i. paper Indikatorpapier *n*, Reagenzpapier *n*, Prüfpapier *n*
i. range Umschlagbereich *m*, Umschlaggebiet *n*, Indikatorbereich *m*
i. solution Indikatorlösung *f*
i. substance *s.* i. ingredient
i. transition Umschlag *m*, Farbumschlag *m* *(eines Indikators)*
indicolite *(Min)* Indigolith *m*, blauer Turmalin *m*
indifferent indifferent, wirkungslos
i. materials indifferente Stoffe *mpl*
indigestibility Unverdaulichkeit *f*
indigestible unverdaulich
indigo[-blue] indigoblau
indigo [blue] Indig[o]blau *n*, Indigo *m(n)*, Indigotin *n*
i. copper *(Min)* Kupferindig[o] *m*, Kovellin *m* *(Kupfer(II)-sulfid)*
i. paper Indigopapier *n*

i. plant Indigopflanze *f*
indigoid dye Indigofarbstoff *m*, indigoider Farbstoff *m*
indigolite *s.* indicolite
indigosol Indigosol *n* *(Leukoküpenfarbstoffester)*
indirect arc[-heated] furnace indirekter Lichtbogenofen *m*, Lichtbogenofen *m* mit indirekter Beheizung, Lichtbogenofen *m* mit reiner Strahlungsbeheizung, Lichtbogenstrahlungsofen *m*
i. cook process *(Pap)* indirekte Kochung *f*
i. drying Kontakttrocknung *f*, Berührungstrocknung *f*
i.-fired indirekt beheizt
i. heating indirektes Heizen (Beheizen) *n*
i. process indirektes Verfahren (Ammoniakgewinnungsverfahren, NH_3-Gewinnungsverfahren) *n*
i. resistance-heated furnace indirekter Widerstandsofen *m*
i. rotary dryer Röhrentrockner *m*; Trommeltrockner *m* mit Kontaktheizung, Röhrentrommeltrockner *m*
i. solvent mittelbarer (latenter) Löser *m*, Hilfslöser *m*
i. system *s.* i. process
indissoluble unlöslich, nicht löslich, unl., unlö, nl
indistinguishable nicht unterscheidbar
indium In Indium *n*
i. cyanide $In(CN)_3$ Indium(III)-zyanid *n*
i. dibromide $InBr_2$ Indiumdibromid *n*, Indium(II)-bromid *n*
i. dichloride $InCl_2$ Indiumdichlorid *n*, Indium(II)-chlorid *n*
i. diiodide InJ_2 Indiumdijodid *n*, Indium(II)-jodid *n*
i. hydroxide $In(OH)_3$ Indium(III)-hydroxid *n*
i. iodate $In(JO_3)_3$ Indium(III)-jodat *n*
i. monobromide $InBr$ Indium[mono]bromid *n*, Indium(I)-bromid *n*
i. monochloride $InCl$ Indium[mono]chlorid *n*, Indium(I)-chlorid *n*
i. monoiodide InJ Indium[mono]jodid *n*, Indium(I)-jodid *n*
i. monosulphide InS Indium[mono]sulfid *n*, Indium(II)-sulfid *n*
i. monoxide InO Indium[mon]oxid *n*, Indium(II)-oxid *n*
i. nitrate $In(NO_3)_3$ Indium(III)-nitrat *n*
i. perchlorate $In(ClO_4)_3$ Indium(III)-perchlorat *n*
i. selenate $In_2(SeO_4)_3$ Indium(III)-selenat *n*
i. sesquioxide *s.* i. trioxide
i. suboxide In_2O Indiumsuboxid *n*, Indium(I)-oxid *n*
i. subsulphide In_2S Indiumsubsulfid *n*, Indium(I)-sulfid *n*
i. sulphate $In_2(SO_4)_3$ Indium(III)-sulfat *n*
i. tribromide $InBr_3$ Indiumtribromid *n*, Indium(III)-bromid *n*
i. trichloride $InCl_3$ Indiumtrichlorid *n*, Indium(III)-chlorid *n*

i. trifluoride InF$_3$ Indiumtrifluorid *n*, Indium(III)-fluorid *n*
i. triiodide InJ$_3$ Indiumtrijodid *n*, Indium(III)-jodid *n*
i. trioxide In$_2$O$_3$ Indiumtrioxid *n*, Indium(III)-oxid *n*
i. trisulphide In$_2$S$_3$ Indiumtrisulfid *n*, Indium(III)-sulfid *n*
individual curing unit *(Gum)* Einzelheizer *m*
i. kiln *(Ker)* Einzelofen *m*
i. milk Einzelmilch *f*
i. vulcanizer s. i. curing unit
indivisibility Unteilbarkeit *f*
indivisible unteilbar; nicht spaltbar
indole Indol *n*, 2,3-Benzopyrrol *n*
i. alkaloid Indolalkaloid *n*
i. test Indolprobe *f*, Indolnachweis *m*
indoleacetic acid Indol[yl]-3-essigsäure *f*, β-Indolylessigsäure *f*, IES *f*
indolebutyric acid Indol[yl]buttersäure *f*
indolepyruvic acid Indol[yl]brenztraubensäure *f*
indophenol Indophenol *n*, Chinonphenolimin *n*
i. paste Indophenolpaste *f*
indophenoloxidase Indophenoloxidase *f (Ferment)*
indoxyl Indoxyl *n*, 3-Hydroxyindol *n*, 3-Indolol *n*
indoxylcarboxylic acid Indoxylkarbonsäure *f*
induce crystallization / to die Kristallisation anregen
induced draught Saugzug *m*
i.-draught cooling tower Ventilatorkühlturm *m*
i.-draught fan Saugzuggebläse *n*
i. enzyme synthesis induzierte Enzymsynthese *f*
i. flux Induktionsfluß *m*
i. oxidation Autoxydation *f*
i. polarization Verschiebungspolarisation *f*
i. radioactivity induzierte (künstliche) Radioaktivität *f*
i. reaction induzierte Reaktion *f*
i.-roll [magnetic] separator Walzenscheider *m*
inductance furnace s. induction furnace
induction *(Bioch)* Induktion *f*
i. effect *(phys Ch)* Induktionseffekt *m*, induktiver Effekt *m*, I-Effekt *m*
i. factor Induktionsfaktor *m*
i. furnace Induktionsofen *m*, induktionsbeheizter Ofen *m*
i.-harden / to induktionshärten
i. hardening Induktionshärten *n*, Induktionshärtung *f*
i.-heated furnace s. i. furnace
i. heater Induktionsheizgerät *n*
i. heating induktives Beheizen *n*, Induktiv[be]heizung *f*, Induktionsheizung *f*, induktive Erwärmung *f*, Induktionserwärmung *f*
i.-heating furnace s. i. furnace
i. period Induktionsperiode *f*
i. surface hardening Induktionsoberflächenhärten *n*, Induktionsoberflächenhärtung *f*
i. time Induktionszeit *f*

inductive agent *(Bioch)* Induktionsstoff *m*, Induktor *m*
i. effect *(phys Ch)* induktiver Effekt *m*, Induktionseffekt *m*, I-Effekt *m*; *(Bioch)* Induktionswirkung *f*
i. reaction *(Bioch)* Induktionsvorgang *m*
inductor *(Bioch)* Induktor *m*, Induktionsstoff *m*; *(Chem)* Induktor *m*
indurate / to härten; indurieren, verhärten *(z.B. von Geweben)*
induration *(Geol)* Erhärtung *f*, Verfestigung *f*, Lithifikation *f*; Induration *f*, Verhärtung *f (z.B. von Geweben)*
industrial alcohol technischer Alkohol *m*, Industriealkohol *m*
i. chemical technische Chemikalie *f*
i. chemist Industriechemiker *m*
i. chemistry industrielle (technische) Chemie *f*, Industriechemie *f*, chemische Technik *f*, Chemietechnik *f*
i. coal Industriekohle *f*, Kohle *f* für Industriezwecke
i. dust Industriestaub *m*
i. gas Industriegas *n*
i. gas burner Industriegasbrenner *m*
i. kiln Industrieofen *m*
i. plant Industrieanlage *f*
i. sewage s. i. wastes
i. spirit s. i. alcohol
i. wastes Industrieabfälle *mpl*; gewerbliches (industrielles) Abwasser *n*, Industrieabwasser *n*, Fabrikabwasser *n*
i. water Brauchwasser *n*, Betriebswasser *n*, Gebrauchswasser *n*, Fabrikationswasser *n*, Nutzwasser *n*
inedibility Ungenießbarkeit *f*
inedible nicht eßbar (genießbar), ungenießbar
i. fat technisches Fett *n*
ineffective unwirksam, wirkungslos
inelastic unelastisch, nicht elastisch
i. collision unelastischer Stoß *m*
i. scattering *(Kch)* unelastische Streuung *f*
inert inert, inaktiv, reaktionsträge, reaktionslos
i. atmosphere Schutzgasatmosphäre *f*
i. black inaktiver Ruß *m*, Inaktivruß *m*
i. filler inaktiver (passiver, inerter) Füllstoff *m*
i. gas inertes Gas *n*, Inertgas *n*, Edelgas *n*; Schutzgas *n*
i.-gas configuration Edelgaskonfiguration *f*
i.-gas core Edelgasrumpf *m*
i.-gas electron configuration s. i.-gas configuration
i.-gas electronic structure s. i.-gas configuration
i.-gas octet Edelgasoktett *n*
i.-gas pressure Edelgasdruck *m*
i.-gas shell Edelgasschale *f*
i.-gas-shielded arc-welding Schutzgas-Lichtbogenschweißen *n*
i.-gas structure s. i.-gas configuration
i. ingredient Beistoff *m*

i. substance indifferenter Stoff *m*
inertia Trägheit *f*, Beharrungsvermögen *n*; *(Foto)* Inertia *f*
i. effect Trägheitsmoment *n*, Beharrungsmoment *n*
i.-free trägheitslos
inertial separator Prall[ab]scheider *m*
inertinite Inertinit *m* *(zusammenfassende Bezeichnung für vier Mazerale der Steinkohle)*
infertility Unfruchtbarkeit *f (des Bodens)*
infiltration Infiltration *f*
infinite dilution unendliche Verdünnung *f*
infl *s.* 1. inflammable; 2. influence
inflame / to anzünden, entzünden; entflammen, sich entzünden
inflammability Entflammbarkeit *f*, Entzündbarkeit *f*, Entzündlichkeit *f*, Brennbarkeit *f*
inflammable entflammbar, entzündbar, entzündlich, brennbar, inflammabel
i. principle *(historisch)* brennbares Prinzip *n*, Feuerstoff *m*, Feuermaterie *f*, Phlogiston *n*
inflammation Inflammation *f*, Entflammung *f*, Entzündung *f*, Zündung *f*
i. time Entflammungszeit *f*
inflating agent Treibmittel *n*, Blähmittel *n*
inflow Einfließen *n*, Einfluß *m*, Einströmen *n*; Zufluß *m*
influence / to einwirken, beeinflussen
influence Einwirkung *f*, Einwirken *n*, Wirkung *f*, Einfluß *m*, Beeinflussung *f*
i. on the stiffness *(Gum)* versteifende Wirkung *f*
influent pipe Zulaufrohr *n*, Einströmrohr *n*, Einlaßrohr *n*
i. well Zulauftauchrohr *n*
infracrustal rock infrakrustales Gestein *n*
infrared infrarot, Infrarot..., IR-..., ultrarot, Ultrarot..., UR-...
infrared *s.* i. radiation
i. absorption Infrarotabsorption *f*
i. absorption spectrum Infrarotabsorptionsspektrum *n*
i. analysis Infrarotanalyse *f*
i. dryer Infrarottrockner *m*
i. drying Infrarottrocknung *f*
i. drying adhesive Infrarotkleber *m*
i. drying oven Infrarottrockenofen *m*
i. frequency Infrarotfrequenz *f*
i. heater Infrarotheizgerät *n*
i. heating Infrarot[be]heizung *f*
i. lamp Infrarotstrahler *m*, Infrarotlampe *f*
i. light Infrarotlicht *n*
i. microscope Infrarotmikroskop *n*
i. photography Infrarotfotografie *f*
i. photometer Infrarotfotometer *n*
i. preheating Infrarotvorwärmung *f*
i. radiation Infrarot *n*, Ultrarot *n*, infrarote (ultrarote) Strahlung *f*, Infrarotstrahlung *f*
i. radiator *s.* i. lamp
i. spectrophotometer Infrarotspektralfotometer *n*, Infrarotspektrometer *n*

i. spectroscopy Infrarotspektroskopie *f*
i. spectrum Infrarotspektrum *n*
i. study Infrarotuntersuchung *f*
i. transmittance Infrarotdurchlässigkeit *f*
i. transmitting infrarotdurchlässig
infusibility Unschmelzbarkeit *f*
infusible unschmelzbar, *(i.e.S.)* schwer schmelzbar
i. white precipitate [$Hg(NH_2)]_nCl_n$ unschmelzbares weißes Präzipitat *n*, Amidoquecksilber(II)-chlorid *n*
infusion *(Pharm)* Aufguß *m*, Infus[um] *n*, Infusionslösung *f*; Infusion *f*, Eingießung *f*, Einfließenlassen *n*
i. mashing (method, process) *(Brauwesen)* Infusionsverfahren *n*, Aufgußverfahren *n*, Auslaugungsverfahren *n*
infusorial earth Infusorienerde *f*, Diatomeenerde *f*, Kieselgur *f*
ingot Block *m*, Gußblock *m*, Rohblock *m*, Ingot *m*; Barren *m*; Substanzbarren *m*, Schmelzbarren *m*, Schmelzling *m*, Barren *m (beim Zonenschmelzen)*
i. casting Blockgießen *n*, Blockguß *m*; Barrenguß *m*
i. iron Flußstahl *m*
i. length Barrenlänge *f*, Länge *f* des Substanzbarrens (Schmelzlings) *(beim Zonenschmelzen)*
i. mould Blockform *f*, Blockkokille *f*, Kokille *f*
i. steel *s.* i. iron
ingrain dye Ingrain-Farbe *f*
i. effect Melangeeffekt *m*
ingredient Ingrediens *n*, Ingredienz *f*, Zutat *f*, Zusatzstoff *m*, Bestandteil *m (z.B. von Arzneien)*
inhalation narcotic Inhalationsnarkotikum *n*
inherent ash innere Asche *f*, Pflanzenasche *f*
i. moisture inneres Wasser *n*, Innenwasser *n*
i. toxicity innewohnende Toxizität *f*
inhibit / to inhibieren, hemmen, verzögern, bremsen, behindern
inhibiter, inhibiting substance *s.* inhibitor
inhibition Inhibition *f*, Inhibitorwirkung *f*, Hemmung *f*, Verzögerung *f*, *(i.e.S.)* negative Katalyse *f*, Antikatalyse *f*
i. curve Hemmungskurve *f*
i. of growth Wachstumshemmung *f*
i. of reaction Reaktionshemmung *f*
inhibitor Inhibitor *m*, Hemmstoff *m*, Verzögerer *m*, Passivator *m*, *(i.e.S.)* negativer Katalysator *m*, Antikatalysator *m*
inhibitory inhibierend, hemmend, verzögernd, bremsend; lindernd
inhomogeneity Inhomogenität *f*, Heterogenität *f*, Verschiedenartigkeit *f*, Ungleichartigkeit *f*
inhomogeneous inhomogen, heterogen, verschiedenartig, ungleichartig zusammengesetzt, innerlich uneinheitlich
initial amount Anfangsmenge *f*
i. boiling point Siedebeginn *m*, SB, Anfangssiedepunkt *m*, Beginn-Siedepunkt *m*, Beginn-Kochpunkt *m*

i. brightness *(Pap)* Weißgehalt *m* (Weißgrad *m*, Weiße *f*) vor der Bleiche
i. charge Ansatz *m*
i. concentration Anfangskonzentration *f*, Ausgangskonzentration *f*
i. deposit Anfangsbelag *m* *(von Pflanzenschutzmitteln)*
i. detonating agent Initialsprengstoff *m*, Primärsprengstoff *m*
i. exposure *(Foto)* Erstbelichtung *f*
i. ignition Initialzündung *f*
i. ionization (ionizing event) Primärionisation *f*
i. material *s.* i. substance
i. mixture Vormischung *f*
i. moisture content Anfangsfeuchte *f*; Anfangsfeuchtebeladung *f*
i. period Vorperiode *f*, Vorversuch *m* *(einer kalorimetrischen Messung)*
i. point Anfangspunkt *m*
i. product Ausgangsprodukt *n*, Vorprodukt *n*
i. rate Anfangsgeschwindigkeit *f*
i. reactant *s.* i. substance
i. retention anfängliche Haftfähigkeit *f* *(von Pflanzenschutzmitteln)*
i. solubility Anfangslöslichkeit *f*
i. solute concentration Ausgangskonzentration *f* des gelösten Stoffes, Ausgangskonzentration *f* der Beimengung (Verunreinigung) *(in der Zonenschmelztheorie)*
i. solution Ausgangslösung *f*
i. spreading coefficient Anfangsspreitungskoeffizient *m*
i. state Ausgangszustand *m*, Anfangszustand *m*
i. substance Ausgangssubstanz *f*, Ausgangsstoff *m*
i. temperature Anfangstemperatur *f*, Initialtemperatur *f*
i. vacuum Vorvakuum *n*
i. value Anfangswert *m*
i. velocity *s.* i. rate
initiate / to initiieren, anregen, einleiten, anfangen, beginnen, in Gang bringen, starten, auslösen
initiated by irradiation strahlungsinitiiert *(z.B. eine Polymerisation)*
initiating agent *s.* initiator
i. electron primäres Elektron *n*, Primärelektron *n*
i. explosive *s.* initiator 1.
i. properties Sprengstoffeigenschaften *fpl*, explosive Eigenschaften *fpl*
i. reaction Startreaktion *f*, Primärreaktion *f*
i. step erste Stufe *f*, Primärstufe *f*
initiation Initiierung *f*, Anregung *f*, Einleitung *f*, Start *m*, Beginn *m* *(z.B. einer Reaktion)*; *(Plast)* Kettenstart *m*
i. of cracks Rißbildung *f*
i. stage Initiierungsstadium *n*, Startreaktionsstadium *n*, Aktivierungsstadium *n*
initiator 1. Initialsprengstoff *m*, Zünd[spreng]stoff *m*, Initialexplosivstoff *m*, initiierender Sprengstoff *m*; 2. Initiator *m*, Aktivator *m*

inject / to [ein]spritzen
i. in a mould in ein Werkzeug einspritzen, in eine Form einspritzen
injection Injektion *f*, Einspritzung *f*
i. apparatus Injektionsgerät *n* *(z.B. für flüssige Düngemittel)*
i. capacity *(Plast)* Spritzvolumen *n*
i. compound *s.* i. moulding compound
i. cure Schnellpökeln *n*, Schnellpökelung *f*
i. cure method Schnellpökelverfahren *n*
i. cylinder *(Plast)* Spritz[gieß]zylinder *m*, Massezylinder *m*
i. defect *(Plast)* Spritzfehler *m*
i. metamorphism *(Geol)* Injektionsmetamorphose *f*
i. mould Spritzgießwerkzeug *n*, Spritzgußwerkzeug *n*, Spritzgußform *f*
i. moulded parison spritzgegossener Vorformling *m*, spritzgegossenes Külbel *n*
i. moulded part Spritzling *m*, Spritzgußteil *n*, Spritzgießteil *n*
i. moulded plastic Spritzgußkunststoff *m*
i. moulder Spritzgießer *m*
i. moulding Spritzgießverfahren *n*, Spritzgußverfahren *n*, Spritzgießen *n*, Spritzguß *m*; Spritzgußteil *n*, Spritzgießteil *n*, Spritzling *m*
i. moulding compound Spritzgußmasse *f*
i. moulding machine Spritzgießmaschine *f*
i. moulding material *s.* i. moulding compound
i. moulding polystyrene Polystyrolspritzgußmasse *f*
i. moulding pressure Spritzdruck *m* beim Spritzgießen
i. piston Spritzkolben *m*
i. plowshare Injektionspflugschar *n* *(für flüssige Düngemittel)*
i. port Einspritzöffnung *f*
i. press Spritzpresse *f*
i. pressure Spritzdruck *m*, Einspritzdruck *m*
i. ram Spritzkolben *m*
i. rate Einspritzrate *f*
i. speed Einspritzgeschwindigkeit *f*
injector Injektor *m*, Druckstrahlpumpe *f*, Dampf[-Druck]strahlpumpe *f*
i. gun Injektor *m* *(zur Bodenentseuchung)*
i. mixer Injektormischer *m*
ink Druckfarbe *f*, [grafische] Farbe *f*; Tinte *f*
i. binder *(Pap)* Druckfarbenbindemittel *n*
i. dye Tintenfarbstoff *m*
i. receptivity *(Pap)* Druckfarbenaufnahmevermögen *n*
i. resistance *(Pap)* Tintenfestigkeit *f*
i. tablet Tintentablette *f*
inleakage Leckluft *f* *(Vakuumtechnik)*
inlet Einlaß *m*, Einlauf *m*, Eintritt *m*, Zuleitung *f*, Zuführung *f*; Maul *n* *(am Backenbrecher)*; *(Plast)* Angußsteg *m*
i. air Zuluft *f*; Frischluft *f*
i. duct Einlaßkanal *m*
i. end Aufgabeende *n* *(einer Herdplatte)*

i. gas temperature Gaseintrittstemperatur f
i. pipe Einleitungsrohr n
i. port Eintrittsöffnung f, Einlauföffnung f
i. sluice Beschickungsschleuse f, Eintragzelle f
i. temperature Eingangstemperatur f, Eintritts-
temperatur f
i. valve Einlaßventil n
i.-valve control unloading Leerlaufregulierung f
*(eines Verdichters durch Offenhalten der Saug-
ventile)*
i. vane Vorleitschaufel f
INN *(Abk. für)* international nonproprietary name
inner complex Innerkomplex m
i. complex anion Innerkomplexanion n
i. complex cation Innerkomplexkation n
i. complex salt inneres Komplexsalz n, In-
nerkomplexsalz n, innere Komplexverbindung f,
innerkomplexe Verbindung f
i. diameter Innendurchmesser m
i. end von der Koksnaht herstammendes Ende n,
Ende n der Koksnaht *(eines Koksstückes)*
i. orbital complex Innenorbitalkomplex m, In-
ner-orbital-Komplex m
i. pipe Kernrohr n
i. quantum number innere Quantenzahl f
i. radius Innenradius m
i. salt inneres Salz n
i. shell innere Schale f, Innenschale f
i. surface innere Oberfläche f, Innenfläche f
i. transition element Übergangselement n *(im
periodischen System)*
i. tube *(Gum)* Luftschlauch m; Schlauchseele f
i. tube compound *(Gum)* Luftschlauchmischung
f; Mischung f für die Schlauchseele, Seelen-
mischung f
i. tube reclaim *(Gum)* Luftschlauchregenerat n
innerliner Innerliner m, Innengummi m
innocuous unschädlich, harmlos
inoculate / to *(eine Person)* impfen; *(Impfmaterial)*
einimpfen; *(z.B. einen Gärtank)* beimpfen
inoculating yeast Mutterhefe f; Anstellhefe f,
Stellhefe f
inoculation Impfen n, Impfung f; Einimpfen n,
Einimpfung f, Inokulation f; Beimpfen n, Beimp-
fung f
inoculum Inokulum n, Impfkultur f; Impfmenge f,
Impfmaterial n
inodorous geruchlos, geruchfrei
inodorousness Geruchlosigkeit f
inorganic anorganisch
i. accelerator *(Gum)* anorganischer Beschleuniger m
i. acid anorganische Säure f
i. benzene $B_3N_3H_6$ anorganisches Benzol n,
Pseudobenzol n, Borazol n, Triborintriamin n
i. chemistry anorganische Chemie f
i. fertilization Mineralstoffdüngung f
i. fertilizer Mineraldüngemittel n
i. fibre anorganischer Faserstoff m; anorganische Faser f

i. polymer anorganisches Polymer[es] n
inosine Inosin n, Hypoxanthosin n, Ino
i.-phosphoric acid s. inosinic acid
inosinic acid Inosinsäure f, Muskelinosinsäure f,
Hypoxanthinribosid-5-phosphorsäure f *(Nukleotid)*
inosite, inositol Inosit[ol] n, Hexahydroxyzyklohexan n
input 1. Aufwand m; 2. *(Gerb)* Einarbeitung f
i. hopper Aufgabetrichter m, Einschütttrichter m,
Beschickungstrichter m
insect attractant Insektenanlockmittel n, In-
sektenlockstoff m
i. bite remedy Insektenstichmittel n
i. dye Insektenfarbstoff m
i. powder Insektenpulver n, Insektenpuder m
i. repellent Insektenabwehrmittel n, Insekten-
schutzmittel n, insektenvertreibendes (insekten-
abschreckendes, insektenabstoßendes) Mittel n
i. wax Insektenwachs n, Chinawachs n, chinesisches Wachs n, Cera chinensis *(durch
Schildläuse abgeschiedenes Wachs von Fraxinus
chinensis Roxb.)*
insecticidal insektizid, insektentötend, Insekten
vernichtend
i. dust insektizides Stäubemittel n
i. efficiency insektizide Wirksamkeit f
i. fog Insektizidnebel m
i. power insektizide Wirkung f, Wirkungskraft f
eines Insektizids, Insektizidaktivität f
i. smoke Insektizidrauch m
insecticide Insektizid n, Insektenbekämpfungs-
mittel n
i. paper Insektenvertilgungspapier n, Giftpapier
n
insectifuge s. insect repellent
insensitive unempfindlich
inseparable un[ab]trennbar
insert / to einführen, einsetzen, einbringen,
[hin]einschieben; einlagern, einbauen, einfügen
(z.B. Atome in Zwischengitterplätze)
insert *(Plast)* Einlage f, Einpreßteil n, Einspritzteil
n
insertion Einführen n, Einführung f, Einsetzen n,
Einsetzung f, Einbringen n, Einbringung f,
Hineinschieben n, Einschiebung f; Einlagerung
f, Einbau m, Einfügen n *(z.B. von Atomen in
Zwischengitterplätze)*
i. funnel Einsatztrichter m
inset *(Geol)* Einsprengling m
inside diameter Innendurchmesser m
i. drum filter Innen[trommel]filter n, Innen-
zellenfilter n
i. header *(beim Verpacken der Papierrolle an
deren Stirnseite eingelegter Papierdeckel)*
insipid ohne Geschmack, geschmacklos, fad[e],
schal, abgestanden
insol s. insoluble
insolubility Unlöslichkeit f, Nichtlöslichkeit f
insolubilize / to unlöslich machen

insoluble unlöslich, nicht löslich, unl., unlö, nl
insoluble unlösliche (nicht lösliche) Substanz *f*, Unlösliches *n*
i. germanium dioxide GeO$_2$ *(in Wasser)* unlösliches Germanium(IV)-oxid (Germaniumdioxid) *n (tetragonale Modifikation)*
i. in alcohol alkoholunlöslich, in Alkohol unlöslich, nicht in Alkohol löslich
i. in alkali alkaliunlöslich, in Alkali unlöslich, nicht in Alkali löslich
i. in benzene benzolunlöslich, in Benzol unlöslich, nicht in Benzol löslich
i. in ether ätherunlöslich, in Äther unlöslich, nicht in Äther löslich
i. in water wasserunlöslich, in Wasser unlöslich, nicht in Wasser löslich, nichtwasserlöslich
i. residue unlöslicher Rückstand *m*
i. sulphur unlöslicher (plastischer) Schwefel *m*
inspection belt Klaubeband *n*, Leseband *n (bei der Kohlenaufbereitung)*
i. door Kontrollöffnung *f*
i. glass Schauglas *n*
i. hatch (port) Schauloch *n*, Schauöffnung *f*
inspissate / to eindampfen, eindunsten, eindicken, verdicken, *(z.B. Milch)* kondensieren; dickflüssig (dick, zähflüssig, zäh) werden
inspissated juice Dicksaft *m*, eingedickter Saft *m*
inspissation Eindampfen *n*, Eindampfung *f*, Eindunsten *n*, Eindicken *n*, Verdicken *n*; Verdickungsmittel *n*
instability Instabilität *f*, Unbeständigkeit *f*, Veränderlichkeit *f*
i. constant [of complex ions] Instabilitätskonstante *f*, Dissoziationskonstante *f*
instable unstabil, instabil, unbeständig, veränderlich
i. dip *(Landw)* Kurznaßbeize *f*
instrumental analysis Instrumentenanalyse *f*
i. error Instrumentenfehler *m*, Gerät[e]fehler *m*
insulate / to isolieren, *(manchmal auch)* dämmen
insulated with rubber gummiisoliert
insulating board Isolierpappe *f*
i. brick Isolierstein *m*
i. compound *(Gum)* Isoliermischung *f*; Isoliermasse *f*, Isolationsmasse *f*
i. medium Isoliermittel *n*, Isolationsmittel *n*
i. oil Isolieröl *n*
i. paper Isolierpapier *n*
i. varnish Isolierlack *m*
insulation Isolieren *n*, Isolierung *f*, Isolation *f*, *(manchmal auch)* Dämmung *f*; Isolation *f*, Isolierung *f*, Isolierstoff *m*, Isoliermaterial *n*, *(manchmal auch)* Wärmeisolierstoff *m*, Wärmedämmstoff *m (oder)* Schalldämmstoff *m*
i. compound Isolationsmasse *f*, Isoliermasse *f*
i. concrete Isolierbeton *m*, Dämmbeton *m*
i. material Isolierstoff *m*, Isoliermaterial *n*, *(manchmal auch)* Wärmeisolierstoff *m*, Wärmedämmstoff *m (oder)* Schalldämmstoff *m*
i. resistance Isolationswiderstand *m*, Isolierfestigkeit *f*

i. stock *(Gum)* Isoliermischung *f*
i. strip Isolierstreifen *m*
i. tape Isolierband *n*
i. tubing Isolierschlauch *m*
insulator Isolator *m*; Isolierstoff *m*, Isoliermaterial *n*, *(manchmal auch)* Wärmeisolierstoff *m*, Wärmedämmstoff *m (oder)* Schalldämmstoff *m*
insulin Insulin *n*
insulinase Insulinase *f (insulinzerstörendes Leberferment)*
insulinize / to mit Insulin behandeln
intaglio ink Tiefdruckfarbe *f*
i. paper Tiefdruckpapier *n*
i. printing Tiefdruck *m*
i. printing ink s. i. ink
i. printing paper s. i. paper
i. printing process Tiefdruckverfahren *n*
intake Einlaß *m*, Eintritt *m*, Einlaufseite *f*, Aufgabeseite *f*; Zuführungsrohr *n*; Anschluß *m*
Intalox saddle Intalox-Sattel[körper] *m (ein Füllkörper)*
integral heat integrale Wärme *f*
i. heat of adsorption integrale Adsorptionswärme *f*
i. heat of dilution integrale Verdünnungswärme (Verdünnungsenthalpie) *f*
i. heat of solution integrale Lösungswärme (Lösungsenthalpie) *f*
i. tannin Austauschgerbstoff *m*
integrated mill Zellstoff- und Papierfabrik *f*
intensely coloured intensiv farbig, farbkräftig
intensification Verstärkung *f*, Intensivierung *f*
i. of colour Farbvertiefung *f*
intensifier (Foto) Verstärker *m*
intensify / to verstärken, intensivieren; *(Farbe)* vertiefen
intensity Intensität *f*, Stärke *f*
i. distribution Intensitätsverteilung *f*
i. of a spectral line Linienintensität *f*, Intensität *f* einer Spektrallinie
i. of light Helligkeit *f*, Lichtintensität *f*, *(i.e.S.)* Lichtstromdichte *f*
i. of radiation Strahlungsintensität *f*
i. rules *(phys Ch)* Intensitätsregeln *fpl*
intensive mixer Banbury-Mischer *m*, Banbury-Innenmischer *m*, Banbury-Kneter *m*
i. property intensive Eigenschaft (Größe) *f*, Intensitätsgröße *f*
inter-ester innerer Ester *m*
i.-glaze decoration *(Ker)* Zwischenglasurdekor *n*, Inglasurdekor *n*
interact / to in Wechselwirkung treten; in Wechselwirkung stehen
interaction Wechselwirkung *f*
interatomic interatomar
i. distance (spacing) Atomabstand *m*, Kernabstand *m*
interbanded mit Streifen durchsetzt, mit streifigen Einlagerungen
intercalary zwischengelagert, eingelagert, eingeschlossen; *(Bot)* interkalar, eingeschaltet

intercalate / **to** zwischenlagern, einlagern, einschließen; *(Krist)* als Schichten einlagern
intercalate compound s. intercalation compound
intercalation Zwischenlagerung f, Einlagerung f, Einschluß m; *(Krist)* schichtförmige Einlagerung f
i. compound schichtförmig ausgebildete Einlagerungsverbindung f, lamellare Verbindung f
i. compound of graphite Einlagerungsverbindung f des Graphits, Graphit-Einlagerungsverbindung f
intercellular pigment Interzellularpigment n
i. space Interzellularraum m, Zwischenzellenraum m, Interzellulare f
intercept *(Krist)* Achsenabschnitt m, Parameter m; *(Math)* Abschnitt m
interchange Austausch m, Wechsel m
i. force Austauschkraft f
i. of energy Energieaustausch m
interchangeability Austauschbarkeit f, Auswechselbarkeit f
interchangeable austauschbar, auswechselbar
interchanger of heat Wärme[aus]tauscher m, Wärmeaustauschapparat m
interconversion gegenseitige Umwandlung f
interconvert / **to** ineinander umwandeln (überführen); sich ineinander umwandeln (verwandeln)
interconvertible ineinander umwandelbar
intercooler Zwischenkühler m
intercrustal *(Geol)* interkrustal
intercrystalline interkristallin, zwischenkristallin
i. corrosion interkristalline Korrosion f
i. cracking (failure, fracture) interkristalliner Bruch m, Korngrenzenbruch m
interelectrode distance Elektrodenabstand m
interest dosage *(Landw)* Verzinsungsgabe f *(von Düngemitteln)*
interesterification Umesterung f
interface Grenzfläche f, Trennungsfläche f, Phasengrenzfläche f
i. line Trennlinie f, Grenzlinie f *(zwischen zwei Komponenten)*
i. liquid-liquid Grenzfläche f flüssig-flüssig, Flüssigkeit-Flüssigkeit-Grenzfläche f
i. radius Steiglochkreisradius m *(beim Separator)*
interfacial Grenzflächen...
i. activity Grenzflächenaktivität f
i. angle Grenzflächenwinkel m; *(Krist)* Flächenwinkel m, Kantenwinkel m
i. energy s. i. surface energy
i. film Grenzflächenfilm m *(in Emulsionen)*
i. free energy freie Energie f der Grenzfläche
i. layer Zwischenschicht f
i. phenomenon Grenzflächenerscheinung f
i. surface energy Grenzflächenenergie f
i. [surface] tension Grenzflächenspannung f
i. work Grenzflächenreibungsarbeit f
interfere with / **to** eingreifen in, beeinflussen, interferieren mit; beeinträchtigen, behindern, störend beeinflussen (einwirken auf)

interference Raumbeanspruchung f, Raumerfüllung f *(z.B. von Substituenten in der Nähe des Reaktionszentrums)*; *(Phys)* Interferenz f
i. colour Interferenzfarbe f
i. figure Interferenzfigur f, Interferenzbild n
i. filter Interferenzfilter n
i. fringe Interferenzstreifen m
i. microscope Interferenzmikroskop n
i. pattern s. i. figure
i. phenomenon Interferenzerscheinung f
i. ring Interferenzring m
interferometry Interferometrie f
interfibre bonding *(Pap)* Zwischenfaserbindung f, Faserverkettung f
interfuse / **to** [ver]mischen, durchmischen, durchsetzen
interfusion Vermischen n, Vermischung f, Durchmischen n, Durchmischung f, Durchsetzung f
intergranular attack (corrosion) interkristalline Korrosion f, Korngrenzenkorrosion f
i. cracking interkristalliner Riß m
intergrown verwachsen
i. material Verwachsenes n
intergrowth *(Krist)* Zusammenwachsen n, Verwachsen n, Verwachsung f; *(Krist)* Durchwachsen n, Durchwachsung f, Penetration f
interhalogen compound Verbindung f von Halogenen untereinander (miteinander), Interhalogenverbindung f
interionic interionisch
i. action interionische Wechselwirkung f, Ionenwechselwirkung f
i. force interionische Wechselwirkungskraft f
interior paint Innenanstrichfarbe f
i. varnish Innenlack m, Lack m für Innenanstriche (innen)
interlaminar strength *(Plast)* Spaltfestigkeit f *(von Schichtstoffen)*
interlattice position Zwischengitterplatz m
interlayer Zwischenschicht f, Zwischenlage f, Trennschicht f
interleakage inneres Lecken n
interleave / **to** als Schichten einlagern
interlining 1. Einlagestoff m; 2. Sperrschicht f; s. interlayer
interlock / **to** zusammenbacken *(von Kristallen)*
intermediate intermediär, Zwischen...
intermediate 1. Zwischenverbindung f, Intermediärverbindung f, intermediäre Verbindung f; Zwischenprodukt n, Zwischengut n, Zwischenstoff m, Zwischenerzeugnis n, Intermediärprodukt n; 2. Zwischenstufe f
i. asphaltic petroleum gemischtbasisch asphaltisches Erdöl n
i. base gemischte Basis f *(eines Roherdöls)*
i. calender s. i. rolls
i. chelate Zwischenchelat n
i. chelate form Zwischenchelatform f
i. chemical chemisches Zwischenprodukt n
i. complex Zwischenkomplex m

i. compound Zwischenverbindung f, Intermediärverbindung f, intermediäre Verbindung f
i. condition Zwischenzustand m
i. constituent s. intermetallic compound
i. crushing Mittelbrechen n
i. cut Zwischenlauf m, Übergangsfraktion f *(bei diskontinuierlicher Destillation)*
i. form intermediäre Form f, Zwischenform f, Übergangsform f
i. fraction Zwischenfraktion f
i. heater Zwischenerhitzer m
i. Herz compound Herzsche Zwischenverbindung (Verbindung) f, Herz-Verbindung f
i. layer Zwischenschicht f
i. massecuite *(Zucker)* B-Füllmasse f, Mittelproduktfüllmasse f
i. material Zwischenfraktion f, Mittelprodukt n, Mittelgut n *(beim Klassieren)*
i. nucleus Zwischenkern m
i. oxide amphoteres Oxid n
i. paraffinic petroleum gemischtbasisch paraffinisches Erdöl n
i. phase 1. Zwischenphase f; 2. s. intermetallic compound
i. position Mittelstellung f
i. product 1. Zwischenprodukt n, Zwischengut n, Zwischenstoff m, Zwischenerzeugnis n, Intermediärprodukt n; 2. *(Lebm)* Mittelproduktzucker m, Mittelprodukt n
i.-product reaction Zwischenstoffreaktion f
i. reaction Zwischenreaktion f
i. rock intermediäres Eruptivgestein (Erstarrungsgestein, Gestein) n, neutrales Gestein n
i. rolls *(Pap)* Feuchtglättwerk n, Feuchtglätte f, Zweiwalzen[feucht]kalander m
i. shed Zwischenproduktenbetrieb m
i. stage Zwischenstufe f, Zwischenstadium n
i. state Zwischenzustand m
i. substance s. i. product
i. zinc *(Zink mit höchstens 0,50% Verunreinigungen)*
intermetallic compound intermetallische Phase (Verbindung) f, metallische intermediäre Phase (Verbindung) f
intermolecular zwischenmolekular, intermolekular
i. force [zwischen]molekulare Kraft f, Molekularkraft f
internal energy innere Energie f
i. indicator interner Indikator m
i. phase innere (disperse, offene) Phase f, disperser Bestandteil m, Dispersum n *(einer Emulsion)*
International Union of Pure and Applied Chemistry Internationale Union f für reine und angewandte Chemie
international unit internationale Einheit f, I.E. *(von Wirkstoffen)*
interstice Zwischenraum m, Lücke f; Zwischengitterplatz m, Zwischengitterstelle f, Gitterhohlraum m, Hohlraum m

interstitial interstitiell, Interstitial..., Einlagerungs...
interstitial Zwischengitteratom n, Einlagerungsatom n, Atom n auf Zwischengitterplatz
i. alloy Einlagerungslegierung f
i. atom s. interstitial
i. carbide Einlagerungskarbid n, interstitielles Karbid n
i.-cell-stimulating hormone zwischenzellenstimulierendes (luteinisierendes, die interstitiellen Zellen stimulierendes) Hormon n, Luteinisierungshormon n, ICSH, LH
i. compound Einlagerungsverbindung f, interstitielle Verbindung f
i. diffusion Zwischengitterplatzdiffusion f
i. hydride Einlagerungshydrid n, interstitielles Hydrid n
i. impurity [atom] Einlagerungsfremdatom n
i. ion Zwischengitterion n, Ion n auf Zwischengitterplatz
i. lattice site Zwischengitterplatz m, Zwischengitterstelle f, Gitterhohlraum m, Hohlraum m
i. pair Zwischengitterpaar n
i. phase Einlagerungsphase f
i. position (site) s. i. lattice site
i. [solid] solution Einlagerungsmischkristall m, Interstitialmischkristall m, interstitielle feste Lösung f, Interstitiallösung f
i. structure Einlagerungsstruktur f
i.-vacancy pair Frenkelsches Fehlstellenpaar n
interval timer Kurzzeitwecker m
intimate innig *(z.B. Mischung)*, eng, innig *(z.B. Kontakt)*
intoxicant Rauschgift n; berauschendes Getränk n
intoxicate / to berauschen, trunken machen; vergiften
intoxicating liquor berauschendes Getränk n
intoxication Intoxikation f, Vergiftung f
intra ester innerer Ester m
intramolecular intramolekular, innermolekular
i. change intramolekulare (innermolekulare) Umlagerung f
intrastratal solution *(Geol)* Auflösung f im Gestein
intrinsic angular momentum innerer Drehimpuls m, Eigendrehimpuls m, Drall m, Spin m *(der Elementarteilchen)*
i. electrical conductivity Eigen[halb]leitfähigkeit f, Intrinsic-Leitfähigkeit f, Eigenleitung f
i. energy innere Energie f
i. factor innerer Faktor m, Hämogenase f *(im Magensaft nachgewiesenes Ferment)*
i. parity *(Kch)* Eigenparität f
i. pressure Kohäsionsdruck m, Binnendruck m
i. semiconduction Eigen[halb]leitung f, Intrinsic-Leitung f, i-Leitung f
i. semiconductivity Eigen[halb]leitfähigkeit f, Intrinsic-Leitfähigkeit f
i. semiconductor Eigenhalbleiter m, Halbleiter m mit Eigenleitfähigkeit
i. viscosity Grundviskosität f, Grenzviskosität f
introduce / to einführen, einbringen, einleiten, eintragen, einarbeiten

i. commercially in den Handel bringen
introduction Einführen *n*, Einbringen *n*, Einleiten *n*, Eintragen *n*, Einarbeiten *n*
i. of casing Futterrohreinbau *m*
introfaction *(Herabsetzung der Grenzflächenspannung durch ein Netzmittel, um das Eindringen in feste Stoffe zu erleichtern)*
introfier *(das Eindringen einer Flüssigkeit erleichterndes Netzmittel)*
intrusion *(Geol)* Intrusion *f*, Injektion *f*, Eindringen *n*; *(Plast)* Fließgußverfahren *n*, Intrusion *f*, Intrudieren *n*, Intrusionsverfahren *n*
intrusive rock Intrusivgestein *n*
intumescence Blähen *n*, Blähung *f*, Aufblähen *n*, Aufblähung *f*
intussusception *(Bio)* Intussuszeption *f* *(Einbau von Stoffen, besonders in wachsende Zellmembranen)*
inulase *s.* inulinase
inulin $(C_6H_{10}O_5)_n$ Inulin *n* *(ein Reservekohlenhydrat)*
inulinase Inulinase *f* *(zu den Polyasen gehörendes Ferment)*
invariable unveränderlich, invariabel
invariant invariant, nonvariant, ohne Freiheit[sgrad]
inverse plasticity *(Koll)* Dilatanz *f* *(Gegensatz zur Thixotropie)*
inversely proportional umgekehrt proportional
inversion Inversion *f*, Invertierung *f*, Umkehrung *f*, Umwandlung *f*
i. axis *(Krist)* Inversionsdrehachse *f*
i. layer Inversionsschicht *f*
i. of cane-sugar Rohrzuckerinversion *f*
i. of configuration Konfigurationsumkehr *f*, Konfigurationswechsel *m*, Inversion *f* [der Konfiguration]
i. of phases Phasenumkehr *f*
i. of sucrose Rohrzuckerinversion *f*
i. point Umwandlungspunkt *m* *(von Modifikationen)*
i. temperature Inversionstemperatur *f*
invert / **to** *(Vorgänge)* umkehren, rückgängig machen; *(Gefäße)* umdrehen, umkehren, *(auch)* umgekehrt halten
invert emulsion Invertemulsion *f* *(Anwendungsform von Pflanzenschutzmitteln)*
i. soap Invertseife *f*, Kationseife *f*
i. sugar Invertzucker *m*
i. syrup Invertzuckersirup *m*
invertase Invertase *f*, Invertin *n*, Sa[c]charase *f* *(rohrzuckerspaltendes Ferment)*
inverted bottle Schauglas *n*, Präparatenglas *n*
i. L type of calender Vierwalzenkalander *m* mit oberer Brustwalze, Vierwalzenbrustkalander *m*, Brustkalander *m*, F-Kalander *m*
i. multiplet *(phys Ch)* verkehrtes Multiplett *n*
i. sugar *s.* invert sugar
invest / **to** umhüllen, umkleiden, umgeben
investigate / **to** untersuchen

investigation of structure Strukturuntersuchung *f*
investigational work Forschungsarbeit *f*
investment Umhüllen *n*, Umkleiden *n*; feuerfester Formstoff *m* (feuerfeste Masse *f*) zum Umhüllen der Modelle *(beim Investmentguß)*
i. casting Investmentguß *m*, Gießen *n* mit verlorener Gußform, Präzisionsguß *m* nach dem Ausschmelzverfahren
i. moulding Formen. *n* mit Ausschmelzmodellen
i. pattern Ausschmelzmodell *n*, ausschmelzbares (verlorenes) Modell *n* *(beim Investmentguß)*
inwall Schachtmauerwerk *n*; Schacht *m* *(eines Hochofens)*
inward flow strainer *(Pap)* Knotenfänger *m* mit Stoffdurchgang von außen nach innen
IOC *(Abk. für)* Institute of Chemistry
iodargyrite *(Min)* Jodargyrit *m* *(Silberjodid)*
iodate M^IJO_3 Jodat *n*
i. method Jodatmethode *f*
iodembolite *(Min)* Jodobromit *m*, Jodbromchlorsilber *n*
iodic Jod..., *(i.e.S.)* Jod(V)-...
i. acid HJO_3 Jodsäure *f*
i. [acid] anhydride J_2O_5 Dijodpentoxid *n*, Jodpentoxid *n*, Jod(V)-oxid *n*
iodide M^IJ Jodid *n*
iodimetry Jodometrie *f*
iodinate / **to** jodieren
iodination Jodieren *n*, Jodierung *f*
iodine J Jod *n*
i. azide JN_3 Jodazid *n*
i. bisulphide *s.* i. disulphide
i.-containing jodhaltig
i. cyanide JCN Jodzyanid *n*, Zyanjodid *n*
i. dioxide JO_2 Joddioxid *n*
i. disulphide J_2S_2 Dijoddisulfid *n*
i. heptafluoride JF_7 Jodheptafluorid *n*, Jod(VII)-fluorid *n*
i. hydroxide JOH Jod(I)-hydroxid *n*, hypojodige (unterjodige) Säure *f*
i. method Jodmethode *f*
i. monobromide JBr Jod[mono]bromid *n*, Jod(I)-bromid *n*
i. monochloride JCl Jod[mono]chlorid *n*, Jod(I)-chlorid *n*
i. number *s.* i. value
i. pentafluoride JF_5 Jodpentafluorid *n*, Jod(V)-fluorid *n*
i. pentoxide J_2O_5 Jodpentoxid *n*, Dijodpentoxid *n*, Jod(V)-oxid *n*
i. solution Jodlösung *f*
i. staining power Jodfärbevermögen *n*
i. tetroxide J_2O_4 Dijodtetroxid *n*
i. tribromide JBr_3 Jodtribromid *n*, Jod(III)-bromid *n*
i. trichloride JCl_3 Jodtrichlorid *n*, Jod(III)-chlorid *n*
i. value Jodzahl *f*, JZ
i. vapour Joddampf *m*
iodization Jodieren *n*, Jodierung *f*

iodize / **to** jodieren
iodized active carbon Jodkohle f
i. salt Jodspeisesalz n
iodoaurate $M^I[AuJ_4]$ Jodoaurat(III) n, Tetrajodoaurat(III) n
iodoazide JN_3 Jodazid n
iodobenzene C_6H_5J Jodbenzol n, Phenyljodid n
iodobismuthate Jodobismutat(III) n
iodobromite (Min) Jodbromit m, Jodbromchlorsilber n
iodoethane CH_3CH_2J Jodäthan n, Äthyljodid n
iodoform CHJ_3 Jodoform n, Trijodmethan n
i. test Jodoformprobe f
iodogorgonic acid $HOC_6H_2(J_2)CH_2CH(NH_2)COOH$ Jodgorgosäure f, 3,5-Dijodtyrosin n, 2-Amino-3-(3,5-dijod-4-hydroxyphenyl)propansäure f
iodometric jodometrisch
iodometry Jodometrie f
iodonium compound Jodoniumverbindung f
iodophosphine, iodophosphonium PH_4J Phosphoniumjodid n
iodoplatinic acid H_2PtJ_6 Hexajodoplatin(IV)-säure f, Platinjodwasserstoffsäure f, Jodoplatinsäure f, Platindihydrogenjodid n
iodoquinoline Jodchinolin n
iodous Jod..., (i.e.S.) Jod(III)-...
iodyrite (Min) Jodyrit m, (veraltet für) Jodargyrit m (Silberjodid)
IOF (Abk. für) Institute of Fuel
iolite (Min) Iolith m, (veraltet für) Kordierit m (Magnesiumaluminiumalumopentasilikat)
ion Ion n
i. absorption Ionenaufnahme f
i. activity Ionenaktivität f
i. activity coefficient Ionenaktivitätskoeffizient m
i.-adsorbing capacity Ionenadsorptionsvermögen n
i. antagonism (Bioch) Ionenantagonismus m
i. association Ionenassoziation f
i. atmosphere Ionenwolke f, Ionenatmosphäre f
i. beam Ionenstrahl m
i. cloud s. i. atmosphere
i. colour Ionenfarbe f
i. conductance Ionenleitfähigkeit f
i. density Ionendichte f
i.-dipole complex Ionendipolkomplex m
i.-dipole forces Ionendipolkräfte fpl
i. dosage (dose) Ionendosis f
i. exchange Ionenaustausch m
i.-exchange bead Austauscherkorn n
i.-exchange chromatography Ionenaustauschchromatografie f, Austauschchromatografie f
i.-exchange compound (Landw) adsorbierender Bodenkomplex m, Austauschmaterial n
i.-exchange equilibrium Ionenaustauschgleichgewicht n
i.-exchange material Ionenaustauscher m
i.-exchange membrane ionenaustauschende Membran f, Ionenaustauschermembran f

i.-exchange method Ionenaustauschmethode f
i.-exchange reaction Ionenaustauschreaktion f
i.-exchange resin Ionenaustausch[er]harz n, Austausch[er]harz n, Ionenaustauscher m auf Kunstharzbasis, Kunstharz[ionen]austauscher m
i.-exchange separation Ionenaustauschtrennung f
i. exchanger Ionenaustauscher m
i.-exclusion process Ionenausschlußverfahren n, Elektrolytvorlaufverfahren n
i. floatation (Abtrennung von Ionen mit Hilfe von Tensiden)
i. focus grid Ionenrichtgitter n (im Spektrometer)
i. hydrate Ionenhydrat n
i.-ion complex Ion-Ion-Komplex m
i.-ion recombination Ion-Ion-Rekombination f
i. lattice (Krist) Ionengitter n
i. mobility Ionenbeweglichkeit f
i. movement Ionenbewegung f
i. pair Ionenpaar n
i. product Ionenprodukt n
i. reaction Ionenreaktion f
i.-resonant spectrometer Ionenresonanz-Hochfrequenzspektrometer n
i.-retardation process Ionenverzögerungsverfahren n
i. separation Ionentrennung f
i. source Ionenquelle f
ionic ionisch, ional, Ionen...; ionoid; ionisiert
i. activity Ionenaktivität f
i. activity coefficient Ionenaktivitätskoeffizient m
i. addition ionoide Addition f (von Halogenen oder Halogenwasserstoffen an Alkene)
i. atmosphere Ionenwolke f, Ionenatmosphäre f
i. balance s. i. equilibrium
i. bond Ionenbeziehung f, Ionenbindung f, heteropolare (polare, elektrovalente, elektrostatische, ionogene) Bindung f, Elektrovalenz f
i. charge Ionenladung f
i. charge number Ionenladungszahl f
i. compound Ionenverbindung f, heteropolare Verbindung f
i. concentration Ionenkonzentration f, (i.e.S.) ionale Konzentration f
i. conductance Ionenleitfähigkeit f
i. conduction Ionenleitung f
i. conductivity Ionenleitfähigkeit f
i. content Ionengehalt m
i. crystal Ionenkristall m
i. crystal lattice (Krist) Ionengitter n
i. electrode Sprühelektrode f
i. equation Ionengleichung f
i. equilibrium Ionengleichgewicht n
i. equivalent Ionen-Äquivalentmasse f
i. forces Ionenkräfte fpl
i. formula Ionenformel f
i. getter pump Ionengetterpumpe f
i. hydration Hydratation f der Ionen, Ionenhydratation f
i. interaction interionische Wechselwirkung f, Ionenwechselwirkung f

i. lattice *(Krist)* Ionengitter *n*
i. linkage *s.* i. bond
i. migration Ionenwanderung *f*
i. mobility Ionenbeweglichkeit *f*
i. molecule Ionenmolekül *n*, Ionenmolekel *f*
i. movement Ionenbewegung *f*
i. polymerization Ionenpolymerisation *f*, ionische Polymerisation *f*
i. product Ionenprodukt *n*
i. propellant Ionentreibstoff *m*
i. radius Ionenradius *m*
i. reaction Ionenreaktion *f*
i. species Ionenart *f*, Ionensorte *f*, Ionengattung *f*
i. speed Ionen[wanderungs]geschwindigkeit *f*, Wanderungsgeschwindigkeit *f*
i. state Ionenzustand *m*
i. strength Ionenstärke *f* *(Hälfte der gesamten ionalen Konzentration)*
i. structure Ionenstruktur *f*
i. surfactant ionogener grenzflächenaktiver Stoff *m*
i. susceptibility Ionensuszeptibilität *f*
i. valence (valency) Ionenwertigkeit *f*
i. velocity *s.* i. speed
i. weight Ionenmasse *f*
i. wind elektrischer Wind *m*, Ionenwind *m*
ionizable ionisierbar
ionization Ionisation *f*, Ionisierung *f*, Ionenbildung *f*
i. by collision Stoßionisation *f*
i. chamber Ionisationskammer *f* *(ein Zählrohr)*
i. constant Dissoziationskonstante *f*, Affinitätskonstante *f*
i. energy Ionisationsenergie *f*, Ionisationsarbeit *f*
i. potential Ionisationspotential *n*, Ionisationsspannung *f* *(in Volt ausgedrückte Ionisationsenergie)*
i. spectrometer Bragg-Spektrometer *n*, Braggsches Spektrometer *n*
ionize / to ionisieren; in Ionen zerfallen
ionized atom ionisiertes Atom *n*, Atomion *n*
ionizer wire Sprühdraht *m*
ionizing *s.* ionization
i. electrode Sprühelektrode *f*
i. potential *s.* ionization potential
i. power Ionisierungsfähigkeit *f*, Ionisationsfähigkeit *f*, Ionisierungsvermögen *n*
i. radiation ionisierende Strahlung *f*
i. region *(Spektrometrie)* Ionisierungsbereich *m*
ionogenic ionogen, ionenerzeugend; ionisierbar
ionomer Ionomer[es] *n*
ionometric ionometrisch
ionophilic ionophil
ionophoresis Ionophorese *f*
ionotropic gel ionotropes Gel *n*
IOP *(Abk. für)* Institute of Petroleum
IOS *(Abk. für)* International Organization for Standardization

IP *(Abk. für)* Institute of Petroleum
IPC 1. *(Abk. für)* Institute of Paper Chemistry; 2. *s.* isopropyl-N-phenylcarbamate
ipecac *s.* ipecacuanha
ipecacuanha Brechwurz *f*, Ipekakuanha *f*, Cephaëlis ipecacuanha (Brot.)A. Rich. und C. acuminata Karsten; *s.*i. root
i. alkaloid Ipekakuanhaalkaloid *n*
i. root *(Pharm)* Brechwurzel *f*, Ipekakuanhawurzel *f* *(von Cephaëlis ipecacuanha (Brot.)A. Rich. und C. acuminata Karsten)*
IPT *(Abk. für)* Institute of Petroleum Technologists
IR *s.* isoprene rubber
I.R.H.D. *(Abk. für)* International Rubber Hardness Degrees
iridescence Irisieren *n*, Schillern *n*
iridescent irisierend
i. paper irisierendes Papier *n*, Irispapier *n*, Perlmutterpapier *n*
iridium Ir Iridium *n*
i.-osmine *s.* iridosmine
i.-potassium chloride $K_2[IrCl_6]$ Kaliumhexachloroiridat(IV) *n*
i.-sodium chloride $Na_2[IrCl_6]$ Natriumhexachloroiridat(IV) *n*
iridosmine *(Min)* Iridosmium *n*
Irish moss *(Pharm, Gerb)* Irländisches Moos *n*, Karrag[h]eenmoos *n*, Knorpeltang *m*, Knopftang *m*, Karrag[h]een *n* *(Droge der Rotalgen Chondrus crispus (L.)Stackh. und Gigartina mamillosa J. Agardh)*
I. moss mucilage Karragheenschleim *m*
iron Fe Eisen *n*
i. alloy Eisenlegierung *f*
i. alum Eisenalaun *m* *(Sammelname für Doppelsalze von Eisen(III)-sulfat mit Alkalisulfaten); (Min)* Halotrichit *m* *(Aluminiumeisen(II)-sulfat-22-Wasser)*
i.-aluminium garnet *(Min)* Eisentongranat *m*, gemeiner Granat *m*, Almandin *m* *(Aluminiumeisen(II)-orthosilikat)*
i.-ammonium alum *s.* i.-ammonium sulphate
i.-ammonium sulphate $(NH_4)Fe(SO_4)_2 \cdot 12H_2O$ Eisenammoniumsulfat-12-Wasser *n*, Ammoniumeisenalaun *m*, Eisenammoniakalaun *m*
i. and ammonium citrate Eisenammoniumzitrat *n* *(komplexe Eisen(III)-ammoniumzitratverbindung)*
i. arsenide Eisenarsenid *n*
i. bacteria Eisenbakterien *npl* *(Leptothrix-, Crenothrix- und Gallionella-Arten)*
i.-base auf Eisenbasis (Eisengrundlage)
i. blast furnace Eisenhochofen *m*
i. boride Eisenborid *n*
i. borings Eisen[bohr]späne *mpl*
i. buff Rostgelb *n*
i. carbide Fe_3C Eisenkarbid *n*
i. catalyst Eisenkatalysator *m*, Eisenkontakt *m*
i. cement Eisenkitt *m*, Rostkitt *m*
i. chelate of indigo Eisenindigochelat *n*

i. chlorosis *(Landw)* Eisenchlorose *f (Bleichsucht von Pflanzen bei Eisenmangel)*
i. chrysolite *(Min)* Neochrysolith *m, (veraltet für)* Fayalit *m (Eisen(II)-orthosilikat)*
i.-containing eisenhaltig
i. content Eisengehalt *m*
i. crucible Eisentiegel *m*
i. dichloride $FeCl_2$ Eisendichlorid *n*, Eisen(II)-chlorid *n*
i. dichromate $Fe_2(Cr_2O_7)_3$ Eisen(III)-dichromat(VI) *n*
i. disulphide FeS_2 Eisen(II)-disulfid *n*
i. filings Eisenfeilspäne *mpl*
i.-free redox system eisenfreies Redoxsystem *n*
i. group Eisengruppe *f*, Gruppe *f* der Eisenmetalle
i. hat *(Geol)* Eiserner Hut *m*
i. liquor *(Farb)* Eisenbeize *f*, Schwarzbeize *f (Eisenazetatlösung)*
i. meteorite Eisenmeteorit *m*
i. mica *(Min)* Eisenglimmer *m, (veraltet für)* Lepidomelan *m*
i. monosulphide FeS Eisen[mono]sulfid *n*, Eisen(II)-sulfid *n*
i. monoxide FeO Eisen[mon]oxid *n*, Eisen(II)-oxid *n*
i. mordant Eisenbeize *f*
i.-nickel core *(Geol)* Eisen-Nickel-Kern *m*, Nickeleisenkern *m*, Nife-Kern *m*, Erdkern *m*
i. nitride Eisennitrid *n*
i. notch *s.* i. tap hole
i. ore Eisenerz *n*
i. oxide red Eisenoxidrot *n*
i. pan *s.* ironpan
i. pentacarbonyl $Fe(CO)_5$ Eisenpentakarbonyl *n*
i. perbromide *s.* i. tribromide
i. perchloride *s.* i. trichloride
i. persulphate *s.* i. trisulphate
i. phosphide Eisenphosphid *n*
i. porphyrin protein Eisenporphyrinprotein *n*
i.-potassium alum *s.* i.-potassium sulphate
i.-potassium sulphate $KFe(SO_4)_2 \cdot 12H_2O$ Kaliumeisen(III)-sulfat-12-Wasser *n*, Kaliumeisenalaun *m*
i. powder Eisenpulver *n*
i. proto compound Eisen(II)-verbindung *f*
i. protochloride *s.* i. dichloride
i. protoiodide FeJ_2 Eisen(II)-jodid *n*
i. protosulphide *s.* i. monosulphide
i. protoxalate $Fe(COO)_2$ Eisen(II)-oxalat *n*
i. pyrite[s] *(Min)* Pyrit *m*, Eisenkies *m*, Schwefelkies *m (Eisen(II)-sulfid)*
i. removal Enteisenung *f*, Enteisenen *n (z.B. des Wassers)*
i. runner Eisenabstichrinne *f*
i. runoff Eisenablauf *m*, Eisenauslauf *m*
i. rust Eisenrost *m*
i.-rust cement *s.* i. cement
i. sesqui compound Eisen(III)-Verbindung *f*
i. sesquichloride *s.* i. trichloride
i. sesquioxide Fe_2O_3 Eisen(III)-oxid *n*

i. sesquisulphate *s.* i. trisulphate
i. silicide Eisensilizid *n*
i. sludge Eisenschlamm *m*
i. speck Eisenfleck *m (Papierfehler)*
i. sponge Eisenschwamm *m*
i. stain Eisenfleck *m (im Holz)*
i. sulphuret *s.* i. monosulphide
i. tap hole Eisen[ab]stichloch *n*, Eisen[ab]stich *m*, Stichloch *n* für Eisen
i. tapping Eisen[ab]stich *m*
i. tapping hole *s.* i. tap hole
i. tersulphate *s.* i. trisulphate
i. tetracarbonyl $Fe(CO)_4$ Eisentetrakarbonyl *n*
i. tray Eisenpfanne *f*
i. tribromide $FeBr_3$ Eisentribromid *n*, Eisen(III)-bromid *n*
i. trichloride $FeCl_3$ Eisentrichlorid *n*, Eisen(III)-chlorid *n*
i. trisulphate $Fe_2(SO_4)_3$ Eisentrisulfat *n*, Eisen(III)-sulfat *n*
i. vitriol $FeSO_4 \cdot 7H_2O$ Eisenvitriol *n*, Eisen(II)-sulfat-7-Wasser *n*; *(Min)* Eisenvitriol *m*, Melanterit *m*
ironpan *(Bodenchemie)* Ortstein *m*
ironstone Eisenstein *m (eisenreiches Sedimentgestein)*
i. clay Toneisenstein *m*
irradiate / to 1. bestrahlen; *(Foto)* belichten; 2. *(z.B. Licht)* ausstrahlen; strahlen; leuchten
irradiation 1. Bestrahlen *n*, Bestrahlung *f*; *(Foto)* Belichtung *f*; 2. Ausstrahlen *n*; Strahlen *n*; Leuchten *n*; *(Phys)* Strahlungsintensität *f*
irreversible irreversibel, nicht umkehrbar
i. sol irreversibles Sol *n*
irrigation field Rieselfeld *n*
irritant Reizmittel *n*, Reizstoff *m*
i. action Reizwirkung *f*
i. timber *(Reizstoffe enthaltendes und Hautentzündungen verursachendes Holz)*
irritate / to reizen
irruptive rock Intrusivgestein *n*, Tiefengestein *n*, plutonisches Gestein *n*, Plutonit *m*
i.s. *s.* insoluble
Isbell vanner *(Aufbereitungstechnik)* Isbell-Vanner *m (eine Setzmaschine)*
isentrope *(phys Ch)* Isentrope *f (Diagrammkurve einer reversiblen adiabatischen Zustandsänderung)*
isentropic *(phys Ch)* isentropisch
iserine, iserite *(Min)* Iserit *m*, *(veraltet für)* Iserin *m*
isethionic acid $CH_2(OH)CH_2SO_2OH$ Isäthionsäure *f*, 2-Hydroxyäthansulfonsäure *f*, Äthanol-(1)-sulfonsäure-(2) *f*
ISIC *(Abk. für)* International Solvay Institute of Chemistry
isinglass 1. Fischleim *m*; Hausenblasenleim *m*, Hausenblase *f*; 2. Glimmer *m*; 3. Agar[-Agar] *m(n)*
ISO *(Abk. für)* International Organization for Standardization
iso acid isomere Säure *f*

isoalkane Isoalkan *n*
isoamyl alcohol $(CH_3)_2CHCH_2CH_2OH$ Isoamylalkohol *m*, 3-Methylbutanol-(1) *n*, Isobutylkarbinol *n*
  i. aldehyde $(CH_3)_2CHCH_2CHO$ Isoamylaldehyd *m*, Isovaleraldehyd *m*, Methylbutyraldehyd *m*, 3-Methylbutanal *n*
  i. isovalerate *s.* i. valerianate
  i. nitrate $(CH_3)_2CHCH_2CH_2ONO_2$ Isoamylnitrat *n*, 3-Methylbutylnitrat *n*, Salpetersäureisoamylester *m*
  i. valer[ian]ate $(CH_3)_2CHCH_2COO \cdot C_5H_{11}$ Valeriansäureisoamylester *m*, Isoamylvalerianat *n*, Apfelessenz *f*, Apfeläther *m*
isoascorbic acid Isoaskorbinsäure *f*
isoatisine Isoatisin *n*
isobar *(phys Ch)* Isobare *f*; *s.* isobare
isobare *(Kch)* Isobar[e] *n*
isobaric isobar
  i. distillation isobare Destillation *f*
isobarometric filler Gegendruckfüller *m*, Gegendruckfüllmaschine *f*, Gegendruckabfüllapparat *m*
isobutane $CH_3CH(CH_3)CH_3$ Isobutan *n*, *i*-Butan *n*, Trimethylmethan *n*, 2-Methylpropan *n*
  i. recycle Isobutankreislauf *m*, Isobutanumlauf *m*, iC_4-Kreislauf *m*, iC_4-Umlauf *m*
isobutanol $(CH_3)_2CHCH_2OH$ Isobutanol *n*, Isobutylalkohol *m*, 2-Methylpropanol-(1) *n*, Isopropylkarbinol *n*
isobutene $CH_2=C(CH_3)_2$ Isobuten *n*, Isobutylen *n*, *asym*-Dimethyläthylen *n*, 2-Methylpropen *n*
isobutyl acetic acid $(CH_3)_2CH(CH_2)_2COOH$ Isobutylessigsäure *f*, Isokapronsäure *f*, 4-Methylpentansäure *f*
  i. alcohol $(CH_3)_2CHCH_2OH$ Isobutylalkohol *m*, Isobutanol *n*, 2-Methylpropanol-(1) *n*, Isopropylkarbinol *n*
  i. carbinol $(CH_3)_2CHCH_2CH_2OH$ Isobutylkarbinol *n*, Isoamylalkohol *m*, 3-Methylbutanol-(1) *n*
  i. mercaptan $(CH_3)_2CHCH_2SH$ Isobutylmerkaptan *n*, 2-Methyl-1-propanthiol *n*
isobutylation Isobutylierung *f*
isobutylene *s.* isobutene
i.-isoprene rubber Isobutylen-Isopren-Kautschuk *m* (Butylkautschuk)
isobutyric acid $(CH_3)_2CHCOOH$ Isobuttersäure *f*, *i*-Buttersäure *f*, Dimethylessigsäure *f*, 2-Methylpropionsäure *f*, 2-Methylpropansäure *f*
isocal, isocalorific line Isokale *f*, Linie (Kurve) *f* gleichen Heizwertes
isocaproic acid $(CH_3)_2CH(CH_2)_2COOH$ Isokapronsäure *f*, 4-Methylpentansäure *f*
isocarbon line Linie (Kurve) *f* gleichen Kohlenstoffgehaltes
isochore *(phys Ch)* Isochore *f*
isocitric acid
  $COOH \cdot CH(OH)CH(COOH)CH_2 \cdot COOH$ Isozitronensäure *f*, 1-Hydroxy-1,2,3-propantrikarbonsäure *f*

i. acid dehydrogenase Isozitratdehydr[ogen]ase *f*, Isozitronensäuredehydr[ogen]ase *f*
isocrotonic acid $CH_3CH=CHCOOH$ Isokrotonsäure *f*, β-Krotonsäure *f*, Allokrotonsäure *f*, *cis*-Buten-(2)-säure-(1) *f*, *cis*-3-Methylakrylsäure *f*
isocyanate $R \cdot N=C=O$ Isozyanat *n*
  i. adhesive Isozyanatkleber *m*
  i. resin Isozyanatplast *m*, Polyurethan *n*
  i. rubber Urethankautschuk *m*, Polyurethankautschuk *m*
isocyanide $R \cdot N \equiv C$ Isozyanid *n*, Isonitril *n*
isocyanine Isozyanin *n*
  i. condensation Isozyaninkondensation *f*
isocyanuric acid Isozyanursäure *f*
isocyclic isozyklisch, karbozyklisch, homozyklisch
  i. compounds Isozyklen *pl*, Karbozyklen *pl*, isozyklische (karbozyklische, homozyklische) Verbindungen *fpl*
isodiametric isodiametrisch
isodimensional isodimensional
isodisperse isodispers
isodurene $C_6H_2(CH_3)_4$ Isodurol *n*, 1,2,3,5-Tetramethylbenzol *n*
isodurylic acid $(CH_3)_3C_6H_2COOH$ Isodurylsäure *f*, Trimethylbenzoesäure *f*
isoelectric isoelektrisch
  i. point isoelektrischer (elektrisch neutraler) Punkt *m*
  i. precipitate *(Koll)* am isoelektrischen Punkt ausgeflocktes Gel *n*
  i. precipition *(Koll)* isoelektrische Fällung *f*
isoelectronic isoelektronisch, isoster
isoerucic acid $C_8H_{17}CH=CHC_{11}H_{22}COOH$ Brassidinsäure *f*, *trans*-13-Dokosensäure *f*
isoeugenol $C_6H_3(OH)(OCH_3)CH=CHCH_3$ Isoeugenol *n*, 4-Hydroxy-3-methoxy-1-propenylbenzol *n*
isoferulic acid $HO(CH_3O)C_6H_3CH=CHCOOH$ Isoferulasäure *f*, Hesperetinsäure *f*, Kaffeesäure-4-methyläther *m*, 3-(3-Hydroxy-4-methoxyphenyl)-propensäure *f*
isoflavone Isoflavon *n*
isogarryfoline Isogarryfolin *n*
isogel *(Koll)* Isogel *n*
isohemipinic acid Isohemipinsäure *f*, 4,5-Dimethoxyisophthalsäure *f*
isohumulone Isohumulon *n (Bitterstoff des Hopfens)*
isohydric von gleichem pH-Wert, mit gleicher Hydroniumionenkonzentration
isoionical point isoionischer Punkt *m*
isolable isolierbar, rein darstellbar
isolate / to isolieren, rein darstellen; isolieren, absondern, abtrennen; abdichten
isolated double bond isolierte Doppelbindung *f*
  i. reaction isolierte Reaktion *f*
isolation Gewinnung *f*, Darstellung *f*, Reindarstellung *f*
isolysergic acid Isolysergsäure *f*
isomer Isomer[es] *n*

m-isomer m-Isomer[es] n, Meta-Isomer[es] n, (Benzolderivat mit zwei Substituenten in 1,3-Stellung)

o-isomer o-Isomer[es] n, Ortho-Isomer[es] n (Benzolderivat mit zwei Substituenten in 1,2-Stellung)

p-isomer p-Isomer[es] n, Para-Isomer[es] n (Benzolderivat mit zwei Subsituenten in 1,4-Stellung)

isomerase Isomerase f (intramolekulare Umlagerung bewirkendes Ferment)

isomeric isomer
  i. compound isomere Verbindung f, Isomer[es] n
  i. pair Isomerenpaar n
  i. polymer isomeres Polymer[es] n

isomeride s. isomer

isomerism Isomerie f

isomerization Isomerisierung f, Isomerisation f
  i. equilibrium Isomerisierungsgleichgewicht n, Isomerisationsgleichgewicht n
  i. process Isomerisierungsverfahren n, Isomerisationsverfahren n
  i. reaction Isomerisierungsreaktion f, Isomerisationsreaktion f

isomerize / to sich isomerisieren, isomerisiert werden; isomerisieren

isomethine Isomethin n

isomorphic isomorph, gleichgestaltig

isomorphism (Krist) Isomorphie f, Isomorphismus m, Gleichgestaltigkeit f

isomorphous s. isomorphic

isoniazid Isoniazid n, Isonikotinsäurehydrazid n, Pyridin-(4)-karbonsäurehydrazid n, INH

isonicotinic acid Isonikotinsäure f, γ-Pyridinkarbonsäure f, Pyridin-(4)-karbonsäure f
  i. acid hydrazide s. isoniazid

isonitrile $R \cdot N \equiv C$ Isonitril n, Isozyanid n

isooctane Isooktan n (Strukturisomeres des Oktans), (i.e.S.) $(CH_3)_2CHCH_2C(CH_3)_3$ Isooktan n, 2,2,4-Trimethylpentan n

isooctyl alcohol Isooktylalkohol m, Isooktanol n, 2-Äthylhexanol-(1) n

isoosmotic isosmotisch, isotonisch

isoparaffin synthesis Isosynthese f (von Isoparaffinen)

isopentane $CH_3CH(CH_3)CH_2CH_3$ Isopentan n, i-Pentan n, 2-Methylbutan n, Äthyldimethylmethan n

isophthalic acid $C_6H_4(COOH)_2$ Isophthalsäure f, m-Phthalsäure f, Benzol-m-dikarbonsäure f, Benzoldikarbonsäure-(1,3) f

isopiestic method isopiestische Methode f (zur Bestimmung der relativen Molekülmasse)
  i. solutions Lösungen fpl gleichen Dampfdrucks

isopoly acid Isopolysäure f

isopolymorphism Isopolymorphie f

isoprene $CH_2 = CHC(CH_3) = CH_2$ Isopren n, 2-Methyl-1,3-butadien n
  i. rubber Isoprenkautschuk m (synthetisches Polyisopren)

isoprenoid aus Isoprenmolekülen aufgebaut

isoprenoids Isoprenoide npl (Sammelname für Terpene und Steroide)

isopropanol s. isopropyl alcohol

isopropyl acetic acid $(CH_3)_2CHCH_2COOH$ Isopropylessigsäure f, 3-Methylbutansäure f, Isovaler[ian]säure f, Isobaldriansäure f
  i. alcohol $CH_3CHOHCH_3$ Isopropylalkohol m, sek-Propylalkohol m, 2-Propanol n, Dimethylkarbinol n
  i. chloride $CH_3CHClCH_3$ Isopropylchlorid n, 2-Chlorpropan n
  i. ether $(CH_3)_2CHOCH(CH_3)_2$ Isopropyläther m, Diisopropyläther m, 2-Isopropoxypropan n
  i. group $(CH_3)_2CH-$ Isopropylgruppe f, Methoäthylgruppe f, Methyläthylgruppe f

isopropyl-N-phenylcarbamate $C_6H_5NHCOO \cdot CH(CH_3)_2$ Isopropyl-N-phenylkarbamat n, Propham n, IPC

isopropylamine $(CH_3)_2CHNH_2$ Isopropylamin n, 2-Aminopropan n

isopropylbenzene $C_6H_5CH(CH_3)_2$ Isopropylbenzol n, 2-Phenylpropan n, Kumol n

isopropylcarbinol $(CH_3)_2CHCH_2OH$ Isopropylkarbinol n, Isobutylalkohol m, Isobutanol n, 2-Methylpropanol-(1) n

isopropylidene acetone $(CH_3)_2C=CHCOCH_3$ Mesityloxid n, 2-Methylpenten-(2)-on n, Isopropylidenazeton n

isoquinoline Isochinolin n, 3,4-Benzopyridin n

isoquinuclidine Isochinuklidin n

isorubber Iso-Kautschuk m

isosmotic s. isotonic

isostere isostere Verbindung f; (phys Ch) Isostere f (eine Kurve)

isosteric isoster, isoelektronisch
  i. compound isostere Verbindung f

isosterism Isosterie f

isosuccinic acid $CH_3CH(COOH)_2$ Isobernsteinsäure f, 2-Methylpropandisäure f

isosynthesis Isosynthese f

isotactic isotaktisch
  i. polymer isotaktisches Polymer[es] n
  i. structure isotaktische Struktur f

isotacticity Isotaktizität f

isoteniscope Isoteniskop n (Gerät zur Bestimmung des Sättigungsdruckes von Flüssigkeiten)

isotherm Isotherme f

isothermal isotherm[isch]
  i. annealing isothermes Glühen n
  i. calorimeter isothermes Kalorimeter n
  i. distillation isotherme Destillation f
  i. expansion isotherme Expansion f
  i. transformation isotherme Umwandlung f, Umwandlung f bei gleichbleibender Temperatur (Unterkühlungstemperatur)
  i. transformation diagram isothermes Zeit-Temperatur-Umwandlungsdiagramm (ZTU-Diagramm) n, Zeit-Temperatur-Diagramm n für isotherme Umwandlung

isotone *(Kch)* Isoton *n*
isotonic isotonisch, isosmotisch
i. solution isotonische (isosmotische) Lösung *f*
isotope Isotop *n*
i. counter Zählrohr *n*
i. dilution procedure Isotopenverdünnungsverfahren *n*
i. effect Isotopieeffekt *m*, Isotopeneffekt *m*
i. separation Isotopentrennung *f*
isotopic isotop, Isotopen..., Isotopie...
i. analysis Isotopenanalyse *f*
i. atomic weight Massenwert *m*, *(bisher)* Isotopengewicht *n*
i. compound Isotopenverbindung *f*
i. dilution analysis (method) Isotopenverdünnungsmethode *f*
i. exchange Isotopenaustausch *m*
i. exchange reaction Isotopenaustauschreaktion *f*
i. indicator *s.* i. tracer
i. mixture Isotopengemisch *n*
i. separation Isotopentrennung *f*
i. tracer Isotopentracer *m*, Isotopenindikator *m*, isotoper Indikator *m*, Indikatorisotop *n*, Leitisotop *n*, Zusatzisotop *n*
isotopism, isotopy Isotopie *f*
isotrope, isotropic *(Krist)* isotrop
isotropy *(Krist)* Isotropie *f*
isotypy *(Krist)* Isotypie *f*
isovaleral, isovaleraldehyde, isovalerial $(CH_3)_2CHCH_2CHO$ Isovaleraldehyd *m*, Isoamylaldehyd *m*, Methylbutyraldehyd *m*, 3-Methylbutanal *n*
isovalerianic acid *s.* isovaleric acid
isovaleric acid $(CH_3)_2CHCH_2COOH$ Isovaler[ian]säure *f*, Isobaldriansäure *f*, Delphinsäure *f*, 3-Methylbutansäure *f*
i. aldehyde *s.* isovaleral
isoviolanthrone Isoviolanthron *n*, Isodibenzanthron *n*
isovol Isovole *f*, Linie (Kurve) *f* gleichen Gehalts an flüchtiger Substanz
isoxylic acid $(CH_3)_2C_6H_3COOH$ Isoxylylsäure *f*, 2,5-Xylylsäure *f*, 2,5-Dimethylbenzoesäure *f*
issue / **to** entweichen, ausströmen, abziehen; aussenden, emittieren, ausströmen lassen
Istrian galls *(Gerb)* *(Gallen von Quercus ilex L.)*
It diagram It-Diagramm *n*, I-t-Diagramm *n*, Wärmeinhalt-Temperatur-Diagramm *n*
itaconic acid $HOOC \cdot C(=CH_2)CH_2 \cdot COOH$ Itakonsäure *f*, Methylenbernsteinsäure *f*, Propylendikarbonsäure *f*
Italian fennel Fenchel *m*, Foeniculum vulgare Mill.
I. red Italienischer Ocker *m* *(Pigmentfarbe aus Eisen(III)-oxid)*
I. senna *(Sennesblätter von Cassia obovata Collad)*
I.U. *s.* international unit
IUB *(Abk. für)* International Union of Biochemistry
IUC *(Abk. für)* 1. International Union of Chemistry; 2. International Union of Crystallography

IUC rule IUC-Regel *f*, Regel *f* der IUC
IUC system IUC-System *n* *(Nomenklatursystem der Internationalen Union für Chemie)*
IUCr *(Abk. für)* International Union of Crystallography
IUPAC *s.* International Union of Pure and Applied Chemistry
IUPAC-Dyson notation IUPAC-Dyson-Notation *f*
IUPAC-Dyson [notation] system IUPAC-Dyson-System *n*
IUPAC name nach den IUPAC-Regeln gebildeter Name *m*
IUPAC rule IUPAC-Regel *f*, Regel *f* der IUPAC
IUPAC system IUPAC-System *n* *(Nomenklatursystem)*
I.V. *s.* iodine value
Ivanov reaction Ivanoff-Reaktion *f* *(Hydroxykarbonsäure-Synthese)*
ivory Elfenbein *n*
i. black Elfenbeinschwarz *n*
i. cardboard Elfenbeinkarton *m*
Izod impact strength *(Plast)* Schlagbiegefestigkeit (Kerbschlagzähigkeit) *f* nach Izod
Izod impact test *(Plast)* Izod-Prüfung *f* *(zur Bestimmung der Schlagbiegefestigkeit)*

J

J acid $C_{10}H_5NH_2OHSO_3H$ J-Säure *f*, I-Säure *f*, 6-Amino-1-naphthol-3-sulfonsäure *f*, 2-Amino-5-naphthol-7-sulfonsäure *f*
J acid urea J-Säureharnstoff *m*, I-Säureharnstoff *m*
J box *(Text)* J-Box *f*, Warenspeicher *m*, J-Gefäß *n*, Stiefel *m*, Bleichstiefel *m*, Freiberger Stiefel *m*
jaborandi oil Jaborandiöl *n* *(aus Blättern von Pilocarpus pennatifolius Lem.)*
jacket / **to** ummanteln, verkleiden, umhüllen
jacket Mantel *m*, Ummantelung *f*, Verkleidung *f*, Umhüllung *f*; Hülle *f*, Kapsel *f*, Hülse *f*
j. cooler Mantelkühler *m*
j. curing press *(Gum)* Einzelheizer *m*
j. steam Heizdampf *m*
jacketed kettle Mantelkessel *m*, Mantelgefäß *n*, Mantelbehälter *m*
j. shelf dryer Heizplattentrockner *m*
jade *(Min)* Jade *f* *(Halbedelstein aus Jadeit oder Nephrit)*
jadeite *(Min)* Jadeit *m* *(Natriumaluminiumdisilikat)*
jag-knotter *(Pap)* Ästefänger *m*, Astfänger *m*, Astfang *m*, Knotenfänger *m*
Jamaica kino Westindisches Kino *n* *(Rindenextrakt von Coccoloba uvifera L.)*
jamb wall Oberbauseitenwand *f* *(im Glasschmelzofen)*
Jamba oil *(Öl von Eruca sativa Mill.)*
jamesonite *(Min)* Jamesonit *m*, Bleiantimonspießglanz *m* *(Blei(II)-antimon(III)-sulfid)*
japan / **to** *(lackieren mit anschließender Hitzehärtung)*

japan Japanlack *m (i.w.S.)*
  j. water *(wäßrige Emulsion von Japanlack)*
Japan acid $HOOC(CH_2)_{19}COOH$ Japansäure *f*, Heneikosandisäure *f*
  J. agar Agar[-Agar] *m(n)*, Gelose *f*
  J. camphor Japankampfer *m*, *d*-Kampfer *m*
  J. isinglass *s*. J. agar
  J. lacquer *s*. japan
  J. leather *s*. japanned leather
  J. paper Japanpapier *n*
  J. tallow (wax) Japantalg *m*, *(oft unkorrekt)* Japanwachs *n (von Rhus succedanea L.* und *Rhus verniciflua Stokes)*
Japanese anise oil Sternanisöl *n (von Illicium verum Hook. fil.)*
  J. chestnut 1. Japanische Kastanie *f*, Castanea crenata Sieb. et Zucc.; 2. *Frucht von* 1.
  J. galls Japanische (Chinesische) Gallen *fpl (von Rhus chinensis Mill.)*
  J. ginger Japanischer Ingwer *m (von Zingiber mioga (Thunb.)Rosc.)*
  J. lacquer Japanischer Firnis *m*, Japanlack *m (i.e.S.) (von Rhus verniciflua Stokes)*
  J. larch Japanische Lärche *f*, Larix leptolepis (Sieb. et Zucc.)Gord.
  J. paper Japanpapier *n*
  J. tissue paper Japanseidenpapier *n*
  J. tung oil Japanisches Tungöl *n (Samenöl von Aleurites cordata (Thunb.)R.Br. ex Steud.)*
  J. turpentine *(Terpentin von Larix gmelini (Rupr.)Kuzenewa)*
japanned leather Lackleder *n*
jar *(irdenes oder gläsernes)* Gefäß *n*; *(Chem)* Becherglas *n*; Glaskolben *m*, Glaszylinder *m*
  j. brush Gläserbürste *f*
jargo[o]n *(Min)* Jargon *m (Zirkoniumorthosilikat)*
jasp-agate *s*. jasper
jasper *(Min)* Jaspis *m (Silizium(IV)-oxid)*
  j. ware *(Ker)* Jaspisware *f*
Java cardamom Java-Kardamom *m(n) (von Amomum maximum Roxb.)*
  J. devilpepper Rauvolfia serpentina (L.)Benth. *(alkaloidhaltige Heilpflanze)*
Javel[le] water Javellesche Lauge *f*, Eau de Javelle *n (wäßrige Kaliumhypochloritlösung)*
jaw Backe *f*; Brechbacke *f*
  j. breaker (crusher) Backenbrecher *m*
  j. type press Maulpresse *f*, einhüftige Presse *f*
jellification Gel[atin]ieren *n*, Gel[atin]ierung *f*, Gallertbildung *f*, *(Koll auch)* Gelbildung *f*
jellify / to, jelly / to gelatinieren, in Gelee überführen; gel[atin]ieren, zu Gelee erstarren
jelly Gelee *n*, Gallerte *f*, Gallert *n*, *(Koll i.w.S. auch)* Gel *n*
  j.-like gallertartig, gallertähnlich, gallertig; *(Koll)* gelartig; salbenartig
jelutong Djelutung *m*, Djelutungharz *n (Kopal von Dyera-Arten und von Parthenium argentatum A.Gray)*
Jensen tower *(Pap)* Jensen-Turm *m*

Jensen tower system *(Pap)* Zweiturmsystem *n (Jensen-Türme)*
Jensen two-tower acid-making system *s*. Jensen tower system
Jequie rubber Jequié-Kautschuk *m (von Manihot dichotoma Ule)*
Jeremiassen crystallizer *(ein Kristallisator mit Strömungssichtung)*
jerva[sic] acid Jervasäure *f*, Chelidonsäure *f*, Pyron-(4)-dikarbonsäure-(2,6) *f*
Jesuit's balsam Kopaivabalsam *m*, *(ungenau für)* Kopaivaterpentin *n(m) (von Copaifera-Arten)*
jet / to ausstoßen, ausschleudern; druckspülen; ausströmen, herausschießen
jet 1. Strahl *m*; Düse *f*; 2. *(Min)* Jet[t] *m(n)*, Gagat *m (Varietät von Braunkohle)*; 3. *(Farb)* Tiefschwarz *n*
  j.-black tiefschwarz, pechschwarz, glänzend schwarz
  j. black Tiefschwarz *n*
  j.-black masterbatch *(Gum) (auf nassem Wege ohne Dispergator hergestellte Vormischung)*
  j. coal *s*. cannel coal
  j. condenser nasser Mischkondensator *m*, Einspritzkondensator *m*
  j. dryer Luftstromtrockner *m*, Düsentrockner *m*
  j. drying Luftstromtrocknung *f*, Düsentrocknung *f*
  j. drying dobbin *(Ker)* Karusselltrockner *m* mit Luftstromtrocknung
  j. dyeing machine Düsenfärbemaschine *f*; Düsenfärbeapparat *m*
  j. fuel Düsenkraftstoff *m*, Düsentreibstoff *m*
  j. mill Strahlmühle *f*
  j. mixer Strahl[apparat]mischer *m*
  j. moulding Spritzpreßverfahren *n (für hitzehärtbare Plaste)*
  j. of liquid Flüssigkeitsstrahl *m*
  j. of steam Dampfstrahl *m*
  j. of water Wasserstrahl *m*
  j. process Düsenblasverfahren *n (für Glasfaserstoffe)*
  j. pump Treibmittelpumpe *f*, Strahlpumpe *f*
  j. velocity Geschwindigkeit *f* an der Düse
Jet-O-Mizer Jet-O-Mizer *m (Strahlmühle mit einem aufrechtstehenden, ovalen Ringraum als Mahlkammer)*
jeweller's borax $Na_2B_4O_7 \cdot 5H_2O$ Juwelierborax *m*, oktaedrischer Borax *m*, Natriumtetraborat-5-Wasser *n*
  j. putty *(Polierpulver mit Zinndioxid als Hauptbestandteil)*
jig / to ständig ruckartig auf- und abbewegen (hin- und herbewegen); *(Aufbereitungstechnik)* setzen, in der Setzmaschine aufbereiten
jig 1. *(Aufbereitungstechnik)* Setzmaschine *f*, Setzapparat *m*; 2. *(Text)* Jigger *m (eine Färbemaschine)*; 3. Schablone *f*
  j. bed Setzbett *n*
  j. padding process *(Text)* Klotzfärbeverfahren *n*

j. screen Setzsieb *n*
j.-wash / to in der [Naß-]Setzmaschine waschen (naßsortieren)
j. washer Naßsetzmaschine *f*
jigger *(Ker)* Drehmaschine *f*, Überdrehmaschine *f*; *(Text)* Jigger *m (eine Färbemaschine)*
jiggering *(Ker)* Überdrehen *n*, Überformen *n*
  j. machine *(Ker)* Drehmaschine *f*, Überdrehmaschine *f*
  j. process *(Ker)* Überdrehverfahren *n*
jigging *(Aufbereitungstechnik)* Setzen *n*, Setzarbeit *f*
  j. compartment Setzraum *m*
  j. screen Setzsieb *n*
Jingan gum *(Pflanzengummi von Odina wodier Roxb.)*
jinking *(Ker)* Abheben *n*, Abschälen *n (der Glasur)*
jj-coupling *(Kch)* jj-Kopplung *f*
job-press ink Akzidenzfarbe *f*
  j. printing Akzidenzdruck *m*
johannite *(Min)* Johannit *m*, Uranvitriol *m*
Johnson's mixture Johnsons Gemisch *n (veraltetes Pflanzenschutzmittel aus Kupfersulfat und Ammoniumkarbonat)*
Johnson's solution *(Bot)* Nährlösung *f* nach Johnson
join / to:
  j. on a polymer to another ein Polymeres auf ein anderes aufpfropfen
  j. together miteinander verbinden
joining power Wertigkeit *f*, Valenz *f*
joint Verbindung *f*; Anschluß *m*
  j. action Verbundwirkung *f*
  j. clamp Halteklemme *f*, Halteschelle *f*, Ligatur *f (für Schliffverbindungen)*
  j. line *(Glas)* Trennfuge *f*, Formennaht *f (Fehler)*, *(Plast)* Gratlinie *f*, Gratnaht *f*
Jojoba oil *(Öl von Simmondsia californica Nutt.)*
jolleying *(Ker)* Eindrehen *n*
jolt-ram machine Rüttelformmaschine *f*
Jones-Bertrams beater *(Pap)* Holländer *m* nach Jones-Bertrams
jordan Kegel[stoff]mühle *f*, *(i.e.S.)* Jordan-Kegel[stoff]mühle *f*, Jordan-Mühle *f*
  j. bars Messer *npl* der Kegelstoffmühle
  j. plug Kern (Kegel, Konus, Rotor, Kegelrotor) *m* der Kegelstoffmühle
  j. refining *(Pap)* Aufschlagen *n* des Stoffs mittels Kegelstoffmühlen
  j. rotor s. j. plug
Jordan engine (mill, refiner) *(Pap)* Jordan-Mühle *f*, Jordan-Kegel[stoff]mühle *f*
joule Joule *n*, J
Joule-Thomson coefficient Joule-Thomson-Koeffizient *m*
Joule-Thomson effect Joule-Thomson-Effekt *m (isenthalpischer Drosseleffekt)*
journal bearing Zapfenlager *n*, Radialgleitlager *n*
juglone Juglon *n*, 5-Hydroxy-1,4-naphthochinon *n*
juice Saft *m*, *(i.e.S.)* Fleischsaft *m (oder)* Gemüsesaft *m (oder)* Obstsaft *m*, Fruchtsaft *m*

j. catcher *(Zucker)* Saftfänger *m*
j. clarification Saftreinigung *f*
j. expressing Entsaften *n*; Keltern *n*
j. heater Safterhitzer *m*
j. pump Saftpumpe *f*
jumbo roll von der Maschine kommende Papierrolle *f*, Maschinenrolle *f*
junction Grenzfläche *f*, Berührungsfläche *f*, Berührungszone *f*; *(phys Ch)* Phasengrenzfläche *f*; Übergang *m (z.B. im Halbleiter)*; Lötstelle *f*, Lötverbindung *f*
juniper Wacholder *m*, *(Gattung)* Juniperus L., *(i.e.S.)* Juniperus communis L.
  j. oil Wacholder[beer]öl *n (von Juniperus communis L.)*
juniperic acid $HOC_{15}H_{30}COOH$ Juniperinsäure *f*, 16-Hydroxypalmitinsäure *f*, 16-Hydroxyhexadekansäure *f*
junk 1. Altmaterial *n*, Altwaren *pl*; Abfall *m*; 2. *(Pap)* grobe Verschmutzungen *fpl*, Grobstoff *m*, Schmutzstoff *m*
  j. remover *(Pap)* Schmutzfang *m*, Schmutzfänger *m*, Schmutzschleuse *f*
Jupiter oil s. juniper oil
Jura turpentine *(Terpentin von Picea abies (L.)Karst.)*
jute Jute *f*, Corchorus L.
  j. fibre Jutefaser *f*
  j. packing Jutedichtung *f*, Jutepackung *f*

K

K acid K-Säure *f*, 1-Amino-8-naphthol-4,6-disulfonsäure *f*
K capture s. K-electron capture
K electron K-Elektron *n*
K-electron capture K-Elektroneneinfang *m*, K-Einfang *m*
K factor *(Plast)* K-Wert *m*, Eigenviskosität *f*
K meson K-Meson *n*, Kaon *n*
K radiation K-Röntgenstrahlung *f*
K-shell K-Schale *f*
K strophanthin k-Strophanthin *n*, Strophanthin K *n*
K value *(Plast)* K-Wert *m*, Eigenviskosität *f*; *(Zucker)* Kupferzahl *f*
kaempferol Kämpferol *n*, 3,5,7,4'-Tetrahydroxyflavon *n*
kainit[e] *(Min)* Kainit *m (Kaliummagnesiumchloridsulfat)*
kaiser green $Cu(CH_3COO)_2 \cdot 3Cu(AsO_2)_2$ Kaisergrün *n*, *(veraltet für)* Schweinfurter Grün *n*, Kupferarsenitazetat *n*
kakoxene *(Min)* Kakoxen *m (Eisen(III)-trihydroxidorthophosphat)*
kalaite *(Min)* Kal[l]ait *m*, Türkis *m (Aluminiumkupfer(II)-oktahydroxidtetraorthophosphat)*

Kaldo process Kaldo-Verfahren *n*, Kalling-Domnarvet-Verfahren *n* *(Sauerstoffaufblaseverfahren)*

kalinite *(Min)* Kalinit *m*, Kali[um]alaun *m* *(Kaliumaluminiumsulfat-12-Wasser)*

kalkammon Kalkammon *n*, *(veraltet für)* Kalkammoniak *n* *(Düngemittel)*

kalksaltpeter *(Min)* Kalksalpeter *m*, Nitrokalzit *m* *(Kalziumnitrat)*

kallidin Kallidin *n* *(ein Peptid, das durch Einwirkung von Kallikrein auf das Blutserum entsteht)*

kallikrein Kallikrein *n* *(hormonartiger Wirkstoff der Bauchspeicheldrüse)*

kamacite Kamazit *m* *(Ni-armes Balkeneisen der Eisenmeteoriten)*

kampylite *(Min)* Kampylit *m*

Kamyr bleacher *(Pap)* Kamyr-Bleichturm *m*

Kamyr grinder *(Pap)* Kamyr-Schleifer *m*

kanaf[f] *s.* kenaf

kaneite *(Min)* Kaneit *m* *(Mangan(III)-arsenid)*

kanya butter Kanyabutter *f* *(von Pentadesma butyraceum Sabine)*

kaolin *(Min)* Kaolin *m*, Porzellanerde *f*, weißer (reiner) Ton *m*, China Clay *m(n)*; *(Tech)* Kaolin *n*, Schlämmkaolin *n*, geschlämmte Porzellanerde *f*

k. deposit Kaolinlagerstätte *f*, Kaolinlager *n*, Kaolinvorkommen *n*

kaoline *s.* kaolin

kaolinite *(Min)* Kaolinit *m* *(Aluminiumhydroxidsilikat)*

kaolinitic kaolinitisch

kaolinization Kaolinisierung *f*, Kaolinisation *f*

kaolinize / to kaolinisieren

kaori *s.* kauri

kapok Kapok *m* *(Kapselwolle des Kapokbaums)*

k. oil Kapoköl *f* *(Samenöl des Kapokbaums)*

karamanni wax *(Gummiharz von Symphonia globulifera L.)*

karaya [gum] Karayagummi *n*, Sterkuliagummi *n*, Indischer (Ostindischer) Tragant *m* *(meist von Sterculia urens Roxb.)*

k. mucilage Karayaschleim *m* *(von Sterculia-Arten)*

Karl-Fischer reagent Karl-Fischer-Reagens *n*, Karl-Fischer-Lösung *f*

karpholite *(Min)* Karpholith *m*

Kassel[er] kiln Kasseler Ofen *m*

Kassner process Kassnersches Verfahren *n* zur Sauerstoffgewinnung

katabolism Katabolismus *m*, Abbau *m*, Zerlegung *f* *(hochmolekularer Stoffe in einfachere Verbindungen)*

katamorphism *(Geol)* Kata[meta]morphose *f*, Katamorphismus *m*

katapleit[e] *(Min)* Katapleit *m* *(Natriumzirkonium(IV)-trisilikat)*

kateera (kateira) gum Kuteragummi · *n* *(von Sterculia-, Cochlospermum- oder Astragalus-Arten)*

kath *(kristalliner Anteil des Extrakts von Katechuholz)*

kathaemoglobin Kathämoglobin *n* *(ein Hämochromogen)*

katharometer Katharometer *n*, Wärmeleitfähigkeitszelle *f*

kathode *s.* cathode

kathodic *s.* cathodic

kation *s.* cation

kaurene Kauren *n*, α-Podokarpren *n*

kauri Kaurifichte *f*, Agathis Salisb.; Kaurikopal *m*, Kauriharz *n*

k.-butanol number (value) Kauri-Butanol-Wert *m*, Kauri-Butanol-Zahl *f*

k. copal (gum) Kaurikopal *m*, Kauriharz *n*, Kaurigum *m*, Cowrikopal *m*

k. oil Kauriöl *n*

k.-reduction test Kaurireduktionsprüfung *f*

k. resin *s.* k. copal

kaurie, kaury *s.* kauri

kava Kawapfeffer *m*, Rauschpfeffer *m*, Piper methysticum G.Forst.; Kava-Kava-Wurzelstock *m*, Kawawurzel *f*, Pfefferwurzel *f*, Rauschpfefferwurzel *f*, Awapfefferwurzel *f*

k. resin Kawaharz *n*

kavahin Kawahin *n*, Methystizin *n*

kavaic acid Kawasäure *f*, β-Methoxy-γ-zinnamylidenkrotonsäure *f*

kavakava, kawa *s.* kava

kawaic acid *s.* kavaic acid

kawain Kawain *n*, 5,6-Dihydro-4-methoxy-6-styryl-pyran-2-on *n*

kawrie, kawry *s.* kauri

Kaysam process Kaysam-Prozeß *m* *(Latex-Verarbeitungsverfahren)*

keep / to:

k. at the boil am Sieden (Kochen) halten

k. for an indefinite period unbegrenzt haltbar sein

k. in solution in Lösung halten

k. in suspension in Suspension halten

k. near the boil nahe am Sieden halten

k. screened from the light lichtgeschützt aufbewahren

keeping quality Haltbarkeit *f*

keesh Garschaum[graphit] *m*, primärer Graphit *m*, Primärgraphit *m*

kefir Kefir *m*

k. grains (seeds) Kefirkörner *npl*

keg beer Faßbier *n*

k. racker Faßfüller *m*, Faßabfüllmaschine *f*

keilhauite *(Min)* Keilhauit *m*, Yttrotitanit *m*

keir *(veraltet für)* kier

Kekulé formula Kekulé-Formel *f*

Kekulé-[-like] structures Kekulé-Strukturen *fpl*, Kekulé-Grenzstrukturen *fpl*

Kellog fluidized synthesis Kellog-Synthese *f*

kelly Mitnehmerstange *f*, Kellystange *f*, Kelly *m(n)*

Kelly filter Kelly-Filter *n*

kelp Kelp *n* *(Tangasche)*

Kelvin degree Kelvin-Grad *m*, °K

Kelvin [temperature] scale Kelvin-Skale *f*

kenaf Kenaf *m(n)*, Kanaff *m*, Kenab *m*, Kenafpflanze *f*, Hibiscus cannabinus L.
k. fibre Kenaffaser *f*, Hanfeibischfaser *f (Faser von Hibiscus cannabinus L.)*
Kent Maxecon mill Maxecon-Kent-Mühle *f*, Maxecon-Kent-Ring[rollen]mühle *f*
kephalin Kephalin *n (ein Phosphatid)*
kephir *s.* kefir
kerasin Kerasin *n (zu den Lipoiden gehörendes Zerebrosid)*
keratin Keratin *n*, Hornsubstanz *f*
keratinization Keratinbildung *f*
kermek *(Gerb)* Kermek *(Wurzeln von Limonium gmelini (Willd.)O. Ktze. und L. latifolium (Sm.)O. Ktze.)*
kermes *s.* 1. k. grains; 2. k. insect; 3. k. mineral; 4. k. scarlet; 5. k. oak
k. berries *s.* k. grains
k. dye *s.* k. scarlet
k. grains Kermeskörner *npl*, Scharlachkörner *npl*, Kermes *m (getrocknete weibliche Kermesschildläuse)*
k. insect Kermesschildlaus *f*
k. mineral [mineralischer] Kermes *m*, [oxidhaltiger] Mineralkermes *m (Doppelverbindung des Antimontrisulfids mit Antimontrioxid)*
k. oak Kermeseiche *f*, Quercus coccifera L.
k. scarlet Kermesscharlach *m*, Kermes[farbstoff] *m*
kermesic acid Kermessäure *f*, 1,3,4,6-Tetrahydroxy-2-methoxy-8-methylanthrachinon-5-karbonsäure *f*
kermesite, kermesome *(Min)* Kermesit *m*, Rotspießglanz *m (Antimon(III)-oxidsulfid)*
kernel Kern *m*; Korn *n*
k. oil Kernöl *n*
k. weight *(Landw)* Tausendkornmasse *f*, *(bisher)* Tausendkorngewicht *n*
kernite Kernit *m (Natriumtetraborat-4-Wasser)*
kerosene *s.* kerosine
kerosine Petroleum *n*, Petrol *n*, Kerosin *n*, Kerosen *n*
k. fraction Petrol[eum]fraktion *f*, Kerosinfraktion *f*
k. shale Kerosinschiefer *m*
Kerr constant Kerr-Konstante *f*
Kerr effect Kerr-Effekt *m*
kessyl alcohol Kessylalkohol *m*
k. ketone Kessylketon *n*
Kesting electrolytic process *(Pap)* Kesting-Verfahren *n*, Münchner Verfahren *n (zur Chlordioxidbleichlaugenherstellung)*
kestose Kestose *f (ein Trisaccharid)*
ketal Ketal *n*, Ketonazetal *n*
ketazine Ketazin *n (Kondensationsprodukt von Ketonen mit Hydrazin)*
keten[e] Keten *n (organische Verbindung mit der Gruppe =C=C=O), (i.e.S.)* $CH_2=C=O$ Keten *n*
k. base Ketenbase *f*
ketimide *s.* ketimine

ketimine Ket[on]imin *n*, Ket[on]imid *n*
k.-enamine tautomerism Ketimin-Enamin-Tautomerie *f*
ketine Ketin *n*, 2,5-Dimethylpyrazin *n*
ketipic acid Ketipinsäure *f*, 3,4-Dioxohexandisäure *f*
ketira gum *s.* kateera gum
keto acid Keto[n]säure *f*, Ketokarbonsäure *f*
k.-acid ester *s.* k. ester
k. alcohol Keto[n]alkohol *m*, Hydroxyketon *n*, *(ungenau)* Ketol *n*
k. aldehyde Keto[n]aldehyd *m*
k. compound Keto[n]verbindung *f*
k.-enol constant Enolkonstante *f*
k.-enol tautomerism Keto-Enol-Tautomerie *f*
k. ester Keto[säure]ester *m*, Ketokarbon[säure]ester *m*
k. form Keto[n]form *f*
k. group Keto[n]gruppe *f*, Ketonkarbonyl *n*, ketonartig gebundene Karbonylgruppe *f*; Keto[n]gruppe *f (i.w.S.)*, Karbonylgruppe *f*
k. lactam Ketolaktam *n*
k. structure Ketonstruktur *f*
ketoacidosis *s.* ketosis
ketoamine Ketoamin *n*
ketobemidone Ketobemidon *n*, 1-Methyl-4-(3'-hydroxyphenyl)piperidyl-(4)-äthylketon *n*
ketocarbonyl [group] Ketonkarbonyl *n*, ketonartig gebundene Karbonylgruppe *f*, Keto[n]gruppe *f (i.e.S.)*
ketocarboxylic acid *s.* keto acid
k.[-acid] ester *s.* keto ester
ketogenesis Ketogenese *f*, Ketonkörperbildung *f*, Ketonkörperentstehung *f*
ketogenic ketogen
ketoglutaric acid Ketoglutarsäure *f*, Oxoglutarsäure *f*, Oxopentandisäure *f*, *(i.e.S.)* $HOOCCH_2CH_2COCOOH$ α-Ketoglutarsäure *f*, 2-Oxoglutarsäure *f*, 2-Oxopentandisäure *f*
ketoheptose Ketoheptose *f (ein Ketozucker)*
ketohexose Ketohexose *f (ein Ketozucker)*
ketoketene $(R)_2C=C=O$ Ketoketen *n*
ketol Ketol *n*, Monohydroxyketon *n*; Keto[n]alkohol *m*, Hydroxyketon *n*, *(ungenau)* Ketol *n*
ketolysis Ketolyse *f*
ketolytic ketolytisch
ketomalonic acid $OC(COOH)_2$ Ketomalonsäure *f*, Mesoxalsäure *f*, Oxopropandisäure *f*
ketonaemia Ketonämie *f (Zustand einer abnormen Anhäufung von Ketonkörpern im Blut)*
ketone $R_1C(=O)R_2$ Keton *n*
k. acetal *s.* ketal
k. alcohol *s.* keto alcohol
k. body Keto[n]körper *m*
k. formation Ketonbildung *f*
k. group *s.* keto group
k. musk Ketonmoschus *m*, Moschusketon *n*, Moschus C *n*, 4-Azetyl-5-(2'-methyl-2'-propyl)-2,6-dinitro-*m*-xylol *n*
k. peroxide Ketonperoxid *n*

k. reaction Ketonreaktion f
k. resin Ketonharz n
ketonemia s. ketonaemia
ketonic Keton...
  k. acid s. keto acid
  k. carbonyl [group] s. ketocarbonyl [group]
  k. cleavage Ketonspaltung f
  k. compound s. keto compound
  k. ester s. keto ester
  k. fission s. k. cleavage
  k. function Keto[n]funktion f
  k. group s. keto group
  k. hydrolysis s. k. cleavage
  k. rancidity Keto[n]ranzigkeit f, Parfümranzigkeit f
  k. resin s. ketone resin
  k. structure s. keto structure
  k. sugar s. ketose
ketonimine s. ketimine
  k. dye[stuff] Ketoniminfarbstoff m, Ketonimidfarbstoff m
ketonize / to ketonisieren
ketonuria Ketonurie f (Ausscheidung von Ketonkörpern im Harn)
ketopentose Ketopentose f (ein Ketozucker)
ketopinic acid Ketopinsäure f, 7,7-Dimethyl-2-oxobizyklo-[1,2,2]-1-heptankarbonsäure f
ketose Ketose f, Keto[n]zucker m
ketoside Ketosid n (Glykosid, das bei der Hydrolyse eine Ketose ergibt)
ketosis Ketosis f, Ketose f (diabetische Azidose)
ketosteroid Ketosteroid n
  17-ketosteroid 17-Ketosteroid n (Steroidhormon, das am 17. Kohlenstoffatom eine Ketogruppe trägt)
γ-ketovaleric acid $CH_3CO(CH_2)_2COOH$ 4-Ketovaleriansäure f, 4-Oxopentansäure f, Pentanon-(4)-säure-(1) f, Lävulinsäure f
ketoxime Ketoxim n
kettle-type reboiler Rührkessel-Destillierblase f
ketyl Ketyl n, Metallketyl n
keuper marl Keupermergel m
key Schlüsselkomponente f
  k. chemical Grundchemikalie f, Grundstoff m, Schlüsselstoff m, Schlüsselprodukt n
  k. component s. key
  k. intermediate wichtiges Zwischenprodukt n
Khari salt (Gerb) Kharisalz n (natürliches Salzgemisch mit Natriumsulfat als Hauptbestandteil)
khellin Khellin n, 5,8-Dimethoxy-2-methyl-4',5',6,7-furanochromon n
kibbled gelatin gekörnte Gelatine f, Körnergelatine f
Kick's law Kicksches Gesetz n, Zerkleinerungsgesetz n nach Kick
kier (Text) Beuchfaß n, Beuchkessel m; (Pap) Kocher m
  k. assistant s. k.-boiling assistant
  k.-boil / to (Text) beuchen
  k. boiling (Text) Beuchen n, Beuche f

  k.-boiling assistant (Text) Beuchhilfsmittel n
kiering s. kier boiling
kieselgu[h]r Kieselgur f, Diatomeenerde f, Infusorienerde f
kieserite (Min) Kieserit m (Magnesiumsulfat-1-Wasser)
kilbrickenite (Min) Kilbrickenit m, (veraltet für) Geokronit m
Kiliani[-Fischer] synthesis Kilianische Synthese f (kohlenstoffreicher Zucker über Zyanhydrine)
kill / to (eine Schmelze) beruhigen; (Gum) übermastizieren, totmastizieren, totwalzen
killed spirit[s] Lötwasser n
  k. steel beruhigter (beruhigt vergossener, ruhig vergossener) Stahl m
killing Beruhigen n (einer Schmelze); (Gum) Übermastizieren n, Totmastizieren n, Totwalzen n
  k. agent (Met) Beruhigungsmittel n
kiln / to [im Ofen] darren, dörren, rösten
kiln Trockenofen m; Trockenkammer f; Röstofen m; Kalzinierofen m; Brennofen m; Darre f, Darrofen m; Kiln m, Bienenkorbofen m; Kiln m, Regenerierofen m, Regenerator m, Katalysatorregenerator m (beim katalytischen Kracken)
  k. car (Ker) Ofenwagen m
  k.-dried ofengetrocknet, ofentrocken; gedarrt, gedörrt
  k.-dried malt s. kilned malt
  k.-dry / to s. to kiln
  k. dryer Trockenofen m; Trockenkammer f; Darre f, Darrofen m
  k. floor Ofensohle f, Ofenboden m; Rösthorde f, Darrhorde f, Horde f
  k. furniture (Ker) Brennhilfsmittel n, Ofenstützmaterial n, Setzhilfsmittel n
  k. loss (Ker) Ofenausschuß m
  k. malt s. kilned malt
  k. temperature (Lebm) Darrtemperatur f, Abdarrtemperatur f
  k. waste heat Ofenabwärme f
  k. wicket Ofeneingang m
kilned malt Darrmalz n
kilning Darren n, Dörren n, Rösten n
kilocalorie Kilokalorie f, kcal, (veraltet) Kilogrammkalorie f
kilogram calorie s. kilocalorie
  k. molarity Kilogramm-Molarität f, kg-Molarität f, Molalität f
kimberlite Kimberlit m (vulkanischer Tuff)
kinase Kinase f
kind of bond (linkage) Bindungsart f
kindle / to anzünden, entzünden; sich entzünden
kindling point (temperature) Entflammungstemperatur f
kinematic viscosity kinematische Viskosität f, Viskositäts-Dichte-Verhältnis n
kinetic kinetisch
  k. activity factor kinetischer Aktivitätsfaktor m
  k. chain length kinetische Kettenlänge f

k. control kinetische Kontrolle f
k. energy kinetische Energie f, Bewegungsenergie f
k. theory of gases kinetische Gastheorie f
kinetin Kinetin n, 6-Furfurylaminopurin n
king's blue Königsblau n, (veraltet für) Kobaltblau n, Kobaltultramarin n, Thénards Blau n (Kobaltaluminat)
k. gold s. k. yellow
k. green $Cu(CH_3COO)_2 \cdot 3Cu(AsO_2)_2$ Königsgrün n, (veraltet für) Schweinfurter Grün n, Kupferarsenitazetat n
k. yellow As_2S_3 Königsgelb n, reines Auripigment n (Arsen(III)-sulfid); Königsgelb n, Auripigment n, Operment n, Rauschgelb n, Chinagelb n, Gelbglas n, gelber Arsenik m, gelbes Schwefelarsen n (technisches Arsen(III)-sulfid)
kinin Kinin n (im tierischen Organismus erzeugtes, pharmakologisch hochaktives Polypeptid); Kinin n, Phytokinin n, Zytokinin n (pflanzlicher Wirkstoff mit wachstumsfördernder Aktivität)
kino [gum] Kino n(m), Kinoharz n, Kinogummi n (eingetrockneter, gerbstoffreicher Saft aus den Rinden verschiedener tropischer Bäume)
kinotannic acid Kinogerbsäure f
Kipp [gas] generator Kippscher Apparat (Gasentwickler) m, Gasentwicklungsapparat m nach Kipp
Kipp's apparatus s. Kipp [gas] generator
Kirchhoff law of radiation Kirchhoffsches Strahlungsgesetz n
Kirschner value Kirschner-Zahl f, Ki-Z
kish Garschaum[graphit] m, primärer Graphit m, Primärgraphit m
Kjeldahl connecting bulb Tropfenfänger m (der Kjeldahl-Apparatur)
Kjeldahl determination Kjeldahl-Bestimmung f, Kjeldahlsche Stickstoffbestimmung f, Stickstoffbestimmung f nach Kjeldahl
Kjeldahl digestion apparatus Kjeldahl-Apparat m, Stickstoffbestimmungsapparat m nach Kjeldahl
Kjeldahl flask Kjeldahl-Kolben m
Kjeldahl [nitrogen] method Kjeldahl-Methode f, Kjeldahl-Verfahren n (Aufschlußverfahren zur Stickstoffbestimmung)
Kjellin furnace Kjellin-Ofen m (ein Induktionsofen)
kliachite (Min) Kliachit m
klockmannite (Min) Klockmannit m (Kupfer(II)-selenid)
Knapp solution Knappsche Lösung f
knapsack duster (Landw) Rückenstäubegerät n, Rücken[ver]stäuber m
k. sprayer (Landw) Rückensprühgerät n, Rückensprüher m; (Landw) Rückenspritze f
knead / to [durch]kneten, anteigen, plasti[fi]zieren
kneader s. kneading machine
k. mixer Knetmischer m, Mischkneter m
kneading Kneten n, Durchkneten n, Anteigen n, Plasti[fi]zieren n
k. machine Knetmaschine f, Knetwerk n, Knetap-

parat m, Kneter m, (oft auch) Mischmaschine f, Mischer m, (i.e.S.) Schaufelkneter m, Doppelarmkneter m, Doppelpaddelmischer m; (Pap) Zerfaserer m, Zerfaserungsmaschine f; (Ker) Masseschlagmaschine f
k. process Knetprozeß m
k. roll Knetwalze f
k. table Tellerkneter m, Knetteller m; (Ker) Masseschlagmaschine f
k. table with fluted roll Riffelkneter m (zur Margarineherstellung)
knee (Foto) Schulter f (der Schwärzungskurve), Gebiet n der Überexposition (Überbelichtung)
knife Messer n; Schaber m, Schabemesser n, Abstreifer m, Abstreifermesser n, Rakel f
k. barker (barking machine) (Pap) Messer[scheiben]entrinder m
k. coater Messerstreichmaschine f, Rakelstreichmaschine f, Rakelauftragmaschine f
k. coating Beschichten (Streichen) n mit Rakel
k. cylinder Messerwalze f
k.-discharge centrifuge Schälzentrifuge f, Schälschleuder f
k. drum s. k. roll
k. edge Schneide f (z.B. an einer Analysenwaage); Schabemesser n, Schaber m, Messer n, Schabevorrichtung f
k. roll (Pap) Messerwalze f, Holländerwalze f, Mahlwalze f
knived roll s. knife roll
knives (Pap) Holländermesser npl, Messer npl
knock Klopfen n (im Vergasermotor)
k.-free klopffrei, nichtklopfend
k. inhibitor Antiklopfmittel n, Klopfbremse f
k.-out bar Ausdrückstange f
k.-out pin Ausdrückstift m, Auswerferstift m
k.-out point Knock-out-Punkt m (bei der Insektizidprüfung)
k. proneness Klopfneigung f, Klopffreudigkeit f, Klopfempfindlichkeit f, Klopfstärke f
k. rating 1. Klopfwertbestimmung f, Klopf[wert]prüfung f, Ermittlung f der Klopffestigkeit; 2. Klopfwert m, Klopffestigkeit f
k.-rating engine Klopfprüfmotor m
k. reducer s. k. inhibitor
k. resistance Klopffestigkeit f
k. suppressor s. k. inhibitor
k. value Klopfwert m, Klopffestigkeit f
knockdown (Insektizidprüfung) Knockdown n, KD (Zusammenbruch der Gleichgewichtshaltung vergifteter Tiere)
k. agent Knockdown-Mittel n (sofort wirkendes Insektizid)
k. effect Knockdown-Effekt m, KD-Effekt m (sofortige Wirkung auf Schadinsekten)
k. poison s. k. agent
k. threshold (Insektizidprüfung) (zum Knockdown-Effekt führender Schwellenwert)
knocked-on atom Anstoßatom n, angestoßenes Atom n

knocker Klopfer m, Klopfwerk n
knocking Klopfen n *(im Vergasermotor)*
k. propensity s. knock proneness
k. property Klopfeigenschaft f
knockless s. knock-free
knockout Ausdrücker m, Auswerfer m
k. plate Ausdrückplatte f
knockproof klopffest
Knoevenagel condensation (reaction) Knoevenagel-Reaktion f, Knoevenagel-Kondensation f, Knoevenagelsche Synthese f
Knoop hardness test Härteprüfung f nach Knoop
knopper, knoppern nut Knopper f *(Eichelgalle)*
Knop's solution *(Landw)* Knopsche Lösung f, Nährlösung f nach Knop
Knorr pyrrole synthesis Knorrsche Pyrrolsynthese f
knot *(Pap)* Knoten m, Faserstoffknoten m, Batzen m; Ast[knoten] m *(im Holz)*; *(Glas)* Knoten m *(Fehler)*
k. screen *(Pap)* Ästefänger m, Astfänger m, Astfang m, Knotenfänger m
knotter [screen] s. knot screen
knotting *(Pap)* Rückhaltung f (Abtrennen n) grober Bestandteile *(Äste, Splitter)* aus Stoffsuspensionen, Vorsortierung f
knurled screw Rändelschraube f
Ko-kneader Ko-Kneter m *(Misch- und Plastiziermaschine)*
Koch acid $C_{10}H_4(NH_2)(SO_3H)_3$ Koch-Säure f, Kochsche Säure f, 1-Naphthylamin-3,6,8-sulfonsäure f
KOH number *(Gum)* KOH-Zahl f
Kohlrausch bridge *(phys Ch)* Kohlrausch-Brücke f
Kohlrausch law Kohlrauschsches Gesetz n *(der unabhängigen Ionenwanderung)*
kojic acid Kojisäure f, 5-Hydroxy-2-(hydroxymethyl)-1,4-pyron n
kok-saghyz Kok-Saghys m, Kok-Saghyz m, Kautschuklöwenzahn m, Taraxacum bicorne Dahlst.
k.-saghyz rubber Kok-Saghys-Kautschuk m
k.-sagyz s. kok-saghyz
kokum butter Kokumbutter f *(von Garcinia indica Choisy)*
kola [nut] Kolanuß f
k. tree Kolabaum m, *(Gattung)* Cola Schott et Endl.
Kolbe electrolysis reaction Kolbesche Kohlenwasserstoffsynthese f (Paraffinsynthese f, Synthese f von Paraffinkohlenwasserstoffen)
Kolbe reaction Kolbe-Reaktion f, Kolbe-Synthese f, Kolbesche Reaktion (Synthese) f
Kolbe reaction of phenols Kolbesche Synthese f von Phenolkarbonsäuren, Kolbesche Salizylsäuresynthese f
Kolbe-Schmitt reaction (synthesis) Kolbe-Schmitt-Reaktion f, Kolbe-Schmitt-Synthese f, Kolbe-Schmittsche Reaktion (Synthese) f
Kolbe synthesis s. Kolbe reaction

Kolbe's electrochemical reaction s. Kolbe electrolysis reaction
kollergang Kollergang m, Kollermühle f
Kolloplex ·mill Kolloplexmühle f *(eine Schlagstiftmühle)*
Konowaloff rule Konowalowsche Regel f
Kopp law Koppsche Regel f, Kopp-Neumannsche Regel f
Koppers oven Koppers-Ofen m, Koppers-Verbund[kreisstrom]ofen m, Koppers-Kreisstrom-Verbundkammerofen m, Kreisstrom-Verbund[koks]ofen m nach Koppers
Koppers process 1. Koppers-Verfahren n *(zur Gasreinigung)*; 2. s. Koppers process for gasification of pulverized coal
Koppers process for gasification of pulverized coal Staubvergasungsverfahren n nach Koppers[-Totzek], Koppers-Totzek-Verfahren n [zur Kohlenstaubvergasung]
koppite *(Min)* Koppit m
kopsin Kopsin n *(Alkaloid)*
Kosey silk Koseiseide f *(aus regeneriertem Fibroin)*
Kostinky effect *(Foto)* Kostinky-Effekt m
koumis[s], koumyss Kumys m, Kumyß m
Kraemer and Sarnow softening temperature Erweichungspunkt m nach Kraemer-Sarnow, Erweichungspunkt KS m
kraft s. k. paper
k. cardboard Kraftkarton m
k. cook Sulfatkochung f für Kraftzellstoff
k. cooking liquor *(Pap)* Sulfat[koch]lauge f
k. digester Kraftzellstoffkocher m, Sulfatzellstoffkocher m
k. paper Kraft[pack]papier n
k. paper machine Kraftpapiermaschine f
k. process Kraftzellstoffverfahren n
k. pulp Kraftzellstoff m *(hochfester Sulfatzellstoff)*
k. pulp mill Sulfatzellstoffabrik f, Sulfatfabrik f
k. semichemical process *(Pap)* Semikraft-Verfahren n
k. wrapping paper s. k. paper
Krämer mill Krämer-Mühle f *(eine Schlägermühle)*
krausen Kräusen pl *(bei der Bierherstellung)*
kraznozem Roterde f
Krebs cycle Krebs-Zyklus m, Szent-Györgyi-Krebs-Zyklus m, Zitronensäurezyklus m, Zitratzyklus m, Trikarbonsäurezyklus m
krennerite *(Min)* Krennerit m (Gold(III)-tellurid)
Kröcker bomb Berthelot-Mahler-Kröcker-Bombe f, Berthelotsche Bombe f
kröhnkite *(Min)* Kröhnkit m *(Natriumkupfersulfat)*
krokidolite, krokydolite, krokydolith Krokydolith m *(Hornblende-Abart)*
Kroll process *(Met)* Kroll-Verfahren n *(ein Reduktionsverfahren)*
Kron effect *(Foto)* Kron-Effekt m
krönkite *(fälschlich für)* kröhnkite
Krupp-Lurgi process Krupp-Lurgi-Verfahren n, Krupp-Lurgi-Schwelverfahren n

Krupp-Renn process *(Met)* Krupp-Rennverfahren *n*
Krupp ring-roll press Ringwalzenpresse *f* nach Krupp-Gruson
kryptocyanine Kryptozyanin *n*, 1,1'-Diäthyl-4,4'-karbozyaninjodid *n*
krypton Kr Krypton *n*
Kubierschky tower Kubierschky-Turm *m*, Kubierschky-Apparat *m* *(zur technischen Gewinnung von Brom aus Bromlaugen)*
Kubierschky's nitrator Kubierschky-Nitrierapparat *m*, Nitrierapparat *m* nach Kubierschky
Kuhn-Roth determination Kuhn-Roth-Bestimmung *f*
Kullgren [lingnosulphonic] acid nach Kullgren bestimmte wasserlösliche Lingninsulfonsäure *f*
kumis[s] *s.* koumis
kupfernickel *(Min)* Kupfernickel *m*, *(veraltet für)* Nickelin *m*, Rotnickelkies *m* *(Nickelarsenid)*
kurchee (kurchi) bark *(Pharm)* Kurchirinde *f* *(von Holarrhena antidysenterica Wall.)*
Kurrol salt Kurrolsches Salz *n*
kussum oil Macassar-Öl *n* *(von Schleichera trijuga Willd.)*
kutch *(Gerb)* Mangroverindenextrakt *m(n)*
kuteera (kutira) gum Kuteragummi *n* *(von Sterculia-, Cochlospermum- oder Astragalus-Arten)*
kvas[s] Kwaß *m*
kyanite *(Min)* Kyanit *m*, Zyanit *m*, Disthen *m* *(Aluminiumoxidorthosilikat)*
kyanize / to kyanisieren *(Holz durch Tränkung mit 0,66%iger Sublimatlösung konservieren)*
kyanizing Kyanisieren *n*, Kyanisierung *f*
kynurenic acid NC$_9$H$_5$(OH)COOH Kynurensäure *f*, 4-Hydroxy-2-chinolinkarbonsäure *f*
kynurenine C$_6$H$_4$(NH$_2$)COCH$_2$CHNH$_2$COOH Kynurenin *n*, o-Aminophenazylaminoessigsäure *f*
kynuric acid HOOCCONHC$_6$H$_4$COOH Kynursäure *f*, Oxanilsäure-2-karbonsäure *f*
kynurine NC$_9$H$_6$OH Kynurin *n*, 4-Hydroxychinolin *n*

L

L-D plant LD-Stahlwerk *n* *(Blasstahlwerk)*
L-D process LD-Verfahren *n*, LD-Blasstahlverfahren *n*, LD-Aufblaseverfahren *n*, Linz-Donawitz-Verfahren *n*
L/D ratio *(Plast)* L/D-Verhältnis *n* *(Länge zu Durchmesser der Schnecke)*
l-form l-Form *f*, (-)-Form *f*, linksdrehende Form *f*, Linksform *f*
L-layer *(Bodenkunde)* Förnahorizont *m* *(Symbol: L)*
L-shell L-Schale *f*
L type of calender L-Kalander *m*, L-Walzenkalander *m*, Vierwalzenkalander *m* mit unten vorliegender Walze
laavenite *s.* lavenite

lab, lab. *s.* laboratory
labdanum La[b]danum *n*, La[b]danharz *n*, La[b]dangummi *n*
l. oil La[b]danumöl *n*
l. resin *s.* labdanum
label / to kennzeichnen, angeben, markieren; indizieren, beziffern; etikettieren, durch eine Aufschrift kennzeichnen; markieren *(eine Substanz durch den Einbau gut nachweisbarer, meist radioaktiver Atome kenntlich machen)*
label Etikett *n*, Aufschrift *f*; Markierung *f*, markiertes Atom *n*
l. clearance amtliche Anerkennung *f (Handelserlaubnis für ein Pflanzenschutzmittel)*
l. direction Gütevorschrift *f*
labelled atom markiertes Atom *n*
l. compound markierte Verbindung *f*
labeller *s.* labelling machine
labelling Kennzeichnen *n*, Kennzeichnung *f*, Markieren *n*, Markierung *f*; Indizierung *f*, Bezifferung *f*; Etikettieren *n*, Etikettierung *f*; Markieren *n*, Markierung *f* *(Kenntlichmachung einer Substanz durch den Einbau gut nachweisbarer, meist radioaktiver Atome)*
l. machine Etikettiermaschine *f*
labile labil; empfindlich *(z.B. gegen Alkalien)*
lability Labilität *f*; Empfindlichkeit *f* *(z. B. gegen Alkalien)*
laboratory labor[atoriums]mäßig, Labor[atoriums]...
laboratory Labor[atorium] *n*
l. abrasion test Labor[atoriums]abriebversuch *m*, Labor[atoriums]abriebprüfung *f*, Labor[atoriums]abnutzungsprüfung *f*
l. apparatus Labor[atoriums]gerät *n*, Labor[atoriums]apparat *m*
l. assistant Laborant *m*, Chemielaborant *m*; Laborantin *f*, Chemielaborantin *f*
l. bench *s.* l. table
l. bench sink Einsatzbecken *n* für Labortische
l. bottle Labor[atoriums]standflasche *f*, Standflasche *f*
l. column Labor[atoriums]kolonne *f*
l. determination Labor[atoriums]bestimmung *f*
l. equipment Labor[atoriums]einrichtung *f*, Labor[atoriums]ausstattung *f*
l. furnace Labor[atoriums]ofen *m*
l. hood Abzug[s]schrank *m*, Abzug *m*
l. instrument *s.* l. apparatus
l. manual Labor[atoriums]tagebuch *n*
l. method Labor[atoriums]methode *f*
l. pH meter Labor[atoriums]-pH-Meter *n*
l. potentiometer Labor[atoriums]potentiometer *n*
l. press Labor[atoriums]presse *f*
l. roller mill *s.* l.-size mill
l.-scale test Prüfung *f* im Labor[atoriums]maßstab
l. sieve Labor[atoriums]sieb *n*
l. sink Labor[atoriums]becken *n*
l.-size mill *(Gum)* Labor[atoriums]walzwerk *n*

l. table Labor[atoriums]tisch *m*
l. test Labor[atoriums]versuch *m*, Labor[atoriums]prüfung *f*, Labor[atoriums]test *m*
l. titrimeter Labor[atoriums]titrimeter *n*
labrador feldspar *(Min)* Labradorfeldspat *m*, Labradorstein *m*, Labradorit *m*
l. hornblende *(Min)* Labradorhornblende *f*, *(veraltet für)* Hypersthen *m*
labradorite, labrodite *s.* labrador feldspar
labyrinth seal Labyrinth[spalt]dichtung *f*
l. trap Labyrinthkondens[at]wasserableiter *m*
lac Rohschellack *m*, Gummilack *m*; Schellack *m*, Schollenlack *m*
l. dye Lac[k]dye *m(n)*, Lac-Dye *m(n)*, roter Lackfarbstoff *m*
l. insect Lack[schild]laus *f*, Carteria lacca
l. lac (lake) Lack-Lack *m*, Lac-Lac *m*, [roter] Färbelack *m*
lacca[in]ic acid Lakkainsäure *f*
laccase Lakkase *f*, Urushioloxydase *f*
lachrymator Augenreizstoff *m*, Tränenreizstoff *m*, Tränengas *n*
lachrymatory tränenreizend, tränenerregend, tränenerzeugend, lakrimogen, augenreizend
lacker *s.* lacquer
lac[k]moid $(OH)_2C_6H_3N = [C_6H_2(OH)_3]_2$ La[c]kmoid *n*, Resorzinblau *n*
lacmus Lackmus *n(m)* *(Flechtenfarbstoff)*
lacquer Lack *m*, *(besonders)* Lackfarbe *f*, pigmentierter Lack *m*
l. dye Lackfarbstoff *m*
l. remover Lackentferner *m*
l. solvent Lacklösungsmittel *n*
lactacidogen Laktazidogen *n*
lactalbumin Laktalbumin *n*, Milchalbumin *n*
lactam Laktam *n*
l.-lactim tautomerism Laktam-Laktim-Tautomerie *f*
β-l. ring β-Laktamring *m*
lactaric acid $CH_3(CH_2)_{16}COOH$ Laktarsäure *f*, Stearinsäure *f*, Oktadekansäure *f*
lactarinic acid Laktarinsäure *f*, ε-Ketostearinsäure *f*, 6-Oxooktadekansäure *f*
lactase Laktase *f*, β-Galaktosidase *f*
lactate Laktat *n*
lactic acid Milchsäure *f*, *(i.e.S.)* $CH_3CH(OH)COOH$ Milchsäure *f*, Äthylidenmilchsäure *f*, 2-Hydroxypropionsäure *f*
l.-acid bacteria Milchsäurebakterien *npl*
l.-acid fermentation Milchsäuregärung *f*, milchsaure Gärung *f*
l. acid of fermentation $CH_3CH(OH)COOH$ Gärungsmilchsäure *f*, DL-Milchsäure *f*
l.-acid organisms Milchsäuremikroben *fpl*, Milchsäurebildner *mpl*, Milchsäureerzeuger *mpl*
l.-acid-producing bacteria *s.* l.-acid bacteria
l. anhydride $[CH_3CH(OH)CO]_2O$ Milchsäureanhydrid *n*, 2-Hydroxypropansäureanhydrid *n*
l. dehydrogenase Laktatdehydr[ogen]ase *f*, Laktikodehydrase *f*, Milchsäuredehydr[ogen]ase *f*, LDH

l. fermentation *s.* l.-acid fermentation
lactide Laktid *n*
lactim Laktim *n*
lactobacillus Laktobazillus *m*, Milchsäurebazillus *m*
lactobionic acid Laktobionsäure *f*, 4-[β-*d*-Galaktosido-]-*d*-glukonsäure *f*
lactobiose *s.* lactose
lactobutyrometer Laktobutyrometer *n*, Milchbutyrometer *n*
lactodensimeter Laktodensimeter *n*, Milchspindel *f*
lactogen *s.* lactogenic hormone
lactogenic laktotrop
l. hormone laktotropes (luteotrophes) Hormon *n*, Laktogen *n*, Laktationshormon *n*, Laktotropin *n*, Prolaktin *n*, Luteotrophin *n*, LTH
lactoglobulin Laktoglobulin *n*
lactometer Laktometer *n*, Galaktometer *n*
lactone Lakton *n*
l. acid Laktonsäure *f* *(Laktonderivat mit Karboxylgruppe am Laktonring)*
l. formation Laktonbildung *f*
l. ring Laktonring *m*
l. rule Laktonregel *f* [der optischen Drehung], Hudsonsche Laktonregel *f*
lactonic Lakton...
l. acid Laktonsäure *f* *(Laktonderivat mit Karboxylgruppe am Laktonring)*; $HOCH_2(CHOH)_4COOH$ *d*-Laktonsäure *f*, *d*-Galaktonsäure *f*
l. linkage Laktonbindung *f*
lactonitrile $CH_3CH(OH)CN$ Laktonitril *n*, Milchsäurenitril *n*, Azetaldehydzyanhydrin *n*, 2-Hydroxypropannitril *n*
lactonization Laktonisierung *f*, Laktonbildung *f*
lactonize / to laktonisieren
lactoperoxidase Laktoperoxydase *f*
lactoscope Laktoskop *n*, Galaktoskop *n*
lactose Lakto[bio]se *f*, Milchzucker *m*
l. fermentation Milchzuckergärung *f*
l. syrup Laktosesirup *m*
lactoserum Milchserum *n*, Molken *m*, Molke *f*, Käswasser *n*
lactosuria Laktosurie *f* *(Auftreten von Milchzucker im Harn)*
lact[o]yl $CH_3CH(OH)CO-$ Lakt[o]yl *n*
lactyllactic acid $CH_3CH(OH)COOCH(CH_3)COOH$ Laktylmilchsäure *f*, 2-(2-Hydroxypropanoyloxy)propansäure *f*
lactylurea Laktylharnstoff *m*
lacustrine *(Geoch)* lakustrisch, limnisch
ladanum *s.* labdanum
ladder polymer Leiterpolymer[es] *n*
ladders *(Glas)* Falten *fpl* *(Oberflächenfehler)*
ladle Gießlöffel *m*; Pfanne *f*; *(Glas)* Schöpfkelle *f*, Kelle *f*, Löffel *m*
l. lining Pfannenauskleidung *f*, Pfannenfutter *n*
ladling Ausschöpfen *n* *(Entnahme des Glases aus dem Schmelzofen mit Kellen oder Löffeln)*
laevo *s.* laevorotatory

l. acid s. laevorotatory acid
l. dichloro chelate l-Dichlorochelat n
l. form s. laevorotatory form
**laevogyrate, laevogyratory, laevogyre, laevo-
gyrous** s. laevorotatory
laevolactic acid Linksmilchsäure f, l-Milchsäure f,
linksdrehende Milchsäure f
laevopimaric acid Lävopimarsäure f, β-Pimarsäure
f (Harzsäure)
laevorotary, laevorotating s. laevorotatory
laevorotation Linksdrehung f
laevorotatory linksdrehend, l-drehend, l
l. acid linksdrehende Säure f, Linkssäure f,
l-Säure f
l. form linksdrehende Form f, Linksform f, l-Form
f, (-)-Form f
l. lactic acid s. laevolactic acid
l. tartaric acid s. laevotartaric acid
laevotartaric acid Linksweinsäure f, l-Weinsäure f,
linksdrehende Weinsäure f, l-2,3-Dihydroxybu-
tandisäure f
laevul[in]ic acid $CH_3CO(CH_2)_2COOH$ Lävulinsäure
f, 4-Oxopentansäure f, 4-Ketovaleriansäure f,
Pentanon-(4)-säure-(1) f
laevulose $C_6H_{12}O_6$ Lävulose f, D-Fruktose f,
Fruchtzucker m
lag / to verkleiden, isolieren
lager [beer] Lagerbier n
l. fermentation Untergärung f
l. yeast untergärige Hefe f, Unterhefe f
lagging Verkleidung f, Isolierung f
lagosin Lagosin n (ein Polyenantibiotikum)
laid paper geripptes Papier n, mit Wasser[zeichen]linien versehenes Papier n, Wasser-
linienpapier n, Vergépapier n
Lainer effect (Foto) Lainer-Effekt m
laitance Zementmilch f, Zementschlamm m, Ze-
mentschlämme f, Zementschlempe f
lake Lack m; (beim Beizen der Gewebe) Farblack
m
l. colour s. l. dye
l. dye Lackfarbstoff m, Lackfarbe f
l. lac Lack-Lack m, Lac-Lac m, [roter] Färbelack
m
lakmus s. lacmus
lambda point (phys Ch) λ-Punkt m
Lambert-Beer law Lambert-Beersches Gesetz n
(der Lichtabsorption)
Lambert law [of absorption] Lambertsches Gesetz
(Absorptionsgesetz) n, Bouguer-Lambertsches
Gesetz (Absorptionsgesetz) n
lamella Lamelle f
lamellar lamellar, blattförmig, blättchenförmig,
blätt[e]rig, tafelförmig, plattenförmig, schichtförmig, geschichtet
l. colloid Neutralkolloid n
l. compound lamellare Verbindung f, schichtförmig ausgebildete Einlagerungsverbindung f
l. compound of graphite Einlagerungsverbindung f des Graphits, Graphit-Einlagerungsverbindung f

laminar laminar, Laminar...
l. flow laminare Strömung (Bewegung) f, Laminarströmung f, Laminarbewegung f, Schichtenströmung f
laminate / to [be]schichten, laminieren; (Pap)
bekleben, kaschieren; (Pap) [zusammen]gaut-
schen
laminate Schicht[preß]stoff m, Laminat n, geschichteter Werkstoff m, (manchmal auch)
Hartgewebe n, Hgw
laminated fabric Hartgewebe n, Hgw
l. film (foil) Verbundfolie f
l. glass s. l. safety [sheet] glass
l. material Schicht[preß]stoff m, Laminat n,
geschichteter Werkstoff m
l. moulding Schichtpreßverfahren n; Schicht-
preßteil n
l. paper laminiertes (befilmtes) Papier n; (Plast)
Hartpapier n, Hp
l. plastic s. l. material
l. preform (Plast) geschichteter Vorformling m
l. product Schichtpreßstofferzeugnis n
l. safety [sheet] glass geschichtetes Sicherheitsglas n, Verbund[sicherheits]glas n, Mehr-
scheiben[sicherheits]glas n, Mehrschichten[si-
cherheits]glas n
l. sheet Schichtpreßstoffplatte f, Schichtfolie f
l. wood Schicht[preß]holz n
laminating (Plast) Schichten n, Schichtung f,
Laminieren n, Laminierung f; (Pap) Kaschieren
n, Kaschierung f; (Pap) Zusammengautschen n
l. resin Laminierharz n
lamination Schichten n, Schichtung f, Laminieren
n, Laminierung f; (Geol) Schichtung f, Schich-
tenbildung f; (Ker) Textur f (Fehler)
lamp-blown (Glas) vor der Lampe geblasen,
lampengeblasen
l. method Lampenmethode f (zur Schwefelbestimmung)
l. oil Lampenpetroleum n, Leuchtöl n, Leuchtpetroleum n
lampblack Lampenschwarz n, Lampenruß m;
(Gum) Flammruß m
lampworked s. lamp-blown
lampworker Glasbläser m vor der Lampe, Lampen-
bläser m
lampworking (Glas) Lampenarbeit f, Lampen-
bläserei f
lamy butter Kanyabutter f (von Pentadesma
butyraceum Sabine)
lanarkite (Min) Lanarkit m (Blei(II)-oxidsulfat)
lanatoside Lanatosid n, Digilanid n
Lancashire boiler Lancashirekessel m (Zwei-
flammrohrkessel)
lance Blaslanze f, Sauerstofflanze f, O_2-Lanze f
(beim Sauerstoffaufblaseverfahren); (Landw)
Stabspritze f
lancet-point dissecting needle Lanzettnadel f, lan-
zettenförmige Präpariernadel f
land [area] (Plast) Abquetschfläche f

Landé g-factor s. Landé splitting factor g
Landé splitting factor g Landéscher g-Faktor m, Landé-Faktor g m, g-Faktor m
Landolt reaction Landoltsche Reaktion f
Landsiedl absorption (potash) bulb Kaliapparat m nach Landsiedl
langite (Min) Langit m (Kupfer(II)-hexahydroxidsulfat)
Langmuir adsorption isotherm Langmuirsche Adsorptionsisotherme f
lanoceric acid $(HO)_2C_{29}H_{57}COOH$ Lanozerinsäure f
lanolin[e] Lanolin n, gereinigtes Wollfett n
lanopalm[in]ic acid $HOC_{15}H_{30}COOH$ Lanopalmitinsäure f
lanostadienol s. lanosterol
lanosterol Lanosterol n, Lanosterin n
lantern ring Zwischenlaterne f (Teil der Stopfbuchspackung)
lanthana La_2O_3 Lanthanerde f, Lanthana f, (veraltet für) Lanthanoxid n
lanthanide s. lanthanoid
lanthanite (Min) Lanthanit m
lanthanoid Lanthanoid n, Lanthanoidenelement n, (bisher) Lanthanid n, Lanthanidenelement n, Element n der Lanthanreihe
l. contraction Lanthanoidenkontraktion f
l. element s. lanthanoid
l. group s. l. series
l. metal Lanthanoidenmetall n
l. series Lanthanoidenreihe f, Lanthanoidengruppe f, (bisher) Lanthan[iden]reihe f, Lanthanidengruppe f
lanthanon s. lanthanoid
lanthanum La Lanthan n
l. bromide $LaBr_3$ Lanthanbromid n
l. carbide LaC_2 Lanthankarbid n, Lanthanazetylid n
l. carbonate $La_2(CO_3)_3$ Lanthankarbonat n
l. chloride $LaCl_3$ Lanthanchlorid n
l. fluoride LaF_3 Lanthanfluorid n
l. hydroxide $La(OH)_3$ Lanthanhydroxid n
l. iodate $La(JO_3)_3$ Lanthanjodat n
l. nitrate $La(NO_3)_3$ Lanthannitrat n
l. salt Lanthansalz n
l. sesquioxide La_2O_3 Lanthansesquioxid n, (veraltet für) Lanthan[tri]oxid n, Lanthan(III)-oxid n
l. sulphate $La_2(SO_4)_3$ Lanthansulfat n
l. trioxide La_2O_3 Lanthan[tri]oxid n, Lanthan(III)-oxid n
lanthionine $S[CH_2CH(NH_2)CO_2H]_2$ Lanthionin n, α,α-Diamino-β,β-dikarboxydiäthylsulfid n, β,β-Thiodialanin n
lap (Glas) Falte f, Verfaltung f (Fehler); (Text) Wickel m
l.-joint flange Bördelflansch m, loser Flansch m
l. pulp s. lapped pulp
lapacho Lapachoholz n
lapachoic acid s. lapachol

lapachol Lapachol n, Lapachosäure f, Tekomin n, Taigusäure f, Grönhartin n, 2-Hydroxy-3-(3-methyl-2-butenyl)-1,4-naphthochinon n
lapilli tuff Lapillituff m
lapis [lazuli] (Min) Lapislazuli m, Lasurit m, Lasurstein m
Laplacian operator Laplace-Operator m, Laplacescher Operator m, Delta-Operator m
lapped pulp (Pap) Halbstoff m in Bogenform (trockenen Bogen), Bogenhalbstoff m, (i.e.S.) Holzschliff m in Bogenform (Pappenform), Holzschliffpappe f, Holzschliffblätter npl (oder) Zellstoff m in Bogenform (Pappenform), Zellstoffpappe f, Zellstoffblätter npl
laps of pulp s. lapped pulp
lard Schweineschmalz n, Schmalz n, Schweinefett n
l. oil Lardöl n, Specköl n, Fettöl n, Schmalzöl n
large bell große Glocke (Verschlußglocke, Gichtglocke) f (eines Hochofens)
l.-bore mit weiter Bohrung
l. calorie größe Kalorie f, Kilogrammkalorie f, (veraltet für) Kilokalorie f, kcal
l. coke Großkoks m
l.-grained grobkörnig
l.-meshed weitmaschig, grobmaschig
l. molecule Makromolekül n, Riesenmolekül n
l.-pored großporig
l.-pressure cooker Autoklav m
l. producer Großproduzent m, Großerzeuger m
l. ring großer Ring m, Makroring m (aus 13 oder mehr Gliedern bestehend)
l.-ring compound Verbindung f mit großem Ring, Ringverbindung f mit großer Gliederzahl, makrozyklische Verbindung f, Großringverbindung f
l.-ring ketone Keton n mit großem Ring, großes Ringketon n, makrozyklisches Keton n
l.-scale groß, im großen [Maße], in großem Umfang (Maßstab), ausgedehnt, Groß..., Massen..., (oft auch) großtechnisch, großindustriell, im großtechnischen Maßstab
l.-scale process [technischer] Großprozeß m, großtechnischer Prozeß m, großtechnisches Verfahren n
l.-scale test Großversuch m
laricic acid $C_{19}H_{36}OH(COOH)_3$ Agarizin n, Agarizinsäure f, Zetylzitronensäure f
laricin Koniferin n, Larizin n, Abietin n, Koniferosid n (β-Glukosid des Koniferylalkohols)
Larmor precession Larmor-Präzession f
larnite (Min) Larnit m (Kalziumorthosilikat)
larvicide Larvizid n, Larvengift n
Lassaigne's test Lassaignesche Probe f, Lassaigne-Probe f, Stickstoffnachweis m nach Lassaigne
last / to:
l. well gut beständig sein
last hypochloride treatment (Pap) Hypochlorit-Endbleiche f, Endhypochloritbleiche f
l. pressed juice Scheitermost m, Nachdruck m

l. wort Nachwürze *f*
late hop Späthopfen *m*
l. wood Spätholz *n*
latensification *(Foto)* Latensifikation *f*, Verstärkung *f* des latenten Bildes
latent crimp *(Text)* latente Kräuselung *f*
  l. heat latente (gebundene) Wärme *f*, Umwandlungswärme *f*
  l. image *(Foto)* latentes Bild *n*
  l.-image intensification *(Foto)* Verstärkung *f* des latenten Bildes
  l.-image regression *(Foto)* Abklingen *n* (Rückgang *m*) des latenten Bildes
  l. solvency latentes Lösungsvermögen (Lösevermögen) *n*
  l. solvent latentes Lösungsmittel *n*, latenter Löser *m*
lateral Ansatzstück *n*
  l. chain Seitenkette *f*
  l. order laterale Ordnung *f*
laterite *(Min)* Laterit *m*
lateritic lateritisch, Laterit...
  l. soil Lateritboden *m*
lateritization Lateritisierung *f*
latex Latex *m*, Milchsaft *m*; Latex *m*, Kunstharzdispersion *f*
  l.-bearing latexführend, milchsaftführend
  l. cell *s.* laticiferous cell
  l.-chlorinated rubber Latex-Chlorkautschuk *m*
  l. compound Latexmischung *f*
  l. compounding Herstellung *f* von Latexmischungen
  l. concentrate Latexkonzentrat *n*
  l. foam [rubber] Latexschaum[gummi] *m*, Schaumgummi *m*
  l. foam sponge Latexschwamm *m*
  l. paint Latex[anstrich]farbe *f*
  l.-sprayed rubber Sprühkautschuk *m*
  l. technology Latextechnologie *f*
  l. thread Latex[gummi]faden *m*
  l. tube *s.* laticiferous tube
  l. vessel *s.* laticiferous vessel
lather / to schäumen
lather Schaum *m*, *(i.e.S. auch)* Seifenschaum *m*
  l. booster Schaumverbesserer *m*
  l. collapse Zusammenbruch *m* (Zusammensinken *n*) des Schaums
  l. shave cream schäumende Rasiercreme *f*
  l. value Schaumwert *m*, Schaumzahl *f*
  l. value in presence of dirt Belastungsschaumzahl *f*
lathering property Schäumeigenschaft *f*
laticiferous latexführend, milchsaftführend
  l. cell Milch[saft]zelle *f*, ungegliederte Milchröhre *f*
  l. tube Milchröhre *f*, Milchsaftschlauch *m*
  l. vessel Milch[saft]gefäß *n*, gegliederte Milchröhre *f*
lattice *(Krist)* Gitter *n*
  l. arrangement Gitter[an]ordnung *f*

l. compound Gitterverbindung *f*
l. constant Gitterkonstante *f*
l. defect Gitter[bau]fehler *m*; Gitterfehlstelle *f*, Fehlstelle *f*
l. disorder Gitterfehlordnung *f*
l. distance Gitterabstand *m*, Gitterperiode *f*, Netzebenenabstand *m*
l. distortion Gitterverzerrung *f*
l.-enclosure compound Gittereinschlußverbindung *f*
l. energy Gitterenergie *f*
l. expansion Gitteraufweitung *f*, Gitterexpansion *f*
l. face Gitterfläche *f*
l. forces Gitterkräfte *fpl*
l. hole *s.* l. vacancy
l. imperfection Gitterstörung *f*, Gitterfehler *m* *(i.w.S.)*
l. orientation Gitterorientierung *f*
l. plane Gitterebene *f*, Netzebene *f*
l. point Gitterpunkt *m*; Gitterplatz *m*, Gitterstelle *f*
l. position (site) Gitterplatz *m*, Gitterstelle *f*
l. spacing *s.* l. distance
l. structure Gitterstruktur *f*, Gitterverband *m*, Gitter[auf]bau *m*
l. type Gittertyp *m*, Kristallgittertyp *m*, Strukturtyp *m*
l. vacancy Gitterleerstelle *f*, Gitterloch *n*, Gitterlücke *f*, Leerstelle *f*, Leerplatz *m*, Loch *n*, Lücke *n*, unbesetzter Gitterplatz *m*
laudanidine Laudanidin *n*, *l*-Laudanin *n*
laudanine Laudanin *n* *(Alkaloid)*
laudanosine Laudanosin *n*, N-Methyltetrahydropapaverin *n*
laudanum *(Pharm)* Laudanum *n* *(Opiumtinktur)*
Laue diagram *s.* Laue pattern
Laue experiment Laue-Versuch *m*
Laue method *(Krist)* Laue-Verfahren *n*, Laue-Methode *f*
Laue pattern (photograph) Laue-Aufnahme *f*, Laue-Bild *n*, Laue-Fotogramm *n*, Laue-Diagramm *n*
Laue photograph method *s.* Laue method
Laue X-ray method *s.* Laue method
laughing gas Lachgas *n*, Lustgas *n* *(Stickstoff(I)-oxid)*
laumon[t]ite *(Min)* Laumon[t]it *m* *(Kalziumbisalumodisilikat)*
launder / to *(Textilien)* waschen
launder Gerinne *n*, Rinne *f*
  l. classifier Stromklassierer *m*
laundering Waschen *n* *(von Textilien)*
  l. fastness Waschechtheit *f*, Waschbeständigkeit *f*
lauraldehyde $CH_3(CH_2)_{10}CHO$ Lauraldehyd *m*, Laurinaldehyd *m*, Dodekanal *n*, Dodezylaldehyd *m*, Aldehyd C 12 [L] *m*
laurate Laurat *n*
laurel leaves Lorbeerblätter *npl*

l. oil Lor[beer]öl *n*, Lorbeerfett *n*, Lorbeerbutter *f*

Laurent's acid $C_{10}H_6NH_2(SO_3H) \cdot H_2O$ Laurent-Säure *f*, Laurentsche Säure *f*, 1-Naphthylamin-5-sulfonsäure *f*

lauric acid $CH_3(CH_2)_{10}COOH$ Laurinsäure *f*, Dodekansäure *f*

l. aldehyde *s.* lauraldehyde

lauroleic acid $C_{11}H_{21}COOH$ Lauroleinsäure *f*, 9-Dodezensäure *f*

laurostearic acid *s.* lauric acid

lauroyl $-C(=O)[CH_2]_{10}CH_3$ Lauroyl *n*, Dodekanoyl *n*

l. peroxide Lauroylperoxid *n*, Dodekanoylperoxid *n*

lauryl $-CH_2[CH_2]_{10}CH_3$ Lauryl *n*, *n*-Dodezyl *n*

l. alcohol $CH_3(CH_2)_{11}OH$ Laurylalkohol *m*, *n*-Dodezylalkohol *m*, 1-Dodekanol *n*, Alkohol C 12 *m*

l. aldehyde *s.* lauraldehyde

l. mercaptan $CH_3(CH_2)_{11}SH$ Laurylmerkaptan *n*, *n*-Dodezylmerkaptan *n*, 1-Dodekanthiol *n*

laurylamine $CH_3(CH_2)_{11}NH_2$ Laurylamin *n*, Dodezylamin *n*, 1-Aminododekan *n*

lauter / to *(Bierwürze)* [ab]läutern *f*

lauter mash Lautermaische *f*, Dünnmaische *f*

l. tub (tun) Läuterbottich *m*, Filterbottich *m*

lautering Läutern *n*, Abläutern *n*

lava Lava *f*

lavandin oil Lavandinöl *n*

lavandulol Lavandulol *n* *(flüssiger Terpenalkohol in Lavendel- und Lavandinöl)*

lavender [flower] oil Lavendelöl *n*

l.-spike oil Spiköl *n*

l. water *(Kosmet)* Lavendelwasser *n*

lavenite *(Min)* Lávenit *m* *(ein zirkonhaltiges Sorosilikat)*

Laves phase Laves-Phase *f*

law of Boyle-Mariotte Boyle-Mariottesches Gesetz *n*

l. of Bunsen-Roscoe Bunsen-Roscoesches Gesetz *n*, Reziprozitätsgesetz *n*

l. of conservation of energy Gesetz *n* (Prinzip *n*, Satz *m*) [von] der Erhaltung der Energie, Erhaltungssatz *m* der Energie, Energieerhaltungssatz *m*, Energieprinzip *n*

l. of conservation of mass (matter) Gesetz *n* von der Erhaltung der Masse, Massenerhaltungssatz *m*, Erhaltungssatz *m* der [wägbaren] Masse (Materie)

l. of constant heat summation *s.* l. of Hess

l. of constant proportions Gesetz *n* der konstanten Proportionen (Gewichtsverhältnisse)

l. of decay [radioaktives] Zerfallsgesetz *n*

l. of definite composition (proportions) *s.* l. of constant proportions

l. of distribution Verteilungsgesetz *n*; Nernstsches Verteilungsgesetz *n*, Nernstscher Verteilungssatz *m*

l. of Dulong and Petit Dulong-Petitsche Regel *f*, Dulong-Petit-Regel *f*

l. of effectivity of growth factors *(Landw)* Wirkungsgesetz *n* der Wachstumsfaktoren

l. of equipartition of energy Gesetz *n* der Gleichverteilung der Energie, Gleichverteilungssatz *m*, Äquipartitionstheorem *n*

l. of equivalent proportions Gesetz *n* der äquivalenten Proportionen

l. of Hess Heßscher Satz *m*, Heßsches Gesetz *n*, Gesetz *n* der konstanten Wärmesummen

l. of independent migration of ions Gesetz *n* der unabhängigen Ionenwanderung, Kohlrauschsches Gesetz *n*

l. of inertia Trägheitsgesetz *n*

l. of intermediate reactions *s.* l. of intermediate stages

l. of intermediate stages Stufenregel *f* von Ostwald, Ostwaldsche Stufenregel *f*

l. of mass action Massenwirkungsgesetz *n*, MWG *n*, Reaktionsisotherme *f*

l. of multiple proportions Gesetz *n* der multiplen Proportionen

l. of partition *s.* l. of distribution

l. of Paschen *(Phys)* Paschensches Gesetz *n*

l. of rational indices *(Krist)* Rationalitätsgesetz *n*, Gesetz *n* der rationalen Parameterverhältnisse, Haüysches Gesetz *n*

l. of rational[ity of] intercepts *(Krist)* Gesetz *n* von der Rationalität der Achsenabschnitte

l. of reciprocal proportions *s.* l. of equivalent proportions

l. of size distribution Körnungsgesetz *n*, Kornverteilungsgesetz *n*

l. of the photochemical equivalent fotochemisches (Stark-Einsteinsches) Äquivalenzgesetz *n*, Quantenäquivalenzgesetz *n*

l. of volumes Volum[en]gesetz *n*

l. of yields *(Landw)* Ertragsgesetz *n*

lawrencium Lr, *(bisher)* Lw Lawrencium *n*

lawsone Lawson *n*, 2-Hydroxy-1,4-naphthochinon *n*

lawsonite *(Min)* Lawsonit *m*

lay-up *(Pap)* Ablegen *n* *(der Bogen)*

l.-up machine *(Gum)* Reifenaufbaumaschine *f*, Reifenerzeugungsmaschine *f*, Reifenwickelmaschine *f*, Konfektioniermaschine *f*

layer Schicht *f*, Lage *f*, *(manchmal auch)* Film *m*; *(Foto)* Emulsionsschicht *f*; *(Gerb)* Versatz *m*; *(Geol)* Lager *n*, Schicht *f*, Flöz *n*; *(Bodenkunde)* Horizont *m*

l. lattice Schicht[en]gitter *n*

l. line Schichtlinie *f*

l. of oxide Oxidschicht *f*, Oxydationsschicht *f*, *(bei Eisen auch)* Zunderschicht *f*

l. of protein Proteinschicht *f*, Proteinhäutchen *n*

l. of rust Rostschicht *f*

l. structure Schichtstruktur *f*

l. thickness Schichtdicke *f*

lazulite *(Min)* Lazulith *m*

lazurite *(Min)* Lasurit *m*, Lasurstein *m*, Lapislazuli *m*

371   **lead**

LCAO s. linear combination of atomic orbitals
LCAO **approximation** s. linear-combination-
of-atomic-orbitals approximation
LCAO **method** s. linear-combination-of-atomic-
orbitals method
LCAO MO s. LCAO molecular orbital
LCAO MO **method** s. LCAO molecular-orbital
method
LCAO MO **theory** s. LCAO molecular-orbital theory
LCAO **molecular orbital** LCAO-Molekülorbital n
LCAO **molecular-orbital method** LCAO-MO-
Methode f, LCAO-Methode f der Molekülorbitale
LCAO **molecular-orbital theory** LCAO-MO-Theorie
f, LCAO-Theorie f der Molekülorbitale
LCAO SCF **calculation** SCF-LCAO-Rechnung f
LCAO **self-consistent method** selbstkonsistente
LCAO-Methode f
LCAO **theory** s. linear-combination-of-atomic-
orbitals theory
LD s. lethal dose
L.D. **polythene** s. low-density polyethylene
Le **Chatelier[-Braun] principle** Le Chatelier-
Braunsches Prinzip n, Prinzip n des kleinsten
Zwanges, Prinzip n des beweglichen Gleich-
gewichtes
Le **Chatelier's principle of least resistance**
(restraint) s. Le Chatelier[-Braun] principle
Lea **[peroxide] value** Lea-Zahl f (Maß für den
Peroxidgehalt von Fetten)
leach **[out] / to** laugen (Erzen oder metallurgischen
Zwischenprodukten die metallischen Anteile
durch geeignete Flüssigkeiten entziehen); auslaugen, auswaschen, herauslösen, ausziehen,
extrahieren
leach 1. Lauge f; 2. Laugen n, Laugung f (Entziehen
der metallurgischen Anteile von Erzen oder metallurgischen Zwischenprodukten durch geeignete
Flüssigkeiten); Auslaugen n, Auslaugung f,
Auswaschen n, Auswaschung f, Herauslösen n,
Ausziehen n, Extrahieren n, Extraktion f; 3.
Auslaug[e]behälter m, Auslaug[e]bottich m, Extraktionsgefäß n
l. **liquor** (Gerb) Extraktionsbrühe f
l. **tank (trough, vat, vessel)** s. leach 3.
leachate Auszug m (aus Drogen oder Böden),
Perkolat n
leached **cossettes** ausgelaugte Rübenschnitzel
(Zuckerrübenschnitzel) npl
leaching s. leach 2., (i.e.S.) Feststoffextraktion f
l. **agent** Laugungsmittel n, Laugemittel n
l. **by agitation** Rührlaugung f
l. **in place** Laugung f in situ, in-situ-Laugung f,
Laugung f in der Grube
l. **solution** Laugenlösung f, Lauge f
l. **tank (trough, vat, vessel)** s. leach 3.
lead / **to** I. [ver]bleien, mit Bleitetraäthyl versetzen
(Kraftstoff); II. leiten, führen
l. **away** abführen, ableiten
lead I. Pb Blei n; II. 1. Zuleitung f; 2. Steigung f (z. B.
einer Schnecke)

24*

l. **accumulator** s. l.-acid accumulator
l. **acetate** Pb(CH$_3$COO)$_2$ Blei(II)-azetat n,
Bleidiazetat n
l.-**acetate paper** Blei[azetat]papier n
l.-**acid accumulator (battery)** Bleiakkumulator m,
Säureakkumulator m, Bleisammler m
l.-**alkali silicate glass** Bleialkalisilikatglas n
l. **alloy** Bleilegierung f
l. **azide (azoimide)** Pb(N$_3$)$_2$ Blei(II)-azid n
l.-**base** auf Bleibasis (Bleigrundlage)
l. **battery** s. l.-acid accumulator
l. **benzoate** Pb(C$_6$H$_5$COO)$_2$ Bleibenzoat n
l. **biorthoarsenate** s. l. hydrogen arsenate
l. **blast furnace** Bleischachtofen m, Blei-
[schmelz]ofen m
l. **block** Bleiblock m, (Sprengstoffprüfung auch)
Trauzl-Block m, Bleiblock m nach Trauzl
l. **block expansion** Bleiblockausbauchung f,
Trauzl-Blockausweitung (Sprengstoffprüfung)
l.-**block expansion test** Bleiblockprobe f (zur
Sprengstoffprüfung)
l. **brick** Bleibaustein m (Schutzbaustein für
Isotopenlaboratorien)
l. **bromate** Pb(BrO$_3$)$_2$ Bleibromat n
l. **caprate** Pb(C$_9$H$_{19}$COO)$_2$ Bleikaprinat n
l. **caproate** Pb(C$_5$H$_{11}$COO)$_2$ Bleikapronat n
l. **caprylate** Pb(C$_7$H$_{15}$COO)$_2$ Bleikaprylat n
l. **carbonate** PbCO$_3$ Blei(II)-karbonat n
l. **chamber** Bleikammer f
l.-**chamber process** Bleikammerverfahren n (zur
Schwefelsäuregewinnung)
l. **chlorate** Pb(ClO$_3$)$_2$ Bleichlorat n
l. **chlorite** Pb(ClO$_2$)$_2$ Bleichlorit n
l. **chromate** PbCrO$_4$ Blei(II)-chromat n
l.-**clad / to** s. to l.-line
l. **cladding** s. l. lining
l.-**coat / to** mit Blei überziehen, verbleien
l. **coating** Überziehen n mit Blei, Bleiüberzug m,
Verbleien n, Verbleiung f; (Gum) Umpressen n
mit einem Bleimantel (zur Vulkanisation)
l. **coil** Blei[rohr]schlange f
l.-**covered** mit Blei überzogen, verbleit
l. **crown glass** Kronflintglas n
l. **crystal glass** Bleikristall[glas] n
l. **cyanate** Pb(OCN)$_2$ Bleizyanat n
l. **cyanide** Pb(CN)$_2$ Bleizyanid n
l. **dibromide** PbBr$_2$ Bleidibromid n, Blei(II)-bro-
mid n
l. **dichloride** PbCl$_2$ Bleidichlorid n, Blei(II)-chlorid
n
l. **dichromate** PbCr$_2$O$_7$ Blei(II)-dichromat(VI) n
l. **difluoride** PbF$_2$ Bleidifluorid n, Blei(II)-fluorid
n
l. **diiodide** PbJ$_2$ Bleidijodid n, Blei(II)-jodid n
l. **dioxide** PbO$_2$ Bleidioxid n, Blei(IV)-oxid n
l.-**dip / to** tauchverbleien, im Tauchverfahren
feuerverbleien
l. **dithionate** PbS$_2$O$_6$ Blei(II)-dithionat n
l. **electrode** Bleielektrode f
l. **fan** Bleiventilator m

l. ferricyanide $Pb_3[Fe(CN)_6]_2$ Blei(II)-hexazyano-ferrat(III) n
l. ferrite $Pb(FeO_2)_2$ Bleiferrat(III) n
l. ferrocyanide $Pb_2[Fe(CN)_6]$ Blei(II)-hexazyano-ferrat(II) n
l. fluorosilicate $Pb[SiF_6]$ Blei(II)-hexafluorosilikat n
l. formate $Pb(HCOO)_2$ Blei(II)-formiat n
l.-free bleifrei
l. frit *(Ker)* Bleifritte f
l. gasket Bleidichtung f
l. glance *(Min)* Bleiglanz m, Galenit m *(Blei(II)-sulfid)*
l. glass Bleiglas n
l. glaze *(Ker)* Bleiglasur f
l.-glazed *(Ker)* bleiglasiert
l. hydrate s. l. hydroxide
l. hydrogen arsenate $PbHAsO_4$ Blei(II)-hydrogenarsenat(V) n
l. hydroxide $Pb(OH)_2$ Blei(II)-hydroxid n
l. hyposulphite s. l. thiosulphate
l. iodate $Pb(JO_3)_2$ Blei(II)-jodat n
l. limit Grenzwert m für Blei
l.-line / to mit Blei auskleiden, ausbleien
l. line Anschlußleitung f, Zuführungsleitung f
l. lining Auskleiden n mit Blei, Bleiauskleidung f, Ausbleien n, Ausbleiung f
l. metaarsenate $Pb(AsO_3)_2$ Blei(II)-metaarsenat(V) n
l. metaarsenite $Pb(AsO_2)_2$ Blei(II)-metaarsenat(III) n
l. metaborate $Pb(BO_2)_2$ Blei(II)-metaborat n
l. metaphosphate $Pb(PO_3)_2$ Blei(II)-metaphosphat n
l. metasilicate $PbSiO_3$ Blei(II)-metasilikat n
l. metatitanate $PbTiO_3$ Blei(II)-metatitanat n
l. metavanadate $Pb(VO_3)_2$ Blei(II)-metavanadat(V) n
l. molybdate *(Min)* Molybdänbleispat m, *(veraltet für)* Wulfenit m *(Bleimolybdat)*; $PbMoO_4$ Blei(II)-molybdat(VI) n
l. monoxide PbO Blei[mon]oxid n, Blei(II)-oxid n
l. nitrate $Pb(NO_3)_2$ Blei(II)-nitrat n
l. nitrite $Pb(NO_2)_2$ Blei(II)-nitrit n
l. ochre Massicot m *(gelbes Blei(II)-oxidpulver)*
l. orthoarsenate $Pb_3(AsO_4)_2$ Blei(II)-[ortho]arsenat(V) n
l. orthophosphate $Pb_3(PO_4)_2$ Blei(II)-[ortho]phosphat n
l. oxalate $Pb(C_2O_4)$ Blei(II)-oxalat n
l. oxychloride $2PbO \cdot PbCl_2$ Blei(II)-oxidchlorid n
l. packing Bleidichtung f, Bleipackung f
l. paint Bleifarbe f
l. pan Bleipfanne f
l. paper Blei[azetat]papier n
l. perchlorate $Pb(ClO_4)_2$ Blei(II)-perchlorat n
l. peroxide s. l. dioxide
l. persulphate PbS_2O_8 Blei(II)-peroxodisulfat n
l. phosphite $PbPHO_3$ Blei(II)-phosphit n
l. phthalocyanine Bleiphthalozyanin n

l. picrate $Pb[OC_6H_2(NO_2)_3]_2$ Bleipikrat n
l. pipe Bleirohr n
l. plate Bleiplatte f
l. poisoning Bleivergiftung f
l. press cure (technique) Bleimantelverfahren n, Bleimantelvulkanisation f, Vulkanisation f unter Blei
l. protoxide s. l. monoxide
l. pyroantimonate $Pb_2Sb_2O_7$ Blei(II)-diantimonat(V) n, Bleipyroantimonat(V) n
l. pyroarsenate $Pb_2As_2O_7$ Blei(II)-diarsenat(V) n, Bleipyroarsenat(V) n
l. pyrophosphate $Pb_2P_2O_7$ Bleidiphosphat n, Bleipyrophosphat n
l. refining Bleiaffination f
l. roll s. leading roll
l. salt Bleisalz n
l. selenate $PbSeO_4$ Blei(II)-selenat n
l. selenide $PbSe$ Blei(II)-selenid n
l. sesquioxide Pb_2O_3 Blei(II,IV)-oxid n, *(bisher)* Bleisesquioxid n
l. silicofluoride s. l. fluorosilicate
l. sludge Bleischlamm m
l. soap Bleiseife f
l. solubility *(Ker)* Bleilässigkeit f, Bleilöslichkeit f
l. stearate $Pb(C_{17}H_{35}COO)_2$ Bleistearat n
l. suboxide $Pb + PbO$ Bleisuboxid n
l. sulphate $PbSO_4$ Blei(II)-sulfat n, Bleivitriol n
l. sulphide PbS Blei(II)-sulfid n
l.-sulphide process Bleisulfidverfahren n, Bender-Prozeß m *(zur Entschwefelung von Erdöldestillaten)*
l.-sulphide sweetening (treating) *(Erdöl)* Bleisulfidsüßen n, Bleisulfidbehandlung f
l. sulphite $PbSO_3$ Blei(II)-sulfit n
l. sulphocyanate s. l. thiocyanate
l. superoxide s. l. dioxide
l. susceptibility Bleiempfindlichkeit f
l. telluride *(Min)* Altait m, Tellurblei n; $PbTe$ Blei(II)-tellurid n
l. tetraacetate $Pb(CH_3COO)_4$ Bleitetraazetat n, Blei(IV)-azetat n
l. tetrachloride $PbCl_4$ Bleitetrachlorid n, Blei(IV)-chlorid n
l. tetraethyl $Pb(C_2H_5)_4$ Bleitetraäthyl n, BTÄ, Tetraäthylblei n
l. tetramethyl $Pb(CH_3)_4$ Bleitetramethyl n, Tetramethylblei n
l. tetroxide Pb_3O_4 Blei(II,IV)-oxid n, Bleimennige f, Mennige f
l. thiocyanate $Pb(SCN)_2$ Blei(II)-thiozyanat n, Blei(II)-rhodanid n
l. thiosulphate PbS_2O_3 Bleithiosulfat n
l.-tin solder Blei-Zinn-Lot n
l. tree Bleibaum m
l. tungstate $PbWO_4$ Blei(II)-wolframat n
l. vanadinate s. l. metavanadate
l. wire Bleidraht m
l. wolframate s. l. tungstate

leaded bronze Bleibronze f
l. **fuel** gebleiter (verbleiter, mit Bleitetraäthyl versetzter) Kraftstoff m
l. **gasoline** gebleites (verbleites, mit Bleitetraäthyl versetztes) Benzin n, Bleibenzin n
leaden egg Bleigefäß n mit halbkugeligem Boden
leader (Gum) Zwischenläufer m, Zwischenleinen n, Mitläufer m
leadhillite (Min) Leadhillit m
leading l. Bleien n, Verbleien n, Verbleiung f, Versetzen n mit Bleitetraäthyl (von Kraftstoffen); ll. Leiten n, Führen n
l. **roll** (Pap) Leitwalze f
l.-**through tape** (Pap) Einführungsseil n, Führungsseil n
leadless bleifrei
leaf Blatt n, Scheibe f (eines Blattfilters)
l. **agitator** Blattrührer m
l. **dressing** Blattdüngung f
l. **fat** Nierenfett n
l. **fibre** Blattfaser f
l. **filter** Blattfilter n, Scheibenfilter n
l. **gold** Blattgold n
l. **green** Blattgrün n, Chlorophyll n
l. **pigment** Blattfarbstoff m
l. **spring** Blattfeder f
leak / to lecken, leck (undicht, defekt) sein; durchsickern, ausströmen, entweichen
leak 1. Leck n, Spalt m, Undichtigkeit f; 2. Lecken n; Durchsickern n, Ausströmen n, Entweichen n
to spring a l. undicht werden
l. **detector** Gasspürgerät n
leakage 1. Lecken n; 2. Leckage f, Leckverlust m, Sickerverlust m; Leckflüssigkeit f; Leckgas n; Leckluft f
l. **curve** s. break-through curve
l. **flow** Leckströmung f
l. **path** Leckweg m, Sickerweg m
leakiness Undichtheit f, Undichtigkeit f
leaking margarine zum Tränen neigende Margarine f
leakproof, leaktight leckfrei, lecksicher, dicht
lean mager, arm
l. **clay** Magerton m
l. **coal** Magerkohle f, magere Kohle f, Halbanthrazit m
l. **concrete** Magerbeton m, magerer Beton m, Sparbeton m, Füllbeton m
l. **fuel gas** mainne Sammelleitung (Zuführungsleitung) f für Schwachgas (eines Koksofens)
l. **gas** Schwachgas n, Armgas n
l.-**mixed concrete** s. l. concrete
l. **oil** Mageröl n, mageres Absorptionsöl n, armes (frisches, regeneriertes) Waschöl n (für die Absorptionskolonne)
l. **ore** geringwertiges (geringhaltiges, armes) Erz n, Magererz n
lear s. lehr
leather Leder n
l. **board** [braune] Lederpappe f, Braunholzpappe

f, Braunschliffpappe f, Braunholzkarton m
l. **cement** Lederkitt m
l. **charcoal** Lederkohle f
l.-**dressing agent** Lederpflegemittel n
l. **glue** Lederleim m, Hautleim m
l. **grease** Lederfett n
l.-**hard** lederhart
leathercloth Kunstleder n, Lederaustauschstoff m
leathering (Gerb) Durchgerbung f, Intensität f der Gerbung
leathery lederartig, lederähnlich
leaven l. to (Teig) [auf]gehen lassen; (Teig) säuern, ansäuern, einsäuern, sauer werden lassen, zur Gärung bringen
leaven Hefe f, Bärme f, Treibmittel n, Triebmittel n, Teiglockerungsmittel n; Sauer[teig] m
leavening 1. Gehenlassen n (von Teig); Säuern n, Ansäuern n, Einsäuern n (von Teig); 2. s. leaven
l. **agent** s. leaven
leaving air Abluft f
Lebedev process Lebedew-Verfahren n (zur Butadiengewinnung)
Leblanc process Leblanc-Verfahren n (zur Sodagewinnung)
Leblanc soda Leblanc-Soda f
lecanoric acid Lekanorsäure f, Diorsellinsäure f, Glabratsäure f
lechatelierite (Min) Lechatelierit m
lecithin Lezithin n
l. **content** Lezithingehalt m
lecithinase Lezithinase f, Phospholipase f (Glyzerin-Fettsäure-Esterbindungen lösendes Ferment)
Leclanché cell Leclanché-Element n
ledererite (Min) Ledererit m, (veraltet für) Gmelinit m
ledger [paper] Bücher[schreib]papier n, Geschäftsbücherpapier n
leer s. lehr
lees Bodenkörper m, Niederschlag m, Bodensatz m, Satz m, Ausscheidung f, Ausscheidungsprodukt n, Ablagerung f, Sediment n; Trub m
left[-hand] circularly polarized linkszirkular polarisiert
l.-**hand pan** linke Waagschale f, Lastschale f (z.B. einer Analysenwaage)
l.-**handed** s. l.-rotating
l.-**handed rotation** Linksdrehung f
l.-**rotating** linksdrehend, l̄-drehend
leg Schenkel m (z.B. eines U-Rohres)
legal chemistry Gerichtschemie f, forensische (gerichtliche) Chemie f
legume bacteria Knöllchenbakterien npl (Rhizobium-Arten)
legumelin Legumelin n
legumine Legumin n
lehr (Glas) Kühlofen m
l. **loader** (Glas) Kühlofenbeschicker m
Leitz microscope Leitz-Erhitzungsmikroskop n
lemon Zitrone f
l. **balm** Zitronenmelisse f, Melissa officinalis L.

l. balm oil Melissenöl n *(aus Melissa officinalis L.)*
l. chrome $PbCrO_4$ Chromgelb n, Zitronengelb n, Blei(II)-chromat n; $BaCrO_4$ Barytgelb n, Ultramaringelb n, gelbes Ultramarin n, Steinbühler Gelb n, Bariumchromat n
l.-grass oil Lemongrasöl n, Indisches Verbenaöl n *(von Cymbopogon-Arten)*
l.-yellow zitronengelb
length-by-length cure absatzweise (stückweise) Vulkanisation f
l. of exposure Expositionszeit f
l. of life Lebensdauer f
l. of path *(Kolorimetrie)* Weglänge f, Schichtdicke f
l. of sheet *(Pap)* Blattlänge f, Bogenlänge f
l. of stroke Hublänge f, Hub m
l. of travel Förderstrecke f, Förderweg m
Lennard-Jones potential Lennard-Jones-Potential n
lens *(Phys)* Linse f; *(Foto)* Objektiv n; Lupe f
l. paper José-Papier n *(Wischpapier für optische Linsen)*
l. surface Linsenfläche f
lenticular in shape linsenförmig
lentisk s. mastic gum
lepidine $CH_3NC_9H_6$ Lepidin n, 4-Methylchinolin n, p-Methylchinolin n, γ-Methylchinolin n
l. ethiodide Lepidinäthyljodid n
lepidoblastic lepidoblastisch, schuppig *(Gefüge)*
lepidocrocite *(Min)* Lepidokrokit m, Rubinglimmer m *(Eisenoxidhydroxid)*
lepidolite *(Min)* Lepidolith m, Lithionglimmer m
lepidomelane *(Min)* Lepidomelan m
leptin Leptin n *(Alkaloidglykosid)*
lepton Lepton n *(leichtes Elementarteilchen)*
less volatile component hochsiedende (höhersiedende, schwer[er]siedende, schwer[er]flüchtige) Komponente f, hochsiedender (höhersiedender, schwer[er]siedender, schwer[er]flüchtiger) Anteil m
Lessing ring Lessing-Ring m *(Raschig-Ring mit einfachem Steg)*
let-off arrangement *(Gum)* Abwickelvorrichtung f
lethal dose letale (tödliche) Dosis f, Tödlichkeitsdosis f, dosis letalis, LD
letter grade of buna s. lettered buna rubber
l. paper Briefpapier n, Ausstattungspapier n, Postpapier n, Billetpapier n
lettered buna rubber Buchstabenbuna m
letterpress ink Buchdruckfarbe f
leucaenine, leucaenol, leucenine s. leucenol
leucenol $C_8H_{10}O_4N_2$ Leuzenol n, Leuzenin n, Mimosin n, 2-Amino-3-[3-hydroxy-4-oxo-1-pyridinyl]propionsäure f
leucine $(CH_3)_2CHCH_2CH(NH_2)COOH$ Leuzin n, 2-Aminoisokapronsäure f, 2-Amino-4-methylpentansäure f
leucite *(Min)* Leuzit m *(Kaliumalumodisilikat)*
leucitophyre *(Geol)* Leuzitophyr m, Leuzitporphyr m

Leuckart reaction Leuckart-Reaktion f *(Aminalkylierung)*
leuco base Leukobase f
l. compound Leukoverbindung f
l. ester Leukoester m
l. form Leukoform f
l. salt Leukosalz n
leucoalizarin Leukoalizarin n
leucoaurin Leukoaurin n
leucocratic (Geol) leukokrat
leucogen Leukogen n *(wäßrige Natriumhydrogensulfitlösung)*
leucoindigo Leuk[o]indigo m
leucoline Chinolin n, 2,3-Benzopyridin n
leucomalachite green Leukomalachitgrün n
leucomanganite *(Min)* Leukomanganit m, *(veraltet für)* Fairfieldit m *(Kalziummangan(II)-orthophosphat)*
leucopararosaniline Leukopararosanilin n, Paraleukorosanilin n
leucophane, leucophanite *(Min)* Leukophan m
leucopyrite *(Min)* Leukopyrit m
leucoquinizarine Leukochinizarin n, 1,4,9,10-Tetrahydroxyanthrazen n
leucorosaniline Leukorosanilin n
leucosulphuric acid ester Leukoschwefelsäureester m
l. acid ester salt Leukoschwefelsäureestersalz n
leucotetrasulphuric acid ester Leukotetraschwefelsäureester m
leucovorin Leukovorin n, Folinsäure f, Formyltetrahydrofolsäure f, Zitrovorumfaktor m
leucoxene *(Min)* Leukoxen m
leukargyrite *(Min)* Leukargyrit m, *(veraltet für)* Freibergit m
leukerin 6-Purinthiol n, 6-Merkaptopurin n
Levant galls *(Gerb)* Aleppogallen fpl, Türkische (Smyrnaer) Gallen fpl *(von Quercus infectoria Oliv.)*
L. storax levantiner (asiatischer) Styrax (Storax) m *(von Liquidambar orientalis Mill.)*
level / to [ein]ebnen, planieren; ausgleichen; egalfärben, egalisieren
l. out ausgleichen
level *(Farb)* egal, gleichmäßig
level *(phys Ch)* Niveau n, Term m; *(Tech)* Füllstand m, Füllhöhe f, Niveau n, Spiegel m, Pegel m; *(Bodenkunde)* Horizont m
l. dyeing Egalfärben n, Egalisieren n, Egalisierung f
l. dyeing assistant s. levelling agent
l. measurement Füllhöhenmessung f, Füllstand[s]messung f, Niveaumessung f
l. of energy Energieniveau n, Energiezustand m, Energiestufe f, Energieterm m, Term m, Elektronenniveau n
l. of sulphur *(Gum)* Schwefeldosierung f
leveller Glattstreicher m
levelling *(Text)* Egalisierung f, Egalisieren n, Egalfärben n

l. agent Egalisier[hilfs]mittel n, Egalisierer m
l. bar Planierstange f, Einebnungsstange f (zum Einebnen der in die Kokskammer eingefüllten Kohle)
l. bottle Niveauflasche f, Niveaugefäß n, Ausgleich[s]gefäß n
l. bulb Niveaukugel f, Ausgleich[s]kolben m, Niveaubirne f
l. device Einebnungsvorrichtung f, Planiervorrichtung f
l. tube Niveaurohr n
l. wire Markenspitze f (eines Redwood-Viskosimeters)
levelness (Farb) Egalität f, Gleichmäßigkeit f, Egalisierung f (Zustand)
lever Hebel m, Hebestange f
l.-sealed plug cock Hahn m (Hahnventil n) mit hebelgelüftetem Küken
Leviathan washer Leviathan m, Leviathan-Wollwaschmaschine f
levisticum oil Liebstöckelöl n (von Levisticum officinale W.D.I. Koch)
levo, levo... s. laevo, laevo...
levyne, levynite (Min) Levyn m (Kalziumdialumotrisilikat)
Lewis acid Lewis-Säure f (Sammelname für Verbindungen mit Säurecharakter im Sinne von Lewis)
Lewis base Lewis-Base f (Sammelname für Verbindungen mit Basecharakter im Sinne von Lewis)
LH s. luteinizing hormone
Libbey-Owens process Libbey-Owens-Verfahren n, Colburn-Verfahren n (Tafelglasziehverfahren)
liberate / to befreien, freimachen, freisetzen, freilegen, in Freiheit setzen, entwickeln, (Wärme) entbinden
liberation Befreien n, Befreiung f, Freimachen n, Freisetzen n, Freilegen n, Entwickeln n, Entwicklung f, Entbinden n, Entbindung f (von Wärme); Freiwerden n
libethenite (Min) Libethenit m (Kupfer(II)-hydroxidorthophosphat)
libi-dibi (Gerb) Dividivi pl (Hülsen von Caesalpinia coriaria (Jacq.)Wild.)
licanic acid Likansäure f
lichen blue Lackmus n(m) (Flechtenfarbstoff)
l. dye Flechtenfarbstoff m
lichenic acid (veraltet für) fumaric acid
lichenin Lichenin n
lick-up (Pap) Bahnabnahme f [mit Oberfilz], Lick-up-Bahnabnahme f
l.-up felt s. l.-up overfelt
l.-up machine Selbstabnahme[papier]maschine f, Yankee-Maschine f
l.-up overfelt (wet felt) (Pap) Selbstabnahme-[ober]filz m, Selbstabnahmeobertuch n, Abnahmefilz m, Lick-up-Filz m
licorice Lakritze f, Lakritzensaft m, Süßholzsaft m (von Glycyrrhiza glabra L.)

lid Deckel m, Klappe f
lidded mit einem Deckel versehen (verschlossen), mit einer Klappe versehen (verschlossen)
Liebermann reaction for phenols Liebermannsche Phenolreaktion f
Liebermann-Storch test Liebermann-Storch-Test m (Nachweis von Naturharzen)
Liebig condenser Liebig-Kühler m, Liebigscher Kühler m
Liesegang phenomenon Phänomen n der Liesegangschen Ringe
Liesegang rings Liesegang-Ringe mpl, Liesegangsche Ringe mpl
lievrite (Min) Lievrit m, Ilvait m
life Lebensdauer f, Dauerhaftigkeit f, Haltbarkeit f, Laufzeit f
l. of the felt (Pap) Filzlaufzeit f, Filzlaufdauer f
l. period Lebensdauer f (z. B. von Ladungsträgern)
lifeless (Gum) tot
lifetime Lebensdauer f (z.B. von Ladungsträgern); Halbwertszeit f
l. of a foam Lebensdauer (Beständigkeit, Haltbarkeit, Stabilität) f eines Schaumes, Schaumdauer f, Schaumbeständigkeit f, Schaumstabilität f
lift / to liften, [in die Höhe] heben, hinaufheben, nach oben befördern
l. out austragen
lift Lift m, Heber m, Hebevorrichtung f, Fördereinrichtung f, Fördervorrichtung f; (Plast) Ausstoß m (bei einer Pressung)
l. air Förderluft f, Luft f für Airlift
l. check valve Hubrückschlagventil n
l. line (pipe) Liftleitung f, Heberleitung f (zum Heben des Katalysators beim katalytischen Kracken); Steigrohr n
l. pot Lifttopf m
l. steam trap Kondensat-Rückleiter m
l. tank Liftbehälter m, Förderbehälter m
lifter Hubflügel m
lifting Liften n, Heben n, Hebung f, Förderung f (z.B. des Katalysators beim katalytischen Kracken); Hochziehen n, Hochgehen n (Anquellen und Loslösung eines Anstrichs durch Lösungsmittel); (Pap) Rupfen n
l. device Hebevorrichtung f
l. flight Hubleiste f, Hebeleiste f, Mitnehmerblech n, Schaufel f (z.B. in einer Trockentrommel)
l. method Liftverfahren n
l. plate s. l. flight
ligancy Koordinationszahl f, Zähligkeit f, koordinative Wertigkeit f
ligand Ligand m
l. concentration Ligandenkonzentration f
l. field theory Ligandenfeldtheorie f
light / to anheizen, anzünden, anbrennen; beleuchten
light I. hell, licht; II. leicht; leicht[er]siedend, niedrigsiedend, tiefsiedend; (Glas) dünnwandig;

(Lebm) leicht, von geringem Alkoholgehalt, alkoholarm, schwach alkoholhaltig (alkoholisch)
light Licht *n*
l.-absorbing lichtabsorbierend, lichtschluckend
l. absorption Lichtabsorption *f*
l. ageing Lichtalterung *f*, Alterung *f* durch Licht
l. benzine *s.* l. gasoline
l. benzol Leichtbenzol *n*
l.-blue hellblau, lichtblau
l. blue Hellblau *n*, Lichtblau *n*
l. camphor oil leichtes Kampferöl *n*
l.-catalyzed lichtkatalysiert, fotokatalytisch
l.-coloured leicht gefärbt, hell [gefärbt]
l.-coloured beer helles Bier *n*, Helles *n*
l.-coloured filler heller Füllstoff *m*
l. component leicht[er]siedende (niedrigsiedende, tiefsiedende) Komponente *f*, Leicht[er]siedendes *n*
l. distillate leichtes Destillat *n*
l.-duty detergent Leichtwaschmittel *n*
l. energy Lichtenergie *f*
l. filter Lichtfilter *n*
l. flint glass Leichtflintglas *n*
l. fraction leichte (niedrigsiedende, tiefsiedende) Fraktion *f*; Leichtgut *n*
l. gasoline Leichtbenzin *n*, leichtes Benzin *n*
l. hydrogen ^{1}H leichter Wasserstoff *m*, Protium *n*
l. key niedrigsiedende Schlüsselkomponente *f*
l. ligroin *s.* l. petroleum
l. liquid paraffin Leichtparaffin *n*, Sprühparaffin *n*, Paraffin *n* für Sprühzwecke
l.-liquor filter *(Zucker)* Dünnsaftfilter *n*
l. malt helles Malz *n*
l. material Leichtgut *n*
l. metal Leichtmetall *n*
l. microscope Lichtmikroskop *n*
l. mineral Leichtmineral *n*
l. naphtha *s.* l. gasoline
l. oil Leichtöl *n*, leichtes Öl *n*
l. petroleum Petrol[eum]äther *m (Siedebereich 40 bis 70 °C)*
l. phase *s.* l. component
l. quant[um] Lichtquant *n*, Photon *n*, Strahlungsquant *n*
l. red silver ore *(Min)* Lichtes Rotgültigerz *n*, Proustit *m (Arsen(III)-silbersulfid)*
l. resistance *s.* lightfastness
l. scattering Lichtstreuung *f*, Streuung (Zerstreuung) *f* des Lichts
l. scattering method Lichtstreuungsmethode *f (zur Molekulargewichtsbestimmung)*
l.-sensitive lichtempfindlich
l. sensitivity Lichtempfindlichkeit *f*
l. source Lichtquelle *f*
l. spirit *s.* l. gasoline
l. spot *(Foto)* Lichtfleck *m*, Leuchtfleck *m*
l. stabilizer Lichtstabilisator *m*
l. test Prüfung *f* auf Lichtechtheit (Lichtbeständigkeit), Lichtechtheitsprüfung *f*

l. transmission Lichtdurchlässigkeit *f*
l. transmittance Lichtdurchlässigkeit *f*, Durchlässigkeitsgrad *m*, Transparenz *f*
l. unwashed benzol ungewaschenes Leichtbenzol *n*
l.-water reactor Leichtwasserreaktor *m*
l. wave Lichtwelle *f*
lighter component *s.* light component
l. flint [tip] Zündstein *m*
l.-than-light component leichtest siedende Komponente *f*
lightfast lichtecht, lichtbeständig
lightfastness Lichtechtheit *f*, Lichtbeständigkeit *f*
lighting gas Leuchtgas *n*
lightly branched schwach verzweigt
lightproof lichtundurchlässig, lichtdicht
lightproofness Lichtundurchlässigkeit *f*, Lichtdichtheit *f*
lighttight *s.* lightproof
lightweight aggregate leichter Zuschlagstoff *m*, Leichtzuschlagstoff *m*
l. brick Leichtstein *m*
l. concrete Leichtbeton *m*
l. expanded clay [aggregate] Blähton *m*, Porensinter *m*
lignan Lignan *n*
ligneous substance Holzsubstanz *f*
lignification Verholzung *f*, Lignifizierung *f*
lignify / to verholzen, lignifizieren
lignin Lignin *n*, Holzstoff *m*
l. coal Ligninkohle *f*
l. content Ligningehalt *m*
l. pitch Ligninpech *n (Sulfitablaugenkonzentrat)*
l. removal *(Pap)* Delignifizierung *f*, Ligninentfernung *f*, Lignin[her]auslösung *f*
l. residues Ligninreste *mpl*, restliches Lignin *n*, Restlignin *n*
ligninase Ligninase *f (Lignin abbauendes Ferment)*
ligninsulphonic acid Ligninsulfo[n]säure *f*, Lignosulfo[n]säure *f*
lignite Braunkohle *f*; verfestigte (erhärtete) Braunkohle *f (nach der Kohlenklassifikation der ASTM)*; Xylit *m(n)*, Braunkohlenxylit *m(n)*, Braunkohlenholz *n*, Hylit *m(n)*, xylitische (holzartige) Braunkohle *f*, *(veraltet)* Lignit *m*, Braunkohlenlignit *m*, lignitische Braunkohle (Kohle) *f*
l. coke Braunkohlenkoks *m*
l. stage Braunkohlenstadium *n*
l. tar Braunkohlenteer *m*
lignitic coal Braunkohle *f (nach der Kohlenklassifikation der ASTM)*
lignitous coal subbituminöse Kohle *f*, bituminöse (steinkohlenähnliche) Braunkohle *f*, Glanz[braun]kohle *f*
lignocellulose Holzzellulose *f*; Holzzellstoff *m*, Zellstoff *m*
lignosulphonic acid *s.* ligninsulphonic acid
ligroin[e] Ligroin *n (Siedebereich 90 bis 120 °C)*; Petrol[eum]äther *m (Siedebereich 40 bis 70 °C)*; Leichtbenzin *n*, leichtes Benzin *n (Siedebereich 20 bis 135 °C)*

like charge gleichnamige Ladung *f*
Lima weed tangartige Färberflechte *f*, Roccella fuciformis (L.)Lam.
L. wood *(Farb)* Coulteriaholz *n*, Coulteria-Rotholz *n (von Caesalpinia tinctoria (H.B.K.)Benth.)*
lime / to *(Landw)* kalken, [mit Kalk] düngen; *(Gerb)* äschern *(Häute mit Erdalkali- oder Alkalilösungen behandeln)*
l. out *(Farb)* kalken
lime Kalk *m*, *(Landw auch)* Kalkdünger *m*, Kalkdüngemittel *n*
l. accumulation horizon *(Bodenkunde)* Kalkanreicherungshorizont *m*
l. acetate $Ca(CH_3COO)_2$ Kalziumazetat *n*
l. balance *(Landw)* Kalkbilanz *f*
l. bath Kalkbad *n*
l. blast *(Gerb)* Kalkflecken *mpl*, Kalkschatten *mpl (auf Häuten nach unsachgemäßem Äschern)*
l. boil *(Text)* Kalkbeuche *f*, Kalkäscher *m*
l. burning Kalkbrennen *n*
l.-carbon-dioxide process *(Zucker)* Kalk-Kohlensäureverfahren *n*, Scheidesaturation *f*
l.-cement hydraulischer Kalk *m*, Zementkalk *m*
l.-chrome garnet *(Min)* Kalkchromgranat *m (Kalziumchrom(III)-orthosilikat)*
l. concrete Kalkbeton *m*
l. cream *(Gerb)* Schwödebrei *m*, Kalkbrei *m*
l. defecation *(Zucker)* Kalkscheidung *f*
l. feldspar *(Min)* Kalkfeldspat *m*, Anorthit *m (Kalziumalumodisilikat)*
l. fertilizer Kalkdüngemittel *n*, Kalkdünger *m*
l. funnel furnace Kalktrichterofen *m (Brennofen)*
l. glass Kalkglas *n*
l. harmotome *(Min)* Kalkharmotom *m*, *(veraltet für)* Phillipsit *m*
l. hydrate $Ca(OH)_2$ Kalkhydrat *n*, Kalziumhydroxid *n*, Löschkalk *m*, gelöschter Kalk *m*
l. hydrating *s.* l. slaking
l. hypophosphite $Ca(H_2PO_2)_2$ Kalziumhypophosphit *n*
l.-induced chlorosis *(Landw)* Kalkchlorose *f*
l.-intolerant *(Landw)* kalkmeidend
l. kiln Kalk[brenn]ofen *m*
l. light Drummondsches Kalklicht *n*
l. liquor Kalkmilch *f*; Kalkwasser *n*; *(Gerb)* Äscherbrühe *f*
l. marl Kalkmergel *m*
l. mica *(Min)* Kalkglimmer *m*, *(veraltet für)* Margarit *m (Kalziumaluminiumdihydroxiddialumodisilikat)*
l. milk Kalkmilch *f (Suspension von Kalziumhydroxid in Wasser)*
l. mortar Kalkmörtel *m*
l. mud Kalkschlamm *m*
l. mud washer Kalkschlammwäscher *m*
l. nitrate $Ca(NO_3)_2$ Kalziumnitrat *n*, Kalksalpeter *m*
l. nitrogen $CaCN_2$ Kalkstickstoff *m*, Kalziumzyanamid *n*
l. paint *s.* l. cream

l. pan *s.* l. accumulation horizon
l. pit *(Gerb)* Äschergrube *f*, Äscher *m*, *(i.e.S.)* Kalkäscher *m*
l. plaster Kalkputz *m*
l. pyrolignite $Ca(CH_3COO)_2$ Kalziumazetat *n*
l. reburning *(Pap)* Rückbrennen *n* des Kalkschlammes zu Branntkalk
l. refractory *(Ker)* Kalziumoxiderzeugnis *n*
l. requirements Kalkbedarf *m*
l. ring furnace Kalkringofen *m (Brennofen)*
l. saccharate Kalksa[c]charat *n*, Kalziumsa[c]charat *n*
l. saltpetre $Ca(NO_3)_2$ Kalksalpeter *m*, Kalziumnitrat *n*; *(Min)* Kalksalpeter *m*, Nitrokalzit *m*
l. slaker Kalklöscher *m*, Kalklöschapparat *m*
l. slaking Kalklöschen *n*
l. sludge *s.* l. mud
l. soap Kalkseife *f*
l.-soda [softening] method Kalk-Soda-Verfahren *n (Wasseraufbereitung)*
l. spar Kalkspat *m*
l. speck *s.* l. blast
l. status *(Landw)* Kalkversorgungsgrad *m*
l. sulphur Schwefelkalk *m*, *(i.e.S.)* Schwefelkalkbrühe *f (Pflanzenschutzmittel)*
l. tunnel furnace Kalkschachtofen *m (Brennofen)*
l. uranite *(Min)* Kalkuranit *m*, *(veraltet für)* Autunit *m (Kalziumbisuranylorthophosphat)*
l. water Kalkwasser *n (klare wäßrige Kalziumhydroxidlösung)*
limed juice *(Zucker)* geschiedener Saft *m*, Scheidesaft *m*
limer *(Zucker)* Scheidepfanne *f*
limerock *s.* limestone
limestone Kalkstein *m*
l. coal Kohle *f* der Kalksteingruppe
l. packing *(Pap)* Kalksteinfüllung *f*
l. tower *(Pap)* Säureturm *m (Sulfitverfahren)*
limewash Kalktünche *f*, Tünche *f*
liming *(Landw)* Kalken *n*, Kalkung *f*; *(Gerb)* Äschern *n*, Äscher *m*, *(i.e.S.)* Kalkäscher *m*; *(Text)* Kalkäscher *m*, Kalkbeuche *f*; *(Zucker)* Defäkation *f*, Scheiden *n*, Scheidung *f*, Kalkung *f*
l. for preservation *(Landw)* Erhaltungskalkung *f*
l. tank *(Zucker)* Scheidepfanne *f*
limit Grenze *f*; Grenzwert *m*
l. dextrin Grenzdextrin *n*, Restdextrin *n*
l. of error Fehlergrenze *f*
l. test Grenzwertbestimmung *f*
limiting concentration Grenzkonzentration *f*
l. crystallization temperature Liquidustemperatur *f*
l. current *(phys Ch)* Grenzstrom *m*, Diffusionsgrenzstrom *m*, maximaler Diffusionsstrom *m*, Sättigungsstrom *m*
l. devitrification temperature *s.* l. crystallization temperature
l. diffusion current *s.* l. current
l. formula Grenzformel *f*
l. frequency Grenzfrequenz *f*

l. ratio Grenzverhältnis *n*
l. state Grenzzustand *m*
l. structure Grenzstruktur *f*
l. value Grenzwert *m*
l. viscosity number Grenzviskositätszahl *f*
limnic limnisch, im Süßwasser entstanden
limonene Limonen *n*, 1,8(9)-*p*-Menthadien *n*, 1-Methyl-4-isopropenylzyklohexen-(1) *n*
***dl*-limonene** Dipenten *n*, *dl*-Limonen *n*, *dl*-1,8(9)-p-Menthadien *n*
limonite *(Min)* Limonit *m*, Brauneisen[erz] *n*, Brauneisenstein *m*
limy kalkhaltig, kalkig; kalkartig
l. soil Kalkboden *m*
linarite *(Min)* Linarit *m*, Bleilasur *m* *(Blei(II)-kupfer-(II)-dihydroxidsulfat)*
lincolnine, lincolnite *(Min)* Lincolnit *m*, *(veraltet für)* Heulandit *m* *(Kalziumdialumohexasilikat)*
linctus *(Pharm)* Lecksaft *m*
Linde process Linde-Verfahren *n* *(zur Luftverflüssigung)*
Lindemann glass Lindemann-Glas *n* *(für Röntgenstrahlen durchlässiges Glas)*
lindgrenite *(Min)* Lindgrenit *m* *(Kupfer(II)-hydroxidmolybdat)*
line / **to** verkleiden, überziehen; plattieren; auskleiden, ausmauern, *(Met auch)* zustellen, [aus]füttern; *(Pap)* bekleben, kaschieren; *(Pap)* [zusammen]gautschen
line 1. Linie *f*; 2. Leitung *f*
l. defect linienhafter Gitterfehler *m*; linienhafte Gitterfehlstelle *f*
l. grating *(phys Ch)* Liniengitter *n*, Strichgitter *n*
l. intensity Linienintensität *f*, Intensität *f* einer Spektrallinie
l. mixer kontinuierlicher Flüssigkeitsmischer *m*
l. of equal volatile matter Linie (Kurve) *f* gleichen Gehalts an flüchtiger Substanz, Isovole *f*
l. of flow Fließlinie *f*
l. of force Kraftlinie *f*
l. pipe nahtloses Stahlrohr *n*
l. spectrum Linienspektrum *n*, Atomspektrum *n*
l. width *(phys Ch)* Linienbreite *f*
linear accelerator linearer Beschleuniger (Teilchenbeschleuniger) *m*, Linearbeschleuniger *m*
l. annulization lineare Anellierung *f* *(kondensierter Ringsysteme)*
l. belt filter Bandfilter *n*, Umlauffilter *n*
l. colloid Linearkolloid *n*
l. combination lineare Kombination *f*, Linearkombination *f*
l. combination of atomic orbitals Linearkombination *f* von Atomorbitalen, LCAO
l.-combination-of-atomic-orbitals approximation LCAO-Näherung *f*
l.-combination-of-atomic-orbitals method LCAO-Methode *f*, LCAO-Verfahren *n*, Methode *f* der linearen Kombination von Atomorbitalen
l.-combination-of-atomic-orbitals theory LCAO-Theorie *f*

l. expansion lineare Ausdehnung *f*, Längsdehnung *f*
l. molecular chain lineares Kettenmolekül *n*
l. molecule Fadenmolekül *n*
l. polarization lineare Polarisation *f*
l. polyethylene Linearpolyäthylen *n*, Niederdruckpolyäthylen *n*
l. polymer lineares (eindimensionales) Polymer[es] *n*
l. polymerization Linearpolymerisation *f*
l. shrinkage *(Ker)* lineare Schwindung *f*
l. thermal expansion coefficient linearer Ausdehnungskoeffizient *m*
linearly polarized linear polarisiert
lined board beklebte (kaschierte) Pappe *f*; gedeckte (gegautschte) Pappe *f*
linen *s.* l. fabric
l. embossed paper *s.* l.-faced paper 2.
l. embossed writing paper Leinenpostpapier *n*
l. fabric Leinen[gewebe] *n*, Leinenstoff *m*, leinener Stoff *m*, Leinwand *f*
l.-faced paper 1. *(ursprünglich)* Gewebepapier *n*, Leinenpapier *n*, Papyrolin *n*, Packgewebe *n* *(Papierbahn mit ein- oder zweiseitiger Gewebebahn)*; 2. *(jetzt)* Leinenpapier *n*, leinengeprägtes Papier *n*, Papier *n* mit Leinenprägung
l. fibre Flachsfaser *f*
l. finish *(Pap)* Leinenprägung *f*
l.-finished paper *s.* l.-faced paper 2.
l. paper Leinenpapier *n* *(aus Leinenlumpen hergestelltes Papier)*; *s.* l.-faced paper
l. rags Leinenhadern *mpl*, Leinenlumpen *mpl*
linenizing *(Pap)* Leinenprägen *n*
l. calender *(Pap)* Leinenprägekalander *m*, Leinenprägepresse *f*
liner *(Erdöl)* Liner *m* *(gelochter Rohrstrang)*; *(Gum)* *s.* lining 3.
liners Beklebepapier *n*, Aufklebepapier *n*, Kaschierpapier *n*, Überzugspapier *n*
lingering period Verweilzeit *f*
liniment *(Pharm)* Liniment *n*
lining 1. Verkleiden *n*, Überziehen *n*; Plattieren *n*; Auskleiden *n*, Ausmauern *n*, *(Met auch)* Zustellen *n*, Füttern *n*; *(Pap)* Bekleben *n*, Kaschieren *n*; *(Pap)* Gautschen *n*; 2. Verkleidung *f*; Plattierung *f*; Auskleidung *f*, Ausmauerung *f*, *(Met auch)* Zustellung *f*, Futter *n*; 3. *(Gum)* Mitläuferstoff *m*, Mitläufergewebe *n*
l. leather Futterleder *n*
l. paper *s.* liners
link / **to** verbinden, verketten, *(Farbstoff an die Unterlage)* binden, *(auf der Unterlage)* fixieren; sich verbinden
l. together miteinander verbinden (verknüpfen); sich [miteinander] verbinden
link 1. Bindung *f*; 2. *(Tech)* Bindeglied *n*, Kettenglied *n*, Zwischenglied *n*
l.-suspended centrifuge Pendelzentrifuge *f*
linkage 1. Verbindung *f*, Verkettung *f*, Verknüpfung *f* *(Vorgang)*; 2. Bindung *f* *(Zustand)*

l. electron Bindungselektron *n*, Bindeelektron *n*
l. energy Bindungsenergie *f*
l. force Bindungskraft *f*, Bindekraft *f*, bindende Kraft *f*
l. system Bindungssystem *n*
l. valence (valency) Bind[ungswert]igkeit *f*, Atombindigkeit *f*, Atomwertigkeit *f*, kovalente Wertigkeit *f*, Kovalenz *f*, Atombindungszahl *f*, Bindungszahl *f*
linked benzene ring system *(Kohlenwasserstoff mit mehreren nicht kondensierten Benzolkernen)*
l. by a double bond zweifach gebunden
l. by a triple bond dreifach gebunden
l. kilns gekoppelte Öfen *mpl*
l. reaction gekoppelte Reaktion *f*
linking *s.* linkage 1.
linn[a]eite *(Min)* Linneit *m*, Kobaltkies *m* *(Kobalt-(II,III)-sulfid)*
linoleic acid
$CH_3(CH_2)_4CH=CHCH_2CH=CH(CH_2)_7COOH$
Linolsäure *f*, Leinölsäure *f*, 9,12-Oktadekadiensäure *f*
linolenic (α-linolenic) acid $C_{17}H_{29}COOH$ Linolensäure *f*, 9,12,15-Oktadekatriensäure *f*
linolic acid *s.* linoleic acid
linoxy[li]n Linoxyn *n* *(Oxydations- und Polymerisationsprodukt des Leinöls)*
linseed Leinsaat *f*, Leinsamen *m*
l. cake Leinkuchen *m*
l. mucilage Leinsamenschleim *m*
l. oil Leinöl *n*
l.-oil varnish Leinöllack *m*; Leinölfirnis *m*
lint cotton Lint[baum]wolle *f*, Lint *m*
linters Linters *mpl*, Baumwollinters *mpl*, Faserflug *m*
l. pulp *(Pap)* Lintershalbstoff *m*
Linz-Donawitz process Linz-Donawitz-Verfahren *n*, Linz-Donawitzer Verfahren *n*, LD-Verfahren *n*, LD-Blasstahlverfahren *n*, LD-Aufblaseverfahren *n*
liparite *(Geol)* Liparit *m*, Rhyolit *m*; *(Min)* Liparit *m*
lipase Lipase *f* *(fettspaltendes oder fettaufbauendes Ferment)*
lipides *s.* lipids
lipids Lipide *npl* *(Sammelname für Fette und Lipoide)*
lipins *s.* 1. lipids; 2. lipoids
α-lipoic acid α-Liponsäure *f*, 6,8-Thioktinsäure *f*, 6,8-Dithio-*n*-oktansäure *f*
lipoids Lipoide *npl*, *(i.w.S.)* *s.* lipids
lipolysis Lipolyse *f*, Fettspaltung *f*
lipolytic lipolytisch, fettspaltend
l. enzyme fettspaltendes Ferment *n*
l. rancidity Fettsäureranzidität *f*, hydrolytische Ranzigkeit *f*
α-liponic acid *s.* α-lipoic acid
lipophile, lipophilic lipophil, fettliebend, sich mit Fett mischend, in Fett löslich
lipoproteid[e], lipoprotein Lipoproteid *n*

liposoluble fettlöslich
lipovitamin fettlösliches Vitamin *n*
lipstick Lippen[schmink]stift *m*
liq. s. liquid
liquating *(Met)* Seigern *n*, Seigerung *f*, Pauschen *n*
liquation *(Met) s.* liquating; *(Geol) s.* l. differentiation
l. differentiation *(Geol)* liquide Entmischung *f*, Liquation *f* *(Entmischung magmatischer Schmelzen)*
l. refining Seigerraffination *f*, Umschmelzen *n* *(zum Entfernen von Verunreinigungen aus Metallschmelzen)*
liquefaction Verflüssigen *n*, Verflüssigung *f*; Flüssigwerden *n*, Verflüssigung *f*; *(Met)* Schmelzen *n*
l. of air Luftverflüssigung *f*
l. of gas Gasverflüssigung *f*
liquefiable verflüssigbar
liquefication *s.* liquefaction
liquefied gas verflüssigtes Gas *n*, Flüssiggas *n*; *(i.e.S.) s.* l. petroleum gas
l. natural gas verflüssigtes Erdgas *n*, LNG
l. petroleum gas verflüssigtes Erdölgas *n*, Flüssiggas *n*, LPG
l. phenol *(Pharm)* verflüssigtes Phenol *n*
liquefier Verflüssiger *m*
liquefy / to verflüssigen; sich verflüssigen, flüssig werden, zerfließen; *(Met)* schmelzen
liquefying *s.* liquefaction
liquid flüssig, tropfbar; klar, durchsichtig
liquid Flüssigkeit *f*; flüssige Phase *f*, Schmelze *f* *(in der Zonenschmelztheorie)*
l. air flüssige Luft *f*
l. air trap Ausfriertasche *f*
l. ammonia flüssiges (verflüssigtes) Ammoniak *n*
l. being pumped Förderflüssigkeit *f*, geförderte Flüssigkeit *f*
l.-bright metal Glanzmetall *n*
l.-buffered seal Sperrflüssigkeitsdichtung *f*
l.-carburize / to badaufkohlen, badzementieren, badeinsetzen, im Salzbad (in flüssigen Mitteln) aufkohlen (zementieren, einsetzen)
l. carburizing Salzbadaufkohlen *n*, Badaufkohlen *n*, flüssige Aufkohlung *f*, Salzbadzementieren *n*, Badzementieren *n*, Badeinsetzen *n*, Aufkohlen (Zementieren, Einsetzen) *n* im Salzbad (in flüssigen Mitteln)
l. chromatography Flüssig[keits]-Chromatografie *f* *(als Gegensatz zur Gaschromatografie)*
l. crotonic acid *s.* isocrotonic acid
l. crystal flüssiger Kristall *m*, kristalline (anisotrope) Flüssigkeit *f*, Fastkristall *m*
l. culture Flüssigkeitskultur *f*
l.-curing method Flüssigkeitsbadvulkanisation *f*, LCM-Vulkanisation *f*
l. cyclone Hydrozyklon *m*, Naßzyklon *m*
l. detergent flüssiges Waschmittel *n*
l. distributor Flüssigkeitsverteiler *m*

l. egg yolk flüssiges Eigelb *n*, Flüssigeigelb *n*
l. expanded film flüssig-expandierter Film *m*
l. expansion thermometer Flüssigkeitsthermometer *n*
l. feed flüssiger Öleinsatz *m (beim Thermofor-Catalytic-Cracking)*
l. fertilizer Flüssigdünger *m*
l. filler (filling machine) Füllmaschine *f*, Abfüllmaschine *f*, Abfüllautomat *m*, Füller *m (für Flüssigkeiten)*
l. film Flüssigkeitsfilm *m*
l. filtration Filtrieren *n* von Flüssigkeiten, Flüssigkeitsfiltration *f*
l. freezing Eintauch[gefrier]verfahren *n*
l. fresh egg yolk Frischeigelb *n*
l.-fuel burner Brenner *m* für flüssigen Brennstoff
l.-fuel system Flüssigbrennstoffsystem *n*
l. gas Flüssiggas *n (verflüssigte Kohlenwasserstoffe)*
l. glass Wasserglas *n*
l. glue flüssiger Leim *m*
l. gold *(Ker)* Glanzgold *n*
l. hold-up Haftinhalt *m*, Ruheinhalt *m (z. B. einer Kolonne)*
l. holding volume Flüssigkeitsfassungsvermögen *n*
l. hydrate Flüssigkeitshydrat *n*
l. hydrate crystal Flüssigkeitshydratkristall *m*
l.-in-glass thermometer Flüssigkeits-Glasthermometer *n*
l. junction potential Flüssigkeitspotential *n*, Diffusionspotential *n*
l. layer Flüssigkeitsschicht *f*
l. level Flüssigkeitsstand *m*, Flüssigkeitsspiegel *m*, Flüssigkeitsniveau *n*
l.-liquid chromatography Flüssig-Flüssig-Chromatografie *f*, Liquidus-Liquidus-Chromatografie *f*, LLC
l.-liquid extraction Solventextraktion *f*, Lösungsmittelextraktion *f*, Flüssig-Flüssig-Extraktion *f*, Extraktion *f* in flüssigen Systemen
l.-liquid interface Flüssigkeit-Flüssigkeit-Grenzfläche *f*, Grenzfläche *f* flüssig-flüssig
l.-liquid partition chromatography Flüssig-Flüssig-Verteilungschromatografie *f*
l.-liquid potential Flüssigkeitspotential *n*, Diffusionspotential *n*
l. manure Jauche *f*
l. metal fuelled reactor *(Kch)* Reaktor *m* mit Flüssigmetallbrennstoff
l. metal reactor fuel *(Kch)* Flüssigmetallbrennstoff *m*
l. meter Durchfluß[mengen]messer (Mengenstrommesser) *m* für Flüssigkeiten
l. milk Trinkmilch *f*
l. mixer Flüssigkeitsmischer *m*
l. mixture Flüssigkeitsgemisch *n*
l.-motivated ejector Flüssigkeitsstrahlpumpe *f*; Flüssigkeitsstrahlgebläse *n*, Flüssigkeitsstrahlverdichter *m*

l. nitrogen flüssiger (verflüssigter) Stickstoff *m*, Flüssigstickstoff *m*
l.-nitrogen bath flüssiges Stickstoffbad *n*
l. operating hold-up Betriebsinhalt *m*, Arbeitsinhalt *m (z. B. einer Kolonne)*
l. oxygen flüssiger (verflüssigter) Sauerstoff *m*, Flüssigsauerstoff *m*
l.-oxygen bath flüssiges Sauerstoffbad *n*
l. paraffin (petrolatum) Paraffinöl *n*; *(Pharm)* flüssiges Paraffin *n*
l. phase flüssige Phase *f*, Flüssig[keits]phase *f*, *(bei der Hochdruckhydrierung meist)* Sumpfphase *f*
l.-phase catalyst Flüssigphasekatalysator *m*, Sumpfphasekatalysator *m*
l.-phase converter Sumpfphase[hydrier]ofen *m*, Sumpfofen *m*
l.-phase cracking Flüssigphasekracken *n*, Kracken *n* in Flüssigphase (flüssiger Phase)
l.-phase gasoline Sumpfphasenbenzin *n*, S-Benzin *n*
l.-phase hydrogenation Sumpfphase[n]hydrierung *f*, Hydrierung *f* in Sumpfphase (flüssiger Phase)
l.-phase isomerization Flüssigphaseisomerisierung *f*, Flüssigphaseisomerisation *f*, Isomerisierung *f* in der Flüssigphase, Isomerisierung *f* in flüssiger Phase
l.-phase petrol s. l.-phase gasoline
l.-phase process Flüssigphaseverfahren *n*, Flüssigphaseprozeß *m*
l.-phase stall Sumpfphasekammer *f (bei der Kohlehydrierung)*
l.-phase transfer unit *(Destillation)* Übergangseinheit *f* in der flüssigen Phase
l.-piston compressor Flüssigkeitsringverdichter *m*, Flüssigkeitsringgebläse *n*, Elmo-Kompressor *m*
l. polymer flüssiges Polymer[es] *n*, Liquid-Polymer *n*
l. powder flüssiger Puder *m*
l. product *(Destillation)* Erzeugnis *n*, Destillat *n*
l. range Flüssigkeitsbereich *m*
l. reflux *(Destillation)* Rücklaufflüssigkeit *f*
l. refrigerant Kühlflüssigkeit *f*
l. resin Tallöl *n*, Kiefernöl *n*, Harzöl *n*
l.-rich flüssigkeitsreich
l. rosin s. l. resin
l.-salt carburizing s. l. carburizing
l. seal gasholder Glockengasbehälter *m*
l. slip *(Ker)* flüssiger Schlicker *m*, Gießschlicker *m*
l. soap flüssige Seife *f*
l.-solid chromatogram Flüssig-Fest-Chromatogramm *n*
l.-solid chromatography Flüssig-Fest-Chromatografie *f*, Liquidus-Solidus-Chromatografie *f*, LSC
l.-solid extraction Feststoffextraktion *f*
l.-solid interface Flüssig-Fest-Grenzfläche *f*, Grenzfläche *f* flüssig-fest

l. thermometer Flüssigkeitsthermometer *n*
l. throughput Flüssigkeitsdurchsatz *m*
l. to be pumped Förderflüssigkeit *f*, zu befördernde Flüssigkeit *f*
l.-to-solid ratio *(Pap)* Flottenverhältnis *n*
l. under examination (investigation) Untersuchungsflüssigkeit *f*
l.-vapour equilibrium diagram Siedediagramm *n*
l.-vapour interface Flüssigkeit-Dampf-Grenzfläche *f*, Grenzfläche *f* flüssig-gasförmig
liquidus curve (line) Liquiduskurve *f*, Liquiduslinie *f*, Flüssiglinie *f*
l. temperature Liquidustemperatur *f*
liquiduster Flüssigstäuber *m* *(zum gemeinsamen Versprühen flüssiger und staubförmiger Pflanzenschutzmittel)*
liquifiable *s.* liquefiable
liquify / to *s.* to liquefy
liquor Flüssigkeit *f*; *(Tech)* Lauge *f*; Brühe *f*; *(Gär)* Brauwasser *n*; *(Zucker)* Kläre *f*; *(Text)* Flotte *f*, Bad *n*; *(Pap)* Kochsäure *f* *(oder)* Kochlauge *f*; Wasser *n* *(bei der Kohlehydrierung)*; wäßriges Kondensat *n* *(bei der Schwelung)*; *(Lebm)* alkoholisches (geistiges) Getränk *n*
l.-circulating line *(Pap)* Säureleitung *f*, Kochsäureleitung *f*; Kochlaugenleitung *f*
l. circulation *(Text)* Flottenzirkulation *f*, Flottenkreislauf *m*
l. flow Flüssigkeitsstrom *m*; *(Text)* Flotten[zu]lauf *m*
l.-making plant *(Pap)* Säurestation *f*, Kochsäureanlage *f*; Laugenstation *f*, Kochlaugenanlage *f*
l. ratio *(Text)* Flottenverhältnis *n*; *(Pap)* *s.* l.-to-wood ratio
l. recovery *(Pap)* Ablaugengewinnung *f*, Ablaugenregeneration *f*, Laugenregeneration *f*
l. tank *(Pap)* Kochlaugenbehälter *m*, Kochlaugenvorratstank *m*
l.-to-wood ratio *(Pap)* Flottenverhältnis *n*, *(i.e.S.)* Säureverhältnis *n* *(oder)* Laugenverhältnis *n*
liquorice Lakritze *f*, Lakritzensaft *m*, Süßholzsaft *m* *(von Glycyrrhiza glabra L.)*
liroconemalachite *(Min)* Lirokonmalachit *m*, *(veraltet für)* Lirokonit *m*, Linsenerz *(Aluminiumkupfer(II)-trihydroxidorthoarsenat)*
liroconite *(Min)* Lirokonit *m*, Linsenerz *n* *(Aluminiumkupfer(II)-trihydroxidorthoarsenat)*
lisoloid Lisoloid *n*, *(veraltet für)* feste Emulsion *f* *(kolloidales System aus fester und flüssiger Phase)*
liter-atmosphere *s.* litre-atmosphere
litharge PbO Bleiglätte *f*, Bleimonoxid *n*, Blei(II)-oxid *n*
lithia Li$_2$O Lithium[mon]oxid *n*
l. mica Lithionglimmer *m*, Lithiumglimmer *m*
l. porcelain lithiumoxidhaltiges Porzellan *n*
lithiate / to *(mit Lithium oder Lithiumverbindungen versetzen oder verbinden)*
lithic acid Harnsäure *f*, 2,6,8-Trihydroxypurin *n*
lithidionite *s.* litidionite

lithification *(Geol)* Verfestigung *f*, Erhärtung *f*, Lithifikation *f*
lithify / to *(Geol)* versteinern, erhärten, sich verfestigen
lithiophilite *(Min)* Lithiophilit *m*
lithiophorite *(Min)* Lithiophorit *m*
lithium Li Lithium *n*
l. acetate CH$_3$COOLi Lithiumazetat *n*
l. acetylide Li$_2$C$_2$ Lithiumazetylid *n*, Lithiumkarbid *n*
l. acid sulphate LiHSO$_4$ Lithiumhydrogensulfat *n*
l. alkyl Lithiumalkyl *n*
l. aluminate LiAlO$_2$ Lithiumaluminat *n*
l. aluminium hydride Li[AlH$_4$] Lithiumalanat *n*, Lithiumaluminiumhydrid *n*
l. amide LiNH$_2$ Lithiumamid *n*
l. arsenate *s.* l. orthoarsenate
l. aryl Lithiumaryl *n*
l. bicarbonate LiHCO$_3$ Lithiumhydrogenkarbonat *n*
l. bichromate *s.* l. dichromate
l. bromide LiBr Lithiumbromid *n*
l. carbide Li$_2$C$_2$ Lithiumkarbid *n*, Lithiumazetylid *n*
l. carbonate Li$_2$CO$_3$ Lithiumkarbonat *n*
l. chlorate LiClO$_3$ Lithiumchlorat *n*
l. chloride LiCl Lithiumchlorid *n*
l. chloroplatinate Li$_2$[PtCl$_6$] Lithiumhexachloroplatinat(IV) *n*
l. chromate Li$_2$CrO$_4$ Lithiumchromat *n*
l. dichromate Li$_2$Cr$_2$O$_7$ Lithiumdichromat(VI) *n*
l. dihydrogen phosphate LiH$_2$PO$_4$ Lithiumdihydrogenphosphat *n*
l. fluoride LiF Lithiumfluorid *n*
l. fluorosilicate Li$_2$[SiF$_6$] Lithiumhexafluorosilikat *n*
l. hydrate *s.* l. hydroxide
l. hydride LiH Lithiumhydrid *n*
l. hydrosulphide LiHS Lithiumhydrogensulfid *n*
l. hydroxide LiOH Lithiumhydroxid *n*
l. iodate LiJO$_3$ Lithiumjodat *n*
l. metasilicate Li$_2$SiO$_3$ Lithiumtrioxosilikat *n*, Lithiummetasilikat *n*
l. metavanadate LiVO$_3$ Lithiumtrioxovanadat(V) *n*, Lithiummetavanadat(V) *n*
l. mica *s.* lithia mica
l. molybdate Li$_2$MoO$_4$ Lithiummolybdat(VI) *n*
l. nitrate LiNO$_3$ Lithiumnitrat *n*
l. nitride Li$_3$N Lithiumnitrid *n*
l. nitrite LiNO$_2$ Lithiumnitrit *n*
l. orthoarsenate Li$_3$AsO$_4$ Lithium[ortho]arsenat(V) *n*
l. orthophosphate *s.* l. phosphate
l. orthosilicate Li$_4$SiO$_4$ Lithiumorthosilikat *n*, Lithiumtetroxosilikat *n*
l. oxide Li$_2$O Lithiumoxid *n*
l. perchlorate LiClO$_4$ Lithiumperchlorat *n*
l. permanganate LiMnO$_4$ Lithiumpermanganat *n*, Lithiummanganat(VII) *n*
l. peroxide Li$_2$O$_2$ Lithiumperoxid *n*

l. phosphate Li_3PO_4 Lithiumphosphat *n*
l. salicylate HOC_6H_5COOLi Lithiumsalizylat *n*
l. selenide Li_2Se Lithiumselenid *n*
l. silicide Li_6Si_2 Lithiumsilizid *n*
l. silicofluoride *s.* l. fluorosilicate
l. sulphate Li_2SO_4 Lithiumsulfat *n*
l. sulphide Li_2S Lithiumsulfid *n*
l. sulphite Li_2SO_3 Lithiumsulfit *n*
l. sulphocyanate *s.* l. thiocyanate
l. thiocyanate LiSCN Lithiumthiozyanat *n*, Lithiumrhodanid *n*
l. tungstate Li_2WO_4 Lithiumwolframat *n*
litho *s.* lithographic paper; *(Ker)* Abziehbild *n*
l. ink *s.* lithographic ink
l. varnish *s.* lithographic varnish
lithocholic acid Lithocholsäure *f*, 3-Monohydroxycholansäure *f*
lithographic crayon Lithografiekreide *f*, lithografische Kreide (Fettkreide) *f*
l. ink Steindruckfarbe *f*, Lithografiefarbe *f*, Druckfarbe *f* für Steindruck
l. limestone Lithografenkalk *m*, Lithografenschiefer *m*, Lithografiestein *m*, lithografischer Stein *m*
l. paper Lithografiepapier *n*, Steindruckpapier *n*, *(heute meist)* Offset[druck]papier *n*
l. printing lithografischer Druck *m*, Steindruck *m*, Lithografie *f*
l. printing ink *s.* l. ink
l. printing paper *s.* l. paper
l. process lithografisches Druckverfahren (Verfahren) *n*
l. tusche lithografische Tusche (Fettusche) *f*
l. varnish Lithografenfirnis *m*, lithografischer Firnis *m*
lithography Lithografie *f*, lithografischer Druck *m*, Steindruck *m*; *(Ker)* Dekorieren *n* (Dekoration *f*) mit Abziehbildern
lithol red Litholrot *n*
lithophile element *(Geol)* lithophiles Element *n*
lithop[h]one Lithopone *f*, Lithopon *n* *(Weißpigment aus ZnS und BaSO$_4$)*
lithosphere *(Geol)* Lithosphäre *f*, Gesteinsmantel *m*, Gesteinshülle *f*, Gesteinskruste *f*
lithotype Lithotype *f*, Hauptstreifenart *f* *(der Steinkohlenlagerung)*
litidionite *(Min)* Litidionit *m*, Neozyanit *m* *(ein Cu-haltiges Phyllosilikat)*
litmus Lackmus *n(m)* *(Flechtenfarbstoff)*
l. [test] paper Lackmuspapier *n*
l. tincture Lackmustinktur *f*
litre-atmosphere Literatmosphäre *f* *(veraltetes Energiemaß)*
Littleton [softening] point *(Glas)* Littleton-Punkt *m*, Erweichungspunkt *m*, Erweichungstemperatur *f*
Littrow spectrograph Littrow-Spektrograf *m*
live steam Frischdampf *m*, Direktdampf *m*, direkter (gespannter) Dampf *m*
l.-steam pipe Rohr *n* für direkten Dampf (Wasserdampf)

liver of lime Kalkschwefelleber *f* *(Gemisch aus Kalziumsulfiden und Kalziumsulfat)*
l. of sulphur Schwefelleber *f*, Hepar sulfuris *(technisches Kaliumsulfid)*
l. oil Leberöl *n*, Lebertran *m*
l. ore *(Min)* Zinnober *m*, Zinnabarit *m* *(Quecksilber(II)-sulfid)*
l. starch Leberstärke *f*, tierische Stärke *f*, Tierstärke *f*, Glykogen *n*
living polymer lebendes Polymer[es] *n*, stöchiometrisches Polymer[es] *n*
l. polymerization stöchiometrische Polymerisation *f*
livingstonite *(Min)* Livingstonit *m* *(Antimon(III)-quecksilber(II)-sulfid)*
lixiviate / to [aus]laugen, auswaschen, herauslösen, ausziehen, extrahieren
lixiviation Auslaugen *n*, Auslaugung *f*, Laugung *f*, Auswaschen *n*, Herauslösen *n*, Ausziehen *n*, Extrahieren *n*, Extraktion *f*
Ljungstrom heater (regenerator) Ljungström-Vorwärmer *m*, Ljungström-Regenerator *m*
L.L.C. *s.* liquid-liquid chromatography
LMFR *s.* liquid metal fuelled reactor
LNG *s.* liquefied natural gas
load / to *(Gefäße)* beladen, beschicken, füllen, chargieren; *(Füllgut)* [auf]laden, aufgeben, einfüllen, einschütten, einlegen, eintragen, einwerfen; *(mit Füllstoffen)* beschweren, füllen
load 1. Ladung *f*, Beladung *f*, Beschickung *f*, Füllmaterial *n*, Füllgut *n*, Füllung *f*, Charge *f*; Füllstoff *m*, Zuschlag *m*, Zusatz *m*, Beschwerungsmaterial *n*; 2. Last *f*, Belastung *f*, *(Phys)* Lastkraft *f*; Beanspruchung *f*
l.-bearing strength Tragfähigkeit *f*, Tragkraft *f*
l. deflection curve *(Plast)* Last-Durchbiegungskurve *f*
l.-elongation curve *(Gum, Text)* Spannungs-Dehnungsdiagramm *n*, Spannungs-Dehnungslinie *f*, Zug-Dehnungsdiagramm *n*, Zug-Dehnungskurve *f*, Kraft-Dehnungskurve *f*, Dehnungs-Spannungskurve *f*, Dehnungs-Spannungslinie *f*, Kraft-Verlängerungsschaubild *n*
l.-extension curve *s.* l.-elongation curve
l. metamorphism *(Geol)* Belastungsmetamorphose *f*
l. resistance Lastwiderstand *m*, Belastungswiderstand *m*
loaded gefüllt, füllstoffhaltig, mit Füllstoffen
loader 1. Beschickungsvorrichtung *f*; 2. *(Pap)* Füllstoff *m*, Papierfüllstoff *m*
loading 1. Laden *n*, Aufladen *n*, Aufgeben *n*, Aufgabe *f*, Einfüllen *n*, Einschütten *n*, Einlegen *n*, Eintragen *n*, Einwerfen *n*; Zuschlag *m*, Zusatz *m*, Dosierung *f* *(von Füllstoffen)*, Füllung *f*; Beschweren *n*, Füllen *n* *(mit Füllstoffen)*; 2. Füllstoff *m*, Zuschlag *m*, Zusatz *m*, Beschwerungsmaterial *n*; Gehalt *m*, Beladung *f* *(eines Gases mit festen Teilchen)*
l. agent *s.* l. material

l. behaviour Belastungsverhalten *n*; Überflutungsverhalten *n (einer Kolonne)*
l. chamber *(Plast)* Füllraum *m*
l. density Ladungsdichte *f*, Ladungskonzentration *f (von Sprengstoffen)*
l. funnel Einfülltrichter *m*
l. gas velocity Flutgeschwindigkeit *f* des Gases *(in einer Kolonne)*
l. grade Füllungsgrad *m*
l. hopper Aufgabetrichter *m*, Einschütttrichter *m*, Beschickungstrichter *m*
l. material Füllstoff *m*, Zuschlag *m*, Zusatz *m*, Beschwerungsmaterial *n*
l. of black *(Gum)* Rußzusatz *m*, Rußzuschlag *m*, Füllung *f* mit Ruß, Rußdosierung *f*
l. of fabrics *(Text)* Gewebeinkrustierung *f*
l. point Belastungsgrenze *f*; *(Destillation)* Staupunkt *m*
l. ratio Flottenverhältnis *n*
l. tray *(Plast)* Füllvorrichtung *f* für Preßwerkzeuge, Fülltablett *n*
loadstone *s.* lodestone
loaf sugar Hutzucker *m*
loam Lehm *m*
l. mould Lehmform *f*
l. moulding Lehmformen *n*, Formen *n* in Lehm
loamification *(Bodenkunde)* Verlehmung *f*
loamy lehmhaltig, lehmig
l. ground Lehmboden *m*
loan paper Wert[titel]papier *n*, Wertzeichenpapier *n*
lobe pump Drehkolbenpumpe *f*, Kapselpumpe *f*
lobelanine Lobelanin *n*, 1-Methyl-2,6-diphenazylpiperidin *n (Alkaloid)*
lobelia alkaloid Lobeliaalkaloid *n*
lobeline Lobelin *n*, 1-Methyl-2-[2-hydroxy-2-phenyläthyl]-6-phenazylpiperidin *n*
l. hydrochloride Lobelinhydrochlorid *n*
loblolly pine Weihrauchkiefer *f*, Pinus taeda L.
local anaesthetic Lokalanästhetikum *n*, Mittel *n* zur örtlichen Schmerzausschaltung
l. analgesic Lokalanalgetikum *n*, örtlich schmerzstillendes Mittel *n*
l. cell (galvanic element) Lokalelement *n (sehr kleines Korrosionselement)*
l. overheating lokale (örtliche) Überhitzung *f*
l. reduction *(Foto)* örtliche (partielle) Abschwächung *f*
localize */ to* lokalisieren
localized bond lokalisierte Bindung *f*
l. corrosion örtliche Korrosion *f*
localization [of position] Lokalisieren *n*, Lokalisation *f*, Ortsbestimmung *f*
locant *(Nomenklatur)* Stellungsziffer *f*
locao Lokao *n*, Chinesisches Grün *n*, Chinagrün *n (Naturfarbstoff aus Rhamnus-Arten)*
locaonic acid *s.* locao
locate */ to (Atome in Zwischenräumen eines Gitters)* unterbringen, *(Atome auf Zwischengitterplätze)* bringen

lock Schleuse *f*, Schleuseneinrichtung *f*
locking device Verriegelung *f*; Feststelleinrichtung *f*, Arretierung *f*, Arretiervorrichtung *f*; Hahnsicherung *f*
l. pressure *(Plast)* Schließdruck *m*, Spanndruck *m*
l. ring Schließring *m*, Klemmring *m*, Spannring *m*, Sperring *m*
locust bean gum Johannisbrotgummi *n (von Ceratonia siliqua L.)*
lode *(Geol)* Gang *m*, Ader *f*; Lager *n*; Flöz *n*
lodestone *(Min)* Magnetit *m*, Magneteisenerz *n*
Loesche coal mill Loesche-Kohlen[staub]mühle *f*
Loesche mill Loesche-Mühle *f (eine Federrollenmühle)*
loess Löß *m*
l. loam Lößlehm *m*
l. soil Lößboden *m*
loevigite *s.* loewigite
loeweite *(Min)* Löweit *m (ein Natriummagnesiumsulfat)*
loewigite Löwigit *m,* *(veraltet für)* Alunit *m (Kaliumaluminiumsulfat)*
loewite *s.* loeweite
LOF-Colburn process Libbey-Owens-Verfahren *n*, Colburn-Verfahren *n (Tafelglasziehverfahren)*
loft-dried *(Pap)* auf dem Dachboden (Trockenboden) getrocknet, luftgetrocknet
log *(Pap)* Langholz *n*
l. book Arbeitstagebuch *n*
l. scale logarithmischer Maßstab *m*
logarithmic paper Logarithmenpapier *n*, logarithmisches Papier *n*
l. viscosity number logarithmische Viskositätszahl *f*
loglog paper doppelt logarithmisch geteiltes Papier *n*
logwasher Doppelflügelwäscher *m (ein Trogläuterapparat)*
logwood Kampescheholz *n*, Campecheholz *n*, Blauholz *n*, Blutholz *n (von Haematoxylum campechianum L.)*
löllingite *(Min)* Löllingit *m (Eisenarsenid)*
lomonite *s.* laumontite
lone electron pair, l. pair of electrons einsames (freies) Elektronenpaar *n*
long-arm centrifuge Langarmzentrifuge *f*
l.-chain langkettig
l.-chain molecule Kettenmolekül *n*
l.-chain polymer langkettiges Polymer[es] *n*
l. direction *(Pap)* Längsrichtung *f*, Laufrichtung *f*, Maschinenrichtung *f*, Arbeitsrichtung *f*
l. felt *(Pap)* Langfilz *m (bei Rundsieb-Kartonmaschinen)*
l.-fibre langfas[e]rig
l.-flame coal langflammige (langflammig brennende) Kohle (Steinkohle) *f*
l. free stock (stuff) *(Pap)* langröscher Stoff *m*
l. glass langes Glas *n*
l. ink lange Druckfarbe *f*

l.-lived radiation langlebige Strahlung f
l. molecule langgestrecktes Molekül n
l.-necked langhalsig
l.-necked flask Langhalskolben m
l. oil s. l.-oil varnish
l.-oil alkyd langöliges Alkyd[harz] n
l.-oil varnish fetter Öllack (Lack) m
l. pepper Langer Pfeffer m (Früchte von Piper longum L.)
l. period lange (große) Periode f, Langperiode f (im Periodensystem)
l.-range order (Krist) weitreichende Ordnung f, Fernordnung f, Fernordnungsgrad m
l.-range preservation (Lebm) Langzeitkonservierung f
l. residue (residuum) langer Destillationsrückstand m, atmosphärischer Rückstand m, Topprückstand m
l. retention time dryer Langzeittrockner m
l.-stemmed funnel Trichter m mit langem Stiel, Bunsentrichter m
l.-term preservation s. l.-range preservation
l.-term treatment Dauerbehandlung f
l.-time burning oil Langzeitbrennöl m, Signalöl m
l.-tube evaporator Langrohrverdampfer m, Kestner-Verdampfer m
l.-tube method Standglasverfahren n (z.B. bei der Bestimmung der Absetzgeschwindigkeit)
l.-tube vertical-film evaporator Langrohr-Vertikalverdampfer m
l. wavelength region Langwellenbereich m; langwelliger Spektralbereich m, langwelliges Spektralgebiet n
l. wet stock (Pap) langschmieriger Stoff m
longifolene Longifolen n (trizyklisches Sesquiterpen)
longitudinal arch kiln Ofen m mit Längsgewölbe
l. conche (Lebm) Längsreibe[maschine] f, Konche f, Conche f
l. covering machine Längsbedeckungsmaschine f, Schermaschine f (zur Kabelherstellung)
l.-fin tube Längsrippenrohr n
l. insulating process Längsbedeckungsverfahren n, Scherverfahren n (zur Kabelherstellung)
l. riffle Längsleiste f
look-through (Pap) Durchsicht f
loop dryer Hänge[band]trockner m, Laufbandtrockner m
loose / to abgeben
loose locker, lose
l. grain (Gerb) loser Narben m (Fehler)
l.-grained (Gerb) losnarbig
l. place (position) (Krist) Lockerstelle f
l. stock (Text) loses Fasergut (Material) n
l. stock dyeing (Text) Färben n in der Flocke
l. water (Pap) Sieb[ab]wasser n
loosen / to [auf]lockern
loosening Lockerung f, Lösung f (z.B. einer chemischen Bindung)
l. off Loslösen n, Loslösung f

Lopulco mill Lopulco-Mühle f (Federrollenmühle nach dem Prinzip der Loesche-Mühle)
loranskite (Min) Loranskit m
Loschmidt number Avogadro-Zahl f, Avogadrosche Zahl f (Anzahl der Moleküle in 1 ml eines idealen Gases bei 0 °C und 1 atm; hingegen: Loschmidtsche Zahl=Avogadro number)
loss Verlust m, Abnahme f, Abgang m, Schwund m, Verarmung f
l. angle Verlustwinkel m
l. by radiation Strahl[ungs]verlust m, Abstrahlungsverlust m
l. factor [di]elektrischer Verlustfaktor m, [di]elektrische Verlustziffer f, Verlustfaktor m
l. in strength Festigkeitsverlust m, Festigkeitsrückgang m, Festigkeitseinbuße f
l. in weight Masseverlust m
l. of activity Wirksamkeitsverlust m
l. of caking power Backfähigkeitsverlust m
l. of calorific value Heizwertverlust m
l. of chemical (Pap) Chemikalienverlust m, (meist) Alkaliverlust m
l. of energy Energieverlust m
l. of entropy Entropieverlust m
l. of heat Wärmeverlust m
l. of liquid Flüssigkeitsverlust m
l. of material Substanzverlust m
l. of moisture Feuchtigkeitsverlust m
l. of rag Stoffverluste mpl (Hadernaufbereitung)
l. of sulphur (Pap) Schwefelverlust m
l. on drying Trocknungsverlust m
l. on ignition Glühverlust m
Lossen rearrangement Lossensche Umlagerung f, Lossenscher Abbau m (von Hydroxamsäure zu primären Aminen)
lost-wax investment pattern (Met) [verlorenes] Wachsmodell n, Wachsausschmelzmodell n
l.-wax moulding (Met) Formen n mit [verlorenen] Wachsmodellen, Formen n mit Wachsausschmelzmodellen
l.-wax process (Met) Wachsausschmelzverfahren n, Methode f des verlorenen Wachsmodells
lot number Fabrikationsnummer f
Lothar Meyer periodic table Tabelle f des Periodensystems nach Lothar Meyer
lotion (Kosmet, Pharm) Lotion f
loup (Met) Luppe f
lousicide Läuse[bekämpfungs]mittel n
lousy silk knötchenhaltige Naturseide f
louver Jalousiedrosselklappe f
l. dryer Jalousietrockner m
l. separator Filtrex-Abscheider m
louvre s. louver
lovage oil Liebstöckelöl n (von Levisticum officinale W.D.I. Koch)
love arrows (Min) Crinis veneris, Venushaar n (faseriger Rutil; Titandioxid)
lovenite (Min) Låvenit m (ein zirkonhaltiges Sorosilikat)
Lovibond tintometer Tintometer n nach Lovibond, Lovibondsches Tintometer n (Farbmeßgerät)

low-acid food (good) *(Lebm) (Konserve mit pH größer als 4,5)*
l.-**alkali glass** alkaliarmes Glas *n*
l.-**alloy steel** niedriglegierter Stahl *m*
l.-**ammonia latex** Latex *m* mit niedrigem Ammoniakgehalt
l.-**analysis fertilizer** *(Am) (Düngemittel mit weniger als 20% Nährstoffgehalt)*
l.-**angle X-ray scattering** *(Koll)* Röntgenkleinwinkelstreuung *f*
l.-**ash** asche[n]arm, mit niedrigem Aschegehalt
l.-**boiled pulp** *(Pap)* harter (hartgekochter, wenig aufgeschlossener) Zellstoff *m*
l. **boiler** Niedrigsieder *m*, niedrigsiedendes Lösungsmittel *n*
l.-**boiling** tiefsiedend, niedrigsiedend, leichtsiedend, niedersiedend
l.-**boiling component** leicht[er]siedende (niedrigsiedende, tiefsiedende) Komponente *f*, Leicht[er]siedendes *n*
l.-**carbon** niedriggekohlt, kohlenstoffarm, mit niedrigem Kohlenstoffgehalt *(z.B. Stahl)*
l.-**chalcocite** *(Min)* Chalkosin *m*, Tiefkupferglanz *m (Kupfer(I)-sulfid)*
l.-**contrast** *(Foto)* kontrastarm
l.-**contrast developer** *(Foto)* weich arbeitender Entwickler *m*
l.-**contrast emulsion** *(Foto)* kontrastarme Emulsion *f*
l.-**density bleaching** *(Pap)* Dünnstoffbleiche *f*
l.-**density chlorination** *(Pap)* Chlorierung *f* bei niedriger Stoffdichte
l.-**density polyethylene** Polyäthylen *n* niedriger Dichte
l.-**density tower bleaching** *(Pap)* Dünnstoff-Turmbleiche *f*
l.-**energy** energiearm, von geringer (niedriger) Energie
l.-**ester pectin** niederverestertes (niedrig verestertes, leichtverestertes, methoxylarmes) Pektin *n*
l. **explosive** Treibstoff *m*, Schießstoff *m*, Schießmittel *n*, Pulversprengstoff *m*
l. **fermentation** Untergärung *f*
l.-**fermentation beer** untergäriges Bier *n*
l. **flow** *(Plast)* schlechtes Fließen *n*
l.-**fusion** niedrigschmelzend, leichtschmelzend, leichtschmelzbar, tiefschmelzend, mit niedrigem Schmelzpunkt
l.-**grade** niederwertig, minderwertig, geringwertig, wertarm, von minderer Qualität; arm *(Erze)*
l.-**grade massecuite** *(Zucker)* C-Füllmasse *f*, Nachproduktfüllmasse *f*
l.-**grade ore** geringwertiges (geringhaltiges, armes) Erz *n*, Magererz *n*
l.-**grade sugar** Nachproduktzucker *m*, drittes Produkt *n*
l. **grinding** Flachmüllerei *f*
l.-**intensity magnetic separation** Schwachfeldscheidung *f*

l. **krausen** *(Gär)* Niederkräusen *pl*, Jungkräusen *pl*
l.-**leaded** schwach verbleit *(Kraftstoffe)*
l.-**load compound** *(Gum)* schwach gefüllte Mischung *f*
l.-**loss body** *(Ker)* Masse *f* mit niedrigem Verlustfaktor
l.-**melting** niedrigschmelzend, leichtschmelzend, tiefschmelzend, mit niedrigem Schmelzpunkt, leichtschmelzbar
l.-**melting alloy** niedrigschmelzende Legierung *f*
l.-**melting-point** *s.* l.-melting
l.-**methoxyl pectin** *s.* l.-ester pectin
l.-**modulus rubber** Kautschuk *m* mit geringem (niederem) Spannungswert
l.-**molecular** niedrigmolekular, niedermolekular
l.-**molecular-weight** von geringer Molekülmasse
l.-**octane** niedrigoktanig, mit niedriger (geringer) Oktanzahl
l.-**oxygen** sauerstoffarm, O-arm, mit niedrigem (geringem) Sauerstoffgehalt
l.-**phosphorous** phosphorarm, P-arm, niedrigphosphorhaltig, mit niedrigem (geringem) Phosphorgehalt
l. **pressure** Niederdruck *m*
l.-**pressure laminate** Niederdruckschichtstoff *m*
l.-**pressure laminating** Niederdrucklaminieren *n*, Niederdruckschichtpressen *n*, Herstellen *n* von Niederdruckschichtstoffen
l.-**pressure line** Niederdruckleitung *f*
l.-**pressure moulding** Niederdruckpreßverfahren *n*; nach dem Niederdruckpreßverfahren hergestelltes Formteil *n*
l.-**pressure [process] polyethylene** Niederdruckpolyäthylen *n*, Linearpolyäthylen *n*
l.-**pressure resin** Niederdruckharz *n*
l.-**pressure sprayer** *(Landw)* Niederdrucksprühgerät *n*
l.-**pressure steam** Niederdruckdampf *m*
l.-**rank** geringinkohlt, niedriginkohlt, niederinkohlt, gering (niedrig, schwach) inkohlt, von niedrigem (niederem) Inkohlungsgrad
l.-**raw sugar** Nachproduktzucker *m*, drittes Produkt *n*
l.-**riser plate (tray)** Flachglockenboden *m*
l.-**strain** spannungsarm
l.-**structure [carbon] black** *(Gum)* Low-structure-Ruß *m*, Niederstruktur-Ruß *m*
l. **sugar** *s.* l. raw sugar
l.-**sulphur** schwefelarm, S-arm, mit niedrigem (geringem) Schwefelgehalt
l.-**temperature** Tieftemperatur...
l.-**temperature behaviour** Kälteverhalten *n*, Verhalten *n* bei Kälte, Tieftemperatureigenschaften *fpl*, Kälteeigenschaften *fpl*
l.-**temperature brittleness** Kältesprödigkeit *f*
l.-**temperature carbonization** Tieftemperaturverkokung *f*, Schwelung *f*, Schwelen *n*, Verschwelung *f*, Verschwelen *n*, Tieftemperaturentgasung *f*

l.-temperature carbonization (carbonizing) plant Schwelanlage f, Schwelwerk n, Schwelerei f
l.-temperature carbonization (carbonizing) process Schwelverfahren n, Tieftemperaturverfahren n; Schwelvorgang m
l.-temperature characteristics s. l.-temperature behaviour
l.-temperature chlorination Tieftemperaturchlorierung f
l.-temperature coke Tieftemperaturkoks m, Schwelkoks m
l.-temperature coking s. l.-temperature carbonization
l.-temperature evaporator Tieftemperaturverdampfer m
l.-temperature flexibility Tieftemperaturflexibilität f, Kälteflexibilität f
l.-temperature form Tieftemperaturform f
l.-temperature hydrogenation Tieftemperaturhydrierung f
l.-temperature insulation Kälteisolierung f
l.-temperature kiln Niedertemperaturofen m
l.-temperature material Werkstoff m für die Tieftemperaturtechnik
l.-temperature pasteurization Dauererhitzungsverfahren n, Dauererhitzung f
l.-temperature polymer Tieftemperaturkautschuk m, Kaltkautschuk m, kalt polymerisierter Kautschuk m, Coldrubber m
l.-temperature polymerization Tieftemperaturpolymerisation f
l.-temperature properties s. l.-temperature behaviour
l.-temperature resistance Tieftemperaturbeständigkeit f, Kältebeständigkeit f, Kältefestigkeit f
l.-temperature retort Schwelretorte f
l.-temperature rubber s. l.-temperature polymer
l.-temperature separation Tieftemperaturabscheidung f
l.-temperature tar Tieftemperaturteer m, TTT, Schwelteer m, Urteer m, Primärteer m; Steinkohlen-Tieftemperaturteer m, Steinkohlenschwelteer m, Steinkohlenurteer m
l.-temperature test Kälte[beständigkeits]prüfung f
l.-tension [electrical] porcelain Niederspannungs-Elektroporzellan n, Niederspannungsporzellan n
l. unsaturation geringe Ungesättigtheit f, geringer Anteil m ungesättigter Gruppen
l. vacuum Grobvakuum n, Niedervakuum n
l.-viscosity niedrigviskos, leichtviskos, wenig viskos, von niedriger Viskosität
l.-volatile bituminous [steam] coal niedrigflüchtige bituminöse Kohle f, niedrigflüchtige (geringbituminöse) Kohle f
l. voltage Niederspannung f
l.-voltage electrophoresis Niederspannungselektrophorese f
l.-volume machinery (Geräte zum Versprühen von Pflanzenschutzmitteln mit wenig Wasser und Preßluft)
l.-volume mist blower (Landw) Sprühnebler m, Sprühblaser m, Atomiseur m
l.-volume spraying (Versprühen von Pflanzenschutzmitteln mit nur 0,5 bis 2 Liter/Ar Wasser als Träger und Luft als Dispersionsmittel)
l. yeast untergärige Hefe f, Unterhefe f
l.-zinc brass Messing n mit niedrigem Zinkgehalt
löweite (Min) Löweit m, Loeweit m (Natriummagnesiumsulfat)
lower / to herabsetzen, [ver]mindern, verringern, verkleinern, reduzieren, erniedrigen, senken
lower bainite unterer Bainit m, Unterbainit m, Gefüge n der unteren Zwischenstufe
l.-boiling niedriger (leichter) siedend, leichtersiedend
l. consolute (critical solution) temperature untere kritische Lösungstemperatur (Mischungstemperatur) f, unterer kritischer Lösungspunkt (Mischungspunkt) m
l. explosion limit untere Explosionsgrenze f
l. fatty acid niedere Fettsäure f
l. felt (Pap) Unterfilz m, unterer Abnahmefilz m (bei Rundsieb-Kartonmaschinen)
l. guide roll (Pap) Siebleitwalze f im rücklaufenden Siebtrum
l. heating value Heizwert m (i.e.S.), H, (veraltet) unterer Heizwert m, H_u
l. millstone Unterstein m, Bodenstein m
l.-nitrated niedrignitriert
l. run of the deckle strap (Pap) unterer Deckelriemen m
l. size untere Grenzkorngröße f
lowering Herabsetzung f, Minderung f, Verminderung f, Verringerung f, Verkleinerung f, Reduzierung f, Erniedrigung f, Senkung f
l. of the freezing point Gefrierpunktserniedrigung f, Gefrierpunktsdepression f
l. of the melting point Schmelzpunktserniedrigung f, Schmelzpunktsdepression f
l. of vapour pressure Dampfdruckerniedrigung f
löwigite (Min) Löwigit m, (veraltet für) Alunit m (Kaliumaluminiumhexahydroxiddisulfat)
lowquartz modification Tiefquarz-Modifikation f (des Germaniums)
L.P. gas s. liquefied petroleum gas
LP steam s. low-pressure steam
L.P.G. s. liquefied petroleum gas
lq. s. liquid
LS coupling (Kch) LS-Kopplung f, Russell-Saunders-Kopplung f
L.S.C. s. liquid-solid chromatography
LSD s. lysergic acid diethylamide
LTP s. low-temperature polymer
lube s. lubricant
l. oil s. lubricating oil
lubricant Schmierstoff m, Schmiermittel n, (i.e.S.) Schmierflüssigkeit f; Gleitmittel n, (Text auch) Schmälzmittel n, Schmälze f

l. exudation *(Text)* Ausschwitzen n des Gleitmittels (Schmälzmittels)
lubricate / to [ab]schmieren; *(Text)* schmälzen
lubricating agent *s.* lubricant
  l. cream *(Kosmet)* Nährcreme f, Hautnährcreme f
  l. film Schmier[mittel]schicht f, *(i.e.S.)* Schmierölfilm m
  l. grease Schmierfett n
  l. oil Schmieröl n
  l.-oil distillate Schmieröldestillat n
  l.-oil extract Schmierölextrakt m
  l.-oil fraction Schmierölfraktion f
  l.-oil refining Schmierölraffination f
  l.-oil stock Schmierölausgangsstoff m
  l. power Schmierfähigkeit f
  l. properties Schmiereigenschaften fpl
  l. stock *s.* l.-oil stock
lubrication Schmieren n, Schmierung f
  l. oil *s.* lubricating oil
lubricity Schmierfähigkeit f; Schlüpfrigkeit f, Fettigkeit f
luciferase Luziferase f
luciferine Luziferin n
ludlamite *(Min)* Ludlamit m *(Eisen(II)-hydroxiddiorthophosphat)*
ludwigite *(Min)* Ludwigit m *(Eisenmagnesiumborat)*
Lugol's solution *(Pharm)* Lugolsche Lösung f *(wäßrige Jodlösung)*
Lukas test Lukas-Test m *(zum Alkoholnachweis)*
Lumbang oil *(Öl von Samen der Aleurites moluccana (L.)Willd.)*
lumen 1. Lumen n, lm; 2. Lumen n, Zellraum m
luminance Leuchtdichte f, L
luminesce / to lumineszieren
luminescence Lumineszenz f, Leuchtanregung f, kaltes Leuchten n
luminescent lumineszent, lumineszierend, Lumineszenz...
  l. analysis Lumineszenzanalyse f
  l. indicator Lumineszenzindikator m
  l. paint Leuchtfarbe f, Lumineszenzfarbe f
  l. substance *s.* luminophore
luminiferous lichtspendend, leuchtend; lichterzeugend; lichtfortpflanzend
luminophore Luminophor m, Luminofor m, Lumineszenzstrahler m, *(i.e.S.)* Leuchtstoff m
luminosity Leuchten n, Glanz m, Helle f; Lichtstärke f, Helligkeit f; Leuchtstärke f, Leuchtkraft f, Luminosität f
luminous glänzend, scheinend, leuchtend, Leucht..., Licht...
  l. arc process Lichtbogenverfahren n *(zur Bindung des Luftstickstoffs)*
  l. bacteria Leuchtbakterien npl
  l. effect Leuchtwirkung f, Lichtwirkung f, Leuchterscheinung f, Lichterscheinung f
  l. energy Lichtenergie f, Strahlungsenergie f
  l. flame leuchtende Flamme f

  l. flux Lichtstrom m
  l.-flux density Lichtstromdichte f
  l. intensity Lichtstärke f, Helligkeit f
  l. paint Leuchtfarbe f, Lumineszenzfarbe f
  l. paper leuchtendes Papier n, *(i.e.S.)* fluoreszierendes Papier n, Tageslicht-Leuchtpapier n *(oder)* phosphoreszierendes Papier n
  l. source Lichtquelle f
lump / to klumpen, Klumpen bilden, klumpig werden, sich zusammenballen
lump Klumpen m, Brocken m, Stück n, *(manchmal auch)* Korn n
  l. coal Stückkohle f
  l. coke Stückkoks m
  l. density Korndichte f *(von Schüttgütern)*
  l. fuel Stückbrennstoff m
  l. lime Branntkalk m, Ätzkalk m, *(i.e.S.)* Stückkalk m
  l. ore Stückerz n, Erzbrocken mpl
  l. sugar Würfelzucker m
lumpbreaker Ballenzerteiler m *(z.B. in Mischern)*
  l. roll *(Pap)* Kittner-Walze f, Abpreßwalze f
lumpiness Klumpigkeit f, Stückigkeit f, klumpige (stückige) Beschaffenheit f
lumpy klumpig, klumpenförmig, Klumpen..., stückig, Stück...
lunar caustic $AgNO_3$ Höllenstein m, Silbernitrat n
  l. dust Mondstaub m
  l. mineral Mondmineral n
  l. rock Mondgestein n
lung injurant Lungengift n
lupine alkaloid Lupinenalkaloid n, Chinolizidinalkaloid n
lupinidine Lupinidin n, Spartein n
lupinine Lupinin n
Lüpke pendulum (resiliometer) Lüpke-Pendel n, Rückprallpendel n nach Lüpke
lupulin Lupulin n, Hopfenmehl n
  l. gland Lupulinbecher m, Drüsenbecher m, Becherdrüse f
α-lupulinic acid α-Lupulinsäure f, α-Hopfenbittersäure f, α-Bittersäure f, Humulon n
β-lupulinic acid β-Lupulinsäure f, β-Hopfenbittersäure f, β-Bittersäure f, Lupulon n
lupulon[e] *s.* β-lupulinic acid
lure Köder m, Lockspeise f
Lurgi gasification plant Lurgi-Druckvergasungsanlage f, Lurgi-Druckvergasungswerk n
Lurgi high-pressure process, Lurgi pressure-gasification process Lurgi-Druckvergasungsverfahren n, Lurgi-Druckgasverfahren n
Lurgi pressure gasifier Lurgi-Druckvergaser m
Lurgi process Lurgi-Verfahren n
Lurgi Spülgas low-temperature carbonization process Lurgi-Spülgas[schwel]verfahren n
lustre Glanz m, Politur f, Schein m, Schimmer m; *(Min, Text, Gerb)* Glanz m
  l. colour *(Ker)* Lüsterfarbe f
  l. finish *(Gerb)* Glanzappretur f, Glanzauftrag m
  l. glaze *(Ker)* Lüsterglasur f

25*

lustreless glanzlos, matt
lustring agent (Text) Glanzausrüstungsmittel n, Lüstriermittel n
lustrous glänzend, Glanz...
 l. carbon Glanzkohlenstoff m
lute / to [ver]kitten, (i.w.S.) verschmieren, dichten
lute Kitt m, Dichtungsmasse f; Gummiring m (für Flaschen usw.)
lutecium s. lutetium
lutein Lutein n, Xanthophyll n, 3,3'-Dihydroxy-α-karotin n (gelber Naturfarbstoff)
luteinize / to luteinisieren
luteinizing hormone luteinisierendes (zwischenzellenstimulierendes) Hormon n, Luteinisierungshormon n, LH, ICSH
luteocobaltic chloride [Co(NH₃)₆]Cl₃ Luteokobaltchlorid n, (veraltet für) Hexamminkobalt(III)-chlorid n
luteolin Luteolin n (ein Flavonfarbstoff)
luteotrophic luteotroph
 l. hormone s. luteotrophin
luteotrophin Luteotrophin n, luteotrophes (laktotropes) Hormon n, Prolaktin n, Laktationshormon n, Laktogen n, Laktotropin n, LTH
luteotropic s. luteotrophic
luteotropin s. luteotrophin
lutetia Lu₂O₃ Lutetiumerde f, (veraltet für) Lutetiumoxid n
lutetium Lu Lutetium n
 l. chloride LuCl₃ Lutetium(III)-chlorid n
 l. oxide Lu₂O₃ Lutetiumoxid n
 l. sulphate Lu₂(SO₄)₃ Lutetiumsulfat n
lutidine (CH₃)₂C₅H₃N Lutidin n, Dimethylpyridin n
luting s. lute
lutoids (Gum) Lutoide npl
Luwesta extractor Luwesta-Zentrifugalextraktor m
luxmasse Lux-Masse f, Luxmasse f
luxury consumption Luxuskonsum m (Aufnahme unnötiger Nährstoffmengen durch Pflanzen)
lycine C₅H₁₁NO₂ Lyzin n, Betain n, Oxyneurin n, Trimethylglykokoll n
lycopene Lykopen n, Dikaroten n (Naturfarbstoff)
lycopodium powder Bärlappsporen fpl, Blitzpulver n, Hexenmehl n (von Lycopodium-Arten)
lycorine Lykorin n (Alkaloid)
lydian stone, lydite (Min) Lydit n
lye Lauge f, Alkalilauge f, (i.e.S.) Kalilauge f (oder) Natronlauge f
 l. phase Laugenphase f
 l.-proof laugenbeständig, alkalibeständig, alkalifest
 l. treating Laugenbehandlung f
Lyman series Lyman-Serie f
lyo-enzyme Lyoferment n, Lyoenzym n (im Zellinhalt gelöstes Ferment)
lyogel Lyogel n
lyolysis Lyolysis f, Solvolyse f
Lyons blue Bleu de Lyon n (Cl-Salz des Triphenylrosanilins)
lyophile, lyophilic lyophil, lösungsmittelanziehend

 l. colloid lyophiles Kolloid n
 l. drying lyophile Trocknung f
lyophilization Lyophilisieren n, Lyophilisierung f, Gefriertrocknen n, Gefriertrocknung f, Sublimationstrocknen n, Sublimationstrocknung f
lyophobe, lyophobic lyophob, lösungsmittelabstoßend
 l. colloid lyophobes Kolloid n
lyosphere [dicke] Flüssigkeitsschicht f
lyotropic order (series) lyotrope Reihe f, Hofmeistersche Ionenreihe (Reihe) f
lysergic acid Lysergsäure f
 l. acid diethylamide Lysergsäurediäthylamid n, LSD n
lysin[e] NH₂(CH₂)₄CH(NH₂)COOH Lysin n, 2,6-Diaminohexansäure f
lysozyme Lysozym n (bakteriolytisches Enzym)

M

M s. monofunctional
M-shell M-Schale f
M unit s. monofunctional unit
MAA s. monomethylarsonic acid
Macaja oil (Öl der Palme Acrocomia sclerocarpa Mart.)
Macassar oil Macassar-Öl n (von Schleichera trijuga Willd. und Sch. oleosa Merr.)
MacDougall furnace MacDougall-Ofen m (Mehretagenofen mit zentraler Rührwelle)
mace Mazis m, Muskatblüte f (von Myristica fragans Houtt.)
maceral Mazeral n, Gefügebestandteil m (der Kohle)
 m. group Mazeralgruppe f
macerate / to mazerieren, einweichen, aufweichen, durchfeuchten, [ein]wässern; (Nahrungsmittel) aufschließen; (Rohtorfstücke) zerreißen, zerkleinern
macerate (Plast) Schnitzelmaterial n (Füllstoff)
 m. moulding (Plast) Verpressen n von Schnitzelpreßmasse; Preßteil n mit Schnitzelfüllstoff
 m. moulding compound (Plast) Schnitzelpreßmasse f
macerated fabric (Plast) Gewebeschnitzel npl
 m. peat zerrissener Torf m
maceration Mazerieren n, Mazeration f, Einweichen n, Aufweichen n, Durchfeuchten n, Wässern n, Einwässern n; Aufschließen n (von Nahrungsmitteln bei der Verdauung); Zerreißen n, Zerkleinern n, Zerkleinerung f (von Rohtorfstücken)
macerator Zerreißmaschine f, Zerreißwerk n, Reißwerk n
machine broke Ausschußpapier n, Ausschuß m, Papierausschuß m, Kollerstoff m
 m. calender stack (Pap) Maschinenkalander m,

Kalander *m*, Papierkalander *m*, Glätt[walzen]werk *n*, Walzenglättwerk *n*, Glättmaschine *f*, Maschinenglättwerk *n*, Trockenglättwerk *n*, Satinierkalander *m*, Satinier[walz]werk *n*
m. chest *(Pap)* Maschinenbütte *f*, Arbeitsbütte *f*, Stoffbütte *f*
m.-coated *(Pap)* maschinengestrichen, in der Maschine gestrichen
m. coating *(Pap)* Streichen *n* in der Papiermaschine, Maschinenstrich *m*
m.-cut peat *s.* m. peat
m. cylinder method *(Glas)* Walzenaushebeverfahren *n*
m. direction *(Pap)* Längsrichtung *f*, Laufrichtung *f*, Maschinenrichtung *f*, Arbeitsrichtung *f*
m.-dried *(Pap)* maschinengetrocknet, auf der Maschine getrocknet, maschinentrocken
m. finish *(Pap)* Maschinenglätte *f*, M'gl
m.-finished *(Pap)* maschinenglatt, m'gl, ·mgl; *(manchmal auch)* ungeglättet
m.-glazed *(Pap)* einseitigglatt, egl
m.-made deckle-edge paper Maschinenbüttenpapier *n*, Rundsieb-Büttenpapier *n*
m.-made paper *s.* m. paper
m. moulding Maschinenformen *n*, maschinelles Formen (Einformen) *n*
m. oil Maschinen[schmier]öl *n*
m. paper Maschinenpapier *n*, Patentpapier *n*
m. peat Maschinentorf *m*, maschinengeformter Torf *m*
m. process maschinelles Verfahren *n*
m. roll *(Pap)* von der Maschine kommende Papierrolle *f*, Maschinenrolle *f*
m. service chest *s.* m. chest
m. speed *(Pap)* Maschinengeschwindigkeit *f*
m. way *s.* m. direction
m. winder *(Pap)* Umroller *m*, Rollenschneidmaschine *f*, Rollenschneider *m*, Längsschneidemaschine *f*, Längsschneider *m*
m. wire Papiermaschinensieb *n*
m.-wire pit *(Pap)* Siebwasserbehälter *m*, Siebwassersammelbecken *n*
machinery oil *s.* machine oil
Mackie line *(Foto)* Mackie-Linie *f*
macroanalysis Makroanalyse *f*
macrochemistry Makrochemie *f*
macroclastics *(Geol)* grobklastische (grobkörnige klastische, makroklastische) Sedimente *npl*
macrocomponent, macroconstituent Makrokomponente *f*, Makrobestandteil *m*, makropetrografischer Bestandteil *m*
macrocrystalline makrokristallin
macrocyclic makrozyklisch
m. compound makrozyklische Verbindung *f*, Verbindung *f* mit großem Ring, Großringverbindung *f*
m. ketone makrozyklisches Keton *n*, Keton *n* mit großem Ring, großes Ringketon *n*
macroelement *(Landw)* Makronährstoff *m*, Hauptnährstoff *m*

macroetch / to makroätzen, makroskopisch ätzen, grobätzen
macroetching Makroätzen *n*, Makroätzung *f*, makroskopische Ätzung *f*, Grobätzung *f*
m. reagent Ätzmittel *n* zur Makroätzung
macrofibrille Makrofibrille *f*
macrolide Makrolid *n*
macromethod Makromethode *f*
macromolecular makromolekular
macromolecule Makromolekül *n*, Riesenmolekül *n*
macronutrient *s.* macroelement
macroring Makroring *m*, großer Ring *m* *(aus 13 oder mehr Gliedern bestehend)*
m. compound *s.* macrocyclic compound
m. ketone *s.* macrocyclic ketone
macroscopic makroskopisch
macrostructure Makrostruktur *f*, Makrogefüge *n*, Grobgefüge *n*
macrotetrolide Makrotetrolid *n*
Madagascar cardamom *(Lebm)* Abessinischer (Madagassischer) Malagetta *m*, Korarima-Malagetta *m* *(von Aframomum angustifolium Schum.)*
M. copal Sansibarkopal *m* *(von Trachylobium verrucosum Oliv. und Tr. hornemannianum Hayne)*
madder Krapp *m*, Rubia L., *(i.e.S.)* Färberkrapp *m*, Färberröte *f*, Rubia tinctorum L.; Krappwurzel *f*; Krappfarbstoff *m*, Krapprot *n*
m. root Krappwurzel *f*
Madelung constant Madelung-Konstante *f*, Madelungsche Konstante *f*
madia oil Madiöl *n* *(von Madia sativa Mol.)*
maf *s.* moisture-and-ash-free
magazine *(Pap)* Holzschacht *m*, Magazin *n* *(eines Schleifers)*
m. grinder *(Pap)* Magazinschleifer *m*
m. paper Zeitschriftenpapier *n*
magenta Magenta *n*, Fuchsin *n*, Rosanilin[chlorhydrat] *n*
magma *(Geol)* Magma *n*; *(manchmal auch)* Masse *f*, Brei *m*, Milch *f*; *(Zucker)* Füllmasse *f*
magmatic magmatisch, Magma...
m. corrosion *(Geol)* Korrosion *f*
m. digestion *(Geol)* [magmatische] Assimilation *f*, Einschmelzung *f*, Aufzehrung *f*
m. hearth Magmaherd *m*
m. intrusion Magmaintrusion *f*
m. rock magmatisches Gestein *n*, Magmagestein *n*, Magmatit *m*
magnesia MgO Magnesia *f*, Bittererde *f*, *(veraltet für)* Magnesiumoxid *n*
m. alba $4MgCO_3 \cdot Mg(OH)_2 \cdot 5H_2O$ Magnesia alba *f*, *(veraltet für)* Magnesiumhydroxidkarbonat-5-Wasser *n*
m. alum *(Min)* Magnesiaalaun *m*, *(veraltet für)* Pickeringit *m* *(Magnesiumaluminiumsulfat-22-Wasser)*
m. cement Magnesiabinder *m*, Magnesitbinder *m*, Magnesiazement *m*

m. magma Magnesiamilch f
m. mixture Magnesiamixtur f
m. usta MgO Magnesia usta f, *(veraltet für)* Magnesiumoxid n
magnesial, magnesian magnesiahaltig, Magnesia...; Magnesium...
magnesinitre *(Min)* Magnesiasalpeter m, Nitromagnesit m *(Magnesiumnitrat)*
magnesiochromite *(Min)* Magnesiochromit m *(Magnesiumchromat(III))*
magnesite *(Min)* Magnesit m, Bitterspat m *(Magnesiumkarbonat)*
m. brick [feuerfester] Magnesitstein m
magnesium Mg Magnesium n
m. acetate $Mg(CH_3COO)_2$ Magnesiumazetat n
m. acid arsenate s. m. hydrogen arsenate
m. acid orthophosphate s. m. hydrogen phosphate
m.-ammonium carbonate $(NH_4)_2CO_3 \cdot MgCO_3$ Ammoniummagnesiumkarbonat n
m.-ammonium sulphate $(NH_4)_2Mg(SO_4)_2$ Ammoniummagnesiumsulfat n
m.-base auf Magnesiumbasis (Magnesiumgrundlage)
m.-base acid (liquor, sulphite liquor) *(Pap)* Magnesiumbisulfitkochsäure f
m. bicarbonate s. m. hydrogen carbonate
m. biphosphate s. m. tetrahydrogen diphosphate
m. bisulphite $Mg(HSO_3)_2$ Magnesiumhydrogensulfit n, *(Tech)* Magnesiumbisulfit n
m. bromate $Mg(BrO_3)_2$ Magnesiumbromat n
m. bromide $MgBr_2$ Magnesiumbromid n
m. carbonate $MgCO_3$ Magnesiumkarbonat n
m. chlorate $Mg(ClO_3)_2$ Magnesiumchlorat n
m. chloride $MgCl_2$ Magnesiumchlorid n
m. chloropalladate $Mg[PdCl_6]$ Magnesiumhexachloropalladat(IV) n
m. chloroplatinate $Mg[PtCl_6]$ Magnesiumhexachloroplatinat(IV) n
m. chlorostannate $Mg[SnCl_6]$ Magnesiumhexachlorostannat(IV) n
m. chromate $MgCrO_4$ Magnesiumchromat n
m. dioxide s. m. peroxide
m. ferrocyanide $Mg_2[Fe(CN)_6]$ Magnesiumhexazyanoferrat(II) n
m. fluoride MgF_2 Magnesiumfluorid n
m. fluorosilicate $Mg[SiF_6]$ Magnesiumhexafluorosilikat n
m. hydrate s. m. hydroxide
m. hydride MgH_2 Magnesiumhydrid n
m. hydrogen arsenate $MgHAsO_4$ Magnesiumhydrogen[ortho]arsenat(V) n
m. hydrogen carbonate $Mg(HCO_3)_2$ Magnesiumhydrogenkarbonat n
m. hydrogen phosphate $MgHPO_4$ Magnesiumhydrogen[ortho]phosphat n
m. hydroxide $Mg(OH)_2$ Magnesiumhydroxid n
m. hypophosphite $Mg(H_2PO_2)_2$ Magnesiumhypophosphit n
m. hyposulphite s. m. thiosulphate

m. iodate $Mg(JO_3)_2$ Magnesiumjodat n
m. iodide MgJ_2 Magnesiumjodid n
m. nitrate $Mg(NO_3)_2$ Magnesiumnitrat n
m. nitride Mg_3N_2 Magnesiumnitrid n
m. orthoarsenite $Mg_3(AsO_3)_2$ Magnesium[ortho]arsenat(III) n
m. orthophosphite $MgPHO_3$ Magnesium[ortho]phosphit n
m. oxide MgO Magnesiumoxid n
m. oxychloride cement Magnesiabinder m, Magnesitbinder m, Magnesiazement m
m. perchlorate $Mg(ClO_4)_2$ Magnesiumperchlorat n
m. permanganate $Mg(MnO_4)_2$ Magnesiumpermanganat n, Magnesiummanganat(VII) n
m. peroxide MgO_2 Magnesiumperoxid n
m. phthalocyanine Magnesiumphthalozyanin n
m.-potassium chloride $KCl \cdot MgCl_2$ Kaliummagnesiumchlorid n
m.-potassium sulphate $K_2SO_4 \cdot 2MgSO_4$ Kaliummagnesiumsulfat n
m. pyrophosphate $Mg_2P_2O_7$ Magnesiumdiphosphat n, Magnesiumpyrophosphat n
m. silicofluoride s. m. fluorosilicate
m. sulphate $MgSO_4$ Magnesiumsulfat n
m. sulphide MgS Magnesiumsulfid n
m. sulphite $MgSO_3$ Magnesiumsulfit n
m. tetrahydrogen [di]phosphate $MgH_4(PO_4)_2$ Magnesiumtetrahydrogenbisphosphat n
m. thiosulphate MgS_2O_3 Magnesiumthiosulfat n
m. turnings Magnesiumdrehspäne mpl
magnetic magnetisch, Magnet...
m. attraction magnetische Anziehung (Anziehungskraft) f, magnetischer Zug m
m. concentration magnetische Aufbereitung f
m. drum Magnettrommelscheider m; Magnettrommel f
m. drum separator Magnettrommelscheider m
m. field magnetisches Feld n, Magnetfeld n
m. filter Magnetfilter n
m. iron [ore] s. magnetite
m. meter elektromagnetischer Durchfluß[mengen]messer m
m. method magnetische Methode (Spürmethode) f *(bei der Erdölsuche)*
m. moment magnetisches Moment n
m. nuclear moment magnetisches Kernmoment n
m. nuclear resonance kernmagnetische Resonanz f, Kernspinresonanz f
m. pulley Magnet[band]rolle f *(in Förderbändern)*
m. pyrites *(Min)* Magnetkies m, Pyrrhotin m *(Eisen(II)-sulfid)*
m. quantum number magnetische (räumliche) Quantenzahl f
m. resonance spectroscopy magnetische Resonanzspektroskopie f
m. rotation Magnetorotation f, magnetische Drehung f [der Polarisationsebene], Faraday-Effekt m

m. **rotatory power** magnetisches Dreh[ungs]vermögen n
m. **separation** Magnetscheiden n, Magnetscheidung f, Magnetsortieren n
m. **separator** Magnetscheider m
m. **stirrer (stirring apparatus)** magnetisches Rührwerk n
m. **tape** Magnetband n
m. **transition point** Curie-Punkt m, Curie-Temperatur f
magnetics magnetisierbares Gut n; magnetisches Gut n
magnetism Magnetismus m
magnetite (Min) Magnetit m, Magneteisen[erz] n, Magneteisenstein m (Eisen(II,III)-oxid)
magnetizable magnetisierbar
magnetization Magnetisierung f
magnetize / to magnetisieren
magnetocaloric effect magnetokalorischer Effekt m
magnetochemistry Magnetochemie f
magnetomotive force magnetomotorische Kraft f, MMK f, magnetische Ringspannung f, Umlaufspannung f
magneton Magneton n (Maß für atomare magnetische Momente)
magnetopyrite (Min) Magnetopyrit m, (veraltet für) Magnetkies m (Eisen(II)-sulfid)
magnochromite (Min) Magnochromit m, (veraltet für) Magnesiachromit m (Magnesiumchromat-(III))
magnolite (Min) Magnolit m (Quecksilbertellurat)
Mahler bomb Berthelot-Mahler-Kröcker-Bombe f, Berthelotsche Bombe f
Mahlke thread thermometer Fadenthermometer n nach Mahlke
Maillard-type browning (Lebm) Bräunung f vom Maillard-Typus, Maillard-Reaktion f, nichtenzymatische Bräunungsreaktion f, Amin-Zucker-Bräunung f
main Hauptleitung f, Hauptstrang m, Hauptrohr n
m. **alkaloid** Hauptalkaloid n
m. **chain** Hauptkette f (eines verzweigten Moleküls)
m. **column** Hauptkolonne f
m. **component** Hauptkomponente f, Grundkomponente f, Hauptbestandteil m
m. **defecation** (Zucker) Hauptscheidung f, Hauptkalkung f
m. **fermentation** Hauptgärung f
m. **fraction** Hauptfraktion f, Hauptlauf m
m. **group** Hauptgruppe f (des Periodensystems)
m. **product** Hauptprodukt n
m. **reaction** Hauptreaktion f
m. **steam pipe** Hauptdampfleitung f
maintain / to [aufrecht]erhalten, (z.B. auf einer Temperatur) halten
m. **constant** konstant halten
maintenance cost[s] Unterhaltungskosten pl, Wartungskosten pl, Instandhaltungskosten pl

m. **dressing** (Landw) Erhaltungsdüngung f
maiolica s. majolica
maize Mais m, Zea mays L.
m. **flakes** Maisflocken fpl
m. **germ oil** s. m. oil
m. **grits** Maisgrieß m
m. **meal** Maismehl n
m. **oil** Mais[keim]öl n
m. **protein** Maisprotein n
m. **protein staple** Zeïnfaser f
m. **starch** Maisstärke f
majolica Majolika f
major alkaloid Hauptalkaloid n
m. **component** Hauptkomponente f, Grundkomponente f, Hauptbestandteil m
m. **element** (Landw) Makronährstoff m, Hauptnährstoff m
m. **ingredient** Hauptbestandteil m
m. **valency bond** Hauptvalenzbindung f
majority carrier Majoritäts[ladungs]träger m, Mehrheitsträger m
Makarev shaft-chain grate Schacht-Kettenrost m von Makarjew
make / to:
m. **a bond** eine Bindung herstellen (bilden)
m. **acidic** [an]säuern
m. **alkaline** alkalisch machen, alkalisieren, (Farb auch) alkalisch stellen
m. **coincide** zur Deckung bringen
m. **into a dough (paste)** anteigen
m. **into a pulp** (Pap) zu Halbstoff aufschließen, (i.e.S.) zu Holzschliff verschleifen (oder) chemisch aufschließen
m. **into a roll** (Pap) aufwickeln, aufrollen
m. **into paper** zu (auf) Papier verarbeiten
m. **neutral** neutralisieren
m. **red-hot** auf Rotglut erhitzen
m. **to measure** (z.B. Polymere) „nach Maß" (gezielt) aufbauen (herstellen), „maßschneidern"
m. **unstable** destabilisieren
m. **up** ausgleichen; auffüllen, füllen (bis...)
m. **up for the loss of chemical** (Pap) die Alkaliverluste decken
make 1. Ausführung f, Bauart f, Typ m; Fabrikat n, Erzeugnis n; Produkt n; Fabrikmarke f; 2. Anfertigung f, Herstellung f, Produktion f, Fabrikation f; 3. Produktion[smenge] f, Ausstoß m; 4. Beschaffenheit f, Zustand m, Verfassung f; 5. Gasen n, Gasung f (bei der Wassergaserzeugung)
m. **part** Produktionsphase f (z.B. eines Schwelgenerators)
maker Hersteller[betrieb] m, Herstellerfirma f, Produzent m; (Pap) Schöpfer m
makeup Ausgleich m; Auffüllung f; Deckung f von Verlusten; Zusatzchemikalie f (zur Deckung von Verlusten); s. m. preparation
m. **base** (Kosmet) Grundlagencreme f, Fondcreme f
m. **chemical** (Pap) zur Deckung der Alkaliverluste

zugesetzte Chemikalie *f*, Zusatzchemikalie *f*, Alkalizusatz *m*

m. gas Frischgas *n*

m. of chemical loss *(Pap)* Deckung *f* der Alkaliverluste

m. preparation *(Kosmet)* Make-up *n*, Verschönerungsmittel *n*, Schönheitsmittel *n*

m. water Zusatzwasser *n*

making direction *(Pap)* Längsrichtung *f*, Laufrichtung *f*, Maschinenrichtung *f*, Arbeitsrichtung *f*

m. of paper Paierherstellung *f*, Papiererzeugung *f*, Papierfabrikation *f*, Papiermachen *n*, Papiermacherei *f*

m. al. s. methyl alcohol

Malabar almond Katappenbaum *m*, Katappa-Terminalie *f*, Indischer Mandelbaum *m*, Terminalia catappa L.

M. kino Malabarkino *n*, Amboinokino *n*, indisches Kino *n* *(Kinoharz von Petrocarpus marsupium Roxb.)*

malachite *(Min)* Malachit *m*, Kupferspat *m* *(Kupfer(II)-dihydroxidkarbonat)*

m. green Malachitgrün *n*, Bittermandelölgrün *n*, Benzalgrün *n*, Benzolgrün *n*, Viktoriagrün *n*, Neugrün *n*; Berggrün *n* *(gemahlener Malachit oder künstlich hergestelltes Pigment)*

malacolite *(Min)* Malakolith *m*, *(veraltet für)* Diopsid *m*

malacon *(Min)* Malakon *m*

malate Malat *n* *(Salz oder Ester der Äpfelsäure)*

Malay[an] camphor Borneokampfer *m*, Sumatrakampfer *m*, Baroskampfer *m* *(von Dryobalanops aromatica Gaertn. fil.)*

maldonite *(Min)* Maldonit *m*, Wismutgold *n*

male fern extract *(Pharm)* Wurmfarnextrakt *m*, Filixextrakt *m*

m. form (mould) *(Plast)* Stempel *m*, Stempelprofil *n*, Preßstempel *m*, Patrize *f*

maleate Male[in]at *n* *(Salz oder Ester der Maleinsäure)*

maleic acid HOOCCH=CHCOOH Maleinsäure *f*, *cis*-Butendisäure *f*, *cis*-1,2-Äth[yl]endikarbonsäure *f*

m. anhydride Maleinsäureanhydrid *n*, *cis*-Butendisäureanhydrid *n*, 2,5-Furandion *n*

m. hydrazide Maleinylhydrazin *n*, *(fälschlich)* Maleinsäurehydrazid *n*

maleinate s. maleate

maleinic acid s. maleic acid

malformation Mißbildung *f*

malic acid HOOCCH(OH)CH$_2$COOH Äpfelsäure *f*, Apfelsäure *f*, Malinsäure *f*, Hydroxybernsteinsäure *f*, Hydroxybutandisäure *f*

m. acid dehydrogenase Malatdehydr[ogen]ase *f*, Äpfelsäuredehydr[ogen]ase *f*

m. enzyme Malathydr[ogen]ase *f*

mallardite *(Min)* Mallardit *m*, Manganvitriol *m* *(Mangan(II)-sulfat-7-Wasser)*

malleable [cast] iron Temperguß *m*

malleabl[e]ize / **to** tempern

malleabl[e]izing Tempern *n*

mallet bark *(Gerb)* Mallett[o]rinde *f* *(von Eucalyptus-Arten, vor allem Eu. occidentalis Endl.)*

mallotoxin Mallotoxin *n*, Rottlerin *n*

malo-lactic fermentation Apfelsäure-Milchsäuregärung *f*

malonate Malonat *n* *(Salz oder Ester der Malonsäure)*

malonic acid HOOCCH$_2$COOH Malonsäure *f*, Methandikarbonsäure *f*, Propandisäure *f*

m. ester H$_5$C$_2$OOC·CH$_2$·COOC$_2$H$_5$ Malonsäure-[diäthyl]ester *m*, Äthylmalonat *n*

m. ester synthesis Malonestersynthese *f*

malonitrile CH$_2$(CN)$_2$ Malonsäuredinitril *n*, Methylendizyanid *n*

malonyl urea Malonylharnstoff *m*, Barbitursäure *f*

malt / **to** malzen, [ver]mälzen

malt Malz *n*

m. amylase Malzamylase *f*

m. analysis Malzanalyse *f*

m. beer Malzbier *n*, Karamelbier *n*

m. coffee Malzkaffee *m*

m. crusher s. m. mill

m. drying kiln s. m. kiln

m. extract Malzextrakt *m*, Maltosesirup *m*

m. factory Malzfabrik *f*, Mälzerei *f*

m. floor Malztenne *f*

m. for brewing Brau[erei]malz *n*

m. hopper Malztrichter *m*, Malzrumpf *m*

m.-house s. m. factory

m.-house floor s. m. floor

m.-house waste Mälzereiabwasser *n*

m. kiln Malzdarre *f*, Darre *f*

m. meal Malzschrot *n(m)*

m. mill Malzzerkleinerungsapparat *m*, Malzzerkleinerungsvorrichtung *f*, *(i.e.S.)* Walzen[schrot]mühle *f*, Schrotmühle *f*, Malzquetsche *f*, Malz[zer]quetscher *m*

m. plough s. m. turning device

m. polisher (polishing apparatus) Malzbürstmaschine *f*, Malzputzmaschine *f*, Malzputz- und -entkeimungsmaschine *f*

m. rootlets Malzkeime *mpl*

m. shovel Wichsschaufel *f*, Wenderschaufel *f* *(zum Wenden des Malzes)*

m. slurry Malzmilch *f*

m. spent grains Malztreber *pl*, Biertreber *pl*, Treber *pl*

m. starch Malzstärke *f*

m. sugar Malzzucker *m*, Maltose *f*

m. turning device Malzwendevorrichtung *f*, Malzwender *m*, Tennenwender *m*

m. vinegar Malzessig *m*

m. wort Malzwürze *f*

maltase Maltase *f* *(in Bierhefe enthaltenes Ferment)*

malted barley Gerstenmalz *n*

m. oat Hafermalz *n*

m. wheat Weizenmalz *n*

malting Mälzen n, Mälzung f, Vermälzung f, Malzbereitung f, Mälzerei f, Mälzungsprozeß m; *(manchmal auch für)* m. plant
  m. barley Malzgerste f, *(i.e.S.)* Braugerste f, BG *(oder)* für Brauzwecke geeignete Sommergerste f, SG
  m. floor Malztenne f
  m. loss Mälzungsschwund m
  m. plant Malzfabrik f, Mälzerei f
  m. process Mälzungsprozeß m, Mälzen n
maltodextrin Malto[se]dextrin n
maltomobile Malzwendevorrichtung f, Malzwender m, Tennenwender m
maltonic acid $CH_2OH(CHOH)_4COOH$ Maltonsäure f, d-Glukonsäure f
maltose Maltose f, Malzzucker m
maltster Mälzer m, Malzmeister m, Obermälzer m
malty malzartig, Malz...
mammalian toxicity Giftigkeit (Toxizität) f für Säugetiere
  m. toxicology Toxikologie f der Säugetiere
mammoth pump Mammutpumpe f, Drucklufttheber m, Lufttheber m, Airlift m
man hole s. manhole
  m.-made fibre Chemiefaser f; Chemiefaserstoff m
  m.-made leather synthetisches Leder n
  m.-made rubber synthetischer (künstlicher) Kautschuk m, Synthesekautschuk m, Kunstkautschuk m, SK
Manchester brown Manchesterbraun n
  M. kiln *(Ker)* Manchester-Ofen m *(deckenbeheizter Ofen mit Längsgewölbe)*
  M. yellow Manchestergelb n, Martiusgelb n, Naphtholgelb n
mandelic acid $C_6H_5CH(OH)COOH$ Mandelsäure f, α-Hydroxyphenylessigsäure f, Phenylglykolsäure f
dl-**mandelic acid** $C_6H_5CH(OH)COOH$ dl-Mandelsäure f, para-Mandelsäure f
Mandelin reagent *(Lebm)* Mandelins Reagens n
mandrel *(Gum, Plast)* Dorn m; *(Glas)* Pfeife f *(beim Röhrenziehverfahren)*
Mangabeira rubber Mangabeirakautschuk m, Pernambukokautschuk m *(von Hancornia speciosa Gomez)*
manganate Manganat n
manganblende *(Min)* Manganblende f, Alabandin m *(Mangan(II)-sulfid)*
manganepidote *(Min)* Mangan-Epidot m *(ein Piemontit)*
manganese Mn Mangan n
  m. acetate $Mn(CH_3COO)_2$ Mangan(II)-azetat n
  m. alumina *(Min)* Mangantongranat m, Spessartin m *(Aluminiummangan(II)-orthosilikat)*
  m.-ammonium-phosphate NH_4MnPO_4 Ammoniummangan(II)-phosphat n
  m.-ammonium-sulphate $(NH_4)_2SO_4 \cdot MnSO_4$ Ammoniummangan(II)-sulfat n
  m. black Manganschwarz n, Zementschwarz n *(Mangandioxid)*

  m. borate $MnB_4O_7 \cdot 8H_2O$ Manganborat n
  m. bronze Manganbronze f
  m. brown Manganbraun n, Manganbister $m(n)$, Bister $m(n)$
  m. carbide Mangankarbid n
  m. dioxide MnO_2 Mangandioxid n, Mangan(IV)-oxid n
  m. disilicide $MnSi_2$ Mangandisilizid n
  m. ferrocyanide $Mn_2[Fe(CN)_6]$ Mangan(II)-hexazyanoferrat(II) n
  m. fluorosilicate $Mn[SiF_6]$ Mangan(II)-hexafluorosilikat n
  m. glance *(Min)* Manganglanz m, *(veraltet für)* Alabandin m *(Mangan(II)-sulfid)*
  m. green Mangangrün n, Kasselergrün n
  m. heptoxide Mn_2O_7 Manganheptoxid n, Mangan(VII)-oxid n
  m. hydrogen phosphate $MnHPO_4$ Mangan(II)-hydrogen[ortho]phosphat n
  m. hydroxide $MnO(OH)_2$ Mangan(IV)-oxiddihydroxid n
  m. hypophosphite $Mn(H_2PO_2)_2$ Mangan(II)-hypophosphit n
  m. monoxide MnO Mangan[mon]oxid n, Mangan(II)-oxid n
  m. nodule *(Geol)* Manganknolle f
  m. ore Manganerz n
  m. perchloride $MnCl_4$ Manganperchlorid n, Mangantetrachlorid n, Mangan(IV)-chlorid n
  m. peroxide s. m. dioxide
  m. phosphide Manganphosphid n
  m. phosphite $MnHPO_3$ Mangan(II)-[ortho]phosphit n
  m. protoxide s. m. monoxide
  m. pyrophosphate $Mn_2P_2O_7$ Mangan(II)-pyrophosphat n, Mangan(II)-diphosphat n
  m. removal Entmanganung f
  m. resinate Manganresinat n
  m. selenate $MnSeO_4$ Mangan(II)-selenat n
  m. selenide $MnSe$ Mangan(II)-selenid n
  m. sesquioxide Mn_2O_3 Dimangantrioxid n, Mangan(III)-oxid n
  m. silicate $MnSiO_3$ Mangan(II)-metasilikat n, Mangan(II)-trioxosilikat n
  m. silicofluoride s. m. fluorosilicate
  m. soap Manganseife f
  m. spar *(Min)* Mangankiesel m, Rhodonit n
  m. steel Mangan[hart]stahl m
  m. thiocyanate $Mn(SCN)_2$ Mangan(II)-thiozyanat n, Mangan(II)-rhodanid n
  m. trioxide MnO_3 Mangantrioxid n, Mangan(IV)-oxid n
manganic Mangan..., *(i.e.S.)* Mangan(III)-... *(s. dagegen* m. acid*)*
  m. acid H_2MnO_4 Mangansäure f
  m. fluoride MnF_3 Mangan(III)-fluorid n, Mangantrifluorid n
  m. hydroxide $Mn(OH)_3$ Mangan(III)-hydroxid n, *(besser)* $MnO(OH)$ Manganoxidhydroxid n *(oder)* $Mn_2O_3 \cdot nH_2O$ Mangan(III)-oxidhydrat n

m. oxalate chelate Mangan(III)-oxalatchelat n
m. oxide Mn_2O_3 Mangan(III)-oxid n, Dimangantrioxid n
m. potassium sulphate $KMn(SO_4)_2$ Kaliummangansulfat n
m. sulphate $Mn_2(SO_4)_3$ Mangan(III)-sulfat n
m. sulphide MnS_2 Mangan(IV)-sulfid n, Mangandisulfid n
manganicyanide $M^I_3[Mn(CN)_6]$ Zyanomanganat(III) n, Hexazyanomanganat(III) n
manganite Manganat(IV) n, Oxomanganat(IV) n; (Min) Manganit m (Mangan(III)-hydroxid)
mangano-manganic oxide Mn_3O_4 Mangan(II,IV)-oxid n, rotes Manganoxid n
manganocyanhydric acid $H_4Mn(CN)_6$ Mangan(II)-zyanwasserstoffsäure f
manganocyanide $M^I_4[Mn(CN)_6]$ Zyanomanganat(II) n, Hexazyanomanganat(II) n
manganosite (Min) Manganosit m (Mangan(II)-oxid)
manganostibiite (Min) Manganostibiit m
manganotantalite (Min) Mangan[o]tantalit m
manganous Mangan..., (i.e.S.) Mangan(II)-...
m. acetate $Mn(CH_3COO)_2$ Mangan(II)-azetat n
m. acid orthophosphate $MnHPO_4$ Mangan(II)-hydrogen[ortho]phosphat n
m. ammonium sulphate $(NH_4)_2SO_4 \cdot MnSO_4$ Ammoniummangan(II)-sulfat n
m. bromide $MnBr_2$ Mangan(II)-bromid n
m. carbonate $MnCO_3$ Mangan(II)-karbonat n
m. chloride $MnCl_2$ Mangan(II)-chlorid n, Mangandichlorid n
m. dithionate MnS_2O_6 Mangan(II)-dithionat n
m. fluoride MnF_2 Mangan(II)-fluorid n, Mangandifluorid n
m. hydroxide $Mn(OH)_2$ Mangan(II)-hydroxid n
m. iodide MnJ_2 Mangan(II)-jodid n, Mangandijodid n
m. metasilicate $MnSiO_3$ Mangan(II)-trioxosilikat n, Mangan(II)-metasilikat n
m. nitrate $Mn(NO_3)_2$ Mangan(II)-nitrat n
m. orthophosphate $Mn_3(PO_4)_2$ Mangan(II)-[ortho]phosphat n
m. orthosilicate Mn_2SiO_4 Mangan(II)-orthosilikat n, Mangan(II)-tetroxosilikat n
m. oxide MnO Mangan(II)-oxid n, Mangan[mon]oxid n
m. pyrophosphate $Mn_2P_2O_7$ Mangan(II)-pyrophosphat n, Mangan(II)-diphosphat n
m. silicide Mn_2Si Mangan(II)-silizid n, Dimangansilizid n
m. sulphate $MnSO_4$ Mangan(II)-sulfat n
m. sulphide MnS Mangan(II)-sulfid n, Mangan[mono]sulfid n
manganowolframite (Min) Manganwolframit m, (veraltet für) Hübnerit m (Mangan(II)-wolframat)
mangantantalite s. manganotantalite
mangle dryer Mangeltrockner m
mangrove bark (Gerb) Mangroverinde f
m. cutch (Gerb) Mangroverindenextrakt m

manhole Mannloch n, Einsteigöffnung f, Einsteigluke f
m. cover Mannlochdeckel m
manicure preparation Nagelpflegepräparat n, Nagelpflegemittel n
manifold Sammelleitung f, Sammelrohr n; Verteiler m, Verteilungsrohr n, Rohrverteiler m
m. paper (Am) Durchschlagpapier n; Durchschreibpapier n
Manila copal Manila-Kopal m (Harz von Agathis-Arten, vorwiegend von Agathis alba (Lam.)Foxw.)
M. elemi (Elemiharz von Canarium-Arten)
M. fibre (hemp) Manilahanf m, Abaka m (von Musa textilis Née)
M. paper s. Manilla paper
Manilla paper Manila[kraft]papier n, Manilapackpapier n
M. rope stock (Pap) Hadernhalbstoff m aus Manilatauen
Manketti nut oil Mankettinußöl n, Munkettinußöl n, Omunketenußöl n (von Ricinodendron rautaneni Schinz)
manna Manna $n(f)$ (Ausscheidungsprodukt der Mannaesche Fraxinus ornus L.)
m. sugar Mannazucker m, Mannit m
mannan Mannan n (ein Polysaccharid)
Mannheim furnace Mannheimer Ofen m (mechanischer Salzsäureofen)
Mannich base Mannich-Base f
Mannich reaction Mannich-Reaktion f (Aminoalkylierung zur Darstellung tertiärer Basen; Aldehydkondensation)
Manning-Shepperd method Methode f von Manning-Shepperd (zur Bestimmung der Paraffine)
mannite s. manna sugar
mannitol $HOCH_2[CH(OH)]_4CH_2OH$ Mannit m (ein Hexit)
mannogalactan Mannogalaktan n (ein Pflanzengummi)
mannose Mannose f (ein Monosaccharid)
manometer Druckmesser m, Manometer n
manometric manometrisch, Manometer..., Druck[messer]...
m. thermometer manometrisches Thermometer n, Druckthermometer n
manool Manool n (bizyklisches Diterpen)
mantle (Tech) Mantel m
manual control nichtselbsttätige Steuerung f, Handsteuerung f; nichtselbsttätige Regelung f, Handregelung f
manually operated handbetätigt, manuell betätigt, handbewegt
manufacture Erzeugung f, Herstellung f, Fabrikation f
m. of chemical pulp Zellstofferzeugung f
m. of mechanical wood pulp Holzschlifferzeugung f
m. of nitric acid Salpetersäureherstellung f
m. of paper Papierherstellung f, Papiererzeugung f, Papierfabrikation f

m. of raw acid *(Pap)* Turmsäureherstellung *f*
manufactured künstlich [umgewandelt], veredelt,
synthetisch *(z.B. Brennstoff)*
   m. aggregate künstlicher Zuschlagstoff *m*
   m. fibre Regeneratfaserstoff *m*
   m. gas technisches (künstlich hergestelltes) Gas
n
   m. ice Kunsteis *n*
manufacturing process Produktionsprozeß *m*,
Produktionsverfahren *n*, Produktionsgang *m*,
Herstellungsverfahren *n*, Fertigungsverfahren *n*;
Verarbeitungsverfahren *n*, Verarbeitungsprozeß
m
manure / to *(mit organischen Düngemitteln)*
düngen
manure Mist *m*, Dung *m*, Dünger *m*, Wirtschaftsdünger *m*, *(i.w.S.)* Düngemittel *n*
   m. reinforcement *(Zugabe von Phosphaten zu
organischem Dünger)*
   m. salt Düngesalz *n*
manurial value Düngerwert *m*
manway Mannloch *n*, Einsteigöffnung *f*,
Einsteigluke *f*
· **many-body forces** *(phys Ch)* Mehrkörperkräfte *fpl*
   m.-body problem *(phys Ch)* Mehrkörperproblem
n
   m.-membered vielgliedrig
   m.-step synthesis Mehrstufensynthese *f*
map paper Landkartenpapier *n*, Kartenpapier *n*;
Seekartenpapier *n*
maple Ahorn *m*, Acer L.
   m. sap Ahornsaft *m*
   m. sugar Ahornzucker *m*
   m. syrup Ahornsirup *m*
mar proofness (resistance) Kratzfestigkeit *f*
   m.-resistant kratzfest
Maracaibo resin *(Harz von Copaifera officinalis L.)*
marajuana *s.* marihuana
marble Marmor *m*; Glaskugel *f (Glasfaserherstellung)*
   m. board Marmorpappe *f*
   m. paper *s.* marbled paper
marbled paper Mamorpapier *n*, Sprenkpapier *n*
marc Treber *pl*, Trester *pl*; Extraktionsrückstand *m*
marcasite *(Min)* Markasit *m (Eisendisulfid)*
marching modulus *(Gum)* ansteigende Span-
nungswertcharakteristik *f*
Marcy [ball] mill Marcy-Mühle *f (Kugelmühle mit
seitlichem Rostaustrag)*
margaric acid $CH_3(CH_2)_{15}COOH$ Margarinsäure *f*,
n-Heptadekansäure *f*
margarin *s.* margarine
margarine Margarine *f*
   m. blend Fettmischung *f* (Fettkomposition *f*,
Fettansatz *m*) zur Margarineherstellung, Margarinerezeptur *f*
   m. blender Mischmaschine *f*, Kneter *m*, *(i.e.S.)*
Vakuumkneter *m*
   m. colouring Margarinefärbung *f (mit Farbstoffen)*; Margarinefarbe *f (Farbstoff)*

   m. emulsion Margarineemulsion *f*
   m. factory Margarinefabrik *f*, Margarinewerk *n*
   m. fat Margarinefett *n*; Schmelzmargarine *f*,
Margarineschmalz *n*
   m. flavour Margarinearoma *n*
   m. for frying Bratmargarine *f*
   m. for leafy (puff) pastry Ziehmargarine *f*
   m. industry Margarineindustrie *f*
   m. made with milk Milchmargarine *f*
   m. made with water Wassermargarine *f*
   m. made with whey mit Molke hergestellte
Margarine *f*, Molkenmargarine *f*
   m.[-making] plant Margarine[herstellungs]an-
lage *f*
   m. preservative Margarinekonservierungsmittel
n
   m. vacuum blender Vakuumkneter *m*
   m. votator Votator *m*, Votatoranlage *f (Gerät zur
kontinuierlichen Margarineherstellung)*
   m. works *s.* **m. factory**
margarite *(Min)* Margarit *m*, Perlglimmer *m
(Kalziumaluminiumdihydroxiddialumodisilikat)*
marginal emulsion Grenzemulsion *f*
   m. watermark *(Pap)* Randwasserzeichen *n*
Margosa oil *(Samenöl von Antelaea azadirachta
(L.)Adelbert)*
marialite *(Min)* Marialith *m*
marihuana, marijuana Marihuana *n*, Haschisch
m(n) (von Cannabis indica Lam.)
marine acid Muriatsäure *f*, *(historisch für)* Salzsäure *f*
   m. animal oil Seetieröl *n*, Marineöl *n*, Seetierfett
n
   m. drilling Unterwasserbohrung *f*
   m. glue Marineleim *m*
   m. luminescence Meeresleuchten *n*
   m. oil *s.* **m. animal oil**
   m. salt Seesalz *n*, Meersalz *n*
Mariotte flask Mariottesche Flasche *f*
Mariotte law Boyle-Mariottesches Gesetz *n*
marjoram oil Majoranöl *n*
mark / to markieren, kennzeichnen, [be]zeichnen
   m. off abflecken, abfärben, ausbluten
mark Marke *f*, Strichmarke *f*, Eichmarke *f*;
Extraktionsrückstand *m*
market / to in den Handel bringen
market grade Handelsqualität *f*
   m. type Handelssorte *f*
marking Markieren *n*, Kennzeichnen *n*, Bezeichnen
n
   m. felt *(Pap)* Markierfilz *m*
   m. ink Glastinte *f*
   m. off Abflecken *n*, Abfärben *n*, Ausbluten *n*
   m. pencil Fettstift *m*
   m. roll *(Pap)* Wasserzeichenwalze *f*
Markovnikov addition Markownikow-Addition *f*,
Anlagerung *f (von Halogenwasserstoffen an
aliphatische Doppelbindungen) nach der Mar-
kownikowschen Regel*
Markovnikov's rule Markownikow-Regel *f*, Mar-
kownikowsche Regel *f*

Markownikoff rule s. Markovnikov's rule
marl Mergel m; (Landw) Wiesenkalk m; (Geol)
(manchmal für) glauconite
m. clay Mergelton m
marly merg[e]lig
marmatite (Min) Marmatit m, Eisenzinkblende f
Marme reagent (solution) Marmes Reagens n
(Alkaloidnachweis)
Marquis solution Marquis Reagens n (Alkaloidnachweis)
Marseilles soap Marseiller Seife f
Marsh arsenic test s. Marsh's test
marsh gas Sumpfgas n
marshite (Min) Marshit m (Kupfer(I)-jodid)
Marsh's test Marshsche Probe f (As-Nachweis)
Marsh's testing apparatus Marshscher Apparat m
Martens heat resistance (Plast) Formbeständigkeit f in der Wärme nach Martens
Martens thermostability s. Martens heat resistance
martensite (Met) Martensit m
martensitic (Met) martensitisch
m. steel s. martensite
martite (Min) Martit m
Martius yellow Martiusgelb n, Manchestergelb n,
Naphtholgelb n
marver / to (Glas) (einen Glasposten auf ebener
Platte) wälzen, marbeln; (Glas) (einen Glasposten
in einer eiförmig ausgehöhlten Form) wulchern,
motzen
marver [plate] (Glas) Marbel[platte] f, Wälzplatte f
marvering (Glas) Wälzen n, Marbeln n (eines
Glaspostens auf ebener Platte); (Glas) Wulchern
n, Motzen n (eines Glaspostens in einer eiförmig
ausgehöhlten Form)
Marzetti plastometer Marzetti-Plastometer n, Aus-
flußplastometer n nach Marzetti
mascagnine, mascagnite (Min) Mascagnin m,
Mascagnit m (Ammoniumsulfat)
mascara Mascara m (Wimperntusche)
mash / to [ein]maischen
mash Maische f
m. copper s. m. tun
m. filter Maischefilter n
m. filter plate Maischefilterplatte f
m. kettle s. m. tun
m. making water Einmaischwasser n
m. system Maischverfahren n, Maischprozeß m
m. tub Maisch[e]bottich m, Maisch[e]kessel m
m. tun Maischekochkessel m, Maischepfanne f;
s. m. tub
masher Maischapparat m, (i.e.S.) Vormaischapparat m, Vormaischer m
mashing Maischen n, Einmaischen n
m. process Maischverfahren n, Maischprozeß m
m. temperature Maischtemperatur f
m. tub s. mash tub
mask / to maskieren (Kationen durch Komplexbildner binden); (eine Reaktion) überdecken; (Gerb)
maskieren
masked liquor (solution) (Gerb) maskierte Brühe f

masking Maskierung f (Bindung von Kationen
durch Komplexbildner)
m. agent Maskierungsmittel n, Maskierungs-
reagens n, maskierendes Mittel n, maskierender
Stoff m
m. power Maskierungsvermögen n (einer Glasur)
masonite (Min) Masonit m
Masonite process (Pap) Masonite-Verfahren n,
Explosionsverfahren n
mass Masse f
m.-absorption coefficient Massenabsorptions-
koeffizient m
m.-action expression (law) Massenwirkungsgesetz n, MWG n, Reaktionsisotherme f
m. concrete Massenbeton m
m. defect Masse[n]defekt m, Massenschwund m
m. density Massendichte f, Dichte f (i.e.S.),
Raumdichte f der Masse
m.-dyed [er]spinngefärbt, düsengefärbt
m. effect Masseneffekt m
m.-energy equivalence Äquivalenz f von Energie
und Masse
m.-energy equivalence principle Masse-
Energie-Äquivalenzprinzip n
m.-energy relation Masse-Energie-Gleichung f,
Masse-Energie-Beziehung f, Einsteinsche Gleichung f
m. equivalent Masse[n]äquivalent n
m.-luminosity relation Masse-Leuchtkraft-
Beziehung f
m. number Massenzahl f, M
m. of a molecule Molekülmasse f, Molekularmasse f
m. of an electron Elektronenmasse f
m. per unit volume Massenkonzentration f (z.B.
Gramm je Liter bei Lösungen)
m.-pigmented s. m.-dyed
m. polymerization Massepolymerisation f, Polymerisation f in Masse (Substanz)
m. ratio Massenverhältnis n
m. ratio of liquid to solid Massenverhältnis n
flüssig zu fest
m. scale Massenskale f, Atommassenskale f
m.-spectrogram Massenspektrogramm n
m. spectrograph Massenspektrograf m
m. spectrometer Massenspektrometer n
m. spectrometric analysis s. m. spectrometry
m. spectrometry Massenspektrometrie f
m. spectroscopy Massenspektroskopie f
m. spectrum Massenspektrum n
m. susceptibility Massensuszeptibilität f,
Grammsuszeptibilität f, spezifische Suszeptibilität f
m. transfer Stoffübergang m
m.-transfer coefficient Stoffübergangszahl f
m. transport Stofftransport m
m. unit Masseneinheit f
massage cream Massagecreme f
massecuite (Zucker) Füllmasse f
Massey [process-coated] paper Massey-Papier n

(nach dem Massey-Verfahren hergestelltes maschinengestrichenes Papier)
massicot PbO Massicot *m (gelbes Pulver aus Blei(II)-oxid)*
master filter valve Steuerkopf *m* eines Filters
  m. mould *(Ker)* Mutterform *f*
masterbatch / to *(Gum)* Masterbatches (Vormischungen) herstellen; mit Masterbatches (Vormischungen) mischen
masterbatch *(Gum)* Masterbatch *m*, Vormischung *f*, Grundmischung *f*
  m. method of mixing Mischen *n* mit Masterbatches (Vormischungen)
masterbatching Herstellung *f* von Masterbatches (Vormischungen); Mischen *n* mit Masterbatches (Vormischungen)
mastic [gum] Mastix *m*, Mastixharz *n (von Pistacia lentiscus L.)*
masticate / to *(Gum)* mastizieren
masticating mill *(Gum)* Walzwerk *n* zum Mastizieren
mastication *(Gum)* Mastikation *f*, Mastizieren *n*
masticator *(Gum)* Mastikator *m*, Mastiziermaschine *f*
mastix *s.* mastic [gum]
masurium Ma Masurium *n*, *(veraltet für)* Tc Technetium *n*
masut *s.* mazout
mat matt, mattiert, glanzlos
mat Masse *f*, Vlies *n*; Papierbahn *f*, Stoffbahn *f*, Papiervlies *n*, Faserfilz *m*; *(Mälzerei)* Greifhaufen *m*
  m. moulding *(Plast)* Mattenpreßverfahren *n*
matairesinol Matairesinol *n (Lignan)*
match / to *(Text)* nach Muster (Farbvorlage) färben
  m. a shade *(Text)* einen Ton (Farbton) treffen
match mark *(Glas)* Formennaht *f*, Trennfuge *f (Fehler)*
  m. wax Zündholzparaffin *n*, Paraffin *n* zur Imprägnierung von Streichhölzern, Match-Paraffin *n*
matched-metal moulding *(Plast)* Formpressen *n* mit aufeinanderpassenden Metallformen
  m.-mould forming *(Plast)* Formpressen *n* mit aufeinanderpassenden Werkzeugen
matching *(Text)* Färben *n* nach Muster (Farbvorlage)
material materiell, stofflich, körperlich, substantiell
material Material *n*, Werkstoff *m*; Stoff *m*, Substanz *f*; Gut *n*
  m. balance Materialbilanz *f*
  m. being blended Mischgut *n*
  m. being compacted Preßgut *n*
  m. being conveyed Fördergut *n*
  m. being crushed Brechgut *n*
  m. being dried trocknendes Gut *n*, Trockengut *n*, Trocknungsgut *n*
  m. being filtered Filtergut *n*
  m. being mixed Mischgut *n*; Knetgut *n*; Rührgut *n*

m. chain *(phys Ch)* Stoffkette *f*
m. cost[s] Materialkosten *pl*
m. of construction Konstruktionswerkstoff *m*
m. of consumption Verbrauchsstoff *m*
m. of sizing *(Pap)* Leim *m*, Leim[ungs]mittel *n*, Leimstoff *m*, Leimungsmaterial *n*
m. particle materielles Teilchen *n*, materielle Partikel *f*, Materieteilchen *n*, Stoffteilchen *n*
m. to be absorbed Absorbend *m*, zu absorbierender Stoff *m*
m. to be adsorbed Adsorbend *m*, zu adsorbierender Stoff *m*
m. to be compacted Preßgut *n*
m. to be dried zu trocknendes Gut *n*, Trockengut *n*, Trocknungsgut *n*, Trockneraufgabegut *n*
m. to be expressed Preßgut *n*
m. to be extracted Extraktionsgut *n*, Lösegut *n*, Abgeber *m*
m. to be filtered Filtergut *n*, Trübe *f*
m. to be ground Mahlgut *n*; Brechgut *n*
m. to be handled [zu behandelndes] Gut *n*
m. to be mixed Mischgut *n*; Knetgut *n*; Rührgut *n*
m. to be pressed *s.* m. to be expressed
m. to be pulverized Mahlgut *n*
m. to be screened Siebgut *n*
m. to be separated Scheidegut *n*, Setzgut *n*
m. to be sintered Sintergut *n*
m. to be sized Klassiergut *n*; Sortiergut *n*; Siebgut *n*
m. to be stirred Rührgut *n*
Mathieson cell Mathieson-Zelle *f (eine Quecksilberzelle)*
matlockite *(Min)* Matlockit *m*
matrix *(Ker)* Grundmasse *f*, Einbettungsmasse *f*, Matrix *f*; *(Geol)* Grundmasse *f*; Gangart *f*, Ganggestein *n*; Muttergestein *n*
m. mechanics Matrizenmechanik *f*, Quantenmechanik *f (von Heisenberg)*
matt matt, mattiert, glanzlos
m. glaze *(Ker)* Mattglasur *f*
Mattauch rule *(Kch)* Mattauchsche Regel *f*
matte Stein *m (künstlich erschmolzenes Gemisch von Metallsulfiden)*
m.-smelt / to auf Stein [ver]schmelzen
m. smelting Steinschmelzen *n*, Schmelzen (Verschmelzen) *n* auf Stein
matter Materie *f*, Stoff *m*, Substanz *f*; Material *n*, Werkstoff *m*, Stoff *m*
m. transport Substanztransport *m (beim Zonenschmelzen)*
m. waves Materiewellen *fpl*
Matthiessen rule *(phys Ch)* Matthiessensche Regel *f*
matting Mattieren *n*, Mattierung *f*; *(Pap)* Verfilzung *f (der Fasern)*
m. of the fibres *(Pap)* Faserverfilzung *f*
m. sand Mattiersand *m (z.B. zur Glasmattierung)*
maturation Reife *f*, Reifen *n*, Reifwerden *n*, Heranreifen *n*, Ausreifen *n*, Ausreifung *f*; Reife *f (Zustand)*

mature / **to** [heran]reifen, ausreifen; reifen lassen, (z.B. Wein) ablagern; (Glas) einbrennen; (Ker) garbrennen

mature reif, ausgereift, vollentwickelt, ganz ausgebildet; abgelagert (z.B. Wein); gealtert

matured glaze (Ker) ausgeschmolzene Glasur f

maturing Reife f, Reifen n; (Ker) Mauken n, Sumpfen n (Lagern in feuchten Räumen); (Ker) Gutbrennen n, Garbrennen n

m. cellar (Ker) Maukkeller m, Massekeller m

m. point (Ker) Garpunkt m

m. range (Ker) Gutbrandbereich m, Garbrandbereich m

m. temperature (Ker) Gutbrandtemperatur f, Gar[brand]temperatur f

m. time (Plast) Nachhärtezeit f

maturity Reife f (Zustand); (Ker) Brennreife f

m. level Reifegrad m

m. test Reifegradbestimmung f

mauve, mauvein[e] Mauvein n, Mauve n (basischer Azinfarbstoff)

mavacurine Mavakurin n (Alkaloid)

maw seed Mohnsamen m

m. seed oil Mohnöl n

maximal dose Maximaldosis f, größte Gabe f

maximum and minimum thermometer Maximum-Minimum-Thermometer n

m. boiling point Maximum-Siedepunkt m, Siedepunktsmaximum n, Siedepunktshöchstwert m

m. bubble pressure maximaler Blasendruck m

m. bubble pressure method Blasen[druck]methode f

m. covalence (covalency) maximale Bindigkeit f

m. density (Foto) Maximalschwärzung f

m. diffusion current maximaler Diffusionsstrom m, Diffusions[grenz]strom m, Grenzstrom m, Sättigungsstrom m

m. efficiency maximaler Wirkungsgrad m; Höchstleistung f

m. initial retention maximale Anfangsretention f (Pflanzenschutz)

m. multiplicity größte (größtmögliche, maximale) Multiplizität f

m.-multiplicity principle (rule) [Hundsches] Prinzip n der größten Multiplizität, [Hundsche] Regel f der größten Multiplizität, Hundsches Multiplizitätsprinzip (Prinzip) n, [erste] Hundsche Regel f

m. negative valence negative Maximalvalenz f, maximale negative Wertigkeit f

m. permissible concentration höchstzulässige Konzentration f, Grenzkonzentration f

m. positive valence positive Maximalvalenz f, maximale positive Wertigkeit f

m. rate maximale Konzentration f

m. tolerance (Toxikologie) duldbare (zulässige, zugelassene) Höchstmenge f, Toleranz[dosis] f, Toleranzwert m

m. useful work (phys Ch) maximale Nutzarbeit (Arbeit) f [der Reaktion]

m. valence Maximalvalenz f, maximale Wertigkeit f

m. work s. m. useful work

maxite (Min) Maxit m, (veraltet für) Leadhillit m

Maxwell-Boltzmann distribution law Maxwell-Boltzmannsches [Impuls-]Verteilungsgesetz n, klassisches Verteilungsgesetz n

Maxwell-Boltzmann statistics Maxwell-Boltzmann-Statistik f, Boltzmann-Statistik f

Maxwell-Boltzmann velocity-distribution law Maxwellsches Geschwindigkeitsverteilungsgesetz (Verteilungsgesetz) n

Mayer reagent [for alkaloids] Mayers Reagens n (Alkaloidnachweis)

mayonnaise Mayonnaise f; Wasser-in-Öl-Emulsion f (z.B. von Pflanzenschutzmitteln)

mazout, mazut Masut m

McDougall furnace McDougall-Ofen m, Mehretagenröstofen m nach McDougall

McFadyen-Stevens reduction McFadyen-Stevenssche Karbonsäure-Reduktion f

McLeod gauge McLeod-Manometer n, McLeod-Vakuummeter n, Kompressionsmanometer n nach McLeod

MCPA MCPA n, 2-Methyl-4-chlorphenoxyessigsäure f (Herbizid)

McQuaid-Ehn test McQuaid-Ehn-Probe f (Prüfverfahren zur Bestimmung der arteigenen Korngröße)

MEA s. monoethanolamine

mead Met m, Honigwein m

meadow green $Cu(CH_3COO)_2 \cdot 3Cu(AsO_2)_2$ Schweinfurter Grün n, Kupferarsenitazetat n

m. limestone Wiesenkalk m

meal [grobes] Mehl (Getreidemehl) n, Schrotmehl n; Mehl n, Pulver n, Staub m (z.B. aus Mineralen, Gesteinen)

mealiness mehlige Beschaffenheit f, Mehligkeit f

mealy mehlig, mehlartig

m. barley milde Gerste f (Qualitätsmerkmal der Braugerste)

mean mittel, Mittel..., durchschnittlich, Durchschnitts...

mean s. m. value

m. activity coefficient mittlerer Aktivitätskoeffizient m

m. free path (phys Ch) mittlere freie Weglänge f

m. free time (phys Ch) mittlere freie Zeit f zwischen zwei Stößen

m. life[time] (phys Ch) mittlere Lebensdauer f

m. molecular velocity mittlere Geschwindigkeit f der Moleküle

m. rate of stressing mittlere Beanspruchungsgeschwindigkeit f

m. value Mittelwert m, Mittel n, Durchschnitt[swert] m

m. velocity mittlere (durchschnittliche) Geschwindigkeit f

measure 1. Maß n; 2. Maßnahme f

to make to m. (z.B. Polymere) „nach Maß" aufbauen

m. of contraction Schwindmaß n
m. of safety Sicherheitsmaßnahme f
m. of shrinkage s. m. of contraction
measured quantity Meßgröße f
measurement Messen n, Messung f
  m. element Meßeinrichtung f, Meßglied n
  m. of conductance Leitfähigkeitsmessung f, Konduktometrie f
  m. of dielectric constant Dielektrometrie f, Dekametrie f, DK-Metrie f, Messung f der Dielektrizitätskonstante
  m. of gas density Gasdichtemessung f
  m. of pH pH-[Wert-]Messung f
  m. of temperature Temperaturmessung f
  m. of the contact angle Randwinkelmessung f, Kontaktwinkelmessung f
  m. of the scattered light Streulichtmessung f, Tyndallometrie f
  m. of viscosity Viskositätsmessung f, Viskosimetrie f
measuring Messen n, Messung f
  m. accuracy Meßungenauigkeit f
  m. cell Meßzelle f
  m. chamber Meßkammer f
  m. cup Meßkelch m
  m. cylinder Meßzylinder m, Maßzylinder m, Mensur f
  m. flask Meßkolben m, Maßkolben m, Meßflasche f
  m. glass Meßglas n
  m. junction Meßstelle f eines Thermoelements
  m. pipette Meßpipette f, Teilpipette f
  m. pot Meßgefäß n
  m. range Meßbereich m
  m. spoon Maßlöffel m, Dosierlöffel m
  m. van Meßwagen m
  m. vessel Meßgefäß n, Meßbehälter m
meat curing Fleischpökeln n, Pökeln n, Einpökeln n
  m.-curing process Pökelverfahren n, Pökelprozeß m; Pökelvorgang m
  m. extract Fleischextrakt m
  m. flavour Fleischaroma n
  m.-infusion agar (Bakt) Fleisch-Pepton-Agar m(n)
  m. juice Fleischsaft m
  m. meal Fleisch[futter]mehl n
Mecca aloë (Aloe-Sorte von Aloe abyssinica Lam.)
  M. balsam Mekkabalsam m (von Commiphora opobalsamum (L.)Engl.)
  M. galls (Gerb) Aleppogallen fpl, Türkische (Smyrnaer) Gallen fpl (von Quercus infectoria Oliv.)
mechanical applicator (Landw) Ausbringungsgerät n (für Chemikalien, bei Pflanzenschutzmitteln auch) Pflanzenschutzgerät n, Applikationsgerät n
  m. barking (Pap) mechanische Entrindung f
  m. breakdown (Gum) mechanischer Abbau m, mechanische Plastizierung f

m. classifier mechanischer Klassierer m
m.-draught cooling tower Ventilatorkühlturm m
m. equivalent of heat mechanisches Wärmeäquivalent n
m. pulp s. m. wood-pulp
m. pulp mill (Pap) Holzschleiferei f, Schleiferei f, Schleifereibetrieb m
m. pulping mechanischer Aufschluß m des Holzes, Holzschlifferzeugung f, Holzschliffgewinnung f
m. rubber goods technische Gummiwaren fpl
m. seal Gleitringdichtung f
m. shaker Schüttelmaschine f
m. washing mechanische Wäsche f
m. water s. mechanically-held water
m. wood[-pulp] [mechanischer] Holzschliff m, [mechanischer] Holzstoff m, Holzmasse f, Schleifmasse f
mechanically agitated extractor Rührwerk[misch]-extrakteur m
  m. foamed plastic physikalisch getriebener Schaumstoff m
  m.-held water (Ker) mechanisch gebundenes Wasser n
mechanism of action Wirkungsmechanismus m, Wirkungsweise f
mechanocaloric effect mechanokalorischer Effekt m
mechanochemical mechanochemisch, mechano-chemisch
  m. reaction mechanochemische (mechano-chemische) Reaktion f
mechanochemistry Mechanochemie f
mechlorethamine Mechloräthamin n, Methyl-bis-(2-chloräthyl)amin n, Chlormethin n
  m. hydrochloride Mechloräthaminhydrochlorid n, Methyl-bis-(2-chloräthyl)aminhydrochlorid n, Methyl-bis-(2-chloräthyl)aminchlorhydrat n
Mecke solution Meckes Reagens n (Alkaloidnachweis)
meconic acid Mekonsäure f, Mohnsäure f, Opiumsäure f, 3-Hydroxy-4-oxo-1,4-pyran-2,6-dikarbonsäure f
median effective dose mittlere effektive Dosis f, ED 50
  m. lethal dose mittlere letale Dosis f, LD 50
medicament Medikament n, Arzneimittel n
medicated paper medizinisches Papier n
  m. soap medizinische Seife f
medicinal medizinisch, arzneilich
  m. oil medizinisches Öl n, Medizin[al]öl n, Öl n für medizinische Zwecke
  m. paraffin flüssiges Paraffin n
  m. soap s. medicated soap
  m. substance Arzneistoff m
medium Medium n, Mittel n, Stoff m, Träger m; Lösungsmittel n; (Bakt) Nährmedium n, Nährboden m, Nährsubstrat n; (Farb) Bindemittel n; (Farb) Bindemittellösung f
  m. accelerator (Gum) mittelstarker (mittelschneller) Beschleuniger m

m. benzine s. m. gasoline
m. boiler Mittelsieder m, mittelsiedendes Lösungsmittel n
m.-caking coal mittelbackende (mäßigbackende, mittelmäßig backende) Kohle f, Kohle f mit mittelmäßigem Backvermögen
m.-carbon mittelgekohlt, mit mittlerem Kohlenstoffgehalt (z.B. Stahl)
m. circuit Trübekreislauf m, Trübeumlauf m (beim Schwerflüssigkeitsverfahren)
m.-density polyethylene Polyäthylen n mittlerer Dichte
m. gasoline Mittel[schwer]benzin n
m.-grained mittel[fein]körnig
m.-grained sand Mittelsand m
m.-heavy gasoline s. m. gasoline
m.-modulus rubber Kautschuk m mit mittlerem Spannungswert
m. naphtha s. m. gasoline
m. of adsorption Adsorptionsmedium n
m.-oil alkyd mittelöliges Alkyd[harz] n
m.-oil varnish halbfetter (mittelfetter) Öllack (Lack) m
m. period mittlere Periode f (des Periodensystems)
m.-pressure synthesis Mitteldrucksynthese f
m. processing channel black Medium-Processing-Channel-Ruß m, MPC-Ruß m, mittelverarbeitbarer (mittelmäßig verarbeitbarer) Kanalruß m
m.-rank von mittlerem Inkohlungsgrad
m. ring mittlerer (mittelgroßer) Ring m (aus 8 bis 12 Gliedern bestehend)
m.-ring compound Verbindung f mit mittlerem (mittelgroßem) Ring, Verbindung f mit mittlerer Ringgröße
m.-size ring s. m. ring
m.-size ring compound s. m.-ring compound
m.-soft pitch mittelhartes Pech n, Mittelpech n
m. solid Trübefeststoff m (bei der Schwerflüssigkeitsaufbereitung)
m. speed accelerator s. m. accelerator
m.-temperature carbonization Mitteltemperaturverkokung f, Mitteltemperaturentgasung f
m.-temperature coke Mitteltemperaturkoks m
m. thermal black Medium-Thermal-Ruß m, MT-Ruß m (Gasruß von mittlerer Teilchengröße)
m.-viscosity mittelviskos, von mittlerer Viskosität
m.-volatile [bituminous] coal mittelflüchtige bituminöse Kohle f, mittelflüchtige (mittelbituminöse) Kohle f
m. volume spraying (Versprühen von Pflanzenschutzmitteln mit 2 bis 6 Liter/Ar Wasser als Träger)
meerschaum (Min) Meerschaum m, Sepiolith m
Meerwein-Ponndorf-Verley reduction Meerwein-Ponndorf-Verley-Reduktion f, Meerwein-Ponndorf-Verley-Karbonylreduktion f, Meerwein-Ponndorfsche Reaktion f
megascopic megaskopisch

megass[e] Bagasse f, Begasse f, Zuckerrohrbagasse f (Zuckerrohrabfälle)
meionite, mejonite (Min) Mejonit m
MEK, M.E.K. s. methylethylketone
MEK/benzene dewaxing plant MEK-Benzol-Entparaffinierungsanlage f
MEK/benzene [dewaxing] process MEK-Benzol-Verfahren n (Entparaffinierungsverfahren mit Methyläthylketon-Benzol als Lösungsmittel)
MEK dewaxing MEK-Entparaffinierung f (Entparaffinierung mit Methyläthylketon als Lösungsmittel)
MEK [dewaxing] process MEK-Verfahren n (Entparaffinierungsverfahren mit Methyläthylketon als Lösungsmittel)
Meker burner Méker-Brenner m
Mekka s. unter Mecca
melaconite (Min) Melakonit m, (veraltet für) Tenorit m (Kupfer(II)-oxid)
melamine $C_3H_6N_6$ Melamin n, Zyanursäureamid n, Triamino-sym-triazin n
m.-formaldehyde resin Melamin-Formaldehydharz n, Melaminharz n
m.-phenolic resin Melamin-Phenolharz n
m. resin s. m.-formaldehyde resin
melampyrite $CH_2OH(CHOH)_4CH_2OH$ Melampyrit m, Melampyrin n, Dulzit m (ein Hexit)
melanite (Min) Melanit m
melanocerite (Min) Melanozerit m
melanochroite (Min) Melanochroit m, (veraltet für) Phönizit m (Blei(II)-oxychromat)
melanoidin[e] Melanoidin n (Farb-und Aromastoff des Malzes)
m. malt Mela[noidin]malz n
melanosiderite (Min) Melanosiderit m
melanterite (Min) Melanterit m, Eisenvitriol m (Eisen(II)-sulfat-7-Wasser)
melassigenic melassebildend
Meldola's blue Meldolas Blau n, Meldolablau n, Echtneublau 3R n, Neublau R n
meletin Meletin n, 3,5,7,3',4'-Pentahydroxyflavon n
melibiose Melibiose f (ein Disaccharid)
melilite (Min) Melilith m
melilotic acid $HOC_6H_4CH_2CH_2COOH$ Melilotsäure f, o-Hydrokumarsäure f
meli[no]phane (Min) Melinophan m
meliphanite (Min) Meliphanit m, (veraltet für) Melinophan m
melissa oil Melissenöl n (von Melissa officinalis L.)
melissic acid $CH_3(CH_2)_{28}COOH$ Melissinsäure f, n-Triakontansäure f
melissyl alcohol $C_{31}H_{63}OH$ Melissylalkohol m, Myrizylalkohol m
mellic acid s. mellitic acid
melliferous honigerzeugend, nektarführend, honigend
mellite (Min) Mellit m, Honigstein m
mellitic acid $C_6(COOH)_6$ Mellit[h]säure f, Honigsteinsäure f, Benzolhexakarbonsäure f
mellow lime liquor (Gerb) milder (fauler) Äscher m, Stinkäscher m

m. tan liquor *(Gerb)* süße Gerbbrühe f
melonite *(Min)* Melonit *m*, Tellurnickel *n* *(Nik-keltellurid)*
melt / to 1. [zer]schmelzen, flüssig werden; sich auflösen; 2. [ein]schmelzen; *(z.B. Zucker)* auflösen
  m. down niederschmelzen, einschmelzen, zusammenschmelzen
  m. together zusammenschmelzen
melt 1. Schmelze *f*, Schmelzfluß *m*; Schmelze *f*, Charge *f*; 2. Schmelzen *n*, Einschmelzen *n*, Schmelzung *f*
  m. extractor Schmelzextraktor *m* *(Bauart eines Torpedos bei Kolbenspritzgußmaschinen)*
  m. extrusion Schmelzspinnen *n*, Spinnen (Erspinnen) *n* aus der Schmelze
  m. [flow] index Schmelzindex *m*
  m. pan Schmelzpfanne *f*, Schmelzkessel *m*, Schmelzgefäß *n*
  m. spinning Schmelzspinnverfahren *n*, Spinnen (Erspinnen) *n* aus der Schmelze
  m. spinning line Schmelzspinnanlage *f*
  m. spinning process s. m. spinning
  m.-spun filament schmelzgesponnener (im Schmelzspinnverfahren ersponnener) Elementarfaden *m*
  m. temperature Schmelztemperatur *f*
  m. viscosity Schmelzviskosität *f*
meltability Schmelzbarkeit *f*
meltable schmelzbar
melted butter zerlassene Butter *f*
m.[-down] fat ausgelassenes Fett *n*, Schmalz *n*
melter Schmelzvorrichtung *f*; Schmelzwanne *f*, Schmelzraum *m*, Schmelzteil *m*; Schmelzer *m*
melting Schmelzen *n*, Einschmelzen *n*, Schmelzung *f*
  m. chamber s. m. end
  m. crucible Schmelztiegel *m*
  m. curve Schmelzkurve *f*
  m. diagram Schmelzdiagramm *n*
  m. dilatation Schmelzausdehnung *f*, Schmelzdilatation *f*
  m. dilatation curve Schmelzausdehnungskurve *f*
  m. dilation s. m. dilatation
  m.-down Niederschmelzen *n*, Einschmelzen *n*, Zusammenschmelzen *n*
  m. end Schmelzteil *m*, Schmelzraum *m*, Schmelzwanne *f*
  m. enthalpy Schmelzenthalpie *f*
  m. furnace Schmelzofen *m*
  m. heat Schmelzwärme *f*
  m. pan Schmelzpfanne *f*, Schmelzkessel *m*, Schmelzgefäß *n*
  m. point Schmelzpunkt *m*, Fließpunkt *m*; *(Gerb veraltet für)* shrinkage temperature
  m.-point apparatus Schmelzpunktapparat *m*
  m.-point bath Schmelzpunktbad *n*
  m.-point capillary Schmelzpunktkapillare *f*, Schmelzpunktröhrchen *n*

  m.-point curve s. m.-point pressure curve
  m.-point diagram s. m. diagram
  m.-point pressure curve Schmelzdruckkurve *f*
  m.-point tube s. m.-point capillary
  m. range Schmelzbereich *m*
  m. tank Schmelzwanne *f*, *(i.e.S.)* Glasschmelzwanne *f*
  m. temperature Schmelztemperatur *f*
  m. viscosity Schmelzviskosität *f*
  m. zone Schmelzzone *f*
meltwater Schmelzwasser *n*
member [einzelnes] Glied *n*, Einzelglied *n*
  m. of a ring Ringglied *n*
membered gegliedert, ...gliedrig
membrane Membran[e] *f*, Wand *f*, Scheidewand *f*, Diaphragma *n*
  m. bag Diaphragmasack *m*
  m. equilibrium Membrangleichgewicht *n*
  m. filter Membranfilter *n*
  m. hydrolysis Membranhydrolyse *f*
  m. potential Membranpotential *n*
memory effect elastische Nachwirkung *f*; *(Kch)* Memory-Effekt *m*, Gedächtniseffekt *m* *(Verbleiben von Restaktivität aus radioaktiven Substanzen in Gefäßen und Meßinstrumenten)*
menaccanite *(Min)* Menaccanit *m*, *(veraltet für)* Ilmenit *m* *(Eisen(II)-metatitanat)*
menadione Menadion *n*, 2-Methyl-1,4-naphthochinon *n*, Vitamin K_3 *n*
mend / to *(Gerb)* *(Äscher)* anschärfen, zubessern, nachbessern
Mendeléeff [periodic] system Mendelejewsches Periodensystem *n* [der Elemente], Mendelejew-System *n*
Mendeléeff [periodic] table Mendelejewsche Tabelle *f* des Periodensystems; *s.* Mendeléeff periodic system
mendelevium Md Mendelevium *n*
Mendheim kiln *(Ker)* Mendheim-Ofen *m* *(gasbefeuerter Kammerringofen nach Mendheim)*
mendipite *(Min)* Mendipit *m*
mendozite *(Min)* Mendozit *m* *(Natriumaluminium-sulfat-11-Wasser)*
meneghinite *(Min)* Meneghinit *m* *(Blei(II)-antimon-(III)-sulfid)*
menhaden oil Menhadenöl *n*
Meni oil Meni-Öl *n* *(von Lophira alata Banks)*
menilite *(Min)* Menilit *m*
meniscus Meniskus *m*; *(Glas)* Zwiebel *f* *(beim Vertikalziehverfahren)*
  m. reader Meniskusvisierblende *f* *(Titration)*
mensuration analysis Maßanalyse *f*, volumetrische Analyse *f*, Titrieranalyse *f*, Volumetrie *f*, Titrimetrie *f*
1,8(9)-p-menthadiene 1,8(9)-*p*-Menthadien *n*, Limonen *n*, 1-Methyl-4-isopropylzyklohexadien-(1,8) *n*
p-menthane *p*-Menthan *n*, 1-Methyl-4-isopropyl-zyklohexan *n*, Hexahydrozymol *n*
menthene Menthen *n*, *(i.e.S.)* δ-3-Menthen *n*

menthol Menthol n, Menthanol-(3) n, Hexahydro-thymol n
menthone Menthon n, p-Menthanon-(3) n
menthyl salicylate HO · C_6H_4 · CO · $OC_{10}H_{19}$ Menthylsalizylat n
Mercapsol process Mercapsol-Verfahren n, Mercapsol-Prozeß m *(zur Entschwefelung von Erdöldestillaten)*
mercaptal Merkaptal n *(Sammelname für Kondensationsprodukte von Merkaptanen mit Aldehyden)*
mercaptan Merkaptan n, Thioalkohol m
  m. conversion Merkaptanumwandlung f
  m. extraction Merkaptanextraktion f
  m. removal Merkaptanentfernung f
  m.-rich merkaptanreich, reich an Merkaptanen
  m. sulphur *(Gum)* Merkapto-Schwefel m
mercaptide Merkaptid n *(Sammelname für Salze der Thioalkohole)*
mercapto group —SH Merkaptogruppe f, Merkaptorest m, Sulfhydrylgruppe f, Thiolgruppe f
mercaptoacetic acid HS·CH_2·COOH Merkaptoessigsäure f, Thioglykolsäure f
1-mercaptododecane $CH_3(CH_2)_{10}CH_2SH$ n-Dodezylmerkaptan n, 1-Dodekanthiol n, Laurylmerkaptan n
2-mercaptoethanoic acid s. mercaptoacetic acid
mercaptol[e] Merkaptol n *(Sammelname für Kondensationsprodukte von Merkaptanen mit Ketonen)*
6-mercaptopurine $C_5H_4N_4S$·H_2O 6-Merkaptopurin n, 6-Purinthiol n
mercerization Merzerisieren n, Merzerisierung f, Merzerisation f, Laugen n *(kurzzeitige Behandlung von Baumwollgarnen und -geweben mit starker kalter Natronlauge)*
  m. with tension Merzerisieren (Laugen) n unter Spannung
  m. without tension spannungsloses Merzerisieren (Laugen) n
mercerize / to merzerisieren, laugen
mercerizing s. mercerization
  m. assistant Merzerisierhilfsmittel n
  m. machine Merzerisiermaschine f
merchant bar Raffinierstahl m, Garbstahl m, Paketstahl m
mercuration s. mercurization
mercurial quecksilberhaltig; Quecksilber...
  m. barometer Quecksilberbarometer n
  m. manometer Quecksilbermanometer n
  m. ointment Quecksilbersalbe f
mercurialism Quecksilbervergiftung f
mercuric Quecksilber..., *(i.e.S.)* Quecksilber(II)-...
  m. acetate $Hg(CH_3COO)_2$ Quecksilber(II)-azetat n
  m. acetylide HgC_2 Quecksilber(II)-azetylid n
  m. arsenate s. m. orthoarsenate
  m. barium iodide HgJ_2·BaJ_2 Quecksilber(II)-bariumjodid n

  m. bromate $Hg(BrO_3)_2$ Quecksilber(II)-bromat n
  m. bromide $HgBr_2$ Quecksilber(II)-bromid n, Quecksilberdibromid n
  m. bromide iodide HgBrJ Quecksilber(II)-bromidjodid n
  m. chlorate $Hg(ClO_3)_2$ Quecksilber(II)-chlorat n
  m. chloride $HgCl_2$ Quecksilber(II)-chlorid n, Quecksilberdichlorid n, Sublimat n
  m. chloride iodide HgClJ Quecksilber(II)-chloridjodid n
  m. chloride paper Sublimatpapier n *(ein Reagenzpapier)*
  m. chromate $HgCrO_4$ Quecksilber(II)-chromat n
  m. cyanide $Hg(CN)_2$ Quecksilber(II)-zyanid n
  m. dioxysulphate $HgSO_4$·2HgO Quecksilber(II)-oxidsulfat n
  m. fluoride HgF_2 Quecksilber(II)-fluorid n, Quecksilberdifluorid n
  m. fluorosilicate $Hg[SiF_6]$ Quecksilber(II)-hexafluorosilikat n
  m. fulminate $Hg(CNO)_2$ Quecksilber(II)-fulminat n, Knallquecksilber n
  m. hydrogen arsenate $HgHAsO_4$ Quecksilber(II)-hydrogenarsenat(V) n
  m. hydroxide $Hg(OH)_2$ Quecksilber(II)-hydroxid n
  m. iodate $Hg(JO_3)_2$ Quecksilber(II)-jodat n
  m. iodide HgJ_2 Quecksilber(II)-jodid n, Quecksilberdijodid n
  m. iodobromide s. m. bromide iodide
  m. iodochloride s. m. chloride iodide
  m. metatellurate $HgTeO_4$ Quecksilber(II)-metatellurat(VI) n, Quecksilber(II)-tetroxotellurat(VI) n
  m. nitrate $Hg(NO_3)_2$ Quecksilber(II)-nitrat n
  m. nitride Hg_3N_2 Quecksilber(II)-nitrid n
  m. orthoarsenate $Hg_3(AsO_4)_2$ Quecksilber(II)-[ortho]arsenat(V) n
  m. orthophosphate $Hg_3(PO_4)_2$ Quecksilber(II)-[ortho]phosphat n
  m. orthotellurate Hg_3TeO_6 Quecksilber(II)-[ortho]tellurat(VI) n, Quecksilber(II)-hexoxotellurat(VI) n
  m. oxide HgO Quecksilber(II)-oxid n
  m. oxybromide $HgBr_2$·3HgO Quecksilber(II)-oxidbromid n
  m. oxychloride 4HgO·$HgCl_2$ Quecksilber(II)-oxidchlorid n
  m. oxycyanide $Hg(CN)_2$·HgO Quecksilber(II)-oxidzyanid n
  m. oxyfluoride HgF_2·HgO Quecksilber(II)-oxidfluorid n
  m. oxyiodide HgJ_2·3HgO Quecksilber(II)-oxidjodid n
  m. phosphate s. m. orthophosphate
  m. potassium cyanide $K_2[Hg(CN)_4]$ Kaliumtetrazyanomerkurat(II) n
  m. rhodanide s. m. thiocyanate
  m. salt Quecksilber(II)-salz n
  m. selenide HgSe Quecksilber(II)-selenid n
  m. silver iodide $Ag_2[HgJ_4]$ Silbertetrajodomerkurat(II) n

m. **subsulphate** s. m. dioxysulphate
m. **sulphate** HgSO$_4$ Quecksilber(II)-sulfat n
m. **sulphide** HgS Quecksilber(II)-sulfid n
m. **sulphocyanate (sulphocyanide)** s. m. thiocyanate
m. **thiocyanate** Hg(SCN)$_2$ Quecksilber(II)-thiozyanat n, Quecksilber(II)-rhodanid n
m. **tungstate** HgWO$_4$ Quecksilber(II)-wolframat-(VI) n
mercuriiodide M$^I_2$[HgJ$_4$] Jodomerkurat(II) n, Tetrajodomerkurat(II) n
mercurimetric merkurimetrisch
mercurimetry Merkurimetrie f (Titration mit Quecksilber(II)-nitratlösung)
mercurisulphite M$^I_2$[Hg(SO$_3$)$_2$] Sulfitomerkurat(II) n, Disulfitomerkurat(II) n
mercurization Merkurierung f (Einführung von Hg in organische Verbindungen)
mercurometric merkurometrisch
mercurometry Merkurometrie f (Titration mit Quecksilber(I)-nitratlösung)
mercurous Quecksilber..., (i.e.S.) Quecksilber(I)-...
m. **acetate** CH$_3$COOHg Quecksilber(I)-azetat n
m. **acid [ortho]arsenate** Hg$_2$HAsO$_4$ Quecksilber(I)-hydrogen[ortho]arsenat(V) n
m. **arsenate** s. m. orthoarsenate
m. **azide** HgN$_3$ Quecksilber(I)-azid n
m. **bromate** HgBrO$_3$ Quecksilber(I)-bromat n
m. **bromide** Hg$_2$Br$_2$ Quecksilber(I)-bromid n
m. **carbonate** Hg$_2$CO$_3$ Quecksilber(I)-karbonat n
m. **chlorate** HgClO$_3$ Quecksilber(I)-chlorat n
m. **chloride** Hg$_2$Cl$_2$ Quecksilber(I)-chlorid n, Kalomel n
m. **chromate** Hg$_2$CrO$_4$ Quecksilber(I)-chromat n
m. **fluoride** Hg$_2$F$_2$ Quecksilber(I)-fluorid n
m. **fluorosilicate** Hg$_2$[SiF$_6$] Quecksilber(I)-hexafluorosilikat n
m. **iodate** HgJO$_3$ Quecksilber(I)-jodat n
m. **iodide** Hg$_2$J$_2$ Quecksilber(I)-jodid n
m. **nitrate** Hg$_2$(NO$_3$)$_2$ Quecksilber(I)-nitrat n
m. **nitrite** Hg$_2$(NO$_2$)$_2$ Quecksilber(I)-nitrit n
m. **orthoarsenate** Hg$_3$AsO$_4$ Quecksilber(I)-[ortho]arsenat(V) n
m. **orthophosphate** Hg$_3$PO$_4$ Quecksilber(I)-[ortho]phosphat n
m. **oxide** Hg$_2$O Quecksilber(I)-oxid n
m. **phosphate** s. m. orthophosphate
m. **salt** Quecksilber(I)-salz n
m. **silicofluoride** s. m. fluorosilicate
m. **sulphate** Hg$_2$SO$_4$ Quecksilber(I)-sulfat n
m. **sulphide** Hg$_2$S Quecksilber(I)-sulfid n
m. **thiocyanate** HgSCN Quecksilber(I)-thiozyanat n, Quecksilber(I)-rhodanid n
m. **tungstate** Hg$_2$WO$_4$ Quecksilber(I)-wolframat n
mercury Hg Quecksilber n
m. **alkyl** Quecksilberalkyl n, Quecksilberalkylverbindung f
m. **arc** Quecksilberlichtbogen m
m. **arc lamp** s. m.-vapour lamp

m. **barometer** Quecksilberbarometer n
m. **bichloride** s. m. dichloride
m.-**binary-system boiler** Quecksilberdampfkessel m mit nachgeschaltetem Wasserdampferzeuger
m. **biniodide** HgJ$_2$ Quecksilberdijodid n, Quecksilber(II)-jodid n
m. **cathode** Quecksilberkatode f
m.-**cathode electrolysis** Elektrolyse f mit Quecksilberkatode
m. **cell** Quecksilberzelle f, Amalgamzelle f
m.-**cell process** Quecksilberverfahren n, Amalgamverfahren n (Elektrolyse)
m. **chloride paper** Sublimatpapier n (ein Reagenzpapier)
m. **clay** Quecksilberton m
m. **column** Quecksilbersäule f
m. **dichloride** HgCl$_2$ Quecksilberdichlorid n, Quecksilber(II)-chlorid n, Sublimat n
m. **diffusion pump** s. m.-vapour pump
m. **discharge lamp** s. m.-vapour lamp
m. **dropping cathode** Quecksilbertropfkatode f
m. **electrode** Quecksilberelektrode f
m. **fulminate** Hg(CNO)$_2$ Quecksilber(II)-fulminat n, Knallquecksilber n
m. **gauge** s. m. manometer
m.-**in-glass thermometer** s. m. thermometer
m. **intensification** (Foto) Quecksilberverstärkung f
m. **intensifier** (Foto) Quecksilberverstärker m
m. **lamp** s. m.-vapour lamp
m. **level** Quecksilberstand m
m. **manometer** Quecksilbermanometer n
m. **microelectrode** Quecksilbermikroelektrode f
m. **monochloride** s. m. protochloride
m. **ore** Quecksilbererz n
m. **perchloride** s. m. dichloride
m. **pernitrate** Hg(NO$_3$)$_2$ Quecksilber(II)-nitrat n
m. **persulphate** HgSO$_4$ Quecksilber(II)-sulfat n
m. **protochloride** Hg$_2$Cl$_2$ Quecksilber(I)-chlorid n, Kalomel n
m. **protoiodide** Hg$_2$J$_2$ Quecksilber(I)-jodid n
m. **protonitrate** Hg$_2$(NO$_3$)$_2$ Quecksilber(I)-nitrat n
m. **pump** s. m.-vapour pump
m. **salt** Quecksilbersalz n
m.-**silver iodide** Ag$_2$[HgJ$_4$] Silbertetrajodomerkurat(II) n
m. **subchloride** s. m. protochloride
m. **thermometer** Quecksilberthermometer n
m. **tong[s]** Quecksilberzange f
m. **trap** Quecksilberfalle f
m. **vapour** Quecksilberdampf m
m.-**vapour lamp** Quecksilber[dampf]lampe f
m.-**vapour pump** Quecksilberdiffusionspumpe f, Diffusionspumpe f mit Quecksilberfüllung, Quecksilberpumpe f
meroxene (Min) Meroxen m
mescaline C$_{11}$H$_{17}$NO$_3$ Meskalin n, Mezkalin n, 3,4,5-Trimethoxy-β-phenäthylamin n
mesenteric fat Mickerfett n, Gekrösefett n

mesh Masche *f*; Siebnummer *f*, Maschenzahl *f* (*Anzahl der Maschen je Zoll linear*)
  m. analysis Siebanalyse *f*
  m. packings Maschendrahtfüllkörper *mpl*
  m. size 1. Maschenweite *f*; 2. (*Korngröße, ausgedrückt durch die Maschenzahl des Siebs*)
mesitine [spar], mesitinspath *s.* mesitite
mesitite (*Min*) Mesitit *m*, Mesitinspat *m*
mesitoic acid $(CH_3)_3C_6H_2COOH$ Mesitylen-2-karbonsäure *f*, 2,4,6-Trimethylbenzoesäure *f*
mesityl oxide $(CH_3)_2C=CHCOCH_3$ Mesityloxid *n*, 2-Methylpenten-(2)-on-(4) *n*
mesitylene $C_6H_3(CH_3)_3$ Mesitylen *n*, 1,3,5-Trimethylbenzol *n*
meso compound Mesoverbindung *f*
  m. form *meso*-Form *f*, Mesoform *f*, Meso-Form *f*
mesocolloid Mesokolloid *n*
mesoionic mesoionisch
  m. compound mesoionische Verbindung *f*
mesolite (*Min*) Mesolith *m* (*Kalziumnatriumtrisdialumotrisilikat*)
mesomeric mesomer, Mesomerie...
  m. effect Mesomerieeffekt *m*, M-Effekt *m*, mesomerer Substituenteneffekt *m*, elektromerer Effekt *m*
  m. energy Mesomerieenergie *f*, Resonanzenergie *f*, Delokalisierungsenergie *f*, Konjugationsenergie *f*
mesomerism Mesomerie *f*, Resonanz *f*, Strukturresonanz *f*
mesomorphic mesomorph
  m. state mesomorpher (liquokristalliner) Zustand *m*
mesomorphous *s.* mesomorphic
meson Meso[tro]n *n* (*mittelschweres Elementarteilchen*)
  μ-**meson** *μ*-Meson *n*, Mu-Meson *n*, Myon *n*, Müon *n*, Muon *n*
  π-**meson** *π*-Meson *n*, Pion *n*
  m. field Mesonenfeld *n*
  m. theory Mesonentheorie *f*
mesonic mesonisch
  m. atom Meso[nen]atom *n*, Mesonatom *n*
mesoperiodate $M^I_3[JO_5]$ Pentoxoperjodat *n*, Mesoperjodat *n*
mesoperrhenate $M^I_3[ReO_5]$ Pentoxorhenat(VII) *n*, Mesoperrhenat *n*
mesosiderite (*Min*) Mesosiderit *m*
mesotartaric acid $HOOC(CHOH)_2COOH$ *meso*-Weinsäure *f*, Mesotartarsäure *f*, Antiweinsäure *f*
mesothorium Mesothorium *n*
mesotype (*Min*) Mesotyp *m*
mesoxalic acid $OC(COOH)_2$ Mesoxalsäure *f*, Oxomalonsäure *f*, Oxopropandisäure *f*
messenger ribonucleic acid Messenger-Ribonukleinsäure *f*, Messenger-RNS *f*, Boten-RNS *f*
  m. RNA *s.* m. ribonucleic acid
mesylate Mesylat *n*, Mesylester *m*, Methansulfonsäureester *m*, Methansulfonat *n*
meta metaständig, *m*-ständig

in m. position in meta-Stellung (*m*-Stellung, 1,3-Stellung), metaständig, *m*-ständig
  to be [located, situated] m. sich in meta-Stellung (*m*-Stellung, 1,3-Stellung) befinden, in meta-Stellung stehen, metaständig (*m*-ständig) sein
m.-aldehyde $(CH_3CHO)_4$ Metaldehyd *m*
m.-anthracite Meta-Anthrazit *m*
m. compound meta-Verbindung *f*, *m*-Verbindung *f*
m.-directing meta-dirigierend, *m*-dirigierend, nach der meta-Stellung dirigierend
m.-directing group meta-dirigierende Gruppe *f*, meta-dirigierender Substituent *m*, Substituent *m* zweiter Ordnung (Klasse)
m. director *s.* m.-directing group
m. isomer meta-Isomer[es] *n*, *m*-Isomer[es] *n*
m. position meta-Stellung *f*, *m*-Stellung *f*, 1,3-Stellung *f*
metaacetaldehyde $(CH_3CHO)_4$ Metaldehyd *m*
metaaluminate $M^I[AlO_2]$ Metaaluminat *n*
metaantimonate M^ISbO_3 Trioxoantimonat(V) *n*, Metaantimonat(V) *n*
metaantimonic acid $HSbO_3$ Metaantimon(V)-säure *f*
metaantimonous acid $HSbO_2$ Metaantimon(III)-säure *f*
metaarsenate M^IAsO_3 Metaarsenat(V) *n*
metaarsenic acid $HAsO_3$ Metaarsen(V)-säure *f*
metaarsenite M^IAsO_2 Metaarsenat(III) *n*
metaaurate $M^I[AuO_2]$ Metaaurat(III) *n*
metabismuthate M^IBiO_3 Trioxobismutat(V) *n*, Metabismutat(V) *n*, Bismutat(V) *n*
metabismuthic acid $HBiO_3$ Wismut(V)-säure *f*, Metawismutsäure *f*
metabisulphite $M^I_2S_2O_5$ Disulfit *n*, Pyrosulfit *n*, (*veraltet*) Metabisulfit *n*
metabituminous coal metabituminöse Kohle *f*
metabolic Stoffwechsel..., den Stoffwechsel (Nahrungsumsatz) betreffend
  m. disturbance Stoffwechselstörung *f*
  m. energy Stoffwechselenergie *f*
  m. process Stoffwechselvorgang *m*, Stoffwechselprozeß *m*
  m. reaction Stoffwechselreaktion *f*
metabolism Stoffwechsel *m*, Metabolismus *m*
metabolite Stoffwechselprodukt *n*; Metabolit *m* (*für Stoffwechselreaktionen unentbehrliche Substanz*)
metaborate M^IBO_2 Dioxoborat *n*, Metaborat *n*
metaboric acid HBO_2 Dioxoborsäure *f*, Metaborsäure *f*
metacarbonate $M^I_2CO_3$ Karbonat *n*
metachrome [dyeing] method Metachromverfahren *n*
metacinnabar[ite] (*Min*) Metazinnabarit *m* (*Quecksilber(II)-sulfid*)
metacolumbate M^INbO_3 Metaniobat *n*
metaformaldehyde $(CH_2O)_3$ *α*-Trioxymethylen *n*, 1,3,5-Trioxan *n*
metahalloysite (*Min*) Metahalloysit *m* (*Aluminiumsilikat*)

metahypophosphate $M^I_4P_2O_6$ Hypophosphat n
metakaolin Metakaolin m
metal Metall n
  m.-**activated** metallaktiviert
  m. **alkyl** Metallalkyl n
  m. **amide** $M^I NH_2$ Metallamid n, Amid n
  m. **atom** Metallatom n
  m. **bath** Metallbad n
  m. **bond** Metallbindung f, metallische Bindung f
  m. **bonding** Metall[ver]kleben n, Metall[ver]klebung f, Metallverleimung f
  m. **buffer** Metallpuffer m
  m. **carbide** Metallkarbid n
  m. **catalyst** Metallkatalysator m
  m.-**catalyzed** metallkatalysiert
  m. **cation** Metallkation n
  m.-**ceramic** keramisch, pulvermetallurgisch; Metall-Keramik-... (z.B. Verbundwerkstoff)
  m. **ceramics** Metallkeramik f, Pulvermetallurgie f
  m. **chelate** s. m.-chelate complex
  m.-**chelate bond** Metallchelatbindung f
  m.-**chelate complex (compound)** Metallchelatkomplex m, Metallchelatverbindung f, Metallchelat n
  m. **colouring** Metallfärbung f
  m. **complex** Metallkomplex m
  m. **compound** intermetallische Phase (Verbindung) f, metallische intermediäre Phase f
  m.-**containing** metallhaltig
  m. **crust** Metallbeschlag m
  m. **deactivator** Metallde[s]aktivator m
  m. **deposition** Metallabscheidung f
  m.-**donor bond (linkage)** Metall-Donatorbindung f
  m. **electrode** Metallelektrode f
  m.-**enzyme bond** Metall-Enzymbindung f
  m. **fabric** Metallgewebe n, Drahtgewebe n
  m.-**filament lamp** Metallfadenlampe f
  m. **flange** Metallflansch m
  m. **foil** Metallfolie f
  m.-**free** metallfrei
  m.-**free phthalocyanine** metallfreies (freies) Phthalozyanin n
  m. **gasket** Metalldichtung f
  m. **gauze** s. m. fabric
  m.-**halogen exchange** Metall-Halogen-Austausch m
  m. **hose** Metallschlauch m
  m.-**hydrogen exchange** Metall-Wasserstoff-Austausch m
  m. **ion** Metallion n
  m. **ketyl** Metallketyl n, Ketyl n
  m.-**metal exchange** Metall-Metall-Austausch m
  m. **mould** Metallform f, metallische Dauergießform (Dauerform, Form) f, Kokille f
  m. **ore** Metallerz n
  m. **packing** Metalldichtung f, Metallpackung f
  m. **paper** Metallpapier n
  m.-**paper capacitor** Metallpapierkondensator m, MP-Kondensator m, MPko m

m. **phthalocyanine** Metallphthalozyanin n
m. **powder** Metallpulver n, metallisches Pulver n
m.-**powder filter** Metallpulverfilter n, pulvermetallurgisches Filter n
m. **screen** Metallsieb n, Drahtsieb n
m. **seal** Metalldichtung f
m. **sol** Metallsol n
m. **spraying** Metallspritzverfahren n, Metallspritzen n, Spritzmetallisieren n, Schoop[is]ieren n
m. **stearate** Metallstearat n
m. **sulphide** Metallsulfid n
m. **thread** Metallfaden m, Metalldraht m
m. **titration** Metalltitration f
m.-**to-rubber bonding** Gummi-Metall-Verbindung f
m.-**valence-bond orbital** Metallvalenzorbital n
metalation Metallierung f
metaldehyde $(CH_3CHO)_4$ Metaldehyd m
metallic metallisch, metallen, Metall...
metallic (Sammelname für Chemiefaserstoffe, die aus Metall, einem mit Kunststoff beschichteten Metall, einem mit Metall beschichteten Kunststoff oder aus einer völlig mit Metall ummantelten Seele bestehen)
m. **alloy** Metallegierung f
m. **atom** Metallatom n
m. **bond** metallische Bindung f, Metallbindung f
m. **copper** metallisches Kupfer n
m. **electrode** Metallelektrode f
m. **fibre** Metallfaser f, MT; Metallfaserstoff m
m. **filament** Metallfaden m, Metalldraht m
m. **ink** Metallfarbe f
m. **ion** Metallion n
m. **lattice** metallisches Gitter n, Metallgitter n
m. **lustre** metallischer Glanz m, Metallglanz m
m. **mordant** Metall[salz]beize f, basische Beize f
m. **ore** Metallerz n
m. **oxide** Metalloxid n
m.-**oxide cure** (Gum) Metalloxidvernetzung f, Vernetzung f mit Metalloxiden
m. **packing** Metalldichtung f, Metallpackung f
m. **paper** Metallicpapier n, Omskriptpapier n, Metallschreibstiftpapier n
α-m. **phosphorus** violetter (Hittorfscher) Phosphor m
β-m. **phosphorus** schwarzer Phosphor m
m. **poisoning** durch Kautschukgifte beschleunigte Autoxydation f
m. **silver** metallisches Silber n
m. **soap** Metallseife f
m. **state** metallischer Zustand m, Metallzustand m
m. **tungsten powder** Wolframmetallpulver n, metallisches Wolframpulver n
m. **yarn** Metallfaden m; metallisierter Faden m; Metallgarn n
metalliferous metallhaltig; metallführend, erzführend
metallization 1. Metallisieren n, Metallisierung f, (i.e.S.) Metallspritzen n, Spritzmetallisieren n.

Metallspritzverfahren *n*, Schoop[is]ieren *n*; 2.
(Geol) Vererzung *f*
metallized paper Metallpapier *n*
metallizing *s.* metallization 1.
metallo catalyst Metallkatalysator *m*, Metallkontakt *m*
m. enzyme Metallenzym *n*
m.-organic metallorganisch, Organometall...
m.-organic compound metallorganische Verbindung *f*, Organometallverbindung *f*, Metallorganyl *n*
metallogen[et]ic metallogen, erzbildend, Erzbildungs...
metallographic metallografisch
metallography Metallografie *f*, Metallbeschreibung *f*
metalloid Metalloid *n*, *(veraltet für)* Nichtmetall *n*
metallurgical metallurgisch, hüttenmännisch
m. coke Hüttenkoks *m*, metallurgischer (hüttenfähiger, verhüttungsfähiger) Koks *m*, Koks *m* für metallurgische Zwecke
metallurgy Metallurgie *f*, Hüttenkunde *f*, Hüttenwesen *n*
metalorganic *s.* metallo-organic
metamer Metamer[es] *n*
metameric metamer
metamerism Metamerie *f* *(eine Form der Strukturisomerie)*
metametal Metametall *n*
metamorphic *(Geol)* metamorph
m. rock metamorphes Gestein *n*, Metamorphit *m*
metamorphism Metamorphose *f*, Metamorphismus *m*
metamorphite, metamorphosed rock *s.* metamorphic rock
metamorphosis *s.* metamorphism
metanilic acid $C_6H_4(NH_2)SO_3H$ Metanilsäure *f*, *m*-Aminobenzolsulfonsäure *f*, Anilin-*m*-sulfonsäure *f*
metaniobate M^INbO_3 Metaniobat *n*
metaperiodate M^IJO_4 Tetroxoperjodat *n*, Metaperjodat *n*
metaperiodic acid HJO_4 Tetroxojod(VII)-säure *f*, Metaperjodsäure *f*
metaperrhenate M^IReO_4 Tetroxorhenat(VII) *n*
metaphosphate M^IPO_3 Metaphosphat *n*
metaphosphite M^IPO_2 Metaphosphit *n*
metaphosphoric acid $(HPO_3)_n$ Metaphosphorsäure *f*
metaplumbate $M^I_2PbO_3$ Trioxoplumbat(IV) *n*, Metaplumbat(IV) *n*
metasilicate $M^I_2SiO_3$ Metasilikat *n*, Trioxosilikat *n*
metasilicic acid H_2SiO_3 Metakieselsäure *f*, Trioxokieselsäure *f*
metastable metastabil
m. state metastabiler Zustand *m*
m. system metastabiles System *n*
metastannate $M^I_2SnO_3$ Trioxostannat(IV) *n*
metastannic α **acid** H_2SnO_3 α-Zinnsäure *f*, a-Zinnsäure *f*, gewöhnliche Zinnsäure *f*

m. β **acid** $H_{10}Sn_5O_{15}$ β-Zinnsäure *f*, b-Zinnsäure *f*, Metazinnsäure *f*
metatantalate M^ITaO_3 Trioxotantalat(V) *n*, Metatantalat(V) *n*
metatellurate $M^I_2TeO_4$ Tetroxotellurat(VI) *n*, Metatellurat(VI) *n*
metathesis doppelte Umsetzung *f*, Wechselzersetzung *f*, Metathese *f*
metathetical metathetisch
m. reaction *s.* metathesis
metathioarsenate M^IAsS_3 Trithioarsenat(V) *n*, Metathioarsenat(V) *n*
metathioarsenite M^IAsS_2 Dithioarsenat(III) *n*, Metathioarsenat(III) *n*
metathiostannate $M^I_2SnS_3$ Trithiostannat(IV) *n*, Metathiostannat *n*
metatitanate $M^I_2TiO_3$ Trioxotitanat(IV) *n*, Metatitanat(IV) *n*
metatungstate $M^I_6[H_2W_{12}O_{40}]$ Dihydrogendodekawolframat(VI) *n*, Metawolframat(VI) *n*
metatungstic acid $H_6[H_2W_{12}O_{40}]$ Dihydrogendodekawolframsäure *f*, Metawolframsäure *f*
metavanadate M^IVO_3 Trioxovanadat(V) *n*, Metavanadat(V) *n*
metazirconate $M^I_2ZrO_3$ Trioxozirkonat(IV) *n*, Metazirkonat(IV) *n*
meteoric iron Meteoreisen *n*, Eisenmeteorit *m*, Siderit *m*
m. stone Meteorstein *m*, Steinmeteorit *m*, Aerolith *m*
meteorite Meteorit *m*
meter / to dosieren, zumessen
meter Meßgerät *n*, Messer *m*, Zähler *m*; Prüfgerät *n*
metering Dosierung *f*, Dosieren *n*, Zumessen *n*
m. pump Dosier[ungs]pumpe *f*, Zumeßpumpe *f*, *(Plast auch)* Titerpumpe *f*, Spinnpumpe *f*
m. roll *(Pap)* Meßwalze *f*
m. screw *(Plast)* Schnecke *f* mit Homogenisier[ungs]zone
m. section (zone) *(Plast)* Homogenisier[ungs]zone *f* *(einer Strangpresse)*
methacrolein *s.* methacrylaldehyde
methacrylaldehyde $CH_2=C(CH_3)CHO$ Methakrylaldehyd *m*, Methakrolein *n*
methacrylate Methakrylat *n*
m. polymer Polymethakrylat *n*
methacrylic acid $CH_2=C(CH_3)COOH$ Methakrylsäure *f*, 2-Methylpropen-(1)-säure-(3) *f*
methaemoglobin Methämoglobin *n*
methallyl chloride $CH_2=C(CH_3)\cdot CH_2Cl$ Methallylchlorid *n*, 3-Chlor-2-methylpropen *n*
methanal HCHO Methanal *n*, Formaldehyd *m*
methanamide $HCONH_2$ Methanamid *n*, Ameisensäureamid *n*, Formamid *n*
methane CH_4 Methan *n*
m.-dicarbonic acid *s.* m.-dicarboxylic acid
m.-dicarboxylic acid $HOOCCH_2COOH$ Methandikarbonsäure *f*, Propandisäure *f*, Malonsäure *f*
m. recovery Methangewinnung *f* *(durch Flözausgasung)*

methanecarboxylic acid s. ethanoic acid
methanoic acid HCOOH Methansäure f, Ameisen-säure f
methanol CH_3OH Methanol n, Methylalkohol m
m. synthesis Methanolsynthese f
methanolic methanolisch
methanolysis Methanolyse f
methemoglobin s. methaemoglobin
methenamine $(CH_2)_6N_4$ Methenamin n, Hexamethylentetramin n, Hexamin n
methene s. methylene
methenyl tribromide $CHBr_3$ Tribrommethan n, Bromoform n
m. trichloride $CHCl_3$ Trichlormethan n, Chloroform n
m. triiodide CHJ_3 Trijodmethan n, Jodoform n
methide Methid n
methine Methin n
m. bridge Methinbrücke f
m. dye Methinfarbstoff m
methionine Methionin n, 2-Amino-4-methylmerkaptobuttersäure f
method Methode f, Verfahren n, Technik f, Arbeitsweise f
m. for preparation s. m. of preparation
m. in the laboratory Labor[atoriums]methode f, Labor[atoriums]verfahren n
m. of application Anwendungsweise f
m. of approximation Näherungsmethode f, Näherungsverfahren n
m. of chlorination Chlor[ier]ungsverfahren n
m. of electron pair s. m. of valence-bond structures
m. of isotope separation Verfahren n zur Isotopentrennung
m. of measurement Meßmethode f, Meßverfahren n
m. of mixture (phys Ch) Mischungsmethode f
m. of molecular beams Molekularstrahlmethode f, Methode f der Molekularstrahlen
m. of molecular orbitals Molekülorbitalmethode f, MO-Methode f, Methode f der Molekülorbitale, Hund-Mulliken-Lennard-Jones-Hückel-Methode f
m. of notation Bezeichnungsweise f
m. of preparation Darstellungsmethode f, Herstellungsverfahren n
m. of reclaiming Regenerierverfahren n
m. of reduction Reduktionsmethode f
m. of resolution Spalt[ungs]methode f
m. of rotating crystals Drehkristallverfahren n, Drehkristallmethode f, Bragg-Verfahren n
m. of separation Trennmethode f, Trennverfahren n
m. of spin states s. m. of valence-bond structures
m. of the self-consistent field Methode f des selbstkonsistenten Feldes, SCF-Methode f, Self-consistent-field-Methode f
m. of valence-bond structures Valenzstrukturmethode f, Methode f der Valenzstrukturen,

Valenzbindungsmethode f, VB-Methode f, Heitler-London-Slater-Pauling-Methode f, HLSP-Methode f, Elektronenpaarmethode f, Spinmethode f
m. of working Arbeitsweise f
methoxide CH_3OM^I Methoxid n, Methylat n
methoxybenzene $CH_3OC_6H_5$ Methoxybenzol n, Methylphenyläther m, Anisol n
p-methoxybenzoic acid $CH_3OC_6H_4COOH$ p-Methoxybenzoesäure f, 4-Methoxybenzolkarbonsäure f, Anissäure f
methoxyl group $—OCH_3$ Methoxy[l]gruppe f, Methoxy[l]rest m
methoxymethane CH_3OCH_3 Methoxymethan n, Dimethyläther m, Methyläther m
methyl Methyl n
m. acetate CH_3COOCH_3 Methylazetat n, Essigsäuremethylester m
m. acrylate Methylakrylat n, Akrylsäuremethylester m
m. alcohol CH_3OH Methylalkohol m, Methanol n
m. bromide CH_3Br Methylbromid n, Monobrommethan n, Brommethyl n
m. cellulose Methylzellulose f
m. chloride CH_3Cl Methylchlorid n, Monochlormethan n, Chlormethyl n
m. α-crotonate $CH_3CH=CHCOOCH_3$ Methyl-α-krotonat n, α-Krotonsäuremethylester m
m. ester Methylester m
m. ether CH_3OCH_3 Methyläther m, Dimethyläther m
m. formate $HCOOCH_3$ Methylformiat n, Ameisensäuremethylester m
m. group $CH_3—$ Methylgruppe f, Methylrest m
m. halide Methylhalogenid n, Halogenmethan n, Halogenmethyl n
m. iodide CH_3J Methyljodid n, Monojodmethan n
m. methacrylate $CH_2=C(CH_3)COOCH_3$ Methylmethakrylat n, Methyl-2-methylpropenoat n, Methakrylsäuremethylester m
m. orange Methylorange n, Goldorange n, Orange III n, Helianthin n
m. radical Methylgruppe f, Methylrest m, Methylradikal n; [freies] Methylradikal n
m. red Methylrot n, p-Dimethylaminoazobenzolo-karbonsäure f, Anthranilsäureazodimethylanilin n
m. rubber Methylkautschuk m
m. salicylate $HOC_6H_4COOCH_3$ Methylsalizylat n, Salizylsäuremethylester m
m. silicone Methylsilikon n
m. silicone fluid (oil) Methylsilikonflüssigkeit f, Methylsilikonöl n
m. silicone resin Methylsilikonharz n
m. silicone rubber Methylsilikongummi m
m. sodium CH_3Na Methylnatrium n, Natriummethyl n
m. substituent Methylsubstituent m
m. sulphate $(CH_3)_2SO_4$ Methylsulfat n, Dimethylsulfat n, Schwefelsäure[di]methylester m

m. sulphoxide $(CH_3)_2SO$ Methylsulfoxid n, Dimethylsulfoxid n, Methylsulfinylmethan n

m. violet Methylviolett n

m. yellow Dimethylgelb n, Buttergelb n, p-Dimethylaminoazobenzol n

methylacetaldehyde CH_3CH_2CHO Methylazetaldehyd m, Propionaldehyd m, Propanal n

methylacetic acid CH_3CH_2COOH Propionsäure f, Propansäure f

methylacetylene $CH_3C{\equiv}CH$ Methyläthin n, Methylazetylen n, Propin n, Allylen n

β-methylacrolein $CH_3CH{=}CHCHO$ 3-Methylakrolein n, Buten-(2)-al-(1) n, Krotonaldehyd m

α-methylacrylic acid $CH_2{=}C(CH_3)COOH$ 2-Methakrylsäure f, 2-Methylpropen-(1)-säure-(3) f

cis-β-methacrylic acid $CH_3CH{=}CHCOOH$ cis-3-Methakrylsäure f, Isokrotonsäure f, β-Krotonsäure f, Allokrotonsäure f, cis-Buten-(2)-säure-(1) f

trans-β-methacrylic acid $CH_3CH{=}CHCOOH$ trans-3-Methakrylsäure f, Krotonsäure f, Buten-(2)-säure-(1) f

methylal $CH_2(OCH_3)_2$ Methylal n, Dimethylformal n, Formaldehyddimethylazetal n, Formal n, Dimethoxymethan n

methylamine CH_3NH_2 Methylamin n, Monomethylamin n, Aminomethan n

p-methylaminophenol $HOC_6H_4NHCH_3$ p-Methylaminophenol n, N-Monomethyl-p-aminophenol n, p-Hydroxymethylanilin n

N-methylaniline $C_6H_5NHCH_3$ N-Methylanilin n, N-Monomethylanilin n, N-Methylaminobenzol n

methylate / to methylieren; mit Methanol vergällen (denaturieren)

methylate CH_3OM^I Methylat n, Methoxid n

methylated spirit vergällter (denaturierter) Alkohol (Spiritus, Branntwein) m, Spiritus denaturatus m

methylation Methylieren n, Methylierung f; Vergällen n, Vergällung f, Denaturieren n, Denaturierung f, Denaturation f (von Alkohol)

methylbenzene $C_6H_5CH_3$ Methylbenzol n, Phenylmethan n, Toluol n

methylbenzoic acid $CH_3C_6H_4COOH$ Methylbenzoesäure f, Methylbenzolkarbonsäure f, Toluylsäure f

methylbenzol s. methylbenzene

methylbenzoylecgonine Methylbenzoylekgonin n, Benzoylmethylekgonin n, Benzoylekgoninmethylester m

α-methylbivinyl $CH_2{=}CHCH{=}CHCH_3$ 1-Methylbutadien n, 1-Methyldivinyl n, 1,3-Pentadien n, Piperylen n

3-methylbutanal $(CH_3)_2CHCH_2CHO$ 3-Methylbutanal n, Methylbutyraldehyd m, Isovaleraldehyd m, Isoamylaldehyd m

2-methylbutane $CH_3CH_2CH(CH_3)_2$ 2-Methylbutan n, Isopentan n, i-Pentan n, Äthyldimethylmethan n

3-methylbutanoic acid $(CH_3)_2CHCH_2COOH$ 3-Methylbutansäure f, Isopropylessigsäure f, Isovaler[ian]säure f, Isobaldriansäure f

2-methyl-1-butanol $CH_3CH_2CH(CH_3)CH_2OH$ 2-Methylbutanol-(1) n, optisch aktiver prim-Amylalkohol m

2-methyl-2-butanol $CH_3CH_2C(CH_3)(OH)CH_3$ 2-Methylbutanol-(2) n, Dimethyläthylkarbinol n, tert-Amylalkohol m

3-methyl-1-butanol $(CH_3)_2CHCH_2CH_2OH$ 3-Methylbutanol-(1) n, Isoamylalkohol m, Isobutylkarbinol n

cis-methylbutenedioic acid $CH_3C(COOH){=}CHCOOH$ Methylbutendisäure f, Zitrakonsäure f, Methylmaleinsäure f

cis-2-methyl-2-butenoic acid $CH_3CH{=}C(CH_3)COOH$ cis-2-Methyl-2-butensäure f, 2-Methylisokrotonsäure f, Angelikasäure f

methyl-tert-butylketone $CH_3COC(CH_3)_2$ Methyl-tert-butylketon n, 2,2-Dimethylbutanon-(3) n, Trimethylazeton n, Pinakolon n

methylcellosolve Methylzellosolve n (Methylglykoläther)

methylchavicol $C_{10}H_{12}O$ Methylchavikol n, 1-Methoxy-4-(2-propenyl)-benzol n

methylchloroform CH_3CCl_3 Methylchloroform n, 1,1,1-Trichloräthan n

2-methyl-4-chlorophenoxyacetic acid $ClC_6H_3(CH_3)OCH_2COOH$ 2-Methyl-4-chlorphenoxyessigsäure f, MCPA n (Herbizid)

methylchlorosilane Methylchlorosilan n

methylcholanthrene Methylcholanthren n

methylconidine Methylkonidin n

methylcoumarin Methylkumarin n

methylcycloheptane $C_7H_{13}CH_3$ Methylzykloheptan n

methylcyclohexane $C_6H_{11}CH_3$ Methylzyklohexan n, Hexahydrotoluol n, Zyklohexylmethan n, Heptanaphthen n

methylcyclopentane $C_5H_9CH_3$ Methylzyklopentan n

N-methylemetine N-Methylemetin n

methylene Methylen n

m. blue Methylenblau n

m.-blue reductase fermentation test Reduktaseprobe f (Schardingersche Reaktion)

m.-blue reduction test s. m.-blue reductase fermentation test

m.-blue test Methylenblau[reduktions]probe f

m. bridge Methylenbrücke f

m. bromide CH_2Br_2 Methylenbromid n, Dibrommethan n

m. chloride CH_2Cl_2 Methylenchlorid n, Dichlormethan n

m. compound Methylenverbindung f

m. dichloride s. m. chloride

m. halide Methylenhalogenid n

m. iodide CH_2J_2 Methylenjodid n, Dijodmethan n

β-methylesculetin β-Methyläskuletin n, 6-Methoxy-7-hydroxykumarin n, Chrysatropasäure f, Gelseminsäure f, Skopoletin n

methylethylcarbinol $CH_3CH_2CHOHCH_3$ Methyläthylkarbinol n, sek-Butylalkohol m, sek-Butanol n, Butanol-(2) n

sym-**methylethylethylene** $CH_3CH_2CH=CHCH_3$ 1-Methyl-2-äthyläth[yl]en n, Penten-(2) n, n-Amylen-(2) n

methylethylketone $CH_3 \cdot CO \cdot C_2H_5$ Methyläthylketon n, Äthylmethylketon n, Butanon-(2) n

methylethylphenanthrene Methyläthylphenanthren n

methylethylsulphide $CH_3SC_2H_5$ Methyläthylsulfid n, Äthylmethylsulfid n, Methylthioäthan n

methylglyoxal CH_3COCHO Methylglyoxal n, 2-Oxopropanal n, Brenztraubensäurealdehyd m, Azetylformaldehyd m

methylisobutylic ketone Methylisobutylketon n

α-**methylisocrotonic acid** $CH_3CH=C(CH_3)COOH$ 2-Methylisokrotonsäure f, Angelikasäure f, cis-2-Methyl-2-butensäure f

methylisopelletierine Methylisopelletierin n

methyllithium CH_3Li Lithiummethyl n

methylmagnesium iodide CH_3MgJ Methylmagnesiumjodid n (eine Grignard-Verbindung)

methylmaleic acid $CH_3C(COOH)=CHCOOH$ Methylmaleinsäure f, Zitrakonsäure f, Methylbutendisäure f

methylnaphthalene $C_{10}H_7CH_3$ Methylnaphthalin n

2-methyl-1,4-naphthoquinone $C_{10}H_5CH_3O$ 2-Methyl-1,4-naphthochinon n, Menadion n, Vitamin K_3 n

methylnitroaniline $CH_3C_6H_3(NO_2)NH_2$ Methylnitroanilin n

methylnonane $C_9H_{19}CH_3$ Methylnonan n

methyloctanoic acid Methyloktansäure f

methylol compound Methylolverbindung f

methylolurea $HOCH_2 \cdot NH \cdot CO \cdot NH_2$ Methylolharnstoff m, Hydroxymethylharnstoff m

methylpelletierine Methylpelletierin n

4-methylpentanoic acid $(CH_3)_2CH(CH_2)_2COOH$ 4-Methylpentansäure f, Isokapronsäure f

2-methyl-3-pentanol $(CH_3)_2CHCHOHCH_2CH_3$ 2-Methylpentanol-(3) n, Äthylisopropylkarbinol n

2-methyl-2-pentene $(CH_3)_2C=CHCH_2CH_3$ 2-Methyl-2-penten n, 2-Äthyl-1,1-dimethyläth[yl]en n

methylpentose Methylpentose f

methylphenol $CH_3C_6H_4OH$ Methylphenol n, Kresol n, Hydroxytoluol n

methylphenyl ether $CH_3OC_6H_5$ Methylphenyläther m, Methoxybenzol n, Anisol n

m. ketone $C_6H_5COCH_3$ Methylphenylketon n, Azetylbenzol n, Azetophenon n

m. silicone Methylphenylsilikon n

m. silicone fluid Methylphenylsilikonflüssigkeit f, Methylphenylsilikonöl n

m. silicone resin Methylphenylsilikonharz n

methylphenylpyrazolone Methylphenylpyrazolon n

2-methylpropanoic acid $(CH_3)_2CHCOOH$ 2-Methylpropansäure f, 2-Methylpropionsäure f, Dimethylessigsäure f, Isobuttersäure f, i-Buttersäure f

2-methyl-1-propanol $(CH_3)_2CHCH_2OH$ 2-Methylpropanol-(1) n, Isobutylalkohol m, Isobutanol n, Isopropylkarbinol n

2-methyl-2-propanol $(CH_3)_2COHCH_3$ 2-Methylpropanol-(2) n, $tert$-Butylalkohol m, $tert$-Butanol n, Trimethylkarbinol n

2-methylpropene $CH_2=C(CH_3)_2$ 2-Methylpropen n

2-methylpropenoic acid $CH_2=C(CH_3)COOH$ 2-Methylpropen-(1)-säure-(3) f, 2-Methakrylsäure f

α-**methylpropionic acid** $s.$ 2-methylpropanoic acid

methylpropyl ketone $CH_3COC_3H_7$ Methylpropylketon n, Pentanon-(2) n

methylpropylacetylene $CH_3C\equiv C(CH_2)_2CH_3$ Methylpropylazetylen n, 2-Hexin n

methylrosaniline Methylrosanilin n

methylstyryl ketone $C_6H_5CH=CHCOCH_3$ Benzalazeton n

methylsuccinic acid $HOOCCH(CH_3)CH_2COOH$ Methylbernsteinsäure f, Methylbutandisäure f, Brenzweinsäure f, Pyroweinsäure f

methylsulphate $(CH_3)_2SO_4$ Methylsulfat n, Dimethylsulfat n, Schwefelsäuredimethylester m

methylsulphinylmethane $s.$ methyl sulphoxide

N-**methyltaurine** $CH_3NHCH_2CH_2SO_3H$ N-Methyltaurin n, 2-(Methylamino)-äthan-1-sulfonsäure f

methyltrichlorosilane CH_3SiCl_3 Methyltrichlorsilan n

methyltrihydroxyanthraquinone Methyltrihydroxyanthrachinon n

methymycin Methymyzin n

methysticin Methystizin n, Kawahin n

metol-hydroquinone (Foto) Metol-Hydrochinon n

m.-hydroquinone developer (Foto) Metol-Hydrochinon-Entwickler m

mevaldic acid $OHCCH_2C(OH)(CH_3)CH_2COOH$ 3-Hydroxy-3-methylglutaraldehydsäure f

mevalonic acid $HOH_2CCH_2C(CH_3)(OH)CH_2COOH$ Mevalonsäure f, 3,5-Dihydroxy-3-methylvaleriansäure f

Mexican hallucinogenic mushroom mexikanischer Rauschpilz m, Psilocybe mexicana Heim

Meyer tangential chamber Tangentialkammer f nach Meyer (bei der Schwefelsäuregewinnung)

mf $s.$ moisture-free

MF, M.F. $s.$ 1. machine finish; 2. machine-finished

M.F.I. $s.$ melt flow index

MG, M.G. $s.$ machine-glazed

M.G. machine Selbstabnahme[papier]maschine f, Yankee-Maschine f

MH $s.$ maleic hydrazide

miargyrite (Min) Miargyrit m (Antimon(III)-silbersulfid)

miarolitic (Geol) miarolitisch, kleindrusig

m. cavity (Geol) miarolitischer Hohlraum (Drusenraum) m

miazine Miazin n, 1,3-Diazin n, Pyrimidin n

MIBK $s.$ methylisobutylic ketone

mica Glimmermineral n, Glimmer m

m. schist Glimmerschiefer m

micaceous glimmerig, Glimmer...

m. hematite s. m. iron ore
m. iron ore Eisenglimmer m
m. sandstone Glimmersandstein m
m. schist Glimmerschiefer m
micalike glimmerartig
Micauba oil (Öl der Palme Acrocomia sclerocarpa Mart.)
micell[a] s. micelle
micellar mizellar, Mizellar...
m. colloid Mizellkolloid n, Assoziationskolloid n
m. hypothesis s. m. theory
m. string Mizellarstrang m
m. structure Mizellarstruktur f
m. theory Mizellartheorie f
micelle (Koll) Mizelle f, Mizell n
Michael condensation (reaction) Michael-Addition f (nukleophile Methylenaddition)
Michler's hydrol Michlersches Hydrol n
Michler's ketone Michlers Keton n (für Farbstoffsynthesen)
micrinite Mikrinit m (Mazeral der Steinkohle)
micro gas analysis Mikrogasanalyse f
m.-glass Dünnglas n
m. length stretch (Text) Mikrodehnung f (vor der Harzbehandlung)
m. Rast s. Rast microprocedure
m. wax Mikrowachs n, Mikroparaffin n, mikrokristallines Wachs (Paraffin, Paraffinwachs) n
microanalysis Mikroanalyse f
microanalytic[al] mikroanalytisch
m. titration Mikrotitration f
microbalance Mikrowaage f
microbes Mikrob[i]en fpl, Klein[st]lebewesen npl, Mikroorganismen mpl
microbial, microbian, microbic mikrobiell
microbicidal mikrobizid, Mikroorganismen abtötend, antimikrobiell
microbicide mikrobizides (Mikroorganismen abtötendes) Mittel n, antimikrobieller Stoff m
microbiologic[al] mikrobiologisch
m. resistance mikrobiologische Resistenz f
microbiology Mikrobiologie f
microbomb Mikrobombe f (Laborgerät)
microburet[te] Mikrobürette f, Feinbürette f, Bankbürette f
microburner Mikrobrenner m
microcalorimeter Mikrokalorimeter n
microcalorimetric mikrokalorimetrisch
microcalorimetry Mikrokalorimetermethode f
microcell (Fotometrie) Mikroküvette f
microcellular mikroporös
m. rubber mikroporöser Gummi m, Moosgummi m, Porengummi m
microchemical mikrochemisch
microchemistry Mikrochemie f
microchromatography Mikrochromatografie f
microcline (Min) Mikroklin m (Kaliumaluminiumsilikat)
microcolorimeter Mikrokolorimeter n
microcomponent Mikrokomponente f, Mikro-

bestandteil m, mikropetrografischer Bestandteil m
microcosmic bead Phosphorsalzperle f
m. salt Na[NH$_4$]HPO$_4$·4H$_2$O Phosphorsalz n, Natriumammoniumhydrogenphosphat-4-Wasser n
microcrystalline mikrokristallin[isch], feinkristallin[isch]
m. wax mikrokristallines Wachs (Paraffin, Paraffinwachs) n, Mikrowachs n, Mikroparaffin n
microdensitometer Mikrodichtemesser m
microdistillation Mikrodestillation f
microelectrode Mikroelektrode f
microelectrophoresis Mikroelektrophorese f
m. on paper Papierelektrophorese f
microelement (Landw) Mikroelement n, Mikronährstoff m, Spurenelement n, Hochleistungselement n
microexamination Mikrountersuchung f, mikroskopische Untersuchung f
microfibril Mikrofibrille f
microfibrillar mikrofibrillär
microgel Mikrogel n
microgram[me] method Mikrogramm-Methode f (Ultramikroanalyse)
microhardness Mikrohärte f
microheterogeneity Mikroheterogenität f
m. theory Kristallittheorie f (Theorie vom kristallitischen Aufbau des Glases)
microhydrogenation Mikrohydrierung f
microlite (Min) Mikrolith m
microlithotype Mikrolithotype f, Streifenart f
micrology Mikrotechnik f, mikroskopische Technik f
micromanipulation Arbeitstechnik f mit dem Mikromanipulator
micromanipulator Mikromanipulator m, Feinmanipulator m (hochempfindliches mikrochemisches Arbeitsgerät)
micromanometer Mikromanometer n
micromethod Mikromethode f
micromolecular mikromolekular
micron Mikron n (mikroskopisch sichtbares Schwebeteilchen)
micronized pigments (Farb) mikronisierte Pigmente npl
Micronizer [fluid-energy mill] s. Micronizer jet mill
Micronizer jet mill Micronizer-Mühle f, Micronizer m (eine Spiralstrahlmühle)
micronutrient (Landw) Mikronährstoff m, Mikroelement n, Spurenelement n, Hochleistungselement n
m. deficiency (Landw) Spurenelementmangel m
microorganisms s. microbes
microparaffin Mikroparaffin n, Mikrowachs n, mikrokristallines Paraffin (Paraffinwachs, Wachs) n
microphysics Mikrophysik f
micropipet[te] Mikropipette f
micropore Mikropore f

microporous mikroporös
micropyrometer Mikropyrometer *n*, optisches Glühfadenpyrometer *n*
microradiometer Mikroradiometer *n*
microscope Mikroskop *n*
  m. slide Mikroskopobjektträger *m*, Objektträger *m*
microscopic mikroskopisch
  m. reversibility mikroskopische Reversibilität *f*
  m. stain Färbemittel *n* für die Mikroskopie
  m. structure Mikrostruktur *f*, Fein[st]struktur *f*, Mikrogefüge *n*, Feingefüge *n*, Kleingefüge *n*
  m. technique mikroskopische Technik *f*, Mikrotechnik *f*
microscopical *s.* microscopic
microscopy Mikroskopie *f*
microsection mikroskopischer Schnitt *m*
microsensor Mikrosensor *m* *(automatischer Gasspurenanzeiger)*
microseparator Magnetscheider *m* für kleine Eisenteilchen
microspore Mikrospore *f*
microstate *(phys Ch)* Mikrozustand *m*
microstructure Mikrostruktur *f*, Fein[st]struktur *f*, Mikrogefüge *n*, Feingefüge *n*, Kleingefüge *n*
microsublimation Mikrosublimation *f*
microtechnic, microtechnique Mikrotechnik *f*, mikroskopische Technik *f*
microtitration Mikrotitration *f*
microtome / to mit dem Mikrotom dünne Schnitte *(für mikroskopische Untersuchungen)* herstellen, Mikrotomschnitte herstellen
microtome Mikrotom *n*
microtomy Herstellung *f* von Mikrotomschnitten
microvitrain Mikrovitrit *m*, Mikrovitrain *m*
microwave Mikrowelle *f*
  m. spectroscope Mikrowellenspektroskop *n*
  m. spectroscopy Mikrowellenspektroskopie *f*
  m. spectrum Mikrowellenspektrum *n*
mid-boiling point mittlerer Siedepunkt *m*
  m.-b.pt. *s.* m.-boiling point
  m. fire Mittelfeuer *n*
  m.-season hop mittelfrüher Hopfen *m*
  m.-wall *s.* midfeather
Midcontinent petroleum Midcontinent-Erdöl *n*, Midcontinent-Öl *n* *(Erdöl der Mittelstaaten der USA)*
middle *(Pap)* Einlage *f* *(beim Triplexkarton)*
  m. chamber (compartment) Mittelraum *m*, Mittelkammer *f* *(z.B. einer Elektrolysezelle)*
  m. component mittlere Komponente *f*
  m. distillate Mitteldestillat *n*, mittleres Destillat *n*
  m. floret Mittelährchen *n* *(z.B. bei der Braugerste)*
  m. fraction Mittelfraktion *f*, mittlere Fraktion *f*
  m. grain Mittelkorn *n*
  m. lamella *(Bio)* Mittellamelle *f*
  m. oil Mittelöl *n*
  m. tones *(Foto)* Mitteltöne *mpl*
middling *s.* middlings product
middlings Grieß *m*, Grießmehl *n*, Mittelgrieß *m*; Nachmehl *n*

m. product Mittelgut *n*, Mittelprodukt *n*, Zwischenfraktion *f*
m. purifier Grießputzmaschine *f*
midfeather 1. *(Glas)* Brennerzunge *f*; 2. *(Pap)* Mittelwand *f*, Zwischenwand *f*, Scheidewand *f* *(des Holländers)*
midriff *s.* midfeather 2.
miemite *(Min)* Miemit *m* *(Kalziummagnesiumkarbonat)*
migma *(Geol)* Migma *n*
migmatite Migmatit *m*, Mischgestein *n*
migmatization *(Geol)* Migmatisierung *f*, Migmabildung *f*
migrate / to wandern, migrieren *(z.B. Ionen, Erdöl)*
  m. in [hin]zuwandern, [hin]einwandern
  m. out fortwandern, [hin]auswandern, wegwandern, abwandern
migration Migration *f*, Wandern *n*, Wanderung *f* *(z.B. von Ionen, Erdöl)*
  m. area *(Kch)* Wanderfläche *f*, Migrationsfläche *f*
  m. chamber Elektrophoresekammer *f*
  m. current Wanderungsstrom *m*, Migrationsstrom *m*
  m. length *(Kch)* Wanderlänge *f*, Migrationslänge *f*
  m. of a halogen atom Halogenwanderung *f*
  m. of an alkyl group Alkylwanderung *f*
  m. of ions Ionenwanderung *f*
  m. of plasticizer Weichmacherwanderung *f*
  m. speed (velocity) Wanderungsgeschwindigkeit *f (z.B. der Ionen)*
migratory staining *(Gum)* Ausbluten *n*
Mikro-Atomizer Mikro-Atomizer *m* *(eine sieblose Feinstaubmühle zur Herstellung von ultrafeinem Pulver)*
Mikro-Pulverizer [hammer mill] Mikro-Pulverizer *m* *(eine Hammermühle zum Feinmahlen)*
mild mild, schonend, gelinde, leicht
  m. ale leichtes (schwach gehopftes) Bier *n*
  m. ale malt dunkles [Münchner] Malz *n*
  m. beer *s.* m. ale
  m. mercury chloride Hg_2Cl_2 Quecksilber(I)-chlorid *n*, Kalomel *n*
  m. oxidation milde Oxydation *f*
  m. steel Weichstahl *m*, weicher Stahl *m*
mildew Schimmel *m* *(von Schimmelpilzen gebildeter Überzug)*; Mehltau *m* *(eine Pflanzenkrankheit)*
  m.-proof *s.* m.-resistant
  m. resistance Schimmelbeständigkeit *f*, Widerstandsfähigkeit *f* gegen Schimmel
  m.-resistant schimmelbeständig, schimmelfest
mildewproofing Schimmelfestappretur *f*, Schimmelfestausrüstung *f*
milfoil oil Schafgarbenöl *n*
milk Milch *f*, *(i.e.S.)* Kuhmilch *f*; Latex *m*, *(i.w.S.)* Milchsaft *m*; Kokosmilch *f*, *(i.w.S.)* Pflanzenmilch *f*
  m. acid $CH_3CH(OH)COOH$ Milchsäure *f*, Lak-

tinsäure f, 2-Hydroxypropansäure f, 2-Hydroxy-
propionsäure f
m. albumin Laktalbumin n, Milchalbumin n
m. bottle capping (filling, sealing) machine
Milchflaschenabfüllautomat m, Milchflaschen-
füll- und -verschließanlage f
m. casein Kasein n
m. catalase Milchkatalase f
m. centrifuge s. m. separator
m. chocolate Milchschokolade f
m. condensery (condensing plant) Kondens-
milchfabrik f
m. constituents Milchbestandteile mpl
m. cooler Milchkühler m
m. defect Milchfehler m
m. enzymes Milchenzyme npl, Milchfermente npl
m. evaporating plant s. m. condensery
m. fat Milchfett n
m. fat synthesis Milchfettsynthese f
m. for cheese[making] Käsereimilch f
m. glass Milchglas n
m. globulin Milchglobulin n, Laktoglobulin n
m. house Milchhaus n
m. lipase Milchlipase f
m. margarine Milchmargarine f
m. mould Milchschimmel m
m. of almonds Mandelmilch f
m. of lime Kalkmilch f
m.-of-lime system (Pap) Kalkmilchsystem n,
Kalkmilchverfahren n
m. of magnesia Magnesiamilch f
m. of sulphur Schwefelmilch f, gefällter Schwefel
m
m. opal (Min) Milchopal m, Hydrophan m
(Silizium(IV)-oxid)
m. pasteurizer Milcherhitzer m, Milcherhitzungsapparat m
m. phase Milchphase f
m. powder Milchpulver n, Trockenmilchpulver n,
Trockenmilch f
m. preserves Milchdauerwaren fpl, Dauermilch-
waren fpl
m. processing Milchverarbeitung f
m. processor Milchverarbeiter m
m. producer Milchproduzent m, Milcherzeuger m
m. product Milchprodukt n, Milcherzeugnis n
m. production Milchproduktion f, Milcherzeu-
gung f
m. protein Milcheiweiß n, Milchprotein n
m. ripener Rahmreifer m, Säuerungsgefäß n,
Säuerungswanne f
m. ripening Milchreifung f (Zeit, in der man Milch
über ihre Dicklegung hinaus stehen läßt)
m. salt Milchsalz n
m. secretion Milchsekretion f, Milchabsonde-
rung f
m. separator Milchseparator m, Milchschleuder
f, Entrahmungszentrifuge f
m. serum Milchserum n, Molken m, Molke f,
Käs[e]wasser n

m. serum protein Milchserumprotein n, Molkenprotein n
m. solids Milchtrockenmasse f
m. soured with starter saure (dickgelegte) Milch
f, mit Säureweckern gesäuerte (versetzte) Milch
f, Sauermilch f, Dickmilch f
m. stone Milchstein m
m. stone remover Milchsteinentferner m
m. storage vessel Milchvorratstank m, Milch-
annahmebehälter m
m. sugar Milchzucker m, Lakto[bio]se f
m. treatment Milchbehandlung f
m. vitaminizing Milchvitamin[is]ierung f
m. yield Milchertrag m, Milchleistung f
milkiness milchige Trübung f
milking grease Melkfett n
m. machine Melkmaschine f
milkstone s. milk stone
milky milchig, milchartig; milchhaltig; milch-
gebend
m. ice Trübeis n
m. juice s. m. sap
m. quartz Milchquarz m
m. sap Milchsaft m, (i.e.S.) Latex m
mill / to [grob]mahlen, vermahlen, zermahlen,
[grob]zerkleinern, brechen, (im Kollergang) zerdrücken, zerreiben, zerquetschen; feinmahlen;
[aus]walzen; auf dem Walzwerk mischen; (Scho-
koladenmasse) konchieren, auf der Konche
(schlagend, reibend, peitschend) vermengen;
(Tuch, Leder) walken; (Körner) enthülsen, schälen
m. to death (Gum) totwalzen, übermastizieren
mill Mühle f (Betrieb); Mühle f (Maschine zum
Zerkleinern fester Stoffe); (Text) Walkmaschine
f, Walke f; (Gum) Walzwerk n; Aufbereitungs-
anlage f (z.B. für Erze)
m. addition (Ker) Mühlenzusatz m
m. board Maschinenpappe f
m. breakdown (Gum) mechanischer Abbau m,
mechanische Plastizierung f (auf Walzwerken)
m. broke Ausschußpapier n, Ausschuß m,
Papierausschuß m, Kollerstoff m
m. cake Preßkuchen m, Ölkuchen m
m. drying Mahltrocknen n, Mahltrocknung f
m.-lined board in der Rolle geklebte (kaschierte)
Pappe f
m.-mix / to (Gum) auf dem Walzwerk mischen
m. mixer Kugelmühle f
m. mixing Mahlen n in der Kugelmühle, Kugeln
n; (Gum) Mischen n auf dem Walzwerk
m.-opening (Gum) Walzenspalt m
m. operator (Gum) Walzwerksarbeiter m
m. pan (Gum) Walzenschiff n (Auffangblech an
Mischwalzwerken)
m. roll (Pap) von der Maschine kommende
Papierrolle f, Maschinenrolle f
m. runner Läufer m, Laufstein m, Laufrolle f
m. scale Walzzunder m, Walz[en]sinter m
m. shell Mahltrommel f

m.-strainer (Gum) Mill-Strainer m
m. wrapper (wrapping) Blaupackpapier n, Packpapier (Einschlagpapier) n für Rollenpapier (Papierrollen); Rieseinschlagpapier n
millable urethane elastomer walzbares Polyurethanelastomer[es] n
mille (Pap) Neuries n (= 1000 Bogen)
milled fibre gemahlene Faser f, Kurzfaser f
  m. peat Frästorf m
  m.-peat process Fräs[torf]verfahren n, Torffräsverfahren n
  m. rice geschälter (entspelzter) Reis m
Miller [crystal] indices (Krist) Miller-Indizes pl, Millersche Indizes pl
millerite (Min) Millerit m, Haarkies m (Nikkel(II)-sulfid)
Miller's tannin (Gerb) (Extrakt von Hemlocktannenrinde)
millibarn Millibarn n (Einheit für den Wirkungsquerschnitt des Atomkerns, 10^{-27} cm^2)
milligram method Milligramm-Methode f (Mikroanalyse)
milling Mahlen n, Vermahlen n, Mahlung f, Vermahlung f; (Gum) Walzen n; (Text, Gerb) Walken n; (Lebm) Müllerei f, (i.e.S.) Mahlmüllerei f (oder) Schälmüllerei f
  m. machine Walkmaschine f, Walke f
  m. product Mahlprodukt n
  m. technique Mahlverfahren n
  m. time Mahlzeit f, Mahldauer f
millipoise Millipoise n, mP (Einheit der Viskosität)
millipore filter Ultrafilter n
Millon's base Millonsche Base f
Millon's reaction Millonsche Reaktion f (auf Eiweißstoffe)
Millon's reagent Millons Reagens n (zum Nachweis von Eiweißstoffen)
Mills-Packard chamber Mills-Packard-Kammer f (Schwefelsäureherstellung)
millstone Mühlstein m, Mahlstein m
mimeograph ink Mimeograf[en]farbe f
mimet[es]ite (Min) Mimetesit m
mimosa bark Mimosarinde f (Gerbrinden mehrerer Acacia-Arten)
Mindanao cinnamon (Zimtsorte von Cinnamomum mindanaense Elm.)
mine Grube f, Bergwerk n, Zeche f, Mine f
  m. gas Grubengas n
  m.-run salt Rohsalz n, (Düngemittelindustrie i.e.S.) Kalirohsalz n
  m. water Grubenwasser n, Schachtwasser n
mineral mineralisch, Mineral...; anorganisch
mineral Mineral n; Erz n
  m. acid Mineralsäure f, anorganische Säure f
  m. aggregate Mineralaggregat n
  m. association Mineralassoziation f, Mineralkombination f, Mineralvergesellschaftung f
  m. butter $SbCl_3$ Antimonbutter f, Antimontrichlorid n
  m. caoutchouc Elaterit m, Erdpech n

  m. charcoal mineralische (fossile) Holzkohle f
  m. coal Mineralkohle f, Naturkohle f, mineralische (fossile, natürliche) Kohle f
  m. colza [oil] mineralisches Colzaöl (Rüböl) n
  m. content Mineralstoffgehalt m
  m. cotton s. m. wool
  m. deposit Erzlagerstätte f
  m. dressing Mineralaufbereitung f, bergbauliche Aufbereitung f
  m. facies Mineralfazies f
  m. fat s. m. jelly
  m. fertilizer Mineraldünger m
  m. fibre mineralische Faser f, Mineralfaser f; Mineralfaserstoff m
  m. filler mineralischer Füllstoff m
  m. jelly Petrolatum n
  m. matter mineralischer Stoff m, mineralische Substanz f, Mineralstoff m, Mineralsubstanz f, mineralisches Material n
  m.-matter-free mineral[stoff]frei
  m. metabolism Mineralstoffwechsel m
  m. nutrition Mineralsalzernährung f, mineralische Ernährung f
  m. oil Mineralöl n, mineralisches Öl n
  m.-oil crude Rohöl n
  m. phosphate Mineralphosphat n
  m. pigment Mineralpigment n, künstliches anorganisches Pigment n
  m. pitch Asphalt m, Erdpech n, Erdharz n
  m. province Mineralprovinz f
  m. rubber Mineralrubber m, geblasenes Bitumen n
  m. salt Mineralsalz n, (i.e.S.) Steinsalz n
  m. seal [oil] mineralisches Robbentran n
  m. sperm [oil] mineralisches Sperm[azeti]öl n, mineralisches Walratöl n
  m. spirit[s] Lösungsbenzin n
  m. spring Mineralquelle f, (i.e.S.) Heilquelle f, Brunnen m
  m. substance s. m. matter
  m. sulphur anorganisch gebundener Schwefel m
  m. tallow (Min) Hatchettin m
  m. tanning Mineralgerbung f
  m. water Mineralwasser n
  m. weathering Mineralverwitterung f
  m. wool Mineralwolle f
  m. yeast Mineralhefe f, Trockenfutterhefe f, Futterhefe f
  m. zoning zonale (zonare) Erzverteilung f (Verteilung f von Erzlagerstätten)
mineralization (Bodenchemie) Mineralisation f, Mineralisierung f
mineralize / to (Bodenchemie) mineralisieren
mineralized lode Mineralgang m
mineralizer Mineralisator m
mineralocorticoid Mineralkortikoid n (Nebennierenrindenhormon)
mineralogy Mineralogie f, Mineralkunde f
minerals beneficiation s. mineral dressing
mingle / to [ver]mischen, zusammenmischen, [ver]mengen

mini-shaker *(Labortechnik)* Kleinschüttler *m*, Kleinschüttelgerät *n*
miniature recorder Kleinschreiber *m*
minicare pflegeleicht
minimize / **to** auf ein Mindestmaß verringern (reduzieren), *(z.B. Verluste)* klein halten
minimum minimal, Mindest..., Kleinst...
minimum Minimum *n*
  m. boiling point Minimum-Siedepunkt *m*, Siedepunktsminimum *n*
  m. dosage (dose) Mindestdosis *f*
  m. effective rate kleinste wirksame Dosis *f*
  m. fat content Mindestfettgehalt *m*
  m. reflux ratio Mindestrücklaufverhältnis *n*
  m. temperature Minimaltemperatur *f*
  m. tray number Mindestbodenzahl *f*, Mindesttrennstufenzahl *f*
  m. vapour pressure Siedeminimum *n*
mining explosive (powder) Bergbausprengstoff *m*
minium Pb_3O_4 Mennige *f*, Bleimennige *f*, Minium *n*, rotes Bleioxid *n*, Blei(II,IV)-oxid *n*
minor element *s.* **m. nutrient element**
  m. mineral Nebenmineral *n*
  m. nutrient element *(Landw)* Spurenelement *n*, Mikronährstoff *m*, Mikroelement *n*, Hochleistungselement *n*
minority carrier Minoritäts[ladungs]träger *m*, Minderheitsträger *m*
Minton oven *(Ker)* Minton-Ofen *m* *(mit überschlagender Flamme arbeitender Brennofen)*
minus material *(Klassierung)* Feinkorn *n*, Unterkorn *n*, *(Siebklassierung auch)* Siebfeines *n*, *(Windsichten auch)* Feingut *n*, Sichtfeines *n*
minutely crystalline kleinkristallin
Mioga ginger Japanischer Ingwer *m*, Zingiber mioga (Thunb.)Rosc.
miotic miotisch, pupillenverenge[r]nd
miotic Miotikum *n*, pupillenverenge[r]ndes Medikament (Mittel) *n*
MIR *s.* maximum initial retention
mirabilite *(Min)* Mirabilit *m*, Glaubersalz *n* *(Natriumsulfat-10-Wasser)*
mirbane oil Mirbanöl *n*, unechtes (künstliches) Bittermandelöl *n*, Nitrobenzol *n*
mirror amalgam Spiegelamalgam *n* *(Spiegelbelag aus 77 % Hg und 23 % Sn)*
  m. galvanometer Spiegelgalvanometer *n*
  m. image Spiegelbild *n*
  m.-image isomer Spiegelbildisomeres *n*, Enantiomeres *n*, optischer Antipode *m*
  m.-image isomerism (relationship) Spiegelbildisomerie *f*, Enantiomorphie *f*, optische Isomerie *f*
  m. plane Spiegelebene *f*
  m. reflection spiegelnde (gerichtete) Reflexion *f*, Spiegelreflexion *f*, Spieg[e]lung *f*
  m. varnishing Spiegellackierung *f*
miscella Miszella *f* *(beladenes Extraktionsmittel)*
  m. filter Miszellaschleuder *f*
miscibility Mischbarkeit *f*

  m. gap Mischungslücke *f*
miscible mischbar
  m. in all proportions beliebig (in jedem Verhältnis) mischbar
  m. oil mischbares Öl *n* *(durch Emulgatorzusatz mit Wasser mischbares, stabile Emulsion bildendes Öl)*
misnomer Fehlbenennung *f*, falscher Name *m*
mispickel *(Min)* Mißpickel *m*, Arsenopyrit *m*, Arsenkies *m*, Giftkies *m* *(Eisenarsensulfid)*
missing bond Lückenbindung *f*
mist leichter Nebel *m*, feuchter Dunst *m*
mitis green $Cu(CH_3COO)_2 \cdot 3Cu(AsO_2)_2$ Mitisgrün *n*, Schweinfurter Grün *n*, Mitisgrün *n*, Schweinfurter Grün *n*, Mitisarsenitazetat *n*
mitomycin Mitomyzin *n* *(Antikrebsmittel)*
mitotic index *(Bio)* Mitoseindex *m*
  m. poison Mitosegift *n*
mitraphylline Mitraphyllin *n*
Mitscherlich cook process *(Pap)* indirekte Kochung *f* nach Mitscherlich
Mitscherlich's law of isomorphism *(Krist)* Mitscherlichs Isomorphieregel *f*
Mittler's green $Cr_2O_3 \cdot xH_2O$ Mittlers Grün *n*, Chromoxidhydratgrün *n*, Guignetgrün *n*, Chromoxidgrün feurig *n*, Viridian *n*
mix / **to** [ver]mischen, zusammenmischen, [ver]mengen, verrühren, anrühren; *(Filtrate)* vereinigen
mix Mischen *n*, Vermischen *n*, Mengen *n*, Vermengen *n*; Mischung *f*, Gemisch *n*; Gemenge *n*
  m.-crystal Mischkristall *m*
  m. formula Mischungsrezept *n*
mixability Mischbarkeit *f*
mixable mischbar
mixed acid Mischsäure *f*, *(meist)* Salpeterschwefelsäure *f*, Nitriersäure *f*
  m. adhesive Mischklebstoff *m*, Zweikomponentenkleber *m*, Reaktionskleber *m*
  m. aniline point Mischanilinpunkt *m*
  m. base gemischte Basis *f* *(eines Roherdöls)*
  m.-base crude [oil] *s.* **m.-base petroleum**
  m.-base petroleum gemischtbasisches Erdöl (Rohöl) *n*, Mischöl *n*, Erdöl *n* auf gemischter Basis
  m. bed Mischbett *n*
  m.-bed [ion] exchanger Mischbettaustauscher *m*, Mischbettfilter *m*
  m. board Mischpappe *f*
  m. catalyst Mischkatalysator *m*, Mehrstoffkatalysator *m*
  m. colour Mischfarbe *f*
  m. complex Mischkomplex *m*
  m. conductor Mischleiter *m*
  m. crystal Mischkristall *m*
  m.-crystal formation Mischkristallbildung *f*
  m. culture Mischkultur *f*
  m. dye gemischter Farbstoff *m*
  m. element Mischelement *n*
  m. ether R' · O · R'' gemischter Äther *m*

m.-feed operation Mischstromführung f (z.B. von Mehrkörperverdampfern)
m. fertilizer Mischdünger m
m.-fertilizer plant Mehrnährstoffdüngerfabrik f, Mischdüngerwerk n
m.-flow pump halbaxiale Pumpe f, Pumpe f in diagonaler Bauart
m. gas Mischgas n
m. glyceride gemischtes (gemischtsäuriges) Glyzerid n
m. indicator Mischindikator m, Indikatorgemisch n
m. ketone R' · CO · R'' gemischtes Keton n
m. lead alkyls gemischte Bleialkyle npl (Antiklopfmittel)
m. melting point Mischschmelzpunkt m, Schmelzpunkt m eines Gemisches
m. milk Mischmilch f
m. molecule Mischmolekül n
m. oil Mischöl n
m. oxide Mischoxid n
m. phase Mischphase f, Gemischtphase f, gemischte Phase f
m.-phase cracking Gemischtphasekracken n, Kracken n in gemischt flüssiger und dampfförmiger Phase
m.-phase process Gemischtphaseverfahren n, Gemischtphaseprozeß m
m. polymer Mischpolymer[es] n, Mischpolymerisat n, Kopolymer[es] n, Kopolymerisat n
m. salt Mischsalz n
m.-salt catalyst Mischsalzkatalysator m, Mischsalzkontakt m
m. solvent gemischtes Lösungsmittel n, Lösungsmittelgemisch n
m. strawboard Strohmischpappe f
m. vegetables Mischgemüse n
mixer Mischer m, Mischapparat m, Mischmaschine f; Rührer m, Rührwerk n, Rührmaschine f; Kneter m, Knetwerk n, Knetmaschine f; Mischer m, Mixer m, Mischgefäß n, Misch[er]behälter m
m. bowl Misch[er]behälter m, Rühr[werk]behälter m, Rühr[werk]kessel m
m.-settler Mixer-Settler-Extraktor m, Mixer-Settler-Apparat m, Misch-Trenn-Behälter m, Mischer-Abscheider m
m.-settler unit Mixer-Settler-Anlage f (Extraktionsanlage aus Mischern und Absetzbehältern)
mixible s. mixable
mixing Mischen n, Vermischen n, Mengen n, Vermengen n
m. arm Mischerschaufel f, Mischflügel m, Rühr[er]schaufel f, Rühr[er]flügel m, Rühr[er]stab m; Knetschaufel f, Knetarm m
m. bin Mischbunker m
m. blade s. m. arm
m. blunger Mischquirl m
m. box s. m. chest
m. chamber Mischkammer f; Knettrog m (eines Innenmischers)

m. chest Mischbütte f, Mischbottich m, Mischbehälter m
m. condenser Mischkondensator m
m. device Mischvorrichtung f
m. element Rührorgan n, Knetorgan n, Schaufelelement n
m. equipment (Gum) Maschinen fpl für die Mischungsherstellung
m. formula Mischungsrezept n
m. impeller Rührerlaufrad n
m. instruction Mischvorschrift f
m. machine Mischmaschine f; Rührmaschine f, Knetmaschine f
m. mill Mischwalzwerk n, Walzenmischer m
m. nozzle Mischdüse f
m. plant Mischanlage f
m. potcher (Pap) Mischholländer m
m. procedure Mischweise f
m. process Mischvorgang m, Mischprozeß m
m. ratio Mischungsverhältnis n
m. rolls s. m. mill
m. screw Mischschnecke f; Knetschnecke f
m. tank Mischtank m, Mischbehälter m, Mischgefäß n, Mischer m
m. time Mischzeit f, Mischdauer f; Rührzeit f, Rührdauer f; Knetzeit f, Knetdauer f
m. vat (vessel) Mischgefäß n, Mischkessel m, Misch[er]behälter m; Rühr[werk]kessel m, Rühr[werk]behälter m
m. water Anmach[e]wasser n (Betonherstellung)
mixite (Min) Mixit m (Wismut(III)-kupfer(II)-oktahydroxidpentaorthoarsenat)
mixture Mischen n, Vermischen n, Mengen n, Vermengen n; Mischung f, Gemisch n; Gemenge n; (Pharm) Mixtur f
m. for freezing Kältemischung f
m. for spraying Spritzmittel n, Spritzflüssigkeit f, Spritzbrühe f
m. of cis and trans isomers cis-trans-Gemisch n
m. of isomers Isomerengemisch n
m. of isotopes Isotopengemisch n
m. of solids Feststoffgemenge n
m. ratio Mischungsverhältnis n
m. yarn Melangegarn n (aus verschiedenfarbigen Fasern); Mischgarn n (aus verschiedenen Einfachgarnen)
mizzonite (Min) Mizzonit m
ML s. micro length stretch
MLA s. mixed lead alkyls
mmf s. mineral-matter-free
mn. s. monoclinic
MO s. molecular orbital
mobile mobil, beweglich, verschiebbar, versetzbar, verstellbar
m. equilibrium bewegliches Gleichgewicht n
m. hydrogen atom bewegliches Wasserstoffatom n
m. phase mobile (bewegliche) Phase f (Chromatografie)

m. solvent Laufmittel *n*, Fließmittel *n* *(Chromatografie)*
mobility Beweglichkeit *f*
m. of ions Ionenbeweglichkeit *f*
mocha stone *(Min)* Mokkastein *m*, Mochhastein *m*, Moosachat *m*
modacrylic fibre Modakrylfaser *f*; Modakrylfaserstoff *m*
modacrylonitrile fibre Modakrylnitrilfaser *f*
mode of action Wirkungsweise *f*
m. of administration Applikationsart *f*, Art *f* der Verabreichung
model Modell *n*
m. compound Modellverbindung *f*
m. oven Versuchskammer *f* *(zur Treibdruckbestimmung)*
moderate / **to** *(Kch)* moderieren, [ab]bremsen, verlangsamen
moderate accelerator *(Gum)* mittelstarker (mittelschneller) Beschleuniger *m*
m. heat mäßige Wärme *f*
m. speed paper machine Papiermaschine *f* [mit] mittlerer Geschwindigkeit
m. strike *(Farb)* mäßiges Aufziehen *n*
moderately coarse mittelgrob
m. fast *(Farb)* mäßig (mittelmäßig) echt
m. fine mittelfein
m. weak acid mittelstarke Säure *f*
moderation *(Kph)* Bremsen *n*, Bremsung *f*, Abbremsen *n*, Abbremsung *f* *(z.B. von Neutronen)*
moderator *(Kph)* Moderator *m*, Moderatorsubstanz *f*, Bremssubstanz *f*, Bremsstoff *m*
modification Modifizierung *f*, Abänderung *f*, Abwandlung *f* *(z.B. eines Verfahrens)*; Modifikation *f* *(Zustandsform eines Elements oder einer Verbindung)*
modified modifiziert, abgeändert, abgewandelt
m. acrylic fibre Modakrylfaser *f*; Modakrylfaserstoff *m*
m. broth dilution method modifizierte Verdünnungsmethode (Bouillonverdünnungsmethode) *f*
modifier Modifikationsmittel *n*; *(Flot)* Regler *m*, regelndes Reagens (Mittel, Schwimmittel) *n*; Modifikator *m*, Modifier *m* *(Zusatz zur Viskose oder zum Viskosespinnbad)*
modify / **to** modifizieren, abändern, abwandeln *(z.B. ein Verfahren)*
modifying agent *s.* modifier
modular principle Baukastenprinzip *n*, Baukastensystem *n*
modulus Modul *m*; *(Gum)* Spannungswert *m*
m. of elasticity Elastizitätsmodul *m*, E-Modul *m*
m. of rupture Biegefestigkeit *f*, Bruchmodul *m*
moellon Degras *m(n)*, Moellon *n* *(Lederfettungsmittel)*
Mogador gum *(Pflanzengummi von Acacia gummifera Willd.)*
Mohler solution for tartaric acid Mohlers Reagens *n* *(zum Nachweis von Weinsäure)*

mohnseed Mohnsamen *m*
Mohr clip *s.* Mohr pinchcock clamp
Mohr litre Mohrsches Liter *n* *(für gasvolumetrische Bestimmungen)*
Mohr measuring pipet[te] Meßpipette *f* nach Mohr
Mohr method Endpunkt[s]bestimmung *f* nach Mohr
Mohr pinchcock clamp Quetschhahn *m* (Schlauchklemme *f*) nach Mohr
Mohr's salt $(NH_4)_2Fe(SO_4)_2 \cdot 6H_2O$ Mohrsches Salz (Doppelsalz) *n*, Ammoniumeisen(II)-sulfat-6-Wasser *n*
Mohs' [hardness] scale, Mohs' scale of hardness Härteskala *f* nach Mohs, Mohssche Härteskala *f*, Mohs-Skala *f*
moiety Teil *m*, Anteil *m*, Komponente *f*
moil *(Glas)* Absprengkappe *f*, Kappe *f*
molre effect Moiré-Effekt *m*
moist feucht; naß
m. chamber culture dish feuchte Kammer *f* *(Laborgerät)*
m. pellet Grünpellet *n*
moisten / **to** 1. anfeuchten, befeuchten; benetzen; annässen, anteigen; 2. feucht werden
moistening Anfeuchten *n*, Anfeuchtung *f*, Befeuchten *n*; Benetzen *n*; Annässen *n*, Anteigen *n*
moisture Feuchte *f*, Feuchtigkeit *f*; Nässe *f*; Feuchtigkeitsgehalt *m*, Feuchtegehalt *m*, Feuchteanteil *m*
m. absorption Feuchteaufnahme *f*
m.-and-ash-free wasser- und aschefrei, waf
m.-carrying capacity Feuchteaufnahmevermögen *n*, Feuchtigkeitsaufnahmevermögen *n*
m. content Feuchtigkeitgehalt *m*, Feuchtegehalt *m*, Feuchteanteil *m*; Feuchtebeladung *f*
m. content dry weight basis Feuchtesatz *m*, absoluter Feuchtegehalt *m*
m. content wet weight basis Feuchtigkeitsgehalt *m*, Feuchtegehalt *m*, Feuchteanteil *m*
m. expansion Feuchtigkeitsausdehnung *f*
m.-free wasserfrei, wf, trocken
m.-free weight *(Pap)* Trockenmasse *f*, Darrmasse *f*, *(bisher)* Darrgewicht *n*
m. gradient Feuchtegefälle *n*, Feuchtigkeitsgefälle *n*
m.-laden air mit Feuchtigkeit beladene Luft *f*, Feuchtluft *f*
m. loss Feuchtigkeitsverlust *m*
m. meter Feuchtigkeitsmesser *m*, Feuchtemesser *m*
m.-repellent feuchtigkeitsabweisend
m. resistance Feuchtigkeitsbeständigkeit *f*, Feuchtebeständigkeit *f*, Feuchtigkeitsfestigkeit *f*, Feuchtefestigkeit *f*
m. resistant feuchtigkeitsbeständig, feuchtigkeitsfest
m.-set ink Moisture-set Ink *f* *(rasch trocknende Druckfarbe)*
m.-vapour resistance Wasserdampfundurchlässigkeit *f*

m.-vapour resistant wasserdampfundurchlässig, wasserdampfdicht
moistureless *s.* moisture-free
moistureproof feuchtigkeitsdicht, *(i.e.S.)* wasserdampfundurchlässig, wasserdampfdicht
moisturizer Feuchthaltemittel *n*
Mojonnier fat test *(Lebm)* Mojonnier-Test *m (zur Bestimmung von Fett- und Wassergehalt)*
Mojonnier method *s.* 1. Mojonnier fat test; 2. Mojonnier solids test
Mojonnier solids test *(Lebm)* Mojonnier-Test *m (zur Trockensubstanzbestimmung)*
Mojonnier test *s.* 1. Mojonnier fat test; 2. Mojonnier solids test
Moka aloe *s.* Mecca aloe
mol *s.* mole
molal molal, gewichtsmolar
  m. boiling-point[-elevation] constant molale (molare, molekulare) Siedepunktserhöhung *f*, ebullioskopische Konstante *f*
  m. concentration *s.* molality
  m. depression constant, m. depression of freezing point *s.* m. freezing-point[-depression] constant
  m. elevation constant, m. elevation of boiling point *s.* m. boiling-point[-elevation] constant
  m. freezing-point[-depression] constant molale (molare, molekulare) Gefrierpunktserniedrigung *f*, kryoskopische Konstante *f*
  m. solution molale (gewichtsmolare) Lösung *f*
molality Molalität *f*, Kilogramm-Molarität *f*, kg-Molarität *f*, Gewichtsmolarität *f*, molale Konzentration *f*
molar [volumen]molar, Mol...
  m. absorbancy index *s.* m. extinction coefficient
  m. absorptivity *s.* m. extinction coefficient
  m. concentration *s.* molarity
  m. conductance molare Leitfähigkeit *f*, molares Leitvermögen *n*
  m. depression constant, m. depression of freezing point *s.* molal freezing-point[-depression] constant
  m. dispersivity Mol[ekul]ardispersion *f*
  m. elevation constant, m. elevation of boiling point *s.* molal boiling-point[-elevation] constant
  m. entropy molare Entropie *f*
  m. extinction coefficient molarer Extinktionskoeffizient *m*
  m. fraction Molenbruch *m*
  m. free energy molare freie Energie *f*
  m. heat [capacity] Mol[ekular]wärme *f*
  m. latent heat of vaporization molare Verdampfungsenthalpie (Verdampfungswärme) *f*
  m. magnetic rotation molares magnetisches Dreh[ungs]vermögen *n*
  m. parachor Parachor *m*
  m. polarizability Mol[ekul]polarisierbarkeit *f*
  m. polarization Mol[ekular]polarisation *f*
  m. ratio Molverhältnis *n*
  m. refraction (refractivity) Mol[ekular]refraktion *f*

m. rotation molare Drehung *f*
m. rotatory power molares Dreh[ungs]vermögen *n*
m. solution [volumen]molare Lösung *f*
m. susceptibility molare Suszeptibilität *f*, Molsuszeptibilität *f*
m. translational energy molare Translationsenergie *f (Translationsanteil der molaren inneren Energie)*
m. volume Molvolumen *n*
molarity Molarität *f*, Liter-Molarität *f*, molare Konzentration (Volumkonzentration) *f*, Mol[ar]konzentration *f*, Volumenmolarität *f*
molasses Melasse *f*; Sirup *m*
  m. fodder Melassefuttermittel *n*
  m. formation Melassebildung *f*
  m.-forming substance Melassebildner *m*
  m. sugar Melassezucker *m*
  m. wort Melassewürze *f*
mold / to *s.* to mould
mold *s.* mould
moldavite *(Min)* Moldavit *m*
mole Mol *n*, Grammol[ekül] *n*
  m. fraction Molenbruch *m*
  m. per cent unsaturation Molprozent (Mol-%) *n* Ungesättigtheit
  m. quantity Molmenge *f*
  m. ratio Molverhältnis *n*
molecular molekular, Molekular..., Molekül...
  m. activation Molekülaktivierung *f*
  m. association Molekülassoziation *f*, Assoziation *f*
  m. asymmetry Molekularasymmetrie *f*, Molekülasymmetrie *f*
  m. attraction Molekularattraktion *f*, Molekelanziehung *f*, intermolekulare (zwischenmolekulare) Anziehung *f*
  m. beam Molekularstrahl *m*, Molekülstrahl *m*
  m.-beam apparatus Molekularstrahlapparatur *f*, Molekularstrahlapparat *m*
  m.-beam experiment Molekularstrahlversuch *m*, Versuch *m* mit Molekularstrahlen
  m.-beam measurement Molekularstrahlmessung *f*
  m.-beam method Molekularstrahlmethode *f*, Methode *f* der Molekularstrahlen
  m.-beam spectroscopy Molekularstrahlspektroskopie *f*
  m.-beam technique *s.* m.-beam method
  m. biology Molekularbiologie *f*
  m. chain Molekülkette *f*
  m. cleavage Molekülspaltung *f*
  m. colloid Molekülkolloid *n*
  m. complex Molekülkomplex *m*
  m. compound Molekularverbindung *f*, Molekülverbindung *f*
  m.-compound formula *s.* m. formula
  m. conductivity molekulare Leitfähigkeit *f*, molekulares Leitvermögen *n*
  m. crystal Molekülkristall *m*

m. depression constant, m. depression of freezing point s. molal freezing-point[-depression] constant
m. diameter Moleküldurchmesser m
m. dissociation Moleküldissoziation f, Dissoziation f
m. dissymetry Molekularasymmetrie f, Molekül-asymmetrie f
m. distillation Molekulardestillation f
m. eigenfunction Moleküleigenfunktion f
m. electron cloud Molekülelektronenwolke f
m. elevation constant, m. elevation of boiling point s. molal boiling-point[-elevation] constant
m. force Molekularkraft f, [zwischen]molekulare Kraft f
m. formula Molekularformel f, Molekülformel f
m. group Molekülgruppe f
m. heat Mol[ekular]wärme f
m. interaction zwischenmolekulare Wechselwirkung f
m. lattice molekulares Gitter n, Molekülgitter n
m. layer molekulare Schicht f, Molekülschicht f
m. model Molekülmodell n
m. motion (movement) Molekularbewegung f
m. orbital Molekülorbital n, Molekularorbital n, molekulares Orbital n, MO n
m.-orbital approximation MO-Näherung f
m.-orbital calculation Molekülorbitalrechnung f, MO-Rechnung f
m.-orbital method Molekülorbitalmethode f, MO-Methode f, Methode f der Molekülorbitale, Hund-Mulliken-Lennard-Jones-Hückel-Methode f
m.-orbital theory Molekülorbitaltheorie f, MO-Theorie f, Theorie f der Molekülorbitale, Hund-Mulliken-Lennard-Jones-Hückel-Theorie f
m. orientation molekulare Orientierung f, Molekülorientierung f
m. phosphorescence Molekülphosphoreszenz f
m. polarizability s. molar polarizability
m. polarization s. molar polarization
m. ray s. m. beam
m. refraction (refractivity) s. molar refraction
m. rotation molekulare Drehung f, Molekulardrehung f, Molekularrotation f
m. sandwich Sandwichmolekül n
m. sieve Molekularsieb n, Molekülsieb n
m. spectrum Molekülspektrum n, Bandenspektrum n
m. spin orbital molekulares Spinorbital n
m. still Molekulardestillierapparat m
m. structure Molekülstruktur f, Molekularstruktur f
m. velocity Molekülgeschwindigkeit f, Geschwindigkeit f der Moleküle
m. volume s. molar volume
m. wave function molekulare Wellenfunktion f, Molekülwellenfunktion f
m. weight relative Molekülmasse f, (bisher) Molekulargewicht n

m.-weight determination Bestimmung f der relativen Molekülmasse, (bisher) Molekulargewichtsbestimmung f
m.-weight distribution Verteilung f der relativen Molekülmassen, (bisher) Mol[ekular]gewichtsverteilung f
molecularity [of reaction] Molekularität f [einer Reaktion], Reaktionsmolekularität f
molecularly disperse molekulardispers
molecule Molekül n, Molekel f
m. of solvent Lösungsmittelmolekül n
moler Moler m, Molererde f
molfraction Molenbruch m
Molisch reaction Molisch-Reaktion f, Molischsche Reaktion f (Nachweis von Kohlenhydraten)
Molisch reagent Reagens n nach Molisch (α-Naphthol in Äthylalkohol)
Molisch test Nachweis m nach Molisch (von Kohlenhydraten)
Mollier chart Mollier-Diagramm n, i-s-Diagramm n, Enthalpie-Entropie-Diagramm n
molten geschmolzen, schmelzflüssig
m. bath Schmelzbad n
m. material Schmelzgut n
m. metal schmelzflüssiges Metall n, Metallschmelze f
m.-metal dyeing Metallbadfärben n, Metallbadfärbung f, Färben n im Metallbad
m.-metal dyeing process Metallbadfärbeverfahren n
m. rock (Geol) Schmelzfluß m, Schmelze f
m. salt Salzschmelze f, geschmolzenes Salz n
m.-salt carburizing Salzbadaufkohlen n, Badaufkohlen n, flüssige Aufkohlung f, Salzbadzementieren n, Badzementieren n, Badeinsetzen n, Aufkohlen n im Salzbad, Aufkohlen n in flüssigen Mitteln
m.-salt extraction Salzschmelzflußextraktion f
m.-salt mixture geschmolzene (flüssige) Salzmischung f, geschmolzenes (flüssiges) Salzgemisch n, geschmolzene Metallsalze npl (beim Houdry-Verfahren)
m.-salt reactor Salzschmelzenreaktor m
m. zone aufgeschmolzene Zone f, Schmelzzone f (beim Zonenschmelzen)
molybdate M'_2MoO_4 Molybdat(VI) n
molybdeniferous molybdänhaltig
molybdenite (Min) Molybdänit m, Molybdänglanz m (Molybdän(IV)-sulfid)
molybdenum Mo Molybdän n
m. anhydride s. molybdic[-acid] anhydride
m. blue $Mo_2O_5 \cdot xMoO_3$ Molybdänblau n, Molybdän(V,VI)-oxid n
m. bromohydroxide s. m. dihydroxytetrabromide
m. carbide Molybdänkarbid n
m. chlorohydroxide s. m. dihydroxytetrachloride
m. dihydroxytetrabromide $Mo_3Br_4(OH)_2$ Molybdän(II)-dihydroxidtetrabromid n
m. dihydroxytetrachloride $Mo_3Cl_4(OH)_2$ Molybdän(II)-dihydroxidtetrachlorid n

m. ore Molybdänerz n
m. oxybromide MoO_2Br_2 Molybdändioxiddibromid n
m. oxytetrachloride $MoOCl_4$ Molybdänoxidtetrachlorid n
m. oxytetrafluoride $MoOF_4$ Molybdänoxidtetrafluorid n
m. powder Molybdänpulver n
m. sesquioxide Mo_2O_3 Molybdänsesquioxid n, (veraltet für) Molybdän(III)-oxid n
m. sesquisulphide Mo_2S_3 Molybdänsesquisulfid n, (veraltet für) Molybdän(III)-sulfid n
m. steel Molybdänstahl m
m. tetrabromohydroxide s. m. dihydroxytetrabromide
m. trioxide s. molybdic oxide
m. trioxyhexachloride $Mo_2O_3Cl_6$ Molybdäntrioxidhexachlorid n
m. trioxypentachloride $Mo_2O_3Cl_5$ Molybdäntrioxidpentachlorid n
molybdic Molybdän...; höherwertigem Molybdän entsprechend
m. acid H_2MoO_4 Molybdänsäure f
m.[-acid] anhydride MoO_3 Molybdän(VI)-oxid n, Molybdäntrioxid n
m. ochre (Min) Molybdänocker m, Molybdit m; (Min) Molybdänocker m, Ferrimolybdit m
m. oxide MoO_3 Molybdän(VI)-oxid n, Molybdäntrioxid n
m. sulphide MoS_2 Molybdän(IV)-sulfid n, Molybdändisulfid n
molybdite (Min) Molybdit m, Molybdänocker m
molybdo-molybdic oxide Mo_2O_3 Molybdän(III)-oxid n
molybdophosphate $M^I_3[PMo_{12}O_{40}]$ Molybdatophosphat n, Dodekamolybdatophosphat n
molybdous Molybdän...; niederwertigem Molybdän entsprechend
moment of discharge Schußmoment n (bei seismischen Verfahren)
m. of inertia Trägheitsmoment n
m. of momentum Bahndrehimpuls m, Drehimpuls m, Impulsmoment n, Drall m
momentum Impuls m, Bewegungsgröße f; (Tech) bewegende Kraft f, Triebkraft f
m. separator Prall[ab]scheider m
m. space Impulsraum m
mon. s. monoclinic
monacid s. monoacid
monacite s. monazite
monactin Monaktin n (Antibiotikum)
monad einwertig
monad einwertiges Element n; einwertige Atomgruppe f
monamide s. monoamide
monamine s. monoamine
monarda oil Monardaöl n (ätherisches Öl von Monarda-Arten)
monatomic einatomig, monoatomar
monatomicity Einatomigkeit f

monazite (Min) Monazit m
m. sand Monazitsand m
Mond carbonyl process Mond-Verfahren n, Mond-Prozeß m, Mond-Niederdruckkarbonylverfahren n
Mond gas Mond-Gas n
Mond gas producer Mond-Gasgenerator m, Mond-Gaserzeuger m, Mond-Generator m
Mond process 1. Mond-Verfahren n, Mond-Prozeß m, Mondsches Verfahren n, Mond-Gas[generator]verfahren n, Mond-Gasprozeß m; 2. s. Mond carbonyl process
Mond producer s. Mond gas producer
money paper Banknotenpapier n
Monnier kiln Monnier-Ofen m (deckenbeheizter Tunnelofen)
mono derivative Monoderivat n
m.-ion einfach geladenes Ion n, einwertiges Ion n
Mono pump Mono-Pumpe f (eine Einspindelpumpe)
monoacetin Monoazetin n, Glyzerinmonoazetat n
monoacid einsäurig (Base); einfachsauer, sekundär, Monohydrogen..., Hydrogen...
monoacid einbasige (einbasische, einwertige) Säure f
m. phosphate $M^I_2HPO_4$ Monohydrogenphosphat n, Hydrogen[ortho]phosphat n, Hydrogenmonophosphat n, sekundäres Phosphat n
m. salt einfachsaures (sekundäres) Salz n, Monohydrogensalz n, Hydrogensalz n (einer dreibasigen Säure)
monoacidic s. monoacid
monoalkylate / to monoalkylieren
monoalkylation Monoalkylierung f
monoalkylbenzene Monoalkylbenzol n
monoamide Mon[o]amid n
monoamine Mon[o]amin n
m. oxidase Monoamin[o]oxydase f
monoatomic s. monatomic
monoaxial orientation einachsige (axiale) Orientierung f
monoazo dye Monoazofarbstoff m
monobasic einbasig, einbasisch
m. acid s. monoacid
m. ammonium phosphate $NH_4H_2PO_4$ Monoammoniumdihydrogenphosphat n, Ammoniumdihydrogen[ortho]phosphat n
m. barium phosphate $BaH_4(PO_4)_2$ Bariumtetrahydrogenbisphosphat n
m. calcium phosphate $Ca(H_2PO_4)_2$ Monokalziumhydrogenphosphat n, Kalziumdihydrogen[ortho]phosphat n
m. lead acetate $Pb_2O(CH_3COO)_2$ Blei(II)-oxidazetat n
m. lead arsenate $PbH_4(AsO_4)_2$ Monoblei[ortho]arsenat(V) n, Blei(II)-tetrahydrogen[ortho]arsenat n
m. lithium phosphate LiH_2PO_4 Lithiumdihydrogen[ortho]phosphat n

m. magnesium phosphate $Mg(H_2PO_4)_2$ Monomagnesiumdihydrogenphosphat n, Magnesiumdihydrogen[ortho]phosphat n

m. manganous phosphate $Mn(H_2PO_4)_2$ Mangan(II)-dihydrogen[ortho]phosphat n

m. phosphate $M^IH_2PO_4$ primäres Phosphat n, Dihydrogen[ortho]phosphat n

m. potassium orthoarsenate KH_2AsO_4 Kaliumdihydrogen[ortho]arsenat(V) n

m. potassium phosphate KH_2PO_4 Monokaliumdihydrogenphosphat n, Kaliumdihydrogen[ortho]phosphat n

m. sodium hypophosphite NaH_2PO_2 Natriumhypophosphit n

m. sodium orthoarsenate NaH_2AsO_4 Natriumdihydrogen[ortho]arsenat(V) n

m. sodium [ortho]phosphate NaH_2PO_4 Mononatriumdihydrogenphosphat n, Natriumdihydrogen[ortho]phosphat n

monoboride Monoborid n

monobromate / to s. to monobrominate

monobromide Monobromid n

monobrominate / to monobromieren

monobrominated camphor Monobromkampfer m, Bromkampfer m

monobromination Monobromierung f

monobromobenzene C_6H_5Br Monobrombenzol n, Brombenzol n

monocalcium phosphate $Ca(H_2PO_4)_2$ Monokalziumdihydrogenphosphat n, Kalziumdihydrogen[ortho]phosphat n

monocarboxylic acid Monokarbonsäure f, einbasige Karbonsäure f

monochloride Monochlorid n

monochlorinate / to monochlorieren

monochlorination Monochlorierung f

monochloro derivative Monochlorderivat n

monochloroacetic acid $CH_2ClCOOH$ Chloressigsäure f, Chloräthansäure f

monochlorobenzene C_6H_5Cl Monochlorbenzol n, Chlorbenzol n

monochloroethane CH_3CH_2Cl Monochloräthan n, Chloräthan n, Chloräthyl n, Äthylchlorid n

monochloromethane CH_3Cl Monochlormethan n, Chlormethan n, Chlormethyl n, Methylchlorid n

monochloroparaffin Monochlorparaffin n, Monochloralkan n, monochlorierter Paraffinkohlenwasserstoff m

monochlorosilane SiH_3Cl Monochlorsilan n, Chlorsilan n

monochromatic monochromatisch, einfarbig, spektralrein

m. illuminator s. monochromator

monochromator Monochromator m

monochrome (Foto) monochromatisch, einfarbig, Schwarzweiß...

monoclinic (Krist) monoklin, monokl.

m. sulphur monokliner Schwefel m, β-Schwefel m

monocrystal Einkristall m

monocyclic monozyklisch

monodentate (Komplexchemie) einzahnig, einzähnig, einzählig (Koordinationswert des Liganden)

monodisperse monodispers, homodispers, isodispers

m. system monodisperses System n

monoester Monoester m

monoethanolamine $CH_2OHCH_2NH_2$ Monoäthanolamin n, Aminoäthylalkohol m, 2-Aminoäthanol-(1) n, Kolamin n

monoethenoid fatty acid einfach ungesättigte Fettsäure f, Fettsäure f mit einer Doppelbindung, Monoolefinkarbonsäure f

monofil, monofilament [yarn] Monofil[garn] n, monofile Seide f

monofluoride Monofluorid n

monofuel Einstoffkraftstoff m, Einstofftreibstoff m, homogener Treibstoff m, Monotreibstoff m

monofunctional monofunktionell

m. unit monofunktionelle Einheit f, M-Einheit f (beim molekularen Aufbau)

monogermane GeH_4 Monogerman n, German[iumtetrahydrid] n

monoglyceride Monoglyzerid n

monohalide Monohalogenid n

monohalogen Monohalogen...

m. derivative Monohalogenderivat n

m. derivative of a paraffin Monohalogenparaffin n, Monohalogenalkan n, Alkylmonohalogenid n

monohalogenate / to monohalogenieren

monohalogenation Monohalogenierung f

monohydrate Monohydrat n

monohydric einwertig, mit einer Hydroxylgruppe (OH-Gruppe); einfachsauer, sekundär, Monohydrogen..., Hydrogen...

m. alcohol einwertiger Alkohol m

m. phenol einwertiges Phenol n

m. phosphate s. monohydrogen phosphate

monohydride Monohydrid n

monohydrogen Monohydrogen..., Hydrogen...

monohydrogen Monowasserstoff m, atomarer (einatomiger) Wasserstoff m

m. phosphate $M^I_2HPO_4$ Monohydrogenphosphat n, Hydrogen[ortho]phosphat n, Hydrogenmonophosphat n, sekundäres Phosphat n

m. potassium orthophosphate K_2HPO_4 Kaliumhydrogen[ortho]phosphat n, Dikaliumhydrogenphosphat n

monohydroxy Monohydroxy...

m. compound Monohydroxyverbindung f

monoiodide Monojodid n

monoiodoethane CH_3CH_2J Äthyljodid n, Jodäthyl n, Jodäthan n

monoisotopic monoisotop

m. element Reinelement n, monoisotopes Element n

monoketone Monoketon n

m. imide Monoketonimid n

monolayer monomolekulare Schicht (Adsorptions-

schicht) f, monomolekularer Film (Oberflächenfilm) m, Monomolekularfilm m
  m. capacity Sättigungswert m bei monomolekularer Bedeckung
monolithic monolithisch
monomer Monomer[es] n, monomere Substanz f;
(Plast) Grundmolekül n
monomeric monomer
monomethylarsonic acid $CH_3 \cdot AsO(OH)_2$ Methyl-
arsonsäure f (Pflanzenschutzmittel)
monomineral monomineralisch
monomolecular monomolekular, unimolekular,
1molekular
  m. film (layer) s. monolayer
  m. reaction monomolekulare (unimolekulare)
Reaktion f
mononitrate / to mononitrieren
mononitrate Mononitrat n
mononitration Mononitrieren n, Mononitrierung f
mononitro Mononitro...
  m. body Mononitrokörper m
  m. compound Mononitroverbindung f
mononitrobenzene $C_6H_5NO_2$ Mononitrobenzol n,
Nitrobenzol n
mononuclear einkernig
mononucleotide Mononukleotid n
  m. flavin enzyme Flavinmononukleotid n, FMN
monoperphthalic acid Monoperphthalsäure f,
Phthalomonopersäure f
monophane (Min) Monophan m, (veraltet für)
Epistilbit m
monophosphate Monophosphat n, Orthophosphat
n, Phosphat n
monophosphide Monophosphid n
monopotassium phosphate KH_2PO_4 Mono-
kaliumphosphat n, Kaliumdihydrogen[ortho]-
phosphat n
monosaccharide $C_nH_{2n}O_n$ Monosa[c]charid n,
Einfachzucker m, einfacher Zucker m
monoselenide Monoselenid n
monosilane SiH_4 Monosilan n, Silan n
monosilicate Monosilikat n, Singulosilikat n
  m. slag Singulosilikatschlacke f
monosodium glutamate
Na $OOCCH_2CH_2CH(NH_2)COOH$ Mononatrium-
glutam[in]at n, Natriumglutam[in]at n
  m. phosphate NaH_2PO_4 Mononatriumphosphat
n, Natriumdihydrogen[ortho]phosphat n
monostearin Monostearin n, Glyzerinmonostearat
n, Glyzerolmonostearat n, Glyzerylmonostearat n
monosubstitute / to monosubstituieren, einfach
substituieren
monosubstitution Monosubstitution f, Einfach-
substitution f
  m. product Monosubstitutionsprodukt n, mono-
substituiertes Produkt n
monosulphide Monosulfid n
  m. bridge (crosslink) Monosulfidbrücke f
monosulphonic acid Monosulfonsäure f
monotelluride Monotellurid n

monoterpene $C_{10}H_{16}$ Monoterpen n
monotropic monotrop
monotropy Monotropie f
monounsaturated einfach ungesättigt
monovalence, monovalency Einwertigkeit f
monovalent einwertig, monovalent
monovariant monovariant, univariant, einfachfrei
  m. system monovariantes System n
monoxide Monoxid n
Mont Cenis process Mont-Cenis-Verfahren n (zur
synthetischen Erzeugung von Ammoniak)
montan wax Montanwachs n
  m.-wax pitch Montanwachspech n
  m.-wax size (Pap) Montanwachsleim m
montanic acid Montansäure f, Oktakosansäure f
montanin wax s. montan wax
montejus Montejus m, Druckgefäß n, Druckfaß n
monticellite (Min) Monticellit m (Kalziummagne-
siumorthosilikat)
montmorillonite (Min) Montmorillonit m (Aluminiumdihydrogentetrasilikat)
monuron $ClC_6H_4NHCON(CH_3)_2$ Monuron f,
3-(4-Chlorphenyl)1,1-Dimethylharnstoff m
(Herbizid)
moon knife (Gerb) Schlichtmond m
Mooney s. Mooney value
Mooney cure test Prüfung f des Anvulkanisations-
verhaltens mit dem Mooney-Plastometer
Mooney instrument s. Mooney viscometer
Mooney plasticity Mooney-Plastizität f
Mooney plastometer s. Mooney viscometer
Mooney scorch [time] Mooney-Anvulkanisations-
zeit f
Mooney unit Mooney-Grad m
Mooney value Mooney-Wert m, Mooney-Zahl f
Mooney viscometer Mooney-Viskosimeter n,
Mooney-Scherscheibenviskosimeter n, Mooney-
Plastometer n
Mooney viscosity Mooney-Viskosität f
moonstone (Min) Mondstein m (Feldspat-Varietät)
moor peat Hochmoortorf m
Moore-Campbell kiln Moore-Campbell-Ofen m
(elektrischer Tunnelofen)
Moore filter Moore-Filter n, Tauchnutsche f
mop polishing Schwabbeln n, Polieren n mit der
Schwabbelscheibe
mor Auflagehumus m, Rohhumus m, Mor
mordant / to beizen
mordant Beizmittel n, Beizstoff m, Beize f
  m. colour s. m. dye
  m. dye Beizenfarbstoff m, adjektiver (beizenfärbender) Farbstoff m
  m. dyeing Beizenfärben n, Beizenfärberei f
  m. dyestuff s. m. dye
  m. rouge (Text) Rotbeize f (Aluminiumazetat in
Essigsäure)
mordanting Beizen n, Beizung f
  m. process Beizverfahren n
more volatile component Leichterflüchtiges n,
Leichtersiedendes n, leichterflüchtige (leichter-

siedende) Komponente f, leichterflüchtiger (leichtersiedender) Anteil m

morenosite *(Min)* Morenosit m, Nickelvitriol m *(Nickelsulfat-7-Wasser)*

Morgan [gas] producer Morgan-Gasgenerator m, Morgan-Gaserzeuger m, Morgan-Generator m

morin Morin n, 3,5,7,2′,4′-Pentahydroxyflavon n

Moroccan ammoniac Afrikanisches Ammoniakgummi n *(Pflanzengummi aus Ferula tingitana L. und Ferula communis L. var. brevifolia Marcz)*

Morocco gum *(Pflanzengummi von Acacia gummifera Willd.)*

morphine Morphin n, 3,6-Dihydroxy-N-methyl-4,5-epoxymorphinen-(7) n

  m. alkaloid Morphinalkaloid n

  m. hydrochloride Morphinhydro[gen]chlorid n

  m. sulphate Morphinsulfat n

morphinelike morphinähnlich

morphol Morphol n, 3,4-Phenanthrendiol n

morpholine Morpholin n, Tetrahydro-1,4-oxazin n

mortar 1. Mörser m, Reibschale f; 2. Mörtel m

  m. of cement Zementmörtel m

  m. structure *(Geol)* Mörtelstruktur f

mosaic bloc *(Krist)* Mosaikblock m, Mosaikblöckchen n

  m. crystal Mosaikkristall m

  m. gold Musivgold n, Mosaikgold n, Maiergold n *(Zinn(IV)-sulfid)*; Musivgold n, Mosaikgold n *(eine Messingsorte)*

  m. structure (texture) *(Krist)* Mosaikstruktur f, Mosaiktextur f

mosandrite *(Min)* Mosandrit m

moscovite *(Min)* Muskovit m *(Kaliglimmer)*

Moseley diagram Moseley-Diagramm n

Moseley law Moseleysches Gesetz n

Mosotti-Clausius equation Clausius-Mosotti-Gleichung f, Clausius-Mosottische Gleichung f

moss agate *(Min)* Moosachat m

  m. green $Cu(CH_3COO)_2 \cdot 3Cu(AsO_2)_2$ Schweinfurter Grün n, Kupferarsenitazetat n

  m. peat Moostorf m, Fasertorf m

Mössbauer effect Mößbauer-Effekt m

most probable [molecular] velocity wahrscheinlichste Geschwindigkeit f [der Moleküle]

mote Nisse f, Knötchen n

moth ball Mottenkugel f

  m. repellent s. mothproofing agent

mother cell Mutterzelle f

  m. culture Mutter[säure]kultur f

  m. juice Muttersaft m, naturreiner Saft m

  m. liquid (liquor) Mutterlauge f, Stammlauge f, Urlauge f

  m. of coal mineralische (fossile) Holzkohle f

  m. of pearl Perlmutter f, Perlmutt n

  m.-of-pearl effect Perlmutteffekt m

  m.-of-pearl paper Perlmutterpapier n, Irispapier n, irisierendes Papier n

  m.-of-pearl sulphur perlmuttartiger Schwefel m *(eine Schwefelmodifikation)*

  m. of vinegar Essigmutter f

m. rock Muttergestein n, Ursprungsgestein n, Ausgangsgestein n, *(manchmal auch)* Erdölmuttergestein n

m. starter s. m. culture

m. stock *(Gum)* Vormischung f, Masterbatch m, Grundmischung f

m. substance Muttersubstanz f, Grundsubstanz m, Stammkörper m, Grundkörper m

m. water s. m. liquid

mothproof / to mottenfest (mottenecht, mottensicher, mottenbeständig) machen

mothproof mottenfest, mottenecht, mottensicher, mottenbeständig

  m. finish Mottenfestappretur f, Mottenfestausrüstung f, Mottenechtappretur f, Mottenechtausrüstung f

  m. paper Mottenpapier n, Naphthalinpapier n

mothproofer s. mothproofing agent

mothproofing Mottenfestmachen n, Mottenechtmachen n, Mottensichermachen n, Mottenbeständigmachen, Mottenschutz m

  m. agent Motten[schutz]mittel n

motion Bewegung f

  m. of rotation Rotation[sbewegung] f, Drehbewegung f, drehende Bewegung f

  m. of the ions Ionenbewegung f

  m. of translation Translation[sbewegung] f, fortschreitende Bewegung f

motivating fluid Treibmittel n *(bei Strahlpumpen)*

motive steam Treibdampf m *(z.B. bei Dampfstrahlpumpen)*

motor benzol Motorenbenzol n

  m. fuel Motor[en]kraftstoff m, Motor[en]treibstoff m

  m. gasoline Motor[en]benzin n, Fahrbenzin n

  m. method Motormethode f, Motorverfahren n, F-2-Methode f *(zur Oktanzahlbestimmung)*

  m. octane number Motor-Oktanzahl f, F 2

  m. oil Motorenöl n

  m. spirit s. m. gasoline

motorized-head pulley angetriebene Kopfrolle (Kopftrommel) f

mottled glaze *(Ker)* gesprenkelte Glasur f

  m. marble Achatmarmorpapier n

  m. paper meliertes (gefasertes) Papier n

  m. pig iron meliertes Roheisen n

mottling *(Pap)* Melierung f

mottramite *(Min)* Mottramit m

mould / to I. formen; form[press]en, *(Gum auch)* in Formen vulkanisieren; *(Pap)* schöpfen; II. [ver-] schimmeln, schimm[e]lig werden

mould I. Form f, Gießform f, Gußform f; *(Plast)* Form f, Werkzeug n, Preßform f, Preßwerkzeug n; *(Gum)* Form f, Vulkanisierform f; *(Pap)* Form f, Schöpfform f, Schöpfrahmen m; II. Schimmel m; Schimmelpilz m

  m.-blown glass in Formen geblasenes Glas n

  m. breaker, m.-breaking jack *(Gum)* Formenbrecher m, Formenöffner m, Brecheisen n *(zum Öffnen von Vulkanisierformen)*

m. cavity Werkzeughohlraum *m*, Werkzeughöhlung *f*, Formhöhlung *f*, Formnest *n*, formgebende Aushöhlung *f*

m. charge Beschickung *f*, Füllgut *n*

m. clamp Formschließeinheit *f*, Werkzeugschließeinheit *f*

m.-clamping force Formschließkraft *f*, Werkzeugzuhaltekraft *f*

m.-clamping stroke Werkzeugschließhub *m*, Werkzeugschließweg *m*

m.-clearing jack *s. m.* breaker

m. closing Formenschluß *m*, Werkzeugschluß *m*, Schließen *n* der Form, Schließen *n* des Werkzeugs

m.-closing time Formschließzeit *f*, Werkzeugschließzeit *f*

m. cracker *s. m.* breaker

m. cure (curing) *(Gum)* Vulkanisation (Heizung) *f* in Formen, Formheizung *f*

m. design Formkonstruktion *f*, Werkzeugkonstruktion *f*

m. designer Formkonstrukteur *m*, Werkzeugkonstrukteur *m*

m. development *s. m.* formation

m. felt *(Pap)* Unterfilz *m*, unterer Abnahmefilz *m* *(bei Rundsieb-Kartonmaschinen)*

m. formation Schimmelbildung *f*

m. fungus Schimmelpilz *m*

m.-locking pressure Werkzeugschließdruck *m*, Schließdruck *m*, Spanndruck *m*

m. lubricant Gleitmittel *n*, *(Kunststoffen oder Kautschuk zugesetztes)* Form[en]trennmittel *n*, Entformungsmittel *n*; *(Glas)* Formenschmiermittel *n*, Formenschmiere *f*

m. machine Rundsieb[papier]maschine *f*

m. mark *(durch das Pressen in der Form verursachte)* Fehl[er]stelle *f*; *(Glas)* Formennaht *f*, Trennfuge *f (Fehler)*

m. opening Formöffnung[shöhe] *f*, [lichte] Einbauhöhe *f*

m.-parting line Trennfuge (Nahtstelle) *f* der Form

m. pigment Schimmelpilzfarbstoff *m*

m. plug Formzapfen *m*

m. pressure Preßdruck *m*

m. release 1. Formentrennung *f*, Entformung *f*; 2. *s. m.*-release agent

m.-release agent (medium) Form[en]einstreichmittel *n*, Form[en]einsprühmittel *n*, *(auf die Form aufgebrachtes)* Form[en]trennmittel *n*, Entformungsmittel *n*

m.-ripened *(Lebm)* schimmelgereift

m. seam *(Glas)* Formennaht *f*, Trennfuge *f (Fehler)*

m. shrinkage Formenschwindmaß *n*, Werkzeugschwindmaß *n*

m. sticking Haften *n* des Teils am Werkzeug *(beim Entformen)*

m. temperature Werkzeugtemperatur *f*

m. vent Entlüftungskanal *m* eines Werkzeugs (einer Form)

m. vulcanization *s. m.* cure

mouldability Formbarkeit *f*, Verformbarkeit *f*

mouldable [ver]formbar

moulded articles (goods) Formartikel *mpl*, Formteile *npl*

m.-in *(Plast)* eingepreßt, eingespritzt

m. laminate Schichtstoffpreßteil *n*, Preßteil *n* aus Schichtstoff

m. laminated section [form]gepreßtes Schichtstoffprofil *n*

m. laminated tube [form]gepreßtes Schichtstoffrohr *n*

m. material Formstoff *m*

m.-on-sole process Preßmethode *f (Herstellung von formgepreßtem Schuhwerk)*

m. parts (products) *s. m.* articles

m. seal Formdichtung *f*

moulder Former *m*; Formmaschine *f*; *(Pap)* Schöpfer *m*

m. and wrapper *s.* moulding and wrapping machine

moulding I. Formen *n*; Form[press]en *n*, Herstellung *f* von Formartikeln (Formteilen), *(Gum auch)* Vulkanisation *f* in Formen; Formartikel *m*, Formteil *n*; II. Schimmeln *n*, Verschimmeln *n*

m. and wrapping machine Form- und Verpackungsmaschine *f*, Form- und Einwickelmaschine *f*

m. chamber Formkammer *f*

m. compound Formmasse *f*

m. cycle Preßzyklus *m*

m. index *(Plast)* Becherfließzahl *f*, Becherschließzeit *f*

m. machine Formmaschine *f*

m. material Formmasse *f (i.w.S.)*

m. plaster Formengips *m*

m. plug Preßstempel *m*, Patrize *f*, Stempel *m*, Stempelprofil *n*

m. powder *(Plast)* pulvrige Preßmasse *f*

m. press *(Plast)* Formpresse *f*, Presse *f*; *(Gum)* Vulkanisierpresse *f*

m. resin Preßharz *n*

m. sand Formsand *m*

m. shrinkage Schrumpfung *f* eines Formteils

m. technique Formtechnik *f*

m. temperature Preßtemperatur *f*

m. time Formzeit *f*, Preßzeit *f*

mouldmade paper Maschinenbüttenpapier *n*, Rundsieb-Büttenpapier *n*

mouldmaker Formenbauer *m*, *(Plast auch)* Werkzeugmacher *m*

mouldmaking Formherstellung *f*, Formenbau *m*, *(Plast auch)* Werkzeugbau *m*

mouldy flavour Schimmelgeschmack *m*

m. smell Schimmelgeruch *m*

m. taste *s. m.* flavour

mountain cork *(Min)* Bergkork *m*

m. leather *(Min)* Bergleder *n*

m. tallow *(Min)* Hatchettin *m*

mounting board Aufziehkarton *m*

m. plate *(Plast)* Aufspannplatte *f*
moura butter (fat, oil) *s.* mowra[h] butter
mouth blowpipe Lötrohr *n*
mouthpiece Mundstück *n*
mouthwash Mundwasser *n*
movable beweglich [angeordnet]
  m. collar drehbare Manschette *f (am Bunsenbrenner)*
  m. platen *(Plast)* bewegliche Formplatte (Werkzeugplatte) *f*
  m.-sieve jig Setzmaschine *f* mit bewegtem Sieb
moving bed Bewegtbett *n*, Wanderbett *n*
  m.-bed adsorption Fließbettadsorption *f*
  m.-bed process Bewegtbettverfahren *n*, Wanderbettverfahren *n*, Verfahren *n* mit bewegtem Katalysatorbett, Verfahren *n* mit sich bewegendem Katalysator
  m.-boundary method *(phys Ch)* Methode *f* der wandernden Grenzflächen
  m.-burden process Verfahren *n* mit umlaufender Beschickung *(Verfahren der I.C.I. zur allothermen Vergasung in der Wirbelschicht)*
  m. catalyst bewegter (sich bewegender, beweglicher) Katalysator (Kontakt) *m*
  m. catalyst bed bewegtes Katalysatorbett *n*
  m.-coil galvanometer Drehspulgalvanometer *n*
  m.-fire kiln Ofen *m* mit wanderndem Feuer
  m. jaw schwingende Brechbacke *f*, Schwingerbrechbacke *f*, Brechschwinge *f (eines Backenbrechers)*
  m. mould half *(Plast)* bewegliche Werkzeughälfte *f*
  m. phase bewegliche (mobile) Phase *f (in der Chromatografie)*
  m. platen *s.* movable platen
  m.-product dryer Rieseltrockner *m*; Trockner *m* mit bewegtem Trockengut, Schwebetrockner *m*
  m.-sieve-type jig Stauchsetzapparat *m*, Stauchsetzmaschine *f*
mowra[h] butter (fat, oil) Mowra[h]butter *f*, Mowra[h]öl *n*
Moyno pump Moyno-Pumpe *f (eine Einspindelpumpe)*
mp, m.p. *s.* melting point
6-MP *s.* 6-mercaptopurine
MP capacitor *s.* metal-paper capacitor
MPC black *s.* medium processing channel black
M.Q. developer *(Foto)* Metall-Hydrochinon-Entwickler *m*
MSO *s.* molecular spin orbital
MT black *s.* medium thermal black
mu meson Mu-Meson *n*, μ-Meson *n*, Muon *n*, Myon *n*, Müon *n*
mucic acid COOH(CHOH)$_4$COOH Schleimsäure *f*, Muzinsäure *f*, Tetrahydroxyadipinsäure *f*
mucilage Schleim *m*, *(meist)* Pflanzenschleim *m*
  m. dressing *(Gerb)* Schleimappretur *f*
mucilaginous schleimig, schleimhaltig
  m. material (substance) Schleimsubstanz *f*, Schleimstoff *m*, schleimige Substanz *f*

mucin Muzin *n (ein Glykoproteid)*
mucobromic acid OHCC(Br)=C(Br)COOH Mukobromsäure *f*, 4-Aldo-2,3-dibrom-2-butensäure *f*
mucochloric acid OHCC(Cl)=C(Cl)COOH Mukochlorsäure *f*, 4-Aldo-2,3-dichlor-2-butensäure *f*
mucoid schleimartig
mucoid Mukoid *n (ein Glykoproteid)*
mucoitin sulphate *s.* mucoitinsulphuric acid
mucoitinsulphuric acid Mukoitinschwefelsäure *f (ein hochmolekulares Mukopolysaccharid)*
muconic acid HOOCCH=CHCH=CHCOOH Mukonsäure *f*, 2,4-Hexadiendisäure *f*
mucopeptide Mukopeptid *n*
mucopolysaccharide Mukopolysa[c]charid *n*
mucoprotein Mukoproteid *n*, Mukoprotein *n*
mucor Köpfchenschimmel *m*, Mucor mucedo
mucus Schleim *m*
mud Schlamm *m*, Schlick *m*; Spülung *f*, Spülschlamm *m*, [schlammartige] Spülflüssigkeit *f*, Bohrspülung *f*, [schlammartige] Bohrflüssigkeit *f*, Bohrschlamm *m*, Mud *m*, Spülmud *m (beim Erdölbohren)*
  m. column Spül[ungs]säule *f (beim Erdölbohren)*
  m. hose Spülschlauch *m (einer Rotary-Bohranlage)*
  m. outlet Ausguß *m (einer Rotary-Bohranlage)*
  m. pit Spülgrube *f*, Spülungstank *m*, Spülungsbehälter *m*, Schlammgrube *f (einer Rotary-Bohranlage)*
  m. pump Schlammpumpe *f*; Spülpumpe *f (einer Rotary-Bohranlage)*
  m. turbine spülungsbetriebene Turbine *f*, mit der Spülung (Bohrlochspülung) angetriebene Turbine *f*, durch die Spülflüssigkeit (den Spülungsstrom) angetriebene Turbine *f*
mudstone *(Geol)* Schlammstein *m*
muff dyeing *(Text)* Hülsenlosfärben *n*
muffle Muffel *f*
  m. furnace (kiln) Muffelofen *m*, gemuffelter Ofen *m*
  m. tunnel kiln Muffeltunnelofen *m*
mule gum Jequié-Kautschuk *m (von Manihot dichotoma Ule)*
muller Laufstein *m*, Läufer *m*, Mahlwalze *f*
  m. mixer Mischkollergang *m*
Müller-Rochow process Müller-Rochow-Verfahren *n*, Rochow-Verfahren *n*, direktes Verfahren *n*, Direktverfahren *n*, direkte Methode *f (zur Herstellung von Chlorsilanen)*
mullicite *(Min)* Mullizit *m*, *(veraltet für)* Vivianit *m (Eisen(II)-orthophosphat)*
mulling machine Kollergang *m*, Kollermühle *f*
mullite *(Min)* Mullit *m (Aluminiumsilikat)*
  m. porcelain Mullitporzellan *n*
multicavity mould *(Plast)* Mehrfachform *f*, Mehrfachwerkzeug *n*
multicell Mehrzellen..., Vielzellen...
  m. dust collector (extractor) Vielzellenentstauber *m*, Vielzellenabscheider *m*

m. floatation machine Mehrzellenflotationsmaschine f, Mehrzellenflotationsgerät n
multicentre Mehrzentren..., polyzentrisch
m. bond Mehrzentrenbindung f
multichamber Mehrkammer...
m. centrifuge Mehrkammerzentrifuge f, Kammerzentrifuge f
m. kiln Mehrkammerofen m
Multiclone Multiklon m (ein Gasgegenstrom-Staubgleichstrom-Zyklon)
multicoil condenser Rohrschlangenkondensator m
multicolour mehrfarbig; Mehrfarben..., Vielfarben...
m. effect Mehrfarbeneffekt m, Vielfarbeneffekt m
m. printing Mehrfarbendruck m, Farbendruck m, Chromodruck m
multicoloured mehrfarbig
m. effect s. multicolour effect
multicompartment Mehrkammer...; Mehrzellen..., Vielzellen...
m. drum filter Trommelzellenfilter n
m. mill Mehrkammer[rohr]mühle f, Verbund[rohr]mühle f
multicompartmented s. multicompartment
multicomponent Mehrkomponenten..., Mehrstoff..., Vielstoff...
m. distillation Destillation f eines Mehrstoffgemischs (Mehrkomponentensystems)
m. mixture Mehrkomponentengemisch n, Mehrstoffgemisch n, Vielstoffgemisch n
m. system Mehrkomponentensystem n, Mehrstoffsystem n
multiconveyor [tunnel] dryer Mehrbandtrockner m
Multicyclone Multizyklon m (ein Gasgegenstrom-Staubgleichstrom-Zyklon)
multicylinder machine (Pap) Mehrrundsiebmaschine f
multidaylight press Mehretagenpresse f, Etagenpresse f
multideck screen Mehrdeckersiebmaschine f
multidentate (Komplexchemie) mehrzahnig, mehrzähnig, vielzähnig, mehrzählig
multifil, multifilament [yarn] Multifil[garn] n, multifile (polyfile) Seide f
multifunctional polyfunktionell
multigrade oil Mehrbereichsöl n
multihearth furnace mehrherdiger Ofen m, (meist) Mehretagenofen m, mehretagiger Ofen m
m. roaster (roasting furnace) mehrherdiger Röstofen m, (meist) Mehretagenröstofen m, mehretagiger Röstofen m
multi-impression mould (Plast) Mehrfachform f, Mehrfachwerkzeug n
multiknife chipper (Pap) Vielmesserhackmaschine f
multilayer Mehrfachschicht f, multimolekulare Schicht (Adsorptionsschicht) f
multimembered vielgliedrig (Ring)
multimembrane electrodialyzer Mehrzellenelektrodialysator m

multimolecular [adsorbed] layer s. multilayer
multinuclear mehrkernig, Mehrkern...
m. compound mehrkernige Verbindung f
multinutrient fertilizer Mehrnährstoffdünger m, Kombinationsdünger m
multipass clarifier bowl Tellerzentrifuge f zur Trennung von Emulsionen
multipassage kiln (Ker) Mehrbahnofen m, Passageofen m, Mehrkanaldurchschubofen m
multiphase mehrphasig, Mehrphasen...
m. system Mehrphasensystem n, mehrphasiges System n
multipiston pump Mehrkolbenpumpe f
multiplate Mehrplatten...
m. beater (Pap) Holländer m mit mehreren Grundwerken
m. filter Etagennutsche f
m. freezer Mehrplatten[schnell]gefrierapparat m
multiplaten press s. multidaylight press
multiple vielfach, mehrfach, multipel, Viel[fach]..., Mehr[fach]...
multiple Vielfaches n, Multiples n; Multieinheit f, Mehrfachapparat m
m. acceleration Vielfachbeschleunigung f
m. accelerator Vielfachbeschleuniger m
m.-belt [tunnel] dryer Mehrbandtrockner m
m. bond Mehrfachbindung f, mehrfache Bindung f
m.-bond system Mehrfachbindungssystem n
m.-cell s. multicell
m. cyclone s. m.-unit cyclone
m.-daylight press s. multidaylight press
m. decay s. m. disintegration
m.-deck screen s. multideck screen
m. disintegration Mehrfachzerfall m, Dualzerfall m, dualer (verzweigter) Zerfall m, [radioaktive] Verzweigung f
m.-effect evaporation Mehrstufenverdampfung f, Mehrkörperverdampfung f
m.-effect evaporator Mehrstufenverdampfer m, Mehrkörperverdampfer m, Mehrfachverdampfer m, Kaskadenverdampfer m
m. electrode mehrfache Elektrode f, Mehrfachelektrode f
m.-function s. multifunctional
m.-hearth furnace s. multihearth furnace
m.-hearth roaster (roasting furnace) s. multihearth roaster
m.-layer insulation Folienisolierung f
m. link[age] s. m. bond
m.-plate s. multiplate
m. polarogram Mehrstufenpolarogramm n
m. proportions multiple Proportionen fpl
m. reflection mehrfache Reflexion f, Mehrfachreflexion f
m.-retort [underfeed] stoker Mehrmulden-Unterschubrost m
m.-roll s. multiroll
m. scattering (phys Ch) Vielfachstreuung f
m.-stage s. multistage

m.-type cyclone s. m.-unit cyclone
m.-unit cyclone Mehrfachzyklon m, Vielfachzyklon m, Zyklonbatterie f
multiplet Multiplett n
  m. level Multipletterm m, Multiplettniveau n
  m. splitting Multiplettaufspaltung f
  m. structure Multiplettstruktur f
multiplex [roll] Multiplexwalze f
  m.-roll plant Multiplexanlage f, Etagenwalzwerk n *(System mit mehreren hintereinandergeschalteten Walzenpaaren)*
multiplication constant *(Kch)* Multiplikationsfaktor m, Vermehrungsfaktor m, Reproduktionsfaktor m
multiplicative numeral (numerical) prefix s. multiplying prefix
multiplicity Multiplizität f
multi-ply mehrschichtig, mehrlagig, Mehrschichten..., Mehrlagen...
multi-ply board Multiplexpappe f, mehrschichtige Pappe f
multiplying prefix multiplizierendes (vervielfachendes) Präfix n, vervielfachende Vorsilbe f, multiplikatives Zahlwort n, Multiplikativzahl f
multipoint recorder Mehrfachpunktschreiber m
multiport plug valve Mehrweg[e]hahn m
multiproduct unit Mehrproduktenapparat m, Mehrgutapparat m
multiroll Mehrwalzen..., mit mehreren Walzen [versehen]
  m. crusher Mehrwalzenbrecher m
  m. mill Mehrwalzenstuhl m; Mehrwalzenmühle f, Mehrpendelmühle f
multiroller s. multiroll
multirotation Multirotation f, Mutarotation f
multiscrew extruder *(Plast)* Mehrschneckenextruder m
multistage mehrstufig, Mehrstufen...
  m. belt dryer Mehrbandtrockner m
  m. bleaching *(Pap)* Mehrstufenbleiche f, Vielstufenbleiche f
  m. compressor mehrstufiger Verdichter (Kompressor) m, Mehrstufenverdichter m, Mehrstufenkompressor m
  m. pump mehrstufige Pumpe f
  m. reciprocating-pusher centrifuge Mehrstufenschubzentrifuge f
  m. separator *(Erdöl)* Mehrstufenseparator m, Stufenseparator m, mehrstufiger Separator (Gas-Öl-Separator, Gasseparator) m, mehrstufige Gastrennanlage f
  m. washer Mehrstufenwäscher m
multistep mehrstufig, Mehrstufen...
multisweep method oszillografische Polarografie f mit Wechselstrom
multivalence, multivalency Mehrwertigkeit f, Polyvalenz f
multivalent mehrwertig, polyvalent
multivat machine *(Pap)* Mehrrundsiebmaschine f
multiwall paper sack mehrlagiger Papiersack m

multiway union Abzweigstück n
mundic *(Min)* Eisenkies m, Schwefelkies m, Pyrit m *(Eisendisulfid)*
mungo Mungo m, Reißwolle f *(aus gewebten oder gewalkten Stoffen und Filzen)*
Munich malt Münchner (dunkles) Malz n
muon Muon n, Myon n, Müon n, Mu-Meson n, μ-Meson n
Murakami's reagent Murakami-Reagens n, Murakami-Ätzmittel n
muramic acid Muraminsäure f, 2-Amino-2-desoxy-3-O-(α-D-karboxyäthyl)-D-glukose f
murexide $NH_4 \cdot C_8H_4N_5O_6 \cdot H_2O$ Murexid n, Ammoniumpurpurat n
  m. reaction Murexidreaktion f
  m. test Murexidprobe f *(Harnsäurenachweis mittels Murexidreaktion)*
muriacite *(Min)* Muriazit m, *(veraltet für)* Anhydrit m
muriate Muriat n, *(historisch für)* Metallchlorid n; *(technisches)* Chlorid n
  m. of potash *(technisches)* Kaliumchlorid n; *(Landw)* Kalidüngesalz n, *(i.e.S.)* 50er Kalidüngesalz n *(mit etwa 50 % K_2O-Gehalt)*
  m. of soda *(technisches)* Natriumchlorid n
muriatic acid Muriatsäure f, *(historisch für)* Salzsäure f; *(technische)* Salzsäure f
murium Murium n *(hypothetisches Element)*
Murphree [plate] efficiency *(Destillation)* Bodenwirkungsgrad m [nach Murphree]
muscarine Muskarin n, 2-Methyl-3-hydroxy-5-dimethylaminomethyl-tetrahydrofuran-hydroxymethyl n
muscle adenylic acid Muskeladenylsäure f, 5'-Adenylsäure f, Adenosin-5'-monophosphorsäure f, Adenosin-5'-monophosphat n
  m. oil *(Kosmet)* Muskelöl n
muscone Muskon n, 3-Methylzyklopentadekanon n
muscovite, muscovy glass *(Min)* Muskovit m, Kaliglimmer m
museum jar Schauglas n, Präparatenglas n, Ausstellungsglas n
mushroom mixer Pilzmischer m
  m.-seated valve Pilzventil n, pilzförmiges Tellerventil n
music paper Notenpapier n, *(i.e.S.)* Notendruckpapier n *(oder)* Notenschreibpapier n
musite *(Min)* Musit m, *(veraltet für)* Parisit m
musk Moschus m
  m. C *(ketone)* Moschusketon n, Ketonmoschus m, Moschus C m, 4-Azetyl-5-(2'-methyl-2'-propyl)-2,6-dinitro-m-xylol n
musklike moschusartig, Moschus...
mussite *(Min)* Mussit m, *(veraltet für)* Diopsid m
must 1. Most m; 2. Moder m, Schimmel m
  m. gauge Mostwaage f, Mostaräometer n, Oechsle-Waage f
mustard 1. Senf m, Sinapis L.; Senf m, Speisesenf m, Mostrich m; 2. s. m. gas
  m. gas Senfgas n, Mustardgas n, Lost m, Yperit n

m. oil s. mustardseed oil
m. paper Senfpapier n, Senfpflaster n
m. therapy Losttherapie f
mustardseed oil Senföl n
mustine Mustin n, Methylbis-(ß-chloräthyl)amin n
mutagen, mutagenic agent Mutagen n, mutationsauslösendes Agens n
mutarotation Mutarotation f, Multirotation f
mutase Mutase f
mutator shaft (Tech) Messerwelle f
Muthmann's liquid Muthmanns Flüssigkeit f (1,1,2,2-Tetrabromäthan)
mutton fat (tallow) Hammeltalg m, Hammelfett n
mutual gegenseitig, wechselseitig, beiderseitig, Wechsel...
m. solubility gegenseitige Löslichkeit f
MVA s. mevalonic acid
M.V.C. s. more volatile component
M.W. s. molecular weight
M.W.P. s. mechanical wood-pulp
mycaminose Mykaminose f, 3,6-Didesoxy-3-dimethyl-amino-D-glukose f
mycarose Mykarose f
mycelium Myzel[ium] n
mycolic acid Mykolsäure f
mycolipenic acid Mykolipensäure f, (+)2,4,6-Trimethyl-tetrakos-2-ensäure f
mycomycin Mykomyzin n, n-Tridekatetraen-(2,4,6,7)-diin-(9,11)-karbonsäure f
mycophenolic acid Mykophenolsäure f (Antibiotikum)
mycosamine Mykosamin n (ein Aminozucker)
mycosterol Mykosterin n
mydriatic mydriatisch, pupillenerweiternd
mydriatic Mydriatikum n, pupillenerweiterndes Medikament (Mittel) n
myelin[e] Myelin n
myeloperoxidase Myeloperoxydase f
mylonite (Geol) Mylonit n
mylonitization (Geol) Mylonitisierung f
myogen Myogen n (am Aufbau des Muskelproteins beteiligter Eiweißbaustein)
myoglobin Myo[hämo]globin n
myokinase Myokinase f, Adenylatkinase f, ADP-Phosphomutase f
myosin Myosin n (zu den Globulinen gehörendes Muskelprotein)
myotic miotisch, pupillenverenge[r]nd
myotic Miotikum n, pupillenverenge[r]ndes Medikament (Mittel) n
myrab[olan] s. myrobalan
myrbane oil Mirbanöl n, unechtes (künstliches) Bittermandelöl n, Nitrobenzol n
myrcene Myrzen n, 2-Methyl-6-methylen-2,7-oktadien n
myrcia oil Bayöl n (von Pimenta racemosa (Mill.)I.W. Moore)
myrica tallow Myrikawachs n, Myrikatalg m, Myrthenwachs n, Bayberrytalg m, Kapbeerenwachs n

myricetin Myrizetin n, 3,5,7,3',4',5'-Hexahydroxyflavon n
myricin s. myricyl palmitate
myricitrin Myrizitrin n (glykosidischer Farbstoff)
myricyl alcohol Myrizylalkohol m, Melissylalkohol m
m. palmitate Myrizyl[alkohol]palmitat n, Palmitinsäuremyrizylester m, Myrizin n, Myrizylhexadekanoat n
myristaldehyde Myrist[in]aldehyd m, Myristylaldehyd m, Tetradekanal n
myristate Myristat n (Salz oder Ester der Myristinsäure)
myristic acid $CH_3(CH_2)_{12}COOH$ Myristinsäure f, Tetradekansäure f
m. aldehyde s. myristaldehyde
myristica Muskatnußbaum m, Myristica fragrans Houtt.; Muskatnuß f, Muskatsamen m
m. oil ätherisches Muskatöl n, Muskatnußöl n
myristicic acid Myristizinsäure f, 5-Methoxy-3,4-methylendioxybenzolkarbonsäure f
myristicin Myristizin n, 1-Methoxy-2,3-methylendioxy-5-(2-propenyl)benzol n
myristicinaldehyde Myristizinaldehyd m, 5-Methoxy-3,4-methylendioxybenzolkarbonal n
myristin Myristin n; Trimyristin n, Myristin n, Glyzerintrimyristat n, Glyzeryltetradekanoat n
myristoleic acid $C_{13}H_{25}COOH$ Myristoleinsäure f, 9-Tetradezensäure f
myristone Myriston n, 14-Heptakosanon n
myristyl alcohol Myristylalkohol m, Myristinalkohol m, 1-Tetradekanol n
myrmekite (Min) Myrmekit m
myrobalan (Gerb) Myrobalane f (Frucht mehrerer Terminalia-Arten)
myronic acid Myronsäure f
myrosin Myrosin n (in Senföl und Meerrettich vorkommendes Enzym)
myrrh gum Myrrhenharz n, Myrrhe f
m. oil Myrrhenöl n (von Commiphora-Arten)
myrtenal Myrtenal n, 6,6-Dimethyl-2-methanoylbizyklo-[1,1,3]-2-hepten n
myrtenic acid Myrtensäure f, 6,6-Dimethylbizyklo-[1,1,3]-2-hepten-2-karbonsäure f
myrtenol Myrtenol n, 6,6-Dimethyl-2-hydroxymethylbizyklo-[1,1,3]-2-hepten n
myrticolorin Myrtikolorin n, Rutin n, Querzetin-3-rutinosid n, Violaquerzitrin n
myrtle oil Myrtenöl n
m. wax Myrtenwachs n, Myrikawachs m, Myrikatalg m, Bayberrytalg m, Kapbeerenwachs n
myrtol Myrtol n (Fraktion des Myrtenöls zwischen 160 und 180 °C)

N

N-O peptidyl shift *(Text)* N-O-Peptidylverschiebung *f*
N-shell *(Kch)* N-Schale *f*
N-terminal N-terminal, N-endständig
na. s. nonagglomerating
nacre Perlmutter *f*, Perlmutt *n*
nacreous Perlmutt[er]..., aus Perlmutt[er] [bestehend], perlmuttern; perlmutt[er]artig, perlmutt[er]ähnlich, perlmutt[er]glänzend
n. effect Perlmutteffekt *m*
n. sulphur perlmutterartiger Schwefel *m (eine Schwefelmodifikation)*
nacrine, nacrite *(Min)* Nakrit *m (ein Phyllosilikat)*
nacrous *s.* nacreous
Nadi reagent Reagens *n* nach Nadi *(aus Dimethyl-p-phenylendiamin, α-Naphthol und Na₂CO₃, zum Nachweis von Indophenoloxydase)*
nadorite *(Min)* Nadorit *m (Blei(II)-antimon(III)-dioxidchlorid)*
nagyagite *(Min)* Nagyagit *m*, Blättertellur *m*
nail lacquer Nagellack *m*
n. lacquer remover Nagellackentferner *m*
n. polish Nagelpoliermittel *n*, Nagelpolitur *f*
naked barlay Nacktgerste *f*
n. flame offene Flamme *f*
nakrite *s.* nacrite
name / to benennen
name reaction Namenreaktion *f*
naming Benennen *n*, Benennung *f*, Namengeben *n*, Namengebung *f*
n. system Benennungssystem *n*
nantokite *(Min)* Nantokit *m (Kupfer(I)-chlorid)*
napalm Napalm *n*
naphtha Benzin *n (besonders für technische Zwecke oder als Reformingstock)*; Schwerbenzin *n*, schweres Benzin *n*, Naphtha *n(f) (Siedebereich 150 bis 210 °C)*; Ligroin *n (Siedebereich 90 bis 120 °C)*
n. fraction Benzinfraktion *f*; Schwerbenzinfraktion *f*, Naphthafraktion *f*
n. furnace Benzinofen *m*
n. recovery Benzinrückgewinnung *f*
n. wash tower Benzinwäscher *m (bei der Bleicherdebehandlung nach dem Perkolationsverfahren)*
naphthadianthrone Naphthadianthron *n*
naphthalane Naphthalan *n*, Perinaphthopyran *n*
naphthalene C₁₀H₈ Naphthalin *n*, Naphthalen *n*
n. carboxylic acid *s.* naphthoic acid
1,8-n. dicarboxylic acid C₁₀H₆(COOH)₂ 1,8-Naphthalindikarbonsäure *f*, Naphthalsäure *f*
n. series Naphthalinreihe *f*
n. sulphonyl chloride C₁₀H₇SO₂Cl Naphthalinsulfo[nyl]chlorid *n*
n. tetrachloride C₁₀H₈Cl₄ Naphthalintetrachlorid *n*, 1,2,3,4-Tetrachlor-1,2,3,4-tetrahydronaphthalin *n*
n. 1,2,3,4-tetrahydride C₁₀H₁₂ Naphthalin-

1,2,3,4-tetrahydrid *n*, 1,2,3,4-Tetrahydronaphthalin *n*
n. tray Naphthalinpfanne *f*
n. vapour Naphthalindampf *m*
α-naphthaleneacetic acid C₁₀H₇CH₂COOH α-Naphthylessigsäure *f*
naphthalenedisulphonic acid C₁₀H₆(SO₃H)₂ Naphthalindisulfo[n]säure *f*
naphthalenesulphonic acid C₁₀H₇SO₃H Naphthalinsulfo[n]säure *f*
naphthalic acid C₁₀H₆(COOH)₂ Naphthalsäure *f*, 1,8-Naphthalindikarbonsäure *f*
naphthalol Naphthalol *n*, Betol *n*, Salizylsäure-β-naphtholester *m*
naphthane C₁₀H₁₈ Naphthan *n*, *(veraltet für)* Dekahydronaphthalin *n*, Bizyklo-(0,4,4)-dekan *n*
naphthenate Naphthenat *n*
naphthene Naphthen *n*, Zykloalkan *n*, Zykloparaffin *n*
n. base Naphthenbasis *f*, naphthenische Basis *f (eines Roherdöls)*
n.-base crude [oil] *s.* n.-base petroleum
n.-base petroleum naphthen[bas]isches Erdöl *n (Rohöl) n*, Naphthenerdöl *n*, Naphthen[basis]öl *n*, Erdöl *n* auf Naphthenbasis
naphthenic naphthenartig, naphthenisch, gesättigt alizyklisch
n. acid Naphthensäure *f*
n.-aromatic petroleum naphthenisch-aromatisches Erdöl *(Rohöl) n*
n.-base crude [oil] *s.* naphthene-base petroleum
n. oil naphthenisches (naphthenhaltiges) Öl *n*
n. petroleum *s.* naphthene-base petroleum
naphthindigo Naphth[alin]indigo *m*, Naphthindolindigo *m*
naphthionic acid NH₂C₁₀H₆SO₃H Naphthionsäure *f*, 1-Naphthylamin-4-sulfonsäure *f*
naphthoic acid C₁₀H₇COOH Naphthoesäure *f*, Naphthalinkarbonsäure *f*
α-naphthol, 1-naphthol C₁₀H₇OH α-Naphthol *n*, 1-Hydroxynaphthalin *n*
α-naphthol orange NaSO₃C₆H₄N=NC₁₀H₆OH α-Naphtholorange *n*, Orange I *n (Natriumsalz der α-Naphtholazobenzolsulfonsäure-(4))*
β-naphthol, 2-naphthol C₁₀H₇OH β-Naphthol *n*, 2-Hydroxynaphthalin *n*, Isonaphthol *n*
naphthol AS Naphthol AS *n*, 2-Hydroxy-3-naphthoesäureanilid *n*, Anilidsäure *f*
n. blue black B Naphtholblauschwarz B *n*, Amidoschwarz 10 B *n*
n. component Naphtholkomponente *f*
n. pitch Naphtholpech *n*
n. salol Naphthalol *n*, Betol *n*, Salizylsäure-β-naphtholester *m*
naphtholdisulphonic acid HOC₁₀H₅(SO₃H)₂ Naphtholdisulfonsäure *f*
β-n. acid R C₁₀H₅(OH)(SO₃H)₂ β-Naphtholdisulfonsäure R *f*, R-Säure *f*, 2-Naphthol-3,6-disulfonsäure *f*
naphtholsulphonic acid HOC₁₀H₆SO₃H Naphtholsulfonsäure *f*

naphthopyrone Naphthopyron n
naphthoquinone $C_{10}H_6O_2$ Naphthochinon n, Dihydrodiketonaphthalin n
naphthostyril Naphthostyril n
naphthyl group $C_{10}H_7$ — Naphthylgruppe f, Naphthylrest m
naphthylamine $C_{10}H_7NH_2$ Naphthylamin n, Aminonaphthalin n
n. sulphate $(C_{10}H_7NH_2)_2H_2SO_4$ Naphthylaminsulfat n
naphthylaminedisulphonic acid $NH_2C_{10}H_5(SO_3H)_2$ Naphthylamindisulfonsäure f, Aminonaphthalindisulfonsäure f
naphthylaminesulphonic acid $C_{10}H_6NH_2(SO_3H)$ Naphthylaminsulfonsäure f, Aminonaphthalinsulfonsäure f
1-naphthylamine-3,6,8-trisulphonic acid $C_{10}H_4(NH_2)(SO_3H)_3$ 1-Naphthylamin-3,6,8-trisulfonsäure f, Kochsche Säure f
Naples yellow Neapelgelb n, Antimongelb n, Bleiantimonat(V) n
napping (Text) Noppen n, Rauhen n, Aufrauhen n
narbomycin Narbomyzin n
narcotic narkotisch [wirkend]
narcotic Narkotikum n, Narkosemittel n
narcotine Narkotin n (Alkaloid)
narrow-meshed engmaschig, kleinmaschig, feinmaschig
n.-mouth bottle s. n.-neck[ed] bottle
n.-neck[ed] bottle Enghalsflasche f
n.-neck[ed] flask Enghalskolben m
nascent naszierend, in statu nascendi
Nash Hytor pump Flüssigkeitsringverdichter m; Flüssigkeitsringpumpe f; Flüssigkeitsringgebläse n; Elmo-Pumpe f
native nativ, natürlich, unverändert; (Min, Met) gediegen, natürlich vorkommend
native s. n. metal
n. asphalt Naturasphalt m, natürlicher Asphalt m
n. copper gediegen[es] Kupfer n, Bergkupfer n
n. mercury gediegenes Quecksilber n, Jungfernquecksilber n
n. metal gediegenes Metall n
n. paraffin Erdwachs n, Bergtalg m, Ozokerit m
n. rubber (auf Kleinplantagen erzeugter Kautschuk)
n. serum Nativserum n, natives Serum n
n. silver (Min) gediegen[es] Silber n
n. starch native Stärke f
n. sulphur gediegener (natürlicher) Schwefel m
natrite s. natron
natrolite (Min) Natrolith m (Natriumdialumotrisilikat)
natron $Na_2CO_3 \cdot 10H_2O$ Kristallsoda f, Natriumkarbonat-10-Wasser n; (Min) Natron n, Soda f
n. cellulose Natronzellulose f
natural:
from n. sources natürlich [vorkommend], Natur...
n. ageing natürliche Alterung f
n. alloy Naturlegierung f

n. asphalt s. native asphalt
n. bleaching (Text) natürliche Bleiche f, Naturbleiche f, Rasenbleiche f
n. camphor natürlicher Kampfer m (von Cinnamomum camphora (L.)Sieb.)
n. cellulose natürliche (native) Zellulose f
n. cement Naturzement m
n. circulation natürlicher Umlauf m, Naturumlauf m
n. clay s. naturally occurring clay
n. coal Naturkohle f, Mineralkohle f, natürliche (mineralische, fossile) Kohle f
n. draught natürlicher Zug m, Naturzug m (z.B. eines Ofens)
n. draught cooling tower Kühlturm m mit natürlichem Zug, selbstbelüfteter Kühlturm m, Kaminkühlturm m
n. dye Naturfarbstoff m, natürlicher Farbstoff m
n. earth naturaktive Erde (Bleicherde) f, Roherde f
n. enzyme originäres Enzym (Ferment) n
n. fat natürliches Fett n, Naturfett n
n. fibre natürliche (native) Faser f, Naturfaser f; Naturfaserstoff m
n. flow freies Ausfließen n (des Erdöls bei der Förderung)
n. frequency Eigenfrequenz f, charakteristische Frequenz f
n. gas Naturgas n, Erdgas n
n. gas pipeline Erdgaspipeline f, Erdgasrohrleitung f
n. gasoline Erdgasbenzin n, Natur[gas]benzin n
n. glass Naturglas n, Gesteinsglas n
n. gum Naturgummi n, Pflanzenschleim m
n. hydrocarbon Naturkohlenwasserstoff m
n. ice Natureis n
n. indigo natürlicher Indigo m, Naturindigo m, Pflanzenindigo m
n. juice naturreiner Saft m, Muttersaft m
n. latex s. n. rubber latex
n. line width natürliche Linienbreite f (von Spektrallinien)
n. mineral water natürliches Mineralwasser n
n. moulding sand Natur[form]sand m, natürlicher (natürlich vorkommender) Sand (Formsand) m
n. musk natürlicher (echter) Moschus m, Naturmoschus m
n. parchment tierisches (animalisches) Pergament n, Hautpergament m, Pergament n
n. pearl essence Perl[en]essenz f, Fischsilber n
n. perfume Naturriechstoff m
n. pigment natürliches Pigment n
n. plastic plastischer Naturstoff m
n. product Naturprodukt n, Naturstoff m
n. resin Naturharz n, natürliches Harz n
n. rubber Naturkautschuk m, NK
n.-rubber compound s. n. rubber mix
n.-rubber graft [polymer] Pfropfpolymerisat (Pfropfpolymeres) n von Naturkautschuk
n. rubber latex Natur[kautschuk]latex m

n. rubber mix (stock) Naturkautschukmischung *f*

n. rubber vulcanizate Naturkautschukvulkanisat *n*

n. sand *s.* n. moulding sand

n. silk Naturseide *f*, reine (echte) Seide *f*

n. souring *(Lebm)* Spontansäuerung *f*

naturally bonded sand *s.* natural moulding sand

n. carbonated water [einfacher] Säuerling *m*, Sauerbrunnen *m*, kohlensaures Wasser *n*

n. occurring natürlich [vorkommend], Natur...

n. occurring clay Naturton *m*, natürlicher Ton *m*; natürliche (naturaktive) Tonerde (Bleicherde) *f*, Naturbleicherde *f*

n. soured milk spontan gesäuerte Milch *f*, Sauermilch *f*, Setzmilch *f*, Schlippermilch *f*, Plundermilch *f*

naumannite *(Min)* Naumannit *m (Silberselenid)*

navigation coal Marinekohle *f*, Schiffskesselkohle *f*

navy blue Marineblau *n*

NBR *s.* 1. nitrile-butadiene rubber; 2. acrylonitrile-butadiene rubber

NBS 1. *(Abk. für)* National Bureau of Standards; 2. *(Abk. für) New British Standard;* 3. *s.* N-bromo-succinimide

NBS abrader *(Gum)* Abriebprüfmaschine *f* des National Bureau of Standards

NCB *(Abk. für)* National Coal Board

N.C.I. powder *(Läusevertilgungsmittel aus Naphthalin, Kreosot und Jodoform)*

NCL *(Abk. für)* National Chemical Laboratory

NCR *s.* nitrile-chloroprene rubber

nd. *(Abk. für)* needles

neamine Neamin *n*, Neomyzin A *n (Antibiotikum)*

near-by position Nachbarstellung *f*, 1,2,3-Stellung *f*

n.-infrared radiation element Infrarot-Hellstrahler *m*

neat unvermengt, unvermischt, rein, unverdünnt *(z.B. Wein)*

neatsfoot oil Rinderklauenöl *n*, Klauenöl *n*, Rinderklauenfett *n*

nebulite Nebulit *m (ein metamorphes Mischgestein)*

nebulization Vernebeln *n*, Verneb[e]lung *f*

nebulize / to vernebeln, aerosolieren, aerolisieren

nebulizer *(Landw)* Nebelblaser *m*, Nebelgerät *n*, Atomiseur *m*

necic acid Nezinsäure *f (Säurekomponente eines Pyrrolizidin-Alkaloids)*

necine Nezin *n*

neck Hals *m (z.B. einer Flasche)*; Zapfen *m*, Hals *m (einer Welle)*; Zapfen *m (einer Walze)*

necking Querschnittsverminderung *f (beim Recken von Elementarfäden)*

neckring *(Glas)* Kopfform *f*

nectariferous plant honigende Pflanze *f*, Honigpflanze *f*, Trachtpflanze *f*, Nektarspender *m*, Nektarlieferant *m*

needle *(Krist)* Nadel *f*; *(Glas)* Plunger *m*, Stempel *m*, Stößel *m*, Treiber *m (eines Speisers)*

n.-bonded fabric *(Text)* Nadelfilz *m*

n. bonding *(Text)* Nadelfilzherstellung *f*

n. chain *(Text)* Nadelkette *f*

n. culture *(Bioch)* Stichkultur *f*

n. felt *s.* n.-bonded fabric

n. lead Nadelblei *n*

n.-like nadelförmig, nadelartig, nadelig, Nadel...

n.-punched felt *s.* n.-bonded fabric

n.-shaped *s.* n.-like

n. tear resistance Nadelausreißwiderstand *m*, Nadelausreißfestigkeit *f*

n. valve Nadelventil *n*

n. vibrator Innenrüttler *m*, Innenvibrator *m*, Tauchrüttler *m*, Tiefenrüttler *m*

n. zeolite *s.* natrolite

negative *(Foto)* Negativ *n*

n. adsorption negative Adsorption *f*

n. auxochromic group negatives Auxochrom *n*

n. catalysis negative Katalyse *f*, Antikatalyse *f*, Inhibition *f*, Hemmung *f*, Verzögerung *f*

n. catalyst negativer Katalysator *m*, Antikatalysator *m*, Inhibitor *m*, Hemmstoff *m*, Verzögerer *m*, Passivator *m*

n. copy *(Foto)* Negativkopie *f*, Schwarzkopie *f*

n. developer *(Foto)* Negativentwickler *m*

n. development *(Foto)* Negativentwicklung *f*

n. electrode negative (negativ geladene) Elektrode *f*, Katode *f*

n. electron *s.* negatron

n. emulsion *(Foto)* Negativemulsion *f*

n. image *(Foto)* negatives Bild *n*, Negativbild *n*

n. ion negatives (negativ geladenes) Ion *n*, Anion *n*

n. material *(Foto)* Aufnahmematerial *n*, Negativmaterial *n*

n. paper Negativpapier *n*

n. proton Antiproton *n*

negatively charged negativ geladen

negatron Negatron *n*, *(normales, negativ geladenes)* Elektron *n*

neglect / to vernachlässigen, unberücksichtigt lassen

negligible zu vernachlässigen[d], vernachlässigbar

neighbouring benachbart, nachbarständig, angrenzend, anliegend, anstoßend, vizinal, Nachbar..., Neben..., Rand...

n. atom Nachbaratom *n*

n. group effect Nachbargruppeneffekt *m*

n. position Nachbarstellung *f*, 1,2,3-Stellung *f*

Nelson cell Nelson-Zelle *f (zur Chloralkali-Elektrolyse)*

nemalite *(Min)* Nemalith *m (Magnesiumhydroxid)*

nematic phase *(Koll)* nematische Phase *f*, pl-Phase *f*

n. state nematischer Aggregatzustand *m*

nematoblastic *(Geol)* nematoblastisch, faserig

nematocide Nematozid *n (Mittel gegen Fadenwürmer)*

nematolith s. nemalite
neo-ceramic glass Glaskeramik f, Keramik f aus Glas, glaskeramischer Stoff m, Vitrokeram n
neoarsphenamine $(NH_2)(OH)C_6H_3As=As-C_6H_3(OH)(NHCH_2OSOH)$ Neoarsphenamin n, m,m'-Diamino-p,p'-dihydroxyarsenobenzol-N-methylensulfoxylsäure f
neochrysolite, neocrisolite (Min) Neochrysolith m, (veraltet für) Fayalit m (Eisen(II)-orthosilikat)
neocyanine Neozyanin n, Allozyanin n
neocyanite (Min) Neozyanit m, Lit[h]idionit m (ein kupferhaltiges Phyllosilikat)
neodymia Nd_2O_3 Neodymerde f, Neodymia f, (veraltet für) Neodymoxid n
neodymium Nd Neodym n
 n. acetate $Nd(CH_3COO)_3$ Neodymazetat n
 n. bromate $Nd(BrO_3)_3$ Neodymbromat n
 n. bromide $NdBr_3$ Neodymbromid n
 n. carbide NdC_2 Neodymkarbid n
 n. chloride $NdCl_3$ Neodymchlorid n
 n. iodide NdJ_3 Neodymjodid n
 n. molybdate $Nd_2(MoO_4)_3$ Neodymmolybdat(VI) n
 n. nitrate $Nd(NO_3)_3$ Neodymnitrat n
 n. nitride NdN Neodymnitrid n
 n. oxide Nd_2O_3 Neodymoxid n
 n. sulphate $Nd_2(SO_4)_3$ Neodymsulfat n
 n. sulphide Nd_2S_3 Neodymsulfid n
neohexane $C_2H_5C(CH_3)_3$ 2,2-Dimethylbutan n
neolite (Min) Neolith m (unreiner Serpentin)
neomagma palingenetisches Magma n
neomethymycin Neomethymyzin n (Antibiotikum)
neomineralization Mineralneubildung f
neomorph (Geol) neomorph
neomycin Neomyzin n (Antibiotikum)
 n. A Neomyzin A n, Neamin f
neon Ne Neon n
 n. discharge tube Neonröhre f
neopentane $C(CH_3)_4$ Neopentan n, Tetramethylmethan n, 2,2-Dimethylpropan n
neoprene Neopren n, Poly-2-chlor-1,3-butadien n
 n. cement Neopren[klebe]zement m
 n. compound Neoprenmischung f
 n. latex Neoprenlatex m
 n. stock s. n. compound
 n. vulcanizate Neoprenvulkanisat n
neotype (Min) Neotyp m, (veraltet für) Bariumkalzit m
neovolcanic neovulkanisch
nep (Text) Nest n, Nisse f, Knötchen n
Nepal cardamom Nepalkardamom m(n), Bengalischer Kardamom m(n) (von Amomum aromaticum Roxb. und A. subulatum Roxb.)
nepheline, nephelite (Min) Nephelin m (ein Tektosilikat)
nephelometer Nephelometer n, Trübungsmesser m
nephelometric nephelometrisch
 n. analysis Nephelometrie f
 n. titration nephelometrische Titration f

nephelometry Nephelometrie f, (i.e.S.) Tyndallometrie f, Streulichtmessung f
nephrite (Min) Nephrit m
nephrotoxic nephrotoxisch, nierenschädigend
neptunium Np Neptunium n
neral $(CH_3)_2C=CHCH_2CH_2C(CH_3)=CHCHO$ Neral n, Zitral B n, cis-Zitral n
Nernst calorimeter Kalorimeter n von Nernst, Nernstsches Metallkalorimeter n
Nernst distribution law Nernstsches Verteilungsgesetz n, Nernstscher Verteilungssatz m
Nernst glower s. Nernst lamp
Nernst heat theorem Nernstsches Wärmetheorem n
Nernst lamp Nernst-Lampe f, Nernst-Brenner m
nerol $(CH_3)_2C=CHCH_2CH_2C(CH_3)=CHCH_2OH$ Nerol n, 3,7-Dimethyl-2,6-oktadien-1-ol n
neroli oil Neroliöl n, Pomeranzenblütenöl n (von Citrus aurantium L. ssp. aurantium)
nerve (Gum) Nerv m
 n. gas Nervengas n
nervonic acid $CH_3(CH_2)_7CH=CH(CH_2)_{13}COOH$ Nervonsäure f
nervy (Gum) nervig; (Plast) strähnig (von Extrudaten bei zu niedriger Düsentemperatur)
NES cure s. nonelemental sulphur cure
Nessler glass (tube) (Kolorimetrie) Neßler-Rohr n, Neßler-Röhre f, Neßler-Gefäß n
nesslerization Stickstoffbestimmung f nach Neßler
nesslerize / to mit Neßlers Reagens versetzen
Nessler's reagent (solution) Neßlers Reagens n (zum Nachweis von Ammoniak)
nest Satz m (gleicher Geräte von verschiedener Größe)
 n. fertilization (Landw) Nest[er]düngung f
net calorific value Heizwert m (i.e.S.), H_u (veraltet) unterer Heizwert m, H_u
 n. coking heat untere Verkokungswärme f, Nettowärme f, Nutzwärme f
 n. heating value s. n. calorific value
 n. plane (Krist) Netzebene f, Gitterebene f
 n. positive suction head Haltedruck m, größtmögliche Saughöhe f
nettle fibre Nesselfaser f
network Netzwerk n
 n. analysis Netzwerkanalyse f
 n. co-former (Glas) Netzwerkbildner und -wandler m
 n. former, n.-forming ion (Glas) Netzwerkbildner m, Netzwerkformer m
 n. intermediate (Glas) Netzwerkzwitter m
 n. modifier, n.-modifying ion (Glas) Netzwerkwandler m, Modifizierer m
Neumann lines (Krist) Neumannsche Linien fpl
neurine $CH_2=CHN(CH_3)_3OH$ Neurin n, Vinyltrimethylammoniumhydroxid n
neurotoxic neurotoxisch, nervenschädigend
neurotoxicity Neurotoxizität f
neutral neutral
 to be n. neutral reagieren

to make n. neutralisieren
n. atom neutrales Atom *n*
n. black neutralschwarz
n. colloid Neutralkolloid *n*
n. fat Neutralfett *n*
n. lard Neutrallard *n*, Neutralschmalz *n*
n. lead arsenate Pb₃(AsO₄)₂ Bleiarsenat *n*
n. magnesium phosphate $Mg_3(PO_4)_2$ Magnesium[ortho]phosphat *n*, Trimagnesiumphosphat *n*
n. mercuric phosphate $Hg_3(PO_4)_2$ Quecksilber(II)-[ortho]phosphat *n*
n. mercurous phosphate Hg_3PO_4 Quecksilber(I)-[ortho]phosphat *n*
n. meson neutrales Meson *n*, Neutretto *n*
n. oil Neutralöl *n*, neutrales Öl *n*
n. point Neutralpunkt *m*
n. potassium phosphate K_3PO_4 Kalium[ortho]phosphat *n*, Trikaliumphosphat *n*
n. [reclaiming] process *(Gum)* Neutral[salz]verfahren *n* *(Regenerierverfahren)*
n. red Neutralrot *n*, Toluylenrot *n* *(ein Redoxindikator)*
n. refractory [material] neutrales feuerfestes (ff.) Material *n*
n. rock neutrales (intermediäres) Gestein (Eruptivgestein, Erstarrungsgestein) *n*
n. salt neutrales (normales) Salz *n*, Neutralsalz *n*, Normalsalz *n*
n.-salt effect Neutralsalzwirkung *f* *(in der Azidimetrie)*
n.-salt error Neutralsalzfehler *m*, Salzfehler *m* *(Meßfehler bei pH-Bestimmungen)*
n. soap neutrale Seife *f*
n. sodium sulphite cooking liquor *(Pap)* Neutralsulfitkochlauge *f*
n. sodium sulphite process *(Pap)* Neutralsulfitverfahren *n*, NSSC-Verfahren *n*
n. sodium sulphite semichemical pulp Neutralsulfit-Halbzellstoff *m*, Neutralsulfitstoff *m*, NSSC-Stoff *m*
n. sodium sulphite semichemical pulping Neutralsulfit- Halbzellstoffaufschluß *m*
n. sodium sulphite waste liquor *(Pap)* Neutralsulfitablauge *f*
n. sulphite process *s.* n. sodium sulphite process
n. sulphite semichemical liquor *(Pap)* Neutralsulfitkochlauge *f*
n. sulphite semichemical process *s.* n. sodium sulphite process
n. sulphite semichemical spent liquor *(Pap)* Neutralsulfitablauge *f*
n. syntan synthetischer Hilfsgerbstoff *m*
n. verdigris $Cu(CH_3COO)_2 \cdot H_2O$ kristallisierter (gereinigter) Grünspan *m*, Kupfer(II)-azetat *n*
neutrality Neutralität *f*
neutralization Neutralisieren *n*, Neutralisation *f*, *(Ölraffination auch)* chemisches Entsäuern *n*, Laugeentsäuerung *f*; Neutralisationsvorgang *m*, Neutralisationsreaktion *f*

n. curve Neutralisationskurve *f*
n. indicator Neutralisationsindikator *m*
n. method Neutralisationsverfahren *n*, Neutralisationsanalyse *f*
n. number Neutralisationszahl *f*
n. reaction Neutralisationsvorgang *m*, Neutralisationsreaktion *f*
n. titration Neutralisationstitration *f*, Säure-Basen-Titration *f*
n. with alkali solutions Laugeentsäuerung *f*, Neutralisationsverfahren *n* mit (durch) Laugen
neutralize / to neutralisieren, *(Ölraffination auch)* entsäuern
neutralizer Neutralisationsmittel *n*, Neutralisierungsmittel *n*, Neutralisator *m*
neutralizing activity Neutralisationsaktivität *f*, Neutralisierungsaktivität *f*
n. agent *s.* neutralizer
neutrals Neutralöle *npl*
neutretto Neutretto *n*, neutrales Meson *n*
neutrino Neutrino *n*, *ν*
neutron Neutron *n*
n. absorber Neutronenabsorber *m*, Neutronenfänger *m*
n.-activation analysis Neutronenaktivierungsanalyse *f*
n.-binding energy Neutronenbindungsenergie *f*, Bindungsenergie *f* des Neutrons
n. bombardment Neutronenbeschuß *m*, Neutronenbestrahlung *f*
n. capture Neutroneneinfang *m*
n.-capture cross section Neutroneneinfangquerschnitt *m*
n. decay Neutronenzerfall *m*
n. density Neutronendichte *f*
n. detection Neutronennachweis *m*
n. diffraction Neutronenbeugung *f*
n. diffraction study Neutronenbeugungsuntersuchung *f*
n. emission Neutronenemission *f*
n. excess Neutronenüberschuß *m*
n. flux Neutronenfluß *m*
n.-irradiated neutronenbestrahlt
n. irradiation Neutronenbestrahlung *f*
n. moderation Neutronenbremsung *f*
n. poison Reaktorgift *n*
n. rest mass Neutronenruh[e]masse *f*
n. source Neutronenquelle *f*
n. spectrum Neutronenspektrum *n*
Nevile-Winther acid $C_{10}H_6(OH)SO_3H$ Nevile-Winther-Säure *f*, 1-Naphthol-4-sulfonsäure *f*
new beer grünes Bier *n*, Jungbier *n*
n. fuchsine Neufuchsin *n*
n. green $Cu(CH_3COO)_2 \cdot 3Cu(AsO_2)_2$ Neugrün *n*, Schweinfurter Grün *n*, Kupferarsenitazetat *n*
n. rubber Frischkautschuk *m*
n. wine Jungwein *m*
New Jersey zinc[-recovery] process New-Jersey-Zinkverfahren *n*
Newcastle kiln Newcastle-Ofen *m* *(Ofen mit horizontaler Flammenführung)*

news s. newsprint paper
n. board Graupappe f
n. ink s. newsprint ink
n. machine s. newsprint paper machine
newsprint Zeitungsdruck m; s. n. paper
n. ink Zeitungs[druck]farbe f
n. paper Zeitungsdruckpapier n, Rotations-
druckpapier n
n. paper machine Zeitungsdruckpapiermaschine
f
newton Newton n, N
Newton law for cooling (heat loss) Newtonsches
Abkühlungsgesetz n
Newton laws of motion Newtonsche Bewegungs-
gesetze npl   —
Newton rings Newtonsche Ringe mpl
Newtonian flow (Plast) Newtonsches Fließen n
Newtonian fluid (liquid) Newtonsche Flüssigkeit f
next-to-end carbon endständiges C-Atom n
ngai camphor Ngaikampfer m (chemisch fast
reines l-Borneol)
ni-carbing Karbonitrieren n, Karbonitrierung f,
Gaszyanieren n, Trockenzyanieren n
niacin C_5H_4NCOOH Niazin n, Nikotinsäure f,
3-Pyridinkarbonsäure f
n. amide $C_5H_4NCONH_2$ Nikotin[säure]amid n,
Niazinamid n, Pyridin-3-karbonsäureamid n
niamide s. niacin amide
Nicaragua ipecacuanha Nikaragua-Brechwurzel f,
Kartagena-Brechwurzel f, Panama-Brechwurzel
f (von Cephaëlis acuminata Karsten)
niccolic s. niccolous
niccolite (Min) Niccolit m, Nickelin m, Rotnickelkies m (Nickelarsenid)
niccolous Nickel...
Nichols-Freeman flash roaster Nichols-Freeman-
Ofen m, NF-Ofen m, Blitzröstofen m nach
Nichols-Freeman
Nichols furnace Nichols-Ofen m, Mehretagenröstofen m nach Nichols
Nicholson blue Nicholson-Blau n, Alkaliblau n
nickel Ni Nickel n
n. acetate $Ni(CH_3COO)_2$ Nickelazetat n
n. acid fluoride $NiF_2 \cdot 5HF$ Nickelhydrogenfluorid
n
n. alloy Nickellegierung f
n. alloy steel nickellegierter Stahl m, Nickelstahl
m
n.-ammonium bromide $[Ni(NH_3)_6]Br_2$ Hexam-
minnickel(II)-bromid n
n.-ammonium chloride $NH_4Cl \cdot NiCl_2 \cdot 6H_2O$ Am-
moniumnickel(II)-chlorid-6-Wasser n
n.-ammonium sulphate $(NH_4)_2SO_4 \cdot NiSO_4 \cdot 6H_2O$
Ammoniumnickel(II)-sulfat-6-Wasser n
n. antimony glance (Min) Nickelantimonglanz m,
(veraltet für) Ullmannit m (Nickelantimonsulfid)
n. arsenate s. n. orthoarsenate
n.-base auf Nickelbasis (Nickelgrundlage)
n. bloom (Min) Nickelblüte f (Nickelarsenat)
n. brass (Am) Neusilber n

n.-cadmium accumulator (cell) Nickel-
Kadmium-Akkumulator m, Kadmium-Nickel-
Sammler m
n. carbide Ni_3C Nickelkarbid n
n. carbonate $NiCO_3$ Nickelkarbonat n
n. carbonyl s. n. tetracarbonyl
n. catalyst Nickelkatalysator m
n. chelate Nickelchelat n
n. chloride hexammine $[Ni(NH_3)_6]Cl_2$ Hexammin-
nickel(II)-chlorid n
n. cyanide $Ni(CN)_2$ Nickelzyanid n, Nik-
kel(II)-zyanid n
n. dimethylglyoxime $Ni[(CH_3)_2(CNO)_2H]_2$ Nik-
keldimethylglyoxim n, Nickeldiazetyldioxim n
n. dithionate NiS_2O_6 Nickel(II)-dithionat n
n. electrode Nickelelektrode f
n. formate $Ni(HCOO)_2$ Nickelformiat n
n.-free nickelfrei, Ni-frei
n. gymnite (Min) Nickelgymnit m
n. iodide hexammine $[Ni(NH_3)_6]J_2$ Hexamminnik-
kel(II)-jodid n
n.-iron battery (cell) Nickel-Eisen-Akkumulator
m, NiFe-Akkumulator m
n.-linn[a]eite (Min) Nickellinneit m, (veraltet für)
Polydymit m (Trinickeltetrasulfid)
n. matte Nickelstein m
n. monosulphide NiS Nickel[mono]sulfid n,
Nickel(II)-sulfid n
n. monoxide NiO Nickel[mon]oxid n, Nik-
kel(II)-oxid n
n. nitrate $Ni(NO_3)_2$ Nickel(II)-nitrat n
n. nitrate tetrammine $[Ni(NH_3)_4](NO_3)_2$ Te-
tramminnickel(II)-nitrat n
n. orthoarsenate $Ni_3(AsO_4)_2$ Nickel[ortho]ar-
senat(V) n
n. orthophosphate $Ni_3(PO_4)_2$ Nickel(II)-[ortho]-
phosphat n
n. pellets Nickelkugeln fpl
n. perchlorate $Ni(ClO_4)_2$ Nickelperchlorat n
n. peroxide s. n. sesquioxide
n. phosphate s. n. orthophosphate
n. phthalocyanine Nickelphthalozyanin n
n.-plated vernickelt
n. plating Vernickeln n, Vernickelung n
n.-plating bath Vernickelungsbad n
n.-potassium cyanide $K_2[Ni(CN)_4]$ Kaliumtetra-
zyanoniccolat(II) n
n. protoxide s. n. monoxide
n. sesquioxide Ni_2O_3 Dinickeltrioxid n, Nik-
kel(III)-oxid n
n. silver Neusilber n
n.-skutterudite (Min) Nickelskutterudit m, Chlo-
antit m, Weißnickelkies m (Nickelarsenid)
n. steel Nickelstahl m
n.-steel lining Auskleidung f aus Nickelstahl
n. stibine (Min) Ullmannit m (Nickelantimonsulfid)
n. subsulphide Ni_2S Nickelsubsulfid n, Di-
nickelmonosulfid n
n. sulphate $NiSO_4$ Nickelsulfat n

n. tetracarbonyl Ni(CO)$_4$ Nickeltetrakarbonyl *n*
n. vitriol *(Min)* Nickelvitriol *m*, Morenosit *m* *(Nickelsulfat-7-Wasser)*
nickelic Nickel..., *(i.e.S.)* Nickel(III)-...
n. hydroxide Ni(OH)$_3$ Nickel(III)-hydroxid *n*
n. oxide *s.* nickel sesquioxide
nickeliferous nickelhaltig
nickelin[e], nickelite *(Min)* Nickelin *m*, Niccolit *m*, Rotnickelkies *m (Nickelarsenid)*
nickelization Vernickeln *n*, Vernickelung *f*
nickelize / *to* vernickeln
nickelling *s.* nickelization
nickelous Nickel..., *(i.e.S.)* Nickel(II)-...
n. arsenate Ni$_3$(AsO$_4$)$_2$ Nickel[ortho]arsenat(V) *n*
n. bromide NiBr$_2$ Nickel(II)-bromid *n*, Nikkeldibromid *n*
n. chloride NiCl$_2$ Nickel(II)-chlorid *n*, Nikkeldichlorid *n*
n. ferrocyanide Ni$_2$[Fe(CN)$_6$] Nickel(II)-hexazyanoferrat(II) *n*
n. fluoride NiF$_2$ Nickel(II)-fluorid *n*, Nickeldifluorid *n*
n. fluorosilicate Ni[SiF$_6$] Nickelhexafluorosilikat *n*
n. hydroxide Ni(OH)$_2$ Nickel(II)-hydroxid *n*
n. iodide NiJ$_2$ Nickel(II)-jodid *n*, Nickeldijodid *n*
n.-nickelic oxide Ni$_3$O$_4$ Nickel(II,III)-oxid *n*
n.-nickelic sulphide Ni$_3$S$_4$ Nickel(II,III)-sulfid *n*, Trinickeltetrasulfid *n*
n. orthophosphate Ni$_3$(PO$_4$)$_2$ Nickel(II)-[ortho]phosphat *n*
n. oxide NiO Nickel(II)-oxid *n*, Nickel[mon]oxid *n*
n. phosphate *s. n.* orthophosphate
n. silicofluoride *s. n.* fluorosilicate
Nicol prism Nicol-Prisma *n*, Nicolsches Prisma *n*, Nicol *m*
nicotinamide *s.* nicotinic acid amide
nicotine Nikotin *n*
n. poisoning Nikotinvergiftung *f*
nicotinic acid C$_5$H$_4$NCOOH Nikotinsäure *f*, 3-Pyridinkarbonsäure *f*, Niazin *n*
n. acid amide Nikotin[säure]amid *n*, 3-Pyridinkarbon[säure]amid *n*
Niger copal *(Kopal von Daniella oblonga Oliv.)*
night cream Nachtcreme *f*
nigrine Nigrin *m (titanreiches Mineral)*
nigrotic acid Nigrotinsäure *f*, 3,5-Dihydroxynaphthoesäure-(2)-sulfonsäure-(7) *f*
nihilum album Nihilum album *n*, weißes Nichts *n*, *(veraltet für)* Zinkblumen *fpl (weißes, wollartiges Zinkoxid)*
nikethamide Nikäthamid *n*, Nizethamid *n*, Pyridin-3-karbonsäurediäthylamid *n*
nil *(Versuchswesen)* Null[variante] *f*, Nullversuch *m*, *(Landw auch)* Nullfläche *f*
ninhydrin reaction (test) Ninhydrinreaktion *f (zum Nachweis von Proteinen und Aminosäuren)*
niobe oil C$_6$H$_5$COOCH$_3$ Niobeöl *n*, Benzoesäuremethylester *m*, Methylbenzoat *n*
niobic Niob..., *(i.e.S.)* Niob(V)-...

n. acid Nb$_2$O$_5$·xH$_2$O Niobpentoxidhydrat *n*, Niobsäure *f*
niobite Niobit *m (Nb- und Ta-haltiges oxidisches Mineral)*
niobium Nb Niob *n*
n. carbide NbC Niobkarbid *n*
n. dioxide NbO$_2$ Niobdioxid *n*, Niob(IV)-oxid *n*
n. hydride NbH Niobhydrid *n*
n. hydroxide Nb(OH)$_5$ Niobhydroxid *n*
n. monoxide NbO Niob[mon]oxid *n*, Niob(II)-oxid *n*
n. pentabromide NbBr$_5$ Niobpentabromid *n*, Niob(V)-bromid *n*
n. pentachloride NbCl$_5$ Niobpentachlorid *n*, Niob(V)-chlorid *n*
n. pentafluoride NbF$_5$ Niobpentafluorid *n*, Niob(V)-fluorid *n*
n. pentoxide Nb$_2$O$_5$ Niobpentoxid *n*, Niob(V)-oxid *n*
niobous Niob..., *(i.e.S.)* Niob(III)-...
nip [of the rolls] Walzenspalt *m*
n. pressure *(Pap)* Liniendruck *m (zwischen den Walzen)*, Walzenanpreßdruck *m*
n. roller Haltewalze *f*, Preßwalze *f*, Andruckwalze *f*
n. rolls *(Pap)* Feuchtglättwerk *n*, Feuchtglätte *f*, Zweiwalzen[feucht]kalander *m*
nipple Nippel *m*
nisinic acid C$_{23}$H$_{35}$COOH Nissinsäure *f*, 4,8,12,15,18,21-Tetrakosahexaensäure *f*
niter *(Am) s.* nitre
niton Nt Niton *n*, *(veraltet für)* Rn Radon *n*
nitramide NH$_2$NO$_2$ Nitramid *n*
nitramine *s.* nitramide; Nitramin *n (Nitramid-Derivat der allgemeinen Formel RHN·NO$_2$ oder R$_1$R$_2$N·NO$_2$)*
nitranilic acid (NO$_2$)$_2$C$_6$O$_2$(OH)$_2$ Nitranilsäure *f*, 2,5-Dihydroxy-3,6-dinitrochinon *n*
nitraniline Nitr[o]anilin *n*, Aminonitrobenzol *n*
nitrate / *to* nitrieren
nitrate Nitrat *n*
n. bacteria Nitratbakterien *npl*, Nitratbildner *mpl*, Nitrifikanten *mpl (Gattung Nitrobacter)*
n. compound Nitratverbindung *f*
n. nitrogen Nitratstickstoff *m*
n. of lime Ca(NO$_3$)$_2$ Kalksalpeter *m*, Kalziumnitrat *n*
n. of potash KNO$_3$ Kaliumnitrat *n*, Kalisalpeter *m*
n. of soda NaNO$_3$ Natriumnitrat *n*, Natronsalpeter *m*
n. paper *s.* nitrated paper
n. pulping *(Pap)* Salpetersäureprozeß *m*, Salpetersäureverfahren *n*, Salpetersäureaufschluß *m*
nitrated paper Nitrierkrepp *m*
nitratine *(Min)* Nitratin *m*, *(veraltet für)* Natronsalpeter *m (Natriumnitrat)*
nitrating acid Nitriersäure *f*, *(Tech)* Mischsäure *f* *(Gemisch aus konzentrierter Salpeter- und Schwefelsäure)*

n. agent Nitrier[ungs]mittel *n*, Nitrierungsagens *n*, nitrierendes Agens *n*

n. department Nitrierabteilung *f*

n. pan Nitriergefäß *n*, Nitrierkessel *m*

n. paper s. nitrated paper

n. unit Nitrier[ungs]anlage *f*

nitration Nitrieren *n*, Nitrierung *f*

n. mixture Nitrier[ungs]gemisch *n*

n. plant Nitrier[ungs]anlage *f*, Nitrierungsbetrieb *m*

n. product Nitrier[ungs]produkt *n*

nitratite s. nitratine

nitrator Nitrierer *m*, Nitrator *m*, Nitrierapparat *m*, Nitriergefäß *n*

nitre Salpeter *m (i.w.S.); (Min)* Niter *m*, Nitronatrit *m*, Natronsalpeter *m (Natriumnitrat)*; C_2H_5ONO Salpetrigsäureäthylester *m*, Äthylnitrit *n; (veraltet für)* natron

n. cake Natriumkuchen *m (Gemisch aus Na_2SO_4 und $NaHSO_4$ als Nebenprodukt der Salpetersäureherstellung); (Am)* $NaHSO_4$ Natriumhydrogensulfat *n*

nitric Stickstoff...; höherwertigem Stickstoff entsprechend, *(meist)* Stickstoff(V)-...

n. acid HNO_3 Salpetersäure *f*

n. acid oxidation Salpetersäureoxydation *f*

n. acid pulping Salpetersäure[zellstoff]verfahren *n*, Holzaufschluß *m* mit Salpetersäure

h. anhydride N_2O_5 Stickstoff(V)-oxid *n*, Distickstoffpentoxid *n*

n. bacteria s. nitrate bacteria

n. ether $C_2H_5NO_3$ Salpetersäureäthylester *m*, Äthylnitrat *n*

n. oxide NO Stickstoff(II)-oxid *n*, Stickstoff[mon]oxid *n*, Stickoxid *n (i.e.S.)*

nitridation s. nitride hardening

nitride / to *(Met)* nitrier[härt]en, durch Nitrierung härten, aufsticken, versticken

nitride Nitrid *n*

n. case s. nitrided case

n. hardening *(Met)* Nitrier[härt]en *n*, Nitrier[härt]ung *f*, Nitridhärten *n*, Nitridhärtung *f*, Stickstoffhärten *n*, Stickstoffhärtung *f*, Aufsticken *n*, Aufstickung *f*, Versticken *n*, Verstickung *f*

nitrided case (layer) *(Met)* nitrierte Randschicht (Schicht) *f*, nitrierter Rand *m*, Nitrierschicht *f*

n. steel nitrier[gehärte]ter Stahl *m*

nitriding s. nitride hardening

n. action *(Met)* Nitrierwirkung *f*, nitrierende Wirkung *f*

n. bath *(Met)* Nitrier[salz]bad *n*

n. box *(Met)* Nitrierkasten *m*

n. depth *(Met)* Nitriertiefe *f*, Tiefe *f* der Nitrierschicht

n. furnace *(Met)* Nitrierofen *m*

n. process *(Met)* Nitrier[härte]verfahren *n*

n. steel Nitrierstahl *m*, Stahl *m* für Nitrierhärtung

n. temperature *(Met)* Nitriertemperatur *f*

n. time *(Met)* Nitrierzeit *f*

nitridize / to 1. mit Stickstoff vereinen; 2. *(die Oxydationsstufe durch Einwirkung von Stickstoff verändern)*

nitrification *(Landw)* Nitrifikation *f*, Nitrifizierung *f*

nitrifiers nitrifizierende Organismen *mpl*

nitrifying bacteria nitrifizierende Bakterien *npl*, Nitrifikationsbakterien *npl*, Nitrobakterien *npl (Sammelname für Nitrit- und Nitratbakterien)*

nitrile R·C≡N Nitril *n*

n.-butadiene rubber Nitrilkautschuk *m*, Butadien-Akrylnitrilkautschuk *m (Butadien-Akrylnitril-Mischpolymerisat)*

n. cement Klebelösung *f* auf Nitrilkautschukbasis

n.-chloroprene rubber Nitril-Chloroprenkautschuk *m (Chlorbutadien-Akrylnitril-Mischpolymerisat)*

n. group —C≡N Nitrilgruppe *f*

n. rubber s. n.-butadiene rubber

n.-silicone rubber Nitrilsilikongummi *m*, Nitrilsilikonkautschuk *m*

n. synthesis Nitrilsynthese *f*

nitrilo-triacetic acid $N(CH_2COOH)_3$ Nitrilotriessigsäure *f*, NTE

nitrite Nitrit *n*

n. bacteria s. n.-forming bacteria

n.-containing nitrithaltig

n.-forming bacteria Nitritbakterien *npl*, Nitritbildner *mpl*, Nitrosebakterien *npl*

nitro body Nitrokörper *m*

n. compound Nitroverbindung *f*

n. derivative Nitroderivat *n*

n. dye Nitrofarbstoff *m*

n. group —NO_2 Nitrogruppe *f*

n.-isonitro tautomerism Nitro-Isonitro-Tautomerie *f*

n. musk Nitromoschus *m*

nitroacetanilide Nitr[o]azetanilid *n*

nitroacetic acid O_2NCH_2COOH Nitroessigsäure *f*

nitroalkane Nitroalkan *n*, Nitroparaffin *n*

nitroaniline $NH_2C_6H_4NO_2$ Nitr[o]anilin *n*, Aminonitrobenzol *n*

nitroanthraquinone Nitroanthrachinon *n*

nitrobacteria s. 1. nitrifying bacteria; 2. *(i.e.S.)* nitrate bacteria

5-nitrobarbituric acid 5-Nitrobarbitursäure *f*, Dilitursäure *f*

nitrobarite *(Min)* Nitrobarit *m*, Nitrobaryt *m*, Barytsalpeter *m (Bariumnitrat)*

nitrobenzaldehyde $O_2NC_6H_4CHO$ Nitrobenzaldehyd *m*

nitrobenzene $C_6H_5NO_2$ Nitrobenzol *n*

p-nitrobenzenesulphonyl chloride $NO_2C_6H_4SO_2Cl$ p-Nitrobenzolsulfo[nyl]chlorid *n*

nitrobenzoic acid $O_2NC_6H_4COOH$ Nitrobenzoesäure *f*, Nitrobenzolkarbonsäure *f*

nitrocalcite *(Min)* Nitrokalzit *m*, Kalksalpeter *m (Kalziumnitrat)*

nitrocellulose Nitrozellulose *f*, *(fälschlich für)* Zellulosenitrat *n*, Nitratzellulose *f*, Zellulosesalpetersäureester *m*

n. lacquer Nitro[zellulose]lack *m*
n. propellant Nitrozellulosepulver *n*, Nc-Pulver *n*, *(fälschlich für)* Zellulosenitratpulver *n*
nitrocementation *(Met)* Gaszyanieren *n*, Trokkenzyanieren *n*, Karbonitrieren *n*
nitrochalk Kalkammonsalpeter *m*
nitrochlorobenzene $C_6H_4Cl(NO_2)$ Nitrochlorbenzol *n*, Chlornitrobenzol *n*
nitrocobaltate Nitrokobaltat *n*
nitrocotton Schieß[baum]wolle *f*
nitroethane $CH_3CH_2NO_2$ Nitroäthan *n*
nitrogation *(Düngung durch Berieseln mit schwach ammoniakalischem Wasser)*
nitrogen N Stickstoff *m*
n. bridge Stickstoffbrücke *f*
n. carrier *(Landw)* Stickstoffträger *m*
n. case hardening Stickstoffhärten *n*, Nitrierhärten *n*, Nitridhärten *n*, Nitrieren *n*, Aufsticken *n*, Versticken *n*
n. case-hardening process Nitrier[härte]verfahren *n*
n. chloride s. n. trichloride
n. content Stickstoffgehalt *m*
n. cycle Stickstoffkreislauf *m*
n. determination Stickstoffbestimmung *f*
n. determination apparatus Stickstoffbestimmungsapparat *m*
n. dioxide NO_2 Stickstoffdioxid *n*
n. donor Stickstoffdon[at]or *m*
n. family Stickstoffgruppe *f*, Stickstoff-Phosphor-Gruppe *f*
n. fertilization Stickstoffdüngung *f*
n. fertilizer Stickstoffdüngemittel *n*, Stickstoffdünger *m*
n. fixation *(Landw)* Stickstoffbindung *f*, Stickstoff-Fixierung *f* *(durch Bodenbakterien aus freiem N_2)*
n.-fixing bacteria stickstoffbindende Bakterien *npl*
n.-free stickstofffrei
n.-gathering plant Stickstoffsammler *m*, Stickstoffmehrer *m*
n. group s. n. family
n. hardening s. n. case hardening
n. iodide s. n. triiodide
n. monoxide N_2O Distickstoff[mon]oxid *n*, Stickstoff(I)-oxid *n*; NO Stickstoff[mon]oxid *n*, Stickstoff(II)-oxid *n*, Stickoxid *n*
n. mustard Stickstoffsenfgas *n*, Stickstofflost *m*, Stickstoffyperit *n*, N-Lost *m*, N-Yperit *n*
n. of the diazonium group Stickstoff *m* der Diazogruppe, Diazostickstoff *m*, Azostickstoff *m*
n. oxide Stickstoffoxid *n*, *(i.e.S.)* s. n. dioxide
n. oxyfluoride NOF Nitrosylfluorid *n*, Stickstoffoxidfluorid *n*
n. pentasulphide N_2S_5 Stickstoffpentasulfid *n*, Distickstoffpentasulfid *n*, Stickstoff(V)-sulfid *n*
n. pentoxide N_2O_5 Stickstoffpentoxid *n*, Distickstoffpentoxid *n*, Stickstoff(V)-oxid *n*
n. peroxide s. n. dioxide

n.-phosphorus fertilizer stickstoffhaltiger Phosphatdünger *m*
n. reservoir *(Landw)* Stickstoffquelle *f*, Stickstoffreservoir *n*
n. tetroxide N_2O_4 Stickstofftetroxid *n*, Distickstofftetroxid *n*
n. trichloride NCl_3 Stickstofftrichlorid *n*, Stickstoff(III)-chlorid *n*
n. trifluoride NF_3 Stickstofftrifluorid *n*, Stickstoff(III)-fluorid *n*
n. triiodide NJ_3 Stickstofftrijodid *n*, Stickstoff(III)-jodid *n*
n. trioxide N_2O_3 Stickstofftrioxid *n*, Distickstofftrioxid *n*, Stickstoff(III)-oxid *n*
n. uptake Stickstoffaufnahme *f*
nitrogenous stickstoffhaltig, Stickstoff enthaltend, Stickstoff...
n. base Stickstoffbase *f*, stickstoffhaltige Base *f*
n. fertilizer (manure) Stickstoffdüngemittel *n*, Stickstoffdünger *m*
nitroglycerine $C_3H_5(ONO_2)_3$ Nitroglyzerin *n*, *(fälschlich für)* Glyzerintrinitrat *n*
n. propellant Nitroglyzerinpulver *n*, *(fälschlich für)* Glyzerintrinitratpulver *n*
nitrohydrochloric acid Königswasser *n*, Kö., Kw.
nitroiridate $M^I_3[Ir(NO_2)_6]$ Hexanitroiridat(III) *n*
nitrojection *(Landw)* Injektion *f* von flüssigem Ammoniak *(als Düngemittel)*, Nitrojektion *f*
nitrolic acid Nitrolsäure *f*
nitrolime Kalkstickstoff *m*, Kalziumzyanamid *n*
nitromagnesite *(Min)* Nitromagnesit *m*, Magnesiasalpeter *m* *(Magnesiumnitrat)*
nitrometer Azotometer *n*, Nitrometer *n*
nitromethane CH_3NO_2 Nitromethan *n*
nitromuriatic acid Königswasser *n*, Kö., Kw.
nitronaphthalene Nitronaphthalin *n*
nitronatrite *(Min)* Nitronatrit *m*, Natronsalpeter *m* *(Natriumnitrat)*
nitronic acid Nitronsäure *f*
nitronickelate $M^I_4[Ni(NO_2)_6]$ Hexanitroniccolat(II) *n*
nitronium s. n. ion
n. chloride NO_2Cl Nitrylchlorid *n*
n. ion NO_2^+ Nitroniumion *n*, Nitrylion *n*
n. perchlorate NO_2ClO_4 Nitroniumperchlorat *n*, Nitrylperchlorat *n*
nitroosmate $M^I_2[Os(NO_2)_5]$ Pentanitroosmat(III) *n*
nitroparaffin Nitroparaffin *n*, Nitroalkan *n*
nitroperbenzoic acid Nitroperbenzoesäure *f*
nitrophosphate Nitrophosphat *n* *(Sammelname für N-haltigen Phosphatdünger)*
nitroprussiate s. nitroprusside
nitroprusside $M^I_2[Fe(CN)_5(NO)]$ Nitroprussid *n*, Nitroprussiat *n*, Pentazyanonitrosylferrat *n*
nitrorhodate $M^I_3[Rh(NO_2)_6]$ Hexanitrorhodat(III) *n*
nitrosamine Nitrosamin *n*
n. red Nitrosaminrot *n*
nitrosate / to nitrosieren, die Nitrosogruppe *(in eine organische Verbindung)* einführen
nitrosate Nitrosat *n*

nitrosating agent Nitrosier[ungs]mittel *n*, nitrosierendes Agens *n*
nitrosation Nitrosierung *f*
nitrosite Nitrosit *n*
nitroso compound Nitrosoverbindung *f*
 n. dye Nitrosofarbstoff *m*
 n. group —N=O Nitrosogruppe *f*
 n.-isonitroso tautomerism Nitroso-Isonitroso-Tautomerie *f*
 n. rubber Nitrosokautschuk *m*
p-**nitrosoaniline** ONC$_6$H$_4$NH$_2$ *p*-Nitrosoanilin *n*, 4-Nitrosoanilin *n*
nitrosobacteria *s.* nitrite-forming bacteria
nitrosobenzene C$_6$H$_5$N=O Nitrosobenzol *n*
nitrosocresol ON·C$_6$H$_3$(CH$_3$)OH Nitrosokresol *n*
nitrosoferricyanide M'$_2$[Fe(CN)$_5$(NO)] Pentazyanonitrosylferrat *n*, Nitroprussid *n*, Nitroprussiat *n*
nitrosonaphthol Nitrosonaphthol *n*
p-**nitrosophenol** NOC$_6$H$_4$OH *p*-Nitrosophenol *n*, 4-Nitrosophenol *n*
nitrososulphuric acid *s.* nitrosyl sulphuric acid
nitrostarch Nitrostärke *f*, Stärkenitrat *n* *(Sprengstoff)*
nitrosulphuric acid *s.* nitrosyl sulphuric acid
nitrosyl chloride NOCl Nitrosylchlorid *n*, Stickstoffoxidchlorid *n*
 n. fluoride NOF Nitrosylfluorid *n*, Stickstoffoxidfluorid *n*
 n. hydrogen sulphate NOHSO$_4$ Nitrosylhydrogensulfat *n*
 n. ion NO$^+$ Nitrosylion *n*
 n. sulphuric acid NOHSO$_4$ Nitrosylschwefelsäure *f*, *(veraltet für)* Nitrosylhydrogensulfat *n*
nitrotoluidine CH$_3$C$_6$H$_3$(NO$_2$)NH$_2$ Nitrotoluidin *n*
nitrous nitros *(Stickoxid enthaltend)*; salpeterhaltig, salpetrig, salpeterartig, Salpeter...; Stickstoff...; niederwertigem Stickstoff entsprechend, *(meist)* Stickstoff(III)-...
 n. acid HNO$_2$ salpetrige Säure *f*
 n. anhydride N$_2$O$_3$ Stickstoff(III)-oxid *n*
 n. bacteria *s.* nitrite-forming bacteria
 n. ether C$_2$H$_5$ONO Salpetrigsäureäthylester *m*, Äthylnitrit *n*
 n. gases nitrose Gase *npl (Gemisch aus Luft und Stickstoffoxiden)*
 n. oxide N$_2$O Stickstoff(I)-oxid *n*, Distickstoff[mon]oxid *n*
 n. vitriol nitrose Säure *f*, Nitrose *f*
nitroxanthic acid C$_6$H$_2$(NO$_2$)$_3$OH Pikrinsäure *f*, 2,4,6-Trinitrophenol *n*
nitroxyl chloride *s.* nitronium chloride
nitroxylene NO$_2$C$_6$H$_3$(CH$_3$)$_2$ Nitroxylol *n*, Dimethylnitrobenzol *n*
nitryl chloride *s.* nitronium chloride
 n. perchlorate *s.* nitronium perchlorate
nix alba Nix alba *f*, *(veraltet für)* Zinkblumen *fpl (weißes, wollartiges Zinkoxid)*
NMR *s.* nuclear magnetic resonance
NMR-spectroscopy *s.* nuclear magnetic resonance spectroscopy

no-bond resonance Hyperkonjugation *f*, Baker-Nathan-Effekt *m*
no-fines concrete entfeinter Beton *m*, Schüttbeton *m*
no-iron finish *(Text)* Bügelfreiausrüstung *f*, Bügelarmausrüstung *f*, No-iron-Ausrüstung *f*, Naßknitterausrüstung *f*
no-strength temperature *(Plast)* NST-Wert *m*
nobelium No Nobelium *n*
noble gas Edelgas *n*
 n. metal edles Metall *n*, Edelmetall *n*
 n. mould Edelfäule *f*
nodal line Knotenlinie *f*
 n. plane Knotenebene *f*, Knotenfläche *f*
 n. point Knotenpunkt *m*, Schwingungsknoten *m*
nodular *(Min)* knollig, kugelig
 n. cast iron Kugelgraphit[grau]guß *m*, sphärolithischer (globularer) Grauguß *m*
 n. graphite Kugelgraphit *m*
nodule-forming bacteria Knöllchenbakterien *npl (Rhizobium-Arten)*
nodulizing Pelletisieren *n*, Pelletisierung *f*, *(bei Hitzeeinwirkung)* Kugelsintern *f*
noil *(Text)* Kämmling *m*
noiseless paper geräuschloses Papier *n*, Programmpapier *n*
nomenclature Nomenklatur *f*
 n. commission Nomenklaturkommission *f*
 n. rule Nomenklaturregel *f*
 n. system Nomenklatursystem *n*
nominal diameter Nenndurchmesser *m*
nonaccelerated *(Gum)* nicht beschleunigt, beschleunigerfrei, ohne Beschleuniger
nonacidic nicht sauer
nonacosane Nonakosan *n*
nonactic acid *s.* nonactinic acid
nonactin Nonaktin *n (Antibiotikum)*
nonactinic acid Nonaktinsäure *f*
nonactivated clay Naturton *m*, natürlicher Ton *m*; natürliche (naturaktive) Tonerde (Bleicherde) *f*, Naturbleicherde *f (zum Raffinieren)*
nonageing nicht alternd, alterungsbeständig
nonagglomerating nicht klumpend
nonaging *s.* nonageing
nonalcoholic alkoholfrei, nichtalkoholisch
nonanedioic acid HOOC(CH)COOH Nonandisäure *f*, 1,7-Heptandikarbonsäure *f*, Azelainsäure *f*, Lepargylsäure *f*
n-**nonanoic acid** CH$_3$(CH$_2$)$_7$COOH *n*-Nonylsäure *f*, *n*-Nonansäure *f*, Pelargonsäure *f*
nonaqueous wasserfrei; nichtwäßrig, nichtwässerig, nicht wäßrig (wässerig)
 n. solution nichtwäßrige Lösung *f*
 n. solvent nichtwäßriges Lösungsmittel *n*
 n. titration Titration *f* in nichtwäßriger Lösung
nonaromatic nichtaromatisch
nonasphaltic petroleum nichtasphaltisches Erdöl *n*
nonassociated liquid normale Flüssigkeit *f*
nonavailable sulphur dioxide *(Pap)* gebundenes SO$_2$ *n*

nonbanded nicht streifig
nonbenzenoid nichtbenzoid
nonbevelled nichtschräggeschliffen, ohne Schrägfläche
nonbiological haze nichtbiologische Trübung *f*
(*z.B. des Bieres*)
nonblack *(Gum)* rußfrei, ohne Ruß
n. filler (pigment) *(Gum)* heller Füllstoff *m*
n. reinforcing filler *(Gum)* heller Verstärkerfüllstoff *m*, heller aktiver Füllstoff *m*
nonblooming *(Gum)* nicht ausblühend (ausschwe-
felnd)
nonboiling zone Vorwärmzone *f (im Verdampfer)*
nonbonded bindungslos
nonbonding nichtbindend
n. electron nichtbindendes (einsames) Elektron
n, Nichtbindungselektron *n*
n. orbital nichtbindendes Orbital *n*
nonbranched unverzweigt
nonbronzing nicht bronzierend
noncaking coal nichtbackende Kohle *f*
noncarbonate hardness Nichtkarbonathärte *f*,
permanente (bleibende) Härte *f*
noncarcinogenic nicht karzinogen (kanzerogen,
krebserzeugend, krebserregend, krebsauslösend)
noncellular filter zellenloses Filter *n*
noncellulosic constituents (materials) *(Pap)* Nicht-
zellulosebestandteile *mpl*
noncentral nichtzentral
nonchelated nicht chelatgebunden
nonclay nichttonig, tonfrei
n. body *(Ker)* tonfreie Masse *f*, tonfreies Gemisch
n
n. casting slip *(Ker)* tonfreier Gießschlicker *m*
noncoking coal nichtkokende Kohle *f*
noncombustible nicht brennbar, un[ver]brennbar
noncombustible [matter] Unverbrennbares *n*, Unverbrennliches *n*
n. paper feuerfestes (feuersicheres, unentflammbares, nicht entflammbares) Papier *n*,
(*i.e.S.*) flammensicheres (flammensicher imprägniertes, schwer entflammbares) Papier *n*
noncomplexing nicht komplexbildend
noncondensability Nichtkondensierbarkeit *f*
noncondensable nicht kondensierbar
nonconducting nichtleitend, dielektrisch
nonconductor Nichtleiter *m*, Dielektrikum *n*
nonconjugated nichtkonjugiert
nonconvertible physikalisch trocknend *(z.B. Anstrichmittel)*
noncorroding nicht korrodierend, korrosionsfest,
korrosionsbeständig
noncreasable *s.* noncreasing
noncrease finish Knitterarmausrüstung *f*, Knitterarmappretur *f*, Knitterfestausrüstung *f*, Knitterechtausrüstung *f*, knitterbeständige Ausrüstung *f*
noncreasing knitterarm, knitterfest, knitterecht,
knitterbeständig

noncrushable *(Am) s.* noncreasing
noncrystalline nichtkristallin[isch], amorph, gestaltlos, formlos, ohne Kristallform
noncumulative nichtkumuliert
noncuring *s.* nonvulcanizing
noncurling paper nichtrollendes (nichtkräuseln-
des) Papier *n*
noncyclic[al] nichtzyklisch, azyklisch
non-Daltonian (non-daltonide) compounds nichtdaltonide (nichtdaltonische, berthollide) Verbindungen *fpl*, Berthollide *npl*, Berthollidverbindungen *fpl*, Verbindungen *fpl* von nichtkonstanter
Zusammensetzung
nondegenerate nichtentartet
nondeliquescent nicht zerfließend (zergehend)
nondestructive test zerstörungsfreie Prüfung *f*,
zerstörungsfreier Test *m*
nondipolar dipolfrei
nondiscolouring nicht verfärbend, farbbeständig
nondrying nichttrocknend
n. oil nichttrocknendes Öl *n*
nondyeing nicht anfärbend
noneclipsed *(Stereochemie)* gestaffelt, auf Lücke,
±anti-periplanar
n. conformation gestaffelte (±anti-periplanare)
Konformation *f*
nonelectrolyte Nichtelektrolyt *m*, Anelektrolyt *m*
n. chelate Nichtelektrolytchelat *n*
nonelemental sulphur cure Vulkanisation *f* mit
Schwefelspendern (schwefelabspaltenden
Verbindungen), Vulkanisation *f* ohne freien
Schwefel
nonenolizable nicht enolisierbar
nonenzymatic nichtenzymatisch
nonequivalent nicht gleichwertig
nonexchangeable nichtaustauschbar
nonexistence Nichtexistenz *f*
nonexplosive nicht explosiv (explosibel)
nonfading *(Pap)* nichtgilbend, lichtecht
nonfat[ty] fettfrei
n. milk solids fettfreie Trockenmasse *f* der Milch
n. solids fettfreie Trockenmasse *f*
nonferrous nichteisenhaltig, eisenfrei, Nicht-
eisen..., NE-...
n. alloy NE-Legierung *f*
n. blast furnace Gebläseschachtofen (Schacht-
schmelzofen) *m* für NE-Metalle
n. metal Nichteisenmetall *n*, NE-Metall *n*
nonfissile, nonfissionable nichtspaltbar
nonflam *s.* nonflammable
nonflammability Nichtentflammbarkeit *f*, Unentflammbarkeit *f*, Flammwidrigkeit *f*
nonflammable nicht entflammbar (entzündbar,
inflammabel), unentflammbar, flammwidrig, un-
entzündbar
nonflocculating yeast Staubhefe *f*
Nongo gum *(Harz von Albizzia brownii Walp.)*
nongreasy nichtfettig, nichtfettend
non-heat-treatable nicht aushärtbar (vergütbar)
nonhydrated nichthydratisiert

nonhydraulic lime Luftkalk *m*, an der Luft erhärtender Kalk *m*
  n. mortar Luftmörtel *m*, an der Luft erhärtender Mörtel *m*
nonhydrolyzable tannin kondensierter Gerbstoff *m*
nonideal nichtideal
1-nonine *s.* 1-nonyne
noninflammable *s.* nonflammable
nonionic nichtionogen, nichtionisch, ioneninaktiv, nichtionisierend
nonionic nichtionogener (nichtionischer) grenzflächenaktiver Stoff *m*
  n. emulsifier nichtionogener Emulgator *m*
  n. substance *s.* nonionic
  n. surfactant nichtionogenes Netzmittel *n*
nonionizing, nonionogenic *s.* nonionic
non-key [component] *(Destillation) (nicht als Schlüsselkomponente fungierender Anteil eines Mehrstoffgemisches)*
nonknocking nichtklopfend, klopffrei
nonlathering nichtschäumend
nonleafing *(Farb)* nicht ausschwimmend, nonleafing
nonlinear nichtlinear, unlinear
nonlinearity Nichtlinearität *f*, Unlinearität *f*
nonlocalized nichtlokalisiert
nonlubricated compressor ölfreier (schmierloser) Verdichter (Kompressor) *m*, Trockenlaufverdichter *m*
nonluminous nichtleuchtend
  n. flame nichtleuchtende Flamme *f*
nonmagnetic nichtmagnetisch, unmagnetisch
nonmagnetics nichtmagnetisches (nichtmagnetisierbares) Gut *n*
nonmechanical classifier Freifallklassierer *m*
nonmetal Nichtmetall *n*, *(veraltet)* Metalloid *n*
nonmetallic nichtmetallisch, Nichtmetall...
  n. element *s.* nonmetal
  n. minerals Steine *mpl* und Erden *fpl*; nichtmetallische Minerale *npl*, nichtmetallische [mineralische] Rohstoffe *mpl*
nonmigrating nichtwandernd, wanderungsbeständig
nonmiscible nicht mischbar
nonmobile unbeweglich, stationär, ortsfest
  n. phase *(Chromatografie)* unbewegliche (stationäre) Phase *f*
nonnatural künstlich [umgewandelt], synthetisch *(z.B. Brennstoff)*
non-Newtonian flow nicht-Newtonsches Fließen *n*
non-Newtonian fluid nicht-Newtonsche Flüssigkeit *f*

nonnitrogenous stickstofffrei, nicht stickstoffhaltig
nonodorous geruchlos
nonoic acid *s.* n-nonanoic acid
nonoxide Nonoxid *n*
nonoxidizing nichtoxydierend
nonpasted board *(Am)* gedeckte (gegautschte) Pappe *f*
nonpathogenic nicht pathogen, apathogen

nonpersistant pesticide nicht persistentes Pflanzenschutzmittel *n*
nonphenolic nichtphenolisch
nonphototropic nicht phototrop
nonpigmented *(Gum)* ungefüllt, füllstofffrei, ohne Füllstoffe
nonplastic nichtplastisch, unplastisch
  n. material *(Ker)* Magerungsmittel *n*
nonpoisonous ungiftig, nicht giftig, nichttoxisch
nonpolar nichtpolar, unpolar, apolar; unpolar, homöopolar, einpolar *(chemische Bindung)*
  n. adsorption apolare Adsorption *f*
  n. bond *s.* n. linkage
  n. compound unpolare (homöopolare) Verbindung *f*
  n. linkage unpolare (homöopolare, einpolare, kovalente, unitarische) Bindung *f*, Elektronenpaarbindung *f*, Atombindung *f*
  n. liquid normale Flüssigkeit *f*
  n. molecule nichtpolares (unpolares) Molekül *n*
nonpolarizable unpolarisierbar, nichtpolarisierbar
nonproprietary name nichtgeschützte (freie) Bezeichnung *f*, nicht wortgeschützter Name *m*, Freiname *m*, freier Warenname *m*
nonprotein nichteiweißartiger Stoff *m*
  n. nitrogen Nichteiweißstickstoff *m*, Nichtproteinstickstoff *m*, Reststickstoff *m*
nonpulsating pulsationsfrei
nonquinonoid nichtchinoid
nonradiative strahlungslos, strahlungsfrei
nonradioactive nichtradioaktiv
  n. isotope stabiles (nichtradioaktives) Isotop *n*
nonreactive reaktionslos, nichtreaktionsfähig
nonreducing nichtreduzierend
nonreflecting glass reflexfreies Glas *n*
nonregenerative nichtregenerativ
nonreinforcing *(Gum)* inaktiv, Inaktiv...
  n. black inaktiver Ruß *m*, Inaktivruß *m*
  n. filler inaktiver Füllstoff *m*
nonrelativistic nichtrelativistisch
nonresidue rückstandslos, rückstandsfrei
  n. cracking rückstandsloses Kracken *n*, Kracken *n* auf Koksrückstand, Kracken *n* mit Kokungsarbeitsweise, Kracken *n* nach der Verkokungsfahrweise
  n. method Krackprozeß *m* mit Kokungsarbeitsweise, Verkokungs[krack]verfahren *n*
nonreturn [flow] valve Rückschlagventil *n*
nonreversible irreversibel, nicht umkehrbar (rückläufig)
nonrigid nicht starr, nichtstarr
  n. plastic weich[gestellt]er Plast *m*
nonrubber [constituent, material, substance] Nichtkautschuksubstanz *f*, Nichtkautschukbestandteil *m*
nonsafety Gesundheitsschädlichkeit *f (z.B. von Lebensmittelzusätzen)*
nonsaponifiable unverseifbar, nicht verseifbar
nonsaponifying nichtverseifend
nonscaling zunderbeständig, zunderfrei

nonselective herbicide nichtselektives (allgemein wirkendes, total wirkendes) Herbizid n, Totalherbizid n

n. treatment Ganzflächenbehandlung f (mit Pflanzenschutzmitteln)

nonshattering glass Sicherheitsglas n

nonshrinking (Text) nichtschrumpfend, krumpffrei, nicht einlaufend; nicht schwindend

n. body (Ker) nicht schwindende Masse f

nonslagging nicht schlackend

nonslip finish Schiebefestappretur f, Schiebefestausrüstung f

nonsludging oil [bei Gebrauch] keinen Schlamm bildendes Öl n

nonsoapy seifenfrei, seifenlos

nonsoluble nicht löslich, unlöslich, nl, unl., unlö

nonsolvent inaktives Lösungsmittel n, inaktiver Löser m, Nichtlöser m

nonstaining nicht verfärbend

nonstick nichtklebend

nonstoichiometric nichtstöchiometrisch

n. compounds nichtdaltonide (nichtdaltonische, berthollide) Verbindungen fpl, Berthollidverbindungen fpl, Berthollide npl, Verbindungen fpl von nichtkonstanter Zusammensetzung

nonsugar zuckerfrei

nonsugar nichtzuckerartiger (zuckerfreier, zuckerfremder) Bestandteil m, Nichtzuckeranteil m, Aglykon n

nonsulphur schwefelfrei

n. cure (vulcanization) schwefelfreie Vulkanisation (Vernetzung) f

nonsuperimposable nicht deckungsgleich

nonsupport of combustion Unbrennbarkeit f, Flammwidrigkeit f, Feuersicherheit f

nonswelling quellfest

n. agent Quellfestmittel n

nonsymbiotic fixation (Landw) nichtsymbiotische Stickstoffbindung f (Fixierung f von Luftstickstoff), Stickstoffbindung f durch freilebende Bakterien

nonsystemic chemical nichtsystemischer Stoff m (nicht über das Gefäßsystem der Pflanze wirkendes Pflanzenschutzmittel)

n. fungicide nichtsystemisches (nicht über das Gefäßsystem der Pflanze wirkendes) Fungizid n

nontan[nin] Nichtgerbstoff m

nontitrative nichttitrimetrisch

nontorque textured (Text) verdrehungsfrei texturiert

nontoxic nichttoxisch, ungiftig, nicht giftig

non-two-sided (Pap) nicht zweiseitig

nonuniform ungleichförmig, ungleichmäßig, uneinheitlich

nonvalent nullwertig

nonvariant nonvariant, invariant, ohne Freiheit[sgrad]

nonviscous nicht viskos

nonvitreous, nonvitrified nicht glasartig

nonvolatile nichtflüchtig

n. matter nichtflüchtiger Stoff m, nichtflüchtige Bestandteile mpl, Nichtflüchtiges n

nonvolatility Nichtflüchtigkeit f

nonvulcanizable, nonvulcanizing nicht vulkanisierbar (vernetzungsfähig, vernetzbar)

nonweathering nicht verwitternd

nonwettable unbenetzbar, nicht benetzbar

nonwetter Nichtnetzer m

nonwoody pulp (Pap) textiler Halbstoff m, (i.e.S.) Hadernhalbstoff m

nonwoven ungewebt, nichtgewebt

n. fabric Faservlies n, Fasergewirre n, Vliesfolie f

nonwovens Textilverbundstoffe mpl, Vliesstoffe mpl, Vlieswaren fpl, ungewebte Textilien pl

nonyellowing vergilbungsbeständig, nicht[ver]gilbend, nicht gelb werdend

n-nonyl alcohol $CH_3(CH_2)_7CH_2OH$ n-Nonylalkohol m, n-Nonanol n, Nonanol-(1) n

n-nonyl aldehyde $CH_3(CH_2)_7CHO$ n-Nonylaldehyd m, Pelargonaldehyd m

n-nonylic acid s. n-nonanoic acid

nonylone $(C_8H_{17})_2CO$ Nonylon n, Dioktylketon n, Pelargon n, 9-Heptadekanon n

1-nonyne $CH≡C(CH_2)_6CH_3$ 1-Nonin n, n-Heptylazetylen n

non-zero spin von Null verschiedener Spin m, nichtverschwindender Spin m

nootkatene Nootkaten n (dreifach ungesättigtes Sesquiterpen)

nootkatin Nootkatin n (Sesquiterpen-Tropolon)

norbornadiene Norbornadien n, Bizyklo-2,2,1-heptadien n

norbornene Norbornen n, Bizyklo-2,2,1-hepten-(2) n

nordhausen acid Nordhäuser Vitriolöl n, (veraltet für) rauchende Schwefelsäure f, Oleum n

norecgonine Norekgonin n, Kokayloxyessigsäure f

Norge nitre Norgesalpeter m, (technisches) Kalziumnitrat n

normal aluminium acetate $Al(CH_3COO)_3$ Aluminiumazetat n

n. antimonyl sulphate $(SbO)_2SO_4$ Antimon(III)-oxidsulfat n

n. atom normales Atom n (bei dem sich alle Elektronen im Grundzustand befinden)

n. beet normale Zuckerrübe f, Normalrübe f, N-Rübe f

n. benzine Normalbenzin n

n. butane $CH_3CH_2CH_2CH_3$ Normalbutan n, n-Butan n

n. calomel electrode Normalkalomelelektrode f, Kalomelnormalelektrode f

n. carbonization Normalverkokung f, Vollverkokung f, Hochtemperaturverkokung f, HT.-Verkokung f, Verkokung f (i.e.S.), Hochtemperaturentgasung f, HT.-Entgasung f

n. complex Normalkomplex m, Anlagerungskomplex m

n. compound Normalverbindung *f*, n-Verbindung *f*

n. conditions Norm[al]bedingungen *fpl*, Standardbedingungen *fpl*

n. density Normdichte *f*

n. developer *(Foto)* Normalentwickler *m*

n. dose Norm[al]dosis *f*

n. electrode Normalelektrode *f*

n. fluid *s.* n. liquid

n. freezing normales Erstarren *n (einer Schmelze)*

n. hydrocarbon Normalkohlenwasserstoff *m*

n. hydrogen electrode Normalwasserstoffelektrode *f*, Wasserstoffnormalelektrode *f*

n. KOH solution Normalkalilauge *f*, normale Kalilauge *f*

n. latex normaler Latex *m*

n. lead orthophosphate $Pb_3(PO_4)_2$ Blei[ortho]phosphat *n*

n. liquid normale Flüssigkeit *f*

n. mercuric phosphate $Hg_3(PO_4)_2$ Quecksilber(II)-[ortho]phosphat *n*

n. mercurous phosphate Hg_3PO_4 Quecksilber(I)-[ortho]phosphat *n*

n. molecule normales (nichtaktiviertes) Molekül *n*

n. multiplet *(phys Ch)* normales (regelrechtes) Multiplett *n*

n. pentane $CH_3(CH_2)_3CH_3$ Normalpentan *n*, *n*-Pentan *n*

n. phosphate $M^I_3PO_4$ neutrales (tertiäres) Phosphat *n*

n. potassium phosphate K_3PO_4 Kalium[ortho]phosphat *n*, Trikaliumphosphat *n*

n. potassium pyrophosphate $K_4P_2O_7$ Kaliumdiphosphat *n*, Kaliumpyrophosphat *n*

n. potential Normalpotential *n*

n. pressure Norm[al]druck *m*, Standarddruck *m*, Normalluftdruck *m*

n. rotatory dispersion normale Rotationsdispersion *f*

n. salt normales (neutrales) Salz *n*, Normsalz *n*, Neutralsalz *n*

n. silver-silver chloride electrode Normal-Silber-Silberchloridelektrode *f*

n. silver sulphate Ag_2SO_4 Silbersulfat *n*

n. sodium pyrophosphate $Na_4P_2O_7$ Natriumdiphosphat *n*, Natriumpyrophosphat *n*

n. solution Normallösung *f*, normale Lösung *f*, n-Lösung *f*; isotonische Lösung *f*

n. spectrum Normalspektrum *n*, Gitterspektrum *n*, Beugungsspektrum *n*

n. state Normalzustand *m*, Grundzustand *m*, Grundniveau *n*, Grundterm *m (eines Atoms)*

n. temperature Norm[al]temperatur *f*, *(i.e.S.)* Zimmertemperatur *f*, Raumtemperatur *f*, gewöhnliche Temperatur *f*

n. temperature and pressure Norm[al]temperatur *f* und -druck *m*, physikalischer Normzustand *m (Standardbedingungen von 0 °C und 760 Torr)*

n. test sieve Standardprüfsieb *n*, Norm[al]prüfsieb *n*, standardisiertes Prüfsieb *n*

n. unit Normaleinheit *f*, Standardeinheit *f*

n. vibration *(phys Ch)* Normalschwingung *f*

n. voltage Normalspannung *f*

n. volume Norm[al]volumen *n*

normality Normalität *f (ein Konzentrationsmaß)*

n. factor Normalitätsfaktor *m*, Korrekturfaktor *m (bei ungenauen Normallösungen)*

normalization *s.* normalizing

normalize / to normalglühen, normalisierend glühen, normalisieren

normalizing Normal[isierungs]glühen *n*, Normal[isierungs]glühung *f*, normalisierendes Glühen *n*, Normalisieren *n*, Normalisierung *f*

n. furnace Normalglühofen *m*, Normalisierofen *m*

n. temperature Normalglühtemperatur *f*, Normalisiertemperatur *f*

normally distributed normal verteilt

nornicotine Nornikotin *n (ein Tabakalkaloid)*

norpseudoecgonine Norpseudoekgonin *n*

norpseudotropine Norpseudotropin *n*

nortropine Nortropin *n*, 3-Nortropanol *n*, Tropigenin *n*, Tropolin *n*

Norway (Norwegian) saltpeter Norgesalpeter *m*, *(technisches)* Kalziumnitrat *n*

nose *(Glas)* Arbeitswanne *f*, Läuterwanne *f*; *s.* n.-piece

n.-piece *(Glas)* Pfeifenkopf *m*, Pfeifenende *n*

nosean[ite], noselite Nosean *m*, Noselith *m (Feldspatvertreter)*

not printed upon *(Pap)* unbedruckt

notate / to bezeichnen

notation Bezeichnung *f*; Bezeichnungsweise *f*; Notation *f (Kurzdarstellung chemischer Strukturen, besonders zur Speicherung in Datenverarbeitungsanlagen)*

n. system Bezeichnungssystem *n*, Bezeichnungsweise *f*; Notationssystem *n*

notch Überfall *m (Mengenstrommessung)*

n. factor Kerbeinflußzahl *f*, Kerbwirkungszahl *f*, Kerbwirkungsfaktor *m*

n. impact resistance (strength) Kerbschlagzähigkeit *f*

n.-sensitive kerbempfindlich

n. sensitivity Kerbempfindlichkeit *f*

note paper Briefpapier *n*, Ausstattungspapier *n*, Postpapier *n*, Billetpapier *n*

noume[a]ite *(Min)* Noumeait *m*, *(veraltet für)* Garnierit *m (Nickelsilikat)*

nourish / to nähren, nahrhaft sein; [er]nähren

nourishing nährend, nahrhaft, Nähr...

n. cream Hautnährcreme *f*, hautnährende Creme *f*, Nährcreme *f*

nourishment Ernährung *f*; Nahrungsmittel *n*, Nahrung *f*

novolak Novolackharz *n*, Novola[c]k *m*

noxious schädlich

n.-smelling übelriechend

nozzle Düse *f*

n. atomization Verdüsen *n*, Zerstäubung (Versprühung) *f* durch Düsen

n. coefficient Verlustbeiwert *m* der Düse, Düsenbeiwert *m*

n. discharge disk centrifuge Tellerzentrifuge *f* mit Düsenaustrag, Ventiltellerzentrifuge *f*

n.-mix burner Leuchtflammenbrenner *m*; Gebläsebrenner *m*

n. mixer Düsenmischer *m*, Mischdüse *f*

n. pulverizer Strahlprallmühle *f*

n. throat Verengung *f* einer Düse, Düsenverengung *f*, Düsenhals *m*

n.-type relief valve Düsensicherheitsventil *n*

NPK fertilizer Volldünger *m*, NPK-Dünger *m*

NPSH *s*. net positive suction head

NQR *s*. nuclear quadrupole resonance

NR *s*. natural rubber

NSR *s*. nitrile-silicone rubber

NSSC process (*Pap*) Neutralsulfitverfahren *n*, NSSC-Verfahren *n*

NTP *s*. normal temperature and pressure

nubby (*Text*) noppig, mit Noppen

nuclear nuklear, Kern...

n. angular momentum Kernspin *m*, Kerndrehimpuls *m*

n. atom [model] Kernmodell *n* des Atoms, Rutherfordsches Atommodell *n*

n. capture Kerneinfang *m*

n. chain reaction nukleare Kettenreaktion *f*, Kernkettenreaktion *f*

n. charge Kernladung *f*

n. charge number Kernladungszahl *f*, Atomnummer *f*, Ordnungszahl *f*, OZ, Z

n. chemistry Kernchemie *f*

n. collision Kern[zusammen]stoß *m*

n. constituent Kernbaustein *m*, Kernteilchen *n*, Nukleon *n*

n. cross section Wirkungsquerschnitt *m* des Atomkerns (Kerns), nuklearer Wirkungsquerschnitt *m*

n. decay Kernzerfall *m*

n. density Kerndichte *f*

n. disintegration *s*. n. decay

n. distance Kernabstand *m*, Atomabstand *m*

n. emulsion Kern[spuren]emulsion *f*, Kernspuremulsion *f*

n. emulsion technique Kern[spur]emulsionstechnik *f*

n. energy nukleare Energie *f*, Kernenergie *f*, Atom[kern]energie *f*

n. energy level Kern[energie]niveau *n*, Kernterm *m*

n. energy level density Kern[energie]niveaudichte *f*

n. equation *s*. n. reaction formula

n. evaporation Kernverdampfung *f*

n. excitation Kernanregung *f*

n. explosive Kernsprengstoff *m*, Atomsprengstoff *m*

n. field Kernfeld *n*, nukleares Feld *n*

n. fission Kernspaltung *f*, Atomkernspaltung *f*

n. fission spectrum Kernspaltungsspektrum *n*

n. forces Kern[feld]kräfte *fpl*

n. fragments Kernbruchstücke *npl*

n. fuel Kernbrennstoff *m*

n. fusion Kernfusion *f*, Kernverschmelzung *f*, Kernsynthese *f*

n. gyromagnetic ratio gyromagnetisches Verhältnis *n* des Kerns (Atomkerns)

n. induction Kerninduktion *f*

n. isobar Kernisobar *n*, Isobar *n*, isobare Atomart *f*

n. isomer Kernisomeres *n*

n. isomerism Kernisomerie *f*

n. magnetic moment magnetisches Kernmoment *n* (Moment *n* des Kerns), kernmagnetisches Moment *n*

n. magnetic resonance kernmagnetische Resonanz *f*, magnetische Kernresonanz *f*, NMR

n. magnetic resonance spectroscopy magnetische Kernresonanzspektroskopie *f*

n. magnetic resonance spectrum magnetisches Kernresonanzspektrum *n*

n. magnetic resonance study magnetische Kernresonanzuntersuchung *f*, NMR-Untersuchung *f*

n. magnetism Kernmagnetismus *m*

n. magneton Kernmagneton *n*, KM

n. mass Kernmasse *f*

n. matter Kernmaterie *f*

n. membrane (*Bioch*) Zellmembran *f*

n. model Kernmodell *n*

n. model of the atom *s*. n. atom [model]

n. number Massenzahl *f*, M

n. paramagnetism Kernparamagnetismus *m*

n. particle Kernbaustein *m*, Kernteilchen *n*, Nukleon *n*

n. photoeffect (photoelectric effect) Kernfotoeffekt *m*

n. physics Kernphysik *f*

n. polarization Kernpolarisation *f*

n. potential Kernpotential *n*

n.-powered mit Kernenergie (Atomkraft) angetrieben (betrieben), mit Atomantrieb, atomgetrieben, atombetrieben, kernenergiegetrieben

n. property Kerneigenschaft *f*

n. quadrupole moment Kernquadrupolmoment *n*, Quadrupolmoment *n* des Kerns

n. quadrupole resonance Kernquadrupolresonanz *f*

n. radiation Kernstrahlung *f*

n. radiation chemistry Kernstrahlenchemie *f*

n. radius Kernradius *m*

n. reaction Kernreaktion *f*

n. reaction equation (formula) Kernreaktionsgleichung *f*, Kernreaktionsformel *f*

n. reactor Kernreaktor *m*, Reaktor *m*

n. resonance Kernresonanz *f*

n. resonance absorption Kernresonanzabsorption *f*

n. resonance spectrograph Kernresonanzspektrograf m
n. resonance spectroscopy Kernresonanzspektroskopie f
n. sap *(Bioch)* Zellsaft m, Nukleoplasma n, Karyolymphe f
n. solution *(Koll)* Keimlösung f
n. spectroscopy Kernspektroskopie f
n. spin Kernspin m, Kerndrehimpuls m
n. spin moment Kernspinmoment n
n. spin quantum number Kernspinquantenzahl f
n. stability nukleare Stabilität f, Atomkernstabilität f, Kernstabilität f
n. statistics Kernstatistik f
n. structure Kernaufbau m, Kernstruktur f
n. substitution product kernsubstituiertes Produkt n
n. temperature Kerntemperatur f
n. theory of Rutherford Kerntheorie f von Rutherford
n. track Kernspur f
n. track emulsion s. n. emulsion
n. track measuring microscope Kernspurmeßmikroskop n
n. transformation Kernumwandlung f
n. transition Kernübergang m
nuclease Nuklease f *(zu den Phosphoesterasen gehörendes Ferment)*
nucleate / **to** als Kristall[isations]keim (Kristallisationskern) wirken; Kristall[isations]keime (Kristallisationskerne) bilden
nucleating agent s. nucleation agent
nucleation Kristall[isations]keimbildung f, Keimbildung f, Kristallkernbildung f, Kernbildung f
n. agent *(Krist)* Keimbildner m, keimbildendes Mittel n, keimbildender Zusatz m
n. centre Kristallisationszentrum n
nucleic acid *(Bioch)* Nukleinsäure f
n. acid synthesis Nukleinsäuresynthese f
nucleon Nukleon n, Kernteilchen n, Kernbaustein m
nucleonic component Nukleonenkomponente f
nucleonics angewandte Kernphysik f, Kerntechnik f
nucleophile Nukleophil n, nukleophiles (anionisches) Reagens n
nucleophilic nukleophil, kernsuchend, kernfreundlich
n. attack nukleophile Attacke f
n. displacement s. n. substitution
n. reaction nukleophile (anionoide) Reaktion f
n. substitution nukleophile (anionoide) Substitution f, S_N-Reaktion f, nukleophiler Austausch m
nucleoprotein Nukleoproteid n
nucleosidase Nukleosidase f *(zu den Amidasen gehörendes Ferment)*
nucleoside Nukleosid n
nucleotidase Nukleotidase f *(zu den Phosphoesterasen gehörendes Ferment)*

nucleotide Nukleotid n
nucleus Nukleus m, Atomkern m, Kern m; *(org Ch)* Kern m, Ring m; *(Bio)* Nukleus m, Zellkern m, Kern m; *(Krist)* s. n. of crystallization
n.-loving s. nucleophilic
n. of crystallization Kristall[isations]kern m, Kristall[isations]keim m, Kristallisationszentrum n, Keim m
nuclide Nuklid n
null-balance instrument Ausgleich[meß]instrument n
n.-balance method Kompensationsmethode f
n. electrode Nullelektrode f
n. instrument Nullinstrument n
n. line Nullinie f *(im Bandenspektrum)*
n. method of measurement Nullmethode f
number / **to** numerieren, beziffern, durchnumerieren
number Zahl f; Ziffer f; Nummer f; Anzahl f, Menge f; *(Pap)* Siebnummer f; *(Text)* Nummer f, Feinheit f, Titer m
n. of actual plates (trays) praktische (wirkliche, tatsächliche) Bodenzahl f *(einer Destillationskolonne)*
n. of cuts *(Pap)* Schnittzahl f
n. of folds *(Pap)* Falz[ungs]zahl f
n. of moles Molzahl f
n. of nuclei Kristall[isations]keimzahl f, Kristall[isations]kernzahl f, Keimzahl f
n. of plates Bodenzahl f *(einer Destillationskolonne)*
n. of revolutions Drehzahl f, Umdrehungszahl f
n. of stages Stufenzahl f
n. of strokes Hubzahl f
n. of theoretical plates (trays) theoretische Bodenzahl f *(einer Destillationskolonne)*
n. of vibrations Schwingungszahl f
n. of zone passes Zonen[durchgangs]zahl f, Anzahl f der Zonendurchgänge *(beim Zonenschmelzen)*
numbered buna rubber Zahlenbuna m
numbering Numerierung f, Bezifferung f, Durchnumerierung f
n. system Numerierungssystem n, Bezifferungssystem n, Durchnumerierungssystem n
numbness Unempfindlichkeit f, örtliche Betäubung f
numeral grade of buna s. numbered buna rubber
numeric subscript unten angehängter Zahlenindex m
n. superscript hochgestellte Zahl f, Hochzahl f, Exponent m
numerical num[m]erisch, zahlenmäßig, Zahlen...
n. aperture numerische Apertur f
n. prefix numerisches Präfix n, Zahlwortpräfix n, Zahlenpräfix n, Zahlenvorsatz m
n. ratio Zahlenverhältnis n
n. value Zahlenwert m
nut gall Gallapfel m
n. oil Nußöl n

nutating-piston meter Scheibenzähler m, Taumelscheibenzähler m
nutmeg Muskatnuß f, Muskatsamen m; Muskatnußbaum m, Myristica fragrans Houtt.
n. butter Muskat[nuß]butter f, Muskatbalsam m, Muskatnußöl n
n. oil ätherisches Muskatöl n, Muskatnußöl n; s. n. butter
nutrient nährend, nahrhaft, Nähr...; Nährstoff...
nutrient Nährstoff m, (bei Nährlösungen) Nährsalz n
  n. absorption Nährstoffaufnahme f
  n. agar Nähragar m(n), Agarnährboden m
  n. agar plate Agar[nähr]platte f
  n. balance Nährstoffbilanz f
  n. broth Nährbouillon f
  n. broth bottle Nährbodenflasche f
  n. carrier Nährstoffträger m
  n. content Nährstoffgehalt m
  n. deficiency Nährstoffmangel m, Nährstoffarmut f
  n. deficiency symptom Nährstoffmangelerscheinung f
  n. demand Nährstoffbedarf m
  n. displacement Nährstoffverlagerung f
  n. element Nähr[stoff]element n
  n. elution Nährstoffauswaschung f
  n. enrichment Nährstoffanreicherung f
  n. lack s. n. deficiency
  n. leaching s. n. elution
  n. line Nährstofflinie f
  n. medium Nährboden m, Nährmedium n, Nährsubstrat n
  n. mobility Beweglichkeit f der Nährstoffe
  n. ratio Nährstoffverhältnis n
  n. remigration Nährstoffrückwanderung f
  n. requirements Nährstoffansprüche mpl, Nährstoffbedürfnis n, Gesamtnährstoffbedarf m
  n. reserve Nährstoffvorrat m, Nährstoffreserve f
  n. salt Nährsalz n
  n. solution Nährlösung f
  n. supply Nährstoffversorgung f, Nährstoffzufuhr f, Nährstoffnachlieferung f
  n. uptake Nährstoffaufnahme f
  n. utilization Nährstoffausnutzung f
  n. withdrawal Nährstoffentzug m
  n. yeast Nährhefe f
nutrition of plants Pflanzenernährung f
  n. science Ernährungswissenschaft f
nutritional die Ernährung betreffend, Ernährungs..., Nahrungs..., Nähr...
  n. investigation Ernährungsforschung f
  n. value Nährwert m
nutritious nährend, nahrhaft, Nähr...; nährstoffreich
nutritive nährend, nahrhaft, Nähr...
nutritive Nährstoff m
  n. humus Nährhumus m
  n. medium (Bioch) Nährboden m, Nährmedium n, Nährsubstrat n

n. value Nährwert m
nutsch[e], nutsch filter Filternutsche f, Nutsche f
nux vomica Brechnuß f, Krähenauge n (von Strychnos nux-vomica L.)
NW acid, NW-acid s. Nevile-Winther acid
Nyassa rubber (Lianenkautschuk von Landolphia kirkii Dyer)
Nylander solution Nylanders Reagens n
nylon dye Nylonfarbstoff m
  n. felt (Pap) Nylonfilz m
  n. salt Hexamethylendiammoniumadipat n

O

o-rh. s. orthorhombic
O-ring Rund[schnur]ring m, O-Ring m (Dichtung)
oak bark (Gerb) Eichenrinde f
  o. gall Eichengallapfel m, Eichengalle f
Oakes frother (Gum) Oakes-Maschine f
oast[-house] Darrhaus n, Darre f, (i.e.S.) Hopfendarre f
oat Hafer m, Avena L., (i.e.S.) Saathafer m, Avena sativa L.
  o. flakes Haferflocken pl
  o. flour Hafermehl n
  o. groats Hafergrütze f
  o. huller Haferschälgang m
  o. malt Hafermalz n
  o. meal s. o. flour
  o. starch Haferstärke f
object to be weighed Wägegut n, abzuwägende Substanz f
objectionable zu beanstanden[d], störend, schädlich; unangenehm, schlecht, aufdringlich, widerlich, übel (Geruch)
oblique head (Plast) Schrägspritzkopf m
  o. stopcock Hahn m mit Bohrung schräg zur Achse, Hahn m mit schräger Bohrung
obliterating power Deckvermögen n, Deckfähigkeit f, Deckkraft f
oblong-mesh cloth Langmaschengewebe n
obnoxious übelriechend, unangenehm (schlecht, widerlich, aufdringlich) riechend
observation tube Beobachtungsrohr n (Laborgerät)
obsidian (Geol) Obsidian m
obtain / to erhalten, gewinnen, darstellen, herstellen
occidentalol Okzidentalol n (Sesquiterpenalkohol)
occlude / to okkludieren, einschließen, absorbieren
occlusion Okklusion f, Einschluß m, Absorption f
  o. compound Einschlußverbindung f
occupation number Besetzungszahl f
occupy / to besetzen, einnehmen
occur / to vor sich gehen, erfolgen, stattfinden, eintreten; vorkommen, auftreten, sich finden
occurrence Vorkommen n, Auftreten n

o. in nature natürliches Vorkommen n

ocher, ochre *(Min)* Ocker m

ochreous ockerhaltig

ochrolite *(Min)* Ochrolith m, *(veraltet für)* Nadorit m *(Blei(II)-antimon(III)-dioxidchlorid)*

ocimene Ozimen n, 3,7-Dimethyl-1,3,6-oktatrien n

oct. *s.* octahedral

octaacetylcarminic acid Oktazetylkarminsäure f

9,12-octadecadienoic acid $CH_3(CH_2)_4CH=CHCH_2CH=CH(CH_2)_7COOH$ Oktadekadien-(9,12)-säure f, Linolsäure f, Leinölsäure f

n-octadecanoic acid $CH_3(CH_2)_{16}COOH$ n-Oktadekansäure f, Zetylessigsäure f, Stearinsäure f

1-octadecanol $CH_3(CH_2)_{16}CH_2OH$ Oktadekanol-(1) n, n-Oktadezylalkohol m, Stearylalkohol m

9,12,15-octadecatrienoic acid $CH_3CH_2CH=CHCH_2CH=CHCH_2CH=CH-(CH_2)_7COOH$ Oktadekatrien-(9,12,15)-säure f, Linolensäure f

***cis*-6-octadecenoic acid** $CH_3(CH_2)_{10}=CH(CH_2)_4COOH$ *cis*-6-Oktadezensäure f, Δ_6-Oktadezensäure f, Petroselinsäure f

***cis*-9-octadecenoic acid** $CH_3(CH_2)_7CH=CH(CH_2)_7COOH$ *cis*-9-Oktadezensäure f, Elainsäure f, Oleinsäure f, Ölsäure f

***trans*-11-octadecenoic acid** $CH_3(CH_2)_5CH=CH(CH_2)_9COOH$ *trans*-11-Oktadezensäure f, Vakzensäure f

n-octadecyl alcohol *s.* 1-octadecanol

n-octadecylic acid *s.* n-octadecanoic acid

6-octadecynoic acid $CH_3(CH_2)_{10}C\equiv C(CH_2)_4COOH$ 6-Oktadezinsäure f, Taririnsäure f

9-octadecynoic acid $CH_3(CH_2)_7C\equiv C(CH_2)_7COOH$ 9-Oktadezinsäure f, Stearolsäure f

octafluoride Oktafluorid n

octahedral *(Krist)* oktaedrisch, achtflächig

o. borax $Na_2B_4O_7 \cdot 5H_2O$ oktaedrischer Borax m, Juwelierborax m, Natriumtetraborat-5-Wasser n

octahedrite *(Min)* Oktaedrit m

octahedron *(Krist)* Oktaeder n, [regelmäßiger, regulärer] Achtflächner m

octahydrate Oktahydrat n

octahydride Oktahydrid n

octahydroisoquinoline Oktahydroisochinolin n

octalin Oktalin n, Oktahydronaphthalin n

octamolybdate $M^I_4Mo_8O_{26}$ Oktamolybdat n

octane C_8H_{18} Oktan n, *(i.e.S.) s.* n-octane

n-octane C_8H_{18} n-Oktan n, Normaloktan n, Oktan n

o. number Oktanzahl f, Oktanziffer f, OZ f

o. rating Oktanzahlbestimmung f, OZ-Bestimmung f; *s.* o. number

o. value *s.* o. number

octanedioic acid $HOOC(CH_2)_6COOH$ Oktandisäure f, Hexandikarbonsäure-(1,6) f, Suberinsäure f, Korksäure f

octanoic acid $CH_3(CH_2)_6COOH$ n-Oktansäure f, n-Kaprylsäure f, Heptankarbonsäure-(1) f, Hexylessigsäure f

1-octanol $CH_3(CH_2)_7OH$ 1-Oktanol n, n-Oktylalkohol m, Heptylkarbinol n, n-Kaprylalkohol m

octatomic achtatomig

octatungstate Oktawolframat n

octavalency Achtwertigkeit f, Oktavalenz f

octavalent achtwertig, oktavalent

octet Oktett n, Achtergruppe f, Achtergruppierung f, Achterschale f

o. gap Oktettlücke f

o. rule Oktettregel f, Oktett-Theorie f

octic acid *s.* octanoic acid

1-octine *s.* 1-octyne

4-octine *s.* 4-octyne

n-octoic acid *s.* octanoic acid

octose Oktose f *(Monosaccharid mit acht Sauerstoffatomen)*

octoxide Oktoxid n

n-octyl alcohol *s.* 1-octanol

n-octylacetylene $HC\equiv C(CH_2)_7CH_3$ n-Oktylazetylen n, 1-Dezin n

n-octylic acid $CH_3(CH_2)_6COOH$ n-Oktylsäure f, n-Oktansäure f, n-Kaprylsäure f, Heptankarbonsäure-(1) f, Hexylessigsäure f

1-octyne $HC\equiv C(CH_2)_5CH_3$ 1-Oktin n, n-Hexylazetylen n, Krypyliden n

4-octyne $CH_3(CH_2)_2C\equiv C(CH_2)_2CH_3$ 4-Oktin n, Dipropylazetylen n

OD *s.* oven-dry

O.D. *s.* outer diameter

odd atomic number ungerade Atomnummer (Ordnungszahl, Kernladungszahl) f

o. electron einsames (unpaares, ungepaartes, unpaariges) Elektron n

o.-even nucleus Ungerade-gerade-Kern m, ug-Kern m

o.-numbered ungeradzahlig, ungerade

o.-odd nucleus Ungerade-ungerade-Kern m, uu-Kern m

Oderberg [colloid] mill Oderberg-Mühle f, Oderberger Kolloidmühle f

odontolite *(Min)* Odontolith m

odorant duftend, wohlriechend, Duft..., Riech...

odorant Odorans n, Odor[is]ierungsmittel n, Odoriermittel n

odoriferous *s.* odorant

odorimeter *s.* olfactometer

odorimetry *s.* olfactometry

odorize / to *(giftige Gase)* odor[is]ieren

odorizing Odor[is]ierung f, Odor[is]ieren n *(von giftigen Gasen)*

odorless *s.* odourless

odorometer *s.* olfactometer

odorous *s.* odorant

o. substance Geruchsstoff m

odour Geruch m, Duft m; Geruchsstoff m

o.-free *s.* odourless

o. improvement Geruchsverbesserung f

odourless geruchlos, geruch[s]frei, nichtriechend

OE-SBR *s.* oil-extended styrene-butadiene rubber

oenanthal $CH_3(CH_2)_5CHO$ Oenanthaldehyd m, n-Heptylaldehyd m, n-Heptanal n

oenanthic acid $CH_3(CH_2)_5COOH$ Oenanthsäure f, n-Heptylsäure f
oenological önologisch, weinfachkundlich
oenology Önologie f, Wein[bau]kunde f
oenometer Önometer n (Aräometer zur Bestimmung des Alkoholgehalts der Weine)
OEP s. oil-extended polymer
oerstedtite (Min) Oerstedtit m (Zirkoniumorthosilikat)
oestradiol Östradiol n (ein Sexualhormon)
  o. **benzoate** Östradiolbenzoat n, Östradiol-3-monobenzoat n
off-centre außermittig, exzentrisch
  o. **colour** Farbfehler m, Verfärbung f
  o. **flavour** Geschmacksfehler m (oder) Geruchsfehler m
  o.-**gas** Abgas n
  o.-**machine coating** (Pap) Streichen n außerhalb der Maschine (Papiermaschine), Separatstreichen n, Separatstrich m
  o.-**machine operation** (Arbeitsgang innerhalb eines technologischen Prozesses, der außerhalb der Hauptmaschine durchgeführt wird)
  o.-**shade / to be** im Farbton abfallen
  o.-**weight** Masseabweichung f, (bisher) Gewichtsabweichung f
offhand glass freihandgeblasenes Glas n
  o. **process** (Glas) Freihandblasen n
official (Pharm) offizinell; offiziell
  o. **name** offizieller Name m (einer organischen Verbindung)
  o. **sample** Vergleichsprobe f
  o. **toxicologist** vereidigter Toxikologe m
offset / to ausgleichen, kompensieren
  o. **the yellow cast** den Gelbstich beseitigen (beheben, auslöschen, kompensieren, neutralisieren)
offset (Werkstoffprüfung) bleibende (dauernde) Dehnung f
  o. **base** s. o. punt
  o. **cardboard** s. o. printing cardboard
  o. **finish** (Glas) versetzte Mündung f (Fehler)
  o. **ink** s. o. printing ink
  o. **paper** s. o. printing paper
  o. **press** (Pap) Offsetpresse f, glättende (filzlose) Presse f
  o. **printing** Offsetdruck m
  o. **printing cardboard** Offsetkarton m
  o. **printing ink** Offset[druck]farbe f, Druckfarbe f für Offsetdruck
  o. **printing paper** Offset[druck]papier n
  o. **process** Offset[druck]verfahren n
  o. **punt** (Glas) versetzter Boden m (Fehler)
  o. **yield strength (stress)** (Plast) Proportionalitätsgrenze f
offtake [pipe] Austragrohr n, Ableitungsrohr n, Austrittsrohr n, Abzugsrohr n
O.F.H.C. copper s. oxygen-free high-conductivity copper
OH-furnace s. open-hearth furnace

OH group OH-Gruppe f, OH-Rest m, Hydroxy[l]gruppe f
OH radical OH-Gruppe f, Hydroxy[l]gruppe f, OH-Rest m, OH-Radikal n; [freies] OH-Radikal n
oil / to ölen; schmieren; (Gerb) fetten, abölen; (Text) schmälzen
oil Öl n
  o. **absorption** Absorption f in Öl, Ölabsorption f, Waschölabsorption f
  o. **accumulation** Ölansammlung f, Ölanreicherung f, Ölakkumulation f
  o.-**base mud** Ölspülung f, Spülschlamm m (Spülung f) auf Erdölbasis
  o.-**base[d] paint** Ölfarbe f
  o. **bath** Ölbad n
  o.-**bearing** [erd]ölführend, [erd]ölhaltig
  o.-**buffered seal** Sperröldichtung f
  o. **burner** Ölbrenner m
  o. **cake** Ölkuchen m, Preßkuchen m
  o.-**cake breaker (crusher)** Ölkuchenbrecher m
  o.-**cake meal** Ölkuchenmehl n
  o./**carbon black masterbatch** Öl-Ruß-Batch m, Vormischung f aus Polymer, Ruß und Öl
  o. **colour** s. o.-base[d] paint
  o.-**containing** ölhaltig
  o. **cracking** Ölkracken n, Ölspaltung f
  o. **cup** Petroleumgefäß n (eines Flammpunktprüfers); Ölbehälter m (eines Viskosimeters)
  o. **deposit** Erdöllagerstätte f, Erdölvorkommen n; Ölkruste f, Ölrückstand m
  o.-**drop method** Öltropfenmethode f
  o. **droplet** Öltröpfchen n
  o.-**extended** (Gum) ölplastiziert, ölgestreckt, ölhaltig
  o.-**extended polymer** s. o.-extended styrenebutadiene rubber
  o.-**extended rubber** ölgestreckter (ölplastizierter, ölhaltiger) Kautschuk m
  o.-**extended styrene-butadiene rubber** ölgestreckter (ölplastizierter, ölhaltiger) Butadien-Styrol-Kautschuk m, OP-Kautschuk m
  o. **extension** (Gum) Ölstreckung f, Ölplastizierung f
  o. **feed** Öleinsatz m
  o. **field** s. oilfield
  o.-**filled** s. o.-extended
  o. **filler** (Farb) Ölspachtel m(f)
  o. **filter** Ölfilter n
  o.-**fired** ölgefeuert, ölbeheizt, mit Öl geheizt (beheizt, betrieben)
  o. **firing** Ölfeuerung f, Öl[be]heizung f
  o. **flame** Ölflamme f
  o. **flax** Öllein m, Linum usitatissimum L.
  o.-**flocculated suspension** Suspension f mit Ölflockung
  o.-**forming** ölbildend
  o.-**free** ölfrei
  o.-**fuel firing** s. o. firing
  o.-**furnace black** (Gum) Öl-Furnace-Ruß m
  o.-**furnace plant** (Gum) Öl-Furnace-Anlage f

o.-furnace process *(Gum)* Öl-Furnace-Verfahren *n*

o. gas Ölgas *n*

o./gas separator Gas-Öl-Separator *m*, Gas-Öl-Trennvorrichtung *f*, Gasseparator *m*, Gas[ab]scheider *m*, Gastrennanlage *f*

o. gasification Ölvergasung *f*

o. genesis Erdölgenesis *f*, Erdölentstehung *f*, Erdölbildung *f*

o. green Cr_2O_3 Chrom[oxid]grün *n*, Ölgrün *n*, grüner Zinnober *m*, Chrom(III)-oxid *n*

o. gun *s.* o. burner

o.-harden / to ölhärten, in Öl härten

o. hardening Ölhärten *n*, Ölhärtung *f*, Härten *n* in Öl

o.-hardening steel ölhärtender Stahl *m*, Öl-härtestahl *m*, Ölhärter *m*

o.-heated *s.* o.-fired

o. heating *s.* o. firing

o. horizon *s.* o. layer

o.-impregnated ölgetränkt, mit Öl getränkt

o. impregnation Öltränkung *f*

o.-in-water emulsion Öl-in-Wasser-Emulsion *f*, Ö/W

o.-in-water type Öl-in-Wasser-Typ *m* *(von Emulsionen)*

o. indication Erdölanzeichen *n*, Erdölindikation *f*

o. industry Erdölindustrie *f*, Ölindustrie *f*

o. jacket Ölbad *n*, ölgespeister Heizmantel *m*

o.-jacketed mit Ölbad

o. layer *(Geol)* Ölschicht *f*, Ölhorizont *m*

o. length *(Farb)* Ölgehalt *m* *(bezogen auf Harz)*, Verhältnis *n* Öl/Harz

o. level Ölspiegel *m*, Ölniveau *n*, Ölstand *m*

o. liberation Öl[ab]scheidung *f*

o. masterbatch *(Gum)* Ölbatch *m*, Ölvor-mischung *f*

o.-masterbatched polymer *s.* o.-extended sty-rene-butadiene rubber

o. mill Ölmühle *f*, Ölschlägerei *f*, Ölfabrik *f*

o. mixture Ölmischung *f*, Ölgemisch *n*

o.-modified ölmodifiziert

o.-modified resin ölmodifiziertes Harz *n*

o. occurrence Erdölvorkommen *n*, Erdöllager-stätte *f*

o. of anise (aniseed) Anisöl *n* *(von Pimpinella anisum L. und Illicium verum Hook. f.)*

o. of bay Bayöl *n* *(von Pimenta racemosa (Mill.)I.W. Moore)*

o. of ben Ben-Öl *n*, Behen-Öl *n* *(von Moringa aptera Gaertn., seltener von Moringa oleifera Lam.)*

o. of bene Sesamöl *n* *(von Sesamum indicum L.)*

o. of birch buds Birkenknospenöl *n*

o. of bitter almonds Bittermandelöl *n*

o. of caraway Kümmelöl *n* *(von Carum carvi L.)*

o. of citronella Zitronellöl *n* *(von Cymbopogon nardus (L.)Rendle)*

o. of cloves Nelkenöl *n*, Gewürznelkenöl *n* *(von Syzygium aromaticum (L.)Merr. et L. M. Perry)*

o. of hartshorn Dippels Tieröl *n* *(aus Knochen oder Knochenteer zum Vergällen von Spiritus)*

o. of juniper berries Wacholder[beer]öl *n* *(von Juniperus communis L.)*

o. of lavender Lavendelöl *n* *(von Lavandula-Arten)*

o. of mirbane $C_6H_5NO_2$ *(Kosmet)* Mirbanöl *n*, Nitrobenzol *n*

o. of mustard Senföl *n*

o. of myrrh Myrrhenöl *n* *(von Commiphora-Arten)*

o. of nutmeg ätherisches Muskatnußöl *n* *(von Myristica fragrans Houtt.)*

o. of roses Rosenöl *n*

o. of sassafras Sassafrasöl *n* *(von Sassafras albidum (Nutt.)Nees)*

o. of thyme Thymianöl *n* *(von Thymus vulgaris L. und Th. zygis L.)*

o. of turpentine Terpentinöl *n*

o. of vitriol Vitriolöl *n* *(i.w.S.)*, *(veraltet für)* Schwefelsäure *f*

o. of wintergreen Wintergrünöl *n* *(von Gaultheria procumbens L.)*

o. paint Ölfarbe *f*

o. parchment Ölpergament *n*

o. plant Ölpflanze *f*, Ölfrucht *f*

o. pool *s.* oilfield

o. preserves Ölpräserven *fpl*

o.-proof *s.* oilproof

o. pump Ölpumpe *f*

o.-quench / to in Öl abschrecken (ablöschen)

o. quenching Ölabschrecken *n*, Ölabschreckung *f*, Ölablöschung *f*, Abschrecken *n* in Öl

o.-reactive ölreaktiv

o.-reactive resin ölreaktives Harz *n*

o. refinery Erdölraffinerie *f*

o. repellency Ölabweisungsvermögen *n*, ölab-weisende Kraft (Eigenschaft) *f*, ölabweisendes Verhalten *n*, ölabweisender Charakter *m*

o./resin ratio *s.* o.-to-resin ratio

o. resistance Ölfestigkeit *f*, Ölbeständigkeit *f*

o.-resistant, o.-resisting ölfest, ölbeständig

o. rock *s.* o.-source rock

o. rubber Ölkautschuk *m*, *(veraltet für)* Faktis *m*

o. sand *(Geol)* Ölsand *m*, ölhaltiger Sand *m*; *(Met)* Ölsand *m* *(reiner Quarzsand mit einem Kernöl als Bindemittel zur Kernherstellung)*

o.-sand core Öl[sand]kern *m*

o. scrubber Ölwäscher *m*, Ölwascher *m*

o. seal ring Manschettendichtung *f*, Wellendicht-ring *m*

o. separator Ölabscheider *m*

o. shale *(Geol)* Ölschiefer *m*, bituminöser Schiefer *m*, Bitumenschiefer *m*

o. show *s.* o. indication

o.-soluble öllöslich

o.-source rock Erdölmuttergestein *n*

o. spot Ölfleck *m*

o. stain *(Farb)* Ölbeize *f*

o. tannage Sämischgerbung *f*, *(i.w.S.)* Fett-gerbung *f*

o.-tanned sämischgegerbt, *(i.w.S.)* fettgar
o. tanning s. o. tannage
o. tar Ölteer *m*
o.-tight s. oiltight
o.-to-resin ratio Öl-Harz-Verhältnis *n*, Verhältnis *n* von Öl zu Harz
o. trace Ölspur *f*
o. trap Ölfalle *f*, Erdölfalle *f*
o. turbine ultracentrifuge Ölturbinen[ultra]zentrifuge *f*
o. varnish Öllack *m*; Ölfirnis *m*
o. washing Ölwäsche *f*, Auswaschen *n* mit[tels] Waschöl
o.-water emulsion Öl-in-Wasser-Emulsion *f*
o. well Ölbohrloch *n*, Ölbohrung *f*, Ölbrunnen *m*, Erdölbohrloch *n*, Erdölbohrung *f*; Ölquelle *f*, Erdölquelle *f*
o. wrapping paper Öl[pack]papier *n*
o. yellow $C_6H_5N=NC_6H_4N(CH_3)_2$ Buttergelb *n*, Dimethylgelb *n*, *p*-Dimethylaminoazobenzol *n*
oilcloth Wachstuch *n*
oiled paper geöltes Papier *n*, Ölpapier *n*
oilfield Ölfeld *n*, Erdölfeld *n*
o. development Ölfeldentwicklung *f*
oiliness 1. Öligkeit *f*, Fettigkeit *f*; 2. altöliger Geschmack *m* *(z.B. verdorbener Fette)*; Peroxidranzigkeit *f*; 3. Schmierfähigkeit *f*, Schmiergüte *f*, Schmierwert *m*, Öligkeit *f*, Schmierergiebigkeit *f* *(eines Schmieröls)*
oiling Ölen *n*; Schmieren *n*; *(Gerb)* Fetten *n*, Abölen *n*; *(Text)* Schmälzen *n*
oilless bearing ölloses (ölfreies) Lager *n*, Öllos-Lager *n*
oilproof öldicht, ölundurchlässig
oiltight öldicht
oily öl[art]ig; ölhaltig, Öl...
o. bitumen Ölbitumen *n*
o.-petrolatum dryer Trockner *m* für öliges Petrolatum
o.-petrolatum tank Tank *m* für öliges Petrolatum
ointment *(Pharm)* Salbe *f*
o. base Salbengrundlage *f*
oiticica oil *(Farb)* Oiticicaöl *n* *(von Licania rigida Benth.)*
ol bridge s. olation bridge
olate / to verolen
olation Verolung *f* *(Bildung von Nebenvalenzbrücken zwischen Metallatomen und OH-Gruppen)*
o. bridge Olbrücke *f*
old fustic echtes Gelbholz *n*, echter (alter) Fustik *m* *(von Chlorophora tinctoria Gaud.)*
o. hop Althopfen *m*
o. paper Altpapier *n*, Abfallpapier *n*, Papierabfälle *mpl*, Makulatur *f*, Ausschußpapier *n*, Ausschuß *m*
o.-paper stock Altpapierstoff *m*, wiederaufbereitetes (aufbereitetes, regeneriertes) Altpapier *n*
o.-rope stock *(Pap)* Hadernhalbstoff *m* aus Tauen (Manilatauen, Hanftauabfällen)

o. tuberculin *(Pharm)* Alttuberkulin *n*, Tuberkulin *n* Koch, AT, A.T. Koch, TOA
oleanoic acid Oleanolsäure *f*, Karyophyllin *n*
oleate Oleat *n*
olefiant gas $H_2C=CH_2$ Äth[yl]en *n*
olefin[e] s. o. hydrocarbon
o.-forming reagent Olefinierungsreagens *n*
o. hydrocarbon Äth[yl]enkohlenwasserstoff *m*, Äth[yl]en *n* *(i.w.S.)*, Alk[yl]en *n*, Olefin *n*, Monoolefin *n*
o.-oxide polymerization Olefinoxidpolymerisation *f*
o. polymerization Olefinpolymerisation *f*
oleic acid $CH_3(CH_2)_7CH=CH(CH_2)_7COOH$ Ölsäure *f*, Oleinsäure *f*, Elainsäure *f*, *cis*-9-Oktadezensäure *f*
olein Olein *n* *(technische Ölsäure)*
oleinic acid s. oleic acid
oleomargarin[e] Oleomargarin *n*, Oleo *n*
oleoresin Oleoresin *n*, Ölharz *n* *(natürlich vorkommende Lösung von Harz in essentiellem Öl)*
o. of male fern *(Pharm)* Wurmfarnextrakt *m*
oleoresinous paint Ölharzfarbe *f*, Harz-Öl-Farbe *f*, Öl-Naturharz-Farbe *f*
o. varnish Ölharzlack *m*, Harz-Öl-Lack *m*, Öl-Naturharz-Lack *m*
oleostearin[e] *(Lebm)* Oleostearin *n*, Oleostock *n*, Preßtalg *m*, Preßling *m*
oleum Oleum *n*, rauchende Schwefelsäure *f*
o.-measuring vessel Meßgefäß *n* für Oleum
oleyl alcohol $C_8H_{17}CH=CH(CH_2)_8OH$ Oleylalkohol *m*, *cis*-9-Oktadezenol-(1) *n*
olfactometer Olfaktometer *n* *(Gerät zur Geruchsprüfung)*
olfactometry Olfaktometrie *f*, Odorimetrie *f*
olibanum Weihrauch *m* *(Gummiharz von Boswellia-Arten)*
o. oil *(Pharm)* Weihrauchöl *n*, Olibanumöl *n*
oligase Oligase *f* *(Ferment)*
oligoclase, oligoclasite *(Min)* Oligoklas *m*
oligomer Oligomer[es] *n*
oligosaccharide Oligosaccharid *n*
olive olivfarben
olive Olive *f* *(von Olea europaea L.)*
o.-green olivgrün
o.-kernel oil Olivenkernöl *n*
o. oil Olivenöl *n*
o.-oil castile soap kastilianische Seife *f*
olivenite *(Min)* Olivenit *n* *(Kupfer(II)-hydroxidorthoarsenat)*
Oliver filter Oliver-Filter *n*
olivine *(Min)* Olivin *m* *(Magnesiumeisen(II)-orthosilikat)*
o. refractory *(Ker)* Olivinerzeugnis *n*
O.L.P. process OLP-Verfahren *n*, Oxygène-Lance-Poudre-Verfahren *n* *(ein Sauerstoffaufblaseverfahren)*
ommatine Ommatin *n* *(Naturfarbstoff)*
ommine Ommin *n* *(Naturfarbstoff)*
ommochrome Ommochrom *n* *(Naturfarbstoff)*

on-centre mittig, zentrisch
on-glaze decoration *(Ker)* Aufglasurdekor *n*
on-glaze painting *(Ker)* Aufglasurmalerei *f*
on-machine coating *(Pap)* Streichen *n* in der Papiermaschine, Maschinenstrich *m*
on-off control Zweipunktregelung *f*
on standing beim Stehen[lassen]
once-fired ware *(Ker)* einmal gebrannte Ware *f*
o.-run benzol vordestilliertes Benzol *n*, Benzolvorprodukt *n*, Benzolvorerzeugnis *n*
o.-run[ning] still Destillationsapparat (Destillierapparat) *m* für Benzolvorprodukt
o.-through[-flow] boiler Zwangdurchlaufkessel *m*, Zwangdurchlaufdampferzeuger *m*
one-and-a-half bond, one-and-one-half bond anderthalbfache Bindung *f*, Anderthalbfachbindung *f*
o.-bath chrome liquor *(Gerb)* Einbadchrombrühe *f*
o.-bath chrome tannage Einbad[chrom]gerbung *f*
o.-bath chroming method *(Text)* Einbadchrom[ier]verfahren *n*
o.-bath method *(Text)* Einbadverfahren *n*
o.-bath tannage *s.* o.-bath chrome tannage
o.-colour indicator einfarbiger Indikator *m*
o.-component system Einkomponentensystem *n*, Einstoffsystem *n*, unitäres System *n*
o.-daylight press *(Gum)* Einetagenpresse *f*
o.-dimensional eindimensional
o.-dimensional paper chromatography eindimensionale Papierchromatografie *f*
o.-electron atom Einelektronenatom *n*
o.-electron bond Einelektronenbindung *f*
o.-electron orbital Einelektronenorbital *n*
o.-electron reduction Einelektronenreduktion *f*
o.-electron state Einelektronenzustand *m*
o.-electron transfer process Einelektronenaustauschreaktion *f (Übergang eines Einzelelektrons bei Radikalreaktionen)*
o.-face centred *(Krist)* einseitig flächenzentriert
o.-floor[ed] kiln Einhordendarre *f*
o.-mark bulb pipet Vollpipette *f* mit einer Marke
o.-pack Eintopf... *(z.B. Washprimer)*; Einkomponenten... *(z.B. Anstrichfarbe)*
o.-pack system *(Farb)* Einkomponentensystem *n*
o.-pan balance Einschalenwaage *f*
o.-pass evaporator Durchlaufverdampfer *m*
o.-phase system Einphasensystem *n*, einphasiges (homogenes) System *n*
o.-pit liming system *(Gerb)* Eingrubenäschersystem *n*
o.-shot process *(Plast)* Einstufenverfahren *n (z.B. Verschäumung)*
o.-side coater *(Pap)* einseitig streichende Streichmaschine *f*, Streichmaschine *f* für einseitige Beschichtung
o.-sided coated *(Pap)* einseitig gestrichen, mit einseitigem Strich
o.-sided coating *(Pap)* einseitiges Streichen *n*, einseitiger Strich *m*, einseitige Beschichtung *f*

o.-stage resin Phenolharz *n* im A-Zustand, Resol *n*, A-Harz *n (i.e.S.)*
o.-step process Einstufenprozeß *m*, Einstufenverfahren *n*
onglaze *(Ker)* Aufglasur *f*
Onia-Gegi process Onia-Gegi-Verfahren *n (thermisch-katalytisches Ölspaltungsverfahren)*
onion *(Glas)* Zwiebel *f (beim Vertikalziehverfahren)*
o. skin Zwiebelhautpapier *n*, Zwiebelschalenpapier *n*, Onionskin *n*; Florpostpapier *n*
onium compound Oniumverbindung *f*
o. structure Oniumstruktur *f*
onofrite *(Min)* Onofrit *m*
Onsager equation Debye-Hückel-Onsagersche Gleichung *f*
onset of vulcanization Vulkanisationseinsatz *m*
onyx *(Min)* Onyx *m (Silizium(IV)-oxid)*
oolite *(Min)* Oolith *m*, Rogenstein *m*
oolitic oolithisch, Oolith[en]...
ooze Schlamm *m*, Schlick *m*
OP *s.* osmotic pressure
opacifier Trübungsmittel *n (z.B. für Glas und Plaste)*
opacifying power (properties) Deckfähigkeit *f*, Deckkraft *f*
opacity Undurchsichtigkeit *f*, Opazität *f*, Lichtundurchlässigkeit *f*; Trübung *f*
opal *(Min)* Opal *m (Silizium(IV)-oxid)*
o. glass *s.* opaque glass
o. jasper *(Min)* Opaljaspis *m*, *(veraltet für)* Jaspopal *m (Silizium(IV)-oxid)*
opalesce / to opaleszieren, opalisieren; getrübt sein
opalescence Opaleszenz *f*, Opaleszieren *n*, Opalisieren *n*; Trübung *f*
o. colour Opaleszenzfarbe *f*
opalescent opaleszierend [getrübt], opalisierend
o. cloud weißer Bruch *m (Weinfehler)*
opaque undurchsichtig, undurchdringlich, nicht durchscheinend, opak, lichtundurchlässig; milchig, trüb; *(Farb)* gut deckend
o. glass Trübglas *n*, opakes Glas *n*, Opalglas *n*, *(i.e.S.)* Milchglas *n*
o. glaze *(Ker)* Opakglasur *f*, opake (deckende, trübe) Glasur *f*, Trübglasur *f*
o. ice Trübeis *n*
o. matter Opaksubstanz *f*, opake Substanz *f*
open up / to *(den Walzenspalt)* weiter stellen; *(Erz)* aufschließen
open-air drying Lufttrocknung *f*, Trocknen *n* an der Luft, atmosphärische Trocknung *f*
o. area offene Fläche *f (z.B. eines Siebes)*
o. assembly time *(Plast)* offene Wartezeit *f (vor dem Kleben)*
o. bubble *(Plast)* offene Blase *f* an der Oberfläche *(Preßfehler)*
o. carbide furnace offener Kalziumkarbidofen *m*
o.-cell offenzellig *(Schaumstoff)*
o.-cell foam offenzelliger Schaum[kunst]stoff *m*
o.-chain offenkettig, mit offener Kette

o.-chain complex salt offenes (offenkettiges) Komplexsalz *n*
o.-chain compound offenkettige Verbindung *f*
o.-chain form offene (gestreckte) Kettenform *f*
o.-chain hydrocarbon offenkettiger (kettenförmiger, azyklischer, aliphatischer) Kohlenwasserstoff *m*
o.-chain system offenkettiges System *n*
o. channel offenes Gerinne *n*
o. cup *s.* o. flash cup
o.-cup flash point *s.* o. flash point
o.-cup flash tester *s.* o. flash tester
o. cure Freivulkanisation *f*, Freiheizung *f*
o.-cut mining (working) Tagebauförderung *f*, Tagebaubetrieb *m*, Förderung *f* im Tagebau
o. fermentation offene Gärung *f*
o. filter offenes (offen arbeitendes) Filter *n*
o.-flame kiln *(Ker)* Brennofen *m* mit offenem (direktem) Feuer
o. flash cup offener Tiegel *m*, o.T. *(eines Flammpunktprüfers)*
o. flash point Flammpunkt *m* im offenen Tiegel, Fl. o. T.
o. flash tester offener Flammpunkt[s]prüfer (Flammpunkt[s]apparat) *m*
o.-gap press *s.* o.-side press
o. grinding Hochmahlverfahren *n*, Hochmüllerei *f*, Grießmüllerei *f*
o.-hearth furnace Siemens-Martin-Ofen *m*, SM-Ofen *m*; Herdofen *m*
o.-hearth process Herdfrischverfahren *n*, Herdfrischprozeß *m*, Siemens-Martin-Verfahren *n*, SM-Verfahren *n*
o.-hearth regenerator Winderhitzer *m* am Siemens-Martin-Ofen
o.-hearth slag Siemens-Martin-Schlacke *f*, SM-Schlacke *f*
o.-hearth steel Herdfrischstahl *m*, herdgefrischter Stahl *m*, Siemens-Martin-Stahl *m*, SM-Stahl *m*
o.-hearth steel plant Siemens-Martin-Stahlwerk *n*
o.-hearth steel scrap Siemens-Martin-Stahlschrott *m*
o. mill *s.* o. roll mill
o.-mill mastication *(Gum)* Mastikation *f* (Mastizieren *n*) auf Walzwerken
o. mould *(Plast)* offenes Werkzeug *n*
o.-pan mixer Trogmischer *m*
o. pot *(Glas)* offener Hafen *m*
o. roll mill Walzwerk *n*
o. sand gut gasdurchlässiger Sand (Formsand) *m*, Sand *m* guter (hoher) Gasdurchlässigkeit
o. setting *(Ker)* offene Setzweise *f*, offenes Setzen *n* (Setzen des Brenngutes im Ofen ohne Kapseln)
o.-side press einhüftige Presse *f*, Maulpresse *f*
o. steam direkter Dampf *m*, Direktdampf *m*, Frischdampf *m*
o.-steam cure (curing) Freidampfvulkanisation *f*, Vulkanisation *f* in offenem Dampf, Freidampfheizung *f*

o.-steam vulcanizer *(Gum)* Vulkanisierkessel *m*, Vulkanisationskessel *m*
o. system *(phys Ch)* offenes System *n*
o.-tank leaf filter Tauch[blatt]filter *n*
o.-type headbox *(Pap)* offener Stoffauflauf[kasten] *m*, Stoffauflauf *m* offener Bauart
o. vulcanization *s.* o. cure
o.-width bleaching *(Text)* Breitbleiche *f*
o.-width boilout *(Text)* Abkochen *n* in breitem Zustand
o.-width washing machine *(Text)* Breitwaschmaschine *f*
opencast work *s.* open-cut mining
opening pressure Öffnungsdruck *m*
operability Betriebsbereitschaft *f*; Arbeitsfähigkeit *f*
operable betriebsbereit; arbeitsfähig
operate / to 1. betreiben, bedienen, betätigen; steuern; antreiben; 2. arbeiten
operating characteristics Betriebsverhalten *n*
o. floor Arbeitsbühne *f*, Bedienungsbühne *f*
o. holdup Betriebsinhalt *m*, Arbeitsinhalt *m* *(z.B. einer Kolonne)*
o. line *(Destillation)* Arbeitslinie *f*, Betriebslinie *f*
o. mechanism Stellantrieb *m* *(z.B. eines Ventils)*
o. platform *s.* o. floor
o. pressure Betriebsdruck *m*, Arbeitsdruck *m*; Steuerdruck *m*
o. speed Betriebsdrehzahl *f*; Arbeitsgeschwindigkeit *f*
o. steady state stationärer Betriebszustand *m*
o. steam Treibdampf *m* *(z.B. einer Dampfstrahlpumpe)*
o. temperature Betriebstemperatur *f*, Arbeitstemperatur *f*
operation Betrieb *m*, Arbeitsweise *f*, Fahrweise *f*, Betriebsmethode *f*; Arbeitsgang *m*, Operation *f*
o. of coke ovens Koksofenbetrieb *m*
operator Stellantrieb *m*, Stellmotor *m* *(Regelungstechnik)*
ophicalcite, ophiolite *(Min)* Ophikalzit *m*
ophite *(Min)* Ophit *m*, *(veraltet für)* Serpentin *m*
opium Opium *n*, Rohopium *n*
o. alkaloid Opiumalkaloid *n*, Mohnalkaloid *n*
o. poppy Schlafmohn *m*, Papaver somniferum L.
Oppenauer oxidation (reaction) Oppenauersche Oxydation *f*, Oppenauer-Reaktion *f* *(Alkohol-Dehydrierung)*
Oppenheimer-Phillips process *(Kch)* Oppenheimer-Phillips-Prozeß *m*
opposed ekliptisch, verdeckt, ±syn-periplanar *(Stereochemie)*
opposing reaction umkehrbare (reversible) Reaktion *f*
optic axis *(Krist)* optische Achse *f*
optical activity optische Aktivität (Drehung, Rotation) *f*, optisches Dreh[ungs]vermögen *n*, Drehung *f* der Polarisationsebene, Rotationspolarisation *f*
o. anomaly optische Anomalie *f*

o. antipode [optischer] Antipode *m*, enantiomorphe Form *f*, Enantiomer[es] *n*, Spiegelbildisomer[es] *n*
o. axis *s.* optic axis
o. bleach *s. o.* bleaching agent
o. bleaching optische Bleiche (Aufhellung) *f*
o. bleaching agent, o. brightener optischer Aufheller *m*, optisches Aufhellungsmittel (Bleichmittel) *n*, Weißtöner *m*
o. density [optische] Dichte *f*, Extinktion *f*, logarithmische Opazität *f*, *(Foto auch)* Schwärzung *f*, Deckung *f*
o. electron optisches Elektron *n*, Leuchtelektron *n*, Valenzelektron *n*, Bindungselektron *n*
o. exaltation optische Exaltation *f*
o. fog *(Foto)* Belichtungsschleier *m*
o. glass optisches Glas *n*, Linsenglas *n*
o. inversion Waldensche Umkehrung *f*, Inversion *f*
o. isomer optisch isomere Verbindung *f*, optisches Isomer[es] *n*, enantiomorphe Form *f*, Enantiomer[es] *n*, Spiegelbildisomer[es] *n*
o. isomerism optische Isomerie *f*, Enantiomorphie *f*, Spiegelbildisomerie *f*
o. opposite *s. o.* antipode
o. photographic density optische Dichte *f*, Schwärzung *f*, Extinktion *f*
o. pyrometer optisches Pyrometer *n*
o. rotation *s. o.* activity
o. rotatory dispersion Rotationsdispersion *f*, R.D.
o. rotatory power *s. o.* activity
o. sensitization *(Foto)* optische Sensibilisierung *f*, Farbensensibilisierung *f*
o. sensitizer *(Foto)* Sensibilisator *m*, Sensibilisierungsfarbstoff *m*
o. sensitizing *s. o.* sensitization
o. spectrum optisches (sichtbares) Spektrum *n*
o. whitening agent *s. o.* bleaching agent
optically active optisch aktiv
o. inactive optisch inaktiv
o. isotropic optisch isotrop
o. uniaxial optisch einachsig
optimum cure optimale Vulkanisation *f*
o. of cure Vulkanisationsoptimum *n*
orange I $NaSO_3C_6H_4N=NC_{10}H_6OH$ Orange I *n*, α-Naphtholorange *n (Natriumsalz der α-Naphtholazobenzolsulfonsäure-(4))*
orange II $NaSO_3C_6H_4N=NC_{10}H_6OH$ Orange II *n*, β-Naphtholorange *n (Natriumsalz der β-Naphtholazobenzolsulfonsäure-(4))*
orange III $NaSO_3C_6H_4N=NC_6H_4N(CH_3)_2$ Orange III *n*, Methylorange *n*, Helianthin *n (Natriumsalz der 4'-Dimethylaminoazobenzolsulfonsäure-(4))*
orange IV (N) $NaSO_3C_6H_4N=NC_6H_4NHC_6H_5$ Orange IV *n*, Tropäolin OO *n*, Diphenylaminorange *n (Natriumsalz der 4'-Anilinoazobenzolsulfonsäure-(4))*
o.-coloured orangefarben, orangefarbig
o.-flower oil Pomeranzenblütenöl *n (von Citrus aurantium L. ssp. aurantium)*

o. lac *s. o.* shellac
o. lead Pb_3O_4 Orangemennige *f*, Saturnzinnober *m*, Pariser Rot *n*
o. peel Pomeranzenschale *f (von Citrus aurantium L. ssp. aurantium)*; Apfelsinenschale *f*, Orangenschale *f (von Citrus sinensis (L.)Osbeck)*; Apfelsinenschale[nhaut] *f*, Apfelsinenschaleneffekt *m*, Orangenschaleneffekt *m*, *(bei Spritzlackierungen auch)* Spritznarben *fpl (Oberflächenfehler)*
o.-peel oil [bitteres] Pomeranzenschalenöl *n*; Apfelsinenschalenöl *n*, Orangenschalenöl *n*, süßes Pomeranzenschalenöl *n*, Portugalöl *n*
o.-red orangerot
o. shellac Orangeschellack *m*
o. wrapper Apfelsineneinwickelpapier *n*
o.-yellow orangegelb
orangite *(Min)* Orangit *m (Thoriumorthosilikat)*
orbit Bahn *f*, Umlaufbahn *f*, Kreisbahn *f*
orbital Orbital *n(m)*
p orbital *p*-Orbital *n*
s orbital *s*-Orbital *n*
sp orbital *sp*-Orbital *n*
π orbital *π*-Orbital *n*
o. angular momentum Orbitaldrehimpuls *m*
o. electron Orbitalelektron *n*
o. electron arrangement Elektronenanordnung *f*, Elektronenkonfiguration *f*
o. [magnetic] moment Orbitalmoment *n*, orbitalmagnetisches Moment *n*
o. motion Orbitalbewegung *f*
o. quantum number azimutale (sekundäre) Quantenzahl *f*, Azimutalquantenzahl *f*, Orbitaldrehimpulsquantenzahl *f*
o. theory Orbitaltheorie *f*
orcein Orzein *n (Farbstoff)*
orcellinic acid $(HO)_2C_6H_2(CH_3)COOH$ Orsellinsäure *f*, 2,4-Dihydroxy-6-methylbenzoesäure *f*
orchil *s.* archil
orcin[ol] $CH_3C_6H_3(OH)_2$ Orzin *n*, 5-Methylresorzin *n*, *m*-Dihydroxytoluol *n*
order Ordnung *f*
o.-disorder transformation (transition) *(Krist)* Ordnung-Unordnung-Umwandlung *f*
o. of adding materials (the compounding ingredients) *(Gum)* Mischfolge *f*
o. of magnitude Größenordnung *f*
o. of milling *(Gum)* Mischvorschrift *f*
o. of reaction Reaktionsordnung *f*
ordering Konditionieren *n*, Konditionierung *f*
ordinal number Ordnungszahl *f*, Atomnummer *f*, Kernladungszahl *f*, OZ, Z
ordinary extract *(Gerb)* unsulfitierter Gerbstoffauszug *m*
o. lactic acid $CH_3CH(OH)COOH$ gewöhnliche Milchsäure *f*, *dl*-Milchsäure *f*, Gärungsmilchsäure *f*, 2-Hydroxypropionsäure *f*
o. malic acid $HOOCCH(OH)CH_2COOH$ natürliche Äpfelsäure *f*, *l*-Äpfelsäure *f*, *l*-Hydroxybutandisäure *f*

o. malt helles Malz *n*
o. pipette Vollpipette *f*
o.-pressure steam Niederdruckdampf *m*
o. ray ordentlicher (ordinärer) Strahl *m*
o. temperature gewöhnliche Temperatur *f*, Zimmertemperatur *f*, Raumtemperatur *f*
ore Erz *n*
o.-bearing erzhaltig, erzführend
o. bed Erzbett *n*
o. beneficiation Erzaufbereitung *f*
o.-beneficiation plant Erzaufbereitungsanlage *f*
o. body Erzkörper *m*
o. box Setzraum *m (einer Kolbensetzmaschine)*
o. bridge Erzbrücke *f*
o. bunker Erzbunker *m*
o. burden Erzmöller *m*
o. cleaning Erzwäsche *f*
o. deposit Erzlagerstätte *f*, Erzlager *n*, Erzvorkommen *n*
o. dressing *s. o.* beneficiation
o.-dressing plant *s.* o.-beneficiation plant
o.-dressing process Erzaufbereitungsprozeß *m*, Erzaufbereitungsverfahren *n*
o. gangue Erzgangart *f*
o. hearth Herd[ofen] *m*
o. mineral Erzmineral *n*
o. preparation Erzvorbereitung *f*
o.-preparation plant Erzvorbereitungsanlage *f*
o. pulp Erztrübe *f*
o. shoot Erzfall *m*
Oregon balsam *(Terpentin von Pseudotsuga menziesi (Mirbel)Franco)*
Orford process Orford-Verfahren *n (Kopf- und Bodenschmelzen zur getrennten Gewinnung von Kupfer und Nickel)*
organic organisch
o. accelerator organischer Beschleuniger *m*
o. acid organische Säure *f*
o. base organische Base *f*
o. chemist Organochemiker *m*, Organiker *m*
o. chemistry organische Chemie *f*
o.-chemistry nomenclature Nomenklatur *f* der organischen Chemie
o. compound organische Verbindung *f*
o. fibre organischer Faserstoff *m*; organische Faser *f*
o. glass organisches Glas *n*
o. group organische Gruppe *f*, Organogruppe *f*
o. matter organische Substanz *f*
o.-modified organisch modifiziert
o. peroxide organisches Peroxid *n*
o. polymer organisches Polymer[es] *n*
o. portion organischer Anteil *m (z.B. eines Silikons)*
o. remains organische Überreste *mpl*
o. rock organogenes (biogenes) Gestein (Sediment) *n*, Biolith *m*
o. silane organisches Silan *n*, Organosilan *n*
o. siloxane organisches Siloxan *n*, Organosiloxan *n*, Organosiliziumoxid *n*, siliziumorganisches Oxid *n*

o. solvent organisches Lösungsmittel *n*
organized ferment geformtes Ferment *n*
organo-clay complex organomineralischer Komplex *m*, Ton-Humus-Komplex *m*
o. portion *s.* organic portion
organoberyllium compound berylliumhaltige organische Verbindung *f, (meist)* berylliumorganische Verbindung *f*, Organoberylliumverbindung *f (mit direkter Be-C-Bindung)*
organoboron compound borhaltige organische Verbindung *f, (meist)* bororganische Verbindung *f*, Organoborverbindung *f (mit direkter B-C-Bindung)*
organocadmium compound kadmiumhaltige organische Verbindung *f, (meist)* kadmiumorganische Verbindung *f*, Organokadmiumverbindung *f (mit direkter Cd-C-Bindung)*
organochlorine compound organische Chlorverbindung *f, (meist)* chlorierter Kohlenwasserstoff *m*
organochlorosilane Organochlorsilan *n*, organisches Chlorsilan *n*
organoderivative organisches Derivat *n, (meist)* Organoderivat *n*
organofunctional organofunktionell, karbofunktionell
o. silicone organofunktionelles Silikon *n*; organofunktionelle Organosiliziumverbindung *f*
organogen Organogen *n (am Aufbau organischer Verbindungen vorwiegend beteiligtes Element)*
organogenic *(Geol)* organogen
organohalo[geno]silane Organohalogensilan *n*, Organosiliziumhalogenid *n*, organisches Halogensilan *n*, siliziumorganisches Halogenid *n*
organoleptic organoleptisch, sinnlich wahrnehmbar
o. estimation (evaluation, rating, test) organoleptische Prüfung *f*, Sinnenprüfung *f*, sensorische Analyse *f (von Lebensmitteln)*, *(i.e.S.)* Geschmacksprüfung *f*
organolithium compound lithiumhaltige organische Verbindung *f, (meist)* lithiumorganische Verbindung *f*, Organolithiumverbindung *f (mit direkter Li-C-Bindung)*
organomercury clay Organoquecksilberton *m*
o. compound quecksilberhaltige organische Verbindung *f, (meist)* quecksilberorganische Verbindung *f*, Organoquecksilberverbindung *f (mit direkter Hg-C-Bindung)*
o. derivative Organoderivat *n* des Quecksilbers, Organoquecksilberverbindung *f*
o. dressing Organoquecksilberbeize *f*
organometallic compound metallhaltige organische Verbindung *f, (meist)* metallorganische (organometallische) Verbindung *f*, Organometallverbindung *f (mit direkter Metall-Kohlenstoff-Bindung)*
organophosphate insecticide *s.* organophosphorus insecticide
organophosphorus compound phosphorhaltige

organische Verbindung f, *(meist)* phosphororganische Verbindung f, Organophosphorverbindung f *(mit direkter P-C-Bindung)*

o. insecticide organisches Phosphorinsektizid n

organopolysiloxane Organopolysiloxan n, Polyorganosiloxan n, organisches Polysiloxan n, polymeres Organosiloxan n, Silikon n

organosilane Organosilan n, organisches Silan n

organosilazane Organosilazan n, organisches Silazan n

organosilicon chemistry Chemie f der siliziumhaltigen organischen Verbindungen, *(meist)* Organosiliziumchemie f, siliziumorganische Chemie f

o. compound siliziumhaltige organische Verbindung f, *(meist)* Organosiliziumverbindung f, siliziumorganische Verbindung f *(mit direkter Si-C-Bindung)*

o. halide siliziumorganisches Halogenid n, Organosiliziumhalogenid n, Organohalogensilan n, organisches Halogensilan n

o. oxide siliziumorganisches Oxid n, Organosiliziumoxid n, organisches Siloxan n, Organosiloxan n

o. polymer organisches Siliziumpolymer[es] n, *(meist)* siliziumorganisches Polymer[es] n, Organosiliziumpolymer[es] n

organosilicone Organosilikon n

organosiloxane Organosiloxan n, organisches Siloxan n, Organosiliziumoxid n, siliziumorganisches Oxid n

o. polymer Organosiloxanpolymer[es] n

organosilyl Organosilyl...

organosol Organosol n

organotin compound zinnhaltige organische Verbindung f, *(meist)* zinnorganische Verbindung f, Organozinnverbindung f *(mit direkter Sn-C-Bindung)*

o. mercaptide Organozinnmerkaptid n

o. stabilizer Organozinnstabilisator m

orient / to orientieren

orient yellow CdS Kadmiumgelb n, Kadmiumsulfid n

oriental emerald *(Min)* Korund m *(α-Aluminiumoxid)*

o. ruby *(Min)* Rubin m

o. storax levantiner (asiatischer) Styrax (Storax) m *(von Liquidambar orientalis Mill.)*

o. sweet gum Orientalischer Amberbaum m, Storaxbaum m, Liquidambar orientalis Mill.; *s. o.* storax

o. topaz *(Min)* orientalischer Topas m, *(veraltet für)* gelber Korund m

orientate / to *s.* to orient

orientation Orientierung f, [räumliche] Ausrichtung f *(von Molekülen)*; Ortsbestimmung f *(für Substituenten)*; *(Krist)* Orientierung f

o. effect Orientierungseffekt m, Richteffekt m

o. polarization Orientierungspolarisation f

orifice kleine (enge) Öffnung f; Mündung f, Austritt m, Auslauföffnung f; Düse f; Mundstück n; Blende f; Bohrung f, Lochweite f, Sieblochung f

o. edge Blenden[einlauf]kante f

o. [flow]meter Meßblende f *(ein Mengenstrommesser)*

o. mixer Blendenmischer m, Mischblende f

o. plate Drosselscheibe f, Meßblende f *(Mengenstrommessung)*

o. ring *(Glas)* Speiserring m

o. trap Düsenkondens[at]ableiter m

orificial Öffnungs...; Mündungs...; Blenden...

origin of the band Bandenmitte f, Nullstelle f *(eines Bandenspektrums)*

original assay Originalbestimmung f

o. container Originalbehälter m

o. cross section ursprünglicher Querschnitt m, Ausgangsquerschnitt m

o. position Ausgangsstellung f

o. starter culture *(Margarineherstellung)* Mutter[säure]kultur f

o. state Ausgangszustand m, Anfangszustand m

o. wort Vorderwürze f; Stammwürze f

ormolu Musivgold n, Mosaikgold n *(eine Messingsorte)*; Musivgold n, Mosaikgold n, Malergold n *(Zinn(IV)-sulfid)*

ornithine $NH_2(CH_2)_3CH(NH_2)COOH$ Ornithin n, α,δ-Diaminovaleriansäure f, 2,5-Diaminopentansäure f

orotic acid Orotsäure f, Urazil-4-karbonsäure f

orpiment [yellow] Auripigment n, Operment n, Königsgelb n, Rauschgelb n, Chinagelb n, Gelbglas n, gelber Arsenik m, gelbes Schwefelarsen n *(Arsen(III)-sulfid)*

Orr's white Barytzinkweiß n, Lithopone f, Lithopon n

Orsat analyzer (apparatus) Orsat-Apparat m, Orsat-Gerät n, Gasanalysenapparat m nach Orsat

Orsat gas [analysis] apparatus *s.* Orsat analyzer

Orsat gas manifold Hahnsystem (Hahnrohr) n des Orsat-Apparates

orsell[in]ic acid *s.* orcellinic acid

orthanilic acid $NH_2C_6H_4SO_3H$ Orthanilsäure f, o-Aminobenzolsulfonsäure f

orthite *(Min)* Orthit m, Allanit m *(zerhaltiges Sorosilikat)*

ortho orthoständig, o-ständig

in o. position in ortho-Stellung (o-Stellung, 1,2-Stellung), orthoständig, o-ständig

to be [located, situated] o. sich in ortho-Stellung (o-Stellung, 1,2-Stellung) befinden, in ortho-Stellung stehen, orthoständig (o-ständig) sein

o.-and-peri-fused *s.* o.-peri-fused

o. compound ortho-Verbindung f, o-Verbindung f

o.-directing ortho-dirigierend, nach der ortho-Stellung dirigierend

o.-fused orthokondensiert, orthoanelliert

o. fusion Orthokondensation f, Orthoanellierung f

o. hydrogen Orthowasserstoff m, ortho-Wasserstoff m

o. **isomer** Ortho-Isomer[es] n, o-Isomer[es] n
o. **molecule** ortho-Molekül n
o. **oil** Ortho-Öl n, o-Öl n
o. **orientation** ortho-Orientierung f
o.-**para-directing** ortho-para-dirigierend, nach der ortho-und para-Stellung dirigierend (lenkend)
o.-**para-directing group** ortho-para-dirigierende Gruppe f, ortho-para-dirigierender Substituent m, Substituent m erster Ordnung
o.-**para director** s. o.-para-directing group
o.-**peri-fused** ortho-peri-kondensiert, ortho-peri-anelliert
o. **position** ortho-Stellung f, o-Stellung f, 1,2-Stellung f
o. **rock** Orthogestein n, Gestein n der Orthoreihe
o. **state** ortho-Zustand m
orthoaluminate $M^I_3AlO_3$ Orthoaluminat n
orthoantimonate $M^I_3SbO_4$ Tetroxo[mono]antimonat(V) n, Orthoantimonat(V) n
orthoantimonic acid H_3SbO_4 Orthoantimonsäure f, Monoantimon(V)-säure f, Tetroxomonoantimon(V)-säure f
orthoantimonous acid H_3SbO_3 (oder) $Sb(OH)_3$ orthoantimonige Säure f, Trioxoantimon(III)-säure f
orthoarsenate Orthoarsenat(V) n, Arsenat(V) n
orthoarsenic acid H_3AsO_4 Arsen(V)-säure f, Arsensäure f
orthoarsenite Orthoarsenat(III) n, Arsenat(III) n
orthobituminous coal orthobituminöse Kohle f
orthoborate $M^I_3BO_3$ Orthoborat n, Trioxoborat n
orthoboric acid $B(OH)_3$ Orthoborsäure f, Monoborsäure f, Trioxoborsäure f, [normale] Borsäure f
orthocarbonate $M^I_4CO_4$ Orthokarbonat n
orthochromatic (Foto) orthochromatisch
orthoclase (Min) Orthoklas m (Kaliumalumotrisilikat)
orthocortex (Text) Orthokortex m
Orthoflow catalytic cracking (Erdöl) Katkracken (katalytisches Kracken) n im Orthoflow-Verfahren
Orthoflow [catalytic cracking] process (Erdöl) Orthoflow-Verfahren n (katalytisches Krackverfahren)
orthohydrogen Orthowasserstoff m, ortho-Wasserstoff m
orthomagmatic stage (Geol) liquidmagmatisches Stadium n
orthoperiodate $M^I_5[JO_6]$ Hexoxoperjodat n, Orthoperjodat n
orthoperiodic acid H_5JO_6 Hexoxojod(VII)-säure f, Orthoperjodsäure f
orthophosphate Orthophosphat n
orthophosphoric acid H_3PO_4 Orthophosphorsäure f, [gewöhnliche] Phosphorsäure f, Monophosphorsäure f
orthoplumbate $M^I_4PbO_4$ Orthoplumbat n, Tetroxoplumbat(IV) n

orthopositronium Ortho-Positronium n (mit entgegengesetztem Spin von Positron und Elektron)
orthorhombic (Krist) [ortho]rhombisch
o. **[crystal] system** (Krist) rhombisches System n
orthosilicate $M^I_4SiO_4$ Tetroxosilikat n, Orthosilikat n
orthosilicic acid H_4SiO_4 Orthokieselsäure f, Tetroxokieselsäure f
orthostannate $M^I_4SnO_4$ Tetroxostannat(IV) n, Orthostannat n
orthostannic acid H_4SnO_4 Tetroxozinnsäure f, Orthozinnsäure f
orthotantalate $M^I_3TaO_4$ Orthotantalat n
orthotellurate Orthotellurat n
orthotelluric acid H_6TeO_6 Orthotellursäure f, Hexoxotellursäure f
orthotitanate $M^I_4TiO_4$ Orthotitanat n
orthotitanic acid H_4TiO_4 Orthotitansäure f
orthotungstate $M^I_2[WO_4]$ normales Wolframat n
orthotungstic acid H_2WO_4 [gelbe] Wolframsäure f
orthovanadate $M^I_3VO_4$ (3 : 1)-Vanadat(V) n, Orthovanadat n, Tetroxovanadat(V) n
Orton cone (Ker) Ortonkegel m (ein Schmelzkörpertyp zur Temperaturüberwachung)
ortstein (Bodenchemie) Ortstein m
osazone Osazon n
o. **formation** Osazonbildung f
oscillate / to oszillieren, schwingen, pendeln, vibrieren; schwingen (pendeln) lassen, hin und her schwenken (bewegen)
oscillating-basket centrifuge Schwingsiebzentrifuge f, Schwingsiebschleuder f
o. **conveyor** Schwingförderrinne f, Förderrutsche f
o. **crystal method** Schwenkmethode f (Abart der Drehkristallmethode)
o. **disk method** Methode f der schwingenden Scheibe (zur Bestimmung der Viskosität von Gasen)
o. **doctor** (Pap) Vibrationsschaber m, traversierender Schaber m
o. **mill** s. oscillatory mill
o. **screen** Flachwurfsieb n, Planschwingsiebmaschine f; s. vibrating screen
o.-**screen centrifuge** Schwingsiebzentrifuge f, Schwingsiebschleuder f
o. **strainer** s. o. screen
oscillation Oszillation f, Schwingung f
oscillator Oszillator m
oscillatory [ball] mill Schwingmühle f, schwingende Kugelmühle f
oscillo-polarographic oszillopolarografisch
oscillographic polarography oszillografische Polarografie f
oscillometry Oszillometrie f
oscine Oszin n, Skopolin n, 3,7-Oxido-6-tropanol n (Alkaloid)
Oslo crystallizer (ein Kristallisator mit Strömungssichtung)
osmate $M^I_2OsO_4$ Osmat(VI) n, Tetroxoosmat(VI) n

osmic Osmium...; höherwertigem Osmium entsprechend
  o. acid H_2OsO_4 Tetroxoosmium(VI)-säure f; Osmiumsäure f; s. o.[-acid] anhydride
o.[-acid] anhydride OsO_4 Osmiumtetroxid n, Osmium(VIII)-oxid n, (oft unkorrekt) Osmiumsäure f
osmious s. osmous
osmium Os Osmium n
  o. dichloride $OsCl_2$ Osmiumdichlorid n, Osmium(II)-chlorid n
  o. dioxide OsO_2 Osmiumdioxid n, Osmium(IV)-oxid n
  o. disulphide OsS_2 Osmiumdisulfid n, Osmium(IV)-sulfid n
  o. hexafluoride OsF_6 Osmiumhexafluorid n, Osmium(VI)-fluorid n
  o. monoxide OsO Osmiummonoxid n, Osmium(II)-oxid n
  o. octofluoride OsF_8 Osmiumoktafluorid n, Osmium(VIII)-fluorid n
  o.-potassium hexachloride(III) $K_3[OsCl_6]$ Kaliumhexachloroosmat(III) n
  o.-potassium hexachloride(IV) $K_2[OsCl_6]$ Kaliumhexachloroosmat(IV) n
  o. sesquioxide Os_2O_3 Osmium(III)-oxid n
  o. tetrachloride $OsCl_4$ Osmiumtetrachlorid n, Osmium(IV)-chlorid n
  o. tetrafluoride OsF_4 Osmiumtetrafluorid n, Osmium(IV)-fluorid n
  o. tetrasulphide OsS_4 Osmiumtetrasulfid n, Osmium(VIII)-sulfid n
  o. tetroxide OsO_4 Osmiumtetroxid n, Osmium(VIII)-oxid n
  o. trichloride $OsCl_3$ Osmiumtrichlorid n, Osmium(III)-chlorid n
osmocyanide $M^I_4[Os(CN)_6]$ Zyanoosmat(II) n
osmometer Osmometer n
osmometry Osmometrie f
osmophilic yeast osmophile Hefe f
osmophore osmophore Gruppe f
osmosis Osmose f
osmotic osmotisch
  o. coefficient osmotischer Koeffizient m
  o. force osmotische Kraft f
  o. pressure osmotischer Druck m
  o. swelling (Gerb) Neutralsalzquellung f
osmous Osmium...; niederwertigem Osmium entsprechend
  o. chloride s. osmium dichloride
  o. sulphite $OsSO_3$ Osmiumsulfit n
osone Oson n
osteolite (Min) Osteolith m (ein Phosphorit)
Ost's solution Ostsche Lösung f (aus $CuSO_4$, Na_2CO_3 und $NaHCO_3$ zum Traubenzuckernachweis)
Ostwald dilution law Ostwaldsches Verdünnungsgesetz n
Ostwald process Ostwald-Verfahren n (zur Gewinnung von NO_2)

Ostwald ripening (Koll) Ostwald-Reifung f
Ostwald rule Ostwaldsche Stufenregel f
Ostwald viscosimeter Ostwald-Viskosimeter n
Ostwald viscosity pipette Kapillarviskosimeter n nach Ostwald
otto of roses Rosenöl n
Otto-Hoffmann oven Otto-Hoffmann-Ofen m, Otto-Hoffmann-Koksofen m
Otto oven Otto-Ofen m, Otto-Zwillingszugofen m, Zwillingsverbundofen m von Otto
Otto-Rummel process Rummel-[Otto-]Verfahren n, Rummel-Schlackenbadverfahren n
ouabain Ouabain n, g-Strophantin n, Strophantin G n (Glykosid)
ouricury wax Ouricury-Wachs n (von Cocos coronata)
outage Schwund m (verlorengegangene Menge)
  o. time Ausfallzeit f, Stillstandszeit f; Nebenzeit f
outbreathing Entweichen n von Dämpfen
outcrop / to (Geol) ausgehen, ausstreichen, zutage treten
outcrop (Geol) Anstehen n, Ausstreichen n, Zutagetreten n, Zutageliegen n; anstehende Ader f, Ausgehendes n
outdoor air Außenluft f
  o. durability Außenbeständigkeit f, Witterungsbeständigkeit f, Wetterbeständigkeit f, Wetterfestigkeit f
  o. weathering (Gum) Außenbewetterung f
  o. weathering test (Gum) Prüfung f durch Außenbewetterung
outer cylinder Außenzylinder m
  o. diameter Außendurchmesser m
  o. electron Außenelektron n, äußeres Elektron n
  o. layer effect (Glas) Rampenbildung f, Wellenbildung f (Fehler)
  o. orbital bond Außenorbitalbindung f
  o.-orbital complex Außenorbitalkomplex m
  o. shell äußer[st]e Schale f, Außenschale f
  o. wall s. outside wall
outermost electron Valenzelektron n, Bindungselektron n, Leuchtelektron n, optisches Elektron n
  o. layer (shell) s. outer shell
outflow Abfluß m, Ablauf m, Ausfluß m, Austritt m; Abflußmenge f; Abfließen n, Ausfließen n, Ausströmen n
  o. tube Ausflußrohr n, Auslaßrohr n
outgas / to entgasen
outgassing Entgasen n, Entgasung f
outgo of pulp (Pap) Stoffaustritt m
outlet Ablaß m, Auslaß m, Ablauf m, Austritt m, Ausfluß m, Abgang m, Austrag m
  o. air temperature Ablufttemperatur f
  o. end (Aufbereitungstechnik) Austragende n (einer Herdplatte)
  o. for drilling fluid Ausguß m (einer Rotary-Bohranlage)
  o. pipe s. o. tube

o. **sluice** Auslaßschleuse f, Austragschleuse f
o. **tube** Ablaßrohr n, Ablaufrohr n, Abflußrohr n
o. **valve** Abgabeschieber m, Austragventil n
output Ausstoß m, Leistung f
o. **fertilization** Leistungsdüngung f
o. **rate** Ausstoßgeschwindigkeit f, Ausstoßrate f
outside diameter s. outer diameter
o. **electron** s. outer electron
o. **header** (beim Verpacken der Papierrolle an deren Stirnseite aufgeklebter Papierdeckel)
o. **indicator** externer (außerhalb verwendeter) Indikator m
o. **wall** Primärwand f (einer Faser)
outward flow strainer (Pap) Knotenfänger m mit Stoffdurchgang von innen nach außen
ouvarovite (Min) Uwarowit m, Kalkchromgranat m (Kalziumchrom(III)-orthosilikat)
oven (technischer) Ofen m; (Ker) Brennofen m; (Labortechnik) Trockenschrank m
o. **ageing** (Gum) Ofenalterung f, Alterung f im Wärmeschrank, Heißluftalterung f [im Geer-Ofen]
o. **coke** Kammerofenkoks m
o.-**dried** s. o.-dry
o.-**dry / to** im Trockenschrank trocknen; in einem (technischen) Ofen trocknen
o.-**dry** im Trockenschrank (bei 100 bis 110 °C) getrocknet, absolut trocken, (Pap auch) atro
o. **drying** Ofentrocknung f
o. **losses** (Lebm) Backverluste mpl
o. **operation** Ofenbetrieb m
ovens coal bunker Kokskohlenbunker m, Kokskohlenturm m, Kohlenturm m, Kohlensilo m (n), Kohlen[füll]bunker m, Kohlenaufgabebunker m
over-burned überbrannt
o.-**develop / to** (Foto) überentwickeln
o.-**development** (Foto) Überentwicklung f
o.-**exposure** (Foto) Überbelichtung f
o.-**riding effect** dominierende Wirkung f
o.-**stimulation** übermäßige Förderung f (z.B. des Wachstums)
o.-**vulcanization** Übervulkanisation f, Übervernetzung f
o.-**vulcanize / to** übervulkanisieren, übervernetzen
overall application (Landw) Ganzflächenapplikation f, ganzflächiges Ausbringen n (von Pflanzenschutzmitteln)
o. **basicity** (Gerb) Gesamtbasizität f
o. **collection efficiency** Gesamtentstaubungsgrad m
o. **column efficiency** (Destillation) mittleres Verstärkungsverhältnis n (Quotient aus theoretischer und praktischer Trennstufenzahl)
o. **contrast** (Foto) Gesamtkontrast m
o. **effect** Gesamtwirkung f
o. **efficiency** Gesamtwirkungsgrad m
o. **fog** (Foto) Allgemeinschleier m
o. **plate efficiency** s. o. column efficiency
o. **rate** Gesamtgeschwindigkeit f
o. **reaction** Gesamtreaktion f

o. **sensitivity** (Foto) Allgemeinempfindlichkeit f
o. **spraying** (Landw) Ganzflächenbesprühung f
o. **thermal efficiency** thermischer Gesamtwirkungsgrad m
o. **treatment** (Landw) Ganzflächenbehandlung f
o. **velocity** s. o. rate
o. **volatility** Gesamtflüchtigkeit f
o. **yield** Gesamtausbeute f
overbleach / to überbleichen
overbleaching Überbleichung f, Überbleiche f
overblowing (Met) Überfrischen n
overburdening (Met) Übermöllerung f
overcarburize / to überkohlen
overcook / to (Pap) überkochen
overcooking (Pap) Überkochung f
overcool / to unterkühlen
overcooling Unterkühlung f
overcure / to (Plast) überhärten; (Gum) übervulkanisieren, übervernetzen
overcure, overcuring (Plast) Überhärten n, Überhärtung f; (Gum) Übervulkanisation f, Übervernetzung f
overdose / to überdosieren
overdose Überdosis f
overdriven centrifuge Hängezentrifuge f
overdry / to übertrocknen, zu lange trocknen
overdrying Übertrocknen n, Übertrocknung f
overdye / to überfärben
overdye fastness Überfärbeechtheit f
overfeed firing Unterwindfeuerung f
overfelt (Pap) Oberfilz m, Obertuch n
overfermentation Vergärung f, Ausgärung f
overfining Überschönung f (des Weins)
overfire air Zweitluft f (Feuerung)
overfiring (Ker) Überfeuern n, Überfeuerung f, Überbrennen n (Fehler)
overflow / to überlaufen, überfließen, überströmen
overflow Überlauf m (Vorrichtung); s. o. product; (Gum) Austrieb m
o. **box** Überlaufkasten m
o. **fraction** Überlauffraktion f, Überlaufklasse f, Überlauf m
o. **launder** Überlaufrinne f
o. **lip** Überlaufwehr n, Überlaufdamm m
o. **pipe** Überlaufrohr n, Überlaufstutzen m, Überströmrohr n
o. **port** Überlauf m, Überlauföffnung f
o. **product** Überlaufprodukt n, Überlaufgut n, Überlauf m; (Hydroklassierung) Klarflüssigkeit f
o. **rate** Überlauf m (Durchsatz), Überlaufmenge f in der Zeiteinheit
o. **tube** s. o. pipe
o. **weir** Überlaufwehr n, Überfallwehr n, Überlaufdamm m, Ablaufwehr n
overglaze (Ker) Aufglasur f
o. **colour** (Ker) Aufglasurfarbe f; Schmelzfarbe f; Emailfarbe f
o. **decoration** (Ker) Aufglasurdekor n
overgrind / to übermahlen, totmahlen
overgrowth (Krist) Überwachsung f

oxaluria

overhead s. o. product
o. application *(Landw)* Ganzflächenapplikation f, ganzflächiges Ausbringen n *(von Pflanzenschutzmitteln)*
o. bin *(Pap)* über dem Kocher angeordneter Hackschnitzelsilo m *(n)*
o. hopper Hochbunker m
o. product *(Destillation)* Kopfprodukt n, Überkopfprodukt n, Topp-Produkt n
o. take-off Destillatabnahme f
o. temperature *(Destillation)* Kopftemperatur f
o. vapour über Kopf abgehender Dampf m
overheat / to überhitzen, überheizen
overheating Überhitzen n, Überhitzung f, Überheizen n
overlaid lumber *(mit Dekorfolie beschichtetes Holz)*
overlap / to *(Bindungslehre)* überlappen
overlap *(Bindungslehre)* Überlappen n, Überlappung f
o. integral Überlappungsintegral n
overlapping s. overlap
overlay *(Holz)* Deckschicht f, Belag m; s. o. paper
o. paper *(Holz)* Dekorfolie f
o. sheet *(Plast)* Oberflächenmatte f *(Vlies als Oberschicht)*
overlime / to überkalken
overliming Überkalkung f
overload / to überlasten
overload alarm Überlastungsmelder m
overmasticate / to *(Gum)* übermastizieren, totwalzen
overmastication *(Gum)* Übermastizieren n, Totwalzen n
overmill / to s. to overmasticate
overmilling s. overmastication
overpole / to *(Kupfer)* überpolen, zu weit polen
overpoling Überpolen n *(von Kupfer)*
overpotential s. overvoltage
overpressure Überdruck m
overs *(Klassierung)* Siebüberlauf m, Überlauf m, Siebübergang m, Siebrückstand m, Siebgrobes n
oversaturated übersättigt
oversaturation Übersättigen n, Übersättigung f, *(Zucker)* Übersaturation f
overshot *(Erdöl)* Fangglocke f, Schraubentube f, Fangmuffe f
oversize s. o. material
o. chips *(Pap)* Grobgut n, Grobspäne mpl
o. material (product) *(Klassierung)* Überkorn n, Grobkorn n, Grobgut n, *(Siebklassierung auch)* Siebgrobes n, *(Windsichten auch)* Sichtgrobes n, Grieß m
oversteeping übermäßiges Einweichen n, Überweiche f, Totweiche f *(des Malzes)*
overstoving *(Farb)* Überbrennen n *(Überschreiten der Einbrennbedingungen)*
overtannage Übergerbung f, Totgerbung f
overtone band Oberschwingungsbande f *(bei Molekülspektren)*

overuse / to übermäßig gebrauchen, überbeanspruchen
overuse übermäßiger Gebrauch m, Überbeanspruchung f
overvoltage [elektrochemische, elektrolytische] Überspannung f
overworked butter überbutterte Butter f
ovicidal ovizid, eiabtötend
ovicide Ovizid n, Eiergift n *(eiabtötendes Schädlingsbekämpfungsmittel)*
ovoid Eierbrikett n, Eiformbrikett n
ovomucoid Ovomukoid n
ovovitellin Ovovitellin n, Vitellin n *(Phosphoproteid aus Eidotter)*
ovulation inhibitor Ovulationshemmer m
O/W emulsion Öl-in-Wasser-Emulsion f
O/W type Öl-in-Wasser-Typ m *(von Emulsionen)*
Owens [bottle] machine Owens-Maschine f *(zur Flaschenherstellung)*
Owens process Owens-Verfahren n *(zur Flaschenherstellung)*
own weight Eigenmasse f, *(bisher)* Eigengewicht n
oxalacetic acid HOOCCH$_2$COCOOH Oxalessigsäure f, Ketobernsteinsäure f
o. carboxylase Oxalessigsäurekarboxylase f
o. ester C$_2$H$_5$OOC·C(OH)=CH·COOC$_2$H$_5$ Oxalessigester m, Oxalessigsäurediäthylester m
oxalaldehyde CHOCHO Oxalaldehyd m, Oxalsäuredialdehyd m, Äthandial n, Glyoxal n
oxalaldehydic acid CHOCOOH Oxalaldehydsäure f, Oxoäthansäure f, Oxoessigsäure f, Äthanalsäure f, Glyoxylsäure f, Glyoxalsäure f
oxalamide s. oxamide
oxalate / to mit Oxalaten behandeln, *(i.e.S.)* *(Blut)* durch Zusatz von Oxalaten ungerinnbar (gerinnungsunfähig) machen
oxalate Oxalat n, *(i.e.S.)* M$^I_2$C$_2$O$_4$ [neutrales] Oxalat n
o. chelate Oxalatchelat n
o. complex Oxalatokomplex m
o. complex of boron Oxalatoboratkomplex m
oxalato-complex s. oxalate complex
oxalatoaluminate Oxalatoaluminat n
oxalatochromate(III) M$^I_3$[Cr(C$_2$O$_4$)$_3$] Trioxalatochromat(III) n
oxalatocobaltate(III) M$^I_3$[Co(C$_2$O$_4$)$_3$] Trioxalatokobaltat(III) n
oxalatoferrate(II) M$^I_2$[Fe(C$_2$O$_4$)$_2$] Dioxalatoferrat(II) n
oxalatoferrate(III) M$^I_3$[Fe(C$_2$O$_4$)$_3$] Trioxalatoferrat(III) n
oxalic acid HOOCCOOH Oxalsäure f, Kleesäure f, Äthandisäure f
o. ester (C$_2$H$_5$COO)$_2$ Oxalsäurediäthylester m
oxalonitrile (CN)$_2$ Dizyan n, Zyan n, Oxalsäurenitril n
oxalsuccinic acid Oxal[o]bernsteinsäure f, 1-Oxopropantrikarbonsäure-(1,2,3) f
o. carboxylase Oxalbernsteinsäurekarboxylase f
oxaluria *(Med)* Oxalurie f

oxaluric acid $NH_2CONHCOCOOH$ Oxalursäure *f*, Äthandisäuremonoureid *n*

oxalyl urea Oxalylharnstoff *m*, Parabansäure *f*, Imidazoltrion *n*

oxamic acid $H_2NCOCOOH$ Oxamidsäure *f*

oxamide $NH_2COCONH_2$ Oxamid *n*, Äthandiamid *n*

oxammonium NH_2OH Hydroxylamin *n*

o. hydrochloride $NH_2OH \cdot HCl$ Hydroxylamin-hydrogenchlorid *n*, Hydroxylammoniumchlorid *n*

o. sulphate $2NH_2OH \cdot H_2SO_4$ Hydroxylammoniumsulfat *n*

oxanilic acid 2-carboxylic acid $HOOCCONHC_6H_4COOH$ Oxanilsäure-2-karbonsäure *f*, Kynursäure *f*

oxazine dye Oxazinfarbstoff *m*

oxetane polymer Polyäther *m* aus Oxazyklobutanderivaten

Oxford India paper Bibeldruckpapier *n*, Dünndruckpapier *n*

Oxford method Zylinder[platten]methode *f*, Oxfordmethode *f (zur Penizillin-Wertbestimmung)*

Oxford unit Oxford-Einheit *f*, OE, Florey-Einheit *f (veraltete biologische Maßeinheit des Penizillins)*

oxidant Oxydans *n*, Oxydationsmittel *n*, Oxydier[ungs]mittel *n*, oxydierendes Mittel *n*

oxidase Oxydase *f*

oxidation Oxydation *f*, Oxydierung *f*

o. bleaching Oxydationsbleiche *f*

o. by air Luftoxydation *f*

o. catalyst Oxydationskatalysator *m*

o. discharge *(Text)* Oxydationsätze *f*, Ätzen *n* mit Oxydationsmitteln

o. inhibitor Oxydationsinhibitor *m*, Oxydationsverhinderer *m*

o. number Oxydationszahl *f*, Oxydationswert *m*

o. of ammonia Ammoniakverbrennung *f*

o. of coal Kohle[n]oxydation *f*

o. potential Oxydationspotential *n*

o. process Oxydationsverfahren *n*, *(Erdöl auch)* oxydatives Süßungsverfahren *n (zur Umwandlung der Merkaptane)*; Oxydationsvorgang *m*

o. product Oxydationsprodukt *n*

o.-reduction catalyst Oxydations-Reduktionskatalysator *m*

o.-reduction electrode Oxydations-Reduktionselektrode *f*, Reduktions-Oxydationselektrode *f*, Redoxelektrode *f*

o.-reduction enzyme Oxydoreduk[t]ase *f*, Redoxase *f*

o.-reduction indicator Redoxindikator *m*

o.-reduction potential Oxydations-Reduktionspotential *n*, Redoxpotential *n*

o.-reduction reaction Oxydations-Reduktionsreaktion *f*, Reduktions-Oxydationsreaktion *f*, Redoxreaktion *f*

o.-reduction system Oxydations-Reduktionssystem *n*, Reduktions-Oxydationssystem *n*, Redoxsystem *n*

o.-reduction titration Oxydations-Reduktionstitration *f*, Redoxtitration *f*

o. resistance Oxydationsbeständigkeit *f*, Oxyda-

tionsstabilität *f*, Beständigkeit *f* gegen oxydative Einflüsse

o. state Oxydationsstufe *f*, Oxydationszustand *m*

o. zone Oxydationszone *f*, Verbrennungszone *f*

oxidative oxydativ

o. breakdown oxydativer Abbau *m*

o. copperization oxydative Kupferung *f*

o. coupling oxydative Kupplung *f*

o. deamination oxydative Desaminierung *f*

o. degradation oxydativer Abbau *m*

o. desamination *s. o.* deamination

o. rancidity Peroxidranzigkeit *f*, Peroxidigkeit *f*

o. stability *s.* oxidation resistance

oxidatively stable oxydationsbeständig, beständig gegen oxydative Einflüsse

oxide Oxid *n*

o. cathode Oxidkatode *f*

o.-ceramic products oxidkeramische Erzeugnisse *npl*, Oxidkeramik *f*

o. ceramics oxidkeramische Erzeugnisse *npl*, Oxidkeramik *f*; oxidkeramische Stoffe *mpl*

o.-coated cathode *s. o.* cathode

o. coating oxidischer Überzug *m*, oxidische Deckschicht *f*

o. electrode Oxidelektrode *f*

o. film Oxidfilm *m*, Oxidhaut *f*, Oxidbelag *m*, dünne Oxidschicht *f*

o. layer Oxidschicht *f*, Oxidationsschicht *f*, *(Met)* Zunderschicht *f*

o. phosphor Oxidphosphor *m*

o.-reduced powder Reduktionspulver *n*

o. sintering *(Ker)* Oxidsinterung *f*

oxidic oxidisch

oxidimetric oxydimetrisch

oxidimetry Oxydimetrie *f*, Redoxanalyse *f*

oxidizability Oxydierbarkeit *f*

oxidizable oxydierbar

oxidize / to oxydieren

o. back zurückoxydieren

oxidized flavour Oxydationsgeschmack *m*

o. mineral (ore) oxidisches Erz *n*

o. zone Oxydationszone *f*

oxidizer Sauerstoffträger *m* *(in Explosivstoffen)*

oxidizing action oxydierende Wirkung *f*, Oxydationswirkung *f*

o. agent Oxydationsmittel *n*, Oxydier[ungs]mittel *n*, oxydierendes Mittel *n*, Oxydans *n*

o. atmosphere oxydierende Atmosphäre *f*

o. bath Oxydationsbad *n*

o. bleach Oxydationsbleiche *f*

o. bleaching agent oxydativ wirkendes Bleichmittel *n*

o. catalyst Oxydationskatalysator *m*

o. flame Oxydationsflamme *f*

o. roasting oxydierendes Rösten *n*, oxydierende Röstung *f*

o. zone Oxydationszone *f*, *(Brennerflamme)* Oxydationsraum *m*

oxido-reductase Oxydoreduk[t]ase *f*, Redoxase *f*

oxime Oxim *n* *(Isonitrosoverbindung)*

oxinate Oxinat n (Komplexverbindung des Oxins)
oxindole C_8H_7NO Oxindol n, 2-Oxoindolin n
  o. alkaloid Oxindolalkaloid n
oxine Oxin n, 8-Hydroxychinolin n
  o. chelate Oxinchelat n
oxirane Oxiran n, Äthylenoxid n, 1,2-Epoxyäthan n
oxo acid Oxosäure f, Sauerstoffsäure f (Säure mit koordinativ gebundenem Sauerstoff)
  o. bridge Oxobrücke f
  o. complex Oxokomplex m
  o. compound Oxoverbindung f
  o.-cyclo-tautomerism Oxo-Zyklo-Tautomerie f
  o. form Oxoform f
  o. function Oxofunktion f
  o. group =O Oxogruppe f
  o. process s. 1. o. reaction; 2. o. synthesis
  o. reaction Oxo[synthese]reaktion f, Roelen-Reaktion f, Oxo[synthese]prozeß m
  o. synthesis Oxosynthese f, Hydroformylierung f, Oxo[synthese]prozeß m
oxoethanoic acid CHOCOOH Oxoäthansäure f, Oxalaldehydsäure f, Äthanalsäure f, Glyoxylsäure f, Glyoxalsäure f, Oxoessigsäure f
oxolation (Bildung von Oxoverbindungen aus verolten Metallsalzkomplexen)
oxomethane HCHO Formaldehyd m, Methanal n, Oxomethan n
oxonium compound Oxoniumverbindung f
  o. ion Hydroniumion n, Oxoniumion n
  o. salt Oxoniumsalz n
3-oxopentanedioic acid O=C(CH$_2$COOH)$_2$ Pentanon-(3)-disäure f, Azetondikarbonsäure f, 3-Ketoglutarsäure f
oxy sauerstoffhaltig, Oxy...; (org Ch meist) hydroxylhaltig, Hydroxy...
  o. acetylene blowpipe (torch) Azetylenbrenner m
  o. acid s. oxo acid
  o. compound Sauerstoffverbindung f
  o.-hydrogen s. unter oxyhydrogen
  o.-phenic acid (veraltet für) pyrocatechol
oxyacetate basisches Azetat n
oxyacetylene cutting s. oxygen-acetylene cutting
  o. welding s. oxygen-acetylene welding
oxybionic, oxybiotic oxybiontisch, aerob
oxycellulose Oxyzellulose f, oxydierte Zellulose f
oxychloride Oxidchlorid n
oxycompound s. oxy compound
oxydase s. oxidase
oxydation s. oxidation
oxyde s. oxide
oxydiacetic (oxydiethanoic) acid HOOC·CH$_2$OCH$_2$·COOH Oxydiessigsäure f, Oxydiäthansäure f, Diglykolsäure f
oxydizable s. oxidizable
oxygen O Sauerstoff m
  o. absorbent Sauerstoffabsorptionsmittel n
  o. absorption test (Gum) Sauerstoffabsorptionsprüfung f (Alterungstest)
  o.-acetylene cutting Azetylen-Sauerstoff-Brennschneiden n, autogenes Brennschneiden n, Autogenbrennschneiden n

  o.-acetylene welding Azetylen-Sauerstoff-Schweißen n, Autogenschweißen n
  o. ageing (Gum) Sauerstoffalterung f
  o. atom Sauerstoffatom n
  o. balance Sauerstoffbilanz f (von Explosivstoffen)
  o. bomb Sauerstoffbombe f
  o. bomb ageing (Gum) Sauerstoffbombenalterung f, Alterung f in der Sauerstoffbombe, Sauerstoffdruckalterung f
  o. carrier Sauerstoff[über]träger m
  o.-carrying sauerstofftragend
  o. compound Sauerstoffverbindung f
  o.-containing sauerstoffhaltig
  o. cutting autogenes Schneiden n, Brennschneiden n
  o. debt (Med) Sauerstoffschuld f
  o. deficiency Sauerstoffmangel m
  o. demand Sauerstoffbedarf m
  o. depletion Verarmung f an Sauerstoff
  o. donor Sauerstoffdon[at]or m
  o. electrode Sauerstoffelektrode f
  o.-enriched sauerstoffangereichert, mit Sauerstoff angereichert
  o. evolution Sauerstoffentwicklung f, Sauerstoffabscheidung f
  o. flask combustion (Analytik) Kolbenverbrennung f
  o.-free sauerstofffrei
  o.-free high-conductivity copper OFHC-Kupfer n, sauerstofffreies Kupfer n hoher Leitfähigkeit
  o. lance (Met) Sauerstofflanze f, O$_2$-Lanze f, Blaslanze f
  o. lance process (Met) Sauerstoff[auf]blas[e]verfahren n, Sauerstoffblasstahlverfahren n
  o.-linked sauerstoffgebunden
  o. of the air Luftsauerstoff m
  o. overvoltage (Ech) Sauerstoffüberspannung f
  o. point (phys Ch) Sauerstoffpunkt m
  o. pressure method s. o. bomb ageing
  o. process of steelmaking Sauerstoff[auf]blas[e]verfahren n, Sauerstoffaufblaskonverterprozeß m, Sauerstofffrischverfahren n, Aufblas[e]verfahren n, Blasstahlverfahren n, Oberwindfrischverfahren n
  o. release Sauerstoffabspaltung f, Freiwerden n von Sauerstoff
  o. source Sauerstoffquelle f
  o. standard (Bezug der relativen Atommassen auf das Sauerstoffatom)
  o. steel plant Sauerstoffblaswerk n, Blasstahlwerk n
  o. uptake Sauerstoffaufnahme f
  o. vulcanization (Gum) Zyklisierung f, Molekülverkettung f, Vernetzung f
oxygenase Oxygenase f
oxygenate / to oxydieren; mit Sauerstoff sättigen; mit Sauerstoff anreichern
oxygenated oxydiert; sauerstoffgesättigt; sauerstoffbeladen, mit Sauerstoff angereichert

o. muriatic acid *s.* oxymuriatic acid

o. water 1. mit Sauerstoff angereichertes *(oder* behandeltes)* Wasser *n;* 2. H_2O_2 Wasserstoffperoxid *n*

oxygenation Oxydation *f,* Oxydierung *f;* Sauerstoffaufnahme *f,* Anreicherung *f* mit Sauerstoff

oxygenerator Sauerstoffentwicklungsapparat *m*

oxygenic sauerstoffhaltig; sauerstoffähnlich; aus Sauerstoff bestehend, Sauerstoff...

oxygenize / to *s.* to oxygenate

oxygenous *s.* oxygenic

oxyhalide Oxidhalogenid *n*

oxyhemocyanin Oxyhämozyanin *n (kupferhaltiger Blutfarbstoff)*

oxyhemoglobin Oxyhämoglobin *n*

oxyhydrogen blowpipe (burner) Knallgasgebläse *n*

  o. cell Knallgaselement *n*

  o. flame Knallgasflamme *f*

  o. gas Knallgas *n (i.e.S.)*

  o. torch *s.* o. blowpipe

  o. welding Wasserstoff-Sauerstoff-Schweißen *n*

oxyliquit Oxyliquit *n,* Sprengluft *f,* Flüssigluftsprengstoff *m*

oxyluminescence *(durch Oxydation bedingte Chemilumineszenz)*

oxymel *(Pharm) (Gemisch aus verdünnter Essigsäure und Honig; Expektorans)*

oxymuriatic acid oxydierte (oxygenierte) Muriatsäure *f, (historisch für)* Chlor *n*

oxyn Oxyn *n (festes Oxydationsprodukt eines trocknenden Öles)*

oxynaphthoic acid *(unkorrekt für)* hydroxynaphthoic acid

oxyneurine Oxyneurin *n,* Betain *n,* Lyzin *n,* Trimethylglykokoll *n*

oxyphil[e], oxyphilic, oxyphilous *(Bio)* azidophil, säureliebend; azidophil *(durch saure Farbstoffe leicht färbbar)*

oxyproline Hydroxyprolin *n,* 4-Hydroxypyrrolidin-2-karbonsäure *f*

oxyquinoline *(unkorrekt für)* hydroxyquinoline

oxysalt 1. Oxosalz *n (Salz einer Oxosäure);* 2. Oxidsalz *n, (i.w.S.)* basisches Salz *n*

oxytetracycline Oxytetrazyklin *n,* 5-Hydroxytetrazyklin *n (ein Antibiotikum aus Streptomyces rimosus)*

oxytocic oxytozisch

  o. activity oxytozische Wirksamkeit *f,* Oxytozinwirksamkeit *f*

oxywelding *s.* oxygen-acetylene welding

ozocerite *s.* ozokerite

ozokerite *(Min)* Ozokerit *m,* Erdwachs *n,* Bergtalg *m*

  o. rock Ozokeritgestein *n*

  o. wax *s.* ozokerite

ozone O_3 Ozon *n,* Trisauerstoff *m*

  o. concentration Ozonkonzentration *f*

  o. crack *(Gum)* Ozonriß *m*

  o. cracking *(Gum)* Ozonrißbildung *f*

  o. cut *s.* o. crack

  o. exposure test *s.* o. test

  o. generator Ozongenerator *m,* Ozonisator *m*

  o. resistance Ozonbeständigkeit *f,* Ozonfestigkeit *f,* Ozonresistenz *f,* Ozonwiderstand *m*

  o.-resistant, o.-resisting ozonbeständig, ozonfest, ozonresistent

  o. test *(Gum)* Ozonprüfung *f,* Ozonalterung *f*

ozonic ozonartig; ozonhaltig; Ozon...

ozonide Ozonid *n*

ozoniferous ozonhaltig

ozonification *s.* ozonization

ozonify / to *s.* to ozonize

ozonization Ozonisierung *f,* Ozonisieren *n,* Ozonisation *f*

ozonize / to ozonisieren, in Ozon verwandeln; mit Ozon behandeln *(oder)* anreichern; sich in Ozon verwandeln

ozonizer *s.* ozone generator

ozonizing *s.* ozonization

ozonolysis Ozonolyse *f,* Ozon[id]spaltung *f,* Ozonabbau *m*

ozonous *s.* ozonic

P

P-branch P-Zweig *m,* negativer Zweig *m (eines Bandenspektrums)*

p-n junction p-n-Übergang *m,* pn-Übergang *m,* Elektronen-Löcher-Übergang *m*

p_z electron *s.* π electron

Paalsgard emulsion oil Paalsgard-Emulsionsöl *n,* Paalsgardsches Öl *n*

PABA *s.* p-aminobenzoic acid

pachnolite *(Min)* Pachnolith *m (Kalziumnatriumfluoroaluminat)*

Pachuca tank *(Met)* Pachuca-Tank *m*

pack / to [ein]packen, verpacken, paketieren; einpacken, verpacken, einbetten, umgeben *(z.B. mit einem Zementationspulver);* [ab]dichten; *(Destillationstechnik)* [mit Füllkörpern] füllen

pack Packen *m,* Ballen *m;* Paket *n;* Packung *f (abgepackte Menge); (Med)* Packung *f; (Kosmet)* Gesichtspackung *f; (Tech)* Einbau *m,* Einbauten *mpl*

  p.-carburize / to in festen Kohlungsmitteln aufkohlen, pulveraufkohlen, in festen Einsatzmitteln *(in festem Einsatz)* zementieren, pulverzementieren, in festen Mitteln einsetzen, pulvereinsetzen

  p. carburizing Aufkohlen *n* in festen Kohlungsmitteln, Pulveraufkohlen *n,* Zementieren *n* in festen Einsatzmitteln *(in festem Einsatz),* Pulverzementieren *n,* Einsetzen *n* in festen Mitteln, Pulvereinsetzen *n*

  p. dyeing *s.* package dyeing

  p. hardening Einsatzhärten *n* mit festen Mitteln

packability Verpackungsfähigkeit *f*

packable verpackungsfähig

package / to [ein]packen, verpacken, paketieren
package Packen *n*, Verpacken *n*, Verpackung *f*; Packung *f (abgepackte Menge)*; Verpackung *f*, Emballage *f*, Verpackungsmaterial *n*; *(Text)* Pack *m*, Wickelkörper *m*, Garnkörper *m*, Aufmachungseinheit *f*
p.-dyed *(Text)* packgefärbt, im (nach dem) Packsystem gefärbt
p. dyeing *(Text)* Packfärberei *f*, Färben *n* im (nach dem) Packsystem, Packfärben *n*
p. dyeing machine *(Text)* Packfärbeapparat *m*
p. paper *(Am)* Verpackungspapier *n*, Packpapier *n*, Einpackpapier *n*
packaging Packen *n*, Verpacken *n*, Verpackung *f*
p. film (foil) Verpackungsfolie *f*
p.-forming varnish Verpackungslack *m*
p. material Verpackungsmaterial *n*, Verpackung *f*, Emballage *f*
p. paper *s.* package paper
packed column Füllkörperkolonne *f*, Füllkörpersäule *f*
p. for prolonged storage haltbar verpackt
p. gland Stopfbuchse *f*, Stopfbuchspackung *f*
p.-gland joint Stopfbuchsverbindung *f (Rohrverbindung)*
p. height Füllkörperhöhe *f (in Destillations- und Absorptionskolonnen)*
p. tower Füll[körper]turm *m*
packer Packer *m (zum Abdichten eines Teils des Bohrlochs oder Futterrohrs gegen einen anderen)*
packing Packen *n*, Verpacken *n*, Verpackung *f*; Verpackung *f*, Verpackungsmaterial *n*, Emballage *f*; Füllung *f*, Füllkörper *mpl*, Füllstoff *m*, Füllmaterial *n (z.B. in Destillationskolonnen)*; Einpacken *n*, Verpacken *n*, Einbetten *n*, Umgeben *n (z.B. mit einem Zementationspulver)*; Dichten *n*, Dichtung *f*, Abdichten *n*, Abdichtung *f*; Liderung *f (von Pumpen)*; Packung *f*, Füllung *f*, Schüttung *f*; *(Krist)* Packung *f*, Kugelpackung *f*
p. absorption tower Füll[körper]turm *m*
p. box *s.* p. gland
p. density Packungsdichte *f*
p. depth Schüttungshöhe *f*
p. effect Packungseffekt *m*
p. fraction Packungsanteil *m*, Packungsbruch *m*
p. gland Stopfbuchse *f*, Stopfbuchspackung *f*
p. machine Verpackungsmaschine *f*
p. material Verpackungsmaterial *n*, Verpackung *f*, Emballage *f*; Einpackmittel *n*, Einbettungsmittel *n*, Einbettungswerkstoff *m*, Einbettungsmasse *f*; Dicht[ungs]werkstoff *m*, Dicht[ungs]stoff *m*, Dicht[ungs]material *n*, Dicht[ungs]mittel *n*; Packungs[werk]stoff *m*, Packungsmaterial *n*; Füllmaterial *n*, Füllung *f*, Füllstoff *m*, Füllkörper *mpl*
p. of particles *(Krist)* Packungsart *f*, Gitteranordnung *f*
p. of spheres (spherical units) *(Krist)* Kugelpackung *f*

p. of the chips *(Pap)* Holzfüllung *f*, Hackschnitzelfüllung *f*, Einbringen *n* der Hackschnitzel
p. paper *s.* wrapping paper
p. ring Dicht[ungs]ring *m*, Packungsring *m*
p. room Packraum *m*
p. set Dicht[ungs]satz *m*
p. strip Dicht[ungs]schnur *f*
p. table Packtisch *m*
p. tissue Packseidenpapier *n*, Einschlagseidenpapier *n*, Einwickelseidenpapier *n*
packings Füllkörper *mpl*, Füllmaterial *n*, Füllstoff *m*, Füllung *f*
pad / to *(Text)* [auf]klotzen, foulardieren
pad Einlage *f*; Unterlage *f*, Kissen *n*, Polster *n*; Platte *f (Masseplatte)*; *(Text)* Foulard *m*, Klotzmaschine *f*, Breitfärbemaschine *f*
p.-acid develop process *(Text)* Säureschockverfahren *n*
p.-batch process *(Text)* Kaltverweil-Färbeverfahren *n*
p. bath *s.* padding bath
p. box *(Text)* Klotztrog *m*, Klotzchassis *n*
p.-dry cure *(Text)* Klotz-Trocken-Kondensierverfahren *n*
p.-dry process *(Text)* Klotz-Trocken-Verfahren *n*
p. dyeing *(Text)* Klotzfärben *n*, Klotzfärbung *f*, Klotzen *n*, Foulardfärbung *f*
p. dyeing process *(Text)* Klotzfärbeverfahren *n*
p. filter Kammerfilterpresse *f*
p.-fix process *(Text)* Klotz-Fixier-Verfahren *n*
p.-jig method (process) *(Text)* Pad-Jig-Verfahren *n*, Foulard-Jigger-Verfahren *n*
p.-roll machine *(Text)* Pad-Roll-Maschine *f*
p.-roll method *(Text)* Klotz-Roll-Verfahren *n*, Pad-Roll-Verfahren *n*
p.-steam process *(Text)* Klotz-Dämpf-Verfahren *n*, Klotz-Dämpf-Färbeverfahren *n*, Klotzdämpfen *n*, Pad-Steam-Verfahren *n*
p.-store process *(Text)* Klotz-Verweil-Verfahren *n*
p. thermofix dyeing *(Text)* Klotz-Thermofixier-Färbeverfahren *n*
padded dyeing *s.* pad dyeing
padder *(Text)* Foulard *m*, Klotzmaschine *f*, Breitfärbemaschine *f*
padding *(Text)* Klotzen *n*, Klotzverfahren *n*, Aufklotzen *n*, Foulardieren *n*, Foulardverfahren *n*; *(Pap)* Oberflächenfärbung *f*; Kalanderfärbung *f*, Oberflächenfärbung *f* im Kalander
p. auxiliary *(Text)* Klotzhilfsmittel *n*
p. bath (liquor) *(Text)* Klotz *m*, Klotzbad *n*, Klotzflotte *f*, Foulardierlösung *f*
p. machine (mangle) *(Text)* Foulard *m*, Klotzmaschine *f*, Breitfärbemaschine *f*
paddle / to *(Gerb)* haspeln, treiben; *(ein Glasstück im Ofen vor dem Pressen)* patschen
paddle Schaufelblatt *n*, Schaufel *f*, Blatt *n*, Flügel *m*, Rührarm *m*, Paddel *n*; *(Gerb)* Haspel *m(f)*, Haspelgeschirr *n*

p. agitator Balkenrührer *m*, Paddelrührer *m*
p. dyeing machine *(Text)* Paddelfärbemaschine *f*, Schaufelradfärbemaschine *f*
p. kneading table Schaufelkneter *m (Margarineherstellung)*
p. liming *(Gerb)* Haspeläscher *m*
p. mixer *s.* p. agitator
p. wheel dyeing machine *s.* p. dyeing machine
paddling *(Gerb)* Haspeln *n*, Treiben *n*, Behandlung *f* im Haspel; Patschen *n (Vorformen eines Glasstückes im Ofen vor dem Pressen)*
paddy *s.* p. rice
p. eliminator (machine) *s.* p. separator
p. rice Paddy[reis] *m*, ungeschälter bespelzter Rohreis *m*
p. separator Paddyausleser *m*, Rohreisausleser *m*
padi *s.* paddy rice
paeonidin Päonidin *n (ein Anthozyanidin)*
paeonin Päonin *n (ein Anthozyan des Päonidins)*
pagodite *(Min)* Pagodit *m*, *(veraltet für)* Agalmatolith *m (Aluminiumdihydrogentetrasilikat)*
paH *(Kennwert der Wasserstoffionenaktivität, negativer dekadischer Logarithmus der Konzentration aktiver Wasserstoffionen)*
pain-reliever schmerzstillendes (schmerzlinderndes, analgetisches) Mittel *n*, Schmerzlinderungsmittel *n*, Analgetikum *n*
paint / to *(Gerb)* schwöden
paint Anstrichfarbe *f*
p. binder Farb[en]bindemittel *n*
p. coat[ing] Farbanstrich *m*
p. drying Anstrichtrocknung *f*
p. extender Streckmittel *n*, Verschnittmittel *n*, Extender *m*
p. film Anstrichfilm *m*
p. pigment Farbpigment *n*, Pigment *n*
p. remover Farbabbeizmittel *n*, Abbeizmittel *n*, Farbentferner *m*
p. spray gun Farbspritzpistole *f*, Spritzpistole *f*
p. sprayer *s.* p. spray gun
p. spraying Farbspritzen *n*
p. spraying technique Farbspritzverfahren *n*
p. system Anstrichsystem *n*, Anstrichaufbau *m*
p. unhairing *(Gerb)* Schwöden *n*, Schwöde *f*
p. vehicle *s.* p. binder
paintability Bemalbarkeit *f*
painter's cardboard Zeichenkarton *m*, Mal[er]pappe *f*, Malkarton *m*, Ölkarton *m*
p. naphtha Lackbenzin *n*
p. putty Glaserkitt *m*, Fensterkitt *m*
painting *(Gerb)* Schwöden *n*, Schwöde *f*
p. board *s.* painter's cardboard
pair / to paarweise zusammentreten, sich paarweise vereinigen; *(z.B. Elektronen)* paaren
pair Paar *n*
p. annihilation Paarvernichtung *f*, Annihilation *f*
p. creation (formation) *s.* p. production
p. of antipodes Antipodenpaar *n*

p. of optical isomers optisch aktives Isomerenpaar *n*
p. of salts Salzpaar *n*
p. of shared electrons gemeinsames Elektronenpaar *n*
p. production Paarbildung *f*, Paarerzeugung *f*, Paarung *f*
p. spectrometer Paarspektrometer *n*
p. spectroscope Paarspektroskop *n*
paired electron gepaartes Elektron *n*
pairing *s.* pair production
p. energy Paarbildungsenergie *f*
pakice Schuppeneis *n*
palateful vollmundig *(z.B. Bier)*
palatefulness Vollmundigkeit *f (z.B. des Biers)*
Palay rubber *(Kautschuk der Liane Cryptostegia grandiflora R. Br.)*
palcotanic acid *(aus der Rinde von Mammutbäumen extrahiertes Gemisch organischer Säuren, Ausgangsstoff für Bindemittelhärter)*
pale ale helles Bier *n*, Helles *n*
p. ale malt helles [Pilsner] Malz *n*
p. beer *s.* p. ale
p.[-dried] malt *s.* p. ale malt
p. oil helles Öl *n*, Pale Oil *n (leicht gefärbtes raffiniertes Schmieröldestillat)*
p.-yellow fahlgelb, blaßgelb, schwachgelb
paleobiochemistry Paläobiochemie *f*
palingenesis *(Bio, Geol)* Palingenese *f*, Palingenesis *f*, Palingenesie *f*
palingen[et]ic magma palingenetisches Magma *n*
Pall ring Pall-Ring *m (ein Füllkörper)*
palladic Palladium..., *(i.e.S.)* Palladium(IV)-...
p. hydroxide $Pd(OH)_4$ Palladium(IV)-hydroxid *n*
p. oxide PdO_2 Palladium(IV)-oxid *n*, Palladiumdioxid *n*
p. potassium chloride $K_2[PdCl_6]$ Kaliumhexachloropalladat(IV) *n*
palladinized asbestos Palladiumasbest *m*
palladium Pd Palladium *n*
p. black Palladiummohr *n*, Palladiumschwarz *n*
p. chloride paper Palladium(II)-chlorid-Papier *n*
p. diammine dihydroxide $[Pd(NH_3)_2](OH)_2$ Diamminpalladium(II)-hydroxid *n*
p. dichloride $PdCl_2$ Palladiumdichlorid *n*, Palladium(II)-chlorid *n*
p. difluoride PdF_2 Palladiumdifluorid *n*, Palladium(II)-fluorid *n*
p. diiodide PdJ_2 Palladiumdijodid *n*, Palladium(II)-jodid *n*
p. dioxide PdO_2 Palladiumdioxid *n*, Palladium(IV)-oxid *n*
p. disulphide PdS_2 Palladiumdisulfid *n*, Palladium(IV)-sulfid *n*
p. gold *(Min)* Palladiumgold *n*, Porpezit *m*
p. hydride Pd_2H Palladiumhydrid *n*
p. monosulphide PdS Palladium[mono]sulfid *n*, Palladium(II)-sulfid *n*
p. monoxide PdO Palladium[mon]oxid *n*, Palladium(II)-oxid *n*

p. suboxide Pd_2O Palladiumsuboxid *n*
p. subsulphide Pd_2S Palladiumsubsulfid *n*
p. tetrammine chloride $[Pd(NH_3)_4]Cl_2$ Tetramminpalladium(II)-chlorid *n*
p. trifluoride PdF_3 Palladiumtrifluorid *n*, Palladium(III)-fluorid *n*
p. tube Palladiumröhre *f*
palladous Palladium..., *(i.e.S.)* Palladium(II)-...
p. bromide $PdBr_2$ Palladium(II)-bromid *n*, Palladiumdibromid *n*
p. chloride $PdCl_2$ Palladium(II)-chlorid *n*, Palladiumdichlorid *n*
p. cyanide $Pd(CN)_2$ Palladium(II)-zyanid *n*
p. dichlorodiammine $PdCl_2(NH_3)_2$ Dichlorodiamminpalladium(II) *n*
p. dihydroxydiammine $[Pd(NH_3)_2](OH)_2$ Diamminpalladium(II)-hydroxid *n*
p. hydroxide $Pd(OH)_2$ Palladium(II)-hydroxid *n*
p. iodide PdJ_2 Palladium(II)-jodid *n*, Palladiumdijodid *n*
p. nitrate $Pd(NO_3)_2$ Palladium(II)-nitrat *n*
p. oxide PdO Palladium(II)-oxid *n*, Palladium[mon]oxid *n*
p. potassium chloride $K_2[PdCl_4]$ Kaliumtetrachloropalladat(II) *n*
p. sodium chloride $Na_2[PdCl_4]$ Natriumtetrachloropalladat(II) *n*
p. sulphate $PdSO_4$ Palladium(II)-sulfat *n*
pallas iron *s.* pallasite
pallasite *(Geol)* Pallasit *m*
pallet *(Tech)* Palette *f*, Stapelplatte *f*
palliative Palliativum *n*, Linderungsmittel *n*
palm butter (grease) *s.* p. oil
p. kernel (nut) oil Palmkernöl *n*, *(i.e.S.)* Palmkernfett *n (raffiniertes Palmkernöl)*
p. oil Palmöl *n*, Palmfett *n*
p. sap Palmensaft *m*
p. starch Palmenstärke *f*, Palm[en]sago *m*, Sagostärke *f*
p. sugar Palmzucker *m*
p. wine Palmwein *m*
palmarosa oil Palmarosaöl *n (von Cymbopogon martini (Roxb.)Stapf)*
Palmer process *(Gum)* Palmer-Verfahren *n*, Hochdruckdampfverfahren *n (Regenerierverfahren)*
palmitate Palmitat *n*
palmitic acid $CH_3(CH_2)_{14}COOH$ Palmitinsäure *f*, Hexadekansäure *f*, Zetylsäure *f*
p. chloride $C_{15}H_{32}COCl$ Palmitinsäurechlorid *n*, Palmitoylchlorid *n*
palmitin $(C_{15}H_{31}COO)_3C_3H_5$ Tripalmitin *n*, Glyzerintripalmitat *n*, Glyzerintripalmitinsäureester *m*
palmitinic acid *s.* palmitic acid
palmitole[in]ic acid $CH_3(CH_2)_5CH=CH(CH_2)_7COOH$ Palmitoleinsäure *f*, Palmito-oleinsäure *f*, Zoomarinsäure *f*, Physetölsäure *f*
palmitoyl chloride *s.* palmitic chloride
palmkernel oil *s.* palm kernel oil

palosine Palosin *n*
palygorskite Palygorskit *m*, Attapulgit *m (Tonmineral)*
pamaquin[e] Pamaquin *n*, 8-(4-Diäthylamino-1-methylbutylamino)-6-methoxychinolin *n*
pan Pfanne *f*, Mulde *f*, Trog *m*, Schüssel *f*, Schale *f*, Wanne *f*; Waagschale *f*, Schale *f*; Siedepfanne *f*, Pfanne *f (zur Speisesalzgewinnung); (Gum)* Kessel *m*, Vulkanisierkessel *m*, Vulkanisationskessel *m*; *(Pap)* Holländertrog *m*, Stoffwanne *f*
p. amalgamation Pfannenamalgamierung *f*, Pfannenamalgamation *f*
p. crusher *s.* p. mill
p. devulcanization *s.* p. [reclaiming] process
p.-film *s.* panchromatic film
p. granulator Granulierteller *m*
p. mill (mixer) Kollergang *m*, Kollermühle *f*, chilenische Mühle *f*
p. of the beater *(Pap)* Holländertrog *m*, Stoffwanne *f*
p. [reclaiming] process *(Gum)* Pan-Prozeß *m*, Pan-Verfahren *n (Regenerierverfahren)*
p.-type roller mill *s.* p. mill
panacea Allheilmittel *n*, Universalmittel *n*, Panazee *f*
Panama bark Panamarinde *f*, Seifenrinde *f*, Quillajarinde *f (von Quillaja saponaria Mol.)*
P. ipecac[uanha] Panama-Brechwurzel *f*, Kartagena-Brechwurzel *f*, Nikaragua-Brechwurzel *f (von Cephaelis acuminata Karsten)*
Panau resin *(Harz von Dipterocarpus vernicifluu Blanco)*
panchromatic panchromatisch
p. emulsion *(Foto)* panchromatische Emulsion *f*
p. film *(Foto)* panchromatischer Film *m*
panchromatize / to *(Foto)* panchromatisch machen (sensibilisieren)
pancreatic bate *(Gerb)* Pankreasbeize *f*
p. juice Pankreassaft *m*
pandermite *(Min)* Pandermit *n*
panel board gehärtete Pappe *f*, Hartpappe *f*
p. of coal Kohlenfeld *n (bei der Untertagevergasung)*
Paneth rule Panethsches Gesetz *n*
panidiomorphic *(Geol)* panidiomorph
pantothenic acid Pantothensäure *f*, N-(2,4-Dihydroxy-3,3-dimethylbutyryl)-β-alanin *n*
p.a.p. *s.* p-aminophenol
papain Papain *n (aus dem Milchsaft des Melonenbaumes gewonnenes Ferment)*
papainase Papainase *f (ein eiweißspaltendes Ferment)*
papaver Mohn *m*, Papaver L.
papaveraldine Papaveraldin *n*, 6,7-Dimethoxy-1-(3,4-dimethoxybenzoyl)isochinolin *n*
papaverine Papaverin *n*, 1-(3',4'-Dimethoxybenzyl)-6,7-dimethoxyisochinolin *n*
p. hydrochloride Papaverinhydrochlorid *n*, 1-(3',4'-Dimethoxybenzyl)-6,7-dimethoxyisochinolinhydrochlorid *n*

papaverinol Papaverinol *n*, 6,7-Dimethoxy-1-(α-hydroxy-3,4-dimethoxybenzyl)isochinolin *n*
paper / to (*Pap*) bekleben, kaschieren; tapezieren; (*mit Sandpapier*) abschleifen
paper Papier *n*
to make into p. zu (auf) Papier verarbeiten
p. base (*Foto*) Papierunterlage *f*
p. birch Papierbirke *f*, Betula papyrifera Marsh.
p. board *s.* paperboard
p. bowl Papier[kalander]walze *f*
p. capacitor Papierkondensator *m*, Pko *m*
p. chromatogram Papierchromatogramm *n*
p. chromatography Papierchromatografie *f*
p. colouring Papierfärbung *f*, Papierfärben *n*, Färben *n* von Papier, Papierfärberei *f*
p. developer (*Foto*) Papierentwickler *m*
p. development (*Foto*) Papierentwicklung *f*
p. dryer Papiertrockenzylinder *m*
p. drying Papiertrocknung *f*
p. electrophoresis Papierelektrophorese *f*
p. factory Papierfabrik *f*
p. filler Papierfüllstoff *m*
p. filter Papierfilter *n*
p. finishing Papierausrüstung *f*, Fertigstellung *f* des Papiers
p. for banknotes Banknotenpapier *n*
p. for calculating machines Additionsrollenpapier *n*, Kassenrollenpapier *n*, Registerrollenpapier *n*, Registrier[kassen]papier *n*
p. for drawing Zeichenpapier *n*
p. for parchmentizing Pergamentrohstoff *m*
p. for technical purposes technisches Papier (Sonderpapier) *n*
p. for wax flowers Wachsblumen[roh]papier *n*
p. grade (*Foto*) Papierhärtegrad *m*, Papiergradation *f*
p. honeycomb core (*Wabenmittellage aus harzgetränktem Kraftpapier für Verbundkonstruktionen*)
p. in sheets Bogenpapier *n*, Formatpapier *n*, auf Format geschnittenes Papier *n*
p. industry Papierindustrie *f*
p. machine Papier[herstellungs]maschine *f*, (*i.e.S.*) Langsieb[papier]maschine *f*
p. machine coating Streichen *n* in der Papiermaschine, Maschinenstrich *m*
p. machine felt Papiermaschinenfilz *m*
p. machine speed Papiermaschinengeschwindigkeit *f*, Arbeitsgeschwindigkeit *f* der Papiermaschine
p. machine wire Papiermaschinensieb *n*
p.-maker *s.* papermaker
p.-making *s.* papermaking
p. mat Papiereinlage *f*
p. mill Papierfabrik *f*; Papiermühle *f*
p. mill slime sich bei der Papierherstellung bildender Schleim *m*
p. mill sludge (*Landw*) Kalkschlamm *m* (*aus Papierfabriken*)
p. mill wastes Papierfabrikabwasser *n*

p.-mould (*Pap*) Schöpfform *f*, Form *f*, Schöpfrahmen *m*
p.-mulberry Papiermaulbeerbaum *m*, Broussonetia papyrifera (L.)L'Hérit.
p.-plastic overlay (*Belag aus phenolharzgetränktem Kraftpapier zur Holzvergütung*)
p. pressing (*Text*) Spanpressen *n*
p. pressing plant (*Text*) Spanpreßanlage *f*
p. print (*Foto*) Papierabzug *m*, Papierbild *n*
p. pulp Papierzellstoff *m*, Zellstoff *m* für die Papierindustrie; *s. p.* stock
p. roll [elastische] Papierwalze *f* (*des Superkalanders*)
p. size Papierformat *n*
p. sizing Papierleimung *f*
p. sizing agent (material) (*Pap*) Leim *m*, Leim[ungs]mittel *n*, Leimstoff *m*, Leimungsmaterial *n*
p. stock Papierrohstoff *m*, Rohstoff *m* für die Papiererzeugung, Faserrohstoff *m*, Papierfaserstoff *m*; [fertiger] Papierstoff *m*, Papiermasse *f*, Papierbrei *m*, Ganzstoff *m*, Ganzzeug *n*, Stoff *m*; Altpapierstoff *m*, wiederaufbereitetes (aufbereitetes, regeneriertes) Altpapier *n*; Faserhalbstoff *m*, Halbstoff *m*, Halbzeug *n*, Stoff *m*
p.-stuff (*veraltet für*) stuff
p. tension Papierzugspannung *f*, Papierbahnspannung *f*, Bahnspannung *f*
p. towelling Abtupfen *n* mit Fließpapier
p. web Papierbahn *f*, Stoffbahn *f*, Papiervlies *n*, Faserfilz *m*
p. weight Papiermasse *f*, (*bisher*) Papiergewicht *n*
p. wool Papierwolle *f*
p. yarn Papiergarn *n*
paperboard Pappe *f*; Karton *m*
p. carton Faltschachtel *f*, Pappschachtel *f*, Pappkarton *m*
p. machine Kartonmaschine *f*
p. mill Kartonfabrik *f*; Pappenfabrik *f*
paperhanger's paste Tapetenkleister *m*
papermaker Papiermacher *m*; Schöpfer *m*
papermaker's alum Papiermacheralaun *m* (*technisches Aluminiumsulfat*)
p. felt Papiermaschinenfilz *m*
p. furnish Papierrohstoffe *mpl*, Rohstoffe *mpl* für die Papiererzeugung
papermaking Papierherstellung *f*, Papiererzeugung *f*, Papierfabrikation *f*, Papiermachen *n*, Papiermacherei *f*
p. by hand Echt-Büttenpapier-Herstellung *f*
p. fibre Papierfaser *f*
p. machine *s.* paper machine
p. pulp *s.* paper pulp
p. stock *s.* paper stock
papyrolin Gewebepapier *n*, Leinenpapier *n*, Papyrolin *n*, Packgewebe *n* (*mit einer Gewebebahn zwischen zwei Papierbahnen*)
para paraständig, *p*-ständig
in p. position in para-Stellung (*p*-Stellung, 1,4-Stellung), paraständig, *p*-ständig

to be [located, situated] p. sich in para-Stellung (*p*-Stellung, 1,4-Stellung) befinden, in para-Stellung stehen, paraständig (*p*-ständig) sein

p.-aminophenol developer *(Foto)* Par[a]aminophenolentwickler *m*, *p*-Aminophenol-Entwickler *m*

p.-bridged para-überbrückt

p. compound para-Verbindung *f*, *p*-Verbindung *f*

p. fuchsine Parafuchsin *n*

p. hydrogen para-Wasserstoff *m*, Parawasserstoff *m*

p. isomer Para-Isomer[es] *n*, *p*-Isomer[es] *n*

p. molecule para-Molekül *n*

p.-nitroaniline Paranitranilin *n*, *p*-Nitranilin *n*, 4-Nitr[o]anilin *n*

p. nut Paranuß *f*, Brasilnuß *f (von Bertholletia excelsa Humb. et Bonpl.)*

p. position para-Stellung *f*, *p*-Stellung *f*, 1,4-Stellung *f*

p. red *s.* paranitraniline red

p.-rock Paragestein *n*, Gestein *n* der Parareihe

p. state para-Zustand *m*

Para [rubber] Parakautschuk *m*, Para[gummi] *m*

P. rubber tree Parakautschukbaum *m (Hevea brasiliensis (H.B.K.) Muell. Arg.)*

paraacetaldehyde (C$_2$H$_4$O)$_3$ Paraazetaldehyd *m*, Paraldehyd *m*, 2,4,6-Trimethyl-1,3,5-trioxan *n*

parabanic acid Parabansäure *f*, Oxalylharnstoff *m*, Imidazoltrion *n*

parabituminous coal parabituminöse Kohle *f*

paracasein Parakasein *n (unlösliches Kasein)*

parachor *(phys Ch)* Parachor *m(n)*, P

p. measurement Parachormessung *f*

paracortex Parakortex *m*

paracrystal *s.* liquid crystal

paracrystalline parakristallin

paradamite *(Min)* Paradamit *m (Zinkarsenat)*

paradichlorbenzene C$_6$H$_4$Cl$_2$ Paradichlorbenzol *n*, *p*-Dichlorbenzol *n*

paraffin 1. Paraffin *n*; Petrol[eum] *n*, Kerosin *n*, Kerosen *n*; 2. *s.* p. hydrocarbon; 3. *s.* p. wax

p. base Paraffinbasis *f*, paraffinische Basis *f (eines Roherdöls)*

p.-base crude [oil] *s.* p.-base petroleum

p.-base petroleum paraffinbasisches (paraffinisches, paraffinöses) Erdöl (Rohöl) *n*, Paraffin[basis]öl *n*, Erdöl *n* auf Paraffinbasis

p.-coated paraffiniert

p. distillate Paraffindestillat *n*

p. hydrocarbon Paraffinkohlenwasserstoff *m*, paraffinischer (gesättigter) Kohlenwasserstoff *m*, Paraffin *n*, Alkan *n*, Grenzkohlenwasserstoff *m*

p. oil Paraffinöl *n*; *(Pharm)* flüssiges Paraffin *n*; Petrol[eum] *n*, Kerosin *n*, Kerosen *n*

p. paper Paraffinpapier *n*, paraffiniertes Papier *n*

p. scale Schuppenparaffin *n*, Paraffinschuppen *fpl*

p. series Paraffinreihe *f*, Alkanreihe *f*, Grenzkohlenwasserstoffreihe *f*

p. slack wax Paraffingatsch *m*

p. wax [festes] Paraffin *n*, Festparaffin *n*, Paraffinwachs *n*, Wachs *n (fester Erdölkohlenwasserstoff vorwiegend aliphatischen Charakters)*

p. wax size *(Pap)* Paraffinleim *m*

p. wax sizing *(Pap)* Paraffinleimung *f*

p.-waxed paper *s.* p. paper

paraffined paper *s.* paraffin paper

paraffinic paraffinisch, paraffinhaltig, Paraffin...

p. alcohol Paraffinalkohol *m*

p.-naphthenic petroleum paraffinisch-naphthenisches Erdöl (Rohöl) *n*

p. oil Paraffinöl *n*

p. petroleum *s.* paraffin-base petroleum

paraffinicity Paraffinität *f*, Paraffinanteil *m*

paraffinity *(Gehalt an gesättigten aliphatischen Kohlenwasserstoffen)*

paraform[aldehyde] (CH$_2$O)$_n$ Paraformaldehyd *m*, Polyoxymethylen *n*

parafuchsin[e] Parafuchsin *n*

paragenesis *(Geol)* Paragenese *f*, Paragenesis *f*

paragenetic *(Geol)* paragenetisch

paragonite *(Min)* Paragonit *m*, Natronglimmer *m*

parahydrogen Parawasserstoff *m*, para-Wasserstoff *m*

paralactic acid Paramilchsäure *f*, Fleischmilchsäure *f*, Rechtsmilchsäure *f*, *d*-Milchsäure *f*

paraldehyde (CH$_3$CHO)$_3$ Paraldehyd *m*, Paraazetaldehyd *m*, 2,4,6-Trimethyl-1,3,5-trioxan *n*

parallel flow Parallelstrom *m*, Gleichstrom *m*

p. plate plastometer Parallelplattenplastometer *n*, Parallelplattendruckgerät *n*

p. roll parallel geführte Walze *f*

p. screening *(kombiniertes Verfahren zur Analyse von Pflanzenschutzmittelrückständen: Tierversuch, Bestimmung der chlororganischen Verbindungen, Messung der Cholinesterase-Hemmung)*

p.-seat gate valve Parallelschieber *m*, Plattenschieber *m*

paralyzer Katalysatorgift *n*, Kontaktgift *n*

paramagnetic paramagnetisch

paramagnetic *s.* p. material

p. compound paramagnetische Verbindung *f*

p. dispersion paramagnetische Dispersion *f*

p. electronic resonance *s.* p. resonance

p. material paramagnetischer Stoff (Körper) *m*, paramagnetische Substanz *f*, Paramagnetikum *n*

p. resonance paramagnetische Resonanz *f*, Elektronenspinresonanz *f*

p. resonance method paramagnetische Resonanzmethode (Resonanzspektroskopie) *f*

p. resonance spectrum paramagnetisches Resonanzspektrum *n*

p. substance *s.* p. material

p. susceptibility paramagnetische Suszeptibilität *f*

paramagnetism Paramagnetismus *m*

paramandelic acid C$_6$H$_5$CH(OH)COOH para-Mandelsäure *f*, *dl*-Mandelsäure *f*

parameter *(Krist)* Parameter *m*, Achsenabschnitt *m*
paramine brown Paraminbraun *n*
paraminophenol $NH_2C_6H_4OH$ *p*-Aminophenol *n*, 4-Amino-1-hydroxybenzol *n*, 4-Aminophenol *n*
  p. developer *(Foto)* Par[a]aminophenolentwickler *m*, *p*-Aminophenol-Entwickler *m*
paramolybdate $M^I_6Mo_7O_{24}$ Paramolybdat *n*
paramorphine Paramorphin *n*, Thebain *n*
paranitraniline Paranitranilin *n*, *p*-Nitranilin *n*, 4-Nitr[o]anilin *n*
  p. red Paranitranilinrot *n*, Pararot *n*, 1-(4-Nitrophenyl)azo-2-naphthol *n*, *p*-Nitrobenzolazo-β-naphthol *n*
paraplasm Paraplasma *n* *(im Zytoplasma enthaltene Stoffwechselprodukte)*
parapositronium Para-Positronium *n* *(mit gleichgerichtetem Spin von Elektron und Positron)*
pararosaniline Pararosanilin *n*
pararosolic acid Pararosolsäure *f*, Aurin *n*, *p*-Rosolsäure *f*
parasitic capture parasitärer Neutroneneinfang *m*
parasorbic acid Parasorbinsäure *f*
parasympathomimetic *(Med)* parasympath[ik]omimetisch
  p. agent *(Med)* Parasympath[ik]omimetikum *n*
paratartaric acid $HOOC(CHOH)_2COOH$ paraWeinsäure *f*, razemische Weinsäure (Säure) *f*, Traubensäure *f*
parathion Parathion *n* *(O,O-Diäthyl-O-p-nitrophenylester der Thiophosphorsäure, Insektizid)*
parathormone Parathormon *n*
parathyroid gland Nebenschilddrüse *f*, Glandula parathyreoidea
paratungstate Parawolframat *n*
parautochthonous *(Geol)* parautochthon
parchment tierisches (animalisches) Pergament *n*, Hautpergament *n*, Pergament *n*; vegetabilisches Pergament *n*, [echtes] Pergamentpapier *n*, Echtpergamentpapier *n*, Säurepergament *n*; Pergamentersatzpapier *n*, Pergamentersatz *m*
  p. cardboard Pergamentkarton *m*
  p. paper [echtes] Pergamentpapier *n*, Echtpergamentpapier *n*, vegetabilisches Pergament *n*, Säurepergament *n*
parchmentization *(Pap)* Pergamentierung *f*
parchmentize /to *(Pap)* pergamentieren
parchmentizing *s.* parchmentization
  p. agent *(Pap)* Pergamentierungsmittel *n*
parchmenty pergamentartig
parent Stamm *m*; *s.* 1. p. element; 2. p. substance
  p. acid Stammsäure *f*
  p. base Stammbase *f*
  p. cell Mutterzelle *f*
  p. compound Stammverbindung *f*
  p. element Mutterelement *n*, Anfangsglied *n*, Ausgangselement *n*
  p. hydrocarbon Grundkohlenwasserstoff *m*, Stammkohlenwasserstoff *m*
  p. mass peak Ausgangslinie *f* (Massenpik *m*) im Massenspektrum

  p. material Ausgangsmaterial *n*
  p. name Stammname *m*
  p. peak *s.* p. mass peak
  p. ring system Stammringsystem *n*, zyklisches Grundgerüst *n*
  p. rock Muttergestein *n*, Ursprungsgestein *n*, Ausgangsgestein *n*
  p. substance Muttersubstanz *f*, Grundsubstanz *f*, Grundstoff *m*, Ursprungssubstanz *f*, Stammkörper *m*, Grundkörper *m*, Ausgangssubstanz *f*, Ausgangsstoff *m*, Ausgangsprodukt *n*
parental magma Stamm-Magma *n*, Muttermagma *n*
parhelium Parhelium *n*
parian cement Pariangips *m*
Paris green $Cu(CH_3COO)_2 \cdot 3Cu(AsO_2)_2$ Pariser Grün *n*, Schweinfurter Grün *n*, Kupferarsenitazetat *n*
  P. white Pariser Weiß *n* *(fein gemahlene Kreide)*
parisite *(Min)* Parisit *m*
parison *(Plast)* Rohling *m*, Vorformling *m*; *(Glas)* Külbel *n*, Kölbel *n*
  p. mould *(Glas)* Vorform *f*
parity *(phys Ch)* Parität *f*
Parker process Parker-[Coalite-]Verfahren *n*
Parker solution Parker-Lösung *f*
parkerize /to parkerisieren
parkerizing Parkerisieren *n* *(Phosphatierungsverfahren)*
Parker's cement *s.* Roman cement
Parkes process Parkes-Verfahren *n* *(zur Entfernung von Edelmetallen aus Blei)*
Parkes reagent Parkes-Reagens *n*
Parr bomb Parr-Bombe *f*
parrot green $Cu(CH_3COO)_2 \cdot 3Cu(AsO_2)_2$ Papageiengrün *n*, Schweinfurter Grün *n*, Kupferarsenitazetat *n*
Parr's basis Parrsche Grundlage *f* *(in Seylers Kohlenklassifizierungsschaubild)*
parsley camphor Petersilienkampfer *m*, Apiol *n*
  p. oil Petersilienöl *n*
part *s.* partial
part Teil *n(m)*
  p. by volume Raumteil *m*, Volumenteil *m*
  p. by weight Masseteil *m*
  p.-hydrogenated halbgehärtet
partial partial, partiell, Partial..., teilweise, Teil...
  p. combustion unvollkommene (unvollständige) Verbrennung *f*, Teilverbrennung *f*
  p. condensation partielle Kondensation *f*, Teilkondensation *f*, Dephlegmation *f*
  p.-condensation head *s.* p. condenser
  p. condenser Teilkondensator *m*, Dephlegmator *m*
  p. glyceride gemischtes (gemischtsäuriges) Glyzerid *n*
  p. heat of dilution differentielle (differentiale) Verdünnungswärme (Verdünnungsenthalpie) *f*
  p. heat of solution differentielle (differentiale) Lösungswärme (Lösungsenthalpie) *f*

p. hydrolysis partielle Hydrolyse *f*, Partialhydrolyse *f*
p. miscibility begrenzte (teilweise) Mischbarkeit *f*
p. molal quantity *s.* p. molar quantity
p. molal volume *s.* p. molar volume
p. molar entropy partielle molare Entropie *f*
p. molar quantity partielle molare Größe *f*
p. molar volume partielles molares Volumen *n*, partielles Molvolumen *n*
p. pressure Partialdruck *m*, Teildruck *m*
p. pressure curve Partialdruckkurve *f*
p. pressure of water vapour Wasserdampfpartialdruck *m*, Wasserdampfteildruck *m*
p. racemate partielles Razemat *n*
p. racemization partielle Razemisierung *f*
p.-radiation pyrometer Teilstrahlungspyrometer *n*, Leuchtdichtepyrometer *n*
p. solution Anlösung *f*
p. sterilization Teilsterilisation *f*
p. valence Partialvalenz *f*
p.-valence theory Partialvalenztheorie *f*, Theorie *f* der Partialvalenzen
p. valency *s.* p. valence
p. vapour pressure Partialdampfdruck *m*, Teildampfdruck *m*
partially miscible teilweise (begrenzt) mischbar
p. racemic compound partielles Razemat *n*
p. soluble teilweise löslich
particle Teilchen *n*, Partikel *f*, *(i.e.S.)* Korpuskel *n*; Korn *n*
α-**particle** α-Teilchen *n*, Alphateilchen *n*, Heliumkern *m*
α-**particle source** α-Strahlenquelle *f*
β-**particle** β-Teilchen *n*, Betateilchen *n*, Elektron *n*
p. acceleration Teilchenbeschleunigung *f*
p. accelerator Teilchenbeschleuniger *m*
p. board *(Holz)* Spanplatte *f*
p. charge Teilchenladung *f*
p. counter Teilchenzähler *m*
p. energy Teilchenenergie *f*
p. fineness Teilchenfeinheit *f*
p. of dirt Schmutzteilchen *n*
p. of grease Fetteilchen *n*
p. of matter Stoffteilchen *n*, Materieteilchen *n*, materielles Teilchen *n*, materielle Partikel *f*
p. production Teilchenerzeugung *f*
p. radiation Teilchenstrahlung *f*, Korpuskularstrahlung *f*
p. shape Teilchenform *f*; Kornform *f*, Korngestalt *f*
p. size Teilchengröße *f*, Partikelgröße *f*; Korngröße *f*, Stückgröße *f*; Tröpfchengröße *f*
p. size analysis Korngrößenanalyse *f*
p. size determination Teilchengrößenbestimmung *f*
p. size distribution Korn[größen]verteilung *f*
p. size distribution curve Kornverteilungskurve *f*, Kornkennlinie *f*, Körnungslinie *f*

p. size range Körnungsstufe *f*
p. spin *(phys Ch)* Teilchenspin *m*
p. structure Teilchenstruktur *f*
p. trajectory Teilchen[flug]bahn *f*
p. velocity Teilchengeschwindigkeit *f*
parting Lösen *n*, Ablösen *n*, Trennen *n*, Trennung *f*; *(Naßmetallurgie)* Scheiden *n*, Scheidung *f*
p. agent (compound) Trennmittel *n*
p. dish Trennplatte *f*
p. line Trennlinie *f*, Trennfuge *f*, Teilfuge *f*, Formennaht *f*
p. plane Trennfläche *f*, Teilungsebene *f*, Ablösungsebene *f*
partition / **to** teilen, trennen; verteilen; abteilen
partition Teilen *n*, Trennen *n*; Verteilen *n*; Abteilen *n*; Trennwand *f*, Scheidewand *f*, Zwischenwand *f*, Membran[e] *f*; Fach *n*, Abteilung *f*
p. chromatography Verteilungschromatografie *f*
p. coefficient Verteilungskoeffizient *m*, Verteilungskonstante *f*; Segregationskonstante *f*, Abscheidungskonstante *f*, Verteilungskoeffizient *m* (in der Zonenschmelztheorie)
p. function *(phys Ch)* Verteilungsfunktion *f*, Zustandssumme *f*
p. function for vibrational energy Schwingungsverteilungsfunktion *f*, Verteilungsfunktion (Zustandssumme) *f* der Schwingungsenergie
p. law Verteilungsgesetz *n*; Nernstsches Verteilungsgesetz *n*, Nernstscher Verteilungssatz *m*
p. wall Trennwand *f*, Scheidewand *f*, Zwischenwand *f*, Membran[e] *f*
partly bleached halbgebleicht
p. preserved food Halbkonserve *f*
p. saponified partiell verseift
p. skimmed milk teilentrahmte Milch *f*
Paschen-Back effect Paschen-Back-Effekt *m*
Paschen series Paschen-Serie *f*
pass / **to** 1. [ein]leiten, [hin]durchleiten; 2. durchfließen *(z.B. vom Strom)*; strömen *(z.B. von einer Flüssigkeit)*; passieren, durchgehen, durchlaufen
p. in[to] einleiten
p. over überleiten; überdestillieren, übergehen
p. the stock endwise through the mill *(Gum)* die Mischung über Kopf stürzen
p. through *(durch ein Sieb)* durchfallen, passieren; *(durch ein Sieb)* hindurchdrücken, hindurchlaufen lassen, seihen, passieren
p. up[wards] nach oben steigen, aufsteigen, emporsteigen; aufwärts (nach oben) führen (transportieren), hochfördern
pass Durchgang *m*, Passage *f* (beim Zonenschmelzen)
passage Durchgang *m*, Durchlauf *m*, Durchfluß *m*; Durchtritt *m*, Durchlaß *m*
p. of current (electricity) Stromdurchgang *m*
Passburg dryer Paßburg-Trockenschrank *m*, Vakuumtrockenschrank *m* nach Paßburg
passing of the sheet Papierlauf *m*, Papierdurchgang *m*

passivate / to *(ein unedles Metall)* passivieren
passivating, passivation Passivierung *f (eines unedlen Metalls)*
passivator Passivierungsmittel *n*
passive passiv, inaktiv; passiviert
p. metal passiviertes Metall *n*
passivity Passivität *f (Eigenschaft einiger unedler Metalle)*
paste / to *(z.B. Kohle bei der Hydrierung)* anpasten, anreiben, anrühren; *(z.B. Farbe)* anteigen; *(Pap)* bekleben, kaschieren; *(z.B. Akkuplatten)* pastieren
  p. together zusammenkleben
paste Paste *f*, Brei *m*, Teig *m*; Kleister *m*, Klebstoff *m*; Fruchtpaste *f*; Zementbrei *m*; *(Ker)* Masse *f*
  to make into a p. *(z.B. Farbe)* anteigen
  p.-dried *(Gerb)* gepastet *(durch Aufkleben auf Glas- oder Metallplatten getrocknet)*
  p. drying *(Gerb)* Klebetrocknen *n*, Klebetrocknung *f*
  p.-like paste[n]artig, pastenförmig, pastös, pastig; salb[enart]ig
  p. mill (mixer) Pastenmischer *m*
  p. mould *(Glas)* graphit[is]ierte Tauchform *f*, Drehkülbelform *f*
  p. resin Pastenharz *n*, Harzpaste *f*
  p. solder Lötpaste *f*
pasteboard geklebter Karton *m*, Klebekarton *m*
pasted board *s.* pasteboard
  p. up angeteigt
pastel shade Pastellton *m*
Pasteur effect Pasteur-Effekt *m*
Pasteur flask Pasteur-Kolben *m*
pasteurization Pasteurisieren *n*, Pasteurisierung *f*, Pasteurisation *f*
  p. flavour Pasteurisierungsgeschmack *m*
  p. section Pasteurisierungsabteilung *f*
  p. temperature Pasteurisier[ungs]temperatur *f*, *(i.e.S.)* Milcherhitzungstemperatur *f*
pasteurize / to *s.* pasteurisieren
pasteurizer Pasteuriseur *m*, Pasteurisierapparat *m*, *(i.e.S.)* Milcherhitzer *m*, Milcherhitzungsapparat *m*
pasteurizing *s.* pasteurization
pastilles *(Pharm)* Pastillen *fpl*
pasting Anpasten *n*, Anrühren *n*, Anreiben *n*, Anreibung *f (z.B. bei der Kohlehydrierung)*; Anteigen *n (z.B. von Farbe)*; *(Pap)* Kleben *n*, Kaschieren *n*; Pastieren *n (z.B. von Akkuplatten)*; *s.* p.*paper
  p. agent Anteigmittel *n*
  p. paper Beklebepapier *n*, Aufklebepapier *n*, Überzugspapier *n*, Kaschierpapier *n*
  p. press Klebepresse *f*
  p. process *(Gerb)* Klebetrockenverfahren *n*
  p.-up Anteigen *n*
pastry fat Backfett *n*, Ziehfett *n*
  p. margarine Backmargarine *f*, Bäckermargarine *f*
pasty paste[n]artig, pastenförmig, pastös, pastig; salb[enart]ig

p. butter überbutterte Butter *f*
patching cement Reifenreparaturkitt *m*
  p. rubber Reparaturplatte *f*
patchouli alcohol Patsch[o]ulialkohol *m*
  p. oil Patsch[o]uliöl *n*
pate dure *(Ker)* Hartmasse *f*
  p. tender *(Ker)* Weichmasse *f*
patent / to *(z.B. Draht)* patentieren
patent flour *(helles feinstes Weizenmehl von niedrigem Ausmahlungsgrad)*
  p. green *s.* patgreen
  p. index Patentregister *n*
  p. leather Lackleder *n*
  p. litigation Patentstreit *m*, Patentverletzungsprozeß *m*
patenting Patentieren *n (Wärmebehandlung von Draht)*
  p. furnace Patentierofen *m*
patgreen $Cu(CH_3COO)_2 \cdot 3Cu(AsO_2)_2$ Patentgrün *n*, Schweinfurter Grün *n*, Kupferarsenitazetat *n*
path Weg *m*, Bahn *f*, Flugbahn *f*; Weglänge *f*; Lauf *m (z.B. seismischer Wellen)*
  p. of the particle Teilchenbahn *f*
pathogen Pathogen *n*, pathogener Organismus *m*
pathogenic pathogen, krankheitserzeugend, krankheitserregend
  p. bacteria pathogene (krankheitserregende) Bakterien *npl*
patina Patina *f*, Edelrost *m*
patrinite *(Min)* Patrinit *m*, Aikinit *m*, Aikenit *m*
patrix Patrize *f*, Stempel *m*, Stempelprofil *n*, Preßstempel *m*
pattern Muster *n*, Probe *f*; Schablone *f*, Lehre *f*; Modell *n*
  p.-draw[ing] machine Abhebeformmaschine *f*
  p. plate Modellplatte *f*, Formplatte *f*
patterned glass Ornamentglas *n*
Patterson name Name *m* nach Patterson
pattersonite *(Min)* Pattersonit *m*
Pattinson process Pattinson-Verfahren *n*, Pattinsonieren *n*
Pattinson's white lead Pattinson-Bleiweiß *n*
Pauli exclusion principle Paulisches Ausschließungsprinzip (Prinzip) *n*, Pauli-Prinzip *n*, Ausschließungsprinzip *n*
Pauli-type electrodialyzer Elektrodialysator *m* nach Pauli
Pauling's electronegativity Elektronegativität *f* nach Pauling
Pauly [protein] reaction Paulysche Diazoreaktion (Reaktion) *f*
Pavy solution Pavys (Pavysche) Lösung *f*
pay ore [ab]bauwürdiges Erz *n*
PBI *s.* polybenzimidazole
PBR *s.* 1. pebble bed reactor; 2. pyridinebutadiene rubber
PBT *s.* polybenzothiazole
PCE, P.C.E. *s.* pyrometric-cone equivalent
pcH *(negativer dekadischer Logarithmus der Wasserstoffionenkonzentration)*

PCP C_6Cl_5OH Pentachlorphenol *n*, PCP *n* *(Herbizid)*; *s.* polychloroprene
p.c.t.f.e., P.C.T.F.E. *s.* polychlorotrifluoroethylene
pd. *s.* powder
P.D. *s.* pressure distillate
PE *s.* 1. pentaerythritol; 2. polyethylene
pea Erbse *f*, Pisum sativum L.; kleines Erzstück *n* *(oder)* Kohlestück *n*
  p. meal Erbsenmehl *n*
  p. nut *s.* peanut
  p. ore Bohnerz *n*
  p. shelling mill Erbsenschälmühle *f*
  p.-size erbsengroß
peach Pfirsich *m*
  p.-kernel oil Pfirsichkernöl *n*; *s.* bitter almond oil
Peachey cure (process) Vulkanisation *f* nach Peachey, Peachey-Prozeß *m*
peacock ore *(Min)* Buntkupfererz *n*, Buntkupferkies *m*, Bornit *m (Eisen(II)-kupfer(II)-sulfid)*
peak Höchstwert *m*, Maximum *n*, Scheitelwert *m*; Scheitelpunkt *m*, Gipfelpunkt *m* *(z.B. einer Kurve)*; Berg *m*, Peak *m*, Pik *m*, Zacken *m*, Zacke *f (z.B. eines Chromatogramms)*
  p. firing temperature Spitzentemperatur *f*, maximale Brenntemperatur *f*
  p. width Bergbreite *f*, Peakbreite *f (z.B. in einem Chromatogramm)*
peaky[-curing] *(Gum)* mit kurzem Vulkanisationsplateau (Plateau) *(mit engem Optimalbereich der Vulkanisation)*
peanut Erdnuß *f*, Arachis hypogaea L.
  p. butter Erdnußbutter *f*
  p. cake Erdnußkuchen *m*
  p. cake meal *s.* oil meal
  p. oil Erdnußöl *n*, Arachisöl *n*
  p. oil meal Erdnußkuchenmehl *n*
pear oil $CH_3COOC_5H_{11}$ Birnenöl *n*, Amylazetat *n*, Essigsäureisoamylester *m*, Birnenäther *m*
pearl ash K_2CO_3 Perlasche *f*, Kaliumkarbonat *n*
  p. barley Perlgraupen *fpl*
  p. essence Perl[en]essenz *f*, Fischsilber *n*; Perl[en]essenz *f*, Essence d'Orient *f*
  p. filler (hardening) *(Pap)* durch Fällung gewonnener Industriegips *m*
  p.-mica *(Min)* Perlglimmer *m*, Margarit *m* *(Kalziumaluminiumdihydroxiddialumodisilikat)*
  p. sinter *(Min)* Perlsinter *m*, *(veraltet für)* Opalsinter *m*
  p. spar *(Min)* Perlspat *m*, Dolomit *m (Kalziummagnesiumkarbonat)*; *(Min)* Perlspat *m*, Aragonit *m (Kalziumkarbonat)*
  p. white Perlweiß *n*, Wismutweiß *n*, Spanischweiß *n (aus Wismutoxidnitrat)*; Perlweiß *n*, Schminkweiß *n (aus Wismutoxidchlorid)*; *s.* p. filler
pearlite *(Met)* Perlit *m (lamellares Aggregat aus Ferrit und Zementit mit perlmutterartigem Glanz)*; *(Min)* Perlit *m*, Perlstein *m*
pearlitic *(Met, Min)* perlitisch
pearly lustre Perlmutterglanz *m*
peat Torf *m*

  p. bank Torfstich *m*
  p. board Torfpappe *f*
  p. bog Torfmoor *n*
  p. briquette Torfbrikett *n*
  p. coke Torfkoks *m*
  p. drying Torftrocknung *f*
  p. dust Torfmull *m*
  p. formation Torfbildung *f*, Vertorfung *f*
  p.-forming torfbildend, torfogen, vertorfend
  p.-forming process Torfbildungsprozeß *m*, Vertorfungsvorgang *m*
  p. gas Gas *n* aus Torf, Torfgas *n*
  p. humus Torfhumus *m*
  p. machine Torfmaschine *f*
  p. moor *s.* p. bog
  p. moss Torfmoos *n*, Sphagnum *n*
  p. soil Torfboden *m*
  p. substance Torfsubstanz *f*, Torfmasse *f*
  p. tar Torfteer *m*
  p. winning Torfgewinnung *f*
peatery *s.* 1. peat bank; 2. peat bog
peaty torfig
pebble Kiesel[stein] *m*; Pebble *n*, Wärmestein *m*, Steinkugel *f (Kugel aus hitzebeständigem Material oder Koks im Wärmesteinerhitzer)*
  p. bed Wärmesteinschüttung *f (im Wärmesteinerhitzer)*
  p. bed reactor *(Kch)* Kugelhaufenreaktor *m*, Pebble-Reaktor *m*
  p. heater Pebble-Heater *m*, Wärmesteinerhitzer *m*, Steinerhitzer *m*
  p.-heater process Pebble-Heater-Verfahren *n*, Steinerhitzerverfahren *n*
  p.-heater pyrolysis Pebble-Heater-Pyrolyse *f*, Pyrolyse *f* nach dem Pebble-Heater-Verfahren, Pyrolyse *f* mit Pebbles
  p. mill Kugelmühle *f (mit Flintstein- oder Hartporzellankugelfüllung)*
  p. reactor *s.* p. bed reactor
  p. stove *s.* p. heater
Pechmann condensation (coumarin synthesis) Pechmannsche Kumarinsynthese *f*
pecktolite *(Min)* Pektolith *m (Kalziumnatriumhydrogentrisilikat)*
Peco process Peco-Verfahren *n (zur Trocknung von Torf)*
pectase *s.* pectinesterase
pectate Pektat *n (Salz oder Ester der Pektinsäure)*
pectic acid Pektinsäure *f*, Poly-D-galakturonsäure *f*
  p. enzyme pektisches Ferment *n*, Pektin[glykosid]ase *f (pektinabbauendes Fermentgemisch)*
  p. substance Pektinstoff *m*, Pektin *n (i.w.S.)*
pectin Pektin *n (i.w.S.)*, Pektinstoff *m*; Pektin *n*, pektinige Säure *f (teilweise oder vollmethoxylierte Poly-D-galakturonsäure)*
  p. chemistry Pektinchemie *f*
  p. grade Geliergrad *m (Geliereinheit f)* des Pektins
  p. jelly Pektingelee *n*, Pektingel *n*

p. preparation Pektinpräparat n
p. proper eigentliches Pektin n
p. solution Pektinlösung f
p. sugar s. pectinose
pectinase Pektin[glykosid]ase f, pektisches Ferment n (pektinabbauendes Fermentgemisch); Pektinase f, Pektinpolygalakturon[id]ase f, Polygalakturon[id]ase f
pectinate Pektinat n (Salz oder Ester der pektinigen Säure)
pectinesterase Pekt[inester]ase f, Pektinmethylesterase f
pectinic acid pektinige Säure f, Pektin n (teilweise oder voll methoxylierte Poly-D-galakturonsäure)
pectinogen Protopektin n
pectinolytic s. pectolytic
pectinose Pektinose f, Pektinzucker m, Arabinose f, Gummizucker m
pectinous substance s. pectic substance
pectolite (Min) Pektolith m (Kalziumnatriumhydrogentrisilikat)
pectolytic pekt[in]olytisch, pektinabbauend, pektinspaltend
p. enzyme pektolytisches Enzym (Ferment) n
pectose Pektose f, (veraltet für) Protopektin n
peculiarity besondere Konfiguration f (z.B. der Elektronenschale)
ped (Bodenkunde) Krümel m(n)
Pedersen process Pedersen-Verfahren n (zur Aluminiumgewinnung)
pedestal Sockel m, Ständer m
p.-mounted pump Grundplattenpumpe f
pedogenesis Bodenbildung f, Bodenentwicklung f, Entstehung f des Bodens
pedologic[al] bodenkundlich, pedologisch
pedologist Bodenkundler m
pedology Bodenkunde f, Pedologie f
peel / to 1. schälen, enthülsen; abschälen, ablösen (z.B. eine Folie); (Pap) schälen, entrinden; 2. abplatzen, abspringen, abbröckeln (von einer Schicht); (Ker) sich abschälen, abblättern (Glasur)
peel bond test (Plast) Schäl- und Haftprüfung f
p. strength (Plast) Schälfestigkeit f, Ablösefestigkeit f
peeling Schälen n, Enthülsen n; Abschälen n, Ablösen n (z.B. einer Folie); (Pap) Schälen n, Schälung f, Entrinden n, Entrindung f; Abplatzen n, Abspringen n, Abbröckeln n (einer Schicht); (Ker) Abschälen n, Abblättern n (der Glasur)
p. machine Schälmaschine f
p. test (Holz) Schälprüfung f, Abhebeprüfung f (auf Festigkeit von Leimverbindungen)
peephole Schauloch n
peganite (Min) Peganit m, (veraltet für) Variscit m (Aluminiumorthophosphat)
pegmatite (Min) Pegmatit m
pegmatitic pegmatitisch
pegmatolite (Min) Pegmatolith m, (veraltet für) Orthoklas m (Kaliumalumotrisilikat)

Pegu catechu (cutch) Braunes Katechu n, Pegukatechu n, Bombaykatechu n (Extrakt aus Acacia catechu Willd.)
Peirce-Smith converter Peirce-Smith-Konverter m, Peirce-Smith-Trommelkonverter m, Trommelkonverter m nach Peirce-Smith
pelagite (Min) Pelagit m
pelargone Pelargon n, Dioktylketon n, Nonylon n, 9-Heptadekanon n
pelargonic acid $CH_3(CH_2)_7COOH$ Pelargonsäure f, n-Nonylsäure f, Nonansäure f
p. aldehyde $CH_3(CH_2)_7CHO$ Pelargonaldehyd m, n-Nonylaldehyd m
Peligot's salt $K[CrO_3Cl]$ Peligotsches Salz n (Kaliumchlorochromat)
pelite (Geol) Pelit m
pellagra-preventive factor Pellagrapräventivvitamin n, Antipellagravitamin n, PP-Faktor m (Pellagraschutzstoff; Nikotinsäureamid)
pellet Pellet n, Kugel f, Kügelchen n, Pille f, Tablette f; Granalie f, Granulum n, Körnchen n, kug[e]liges Granulatkorn n; (Gum) Pellet n, Krümel m(n); (Gum) Rußperle f, Korn n
pellet / to s. to pelletize
p. mill s. pelletizing machine
pelletierine Pelletierin n, β-[2-Piperidyl]-propionaldehyd m (Punikaalkaloid)
pelleting s. pelletizing
pelletize / to pellet[is]ieren, zu Kügelchen formen, tablettieren; (Erz) pellet[is]ieren, kugelsintern; (Ruß) körnen, perlen; (Gummimischung) zu Pellets (Krümeln) formen (verarbeiten)
pelletized black geperlter Ruß m, Perlruß m
p. catalyst tablettierter (pillenförmiger, in Pillenform gepreßter) Katalysator m, Katalysator m in Pillenform
pelletizer Pellet[is]iermaschine f, Tablettiermaschine f; (Gum) Pelletizer m
pelletizing Pellet[is]ieren n, Pellet[is]ierung f, Tablettieren n, Aufbaugranulieren n; Pellet[is]ieren n, Kugelsintern n, Kugelsinterung f (von Erz); (Gum) Körnen n, Perlen n (von Ruß); (Gum) Verarbeitung f zu Pellets (Krümeln)
p. cone Pellet[is]ierkonus m
p. device Pelletformeinrichtung f
p. disk Pellet[is]ierteller m
p. drum Pellet[is]iertrommel f
p. equipment (Gum) Verdichter m, Perlanlage f, Granulator m
p. machine Pellet[is]iermaschine f, Tablettiermaschine f
p. plant Pelletisier[ungs]anlage f
p. process Pelletisier[ungs]verfahren n
pellotine Pellotin n (Anhaloniumalkaloid)
pelt (Gerb) Blöße f, enthaarte Haut f
Peltier effect Peltier-Effekt m
pen-and-ink recorder Tintenschreiber m
penaldic acid Penald[in]säure f (Sammelname für Penizillinspaltprodukte der allgemeinen Formel $RCONH \cdot CH(COOH)CHO$)

p.-F acid Penaldin-F-Säure *f*, 2-Pentenylpenaldinsäure *f*
p.-G acid Penaldin-G-Säure *f*, Benzylpenaldinsäure *f*
p.-K acid Penaldin-K-Säure *f*, Heptylpenaldinsäure *f*
p.-X acid Penaldin-X-Säure *f*, p-Hydroxybenzylpenaldinsäure *f*
pencil stone *(Min)* Bildstein *m (Aluminiumdihydrogentetrasilikat)*
pendulum method *(Gum)* Pendelfallmethode *f*
p. roller mill Pendel[rollen]mühle *f*, Fliehkraftpendelmühle *f*, Fliehkraftwalzenmühle *f*
penetrability *(Text)* Durchdringbarkeit *f*, Durchfärbbarkeit *f*
penetrate / to [durch]tränken, imprägnieren; penetrieren, durchdringen, durchwandern, passieren; *(Text)* durchdringen, eindringen, durchfärben; *(Gerb)* durchbeißen *(von einem Gerbstoff)*; *(Gerb)* durchschlagen *(von einem Farbstoff)*
penetrating durchdringend *(z.B. Geruch)*
penetrating *(Text)* Durchdringen *n*, Eindringen *n*, Durchfärben *n*, Durchfärbung *f*, Penetration *f*
p. agent *(Text)* Durchdringungsmittel *n*, Eindringungsmittel *n*, Durchfärbemittel *n*
p. power Durchdringungsvermögen *n*, Durchdringungsfähigkeit *f*, Durchdringungskraft *f*
penetration Tränkung *f*, Durchtränkung *f*, Imprägnierung *f*; Penetration *f*, Durchdringung *f*, Durchsetzung *f*; *(Text)* Durchdringen *n*, Eindringen *n*, Durchfärben *n*, Durchfärbung *f*, Penetration *f*
p. capacity *s.* penetrating power
p. complex Durchdringungskomplex *m*
p. depth Eindringtiefe *f*
p. dyeing Durchfärbung *f*
p. of cooking acid *(Pap)* Kochsäuredurchtränkung *f*
p. of sulphite cooking liquor *s.* p. of cooking acid
p. of wood *(Pap)* Holzdurchtränkung *f*, Hackschnitzeldurchtränkung *f*, Hackschnitzelimprägnierung *f*
p. period *(Pap)* Durchtränkungszeit *f*, Imprägnierungsperiode *f*; *(Pap)* Ankochperiode *f*
p. probability Durchdringwahrscheinlichkeit *f*
p. time *(Pap)* Durchtränkungszeit *f*, Imprägnierungsperiode *f*
p. twins *(Krist)* Durchdringungszwillinge *mpl*, Verwachsungszwillinge *mpl*
p. under pressure *(Pap)* Druckdurchtränkung *f*, Druckimprägnierung *f*
penetrometer Penetrometer *n*
p. method Penetrometerverfahren *n*
penicillamine $(CH_3)_2C(SH) \cdot CH(NH_2) \cdot COOH$ Penizillamin *n*, β,β-Dimethylzystein *n*, β-Thiolvalin *n*
penicillaminic acid $(CH_3)_2C(SO_3H) \cdot CH(NH_2) \cdot COOH$ Penizillaminsäure *f*

penicillanic acid Penizillansäure *f (Baustein der Penizilline)*
penicillase *s.* penicillinase
penicillic acid Penizill[in]säure *f*, 3-Methoxy-5-methyl-4-oxo-2,5-hexadiensäure *f*
penicillin Penizillin *n (Sammelname für von verschiedenen Schimmelpilzen erzeugte Antibiotika)*
p. I *s.* p. F
p. II *s.* p. G
p. III *s.* p. X
p. IV *s.* p. K
p. activity Penizillinwirksamkeit *f*
p. amidase Penizillinamidase *f*
p. F Penizillin F *n*, Penizillin I *n*, Δ^2-Pentenylpenizillin *n*
p. G Penizillin G *n*, Penizillin II *n*, Benzylpenizillin *n*
p. K Penizillin K *n*, Penizillin IV *n*, n-Heptylpenizillin *n*
p. N Penizillin N *n*, δ-Amino-δ-karboxybutylpenizillin *n*
p. nucleus Grundkörper *m* der Penizilline
p. O Penizillin O *n*, Allylmerkaptomethylpenizillin *n*
p.-resistant penizillinresistent
p.-sensitive penizillinempfindlich
p. V Penizillin V *n*, Phenoxymethylpenizillin *n*
p. X Penizillin X *n*, Penizillin III *n*, p-Hydroxybenzylpenizillin *n*
penicillinase Penizill[in]ase *f (penizillinzerstörendes Enzym)*
penicilloate Penizilloat *n (bei der Alkali-Inaktivierung entstehendes Abbauprodukt eines Penizillins)*
penicilloic acid Penizillo[in]säure *f (Sammelname für Amidodikarbonsäuren der allgemeinen Formel $RCONH(C_6H_{10}NS)(COOH)_2$)*
p.-F acid Penizilloin-F-Säure *f*, 2-Pentenylpenizilloinsäure *f*
p.-G acid Penizilloin-G-Säure *f*, Benzylpenizilloinsäure *f*
p.-K acid Penizilloin-K-Säure *f*, Heptylpenizilloinsäure *f*
penillic acid Penillsäure *f (ein Inaktivierungsprodukt eines Penizillins in saurer Lösung)*
p.-F acid Penill-F-Säure *f*, 2-Pentenylpenillsäure *f*
p.-G acid Penill-G-Säure *f*, Benzylpenillsäure *f*
p.-K acid Penill-K-Säure *f*, Heptylpenillsäure *f*
p.-X acid Penill-X-Säure *f*, p-Hydroxybenzylpenillsäure *f*
penilloaldehyde $RCONHCH_2CHO$ Penilloaldehyd *m*
p. F Penilloaldehyd F *m*, 2-Pentenylpenilloaldehyd *m*
p. G Penilloaldehyd G *m*, Benzylpenilloaldehyd *m*
p. K Penilloaldehyd K *m*, Heptylpenilloaldehyd *m*
p. X Penilloaldehyd X *m*, p-Hydroxybenzylpenilloaldehyd *m*

penilloic acid Penillosäure f
 p.-G acid Penillo-G-Säure f, Benzylpenillosäure f
 p.-K acid Penillo-K-Säure f, Heptylpenillosäure f
 p.-X acid Penillo-X-Säure f, p-Hydroxybenzyl-penillosäure f
pennine, penninite (Min) Pennin m
Pennvernon process s. Pittsburgh (sheet) process
Pensky-Martens apparatus s. Pensky-Martens flash-point apparatus
Pensky-Martens closed apparatus (tester) geschlossener Flammpunktprüfer m nach Pensky-Martens, geschlossenes Pensky-Martens-Gerät n, geschlossener Pensky-Martens-Apparat m
Pensky-Martens flash-point apparatus (tester) Flammpunktprüfer m nach Pensky-Martens, Pensky-Martens-Gerät n, Pensky-Martens-Apparat m
Pensky-Martens [flash] tester s. Pensky-Martens flash-point apparatus
pentabasic fünfbasig, fünfbasisch
pentaborane B_5H_9 Pentaboran n
pentaborate Pentaborat n
pentabromide Pentabromid n
pentacarbonyl Pentakarbonyl n
pentacene Pentazen n, 2,3,6,7-Dibenzanthrazen n
pentachloride Pentachlorid n
pentachlorophenol C_6Cl_5OH Pentachlorphenol n (Herbizid)
pentacid fünfsäurig
pentacontane Pentakontan n (ein Alkan mit 50 C-Atomen)
pentacosane Pentakosan n (ein Alkan mit 25 C-Atomen)
pentacovalent pentakovalent
pentacyclic pentazyklisch
pentad fünfwertig
pentad fünfwertiges Element n; fünfwertige Atomgruppe f
pentadecane Pentadekan n (ein Alkan mit 15 C-Atomen), (i.e.S.) n-Pentadekan n, Pentadekan n
pentadecanoic acid Pentadekansäure f, Pentadezylsäure f
pentadecanol Pentadekanol n, (i.e.S.) 1-Pentadekanol n, Pentadezylalkohol m
pentadecyl Pentadezyl n (Atomgruppe)
 p. alcohol Pentadezylalkohol m, 1-Pentadekanol n
pentadecylic acid s. pentadecanoic acid
pentadentate (Komplexchemie) fünfzahnig, fünfzähnig, fünfzählig
pentadiene Pentadien n, (i.e.S.) $CH_2=CHCH=CHCH_3$ 1,3-Pentadien n, 1-Methylbutadien n, 1-Methyldivinyl n, Piperylen n
2,4-pentadienoic acid $CH_2=CHCH=CHCOOH$ 2,4-Pentadiensäure f, 3-Vinylakrylsäure f
pentadigalloylglucose Pentadigalloylglukose f (mit Digallussäure veresterte Glukose)
pentaerythrite s. pentaerythritol

pentaerythritol $C(CH_2OH)_4$ Pentaerythrit m, 2,2-Bis-(hydroxymethyl)-1,3-propandiol n
 p. tetranitrate $C(CH_2ONO_2)_4$ Pentaerythrittetranitrat n, Pentrit n, PETN (Sprengstoff)
pentaglycerol $CH_3C(CH_2OH)_3$ Pentaglyzerin n, Pentaglyzerol n, 1,1,1-Trimethyloläthan n, 2-(Hydroxymethyl)-2-methyl-1,3-propandiol n
pentahydrate Pentahydrat n
pentahydric fünfwertig, mit 5 Hydroxylgruppen (OH-Gruppen) (Alkohol, Phenol)
pentahydroxy Pentahydroxy...; s. pentahydric
pentaiodide Pentajodid n
pentamethonium $(CH_3)_3N(CH_2)_5N(CH_3)_3$ Pentamethonium n, Pentamethylen-1,5-bis-(trimethylammonium) n
pentamethylbenzene Pentamethylbenzol n
pentamethylbenzoic acid Pentamethylbenzoesäure f, Pentamethylbenzolkarbonsäure f
pentamethylene Pentamethylen n, Zyklopentan n; Pentamethylen n (Atomgruppe)
pentamethylenediamine $H_2N(CH_2)_5NH_2$ Pentamethylendiamin n, Kadaverin n
pentamethyleneimine Pentamethylenimin n, Hexahydropyridin n, Piperidin n
pentamidine Pentamidin n, 4,4'-Diamidinodiphenoxypentan n
pentammine Pentammin n (ein Ammin mit 5 Ammoniakmolekülen)
pentanal $CH_3(CH_2)_3CHO$ Pentanal n, Valeraldehyd m, n-Amylaldehyd m
pentane Pentan n (ein Alkan mit 5 C-Atomen), (i.e.S.) $CH_3(CH_2)_3CH_3$ n-Pentan n, Normalpentan n, Pentan n
pentanecarboxylic acid Pentankarbonsäure f
pentanedial $OCH(CH_2)_3CHO$ Pentandial n, Glutar[säure]dialdehyd m, Glutaraldehyd m
pentanedicarboxylic acid Pentandikarbonsäure f
pentanedioic acid $HOOC(CH_2)_3COOH$ Pentandisäure f, Glutarsäure f
pentanediol Pentandiol n (ein zweiwertiger Alkohol eines Pentans)
pentanedione Pentandion n (ein Diketon des n-Pentans)
pentanoic acid Pentansäure f, Valeriansäure f, (i.e.S.) $CH_3(CH_2)_3COOH$ n-Pentansäure f, n-Valeriansäure f, Baldriansäure f, Pentansäure f
pentanol Pentanol n, Pentylalkohol m, Amylalkohol m, (i.e.S.) $CH_3(CH_2)_3CH_2OH$ 1-Pentanol n, n-Pentylalkohol m, n-Amylalkohol m
pentanone Pentanon n (ein Keton des n-Pentans)
pentapeptide Pentapeptid n (Peptid mit 5 Aminosäureeinheiten)
pentaquin Pentaquin n, 8-(5-Isopropylaminopentyl-amino)-6-methoxychinolin n
pentaselenide Pentaselenid n
pentasilane Si_5H_{12} Pentasilan n
pentasodium triphosphate $Na_5P_3O_{10}$ Pentanatriumtriphosphat n, Natriumtriphosphat n
pentasulphide Pentasulfid n
pentathionate $M^I_2S_5O_6$ Penthionat n

pentathionic acid $H_2S_5O_6$ Pentathionsäure f
pentatriacontane Pentatriakontan n *(ein Alkan mit 35 C-Atomen), (i.e.S.)* n-Pentatriakontan n, Pentatriakontan n
18-pentatriacontanone 18-Pentatriakontanon n, Diheptadezylketon n, Stearon n
pentatungstate Pentawolframat n
pentavalence, pentavalency Fünfwertigkeit f, Pentavalenz f
pentavalent fünfwertig, pentavalent
pentazolyl Pentazolyl n *(Atomgruppe)*
pentene Penten n, Amylen n *(ein Alken mit 5 C-Atomen)*
pentenyl Pentenyl n *(Atomgruppe)*
Δ^2-**pentenylpenicillin** Δ^2-Pentenylpenizillin n, Penizillin F n, Penizillin I n
pentetrazol Pentetrazol n, 1,5-Pentamethylentetrazol n
penthrite s. pentaerythritol tetranitrate
pentine s. pentyne
pentite Pentit m *(ein fünfwertiger Alkohol)*
pentlandite *(Min)* Pentlandit m, Eisennickelkies m
pentosan $(C_5H_8O_4)_n$ Pentosan n *(ein aus Pentosen aufgebautes Polysaccharid)*
pentose $C_5H_{10}O_5$ Pentose f *(Monosaccharid mit 5 Sauerstoffatomen)*
pentoxide Pentoxid n
pentyl Pentyl n, Amyl n
  p. alcohol s. pentanol
pentyne Pentin n *(ein Alkin mit 5 C-Atomen)*
P.E.O. s. Paalsgard emulsion oil
peonidin s. paeonidin
peonin s. paeonin
pepper oil Pfefferöl n *(von Piper nigrum L.)*
pepsin Pepsin n *(Ferment des Magensaftes)*
pepsinogen Pepsinogen n, Propepsin n *(inaktive Vorstufe des Pepsins)*
peptic digestion Pepsinverdauung f
peptidase Peptidase f *(auf die Spaltung von Peptiden spezialisierte Protease)*
peptide Peptid n
  p. bond Peptidbindung f, peptidische Verknüpfung f, peptidartige Bindung f
  p. chain Peptidkette f
  p. group Peptidgruppe f
  p. link[age] s. p. bond
  p. moiety (portion) Peptid[an]teil m, Peptidkomponente f
  p. synthesis Peptidsynthese f
peptisize / to s. to peptize
peptization Peptisierung f, Peptisation f, Gel-Sol-Umwandlung f; *(Gum)* Peptisierung f, Peptisation f, Plastizierung f (Abbau m, Erweichung f) mit Peptisiermitteln
peptize / to peptisieren, *(Gum auch)* mit Peptisiermitteln erweichen (abbauen, plastizieren)
peptizer Peptisier[ungs]mittel n, Peptisationsmittel n, Peptisator m; *(Gum)* Peptisier[ungs]mittel n, chemisches Plastizier[ungs]mittel (Abbaumittel) n

peptizing agent s. peptizer
  p. effect *(Gum)* Abbauwirkung f, erweichende Wirkung f
peptolysis Peptolyse f
peptolytic peptolytisch
peptone Pepton n *(ein hochmolekulares Spaltprodukt aus Eiweißstoffen)*
peptonization Peptonisierung f
peptonize / to peptonisieren
peptonizing rest Eiweißrast f
peptotoxine Peptotoxin n
per capita usage Pro-Kopf-Verbrauch m
per compound Perverbindung f
  p. substitution Persubstitution f
peracetic acid s. peroxyacetic acid
peracid Persäure f, Übersäure f, Azylhydroperoxid n
perbituminous coal perbituminöse Kohle f
perborate s. peroxyborate
perboric acid s. peroxyboric acid
percarbonate s. peroxycarbonate
percarbonic acid s. peroxycarbonic acid
percent concentration Konzentration f in Prozent, prozentuelle Konzentration f
percentage prozentualer Anteil m; Prozentgehalt m, prozentualer Gehalt m, %-Gehalt m
  p. by volume Volum[en]prozent n, Vol.%
  p. by weight Masseprozent n, Masse%, *(bisher)* Gewichtsprozent n, Gew.%, Gewichtsanteile mpl
  p. composition prozentuale Zusammensetzung f
  p. humidity relative Feuchte (Feuchtigkeit) f
  p. purity Reinheitsgrad m
perch / to *(Gerb)* schlichten
perch *(Gerb)* Schlichtbaum m
perchlorate $M^I ClO_4$ Perchlorat n
perchloric acid $HClO_4$ Perchlorsäure f, Überchlorsäure f
perchloroethane CCl_3CCl_3 Perchloräthan n, Hexachloräthan n
perchloromethane CCl_4 Perchlormethan n, Tetrachlormethan n, Tetrachlorkohlenstoff m, Kohlenstofftetrachlorid n
perchromate s. peroxychromate
perchromic acid s. peroxychromic acid
percolate / to durchsickern, durchlaufen; durchsickern (durchlaufen) lassen; perkolieren, durchseihen
percolate Perkolat n
percolation Durchsickern n, Durchlaufen n; Durchsickernlassen n; Perkolieren n, Perkolation f, *(Bodenkunde auch)* Durchschlämmung f
  p. method (process) Perkolationsmethode f, Perkolationsverfahren n, Seiherverfahren n, Filtrationsmethode f *(zur Bleicherderaffination)*; Perkolationsmethode f, Perkolationsverfahren n, Filterverfahren n *(bei der Untertagevergasung)*
percolator Perkolator m
percrystalline perkristallin
percussion mortar Diamantmörser m, Stahlmörser m, Mineralmörser m

p. system Schlagbohrverfahren n *(zum Erdöl-bohren)*

perester Perester m

perezinone Perezinon n, 2-Hexyl-3-hydroxy-5,6-(1-propylen)-1,4-zyklohexadiendion n

perezone Perezon n, Pipitzahoinsäure f, 2-Hexyl-3-oxy-5-propenyl-*p*-benzochinon n, 2-Hexyl-3-hydroxy-5-(1-propenyl)-1,4-zyklohexadiendion n

perfect cleavage *(Min)* vollkommene Spaltbarkeit f

p.-discharge elevator Schwerkraftbecherwerk n mit Ablenkrollen

p. gas ideales (vollkommenes) Gas n

p. paper einwandfreies Papier n, Papier n erster Wahl

p. plate idealer (theoretischer) Boden m, *(i.w.S.)* theoretische Trennstufe f

perfecting engine *(Pap)* Kegel[stoff]mühle f, *(i.e.S.)* Jordan-Kegel[stoff]mühle f, Jordan-Mühle f; *(Pap)* Refiner m, Stoffaufschläger m, Aufschläger m, Ganzstoffmahlmaschine f

perfluorinate / to perfluorieren

perfluorination Perfluorierung f

perfluoroethylene $CF_2=CF_2$ Perfluoräth[yl]en n, Tetrafluoräth[yl]en n

perforate / to perforieren, lochen

perforated basket Lochtrommel f, Siebtrommel f

p. bottom Siebboden m *(eines Gooch-Tiegels)*

p. brick Loch[ziegel]stein m

p. edge-runner mill Kollergang m mit perforierter Mahlbahn

p. metal gelochtes Metall n, gelochte Metallplatte f, Lochplatte f

p.-metal screen Lochblechsieb n

p. plate Lochplatte f, Lochblech n, gelochtes Blech n, Siebblech n; *(Destillation)* Siebboden m, gelochter Austauschboden m; Siebplättchen n *(eines Gooch-Tiegels)*

p.-plate column (tower) *(Destillation)* Siebbodenkolonne f

p. roll Lochwalze f

p. tray *(Destillation)* Siebboden m, gelochter Austauschboden m

perforation Perforation f, Perforieren n, Perforierung f; Sieblochung f, Bohrung f, Lochwerte mpl, Öffnung f *(eines Siebes)*

perforator Perforationsgerät n

performance *(Tech)* Leistung f, Arbeitsleistung f, Gebrauchsleistung f

p. coefficient Leistungszahl f

p. curve Leistungskurve f, Leistungskennlinie f

p. number Leistungszahl f, LZ

performic acid s. peroxyformic acid

perfume / to parfümieren

perfume Parfüm n, Riechstoff m, Geruchsstoff m

p. carrier Dioftträger m

perfumed paper parfümiertes Papier n, Parfümpapier n

perfuse / to durchströmen, durchtränken; benetzen

pergamyn Pergamin[papier] n, Pergamyn n, hochsatiniertes Pergamentersatzpapier n

perhydrate s. peroxyhydrate

perhydrogenate / to perhydrieren

perhydrogenation Perhydrierung f

perhydrogenize / to s. to perhydrogenate

perhydrous coal wasserstoffreiche (perhydrierte) Kohle f, Kohle f mit erhöhtem Wasserstoffgehalt

peri acid $H_2NC_{10}H_6SO_3H$ Perisäure f, 1-Naphthylamin-8-sulfonsäure f

p. bridge Peribrücke f

p.-fused perikondensiert, perianelliert

p. fusion Perikondensation f, Perianellierung f

p. position peri-Stellung f, Peristellung f

periclas[it]e *(Min)* Periklas m *(Magnesiumoxid)*

pericline *(Min)* Periklin m

peridot *(Min)* Peridot m

peridotite *(Min)* Peridotit m

perilla alcohol Perillaalkohol m, 4-Isopropenyl-1-zyklohexen-1-ol n

p. aldehyde s. perillaldehyde

p. oil Perillaöl n *(von Perilla frutescens (L.)Britt. und P. arguta Benth.)*

perillaldehyde Perillaldehyd m, 4-Isopropenyl-1-zyklohexen-1-karbonal n

perillic acid Perillasäure f, 4-Isopropenyl-1-zyklohexen-1-karbonsäure f

p. alcohol s. perilla alcohol

perillyl alcohol s. perilla alcohol

p. aldehyde s. perillaldehyde

period Periode f *(z.B. im Periodensystem)*; Periode f, Dauer f, Zeitabschnitt m, Zeit f

p. of average life *(phys Ch)* mittlere Lebensdauer f

p. of drying Trocknungsabschnitt m

p. of half-change, p. of half-life *(phys Ch)* Halbwert[s]zeit f

p. of induction Induktionsperiode f; s. p. of latency

p. of latency Latenzzeit f

p. of retention *(Pap)* Durchgangszeit f, Verweilzeit f

periodate Perjodat n; $M^I JO_4$ Perjodat n, Metaperjodat n; $M^I_5 JO_6$ Orthoperjodat n

periodic acid Perjodsäure f, Überjodsäure f; HJO_4 Perjodsäure f, Metaperjodsäure f; $H_5 JO_6$ Orthoperjodsäure f

periodic arrangement [of the elements] periodische (natürliche) Anordnung f der Elemente

p. chart [of the elements] s. p. table [of the elements]

p. function periodische Funktion f

p. kiln periodischer (periodisch arbeitender, intermittierender) Ofen (Brennofen) m

p. law Periodengesetz n, Gesetz n der Periodizität f, *(bisher auch)* periodisches Gesetz n [der Elemente]

p. property periodische (sich periodisch wiederholende) Eigenschaft f *(der Elemente)*

p. reaction periodische (rhythmische) Reaktion f (Fällungsreaktion, Niederschlagsreaktion) f

p. system [of the elements] Periodensystem *n* [der Elemente], PSE *n*, natürliches System *n* [der Elemente], *(bisher auch)* periodisches System *n* [der Elemente]
p. table [of the elements] Tabelle (Tafel) *f* des Periodensystems [der Elemente], *(bisher auch)* periodische Tabelle *f*, Tabelle *f* des periodischen Systems; *s.* p. system
periodicity Periodizität *f*, periodischer Charakter *m*, periodische Wiederholung *f*
peripheral port Öffnung *f* am Umfang
p. pressure Druck *m* am Umfang
p. solids-discharge centrifuge Zentrifuge *f* mit Umfangsaustrag, Zentrifuge *f* mit seitlicher Entleerung
p. speed Umfangsgeschwindigkeit *f*
periphery Umfang *m*, Peripherie *f*
perishability Verderblichkeit *f*
perishable [leicht] verderblich
peristerite *(Min)* Peristerit *m* *(Natriumalumotrisilikat)*
peritectic peritektisch
p. point peritektischer Punkt *m*
p. system Peritektikum *n*
p. temperature peritektische Temperatur *f*
Perkin condensation (reaction) Perkin-Reaktion *f*, Perkinsche Reaktion (Synthese) *f*, Perkin-Kondensation *f*
Perkin's mauve (purple, violet) Perkin[sche]s Mauve[in] *n*, Perkin-Violett *n*
perlite *(Min)* Perlit *m*, Perlstein *m*; *(Met)* Perlit *m* *(lamellares Aggregat aus Ferrit und Zementit mit perlmutterartigem Glanz)*
perlitic *(Min, Met)* perlitisch
permanent permanent, bleibend
p. blue Permanentblau *n*, Waschblau *n*, Neublau *n*
p. compression set Druckverformungsrest *m*, Zusammendrückungsrest *m*, [bleibende] Druckverformung *f*, bleibende Verformung *f* nach Druckbelastung (Druckeinwirkung), Formänderungsrest *m* bei Druckbeanspruchung
p. deformation Verformungsrest *m*, bleibende Verformung (Deformation, Gestaltsänderung) *f*, Formänderungsrest *m*
p. elongation *s.* p. set at elongation
p. gas permanentes Gas *n*
p. green Permanentgrün *n* *(Mischpigment aus Bariumsulfat und Chromoxidhydratgrün)*
p. hair dye chemisch wirkende Haarfarbe *f*
p. hardness permanente (bleibende) Härte *f*, Nichtkarbonathärte *f*, Permanenthärte *f*
p. magnet Dauermagnet *m*, Permanentmagnet *m*
p. metal mould metallische Dauergießform (Dauerform, Form) *f*, Metallform *f*, Kokille *f*
p. mould Dauer[gieß]form *f*, *(meist)* metallische Dauergießform (Dauerform, Form) *f*, Kokille *f*
p.-mould casting Dauerformguß *m*, [normales] Kokillengießen *n*, [normaler] Kokillenguß *m*
p.-mould casting process Dauerformgießver-

fahren *n*, Dauerformgußverfahren *n*, [normales] Kokillengußverfahren *n*
p. oil nichttrocknendes Öl *n*
p. press *(Text)* Permanentappretur *f*, Permanentausrüstung *f*
p. red Permanentrot *n* *(ein Azofarbstoff)*
p. set Verformungsrest *m*, bleibende Verformung (Deformation, Gestaltsänderung) *f*, Formänderungsrest *m*, *(i.e.S.)* Dehnungsrest *m*, Zugverformungsrest *m*, Formänderungsrest *m* bei Dehnungsbeanspruchung; *(Text)* permanente Fixierung *f*, Dauerfixierung *f*
p. set at elongation Dehnungsrest *m*, Zugverformungsrest *m*, Formänderungsrest *m* bei Dehnungsbeanspruchung
p.-set test in compression Messung *f* des Formänderungsrestes bei Druckbeanspruchung
p.-set test in tension Messung *f* des Formänderungsrestes bei Dehnungsbeanspruchung
p. waving *(Kosmet)* Dauerwellung *f*, Dauerwellherstellung *f*
p.-waving solution Dauerwellflüssigkeit *f*
p. white Permanentweiß *n*, Blanc fixe *n*, Blankfix *n*, Barytweiß *n* *(gefälltes Bariumsulfat)*
permanganate M^IMnO_4 Permanganat *n*, Manganat(VII) *n*
p. ion Permanganat-Ion *n*
p. number Permanganatzahl *f*
p. oxidation Permanganatoxydation *f*
p. test Permanganatprobe *f*, Prüfung *f* mit Permanganat
p. titration Titration *f* mit Permanganat, manganometrische Titration *f*
permanganatometry *s.* permanganometry
permanganic acid $HMnO_4$ Permangansäure *f*, Übermangansäure *f*, Mangan(VII)-säure *f*
p. anhydride Mn_2O_7 Manganheptoxid *n*, Mangan(VII)-oxid *n*
permanganometry Permanganometrie *f*, Manganometrie *f*
permanganyl —MnO_3 Permanganyl *n*
permeability Permeabilität *f*, Durchlässigkeit *f*; *(Met)* Gasdurchlässigkeit *f*, Durchlässigkeit *f* *(eines Formsandes)*
p. number *(Met)* Gasdurchlässigkeitszahl *f*, Gasdurchlässigkeit *f*, Gd *(eines Formsandes)*
p. test *(Met)* Gasdurchlässigkeitsprüfung *f* *(bei Formsand)*
p. to air Luftdurchlässigkeit *f*
p. to gas[es] Gasdurchlässigkeit *f*
p. to vapour Wasserdampfdurchlässigkeit *f*
permeable permeabel, durchlässig
permeate / to permeieren, durchdringen
permeation Permeation *f*, Durchdringung *f*
permeativity Permeiervermögen *n*, Durchdringungsvermögen *n*; Eindringfähigkeit *f* *(z. B. von Pflanzenschutzmitteln)*
permethanoic acid *s.* peroxyformic acid
permissible [explosive] Sicherheitssprengstoff *m*
permitted zulässig

p. load Höchstlast f
p. transition (phys Ch) erlaubter Übergang m
permittivity Dielektrizitätskonstante f, DK
pernigraniline Pernigranilin n (Farbstoff)
pernitrate s. peroxynitrate
pernitric acid s. peroxynitric acid
pernitride Pernitrid n
pernitrite s. peroxynitrite
pernitrous acid s. peroxynitrous acid
perofskite s. perovskite
perosmic acid s. p. oxide
p.[-acid] anhydride s. p. oxide
p. oxide Osmiumperoxid n, Überosmiumsäure-anhydrid n, Überosmiumsäure f, Perosmium-säure f, Osmiumsäure f, (veraltet für) Os-miumtetroxid n, Osmium(VIII)-oxid n
perovskite, perowskite (Min) Perowskit m (Kalziummetatitanat)
peroxidase Peroxydase f (eine Oxydoreduktase, die die Oxydation von Substraten unter Verwendung von Wasserstoffperoxid katalysiert)
p. test Peroxydasetest m, Peroxydaseprobe f
peroxidate / to s. to peroxidize
peroxidatic peroxydatisch, Peroxydase...
peroxidation Peroxydieren n, Peroxydierung f, Peroxydation f; Epoxydieren n, Epoxydierung f, Epoxydation f
peroxide Peroxid n; H_2O_2 Wasserstoffperoxid n, Hydrogenperoxid n
p. bleaching Peroxidbleiche f
p. bleaching solution Peroxidbleichlösung f
p. crosslinking peroxidische Vernetzung f, Peroxidvernetzung f
p. cure (Gum) Peroxidvulkanisation f, Vulkanisation f mit Peroxid
p.-cured mit Peroxid vulkanisiert
p. effect Peroxideffekt m
p.-initiated addition durch Peroxide initiierte (ausgelöste) Addition f, Anti-Markownikow-Addition f, Addition f im Anti-Markownikow-Sinn
p. link Peroxidbindung f
p. number Peroxidzahl f, Peroxidwert m
p. rancidity Peroxid[ranz]igkeit f
p. value s. p. number
p. vulcanization s. p. cure
peroxidic peroxidisch, Peroxid...
peroxidize / to peroxydieren; epoxydieren
peroxo acid Peroxosäure f
p. compound Peroxoverbindung f
p. derivative Peroxoderivat n
p. group Peroxogruppe f
p. salt Peroxosalz n
peroxoborate Peroxoborat n
peroxoboric acid Peroxoborsäure f
peroxocarbonate Peroxokarbonat n
peroxocarbonic acid Peroxokohlensäure f
peroxochromate Peroxochromat n
peroxocobaltic ammine Peroxokobalt(III)-ammin n
peroxohydrate Peroxohydrat n
peroxonitrate $M^I[NO_4]$ Peroxonitrat n

peroxonitric acid HNO_4 Peroxosalpetersäure f
peroxonitrite $M^I[OON = O]$ Peroxonitrit n
peroxonitrous acid HOON=O peroxosalpetrige Säure f
peroxophosphate Peroxophosphat n; $M^I_3[PO_5]$ Peroxo[mono]phosphat n
peroxophosphoric acid Peroxophosphorsäure f; H_3PO_5 Peroxo[mono]phosphorsäure f
peroxosulphate Peroxosulfat n; $M^I_2SO_5$ Peroxo[mono]sulfat n
peroxosulphuric acid Peroxoschwefelsäure f; H_2SO_5 Peroxo[mono]schwefelsäure f
peroxouranate $M^I_2UO_6$ Peroxouranat n
peroxovanadate Peroxovanadat n
peroxovanadic acid Peroxovanadinsäure f
peroxy acid Peroxysäure f; (Komplexchemie) Peroxysäure f, (veraltet für) Peroxosäure f
p. compound Peroxyverbindung f; (Komplexchemie) Peroxyverbindung f, (veraltet für) Peroxoverbindung f
p. ester Peroxyester m
p. group Peroxygruppe f; (Komplexchemie) Peroxygruppe f, (veraltet für) Peroxogruppe f
peroxyacetic acid $CH_3CO \cdot O \cdot OH$ Peroxy[mono]essigsäure f, Peressigsäure f, Peräthansäure f, Azetylwasserstoffperoxid n, Azetylhydroperoxid n
peroxybenzoic acid $C_6H_5CO \cdot O \cdot OH$ Peroxy[mono]benzoesäure f, Perbenzoesäure f, Benzopersäure f, Perbenzolkarbonsäure f, Benzolhydroperoxid n
peroxyborate Per[oxy]borat n, (veraltet für) Peroxoborat n; Boratperoxyhydrat n, Perborat n, (veraltet für) Boratperoxohydrat n
peroxyboric acid Per[oxy]borsäure f, Überborsäure f, (veraltet für) Peroxoborsäure f
peroxycarbonate Per[oxy]karbonat n, (veraltet für) Peroxokarbonat n; Karbonatperoxyhydrat n, Perkarbonat n, (veraltet für) Karbonatperoxohydrat n
peroxycarbonic acid Per[oxy]kohlensäure f, Überkohlensäure f, (veraltet für) Peroxokohlensäure f
peroxycarboxylic acid Peroxykarbonsäure f
peroxychromate Per[oxy]chromat n, (veraltet für) Peroxochromat n
peroxychromic acid Per[oxy]chromsäure f, Überchromsäure f, (veraltet für) Peroxochromsäure f
peroxyformic acid HCO·O·OH Peroxyameisensäure f, Perameisensäure f, Permethansäure f, Formylhydroperoxid n
peroxyhydrate Per[oxy]hydrat n, (veraltet für) Peroxohydrat n
peroxyhydrated peroxycarbonate Karbonatperoxyhydrat n, (veraltet für) Karbonatperoxohydrat n
peroxynitrate $M^I[NO_4]$ Per[oxy]nitrat n, (veraltet für) Peroxonitrat n
peroxynitric acid HNO_4 Per[oxy]salpetersäure f, Übersalpetersäure f, (veraltet für) Peroxosalpetersäure f

peroxynitrite $M^I[OON=O]$ Per[oxy]nitrit n, *(veraltet für)* Peroxonitrit n

peroxynitrous acid $HOON=O$ per[oxy]salpetrige Säure f, übersalpetrige Säure f, *(veraltet für)* peroxosalpetrige Säure f

peroxyphosphate Per[oxy]phosphat n, *(veraltet für)* Peroxophosphat n; Phosphatperoxyhydrat n, Perphosphat n, *(veraltet für)* Phosphatperoxohydrat n

peroxyphosphoric acid Per[oxy]phosphorsäure f, Überphosphorsäure f, *(veraltet für)* Peroxophosphorsäure f

peroxysulphate Peroxysulfat n, *(veraltet für)* Peroxosulfat n; $M^I_2S_2O_8$ Per[oxy]disulfat n, Persulfat n, *(veraltet für)* Peroxodisulfat n

peroxysulphuric acid Peroxyschwefelsäure f, *(veraltet für)* Peroxoschwefelsäure f; $H_2S_2O_8$ Per[oxy]dischwefelsäure f, Perschwefelsäure f, Überschwefelsäure f, *(veraltet für)* Peroxodischwefelsäure f

peroxyuranate $M^I_2UO_6$ Per[oxy]uranat n, *(veraltet für)* Peroxouranat n

peroxyvanadate Per[oxy]vanadat n, *(veraltet für)* Peroxovanadat n

peroxyvanadic acid Per[oxy]vanadinsäure f, *(veraltet für)* Peroxovanadinsäure f

perpendicular position Senkrechtstellung f

perpetual motion of the first kind *(phys Ch)* Perpetuum mobile n erster Art

p. motion of the second kind *(phys Ch)* Perpetuum mobile n zweiter Art

perphosphate s. peroxyphosphate

perphosphoric acid s. peroxophosphoric acid

perrhenate Perrhenat n, Rhenat(VII) n

perrhenic acid $HReO_4$ Perrheniumsäure f, Überrheniumsäure f, Rhenium(VII)-säure f, Rheniumsäure f

Perrin process Perrin-Verfahren n, Schlackenreaktionsverfahren n, Waschverfahren n

persalt Persalz n

Persian ammoniac Ammoniakgummi[harz] n *(von Dorema ammoniacum Don und Verwandten)*

P. bark Cascara-sagrada-Rinde f, Kaskararinde f, Sagrada[rinde] f, Amerikanische Faulbaumrinde f *(von Rhamnus purshianus DC.)*

P. berry Persische Gelbbeere f

P. red $Pb(OH)_2 \cdot PbCrO_4$ Persischrot n, Chromrot n, Chromzinnober m

P. tragacanth Persischer Tragant m *(von Astragalus-Arten)*

persic oil Pfirsichkernöl n

persist / to beständig (stabil, persistent) sein *(z.B. von einem Schädlingsbekämpfungsmittel)*; bestehen bleiben

persistence *(phys Ch)* Persistenz f; Beständigkeit f, Stabilität f, Persistenz f *(z.B. eines Schädlingsbekämpfungsmittels)*

p. of a foam Schaumbeständigkeit f, Schaumstabilität f, Schaumdauer f, Beständigkeit (Haltbarkeit, Lebensdauer, Stabilität) f eines Schaums

persistent beständig, stabil, persistent *(z.B. Schädlingsbekämpfungsmittel)*; anhaltend, nachhaltig

p. lines *(Spektralanalyse)* letzte (beständige) Linien fpl, Restlinien fpl, Grundlinien fpl, Nachweislinien fpl

personal error persönlicher (subjektiver) Fehler m

Persoz's reagent *(Text)* Persoz-Reagens n

perspective formula perspektivische Formel f

perspiration check Schweiß[hemmungs]mittel n, Antischweißmittel n, Antitranspirationsmittel n, Antiperspirant n, Antihidrotikum n, schweißhemmendes (schweißhinderndes, antitranspirierendes) Mittel n

p. fastness *(Text)* Schweißechtheit f

p. resistance *(Gerb)* Schweißechtheit f *(des Leders)*

p.-resistant *(Gerb)* schweißecht

persulphate s. peroxysulphate

persulphide Persulfid n, Übersulfid n

persulphuric acid s. peroxysulphuric acid

pertechnate M^ITcO_4 Pertechnetat n, Technetat(VII) n

pertechnetic acid $HTcO_4$ Pertechnetiumsäure f, Technetium(VII)-säure f

perthiocarbonate $M^I_2CS_4$ Perthiokarbonat n

perthiocarbonic acid H_2CS_4 Perthiokohlensäure f

perthite *(Min)* Perthit m

Peru balsam Perubalsam m, Peruanischer (Indischer) Balsam m *(von Myroxylon balsamum (L.)Harms var. pereirae)*

peruranate s. peroxyuranate

Peruvian balsam s. Peru balsam

pervanadate s. peroxyvanadate

pervanadic acid s. peroxyvanadic acid

perylene Perylen n, peri-Dinaphthylen n

p. ring system Perylenringsystem n

perylenecarboxylic acid Perylenkarbonsäure f

perylenequinone Perylenchinon n

pest Schädling m, Schaderreger m, Schadorganismus m

p. control Schädlingsbekämpfung f

p.-control material s. pesticide

pesticidal chemical s. pesticide chemical

pesticide Pestizid n, Schädlingsbekämpfungsmittel n

p. chemical chemisches Schädlingsbekämpfungsmittel n

p.-chemical petition Anmeldung f eines Schädlingsbekämpfungsmittels zur Zulassung *(oder)* Anerkennung

p. residue Pestizidrückstand m, Schädlingsbekämpfungsmittelrückstand m

pestle Stößel m, Pistill n, Mörserkeule f

pet s. petrolatum

p. cock Kondenswasserhahn m, Kompressionshahn m, Probierhahn m; Wasserablaßhahn m

petalite Petalit m *(Lithiummineral)*

Petersen process Petersen-Verfahren n, Petersen-Turmverfahren n *(zur Schwefelsäureherstellung)*

pethidine Pethidin *n*, 1-Methyl-4-phenylpiperi-din-4-karbonsäureäthylester *m*
p. hydrochloride Pethidinhydrochlorid *n*
PETN *s.* pentaerythritol tetranitrate
Petri [culture] dish Petrischale *f*
Petri dish bottom Petri-Unterschale *f*
Petri dish box Petrischalenbüchse *f*, Sterilisier-büchse *f*
Petri dish top Deckel *m* (Oberschale *f*) der Petrischale
Petri plate *s.* Petri dish
petrifaction Petrifikation *f*, Versteinerung *f*
petrify / to petrifizieren, versteinern
petrochemical petrochemisch, gesteinschemisch; petrolchemisch, *(fälschlich)* petrochemisch
petrochemical Petrolchemikalie *f*, Erdölchemikalie *f*, chemisches Produkt *n* auf Erdölbasis, *(fälschlich)* Petrochemikalie *f*
petrochemistry Petrochemie *f*, Gesteinschemie *f*; Petrolchemie *f*, Erdölchemie *f*, *(fälschlich)* Petrochemie *f*
petrofabric *(Geol)* Gefüge *n*
p. analysis *(Geol)* Gefügeanalyse *f*
petrofabrics *(Geol)* Gefügekunde *f*
petrogenesis Petrogenese *f*, Petrogenesis *f*
petrogenetic petrogenetisch
petrogenic petrogen; petrogenetisch
petrographic[al] petrografisch
petrography Petrografie *f*, Gesteinskunde *f*
petrol Benzin *n (besonders als Vergaserkraftstoff)*
p. vapour Benzindampf *m*
petrolatum Petrolat[um] *n*, Rohvaseline *f*, Roh-vaselin *n*
p. tank Tank *m* für Petrolatum
petroleum Erdöl *n*
p. asphalt Erdölasphalt *m*, Petrolasphalt *m*
p.-bearing *s.* petroliferous
p. benzin Petroleumbenzin *n*
p. ceresin Erdölzeresin *n*
p. coke Erdölkoks *m*, Petrolkoks *m*
p.-derived aus Erdöl erzeugt
p. distillate Erdöldestillat *n*
p. distillation Erdöldestillation *f*
p. distillation residue Erdöldestillationsrück-stand *m*
p. ether Petrol[eum]äther *m (Siedebereich 40 bis 70 °C)*; Leichtbenzin *n*, leichtes Benzin *n (Siedebereich 20 bis 135 °C)*
p. fraction Erdölfraktion *f*
p. fuel oil Heizöl *n* auf Erdölbasis
p. gas Erdölgas *n*
p. genesis Erdölgenesis *f*, Erdölentstehung *f*, Erdölbildung *f*
p. hydrocarbon Erdölkohlenwasserstoff *m*
p. industry Erdölindustrie *f*
p. jelly *s.* petrolatum
p. lubricating oil Schmieröl *n* auf Erdölbasis
p. naphtha Erdölschwerbenzin *n*, Petrolnaphtha *n(f)*
p. oil Öl *n* aus Erdöl, Öl *n* auf Erdölbasis

p. pitch Erdölpech *n*, Petrolpech *n*
p. product Erdöl[folge]produkt *n*
p. refining Erdölverarbeitung *f*
p. residue Erdölrückstand *m*, Erdölresiduum *n*
p. resin Erdölharz *n*, Petrolharz *n*
p. spirit[s] Lösungsbenzin *n*
p. tar Erdölteer *m*
p. wash Petroleumemulsion *f (zur Schädlings-bekämpfung)*
p. wax Erdölwachs *n*, Erdölparaffin *n*, Paraffin-wachs *n* aus Erdöl
p. well Erdölbohrloch *n*, Erdölbohrung *f*, Ölbohrloch *n*, Ölbohrung *f*, Ölbrunnen *m*; Erdölquelle *f*, Ölquelle *f*
p. white oil Weißöl *n (Petroleum der höchsten Reinheitsstufe)*
petroliferous [erd]ölführend, [erd]ölhaltig
petrologic[al] petrologisch
petrology Petrologie *f*
petroselinic acid $CH_3(CH_2)_{10}=CH(CH_2)_4COOH$ Petroselinsäure *f*, *cis*-6-Oktadezensäure *f*
petun[t]se, petun[t]ze Pe-tun-tse *m (chinesischer Feldspatpegmatit)*
petzite *(Min)* Petzit *m*
PF *s.* phenolformaldehyde
P.F.D. *s.* primary flash distillate
PFEP *s.* polyfluoroethylene propylene
Pfitzinger reaction Pfitzingersche Reaktion *f (zur Chinolinsynthese)*
Pfund series Pfund-Serie *f (fünfte Spektralserie des Wasserstoffatoms)*
pH concept pH-Begriff *m*
pH control pH-[Wert-]Regelung *f*; pH-Kontrolle *f*, Kontrolle (Überwachung) *f* des pH-Wertes
pH-control system pH-[Wert-]Regelsystem *n*
pH controller pH-Regler *m*, pH-Regelgerät *n*
pH decrease pH-Abnahme *f*, pH-Abfall *m*
pH determination pH-[Wert-]Bestimmung *f*, pH-[Wert-]Ermittlung *f*
pH-determination apparatus pH-Bestimmungs-gerät *n*, pH-Bestimmungsapparat *m*
pH drop *s.* pH decrease
pH indicator pH-Indikator *m*; pH-Anzeigegerät *n*
pH instrument *s.* pH meter
pH interval pH-Intervall *n*
pH measurement pH-[Wert-]Messung *f*
pH-measurement system *s.* pH-measuring system
pH-measuring system pH-[Wert-]Meßsystem *n*
pH meter pH-[Wert-]Messer *m*, pH-[Wert-]Meßgerät *n*, pH-Meter *n*
pH number *s.* pH value
pH range pH-Bereich *m*, pH-Gebiet *n*
pH regulator *(Flot)* pH-Wert-Regler *m*, Mittel *n* zur pH-Regelung, pH-regelnder Zusatz *m*
pH scale pH-Skale *f*
pH standard pH-Standard *m*
pH unit pH-Einheit *f*
pH value pH-Wert *m*, pH *m(n)*, pH-Zahl *f*, Wasserstoff[ionen]exponent *m*
phacolite *(Min)* Phakolith *m*

phene

phage Phage *m*, Bakteriophag[e] *m*
phanerocrystalline phanerokristallin
pharmaceutic[al] pharmazeutisch
pharmaceutic[al] Pharmazeutikum *n*, pharmazeutisches Mittel *n*, Arzneimittel *n*
  p. chemical pharmazeutische Chemikalie *f*
  p. chemist pharmazeutischer Chemiker *m*, Pharmakochemiker *m*; Apotheker *m*
  p. chemistry pharmazeutische Chemie *f*, Pharmakochemie *f*
  p. emulsion pharmazeutische Emulsion *f*
  p. industry pharmazeutische Industrie *f*, Arzneimittelindustrie *f*
  p. preparation pharmazeutisches Präparat *n*, Arzneipräparat *n*, Arzneizubereitung *f*
pharmaceutics Pharmazeutik *f*
pharmac[eut]ist Pharmazeut *m*; Apotheker *m*
pharmacodynamic pharmakodynamisch
pharmacodynamics Pharmakodynamik *f*, Pharmakodynamie *f*
pharmacognosia *s.* pharmacognosy
pharmacognostic pharmakognostisch
pharmacognosy Pharmakognosie *f*, Drogenkunde *f*
pharmacolite *(Min)* Pharmakolith *m* *(Kalziumhydrogenorthoarsenat)*
pharmacologia *s.* pharmacology
pharmacologic[al] pharmakologisch
pharmacology Pharmakologie *f*
pharmacon Pharmakon *n*
pharmacopeia, pharmacopoeia Pharmakopöe *f*, Arzneibuch *n*
  p. commission Arzneibuchkommission *f*
pharmacopoeial Pharmakopöe..., Arzneibuch..., offizinell
  to be of p. quality den Anforderungen des Arzneibuches entsprechen
pharmacosiderite *(Min)* Pharmakosiderit *m*, Würfelerz *n* *(Eisen(III)-hexahydroxidtriorthoarsenat)*
pharmacotherapeutic[al] pharmakotherapeutisch
pharmacotherapeutics Pharmakotherapeutik *f*
pharmacotherapy Pharmakotherapie *f*, medikamentöse Therapie *f*, Arzneibehandlung *f*
pharmacy Pharmazie *f*
phase Phase *f*
  p. angle Phasenwinkel *m*
  p. boundary Phasengrenze *f*
  p.-boundary potential Phasengrenzpotential *n*
  p. change Phasenänderung *f*, Phasenumwandlung *f*, Phasenübergang *m*
  p.-contrast microscope Phasenkontrastmikroskop *n*
  p. diagram Phasendiagramm *n*, Zustandsdiagramm *n*
  p. difference Phasendifferenz *f*, Phasenunterschied *m*
  p. equilibrium Phasengleichgewicht *n*, Gleichgewicht *n* der Phasen
  p. inversion Phasenumkehr *f*
  p.-inversion point Inversionspunkt *m*, Staupunkt *m* *(Belastung von Kolonnen)*

  p. relation Phasenbeziehung *f*
  p. rule Phasenregel *f*, Phasengesetz *n*
  p. rule of Gibbs Phasenregel *f* von Gibbs, Gibbssches Phasengesetz *n*
  p. separation Phasentrennung *f*
  p. shift Phasen[ver]schiebung *f*, Phasendrehung *f*
  p. stability Phasenstabilität *f*
  p. transition *s.* p. change
  p. velocity Phasengeschwindigkeit *f*
phellandral Phellandral *n*, 4-Isopropyl-1-zyklohexen-1-karbonal *n*
phellandrene Phellandren *n* *(ein optisch aktives monozyklisches Terpen)*
phellonic acid Phellonsäure *f*, 22-Hydroxydokosansäure *f*
phenacetin Phenazetin *n*, 4-Äthoxyazetanilid *n*
phenaceturic acid $C_6H_5CH_2CONHCH_2COOH$ Phen[yl]azetursäure *f*, Phenazetylglykokoll *n*, Azetylphenylglyzin *n*
phenacetylurea *s.* phenylacetylurea
phenacite *s.* phenakite
phenacyl $C_6H_5COCH_2-$ Phenazyl *n*
  p. alcohol $C_6H_5COCH_2OH$ Phenazylalkohol *m*, α-Hydroxyazetophenon *n*
  p. chloride $C_6H_5COCH_2Cl$ Phenazylchlorid *n*, α-Chlorazetophenon *n*, ω-Chlorazetophenon *n*
phenacylidene $C_6H_5COCH=$ Phenazyliden *n*
phenakite *(Min)* Phenakit *m* *(Berylliumorthosilikat)*
phenanthraquinone *s.* phenanthrenequinone
phenanthrene Phenanthren *n* *(ein aromatischer Kohlenwasserstoff)*
  p. ring system Phenanthrenringsystem *n*
phenanthrenequinone Phenanthrenchinon *n* *(ein Chinon der Phenanthrenreihe), (i.e.S.)* Phenanthrenchinon *n*, 9,10-Phenanthrenchinon *n*, 9,10-Dihydro-9,10-dioxophenanthren *n*
phenanthridine Phenanthridin *n*, 3,4-Benzochinolin *n*
phenanthrin *s.* phenanthrene
phenanthroindolizidine alkaloid Phenanthroindolizidinalkaloid *n*
phenanthrol Phenanthrol *n*, Hydroxyphenanthren *n*
phenanthroline Phenanthrolin *n* *(eine vom Phenanthren abgeleitete heterozyklische Verbindung)*
phenanthroquinolizidine alkaloid Phenanthrochinolizidinalkaloid *n*
phenanthryl Phenanthryl *n* *(Atomgruppe)*
phenanthrylene Phenanthrylen *n* *(Atomgruppe)*
phenarsazine Phenarsazin *n*, Phenazarsin *n*
  p. chloride Phenarsazinchlorid *n*, Diphenylaminchlorarsin *n*, 10-Chlor-5,10-dihydrophenarsazin *n*
phenarsazinic acid Phenarsazinsäure *f*, Phenazarsinsäure *f*, 5,10-Dihydro-10-hydroxyphenarsazin-10-oxid *n*
phenate *s.* phenolate

phenazarsine s. phenarsazine
phenazarsinic acid s. phenarsazinic acid
phenazine Phenazin n, Dibenzopyrazin n, Azophenylen n
phenazocine Phenazozin n, 2'-Hydroxy-2-phenyl-äthyl-5,9-dimethyl-6,7-benzomorphan n
phenazone Phenazon n, 1-Phenyl-2,3-dimethyl-5-pyrazolon n
phene Phen n, *(veraltet für)* Benzol n
phenenyl $C_6H_3\equiv$ Phenenyl n
phenethicillin Phenethizillin n, α-Phenoxyäthyl-penizillin n
phenethyl $C_6H_5CH_2CH_2-$ Phenäthyl n
  p. alcohol Phenäthylalkohol m, β-Phenyläthyl-alkohol m, 2-Phenyläthanol n, Benzylkarbinol n
phenethylamine s. phenylethylamine
phenetidine Phenetidin n, Äthoxyanilin n, Amino-phenetol n, Aminophenoläthyläther m
phenetidino $C_2H_5OC_6H_4NH-$ Phenetidino...
phenetole $C_6H_5OC_2H_5$ Phenetol n, Äthoxybenzol n, Äthylphenyläther m
phenetyl $C_2H_5OC_6H_4-$ Phenetyl n, Äthoxyphenyl n
phenformin Phenformin n, N'-Phenyläthyldiguanid n
phengite *(Min)* Phengit m
phenglutarimide Phenglutarimid n, α-2-Diäthyl-aminoäthyl-α-phenylglutarimid n
phenic acid C_6H_5OH Phen[yl]säure f, *(veraltet für)* Phenol n
phenicin s. phoenicine
phenidone $C_9H_{10}N_2O$ Phenidon n, 1-Phenyl-3-pyrazolidon n
phenindamine Phenindamin n *(Antihistaminikum)*
phenindione Phenindion n, 2-Phenyl-1,3-indan-dion n
pheniramine Pheniramin n, 3-Diäthylamino-1-phenyl-1,2'-pyridylpropan n
phenmetrazine Phenmetrazin n, 2-Phenyl-3-me-thyltetrahydro-1,4-oxazin n
phenobarbital Phenobarbital n, Phenobarbiton n, 5-Phenyl-5-äthylbarbitursäure f
  p. sodium Phenobarbitalnatrium n
phenobarbitone s. phenobarbital
phenocryst *(Geol)* Einsprengling m
phenol Phenol n, *(i.e.S.)* C_6H_5OH Phenol n, Monohydroxybenzol n
  p. acid s. phenolic acid
  p. coefficient Phenolkoeffizient m
  p.-furfural resin Phenolfurfuralharz n
  p. nucleus s. p. ring
  p. oxidase Phenol[oxyd]ase f
  p. red s. phenolsulphonephthalein ·
  p. ring Phenolring m, Phenolkern m
  p. solution Phenollösung f
  p. test Prüfung f auf Phenol; Phenolprobe f, Prüfung f mit Phenol
  p. trinitrate $C_6H_2(NO_2)_3OH$ Trinitrophenol n
phenolase s. phenol oxidase
phenolate / to phenolisieren

phenolate Phen[ol]at n, Phenoxid n
  p. process Phenolatverfahren n
phenolformaldehyde Phenolformaldehyd m, Phenolformaldehydkondensat n, Phenol-Form-aldehyd-Kondensationsprodukt n, PF
  p. condensation Phenolformaldehydkondensa-tion f
  p. resin Phenolformaldehydharz n, PF-Harz n
phenolic phenolisch, Phenol...
phenolic Phenolharz n; Phenoplast m, Phenolharz-kunststoff m
  p. acid Phenolsäure f
  p. adhesive Phenolharzklebstoff m, Phenol-kleber m, Phenol[harz]kitt m
  p. alcohol Phenolalkohol m
  p. aldehyde Phenolaldehyd m
  p. cement s. p. adhesive
  p. ester Phenolester m
  p. ether Phenoläther m
  p. foam Phenol[harz]schaum m, Phenolharz-schaumstoff m
  p. hydroxyl phenolisches Hydroxyl (OH) n, phenolische Hydroxylgruppe f, Phenolhydroxyl n
  p. laminate Phenolharzlaminat n
  p. moulding compound Phenolharzpreßmasse f
  p. plastic s. phenoplast
  p. resin Phenolharz n
  p.-resin foam s. p. foam
  p. taste Phenolgeschmack m
  p. varnish Phenolharzlack m
  p. wastes phenolhaltiges Abwasser n, Phenolab-wasser n
phenolize / to s. to phenolate
phenolphthalein Phenolphthalein n, 3,3-Bis-(p-hy-droxyphenyl)phthalid n
phenolsulphonephthalein Phenolsulfonphthalein n, Phenolrot n
phenolsulphonic acid Hydroxybenzolsulfonsäure f
phenolsulphonphthalein s. phenolsulphone-phthalein
phenomenon Phänomen n, Erscheinung f
phenonium ion Phenoniumion n, Benzeniumion n
phenoplast Phenoplast m, Phenolharzkunststoff m
phenosafranine Phenosafranin[chlorid] n *(ein Desensibilisator)*
phenoxide s. phenolate
phenoxy C_6H_5O- Phenoxy...
  p. group Phenoxygruppe f, Phenoxyrest m
  p. resin Phenoxyharz n
phenoxyacetic acid $C_6H_5O \cdot CH_2COOH$ Phenoxyes-sigsäure f, O-Phenylglykolsäure f
phenoxymethylpenicillin Phenoxymethylpenizillin n, Penizillin V n
phenyl C_6H_5- Phenyl n
  p. acetate $CH_3COOC_6H_5$ Phenylazetat n, Es-sigsäurephenylester m
  p. benzoate $C_6H_5COOC_6H_5$ Phenylbenzoat n, Benzoesäurephenylester m, Phenylbenzolkarb-oxylat n
  p. bromide C_6H_5Br Phenylbromid n, *(veraltet für)* Brombenzol n

p. chloride C_6H_5Cl Phenylchlorid *n*, *(veraltet für)* Chlorbenzol *n*

p. disulphide $C_6H_5SSC_6H_5$ Phenyldisulfid *n*, Diphenyldisulfid *n*, Phenyldithiobenzol *n*

p. ester Phenylester *m*

p. ether $C_6H_5OC_6H_5$ Phenyläther *m*, Diphenyläther *m*, Diphenyloxid *n*

p. gamma acid $HOC_{10}H_5(NHC_6H_5)(SO_3H)$ Phenyl-γ-Säure *f*, 7-Phenylamino-1-naphthol-3-sulfonsäure *f*

p. group Phenylgruppe *f*, Phenylrest *m*

p. hydrate (hydroxide) C_6H_5OH Phenylhydrat *n*, *(veraltet für)* Phenol *n*

p. isocyanide *s.* phenylcarbylamine

p. J acid $HOC_{10}H_5(NHC_6H_5)SO_3H$ Phenyl-J-Säure *f*, Phenyl-I-Säure *f*, 2-Phenylamino-5-naphthol-7-sulfonsäure *f*

p. ketone $C_6H_5COC_6H_5$ Phenylketon *n*, Diphenylketon *n*, Benzophenon *n*, Benzoylbenzol *n*

p. peri acid $C_{16}H_{13}NO_3S$ Phenylperisäure *f*, Phenyl-1-naphthylamin-8-sulfonsäure *f*, 1-Phenylaminonaphthalin-8-sulfonsäure *f*

p. radical Phenylgruppe *f*, Phenylrest *m*, Phenylradikal *n*; [freies] Phenylradikal *n*

p. salicylate $C_6H_4(OH)COOC_6H_5$ Phenylsalizylat *n*, Salizylsäurephenylester *m*, Salol *n*

p. silicone Phenylsilikon *n*

p. silicone fluid Phenylsilikonflüssigkeit *f*, Phenylsilikonöl *n*

N-phenylacetamide $C_6H_5NHCOCH_3$ *N*-Phenylazetamid *n*, Azetanilid *n*

phenylacetic acid $C_6H_5CH_2COOH$ Phenylessigsäure *f*, 1'-Tolylsäure *f*, 1'-Toluolkarbonsäure *f*

phenylaceturic acid *s.* phenaceturic acid

phenylacetyl $C_6H_5CH_2CO-$ Phenylazetyl *n*

phenylacetylene $C_6H_5C≡CH$ Phenylazetylen *n*, Azetylenylbenzol *n*, Äthinylbenzol *n*

phenylacetylglycine *s.* phenaceturic acid

phenylacetylurea Phen[yl]azetylharnstoff *m*

phenylacrylic acid Phenylakrylsäure *f*, *(i.e.S.)* $C_6H_5CH=CHCOOH$ Phenylakrylsäure *f*, *trans*-β-Phenylakrylsäure *f*, [gewöhnliche] Zimtsäure *f*, *trans*-Zimtsäure *f*, Zinnamsäure *f*, *trans*-3-Phenylpropensäure *f*

phenylalanine Phenylalanin *n*, *(i.e.S.)* *s.* β-phenylalanine

α-**phenylalanine** $CH_3C(NH_2)(C_6H_5)COOH$ α-Phenylalanin *n*, 2-Amino-2-phenylpropionsäure *f*

β-**phenylalanine** $C_6H_5CH_2CH(NH_2)COOH$ β-Phenylalanin *n*, 2-Aminohydrozimtsäure *f*, 2-Amino-3-phenylpropionsäure *f*

phenylalkane Phenylalkan *n*

phenylallyl alcohol Phenylallylalkohol *m*

phenylamine $C_6H_5NH_2$ Phenylamin *n*, Aminobenzol *n*, Anilin *n*

phenylate / to phenylieren

phenylation Phenylierung *f*

phenylazo $C_6H_5N=N-$ Phenylazo...

phenylbenzene $C_6H_5·C_6H_5$ Phenylbenzol *n*, *(veraltet für)* Biphenyl *n*, Diphenyl *n*

phenylbenzoylcarbinol $C_6H_5CH(OH)COC_6H_5$ Phenylbenzoylkarbinol *n*, Benzoylphenylkarbinol *n*, Benzoin *n*, 2-Hydroxy-2-phenylazetophenon *n*

phenylbenzylcarbinol $C_6H_5CH(OH)CH_2C_6H_5$ Phenylbenzylkarbinol *n*, Benzylphenylkarbinol *n*, 1,2-Diphenyläthanol *n*

phenylboric acid $C_6H_5B(OH)_2$ Phenylborsäure *f*, Benzolboronsäure *f*

phenylbutyric acid Phenylbuttersäure *f*, Phenylbutansäure *f*

phenylcarbamic acid $C_6H_5NHCOOH$ Phenylkarbaminsäure *f*, Phenylkarbamidsäure *f*, Karbanilsäure *f*

phenylcarbamide *s.* phenylurea

phenylcarbamido *s.* phenylureido

phenylcarbinol $C_6H_5CH_2OH$ Phenylkarbinol *n*, Benzylalkohol *m*

phenylcarbylamine C_6H_5NC Phenylkarbylamin *n*, Phenylisozyanid *n*

phenylchloroform $C_6H_5CCl_3$ Phenylchloroform *n*, α-Trichlortoluol *n*, ω-Trichlortoluol *n*, ω-Trichlormethylbenzol *n*

phenylchlorosilane Phenylchlorsilan *n*

phenylene $-C_6H_4-$ Phenylen *n*

p. blue Phenylenblau *n* *(ein Indaminfarbstoff)*

p. group Phenylengruppe *f*, Phenylenrest *m*

phenylenebisazo $-N=NC_6H_4N=N-$ Phenylenbisazo...

phenylenediamine $C_6H_4(NH_2)_2$ Phenylendiamin *n*, Diaminobenzol *n*

phenylenedisazo *s.* phenylenebisazo

phenylethane $C_6H_5C_2H_5$ Phenyläthan *n*, Äthylbenzol *n*

phenylethanol Phenyläthanol *n*, Phenyläthylalkohol *m*

phenylethyl $C_6H_5C_2H_4-$ Phenyläthyl..., *(i.e.S.)* $C_6H_5CH_2CH_2-$ Phenäthyl..., β-Phenyläthyl...

p. alcohol Phenyläthylalkohol *m*, Phenyläthanol *n*, *(i.e.S.)* Phenäthylalkohol *m*, β-Phenyläthylalkohol *m*, 2-Phenyläthanol *n*, Benzylkarbinol *n*

p. ether *s.* phenetole

phenylethylamine Phenyläthylamin *n*, Aminoäthylbenzol *n*

phenylethylene $C_6H_5·CH=CH_2$ Phenyläthylen *n*, Vinylbenzol *n*, Styrol *n*

phenylformic acid C_6H_5COOH Phenylameisensäure *f*, *(veraltet für)* Benzoesäure *f*, Benzolkarbonsäure *f*

phenylglycine $C_6H_5·NH·CH_2·COOH$ Phenylglyzin *n*, Phenylglykokoll *n*, Anilidoessigsäure *f*

phenylglyoxylic acid $C_6H_5COCOOH$ Phenylglyoxylsäure *f*, Benzoylameisensäure *f*

phenylhydrazine $C_6H_5NHNH_2$ Phenylhydrazin *n*

phenylhydrazone Phenylhydrazon *n* *(ein Substitutionsprodukt von Phenylhydrazin)*

phenylic acid *s.* phenic acid

phenylidene $C_6H_6=$ Phenyliden n, Zyklohexadienyliden n

phenylketonuria Phenylketonurie f

phenyllactic acid Phenylmilchsäure f, Hydroxyphenylpropansäure f

phenylmagnesium bromide C_6H_5MgBr Phenylmagnesiumbromid n

phenylmercuric acetate $CH_3 \cdot COOHg \cdot C_6H_5$ Phenylquecksilberazetat n, PMAS n, PMA n

phenylmercury urea Phenylquecksilberharnstoff m

phenylmethane $C_6H_5CH_3$ Phenylmethan n, Methylbenzol n, Toluol n

phenylnaphthylamine Phenylnaphthylamin n

phenylnitromethane $C_6H_5CH_2NO_2$ Phenylnitromethan n, ω-Nitrotoluol n

phenylphenazonium Phenylphenazonium n
p. salt Phenylphenazoniumsalz n

phenylphosphinic acid $C_6H_5PO_2H_2$ Phenylphosphinsäure f, prim-Phenylphosphinsäure f, Benzolphosphinsäure f, prim-Benzolphosphinsäure f

phenylphosphonic acid $C_6H_5PO_3H_2$ Phenylphosphonsäure f, Benzolphosphonsäure f

phenylpropiolic acid s. phenylpropynoic acid

phenylpropionic acid Phenylpropionsäure f, Phenylpropansäure f

phenylpropynoic acid $C_6H_5C≡CCOOH$ Phenylpropynsäure f, Phenylpropiolsäure f

phenylpyruvic acid $C_6H_5CH_2COCOOH$ Phenylbrenztraubensäure f, Benzylglyoxylsäure f

phenylsuccinic acid $HOOC \cdot CH(C_6H_5)CH_2 \cdot COOH$ Phenylbernsteinsäure f

phenylsulphamic acid $C_6H_5NHSO_3H$ Phenylsulfamidsäure f, Phenylsulfaminsäure f

phenylsulphamyl $C_6H_5NHSO_2-$ Phenylsulfam[o]yl n

phenylsulfonamido $C_6H_5SO_2NH-$ Phenylsulfonamido..., Phenylsulfonylamino...

phenylsulphonic acid Phenylsulfonsäure f, (veraltet für) Benzolsulfonsäure f, (i.e.S.) $C_6H_5SO_3H$ Phenyl[mono]sulfonsäure f, (veraltet für) Benzol[mono]sulfonsäure f

phenylsulphuric acid $C_6H_5OSO_3H$ Phenylschwefelsäure f

phenylurea $C_6H_5NHCONH_2$ Phenylharnstoff m, Phenylkarbamid n

phenylureido $C_6H_5NHCONH-$ Phenylureido..., Phenylkarbamido...

phenytoin Phenytoin n, 5,5-Diphenylhydantoin n

pheromone Pheromon n, Ektohormon n

pheron Pheron n, Träger m (kolloider Träger eines Ferments)

phillipite (Min) Phillipit m (Eisen(III)-kupfer(II)-sulfat)

Phillips process Phillips-Verfahren n (Mitteldruckpolymerisationsverfahren für Polyäthylen)

phillipsine, phillipsite (Min) Phillipsit m

philosopher's wool Lana philosophica f, Philosophenwolle f, (veraltet für) Zinkblumen fpl (weißes, wollartiges Zinkoxid)

phlobaphenes Phlobaphene npl (Oxydationsprodukte von Gerbstoffen)

phloem (Bio) Phloem n, Leitgewebe n

phlogistian s. phlogistonist

phlogisticated air (historisch) phlogistierte Luft f

phlogiston (historisch) Phlogiston n, brennbares Prinzip n, Feuerstoff m, Feuermaterie f
p. theory Phlogistontheorie f

phlogistonated muriatic acid (historisch) phlogistierte Muriatsäure f

phlogistonist Phlogistiker m (Anhänger der Phlogistontheorie)

phlogolite, phlogopite (Min) Phlogopit m

phlori[d]zin
$HOC_6H_4CH_2CH_2COC_6H_2(OH)_2O \cdot C_6H_{11}O_5$ Phloridzin n, Phlorrhizin n (Glukosid)

phloroglucin[ol] $C_6H_3(OH)_3$ Phlorogluzin n, 1,3,5-Trihydroxybenzol n

phlorrhizin s. phlori[d]zin

phloxin Phloxin n (ein Farbstoff)

Phoenician purple Antiker (Tyrischer, Byzantinischer) Purpur m, Purpur m der Alten, 6,6'-Dibromindigo m

phoenicin[e] Phönizin n, 2,2'-Dihydroxy-4,4'-ditoluchinon n

phoenicite, phoenicochroite s. phönicit

phönicit (Min) Phönizit m, Phönikochroit m (Blei(II)-oxidchromat)

phonolite, phonolyte (Min) Phonolith m, Klingstein m

phormium Neuseelandflachs m, Neuseeländer Flachs m, Phormium tenax J.R. et G. Forst

phorone $(CH_3)_2C=CHCOCH=C(CH_3)_2$ Phoron n

phosgenation Phosgenieren n, Phosgenierung f

phosgene $COCl_2$ Phosgen n, Kohlenoxidchlorid n, Karbonylchlorid n, Kohlensäuredichlorid n

phosgenite (Min) Phosgenit m, Bleihornerz n, Hornblei n

phosphamic acid $NH_2P(=O)(OH)_2$ Phosphamsäure f, Amidophosphorsäure f

phosphatase Phosphatase f (zu den Hydrolasen gehörendes Ferment)
p. test (Lebm) Phosphataseprobe f

phosphate / to phosphatieren (eine Phosphatschicht als Rostschutz aufbringen)

phosphate Phosphat n, (i.e.S.) $M^I_3PO_4$ Orthophosphat n
p. buffer Phosphatpuffer m
p.-buffered phosphatgepuffert
p. carrier (Landw) Phosphatträger m
p. fertilizer Phosphatdüngemittel n
p. furnace Phosphorofen m
p. glass Phosphatglas n
p. of lime (Landw) Fluorapatit m
p. plasticizer Phosphatweichmacher m
p. process Phosphat[entschwefelungs]verfahren n
p. rock Phosphaterz n

p. slag Kalziumsilikatschlacke f *(bei der Darstellung von Phosphor im Elektroofen)*
p. solubility *(Landw)* Phosphatlöslichkeit f
phosphatide Phosphatid n, Phospholip[o]id n
phosphatidylcholine Phosphatidylcholin n, Cholinphosphatid n
phosphatidylethanolamine Phosphatidyläthanolamin n, Äthanolaminphosphatid n, Phosphatidylkolamin n
phosphatidylserine Phosphatidylserin n, Serinphosphatid n
phosphating, phosphation Phosphatierung f *(Aufbringen einer Phosphatschicht als Rostschutz)*
phosphide Phosphid n
phosphine Phosphin n, Phosphor(III)-hydrid n, Phosphorwasserstoff m, *(i.e.S.)* PH_3 Monophosphin n, gasförmiger Phosphorwasserstoff m
phosphinic acid RPH=O(OH) Phosphinsäure f
phosphinous acid RPH(OH) Phosphensäure f
phosphite Phosphit n
phosphocalcite s. phosphorochalcite
phosphocerite *(Min)* Phosphozerit m, *(veraltet für)* Monazit m
phosphochalcite s. phosphorochalcite
phosphoglyceric acid Phosphoglyzerinsäure f
phosphokinase Phosphokinase f
phospholipid[e] s. phosphatide
phosphomolybdate $M'_3[P(Mo_3O_{10})_4]$ Tetrakistrimolybdatophosphat n, Dodekamolybdatophosphat n, Molybdatophosphat n
phosphomolybdic acid $H_3[P(Mo_3O_{10})_4]$ Tetrakistrimolybdatophosphorsäure f, Dodekamolybdatophosphorsäure f, Molybdatophosphorsäure f
phosphonic acid $RP=O(OH)_2$ Phosphonsäure f, Alkylphosphonsäure f
phosphonium Phosphonium n
p. base Phosphoniumbase f
p. bromide $[PH_4]Br$ Phosphoniumbromid n
p. chloride $[PH_4]Cl$ Phosphoniumchlorid n
p. iodide $[PH_4]J$ Phosphoniumjodid n
p. ion $[PH_4]^+$ Phosphoniumion n
p. salt Phosphoniumsalz n
p. sulphate $[PH_4]_2SO_4$ Phosphoniumsulfat n
phosphonous acid $RP(OH)_2$ Phosphonigsäure f, phosphonige Säure f
phosphoproteid[e], phosphoprotein Phospho[r]proteid n
phosphor Phosphor m *(phosphoreszenzfähiger Stoff)*
p. bronze Phosphorbronze f
phosphoresce / **to** phosphoreszieren
phosphorescence Phosphoreszenz f, *(i.e.S.)* Chemilumineszenz f, Chemolumineszenz f
p. spectrum Phosphoreszenzspektrum n
phosphorescent phosphoreszierend
p. paint Phosphoreszenzfarbe f, nachleuchtende (phosphoreszierende) Farbe (Leuchtfarbe) f
phosphoretted hydrogen PH_3 Monophosphin n, Phosphin n, gasförmiger Phosphorwasserstoff m
phosphorfluoric acid $H[PF_6]$ Hexafluorophosphorsäure f

phosphoric Phosphor...; höherwertigem Phosphor entsprechend, *(meist)* Phosphor(V)-...; phosphoreszierend
p. acid Phosphorsäure f, *(i.e.S.)* H_3PO_4 Orthophosphorsäure f, Monophosphorsäure f
p.-acid polymerization process Phosphorsäurepolymerisationsverfahren n
p.-acid process Phosphorsäureverfahren n *(bei der katalytischen Polymerisation)*
p. anhydride s. p. oxide
p. bromide PBr_5 Phosphor(V)-bromid n, Phosphorpentabromid n
p. chloride PCl_5 Phosphor(V)-chlorid n, Phosphorpentachlorid n
p. oxide P_2O_5 Phosphor(V)-oxid n, Phosphorpentoxid n
p. perbromide s. p. bromide
p. perchloride s. p. chloride
phosphorite *(Min)* Phosphorit m
phosphorochalcite *(Min)* Phosphorochalzit m, Pseudomalachit m *(Kupfer(II)-trihydroxidorthophosphat)*
phosphorofluoric acid HPF_6 Hexafluorophosphorsäure f
phosphorous Phosphor...; niederwertigem Phosphor entsprechend, *(meist)* Phosphor(III)-...; phosphoreszierend
p. acid H_2PHO_3 [ortho]phosphorige Säure f, monophosphorige Säure f
p. bromide PBr_3 Phosphor(III)-bromid n, Phosphortribromid n
p. chloride PCl_3 Phosphor(III)-chlorid n, Phosphortrichlorid n
p. sulfide P_2S_3 Phosphor(III)-sulfid n, Diphosphortrisulfid n
phosphorus P Phosphor m
p.-bearing phosphorhaltig
p. chloride nitride s. p. dichloride nitride
p. chlorofluoride s. p. dichloride trifluoride
p. dibromide trichloride PBr_2Cl_3 Phosphor(V)-dibromidtrichlorid n
p. dibromide trifluoride PBr_2F_3 Phosphor(V)-dibromidtrifluorid n
p. dibromonitride $PNBr_2$ Phosphornitriddibromid n
p. dibromotrichloride s. p. dibromide trichloride
p. dichloride PCl_2 Phosphordichlorid n
p. dichloride nitride $(PNCl_2)_n$ Phosphornitriddichlorid n, Phosphornitridkautschuk m, anorganischer Kautschuk m
p. dichloride trifluoride PCl_2F_3 Phosphor(V)-dichloridtrifluorid n
p. diiodide P_2J_4 Diphosphortetrajodid n
p. fixation *(Landw)* Phosphorfestlegung f
p. furnace Phosphorofen m
p. halide Phosphorhalogenid n
p. heptasulphide P_4S_7 Phosphorheptasulfid n, Tetraphosphorheptasulfid n, Phosphor(III,IV)-sulfid n
p. hydride Phosphor(III)-hydrid n, Phosphin n,

Phosphorwasserstoff *m*, *(i.e.S.)* PH_3 Monophosphin *n*, gasförmiger Phosphorwasserstoff *m*
p. metabolism Phosphorstoffwechsel *m*
p. monoselenide P_2Se Phosphor[mono]selenid *n*, Diphosphor[mono]selenid *n*
p. needs *(Landw)* Phosphorbedarf *m*
p. nitride P_3N_5 Phosphornitrid *n*, Triphosphorpentanitrid *n*, Phosphorstickstoff *m*
p. oxybromide $POBr_3$ Phosphor(V)-oxidbromid *n*, Phosphoryl(V)-bromid *n*
p. oxychloride $POCl_3$ Phosphor(V)-oxidchlorid *n*, Phosphoryl(V)-chlorid *n*
p. oxyfluoride POF_3 Phosphor(V)-oxidfluorid *n*, Phosphoryl(V)-fluorid *n*
p. pentabromide PBr_5 Phosphorpentabromid *n*, Phosphor(V)-bromid *n*
p. pentachloride PCl_5 Phosphorpentachlorid *n*, Phosphor(V)-chlorid *n*
p. pentafluoride PF_5 Phosphorpentafluorid *n*, Phosphor(V)-fluorid *n*
p. pentaselenide P_2Se_5 Phosphorpentaselenid *n*, Diphosphorpentaselenid *n*
p. pentoxide P_2O_5 Phosphorpentoxid *n*, Phosphor(V)-oxid *n*
p. perbromide *s.* p. pentabromide
p. perchloride *s.* p. pentachloride
p. salt $Na(NH_4)HPO_4 \cdot 4H_2O$ Phosphorsalz *n*, Natriumammoniumhydrogenphosphat-4-Wasser *n*
p. sesquisulphide P_4S_3 Phosporsesquisulfid *n*, *(veraltet für)* Tetraphosphortrisulfid *n*
p. subselenide P_4Se Phosphorsubselenid *n*, *(veraltet für)* Tetraphosphormonoselenid *n*
p. tetrabromotrichloride PBr_4Cl_3 Phosphor(V)-tetrabromidtrichlorid *n*
p. tetroxide P_2O_4 Phosphortetroxid *n*
p. thioamide $PS(NH_2)_3$ Thiophosphorylamid *n*, Thiophosphorsäuretriamid *n*
p. thiobromide $PSBr_3$ Thiophosphoryl(V)-bromid *n*
p. thiobromochloride $PSBrCl_2$ Thiophosphorylbromiddichlorid *n*
p. thiochloride $PSCl_3$ Thiophosphoryl(V)-chlorid *n*
p. thiochlorobromide $PSClBr_2$ Thiophosphoryldibromidchlorid *n*
p. thiocyanate $P(SCN)_3$ Phosphor(III)-thiozyanat *n*, Phosphor(III)-rhodanid *n*
p. thiofluoride PSF_3 Thiophosphoryl(V)-fluorid *n*
p. triamide $OP(NH_2)_3$ Phosphoryltriamid *n*, Phosphoroxidtriamid *n*, Phosphorsäuretriamid *n*, Triamidophosphorsäure *f*
p. tribromide PBr_3 Phosphortribromid *n*, Phosphor(III)-bromid *n*
p. trichloride PCl_3 Phosphortrichlorid *n*, Phosphor(III)-chlorid *n*
p. trifluoride PF_3 Phosphortrifluorid *n*, Phosphor(III)-fluorid *n*
p. triiodide PJ_3 Phosphortrijodid *n*, Phosphor(III)-jodid *n*

p. trioxide P_2O_3 Phosphortrioxid *n*, Phosphor(III)-oxid *n*
p. triselenide P_2Se_3 Phosphortriselenid *n*, Diphosphortriselenid *n*
phosphoryl PO Phosphorylradikal *n*
p. amide *s.* phosphorus triamide
p. bromide $POBr_3$ Phosphor(V)-oxidbromid *n*, Phosphoryl(V)-bromid *n*
p. chloride $POCl_3$ Phosphor(V)-oxidchlorid *n*, Phosphoryl(V)-chlorid *n*
phosphorylase Phosphorylase *f* *(Ferment)*
phosphotungstate $M^I_3[P(W_3O_{10})_4]$ Tetrakiswolframatophosphat *n*, Dodekawolframatophosphat *n*, Wolframatophosphat *n*
phosphotungstic acid $H_3[P(W_3O_{10})_4]$ Tetrakistriwolframatophosphorsäure *f*, Dodekawolframatophosphorsäure *f*, Wolframatophosphorsäure *f*
phosphouranylite *(Min)* Phosphuranylit *m*
phosphowolframate *s.* phosphotungstate
phosphowolframic acid *s.* phosphotungstic acid
phosphuranylite *s.* phosphouranylite
phosphuretted hydrogen *s.* phosphoretted hydrogen
phosvitin Phosvitin *n* *(ein Phosphoproteid)*
photo paper Fotopapier *n*, fotografisches Papier *n*
photoactive lichtempfindlich; fotoaktiv, lichtelektrisch aktiv
photoactivity Lichtempfindlichkeit *f*; Fotoaktivität *f*, lichtelektrische Aktivität *f*
photoanalysis Fotoanalyse *f*
photobarrier cell *s.* photovoltaic cell
photocatalysis Fotokatalyse *f*
photocatalyst Fotokatalysator *m*
photocathode Fotokatode *f*
photocell Foto[emissions]zelle *f*, fotoelektrische (lichtelektrische) Zelle *f*
photochemical fotochemisch
p. cell *s.* photovoltaic cell
p. chlorination fotochemische Chlorierung *f*, Fotochlorierung *f*, Lichtchlorierung *f*
p. destruction fotochemischer Abbau *m*
p. equilibrium fotochemisches Gleichgewicht *n*
p. equivalent fotochemisches Äquivalent *n*
p. radiation fotochemische Strahlung *f*
p. reaction fotochemische Reaktion (Umsetzung) *f*, Fotoreaktion *f*, Lichtreaktion *f*
p. state *s.* equilibrium
photochemistry Fotochemie *f*
photocolorimetric method lichtelektrisches (objektives) kolorimetrisches Verfahren *n*, lichtelektrische (objektive) kolorimetrische Methode *f*
photocolorimetry lichtelektrische (objektive) Kolorimetrie *f*
photoconductive cell Fotowiderstandszelle *f*, Fotowiderstand *m*, Widerstands[foto]zelle *f*, Lichtwiderstand *m*, lichtelektrischer (fotoelektrischer) Widerstand *m*
p. effect Fotoleitungseffekt *m*, Halbleiterfotoeffekt *m*, innerer Fotoeffekt *m*, innerer lichtelektrischer (fotoelektrischer) Effekt *m*

photoconductivity Fotoleitfähigkeit *f*, lichtelektrische (fotoelektrische) Leitfähigkeit *f*
photocurrent Fotostrom *m*, fotoelektrischer Strom *m*
photodecomposition, photodegradation *s.* photolysis
photodissociation Fotodissoziation *f*
photoelasticity Fotoelastizität *f*, Spannungsoptik *f*
photoelectric fotoelektrisch, lichtelektrisch
  p. cell fotoelektrische (lichtelektrische) Zelle *f*, Foto[emissions]zelle *f*
  p. colorimeter lichtelektrisches (objektives) Kolorimeter *n*
  p. effect fotoelektrischer (lichtelektrischer) Effekt *m*, Fotoeffekt *m*
  p. exposure meter fotoelektrischer Belichtungsmesser *m*
  p. photometer lichtelektrisches (objektives) Fotometer *n*
  p. pyrometer fotoelektrisches Pyrometer *n*
photoelectricity Fotoelektrizität *f*
photoelectrolytic cell *s.* photovoltaic cell
photoelectron Fotoelektron *n*
photoemission Fotoemission *f*, lichtelektrische Emission *f*
photogelatin Lichtdruckgelatine *f*
  p. ink Lichtdruckfarbe *f*, Druckfarbe *f* für Lichtdruck
  p. process Lichtdruckverfahren *n*
photographic base paper Fotorohpapier *n*, fotografisches Rohpapier *n*
  p. chemical fotografische Chemikalie *f*, Fotochemikalie *f*
  p. chemistry fotografische Chemie *f*, Chemie *f* der Fotografie (fotografischen Prozesse)
  p. density optische Dichte *f*, Extinktion *f*, Schwärzung *f*
  p. developer fotografischer Entwickler *m*
  p. emulsion fotografische (lichtempfindliche) Emulsion *f*, Fotoemulsion *f*
  p. paper fotografisches Papier *n*, Fotopapier *n*
  p. plate fotografische Platte *f*, Fotoplatte *f*
  p. record fotografische Aufzeichnung (Registrierung) *f*
  p. recording polarograph Polarograf *m* mit fotografischer Registrierung
photogravure Heliogravüre *f*, Fotogravüre *f*, Heliografie *f*
  p. ink Druckfarbe *f* für Heliogravüre
photohalide Fotohalogenid *n*
photoionization Fotoionisation *f*
photoluminescence Fotolumineszenz *f*
photolysis Fotolyse *f*, Zersetzung *f* durch Licht
photolytic fotolytisch
photomacrograph Makrofotografie *f*, Makrofoto *n*, Makroaufnahme *f*, makrofotografische Aufnahme *f*
photomacrography Makrofotografie *f*
photometer Fotometer *n*
  p. head Fotometerkopf *m*

photometric fotometrisch, Fotometer...
  p. analysis fotometrische Analyse *f*, Fotometrie *f*, Lichtstärkemessung *f*, Strahlungsmessung *f*
photometry Fotometrie *f*, Lichtstärkemessung *f*, Strahlungsmessung *f*
photomicrograph Mikrofotografie *f*, Mikrofoto *n*, Mikroaufnahme *f*, mikrofotografische Aufnahme *f*
photomicrography Mikrofotografie *f*
photomultiplier Foto[elektronen]vervielfacher *m*, Fotosekundärelektronenvervielfacher *m*, Sekundärelektronenvervielfacher *m*, SEV *m*, Fotomultiplier *m*, Multiplier *m*, Vervielfacherfotozelle *f*
photon Photon *n*, Lichtquant *n*, Strahlungsquant *n*
photoneutron Fotoneutron *n*
photonuclear reaction Kernfotoeffekt *m*
photooxidation Fotooxydation *f*, fotochemische Oxydation *f*
photophoresis Fotophorese *f*
photopolymer Fotopolymer[es] *n*
photopolymerization Fotopolymerisation *f*
photoproton Fotoproton *n*
photoreaction fotochemische Reaktion (Umsetzung) *f*, Fotoreaktion *f*, Lichtreaktion *f*
photoresistor Fotowiderstand *m*, Fotowiderstandszelle *f*, Widerstands[foto]zelle *f*, Lichtwiderstand *m*, lichtelektrischer (fotoelektrischer) Widerstand *m*
photosensitive lichtempfindlich
photosensitiveness Lichtempfindlichkeit *f*
photosensitization Fotosensibilisation *f*, Fotosensibilisierung *f*, Sensibilisierung *f*
photosensitize / to [foto]sensibilisieren
photosensitizer Sensibilisator *m*
photosynthate Assimilat *n*
photosynthesis Fotosynthese *f*
photosynthetic fotosynthetisch
phototropism, phototropy Fototropie *f*
phototube Foto[emissions]zelle *f*, fotoelektrische (lichtelektrische) Zelle *f*
phototype paper Lichtdruckpapier *n*
phototyping cardboard Lichtdruckkarton *m*
photovalve *s.* phototube
photovoltaic cell Sperrschicht[foto]zelle *f*, Sperrschicht[foto]element *n*, Fotoelement *n*
  p. effect Sperrschicht[foto]effekt *m*, Foto-Volta-Effekt *m*, Fotovolteffekt *m*
phragmites peat Schilftorf *m*
phrenosin Phrenosin *n*, Zerebron *n*
phthalamic acid $C_6H_4(CONH_2)COOH$ Phthalamidsäure *f*, Phthalaminsäure *f*, Phthalsäuremonamid *n*
phthalamide $C_6H_4(CONH_2)_2$ Phthalamid *n*, Phthalsäurediamid *n*
phthalandione *s.* phthalic anhydride
phthalate Phthalat *n*
  p. plasticizer Phthalatweichmacher *m*
phthalein Phthalein *n*
  p. group Phthaleingruppe *f*
phthalic acid *s.* *o*-p. acid

m-p. acid $C_6H_4(COOH)_2$ Isophthalsäure *f*, Benzol-*m*-dikarbonsäure *f*, Benzoldikarbonsäure-(1,3) *f*

o-p. acid $C_6H_4(COOH)_2$ Phthalsäure *f*, Benzol-*o*-dikarbonsäure *f*, Benzoldikarbonsäure-(1,2) *f*

p-p. acid $C_6H_4(COOH)_2$ Terephthalsäure *f*, Benzoldikarbonsäure-(1,4) *f*

p. anhydride $C_6H_4(CO)_2O$ Phthalsäureanhydrid *n*, Phthalandion *n*, 1,2-Benzoldikarbonsäureanhydrid *n*, 1,3-Dioxophthalan *n*

p. diamide *s.* phthalamide

p. glyceride resin Glyptalharz *n*

phthalimide $C_6H_4(CO)_2NH$ Phthalimid *n*, Phthalsäureimid *n*, 1,3-Isoindoldion *n*

phthalocyanine Phthalozyanin *n*

p. dye Phthalozyaninfarbstoff *m*

p. pigment Phthalozyaninpigment *n*

p. series Phthalozyaninreihe *f*

phthalogen brilliant blue Phthalogenbrillantblau *n*

phthalonitrile $C_6H_4(C{\equiv}N)_2$ Phthalsäuredinitril *n*, Phthalodinitril *n*, 1,2-Dizyanobenzol *n*

phthalophenon Phthalophenon *n*, 3,3-Diphenylphthalid *n*

phthalyl chloride $C_6H_4(COCl)_2$ Phthalylchlorid *n*

phthalylsulphathiazole Phthalylsulfathiazol *n*, 2-(N^4-Phthalylsulfanilamido)thiazol *n*

phthiocol Phthiokol *n*, 2-Hydroxy-3-methyl-1,4-naphthochinon *n*

Phurnacite Phurnacite *m (Warenzeichen für einen rauchlosen Brennstoff)*

Phurnacite process Phurnacite-Verfahren *n*

phycitol $(HOCH_2CHOH)_2$ Phyzit *m, (veraltet für)* Erythrit *m*, Threit *m*, Butantetrol *n*

phycocolloid Phykokolloid *n*

phycocyan[in] Phykozyan *n (Farbstoff der blaugrünen Algen)*

phycoerythrin Phykoerythrin *n (Farbstoff der Rotalgen)*

phyllite *(Geol)* Phyllit *m*

phyllonite *(Geol)* Phyllonit *m*

phylloquinone Phyllochinon *n*, Vitamin K$_1$ *n*, Koagulationsvitamin *n*, antihämorrhagisches Vitamin *n*

physalite *(Min)* Physalit *m*, Pyrophysalit *m*

physic nut Purgiernuß *f*, Jatropha curcas L.

physical physikalisch; physisch

p. adsorption physikalische (van-der-Waalssche) Adsorption *f*

p. chemistry physikalische Chemie *f*

p. decolorization *(Glas)* physikalische Entfärbung *f*

p. property physikalische Eigenschaft *f*

p. tracer physikalischer Tracer *m*

physicochemical physikalisch-chemisch, physikochemisch

physiologic[al] physiologisch

p. chemistry physiologische Chemie *f*

p. inertness physiologische Indifferenz (Reizlosigkeit) *f*

p. saline (salt solution) physiologische Kochsalzlösung *f*

physiologically inert physiologisch indifferent (inert, reizlos)

physostigmine Physostigmin *n*, Eserin *n (Alkaloid)*

p. salicylate Physostigminsalizylat *n*

p. sulphate Physostigminsulfat *n*

physostigmol Physostigmol *n (Alkaloid)*

phytase Phytase *f (zu den Phosphoesterasen gehörendes Ferment)*

phyteral Phyteral *n (pflanzliches Aufbauelement der Kohle)*

phytic acid Phytinsäure *f*

phytin Phytin *n*

phytoalexin Phytoalexin *n (von Pflanzen auf Parasitenbefall hin gebildeter Abwehrstoff)*

phytochemistry Phytochemie *f*

phytocidal pflanzentötend

phytol Phytol *n*

phytosterol Phytosterin *n*, pflanzliches Sterin *n*, Pflanzensterin *n*

p.-acetate test Phytosterinazetatprobe *f*

phytotoxic phytotoxisch, pflanzenschädigend

phytotoxicity Phytotoxizität *f*, Giftigkeit *f* für Pflanzen

PI *s.* polyimide

pi bond π-Bindung *f*, Pi-Bindung *f*, π-Elektronenpaarbindung *f*

pi meson π-Meson *n*, Pion *n*

pi orbital π-Orbital *n*

piceatannol Pizeatannol *n*, 2,5,6,3',4'-Pentahydroxy-3,4-tetramethylstilben *n (Gerbstoff)*

picein Pizein *n*

picene Pizen *n*

pick / to lesen, ernten, sammeln, pflücken; [aus]wählen, sortieren, [aus]lesen, [aus]klauben; *(Pap)* rupfen

p. off by hand [von Hand] ausklauben (klauben, auslesen) *(bei der Kohlenaufbereitung)*

p. out herauslösen

p. up *(z.B. Wellen durch ein Geofon)* erfassen; *(Flüssigkeiten)* aufnehmen; *(Papierbahn vom Sieb)* abnehmen

pickeringite *(Min)* Pickeringit *m (Magnesiumaluminiumsulfat-22-Wasser)*

picking Lesen *n*, Ernten *n*, Sammeln *n*, Pflücken *n*; Auswählen *n*, Wählen *n*, Sortieren *n*, Auslesen *n*, Ausklauben *n*, Klauben *n*; *(Pap)* Rupfen *n*

p. resistance *(Pap)* Rupfwiderstand *m*

p.-resistance testing *(Pap)* Rupffestigkeitsprüfung *f*

pickle / to 1. *(Lebm)* naßpökeln; [ein]säuern, *(bei Fisch auch)* marinieren; 2. *(Gerb)* pickeln; 3. *(Saatgut, Metall)* [ab]beizen

pickle Pökellake *f*, Pökelsalzlösung *f*; Marinade *f*, Aufguß *m (z.B. gewürzte Salz-Essig-Lösung)*; 2. *(Gerb)* Pickel *m*; 3. Beize *f*, Beizmittellösung *f*

p. cure *s.* pickling 1.

p.-cured naßgepökelt; [ein]gesäuert, *(bei Fisch auch)* mariniert

p. curing s. pickling 1.
p. liquor (solution) s. pickle
pickled s. pickle-cured
p. fish marinierter Fisch m, Fischmarinade f
p. meat Pökelfleisch n
pickles Essiggemüse n; Essigfrüchte fpl
pickling 1. Naßpökeln n, Naßpökelung f; Säuern n, Säuerung f, Einsäuern n, (bei Fisch auch) Marinieren n; 2. (Gerb) Pickeln n, Pickel m; 3. Beizen n, Abbeizen n, Beize f
p. method (Landw) Tauchverfahren n (Saatgutbeize)
p. solution s. pickle
p. wastes Beizereiabwasser n
pickup Erfassen n (z.B. von Wellen durch ein Geofon); Aufnahme f (z.B. von Leim durch Papier); (Text) Flüssigkeitsaufnahme f, Flottenaufnahme f; (Ker) Glasuraufnahme f; (Tech) Auffangrinne f, Rinne f; (Pap) Abnahme f (der Papierbahn vom Sieb durch Walzen)
p. felt (Pap) Abnahmefilz m
p. of size (Pap) Leimaufnahme f
p. roll (Pap) Pickup-Walze f, Selbstabnahmewalze f
picoline Pikolin f
picolinic acid C_5H_4NCOOH Pikolinsäure f, α-Pyridinkarbonsäure f, Pyridinkarbonsäure-(2) f
picotite (Min) Pikotit m, Chromspinell m
picram[in]ic acid $C_6H_2(NO_2)_2(NH_2)OH$ Pikraminsäure f, 2-Amino-4,6-dinitrophenol n
picrate Pikrat n
picric acid $C_6H_2(NO_2)_3OH$ Pikrinsäure f, 2,4,6-Trinitrophenol n
picrochromite (Min) Pikrochromit m, (veraltet für) Magnesiochromit m (Magnesiumchromat(III))
picromycin Pikromyzin n (Antibiotikum)
picronitric acid s. picric acid
picrotoxin Pikrotoxin n
Pictet-Spengler reaction Pictet-Spenglersche Reaktion f (Isochinolin-Ringschluß)
Pidgeon [vacuum] process Pidgeon-Verfahren n (Retortenverfahren zur Magnesiumgewinnung)
piece dyeing Stückfärben n, Färben n im Stück
p.-dyeing machine Stückfärbemaschine f
pie[d]montite (Min) Piemontit m
piezo effect s. piezoelectric effect
piezochemistry Piezochemie f
piezochromism Piezochromie f
piezocrystallization Piezokristallisation f, Druckkristallisation f
piezoelectric piezoelektrisch
p. ceramics piezoelektrische Keramik f
p. coupling coefficient piezoelektrischer Kopplungskoeffizient m
p. crystal piezoelektrischer Kristall m, Piezokristall m
p. effect piezoelektrischer Effekt m, Piezoeffekt m
piezoelectricity Piezoelektrizität f
piezometer Piezometer n, piezoelektrischer Druckmesser m, Piezodruckmesser m

pig Massel f
p. bed Gießbett n, Masselbett n, Masselbeet n
p.-casting machine Masselgießmaschine f
p. fat Schweinefett n, Schweineschmalz n, Schmalz n
p. iron Roheisen n, Masseleisen n
p. lead Blockblei n, Rohblei n, Werkblei n
pigment / to pigmentieren, färben; (Gum) füllen, Füllstoffe zusetzen
pigment Pigment n, Körperfarbe f; Pigmentfarbstoff m; (Gum) Füllstoff m
p.-binder ratio Pigment-Bindemittel-Verhältnis n
p. chlorine 2G Pigmentchlorin GG n
p. dispersion Pigmentdispergierung f, Pigmentdispersion f
p. dye Pigmentfarbstoff m
p. floating (flooding) Ausschwimmen n (sichtbares Entmischen der Pigmente in Anstrichstoffen)
p. grinding Pigmentanreibung f
p. loading (Gum) Zusatz (Zuschlag) m von Füllstoffen, Füllung f, Füllstoffdosierung f
p. migration Pigmentwanderung f
p. padding (Text) Pigmentklotzverfahren n, Pigmentklotzung f
p. printing Pigmentdruck m
p.-producing lichen Färberflechte f
p. settling Absetzen n der Pigmente
p. volume concentration Pigmentvolumenkonzentration f, PVK
pigmentation Pigmentation f, Färbung f; (Gum) Füllung f, Zusatz (Zuschlag) m von Füllstoffen, Füllstoffdosierung f
pigmented (Gum) gefüllt, füllstoffhaltig, mit Füllstoffen
pikromycin s. picromycin
pile (Tech) Haufen[speicher] m; (Pap) Papierstapel m; (Kch) Kernreaktor m, Reaktor m, Atommeiler m, Pile m; Meiler m (Holzverkohlung)
p. curing (Gum) Anvulkanisation f, Anvulkanisieren n, Anbrennen n, Anspringen n
p.-on property (Text) (Am) Aufziehvermögen n, Aufzieheigenschaft f
pileless finish (Text) Kahlappretur f
pili nut oil Pilinuß-Öl n (von Canarium-Arten)
Pilkington process Pilkington-Verfahren n (Spiegelglasherstellung)
pill / to pillieren, zu Kügelchen formen
pill (Plast) Vorformling m; (Pharm) Pille f
pilling (Text) Pilling n, Pillbildung f, Pillingeffekt m
pilocarpine Pilokarpin n (Alkaloid)
p. nitrate Pilokarpinnitrat n
pilot-controlled mit Hilfssteuerung; vorgesteuert (Hydraulik- und Pneumatikventile)
p.-controlled valve Ventil n mit Hilfssteuerung; vorgesteuertes Ventil n (Hydraulik und Pneumatik)
p.-operated s. p.-controlled
p. plant Pilotanlage f, Halbbetriebsanlage f, halbtechnische Versuchsanlage f; Versuchsbetrieb m

p.-plant-scale production Produktion *f* in halbtechnischem Maßstab
p.-plant stage Versuchsstadium *n*
p. production s. p.-plant-scale production
p.-supply line Hilfssteuerleitung *f*; Vorsteuerleitung *f (Hydraulik und Pneumatik)*
p. valve Hilfssteuerventil *n*; Vorsteuerventil *n (Hydraulik und Pneumatik)*
pilotaxitic pilotaxitisch *(Gesteinsstruktur)*
Pilsen malt *(helles)* Pilsner Malz *n*
d-**pimaric acid** Dextropimarsäure *f*, α-Pimarsäure *f*, *d*-Pimarsäure *f (Phenanthrenderivat)*
pimaricin Pimarizin *n (Antibiotikum)*
pimelic acid COOH(CH$_2$)$_5$COOH Pimelinsäure *f*, 1,5-Pentandikarbonsäure *f*, 1,7-Heptandisäure *f*
pimelite *(Min)* Pimelit *m*
pimple *(Plast)* Pickel *m (Preßfehler)*
pin *(Tech)* Stift *m*, Bolzen *m*, Dorn *m*; *(Ker)* Pinne *f*, Brennstütze *f*
p.-disk mill s. pinned-disk disintegrator
p. frame *(Text)* Nadel[spann]rahmen *m*
p.[-type] mill s. pinned-disk disintegrator
pinacoid *(Krist)* Pinakoid *n*, Zweiflächner *m*
pinacoidal *(Krist)* pinakoidal
pinacol (CH$_3$)$_2$C(OH)COH(CH$_3$)$_2$ Pinakol *n*, Tetramethyläthylenglykol *n*
p. rearrangement Pinakol-Pinakolon-Umlagerung *f*
pinacoline, pinacolone CH$_3$COC(CH$_3$)$_3$ Pinakolin *n*, *(veraltet für)* Pinakolon *n*, 2,2-Dimethylbutanon-(3) *n*, Trimethylazeton *n*, Methyl-*tert*-butylketon *n*
pinacone (CH$_3$)$_2$C(OH)COH(CH$_3$)$_2$ Pinakon *n*, *(veraltet für)* Pinakol *n*, Tetramethyläthylenglykol *n*
pinacryptol green Pinakryptolgrün *n*
p. yellow Pinakryptolgelb *n*
pinaverdol Pinaverdol *n*, 1,1′,6-Trimethylisozyaninjodid *n*
pinc rubber *(Lianenkautschuk von Landolphia kirkii Dyer)*
pinch Prise *f*; *(Kch)* Faden *m*, Pinch *m (des Ionenstromes)*
p. clamp Halteklemme *f (z.B. für Kugelschliffverbindungen)*
p. cock Quetschhahn *m*, Quetschklemme *f*, Schlauchklemme *f*
p. valve Schlauchventil *n*, Gummitaschenventil *n*
pine Kiefer *f*, *(Gattung)* Pinus L.
p.-needle oil Kiefernnadelöl *n*; *(oft auch)* Fichtennadelöl *n (oder)* Edeltannennadelöl *n*
p. oil Pine Oil *n*, Kienöl *n (von Pinus-Arten)*, *(auch)* Kiefernnadelöl *n (oder)* Kiefernzapfenöl *n*
p. resin (rosin) Kiefernharz *n*, *(i.e.S.)* Terpentinharz *n*, Balsamharz *n*, Harz *n* aus Rohterpentin, Kolophonium *n*, Balsamkolophonium *n*, Geigenharz *n*
p. splint test *(Ligninnachweis mittels Phlorogluzin und Salzsäure)*
p. tar Kiefern[holz]teer *m*

pinene C$_{10}$H$_{16}$ Pinen *n (bizyklisches Monoterpen)*
pinewood Kiefernholz *n*
piney tallow Pineytalg *m*, Vateriafett *n*, Malabartalg *m (Samenfett von Vateria indica L.)*
ping Klingeln *n*, Pinken *n (im Vergasermotor)*
pinhole *(Ker)* Nadelstich *m (Fehler)*; *(Plast)* Loch *n (Fehler)*
p. gate s. pinpoint gate
pinholing Porenbildung *f (bei Anstrichfilmen)*
pinite *(Min)* Pinit *m*
pinitol C$_6$H$_6$(OH)$_5$OCH$_3$ *d*-Pinit *n*, Matezit *n*
pink rosa[farben], rosafarbig
pink Rosa *n (Farbempfindung)*; rosafarbener Farbstoff *m*, Rosa *n*
p. champagne Rotsekt *m*
p. salt (NH$_4$)$_2$[SnCl$_6$] Pinksalz *n*, Ammoniumchlorostannat(IV) *n*
pinking Pinken *n*, Klingeln *n (im Vergasermotor)*
pinkish blaßrosa
pinned-disk disintegrator (mill) Stift[scheiben]mühle *f*, Schlagstiftmühle *f*, Dismembrator *m*
pinning forceps Präparierpinzette *f*
pinobanksin Pinobanksin *n (Flavanon)*
pinocembrin Pinozembrin *n (Flavanon, in Kiefernholz)*
pinolene s. pinolin[e]
pinolin[e] Pinolin *n*, Harzgeist *m*, Harzessenz *f*, Harzsprit *m*, Terpentinessenz *f*
pinoresinol Pinoresinol *n (Derivat des Koniferylalkohols)*
pinostrobin Pinostrobin *n (Flavonderivat, in Kiefernholz)*
pinosylvin (HO)$_2$C$_6$H$_3$CH=CHC$_6$H$_5$ Pinosylvin *n*, 1-(3,5-Dihydroxyphenyl)-2-phenyläthen *n*, 3,5-Dihydroxystilben *n*
pinpoint gate *(Plast)* Nadelpunktanguß *m*
Pintsch gas Pintsch-Gas *n*, Ölgas *n*
Pintsch-Hillebrand process Pintsch-Hillebrand-Verfahren *n*, Wälzgasverfahren *n* nach Pintsch-Hillebrand
pioscope Pioskop *n (zur kolorimetrischen Bestimmung des Fettgehaltes von Milch)*
pipe / to Rohre legen, mit Rohren (Rohrleitung) versehen, durch Rohrleitung verbinden; [in Rohrleitungen] leiten (führen)
pipe 1. Rohr *n*, Röhre *f*, Leitung *f*; 2. Lunker *m (im Gußblock)*
p. coil Rohrschlange *f*, Rohrspirale *f*
p. cooler Röhrenkühler *m*
p. dope Dichtungsmaterial *n* für Rohrgewindeverbindungen
p. flange Rohr[leitungs]flansch *m*
p. friction Rohrreibung *f*
p. furnace Röhrenofen *m*, Röhrenerhitzer *m*
p. gamboge Röhrengutti *n (Gummiguttsorte)*
p. hanger Rohrschelle *f*, Rohraufhänger *m*
p. heater s. p. furnace
p. joint Rohrverbindung *f*, Rohrverbinder *m*
p. line s. pipeline

p. **machine** *(Ker)* Röhrenpresse f
p. **precipitator** Röhrenelektrofilter *n*, Röhren[elektro]abscheider *m*
p. **press** *(Ker)* Röhrenpresse f, Rohrpresse f
p. **scale** Rohrzunder *m*
p. **still** *s.* 1. p.-still plant; 2. p. furnace
p.-**still distillation** Röhren[ofen]destillation f
p.-**still furnace [heater]** *s.* p. furnace
p.-**still plant (unit)** Röhrenofenanlage f, Pipestill-Anlage f
p. **thread** Rohrgewinde f
p. **wall** Rohrwand[ung] f
pipeclay Pfeifenton *m*
p. **triangle** Tondreieck *n*, Porzellandreieck *n*, Dreieck *n* mit Porzellanröhren
pipecolic acid Pipekolinsäure f *(Piperidinderivat)*
pipeline Pipeline f, Überlandrohrleitung f, Fernleitung f, Rohrleitung f, Leitung f
p. **agitator** Rohrleitungsrührwerk *n*
p. **milkers** Rohrmelkanlage f, Melkrohrsystem *n*
p. **tracer** Rohrleitungsheizschlange f; Rohrleitungskühlschlange f
piperazine $HN(CH_2)_4NH$ Piperazin *n*, Diäthylendiamin *n*
piperic acid $CH_2(O_2)C_6H_3CH=CHCH=CHCOOH$ Piperinsäure f
piperidine $(CH_2)_5NH$ Piperidin *n*, Hexahydropyridin *n*, Pentamethylenimin *n*
2-**piperidone** C_5H_8ONH 2-Piperidon *n*, α-Piperidon *n*, δ-Valerolaktam *n*, 5-Aminovaleriansäure-δ-laktam *n*
piperine Piperin *n*, 1-Piperylpiperidin *n*
piperolidine Piperolidin *n*, δ-Konizein *n*, 1,2-Trimethylenpiperidin *n*
piperonal $CH_2(O_2)C_6H_3CHO$ Piperonal *n*, Heliotropin *n*
piperonylic acid $CH_2(O_2)C_6H_3COOH$ Piperonylsäure f
piperylene $CH_2=CHCH=CHCH_3$ Piperylen *n*, 1-Methylbutadien *n*, Pentadien-(1,3) *n*, 1-Methyldivinyl *n*
pipet *s.* pipette
pipette / to pipettieren
p. **off** abpipettieren
pipette Pipette f
p. **brush** Pipettenbürste f
p. **rack (stand, support)** Pipettenständer *m*, Pipettengestell *n*
p. **tip** Pipettenspitze f
p. **with graduated scale** Meßpipette f
pipetting Pipettieren *n*, Pipettierung f
p. **method** Pipettmethode f
pipey *(Gerb)* rinnend *(Narben)*; losnarbig *(Leder)*
p. **grain** *(Gerb)* rinnender (loser) Narben *m*
piping 1. Leiten (Führen) *n* in Rohrleitungen; 2. Rohrmaterial *n*; 3. Lunkern *n*, Lunkerung f, Lunker[aus]bildung f *(im Gußblock)*
Pirani gauge Pirani-Manometer *n*, Pirani-Vakuummeter *n*, Wärmeleitvakuummeter *n*
pisolite *(Min)* Pisolith *m*, Erbsenstein *m* *(Kalziumkarbonat)*

pisolitic tuff Pisolithtuff *m*
pistachia galls Mastix *m*, *(körniges)* Mastixharz *n*
pistachio nut Pistazie f, Pistazienmandel f, Alepponuß f *(von Pistacia vera L.)*
pistacite *(Min)* Pistazit *m*, *(veraltet für)* Epidot *m*
piston Kolben *m*, Stempel *m*, Schieber *m*
p.-**actuated** kolbenbetätigt
p. **area** Kolbenfläche f
p. **pump** Kolbenpumpe f *(i.e.S.)*
p. **ring** Kolbenring *m*
p. **rod** Kolbenstange f
p.-**rod packing** Kolbenstangendichtung f, Kolbenstangenpackung f
p.-**type diecasting machine** Kolben[druck]gießmaschine f
pit Grube f; *(Gerb)* Gerbgrube f, Farbgrube f; *(Plast)* Grübchen *n*, Loch *n* *(Preßfehler)*
p.-**cast pipe** Sandgußrohr *n*
p. **lime** *(Gerb)* Grubenäscher *m*
p.-**limed** *(Gerb)* in der Grube geäschert
p. **liming** *(Gerb)* Grubenäscher *m*
p. **tannage** Grubengerbung f
pitch / to 1. teeren, pichen; 2. *(Hefe)* anstellen
pitch 1. Teeren *n*, Pichen *n*; Pech *n*; *(Pap)* [schädliches] Harz *n*; 2. Anstellen *n* *(der Hefe)*; 3. Teilung f *(von Ketten, Zahnrädern usw.)*; Steigung f *(Gewinde)*
p.-**bound briquette** pechgebundenes (mit Pech gebundenes) Brikett *n*, Kohlepechbrikett *n*
p. **coal** Pechkohle f
p. **lake** Asphaltsee *m*, Pechsee *m*
p.-**like** pechartig
p. **of buckets** Becherteilung f
p. **paper** Bitumenpapier *n*, bituminiertes Papier *n*, Teerpapier *n*, Asphaltpapier *n*; Doppelpechpapier *n*
p. **pine** Pitchpine f *(n)*, Pitchpineholz *n*
p. **polishing** *(Glas)* Pechpolitur f
p. **tip** *(Text)* Pechspitze f *(Wolle)*
p. **trouble[s]** *(Pap)* Harzschwierigkeiten *fpl*
pitchblende *(Min)* Pechblende f *(Uran(IV)-oxid)*
pitchers *(Ker)* Scherben *mpl*, Massescherben *mpl*; Glattscherben *n*
pitching Teeren *n*, Pichen *n*; Anstellen *n* *(der Hefe)*
p. **temperature** *(Gär)* Anstelltemperatur f
p. **vessel** *(Gär)* Anstellbottich *m*
p. **yeast** Anstellhefe f, Stellhefe f; Mutterhefe f
pitchy pechartig; *(Pap)* harzhaltig *(Zellstoff)*
pith Mark *n* *(des Holzes)*
pitman Schubstange f *(eines Wilfley-Herdes)*; Zugstange f *(eines Backenbrechers)*
Pitot tube Pitot-Rohr *n*, Staurohr *n*, Pitotsche Röhre f
pitted narbig, löcherig
pitticite *(Min)* Pittizit *m*, Arseneisensinter *m*
pitting Lochfraß *m*, punktförmige Anfressung f, Grübchenbildung f, *(veraltet)* Pitting *m*
p. **corrosion** Lochfraßkorrosion f, *(veraltet)* Pitting-Bildung f
pittizite *s.* pitticite

Pittsburgh [sheet] process Pittsburgh-Verfahren *n* (*Tafelglasziehverfahren*)
pituitary Hypophysenhinterlappenpulver *n*
pivaldehyde (CH₃)₃CCHO Pivalaldehyd *m*, Trimethylazetaldehyd *m*
pivalic acid (CH₃)₃CCOOH Pivalinsäure *f*, Dimethylpropansäure *f*, Trimethylessigsäure *f*
pivoted-bucket conveyor Pendelbecherwerk *n*; Schaukelbecherwerk *n*, Schaukelförderer *m*
place / to (*z.B. Chemikalien in Gefäße*) eintragen, [ein]bringen; (*Düngemittel*) ausbringen; (*Beton*) einbringen; (*Brenngut in den Ofen*) setzen
place Stelle *f*, Ort *m*
  p. isomerism Stellungsisomerie *f*, Ortsisomerie *f*
placer [deposits] Seife *f*, Seifenerz *n*, Wascherz *n*
  p. gold Seifengold *n*, Waschgold *n*
placing Eintragen *n*, Einbringen *n* (*z.B. von Chemikalien in Gefäße*); Ausbringen *n* (*von Düngemitteln*); Einbringen *n* (*von Beton*); Setzen *n* (*von Brenngut in den Ofen*)
  p. of concrete Betoneinbringen *n*, Betoneinbau *m*
  p. sand (*Ker*) Streusand *m*
plagioclase (*Min*) Plagioklas *m*
plagionite (*Min*) Plagionit *m* (*Blei(II)-antimon(III)-sulfid*)
plain 1. eben, flach, plan; glatt; 2. rein, unlegiert (*Stahl*)
  p. air condenser Kühlrohr *n*
  p. bearing Gleitlager *n*
  p. carbon steel reiner (unlegierter) Kohlenstoffstahl *m*
  p. core pin Lochstift *m*, Kernlochstift *m*
  p. dressing Flächendüngung *f*
  p. glass funnel Trichter *m* mit glatter Wandung
plaining (*Glas*) Läuterung *f*
  p. end (*Glas*) Läuterwanne *f*
plait point kritischer Punkt *m*, (*i.e.S.*) kritischer Entmischungspunkt *m*
plan paper Landkartenpapier *n*, Kartenpapier *n*; Seekartenpapier *n*
planar eben[flächig], plan[ar]
planchéite (*Min*) Plancheit *m*
Planck constant Plancksches Wirkungsquantum *n*, Plancksche Konstante *f*
Planck quantum of action *s.* Planck constant
Planck radiation formula (law) Plancksches Strahlungsgesetz *n*, Plancksche Strahlungsformel *f*
plane flach, eben, plan
plane Ebene *f*, [ebene] Fläche *f*
  p. face (*Krist*) ebene Fläche *f*
  p. of evaporation Trockenspiegel *m*
  p. of polarization Polarisationsebene *f*
  p. of symmetry (*Krist*) Symmetrieebene *f*, Spiegelebene *f*
  p. of the boiling-point diagram Siedefläche *f*
  p. of the page (paper) Papierebene *f*, Schreibebene *f*
  p.-parallel planparallel

p. polarization lineare Polarisation *f*
  p.-polarized linear polarisiert
planet stirrer *s.* planetary stirrer
planetary gear train Planetenradgetriebe *n*
  p. stirrer Planetenrührwerk *n*, Planetenrührer *m*, Planetenmischer *m*
plankton Plankton *n*
planktonic planktonisch
planographic ink Flachdruckfarbe *f*, Druckfarbe *f* für Flachdruck
  p. printing Flachdruck *m*
  p. printing ink *s.* p. ink
  p. process Flachdruckverfahren *n*
planography Flachdruck *m*
plansifter Plansichter *m*
plant 1. Anlage *f*, Betriebsanlage *f*; Betrieb *m*, Werk *n*, Fabrik *f*; 2. Pflanze *f*
  p. biochemistry Pflanzenbiochemie *f*
  p. blank value Pflanzenblindwert *m* (*Rückstandsanalytik*)
  p.-constructed industriell gefertigt (hergestellt)
  p. debris *s.* p. remains
  p.-derived pflanzlicher Herkunft, pflanzlichen Ursprungs
  p. entity Pflanzensubstanz *f*
  p. extract Pflanzenextrakt *m*, Pflanzenauszug *m*
  p.-fibre gasket Pflanzenfaserdichtung *f*
  p. filter Industriefilter *n*
  p. food element (*Landw*) Nährstoffelement *n*
  p. material pflanzlicher Rohstoff *m*
  p. nutrient Pflanzennährstoff *m*
  p. nutrition Pflanzenernährung *f*
  p. pigment Pflanzenfarbstoff *m*
  p. poison pflanzliches Gift *n*, Pflanzengift *n*
  p. protective protektives (protektiv wirkendes, vorbeugend wirkendes) Pflanzenschutzmittel *n*
  p. remains Pflanzen[über]reste *mpl*, pflanzliche Reste *mpl*
  p. resistance Pflanzenresistenz *f*, pflanzliche Resistenz *f*
  p. shutdown Betriebsstillegung *f*
  p. source pflanzlicher Rohstoff *m*
  p. spray flüssiges Pflanzenschutzmittel *n*, (*i.e.S.*) Sprühmittel *n*, Sprühflüssigkeit *f*, Nebelmittel *n* (*oder*) Spritzmittel *n*, Spritzflüssigkeit *f*
  p. sterol pflanzliches Sterin *n*, Pflanzensterin *n*, Phytosterin *n*
plantation latex Plantagenlatex *m*
  p. rubber Plantagenkautschuk *m*, Pflanzungskautschuk *m*
plasma (*phys Ch*) Plasma *n*; (*Min*) Plasma *n* (*Varietät von Chalzedon*); (*Med*) Blutplasma *n*, Plasma *n*; (*Bio*) Protoplasma *n*, Plasma *n*
  p. burner Plasmabrenner *m*
  p. chemistry Plasmachemie *f*
  p. electron oscillations Plasmaschwingungen *fpl*
  p. membrane Plasmamembran *f*
  p. of gas discharge Gasentladungsplasma *n*
  p. oscillations Plasmaschwingungen *fpl*
  p. physics Plasmaphysik *f*

p. state Plasmazustand *m*
plasmolysis Plasmolyse *f*, Plasmaablösung *f*
plaster / to *(Bau)* [ver]putzen, bewerfen; vergipsen, mit Gips überziehen; überziehen, bekleiden; *(Gär)* gipsen, mit Gips behandeln
plaster 1. *(Med)* Pflaster *n*; 2. s. p. mortar
p. mortar Putzmörtel *m*; Gipsmörtel *m*
p. mould Gipsform *f*
p. of Paris gebrannter Gips *m* *(im wesentlichen $CaSO_4 \cdot 0,5H_2O$); (manchmal auch)* [kristalliner] Gips *m (Kalziumsulfat-2-Wasser)*
plastic plastisch, verformbar, bildsam
plastic Plast *m*, [organischer] Kunststoff *m*
p. bag Plastsack *m*, Kunststoffsack *m*
p. body *(Ker)* plastische (bildsame) Masse *f*
p. clay plastischer (fetter) Ton *m*
p. composition Plastmasse *f*, Plastansatz *m*, Plastmischung *f*
p. concrete plastischer (weicher) Beton *m*
p. covering Plastüberzug *m*, Kunststoffüberzug *m*
p. deformation bleibende (irreversible, plastische) Verformung (Deformation) *f*
p. fat halbgehärtetes Fett (Öl) *n*
p. film s. p. foil
p. flow Fließen *n (eine Art der plastischen Verformung)*
p. foil Plastfolie *f*, Kunststoffolie *f*
p. foil welder Plastfolienschweißmaschine *f*, Plastfolienschweißgerät *n*
p. material plastische Masse *f*; Plast *m*, [organischer] Kunststoff *m*
p. melt Plastschmelze *f*
p. paint plastische Anstrichfarbe (Farbe) *f*
p. pipe Plastrohr *n*, Kunststoffrohr *n*
p. piping Plastrohrmaterial *n*, Kunststoffrohrmaterial *n*; s. p. pipe
p. pressing *(Ker)* plastisches Pressen *n*, Naßpressen *n*
p. product Plasterzeugnis *n*, Kunststofferzeugnis *n*
p.-proofed kunststoffimprägniert
p. range plastischer Bereich *m*, plastisches Gebiet *n*, Plastizitätsbereich *m*, Erweichungsbereich *m*
p. state plastischer Zustand *m*
p. steel knetbarer Stahl *m*
p. sulphur plastischer Schwefel *m*
p. yield Wärmebeständigkeit *f*, Formbeständigkeit *f* in der Wärme
plasticate / to s. to plasticize
plasticating cylinder *(Plast)* Massezylinder *m*, Spritz[gieß]zylinder *m*
plastication Plasti[fi]zieren *n*, Plasti[fi]zierung *f*, Weichmachen *n*, Erweichen *n*
plasticator *(Plast)* Plastikator *m*, Plastiziermaschine *f*; *(Gum)* Mastikator *m*, Mastiziermaschine *f*, Plastikator *m*
plasticise / to s. to plasticize
plasticity Plastizität *f*, Verformbarkeit *f*, Bildsamkeit

p. measurement Plastizitätsmessung *f*
p. test Plastizitätsprüfung *f*
plasticization s. plastication
plasticize / to plasti[fi]zieren, weichmachen, erweichen
plasticized PVC Weich-PVC *n*, PVC-weich *n*, PVC-W
plasticizer Weichmacher *m*, Weichmachungsmittel *n*, Plasti[fi]zier[ungs]mittel *n*, Plastifikationsmittel *n*, weichmachender (plastischmachender) Zusatz *m*, Plasti[fi]kator *m*, *(Gum auch)* [chemisches] Abbaumittel *n*, Peptisier[ungs]mittel *n*
plasticizing agent s. plasticizer
p. capacity Plastizierleistung *f*
plastics processing Plastverarbeitung *f*
p. processor Plastverarbeiter *m*
p. technology Technologie *f* der Plaste (Kunststoffe)
plastify / to s. to plasticize
plastigel Plastigel *n*
plastimeter s. plastometer
plastograph Plastograf *m (Gerät zur Plastizitätsbestimmung)*
plastomer Plastomer[es] *n*
plastometer Plastometer *n*, Plastizitätsprüfgerät *n*
plate / to beschichten, *(Met, Text)* plattieren, *(Ech)* galvanisieren, elektroplattieren; *(Pap)* kalandern, kalandrieren, glätten, satinieren
plate Platte *f*, Blech *n*; Teller *m*; Tafel *f*; *(Destillation)* Boden *m*, Austauschboden *m*; *(Galvanotechnik)* Überzug *m*; *(Krist)* Tafel *f*; *(Ech)* (flache) Elektrode *f*, Platte *f*; *(Pap)* Grundwerk *n*, Messerwerk *n*, Messerblock *m (eines Holländers)*; *(bei Waagen)* Platte *f*, Pfanne *f*, *(bei Feinwaagen)* Planlager *n*
p.-and-frame filter Rahmenfilter *n*; s. p.-and-frame filter press
p.-and-frame [filter] press Rahmen[filter]presse *f*
p.-and-ring filter press s. p.-and-frame filter press
p. column Bodenkolonne *f*, Bodensäule *f*
p. cooler Plattenkühler *m*, Plattenkühlapparat *m*, Plattenkühlmaschine *f*
p. current Anodenstrom *m*
p. efficiency [factor] Bodenwirkungsgrad *m (einer Destillationskolonne)*
p. etching *(Glas)* Schablonenätzen *n*, Schablonenätzung *f*
p.-fin heat exchanger berippter Plattenwärmeaustauscher (Plattenwärmeübertrager) *m*, Wirbelzellenwärmeaustauscher *m*, Wirbelzellenwärmeübertrager *m*
p. freezer (froster) Platten[schnell]gefrierapparat *m*
p. glass Spiegelglas *n*
p.-glass furnace Spiegelglaswanne *f*
p. heat exchanger Plattenwärmeaustauscher *m*, Plattenwärmeübertrager *m*, Kammerwärmeaustauscher *m*, Kammerwärmeübertrager *m*
p. ice Platteneis *n*

plate 492

p.-jiggering machine *(Ker)* Tellerdrehmaschine *f*
p. method *(Pharm)* Plattenmethode *f*
p. Murphree efficiency *(Destillation)* Bodenwirkungsgrad *m* [nach Murphree]
p. number Bodenzahl *f* *(einer Destillationskolonne)*
p. paper Kupferdruckpapier *n*
p. pasteurizer *(Lebm)* Plattenerhitzer *m*, Platten-[erhitzungs]apparat *m*
p. precipitator Plattenelektrofilter *n*, Platten-[elektro]abscheider *m*
p. press Schichtenfilter *n*
p. separator Freifallscheider *m* *(zum Elektroscheiden)*
p. theory *(Chromatografie)* Theorie *f* der Böden
p. tower Bodenkolonne *f*, Bodensäule *f*
p.-type heat exchanger *s.* p. heat exchanger
p. voltage Anodenspannung *f*
plateau effect *(Gum)* Plateaueffekt *m*, Plateau *n*, Vulkanisationsplateau *n*
platen *(Gum, Plast)* Heizplatte *f*; Pressentisch *m* *(einer Etagenpresse)*
p. mark *(Plast)* Markierung *f* durch die Heizplatte *(Fehler)*
p. press Etagenpresse *f*, Packpresse *f*, Plattenpresse *f*, *(Gum auch)* Etagenvulkanisierpresse *f*
platformate *(Erdöl)* Platformat *n*, Platformierungsprodukt *n*, Platformerprodukt *n*
platformer *(Erdöl)* Platformer *m*, Platforming-Anlage *f*
platforming *(Erdöl)* Platforming *n*, Platformen *n*, Platin-Reforming *n*, Reformieren *n* an Platinkatalysatoren (Platinkontakten)
p. process *(Erdöl)* Platforming-Verfahren *n*, Platin-Reforming-Prozeß *m*, Reformierungsprozeß *m* an Platinkatalysatoren (Platinkontakten)
p. unit *s.* platformer
platina mohr *s.* platinum black
platinammine *s.* platinum ammine
platinate / to *s.* to platinize
platinate Platinat *n*
plating Beschichten *n*, *(Met, Text)* Plattieren *n*, *(Ech)* Galvanisieren *n*, Elektroplattieren *n*; Schutzschicht *f*, *(Met, Text)* Plattierung *f*, *(Ech)* galvanischer Überzug *m*
p. bath Bad *n*, Galvanisierbad *n*
p. rack *(Ech)* Einhängegestell *n*, Galvanisiergehänge *n*
p. solution *s.* p. bath
platinibromide $M^I_2[PtBr_6]$ Bromoplatinat(IV) *n*, Hexabromoplatinat(IV) *n*
platinic Platin..., *(i.e.S.)* Platin(IV)...
p. ammine Polyamminplatin(IV)-verbindung *f*
p. ammonium chloride $(NH_4)_2[PtCl_6]$ Ammoniumhexachloroplatinat(IV) *n*
p. chloride $PtCl_4$ Platin(IV)-chlorid *n*, Platintetrachlorid *n*
p. oxide PtO_2 Platin(IV)-oxid *n*, Platindioxid *n*
p. potassium bromide $K_2[PtBr_6]$ Kaliumhexabromoplatinat(IV) *n*

p. potassium chloride $K_2[PtCl_6]$ Kaliumhexachloroplatinat(IV) *n*
p. potassium iodide $K_2[PtJ_6]$ Kaliumhexajodoplatinat(IV) *n*
p. pyrophosphate PtP_2O_7 Platindiphosphat *n*, Platinpyrophosphat *n*
p. sal ammoniac *s.* p. ammonium chloride
p. sodium chloride $Na_2[PtCl_6]$ Natriumhexachloroplatinat(IV) *n*
p. sulphate $Pt(SO_4)_2$ Platin(IV)-sulfat *n*
p. sulphide PtS_2 Platin(IV)-sulfid *n*, Platindisulfid *n*
platinichloride $M^I_2[PtCl_6]$ Chloroplatinat(IV) *n*, Hexachloroplatinat(IV) *n*
platiniferous platinhaltig
platinization Platinieren *n*
platinize / to platinieren, mit Platin überziehen
platinized asbestos Platinasbest *m*
platinizing *s.* platinization
platinochloride $M^I_2[PtCl_4]$ Chloroplatinat(II) *n*, Tetrachloroplatinat(II) *n*
platinocyanide $M^I_2[Pt(CN)_4]$ Zyanoplatinat(II) *n*, Tetrazyanoplatinat(II) *n*
platinotype paper *(Foto)* Platinpapier *n*
platinous Platin..., *(i.e.S.)* Platin(II)-...
p. ammine Polyamminplatin(II)-verbindung *f*
p. ammonium chloride $(NH_4)_2[PtCl_4]$ Ammoniumtetrachloroplatinat(II) *n*
p. ammonium cyanide $(NH_4)_2[Pt(CN)_4]$ Ammoniumtetrazyanoplatinat(II) *n*
p. chloride $PtCl_2$ Platin(II)-chlorid *n*, Platindichlorid *n*
p. cyanide $Pt(CN)_2$ Platin(II)-zyanid *n*
p. oxide PtO Platin(II)-oxid *n*, Platinmonoxid *n*
p.-platinic oxide Pt_3O_4 Platin(II,IV)-oxid *n*
p. potassium chloride $K_2[PtCl_4]$ Kaliumtetrachloroplatinat(II) *n*
p. potassium cyanide $K_2[Pt(CN)_4]$ Kaliumtetrazyanoplatinat(II) *n*
p. sal ammoniac *s.* p. ammonium chloride
p. sodium chloride $Na_2[PtCl_4]$ Natriumtetrachloroplatinat(II) *n*
p. sulphide PtS Platin(II)-sulfid *n*, Platin[mono]sulfid *n*
platinum Pt Platin *n*
p. ammine Ammoniakkomplexverbindung *f* des Platins, Platiak *n*
p.-bearing platinhaltig
p. black Platinmohr *n*, Platinschwarz *n*
p. boat Platinschiffchen *n*
p. catalyst Platinkatalysator *m*, platinhaltiger Katalysator *m*, Platinkontakt *m*
p. crucible Platintiegel *m*
p. dichloride $PtCl_2$ Platindichlorid *n*, Platin(II)-chlorid *n*
p. dioxide PtO_2 Platindioxid *n*, Platin(IV)-oxid *n*
p. dish Platinschale *f*
p. electrode Platinelektrode *f*
p. microelectrode Platinmikroelektrode *f*
p. mohr *s.* p. black

p. monosulphide PtS Platin[mono]sulfid n, Platin(II)-sulfid n
p. monoxide PtO Platinmonoxid n, Platin(II)-oxid n
p. oxide catalyst Platinoxidkatalysator m
p. plating Platinieren n
p. pyrophosphate PtP_2O_7 Platindiphosphat n, Platinpyrophosphat n
p. reforming s. platforming
p. resistance thermometer Platinwiderstandsthermometer n
p. sesquisulphide Pt_2S_3 Platin(III)-sulfid n
p. sol Platinsol n
p. sponge Platinschwamm m
p. tetrachloride $PtCl_4$ Platintetrachlorid n, Platin(IV)-chlorid n
p.-tipped forceps Tiegelzange f mit Platinschuhen
p. toning (Foto) Platintonung f
p. trichloride $PtCl_3$ Platintrichlorid n, Platin(III)-chlorid n
p. trioxide PtO_3 Platintrioxid n, Platin(VI)-oxid n
p. wire Platindraht m
platosammines (Sammelname für Polyamminplatin(II)-verbindungen der allgemeinen Formel $X_2(NH)_3Pt$)
platy (Krist) tafelig, tafelförmig
Plauson [colloid] mill Plauson-Mühle f, Plausonsche Kolloidmühle f
playing cardboard Spielkartenkarton m
pleasant-smelling angenehm riechend, von angenehmem Geruch
plenum chamber Trockenkammer f, Luftkammer f, (Plast auch) Spritzkammer f, Vorformkammer f, Massekammer f; (Plast) Absaugkasten m
pleochro[mat]ic (Krist) pleochroitisch
pleochro[mat]ism (Krist) Pleochroismus m, Mehrfarbigkeit f
pleonaste (Min) Pleonast m (ein eisen(II)-haltiger Spinell)
plerotic water Grundwasser n
plessite (Min) Plessit m (ein nickelhaltiges gediegenes Eisen aus Meteoriten)
pliability Biegsamkeit f, Geschmeidigkeit f
pliable, pliant biegsam, geschmeidig
plot / to (in ein Koordinatensystem) eintragen
plough Austragspflug m, Abstreicher m (an Stetigförderern)
p. bar Streichstange f
p.-type agitator Flügelrührer m
plow (Am) s. plough
pluck / to (Pap) rupfen
plucking (Pap) Rupfen n
p. resistance (Pap) Rupfwiderstand m
plug / to verstopfen, zustöpseln, verschließen; sich verstopfen, sich zusetzen
plug Stopfen m, Pfropfen m, Stöpsel m, (an Fässern) Zapfen m, Spund m; Küken n, Hahnküken n; (Glas) Stempel m; Kern m, Kegel m, Konus m, Rotor m, Kegelrotor m (der Kegelstoffmühle)

p.-and-ring forming (Plast) Streckformen n mit Ring, Ringverformung f
p.-assist forming (Plast) Vakuumformen n mit mechanischer Vorstreckung
p.-assist pressure forming (Plast) Druckluftformen n mit Vorstreckung
p.-assist vacuum forming (Plast) Vakuumformen (Vakuumziehverfahren) n mit Vorstreckung
p. bib s. p. cock
p. cock Hahn m, Hahnventil n; Zapfen m, Spund m (an Fässern)
p. of cotton wool Wattebausch m
p. pressure Mahldruck m, Anpreßdruck m (bei Kegelstoffmühlen)
p.-up (Plast) Pfropfenbildung f (Spritzgießfehler)
p. valve Hahn m, Hahnventil n
plugging Verstopfen n, Verlegen n, Zusetzen n
plumbagin Plumbagin n, 2-Methyl-5-hydroxy-1,4-naphthochinon n
plumbago (Min) Plumbago f (veraltet für: 1. Naturgraphit; 2. Bleiglanz)
p. crucible Graphittiegel m
plumbane PbH_4 Plumban n, Bleiwasserstoff m
plumbate Plumbat n
plumber's solder Blei-Zinn-Weichlot n, Lötzinn n
plumbic Blei..., (i.e.S.) Blei(IV)-...
plumbiferous bleihaltig
plumbite $M^I_2PbO_2$ Plumbat(II) n
p. [sweetening] process (Erdöl) Plumbitverfahren n
plumbo-plumbic oxide Pb_3O_4 Blei(II,IV)-oxid n, Bleimennige f, Mennige f
plumbobinnite (Min) Plumbobinnit m, (veraltet für) Dufrenoysit m (Blei(II)-arsen(III)-sulfid)
plumbogummite, plumboresinite (Min) Plumbogummit m (Aluminiumblei(II)-hydrogenhexahydroxiddiorthophosphat)
plumbous Blei..., (i.e.S.) Blei(II)-...
p. oxide PbO Blei(II)-oxid n, Bleimonoxid n
p. sulphide PbS Blei(II)-sulfid n
plummet Senkgewicht n, Senklot n, Senkblei n; Schwebekörper m (in Durchflußmessern)
plump / to (Gerb) (Blößen) schwellen; (Gerb) aufgehen
plumper s. plumping agent
plumping (Gerb) Schwellen n (der Blößen)
p. agent (Gerb) Schwellmittel n
p. power Schwellwirkung f
p. tannage Füllgerbung f
plunge / to [ein]tauchen
plunger Tauchkolben m, Plunger[kolben] m, Kolben m, Stempel m, Stößel m
p. compartment Kolbenraum m
p. jig (Aufbereitungstechnik) Kolbensetzmaschine f, Kolbensetzapparat m
p. moulding (Plast) Spritzpressen n mit Doppelkolbenpresse
p. pump Tauchkolbenpumpe f, Plungerpumpe f
p.-type fixed-sieve jig s. p. jig
p.-type injection machine (Plast) Kolbenspritzgußmaschine f

p.-type jig s. p. jig
plural scattering (phys Ch) Mehrfachstreuung f
plus material (mesh) Überkorn n (beim Sieben)
plush copper (Min) Chalkotrichit m (nadeliger Kuprit, Kupfer(I)-oxid)
plutonic consolidation (Geol) intratellurische Erstarrung f, Erstarrung f in der Tiefe
p. **rock** s. plutonite
plutonite (Geol) Plutonit m, plutonisches Gestein n, Tiefengestein n
plutonium Pu Plutonium n
ply Lage f, Schicht f; Zwischenlage f; Furnierplatte f
p. **separation** (Gum) Lagenlösung f
p. **yarn** Mehrfachzwirn m
plywood Sperrholz n
p. **adhesive** Klebstoff m für Sperrholz
p. **board** Sperrholzpappe f
PMA s. 1. phenylmercuric acetate; 2. phosphomolybdic acid
PMR s. paramagnetic resonance
PNA s. para-nitroaniline
pneumatic pneumatisch; Druckluft..., Preßluft...
pneumatic s. p. tyre
p. **atomizer** Druckluftzerstäuber m, Druckluftdüse f
p. **cleaning** Luftwäsche f, Luftaufbereitung f, Trockenaufbereitung f, pneumatische (trockene) Aufbereitung f, Aufbereitung f auf trockenem Wege
p. **[conveying] dryer** Stromtrockner m, pneumatischer Trockner m
p. **malting** pneumatische (mechanische) Mälzerei f
p. **mill** Strahlprallmühle f
p. **nozzle atomization** Zweistoffzerstäubung f, Druckluftzerstäubung f
p. **process** (Met) Blasverfahren n
p. **sprayer** (Landw) Sprühblaser m
p. **spraying** Zweistoffversprühung f, Druckluftversprühung f
p. **trough** pneumatische Wanne f
p. **tyre** Luftreifen m, Pneumatikreifen m
p. **vibrator** Preßlufttrüttler m, Drucklufttrüttler m, Druckluftvibrator m
poach / to s. to potch
poacher s. potcher
poaching engine s. potching engine
pocher s. potcher
pocket (Pap) Tasche f, Preßtasche f, Preßkasten m, Schacht m, (i.w.S.) Presse f (eines Holzschleifers)
p. **battery** Taschenlampenbatterie f
p. **electrode** Fangtaschenelektrode f
p. **grinder** (Pap) Pressenschleifer m, Mehrpressenschleifer m
p. **lamp battery** s. p. battery
p. **of magma** Magmaherd m, Magmakammer f
p. **spectroscope** Taschenspektroskop n
p. **sprayer** Sprühdose f

p.-type grinder s. p. grinder
Podbielniak centrifugal [countercurrent] contactor s. Podbielniak contactor
Podbielniak contactor Podbielniak-Kontaktor m, Podbielniak-Extraktor m, Podbielniak-Zentrifugalextraktor m, Podbielniak-Zentrifuge f, Gegenstromextraktionsapparat m nach Podbielniak
Podbielniak machine s. Podbielniak contactor
podocarprene Podokarpren n (ein Diterpenkohlenwasserstoff)
podsol s. podzol
podzol [soil], podzolic soil Podsol[boden] m, Bleicherde f, Bleicherdeboden m, Bleichsand m
podzolization Podsolierung f
Poggendorff compensation method Poggendorffsche Kompensationsmethode f
poikilitic poikilitisch (Gesteinsstruktur)
poikiloblastic poikiloblastisch (Gefüge)
point defect punktförmiger Gitterfehler m; punktförmige Fehlstelle f
p. **group** (Krist) Punkt[symmetrie]gruppe f
p. **of attachment** (Nomenklatur) Verknüpfungsstelle f
p. **of congelation** Gefrierpunkt m, Gefriertemperatur f; Erstarrungspunkt m, Koagulierungspunkt m; Stockpunkt m (bei Ölen)
p. **of criticality** kritischer Punkt m
p. **of entry** Eintrittsstelle f, Einlaufstelle f, Einspritzstelle f, Eintritt m, Einlauf m
p. **of fusion** Kondensationsstelle f, Anellierungsstelle f
p. **of inflection** (Titration) Umschlag[s]punkt m; Wendepunkt m (einer Kurve)
p. **of intersection** Schnittpunkt m
p. **of linkage** Verknüpfungsstelle f, Verbindungsstelle f
p. **of neutrality** Neutralpunkt m, Äquivalenzpunkt m
p. **of rupture** Zerreißpunkt m, Bruchpunkt m
pointer Zeiger m, Zunge f, Nadel f (z.B. eines Meßgerätes)
poise Poise n, P (Maßeinheit der dynamischen Viskosität)
Poiseuille's equation (Pap) Gleichung f von Poiseuille (Filtrationsvorgang bei der Blattbildung)
poison / to vergiften
poison Gift n
p. **bait** Giftköder m
p. **by contact** (veraltet für) contact poison
p. **by food** (veraltet für) stomach poison
p. **for baits** Ködergift n
p. **for traps** Fallengift n
p. **gas** Giftgas n
p. **hemlock** Fleckenschierling m, Gefleckter Schierling m, Giftschierling m, Conium maculatum L.
poisoned grain Giftgetreide n
p. **paper** Giftpapier n, Insektenvertilgungspapier n

poisoning Vergiftung f, Intoxikation f; Verseuchung f
  p. effect Giftwirkung f
poisonous giftig, toxisch
  p. gas Giftgas n
Poisson's ratio Poissonsche Zahl (Konstante) f, Querzahl f (Kennzahl für die Querkontraktion)
pokehole Stoch[er]loch n, Stoch[er]öffnung f
poker Stoch[er]vorrichtung f, Schürvorrichtung f
  p. vibrator Tauchrüttler m, Tiefenrüttler m, Innenrüttler m, Innenvibrator m
poking hole s. pokehole
polar polar
  p. adsorption polare Adsorption f
  p. bond polare (heteropolare, ionogene, elektrostatische, elektrovalente) Bindung f, Ionenbindung f, Ionenbeziehung f, Elektrovalenz f
  p. compound [hetero]polare Verbindung f, Ionenverbindung f
  p. link s. p. bond
  p. linkage polare (heteropolare, elektrostatische, ionogene, elektrovalente) Verknüpfung f (als Vorgang); s. p. bond
  p. liquid polare Flüssigkeit f
  p. molecule polares Molekül n
  p. solvent polares Lösungsmittel n
polarimeter Polarimeter n, Polarisationsgerät n, Polarisator m
polarimetry Polarimetrie f
polariscope s. polarimeter
polarise / to s. to polarize
polaristrobometer Polaristrobometer n
polarity Polarität f
  p. paper (Am) Pol[reagenz]papier n
  p. tester Polprüfer m
polarizability Polarisierbarkeit f
  p. of molecule molekulare Polarisierbarkeit f
polarizable polarisierbar
polarization Polarisation f, Polarisierung f
  p. current Polarisationsstrom m
  p. effect Polarisationseffekt m, Polarisationswirkung f
  p. electrode Polarisationselektrode f
  p. ellipsoid Polarisationsellipsoid n, Polarisierbarkeitsellipsoid n
  p. plane Polarisationsebene f
  p. potential s. p. voltage
  p. resulting from deformation (induction) Verschiebungspolarisation f
  p. saturation Polarisationssättigung f
  p. voltage Polarisationsspannung f
polarize / to polarisieren
polarizer Polarisator m (Gerät); polarisierende Substanz f
polarizing current Polarisationsstrom m
  p. effect Polarisationseffekt m, Polarisationswirkung f
  p. microscope Polarisationsmikroskop n
  p. power Polarisierungskraft f
polarogram Polarogramm n, polarografische Kurve f

polarograph Polarograf m
polarographic polarografisch
  p. analysis polarografische Analyse f
  p. curve polarografische Kurve f, Polarogramm n
  p. wave polarografische Stufe (Welle) f
polarography Polarografie f
pole / to (Kupfer) polen
pole fabric (Text) Polgewebe n
  p.[-finding] paper, p. reagent paper Pol[reagenz]papier n
Polenske number (value) Polenske-Zahl f, Po-Z, PZ (Kennzahl der Fette und fetten Öle)
polianite (Min) Polianit m (idiomorpher Pyrolusit, Mangan(IV)-oxid)
policeman Gummiwischer m (für Rührstäbe)
poling Polen n (von Kupfer), (i.e.S.) Zähpolen n (zum Entfernen überschüssigen Sauerstoffs)
  p.-down Dichtpolen n (von Kupfer, zum Austreiben von Schwefel)
polish / to polieren, glätten; (Edelsteine) schleifen; blankfiltrieren, polierfiltrieren
polish Politur f
polished face polierte Anschliffffläche f, Anschliff m
  p.-face method Anschliffmethode f, Anschliffverfahren n
  p. plate glass Spiegelglas n
  p. rice polierter Reis m, Kochreis m
  p. section Anschliff m (in der Auflichtmikroskopie)
  p.-specimen technique s. p.-surface technique
  p. surface polierte Oberfläche f; s. p. section
  p.-surface technique Anschlifftechnik f
  p. thin section polierter Dünnschliff m
polisher block (Glas) Poliertisch m, Polierstein m
polishing Polieren n, Glätten n; Schleifen n (von Edelsteinen); s. p. filtration
  p. agent Poliermittel n
  p. compound Poliermasse f
  p. drum Poliertrommel f
  p. filter Klärfilter n, Blankfilter n, Polierfilter n
  p. filtration Blankfiltration f, Polierfiltration f, Klärfiltration f, Klarfiltration f, Feinfiltration f
  p. material s. p. agent
  p. oil Polieröl n
  p. pitch Polierpech n
  p. plate Polierblech n
  p. rouge Polierrot n, Englischrot n (Eisen(III)-oxid)
  p. wheel Polierscheibe f
pollucite (Min) Polluzit m (zäsiumhaltiges Tektosilikat)
pollution load Belastung f mit Schmutzstoffen, Abwasserlast f
  p. of the environment Umweltverschmutzung f
pollux s. pollucite
polonium Po Polonium n
polyacetal Polyazetal n
polyacid mehrsäurig (Base)
polyacid mehrbasige Säure f

polyacrylate Polyakrylat *n*, Polyakrylsäureester *m*
p. elastomer Polyakrylatelastomer[es] *n*, Akryl-elastomer[es] *n*
p. rubber Akrylatkautschuk *m*, Akryl-Butadien-Kautschuk *m* *(Butadien-Akrylsäureester-Misch-polymerisat)*
polyacrylic acid Polyakrylsäure *f*   •
polyacrylonitrile Polyakrylnitril *n*, PAN
p. fibre Polyakrylnitrilfaser *f*, Akrylfaser *f*; Polyakrylnitrilfaserstoff *m*, PVY, Akrylfaserstoff *m*
polyad mehrwertig, vielwertig, polyad
polyad mehrwertiges Element *n*; mehrwertige Atomgruppe *f*
polyaddition Polyaddition *f*, additive Polykondensation *f*
polyadic *s.* polyad
polyaffinity theory Polyaffinitätstheorie *f*
polyalcohol Polyalkohol *m*, Polyol *n*, mehrwertiger Alkohol *m*
polyallomer *(Plast)* Polyallomer[es] *n*
polyamide Polyamid *n*, polymeres Amid *n*, PA
p. fibre Polyamidfaser *f*; Polyamidfaserstoff *m*, PA
polyampholyte Polyampholyt *m* *(sowohl mit Säuren als auch mit Basen reagierendes Polymeres)*
polyanion Polyanion *n*
polyargyrite *(Min)* Polyargyrit *m*
polyaryl ether Polyaryläther *m*
polyatomic mehratomig, vielatomig, polyatomar
p. molecule mehratomiges (vielatomiges) Molekül *n*
polybasic acid mehrbasige *(oft unkorrekt mehrbasische)* Säure *f*
polybasite *(Min)* Polybasit *m* *(ein Silberspießglanz)*
polybenzimidazole Polybenzimidazol *n*
polybenzothiazole Polybenzothiazol *n*
polyblend Polyblend *n*, Polymermischung *f* *(Gemisch mehrerer Thermoplaste)*
polybutadiene Polybutadien *n*
polybut[yl]ene Polybut[yl]en *n*
polycaprolactam Polykaprolaktam *n*
polycarbonate Polykarbonat *n*, PC
p. fibre Polykarbonatfaser *f*; Polykarbonatfaserstoff *m*, PK
p. resin Polykarbonatharz *n*
polycarboxylic acid Polykarbonsäure *f*
polycentric molecular orbital polyzentrisches Molekülorbital *n*, Mehrzentrenmolekülorbital *n*
polychlorethane Polychloräthan *n*
polychloroprene Polychloropren *n*
polychlorotrifluoroethylene Polytrifluorchloräthylen *n*, PCTFE
polychroism *(Krist)* Pleochroismus *m*, Mehrfarbigkeit *f*
polycondensate Polykondensat *n*
p. fibre Polykondensatfaser *f*; Polykondensatfaserstoff *m*
polycondensation Polykondensation *f*
polycrase, polycrasite Polykras *m* *(oxidisches uranhaltiges Seltenerdmineral)*

polycyclic polyzyklisch
p. hydrocarbon polyzyklischer Kohlenwasserstoff *m*
p. quinone Polyzyklochinon *n*
polycyclohydrocarbon *s.* polycyclic hydrocarbon
polydentate *(Komplexchemie)* mehrzahnig, mehrzähnig, vielzähnig, mehrzählig
polydichlorstyrene Polydichlorstyrol *n*
polydisperse polydispers
p. system polydisperses System *n*
polydispersion *s.* polydispersity
polydispersity Polydispersität *f*
polydymite *(Min)* Polydymit *m* *(Trinickeltetrasulfid)*
polyelectrode *(Ech)* mehrfache Elektrode *f*, Mehrfachelektrode *f*
polyelectrolyte *(Ech)* Polyelektrolyt *m*
polyene Polyen *n* *(Verbindung mit mehreren konjugierten Kohlenstoffdoppelbindungen)*
p. antibiotic Polyenantibiotikum *n*
polyenoid system *(konjugiertes Ringsystem mit eingebauter Polyenkette)*
polyepoxide Polyepoxid *n*
polyester Polyester *m*
p. fibre Polyesterfaser *f*; Polyesterfaserstoff *m*, PE
p. resin Polyesterharz *n*, Polyester *m*
p. rubber Polyesterkautschuk *m*
p. urethane auf Polyesterbasis aufgebautes Urethanelastomer[es] *n*
polyesteramide rubber (urethane) auf Polyesteramid aufgebautes Urethanelastomer[es] *n*
polyethenoid fatty acid mehrfach ungesättigte Fettsäure *f*
polyether urethane auf Polyätherbasis aufgebautes Urethanelastomer[es] *n*
polyethylene Polyäth[yl]en *n*, PE
p. adipate Polyäthylenadipat *n*
p. derivative Polyäthylenderivat *n*
p. fibre Polyäthylenfaser *f*; Polyäthylenfaserstoff *m*, PT
p. foam Polyäthylenschaum[stoff] *m*
p. glycol *s.* p. oxide
p. imine Polyäthylenimin *n*
p. oxide $HOCH_2(CH_2OCH_2)_x \cdot CH_2OH$ Polyäthylenoxid *n*, Polyäthylenglykol *n*, Polyglykol *n*
p. terephthalate Polyäthylenterephthalat *n*, PETP
polyfluoroethylene propylene Polyfluoräthylenpropylen *n*
polyform process *(Erdöl)* Polyform[ing]-Verfahren *n*
polyformaldehyde Polyformaldehyd *m*
polyforming *(Erdöl)* Polyformen *n*, Polyforming *n* *(thermisches Reformieren mit Gasrückführung)*
polyfunctional polyfunktionell, mehrfunktionell
p. alcohol mehrwertiger Alkohol *m*, Polyalkohol *m*
polygalacturonic acid Polygalakturonsäure *f*
polygalite *(fälschlich für)* polyhalite
polyglycerol Polyglyzerin *n*

polygonal polygonal
polyhalite *(Min)* Polyhalit *m (Kaliumkalziummagne-siumsulfat)*
polyhalogenated höherhalogeniert, mehrfach halogeniert
polyhedral *(Krist)* polyedrisch, vielflächig
polyhedron *(Krist)* Polyeder *n*, Vielflach *n*, Vielflächner *m*
polyheteroatomic ring *(fünf- oder sechsgliedrige)* heterozyklische Verbindung *f* mit mehreren Heteroatomen
polyhexamethylene adipamide Polyhexamethylenadipinsäureamid *n*
polyhydrate Polyhydrat *n*
polyhydric mehrere Hydroxylgruppen enthaltend
  p. alcohol *s.* polyol
  p. phenol mehrwertiges Phenol *n*
polyhydrocarbon polymerisierter Kohlenwasserstoff *m*
polyhydroxy alcohol *s.* polyol
  p. aldehyde Polyhydroxyaldehyd *m*
  p. compound Polyhydroxyverbindung *f*
  p. ketone Polyhydroxyketon *n*
polyhydroxyanthraquinone Polyhydroxyanthrachinon *n*
polyimide Polyimid *n*
polyiodide Polyjodid *n*
polyion Polyion *n*
polyisobut[yl]ene Polyisobut[yl]en *n*, PIB
polyisoprene Polyisopren *n*
polyketone Polyketon *n*
polymembered vielgliedrig
polymer Polymer[es] *n*, Polymerisat *n*
  p. binder Polymerisatbinder *m*
  p. blend *s.* polyblend
  p. chain Polymerkette *f*
  p. chemistry Chemie *f* der Hochpolymeren, makromolekulare Chemie *f*
  p. gasoline Polymer[isations]benzin *n*
  p.-homologous polymerhomolog, polymereinheitlich
  p. homologue Polymerhomologes *n*
polymeric polymer
  p. organosiloxane polymeres Organosiloxan *n*, organisches Polysiloxan *n*, Polyorganosiloxan *n*, Organopolysiloxan *n*
  p. siloxane polymeres Siloxan *n*, Polysiloxan *n*
  p. tannage *s.* polymerization tannage
polymeride *s.* polymer
polymerizability Polymerisierbarkeit *f*
polymerizable polymerisierbar, polymerisationsfähig
  p. plasticizer innerer Weichmacher *m*
polymerizate Polymerisat *n*
polymerization Polymerisation *f*, Polymerisieren *n*, Polymerisierung *f*
  p. initiator Polymerisationsinitiator *m*, Polymerisationserreger *m*, Starter *m*, Aktivator *m*
  p. kettle Polymerisationskessel *m*
  p. product Polymerisationsprodukt *n*

  p. recipe Polymerisationsansatz *m*
  p. tannage Polymerisationsgerbung *f*
  p. temperature Polymerisationstemperatur *f*
polymerize / to polymerisieren
polymetamorphic rock polymetamorphes Gestein *n*
polymetamorphism *(Geol)* Polymetamorphose *f*
polymethacrylate Polymethakrylat *n*, Polymethakrylsäureester *m*, Polymethylmethakrylat *n*, PMMA
polymethacrylic acid Polymethakrylsäure *f*
polymethyl acrylate (methacrylate) *s.* polymethacrylate
polymethylene Polymethylen *n*
polymignite, polymignyte Polymignit *m (oxidisches Seltenerdmineral)*
polymineral polymineralisch
polymixin *s.* polymyxin
polymolecular hochmolekular, höhermolekular
polymorphic polymorph, vielgestaltig
polymorphism Polymorphie *f*, Polymorphismus *m*, Vielgestaltigkeit *f*
polymorphous *s.* polymorphic
polymyxin *(Eiweißchemie)* Polymyxin *n*
polyneuridine Polyneuridin *n*
polynitro derivative Polynitroderivat *n*
polynosic fibre polynosische Faser *f*, Polynosefaser *f*; polynosischer Faserstoff *m*, Polynosefaserstoff *m*
polynuclear mehrkernig
polyol Polyol *n*, Polyalkohol *m*, mehrwertiger Alkohol *m*
polyolefin Polyolefin *n*, PO
  p. fibre Polyolefinfaser *f*; Polyolefinfaserstoff *m*, PO
  p. rubber Polyolefinkautschuk *m*.
polyorganosiloxane Polyorganosiloxan *n*, Organopolysiloxan *n*, polymeres Organosiloxan *n*, organisches Polysiloxan *n*, Silikon *n*
polyose Polyose *f*
polyoxacyclobutane Polyoxazyklobutan *n*
polyoxymethylene $(CH_2O)_n$ Polyoxymethylen *n*, POM, Paraformaldehyd *m*, Paraform *f*
polypeptide Polypeptid *n*
  p. antibiotic Polypeptidantibiotikum *n*
  p. chain Polypeptidkette *f*
polyphenol oxydase, polyphenolase Polyphenol[oxyd]ase *f*
polyphenyl sulfide Polyphenylsulfid *n*
polyphenylene oxide Polyphenylenoxid *n*, PPO
polyphosphate Polyphosphat *n*
polyporic acid Polyporsäure *f*, 2,5-Diphenyl-3,6-dihydroxybenzochinon *n*
polypropylene Polypropylen *n*, PP
  p. fibre Polyprop[yl]enfaser *f*; Polyprop[yl]enfaserstoff *m*, PP
  p. foam Polypropylenschaum[stoff] *m*
polyprotic acid mehrbasige Säure *f*
polyreaction Polyreaktion *f (Polymerisations-, Polykondensations- oder Polyadditionsreaktion)*

polysaccharide Polysa[c]charid *n*
polysaccharose Polyose *f*
polysilicic acid Polykieselsäure *f*
polysiloxane $(H_2SiO)_n$ Polysiloxan *n*, polymeres Siloxan *n*
polystyrene $(C_6H_5CH = CH_2)_n$ Polystyrol *n*, PS
  p. fibre Polystyrolfaser *f*, Polyvinylbenzolfaser *f*; Polystyrolfaserstoff *m*, Polyvinylbenzolfaserstoff *m*
  p. foam Polystyrolschaum[stoff] *m*
polysubstituted polysubstituiert, mehrfach substituiert
polysubstitution Polysubstitution *f*, Mehrfachsubstitution *f*
polysulphide Polysulfid *n*
  p. bridge (cross-link, link) Polysulfidbrücke *f*, polysulfidische Brückenbindung *f*
  p. liquid polymer flüssiges Polymer[es] *n*, Liquid-Polymer *n*
  p. rubber Polysulfidkautschuk *m*, Thioplast *m*
polysulphidic polysulfidisch
  p. cross-link *s.* polysulphide bridge
  p. sulphur polysulfidisch gebundener Schwefel *m*
polysulphone Polysulfon *n*
  p. resin Polysulfonharz *n*
polysynthetic twinning polysynthetische Zwillingsbildung (Verzwillingung) *f*
polyterpene $(C_5H_8)_n$ Polyterpen *n*
polytetrafluoroethylene Polytetrafluoräth[yl]en *n*, PTFE
  p. fibre Polytetrafluoräth[yl]enfaser *f*; Polytetrafluoräth[yl]enfaserstoff *m*, PFT
polythene *s.* polyethylene
polythionate $M'_2[S_xO_6]$ Polythionat *n*
polythionic acid Polythionsäure *f*
polythiourea Thioharnstoffharz *n*, Polythiokarbamid *n*
polytriazole Polytriazol *n*
polytropic *(phys Ch)* polytropisch
polyunsaturated mehrfach ungesättigt
polyurea Harnstoffharz *n*, Polyharnstoff *m*, Polykarbamid *n*
  p. fibre Polyharnstoffaser *f*; Polyharnstofffaserstoff *m*, PH
polyurethane Polyurethan *n*, PUR
  p. elastomer Polyurethanelastomer[es] *n*, Urethanelastomer[es] *n*
  p. fibre Polyurethanfaser *f*; Polyurethanfaserstoff *m*, PU
  p. foam Polyurethanschaum[stoff] *m*
  p. resin Polyurethanharz *n*
  p. rubber Polyurethankautschuk *m*, Urethankautschuk *m*
polyuronic acid Polyuronsäure *f*
polyuronid[e] Polyuronid *n*
polyvalency Mehrwertigkeit *f*, Polyvalenz *f*
polyvalent mehrwertig, polyvalent
polyvinyl acetal Polyvinylazetal *n*
  p. acetate Polyvinylazetat *n*, PVAC

p. alcohol Polyvinylalkohol *m*, PVA
p. alcohol fibre Polyvinylalkoholfaser *f*, Vinylalfaser *f*; Polyvinylalkoholfaserstoff *m*, Vinylalfaserstoff *m*
p. butyral Polyvinylbutyral *n*, PVB
p. carbazole Polyvinylkarbazol *n*, PCV
p. chloride Polyvinylchlorid *n*, PVC
p. chloride acetate Polyvinylchloridazetat *n*, PVCA
p. chloride fibre Polyvinylchloridfaser *f*; Polyvinylchloridfaserstoff *m*, PVC
p. chloride-vinyl acetate copolymer *s.* p. chloride acetate
p. cyclohexane Polyvinylzyklohexan *n*
p. ether Polyvinyläther *m*
p. ethyl ether Polyvinyläthyläther *m*
p. fibre Polyvinylfaser *f*; Polyvinylfaserstoff *m*, PV
p. fluoride Polyvinylfluorid *n*
p. formal Polyvinylformal *n*, Polyvinyl-Formaldehydazetal *n*
p. methyl ether Polyvinylmethyläther *m*
p. propionate Polyvinylpropionat *n*
p. propionate resin Polyvinylpropionatharz *n*
p. toluene Polyvinyltoluol *n*
polyvinylidene chloride Polyvinylidenchlorid *n*, PVDC
p. chloride fibre Polyvinylidenchloridfaser *f*; Polyvinylidenchloridfaserstoff *m*, PVD
p. chloride resin Polyvinylidenchloridharz *n*
p. cyanide fibre Polyvinylidenzyanidfaser *f*, Dinitrilfaser *f*; Polyvinylidenzyanidfaserstoff *m*, Dinitrilfaserstoff *m*
p. dinitrile fibre Polyvinylidendinitrilfaser *f*; Polyvinylidendinitrilfaserstoff *m*
p. fluoride Polyvinylidenfluorid *n*
polyvinylpyrrolidone Polyvinylpyrrolidon *n*, Polyvinylpyrrolidinon *n*, PVP
pomace 1. Mus *n*, Pulpe *f*, Pülpe *f*, Fruchtmasse *f*, Fruchtbrei *m*; Maische *f*, Treber *pl*, Trester *pl*; Preßrückstand *m*, Preßkuchen *m*, *(i.e.S.)* Rizinussaatkuchen *m*; 2. organischer Dünger *m*
pomade, pomatum Pomade *f*
Pomilio process *(Pap)* Pomilio-Verfahren *n* *(Chloraufschluß)*
pommel / to *(Gerb)* krispeln, aufkrausen
pommel Kolben *m* *(z.B. einer Schneckenpresse)*; *(Gerb)* Krispelholz *n*
Pompeian red Pompejanischrot *n* *(rotbraune Anstrichfarbe aus feinpulvrigem Fe_2O_3)*
pond 1. *(Tech)* Sumpf *m*; 2. *(Tech)* Teich *m*, Becken *n*
p. cooler Tauchkühler *m*
p. depth *(Tech)* Sumpftiefe *f*
Pongam oil Kurrunjeöl *n* *(von Pongamia pinnata (L.)Merr.)*
pontianac, pontianak [gum] *(halbfossiler)* Manila-Kopal *m* *(von Agathis alba (Lam.) Foxw.)*; Pontianak *m*, *(veraltet für)* Djelutung *n*, Djelutungharz *n* *(Kopal von Dyera-Arten und Parthenium argentatum A. Gray)*

pony press *(Pap)* Babypresse *f*, Vorpresse *f*
p. roll *(Pap)* Vorpreßwalze *f*
pool 1. erdölführendes *(oder)* erdgashaltiges Speichergestein *n*; 2. Becken *n*
p. cathode flüssige Katode *f*, Flüssigkeitskatode *f*
poor arm; schlecht, gering[wertig]
  p. concrete magerer Beton *m*, Magerbeton *m*, Sparbeton *m*, Füllbeton *m*
  p. conductor schlechter Leiter *m*
  p. in hydrogen wasserstoffarm
  p. lime Magerkalk *m*
  p. solvent schlechtes Lösungsmittel *n*
  p.-venting sand wenig gasdurchlässiger Sand (Formsand) *m*, Sand *m* geringer Gasdurchlässigkeit
poorly absorbing schwach absorbierend
  p. conducting schlecht leitend
P.O.P. *s.* printing-out paper
pop / **to** *(z.B. Getreidekörner)* puffen
popcorn Puffmais *m*, gepuffter Mais *m*
poppy Mohn *m*, Papaver L., *(i.e.S.)* Schlafmohn *m*, Papaver somniferum L.
  p. capsule (head) Mohnkapsel *f*
  p. oil *s.* p.-seed oil
  p. seed Mohnsamen *m*
  p.-seed oil Mohnöl *n*
populate / **to** *(z.B. ein Energieband mit Elektronen)* besetzen
population Besetzung *f (z.B. eines Energiebandes mit Elektronen)*
  p. equivalent *(Wasserchemie)* Einwohnergleichwert *m*
  p. inversion *(phys Ch)* Besetzungsinversion *f (des Energieniveaus von Atomen)*
porcelain 1. Porzellan *n*, *(i.e.S.)* Hartporzellan *n*; Weichporzellan *n*; 2. Porzellan *n*, Porzellanwaren *pl*
  p. basin Porzellan[abdampf]schale *f*
  p. boat Porzellanschiffchen *n*
  p. body Porzellanscherben *m*
  p. casserole Porzellankasserolle *f*
  p. cement Porzellankitt *m*
  p. clay Porzellanerde *f*, weißer (reiner) Ton *m*, Kaolin *m*, China Clay *m(n)*
  p. crucible Porzellantiegel *m*
  p. dish *s.* p. basin
  p. earth *s.* p. clay
  p. enamel Email *n*, Emaille *f*
  p. evaporating basin (dish) *s.* p. basin
  p. filter Porzellanfilter *n*
  p. insulator Porzellanisolator *m*
  p. jasper Porzellanjaspis *m (natürlich gefritteter Ton)*
  p. packing Füllkörper *mpl* aus Porzellan
  p. pipe Porzellanrohr *n*, Porzellanröhre *f*
  p. plate Porzellanplatte *f*
  p. process Porzellanverfahren *n*
  p. tank Porzellanbehälter *m*
  p. tile Porzellanfliese *f*

p. tissue paper Porzellanseidenpapier *n (ein Einwickelpapier)*; Porzellandruck-Seidenpapier *n*
p. triangle Porzellandreieck *n*, Dreieck *n* mit Porzellanröhren
porcelainization Porzellanherstellung *f*, Porzellanbereitung *f*
porcelainize / **to** zu Porzellan brennen, Porzellan herstellen; porzellanartige Überzüge herstellen
porcelainous, porcel[i]aneous, porcel[i]anous aus Porzellan, porzellanen; porzellanartig, porzellanähnlich
pore Pore *f*
  p. diameter Porendurchmesser *m*
  p. filler Porenfüllmittel *n*, Porenfüller *m*
  p.-forming porenbildend
  p. size Porenweite *f*, Porengröße *f*
  p.-size distribution Porengrößenverteilung *f*
  p. space Porenraum *m*
  p. volume Porenvolumen *n*
  p. water Porenwasser *n*
poriferous porös, porig, mit Poren versehen
porose *s.* porous
porosimeter Porosimeter *n (Gerät zur Messung der Porosität)*
porosity Porosität *f*, Porigkeit *f*; relatives Porenvolumen *n*, relativer Porenraum *m*
porous porös, porig, mit Poren versehen; löcherig; blasig, schwammig; undicht, durchlässig
  p. cell (cup) *s.* p. pot
  p. diaphragm Diaphragma *n*, poröse Scheidewand *f*
  p. filter medium poröse Filtermasse *f*
  p.-glass filter poröses Glasfilter *n*
  p. material poröser (poriger) Werkstoff (Sinterwerkstoff) *m*
  p. membrane *s.* p. diaphragm
  p.-metal filter Metallfilter *n*
  p. plug Drosselpfropfen *m*
  p. pot poröser Tonzylinder *m*, poröse Tonzelle *f*, Tondiaphragma *n*
porpezite *(Min)* Porpezit *m*, Palladiumgold *n*
porphin[e] Porphin *n (ein Pyrrolfarbstoff)*
  p. ring Ringsystem *n* des Porphins, Porphinskelett *n*
porphobilinogen Porphobilinogen *n*
porphyrin Porphyrin *n (Porphinderivat)*
  p. chelate Porphyrinchelat *n*
porphyrite Porphyrit *m*
porphyritic porphyrisch
porphyroblast *(Geol)* Porphyroblast *m*
porphyroblastic *(Geol)* porphyroblastisch
porphyry Porphyr *m*
  p. roll Porphyrwalze *f*
port 1. Öffnung *f*, Stutzen *m*, Mund *m*, Anschluß *m*; 2. Portwein *m*
  p. mouth *(Glas)* Brennermaul *n*, Brennermündung *f*
  p. neck *(Glas)* Brennerhals *m*
  p. opening *s.* p. mouth
portable tragbar; transportabel, ortsbeweglich

p. mixer Anklemmrührer *m*
portion Teil *m*, Anteil *m*, Portion *f*, [abgemessene] Menge *f*, Quantum *n*
 in portions portionsweise, portionenweise
portioning device Dosierapparat *m*, Zuteiler *m*
portland blast-furnace [slag] cement Hochofenzement *m*, HOZ, Hüttenzement *m*
 p. cement Portlandzement *m*, PZ
 p. cement clinker Portlandzementklinker *m*
position / to einstellen, in die richtige Lage (Stellung) bringen
position Lage *f*, Stellung *f*, Position *f*
 in meta p. in meta-Stellung (*m*-Stellung, 1,3-Stellung), metaständig, *m*-ständig
 in ortho p. in ortho-Stellung (*o*-Stellung, 1,2-Stellung), orthoständig, *o*-ständig
 in para p. in para-Stellung (*p*-Stellung, 1,4-Stellung), paraständig, *p*-ständig
 p. isomer Stellungsisomer[es] *n*
 p. isomerism Stellungsisomerie *f*, Ortsisomerie *f*
 p. of equilibrium Gleichgewichtslage *f*
 p. of fusion Kondensationsstelle *f*, Anellierungsstelle *f*
positioner Positioner *m*, Stellwerk *n*
positive positiv
positive *(Foto)* Positiv *n*
 p. adsorption positive Adsorption *f*
 p. auxochromic group positives Auxochrom *n*, basische auxochrome Gruppe *f*
 p. birefringence positive Doppelbrechung *f*
 p. catalysis [positive] Katalyse *f*
 p. catalyst [positiver] Katalysator *m*
 p. column *(phys Ch)* positive Säule *f*
 p. crystal optisch positiver Kristall *m*
 p. developer *(Foto)* Positiventwickler *m*
 p.-displacement blower Umlaufkolbengebläse *n*, Kapselgebläse *n*
 p.-displacement flowmeter Verdrängungs[volumen]zähler *m*
 p.-displacement pump Verdrängerpumpe *f*
 p. electricity positive Elektrizität *f*
 p. electrode positive (positiv geladene) Elektrode *f*, Anode *f*
 p. electron positives Elektron *n*, Positron *n*
 p. emulsion *(Foto)* Positivemulsion *f*
 p. film Positivfilm *m*
 p. image positives Bild *n*, Positivbild *n*
 p. ion positives (positiv geladenes) Ion *n*, Kation *n*
 p. material *(Foto)* Positivmaterial *n*
 p. mould Füll[raum]form *f*, Füllraumwerkzeug *n*, Positivform *f*
 p.-negative electron pair Elektron-Positron-Paar *n*, Positron-Elektron-Paar *n*
 p. plate positive Platte *f*, Plusplatte *f*
 p. print positive Kopie *f*
 p. ray positiver Strahl *m*, Kanalstrahl *m*
 p.-ray analysis Kanalstrahlanalyse *f*
 p. rotary blower s. p. displacement blower

p. valency positive [elektrochemische] Wertigkeit *f*, positive Elektrovalenz *f*
positively charged positiv geladen
positon *s.* positron
positron Positron *n*, positives Elektron *n*
 p. decay (disintegration) Positronenzerfall *m*
 p.-electron pair Positron-Elektron-Paar *n*, Elektron-Positron-Paar *n*
 p. emission Positronenemission *f*
 p. formation Positronenbildung *f*
 p. radiation Positronenstrahlung *f*
 p. radiator Positronenstrahler *m*
positronium Positronium *n* *(aus Positron und Elektron bestehendes System)*
post *(Ker)* Brennstütze *f*, Stütze *f* *(Brennhilfsmittel)*
postage stamp paper Briefmarkenpapier *n*
postblossom spray *(Landw)* Nachblütenspritzung *f*
postchlorinate / to nachchlorieren; nachchloren
postchlorinating, postchlorination Nachchlorieren *n*, Nachchlorierung *f*; Nachchloren *n*, Nachchlorung *f*
postcrystalline deformation *(Geol)* postkristalline Deformation *f*
postcrystallization Nachkristallisation *f*
postcure, postcuring *(Gum)* *s.* postvulcanization; *(Plast)* Nachhärten *n*, Nachhärtung *f*
postdefecation *(Zucker)* Nachscheidung *f*
postemergence herbicide Nachauflaufherbizid *n*
 p. treatment Nachauflaufbehandlung *f* *(Einbringen von Herbiziden in den Boden nach dem Keimen der Kulturpflanzensamen)*
poster paper Affichenpapier *n*, Anschlagpapier *n*, Plakatpapier *n*
posterior pituitary Hypophysenhinterlappenpulver *n*
 p.-pituitary extract Hypophysenhinterlappenextrakt *m*
posters *s.* poster paper
posteruption phenomena postvulkanische (nachvulkanische) Erscheinungen *fpl*, Nachklangerscheinungen *fpl*, Nachhallerscheinungen *fpl*
postexposure *(Foto)* Nachbelichtung *f*
postformed laminated section nachgeformtes Schichtstoffprofil *n*
 p. moulding *(Plast)* nachgeformtes Formteil *n*
postforming nachträgliches Formen *n*, nachträgliche Formung *f*
postharvest maturation Nachreifen *n*, Nachreifung *f*
 p. maturation in the stack (stock) Schwitzen *n* auf dem Stock, Ausschwitzen *n*, Nachreifen *n* *(der in die Mälzerei angelieferten Rohgerste)*
postheat treatment Wärmenachbehandlung *f*
postliming *s.* postdefecation
postmoulding deformation Nachverformung *f*
postoven cure Nachvulkanisation *f* [im Ofen], Nachheizung *f* [im Ofen]
postperiod Nachperiode *f*, Nachversuch *m* *(einer kalorimetrischen Messung)*

postpolymerization Nachpolymerisation *f*
poststretching *(Text)* Nachverstrecken *n*
posttectonic recrystallization posttektonische (postkinematische) Umkristallisation *f*
postvolcanic postvulkanisch, nachvulkanisch
  p. phenomena *s.* posteruption phenomena
postvulcanization Nachvulkanisation *f*, Nachheizung *f*; Vulkanisation *f* nach dem Trocknen von Latex
pot / **to** *(z.B. Eier, Fleisch usw.)* einlegen, einmachen
pot Topf *m*; Tiegel *m*; Krug *m*, Kanne *f*; Hafen *m*, Glashafen *m*; Blase *f (eines Destillierapparates)*; *(Plast)* Spritztopf *m*, Füllzylinder *m*, Füllraum *m*; *(Landw)* Kulturgefäß *n*; Farbkessel *m (in der Farbspritztechnik)*; *s.* p. heater [vulcanizer]
  p. ale Schlempe *f*
  p.-anneal / **to** topfglühen
  p. annealing Topfglühen *n*, Topfglühung *f*
  p.-annealing furnace Topfglühofen *m*
  p. arch *(Glas)* Hafentemperofen *m*
  p. barley Gerstengraupen *pl*, Graupen *pl*
  p. cheese Topfkäse *m*, Kochkäse *m*
  p. curare Topfkurare *n*
  p. experiment *(Bio)* Gefäßversuch *m*
  p. furnace *(Glas)* Hafenofen *m*
  p. glass Hafenglas *n*, im Hafen geschmolzenes Glas *n*
  p. heater [vulcanizer] Autoklavheizpresse *f*, Autoklav[en]presse *f*, Autoklav *m*
  p. life Topfzeit *f*, Standzeit *f*, Potlife *n (Zeit, in der Zwei- oder Mehrkomponentenlacke verarbeitungsfähig bleiben, ohne daß Verdickung eintritt)*
  p.-melted glass *s.* p. glass
  p. melting *(Glas)* Hafenschmelze *f*
  p. plant Mehrkörperextraktionsanlage *f*
  p. plunger Spritz[preß]kolben *m*
  p. press Kachelpresse *f*, Trogpresse *f*, Ring-, presse *f*, Schachtelpresse *f*
  p. spinning Topfzentrifugenspinnverfahren, *n*, Spinntopfverfahren *n*, Zentrifugen[spinn]verfahren *n*, Zentrifugenspinnen *n*
  p. still Schalendestillierapparat *m*, Blasendestillierapparat *m*
  p. study (test) *s.* p. experiment
  p. valve Tellerventil *n (Kegelsitzventil)*
potability Trinkbarkeit *f*
potable trinkbar
potable Getränk *n*
  p. spirit Trinkbranntwein *m*
  p. water Trinkwasser *n*
potash K_2CO_3 Pottasche *f*, Kaliumkarbonat *n*; KOH Kaliumhydroxid *n*, Ätzkali *n*; K_2O Kaliumoxid *n*, Kali *n*; Kalisalz *n*, Kali *n*
  p. alum $KAl(SO_4)_2 \cdot 12H_2O$ Kalium[aluminium]-alaun *m*, Kalialaun *m*, Alaun *m*, Kaliumaluminiumsulfat-12-Wasser *n*
  p. brine *(natürlich vorkommende)* kalihaltige Lauge *f*
  p. bulb Kaliapparat *m (ein Absorptionsgefäß)*

p. chrome alum $KCr(SO_4)_2 \cdot 12H_2O$ Kaliumchromalaun *m*, Chromalaun *m*, Kaliumchrom(III)-sulfat-12-Wasser *n*
p. deposit Kali[salz]lager *n*, Kalisalzlagerstätte *f*
p. feldspar *(Min)* Kalifeldspat *m (Kaliumalumotrisilikat)*
p. fertilizer Kalidünger *m*, Kalidüngemittel *n*
p. fusion Kalischmelze *f*
p. glass Kaliglas *n*
p. industry Kaliindustrie *f*
p. lye Kalilauge *f*, Kaliumhydroxidlösung *f*
p. magnesia $K_2SO_4 \cdot MgSO_4$ Kalimagnesia *f*, Kaliummagnesiumsulfat *n*
p. melt *s.* p. fusion
p. mica *(Min)* Kaliglimmer *m*, Muskovit *m*
p. mine wastes Kaliabwasser *n*
p. ore Kalirohsalz *n*
p. salt Kalisalz *n*, Kali *n*
p. [soft] soap Kali[um]seife *f*
p. water glass Kaliwasserglas *n*
potassa *s.* 1. potash; 2. potassium hydroxide
potassamide *s.* potassium amide
potassic kaliumhaltig, Kalium...; kalihaltig, Kali...
  p. fertilizer Kalidünger *m*, Kalidüngemittel *n*
potassiferous kalihaltig, Kali...
  p. salt Kalisalz *n*, Kali *n*
potassium K Kalium *n*
  p. acetate CH_3COOK Kaliumazetat *n*
  p. acid carbonate *s.* p. hydrogen carbonate
  p. acid fluoride *s.* p. hydrogen fluoride
  p. acid oxalate *s.* p. hydrogen oxalate
  p. acid phosphate *s.* p. dihydrogen phosphate
  p. acid phthalate *s.* p. hydrogen phthalate
  p. acid sulphate *s.* p. hydrogen sulphate
  p. acid sulphite *s.* p. hydrogen sulphite
  p. acid tartrate *s.* p. hydrogen tartrate
  p. alum $KAl(SO_4)_2 \cdot 12H_2O$ Kalium[aluminium]-alaun *m*, Kalialaun *m*, Alaun *m*, Kaliumaluminiumsulfat-12-Wasser *n*
  p. aluminate $K(AlO_2)$ Kaliumaluminat *n*
  p.-aluminium sulphate *s.* p. alum
  p. amalgam Kaliumamalgam *n*
  p. amide KNH_2 Kaliumamid *n*
  p. ammine pentachloroplatinate $K_2[Pt(NH_3)Cl_5]$ Kaliumpentachloroamminplatinat(IV) *n*
  p.-ammonium tartrate $NH_4KC_4H_4O_6$ Kaliumammoniumtartrat *n*
  p. anhydrosulphate *s.* p. pyrosulphate
  p. antimonyl tartrate $K[C_4H_2O_6Sb(OH)_2]$ Kaliumantimonyltartrat *n*, *(besser)* Kaliumantimonotartrat *n*
  p. argentocyanide *s.* p.-silver cyanide
  p. aurate $KAuO_2$ Kaliumaurat *n*
  p. auricyanide $K[Au(CN)_4]$ Kaliumtetrazyanoaurat(III) *n*
  p. aurocyanide *s.* p. cyanaurite
  p. azide KN_3 Kaliumazid *n*
  p. bicarbonate *s.* p. hydrogen carbonate
  p. bichromate *s.* p. dichromate
  p. bifluoride *s.* p. hydrogen fluoride

p. binoxalate *s.* p. hydrogen oxalate
p. biphosphate *s.* p. dihydrogen phosphate
p. biphthalate *s.* p. hydrogen phthalate
p. bisulphate *s.* p. hydrogen sulphate
p. bisulphide *s.* p. hydrogen sulphide
p. bisulphite *s.* p. hydrogen sulphite
p. bitartrate *s.* p. hydrogen tartrate
p. borotartrate $KC_4H_4BO_7$ Kaliumborotartrat *n*
p. bromate $KBrO_3$ Kaliumbromat *n*
p. bromaurate $K[AuBr_4]$ Kaliumtetrabromoaurat(III) *n*
p. bromide KBr Kaliumbromid *n*
p. bromoplatinate $K_2[PtBr_6]$ Kaliumhexabromoplatinat(IV) *n*
p. bromoplatinite $K_2[PtBr_4]$ Kaliumtetrabromoplatinat(II) *n*
p. carbolate C_6H_5OK Phenolkalium *n*
p. carbonate K_2CO_3 Kaliumkarbonat *n*, Pottasche *f*
p. carbonyl $K_6C_6O_6$ Kaliumkarbonyl *n*
p. chlorate $KClO_3$ Kaliumchlorat *n*
p. chloraurate $K[AuCl_4]$ Kaliumtetrachloroaurat(III) *n*
p. chloride KCl Kaliumchlorid *n*
p. chlorite $KClO_2$ Kaliumchlorit *n*
p. chloroplatinate $K_2[PtCl_6]$ Kaliumhexachloroplatinat(IV) *n*
p. chloroplatinite $K_2[PtCl_4]$ Kaliumtetrachloroplatinat(II) *n*
p. chromate K_2CrO_4 Kaliumchromat *n*
p.-chrome alum *s.* p. chromium sulphate
p.-chromium sulphate $KCr(SO_4)_2 \cdot 12H_2O$ Kaliumchrom(III)-sulfat-12-Wasser *n*, Kaliumchromalaun *m*, Chromalaun *m*
p. citrate $K_3C_6H_5O_7$ Kaliumzitrat *n*
p. cobaltinitrite *s.* p. hexanitrocobaltate(III)
p. cobaltocyanide $K_4[Co(CN)_6]$ Kaliumhexazyanokobaltat(II) *n*
p.-containing kaliumhaltig
p. cyanate KOCN Kaliumzyanat *n*
p. cyanaurite $K[Au(CN)_2]$ Kaliumdizyanoaurat(I) *n*, Kaliumgold(I)-zyanid *n*
p. cyanide KCN Kaliumzyanid *n*, Zyankali *n*
p. cyanoplatinite $K_2[Pt(CN)_4]$ Kaliumtetrazyanoplatinat(II) *n*
p. deficiency Kalimangel *m*
p. diborane (diboranide) $K_2B_2H_6$ Dikaliumhexahydridoborat *n*
p. dichromate $K_2Cr_2O_7$ Kaliumdichromat(VI) *n*
p. dihydrogen arsenate KH_2AsO_4 Kaliumdihydrogen[ortho]arsenat *n*
p. dihydrogen [ortho]phosphate KH_2PO_4 Kaliumdihydrogen[ortho]phosphat *n*, Monokaliumdihydrogenphosphat *n*
p. dioxide K_2O_2 Kaliumperoxid *n*
p. diphosphate *s.* p. dihydrogen [ortho]phosphate
p. dysentery *(Med)* Kaliruhr *f*
p. ethyl dithiocarbonate *s.* p. xanthogenate
p. ethyl xanthate (xanthogenate) *s.* p. xanthogenate

p. ferric sulphate $KFe(SO_4)_2 \cdot 12H_2O$ Kaliumeisen(III)-sulfat-12-Wasser *n*, Kaliumeisenalaun *m*
p. ferricyanide *s.* p. hexacyanoferrate(III)
p. ferrocyanide *s.* p. hexacyanoferrate(II)
p. fluoride KF Kaliumfluorid *n*
p. fluoroborate $K[BF_4]$ Kaliumtetrafluoroborat *n*
p. fluorosilicate $K_2[SiF_6]$ Kaliumhexafluorosilikat *n*
p. fluorozirconate $K_2[ZrF_6]$ Kaliumhexafluorozirkonat(IV) *n*
p. formate HCOOK Kaliumformiat *n*
p. hexachloroplatinate $K_2[PtCl_6]$ Kaliumhexachloroplatinat(IV) *n*
p. hexacyanoferrate(II) $K_4[Fe(CN)_6]$ Kaliumhexazyanoferrat(II) *n*, gelbes Blutlaugensalz *n*
p. hexacyanoferrate(III) $K_3[Fe(CN)_6]$ Kaliumhexazyanoferrat(III) *n*, rotes Blutlaugensalz *n*
p. hexanitrocobaltate(III) $K_3[Co(NO_2)_6]$ Kaliumhexanitrokobaltat(III) *n*
p. hexylxanthogenate $C_6H_{13}OCSSK$ Kaliumhexylxanthogenat *n*
p. hydrate *s.* p. hydroxide
p. hydride KH Kaliumhydrid *n*
p. hydrogen carbonate $KHCO_3$ Kaliumhydrogenkarbonat *n*
p. hydrogen fluoride KHF_2 Kaliumhydrogenfluorid *n*
p. hydrogen oxalate KHC_2O_4 Kaliumhydrogenoxalat *n*, Monokaliumoxalat *n*
p. hydrogen phosphate K_2HPO_4 Kaliumhydrogen[ortho]phosphat *n*, Dikaliumhydrogenphosphat *n*
p. hydrogen phthalate $C_6H_4(COOH)(COOK)$ Kaliumhydrogenphthalat *n*
p. hydrogen sulphate $KHSO_4$ Kaliumhydrogensulfat *n*
p. hydrogen sulphide KSH Kaliumhydrogensulfid *n*
p. hydrogen sulphite $KHSO_3$ Kaliumhydrogensulfit *n*
p. hydrogen tartrate $KHC_4H_4O_6$ Kaliumhydrogentartrat *n*
p. hydrosulphide *s.* p. hydrogen sulphide
p. hydrotartrate *s.* p. hydrogen tartrate
p. hydroxide KOH Kaliumhydroxid *n*, Ätzkali *n*
p. hydroxide fusion Kaliumhydroxidschmelze *f*
p. hydroxide solution Kaliumhydroxidlösung *f*, Kalilauge *f*
p. hyperchlorate *s.* p. perchlorate
p. hypochlorite KOCl Kaliumhypochlorit *n*
p. hypophosphite KH_2PO_2 Kaliumhypophosphit *n*
p. hyposulphite *s.* p. thiosulphate
p. iodate KJO_3 Kaliumjodat *n*
p.-iodate-starch paper Kaliumjodatstärkepapier *n*
p. iodide KJ Kaliumjodid *n*
p.-iodide-starch indicator Kaliumjodidstärkeindikator *m*

p.-iodide-starch paper Kaliumjodidstärkepapier n, Jodkaliumstärkepapier n, Ozonpapier n
p. lactate $KC_3H_5O_3$ Kaliumlaktat n
p. line Spektrallinie f des Kaliums, Kaliumlinie f
p.-magnesium sulphate $K_2SO_4 \cdot 2MgSO_4$ Kaliummagnesiumsulfat n
p. manganate K_2MnO_4 Kaliummanganat n
p. mercuric cyanide $K_2[Hg(CN)_4]$ Kaliumtetrazyanomerkurat(II) n
p. mercuric iodide $K_2[HgJ_4]$ Kaliumquecksilberjodid n
p. metaborate KBO_2 oder $K_2B_2O_4$ Kaliummetaborat n
p. metaphosphate $(KPO_3)_n$ Kaliummetaphosphat n
p. metaplumbate K_2PbO_3 Kaliumtrioxoplumbat-(IV) n, Kaliummetaplumbat(IV) n
p. metasilicate K_2SiO_3 Kaliumtrioxosilikat n, Kaliummetasilikat n
p. metatungstate $K_2W_4O_{13}$ Kaliummetawolframat(VI) n
p. methylate $KOCH_3$ Kaliummethylat n, Kaliummethoxid n
p. mica *(Min)* Kaliglimmer m, Muskovit m
p. monophosphate s. p. hydrogen phosphate
p. monosulphide K_2S Kalium[mono]sulfid n
p. muriate s. p. chloride
p.-nickel cyanide $K_2[Ni(CN)_4]$ Kaliumtetrazyanoniccolat n
p. nitrate KNO_3 Kaliumnitrat n
p. nitride K_3N Kaliumnitrid n
p. nitrite KNO_2 Kaliumnitrit n
p. nitroprusside $K_2[Fe(CN)_5(NO)]$ Kaliumnitroprussid n, Kaliumnitroprussiat n, Dikaliumpentazyanonitrosylferrat(II) n
p. oleate Kaliumoleat n
p. orthoarsenite K_3AsO_3 Kalium[ortho]arsenat-(III) n
p. orthophosphate K_3PO_4 Kalium[ortho]phosphat n, Trikaliumphosphat n
p. orthotungstate s. p. tungstate
p. osmate K_2OsO_4 Kaliumosmat(VI) n
p. oxalate $K_2C_2O_4$ Kaliumoxalat n
p. oxide K_2O Kalium[mon]oxid n
p. oxymuriate s. p. chlorate
p. palmitate Kaliumpalmitat n
p. paratungstate $K_6W_7O_{24}$ Kaliumparawolframat n
p. pentaborane (pentaboranide) $K_2B_5H_9$ Dikaliumnonahydridopentaborat n
p. pentasulphide K_2S_5 Kaliumpentasulfid n
p. pentathionate $K_2S_5O_6$ Kaliumpentathionat n
p. perborate KBO_3 Kaliumperoxoborat n
p. percarbonate $K_2C_2O_6$ Kaliumperoxo[di]karbonat n
p. perchlorate $KClO_4$ Kaliumperchlorat n
p. perchromate K_3CrO_8 Kaliumperoxochromat n, Kaliumtetraperoxochromat(V) n
p. periodate KJO_4 Kaliummetaperjodat n, Kaliumtetroxojodat n

p. permanganate $KMnO_4$ Kaliumpermanganat n, Kaliummanganat(VII) n
p. peroxide K_2O_2 Kaliumperoxid n; s. p. superoxide
p. peroxychromate s. p. perchromate
p. peroxydisulphate $K_2S_2O_8$ Kaliumperoxodisulfat n
p. peroxysulphate s. p. peroxydisulphate
p. perrhenate $KReO_4$ Kaliumperrhenat n, Kaliumtetraoxorhenat(VII) n
p. persulphate s. p. peroxydisulphate
p. phenate C_6H_5OK Phenolkalium n
p. platinic chloride s. p. hexachloroplatinate
p. platinichloride s. p. hexachloroplatinate
p. platinochloride s. p. chloroplatinite
p. platinocyanide s. p. cyanoplatinite
p.-platinum cyanide s. p. cyanoplatinite
p. press Natriumpresse f
p. pyroborate s. p. tetraborate
p. pyrophosphate $K_4P_2O_7$ Kaliumdiphosphat n, Kaliumpyrophosphat n
p. pyrosulphate $K_2S_2O_7$ Kaliumdisulfat n, Dikaliumdisulfat n, Kaliumpyrosulfat n
p. pyrosulphite $K_2S_2O_5$ Kaliumdisulfit n, Kaliumpyrosulfit n
p. quadroxalate s. p. tetroxalate
p. rhodanide s. p. thiocyanate
p. selenate K_2SeO_4 Kaliumselenat n
p. selenide K_2Se Kaliumselenid n
p. selenite K_2SeO_3 Kaliumselenit n
p. selenocyanide $KSeCN$ Kaliumselenozyanat n
p. silicate s. p. metasilicate
p. silicofluoride s. p. fluorosilicate
p. silicotungstate $K_4[SiW_{12}O_{40}]$ Kaliumsilikododekawolframat n
p.-silver cyanide $K[Ag(CN)_2]$ Kaliumsilberzyanid n, Kaliumdizyanoargentat n
p. soap Kali[um]seife f
p.-sodium carbonate $KNaCO_3$ Kaliumnatriumkarbonat n
p.-sodium cobaltinitrite $K_2Na[Co(NO_2)_6]$ Kaliumnatriumhexanitrokobaltat(III) n
p.-sodium tartrate $KNaC_4H_4O_6 \cdot 4H_2O$ Kaliumnatriumtartrat n, Seignettesalz n
p. stannate K_2SnO_3 Kaliumtrioxostannat(IV) n
p. stearate Kaliumstearat n
p. sulphate K_2SO_4 Kaliumsulfat n
p. sulphhydrate s. p. hydrogen sulphide
p. sulphite K_2SO_3 Kaliumsulfit n
p. sulphocarbonate s. p. thiocarbonate
p. sulphocyanate s. p. thiocyanate
p. sulphocyanide s. p. thiocyanate
p. superoxide KO_2 Kaliumhyperoxid n
p. tartrate $K_2C_4H_4O_6$ Kaliumtartrat n
p. tetraborate $K_2B_4O_7$ Kaliumtetraborat n
p. tetrasilicate $K_2Si_4O_9$ Kaliumtetrasilikat n
p. tetrasulphide K_2S_4 Kaliumtetrasulfid n
p. tetrathionate $K_2S_4O_6$ Kaliumtetrathionat n
p. tetroxalate $KHC_2O_4 \cdot H_2C_2O_4$ Kaliumtetroxalat n, Kaliumtrihydrogenoxalat n

p. tetroxide KO_2 Kaliumhyperoxid *n*
p. thioarsenate K_3AsS_4 Kaliumthioarsenat(V) *n*
p. thioarsenite K_3AsS_3 Kaliumthioarsenat(III) *n*
p. thiocarbonate K_2CS_3 Kaliumthiokarbonat *n*
p. thiocyanate. KSCN Kaliumthiozyanat *n*, Kaliumrhodanid *n*
p. thiocyanate paper Kaliumthiozyanatpapier *n*
p. thiosulphate $K_2S_2O_3$ Kaliumthiosulfat *n*
p. tongs Natriumzange *f*
p. trithiocarbonate *s.* **p. thiocarbonate**
p. tungstate K_2WO_4 Kaliumwolframat *n*
p. water glass Kaliwasserglas *n*
p. xanthogenate C_2H_5OCSSK Kaliumxanthogenat *n*, Kaliumäthylxanthogenat *n*
p. zirconifluoride *s.* **p.** fluorozirconate
potato alcohol Kartoffelspiritus *m*, Kartoffelsprit *m*
p. flour Kartoffelmehl *n*, Stärkemehl *n*; *s.* **p.** starch
p. mash Kartoffelmaische *f*, Kartoffelbrei *m*
p. meal Kartoffelwalzmehl *n*
p. oil Fuselöl *n*
p. slump Kartoffelschlempe *f*
p. spirits *s.* **p.** alcohol
p. starch Kartoffelstärke *f*
potch / to (Pap) aufschlagen; (Pap) (Stoff im Bleichholländer mit Bleichlauge) mischen
potcher, potching engine (Pap) Bleichholländer *m*; Mischholländer *m*; Waschholländer *m*
potency Wirkkraft *f*, Wirkungsstärke *f*, Wirksamkeit *f*, Potenz *f*; (Homöopathie) Potenz *f*, [hohe] Verdünnung *f*
potent stark wirkend, wirksam
potential potentiell, Potential...
potential Potential *n*
p. barrier Potentialwall *m*, Potentialschwelle *f*, Potentialberg *m*, Potentialbarriere *f*
p. difference Potentialdifferenz *f*, Potentialunterschied *m*
p. drop Potentialabfall *m*, Potentialgefälle *n*, Potentialrückgang *m*
p. energy potentielle Energie *f*, Lageenergie *f*, Zustandsenergie *f*
p.-energy barrier *s.* **p.** barrier
p.-energy curve Potentialkurve *f*
p.-energy equation *s.* **p.** equation
p.-energy surface Potentialfläche *f*, Äquipotentialfläche *f*, Niveaufläche *f*
p. equation Potentialgleichung *f*
p. field Potentialfeld *n*
p.-forming potentialbildend
p. function Potentialfunktion *f*
p. gradient Potentialgradient *m*
p. gum potentieller (möglicher) Gum *m*, potentielles (mögliches) Harz *n*, Harzneubildung *f*, Potential Gum *m*
p. hill *s.* **p.** barrier
p. hole *s.* **p.** well
p. jump Potentialsprung *m*
p. measurement Potentialmessung *f*
p. mediator Potentialvermittler *m*
p. of a single electrode Einzelpotential *n*, E

p. of an electrode Elektrodenpotential *n*
p. scattering Potentialstreuung *f*
p. temperature potentielle Temperatur *f*
p. threshold (wall) *s.* **p.** barrier
p. well Potentialtopf *m*, Potentialmulde *f*, Potentialkasten *m*
potentiate / to potenzieren, steigern, verstärken; (Homöopathie) potenzieren, verdünnen
potentiation Potenzierung *f*, Steigerung *f*, Verstärkung *f*; Potenzieren *n*, Verdünnen *n* (von homöopathischen Mitteln)
potentiative potenzierend, steigernd, verstärkend
potentiometer Potentiometer *n*
potentiometric potentiometrisch
p. analysis potentiometrische Analyse *f*, (i.e.S.) potentiometrische (elektrometrische) Maßanalyse (Titration) *f*
p. method potentiometrische Methode *f*, Potentiometermethode *f*, potentiometrisches Verfahren *n*, Potentiometerverfahren *n*
p. titration potentiometrische (elektrometrische) Titration (Maßanalyse) *f*
potentiometry Potentiometrie *f*
potette (Glas) Stiefel *m* (Durchlaß zwischen Schmelz- und Arbeitswanne)
p. tank (Glas) Stiefelwanne *f*
potherbs Küchenkräuter *pl*
potstone (Min) Topfstein *m*
Pott-Broche process Pott-Broche-Verfahren *n*, Verfahren *n* von Pott-Broche, Pott-Broche-Druckextraktionsverfahren *n*
potter Töpfer *m*
potter's clay Töpferton *m*
p. wheel Töpferscheibe *f*, Drehscheibe *f*
pottery Töpferei *f*; Töpferware *f*, (i.e.S.) Steingut *n*
p. tissue paper Porzellanseidenpapier *n* (ein Einwickelpapier); Porzellandruck-Seidenpapier *n*
potting (Text) Potting *n*, Potten *n*, Topffärben *n*; (Plast) Einbetten *n*
pounce Bimssteinpulver *n*
pour / to gießen, schütten; fließen, strömen, rinnen
p. concrete betonieren, Beton einbringen
p. off abgießen, dekantieren
p. steel Stahl gießen
pour-on method (Landw) Rückenbegießung *f* (zur chemischen Bekämpfung von Dasellarven)
p. point Fließpunkt *m*, Pourpoint *m* (niedrigste, 2-5 °C über dem Stockpunkt liegende Temperatur, bei der ein Kraftstoff beim Abkühlen gerade noch fließt), Stockpunkt *m* in Form des Pourpoints
p.-point depressant Fließpunktserniedriger *m*, Pourpoint-Depressor *m*
p.-point test Fließpunkt[s]prüfung *f*
pourability Gießbarkeit *f*, Vergießbarkeit *f*
pourable [ver]gießbar
poured-in-place concrete Ortbeton *m*
p. joint ausgegossene Verbindung *f* (Rohrverbindung)

505

praseolite

powder / to 1. pulver[isiere]n, zerpulvern; zu Pulver zerstäuben; 2. [ein]pudern
p. together gemeinsam verreiben
powder Pulver *n*, Mehl *n*; Puder *m*
in fine p. feingepulvert, feinpulverisiert, feinpulverig
p. adhesive Klebstoffpulver *n*
p. blue Smalte *f*, Schmalte *f*, Blaufarbenglas *n* *(Kobalt(II)-kaliumsilikat)*
p. camera Pulverkamera *f*
p. density Schüttdichte *f* von Pulver
p. diagram *(Krist)* Pulverdiagramm *n*, Debye-Scherrer-Diagramm *n*, Debye-Scherrer-Aufnahme *f*
p. extinguisher Trockenlöscher *m*
p. funnel Pulvertrichter *m*
p. insulation Füllstoffisolierung *f*, Stopfisolierung *f*
p. metal gepulvertes Metall *n*, Metallpulver *n*, metallisches Pulver *n*
p. metal compact Metallpulverpreßling *m*, Pulverpreßling *m*, Pulverpreßkörper *m* *(in der Pulvermetallurgie)*
p.-metallurgical pulvermetallurgisch
p.-metallurgical method (process) pulvermetallurgisches Verfahren *n*, pulvermetallurgischer Prozeß *m*, pulvermetallurgische Methode *f*
p. metallurgy Pulvermetallurgie *f*
p.-metallurgy process (technique) s. p.-metallurgical method
p. method [of analysis] *(Krist)* Pulververfahren *n*, Pulvermethode *f*, Debye-Scherrer-Verfahren *n*, Debye-Scherrer-Methode *f*
p. moulding *(Plast)* Pulversinterverfahren *n*
p. of algaroth Sb$_4$O$_5$Cl$_2$ Algarotpulver *n*, Antimonoxidchlorid *n*
p. paper Puderpapier *n*
p. pattern (photograph) *(Krist)* Pulveraufnahme *f*, Debye-Scherrer-Aufnahme *f*
p. rolling Pulverwalzen *n*
p. sintering process s. p. moulding
p. test *(Glas)* Grießprobe *f*
p.-type fire extinguisher s. p. extinguisher
powdered pulv[e]rig, gepulvert, pulverisiert; bepudert, bestäubt; bestreut
p. brown coal Braunkohlen[brenn]staub *m*
p. buttermilk Buttermilchpulver *n*
p. catalyst pulverisierter (gepulverter, pulvriger, pulverförmiger, staubförmiger, staubfeiner) Katalysator *m*, Katalysator *m* in Pulverform, Katalysatorpulver *n*, Katalysatorstaub *m*, Kontaktstaub *m*
p. coal Kohlenstaub *m*, Kohlepulver *n*, pulverisierte (gepulverte) Kohle *f*; Steinkohlenstaub *m*
p. fuel Brenn[stoff]staub *m*
p. glass Glaspulver *n*
p. metal s. powder metal
p.-metal process s. powder-metallurgical method
p. milk Milchpulver *n*, Trockenmilch *f*

p. opium Opiumpulver *n*
p. pectin Trockenpektin *n*
p. porcelain Porzellanmehl *n*
p. reclaim *(Gum)* Regeneratpulver *n*
p. strychnos seed *(Pharm)* Brechnußpulver *n* *(von Strychnos nux-vomica L.)*
p. sugar Puderzucker *m*, Staubzucker *m*
p. whey Trockenmolke *f*, Molkenpulver *n*
powdering agent Pudermittel *n*
powdery pulv[e]rig, pulverartig, pulverförmig, Pulver...; bepudert, bestäubt
power / to *(z.B. einen Stromkreis)* speisen
power Kraft *f*, Stärke *f*; Energie *f*, Leistung *f*; Leistungsvermögen *n*, Leistungsfähigkeit *f*; *(Math)* Potenz *f*
p. alcohol Kraftsprit *m*, Kraftspiritus *m*, Treibspiritus *m*
p. breeder reactor *(Kph)* Energiebrutreaktor *m*, Leistungsbrutreaktor *m*
p. duster *(Landw)* Motorverstäuber *m*
p. factor [dielektrischer] Leistungsfaktor *m*
p. fuel Kraftstoff *m*, Motor[en]kraftstoff *m*, Treibstoff *m*, Motor[en]treibstoff *m*, Motorenbetriebsstoff *m*
p. gas Kraftgas *n*
p. kerosine Motorenpetroleum *n*
p. number spezifische Leistung *f*
p. of holding dirt Schmutzrückhaltevermögen *n*
p.-operated sprayer s. p. sprayer
p. reactor Energiereaktor *m*, Leistungsreaktor *m*
p. sprayer *(Landw)* Druckluftvernebler *m*
p. vaporizing oil s. p. kerosine
pozzolana Puzzolanerde *f*, Pozz[u]olanerde *f*
pozzolanic cement Puzzolanzement *m*
PP s. permanent press
p.p.d. s. 1,4-diaminobenzene
PP factor s. pellagra-preventive factor
PPO s. polyphenylene oxide
practicum Praktikum *n* *(in der Studentenausbildung)*
prase[m] *(Min)* Prasem *m* *(Silizium(IV)-oxid)*
praseo salt *(Komplexchemie)* Praseosalz *n*
praseodymia Pr$_2$O$_3$ Praseodymerde *f*, Praseodymia *f*, *(veraltet für)* Praseodym(III)-oxid *n*
praseodymium Pr Praseodym *n*
p.-ammonium sulphate (NH$_4$)$_2$SO$_4$·Pr$_2$(SO$_4$)$_3$ Ammoniumpraseodymsulfat *n*
p. carbonate Pr$_2$(CO$_3$)$_2$ Praseodymkarbonat *n*
p. chloride PrCl$_3$ Praseodym(III)-chlorid *n*
p. dioxide PrO$_2$ Praseodymdioxid *n*, Praseodym(IV)-oxid *n*
p. sesquioxide s. p. trioxide
p. sulphate Pr$_2$(SO$_4$)$_3$ Praseodymsulfat *n*
p. sulphide Pr$_2$S$_3$ Praseodymsulfid *n*
p. tetroxide PrO$_4$ Praseodymtetroxid *n*, Praseodym(IV)-peroxid *n*
p. trioxide Pr$_2$O$_3$ Praseodymtrioxid *n*, Praseodym(III)-oxid *n*
praseolite, prasiolite *(Min)* Praseolith *m* *(Magnesiumaluminiumalumopentasilikat)*

Prayon [continuous] filter Prayon-Filter *n (Band-zellenfilter aus vielen einzelnen Planfilterzellen, die auf einer kreisförmigen Rollenbahn umlaufen)*
prebaked electrode vorgebackene (vorgebrannte) Elektrode *f*, vorgebrannte Elektrodenkohle (Kohle) *f*
preblend / to vormischen
preblend Vormischung *f*
prebloom spray *(Landw)* Vorblütenspritzung *f*
prebreaker Vorbrecher *m*
precarbonization Vorschwelen *n*, Vorschwelung *f*
precarbonize / to vorschwelen
precast / to vorfertigen
precast concrete vorgefertigter Beton *m*
precession *(phys Ch)* Präzession *f*
prechill / to vorkühlen
prechill Vorkühlen *n*, Vorkühlung *f*
prechlorinate / to *(Pap)* vorchlorieren, vorbleichen; *(Wasser)* vorchloren
prechlorination *(Pap)* Vorchlorierung *f*, Vorbleiche *f*; *(Wasseraufbereitung)* Vorchloren *n*, Vorchlorung *f*
prechrome / to vorchromieren
prechrome process *(Text)* Chrombeizverfahren *n*
prechurning *(Lebm)* Anbuttern *n*
precious metal Edelmetall *n*
p. stone Edelstein *m*
precip. *(Abk. für)* precipitated
precipitability Fällbarkeit *f*, Abscheidbarkeit *f*
precipitable [aus]fällbar, abscheidbar, niederschlagbar
precipitant Fällungsmittel *n*, Fällungs[re]agens *n*, Ausfällungsmittel *n*, Abscheidungsmittel *n*, Niederschlagsmittel *n*, Präzipitiermittel *n*
p. power Fällungsvermögen *n*
precipitate / to [aus]fällen, abscheiden, ausscheiden, niederschlagen, präzipitieren; *(Dämpfe)* kondensieren; ausfallen, sich abscheiden; *(von Dämpfen)* sich kondensieren, sich niederschlagen
precipitate [chemischer] Niederschlag *m*, Bodenkörper *m*, Fällung *f*
precipitated barium sulphate gefälltes Bariumsulfat *n (Permanentweiß)*
p. calcium carbonate gefälltes (präzipitiertes) Kalziumkarbonat *n*, gefällte (präzipitierte) Kreide *f*
p. calcium sulphate gefälltes (präzipitiertes) Kalziumsulfat *n*, Annalin *n*
p. chalk s. p. calcium carbonate
p. copper Zementkupfer *n*
p. form Fällungsform *f (Gewichtsanalyse)*
p. gypsum s. p. calcium sulphate
p. sulphur gefällter Schwefel *m*, Schwefelmilch *f*
p. whiting s. p. calcium carbonate
precipitating s. precipitation
p. action Fällungswirkung *f*
p. agent s. precipitant

p. bath Fällbad *n*
p. electrode Niederschlagselektrode *f*
precipitation Ausfällen *n*, Fällen *n*, Niederschlagen *n*, Abscheiden *n*, Präzipitieren *n*, Präzipitation *f*
p. analysis Fällungs[maß]analyse *f*, Fällungstitration *f*, Fällungsmethode *f*
p. bath s. precipitating bath
p. box *(Met)* Fällkasten *m*
p. by electrolytes Elektrolytfällung *f*
p. figure test *(Gerb)* Basizitätsbestimmung *f (bei Chrombrühen)*
p. hardening Ausscheidungshärten *n*, Aushärten *n*
p. naphtha Normalbenzin *n (als Fällungsmittel)*
p. polymerization Fällungspolymerisation *f*
p. rate Absetzgeschwindigkeit *f*
p. reaction Fällungsreaktion *f*, Niederschlagsreaktion *f*
p. tank Absetzbecken *n*, Absetzbehälter *m*, Absitzbehälter *m*, Abscheider *m*, Fällkasten *m*, Klärtank *m*, Klärbehälter *m*
p. titration s. p. analysis
precipitative [aus]fällend, präzipitierend
precipitator s. 1. precipitant; 2. precipitation tank; 3. dry precipitator
precipitin *(Med)* Präzipitin *n (Antikörper)*
precision Genauigkeit *f*, Präzision *f*
p. balance Präzisionswaage *f*
p. casting Präzisionsguß *m*, Feinguß *m*, Genauguß *m*
p. distillation Feindestillation *f*
p. investment casting Investmentguß *m*, Gießen *n* mit verlorener Gußform, Präzisionsguß *m* nach dem Ausschmelzverfahren
p. photometry Präzisionsfotometrie *f*
p. polarimeter Präzisionspolarimeter *n*
precleaner Vorreiniger *m*
precoat [bed] *(Filtration)* Anschwemmschicht *f*, Anschwemmgut *n*, Precoatschicht *f*
p. clarifier Anschwemmklärfilter *n*
p. filter Anschwemmfilter *n*, Precoatfilter *n*
p. filtering medium angeschwemmtes Filtermedium (Filtriermaterial, Filtermaterial) *n*, Filterhilfsschicht *f*
p. layer s. precoat [bed]
precoated filter s. precoat filter
precoating material s. precoat filtering medium
precompounded *(Plast)* vorbeharzt, vorimprägniert
precondensate *(Plast)* Vorkondensat *n*, Vorkondensationsprodukt *n*
precondition / to vorbehandeln
precondition Vorbedingung *f*
preconditioning Vorbehandlung *f*
p. screen Vorsieb *n*
precool / to vorkühlen
precooler Vorkühler *m*
precooling Vorkühlen *n*, Vorkühlung *f*
precrusher Vorbrecher *m*
precrystalline deformation *(Geol)* präkristalline Deformation *f*

precrystallization Vorkristallisation *f*
precrystallizer Vorkristallisator *m*, Vorkristaller *m*
precure / to *s.* to prevulcanize
precure, precuring *(Plast)* Vorhärten *n*; *(Gum)* s. prevulcanization; *(Text)* Vorkondensation *f*
precursor Vorläufer *m*, Vorstufe *f*, Präkursor *m*
  p.-free vorstufenfrei, präkursorfrei
predefecated juice *(Zucker)* vorgeschiedener Saft *m*
predefecation *(Zucker)* Vorscheidung *f*, Vorkalkung *f*
  p. juice *s.* predefecated juice
predigestion *(Pap)* Vorkochung *f*
predissociation *(phys Ch)* Prädissoziation *f*
  p. level Prädissoziationszustand *m*
  p. spectrum Prädissoziationsspektrum *n*
predistillation vorhergehende Entgasung *f*, Vorschwelung *f* *(bei Doppelgasgeneratoren)*
  p. [gas] producer Schwelgenerator *m*, Doppelgasgenerator *m*, Doppelgaserzeuger *m*
predosage level *(Toxikologie)* Blindwert *m*
predry / to vortrocknen
predryer Vortrockner *m*, *(Pap auch)* Vortrockenzylinder *m*
predrying Vortrocknen *n*, Vortrocknung *f*
  p. shaft Schachttrockner *m* *(zur Vortrocknung)*
pre-emergence herbicide Vorauflaufherbizid *n*
pre-emergence treatment *(Landw)* Vorauflaufbehandlung *f* *(Ausbringen von Herbiziden vor dem Keimen der Kulturpflanzensamen)*
pre-emergent herbicide *s.* pre-emergence herbicide
pre-evaporator Vorverdampfer *m*
pre-expander *(Plast)* Vorexpansionseinrichtung *f*
pre-expose / to *(Foto)* vorbelichten
pre-exposure *(Foto)* Vorbelichtung *f*
preferential adsorption bevorzugte (selektive) Adsorption *f*
  p. oxidation selektive Oxydation *f* *(z.B. bei der Feuerraffination)*
  p. retention Vorzugsretention *f* *(von Komponenten in Wirkstoffgemischen)*
  p. wetting Umnetzung *f*
pre-fermentation Vorgärung *f*, Angärung *f*
preferred orientation Vorzugsorientierung *f*
prefilt [feed, slurry] Aufgabegut *n* *(beim Filtrieren)*, Filtergut *n*, Trübe *f*
prefiltration Vorfiltern *n*, Vorfiltration *f*
prefire *(Ker)* Vorfeuer *n*, Schmauchfeuer *n*
prefiring *(Ker)* Vorbrennen *n*
prefix / to *(Kosmet)* vorfixieren; *(Nomenklatur)* *(als Präfix)* voranstellen
prefoam / to *(Plast)* vorschäumen, aufschäumen
prefoaming *(Plast)* Vorschäumen *n*, Aufschäumen *n*
  p. temperature *(Plast)* Vorschäumtemperatur *f*, Aufschäumtemperatur *f*
preform / to vorformen
preform *(Gum)* vorkonfektionierter (vorgeformter) Rohling *m*; *(Plast)* Vorformling *m*, Vorpreßling *m*

p. machine Vorformmaschine *f*
p. moulding (process) Vorformverfahren *n*
p. screen Vorformschirm *m*, Saugform *f* *(für glasfaserverstärkte Plaste)*
preformed vorgeformt; als Vorprodukt gebildet
  p. gum vorgebildeter (vorhandener, aktueller) Gum *m*, vorgebildetes (vorhandenes, aktuelles) Harz *n*, aktuelle Verharzungsprodukte *npl*
preformer Vorformmaschine *f*
preforming Vorformen *n*
  p. press Vorformpresse *f*
  p. tool Vorformwerkzeug *n*
prefractionate / to vorfraktionieren
prefractionator Vorfraktionierturm *m*
prefuse / to vorschmelzen
preharvest interval Karenzzeit *f*, Wartezeit *f* *(nach dem Einsatz von Pflanzenschutzmitteln)*
preheat / to vorwärmen, anwärmen, vorerhitzen; vorheizen
preheat time Vorwärmzeit *f*
preheater Vorerhitzer *m*, Vorheizer *m*, Vorheizofen *m*, Vorwärmer *m*, Vorwärmgerät *n*, *(Plast auch)* Vorwärmwalze *f*, *(Pebble-Heater-Verfahren auch)* Vorheizkammer *f*
preheating Vorwärmen *n*, Anwärmen *n*, Vorerhitzen *n*; Vorheizen *n*
  p. compartment *s.* p. zone
  p. dryer *(Ker)* Vortrockner *m*
  p. mill *(Gum)* Vorwärmwalzwerk *n*
  p. temperature Vorwärmtemperatur *f*, Vorerhitzungstemperatur *f*
  p. zone Vorwärmzone *f*, Vorheizzone *f*, Anwärmzone *f*
prehnite *(Min)* Prehnit *m* *(ein Sorosilikat)*
prehnitene $C_6H_2(CH_3)_4$ Prehnitol *n*, 1,2,3,4-Tetramethylbenzol *n*
prehnitic acid $C_6H_2(COOH)_4$ Prehnitsäure *f*, 1,2,3,5-Benzoltetrakarbonsäure *f*
prehnitylic acid $(CH_3)_3C_6H_2COOH$ Prehnitylsäure *f*, 2,3,4-Trimethylbenzoesäure *f*
preignition Frühzündung *f*
preimpregnate / to vorimprägnieren, *(Plast auch)* vorbeharzen
preimpregnation Vorimprägnierung *f*, *(Pap auch)* Vorhydrolyse *f*, vorhergehende Hydrolyse *f* *(der Hackschnitzel)*
preionization Präionisation *f*, Autoionisation *f*, Selbstionisation *f*
preirradiation *(Plast)* Vorbestrahlung *f*
preknotter *(Pap)* Vorsortierer *m*
preknotting *(Pap)* Vorsortierung *f*
prelimed juice *(Zucker)* vorgeschiedener Saft *m*
preliminary concentration Vorkonzentrierung *f*, Voreindickung *f*
  p. cooling Vorkühlung *f*
  p. degassing Vorentgasung *f*
  p. drying Vortrocknung *f*
  p. emulsion Voremulsion *f*
  p. filtration Vorfiltration *f*, Vorfiltern *n*
  p. impregnation *s.* preimpregnation

p. **neutralization** Vorneutralisation *f*
p. **penetration** *s.* preimpregnation
p. **purification** Vorreinigung *f*
p. **screening (sizing)** Vorklassierung *f*
p. **test** Vorprüfung *f*, Voruntersuchung *f*, Vorversuch *m*, Vorprobe *f*; orientierende Prüfung *f* *(von Pflanzenschutzmitteln)*
p. **treatment** Vorbehandlung *f*
preliming *(Zucker)* Vorscheidung *f*, Vorkalkung *f*
premash / to vormaischen
premasher Vormaischer *m*, Vormaischapparat *m*
premasticate / to *(Gum)* vormastizieren
premastication *(Gum)* Vormastikation *f*
premature curing *(Plast)* vorzeitige Härtung *f*, vorzeitiges Aushärten *n* *(Preßfehler)*
premetallize / to vormetallisieren
Premier [colloid] mill Premier-Mühle *f*, Premier-Kolloidmühle *f*
premier jus Premier jus *m* *(feiner Speisetalg)*
premilling Replastizieren *n* *(von Silikonkautschukmischungen)*
premium fuel (gasoline) Premium-Kraftstoff *m*, Superkraftstoff *m*, Premium-Benzin *n*, Superbenzin *n*
p. **motor fuel (spirit)** *s.* p. fuel
p. **oil** Premium-Öl *n* *(mit Additiven versetztes Schmieröl)*
p. **spirit** *s.* p. fuel
premix / to vormischen
premix Vormischung *f*; *(Plast)* vorgemischtes harzgetränktes Glasfasermaterial *n*
p. **burner** Treibdüsenbrenner *m*
premixer Vormischgerät *n*, Vormischgefäß *n*
preneutralize / to vorneutralisieren
prenitol *s.* prehnitene
prepacked concrete vorgepackter Beton *m*, Prepaktbeton *m*, Schlämmbeton *m*
preparation 1. Herstellung *f*, Darstellung *f*, Ansetzen *n*, Bereitung *f*, Zubereitung *f*; Vorbereiten *n*, Vorbereitung *f*; Aufbereiten *n*, Aufbereitung *f*; Vorbehandlung *f*, Präparieren *n*; Haltbarmachen *n*, Präparieren *n*; 2. Präparat *n*, *(Med auch)* Mittel *n*, Arzneimittel *n*, Arzneizubereitung *f*, Fertigpräparat *n*
p. **dish** Präparatenglas *n*
preparative präparativ
p. **chemistry** präparative Chemie *f*, Präparatenchemie *f*
p. **method** Darstellungsmethode *f*, Herstellungsmethode *f*
prepare / to herstellen, darstellen, ansetzen, [zu]bereiten; vorbereiten; aufbereiten; vorbehandeln; präparieren; haltbar machen, präparieren, konservieren
prepared calcium carbonate, p. chalk *(gemahlene)* Schlämmkreide *f*, präparierte (geschlämmte und gemahlene) Kreide *f*
p. **fuel** künstlicher (veredelter, synthetischer) Brennstoff *m*
p. **mustard** Speisesenf *m*, Senf *m*, Mostrich *m*

p. **tar** präparierter (destillierter) Teer *m*
preparing salt $Na_2[Sn(OH)_6]$ Präpariersalz *n*, Natriumhexahydroxostannat(IV) *n*
prepd. *(Abk. für)* prepared
pre-period Vorperiode *f*, Vorversuch *m* *(einer kalorimetrischen Messung)*
prephenic acid $HOC_6H_5(COOH)CH_2COCOOH$ Prephensäure *f* *(Zyklohexanderivat)*
preplanting application *(Landw)* Applikation *f* (Ausbringen *n*) vor dem Auspflanzen
p. **treatment** Behandlung *f* *(des Bodens mit Pflanzenschutzmitteln)* vor dem Auspflanzen
preplasticating *s.* preplastication
p. **system** *(Plast)* Vorplastiziersystem *n*
preplastication *(Plast)* Vorplasti[fi]zieren *n*, Vorplasti[fi]zierung *f*
preplasticator cylinder *(Plast)* Vorplastizierzylinder *m*
preplasticizing *s.* preplastication
prepn. *s.* preparation
prepolymer Vorpolymer[es] *n*, Vorpolymerisat *n*
p. **process** *(Plast)* Prepolymer-Verfahren *n* *(Verschäumung)*
prepreg Prepreg *n*, vorimprägniertes Glasfasermaterial *n*
preroast / to vorrösten
prescreen / to *(Pap)* vorsortieren
prescreening *(Pap)* Vorsortierung *f*
prescription Vorschrift *f*, *(Med)* Rezept *n*
p. **balance** Präzisionsapothekerwaage *f*
preservation Konservieren *n*, Konservierung *f*, Erhaltung *f*, *(i.e.S.)* Einkochen *n*, Einmachen *n*
.p. **jar** Präparatenglas *n*, Schauglas *n*
preservative konservierend
preservative [agent] Konservierungsmittel *n*, Konservierungsstoff *m*, Fäulnisverhütungsmittel *n*; *(Foto)* Entwicklerkonservierungsmittel *n*, Entwicklerschutzmittel *n*; *(Gum)* Stabilisierungsmittel *n*, Konservierungsmittel *n*
p. **for margarine** Margarinekonservierungsmittel *n*
p. **paper** konservierendes Papier *n*
preserve / to konservieren, haltbar machen, erhalten, *(z.B. Obst)* einmachen, einkochen, einlegen
p. **in brine** naßpökeln
p. **with salt** einsalzen, trockenpökeln
preserve 1. Konserve *f*, *(i.e.S.)* Obstkonserve *f*; Präserve *f*, Halbkonserve *f*; Marmelade *f*; 2. Arbeitsschutzmittel *n*, *(i.e.S.)* Schutzbrille *f*
p. **can** Konservendose *f*, Dose *f*, Konservenbüchse *f*, Büchse *f*
p. **jar** Einmachglas *n*, Konservenglas *n*
p. **tin** *s.* p. can
preserved milk Dauermilch *f*
p.-**milk factory** Dauermilchwerk *n*
preserving agent *s.* preservative [agent]
p. **jar** *s.* preserve jar
p. **paper** *s.* preservative paper
p. **plant** Konservenfabrik *f*

presetting *(Text)* Vorstabilisieren *n*, Vorstabilisierung *f*, Vorfixieren *n*, Vorfixierung *f*, Vordämpfen *n*

pre-shave lotion vorbehandelndes Rasierwasser *n*, Pre-shave-Lotion *f*

preshrinking Vorschrumpfen *n*, Vorschrumpfung *f*

preshrunk vorgeschrumpft

presinter / to vorsintern

presintering Vorsinterung *f*

presoak / to *(Pap)* vorweichen, vorimprägnieren; *(Gerb)* anweichen, vorweichen

presoaking *(Pap)* Vorweichen *n*, Vorimprägnierung *f* *(der Hackschnitzel vor der Kochung)*; *(Gerb)* Vorweiche *f*

presowing application Vorsaatanwendung *f* *(von Herbiziden)*

　p. herbicide Vorsaatherbizid *n*

　p. treatment Vor[aus]saatbehandlung *f*

press / to pressen; zusammenpressen; abpressen; auspressen; *(Met)* verpressen; *(Pap)* pressen, grainieren; *(Text)* bügeln; *(z.B. Hebel)* [nieder]drücken

　p. off abpressen

press Presse *f*, Scheidepresse *f*, Quetsche *f*; *(Gär)* Kelter *f*, Traubenpresse *f*, Traubenquetsche *f*

　p.-and-blow machine *(Glas)* Preßblas[e]maschine *f*

　p.-and-blow process *(Glas)* Preßblas[e]verfahren *n*

　p. arrangement Pressenanordnung *f*

　p. body *(Ker)* Preßmasse *f*

　p. cake Preßkuchen *m*

　p. cloth Preßtuch *n*

　p. copy paper Durchschlagpapier *n*

　p. cure (curing) Vulkanisation *f* in der Presse (Preßform), Preßvulkanisation *f*, Pressenheizung *f*

　p. doctor *s.* p. roll doctor

　p. dust *(Ker)* Preßstaub *m*

　p. felt *(Pap)* Preßfilz *m*

　p. liquor Preßlauge *f*

　p. mark *(Pap) (Am)* Molettewasserzeichen *n*

　p. mix *(Ker)* Preßmasse *f*

　p. mould Preßform *f*, *(Plast auch)* Preßwerkzeug *n*

　p. must *(Lebm)* abgepreßter Most *m*

　p. part *(Pap)* Pressenpartie *f*, Naßpressenpartie *f*

　p. platen *(Plast)* Pressentisch *m*

　p. roll *(Pap)* Preßwalze *f* *(der Naßpresse)*; Formatwalze *f*, Formatzylinder *m* *(der Pappen-Rundsiebmaschine)*

　p. roll doctor *(Pap)* Schaber *m* an der Preßwalze

　p. section *s.* p. part

　p. vulcanization *s.* p. cure

　p. water Preßwasser *n*

pressafiner *(Pap)* Pressafiner *m* *(eine Schneckenpresse)*

pressboard Preßspan *m*

presse-pâte, p.-pate *(Pap)* Holzschliffentwäs-

serungsmaschine *f*; Zellstoffentwässerungsmaschine *f*

pressed cake Preßkuchen *m*

　p. distillate *(Erdöl)* abgepreßtes Öl *n*, Preßöl *n* *(bei der Entparaffinierung vor der Aufarbeitung zu Neutralöl erhaltenes Öl)*

　p. glass Preßglas *n*

　p. sheet *(Plast)* gepreßte Folie *f*

pressing 1. Pressen *n*, Pressung *f*, Abpressen *n*; Auspressen *n*; *(Met)* Verpressen *n*; *(Pap)* Pressen *n*, Grainieren *n*; *(Text)* Bügeln *n*; 2. *(Met)* Preßteil *n*

　p. board *s.* pressboard

　p. dust *s.* press dust

　p. felt *s.* press felt

　p. method Preßverfahren *n*

　p. pressure Preßdruck *m*

　p. process Preßverfahren *n*; Preßvorgang *m*

　p. time *(Plast)* Preßzeit *f*, Standzeit *f*

　p. tool Preßwerkzeug *n*

pressor pressorisch, blutdrucksteigernd

press[s]pahn Preßspan *m*

pressure Druck *m*

　p. accumulator *s.* p. container

　p. at the roll nips *(Pap)* Liniendruck *m* *(zwischen den Walzen)*, Walzenanpreßdruck *m*

　p. atomization Druckzerstäubung *f*, Drucksprühung *f*

　p. atomizer Druckzerstäuber *m*, Druckversprüher *m*

　p. bag *(Plast)* Drucksack *m*

　p. bag moulding *(Plast)* Gummisackverfahren *n*, Niederdruckpreßverfahren *n*

　p. bar Anreibbarren *m*, Barren *m* *(eines Walzenstuhls)*

　p. boil[ing] *(Text)* Druckkochen *n*, Beuchen *n*

　p. bulb Druckball *m*

　p. case Druckkörper *m* *(Filtration)*

　p. casting Druckgießen *n*, Druckguß *m*, Gießen *n* unter Druck

　p.-casting die Druckgießform *f*, Druckgußform *f*

　p. chamber Druckkammer *f*

　p.-compensating vessel Druckausgleichsgefäß *n*

　p. container *(Pap)* Druck[säure]behälter *m*, Kochsäuredruckspeicher *m*, Rezipient *m*

　p. cooler Druckkühler *m*

　p. cracking *(Ker)* Preßrißbildung *f* *(Fehler)*

　p. cylinder Druck[gas]flasche *f*, Bombe *f*; Druckkörper *m* *(eines Filters)*; *(Plast)* Druckzylinder *m*

　p. decatizing *(Text)* Preßglanzdekatur *f*

　p.-diecast / to druckgießen, unter Druck gießen, durch Druck (auf Druckgießmaschinen) vergießen, im Druckguß herstellen

　p. diecasting Druckgießen *n*, Druckguß *m*, Gießen *n* unter Druck; Druckgußstück *n*, Druckgußteil *n*

　p.-diecasting machine Druckgießmaschine *f*, Druckgußmaschine *f*

p. difference Druckdifferenz f. Druckgefälle n
p. differential Druckdifferential n; s. p. difference
p. distillate Druckdestillat n
p. distillation Druckdestillation f
p. drop Druckabfall m, Druckminderung f; Druckgefälle n
p. dyeing Druckfärben n, Färben n unter Druck
p. energy Druckenergie f
p. evaporator Druckverdampfer m
p. filter Druckfilter n, Überdruckfilter n
p. filtration Druckfiltrieren n, Druckfiltration f, Druckfiltern n
p. filtration funnel s. p. filter
p. flow Druckströmung f
p. foot (Pap) Preßplatte f (eines Pressenschleifers)
p. forming (Plast) Formstanzen n
p. gauge Manometer n, Druckmesser m
p. head Druckhöhe f, Druck m
p. headbox (Pap) Hochdruckstoffauflauf m
p. impregnation (Pap) Druckdurchtränkung f, Druckimprägnierung f
p. increase Druckanstieg m, Druckerhöhung f, Drucksteigerung f
p. line Druckleitung f
p. nozzle Flüssigkeitsdruckdüse f, Druckdüse f, Druckzerstäuber m
p. nutsche Druck[filter]nutsche f
p. of swelling Quellungsdruck m
p. pad (Plast) Druckscheibe f, Druckleiste f, Druckaufnahmefläche f
p. port Druckeintritt m, Druckanschluß m
p. pot Druckkessel m (in der Farbspritztechnik)
p. pump Druckpumpe f
p. relief device Überdrucksicherung f
p. relief valve Überdruckventil n, Druckbegrenzungsventil n, Sicherheitsventil n
p. roll Quetschwalze f, Preßwalze f, Druckwalze f
p.-seal joint druckgespannte (selbstdichtende) Verbindung f
p.-sensitive adhesive Selbstkleber m, Kleb[e]streifen m; Haftkleber m
p. shell Druckkörper m, Druckgehäuse n
p. tank Druckbehälter m, Druckspeicher m
p. tap Druckentnahmestelle f
p. thermometer Druckthermometer n, manometrisches Thermometer n
p.-tight druckdicht
p. tower Druckturm m
p. tubing Druckschlauch m; Druckschlauchmaterial n, Druckschläuche mpl
p.-type filling machine Gegendruckfüllmaschine f, Gegendruckfüller m
p. vessel Druckgefäß n, Druckbehälter m, Autoklav m
pressureless sintering druckloses Sintern n, Sintern n ohne Druckanwendung, Normalsintern n
pressurize / to unter Druck setzen

pressurized-gas producer Druckgasgenerator m, Druckgaserzeuger m, Druckvergaser m
p. headbox (Pap) Hochdruckstoffauflauf m
p. heavy-water reactor (Kch) Schwerwasserdruckreaktor m
p.-water reactor Preßwasserreaktor m, Druckwasserreaktor m
presteam / to (Hackschnitzel vor der Kochung) dämpfen, vordämpfen
presteaming (Pap) Dämpfverfahren n, Dämpfung f (der Hackschnitzel vor der Kochung), Vordämpfen n
prestress / to vorspannen
prestressed concrete vorgespannter Beton m, Spannbeton m
p. [safety] glass vorgespanntes Glas n
pretan / to angerben, vorgerben
pretannage Vorgerbung f
pretreat / to vorbehandeln
pretreatment Vorbehandlung f
prevention of fog (Foto) Schleierverhinderung f, Schleierverhütung f
prevulcanization (Gum) Anvulkanisation f, Anvulkanisieren n, Anspringen n, Anbrennen n; (Gum) Vorvulkanisation f, Vorheizung f
prevulcanize / to (Gum) anvulkanisieren, anspringen, anbrennen; (Gum) vorvulkanisieren, vorheizen
priceite (Min) Priceit m
prill / to sprühkristallisieren, prillen (eine Schmelze oder konzentrierte Lösung in Granulat verwandeln)
prill 1. Metallkönig m, Regulus m; 2. (durch Sprühkristallisation erzeugte Granalie)
prilling Sprühkristallisieren n, Sprühkristallisation f, Prillen n
p. tower Prillturm m (Düngemittelindustrie)
prill[i]on Schlackenzinn n, Schlackenmetall n
primaquine Primaquin n, 8-(4'-Amino-1'-methylbutylamino)-6-methoxychinolin n
primary primär
p. accelerator (Gum) Primärbeschleuniger m
p. acetate (Text) Primärazetat n, Triazetat n, Zellulosetriazetat n
p. air Primärluft f, Erstluft f
p. alcohol primärer Alkohol m
p. aluminium pig Hüttenaluminium n
p. amine primäres Amin n
p. ammonium phosphate $NH_4H_2PO_4$ Monoammoniumdihydrogenphosphat n, Ammoniumdihydrogen[ortho]phosphat n
p. battery Primärbatterie f
p. beam Primärstrahl m
p. bond s. p. valency bond
p. calcium phosphate $Ca(H_2PO_4)_2$ Monokalziumdihydrogenphosphat n, Kalziumdihydrogen[ortho]phosphat n
p. carbon atom primäres Kohlenstoffatom n
p. cell Primärelement n
p. cellulose acetate s. p. acetate

p. centrifuge Primärzentrifuge *f*
p. clay Primärton *m*, primärer (eluvialer) Ton *m*, Verwitterungston *m*, Residualton *m*
p. colour *(Farbentheorie)* Grundfarbe *f*
p. column Vordestillationskolonne *f*
p. component *(Farb)* Erstkomponente *f*, aktive Komponente *f*, Diazo[tierungs]komponente *f*
p. condition Anfangszustand *m*
p. cooler Vorkühler *m*
p. creep Anfangskriechen *n*
p. crusher Vorbrecher *m*
p. deposit Primärlagerstätte *f*, primäre Lagerstätte *f*
p. dilute charge zusätzlich [mit Öllösung aus Sekundärzentrifuge] verdünnte Primärcharge *f* *(beim Entwachsungsprozeß)*
p. distillation Primärdestillation *f*, Erstdestillation *f*, primäre (erste) Destillation *f*
p. electron primäres Elektron *n*, Primärelektron *n*
p. element *s.* p. nutrient
p. explosive Initialsprengstoff *m*, Zünd[spreng]stoff *m*, Initialexplosivstoff *m*
p. fermentation Vorgärung *f*, Angärung *f*; Hauptgärung *f*
p. flash distillate primäres Flash-Destillat *n* *(Kopfprodukt aus der ersten Rohöldestillation)*
p. fraction Grundfraktion *f*
p. layer Primärschicht *f*
p. magnesium phosphate $Mg(H_2PO_4)_2$ Monomagnesiumdihydrogenphosphat *n*, Magnesiumdihydrogen[ortho]phosphat *n*
p. metal Primärmetall *n*
p. mineral primäres Mineral *n*
p. nutrient Kernnährstoff *m*
p. ore primäres (protogenes) Erz *n*
p. phosphate $M^IH_2PO_4$ primäres Phosphat *n*, Dihydrogen[ortho]phosphat *n*
p. photochemical reaction, p. photoreaction fotochemische Primärreaktion *f*, primäre fotochemische Reaktion *f*
p. pipe *(Met)* [offener] Primärlunker *m*, primärer Lunker *m* *(im Gußblock)*
p. product Primärprodukt *n*, primäres Produkt *n*
p. ray Primärstrahl *m*
p. reaction Primärreaktion *f*, Primärvorgang *m*, Primärakt *m*
p. reference fuel Urbezugskraftstoff *m*, Primärbezugskraftstoff *m*
p. salt primäres Salz *n* *(einer mehrbasigen Säure)*
p. salt effect primärer Salzeffekt *m*
p. sodium orthophosphate NaH_2PO_4 Mononatriumdihydrogenphosphat *n*, Natriumdihydrogen[ortho]phosphat *n*
p. solution Grundlösung *f*
p. standard *(Maßanalyse)* Urtitersubstanz *f*; *(pH-Messung)* Primärstandard *m*
p. valency Hauptvalenz *f*
p. valency bond Hauptvalenzbindung *f*
p. wall Primärwand *f* *(einer Faser)*
p. zinc Rohzink *n*

prime / to 1. vorbereiten, in Betrieb setzen; 2. füllen, laden; *(Pumpen)* [vor]füllen; 3. zur Zündung vorbereiten; zünden, als Zünder dienen für *(von Initialsprengstoffen)*; 4. *(Farb)* grundieren, grundierend bestreichen; 5. spucken *(von Kesseln)*; 6. *(Destillation)* übertreiben
prime coking coal bituminöse Spezialkokskohle *f*
  p. paint *s.* priming paint
  p. strong pulp sehr harter Zellstoff *m*
primer 1. Sprengkapsel *f*, Zündhütchen *n*, Zünder *m*; *s.* primary explosive; 2. Grundiermittel *n*, Grund[ier]anstrichmittel *n*
  p. coat *s.* priming coat
priming 1. Vorbereitung *f*, Inbetriebnahme *f*; 2. Füllen *n*, Laden *n*; 3. Stoßzündung *f*; *s.* primary explosive; *s.* primer 1.; 4. Grundieren *n*, Grundierung *f*; *s.* p. coat; 5. Spucken *n* *(von Kesseln)*; 6. *(Destillation)* Übertreiben *n*
  p. coat Grund[ier]anstrich *m*, Untergrundanstrich *m*, Grundierung *f*
  p. composition Zündsatz *m*
  p. paint Grund[anstrich]farbe *f*, Grundierfarbe *f*
primuline Primulin *n* *(Anthrachinonderivat)*
  p. red Primulinrot *n*
  p.-type base Primulinbase *f*
  p. yellow Primulingelb *n*
principal alkaloid Hauptalkaloid *n*
  p. colouring material Hauptfarbstoff *m*
  p. metal Grundmetall *n*
  p. quantum number Hauptquantenzahl *f*
  p. series *(phys Ch)* Hauptserie *f*
  p. valence (valency) Hauptvalenz *f*
principle Prinzip *n*, Grundsatz *m*, Lehrsatz *m*, Satz *m*, Gesetz *n*, Regel *f*; Grundbestandteil *m*, Prinzip *n*
  p. of equipartition [of energy] Äquipartitionsprinzip *n* (Äquipartitionstheorem *n*, Gleichverteilungssatz *m*) [der Energie]
  p. of least motion Prinzip *n* der möglichst geringen Konfigurationsänderungen
  p. of least resistance (restraint) Prinzip *n* des kleinsten Zwanges, Prinzip *n* des beweglichen Gleichgewichtes, Le Chatelier-Braunsches Prinzip *n*
  p. of mass action Massenwirkungsgesetz *n*
  p. of maximum multiplicity [Hundsches] Prinzip *n* der größten Multiplizität, [Hundsche] Regel *f* der größten Multiplizität, Hundsches Multiplizitätsprinzip (Prinzip) *n*, [erste] Hundsche Regel *f*
  p. of mobile equilibrium *s.* p. of least resistance
Prins reaction Prins-Reaktion *f* *(Aldehydkondensation)*
print / to drucken; bedrucken; *(Foto)* kopieren, abziehen; lichtpausen
print Druck *m*; Abdruck *m*; *(Foto)* Kopie *f*, Abzug *m*; Lichtpause *f*
  p. developer *(Foto)* Entwickler *m* für Papiere
  p.-out paper *s.* printing-out paper
  p. paper *s.* printing paper

p. paste s. printing paste
p. washer *(Foto)* Wässerungstank *m*
printability *(Pap)* Bedruckbarkeit *f*, Druckfähigkeit *f*
printer's acetate $Al_2O(CH_3COO)_4 \cdot 4H_2O$ basisches Aluminiumazetat *n*
printing Drucken *n*; Bedrucken *n*; *(Foto)* Kopieren *n*, Abziehen *n*; Lichtpausen *n*
p. additive Druckereihilfsmittel *n*
p. ink Druckfarbe *f*, grafische Farbe *f*, *(i.e.S.)* Druckerschwärze *f*
p. machine Druckmaschine *f*
p.-out paper *(Foto)* Auskopierpapier *n*
p. paper Druckpapier *n*; *(Foto)* Kopierpapier *n*, Abzugspapier *n*; Lichtpauspapier *n*
p. paste Druckpaste *f*
p. process Druckverfahren *n*
p. properties *(Pap)* Druckeigenschaften *fpl*, Bedruckbarkeit *f*, Druckfähigkeit *f*
p. roller Druckwalze *f*
p. thickener *(Text)* Druckverdickungsmittel *n*, Druckverdickung *f*
prism *(Krist)* Prisma *n*, Säule *f*
p. spectrograph Prismenspektrograf *m*
p. spectroscope Prismenspektroskop *n*
probability Wahrscheinlichkeit *f*
p. amplitude Wahrscheinlichkeitsamplitude *f*
p. density Wahrscheinlichkeitsdichte *f*
p. distribution Wahrscheinlichkeitsverteilung *f*
p. factor Wahrscheinlichkeitsfaktor *m*, sterischer Faktor *m*
p. of collision Stoßwahrscheinlichkeit *f*
p. of disintegration Zerfallswahrscheinlichkeit *f*
p. of excitation Anregungswahrscheinlichkeit *f*
p. of ionization Ionisationswahrscheinlichkeit *f*
p. of transition Übergangswahrscheinlichkeit *f*
probit mortality Probitmortalität *f* *(Toxikologie)*
procaine Prokain *n*, *p*-Aminobenzoesäure-β-diäthylaminoäthylester *m*
procedure Arbeitsweise *f*, Methode *f*, Verfahren *n*, Technik *f*
proceed / **to** fortschreiten; vonstatten gehen, verlaufen *(z.B. von Reaktionen)*; verfahren
process / **to** bearbeiten, behandeln; fertigen, herstellen; verarbeiten; veredeln
process Prozeß *m*, Verfahren *n*, Arbeitsmethode *f*; Ablauf *m*, Verlauf *m*, Vorgang *m*, Prozeß *m*; *(Regelungstechnik)* Steuerstrecke *f*, Regelstrecke *f*, Strecke *f*
p. annealing *(Met)* Zwischenglühen *n*, Zwischenglühung *f*
p. cheese Schmelzkäse *m*
p. control Prozeßsteuerung *f*, Verfahrenssteuerung *f*, Prozeßregelung *f*, Verfahrensregelung *f*
p. engineer Verfahrenstechniker *m*, Verfahrensingenieur *m*
p. engineering Verfahrenstechnik *f*
p. industry Verfahrensindustrie *f*
p. ink Illustrations[druck]farbe *f*
p. material *(Foto)* Reproduktionsmaterial *n*, Repro-Material *n*

p. metallurgy Hüttentechnik *f*
p. oil *(Gum)* Weichmacheröl *n*
p. pump Prozeßpumpe *f*; Chemiepumpe *f*; Säurepumpe *f*
p. steam Betriebsdampf *m*
p. temperature Verarbeitungstemperatur *f*
p. variable Verfahrensgröße *f*, Verfahrensveränderliche *f*
p. water Brauchwasser *n*, Gebrauchswasser *n*, Betriebswasser *n*, Fabrikationswasser *n*, Nutzwasser *n*
p. worker Produktionsarbeiter *m*
processability Verarbeitbarkeit *f*, Verarbeitungsfähigkeit *f*
processable verarbeitbar, verarbeitungsfähig
processed cheese Schmelzkäse *m*
p. eggs Eipulver *n*, Trockenei *n*
p. paper veredeltes Papier *n*
p. pulp Edelzellstoff *m*
processibility s. processability
processible s. processable
processing Bearbeitung *f*, Behandlung *f*; Fertigung *f*, Herstellung *f*; Verarbeitung *f*; Veredelung *f*, *(Text auch)* Ausrüstung *f*; Arbeitsweise *f*, Betrieb *m*, Fahrweise *f*, Betriebsmethode *f*
p. aid Verarbeitungshilfsmittel *n*
p. characteristics Verarbeitungseigenschaften *fpl*
p. oil s. process oil
p. property Verarbeitungseigenschaft *f*
p. safety Verarbeitungssicherheit *f*, Fabrikationssicherheit *f*
p. technology Verarbeitungstechnik *f*; Verarbeitungstechnologie *f*
p. temperature Verarbeitungstemperatur *f*
prochlorite *(Min)* Prochlorit *m*, Rhipidolith *m* *(ein Phyllosilikat)*
pro-coating s. protective coating
Proctor dryer *(Ker)* Proctor-Trockner *m* *(Tunneltrockner mit Luftumwälzung)*
prodigiosin Prodigiosin *n* *(hochroter Farbstoff, ein Tripyrrol)*
producer 1. Hersteller *m*, Produzent *m*; 2. Generator *m*
p. coal Generatorkohle *f*
p. gas Generatorgas *n*
p.-gas equilibrium Generatorgasgleichgewicht *n*, Boudouard-Gleichgewicht *n*, Boudouardsches Gleichgewicht *n*
p.-gas reaction Vergasungsreaktion *f*
p. operation Generatorbetrieb *m*
producing formation produzierende Schicht *f*, ölführende Formation (Schicht) *f*, Ölträger *m* *(bei der Erdölförderung)*
product Produkt *n*, Erzeugnis *n*; Gut *n*
p. being dried trocknendes Gut *n*, Trockengut *n* *(während der Trocknung)*
p. being mixed Mischgut *n*; Knetgut *n*; Rührgut *n*
p. collector Gutabscheider *m*

p. loss Gutverlust *m*

p. moisture Gutfeuchte *f*

p. of combustion Verbrennungsprodukt *n*, Verbrennungserzeugnis *n*

p. of decomposition Zersetzungsprodukt *n*, Zersetzungserzeugnis *n*

p. of fusion Schmelzprodukt *n*

p. of roasting Röstprodukt *n*

p. take-off *(Destillation)* Destillatabnahme *f*

p. to be dried zu trocknendes Gut *n*, Trockengut *n*, Trocknungsgut *n*, Trockneraufgabegut *n*

p. to be handled [zu behandelndes] Gut *n*

p. to be separated Scheidegut *n*, Setzgut *n*

p. to be stirred Rührgut *n*

p. to be sublimed Sublimationsgut *n*

production Herstellung *f*, Erzeugung *f*, Produktion *f*; Bildung *f*

p. of flame Flammenbildung *f*

p. of heavy water Schwerwasserherstellung *f*

p. of low temperatures Tieftemperaturerzeugung *f*

p. on a large scale großtechnische Herstellung *f*

p. output Ausstoß *m*, Produktionsleistung *f*, Fertigungsmenge *f*, Produktionszahl *f*

productive capacity *(Landw)* Produktionskraft *f*, Ertragsfähigkeit *f*

proenzyme, proferment Proferment *n*, Proenzym *n*

profile paper Profilpapier *n*, *(i.e.S.)* Millimeterpapier *n* *(oder)* Diagrammpapier *n*; *(Am)* Silhouettenpapier *n*

profiling Formgebung *f*, Fassonieren *n*; *(Gum)* Ziehen *n* von Profilen

p. calender Profilkalander *m*

proflavine Proflavin *n*, 3,6-Diaminoakridin *n*

progesterone, progestin Progesteron *n* *(Hormon)*

programme paper Programmpapier *n*, geräuschloses Papier *n*

programmed temperature gas chromatography Gaschromatografie *f* mit Temperaturprogramm (programmierter Temperatur)

progressive predefecation (preliming) *(Zucker)* progressive Vorscheidung *f*

proguanil Proguanil *n*, Chlorguanid *n*, N^1-*p*-Chlorphenyl-N^5-isopropylbiguanid *n*

p. hydrochloride Proguanilhydrochlorid *n*

pro-ignition zündbeschleunigend

pro-ignition dope Zündbeschleuniger *m*, zündbeschleunigender Zusatz *m*

projected area *(Plast)* Preß[lings]fläche *f*

projection formula Projektionsformel *f*

p. galvanometer Projektionsgalvanometer *n*

prolactin Prolaktin *n*, laktotropes (luteotrophes) Hormon *n*, Laktogen *n*, Laktationshormon *n*, Laktotropin *n*, Luteotrophin *n*, LTH

prolamine Prolamin *n* *(Eiweißbestandteil des Getreidemehls)*

prolidase Prolidase *f* *(zu den Dipepsidasen gehörendes Ferment)*

prolinase Prolinase *f* *(zu den Dipeptidasen gehörendes Ferment)*

proline $C_4H_8N \cdot COOH$ Prolin *n*, Pro, Pyrrolidin-(2)-karbonsäure *f*

prolonged development *(Foto)* ausgedehnte (verlängerte) Entwicklung *f*

p. exposure *(Foto)* lange Belichtung *f*

promethium Pm Promethium *n*

promote *I* **to** *(z.B. eine Reaktion)* fördern, beschleunigen; *(phys Ch)* (ein Elektron von einem Orbital in ein anderes heben)

promoted heat treatment thermische Behandlung *f* mit Promotor *(Herstellung von Butylkautschukmischungen mit Promotor)*

promoter Promotor *m*, Aktivator *m*, Beschleuniger *m*, synergetischer Verstärker *m*; *(Flot)* Sammler *m*, Kollektor *m*

p. action Promotorwirkung *f*

promoting agent *s.* promoter

promotion Fördern *n*, Förderung *f*, Beschleunigen *n*, Beschleunigung *f* *(z.B. einer Reaktion)*; *(phys Ch)* Promovierung *f*, Promotion *f*

p. energy *(phys Ch)* Promotionsenergie *f*

prone to knocking klopffreudig, klopfempfindlich, klopfstark

proof *I* **to** 1. beweisen; 2. beständig (dicht, undurchlässig, undurchdringlich) machen; imprägnieren; gummieren; *(Pap)* abziehen, andrucken

proof 1. beständig, dicht, undurchlässig, undurchdringlich; 2. normalstark, probehaltig *(alkoholische Flüssigkeit)*

proof 1. Beweis *m*; 2. Normalstärke *f (alkoholischer Getränke)*; 3. Reagenzglas *n*

p. of configuration Konfigurationsbeweis *m*

p. of structure Strukturbeweis *m*, Konstitutionsbeweis *m*

p. paper *s.* proofing paper

p. spirit Normalalkohol *m*, Normalweingeist *m*, Probeweingeist *m*

proofed fabric gummiertes Gewebe *n*, gummierter Stoff *m*

proofing 1. Beständigmachen *n*, Dichtmachen *n*, Undurchlässigmachen *n*, Undurchdringlichmachen *n*; Imprägnieren *n*, Imprägnierung *f*; Gummieren *n*, Gummierung *f*; 2. Imprägnierungsmittel *n*, Dichtungsmittel *n*; 3. gummierter Stoff *m*, gummiertes Gewebe *n*

p. paper Abziehpapier *n*, Abzug[s]papier *n*, Probeabzug[s]papier *n*, Andruckpapier *n*

prooxidant Prooxygen *n (Stoff, der schon in Spuren Oxidationsprozesse fördert)*

prop *(Ker)* Brennstütze *f*, Stütze *f (Brennhilfsmittel)*

propadiene $CH_2=C=CH_2$ Propadien *n*, Allen *n*

propagate *I* **to** sich ausbreiten (fortpflanzen, vermehren); weitertragen; fortpflanzen, vermehren

propagation Ausbreitung *f*, Fortpflanzen *n*, Fortpflanzung *f*, Vermehrung *f*; Weitertragen *n*; Fortpflanzungsreaktion *f*, Wachstumsreaktion *f*, Kettenwachstum *n*, Kettenfortpflanzung *f* *(z.B. bei der Polymerisation)*

p. stage Fortpflanzungsstadium *n*, Wachstumsstadium *n* (*z.B. bei der Polymerisation*)
p. velocity Fortpflanzungsgeschwindigkeit *f*, Ausbreitungsgeschwindigkeit *f*
propanal CH_3CH_2CHO Propanal *n*, Propionaldehyd *m*
propane C_3H_8 Propan *n*
p. deasphalting Propanentasphaltierung *f*, Entasphaltierung *f* (Entasphaltieren *n*) mit Propan, Propanextraktion *f* von Asphalt
p.-deasphalting plant Propanentasphaltierungsanlage *f*, Anlage *f* zur Propanentasphaltierung (Entasphaltierung mit Propan)
p. dewaxing Propanentparaffinierung *f*, Entparaffinierung *f* (Entparaffinieren *n*) mit Propan
p.-dewaxing plant Propanentparaffinierungsanlage *f*, Anlage *f* zur Propanentparaffinierung (Entparaffinierung mit Propan)
p. nitrile CH_3CH_2CN Propannitril *n*, Propionitril *n*, Propionsäurenitril *n*, Äthylzyanid *n*
p. process Propanverfahren *n*
1,2-propanediamine $CH_3CH(NH_2)CH_2NH_2$ 1,2-Propandiamin *n*, 1,2-Diaminopropan *n*, Propylendiamin *n*
propanedioic acid $HOOCCH_2COOH$ Propandisäure *f*, Methandikarbonsäure *f*, Malonsäure *f*
1,2-propanediol $HOCH_2CH(OH)CH_3$ 1,2-Propandiol *n*, Propylenglykol *n*
1-propanethiol C_3H_7SH 1-Propanthiol *n*, Propylmerkaptan *n*, Thiopropylalkohol *m*
1,2,3-propanetriol $CH_2OHCHOHCH_2OH$ 1,2,3-Propantriol *n*, Trihydroxypropan *n*, Glyzerin *n*, Ölsüß *n*
propanoic acid CH_3CH_2COOH Propansäure *f*, Propionsäure *f*
1-propanol $CH_3CH_2CH_2OH$ 1-Propanol *n*, Propylalkohol *m*
2-propanol $CH_3CHOHCH_3$ 2-Propanol *n*, Isopropylalkohol *m*, *sek*-Propylalkohol *m*
propanolysis Propanolyse *f*
2-propanone CH_3COCH_3 2-Propanon *n*, Dimethylketon *n*, Azeton *n*
propargyl alcohol $HC≡CCH_2OH$ Propargylalkohol *m*, 2-Propin-1-ol *n*
propargylic acid $HC≡CCOOH$ Propargylsäure *f*, Propiolsäure *f*
propellant 1. Treibstoff *m*, Treibmittel *n*, Schießstoff *m*, Schießmittel *n*, Pulversprengstoff *m*; 2. Treibgas *n* (*für Aerosole*)
p. charge Treibladung *f*
p. explosive *s.* propellant 1.
p. gas *s.* propellant 2.
p. powder Treibladungspulver *n*
propellent *s.* propellant
propeller Propeller *m*, Schraube *f*
p. agitator *s.* p. mixer
p. blunger Schraubenquirl *m*
p.-calandria evaporator Vertikalrohrverdampfer *m* mit Innenheizkammer und Axialförderrad im Fallrohr

p. fan Flügelradlüfter *m*, Propellerventilator *m*, Axialventilator *m*, Axiallüfter *m*
p. mixer Propellerrührer *m*, Propellerrührwerk *n*, Schraubenrührer *m*, Propellermischer *m*, Propellermischgerät *n*
p. pump Propellerpumpe *f*, Axialpumpe *f*
p.-type agitator 1. *s.* p. mixer; 2. (*Pap*) Umtriebpropeller *m*, Stofftreiber *m* (*beim Holländer*)
propenal $CH_2=CHCHO$ Propenal *n*, Allylaldehyd *m*, Akrylaldehyd *m*, Akrolein *n*
propene $CH_3CH=CH_2$ Prop[yl]en *n*, Methyläthen *n*
2-propene-1-ol CH_2CHCH_2OH 2-Propen-1-ol *n*, Propenol-(3) *n*, Allylalkohol *m*, Vinylkarbinol *n*
1,2,3-propenetricarboxylic acid $C_3H_3(COOH)_3$ Propentrikarbonsäure-(1,2,3) *f*, β-Karboxyglutakonsäure *f*, Akonitsäure *f*
propenoic acid $CH_2=CHCOOH$ Propensäure *f*, Vinylkarbonsäure *f*, Äthenkarbonsäure *f*, Akrylsäure *f*
propenyl Propenyl *n*
p. bromide $CH_3CH=CHBr$ Propenylbromid *n*, 1-Brompropen *n*
proper frequency Eigenfrequenz *f*, charakteristische Frequenz *f*
p. state (*phys Ch*) Eigenzustand *m*
p. value Eigenwert *m*
property Eigenschaft *f*, Fähigkeit *f*, Vermögen *n*
prophylactic prophylaktisch, vorbeugend, verhütend
prophylactic Prophylaktikum *n*, vorbeugendes Mittel *n*
propine *s.* propyne
β-propiolactone β-Propiolakton *n*, β-Hydroxypropionsäurelakton *n*, Propanolid-(3,1) *n*
propiolic acid $HC≡CCOOH$ Propiolsäure *f*, Propargylsäure *f*
p. alcohol $HC≡CCH_2OH$ Propiolalkohol *m*, Propargylalkohol *m*, 2-Propin-1-ol *n*
propionaldehyde CH_3CH_2CHO Propionaldehyd *m*, Propanal *n*
propionate Propionat *n*
propionic acid CH_3CH_2COOH Propionsäure *f*, Propansäure *f*, Methylessigsäure *f*
p.-acid fermentation Propionsäuregärung *f*
p. aldehyde *s.* propionaldehyde
propionitrile CH_3CH_2CN Propionitril *n*, Propannitril *n*, Propionsäurenitril *n*, Äthylzyanid *n*
propionyl group $C_2H_5CO—$ Propionylgruppe *f*, Propionylrest *m*
propiophenone $C_6H_5 \cdot CO \cdot CH_2 \cdot CH_3$ Propiophenon *n*, Äthylphenylketon *n*, Propionylbenzol *n*
propolis Propolis *f*, Bienenharz *n*, Bienenvorwachs *n*, Kittharz *n*, Vorwachs *n*, Stopfwachs *n*, Halbwachs *n*
proportion / to dosieren, zuteilen, zumessen
proportion Proportion *f*, Verhältnis *n*; Anteil *m*
in molecular proportions in molekularem Verhältnis
miscible in all proportions in jedem Verhältnis mischbar

p. by atoms Atomverhältnis n, Atomverhältniszahl f
p. by volume Volumverhältnis n, Raumverhältnis n
proportional counter Proportional[itäts]zähler m, Proportionalzählrohr n
p. intensifier *(Foto)* proportionaler Verstärker m
p. limit Proportionalitätsgrenze f
p. reducer *(Foto)* proportionaler (proportional wirkender) Abschwächer m
p. region Proportionalitätsbereich m
proportionality constant Proportionalitätsfaktor m
proportioning Dosierung f, Zuteilung f, Zumessung f
p. apparatus Dosiergerät n, Dosierapparat m, Zuteileinrichtung f, Zuteilungsvorrichtung f
p. pump Dosier[ungs]pumpe f, Zumeßpumpe f
p. screw Dosierschraube f
proprietary Eigentums...; *(gesetzlich oder warenzeichenrechtlich)* geschützt
p. name geschützte Bezeichnung f, wortgeschützter Name, m
propyl Propyl n
n-p. acetate $CH_3COOC_3H_7$ n-Propylazetat n, Essigsäure-n-propylester m
p. acetic acid $CH_3(CH_2)_3COOH$ Propylessigsäure f, n-Pentansäure f, Baldriansäure f
n-p. acetylene $CH_3CH_2CH_2C{\equiv}CH$ n-Propylazetylen n, 1-Pentin n
n-p. alcohol $CH_3CH_2CH_2OH$ Propylalkohol m, 1-Propanol n
sec-p. alcohol $CH_3CHOHCH_3$ sek-Propylalkohol m, Isopropylalkohol m, 2-Propanol n
p. aldehyde CH_3CH_2CHO Propionaldehyd m, Propanal n
p. carbinol $CH_3(CH_2)_2CH_2OH$ Propylkarbinol n, n-Butylalkohol m, 1-Butanol n
n-p. chloride $CH_3CH_2CH_2Cl$ [n-]Propylchlorid n, 1-Chlorpropan n
p. hydride C_3H_8 Propan n
p. mercaptan C_3H_7SH Propylmerkaptan n, 1-Propanthiol n, Thiopropylalkohol m
propylamine $C_3H_7NH_2$ Propylamin n
propylene $CH_3CH{=}CH_2$ Prop[yl]en n, Methyläthen n
p. aldehyde $CH_3CH{=}CHCHO$ Propylenaldehyd m, Krotonaldehyd m, 3-Methylakrolein n, Buten-(2)-al-(1) n
p. bromide (dibromide) $CH_3CHBrCH_2Br$ Propylen[di]bromid n, 1,2-Dibrompropan n
1,2-p. glycol $HOCH_2CH(OH)CH_3$ Propylenglykol n, 1,2-Propandiol n
p. oxide Propylenoxid n
p.-oxidic propylenoxidisch
propylenediamine $CH_3CH(NH_2)CH_2NH_2$ Propylendiamin n, 1,2-Propandiamin n, 1,2-Diaminopropan n
propylethylene $CH_3CH_2CH_2CH{=}CH_2$ n-Propyläth[yl]en n, n-Amylen-(1) n, Penten-(1) n
propylite *(Geol)* Propylit m

propylitization *(Geol)* Propylitisierung f, Propylitisation f, Propylitbildung f
n-propylmagnesium bromide C_3H_7MgBr n-Propylmagnesiumbromid n
d-2-propylpiperidine $C_8H_{17}N$ d-2-Propylpiperidin n, d-Koniin n
propyne $CH_3C{\equiv}CH$ Propin n, Allylen n, Methyläthin n, Methylazetylen n
2-propyne-1-ol $HC{\equiv}CCH_2OH$ 2-Propin-1-ol n, Propargylalkohol m, Propiolalkohol m
propynoic acid $HC{\equiv}CCOOH$ Propiolsäure f, Propargylsäure f
Prosize process *(Pap)* Prosize-Verfahren n
prosopite *(Min)* Prosopit m
prospect / to prospektieren, schürfen, aufspüren, suchen, *(nach Erz)* graben, *(nach Öl)* bohren
prospecting Prospektierung f, Prospektion f, Schürfung f, Aufspürung f, Suche f
prosthetic prosthetisch
p. group prosthetische (aktive) Gruppe f, Wirk[ungs]gruppe f, Koferment n, Koenzym n, Agon n
protactinium Pa Protaktinium n
protaminase Protaminase f *(zu den Karboxypeptidasen gehörendes Ferment)*
protamine Protamin n *(eine Eiweißart)*
protease Protease f, proteolytisches (proteinspaltendes, eiweißabbauendes, eiweißspaltendes, eiweißverdauendes) Ferment (Enzym) n
protected from light lichtgeschützt, vor Licht geschützt
p. rosin size *(Pap)* stabilisierter Freiharzleim m, stabilisierte Harzemulsion f
protecting film s. protective film
p. group Schutzgruppe f
p. ingredient s. protective agent
p. layer s. protective coating 1.
p. tube Schutzrohr n, Schutzhülse f
protection Schutz m
p. from light Lichtschutz m
p. layer s. protective coating 1.
p. tube s. protecting tube
protective schützend, Schutz...
protective protektives (protektiv wirkendes, vorbeugend wirkendes) Mittel n
p. action Schutzwirkung f
p. agent *(Gum)* Alterungsschutzmittel n
p. atmosphere Schutz[gas]atmosphäre f
p. capacity Schutzfähigkeit f
p. clothing Schutz[be]kleidung f
p. coating 1. schützender Überzug m, Schutzüberzug m; Schutzanstrich m; 2. Schutz[be]kleidung f
p. colloid Schutzkolloid n
p. contact insecticide protektives Kontaktinsektizid n
p. engobe *(Ker)* Schutzengobe f
p. film Schutzfilm m, Schutzhaut f, dünne Schutzschicht f
p. fungicide Residualfungizid n, protektives Fungizid n

p. group s. protecting group
p. insecticide protektives (protektiv wirkendes) Insektizid n
p. lacquer Schutzlack m
p. layer Schutzschicht f
p. sheath Schutzhülle f
p. toxicant protektives (protektiv wirkendes, vorbeugend wirkendes) Gift (Pflanzenschutzmittel) n
protector kolloidaler Träger m, Trägersubstanz f, Apoferment n, Apoenzym n, Pheron n
proteic substance Eiweißstoff m, Eiweißkörper m, Eiweiß n
proteid[e] Eiweißstoff m, Eiweißkörper m, Eiweiß n; Proteid n, zusammengesetzter (konjugierter) Eiweißstoff (Eiweißkörper) m
protein proteinartig; proteinhaltig, Protein...
protein Eiweißstoff m, Eiweißkörper m, Eiweiß n, Protein n (i.w.S.); Protein n, einfacher Eiweißstoff (Eiweißkörper) m
p. adhesive Eiweißleim m
p. breakdown s. p. degradation
p. chemist Eiweißchemiker m
p. chemistry Eiweißchemie f
p. cleavage s. p. degradation
p. component Eiweißkomponente f
p. deficiency Eiweißmangel m
p. degradation Eiweißabbau m, Proteinabbau m
p.-degrading enzyme s. proteolytic enzyme
p. denaturation Proteindenaturierung f, Proteindenaturation f
p.-digesting s. proteolytic
p. digesting enzyme s. proteolytic enzyme
p. fibre Eiweißfaser f, Proteinfaser f; Eiweiß[chemie]faserstoff m, CE, Proteinfaserstoff m
p.-free eiweißfrei
p. haze Eiweißtrübung f (z.B. des Bieres)
p. inclusion Eiweißeinschluß m
p. material (matter) s. proteic substance
p. metabolism Eiweißstoffwechsel m
p.-metal turbidity Metalleiweißtrübung f
p. moiety Protein[an]teil m, Proteinkomponente f
p. nitrogen Eiweißstickstoff m
p. plastic Proteinkunststoff m, Kunsthorn n
p. sol Proteinsol n
p. solution Proteinlösung f
p.-splitting s. proteolytic
p.-splitting bacteria s. proteolytic bacteria
p. synthesis Eiweißsynthese f
p.-tannin turbidity Eiweißgerbstofftrübung f
p. turbidity s. p. haze
proteinaceous proteinartig; proteinhaltig, Protein...
proteinase Proteinase f
proteinized (Text) animalisiert
proteoclastic s. proteolytic
proteohormone Proteohormon n
proteolysis Proteolyse f, Eiweißspaltung f

proteolytic proteolytisch, proteinspaltend, eiweißabbauend, eiweißspaltend, eiweißverdauend
p. bacteria Proteolyten mpl, Eiweißzersetzer mpl
p. enzyme proteolytisches (proteinspaltendes, eiweißabbauendes, eiweißspaltendes, eiweißverdauendes) Ferment (Enzym) n, Protease f
proteose Proteose f, Albumose f (Spaltprodukt des Proteins)
prothrombin Prothrombin n (Glykoproteid)
protium ^{1}H Protium n, leichter Wasserstoff m
protoactinium s. protactinium
protocatechualdehyde $(HO)_2C_6H_3CHO$ Protokatechualdehyd m
protocatechuic acid $(HO)_2C_6H_3COOH$ Protokatechusäure f, 3,4-Dihydroxybenzoesäure f
p. aldehyde s. protocatechualdehyde
protochloride (Chlorid der niedrigsten Oxydationsstufe)
protoenstatite (Min) Protoenstatit m (Magnesiumsilikat)
protogenic protonogen, protonenabspaltend, Protonen abspaltend (abgebend, liefernd), H^+-abgebend
protoiodide (Jodid der niedrigsten Oxydationsstufe)
protolysis Protolyse f
protolyte Protolyt m
protolytic protolytisch
p. reaction protolytische Reaktion f
proton Proton n, p
p. acceptor Protonenakzeptor m, Emprotid n
p. affinity Protonenaffinität f
p. binding energy Protonenbindungsenergie f, Bindungsenergie f des Protons
p. catcher Protonenfänger m
p. don[at]or Protonendon[at]or m, Dysprotid n
p.-fissioning s. protogenic
p.-free protonenfrei
p. loss Protonenabgabe f
p. rest mass Protonenruh[e]masse f
p. shift Protonenverschiebung f
p. transfer Protonenübertragung f
protonation Proton[is]ierung f, Addition f eines Protons
protonic acid Proton[en]säure f, Protonendonator m, Dysprotid n
protonitrate (Nitrat der niedrigsten Oxydationsstufe)
protopectin Pro[to]pektin n (wasserunlösliches natives Pektin)
protophilic protonophil, Protonen aufnehmend (anlagernd), H^+-aufnehmend
protopine Protopin n, Fumarin n
protoplasm[a] Protoplasma n
protoporphyrin Protoporphyrin n
protosalt (Metallsalz der niedrigsten Oxydationsstufe)
protosulphate (Sulfat der niedrigsten Oxydationsstufe)
protosulphide (Sulfid der niedrigsten Oxydationsstufe)

prototropic prototrop
prototropism, prototropy Prototropie f (auf Protonenwanderung basierende Tautomerie)
protoxide (Oxid der niedrigsten Oxydationsstufe)
protruded metal Streckmetall n
protruding-type insert (Plast) an einer Seite herausragendes Einlageteil (Einpreßteil) n
proustite (Min) Proustit m, lichtes Rotgültigerz n (Arsen(III)-silbersulfid)
Prout hypothesis Proutsche Hypothese f
Prout law Gesetz n der konstanten Proportionen
provisions industry Lebensmittelindustrie f, Nahrungsmittelindustrie f
provitamin Provitamin n
proximate analysis Schnellanalyse f, Rapidanalyse f, Schnellbestimmung f
prussate s. prussiate
Prussian blue $Fe^{III}_4[Fe^{II}(CN)_6]_3$ [unlösliches] Berliner Blau n, Preußischblau n
prussiate 1. $M^I CN$ Zyanid n; 2. $M^I_4[Fe(CN)_6]$ Hexazyanoferrat(II) n, Zyanoferrat(II) n; $M^I_3[Fe(CN)_6]$ Hexazyanoferrat(III) n, Zyanoferrat(III) n; 3. Prussid n, Prussiat n
prussic acid HCN Blausäure f, Zyanwasserstoffsäure f
prussite $(CN)_2$ Dizyan n, Zyan n
psammite (Geol) Psammit m
psammitic limestone Kalksandstein m, sandiger Kalkstein m
Pschorr reaction (synthesis) Pschorrsche Phenanthrensynthese f, Pschorr-Synthese f
psephite (Geol) Psephit m
pseudo acid Pseudosäure f
p. base Pseudobase f
pseudoalbite (Min) Pseudoalbit m, (veraltet für) Andesin m
pseudoanthracite Pseudoanthrazit m
pseudoaromatics Pseudoaromaten pl
pseudoasymmetric[al] pseudoasymmetrisch
pseudoasymmetry Pseudoasymmetrie f
pseudobreccia (Geol) Pseudobrekzie f
pseudobrookite (Min) Pseudobrookit m
pseudobutylene glycol $CH_3CHOHCHOHCH_3$ 2,3-Butylenglykol n, 2,3-Dihydroxybutan n, 2,3-Butandiol n
pseudocannel [coal] Pseudokannelkohle f
pseudocatalysis Pseudokatalyse f
pseudocellulose Hemizellulose f
pseudochlorogenic acid Pseudochlorogensäure f
pseudocrystalline pseudokristallin
pseudocumene $C_6H_3(CH_3)_3$ Pseudokumol n, 1,2,4-Trimethylbenzol n, asym-Trimethylbenzol n
pseudocyanine Pseudozyanin n
pseudoecgonine Pseudoekgonin n, ψ-Ekgonin n
pseudohalogen Pseudohalogen n
pseudomalachite (Min) Pseudomalachit m, Phosphorochalzit m (Kupfer(II)-trihydroxidorthophosphat)
pseudomauveine Pseudomauvein n
pseudomonomolecular pseudomonomolekular

pseudomorph pseudomorpher Kristall m, Pseudomorphose f, Afterkristall m
pseudomorphic pseudomorph
pseudomorphism Pseudomorphie f
pseudomorphosis Pseudomorphosierung f
pseudomorphous s. pseudomorphic
pseudopelletierine $C_9H_{15}NO$ Pseudopelletierin n, ψ-Pelletierin n, N-Methylgranatonin n
pseudoplastic quasiplastisch
p. flow quasiplastisches Fließen n, Quasifließen n
pseudoplasticity 1. s. pseudoplastic flow; 2. (Koll) Thixotropie f
pseudoracemic pseudorazemisch ·
pseudosalt Pseudosalz n
pseudoscopine Pseudoskopin n
pseudosolution Pseudolösung f, kolloid[al]e Lösung f
pseudostable pseudostabil
pseudosymmetric[al] pseudosymmetrisch
pseudosymmetry Pseudosymmetrie f
pseudotachylyte (Geol) Pseudotachylyt m
pseudotropine Pseudotropin n, ψ-Tropin n
pseudounimolecular reaction pseudomonomolekulare Reaktion f
pseudowollastonite (Min) Pseudowollastonit m (Kalziummetasilikat)
psilocybin Psilozybin n, 4-Phosphoryloxy-N,N-dimethyltryptamin n
psilomelane, psilomelanite (Min) Psilomelan m (Mangan(IV)-oxid)
psychochemical Psychochemikalie f
psycho[to]mimetic psychomimetisch
psycho[to]mimetic Psychomimetikum n, psychomimetisches Agens n
psychotrine Psychotrin n
psychotropic agent Psychopharmakon n, psychotropes Pharmakon n
psychrometer Psychrometer n, Luftfeuchtigkeitsmesser m, Verdunstungsmesser m
psychrometric chart Feuchtigkeitsdiagramm n, Feuchtigkeitstafel f
psychrophile, psychrophilic psychrophil, kälteliebend
psyllium mucilage Flohsamenschleim m, Psylliumschleim m (von Plantago psyllium L.)
pteridine Pteridin n (unsubstituierter Grundkörper der Pterine)
pterin Pterin n
pteropod ooze Pteropodenschlamm m
pterostilbene $(CH_3O)_2C_6H_3CH=CHC_6H_4OH$ Pterostilben n
p.t.f.e., PTFE, P.T.F.E. s. polytetrafluoroethylene
ptomaine Ptomain n, Leichengift n
ptyalin Ptyalin n (im Speichel enthaltene Amylase)
pucherite (Min) Pucherit m (Wismut(III)-orthovanadat)
pucker / to verbiegen
puckered configuration geknickte Konfiguration f
p. ring nicht ebener Ring m

puddle / to umrühren, puddeln
puddle *(Landw)* Gülle f
  p. iron Puddeleisen n
  p. steel Puddelstahl m
puddling Umrühren n, Puddeln n
  p. furnace Puddelofen m
  p. process Puddelverfahren n, Puddelprozeß m
  (zur Herstellung von Schweißstahl)
puering *(Gerb) (historisch)* Hundekotbeize f
puff / to [auf]blähen; *(Glas)* vorblasen
puff *(Glas)* Blasluft f
  p. drying Verdampfungstrocknung f im Vakuum
  p. paste Blätterteig m
  p. pastry margarine Ziehmargarine f
puffing Blähen n, Aufblähen n; *(Glas)* Vorblasen n
pug / to *(eine plastische Masse in einem Knetwerk bearbeiten)*
pug Knetwerk n; Dampfknetwerk n
  p. mill *(Ker)* Tonschneider m, Tonknetmaschine f, Tonkneter m
pugging *(Bearbeitung einer plastischen Masse in einem Knetwerk)*
pulegone Pulegon n *(ein Terpenketon)*
Pulfrich refractometer Pulfrich-Refraktometer n
pull / to ziehen; anziehen; *(Glas)* ziehen
  p. up *(Pap)* rupfen
  p. upward *(Glas)* emporziehen
pull Anziehung f; Zug m *(z.B. Schornsteinzug)*
  p.-back ram *(Plast)* Rückzugkolben m
  p. of the web Papierzugspannung f, Papierbahnspannung f, Bahnspannung f
  p. rod *(Plast)* Ausdrückstange f *(einer Presse)*
pulled surface *(Plast)* rauhe Oberfläche, Rauhigkeit f *(Fehler)*
pulp / to 1. *(Flot)* aufschwemmen; 2. *(Pap)* zu Halbstoff aufschließen, *(i.e.S.)* zu Holzschliff verschleifen *(oder)* chemisch aufschließen
pulp 1. Brei m, breiige Masse f, Pulpe f, Pülpe f, Pulp m; *(Lebm)* Pülpe f, *(i.e.S.)* Fruchtfleisch n, Fruchtmark n, Obstpülpe f; 2. *(Flot)* Trübe f, *(beim Sedimentieren auch)* Suspension f; 3. *(Pap)* Faser[stoff]brei m, Stoffbrei m, Fasermasse f, Fasersuspension f, Stoff m; Faserhalbstoff m, Halbstoff m, Stoff m, Halbzeug n, *(i.e.S.)* [mechanischer] Holzschliff m, [mechanischer] Holzstoff m, Holzmasse f, Schleifmasse f *(oder)* Holzzellstoff m, Zellstoff m; Zellstoffsuspension f, Zellstoffbrei m
to make into [a] p. s. to pulp 2.
  p. bleaching Zellstoffbleiche f
  p.-bleaching plant *(Pap)* Bleich[erei]anlage f, Bleicherei f
  p. board *(Pap)* Halbstoff m in Bogenform (in trockenen Bogen), *(i.e.S.)* Holzschliff m in Bogenform (Pappenform), Holzschliffpappe f, Holzschliffblätter npl *(oder)* Zellstoff m in Bogenform (Pappenform), Zellstoffpappe f, Zellstoffblätter npl
  p. brightness *(Pap)* Stoffweiße f, Weißgehalt (Weißgrad) m des Zellstoffs

  p. cardboard Holzkarton m *(aus Holzschliff)*; Zellstoffkarton m
  p. chest *(Pap)* Maschinenbütte f, Arbeitsbütte f, Stoffbütte f
  p.-coloured *(Pap)* in der Masse (im Stoff, im Holländer) gefärbt
  p. consistency (density) *(Pap)* Stoffdichte f, Stoffkonzentration f, Stoffkonsistenz f
  p. disk Masseplatte f, Massekuchen m
  p. dryer *(Zucker)* Schnitzeltrockner m, Schnitzeltrocknungsanlage f
  p. drying *(Zucker)* Schnitzeltrocknung f; *(Pap)* Zellstofftrocknung f
  p.-drying machine *(Pap)* Holzschliffentwässerungsmaschine f; Zellstoffentwässerungsmaschine f
  p. engine *(Pap)* Holländer m, Mahlholländer m, Messerholländer m, Ganzzeugholländer m
  p. furnish *(Pap)* eingetragener Faserhalbstoff m, Mahlgut n
  p. grinder s. p. engine
  p. industry Zellstoffindustrie f
  p. inlet *(Flot)* Trübezuleitung f
  p. machine s. 1. p. engine; 2. p.-drying machine
  p. making (manufacture) Holzschlifferzeugung f; Zellstofferzeugung f
  p. mill *(Pap)* Holzschleiferei f, Schleiferei f, Schleifereibetrieb m; Zellstoffwerk n, Zellstoffabrik f
  p. moulding *(Plast)* Pressen n von Faserbreiformteilen; Faserbreipreßteil n
  p. of ore Erztrübe f, Stofftrübe f, Trübe f
  p. press *(Zucker)* Schnitzelpresse f
  p.-press water *(Zucker)* Schnitzelpreßwasser n
  p. pump *(Zucker)* Schnitzelpumpe f
  p. purification Zellstoffveredelung f, Zellstoffreinigung f
  p. quality *(Pap)* Stoffqualität f
  p. refining s. p. purification
  p. roller *(veraltet für)* dandy roll
  p. saver *(Pap)* Stoffrückgewinnungsanlage f, Faserrückgewinnungsanlage f, Fangstoffanlage f, Stoffang m, Stoffänger m
  p. screen *(Pap)* Halbstoffsortierer m
  p. sheet Prüfbogen m, Handmuster n, Schöpfpapiermuster n, Laborpapierblatt n; Zellstoffbogen m
  p.-sized *(Pap)* massegeleimt, stoffgeleimt
  p.-sizing *(Pap)* Leimung f im Stoff, Leimung f in der Masse, Masseleimung f
  p. slurry (stock) *(Pap)* Faser[stoff]brei m, Stoffbrei m, Fasermasse f, Fasersuspension f, Stoff m
  p. strength Zellstofffestigkeit f
  p. suspension *(Pap)* Stoffsuspension f, Stoffwasser n
  p. wadding *(Am)* Zellstoffwatte f
  p. washer Zellstoffwäscher m
  p. washing Zellstoffwäsche f
  p. water *(Pap)* Sieb[ab]wasser n; *(Zucker)* Diffusionsabwasser n

p. **wood** (Pap) Papierholz n, Faserholz n, (i.e.S.) Schleifholz n (oder) Zellstoffholz n, Chemieholz n

p. **wood storage** (Pap) Holzlagerung f

p. **yield** Zellstoffausbeute f

pulper (Pap) Pulper m, Stoff[auf]löser m

pulping (Pap) Aufschließen n, Aufschließung f, Aufschluß m

p. **agent** (Pap) Aufschlußmittel n, Aufschluß-chemikalie f, Aufschließungsreagens n

p. **engine** s. pulper

p. **liquor** (Pap) Aufschlußlösung f, Kochflüssig-keit f, Kochlösung f

p. **machine** (Lebm) Passiermaschine f

p. **method** s. p. process

p. **of cereal straw** Strohaufschluß m

p. **of rags** Hadernaufschluß m, Hadernauf-bereitung f

p. **of wood** Holzaufschluß m

p. **period** (Pap) Fertigkochperiode f

p. **process** (Pap) Aufschlußverfahren n

p. **with chlorine** (Pap) Chloraufschluß m

p. **with nitric acid** Salpetersäure[zellstoff]ver-fahren n, Holzaufschluß m mit Salpetersäure

p. **with sodium sulphite** (Pap) Monosulfitauf-schluß m

pulpmaker Zellstoffhersteller m, Zellstoffproduzent m

pulpstone (Pap) Schleif[er]stein m

pulpwater (Pap) Sieb[ab]wasser n

pulpwood grinder (Pap) Holzschleifer m, Holz-schleifmaschine f, Holzzerfaserungsmaschine f, Defibrator m

pulpy breiig, breiartig; fleischig

p. **mass [of fibres]** (Pap) Fasermasse f, Faserbrei m

pulsate / to pulsieren

pulsation Pulsation f, pulsierende Bewegung f, Pulsieren n

p. **damper (snubber)** Pulsationsdämpfer m, Druckschwingungsdämpfer m

pulse column Pulsationskolonne f, pulsierte Ko-lonne f

pulsed sieve-plate column Siebbodenkolonne f mit Pulsation, pulsierte Siebbodenkolonne f

p. **tower** s. pulse column

pulser Pulsator m

pulsing s. pulsation

p. **device** s. pulser

pulsometer [pump] Pulsometer n

pulverization s. pulverizing

pulverize / to pulverisieren, [zer]pulvern, auf Staubfeinheit mahlen; (harte bis mittelharte Stoffe) [ver]mahlen; (Flüssigkeiten) zerstäuben

pulverized brown coal Braunkohlen[brenn]staub m

p. **coal** Kohlenstaub m, Kohlepulver n, pul-verisierte (gepulverte) Kohle f; Steinkohlenstaub m

p.-**coal burner** Kohlenstaubbrenner m, Staub-

brenner m; Steinkohlenstaubbrenner m

p.-**coal mill** Kohlenstaubmühle f, Kohle[n]mühle f; Steinkohlenmühle f

p. **fuel** Brenn[stoff]staub m

p.-**fuel-fired** [brenn]staubgefeuert

pulverizer s. pulverizing mill

p. **disk** Mahlscheibe f (einer Scheibenmühle)

pulverizing Pulverisieren n, Pulverisierung f, Pulvern n, Zerpulvern n, Mahlen n auf Staubfeinheit; Mahlen n, Vermahlen n (harter bis mittelharter Stoffe); Zerstäuben n, Zerstäubung f (von Flüssigkeiten)

p. **chamber** Mahlkammer f, Mahlraum m

p. **equipment** Pulverisieranlage f, Fein[st]mahl-anlage f

p. **mill** Mühle f (für harte bis mittelharte Stoffe); Staubmühle f, Pulverisiermühle f, Fein[st]mühle f

pulverulent pulv[e]rig, pulverförmig; staubig, staubbedeckt; brüchig, [leicht] zerbröckelnd, zerkrümelnd

pumace s. pomace

pumice Bimsstein m

p. **concrete** Bims[kies]beton m

p. **powder** Bimsstaub m

p. **stone** s. pumice

p. **tuff** Bimssteintuff m

pumiceous bimssteinartig

pummace s. pomace

pummel / to (Gerb) (Häute im Weichprozeß mechanisch) durcharbeiten

pump / to pumpen

p. **off** abpumpen

pump Pumpe f; Spülpumpe f (einer Rotary-Bohranlage)

p. **box** (Pap) Sauger m, Saug[er]kasten m

p. **casing** Pumpengehäuse n, Pumpenkörper m

p. **efficiency** Pumpenwirkungsgrad m

pumpable pumpfähig

pumped concrete Pumpbeton m

pumping Pumpen n

p. **concrete** s. pumped concrete

p. **fluid** Treibmittel n (Strahlpumpe)

p. **limit (point)** Pumpgrenze f (Verdichter)

p. **well** pumpender Ölbrunnen m

pumpkin [seed] oil Kürbiskernöl n (von Cucurbita pepo L.)

punch Stempel m, Prägestempel m, Pfaff m

p. **press** Lochpresse f, Lochstanze f, Formstanze f

punched plate Lochblech n

p.-**plate screen** Lochblechsieb n

punching Lochen n, Stanzen n, Ausschneiden n, (Plast) Prägen n (der Formteilkonturen)

puncture Durchschlag m, Durchschlagen n, Durch-bruch m

p.-**proof** durchschlagfest

p. **strength** Durchschlag[s]festigkeit f

pungent beißend, stechend [riechend]; scharf (im Geschmack)

p. smell beißender (stechender) Geruch *m*
p.-smelling stechend riechend
punty [iron] *(Glas)* Anfangeisen *n*, Hefteisen *n*
puppet *(Gum)* Puppe *f*, Rolle *f*
pure rein
  p. aluminium reines Aluminium *n*, Reinaluminium *n*
  p. benzene Reinbenzol *n*
  p.-blue reinblau
  p. brine Reinsole *f*
  p. coal 1. *s*. p. coal material; 2. Reinkohle *f*, reine Kohle *f* *(aschenärmste Kohle in der Aufbereitungstechnik)*
  p. coal material (substance) wasser- und aschefreie Kohlensubstanz (Kohle, Substanz) *f*, waf-Kohle *f*, Reinkohlensubstanz *f*, Reinkohle *f*, reine Kohlensubstanz (Kohle) *f*
  p. copper reines Kupfer *n*, Reinkupfer *n*
  p. culture Reinkultur *f*, Reinzucht *f*
  p. element Reinelement *n*, monoisotopes Element *n*
  p. flake naphthalene *(durch Sublimation)* hochgereinigtes Naphthalin *n*
  p. growth *s*. p. culture
  p. gum compound (mix, stock) *(Gum)* ungefüllte Mischung *f*
  p. gum vulcanizate *(Gum)* Reinvulkanisat *n*
  p. lecithin Reinlezithin *n*
  p. licorice Lakritzensaft *m*, Lakritze *f*, Süßholzsaft *m* *(von Glycyrrhiza glabra L.)*
  p. metal reines Metall *n*, Reinmetall *n*
  p. product Reinprodukt *n*, Reinerzeugnis *n*
  p. protein Reinprotein *n*
  p. research reine Forschung *f*
  p. substance reine Substanz *f*, Reinsubstanz *f*, reiner Stoff *m*
  p. toluene Reintoluol *n*
  p. xylene Reinxylol *n*
pures Reinprodukte *npl*
purgative Purgativum *n*, *(mäßig starkes)* Abführmittel *n*
purge / to entlüften, *(Gase, Luft)* abführen
purge gas Spülgas *n* *(Gaschromatografie)*
  p. steam Spüldampf *m* *(z.B. beim Thermofor-Catalytic-Cracking)*
purging flax Purgierlein *m*, Linum catharticum L.
  p. liquid Spülflüssigkeit *f*
  p. nut Purgiernuß *f*, Jatropha curcas L.
purification Reinigung *f*, Klärung *f*; Raffination *f*, Raffinieren *n*; Läuterung *f*, Läutern *n*; Purifikation *f*; Reindarstellung *f*
purified wood pulp Edelzellstoff *m*
purifier Reiniger *m*, Reinigungsapparat *m*; Reinigungsmittel *n*
purify / to reinigen, klären; raffinieren; läutern; reindarstellen
purifying effect Reinigungswirkung *f*, Waschwirkung *f*
purine Purin *n*
  p. base Purinbase *f*

p. group Puringruppe *f*
6-purinethiol 6-Purinthiol *n*, 6-Merkaptopurin *n*
purity Reinheit *f*
  p. coefficient *s*. p. quotient
  p. degree Reinheitsgrad *m*
  p. quotient *(Zucker)* Reinheitsquotient *m*
Purkinje effect *(Foto)* Purkinje-Phänomen *n*, Purkinjesches Phänomen *n*
purple purpurfarben, purpurfarbig, purpurn
  p. blende *(Min)* Purpurblende *f*, *(veraltet für)* Kermesit *m* *(Antimon(III)-oxidsulfid)*
  p. copper [ore] *(Min)* Buntkupfererz *n*, Buntkupferkies *m*, Bornit *m* *(Eisen(II)-kupfer(II)-sulfid)*
  p. of Cassius Cassiusscher (Kassiusscher) Goldpurpur *m* *(purpurfarbene kolloidale Goldlösung)*
  p. ore Purpurerz *n* *(gelaugte Rückstände bei der Verarbeitung Cu-haltiger Kiesabbrände)*
  p.-red purpurrot
  p. salt $KMnO_4$ Kaliumpermanganat *n*, Kaliummanganat(VII) *n*
  p. snail Purpurschnecke *f*
purplish-red *s*. purple-red
purpureate chelate Purpursäurechelat *n*
purpureo cobaltic chloride $[Co(NH_3)_5Cl]Cl_2$ Chloropentamminkobalt(III)-chlorid *n*
purpuric acid Purpursäure *f*
purpurin Purpurin *n*, 1,2,4-Trihydroxy-9,10-anthrachinon *n*
push out / to [her]ausstoßen, [aus]drücken *(z.B. Koks)*; verdrängen *(z.B. Luft aus einem Gefäß)*
push button Druckknopf *m*, Drucktaste *f*
  p.-button switch Druckknopfschalter *m*, Knopfschalter *m*
  p.-type centrifuge Schubzentrifuge *f*, Schubschleuder *f*
pushed-bat kiln *(Ker)* Durchschubofen *m*
pusher *s*. p.-type kiln
  p. disk *s*. p. plate
  p. machine Ausdrückmaschine *f*, Ausstoßmaschine *f*, Koks[aus]drückmaschine *f*, Koksausstoßmaschine *f*
  p. plate Schubboden *m*
  p. ram Druckstange *f*, Koksausdrückstange *f*, Koksdruckstange *f*, Stoßstange *f* *(zum Ausstoßen des Kokses)*
  p. side Stoßmaschinenseite *f*, Maschinenseite *f* *(am Koksofen)*
  p.-side bench maschinenseitige Bedienungsbühne *f* *(am Koksofen)*
  p.-type kiln *(Ker)* Stoßofen *m*
put into solution / to in Lösung bringen, [auf]lösen
  p. on *(Pap)* *(ein Sieb)* einziehen
  p. up the batch *(Gum)* die Mischung zusammenstellen
putrefacient *s*. putrefactive
putrefaction Faulen *n*, Fäule *f*, Fäulnis *f* *(Zersetzung durch Mikroorganismen bei gehemmtem Sauerstoffzutritt)*
putrefactive fäulniserregend, Fäulnis...

p. bacteria Fäulnisbakterien *npl*, Putride *npl*
p. fermentation Fäulnisgärung *f*
putrefy / to [ver]faulen
putrescent [ver]faulend; Fäulnis...
putrescibility Faulfähigkeit *f*, Fäulnisfähigkeit *f*
putrescible faulfähig, fäulnisfähig
putrescine $H_2N(CH_2)_4NH_2$ Putreszin *n*, Diaminobutan *n*, Tetramethylendiamin *n*
putrid faul[ig], angefault
putridity Fäule *f*, Fäulnis *f*
putty / to [ver]kitten
putty 1. Kitt *m*, Klebkitt *m*, *(i.e.S.)* Glaserkitt *m*, Fensterkitt *m*; 2. *s*. **p. powder**
p. powder *(Polierpulver mit Zinndioxid als Hauptbestandteil)*
Puzzolan cement Puzzolanzement *m*
p.v.a., PVA, P.V.A. *s*. polyvinyl acetate
p.v.c., PVC, P.V.C. *s*. 1. polyvinyl chloride; 2. pigment volume concentration
p.v.f., PVF, P.V.F. *s*. polyvinyl fluoride
p.v.p., PVP, P.V.P. *s*. polyvinylpyrrolidone
PWR *s*. pressurized-water reactor
pycnometer Pyknometer *n*, Wägefläschchen *n*
pycnometric pyknometrisch, Pyknometer...
pyocyanase Pyozyanase *f*
pyocyanin[e] Pyozyanin *n* *(Antibiotikum)*
pyr. *s*. pyridine
pyramidal pyramidal
pyran Pyran *n*
pyranose Pyranose *f* *(ringförmiges Monosaccharid)*
p. ring Pyranosering *m*
p. structure Pyranosestruktur *f*
pyranoside Pyranosid *n*
pyranthrone Pyranthron *n*
pyrargyrite *(Min)* Pyrargyrit *m*, dunkles Rotgültigerz *n* *(Antimon(III)-silbersulfid)*
pyrazine Pyrazin *n*, 1,4-Diazin *n*, Piazin *n*
pyrazolanthrone Pyrazolanthron *n*
pyrazole Pyrazol *n*, 1,2-Diazol *n*
pyrazoline Pyrazolin *n*
pyrazolone Pyrazolon *n*
p. dye Pyrazolonfarbstoff *m*
pyrene Pyren *n*
pyreneite *(Min)* Pyreneit *m* *(Kalziumaluminiumorthosilikat)*
pyrethrin Pyrethrin *n* *(natürliches Insektizid)*
pyrethrinization Pyrethrineinwirkung *f*, Pyrethrinvergiftung *f* *(Schädlingsbekämpfung)*
pyrheliometer Pyrheliometer *n*
pyrichrolite *(Min)* Pyrichrolith *m*, *(veraltet für)* Pyrostilpnit *m*
pyridazine Pyridazin *n*, 1,2-Diazin *n*, Oiazin *n*
pyridine C_5H_5N Pyridin *n*
p.-butadiene rubber Pyridin-Butadienkautschuk *m* *(Butadien-Vinylpyridin-Mischpolymerisat)*
3-p. carboxylic acid C_5H_4NCOOH 3-Pyridinkarbonsäure *f*, Nikotinsäure *f*, Niazin *n*
p. extraction Pyridinextraktion *f*
p. ring Pyridinring *m*

pyridinediol $C_5H_3N(OH)_2$ Pyridindiol *n*, Dihydroxypyridin *n*
pyrid[in]ol, pyridone C_5H_5NO Pyridon *n*, Hydroxypyridin *n*
pyridoxal Pyridoxal *n*
p. phosphate Pyridoxalphosphat *n* *(Koenzym in der Dekarboxylase)*
pyridoxine Pyridoxin *n*, Vitamin B 6 *n*, Adermin *n*
pyridylamine $C_5H_6N_2$ Pyridylamin *n*, Aminopyridin *n*
pyrimethamine Pyrimethamin *n*, 2,4-Diamino-5-p-chlorphenyl-6-äthylpyrimidin *n*
pyrimidine Pyrimidin *n*, 1,3-Diazin *n*, Miazin *n*
pyrite *(Min)* Pyrit *m*, Eisenkies *m*, Schwefelkies *m* *(Eisendisulfid)*
p. concentrate Pyritkonzentrat *n*
pyritic smelting pyritisches Schmelzen *n*, Pyritschmelzen *n*
pyro *s*. pyrogallol
pyroacetic ether CH_3COCH_3 Azeton *n*, 2-Propanon *n*, Dimethylketon *n*
pyroacid Pyrosäure *f*
pyroantimonate Diantimonat(V) *n*, Pyroantimonat(V) *n*
pyroantimonic acid $H_4Sb_2O_7$ Diantimon(V)-säure *f*, Pyroantimonsäure *f*
pyroarsenate Diarsenat(V) *n*, Pyroarsenat(V) *n*
pyroarsenic acid $H_4As_2O_7$ Diarsen(V)-säure *f*, Pyroarsensäure *f*
pyroarsenite Diarsenat(III) *n*, Pyroarsenat(III) *n*
pyroborate $M'_2B_4O_7$ Tetraborat *n*, Heptoxotetraborat *n*, Pyroborat *n*
pyroboric acid $H_2B_4O_7$ Tetraborsäure *f*, Heptoxotetraborsäure *f*, Pyroborsäure *f*
pyrobutanedioic acid *s*. pyrotartaric acid
pyrocatechin, pyrocatechinic acid *s*. pyrocatechol
pyrocatechol $C_6H_4(OH)_2$ Pyrokatechol *n*, Brenzkatechin *n*, 1,2-Dihydroxybenzol *n*, o-Dihydroxybenzol *n*
p. monoethyl ether $C_2H_5OC_6H_4OH$ Brenzkatechinmonoäthyläther *m*, o-Äthoxyphenol *n*, Guäthol *n*
pyrocellulose Zellulosenitrat *n*, Nitratzellulose *f*, *(fälschlich auch)* Nitrozellulose *f*
pyrochlore, pyrochlorite Pyrochlor *m* *(niob- und tantalhaltiges Mineral)*
pyrochroite *(Min)* Pyrochroit *m* *(Mangan(II)-hydroxid)*
pyroclastic rock pyroklastisches Gestein *n*
pyroconite *(Min)* Pyrokonit *m*, *(veraltet für)* Pachnolith *m* *(Kalziumnatriumfluoroaluminat)*
pyroelectricity Pyroelektrizität *f*
pyrogallol $C_6H_3(OH)_3$ Pyrogallol *n*, Pyrogallussäure *f*, 1,2,3-Trihydroxybenzol *n*
p. tan Pyrogallolgerbstoff *m*
pyrogen test Prüfung *f* auf pyrogene Substanzen
pyrogenetic *s*. pyrogenic
pyrogenic pyrogen
p. distillation trockene Destillation *f*, Trockendestillation *f*, Entgasung *f*

p. ore mineral pyrogenes Erzmineral n
pyroligneous acid Pyroligninsäure f, Holzessig m *(rohe Essigsäure aus trockener Holzdestillation)*
p. alcohol CH_3OH Methylalkohol m, Methanol n *(Rohprodukt aus trockener Holzdestillation)*
pyrolignite of lime $Ca(CH_3COO)_2$ Kalziumazetat n
pyrolusite *(Min)* Pyrolusit m *(Mangan(IV)-oxid)*
pyrolysis Pyrolyse f
pyrolytic pyrolytisch
pyrolyzate Pyrolyseprodukt n
pyrolyze / to pyrolysieren
pyromellitic acid $C_6H_2(COOH)_4$ Pyromellit[h]säure f, 1,2,4,5-Benzoltetrakarbonsäure f
p. dianhydride Pyromellit[h]säuredianhydrid n
pyrometallurgical pyrometallurgisch
pyrometallurgy Pyrometallurgie f
pyrometamorphism *(Geol)* Pyrometamorphose f, kaustische Metamorphose f
pyrometer Pyrometer n, Strahlungstemperaturmesser m, Strahlungsthermometer n
pyrometric pyrometrisch
p. cone *(Ker)* pyrometrischer Kegel m, Schmelzkegel m *(Temperaturkennkörper)*
p.-cone equivalent *(Ker)* Schmelzkegeläquivalent n, PCE
pyrometry Pyrometrie f, Messung f hoher Temperaturen
pyromorphite *(Min)* Pyromorphit m, Braunbleierz n, Grünbleierz n, Buntbleierz n
pyromucic acid $C_4H_3O \cdot COOH$ Brenzschleimsäure f, Pyroschleimsäure f, Furan-2-karbonsäure f
p. aldehyde $C_4H_3O \cdot CHO$ 2-Fur[fur]aldehyd m, 2-Furfurylaldehyd m, 2-Furankarbonal n, Fur[fur]al n
pyrone Pyron n *(Sammelname für eine Gruppe ketonartiger heterozyklischer Verbindungen)*
pyronine Pyronin n *(ein Xanthenfarbstoff)*
pyrope *(Min)* Pyrop m *(Magnesiumaluminiumorthosilikat)*
pyrophanite *(Min)* Pyrophanit m *(Manganmetatitanat)*
pyrophillite s. pyrophyllite
pyrophoric pyrophor, selbstentzündlich
p. alloy pyrophore Legierung f
pyrophorous s. pyrophoric
pyrophosphatase Pyrophosphatase f *(zu den Anhydridphosphatasen gehörendes Ferment)*
pyrophosphate Diphosphat n, Pyrophosphat n
pyrophosphite Diphosphit n, Pyrophosphit n
pyrophosphoric acid $H_4P_2O_7$ Diphosphorsäure f, Pyrophosphorsäure f
pyrophosphorous acid $H_4P_2O_5$ diphosphorige (pyrophosphorige) Säure f
pyrophyllite *(Min)* Pyrophyllit m *(Aluminiumdihydrogentetrasilikat)*
pyrophysalite *(Min)* Pyrophysalit m, Physalit m
pyroracemic acid $CH_3COCOOH$ Brenztraubensäure f, Pyr[o]uvinsäure f, 2-Oxopropansäure f, 2-Oxypropionsäure f, Azetylkarbonsäure f
pyroscope *(Ker)* Pyroskop n

pyrosol *(Koll)* Pyrosol n *(schmelzflüssiges Sol)*
pyrostibine, pyrostibite *(Min)* Pyrostibit m, Kermesit m *(Antimon(III)-oxidsulfid)*
pyrostilpnite *(Min)* Pyrostilpnit m *(Antimon(III)-silbersulfid)*
pyrosulphate $M^I_2S_2O_7$ Disulfat n, Pyrosulfat n
pyrosulphite $M^I_2S_2O_5$ Disulfit n, Pyrosulfit n
pyrosulphuric acid $H_2S_2O_7$ Dischwefelsäure f, Pyroschwefelsäure f
pyrosulphurous acid $H_2S_2O_5$ dischweflige (pyroschweflige) Säure f
pyrosulphuryl chloride $S_2O_5Cl_2$ Disulfurylchlorid n, Pyrosulfurylchlorid n, Dischwefelpentoxiddichlorid n
pyrotantalate $M^I_4Ta_2O_7$ Ditantalat n
pyrotartaric acid $HOOCCH(CH_3)CH_2COOH$ Pyroweinsäure f, Brenzweinsäure f, Methylbernsteinsäure f, Methylbutandisäure f
pyrotechnic pyrotechnisch
p. mixture pyrotechnischer Satz m
p. smoke generator pyrotechnischer Rauchgenerator m
pyrotechnical s. pyrotechnic
pyrotechnician s. pyrotechnist
pyrotechnics 1. Pyrotechnik f, Feuerwerkerei f, Feuerwerkskunst f; 2. pyrotechnische Erzeugnisse npl
pyrotechnist Pyrotechniker m, Feuerwerker m
pyrotechny s. pyrotechnics 1.
pyrotellurate $M^I_2Te_2O_7$ Ditellurat n
pyrothioarsenate $M^I_4As_2S_7$ Dithioarsenat(V) n
pyrothioarsenite $M^I_4As_2S_5$ Dithioarsenat(III) n
pyrouranate $M^I_4U_2O_7$ Diuranat n
pyrovanadate $M^I_4V_2O_7$ Heptoxodivanadat(V) n, Divanadat(V) n, Pyrovanadat(V) n
pyroxene *(Min)* Pyroxen n
p. group *(Min)* Pyroxengruppe f, Pyroxenfamilie f
pyroxylin[e] Kollodiumwolle f, Kolloxylin n, Nitrozellulose f
pyrrhosiderite *(Min)* Pyrrhosiderit m, *(veraltet für)* Goethit m *(Eisenoxidhydroxid)*
pyrrhotine, pyrrhotite *(Min)* Pyrrhotin m, Magnetkies m *(Eisen(II)-sulfid)*
pyrrole $(CH=CH)_2=NH$ Pyrrol n, Imidol n, Azol n
p. pigment Pyrrolfarbstoff m
p. ring Pyrrolring m
pyrrolidine Pyrrolidin n, Tetrahydropyrrol n, Tetramethylenimin n
pyrrolylene Pyrrolylen n, 1,3-Butadien n
pyruvaldehyde s. pyruvic aldehyde
pyruvate Pyruvat n *(Salz oder Ester der Brenztraubensäure)*
pyruvic acid $CH_3COCOOH$ Brenztraubensäure f, Pyr[o]uvinsäure f, 2-Oxopropansäure f, 2-Oxypropionsäure f, Azetylkarbonsäure f
p. aldehyde CH_3COCHO Pyruvinaldehyd m, Brenztraubensäurealdehyd m, Methylglyoxal n, 2-Oxopropanal n, Azetylformaldehyd m

Q

Q-branch Q-Zweig *m*, Nullzweig *m (eines Bandenspektrums)*
Q-electron Q-Elektron *n*
Q-shell Q-Schale *f*
Q unit Q-Einheit *f*, tetrafunktionelle Einheit *f (Struktureinheit)*
Q value *(phys Ch)* Q-Wert *m (einer Kernreaktion)*
quadravalent *s.* quadrivalent
quadribasic vierbasig
quadricovalent tetrakovalent, vierbindig
quadridentate *(Komplexchemie)* vierzahnig, vierzähnig, vierzählig *(Koordinationswert des Liganden)*
quadrimolecular tetramolekular
quadrivalence, quadrivalency Vierwertigkeit *f*, Tetravalenz *f*
quadrivalent vierwertig, tetravalent
quadruple point *(phys Ch)* Quadrupelpunkt *m*
quadruply connected vierfach gebunden
quadrupole Quadrupol *m*
  q. forces Quadrupolkräfte *fpl*
  q. moment Quadrupolmoment *n*, *(i.e.S.)* Quadrupolmoment *n* des Kerns, Kernquadrupolmoment *n*
  q. radiation Quadrupolstrahlung *f*
qualitative analysis qualitative Analyse *f*
quality Qualität *f*, Beschaffenheit *f*, Eigenschaft *f*; Güte *f*; *(Gum)* Mischung *f*
  q. control Qualitätskontrolle *f*
  q. demands Qualitätsanforderungen *fpl*
  q. fertilization Qualitätsdüngung *f*
  q. test Qualitätsprüfung *f*
quant *(phys Ch)* Quant *n*
  q. response *(Versuchswesen)* gequantelte Reaktion *f*
quantisation *s.* quantization
quantitative analysis quantitative Analyse *f*
quantity Quantität *f*, Menge *f*, Anzahl *f*
  q. meter Mengenmesser *m*
  q. of caustic *(Pap)* Alkalimenge *f*
  q. of electricity Elektrizitätsmenge *f*
  q. of heat Wärmemenge *f*
  q. of light Lichtmenge *f*
  q. of precipitate Bodenkörpermenge *f*, Niederschlagsmenge *f*
  q. of radiation Strahlungsintensität *f*
  q. ratio Mengenverhältnis *n*
quantization Quanteln *n*, Quantelung *f*, Quantisieren *n*, Quantisierung *f*
  q. of direction Richtungsquantelung *f*, Raumquantelung *f*, räumliche Quantelung (Quantisierung) *f*
quantize / to quanteln, quantisieren
quantometer Quantometer *n (ein Analysengerät)*
quantum 1. Quantum *n*, Menge *f*, Anzahl *f*; Betrag *m*, Summe *f*; 2. *(phys Ch)* Quant *n*
  q. chemistry Quantenchemie *f*
  q. condition Quantenbedingung *f*

  q. counter Quantenzähler *m*
  q. efficiency Quantenausbeute *f*
  q. electrodynamics Quantenelektrodynamik *f*
  q. energy Quantenenergie *f*
  q. equivalence Quantenäquivalenz *f*
  q. jump Quantensprung *m*
  q. liquid Quantenflüssigkeit *f*
  q.-mechanical quantenmechanisch
  q.-mechanical resonance quantenmechanische Resonanz *f*
  q. mechanics Quantenmechanik *f*, *(i.w.S.)* Quantenmechanik (Matrizenmechanik) *f* von Heisenberg
  q. number Quantenzahl *f*
  q. of action Plancksches Wirkungsquantum *n*, Plancksche Konstante *f*
  q. of energy Energiequant[um] *n*
  q. of light Lichtquant *n*, Photon *n*, Strahlungsquant *n*
  q. of radiation Strahlungsquant *n*
  q. of X-rays Röntgenquant *n*
  q. state Quantenzustand *m*
  q. statistics Quantenstatistik *f*
  q. theory Quantentheorie *f*
  q. transition *s.* q. jump
  q. yield *s.* q. efficiency
quarrying Tagebauförderung *f*, Tagebaubetrieb *m*, Förderung *f* im Tagebau
quartation Quartation *f (Trennung von Au und Ag durch heiße HNO_3)*
quarter-sized *(Pap)* viertelgeleimt, 1/4geleimt
quartz *(Min)* Quarz *m (Siliziumdioxid)*
  q. boat Quarzboot *n*, Quarzschiffchen *n*, Substanzschiffchen *n* aus Quarz *(beim Zonenschmelzen)*
  q. clock Quarzuhr *f*
  q. crystal Quarzkristall *m*
  q.-crystal clock *s.* q. clock
  q. diorite *(Min)* Quarzdiorit *m*
  q. filter funnel Quarzfilternutsche *f*
  q. glass Quarzglas *n*, durchsichtiges (klares) Kieselglas *n*
  q. grains Quarzkörner *npl*
  q. lamp Quarzlampe *f*
  q. porphyry Quarzporphyr *m*
  q. rock *s.* quartzite
  q. sand Quarzsand *m*
  q. spectrograph Quarzspektrograf *m*
  q. vessel Quarzgefäß *n*
  q. wedge Quarzkeil *m*
quartzic, quartziferous quarzhaltig, Quarz enthaltend, quarzführend, Quarz...
quartzite *(Min)* Quarzit *m*
quartzitic quarzitisch
quartzose, quartzous, quartzy quarzhaltig, quarzführend, Quarz enthaltend, Quarz...; aus Quarz [bestehend], quarzig; quarzähnlich, quarzartig, wie Quarz beschaffen, quarzig
quas Kwaß *m*
quasi-crystalline quasikristallin

q.-emulsifyer Quasi-Emulgator *m*
q.-Fermi level Quasi-Fermi-Niveau *n*
q.-racemate Quasirazemat *n*
quasistable quasistabil
quasistatic quasistatisch
quasistationary quasistationär
quasiviscous quasiviskos
quass *s.* quas
quassia Quassiaholz *n*, Bitterholz *n*, Fliegenholz *n*
quaternary quaternär, quartär
quaternary *s.* q. ammonium compound
  q. alloy quaternäre (quartäre) Legierung *f*
  q. ammonium base quaternäre (quartäre) Ammoniumbase *f*
  q. ammonium compound quaternäre (quartäre) Ammoniumverbindung *f*
  q. ammonium hydroxide quaternäres (quartäres) Ammoniumhydroxid *n*
  q. salt Quartärsalz *n*, quaternäres (quartäres) Salz *n*
quatropulper *(Pap)* Quatropulper *m*, Pulper (Stofflöser) *m* mit vier Auflösescheiben
quebrachamine Quebrachamin *n*
quebrachine Quebrachin *n*, Yohimbin *n* *(Alkaloid)*
quebracho Quebrachobaum *m* *(Quebrachia lorentzi Griseb. und Schinopsis balansae Engl.)*; Quebrachoholz *n*; *s.* q. bark
  q. bark Quebrachorinde *f*
  q. extract Quebrachoextrakt *m*
  q. ordinary *(Gerb)* warmlöslicher Quebrachoextrakt *m*
queen-bee's nutrient jelly Weisel[zellen]futtersaft *m*, Gelée royale *f*
  q. substance Königinnensubstanz *f (biologisch wirksame Substanz in den Speicheldrüsen der Bienenkönigin)*
quench / to *(Stahl)* rasch abkühlen, abschrecken, *(i.e.S.)* abschreckhärten, [durch Abschrecken] härten; *(Glas, Ker)* abschrecken, [ab]kühlen; *(Koks)* [ab]löschen; *(Lichtbogen)* [aus]löschen
quench *s.* quenching
  q. ag[e]ing *(Met)* Abschreckalterung *f*
  q. gas *s.* quenching gas
  q.-harden / to abschreckhärten, [durch Abschrecken] härten
  q. hardening Abschreckhärten *n*, Abschreckhärtung *f*, Härten *n*
  q. oil *s.* quenching oil
quenched lime Löschkalk *m*, gelöschter Kalk *m*
quenching rasches Abkühlen *n*, Abschrecken *n*, Abschreckung *f*; Abkühlen *n*, Abkühlung *f*, Kühlen *n*; Löschen *n*, Ablöschen *n*, Ablöschung *f (z.B. des Kokses)*; Löschen *n*, Auslöschen *n (des Lichtbogens)*
  q. agent *s.* q. medium
  q. bath Abschreckbad *n*
  q. car Löschwagen *m*, Koks[lösch]wagen *m*
  q. gas Löschgas *n*
  q. medium Abschreckmittel *n*; Ablöschmittel *n*
  q. of orbital angular momentum *(phys Ch)* Spinauslöschung *f*

q. oil Abschrecköl *n*, Quenchöl *n*
q. station Löschstation *f*, Löschanlage *f*
q. tower Löschturm *m*, Kokslöschturm *m*
quercetin Querzetin *n*, Meletin *n*, Sophoretin *n*, 3,5,7,3',4'-Pentahydroxyflavon *n*
quercetin-3-rutinoside Querzetin-3-rutinosid *n*, Violaquerzitrin *n*, Rutin *n*, Myrtikolorin *n*
quercitin *s.* quercetin
quercitol Querzitol *n*
quercitrin Querzitrin *n (Rhamnosid des Querzetins, Farbstoff)*
quercitron [bark] Querzitron *n*, Querzitronrinde *f*
quick ash Flugasche *f*
  q. assay Schnelltest *m*
  q.-assay method Schnellmethode *f*, Kurzzeitmethode *f*
  q. bleach Schnellbleiche *f*
  q.-bleaching schnellbleichend
  q.-breaking emulsion schnellbrechende Emulsion *f*
  q. clearance *(Anerkennung z.B. eines Pflanzenschutzmittels ohne Prüfverfahren)*
  q. cook *(Pap)* Schnellkochung *f*, direkte Kochung *f*
  q.-curing *(Plast)* schnellhärtend; *(Gum)* rasch (schnell) vulkanisierend, schnellvulkanisierend, rasch (schnell) heizend
  q. curing Schnellpökeln *n*, Schnellpökelung *f*
  q.-curing moulding compound *(Plast)* Schnellpreßmasse *f*
  q.-drying schnelltrocknend, schnell trocknend
  q. drying Schnelltrocknen *n*, Schnelltrocknung *f*
  q. fire *(Ker)* Scharffeuer *n*
  q.-freeze / to schnell gefrieren, schnellfrieren, schnell gefrieren lassen
  q.-freeze[r] Schnellgefrierapparat *m*
  q. freezing schnelles Gefrieren (Einfrieren) *n*, Schnellgefrieren *n*, Schnellgefrierverfahren *n*
  q. freezing plant Schnellgefrieranlage *f*
  q. hardener Schnellerhärter *m*, Schnellbinder *m*, Abbenderegler *m*
  q.-operating valve Schnellschlußventil *n*
  q.-setting agent *s.* q. hardener
  q. vinegar process Schnellessigverfahren *n*
quicklime CaO gebrannter (ungelöschter) Kalk *m*, Branntkalk *m (Kalziumoxid)*
quicksilver Quecksilber *n (Zusammensetzungen s. unter mercury)*
quiescent tank *(Abwasserreinigung)* Absitztank *m*
quinaldine Chinaldin *n*, 2-Methylchinolin *n*
  q. alkiodide Chinolinalkyljodid *n*
  q. chelate Chinaldinchelat *n*
  q. ethiodide Chinaldinäthyljodid *n*
quinalizarin[e] Chinalizarin *n*, Alizarinbordeaux *n*, 1,2,5,8-Tetrahydroxy-9,10-anthrachinon *n*
quinazoline Chinazolin *n*, 1,3-Benzodiazin *n*, Benzopyrimidin *n*
quince seed Quittensamen *m*, Quittenkerne *mpl*
  q. seed mucilage Quitten[samen]schleim *m*
quinhydrone Chinhydron *n*, *(i.e.S.)* $C_6H_4O_2 \cdot C_6H_4(OH)_2$ Chinhydron *n*

q. electrode Chinhydronelektrode f
quinic acid $(HO)_4C_6H_7COOH$ Chinasäure f, 1,3,4,5-Tetrahydroxyzyklohexankarbonsäure f
quinidine Chinidin n, β-Chinin n, Konchinin n
α-**quinidine** α-Chinidin n, Zinchonidin n
q. sulphate Chinidinsulfat n
quinine Chinin n, [5-Vinylchinuklidyl-(2)]-[6′-methoxychinolyl-(4′)]karbinol n
q. hydrochloride Chininhydrochlorid n
q. sulphate Chininsulfat n
quinizarin Chinizarin n, 1,4-Dihydroxy-9,10-anthrachinon n
q. condensation Chinizarinkondensation f
quinoid chinoid
q. cure (Gum) Chinondioximvulkanisation f, Chinondioximvernetzung f
quinol $C_6H_4(OH)_2$ Hydrochinon n, 1,4-Dihydroxybenzol n
q. clathrate Hydrochinonklathrat n
q. rearrangement Chinolumlagerung f
quinolate Chinolat n (Salz oder Ester der Chinolinsäure)
quinoline Chinolin n, 2,3-Benzopyridin n
q. dye Chinolinfarbstoff m
q. ethiodide Chinolinäthyljodid n
quinolinic acid $C_5H_3N(COOH)_2$ Chinolinsäure f, Pyridindikarbonsäure-(2,3) f
quinolizidine Chinolizidin n, Oktahydropyridokolin n, Norlupinan n, 1-Azabizyklo-[4,4,0]-dekan n
q. ring Chinolizidinring m
quinolylamine Chinolylamin n, Aminochinolin n
quinone Chinon n, (i.e.S.) $C_6H_4O_2$ Chinon n, p-Chinon n, p-Benzochinon n, 1,4-Zyklohexadiendion n
q. imine dye Chinoniminfarbstoff m
q. methide Chinonmethid n
q. monoxime $C_6H_4O(NOH)$ Chinonmonoxim n
q. ring Chinonring m
quinonediimine Chinondiimin n
quinonoid chinoid
q. dye Chinonfarbstoff m
q. form chinoide Form f
quinoxaline Chinoxalin n, Benzopyrazin n, 1,4-Benzodiazin n, 1,4-Diazanaphthalin n
quinquevalence, quinquevalency Fünfwertigkeit f, Pentavalenz f
quinquevalent fünfwertig, pentavalent
quintuply connected fünffach gebunden
quinuclidine Chinuklidin n (bizyklisches Zinchoninderivat)
quotient Quotient m
q. of purity Reinheitsquotient m (des Zuckerrübensaftes)

R

R acid $C_{10}H_5(OH)(SO_3H)_2$ R-Säure f, β-Naphtholdisulfonsäure R f, 2-Naphthol-3,6-disulfonsäure f

R-branch R-Zweig m, positiver Zweig m (eines Bandenspektrums)
R-M number s. Reichert-Meissl number
R salt R-Salz n, Dinatriumsalz n der 2-Naphthol-3,6-disulfonsäure, 2-naphthol-3,6-disulfonsaures Natrium n
rabbit fat Kaninchenfett n
r. hair Kanin[chen]haar n
rabble / to krählen, [um]rühren; kratzen; (Met) abschöpfen
rabble [arm] Krählarm m, Mischarm m, Rührarm m, Krähler m, Kratze f, Kratzer m, Krücke f
r. blade Rührzahn m
rabbler s. rabble [arm]
rabbling Krählen n, Rühren n, Umrühren n, Kratzen n
Rabi method (Kch) Methode f von Rabi, Rabi-Methode f
rac. s. racemic
race / to zu hoch steigen (Temperatur)
racemase Razemase f
racemate 1. Razemat n, razemisches Gemisch n, razemische Modifikation f (optisch inaktive 1 : 1-Mischung zweier optischer Antipoden); 2. Razemat n, DL-Tartrat n (Salz oder Ester der Traubensäure)
racemation s. racemization
racemic razemisch
r. acid $HOOC(CHOH)_2COOH$ razemische Weinsäure (Säure) f, para-Weinsäure f, DL-Weinsäure f, Traubensäure f
r. compound razemische Verbindung f, Razematverbindung f
r. form Razemform f
r. material s. r. substance
r. mixture (pair) s. racemate 1.
r. substance razemische Substanz f
racemisation s. racemization
racemizability Razemisierbarkeit f
racemizable razemisierbar
racemization Razemisierung f
racemize / to razemisieren
racemoid s. racemate
rachelle rachel, gelbbraun
racing fuel Rennkraftstoff m
rack / to (Bier auf Fässer) abfüllen; (Wein auf Fässer) abziehen
rack 1. Gestell n, Ständer m, Rahmen m; 2. Arrak m
racker s. racking machine
racking Abfüllen n (des Biers) auf Fässer, Faßabfüllung f, Abziehen n (des Weins) auf Fässer
r. cock Abfüllhahn m; Abziehhahn m
r. hose Abfüllschlauch m
r. machine Faßabfüllmaschine f, Faß[ab]füller m
rad s. radiation absorbed dose
raddle s. ruddle
radial radial, Radial...
r. blade Radialschaufel f

r. distribution function radiale Verteilungsfunktion f
r.-piston pump Radialkolbenpumpe f
r. probability [distribution] function s. r. distribution function
r. quantum number radiale Quantenzahl f
radiant strahlend
   r. density Strahlungsdichte f
   r. energy Strahlungsenergie f
   r. energy bombardment Anregung f durch radioaktive Strahlen (Strahlung)
   r. energy density Strahlungsenergiedichte f
   r. flux Strahlungsfluß m, Strahlungsleistung f
   r. flux density Strahlungsflußdichte f
   r. heat Strahlungswärme f
   r. heat dryer s. r.-heating dryer
   r. heater Heizstrahler m
   r.-heating dryer Strahlungtrockner m
   r. intensity Strahlungsintensität f
   r. power Strahlungsleistung f, Strahlungsfluß m
radiate / **to** strahlen, leuchten; [aus]strahlen, abstrahlen, emittieren, (Strahlen, Wellen) aussenden; bestrahlen
radiating element Strahler m
radiation Strahlen n, Strahlung f, Leuchten n; Ausstrahlen n, Ausstrahlung f, Abstrahlen n, Abstrahlung f, Radiation f, Emission f; Bestrahlen n, Bestrahlung f
   r. absorbed dose Rad-Einheit f, Rad n, rd (Einheit der Strahlendosis)
   r. absorption Strahlungsabsorption f
   r. biology Strahlenbiologie f, Radiobiologie f
   r. buffer Strahlungspuffer m
   r. catalysis Strahlungskatalyse f
   r.-chemical strahlenchemisch, strahlungschemisch, radiationschemisch
   r. chemistry Strahlenchemie f, Strahlungschemie f, Radiationschemie f, Kernstrahlenchemie f
   r. concrete Strahlenschutzbeton m, Abschirmbeton m
   r. constant Strahlungskonstante f
   r. counter Strahlenzähler m, Strahlenzählrohr n, Strahlungszähler m
   r. cure (curing) s. r. vulcanization
   r. damage Strahlungsschaden m, Strahlenschaden m
   r. damping Strahlungsdämpfung f
   r. density Strahlungsdichte f
   r. detector Strahlendetektor m
   r. dosage (dose) Strahlungsdosis f, Strahlendosis f
   r. dosimetry Strahlendosimetrie f
   r. dryer Strahlungstrockner m
   r. drying Strahlungstrocknung f
   r. effect Strahlungswirkung f, Strahlenwirkung f
   r. energy Strahlungsenergie f
   r. field Strahlungsfeld n
   r. filter Strahlenfilter n
   r. formula Strahlungsformel f
   r. hazard Strahlengefährdung f, Strahlengefahr f

r. heating Strahlungsheizung f
r.-induced strahlungsinduziert, strahleninduziert, strahlungsangeregt
r. intensity Strahlungsintensität f
r. law Strahlungsgesetz n
r. length Strahlungslänge f
r. loss Strahl[ungs]verlust m, Abstrahlungsverlust m, Ausstrahlungsverlust m
r. measurement Strahlungsmessung f, Strahlenmessung f
r.-measuring device Strahlungsmeßgerät n
r. monitor Strahlungsüberwachungsgerät n, Strahlenüberwachungsgerät n, Strahlenwarngerät n; radioaktive Strahlung registrierende Luftüberwachungsanlage f
r. monitoring Strahlenüberwachung f
r. of heat Wärmestrahlung f
r. oven Strahlungsofen m
r. pasteurization Strahlenpasteurisierung f
r. pressure Strahlungsdruck m, (i.e.S.) Lichtdruck m
r. protection Strahlungsschutz m, Strahlenschutz m
r. pyrometer Strahlungspyrometer n; Teilstrahlungspyrometer n, Leuchtdichtepyrometer n
r. reaction Strahlungsreaktion f, Strahlenreaktion f
r. resistance Strahlungsbeständigkeit f, Strahlungswiderstand m
r. shielding Strahlenabschirmung f
r. shielding concrete s. r. concrete
r. shielding glass Röntgenschutzglas n
r. sickness Strahlenkrankheit f, Strahlensyndrom n
r. source Strahlungsquelle f, Strahlenquelle f
r. spectrum Strahlungsspektrum n
r. survey s. r. monitoring
r. syndrome s. r. sickness
r. temperature Strahlungstemperatur f
r. therapy Strahlentherapie f
r.-type drying s. r. drying
r. vulcanization Vulkanisation (Vernetzung) f durch [radioaktive] Bestrahlung
radiationless strahlungslos, strahlungsfrei
   r. annihilation strahlungslose Vernichtung f
   r. transition strahlungsloser (strahlungsfreier) Übergang m
radiative strahlend, Strahlungs...
   r. capture Strahlungseinfang m
   r. equilibrium Strahlungsgleichgewicht n
   r. transition strahlender Übergang m, Strahlungsübergang m
radiator Strahler m
radical radikalisch
radical Gruppe f, Rest m, Radikal n; freies Radikal n, Radikal n (i.e.S.)
   r. chain polymerization Radikalkettenpolymerisation f
   r. formation Radikalbildung f

r. ion Radikalion *n*
r.-like radikalartig
r. migration Radikalwanderung *f*
r. name Radikalname *m*, Radikalbenennung *f*
r. polymerization Radikalpolymerisation *f*, radikalische Polymerisation *f*
r. reaction Radikalreaktion *f*
r. transfer Radikalübertragung *f*
r. vinegar Eisessig *m*
radicofunctional name radikofunktioneller Name *m* *(aus einem Radikal und einer funktionellen Gruppe zusammengesetzter Name)*
radio-frequency curing *(Gum)* Hochfrequenzheizung *f*
r.-frequency drying Hochfrequenztrocknung *f*, dielektrische Trocknung *f*
r.-frequency heating Hochfrequenz[be]heizung *f*, dielektrisches Beheizen *n*, Dielektro[be]heizung *f*
r.-frequency moulding *(Plast)* Formpressen *n* mit Hochfrequenzvorwärmung (Hochfrequenzheizung)
r.-frequency plastic welding Hochfrequenzschweißen *n* (Hochfrequenzschweißung *f*) von Plasten
r.-frequency preheating Hochfrequenzvorwärmung *f*
r.-labelled radioaktiv markiert, radioindiziert
r. radiation Radiostrahlung *f*
radioactinium Radioaktinium *n*
radioactivate / to radioaktiv machen, aktivieren
radioactivation Aktivieren *n*, Aktivierung *f* *(Methode der Radiochemie)*
r. analysis Aktivierungsanalyse *f*, Aktivitätsanalyse *f*
radioactive radioaktiv
r. carbon radioaktiver Kohlenstoff *m*, *(i.e.S.)* ^{14}C Radiokohlenstoff *m*, Kohlenstoff-14 *m*
r. chain *s.* r. decay series
r. change *s.* r. decay
r. cobalt radioaktives Kobalt *n*, *(i.e.S.)* ^{60}Co Radiokobalt *n*, Kobalt-60 *n*
r. constant Zerfallskonstante *f*
r. contamination radioaktive Verseuchung *f*
r. dating radioaktive (absolute, physikalische, physikalisch-chemische) Altersbestimmung *f*, Altersbestimmung *f* durch Radionuklide
r. decay radioaktiver Zerfall *m*, Atomzerfall *m*, Kernzerfall *m*
r. decay constant *s.* r. constant
r. decay law radioaktives Zerfallsgesetz *n*
r. decay series radioaktive Zerfallsreihe (Stammreihe, Reihe) *f*, Stammbaum *m*
r. decontamination radioaktive Entseuchung *f*
r. disintegration *s.* r. decay
r. disintegration constant *s.* r. constant
r. disintegration series *s.* r. decay series
r. displacement law radioaktiver Verschiebungssatz *m*, radioaktives Verschiebungsgesetz *n*
r. element radioaktives Element *n*, Radioelement *n*

r. emanation Emanation *f* *(radioaktive gasförmige Ausscheidung)*
r. equilibrium radioaktives Gleichgewicht *n*
r. fallout *(atmosphärischer)* radioaktiver Niederschlag *m* *(als festes Sediment)*, Fallout *m*
r. family *s.* r. decay series
r. heat radiogene Wärme *f*
r. indicator Radioindikator *m*, radioaktiver Indikator (Tracer) *m*
r. isotope radioaktives (instabiles) Isotop *n*, Radioisotop *n*
r. logging Messung *f* der Radioaktivität in Bohrlöchern
r. metal *s.* r. element
r. nuclide radioaktives Nuklid *n*, Radionuklid *n*
r. phosphorescent material radioaktive Leuchtmasse *f*
r. series *s.* r. decay series
r. standard radioaktives Standardpräparat *n*, radioaktiver Standard *m*
r. tracer *s.* r. indicator
r. transformation radioaktive Umwandlung *f*, radioaktiver Übergang *m*
r. units radioaktive Einheiten *fpl*
r. waste radioaktiver Abfall *m*, Atommüll *m*
radioactivity Radioaktivität *f*
r. logging *s.* radioactive logging
r. standard *s.* radioactive standard
radioadap[ta]tion Radioadaption *f*
radioautogram Autoradiogramm *n*, autoradiografische Aufnahme *f*
radioautographic autoradiografisch
radioautography Radioautografie *f*, Autoradiografie *f*
radiobiologic[al] strahlenbiologisch
radiobiology Strahlenbiologie *f*, Radiobiologie *f*
radiocarbon ^{14}C Radiokohlenstoff *m*, Kohlenstoff-14 *m*, *(i.w.S.)* radioaktiver Kohlenstoff *m*
r. age Kohlenstoffalter *n*, ^{14}C-Alter *n*
r. dating (method) Radiokohlenstoffmethode *f*, Radiokarbonmethode *f*, Radiokarbonverfahren *n*, Radiokohlenstoffdatierung *f*, Kohlenstoff-14-Methode *f*
radiochemical radiochemisch, strahlenchemisch
r. laboratory radiochemisches Laboratorium *n*
radiochemistry Radiochemie *f*; *s.* radiation chemistry
radiochromatography Radiochromatografie *f*
radiocobalt ^{60}Co Radiokobalt *n*, Kobalt-60 *n*, *(i.w.S.)* radioaktives Kobalt *n*
radiocolloid Radiokolloid *n*
radioelement Radioelement *n*, radioaktives Element *n*
radiogenetics Strahlengenetik *f*
radiogenic radiogen, durch radioaktiven Zerfall entstanden
r. heat radiogene Wärme *f*
radiogram *s.* radiograph
radiograph Röntgenaufnahme *f*, Röntgenbild *n*, Röntgenogramm *n*

radiographic radiografisch, Radiografie...; röntgenografisch, Röntgen...

radiography Radiografie *f (Verfahren zur Werkstoffprüfung)*; Röntgenografie *f*, Röntgenfotografie *f*

radioiodine ^{131}J Radiojod *n*, Jod-131 *n*, *(i.w.S.)* radioaktives Jod *n*

radioisotope radioaktives (instabiles) Isotop *n*, Radioisotop *n*

radiolarian ooze Radiolarienschlamm *m*

radiolarite *(Geol)* Radiolarit *m*

radiology Radiologie *f*, Strahlenkunde *f*

radiolucent strahlendurchlässig, *(i.e.S.)* durchlässig für Röntgenstrahlen

radioluminescence Radiolumineszenz *f*

radiolysis Radiolyse *f*, Strahlungsdissoziation *f*, strahlenchemische Zersetzung *f*

radiometer Radiometer *n*

radiometric radiometrisch

  r. adsorption analysis radiometrische Adsorptionsanalyse *f*

  r. analysis radiometrische Analyse *f*, Radioreagenzverfahren *n*, Radioreagenzmethode *f*, Analyse *f* mit radioaktiven Reagenzien

radiometry Radiometrie *f*

radiomicrometer Mikroradiometer *n*

radiomimetic *(Med)* radiomimetisch

radiomimetic *(Med)* Radiomimetikum *n*

radionuclide Radionuklid *n*, radioaktives Nuklid *n*

radiopacity Strahlenundurchlässigkeit *f*, *(i.e.S.)* Undurchlässigkeit *f* für Röntgenstrahlen

radiopaque strahlenundurchlässig, *(i.e.S.)* undurchlässig für Röntgenstrahlen

radiopasteurization Strahlenpasteurisierung *f*

radiophosphorus Radiophosphor *m*, Phosphor-32 *m*, *(i.w.S.)* radioaktiver Phosphor *m*

radiophotographic chemicals Röntgenchemikalien *fpl*

radiosensitive strahlungsempfindlich, strahlenempfindlich

radiosensitivity Strahlungsempfindlichkeit *f*, Strahlenempfindlichkeit *f*

radiosodium Radionatrium *n*, Natrium-24 *n*, *(i.w.S.)* radioaktives Natrium *n*

radiostrontium Radiostrontium *n*, Strontium-90 *n*, *(i.w.S.)* radioaktives Strontium *n*

radiosulphur Radioschwefel *m*, Schwefel-35 *m*, *(i.w.S.)* radioaktiver Schwefel *m*

radiotherapy Strahlentherapie *f*

radiothorium Radiothorium *n*, Thorium-228 *n*

radiotoxicity Radiotoxizität *f*

radiotracer Radioindikator *m*, radioaktiver Tracer (Indikator) *m*

radium Ra Radium *n*

  r.-beryllium neutron source *(Kch)* Radium-Beryllium-Quelle *f*

  r. bromide RaBr$_2$ Radiumbromid *n*

  r. carbonate RaCO$_3$ Radiumkarbonat *n*

  r. chloride RaCl$_2$ Radiumchlorid *n*

  r. content Radiumgehalt *m*

  r. emanation EmRa, RaEm *(oder)* ^{222}Rn Radiumemanation *f*, *(veraltet für)* Radon *n*

  r. iodate Ra(JO$_3$)$_2$ Radiumjodat *n*

  r. isotope Radiumisotop *n*

  r. mould *(Med)* Radiummoulage *f*

  r. needle *(Med)* radiumhaltige Hohlnadel *f*, Radiumnadel *f*

  r. pack Radiumeinlage *f*, Packung *f* mit radioaktivem Heilschlamm, radioaktive Heilschlammpackung *f*

  r. preparation Radiumpräparat *n*

  r. series Radiumreihe *f*, Uran-Radium-Reihe *f*, Uran-Radium-Zerfallsreihe *f*

  r. sulphate RaSO$_4$ Radiumsulfat *n*

  r. therapy Radiumtherapie *f*, Radiumbehandlung *f*

radon Rn Radon *n*

  r. fluoride Radonfluorid *n*

raffia Raffiabast *m*

Raffia palm Raffiapalme *f*, Raphia farinifera (Gaertn.)Hyl.

raffinase Raffinase *f (ein Raffinose zersetzendes Ferment)*

raffinate Raffinat *n*; Raffinat *n*, Solvat *n (im Lösungsmittel unlösliche Anteile bei der Solventextraktion)*

  r. end Raffinatende *n*, Raffinatseite *f*

  r. evaporator Raffinatverdampfer *m*

  r. phase Raffinatphase *f*

  r. solvent evaporator Verdampfer *m* für Raffinatlösung

  r. stripper Raffinatstripper *m*

raffinose Raffinose *f*, Melitose *f*, Melitriose *f (ein Trisaccharid)*

rag board Hadernpappe *f*

  r. boiler *(Pap)* Hadernkocher *m*

  r. breaker s. r. engine

  r. catcher *(Pap)* Zopfwinde *f*

  r. chopper s. r. cutter

  r. content *(Pap)* Haderngehalt *m*

  r. content paper hadernhaltiges Papier *n*

  r. cutter *(Pap)* Hadernschneider *m*, Lumpenschneider *m*

  r. duster *(Pap)* Hadernstäuber *m*

  r. engine *(Pap)* Halbzeugholländer *m*, Halbstoffholländer *m*

  r. felt Wollfilzpappe *f*

  r. fibre *(Pap)* Hadernfaser *f*

  r. mill Hadernaufbereitungsanlage *f*; Hadernpapierfabrik *f*

  r. mix *(Gum)* Ragmischung *f*

  r. paper Hadern[halbstoff]papier *n*, Lumpenpapier *n*

  r. paper making Hadernpapierherstellung *f*, Hadernpapierfabrikation *f*

  r. pulp *(Pap)* Hadern[halb]stoff *m*, Lumpenhalbstoff *m*, Hadernhalbzeug *m*

  r. roll *(Pap)* Holländerwalze *f*, Messerwalze *f*, Mahlwalze *f*

  r.-sorting room *(Pap)* Sortiersaal *m (bei der Hadernaufbereitung)*

r. stock *(Pap)* Hadern[halb]stoff *m*, Lumpenhalbstoff *m*, Hadernhalbzeug *n*; *(Gum)* Ragmischung *f*
r. stuff *s*. r. pulp
r. thrasher *s*. r. willow
r. tissue paper Hadernseidenpapier *n*
r. washing *(Pap)* Waschen *n* der gekochten Hadern
r. willow *(Pap)* Haderndrescher *m*
ragger *(Pap)* Zopfwinde *f*
rags *(Pap)* Hadern *mpl*, Lumpen *mpl*
raies ultimes *(Spektralanalyse)* letzte (beständige) Linien *fpl*, Restlinien *fpl*, Grundlinien *fpl*, Nachweislinien *fpl*
rail tank [car] Eisenbahnkesselwagen *m*, Eisenbahntankwagen *m*, Kesselwagen *m* [der Eisenbahn]
rain-making Erzeugung *f* von künstlichem Regen
r. test *(Text)* Beregnungsprüfung *f*, Beregnungsversuch *m*
r. water Regenwasser *n*
rainout 1. Rainout *n (Ausfallen radioaktiver Schwebstoffe als Kondensationskerne, atmosphärischer Niederschlagteilchen)*; 2. Rainout *m (Produkt von 1.)*
rainproof regendicht, wasserdicht
raise / to *(z.B. die Temperatur)* steigern, erhöhen; *(Brot, Teig)* [auf]gehen lassen, zum Gehen bringen; *(Farben)* aufhellen; *(Tuch)* [auf]rauhen
r. the basicity *(Gerb)* *(Chrombrühe)* abstumpfen
r. the flame die Flamme verstärken
r. to the boil zum Kochen bringen
raisin Rosine *f*
r.-seed oil Weintraubenkernöl *n*, Traubenkernöl *n*
raising of the boiling point Siedepunktserhöhung *f*
rake / to durch Kratzen fördern; auskratzen, ausschaben
rake Rechen *m*
r. arm *s*. raking arm
r. classifier Rechenklassierer *m*
raking arm Krählarm *m*, Krähler *m*
r. blade Schaberblech *n*, Schlammschild *m*
r. mechanism Krählwerk *n*, Rührwerk *n*
ralstonite *(Min)* Ralstonit *m*
ram / to rammen, stampfen
ram Kolben *m*, Stempel *m*; Pressentisch *m*; Druckstange *f (zum Ausstoßen des Kokses)*
r. extruder Kolbenstrangpresse *f*
r. extrusion Kolbenstrangpressen *n*
r. force Kolbendruck *m*
r. pump Tauchkolbenpumpe *f*, Plungerpumpe *f*
r.-type preplastication Kolbenvorplastizierung *f*
Raman active ramanaktiv
Raman effect Raman-Effekt *m*, Smekal-Raman-Effekt *m*
Raman frequency Raman-Frequenz *f*
Raman inactive ramaninaktiv
Raman line Raman-Linie *f*

Raman scattering Raman-Streuung *f*
Raman shift Raman-Verschiebung *f*
Raman spectroscopy Raman-Spektroskopie *f*
Raman spectrum Raman-Spektrum *n*
Raman study Raman-Untersuchung *f*
ramee *s*. ramie
ramie Ramie *f*, Chinagras *n*, Boehmeria nivea (L.)Gaudich
r. fibre Ramiefaser *f*
r. packing Ramiedichtung *f*, Ramiepackung *f*
Rammelkamp method Rammelkamp-Methode *f (zur Penizillinbestimmung)*
rammer Ramme *f*, Rammgerät *n (z.B. zur Formsandprüfung)*
ramming Rammen *n*, Stampfen *n*
r. mass (material, mix, mixture) Stampfmasse *f*, Stampfgemisch *n*, Stampfmischung *f*
Ramsauer effect Ramsauer-Effekt *m*
Ramsay-Young rule Ramsay-Youngsche Regel *f*
Ramsbottom carbon residue Verkokungsrückstand (Koksrückstand) *m* nach Ramsbottom, Ramsbottom-Carbon *n*
Ramsbottom carbon residue method *s*. Ramsbottom method
Ramsbottom coke number *s*. Ramsbottom value
Ramsbottom coking method *s*. Ramsbottom method
Ramsbottom method Ramsbottom-Methode *f (zur Bestimmung der Verkokungsneigung)*
Ramsbottom test Ramsbottom-[Carbon-]Test *m*, Verkokungstest *m* nach Ramsbottom
Ramsbottom value Ramsbottom-[Carbon-]Wert *m*, Ramsbottom-Verkokungswert *m*, Ramsbottom-Verkokungszahl *f*
Ramtilla oil Nigeröl *n (von Guizotia abyssinica (L.f.)Cass.)*
rancid ranzig
rancidification Ranzigwerden *n*
rancidify / to ranzig werden; ranzig machen (werden lassen)
rancidity Ranzidität *f*, Ranzigkeit *f*, ranziger Geschmack *m (oder)* ranziger Geruch *m*
r. test Ranziditätsprüfung *f*, Methode *f* zur Bestimmung der Schnelligkeit des Ranzigwerdens
r. with formation of aldehydes [freie] Aldehydranzigkeit *f*, Freialdehydigkeit *f*, Aldehydigkeit *f*
r. with formation of epihydrinaldehyde Epihydrinaldehydranzigkeit *f*
rancidness *s*. rancidity
random regellos, wahllos, planlos, ziellos, zufällig, ungeordnet, chaotisch
to be oriented in a r. manner to each other wirr durcheinanderliegen
r. copolymer statistisches Kopolymer[es] *n*
r. copolymerization statistische Kopolymerisation *f*
r. distribution *(Krist)* zufällige (regellose, statistische) Verteilung *f*

r. dyeing Buntfärbeverfahren *n* zum Erzielen von Mehrfarbeneffekten

r. error Zufallsfehler *m*, zufälliger Fehler *m*

r. fashion *(phys Ch)* ungeordnete Lage *f*, Zufallslage *f*

r. network *(Glas)* ungeordnetes Netz *n*

r. orientation *(Krist)* Zufallsorientierung *f*, regellose (nichtbevorzugte) Orientierung *f*

r. sample Stichprobe *f*

randomly distributed *(Krist)* zufällig (regellos, statistisch) verteilt

randomness *(Krist)* Unordnung *f*

Raney nickel Raney-Nickel *n*, Raney-Katalysator *m* aus Nickel

range Bereich *m*, Gebiet *n*; Abstand *m*, Entfernung *f*; Reichweite *f*; Umfang *m*, Spanne *f*

r.-energy relation Reichweite-Energie-Beziehung *f*, Energie-Reichweite-Beziehung *f*

r. of contrast *(Foto)* Kontrastumfang *m*

r. of dyes Farbstoffsortiment *n*

r. of electrons Elektronenreichweite *f*

r. of exposure *(Foto)* Belichtungsbereich *m*

r. of nuclear forces Reichweite *f* der Kernkräfte

r. of pH pH-Bereich *m*, pH-Gebiet *n*

r. of screen sizes Korngrößenbereich *m*, Kornspanne *f*, Teilchengrößenbereich *m*

r. of sensitivity *(Foto)* Empfindlichkeitsbereich *m*

rank Rang *m*, Rangstufe *f*, Inkohlungsgrad *m*

by r. nach dem Inkohlungsgrad

r. classification of coals Rangklassifikation *f* der Kohlen, Kohlenklassifikation *f* (Einteilung *f* der Kohlenarten) nach ihrem Inkohlungsgrad

r. parameter Inkohlungsparameter *m*, Inkohlungsmaßstab *m*

Rankine temperature scale Rankine-Skale *f*

Raoult's law Raoultsches Gesetz *n*

rap / to [be]klopfen, rütteln

rape 1. Raps *m*, Ölraps *m*, Brassica napus L.; 2. Trester *pl*; Treber *pl*; 3. Standfaß *n (bei der Essigherstellung)*, Essigbildner *m*

r. cake Rapskuchen *m*

r.[-seed] oil Kolzaöl *n*, Kohlsaatöl *n*, *(i.e.S.)* Rapsöl *n (aus Brassica napus L. em. Metzg.)* *(oder)* Rüb[sen]öl *n (aus Brassica rapa L. em. Metzg.)*

r. seed-oil green *(dem Schweinfurter Grün analoge Komplexverbindung aus Kupferarsenit und verseiftem Rapsöl) (Insektizid)*

raphia *s.* raffia

Raphia palm *s.* Raffia palm

rapid accelerator *(Gum)* schnellwirkender (schneller, starker) Beschleuniger *m*

r. acetification Schnellessigfabrikation *f*

r. cementing agent Schnellbinder *m*, Schnellerhärter *m*, Abbinderegler *m*

r. cooling Intensivkühlung *f*

r. determination Schnellbestimmung *f*

r. drying Schnelltrocknen *n*, Schnelltrocknung *f*

r. filter Schnellfilter *n*

r. filtration Schnellfiltrierung *f*, Schnellfiltration *f*

r. fixer *(Foto)* Schnellfixierbad *n*

r. fixing *(Foto)* Schnellfixierung *f*

r. freezing schnelles Gefrieren (Einfrieren) *n*, Schnellgefrieren *n*, Schnellgefrierverfahren *n*

r. hardening schnellbindend; schnellhärtend

r. sand filter Schnellsandfilter *n*

r. sand filtration Schnellsandfilterung *f*, Schnellsandfiltration *f*

r. strike schnelles Aufziehen *n (von Farbstoffen)*

r. system of digesting *(Pap)* Schnellkochverfahren *n*

r. tannage Schnellgerbung *f*

r. test *s.* r. determination

r. wetting agent Rapidnetzer *m*

rapper Klopfvorrichtung *f*, Rüttelvorrichtung *f*, Erschütterungsvorrichtung *f*, Klopfer *m*

rare selten; dünn *(z.B. Luft)*; nicht (halb) gar, halbgar *(z.B. Fleisch)*

r.-earth elements (metals) Seltenerdmetalle *npl*, Metalle *npl* der Seltenerden

r. earths seltene Erden *pl*, Seltenerden *pl*; *s.* r.-earth elements

r. gas Edelgas *n*

r.-gas configuration Edelgaskonfiguration *f*

r.-gas shell Edelgasschale *f*

rarefaction Verdünnen *n*; Verdünnung *f (Zustand)*

rarefactive verdünnend, Verdünnungs...

rarefiable verdünnbar

rarefy (rarify) / to verdünnen

Rasamala resin (wood oil) Rasamalaharz *n (von Altingia excelsa Noron.)*

Raschig hydrazine synthesis *s.* Raschig synthesis

Raschig process Raschig-Verfahren *n (Herstellung von Phenol aus Benzol)*

Raschig ring Raschig-Ring *m (ein Füllkörper)*

Raschig synthesis Raschig-Synthese *f*, Raschig-Prozeß *m (Hydrazindarstellung)*

Raschig's method Raschig-Verfahren *n (zur katalytischen Umsetzung von Chlorbenzol)*

Rast method (micromethod) Mikromethode *f* von Rast, Rast-Methode *f*, Kampfermethode *f* [nach Rast] *(zur Bestimmung der relativen Molekülmasse)*

Rast microprocedure Bestimmung *f* der relativen Molekülmasse nach [der Mikromethode von] Rast

Rast molecular weight method *s.* Rast method

Rast's camphor method *s.* Rast method

rat destruction Rattenvertilgung *f*, Rattenbekämpfung *f*

r. poison Rattengift *n*

rate Geschwindigkeit *f*; Menge *f* je Zeiteinheit; Anzahl *f* je Zeiteinheit

r. action Vorhaltwirkung *f*, differenzierende Wirkung *f (Regelungstechnik)*

r. constant Geschwindigkeitskonstante *f*

r.-determining geschwindigkeitsbestimmend

r. meter Durchfluß[mengen]messer *m*, Durchfluß[mengen]meßgerät *n*, Mengenstrommesser *m*, Flußmesser *m*

r. of absorption Absorptionsgeschwindigkeit *f*; Aufziehgeschwindigkeit *f (von Farbstoffen)*
r. of ag[e]ing Alterungsgeschwindigkeit *f*
r. of beating *(Pap)* Mahlgeschwindigkeit *f*
r. of bromination Bromierungsgeschwindigkeit *f*
r. of burning Brenngeschwindigkeit *f*
r. of change Umwandlungsgeschwindigkeit *f*
r. of cleavage Spalt[ungs]geschwindigkeit *f*
r. of combustion Verbrennungsgeschwindigkeit *f*, Brenngeschwindigkeit *f*
r. of condensation Kondensationsgeschwindigkeit *f*
r. of corrosion Korrosionsgeschwindigkeit *f*
r. of coupling Kupplungsgeschwindigkeit *f*
r. of crystallization Kristallisationsgeschwindigkeit *f*
r. of cure Anvulkanisationsgeschwindigkeit *f*; Vulkanisationsgeschwindigkeit *f*, Heizgeschwindigkeit *f*
r. of cyclization Zyklisierungsgeschwindigkeit *f*
r. of decay Zerfallsgeschwindigkeit *f*
r. of deformation Deformationsgeschwindigkeit *f*
r. of delignification *(Pap)* Auslösungsgeschwindigkeit *f* des Lignins
r. of delivery Fördergeschwindigkeit *f (z.B. einer Pumpe); (Text)* Liefergeschwindigkeit *f*; Lieferstrom *m*, Förderstrom *m*
r. of deposition Ablagerungsgeschwindigkeit *f*, Kuchenbildungsgeschwindigkeit *f*
r. of descent Sinkgeschwindigkeit *f*
r. of detonation Detonationsgeschwindigkeit *f*
r. of development *(Foto)* Entwicklungsgeschwindigkeit *f*, Entwicklungstempo *n*
r. of diffusion Diffusionsgeschwindigkeit *f*; Transfusionsgeschwindigkeit *f*
r. of discharge Lieferstrom *m*, Förderstrom *m*
r. of disintegration Zerfallsgeschwindigkeit *f*, Zerfallsrate *f*
r. of dissolution Lösegeschwindigkeit *f*, Lösungsgeschwindigkeit *f*, Auflösungsgeschwindigkeit *f*, Auflösungstempo *n*
r. of distillation Destillationsgeschwindigkeit *f*
r. of drainage *(Pap)* Entwässerungsgeschwindigkeit
r. of drying Trockengeschwindigkeit *f*, Trocknungsgeschwindigkeit *f*, spezifische Austauschfeuchtemenge *f*
r. of dye fixation Farbstoffixierungsgeschwindigkeit *f*
r. of dyeing Färbegeschwindigkeit *f*
r. of effusion Effusionsgeschwindigkeit *f*, Ausströmgeschwindigkeit *f*
r. of elution Elutionsgeschwindigkeit *f*
r. of esterification Veresterungsgeschwindigkeit *f*
r. of evaporation Verdampfungsgeschwindigkeit *f*; Verdunstungsgeschwindigkeit *f*
r. of exchange Austauschgeschwindigkeit *f*
r. of extraction Extraktionsgeschwindigkeit *f*

r. of fall Fallgeschwindigkeit *f*; Sinkgeschwindigkeit *f*
r. of fall under gravity Sinkgeschwindigkeit *f* im Gravitationsfeld
r. of feed[ing] Eintragmenge *f* in der Zeiteinheit; Durchsatz *m*; Vorschub *m*, Vorschubgeschwindigkeit *f*
r. of filling [of the chips with liquor] s. r. of penetration
r. of filtration Filtriergeschwindigkeit *f*, Filtergeschwindigkeit *f*
r. of fixing *(Foto)* Fixiergeschwindigkeit *f*
r. of flow Fließgeschwindigkeit *f*, Strömungsgeschwindigkeit *f*, Durchflußgeschwindigkeit *f*, Durchlaufgeschwindigkeit *f*; Durchfluß *m*, Durchsatz *m*, Förderstrom *m*, Strom *m*, Durchflußstrom *m*
r. of formation Bildungsgeschwindigkeit *f*
r. of formation of centres of crystallization s. r. of nuclei forming
r. of formation of nuclei s. r. of nuclei forming
r. of formation of the condensation centres s. r. of nuclei forming
r. of growth Wachstumsgeschwindigkeit *f*
r. of growth of the crystals Kristallwachstumsgeschwindigkeit *f*
r. of heat flow Wärmestrom *m*
r. of heating Erwärmungsgeschwindigkeit *f*, Aufheiz[ungs]geschwindigkeit *f*, Erhitzungsgeschwindigkeit *f*
r. of hydrogenation Hydrierungsgeschwindigkeit *f*
r. of hydrolysis Hydrolyse[n]geschwindigkeit *f*
r. of inflow Einström[ungs]geschwindigkeit *f*
r. of migration Wanderungsgeschwindigkeit *f*
r. of mixing Mischgeschwindigkeit *f*
r. of movement s. r. of migration
r. of nitration Nitrier[ungs]geschwindigkeit *f*
r. of nuclei forming *(Krist)* Keimbildungsgeschwindigkeit *f*
r. of oxidation Oxydationsgeschwindigkeit *f*
r. of penetration *(Pap)* Durchtränkungsgeschwindigkeit *f*, Imprägniergeschwindigkeit *f* *(der Hackschnitzel)*
r. of polymerization Polymerisationsgeschwindigkeit *f*
r. of racemization Razemisierungsgeschwindigkeit *f*
r. of reaction Reaktionsgeschwindigkeit *f*
r. of rotation Winkelgeschwindigkeit *f*; Drehzahl *f*
r. of saponification Verseifungsgeschwindigkeit *f*
r. of sedimentation (settling) Absetzgeschwindigkeit *f*, Sedimentationsgeschwindigkeit *f*, Klärgeschwindigkeit *f*, Absinkgeschwindigkeit *f*, Sinkgeschwindigkeit *f*
r. of stirring Rührgeschwindigkeit *f*
r. of sublimation Sublimationsgeschwindigkeit *f*
r. of swelling Quellungsgeschwindigkeit *f*

r. of travel Laufgeschwindigkeit *f (z.B. einer Welle)*

r. of vaporization Verdampfungsgeschwindigkeit *f*; Verdunstungsgeschwindigkeit *f*

r. of vertical decent Fallgeschwindigkeit *f*

r. of vulcanization Vulkanisationsgeschwindigkeit *f*, Heizgeschwindigkeit *f*

r. of water removal *s.* r. of drainage

r. of withdrawal *(Destillation)* Entnahmeverhältnis *n*

r. of zone travel Geschwindigkeit *f* der Zonenwanderung, Wanderungsgeschwindigkeit *f* der Zone (Schmelzzone), Zonengeschwindigkeit *f (beim Zonenschmelzen)*

ratholite *(Min)* Ratholith *m, (veraltet für)* Pektolith *m (Kalziumnatriumhydrogentrisilikat)*

raticide Rattenbekämpfungsmittel *n*

ratio Verhältnis *n*

r. of chemical-to-wood *(Pap)* Chemikalienverhältnis *n, (i.e.S. meist)* Alkaliverhältnis *n*

r. of concentrations Konzentrationsverhältnis *n*

r. of cooking liquor to wood weight *(Pap)* Flottenverhältnis *n, (i.e.S.)* Säureverhältnis *n (oder)* Laugenverhältnis *n*

r. of overburden to coal Verhältnis *n* Deckgebirge zu Kohle, Verhältnis *n* zwischen Deckgebirge und Kohlenschicht, Decke-Kohle-Verhältnis *n*, D : K-Verhältnis *nn*

r. of specific heats Verhältnis *n* der spezifischen Wärmen

r. of submergence Eintauchverhältnis *n*

r. of the intercepts *(Krist)* Achsenverhältnis *n*

rational index law *(Krist)* Rationalitätsgesetz *n*, Gesetz *n* der rationalen Parameterverhältnisse, Haüysches Gesetz *n*

r. intercept *(Krist)* rationaler Achsenabschnitt *m*

rattle *(Pap)* Klang *m*

raw roh, Roh...; unbearbeitet; *(Lebm)* ungekocht, unzubereitet; *(Ker)* ungebrannt; *(Foto)* unbelichtet

r. acid *(Pap)* Rohsäure *f*, Turmsäure *f*

r. acid liquor making *s.* r. acid production

r. acid production *(Pap)* Turmsäureherstellung *f*

r. barley Rohgerste *f*

r. batch *(Glas)* scherbenfreies Gemenge *n*, Rohgemenge *n*

r. beet sugar Rübenrohzucker *m*

r. body *(Ker)* ungebrannte Masse *f*

r. cane sugar Rohrrohzucker *m*

r. clay Rohton *m*; Rohkaolin *m*

r. coal Rohkohle *f*

r. coffee Rohkaffee *m*

r. cotton Rohbaumwolle *f*

r. cullet *(Glas)* Scherbengemenge *n*

r. fat Rohfett *n*

r. fibrous material *(Pap)* Faserrohstoff *m*

r. gas Rohgas *n*

r. gasoline Rohbenzin *n*

r. glaze *(Ker)* Rohglasur *f*

r. hide Rohhaut *f*, ungegerbte Haut *f*

r. humus Rohhumus *m*, Auflagehumus *m*, Mor

r. juice Rohsaft *m*; *(Zucker)* Diffusionssaft *m*, Rohsaft *m*

r. juice pump *(Zucker)* Rohsaftpumpe *f*

r. kaolin Rohkaolin *m*

r. lignite Rohbraunkohle *f*, Förderbraunkohle *f*

r. material Rohmaterial *n*, Rohstoff *m*, Ausgangsmaterial *n*, Ausgangsstoff *m*

r.-material cost[s] Rohstoffkosten *pl*, Ausgaben *fpl* für Rohstoffe

r. milk Rohmilch *f*

r. mixture Rohmischung *f*

r. opium Rohopium *n*

r. ore Roherz *n*, Fördererz *n*

r. paper Rohpapier *n*

r. papermaking material Papierrohstoff *m*, Rohstoff *m* für die Papiererzeugung, Faserrohstoff *m*, Papierfaserstoff *m*

r. peat Rohtorf *m*

r. product Rohprodukt *n*

r. rubber Rohkautschuk *m*

r. sewage rohes Abwasser *n*, Rohabwasser *n*

r. silk Rohseide *f*

r. spirits Rohspiritus *m*, Rohsprit *m*

r. starch Rohstärke *f*, Grünstärke *f*

r. stock Rohcharge *f*; *s.* r. papermaking material

r. sugar Rohzucker *m*

r. sulphite cooking acid *s.* r. acid

r. tallow Rohtalg *m*

r. water Rohwasser *n*; Frischwasser *n*

r. wool Rohwolle *f*, Schweißwolle *f*, Schmutzwolle *f*

r. wool scouring Waschen *n* der Rohwolle (Schweißwolle, Schmutzwolle)

rawhide Rohhaut *f*, ungegerbte Haut *f*

ray Strahl *m*

β-r. disintegration Beta-Zerfall *m*, β-Zerfall *m*, Beta-Umwandlung *f*

β-r. ionization detector β-Strahlenionisationsdetektor *m*

r. of light Lichtstrahl *m*

r. tracing Strahlenspurverfolgung *f*

r. tube Katodenstrahlröhre *f*, Elektronenstrahlröhre *f*

Rayleigh disk *(Phys)* Rayleigh-Scheibe *f*

Rayleigh-Jeans law *(Phys)* Rayleigh-Jeanssches Strahlungsgesetz *n*

Rayleigh law *(Phys)* Rayleighsches Gesetz *n*

Rayleigh refractometer Rayleigh-Interferometer *n (nach Haber und Löwe)*

Rayleigh scattering Rayleigh-Streuung *f*

Raymond bowl mill Raymond-Schüsselmühle *f*

Raymond Imp [hammer] mill Raymond-Imp-Mühle *f (eine Hammermühle)*

Raymond mill Raymond-Mühle *f*, Raymond-Pendel[rollen]mühle *f*

Raymond ring-roll[er] mill *s.* Raymond mill

Raymond roller mill *s.* Raymond mill

rayon Kunstseide *f*, Reyon *n(m)*

r. industry Kunstseidenindustrie *f*, Reyonindustrie *f*

r. mill wastes Kunstseidenfabrikabwasser n, Reyonfabrikabwasser n

r. pulp Textilzellstoff m, Kunstfaserzellstoff m, Chemiezellstoff m, Zellstoff m für die Chemiefaserindustrie, Reyonzellstoff m

α-**rays** α-Strahlen mpl, Alphastrahlen mpl

β-**rays** β-Strahlen mpl, Betastrahlen mpl

γ-**rays** γ-Strahlen mpl, Gammastrahlen mpl

razorite (Min) Rasorit m (Natriumtetraborat-4-Wasser)

RBE s. relative biological effectiveness

Re. s. Reynolds number

re-absorb / to wieder absorbieren (aufsaugen, einsaugen, aufnehmen); s. to resorb

reacidify / to erneut ansäuern

react / to reagieren, sich umsetzen, [chemisch] aufeinander einwirken; zur Reaktion bringen, aufeinander einwirken lassen; rückwirken

  r. to alkaline alkalisch reagieren

reactant Reaktionspartner m, Reaktionsteilnehmer m, Reaktant m, reagierender Stoff m; (Enzymchemie) Substrat n

  r.-type resin Reaktantharz n

reacting column Reaktionsturm m

  r. substance s. reactant

  r. tower Reaktionsturm m

reaction Reaktion f, Vorgang m

  r. catalyst Katalysator m, (Polymerisation auch) Aktivator m

  r. chain Reaktionskette f

  r. chamber Reaktionskammer f, Reaktionsraum m

  r. coordinate Reaktionskoordinate f

  r. course Reaktionsverlauf m, Reaktionsablauf m

  r. curve Reaktionskurve f

  r. energy Reaktionsenergie f

  r. entropy Reaktionsentropie f

  r. equation Reaktionsgleichung f, Umsatzgleichung f

  r. gas Reaktionsgas n

  r. gas chromatography Reaktionsgaschromatografie f

  r.-granulation drum Ammonisier-Granuliertrommel f (Düngemittelindustrie)

  r. in solution Lösungsreaktion f

  r. inhibition Reaktionshemmung f

  r. intermediate Reaktionszwischenstufe f, Intermediärprodukt n

  r. isochore Reaktionsisochore f

  r. isotherm Reaktionsisotherme f, Massenwirkungsgesetz n, MWG n

  r. kinetics Reaktionskinetik f, chemische Kinetik (Reaktionskinetik) f

  r. mechanism Reaktionsmechanismus m

  r. mixture Reaktionsgemisch n

  r. of first order Reaktion f erster Ordnung

  r. of second order Reaktion f zweiter Ordnung

  r. of zero order Reaktion f nullter Ordnung

  r. order Reaktionsordnung f

  r. pan (flaches) Reaktionsgefäß n

r. paper (Am) Prüfpapier n, Reagenzpapier n, Indikatorpapier n

r. period Hauptperiode f, Hauptversuch m (einer kalorimetrischen Messung)

r. primer (Farb) Reaktionsgrundierung f, Reaktionsprimer m, Wash-Primer m (Haftgrundmittel für Metalloberflächen)

r. product Reaktionsprodukt n, Endprodukt n, Endstoff m

r. rate Reaktionsgeschwindigkeit f, Umsetzungsgeschwindigkeit f

r.-rate constant Reaktionsgeschwindigkeitskonstante f

r. scheme Reaktionsschema f

r. sequence Reaktionsablauf m, Reaktionsfolge f

r. space Reaktionsraum m

r. temperature Reaktionstemperatur f

r. tower Reaktionsturm m; (Pap) Säureturm m (Sulfitverfahren); (Pap) Laugenturm m, Alkaliturm m (Chloraufschluß); (Pap) Chlor[ierungs]turm m (Zellstoffbleiche)

r. velocity s. r. rate

r.-velocity constant s. r.-rate constant

r. vessel Reaktionsgefäß n, Reaktionskessel m, Reaktionsbehälter m

r. zone Reaktionszone f

reactivate / to reaktivieren

reactivation Reaktivierung f

reactive reaktiv, reaktionsfähig, reaktionsfreudig

  r. diluent (Plast) reaktionsfähiger Verdünner m, reaktionsfähiges Lösungsmittel n

  r. dye Reaktivfarbstoff m

  r. group reaktive (reaktionsfähige) Gruppe f, Reaktivgruppe f

reactivity Reaktionsfähigkeit f, Reaktionsvermögen n, Reaktivität f, Reaktionsfreudigkeit f

  r. index Reaktionsfähigkeitsindex m

  r. to (with) oxygen Reaktionsfähigkeit (Reaktivität) f gegen[über] Sauerstoff

reactor 1. Kernreaktor m, Reaktor m; 2. Reaktor m, Reaktionsapparat m, Stoffumsetzer m, Umsetzer m; Reaktionsofen m; Reaktionsraum m, Reaktionskammer f

  r.-clarifier [kombinierter] Flockungsklärapparat m

  r. product Reaktorprodukt n

  r. shielding (Kch) Reaktorabschirmung f

  r. vessel s. reaction vessel

readily oxidizable leicht oxydierbar

  r. soluble leichtlöslich, leicht (gut) löslich

  r. volatile leichtflüchtig

reading Ablesen n, Ablesung f (z.B. eines Skalenwertes); Skalenwert m, Stand m, Anzeige f

to show a constant r. einen konstanten Wert aufweisen, konstant sein

to take a r. of ablesen

  r. device Ablesevorrichtung f

  r. error Ablesefehler m

r. microscope Ablesemikroskop *n*
readjust / to wieder einstellen *(auf einen Wert)*
ready for packing packfertig
  r. for use gebrauchsfertig; betriebsfertig
  r.-mixed concrete Fertigbeton *m*, Transport-
beton *m*, Lieferbeton *m*, Frischbeton *m*
  r.-to-cook kochfertig
  r.-to-drink trinkfertig
  r.-to-use *s.* r. for use
reagent Reagens *n*, [chemisches] Nachweismittel
n, Prüfstoff *m*, Prüf[ungs]mittel *n*
  of r. purity *s.* r. grade
  r. bottle Reagenzienflasche *f*
  r. grade analysenrein, zur Analyse, p.a.
  r. room Reagenzienraum *m*
  r. solution Reagenslösung *f*
real gas reales Gas *n*
  r. solution echte Lösung *f*
realgar *(Min)* Realgar *m (Tetrarsentetrasulfid)*
ream *(Pap)* Ries *n*
  r. paper Bogenpapier *n*, Formatpapier *n*, auf
Format geschnittenes Papier *n*
  r. wrapper Rieseinschlagpapier *n*
rear heat zone *(Plast)* hintere Heizzone *f*
  r.-phase chromatography Chromatografie *f* mit
Phasenumkehr (umgekehrter Phase)
rearrange / to umlagern, umgruppieren, umstellen;
sich umlagern (umgruppieren)
rearrangement Umlagerung *f*, Umgruppierung *f*
  r. of esters Esterumlagerung *f*
  r. polymerization Umlagerungspolymerisation *f*
  r. reaction Umlagerungsreaktion *f*
reassay Neubestimmung *f*, nochmalige Bestimmung *f*
  r. at source vom Hersteller vorgenommene
Neubestimmung *f*
reboil / to [wieder] aufkochen; *(Glas)* aufkochen,
aufschäumen; nachschäumen
reboiler 1. Destillationsgefäß *n*, Destillationsblase
f, Destillierblase *f*, Blase *f*, *(Tech auch)* Verdampfungsofen *m*; 2. *(Erdöl)* Aufkocher *m*,
Aufkochofen *m*, Rückverdampfer *m*, Wieder-
aufkocher *m*, Reboiler *m*
  r. furnace *s.* reboiler 2.
  r. liquid *(Destillation)* Blasenflüssigkeit *f*
rebound / to zurückspringen, zurückprallen;
zurückspringen (zurückprallen) lassen; reflek-
tiert werden
rebound Rückprall *m*, Rückstoß *m*, Zurückspringen
n; *(Gum)* [elastische] Erholung *f*, Rückverformung *f*, Rückfederung *f*, elastischer Anteil *m*;
s. r. elasticity
  r. elasticity *(Gum)* Rückprallelastizität *f*, Stoßelastizität *f*, elastischer Wirkungsgrad *m* bei
Stoß-Druckbeanspruchung
  r. height *(Gum)* Rückprallhöhe *f*
  r. resilience *s.* r. elasticity
  r. test *(Gum)* Rückpralltest *m*, Rückprallversuch
m
rebromination *(Foto)* Rebromierung *f*, Rehalo-
gen[is]ierung *f*

reburning *(Pap)* Rückbrennen *n*
recalescence *(Krist)* Rekaleszenz *f*
recap / to *(Reifen)* runderneuern
recapping Runderneuerung *f (von Reifen)*
recapture Wiedereinfangen *n*
recarbonize / to *s.* to recarburize
recarburization *(Met)* Wiederaufkohlen *n*, Wieder-
aufkohlung *f*
recarburize / to *(Met)* wiederaufkohlen
recarburizer, recarburizing agent *(Met)* Aufkohlungsmittel *n*
recausticizing *(Pap)* Kausti[fi]zieren *n*, Kausti[fi]zierung *f*, Aussüßen *n*
receding contact angle Rückzugsrandwinkel *m*,
Kontraktionsberührungswinkel *m* *(Prüfung
grenzflächenaktiver Stoffe)*
receive / to auffangen, sammeln
receiver Sammelbehälter *m*, Sammelgefäß *n*,
Auffangbehälter *m*, Auffanggefäß *n*, Auffänger
m; Auffangschale *f*; *(Destillation)* Vorlage *f*,
Destilliervorlage *f*, Destillat[ions]vorlage *f*, Destillatsammler *m*; *(Glas)* Auffangrinne *f*
  r. tube Vorstoß *m*
receiving dryer *(Pap)* Vortrockenzylinder *m*,
Vortrockner *m*
  r. electrode Niederschlagselektrode *f*, Sammelelektrode *f*
  r. flask Vorlage *f (der Destillationsapparatur)*
  r. tank Sammelbehälter *m*, Auffangbehälter *m*,
Auffänger *m*; *(Pap)* Stoffgrube *f*, Stoffkasten *m*,
Kochergrube *f*, Kocherbütte *f*, Blastank *m*,
Ausblasbehälter *m*
recent frisch, neu, rezent; ständig nachwachsend
(Brennstoff)
receptacle Behälter *m*, Sammelbehälter *m*, Gefäß
n, Sammelgefäß *n*
recessed-plate [filter] press Kammerfilterpresse *f*
recheck / to nachprüfen, nochmals [über]prüfen
rechipper *(Pap)* Desintegrator *m*, Desintegratormühle *f*, Schlagmühle *f*, Schlägermühle *f*,
Schleudermühle *f*
recipe Rezept *n*; Rezeptur *f*; Gebrauchsanweisung
f
  r. of mix Mischungsrezept *n*
reciprocal Kehrwert *m*
  r. relative dispersion reziproke relative Dispersion *f*, Abbesche Zahl *f*, *v*-Wert *m*
reciprocate / to hin- und herbewegen; hin- und
hergehen, sich hin- und herbewegen
reciprocating compressor Kolbenverdichter *m*,
Kolbenkompressor *m*
  r.-conveyor centrifuge Schubzentrifuge *f*,
Schubschleuder *f*
  r. impeller agitator Vibrationsrührer *m*, Vi-
bromischer *m*, Vibrationsmischer *m*
  r. plate column Kolonne *f* mit schwingenden
Siebböden
  r. pump Hubkolbenpumpe *f*, Kolbenpumpe *f* mit
hin- und hergehendem Kolben
  r.-pusher centrifuge *s.* r. conveyor centrifuge

r. screen Vibratorsieb *n*, Vibrationssieb *n*, Zittersieb *n*; Steilwurfsieb *n*
r. screw machine *(Plast)* Schneckenmaschine *f* mit oszillierender Schnecke
reciprocity failure *s.* r. law failure
r. law *(Foto)* Reziprozitätsgesetz *n*, Reziprozitätsregel *f*, Bunsen-Roscoesches Gesetz *n*
r. law failure *(Foto)* Abweichung *f* vom Reziprozitätsgesetz, Schwarzschild-Effekt *m*
recirculate / to im Kreislauf führen, [im Kreislauf] zurückführen, umpumpen; [in den Kreislauf] zurückführen, dem Kreislauf wieder zuführen (zusetzen), rezirkulieren
recirculation Umlauf *m*, Zirkulation *f*
reclaim / to wiedergewinnen, [zu]rückgewinnen, rückführen; wiederaufbereiten, *(Gum, Plast meist)* regenerieren
reclaim *(Gum)* Regenerat *n*, regenerierter Kautschuk *m*, Regenerativgummi *m*; *(Plast)* Regenerat *n*
r. compound Regeneratmischung *f*
r. dispersion *s.* r. rubber dispersion
r. mix *s.* r. compound
r. mixing mill Regeneratmischwalzwerk *n*
r. rubber dispersion Regeneratdispersion *f*
reclaimed rubber Regenerativgummi *m*, regenerierter Kautschuk *m*, Regenerat *n*
r. wool Reißwolle *f*, *(manchmal auch)* Regeneratwolle *f*
reclaimer Regenerathersteller *m*
reclaiming Regenerieren *n*, Regenerierung *f*
r. agent Regeneriermittel *n*
r. method (process) Regenerierverfahren *n*
r. tank *(Pap)* Rücklaufbehälter *m*
reclamation *s.* reclaiming
r. disease *(Landw)* Heidemoorkrankheit *f* *(Mangelkrankheit)*
reclamator process *(Gum)* Reclamator-Verfahren *n* *(Regenerierverfahren)*
recoil Rückstoß *m*, Rückschlag *m*, Rückprall *m*
r. atom Rückstoßatom *n*
r. electron Rückstoßelektron *n*, Compton-Elektron *n*
r. ion Rückstoßion *n*
r. nucleus Rückstoßkern *m*
r. particle Rückstoßteilchen *n*
r. radiation Rückstoßstrahlung *f*
recombination Rekombination *f*, Wiedervereinigung *f*
r. of carriers Ladungsträgerrekombination *f*
r. of ions Ionenrekombination *f*
r. rate Rekombinationsgeschwindigkeit *f*, Wiedervereinigungsgeschwindigkeit *f*, Rekombinationsrate *f*
r. trap Rekombinationszentrum *n*, tiefe (tiefliegende) Haftstelle *f*
r. velocity *s.* r. rate
recombine / to wiedervereinigen, rekombinieren; sich wiedervereinigen (rekombinieren)
reconcentrate / to aufsättigen

recondense / to wieder kondensieren (niederschlagen, verdichten); [sich] wieder kondensieren
reconstitute / to zur ursprünglichen Konzentration lösen
reconstituted milk auf Milchkonzentration verdünnte Trockenmilch *f*, rekonstituierte *(aus aufgelöstem Milchpulver erhaltene)* Milch *f*, Milchpulvermilch *f*
reconvert / to zurückverwandeln, zurückbilden; sich zurückverwandeln (zurückbilden)
recooking nochmaliges Kochen *n*
recorder Schreiber *m*, Schreibwerk *n*
r. chart Registrierpapier *n*, Diagrammpapier *n*
recording Registrierung *f*, Aufzeichnung *f*; Schreibung *f*
r. instrument registrierendes Gerät (Instrument) *n*, Registriergerät *n*, Registrierinstrument *n*; [selbst]schreibendes Gerät (Instrument) *n*, Schreibgerät *n*, Schreibinstrument *n*, Selbstschreiber *m*, Schreiber *m*
recover / to 1. wiedergewinnen, [zu]rückgewinnen, rückführen; [wieder]aufbereiten, *(Gum, Plast meist)* regenerieren; 2. gewinnen, darstellen; 3. sich nachbilden *(z.B. von Isotopen)*; 4. *(Gum)* rückfedern, sich erholen, zurückspringen
recoverable wiedergewinnbar
recovered material *s.* r. stock
r. oil zurückgewonnenes Öl *n* *(bei der Altöl-Regenerierung)*; Öl *n* aus Rückstandsaufarbeitung *(bei der Kohlehydrierung)*
r.-oil tank Tank *m* für zurückgewonnenes Öl
r. stock *(Pap)* Fangstoff *m*, zurückgewonnener (wiedergewonnener) Stoff *m*
r. waste paper Altpapierstoff *m*, wiederaufbereitetes (aufbereitetes, regeneriertes) Altpapier *n*
r. wool Reißwolle *f*, *(manchmal auch)* Regeneratwolle *f*
recovery 1. Wiedergewinnung *f*, Rückgewinnung *f*, Rückführung *f*; Aufbereitung *f*, *(Gum, Plast meist)* Regenerierung *f*, Regeneration *f*; 2. Gewinnung *f*, Darstellung *f*, Ausbringen *n*; 3. Ausbeute *f*; 4. Nachbildung *f* *(z.B. von Isotopen)*; 5. *(Gum)* [elastische] Erholung *f*, Rückverformung *f*, Rückfederung *f*, Zurückspringen *n*, elastischer Anteil *m*; *(Text)* zeitabhängige Erholung *f*
r. equipment Rückgewinnungsanlage *f*
r. furnace *(Pap)* Schmelzofen *m* zur Verbrennung der Schwarzlauge
r. of chemicals Chemikalienrückführung *f*, Chemikalienrückgewinnung *f*
r. of coal chemicals Kohlenwertstoffgewinnung *f*
r. of heat Wärmerückgewinnung *f*
r. of papermaking fibres *(Pap)* Faserrückgewinnung *f*, Faserwiedergewinnung *f*, Stoffrückgewinnung *f*, Stoffwiedergewinnung *f*
r. of waste paper Wiederaufbereitung *f* von

Altpapier, Altpapieraufbereitung *f*, Altpapier-regeneration *f*

r. plant Rückgewinnungsanlage *f*, Wieder-gewinnungsanlage *f*

recrystallization Umkristallisieren *n (Tätigkeit)*; Rekristallisation *f (Vorgang)*

r. annealing Rekristallisationsglühen *n*, re-kristallisierendes Glühen *n*

r. cylinder (tube) Ruherohr *n (Margarineherstel-lung)*

recrystallize / to umkristallisieren, wieder aus-kristallisieren

rect plant *s.* rectifying plant

rectangular basin rechteckiges Becken *n*, Längs-becken *n*

r. basket extractor kombinierter Becherwerkex-traktor *m*

r. kiln Rechteckofen *m*

r. notch rechteckiger Überfall *m (Mengenstrom-messung)*

r.-opening cloth Langmaschengewebe *n*

r. tank *s.* r. basin

rectification 1. Rektifikation *f*, Rektifizierung *f*, Gegenstromdestillation *f*; 2. Justieren *n*, Ju-stierung *f*, Justage *f*, [richtige] Einstellung *f (eines Instruments)*

r. column *s.* rectifying column

r. still Rektifizierapparat *m*, Rektifikationsapparat *m*, Rektifizieranlage *f*

rectifier *s.* rectifying apparatus

rectify / to *(Destillation)* rektifizieren; *(Instrumente)* justieren, [richtig] einstellen

rectifying apparatus Rektifikationsapparat *m*, Rektifizierapparat *m*, Rektifiziervorrichtung *f*

r. column Rektifikationssäule *f*, Rektifikations-kolonne *f*, Rektifiziersäule *f*, Rektifizierkolonne *f*, Austauschsäule *f*, Austauschkolonne *f*, Trenn-säule *f*, Trennkolonne *f*

r. operating line Verstärkungsgerade *f*

r. plant Rektifikationsanlage *f*, Rektifizieranlage *f*

r. section Rektifizierteil *m*, Rektifikationsteil *m*, Rektifizierzone *f*, Rektifikationszone *f*, Ver-stärkungsteil *m*

r. still *s.* r. apparatus

recuperative rekuperativ

r. firing Rekuperativfeuerung *f*

r. furnace Rekuperativofen *m*

recuperator Rekuperator *m*, rekuperativer Vor-wärmer *m*

recycle / to im Kreislauf führen, [im Kreislauf] zurückführen, umpumpen; [in den Kreislauf] zurückführen, dem Kreislauf wieder zuführen (zusetzen), rezirkulieren

recycle gas Kreislaufgas *n*, Umlaufgas *n*, Zir-kulationsgas *n*

r. hydrogen Kreislaufwasserstoff *m*

r. hydrogen chloride Kreislaufchlorwasserstoff *m*

r. oil Rücklauföl *n*, Rückkreisöl *n*, Rückführöl *n*

r. pump Umlaufpumpe *f*, Umwälzpumpe *f*

r. stock *s.* r. oil

recycling 1. Kreislaufführung *f*; 2. Recycling *n (Wiedereinpressen von im Überschuß vor-handenen Raffinerieerzeugnissen in die Erdöllagerstätte)*

red Rot *n (Farbempfindung)*; rot[färbend]er Farb-stoff *m*, Rot *n*

r. acaroid resin rotes Akaroidharz *n (von Xanthorrhoea-Arten)*

r. acetate *,s.* r. mordant

r. acid 1,5-Dihydroxynaphthalin-3,7-disulfon-säure *f*

r. algae Rotalgen *fpl*, Rottange *mpl*, Rhodophyceae

r. antimony *(Min)* Rotspießglanz *m*, Kermesit *m (Antimon(III)-oxidsulfid)*

r. arsenic 1. *s.* r. arsenic sulphide; 2. *(Min)* Rote Arsenblende *f*, *(veraltet für)* Realgar *m (Te-trarsentetrasulfid)*; 3. *(Gerb)* Rotes Arsenik *n (Gemisch aus Arsensulfiden)*

r. arsenic glass As_4S_4 rotes Arsenikglas *n*, Rotglas *n (glasiges Arsensulfid)*

r. arsenic sulphide As_4S_4 Tetrarsentetrasulfid *n*

r. bark Chinarinde *f*, Fieberrinde *f (von Cinchona-Arten, i.e.S. von Cinchona succirubra Pavon)*

r. bole roter Bolus *m*, Rötel *m*, Rotocker *m*, Rotstein *m*, Eisenrot *n*

r. brass Rotguß *m*, Rotmetall *n*

r. colouration Rotfärbung *f*

r. copper ore *(Min)* Rotkupfererz *n*, Kuprit *m (Kupfer(I)-oxid)*

r. copper oxide Cu_2O Kupfer(I)-oxid *n*

r. glow *s.* r. heat

r. gum 1. amerikanischer Styrax (Storax) *m (Balsamharz von Liquidambar styraciflua L.)*; 2. *s.* eucalyptus gum; 3. *s.* r. acaroid resin; 4. Rhodiumholz *n (Kernholz von Liquidambar styraciflua L.)*

r. heat Rotglut *f*

r. heat discolouration *(Gerb)* rote Erhitzung *f (durch Bakterien verursachte rote Verfärbung an Häuten)*

r. hematite *(Min)* Hämatit *m*, Haematit *m (Eisen(III)-oxid)*

r.-hot rotglühend

r. iron ore *(Min)* Roteisenerz *n*, Roteisenstein *m*, Blutstein *m (Eisen(III)-oxid)*

r. ironbark *(Sammelname für mehrere Gerb-rinden liefernde Eucalyptus-Arten, besonders Eucalyptus sideroxylon A. Cunn.)*

r. lead Pb_3O_4 Mennige *f*, Bleimennige *f*, rotes Bleioxid *n*, Bleioxidrot *n*, Blei(II)-orthoplumbat *n*, Blei(II,IV)-oxid *n*

r. lead ore *(Min)* Rotbleierz *n*, Krokoit *m (Blei(II)-chromat)*

r. lead oxide *s.* r. lead

r.-lead paste Mennigepaste *f*

r. liquor *(Pap)* Sulfitablauge *f*, Ablauge *f*, Urlauge *f*; *s.* r. mordant

r. **mercuric iodide** HgJ_2 rotes Quecksilber(II)-jodid n
r. **mercuric oxide** s. r. precipitate
r. **mercuric sulphide** HgS Zinnober m, rotes Quecksilbersulfid n
r. **mordant** (Text) Rotbeize f (Aluminiumazetat in Essigsäure)
r. **mud** (Met) Rotschlamm m
r. **ochre** Roter Ocker m, Eisenmennige f
r. **oil** Red Oil n (rot bis rotorange gefärbtes raffiniertes Schmieröldestillat)
r. **orpiment** As_4S_4 Rauschrot n, Rubinschwefel m, Tetrarsentetrasulfid n
r. **oxide** (Farb) Oxidrot n, Eisenoxidrot n
r. **phosphorus** roter Phosphor m (des Handels)
r. **pine** Rotkiefer f, Harzkiefer f, Pinus resinosa Ait.
r. **potassium prussiate** s. r. prussiate of potash
r. **precipitate** HgO rotes Präzipitat n, rotes Quecksilber(II)-oxid n
r. **prussiate of potash** $K_3[Fe(CN)_6]$ rotes Blutlaugensalz n, Kaliumhexazyanoferrat(III) n
r. **prussiate of soda (sodium)** $Na_3[Fe(CN)_6]$ Natriumhexazyanoferrat(III) n, Natriumeisen(III)-zyanid n
r. **rot** Rotfäule f
r. **rudd** s. r. ochre
r. **shade** Rotton m, roter Farbton m, rote Farbnuance f
r. **silver ore** (Min) dunkles Rotgültigerz n, Pyrargyrit m (Antimon(III)-silbersulfid)
r. **spruce** Rotfichte f, Picea rubens Sarg.
r. **squill** Meerzwiebel f, Urginea maritima (L.)Bak.
r. **stain** (Glas) Rotbeize f, Rotätze f
r. **wine** Rotwein m
r. **wood** s. redwood
r. **zinc ore** (Min) Rotzinkerz n, Zinkit m (Zinkoxid)
reddingite (Min) Reddingit m (eisen(II)-haltiges Mangan(II)-orthophosphat)
reddle (Min) Rötel m
redeposit / **to** sich wieder absetzen (ablagern, niederschlagen)
redeposition of soil (Text) Wiederaufziehen n des Schmutzes, Rückvergrauen n, Rückvergrauung f, Wiederanschmutzen n
redetermination Neubestimmung f, nochmalige Bestimmung f
redevelop / **to** (Foto) nachentwickeln
redevelopment (Foto) Nachentwicklung f
rediazotize / **to** erneut (weiter) diazotieren, nachdiazotieren
redissolve / **to** wiederauflösen, wieder lösen
redissolved milk powder auf Magermilchkonzentration verdünntes Milchpulver n
redistil / **to** nochmals (erneut, wiederholt) destillieren, umdestillieren, redestillieren
redistillation nochmalige (erneute) Destillation f, Zweitdestillation f, Redestillation f, Redestillieren n
redistribution Neuverteilung f
r. **reaction** Neuverteilungsreaktion f

redistributor Boden m zur Neuverteilung (in Füllkörperkolonnen)
Redler conveyor Redler-Kettenförderer m, Redler-Förderer m, Redler-Band n, Redler m, Trogkettenförderer m
redness Rotglut f
redox electrode Redoxelektrode f, Reduktions-Oxydations-Elektrode f
r. **equilibrium** Redoxgleichgewicht n
r. **exchanger** Redoxaustauscher m
r. **indicator** Redoxindikator m
r. **ion exchanger (resin)** Redox-Ionenaustauscher m
r. **polymer** Redoxpolymerisat n
r. **potential** Redoxpotential n, Reduktions-Oxydations-Potential n, Oxydations-Reduktions-Potential n
r. **reaction** Redoxreaktion f, Reduktions-Oxydations-Reaktion f, Oxydations-Reduktions-Reaktion f, Oxydoreduktion f
r. **resin** Redoxharz n, (i.w.S.) Redox-Ionenaustauscher m
r. **system** Redoxsystem n, Reduktions-Oxydations-System n, Oxydations-Reduktions-System n
r. **titration** Redoxtitration f, Reduktions-Oxydations-Titration f, Oxydations-Reduktions-Titration f
redoxase Redoxase f, Oxydoreduk[t]ase f (Ferment)
redruthite Redruthit m, (veraltet für) Chalkosin m, Kupferglanz m (Mineralgruppe)
reds (Gerb) Gerbstoffrote npl, Phlobaphene npl
reduce / **to** 1. reduzieren, [ver]mindern, verringern, herabsetzen, senken; zerkleinern; (Foto) abschwächen; (Chem) reduzieren; 2. sich vermindern (verkleinern, verringern), [ab]sinken, abnehmen, zurückgehen, nachlassen, [ab]fallen; (Chem) reduziert werden, sich reduzieren lassen
r. **to ashes** veraschen
r. **to fibres** defibrieren, zerfasern, in Einzelfasern zerlegen
r. **to powder** pulverisieren, pulvern
r. **to pulp** (Pap) zu Halbstoff aufschließen, (i.e.S.) zu Holzschliff verschleifen (oder) chemisch aufschließen
r. **to standard (type strength)** typkonform stellen
reduced crude [oil] reduziertes (getopptes, abgetopptes) Rohöl (Öl) n
r. **equation of state** reduzierte Zustandsgleichung f
r. **mass** (phys Ch) reduzierte Masse f
r. **oil** s. r. crude [oil]
r. **pressure** reduzierter (verminderter) Druck m
reducer 1. Übergangsstück n, Reduzierstück n; 2. (Foto) Abschwächer m; 3. s. reductant
r. **tee** T-Stück n mit Reduzierung
reducible reduzierbar
reducing action reduzierende Wirkung f, Reduktionswirkung f

r. adapter Übergangsstück n (mit kleiner Hülse auf großem Kern)
r. agent s. reductant
r. atmosphere reduzierende Atmosphäre f
r. bath Reduktionsbad n
r. flame Reduktionsflamme f
r. power Reduktionsvermögen n
r. roasting reduzierendes Rösten n, reduzierende Röstung f
r. sugar reduzierender Zucker m
r. valve Reduzierventil n, Druckminder[ungs]ventil n, Druckminderer m
r. zone Reduktionszone f
reductant Reduktionsmittel n, reduzierendes Mittel n, Reduktor m, Desoxydationsmittel n
reductase Reduktase f (Ferment)
r. test Reduktaseprobe f (Schardingersche Reaktion)
reduction Reduktion f, Reduzierung f, Verminderung f, Minderung f, Verringerung f, Herabsetzung f, Senkung f; Sinken n, Absinken n, Abnahme f, Rückgang m, Nachlassen n, Abfallen n, Abfall m, Fallen n; Zerkleinerung f; (Chem) Reduktion f; (Foto) Abschwächung f
r. bleaching Reduktionsbleiche f
r. clearing reduktive Nachbehandlung f
r. flask Reduktionskolben m
r. in image contrast (Foto) Kontrastverminderung f
r. liquor Reduktionslauge f, Reduktionslösung f, Reduktionsbrühe f, Reduktionsflüssigkeit f
r. method Reduktionsmethode f
r. milling (Lebm) Hochmahlverfahren n, Hochmüllerei f, Grießmüllerei f
r. mixture Reduktionsgemisch n, Reduktionsmischung f, reduzierendes Gemisch n
r. of caking power Backfähigkeitsverminderung f, Rückgang m des Backvermögens
r.-oxidation potential s. redox potential
r. pan Reduziergefäß n, Reduktor m
r. plant Reduktionsanlage f, Reduktionsbetrieb m
r. potential Reduktionspotential n
r. process Reduktionsverfahren n; Reduktionsvorgang m
r. ratio Zerkleinerungsgrad m
r. technique Reduktionsverfahren n
r. unit Reduktionsanlage f, Reduktionsapparatur f
r. zone Reduktionszone f
reductive reduzierend, reduktiv
reductive s. reductant
r. capacity Reduktionskraft f
redwood (Pap) Mammutbaum m, Sequoia sempervirens (D. Don) Endl.; (Farb) Rotholz n (Sammelname für zahlreiche Farbhölzer)
Redwood second Redwood-Sekunde f, Redwood-Zahl f (Maßeinheit für die Viskosität)
Redwood visco[si]meter Redwood-Viskosimeter n
redye / to (Text) umfärben, auffärben, nachfärben
Reed reaction Reedsche Reaktion f (durch Licht katalysierte Sulfochlorierung von Paraffinkohlenwasserstoffen)
reef lime Riffkalk m
reel / to (Pap, Text) aufwickeln, aufrollen; (Text) haspeln, winden
r. off (Rohseide) abhaspeln
r. up s. to reel
reel 1. s. reeling machine; 2. s. reel of paper; 3. Rollstange f
r. cylinder s. r. drum
r. drum (Pap) Aufrolltrommel f, Papieraufrolltrommel f, Aufwickeltrommel f, Tambour m, Tambourwalze f
r. of paper Papierrolle f, Papiertambour m, Tambour m, Tambourrolle f
r.-off stand (Pap) Abrollgestell n
r.-packing machine (Pap) Rollenpackmaschine f
r.-slitting machine (Pap) Umroller m, Rollenschneidmaschine f, Rollenschneider m, Längsschneidemaschine f, Längsschneider m
r. spinning Streckspinnen n, Haspelspinnen n
r. stand (Pap) Abrollgestell n
r.-up drum s. r. drum
reeled paper Rollenpapier n
reeler s. reeling machine
reeling (Pap) Aufrollen n, Aufwickeln n, Aufrollung f, Aufwicklung f, Papieraufrollung f, Rollenaufwicklung f; (Text) Haspeln n, Winden n
r. drum s. reel drum
r. machine (Pap) Rollapparat m, Aufrollapparat m, Rollmaschine f, Roller m; (Text) Haspelmaschine f
r.-off stand (Pap) Abrollgestell n
r.-up drum s. reel drum
re-enforced paper s. reinforced paper
re-enrich / to wiederanreichern
re-evaporation Rückverdampfung f
re-exposure (Foto) Nachbelichtung f
referee check (forensische Chemie) Schiedsanalyse f
reference beam Vergleichsstrahl m
r. cell Bezugszelle f
r. electrode Bezugselektrode f, Vergleichselektrode f
r. fuel Bezugskraftstoff m, Bezugstreibstoff m, Eichkraftstoff m, Eichtreibstoff m
r. input Führungsgröße f (Regelungstechnik)
r. junction Ableitungselektrode f (in pH-Meßketten)
r. standard Bezugsstandard m, Vergleichsstandard m
r. substance Bezugssubstanz f, Vergleichssubstanz f
r. system Bezugssystem n
r. unit Bezugseinheit f
r. variable Führungsgröße f (Regelungstechnik)
r. voltage Bezugsspannung f, Vergleichsspannung f, Eichspannung f
refill / to wieder füllen, nachfüllen, auffüllen
refilter / to erneut (nochmals) filtern

refine / to 1. raffinieren, läutern, veredeln, reinigen, verfeinern, feinen; 2. *(Plast)* abläutern, abklären; 3. *(Gum)* refinern; 4. *(Pap)* aufschlagen; *(Pap)* mahlen; *(Pap)* feinmahlen, fertigmahlen
refined bar Raffinierstahl *m*, Gärbstahl *m*, Paketstahl *m*
  r. by the cold process *(Pap)* kaltveredelt, kalt veredelt
  r. by the hot process *(Pap)* heißveredelt, heiß veredelt
  r. copper Raffinatkupfer *n*, Garkupfer *n*
  r. iron *s.* r. bar
  r. lead raffiniertes Blei *n*, Raffinatblei *n*
  r. liquor *(Zucker)* Raffinadekläre *f*
  r.-product pipeline Pipeline (Rohrleitung) *f* für Fertigerzeugnisse (Fertigprodukte, Produkte), Produktenleitung *f*
  r. sugar Raffinadezucker *m*, Raffinade *f*
refinement *s.* refining
refiner 1. *(Pap)* Kegel[stoff]mühle *f*, *(i.e.S.)* Jordan-Kegel[stoff]mühle *f*, Jordan-Mühle *f*; *(Pap)* Refiner *m*, Stoffaufschläger *m*, Aufschläger *m*, Ganzstoffmahlmaschine *f*; 2. *(Glas)* Läuterwanne *f*, Arbeitswanne *f*; 3. *s.* r. mill
  r. disk Refinerscheibe *f*
  r. mill *(Gum)* Refiner-Walzwerk *n*, Refiner *m*
refinery Raffinerie *f*, Raffinationsanlage *f*; *(Met)* Hütte *f*
  r. distillation Raffination *f*
  r. gas Raffineriegas *n*, Raffiniergas *n*
  r. molasses *(Zucker)* Raffineriemelasse *f*
  r. residue Raffinationsrückstand *m*
refining 1. Raffination *f*, Läuterung *f*, Veredelung *f*, Reinigung *f*, Verfeinerung *f*, Säuberung *f*; 2. *(Gum)* Refinern *n*; 3. *(Erdöl)* Raffination *f* *(i.w.S.)*, Verarbeitung *f*; 4. *(Pap)* Veredelung *f*; *(Pap)* Aufschlagen *n*; *(Pap)* Mahlen *n*; *(Pap)* Feinmahlen *n*, Fertigmahlen *n*
  r. agent Läuter[ungs]mittel *n*
  r. chamber (end) *(Glas)* Läuterwanne *f*, Arbeitswanne *f*
  r. engine *s.* refiner 1.
  r. lye Garlauge *f* *(Düngemittelindustrie)*
  r. machine *s.* refiner 1.
  r. mill *s.* refiner mill
  r. treatment *s.* refining
  r. zone *(Glas)* Läuterzone *f*
refire / to *(Ker)* nachbrennen
refiring *(Ker)* Nachbrand *m*
reflect [back] / to reflektieren, zurückwerfen, zurückstrahlen, spiegeln
reflectance Reflexionsvermögen *n*; Reflexionsgrad *m*, Reflexionszahl *f*
  r. measurement Reflexionsmessung *f*
reflecting position Reflexionslage *f*
  r. power *s.* reflectivity
reflection Rückstrahlung *f*, Reflexion *f*, Reflektierung *f*, Spiegelung *f*
  r. angle Reflexionswinkel *m*
  r. coefficient *s.* reflectance

  r. density *(Foto)* Reflexionsdichte *f*
  r. factor *s.* reflectance
  r. goniometer *(Krist)* Reflexionsgoniometer *n*
  r. measurement Reflexionsmessung *f*
  r. method Reflexionsmethode *f*, Reflexionsverfahren *n*
  r. mill Prallmühle *f*
  r. shooting *(Erdöl)* Reflexionsseismik *f*
  r. turbidity *(Koll)* Trübung *f* durch Reflexion
reflectivity Reflexionsvermögen *n*, Reflexionsfähigkeit *f*, Reflexionskraft *f*, Rückstrahlungsvermögen *n*
  r. method *(Geol)* Anschliffmethode *f*
re-flesh / to *(Gerb)* nachentfleischen
reflex copying *(Foto)* Reflexkopierverfahren *n*
reflux / to am Rückflußkühler kochen, unter Rückfluß[kühlung] erhitzen (kochen)
reflux Rückfluß *m*, Rücklauf *m*, Reflux *m*
  r. condenser *(Labortechnik)* Rückflußkühler *m*; *(als Teil einer Rektifikationsapparatur)* Rücklaufkondensator *m*
  r. divider *(Destillation)* Rücklaufteiler *m*, Rückflußteiler *m*
  r. liquid (liquor) *(Destillation)* Rücklaufflüssigkeit *f*, Rückfluß *m*
  r. ratio *(Destillation)* Rücklaufverhältnis *n*, Rückflußverhältnis *n*
  r. separator Rückflußabscheider *m*
  r. stream *s.* reflux
  r. tank Rückflußtank *m*, Rückfluß[sammel]behälter *m*
refluxing Rückflußkochen *n*
reform / to umwandeln; *(Benzinkohlenwasserstoffe)* reformieren
re-form / to zurückbilden
reformate Reformat *n*, reformiertes Produkt *n*
re-formation Rückbildung *f*
Reformatsky reaction Reformatsky-Reaktion *f*, Reformatskysche Reaktion (Synthese) *f* *(von β-Hydroxykarbonsäureestern)*
reformed gasoline reformiertes Benzin *n*, Reform[ier]benzin *n*, Reforming-Benzin *n*
reformer *s.* reforming plant
  r. feedstock Reform[ing]stock *m*, zu reformierendes Material *n*
reforming Reform[ier]en *n*, Reformierung *f*, Reforming *n* *(Änderung der Molekülstruktur und -größe von Benzinkohlenwasserstoffen)*
  r. plant Reform[ier]anlage *f*, Reformierungsanlage *f*, Reforming-Anlage *f*, Reformer *m*
  r. process Reformier[ungs]verfahren *n*, Reforming-Verfahren *n*
  r. reaction Reform[ierungs]reaktion *f*, Reforming-Reaktion *f*
refract / to *(phys Ch)* brechen
refraction Brechung *f*, Refraktion *f*
  r. index *s.* refractive index
  r. method Refraktionsmethode *f*, Refraktionsverfahren *n*
  r. shooting *(Erdöl)* Refraktionsseismik *f*

refractive brechend; lichtbrechend
  r. dispersion Brechungsdispersion f
  r. index Brechungsindex m, Brech[ungs]zahl f, Brechungsquotient m, Brechungskoeffizient m
refractivity Brechungsvermögen n, Refraktionsvermögen n
refractometer Refraktometer n, Brechzahlmesser m, Brechungsmesser m
refractometric refraktometrisch, Refraktometer...
refractometry Refraktometrie f
refractoriness Feuerfestigkeit f, Feuerbeständigkeit f
  r. in service Feuerfestigkeit f im Betrieb
  r. under load Druckfeuerbeständigkeit f, DFB
refractory feuerfest, ff., feuerbeständig; Schamotte...
refractory feuerfestes Material n; feuerfester Stein m; Schamottestein m; feuerfestes Erzeugnis (Produkt) n
  r. brick feuerfester Stein m; Schamottestein m, Schamotteziegel m
  r. ceramics Feuerfestkeramik f
  r. clay feuerfester Ton m, Feuerfestton m
  r. coating feuerfester Überzug m
  r. concrete feuerfester Beton m, Feuer[fest]beton m
  r. lining feuerfeste Auskleidung f
  r. material feuerfestes Material n
  r. mortar feuerfester Mörtel m
  r. product feuerfestes Erzeugnis (Produkt) n
  r. ramming material feuerfeste Stampfmasse f, feuerfestes Stampfgemisch n
refreeze / to wieder erstarren; wieder erstarren lassen
refreshment drink Erfrischungsgetränk n
refrigerant kühlend
refrigerant Kälteträger m, Kältemittel n, Kühlmittel n, Abkühlmittel n
  refrigerant 12 F 12 n (Kältemittel, CF_2Cl_2)
  r. vapour Kältemitteldampf m
refrigerate / to kühlen
refrigerated brine s. refrigerating brine
  r. centrifuge Kühlzentrifuge f
  r. egg Kühlhausei n
  r.-tank truck Kühltankwagen m
refrigerating brine Kühlsole f
  r. coil Kühlschlange f
  r. engineering Kältetechnik f
  r. machine Kältemaschine f
  r. medium s. refrigerant
  r. plant Kälte[erzeugungs]anlage f, Kältemaschinenanlage f, Kühlanlage f
refrigeration Kühlen n, Kühlung f (insbesondere auf Temperaturen unterhalb Atmosphärentemperatur); Kälteerzeugung f
  r. performance Kälteleistung f, Kühlvermögen n
  r. system Kühlsystem n
refrigerator Kältemaschine f; Kühlschrank m
  r. car Kühlwagen m, Kühlwaggon m, Isolierwagen m

  r. egg Kühlhausei n
  r. truck Kühlauto n
  r. wag[g]on s. r. car
re-fuse / to (Met) nochmals (wieder) schmelzen, umschmelzen
refuse Müll m; Berge pl, Abgänge mpl
  r. compost (Landw) Müllkompost m
  r. destructor Müllverbrennungsofen m
  r. discharge Bergeaustrag m
  r.-discharge port Bergeaustragsöffnung f
  r. extraction s. r. discharge
regain / to wiedergewinnen
regenerant Wiederbelebungsmittel n
regenerate / to regenerieren, wiederaufbereiten, wiedergewinnen; regenerieren, erneuern; (eines Sorbens) wiederbeleben, regenerieren
regenerate Regenerat n
regenerated cellulose regenerierte Zellulose f, Hydratzellulose f
  r. cellulose fibre Regeneratzellulosefaser f, Zelluloseregeneratfaser f; Regeneratzellulosefaserstoff m, RZ, Zelluloseregeneratfaserstoff m
  r. naturally occurring protein regenerierter natürlicher Eiweißkörper m
regeneration Regenerierung f, Wiederaufbereitung f, Wiedergewinnung f; Regenerierung f, Erneuerung f; Wiederbelebung f, Regenerierung f (eines Sorbens); Regeneration f, Wärmeregeneration f (z.B. beim Kokereibetrieb)
  r. section regenerative Wärmeaustauschabteilung (Austauschabteilung) f (eines Plattenerhitzers)
regenerative regenerativ, Regenerativ...
  r. firing Regenerativfeuerung f
  r. furnace Regenerativofen m
  r. melting furnace Regenerativschmelzofen m
  r. principle Regenerativprinzip n
  r. system Regenerativsystem n
regenerator Regenerator m, Wärmeregenerator m, regenerativer Vorwärmer m; Regenerator m, Kiln m, Katalysatorregenerator m, Regenerierofen m (beim katalytischen Kracken)
  r. chamber Regeneratorkammer f, Regeneratorraum m, Regenerativkammer f, Vorwärmkammer f
region Bereich m, Gebiet n, Zone f
  r. of correct (normal) exposure (Foto) Gebiet n der richtigen (normalen) Exposition, Gebiet n der Normalexposition (Normalbelichtung), geradliniger Teil m (der Schwärzungskurve), geradliniges Stück (Gebiet) n
  r. of overexposure (Foto) Gebiet n der Überexposition (Überbelichtung), Schulter f (der Schwärzungskurve)
  r. of underexposure (Foto) Gebiet n der Unterexposition (Unterbelichtung), Durchhang m, Kurvendurchhang m, durchängender Teil m (der Schwärzungskurve)
regional metamorphism (Geoch) Regionalmetamorphose f, regionaler Metamorphismus m

register / to angeben *(z.B. Temperatur)*
registered trademark eingetragenes (registriertes) Warenzeichen *n*
registrant Warenzeicheninhaber *m*
regrind / to wieder mahlen (aufarbeiten) *(z.B. Plastabfälle)*
reground material *(Plast)* aufgearbeiteter Abfall *m*, Regenerat *n*
regrouping Umgruppierung *f*, Umlagerung *f*, Umstellung *f*
regular regulär, regelmäßig, normal
  r. boiling konstantes Sieden *n*, Siedekonstanz *f*
  r.-dyeing normal anfärbend
  r. gasoline (motor spirit) [normales] Fahrbenzin *n*
  r. multiplet *(phys Ch)* regelrechtes (normales) Multiplett *n*
  r. spirit *s.* r. gasoline
  r. system *(Krist)* reguläres (kubisches) System *n*
regulating box *(Pap)* Stoffregulierkasten *m*
regulation *(Pharm)* Vorschrift *f*
regulator Regler *m*, Reglersubstanz *f* *(bei der Polymerisation)*
regulatory agency Überwachungsbehörde *f* *(z.B. für Pflanzenschutzmittelrückstände in Lebensmitteln)*
  r. commodity *(einschränkenden Beförderungsvorschriften unterliegender Stoff)*
  r. laboratory Konfektionierlaboratorium *n*
regulus *(Met)* Regulus *m*, Metallkönig *m*
  r. metal Regulusmetall *n*, Hartblei *n* *(Legierung aus 90% Pb, 8% Sb, 2% Sn)*
  r. of antimony Antimonregulus *m*
reheat / to erneut erwärmen; *(Glas)* rückerwärmen, von innen heraus durchwärmen *(beim Külbel)*
reheat erneute Erwärmung *f*; *(Glas)* Rückerwärmung *f*, Durchwärmen *n (des Külbels von innen heraus)*, Temperaturausgleich *m*
rehydrate / to *(Gerb)* erneut (wieder) einweichen
Reichert-Meissl number (value) Reichert-Meissl-Zahl *f*, Reichert-Meisslsche Zahl *f*, R-M-Z, RMZ
Reid apparatus Reidsche Bombe *f*
Reid vapour pressure Reid-Dampfdruck *m*, Dampfdruck *m* nach Reid
re-ignite / to nochmals glühen
Reimer-Tiemann reaction (synthesis) Reimer-Tiemann-Synthese *f*, Reimer-Tiemannsche Synthese (Phenolaldehydsynthese) *f*
reineckate Reineck[e]at *n (vom Reinecke-Salz sich ableitende Verbindung)*
Reinecke salt $NH_4[Cr(SCN)_4(NH_3)_2]$ · H_2O Reinecke-Salz *n*, Ammoniumtetrathiozyanatodiamminchromat(Ⅲ)-1-Wasser *n*
reinforce / to verstärken; Verstärkerwirkung haben; armieren, bewehren
reinforced concrete armierter (bewehrter) Beton *m*, Stahlbeton *m*
  r. hose Schlauch *m* mit Gewebeeinlage
  r. paper textilverstärktes Papier *n*; Gewebepapier *n*, Leinenpapier *n*, Papyrolin *n*, Packgewebe *n*

  r. plastic verstärkter Plast (Kunststoff) *m*
  r. thermoplastic verstärkter Thermoplast *m*
reinforcement Verstärkung *f*; Armierung *f*, Bewehrung *f*
  r. size haftmittelhaltige Schlichte *f*
reinforcing action verstärkende Wirkung *f*, Verstärkerwirkung *f*, Verstärkungseffekt *m*
  r. agent Verstärkungsmittel *n*, Verstärkungsmaterial *n*, Verstärker *m*
  r. black aktiver Ruß *m*, Aktivruß *m*
  r. effect *s.* r. action
  r. element Verstärkungselement *n*
  r. filler (ingredient) Verstärkerfüllstoff *m*, verstärkender (verstärkend wirkender, aktiver) Füllstoff *m*
  r. material *s.* r. agent
  r. pigment *s.* r. filler
  r. white filler heller Verstärkerfüllstoff (aktiver Füllstoff) *m*
reinite *(Min)* Reinit *m (Eisen(II)-wolframat)*
Reinsch test [for arsenic] Reinschsche Probe *f*
reinvert / to umdrehen
reject / to *(Substanzen als unbrauchbar)* verwerfen; *(Namen)* ablehnen, verwerfen
reject Ausschuß *m*
rejected stock *(Pap)* Grobstoff *m*, Spuckstoff *m*, „Sauerkraut" *n*
rejects from the screening operation Siebrückstand *m*, *(bei der Holzschliffsortierung auch)* Grobstoff *m*, Spuckstoff *m*, „Sauerkraut" *n*
relation Beziehung *f*, Verhältnis *n*, Verbindung *f*, Relation *f*, Zusammenhang *m*
relative atomic weight relative Atommasse *f*, *(bisher)* relatives Atomgewicht *n*
  r. biological effectiveness relative biologische Wirksamkeit *f*, RBW *f*
  r. centrifugal force Beschleunigungsverhältnis *n*, Zentrifugenzahl *f*, Schleuderzahl *f*, Trennfaktor *m*, Schleudereffekt *m*
  r. configuration relative Konfiguration *f*
  r. humidity relative Feuchte (Feuchtigkeit) *f*
  r. lowering of vapour pressure relative Dampfdruckerniedrigung *f*
  r. photometry Relativfotometrie *f*
  r. viscosity relative Viskosität *f*
  r. volatility relative Flüchtigkeit *f*
relativistic relativistisch
relativity Relativität *f*
  r. theory Relativitätstheorie *f*
relaxation Relaxation *f*
  r. effect Relaxationseffekt *m*
  r. process Relaxationsverfahren *n*
  r. rate Relaxationsgeschwindigkeit *f*
  r. time Relaxationszeit *f*, Relaxationsperiode *f*
relaxin Relaxin *n (während der Schwangerschaft gebildetes Peptidhormon)*
release / to freimachen, freisetzen, in Freiheit setzen, entwickeln, entbinden; entspannen, die Spannung aufheben; [ab]trennen, ablösen, *(Vulkanisate aus der Form)* herausnehmen

r. electrons Elektronen abgeben
release Freimachen n, Freisetzen n, Freisetzung f, Entwickeln n, Entbinden n; Freiwerden n, Entweichen n; Entspannen n, Aufheben n der Spannung; Trennen n, Abtrennen n, Ablösen n, Herausnehmen n *(der Vulkanisate aus der Form)*; Trennvermögen n; *(Pap)* Kocherabgas n, Abgas n, Übertriebgas n; *(Pap)* Übertriebsäure f, Übertrieb m, Rücklauge f
r. action Trennwirkung f, Trenneffekt m, Ablösungswirkung f, Ablösungseffekt m
r. agent Trennmittel n; Form[en]einstreichmittel n, Form[en]einsprühmittel n, *(auf die Form aufgebrachtes)* Form[en]trennmittel n, Entformungsmittel n, *(Plast auch)* Werkzeugtrennmittel n
r. effect s. r. action
r. of electrons Elektronenauslösung f, Elektronenloslösung f, Elektronenablösung f
r. of liquid Flüssigkeitsabgabe f
relict Relikt[mineral] n
relief *(Pap)* Abgasen n; *(manchmal auch)* Abgas n
r. gas [from the digester] *(Pap)* Kocherabgas n, Abgas n, Übertriebgas n
r. line *(Pap)* Abgasleitung f
r. liquor *(Pap)* Übertriebsäure f, Übertrieb m, Rücklauge f
r. of gases *(Pap)* Abgasen n
r. printing Hochdruck m
r. valve Sicherheitsventil n, Druckbegrenzungsventil n; *(Pap)* Abgasventil n
relieve / to entlasten; *(Kocher)* entlüften, entspannen, abgasen
r. down *(den Druck)* vermindern, herabmindern, herabsetzen
relieving pressure Abblasedruck m
reluctance to crystallize geringe Kristallisationsneigung f
rem s. roentgen equivalent man
remain / to zurückbleiben, [übrig]bleiben
remainder Rest m, Überrest m, Rückstand m
remanufacture Wiederaufbereitung f, Regeneration f *(z.B. von Altpapier)*
remelt / to nochmals (wieder) schmelzen, umschmelzen
remnant Rest m, Spur f, Residuum n
remodel / to rekonstruieren
remolinite *(Min)* Remolinit m, *(veraltet für)* Atakamit m
remote control Fernsteuerung f
r. indication Fernanzeige f
r. operation Fernbetätigung f, Fernbedienung f
removable Fourdrinier [section] *(Pap)* ausfahrbare Siebpartie f
removal Entfernen n, Entfernung f, Beseitigung f, Entzug m, Abspaltung f *(z.B. von Wasser)*, Ablassen n *(z.B. von Luft)*, Abtrennung f *(z.B. einer Säure)*, Abscheidung f *(z.B. von Ammoniak)*; Austragen n, Austrag m *(eines Gutes)*
r. of colours *(Text)* Abziehen n, Ablösen n *(der Farbe)*, Entfärben n

r. of copper Entkupfern n, Entkupferung f
r. of dust Entstaubung f
r. of iron Enteisenen n, Enteisenung f *(z.B. des Wassers)*
r. of lignin *(Pap)* Ligninentfernung f, Lignin[her]auslösung f
r. of lignin residues Ligninrestentfernung f
r. of water Wasserentzug m, Entwässerung f; Wasserabspaltung f
remove / to entfernen, beseitigen, entziehen, abspalten *(z.B. Wasser)*, ablassen *(z.B. Luft)*, abtrennen *(z.B. eine Säure)*, abscheiden *(z.B. Ammoniak)*; *(ein Gut)* austragen
r. a mash Maische abziehen
r. by distillation abdestillieren
r. by hand for discard [von Hand] ausklauben (auslesen) *(bei der Kohlenaufbereitung)*
r. copper entkupfern
r. dust entstauben
r. iron enteisenen
r. the bark *(Pap)* entrinden, schälen
r. the water Wasser entziehen, entwässern; Wasser abspalten
remover Entferner m
render / to 1. *(Foto)* wiedergeben; 2. auslassen, [aus]schmelzen, durch Schmelzen extrahieren (auswaschen) *(zur Fettgewinnung)*
r. inert inert[is]ieren
r. passive *(phys Ch)* passivieren
r. soluble in Lösung bringen
rendered butter Butterschmalz n, Schmelzbutter f, Butterfett n
r. fat ausgelassenes Fett n, Schmalz n
r. margarine Schmelzmargarine f
rendering Auslassen n, Ausschmelzen n, Schmelzen n, Schmelzung f, Schmelze f *(zur Fettgewinnung)*
r. method Schmelzverfahren n
rendzina *(Landw)* Rendzina f, Humuskarbonatboden m
reniala oil *(Samenöl von Adansonia digitata L.)*
renierite Renierit m *(germaniumhaltiges Mineral)*
renin Renin n *(in der Niere gebildetes Enzym)*
rennase s. rennin
rennet / to [ver]laben, Lab zusetzen
rennet Labmagenschleimhaut f; Lab n *(aus zerkleinerten Labmägen gewonnenes Produkt)*; s. rennin
r. bag s. r. stomach
r. casein Labkasein n
r. cheese Labkäse m
r. clotting Labgerinnung f, Labkoagulation f
r.-clotting time Lab[gerinnungs]zeit f
r. coagulability Labgerinnungsfähigkeit f
r. coagulation s. r. clotting
r. curd Lab[käse]bruch m; Labquark m
r. extract Labextrakt m, Labessenz f
r. extraction from vells by soaking Labmagenmazerisierung f
r. ferment s. rennin

r. fermentation Labgärung f
r.-fermentation test Labgärprobe f
r. powder Labpulver n
r.-precipitated casein s. r. casein
r. stomach Labmagen m
r. strength Labstärke f
r. test s. r.-fermentation test
r.-treated labbehandelt (z.B. Kasein)
rennetability Lab[ungs]fähigkeit f
renneting Labung f, Verlabung f, Zusetzen n von
Lab, Labzusatz m
 r. ability s. rennetability
 r. temperature Einlabungstemperatur f
 r. time s. rennet-clotting time
rennin Rennin n, Lab[ferment] n, Chymosin n,
Chymase f
 r. clotting s. rennet clotting
rensselaerite (Min) Rensselaerit m (Magnesiumdihydrogentetrasilikat)
reoxidation Reoxydation f, Wiederoxydation f,
Rückoxydation f, Zurückoxydieren n
rep [unit] s. roentgen equivalent physical
repair patch (Gum) Reparaturplatte f
repasteurization wiederholtes (nochmaliges) Pasteurisieren n, wiederholte (nochmalige) Pasteurisation f
repasteurize / to wiederholt pasteurisieren
repeatability Wiederholbarkeit f, Wiederholstreu-
bereich m
repeatable wiederholbar
repeated flexural strength Dauerbiegefestigkeit f
 r. flexural stress Dauerbiegespannung f,
Dauerbiegebeanspruchung f
 r. stress Dauerbeanspruchung f
repeating unit Grundmolekül n, Struktureinheit f,
Staudinger-Einheit f, Grundeinheit f
repel / to abweisen, abstoßen, zurückstoßen;
vertreiben (z.B. Tiere durch Abschreckmittel)
repellency Abweisungsvermögen n, abweisende
Kraft (Eigenschaft) f, abweisendes Verhalten n,
abweisender Charakter m; Abschreckwirkung f
repellent abweisend, abstoßend; abschreckend
repellent Abstoßungsmittel n, Schutzmittel n mit
abstoßender Wirkung; Abschreckmittel n, Ab-
schreckstoff m, Abwehrmittel n, Abwehrstoff m,
Repellentstoff m, Repellent n
 r. finish Abweisendausrüstung f
replace / to ersetzen, substituieren, austauschen
replaceable ersetzbar, substituierbar, austauschbar
replacement Ersetzen n, Ersatz m, Substitution f,
Austausch m
 r. name Ersetzungsname m, Austauschname m,
Verdrängungsname m, Matrizenname m
 r. reaction Verdrängungsreaktion f, Substitutionsreaktion f, Austauschreaktion f
 r. [syn]tan [synthetischer] Austauschgerbstoff m,
synthetischer Vollgerbstoff m, Vollgerbstoff-
syntan n
replacing power (Bodenkunde) Eintauschstärke f

replenish / to nachfüllen, auffüllen, ergänzen
replenisher [solution] (Foto) Nachfüllösung f,
Regeneratorlösung f, Regenerator m
replenishment Nachfüllung f, Auffüllung f, Ergänzung f
replica (Plast) Abdruck m
Reppe chemistry Reppe-Chemie f
Reppe process Reppe-Verfahren n (z.B. ein
Butadienverfahren)
reprecipitate / to umfällen
reprecipitation Umfällung f
re-press / to nochmals pressen, nachpressen
re-press [machine] (Ker) Nachpresse f, Nach-
preßmaschine f
re-pressing nochmaliges Pressen n, Nachpressen
n
repressur[iz]ing Repressuring n (Sekundärverfahren zur Erdölförderung)
reprocess / to wiederverarbeiten, nachverarbeiten,
wiederaufarbeiten
reprocessing Wiederverarbeitung f, Nachver-
arbeitung f, Wiederaufarbeitung f
reproducibility Reproduzierbarkeit f, Vergleich-
barkeit f, Vergleichstreubereich m
reproducible reproduzierbar
reproduction (Foto) Abzug m, Reproduktion f;
Vermehrung f (von Zellen)
 r. of yeast Hefevermehrung f, Vermehrung f der
Hefe[zellen]
reprography Reproduktionsfotografie f
repulp / to wiederaufschwemmen, aufrühren;
auflösen, aufschließen (z.B. Altpapier)
repulped stock (waste paper) Altpapierstoff m,
wiederaufbereitetes (aufbereitetes, regeneriertes) Altpapier f
repulsion Abstoßung f, Zurückstoßung f, Repulsion
f
repulsive abstoßend, Abstoß...
 r. effect Abstoßeffekt m
 r. force abstoßende Kraft f, Abstoßungskraft f
requirements of raw materials Rohstoffbedarf m
rereel / to (Pap) umrollen, umwickeln
rereeling (Pap) Umrollen n, Umwickeln n
 r. machine (Pap) Umroller m, Rollenschneidmaschine f, Rollenschneider m, Längsschneidemaschine f, Längsschneider m
rerun / to redestillieren, nochmals (erneut, wiederholt) destillieren, umdestillieren
rerun s. rerunning
 r. oil Redestillat n (bei der Redestillation von
Schmieröl)
 r. tower Redestillationskolonne f
rerunning Redestillieren n, Redestillation f, Zweitdestillation f, nochmalige (erneute) Destillation f
resalt / to (Gerb) nachsalzen
resaturate / to aufsättigen
Resazurin [reduction] test Resazurinprobe f
(Schnellreduktionsprobe zur Milchuntersuchung)
rescidine Reszidin n (Alkaloid von Rauvolfia
vomitoria Afzel.)

rescinnamine Reszinnamin *n (3,4,5-Trimethoxy-zimtsäureester des Methylreserpats)*
rescreen / to nachklassieren, nachsichten
research equipment wissenschaftliche Geräte *npl*
  r. institute Forschungsinstitut *n*
  r. laboratory Forschungslabor[atorium] *n*
  r. method Research-Methode *f*, Research-Verfahren *n*, F-1-Methode *f (zur Oktanzahl-bestimmung)*
  r. reactor Forschungsreaktor *m*, Versuchsreaktor *m*
  r. station Forschungsstation *f*, Forschungsstelle *f*
reserpic acid Reserp[in]säure *f*, 11,17-Dimethoxy-18-oxy-epialloyohimban-16-karbonsäure *f*
reserpine Reserpin *n (3,4,5-Trimethoxybenzoe-säureester des Methylreserpats)*
reservation *(Text)* Reservierung *f*, Reservieren *n*
reserve / to aufheben, aufsparen; *(Text)* reservieren
reserve *(Text)* Reservierungsmittel *n*, Reserve-mittel *n*, Reserve *f*
  r. acidity *(Bodenkunde)* potentielle Azididät *f*
  r. carbohydrate Reservekohle[n]hydrat *n*
  r. fertilization Vorratsdüngung *f*
  r. print *(Text)* Reserv[ag]edruck *m*
reservoir Reservoir *n*, Speicher *m*; [offener] Behälter *m*; Tank *m*
  r. rock Speichergestein *n*, Erdölspeichergestein *n*
resid *(Erdöl)* Rückstand *m*, Residuum *n*
  to run *n* **r.** Rückstände (Topprückstände) aufarbeiten
  r. operation Rückstandskracken *n (Aufarbeitung schwerer Topprückstände)*
residence time Verweilzeit *f*, Aufenthaltszeit *f*, Retentionszeit *f*, Rückhaltezeit *f*, Haltezeit *f*, Standzeit *f*, Stehzeit *f*, Durchgangszeit *f*, Durchlaufzeit *f*
residual action Residualwirkung *f*, Rückstandswirkung *f*; Nachwirkung *f*
  r. affinity Restaffinität *f*, Residualaffinität *f*
  r. alkalinity *(Pap)* Restalkaligehalt *m*
  r. asphalt Rückstandsasphalt *m*
  r. clay Residualton *m*, Verwitterungston *m*, Primärton *m*, primärer (eluvialer) Ton *m*
  r. contact insecticide Kontaktinsektizid *n* mit Dauerwirkung
  r. current *(phys Ch)* Reststrom *m*, Grundstrom *m*, Diffusionsstrom *m*
  r. deformation bleibende Verformung (Deformation, Gestaltsänderung) *f*, Verformungsrest *m*, Formänderungsrest *m*
  r. deposits *(Geol)* Restablagerung *f*, Restlagerstätte *f*
  r. dextrin Restdextrin *n*, Grenzdextrin *n*
  r. effect *s.* **r. action**
  r. elongation Dehnungsrest *m*, Zugverformungsrest *m*, Formänderungsrest *m* bei Dehnungsbeanspruchung

r. fuel oil Rückstandsheizöl *n*
r. gas Rückstandsgas *n*, Restgas *n*, restliches Gas *n*
r. hardness Resthärte *f*
r. lignin *(Pap)* Ligninreste *mpl*, restliches Lignin *n*, Restlignin *n*
r. liquor Restflüssigkeit *f*
r. loam Verwitterungslehm *m*
r. lubricating oil Rückstandsschmieröl *n*
r. milk Restmilch *f*
r. moisture Restfeuchte *f*, Restfeuchtigkeit *f*
r. moisture content Restfeuchteanteil *m*, Restfeuchtegehalt *m*; Restfeuchtesatz *m*; Restfeuchtebeladung *f*
r. nitrogen Reststickstoff *m*, Rest-N *n*, Nichteiweißstickstoff *m*, Nichtproteinstickstoff *m*
r. oil Rückstandsöl *n*, Restöl *n*, Residualöl *n*
r. paramagnetism Restparamagnetismus *m*
r. product Rückstand *m*
r. radiation Reststrahlung *f*
r. rays Reststrahlen *mpl*
r. resistance Restwiderstand *m*
r. set *s.* **r. deformation**
r. shrinkage Restschrumpfung *f*, Restschrumpf *m*, Restkrumpfung *f*
r. solid matter fester Rückstand *m*
r. stock *s.* **r. oil**
r. stress Restspannung *f*
r. sugar Restzucker *m*, Zuckerrest *m*, Restsüße *f*
r. titration Rücktitration *f*, Rücktitrieren *n*, Zurücktitrieren *n*
r. toxicity Rückstandstoxizität *f (z.B. von Insektiziden)*
r. valence (valency) Restvalenz *f*
residue Rückstand *m*, Rest *m*, Überrest *m*, Residuum *n*; Rest *m*, Gruppe *f*
r. chemist Rückstandsanalytiker *m*
r. cracking Kracken *n* auf flüssigen Rückstand, Kracken *n* mit Rückstandsarbeitsweise, Kracken *n* nach der Entspannungsfahrweise
r. from distillation Destillationsrückstand *m*
r. fuel Rückstandsbrennstoff *m*
r. gas Rückstandsgas *n*, Restgas *n*, restliches Gas *n*
r.-life 50 percent *(Toxikologie)* Rückstands-Halbwertszeit *f*, RL_{50}
r. method Krackprozeß *m* mit Rückstandsarbeitsweise, Entspannungs[krack]verfahren *n*
r. of combustion Verbrennungsrückstand *m*
r. on ignition Glührückstand *m*
r. separator Rückstandsabscheider *m*, Teerabscheider *m*, Abscheider *m (beim katalytischen Kracken)*
r. tolerance duldbare Rückstandsmenge (Restmenge) *f (z.B. eines Pflanzenschutzmittels)*
residuum *(Erdöl)* Rückstand *m*, Residuum *n*
resilience Verformungsarbeit *f*; elastischer Wirkungsgrad *m*, *(i.e.S.)* elastischer Wirkungsgrad *m* bei Stoß-Druck-Beanspruchung, Rückprallelastizität *f*, Stoßelastizität *f*

r. meter s. resiliometer
r. test Rückpralltest m, Rückprallversuch m
resiliency s. resilience
resilient roll (Pap) [elastische] Papierwalze f, elastische Walze f (des Superkalanders)
r. seat Weichsitz m
resiliometer Resiliometer n (Gerät zur Bestimmung der Rückprallelastizität)
resin / to (Plast) beharzen, mit Harz tränken
resin (natürliches oder synthetisches) Harz n
r. acid Harzsäure f, Resinosäure f
r. bead Harzkorn n (bei Austauschverfahren)
r. binder Harzträger m
r.-bonded plywood Kunstharzsperrholz n
r. canal (Holz) Harzgang m, Harzkanal m
r. constituent Harzkomponente f
r. content Harzgehalt m
r. cure (Gum) Harzvulkanisation f, Harzvernetzung f
r. finish (Gerb) Harzappretur f
r.-like harzähnlich
r. milk (Pap) Harzmilch f, Harzemulsion f, Harzlösung f, Leimmilch f
r. of hops Hopfenharz n
r. of ipomoea Skammoniumharz n, Skammoniaharz n (von Ipomoea orizabensis Ledanois)
r. oil Harzöl n
r./oil ratio s. r.-to-oil ratio
r. pitch Harzpech n
r. pocket (Plast) Harznest n, Harzeinschluß m, Harztasche f (Preßfehler)
r. powder Kunstharzpulver n
r. size (Pap) Harzleim m
r. soap Harzseife f (Salz einer Harzsäure)
r. solution (Farb) Harzlösung f
r. spirit Harzgeist m, Harzessenz f, Harzsprit m, Terpentinessenz f, Pinolin n
r. tannage Harzgerbung f
r.-to-oil ratio Harz-Öl-Verhältnis n, Verhältnis n von Harz zu Öl
r. treatment Kunstharzbehandlung f, Kunstharzausrüstung f, Kunstharzappretur f
resinate Resinat n (Harzseife oder Harzester)
resinifiable verharzbar
resinification Verharzung f
resinified / to become verharzen
resinify / to verharzen
resinoid harzartig
resinoid harzartiger Bestandteil m; Resinoid n
resinous harz[halt]ig, harzreich (z.B. Holz); harzartig
r. binder Harzträger m
r. body s. r. matter
r. exchanger Kunstharzaustauscher m
r. matter (substance) Harzstoff m, Harzkörper m, Harzmasse f, harzige Masse f
r. varnish Harzlack m
resinter / to nochmals sintern, nachsintern
resintering nochmaliges Sintern n, Nachsintern n
resiny s. resinous

resist (Text) Reservierungsmittel n, Reservemittel n, Reserve f
r. paste (Text) Reservierungspaste f
r. printing (Text) Reserv[ag]edruck m
resistance Widerstand m, Beständigkeit f, Stabilität f, Festigkeit f, Widerstandsfähigkeit f, Resistenz f; [elektrischer] Widerstand m, Widerstandswert m, Ohmwert m
r. against chemical agents s. r. to chemicals
r. bridge elektrische Widerstandsbrücke f
r. capacity Widerstandskapazität f
r. coil Widerstandswicklung f
r. furnace [elektrischer] Widerstandsofen m, widerstandsbeheizter Ofen m, Ofen m mit Widerstandserhitzung
r. glass Hartglas n
r.-glass flask Hartglaskolben m
r.-heated furnace s. r. furnace
r. heater Widerstandsheizgerät n
r. heating [elektrische] Widerstandserhitzung f, [elektrische] Widerstands[be]heizung f
r. oven s. r. furnace
r. thermometer Widerstandsthermometer n
r. to abrasion Abriebwiderstand m, Abriebbeständigkeit f, Abriebfestigkeit f, Widerstandsfähigkeit f gegen Abrieb, Abnutzungswiderstand m, Abnutzungsbeständigkeit f, Verschleißwiderstand m, Verschleißfestigkeit f, (Text meist) Scheuerbeständigkeit f, Scheuerfestigkeit f
r. to acid[s] Säurebeständigkeit f, Säurefestigkeit f, Säurewiderstandsfähigkeit f, Säureresistenz f
r. to ageing Alterungsbeständigkeit f, Alterungswiderstand m
r. to alkali[es] Alkalibeständigkeit f, Laugenbeständigkeit f
r. to boiling Kochbeständigkeit f, Kochfestigkeit f
r. to breakage by impact Widerstandsfähigkeit f gegen Bruch durch Stoß, Stoßfestigkeit f
r. to chemical attack chemische Unangreifbarkeit (Beständigkeit, Stabilität, Festigkeit, Widerstandsfähigkeit, Resistenz) f, Beständigkeit f gegen chemische Einwirkungen; s. r. to chemicals
r. to chemicals Chemikalienbeständigkeit f, Chemikalienfestigkeit f, Widerstandsfähigkeit (Resistenz) f gegen Chemikalien
r. to cold Kältebeständigkeit f, Tieftemperaturbeständigkeit f, Kältefestigkeit f
r. to corona [discharge] Koronabeständigkeit f, Beständigkeit f gegen den Koronaeffekt
r. to corrosion Korrosionsbeständigkeit f, Korrosionsfestigkeit f, Korrosionswiderstand m
r. to crack growth Widerstand m gegen Rißwachstum
r. to crystallization Widerstand m gegen Kristallisation, Kristallisationswiderstand m
r. to cut growth Widerstand m gegen Schnittwachstum
r. to damp storing Feuchtlagerbeständigkeit f

r. **to degradation by abrasion** s. r. to abrasion
r. **to fatigue** Ermüdungsbeständigkeit f
r. **to fire** Feuerbeständigkeit f
r. **to flex cracking** Biegungsrißwiderstand m, Widerstand m gegen Biegerißbildung, Biegerißfestigkeit f
r. **to flow** (Filtration) Durchflußwiderstand m; (phys Ch) Fließfestigkeit f, Fließverfestigung f
r. **to fluid flow** Strömungswiderstand m
r. **to freezing** Frostsicherheit f
r. **to friction** Reibungswiderstand m
r. **to fumes** Widerstand m gegen Rauchgase, Rauchgasbeständigkeit f, Rauchgasresistenz f
r. **to further tearing** Weiterreißfestigkeit f, Weiterreißwiderstand m, Durchreißfestigkeit f, Fortreißfestigkeit f
r. **to gases** Widerstandsfähigkeit f gegen Gase, Gasbeständigkeit f
r. **to glow heat** Glutbeständigkeit f, Glutfestigkeit f
r. **to grease** Fettdichtigkeit f, Fettundurchlässigkeit f
r. **to hard water** Hartwasserbeständigkeit f, Beständigkeit f gegen hartes Wasser
r. **to heat** Wärmebeständigkeit f, Wärmefestigkeit f, Hitzebeständigkeit f, Erhitzungsbeständigkeit f, Hitzefestigkeit f, thermische Beständigkeit (Stabilität, Festigkeit, Widerstandsfähigkeit) f, Thermostabilität f
r. **to high temperature[s]** Hochtemperaturbeständigkeit f, Hochhitzebeständigkeit f, Beständigkeit (Widerstandsfähigkeit) f gegen hohe Temperaturen
r. **to hydrolysis** Hydrolysebeständigkeit f
r. **to light** Lichtbeständigkeit f, Lichtechtheit f
r. **to moth** Mottenechtheit f
r. **to oil** Ölfestigkeit f, Ölbeständigkeit f
r. **to oxidation** Oxydationsbeständigkeit f, Oxydationsstabilität f, Beständigkeit f gegen oxydative Einflüsse
r. **to ozone** Ozonbeständigkeit f, Ozonfestigkeit f, Ozonresistenz f, Ozonwiderstand m
r. **to picking (plucking)** (Pap) Rupfwiderstand m
r. **to rusting** Rostbeständigkeit f
r. **to scratching** Widerstandsfähigkeit f gegen Ritzen, Ritzhärte f, Ritzfestigkeit f
r. **to sea water** Seewasserfestigkeit f
r. **to shatter** Sturzfestigkeit f, Koks-Sturzfestigkeit f, Widerstand m gegen das Zerspringen beim Stürzen (des Kokses); (Glas) Splitterfestigkeit f, Splittersicherheit f
r. **to shock** Schlagfestigkeit f
r. **to slagging** Verschlackungsbeständigkeit f, Widerstandsfähigkeit f gegen Verschlackung
r. **to storage** Lagerfähigkeit f, Lagerbeständigkeit f
r. **to sunlight** Sonnenlichtbeständigkeit f, Tageslichtbeständigkeit f
r. **to swelling** Quellbeständigkeit f
r. **to tearing** Zerreißfestigkeit f, Reißfestigkeit f,

Rißbeständigkeit f; Weiterreißfestigkeit f, Durchreißfestigkeit f, Fortreißfestigkeit f
r. **to the boil** s. r. to boiling
r. **to thermal spalling** (Ker) Temperaturwechselbeständigkeit f
r. **to tropical conditions** Tropenbeständigkeit f, Tropenfestigkeit f
r. **to wear** s. r. to abrasion
r. **to weathering [agencies]** Witterungsbeständigkeit f, Wetterbeständigkeit f, Wetterfestigkeit f
r. **welding** Widerstandsschweißen n, Widerstandsschweißung f
r. **winding** Widerstandswicklung f
r. **wire** Widerstandsdraht m
resistant beständig, fest, widerstandsfähig, resistent
to be r. to the action of nicht angegriffen werden von
r. **to abrasion** abriebbeständig, abriebfest, abnutzungsbeständig, verschleißbeständig, verschleißfest, (Text meist) scheuerbeständig, scheuerfest
r. **to acid[s]** säurebeständig, beständig gegen Säuren, säurefest
r. **to ageing** alterungsbeständig
r. **to alkali[es]** alkalibeständig, laugenbeständig
r. **to bleach** bleichecht
r. **to boiling** kochbeständig, kochfest
r. **to cold** kältebeständig, tieftemperaturbeständig, kältefest
r. **to corrosion** korrosionsbeständig, korrosionsfest
r. **to fire** feuerbeständig, feuerfest
r. **to grease** fettdicht, fettundurchlässig
r. **to heat** wärmebeständig, wärmefest, hitzebeständig, erhitzungsbeständig, hitzefest
r. **to high temperature[s]** hochtemperaturbeständig, hochhitzebeständig
r. **to oil** ölfest, ölbeständig
r. **to oxidation** oxydationsbeständig, beständig gegen oxydative Einflüsse
r. **to ozone** ozonbeständig, ozonfest, ozonresistent
r. **to the penetration of greases** s. r. to grease
r. **to the penetration of oils** s. r. to oil
r. **to wear** s. r. to abrasion
r. **to weathering** witterungsbeständig, wetterbeständig, wetterfest
resisting s. resistant
r. **agent** (Text) Reservierungsmittel n, Reservemittel n, Reserve f
resistive heater Widerstandsheizelement n, Widerstandsheizkörper m
resistivity [spezifischer] Widerstand m
r. **curve** Widerstandskurve f
resistor Widerstand m (als Bauelement); Widerstandskörper m, Heizwiderstand m
r. **furnace** s. resistance furnace
r. **material** Widerstandsmaterial n, Widerstandswerkstoff m

r. oven s. resistance furnace
resite Resit n, Phenolharz n im C-Zustand, C-Harz n (i.e.S.)
resitol Resitol n, Phenolharz n im B-Zustand, B-Harz n (i.e.S.)
resizing Nachappretieren n, Nachappretur f, Neuappretieren n, Neuappretur f
reslurry / to wiederaufschwemmen, aufrühren
resol Resol n, Phenolharz n im A-Zustand, A-Harz n (i.e.S.)
resolidification Wiederverfestigung f
  r. temperature Wiederverfestigungstemperatur f, Temperatur f der Wiederverfestigung
resolution Auflösung f; Trennung f, Spaltung f, Aufspaltung f; Auflösung f, Auflösungsvermögen n
  r. of racemates Razematspaltung f, Razemattrennung f
resolvability Auflösbarkeit f; Spaltbarkeit f
resolvable [auf]lösbar; spaltbar
resolve / to auflösen; trennen, [auf]spalten
resolving power (Foto) Auflösungsvermögen n, Auflösung f
resonance Resonanz f; (org Ch) Mesomerie f, Resonanz f, Strukturresonanz f
  r. absorption Resonanzabsorption f
  r. capture Resonanzeinfang m
  r. compound mesomere Verbindung f
  r. effect Resonanzeffekt m, mesomerer Effekt m
  r. energy Resonanzenergie f, Mesomerieenergie f, Dislokalisierungsenergie f, Konjugationsenergie f, Sonderenergie f
  r. escape probability (Kch) Resonanzentkommwahrscheinlichkeit f
  r. fluorescence Resonanzfluoreszenz f, Resonanzstrahlung f
  r. formula Resonanzformel f, mesomere Grenzformel f
  r.-free mesomeriefrei
  r. frequency Resonanzfrequenz f
  r. hybrid Resonanzhybrid n, Resonanzbastard m
  r. integral (Quantenchemie) Resonanzintegral n
  r. interaction Resonanzwechselwirkung f
  r. level Resonanzniveau n
  r. line Resonanzlinie f (im Spektrum)
  r. method Resonanzmethode f, Resonanzverfahren n
  r. neutron Resonanzneutron n
  r. phenomenon Resonanzerscheinung f, Resonanzphänomen n
  r. potential Resonanzpotential n
  r. radiation Resonanzstrahlung f, Resonanzfluoreszenz f
  r. region Resonanzbereich m
  r. scattering Resonanzstreuung f
  r. spectral line s. r. line
  r. spectrum Resonanzspektrum n
  r. stabilization Resonanzstabilisierung f
  r.-stabilized resonanzstabilisiert
  r. state s. r. level

  r. structure Resonanzstruktur f, mesomere Grenzstruktur f
  r. theory Resonanztheorie f
  r. transition Resonanzübergang m
resonant frequency Resonanzfrequenz f
resonating double bond mesomeriefähige Doppelbindung f
  r. structure Resonanzstruktur f, mesomere Grenzstruktur f
  r. valence bond system Resonanzvalenzbindungssystem n
resorb / to resorbieren, aufnehmen, einsaugen, aufsaugen; resorbiert werden
resorcin[ol] $C_6H_4(OH)_2$ Resorzin n, 1,3-Dihydroxybenzol n, m-Dihydroxybenzol n
  r. blue Resorzinblau n, La[c]kmoid n
  r.-formaldehyde resin Resorzinharz n
  r. monoethyl ether $C_2H_5OC_6H_4OH$ Resorzinmonoäthyläther m, m-Äthoxyphenol n
  r. yellow Resorzingelb n, Tropäolin O n, Tropäolin R n
resorption Resorption f, Aufnahme f, Einsaugung f, Aufsaugung f
respiration Respiration f, Atmung f
  r. inhibitor Atmungsinhibitor m
respirator Atemschutzmaske f
respiratory respiratorisch, Atmungs..., Atem...
  r. catalyst Atmungskatalysator m
  r. enzyme Atmungsferment n
  r. poison Atemgift n
  r. protection apparatus Atemschutzgerät n
  r. quotient (ratio) respiratorischer Quotient (Koeffizient) m, Atmungsquotient m, Atmungskoeffizient m, RQ
respond to manuring / to (Landw) auf Düngung ansprechen (reagieren)
response Wirkung f
responsible / to be die Ursache sein, der Träger sein, bewirken
rest / to [auf]sitzen, lagern, ruhen
rest energy Ruh[e]energie f
  r. mass Ruh[e]masse f
  r. nitrogen Reststickstoff m, Rest-N n, Nichteiweißstickstoff m, Nichtproteinstickstoff m
  r. period s. resting period
  r. position Ruhestellung f
resting cylinder s. r. tube
  r. period (stage) Ruheperiode f, Ruhezustand m
  r. tube Ruherohr n (Margarineherstellung)
restore / to wiederherstellen
restoring force rücktreibende Kraft f, Rückstellkraft f
restrainer, restraining agent Verzögerer m
restricted combustion unvollständige Verbrennung f, Verbrennung f unter vermindertem Sauerstoffzutritt, Verbrennung f bei ungenügender Luftzufuhr
restriction Drossel[stelle] f, Drosselung f, Widerstand m (gegen Strömung), Verengung f
reststrahlen Reststrahlen mpl

resulphurization *(Met)* Rückschwefelung *f*
resulphurize / to *(Met)* rückschwefeln
resultant conductance Gesamtleitfähigkeit *f*, Gesamtleitvermögen *n*
  r. orbital angular momentum Gesamtorbitaldrehimpuls *m*
  r. spin angular momentum Gesamtspindrehimpuls *m*
resulting product Endprodukt *n*, Enderzeugnis *n*, Finalprodukt *n*
ret / to *(Flachs)* rösten, rotten
ret Rösten *n*, Rotten *n*, Röste *f*, Rotte *f*, *(i.e.S.)* Flachsröste *f*, Flachsrotte *f*
retail pack Klein[handels]packung *f*
retainer plate *(Plast)* Gesenkplatte *f*, Matrizenplatte *f*
retan / to nachgerben
retannage Nachgerben *n*, Nachgerbung *f*
retard / to verzögern, verlangsamen, bremsen, hemmen
retardation Verzögerung *f*, Verlangsamen *n*, Bremsen *n*, Hemmen *n*
  r. column *(Chromatografie)* Verzögerungssäule *f*, Verzögerungskolonne *f*
  r. factor *(Chromatografie)* Rf-Wert *m*, rf-Wert *m*, Verzögerungsfaktor *m*, Rückhaltefaktor *m*
  r. of cure (vulcanization) Vulkanisationsverzögerung *f*
retarded potential retardiertes Potential *n*
retarder 1. Verzögerungsmittel *n*, Verzögerer *m*, Hemmstoff *m*, Inhibitor *m*, Bremsmittel *n*, Retardiermittel *n*, Retarder *m*; *(Bau)* Abbindeverzögerer *m* *(z.B. für Zement)*; *(Gum)* Vulkanisationsverzögerer *m*, Verzögerer *m*, Antiscorcher *m*; 2. negativer Katalysator *m*, Antikatalysator *m*, Passivator *m*
retarding *s.* retardation
  r. admix[ture] *(Bau)* Abbindeverzögerer *m*
  r. agent (material) *s.* retarder 1.
  r. of the set *(Bau)* Abbindeverzögerung *f*
retene Reten *n* *(Abbauprodukt der Abietinsäure)*
retention Beibehaltung *f*, Erhaltung *f*, Zurückhaltung *f*, Retention *f*
  r. factor *(Chromatografie)* Rf-Wert *m*, rf-Wert *m*, Rückhaltefaktor *m*, Verzögerungsfaktor *m*
  r. of configuration Retention *f* der Konfiguration, Konfigurationserhaltung *f*, Konfigurationserhalt *m*
  r. period *s.* r. time
  r. power *(Chromatografie)* Retentionsvermögen *n*
  r. tank *(Pap)* Reaktionsbehälter *m* *(Tank, in dem ein Stoff im Verlaufe einer bestimmten Durchgangszeit behandelt wird)*
  r. time Verweilzeit *f*, Aufenthaltszeit *f*, Retentionszeit *f*, Rückhaltezeit *f*, Haltezeit *f*, Stehzeit *f*, Standzeit *f*, Durchgangszeit *f*, Durchlaufzeit *f*; *(Milchpasteurisation)* Heißhaltezeit *f*
  r. tower *(Pap)* Reaktionsturm *m*, Stoffturm *m* *(Turm, in dem der Stoff im Verlauf einer bestimmten Durchgangszeit behandelt wird)*

  r. volume Retentionsvolumen *n*, Rückhaltevolumen *n*
reticulation Netzstruktur *f*; Netzbildung *f*, Vernetzung *f*; *(Foto)* Runzelkorn *n*
retinite *(Min)* Retinit *m*
retort / to in der Retorte erhitzen, *(z.B. Ölschiefer)* in der Retorte schwelen
retort Retorte *f*
  r. battery Retortenbatterie *f*; *s.* r. bench
  r. bench Koks[ofen]batterie *f*, Verkokungsbatterie *f*
  r. carbon Retortenkohle *f*
  r. coke Retortenkoks *m*
  r. furnace Retortenofen *m*, Destillierofen *m*
  r. graphite Retortengraphit *m*
  r. neck Retortenhals *m*
  r. oven *s.* r. furnace
  r. setting Retorteneinheit *f*
  r. stand Stativ *n*
  r. stand base Stativfuß *m*
  r. stand rod Stativstab *m*
retorting Retortenschwelen *n* *(von Ölschiefer)*
retractable abnehmbar; abklappbar, abschwenkbar
retread / to *(Reifen)* runderneuern
retreader Vulkaniseur *m*
retreading Runderneuerung *f* *(von Reifen)*
re-treat / to nochmals (wiederholt) behandeln, nachbehandeln
re-treatment nochmaliges (wiederholtes) Behandeln *n*, nochmalige (wiederholte) Behandlung *f*, Nachbehandeln *n*, Nachbehandlung *f*
retrogradation Retrogradation *f* *(von Stärkelösungen)*
retrograde retrograd, rückläufig
  r. condensation Rückkondensation *f*, retrograde Kondensation *f*
retronecine Retronezin *n* *(Alkaloid)*
retropinacolin rearrangement Retropinakolinumlagerung *f*
rettery Flachsrösterei *f*, Flachsröste *f* *(Anlage)*
retting Rösten *n*, Rotten *n*, Röste *f*, Rotte *f*, *(i.e.S.)* Flachsröste *f*, Flachsrotte *f*
  r. bacteria Röstbakterien *npl*, Röstorganismen *mpl*
  r. maturity Röstreife *f*
  r. method Röstverfahren *n*
  r. vat Röstbassin *n*
  r. water Röstwasser *n*
rettory *s.* rettery
return / to zurückführen, wieder zuführen; zurückfüllen; zurückfließen
  r. to the circuit in den Kreislauf zurückführen, dem Kreislauf wieder zuführen (zusetzen), rezirkulieren
return Rückführung *f*, Rückleitung *f*; Rücklauf *m*, Rückfluß *m*
  r. bend Doppelkrümmer *m*
  r. condenser Rückflußkühler *m*
  r. felt run *(Pap)* rücklaufendes Filztrum *n*

r. flow Rückfluß *m*
r. journey *(Pap)* Rücklauf *m (z.B. des Siebes)*
r. oil Rücklauföl *n*
r. pin *(Plast)* Rückdrückstift *m*, Rückstoßstift *m*
r. pipe Rücklaufrohr *n*, Rückleitung *f*
r. roll Tragrolle *f* am Leertrum (Untertrum)
r. spring *(Plast)* Rückzugfeder *f*, Ausdrückbolzenfeder *f*
r. stroke Rückhub *m*, Einfahrhub *m*
r. valve Rückschlagventil *n*, Rückflußventil *n*
r. wire run *(Pap)* rücklaufendes (rückläufiges, unteres) Siebtrum *n*
retzbanyite *(Min)* Retzbanyit *m*
reusable wiederverwendbar, wiederbenutzbar
reuse / to wiederverwenden, von neuem verwenden
reuse Wiederverwendung *f*, Wiederbenutzung *f*, Wiedergebrauch *m*
reverberatory furnace (kiln) Flamm[en]ofen *m*
reversal Umstellen *n*, Umstellung *f*, Umkehr[ung] *f*, Wechsel *m*; Umklappen *n (von Bindungen bei der Waldenschen Umkehrung)*
r. developer *(Foto)* Umkehrentwickler *m*
r. development *(Foto)* Umkehrentwicklung *f*
r. emulsion *(Foto)* Umkehremulsion *f*
r. film Umkehrfilm *m*
r. material[s] *(Foto)* Umkehrmaterial *n*
r. of flow *(Text)* Wechsel *m* der Flottenrichtung
r. of phases Phasenumkehr *f*
r. process *(Foto)* Umkehrverfahren *n*, Direkt-Positiv-Prozeß *m*
r. processing *(Foto)* Umkehrentwicklung *f*
reverse / to umkehren, umdrehen, umlegen; umstellen, umschalten; *(bei Emulsionen)* umschlagen
reverse-jet filter Faservliesfilter *n* mit Blasring
r. press *(Pap)* Steigpresse *f*, Wendepresse *f*
r. press felt *(Pap)* Steigpreßfilz *m*, Wendefilz *m*
r. process (reaction) Rückreaktion *f*, Gegenreaktion *f*
reversed-phase chromatography Chromatografie *f* mit Phasenumkehr (umgekehrter Phase), Umkehrphasenchromatografie *f*
r.-phase partition *(Chromatografie)* Verteilung *f* mit Phasenumkehr
r.-phase partition chromatography Verteilungschromatografie *f* mit Phasenumkehr (umgekehrter Phase)
reversibility Reversibilität *f*, Umkehrbarkeit *f*
reversible reversibel, umkehrbar, in beiden Richtungen verlaufend
r. adsorption reversible (physikalische) Adsorption *f*
r. cell reversibles (umkehrbares) Element *n*
r. colloid reversibles (resolubles) Kolloid *n*
r. electrode reversible (umkehrbare) Elektrode *f*
r. element *s.* r. cell
r. elongation reversible (elastische) Dehnung *f*
r. first order reaction umkehrbare Reaktion *f* erster Ordnung

r. gel reversibles (resolubles) Gel *n*
r. process reversibler (umkehrbarer) Vorgang *m*
r. reaction reversible (umkehrbare) Reaktion *f*
r. sol reversibles (resolubles) Sol *n (lyophiles Sol)*
reversing pan dryer Schaukeltrockner *m*, Kipphordenumlauftrockner *m*
r. valve Reversierventil *n*, Umstellventil *n*, Wechselventil *n*
reversion Umkehrung *f*; Umschlagen *n (z.B. einer Emulsion); (Gum)* Reversion *f*
r. gas chromatography Reversions-Gaschromatografie *f*
r. tendency *(Gum)* Reversionsneigung *f*, Reversionstendenz *f*
revive / to *(Text)* avivieren, auffrischen, schönen
revivificate / to *s.* to revivify
revivification Regenerierung *f*, *(bei Aktivkohle auch)* Wiederbelebung *f*, Reaktivierung *f*
revivifier Regenerator *m*
revivify / to regenerieren, *(Aktivkohle)* wiederbeleben, reaktivieren
reviving *(Text)* Avivieren *n*, Avivage *f*, Schönen *n*
r. agent *(Text)* Aviviermittel *n*
revolver press *(Ker)* Revolverpresse *f*
revolving boiler (digester) *(Pap)* rotierender Kocher *m*, Drehkocher *m*
r. disk Mahlscheibe *f*
r.-disk feeder Tellerspeiser *m*
r. distributor Drehverteiler *m*, drehbarer Gichtverteiler *m*
r. drum strainer *(Pap)* Drehknotenfänger *m*, rotierender Knotenfänger *m*
r. filter Drehfilter *n*, Trommelfilter *n*
r. grate Drehrost *m*
r. knife *(Pap)* Obermesser *n (des Querschneiders)*
r.-knife cutting-machine *(Pap)* Querschneider *m* mit rotierenden Messern, rotierender Querschneider *m*, Messerwellenquerschneider *m*, Rotationsquerschneider *m*
r. paddles *(Pap)* Stofftreiber *m*, Umtriebpropeller *m (beim Holländer)*
r. pot *(Glas)* Drehwanne *f*, Drehtank *m*
r. screen Rundsieb *n*, Siebtrommel *f*, Trommelsieb *n*
r. strainer *s.* drum strainer
r. tube Drehrohr *n*
r. tubular kiln (oven) Drehrohrofen *m*
r.-type syphon *(Pap)* rotierender (umlaufender) Siphon *m (Kondensatableitung)*
rewash / to nachwaschen; *(Foto)* nachwässern
rewet / to wieder anfeuchten
rewind / to *(Pap)* umrollen, umwickeln
rewind shaft *(Pap)* Rollstange *f*, Aufrollstange *f*, Papierrollstange *f*
rewinder *s.* rewinding machine
rewinding *(Pap)* Längsschneiden *n* und Umrollen *n*, *(i.e.S. nur)* Umrollen *n*, Umwickeln *n*
r. machine *(Pap)* Umroller *m*, Rollenschneidmaschine *f*, Rollenschneider *m*, Längsschneidemaschine *f*, Längsschneider *m*

rework / to [wieder]aufarbeiten
rexanthation Umxanthogenierung *f*
Reynolds number *(phys Ch)* Reynoldssche Zahl *f*
Rf value *(Chromatografie)* Rf-Wert *m*, rf-Wert *m*, Rückhaltefaktor *m*, Verzögerungsfaktor *m*
rh [factor] *s.* rhesus factor
rh-negative Rh-negativ
rh-positive Rh-positiv
rH [value] rH-Wert *m*
R.H. *s.* 1. red heat; 2. relative humidity; 3. Rockwell hardness
rhabdophane, rhabdophanite *(Min)* Rhabdophan *m*
rhamnazin Rhamnazin *n (natürlicher Farbstoff)*
rhamnetin Rhamnetin *n (natürlicher Farbstoff)*
rhamnose Rhamnose *f*, L-Mannomethylose *f*
RHC *s.* rubber hydrocarbon content
rheic acid *s.* rhein
rhein Rhein *n (Bestandteil des Rhabarbers)*
rhenate M'_2ReO_4 Rhenat(VI) *n*
rhenic acid $HReO_4$ Perrheniumsäure *f*, Rhenium(VII)-säure *f*
rhenite M'_2ReO_3 Rhenat(IV) *n*
rhenium Re Rhenium *n*
  r. compound Rheniumverbindung *f*
  r. dioxide ReO_2 Rheniumdioxid *n*, Rhenium(IV)-oxid *n*
  r. heptoxide Re_2O_7 Rheniumheptoxid *n*, Dirheniumheptoxid *n*, Rhenium(VII)-oxid *n*
  r. hexachloride $ReCl_6$ Rheniumhexachlorid *n*, Rhenium(VI)-chlorid *n*
  r. pentachloride $ReCl_5$ Rheniumpentachlorid *n*, Rhenium(V)-chlorid *n*
  r. peroxide Re_2O_8 Rheniumperoxid *n*
  r. sesquioxide Re_2O_3 Dirheniumtrioxid *n*, Rhenium(III)-oxid *n*
  r. trioxide ReO_3 Rheniumtrioxid *n*, Rhenium(VI)-oxid *n*
rheological rheologisch
  r. behaviour Fließverhalten *n*, rheologisches Verhalten *n*
rheology Rheologie *f*, Fließkunde *f*
rheomorphism *(Geol)* Rheomorphose *f*
rheopexy *(Koll)* Rheopexie *f*, thixogene Koagulation *f*, Fließverfestigung *f*
rhesus antigen *s.* r. factor
  r. factor Rhesus-Faktor *m*, Rh-Faktor *m*
rhodamine Rhodamin *n*
rhodanate $M'SCN$ Rhodanid *n*, Thiozyanat *n*
rhodanese Rhodanese *f (zu den Transferasen gehörendes Ferment)*
rhodanic acid HSCN Rhodanwasserstoffsäure *f*, Thiozyansäure *f*
  r. value Rhodanzahl *f*, Rh Z
rhodanide *s.* rhodanate
rhodanize / to rhodinieren, mit einer [dünnen] Rhodiumschicht überziehen
rhodanizing [galvanisches] Rhodinieren *n*, Überziehen *n* mit einer [dünnen] Rhodiumschicht
rhodanometry Rhodanometrie *f*

rhodate Rhodat *n*
rhodicite *s.* rhodizite
rhodinic acid Rhodinsäure *f*, Dihydrogeraniumsäure *f*, 3,7-Dimethyl-7(6)-oktensäure *f*
rhodium Rh Rhodium *n*
  r. compound Rhodiumverbindung *f*
  r. dioxide RhO_2 Rhodiumdioxid *n*, Rhodium(IV)-oxid *n*
  r. hydrosulphide $Rh(SH)_3$ Rhodium(III)-hydrogensulfid *n*
  r. monosulphide RhS Rhodium[mono]sulfid *n*, Rhodium(II)-sulfid *n*
  r. monoxide RhO Rhodium[mon]oxid *n*, Rhodium(II)-oxid *n*
  r. nitrate $Rh(NO_3)_3$ Rhodium(III)-nitrat *n*
  r.-plated rhodiniert, mit einer [dünnen] Rhodiumschicht überzogen
  r. plating [galvanisches] Rhodinieren *n*, Überziehen *n* mit einer [dünnen] Rhodiumschicht
  r. sesquioxide Rh_2O_3 Dirhodiumtrioxid *n*, Rhodium(III)-oxid *n*
  r. sesquisulphide Rh_2S_3 Rhodium(III)-sulfid *n*
  r. sulphate $Rh_2(SO_4)_3$ Rhodium(III)-sulfat *n*
  r. sulphite $Rh_2(SO_3)_3$ Rhodium(III)-sulfit *n*
  r. tetrahydroxide $Rh(OH)_4$ Rhodiumtetrahydroxid *n*, Rhodium(IV)-hydroxid *n*
  r. trichloride $RhCl_3$ Rhodiumtrichlorid *n*, Rhodium(III)-chlorid *n*
  r. trifluoride RhF_3 Rhodiumtrifluorid *n*, Rhodium(III)-fluorid *n*
  r. trihydroxide $Rh(OH)_3$ Rhodiumtrihydroxid *n*, Rhodium(III)-hydroxid *n*
  r. trioxide RhO_3 Rhodiumtrioxid *n*, Rhodium(VI)-oxid *n*
rhodizite *(Min)* Rhodizit *m*
rhodochrosite *(Min)* Rhodochrosit *m*, Manganspat *m (Mangan(II)-karbonat)*
rhodommatine Rhodommatin *n (tierischer Farbstoff)*
rhodomycin Rhodomyzin *n (Antibiotikum)*
rhodomycinone Rhodomyzinon *n (Farbstoff)*
rhodonite *(Min)* Rhodonit *m*, Mangankiesel *m*
rhodopsin Rhodopsin *n*, Sehpurpur *m*
rhodosamine Rhodosamin *n*, 2,3,6-Tridesoxy-3-dimethylamino-L-lyxohexose *f*
rhomb. *s.* rhombic
rhomb-spar *(Min)* Rautenspat *m*, *(veraltet für)* Dolomit *m (Kalziummagnesiumkarbonat)*
rhombarsenite *(Min)* Rhombarsenit *m*, *(veraltet für)* Claudetit *m (Arsen(III)-oxid)*
rhombic *(Krist)* rhombisch, orthorhombisch
  r. crystal system *(Krist)* rhombisches System *n*
  r.-pyramidal *(Krist)* rhombisch-pyramidal
  r. sulphur rhombischer Schwefel *m*, α-Schwefel *m*
  r. system *s.* r. crystal system
rhombohedral *(Krist)* rhomboedrisch
rhombohedron *(Krist)* Rhomboeder *n*
rhyolite *(Geol)* Rhyolit *m*, Liparit *m*
rhyotaxitic texture *(Geol)* Fluidaltextur *f*, Fließtextur *f*, Fließgefüge *n*

rib tile Rippenziegel *m*
ribbed gerippt, gerillt, geriffelt
  r. felt *(Pap)* Markierfilz *m*, gerippter Filz *m*, Rippfilz *m*
  r. glass geripptes (geriffeltes) Glas *n*, Rippenglas *n*
ribbing felt *s.* ribbed felt
ribbon-blade agitator Bandrührer *m*
  r. blender Band[schnecken]mischer *m*
  r. flight Bandschnecke *f (Schneckenförderer)*
  r. gum kino Eukalyptuskino *n (Kinoharz von Eucalyptus-Arten)*
  r. mixer Band[schnecken]mischer *m*
  r. of glass Glasband *n*
ribbonization Ribbonisation *f (bei glasfaserverstärkten Plasten)*
riboflavin[e] Riboflavin *n*, Vitamin B_2 *n*, 6,7-Dimethyl-9-(*d*-1′-ribityl)-*iso*-alloxazin *n*
  r. phosphate Riboflavinphosphat *n*, Flavinmononukleotid *n*, FMN
ribonuclease Ribonuklease *f (zu den Phosphatasen zählendes Ferment)*
ribonucleate Ribonukleat *n (Salz der Ribonukleinsäure)*
ribonucleic acid Ribonukleinsäure *f*, RNS
ribonucleoprotein Ribonukleoproteid *n*
ribose Ribose *f (zu den Pentosen gehörendes Monosaccharid)*
RIC *(Abk. für)* Royal Institute of Chemistry
rice beer Reisbier *n*
  r. bran Reiskleie *f*
  r. dust *s.* r. polish
  r. flakes Reisflocken *fpl*
  r. flour Reismehl *n*
  r. grits Reisgrieß *m (zur Bierherstellung)*
  r. meal Reisfuttermehl *n*
  r. mill Reismühle *f*
  r. milling Reismüllerei *f*
  r. oil Reisöl *n*
  r. paper Reispapier *n*, Chinesisches Reispapier (Markpapier) *n (aus dem Mark der Aralia papyrifera)*
  r. polish Reisschleifmehl *n (Abfall beim Reispolieren)*
  r. polishings Reiskleie *f*
  r. powder Reispuder *m*
  r. sheller Reisschälmaschine *f*
  r. starch Reisstärke *f*
rich coal gas Reichgas *n*
  r. concrete fetter Beton *m*
  r. fuel gas main Sammelleitung (Zuführungsleitung) *f* für Starkgas *(eines Koksofens)*
  r. gas Reichgas *n*, reiches Gas *n (oberer Heizwert 7500 bis 8500 kcal/m³)*; Starkgas *n (oberer Heizwert 3600 bis 5500 kcal/m³)*
  r. in carbon kohlenstoffreich
  r. in fat fettreich
  r. in flavour vollmundig *(z.B. Bier)*
  r. in lignin ligninreich
  r. in methane methanreich

  r. in ore erzreich
  r. in water wasserreich
  r. lime Fettkalk *m*, Weißkalk *m*
  r. milk Vollmilch *f*
  r. mortar fetter Mörtel *m*
  r. oil reiches (beladenes) Waschöl *n*, fettes Öl *n (beim Absorptionsverfahren)*
  r.-oil heater Reichölerhitzer *m*
  r. white water *(Pap)* faser- und füllstoffreiches Abwasser (Rückwasser) *n*
Richardson effect *(Phys)* Richardson-Effekt *m*
richellite *(Min)* Richellit *m*
ricidine *s.* ricinine
ricin Rizin *n (Eiweißstoff aus dem Rizinussamen)*
ricinelaidic acid Rizinelaidinsäure *f*
ricinine Rizinin *n (Alkaloid des Rizinussamens)*
ricinoleate Rizinoleat *n*
ricinol[e]ic acid $C_{17}H_{32}(OH)COOH$ Rizinol[ein]säure *f*, Rizinus[öl]säure *f*, Hydroxyölsäure *f*
rick Gradierwerk *n*
rickardite *(Min)* Rickardit *m (Kupfertellurid)*
riddle Schwingsieb *n*, Rüttelsieb *n*, Rätter *m*
rider Reiterchen *n*, Reiter *m*, Reiterwägestück *n*
  r. bar (carrier) Reiterlineal *n (an der Analysenwaage)*
riding roll *(veraltet für)* dandy roll
riebeckite *(Min)* Riebeckit *m*
Riegler reagent Rieglers Reagens *n*
riemer Auftreiber *m (Laborgerät)*
Riesenfeld reaction Riesenfeld-Reaktion *f*
Riesenfeld test Riesenfeld-Probe *f*
riffle Rille *f*; Riffel *f*, Leiste *f*
right-angle elbow Krümmer (Rohrkrümmer) *m* mit rechtwinkliger Ablenkung
  r.-angle stopcock Schwanzhahn *m*
  r.-circularly polarized rechtszirkular polarisiert
  r.-hand circularly polarized *s.* r.-circularly polarized
  r.-hand pan rechte Waagschale *f*, Gewichtsschale *f (z.B. einer Analysenwaage)*
  r.-handed rechtshändig; rechtsdrehend, *d*-drehend, dextrogyr, *d*-; *(Krist)* rechtsdrehend
  r.-handed quartz rechter Quarz *m*, Rechtsquarz *m*
  r.-handed rotation Rechtsdrehung *f*
  r.-rotating rechtsdrehend, *d*-drehend, dextrogyr, *d*-
rigid starr, steif; fest, stabil; hart
  r. arm elevator Tragkettenförderer *m*
  r. chain molecule starres Kettenmolekül *n*
  r. foam fester Schaum *m*; harter Schaumstoff *m*
  r.-foam insulation Schaumstoffisolierung *f*, Schaumstoffdämmung *f*
  r. gel unelastisches Gel *n*
  r. plastic harter Plast (Kunststoff) *m*
  r. PVC Hart-PVC *n*, PVC-hart *n*
rigidity Starrheit *f*, Steifigkeit *f*, Steifheit *f*; Festigkeit *f*, Stabilität *f*; Härte *f*
rimmed (rimming) steel unberuhigter (unberuhigt vergossener, unruhig vergossener) Stahl *m*

rind Rinde *f*, Schale *f*, Kruste *f*; Grat *m*
rindless cheese Käse *m* ohne Rinde, rindenloser Käse *m*
ring Ring *m*
  r. aggregate *(org Ch)* Ringkomplex *m*
  r.-and-ball apparatus Ring[-und]-Kugel-Gerät *n* *(zur Bestimmung des Erweichungspunktes)*
  r.-and-ball method Ring[-und]-Kugel-Methode *f*, Ring[-und]-Kugel-Verfahren *n*, R. u. K., RuK *(zur Bestimmung des Erweichungspunktes)*
  r.-and-ball mill Kugelringmühle *f*
  r.-and-ball softening point Erweichungspunkt *m* „Ring und Kugel", Erweichungspunkt *m* RuK
  r.-and-ball tester *s.* r.-and-ball apparatus
  r. assembly *(org Ch)* Ringsequenz *f*
  r. atom Ringatom *n*
  r.-ball mill *s.* r.-and-ball mill
  r.-branching position Ringverzweigungsstelle *f*
  r. burner Ringbrenner *m*, Gasheizkranz *m*
  r.-carbon atom Ringkohlenstoffatom *n*
  r.-chain tautomerism Ring-Ketten-Tautomerie *f*, Ringkettentautomerie *f*, zyklisch-offene Tautomerie *f*
  r. chelate Ringchelat *n*
  r. closure Ringschluß *m*, Zyklisierung *f*
  r. closure reaction Ringschlußreaktion *f*
  r. compound Ringverbindung *f*, ringförmige (zyklische) Verbindung *f*
  r. contraction Ringverengung *f*
  r. enlargement Ringerweiterung *f*
  r. fission Ring[auf]spaltung *f*, Ringöffnung *f*, Ringsprengung *f*
  r. flip Umklappen *n* des Ringes
  r. formation Ringbildung *f*
  r. gate *(Plast)* ringförmiger Anguß *m*
  r.-halogenated kernhalogeniert
  r. halogenation Halogenierung *f* des Benzolrings (Benzolkerns)
  r. kiln Ringofen *m*
  r. link (member) Ringglied *n*
  r. method *(Ker)* Ringmethode *f*
  r.-methyl group ringständige Methylgruppe *f*
  r. mill *s.* r.-roll mill
  r. mould *(Glas)* Kopfform *f*
  r. nozzle Ringdüse *f*
  r. of carbon atoms Kohlenstoffring *m*
  r. opening *s.* r. fission
  r. packing Ringdichtung *f*
  r. plane Ringebene *f*
  r. polymer Ringpolymer[es] *n*
  r.-roll mill Ringrollenmühle *f (i.w.S.)*, Ringwalzenmühle *f (i.w.S.)*, Fremdkraftwälzmühle *f*, Fremdkraftrollenmühle *f*, Fremdkraftringmühle *f*
  r.-roll mill *(with horizontal grinding ring)* Horizontalringrollenmühle *f*, Horizontalringwalzenmühle *f*; Pendel[rollen]mühle *f*, Fliehkraftpendelmühle *f*, Fliehkraftwalzenmühle *f*; Schüssel-Kegel-Mühle *f*
  r.-roll mill *(with vertical grinding ring)* Vertikalringrollenmühle *f*, Vertikalringwalzenmühle

f, Ringrollenmühle *f*, Ringwalzenmühle *f*, Walzenringmühle *f*, Ringmühle *f (i.e.S.)*
  r.-roll press Ringwalzenpresse *f*
  r.-roller mill *s.* r.-roll mill
  r. sample *s.* r. test piece
  r. scission *s.* r. fission
  r. seal *s.* r. packing
  r.-shaped ringförmig, Ring...
  r.-shaped die *s.* r. nozzle
  r. size Ringgröße *f*
  r. stand Kolbenträger *m*
  r. strain Ringspannung *f*
  r. structure Ringstruktur *f*
  r. system Ringsystem *n*
  r. test Ringprüfung *f (bei Glasuren)*
  r. test piece ringförmiger Probekörper *m*, ringförmige Probe *f*, Ringprobe *f*
Ring Index Ring-Index *m (Verzeichnis der Ringsysteme von A.M. Patterson, L.T. Capell und D.F. Walker)*
  R. Index name nach dem Ring-Index gebildeter Name *m*
  R. Index system Ring-Index-System *n*, System *n* des Ring-Index *(zur Benennung und Bezifferung zyklischer Verbindungen)*
Ringer artificial serum *s.* Ringer's solution
Ringer's solution *(Med)* Ringer-Lösung *f*, Ringersche Lösung *f*
rinkite *(Min)* Rinkit *m*
Rinman[n]'s green Rinmanns Grün *n*, Kobaltgrün *n*
rinse / to [ab]spülen, ausspülen, nachspülen; [ab]waschen, auswaschen; wässern
  r. down herabspülen
  r. off abspülen, wegspülen
  r. out [her]ausspülen
rinse 1. Spülen *n*, Abspülen *n*, Ausspülen *n*, Nachspülen *n*; Waschen *n*; Wässern *n*, Wässerung *f*; 2. Spülmittel *n*, Nachspülmittel *n*
  r. amount Spülmittelmenge *f*
  r. liquor Spülflüssigkeit *f*
  r. screen Entbrühungssieb *n*, Entwässerungssieb *n*, Waschsieb *n*
  r. valve Spülventil *n*, Spülflüssigkeitsschieber *m*
  r. water Spülwasser *n*, Waschwasser *n*
rinsed milk Spülmilch *f*
rinser Spülapparat *m*, Spüler *m*, Ausspritzapparat *m*
rinsing Spülen *n*, Spülung *f*, Abspülen *n*, Ausspülen *n*, Nachspülen *n*; Waschen *n*, Wässern *n*, Wässerung *f*; Spülflüssigkeit *f*
  r. agent Spülmittel *n*, Nachspülmittel *n*
  r. bath Spülbad *n*
  r. water Spülwasser *n*, Waschwasser *n*
rinsings Spülflüssigkeit *f*
Rio ipecac[uanha] Rio-Brechwurzel *f*, Rio-Ipekakuanha *f*, brasilianische Brechwurzel *f (von Cephaëlis ipecacuanha (Brot.)A. Rich.)*
ripen / to 1. reifen [lassen]; *(Milch)* säuern, reifen lassen; 2. reifen, reif werden; *(Milch)* sauer werden, reifen

ripened butter Sauerrahmbutter f
  r. cheese gereifter Käse m
  r. filter eingearbeitetes Filter n *(Wasserauf-bereitung)*
  r. milk gesäuerte (gereifte) Milch f
ripener Rahmreifer m, Säuerungsgefäß n, Säuerungswanne f
ripeness Reife f *(Zustand)*
  r. figure *(Text)* Reifezahl f
ripening Reifen n, Reifung f; Einarbeiten n, Einarbeitung f, Reifung f *(eines Filters)*
  r. agent Reifebeschleuniger m, Reifungsbeschleuniger m *(z.B. Äthylen)*
  r. process Reifungsvorgang m, Reifungsprozeß m, Reifeprozeß m, Reifung f, *(bei der Milchbehandlung auch)* Säuerungsprozeß m, Säuerungsvorgang m
  r. tank (vat) Rahmreifer m, Säuerungsgefäß n, Säuerungswanne f
ripidolite *(Min)* Rhipidolith m
ripper, ripping apparatus *(Pap)* Umroller m, Rollenschneidmaschine f, Rollenschneider m, Längsschneidemaschine f, Längsschneider m
rise / to 1. sich erhöhen, [an]steigen, zunehmen; aufsteigen *(von Gasen)*; aufgehen *(von der Hefe)*; 2. steigern, erhöhen
rise Erhöhung f, Steig[er]ung f, Zunahme f, Anstieg m
  r. in temperature Temperaturerhöhung f, Temperatursteigerung f, Temperaturanstieg m
  r. of boiling point Siedepunktserhöhung f
  r. of temperature s. r. in temperature
riser Steigrohr n; *(Destillation)* Kamin m, Dampfkamin m, Kaminstummel m, Dämpfestutzen m, Dampfhals m *(einer Bodenkolonne)*
  r. cracking Kracken n nach dem Airliftverfahren, Airliftkracken n
  r. tube s. riser
rising-film evaporator Kletterfilmverdampfer m
  r. main Steigleitung f
  r. of cream Aufrahmen n, Aufrahmung f
  r. pipe Steigrohr n
  r. stream Aufstrom m
Ritter-Kellner cook process *(Pap)* direkte Kochung f nach Ritter-Kellner
Rittinger's law Rittinger-Gesetz n *(Zerkleinerung)*
Ritz combination principle s. Ritz-Rydberg combination principle
Ritz formula *(phys Ch)* Ritzsche Formel f
Ritz principle s. Ritz-Rydberg combination principle
Ritz-Rydberg combination principle [Rydberg-]Ritzsches Kombinationsprinzip n
river gravel Flußkies m
  r. pollution Flußverunreinigung f
  r. water Flußwasser n
rivotite *(Min)* Rivotit m
RNA s. ribonucleic acid
roach powder gegen Schaben wirksames Insektenpulver n

road binder Bindemittel n im Straßenbau
  r. marking paint Straßenmarkierungsfarbe f
  r. oil Straßenöl n, Roadoil n *(Öl für Straßenbehandlung zur Staubverhütung, Oberflächenbefestigung und Wasserabdichtung)*
  r. tank wag[g]on, r. tanker Straßentankwagen m, Tankauto n, Autokesselwagen m
  r. tar Straßenteer m
  r. test *(Gum)* Straßenprüfung f
roadline paint s. road marking paint
roaring flame rauschende Flamme f
roast / to [ab]rösten
roast reaction Röstreaktion f
  r.-reaction process Röstreaktionsverfahren n, Röstreaktionsarbeit f
  r.-reduction process Röstreduktionsverfahren n, Röstreduktionsarbeit f
roasted coffee Röstkaffee m
  r. malt geröstetes Malz n, Farbmalz n
  r. material Röstgut n
  r. product Röstprodukt n
  r. pyrites Kiesabbrand m
  r. starch Röststärke f
roaster Röstofen m
  r. gas Röstgas n
roasting Rösten n, Röstung f, Abrösten n, Abröstung f
  r.-and-reaction process s. roast-reaction process
  r.-and-reduction process s. roast-reduction process
  r. cylinder Rösttrommel f
  r. dish Röstschale f
  r. furnace (kiln) Röstofen m
  r. of pyrites Pyritröstung f, Kiesabröstung f
  r. oven s. r. furnace
  r. plant Röst[ofen]anlage f
  r. process Röstprozeß m, Röstvorgang m
  r. temperature Rösttemperatur f
  r. time Röstzeit f, Röstbetriebsdauer f
Robert evaporator Vertikalrohrverdampfer m mit Innenheizkammer, Robert-Verdampfer m
Roberts grinder *(Pap)* Roberts-Ringschleifer m, Roberts-Schleifer m
robinin Robinin n *(Glykosid aus Robinia pseudoacacia)*
Robison-Embden ester Robison-[Embden-]Ester m *(Glukose-6-phosphat)*
roborant kräftigend, roborierend
roborant Kräftigungsmittel n, Roborans n
Rochelle powders s. Seidlitz powders
Rochelle salt $KNaC_4H_4O_6 \cdot 4H_2O$ Rochellesalz n, Seignettesalz n, Natronweinstein m, Kaliumnatriumtartrat n
rock / to schütteln; wiegen; schaukeln; *(Gerb)* *(Blößen im Schaukelrahmen)* hin- und herbewegen
rock Gestein n
  r. bit Meißel m für hartes Gestein (Gebirge), Meißel m für harte Formationen *(zum Erdölbohren)*

r. candy Kandis[zucker] *m*, Zuckerkant *m*, Zuckerkand[is] *m*, Kandelzucker *m*
r. catcher Steinfänger *m*, Steinfang *m*
r. crystal *(Min)* Bergkristall *m (Quarzvarietät)*; Kristallglas *n*
r. dammar *(Dammarharz von Hopea odorata Roxb.)*
r. debris Gesteinsstückchen *n*, Gesteinsabrieb *m*
r. explosive Gesteinssprengstoff *m*
r.-forming mineral gesteinsbildendes Mineral *n*
r. maple Zuckerahorn *m*, Acer saccharum Marsh.
r. milk *(Min)* Bergmilch *f (Kalziumkarbonat)*
r. oil Steinöl *n*, Bergöl *n*, *(veraltet für)* Erdöl *n*
r. phosphate Mineralphosphat *n*, Rohphosphat *n*
r. salt Steinsalz *n*
r. salt plate Steinsalzplättchen *n*
r. type Lithotype *f*
r. waste Verwitterungsschutt *m*, Gesteinsschutt *m*, Gesteinsgrus *m*
r. wool Steinwolle *f*, Gesteinswolle *f*, Mineralwolle *f*
rocker *(Gerb)* Hängeäscher *m*; *s.* r. frame
r. frame Schaukelrahmen *m*, Wipprahmen *m* *(zum Bewegen der Blößen in Gerbgruben)*
rocket propellant Raketentreibstoff *m*
rocking furnace Schaukelofen *m*
Rockwell hardness Rockwellhärte *f*, HR
Rockwell hardness test Rockwell-Härteprüfung *f*, Rockwellverfahren *n*
rocky krausen hohe Kräusen *pl*, Hochkräusen *pl* *(Bierherstellung)*
Rocky Mountain Fir Westliche Balsamtanne *f*, Abies lasiocarpa (Hook.) Nutt.
rod Stange *f*, Stab *m*
r.-curtain electrode Schlitzkastenelektrode *f*
r. deck Stabrost *m*, Stangenrost *m*, Runddrahtrost *m (einer Rostsiebmaschine)*
r.-like stäbchenförmig
r. mill Stabmühle *f*
r.-shaped bacteria Stäbchenbakterien *npl*, stäbchenförmige Bakterien *npl*
r. wax Röhrenwachs *m*
rodent bait Giftköder *m* gegen Nagetiere
r. repellent Abschreckmittel (Repellent) *n* gegen Nagetiere
rodenticide Rodentizid *n*, Nagetiergift *n*
rodlike stäbchenförmig
Roe chloride number *(Pap)* Roè-Zahl *f*
Roelig hysteresis apparatus Dämpfungsmesser *m* (Dämpfungsgerät *n*) nach Roelig, Roelig-Maschine *f*, Fliehkraftmaschine *f* nach Roelig
Roelig machine *s.* Roelig hysteresis apparatus
roentgen Röntgen-Einheit *f*, Röntgen *n*, R
r. equivalent Röntgenäquivalent *n*
r. equivalent man Roentgen-equivalent-man *n*, Rem-Einheit *f*, Rem *n*, rem *(biologisches Röntgenäquivalent)*
r. equivalent physical Roentgen-equivalent-physical *n*, Rep-Einheit *f*, Rep *n*, rep *(physikalisches Röntgenäquivalent)*

r. rays Röntgenstrahlen *mpl*
r. spectrum Röntgenspektrum *n*
roentgenize / **to** röntgen *(mit Röntgenstrahlen durchleuchten)*
roentgenologic[al] röntgenologisch
roentgenology Röntgenologie *f*, Röntgen[strahlen]kunde *f*
roentgenometry *(Krist)* Röntgenometrie *f*
roentgenotherapy Röntgentherapie *f*
Roga index Roga-Index *m*, Roga-Backzahl *f*, Backzahl *f* nach Roga
Roga method Roga-Methode *f (zur Bestimmung der Backfähigkeit von Kohle)*
Roga test Roga-Test *m (zur Bestimmung der Backfähigkeit von Kohle)*
roll / **to** rollen; walzen; *(Kautschukmischung)* aufrollen
r. up zusammenrollen
roll Walze *f*, Rolle *f*; Walzenpresse *f*, Walzmaschine *f*; *(Pap)* Holländerwalze *f*, Messerwalze *f*, Mahlwalze *f*; Kalanderwalze *f*, Glättwerkswalze *f*; *s.* r. of paper; *(s.a. unter roller)*
to make into a r. *(Pap)* aufwickeln, aufrollen
r. bars (blades) *(Pap)* Walzenmesser *npl (beim Holländer)*
r. boiling *(Text)* Naßdekatur *f*, Heißwasserdekatur *f*
r. coater Walzenstreichmaschine *f*, Walzenauftragmaschine *f*
r. compound *(Gum)* Walzenmischung *f*, Mischung *f* für Walzen
r. crown Walzenbombage *f*, Bombage (Balligkeit) *f* der Walze
r. crusher Walzenbrecher *m*
r. discharge Walzenabnahme *f*
r. doctor Schaber *m*, Abnahmeschaber *m (an einer Walze)*
r. feeder Walzenspeiser *m*, Aufgabewalze *f*
r. film Rollfilm *m*
r. finished paper Rollenpapier *n*
r. loading Walzenbelastung *f*
r. nip Berührungslinie *f* der Walzen
r. of paper Papierrolle *f*, Papiertambour *m*, Tambour *m*, Tambourrolle *f*
r. pressure Walz[en]druck *m*
r. scale Walz[en]zunder *m*, Walzsinter *m*
r.-slitting machine *(Pap)* Umroller *m*, Rollenschneidmaschine *f*, Rollenschneider *m*, Längsschneidemaschine *f*, Längsschneider *m*
r. stock Papierrollen *fpl*
r. sulphur Stangenschwefel *m*
r. surface Walzenoberfläche *f*
r. train Rollgang *m (im Walzwerk)*
r.-type briquette (briquetting) machine Walzenbrikett[ier]presse *f*, Brikettwalzenpresse *f*
r. wrapping *(Pap)* Verpackung *f* der Papierrollen
rolled glass Walzglas *n*, Gußglas *n*
r. laminated tube gewickeltes Schichtstoffrohr *n*
r. oats Haferflocken *fpl*
r. sheet *(Plast)* Walzfell *n*

r. tube gewickeltes Rohr *n*
roller Walze *f*, Rolle *f*; Zylinder *m*; *(Glas)* Fensterglaszylinder *m* *(beim Walzenblasverfahren)*; *(s.a. unter roll)*
r.-and-bowl mill Schüsselmühle *f*
r. bearing Rollenlager *n*, Walzenlager *n*
r. bit Rollenmeißel *m* *(besonders zum Erdölbohren)*
r. chain Rollenkette *f*
r.-dried milk powder Walzenmilchpulver *n*, walzengetrocknete Milch *f*
r. dryer Walzentrockner *m*, Zylindertrockner *m*, Kurzschleifentrockner *m*
r. drying Walzentrocknung *f*
r. gin *(Text)* Walzenegreniermaschine *f*
r. lehr *(Glas)* Rollenkühlofen *m*
r. milk powder *s.* r.-dried milk powder
r. mill 1. Walzenmühle *f*, Walzenstuhl *m*; 2. Wälzmühle *f*, Rollmühle *f*, Ringmühle *f* *(i.w.S.)*; 3. *s.* ring-roll mill
r. pair Walzenpaar *n*
r. press Walzenpresse *f*
r. printing *(Text)* Rouleauxdruck *m*, Walzendruck *m*
r. printing machine *(Text)* Rouleauxdruckmaschine *f*, Walzendruckmaschine *f*
r. stretching machine *(Plast)* Rollenreckmaschine *f*
r.-type churn Rollbutterfertiger *m*
r. vat Rollenkufe *f*
rolling Rollen *n*; Walzen *n*; Aufrollen *n* *(einer Kautschukmischung)*
r. bearing Wälzlager *n*
r. crusher Walzenbrecher *m*
r. machine Kalander *m*; *(Glas)* Walzmaschine *f*
r. mill Walzwerk *n*
r.-mill stand Walzgerüst *n*
Roman cement hochhydraulischer Kalk *m*, Romanzement *m*, Romankalk *m*
roméine, roméite *(Min)* Roméit *m*
röntgen *s.* roentgen
röntgenize / to *s.* to roentgenize
roof Deckschicht *f*, Decke *f*
r. rock Deckgebirge *n*
roofing board (felt, paper) Dachpappe *f*
r. tile Dachziegel *m*
room cure *(Gum)* Vulkanisation *f* bei Raumtemperatur, Raumtemperaturvernetzung *f*
r. temperature Zimmertemperatur *f*, Raumtemperatur *f*, ZT
r.-temperature vulcanization *s.* r. cure
r.-temperature-vulcanizing bei Zimmertemperatur vulkanisierend, kaltvulkanisierend, kalthärtend
root diffusate Wurzeldiffusat *n*, Wurzelabscheidung *f*
r. excretion (exudate) Wurzelausscheidung *f*
r. hair Wurzelhaar *n*
r.-mean-square velocity mittlere quadratische Geschwindigkeit *f*, Wurzel *f* aus dem mittleren Geschwindigkeitsquadrat

r. rot Wurzelfäule *f*, Wurzelschwamm *m*
Roots blower Roots-Gebläse *n*, Wälzkolbengebläse *n*
rope *(Text)* Strang *m*
r. brown *s.* r. wrapping
r. carrier *(Pap)* Einführungsseil *n*, Führungsseil *n*
r. dyeing *(Text)* Strangfärben *n*, Färben *n* im Strang (in Strangform)
r. form *(Text)* Strangform *f*
r. saturator *(Text)* Strangimprägniermaschine *f*
r. scouring *(Text)* Strangwäsche *f*, Waschen *n* im Strang (in Strangform)
r.-scouring machine *(Text)* Strangwaschmaschine *f*
r. wrapping Tauen[pack]papier *n*
roper Neoprenverseilmaschine *f*, Verseilmaschine *f*
ropiness Klebrigkeit *f*, Zähigkeit *f*, Zähflüssigkeit *f*, Dickflüssigkeit *f*
ropy klebrig, zäh[e], fadenziehend, leimig, schleimig; kahmig *(Wein)*
r. fermentation schleimige Gärung *f*, Schleimgärung *f*
r. wine zäher (kahmiger) Wein *m*
rosaniline Rosanilin *n*, 4,4′,4″-Triamino-3-methyl-triphenylkarbinol *n*; *s.* r. hydrochloride
r. dye Rosanilinfarbstoff *m*, Fuchsinfarbstoff *m*
r. hydrochloride Rosanilinhydrochlorid *n*, Rosanilin[chlorhydrat] *n*, Fuchsin *n*, Magenta *n*
roscoelite *(Min)* Roscoelit *m*, Vanadinglimmer *m*
rose dammar Rosendammar *n* *(von Vatica rassak Blume)*
r. oil Rosenöl *n*
r. quartz *(Min)* Rosenquarz *m* *(Silizium(IV)-oxid)*
r. water Rosenwasser *n*
r. wine Schillerwein *m*, Schieler *m*
Rose crucible Schmelztiegel *m* nach Rose, Rose-Tiegel *m*
roselite *(Min)* Roselith *m* *(Kalziumkobalt(II)-orthoarsenat)*
rosellane *(Min)* Rosellan *m*
Rosenmund reaction (reduction) Rosenmund-Reaktion *f*, Rosenmund-Saizew-Reduktion *f*, Rosenmund-Saizews Säurechloridreduktion *f*
Rosenstiehl's green Rosenstiehls (Kasseler, Böttgers) Grün *n*, Mangangrün *n* *(Bariummanganat)*
roseocobaltic chloride $[Co(NH_3)_5H_2O]Cl_3$ Roseokobaltchlorid *n*, Aquopentamminkobalt(III)-chlorid *n*
Rose's alloy *s.* Rose's metal
Rose's crucible *s.* Rose crucible
Rose's metal Rose[sche]s Metall *n*, Legierung *f* nach Rose
rosin Terpentinharz *n*, Geigenharz *n*, Kolophonium *n*
r. acid Harzsäure *f*
r. oil Terpentinharzöl *n*, Harzöl *n* *(aus trockener Destillation von Kolophonium)*
r. size *(Pap)* Harzleim *m*

r. sizing *(Pap)* Harzleimung *f*
r. soap Harzseife *f*
r. spirit Harzgeist *m*, Harzessenz *f*, Harzsprit *m*, Terpentinessenz *f*, Pinolin *n*
r.-wax emulsion *(Pap)* Harz-Paraffin-Emulsion *f*
r.-wax size *(Pap)* Harz-Paraffin-Leim *m*
Rosin-Rammler exponential law Rosin-Rammlersches Verteilungsgesetz (Kornverteilungsgesetz) *n*
rosinol *s.* rosin oil
rosite *(Min)* Rosit *m*
rosolic acid Rosolsäure *f (Methylderivat des Aurins)*; Aurin *n*, Rosolsäure *f*
ross / to *(Pap) (Am)* schälen, entrinden
Ross effect *(Foto)* Ross-Effekt *m*
rossing *(Pap) (Am)* Schälen *n*, Schälung *f*, Entrinden *n*, Entrindung *f*
rossite *(Min)* Rossit *m*
rot / to verrotten; *(Geol)* verwittern; *(Flachs)* rösten, rotten
rot resistance Verrottungsbeständigkeit *f*, Verrottungsfestigkeit *f*, Fäulnisbeständigkeit *f*, Fäulnisfestigkeit *f*, Fäulniswidrigkeit *f*
r.-resistant verrottungsbeständig, verrottungsfest, fäulnisbeständig, fäulnisfest, fäulniswidrig
r.-resistant finish Verrottungsfestappretur *f*
r. steeping Eintauchen *n* von Baumwollgeweben in Wasser *(zum Beseitigen von Verunreinigungen vor dem Bleichen)*
rotameter Schwebekörpermesser *m*, Schwimmermesser *m*, Rotameter *n*, Rotamesser *m (Durchflußmesser)*
rotary *s.* r. kiln
r. annular column (tower) Kolonne *f* mit rotierendem Zylinder
r. atomizer *s.* rotating atomizer
r. blower Rotationsgebläse *n*, Umlaufkolbengebläse *n*
r. boiler *s.* r. digester
r. burner rotierender Verbrennungsofen *m*, Dreh[rohr]ofen *m*
r. cement kiln Zementdrehofen *m*
r. column Rotationskolonne *f*
r. compressor Rotationsverdichter *m*, Rotationskompressor *m*, rotierender Verdichter *m*
r. crusher Glockenmühle *f*, Kegelmühle *f*
r.-cup atomizer Sprühkorb *m*, Zentrifugalkorb *m*, Düsenkorb *m*
r. cutter Rotorschneidmaschine *f*, Schneidegranulator *m*, Hackapparat *m*; *(Pap)* Querschneider *m* mit rotierenden Messern, rotierender Querschneider *m*, Messerwellenquerschneider *m*, Rotationsquerschneider *m*
r. cutting *(Pap)* Querschneiden *n* mit rotierenden Messern
r. digester *(Pap)* rotierender Kocher *m*, Drehkocher *m*
r.-disk contactor (extractor, tower) *s.* rotating-disk contactor
r. drilling Drehbohren *n*, Rotarybohren *n*

r.-drilling installation Rotary[bohr]anlage *f*
r.-drilling method Drehbohrverfahren *n*, Rotary[bohr]verfahren *n*
r. drum Rotationstrommel *f*
r.-drum ammoniator Ammonisier-Granuliertrommel *f (Düngemittelindustrie)*
r. dryer Drehtrommeltrockner *m*, Trommeltrockner *m*, Röhrentrockner *m*
r. filter Drehfilter *n*
r. furnace *s.* r. kiln
r. grinder *(Pap)* Ringschleifer *m*
r. hand duster *(Landw)* tragbarer Verstäuber *m (mit handbetriebenem Ventilator)*
r. hose Spülschlauch *m (einer Rotarybohranlage)*
r. jacketed-shelf dryer Heiztellertrockner *m*
r. joint Drehdurchführung *f*
r. kiln Drehofen *m*, rotierender Ofen *m*, Reaktionsdrehofen *m*; Dreh[rohr]ofen *m*
r.-kiln shell Drehofenmantel *m*, Ofenmantel *m (des Drehofens)*
r. kneading table Tellerkneter *m*, Knetteller *m*
r. knife cutter *s.* r. cutter
r. machine Rotarybohrtisch *m*, Drehtisch *m*
r.-machine drive Drehtischantrieb *m*
r. method *s.* r.-drilling method
r. milling machine *(Text)* Zylinderwalke *f*, Rotationswalkmaschine *f*
r. motion rotierende (drehende) Bewegung *f*, Drehbewegung *f*
r.-press ink Rotationsdruckfarbe *f*
r. pump Rotationspumpe *f*, rotierende Pumpe *f*, Umlaufkolbenpumpe *f*, Kolbenpumpe *f* mit rotierendem Kolben
r. rig *s.* r.-drilling installation
r. screen Rundsortierer *m*, Rundsieb *n*, Drehsieb *n*; *(Pap)* Drehknotenfänger *m*
r. screen printing *(Text)* Rotationsfilmdruck *m*
r. still Rotationskolonne *f*, Destillierapparat *m* mit rotierender Verdampferfläche
r. strainer *(Pap)* Drehknotenfänger *m*, rotierender Knotenfänger *m*
r. sulphur burner rotierender Schwefel[verbrennungs]ofen *m*
r. system Rotarysystem *n*
r. table Rotarybohrtisch *m*, Drehtisch *m*
r.-table press *(Ker)* Drehtischpresse *f*
r. vacuum drum-type save-all *(Pap)* Vakuumzellenfilter *n*, Saugzellenfilter *n (Filterstoffänger)*
r. vacuum dryer Vakuumschaufeltrockner *m*
r. vacuum filter Vakuumdrehfilter *n*, Vakuumtrommelfilter *n*
r.-vane feeder Zellenradzuteiler *m*, Zellenradzuteileinrichtung *f*
rotatability Drehbarkeit *f*
rotatable drehbar
rotate / to 1. drehen; schwenken; 2. sich drehen, rotieren, umlaufen
rotating atomizer rotierender Zerstäuber *m*, Rotationszerstäuber *m*; rotierender Versprüher *m*, Rotationsversprüher *m*, rotierende Düse *f*

r. converter rotierender (umlaufender, drehbarer) Konverter *m*
r.-core column Kolonne *f* mit rotierendem Zylinder
r. crystal rotierender (gedrehter) Kristall *m*, Drehkristall *m*
r.-crystal camera Drehkristallkamera *f*
r.-crystal diagram *s.* r.-crystal photograph
r.-crystal method Drehkristallverfahren *n*, Drehkristallmethode *f*
r.-crystal photograph Drehkristallaufnahme *f*, Drehkristalldiagramm *n*
r. cylinder rotierender Zylinder *m*; Drehrohr *n*, Drehtrommel *f (eines Trommeltrockners)*
r.-cylinder method rotationsviskosimetrisches Verfahren *n*, Rotationsviskosimeterverfahren *n*
r.-cylinder viscosimeter Rotationsviskosimeter *n*
r.-cylinder viscosimeter of Couette Couette-Viskosimeter *n*, Couette-Apparat *m*, Rotationsviskosimeter *n* nach Couette
r. disk rotierende Scheibe *f*, Drehscheibe *f*
r.-disk atomizer Sprühscheibe *f*, Zerstäuberscheibe *f*, Zerstäubungsscheibe *f*, Zentrifugalteller *m*
r.-disk contactor (extractor) Kolonne (Säule) *f* mit rotierenden Scheiben, Drehscheibenkolonne *f*, Drehscheibenextraktor *m*, Drehscheibenextrakteur *m*
r. distributor Drehverteiler *m*, drehbarer Gichtverteiler *m*
r. grate Drehrost *m*
r. kiln *s.* rotary kiln
r. nozzle *s.* r. atomizer
r. spreader *(Plast)* rotierender Torpedo *m (bei Kolbenspritzgußmaschinen)*
r. strainer *s.* rotary strainer
r.-strip column Drehbandkolonne *f*
r. vacuum dryer Vakuumtaumeltrockner *m*
rotation Drehung *f*, Drehen *n*, Umdrehung *f*, Rotation *f*, Drehbewegung *f*
r. conche *(Lebm)* Umlaufkonche *f*
r. crystal *s.* rotating crystal
r.-disk contactor *s.* rotating-disk contactor
r. moulding *s.* rotational moulding
r. quantum number *s.* rotational quantum number
r. spectrum *s.* rotational spectrum
r. value Drehwert *m*
r.-vibration[al] spectrum Rotationsschwingungsspektrum *n*
rotational casting *(Plast)* Rotationsgießen *n*
r. dispersion Rotationsdispersion *f*, R.D.
r.-dispersion curve Rotationsdispersionskurve *f*
r. energy Rotationsenergie *f*
r. fine structure Rotationsfeinstruktur *f*
r. frequency Rotationsfrequenz *f*, Umlaufsfrequenz *f*
r. heat Rotationswärme *f*
r. isomer Rotationsisomer[es] *n*
r. isomerism Rotationsisomerie *f*, Konformationsisomerie *f*, Konstellationsisomerie *f*

r. level Rotationsniveau *n*
r. motion *s.* rotary motion
r. moulding *(Plast)* Rotationsformen *n*, Rotationspressen *n*
r. partition function Rotationsverteilungsfunktion *f*
r. quantum number Drehimpulsquantenzahl *f*, Rotationsquantenzahl *f*
r. spectrum Rotationsspektrum *n*
r. state Rotationszustand *m*
r. transition Rotationsübergang *m*
r. viscosimeter *s.* rotating-cylinder viscosimeter
rotatory contribution Rotationsanteil *m*, Drehwertanteil *m*
r. dispersion *s.* rotational dispersion
r. dryer *s.* rotary dryer
r. power Dreh[ungs]vermögen *n*
r. value *s.* rotation value
rotenone Rotenon *n (Insektizid)*
rotenonic acid Rotenonsäure *f*
rothoffite *(Min)* Rothoffit *m (Kalziumeisen(III)-triorthosilikat)*
rotogravure ink Rotationstiefdruckfarbe *f*, Druckfarbe *f* für Rotationstiefdruck
rotomoulding *s.* rotational moulding
rotor Rotor *m*; Rotor *m*, Kern *m*, Kegel *m*, Konus *m*, Kegelrotor *m (der Kegelstoffmühle)*; Knetschaufel *f (eines Innenmischers)*
r. disk Rotorscheibe *f (einer Drehscheibenkolonne)*
r. knife Rotormesser *n (einer Rotorschneidmaschine)*
r. separator Walzenscheider *m (zum Elektroscheiden)*
Rotor process [Oberhausener] Rotorverfahren *n*, Rotor-Stahlschmelzverfahren *n*, Rotor-Blasstahlverfahren *n*
rotproof verrottungsbeständig, verrottungsfest, fäulnisbeständig, fäulnisfest, fäulniswidrig
rotproofing agent Verrottungsschutzmittel *n*, Fäulnisverhütungsmittel *n*
rotproofness Verrottungsbeständigkeit *f*, Verrottungsfestigkeit *f*, Fäulnisbeständigkeit *f*, Fäulnisfestigkeit *f*, Fäulniswidrigkeit *f*
rotten lime *(Gerb)* Fauläscher *m*, fauler (toter) Äscher *m*, Stinkäscher *m*
rotting Verrotten *n*, Verrottung *f*; Rösten *n*, Rotten *n (von Flachs)*
röttisite *(Min)* Röttisit *m*
rouge Rouge *n*, Schminkrot *n*; *(Glas)* Polierrot *n*
rough rauh; herb *(Wein)*; grob *(Material)*
r.-cast glass Rohglas *n*
r.-cast plate Rohglasplatte *f*
r. glass *s.* r.-cast glass
r.-grind / to schroten *(Malz, Getreide)*
r. wood Rohholz *n*
roughen / to anrauhen, aufrauhen, rauh machen
rougher *(Flotationszelle zur Grobtrennung)*
roughing filter Grobfilter *n*, Vorfilter *n*
roughness Rauhigkeit *f*, Rauheit *f*
round-bottom flask Rundkolben *m*

r.-bottom four-neck flask Vierhalsrundkolben *m*
r.-bottom long-neck flask Langhalsrundkolben *m*
r.-bottom short-neck flask Kurzhalsrundkolben *m*
r.-bottom three-neck flask Dreihalsrundkolben *m*
r.-bottom two-neck flask Zweihalsrundkolben *m*
r.-bottom wide-neck flask Weithalsrundkolben *m*
r.-bottomed mit rundem (halbkugeligem) Boden
r.-bottomed flask s. r.-bottom flask
r. cell Zylinderzelle *f*, zylindrische Zelle *f*
r. clarifier Rundklärer *m*
r. filter Rundfilter *n*
r. kiln Rundofen *m*
r. of handlers (Gerb) Farbengang *m*
r. of lime (Gerb) Äschergang *m*
r. off / to [ab]runden
rouse / to [um]rühren; (z.B. Heringe) einsalzen
rouser Rührapparat *m*, Rührwerk *n*
routine analysis Routineanalyse *f*, Reihenanalyse *f*, Reihenuntersuchung *f*
Roux bottle Roux-Flasche *f*, Roux-Kolben *m*
Roux culture bottle (Bioch) Kulturkolben *m* nach Roux
roving 1. Roving *m*, Glaseidenstrang *m*; Roving *m*, Vorgarn *n*; 2. Vorspinnen *n*
r. fabric Rovinggewebe *n*
row dressing (Landw) Reihendüngung *f*
r. fertilizer distributor (Landw) Reihendüngerstreuer *m*
r. of cells Zellenreihe *f*
rowlandite (Min) Rowlandit *m*
royal jelly Weisel[zellen]futtersaft *m*, Gelée royale *f*
r. yellow As$_2$S$_3$ Königsgelb *n*, reines Auripigment *n*, Arsen(III)-sulfid *n*; Königsgelb *n*, Auripigment *n*, Operment *n*, Rauschgelb *n*, Chinagelb *n*, Gelbglas *n*, gelber Arsenik *m*, gelbes Schwefelarsen *n* (technisches Arsen(III)-sulfid)
RR acid RR-Säure *f*, 7-Amino-1-naphthol-3,6-disulfonsäure *f*
RTV s. room-temperature-vulcanizing
rub fastness Reibechtheit *f*
rubber / to gummieren
rubber Kautschuk *m(n)*; Gummi *m*
r. accelerator Vulkanisationsbeschleuniger *m*
r. annulus s. r. ring
r. articles s. r. goods
r. bag (Plast) Gummisack *m*
r.-bag moulding (Plast) Pressen *n* mit Gummisack
r. band Gummiband *n*; (Gum) Kautschukfell *n*
r.-bearing plant s. r.-yielding plant
r. belt Gummi[treib]riemen *m*; (endloses) Gummiband *n*
r. belt conveyor Gurtbandförderer *m* mit Gummigurt
r. black s. r. carbon black

r.-brass bond (Gum) Messingbindung *f*
r. bung Gummistopfen *m*
r. carbon black Ruß *m* für kautschuktechnische Zwecke
r. carbon gel [complex] Rußgel *n*
r. cement Gummilösung *f*; Kautschuklösung *f*
r. chemist Kautschukchemiker *m*; Gummichemiker *m*
r. clay Kaolin *n* für kautschuktechnische Zwecke
r.-coat / to s. to r.-cover
r.-coated fabric gummiertes Gewebe *n*, gummierter Stoff *m*
r.-combined sulphur (Gum) gebundener Schwefel *m*
r. compound s. r. stock
r.-cover / to gummieren, mit Gummi überziehen (beziehen, bekleiden, beschichten), mit einem Gummimantel (Gummibezug) verkleiden
r. cover Deckgummi *m*, Gummideckplatte *f*
r.-covered roll gummierte Walze *f*, Gummiwalze *f*
r. derivative Kautschukderivat *n*
r. dibromide (C$_5$H$_8$Br$_2$)$_n$ Kautschukdibromid *n*
r. elasticity Kautschukelastizität *f*; Gummielastizität *f*
r. factory Gummiwerk *n*, Gummibetrieb *m*
r. filament Gummielementarfaden *m*
r.-filler gel Kautschuk-Füllstoff-Gel-Komplex *m*
r.-filler mixture (stock) Kautschuk-Füllstoff-Mischung *f*
r. from smallholdings (auf Kleinplantagen erzeugter Kautschuk)
r. gasket Gummidichtung *f*
r. goods Gummiwaren *fpl*, Gummiartikel *mpl*, Gummifabrikate *npl*
r. hose Gummischlauch *m*
r. hydrocarbon Kautschukkohlenwasserstoff *m*, Kautschuk-KW *m*
r.-hydrocarbon content Kautschukgehalt *m*, Gehalt *m* an Kautschukkohlenwasserstoff
r. hydrochloride Kautschukhydrochlorid *n*, Hydrochlorkautschuk *m*
r. hydrofluoride Kautschukhydrofluorid *n*, Hydrofluorkautschuk *m*
r. industry Kautschukindustrie *f*; Gummiindustrie *f*
r.-insulated gummiisoliert
r. latex Kautschuklatex *m*, Kautschukmilchsaft *m*, Kautschukmilch *f*
r.-like kautschukartig, kautschukähnlich
r. machinery Kautschukmaschinen *fpl*
r.-manufacturing industry s. r. industry
r.-manufacturing plant s. r. factory
r. milk s. r. latex
r. mix[ture] s. r. stock
r. molecule Kautschukmolekül *n*
r. packing Gummidichtung *f*, Gummipackung *f*
r.-pigment mixture s. r.-filler mixture
r. plant Gummiwerk *n*, Gummibetrieb *m*; s. r.-yielding plant

r. plantation Kautschukplantage *f*
r. poison Kautschukgift *n*
r. policeman Gummiwischer *m (für Rührstäbe)*
r. powder Kautschukpulver *n*
r.-producing plant *s.* r.-yielding plant
r. products *s.* r. goods
r. ring Gummiring *m*, Gummimanschette *f*
r.-roll fleshing machine *(Gerb)* Walzenent-
fleischmaschine *f*
r. scrap Abfallgummi *m*, Altgummi *m*, Gummiabfälle *mpl*, Vulkanisatabfälle *mpl*; *(unvulkani-*
sierte) Fabrikationsabfälle *mpl*
r. seal *s.* r. packing
r. sheet Kautschukfell *n*
r.-soluble kautschuklöslich
r. solution Kautschuklösung *f*; Gummilösung *f*
r. solvent Kautschuklösungsmittel *n*; Gummi-
lösungsmittel *n*
r. stamp *(Ker)* Gummistempel *m*
r.-stamp mark *(Pap)* Molettewasserzeichen *n*
r. stock Kautschukmischung *f*
r. stopper Gummistopfen *m*, Gummistöpsel *m*
r. substitute Ölkautschuk *m*, *(veraltet für)* Faktis
m
r.-sulphur blend (compound, mixture, stock)
Kautschuk-Schwefel-Mischung *f*
r. technologist Kautschuktechnologe *m*;
Gummitechnologe *m*
r. technology Kautschuktechnologie *f*; Gummitechnologie *f*
r. thread Gummifaden *m*; Gummifäden *mpl*
r.-tipped glass rod [auf einen Glasstab ge-
steckter] Gummiwischer *m (Laborgerät)*
r.-to-metal bonding (weld) Gummi-Metall-
Verbindung *f*
r. tree Kautschukbaum *m*; Parakautschukbaum
m, Hevea brasiliensis (H.B.K.)Muell. Arg.
r. tube Gummirohr *n*, Gummischlauch *m*
r. tubing Gummischlauch *m*; Gummischlauch-
material *n*, Gummischläuche *mpl*
r.-tubing clamp Schlauchschelle *f*
r. vulcanizate Kautschukvulkanisat *n*
r. waste *s.* r. scrap
r.-yielding plant kautschukliefernde (kautschuk-
bildende, kautschukführende, kautschukhaltige)
Pflanze *f*, Kautschukpflanze *f*, Kautschukge-
wächs *n*, Kautschukträger *m*
rubbered fabric *s.* rubberized fabric
rubberize / to gummieren
rubberized fabric gummiertes Gewebe *n*, gummierter Stoff *m*
r. hair Gummihaar *n*
rubberizing Gummieren *n*, Gummierung *f*
rubbery gummiartig
rubbing fastness Reibechtheit *f*
rubeanic acid $H_2NSCCSNH_2$ Rubeanwasserstoffsäure *f*, Rubeanwasserstoff *m*, Dithiooxalsäurediamid *n*, Dithiooxamid *n*
rubellite *(Min)* Rubellit *m*
ruberythric acid Ruberythrinsäure *f*, Alizarinprim-
verosid *n*

rubiadin Rubiadin *n*, 1,3-Dihydroxy-2-methylan-
thrachinon *n*
rubicelle *(Min)* Rubicell *m (Magnesiumaluminat)*
rubicene Rubizen *n*, Benz[a]indeno[1,2,3-hi]azean-
thrylen *n*
rubidic Rubidium...
rubidium Rb Rubidium *n*
  r. acid carbonate $RbHCO_3$ Rubidiumhydrogenkarbonat *n*
  r. acid sulphate $RbHSO_4$ Rubidiumhydrogensul-
fat *n*
  r. alum *s.* r.-aluminium sulphate
  r.-aluminium sulphate $RbAl(SO_4)_2 \cdot 12H_2O$ Ru-
bidiumaluminiumsulfat-12-Wasser *n*, Rubi-
dium[aluminium]alaun *m*
  r. bicarbonate $RbHCO_3$ Rubidiumbikarbonat *n*,
(veraltet für) Rubidiumhydrogenkarbonat *n*
  r. carbonate Rb_2CO_3 Rubidiumkarbonat *n*
  r. chlorate $RbClO_3$ Rubidiumchlorat *n*
  r. chloride RbCl Rubidiumchlorid *n*
  r. chloroplatinate $Rb_2[PtCl_6]$ Rubidiumhexa-
chloroplatinat(IV) *n*
  r. chromate Rb_2CrO_4 Rubidiumchromat *n*
  r. dichromate $Rb_2Cr_2O_7$ Rubidiumdichromat(VI)
n
  r. fluoride RbF Rubidiumfluorid *n*
  r. hydrate *s.* r. hydroxide
  r. hydroxide RbOH Rubidiumhydroxid *n*
  r. iodate $RbJO_3$ Rubidiumjodat *n*
  r. iodide RbJ Rubidiumjodid *n*
  r. nitrate $RbNO_3$ Rubidiumnitrat *n*
  r. perchlorate $RbClO_4$ Rubidiumperchlorat *n*
  r. periodate $RbJO_4$ Rubidium[meta]perjodat *n*,
Rubidiumtetroxoperjodat *n*
  r. permanganate $RbMnO_4$ Rubidiumpermanga-
nat *n*, Rubidiummanganat(VII) *n*
  r. peroxide Rb_2O_2 Rubidiumperoxid *n*
  r. salt Rubidiumsalz *n*
  r. sulphate Rb_2SO_4 Rubidiumsulfat *n*
rubin[e] Rubin[rot] *n*
  r. number Rubinzahl *f (zur Messung der*
Schutzkolloidwirkung)
rubremetine Rubremetin *n*, Rubremetinium-
hydroxid *n*, Dehydroemetin *n*, Pseudobase *f* der
Rubremetiniumsalze
rubremetinium Rubremetinium *n*
  r. chloride Rubremetiniumchlorid *n*
  r. salt Rubremetiniumsalz *n*
ruby *s.* ruby-red
ruby *(Min)* Rubin *m*; Rubin[rot] *n*
  r. arsenic As_2S_2 Rubinschwefel *m*, Rauschrot *n*,
Sandarach *n*, Arsen(II)-sulfid *n*
  r. glass Rubinglas *n*
  r. number *s.* rubin[e] number
  r.-red rubinrot, rubinfarben, rubinfarbig
  r. red Rubin[rot] *n*
  r. spinel *(Min)* Rubinspinell *m*
ruddle *(Min)* Rötel *m*
Ruff degradation Ruffscher Abbau *m (der Zucker)*
ruffle fat Gekrösefett *n*, Nickerfett *n*

rule of maximum multiplicity [Hundsche] Regel *f* der größten Multiplizität, [Hundsches] Prinzip *n* der größten Multiplizität, Hundsches Multiplizitätsprinzip (Prinzip) *n*, [erste] Hundsche Regel *f*
  r. of thumb Faustregel *f*
rumble Rumpeln *n*, Rumbling *n* (*rumpelndes Verbrennungsgeräusch in großvolumigen Motoren*)
run / to 1. fließen, strömen, laufen, rinnen; auslaufen (*Farbe*); schmelzen, sich verflüssigen (*Metall*), tauen (*Eis*); laufen, undicht (leck) sein (*Gefäß*); laufen, in Gang (Betrieb) sein, arbeiten (*Maschine*); gehen, funktionieren (*Gerät*); 2. (*eine Flüssigkeit*) laufen (fließen, ausströmen) lassen; (*z.B. eine Nitrierung*) durchführen; (*einen Versuch*) anstellen, laufen lassen; destillieren; (*Gerbflüssigkeit in den Gruben*) umpumpen; (*ein Metall*) schmelzen; (*z.B. Wasser*) führen (*von einer Leitung*); (*eine Maschine*) fahren, laufen lassen, in Gang halten, bedienen; (*einen Betrieb*) führen, leiten
  r. a blank eine Blindprobe anstellen
  r. back zurücklaufen
  r. down herunterrieseln
  r. dry leerlaufen lassen
  r. in zulaufen (einlaufen, zufließen, einfließen) lassen, zulassen; zutropfen lassen
  r. off ablaufen lassen, ablassen; abziehen (*z.B. eine Schicht*)
  r. on aufgeben
  r. on to ice auf Eis geben
  r. out auslaufen (*z.B. von Farbe*)
  r. wild außer Kontrolle geraten
run Lauf *m* (*z.B. des Papiermaschinensiebs*); Versuch *m*; Destillation *f*; Gasen *n*, Gasung *f*, Kaltblasen (*bei der Wassergasgewinnung*); (*Gerb*) Elastizität *f*, Nachgiebigkeit *f* (*des Leders*); Strang *m* (*eines endlosen Riemens*)
  r. honey Schleuderhonig *m*
  r.-of-mine coal Grubenkohle *f*, Förderkohle *f*, Rohförderkohle *f*
  r.-of-mine ore Fördererz *n*, [aus dem Bergwerk kommendes] Roherz *n*
  r. of the wire (*Pap*) Sieblauf *m*
  r. part Produktionsphase *f* (*eines Schwelgenerators*)
rundown pipe Ablaufrohr *n*
runnability Laufeigenschaften *fpl*, Verarbeitbarkeit *f* (*des Papiers*)
runner (*Met*) Abstichrinne *f*, Sandrinne *f*, Rinne *f*; (*Plast*) Angußverteiler *m*; s. r. stone
  r. stone Koller[gang]stein *m*, Läufer[stein] *m*
runnerless mould (*Plast*) Spritzgießwerkzeug *n* für angußloses Spritzziehen, angußloses Spritzgießwerkzeug *n*
running Lauf *m*, Laufen *n*; Destillat *n*, Destillationsprodukt *n*
  r. direction (*Pap*) Laufrichtung *f*
  r. of writing ink (*Pap*) Auslaufen *n* der Tinte

r.-out machine (*Ker*) Strangpresse *f* (*z.B. zur Herstellung von Röhren*)
r. time Laufzeit *f* (*z.B. eines Chromatogramms*)
r. water fließendes Wasser *n*
runoff Ablauf *m*, Auslauf *m*, Abfluß *m*
  r. cock Ablaßhahn *m*
  r. gutter for wet cooling Duschrinne *f* (*Margarineherstellung*)
  r. line Ablaufleitung *f*
  r. pipe Ablaufrohr *n*; Ablaufstutzen *m*, Entleerungsstutzen *m*
  r. syrup (*Zucker*) Ablauf *m*
rupture / to [auf]sprengen, spalten
rupture Sprengung *f*, Aufsprengung *f*, Spaltung *f*, Bruch *m*
  r. disk Berstscheibe *f*, Reißscheibe *f*, Platzscheibe *f*
  r. of the web of paper (*Pap*) Reißen *n* der Papierbahn, Bahnriß *m*
Russel effect (*Foto*) Russel-Effekt *m*
Russell-Saunders coupling Russell-Saunderssche Kopplung *f*, Russell-Saunders-Kopplung *f*, LS-Kopplung *f*
Russian dandelion Kok-Saghys *m*, Kautschuklöwenzahn *m*, Taraxacum bicorne Dahlst.
  R. leather Juchtenleder *n*
rust Rost *m*
  r. cement Eisenkitt *m*, Rostkitt *m*
  r. film Rostfilm *m*, dünne Rostschicht *f*
  r. formation Rostbildung *f*
  r.-inhibiting rostinhibierend, rostschützend, rostverhindernd, rostverhütend
  r.-inhibiting grease rostschützendes Fett *n*, Rostschutzfett *n*
  r.-inhibiting oil rostschützendes Öl *n*, Rostschutzöl *n*
  r. inhibition Rostschutz *m*, Rostverhinderung *f*, Rostverhütung *f*
  r. inhibitor Rostinhibitor *m*, Rostschutzmittel *n*, Rostverhinderer *m*, Rostverhütungsmittel *n*
  r. layer Rostschicht *f*
  r. preventer s. r. inhibitor
  r.-preventing s. r.-inhibiting
  r.-preventing additive Rostschutzadditiv *n*, rostverhinderndes (rostverhütendes) Additiv *n*, Antirostadditiv *n*
  r.-preventing agent (medium) s. r. inhibitor
  r. prevention s. r. inhibition
  r.-preventive s. r.-inhibiting
  r. preventive s. r. inhibitor
  r. protection s. r. inhibition
  r.-protective s. r.-inhibiting
  r.-protective paint Rostschutzfarbe *f*, rostschützende Anstrichfarbe *f*
  r. remover s. r.-removing agent
  r.-removing rostentfernend
  r.-removing agent Rostentfernungsmittel *n*, Entrostungsmittel *n*, Rostentferner *m*
  r. resistance Rostbeständigkeit *f*
  r.-resistant, r.-resisting rostbeständig

r.-resisting paint s. r.-protective paint
r. spot Rostfleck m, (Met auch) Roststelle f
rustless rostfrei, nichtrostend, rostbeständig, rostsicher (z.B. Stahl)
rustling finish knirschende Appretur f (mit Seidengriff), Knirschgriffappretur f
ruthenate Ruthenat n
ruthenic Ruthenium...; höherwertigem Ruthenium entsprechend
  r. chloride RuCl$_4$ Ruthenium(IV)-chlorid n, Rutheniumtetrachlorid n
  r. oxide s. ruthenium dioxide
ruthenious Ruthenium...; niederwertigem Ruthenium entsprechend
  r. chloride RuCl$_2$ Ruthenium(II)-chlorid n, Rutheniumdichlorid n
ruthenium Ru Ruthen[ium] n
  r. dioxide RuO$_2$ Rutheniumdioxid n, Ruthenium(IV)-oxid n
  r. hydroxide Ru(OH)$_3$ Ruthenium(III)-hydroxid n
  r. nonoxide Ru$_4$O$_9$ Rutheniumnonoxid n, Ruthenium(IV,V)-oxid n
  r. red Ru$_2$(OH)$_2$Cl$_4$·7NH$_3$·3H$_2$O Ruthen[ium]rot n, Hydroxoamminrutheniumchlorid n
  r. sesquioxide Ru$_2$O$_3$ Ruthenium(III)-oxid n
ruthenous s. ruthenious
Rutherford atom [model] Rutherfordsches Atommodell n, Kernmodell n des Atoms
rutherfordine (Min) Rutherfordin m (Uranylkarbonat)
ruthile s. rutile
rutile (Min) Rutil m (Titan(IV)-oxid)
  r. body (Ker) Rutilmasse f
  r. modification nach dem Rutiltyp kristallisierende Modifikation f
  r. porcelain Rutilporzellan n
rutilite s. rutile
rutin Rutin n, Violaquerzitrin n, Querzetin-3-rutinosid n, Myrtikolorin n
rutinose Rutinose f, 6-(β-L-Rhamnopyranosyl)-D-glukopyranose f
R.V.P. s. Reid vapour pressure
Rydberg constant (number) Rydberg-Konstante f, Rydbergsche Konstante f, Rydberg-Zahl f
rye ergot Roggenmutterkorn n
  r. flour (meal) Roggenmehl n
  r. middlings Roggenkleie f
  r. starch Roggenstärke f

S

s. s. 1. soluble; 2. solubility
s-RNA s. soluble ribonucleic acid
S. s. sized
S acid S-Säure f, 1-Amino-8-naphthol-4-sulfonsäure f
2S acid 2S-Säure f, SS-Säure f, Chicagosäure f, 1-Amino-8-naphthol-2,4-disulfonsäure f

S curve S-Kurve f (im Zeit-Temperatur-Umwandlungsdiagramm)
S finishing (Text) S-Finish n, oberflächliches Verseifen n
S$_N$ reaction S$_N$-Reaktion f, nukleophile (anionoide) Substitution f
S$_N$1 reaction S$_N$1-Reaktion f (monomolekulare nukleophile Substitution)
S$_N$2 reaction S$_N$2-Reaktion f (bimolekulare nukleophile Substitution)
S twist (Gum) S-Drehung f, Linksdraht m
S value (Bodenkunde) S-Wert m (Summe austauschbarer Basen)
sabadilla seeds Sabadillsamen mpl (Insektizid, von Schoenocaulon officinale (Schl. et Ch.)A. Gray)
Sabatier-Senderens reaction (reduction) Sabatier-Senderenssche Reaktion (Reduktion) f
Sabattier effect (Foto) Sabattier-Effekt m
sabinaketone Sabinaketon n, 1-Isopropylbizyklo-[0,1,3]-4-hexanon n
sabinane Sabinan n, Thujan n, 1-Isopropyl-4-methylbizyklo-[0,1,3]-hexan n
sabinene Sabinen n, 1-Isopropyl-4-methylenbizyklo-[0,1,3]-hexan n
sabinic acid Sabininsäure f, 12-Hydroxydodekansäure f
sabinol Sabinol n, 1-Isopropyl-4-methylen-bizyklo-[0,1,3]-3-hexanol n
SAC (Abk. für) Society for Analytical Chemistry
saccharase Sa[c]charase f, Invertase f, Invertin n (rohrzuckerspaltendes Ferment)
saccharate Sa[c]charat n
  s. of lime Kalksa[c]charat n, Kalziumsa[c]charat n
  s. process (Zucker) Sa[c]charatverfahren n
saccharic acid Zuckersäure f; HOOC(CHOH)$_4$COOH Zuckersäure .f, Glukozuckersäure f, (i.e.S.) D-Zuckersäure f, Zuckersäure f, D-Glukozuckersäure f
saccharide Sa[c]charid n, Kohle[n]hydrat n
sacchariferous zuckerhaltig, zuckerliefernd, zuckerspeichernd, zuckererzeugend
saccharification Verzuckerung f, Sa[c]charifikation f, Sa[c]charifizierung f
  s. of starch Stärkeverzuckerung f (durch Diastase)
  s. period (rest, time) Verzuckerungszeit f, Verzuckerungsrast f
saccharifier Verzuckerungsgerät n, Verzuckerungsvorrichtung f, Verzuckerungsapparat m
saccharify / to in Zucker verwandeln, verzuckern, sa[c]charifizieren; zuckern, süßen
saccharimeter Sa[c]charimeter n (ein Polarimeter zur Bestimmung des Zuckergehalts)
saccharimetry Sa[c]charimetrie f
saccharin Sa[c]charin n, o-Benzoesäuresulfimid n, 1,2-Benzisothiazolin-3-on-1,1-dioxid n
saccharine zuckerhaltig, zuckerartig, Zucker...; zuckern, zuck[e]rig, zuckersüß
saccharine s. saccharin

saccharinic acid Saccharinsäure *f*, *(i.e.S.)* $H_2COH \cdot CH(OH) \cdot CH(OH) \cdot C(OH)(CH_3) \cdot COOH$ Saccharinsäure *f*, 2,3,4,5-Tetrahydroxyhexandisäure *f*
saccharization *s.* saccharification
saccharize / to *s.* to saccharify
saccharobiose *s.* saccharose
saccharogenic amylase Saccharogenamylase *f*
saccharolactic acid HOOC(CHOH)$_4$COOH Galaktozuckersäure *f*, Galaktarsäure *f*, Schleimsäure *f*, Muzinsäure *f*, Tetrahydroxyadipinsäure *f*
saccharometer Sa[c]charometer *n* *(eine Senkspindel zur Bestimmung der Dichte einer Zuckerlösung)*
s. for grapes Mostwaage *f*, Mostaräometer *n*, Oechsle-Waage *f*
saccharometry Sa[c]charometrie *f*
saccharomycete Saccharomyzet *m* *(eine askosporenbildende Hefe)*
saccharose Sa[c]charose *f*, Rohrzucker *m*, Rübenzucker *m*, Sucrose *f*
Sachse-Mohr concept of strainless rings Sachse-Mohrsche Theorie *f* [spannungsfreier Ringe], Sachse-Mohrsche Spannungstheorie *f*, Sachse-Mohr-Theorie *f*
Sachsse solution Sachssesche Lösung *f* *(zur oxydimetrischen Zuckerbestimmung)*
sacred bark *(Pharm)* Cascara-sagrada-Rinde *f*, Kaskararinde *f*, Sagrada[rinde] *f*, Amerikanische Faulbaumrinde *f* *(von Rhamnus purshianus DC.)*
sacrificial anode Opferanode *f*, Aktivanode *f*
s. protection katodischer Schutz (Korrosionsschutz) *m* durch (mit) Opferanoden
saddle Sattel[füll]körper *m*; *(Ker)* Dreikant *m*, dreieckige Leiste *f* *(Brennhilfsmittel)*
s. packing Sattel[füll]körper *m*
SAF black *s.* super abrasion furnace black
safe for factory processing verarbeitungssicher
safeguard Sicherung *f*, Schutz *m*, Wächter *m*
safelight Sicherheitslicht *n*
safener *(Zusatzstoff, der Reaktionen zwischen Bestandteilen von Wirkstoffgemischen verhindert)*
safety Sicherheit *f*; Unschädlichkeit *f*
s. cheque paper Sicherheitspapier *n*, Scheckpapier *n*
s. cut-off Sicherheitsabsperrventil *n*
s. engineer Sicherheitsingenieur *m*
s. explosive Sicherheitssprengstoff *m*, *(Bgb auch)* Wettersprengstoff *m*
s. factor Sicherheitsbeiwert *m*, Sicherheitsfaktor *m*
s. fuse Zündschnur *f*
s. glass Sicherheitsglas *n*, Schutzglas *n*
s.-glass interleaver Sicherheitsglas-Zwischenschicht *f*, Zwischenlage *f* im Sicherheitsglas
s. goggles Schutzbrille *f*
s. ink Sicherheitsfarbe *f* *(Druckfarbe für Sicherheitspapiere)*

s. match Sicherheitszündholz *n*
s. package Sicherheitspackung *f*
s. paper *s. s.* cheque paper
s. pipette Sicherheitspipette *f*
s. regulation Sicherheitsbestimmung *f*, Arbeitsschutzanordnung *f*
s.-relief valve *s. s.* valve
s. shower Notbrause *f*, Löschbrause *f*, Feuerlöschbrause *f*
s. valve Sicherheitsventil *n*
s. warning warnende Aufschrift (Belehrung) *f*, *(auf Giftbehältern aufgedruckte)* Vorsichtsmaßregel *f*
safflor *s.* safflower
s. red Saflorrot *n*, Karthamin *n*
safflorite *(Min)* Safflorit *m* *(Kobaltarsenid)*
safflower Saflor *m*, Färberdistel *f*, Carthamus tinctorius L.; Saf[f]lor *m*, Färberdistelblüten *fpl*; Saflorrot *n*, Karthamin *n*
s. oil Safloröl *n*, Carthamusöl *n* *(aus den Samen von Carthamus tinctorius L.)*
saffranine *s.* safranine
saffron Safran *m* *(von Crocus sativus L.)*
safranin[e] Safranin *n* *(ein Azinfarbstoff)*, *(i.e.S.) s.* safranin[e] T
safranin[e] O *s.* safranin[e] T
safranin[e] T Safranin T *n*, Tolusafranin *n*, Safranin *n*
safrol[e] Safrol *n*, 1,2-Methylendioxy-4-(2-propenyl)benzol *n*, 4-Allyl-1,2-methylendioxybenzol *n*
sag / to durchhängen, sich durchbiegen, durchsacken, absacken; [ab]laufen *(von Anstrichmitteln)*
sag Durchhang *m*, Durchbiegung *f*
sagenite *(Min)* Sagenit *m*
saggar *(Ker)* Kapsel *f*, Brennkapsel *f*, Schamottekapsel *f*
s. clay *(Ker)* Kapselton *m*
sagger *s.* saggar
sagging Durchhängen *n*, Durchhang *m*, Durchbiegen *n*, Durchbiegung *f*, Durchsacken *n*, Absacken *n*; Ablaufen *n*, Laufen *n*, Läuferbildung *f*, Gardinenbildung *f*, Vorhangbildung *f* *(von Anstrichmitteln)*
sago Sagopalme *f*, Metroxylon Rottb.; Sagostärke *f*, Palm[en]sago *m*, Palmenstärke *f*
sahlite *(Min)* Sahlit *m*, *(veraltet für)* Salit *m*
sal acetosella *s.* salt of sorrel
s. ammoniac *s.* salmiac
s. polychrestum K$_2$SO$_4$ Polychrestsalz *n*, *(veraltet für)* Kaliumsulfat *n*
s. soda *s.* salt of soda
s. volatile [NH$_4$]HCO$_3 \cdot$ [NH$_4$][CO$_2$NH$_2$] flüchtiges (englisches) Salz *n*, Laugensalz *n*, Geistersalz *n*, *(veraltet für)* Hirschhornsalz *n*, Ammoniumkarbonat *n* [des Handels]
salad oil Salatöl *n*
Saladin malting Kastenmälzerei *f* mit Saladin-Keimkasten, Saladin-Mälzerei *f*

Saladin steep Saladin-Weiche f *(bei der Kasten-mälzerei)*
salammonite *(Min)* Salammonit m, *(veraltet für)* Salammoniak n, [natürlicher] Salmiak m *(Ammoniumchlorid)*
salicin Salizin n, o-Hydroxybenzyl-β-glukosid n
salicyl s. salicyloyl
  s. alcohol s. saligenin
salicylaldehyde $C_6H_4(OH)CHO$ Salizylaldehyd m, o-Hydroxybenzaldehyd m, 2-Phenolmethylal n
salicylamide $HOC_6H_4CONH_2$ Salizyl[säure]amid n, o-Hydroxybenzamid n, 2-Hydroxybenzolkarbon-amid n
salicylanilide $HOC_6H_4CONHC_6H_5$ Salizyl[säure]-anilid n, N-Phenylsalizylamid n, 2-Hydroxy-N-phenylbenzolkarbonamid n
salicylate Salizylat n *(Salz oder Ester der Salizylsäure)*
salicylic acid $C_6H_4(OH)COOH$ Salizylsäure f, Spir[oyl]säure f, 2-Hydroxybenzoesäure f
  s.-acid test Prüfung f auf Salizylsäure
  s. aldehyde s. salicylaldehyde
salicylidene $=CHC_6H_4OH$ Salizyliden n
salicyloyl $-COC_6H_4(OH)$ Salizyloyl n, Salizyl[yl] n
O-salicyloylsalicylic acid $HOC_6H_4COOC_6H_4COOH$ O-Salizyloylsalizylsäure f, o-(Salizyloyloxy)-benzoesäure f
salicylsulfonic acid Salizylsäure-5-sulfonsäure f, Sulfosalizylsäure f
salicylyl s. salicyloyl
saliferous salzhaltig, salzführend
salifiable salzbildungsfähig
salification Salzbildung f
salify / to ein Salz (oder Salze) bilden; *(eine Säure oder Base)* in ein Salz überführen
saligenin $C_6H_4(OH)CH_2OH$ Saligenin n, Salizyl-alkohol m, 2-Hydroxybenzylalkohol m
salimeter s. salinimeter
salina Saline f, Salzsiederei f, Salzwerk n; Salzlager n; Salzsee m
salination Salzen n, Einsalzen n, Salzung f
saline salzig, salzhaltig, Salz...; salzartig; *(Med)* salinisch
saline Saline f, Salzsiederei f, Salzwerk n; Salzlager n; Salzsee m; *(Pharm)* physiologische Koch-salzlösung f
  s. hydride salzartiges Hydrid n
  s. soil *(nicht stark alkalischer)* Salzboden m
  s. solution Salzlösung f; Kochsalzlösung f
  s. water Salzwasser n, Sole f, salzhaltiges Wasser n
salinimeter Salz[gehalt]messer m, Salzwaage f, Halometer n *(Aräometer für Salzlösungen)*
salinity Salzhaltigkeit f; Salzigkeit f; Salzartigkeit f
salinization *(Bodenkunde)* Versalzung f
salinize / to *(Bodenkunde)* versalzen
salite *(Min)* Salit m
saliva Speichel m
salivary gland Speicheldrüse f

salmiac NH_4Cl Salmiak m, Salmiaksalz n, Ammoniumchlorid n; *(Min)* [natürlicher] Salmiak m, Salammoniak n *(Ammoniumchlorid)*
salmine Salmin n *(ein Protamin, das in Lachs-spermien in Verbindung mit Nukleinsäuren vorkommt)*
salmon oil Lachstran m, Lachsöl n, Lachsfett n
salol $C_6H_4(OH)COOC_6H_5$ Salol n, Phenylsalizylat n, Salizylsäurephenylester m
salsoline Salsolin n, 6-Hydroxy-7-methoxy-1-me-thyl-1,2,3,4-tetrahydroisochinolin n
salt / to salzen
  s. out aussalzen
salt Salz n, *(i.e.S.)* Kochsalz n, Salz n
  s. accumulation *(Landw)* Salzanreicherung f
  s. bath *(Met)* Salzbad n, Salzschmelze f
  s.-bath casehardening *(Met)* Salzbadeinsatz-härtung f
  s. bath chromizing *(Met)* Salzbad[in]chromieren n, Salzbadinkromieren n, Inchromieren (In-kromieren, Chromieren) n aus der flüssigen Phase
  s.-bath hardening s. s.-bath casehardening
  s. bridge Salzbrücke f, Elektrolytbrücke f, Elektrolytschlüssel m, elektrolytischer Strom-schlüssel (Heber) m, Flüssigkeitsheber m
  s. brine Salzlake f, Salzlauge f, Salzlösung f, Sole f, Lake f
  s. cake Rohsulfat n, rohes Glaubersalz n, technisches Natriumsulfat n
  s.-cake makeup *(Pap)* zur Deckung der Alkaliver-luste zugesetztes Rohsulfat n, Rohsulfatzusatz m
  s. content Salzgehalt m
  s. crystal Salzkristall m
  s. dissolver Salzlöser m
  s. dome Salzdom m, Salzhut m
  s.-dome trap Salzdomfalle f
  s. effect Salzeffekt m
  s. error Salzfehler m, Neutralsalzfehler m *(Meßfehler bei pH-Bestimmungen)*
  s. figure Reifezahl f *(bei Viskosespinnlösungen)*
  s. fish Salzfisch m, gesalzener Fisch m
  s. flavour s. s. taste
  s. formation Salzbildung f
  s.-forming salzbildend
  s. garden Salzgarten m, Saline f
  s.-glaze / to *(Ker)* salzglasieren
  s. glaze *(Ker)* Salzglasur f
  s. glazing *(Ker)* Salzglasieren n
  s.-ice mixture Kältemischung f aus Salz und Eis
  s. lake Salzsee m
  s.-like salzartig, salzähnlich
  s. link Salzbindung f *(Ionenbeziehung)*
  s. meadow Salzbeet n, *(i.w.S.)* Salzgarten m
  s. meat Salzfleisch n, gesalzenes Fleisch n
  s. melt Salzschmelze f
  s. mixture Salzmischung f, Salzgemisch n
  s. of an amine Aminsalz n
  s. of hartshorn Hirschhornsalz n, [gewöhnliches] Ammoniumkarbonat n *(Gemisch aus Am-*

moniumhydrogenkarbonat und Ammoniumkarbamat)
s. of lemon *s.* s. of sorrel
s. of mercury Quecksilbersalz *n*
s. of phosphorus $Na[NH_4]HPO_4 \cdot 4H_2O$ Phosphorsalz *n*, Natriumammoniumhydrogenphosphat-4-Wasser *n*
s. of Saturn $Pb(CH_3COO)_2$ Bleizucker *m*, Blei(II)-azetat *n*
s. of soda $Na_2CO_3 \cdot 10H_2O$ Kristallsoda *f*, Waschsoda *f*, Soda *f*, Natriumkarbonat-10-Wasser *n*
s. of sorrel Kleesalz *n*, Sauerkleesalz *n*, Bitterkleesalz *n* *(Kaliumtetraoxalat oder ein Gemisch von Kaliumhydrogenoxalat und Kaliumtetraoxalat)*
s. of tartar K_2CO_3 Weinsteinsalz *n*, Kaliumkarbonat *n*
s. pair Salzpaar *n*
s. plug *s.* s. dome
s. solution Salzlösung *f*; Kochsalzlösung *f*
s. spray test Salzsprühversuch *m*, Salzsprühtest *m (zur Korrosionsprüfung)*
s. spue *(Gerb)* Salzfleck *m*, Salzflecken *mpl*
s. stain Salzfleck *m*
s. taste Salzgeschmack *m*
s. tolerance Salztoleranz *f*
s. tower Salzturm *m*
s. water Salzwasser *n*, Sole *f*, salzhaltiges Wasser *n*; *(Glas)* Galle *f*, Glasgalle *f*
s. works Salzwerk *n*, Saline *f*, Salzsiederei *f*
salted almond Salzmandel *f*
saltern Saline *f*, Salzwerk *n*, Salzsiederei *f*; Salzgarten *m*, Saline *f*
saltiness Salzigkeit *f (Geschmacksempfindung)*
salting Salzen *n*, Salzung *f*; Salzablagerung *f*
s. in Einsalzen *n*
s.-in effect Einsalzeffekt *m*
s. in solid salt Trockensalzen *n*, *(bei Fleisch auch)* Trockenpökeln *n*, Trockenpökelung *f*
s. out Aussalzen *n*, Aussalzung *f*
s.-out chromatography Aussalzchromatografie *f*
s.-out effect Aussalzeffekt *m*
s.-out evaporator Verdampfungskristallisator *m*
s. tub Pökelfaß *n*
saltish etwas salzig
saltpeter *s.* saltpetre
saltpetre Salpeter *m*, *(i.e.S.)* Kalisalpeter *m*; *(Min)* Salpeter *m*, Niter *m*, Kalisalpeter *m*, Nitrokalit *m* *(Kaliumnitrat)*
s. refinery Salpetersiederei *f*
salts of lemon (sorrel) *s.* salt of sorrel
salty salzig; salzhaltig
s. flavour (taste) *s.* salt taste
sam / to *s.* to sammy
samaria Sm_2O_3 Samariumerde *f*, Samaria *f*, *(veraltet für)* Samarium(III)-oxid *n*
samaric Samarium..., *(i.e.S) Samarium(III)-...*
s. chloride $SmCl_3$ Samarium(III)-chlorid *n*, Samariumtrichlorid *n*
s. hydroxide $Sm(OH)_3$ Samarium(III)-hydroxid *n*
s. iodide SmJ_3 Samarium(III)-jodid *n*, Samariumtrijodid *n*
samarium Sm Samarium *n*
s. dichloride $SmCl_2$ Samariumdichlorid *n*, Samarium(II)-chlorid *n*
s. diiodide SmJ_2 Samariumdijodid *n*, Samarium(II)-jodid *n*
s. oxide Sm_2O_3 Samarium(III)-oxid *n*
s. peroxide Sm_4O_9 Samariumperoxid *n*
s. sesquioxide *s.* s. oxide
s. trichloride $SmCl_3$ Samariumtrichlorid *n*, Samarium(III)-chlorid *n*
s. triiodide SmJ_3 Samariumtrijodid *n*, Samarium(III)-jodid *n*
samarous Samarium..., *(i.e.S.)* Samarium(II)-...
s. chloride $SmCl_2$ Samarium(II)-chlorid *n*, Samariumdichlorid *n*
s. iodide SmJ_2 Samarium(II)-jodid *n*, Samariumdijodid *n*
samarskite *(Min)* Samarskit *m*
samming *s.* sammying
sammy / to *(Gerb)* abwelken *(vorentwässern)*
sammying *(Gerb)* Abwelken *n (Vorentwässern)*
s. machine *(Gerb)* Abwelkpresse *f*
sample / to Proben entnehmen
sample Probe *f*, Muster *n*, Probestück *n*, Warenprobe *f*; Probe *f*, Substanzprobe *f*
s.-dye jig *s.* s.-dyeing jig
s. dyeing Musterfärben *n*, Musterfärbung *f*
s.-dyeing jig Musterfärbejigger *m*
s. of soil Bodenprobe *f*
s. tube Probenrohr *n*
sampler Probenehmer *m*
sampling Probe[ent]nahme *f*, Probenehmen *n*; *(Pap)* statistische Sortierung *f*
s. cock Probenehmerhahn *m*, Probierhahn *m*
s. device (tool) Probenehmer *m*, Probenahmegerät *n*
sand / to *(Ker) (eine Form)* besanden
sand 1. Sand *m*; 2. Sand *m*, Grobgut *n*, Rückstand *m*, Schlamm *m (beim Klassieren)*
s. acid $H_2[SiF_6]$ Fluorokieselsäure *f*, Hexafluorokieselsäure *f*, Kieselfluorwasserstoffsäure *f*, Kieselflußsäure *f*, Siliziumfluorwasserstoffsäure *f*, Silikofluorwasserstoffsäure *f*
s. bath Sandbad *n*
s. box *s.* s. trap
s. carving *(Glas)* Sandstrahlmattieren *n*
s.-cast / to im Sandguß herstellen
s. casting Sandguß *m*; Sandgußstück *n*
s. catcher *s.* s. trap
s. coal Sandkohle *f*
s. discharge Sandaustrag *m*, Grobgutaustrag *m* *(eines Klassierers)*
s. filter Sandfilter *n*; Kiesfilter *n*
s. floatation Sandflotation *f*
s.-floatation process Sandflotationsprozeß *m*, Sandschwimmprozeß *m*, Sandschwimmverfahren *n*

s. grate s. s. trap
s. mould Sandform f
s.-paper / to (mit Sandpapier) abschleifen
s. paper Sandpapier n
s. product Sand m, Grobgut n, Rückstand m,
Schlamm m (beim Klassieren)
s. sifter s. s. trap
s. sugar Sandzucker m
s. table s. s. trap
s. trap (well) Sandfang m, Sandfänger m
sandalwood Sandelholz n, Santalholz n
s. oil Sandelholzöl n
sandblast / to sandstrahlen; (Glas) sandstrahlen,
mit Sandstrahl mattieren
sandblasting Sandstrahlen n, Sandstrahlreinigung
f; (Glas) Sandstrahlen n, Mattieren n mit
Sandstrahl
sander (Am) Schleifpapier n
sanding (Ker) Sanden n, Besandung f
Sandmeyer [diazo] reaction Sandmeyer-Reaktion
f, Sandmeyersche Reaktion f (Ersatz der Diazogruppe durch andere Reste)
sandslinger (Met) Schleuder[form]maschine f,
Sandschleudermaschine f
sandstone Sandstein m
sandwich (Gum) (aus mehreren Schichten
bestehender) Prüfling m
s. bond Sandwichbindung f, Doppelkegelbin-
dung f
s.-bonded compound s. s. compound
s. complex Sandwichkomplex m
s. compound Sandwichverbindung f
s. construction (Plast) Sandwichbauweise f,
Verbund[platten]bauweise f, Stützstoffbauweise
f, Schichtstoffbauweise f, Wabenbauweise f
s. kiln Sandwichofen m
s. molecule Sandwichmolekül n
s. panel Sandwichplatte f, Verbundplatte f
s. structure Sandwichstruktur f, Doppelkegel-
struktur f
sandy limestone Kalksandstein m
sanforize / to (Text) sanforisieren
sanforizing (Text) Sanforisieren n, Sanforisierung
f
Sanger's reagent Sangersches Reagens n (2,4-Di-
nitrofluorbenzol zum Nachweis von Aminosäuren
und Proteinen)
sanidine (Min) Sanidin m
sanitary cotton Verbandwatte f
s. tissue Toilettenseidenpapier n
sant (Gerb) (Hülsen von Acacia nilotica (L.)Del.)
santal Santal n, 3',4',5-Trihydroxy-7-methoxyisofla-
von n
santalbic acid
$CH_3(CH_2)_5CH=CHC\equiv C(CH_2)_7COOH$ Santalbsäure f, Ximenynsäure f, Oktadezen-(11)-in-(9)-
säure f
santalene Santalen n (ein Sesquiterpen aus
ostindischem Sandelholzöl)
santalic acid Santalsäure f

santalin Santalin n, 1,3-Dioxy-2-methoxy-5,6,7,8-te-
trahydroanthrachinon n
santalol Santalol n (ein Sesquiterpenalkohol aus
ostindischem Sandelholzöl)
santalwood s. sandalwood
santene Santen n, 2,3-Dimethyl-2-norbornen n
s. hydrate Santenhydrat n, β-Santenol n
santenic acid Santensäure f, 1,2-Dimethylzyklo-
pentan-1,3-dikarbonsäure f
santenol Santenol n (ein vom Santen abgeleiteter
Terpenalkohol)
santenone Santenon n (ein vom Santen ab-
geleitetes bizyklisches Keton)
s. alcohol Santenonalkohol m
santonan Santonan n, Tetrahydrosantonin n
santonanic acid Santonansäure f, Tetrahydrosan-
toninsäure f
santonic acid Santonsäure f
s. lactone s. santonin
santonin Santonin[lakton] n
santoninic acid Santoninsäure f
santonous acid santonige Säure f, Santonigsäure
f
São Francisco rubber São-Francisco-Kautschuk m
(von Manihot heptaphylla Ule)
S.A.P. s. sintered aluminium powder
sap / to entsaften
sap Saft m
sap. value s. saponification value
sapanwood s. sappan[wood]
saphirine s. sapphirine
sapid substance Geschmacksstoff m
saponifiable verseifbar
s. matter Verseifbares n
saponification Verseifung f, Saponifikation f
s. flask Verseifungskolben m
s. number (value) Verseifungszahl f, VZ
saponified acetate filament verseifter Azetat-
elementarfaden m
saponify / to verseifen
saponin Saponin n
sappan[wood] Sappanholz n, Indisches Rotholz n
(von Caesalpinia sappan L.)
sapphire (Min) Saphir m (Aluminiumoxid)
sapphirine aus Saphir; saphirartig
sapphirine (Min) Saphirin m (ein Neso-Subsilikat)
saprogenic, saprogenous saprogen, fäulniserregend
saprolite s. sapropelite
sapropel Sapropel n(m), Faulschlamm m
s. wax Sapropelwachs n
sapropelic coal Sapropel[it]kohle f, Faulschlamm-
kohle f, sapropelitische Kohle f
sapropelite Sapropelit m, Faulschlammgestein n,
Sapropelgestein n
saprophyte Saprophyt m, Fäulnisbewohner m
saprophytic saprophytisch, fäulnisbewohnend
s. organism s. saprophyte
sapwood Splint m, Splintholz n
sarcine Sarkin n, Hypoxanthin n, 6-Hydroxypurin
n, 6(1)-Purinon n

sarcolactic acid $CH_3CH(OH)COOH$ Fleischmilchsäure f, Paramilchsäure f, Rechtsmilchsäure f, d-Milchsäure f, d-2-Hydroxypropansäure f
sarcolite (Min) Sarkolith m, (veraltet für) Gmelinit m (ein Tektosilikat)
sarcosine CH_3NHCH_2COOH Sarkosin n, Methylaminoessigsäure f, N-Methylglykokoll n
sardine oil Sardinentran m, Sardinenöl n
sardonyx (Min) Sardonyx m (schwarzweißer Bandachat)
sarin Sarin n (Methylfluorphosphonsäureisopropylester; Kampfstoff)
SAS, S.A.S. s. sodium-aluminium sulphate
sassafras oil Sassafrasöl n (von Sassafras albidum (Nutt.)Nees)
sassolin[e], sassolite (Min) Sassolin m (Orthoborsäure)
satd. (Abk. für) saturated
satin glaze (Ker) Velinglasur f
s. spar 1. (Min) Atlasspat m (faseriger Kalzit, Aragonit oder Gips); 2. s. s. white
s.-vellum glaze s. s. glaze
s. white (Farb) Satinweiß n, Gips m; (Pap, Farb) Satinweiß n, Glanzweiß n (aus Gips, Tonerde und Kalziumoxid)
satine / to (Gerb) glänzen, satinieren
satn. s. saturation
saturant Imprägniermittel n
saturate / to (Lösungen) sättigen; (Valenzen) absättigen
saturated hydrocarbon gesättigter (paraffinischer) Kohlenwasserstoff m, Paraffinkohlenwasserstoff m, Paraffin n, Alkan n, Grenzkohlenwasserstoff m
s. solution gesättigte Lösung f
s. steam (vapour) gesättigter Dampf m, Sattdampf m
s. vapour pressure Sättigungs[dampf]druck m
saturation Sättigung f (von Lösungen); (Bindungslehre) Absättigung f; (Zucker) Saturieren n, Saturation f
s. concentration Sättigungskonzentration f
s. current Sättigungsstrom m, Grenzstrom m
s. gas (Zucker) Saturationsgas n
s. humidity Sättigungsfeuchte f, Sättigungsfeuchtigkeit f; Sättigungsfeuchtebeladung f; Sättigungswassergehalt m
s. juice (Zucker) Schlammsaft m; Saturationssaft m
s. limit Sättigungsgrenze f
s. mud (Zucker) Saturationsschlamm m
s. pan (Zucker) Saturationspfanne f
s. paper Saturationspapier n
s. point Sättigungspunkt m
s. scum s. s. mud
s. tank s. s. pan
s. temperature Taupunkt m
s. value Sättigungswert m
s. vapour pressure Sättigungs[dampf]druck m
saturator Sättiger m, Sättigungsgefäß n, Sätti-

gungsvorrichtung f, Sättigungsapparat m, Satureur m, Saturator m
Saturn salt $Pb(CH_3COO)_2$ Bleizucker m, Blei(II)-azetat n
saturnine Blei...; Bleivergiftungs..., durch Bleivergiftung hervorgerufen (oder) Bleivergiftung betreffend, (Med auch) saturnin
saturnism Saturnismus m, Bleivergiftung f, Bleikrankheit f
saturnium s. protactinium
sausage flask Säbelkolben m [nach Anschütz], Schwertkolben m
saussurite (Min) Saussurit m (ein Tektosilikat)
saussuritization (Geol) Saussuritisierung f, Saussuritisation f, Saussuritbildung f
Savalle's still Destillierapparat m nach Savalle
save-all (Pap) Stoffrückgewinnungsanlage f, Faserrückgewinnungsanlage f, Fangstoffanlage f, Stoffang m, Stoffänger m
s.-all tray (Pap) Siebwasserschiff n; s. s.-all
s.-all water (Pap) Sieb[ab]wasser n
savin oil Sadebaumöl n (von Juniperus sabina L.)
saw gin (Text) Sägeegreniermaschine f, Sägengin m
Sawari fat (Fett von Caryocar amygdaliferum Mutis)
sawdust / to (Gerb) einspänen, in Sägespänen anfeuchten
sawdust Sägemehl n, Sägespäne mpl
sawtooth crusher Daumenbrecher m
saxifrax oil s. sassafras oil
saxoline (veraltet für) petrolatum
Saxony blue Sächsischblau n, (veraltet für) Smalte f, Schmalte f (Kobalt(II)-kaliumsilikat)
Saybolt distilling flask Saybolt-Kolben m
Saybolt visco[si]meter (Erdöl) Sayboltsches Viskosimeter n, Saybolt-Universalviskosimeter n
Saytzeff rule Saizew-Regel f
SBA s. sec-butyl alcohol
SBR s. styrene-butadiene rubber
SC s. spreading coefficient
scab Sulfatblase f (Glasfehler)
scabicide, scabieticide (Pharm) Antiskabiosum n, Mittel n gegen Krätze
scald / to (Gerb) verbrühen (von Blößen in der Beize)
scale / to 1. mit genauer Teilung (Einteilung) versehen, in Grade [ein]teilen, graduieren; 2. (einen Gegenstand) verkrusten, (i.e.S.) mit Kesselstein bedecken; verkrusten, (i.e.S.) Kesselstein ansetzen; (Met) [ver]zundern; entkrusten, (i.e.S.) Kesselstein entfernen; (Met) entzundern; 3. s. to s. off
s. off abblättern, abschuppen, abplatzen, abspringen
scale 1. Skala f, Stufenfolge f; (an Meßgeräten) Skale f, Maßeinteilung f, Gradeinteilung f; 2. dünne Schicht f, Belag m, Ablagerung f, Niederschlag m; (Met) Zunder m, Sinter m; Kesselstein m; Zahnstein m; 3. Schuppe f; 4. Waageschale f; Waage f

s. car *(Met)* Möllerwagen *m (für Kippkübel-begichtung)*
s. division Skalen[ein]teilung *f*; Skalenteil *m*, Teilstrichabstand *m*; Skalen[teil]strich *m*
s. formation Kesselsteinbildung *f*; *(Met)* Zunder-bildung *f*
s. graduated in mm mm-Skale *f*
s. layer *(Met)* Zunderschicht *f*
s. length Skalenlänge *f*
s. of hardness Härteskala *f*
s. of temperature Temperaturskale *f*
s. paper Profilpapier *n, (i.e.S.)* Millimeterpapier *n (oder)* Diagrammpapier *n*
s. span Anzeigebereich *m (eines Meßgeräts)*
s. wax Schuppenparaffin *n*, Paraffinschuppen *fpl*
scalicide Schildlausbekämpfungsmittel *n*
scaling 1. Messung *f (oder)* Einstellung *f* nach einer Skale; 2. Verkrusten *n*, Verkrustung *f*, *(i.e.S.)* Kesselsteinbildung *f*; *(Met)* Zundern *n*, Verzun-derung *f*; Entfernen *n* von Kesselstein; *(Met)* Entzundern *n*; 3. Abschuppen *n*, Abblättern *n*, Abplatzen *n*, Abspringen *n*
scalp [out] / **to** *(übergroße Stücke)* absieben, entfernen
scaly schuppig
s. glass Schuppenglas *n*
scammony [resin] Skammoniumharz *n*, Skam-moniaharz *n (von Ipomoea orizabensis Leda-nois)*
scan / **to** abtasten, [genau] prüfen
scandia Sc_2O_3 Skandia *f*, Skandiumerde *f*, *(veraltet für)* Skandiumoxid *n*
scandium Sc Skandium *n*
s. bromide $ScBr_3$ Skandiumbromid *n*
s. carbonate $Sc_2(CO_3)_3$ Skandiumkarbonat *n*
s. chloride $ScCl_3$ Skandiumchlorid *n*
s. fluoride ScF_3 Skandiumfluorid *n*
s. hydroxide $Sc(OH)_3$ Skandiumhydroxid *n*
s. hydroxynitrate $Sc(OH)(NO_3)_2$ Skandium-hydroxidnitrat *n*
s. iodide ScJ_3 Skandiumjodid *n*
s. nitrate $Sc(NO_3)_3$ Skandiumnitrat *n*
s. oxide Sc_2O_3 Skandiumoxid *n*
s. oxysulphate $(Sc_2O)(SO_4)_2$ Skandiumoxid-sulfat *n*
s. sulphate $Sc_2(SO_4)_3$ Skandiumsulfat *n*
s. sulphide Sc_2S_3 Skandiumsulfid *n*
scanning Abtastung *f*
s. microscope Meßmikroskop *n*
scapolite *(Min)* Skapolith *m (ein Tektosilikat)*
scar *(Met)* Gußfehler *m*
scarlet Scharlach *m*, scharlachroter Farbstoff *m*
s. corns Kermeskörner *npl*, Scharlachkörner *npl*, Kermes *m (getrocknete weibliche Kermesschild-läuse)*
s.-red scharlachrot
s. red *(Farb)* Biebricher Scharlach *m*, Doppel-scharlach *m*, Neurot *n*
scatter / **to** [zer]streuen
scatter Streuen *n*, Zerstreuen *n*, Streuung *f*, Zerstreuung *f*

scattered electron Streuelektron *n*, gestreutes Elektron *n*
s. light gestreutes (zerstreutes, diffuses) Licht *n*, Streulicht *n*, Beugungslicht *n*
s. neutron Streuneutron *n*, gestreutes Neutron *n*
s. radiation Streustrahlung *f*
scattering Streuung *f*, Zerstreuung *f*
s. amplitude Streuamplitude *f*
s. angle Streu[ungs]winkel *m*
s. coefficient Streu[ungs]koeffizient *m*
s. cross section Streu[ungs]querschnitt *m*
s. factor Streufaktor *m*
s. frequency Streufrequenz *f*
s. of light Lichtstreuung *f*
s. power Streuvermögen *n*
s. process Streuprozeß *m*
scavenge / **to** reinigen, säubern, spülen; *(Roherz mechanisch)* läutern; *(Met) (Schmelzen durch Chemikalienzusatz)* desoxydieren
scavenger Reinigungsmittel *n*, Spülmittel *n*; *(Met)* Desoxydationsmittel *n*
s. plate Spülboden *m*
scavenging air Spülluft *f*
scent attractant Duftlockstoff *m*
s.-spray principle Zerstäuberprinzip *n*
s.-tight aromadicht
SCF *s.* self-consistent field
Schäffer acid $C_{10}H_6(OH)SO_3H$ Schäffer-Säure *f*, Schäffersche Säure (β-Säure) *f*, 2-Naphthol-6-sulfonsäure *f*
Schäffer salt Schäffer-Salz *n (Na-Salz der Schäffer-Säure)*
schappe Schappe *f*, Schapp[e]seide *f*
Schardinger dextrin Schardinger-Dextrin *n*
scheduled [amtlich geprüft und] anerkannt *(Schädlingsbekämpfungsmittel)*
Scheele's green Scheeles (Scheelesches) Grün *n (Kupferarsenit)*
scheelite *(Min)* Scheelit *m (Kalziumwolframat)*
Scheibel column Scheibel-Kolonne *f*, Rührkolonne *f* nach Scheibel
Scheibler desiccator Scheibler-Exsikkator *m*, Exsikkator *m* nach Scheibler
Scheibler reagent Scheiblers Reagens *n (Phos-phorwolframsäure)*
Schellbach burette Schellbach-Bürette *f*
scheme of energy levels Termschema *n*
Schiemann reaction Schiemann-Reaktion *f (zur Darstellung von Arylfluoriden)*
Schiff base Schiffsche Base *f (Azomethin)*
Schiff reagent Schiffsches (Schiffs) Reagens *n (meist fuchsinschweflige Säure zum Aldehyd-nachweis)*
Schiff's test Schiffsche Reaktion *f*, Reaktion *f* nach Schiff *(zum Aldehydnachweis)*
schiller spar *(Min)* Schillerspat *m*, Bastit *m (ein Inosilikat)*
schist *(Geol)* Schiefer *m*
schistosity *(Geol)* Schieferung *f*, Schieferigkeit *f*
schistous schieferig, schieferartig

Schlippe's salt $Na_3SbS_4 \cdot 9H_2O$ Schlippesches Salz n, Natriumthioantimonat(V)-9-Wasser n

Schlumberger method Schlumberger-Methode f *(zur elektrischen Bohrlochvermessung)*

Schmidt reaction Schmidt-Reaktion f, Schmidtsche Reaktion f *(Karbonylabbau mit Stickstoffwasserstoffsäure)*

Schöllkopf's acid $HOC_{10}H_5(SO_3H)_2$ Schöllkopf-Säure f, α-Naphtholdisulfonsäure Sch. f, 1-Naphthol-4,8-disulfonsäure f; $C_{10}H_6NH_2SO_3H$ Schöllkopf-Säure f, 1-Naphthylamin-8-sulfonsäure f; *(Am manchmal auch)* 1-Naphthylamin-4,8-disulfonsäure f

Schönherr process Schönherr-Verfahren n *(zur Bindung von Luftstickstoff)*

Schoop metallizing (process) Schoop[is]ieren n, Metallspritzen n, Spritzmetallisieren n, Metallspritzverfahren n

Schopper densometer *(Pap)* Luftdurchlässigkeitsprüfer m nach Schopper

Schopper machine *(Gum)* Zug-Dehnungsprüfgerät n Bauart Schopper, Schopper-Dalen-Maschine f; Abriebprüfgerät n Bauart Schopper

Schopper-Riegler apparatus *(Pap)* Apparat (Mahlungsgradprüfer) m nach Schopper-Riegler

schorl *(Min)* Schörl m *(borhaltiges Zyklosilikat der Turmalin-Reihe)*

Schorlemmer basicity *(Gerb)* Basizität f *(einer Chromlösung)* nach Schorlemmer

schorlite *(Min)* Schorlit m, *(veraltet für)* Schörl m

schorlomite *(Min)* Schorlomit m

Schotten-Baumann reaction Schotten-Baumann-Reaktion f, Schotten-Baumannsche Reaktion f

Schrödinger [wave] equation Schrödinger-Gleichung f, Schrödingersche Wellengleichung (Gleichung) f

Schuller process Schuller-Verfahren n, Aufwärtsziehverfahren n *(Glasrohrherstellung)*

Schumann plate Schumann-Platte f *(UV-Fotografie)*

Schürmann's rule *(Kohlechemie)* Schürmannsche Regel f

Schwarz reaction Schwarzsche Reaktion f *(Naphthalinnachweis; Chloroformnachweis)*

Schwarzschild effect *(Foto)* Schwarzschild-Effekt m

schwazite *(Min)* Schwazit m, Quecksilberfahlerz n

Schweinfurth green $Cu(CH_3COO)_2 \cdot 3Cu(AsO_2)_2$ Schweinfurter Grün n *(Kupferarsenitazetat)*

Schweizer's reagent Schweizer-Reagens n, Schweizers Reagens n, Schweizersche Flüssigkeit f *(Kupferoxidammoniak-Lösung)*

SCI *(Abk. für)* Society of Chemical Industry

science of flow Rheologie f, Fließkunde f

s. of nutrition Ernährungswissenschaft f

scintillate / to szintillieren, aufblitzen, aufleuchten, flimmern, funkeln

scintillation Szintillieren n, Szintillation f, Aufblitzen n, Aufleuchten n, Flimmern n, Funkeln n

s. counter Szintillationszähler m

s. method Szintillationsmethode f

s. spectrometer Szintillationsspektrometer n

scissor compound Scherenverbindung f, Chelatverbindung f, Chelatkomplex m, Chelat n

sclerometer Sklerometer n, Härtemesser m, Ritzhärteprüfer m

scleroprotein Skleroprotein n, Gerüsteiweiß n

scleroscope Skleroskop n, Härtemesser m, Fallhärteprüfer m, Rücksprunghärteprüfer m

scolecite *(Min)* Skolezit m *(Kalziumdialumotrisilikat)*

scoop Schöpfgefäß n; Becher m *(eines Becherwerks)*; Greifer m; Schaufel f; Rührflügel m

scooping bucket elevator Schöpf[becher]werk n

scopine Skopin n *(Alkaloid)*.

scopolamine Skopolamin n *(Alkaloid)*

s. hydrobromide $(C_{17}H_{21}O_4N)HBr \cdot 3H_2O$ Skopolaminhydrobromid n, Tropasäure-6,7-epoxytropylesterhydrobromid n

scopoletin Skopoletin n, Chrysatropasäure f, Gelseminsäure f, β-Methyläskuletin n, 6-Methoxy-7-hydroxykumarin n

scopoline Skopolin n, Oszin n, 3,7-Oxido-6-tropanol n

scorch / to versengen, verbrennen; *(Pflanzenteile durch Chemikalien)* verätzen, verbrennen; *(Gum)* anvulkanisieren, anbrennen, anspringen

scorch Brandfleck m, Sengfleck m; Verätzung f, Verbrennung f *(an Pflanzen durch Chemikalien)*; *(Gum)* Anvulkanisation f, Anvulkanisieren n, Anspringen n, Anbrennen n, Scorch

s. characteristic *(Gum)* Scorchcharakteristik f, Anvulkanisationscharakteristik f, Anspringcharakteristik f, Anbrenncharakteristik f

s. curve *(Gum)* Anvulkanisationskurve f, Anspringkurve f, Anbrennkurve f, Scorchkurve f

s. delay *(Gum)* verzögerter (später) Vulkanisationseinsatz m

s. period *(Gum)* Anvulkanisationsperiode f, Anspringperiode f, Anbrennperiode f, Scorchperiode f

s. point *(Gum)* Anvulkanisationspunkt m, Anspringpunkt m, Anbrennpunkt m, Scorchpunkt m

s. resistance *(Gum)* Scorchbeständigkeit f, Scorchresistenz f

s.-resistant *(Gum)* scorchresistent, scorchbeständig

s. tendency *(Gum)* Anvulkanisationstendenz f, Anvulkanisationsneigung f, Scorchneigung f, Scorchtendenz f, Anbrenntendenz f

s. test *(Gum)* Prüfung f der Anvulkanisation, Prüfung f des Anspringens (Anbrennens), Scorchprüfung f

s. time *(Gum)* Anvulkanisationszeit f, Scorchzeit f, Anspringzeit f, Anbrennzeit f

scorchiness s. scorch tendency

scorching s. scorch

scorchy *(Gum)* scorchanfällig

score [method of] cutting *(Pap)* Längsschneiden n nach dem Druckprinzip

scoria *(Met, Geol)* Schlacke *f*
scoriaceous schlackenartig, schlackenähnlich, schlackig
scorification *(Met)* Verschlacken *n*, Verschlackung *f*; Schlackenbildung *f*
scorifier *(Met)* Schlackenscherben *m*, Schlackenkegel *m*
scorify / to [ver]schlacken, einschlacken, Schlacke bilden
scorodite *(Min)* Skorodit *m* *(Eisen(III)-orthoarsenat)*
Scotch fir *s.* S. pine; Kiefernholz *n (von Pinus sylvestris L.)*
S. fir oil Kiefernnadelöl *n (von Pinus sylvestris L.)*
S. pine Gemeine Kiefer *f*, Föhre *f*, Pinus sylvestris L.
scour / to reinigen; scheuern, blank putzen (reiben); auswaschen, [aus]spülen; abbeizen; *(Met)* entzundern; *(Met)* *(das Ofenfutter)* ausfressen; *(Wolle)* entfetten, entschweißen; *(Seide)* entbasten, degummieren
scour nozzle Spüldüse *f*, Waschdüse *f*
s. valve Spülventil *n*, Spül[flüssigkeits]schieber *m*
scouring Reinigen *n*, Reinigung *f*; Scheuern *n*; Auswaschen *n*, Spülen *n*, Ausspülen *n*; Abbeizen *n*; *(Met)* Entzundern *n*, Beizen *n*; *(Met)* Ausfressung *f (im Ofenfutter)*; *(Text)* Entfetten *n*, Entschweißen *n (von Wolle)*, Entbasten *n*, Degummieren *n*, Entschälen *n (von Seide)*
s. action *(Text)* Waschwirkung *f*
s. agent Reinigungsmittel *n*, Reiniger *m*; Putzmittel *n*; Waschmittel *n*, Spülmittel *n*; Abbeizmittel *n*; *(Met)* Beize *f*; Entschweißungsmittel *n (für Wolle)*; Entbastungsmittel *n (für Seide)*
s. bath Reinigungsbad *n*, Waschbad *n*; Spülbad *n*; *(Met)* Beizbad *n*; Entschweißbad *n (für Wolle)*; Entbastungsbad *n (für Seide)*
s. liquor *(Text)* Waschlauge *f*, Waschflotte *f*; Spülflüssigkeit *f*
s. machine *(Text)* Waschmaschine *f*
s. soap Reinigungsseife *f*, Fleck[en]seife *f*; Scheuerseife *f*
s. train *(Text)* Waschanlage *f*, Waschaggregat *n*
SCR *s.* styrene-chloroprene rubber
scrap Abfall *m*; Ausschuß *m*; *(Met)* Schrott *m*, Bruch *m*, Altmetall *n*; *(Gum)* Scrap *m (minderwertige Kautschuksorte)*
s. iron Eisenschrott *m*, Alteisen *n*
s. lead Bleiabfälle *mpl*, Altblei *n*
s. metal Schrott *m*, Altmetall *n*
s. paper *(veraltet für)* waste paper
s. reclaiming Wiederaufarbeiten *n* von Abfällen
s. rubber Abfallgummi *m*, Altgummi *m*, Gummiabfälle *mpl*, Vulkanisatabfälle *mpl*;*(Gum)* *(un-vulkanisierte)* Fabrikationsabfälle *mpl*
scrape / to kratzen; reiben
s. off abstreifen, abkratzen, abschaben
scraped-pipe crystallizer Kratz-Rohrkristallisator *m*

scraper Abstreifer *m*, Schaber *m*, Ausräumer *m*; *(Ker)* Schrapper *m*, Schabeisen *n*
s. blade *s.* s. knife
s. board Schabkarton *m*
s. chain *(Fördertechnik)* Kratzerkette *f*
s.-chain conveyor Kratzerförderer *m (i.w.S.)*, Stegkettenförderer *m (i.w.S.)*
s. conveyor *s.* s.-chain conveyor
s. flight conveyor Kratzerförderer *m (i.e.S.) (Kette oberhalb des Fördergutes laufend)*
s. knife Abstreif[er]messer *n*, Abstreifmeißel *m*, Abstreifer *m*, Abschabemesser *n*, Schabemesser *n*
scraping Abstreifen *n*, Abkratzen *n*, Abschaben *n*
scratch / to ritzen, kratzen
scratch Kratzer *m*, Schramme *f*
s. board Schabkarton *m*
s. coat Unterputz *m*
s. hardness Ritzhärte *f*, Ritzfestigkeit *f*, Kratzfestigkeit *f*
s.-proof kratzfest
s. resistance *s.* s. hardness
s.-resisting glaze kratzfeste Glasur *f*
s. test Ritzprüfung *f*, Ritzprobe *f*, Kratzprobe *f*
scratching Ritzen *n*, Kratzen *n*
scratted paper gespritztes Papier *n (ein Buntpapier)*
screen / to 1. [durch]sieben, absieben, aussieben, siebklassieren; 2. *(Kch)* [ab]schirmen; *(Licht)* abblenden
s. out aussieben, ausscheiden
screen 1. Sieb *n*, Klassiersieb *n*; Siebapparat *m*, Siebmaschine *f*, Siebvorrichtung *f*; 2. Schirm *m*; Röntgenschirm *m*; Bildschirm *m*; Projektionswand *f*; 3. Filter *n*, Blende *f*
s. analysis Siebanalyse *f*
s. aperture Sieböffnung *f*, Maschenweite *f*
s. area Siebfläche *f*
s. centrifuge Siebzentrifuge *f*, Siebschleuder *f*
s. classification Siebklassieren *n*, Siebklassierung *f*
s. classifier Siebapparat *m*, Siebmaschine *f*, Siebvorrichtung *f*
s.-conveyor centrifuge Siebschnecken[austrag]zentrifuge *f*, Siebschneckenschleuder *f*
s.-covered *(Filtration)* mit Metallgewebe (Siebgewebe) abgedeckt
s. deck Siebboden *m*
s. discharge Siebaustrag *m*
s. efficiency *(Klassierung)* Siebgütegrad *m*, Siebwirkungsgrad *m*, Sieberfolg *m*
s. feed Siebgut *n*
s. filter Siebfilter *n*; Rahmenfilter *n*, Flächenabscheider *m*
s. fines *(Klassierung)* Feinkorn *n*, Unterkorn *n*
s. frame *(Klassierung)* Siebkasten *m*
s. head Siebkopf *m*, *(Gum auch)* Strainerkopf *m*
s. head extruder Siebpresse *f*, Siebkopf-Spritzmaschine *f*, Strainer *m*
s. oversize *(Klassierung)* Siebgrobes *n*, Grobkorn *n*, Überkorn *n*

s. pack *(Plast)* Siebplatte f, Filterplatte f, Siebeinsatz m

s. plate Sortierplatte f, Sortierblech n

s. printing s. s.-process printing

s.-process ink Siebdruckfarbe f, Druckfarbe f für Siebdruck

s.-process printing Siebdruck m, Seidenrasterdruck m

s. rejects *(Holzschliffsortierung)* Grobstoff m, Spuckstoff m, „Sauerkraut" n

s. size Siebweite f; Korngröße f, Körnung f, Teilchengröße f *(beim Siebklassieren)*

s.-size opening Sieböffnung f

s. slot *(Pap)* Knotenfängerschlitz m

s. strainer Siebfilter n

s. surface Sieb[ober]fläche f

s.-type mill Siebmühle f

s. undersize s. s. fines

s. wall Schirmwand f

screened from the light lichtgeschützt

s. stock *(Pap)* Feinstoff m, Gutstoff m, büttenfertiger Stoff m

screening 1. Sieben n, Absieben n, Durchsieben n, Aussieben n, Siebklassieren n; 2. *(Kch)* Abschirmen n, Abschirmung f *(durch sterische Hinderung)*; 3. Vorauswahl f *(von wertvoll scheinenden Neuentwicklungen)*

s. action Siebwirkung f

s. area Siebfläche f

s. capacity Siebleistung f

s. constant *(Kph)* Abschirmungskonstante f, Abschirmungszahl f

s. efficiency s. screen efficiency

s. machine Siebapparat m, Siebmaschine f, Sieb n

s. plant Sieb[erei]anlage f

s. station Siebstation f, Sieberei f

s. surface Sieb[ober]fläche f

screenings Schrenzpapier n; Schrenzkarton m; s. screen rejects

screens s. screen rejects

screw / to verschrauben

s. together zusammenschrauben

screw Schraube f, Schraubenspindel f; Schnecke f *(Fördereinrichtung)*

s. axis *(Krist)* Schraubenachse f

s. clamp Schraubzwinge f; s. s. compressor clamp

s. clip s. s. compressor clamp

s. compressor clamp Schraubenquetschhahn m, Schraubklemme f [nach Hoffmann]

s. conveyor Förderschnecke f, Schneckenförderer m, Transportschnecke f, Schnecke f

s.-conveyor dryer Schneckentrockner m, Trokkenschnecke f

s.-conveyor extractor Extraktor (Extrakteur) m mit Förderschnecken

s. dislocation *(Krist)* Schraubenversetzung f, Burgers-Versetzung f

s. extruder s. s. press

s. flight Schneckengang m

s. injection [moulding] machine *(Plast)* Schneckenspritzgießmaschine f, Spritzgießmaschine f mit Schneckenkolben

s. mixer s. s.-type mixer

s.-on-type cap Schraubenkappe f

s.-operated spindelbetätigt, Spindel...

s.-preplastication *(Plast)* Schneckenvorplastizierung f, Schneckenvorplastizieren n

s. preplasticizing machine *(Plast)* Spritzgießmaschine f mit Schneckenvorplastizierung

s. press Schneckenpresse f, Extruder m

s. pump Schraubenpumpe f, Schneckenpumpe f, Spindelpumpe f

s. speed Schneckendrehzahl f

s.-type compressor Schraubenverdichter m, Schraubenkompressor m, Schneckenverdichter m, Schneckenkompressor m

s.-type mixer Mischschnecke f, Schneckenmischer m, Schneckenrührer m; Schneckenkneter m, Schraubenkneter m

screwed fitting Schraubenformstück n, Schraubenverbindungsstück n, Schraubfitting m(n), Gewindeformstück n, Gewindeverbindungsstück n, Gewindefitting m(n), Gewinde[rohr]verbindung f, Rohrverschraubung f

screwless extruder schneckenloser Extruder m

scriptural reed Riesenschilf n, Pfahlrohr n, Arundo donax L.

scroll Schnecke f, Spirale f

s. conveyor Förderschnecke f, Schneckenförderer m, Transportschnecke f

scroop / to *(Text)* knirschend ausrüsten (appretieren)

scroop *(Text)* Knirschen n, Krachen n, Seidenschrei m

scroopy *(Text)* knirschend, krachend *(Griff)*

scrub / to schrubben, scheuern, *(mit einer Bürste)* reinigen; *(Gase)* waschen

scrub marks senkrechte Falten fpl *(Glasfehler)*

scrubber Wäscher m, Wascher m, Naßabscheider m, Skrubber m; s. scrubbing tower

s. column *(Erdöl)* Waschkolonne f, Auswaschkolonne f

scrubbing Schrubben n, Scheuern n, Reinigen n; Waschen n *(von Gasen)*

s. brush Scheuerbürste f

s. liquid Waschflüssigkeit f *(für Gase beim Naßabscheiden)*

s. oil Waschöl n

s. tower Turmwascher m, Waschturm m, Gaswäscher m, Wäscher m, Wascher m, Skrubber m

scud / to *(Gerb)* streichen, glätten, reinmachen

scud *(Gerb)* Gneist m, Grund m *(Fettreste und Kalkseifen auf Häuten)*

scudding *(Gerb)* Streichen n, Glätten n, Reinmachen n *(Entfernen von Fett- und Kalkseifenresten auf Häuten)*

scuff resistance Abriebfestigkeit f, Abriebbestän-

digkeit f *(bei gleitender oder rollender Reibung und Sand oder Staub)*
scum Schaum *m*, Abschaum *m*; *(Met)* Schlacke *f*; *(Glas)* Schaum *m*; *(Zucker)* Scheideschlamm *m*
  s. juice *(Zucker)* erster Schlammsaft *m*; erster Saturationssaft *m*
  s. pump *(Zucker)* Schlammpumpe *f*
  s. skimmer Schaumabstreifer *m*, Schaumlöffel *m*
  s. thickener Schaumeindicker *m*
  s. yeast Kahmhefe *f*, Kahmpilz *m*, Mycoderma
scumming *(Ker) (beim Brennen entstandene)* Ausblühung *f*; *(Glas)* Schaum *m*
scutch / to *(Text) (Flachs)* schwingen
scutching mill *(Text)* Schwingmaschine *f*
sea foam *(Min)* Meerschaum *m*, Sepiolith *m (ein Phyllosilikat)*
  s. salt Seesalz *n*, Meersalz *n*
  s. water Seewasser *n*, Meer[es]wasser *n*
seal / to [ver]schließen; [ab]dichten; zuschmelzen, verschmelzen, [ver]siegeln; *(Anstrichtechnik) (den Untergrund)* [ab]sperren; *(Met)* nachverdichten *(bei der anodischen Oxydation)*; *(Plastfolien)* siegeln, [ver]schweißen
  s. up zuschmelzen, abschmelzen
seal Verschluß *m*, Abschluß *m*, Dichtung *f*, Abdichtung *f*, Sperre *f*, Packung *f*
  s. face Dicht[ungs]fläche *f*
  s. oil Robbentran *m*, Seehundstran *m*, Robbenöl *n*
  s. ring Dicht[ungs]ring *m*, Packungsring *m*
  s.-weld / to dichtschweißen
sealant Dichtungsmittel *n*
sealed tube Bombenrohr *n*, Einschmelzrohr *n*, Schießrohr *n*
  s.-tube decomposition Aufschluß *m* im Bombenrohr (Einschmelzrohr, Schießrohr)
sealer *(Anstrichtechnik)* Absperrmittel *n*, Sperrgrund *m*, Porenfüller *m*, Porenschließer *m*; *(Plast)* Siegler *m*
sealing Verschließen *n*, Schließen *n*; Abdichten *n*, Dichten *n*; Zuschmelzen *n*, Verschmelzen *n*, Versiegeln *n*, Siegeln *n*; *(Anstrichtechnik)* Absperren *n*, Absperrung *f (des Untergrundes)*; *(Met)* Nachverdichten *n (bei der anodischen Oxydation)*; Siegeln *n*, Schweißen *n*, Verschweißen *n (von Plastfolien)*
  s. face *s.* seal face
  s. fluid Sperrflüssigkeit *f*, Dichtungsflüssigkeit *f*
  s. glass Lötglas *n*
  s. liquid *s. s.* fluid
  s. material Dicht[ungs]werkstoff *m*, Dicht[ungs]stoff *m*, Dicht[ungs]material *n*, Dicht[ungs]mittel *n*
  s. paint *s.* sealer
  s. pressure Dicht[ungs]druck *m*
  s. ring *s.* seal ring
  s. strip Dicht[ungs]schnur *f*
  s. surface *s.* seal face
  s. unit *(Plast)* Schweißgerät *n*, Siegelgerät *n*
  s. wax Siegellack *m*

seam Saum *m*, Naht *f*; *(Geol)* Flöz *n*, Lager *n*
  s. mark Markierung *f* der Siebnaht *(Papierfehler)*
  s. of the machine wire *(Pap)* Siebnaht *f*
seamless pipe (tube) nahtloses Rohr *n*
season / to 1. altern; *(Holz)* trocknen, ablagern; 2. *(Gerb)* appretieren, *(i.e.S.)* glänzen; *(Text)* appretieren; *(Lebm)* würzen
season *(Gerb)* Appretur *f*, *(i.e.S.)* Glanzauftrag *m*
seasoning 1. Altern *n*, Alterung *f*; *(Holz)* Trocknen *n*, Trocknung *f*, Ablagern *n*; 2. *(Text)* Appretieren *n*, Appretur *f*; *(Lebm)* Würzen *n*; *(Lebm)* Würze *f*, Gewürz *n*
  s. machine Appreturmaschine *f*
  s. matter Würzmittel *n*
seat Sitz *m (z.B. eines Ventils)*
  s. ring Sitzring *m*
seatang Seetang *m*, Tang *m*
seaweed Meeresgewächs *n*, Meerespflanze *f*, *(i.e.S.)* Meeresalge *f*
sebaceous talgig, talgartig, Talg...
  s. gland Talgdrüse *f*
sebacic acid $HOOC(CH_2)_8COOH$ Sebazinsäure *f*, Sebazylsäure *f*, Dekandisäure *f*, Oktandikarbonsäure-(1,8) *f*
sec. Redwood *s.* Redwood second
secalonic acid Sekalonsäure *f*, Chrysergonsäure *f*
second carbonation juice *(Zucker)* zweiter Schlammsaft *m*; zweiter Saturationssaft *m*
  s. cure *(Plast)* Nachhärten *n*, Nachhärtung *f*
  s. fillmass *(Zucker)* B-Füllmasse *f*, Mittelproduktfüllmasse *f*
  s. law of thermodynamics zweiter Hauptsatz *m* der Thermodynamik, Entropiesatz *m*
  s. molasses *(Zucker)* Weißsirup *m*, Weißablauf *m*
  s.-order reaction Reaktion *f* zweiter Ordnung
  s.-order transition *(Plast, Gum)* Umwandlung *f* (Phasenumwandlung *f*, Phasenübergang *m*) zweiter Ordnung, *(Gum auch)* Glasumwandlung *f*, Gamma-Umwandlung *f*, α-Anomalie *f*
  s.-order transition point (temperature) *(Gum)* Glasumwandlungstemperatur *f*, Umwandlungspunkt *m* (Umwandlungstemperatur *f*) zweiter Ordnung; *(Plast)* Einfrierpunkt *m*, Einfriertemperatur *f*
  s. raw sugar Rohzuckernachprodukt *n*, Rohzucker II *m*, Nachproduktzucker *m*
  s. run[nings] *(Gär)* Nachlauf *m*
  s. salt *(Gerb) (bereits zur Häutekonservierung verwendetes, wiedergewonnenes Salz)*
  s. screen *s.* secondary screen
  s.-stage floatation Nachflotation *f*
  s. stuff *(veraltet für)* whole stuff
  s. substituent Zweitsubstituent *m*
secondary sekundär
  s. accelerator *(Gum)* Sekundärbeschleuniger *m*, Zweitbeschleuniger *m*, Zusatzbeschleuniger *m*
  s. acetate *s. s.* cellulose acetate
  s. air Sekundärluft *f*, Zweitluft *f*, Zusatzluft *f*
  s. alcohol sekundärer Alkohol *m*
  s. amine $RC \cdot NH \cdot CR$ sekundäres Amin *n*

s. ammonium phosphate $(NH_4)_2HPO_4$ Diammoniumhydrogenphosphat *n*, Ammoniumhydrogen[ortho]phosphat *n*
s. battery Sekundärbatterie *f*, Akkumulator *m*
s. bond *s. s.* valence bond
s. calcium phosphate $CaHPO_4$ Dikalziumhydrogenphosphat *n*, Kalziumhydrogen[ortho]phosphat *n*
s. cell Sekundärelement *n*, sekundäre Kette *f*
s. cellulose acetate Sekundär[zellulose]azetat *n*, Zellulosediazetat *n*, Hydrozelluloseazetat *n*
s. centrifuge Sekundärzentrifuge *f*
s. component *(Farb)* Kupplungskomponente *f*, passive Komponente *f*, Zweitkomponente *f*, Kupplungskörper *m*
s. cooling Nachkühlung *f*
s. coulometry coulometrische Titration *f*
s. creep bleibende Längenänderung *f*, bleibende (konstantzeitabhängige, irreversible) Dehnung *f*
s. crusher Nachbrecher *m*
s. deposit sekundärer Belag *m (von Pflanzenschutzmitteln)*
s. digestion tank *(Wasserchemie)* Nachfaulraum *m*, Nachfaulbecken *n*, Nachfaulbehälter *m*
s. effect Sekundäreffekt *m*
s. electron sekundäres Elektron *n*, Sekundärelektron *n*, SE *n*
s. emission Sekundäremission *f*
s. fermentation Nachgärung *f*, Lagerung *f*; Flaschengärung *f*
s. magnesium phosphate $MgHPO_4$ Magnesiumhydrogen[ortho]phosphat *n*
s. manganous phosphate $MnHPO_4$ Mangan(II)-hydrogen[ortho]phosphat *n*
s. mineral sekundäres Mineral *n*; allothigenes (nicht am Fundort entstandenes) Mineral *n*
s. phosphate $M_2^IHPO_4$ sekundäres Phosphat *n*, Hydrogen[ortho]phosphat *n*
s. photochemical reaction fotochemische Sekundärreaktion *f*
s. pipe Sekundärlunker *m*, sekundärer Lunker *m*, V-Lunker *m (im Gußblock)*
s. quantum number sekundäre (azimutale) Quantenzahl *f*, Azimutalquantenzahl *f*, Orbitaldrehimpulsquantenzahl *f*
s. reaction Nebenreaktion *f*, Sekundärreaktion *f*
s. recovery *(Erdöl)* Sekundärförderung *f*, sekundäre Ausbeutung *f*
s. recovery method *(Erdöl)* sekundäres Förderverfahren (Ausbeutungsverfahren, Verfahren) *n*, Sekundärverfahren *n*
s. reference fuel Unterbezugskraftstoff *m*
s. salt sekundäres (einfachsaures) Salz *n*, Monohydrogensalz *n*, Hydrogensalz *n (einer dreibasigen Säure)*
s. salt effect *(phys Ch)* sekundärer Salzeffekt *m*
s. screen Nachsortierer *m*, Feinsortierer *m*, zweite Sortierstufe *f*
s. screening Nachklassierung *f*, Nachsortierung *f*, Feinsortierung *f*

s. sedimentation basin (tank) Nachklärbecken *n*
s. sodium orthophosphate Na_2HPO_4 Natrium[mono]hydrogenphosphat *n*, Dinatriumhydrogenphosphat *n*
s. standard Sekundärstandard *m*, sekundärer Standard *m*
s. standard electrode sekundäre Standardelektrode *f*
s. structure *(Krist)* Mosaikblöckchen *n*
s. treatment Nachbehandeln *n*, Nachbehandlung *f*
s. valence Nebenvalenz *f*
s. valence bond Nebenvalenzbindung *f*
s. valency *s. s.* valence
s. wall Sekundärwand *f (einer Faser)*
secret ink Geheimtinte *f*, sympathetische Tinte *f*
secretion Sekretion *f*, Absonderung *f*
secretory sekretorisch, den Sekretionsvorgang betreffend, Sekretions...
s. process Sekretionsvorgang *m*
section Abschnitt *m*, Teil *m*; *(Mikroskopie)* Schnitt *m*, *(bei Gesteinen auch)* Schliff *m*; *(Plast)* Profil *n*
s. roller Gliederwalze *f*
sectional paper Profilpapier *n*, *(i.e.S.)* Millimeterpapier *n (oder)* Diagrammpapier *n*
security paper Sicherheitspapier *n*
sedative *(Pharm)* Sedativum *n*, Beruhigungsmittel *n*
sedge peat Seggentorf *m*, Riedgrastorf *m*
sediment / to 1. *(aus Flüssigkeiten)* abscheiden, niederschlagen, ausfällen, absitzen lassen, sedimentieren; ablagern, ausscheiden *(von der Flüssigkeit)*; einen Bodenkörper (Niederschlag) bilden, absetzen; 2. sich setzen (absetzen, abscheiden, ausscheiden, ablagern, niederschlagen), ausfallen, [ab]sinken, absitzen, zur Ausscheidung gelangen, sedimentieren
sediment Sediment *n*, Bodenkörper *m*, Niederschlag *m*, Bodensatz *m*, Satz *m*, Ausscheidung *f*, Ausscheidungsprodukt *n*, Ablagerung *f*, *(Lebm auch)* Trub *m*; *(Geol)* s. sedimentary rock
s. chamber Absetzraum *m*
sedimentary sedimentär, Sediment[ations]..., Niederschlags..., Absatz..., Absetz...
s. deposit *(Geol)* Ablagerung *f*
s. fermentation Untergärung *f*
s. petrography Sedimentpetrografie *f*
s. rock Sedimentgestein *n*, Absatzgestein *n*, Schichtgestein *n*
sedimentate / to *(aus Flüssigkeiten)* abscheiden, niederschlagen, ausfällen, absitzen lassen, sedimentieren
sedimentation Abscheiden *n*, Niederschlagen *n*, Ausfällen *n*, Absitzenlassen *n*, Sedimentieren *n*, Sedimentation *f*; Ablagern *n*, Ausscheiden *n*; Absetzen *n*, Bildung *f* eines Bodenkörpers (Niederschlags); Absitzen *n*, Ausfallen *n*, Sinken *n*, Absinken *n*, Sedimentation *f*; *(Geol)* Sedimentation *f*, Ablagerung *f*

s. analysis Sedimentationsanalyse f
s. area Absetzfläche f
s. balance Sedimentationswaage f
s. basin Absetzbehälter m, Absetzgefäß n, Absetztank m, Absetzbecken n, Absetzer m, Absitzgefäß n, Absitzbehälter m, Absitzbecken n, Klärbecken n, Klärbehälter m, Klärbassin n, Klärgefäß n, Klärtank m
s. centrifuge s. s.-type centrifuge
s. constant Sedimentationskonstante f
s. distance Absetzabstand m, Absetzweg m
s. equilibrium Sedimentationsgleichgewicht n
s. inhibitor (Farb) Absetzverhinderungsmittel n
s. performance Absetzleistung f
s. potential Sedimentationspotential n
s. rate Absinkgeschwindigkeit f, Sinkgeschwindigkeit f, Sedimentationsgeschwindigkeit f, Absetzgeschwindigkeit f, Klärgeschwindigkeit f
s. save-all (Pap) Sedimentationsstoffänger m, Absetzstoffänger m, nach dem Sedimentationsprinzip (Absetzprinzip) arbeitender Stoffänger m
s. tank s. s. basin
s. test Sedimentationstest m
s.-type centrifuge Absetzzentrifuge f, Absetzschleuder f, Sedimentierzentrifuge f, Vollmantelzentrifuge f, Vollmantelschleuder f
s. velocity s. s. rate
sedimentography s. sedimentary petrography
sedoheptulose Sedohept[ul]ose f (eine Ketoheptose)
see-saw development (Foto) kontrollierte Entwicklung f, Handentwicklung f
seed / to (Lösungen mit Kristallkeimen) impfen
seed Samen m; (Glas) Gispe f, Gisbe f, Gäse f, Gasbläschen n (Fehler)
s. crystal Impfkristall m, Keimkristall m, Saatkristall m, Kristallkeim m
s. disinfectant Saat[gut]beizmittel n
s. fat Samenfett n
s. fibre Samenfaser f
s. hair Samenhaar n
s. lac Samenlack m, Körnerlack m
s. oil Samenöl n
s. protectant Saat[gut]beizmittel n
s. protection Samenschutz m, Samenbeize f, Saatgutbeize f
s. treatment Saatgutbehandlung f
s. yeast Mutterhefe f
seeded solution geimpfte Lösung f
seeding Impfen n (zur Anregung der Kristallisation)
s. material (Lebm) Säurewecker m
seedy glass gispiges Glas n
seepage 1. Versickern n; Einsickern n; Durchsickern n; Lecken n; 2. Leckflüssigkeit f; 3. (Erdöl) Ausbiß m (Ölaustritt an der Oberfläche)
Segas process Segas-Verfahren n, Stanier-MacKean-Verfahren n (thermisch-katalytisches Ölspaltungsverfahren)
Seger cone (Ker) Seger-Kegel m, Brennkegel (Schmelzkegel) m nach Seger

Seger formula (Ker) Seger-Formel f
Seger's porcelain Seger-Porzellan n
segregate / to 1. absondern, entmischen, trennen, [ab]scheiden, isolieren; (Met) seigern; 2. sich absondern (entmischen); (Met) seigern
segregation Absonderung f, Entmischung f, Trennung f, Scheidung f, Abscheidung f, Segregation f; (Met) Seigerung f
s. coefficient Segregationskonstante f, Abscheidungskonstante f, Verteilungskoeffizient m (in der Zonenschmelztheorie)
Seidlitz mixtures (powders) (Am) (Brausepulver aus Kaliumnatriumtartrat, Natriumhydrogenkarbonat und gesondert verpackter kristalliner Weinsäure)
Seignette salt $KOOC(CHOH)_2COONa \cdot 4H_2O$ Seignettesalz n, Natronweinstein m, Kaliumnatriumtartrat n
seismic method seismisches Verfahren n, seismische Methode (Spürmethode) f (bei der Erdölsuche)
Seitz [germ-proofing] filter Seitz-Filter n, Seitz-Entkeimungsfilter n
seize / to sich festfressen (festklemmen)
seizing, seizure Festfressen n
selacholeic acid $C_{23}H_{45}COOH$ Selachensäure f, Nervonsäure f, 15-Tetrakosensäure f
selachyl alcohol $C_3H_5(OH)_2 \cdot C_{18}H_{35}$ Selachylalkohol m, 1-Oleylglyzeryläther m
select / to [aus]wählen; auslesen, selektieren
selectifier [screen] (Pap) Selectifier m, Druckseparator m (ein Durchflußsortierer)
selection Auswählen n, Auswahl f, Wahl f; Auslese f, Selektion f
s. principle (rule) Auswahlregel f
selective selektiv [wirkend], auswählend
s. adsorption selektive Adsorption f
s. corrosion selektive Korrosion f
s. cracking selektive Krackung (Spaltung) f, Selektivkrackung f
s.-cracking process Selektivkrackprozeß m, selektives Spaltverfahren n, selektiver Krackprozeß m
s. exchanger Selektivaustauscher m
s. floatation selektive (differentielle) Flotation f, Selektivflotation f, Differentialflotation f
s. growth promoter selektiver Wuchsstoff m
s. herbicide Selektivherbizid n, selektives (selektiv wirkendes) Herbizid n
s. hydrogenation selektive Hydrierung f, (i.e.S.) selektive Härtung f (von Fetten)
s. medium Selektivnährboden m
s. resin Selektivaustauscherharz n
s. solvent Selektivlösungsmittel n, Selektivlösemittel n, selektives (selektiv wirkendes) Lösungsmittel (Lösemittel, Solvent) n
selectivity Selektivität f, selektive Wirkung f, Trennschärfe f
selecto Selekto n (beim Duosolverfahren verwendetes Phenol-Kresol-Gemisch)

selenate $M^I_2SeO_4$ Selenat n
selenic Selen...; höherwertigem Selen entsprechend, *(meist)* Selen(VI)-...
  s. acid H_2SeO_4 Selensäure f
  s. silver *(Min)* Selensilber n, Selensilberglanz m, *(veraltet für)* Naumannit m *(Silberselenid)*
selenide M^I_2Se Selenid n
selenious Selen...; niederwertigem Selen entsprechend, *(meist)* Selen(IV)-...
  s. acid H_2SeO_3 selenige Säure f
  s. [acid] anhydride s. s. oxide
  s. oxide SeO_2 Selen(IV)-oxid n, Selendioxid n
selenite $M^I_2SeO_3$ Selenit n; *(Min)* Selenit m *(Kalziumsulfat-2-Wasser)*
selenium Se Selen n
  s. cell Selenzelle f
  s. dioxide s. selenious oxide
  s. disulphide SeS_2 Selendisulfid n, Selen(IV)-sulfid n
  s. glass Selenfilter n
  s. halide Selenhalogenid n
  s. hexafluoride SeF_6 Selenhexafluorid n, Selen(VI)-fluorid n
  s. monobromide Se_2Br_2 Diselendibromid n, Selen(II)-bromid n
  s. monochloride Se_2Cl_2 Diselendichlorid n, Selen(II)-chlorid n
  s. monoiodide Se_2J_2 Diselendijodid n, Selen(II)-jodid n
  s. monosulphide SeS Selen[mono]sulfid n, Selen(II)-sulfid n
  s. nitride Se_4N_4 Selennitrid n, Tetraselentetranitrid n, Selenstickstoff m
  s. oxybromide $SeOBr_2$ Selenoxidbromid n
  s. oxychloride $SeOCl_2$ Selenoxidchlorid n
  s. oxyfluoride $SeOF_2$ Selenoxidfluorid n
  s. rectifier Selengleichrichter m
  s. ruby glass Selenrubinglas n
  s. sulphotrioxide (sulphoxide) $SeSO_3$ Selensulfit n
  s. tetrabromide $SeBr_4$ Selentetrabromid n, Selen(IV)-bromid n
  s. tetrachloride $SeCl_4$ Selentetrachlorid n, Selen(IV)-chlorid n
  s. tetrafluoride SeF_4 Selentetrafluorid n, Selen(IV)-fluorid n
  s. trioxide SeO_3 Selentrioxid n, Selen(VI)-oxid n
selenonium compound Selenoniumverbindung f
selenous s. selenious
selensilver *(Min)* Selensilber n, Selensilberglanz m, *(veraltet für)* Naumannit m *(Silberselenid)*
self-absorption Selbstabsorption f, Eigenabsorption f
  s.-adherent selbstklebend
  s.-adhesive film Selbstklebefolie f
  s.-adhesive product Selbstklebeerzeugnis n
  s.-alkylation Selbstalkylierung f
  s.-baking electrode selbstbackende (selbst[ein]brennende) Elektrode f, Dauerelektrode f
  s.-catalysis Autokatalyse f

s.-cleansing s. s.-purification
s.-coincidence *(Krist)* Deckung f
s.-colour Eigenfarbe f, Unifarbe f; Selbstfarbstoff m, reiner (ungemischter) Farbstoff m
s.-condensation Selbstkondensation f, extramolekulare Kondensation f
s.-consistent field selbstkonsistentes Feld n
s.-consistent-field calculation SCF-Rechnung f
s.-consistent-field method Methode f des selbstkonsistenten Feldes, SCF-Methode f, self-consistent-field-Methode f
s.-consistent-field molecular orbital SCF-Molekülorbital n
s.-consistent method selbstkonsistente Methode f, selbstkonsistentes Verfahren n
s.-consistent wave function selbstkonsistente Wellenfunktion f
s.-contained press Presse f mit Einzelantrieb
s.-controlled selbsttätig gesteuert, automatisch
s.-curing 1. *(Gum)* selbstvulkanisierend; 2. *(Plast)* selbsthärtend, eigenhärtend
s.-curing cement *(Gum)* selbstvulkanisierende Lösung f
s.-diffusion Selbstdiffusion f, Eigendiffusion f
s.-emulsifying selbstemulgierend
s.-etch pretreatment primer, s.-etching primer Reaktionsgrundierung f, Haftgrundierung f, Haftgrundmittel n, Reaktions-Primer m, Wash-Primer m
s.-extinguishing selbst[ver]löschend, selbstauslöschend
s.-fluxing selbstgehend, selbstgängig *(Erz)*
s.-heating Selbsterhitzung f
s.-ignition Selbstentzündung f; Selbstzündung f
s.-ignition temperature Selbstentzündungstemperatur f
s.-lubricating selbstschmierend
s.-lubrication Selbstschmierung f
s.-oxidation and self-reduction Disproportionierung f, Dismutation f *(z.B. bei der Cannizzaroschen Reaktion)*
s.-potential Eigenpotential n
s.-potential curve Eigenpotentialkurve f
s.-priming selbstansaugend
s.-purification Selbstreinigung f
s.-quenched (s.-quenching) counter selbstlöschendes Zählrohr n, selbstlöschender Zähler m, Auslösezähler m
s.-reinforced elastomer hochstyrolhaltiges Polymerisat (Polymeres) n, Mischpolymerisat n mit hohem Styrolgehalt
s.-smoothing fabric selbstglättendes Gewebe n
s.-stress Leerlaufspannung f *(in der Trommelwand einer Zentrifuge)*
s.-supporting film Folie f, trägerloser (selbsttragender) Film m
s.-sustained, s.-sustaining spontan *(z.B. Verbrennung)*
s.-toning *(Foto)* selbsttonend
s.-vulcanizing s. s.-curing 1.

s.-vulcanizing cement s. s.-curing cement
Seliwanoff reaction Seliwanow-Reaktion f, Seliwanowsche Reaktion f *(zum Nachweis von Hexosen)*
sellaite *(Min)* Sellait m *(Magnesiumfluorid)*
selter, selters water, seltzer [water] Selterswasser n
selwynite *(Min)* Selwynit m *(Cr-haltiger Bol)*
Semet-Solvay oven Semet-Solvay-Ofen m *(Kokereiofen)*
semianthracite [coal] Semianthrazit m, Halbanthrazit m, Magerkohle f
semiautomatic halbautomatisch
semibatch halbkontinuierlich
semibituminous coal semibituminöse (halbbituminöse) Kohle f
semibleached halbgebleicht *(Zellulose)*
semicarbazide $H_2NCONHNH_2$ Semikarbazid n
semicarbazone Semikarbazon n *(Reaktionsprodukt von Aldehyden oder Ketonen mit Semikarbazid)*
semicarbonization s. semicoking
semichemical halbchemisch
s. plant Halbzellstoffwerk n, Halbzellstoffanlage f
s. pulp Halbzellstoff m, Semichemical-Zellstoff m
s.-pulp mill s. s. plant
s. pulping Halbzellstoffaufschluß m, halbchemischer Aufschluß m; Neutralsulfitverfahren n
s. pulping liquor *(Pap)* Neutralsulfitkochlauge f
s. spent liquor *(Pap)* Neutralsulfitablauge f
semichrome tannage Semichromgerbung f *(mit pflanzlichen Gerbstoffen und anschließend Chromsalzen)*
semicoke Halbkoks m
semicoking Halbverkokung f
semicolloid Semikolloid n, Halbkolloid n
semicommercial[-scale] halbtechnisch, im halbtechnischen Maßstab
s.-scale test halbtechnischer Versuch m
semiconducting halbleitend, Halbleiter...
s. compound halbleitende Verbindung f, Halbleiterverbindung f
s. layer Halbleiterschicht f
s. material Halbleitermaterial n, halbleitendes Material n, halbleitender Stoff (Werkstoff) m, Halbleiter m
semiconductive s. semiconducting
semiconductor Halbleiter m
s. junction Halbleiterübergang m, Halbleitersperrschicht f
s. particle counter Halbleiterteilchenzähler m
s. photocell Halbleiterfotozelle f
s. resistor Halbleiterwiderstand m
s. strain gauge Halbleiterdehnungsmesser m
s. thermoelement Halbleiterthermoelement n
semicontinuous halbkontinuierlich
s. distillation halbkontinuierliche Destillation f
semicure *(Gum)* Vorvulkanisation f, Vorheizung f
semicyclic semizyklisch

s. bond semizyklische Bindung (Doppelbindung) f
semicylinder Halbzylinder m
semidecomposed halbzersetzt
semidine $C_6H_5NHC_6H_4NH_2$ Semidin n
s. rearrangement (transformation) Semidinumlagerung f, halbseitige (halbe) Benzidinumlagerung f
semidirect halbdirekt
s. process (system) halbdirektes Ammoniakgewinnungsverfahren n, halbdirektes Verfahren n zur Ammoniakgewinnung
semidry halbtrocken, halbnaß
s. pressing *(Ker)* Pressen n in halbtrockenem Zustand m, Halbtrockenpressen n
s. wine halbtrockener Wein m
semidrying halbtrocknend
s. oil halbtrocknendes (mäßig trocknendes, schwach trocknendes) Öl n
semidull halbmatt
semiebonite Semiebonit n, Halbhartgummi m, halbharter Gummi m
semienclosed impeller halboffenes Laufrad n *(einer Kreiselpumpe)*
semifluid s. semiliquid
semifusinite Semifusinit m, Halbfusinit m *(ein Mazeral der Steinkohle)*
semigloss[y] *(Anstrichtechnik)* Halbglanz...; *(Foto)* halbmatt
semikilled steel halbberuhigter Stahl m
semiliquid halbflüssig, dickflüssig, zähflüssig
s. manure *(Landw)* Gülle f
s. manure management *(Landw)* Güllewirtschaft f
semilog[arithmic] paper halblogarithmisches Papier n, einfach logarithmisch geteiltes Papier n, Exponentialpapier n
semimat halbmatt
s. glaze *(Ker)* Halbmattglasur f
semimatt[e] s. semimat
semimetal Halbmetall n, Übergangsmetall n
semimetallic halbmetallisch
s. packing halbmetallische Dichtung (Packung) f, Metall-Weichstoffdichtung f
semimicro Kjeldahl method Halbmikromethode f von Kjeldahl
s. method Halbmikromethode f, Halbmikroverfahren n
s. Rast method Halbmikromethode f von Rast *(Bestimmung der relativen Molekülmasse)*
s. torsion balance Halbmikro-Torsionswaage f
semimicroanalysis Halbmikroanalyse f, Semimikroanalyse f
semimicrodetermination Halbmikrobestimmung f
semimicromethod s. semimicro method
semimuffle kiln Halbmuffelofen m, halbgemuffelter Ofen m
semiopen impeller s. semienclosed impeller
semipermeability Semipermeabilität f
semipermeable semipermeabel, halbdurchlässig, einseitig durchlässig

s. membrane (partition) semipermeable (halbdurchlässige) Membran[e] (Scheidewand, Trennwand, Wand, Haut) f
semipolar semipolar, halbpolar
s. [double] bond semipolare (halbpolare) Doppelbindung f, koordinative Bindung f, dative Bindung f, Donator-Akzeptor-Bindung f
semiporcelain Halbporzellan n
semipositive mould (Plast) kombiniertes Abquetsch- und Füllpreßwerkzeug n, kombinierte Abquetsch- und Füllform f, Preßwerkzeug n mit vertieft liegendem Abquetschrand
semiprecious stone Halbedelstein m
semiproducer gas Wassergas n
semiquantitative halbquantitativ
semiquinone Semichinon n, Halbchinon n, Merichinon n
semirefractory halbfeuerfest
semireinforcing black (Gum) halbverstärkender (halbaktiver) Ruß m
s. furnace black (Gum) Semi-Reinforcing-Furnace-Ruß m, SRF-Ruß m (halbaktiver Ofenruß)
semirigid foam (Plast) halbharter Schaum m
semisilica brick SiO_2-reicher Schamottestein m
s. refractory SiO_2-reiches Schamotteerzeugnis n (mehr als 72 % Siliziumdioxid)
semisiliceous refractory SiO_2-reiches Schamotteerzeugnis n (weniger als 93 % Siliziumdioxid)
semisolid halbfest, halbhart, semifest
semisynthetic fibre Regeneratfaserstoff m; Regeneratfaser f
semisystematic name halbsystematischer (halbtrivialer) Name m, Halbtrivialname m, Stammsubstanzname m
semitechnical halbtechnisch
semitransparent halbdurchscheinend
semitrivial name s. semisystematic name
semiultra accelerator (Gum) Halbultrabeschleuniger m
semivitreous, semivitrified (Ker) halbverglast, halbglasartig
semivulcanized board Vulkanfiberersatz m
semiwater gas Halbwassergas n
semiworks Halbbetriebsanlage f, Pilotanlage f, Versuchsanlage f, Versuchsbetrieb m
sempervirine Sempervirin n (Alkaloid)
senarmontite (Min) Senarmontit m (Antimon(III)-oxid)
senecic acid $(CH_3)_2C=CHCOOH$ Seneziosäure f, 3-Methyl-2-butensäure f, 2-Methyl-1-propen-1-karbonsäure f, Isopropylidenäthansäure f, 3,3-Dimethylakrylsäure f
senecio alkaloid Senezioalkaloid n, Pyrrolizidinalkaloid n
senecionine Senezionin n
senior (Nomenklatur) vorrangig
seniority (Nomenklatur) Vorrang m, Rangordnung f
senna Sennesblätter npl (von Cassia-Arten)

Sennaar gum Sennaar-Gummi n (vorwiegend von Acacia senegal)
sensible heat fühlbare Wärme f
sensing device Meßfühler m, Fühler m
sensitive empfindlich
s. emulsion fotografische (lichtempfindliche) Emulsion f, Fotoemulsion f
s. ink Sicherheitsfarbe f (Druckfarbe für Sicherheitspapiere)
s. layer fotografische (lichtempfindliche) Schicht f
s. material fotografisches (lichtempfindliches) Material n, Fotomaterial n
s. to acid säureempfindlich
s. to impact schlagempfindlich, stoßempfindlich
s. to light lichtempfindlich
s. to penicillin penizillinempfindlich
sensitiveness s. sensitivity 1.
s. to friction Reibungsempfindlichkeit f, Empfindlichkeit f gegen Reibung
s. to impact Schlagempfindlichkeit f, Stoßempfindlichkeit f
s. to initiation Initiierbarkeit f, Detonationsfähigkeit f
s. to shock s. s. to impact
sensitivity 1. Empfindlichkeit f, Sensibilität f; 2. Kraftstoffempfindlichkeit f, Sensitivity f (Verhältnis von Research- zu Motor-Oktanzahl)
s. to light Lichtempfindlichkeit f
sensitization Sensibilisierung n, (Foto auch) Empfindlichmachen n, Lichtempfindlichmachen n
sensitize / **to** sensibilisieren, (Foto auch) [licht]empfindlich machen
sensitized fluorescence sensibilisierte Fluoreszenz f
s. material [licht]empfindliches (fotografisches) Material n, Fotomaterial n
sensitizer Sensibilisator m
sensitizing s. sensitization
s. dye Sensibilisierungsfarbstoff m, sensibilisierender Farbstoff m
sensitometer Sensitometer n, Empfindlichkeitsmesser m
sensitometry Sensitometrie f
sensor s. sensing device
sensory evaluation (test) sensorische Analyse f, organoleptische Prüfung f, Sinnenprüfung f der Lebensmittel, (i.e.S.) Geschmacksprüfung f
separability Trennbarkeit f, Abtrennbarkeit f, Abscheidbarkeit f
separable [ab]trennbar, abscheidbar
separate / **to** 1. separieren, [ab]trennen, abspalten; zerlegen; absondern, [ab]scheiden, ausscheiden; klassieren; sortieren; 2. sich ausscheiden (abscheiden), zur Ausscheidung gelangen
s. out sich ausscheiden (abscheiden), ausfallen
separate coating (Pap) Streichen n außerhalb der Maschine, Separatstreichen n, Separatstrich m
s. pot mould (Plast) Mehrfachwerkzeug n (Mehrfachform f) mit getrennten Füllräumen

s. [sewerage] system Trennsystem *n*, Trennent-
wässerung *f*
separated milk entrahmte Milch *f*, Magermilch *f*
separating *s.* separation
   s. agent Trennmittel *n*, Scheidemittel *n*; Zusatzstoff *m*, Schleppmittel *n*, Mitnehmer *m* *(bei
der Destillation)*
   s. diffusion Trenndiffundieren *n*, Trenndiffusion
f
   s. efficiency Entrahmungsschärfe *f (Fettentzug)*
   s. fluid Trennmedium *n*, Trennflüssigkeit *f*,
Trennmittel *n*
   s. funnel Scheidetrichter *m*, Trenntrichter *m*,
Schütteltrichter *m*
   s. liquid *s.* s. fluid
   s. temperature Trenntemperatur *f*, *(i.e.S.)* Ent-
rahmungstemperatur *f*
   s. tube Trennrohr *n*
   s. vessel Scheidegefäß *n*, Scheidebehälter *m*
separation 1. Separieren *n*, Separation *f*, Trennen
n, Trennung *f*, Abtrennung *f*, Abspaltung *f*,
Zerlegung *f*; Absonderung *f*, Scheiden *n*,
Abscheiden *n*, Ausscheidung *f*; Klassierung *f*;
Sortierung *f*; 2. Abstand *m* *(z.B. im Kristallgitter)*
   s. column Trennsäule *f*
   s. effect Trenneffekt *m*, Trennwirkung *f*
   s. efficiency Trennleistung *f*, Trennschärfe *f*,
Trenn[ungs]grad *m*; Abscheidegrad *m*, Ausscheidungsgrad *m*, Entstaubungsgrad *m*
   s. factor Trennfaktor *m*
   s. in flocks Ausflocken *n*, Ausflockung *f*
   s. method Trennmethode *f*, Trennverfahren *n*,
Trenntechnik *f*
   s. of charge Ladungstrennung *f*
   s. of ions Ionentrennung *f*
   s. of isotopes Isotopentrennung *f*
   s. of isotopes by centrifuge Isotopentrennung *f*
mit der Zentrifuge, Schleuderverfahren *n*
   s. of isotopes by diffusion pumps Isotopentrennung *f* durch Diffusionspumpen
   s. of isotopes by diffusion through gas
Isotopentrennung *f* durch Gasdiffusion
   s. of milk Entrahmen *n*, Entrahmung *f*
   s. process (technique) *s.* s. method
separator Abscheider *m*, Scheider *m*, Separator *m*,
(i.e.S.) Trennzentrifuge *f*; Entstauber *m*; Sichter
m; *(Erdöl)* Separator *m*, Trenntank *m*, Gas-
Öl-Separator *m*, Gasseparator *m*, Gastrennanlage *f*; Scheidetrichter *m*
   s. tank Absetztank *m*, Absetzbehälter *m*, Absetzgefäß *n*, Absitzgefäß *n*, Absetzbecken *n*,
Absetzer *m*, Absitzbehälter *m*, Absitzbecken *n*,
Klärbecken *n*, Klärbehälter *m*, Klärbassin *n*,
Klärgefäß *n*, Klärtank *m*
Separator-Nobel process Separator-Nobel-Ver-
fahren *n*, SN-Verfahren *n* *(Entparaffinierungsverfahren mit Trichloräthylen als Lösungsmittel)*
separatory funnel *s.* separating funnel
   s. vessel *s.* separating vessel
sepia toning *(Foto)* Sepiatonung *f*, Brauntonung *f*

sepiolite *(Min)* Sepiolith *m*, Meerschaum *m*
septavalence, septavalency Siebenwertigkeit *f*,
Heptavalenz *f*
septavalent siebenwertig, heptavalent
septic tank Faulgrube *f*, Faulbecken *n*, Faulbehälter *m*
   s.-tank method Faulverfahren *n*
septivalence, septivalency *s.* septavalence
septivalent *s.* septavalent
septum Scheidewand *f*
sequence Sequenz *f*, Reihenfolge *f*, Folge *f*
   s. of amino-acid residues Aminosäuresequenz *f*,
Aminosäure[reihen]folge *f*, Reihenfolge *f* [der
Verknüpfung] der Aminosäuren
sequester / to maskieren
sequestering action Maskierung *f*
   s. agent Maskierungsmittel *n*, Maskierungs-
reagens *n*, maskierender Stoff *m*, Komplexbildner *m*, Sequestiermittel *n*
sequestration Maskierung *f*, Komplexbildung *f*
seralbumin *s.* serum albumin
serglobulin *s.* serum globulin
serial dilution Reihenverdünnung *f*
   s.-dilution method Reihenverdünnungsmethode
f (zur biologischen Bestimmung von Antibiotika)
sericin Serizin *n*, Seidenbast *m*, Seidenleim *m*
(Eiweißkörper der Naturseide)
sericinase Serizinase *f (proteolytisches Ferment)*
sericite *(Min)* Serizit *m*
series Serie *f*, Reihe *f*, Folge *f*, Gruppe *f*
   ***D*-series** *D*-Reihe *f*
   ***L*-series** *L*-Reihe *f*
   s. limit Seriengrenze *f (einer Spektralserie)*
   s. of bands Bandengruppe *f*
   s. of lines *s.* s. of spectrum lines
   s. of measurements Meßreihe *f*
   s. of reactions Reaktionskette *f*, *(i.e.S.)* Reak-
tionszyklus *m*, Kettenfortpflanzungsreaktionen
fpl
   s. of salts Salzreihe *f*
   s. of screens (sieves) Siebreihe *f*, Siebskala *f*
   s. of spectrum lines Linienserie *f*, Serie *f* von
Spektrallinien
   s. of tensions Spannungsreihe *f*
   s. of test[ing] sieves Prüfsiebreihe *f*
   s. term Spektralterm *m*
serine $HOCH_2CH(NH_2)COOH$ Serin *n*, Ser,
2-Amino-3-hydroxypropionsäure *f*
serotonin Serotonin *n*, Enteramin *n*, 5-
Hydroxytryptamin *n*, 5-HT
serpentine *(Min)* Serpentin *m*
   s. asbestos *(Min)* Serpentinasbest *m*, Chrysotil-
asbest *m*
   s. cooler Rieselkühler *m*
serrated rubber Glocke *f* mit gezacktem Rand
(einer Glockenbodenkolonne)
serum Serum *n*, Blutserum *n*; Molke *f*
   s. albumin Serumalbumin *n*, Blutserumalbumin
n, Blutwasseralbumin *n*
   s. butter Molkenbutter *f*

s. globulin Serumglobulin *n*
s. protein Serumprotein *n*
service chest *(Pap)* Maschinenbütte *f*, Arbeitsbütte *f*, Stoffbütte *f*
s. temperature Gebrauchstemperatur *f*
s. water Brauchwasser *n*, Gebrauchswasser *n*, Betriebswasser *n*, Fabrikationswasser *n*, Nutzwasser *n*
sesame oil Sesamöl *n (von Sesamum indicum L.)*
Sesci furnace Sesci-Ofen *m (kohlenstaubbeheizter Trommelofen)*
sesquibenihiol Kostol *n (bizyklisches Terpen)*
sesquioxide Sesquioxid *n (veraltet, besser durch Stocksche Benennung zu ersetzen)*
sesquiterpene $C_{15}H_{24}$ Sesquiterpen *n*
set / to 1. aufstellen, aufbauen *(eine Apparatur)*; [aus]fällen, abscheiden, niederschlagen, absitzen lassen, sedimentieren; *(Text)* fixieren; *(Ker)* [ein]setzen; 2. erstarren, erhärten, aushärten, hart (fest) werden, sich [ver]festigen, [ab]binden; sich setzen (absetzen, abscheiden, niederschlagen), absitzen *(Trübstoffe)*; sich klären *(Flüssigkeit)*
s. aside stehen lassen
s. free freimachen, freisetzen, befreien
s. in brickwork einmauern
s. off abschmieren, ablegen, abschmutzen, sich abdrucken *(Farbe auf frischen Druckbogen)*
s. out *(Gerb)* ausstoßen, ausrecken, plattieren
s. to N *(eine Lösung)* auf n einstellen
s. up 1. erhärten; zusammenbacken, verbacken; 2. *(Gum)* anvulkanisieren, anspringen, anbrennen
set 1. Satz *m*; Block *m*, Einheit *f*, Gruppe *f*; 2. Erstarren *n*, Erstarrung *f*, Erhärtung · *f*, Aushärtung *f*, Hartwerden *n*, Festwerden *n*, Verfestigung *f*, Abbinden *n*; 3. bleibende Verformung (Deformation, Gestaltsänderung) *f*, Formänderungsrest *m*, Verformungsrest *m*
s. cure *(Gum)* Vorvulkanisation *f*, Vorheizung *f*
s.-cured *(Gum)* vorvulkanisiert, vorgeheizt
s. milk Setzmilch *f*, Dickmilch *f*, Sauermilch *f*, Schlippermilch *f*, Plundermilch *f*
s. of mouldings *(Plast)* Ausstoß *m (bei einer Pressung)*
s. of rolls Walzensatz *m (z.B. eines Kalanders)*; *(Pap)* Rollensatz *m (Papierrollen)*
s. of screens (sieves) Siebsatz *m*
s. of weights Gewichtssatz *m*
s.-off paper Abschmutzmakulatur *f*, Abschmutzpapier *n*, Antimaculepapier *n*, Zwischenlagepapier *n*
s.-off sheet Einschießbogen *m*, Zwischenlagebogen *n*
s. point Sollwert *m*
s.-up 1. Erhärten *n*; Zusammenbacken *n*, Verbacken *n*; 2. *(Gum)* Anspringen *n*, Anvulkanisieren *n*, Anvulkanisation *f*, Anbrennen *n*; 3. *(fertiger)* Aufbau *m (einer Apparatur)*
s.-up time s. setting time

sett grease Starrschmiere *f*
setter *(Ker)* Einsatzbehälter *m (Brennhilfsmittel)*
setting 1. Aufstellen *n*, Aufbau *m (einer Apparatur)*; 2. Ausfällen *n*, Abscheiden *n*, Niederschlagen *n*, Absitzenlassen *n*, Sedimentieren *n*, Sedimentation *f*; 3. *(Text)* Fixieren *n*, Fixierung *f*; 4. *(Ker)* Setzen *n*, Einsetzen *n*, Besatz *m*; Charge *f*, Besatz *m*; 5. *s.* set 2.; 6. Absitzen *n (von Trübstoffen)*; 7. Klären *n*, Klärung *f (einer Flüssigkeit)*
s. accelerator Abbindebeschleuniger *m*
s. base *(Ker)* Setzunterlage *f*
s. behaviour Abbindeverhalten *n*
s. height *(Ker)* Besatzhöhe *f*, Setzhöhe *f*
s.-out cylinder *(Gerb)* Reckerwalze *f (der Ausreck- oder Ausstoßmaschine)*
s. point Erstarrungspunkt *m*, EP ; Stockpunkt *m*, Setting Point *m (bei Ölen)*; Gerinnungspunkt *m (z.B. bei Milch)*
s. retarder Abbindeverzögerer *m*
s. space *(Ker)* Besatzfläche *f*, Besatzraum *m*, Nutzraum *m (im Brennofen)*
s. time Erstarrungszeit *f*, Härtungszeit *f*, Härtezeit *f*, Abbindezeit *f*, Abbindedauer *f*
s.-up Anvulkanisieren *n*, Anvulkanisation *f*, Anspringen *n*, Anbrennen *n*
settle / to 1. *(aus Flüssigkeiten)* abscheiden, niederschlagen, ausfällen, absitzen lassen, sedimentieren; ablagern, ausscheiden *(von der Flüssigkeit)*; einen Bodenkörper (Niederschlag) bilden, absetzen; 2. sich setzen (absetzen, abscheiden, ausscheiden, ablagern, niederschlagen), ausfallen, [ab]sinken, absitzen, zur Ausscheidung gelangen, sedimentieren
s. down sich absetzen (abscheiden); *(Gerb)* abbinden
s. out *s.* to settle
settle blow *(Glas)* Niederblasen *n*, Festblasen *n*
s. mark *(Glas)* Runzel *f*, Kühlfalte *f (Oberflächenfehler)*
settled material Sinkgut *n*
settlement chamber *s.* settling chamber
settler 1. Absetzbecken *n*, Absetzbehälter *m*, Absetzgefäß *n*, Absetztank *m*, Absetzer *m*, Absitzbecken *n*, Absitzbehälter *m*, Absitzgefäß *n*, Klärbecken *n*, Klärbassin *n*, Klärbehälter *m*, Klärgefäß *n*, Klärtank *m*; 2. Klärapparat *m*, Absetzapparat *m*; Klärwanne *f*, Klärgrube *f*; Scheidebehälter *m*, Abscheider *m*
settling 1. Abscheiden *n*, Niederschlagen *n*, Ausfällen *n*, Absitzenlassen *n*, Sedimentieren *n*, Sedimentation *f*; Ablagern *n*, Ausscheiden *n*, Absetzen *n*, Bildung *f* eines Bodenkörpers (Niederschlags); Absitzen *n*, Ausfallen *n*, Sinken *n*, Absinken *n*, Sedimentation *f*; 2. Bodenkörper *m*, Niederschlag *m*, Bodensatz *m*, Satz *m*, Ausscheidung *f*, Ausscheidungsprodukt *n*, Ablagerung *f*, Sediment *n*, Trub *m*
s. apparatus Schlammapparat *m*, Absetzapparat *m*, Dekantierapparat *m*, Klärapparat *m*
s. basin *s.* settler 1.

s. chamber Absetzkammer *f*, Abscheidekammer *f*, Staubkammer *f*; Kammerabscheider *m*; Absetzgefäß *n*, Beruhigungskammer *f* *(beim Pebble-Heater-Verfahren)*
s. cone Absetzkonus *m*; *(Pap)* Trichterstoffänger *m*
s. curve Absetzkurve *f*
s. pit Absetzgrube *f*, Absetzbecken *n*, Absitzbecken *n*
s. rate *s. s.* velocity
s. table *(Pap)* Sandfang *m*, Sandfänger *m*
s. tank *s.* settler 1.
s. velocity Absinkgeschwindigkeit *f*, Sinkgeschwindigkeit *f*, Absetzgeschwindigkeit *f*, Sedimentationsgeschwindigkeit *f*, Klärgeschwindigkeit *f*
settlings *s.* settling 2.
seven-carbon ring *s. s.*-membered ring
s.-membered siebengliedrig, 7gliedrig
s.-membered ring siebengliedriger Ring (Kohlenstoffring) *m*, Sieben[er]ring *m*, 7-Ring *m*
s.-point-six temperature *(Glas)* Erweichungstemperatur *f*, Erweichungspunkt *m*, Littleton-Punkt *m* *(bei einer Viskosität von* $10^{7,6}$ *P)*
s.-stage bleaching *(Pap)* Siebenstufenbleiche *f*
severe conditions of cracking scharfe Krackbedingungen *fpl*
severity Severity *f* *(Strenge der Bewertung von Kraftstoffen)*
s. factor Severity Factor *m* *(Maß für die Krackintensität)*
sewage Abwasser *n*, Schmutzwasser *n*
s. disposal Abwasserbeseitigung *f*
s. disposal plant (works) *s. s.* treatment plant
s. farm Rieselfeld *n*
s. fungus Abwasserpilz *m*
s. plant *s. s.* treatment plant
s. pump Abwasserpumpe *f*, Schmutzwasserpumpe *f*
s. purification Abwasserreinigung *f*
s. sludge Abwasserschlamm *m*, Klärschlamm *m*, Klärrückstand *m*
s. treatment plant (works) Abwasserreinigungsanlage *f*, Abwasserbeseitigungsanlage *f*, Kläranlage *f*, Klärwerk *n*
s. works *s. s.* treatment plant
sewer / to *(Flüssigkeiten)* ablassen
sewer Abwasserkanal *m*, Ablaufkanal *m*, Ablaßkanal *m*, Schleuse *f*
s. pipe Abwasserrohr *n*
sewerage 1. Kanalisation *f*, Kanalisierung *f*, Beschleusung *f*; 2. Abwasserbeseitigung *f*; Stadtentwässerung *f*, Ortsentwässerung *f*
sex attractant Sexuallockstoff *m*
s. hormone Sexualhormon *n*
sexadentate sechszähnig, sechszählig *(Komplexchemie, Koordinationswert des Liganden)*
sexavalence, sexavalency Sechswertigkeit *f*, Hexavalenz *f*
sexavalent sechswertig, hexavalent

sexivalence, sexivalency *s.* sexavalence
sexivalent *s.* sexavalent
sextuple evaporator Sechsstufenverdampfer *m*, Sechskörperverdampfer *m*
seybertine, seybertite *(Min)* Seybertit *m*, *(veraltet für)* Clintonit *m*
Seyler's [coal] chart Seylers Kohlenklassifizierungsschaubild *n*
S.G. *s.* specific gravity
shade / to schattieren, nuancieren, [ab]tönen
shade Schattierung *f*, Nuancierung *f*, Nuance *f*, Farbnuance *f*, Ton *m*, Farbton *m*, Farbtönung *f*, Stich *m*
s. change Farbtonumschlag *m*
shaded mark *(Pap)* Wasserzeichen *n* mit Schattierungen
shading Schattieren *n*, Schattierung *f*, Nuancieren *n*, Nuancierung *f*, Tönen *n*, Tönung *f*, Abtönen *n*
shadow detail *(Foto)* Durchzeichnung *f* der Schatten, Schattenzeichnung *f*
s. wall Schattenwand *f* *(im Glasschmelzofen)*
shaft Welle *f*; Stange *f*; Schacht *m*
s.-chain grate Schacht-Kettenrost *m*
s. dryer Schachttrockner *m*, Turmtrockner *m*
s. furnace Schachtofen *m*
s. horsepower Wellenleistung *f*, Kupplungsleistung *f*
s. kiln *s. s.* furnace
shagreen Chagrinleder *n*
shake / to [durch]schütteln; ausschütteln
s. out with ether ausäthern, mit Äther extrahieren, durch (mit) Äther ausschütteln
shake 1. Schütteln *n*, Durchschütteln *n*, Schüttelung *f*; Ausschütteln *n*; 2. Schüttelapparat *m*, Schütteleinrichtung *f*, Schüttelbock *m*
s. apparatus *s.* shake 2.
s. culture Schüttelkultur *f*
s. of the wire *(Pap)* Siebschüttelung *f*
s. rails *(Pap)* Registerschienen *fpl*
shaker *s.* shaking machine
s. autoclave Schüttelautoklav *m*
s. flask Schüttelgefäß *n*
s. screen Schüttelsieb *n*, Schüttelsortierer *m*
shaking *s.* shake 1.
s. bottle Schüttelflasche *f*
s. chute Schüttelrutsche *f*, Schüttelrinne *f*
s. cylinder Schüttelzylinder *m*
s. flask *s.* shaker flask
s. gutter *s. s.* chute
s. machine Schüttelmaschine *f*, Schüttelapparat *m*
s. screen (sieve) *s.* shaker screen
s. table Schüttelherd *m*; Stoßherd *m*, Wurfherd *m*
shale *(Geol)* Schiefer *m*
s. clay Schieferton *m*
s. oil Schieferöl *n*
s. tar Schieferteer *m*
shallow flach; seicht
s. bath Flachtrog *m*

s. trap *(Halbleitertechnik)* flache Haftstelle *f*
shaly schief[e]rig, schieferartig, geschiefert; schieferhaltig
shammy, shamoy Sämischleder *n*
shampoo Schampun *n*, Shampoo[n] *n*
shape / to formen, gestalten; profilieren
shape Form *f*, Gestalt *f*; Profil *n*; Formteil *n*
  s. determination Formbestimmung *f*
  s. factor Formfaktor *m*
  s. of greatest stability Zustand *m* der größten Stabilität
shapeability Formbarkeit *f*
shaped coke Formkoks *m*
shaping Formen *n*, Gestalten *n*, Formgebung *f*
  s. machine *(Ker)* Formgebungsmaschine *f*
share / to. [auf]teilen; teilhaben, beteiligt sein
shared electron gemeinsames (anteiliges, aufgeteiltes) Elektron *n*
  s.-electron-pair bond kovalente (unpolare, homöopolare, einpolare, unitarische) Bindung *f*, Kovalenz *f*, Elektronenpaarbindung *f*, Atombindung *f*
  s. electrons gemeinsames Elektronenpaar *n*
sharing electron *s.* shared electron
  s. electrons *s.* shared electrons
shark-liver oil Haifisch[leber]tran *m*
sharp-edged scharfkantig
  s. freezing schnelles Gefrieren (Einfrieren) *n*, Schnellgefrieren *n*, Schnellgefrierverfahren *n*
  s. series *(phys Ch)* scharfe (zweite) Nebenserie *f*
sharpen / to *(Gerb)* anschärfen; *(Text)* [vor]schärfen
sharpened lime *(Gerb)* angeschärfter Äscher *m*, Kalkschwefelnatriumäscher *m* *(mit Natriumsulfid versetzte Kalkmilch)*
sharpener *(Gerb)* Anschärf[ungs]mittel *n* *(meist Natriumsulfid)*
sharpening *(Gerb)* Anschärfen *n*; *(Text)* Schärfen *n*, Vorschärfen *n*
  s. agent *s.* sharpener
Sharples two-stage dewaxing process Sharples-Verfahren *n*, zweistufiger Entwachsungsprozeß *m* der Sharples Corp.
sharpness Schärfe *f*; Schärfe *f*, Deutlichkeit *f*
  s. in print outline *(Text)* Druckschärfe *f*, Konturenschärfe *f*
  s. of separation Trennschärfe *f*
  s. of the vat *(Text)* Schärfe *f* (scharfer Stand *m*) der Küpe
shatter / to zertrümmern; zerspringen, [zer]splittern
shatter Zertrümmern *n*, Zertrümmerung *f*; Zerspringen *n*, Zersplittern *n*, Splittern *n*
  s. resistance (strength) Sturzfestigkeit *f*, Widerstand *m* gegen das Zerspringen *(des Kokses beim Stürzen)*; *(Glas)* Splitterfestigkeit *f*, Splittersicherheit *f*
  s. test Sturzprüfung *f*, Sturzversuch *m*, Fallprüfung *f*, Shatter-Test *m*, Drop-Shatter-Test *m* *(bei Koks)*

s.-test machine Gerät *n* zur Bestimmung der Sturzfestigkeit *(von Koks)*
shattering *s.* shatter
  s. power Sprengkraft *f*, Brisanz *f*
shatterproof splitterfest, splittersicher, splitterfrei, bruchsicher
  s. glass Sicherheitsglas *n*
shattuckite *(Min)* Shattuckit *m*
shave / to *(Gerb)* falzen *(Leder durch Abschaben auf gleichmäßige Dicke bringen)*
shaved ice Splittereis *n*
shaving *(Gerb)* Falzen *n*
  s. cream Rasiercreme *f*
  s. machine *(Gerb)* Falzmaschine *f*
  s. preparation Rasier[hilfs]mittel *n*
shavings Späne *mpl*, Blätter *npl*, Seifenspäne *mpl*, Seifenblätter *npl* *(bei der Seifenherstellung)*; Papierschnitzel *npl* *(mpl)*, Abschnitte *mpl*, Späne *mpl*
SHE *s.* standard hydrogen electrode
shea butter Schibutter *f*, Galambutter *f*, Karitebutter *f* *(von Butyrospermum parkii (Don)Kotschy)*
shear / to [ab]scheren, [ab]schneiden; scherzerkleinern; verschieben, einer Schubwirkung (Scherung) aussetzen; *(Glas)* abnabeln *(einen Tropfen am Speiser)*
shear 1. Schere *f*; 2. Scheren *n*, Abscheren *n*, Schneiden *n*, Abschneiden *n*; Scherzerkleinern *n*; Scherung *f*, Verschiebung *f*, Schub *m*; *(Glas)* Abnabeln *n* *(von Tropfen am Speiser)*
  s.-bar mixer Gatterrührer *m*
  s. cake *(Glas)* Vorsatzkuchen *m*, Kuchen *m*
  s. cut slitter, s. cutter *(Pap)* nach dem Scherenprinzip arbeitende Längsschneid[e]einrichtung *f* *(einer Rollenschneid[e]maschine)*
  s. cutting [method] *s.* s. method of cutting
  s. mark Messernarbe *f*, Schnittnarbe *f* *(Glasfehler)*
  s. method of cutting *(Pap)* Längsschneiden *n* nach dem Scherenprinzip
  s. pin Scherstift *m*
  s. stability *s.* s. strength
  s. strength Scherfestigkeit *f*
shearer *s.* shearing machine
shearing *s.* shear 2.
  s. action *s.* s. effect
  s. cylinder Scherzylinder *m*
  s.-disk viscometer Scherscheibenviskosimeter *n*, Plastometer *n* mit Scherbeanspruchung
  s. effect Scherwirkung *f*, Abscherwirkung *f*
  s. machine Schermaschine *f*
  s. rate (speed) Schergeschwindigkeit *f*
  s. strength *s.* shear strength
shears Schere *f*, Messer *npl*
sheath Hülle *f*, Mantel *m*; *(Text)* Fasermantel *m*, Faserhaut *f*; Mantel *m* *(eines Kernmantelfadens)*
  s. compound *s.* sheathing compound
  s.-core bicomponent fibre *(Text)* Kernmantelfaden *m*

s. electron Hüllenelektron n
s. in lead / to mit einem Bleimantel überziehen, mit Blei ummanteln
s. of electrons Elektronenhülle f
s. of solvent molecules Solvathülle f
sheathing compound (Gum) Mantelmischung f
sheen Glanz m, Schimmer m
sheep milk Schafmilch f
sheet / to 1. (Gum) zu einem Fell auswalzen (ausziehen), bis zur Fellbildung verarbeiten; (Gum) (auf dem Kalander) Platten auswalzen (ausziehen, ziehen); 2. (Pap) in Bogen (Format) schneiden
s. out s. to sheet 1.
sheet Schicht f; (Met) Blech n; (Pap) Bogen m; (Gum) Sheet n, Fell n; (Gum) Platte f; (Plast) Folie f (als Stück; Dicke über 0,01 inch = 0,25 mm); (Geol) Gesteinsschicht f
s. aluminium Aluminiumblech n
s. blow moulding (Plast) Blasformen n von Folienhalbzeug
s. break (Pap) Bahn[ab]riß m, Papierbahn[ab]riß m, Abreißen n der Papierbahn
s. calender (Pap) Bogenkalander m
s.-calendered (Pap) bogengeglättet, bogensatiniert
s. calendering (Pap) Glätten (Satinieren) n von Bogenpapieren, Bogensatinage f; (Gum) Ausziehen (Ziehen, Auswalzen) n von Platten, Plattenziehen n (am Kalander)
s. coater (Pap) Bogenstreichmaschine f
s. cooler Kühler m (bei der Tafelglasherstellung)
s. counter, s.-counting device (Pap) Bogenzählgerät n, Bogenzähler m
s. cutter (Pap) Querschneider m
s.-drawing process Tafelglasziehverfahren n
s. extrusion (Plast) Folienstrangpressen n
s.-fed press (Pap) Bogendruckmaschine f
s. film (Foto) Planfilm m
s. filter Schichtenfilter n
s. formation (Pap) Blattbildung f
s.-forming (Pap) blattbildend
s. forming (Plast) Folienformung f, Plattenformung f
s. gelatin Blattgelatine f, Blättergelatine f
s. glass Tafelglas n
s.-lined board im Format geklebte (kaschierte) Pappe f
s.-metal box Metallbüchse f (der Verkokungsapparatur nach Conradson)
s. of board Pappebogen m
s. of glass [dünne] Glasplatte f
s. of paper Papierblatt n, Papierbogen m
s. paper Bogenpapier n, Formatpapier n, auf Format geschnittenes Papier n
s. rubber Räucherkautschuk m, Smoked Sheet (geräucherte Rohkautschukplatte)
s. sorting (Pap) Bogensortierung f, Sortierung f des Bogenpapiers, Papiersortierung f
s. tension (Pap) Papierzugspannung f, Papierbahnspannung f, Bahnspannung f

sheeted paper s. sheet paper
sheeter (Pap) Querschneider m
s. lines (Plast) Schneidriefen fpl
sheeting 1. (Pap) Querschneiden n, Bogenschneiden n, Fertigstellung f (Schneiden n) in Bogen; 2. (Gum) Ausziehen (Auswalzen) n zu Fellen, Verarbeitung f bis zur Fellbildung; (Gum) Ziehen (Ausziehen, Auswalzen) n von Platten, Plattenziehen n; 3. (Plast) Folie f (als Bahn; Dicke über 0,01 inch = 0,25 mm)
s. calender Kalander m zum Ziehen von Platten
s. dryer (Pap, Text) Bahnentrockner m
s. mill (Gum) Sheet-Mangel f
s.-out s. sheeting 2.
sheets of chemical wood pulp (Pap) Zellstoff m in Bogenform (Pappenform), Zellstoffpappe f, Zellstoffblätter npl
s. of mechanical wood pulp (Pap) Holzschliff m in Bogenform (Pappenform), Holzschliffpappe f, Holzschliffblätter npl
s. of pulp (Pap) Halbstoff m in Bogenform (trockenen Bogen), (i.e.S.) s. s. of chemical wood pulp (oder) s. of mechanical wood pulp
s. of wood pulp s. s. of chemical wood pulp (oder) s. of mechanical wood pulp
shelf dryer Plattentrockner m
s. life Haltbarkeitszeit f, Lagerbeständigkeit f
shell Schale f, Hülle f, Mantel m; Gehäuse n; (Text) Fasermantel m, Faserhaut f
s.-and-tube heat exchanger Röhrenwärmeaustauscher (Röhrenwärmeübertrager) m mit Mantel
s. bars Gehäusemesser npl, Mantelmesser npl (der Kegelstoffmühle)
s. electron Schalenelektron n
s. flour Schalenmehl n (Plastfüllstoff)
s. freezing Roll-Schicht-Frosten n (ein Gefriertrocknungsverfahren)
s. lac s. shellac
s. lime Muschelkalk m
s. limestone Muschelkalkstein m
s. model Schalenmodell n
s. mould (Met) Maskenform f, Formmaske f, Croning-Formmaske f
s. moulding Maskenformen n
s.-moulding process Maskenformverfahren n, Formmaskenverfahren n [nach Croning], Croning-Formmaskenverfahren n, Croning-Verfahren n, Shell-Moulding-Verfahren n
s.-side medium Mantelraummedium n
s. structure Schalenstruktur f, Schalenaufbau m
s.-type boiler Mantelkessel m, Kessel m in Mantelform
Shell hydrodesulphurization process Shell-Hydrodesulphurization-Verfahren n
Shell liquid-phase process Shell-Flüssigphaseverfahren n, Flüssigphaseisomerisierungsverfahren n der Shell Development Co.
Shell phosphate process Shell-Phosphatverfahren n, Phosphatentschwefelungsverfahren n der Shell Development Co.

Shell vapour-phase process Shell-Dampfphaseverfahren *n*, Shell-Gasphaseverfahren *n*, Dampfphaseisomerisierungsverfahren *n* der Shell Development Co.

shellac Schellack *m*, Schollenlack *m*

s. varnish Schellacklösung *f* (*als Spirituslack*)

sheller Schälmaschine *f*, Schälvorrichtung *f*

shelling Schälen *n*, Enthülsen *n*; (*Ker*) Abschälen *n*, Abblättern *n* (*der Glasur*)

shellolic acid Schellolsäure *f* (*im Schellack enthaltene Dikarbonsäure*)

shelly limestone *s.* shell limestone

sherardize / **to** sherardisieren (*nach dem Diffusionsverfahren verzinken*)

sherardizing Sherardisieren *n*, Sherardisierung *f* (*Diffusionsverzinkung*)

shield / **to** abschirmen; sich abschirmen

shielding Abschirmen *n*, Abschirmung *f*, Schirmung *f*

s. concrete Abschirmbeton *m*, Strahlenschutzbeton *m*

shift / **to** verschieben; sich verschieben, wandern

s. back zurückwandern

shift Verschiebung *f*

1,2-shift 1,2-Umlagerung *f*

1,2-s. of alkyl 1,2-Alkylkarboniumumlagerung *f*

s. to the red Rotverschiebung *f*, Verschiebung *f* nach Rot

shifting *s.* shift

shikimic acid Shikimisäure *f*, 3,4,5-Trihydroxyzyklohexen-1-karbonsäure *f*

shikonin Shikonin *n* (*rechtsdrehendes Alkannin aus Shikonwurzeln*)

shim screen (*Pap*) Splitterfänger *m*

shine / **to** glänzen

shiny glänzend

s. effect Glanzeffekt *m*

s. pulp (stock, stuff) (*Pap*) schmieriger Stoff *m*

ship-bottom paint Schiffsbodenfarbe *f*

shive (*Text*) Schäbe *f*

shivering (*Ker*) Abblättern *n*, Abschälen *n* (*der Glasur*)

shock wave (*Sprengtechnik*) Stoßwelle *f*

shockproof stoßfest, stoßsicher

shoddy Shoddy *n*, Shoddywolle *f* (*Reißwolle aus Maschenware*); (*Gum*) Regenerat *n*, regenerierter Kautschuk *m*, Regenerativgummi *m*

shoe polish Schuhkrem *f*

shogaol Shogaol *n*, [4-Oxy-3-methoxyphenyl]äthyl-n-α,β-heptenylketon *n*

shonanic acid Dihydrothujasäure *f* (*Tropolonderivat*)

shooting Torpedieren *n* (*zur Erleichterung des Ölzuflusses zu einer Bohrung*)

s. cylinder (*Plast*) Spritz[gieß]zylinder *m*, Massezylinder *m*

shop-erected vorgefertigt, vormontiert

Shore durometer Shore-Härteprüfer *m*, Shore-Härtemesser *m*

Shore hardness Shore-Härte *f*.

shorl *s.* schorl

shorn wool Schurwolle *f*

short (*Ker*) kurz, wenig bildsam (*Masse*); (*Gum*) stramm

s.-chain kurzkettig

s.-circuit / **to** kurzschließen

s.-cut method Schnellverfahren *n*, abgekürztes Verfahren *n*

s.-cycle kiln (*Ker*) Ofen *m* mit rascher Brandfolge

s.-flame coal kurzflammige (kurzflammig brennende) Kohle (Steinkohle) *f*

s. free stock (stuff) (*Pap*) kurzröscher Stoff *m*

s. glass kurzes Glas *n*

s.-head cone crusher Kegelgranulator *m*, Flachkegelgranulator *m*

s.-hold pasteurization *s.* s.-time high-temperature pasteurization

s. ink kurze Druckfarbe *f*

s.-lived (*phys Ch*) kurzlebig

s.-loop dryer Kurzschleifentrockner *m*

s. moulding (*Plast*) unvollständiges (nichtausgeformtes) Formteil *n* (*Fehler*)

s.-neck flat-bottom flask Kurzhalsstehkolben *m*

s.-neck round-bottom flask Kurzhalsrundkolben *m*

s.-oil alkyd kurzöliges Alkyd[harz] *n*

s.-oil varnish magerer Öllack (Lack) *m*

s. paste Mürbeteig *m*

s.-path [high-vacuum] distillation Kurzwegdestillation *f*

s.-path still Kurzwegdestillierapparat *m*

s. period kurze (kleine) Periode *f*, Kurzperiode *f*

s. pipe kurzes Rohrstück *n*, kurzer Rohrstutzen *m*

s.-range order (*Krist*) Nahordnung *f*, Nahordnungsgrad *m*

s. residue (residuum) kurzer Destillationsrückstand *m*, Vakuumrückstand *m*, Rückstand *m* einer Vakuumkolonne, Short Residuum *n*

s.-retention-time dryer Kurzzeittrockner *m*

s. run (*Plast*) kleines Fertigungslos *n*, Kleinreihe *f*, Kleinserie *f*

s. shot (*Plast*) ungenügende Füllung *f*, unvollständiges Spritzteil *n*

s.-stem[med] funnel Trichter *m* mit kurzem Stiel (Rohr)

s.-time Kurzzeit..., Schnell...

s.-time assay *s.* s.-time test

s.-time heat processing (*Lebm*) Kurzzeiterhitzung *f*, Hochkurzerhitzung *f*, HTST-Erhitzung *f*

s.-time high-temperature method (*Lebm*) Kurzzeiterhitzungsverfahren *n*, Verfahren *n* der Kurzzeiterhitzung, HTST-Verfahren *n*

s.-time high-temperature pasteurization (*Lebm*) Kurzzeitpasteurisation *f*, Hochkurzpasteurisation *f*, HTST-Pasteurisation *f*, Kurzzeiterhitzung *f*, Hochkurzerhitzung *f* (*zur Pasteurisation*)

s.-time high-temperature process *s.* s.-time high-temperature method

s.-time pasteurization s. s.-time high-temperature pasteurization
s.-time pasteurizer (Lebm) Kurzzeitpasteurisierapparat m, Kurzzeiterhitzer m
s.-time test Kurzzeitprüfung f, Schnelltest m
s.-tube evaporator Kurzrohrverdampfer m
s.-tube venturi Kurzventuridüse f
s.-wave kurzwellig, Kurzwellen...
s.-wave-length region Kurzwellenbereich m; kurzwelliger Spektralbereich m, kurzwelliges Spektralgebiet n
s. wet stock (Pap) kurzschmieriger Stoff m
shortage of electrons Elektronenmangel m
shortening Speisehartfett n (für Backzwecke)
shorts feine Weizenkleie (Kleie) f
shortstop / to (Gum) (die Polymerisation) abstoppen, inhibieren, abbrechen
shortstop, shortstopping agent (Gum) Abstoppmittel n, Polymerisationsabstoppmittel n, Polymerisationsstopper m, Inhibitor m
shot / to verperlen; granulieren
shot 1. Schuß m, Sprengung f, Explosion f; 2. (Plast) Schuß m; 3. verperltes Produkt n; Granulat n; (Glas) Perlen fpl (Fehler)
s. capacity (Plast) maximale Schußmasse f
s. point Schießpunkt m, Schußpunkt m, Sprengpunkt m, Sprengstelle f, Explosionspunkt m (bei der Refraktionsseismik)
s. size (Plast) Schußmasse f
s. weight s. s. size
shotting Verperlen n; Granulieren n, Granulierung f
shoulder (Foto) Schulter f (der Schwärzungskurve), Gebiet n der Überexposition (Überbelichtung)
shovel Schaufelblatt n, Schaufel f, Blatt n, Flügel m, Rührarm m
show bottle Schauglas n, Präparatenglas n
s. proof / to nachweisen
s. through / to durchscheinen (Druckfarben)
s.-through Durchscheinen n (Druckfarben)
shower / to rieseln lassen
shower (Kch) Schauer m
s. deck (Erdöl) Showerdeck n
s.-deck tray Sprühkörper m
s. of electrons Elektronenschauer m
s. particle Schauerteilchen n
s.-proof regendicht, wasserdicht
s. pump (Pap) Spritzwasserpumpe f
s. roasting (Met) Blitzröstung f
s. supply tank (Pap) Spritzwasserbehälter m
s. tray Rieselblech n, Rieselboden m
s. water (Pap) Spritzwasser n
shred / to zerfasern
shredder Zerfaserer m
shredding Zerfasern n
s. machine s. shredder
shreds Schnitzel npl, Rübenschnitzel npl
shrend / to (Glasscherben) fritten
shrink / to schrumpfen, schwinden, (Text auch) krumpfen, einlaufen, eingehen

shrink fixture (Plast) Schrumpfvorrichtung f, Abkühlungsvorrichtung f (für Preßteile)
s. ratio Schrumpfverhältnis n
s.-resist finish (Text) Schrumpffestausrüstung f, Krumpffestausrüstung f, schrumpffreie Ausrüstung (Appretur) f
s. resistance (Text) Schrumpffestigkeit f, Krumpfbeständigkeit f, Krumpffestigkeit f
s.-resistant (Text) schrumpffest, schrumpfbeständig, krumpffest, krumpfbeständig, nichtkrumpfend, nichtschrumpfend, nichteinlaufend
shrinkage Schrumpfen n, Schrumpfung f, Schwinden n, Schwindung f, (Text auch) Krumpfen n, Krumpfung f, Einlaufen n, Eingehen n
s. cavity Schwindungshohlraum m
s. crack Schrumpfriß m, Schwind[ungs]riß m, Schwundriß m
s. jig (Plast) Schrumpfvorrichtung f, Abkühlungsvorrichtung f (für Preßteile)
s. meter (Gerb) (Gerät zum Messen des Schrumpfungsgrades)
s. temperature (Gerb) Schrumpfungstemperatur f
s. test (Gerb) Kochprobe f (zur Prüfung des Erfolges von Chromgerbungen)
shrinking s. shrinkage
s. machine (Text) Krumpfmaschine f, Gewebekrumpfmaschine f
shrinkproof s. shrink-resistant
shrinkproofing (Text) Schrumpffestmachen n, Krumpffestmachen n
shroud ring Leitrohr n (am Rührerlaufrad)
shrouded impeller geschlossenes Laufrad n
shut [down] / to abstellen, zum Stillstand bringen, stillsetzen, außer Betrieb setzen, stillegen
s. off absperren; (einen Hahn) schließen
shut Abstellen n, Stillsetzen n, Stillsetzung f, Außerbetriebsetzen n, Außerbetriebsetzung f, Stillegen n, Stillegung f
s.-off nozzle (Plast) Verschlußdüse f, Abschlußdüse f
Siak fat (tallow) Siaktalg m (von Palaquium oleiferum Blanco)
sial (Geoch) Sial n, Sialzone f
sialic acid Sialsäure f (eine azylierte Neuraminsäure)
Siam benzoin Siambenzoe f (von Styrax-Arten)
S. cardamon Siamkardamom m(n), runder Kardamom m (von Elettaria cardamomum (L.)White et Maton)
siaresinolic acid Siaresinolsäure f (aus Siambenzoe)
siberite (Min) Siberit m, (veraltet für) Rubellit m
siccative Sikkativ n, Trockenstoff m, Trockner m
side-blow converter s. s.-blown converter
s. blowing (Met) seitliches Blasen (Einblasen) n
s.-blown converter seitlich blasender Konverter m, Seitenwindkonverter m
s. chain Seitenkette f
s.-chain chlorination Seitenkettenchlorierung f, Chlorierung f in der Seitenkette

s.-chain halogenation Seitenkettenhalogenierung f, Halogenierung f in der Seitenkette
s.-chain metalation Seitenkettenmetallierung f
s. effect (Pharm) Nebenwirkung f
s.-entering seitlich eingebaut
s.-fired furnace (Glas) Wannenofen m mit Querfeuerung (Querbrennern, querziehender Flamme), Querflammenwanne f, querbeheizter Wannenofen m
s.-port furnace s. s.-fired furnace
s. product (Destillation) Seitenprodukt n
s. reaction Nebenreaktion f
s. stream Seitenstrom m
s. stripper Seitenkolonne f, Abstreiferkolonne f, Abtreibkolonne f, Dampfabstreiferkolonne f, Ausdämpf[er]kolonne f, Stripperkolonne f, Strippingkolonne f, Abstreifer m, Abtreiber m, Dampfabstreifer m, Ausdämpfer m, Stripper m
s. tap seitlich angebrachter Hahn m
s.-to-centre shading (Text) Farbablauf m, Seitenablauf m
s. tube seitliches Abzugsrohr n (eines Fraktionierkolbens)
sideaerolite s. siderolite
sideramine Sideramin n (ein eisenhaltiger Bakterienwuchsstoff)
siderazot[e], siderazotite (Min) Siderazot m, Silvestrit m
sideretine (Min) Sideretin m, (veraltet für) Pittizit m
siderin yellow $Fe_2(CrO_4)_3$ Sideringelb n, Eisen(III)-chromat n
siderite Siderit m, Eisenmeteorit m, Meteoreisen n; (Min) Siderit m, Eisenspat m, Spateisenstein m (Eisen(II)-karbonat)
siderolite Siderolith m
sideromycin Sideromyzin n (ein eisenhaltiges Antibiotikum)
siderophile element siderophiles (eisenfreundliches) Element n
sideroschisolite (Min) Sideroschisolith m, (veraltet für) Cronstedtit m
siderosphere Siderosphäre f, Barysphäre f, Ni-Fe-Kern m, Erdkern m
sidewall Seitenwand f
s. compound (Gum) Seitenstreifenmischung f
s. coring Ziehen n von Seitenkernen (beim Erdölbohren)
s. stock s. s. compound
siege (Glas) Bank f, Hafenbank f, Gesäß n, Ofensohle f (eines Hafenofens)
Siemens-Martin furnace Siemens-Martin-Ofen m, SM-Ofen m
Siemens-Martin process Siemens-Martin-Verfahren n, SM-Verfahren n, Herdfrischverfahren n
Siemens-Martin steel Siemens-Martin-Stahl m, SM-Stahl m, Herdfrischstahl m, herdgefrischter Stahl m
Siemens-Martin steel plant Siemens-Martin-Stahlwerk n
Siemens process s. Siemens-Martin process

sienna Sienaerde f (Eisenoxidhydrat)
Sierra Leone butter Kanyabutter f (von Pentadesma butyraceum Sabine)
sieve / to sieben
sieve Sieb n
s. analysis Siebanalyse f
s. area Siebfläche f
s. bottom Siebboden m
s. fineness Siebfeinheit f
s. plate Siebboden m, gelochter Austauschboden m (bei der Destillation)
s.-plate column (tower) Siebbodenkolonne f
s. residue Siebrückstand m
s. scale (series) Siebskala f, Siebreihe f
s. shaker Siebrüttelmaschine f
s. tray s. s. plate
sieving Sieben n, Siebung f
sift / to [durch]sieben, [durch]beuteln
sifter Siebapparat m, Siebmaschine f, Siebvorrichtung f, Sieb n
sifting Sieben n, Durchsieben n, Beuteln n, Durchbeuteln n
sight glass Schauglas n, Sichtfenster n
s. hole Schauloch n
sigma blade Z-förmiger Knetarm m
s. bond Sigmabindung f, σ-Bindung f, σ-Elektronenpaarbindung f
s. electron Sigmaelektron n, σ-Elektron n
s. phase (Met) Sigmaphase f
signal oil Signalöl n, Langzeitbrennöl n
significant signifikant, [statistisch] gesichert
silane Si_nH_{2n+2} Silan n, Siliziumwasserstoff m, Siliziumhydrid n, (i.e.S.) SiH_4 Silan n, Monosilan n
silanol Silanol n (eine Siliziumverbindung mit OH-Gruppen), (i.e.S.) H_3SiOH Silanol n
silazane $H_3Si(NHSiH_2)_nNHSiH_3$ Silazan n
silcrete (Min) Silkret m (Quarzitart)
silica SiO_2 Kieselerde f, (veraltet für) Siliziumdioxid n, Silizium(IV)-oxid n; Silika[masse] f, Silikamaterial n
s. brick Silika[bau]stein m
s. cement Silikamörtel m
s. gel s. silicic-acid gel
s. glass Kieselglas n
s. refractory SiO_2-reiches Schamotteerzeugnis n (mehr als 72% Siliziumdioxid)
s. removal Entkieselung f
s. rock s. siliceous rock
s. sand Silikasand m
s. sol Kiesel[säure]sol n
silicane Silikan n, (veraltet für) Silan n
silicate Silikat n
s. bond Silikatbindung f, silikatische Bindung f
s. glass Silikatglas n
s. phosphor Silikatphosphor m
s. rock Silikatgestein n
s.-sized (Pap) mit Natronwasserglas geleimt
s. sizing (Pap) Leimung f mit Natronwasserglas
s. slag Silikatschlacke f

siliceous Siliziumdioxidhaltig
s. clay kieselsäurereicher Ton *m*
s. dust kieselsäurehaltiger Staub *m*
s. refractory SiO_2-reiches Schamotteerzeugnis *n* *(weniger als 93% Siliziumdioxid)*
s. rock Kieselgestein *n*
s. sediment Kieselsediment *n*
s. sinter Kieselsinter *m*, Opalsinter *m*, Geyserit *m* *(als Absatz von Geysiren)*
silicic acid Kieselsäure *f*, *(i.e.S.)* H_4SiO_4 Orthokieselsäure *f*, Tetroxokieselsäure *f*, Kieselsäure *f*
s.-acid compound Kieselsäureverbindung *f*
s.-acid ester Kieselsäureester *m*, Siliziumester *m*
s.-acid gel Kiesel[säure]gel *n*
s. anhydride Siliziumdioxid *n*, Silizium(IV)-oxid *n*
silicide Silizid *n*
silicified wood verkieseltes Holz *n*
silicify / to verkieseln
silicious *s.* siliceous
silicoacetic acid CH_3SiOOH Silikoessigsäure *f*
silicobenzoic acid C_6H_5SiOOH Silikobenzoesäure *f*, Silikobenzolkarbonsäure *f*
silicobromoform $SiHBr_3$ Silikobromoform *n*, Siliziumbromoform *n*, *(veraltet für)* Tribrom[mono]silan *n*
silicochloroform $SiHCl_3$ Silikochloroform *n*, Siliziumchloroform *n*, *(veraltet für)* Trichlor[mono]silan *n*
silicoethane Si_2H_6 Silikoäthan *n*, *(veraltet für)* Disilan *n*
silicofluoric acid $H_2[SiF_6]$ Fluorokieselsäure *f*, Hexafluorokieselsäure *f*, Siliziumfluorwasserstoffsäure *f*, Silikofluorwasserstoffsäure *f*, *(veraltet)* Kieselfluorwasserstoffsäure *f*, Kieselflußsäure *f*
silicofluoride Fluorosilikat *n*, Silikofluorid *n*, Hexafluorosilikat *n*
silicofluoroform $SiHF_3$ Silikofluoroform *n*, Siliziumfluoroform *n*, *(veraltet für)* Trifluor[mono]silan *n*
silicoformic acid HSiOOH Silikoameisensäure *f*
s. anhydride $H_2Si_2O_3$ Silikoameisensäureanhydrid *n*, Dioxodisiloxan *n*
silicoiodoform $SiHJ_3$ Silikojodoform *n*, Siliziumjodoform *n*, *(veraltet für)* Trijod[mono]silan *n*
silicoketone [—R_2SiO—]$_n$ Silikoketon *n*, *(selten für)* Silikon *n*
silicomanganese steel Mangansiliziumstahl *m*
silicomesoxalic acid $H_4Si_3O_6$ Silikomesoxalsäure *f*
silicomethane SiH_4 Silikomethan *n*, *(veraltet für)* Monosilan *n*
silicomolybdate Silikomolybdat *n*
silicomolybdic acid $H_4SiMo_{12}O_{40} \cdot 14H_2O$ Silikomolybdänsäure *f*
silicon Si Silizium *n*
s. bromodichloride $SiBr_2Cl_2$ Dibromdichlorsilan *n*
s. bromoform *s.* silicobromoform
s. bromohydride SiH_3Br Brom[mono]silan *n*

s. bromotrichloride $SiBrCl_3$ Bromtrichlorsilan *n*
s. bronze Siliziumbronze *f*
s. carbide SiC Siliziumkarbid *n*
s.-carbon bond Silizium-Kohlenstoff-Bindung *f*, Kohlenstoff-Silizium-Bindung *f*
s. cast iron siliziumlegiertes Gußeisen *n*, Siliziumguß *m*
s. chloroform *s.* silicochloroform
s. chlorohydride SiH_3Cl Chlor[mono]silan *n*, Monochlorsilan *n*
s. dioxide SiO_2 Siliziumdioxid *n*, Silizium(IV)-oxid *n*
s. disulphide SiS_2 Siliziumdisulfid *n*, Silizium(IV)-sulfid *n*
s. ester *s.* silicic-acid ester
s. fluoroform *s.* silicofluoroform
s.-functional siliziumfunktionell
s. hexachloride Si_2Cl_6 Siliziumhexachlorid *n*, Disiliziumhexachlorid *n*, Siliziumtrichlorid *n*, Hexachlordisilan *n*
s. hydride *s.* silane
s. iodoform *s.* silicoiodoform
s. iron Siliziumeisen *n*
s.-manganese steel *s.* silicomanganese steel
s. mesoxalic acid *s.* silicomesoxalic acid
s. oxalic acid *s.* silicooxalic acid
s. oxychloride Si_2OCl_6 Siliziumoxidchlorid *n*, Hexachlordisiloxan *n*
s. rectifier Siliziumgleichrichter *m*
s. tetrabromide $SiBr_4$ Siliziumtetrabromid *n*, Silizium(IV)-bromid *n*, Tetrabromsilan *n*
s. tetrachloride $SiCl_4$ Siliziumtetrachlorid *n*, Silizium(IV)-chlorid *n*, Tetrachlorsilan *n*
s. tetraethyl $Si(C_2H_5)_4$ Siliziumtetraäthyl *n*, Silikononan *n*, *(veraltet für)* Tetraäthylsilan *n*
s. tetrafluoride SiF_4 Siliziumtetrafluorid *n*, Silizium(IV)-fluorid *n*, Tetrafluorsilan *n*
s. tetraiodide SiJ_4 Siliziumtetrajodid *n*, Silizium(IV)-jodid *n*, Tetrajodsilan *n*
s. tetramethyl $Si(CH_3)_4$ Siliziumtetramethyl *n*, *(veraltet für)* Tetramethylsilan *n*
s. trichloride *s. s.* hexachloride
siliconate Silikonat *n*
silicone [—R_2SiO—]$_n$ Silikon *n*, Polyorganosiloxan *n*, Organopolysiloxan *n*; Organosiliziumverbindung *f*, siliziumorganische Verbindung *f* *(organische Siliziumverbindung mit direkter Si—C-Bindung)*
s. adhesive Silikonklebstoff *m*
s. antifoam Silikonantischaummittel *n*
s.-base[d] auf Silikonbasis (Silikongrundlage)
s. chemistry Silikonchemie *f*, Polyorganosiloxanchemie *f*
s.-coated silikonüberzogen, silikonbeschichtet
s. coating Silikonüberzug *m*, Schutzüberzug *m* aus Silikon
s. compound Silikonverbindung *f*; Silikon-Compound *n*, Silikonpaste *f*
s. elastomer Silikonelastomer[es] *n*, elastomeres Silikon *n*

s. emulsion Silikonemulsion f
s. filler Silikonfüllstoff m
s. film Silikon[schutz]film m, silikonhaltiger Film m, Silikonhaut f
s. finish Silikonausrüstung f
s. fluid Silikonflüssigkeit f, Silikonfluid n, Silikonöl n
s. grease Silikonfett n
s. gum Silikon[roh]kautschuk m
s. liquid s. s. fluid
s. lubricant Silikonschmiermittel n
s.-modified silikonmodifiziert
s. mould-release agent Silikonform[en]trennmittel n
s. oil s. s. fluid
s. paint Silikon[anstrich]farbe f
s. plant Silikonanlage f, Silikonwerk n
s. polymer Silikonpolymer[es] n, Organopolysiloxanpolymer[es] n
s. putty Silikonkitt m
s. release agent Silikontrennmittel n
s. resin Silikon[kunst]harz n, Organopolysiloxanharz n
s. resin solution Silikonharzlösung f
s. rubber Silikongummi m (Silikonkautschukvulkanisat)
s. rubber compound Silikongummimischung f, Silikonkautschukmischung f
s. rubber gum s. s. gum
s. rubber mixture s. s. rubber compound
s. rubber paste Silikongummipaste f
s. [rubber] stock s. s. rubber compound
s. stopcock grease Silikonhahnfett n
s.-treat / to mit Silikonen behandeln, silikonisieren
s.-treated silikonbehandelt, silikonisiert
s. treatment Silikonbehandlung f, Behandeln n mit Silikonen, Silikonisieren n, Silikonisierung f
s. varnish Silikonlack m
siliconic s. silicono
siliconization Silikonisieren n, Silikonisierung f, Silikonbehandlung f, Behandeln n mit Silikonen; (Met) Insilizieren n, Insilizierung f, Silizieren n, Silizierung f (Zementation mit Silizium)
siliconize / to silikonisieren, mit Silikonen behandeln; (Met) [in]silizieren (mit Silizium zementieren)
siliconizing s. siliconization
silicono (HO)OSi— Silikono...
siliconononane s. silicon tetraethyl
silicooxalic acid $H_2Si_2O_4$ Silikooxalsäure f
silicopropane Si_3H_8 Silikopropan n, (veraltet für) Trisilan n
silicosis (Med) Silikose f
silicospiegel Silikospiegel m (eine Eisen-Silizium-Mangan-Legierung)
silicothermic silikothermisch
s. process silikothermisches Verfahren n, silikothermischer Prozeß m
silicotungstate Silikowolframat n

silicotungstic (silicowolframic) acid $H_4Si(W_{12}O_{40})$ Wolframatokieselsäure f
silicyl $H_3Si—$ Sil[iz]yl n
s. oxide $(SiH_3)_2O$ Silizyloxid n, Disiloxan n, Silikomethyläther m
silicylene $H_2Si=$ Sil[iz]ylen n
silification Verkieselung f, Sili[zi]fizierung f, Silizifikation f
Silit furnace Silitstabofen m, Silitrohrofen m
S. rod Silitstab m
silk Seide f
s. fabric Seidengewebe n, Seidenstoff m
s. fibroin Seidenfibroin n
s. gum Seidenbast m; Seidenleim n
s. pupa Seidenpuppe f
s. rubber Seidenkautschuk m (von Funtumia elastica Stapf)
s. screen Seidensieb n, Seidenraster m
s.-screen ink Druckfarbe f für Silk-Screen-Druck, Siebdruckfarbe f
s.-screen printing Silk-Screen-Druck m, Seidenrasterdruck m, Siebdruck m
s.-screen process Siebdruckverfahren n
silkworm Seidenraupe f
silky seidenartig, seidig
s. lustre (Min) Seidenglanz m
sillimanite (Min) Sillimanit m (Aluminiumalumoorthosilikat)
s. refractory (Ker) Sillimaniterzeugnis n
s. refractory brick [feuerfester] Sillimanitstein m
silo Silo m (n), Bunker m, Behälter m, Zellenspeicher m
siloed beet pulp eingesäuerte Zuckerrübenschnitzel npl
siloxane $H_3Si(OSiH_2)_nOSiH_3$ Siloxan n
s. bond (link, linkage) Siloxanbindung f
s. polymer Siloxanpolymer[es] n
s. unit Siloxaneinheit f
siloxen[e] $Si_6O_3H_6$ Siloxen n
siloxy $H_3SiO—$ Siloxy...
silt Schluff m (Korngröße 0,02 bis 0,002 mm); (Geol) Silt m, Feinsand m
silthiane $H_3Si(SSiH_2)_nSSiH_3$ Silthian n
silvan Silvan n, 2-Methylfuran n
silvan / to versilbern
silver Ag Silber n
s. acetylide C_2Ag_2 Silberazetyl[en]id n, Azetylensilber n, Silberkarbid n
s. bromide AgBr Silberbromid n, (Foto auch) Bromsilber n
s.-bromide crystals (Foto) Silberbromidkristalle mpl, Bromsilberkristalle mpl
s.-bromide grains (Foto) Silberbromidkörner npl
s.-bromide paper (Foto) Bromsilberpapier n
s. carbide s. s. acetylide
s. chloride AgCl Silberchlorid n, (Foto auch) Chlorsilber n
s.-chloride electrode Silberchloridelektrode f
s.-chloride paper (Foto) Chlorsilberpapier n, Kontaktpapier n

s. chromate Ag_2CrO_4 Silberchromat n
s.-clad / to silberplattieren, mit Silber plattieren
s. cladding Silberplattieren n, Silberplattierung f
s. content Silbergehalt m
s. coulometer Silbercoulometer n
s. deposit (Foto) Silberniederschlag m, Silberabscheidung f, Silberbelag m
s. electrode Silberelektrode f
s. ferricyanide $Ag_3[Fe(CN)_6]$ Silberhexazyanoferrat(III) n
s. ferrocyanide $Ag_4[Fe(CN)_6]$ Silberhexazyanoferrat(II) n
s. fluoride Silberfluorid n, (i.e.S.) AgF Silber(I)-fluorid n, Silber[mono]fluorid n
s. fluorosilicate $Ag_2[SiF_6]$ Silberhexafluorosilikat n
s. fulminate CNOAg Silberfulminat n, Knallsilber n
s. glance (Min) Silberglanz m (Silbersulfid)
s. grain (Foto) Silberkorn n
s. halide Silberhalogenid n, (Foto auch) Halogensilber n
s.-halide crystals (Foto) Silberhalogenidkristalle mpl
s.-halide emulsion (Foto) Silberhalogenidemulsion f, Halogensilberemulsion f
s.-halide grains (Foto) Silberhalogenidkörner npl
s. hydrogen phosphate Ag_2HPO_4 Silberhydrogen[ortho]phosphat n, Disilberhydrogenphosphat n
s. image (Foto) Silberbild n
s. intensifier (Foto) Silberverstärker m
s. iodide AgJ Silberjodid n, (Foto auch) Jodsilber n
s. marking (Ker) Silberstrichbildung f (Fehler)
s. mercuric iodide $Ag_2[HgJ_4]$ Silbertetrajodomerkurat(II) n
s. molybdate Ag_2MoO_4 Silbermolybdat(VI) n
s. nitrate $AgNO_3$ Silbernitrat n
s.-nitrate paper Silbernitratpapier n
s.-nitrate solution Silbernitratlösung f
s. nitroprusside $Ag_2[Fe(CN)_5NO]$ Silbernitroprussid n, Silbernitroprussiat n, Disilberpentazyanonitrosylferrat(II) n
s. nucleus (Foto) Silberkeim m
s. orthoarsenate Ag_3AsO_4 Silber[ortho]arsenat(V) n
s. orthoarsenite Ag_3AsO_3 Silber[ortho]arsenat(III) n
s. orthophosphate Ag_3PO_4 Silber[ortho]phosphat n
s. oxide Silberoxid n, (i.e.S.) Ag_2O Silber(I)-oxid n, Silberoxid n
s. paper Silberpackpapier n; Silberpapier n, Aluminiumpapier n
s. particle (Foto) Silberteilchen n
s.-plate / to galvanisch versilbern
s. plating galvanisches Versilbern n, galvanische Versilberung f

s.-plating bath Versilberungsbad n, Silberbad n
s.-potassium nitrate $KNO_3 \cdot AgNO_3$ Kaliumsilbernitrat n
s. protein (Pharm) Proteinsilber n, Albumosesilber n
s. pyrophosphate $Ag_4P_2O_7$ Silberpyrophosphat n, Silberdiphosphat n
s. recovery (Foto) Silberrückgewinnung f
s. ruby (Min) dunkles Rotgültigerz n, Pyrargyrit m (Antimon(III)-silbersulfid)
s. salt Silbersalz n (eine Verbindung des Silbers mit einer Säure); Silbersalz n, Natriumanthrachinon-2-sulfonat n (Natriumsalz der 9,10-Dihydro-9,10-dioxoanthrazen-2-sulfonsäure)
s.-salt diffusion (Foto) Silbersalzdiffusion f
s.-silver chloride electrode Silber-Silberchlorid-Elektrode f
s. solder Silberlot n
s. stain (Glas) Silberbeize f, Silberätze f, Gelbbeize f, Gelbätze f
s. subfluoride Ag_2F Silbersubfluorid n, Disilberfluorid n
s. sulphate Ag_2SO_4 Silbersulfat n
s. thiocyanate AgSCN Silberthiozyanat n, Silberrhodanid n
s.-white silberweiß
silvering Versilbern n, Versilberung f
silvery silb[e]rig, silberglänzend
s.-white silbrigweiß
silvestrite (Min) Silvestrit m, Siderazot m
silyl H_3Si- Sil[iz]yl n
s. compound Silylverbindung f
silylamino H_3SiNH- Silylamino...
silylene $H_2Si=$ Sil[iz]ylen n
silylidyne $HSi\equiv$ Silylidyn n
sima (Geoch) Sima n, Simazone f
simazine Simazin n, 6-Chlor-2,4-bis-äthylamino-1,3,5-triazin n (Herbizid)
similar joint action (Toxikologie) Ähnlichkeitsverbundwirkung f
simple batch distillation diskontinuierliche Gleichstromdestillation f, mehrmalige einfache Destillation f
s. brewhouse Sudhaus n mit einfachem Sudwerk (Sudzeug), Sudhaus n mit Zweigerätesudwerk (2-Geräte-Sudzeug)
s. continuous distillation kontinuierliche Gleichstromdestillation f, einfache kontinuierliche Destillation f
s. cube (Krist) einfacher Elementarwürfel m
s. decomposition einfache Umsetzung f, Ersetzung f, Substitution f
s. distillation einfache (gewöhnliche) Destillation f, Gleichstromdestillation f, Abtrieb m, Abtreiben n
s. protein Protein n, einfacher Eiweißstoff (Eiweißkörper) m
s. reaction einfache Reaktion f
s. replacement s. s. decomposition
s. salt einfaches Salz n, Einfachsalz n

s. sublimation einfache Sublimation f

s. sugar $C_nH_{2n}O_n$ einfacher Zucker m, Einfachzucker m, Monosa[c]charid n, Monose f

simplest [possible] formula einfachst mögliche Formel f, einfachste Formel (Summenformel, Bruttoformel) f, empirische Formel f (i.e.S.), Verhältnisformel f, stöchiometrische Formel (Grundformel) f, Substanzformel f

simplex pump Simplexpumpe f

simultaneous gleichzeitig

s. reaction Simultanreaktion f

s. side reaction Nebenreaktion f

sinalbin Sinalbin n (Glukosid aus Sinapis alba L.)

sinapic acid $HO(CH_3O)_2C_6H_2CH=CHCOOH$ Sinapinsäure f, Hydroxydimethoxyzimtsäure f

s. alcohol $HO(CH_3O)_2C_6H_2CH=CHCH_2OH$ Sinapinalkohol m

sinapin[e] Sinapin n (Cholinester der Sinapinsäure)

singe / to (Text) sengen, abbrennen, abflammen

singeing (Text) Sengen n, Abbrennen n, Abflammen n

s. machine (Text) Sengmaschine f, Senge f

singer s. singeing machine

single-acting einfachwirkend

s.-action compacting einseitige Verdichtung f

s.-base powder einbasiges Pulver (Treibladungspulver) n, reines Nitrozellulosepulver n

s.-bath method Einbadverfahren n

s.-beam instrument Einstrahlinstrument n

s. bond Einfachbindung f

s.-bond orbital Einfachbindungsorbital n

s.-bore stopcock Einweghahn m

s.-cavity mould (tool) s. s.-impression mould

s. cell Einzelzelle f

s. coater (Pap) einseitig streichende Streichmaschine f, Streichmaschine f für einseitige Beschichtung

s.-compartment drum filter zellenloses Trommelfilter n

s.-compartment thickener Einkammereindicker m

s.-component body (Ker) Einstoffmasse f; Einstoffscherben m

s.-conveyor dryer Einbandtrockner m

s. crystal Ein[zel]kristall m, Einling m

s.-crystal fibre Einkristallfaden m

s.-cylinder machine Selbstabnahme[papier]maschine f, Yankee-Maschine f

s.-cylinder pump Einzylinderpumpe f

s.-daylight press Einetagenpresse f

s.-deck oven Eindeckofen m, Einetagenofen m

s.-deck screen Eindeckersiebmaschine f

s.-disk refiner (Pap) Einscheibenrefiner m

s. dose (Pharm) Einzeldosis f

s.-dose container (Pharm) Einzeldosenbehälter m

s.-drum dryer Einwalzentrockner m

s.-drum system Kühlsystem n mit einer gekühlten Walze

s.-effect evaporator Einkörperverdampfer m, Einstufenverdampfer m

s.-electrode potential Einzelpotential n, E

s.-electron bond Einelektron[en]bindung f, Singulettbindung f

s. fertilizer Einzeldünger m

s. fire (firing) (Ker) Ein[mal]brand m

s.-flash curve Gleichgewichtssiedekurve f, Siedekurve f bei geschlossener Verdampfung, Flashkurve f

s.-fluid atomization Zerstäubung f ohne Hilfsstoff, Einstoffzerstäubung f; Versprühung f ohne Hilfsstoff, Einstoffversprühung f

s.-grained structure (Bodenkunde) Einzelkornstruktur f, Einzelkorngefüge n

s.-impression mould (Plast) Einfachwerkzeug n, Einfachform f (bei Preß- und Spritzwerkzeugen)

s.-layer jet stenter (Text) Düsenplanrahmen m

s. line Einzellinie f (im Spektrum)

s.-lined (Pap) einseitig beklebt; einseitig gedeckt

s.-mash process Einmaischverfahren n

s.-material body s. s.-component body

s.-metal powder Metallpulver n mit einer Komponente

s.-pan analytical balance Einschalenanalysenwaage f

s.-pass evaporator Durchlaufverdampfer m

s. potential s. s.-electrode potential

s.-retort [underfeed] stoker Einmuldenunterschubrost m

s. ring Einzelring m, einzelner Ring m

s.-roll crusher Einwalzenbrecher m

s.-roll mill Einwalzenmühle f, Einwalzenstuhl m, Einwalze f; Einfachwalzwerk n; Einpendelmühle f

s. salt einfaches Salz n, Einfachsalz n

s. scattering (phys Ch) Einfachstreuung f

s.-screw extruder (extruding machine) Einschneckenextruder m

s.-seat[ed] valve einsitziges Ventil n, Einsitzventil n

s.-shearing machine (Plast) Einzylinderschermaschine f

s.-sided coating (Pap) einseitiges Streichen n, einseitiger Strich m, einseitige Beschichtung f

s.-sized concrete Einkornbeton m, Beton m mit Haufwerksporosität

s.-solvent process Ein-Lösungsmittel-Verfahren n, Monosolverfahren n

s.-stage einstufig, Einstufen...

s.-stage bleaching (Pap) Einstufen[holländer]bleiche f

s.-stage compressor einstufiger Verdichter (Kompressor) m

s.-stage electrical precipitator Elektrofilter n (Elektroabscheider m) in Einzonenanordnung

s.-stage homogenization Einstufenhomogenisierung f

s.-stage pump einstufige Pumpe f

s.-step reaction Einstufenreaktion f, Einzelreaktion f

s.-strand einsträngig
s.-strand chain Einstrangkette *f*, einsträngige Kette *f*
s.-strand flight conveyor einsträngiger Kratzer[förderer] *m*
s.-strength glass Fensterglas *n* einfacher Dicke
s.-suction pump Pumpe *f* mit einseitigem Flüssigkeitseintritt, einseitig saugende Pumpe *f*
s. thickness *(Glas)* einfache Dicke *f*, ED
s. tyre press *(Gum)* Reifeneinzelheizer *m*
s. yarn einfacher Faden *m*; einfädiges Garn *n*, Einfachgarn *n*
singlet Singulett *n*
s. link[age] *s.* single-electron bond
s. state Singulettzustand *m*
singly bonded einfach gebunden
s. charged einfach geladen
s. ionized einfach ionisiert
s. linked *s. s.* bonded
sinigrin Sinigrin *n*, myronsaures Kalium *n* *(ein Thioglykosid)*
sink / to [ab]sinken, untersinken; [ab]senken; versenken
s. a bore *(Erdöl)* eine Bohrung niederbringen
s. in einsinken
sink Abwaschbecken *n*, Abflußbecken *n*, Spülbecken *n*, Ausgußbecken *n*, Ausguß *m*; Senkgrube *f*, Abzugsschleuse *f*; Abwasserrohr *n*, Abzugsrohr *n*, Abfluß *m*
s.-and-float process Schwimm[-und]-Sink-Verfahren *n*, Schwerflüssigkeitsverfahren *n*
s. curve Sinkkurve *f*
s.-float method *s. s.*-and-float process
s. mark *(Plast)* Einfallstelle *f (Preßfehler, Spritzgießfehler)*
s. material (product) *s.* sinking material
sinking-body viscometer Sinkerviskosimeter *n*
s. fraction Sinkfraktion *f*, Sinkanteil *m*, Sinkgut *n*
s. material Sinkgut *n*
s. paper *(veraltet für)* blotting paper
s. product *s. s.* material
sinks Sinkgut *n*
sinomenine $C_{16}H_{13}O(OCH_3)_2(NCH_3)OH$ Sinomenin *n (Alkaloid)*
sinter / to sintern, *(Ker) (bei ungeformtem Material)* fritten
sinter Sinter *m*, Sintergut *n*; *(Geol)* Sinter *m*
s. cake Rostbelag *m (eines Sinterröstapparates)*
s. firing *(Ker)* Sinterbrand *m*
s. strand Sinterband *n*
sintered alumina *(Ker)* Sintertonerde *f*, Sinterkorund *m*
s. aluminium powder gesintertes Aluminiumpulver *n*, Sinteraluminiumpulver *n*, Sinteraluminiumprodukt *n*, SAP *n*
s. beryllia *(Ker)* Sinterberyllerde *f*
s. carbide Sinterkarbid *n*, gesintertes Karbid *n*
s. compact gesinterter Preßling *m*, Sinterkörper *m*

s. dolomite *(Ker)* Sinterdolomit *m*
s. fly ash Aschensinter *m*
s.-glass crucible Glassintertiegel *m*
s. hard carbide *s. s.* carbide
s. iron Sintereisen *n*
s. material Sinterwerkstoff *m*
s. metal Sintermetall *n*
s. metal carbide *s. s.* carbide
s.-powder metal Sintermetall *n*
s.-powder metal compact gesinterter Preßling *m*, Sinterkörper *m*
s. product Sintererzeugnis *n*
s. steel Sinterstahl *m*
sintering Sintern *n*, Sinterung *f*, *(Ker) (bei ungeformtem Material)* Fritten *n*
s. coefficient Schwindungskoeffizient *m*
s. furnace Sinterofen *m*
s. grate *(Met)* Verblaserost *m*
s. machine Sinterapparat *m*, Sintermaschine *f*
s. plant Sinteranlage *f*
s. point Erweichungspunkt *m*
s. process Sinterverfahren *n*, Sintermethode *f*; Sintervorgang *m*
s. strand Sinterband *n*
s. technique Sintertechnik *f*, Sinterverfahren *n*, Sintermethode *f*
s. temperature Sinter[ungs]temperatur *f*
s. tray *(Ker)* Sinterwanne *f*
s. under pressure Drucksintern *n*, Preßsintern *n*, Heißpressen *n*, Heißsintern *n*
s. zone Sinterzone *f*
siphon / to [ab]hebern, aushebern, *(i.w.S.)* absaugen, ablaufen lassen, entleeren
s. off abheber, *(i.w.S.)* absaugen
siphon Heber *m*, Saugheber *m*; Siphon *m*; Saugstrahlpumpe *f*, Ejektor *m*
s. feed Dükerzulauf *m*
s. pipe Siphonrohr *n*
SIR *s.* styrene-isoprene rubber
sirup *s.* syrup
sisal [hemp] 1. Sisal *m*, Sisalpflanze *f (verschiedene Agavearten, vorzugsweise Agave sisalana Perr.)*; 2. Sisal *m*, Sisalfaser *f (Blattfaser von 1.)*
S.I.T. *s.* spontaneous-ignition temperature
site of injection Injektionsstelle *f*
six-carbon-atom ring *s. s.*-membered ring
s.-carbon chain C_6-Kette *f*
s.-carbon nucleus Sechskohlenstoffkern *m*
s.-coordinate sechsfach koordiniert
s.-membered sechsgliedrig, 6gliedrig
s.-membered ring sechsgliedriger Ring (Kohlenstoffring) *m*, Sechs[er]ring *m*, 6-Ring *m*
s.-stage bleaching *(Pap)* Sechsstufenbleiche *f*
sixfold axis of symmetry sechszählige Symmetrieachse *f*
sizability *(Pap)* Leimbarkeit *f*
size / to 1. kalibrieren; klassieren, nach Korn[größen]klassen trennen (einteilen, zerlegen); 2. *(Text)* schlichten; *(Pap)* leimen; *(Ker)* mit Abziehlack (Klebelack) bestreichen

size 1. Größe *f*, Umfang *m*, Format *n*, Weite *f*, Abmessung *f*; 2. *(Text)* Dicke *f (von Faserstoffen)*, Garnnummer *f*; *s.* sizing material
s. analysis Siebanalyse *f*, *(i.w.S.)* Körnungsanalyse *f*
s.-bath *(Pap)* Leimbad *n*
s. classification Klassieren *n*, Klassierung *f*, Trennen *n* nach Korn[größen]klassen
s. cooker *(Text)* Schlichtekocher *m*; *(Pap)* Leimkocher *m*
s. determination Größenbestimmung *f*
s. distribution Größenverteilung *f* (z.B. von Körnern), Weitenverteilung *f* (z.B. von Poren)
s. emulsion *(Pap)* Harzemulsion *f*, Harzmilch *f*, Harzlösung *f*, Leimmilch *f*
s. fraction Fraktion *f*, Kornklasse *f*, Kornfraktion *f*, Korngruppe *f*
s.-frequency analysis Körnungsanalyse *f*
s. grading *s. s.* classification
s. milk *s. s.* emulsion
s. of aperture Lochmaß *n* (z.B. von Sieböffnungen)
s. of paper Papierformat *n*
s. of particles Teilchengröße *f*, Partikelgröße *f*
s. of separation Trennkorngröße *f*, Kornscheide *f*
s. of the atom Atomgröße *f*
s. of the ion Ionengröße *f*
s. press *s.* sizing press
s. range Kornspanne *f*, Korngrößenbereich *m*
s. reduction Zerteilen *n*, Zerteilung *f*, *(bei festen Stoffen)* Zerkleinern *n*, Zerkleinerung *f*
s. reduction of hard materials Hartzerkleinerung *f*
s. reduction of medium-hard materials Mittelhartzerkleinerung *f*
s. reduction of soft materials Weichzerkleinerung *f*, Weichmüllerei *f*
s. reduction of solids Zerkleinern *n*, Zerkleinerung *f*
s.-reduction ratio Zerkleinerungsgrad *m*
s. requirements *(Krist)* Platzbedarf *m*
s. roll *(Pap)* Tauchwalze *f*, Leitwalze *f (Oberflächenleimung)*
s. separation *s. s.* classification
s. solution *(Pap)* Leimlösung *f*, Leimflüssigkeit *f*, Leimbrühe *f*
s. speck (spot) Leimfleck *m (Papierfehler)*
s. tub *(Pap)* Leimtrog *m*, Leimbütte *f*, Leimwanne *f*
s.-tub treatment *(Pap)* Oberflächenleimung *f*
s. vat *s. s.* tub
s. water *(Pap) (veraltet für)* backwater
sized *(Pap)* geleimt
1/4 sized *(Pap)* viertelgeleimt, 1/4geleimt
1/2 sized *(Pap)* halbgeleimt, 1/2geleimt, mit mittlerer Leimung
s. in the engine (stuff) *(Pap)* massegeleimt, stoffgeleimt
s. with rosin size *(Pap)* [harz]geleimt

sizing 1. Kalibrieren *n*, Kalibrierung *f*; *s.* size classification; 2. *(Text)* Schlichten *n*; *(Pap)* Leimung *f*, Leimen *n*, Papier[stoff]leimung *f*; *s.* s. material
s. agent *s. s.* material
s. bath *(Text)* Schlichtebad *n*, Schlichteflotte *f*
s. chemical *s. s.* material
s. in the engine (stuff) *(Pap)* Leimung *f* im Stoff, Leimung *f* in der Masse, Masseleimung *f*
s. machine *(Text)* Schlichtmaschine *f*, Schlichtanlage *f*
s. material Grundiermasse *f*; *(Text)* Schlichte *f*, Schlichtemittel *n*, Appretur *f*; *(Pap)* Leimungsmaterial *n*, Leim[stoff] *m*, Leim[ungs]mittel *n*
s. of paper Papierleimung *f*
s. pad *(Text)* Auftragsgalette *f*, Präparationsgalette *f*; Schlichtauftragvorrichtung *f (Glasseide)*
s. press *(Pap)* Leimpresse *f*
s. press roll *(Pap)* Leim[pressen]walze *f*
s. screen Klassiersieb *n*
s. solution *s.* size solution
s. substance *s. s.* material
s. varnish Kleblack *m*
s. vat *s.* size tub
s. with Montan wax *(Pap)* Montanwachsleimung *f*
s. with starch *(Pap)* Stärkeleimung *f*
s. with wax emulsions *(Pap)* Paraffinleimung *f*
skarn Skarn *m (kontaktmetasomatischer, eisenreicher Hornfels)*
skatole Skatol *n*, 3-Methylindol *n*
skein *(Text)* Strang *m*, Strähne *f*, Strahn *m*, Docke *f*
s. dyeing machine *(Text)* Strang[garn]färbemaschine *f*
skeletal substratum *(Landw)* Skelettboden *m*
s. vibration Gerüstschwingung *f*
skeleton Skelett *n*, Gerüst *n*
s. catalyst Skelettkatalysator *m*
s.-flight conveyor Trogkettenförderer *m*, Redler[-Förderer] *m*
skew form *(Stereochemie)* schiefe (windschiefe, syn-clinale) Form *f*, syn-Form *f*
s. mounting Walzenschränkung *f*
skid *(Plast)* Markierung *f* durch überfließendes Material *(Fehler an Spritzlingen)*
Skidmore iron crucible Skidmore-Eisentiegel *m*
skim / to abstreifen, abstreichen, entschäumen, abschöpfen, *(Glas)* abschäumen, abfeimen, abfehmen, *(Milch)* entrahmen, abrahmen; *(Erdöl)* [ab]toppen, skimmen; [ab]schälen (z.B. in der Schälzentrifuge); *(Gum) (friktioniertes Gewebe)* skimmen, belegen
s. off abstreifen, abstreichen, entschäumen, abschöpfen
skim *(Gum)* Skim *m*, Serum *n*; *(Pap)* Stift *m*, Splitter *m*
s. coating *(Gum)* Skimmen *n*, Belegen *n*
s.-coating calender *(Gum)* Kalander *m* zum Belegen von Geweben

s. latex *(Gum)* Skimlatex *m*
s. milk entrahmte Milch *f*, Magermilch *f*
s.-milk cheese Magerkäse *m*
s.-milk powder Magermilchpulver *n*, Trockenmagermilch *f*
s. rubber Skimkautschuk *m*
skimcoat / to *(Gum)* *(friktioniertes Gewebe)* skimmen, belegen
skimmed milk *s.* skim milk
skimmer Abstreifer *m*, Abstreicher *m*; Abschäumer *m*, Abschäumlöffel *m*, Schaumlöffel *m*, Schaumkelle *f*; *(Glas)* Abschäumer *m*, Abfehmer *m*; *(Lebm)* Rahmkelle *f*, Rahmlöffel *m*; Schäler *m*; Schällöffel *m*; Schälmesser *n*; Schälrohr *n*; Schälteller *m*; *(Met)* Überlauf *m*; Fuchs *m* *(zur Trennung des flüssigen Eisens von der mitfließenden Schlacke)*
s. centrifuge Schälzentrifuge *f*, Schälschleuder *f*; Schälrohrzentrifuge *f*, Schälrohrschleuder *f*
s. dam *(Met)* Schlackendamm *m*
s. pipe Schälrohr *n* *(einer Schälzentrifuge)*
skimmilk *s.* skim milk
skimming calender *s.* skim-coating calender
s. efficiency *(Lebm)* Entrahmungsschärfe *f*
s. machine Entrahmungsschleuder *f*, Milchseparator *m*, Milchzentrifuge *f*
s. nozzle Austragdüse *f* *(einer Zentrifuge)*
s. plant *(Erdöl)* Toppanlage *f*
s. tube Schälrohr *n* *(einer Schälzentrifuge)*
skimmings Abgeschäumtes *n*, Schaum *m*, Abgeschöpftes *n*; *(Met)* Schlacke *f*
skin Haut *f*; *(Met)* Gußhaut *f*; *(Met)* Walzhaut *f*; *(Text)* Fasermantel *m*, Faserhaut *f*; *(Text)* Mantel *m* *(eines Kernmantelfadens)*
s. absorption Aufnahme (Absorption) *f* durch die Haut
s. blister Oberflächenblase *f* *(Glasfehler)*
s.-core structure *(Text)* Kernmantelstruktur *f*
s. food Hautnährcreme *f*, hautnährende Creme *f*, Nährcreme *f*
s. formation Hautbildung *f*
s. glue Hautleim *m*, Lederleim *m*
s.-irritant hautreizend
s. irritant hautreizender Stoff *m*, Hautreizstoff *m*
s. lotion Hautlotion *f*
s. parchment tierisches (animalisches) Pergament *n*, Hautpergament *n*, Pergament *n*
s. softener Hauterweichungsmittel *n*, hauterweichendes Mittel *n*
s. tonic Hauttonikum *n*
skip Förderkübel *m*, Transportkübel *m*, Kippkübel *m*, Fördergefäß *n*, Skip *m*, Skipgefäß *n*; Förderkorb *m*
s. bridge Schrägbrücke *f* *(eines Hochofens)* mit Kippkübel
s. car *s.* skip
s. hoist Schrägaufzug *m*, Kübelaufzug *m*, Kippaufzug *m*
Skraup quinoline synthesis, Skraup reaction (synthesis) Skraupsche Chinolinsynthese (Synthese) *f*, Skraup-Synthese *f*

skull *(Met)* Pfannenbär *m*, *(im Konverter)* Mündungsbär *m*
skylight *(Spiegelglas schlechter Qualität)*
slab Platte *f*, Tafel *f*, Fliese *f*; *(Met)* Bramme *f*; *(Gum)* Platte *f*, Kautschukplatte *f*, Kautschukkuchen *m*; Slab *m* *(Rohkautschukhandelssorte)*
s. ingot *(Met)* Rohbramme *f*
s. of coagulum *(Gum)* Koagulatplatte *f*
s. off / to *(Kautschukfell)* von der Walze schneiden
s. stock *(Plast)* tafelförmiger Rohstoff *m*
slack belt Schlappgurt *m*
s. mercerization *(Text)* Slack-Merzerisation *f*, spannungsloses Merzerisieren *n*
s. side Leertrum *n(m)* *(unbelastete Seite einer umlaufenden Fördereinrichtung)*
s. sized *(Pap)* schwachgeleimt, mit schwacher Leimung
s. wax Paraffingatsch *m*, Paraffinkuchen *m*
slacking *(Zerfall der Kohle bei abwechselndem Befeuchten und Trocknen)*
slacklime *s.* slaked lime
slag / to verschlacken; ausschlacken
slag Schlacke *f*; *(Glas)* Herdglas *n*
s. attack *(Met)* Schlackenangriff *m*
s. brick Schlackenstein *m*, Schlackenziegel *m*
s. cement Hüttenzement *m*, Schlackenzement *m*, Hochofenzement *m*
s.-forming schlackenbildend
s. hole Schlacken[ab]stichloch *n*, Schlakken[ab]stich *m*, *(i.e.S.)* Schlackenform *f*
s. ladle Schlackenpfanne *f*
s. lead Krätzblei *n*
s. line *(Met)* Schlackenebene *f*
s. notch *s. s.* hole
s. resistance Schlackenbeständigkeit *f*
s. runner (spout) Schlacken[abstich]rinne *f*
s. stone *s. s.* brick
s. sulphate cement Sulfathüttenzement *m*
s. tap Schlackenabstich *m*
s.-tap furnace Schmelzfeuerung *f* *(Feuerung mit flüssigem Schlackenabzug)*
s.-tap hole *s. s.* hole
s. wool Schlackenwolle *f*, Schlackenfaser *f*
slagging Verschlackung *f*, Schlackenbildung *f*; Entschlackung *f*
s. [ash] producer Abstichgenerator *m*, Abstichgaserzeuger *m*, Schlackenabstichgenerator *m*, Schlackenabstichgaserzeuger *m*
slaggy schlackenähnlich, schlackenartig; schlakkenreich, schlackig
slake / to *(Kalk)* [ab]löschen; zerbröckeln, zerfallen *(z. B. Kohle)*
slaked lime Löschkalk *m*, gelöschter Kalk *m* *(Kalziumhydroxid)*
slaker Kalklöscher *m*, Kalklöschapparat *m*
slaking Löschen *n*, Ablöschen *n* *(von Kalk)*
s. tower Kalklöschturm *m*
slant culture *(Biol)* Schräg[agar]kultur *f*
slanting schräg, schief, geneigt

slash / to *(Text)* schlichten
slasher *(Text)* Schlichtmaschine f, Schlichteanlage f
slashing *(Text)* Schlichten n
slat conveyor Stegkettenförderer m *(i.e.S.) (Kette auf der Bahn gleitend)*
slate Schiefer m, Tonschiefer m; Schieferplatte f
  s. clay Schieferton m
  s. flour Schiefermehl n
  s. paper Schieferpapier n
slatey s. slaty
slatted drum *(Gerb)* Lattentrommel f
slaty schief[e]rig, schieferartig, schieferähnlich; schieferhaltig
sledgehammer quantity überreichliche Menge f, Übermaß n, Überschuß m
sleeper block *(Glas)* Durchlaß-Seitenstein m
slice Scheibe f; *(Pap)* Austrittsspalt m, Aufflußspalt m, Ausflußschlitz m *(des Stoffauflaufs)*
  s. ice Platteneis n
  s. nozzle *(Pap)* Ausflußdüse f, Auslaufdüse f
sliced film (sheet) Schälfolie f, geschälte Folie f
slicer Schneidmaschine f, Schnitzelmaschine f
  s. knife Schnitzelmesser n
slicing knife s. slicer knife
  s. machine *(Plast)* Folienschneidmaschine f
slicker *(Gerb)* Stoßeisen n, Recker m, Schlicker m
slide / to gleiten, ziehen *(z.B. Suspension)*
slide Reiter m, Reiterchen n *(der Analysenwaage)*; *(Foto)* Diapositiv n; *(Mikroskopie)* Objektträger m
  s. cell Objektträgerzelle f
  s.-cell method Objektträgerzellenmethode f *(zur Penizillinbestimmung)*
  s. cover glass Deckglas n *(für Objektträger)*
  s. staining rack (tray) Färbegestell n *(für Objektträger)*
sliding-bat kiln *(Ker)* Durchschubofen m
  s. bearing Gleitlager n
  s. carriage *(Ker)* Füllschlitten m
  s. joint *(Plast)* sanftes Werkzeugschließen n
  s. punch *(Plast)* Verschiebestempel m, Schiebestempel m
  s.-vane compressor Drehschieberverdichter m, Vielzellenverdichter m, Zellenverdichter m
slightly grown malt *(Lebm)* Kurzmalz n, Spitzmalz n
  s. soluble schwer (wenig) löslich, schwerlöslich
slime Schleim m; Schlamm m, Schlick m, *(Aufbereitung auch)* Feingut n
  s. fermentation schleimige Gärung f, Schleimgärung f
  s. formation *(Pap)* Schleimbildung f
  s. overflow Schlammüberlauf m, Feingutüberlauf m, Überlauf m für Feingut
  s. spot Schleimfleck m *(Papierfehler)*
slimicide *(Pap)* Schleimverhütungsmittel n
sliminess Schleimigkeit f; Schlammigkeit f
slimy schleimig, schmierig, mukös, mukos, Schleim...; schlammig
slinger Schleuderprallmühle f *(eines Trockners)*

slip 1. Gleiten n, Rutschen n; Schlupf m; Gleitfähigkeit f; 2. *(Ker)* Schlicker m, Schlempe f; 3. s. slippage loss
  s. additive (agent) Gleitmittel n
  s. casting Schlickergießen n, Schlickerguß m
  s. clay *(Ker)* Glasurton m, Glasurlehm m; Begußton m, Engobeton m
  s. coating *(Ker)* Schlickerüberzug m
  s. forming *(Plast)* Streckformverfahren n, Tiefziehen n mit Gleitvorrichtung (gleitendem Niederhalter)
  s. glaze *(Ker)* Schlickerglasur f, Lehmglasur f
  s. kiln *(Ker)* Schlickerofen m
  s.-on flange loser Flansch m
  s. point Fließschmelzpunkt m
  s. process *(Ker)* Schlickerverfahren n, Naßverfahren n
  s.-resistant *(Text)* schiebefest
  s. ring forming *(Plast)* Tiefziehen n mit Ziehring
  s. sheet *(Am)* Einschießbogen m, Zwischenlagebogen m
  s.-sheet paper *(Am)* Zwischenlage[n]papier n, Zwischenlegepapier n
  s.-type expansion joint Gleit[dehnungs]ausgleicher m
slipe wool *(Text)* Gerberwolle f, Kalkäscherwolle f
slippage Schlupf m
  s. loss Leckverlust m, Sickerverlust m
slipperiness Schlüpfrigkeit f, Glätte f; *(Pap)* Rutschvermögen n
slippery schlüpfrig, glatt, glitschig, schmierig
slit / to aufschlitzen, aufschneiden; *(Pap)* schneiden, längsschneiden, längstrennen; einen Riß bekommen, reißen
slit Schlitz m, Spalt m
  s. die *(Plast)* Schlitzdüse f
  s. film *(Text)* Folienfaden m, Folienbändchen n
  s. orifice *(Plast)* Schlitzdüse f
  s. ultramicroscope Spaltultramikroskop n
slitter s. slitting machine; *(Pap)* Tellermesser n, Kreismesser n
  s. break *(Pap)* Bahnriß m in der Rollenschneidmaschine
  s. shaft *(Pap)* Messerwelle f *(der Längsschneideeinrichtung)*
slitting *(Pap)* Schneiden n, Längsschneiden n, Längstrennen n, Längstrennung f
  s. and [re]winding machine s. s. machine
  s. device s. s. machine
  s. machine Längsschneid[e]maschine f, Längsschneider m, Rollenschneid[e]maschine f, Rollenschneider m, Streifenschneid[e]maschine f, Streifenschneider m, *(Pap auch)* Umroller m
sliver Splitter m, Span m, *(Pap auch)* Stift m; *(Text)* Faserband n, Krempelband n; *(Glas)* Lunte f, Vorgarn n
  s. screen *(Pap)* Splitterfänger m
slop Schmutzwasser n; *(Lebm)* Schlempe f, Brennereischlempe f; *(Ker)* Schlicker m
  s. pad liquor *(Text)* Klotzflotte f, Klotzbad n

s.-padding *(Text)* Klotzen *n*, Pflatschen *n*, Foulardieren *n*, Foulardbehandlung *f*

s. wax Slop-Wax *n (nicht preßbares Wachs aus schweren Krackdestillaten)*

slope Neigung *f*, Gefälle *n*; schiefe Ebene *f*, Schräge *f*

sloping grate Schrägrost *m*

slot / to schlitzen, mit Schlitzen versehen

slot Schlitz *m*; *(Destillation)* Dampfdurchtrittsschlitz *m*; *(Pap)* Austrittsspalt *m*, Ausflußspalt *m*, Ausflußschlitz *m (des Stoffauflaufs)*

s. die *(Plast)* Schlitzform *f*, Breitschlitzdüse *f (zur Folienherstellung)*

s. nozzle Schlitzdüse *f*, Flachstrahldüse *f*

s. width Schlitzbreite *f*

slotted cap Schlitzglocke *f*

slow (s.-acting, s.-action) accelerator *(Gum)* schwacher (langsamer, schwach wirkender) Beschleuniger *m*

s. beating *(Pap)* schmierige Mahlung *f*, Schmierigmahlung *f*

s. combustion stille Verbrennung *f*

s. cook *(Pap)* Langsamkochung *f*, indirekte Kochung *f*

s.-curing *(Gum)* langsam vulkanisierend (heizend)

s. curing *(Gum)* langsame Vulkanisation *f*

s. down / to verlangsamen, [ab]bremsen, *(die Geschwindigkeit)* drosseln, herabsetzen, vermindern, *(Kph auch)* moderieren; sich verlangsamen

s. emulsion *(Foto)* geringempfindliche (niedrigempfindliche) Emulsion *f*, Emulsion *f* niedriger Empfindlichkeit

s.-evaporating langsamflüchtig, schwerflüchtig

s. fire Niedrigfeuer *n*

s. freezing langsames Gefrieren (Einfrieren) *n*, Langsamgefrierverfahren *n*

s. neutron langsames Neutron *n*

s. paper machine *s.* s.-speed paper machine

s. pulp *s.* s. stock

s.-renneting *(Lebm)* labträge

s. sand filter Langsamfilter *n*

s. sand filtration Langsamfilterung *f*, Langsamfiltration *f*

s.-speed paper machine Papiermaschine *f* mit kleiner Geschwindigkeit, Langsamläufer *m*

s. stock *(Pap)* schmieriger Stoff *m*

s. strike *(Text)* langsames Aufziehen *n* *(von Farbstoffen)*

s.-striking *(Text)* langsamziehend *(Farbstoffe)*

slowness Trägheit *f*; *(Pap)* Schmierigkeit *f*

slubbing *(Text)* Kammzug *m*; Faserbändchen *n*, Spinnband *n*

sludge Schlamm *m*, Abschlamm *m*, Sinkstoff *m*

s. activation Schlammbelebung *f*

s. concentration Schlammeindickung *f*, Schlammentwässerung *f*; Schlammkonzentration *f*, Schlammeindickungsgrad *m*

s. concentrator Schlammeindicker *m*, Schlammentwässerungsanlage *f*

s. digestion Schlammfaulung *f*

s.-digestion chamber (compartment) Schlammfaulraum *m*, Faulraum *m*

s. discharge Schlammaustrag *m*

s.-discharge line Leitung *f* für den Schlammaustrag

s. fertilizer Klärschlamm *m (als Düngemittel)*

s. gate Schlammschleuse *f*

s. line Suspensionsgrenze *f*

s. pump Schlammpumpe *f*, Dickstoffpumpe *f*, Schmutzwasserpumpe *f*

s.-raking arm Krählarm *m*

s.-raking mechanism Krählwerk *n*, Rührwerk *n*

s. recovery Rückstandsaufarbeitung *f (bei der Kohlehydrierung)*

s. washing *(Pap)* Waschen *n* des Kalkschlammes

s. zone Schlammzone *f*, Dickschlammzone *f*, Rückstandszone *f*

slug Perlen *fpl (Fehler in Glasfaserprodukten)*

sluice Schleuse *f*; Schütz *n (einer Schleuse)*

sluicing nozzle Spüldüse *f*

slurry Schlamm *m*, Schlämme *f*, Aufschlämmung *f*, *(dünner)* Brei *m*; *(Ker)* Schlicker *m*; *(Filtration)* Trübe *f*; *(Zementherstellung)* Rohschlamm *m*

s. method of seed treatment *s.* s. treatment

s. oil *(Erdöl)* Schlammöl *n*

s. preparation *(Filtration)* Vorbehandlung *f* der Trübe

s. settler Schlammabscheider *m*

s. treatment Schlammbeize *f (für Saatgut)*

slush Schlamm *m*, Morast *m*; Schmiere *f*, Schmiermittel *n*

s. moulding *(Plast)* Pastengießen *n*, Ausgießverfahren *n*, Sturzgießverfahren *n (für Hohlkörper aus PVC-Paste)*

s. of stock *(Pap)* Stoffsuspension *f*, Stoffbrei *m*, Faser[stoff]brei *m*, Fasermasse *f*

s. pulp *(Pap)* Dickstoff *m*

s. pulp storage *(Pap)* Dickstoff[vorrats]behälter *m*

s. stock *(Pap)* Dickstoff *m*; *s.* s. of stock

slushing oil Rostschutzöl *n*

small-angle scattering *(phys Ch)* Kleinwinkelstreuung *f*

s.-angle X-ray scattering Röntgenkleinwinkelstreuung *f*

s. bell kleine Glocke (Verschlußglocke, Gichtglocke) *f (eines Hochofens)*

s. coal Kleinkohle *f*, Kohle[n]klein *n*; Steinkohlenklein *n*

s. flame kleine Flamme *f*, Sparflamme *f*

s. formulator *(Kleinhersteller von Wirkstoffzubereitungen)*

s. ice kleinstückiges Eis *n*

s. manufacturer Kleinhersteller *m*

s. period kurze (kleine) Periode *f*, Kurzperiode *f (im Periodensystem)*

s. ring kleiner Ring *m (3- und 4gliedrig)*

s.-ring compound Verbindung *f* mit kleinem Ring, Kleinringverbindung *f*

s.-scale equipment Minimalausstattung f, Mindestausstattung f, Grundausstattung f
s.-scale experiment Kleinversuch m
s.-scale test Vorversuch m mit kleinen Substanzmengen
s.-sized feinstückig, kleinstückig
s.-sized coal s. s. coal
smallholder (Gum) Kleinpflanzer m
smallholding (Gum) kleine Pflanzung f
s. rubber (auf Kleinplantagen erzeugter Kautschuk)
smalls s. small coal
smalt Smalte f, Schmalte f, Blaufarbenglas n (Kobalt(II)-kaliumsilikat)
smaltine, smaltite (Min) Smaltin m, Skutterudit m, Speiskobalt m (Kobalttriarsenid)
smaragdochalcite (Min) Smaragdochalzit m, (veraltet für) Atakamit m (Kupfer(II)-trihydroxidchlorid) (oder) Dioptas m (ein Kupfersilikat)
smaze (Am) Rauchnebel m (Luftverunreinigungstyp)
smear / to 1. [ein]schmieren, bestreichen, (Haut) einsalben; beschmieren, beschmutzen; (Schmiermittel) aufstreichen, aufschmieren; 2. schmieren, schmierig sein
smear 1. Schmiere f; klebrige Substanz f; 2. (veraltet für) ointment; 3. Fettfleck m, Schmutzfleck m; 4. (Ker) Salzglasur f; 5. (Glas) Oberflächenriß m (an einem Flaschenhals)
smectic smektisch
s. phase (Koll) smektische Phase f, bz-Phase f
s. state smektischer Aggregatzustand m (kristalliner Flüssigkeiten)
smell / to riechen
smell Geruch m, Duft m; übler (schlechter) Geruch m, Gestank m
smelling salt Riechsalz n
smelt / to schmelzen, (Metalle) erschmelzen, (Erze) verhütten, (Schrott) einschmelzen
smelt Schmelze f, Schmelzfluß m; (Pap) Schmelze f, Sodaschmelze f, Schmelzsoda f
s. dissolving tank (Pap) Schmelzlöser m, Lösetank m, Lösebehälter m
smelter Schmelzofen m
smelting Schmelzen n; Erschmelzen n (von Metallen), Verhütten n, Verhüttung f (von Erzen), Einschmelzen n (von Schrott)
s. flux electrolysis Schmelz[fluß]elektrolyse f
s. furnace Schmelzofen m
s. lime (Landw) Hüttenkalk m
s. plant Hüttenwerk n, Schmelzhütte f, Hütte f
s. pot Schmelztiegel m
s. zone Schmelzzone f
smithsonite (Min) Smithsonit m, Zinkspat m (Zinkkarbonat)
smog Smog m, Stadtnebel m (Luftverunreinigungstyp)
smoke / to 1. räuchern; ausräuchern; vernebeln; 2. rauchen
smoke Rauch m

s. chamber Räucherkammer f, Rauchkammer f
s. curing (Lebm) Räuchern n, Räucherung f, Räucherei f
s.-dried geräuchert, Räucher...
s. drying s. s. curing
s. generator Rauchgenerator m
s. meter Rauchprobenehmer m
s. monitor Rauchmeldeanlage f
s. point (Erdöl) Rußpunkt m
s. screen Rauchschirm m
s. signal Rauchsignal n
s. stove s. smoking kiln
smoked sheet [rubber] Smoked Sheet, Räucherkautschuk m (geräucherte Rohkautschukplatte)
smokehouse Räucherhaus n, Räucherei f
smokeless fuel rauchloser (rauchfreier, rußfreier) Brennstoff m
s. powder rauchloses (rauchschwaches) Pulver n
smokery Räucher[ungs]anlage f, Räucherei f
smokestack Schornstein m, Esse f, Schlot m, Kamin m
smoking s. smoke curing
s. kiln Räucherofen m
s. tobacco Rauchtabak m
smoky flame rußende Flamme f
s. quartz (Min) Rauchquarz m (Silizium(IV)-oxid)
smooth / to (Pap) glätten
smooth glatt, gleichmäßig (z. B. auch Reaktionsablauf)
s.-drying finish (Text) glatttrocknende Ausrüstung (Appretur) f, Trockenglatt-Ausrüstung f, Naßknitterarm-Ausrüstung f
s. fluidization gleichmäßiges Fließen n
s. porphyry roll Porphyrglattwalze f
s. roll Glattwalze f
s.-roll crusher Walzenbrecher m mit glatten Walzen, Glattwalzenbrecher m
s.-surfaced mit glatter Oberfläche, poliert
s.-surfaced roll Glattwalze f
smoothing Glätten n, Glättung f
s. press (Pap) glättende (filzlose) Presse f, Offsetpresse f
s. roll (Pap) Glättwalze f
smoothness Glätte f, Gleitfähigkeit f
s. number (Pap) Glättezahl f
s. tester (Pap) Glätteprüfgerät n, Glätteprüfer m
smoothstone (Pap) Glättstein m
smoulder / to schwelen, glimmen
smoulder-proof nicht nachglimmend
smudge / to beschmutzen, verschmieren
smudge Schmutz m; Schmutzfleck m; (Plast) Schmierstelle f (an Spritzlingen)
smust (aus Rauch und Staub bestehende Luftverunreinigung)
Smyrna galls (Gerb) Smyrnaer (Türkische) Gallen fpl, Aleppogallen fpl (von Quercus infectoria Oliv.)
S.N.A. s. sodium naphthalene acetate

snake poison Schlangengift n
snakeskin glaze (Ker) Schlangenhautglasur f
snap (Gum) Nerv m
  s.-back fibre (Am) Elastomerfaserstoff m; Elastomerfaser f
snappiness (Pap) Klang m
snappy (Gum) nervig
S.N.F. s. solids-not-fat
snub pulley Ablenktrommel f, Ablenkrolle f, Leittrommel f, Leitrolle f
snubber Pulsationsdämpfer m, Druckschwingungsdämpfer m
snuff[ing] tobacco Schnupftabak m
SO s. spin orbital
soak / to 1. aufsaugen; tränken, einweichen; durchdringen, durchtränken (von Flüssigkeiten); [hindurch]sickern; sich vollsaugen; 2. (Met) ausgleichglühen
soak liquor (Gerb) Weichwasser n
  s. pit s. soaking pit
soaked clay gesumpfter Ton m
soaker (Erdöl) Soaker m, Reaktionskammer f, Soaking Drum f (beim Tube-and-Tank-Verfahren)
  s. bottle washing machine (Lebm) Flaschenweichmaschine f, Weichmaschine f
  s.-hydro (s.-sprayer) bottle washing machine (Lebm) Weich- und Spritzmaschine f
soaking (Gerb) Weiche f, Weichen n, Wässern n
  s. chamber (drum) s. soaker
  s. pit (Ker) Sumpfgrube f; (Met) Wärme[ausgleich]grube f, Tiefofen m
  s. section (Erdöl) Einweichsektion f, Soak-Sektion f (eines Reformofens)
  s. temperature (Ker) Garbrandtemperatur f
  s. zone Durchweichzone f (eines Glühofens)
soap Seife f
  s. bark Seifenrinde f, Quillajarinde f, Panamarinde f (von Quillaja saponaria Mol.)
  s. bubble Seifenblase f
  s. chippings Seifenschnitzel npl
  s. flakes Seifenflocken fpl
  s. manufacture Seifenherstellung f, Seifenfabrikation f
  s. micelle Seifenmizell n, Seifenmizelle f
  s. paper Seifenpapier n
  s. powder Seifenpulver n
  s. shampoo Seifenschampun n, Seifenshampoo[n] f
  s.-stabilized o/w emulsion O/W-Seifenemulsion f (durch Seife stabilisierte O/W-Emulsion)
  s. stock s. soapstock
  s. tissue s. s. paper
soaping Seifen n, Abseifen n
  s. aftertreatment (Text) Nachseifen n, Seifennachbehandlung f
  s. bath Seifenbad n
soapless soap synthetisches Waschmittel n (oder) Reinigungsmittel n, Detergens n, Detergent n
soapstock Soapstock m, Seifenfluß m, Seifenstock m

soapstone / to (Gum) talk[um]ieren, mit Talkum pudern (bestäuben)
soapstone (Min) Speckstein m, Steatit m (Talk-Varietät)
soapstoning (Gum) Talk[um]ieren n, Pudern (Bestäuben) n mit Talkum
socket Hülse f, (bei Kegelschliffen auch) Schliffhülse f; Schliffpfanne f, Schliffschale f (einer Kugelschliffverbindung); Muffe f; Flansch m; Rohrstutzen m
  s. fitting Muffenverbindung f, Muffenverbindungsstück n, Muffenverbinder m
  s. pipe Muffenrohr n
  s.-to-ball adapter Halteklemme f für Kugelschliffverbindungen
  s.-to-cone adapter Halteschelle f für Kegelschliffverbindungen, Ligatur f
  s.-weld joint Muffenschweißverbindung f (Rohrverbindung)
socketed pipe s. socket pipe
Socotra aloe (Pharm) Sokotra-Aloe f (von Aloe perryi Baker)
sod Sode f, Torfsode f (getrocknetes Stechtorfstück in Ziegelform)
  s. oil (Gerb) Sämischgerber-Degras m, Weißgerber-Degras m, deutscher Degras m (Nebenprodukt der Sämischgerbung)
soda 1. Na_2CO_3 Soda f(n), Natriumkarbonat n, (i.e.S.) $Na_2CO_3 \cdot 10H_2O$ Kristallsoda f, Natriumkarbonat-10-Wasser n; 2. $NaHCO_3$ Natron n, Natriumhydrogenkarbonat n; 3. $NaOH$ Ätznatron n, Natriumhydroxid n; 4. Na_2O Natriumoxid n; 5. s. s. water; 6. (Min) Soda f, Natron n, Natrit m (Natriumkarbonat-10-Wasser)
  s. alum $NaAl(SO_4)_2 \cdot 12H_2O$ Sodaalaun m, Natronalaun m, Natriumalaun m, Natriumaluminiumsulfat-12-Wasser n
  s.-asbestos tube Natronasbeströhrchen n
  s. ash Na_2CO_3 [kristall]wasserfreie (kalzinierte) Soda f, wasserfreies Natriumkarbonat n
  s.-base liquor (Pap) Natriumbisulfitkochsäure f
  s.-boiled in Sodalauge ausgekocht
  s. bordeaux Burgunder Brühe f, Kupfersodabrühe f (Pflanzenschutzmittel)
  s. cellulose (Am) Natronzellulose f, Alkalizellulose f
  s. chabazite (Min) Natronchabasit m, (veraltet für) Gmelinit m (ein Tektosilikat)
  s. cook (Pap) Sodakochung f, Natronkochung f
  s. crystals $Na_2CO_3 \cdot 10H_2O$ Kristallsoda f, Natriumkarbonat-10-Wasser n
  s. digester (Pap) Natronzellstoffkocher m
  s. digestion liquor s. s. liquor
  s. extraction Sodaauszug m
  s. feldspar (Min) Natronfeldspat m (Natriumalumotrisilikat)
  s. glass Natronglas n
  s. lime Natronkalk m (Ätznatron-Ätzkalk-Gemisch)
  s.-lime glass Sodakalkglas n, Kalknatronglas n, Natronkalkglas n

s.-lime-silica glass Natron-Kalk-Kieselsäureglas n

s. lime tube Natronkalkrohr n

s. liquor (lye) *(Pap)* Natronkochlauge f

s. mica *(Min)* Natronglimmer m, Paragonit m

s. mill *(Pap)* Natronzellstoffabrik f, Natronwerk n

s. nitre $NaNO_3$ Natronsalpeter m, Natriumnitrat n, *(Min meist)* Nitronatrit m

s. process *(Pap)* Natronverfahren n, Sodaverfahren n

s. pulp Natron[zell]stoff m, Sodazellstoff m

s. pulp process s. s. process

s. pulping *(Pap)* Natronaufschluß m, Sodaaufschluß m

s. recovery *(Pap)* Alkalirückgewinnung f

s.-silica glass Natriumsilikatglas n

s. smelt *(Pap)* Sodaschmelze f, Schmelze f, Schmelzsoda f

s. soap Natronseife f, Natriumseife f

s. softening *(Wasserchemie)* Sodaenthärtung f

s. solution Sodalösung f

s. washing Sodawäsche f

s. water Sodawasser n, kohlensäurehaltiges Wasser n, Selterswasser n *(i.w.S.)*

s. water glass Natronwasserglas n

s. wood pulp process s. s. process

sodalite *(Min)* Sodalith m *(ein Tektosilikat)*

sodamide $NaNH_2$ Natriumamid n

Söderberg cell Elektrolysezelle f mit Söderberg-Elektrode

Söderberg [continuous] electrode Söderberg-Elektrode f, selbstbrennende Elektrode f, Dauerelektrode f

Söderberg self-baking electrode s. Söderberg [continuous] electrode

sodic Natrium...

sodio derivative *(org Ch)* Natriumderivat n

sodioacetoacetic ester $CH_3COCHCOOC_2H_5Na$ Natriumacetessigester m

sodiomalonic ester $CH(COOC_2H_5)_2Na$ Natriummalonsäurediäthylester m, Natriummalonester m

sodium Na Natrium n

s. acetate CH_3COONa Natriumazetat n

s. acetylide Na_2C_2 Natriumazetylid n, Natriumkarbid n

s. acid arsenite s. s. hydrogen arsenite

s. acid carbonate s. s. hydrogen carbonate

s. acid fluoride s. s. hydrogen fluoride

s. acid phosphate s. 1. s. dihydrogen phosphate; 2. s. hydrogen phosphate

s. acid pyrophosphate $Na_2H_2P_2O_7$ Dinatriumdihydrogendiphosphat n, Natriumdihydrogenpyrophosphat n

s. acid sulphate s. s. hydrogen sulphate

s. acid sulphite s. s. hydrogen sulphite

s. acid tartrate s. s. hydrogen tartrate

s. alginate Natriumalginat n

s. alkoxide NaOR Natriumalkoholat n

s. alkyl Natriumalkyl n

s. aluminate Natriumaluminat n

s. aluminate solution [Natronlauge-]Aluminatlösung f, Aluminatlauge f *(Bayer-Verfahren)*

s.-aluminium fluoride $Na_3[AlF_6]$ Natriumfluoroaluminat n, Natriumaluminiumfluorid n

s.-aluminium sulphate $NaAl(SO_4)_2$ Natriumaluminiumsulfat n

s. amalgam Natriumamalgam n

s. amide $NaNH_2$ Natriumamid n

s.-ammonium hydrogen phosphate $Na[NH_4]HPO_4 \cdot 4H_2O$ Natriumammoniumhydrogenphosphat-4-Wasser n, Phosphorsalz n

s.-ammonium sulphate $Na_2SO_4 \cdot (NH_4)_2SO_4$ Natriumammoniumsulfat n

s. amylate Natriumamylat n

s. arsenite Na_3AsO_3 Natriumarsenit n

s. ascorbate Natriumaskorbat n

s. aurothiosulphate $Na_3[Au(S_2O_3)_2]$ Natriumdithiosulfatoaurat(I) n, Natriumgoldthiosulfat n

s. azide NaN_3 Natriumazid n

s.-base acid *(Pap)* Natriumbisulfitkochsäure f

s.-base [sulphite] liquor s. soda-base liquor

s. benzene sulphonate Natriumbenzolsulfonat n, benzolsulfonsaures Natrium n

s. benzoate C_6H_5COONa Natriumbenzoat n

s. bicarbonate s. s. hydrogen carbonate

s. bichromate s. s. dichromate

s. bifluoride s. s. hydrogen fluoride

s. biphosphate s. s. dihydrogen phosphate

s. bismuthate $NaBiO_3$ Natriumtrioxobismutat(V) n

s. bisulphate s. s. hydrogen sulphate

s. bisulphite s. s. hydrogen sulphite

s. bisulphite bleaching *(Pap)* Natriumbisulfitbleiche f

s. bisulphite liquor Natriumhydrogensulfitlösung f, Natriumbisulfitlösung f

s. bitartrate s. s. hydrogen tartrate

s. boranate s. s. borohydride

s. borohydride $Na[BH_4]$ Natriumboranat n, Natriumborwasserstoff m, Natriumborhydrid n, Natrium[tetra]hydridoborat n

s. bromate $NaBrO_3$ Natriumbromat n

s. bromide NaBr Natriumbromid n

s.-bromite desizing *(Text)* Bromitentschlichtung f

s.-butadiene rubber Natrium-Butadienkautschuk m *(Butadien-Na-Polymerisat)*

s. butyrate $CH_3CH_2CH_2COONa$ Natriumbutyrat n

s. cacodylate Natriumkakodylat n

s. carbide s. s. acetylide

s. carbonate Na_2CO_3 Natriumkarbonat n

s. carbonate decahydrate $Na_2CO_3 \cdot 10H_2O$ Natriumkarbonat-Dekahydrat n, Natriumkarbonat-10-Wasser n

s. carboxymethyl cellulose Natriumkarboxymethylzellulose f

s.-catalyzed natriumkatalysiert

s.-catalyzed polymerization s. s. polymerization

s. cellulose xanthate Natriumzellulosexanthogenat n

s. chlorate NaClO$_3$ Natriumchlorat *n*

s. chloraurate Na[AuCl$_4$] Natriumtetrachloroaurat(III) *n*, Natriumgold(III)-chlorid *n*

s. chloride NaCl Natriumchlorid *n*

s. chloride crystal Natriumchloridkristall *m*, Steinsalzkristall *m*, Kochsalzkristall *m*

s. chloride lattice Natriumchloridgitter *n*, Steinsalzgitter *n*, Kochsalzgitter *n*

s. chlorite NaClO$_2$ Natriumchlorit *n*

s. chlorite bleaching *(Pap)* Natriumchloritbleiche *f*

s. chlorite bleaching liquor *(Pap)* Natriumchloritbleichlauge *f*

s. chloroaurate *s.* s.-gold chloride

s. chloroiridate *s.* s.-iridium chloride

s. chloroiridite Na$_3$[IrCl$_6$] Natriumhexachloroiridat(III) *n*

s. chloroosmate Na$_2$[OsCl$_6$] Natriumhexachloroosmat(IV) *n*

s. chloropalladite Na$_2$[PdCl$_4$] Natriumtetrachloropalladat(II) *n*

s. chloroplatinate Na$_2$[PtCl$_6$] Natriumhexachloroplatinat(IV) *n*

s. chloroplatinite Na$_2$[PtCl$_4$] Natriumtetrachloroplatinat(II) *n*

s. chlororhodite Na$_3$[RhCl$_6$] Natriumhexachlororhodat(III) *n*

s. chromate Na$_2$CrO$_4$ Natriumchromat *n*

s. citrate NaOOC·C(OH)(CH$_2$COONa)$_2$ Natriumzitrat *n*

s. cobaltinitrite *s.* s. hexanitrocobaltate(III)

s. condensation Natriumkondensation *f*, Kondensation *f* mittels Natriums

s.-cooled reactor natriumgekühlter Reaktor *m*

s. cyanate NaOCN Natriumzyanat *n*

s. cyanide NaCN Natriumzyanid *n*

s. D line Natrium-D-Linie *f*, Na-D-Linie *f*

s. decamolybdate Na$_2$Mo$_{10}$O$_{31}$ Natriumdekamolybdat *n*

s. dichromate Na$_2$Cr$_2$O$_7$ Natriumdichromat(VI) *n*

s. dihydrogen phosphate NaH$_2$PO$_4$ Natriumdihydrogen[ortho]phosphat *n*, Mononatriumdihydrogenphosphat *n*

s. dinitrophenate Natriumdinitrophenolat *n*

s. dioxide *s.* s. peroxide

s. disilicate Na$_2$Si$_2$O$_5$ Natriumdisilikat *n*

s. dithionate Na$_2$S$_2$O$_6$ Natriumdithionat *n*

s. dithionite Na$_2$S$_2$O$_4$ Natriumdithionit *n*, *(Farb auch)* Dithionit *n*

s. ethoxide C$_2$H$_5$ONa Natriumäthoxid *n*, Natriumäthylat *n*, Natriumalkoholat *n* *(i.e.S.)*

s. ethyl xanthate Natriumäthylxanthogenat *n*

s. ethylate *s.* s. ethoxide

s. feldspar *(Min)* Natronfeldspat *m*, Albit *m* (Natriumalumotrisilikat)

s. ferricyanide Na$_3$[Fe(CN)$_6$] Natriumhexazyanoferrat(III) *n*

s. ferrite NaFeO$_2$ Natriumferrat(III) *n*

s. ferrocyanide Na$_4$[Fe(CN)$_6$] Natriumhexazyanoferrat(II) *n*

s. flame Natriumflamme *f*

s. fluoride NaF Natriumfluorid *n*

s. fluoroaluminate *s.* s. aluminium fluoride

s. fluoroantimonate Na[SbF$_6$] Natriumfluoroantimonat(V) *n*

s. fluoroberyllate Na$_2$[BeF$_4$] Natriumfluoroberyllat *n*, Natriumberylliumfluorid *n*

s. fluoroborate Na[BF$_4$] Natriumtetrafluoroborat *n*

s. fluorosilicate Na$_2$[SiF$_6$] Natriumhexafluorosilikat *n*

s. formaldehyde sulphoxylate NaHSO$_2$·HCHO·2H$_2$O Natriumformaldehydsulfoxylat *n*, Natriumsulfoxylatformaldehyd *m*

s. formate HCOONa Natriumformiat *n*

s. glutamate Natriumglutamat *n*

s.-gold chloride Na[AuCl$_4$] Natriumtetrachloroaurat(III) *n*, Natriumgold(III)-chlorid *n*

s.-gold thiosulphate Na$_3$[Au(S$_2$O$_3$)$_2$] Natriumdithiosulfatoaurat(I) *n*, Natriumgoldthiosulfat *n*

s. graphite reactor Natrium-Graphit-Reaktor *m*, natriumgekühlter Graphitreaktor *m*

s. hexahydroxostannate Na$_2$[Sn(OH)$_6$] *oder* Na$_2$SnO$_3$·3H$_2$O Natriumhexahydroxostannat(IV) *n*

s. hexametaphosphate Natriumhexametaphosphat *n*, Graham-Salz *n*

s. hexanitrocobaltate(III) Na$_3$[Co(NO$_2$)$_6$] Natriumhexanitrokobaltat(III) *n*

s. hydrate *s.* s. hydroxide

s. hydride NaH Natriumhydrid *n*

s. hydrogen arsenite Na$_2$HAsO$_3$ Dinatriumhydrogen[ortho]arsenat(III) *n*

s. hydrogen carbonate NaHCO$_3$ Natriumhydrogenkarbonat *n*

s. hydrogen fluoride NaHF$_2$ Natriumhydrogenfluorid *n*

s. hydrogen phosphate Na$_2$HPO$_4$ Natriumhydrogen[ortho]phosphat *n*, Dinatriumhydrogenphosphat *n*

s. hydrogen sulphate NaHSO$_4$ Natriumhydrogensulfat *n*, Mononatriumsulfat *n*

s. hydrogen sulphide NaHS Natriumhydrogensulfid *n*, *(Gerb)* Natriumsulfhydrat *n*

s. hydrogen sulphite NaHSO$_3$ Natriumhydrogensulfit *n*

s. hydrogen tartrate NaOOC(CHOH)$_2$COOH Natriumhydrogentartrat *n*

s. hydroperoxide NaOOH Natriumhydrogenperoxid *n*

s. hydrosulphide *s.* s. hydrogen sulphide

s. hydrosulphite *s.* 1. s. hydrogen sulphite; 2. s. dithionite

s. hydrosulphite bleaching *(Pap)* Hydrosulfitbleiche *f*, Dithionitbleiche *f*

s. hydrotartrate *s.* s. hydrogen tartrate

s. hydroxide NaOH Natriumhydroxid *n*, Ätznatron *n*

s. hydroxide solution Natriumhydroxidlösung *f*, Natronlauge *f*

s. hypochlorite NaOCl Natriumhypochlorit *n*
s. hypochlorite bleach liquor *(Pap)* Natriumhypochloritbleichlauge *f*
s. hypochlorite process Natriumhypochloritverfahren *n*, NaOCl-Verfahren *n*
s. hyponitrite $Na_2N_2O_2$ Natriumhyponitrit *n*
s. hypophosphate $Na_4P_2O_6$ Natriumhypophosphat *n*
s. hypophosphite NaH_2PO_2 Natriumhypophosphit *n*
s. hyposulphate *s.* s. dithionate
s. hyposulphite *s.* 1. s. thiosulphate; 2. s. dithionite
s. iodate $NaJO_3$ Natriumjodat *n*
s. iodide NaJ Natriumjodid *n*
s. iodoplatinate $Na_2[PtJ_6]$ Natriumhexajodoplatinat(IV) *n*
s.-iridium chloride $Na_2[IrCl_6]$ Natriumhexachloroiridat(IV) *n*
s. lactate $CH_3CHOHCOONa$ Natriumlaktat *n*
s. lamp *s.* s.-vapour [discharge] lamp
s. lauryl sulphate $C_{12}H_{25}SO_4Na$ Natriumlaurylsulfat *n*, Natriumdodezylsulfat *n*
s. light Natriumlicht *n*
s. manganate Na_2MnO_4 Natriummanganat(VI) *n*
s. metaantimonate $NaSbO_3$ Natriumtrioxoantimonat(V) *n*, Natriummetaantimonat(V) *n*
s. metaarsenate $NaAsO_3$ Natriummetaarsenat(V) *n*
s. metaarsenite $NaAsO_2$ Natriummetaarsenat(III) *n*
s. metabisulphite *s.* s. pyrosulphite
s. metaborate $NaBO_2$ Natriummetaborat *n*
s. metaperiodate $NaJO_4$ Natriumtetroxoperjodat *n*, Natriummetaperjodat *n*
s. metaphosphate $(NaPO_3)_n$ Natriummetaphosphat *n*
s. metasilicate Na_2SiO_3 Natriummetasilikat *n*
s. metavanadate $NaVO_3$ Natriumtrioxovanadat(V) *n*, Natriummetavanadat(V) *n*
s. methoxide *s.* s. methylate
s. methylate $NaOCH_3$ Natriummethylat *n*, Natriummethoxid *n*
s. methylsiliconate $CH_3SiOONa$ oder $CH_3Si(OH)_2ONa$ Natriummethylsilikonat *n*
s. mineral Natriummineral *n*
s. molybdate Na_2MoO_4 Natriummolybdat(VI) *n*
s. molybdate crystals $Na_2MoO_4 \cdot 2H_2O$ Natriummolybdat-2-Wasser *n*
s. monohydrogen phosphate *s.* s. hydrogen phosphate
s. monoxide Na_2O Natrium[mon]oxid *n*
s. myristate $CH_3(CH_2)_{12}COONa$ Natriummyristat *n*
s. naphthalene acetate $C_{10}H_7CH_2COONa$ Natriumsalz *n* der Naphthalin-1-essigsäure, Natrium-α-naphthylazetat *n*
s. 2-naphthol-3,6-disulphonate Dinatriumsalz *n* der 2-Naphthol-3,6-disulfonsäure, 2-naphthol-3,6-disulfonsaures Natrium *n*, R-Salz *n*

s. nitrate $NaNO_3$ Natriumnitrat *n*
s. nitride Na_3N Natriumnitrid *n*
s. nitrite $NaNO_2$ Natriumnitrit *n*
s. nitroferricyanide *s.* s. nitroprusside
s. nitroprussiate *s.* s. nitroprusside
s. nitroprusside $Na_2[Fe(CN)_5(NO)]$ Nitroprussidnatrium *n*, Natriumnitroprussid *n*, Natriumnitroprussiat *n*, Dinatriumpentazyanonitrosylferrat(II) *n*
s. nitroprusside paper Nitroprussidnatriumpapier *n*
s. nitrosoferricyanide *s.* s. nitroprusside
s. oleate $C_{17}H_{33}COONa$ Natriumoleat *n*
s. orthoantimonate Na_3SbO_4 Natriumorthoantimonat(V) *n*, Natriumtetroxoantimonat(V) *n*
s. orthosilicate Na_4SiO_4 Natriumorthosilikat *n*, Natriumtetroxosilikat *n*
s. orthovanadate Na_3VO_4 Natriumorthovanadat(V) *n*, Trinatriumtetroxovanadat(V) *n*
s. oxalate $(COONa)_2$ Natriumoxalat *n*
s. oxide *s.* s. monoxide
s. palmitate $C_{15}H_{31}COONa$ Natriumpalmitat *n*
s. perborate $NaBO_3$ Natriumperoxoborat *n*
s. perborate tetrahydrate $NaBO_2 \cdot H_2O_2 \cdot 3H_2O$ Natriummetaborat-Wasserstoffperoxid-3-Wasser *n*
s. perchlorate $NaClO_4$ Natriumperchlorat *n*
s. perchromate Na_3CrO_8 Natriumperoxochromat *n*
s. periodate *s.* s. metaperiodate
s. permanganate $NaMnO_4$ Natriumpermanganat *n*, Natriummanganat(VII) *n*
s. peroxide Na_2O_2 Natriumperoxid *n*
s. peroxydisulphate $Na_2S_2O_8$ Natriumperoxodisulfat *n*
s. perrhenate $NaReO_4$ Natriumperrhenat *n*, Natriumtetroxorhenat(VII) *n*
s. persulphate *s.* s. peroxydisulphate
s. phenate $NaOC_6H_5$ Natriumphenolat *n*, Phenolnatrium *n*
s. phenoxide *s.* s. phenate
s. phosphate bead Natriummetaphosphatperle *f*, Phosphorsalzperle *f*
s. phosphide Na_3P Natriumphosphid *n*
s. phosphite Na_3PO_3 Natriumphosphit *n*
s. platinichloride *s.* s. chloroplatinate
s. platinochloride *s.* s. chloroplatinite
s. polybutadien Natriumpolybutadien *n*
s. polymerization Natriumpolymerisation *f*
s.-polymerized mit Natrium als Katalysator polymerisiert
s. polysulphide Natriumpolysulfid *n*
s. press Natriumpresse *f*
s. propionate C_2H_5COONa Natriumpropionat *n*
s. pyroborate *s.* s. tetraborate
s. pyrophosphate $Na_4P_2O_7$ Natriumdiphosphat *n*, Natriumpyrophosphat *n*
s. pyrosulphate $Na_2S_2O_7$ Natriumdisulfat *n*, Natriumpyrosulfat *n*
s. pyrosulphite $Na_2S_2O_5$ Natriumdisulfit *n*, Natriumpyrosulfit *n*

s. pyrovanadate $Na_4V_2O_7$ Natriumpyrovanadat(V) n, Natrium[heptoxo]divanadat(V) n
s. reactor s. s.-cooled reactor
s. rhodanate (rhodanide) s. s. thiocyanate
s. ricinoleate $C_{17}H_{32}COONa$ Natriumrizinoleat n, Sorizin n
s. salicylate HOC_6H_4COONa Natriumsalizylat n
s. salt Natriumsalz n
s. selenate Na_2SeO_4 Natriumselenat n
s. selenide Na_2Se Natriumselenid n
s. selenite Na_2SeO_3 Natriumselenit n
s. sesquicarbonate $Na_2CO_3 \cdot NaHCO_3$ Trinatriumhydrogendikarbonat n
s. sesquisilicate Natriumsesquisilikat n
s. silicate Natriumsilikat n, Natronwasserglas n
s. silicofluoride s. s. fluosilicate
s. soap Natriumseife f, Natronseife f
s.-spectrum lamp Natriumspektralleuchte f
s. spoon Natriumlöffel m
s. stannate s. s. hexahydroxostannate
s. stearate $CH_3 \cdot (CH_2)_{16} \cdot COONa$ Natriumstearat n
s. subsulphite s. s. thiosulphate
s. succinate $NaOOC(CH_2)_2COONa$ Natriumsukzinat n
s. sulphantimonate s. s. thioantimonate
s. sulphate Na_2SO_4 Natriumsulfat n
s. sulphide Na_2S Natriumsulfid n
s. sulphite Na_2SO_3 Natriumsulfit n
s. sulphocarbonate s. s. thiocarbonate
s. sulphocyanate (sulphocyanide) s. s. thiocyanate
s. sulphonate Natriumsulfonat n
s. sulphoxylate Na_2SO_2 Natriumsulfoxylat n
s. sulphydrate s. s. hydrogen sulphide
s. superoxide s. s. peroxide
s. tartrate $NaOOC(CHOH)_2COONa$ Natriumtartrat n
s. tetraborate $Na_2B_4O_7$ Natriumtetraborat n
s. tetrahydr[id]oborate s. s. borohydride
s. tetrasulphide Na_2S_4 Natriumtetrasulfid n
s. tetrathionate $Na_2S_4O_6$ Natriumtetrathionat n
s. thioantimonate Na_3SbS_4 Natriumthioantimonat(V) n
s. thiocarbonate Na_2CS_3 Natriumthiokarbonat n
s. thiocyanate $NaSCN$ Natriumthiozyanat n, Natriumrhodanid n
s. thiosulphate $Na_2S_2O_3$ Natriumthiosulfat n
s. triphosphate s. 1. tribasic sodium orthophosphate; 2. s. tripolyphosphate
s. tripolyphosphate $Na_5P_3O_{10}$ Natriumtriphosphat n, Pentanatriumtriphosphat n
s. tungstate Na_2WO_4 Natriumwolframat n
s. uranate Na_2UO_4 Natrium[meta]uranat n
s. uranylacetate $NaUO_2(CH_3COO)_3$ Natriumuranylazetat n
s.-vapour [discharge] lamp Natrium[dampf]lampe f
s. wolframate s. s. tungstate
soft weich; alkoholfrei

s.-anneal / to weichglühen
s. annealing Weichglühen n, Weichglühung f
s. asphalt Weichasphalt m
s.-boiled weichgekocht
s. brick (Ker) Schwachbrandstein m, schwach gebrannter Stein m
s. cheese Weichkäse m
s. clay weicher Kaolin m
s. coal bituminöse Kohle f, Weichkohle f
s. contrast (Foto) weicher Kontrast m
s. corn Weichmais m, Stärkemais m, Zea mays L. var. amylacea
s.-curd milk weichgerinnende Milch f
s. drink alkoholfreies Getränk n
s. ferrite (Ker) Weichferrit m
s. fire Rauchfeuer n
s.-fired brick s. s. brick
s.-formation bit (Erdöl) Meißel m für weiche Formationen, Meißel m für lockeres Gestein (Gebirge)
s. glass weiches (leicht schmelzbares) Glas n, Weichglas n
s. handle (Text) weicher Griff m, Weichgriffigkeit f
s. lead Weichblei n
s.-magnetic material weichmagnetischer Werkstoff m
s. maize s. s. corn
s.-mud process (Ker) weichplastisches Verfahren n
s.-nitriding Weichnitrieren n (ein Nitrierhärteverfahren)
s. packing Weichdichtung f, Weichpackung f
s. paraffin [wax] Weichparaffin n, weiches Paraffin n
s. paste (Ker) Weichmasse f
s.-paste porcelain s. s. porcelain
s. plastic body (Ker) weichplastische Masse f
s. plate paper Kupferdruckpapier n
s. porcelain Weichporzellan n
s. pulp weicher (weichgekochter, weit heruntergekochter) Zellstoff m; s. s. stock
s. rays weiche Strahlen mpl
s. resin Weichharz n
s. rubber Weichgummi m
s. rubber clay s. s. clay
s. rubber-covered roll Weichgummiwalze f
s. sealing glass s. s. glass
s.-sized (Pap) schwachgeleimt, mit schwacher Leimung
s. soap weiche Seife f, Schmierseife f
s.-solder / to weichlöten
s. solder Weichlot n
s. soldering Weichlöten n
s. steel Weichstahl m, weicher Stahl m
s. stock (stuff) (Pap) schmieriger Stoff m
s. water weiches Wasser n
s. wax Weichwachs n
s. wheat Weichweizen m, Saatweizen m, Brotweizen m, Triticum aestivum L.

s. wood Weichholz n
s.-working developer (Foto) weich arbeitender Entwickler m
soften / to 1. weich (geschmeidig, biegsam, plastisch) machen, erweichen, aufweichen; (Wasser) weich machen, enthärten; 2. weich werden, erweichen, aufweichen
softener Weichmacher m, erweichendes Mittel n, Erweichungsmittel n; Enthärter m, Enthärtungsmittel n
softening Weichmachen n, Erweichen n, Erweichung f, Aufweichen n; Enthärten n, Enthärtung f (des Wassers); Weichwerden n
s. agent s. softener
s. anneal Weichglühen n, Weichglühung f
s. installation s. s. plant
s. material s. softener
s. plant Enthärtungsanlage f
s. point Erweichungspunkt m, Erweichungstemperatur f
s. range Erweichungsbereich m, Erweichungsintervall n, Erweichungszone f
s. stage Erweichungszustand m, plastischer Zustand m
s. temperature s. s. point
softness (Pap) Schmierigkeit f
s. index (number) (Plast) Weichheitszahl f
softwood Nadelholz n
s. pulp Nadelholzzellstoff m
sogasoid (disperses System aus gasförmigem Dispersionsmittel und festem dispersen Anteil)
soil / to anschmutzen, beschmutzen, schmutzig machen; schmutzen, schmutzig werden
soil 1. Boden m, Grund m, Erdreich n, Erde f; 2. Verschmutzen n, Beschmutzen n; Schmutz m; Schmutzfleck m
s. acidity Bodenazidität f
s. adherence Schmutzhaftung f
s. aggregation Bodenkrümelung f
s. air Bodenluft f, Grundluft f
s. ameliorant s. s. conditioner
s. amendment (nicht total wirkendes) Bodendesinfektionsmittel n; s. 1. s. conditioner; 2. s. conditioning
s. analysis Bodenanalyse f
s. burial test (Text) Erdfaulversuch m, Erdverrottungstest m, Eingrabtest m
s.-cement Bodenbeton m, Bodenzement m, Zementtonbeton m
s. chemistry Bodenchemie f, Chemie f des Bodens
s. colloid Bodenkolloid n
s. compaction Bodenverdichtung f
s. condition Bodenzustand m
s. conditioner Boden[struktur]verbesserungsmittel n
s. conditioning Boden[struktur]verbesserung f, Bodenmelioration f
s. corrosion Bodenkorrosion f, Korrosion f durch das Erdreich

s. exhaustion Bodenmüdigkeit f
s. fertility Bodenfruchtbarkeit f
s. formation Bodenbildung f
s.-forming factor Bodenbildungsfaktor m
s. fumigant Bodendesinfektionsmittel n
s. fumigation Bodendesinfektion f, Bodenentseuchung f (durch Gase und verdampfende Flüssigkeiten)
s. horizon Bodenhorizont m
s. injector Düngelanze f, Bodenlanze f
s. moisture Bodenfeuchtigkeit f
s. nutrient Bodennährstoff m
s. pH s. s. reaction
s. phosphate Bodenphosphat n
s. profile Bodenprofil n
s. reaction Bodenreaktion f
s. redeposition Wiederaufziehen n des Schmutzes, Wiederanschmutzen n, Rückvergrauen n, Rückvergrauung f
s. release (Text) schmutzfreigebende Ausrüstung f, Schmutzauswaschbarkeit f
s. removing capacity (Text) Schmutzlösevermögen n, Schmutzentfernungsvermögen n
s.-repellent (Text) schmutzabweisend, schmutzabstoßend
s. residue 1. Rückstand m (von Herbiziden und Pflanzenschutzmitteln) im Boden; 2. (Text) Restschmutz m
s. respiration Bodenatmung f
s. science Bodenkunde f, Pedologie f
s. sickness Bodenmüdigkeit f
s. solution Bodenlösung f, Bodenflüssigkeit f, Bodenwasser n
s. sourness Bodenversauerung f
s. stabilizer (stabilizing agent) Bodenstabilisator m, Bodenerhärtungsmittel n
s. sterilant 1. Totalherbizid n; 2. Bodendesinfektionsmittel n
s. sterilization Bodendesinfektion f, Bodenentseuchung f
s. structure Bodenstruktur f, Bodengefüge n
s. suspending power (Text) Schmutztragevermögen n
s. water s. s. solution
soiling Anschmutzen n, Beschmutzen n; Schmutzfleck m, Verschmutzung f
s. operation (procedure) (Text) Anschmutzen n, Anschmutzung f
soja s. soybean
sojourn time Verweilzeit f
sol s. 1. soluble; 2. solution
sol Sol n
s.-gel transformation Sol-Gel-Übergang m
s. of hydrated oxide Oxidhydratsol n
solamargine Solamargin n (Alkaloid aus Solanum marginatum L. f.)
solanidine Solanidin n (Aglykon des Solanins)
solanin[e] Solanin n (Steroidalkaloid der Solanazeen)
solar atmosphere Sonnenatmosphäre f

s. chromosphere Chromosphäre f
s. constant Solarkonstante f
s. corona Sonnenkorona f
s. distillate *(Paraffinölsorte mit Siedepunkts-bereich etwa 240 bis 350 °C)*
s. drying Trocknen n in der Sonne, Sonnen-trocknung f
s. energy Sonnenenergie f
s. evaporation solare Eindunstung f
s. furnace Sonnenofen m
s. oil Solaröl n
s. radiation Sonnenstrahlung f
s. salt *(durch Eindunsten von Meerwasser gewonnenes Salz)*
s. spectrum Sonnenspektrum n
s. stearin Schmalzstearin n, Solarstearin n
solarization *(Foto, Glas)* Solarisation f
solate / to von einem Gel zu einem Sol übergehen, ein Sol bilden
solation Gel-Sol-Übergang m
solder / to löten; gelötet werden
s. hard hartlöten
s. soft weichlöten
solder Lot n, Weichlot n, Lötmittel n, Lötmetall n
s. alloy Lotlegierung f
s. bath Lötbad n
s. glass Lötglas n
s. sealing glass *s.* s. glass
s. wire Lötdraht m, Lotdraht m
solderable lötbar, lötfähig
soldered joint Lötverbindung f, Lötstelle f
soldering Löten n, Lötung f; Lötverbindung f, Lötstelle f
s. acid Lötsäure f
s. agent Löt[hilfs]mittel n
s. fluid Lötwasser n
s. liquid *s.* s. fluid
s. paste Lötpaste f, Lötfett n
s. tin Lötzinn n
sole crepe Sohlenkrepp m
s.-flue oven Ofen m mit Sohlebeheizung f, Knab-Ofen m
s. impregnating agent Sohlenimprägniermittel n
solfatarite *(Min)* Solfatarit m
solid fest; dicht, kompakt; massiv
solid fester Körper (Stoff) m, feste Substanz (Materie) f, Festkörper m, Feststoff m, Fest-substanz f; Trockenmasse f, Trockensubstanz f; Festkörper m, feste Phase f, Kristallisat n *(beim Zonenschmelzen)*
s. bitumen Festbitumen n
s.-bowl centrifuge Vollmantelzentrifuge f, Voll-mantelschleuder f
s. carbon dioxide Trockeneis n, festes Koh-lendioxid n
s. carburizing Aufkohlen n in festen Kohlungs-mitteln, Pulveraufkohlen n, Zementieren n in festen Einsatzmitteln (in festem Einsatz), Pul-verzementieren n, Einsetzen n in festen Mitteln, Pulvereinsetzen n

s. casting *(Ker)* Vollguß m, Kernguß m, Massiv-guß m
s. catalyst fester Katalysator m
s. colloid festes Kolloid n
s.-cone nozzle Voll[sprüh]kegeldüse f
s. constituent Trockenbestandteil m
s. content Feststoffgehalt m, Feststoffanteil m; *(Pap)* Gehalt m *(der Ablauge)* an Trockensub-stanz, Trockengehalt m, Festgehalt m
s. crotonic acid *s.* crotonic acid
s. diffusion Festkörperdiffusion f, Diffusion f von Festkörpern
s. electrode massive Elektrode f
s. expansion thermometer Metallausdehnungs-thermometer n
s. fat Hartfett n, gehärtetes (hydriertes) Fett n; festes Fett n, Talg m
s. fuel fester Brennstoff m, Festbrennstoff m; fester Kraftstoff (Treibstoff) m, Festkraftstoff m, Festtreibstoff m
s.-fuel rocket *s.* s. rocket
s. green Solidgrün n
s.-liquid boundary Phasengrenze f fest-flüssig
s.-liquid chromatography Fest-Flüssig-Chromatografie f
s.-liquid interface Grenzfläche f fest-flüssig, Fest-Flüssig-Grenzfläche f
s. material *(Gum)* Kautschuktrockensubstanz f, Gesamt-Festsubstanz f
s. matter Feststoff m, Festsubstanz f, Festkörper m, feste Materie (Substanz) f, fester Stoff (Körper) m; Trockensubstanz f
s. medium *(Bakt)* fester Nährboden m, festes Nährsubstrat n
s.-pack carburizing *s.* s. carburizing
s. paraffin Festparaffin n, festes Paraffin n, Hartparaffin n
s. phase Bodenkörper m; feste Phase f
s.-phase feeding *(Landw)* direkte Nährstoffauf-nahme f *(aus festen Bodenpartikeln)*
s.-phase rule *(Koll)* Bodenkörperregel f
s. point Stockpunkt m
s. propellant fester Kraftstoff (Treibstoff) m, Festkraftstoff m, Festtreibstoff m
s.-propellant rocket *s.* s. rocket
s. pulp board *(Pap)* Halbstoff m in Bogenform (in trockenen Bogen), *(i.e.S.)* Holzschliff m in Bogenform (Pappenform), Holzschliffpappe f, Holzschliffblätter npl *(oder)* Zellstoff m in Bogenform (Pappenform), Zellstoffpappe f, Zell-stoffblätter npl
s. racemate kristallisiertes Razemat n
s. rocket Feststoffrakete f
s. rubber Festkautschuk m
s.-rubber tyre Vollgummireifen m, Massivreifen m
s. shaft Vollwelle f
s.-solid interface Grenzfläche f fest-fest, Fest-Fest-Grenzfläche f

s.-solid mixing Vermengen n, Mischen n von Feststoffkomponenten

s. solubility Festkörperlöslichkeit f, Löslichkeit f in festem Zustand

s. solution feste Lösung f; Mischkristall m

s. spirit Hartspiritus m

s. spray cone Vollsprühkegel m

s. state fester Zustand m, Festkörperzustand m

s.-state physics Festkörperphysik f

s.-state reaction Festkörperreaktion f

s. stopper Vollküken n, massives Küken n *(eines Glashahnes)*

s. surface Feststoffoberfläche f

s. tyre s. s.-rubber tyre

s. wall centrifuge Vollmantelzentrifuge f, Vollmantelschleuder f

solidensing Kondensation f unmittelbar in die feste Phase, festes Kondensieren n, Solidensieren n *(direkte Überführung eines Gases in den festen Aggregatzustand)*

solidification Festwerden n, Erstarren n, Erstarrung f, Verfestigung f

s. point Erstarrungspunkt m, Erstarrungstemperatur f; Stockpunkt m *(bei Ölen)*

s. range Erstarrungsintervall n

solidify / to verfestigen, erstarren lassen, fest (hart, starr) werden lassen; sich verfestigen, erstarren, fest (hart, starr) werden, verhärten, erhärten

solidifying point s. solidification point

solids Trockenmasse f, Trockensubstanz f

s.-bearing feststoffhaltig

s. concentration Feststoffkonzentration f

s. content Feststoffgehalt m, Feststoffanteil m; *(Pap)* Gehalt m *(der Ablauge)* an Trockensubstanz, Trockengehalt m, Festgehalt m

s.-free frei von Feststoffen

s.-handling capacity Feststoffdurchsatzleistung f

s.-holding capacity Feststoffaufnahmevermögen n

s.-liquid separation Trennen n von festen Stoffen und Flüssigkeiten, Trennung f flüssig-fest

s. loading Feststoffgehalt m, Feststoffanteil m

s.-not-fat fettfreie Trockenmasse f

s. recovery filtration Trennfiltration f

solidus [curve, line] Soliduskurve f, Soliduslinie f, Festlinie f

s. point Soliduspunkt m

soling calender *(Gum)* Sohlenkalander m

s. compound *(Gum)* Sohlenmischung f

soliquid *(disperses System aus flüssigem Dispersionsmittel und festem dispersen Anteil, z.B. Suspension)*

soln. s. solution

solonchak *(Bodenkunde)* Solontschak m, Salzboden m *(i.e.S.)*

solonetz *(Bodenkunde)* Solonetz m, Alkaliboden m, Schwarzalkaliboden m

solubility Löslichkeit f

to give s. to löslich machen

s. behaviour Löslichkeitsverhalten n

s. coefficient Löslichkeitskoeffizient m, Löslichkeitskonstante f

s. curve Löslichkeitskurve f, Löslichkeitsdiagramm n

s. in chloroform Chloroformlöslichkeit f

s. of gas Gaslöslichkeit f

s. product Löslichkeitsprodukt n

s. promoter s. solutizer

s. properties Löslichkeitseigenschaften fpl

solubilization Solubilisierung f, Solubilisation f, Löslichmachung f

s. chromatography Lösungschromatografie f

solubilize / to löslich machen

solubilizer s. solutizer

solubilizing Löslichmachen n

s. effect lösungsvermittelnde Wirkung f

s. group löslichmachende Gruppe f

s. power s. solvency

soluble löslich; dispergierbar, emulgierbar *(bei Ölen)*

s. coffee Kaffee-Extrakt m, Kaffee-Extraktpulver n

s. cotton s. s. nitrocellulose

s. cutting oil wasserlösliches Schneidöl n

s. germanium dioxide $GeO_2 \cdot xH_2O$ Germanium(IV)-oxid-Hydrat n, Germanium(IV)-säure f

s. glass Wasserglas n

s. gluside s. s. saccharin

s. guncotton s. s. nitrocellulose

s. in acids säurelöslich

s. in alcohol alkohollöslich

s. in alkalies alkalilöslich

s. in chloroform chloroformlöslich

s. in ether ätherlöslich

s. in fat fettlöslich

s. in water wasserlöslich

s. nitrocellulose Kolloxylin n, Kollodiumwolle f, Nitrozellulose f

s. oil lösliches Öl n; emulgierbares (wasserlösliches) Öl n *(durch Emulgatorzusatz mit Wasser mischbares Öl)*

s. potash glass Kaliwasserglas n

s. resin lösliches Harz n

s. ribonucleic acid lösliche Ribonukleinsäure f, Transfer-Ribonukleinsäure f

s. RNA s. s. ribonucleic acid

s. saccharin Kristallose f *(Natriumsalz des Saccharins)*

s. starch lösliche Stärke f, Dextrinstärke f

s. water glass s. s. potash glass

solute [auf]gelöster Stoff m, Gelöstes n; gelöster Stoff m, Beimengung f, Verunreinigung f *(beim Zonenschmelzen)*

s. concentration Konzentration f des gelösten Stoffes, Konzentration f der Beimengung (Verunreinigung), Fremdstoffkonzentration f *(beim Zonenschmelzen)*

solution Lösung f, Lsg.

to bring into s. in Lösung bringen, lösen

to go into s. in Lösung gehen, sich lösen

s. adhesive Lösungsmittelkleber *m*, Klebstofflösung *f*, flüssiger Klebstoff *m*
s. assistant Lösungsvermittler *m*, lösendes Hilfsmittel *n*
s. behaviour Lösungsverhalten *n*
s.-casting machine Foliengießmaschine *f*
s. coating *(Plast)* Beschichten *n* aus Lösungen
s.-dyed *(Text)* [er]spinngefärbt, düsengefärbt
s. equilibrium Lösungsgleichgewicht *n*
s. for injection Injektionslösung *f*
s. for silver plating Versilberungsbad *n*
s. gas drive *(Erdöl)* Gasentlösungsdruck *m*, Expansionsdruck *m* des im Öl gelösten Gases
s. gas-drive reservoir *(Erdöl)* Gasentlösungslagerstätte *f*, Lagerstätte *f* mit Gasentlösungsdruck
s.-grown in Nährlösung kultiviert
s. heat treatment *s. s.* treatment
s. method Lösungsverfahren *n*
s. of burnt sugar Karamellösung *f*
s. of caustic soda Natronlauge *f*, Natriumhydroxidlösung *f*
s. of common salt Kochsalzlösung *f*
s. of diazonium salt Diazonium[salz]lösung *f*
s. of electrolyte Elektrolytlösung *f*
s. of formaldehyde Formaldehydlösung *f*
s. of lead salts Bleisalzlösung *f*
s. of potassium arsenite Kaliumarsenitlösung *f*
s. of potassium hydroxide Kaliumhydroxidlösung *f*
s. of sodium chloride *s. s.* of common salt
s. polymerization Lösungs[mittel]polymerisation *f*
s. pressure Lösungsdruck *m*, Lösungstension *f*
s. reclaiming process Lösungsregenerierverfahren *n*
s. spinning Erspinnen (Spinnen) *n* aus Lösungen, Lösungsspinnverfahren *n* *(Herstellung von Chemiefaserstoffen)*
s. temperature Lösungstemperatur *f*
s. tension *s. s.* pressure
s. treatment Lösungsglühen *n*
solutizer Lösungsverbesserer *m*, Lösungsvermittler *m*, Lösungsbeschleuniger *m*, Löslichkeitsverbesserer *m*, Löslichkeitsvermittler *m*, Solutizer *m*
s. process Solutizerverfahren *n*, Solutizerprozeß *m* *(zur Entschwefelung von Erdöldestillaten)*
s. solution Solutizerlösung *f*
solutizing agent *s.* solutizer
solvate / to solvatisieren
solvate Solvat *n*
solvation Solvatation *f*, Solvatisierung *f*, Solvatisieren *n*
s. effect Solvatationseffekt *m*
s. energy Solvatationsenergie *f*
s. reaction Solvatationsreaktion *f*
solvatochromism Solvatochromie *f*
Solvay cell Solvay-Zelle *f* *(eine Quecksilberzelle)*
Solvay process Solvay-Verfahren *n*, Ammoniak-Soda-Verfahren *n*

Solvay's ammonia soda process *s.* Solvay process
solve / to [auf]lösen
solvency Lösungsvermögen *n*, Lösevermögen *n*, Lösungsfähigkeit *f*, Lösefähigkeit *f*, Lösungskraft *f*, Lösekraft *f*
solvend [auf]gelöster Stoff *m*, Gelöstes *n*
solvent [auf]lösend
solvent Lösungsmittel *n*, Lm., Lsgm., Lösemittel *n*, Löser *m*, Solvens *n*
s. assistant Lösungshilfsmittel *n*
s. balance Lösungsmittelgleichgewicht *n*
s.-based adhesive Lösungsmittelkleber *m*, Klebstofflösung *f*, flüssiger Klebstoff *m*
s. cement *s.* s.-based adhesive
s.-deoiling process Solventöentölungsverfahren *n*, Solventöentölungsprozeß *m*
s. dewaxing Lösungsmittelentparaffinierung *f*, Solvententwachsung *f*, Entparaffinierung (Entwachsung) *f* mit Lösungsmitteln
s.-dewaxing plant Anlage *f* zur Lösungsmittelentparaffinierung (Entparaffinierung mit Lösungsmitteln)
s.-dewaxing process Solvententparaffinierungsverfahren *n*, Lösungsmittelentparaffinierungsverfahren *n*, Solvententwachsungsprozeß *m*
s. dish Lösungsmittelschale *f*
s. drying Trocknen *n* mittels Lösungsmitteln, Solventtrocknung *f*
s. dyeing Färben *n* in Gegenwart von Lösungsmitteln
s. effect Lösungsmitteleffekt *m*
s. extraction Solventextraktion *f*, Lösungsmittelextraktion *f*, Lösemittelextraktion *f*
s.-extraction process (technique) Solventextraktionsverfahren *n*, Lösungsmittelextraktionsverfahren *n*
s. front *(Chromatografie)* Lösungsmittelfront *f*, Laufmittelfront *f*, Fließmittelfront *f*
s.-hating lösungsmittelabstoßend, kein Lösungsmittel aufnehmend, lyophob
s.-loving lösungsmittelanziehend, Lösungsmittel aufnehmend, lyophil
s. molecule Lösungsmittelmolekül *n*
s. moulding *(Plast)* Tauchen *n* in Lösungen
s. naphtha Solventnaphtha *n(f)* *(Lösungsbenzol oder Schwerbenzin als Steinkohlenteer- bzw. Erdölfraktion)*
s. partition Ausschütteln *n*
s. phase Lösungsmittelphase *f*, Lösemittelphase *f*
s. polymerization Lösungs[mittel]polymerisation *f*
s. power Lösungsvermögen *n*, Lösevermögen *n*, Lösungsfähigkeit *f*, Lösefähigkeit *f*, Lösungskraft *f*, Lösekraft *f*
s. process *(Text)* Lösungsmittelverfahren *n*
s. recovery Lösungsmittelrückgewinnung *f*, Lösungsmittelwiedergewinnung *f*
s.-recovery plant Lösungsmittelrückgewinnungsanlage *f*

s.-recovery section Lösungsmittelrückgewinnungsteil *m*, Lösungsmittelrückgewinnung *f (als Anlageteil)*

s.-refined lösungsmittelraffiniert, mit Lösungsmittel raffiniert

s. refining Solventraffination *f*, Lösungs[mittel]raffination *f*

s.-refining process Solventraffinationsverfahren *n*, Solvent[raffinations]prozeß *m*

s. resistance Lösungsmittelbeständigkeit *f*, Widerstandsfähigkeit (Beständigkeit) *f* gegen Lösungsmittel

s.-resisting lösungsmittelbeständig, lösungsmittelfest

s. retention Lösungsmittelretention *f*

s. scouring *(Text)* Lösungsmittelwäsche *f*, Extraktionswäsche *f*, Trockenwaschverfahren *n*

s. soap Lösungsmittelseife *f*

s. spectrum Lösungs[mittel]spektrum *n*

s. spinning Erspinnen (Spinnen) *n* aus Lösungen, Lösungsspinnverfahren *n (Herstellung von Chemiefaserstoffen)*

s. system Lösungsmittelsystem *n*; *(Chromatografie)* Laufmittelgemisch *n*

s. tannage Lösungsmittelgerbung *f*

s.-thinned paint lösungsmittelverdünnbare (mit Lösungsmittel verdünnbare) Farbe *f (als Gegensatz zur wasserverdünnbaren Farbe)*

s. treatment Lösungsmittelbehandlung *f*

s. trough *(Chromatografie)* Lösungsmitteltrog *m*, Fließmitteltrog *m*

s. vapour Lösungsmitteldampf *m*

s. welding Quellschweißen *n*, Lösungs[mittel]-schweißen *n*, Kleben (Schweißen) *n* durch Anlösen (Anquellen)

solventless lösungsmittelfrei

solvolysis Solvolyse *f*, Lyolysis *f*

solvolyze / to *(gelöste Stoffe)* durch Solvolyse aufspalten (zersetzen)

Somali gum Nilgummi *n (von mehreren Acacia-Arten)*

soman Soman *n (Methylfluorphosphonsäurepinakolylester; Kampfstoff)*

somatotropic hormone somatotropes Hormon *n*, Somatropin *n*, STH

somatotropin s. somatotropic hormone

Sommelet reaction Sommelet-Umlagerung *f (Isomerisierung quartärer Benzylammoniumverbindungen)*

Sommerfeld fine structure constant Sommerfeldsche Feinstrukturkonstante *f*

somnifacient, somnificant Schlafmittel *n*, Hypnotikum *n*

sonde Sonde *f (bei der elektrischen Bohrlochvermessung)*

sonic agglomerator Hochfrequenzsirene *f*, akustischer Staubabscheider *m*

s. cyclone Schallaerozyklon *m*

s. energy Schallenergie *f*

s. luminescence s. sonoluminescence

s. precipitation akustisches Abscheiden *n*

sonoluminescence Sonolumineszenz *f*

soot Ruß *m*

sooty rußig; geschwärzt

sophorine Sophorin *n*, Zytisin *n*, Baptitoxin *n*, Ulexin *n (Lupinenalkaloid)*

soporific Schlafmittel *n*, Hypnotikum *n*

sorb / to sorbieren

sorbate Sorptiv *n*, Sorbend *m*, sorbierter (aufgenommener) Stoff *m*, sorbierte Substanz *f*

sorbent [material] Sorbens *n*, Sorptionsmittel *n*

s. solid Feststoffsorptionsmittel *n*, Feststoffsorbens *n*

sorbic acid $CH_3CH=CHCH=CHCOOH$ Sorbinsäure *f*, 2,4-Hexadiensäure *f*

sorbite *(Met)* Sorbit *m*

d-sorbite s. sorb[it]ol

sorbitic *(Met)* sorbitisch

sorb[it]ol $CH_2OH(CHOH)_4CH_2OH$ Sorbit *m (ein Zuckeralkohol)*

sorbose Sorbose *f*

Sorel cement Sorelzement *m*, Magnesiazement *m*

Sørenson's formol titration Sörensen-Titration *f*, Formoltitration *f* nach Sörensen

sorptate s. sorbate

sorption Sorption *f*, Sorbieren *n*

s. balance Sorptionswaage *f*

s. curve Sorptionskurve *f*

sorrel salt Kleesalz *n*, Bitterkleesalz *n*, Sauerkleesalz *n (Kaliumtetraoxalat oder ein Gemisch von Kaliumhydrogenoxalat und Kaliumtetraoxalat)*

sort / to sortieren

sorter *(Pap)* Sortiererin *f*, *(i.e.S.)* Papiersortiererin *f (oder)* Hadernsortiererin *f*

sorting Sortieren *n*, Sortierung *f*

s. by hand Handsortierung *f*

s. girl *(Pap)* Sortiererin *f*

s. room *(Pap)* Sortiersaal *m*

sound 1. Schall *m*; Ton *m*, Laut *m*, Klang *m*; 2. Sonde *f*

s. field Schallfeld *n*

s. insulation board Schalldämmplatte *f*

s. wave Schallwelle *f*

souple silk Soupleseide *f*, halbentbastete (souplierte) Seide *f*, Weichseide *f*, Souple *m*

soupling *(Text)* Souplieren *n*, Halbentbasten *n*

sour / to [an]säuern, sauer machen; sauer werden, gerinnen *(Milch)*; stichig (säuerlich, schal) werden *(Wein)*; *(Gerb)* pickeln *(mit Säure-Salz-Lösung behandeln)*; *(Ker)* mauken, faulen, lagern

s. the wire *(Pap)* Siebgewebe mit verdünnter Säure reinigen

sour sauer, herb, scharf; bitter; gegoren; ranzig; sauer, naß, kalkarm *(Boden)*; säuerlich, stichig; sauer, doktorpositiv, mit positivem Doktortest *(Erdöldestillat)*

sour Sauer[teig] *m*

s. beer Sauerbier *n*

s. cream Sauerrahm *m*

s. gasoline saures Benzin *n*

s. gum Tupelobaum *m*, Wasser-Tupelo *m*, Nyssa sylvatica Marsh.
s. milk *s.* soured milk
s.-milk cheese Sauermilchkäse *m*
s. oil Saueröl *n*, saures (gesäuertes) Öl *n* *(mit Säure behandeltes Öl vor der Neutralisation)*; saures Erdölprodukt (Erdöldestillat) *n*, doktor-positives Öl *n*, Öl *n* mit positivem Doktortest
source Ausgangsstoff *m*; Quelle *f*; Herkunft *f*; Hersteller *m*
s. of current Stromquelle *f*
s. of electrons Elektronendon[at]or *m*, Elektro-nenspender *m*
s. of energy Energiequelle *f*, Kraftquelle *f*
s. of H⁺ ions H^+-Ionen-Lieferant *m*
s. of heat Wärmequelle *f*
s. of raw material Rohstoffquelle *f*
s. rock Ursprungsgestein *n*, Ausgangsgestein *n*, Muttergestein *n*
sourdough Sauer[teig] *m*
soured milk saure (dickgelegte) Milch *f*, mit Säureweckern gesäuerte (versetzte) Milch *f*, Sauermilch *f*; spontan gesäuerte Milch *f*, Sauermilch *f*, Setzmilch *f*, Schlippermilch *f*, Plundermilch *f*
souring Säuerung *f*, Säuern *n*, Ansäuern *n*; Sauerwerden *n*, Gerinnen *n* *(z.B. der Milch)*; Stichigwerden *n*, Säuerlichwerden *n*, Schalwer-den *n* *(z.B. des Weins)*; *(Gerb)* Pickeln *n*, Pickel *m* *(Behandlung mit Säure-Salz-Lösung)*; *(Ker)* Mauken *n*, Faulen *n*, Lagern *n*
s. of milk Milchsäuerung *f*, Säuerung *f* (Säuern *n*, Dicklegung *f*) der Milch
s. process Säuerungsvorgang *m*, Säuerungs-prozeß *m*
s. vat Säuerungswanne *f*, Säuerungsgefäß *n*, Rahmreifer *m*
sourish säuerlich, angesäuert
sourness Säure *f*, Herbheit *f*
souse / to naß pökeln, [ein]pökeln; [ein]säuern, *(bei Fisch auch)* marinieren
souse Pökellake *f*, Pökelsalzlösung *f*; Marinade *f*, Aufguß *m* *(z.B. gewürzte Salz-Essig-Lösung)*; Pökelfleisch *n*
Soxhlet [extractor] Soxhlet-Extraktor *m*, Soxhlet-Apparat *m*, Soxhletscher Extraktionsapparat *m*, Soxhlet *m*
Soxhlet thimble Extraktionshülse *f* nach Soxhlet
soy *s.* 1. soybean; 2. soybean oil
s. milk *s.* soybean milk
s. oil *s.* soybean oil
soya[-bean] *s.* soybean
soybean Sojabohne *f*, Ölbohne *f*, Fettbohne *f*, Glycine max (L.)Merr.
s. fibre Sojaeiweißfaser *f*; Sojaeiweißfaserstoff *m*
s. flour Soja[bohnen]mehl *n*
s. glue Sojaeiweißleim *m*
s. lecithin Sojalezithin *n*
s. milk Sojamilch *f*
s. oil Soja[bohnen]öl *n*

s.-oil green *(Komplexverbindung aus Kupfer-arsenit und verseiftem Sojabohnenöl; Insektizid)*
s. protein Sojaprotein *n*, Sojaeiweiß *n*
s.p. *s.* setting point
sp. gr. *s.* specific gravity
sp hybrid sp-Hybrid *n*
sp hybridization sp-Hybridisierung *f*
S.P. rubber *s.* superior processing rubber
space Raum *m*; Zwischenraum *m*, Intervall *n*, Abstand *m*; Entfernung *f*
s.-centred *(Krist)* raumzentriert, innenzentriert
s. charge Raumladung *f*
s. chemistry 1. Stereochemie *f*; 2. Kosmochemie *f*
s. diagonal Raumdiagonale *f*
s. dyeing Space-Dyeing *n* *(Verfahren zum Erzielen von Mehrfarbeneffekten auf synthe-tischen Texturseiden)*
s. formula Raumformel *f*, Stereoformel *f*, Konfigurationsformel *f*, geometrische Struktur-formel *f*
s. group *(Krist)* Raum[symmetrie]gruppe *f*
s. isomerism räumliche (stereochemische) Isomerie *f*, Raumisomerie *f*, Stereoisomerie *f*
s. lattice Kristallgitter *n*, Raumgitter *n*, *(i.e.S.)* Translationsgitter *n*, Elementargitter *n*
s. model Raummodell *n*
s. network räumliches Netzwerk *n*
s.-network polymer räumlich vernetztes Poly-mer[es] *n*
s.-time yield Raum-Zeit-Ausbeute *f*
s. velocity Raumgeschwindigkeit *f* *(Maß für den Durchsatz einer Krackanlage)*
spaced-bucket elevator Becherwerk *n* mit Einzel-bechern
spacer Zwischenstück *n*, Distanzstück *n*; *(Pap)* Holzzwischenleiste *f* *(für Bemessung von Mahlmaschinen)*
spacing 1. Abstand *m*, Zwischenraum *m*, Intervall *n*; Entfernung *f*; 2. Einteilen *n* *(in Abstände)*; 3. *s.* s. of the planes
s. between bars *(Pap)* Messerabstand *m*
s. of the planes *(Krist)* Netzebenenabstand *m*, Gitterabstand *m*, Gitterperiode *f*
spall / to abspalten; zertrümmern; abplatzen; abblättern; [zer]splittern
spallation Abspaltung *f*; Spallation *f*, Kernzersplit-terung *f*, Kernzertrümmerung *f*
spalling Abspalten *n*, Abspaltung *f*; Zertrümme-rung *f*; Abplatzen *n*, Abblättern *n*; Splittern *n*, Zersplittern *n*
s. resistance Widerstandsfähigkeit *f* gegen Abplatzen (Abblättern); Temperaturwechsel-beständigkeit *f*
spangle *(Min)* Schüppchen *n*, Flitter *m*
Spanish grass Espartogras *n*, Esparto *m*, Halfagras *n*, Stipa tenacissima L.
S. lavender oil *(Pharm)* Spiköl *n* *(von Lavandula latifolia (L. fil.)Medik.*
S. white Spanischweiß *n*, Perlweiß *n*, Wismutweiß *n* *(aus Wismutoxidnitrat)*

sparge / to *(Gär)* anschwänzen, überschwänzen

sparge liquor *(Gär)* Nachguß *m*

s. steam *(durch eine Flüssigkeit)* perlender Dampf *m*

s. water *(Gär)* Anschwänzwasser *n*, Überschwänzwasser *n*

sparger *(Gär)* Anschwänzvorrichtung *f*

sparging *(Gär)* Anschwänzen *n*, Überschwänzen *n*

sparingly soluble schwerlöslich, schwer (wenig) löslich, slö, wl.

spark / to [ent]zünden; funkeln; Funken sprühen, funken

spark Funke[n] *m*

s. discharge Lichtbogenentladung *f*, Bogenentladung *f*, Funkenentladung *f*

s. spectrum Funkenspektrum *n*

sparking potential kritische Spannung *f*, Zündspannung *f* *(z.B. beim Elektroabscheiden)*

sparkle / to 1. *s.* to spark; 2. schäumen, perlen, moussieren *(Wein)*

sparkler *(sprühendes)* Feuerwerk *n*; Wunderkerze *f*

sparklet bulb Kohlensäurepatrone *f*.

sparkling water Sprudel *m* *(Tafelwasser natürlicher Herkunft)*

s. wine Schaumwein *m*, Sekt *m*

spartalite *(Min)* Spartalith *m*, *(veraltet für)* Zinkit *m* *(Zinkoxid)*

spartalupine Spartalupin *n* *(Alkaloid)*

sparteine Spartein *n*, Lupinidin *n* *(Alkaloid)*

spathic *(Min)* spatig, spatartig, blätterig

s. iron [ore] *(Min)* Eisenspat *m*, Siderit *m*, Spateisenerz *n*, Spateisenstein *m* *(Eisen(II)-karbonat)*

spathiopyrite *(Min)* Spathiopyrit *m*, *(veraltet für)* Safflorit *m* *(Kobaltarsenid)*

spathose *s.* spathic

spatial räumlich, Raum..., Stereo...

s. arrangement (characteristics) räumliche Anordnung *f*, Konfiguration *f*

s. relationship Raumverhältnis *n*

spatter / to [ver]spritzen; *(Met)* spratzen

spatter board *(Pap)* Spritzbrett *n*, Schutzbrett *n*

spattering Spritzen *n*, Verspritzen *n*; *(Met)* Spratzen *n*

spatula Spatel *m*

SPB spirit *s.* special boiling-point spirit

spear pyrite *(Min)* Speerkies *m* *(Markasit-Varietät)*

special board Sonderpappe *f*, Spezialpappe *f*

s. boiling-point spirit Siedegrenzenbenzin *n*

s. cement Sonderzement *m*, Spezialzement *m*

s. coke Sonderkoks *m*, Spezialkoks *m*

s. malt Spezialmalz *n*

s. paper Spezialpapier *n*

s. pitch Spezialpech *n*

s.-purpose Spezial..., Sonder...

s.-purpose tile Fliese *f* für Sonderzwecke

s. reagent spezifisches Reagens *n*

species of coal Kohlenart *f*, Kohlengattung *f*

specific acid catalysis spezifische Säurekatalyse *f*

s. conductance (conductivity) spezifische Leitfähigkeit *f*, spezifisches Leitvermögen *n*

s. dispersivity spezifische Dispersion *f*

s. gravity *(Verhältnis der Masse eines gegebenen Volumens irgendeiner Substanz bei der Temperatur t_1 zur Masse des gleichen Volumens reinen Wassers bei der Temperatur t_2),* Dichteverhältnis *n* *(bei $t_1 = t_2$)*; Dichtezahl *f* *(bei $t_1 \neq t_2$)*

s. gravity balance Mohr-Westphalsche Waage *f* *(zur Dichtebestimmung)*

s. gravity bottle (flask) Wägefläschchen *n* für Dichtemessungen, Pyknometer *n*

s. heat [capacity] spezifische Wärme[kapazität] *f*

s. humidity spezifische Feuchtigkeit *f*

s. ionization spezifische (differentielle) Ionisation *f*, Ionisierungsstärke *f*

s. pressure *(Sprengtechnik)* Explosionsdruck *m*

s. reaction rate (velocity) spezifische Reaktionsgeschwindigkeit *f*, Reaktionsgeschwindigkeitskonstante *f*, Geschwindigkeitskonstante *f*

s. refraction (refractive index, refractivity) spezifische Refraktion *f*

s. resistance spezifischer Widerstand *m*

s. rotary power spezifische Drehung *f*, spezifisches Dreh[ungs]vermögen *n*

s. rotation *s.* s. rotary power

s. speed spezifische Drehzahl *f*; Schnelläufigkeit *f*, Schnellaufzahl *f* *(bei Strömungsmaschinen)*

s. surface spezifische Oberfläche *f*

s. surface activity spezifische Oberflächenaktivität *f*

s. susceptibility spezifische Suszeptibilität *f*, Massensuszeptibilität *f*, Grammsuszeptibilität *f*

s. thermal capacity *s.* s. heat [capacity]

s. thermal conductivity [spezifisches] Wärmeleit[ungs]vermögen *n*, Wärmeleitfähigkeit *f*, Wärmeleitzahl *f*

s. viscosity spezifische Viskosität *f*

s. volume spezifisches Volumen *n*

s. weight Wichte *f*, spezifisches Gewicht *n*

specification Vorschrift *f*, Prüfungsvorschrift *f*; Spezifikation *f* *(festgelegte Eigenschaften für ein bestimmtes Produkt oder dessen Verwendung)*

specify / to vorschreiben

specimen Probe *f*; Probekörper *m*, Prüfkörper *m*, Prüfling *m*, *(i.e.S.)* Probestab *m*, Prüfstab *m*

s. jar Schauglas *n*, Präparatenglas *n*

s. of salt Salzprobe *f*

speck Fleck[en] *m*, Unreinheit *f*; *(Text)* Stippe *f*

s.-free dyeing *(Text)* stippenfreie Färbung *f*

specking Fleckenbildung *f*

spectral spektral, Spektral...

s. analysis Spektralanalyse *f*

s. colour Spektralfarbe *f*

s. dispersion spektrale Zerlegung *f*

s. line Spektrallinie *f*

s. series Spektralserie *f*, Serie *f*

s. term Spektralterm *m*

spectrochemical spektrochemisch
spectrogram Spektrogramm n
spectrograph Spektrograf m
spectrographic spektrografisch
spectrometer Spektrometer n
spectrometry Spektrometrie f
spectrophotometer Spektralfotometer n, Spektrofotometer n
spectrophotometric spektralfotometrisch, spektrofotometrisch
s. analysis s. 1. spectrophotometry; 2. spectral analysis
spectrophotometry Spektralfotometrie f, Spektrofotometrie f
spectropolarimeter Spektropolarimeter n
spectroscope Spektroskop n
spectroscopic spektroskopisch
s. analysis s. spectral analysis
s. displacement law of Kossel and Sommerfeld Sommerfeld-Kosselscher Verschiebungssatz m
spectroscopical s. spectroscopic
spectroscopy Spektroskopie f
spectrum Spektrum n
s. analysis s. spectral analysis
s. line s. spectral line
s. of hydrogen Wasserstoff[atom]spektrum n
specular iron [ore], specularite (Min) Eisenglanz m, Spekularit m (gut kristallisierter Hämatit)
speed 1. Geschwindigkeit f; Drehzahl f; 2. (Foto) Empfindlichkeit f
s. of absorption Absorptionsgeschwindigkeit f; Aufziehgeschwindigkeit f (von Farbstoffen)
s. of cure (Gum) Anvulkanisationsgeschwindigkeit f; Vulkanisationsgeschwindigkeit f, Heizgeschwindigkeit f; (Farb, Plast) Härtungsgeschwindigkeit f, Aushärtungsgeschwindigkeit f
s. of extrusion Spritzgeschwindigkeit f
s. of ions Ionen[wanderungs]geschwindigkeit f, Wanderungsgeschwindigkeit f
s. of nitration Nitrier[ungs]geschwindigkeit f
s. of reaction Reaktionsgeschwindigkeit f
s. of sound Schallgeschwindigkeit f
s. of the stock (Pap) Stoffgeschwindigkeit f, Ausströmungsgeschwindigkeit (Auslaufgeschwindigkeit) f des Stoffs
s. of the wire (Pap) Siebgeschwindigkeit f
s. up / to beschleunigen
spegazzinidine Spegazzinidin n (Alkaloid)
speise Speise f (Zwischenerzeugnis beim Verschmelzen von schwefel-, arsen- bzw. antimonhaltigen Erzen)
spent verbraucht, gebraucht (z.B. Lösung)
s. ammonia liquor Ammoniakabwasser n, Kolonnenablauf m (eines Gaswerks)
s. bark (Gerb) ausgelaugte Rinde f
s. bark liquor (Gerb) ausgelaugte Lohbrühe f
s. brine ablaufende (verbrauchte) Sole f
s. catalyst verbrauchter (gebrauchter) Katalysator m

s. cooking liquor s. s. liquor
s. gas liquor s. s. ammonia liquor
s. hops Hopfentreber m
s. liquor (Pap) Ablauge f
s.-liquor recovery (Pap) Ablaugengewinnung f, Ablaugenregeneration f, Laugenregeneration f
s.-liquor solids (Pap) Ablaugetrockensubstanz f, ATS
s. mash Schlempe f, Brennereischlempe f
s. soda Sodaablauge f
s. sulphite liquor (Pap) Sulfitablauge f, Ablauge f, Urlauge f
s. sulphite liquor solids s. s. liquor solids
s. sulphite liquor yeast Sulfitablaugehefe f
s. tan ausgelaugte Gerberlohe f
s. wash s. s. mash
sperm oil (whale oil) Spermöl n, Spermazet[i]öl n, Wal[rat]öl n
spermaceti [wax] Walrat m(n)
spessartine, spessartite (Min) Spessartin m, Mangantongranat m (Aluminiummangan(II)-orthosilikat)
spew (Gum) Austrieb m (Fehler)
s. line Gratlinie f
sphagnum peat Sphagnumtorf m, Bleichmoostorf m
sphalerite (Min) Sphalerit m, Zinkblende f, Blende f (Zinksulfid)
sphene (Min) Sphen m (Titanit-Varietät)
sphere 1. Sphäre f, Bereich m; 2. Kugel f
s. packing (Krist) Kugelpackung f
spheric[al] sphärisch, kugelförmig; Kugel...
s. aberration sphärische Aberration f, Öffnungsfehler m
s. boiler (cooker, digester, rotary cooker) (Pap) Kugelkocher m
s. shell Kugelschale f
s. symmetry Kugelsymmetrie f
spherically symmetrical kugelsymmetrisch
sphericity Sphärizität f, Kugelförmigkeit f
spherocolloid Sphärokolloid n
spherodizer (Düngemittelindustrie) Spherodizer m (Spezialform einer Granuliertrommel)
spheroidal sphäroidal
s. graphite Kugelgraphit m
s. state sphäroidaler Zustand m
spheroidization s. spheroidizing [anneal]
spheroidize / to weichglühen
spheroidizing [anneal] Weichglühen n, Weichglühung f
spheronizing Spheronizing-Prozeß m (Krümelung von pulverförmigen Stoffen ohne Flüssigkeitszusatz)
spherulite (Geol) Sphärolith m
spherulitic (Geol) sphärolithisch
sphingolipid[e] Sphingolipoid n
sphingomyelin Sphingomyelin n (ein glyzerinfreies Phosphatid)
sphingosine Sphingosin n
SPI (Abk. für) Society of the Plastics Industry

spice / to würzen

spice Gewürz *n*, Würzmittel *n*, Würze *f*

s. plant Gewürzpflanze *f*

spiced gewürzt, würzig, aromatisch

spicy s. spiced

spider *(Plast)* Werkzeugaufnahmegestell *n (beim Schleudergießverfahren)*; Drehhalter *m*, Drehkreuz *n*

spiegel [iron], spiegeleisen Spiegeleisen *n*

spigot Austragsöffnung *f*

spike oil *(Pharm)* Spiköl *n (von Lavandula latifolia (L. fil.)Medik.)*

spill / to verschütten, verspritzen

spin / to 1. [ab]schleudern; 2. *(Garn)* spinnen; *(Spinnlösungen)* verspinnen; *(Chemiefaserstoffe)* erspinnen; 3. sich drehen, kreisen, umlaufen, rotieren

spin Spin *m*, Eigendrehimpuls *m*, Drall *m*, innerer Drehimpuls *m*

s.-dependent spinabhängig

s.-dyed [er]spinngefärbt, düsengefärbt

s. dyeing Spinnfärbung *f*, Erspinnfärbung *f*, Düsenfärbung *f*

s. freezing *(mit rotierenden Ampullen arbeitendes Gefriertrocknungsverfahren)*

s. method Spinmethode *f*, Valenzbindungsmethode *f*, Valenzstrukturmethode *f*, VB-Methode *f*, Heitler-London-Slater-Pauling-Methode *f*, HLSP-Methode *f*, Elektronenpaarmethode *f*

s. moment Spinmoment *n*

s. multiplicity Spinmultiplizität *f*

s. operator Spinoperator *m*

s. orbital Spinorbital *n*

s. quantum number Spinquantenzahl *f*

s.-spin interaction Spin-Spin-Wechselwirkung *f*

s.-state method s. s. method

s. welding Reib[ungs]schweißen *n*

Spinco ultracentrifuge Spinco-Ultrazentrifuge *f*

spindle oil Spindelöl *n*

spinel *(Min)* Spinell *m*, *(i.e.S.)* Magnesiospinell *m (Magnesiumaluminat)*

s. ruby *(Min)* Rubinspinell *m*

spinellane *(Min)* Spinellan *m*, *(veraltet für)* Nosean *m*

spinelle s. spinel

spinnability Spinnfähigkeit *f*; Verspinnbarkeit *f*, technischer Spinnwert *m*

spinneret *(Text)* Spinndüse *f*, Spinnbrause *f*

s. hole *(Text)* Düsenbohrung *f*

spinning 1. Schleudern *n*, Abschleudern *n*; 2. Spinnen *n (von Garn)*; Verspinnen *n (von Spinnlösungen)*; Erspinnen *n (von Chemiefaserstoffen)*; 3. Drehen *n*, Drehung *f*, Kreisen *n*, Umlaufen *n*, Rotieren *n*, Rotation *f*

s. atomizer rotierender Zerstäuber *m*, Rotationszerstäuber *m*, rotierende Sprüheinrichtung (Düse) *f*

s. band column *(Destillation)* Drehbandkolonne *f*

s. bath Spinnbad *n*, Fällbad *n*

s. bulb Spinnzwiebel *f*

s. cabinet (cell) Spinnschacht *m*

s. disk [atomizer] Fliehkraftzerstäuber *m*, Zentrifugalscheibe *f*, Zentrifugalteller *m*, Sprühscheibe *f*, Zerstäuberscheibe *f*, Zerstäubungsscheibe *f*

s. fibre Spinnfaser *f*

s. head Spinnkopf *m*

s. jet s. spinneret

s. machine Spinnmaschine *f*

s. oil *(Text)* Schmälzöl *n*, Schmälze *f*, Schmälzmittel *n*

s. paper Spinnpapier *n (zur Herstellung von Papiergarnen und -bindfäden)*

s. pot Spinntopf *m*, Spinnzentrifuge *f*

s. pump Spinnpumpe *f*, Dosierpumpe *f*, Titerpumpe *f*

s. shower s. spinneret

s. solution Spinnlösung *f*, Erspinnlösung *f*

s. stretch Spinnstrecken *n*

s. table Spinntisch *m*

s. time Schleuderzeit *f*

s. tube s. s. cabinet

s. vessel s. s. pot

spinthariscope Spinthariskop *n*, Spintheriskop *n*

spinulosin Spinulosin *n*, 3,6-Dihydroxy-4-methoxy-2,5-toluchinon *n (Antibiotikum)*

spiral / to sich schraubenförmig drehen (bewegen)

spiral spiralförmig, spiralig, schraubenförmig, Spiral...

s. chute Wendelrutsche *f*

s. classifier Spiralklassierer *m*, Schraubenklassierer *m*

s. coil Spiralrohrschlange *f*

s. condenser Schlangenkühler *m*

s. conveyor Förderschnecke *f*, Schneckenförderer *m*, Transportschnecke *f*

s. exchanger Spiralwärme[aus]tauscher *m*, Spiralwärmeübertrager *m*

s. flow test *(Plast)* Spiraltest *m*

s. granulator Schneckengranulator *m (z.B. für Düngemittel)*

s.-plate heat exchanger s. s. exchanger

s. separator Wendelscheider *m*

s. tile Spirale *f (Füllkörperform)*

s.-tube heat exchanger Spiralrohrbündelwärme[aus]tauscher *m*, Spiralrohrbündelwärmeübertrager *m*

spiran[e] Spiran *n*, spirozyklische Verbindung *f*, Spiranverbindung *f*, Spiroverbindung *f*

spirit Spiritus *m*, Sprit *m*, alkoholisches (geistiges) Getränk *n*, *(i.e.S.)* Alkohol *m*, Branntwein *m*; Benzin *n (besonders als Vergaserkraftstoff)*

s. blue Spritblau *n*, [spritlösliches] Anilinblau *n (Chlorhydrat des Triphenylrosanilins)*

s. of hartshorn Hirschhorngeist *m*, *(veraltet für)* Ammoniakwasser *n*, NH_3-Wasser *n*

s. of turpentine Terpentinöl *n*

s. of wine s. spirits of wine

s.-soluble spritlöslich, alkohollöslich, *(i.w.S.)* löslich in organischen Lösungsmitteln
s. stain *(Farb)* Spiritusbeize *f*, Spritbeize *f*
s. varnish Spirituslack *m*
spirits alkoholische (geistige) Getränke *npl*, Spirituosen *pl*
s. of wine C_2H_5OH Weingeist *m*, Weinspiritus *m*, Äthanol *n*, Äthylalkohol *m*
spirituous spirituos, spirituös, geistig, alkoholisch, Spiritus..., Sprit..., Alkohol...
s. liquor alkoholisches (geistiges) Getränk *n*
spiro compound s. spiran
s. junction s. s. union
s. position Spirostellung *f*
s.-ring system Spiranringsystem *n*
s. union Spirobindung *f*, Spiroverknüpfung *f*
spit / to *(Met)* spratzen
spit-out *(Ker)* Ausspritzer *m* *(Fehler)*
spitting *(Met)* Spratzen *n*
spitzkasten *(Aufbereitungstechnik)* Spitzkasten *m*
splash / to spritzen; verspritzen
s. on aufspritzen *(z.B. Naßgut auf Walzentrockner)*
splash Spritzer *m*; *(Plast)* spritzerförmige Markierung *f* *(Fehler an Spritzlingen)*
s. feed Spritzauftrag *m* *(Naßgutaufgabe in Walzentrocknern)*
s. plate Sprühteller *m*, Spritzteller *m*
splashing loss Spritzverlust *m*
splice / to [zusammen]spleißen; *(Text)* verstärken, verdoppeln
splint coal Splintkohle *f*, Splitterkohle *f*
splinter Splitter *m*, Span *m*
split / to [ab]spalten; aufspalten, zerlegen, teilen; sich spalten (teilen); sich zersetzen, zerfallen
s. off (out) abspalten
s. up aufspalten, zerlegen, teilen; sich spalten (teilen); sich zersetzen, zerfallen, *(i.e.S.)* dissoziieren
split acid s. chebulic acid
s.-body valve Ventil *n* mit geteiltem Gehäuse
s. fibre Foliefaden *m*, Spaltfaser *f*
s. mould *(Plast)* zusammengesetztes Preßwerkzeug (Werkzeug) *n*, Mehrfachwerkzeug *n*; *(Glas)* zweiteilige (geteilte) Form *f*; *(Glas)* mehrteilige Form *f*
s. plot design *(Landw)* Spaltanlage *f* *(z.B. für gestaffelte Düngeversuche)*
splits of mould *(Plast)* Werkzeugeinsätze *mpl*
splitter *(Gum)* Ballenspalter *m*, Kautschukspalter *m*
splitting Abspaltung *f*, Spaltung *f*; Aufspaltung *f*, Zerlegung *f*, Teilung *f*; Zersetzung *f*, Zerfall *m*
s. equipment s. splitter
s. factor g Landéscher g-Faktor *m*, Landé-Faktor g *m*, g-Faktor *m*
s. machine *(Gerb)* Spaltmaschine *f*
s.-off, s.-out Abspalten *n*, Abspaltung *f*
s. pattern *(phys Ch)* Aufspaltungsbild *n*
s.-up Aufspalten *n*, Aufspaltung *f*, Spaltung *f*, Zerlegung *f*; Teilung *f*; Zersetzung *f*, Zerfall *m*, *(i.e.S.)* Dissoziation *f*

spodium Spodium *n*, Knochenkohle *f*
spodumene *(Min)* Spodumen *m* *(Lithiumaluminiumdisilikat)*
spoil / to verderben, unbrauchbar machen; verderben, schlecht (unbrauchbar) werden *(Lebensmittel)*
spoilage Verderb *m*, Verderben *n*, Schlechtwerden *n* *(von Lebensmitteln)*
sponge / to *(Ker)* schwammverputzen, verschwammen
sponge glass Schaumglas *n*
s. iron Schwammeisen *n*, Eisenschwamm *m*
s.-iron powder Schwammeisenpulver *n*
s. plastic Schwammkunststoff *m*
s. process Schwammverfahren *n*, Schwammmethode *f* *(zum Handpressen von ätherischen Ölen)*
s. rubber Schwammgummi *m*
spongelike s. spongy
sponging *(Ker)* Schwammverputzen *n*, Verschwammen *n*
s. agent *(Gum, Plast)* Treibmittel *n*, Blähmittel *n*
spongy schwamm[art]ig, locker, porös
s. nickel Nickelschwamm *m*
s.-nickel catalyst Nickelschwammkatalysator *m*
s. platinum Platinschwamm *m*
spontaneous spontan, freiwillig, von selbst ablaufend, eigenmächtig
s. combustion s. s. ignition
s. fermentation Spontangärung *f*
s. heating Selbsterwärmung *f*, Selbsterhitzung *f*
s. ignition Selbst[ent]zündung *f*, Spontanzündung *f*
s.-ignition temperature Selbstzünd[ungs]temperatur *f*, Selbstentzündungstemperatur *f*, SZT, Selbstzündpunkt *m*, Zündpunkt *m*
s. oxidation spontane Oxydation *f*, Autoxydation *f*
s. souring Spontansäuerung *f* *(der Milch)*
spontaneously spontan, freiwillig, von selbst
s. soured milk spontan gesäuerte Milch *f*, Sauermilch *f*, Setzmilch *f*, Schlippermilch *f*, Plundermilch *f*
spoon proof *(Glas)* Schöpfprobe *f*
spore Spore *f*
s. exine Sporenexine *f*
s. formers s. s.-forming bacteria
s.-forming sporenbildend
s.-forming bacteria, s.-producing bacteria sporenbildende Bakterien *npl*, Sporenbildner *mpl*
sporinite Sporinit *m* *(Mazeral der Steinkohle)*
sporulation Sporenbildung *f*, Sporulation *f*
spot / to tüpfeln
spot Fleck[en] *m*, Unreinheit *f*, Klecks *m*, Tupfen *m*
s. analysis Tüpfelanalyse *f*, Tüpfelprobe *f*
s. coating *(Plast)* Punktbeschichten *n*, punktweises Auftragen *n* der Beschichtung
s. plate Tüpfelplatte *f*
s. remover Fleck[en]entfernungsmittel *n*, Fleckenentferner *m*, Fleckenreinigungsmittel *n*

s. sample Stichprobe *f*
s. test[ing] *s.* s. analysis
s. treatment *(Landw)* Nesterbehandlung *f (z.B. mit Herbiziden)*
s. welding Punktschweißen *n*
spotting agent *(Text)* Detachiermittel *n*
s. plate *s.* spot plate
spout Schnabel *m*, Schnauze *f*, Ausguß *m (eines Laborgeräts)*; Abflußrohr *n*, Speirohr *n; (Glas)* Speiserbecken *n*, Speiserkopf *m*
spouting velocity *(Pap)* Ausströmungsgeschwindigkeit *f*, Auslaufgeschwindigkeit *f (des Stoffs)*
spray / to *(Sprühmittel)* [ver]sprühen; spritzen; *(Flüssigkeiten)* verstäuben, zerstäuben; verdüsen; *(mit Sprühmitteln)* besprühen, übersprühen; bespritzen; *(sortiertes Gut)* bebrausen, abbrausen, [mit einer Brause] abspülen *(Aufbereitungstechnik)*; spritzen *(Farbspritztechnik)*
s. in einspritzen
s. on aufsprühen *(z.B. Naßgut auf Walzentrockner)*
spray 1. Sprühmittel *n*, Sprühflüssigkeit *f*, Nebelmittel *n*; Spritzmittel *n*, Spritzflüssigkeit *f; (Kosmet)* Spray *m*, Sprühmittel *n*; 2. Sprühnebel *m*, Zerstäubungsnebel *m*; 3. Sprühdüse *f*, Bebrausungsdüse *f*, Brause *f*
s. boom Feldspritzrohr *n (Pflanzenschutz)*
s. booth *s.* s. chamber
s. burner Zerstäubungsbrenner *m*
s. chamber Spritzkabine *f*
s. coat Spritzauftrag *m*
s. coater *(Pap)* Spritzstreichmaschine *f*
s. coating *(Pap)* Spritzstreichen *n (Streichen durch Aufsprühen mit Spritzpistolen)*
s. column *s.* s. tower
s. concentrate Spritzkonzentrat *n*
s. cone Sprühkegel *m*
s. cooler Rieselkühler *m*, Berieselungskühler *m*
s. damper *(Pap)* Spritzrohrfeuchter *m*
s.-dried im Sprühverfahren (Zerstäubungsverfahren) getrocknet
s.-dried egg yolk Sprüheigelb *n*, Trockeneigelb *n*
s.-dried milk [powder] Sprühmilchpulver *n*, Zerstäubungsmilchpulver *n*
s. drift *(Landw)* Abdrift (Verwehung) *f* des Sprühmittels
s.-dry / to im Sprühverfahren (Zerstäubungsverfahren) trocknen, mittels Sprühtrocknung (Zerstäubungstrocknung) herstellen
s. dryer Sprühtrockner *m*, Zerstäubungstrockner *m*
s. drying Sprühtrocknung *f*, Zerstäubungstrocknung *f*
s. dust Spritznebel *m (Farbspritztechnik)*
s. dyeing *(Gerb)* Spritzfärbung *f*
s. freezing Berieselungsgefrierverfahren *n*
s. gun Spritzpistole *f*, Sprühpistole *f*, *(Farb auch)* Farbspritzpistole *f; (Pap)* Druckluftzerstäuber *m (Schwefeldioxidherstellung)*

s. injury *(durch Pflanzenschutzmittel verursachter)* Spritzschaden *m*
s. load *s.* s. residue
s. machine *(Landw)* Sprühgerät *n*; Spritzgerät *n*
s. metallization Spritzmetallisieren *n*, Spritzmetallisierung *f*, Metallspritzen *n*, Metallspritzverfahren *n*, Schoop[is]ieren *n*
s. nozzle Sprühdüse *f*, Zerstäuberdüse *f*
s. of water Wasserschleier *m*
s. oil Spritzöl *n*
s. painting Farbspritzen *n*, Spritzen *n*
s. paraffin Sprühparaffin *n*, Paraffin *n* für Sprühzwecke
s. pipe Sprührohr *n*, Verteilerrohr *n*
s. powder milk *s.* s.-dried milk
s. reagent Sprühreagens *n*
s. residue Sprührückstand *m*, Spritzrückstand *m*, Spritzbelag *m*, Spritzniederschlag *m*
s. rotation *(Landw)* Spritzwechsel *m (Wechsel der Pflanzenschutzmittel in systematischer Reihenfolge)*
s. spreader Netzmittel *n (Spritzhilfsmittel)*
s. strength Spritzkonzentration *f*
s. supplement Spritzhilfsmittel *n*
s. swathe Sprühschleier *m*
s. tank Sprühtank *m*, Brühebehälter *m*
s. tower Sprühturm *m*
s.-type ammonia absorber Sprühsättiger *m (zur Ammoniakrückgewinnung)*
s.-type burner *(Pap)* Druckluftschwefelverbrennungsofen *m*, Zerstäubungsschwefelverbrennungsofen *m*
s.-type cooler Sprühdüsenkühler *m*
s. washer *s.* s. tower
s. water Sprühwasser *n*, eingesprühtes (eingespritztes) Wasser *n*
sprayed rubber Sprühkautschuk *m*
sprayer 1. *s.* spray machine; 2. Zerstäuber *m*, Flüssigkeitszerstäuber *m*
spraying Versprühen *n*, Sprühen *n (von Sprühmitteln)*; Spritzen *n*, *(i.w.S.)* Ausbringen *n (von Pflanzenschutzmitteln)* in flüssiger Form; Verstäuben *n*, Zerstäuben *n*, Zerstäubung *f*; Verdüsen *n (von Flüssigkeiten)*; Besprühen *n*, Übersprühen *n (mit Sprühmitteln)*; Bespritzen *n*; Bebrausen *n*, Abbrausen *n*, Abbrausung *f*, Abspülen *n* [mit einer Brause] *(von sortiertem Gut in der Aufbereitungstechnik)*; Spritzen *n*, Farbspritzen *n*, *(manchmal auch)* Spritzlackieren *n*, Spritzlackierung *f (Farbspritztechnik)*
s. gun *s.* spray gun
s. of webs Sprühen *n* zur Verfestigung von Faservliesen
s. water Sprühwasser *n*
spread / to 1. ausbreiten, verteilen; ablegen, auslegen; *(unlöslichen oder schwerlöslichen Stoff auf der Oberfläche von Flüssigkeiten zu einer monomolekularen Schicht)* spreiten; 2. *(Gum)* [be]streichen, auf der Streichmaschine gummieren; sich ausbreiten (ausdehnen)

s. out *s.* to spread 1.
spread coating Überziehen (Beschichten) *n* durch Streichen, Streichverfahren *n*
spreadability Spreitbarkeit *f; (Lebm)* Streichfähigkeit *f*
spreadable *(Lebm)* streichfähig
spreader 1. Streichmaschine *f;* Streichgerät *n;* Leimauftragmaschine *f;* Glättwalze *f (in Walzentrocknern); (Plast)* Angußverteiler *m (im Spritzgießwerkzeug);* 2. *s.* spreading agent
  s. knife Streichmesser *n*
spreading Ausbreiten *n,* Verteilen *n;* Ablegen *n,* Auslegen *n;* Spreiten *n,* Spreitung *f (eines unlöslichen oder schwerlöslichen Stoffs auf der Oberfläche von Flüssigkeiten zu einer monomolekularen Schicht); (Gum)* Streichen *n,* Bestreichen *n,* Gummieren *n* auf der Streichmaschine
  s. ability Spreitungsvermögen *n*
  s. agent Netzmittel *n*
  s. coefficient Ausbreitungskoeffizient *m,* Spreitungskoeffizient *m;* Benetzungskoeffizient *m*
  s. efficiency Verteilerwirksamkeit *f*
  s. factor Hyaluronidase *f (als Diffusionsfaktor wirkendes Ferment)*
  s. knife *s.* spreader knife
  s. machine Streichmaschine *f*
  s. method *s. s.* process
  s. mix[ture] Streichmischung *f*
  s. of writing ink *(Pap)* Auslaufen *n* der Tinte
  s. pressure Spreitungsdruck *m*
  s. process Streichverfahren *n*
  s. properties Verteilungseigenschaften *fpl*
Sprengel pump Sprengel-Pumpe *f*
spring 1. Quelle *f,* Brunnen *m;* 2. Feder *f*
  s.-actuated relief valve federbelastetes Sicherheitsventil *n*
  s. balance Federwaage *f*
  s. barley Sommergerste *f*
  s. centre Federhülse *f*
  s. clip Schlauchklemme *f,* Quetschklemme *f,* Quetschhahn *m*
  s. ejector *(Plast)* mit Federkraft betätigter Ausdrückstift *m*
  s.-loaded mit Federdruck, federbelastet
  s. pad *(Plast)* federnder Niederhalter *m*
  s.[-seated] valve Federventil *n*
  s. water Quellwasser *n*
  s. wood Früh[jahrs]holz *n,* Frühlingsholz *n,* Weitholz *n*
sprinkle / to streuen; spritzen; sprühen; besprengen, beregnen, berieseln
sprinkled paper gespritztes Papier *n (ein Buntpapier)*
Sprout-Waldron-refiner Sprout-Waldron-Refiner *m,* Sprout-Waldron-Mühle *f*
spruce Fichte *f,* Picea A. Dietr.
  s. bark Fichtenrinde *f*
  s. gum *(Harz von Picea mariana (Mill.)B.S.P.; Kaumittel)*

s. turpentine Fichtenöl *n*
sprue *(Plast)* Anguß[kegel] *m*
  s. bush[ing] *(Plast)* Angußbuchse *f*
  s. ejector *(Plast)* Angußausdrückstift *m*
  s. puller *(Plast)* Angußzieher *m*
spue *s.* spew
Spülgas process Spülgas[schwel]verfahren *n*
spume / to schäumen, sich mit Schaum bedecken
spume Schaum *m*
spumescent schaum[art]ig, schäumend
spumous *(Geol)* schaumig
spun-bonded product *(Text)* nach dem Schmelzspinnverfahren hergestellter Vliesstoff *m,* Schmelzspinnverbundstoff *m*
  s.-coloured *s.* s.-dyed
  s. concrete Schleuderbeton *m*
  s.-dyed [er]spinngefärbt, düsengefärbt
  s. fibre Spinnfaser *f,* gesponnene Faser *f*
  s. glass Glasgespinst *n*
  s.-pigmented *s.* s.-dyed
  s. roving *(Text)* Spinnroving *m*
  s. silk gesponnene Seide *f,* Garn *n* aus Rohseidenabfällen
spur *(Ker)* Hahnenfuß *m (Brennhilfsmittel)*
sputter / to [ver]spritzen, [ver]sprühen; zerstäuben; *(Met)* spratzen
sputtering Spritzen *n,* Verspritzen *n,* Sprühen *n,* Versprühen *n,* Versprühung *f;* Zerstäuben *n; (Met)* Spratzen *n;* Aufspritzen *n;* Metallspritzen *n,* Spritzmetallisieren *n,* Schoop[is]ieren *n*
spy hole Schauloch *n*
squalene Squalen *n (aliphatisches Triterpen)*
square basin (tank) Quadratbecken *n*
squash Fruchtbrei *m,* Mus *n,* Pulpe *f,* Pülpe *f;* ausgepreßter Fruchtsaft *m*
squeegee Quetschwalze *f,* Gummiquetscher *m;* Rakel *f (für Siebdruck)*
squeeze / to [aus]quetschen, abquetschen, [aus]drücken, [aus]pressen, abpressen; einzwängen *(z.B. Atome in eine Gitterlücke)*
  s. through durchpressen
squeeze Quetschen *n,* Ausquetschen *n,* Abquetschen *n,* Drücken *n,* Ausdrücken *n,* Pressen *n,* Auspressen *n,* Abpressen *n*
  s. bottle Spritzflasche *f* aus Weichplast
  s. roll Quetschwalze *f,* Abquetschwalze *f,* Preßwalze *f,* Abpreßwalze *f; (Pap)* Walze *f* der Zugpresse (Transportpresse, Greiferpresse, Vorzugspresse)
squeezer Presse *f,* Quetsche *f;* Saftpresse *f,* Entsafter *m,* Fruchtpresse *f*
  s. machine Preßformmaschine *f*
squeezing roll *s.* squeeze roll
Squibb burette Bürette *f* nach Squibb
squirrel-cage disintegrator (mill) Schlagkorbmühle *f,* Desintegrator *m (i.e.S.),* Desintegratormühle *f (i.e.S.),* Schleudermühle *f*
squirt trim *(Pap)* Randabspritzeinrichtung *f,* Randspritzer *m; (Pap)* Randspritzstoff *m*
  s.-trim water *(Pap)* Siebrandspritzwasser *n*

s.-trimmed stock *(Pap)* Randspritzstoff *m*
SR *s.* soil release
S.R.B. *s.* straight-run benzine
SRF black *s.* semireinforcing furnace black
s.s. *s.* slightly soluble
S.S. *s.* soft-sized
SS acid SS-Säure *f*, 2S-Säure *f*, Chicagosäure *f*, 1-Amino-8-naphthol-2,4-disulfonsäure *f*
stab culture Stichkultur *f*
stability Stabilität *f*, Beständigkeit *f*, Festigkeit *f*, Widerstandsfähigkeit *f*, Resistenz *f*; Stabilität *f*, Haltbarkeit *f*
 s. at high temperatures Hochtemperaturbeständigkeit *f*, Beständigkeit (Widerstandsfähigkeit) *f* gegen hohe Temperaturen (bei hohen Temperaturen)
 s. constant *(Ech)* Stabilitätskonstante *f*, Komplexbildungskonstante *f*
 s. factor Stabilitätsfaktor *m*, Festigkeitsfaktor *m*
 s. in storage Lagerbeständigkeit *f*, Lagerfähigkeit *f*, Lagerhaltbarkeit *f*
 s. of a foam Stabilität (Beständigkeit, Haltbarkeit, Lebensdauer) *f* eines Schaumes, Schaumstabilität *f*, Schaumdauer *f*, Schaumbeständigkeit *f*
 s. of colour Farb[ton]beständigkeit *f*, Farbenbeständigkeit *f*, Farb[ton]echtheit *f*
 s. of emulsions Emulsionsstabilität *f*, Emulsionsbeständigkeit *f*
 s. to acid[s] Säurebeständigkeit *f*, Säurefestigkeit *f*, Säurewiderstandsfähigkeit *f*, Säureresistenz *f*
 s. to alkali[es] Alkalibeständigkeit *f*, Laugenbeständigkeit *f*
 s. to chemical attack chemische Unangreifbarkeit (Beständigkeit, Stabilität, Festigkeit, Widerstandsfähigkeit, Resistenz) *f*, Beständigkeit *f* gegen chemische Einwirkungen; Chemikalienbeständigkeit *f*, Chemikalienfestigkeit *f*, Widerstandsfähigkeit (Resistenz) *f* gegen Chemikalien
 s. to storage *s.* s. in storage
 s. zone Stabilitätszone *f*
stabilization Stabilisierung *f*, Stabilisieren *n*, Stabilisation *f*
 s. through resonance Resonanzstabilisierung *f*
stabilize / to stabilisieren; *(Met)* stabilglühen, stabilisierend glühen
stabilized gasoline stabil[isiert]es Benzin *n*, Stabilbenzin *n*
stabilizer Stabilisier[ungs]mittel *n*, Stabilisator *m*; Stabilisator *m*, Stabilisierkolonne *f*, Stabilisationskolonne *f*, Stabilizer *m*
 s. plant Stabilisieranlage *f*, Stabilisationsanlage *f*
stabilizing Stabilisieren *n*, Stabilisierung *f*, Stabilisation *f*; *s.* s. anneal
 s. agent Stabilisier[ungs]mittel *n*, Stabilisator *m*
 s. anneal *(Met)* Stabil[isierungs]glühen *n*, Stabil[isierungs]glühung *f*, stabilisierendes Glühen *n*
 s. bath Stabilisierungsbad *n*

stable stabil, beständig, fest, widerstandsfähig, resistent; stabil, haltbar, dauerhaft
 s. compound stabile Verbindung *f*
 s. gasoline *s.* stabilized gasoline
 s. humus Dauerhumus *m*
 s. in air luftbeständig, an der Luft beständig
 s. in light lichtbeständig
 s. in storage lagerbeständig, lagerfähig
 s. isotope stabiles (nichtradioaktives) Isotop *n*
 s. orbit stabile (stationäre) Bahn *f*, Sollbahn *f*
 s. to acid[s] säurebeständig, säurestabil, säurefest, beständig gegen Säurewirkung
 s. to alkali[es] alkalibeständig, laugenbeständig
 s. to light *s.* s. in light
 s. to oxidation oxydationsbeständig, beständig gegen oxydative Einflüsse
 s. to water wasserbeständig, wasserfest, beständig gegen[über] Wasser
 s. towards acids *s.* s. to acid[s]
 s. towards water *s.* s. to water
stachydrine Stachydrin *n* *(ein Pyrrolidinalkaloid)*
stachyose Stachyose *f* *(ein Tetrasaccharid)*
stack / to [auf]stapeln
stack 1. Kamin *m*, Schornstein *m*, Esse *f*; Schacht *m* *(eines Hochofens)*; 2. Stapel *m*; Satz *m* *(z.B. von Walzen)*; *(Pap)* Kalander[walzen]satz *m*, *(i.w.S.)* Kalander *m*, Papierkalander *m*, Glätt[walzen]werk *n*, Walzenglättwerk *n*, Glättmaschine *f*, Maschinenglättwerk *n*, Trockenglättwerk *n*, Maschinenkalander *m*, Satinierkalander *m*, Satinier[walz]werk *n*
 to be lost at the s. durch den Schornstein abgehen
 s. gas Abgas *n*
 s. loss Flugstaubverlust *m*, Rauchgasverlust *m*
 s. of rolls Walzensatz *m*
 s. temperature Kamintemperatur *f*
 s. valve Kaminklappe *f*
stacker *(Glas)* Kühlofenbeschicker *m*
stacking Stapeln *n*, Aufstapeln *n*; *(Krist)* Stapelung *f*
 s. fault *(Krist)* Stapelfehler *m*
 s. machine *(Ker)* Stapelmaschine *f*
Staffordshire cone *s.* Staffordshire Seger cone
Staffordshire kiln *(Ker)* Staffordshire-Ofen *m* *(kontinuierlicher Brennofen mit von oben beheizten einzelnen Brennkammern)*
Staffordshire Seger cone *(Ker)* Staffordshire-Seger-Kegel *m*
stage Stufe *f*, Stadium *n*, Phase *f*, Schritt *m*
 s. of carbonification Inkohlungsstadium *n*, Inkohlungsstufe *f*
 s. of plasticity plastischer Zustand *m*, Erweichungszustand *m*
 s. of the bleaching process Bleichstufe *f*
stagger / to gegeneinander versetzen, versetzt anordnen
staggered gegeneinander versetzt, versetzt angeordnet; *(Stereochemie)* gestaffelt, auf Lücke; ±anti-periplanar

s. configuration (conformation) gestaffelte (±anti-periplanare) Konformation *f*
s. form gestaffelte Form *f*, Staffelform *f*
staghorn sumac *(Gerb)* Hirschkolbensumach *m*, Rhus typhina L.
stagnation pressure Staudruck *m*, *(i.e.S.)* Gesamtdruck *m*, Ruhedruck *m*
stain / to beflecken, anflecken, beschmutzen, anschmutzen; [an]färben; *(Gum)* verfärben; beizen
stain Fleck *m*; Färbemittel *n*; Beize *f*, Holzbeize *f*; *(Glas)* Beize *f*, Farbbeize *f*
  s. of arsenic Arsenspiegel *m*
  s. removal Fleck[en]entfernung *f*, Fleckenreinigung *f*, Detachur *f*
  s. remover Fleck[en]entfernungsmittel *n*, Flekkenentferner *m*, Fleckenreinigungsmittel *n*
  s. resistance *(Plast)* Fleckenunempfindlichkeit *f*
staining Beflecken *n*, Anflecken *n*, Beschmutzen *n*, Anschmutzen *n*; Färben *n*, Anfärben *n*; *(Gum)* Verfärben *n*; Beizen *n*
  s. dish (trough) Färbeküvette *f*, Färbekasten *m*
  s. well Blockschälchen *n*
stainless nichtrostend, rostbeständig, rostfrei, rostsicher *(z.B. Stahl)*
stake / to *(Gerb)* stollen
staker, staking machine *(Gerb)* Stollmaschine *f*
stalactite *(Min)* Stalaktit *m*
stalactitic *(Min)* stalaktitisch
stalagmite *(Min)* Stalagmit *m*
stalagmitic *(Min)* stalagmitisch
stalagmometer Stalagmometer *n*
stale schal, abgestanden, fad[e] *(Flüssigkeit)*; altbacken *(Brot)*
  s. flavour (taste) Altgeschmack *m*
stalk fibre Stengel[bast]faser *f*
stall Kammer *f* *(bei der Kohlehydrierung)*
stamp battery Pochwerk *n*
  s. mill Stampfmühle *f*
  s. paper Briefmarkenpapier *n*
stamper Stampfmaschine *f*; *s.* stamping mill
stamping ink Stempel[kissen]farbe *f*
  s. mill *(Pap)* Stampfwerk *n*, „Deutsches Geschirr" *n*
stamps *s.* stamping mill
stand Stativ *n*
  to allow to s. stehen lassen
  s. development *(Foto)* Standentwicklung *f*
  s. oil Standöl *n*
standard Standard *m*; Normal *n*; Standardsubstanz *f*
  to bring to s. strength *(einen Farbstoff)* auf Typ bringen
  s. acid standardisierte (eingestellte) Säure *f*, Maßlösung *f* einer Säure, *(i.e.S.)* Normalsäure *f*, normale Säure *f*, n-Säure *f*
  s. alkali standardisiertes (eingestelltes) Alkali *n*, *(i.e.S.)* Normalalkali *n*, normales Alkali *n*, n-Alkali *n*
  s. ASTM method ASTM-Standardmethode *f*

s. atmosphere Normalatmosphäre *f*, physikalische Atmosphäre *f*, atm
s. base standardisierte (eingestellte) Lauge *f*, Maßlösung *f* einer Lauge, *(i.e.S.)* Normallauge *f*, normale Lauge *f*, n-Lauge *f*
s. brick Normalziegel *m*, Normalstein *m*
s. buffer [solution] Standardpufferlösung *f*, standardisierte Pufferlösung *f*, Standardpuffer *m*
s. calomel electrode Standardkalomelelektrode *f*
s. catalyst Standardkatalysator *m*
s. cell Standard[normal]element *n*, Standardzelle *f*, ungesättigtes Normalelement *n*
s. chemical potential chemisches Standardpotential *n*
s. coal Testkohle *f.*
s. colour Standardfarbe *f*
s. conditions Standardbedingungen *fpl*, Norm[al]bedingungen *fpl*, *(Text auch)* Norm[al]klima *n*
s. cone *(Ker)* standardisierter Kegel *m*
s. cone crusher Kegelbrecher *m* in Standardausführung
s. depth [of shade] *(Text)* Normaltontiefe *f*, Richttyptiefe *f*
s. deviation *(Statistik)* Standardabweichung *f*, mittlere quadratische Abweichung *f*, mittlerer quadratischer Fehler *m*
s. dropper Normaltropfenzähler *m*
s. dyeing *(Text)* Typfärbung *f*
s. electrode Standard[bezugs]elektrode *f*
s. electrode potential Standardelektrodenpotential *n*, Standardpotential *n* [der Elektrode]
s. electromotive force Standard-[Bezugs-]EMK *f*
s. emf *s. s.* electromotive force
s. enthalpy Standardenthalpie *f*
s. enthalpy of reaction Standardreaktionsenthalpie *f*
s. entropy Standardentropie *f*
s. equipment Standardausrüstung *f*, Normalausrüstung *f*
s. evaporator Vertikalrohrverdampfer *m* mit Innenheizkammer, Normalverdampfer *m*
s. gold Standardgold *n* *(Münzgold)*
s. gold sol Standardgoldsol *n*
s. ground joint Standardschliffverbindung *f*, Standardschliff *m*, Norm[al]schliff *m*
s. ground stopcock Standardschliffhahn *m*, Normschliffhahn *m*
s. half-cell Standardhalbelement *n*, Standardhalbzelle *f*
s.-half-cell potential Standardpotential *n* des Halbelements
s. heat-content change Änderung *f* der Standardenthalpie
s. heat of formation Standardbildungswärme *f*, Standardbildungsenthalpie *f*
s. hydrogen electrode Standardwasserstoffelektrode *f*
s. IP method IP-Standardmethode *f*

s. method Standardmethode *f*, Standardverfahren *n*, standardisierte Methode *f*, standardisiertes Verfahren *n*

s. mineral Standardmineral *n*

s. nozzle Normdüse *f (Mengenstrommessung)*

s. of emf EMK-Normal *n*, Urspannungsnormal *n*

s. opalescence Standardopaleszenz *f*, Standardtrübung *f*

s. orifice Normblende *f (Mengenstrommessung)*

s. oxidation potential Standardoxydationspotential *n*

s. oxidation-reduction potential *s. s.* redox potential

s. penicillin Standardpenizillin *n*

s. plane *(Krist)* Hauptsymmetrieebene *f*

s. potential Standard[normal]potential *n*; *s. s.* electrode potential

s. preparation Standardpräparat *n*

s. pressure Standarddruck *m*, Norm[al]druck *m*, Normalluftdruck *m*

s. procedure *s. s.* method

s. radius Standardradius *m*

s. reagent Standardreagens *n*

s. redox potential Standardredoxpotential *n*, Standardpotential *n* des Redoxsystems

s. reduction potential Standardreduktionspotential *n*

s. reference emf *s. s.* electromotive force

s. reference voltage *s. s.* voltage

s. resistance Normalwiderstand *m*

s. resistance thermometer Standardwiderstandsthermometer *n*, Normalwiderstandsthermometer *n*

s. sample Standardprobe *f*, Normalprobe *f*

s. screen *s. s.* sieve

s. series of screens (sieves) *s. s.* sieve scale

s. shade Normalton *m*

s. sieve Standardsieb *n*, Norm[al]sieb *n*, standardisiertes Sieb *n*

s. sieve scale (series) Standardsiebskala *f*, Normalsiebskala *f*, Standardsiebreihe *f*, Normalsiebreihe *f*

s. silver Standardsilber *n (Münzsilber)*

s. size *(Pap)* Normformat *n*

s. soap solution Normalseifenlösung *f*

s. soil *(Text)* standardisierter Schmutz *m*

s. solution Standardlösung *f*, standardisierte (eingestellte) Lösung *f*, Maßlösung *f*, Titerlösung *f*, *(i.e.S.)* Normallösung *f*, normale Lösung *f*, n-Lösung *f*

s. solution of sodium nitrite *(Farb)* Normal-Nitritlösung *f*, n-NaNO$_2$-Lösung *f*

s. specification Standardvorschrift *f*

s. stain Standardfleck *m*

s. state Standardzustand *m*

s. substance Standard[bezugs]substanz *f*, standardisierte Substanz *f*

s. technique *s. s.* method

s. temperature Standardtemperatur *f*, Norm[al]temperatur *f*

s. temperature and pressure Norm[al]temperatur *f* und -druck *m*, physikalischer Normzustand *m (Standardbedingungen von 0°C und 760 Torr)*

s. tension *s. s.* voltage

s. test Standardprüfung *f*, Standardversuch *m*

s. test method standardisierte Prüfmethode (Untersuchungsmethode) *f*

s. test sieve Standardprüfsieb *n*, Norm[al]prüfsieb *n*, standardisiertes Prüfsieb *n*

s. testing method *s. s.* test method

s. testing screen (sieve) *s. s.* test sieve

s. thermocouple Standardthermopaar *n*, standardisiertes Thermopaar *n*

s. thermometer Standardthermometer *n*, Normalthermometer *n*

s. titrant (titrimetric substance) Urtitersubstanz *f*, Urtiterstoff *m*

s. turbidity *s. s.* opalescence

s.-type injection nozzle normale Spritzdüse *f*

s. unit Standardeinheit *f*, Normaleinheit *f*

s. value Standardwert *m*, Norm[al]wert *m*

s. voltage Standard[bezugs]spannung *f*

s. volume Norm[al]volumen *n*

s. Weston [cadmium] cell *s. s.* Weston normal cadmium cell

s. Weston normal cadmium cell Standard-Weston-Normalelement *n*, Standard-Weston-Element *n*, Weston-Standardelement *n*, ungesättigtes Weston-Normalelement (Weston-Element) *n*

s. work of reaction Standardreaktionsarbeit *f*

standardization Standardisierung *f*, *(bei Chemikalien auch)* Einstellen *n*, Einstellung *f*

standardize / to standardisieren, *(bei Chemikalien auch)* einstellen

standardized conditions *s.* standard conditions

s. powdered opium *(Pharm)* eingestelltes Opiumpulver *n*

Standfast molten-metal machine *(Text)* Standfast-[Metal-]Maschine *f*, Standfast-Färbemaschine *f*

Standfast molten-metal method (process) *(Text)* Standfast-[Molten-]Metal-Prozeß *m*, Molten-Metal-Verfahren *n*, Metallbadverfahren *n*

standing bath *(Text)* stehendes (altes) Bad *n*, Standbad *n*

s.-bath dyeing *(Text)* Färben *n* auf stehendem Bad

standpipe Standrohr *n*

stannane *s.* stannic hydride

stannate Stannat *n*

stannic Zinn..., *(i.e.S.)* Zinn(IV)-...

s. acid SnO$_2$·xH$_2$O Zinnsäure *f*

s. ammonium chloride (NH$_4$)$_2$[SnCl$_6$] Ammoniumhexachlorostannat(IV) *n*, Pinksalz *n*

s. anhydride *s. s.* oxide

s. bromide SnBr$_4$ Zinn(IV)-bromid *n*, Zinntetrabromid *n*

s. chloride $SnCl_4$ Zinn(IV)-chlorid n, Zinntetrachlorid n

s. compound Zinn(IV)-verbindung f

s. fluoride SnF_4 Zinn(IV)-fluorid n, Zinntetrafluorid n

s. hydride SnH_4 Zinnhydrid n, Zinn(IV)-hydrid n, Zinnwasserstoff m, Stannan n

s. iodide SnJ_4 Zinn(IV)-jodid n, Zinntetrajodid n

s. nitrate $Sn(NO_3)_4$ Zinn(IV)-nitrat n, Zinntetranitrat n

s. oxide SnO_2 Zinn(IV)-oxid n, Zinndioxid n

s. sulphate $Sn(SO_4)_2$ Zinn(IV)-sulfat n, Zinndisulfat n

s. sulphide SnS_2 Zinn(IV)-sulfid n, Zinndisulfid n

stannine (Min) Stannin m, Zinnkies m

stannite M'_2SnO_2 Stannat(II) n; s. stannine

stannolite (Min) Stannolith m, (veraltet für) Kassiterit m, Zinnstein m (Zinn(IV)-oxid)

stannous Zinn..., (i.e.S.) Zinn(II)-...

s. acetate $Sn(CH_3COO)_2$ Zinn(II)-azetat n

s. bromide $SnBr_2$ Zinn(II)-bromid n, Zinndibromid n

s. chloride $SnCl_2$ Zinn(II)-chlorid n, Zinndichlorid n

s. dihydrogen orthophosphate $Sn(H_2PO_4)_2$ Zinn(II)-dihydrogen[ortho]phosphat n

s. fluoride SnF_2 Zinn(II)-fluorid n, Zinndifluorid n

s. hydroxide $Sn(OH)_2$ Zinn(II)-hydroxid n

s. iodide SnJ_2 Zinn(II)-jodid n

s. monohydrogen orthophosphate $SnHPO_4$ Zinn(II)-hydrogen[ortho]phosphat n

s. nitrate $Sn(NO_3)_2$ Zinn(II)-nitrat n, Zinndinitrat n

s. orthophosphate $Sn_3(PO_4)_2$ Zinn(II)-[ortho]-phosphat n

s. oxalate $Sn(COO)_2$ Zinn(II)-oxalat n

s. pyrophosphate $Sn_2P_2O_7$ Zinn(II)-diphosphat n, Zinn(II)-pyrophosphat n

s. salt Zinn(II)-salz n

s. sulphate $SnSO_4$ Zinn(II)-sulfat n, Zinn[mono]sulfat n

s. sulphide SnS Zinn(II)-sulfid n, Zinn[mono]sulfid n

stannyl H_3Sn- Stannyl n

staple / to (Text) Elementarfäden auf Stapel (bestimmte Länge) schneiden (reißen)

staple s. 1. s. fibre; 2. s. length

s. fibre Faser f, Spinnfaser f, (bei Chemiefaserstoffen) Stapelfaser f

s.-fibre glass yarn Glasfasergarn n

s. for filling Füllfaser f, Füllmaterial n

s. length Faserlänge f, Stapellänge f, Stapel m

star anise Sternanis m, Illicium verum Hook. fil.

s. aniseed Sternanisfrüchte fpl, Sternanis m (Sammelfrüchte von Illicium verum Hook. fil.)

s. feeder Zellenrad n (Zuteilvorrichtung)

s. frame (Text) Sternrahmen m, Sternreifen m, Färbestern m

s. steamer (Text) Sterndämpfer m

s. valve (wheel) Zellradschleuse f, Zellenschleuse f, Zellrad n

starch / to (Text) stärken, steifen

starch $(C_6H_{10}O_5)_n$ Stärke f

s. adhesive Stärkeleim m

s.-bearing plant s. s.-yielding plant

s. breakdown Stärkeabbau m

s. chain Stärkekette f

s. chemistry Stärkechemie f

s.-containing stärkehaltig, stärkeführend, stärkereich

s. content Stärkegehalt m

s. conversion Stärkeverzuckerung f (durch Diastase)

s.-converting enzyme stärkespaltendes Ferment n, Amylase f

s. equivalent (Landw) Stärkewert m, St.W.

s. factory Stärkefabrik f

s. flour Stärkemehl n

s. glue Stärkeleim m, Stärkekleister m

s. grain (granule) Stärkekorn n, Stärkekörnchen n

s. gum Stärkegummi n, Dextrin n, Dampfgummi n, Kristallgummi n

s. industry Stärkeindustrie f

s.-iodate paper Stärke-Jodat-Papier n

s. iodide paper Jodstärkepapier n, mit Jodstärke getränktes Papier n

s. kernel s. s. grain

s. milk Stärkemilch f

s. nitrate Stärkenitrat n, Nitrostärke f (Sprengstoff)

s. paper Stärkepapier n

s. paste s. s. glue

s.-reducing enzyme s. s.-converting enzyme

s. sizing (Pap) Stärkeleimung f

s. slurry Stärkemilch f

s. solution Stärkelösung f

s.-splitting stärkespaltend

s.-splitting enzyme s. s.-converting enzyme

s. sugar Stärkezucker m

s. syrup Stärkesirup m, Kapillärsirup m, Bonbonsirup m

s. turbidity Kleistertrübung f

s.-yielding plant stärkeliefernde Pflanze f, Stärkelieferant m

starching (Text) Stärken n, Steifen n, Steifung f

starchy stärkehaltig, stärkeführend, stärkereich; Stärke...; gestärkt, gesteift

s. content s. starch content

s. finish (Text) stärkehaltige Appretur f

Stark-Einstein law Stark-Einsteinsches Äquivalenzgesetz n, Quantenäquivalenzgesetz n, fotochemisches Äquivalenzgesetz n

start / to 1. initiieren, anregen, einleiten, anfangen, beginnen, in Gang bringen; (eine Maschine) anfahren; 2. anfangen, beginnen; einsetzen (Reaktion)

s. up in Betrieb nehmen (setzen), in Gang setzen, anfahren, einfahren, anlassen, (bei Öfen auch) anheizen

s. with ausgehen von
start of transformation Umwandlungsbeginn *m*, Umwandlungsanfang *m*
s. reaction Startreaktion *f*, Primärreaktion *f*, Primärakt *m*
s.-up Inbetriebnahme *f*, Ingangsetzen *n*, Anfahren *n*, Einfahren *n*, Anlassen *n*, *(bei Öfen auch)* Anheizen *n*
starter *(Lebm)* Säurewecker *m*; *(Landw)* Startdünger *m*
  s. can Muttersäurekulturgefäß *n*
  s. culture Säureweckerkultur *f*
  s. distillate Säureweckerdestillat *n*
  s. fertilizer Startdünger *m*
  s. heater Säureweckerapparat *m*
  s.-ripened butter Sauerrahmbutter *f*
starting burner Brenner *m* zum Anlassen (Anheizen) *(eines Ofens)*
  s. material Ausgangsmaterial *n*, Ausgangsstoff *m*, Ausgangssubstanz *f*, Ausgangsgut *n*, Ausgangsprodukt *n*
  s. of cure *(Gum)* Vulkanisationseinsatz *m*
  s. point Ausgangspunkt *m*
  s. sheet Unterlage *f*, Startblech *n*
  s. stack Esse *f* (Schornstein *m*) zum Anlassen (Anbrennen) *(eines Fluo-Solids-Ofens)*
  s.-up *s.* start-up
starvation *(Landw)* Mangel *m*
starved:
  to be s. of liquid abschnappen *(von einer Pumpe)*
Stassfurt [salt] deposits Staßfurter Salze (Abraumsalze) *npl*
state Zustand *m*
  s. of aggregation Aggregatzustand *m*, Formart *f*; Anordnung *f (von Atomen, Atomgruppen oder Molekülen)*
  s. of bonding Bindungszustand *m*
  s. of cure *s.* s. of vulcanization
  s. of division Verteilungszustand *m*
  s. of equilibrium Gleichgewichtszustand *m*
  s. of matter Aggregatzustand *m*, Formart *f*
  s. of oxidation Oxydationsstufe *f*, Oxydationszustand *m*
  s. of vulcanization Vulkanisationszustand *m*
static bed Festbett *n*, ruhendes (statisches) Bett (Feststoffbett) *n*, ruhende (statische) Schüttung *f*
  s. bed of catalyst festliegendes (festes, ruhendes, stationäres) Katalysatorbett (Kontaktbett) *n*, festliegende Katalysatorschicht *f*
  s. carbonizing plant Koksofen *m* mit ruhender Ladung (Beschickung)
  s. catalyst Festbettkatalysator *m*, Festbettkontakt *m*, festliegender (festangeordneter, fester, ruhender) Katalysator (Kontakt) *m*
  s. charge ruhende Ladung (Beschickung) *f (eines diskontinuierlich betriebenen Gaswerksofens)*
  s. electrification statische Aufladung *f*
  s. eliminator Antistatikum *n*, *(manchmal auch)* Entelektrisierungsgerät *n*

s. method statische Methode *f (zur Messung von Dampfdrücken)*
  s. retort Retorte *f* mit ruhender Ladung (Beschickung)
  s. seal Dichtung *f (für nicht gegeneinander bewegte Teile)*
  s. stress statische Beanspruchung (Belastung) *f*
  s. submergence *(Destillation)* Dampfdurchdringtiefe *f*, Eintauchtiefe *f*
  s. test statische Prüfung *f*, statischer Versuch *m*
stationary mould half *(Plast)* feststehende Werkzeughälfte *f*
  s. phase *(Chromatografie)* stationäre (unbewegliche) Phase *f*
  s. platen *(Plast)* feststehende Werkzeugplatte (Formplatte) *f*
  s. state stationärer (stabiler) Zustand *m*, Beharrungszustand *m*
  s.-type syphon *(Pap)* [fest]stehender Siphon *m* *(Kondensatableitung)*
statistical mechanics *s.* s. thermodynamics
  s. thermodynamics statistische Thermodynamik *f*
  s. weight *(phys Ch)* statistisches Gewicht *n*
stator ring Statorring *m* *(einer Drehscheibenkolonne)*
Staudinger molecular weight relative Molekülmasse *f* nach Staudinger
staurolite *(Min)* Staurolith *m*
stay on / to haften
staying-on power Haftvermögen *n*, Haftfähigkeit *f*
Stead's brittleness Steadsche Sprödigkeit *f*
Stead's reagent Steadsches Reagens (Ätzmittel) *n*
steady state stationärer (stabiler) Zustand *m*, Beharrungszustand *m*; dynamisches Gleichgewicht *n*
steam / to dämpfen, mit Dampf (Wasserdampf) behandeln; ausdämpfen, mit Wasserdampf desorbieren; gasen
  s. downwards abwärts (in absteigender Richtung, von oben) gasen
  s. upwards aufwärts (in aufsteigender Richtung, von unten) gasen
steam Dampf *m*
  s. ager *(Text)* Dämpfer *m*
  s. atomization Dampfzerstäubung *f*, Dampfversprühung *f*, Einblasen *n* mittels Dampfes
  s.-atomizing burner Treibdampfbrenner *m*
  s. autoclave Dampfautoklav *m*, Dampfgefäß *n*
  s. bath Dampfbad *n*
  s. battery Dampfregister *n*
  s. blowing *(Text)* Dampfdekatur *f*
  s.-blowing process *(Glas)* Dampfdüsenblasverfahren *n*, Düsenblasverfahren *n*
  s. boiler Dampfkessel *m*
  s. calorimeter Dampfkalorimeter *n*, Kondensationskalorimeter *n*
  s.-carbon reaction Reaktion *f* des Kohlenstoffs mit Wasserdampf, Wassergasreaktion *f*
  s. chamber Dampfkammer *f*, Dampfraum *m*

s. channel *(Plast)* Dampfkanal *m*
s. chest *s.* s. chamber
s. chip distributor *(Pap)* Dampffüllapparat *m (für Hackschnitzel)*
s. coal Dampf[kessel]kohle *f*, Kesselkohle *f*
s. coil Dampf[heiz]schlange *f*
s. collector Dampfsammler *m*
s. condenser Kühler *m (für Dämpfe)*, Kondensator *m*
s. consumption Dampfverbrauch *m*
s. cure *s.* s. curing
s.-cured dampfbehandelt, dampferhärtet, mit Dampf nachbehandelt *(Beton)*
s. curing Dampfbehandlung *f*, Dampferhärtung *f*, Dampfhärten *n (von Beton)*; *(Gum)* Dampfvulkanisation *f*, Vulkanisation *f* in Dampf
s. cylinder Dampfzylinder *m*
s.-cylinder lubricating oil, s.-cylinder stock Dampfzylinderöl *n*
s. demand Dampfbedarf *m*
s. deodorization Dämpfen *n*, Desodorieren *n*, Verdampfen *n* von Geruchsstoffen *(der Fette)*
s.-distil / to mit Dampf (Trägerdampf) destillieren, *(i.e.S.)* mit Wasserdampf destillieren
s.-distillable mit Dampf (Trägerdampf) destillierbar (flüchtig), *(i.e.S.)* mit Wasserdampf destillierbar (flüchtig)
s. distillation Dampfdestillieren *n*, Dampfdestillation *f*, Trägerdampfdestillation *f*, *(i.e.S.)* Wasserdampfdestillation *f*
s.-dried dampfgetrocknet; *(Pap)* maschinengetrocknet, auf der Maschine getrocknet, maschinentrocken
s.-driven dampfgetrieben, durch Dampfdruck betrieben, mit Dampfantrieb
s.-driven turboblower dampfgetriebenes Turbogebläse *n*, Dampfturbogebläse *n*
s. dryer (drying apparatus) Dampftrockner *m*, Dampftrockenapparat *m*
s. drying oven Dampftrockenschrank *m*
s. economy *(Kehrwert des spezifischen Heizdampfbedarfs des Verdampfers)*
s. ejector *s.* s.-jet ejector
s. entrance Dampfeintritt *m*, Dampfeinlaß *m*
s. fat melting Naßschmelze *f* auf Dampf
s. flowmeter Dampfdurchflußmesser *m*
s. generator Dampferzeuger *m*, Dampfentwickler *m*
s.-heated dampfbeheizt, mit Dampf beheizt, Dampf...
s.-heated funnel Dampftrichter *m*
s. heating Dampf[be]heizung *f*
s. injector *s.* s.-jet injector
s. jacket Dampfmantel *m*
s.-jacketed mit Dampfmantel
s. jet Dampfstrahl *m*
s.-jet apparatus Dampfstrahlapparat *m*
s.-jet ejector Dampfstrahlejektor *m*, Dampfstrahlsauger *m*
s.-jet injector Dampfstrahlinjektor *m*

s.-jet refrigerating machine Dampfstrahl[kälte]maschine *f*, Dampfstrahlkühlanlage *f*
s.-jet refrigeration Dampfstrahlkühlung *f*
s. joint Dampfanschluß *m*; *(Pap)* Dampf[einlaß]kopf *m (des Trockenzylinders)*
s. lard Dampfschmalz *n*, Steam Lard *n (aus dem Rohfett mit Dampf ausgeschmolzen)*
s. line Dampfleitung *f*
s. meter Dampfmesser *m*
s.-motivated ejector *s.* s.-jet ejector
s.-moulding process *(Plast)* Dampfstoßverfahren *n (Schäumen)*
s.-operated ejector *s.* s.-jet ejector
s. pipe Dampf[leitungs]rohr *n*; Dampfheizungsrohr *n*
s. pressure Dampfdruck *m*
s. process *(Gum)* Heißdampfverfahren *n (Regenerierverfahren)*
s. pump Dampfpumpe *f*
s. raising Dampferzeugung *f*
s.-raising coal Dampf[kessel]kohle *f*, Kesselkohle *f*
s. reclaim *(Gum)* Heißdampfregenerat *n*
s. rendering Naßschmelze *f* auf Dampf
s. requirement Dampfbedarf *m*
s.-set ink Steam-set-Ink *f (rasch trocknende Druckfarbe)*
s. sterilization Dampfsterilisation *f*
s.-stripping still Ausdämpfkolonne *f*, Abtreiberkolonne *f*, Austreibekolonne *f*
s. superheater Dampfüberhitzer *m*
s. supply Dampfzufuhr *f*
s.-supply line *s.* s. line
s. tempering *(Ker)* Dampfaufbereitung *f*, Heißaufbereitung *f*
s.-tight dampfdicht, dampfundurchlässig
s. trap Kondens[at]wasserableiter *m*, Kondenstopf *m*
s. tube *s.* s. pipe
s.-tube rotary dryer dampfbeheizter Röhrentrockner *m*
s.-turbine oil Dampfturbinenöl *n*
s. valve Dampfventil *m*
s.-volatile dampfflüchtig, *(i.e.S.)* wasserdampfflüchtig
s. vulcanization *(Gum)* Dampfvulkanisation *f*, Vulkanisation *f* in Dampf
steamer Dämpfer *m*
steamfit *(Pap)* Dampf[einlaß]kopf *m (des Trockenzylinders)*
steaming 1. Dämpfen *n*, Dampfbehandlung *f*; *(Farb)* Dämpfen *n*, Dämpferpassage *f*; 2. Dampfen *n*, Dampfung *f*, Naßbetrieb *m (bei der Gaserzeugung)*; Gasen *n*, Gasung *f*, Kaltblasen *n*
steapsin Steapsin *n (fettspaltendes Ferment der Bauchspeicheldrüse)*
stearaldehyde $CH_3(CH_2)_{16}CHO$ Stear[in]aldehyd *m*, Oktadekanal *n*

stearamide $CH_3(CH_2)_{16}CONH_2$ Stear[insäure]amid n, Oktadekanamid n

stearanilide $CH_3(CH_2)_{16}CONHC_6H_5$ Stearanilid n, N-Phenylstearamid n, Oktadekananilid n

stearate Stearat n

stearic acid $CH_3(CH_2)_{16}COOH$ Stearinsäure f, Oktadekansäure f

s. aldehyde s. stearaldehyde

stearin Stearin n, (i.e.S.) Stearin n, Tristearin n, Glyzerintristearat n, Glyzeroltristearat n, Glyzeryloktadekanoat n; Stearin n (technische Stearinsäure)

s. pitch Stearinpech n

s. soap Stearinseife f

stearinic acid s. stearic acid

stearolic acid $CH_3(CH_2)_7C{\equiv}C(CH_2)_7COOH$ Stearolsäure f, 9-Oktadezinsäure f

stearone $(C_{17}H_{35})_2CO$ Stearon n, Diheptadezylketon n, 18-Pentatriakontanon n

stearonitrile $CH_3(CH_2)_{16}CN$ Stearonitril n, Oktadekannitrol n

stearophanic acid $CH_3(CH_2)_{16}COOH$ Stearophansäure f, (veraltet für) Stearinsäure f, Oktadekansäure f

stearophenone $CH_3(CH_2)_{16}COC_6H_5$ Stearophenon n, 1-Phenyl-1-oktadekanon n

stearoxylic acid $CH_3(CH_2)_7COCO(H_2)_7COOH$ Stearoxylsäure f, 9,10-Dioxooktadekansäure f

stearoyl $CH_3(CH_2)_{16}CO-$ Stearoyl n, Oktadekanoyl n, Stearyl n

s. chloride $CH_3(CH_2)_{16}COCl$ Stearoylchlorid n, Stearinsäurechlorid n, Oktadekanoylchlorid n, Stearylchlorid n

stearyl $CH_3(CH_2)_{16}CH_2-$ Stearyl n; s. stearoyl

s. alcohol $CH_3(CH_2)_{16}CH_2OH$ Stearylalkohol m, Oktadezylalkohol m, Oktadekanol n

stearylamine $CH_3(CH_2)_{16}CH_2NH_2$ Stearylamin n, 1-Aminooktadekan n

steatite (Min) Steatit m, Speckstein m

s. ceramics Steatitkeramik f

s. porcelain Steatitporzellan n

s. whiteware Steatitweißware f

steel Stahl m

s. armouring Stahlarmierung f

s. autoclave Stahlautoklav m

s.-blue stahlblau

s.-cased stahlummantelt

s. casting Stahlgußstück n; Stahlguß m

s. cylinder Stahlflasche f, Stahlbombe f

s. engraving s. s.-plate engraving

s. floor Stahlplatte f (bei der Sturzprüfung des Kokses)

s.-grey stahlgrau

s. ingot Stahlblock m, Rohstahlblock m

s. jacket Stahlmantel m

s. pipe Stahlrohr n, Stahlröhre f

s. plate Stahlblech n

s.-plate engraving Stahlstich m

s.-plate-engraving ink Stahlstich[druck]farbe f, Druckfarbe f für Stahlstich, Stahldruckfarbe f

s.-plate fan Lüfter (Ventilator) m mit geraden Schaufeln

s.-plate ink s. s.-plate-engraving ink

s.-plate printing Stahl[stich]druck m, Siderografie f

s. ribbon Stahlband n

s. roll Stahlwalze f

s. scrap Stahlschrott m

s. shell Stahlpanzer m

s. substructure Stahlunterbau m, stählerner Unterbau m

steeling (Pap) Totmahlen n

steelmaking Stahlerzeugung f, Stahlherstellung f, Stahlgewinnung f

s. furnace Stahlschmelzofen m, Stahlwerksofen m

s. pig iron Stahl[roh]eisen n

s. process Stahlerzeugungsverfahren n, Stahlherstellungsverfahren n, Stahlgewinnungsverfahren n

steely glasig (z.B. Braumalz)

s. barley glasige (speckige) Gerste f (Braugerste)

s. malt glasiges Malz n

steep / to [ein]weichen, [ein]wässern, tränken, quellen; rösten, rotten

steep Weichen n, Einweichen n, Wässern n, Einwässern n, Tränken n, Quellen n; Einweichflüssigkeit f

s. tank (Bier) Quellstock m, Quellbottich m, Weichstock m, Weiche f

s. water Einweichwasser n, Weichwasser n

steepage s. steeping

steeper s. steep tank

steeping Weichen n, Einweichen n, Einwässern n, Tränken n, Quellen n; (Pap) Vorweichen n, Vorimprägnierung f (der Hackschnitzel vor der Kochung)

s. degree (Gär) Weichgrad m, Quellreife f

s. liquor Einweichflüssigkeit f, Tränkflüssigkeit f

s. method Benetzungsverfahren n (Saatgutbeize)

s. pan Einweichtrog m, Einweichkufe f, Tränktrog m, Netzbottich m

s. press Tauchpresse f

s. water s. steep water

Steffen molasses (Zucker) Steffen-Melasse f

Steffen [separation] process Steffensches Ausscheidungsverfahren n, Kalksaccharatverfahren n nach Steffen, Steffen-Verfahren n (zur Melasseentzuckerung)

Steffen waste water (Zucker) Steffen-Abwasser n

Steinbühl yellow $BaCrO_4$ Steinbühler Gelb n, Barytgelb n, Ultramaringelb n, gelbes Ultramarin n, Bariumchromat n

steinheilite (Min) Steinheilit m, (veraltet für) Cordierit m (Magnesiumaluminiumalumopentasilikat)

Stelzner method Stelzner-Methode f, Stelznersche Methode f (der Benennung von Heterozyklen)

stelznerite (Min) Stelznerit m, (veraltet für) Antherit m (Kupfer(II)-tetrahydroxidsulfat)

stem Stengel *m*; Ansatzrohr *n (einer Gasbürette)*;
  Stiel *m (eines Trichters)*; Spindel *f (eines Ventils)*
  s. fibre Stengel[bast]faser *f*
stencil Schablone *f*
  s. printing Siebdruck *m*, Seidenrasterdruck *m*
  s. silk Schablonenseide *f*
Stengel process Stengel-Verfahren *n (zur Herstel-*
  lung von Ammoniumnitrat)
stent roll *(Pap)* Spannwalze *f*
stenter *(Text)* Spannrahmen *m*, Rahmenspannmaschine *f*
stenting roll *s.* stent roll
step-by-step stufenweise, schrittweise
  s. cure *(Gum)* Stufenheizung *f*
  s. grate Treppenrost *m*, Stufenrost *m*
  s.-grate producer Treppenrostgenerator *m*,
  Gaserzeuger *m* mit Treppenrost
  s. reaction Stufenreaktion *f*
  s.-up cure *s. s.* cure
  s. wedge *(Foto)* Stufenkeil *m*
stephanite *(Min)* Stephanit *m (Antimon(III)-silber-*
  sulfid)
stepwise stufenweise, schrittweise
  s. reaction *s.* step reaction
stercobilin Sterkobilin *nn (Abbauprodukt des Gal-*
  lenfarbstoffs Bilirubin)
sterculia gum Sterkuliagummi *n*, Karayagummi *n*,
  Indischer (Ostindischer) Tragant *m (meist von*
  Sterculia urens Roxb.)
stereochemical stereochemisch
stereochemistry Stereochemie *f*, Raumchemie *f*
stereoisomer Stereoisomer[es] *n*, Stereomer[es] *n*,
  Raumisomer *n*
stereoisomeric stereoisomer, raumisomer
stereoisomerism Stereoisomerie *f*, stereochemische (räumliche) Isomerie *f*, Raumisomerie *f*
stereomer *s.* stereoisomer
stereometer Stereometer *n*, Volum[en]ometer *n*
sereometry Stereometrie *f*
stereoregular stereoregulär, stereoreguliert,
  sterisch regelmäßig
  s. polymer stereoreguläres Polymer[es] *n*
stereoregularity sterische Regelmäßigkeit *f*
stereoselectivity *s.* stereospecificity
stereospecific stereospezifisch
  s. catalysis stereospezifische Katalyse *f*
  s. catalyst stereospezifischer (stereospezifisch
  wirksamer) Katalysator *m*
  s. polymer stereospezifisches Polymer[es] *n*
  s. polymerization stereospezifische Polymerisation *f*
  s. reaction stereospezifische Reaktion *f*
stereospecificity Stereospezifität *f*, stereochemische Spezifität *f*, Stereoselektivität *f*
stereospectrogram Stereospektrogramm *n*
steric sterisch
  s. compression sterische Spannung *f*
  s. effect sterischer Effekt *m*
  s. factor sterischer Faktor *m*, Wahrscheinlichkeitsfaktor *m*
  s. hindrance (inhibition) sterische Hinderung

(Behinderung) *f*
sterically feasible sterisch möglich
sterile steril, keimfrei
  s. milk *s.* sterilized milk
sterility Sterilität *f*, Keimfreiheit *f*
  s. test Sterilitätsprüfung *f*, Sterilitätsprobe *f*,
  Prüfung *f* auf Sterilität
sterilization Sterilisation *f*, Sterilisieren *n*, Sterili-
  sierung *f*, Entkeimen *n*, Entkeimung *f*, Keimtö-
  tung *f*, Keimfreimachung *f*
  s. by filtration Sterilfiltration *f*
sterilize */ to* sterilisieren, entkeimen, keimfrei
  machen
sterilized cream Sterilsahne *f*, sterilisierte Sahne
  f
  s. milk Sterilmilch *f*, sterilisierte Milch *f*
sterilizer Sterilisator *m*, Sterilisationsapparat *m*,
  Entkeimungsapparat *m*
sterilizing *s.* sterilization
Stern-Gerlach experiment Stern-Gerlach-Versuch
  m, Stern-Gerlachscher Versuch *m*
sternbergite *(Min)* Sternbergit *m* *(Eisen(II,III)-sil-*
  bersulfid)
sternutator *(Pharm)* Niesmittel *n*, Sternutatorium *n*
steroid Steroid *n*
  s. alkaloid Steroidalkaloid *n*
  s. compound *s.* steroid
  s. hormone Steroidhormon *n*
sterol Sterin *n*, Sterol *n*
sterrettite *(Min)* Sterrettit *m*, Eggonit *m (Skan-*
  diumphosphat)
Stevens rearrangement Stevens-Umlagerung *f*,
  Stevenssche Umlagerung *f*
STH *s.* somatotropic hormone
stibial antimonhaltig, Antimon...
stibic Antimon..., *(i.e.S.)* Antimon(V)-...
  s. anhydride *s.* antimony pentoxide
stibiconite *(Min)* Stibikonit *m (ein oxidisches*
  Antimonerz)
stibilite *(Min)* Stibilith *m*, *(veraltet für)* Stibikonit *m*
stibine SbH_3 Stibin *n*, Antimonwasserstoff *m*,
  Antimon(III)-hydrid *n*, *(i.w.S.)* SbR_3 [organisches]
  Stibin *n*
stibious Antimon..., *(i.e.S.)* Antimon(III)-...
stiblite *(Min)* Stiblith *m*, *(veraltet für)* Stibikonit *m*
stibnite *(Min)* Stibnit *m*, Antimonit *m*, Antimonglanz
  m, Grauspießglanz *m (Antimon(III)-sulfid)*
stick */ to* 1. stecken; [an]kleben; verkleben,
  verschmieren, verpichen; beschmieren; durch-
  stechen, durchbohren; 2. haften, kleben; klem-
  men, sich verklemmen, hängenbleiben, stecken-
  bleiben
  s. together zusammenkleben, *(von stückigem*
  Gut) zusammenbacken
  s. up *(Ker)* [an]garnieren
stick Stock *m*; Stange *f*; Stift *m*
  s. bark *(Gerb)* getrocknete Akazienrinde *f*, *(meist*
  unkorrekt) Mimosenrinde *f*
  s. culture *(Bakteriologie)* Stichkultur *f*
  s. downtake *(Pap)* Stababgabe *f (beim Hänge-*
  trockner)

s. lac Stocklack *m*
s.-shaped stangenförmig; stiftförmig
s. uptake *(Pap)* Stabaufnahme *f (beim Hänge-trockner)*
sticker Haftmittel *n*, Haftstoff *m*
stickiness Klebrigkeit *f*
sticking agent *s.* sticker
sticky klebrig, klebend
stiff steif, starr
s.-plastic body *(Ker)* halbplastische Masse *f*
stiffen / to [ver]steifen, *(Gum auch)* verstrammen
stiffening Versteifen *n*, Versteifung *f*, *(Text auch)* griffgebende Appretur *f*, Griffappretur *f*, *(Gum auch)* Verstrammen *n*
stiffness Steifigkeit *f*, *(Gum auch)* Strammheit *f*
s. in bend (flexure) Biegesteifigkeit *f*
s. in torsion Verdrehungssteifigkeit *f*, Torsions-steifigkeit *f*
stilbene $C_6H_5CH=CHC_6H_5$ Stilben *n*, *trans*-1,2-Diphenyläthen *n*
s. diammine $NH_2C_6H_4CH=CHC_6H_4NH_2$ Diamino-stilben *n*
s. dye Stilbenfarbstoff *m*
stilbite *(Min)* Stilbit *m*, Desmin *m* *(ein Tektosilikat)*
still Destillationsapparat *m*, Destillierapparat *m*, Destillationsanlage *f*, Destillieranlage *f*
s. body *s. s.* pot
s. coke Erdölkoks *m*, Petrolkoks *m*
s. dome *s. s.* head
s. head Destillationskopf *m*, Destillierkopf *m*, Destillationsaufsatz *m*, Destillieraufsatz *m*, Destillierhelm *m*, Destillationsdom *m*, Helm *m*, Dom *m*
s. house Destillationshaus *n*, Destillierhaus *n*
s. pot Destillationsblase *f*, Destillierblase *f*, Blase *f*, Destillationsgefäß *n*, *(Tech auch)* Verdampfungsofen *m*
s. receiver Destilliervorlage *f*, Destillat[ions]vor-lage *f*, Destillatsammler *m*, Vorlage *f*
s. residue Destillationsrückstand *m*, Blasen-rückstand *m*
s. wine Stillwein *m*
Still oven Still-Ofen *m*, Still-Koksofen *m*, Still-Regenerativofen *m*, Verbundofen *m* (Regenerativ-Verbund[koks]ofen) *m* nach Still
stillage 1. *(Gär)* Schlempe *f*; 2. Gestell *n*, Stellage *f*
stillhead *s.* still head
stillingia oil Stillingia-Öl *n*, Stillingia-Samenöl *n*, Mong Yu *n (von Sapium sebiferum (L.) Roxb.)*
stillingic acid $CH_3(CH_2)_4(CH=CH)_2COOH$ Deka-dien-(2,4)-säure *f*
stilpnomelane *(Min)* Stilpnomelan *m* *(ein Phyllo-silikat)*
stilpnosiderite *(Min)* Stilpnosiderit *m* *(ein hydroxi-disches Eisenerz)*
stilt *(Ker)* Dreifuß *m*, Dreifüßchen *n* *(Brennhilfs-mittel)*
stimulant Stimulans *n*, Anregungsmittel *n*, Reiz-

mittel *n*
stimulatory drug, stimulus *s.* stimulant
sting out / to *(Glas)* ausflammen
sting-out loss *(Glas)* Ausflammverlust *m*
stir / to rühren
s. in einrühren
s. together zusammenrühren
stir-in resin *(durch einfaches Rühren dispergier-bares PVC-Harz zur Plastisol- und Organosol-herstellung)*
stirrable rührbar
stirred crystallizer Rührkristallisator *m*, Rühr-kristaller *m*
s. [pot] still Rührkolonne *f*
stirrer 1. Rührer *m*, Rührwerk *n*, Rührapparat *m*, Rührvorrichtung *f*; 2. Rührarm *m*
s. guide Rührerführung *f*, Führungslager *n* *(des Rührwerkes)*
s. motor Rührmotor *m*
stirring Rühren *n*
s. arm Rührarm *m*
s. bar Rühranker *m* *(eines magnetischen Rührwerkes)*
s. device *s.* stirrer 1.
s. rate Rührgeschwindigkeit *f*
s. rod Rührstab *m*
stirrup Schalengehänge *n*, Gehänge *n*, Waag-schalenaufhängung *f*
stitch tear strength Stichausreißfestigkeit *f*, Nadel-ausreißfestigkeit *f* Nadelausreißwiderstand *m*
s.-tear test Bestimmung *f* der Stichausreißfestig-keit, Nadelausreißversuch *m*, Nadelausreißprü-fung *f*
s. welding *(Plast)* Heftschweißen *n*
Stobbe condensation (reaction) Stobbe-Reaktion *f*, Stobbe-Kondensation *f* *(von Bernsteinsäure-ester)*
stock 1. Grundwerkstoff *m*, Ausgangsmaterial *n*, zu verarbeitendes Material *n*; 2. *(Gum)* Mischung *f*; 3. *(Met)* Gicht *f*; 4. *(Pap)* Papierrohstoff *m*, Rohstoff *m* für die Papiererzeugung, Faserroh-stoff *m*, Papierfaserstoff *m*; Faser[stoff]brei *m*, Stoffbrei *m*, Fasermasse *f*, Fasersuspension *f*, Stoff *m*; Faserhalbstoff *m*, Halbstoff *m*, Stoff *m*, Halbzeug *n*, [fertiger] Papierstoff *m*, Ganzstoff *m*, Ganzzeug *n*, Stoff *m*; 5. *(Farb)* Druckträger *m*, Druckgrund *m*, Bedruckstoff *m*; 6. *(Geol)* Stock *m*; 7. Lagervorrat *m*, Vorrat *m*
s. beer Lagerbier *n*
s. bin Vorratsbehälter *m*, Vorratsbunker *m*
s. blender *(Gum)* Stockblender *m*
s. circulation *(Pap)* Stoffbewegung *f*, Stoff-umtrieb *m* *(im Holländer)*
s. cleaning (cleanup) *(Pap)* Stoffreinigung *f*, Ganzstoffreinigung *f*
s. column Beschickungssäule *f*
s. deaerator *(Pap)* Dekulator *m*, Stoffentlüfter *m*
s. density *(Pap)* Stoffdichte *f*, Stoffkonzentration *f*, Stoffkonsistenz *f*

s. disintegration (Pap) Stoffmahlung f
s.-dyed (Text) flocke[n]gefärbt, in der Flocke (Wolle) gefärbt
s. dyeing (Text) Färben n in der Flocke (Wolle)
s. emulsion Stammemulsion f
s. fertilization (Landw) Vorratsdüngung f
s. flow (Pap) Stofffluß m, Stoffstrom m, Stofflauf m
s. flow to the wire (Pap) Stoffauflauf m
s. from the squirt trim (Pap) Randspritzstoff m
s. house Lagerhalle f, Lager[haus] n, Rohstofflager n
s. inlet (Pap) Stoffauflauf m
s. level Gichthöhe f, Beschickungshöhe f (eines Hochofens)
s. level indicator Gichtsonde f, Möllersonde f (eines Hochofens)
s. line (Pap) Stoffleitung f (z.B. in einer Stoffrückgewinnungsanlage); (Pap) Sammelrinne f, Stoffrinne f (Holzschliffherstellung); (Met) Beschickungsoberfläche f, Beschickungsoberkante f (eines Schachtofens)
s. line indicator s. s. level indicator
s. liquor (Text) Stammflotte f, Stammansatz m
s. loss (Pap) Faserverlust m
s. preparation (Pap) Ganzstoffaufbereitung f, Ganzzeugbereitung f, Stoffaufbereitung f
s. proportioner (Pap) Stoffmengenregler m, Stoffzuteiler m
s. pump (Pap) Stoffpumpe f
s. room Lagerraum m, Vorratsraum m, Lager n
s. screen (Pap) Ganzstoffsortierer m
s. separation (Pap)Stoffsortierung f
s. sewer (Pap) Sammelrinne f, Stoffrinne f (Holzschliffherstellung)
s. solution Vorratslösung f, (besonders bei Titrationen) Stammlösung f
s. tank (Pap) Stoffbehälter m
s. temperature Lagertemperatur f; (Plast) Schmelztemperatur f (im Extruder)
s. vat (Farb) Stammküpe f
s. velocity (Pap) Stoffgeschwindigkeit f, Ausströmungsgeschwindigkeit (Auslaufgeschwindigkeit) f des Stoffs
s. washing Zellstoffwäsche f
Stock nomenclature Stocksche Nomenklatur f
Stock notation Stocksche Bezeichnung (Wertigkeitsbezeichnung) f; Stocksche Bezeichnungsweise f
Stock system Stocksches System n
Stockholm pitch Holzpech n
  S. tar Stockholmer Teer m, schwedischer Teer m (aus Skandinavien und Finnland kommender Meilerteer aus harzreichem Nadelholz)
stockmaker (Pap) Stoffmengenregler m, Stoffzuteiler m
stockpiling Lagern n, Lagerung f, Vorratshaltung f
stocks (Pap) Stampfwerk n, „Deutsches Geschirr" n

Stoddard solvent (Am) Stoddard-Solvent n (als Reinigungs- und Lösungsmittel benutzte Erdölfraktion)
stoichiometric stöchiometrisch
s. formula stöchiometrische Formel (Grundformel) f, empirische Formel f (i.e.S.), einfachst mögliche Formel f, einfachste Formel (Summenformel, Bruttoformel) f, Verhältnisformel f, Substanzformel f
s. lattice defect stöchiometrische Störstelle (Gitterstörstelle) f
s. valency stöchiometrische Wertigkeit f
stoichiometry Stöchiometrie f
stoke Stokes n (Maßeinheit der kinematischen Viskosität)
stoker Beschickungseinrichtung f, Beschicker m (einer Feuerung)
Stokes fluorescence Stokessche Fluoreszenz f, Stokes-Fluoreszenz f
Stokes law Stokessches Gesetz n, Stokes-Gesetz n
Stokes line Stokessche Linie f (Raman-Spektroskopie)
Stokes rule Stokessche Regel f, Stokessches Fluoreszenzgesetz n
stolzite (Min) Stolzit m (Bleiwolframat)
stomach insecticide Fraßgift n für Insekten
s. poison Fraßgift n, Magengift n
stomachic Stomachikum n, appetitanregendes und verdauungsförderndes Mittel n
stone Stein m; (Glas) Stein m, Steinchen n (Fehler)
s. burnisher (Pap) Achatsteinglätteinrichtung f, Achatsteinglätte f, Steinglätte f
s. catcher Steinfänger m, Steinfang m
s. charging (Pap) Kalksteineinlauf m, Füllung f mit Kalkstein
s. flax (Min) Bergflachs m (Varietät von Hornblendeasbest)
s. green (Gemisch aus Grünerde und Bolus alba)
s. lime Branntkalk m, Ätzkalk m
s. meteorite (Geol) Steinmeteorit m, Meteorstein m, Aerolith m
s. red Roter Ocker m, Eisenmennige f
s. roll Steinwalze f
s. speed (Pap) Umlaufgeschwindigkeit f der Schleif[er]steine, Steinumfangsgeschwindigkeit f
s. surface (Pap) Steinoberfläche f
stoneware (Ker) Steinzeug n
s. clay Steinzeugton m
s. packing Steinzeugfüllkörper m
s. pipe Steinzeugrohr n
stoning machine Steinausleser m; (Lebm) Auskernmaschine f
stonite roll Stonitewalze f
stony meteorite s. stone meteorite
stop bath (Foto) Unterbrecherbad n, Unterbrechungsbad n, Stoppbad n
s.-cock s. stopcock
s. down / to (einen Spektralbereich) ausblenden

s. probability Aufenthaltswahrscheinlichkeit f
s. valve Absperrventil n
stopcock Absperrhahn m, Sperrhahn m, (i.w.S.) Hahn m
s. grease Hahnfett n
s. with tail Schwanzhahn m
stopper / to [mit einem Stopfen] verschließen, zustöpseln, verstöpseln, (Fässer) zuspunden
stopper 1. Stopfen m, Stöpsel m, Pfropfen m, (am Faß) Spund m; (Am) Hahnküken n; 2. Abstoppmittel n, Polymerisationsabstoppmittel n, Polymerisationsstopper m, Inhibitor m
stopping-off Abdecken n (von Oberflächenteilen zum Schutz vor chemischem Angriff)
s.-power Absorptionsvermögen n; Bremsvermögen n
stopple s. stopper 1.
storability Lagerfähigkeit f
storage Lagerung f, Einlagerung f, Aufbewahrung f; s. 1. s. tank; 2. store room; 3. storehouse
s. acid (Pap) Turmsäure f, Rohsäure f
s. battery Akku[mulator] m, Sammler m, Batterie f; s. s. cell 2.
s. battery acid Akkumulator[en]säure f, Akku-Säure f
s. bin Vorratsbehälter m, Vorratsbunker m, Vorratssilo m(n)
s. bottle Vorratsflasche f, (im Labor auch) Standflasche f
s. cell 1. (Pap) Vorratskasten m, Stoffkasten m; 2. (Ech) Akkumulatorzelle f
s. cellar Lagerkeller m
s. chest (Pap) Vorratsbehälter m, Vorratstank m, Vorratsbütte f
s. hopper Vorratstrichter m, Fülltrichter m
s. jar Vorratsgefäß n
s. life Lagerfähigkeit f, Lagerbeständigkeit f
s. period Lager[ungs]zeit f
s. quality (Lebm) Haltbarkeit f
s. regulation Lagerungsvorschrift f
s. resistance (stability) Lagerfähigkeit f, Lagerbeständigkeit f
s. tank Lagertank m, Lagerbehälter m, Lagergefäß n, Vorratstank m, Vorratsbehälter m, Speicherbehälter m, (Pap auch) Vorratsbütte f
s. temperature Lager[ungs]temperatur f, Aufbewahrungstemperatur f
s. tissue (Bio) Speichergewebe n
s. vessel s. s. tank
storax Styrax m (Balsamharz von Liquidambar orientalis Miller und L. styraciflua L.)
s. oil (Kosmet) Styraxöl n, Storaxöl n
store / to lagern, aufbewahren
store Bestand m, Vorrat m, Reserve f; s. 1. s. room; 2. storehouse
s. chest s. storage chest
s. room Lagerraum m, Vorratsraum m, Lager n
s. tank s. storage tank
storehouse Lagerhalle f, Lager[haus] n, Speicher m, Magazin n, Depot n

storing Lagerung f, Einlagerung f, Aufbewahrung f, Vorratshaltung f
stout Stout m (Biertyp)
stove / to erwärmen, erhitzen, heiß machen; warmhalten; (durch Hitze) trocknen, (Farb auch) einbrennen; (Gießharz) härten; (Text) mit Schwefeldioxid bleichen, schwefeln
stove Ofen m; (Tech) Brennofen m; (Farb) Einbrennofen m (für Lacke); Trockenkammer f, Trockenraum m, Darre f
s. enamelling Ofenlackierung f, Einbrennlackierung f
s. for heating air blast Winderhitzer m, Hochofenwinderhitzer m
s. tile Ofenkachel f
stoved salt Siedesalz n
stoving Erwärmen n, Erhitzen n; Warmhalten n; thermisches Trocknen n, Ofentrocknen n; (Farb) Aufbrennen n, Festbrennen n, Einbrennen n, Einbrand m; (Plast) Härten n (von Gießharzen); (Text) Bleichen n mit Schwefeldioxid, Schwefeln n
s. enamel ofentrocknender Emaillack m, Einbrennemaillelack m, Einbrennemaille f
s. lacquer ofentrocknender Lack m, Einbrennlack m, Ofenlack m
s. paint ofentrocknende Anstrichfarbe f, Einbrennfarbe f
s. process (Farb) Einbrennverfahren n
s. section (Farb) Trockenofen m (beim Flow-Coating-Verfahren)
s. synthetic s. s. lacquer
s. temperature (Farb) Einbrenntemperatur f
s. time Härtezeit f (bei Gießharzen); Einbrennzeit f (bei Lacken)
s. varnish s. s. lacquer
St.P. s. strain point
S.T.P. s. standard temperature and pressure
strahlite (Min) Strahl[en]stein m, Aktinolith m (ein Inosilikat)
straight-arm paddle agitator Balkenrührer m
s.-blade fan Lüfter (Ventilator) m mit geraden Schaufeln
s. carbon steel reiner (unlegierter) Kohlenstoffstahl m
s.-chain geradkettig
s. chain geradlinige (gerade, unverzweigte) Kette f
s.-chain formula geradkettige Formel f
s.-chain hydrocarbon Kohlenwasserstoff m mit gerader Kette, geradkettiger Kohlenwasserstoff m
s.-chain structure geradkettige Struktur f (eines Kohlenwasserstoffs)
s. distillation direkte (einfache) Destillation f, Straightrun-Destillation f
s. fertilization Einzeldüngung f
s. fertilizer Einzelnährstoffdüngemittel n
s. lime liquor (Gerb) reiner Kalkäscher m, Weißkalkäscher m

s. line *s.* s.-line portion

s.-line portion *(Foto)* geradliniger Teil *m (der Schwärzungskurve)*, geradliniges Stück (Gebiet) *n*, Gebiet *n* der richtigen (normalen) Exposition, Gebiet *n* der Normalexposition (Normalbelichtung)

s. liquor *(Gerb)* unmaskierte Brühe *f*

s.-lobe compressor Roots-Gebläse *n*, Wälzkolbenverdichter *m*

s.-pipe thread gerades Rohrgewinde *n*

s. portion *s.* s.-line portion

s.-run benzine *s.* s.-run gasoline

s.-run gasoline Straightrun-Benzin *n*, Destillat[ions]benzin *n*, direkt herausdestilliertes Benzin *n*, SR-Benzin *n*

s.-run naphtha (spirit) *s.* s.-run gasoline

s. soap *(Seife mit weniger als 4 % Anteilen von NaCl, freiem Alkali und alkoholunlöslichen Substanzen)*

s. stopcock Hahn *m* mit Bohrung senkrecht zur Achse

s. throat *(Glas)* normaler (bodengleicher) Durchlaß *m*

s.-through press *(Pap)* Liegepresse *f*

s. vacuum forming *(Plast)* Vakuumsaugverfahren *n*, Vakuumtiefziehen *n*

strain */ to* 1. [durch]seihen, *(Suspensionen)* [durch]sieben, *(ohne Druckanwendung)* filtrieren, filtern, *(Gum auch)* strainern; durchsickern, durchlaufen *(vom Filtrat)*; 2. spannen, [straff] anziehen; verformen, deformieren, *(i.e.S.)* dehnen; überdehnen; verbiegen

strain 1. Spannung *f*, Beanspruchung *f*; Form[ver]änderung *f*, Gestalts[ver]änderung *f*, Verformung *f*, Deformation *f*, *(i.e.S.)* Längenänderung *f*; *(Glas)* Verspannung *f*; 2. *(Bioch)* Stamm *m*, Rasse *f*, Kulturstamm *m (von Mikroorganismen)*

s. at break Bruchdehnung *f*, Zerreißdehnung *f*, Reißdehnung *f*

s. disk *(Glas)* Spannungsscheibe *f*

s.-free spannungsfrei

s.-point [temperature] *(Ker)* unterer Kühlpunkt *m*, untere Kühltemperatur *f*

s. test *(Gum)* Straintest *m (Bestimmung der Dehnung unter konstanter Last)*

s. tester *(Gum)* Straintester *m*

s. theory Spannungstheorie *f*

strained */ to be* Spannung besitzen

strainer Seiher *m*, [feines] Sieb *n*, Filter *n*; *(Plast, Gum)* Siebpresse *f*, Siebkopf-Spritzmaschine *f*, Strainer *m*; Lochplatte *f*, Siebeinsatz *m*, Siebeinlage *f*; *(Gär)* s. s. bottom

s. bottom *(Gär)* Läuter[bottich]boden *m*, Siebboden *m*

straining 1. Durchseihen *n*, Sieben *n*, *(i.w.S.)* Filtrieren *n*, *(Gum auch)* Strainern *n*; 2. *s.* strain 1.

s. cloth Siebgewebe *n*, Siebtuch *n*, Seihtuch *n*

s. head *(Gum)* Strainerkopf *m*, Siebkopf *m*

s. machine *(Plast, Gum)* Siebpresse *f*, Siebkopf-Spritzmaschine *f*, Strainer *m*

strainless spannungsfrei

stralite *s.* strahlite

stramonium herb Stechapfelkraut *n (von Datura stramonium L.)*

strand Strang *m*; *(Text)* Strahn *m*, Faserbündel *n*, Spinnfaden *m*, Elementarfadenbündel *n*; Band *n*, Trum *n(m) (eines Bandförderers)*

s. of chain *(Fördertechnik)* Kettenstrang *m*

s. of margarine Margarinestrang *m*

strap pulley *(Pap)* Deckelriemenrolle *f*, Deckelriemenführungsrad *n*, Deckelleitrad *n*

strass Straß *m (Glassorte mit hohem Bleigehalt)*

stratify */ to* schichten, in Schichten auseinanderziehen, auffächern *(z.B. das Aufgabegut bei der Herdaufbereitung)*; Schichten bilden

stratigraphic trap *(Erdöl)* stratigrafische Falle *f*

stratum Schicht *f*; *(Geol)* Schicht[platte] *f*, Formation *f*, Bank *f*, Lage *f*, Flöz *n*

straw *(Pap)* Stroh *n*, Gelbstroh *n*

s. cardboard Strohpappe *f*; Strohkarton *m*

s. cellulose [vollaufgeschlossener] Stroh[zell]-stoff *m*

s. chopper (cutter) *(Pap)* Strohhäckselmaschine *f*

s.-like *(Text)* strohähnlich *(Chemiefaserstoffe)*

s. oil *(hochsiedende Erdölfraktion zur Gasreinigung)*

s. paper Strohpapier *n*

s. pulp [gelber] Strohstoff *m*, Gelbstrohstoff *m*; [vollaufgeschlossener] Stroh[zell]stoff *m*

s. wrapping paper Strohpackpapier *n*

s. yarn querschnittsmodifizierter Chemiefaserstoff *m*

strawboard Strohpappe *f*

stray electron vagabundierendes Elektron *n*, Fremdelektron *n*

s. light Streulicht *n*

s. neutron vagabundierendes Neutron *n*

streak Streifen *m*; Schliere *f*; Lage *f*, dünne Schicht *f*; *s.* s. colour

s. colour *(Min)* Strichfarbe *f*, Strich *m*

s. culture *(Mikrobiologie)* Strichkultur *f*

streaking *(Glas)* Schlierenbildung *f (Fehler)*

stream Strom *m (z.B. von Aufgabegut)*; Strömung *f*

s. barker *(Pap)* hydraulischer Entrinder *m*, Wasserstrahlentrinder *m*, Streambarker *m*

s. gold Waschgold *n*, Seifengold *n*

s. meter Durchfluß[mengen]messer *m*, Mengenstrommesser *m*

s. method Strömungsverfahren *n (bei der Untertagevergasung von Kohle)*

s. of gas Gasstrom *m*

s. of oxygen Sauerstoffstrom *m*

s. of upward moving gas steigender Gasstrom *m*

s. of water Wasserstrom *m*

s. pollution Flußverunreinigung *f*

s. tin Seifenzinn *n*

streaming birefringence (double refraction) Strömungsdoppelbrechung *f*
s. potential Strömungspotential *n*
streamline filter Stromlinienfilter *n*
streamlined flow laminare Strömung (Bewegung) *f*, Laminarströmung *f*, Schichtenströmung *f*, Laminarbewegung *f*
Strecker reaction Strecker-Synthese *f*, Streckersche Synthese *f* *(von Aminosäuren)*
strengite *(Min)* Strengit *m* *(Eisen(III)-orthophosphat)*
strength 1. Stärke *f*, Härte *f*, Festigkeit *f*, Dauerhaftigkeit *f*, Haltbarkeit *f*; 2. Stärke *f*, Wirkungskraft *f*, Wirkungsgrad *m*; *(Pharm)* Wirkungsstärke *f*; 3. Stärke *f*, Intenstät *f*, Echtheit *f* *(z.B. einer Farbe)*; 4. Dichte *f*, Konzentration *f* *(einer Lösung)*; 5. Stärke *f*, Dicke *f* *(von Flachglas)*
s. in compression Druckfestigkeit *f*
s. in tension Zugfestigkeit *f*
s. of coke Koksfestigkeit *f*
s. of concrete Betonfestigkeit *f*
s. of current *(ET)* Stromstärke *f*
s. of field *(ET)* Feldstärke *f*
s. of the clot (curd) *(Lebm)* Bruchfestigkeit *f*
s.-testing machine Festigkeitsprüfer *m*
s.-to-weight ratio Festigkeits-Masse-Verhältnis *n*, Reißlänge *f*
strengthen / to *(Lösungen)* verstärken; *(Gerb)* *(den Äscher)* anschärfen; *(Glas)* härten
strengthening action verstärkende Wirkung *f*, Verstärkerwirkung *f*, Verstärkungseffekt *m*
streptidine Streptidin *n*, 1,3-Diguanidino-2,4,5,6-tetrahydroxyzyklohexan *n*
streptomycin Streptomyzin *n* *(Antibiotikum)*
streptose $C_6H_{10}O_5$ Streptose *f* *(ein Monosaccharid)*
stress / to beanspruchen, belasten; spannen
stress Beanspruchung *f*, Belastung *f*; *(mechanische)* Spannung *f*
s. corrosion Spannungskorrosion *f*
s. cracking Spannungsrißbildung *f*
s.-optical spannungsoptisch
s. relationship Spannungsverhältnis *n*
s. relaxation Spannungsrelaxation *f*
s. relaxation method (test) Spannungsrelaxationsverfahren *n*
s. relief Entspannen *n*, Entspannung *f*; *(Met)* s. s. relief anneal[ing]
s. relief anneal[ing] Spannungsfreiglühen *n*, Spannungsfreiglühung *f*, Entspannungsglühen *n*, Entspannungsglühung *f*, entspannendes Glühen *n*
s. relieving s. s. relief
s.-relieving anneal s. s.-relief anneal[ing]
s.-relieving furnace Entspannungsofen *m*
s.-strain curve Formänderungs-Spannungslinie *f*, *(i.e.S.)* Spannungs-Dehnungsdiagramm *n*, Spannungs-Dehnungslinie *f*, Zug-Dehnungsdiagramm *n*, Zug-Dehnungskurve *f*, Kraft-

Dehnungskurve *f*, Dehnungs-Spannungskurve *f*, Dehnungs-Spannungslinie *f*, Kraft-Verlängerungsschaubild *n*
s.-strain properties Zugdehnungseigenschaften *fpl*, Zugverformungseigenschaften *fpl*, Zug-Dehnungs-Verhalten *n*
stretch / to längen, dehnen, strecken, recken, ausziehen, spannen; [aus]weiten
stretch [elastische] Dehnung *f*, Strecken *n*, Streckung *f*, Verstrecken *n*, Recken *n*, Reckung *f*, *(Text auch)* Stretch *m* *(von Elementarfäden)*; *(Gerb)* Zug *m*, Zügigkeit *f*
s. factor *(Text)* Stretchfaktor *m*
s. forming *(Plast)* Streckformen *n*
s. ratio *(Text)* Gesamtlängenverhältnis *n* *(Verhältnis der Feinheit des ungereckten zu der des gereckten Fadens)*
s. resistance *(Text)* Reckfestigkeit *f*, Verstreckwiderstand *m*
s. roll *(Pap)* Spannwalze *f*
s. spinning *(Text)* Streckspinnen *n*
s. yarn Stretchgarn *n*
stretching Strecken *n*, Streckung *f*, Verstrecken *n*, Recken *n*, Reckung *f*, *(Text auch)* Stretch *m* *(von Elementarfäden)*
s. frequency Valenzfrequenz *f*
s. roll *(Pap)* Spannwalze *f*
s. vibration Valenzschwingung *f*
stria *(Glas, Plast)* Schliere *f* *(Fehler)*
striation *(Glas, Plast)* Schlierenbildung *f*; *(Geol)* Schichtung *f*, Streifung *f*, Streifenbildung *f*
strike *(Text)* Anfangsgeschwindigkeit *f* des Färbens
s.-anywhere match Überallzünder *m* *(Zündholzsorte)*
s. back / to zurückschlagen *(von Flammen)*
s. bath Vorgalvanisierbad *n*
s. out / to *(Gerb)* ausstoßen, ausrecken, plattieren; flach aufspannen
s. pan Vakuumapparat *m* mit Heizschlangen
s. solution s. s. bath
s. solution for silver plating Vorversilberungsbad *n*
s. through / to durchschlagen *(von Farben)*
s.-through Durchschlagen *n* *(von Farben)*
striking-back Zurückschlagen *n* *(der Flamme)*
string *(Glas)* Faden *m* *(fadenförmige Schliere)*
s. catcher *(Pap)* Draht- und Schnurfang *m*
s. discharge Schnürenabnahme *f*, Schnurabnahme *f* *(des Filterkuchens)*
s. galvanometer Saitengalvanometer *n*, Fadengalvanometer *n*
s.-proof test *(Zucker)* Fadenprobe *f*
stringing *(Farb)* Netzbildung *f* *(unerwünschtes Ausschwimmen von Pigmenten)*
strip / to 1. austreiben, desorbieren; *(Destillation)* abstreifen, abtreiben, strippen, ausdämpfen; 2. *(Farb)* abziehen, ablösen, entfärben; 3. *(Leder)* entfärben *(den Gerbstoff entfernen)*
s. off (out) s. to strip 1.
strip action column s. stripper column

s. chart Band *n*, Diagrammband *n*, Diagramm-
streifen *m (des Bandschreibers)*
s.- chart recorder Bandschreiber *m*, Streifen-
schreiber *m*
s. heater Heizband *n*, Bandheizkörper *m*
s. heating Bandheizung *f*
s. of filter paper Filterpapierstreifen *m*
stripiness Streifigkeit *f*
strippable coating abstreifbarer Überzug *m*
stripped atom hochionisiertes Atom *n*
stripper *(Destillation)* Abstreifer *m*, Abtreiber *m*,
Dampfabstreifer *m*, Ausdämpfer *m*, Stripper *m*;
s. s. column
　s. column *(Destillation)* Abstreiferkolonne *f*,
Abtreibkolonne *f*, Dampfabstreiferkolonne *f*,
Ausdämpf[er]kolonne *f*, Abtriebssäule *f*
s. plate Abstreifplatte *f*
stripping Austreiben *n*, Desorption *f (Entfernen
sorbierter Gase von Sorptionsmitteln)*; *(Destillation)* Abstreifen *n*, Ausdämpfen *n*, Strippen *n*,
Strippung *f*; Entmetallisieren *n (Ablösen von
Metallüberzügen vom Grundmetall)*; *(Farb)* Abziehen *n*, Ablösen *n*, Entfärben *n*; *(Ker)* Abheben
n (der Form); *(Toxikologie)* oberflächliches Extrahieren *n (z. B. zum Herunterlösen von Pflanzenschutzmittelrückständen)*
　s. agent (assistant) *(Farb)* Abzieh[hilfs]mittel *n*,
Entfärbungs[hilfs]mittel *n*, Entfärber *m*
s. column *s.* stripper column
s. of a mould *(Plast)* Abspannen *n* eines
Werkzeugs
s. oil Abstreiferöl *n*, Waschöl *n*
s. operating line Abtriebsgerade *f*
s. paper *(Foto)* Abziehpapier *n*, abziehbares
Negativpapier *n*
s. section Abstreiferteil *m*, Abtreibeteil *m*,
Abtrieb[s]teil *m*, Abstreiferzone *f*, Abstrippzone *f*,
Ausdämpfsektion *f*, Ausdämpfungsteil *m (einer
Fraktionierkolonne)*
s. steam Strippdampf *m*; *(Erdöl)* Spüldampf *m*
(beim Thermofor-Catalytic-Cracking)
s. still *s.* stripper column
stroke Hub *m*; *s. s.* length
s. length Hublänge *f*, Hub *m*
stromeyerite *(Min)* Stromeyerit *m (Kupfer(I)-silber-
sulfid)*
strong acid starke Säure *f (nach dem Dissoziationsgrad)*; konzentrierte Säure *f*; *(Pap)*
Starksäure *f*, hochkonzentrierte (hochprozentige) Kochsäure *f*
　s.-acid tower *s. s.* tower
s. aqua reiche Lösung *f (Kältetechnik)*
s. electrolyte starker Elektrolyt *m*
s. gas Starkgas *n*
s. glazing *(Pap)* scharfe Satinage *f*
s. liquor konzentrierte Flüssigkeit *f*
s. paper Kraftpapier *n*
s. pulp *(Pap)* harter (hartgekochter, wenig
aufgeschlossener) Zellstoff *m*
s.-smelling stark duftend; von durchdringendem
(intensivem) Geruch

s. soap solution *(Schwerbenzin mit hohem
Seifengehalt für Zweibadverfahren der Chemischreinigung)*
s. tower *(Pap)* Vorturm *m (eines Zweiturmsystems)*
strongly acid stark sauer
　s. basic stark basisch (alkalisch)
s. caking coal starkbackende Kohle *f*
s. sized *(Pap)* starkgeleimt, vollgeleimt, mit
starker Leimung
strontia SrO Strontian *n*, Strontianerde *f*, *(veraltet
für)* Strontiumoxid *n*
s. process Strontianverfahren *n (zur Melasseentzuckerung)*
strontianite *(Min)* Strontianit *m (Strontiumkarbonat)*
strontic Strontium...
strontium Sr Strontium *n*
s. acetate $Sr(CH_3COO)_2$ Strontiumazetat *n*
s. acid orthophosphate $SrHPO_4$ Strontium-
hydrogen[ortho]phosphat *n*
s. acid sulphate *s. s.* hydrogen sulphate
s. bisulphate *s. s.* hydrogen sulphate
s. bromide $SrBr_2$ Strontiumbromid *n*
s. carbide SrC_2 Strontiumkarbid *n*
s. carbonate $SrCO_3$ Strontiumkarbonat *n*
s. chlorate $Sr(ClO_3)_2$ Strontiumchlorat *n*
s. chloride $SrCl_2$ Strontiumchlorid *n*
s. chromate $SrCrO_4$ Strontiumchromat *n*
s. ferrocyanide $Sr_2[Fe(CN)_6]$ Strontiumhexa-
zyanoferrat(II) *n*
s. fluoride SrF_2 Strontiumfluorid *n*
s. fluosilicate $Sr[SiF_6]$ Strontiumhexafluorosili-
kat *n*
s. hydrate *s. s.* hydroxide
s. hydrogen sulphate $Sr(HSO_4)_2$ Strontium-
hydrogensulfat *n*
s. hydrosulphide $Sr(SH)_2$ Strontiumhydrogensul-
fid *n*
s. hydroxide $Sr(OH)_2$ Strontiumhydroxid *n*
s. iodide SrJ_2 Strontiumjodid *n*
s. nitrate $Sr(NO_3)_2$ Strontiumnitrat *n*
s. nitrite $Sr(NO_2)_2$ Strontiumnitrit *n*
s. orthoarsenite $Sr_3(AsO_3)_2$ Strontium[ortho]ar-
senat(III) *n*
s. oxide SrO Strontiumoxid *n*
s. peroxide SrO_2 Strontiumperoxid *n*
s. saccharate (sucrate) Strontiumsa[c]charat *n*
s. sulphate $SrSO_4$ Strontiumsulfat *n*
strophanthin Strophanthin *n*
strophanthin-G Strophanthin G *n*, g-Strophan-
thin *n*, Ouabain *n*
structural strukturell, Struktur..., Bau..., Gefüge...;
konstruktiv, Konstruktions...
s. board Bauplatte *f*
s. ceramics Baukeramik *f*
s. chemistry Strukturchemie *f*
s. constituent Gefügebestandteil *m*
s. effect Struktureffekt *m*
s. element (entity) Strukturelement *n*, Bau-

element n, Aufbauelement n, Struktureinheit f, Baueinheit f, Baustein m, Baubestandteil m, Aufbaubestandteil m; (Plast) Grundmolekül n, Staudinger-Einheit f

s. formula Strukturformel f, Konstitutionsformel f

s. glass Bauglas n

s. isomer Strukturisomer[es] n

s. isomerism Strukturisomerie f

s. parachor Bindungsparachor m

s. study Strukturuntersuchung f

s. trap (Erdöl) strukturelle (strukturgebundene, tektonische) Falle f

s. unit s. s. element

s. viscosity Strukturviskosität f, Binghamsches Fließen n

structurally isomeric strukturisomer

s. viscose strukturviskos

structure Struktur f, Gefüge n, Aufbau m, Bau m, Konstitution f

with s. strukturzeigend, strukturiert, mit Struktur (Gefüge)

s. amplitude (Krist) Strukturamplitude f

s. analysis Strukturanalyse f

s.-chemical strukturchemisch

s. chemistry Strukturchemie f

s. determination Strukturbestimmung f

s. elucidation Strukturaufklärung f, Konstitutionsaufklärung f, Konstitutionsermittlung f, Konstitutionserforschung f

s. factor (Krist) Strukturfaktor m

s. index (Ölzahl einer Kohleschwarz-Pigmentfarbe in Prozent des Normalwertes als Ausdruck der Mahlfeinheit)

s. type Strukturtyp m, Kristallgittertyp m

structureless strukturlos, gefügelos, unstrukturiert

struv[e]ite (Min) Struvit m (Ammoniummagnesiumorthophosphat)

strychnine Strychnin n (Alkaloid)

s. nitrate Strychninnitrat n

strychnos alkaloid Strychnosalkaloid n

s. seed (Pharm) Brechnußamen mpl, Krähenaugen npl (von Strychnos nux-vomica L.)

S.T.T. s. short-time test

Stuart model Stuart-Modell n, Stuart-Kalotte f (zur Darstellung von Molekülstrukturen)

stucco Stuckgips m

stuff / to (Gerb) fetten

stuff (Pap) [fertiger] Papierstoff m, Ganzstoff m, Ganzzeug n, Stoff m; [mechanischer] Holzschliff m, [mechanischer] Holzstoff m, Holzmasse f, Schleifmasse f

s. box (Pap) Stoffauflauf[kasten] m, Auflaufkasten m; s. s. chest

s. catcher (Pap) Stoffrückgewinnungsanlage f, Faserrückgewinnungsanlage f, Fangstoffanlage f, Stoffang m, Stoffänger m

s. chest (Pap) Maschinenbütte f, Arbeitsbütte f, Stoffbütte f

s. engine (Pap) Holländer m, Mahlholländer m, Messerholländer m, Ganzzeugholländer m

s. pump (Pap) Stoffpumpe f

stuffer box crimping (Text) Stauchkammertexturieren n

stuffing (Pap) Kalanderfärbung f, Oberflächenfärbung f im Kalander; (Gerb) Fetten n, Fettung f, Schmieren n (oder) Abölen n; Lederschmiere f

s. box Stopfbuchse f, Stopfbuchspackung f; (Text) s. s. tube

s. box crimping method (Text) Stauchkammertexturierverfahren n

s. mixture (Gerb) Tafelschmiere f, Fettschmiere f, Aasschmiere f (Gemisch aus Lebertran und Talg)

s. tube (Text) Stauchkammer f

stum / to die Gärung verhindern; (Am) wieder (erneut) gären lassen

stum Most m

stupp Stupp f (bei der Quecksilberraffination anfallende Hg-reiche schlammige Masse)

Sturtevant mill Sturtevant-Ring[rollen]mühle f

Stutzer reagent (Suspension von Kupfer(II)-hydroxid zum Proteinnachweis)

styphnic acid $C_6H(OH)_2(NO_2)_3$ Styphninsäure f, 2,4,6-Trinitroresorzin n, 1,3-Dihydroxy-2,4,6-trinitrobenzol n

styptic styptisch, adstringierend, zusammenziehend, (i.e.S.) hämostyptisch, blutstillend; antidiarrhoisch, stopfend

styptic Styptikum n, Adstringens n, adstringierendes (zusammenziehendes) Mittel n, adstringierende Substanz f, (i.e.S.) blutstillendes Mittel n, Blutstillungsmittel n, Hämostyptikum n; Antidiarrhoikum n, stopfendes Mittel n, Mittel n gegen Durchfall

s. pencil Blutstillstift m, Rasierstein m

stypticity Adstringenz f, adstringierende Wirkung f

styrax Styrax m (Balsamharz von Liquidambar orientalis Miller und L. styraciflua L.)

s. oil (Kosmet) Styraxöl n, Storaxöl n

styrenate / to styrolisieren

styrenated alkyd styrolisiertes Alkydharz n, Styrolalkydharz n

styrene $C_6H_5CH=CH_2$ Styrol n, Vinylbenzol n, Phenyläthen n, Phenyläthylen n

s.-butadiene rubber Styrol-Butadien-Kautschuk m, Butadien-Styrol-Kautschuk m, Styrolkautschuk m, SBK (Butadien-Styrol-Mischpolymerisat)

s.-chloroprene rubber Styrol-Chloropren-Kautschuk m (Chlorbutadien-Styrol-Mischpolymerisat)

s. dibromide $C_6H_5CHBrCH_2Br$ Styroldibromid n, 1,2-Dibrom-1-phenyläthan n

s.-isoprene rubber Styrol-Isopren-Kautschuk m (Isopren-Styrol-Mischpolymerisat)

styrol[ene] s. styrene

styrone $C_6H_5CH=CHCHO$ Zimtaldehyd m, Zinnamal n, 3-Phenylpropenal n

styryl alcohol (carbinol) $C_6H_5CH=CHCH_2OH$ Zimt-

alkohol *m*, Zinnamylalkohol *m*, Styron *n*, Styryl-3-phenylpropen-2-ol *n*
styrylic alcohol *s.* styryl alcohol
S.U. *s. sunshine unit*
Suakin gum Suakingummi *n* (*Arabisches Gummi von Acacia stenocarpa Hochst.*)
Suari fat (*Fett von Caryocar amygdaliferum Mutis*)
sub *s.* substitute 2.
subacetate basisches Azetat *n*, (*unkorrekt für*) Oxidazetat *n* (*oder*) Hydroxidazetat *n*
subacid schwach (etwas) sauer, säuerlich; (*Med*) subazid
subacidity (*Med*) Subazidität *f*, Hyp[o]azidität *f*
sub-aeration floatation cell Unterluftzelle *f*
subalkaline schwach (etwas) alkalisch
subalpine rendzina Tangelrendzina *f* (*Bodentyp*)
subaquatic, subaqueous subaquatisch, Unterwasser...
   s. corrosion Unterwasserkorrosion *f*
subatomic subatomar
subbituminous subbituminös
   s. coal subbituminöse Kohle *f*, Glanz[braun]kohle *f*, bituminöse (steinkohleähnliche) Braunkohle *f*
subcarbonate basisches Karbonat *n*, (*unkorrekt für*) Oxidkarbonat *n* (*oder*) Hydroxidkarbonat *n*
subcool / to unterkühlen
subcooler Unterkühler *m*
subcooling Unterkühlen *n*, Unterkühlung *f*
subcritical (*Kch*) unterkritisch
   s. reactor (*Kch*) unterkritischer Reaktor *m*
subculture / to (*Mikroorganismen*) abimpfen
subculture Subkultur *f*, Nachkultur *f*, Abimpfung *f*; Abimpfen *n*
subcutaneous blowhole Randblase *f*, äußere Blase (Gasblase) *f* (*im Guß*)
subcuticular residue subkutikulärer Rückstand *m* (*von Pflanzenschutzmitteln*)
subcutis (*Gerb*) Subkutis *f*, Unterhaut *f*
subdivide / to in Untergruppen gliedern (unterteilen)
suberane C_7H_{14} Suberan *n*, Heptamethylen *n*, Zykloheptan *n*
subereous korken, aus Kork, Kork...; korkartig, korkähnlich
suberic acid $HOOC(CH_2)_6COOH$ Suberinsäure *f*, Korksäure *f*, Oktandisäure *f*, Hexandikarbonsäure-(1,6) *f*
suberification *s.* suberization
suberin Suberin *n* (*hochmolekulare Korksubstanz*)
suberinic acid *s.* suberic acid
suberization (*Bot*) Verkorkung *f*
suberol *s.* suberyl alcohol
suberone $C_7H_{12}O$ Suberon *n*, Zykloheptanon *n*
suberose, suberous *s.* subereous
suberyl alcohol Suberylalkohol *m*, Suberol *n*, Zykloheptanol *n*
subgroup Untergruppe *f*; Nebengruppe *f* (*des Periodensystems*)
   s. number Untergruppennummer *f*, Unter-

gruppenziffer *f*, Kodenummer *f* der Untergruppe, Dritte Kodeziffer *f* (*des internationalen Kohlenklassifikationssystems*)
subhedral (*Min*) hypidiomorph
subhydrous coal wasserstoffarme (subhydrierte) Kohle *f*, Kohle *f* mit einem geringeren Wasserstoffgehalt
sub-image (*Foto*) Subbild *n*
sub-image speck (*Foto*) Subkeim *m*
subject contrast (*Foto*) Objektkontrast *m*
   s. index Sach[wort]register *n*, Sachverzeichnis *n*
subl (*Abk. für*) sublimes
sublethal dose subletale Dosis *f*
sublimability Sublimierbarkeit *f*
sublimable sublimierbar
sublimand Sublimand *m*, Sublimiergut *n*, zu sublimierender Stoff *m*
sublimate / to sublimieren
sublimate Sublimat *n* (*Endprodukt der Sublimation*); Sublimat *n*, Quecksilber(II)-chlorid *n*
   s. test (*Med*) Sublimatprobe *f*
sublimation Sublimation *f*, Sublimieren *n*
   s. curve Sublimations[druck]kurve *f*
   s. heat Sublimationswärme *f*
   s. in steam Dampfsublimation *f*
   s. point Sublimationspunkt *m*, Sbp., Sublimationstemperatur *f*
   s. pressure Sublimationsdruck *m*
   s. process Sublimationsprozeß *m*
   s. rate Sublimationsgeschwindigkeit *f*
   s. temperature *s. s.* point
sublime / to *s.* to sublimate
sublimed sulphur sublimierter Schwefel *m*, Schwefelblüte *f*, Schwefelblume *f*
sublimer Sublimierblase *f*
submarine gate (*Plast*) Tunnelanguß *m*
   s. throat (*Glas*) versenkter (tiefliegender, tiefer) Durchlaß *m*
submerge / to [ein]tauchen, untertauchen
submerged submers, untergetaucht
   s. burner Tauchbrenner *m*
   s.-coil condenser Tauchrohrverflüssiger *m*, Tauchrohrkondensator *m*
   s. combustion Tauchverbrennung *f*
   s. combustion burner *s. s.* burner
   s. combustion-liquid heater Tauchbrenner *m* zur Beheizung von Flüssigkeiten
   s. culture Submerskultur *f*, submerse Kultur *f*, Tiefkultur *f*
   s. culture method Tieftankmethode *f*, Tieftankverfahren *n*, Submersverfahren *n*, Submerged-culture-Verfahren *n*
   s. pump Tauchpumpe *f*; Unterwasserpumpe *f*
   s. throat *s.* submarine throat
submergence Eintauchtiefe *f*, Tauchtiefe *f*, (*Destillation auch*) Dampfdurchdringtiefe *f*
   s. line Eintauchrohr *n*
   s. ratio Eintauchverhältnis *n*
submersed *s.* submerged
submersion process *s.* submerged culture method

submetallic lustre *(Min)* Halbmetallglanz *m*
submicrogram[me] analysis Submikrogramm-
analyse *f*
submicron Submikron *n*
submicroscopic submikroskopisch
s. morphology submikroskopische Morphologie
f
submicrostructure Submikrostruktur *f*, Submikro-
gefüge *n*
submolecule Submolekül *n*
subnitrate basisches Nitrat *n*, *(unkorrekt für)*
Oxidnitrat *n (oder)* Hydroxidnitrat *n*
subnuclear particles Elementarteilchen *npl (außer
Photonen)*
suboxide Suboxid *n*
subproportional reducer s. subtractive reducer
subsalt basisches Salz *n*, *(unkorrekt für)* Oxidsalz
n (oder) Hydroxidsalz *n*
subsaturated untersättigt
subsaturation Untersättigung *f*
subscript unterer (tiefgesetzter) Index *m*
s. numeral unten angehängter Zahlenindex *m*
subshell *(Kch)* Unterschale *f*, Teilschale *f*
subside / to 1. [ab]sinken, sich setzen (absetzen,
abscheiden, ausscheiden, ablagern, niederschlagen), absitzen, ausfallen, zur Abscheidung
gelangen, sedimentieren; 2. abklingen, nachlassen
subsidence 1. Sinken *n*, Setzen *n*, Setzung *f*; 2.
Bodensatz *m*, Satz *m*, Niederschlag *m*, Sediment
n; 3. Abklingen *n*, Nachlassen *n*
subsidiary quantum number Nebenquantenzahl *f*,
Drehimpulsquantenzahl *f*
s. standard Hilfsstandard *m*
subsilicate basisches Silikat *n*, *(unkorrekt für)*
Oxidsilikat *n (oder)* Hydroxidsilikat *n*
s. slag Subsilikatschlacke *f*
subsilicic rock basisches Gestein (Erstarrungsgestein) *n*, Basit *m*
subsoil *(Landw)* Unterboden *m*, Untergrund *m*
s. water Grundwasser *n*
substage Beleuchtungsapparat *m* unter dem
Mikroskoptisch
s. condenser Kondensor *m* unter dem Mikroskoptisch
substance 1. Substanz *f*, Stoff *m*, Materie *f*, Masse
f; 2. *(Pap)* Masse *f* je Flächeneinheit, *(bisher)*
Flächengewicht *n*, Quadratmetergewicht *n*, Basisgewicht *n*; 3. Dicke *f (beim Flachglas)*
s. number s. substance 2.
s. to be dried Trockengut *n*, Trocknungsgut *n*,
Trockneraufgabegut *n*, zu trocknendes Gut *n*
s. to be extracted Extraktionsgut *n*, Lösegut *n*,
Abgeber *m*
s. to be filtered Filtergut *n*, Trübe *f*
s. to be handled [zu behandelndes] Gut *n*
s. to be mixed Mischgut *n*, Knetgut *n*; Rührgut
n
s. under investigation Versuchssubstanz *f*,
Untersuchungssubstanz *f*, Probesubstanz *f*

s. weight s. substance 2.
substantially free praktisch frei
s. pure sehr rein
s. soluble größtenteils löslich
substantive *(Farb)* direktziehend, sub iv
s. dye Direktfarbstoff *m*, direktziehender (substantiver) Farbstoff *m*, Substantivfarbstoff *m*
substantivity *(Farb)* Substantivität *f*
substituent Substituent *m*
s. effect Substituenteneffekt *m*
substitute / to substituieren, ersetzen, austauschen, auswechseln
substitute 1. Ersatz *m*, Ersatzmittel *i* ...aizstoff *m*,
Surrogat *n*, Austauschstoff *m*; 2. Ölkautschuk *m*,
(veraltet für) Faktis *m*
m-substituted *m*-substituiert, metasubstituiert,
1,3-substituiert
o-substituted *o*-substituiert, orthosubstituiert,
1,2-substituiert
p-substituted *p*-substituiert, parasubstituiert,
1,4-substituiert
substituted group verdrängter (ausgewechselter)
Substituent *m*
s. in the ring kernsubstituiert
substitution Substitution *f*, Ersatz *m*, Austausch *m*
s.-deactivating group die Substitution des-
aktivierende Gruppe *f*
s. in the ring Kernsubstitution *f*
s. isomerism Substitutionsisomerie *f*
s. method in photometry fotometrische Sub-
stitutionsmethode *f*
s. product Substitutionsprodukt *n*
s. reaction Substitutionsreaktion *f*, Verdrängungsreaktion *f*, Austauschreaktion *f*
s. tautomerism Substitutionstautomerie *f*
substitutional alloy Substitutionslegierung *f*
s. isomerism s. substitution isomerism
s. solid solution Substitutionsmischkristall *m*
substitutive name Substitutionsname *m*
substrate Substrat *n*, *(i.e.S.)* Nährboden *m*,
Nährsubstrat *n*; *(Farb)* Substrat *n*, Träger *m*,
Farbträger *m*; Schichtträger *m*, Adhärend *n*,
Packstoff *m (zu verklebender Stoff)*
substructure *(Tech)* Unterbau *m*; *(Krist)* Unter-
struktur *f*
subsubmicron Amikron *n (mikroskopisch unsichtbares Schwebeteilchen)*
subsurface pump Tiefpumpe *f*
s. water Grundwasser *n*
subtilin Subtilin *n (Antibiotikum aus Bacillus
subtilis)*
subtilisin, subtilysine Subtilisin *n (von Bacillus
subtilis ausgeschiedenes Ferment)*
subtraction solid solution Subtraktionsmisch-
kristall *m*
subtractive name Subtraktionsname *m*, Sub-
traktivname *m*
s. reducer *(Foto)* subtraktiver (subtraktiv wirkender) Abschwächer *m*
succession Folge *f*, Reihenfolge *f*

successive experiment Reihenversuch *m*
s. reaction Folgereaktion *f*; Stufenreaktion *f*
s. reactions law Ostwaldsche Stufenregel *f*, Stufenregel *f* von Ostwald
succinaldehyde *s.* succindialdehyde
succinamic acid $NH_2COCH_2CH_2COOH$ Sukzinamidsäure *f*, Bernsteinsäuremon[o]amid *n*, Butandisäuremon[o]amid *n*
succinate Sukzinat *n* *(Salz oder Ester der Bernsteinsäure)*
succindialdehyde $OHCCH_2CH_2CHO$ Sukzin[di]aldehyd *m*, Bernsteinsäuredialdehyd *m*, Butandial-(1,4) *n*
succinic acid $HOOCCH_2CH_2COOH$ Sukzinsäure *f*, Bernsteinsäure *f*, Äthandikarbonsäure-(1,2) *f*, Butandisäure *f*
s. acid dehydrogenase *s.* succinodehydrogenase
s. acid oxidase *s.* succinoxidase
s. dehydrogenase *s.* succinodehydrogenase
s. oxidase *s.* succinoxidase
succinimide Sukzinimid *n*, Bernsteinsäureimid *n*
succinite Sukzinit *m*, Bernstein *m*
succinodehydrogenase Sukzinodehydrogenase *f*, Bernsteinsäuredehydrogenase *f*
succinoxidase Sukzinoxydase *f*, Bernsteinsäureoxydase *f*
succinyl chloride Sukzinylchlorid *n*, Bernsteinsäuredichlorid *n*
succus Succus *m*, Saft *m*, *(i.e.S.)* Pflanzensaft *m*
suck / to [an]saugen, einsaugen
s. back zurücksaugen
s. in einsaugen
s. off absaugen
s. out aussaugen
s. up aufsaugen, einsaugen
suck-and-blow machine *(Glas)* Saugblas[e]maschine *f*
s.-and-blow process *(Glas)* Saugblas[e]verfahren *n*
sucker-rod wax Röhrenwachs *n*
s. rods Pumpgestänge *n* *(einer Gestängetiefpumpe)*
sucrase Sa[c]charase *f*, Invertase *f*, Invertin *n* *(rohrzuckerspaltendes Ferment)*
sucrate Sa[c]charat *n*
sucrochemistry Sucrochemie *f*
sucrol $C_2H_5OC_6H_4NHCONH_2$ Sucrol *n*, Dulzin *n*, 4-Äthoxyphenylharnstoff *m*
sucrose $C_{12}H_{22}O_{11}$ Sa[c]charose *f*, Sucrose *f*
s. determination Sa[c]charosebestimmung *f*
s. octaacetate Sa[c]charoseoktaazetat *n*
s. phosphate Sa[c]charosephosphat *n*
s. phosphorylase Sa[c]charosephosphorylase *f*
s. solution Sa[c]charoselösung *f*
s. splitting Sa[c]charoseumwandlung *f* durch Mikroorganismen, Sa[c]charosespaltung *f*
s. synthesis Sa[c]charosesynthese *f*
suction Saugen *n*, Saugung *f*, Ansaugen *n*, Ansaugung *f*; Absaugen *n*; Einsaugen *n*; Saugzug *m*, Zug *m*

s. area *(Pap)* Saugzone *f*
s. box *(Pap)* Sauger *m*, Saug[er]kasten *m*
s. box cover *(Pap)* Saug[er]kastendeckel *m*
s. conveyor Saugförderer *m*
s. couch [roll] *(Pap)* Siebsaugwalze *f*, Saugwalze *f*, Sauggautsche *f*
s.-drum dryer Saugtrommeltrockner *m*
s. dryer Saugtrockner *m*
s. extractor Absaugvorrichtung *f*, Saugapparat *m*
s. feeder *(Glas)* Saugspeiser *m*
s. feeding *(Glas)* Saugspeisung *f*
s. filter Saugfilter *n*, Vakuumfilter *n*, Unterdruckfilter *n*
s. flask Saugflasche *f*
s. force *s. s.* pressure 2.
s. friction head Saugreibungshöhe *f*
s. gas producer, s. generator Sauggasgenerator *m*
s. head Saughöhe *f*
s. line (pipe) Saugrohr *n*, Saugleitung *f*, Ansaugrohr *n*, Ansaugleitung *f*
s. pipe felt cleaner *(Pap)* Rohrsaug[er]filzwäsche *f*
s. pipette Saugpipette *f*
s. pit Sauggrube *f*
s. point Saugstelle *f*
s. port Ansaugstutzen *m*, Saugstutzen *m*; Saugmund *m*
s. pot Saugtopf *m*
s. press *(Pap)* Saugpresse *f*
s. press roll *(Pap)* Saugpreßwalze *f*
s. pressure 1. Saugdruck *m*, Ansaugdruck *m*; 2. *(Bio)* Sog *m*, Saugkraft *f*
s. process *(Glas)* Saugblas[e]verfahren *n*
s. pump Saugpumpe *f*
s. roll *s. s.* couch [roll]
s. side Saugseite *f*
s. stroke Saughub *m*
s. tank Saugbehälter *m*, Saugbassin *n*
s. tension *s. s.* pressure 2.
s.-type machine *(Glas)* Saugblas[e]maschine *f*
s. valve Saugventil *n*, Fußventil *n*
s. water *(Pap)* Saugerwasser *n*
sudorific diaphoretisch, schweißtreibend
sudorific Diaphoretikum *n*, schweißtreibendes Mittel *n*
suede leather Wildleder *n*, Velourleder *n*
suet Nierenfett *n*, Talg *m*, *(i.e.S.)* Hammeltalg *m* *(oder)* Rindertalg *m*, Rohkern *m*
suety talgig, Talg...
sugar Zucker *m* *(i.e.S.)*, Sa[c]charose *f*, Sucrose *f*; Zucker *m* *(i.w.S.)* *(Monosa[c]charid oder Oligosa[c]charid)*
s. acid Zuckersäure *f*
s. alcohol Zuckeralkohol *m*
s. anhydride Zuckeranhydrid *n*
s. beet Zuckerrübe *f*, Beta vulgaris L. var. altissima Doell
s. beet chips Zuckerrübenschnitzel *npl*, Rübenschnitzel *npl*

s. beet mill Rübenzuckerfabrik f
s. beet pulp ausgelaugte Zuckerrübenschnitzel (Rübenschnitzel) npl
s. beet slices s. s. beet chips
s. candy Zuckerkand[is] m, Zuckerkant m, Kandis[zucker] m, Kandelzucker m
s. cane Zuckerrohr n, Saccharum officinarum L.
s. cane bagasse Zuckerrohrbagasse f, Bagasse f
s. cane molasses Zuckerrohrmelasse f, Rohrmelasse f
s. cane wax Zuckerrohrwachs n
s. charcoal Zuckerkohle f
s. coating Zuckerguß m
s. colouring Zuckercouleur f, Kulör f, Zuckerfarbe f
s. content Zuckergehalt m
s. crusher Rohrmühle f, Zuckerrohrquetsche f
s. crystal Zuckerkristall m
s. derivative Zuckerderivat n, Zuckerabkömmling m
s. determination Zuckerbestimmung f
s. dye s. s. colouring
s. ester Zuckerester m
s. ether Zuckeräther m
s. factory Zuckerfabrik f
s.-factory lime Scheideschlamm m, Scheidekalk m, Saturationsschlamm m
s.-factory waste Zuckerfabrikabwasser n
s. fragment Zuckerzerfallsprodukt n, Zuckerbruchstück n
s. fragmentation Zuckerzerfall m
s. from beet Rübenzucker m
s. house Zuckerhaus n, (i.w.S.) Zuckerfabrik f
s. house molasses Melasse f, Melassesirup m, (dicker schwarzbrauner) Sirup m
s. incinerating dish Zuckerveraschungsschale f
s. industry Zuckerindustrie f
s. loaf Zuckerhut m
s. maple Zuckerahorn m, Acer saccharum Marsh.
s. mill s. 1. s. factory; 2. s. crusher
s. of lead $Pb(CH_3COO)_2$ Bleizucker m, Blei(II)-azetat n
s. of milk Milchzucker m, Lakto[bio]se f
s. off / to (Ahornsaft) auf Korn kochen, (Ahornsirup) zur Kristallisation eindicken
s. palm Zuckerpalme f, Arenga Labill.
s. plant Zuckerpflanze f
s. refiner Zuckerraffineur m
s. refinery Zuckerraffinerie f
s. refining Raffinieren n (Raffination f) des Zuckers, Zuckerraffination f
s. series Zuckerreihe f
s. solution Zuckerlösung f
s. substitute Zuckerersatzstoff m
sugarcane s. sugar cane
sugarhouse s. sugar house
sugarloaf s. sugar loaf
sugarworks Zuckerfabrik f
sugary zuck[e]rig, zuckerhaltig, [zucker]süß

suint Wollschweiß m
s. scouring Schweißwäsche f der Wolle
s. vat Wollschweißküpe f
suitable for packing packfertig
sulf... s. sulph...
sulpha drug Sulfonamidpräparat n, Arzneimittel n auf Sulfonamidbasis
s. preparation Sulfapräparat n (den Sulfonamiden ähnlich aufgebautes Arzneimittel)
sulphacetamide Sulfazetamid n (ein Sulfonamid)
sulphacid s. sulphonic acid
sulphadiazine Sulfadiazin n, 2-(p-Aminobenzolsulfonamido)-pyrimidin n
sulphaethidole Sulfaethidol n (ein Sulfonamid)
sulphaguanidine Sulfaguanidin n, p-Aminobenzolsulfonylguanidin n
sulphamate Sulfamat n
sulphamerazine Sulfamerazin n, 2-(p-Aminobenzolsulfonamido)-4-methylpyrimidin n
sulphamethazine Sulfamethazin n, Sulfadimidin n, 2-(4'-Aminobenzolsulfonamido)-4,6-dimethylpyrimidin n
sulphamethoxypyridazine Sulfamethoxypyridazin n, 3-Sulfanilamido-6-methoxypyridazin n
sulphamic acid H_2NSO_3H Sulfaminsäure f, Sulfamidsäure, f, Amidosulfonsäure f
sulphamidyl s. sulphanilamide
sulphane Sulfan n (Wasserstoffverbindung des Schwefels)
sulphanilamide $H_2NC_6H_4SO_2NH_2$ Sulfanilamid n, p-Aminobenzolsulfonamid n, p-Aminophenylsulfonamid n
sulphanilic acid $NH_2C_6H_4SO_3H \cdot H_2O$ Sulfanilsäure f, p-Aminobenzolsulfonsäure f, Anilin-p-sulfonsäure f
sulphanilylguanidine s. sulphaguanidine
sulphantimonate $M_3^ISbS_4$ Thioantimonat(V) n
sulphantimonite $M_3^ISbS_3$ Thioantimonat(III) n
sulphaphenazole Sulfaphenazol n, 5-(4-Aminobenzolsulfonamido)-1-phenylpyrazol n
sulphapyridine Sulfapyridin n, 2-(p-Aminobenzolsulfonamido)-pyridin n
sulphaquinoxaline Sulfachinoxalin n, 2-(p-Aminobenzolsulfonamido)-chinoxalin n
sulpharsenide $M^{II}AsS$ Arsenosulfid n
sulpharsphenamine Sulfarsphenamin n, m,m'-Diamino-p,p'-dihydroxyarsenobenzol-N,N'-bis-methansulfonsäure f
sulphatase Sulfatase f (zu den Esterasen gehörendes Ferment)
sulphate / to 1. sulf[at]ieren (Alkohole mit Schwefelsäure verestern); 2. sulfatisieren (einen kristallinen Blei(II)-sulfat-Niederschlag im Bleiakku bilden)
sulphate $M_2^ISO_4$ Sulfat n
s. attack Sulfatangriff m
s. cook (Pap) Sulfatkochung f
s. cooking liquor s. s. liquor
s. digester Sulfatzellstoffkocher m
s. digestion liquor s. s. liquor

sulphonate

s. ion SO_4^{2-} Sulfat-Ion n, Sulfatanion n
s. lignin (Pap) Sulfatlignin n
s. liquor (Pap) Sulfat[koch]lauge f
s. mill Sulfatzellstoffabrik f, Sulfatfabrik f
s. of ammonia $(NH_4)_2SO_4$ Ammoniumsulfat n
s. of potash schwefelsaures Kali n (Düngemittel mit Kaliumsulfat als Hauptbestandteil); s. potassium sulphate
s. paper Sulfatpapier n
s. pitch Sulfatpech n
s. process (Pap) Sulfatprozeß m, Sulfatverfahren n
s. pulp Sulfatzellstoff m
s. pulping (Pap) Sulfataufschluß m
s. reducers desulfurierende Bakterien npl
s. reductase Sulfatreduktase f
s.-resisting cement sulfatbeständiger Zement m
s. resistivity Sulfatbeständigkeit f
s. scab Sulfatblase f (Glasfehler)
s. soap (Pap) Sulfatseife f, Rohseife f
s. turpentine (Pap) Sulfatterpentin n
s. wood rosin Tallölkolophonium n
sulphated castor oil sulfatiertes (sulfiertes) Rizinusöl n, Türkischrotöl n
s. oil sulf[at]iertes Öl n; s. sulphonated oil
sulphathiazol[e] Sulfathiazol n, 2-(4-Aminobenzolsulfonamido)-thiazol n
sulphating roasting s. sulphatizing roasting
sulphation 1. Sulf[at]ieren n, Sulf[at]ierung f (Veresterung von Alkoholen mit Schwefelsäure); 2. Sulfatisieren n, Sulfatisierung f (Bildung von kristallinem Blei(II)-sulfat im Bleiakku)
sulphatize / to sulfatisieren (Metallsulfide durch Rösten in Sulfate umwandeln); s. to sulphate 1.
sulphatizing s. sulphation 1.
s. roasting sulfatisierendes Rösten n, sulfatisierende Röstung f
sulphenamide Sulfenamid n
s. accelerator (Gum) Sulfenamidbeschleuniger m
sulphenic acid Sulfensäure f
sulphethylic acid $C_2H_5HSO_4$ Äthylschwefelsäure f, Äthylhydrogensulfat n
sulphide / to s. to sulphidize
sulphide M^I_2S Sulfid n
s. colour s. s. dye[stuff]
s. dye[stuff] Schwefelfarbstoff m, Schwefelfarbe f, Sulfinfarbe f
s. lime (Gerb) Sulfidäscher m, Schwefelnatriumäscher m
s. ore sulfidisches (schwefelhaltiges) Erz n, Sulfiderz n, Schwefelerz n
s.-oxide shell (Geol) Sulfidoxidschale f, Sulfidoxidzone f
s.-painted wool Schwödewolle f
s. phosphor Sulfidphosphor m
s. sulphur Sulfidschwefel m
s. toning (Foto) Schwefeltonung f
s.-type phosphor s. s. phosphor
sulphidic sulfidisch, schwefelhaltig, schwefelführend, Schwefel...

sulphidity Sulfidität f, Sulfidgehalt m
sulphidize / to sulfidieren, xanthogenieren
sulphidizing Sulfidieren n, Sulfidierung f, Xanthogenieren n, Xanthogenierung f
sulphinic acid Sulfinsäure f
sulphisomidine Sulfisomidin n (ein Sulfonamid)
sulphisoxazole Sulfisoxazol n, Sulfafurazol n, 3,4-Dimethyl-5-sulfanilamidoisoxazol n
sulphitation (Zucker) Sulfitieren n, Sulfitierung f, Sulfitation f, Schwefeln n, Schwefelung f, Schwefelsaturation f, Behandlung (Saturation) f mit schwefliger Säure
sulphite / to (Zucker) sulfitieren, schwefeln, mit schwefliger Säure behandeln (saturieren)
sulphite $M^I_2SO_3$ Sulfit n
s. acid liquor s. s. cooking liquor
s. cook (Pap) Sulfitkochung f
s. cooking liquor (Pap) Sulfit[koch]säure f, Kochsäure f
s. digester Sulfitzellstoffkocher m
s. ester Schwefligsäureester m
s. kraft paper Sulfitkraftpapier n
s. laps Sulfitzellstoff m in Bogenform (Pappenform), Sulfitzellstoffblätter npl
s. liquor s. s. cooking liquor
s. lye s. s. spent liquor
s. mill s. s. pulp mill
s. paper Sulfitpapier n
s. pulp Sulfit[zell]stoff m, Bisulfitzellstoff m
s. pulp industry Sulfitzellstoffindustrie f
s. pulp mill Sulfit[zellstoff]werk n, Sulfitzellstoffabrik f
s. pulping [process] (Pap) Sulfitprozeß m, Sulfitverfahren n, Sulfitaufschluß m
s. spent liquor (Pap) Sulfitablauge f, Ablauge f, Urlauge f
s. turpentine (Pap) Fichtenöl n
s. waste liquor s. s. spent liquor
sulphitocobaltate $M^I_3[Co(SO_3)_2]$ Sulfitokobalt(III) n
sulphitomercurate $M^I_2[Hg(SO_3)_2]$ Sulfitomerkurat(II) n, Disulfitomerkurat(II) n
sulpho acid s. sulphonic acid
s. group s. sulphonic acid group
sulphocarbanilide $CS(NHC_6H_5)_2$ Thiokarbanilid n, N,N'-Diphenylthioharnstoff m
sulphocarbolic acid Sulfokarbolsäure f, Karbolschwefelsäure f, (veraltet für) Phenolsulfonsäure f, Hydroxybenzolsulfonsäure f
sulphocarbonate $M^I_2CS_3$ Thiokarbonat n
sulphochlorinate / to sulfochlorieren, chlorsulfonieren
sulphochlorination Sulfochlorieren n, Sulfochlorierung f, Chlorsulfonierung f
sulphocyanate, sulphocyanide M^ISCN Thiozyanat n, Rhodanid n
sulphonal Sulfonal n, Diäthylsulfondimethylmethan n
sulphonamide Sulf[on]amid n, Sulfonsäureamid n
sulphonate / to sulfonieren, sulfurieren; (unkorrekt für) to sulphate 1.

sulphonate Sulfonat n
sulphonated castor oil s. sulphated castor oil
s. oil sulfoniertes (sulfuriertes) Öl n
sulphonating agent Sulfonierungsmittel n, Sulfonierungsagens n, Sulfurierungsmittel n, sulfonierendes Agens (Mittel) n
s. pan s. sulphonation pan
s. reagent s. s. agent
sulphonation Sulfonieren n, Sulfonierung f, Sulfurieren n, Sulfurierung f
s. mixture Sulfonierungsgemisch n, Sulfurierungsgemisch n
s. pan Sulfonierungskessel m, Sulfurierungskessel m, Sulfierkessel m, Sulfurationskessel m, Sulfiergefäß n
s. plant Sulfonier[ungs]anlage f, Sulfier[ungs]anlage f, Sulfierungsbetrieb m
s. temperature Sulfonierungstemperatur f, Sulfurierungstemperatur f
s. unit s. s. plant
sulphonator Sulfonierer m, Sulfonator m, Sulfier[ungs]apparat m, Sulfurationsapparat m
sulphone Sulfon n
sulphonic acid Sulfo[n]säure f
s. acid group —SO₃H Sulfo[n]gruppe f, Sulfo[n]säuregruppe f
sulphonium compound Sulfoniumverbindung f
sulphonyl chloride RSO₂Cl Sulfo[nyl]chlorid n, Sulfonsäurechlorid n
s. group =SO₂ Sulfonylgruppe f
sulphosalicylic acid Sulfosalizylsäure f, Salizylsäure-5-sulfonsäure f
sulphourea H₂NCSNH₂ Thioharnstoff m, Thiokarbamid n, Sulfokarbamid n, Schwefelharnstoff m
sulphovinic acid C₂H₅HSO₄ Äthylschwefelsäure f, Äthylhydrogensulfat n
sulphoxalic acid HSO₂H oder S(OH)₂ Sulfoxylsäure f, Schwefel(II)-hydroxid n
sulphoxide R₂SO Sulfoxid n
sulphoxylate Sulfoxylat n
sulphur / to schwefeln, mit Schwefel (oder Schwefeldioxid) behandeln, (i.e.S.) mit Schwefel (oder Schwefeldioxid) ausräuchern; mit Schwefel räuchern, Schwefel abbrennen
sulphur S Schwefel m
α-sulphur α-Schwefel m, rhombischer Schwefel m
β-sulphur β-Schwefel m, monokliner Schwefel m
λ-sulphur λ-Schwefel m, zyklo-Oktaschwefel m
μ-sulphur flüssiger μ-Schwefel m, katena-Polyschwefel m
sulphur 35 Schwefel-35 m, Radioschwefel m
s. bacteria Schwefelbakterien npl, Thiobakterien npl
s.-bearing schwefelhaltig, schwefelführend
s. bichloride s. s. dichloride
s. black Schwefelschwarz n, Schwefelschwarzfarbstoff m
s. black paste Schwefelschwarzpaste f
s. black plant Schwefelschwarzfabrik f

s. black powder Schwefelschwarzpulver n
s. bloom (Gum) Schwefelausblühung f
s. blooming (Gum) Ausblühen n von Schwefel, Ausschwefeln n
s. blue Schwefelblau n, Schwefelblaufarbstoff m
s. bridge Schwefel[vernetzungs]brücke f, Schwefelvernetzungsstelle f
s. brown Schwefelbraun n
s. burner Schwefel[verbrennungs]ofen m
s.-burner gas (Pap) Verbrennungsgas n eines Schwefelverbrennungsofens
s. colour s. s. dye
s. compound Schwefelverbindung f
s.-containing schwefelhaltig
s. content Schwefelgehalt m
s. cross-link[age] s. s. bridge
s.-curable mit Schwefel vulkanisierbar (vernetzbar)
s. cure Schwefelvulkanisation f, Vernetzung f mit Schwefel
s.-cured mit Schwefel vulkanisiert (vernetzt)
s.-curing s. s.-curable
s. dichloride SCl₂ Schwefeldichlorid n, Schwefel(II)-chlorid n
s. dioxide SO₂ Schwefeldioxid n, Schwefel(IV)-oxid n
s. dioxide-benzol process Benzol-Schwefeldioxid-Verfahren n, SO₂-Benzol-Verfahren n (Entparaffinierungsverfahren)
s. dioxide digester relief gas (Pap) SO₂-Gas n (Kocherabgas)
s. dioxide loss (Pap) SO₂-Verlust m
s. dioxide meter (Pap) SO₂-Meßgerät n
s. dioxide recovery (Pap) SO₂-Rückgewinnung f
s. donor Schwefeldonator m, Schwefelspender m
s. donor cure Vulkanisation f mit Schwefelspendern (schwefelabspaltenden Verbindungen), Vulkanisation f ohne freien Schwefel
s. dosage Schwefeldosierung f
s. dye Schwefelfarbstoff m, Schwefelfarbe f, Sulfinfarbe f
s. flowers Schwefelblüte f, Schwefelblume f, sublimierter Schwefel m
s.-free schwefelfrei
s. heptoxide S₂O₇ Schwefelheptoxid n, Dischwefelheptoxid n, Schwefel(VI)-peroxid n
s. hexafluoride SF₆ Schwefelhexafluorid n, Schwefel(VI)-fluorid n
s. hexaiodide SJ₆ Schwefelhexajodid n, Schwefel(VI)-jodid n
s. in extract (Gum) extrahierbarer Schwefel m
s. indigo blue Schwefelindigoblau n
s. level s. s. dosage
s. link s. s. bridge
s. melter (Pap) Vorschmelzkammer f (eines Schwefelverbrennungsofens)
s.-modified schwefelmodifiziert
s. monobromide S₂Br₂ Schwefel[mono]bromid n, Dischwefeldibromid n

s. monochloride S_2Cl_2 Schwefel[mono]chlorid n, Dischwefeldichlorid n

s. monofluoride S_2F_2 Schwefel[mono]fluorid n, Dischwefeldifluorid n

s. monoiodide S_2J_2 Schwefel[mono]jodid n, Dischwefeldijodid n

s. monoxide SO Schwefel[mon]oxid n, Schwefel(II)-oxid n

s. mustard Schwefellost m, Schwefelyperit n, S-Lost m, S-Yperit n

s. nitride S_4N_4 Schwefelnitrid n, Tetraschwefeltetranitrid n

s. [olive] oil Sulfuröl n *(Olivenöl geringer Qualität)*

s. pentoxydichloride $S_2O_5Cl_2$ Dischwefelpentoxiddichlorid n, Disulfurylchlorid n, Pyrosulfurylchlorid n

s. print Schwefelabdruck m, Baumann-Abdruck m *(zum Nachweis von Schwefel)*

s.-reducing bacteria desulfurierende Bakterien npl

s. scald *(Landw)* Schwefelverbrennungen fpl, Verbrennungsschäden mpl durch Schwefel

s. sensitizer *(Foto)* Schwefelsensibilisator m

s. sesquioxide S_2O_3 Dischwefeltrioxid n

s.-shy *(Landw)* empfindlich gegen Schwefel *(oder)* Schwefelpräparate, schwefelscheu

s. sol Schwefelsol n

s. solution Schwefellösung f

s. subchloride s. s. monochloride

s. tetrachloride SCl_4 Schwefeltetrachlorid n, Schwefel(IV)-chlorid n

s. tetrafluoride SF_4 Schwefeltetrafluorid n, Schwefel(IV)-fluorid n

s. tetroxide SO_4 Schwefeltetroxid n

s. toning *(Foto)* Schwefeltonung f

s. trioxide SO_3 Schwefeltrioxid n, Schwefel(VI)-oxid n

s. vapour Schwefeldampf m

s.-vulcanizable mit Schwefel vulkanisierbar (vernetzbar)

s. vulcanizate Schwefelvulkanisat n

s. vulcanization Schwefelvulkanisation f, Vernetzung f mit Schwefel

s. vulcanization system Schwefelvulkanisationssystem n, Schwefelvernetzungssystem n

s. yellow Schwefelgelb n

sulphurated lime Kalkschwefelleber f *(Gemisch aus Kalziumsulfiden und Kalziumsulfat)*

sulphureous aus Schwefel bestehend; schwefelhaltig, schweflig; schwefelartig; schwefelfarben

sulphuretted hydrogen H_2S Sulfan n, Monosulfan n, Schwefelwasserstoff m, Monoschwefelwasserstoff m, Wasserstoffsulfid n

sulphuric Schwefel...; höherwertigem Schwefel entsprechend, *(meist)* Schwefel(VI)-...

s. acid H_2SO_4 Schwefelsäure f

s.-acid alkylation Schwefelsäurealkylierung f, Alkylierung f mit Schwefelsäure

s.-acid alkylation process Schwefelsäurealkylie-

rungsverfahren n, Alkylierungsverfahren n mit Schwefelsäure

s.-acid industry Schwefelsäureindustrie f

s.-acid polymerization process Schwefelsäurepolymerisationsverfahren n

s.-acid process Schwefelsäureverfahren n *(bei der katalytischen Polymerisation)*

s.-acid refining Schwefelsäureraffinage f, Schwefelsäurebehandlung f, Behandeln n mit Schwefelsäure

s.-acid test *(Farb)* Schwefelsäurewaschprobe f, Schwefelsäurereaktion f

s. anhydride s. sulphur trioxide

s. chlorohydrin $ClSO_3H$ Schwefelsäurechlorhydrin n, Chloroschwefelsäure f

sulphuring-up s. sulphur blooming

sulphurization Schwefeln n, Schwefelung f

sulphurize / **to** schwefeln, mit Schwefel *(oder)* Schwefelverbindungen) verbinden *(oder)* imprägnieren

sulphurless schwefelfrei

s. cure (vulcanization) schwefelfreie Vulkanisation (Vernetzung) f

sulphurous Schwefel...; niederwertigem Schwefel entsprechend, *(meist)* Schwefel(IV)-...

s. acid H_2SO_3 schwef[e]lige Säure f

s.[-acid] anhydride s. sulphur dioxide

s. oxybromide $SOBr_2$ Thionylbromid n

s. oxychloride $SOCl_2$ Thionylchlorid n

s. oxyfluoride SOF_2 Thionylfluorid n

sulphuryl chloride Cl_2SO_2 Sulfurylchlorid n

s. fluoride SO_2F_2 Sulfurylfluorid n

sulphydrate $M'HS$ Hydrogensulfid n, primäres (saures) Sulfid n

sulphydryl [group] —SH Sulfhydrylgruppe f, Thiolgruppe f, Merkaptogruppe f, Merkaptorest m

sumac *(Gerb)* Sumach m *(Blätter von Rhus-Arten, besonders von Rhus coriaria L.)*

s. wax Japanisches (vegetabilisches) Wachs n *(Fett von Rhus succedanea L.)*

Sumatra benzoin [gum] Sumatrabenzoe f *(von Styrax benzoin Dryander)*

S. camphor Sumatrakampfer m, Borneokampfer m, Baroskampfer m *(von Dryobalanops aromatica Gaertn. f.)*

sumbul oil Sumbulöl n, Moschusöl n *(von Ferula sumbul Hook.)*

summer butter Sommerbutter f

s. cream Sommerrahm m

s. wood Spätholz n

sump Anstrichmittelwanne f, Lack[vorrats]behälter m *(beim Flow-Coating-Verfahren)*

s. pump *(kleine)* Wasserhaltungspumpe f, Lenzpumpe f

s. throat *(Glas)* versenkter (tiefliegender, tiefer) Durchlaß m

sun-dried brick sonnengetrockneter (an der Sonne getrockneter, lufttrockneter) Ziegel m

s.-dry / **to** an der Sonne trocknen, durch Sonnenwärme trocknen

s. drying Sonnentrocknung f, Trocknen n an der Sonne, Trocknen n durch Sonnenwärme

s. hemp s. sunn [hemp]

sunflower cake Sonnenblumenkuchen m

s. meal Sonnenblumenkuchenmehl n

s. oil Sonnenblumenöl n

s. oil cake s. s. cake

s. seed oil s. s. oil

sunk spot *(Plast)* Einsackstelle f, Einfallstelle f *(Spritzgußfehler)*

sunlight Sonnenlicht n

s. flavour Sonnenlichtgeschmack m

s. resistance (stability) Sonnenlichtbeständigkeit f, Tageslichtbeständigkeit f

sunn [hemp] Sunnhanf m, Madrashanf m

sunproofing agent Ozonschutzmittel n, Antiozonant m, Antiozonisator m

sunscreen [agent] Sonnenschutzmittel n

s. oil Sonnenschutzöl n

sunshine unit *(Bio)* Strontium-Einheit f *(Aktivität von 1 pc ^{90}Sr/g Ca)*

sunstone *(Min)* Sonnenstein m

suntan make-up, s. preparation Sonnenbräunungsmittel n, Hautbräunungsmittel n

super abrasion furnace black *(Gum)* Super-Abrasion-Furnace-Ruß m, SAF-Ruß m *(besonders hochabriebfester Ölruß)*

s. accelerator *(Gum)* Ultrabeschleuniger m

s. finish (glaze) *(Pap)* Hochglanz m

s.-glycerinated fat *(Backfett mit hohem Gehalt an Mono- und Diglyzeriden)*

s.-heavy concrete Schwerstbeton m

s.-sulphated cement Sulfathüttenzement m, SHZ

s. temperature Supertemperatur f *(von über 5000 °C)*

superacid [sehr] stark sauer

superacidity Superazidität f, Hyperazidität f

superadditivity *(Foto)* Superadditivität f

superalkaline [sehr] stark alkalisch

superalloy Superlegierung f, Superalloy n

supercalender / to *(Pap)* satinieren, glätten

supercalender *(Pap)* Superkalander m, Hochleistungskalander m

supercalendered paper satiniertes (sat) Papier n

supercalendering *(Pap)* Satinieren n, Satinierung f, Satinage f, Glätten n, Glättung f

supercarburization *(Met)* Überkohlung f

supercarburize / to *(Met)* überkohlen

supercentrifuge Superzentrifuge f, Ultrazentrifuge f

superchlorinate / to hochchloren, überchloren

superchlorination Hochchlor[ier]en n, Hochchlorung f, Über[schuß]chlorung f

supercoat Überzug m

superconduct / to Supraleitung (Supraleitfähigkeit) zeigen

superconducting supraleitend, supraleitfähig

s. electron supraleitendes Elektron n, Supraleitungselektron n

s. state supraleitender Zustand m

s. substance supraleitende Substanz f

s. transition Übergang m von der Normal- zur Supraleitung, Übergang m Normalleitung-Supraleitung

superconduction s. superconductivity

superconductive s. superconducting

s. temperature Sprungtemperatur f, Sprungpunkt m in Supraleitern

superconductivity Supraleitfähigkeit f, Supraleitung f

superconductor Supraleiter m

supercool / to unterkühlen

supercooling Unterkühlen n, Unterkühlung f

supercritical *(Kch)* überkritisch

s. reactor *(Kch)* überkritischer Reaktor m

supercrustal rock superkrustales (suprakrustales) Gestein n

superfluid supraflüssig, superflüssig, superfluid, suprafluid, überflüssig

s. state s. superfluidity

superfluidity Supraflüssigkeit f, supraflüssiger (superflüssiger, superfluider, suprafluider, überflüssiger) Zustand m, Suprafluidität f, Überflüssigkeit f

superfuse / to s. to supercool

superheat / to überhitzen

superheat Überhitzungswärme f; Überhitzungstemperatur f

superheated steam überhitzter Dampf m, Heißdampf m

superheater Überhitzer m

superheating Überhitzen n, Überhitzung f

superimposable deckungsgleich

superimpose / to übereinanderlagern, übereinanderlegen; überlagern; *(Reagenzien)* aufbringen

superimposed carbonizing chamber aufgesetzter Schwelschacht m, Schwelaufsatz m, Schwelaufbau m

superinsulation Folienisolierung f

superior processing rubber SP-Kautschuk m, Perfektkautschuk m

superlattice *(Krist)* Übergitter n, Überstruktur f

superliquid s. superfluid

supermolecular structure übermolekulare Struktur f

supernatant überstehend, darüber stehend

supernatant [liquid, liquor] überstehende Flüssigkeit f

superoxide $M^I O_2$ Hyperoxid n

superphosphate Superphosphat n *(Düngemittel)*

superproportional intensifier *(Foto)* superproportionaler (überproportionaler) Verstärker m

s. reducer *(Foto)* superproportionaler (überproportionaler) Abschwächer m

superpure reinst, von höchster Reinheit

s. aluminium Reinstaluminium n

superrefining extreme Reinigung f, Hochreinigung f

superrefractory hochfeuerfest

supersaturate / to übersättigen
supersaturated solution übersättigte Lösung f
supersaturation Übersättigen n, Übersättigung f
superscript [numeral] hochgestellte Zahl f, Hochzahl f, Exponent m
supersensitizer (Foto) Übersensibilisator m
supersolubility curve Übersättigungskurve f
supersonic testing Ultraschallprüfung f (Werkstoffprüfung)
superstructure (Krist) Überstruktur f, Übergitter n
supplementary component (material) Hilfsstoff m
s. spray material Spritzhilfsmittel n
supply chest s. s. tank
s. line Speiseleitung f
s. tank (vat) (Pap) Maschinenbütte f, Arbeitsbütte f, Stoffbütte f
supplying power (Landw) nachschaffende Kraft f (des Bodens)
support / to halten; (z.B. die Verbrennung) unterhalten
support Träger m, Trägersubstanz f, Trägermaterial n (für Katalysatoren); (Foto) Unterlage f, Emulsionsunterlage f, Filmunterlage f, Emulsionsträger m, Filmschichtträger m, Schichtträger m, Träger m
s. roll (Pap) Tragwalze f; Transportwalze f
s. screen Stützsieb n
s. span Spannweite f, Stützweite f
s. stand Stativ n
supported catalyst Trägerkatalysator m
supporting electrolyte (Polarografie) Leitelektrolyt m, Leitsalz n
s. material Trägersubstanz f, Trägermaterial n, Träger m (für Katalysatoren)
s. roll s. support roll
suppository (Pharm) Stuhlzäpfchen n, Suppositorium n
suppress / to unterdrücken; (z.B. eine Blutung) stillen
supraconductivity s. superconductivity
suramin (Pharm) Suramin n
surface / to die Oberfläche behandeln
surface Oberfläche f; Fläche f
s.-active oberflächenaktiv, grenzflächenaktiv, oberflächenwirksam, kapillaraktiv
s.-active agent (compound, substance) oberflächenaktiver Stoff m, oberflächenaktive Substanz (Verbindung) f, oberflächenaktives Mittel n
s. activity Oberflächenaktivität f, Grenzflächenaktivität f
s. aeration Oberflächen[be]lüftung f
s. anaesthetic Oberflächenanästhetikum n
s. atom Oberflächenatom n
s. attachment Oberflächenhaftung f
s. catalysis Oberflächenkatalyse f, Kontaktkatalyse f, heterogene Katalyse f
s. chemistry Oberflächenchemie f
s. coating Überzug m; Anstrich m; Überzugsmittel n; Anstrichstoff m, Anstrichmittel n
s.-coloured (Pap) oberflächengefärbt

s. colouring (Pap) Oberflächenfärbung f
s. combustion Oberflächenverbrennung f (Oxydation)
s. compaction Oberflächenverdichtung f
s. compound Oberflächenverbindung f
s. concentration Oberflächenkonzentration f
s. condenser Oberflächenkondensator m, Oberflächenverdichter m
s. conductivity Oberflächenleitfähigkeit f, Oberflächenleitvermögen n
s. consolidation (Geol) extratellurische Erstarrung f
s. constitution Oberflächenbeschaffenheit f, Oberflächenstruktur f
s. cord (Glas) Oberflächenschliere f (Fehler)
s.-covered überzogen (z.B. mit einem Sieb)
s. crack Oberflächenriß m
s. culture Oberflächenkultur f
s.-culture method Oberflächen[kultur]verfahren n, Oberflächenkulturmethode f
s. defect flächenhafter Gitterfehler m; flächenhafte Gitterfehlstelle f
s. developer (Foto) Oberflächenentwickler m
s. development (Foto) Oberflächenentwicklung f
s. diffusion Oberflächendiffusion f
s. drier (Farb) Oberflächentrockner m
s. dyeing s. s. colouring
s. effect Oberflächeneffekt m, Oberflächenwirkung f
s. energy Oberflächenenergie f
s. enthalpy Oberflächenenthalpie f
s. entropy Oberflächenentropie f
s. evaporation Oberflächenverdunstung f
s. exchanger Oberflächenwärmeaustauscher m, indirekter Austauscher m
s. extent Oberflächenausdehnung f, Oberflächengröße f
s. fermentation Obergärung f
s. film Oberflächenfilm m, Oberflächenhaut f
s. filter Oberflächenfilter n
s. filtration Oberflächenfiltration f
s. force Oberflächenkraft f
s. growth Oberflächenwachstum n
s.-harden / to oberflächenhärten, oberflächlich härten
s. hardening Oberflächenhärten n, Oberflächenhärtung f, (i.e.S.) Oberflächenhärtung f [mit Diffusion]
s. hardness Oberflächenhärte f
s. haze Oberflächentrübung f
s. image (Foto) Oberflächenbild n
s. latent image (Foto) latentes Oberflächenbild n, oberflächiges (äußeres) latentes Bild n
s. layer Oberflächenschicht f, (Met auch) Oberflächenzone f, Randschicht f, Randzone f, Rand m
s. level Oberflächenniveau n, (phys Ch auch) Oberflächenterm m
s. level of the medium Trübeniveau n,

Schwertrübespiegel m (beim Schwerflüssigkeitsverfahren)
s. lifetime Oberflächenlebensdauer f, Lebensdauer f an der Oberfläche (eines Halbleiterkristalls)
s. of contact Berührungsfläche f, Kontaktfläche
f
s. of cut Schnittfläche f
s. of evaporation Verdunstungsfläche f
s. of separation Trennungsfläche f, Grenzfläche
f, Phasengrenzfläche f
s. oxidation oberflächliche Oxydation f
s. phenomenon Oberflächenerscheinung f
s. potential Oberflächenpotential n
s. pressure Oberflächendruck m
s. property Oberflächeneigenschaft f
s. protection Oberflächenschutz m
s. reaction [chemische] Oberflächenreaktion f,
Grenzflächenreaktion f
s. replica Oberflächenabdruck m
s. resistance Oberflächenwiderstand m
s. resistivity spezifischer Oberflächenwiderstand
m
s. rewind method (Pap) Rollenwicklung f auf
Tragwalzen
s. roughness Oberflächenrauh[igk]eit f
s. saponification (Text) oberflächliches Verseifen n, S-Finish n
s. seepage (Erdöl) Ausbiß m, Austritt m an der
Oberfläche
s. size press (Pap) Leimpresse f
s.-sized (Pap) oberflächengeleimt
s.-sized with animal glue (Pap) mit Gelatine
(Tierleim) oberflächengeleimt
s.-sized with starch (Pap) mit Stärke oberflächengeleimt
s. sizing (Pap) Oberflächenleimung f
s.-sizing agent (Pap) Oberflächenleim m, Oberflächenleim[ungs]mittel n
s.-sizing machine (Pap) Oberflächenleimmaschine f
s. sizing with animal glue (Pap) Gelatineleimung f, Oberflächenleimung (Leimung) f mit
Gelatine (Tierleim)
s. sizing with size press (Pap) Oberflächenleimung f in der Leimpresse
s. sizing with size tub (Pap) Oberflächenleimung
f im Leimbadtauchverfahren
s. sizing with starch (Pap) Oberflächenleimung
f mit Stärke
s. skimming device Schwimmdeckenabstreifer
m
s. solution Oberflächenlösung f
s. staining s. s. colouring
s. temperature Oberflächentemperatur f
s. tension Oberflächenspannung f
s.-tension-depressing oberflächenspannungs-
vermindernd
s.-tension difference Unterschied m in der
Oberflächenspannung

s. texture Oberflächentextur f
s. tinting s. s. colouring
s. treating Oberflächenbehandlung f
s. vibration Oberflächenrüttlung f
s. vibrator Oberflächenrüttler m
s. water Oberflächenwasser n
s. waviness (Plast) Schlieren fpl auf der
Oberfläche (Fehler)
s. weathering Oberflächenverwitterung f
surfaced paper Streichpapier n, gestrichenes
Papier n
surfacer Spachtelmasse f, Spachtel m(f)
surfacing Oberflächenbehandlung f
s. end (Pap) Glättwerkspartie f
s. mat (Plast) Oberflächenmatte f (Vlies als
Oberschicht)
s. sheet (Plast) Oberflächenbogen m, Deck-
bogen m
surfactant s. surface-active agent
surge / to [auf]wallen
surge chamber Windkessel m
s. hopper Ausgleichsbunker m
s. tank Puffer m (beim Bleicherde-Kontaktverfahren)
surinamine $HOC_6H_4CH_2CH(NHCH_3)COOH$
Surinamin n, Ratanhin n, Andirin n, N-Me-
thyltyrosin n, 3-(4-Hydroxyphenyl)-2-methyl-
aminopropansäure f
suroxide Peroxid n
surplus Überschuß m
s. gas Überschußgas n, überschüssiges Gas n
s. slip (Ker) Schlickerüberschuß m, Schlicker-
überfluß m
survival time (Toxikologie) Überlebensdauer f
susceptibility Empfindlichkeit f, Suszeptibilität f,
(bei Kraftstoffen auch) Bleiempfindlichkeit f
s. per gram Grammsuszeptibilität f, Massensuszeptibilität f, spezifische Suszeptibilität f
s. per gram atom Atomsuszeptibilität f
s. per gram ion Ionensuszeptibilität f
s. per gram mole molare Suszeptibilität f,
Molsuszeptibilität f
s. to substitution Substitutionsbereitschaft f
s. to vulcanization Vulkanisationsfreudigkeit f
suspend / to suspendieren, aufschwemmen,
aufschlämmen; aufhängen
suspended-cable idler Vielscheibenseilrolle f,
Girlandenrolle f
s. centrifuge Hängezentrifuge f
s. magnet Hängemagnet m
s. material (matter) suspendierter Stoff m,
Schweb[e]stoff m, Schwebeteilchen npl
suspender pit Gerbgrube f, Farbgrube f
s. set Farbengang m (Gerbgrubensystem)
suspending agent Suspendiermittel n, Sus-
pensionsmittel n, Suspensionsmedium n, Stabilisator m (für Suspensionen)
s. capacity Suspendiervermögen n
s. medium s. s. agent
suspensibility Schwebefähigkeit f (einer Suspension)

suspension Suspension *f*, Aufschwemmung *f*, Aufschlämmung *f*
s. clamp *(Labortechnik)* Stativhaken *m* mit Muffe
s. firing [of coal] Kohlenstaubfeuerung *f*, Staubfeuerung *f*
s. of fibres *(Pap)* Fasersuspension *f*
s. polymerization Suspensionspolymerisation *f*, Perlpolymerisation *f*, Kornpolymerisation *f*
s. polyvinylchloride Suspensionspolyvinylchlorid *n*, S-PVC *n*
s. roasting Suspensionsröstung *f*, *(i.w.S.)* Schweberöstung *f*
suspensoid Suspensoid *n*, Suspensionskolloid *n*, lyophobes Kolloid (Sol) *n*
s. catalytic cracking Suspensoid-Kracken *n*, Kracken *n* mit suspendiertem Katalysator
s.-catalytic-cracking process Suspensoid-Krackverfahren *n*, Suspensoid-Verfahren *n*
s. colloid *s.* suspensoid
s. cracking *s. s.* catalytic cracking
s. process *s. s.*-catalytic-cracking process
sussexite *(Min)* Sussexit *m* *(Mangan(II)-hydrogenorthoborat)*
S.V. s ...onification value
swab ... ...ischen, aufwischen; auftragen
swage / to *(Plast)* tiefziehen
swaging *(Plast)* Tiefziehen *n*
swarm *(phys Ch)* Schwarm *m*
s. of molecules Molekülschwarm *m*, Schwarm *m*
swathe Schwaden *m*
swealing *(Text)* Farbwanderung *f (während des Trocknens eines Gewebes)*
sweat / to schwitzen; ausschwitzen
s. off *(eine Fraktion)* langsam übertreiben
sweat Schweiß *m*
s. cylinder *s. s.* roll
s. pan *s.* sweating pan
s. roll *(Pap)* Kühlzylinder *m*
sweated wool Schwitzwolle *f*
sweater room *s.* sweating room
sweating Schwitzen *n*, Schwitzung *f (Entparaffinierungsverfahren)*; *(Gerb)* Schwitze *f*
s. chamber *s. s.* room
s. pan Schwitzwanne *f*, Schwitztasse *f*, Schwitzpfanne *f*
s. process Schwitzverfahren *n*, Schwitzprozeß *m* *(Entparaffinierungsverfahren)*
s. room Schwitzraum *m*, Schwitzkammer *f*
s. stove Schwitzapparat *m*, Schwitzanlage *f*
s. tray *s. s.* pan
sweats [oil] Schwitzöl *n*
Swedish iron schwedisches Holzkohlen[roh]eisen *n*
sweep forward / to mitreißen, mitführen
s. out säubern; *(Gase)* austreiben
sweet süß; *(Erdöl)* süß, gesüßt, doktor-negativ, mit negativem Doktortest
s. almond süße Mandel *f*
s. balm Zitronenmelisse *f*, Melissa officinalis L.
s.-birch oil Birkenöl *n*

s. butter *s.* s.-cream butter
s. cider Apfelmost *m*, *(i.w.S.)* Süßmost *m*
s. cream Süßrahm *m*
s.-cream butter Süßrahmbutter *f*
s.-cream buttermilk Süßrahmbuttermilch *f*
s. flag Kalmus *m*, Acorus calamus L.
s. gasoline süßes (gesüßtes) Benzin *n*
s. glass leicht verarbeitbares Glas *n*
s. gum 1. Amerikanischer Amberbaum *m*, Liquidambar styraciflua L.; 2. Amberholz *n*, Satin-Walnußholz *n* *(Kernholz von 1.)*; 3. Amerikanischer Styrax (Storax) *m (Balsamharz von 1.)*
s. mash süße Maische *f*, Süßmaische *f*
s. oil süßes (gesüßtes) Erdölprodukt (Erdöldestillat) *n*, doktor-negatives Öl *n*, Öl *n* mit negativem Doktortest
s. orange-peel oil Apfelsinenschalenöl *n*, Orangenschalenöl *n*, süßes Pomeranzenschalenöl *n*, Portugalöl *n (von Citrus sinensis (L.)Osbeck)*
s.-smelling süßlich riechend, mit süßlichem Geruch
s. soil neutraler *(nicht versauerter)* Boden *m*
s. water Süßwasser *n*; *(Zucker)* Absüßwasser *n*
s.-water pump *(Zucker)* Absüßpumpe *f*
s. whey süße Molke *f*, Süßmolke *f*
s.-whey powder Süßmolkenpulver *n*
s. wine Süßwein *m*
sweeten / to [ab]süßen, versüßen; *(Butter)* wieder frisch machen; *(Erdöldestillate)* süßen *(um Schwefelverbindungen zu entfernen bzw. in Disulfide zu überführen)*
sweetened [ab]gesüßt, versüßt; gesüßt, süß, doktornegativ, mit negativem Doktortest *(Erdöldestillat)*
s. condensed milk gezuckerte Kondensmilch *f*
sweetener Süßstoff *m*; Adsorptions[klär]mittel *n (für Chemischreinigung)*
sweetening Süßen *n*, Absüßen *n*, Versüßen *n*; Süßen *n*, Süßung *f*, Sweetening *n (von Erdöldestillaten)*
s. agent Süßstoff *m*; *(Erdöl)* Süßungsmittel *n*
s. plant *(Erdöl)* Anlage *f* zum Süßen
s. process *(Erdöl)* Süß[ungs]verfahren *n*, Süßungsprozeß *m*, Sweetening-Prozeß *m*
s. reaction *(Erdöl)* Süßungsreaktion *f*
sweetish süßlich
Sweetland filter, Sweetland [filter] press Sweetland-Filter *n*, Sweetland-Presse *f*
sweetness Süße *f (Geschmacksempfindung)*
swell [up] / to [an]schwellen, [auf]quellen
swell-resistant quellfest
swelling Schwellen *n*, Anschwellen *n*, Anschwellung *f*, Quellen *n*, Quellung *f*, Aufquellen *n*, Aufquellung *f*; Blähen *n*, Blähung *f*, Aufblähen *n*, Aufblähung *f (der Kohle)*
s. after extrusion *s.* s. from the die
s. agent Quellungsmittel *n*
s. behaviour Quellverhalten *n*
s. capacity Quellfähigkeit *f*, Quellvermögen *n*

s. effect Quellwirkung *f*
s. from the die Spritzquellung *f*
s. index (number) Blähindex *m*, Blähzahl *f*, Blähgrad *m*, Swellingindex *m* *(der Kohle)*
s. power Quellvermögen *n*, Quellfähigkeit *f*; Blähvermögen *n* *(der Kohle)*
s. pressure Quellungsdruck *m*; Treibdruck *m* *(der Kohle)*
s.-pressure test Treibversuch *m* *(bei Kohle)*
s. properties *s. s.* behaviour
s. resistance Quellbeständigkeit *f*
s. test Quell[ungs]prüfung *f*, Quellungsversuch *m*; Blähprobe *f* *(bei Kohle)*
s. water Quellungswasser *n*
Swenson-Walker crystallizer Swenson-Walker-Kristallisator *m*
swill / to [ab]spülen
s. out ausspülen
swimming roller *(Text)* schwimmende Walze *f*
swing schwenkbar, Schwenk..., Schwing...
s.-bucket elevator Schaukelbecherwerk *n*, Pendelbecherwerk *n*, Schaukelförderer *m*
s.-bucket rotor ausschwingender Rotor *m* *(einer Zentrifuge)*, Rotor *m* mit frei ausschwingenden Bechern
s. check valve Klappenrückschlagventil *n*, Rückschlagklappe *f*
s. jaw Brechschwinge *f*
s. open / to ausschwingen
s. rake Schwingrechen *m*
s.-rake machine *(Text)* Waschmaschine *f* mit Schwingrechen
s.-tray elevator *s. s.*-bucket elevator
swingfog *(Pflanzenschutz)* Schwingfeuer[nebel]-gerät *n*
swinging-bucket elevator *s.* swing-bucket elevator
s.-bucket rotor *s.* swing-bucket rotor
s. jaw *s.* swing jaw
swingout centrifuge head Rotor *m* einer Horizontalzentrifuge
swirl / to umschwenken
swirl:
to give a s. to in einen Wirbel versetzen
s. [plate] nozzle Kreiselkraftdüse *f*, Dralldüse *f*, Kegelstrahldüse *f*
switch Schalter *m*
s. oil Schalteröl *n*
swivel *(Erdöl)* Spülkopf *m*
s. clamp holder Universaldoppelmuffe *f* *(drehbare Doppelmuffe für Stativklemmen)*
s.-roll machine Kalander *m* mit Walzenschränkung
sydnone Sydnon *n* *(mesoionische Verbindung)*, *(i.e.S.)* Sydnon *n*, N-Phenylsydnon *n*
syenite Syenit *m*
sylvan Sylvan *n*, 2-Methylfuran *n*
sylvanite *(Min)* Sylvanit *m* *(Silbergoldtellurid)*
sylvestrene Sylvestren *n* *(ein zyklisches Terpen)*
sylvic acid Sylvinsäure *f*, *(veraltet für)* Abietinsäure *f*

sylvin[e] *(Min)* Sylvin *m* *(Kaliumchlorid)*
sylvite *(Min)* Sylvit *m*, *(veraltet für)* Sylvin *m*
symbiotic fixation *(Landw)* symbiotische Fixierung (Bindung) *f* *(von Luftstickstoff)*
symbol Symbol *n*, Zeichen *n*
symbolic language Formelsprache *f*
symbolism Symbolik *f*
symmetrical symmetrisch
s. ether R·O·R einfacher (symmetrischer) Äther *m*
symmetry Symmetrie *f*
s. element Symmetrieelement *n*
s. plane *(Krist)* Symmetrieebene *f*, Spiegelebene *f*
Symons cone crusher Symons-Kegelbrecher *m*, Symons-Flachkegelbrecher *m*
Symons disk crusher Symons-Tellerbrecher *m*, Symons-Scheibenbrecher *m*
Symons short-head cone crusher Symons-Kegelgranulator *m*, Symons-Granulator *m*
Symons standard cone crusher Symons-Kegelbrecher *m* in Standardausführung
sympathetic ink sympathetische Tinte *f*, Geheimtinte *f*
s. reaction induzierte Reaktion *f*
sympatheticolytic *s.* sympatholytic
sympatheticomimetic *s.* sympathomimetic
sympathicolytic *s.* sympatholytic
sympathicomimetic *s.* sympathomimetic
sympatholytic sympath[ik]olytisch
sympatholytic Sympath[ik]olytikum *n*
sympathomimetic sympath[ik]omimetisch
sympathomimetic [drug] Sympath[ik]omimetikum *n*
symplektite *(Geol)* Symplektit *m*
symplesite *(Min)* Simplesit *m*
syn syn-..., in syn-Stellung befindlich
to be s. in syn-Stellung stehen
s.-anti isomerism syn-anti-Isomerie *f*
s. form syn-Form *f*
s. isomer syn-Isomer[es] *n*
s. position syn-Stellung *f*
synaeresis Synärese *f*, Synaerese *f*
synantetic *(Geol)* synantetisch
synchrocyclotron Synchrozyklotron *n*, frequenzmoduliertes Zyklotron *n*, FM-Zyklotron *n*
synchrotron Synchrotron *n*
syndet *s.* synthetic detergent
syndiotactic syndiotaktisch, syndyotaktisch
s. polymer syndiotaktisches Polymer[es] *n*
s. structure syndiotaktische Struktur *f*
syndyotactic *s.* syndiotactic
syneresis *s.* synaeresis
synergism Synergismus *m*, synergistische Wirkung *f*, synergistischer Effekt *m*
synergist Synergist *m*
synergistic synergistisch
s. action (effect) *s.* synergism
syngenite *(Min)* Syngenit *m*
synionism Synionie *f* *(Mesomerie im Anion tautomerer Verbindungen)*

synkinematic s. syntectonic

synorogenic (Geol) synorogen

synovial fluid (Med) Synovialflüssigkeit f, Synovia f, Gelenkschmiere f

syntactic s. syndiotactic

syntan Syntan n, synthetischer (künstlicher) Gerbstoff m

s. tannage Syntangerbung f, synthetische Gerbung f, Gerbung f mit synthetischen Gerbstoffen

syntectonic (Geol) syntektonisch, synkinematisch

syntexis (Geol) Syntexis f

synthesis Synthese f, Aufbau m, Vereinigung f

s. gas Synthesegas n

synthesize / to synthetisieren, synthetisch (durch Synthese) herstellen

synthetic synthetisch, Synthese...

synthetic synthetisch gewonnene Substanz f, Syntheseprodukt n; (Text) s. s. fibre; (Gum) s. s. rubber

s. ammonia synthetisches Ammoniak n, Syntheseammoniak n

s. detergent Detergens n, Detergent n, Syndet n, synthetisches Reinigungsmittel n (oder) Waschmittel n

s. fat Synthesefett n, Kunstfett n

s. fibre Synthesefaser f, synthetische Faser f; Synthesefaserstoff m, synthetischer Faserstoff m

s. latex synthetischer Kautschuklatex m, künstlicher Latex m, Synthese[kautschuk]latex m, Kunstkautschuklatex m

s. medium (Bakt) synthetischer Nährboden m

s. mordant (fälschlich für) neutral syntan

s. moulding sand s. s. sand

s. nitrogen fertilizer (durch Nutzung von Luftstickstoff gewonnener Stickstoffdünger)

s. oil of bitter almonds C_6H_5CHO künstliches Bittermandelöl n, Bittermandelessenz f, Benzaldehyd m

s. reagent für Synthesen gebrauchtes Reagens n, Reagens n für Synthesen, Synthesereagens n

s. resin Kunstharz n

s.-resin adhesive Kunstharzkleber m, Kunstharzklebstoff m

s.-resin bonded-fabric sheet Hartgewebe n, Hgw

s.-resin bonded-paper sheet Hartpapier n, Hp

s.-resin cement Kunstharzkitt m

s.-resin varnish Kunstharzlack m

s. rubber synthetischer (künstlicher) Kautschuk m, Synthesekautschuk m, Kunstkautschuk m, SK

s.-rubber adhesive Synthesekautschukkleber m, Kunstkautschukklebstoff m

s.-rubber compound s. s.-rubber mix

s.-rubber latex s. s. latex

s.-rubber mix (stock) Synthesekautschukmischung f, Kunstkautschukmischung f, SK-Mischung f

s. sand synthetischer Sand (Formsand) m

s. tannin (tanning agent) s. syntan

synthetical s. synthetic

synthetize / to s. to synthesize

synthin[e] Synthin n (durch Druckerhitzung aus Synthol erhaltenes Kohlenwasserstoffgemisch)

synthol Synthol n (Gemisch von Alkoholen, Ketonen, Säuren und etwa 2% Kohlenwasserstoffen)

s. process Syntholverfahren n, Syntholprozeß m

syphon s. siphon

syringa-aldehyde $(CH_3O)_2C_6H_2(OH)CHO$ Syringaaldehyd m, Syringaldehyd m, 4-Hydroxy-3,5-dimethoxybenzaldehyd m

syringic acid $(CH_3O)_2C_6H_2(OH)COOH$ Syringasäure f, 4-Hydroxy-3,5-dimethoxybenzoesäure f

s. aldehyde s. syringa-aldehyde

syringin Syringin n, 5-Methoxykoniferin n

syrup 1. Sirup m (dickflüssiger Obst- oder Zuckersaft), (i.e.S.) Speisesirup m; (Pharm) Sirup m; 2. Sirup m, sirupdicke Masse f; Melasse f, Melassesirup m, (dicker schwarzbrauner) Sirup m

syruplike, syrupy sirupartig, sirupös, dickflüssig, klebrig

system of ciphering Chiffrier[ungs]system n, Chiffresystem n

s. of classification Einteilungssystem n, Klassifikationssystem n

s. of crystals Kristallsystem n, kristallografisches System n

s. of enumeration Zählweise f

s. of more components Mehrkomponentensystem n, Mehrstoffsystem n

s. of nomenclature Nomenklatursystem n

s. of notation Bezeichnungssystem n, Bezeichnungsweise f; Notationssystem n

s. of numbering Numerierungssystem n, Bezifferungssystem n, Durchnumerierungssystem n

s. of one component Einkomponentensystem n, Einstoffsystem n, unitäres System n

s. of three components Dreikomponentensystem n, Dreistoffsystem n, ternäres System n

s. of two components Zweikomponentensystem n, Zweistoffsystem n, binäres System n

systematic error systematischer Fehler m

s. name systematischer (rationeller, systematisch gebildeter) Name m, systematische Benennung f

systemic (Pflanzenschutz) systemisch, innertherapeutisch, translokal

systemic [chemical] systemisches (innertherapeutisches) Mittel n, Innertherapeutikum n

s. fungicide systemisches Fungizid n

s. herbicide systemisches Herbizid n

s. insecticide systemisches Insektizid n, Systeminsektizid n

s. property systemische Wirksamkeit f (eines Pflanzenschutzmittels)

Szilard-Chalmers detector Szilard-Chalmers-Detektor m

Szilard-Chalmers effect Szilard-Chalmers-Effekt m

Szilard-Chalmers method Szilard-Chalmers-Ver-

fahren n, Szilard-Chalmers-Methode f (zur
Abtrennung von Radionukliden von ihrer
isotopen Ausgangssubstanz)
Szilard-Chalmers reaction Szilard-Chalmers-Reaktion f

T

T s. 1. transmittance; 2. transport number; 3.
trifunctional
2,4,5-T s. 2,4,5-trichlorophenoxyacetic acid
T-bore stopcock Dreiweghahn m mit Bohrung
senkrecht zur Achse
T-piece s. T-shape connecting tube
T-shape 120° bore stopcock Dreiweghahn m nach
Czako, Czako-Hahn m, Karlsruher Hahn m
T-shape connecting tube T-Stück n, T-förmiges
Zwischenabzweigstück n
T-50 test T-50-Test m (Methode zur Bestimmung
des Vulkanisationszustands)
T-type connector s. T-shape connecting tube
T unit s. trifunctional unit
T-50 value T-50-Wert m (zur Bestimmung des
Vulkanisationszustands)
tab gate (Plast) Vorkammeranguß m
tabernanthine Tabernanthin n, 13-Methoxyibogamin n (Alkaloid aus Tabernanthe iboga)
table Herd m, Setzherd m (Aufbereitungstechnik)
  t. casting process s. t. process
  t. filter Planfilter n
  t. floatation Herdflotation f
  t. margarine Tafelmargarine f
  t. of atomic weights Tabelle f der relativen
Atommassen, (bisher) Atomgewichtstabelle f
  t. oil Tafelöl n
  t. porcelain Geschirrporzellan n
  t. press Pressentisch m
  t. process (Glas) Tischverfahren n
  t. roll (Pap) Registerwalze f
  t.-roll section (Pap) Registerteil m, Registerpartie
f
  t. salt Tafelsalz n, Speisesalz n
  t. syrup Speisesirup m
  t. vibrator Tischrüttler m
  t. vinegar Speiseessig m, Tafelessig m
  t. water Tafelwasser n
tablet Tablette f
  t.-compressing machine, t. press s. tabletting
machine
tabletting Tablettieren n, Tablettierung f
  t. machine (press) Tablettiermaschine f, Tablettenpresse f
tabling Herdarbeit f, Herdsortieren n
tabular tafel[förm]ig
  t. spar (Min) Tafelspat m, (veraltet für)
Wollastonit m (Kalziumtrisilikat)
tabulating paper Additionsrollenpapier n, Kassenrollenpapier n, Registerrollenpapier n, Registrier[kassen]papier n

tabun Tabun n (Dimethylaminozyanphosphorsäureäthylester; Kampfstoff)
tacamahac[a] [gum] 1. Chibonharz n, Gomartharz
n (von Bursera gummifera L.); 2. (Balsam von
Populus candicans Ait.)
tach[h]ydrite, tachyhydrite (Min) Tachyhydrit m,
Tach[h]ydrit m (Kalziummagnesiumchlorid)
tachysterol Tachysterin n (Zwischenprodukt bei
der Bildung von Vitamin D_2 aus Ergosterin) .
tack Klebrigkeit f; Zügigkeit f (der Druckfarbe)
  t.-free kleb[e]frei, nicht klebrig (klebend)
  t.-producing agent Klebrigmacher m
tackifier s. tack-producing agent
tackify / to klebrig machen
tackiness s. tack
tackle (Pap) Bemesserung f, Messergarnierung f
(beim Holländer)
tacky klebrig; zügig (Druckfarbe); (Gum) klebefreudig
taconite (Geol) Takonit m
tacticity Taktizität f
tactoid (Koll) Taktoid n
tactosol (Koll) Taktosol n
taenite (Min) Taenit m, Tänit m (Ni-reiches
Bandeisen)
tag / to markieren (eine Substanz durch den Einbau
gut nachweisbarer, meist radioaktiver Atome
kenntlich machen)
tag [card]board Anhänger[etiketten]karton m
  t. roll (Pap) Walze f der Zugpresse (Transportpresse, Greiferpresse, Vorzugspresse)
Tag closed tester s. Tagliabue closed tester
tagatose Tagatose f (eine Ketose)
tagetone Tageton n, 2-Methylmethylen-6-okten-7-
on-(4) n
tagged atom markiertes Atom n
  t. compound markierte Verbindung f
Tagliabue closed tester geschlossener Flammpunktprüfer m nach Tagliabue, geschlossener
Tagliabue-Prüfer m
taiguic acid Taigusäure f, Lapachol n, Lapachosäure f, Grönhartin n, Tekomin n, 2-Hydroxy-3-
(3-methyl-2-butenyl)-1,4-naphthochinon n
tail Schwanz m (eines Monomeren); (Chromatografie) Schwanz m, Schweif m; (Aufbereitungstechnik) Abgänge mpl, Berge pl; (Destillation)
Nachlauf m; (Gerb) (letztes Gefäß eines
Gerbsystems)
  t. gas Abgas n
  t.-gas tower (Pap) Abgasturm m
  t. oil Schwanzöl n, Schwanzfett n
  t. pulley Hecktrommel f, Heckrolle f
  t.-to-tail polymerization Schwanz-Schwanz-Polymerisation f
  t.-to-tail structure Schwanz-Schwanz-Struktur f
tailed stopcock Schwanzhahn m
tailing 1. Schwanzbildung f, Schweifbildung f (z.B.
in Chromatogrammen); 2. s. tailings
  t. discharge Bergeaustrag m
tailings (Aufbereitungstechnik) Abgänge mpl,
Berge pl, (Flot auch) Flotationsberge pl; (Klas-

sierung) Überlaufprodukt *n*, Überlaufgut *n*, Überlauf *m*, *(bei Holzschliff auch)* Grobstoff *m*, Spuckstoff *m*, „Sauerkraut" *n*; *(Korngrößeneinteilung)* Überkorn *n*, Grobkorn *n*, Grobgut *n*, *(Siebklassierung auch)* Siebgrobes *n*, *(Windsichten auch)* Sichtgrobes *n*; *(Destillation)* Nachlauf *m*; *(Gerb)* Restbrühe *f*, ausgezehrte Farbe *f*
tailkey stopcock *s.* tailed stopcock
tailor[-make] / to *(z.B. Polymere)* „nach Maß" (gezielt) aufbauen (herstellen), „maßschneidern"
tails *s.* tailings
take / to *(eine Ladung)* fassen
　t. off wegnehmen, abnehmen; *(Destillation)* abnehmen, entnehmen
　t. off the mould *(Ker)* entformen
　t. out entnehmen; entziehen *(z.B. Wasser)*
　t. the dye *(Text)* Farbstoff (Farbe) annehmen (aufnehmen)
　t. up *(einen Stoff)* aufnehmen
　t. up the web *(Pap)* die Bahn abnehmen (abgautschen)
　t. up water Wasser aufnehmen
take-off Wegnehmen *n*, Abnehmen *n*, Abnahme *f*; *(Destillation)* Entnahme *f*, Abnahme *f*
　t.-off equipment Abnahmevorrichtung *f*
　t.-out *(Glas)* Entnahme *f*; *s.* t.-out mechanism
　t.-out jaw *s.* t.-out tongs
　t.-out mechanism *(Glas)* Entnahmevorrichtung *f*
　t.-out tongs *(Glas)* Entnahmegreifer *m*
　t.-up Aufnahme *f* *(von Stoffen)*
takut galls *(Gerb)* *(Gallen von Tamarix articulata Vahl)*
Talalay process Talalay-Treibverfahren *n*, Talalay-Prozeß *m* *(Schaumgummiherstellung)*
talc *(Min)* Talk *m*, Talkum *n*, Talkstein *m* *(Magnesiumdihydrogentetrasilikat)*; *(Min)* Muskovit *m*, Kaliglimmer *m* *(ein Phyllosilikat)*; $Mg_3[Si_4O_{10}](OH)_2$ *(synthetischer)* Talk *m*; Talkum *n*, Talk *m* *(pulvrige Form des Talks)*
　t. schist (slate) Talkschiefer *m*
talca gum *s.* talh[a] gum
talcky *s.* talcose
talcose talkig, Talk...
　t. schist (slate) *s.* talc schist (slate)
talcum *s.* talc
　t. powder Talk[um]puder *m*
talh[a] gum Talh[a]-Gummi *n*, Talca-Gummi *n* *(Arabisches Gummi, besonders von Acacia stenocarpa Hochst.)*
talki gum *s.* talh[a] gum
tall oil Tallöl *n*, Kiefernöl *n*, Harzöl *n*
　t.-oil rosin Tallölkolophonium *n*
　t.-oil soap *(Pap)* Tallölseife *f*, Sulfatseife *f*, Rohseife *f*
tallingite *(Min)* Tallingit *m*, Konnellit *m*, Connellit *m*
tallo[e]l, tallöl *s.* tall oil
tallow Talg *m*
　t.-seed oil Stillingia-Samenöl *n*, Stillingia-Öl *n*, Mong Yu *n* *(von Sapium sebiferum (L.)Roxb.)*

tallowiness Talgigwerden *n* *(z.B. der Fette)*; talgiger Geschmack *m*; Peroxidranzigkeit *f*
tallowy talg[art]ig
talomethylose Talomethylose *f* *(eine 6-Desoxyhexose)*
talomucic acid Taloschleimsäure *f*
talonic acid Talonsäure *f* *(eine Polyhydroxymonokarbonsäure)*
talose Talose *f* *(eine Hexose)*
tamarind[o] Tamarinde *f*, Tamarindenbaum *m*, Tamarindus indica L.; Tamarinde *f*, Tamarindenfrucht *f*; Tamarindenmus *n*
　t. pulp Tamarindenmus *n*
tamp / to [fest]stampfen
tamp *(Ker)* Stampfer *m*
tamping *(Ker)* Stampfen *n*
tan / to gerben
tan *s.* 1. tanning material; 2. tanning agent
　t. content Gerbstoffgehalt *m*
　t. fixation Gerbstoffixierung *f*, Fixierung *f* des Gerbstoffs
　t. liquor *s.* tanning liquor
　t. oak *s.* tanbark oak
　t. pit (vat) *s.* tanning vat
TAN *s.* total acid number
tanacetin Tanazetin *n* *(Bitterstoff von Chrysanthemum vulgare (L.) Bernh.)*
tanacetketone Tanazetketon *n*, 6-Methyl-5-methylen-2-heptanon *n*
tanacetone Tanazeton *n*, β-Thujon *n*, d-Isothujon *n*, 1-Isopropyl-4-methylbizyklo-[0,1,3]-3-hexanon *n*
tanacetyl alcohol Tanazetylalkohol *m*, β-Thujylalkohol *m*, 1-Isopropyl-4-methylbizyklo-[0,1,3]-hexanol *n*
tanbark Gerbrinde *f*
　t. oak Gerbereiche *f*, Lithocarpus densiflora Rehd.
tandem calender *(Gum)* Tandemanlage *f*, Kalander *m* in Tandemanordnung
tangential acceleration Tangentialbeschleunigung *f*
　t. firing Eckenfeuerung *f*
tanghinin Tanghinin *n* *(Glykosid aus Tanghina venenifera)*
tank Tank *m*, [geschlossener] Behälter *m*; Wanne *f*, Trog *m*; Pfanne *f*, Mulde *f*; *(Gum)* Kessel *m*, Vulkanisierkessel *m*, Vulkanisationskessel *m*; *(Glas)* Wanne *f*, Schmelzwanne *f*; Sammler *m*, Tank *m* *(beim Tube-and-Tank-Verfahren)*
　t. block *(Glas)* Wannenstein *m*
　t. bottom Tankboden *m*, Behälterboden *m*
　t.-bottom coil Bodenschlange *f*
　t.-bottom wax Tankbodenwachs *n*, Tankbodenparaffin *n*, Tankrückstandswachs *n*, Tankrückstandsparaffin *n*
　t. bottoms Tankbodenrückstände *mpl*
　t. car Eisenbahnkesselwagen *m*, Eisenbahntankwagen *m*
　t. coil Innenrohrschlange *f*

t. crystallizer offener feststehender Kristallisator *m*, Kastenkristallisator *m*, Pfannenkristallisator *m*, Kristallisierpfanne *f*
t. developer *(Foto)* Tankentwickler *m*
t. development *(Foto)* Tankentwicklung *f*
t. farm Tanklager *n*, „Tankfarm" *f*
t. glass Wannenglas *n*, in der Wanne geschmolzenes Glas *n*
t.-mix method *(Pflanzenschutz)* Tankmischmethode *f*
t. retting *(Text)* Röste *f* im Wasserbehälter
t. ship Tankschiff *n*, Tanker *m*
t. sludge Tankschlamm *m*
t. steamer Tankdampfer *m*
t. trailer Tank[an]hänger *m* *(für LKW)*
t. truck Tankauto *n*, Autokesselwagen *m*, Straßentankwagen *m*
t.-type dialyzer Tankdialysator *m*
t. wag[g]on *s.* t. car
tankage Fleischmehl *n*; Tierkörpermehl *n*, Kadavermehl *n*
tanker Tanker *m*, Tankschiff *n*; Straßentankwagen *m*, Tankauto *n*, Autokesselwagen *m*; Tankflugzeug *n*
tanking *(Farb)* Lagern (Ablagern) *n* im Tank
tannage Gerben *n*, Gerbung *f*, Gerberei *f*
tannase Tannase *f* *(gerbstoffspaltendes Ferment)*
tannate Tannat *n* *(Salz oder Ester des Tannins)*
tannenite *(Min)* Tannenit *m*, *(veraltet für)* Emplektit *m*
tanner Gerber *m*
tanner's bark *s.* tanbark
t. liquor *s.* tanning liquor
t. red Gerbstoffrot *n*, Gerberrot *n*, Phlobaphen *n*
t. sumac Gerbersumach *m*, Rhus coriaria L.
tannery Gerberei *f*, Gerbanlage *f*; Gerberei *f*, Gerben *n*, Gerbung *f*
t. stock Gerbmittelvorrat *m*
tannic acid *s.* tannin
t. acid of commerce Handelstannin *n (Gallussäure-Glukoseester mit Penta-m-digalloyl-β-glukose als Hauptbestandteil)*
t. acid proper eigentliches (gewöhnliches) Tannin *n*, eigentliche Gerbsäure *f*
t.-acid solution Tanninlösung *f*, Gerbsäurelösung *f*
tanniferous tanninhaltig, gerbsäurehaltig; gerbstoffhaltig
t. plant Gerbstoffpflanze *f*, gerbstoffhaltige Pflanze *f*
tannin Tannin *n*, Gallusgerbsäure *f*, Gerbsäure *f (1. m-Digalussäure; 2. Gallussäure-Glukoseester mit Penta-m-digalloyl-β-glukose als Hauptbestandteil)*; Gerbstoff *m (pflanzlicher Herkunft)*, Tannin *n (i.w.S.)*, *(ungenau)* Gerbsäure *f (i.w.S.)*
t. extract *s.* tanning extract
t.-mordanted tanningebeizt, tanniert, mit Tannin gebeizt
t. reaction Gerbstoffreaktion *f*
t. solution Tanninlösung *f*, Gerbsäurelösung *f*; Gerb[stoff]lösung *f*

t.-solutizer process Tannin-Solutizerverfahren *n* *(zur Entschwefelung von Erdöldestillaten)*
tanning Gerben *n*, Gerbung *f*, Gerberei *f*; Bräunung *f (der Haut)*; *(Foto)* Gerbung *f*
t. action Gerbwirkung *f*, gerbende (gerberische) Wirkung *f*, Gerbeffekt *m*
t. agent Gerbstoff *m*, gerbender Stoff *m*, gerbende Substanz *f (gerbender Anteil des Gerbmittels)*
t. bark *s.* tanbark
t. developer *(Foto)* gerbender Entwickler *m*
t. drum Gerbfaß *n*, Gerbtrommel *f*
t. effect *s.* t. action
t. extract Gerbmittelauszug *m*, Gerb[stoff]extrakt *m*, Gerbstoffauszug *m*
t. liquor Gerb[stoff]brühe *f*
t. material Gerbmittel *n*, Gerbmaterial *n*
t. matter *s.* t. agent
t. method Gerb[erei]methode *f*, Gerbverfahren *n*
t. oil Gerböl *n*
t. pit *s.* t. vat
t. plant 1. Gerbanlage *f*, Gerberei *f*; 2. Gerbstoffpflanze *f*, gerbstoffhaltige Pflanze *f*
t. power Gerbvermögen *n*
t. process Gerbverfahren *n*, Gerb[erei]methode *f*; Gerbprozeß *m*, Gerbvorgang *m*
t. proper eigentliches Gerben *n*, eigentliche Gerbung (Gerberei) *f*
t. solution Gerb[stoff]lösung *f*
t. substance *s.* t. agent
t. sumac *s.* tanner's sumac
t. value Gerbwert *m*
t. vat Gerbgrube *f*
tannometer Gerbsäuremesser *m*
tansy oil Rainfarnöl *n*, Tanazetöl *n (aus Chrysanthemum vulgare (L.) Bernh.)*
tantalate Tantalat *n*
tantalic Tantal...
t. acid $Ta_2O_5 \cdot xH_2O$ Tantalsäure *f*
t.-acid anhydride *s.* tantalum pentoxide
t. bromide $TaBr_5$ Tantal(V)-bromid *n*, Tantalpentabromid *n*
t. chloride $TaCl_5$ Tantal(V)-chlorid *n*, Tantalpentachlorid *n*
t. fluoride TaF_5 Tantal(V)-fluorid *n*, Tantalpentafluorid *n*
tantalite *(Min)* Tantalit *m*
tantalous Tantal...
t. bromide $TaBr_3$ Tantal(III)-bromid *n*, Tantaltribromid *n*
t. chloride $TaCl_3$ Tantal(III)-chlorid *n*, Tantaltrichlorid *n*
tantalum Ta Tantal *n*
t. carbide Tantalkarbid *n*
t. hydroxide $Ta(OH)_5$ Tantal(V)-hydroxid *n*
t. pentoxide Ta_2O_5 Tantal[pent]oxid *n*, Tantal(V)-oxid *n*
t. sulphide Ta_2S_4 *oder* TaS_2 Tantal(IV)-sulfid *n*
tanwood Gerbholz *n*, gerbstoffhaltiges Holz *n*
tanyard Gerberei *f*, Gerbanlage *f*

tap / to [an]zapfen; *(Schlacke)* abziehen; *(schmelz-flüssiges Metall)* abstechen, ablassen; *(Glas) (eine Wanne)* ablassen, abstechen, anstechen; *(einen Hafen)* ausschöpfen
t. out *(schmelzflüssiges Metall)* abstechen, ablassen
tap Hahn *m*; Zapfen *m*, Spund *m*
t. funnel Tropftrichter *m*
t. grease (lubricant) Hahnfett *n*, Hahnschmiermittel *n*
t. water Leitungswasser *n*
tape Band *n*, Streifen *m*; Bandage *f*, Binde *f*
t. delivery table *(Pap)* Ablegetisch *m*
taper / to sich [konisch] erweitern (verjüngen), konisch zulaufen, konisch [gebaut] sein, konisch erweitert sein, verjüngt sein, Konizität aufweisen
taper konisch [zulaufend], konisch gebaut (erweitert), verjüngt, kegelförmig, keg[e]lig
taper Konizität *f*, Verjüngung *f*, Erweiterung *f*; *(Plast)* Werkzeugneigung *f*, Werkzeugkonizität *f*
t.-joint mit Kegelschliff [versehen]
t. pipe thread kegliges Rohrgewinde *n*
tapering *s.* taper
taphole *(Met)* Abstichloch *n*, Stichloch *n*, Abstichöffnung *f*, Abstich *m*, Stich *m*
t. for matte Steinstichloch *n*, Stein[ab]stich *m*
t. for the molten iron Eisenstichloch *n*, Eisen[ab]stich *m*
tapioca [starch] Tapioka[stärke] *f* *(von Manihot utilissima Pohl)*
tapiolite *(Min)* Tapiolit *m*
tapping Anzapfen *n*, Zapfen *n*, Zapfung *f*; Abziehen *n* *(z.B. von Schlacke)*; Abstechen *n*, Abstich *m*, Ablassen *n* *(von schmelzflüssigem Metall)*
t. cup *(Gum)* Sammelbecher *m*, Latexbecher *m*
t. cut *(Gum)* Zapfschnitt *m*
t. electrode Hilfselektrode *f* zum Öffnen des Abstichlochs
t. hole *s.* taphole
t. incision *s.* t. cut
t. knife *(Gum)* Zapfmesser *n*
tapuru *(Kautschuk von Sapium taburu Ule)*
tar / to teeren
tar Teer *m*
t. acid Teersäure *f*
t. board *s.* tarred board
t. camphor $C_{10}H_8$ Steinkohlenkampfer *m*, *(veraltet für)* Naphthalin *n*
t. cancer Teerkrebs *m*
t. extractor *s.* t. separator
t. filter Teerfilter *n*
t. fraction Teerfraktion *f*
t. oil Teeröl *n*
t. paper *s.* tarred [brown] paper
t. pitch Teerpech *n*
t. sand Teersand *m*, Asphaltsand *m*
t. scrubber Teerwäscher *m*, Teerwascher *m*
t. separation Teer[ab]scheidung *f*, Entteerung *f*
t. separator Teer[ab]scheider *m*, Entteerer *m*
t. vapour Teerdampf *m*

t. white *s.* t. camphor
tara *(Gerb)* Tara *(Hülsen von Caesalpinia tinctoria Domb.)*
tarapacaite *(Min)* Tarapacait *m* *(Kaliumchromat)*
taraxanthin Taraxanthin *n* *(ein Karotinoid)*
taraxasterol Taraxasterin *n* *(Phytosterin von Taraxacum officinale Web.)*
tare / to [aus]tarieren
tarelaidic acid Tarelaidinsäure *f*, *trans*-6-Oktadezensäure *f*
tarfree teerfrei
target Target *n*, Auffänger *m*, Auffangfläche *f*, Auftreffplatte *f*, Treffplatte *f*, Antikatode *f* *(z.B. einer Röntgenröhre)*
tariric acid $CH_3(CH_2)_{10}C\equiv C(CH_2)_4COOH$ Taririnsäure *f*, 6-Oktadezinsäure *f*
tarnish / to anlaufen, blind werden
tarnish Anlaufen *n*, Blindwerden *n*; *(Glas)* Beschlag *m* *(Fehler)*
t. decolourization Auftreten *n* von Anlauffarben
tarnishing Anlaufen *n*, Blindwerden *n*
taroxylic acid Taroxylsäure *f*, 6,7-Dioxooktadekansäure *f*
tarpaulin Persenning *f*, geteertes Leinwandgewebe *n*
tarragon oil Estragonöl *n* *(aus Artemisia dracunculus L.)*
t. vinegar Estragonessig *m*
tarred board Teer[dach]pappe *f*
t. [brown] paper Bitumenpapier *n*, bituminiertes Papier *n*, Teerpapier *n*, Asphaltpapier *n*; Doppelpechpapier *n*
tarry teer[art]ig; nach Teer riechend
t.-smelling nach Teer riechend
tart. a. *s.* tartaric acid
tartar Weinstein *m*, Kaliumhydrogentartrat *n*
t. emetic $K[C_4H_2O_6Sb(OH)_2]\cdot 0{,}5H_2O$ Brechweinstein *m*, *(veraltet für)* Kaliumantimonotartrat-0,5-Wasser *n*
tartaric acid $HOOC(CHOH)_2COOH$ Wein[stein]-säure *f*, Tartrasäure *f*, Dihydroxybernsteinsäure *f*, 2,3-Dihydroxybutandisäure *f*
d-tartaric acid $HOOC(CHOH)_2COOH$ *d*-Weinsäure *f*, Rechtsweinsäure *f*, rechtsdrehende Weinsäure *f*, [gewöhnliche] Wein[stein]säure *f*, *d*-2,3-Dihydroxybutandisäure *f*
dl-tartaric acid $HOOC(CHOH)_2COOH$ *dl*-Weinsäure *f*, razemische Weinsäure (Säure) *f*, Paraweinsäure *f*, Traubensäure *f*
l-tartaric acid $HOOC(CHOH)_2COOH$ *l*-Weinsäure *f*, Linksweinsäure *f*, linksdrehende Weinsäure *f*
t. monoamide *s.* tartramidic acid
t. monoanilide *s.* tartranilic acid
tartramide $NH_2COCH(OH)CH(OH)CONH_2$ Tartramid *n*, Weinsäurediamid *n*, 2,3-Dihydroxybutandiamid *n*
tartramidic acid $NH_2COCH(OH)CH(OH)COOH$ Tartramidsäure *f*, Weinsäuremonoamid *n*, 2,3-Dihydroxybutandisäuremonoamid *n*

tartranilic acid $C_6H_5NHCOCH(OH)CH(OH)COOH$ Tartranilsäure f, Weinsäuremonoanilid n, 2,3-Dihydroxybutandisäuremonoanilid n

tartrate Tartrat n *(Salz oder Ester der Weinsäure)*

t. complex Tartratokomplex m

tartrated antimony $K[C_4H_2O_6Sb(OH)_2]$ Kaliumantimonotartrat n

tartrazine Tartrazin n, Hydrazingelb O n, Echtwollgelb n, Säuregelb n, Echtlichtgelb n, Flavazin T n

tartronic acid $HOOCCH(OH)COOH$ Tartronsäure f, Hydroxymalonsäure f, Hydroxypropandisäure f

tartronylurea Tartronylharnstoff m, 5-Hydroxybarbitursäure f

taste / to auf Geschmack prüfen; schmecken, einen bestimmten Geschmack haben

taste Geschmack m; Verkosten n, Kosten n, Schmecken n, Probieren n *(einer Speise)*
to have an acid t. sauer schmecken

t.-free s. tasteless

t.-impairing geschmacksbeeinträchtigend

t. improver Geschmacksverbesserer m

tasteless geschmacklos, geschmackfrei ·

taster *(berufsmäßiger)* Schmecker m, Koster m, Prüfer m; Probiergläschen n *(z.B. für Wein)*

tastiness Schmackhaftigkeit f

tasty schmackhaft, wohlschmeckend

taurine $CH_2NH_2CH_2SO_2OH$ Taurin n, 2-Aminoäthan-1-sulfonsäure f

taurocarbamic acid $H_2NCONHCH_2CH_2SO_3H$ Taurokarbamidsäure f, 2-Ureidoäthan-1-sulfonsäure f

taurocholic acid Taurocholsäure f *(aus Cholsäure und Taurin bestehende gepaarte Gallensäure)*

tauryl $H_2NCH_2CH_2SO_2—$ Tauryl n

tautomer Tautomer[es] n

tautomeric tautomer

t. equilibrium Tautomeriegleichgewicht n, Tautomerengleichgewicht n

t. form tautomere Form f

tautomerism Tautomerie f

tautomerizable tautomerisierbar

tautomerization Tautomerisierung f

t. constant Tautomeriekonstante f, Tautomerenkonstante f

tautomerize / to tautomerisieren

tautomery s. tautomerism

taw / to weißgerben, alaungerben

tawing Weißgerbung f, Alaungerbung f, Weißgerberei f

taxifolin Taxifolin n, 2,3-Dihydroquerzeton n, 3,5,7,3′,4′-Hydroxy-2,3-dihydroflavonol n

taxine Taxin n *(Alkaloidgemisch aus Taxus baccata L.)*

tazettine Tazettin n *(Alkaloid aus Narcissus tazetta L.)*

2,3,6-TBA s. 2,3,6-trichlorobenzoic acid

T.C. rubber s. technically classified rubber

TCA s. 1. tricarboxylic acid; 2. trichloroacetic acid

TCA cycle s. tricarboxylic-acid cycle

TCC s. Thermofor catalytic cracking

TCC process s. Thermofor catalytic-cracking process

TCP s. 1. Thermofor continuous percolation; 2. tricresyl phosphate

TDI s. toluene diisocyanate

tea oil s. teaseed oil

t. tannin Teegerbstoff m

teal oil s. teel oil

tear / to [zer]reißen

tear 1. Träne f, Tropfen m, Lacktropfen m *(beim Tauchlackieren)*; 2. Reißen n, Zerreißen n; Einriß m; *(Glas)* Klebestelle f *(Fehler)*

t. gas Tränengas n

t. initiation strength Einreißfestigkeit f, Einreißwiderstand m, Widerstand m gegen Einreißen

t.-producing tränenerzeugend

t. propagation strength s. t. resistance 2.

t. propagation test s. t. test

t. resistance (strength) 1. Zerreißfestigkeit f, Reißfestigkeit f, Rißbeständigkeit f; 2. Weiterreißfestigkeit f, Weiterreißwiderstand m, Durchreißfestigkeit f, Fortreißfestigkeit f

t. test Weiterreiß[festigkeits]prüfung f, Durchreißfestigkeitsprüfung f, Fortreißfestigkeitsprüfung f

tearing resistance (strength) s. tear resistance

t. test s. tear test

teartness *(durch Molybdänüberschuß verursachte Pflanzenkrankheit)*

teaseed oil Tee[samen]öl n

teaser *(Glas)* Schmelzer m

teat pipette Mikropipette f, Nadelpipette f

technate Didym n *(Nd-Pr-Gemisch)*

technetium Tc Technetium n

technical aniline technisches Anilin n, Anilinöl n

t. ceramics technische Keramik f

t. chemistry technische (industrielle) Chemie f, chemische Technik f, Chemietechnik f, Industriechemie f

t. paper technisches Papier (Sonderpapier) n

t. white oil technisches Weißöl n, Weißöl n für technische Zwecke

technically classified rubber technisch klassifizierter Kautschuk m, TC-Kautschuk m

t. pure technisch [rein], techn.

technique Verfahren n, Methode f, Technik f, Arbeitsweise f

technology Technik f; Technologie f

t. of rubber Kautschuktechnologie f, Gummitechnologie f

tecoretin Tecoretin m, *(veraltet für)* Fichtelit m *(terpenartiges organisches Mineral)*

tectoquinone Tektochinon n, 2-Methyl-9,10-anthrachinon n

tee [connector] T-Stück n, T-förmiges Zwischenabzweigstück n

teel oil Sesamöl n *(von Sesamum indicum L.)*

teem / to *(Glas)* den Hafen ausgießen

teeth *(Pap)* Holländermesser npl, Messer npl

Teggant galls *(Gerb) (Gallen von Tamarix articulata Vahl)*
teinochemistry Teinochemie *f*
tektite *(Geol)* Tektit *m*
TEL *s.* tetraethyl lead
telethermometer Fernthermometer *n*
tel[i]inite Telinit *m (Einzelmazeral der Kohle)*
tellurate Tellurat *n*
telluric Tellur...; höherwertigem Tellur entsprechend, *(meist)* Tellur(VI)-...
t. acid H_6TeO_6 Tellursäure *f*, Orthotellursäure *f*, Hexoxotellursäure *f*
t.-acid anhydride *s. t.* oxide
t. bismuth *(Min)* Tellurwismut *m*, *(veraltet für)* Tetradymit *m (Wismuttellursulfid)*
t. bromide $TeBr_4$ Tellur(IV)-bromid *n*, Tellurtetrabromid *n*
t. chloride $TeCl_4$ Tellur(IV)-chlorid *n*, Tellurtetrachlorid *n*
t. iodide TeJ_4 Tellur(IV)-jodid *n*, Tellurtetrajodid *n*
t. ochre *s.* tellurite 2.
t. oxide TeO_3 Tellur(VI)-oxid *n*, Tellurtrioxid *n*
telluride M^I_2Te Tellurid *n*
tellurite 1. $M^I_2TeO_3$ Tellurit *n*; 2. *(Min)* Tellurit *m*, Tellurocker *m (Tellur(IV)-oxid)*
tellurium Te Tellur *n*
t. dibromide $TeBr_2$ Tellurdibromid *n*, Tellur(II)-bromid *n*
t. dichloride $TeCl_2$ Tellurdichlorid *n*, Tellur(II)-chlorid *n*
t. dioxide TeO_2 Tellurdioxid *n*, Tellur(IV)-oxid *n*
t. hexafluoride TeF_6 Tellurhexafluorid *n*, Tellur(VI)-fluorid *n*
t. hydride H_2Te Tellurwasserstoff *m*, Tellurhydrid *n*
t. monoxide TeO Tellur[mon]oxid *n*, Tellur(II)-oxid *n*
t. sulphotrioxide $TeSO_3$ Tellursulfit *n*
t. tetrabromide $TeBr_4$ Tellurtetrabromid *n*, Tellur(IV)-bromid *n*
t. tetrachloride $TeCl_4$ Tellurtetrachlorid *n*, Tellur(IV)-chlorid *n*
t. tetrafluoride TeF_4 Tellurtetrafluorid *n*, Tellur(IV)-fluorid *n*
t. tetraiodide TeJ_4 Tellurtetrajodid *n*, Tellur(IV)-jodid *n*
t. trioxide TeO_3 Tellurtrioxid *n*, Tellur(VI)-oxid *n*
tellurous Tellur...; niederwertigem Tellur entsprechend, *(meist)* Tellur(IV)-...
t. acid H_2TeO_3 tellurige Säure *f*
t.-acid anhydride *s. t.* oxide
t. bromide $TeBr_2$ Tellur(II)-bromid *n*, Tellurdibromid *n*
t. chloride $TeCl_2$ Tellur(II)-chlorid *n*, Tellurdichlorid *n*
t. oxide TeO_2 Tellur(IV)-oxid *n*, Tellurdioxid *n*
telomer Telomer[es] *n*, Telomerisat *n*
telomerization Telomerisation *f*
t. reaction Telomerisationsreaktion *f*

Telsmith breaker *s.* Telsmith gyratory crusher
Telsmith cone crusher Telsmith-Kegelbrecher *m*
Telsmith gyratory crusher Telsmith-Kreiselbrecher *m*
temper / **to** *(Metalle)* anlassen; *(Plast)* tempern, spannungsfrei machen; *(Glas) (absichtlich)* verspannen, vorspannen, härten, abschrecken; *(z.B. Mörtel)* mischen, anmachen; *(z.B. Getränke)* verdünnen
tempera paint Temperafarbe *f*
temperance drink alkoholfreies Getränk *n*
temperature Temperatur *f*
7.6 temperature *(Glas)* Erweichungstemperatur *f*, Erweichungspunkt *m*, Littleton-Punkt *m (bei einer Viskosität von $10^{7,6}$ P)*
13.0 temperature *(Glas)* obere Kühltemperatur *f*, oberer Kühlpunkt *m (bei einer Viskosität von 10^{13} P)*
t. build-up *(Gum)* Temperaturentwicklung *f*, Wärmeentwicklung *f (bei dynamischer Beanspruchung)*
t. change Temperaturänderung *f*
t. coefficient Temperaturkoeffizient *m*, Temperaturbeiwert *m*
t. coefficient of viscosity Viskositäts-Temperatur-Koeffizient *m*, Temperaturkoeffizient *m* der Viskosität (Zähigkeit) , VTC
t. compensation Temperaturkompensation *f*, Temperaturausgleich *m*
t. control Temperaturregelung *f*
t. crayon Temperaturmeßfarbstift *m*
t. dependence Temperaturabhängigkeit *f*
t.-dependent temperaturabhängig
t. difference Temperaturdifferenz *f*, Temperaturunterschied *m*
t. drop Temperaturabfall *m*; Temperaturgefälle *n*
t.-entropy chart Temperatur-Entropie-Diagramm *n*, T-s-Diagramm *n*
t. equalization Temperaturausgleich *m*
t. equilibrium Temperaturgleichgewicht *n*
t. gradient Temperaturgradient *m*
t. in the grinder pit *(Pap)* Trogtemperatur *f*
t.-independent temperaturunabhängig
t. limit Temperaturgrenze *f*
t. measurement Temperaturmessung *f*
t. of cooking Kochtemperatur *f*
t. of cure *(Gum)* Vulkanisationstemperatur *f*, Heiztemperatur *f*; *(Farb, Plast)* Härtetemperatur *f*
t. of distillation Destillationstemperatur *f*
t. of explosion Explosionstemperatur *f*
t. of mastication Mastikationstemperatur *f*, Mastiziertemperatur *f*
t. of operation Betriebstemperatur *f*, Arbeitstemperatur *f*
t. of ripening Reifungstemperatur *f*
t. of the stock *(Pap)* Stofftemperatur *f*
t. of vulcanization Vulkanisationstemperatur *f*, Heiztemperatur *f*
t. range Temperaturbereich *m*

t. recorder Temperaturschreiber *m*
t.-resisting temperaturbeständig, temperaturstabil
t. rise Temperaturerhöhung *f*, Temperatursteigerung *f*, Temperaturanstieg *m*
t. scale Temperaturskale *f*
t.-sensitive temperaturempfindlich
t. sensitivity Temperaturempfindlichkeit *f*
t.-stable *s.* t.-resisting
tempered glass vorgespanntes Glas *n*, *(i.e.S.)* Einscheibensicherheitsglas *n*, Einschichtensicherheitsglas *n*
tempering Anlassen *n (von Metallen); (Plast)* Tempern *n*, Spannungsfreimachen *n; (Glas) (absichtliches)* Verspannen *n*, Vorspannen *n*, Vorspannung *f*, Härten *n*, Abschrecken *n*; Mischen *n*, Anmachen *n (z.B. von Mörtel)*; Verdünnen *n*, Verdünnung *f (z.B. von Getränken); (Gerb)* Dampfmachen *n*
t. oil Anlaßöl *n*
t. tank (vat, vessel) Temperiergefäß *n*, Temperierkessel *m*
t. water Anmach[e]wasser *n*
temporary hardness temporäre (vorübergehende, schwindende) Härte *f*, Karbonathärte *f (des Wassers)*
t. preservation Haltbarmachen *n* für beschränkte Zeit (für kürzere Zeiträume)
t. soil-sterilant Bodendesinfektionsmittel *n*
tenacity Festigkeit *f*, Zugfestigkeit *f*, Reißfestigkeit *f*; Zähigkeit *f*; Haftfestigkeit *f*
tend to revert / to *(Gum)* zur Reversion neigen
tendency Tendenz *f*, Neigung *f*, Bestreben *n*
t. to agglomerate Neigung *f* zur Klumpenbildung
t. to crystallize Kristallisationstendenz *f*, Kristallisationsneigung *f*
t. to scorch Anvulkanisationstendenz *f*, Anvulkanisationsneigung *f*, Scorchneigung *f*, Scorchtendenz *f*, Anbrenntendenz *f*
tender / to *(Text)* schwächen, schädigen, morsch (mürbe) machen; *(Text)* schwach (morsch, mürbe) werden
tender clay empfindlicher Ton *m (beim Trocknen und Brennen zum Reißen neigend)*
tendering *(Text)* Schwächung *f*, Schädigung *f*; *(Text)* Morschwerden *n*, Mürbewerden *n*
t. effect *(Text)* Schädigungseffekt *m*
tengerite *(Min)* Tengerit *m*
tennantite *(Min)* Tennantit *m*
tenor of ore Metallgehalt *m* des Roherzes
tenorite *(Min)* Tenorit *m (Kupfer(II)-oxid)*
tenside Tensid *n (die Oberflächenspannung erniedrigender Stoff)*
tensile *s.* t. strength
t. characteristics *s.* t. stress-strain characteristics
t. elongation Zugdehnung *f*
t. failure Zugbruch *m*
t. force Zugkraft *f*
t. hysteresis loop *(Gum)* Dämpfungsschleife *f*, Hysteresiskurve *f*

t. modulus *(Gum)* Modul *m*, Zwischenzahl *f*, Spannungswert *m*
t. properties *s.* t. stress-strain properties
t. set Dehnungsrest *m*, Zugverformungsrest *m*, Formänderungsrest *m* bei Dehnungsbeanspruchung
t. strength Zugfestigkeit *f*, Reißfestigkeit *f*, Zerreißfestigkeit *f*
t.-strength test Zugversuch *m*, Zerreißprüfung *f*
t.-strength tester (testing machine) Zugfestigkeitsprüfgerät *n*, Zerreißmaschine *f*, Festigkeitsprüfer *m*, Festigkeitsprüfmaschine *f*, Zug-Dehnungsprüfgerät *n*
t. stress Zugspannung *f*
t. stress-strain characteristics (properties) Zugdehnungseigenschaften *fpl*, Zugverformungseigenschaften *fpl*, Zug-Dehnungs-Verhalten *n*
t. stretch *(Pap)* Bruchdehnung *f (Zugfestigkeitsprüfung)*
t. test *s.* t.-strength test
t. tester (testing machine) *s.* t.-strength tester
tension Spannung *f*, Zugspannung *f*
t. depressor grenzflächenaktiver (oberflächenaktiver, die Oberflächenspannung erniedrigender) Stoff *m*
t. of the web Papierzugspannung *f*, Papierbahnspannung *f*, Bahnspannung *f*
t. on the felt *(Pap)* Filzspannung *f*
t. pulley Spannrolle *f*, Spanntrommel *f*
t. roll *(Pap)* Spannwalze *f*
tensioning device Spannvorrichtung *f*, Spanneinrichtung *f*
tent fumigation *(Landw)* Zeltbegasung *f*
tenter Spannrahmen *m*, *(Text auch)* Rahmenspannmaschine *f*
t. dryer Spannrahmentrockner *m*
t. frame *s.* tenter
t. frame dryer *s.* t. dryer
tenth-metre *s.* angstrom
tephroite *(Min)* Tephroit *m*
ter Meer's nitrator Nitrierapparat *m* nach Weiler ter Meer
teratolite *(Min)* Teratolith *m*, sächsische Wundererde *f*
terbia Tb_2O_3 Terbinerde *f*, Terbia *f*, *(veraltet für)* Terbium(III)-oxid *n*
terbium Tb Terbium *n*
t. chloride $TbCl_3$ Terbium(III)-chlorid *n*
t. fluoride TbF_3 Terbium(III)-fluorid *n*
t. nitrate $Tb(NO_3)_3$ Terbium(III)-nitrat *n*
t. oxide Tb_2O_3 Terbium(III)-oxid *n*
t. peroxide Tb_4O_7 Tetraterbiumheptoxid *n*
t. sulphate $Tb_2(SO_4)_3$ Terbium(III)-sulfat *n*
terbromide Tribromid *n*
terchloride Trichlorid *n*
tereb[in]ic acid Terebinsäure *f*, 2,2-Dimethylparakonsäure *f*
terephthalic acid $C_6H_4(COOH)_2$ Terephthalsäure *f*, Benzoldikarbonsäure-(1,4) *f*
teri pods *(Gerb)* Teri *pl (Hülsen von Caesalpinia digyna Rottl.)*

term Term *m*, Energieterm *m*, Energiezustand *m*,
Energieniveau *n*, Elektronenniveau *n*, Energiestufe *f*; *(Math)* Term *m*; Fachwort *n*, Fach-
ausdruck *m*, Terminus [technicus] *m*
  t. scheme Termschema *n*
  t. signature (symbol) Termsymbol *n*, Term-
bezeichnung *f*
  t. system Termsystem *n*
terminal endständig, End..., terminal
  t. atom Endatom *n*, endständiges Atom *n*
  t. falling velocity Endfallgeschwindigkeit *f*,
Gleichgewichts-Sinkgeschwindigkeit *f*
  t. group endständige (terminale) Gruppe *f*,
Endgruppe *f*
  t. knife edge Endschneide *f*, Seitenschneide *f*
(z.B. an Analysenwaagen)
  t. velocity *s.* t. falling velocity
terminate / to beenden, abbrechen, abschließen;
begrenzen
termination [of chains] Kettenabbruch *m*, Been-
digung *f* des Kettenwachstums
  t. reaction Abbruchreaktion *f*
  t. stage Abbruchstadium *n*
termolecular trimolekular, 3molekular
  t. reaction trimolekulare Reaktion *f*, Dreier-
reaktion *f*
ternary ternär, dreifach
  t. alloy ternäre Legierung *f*, Dreistofflegierung *f*
  t. collision Dreierstoß *m*
  t. compound ternäre Verbindung *f*
  t. mixture Dreistoffgemisch *n*
  t. system ternäres System *n*, Dreistoffsystem *n*,
Dreikomponentensystem *n*
ternitrate Trinitrat *n*
terpene Terpen *n*
  t. hydrocarbon Terpenkohlenwasserstoff *m*
terpenoid hydrocarbon *s.* terpene hydrocarbon
terpenylic acid Terpenylsäure *f*
terphenyl $(C_6H_5)_2C_6H_4$ Terphenyl *n*, Diphenylbenzol *n*, Diphenylphenylen *n*
terpin Terpin *n*, 1,8-Terpin *n*
terpinene Terpinen *n*
terpineol Terpineol *n* *(ungesättigter Terpenalkohol)*
terpinolene $C_{10}H_{16}$ Terpinolen *n*, 1,4(8)-*p*-Mentha-
dien *n*
terpolymer Terpolymer[es] *n*
terra cotta *(Ker)* Terrakotta *f*, Terrakotte *f*
  t. rossa *(Ker)* Terra rossa *f*, Roterde *f*
  t. sigillata *(Ker)* Terra sigillata *f*
terracotta *s.* terra cotta
terre verte *(Min)* Grünerde *f*, Seladonit *m*
terreic acid Terreinsäure *f*, 2,3-Epoxy-5-hy-
droxy-6-methyl-1,4-benzochinon *n*, 5,6-Epoxy-3-
hydroxy-toluchinon *n*
terrestrial peat terrestrischer Torf *m*
tersulphate Trisulfat *n*
tersulphide Trisulfid *n*
tertiary tertiär, tert.
  t. air Tertiärluft *f*, Drittluft *f*

  t. alcohol tertiärer Alkohol *m*
  t. amine tertiäres Amin *n*
  t. nitrocompound tertiäre Nitroverbindung *f*
  t. phosphate $M^I_3PO_4$ tertiäres (neutrales) Phosphat *n*
  t. salt tertiäres (neutrales) Salz *n* *(einer dreibasigen Säure)*
  t. system ternäres System *n*, Dreistoffsystem *n*,
Dreikomponentensystem *n*
tervalency Dreiwertigkeit *f*, Trivalenz *f*
tervalent dreiwertig, trivalent
test / to testen, prüfen, untersuchen
  t. by spotting tüpfeln
test Test *m*, Prüfung *f*, Versuch *m*, Prüfversuch *m*,
Untersuchung *f*; Nachweis *m*
under t. zu prüfen[d]
  t. animal Versuchstier *n*
  t. bottle Untersuchungsflasche *f*
  t. cock Kontrollhahn *m*, Kontrollhahnventil *n*
  t. compound *(Gum)* Testmischung *f*, Prüf-
mischung *f*
  t. cup Prüftiegel *m* *(eines Flammpunktprüfers)*
  t. dose Prüfungsdosis *f*
  t. for acidity Säuregradbestimmung *f*
  t. for carbon dioxide Prüfung *f* auf Kohlendioxid
  t. for fat content Fett[gehalts]bestimmung *f*
  t. for identification Identitätsprüfung *f*
  t. for potency Bestimmung *f* der Wirksamkeit
  t. for sensivity Empfindlichkeitsprüfung *f*
  t. for sterility Prüfung *f* auf Sterilität, Sterilitätsprüfung *f*
  t. for water Wassernachweis *m*
  t. glass Prüfglas *n*
  t. material Testmaterial *n*, Untersuchungs-
material *n*
  t. method Testverfahren *n*, Prüfverfahren *n*,
Prüfmethode *f*, Untersuchungsverfahren *n*, Untersuchungsmethode *f*
  t. mix[ture] *s.* t. compound
  t. on animals Tierversuch *m*
  t. organism Testorganismus *m*, Testkeim *m*
  t. paper Reagenzpapier *n*, Prüfpapier *n*, Indikatorpapier *n*
  t. piece Prüfkörper *m*, Probe *f*, Probekörper *m*,
Prüfling *m*, Testmuster *n*, Prüfmuster *n*
  t. practice Versuchsdurchführung *f*
  t. reaction Nachweisreaktion *f*, Identifikations-
reaktion *f*
  t. sheet Prüfbogen *m*, Handmuster *n*, Schöpfpapiermuster *n*, Laborpapierblatt *n*
  t. sieve Prüfsieb *n*
  t. sieve series Prüfsiebreihe *f*
  t. solution Prüf[lings]lösung *f*, Untersuchungs-
lösung *f*, zu prüfende (untersuchende) Lösung *f*
  t. specimen *s.* t. piece
  t. strip stabförmige Probe *f*, Stabprobe *f*
  t. substance Versuchssubstanz *f*, Probesubstanz
f, Untersuchungssubstanz *f*
  t. toxin Testtoxin *n*, Prüfungstoxin *n*
  t. tube Reagenzglas *n*, Probierglas *n*

t.-tube basket Sterilisierdrahtkorb *m*

t.-tube brush Reagenzglasbürste *f*, Tüllenbürste *f*

t.-tube centrifuge Flaschenzentrifuge *f*, Flaschenschleuder *f*

t.-tube clamp Reagenzglasklemme *f*, Reagenzglasklammer *f*

t.-tube experiment Reagenzglasversuch *m*

t.-tube holder Reagenzglashalter *m*, Probierglashalter *m*

t.-tube method (*Gum*) Alterung *f* im Zellenofen

t.-tube rack (stand, support) Reagenzglasgestell *n*, Probierglasgestell *n*

tester Prüfer *m*, Prüfender *m*; Prüfgerät *n*, Prüfvorrichtung *f*, Prüfer *m*, Tester *m*

testing Prüfung *f*, Untersuchung *f*, Test *m*; Nachweis *m*

t. method *s.* test method

t. sieve *s.* test sieve

testosterone Testosteron *n* (*Sexualhormon*)

tet. *s.* tetragonal

tetanus toxin Tetanustoxin *n*

tetartohedral (*Krist*) tetartoedrisch

tetr. *s.* tetragonal

tetraacetate Tetraazetat *n*

tetraammine *s.* tetrammine

tetrabasic vierbasig, vierbasisch (*Säure*)

tetraborane B$_4$H$_{10}$ Tetraboran *n*

tetraborate M$'_2$B$_4$O$_7$ Tetraborat *n*, Heptoxotetraborat *n*

tetraboric acid H$_2$B$_4$O$_7$ Tetraborsäure *f*, Heptoxotetraborsäure *f*

tetraboride Tetraborid *n*

tetrabromide Tetrabromid *n*

tetrabromoborate Tetrabromoborat *n*

tetrabromofluorescein Tetrabromfluoreszein *n*, Eosin *n*

tetrabromoindigo Tetrabromindigo *m*, 5,7,5′,7′-Tetrabromindigo *m*

tetrabromomethane CBr$_4$ Tetrabrommethan *n*, Tetrabromkohlenstoff *m*, Kohlenstofftetrabromid *n*

tetrabromophenolsulphonphthalein Tetrabromphenolsulfonphthalein *n*, Bromphenolblau *n* (*pH-Indikator*)

tetrabromosilicane SiBr$_4$ Tetrabromsilan *n*, Siliziumtetrabromid *n*, Silizium(IV)-bromid *n*

tetracaine hydrochloride Tetrakainhydrochlorid *n*, 4-Butylaminobenzoesäure-β-dimethylamino-äthylesterhydrochlorid *n*

tetracarbonyl Tetrakarbonyl *n*

tetracarboxyimide dye Tetrakarboximidfarbstoff *m*, Naphthalintetrakarbonsäure-Farbstoff *m*

tetracene 1. Tetrazen *n*, Naphthazen *n*; 2. Tetrazen *n*, 1-(5-Tetrazolyl)-4-guanyltetrazenhydrat *n* (*Initialsprengstoff*)

tetrachloride Tetrachlorid *n*

tetrachloroanthraquinone Tetrachloranthrachinon *n*

tetrachloroborate Tetrachloroborat *n*

tetrachlorodiammine-platinum [Pt(NH$_3$)$_2$Cl$_4$] Tetrachlorodiamminplatin(IV) *n*

1,1,2,2-tetrachloroethane CHCl$_2$CHCl$_2$ 1,1,2,2-Tetrachloräthan *n*, *sym*-Tetrachloräthan *n*, Azetylentetrachlorid *n*

tetrachloroethylene C$_2$Cl$_4$ Tetrachloräthylen *n*, Tetrachloräthen *n*, Perchloräthylen *n*

tetrachloromethane CCl$_4$ Tetrachlormethan *n*, Tetrachlorkohlenstoff *m*, Kohlenstofftetrachlorid *n*

tetrachloroquinone C$_6$Cl$_4$O$_2$ Tetrachlor-*p*-benzochinon *n*, Chloranil *n*

tetracid viersäurig (*Base*)

tetracoordinated vierfach koordiniert

tetracosa-4,8,12,15,18,21-hexaenoic acid C$_{23}$H$_{35}$COOH 4,8,12,15,18,21-Tetrakosahexaensäure *f*, Nissinsäure *f*

tetracosanoic acid C$_{23}$H$_{47}$COOH Tetrakosansäure *f*, Lignozerinsäure *f*, Kerasinsäure *f*

tetracos-15-enoic acid C$_{23}$H$_{45}$COOH 15-Tetrakosensäure *f*, Selachensäure *f*, Nervonsäure *f*

tetracovalent tetrakovalent, vierbindig

tetracycline Tetrazyklin *n*

tetrad (*Krist*) vierzählig, 4zählig

tetrad vierwertiges Element *n*; vierwertige Atomgruppe *f*

tetradecanal Tetradekanal *n*, Myrist[in]aldehyd *m*, Myristylaldehyd *m*

tetradecanoic acid C$_{13}$H$_{27}$COOH Tetradekansäure *f*, Myristinsäure *f*

tetradec-9-enoic acid C$_{13}$H$_{25}$COOH 9-Tetradezensäure *f*, Myristoleinsäure *f*

tetradecylaldehyde *s.* tetradecanal

tetradentate (*Komplexchemie*) vierzahnig, vierzähnig, vierzählig (*Koordinationswert des Liganden*)

tetradymite (*Min*) Tetradymit *m* (*Wismuttellursulfid*)

tetraene Tetraen *n*

tetraethyl lead Pb(C$_2$H$_5$)$_4$ Bleitetraäthyl *n*, Tetraäthylblei *n*, BTÄ *n*

t. rhodamine [*N,N,N′,n′*-]Tetraäthylrhodamin *n*

tetrafluoride Tetrafluorid *n*

tetrafluoroboric acid H[BF$_4$] Tetrafluoroborsäure *f*, Fluoroborsäure *f*

tetrafluoroethylene CF$_2$=CF$_2$ Tetrafluoräthen *n*, Tetrafluoräthylen *n*, Perfluoräthen *n*, Perfluoräthylen *n*

tetrafluorosilicane SiF$_4$ Tetrafluorsilan *n*, Siliziumtetrafluorid *n*, Silizium(IV)-fluorid *n*

tetrafunctional tetrafunktionell

t. unit tetrafunktionelle Einheit *f*, Q-Einheit *f* (*Struktureinheit*)

tetragonal (*Krist*) tetragonal, tetr., viereckig

t. system tetragonales Kristallsystem *n*

tetrahalide Tetrahalogenid *n*

tetrahedral tetraedrisch, Tetraeder..., vierflächig

t. angle Tetraederwinkel *m*

t. arrangement Tetraederanordnung *f*

t. formula Tetraederformel *f*

t. hybridization tetraedrische Hybridisierung (Bastardisierung) *f*
t. model Tetraedermodell *n*
t. orbital Tetraederorbital *n*
t. valency Tetraedervalenz *f*, tetraedrische Valenz *f*, *q*-Valenz *f*
tetrahedrite *(Min)* Tetraedrit *m*
tetrahedron *(Krist)* Tetraeder *n*, Vierflächner *m*
tetrahydrate Tetrahydrat *n*, 4-Hydrat *n*
tetrahydric alcohol vierwertiger Alkohol *m*, Tetrit *m*
tetrahydride Tetrahydrid *n*
1,2,3,4-tetrahydrobenzene 1,2,3,4-Tetrahydrobenzol *n*, Zyklohexen *n*
tetrahydroborate $M^I_n[BH_4]_n$ Boranat *n*, Metallborwasserstoff *m*, Tetrahydridoborat *n*, Metallborhydrid *n*, Hydridoborat *n*
tetrahydrocytisine Tetrahydrozytisin *n*
tetrahydrofuran Tetrahydrofur[fur]an *n*, Diäthylenoxid *n*, Tetramethylenoxid *n*
tetrahydroisoquinoline Tetrahydroisochinolin *n*
t. base Tetrahydroisochinolinbase *f*
tetrahydronaphthalene $C_{10}H_{12}$ Tetrahydronaphthalin *n*
tetrahydropyrrole Tetrahydropyrrol *n*, Pyrrolidin *n*, Tetramethylenimin *n*
tetrahydroxide Tetrahydroxid *n*
tetraiodide Tetrajodid *n*
tetraiodoborate Tetrajodoborat *n*
tetraiodosilicane SiJ_4 Tetrajodsilan *n*, Siliziumtetrajodid *n*, Silizium(IV)-jodid *n*
tetrakisazo dye Tetrakisazofarbstoff *m*
tetralactone Tetralakton *n*
tetramer Tetramer[es] *n*
tetramethyl biarsine (diarsyl) $(CH_3)_2AsAs(CH_3)_2$ Tetramethylbiarsyl *n*, Tetramethyl[di]arsin *n*, *bis*-Dimethylarsyl *n*, Kakodyl *n*
t. lead $Pb(CH_3)_4$ Bleitetramethyl *n*, Tetramethylblei *n*
tetramethylation Tetramethylierung *f*
1,2,3,4-tetramethylbenzene $C_6H_2(CH_3)_4$ 1,2,3,4-Tetramethylbenzol *n*, Prehnitol *n*
1,2,3,5-tetramethylbenzene $C_6H_2(CH_3)_4$ 1,2,3,5-Tetramethylbenzol *n*, Isodurol *n*
1,2,4,5-tetramethylbenzene $C_6H_2(CH_3)_4$ 1,2,4,5-Tetramethylbenzol *n*, Durol *n*
tetramethylene Tetramethylen *n*, Zyklobutan *n*
t. oxide Tetramethylenoxid *n*, Diäthylenoxid *n*, Tetrahydrofuran *n*, Tetrahydrofurfuran *n*
t. sulphone Tetramethylensulfid *n*, Thiophen *n*, Tetrahydrothiophen *n*
tetramethylen[e]imine Tetramethylenimin *n*, Tetrahydropyrrol *n*, Pyrrolidin *n*
tetramethylethylene glycol $(CH_3)_2C(OH)COH(CH_3)_2$ Tetramethyläthylenglykol *n*, Pinakol *n*
tetramethylglucose Tetramethylglukose *f*
tetramethylmethane $C(CH_3)_4$ Tetramethylmethan *n*, 2,2-Dimethylpropan *n*, Neopentan *n*
tetrammine Tetrammin *n*, Tetramminkomplex *m*
t. salt Tetramminsalz *n*

tetramminecopper(II) sulphate $[Cu(NH_3)_4]SO_4$ Tetramminkupfer(II)-sulfat *n*
tetramolybdate Tetramolybdat *n*
tetranitride Tetranitrid *n*
tetraoxide *s.* tetroxide
tetrapeptide Tetrapeptid *n* *(Peptid mit vier Aminosäureeinheiten)*
tetraphosphorus Tetraphosphor *m*, weißer Phosphor *m*
t. heptasulphide P_4S_7 Tetraphosphorheptasulfid *n*
t. pentasulphide P_4S_5 Tetraphosphorpentasulfid *n*
t. trisulphide P_4S_3 Tetraphosphortrisulfid *n*
tetrapotassium pyrophosphate $K_4P_2O_7$ Kaliumdiphosphat *n*, Kaliumpyrophosphat *n*
tetraquo-iron(II) ion $[Fe(H_2O)_4]^{2+}$ Tetraquoeisen(II)-Ion *n*
tetrasil[ic]ane Si_4H_{10} Tetrasilan *n*
tetrasiloxane $Si_4H_{10}O_3$ Tetrasiloxan *n*
tetrasodium pyrophosphate $Na_4P_2O_7$ Natriumdiphosphat *n*, Natriumpyrophosphat *n*
t. salt Tetranatriumsalz *n*
tetrasubstituted tetrasubstituiert
tetrasubstitution Tetrasubstitution *f*
t. product Tetrasubstitutionsprodukt *n*, tetrasubstituiertes Produkt *n*
tetrasulphide Tetrasulfid *n*
tetrasulphur tetranitride S_4N_4 Tetraschwefeltetranitrid *n*, Schwefelnitrid *n*
tetrathioarsenite $M^I_6As_4S_9$ Tetrathioarsenat(III) *n*
tetrathiocyanate Tetrathiozyanat *n*, Tetrarhodanid *n*
tetrathionate $M^I_2S_4O_6$ Tetrathionat *n*
tetratomic 1. vieratomig; 2. *s.* tetravalent
tetravalency Tetravalenz *f*, Vierwertigkeit *f*
tetravalent tetravalent, vierwertig
tetrazine Tetrazin *n* *(heterozyklischer Sechsring mit vier Stickstoffatomen)*
tetrazo compound Tetrazoverbindung *f*
tetrazole Tetrazol *n*
tetrazotization Tetrazotierung *f*, Tetrazotieren *n*
tetrazotize / to tetrazotieren; sich tetrazotieren lassen
tetrose Tetrose *f* *(Monosaccharid mit vier Kohlenstoffatomen)*
tetroxalate Tetr[a]oxalat *n*
tetroxide Tetroxid *n*
texasite *(Min)* Texasit *m*, *(veraltet für)* Zaratit *m* *(Nickeltetrahydroxidkarbonat)*
text paper *(Am)* Werkdruckpapier *n*, Buchdruckpapier *n*
textile Gewebe *n*, Stoff *m*
t.-and-rubber goods *s.* t.-rubber composites
t. auxiliary Textilhilfsmittel *n*, Textilveredlungsmittel *n*
t. casing *s.* t. insertion
t. chemist Textilchemiker *m*
t. chemistry Textilchemie *f*
t. cleanser Textilreinigungsmittel *n*

t. fabric textiles Flächengebilde *n*, Gewebe *n*
t. fibre Textilfaser *f*; Textilfaserstoff *m*
t. finishing Textilveredlung *f*, Textilausrüstung *f*, Appretieren *n*
t. insertion *(Gum)* Textileinlage *f*, Gewebeeinlage *f*
t. materials textile Rohstoffe *mpl*, Rohstoffe *mpl* der Textilindustrie
t. oil Textilöl *n*, Appreturöl *n*
t. printing Textildruck *m*, Stoffdruck *m*, Gewebedruck *m*
t.-rubber composites Erzeugnisse *npl* der Stoffgummierung
t. size Textilschlichte *f*, textile (haftmittelfreie) Schlichte *f*
textiles Textilien *pl*, Textilerzeugnisse *npl*
texture Textur *f*; *(Holz)* Textur *f*, Zeichnung *f*, Maserung *f*
textured texturiert
T.F.E. *s.* tetrafluoroethylene
Tg [point] *s.* glass transition temperature
TGA *s.* thermogravimetric analysis
thalenite *(Min)* Thalenit *m* *(Yttriumpyrosilikat)*
thallic Thallium..., *(i.e.S.)* Thallium(III)-...
t. acetate $Tl(CH_3COO)_3$ Thallium(III)-azetat *n*
t. bromide $TlBr_3$ Thallium(III)-bromid *n*, Thalliumtribromid *n*
t. chloride $TlCl_3$ Thallium(III)-chlorid *n*, Thalliumtrichlorid *n*
t. fluoride TlF_3 Thallium(III)-fluorid *n*, Thalliumtrifluorid *n*
t. iodide TlJ_3 Thallium(III)-jodid *n*, Thalliumtrijodid *n*
t. nitrate $Tl(NO_3)_3$ Thallium(III)-nitrat *n*
t. oxide Tl_2O_3 Thallium(III)-oxid *n*
t. sulphate $Tl_2(SO_4)_3$ Thallium(III)-sulfat *n*
t. sulphide Tl_2S_3 Thallium(III)-sulfid *n*
thallium Tl Thallium *n*
t. alum $TlAl(SO_4)_2 \cdot 12H_2O$ Thalliumalaun *m*, Thalliumaluminiumsulfat-12-Wasser *n*
t. chloroplatinate $Tl_2[PtCl_6]$ Thalliumhexachloroplatinat(IV) *n*
t. chromate Tl_2CrO_4 Thallium(I)-chromat *n*
t. dihydrogen orthophosphate TlH_2PO_4 Thallium(I)-dihydrogen[ortho]phosphat *n*
t. ferrocyanide $Tl_4[Fe(CN)_6]$ Thallium(I)-hexazyanoferrat(II) *n*
t. fluorosilicate $Tl_2[SiF_6]$ Thalliumhexafluorosilikat *n*
t. monobromide $TlBr$ Thallium[mono]bromid *n*, Thallium(I)-bromid *n*
t. monochloride $TlCl$ Thallium[mono]chlorid *n*, Thallium(I)-chlorid *n*
t. monofluoride TlF Thallium[mono]fluorid *n*, Thallium(I)-fluorid *n*
t. monoiodide TlJ Thallium[mono]jodid *n*, Thallium(I)-jodid *n*
t. monoxide Tl_2O Thallium[mon]oxid *n*, Dithalliummonoxid *n*, Thallium(I)-oxid *n*
t. orthophosphate Tl_3PO_4 Thallium(I)-[ortho]phosphat *n*, Trithalliumphosphat *n*

t. perchlorate $TlClO_4$ Thallium(I)-perchlorat *n*
t. peroxide *s.* t. trioxide
t. sesquioxide *s.* t. trioxide
t. silicofluoride *s.* t. fluorosilicate
t. tribromide $TlBr_3$ Thalliumtribromid *n*, Thallium(III)-bromid *n*
t. trichloride $TlCl_3$ Thalliumtrichlorid *n*, Thallium(III)-chlorid *n*
t. trifluoride TlF_3 Thalliumtrifluorid *n*, Thallium(III)-fluorid *n*
t. triiodide TlJ_3 Thalliumtrijodid *n*, Thallium(III)-jodid *n*
t. trioxide Tl_2O_3 Thalliumtrioxid *n*, Thallium(III)-oxid *n*
t. trisulphide Tl_2S_3 Thalliumtrisulfid *n*, Thallium(III)-sulfid *n*
thallous Thallium..., *(i.e.S.)* Thallium(I)-...
t. acetate CH_3COOTl Thallium(I)-azetat *n*
t. bromate $TlBrO_3$ Thallium(I)-bromat *n*
t. bromide $TlBr$ Thallium(I)-bromid *n*, Thallium[mono]bromid *n*
t. carbonate Tl_2CO_3 Thallium(I)-karbonat *n*
t. chloride $TlCl$ Thallium(I)-chlorid *n*, Thallium[mono]chlorid *n*
t. cyanide $TlCN$ Thallium(I)-zyanid *n*
t. fluoride TlF Thallium(I)-fluorid *n*, Thallium[mono]fluorid *n*
t. hydroxide $TlOH$ Thallium(I)-hydroxid *n*
t. iodide TlJ Thallium(I)-jodid *n*, Thallium[mono]jodid *n*
t. nitrate $TlNO_3$ Thallium(I)-nitrat *n*
t. oxide Tl_2O Thallium(I)-oxid *n*, Dithalliummonoxid *n*, Thallium[mono]oxid *n*
t. sulphate Tl_2SO_4 Thallium(I)-sulfat *n*
t. sulphide Tl_2S Thallium(I)-sulfid *n*
thaw / to [auf]tauen, abtauen, entfrosten, schmelzen
thawing Auftauen *n*, Tauen *n*, Abtauen *n*, Entfrosten *n*, Schmelzen *n*
thebaine Thebain *n*, Paramorphin *n* *(Alkaloid)*
theine Tein *n*, Thein *n*, Koffein *n*, Kaffein *n*, 1,3,7-Trimethylxanthin *n*
Theisen disintegrator Theisen-Desintegrator *m*, Desintegrator[gas]wäscher *m* nach Theisen
thelephoric acid Thelephorsäure *f*
Thénard blue $Al_2[CoO_4]$ Thénards Blau *n*
thenardite *(Min)* Thenardit *m* *(Natriumsulfat)*
theobromine Theobromin *n*, 3,7-Dimethylxanthin *n* *(Alkaloid)*
t. sodium Theobrominnatrium *n*
t. sodium [and sodium] acetate Theobrominnatrium-Natriumazetat *n*, Theobrominnatrium-azetat *n*
t. sodium [and sodium] salicylate Theobrominnatrium-Natriumsalizylat *n*, Theobrominnatrium-salizylat *n*
theophylline Theophyllin *n*, 1,3-Dimethylxanthin *n* *(Alkaloid)*
t. sodium Theophyllinnatrium *n*
t. sodium [and sodium] acetate Theophyllinnatrium-Natriumazetat *n*, Theophyllinnatrium-azetat *n*

theorem Theorem *n*, Lehrsatz *m*, Satz *m*
theoretical:
  5% in excess of the t., 5% over the t. 5% mehr als der Theorie entspricht, 5% über dem theoretischen Wert
  t. plate *(Destillation)* theoretischer (idealer) Boden *m*, *(i.w.S.)* theoretische Trennstufe *f*
  t. plates *(Destillation)* theoretische Bodenzahl *f*
  t. quantity theoretische (theoretisch erforderliche, theoretisch nötige) Menge *f*
  t. tray *s.* t. plate
theory Theorie *f*, Lehre *f*
  in slight excess of t. etwas mehr als der Theorie entspricht, etwas über dem theoretischen Wert
  t. of absolute reaction rate Theorie *f* der absoluten Reaktionsgeschwindigkeit
  t. of electrolytes Elektrolyttheorie *f*
  t. of partial valence[s] Theorie *f* der Partialvalenzen, Partialvalenztheorie *f*
  t. of radioactive disintegration Desintegrationstheorie *f*, Theorie *f* des radioaktiven Zerfalls
  t. of relativity Relativitätstheorie *f*
  t. of special relativity spezielle Relativitätstheorie *f*
  t. of strain Spannungstheorie *f*
  t. of valency Valenztheorie *f*, Valenzlehre *f*, Wertigkeitstheorie *f*
therapeutic therapeutisch [wirksam]
  t. agent Therapeutikum *n*, therapeutischer Wirkstoff *m*, therapeutisch wirksamer Stoff *m*, Heilmittel *n*
  t. dose therapeutische Dosis *f*, Heildosis *f*, heilende Menge *f*
  t. index chemotherapeutischer Index (Quotient) *m*
thermal thermisch, Wärme...
  t. agitation *s.* t. motion
  t. alkylation thermische Alkylierung *f*
  t. analysis thermische Analyse *f*, Thermoanalyse *f*
  t. balance Wärmebilanz *f*, Wärmehaushalt *m*
  t. barrier Hitzebarriere *f*, Hitzemauer *f*, Hitzegrenze *f*
  t. black [thermischer] Spaltruß *m*, Thermalruß *m*
  t. capacity Wärmekapazität *f*
  t. carbon black *s.* t. black
  t.-catalytic cracking thermisch-katalytisches (thermokatalytisches) Kracken *n*, thermischkatalytische (thermokatalytische) Krackung *f*
  t. conduction Wärmeleitung *f*
  t. conductivity Wärmeleitfähigkeit *f*, [spezifisches] Wärmeleit[ungs]vermögen *n*, thermische Leitfähigkeit *f*, Wärmeleitzahl *f*
  t.-conductivity cell Wärmeleitfähigkeitszelle *f*
  t. convection Thermokonvektion *f*, Wärmekonvektion *f*, Wärmemitführung *f*
  t. cracking thermisches (radikalisches) Kracken (Spalten) *n*, thermische (radikalische) Krackung (Spaltung) *f*, Kracken *n* über Radikale
  t.-cracking process thermisches Krackverfahren (Spaltverfahren) *n*

t. decomposition thermische Zersetzung (Spaltung) *f*, thermischer Zerfall (Abbau) *m*, Hitzezersetzung *f*, Hitzespaltung *f*, Hitzezerlegung *f*
  t.-decomposition black *s.* t. black
  t. degradation *s.* t. decomposition
  t. diffusion Thermodiffusion *f*
  t.-diffusion column Trennrohr *n* *(zur Isotopentrennung)*
  t.-diffusion method Thermodiffusionsverfahren *n*, [Clusiussches] Trennrohrverfahren *n*, Clusius-Dickel-Verfahren *n*
  t. diffusivity Temperaturleitfähigkeit *f*, Temperaturleitvermögen *n*, Temperaturleitzahl *f*
  t. dissociation thermische Dissoziation *f*
  t. drying thermisches Trocknen *n*, thermische Trocknung *f*
  t. effect Wärmeeffekt *m*
  t. efficiency thermischer Wirkungsgrad *m*, Wärmewirkungsgrad *m*
  t. endurance *s.* t. stability
  t. energy thermische Energie *f*, Wärmeenergie *f*
  t. equilibrium thermisches (thermodynamisches) Gleichgewicht *n*
  t. excitation thermische Anregung *f*, Temperaturanregung *f*
  t. expansion thermische Ausdehnung *f*, Wärme[aus]dehnung *f*
  t. history thermische Vorgeschichte *f*
  t.-impulse heat sealing Wärmeimpulssiegeln *n* *(von Folien)*
  t. inactivation Hitzeinaktivierung *f*
  t. insulation Wärmeisolation *f*, Wärmeisolierung *f*, Wärmedämmung *f*, Wärmeschutz *m*
  t. insulator Wärmeisolierstoff *m*, Wärmedämmstoff *m*, Wärmeschutzstoff *m*
  t. ionization thermische Ionisierung (Ionisation) *f*
  t. liquid flüssiges Wärmeübertragungsmittel *n*, flüssiger Wärmeträger *m*, Heizflüssigkeit *f*
  t. luminescence Thermolumineszenz *f*
  t. motion Wärmebewegung *f*
  t. neutron thermisches Neutron *n*
  t.-oxidative thermooxydativ, wärmeoxydativ
  t. phosphate Glühphosphat *n*
  t. plasticization *s.* t. softening
  t. process *(Gum)* Heißdampfverfahren *n* *(Regenerierverfahren)*; *(Gum)* Thermalspaltprozeß *m* *(Rußherstellung)*
  t. property thermische Eigenschaft *f*, Wärmeverhalten *n*
  t. radiation Wärme[aus]strahlung *f*, thermische Strahlung *f*
  t. reforming thermisches Reform[ier]en *n*, thermische Reformierung *f*, thermisches Reforming *n*
  t. resistance thermischer Widerstand *m*, Wärme[leit]widerstand *m*
  t. sealing *(Plast)* Heißsiegeln *n*, Heißkleben *n*, Heißverschweißen *n*
  t. shock Abschreckung *f*

t.-shock resistance Temperaturwechselbeständigkeit f, Widerstand m gegen Temperaturwechsel
t.-shock resistant temperaturwechselbeständig
t. siphoning s. t. convection
t. softening thermische Erweichung (Plastizierung) f
t. spalling resistance s. t.-shock resistance
t. stability thermische Beständigkeit (Stabilität, Festigkeit, Widerstandsfähigkeit) f, Thermostabilität f, Wärmebeständigkeit f, Wärmefestigkeit f, Hitzebeständigkeit f, Hitzefestigkeit f, Hitzeresistenz f; (Ker) Temperaturwechselbeständigkeit f
t. transmission Wärmeübertragung f, Wärmeaustausch m, (i.e.S.) Wärmeübergang m
t. value Heizwert m
t. vibration (Krist) Wärmeschwingung f
t. vulcanization Thermovulkanisation f
thermally reversible thermisch reversibel, temperaturreversibel
t. stable thermisch beständig (stabil, widerstandsfähig), thermostabil, thermoresistent, wärmebeständig, wärmefest, wärmestabil, hitzebeständig, hitzefest, hitzestabil, hitzeresistent
thermatomic black (Gum) thermatomischer Ruß m
thermion Thermion n
thermionic thermionisch, Thermionen..., glühelektrisch, Glühelektronen...
t. electron Thermoelektron n, Glühelektron n
t. emission Thermionenemission f; thermische Elektronenemission (Emission) f, thermischer Elektronenaustritt m, Glüh[elektronen]emission f
t. emitter Glühkatodenemitter m, thermischer Elektronensender m
thermit Thermit n
t. process Thermitverfahren n, aluminothermisches Verfahren n
thermoadhesive thermoadhesives Material n
thermobalance Thermowaage f
thermochemical thermochemisch
t. calorie thermochemische Kalorie f
t. mixture thermochemisches Gemisch n
thermochemistry Thermochemie f, chemische Thermodynamik f
thermochromism Thermochromie f
thermocompression Thermokompression f
t. evaporator Verdampfer m mit Thermokompression (Brüdenverdichtung)
thermocompressor Thermokompressor m
thermocouple Thermo[element]paar n, (i.w.S.) Thermoelement n
thermodiffusion Thermodiffusion f
thermoduric s. thermostable
thermodynamic thermodynamisch
t. cycle thermodynamischer Kreisprozeß m
t. equilibrium thermodynamisches Gleichgewicht n
t. function thermodynamische Funktion f

t. potential thermodynamisches Potential n
t. probability thermodynamische Wahrscheinlichkeit f
t. process (Gum) Banbury-Lancaster-Verfahren n (Regenerierverfahren)
t. stability conditions thermodynamische Gleichgewichtsbedingungen fpl
t. temperature scale thermodynamische (absolute) Temperaturskale f
thermodynamical s. thermodynamic
thermodynamics Thermodynamik f
thermoelastic thermoelastisch, wärmeelastisch
thermoelasticity Thermoelastizität f, Wärmeelastizität f
thermoelectric thermoelektrisch, thermisch-elektrisch
t. current Thermostrom m
t. effect thermoelektrischer Effekt m
t. energy transformation thermoelektrische Energieumwandlung f
t. phenomena thermoelektrische Erscheinungen fpl
t. power Thermokraft f, Thermospannung f
t. pyrometer thermoelektrisches Pyrometer n
thermoelectricity Thermoelektrizität f
thermoelectron Thermoelektron n, Glühelektron n
thermoelement Thermoelement n
Thermofor catalytic cracking Thermofor-Catalytic-Cracking n
Thermofor catalytic-cracking process Thermofor-[Catalytic-Cracking-]Verfahren n, TCC-Verfahren n (Krackverfahren mit bewegtem Katalysatorbett)
Thermofor clay-burning kiln Thermofor-Kiln m (Regenerierofen besonderer Konstruktion beim Thermofor-Catalytic-Cracking)
Thermofor continuous percolation Thermofor-Continuous-Percolation f, TCP (kontinuierliche Perkolation nach dem TCP-Verfahren)
Thermofor continuous-percolation process Thermofor-Continuous-Percolation-Verfahren n, TCP-Verfahren n (kontinuierlicher Perkolationsprozeß)
Thermofor kiln s. Thermofor clay-burning kiln
Thermofor process s. Thermofor catalytic-cracking process
thermoforming Warmformen n, Warmformung f, Thermoformung f (von thermoplastischem Halbzeug)
t. machine (Plast) Warmformmaschine f, Thermoformmaschine f
thermogalvanometer Thermogalvanometer n
thermograph Thermograf m, Temperaturschreiber m
thermogravimetric thermogravimetrisch
t. analysis thermogravimetrische Analyse f
thermogravimetry Thermogravimetrie f
thermolabile thermolabil, wärmeunbeständig
thermoluminescence Thermolumineszenz f
thermolysis Thermolyse f
thermomagnetic thermomagnetisch

t. effect thermomagnetischer Effekt *m*
thermomechanical thermomechanisch
t. process *(Gum)* thermomechanisches (mechanisch-thermisches) Verfahren *n*
thermometer Thermometer *n*
t. error Thermometerfehler *m*
t. glass Thermometerglas *n*
t. pipe *s.* t. tube
t. pocket Thermometerstutzen *m*
t. scale Thermometerskale *f*
t. tube Thermometerrohr *n*
thermometric thermometrisch
t. scale Thermometerskale *f*
t. substance Thermometersubstanz *f*
t. titration thermometrische Titration *f*
thermometrical *s.* thermometric
thermometry Thermometrie *f*, Temperaturmessung *f* [mit Hilfe des Thermometers]
thermomolecular pressure difference thermomolekulare Druckdifferenz *f*
thermonatrite, thermonitrite *(Min)* Thermonatrit *m* *(Natriumkarbonat-1-Wasser)*
thermonuclear thermonuklear
t. reaction thermonukleare Reaktion *f*
t. reactor thermonuklearer Reaktor *m*
thermophilic thermophil, wärmeliebend
t. bacteria thermophile (wärmeliebende) Bakterien *npl*
t. digestion thermophile Faulung *f*
thermophillin Thermophillin *n* *(Antibiotikum)*
thermophilous *s.* thermophilic
thermopile Thermosäule *f*
thermoplastic thermoplastisch
thermoplastic Thermoplast *m*, thermoplastischer Kunststoff *m*
t. adhesive thermoplastischer Klebstoff *m*
t. material *s.* thermoplastic
t. yarn thermoplastischer Faden *m*
thermoplasticity Thermoplastizität *f*, Wärmeplastizität *f*
thermoprene Thermopren *n*
thermoreactive *s.* thermosetting
thermoregulator Thermoregulator *m*, Temperaturregler *m*, Temperaturregulator *m*, Wärmeregler *m*
thermoresistant *s.* thermostable
thermoscope Thermoskop *n*
thermoset [resin] *s.* thermosetting plastic
thermosetting hitzehärtbar, wärmehärtbar
t. adhesive hitzehärtbarer (wärmehärtbarer) Klebstoff *m*
t. ink Thermosetting-Ink *f* *(rasch trocknende Druckfarbe)*
t. material (moulding compound) *(Plast)* hitzehärtbare (wärmehärtbare) Formmasse *f*
t. plastic (resin) Duroplast *m*, hitzehärtbarer (wärmehärtbarer) Plast *m*
t. synthetic-resin adhesive *s.* t. adhesive
thermosol method *(Farb)* Thermosolverfahren *n*
t. pad-steam process *(Farb)* Thermosol-

Pad-Steam-Verfahren *n*, Thermosol-Klotz-Dämpfverfahren *n*
thermostability Thermostabilität *f*, thermische Stabilität (Beständigkeit, Festigkeit, Widerstandsfähigkeit) *f*, Wärmebeständigkeit *f*, Wärmefestigkeit *f*, Hitzebeständigkeit *f*, Hitzefestigkeit *f*, Hitzeresistenz *f*
thermostable thermisch beständig (stabil, widerstandsfähig), thermostabil, thermoresistent, wärmebeständig, wärmefest, wärmestabil, hitzebeständig, hitzefest, hitzestabil, hitzeresistent
thermostat / to thermostat[is]ieren, temperieren
thermostat Thermostat *m*
THF *s.* tetrahydrofuran
thiacetic acid *s.* thioacetic acid
thiamin *s.* thiamine
thiaminase Thiaminase *f* *(Vitamin B_1 spaltendes Ferment)*
thiamine Thiamin *n*, Vitamin B_1 *n*
t. hydrochloride Thiaminhydrochlorid *n*
t. pyrophosphate Thiaminpyrophosphat *n*, TPP, Karboxylase *f*
thiazine dye Thiazinfarbstoff *m*
thiazole accelerator *(Gum)* Thiazolbeschleuniger *m*
t. dye Thiazolfarbstoff *m*
t. yellow Thiazolgelb *n*, Titangelb *n*
thiazolidine Thiazolidin *n*
thiaz[ol]ylamine Thiazolylamin *n*, Aminothiazol *n*
thick dick; dicht; dickflüssig, zähflüssig, strengflüssig, viskos; trüb[e]; *(Geol)* mächtig
t. black liquor *(Pap)* Dick[ab]lauge *f*, eingedickte Schwarzlauge *f*
t. glass *s.* t. sheet glass
t. juice *(Zucker)* Dicksaft *m*
t.-juice blowup *(Zucker)* Dicksaftkocher *m*
t.-juice filter *(Zucker)* Dicksaftfilter *n*
t.-juice filtration *(Zucker)* Dicksaftfiltration *f*
t.-juice heater *(Zucker)* Dicksaft[vor]wärmer *m*
t.-juice pump *(Zucker)* Dicksaftpumpe *f*
t.-juice saturation *(Zucker)* Dicksaftsaturation *f*
t. mash Dickmaische *f*
t. sheet glass Dickglas *n*
t. slime Dickschlamm *m*
t. stock *(Pap)* Dickstoff *m*
t.-stock pump *(Pap)* Dickstoffpumpe *f*
t.-walled dickwandig
thicken / to 1. *(Flüssigkeiten)* eindicken, verdicken, dick (dicker, dickflüssig, zähflüssig, viskos) machen; verdichten; trüben; entwässern; 2. eindicken, dick (dicker, dickflüssig, zähflüssig, viskos) werden; dichter werden, sich verdichten
thickened liquor Dickschlamm *m*
t.-liquor outlet Dickschlammaustrag *m*, Schlammaustrag *m*
t. oil eingedicktes Öl *n*
t. sludge *s.* t. liquor
thickener 1. Eindicker *m*, Eindickmittel *n*, Verdicker *m*, Verdickungsmittel *n*, Verdickungszusatz *m*; 2. Eindicker *m*, Eindickapparat *m*, Eindickzylinder

m, Eindickmaschine f, Entwässerungsmaschine f

t. pulp konzentrierter Schlamm m, Schlammkonzentrat n, Dickschlamm m (als Austrag eines Eindickers)

thickening Eindicken n, Eindickung f, Verdicken n; Verdichten n; Dickwerden n (z.B. der Milch)

t. agent s. thickener 1.

t. paste Verdickungspaste f, Verdickungsmasse f

thickness Dicke f; Dichte f; Dickflüssigkeit f, Zähflüssigkeit f, Viskosität f; (Geol) Mächtigkeit f

t. fault (Text) fehlerhafte Dickstelle f

t. gauge Dickenmesser m

t. of the layer Schichtdicke f

t. tester Dickenmesser m

Thiele addition Thiele-Addition f

Thiele reaction Thiele-Reaktion f

thimble 1. (Ker) Fingerhut m (Brennhilfsmittel); 2. Rührer m (für im Hafen geschmolzenes Glas); 3. Extraktionshülse f

thin / to verdünnen

thin dünn; dünnflüssig; mager, arm (Boden); (Foto) kontrastarm, unterbelichtet

t. emulsion film Dünnschichtfilm m, dünnschichtiger Film m

t. film Dünnfilm m, Dünnschicht f

t. glass s. t. sheet glass

t. glassine Dünnpergamin[-Papier] n

t. juice (Zucker) Dünnsaft m

t.-juice filter (Zucker) Dünnsaftfilter n

t.-juice filtration (Zucker) Dünnsaftfiltration f

t.-juice heater (Zucker) Dünnsaftvorwärmer m

t. layer dünne Schicht f, Dünnschicht f

t.-layer chromatogram Dünnschichtchromatogramm n

t.-layer chromatography Dünnschichtchromatografie f

t.-layer distillation s. t.-layer method

t.-layer distillator Dünnschichtdestillator m

t.-layer electrophoresis Dünnschichtelektrophorese f

t.-layer evaporator Dünnschichtverdampfer m

t.-layer method Dünnschichtverfahren n, Dünnschichtdestillation f

t. letter paper s. t. post paper

t.-polished section (Min) Dünnschliff m

t. post paper Dünnpostpapier n

t. section (Min) Dünnschliff m; Dünnschnitt m

t.-section method Dünnschliffmethode f, Dünnschliffverfahren n

t.-section technique Dünnschlifftechnik f

t. sheet glass Dünnglas n

t.-walled dünnwandig

thinner, thinning agent Verdünner m, Verdünnungsmittel n, Abschwächer m, Abschwächungsmittel n, Streckmittel n

thio derivative Thioderivat n

thioacetic acid CH₃COSH Thioessigsäure f

thioacid Thiosäure f

thioalcohol Thioalkohol m, Merkaptan n

thioantimonate M$^I_3$SbS₄ Thioantimonat(V) n

thioantimonite M$^I_3$SbS₃ Thioantimonat(III) n

thioarsenate M$^I_3$AsS₄ Thioarsenat(V) n

thioarsenite M$^I_3$AsS₃ Thioarsenat(III) n

thiocarbamide H₂NCSNH₂ Thiokarbamid n, Thioharnstoff m

thiocarbanilide CS(NHC₆H₅)₂ Thiokarbanilid n, N,N'-Diphenylthioharnstoff m

thiocarbonate M$^I_2$[CS₃] Thiokarbonat n

thiocarbonic acid Thiokohlensäure f

thiocarbonyl chloride CSCl₂ Thiokarbonyldichlorid n, Thiophosgen n

thiocyanate M^{I}SCN Thiozyanat n, Rhodanid n

t. method Rhodanometrie f

thiocyanatoaurate Thiozyanatoaurat n, Rhodanoaurat n

thiocyanatochromate Thiozyanatochromat n, Rhodanochromat n

thiocyanatocobaltate Thiozyanatokobaltat n, Rhodanokobaltat n

thiocyanatoferrate Thiozyanatoferrat n, Rhodanoferrat n

thiocyanatomolybdate Thiozyanatomolybdat n, Rhodanomolybdat n

thiocyanatonickelate Thiozyanatoniccolat n, Rhodanoniccolat n

thiocyanatotitanate Thiozyanatotitanat n, Rhodanotitanat n

thiocyanatotungstate Thiozyanatowolframat n, Rhodanowolframat n

thiocyanatovanadate Thiozyanatovanadat n, Rhodanovanadat n

thiocyanic acid HSCN Thiozyansäure f, Rhodanwasserstoffsäure f

thiocyanide s. thiocyanate

thiocyanogen number (value) Rhodanzahl f, rhodanometrische Jodzahl f, RaZ (Kennzahl der Fette und fetten Öle)

thioether Thioäther m

thioethyl alcohol C₂H₅SH Äthylthioalkohol m, Äthanthiol n, Thioäthanol n, Äthylmerkaptan n, Äthylhydrosulfid n

thiofuran C₄H₄S Thiofuran n, Thiophen n, Thiaphen n, Divinylensulfid n

thioglycerol Thioglyzerin n

thioglycolic acid HSCH₂COOH Thioglykolsäure f, Merkaptoessigsäure f, Sulfhydrylessigsäure f, Merkaptoäthansäure f

thiohypophosphate M$^I_4$P₂S₆ Thiohypophosphat n

thioindigo Thioindigo m

thioindoxyl Thioindoxyl n, 3-Hydroxythionaphthen n

thiol acetic acid s. thioacetic acid

thiolignin Thiolignin n

thio-β-naphthol C₁₀H₇·SH Thio-β-naphthol n, β-Thionaphthol n

thionation Schwefeln n, Schwefelung f, Verschwefeln n, Verschwefelung f

thionyl Thionyl... *(=SO-Gruppe in Halogenverbindungen)*

t. bromide $SOBr_2$ Thionylbromid *n*

t. chloride $SOCl_2$ Thionylchlorid *n*

t. fluoride SOF_2 Thionylfluorid *n*

thiooxalate $M^I_2C_2S_2O_2$ Thiooxalat *n*

thiopental Thiopental *n*, 5-Äthyl-5-(1'-methylbutyl)-2-thiobarbitursäure *f*

t. sodium Thiopentalnatrium *n*

thiophene C_4H_4S Thiophen *n*, Thiaphen *n*, Thiofuran *n*, Divinylensulfid *n*

t.-free thiophenfrei

t. test Thiophenprobe *f*

thiophosgene $CSCl_2$ Thiophosgen *n*, Thiokarbonyldichlorid *n*

thiophosphate Thiophosphat *n*

thiophosphoric anhydride Phosphor(V)-sulfid *n* *(wechselnder Zusammensetzung)*

thiophosphorous anhydride P_2S_3 Phosphor(III)-sulfid *n*, Diphosphortrisulfid *n*

thiosalicylic acid $C_6H_4(SH)COOH$ Thiosalizylsäure *f*, 2-Merkaptobenzoesäure *f*

thiosemicarbazide $H_2NCSNHNH_2$ Thiosemikarbazid *n*, Hydrazinthiokarbonsäureamid *n*

thiostannic acid H_2SnS_3 Thiozinn(IV)-säure *f*

thiostannous acid H_2SnS_2 Thiozinn(II)-säure *f*

thiosulphate $M^I_2S_2O_3$ Thiosulfat *n*

thiosulphatoargentate Thiosulfatoargentat *n*

thiosulphatoaurate Thiosulfatoaurat *n*

thiosulphite $M^I_2S_2O_2$ Thiosulfit *n*

thiotolene C_5H_6S Thiotolen *n*, Methylthiophen *n*

thiourea H_2NCSNH_2 Thioharnstoff *m*, Thiokarbamid *n*

t.-formaldehyde resin Thioharnstoff-Formaldehydharz *n*

t. resin Thioharnstoffharz *n*

third law of thermodynamics dritter Hauptsatz *m* der Thermodynamik, Nernstscher Wärmesatz *m*

t.-order reaction Reaktion *f* dritter Ordnung

thistle funnel (tube) Glockentrichter *m*, Trichterrohr *n*, Trichterröhre *f*

thiuram cure Thiuramvulkanisation *f*, Thiuramvernetzung *f*

t. disulphide cure Vulkanisation *f* mit Thiuramdisulfiden, Thiuramdisulfidvernetzung *f*

thixo arm *(Krählarm mit geringem Strömungswiderstand)*

thixotrope *(Koll)* thixotropes Gel *n*

thixotropic thixotrop

t. agent Thixotropier[ungs]mittel *n*, thixotroper (Thixotropie erzeugender) Stoff *m*

t. paint thixotrope Anstrichfarbe *f*

thixotroping agent *s.* thixotropic agent

thixotropy Thixotropie *f*

Thomas converter Thomas-Konverter *m*, Thomas-Birne *f*

Thomas-Gilchrist process *s.* Thomas process

Thomas meal Thomas-Mehl *n*, Thomas-Phosphat *n*

Thomas meter Thomas-Gasmesser *m*

Thomas phosphate *s.* Thomas meal

Thomas process Thomas-Verfahren *n*, Thomas-Konverterverfahren *n*, basisches Windfrischverfahren (Bessemer-Verfahren) *n*

Thomas slag Thomas-Schlacke *f*

Thomas steel Thomas-Stahl *m*, Thomas-Konverterstahl *m*

Thomsen-Berthelot principle Berthelot-Thomsensches Prinzip *n*, Berthelot-Prinzip *n*, Berthelotsches Prinzip *n*

thomsenolite *(Min)* Thomsenolith *m* *(Kalziumnatriumfluoridoaluminat)*

Thomson [thermoelectric] effect Thomson-Effekt *m*

thomsonite *(Min)* Thomsonit *m*

thorate $M^I_2ThO_3$ Thorat *n*

thoria ThO_2 Thorerde *f*, Thoria *f*, *(veraltet für)* Thorium[di]oxid *n*

thoriate / **to** thorieren, mit Thorium überziehen

thoric Thorium...

thorite *(Min)* Thorit *m* *(Thoriumorthosilikat)*

thorium Th Thorium *n*

t. anhydride *s.* t. dioxide

t. bromide $ThBr_4$ Thoriumbromid *n*

t. carbide ThC_2 Thoriumkarbid *n*

t. chloride $ThCl_4$ Thoriumchlorid *n*

t. decay series *s.* t. series

t. dioxide ThO_2 Thorium[di]oxid *n*

t. emanation Tn Thorium-Emanation *f*, Thoron *n* *(Radonisotop 220)*

t. hydroxide $Th(OH)_4$ Thoriumhydroxid *n*

t. iodide ThJ_4 Thoriumjodid *n*

t. metaphosphate $Th(PO_3)_4$ Thoriummetaphosphat *n*

t. nitrate $Th(NO_3)_4$ Thoriumnitrat *n*

t. orthosilicate $ThSiO_4$ Thoriumorthosilikat *n*, Thoriumtetroxosilikat *n*

t. oxalate $Th(C_2O_4)_2$ Thoriumoxalat *n*

t. oxide *s.* t. dioxide

t. series Thorium[zerfalls]reihe *f*

t. sulphate $Th(SO_4)_2$ Thoriumsulfat *n*

t. sulphide Thoriumsulfid *n*

Thorne barker *(Pap)* Reibungsentrinder *m* System Thorne, Thorne-Entrindungsmaschine *f*, Thorne-Maschine *f*, Thorne-Anlage *f*

Thorne bleacher *(Pap)* Thorne-Bleichturm *m*

thoron Tn Thoron *n*, Thorium-Emanation *f* *(Radonisotop 220)*

thorough mixing Durchmischen *n*

t. washing Durchwaschen *n*, Durchspülen *n*, Durchspülung *f*

thoruranin[ite] *(Min)* Thoruranin *m*, *(veraltet für)* Bröggerit *m*

thousand-corn weight *(Landw)* Tausendkornmasse *f*, *(bisher)* Tausendkorngewicht *n*

thread 1. Faden *m*; 2. Gewinde *n*; Gewindegang *m*

t. molecule *s.* threadlike molecule

t. thermometer Fadenthermometer *n*

threaded joint Gewindeverbindung *f*, Verschraubung *f* *(Rohrverbindung)*

t. pipe Gewinderohr *n*

threadlike fadenförmig, Faden...
t. molecule Fadenmolekül *n*
t. structure Fadenstruktur *f*
three-banded spectrum Dreibandenspektrum *n*
t.-body collision Dreierstoß *m*
t.-bowl calender Dreiwalzenkalander *m*
t.-carbon acid C_3-Karbonsäure *f*, Fettsäure *f* mit drei C-Atomen
t.-carbon-atom tautomerism Dreikohlenstofftautomerie *f*
t.-carbon chain C_3-Kette *f*
t.-centre bond Dreizentrenbindung *f*
t.-centre orbital Dreizentrenorbital *n*
t.-chamber cell *s.* t.-compartment electrodialyzing device
t.-colour process Dreifarbendruckverfahren *n*
t.-compartment electrodialyzing device Dreizellenapparat *m* zur Elektrodialyse
t.-component system Dreikomponentensystem *n*, Dreistoffsystem *n*, ternäres System *n*
t.-coordinate dreifach koordiniert
t.-dimensional dreidimensional
t.-electron bond (linkage) Drei[er]elektronenbindung *f*
t.-fold dreifach; *(Krist)* dreizählig
t.-fold axis of symmetry *(Krist)* dreizählige Achse (Drehachse) *f*
t.-fold collision Dreierstoß *m*
t.-halves bond anderthalbfache Bindung *f*, Anderthalbfachbindung *f*
t.-layer board Dreischichtplatte *f (Spanplattentyp)*
t.-mash method Dreimaischverfahren *n*
t.-membered dreiglied[e]rig, 3glied[e]rig
t.-membered [carbon] ring dreigliedriger Ring (Kohlenstoffring) *m*, Kohlenstoffdrei[er]ring *m*, Drei[er]ring *m*, 3-Ring *m*
t.-membered ring hypothesis Drei[er]ringhypothese *f*
t.-neck[ed] dreihalsig, Dreihals...
t.-neck[ed] adapter Dreifachaufsatz *m*
t.-neck[ed] bottle (flask) Dreihalskolben *m*
t.-piston pump Dreikolbendruckpumpe *f*
t. pit [liming] system *(Gerb)* Drei[gruben]äschersystem *n*
t.-pocket grinder *(Pap)* Dreipressenschleifer *m*, 3-Pressen-Schleifer *m*
t.-product separator Dreigutscheider *m*, Dreiprodukt[en]scheider *m*
t.-product unit Dreigutapparat *m*, Dreiprodukt[en]apparat *m*
t.-roll calender *s.* t.-bowl calender
t.-roll mill Dreiwalzenmühle *f*, Dreiwalzenstuhl *m*, Dreiwalze *f*
t.-stage bleaching *(Pap)* Dreistufenbleiche *f*
t.-stage countercurrent washing Dreistufen-Gegenstromwäsche *f*
t.-stage pump dreistufige Pumpe *f*
t.-stage washing Dreistufenwäsche *f*
t.-throw pump Triplexpumpe *f*

t.-way cock Dreiweg[e]hahn *m*
t.-way valve Dreiwegeventil *n*, Dreiweg[e]hahn *m*
threefold *s.* three-fold
threemembered *s.* three-membered
threonine Threonin *n*, Thr
threose Threose *f (ein Monosaccharid)*
threshold 1. *(Phys)* Schwelle *f*; 2. *(Foto)* Schwelle *f*, Schwärzungsschwelle *f*; 3. *s.* t. energy
t. energy Schwellenenergie *f*
t. point *s.* threshold 2.
t. treatment Schwellen[wert]verfahren *n*, Schwellenwertbehandlung *f*, Thresholdverfahren *n*, Phosphatimpfverfahren *n*
t. value Schwellenwert
throat 1. *(Glas)* Durchlaß *m*, Durchfluß *m*; 2. Gicht *f*, Begichtungsöffnung *f*, Hochofengicht *f*
t. cheek *(Glas)* Durchlaß-Seitenstein *m*
t. cover *(Glas)* Durchlaß-Abdeckstein *m*
thrombin Thrombin *n (Gerinnungsferment aus Prothrombin)*
thrombokinase Thrombokinase *f*
throttle / to drosseln
throttle Drossel[ung] *f*, Drosselpfropfen *m*
throttling Drosseln *n*, Drosselung *f*
through [air] circulation Durch[be]lüftung *f*
t. mixer Knetmischer *m*, Knetmaschine *f*
t. washing Durchwaschen *n*, Durchspülen *n*, Durchspülung *f*
throughput Durchfluß[strom] *m*, Durchsatz[strom] *m*, Strom *m*, Förderstrom *m*
throw-away cartridge Wegwerfpatrone *f*
t. down / to [aus]fällen
throwing *(Ker)* Freidrehen *n*
t. power *(Galvanotechnik)* Streuvermögen *n*
thrust bearing Axiallager *n*
thuja oil Thujaöl *n*, Zedernblätteröl *n (meist von Thuja occidentalis L.)*
thujaketone Thujaketon *n*, 6-Methyl-5-methylen-2-heptanon *n*
thujane Thujan *n*, Sabinan *n*, 1-Isopropyl-4-methylbizyklo-[0,1,3]-hexan *n*
thujaplicin $C_{10}H_{12}O_2$ Thujaplizin *n (Isopropyltropolon)*
thujic acid Thujasäure *f*, Dehydroperillsäure *f (Monoterpenderivat)*
thujol Thujol *n*, Thujylalkohol *m*, 1-Isopropyl-4-methylbizyklo-[0,1,3]-3-hexanol *n*
thujone Thujon *n*, 1-Isopropyl-4-methylbizyklo-[0,1,3]-3-hexanon *n*
thujopsene *s.* widdrene
thujyl Thujyl *n (Atomgruppe)*
t. alcohol *s.* thujol
thulite *(Min)* Thulit *m*
thulium Tm Thulium *n*
t. oxalate $Tm_2(C_2O_4)_3$ Thulium(III)-oxalat *n*
t. oxide Tm_2O_3 Thulium(III)-oxid *n*
t. salt Thuliumsalz *n*
thuringite *(Min)* Thuringit *m*
thyme Thymian *m*, Thymus L., *(i.e.S.)* Gartenthymian *m*, Thymus vulgaris L.

t. camphor $CH_3C_6H_3(C_3H_7)OH$ Thymiankampfer *m*, Thymiansäure *f*, Thymol *n*, 2-Hydroxy-4-isopropyltoluol *n*

t. oil Thymianöl *n (von Thymus vulgaris L. und Th. zygis L.)*

thymic acid *s.* thyme camphor

thymidine *s.* thymine-2-desoxyriboside

thymine Thymin *n*, 5-Methylurazil *n*

thymine-2-desoxyriboside Thymindesoxyribosid *n*, Thymidin *n*

thymol *s.* thyme camphor

t. blue Thymolblau *n*, Thymolsulfophthalein *n (pH-Indikator)*

thymolphthalein Thymolphthalein *n (pH-Indikator)*

thymolsulphon[e]phthalein *s.* thymol blue

thymoquinone $(CH_3)_2CHC_6H_2(CH_3)O_2$ Thymochinon *n*, Karvakrolchinon *n*, 3,6-*p*-Menthadien-2,5-dion *n*

thymus nucleic acid Thymusnukleinsäure *f*, Thymonukleinsäure *f*, Desoxyribonukleinsäure *f*

thyroglobuline Thyreoglobulin *n*

thyroid hormone Schilddrüsenhormon *n*

t.-stimulating hormone *s.* thyrotrop[h]ic hormone

thyrotrop[h]ic thyrotrop

t. hormone thyreotropes Hormon *n*, Thyreotropin *n*, TSH

thyrotrop[h]in *s.* thyrotrop[h]ic hormone

thyroxin[e] Thyroxin *n*, Tetrajodthyronin *n*

Thyssen-Gálocsy process Thyssen-Gálocsy-Verfahren *n (auf Sauerstoffbasis arbeitendes Vergasungsverfahren)*

ticket board Fahrkartenkarton *m*, Billetkarton *m*

tie dyeing Buntfärbeverfahren *n* zum Erzielen von Mehrfarbeneffekten

tiemannite *(Min)* Tiemannit *m*, Selenquecksilber *n (Quecksilber(II)-selenid)*

Tiffeneau rearrangement Tiffeneau-Umlagerung *f*

tiger's eye *(Min)* Tigerauge *n*

tight dicht, fest; undurchlässig, impermeabel; wasserdicht, wasserundurchlässig; dicht schließend

t. dough *(Lebm)* harter (fester, zäher) Teig *m*

t. emulsion beständige Emulsion *f*

t.-fitting dicht [ab]schließend, dicht anliegend

t. mill enggestelltes Walzwerk *n*

t. winding *(Pap)* dichtes (hartes, klanghartes) Aufrollen (Aufwickeln) *n*

tightener [roll] *(Pap)* Spannwalze *f*

tiglic acid $CH_3CH=C(CH_3)COOH$ Tiglinsäure *f*, *trans*-2-Methyl-2-butensäure *f*

tiglium oil Krotonöl *n (von Croton tiglium L.)*

tile Fliese *f*, Platte *f*; Dachziegel *m*; Hohlziegel *m*; Formkörper *m (Füllkörper)*

t. clay Ziegelton *m*

t. dust Ziegelmehl *n*

t. ore *(Min)* Ziegelerz *n*

till Geschiebemergel *m*

tilt / to kippen, schrägstellen, neigen

tilter Kippvorrichtung *f*

tilth Gare *f*, Bodengare *f*

tilting-drum mixer Kipptrommelmischer *m*

t. head press kippbare Presse *f*, Kipppresse *f*

t. hopper *(Met)* Gießpfanne *f*

t. pan dryer Schaukeltrockner *m*, Kipphordenumlauftrockner *m*

t. pan filter Prayon-Filter *n (Bandzellenfilter aus einzelnen Planfilterzellen, die auf einer kreisförmigen Rollenbahn umlaufen)*

t. plate method *(phys Ch)* Methode *f* der geneigten Platte *(Randwinkelmessung)*

timber proofing Holzschutz *m*, Holzkonservierung *f*

time constant Zeitkonstante *f*

t.-gamma curve *(Foto)* Gamma-Zeit-Kurve *f*, Zeit-Gamma-Kurve *f*

t. of application Ausbringungszeit *f (z.B. für Düngemittel)*

t. of beating *(Pap)* Mahldauer *f*

t. of contact Kontaktzeit *f*, Kontaktdauer *f*, Berührungszeit *f*

t. of cook[ing] Kochzeit *f*, Kochdauer *f*

t. of cure *(Gum)* Vulkanisationszeit *f*, Heizzeit *f*, Gesamtheizzeit *f*

t. of development *(Foto)* Entwicklungszeit *f*, Entwicklungsdauer *f*

t. of digestion *s.* t. of cook[ing]

t.-of-flight spectrometer Laufzeitspektrometer *n*, Flugzeitspektrometer *n*

t. of flow Auslaufzeit *f*, Ausflußzeit *f*

t. of half-dyeing Halbfärbezeit *f*

t. of liberation *(Kph)* Auslösezeit *f*

t. of lingering *s.* t. of residence

t. of mixing Mischzeit *f*; Rührzeit *f*, Rührdauer *f*

t. of penetration *(Pap)* Durchtränkungszeit *f*

t. of relaxation Relaxationszeit *f*

t. of rennet coagulation Lab[gerinnungs]zeit *f*

t. of residence Verweilzeit *f*, Aufenthaltszeit *f*, Retentionszeit *f*, Rückhaltezeit *f*, Haltezeit *f*, Standzeit *f*, Stehzeit *f*, Durchgangszeit *f*, Durchlaufzeit *f*

t. of settling Absetzzeit *f*

t. of stirring Rührzeit *f*, Rührdauer *f*

t. of vulcanization *(Gum)* Vulkanisationszeit *f*, Heizzeit *f*, Gesamtheizzeit *f*, *(i.e.S.)* Ausvulkanisationszeit *f*

t. reaction Zeitreaktion *f*

t.-temperature-transformation diagram Zeit-Temperatur-Umwandlungsdiagramm *n*, Zeit-Temperatur-Umwandlungsschaubild *n*, ZTU-Diagramm *n*, ZTU-Schaubild *n*

timing Festlegung (Bestimmung, Wahl) *f* des Behandlungszeitpunktes *(z.B. mit Pflanzenschutzmitteln)*

t. device Kurzzeitmesser *m*

tin / to 1. verzinnen; 2. eindosen, in Dosen konservieren (einmachen); in Dosen verpacken

tin 1. Sn Zinn *n*; 2. [verzinnte] Blechbüchse (Blechdose, Konservendose) *f*, Weißblechbüchse *f*, Weißblechdose *f*

t. **alloy** Zinnlegierung f
t. **anhydride** s. t. dioxide
t. **ash[es]** SnO$_2$ Zinnasche f, Zinn(IV)-oxid n
t.**-base** auf Zinnbasis (Zinngrundlage)
t. **bath** Verzinnungsbad n, Zinnbad n
t.**-bearing** zinnhaltig
t. **bronze** Zinnbronze f
t. **can** s. tin 2.
t. **coating** s. t. plating
t. **cry** Zinngeschrei n, Zinnschrei m
t. **crystals** s. t. salt
t. **dichloride** SnCl$_2$ Zinndichlorid n, Zinn(II)-chlorid n
t. **difluoride** SnF$_2$ Zinndifluorid n, Zinn(II)-fluorid n
t. **diiodide** SnJ$_2$ Zinndijodid n, Zinn(II)-jodid n
t. **dioxide** SnO$_2$ Zinndioxid n, Zinn(IV)-oxid n
t. **disease** s. t. pest
t. **disulphide** SnS$_2$ Zinndisulfid n, Zinn(IV)-sulfid n
t. **foil** Zinnfolie f, Stanniolfolie f, Stanniol n, Zinnstanniol n, Blattzinn n; Stanniolpapier n, Zinnfolienpapier n
t. **glaze** (Ker) Zinnglasur f
t. **hydride** SnH$_4$ Zinnwasserstoff m, Stannan n, Zinn(IV)-hydrid n
t. **monosulphide** SnS Zinn[mono]sulfid n, Zinn(II)-sulfid n
t. **oil** Zinnöl n
t. **ore** Zinnerz n
t. **peroxide** s. t. dioxide
t. **pest** Zinnpest f
t. **plague** s. t. pest
t.**-plate** / to verzinnen
t. **plate** Weißblech n, verzinntes Eisenblech n
t. **plating** Verzinnen n, Verzinnung f
t. **pyrites** (Min) Zinnkies m, Stannin m
t. **salt** SnCl$_2$·2H$_2$O Zinnsalz n, Zinn(II)-chlorid-2-Wasser n
t. **stone** (Min) Zinnstein m, Kassiterit m (Zinn(IV)-oxid)
t. **tetrabromide** SnBr$_4$ Zinntetrabromid n, Zinn(IV)-bromid n
t. **tetrachloride** SnCl$_4$ Zinntetrachlorid n, Zinn(IV)-chlorid n
t. **tetrafluoride** SnF$_4$ Zinntetrafluorid n, Zinn(IV)-fluorid n
t. **tetraiodide** SnJ$_4$ Zinntetrajodid n, Zinn(IV)-jodid n
tincal (Min) Tinkal m, Borax m (Natriumtetraborat-10-Wasser)
tinctorial power Anfärbevermögen n, Färbevermögen n, Färbekraft f
t. **strength** Farbstärke f
tincture (Pharm) Tinktur f
t. **of digitalis** Fingerhuttinktur f, Digitalistinktur f
t. **of iodine** Jodtinktur f
t. **of kino** Kinotinktur f
t. **of opium** Opiumtinktur f
tinfoil paper Stanniolpapier n, Zinnfolienpapier n

tinge / to färben
tinge Anflug m, Stich m, Tönung f
tinman's solder Zinn-Blei-Lot n, Blei-Zinn-Lot n
tinned sheet iron verzinntes Eisenblech n, Weißblech n
t. **steel** verzinnter Stahl m
tinning Verzinnen n, Verzinnung f
t. **process** Verzinnungsverfahren n
tinsel (Glas) Flitter m
tint / to [ab]tönen, nachtönen, nuancieren
tint Farbton m, Farbtönung f, Farbschattierung f, Nuance f, Stich m
tinted cardboard Buntpappe f
tinting Tönen n, Tönung f, Abtönen n, Nachtönen n, Nuancieren n, Nuancierung f
t. **agent** Abtönmittel n, Abtöner m
tintometer Kolorimeter n, Tintometer n
tip (Glas) Pegel m
tippiness Spitzigkeit f (der Färbung von Wollpartien)
tippy dyeing Spitzigfärben n, Schipprigfärben n
tire (Am) s. tyre
tirolite (Min) Tirolit m, Kupferschaum m
tissue (Bio) Gewebe n; Gewebe n, [textiler] Stoff m, Textilgewebe n; s. t. paper
t. **culture** Gewebekultur f, Gewebskultur f; Gewebezüchtung f, Gewebszüchtung f
t.**-culture medium** Gewebskulturmedium n
t.**-culture method** Gewebskulturmethode f
t. **extract** Gewebeextrakt m
t. **fluid** Gewebeflüssigkeit f
t. **hormone** Gewebshormon n
t. **paper** Seidenpapier n (Masse je Flächeneinheit unter 25 g/m^2)
t. **slice** Gewebeschnitt m
t. **wrapper (wrapping)** Packseidenpapier n, Einschlagseidenpapier n, Einwickelseidenpapier n
tit Zipfel m, Zapfen m (Glasfehler)
titan yellow Titangelb n, Thiazolgelb n
titanate Titanat n
titania TiO$_2$ Titanerde f, (veraltet für) Titan(IV)-oxid n, Titandioxid n
t. **porcelain** Titanporzellan n
t. **whiteware** (Ker) Titanweißware f
titanic Titan..., (i.e.S.) Titan(IV)-...
t. **acid** Titansäure f
t.**[-acid] anhydride** s. t. oxide
t. **chloride** TiCl$_4$ Titan(IV)-chlorid n, Titantetrachlorid n
t. **iron** (Min) Titaneisen[erz] n, (veraltet für) Ilmenit m (Eisen(II)-metatitanat)
t. **oxide** TiO$_2$ Titan(IV)-oxid n, Titandioxid n
t. **sulphate** Ti(SO$_4$)$_2$ Titan(IV)-sulfat n
titaniferous titanhaltig, titanführend
titanite (Min) Titanit m (Kalziumtitan(IV)-oxidorthosilikat)
titanium Ti Titan
t. **carbide** TiC Titan[mono]karbid n
t. **chelate** Titanchelat n

t. diboride TiB$_2$ Titandiborid n
t. dibromide TiBr$_2$ Titandibromid n, Titan(II)-bromid n
t. dichloride TiCl$_2$ Titandichlorid n, Titan(II)-chlorid n
t. diiodide TiJ$_2$ Titandijodid n, Titan(II)-jodid n
t. dioxide TiO$_2$ Titandioxid n, Titan(IV)-oxid n
t. disulphide TiS$_2$ Titandisulfid n, Titan(IV)-sulfid n
t. hydride TiH$_2$ Titanhydrid n
t. hydroxide Ti(OH)$_4$ Titan(IV)-hydroxid n
t. monosulphide TiS Titan[mono]sulfid n, Titan(II)-sulfid n
t. monoxide TiO Titan[mon]oxid n, Titan(II)-oxid n
t. nitride TiN Titannitrid n, Titan(III)-nitrid n
t.-potassium fluoride K$_2$[TiF$_6$] Kaliumhexafluorotitanat(IV) n
t. pyrophosphate TiP$_2$O$_7$ Titandiphosphat n, Titanpyrophosphat n
t. sesquioxide Ti$_2$O$_3$ Dititantrioxid n, Titan(III)-oxid n
t. sesquisulphide Ti$_2$S$_3$ Dititantrisulfid n, Titan(III)-sulfid n
t. tetrabromide TiBr$_4$ Titantetrabromid n, Titan(IV)-bromid n
t. tetrachloride TiCl$_4$ Titantetrachlorid n, Titan(IV)-chlorid n
t. tetrafluoride TiF$_4$ Titantetrafluorid n, Titan(IV)-fluorid n
t. tetraiodide TiJ$_4$ Titantetrajodid n, Titan(IV)-jodid n
t. trichloride TiCl$_3$ Titantrichlorid n, Titan(III)-chlorid n
t. trifluoride TiF$_3$ Titantrifluorid n, Titan(III)-fluorid n
t. white Titanweiß n
titanometry Titanometrie f
titanous Titan..., (i.e.S.) Titan(III)-...
t. chloride TiCl$_3$ Titan(III)-chlorid n, Titantrichlorid n
t.-chloride method of E. Knecht Titanchloridmethode (Reduktionsmethode) f von Edmund Knecht
titanyl sulphate TiOSO$_4$ Titanoxidsulfat n
titer s. titre
titrable s. titratable
titrant Titrans n, Titersubstanz f
titratable titrierbar
titrate / to titrieren
titrating apparatus s. titration apparatus
t. beaker Titrierbecher m
t. coulometer Titrationscoulometer n
titration Titration f, Titrieren n, Titrierung f
t. apparatus Titrationsapparat m, Titrierapparat m
t. cell Titrationszelle f
t. curve Titrationskurve f
t. error Titrierfehler m
t. flask Titrierkolben m

t. vessel Titrationsgefäß n
titrator Titriergerät n, Titriervorrichtung f, Titrator m
titre Titer m (Gehalt einer Maßlösung); s. t. value; (Text) Titer m (Kennzeichnung der Feinheit von Garnen)
t. of a soap Titer m einer Seifenlösung, Trübungspunkt m
t. test Titer-Test m (zur Bestimmung des Titerwertes von geschmolzenen Fetten oder fetten Ölen)
t. value Titer[wert] m (Temperatur, bei der die Schmelze eines Fettes oder ein fettes Öl erstarrt)
titrimeter Titrimeter n
titrimetric titrimetrisch
t. analysis Titrieranalyse f, Titrimetrie f, Maßanalyse f, Volumetrie f, volumetrische Analyse f
t. factor Faktor m, Korrekturfaktor m, Normalitätsfaktor m (Größe, um die der Titer von der Normalitätsangabe abweicht)
t. standard Ur[titer]substanz f
titrimetry s. titrimetric analysis
tizerah Tizerahholz n (von Rhus pentaphylla Desf.)
t. extract (Gerb) Tizerahextrakt m (aus Holz und Wurzeln von Rhus pentaphylla Desf.)
tizra s. tizerah
TLC s. thin-layer chromatography
TML s. tetramethyl lead
T.N.T. s. trinitrotoluene
toad poison [venom] Krötengift n
tobacco Tabak m, Nicotiana L., (i.e.S. meist) Virginischer (Großblättriger) Tabak m, Nicotiana tabacum L.; Tabak m (als Produkt)
t. alkaloid Tabakalkaloid n
t. liquor Tabaklauge f
t. oil s. tobaccoseed oil
t. paper Tabakpapier n
t. saucing Tabakaromatisierung f
t. water s. t. liquor
tobaccoseed oil Tabaksamenöl n
tobermorite (Min) Tobermorit m
Tobias acid C$_{10}$H$_6$(NH$_2$)(SO$_3$H) Tobias-Säure f, 2-Aminonaphthalin-1-sulfonsäure f; (manchmal auch) C$_{10}$H$_6$(OH)(SO$_3$H) Oxy-Tobias-Säure f, 2-Oxynaphthalin-1-sulfonsäure f
tocopherol Tokopherol n
toe (Foto) Durchhang m, Kurvendurchhang m, durchhängender Teil m (der Schwärzungskurve), Gebiet n der Unterexposition (Unterbelichtung)
toggle Kniehebel m
t. crusher Kniehebelbackenbrecher m
t. [lever] press Kniehebelpresse f
t. system Kniehebelantrieb m
toilet preparation Toilettepräparat n
t. soap Toilette[n]seife f, Feinseife f
t. tissue Toilettenseidenpapier n
t. water Toilettewasser n
tol. s. toluene
tolan C$_6$H$_5$C≡CC$_6$H$_5$ Tolan n, Diphenyläthin n, Diphenylazetylen n

tolerance Toleranz *f*, zulässige Abweichung *f*; *(Toxikologie)* Toleranz[dosis] *f*, Toleranzwert *m*, duldbare (zulässige, zugelassene) Höchstmenge (Restmenge) *f*

t. to garbage *(Landw)* Müllverträglichkeit *f*

tolidine Tolidin *n*, Dimethylbenzidin *n*

Tollens' reagent Tollens-Reagens *n (ammoniakalische Silbernitratlösung)*

Tolu balsam Tolubalsam *m (von Myroxylon balsamum (L.) Harms var. balsamum)*

toluene $C_6H_5CH_3$ Toluol *n*, Methylbenzol *n*

t. diisocyanate Toluoldiisozyanat *n*, TDI

t. hexahydride $(C_6H_{11})CH_3$ Hexahydrotoluol *n*, Methylzyklohexan *n*, Zyklohexylmethan *n*, Heptanaphthen *n*

toluenediamine *s.* toluylenediamine

p-toluenesulphonate *p*-Toluolsulfonat *n*, *p*-Toluolsulfonsäureester *m*, Tosylester *m*, Tosylat *n*

p-toluenesulphonic acid $CH_3C_6H_4SO_3H$ *p*-Toluolsulfonsäure *f*, 1-Methylbenzolsulfonsäure *f*

p-toluenesulphonyl chloride $CH_3C_6H_4SO_2Cl$ *p*-Toluolsulfonylchlorid *n*, *p*-Toluolsulfo[nsäure]chlorid *n*, Tosylchlorid *n*

p-toluenesulphonyl group $CH_3C_6H_4SO_2-$ *p*-Toluolsulfonylgruppe *f*, Tosylgruppe *f*

toluic acid $CH_3C_6H_4COOH$ Toluylsäure *f*, Methylbenzoesäure *f*

α-**t. acid** $C_6H_5CH_2COOH$ 1′-Tolylsäure *f*, 1′-Toluolkarbonsäure *f*, Phenylessigsäure *f*

toluidine $CH_3C_6H_4NH_2$ Toluidin *n*, Aminotoluol *n*

α-**tolunitril** $C_6H_5CH_2CN$ Phenylessigsäurenitril *n*, Phenylazetonitril *n*, Benzylzyanid *n*

toluol Toluol *n (als Handelsprodukt)*; *s.* toluene

toluylene $C_6H_5CH=CHC_6H_5$ Toluylen *n*, syn-1,2-Diphenyläthen *n*

t. red Toluylenrot *n*, Neutralrot *n (Redoxindikator)*

tol[u]ylenediamine $CH_3C_6H_3(NH_2)_2$ Toluylendiamin *n*, Tolylendiamin *n*, Diaminotoluol *n*

tombozine Tombozin *n*, 10-Desoxysarpagin *n*

tonca bean *s.* tonka bean

tone / to [ab]tönen, nachtönen, nuancieren; *(Foto)* tonen; *(Pharm)* tonisieren

t. down *(eine Farbe)* abschwächen

tone Farbton *m*, Farbtönung *f*, Farbschattierung *f*, Nuance *f*, Stich *m*

t.-in-tone dyeing Ton-in-Ton-Färbung *f*

t. range *(Foto)* Tonumfang *m*

t. rendering (reproduction) *(Foto)* Ton[wert]wiedergabe *f*

tonga bean *s.* tonka bean

tongue *(Glas)* Brennerzunge *f*

t. tear strength *(Gerb)* Zungenreißfestigkeit *f*, Weiterreißfestigkeit *f*

t. tile *s.* tongue

tonic tonisch, tonisierend, kräftigend, stärkend

tonic Tonikum *n*, Kräftigungsmittel *n*, tonisierendes (kräftigendes, stärkendes) Mittel *n*

toning *(Foto)* Tonen *n*, Tonung *f*

t. agent *(Foto)* Toner *m*

t. lotion tonische (tonisierende, kräftigende, stärkende) Lösung *f*

tonka bean Tonkabohne *f (von Dipteryx-Arten)*

t.-bean camphor Tonkabohnenkampfer *m*, Kumarin *n*, 1,2-Benz[o]pyron *n*, 5,6-Benzokumalin *n*

tonqua bean *s.* tonka bean

tool joint Gestängeverbinder *m (einer Rotary-Bohranlage)*

t. room *(Plast)* Werkzeugmacherei *f*

t. steel Werkzeugstahl *m*

tooling resin *(Plast)* Werkzeugharz *n*

toothed-disk mill Zahnscheibenmühle *f*

t.-roll crusher Zahnwalzenbrecher *m*, Stachelwalzenbrecher *m*

toothpaste Zahnpaste *f*

top / to 1. herausdestillieren; *(Erdöl)* toppen; 2. *(Text)* nachfärben, überfärben, [nach]decken, schönen; *(Gerb)* (sauer gefärbtes Leder mit basischen Farbstoffen) übersetzen

top Oberteil *n*; Spitze *f*; Deckel *m*; *(Met)* Kopf *m (beim Kopf- und Bodenschmelzen)*; Gicht *f (des Hochofens)*; *(Text)* Kammzug *m*; Faserbändchen *n*, Spinnband *n*

t.-and-bottom process (smelting) Kopf- und Bodenschmelzen *n (zur getrennten Gewinnung von Kupfer und Nickel)*, Orford-Verfahren *n*

t.-blown basic oxygen converter (furnace) Sauerstoff[auf]blaskonverter *m*, Aufblas[e]konverter *m*, Blasstahlkonverter *m*, Oberwindkonverter *m*

t.-blown oxygen converter process Sauerstoff[auf]blasverfahren *n*, Sauerstoffaufblas-Konverterprozeß *m*, Sauerstoff-Frischverfahren *n*, O_2-Aufblasverfahren *n*, Aufblas[e]verfahren *n*, Blasstahlverfahren *n*, Oberwindfrischverfahren *n*

t. casting *(Met)* fallender Guß *m*

t. centrifugal atomizer Fliehkraftzerstäuber *m*, Zentrifugalscheibe *f*

t. chest *(Pap)* Stoffverteilerkasten *m*

t. chrome dye Nachchromierfarbstoff *m*

t. chrome [dyeing] method Nachchromier[ungs]verfahren *n*

t. chroming Nachchromieren *n*, Nachchromierung *f*

t. coat[ing] Deckanstrich *m*

t. dressing *s.* topdressing

t. dyeing *(Text)* Färben *n* von Kammzug, Kammzugfärben *n*

t.-dyeing machine *(Text)* Kammzugfärbeapparat *m*

t. ejection *(Plast)* Abdrücken *n* von oben, Abstreifen *n* vom Stempel

t. feed Obenaufgabe *f*; Sumpfaufgabe *f (am Zweiwalzentrockner)*

t.-feeding filter Filter *n* mit Obenaufgabe

t. felt *(Pap)* Oberfilz *m*, oberer Abnahmefilz *m (bei Rundsieb-Kartonmaschinen)*

t. fermentation Obergärung *f*

t.-fermentation beer obergäriges (obergängiges) Bier *n*

t.-fermentation yeast obergärige Hefe f, Oberhefe f

t.-fired kiln deckenbeheizter (von oben beheizter) Ofen m

t. force (Plast) Oberstempel m

t. fraction (Erdöl) Kopffraktion f

t.-hat kiln (Ker) Haubenofen m

t. layer obere Schicht f, Oberschicht f

t. liming (Landw) Kopfkalkung f

t. of the column (Destillation) Kolonnenkopf m

t. of the digester (Pap) Gasraum m (des Kochers)

t. pouring s. t. casting

t. press roll doctor (Pap) Schaber m an der [oberen] Preßwalze

t. printing (Text) Vigoureuxdruck m, Druck m auf Kammzug, Überdrucken n

t. product (Destillation) Kopfprodukt n

t. punch (Ker) Oberstempel m

t. ram (Plast) Oberstempel m

t. ram press (Plast) Oberkolbenpresse f, Oberdruckpresse f

t. roll Oberwalze f; (Pap veraltet für) dandy roll

t. roll of the couch press (Pap) obere Gautschwalze f

t. shell (gewölbter) Deckel m

t. side Oberseite f, Filzseite f, Schönseite f (des Papiers)

t. size press roll (Pap) Oberwalze f einer Leimpresse

t.-sized (Pap) oberflächengeleimt

t.-sizing (Pap) Oberflächenleimung f

t.-suspended centrifuge Hängezentrifuge f

t. tub (vat) (Farb) oberster Bottich m

t. yeast obergärige Hefe f, Oberhefe f

topaz (Min) Topas m

topazolite (Min) Topazolith m (Kalziumeisen(III)-triorthosilikat)

topdress / to (Landw) mit Kopfdünger behandeln

topdressing (Landw) Kopfdüngung f; Kopfdünger m

Topham box Spinnzentrifuge f, Spinntopf m

topochemical topochemisch

topochemistry Topochemie f

topped beet geköpfte Rübe (Zuckerrübe) f

t. crude [petroleum] getopptes (abgetopptes, reduziertes) Rohöl (Öl) n, Topped Crude n, Reduced Oil n

topping 1. Toppen n, Toppung f, Toppdestillation f, atomosphärische (unter atmosphärischem Druck arbeitende) Destillation f, Normaldruckdestillation f; 2. (Text) Überfärben n, Überfärbung f, Nachfärben n, Nachfärbung f (einer Komponente in Faserstoffmischungen); (Gerb) Übersetzen n (sauer gefärbter Leder mit basischen Farbstoffen)

t. plant (Erdöl) Toppanlage f

topple over / to umkippen, umfallen

tops (Erdöl) Toppprodukt n

topsoil Ackerkrume f, Krume f, Mutterboden m, Muttererde f

torbanite (Min) Torbanit m, (früher gesonderte Varietät von) Bituminit m, Bogheadkohle f

torber[n]ite (Min) Torbernit m (Kupfer(II)-bisuranylorthophosphat)

torch Brenner m; Lötlampe f; Schweißbrenner m

törnebohmite (Min) Törnebohmit m (zer- und lanthanhaltiges Neso-Subsilikat)

Törnebohm's minerals (Sammelname für die im Portlandzementklinker auftretenden Kristallarten Alit, Belit, Brownmillerit, Felit)

torpedo (Plast) Torpedo n, Verdrängungskörper m

torque Drehmoment n, Moment n

torrefaction (Lebm) Rösten n, Röstung f

    t. dextrin Röstdextrin n

torrefy / to (Lebm) rösten

Torricellian vacuum Torricellisches Vakuum n

torsion Torsion f, Verdrehung f, Drillung f, Verwindung f

    t. head Torsionskopf m

    t. loop (string) galvanometer Schleifengalvanometer n

    t. wire Torsionsdraht m

torula yeast Torulahefe f

tosyl $CH_3C_6H_4SO_2-$ Tosylgruppe f, p-Toluolsulfonylgruppe f

    t. chloride $CH_3C_6H_4SO_2Cl$ Tosylchlorid n, p-Toluolsulfonylchlorid n, p-Toluolsulfo[nsäure]chlorid n

tosylate Tosylat n, Tosylester m, p-Toluolsulfonsäureester m, p-Toluolsulfonat n

tosylation Tosylierung f (Veresterung von Hydroxylverbindungen mittels p-Toluolsulfonylchlorids)

total acid number (Maßzahl für die freie Gesamtsäure in mg KOH/1g Probesubstanz)

    t. acidity (Bodenkunde) Titrationsazidität f, potentielle Azidität f

    t. alkali Gesamtalkali n, Gesamtalkaligehalt m (aktives Alkali + Na_2CO_3)

    t. alkalinity (Pap) Gesamtalkaligehalt m

    t. alkaloids Gesamtalkaloide pl

    t. amount of alkali s. t. alkali

    t. amount of entropy Gesamtentropie f

    t. base number (Maßzahl für den Alkaligehalt, ausgedrückt in mg KOH als Äquivalent für die zur Neutralisation aufgewandte Säuremenge / 1 g Probesubstanz)

    t. charge Gesamtladung f

    t. chemical (Pap) s. t. alkali

    t. concentration Gesamtkonzentration f

    t. condensation (Destillation) Totalkondensation f

    t. conductance Gesamtleitfähigkeit f, Gesamtleitvermögen n

    t. current Gesamtstrom m

    t. emissive power s. t. emissivity

    t. emissivity Gesamtemissionsvermögen n

    t. energy Gesamtenergie f

    t. entropy Gesamtentropie f

    t. extractable sulphur extrahierbarer Schwefel m

t. **gasification** vollständige (restlose, vollkommene, rückstandslose) Vergasung f
t. **hardness** Gesamthärte f
t. **hardness of water** s. t. water hardness
t. **heat of solution** ganze (totale) Lösungswärme (Lösungsenthalpie) f
t. **immersion test** Tauchversuch m, Tauchtest m, Tauchprüfung f
t. **ionic concentration** [gesamte] ionale Konzentration f
t. **load** Gesamtbelastung f (z.B. einer Waage)
t. **make-up** (Pap) Gesamtmenge f der zur Deckung der Alkaliverluste zugesetzten Chemikalien, Gesamtzusatzmenge f
t. **molarity** Gesamtmolarität f
t. **molecular formula** Bruttoformel f, Summenformel f
t. **nitrogen** Gesamtstickstoff m
t. **number** Gesamtmenge f (z.B. von Einheiten)
t. **number of moles** Gesamtmolzahl f
t. **polarization** Gesamtpolarisation f
t. **pressure** Gesamtdruck m
t. **quantum number** Hauptquantenzahl f
t.-**radiation pyrometer** Gesamtstrahlungspyrometer n
t. **reaction** Gesamtreaktion f
t. **reflection** Totalreflexion f
t. **reflux** totaler Rücklauf (Rückfluß) m
t. **reflux operation** Fahrweise (Arbeitsweise) f mit totalem Rücklauf, Fahrweise f ohne Destillatabnahme
t. **shrinkage** (Ker) Gesamtschwindung f
t. **solids** Gesamttrockensubstanz f, Gesamttrockenmasse f; (Gum) Gesamtfestsubstanz f, Kautschuktrockensubstanz f
t.-**solids content** Gehalt m an Gesamttrockensubstanz (Gesamttrockenmasse)
t. **sulphur** totaler Schwefel m, Gesamtschwefel m
t. **sulphur dioxide** (Pap) Gesamtgehalt m an SO_2, Gesamt-SO_2-Gehalt m, Gesamt-SO_2 n
t. **surface** Gesamt[ober]fläche f
t. **swelling** Gesamtquellung f
t. **symmetry** Gesamtsymmetrie f
t. **synthesis** Totalsynthese f
t. **transition probability** Gesamt[übergangs]wahrscheinlichkeit f
t. **valency** Gesamtwertigkeit f
t. **volume** Gesamtvolumen n
t. **water hardness** Gesamthärte f des Wassers
t. **weed control** (Landw) Ganzflächenbehandlung f (zur Unkrautbekämpfung)
t. **weight** Gesamtgewicht n; Gesamtmasse f
t. **work** gesamte Arbeit f, Gesamtarbeit f
t. **yield** Gesamtausbeute f
totaquina, totaquine (Chinin und andere China-Alkaloide enthaltendes Malariamittel aus Cinchona-Rinde)
touchstone Probierstein m; (Min) Lydit m (schwarzer Kieselschiefer)
tough zäh, kräftig, widerstandsfähig; (Text) tragecht, tragfest

t.-**pitch condition** zähgepolter Zustand m
t.-**pitch copper** zähgepoltes (hammergares) Kupfer n, Garkupfer n
toughen / to zäh machen; zäh werden
t. **by poling** (Met) zähpolen
toughened glass vorgespanntes Glas n, (i.e.S.) Einscheibensicherheitsglas n, Einschichtensicherheitsglas n
toughness Zähigkeit f
tourill Tourill n, Turille f
tourmaline, tourmalinite (Min) Turmalin m
tourmalinization Turmalinisierung f, Turmalinisation f
tow (Text) Elementarfadenkabel n, Spinnkabel n, Kabel n aus endlosem Material; Werg n (Bastfasern)
t.-**to-top breaking system** (Text) Reißkonverterverfahren n
t.-**to-top cutting system** (Text) Schneidkonverterverfahren n
t.-**to-top process** (Text) Konverterverfahren n
towanite (Min) Towanit m, (veraltet für) Kupferkies m
towelling paper Handtuchpapier n
tower Turm m, (Destillation auch) Säule f, Kolonne f
t. **acid** (Pap) Turmsäure f, Rohsäure f
t. **bleaching** (Pap) Turmbleichverfahren n, Turmbleiche f
t. **dryer** Brüdenschlottrockner m, Schachttrockner m
t. **packing** Füllkörper m, Füllmaterial n
t. **purifier** Turmreiniger m, Reinigungsturm m
t. **steamer** (Text) Turmdämpfer m
t. **system** (Pap) Turmsystem n, Turmverfahren n
town gas Stadtgas n
t. **refuse** Stadtmüll m
t. **sewage** städtisches Abwasser n
toxalbumin Toxalbumin n
toxic toxisch, giftig
toxic s. toxicant
t. **action** toxische Wirkung f, Giftwirkung f
t. **dose** toxische Dosis f, Giftdosis f, schädigende Menge f
t. **effect** s. t. action
t. **principle** toxisches Prinzip n, giftiger Bestandteil m
t. **substance** s. toxicant
toxical toxisch, giftig
toxicant Giftstoff m, Gift n, giftiger Wirkstoff m, giftige Substanz f, Toxikum n
toxicity Toxizität f, Giftigkeit f
toxicogenic Gift erzeugend; durch Gift (Vergiftung) entstanden (bedingt), (Med auch) toxigen
toxicologic[al] toxikologisch
toxicologist Toxikologe m, Giftkundiger m, Giftsachverständiger m
toxicology Toxikologie f
toxiferine Toxiferin n (ein Kalebassen-Alkaloid)
toxiferous gifthaltig; Gift erzeugend

toxify / to vergiften
toxigenic *(Bio)* Toxine erzeugend; *s.* toxicogenic
toxilic acid *(veraltet für)* maleic acid
toxin Toxin *n*
toxoid *(Med)* Toxoid *n*
toxophore *(Med)* toxophore Gruppe *f (die Giftwirkung auslösender Anteil eines Toxins)*
Tozer process Tozer-Verfahren *n*, Schwelverfahren *n* von Tozer
2,4,5-TP *s.* 2-(2,4,5-trichlorophenoxy)propionic acid
TPN *s.* triphosphopyridine nucleotide
trace 1. Spur *f*, geringe Menge *f*; 2. Schreibspur *f*, Spur *f*, Kurve *f*
  t. amount Spurenmenge *f*
  t. analysis Spurenanalyse *f*
  t. element Spurenelement *n*, *(Bot auch)* Mikronährstoff *m*
  t. metal Spurenmetall *n*
tracer Tracer *m*, Indikatorisotop *n*, Indikatorsubstanz *f*, Leitisotop *n*, Zusatzisotop *n*, Spurenmaterial *n*, Spurensucher *m*
  t. atom Indikatoratom *n*
  t. chemistry Indikator[en]chemie *f*, Tracerchemie *f*
  t. element Indikatorelement *n*, Markierungselement *n*
  t. experiment *(Bioch)* Tracerversuch *m*, *(i.e.S.)* Versuch *m* mit Leitisotopen
  t. method Tracermethode *f*, Indikatormethode *f*
  t. nitrogen Stickstoffisotop N^{15} *n (als Leitisotop)*
  t. technique Tracertechnik *f*
tracing cloth Pausleinen *n*
  t. paper Transparent[zeichen]papier *n*; Durchzeichenpapier *n*, Pauspapier *n*
track resistance *(Plast)* Kriechstromfestigkeit *f*
tracking *(Plast)* Kriechwegbildung *f*
  t. resistance *s.* track resistance
traction thickener Eindicker *m* mit Randantrieb
tractor fuel Traktorenkraftstoff *m*, Traktorentreibstoff *m*, Schlepperkraftstoff *m*
  t. [vaporizing] oil Traktorenpetrol[eum] *n*, Traktorenkerosin *n*, Traktorenöl *n*
trade name Handelsname *m*, Handelsbezeichnung *f*, kommerzieller Name *m*
  t. sample Handelsmuster *n*
  t. term *s.* t. name
  t. waste Schmutzwasser *n*, gewerbliches Abwasser *n*
trademark Warenzeichen *n*, Wz.
tragacanth Tragant[h]pflanze *f*, Tragant *m*, Astragalus L.; *s.* 1. tragacanth gum; 2. African tragacanth
  t. adhesive Tragantleim *m*
  t. gum Tragantgummi *n*, Tragant *m (von Astragalus-Arten)*
  t. mucilage Tragantschleim *m*
trailing *(Papierchromatografie)* Schwanzbildung *f*, Schweifbildung *f*; *(Ker)* Schlickermalerei *f*
  t. blade coater *(Pap)* Schleppprakelstreich-

maschine *f*, Schleppklingenstreichmaschine *f*, Glättschaberstreichmaschine *f*
train oil Walöl *n*, Waltran *m*, *(i.w.S.)* Seetieröl *n*
trajectory Bahn *f*, Weg *m*; *(in Diagrammen)* Trajektorie *f*
tram silk Trame[seide] *f*, Schuß[roh]seide *f*
tramp iron schädliche Eisenteile *npl*, Fremdeisen *n*
  t.-iron magnet[ic separator] Eisenausscheider *m*, Eisenabscheider *m*, Schutzmagnet *m*
  t. material Fremdgut *n*
tranquilite Tranquilit *m (Mondmineral)*
tranquil[l]izer, tranquillizing drug *(Pharm)* Tranquil[l]izer *m*, Ataraktikum *n*, tranquilisierender Stoff *m*
trans trans-..., trans-ständig, in trans-Stellung befindlich
  to be t. trans (trans-ständig, in trans-Stellung) stehen, trans-ständig angeordnet sein
  t. addition trans-Addition *f*
  t. form trans-Form *f*
  t. isomer trans-Isomer[es] *n*
  t. position trans-Stellung *f*, trans-Lage *f*
  t.-trans hydrocarbon trans-trans-Kohlenwasserstoff *m*
transaminase Transaminase *f*, Aminotransferase *f*
transamination Transaminierung *f*
transannular transannular
transcrystalline fracture *(Met)* intrakristalliner Bruch *m*
transesterification Umesterung *f*
transfer / to 1. *(Elektronen, Ionen)* übertragen, abgeben, transportieren; *(Substanzen in ein anderes Gefäß)* überführen, übertragen, bringen, umfüllen; *(Wärme)* übertragen, transportieren; *(Güter)* transportieren; 2. übergehen *(Elektronen)*
transfer Übertragung *f*, Abgabe *f*, Transport *m (von Elektronen, Ionen)*; Überführen *n*, Übertragen *n*, Umfüllen *n (von Substanzen)*; Übertragung *f*, Transport *m (von Wärme)*; Transport *m (von Gütern)*; Abziehbild *n*
  t. base paper Abziehbilderrohpapier *n*
  t. car Zubringe[r]wagen *m*
  t. function Übertragungsfunktion *f*
  t. glass Hafenrohglas *n* für optische Zwecke
  t. hydrogenation Wasserstoffübertragung *f*
  t. mould *(Plast)* Preßspritzwerkzeug *n*, Spritzpreßwerkzeug *n*, Transferpreßwerkzeug *n*
  t. moulder *(Plast)* Spritzpreßmaschine *f*, Spritzpresse *f*
  t. moulding *(Plast)* Preßspritzen *n*, Preßspritzverfahren *n*, Spritzpressen *n*, Spritzpreßverfahren *n*, Transferpressen *n*; *(Plast)* Spritzpreßteil *n*; *(Gum)* Transferformung *f*, Transfer-Verfahren *n*, Fließformen *n (zur Herstellung von Gummiformteilen)*
  t. moulding pressure *(Plast)* Preßdruck *m* beim Preßspritzen (Spritzpressen)
  t. of heat Wärmeübertragung *f*

t. paper Umdruckpapier *n*; Abziehbilderpapier *n*
t. pipette Vollpipette *f*
t. pot *(Plast)* Spritztopf *m (beim Preßspritzen)*
t. reaction *(Plast)* Übertragungsreaktion *f*
t. ribonucleic acid Transfer-Ribonukleinsäure *f*, Transfer-RNA *f*, lösliche Ribonukleinsäure *f*
t. RNA *s.* t. ribonucleic acid
t. unit *(phys Ch)* Übergangseinheit *f*
transference Überführung *f*, Übertragung *f*; Übergang *m*
t. number *(Ech)* [Hittorfsche] Überführungszahl *f*
t. number of the cation Kation-Überführungszahl *f*, Kation[en]überführungszahl *f*
t. of electrons Elektronenübertragung *f*; Elektronenübergang *m*
transferring enzyme Transferase *f*, Transferenzym *n*
transform / to umwandeln
transformability Umwandelbarkeit *f*
transformation Umwandlung *f*, Überführung *f*; Übergehen *n* Übergang *m*, Verwandlung *f*, Umwandlung *f*; *(Kch)* Umwandlung *f*, Kernumwandlung *f*; *(phys Ch)* Transformation *f (von Energieformen)*; *(manchmal auch)* Umlagerung *f (von Atomen oder Atomgruppen)*
t. constant *(Kch)* Zerfallskonstante *f*
t. point *(Glas)* Transformationspunkt *m*
t. product Umwandlungsprodukt *n*
t. range *(Glas)* Transformationsintervall *n*
t. temperature Umwandlungstemperatur *f*
t.-temperature-time diagram Zeit-Temperatur-Umwandlungsdiagramm *n*, Zeit-Temperatur-Umwandlungsschaubild *n*, ZTU-Diagramm *n*, ZTU-Schaubild *n*
transformer oil Transformatorenöl *n*, Trafoöl *n*
transforming *s.* transformation
transglycosylation Transglykosylierung *f*, Transglykosidierung *f (enzymatische Übertragung einer Glykosylgruppe)*
transgranular transkristallin
t. cracking transkristalliner Riß *m*
transient unbeständig, instabil *(z.B. Radikale)*
t. creep Übergangskriechen *n*
t. flow Übergangsfließen *n*
transition Übergehen *n*, Übergang *m*, Umwandlung *f*, Metamorphose *f*
t. element Übergangselement *n*, Übergangsmetall *n*
t. interval *(phys Ch)* Transformationsintervall *n*; *(Plast)* Einfriergebiet *n*; Umschlagsintervall *n*, Umschlagsgebiet *n*, Übergangsbereich *m (eines Farbindikators)*
t. metal *s.* t. element
t.-metal binary hydride binäres Übergangsmetallhydrid *n*, metallartiges (metallisches) Hydrid *n*
t.-metal hydride Übergangsmetallhydrid *n*
t.-metal salt Übergangsmetallsalz *n*
t. point *(Krist, phys Ch)* Umwandlungspunkt *m*;

(bei inkongruent schmelzenden Verbindungen) peritektischer Punkt *m*
t. probability *(Kch)* Übergangswahrscheinlichkeit *f*
t. region Übergangsbereich *m*, Übergangsgebiet *n*, Übergangszone *f*
t. section *s.* t. region
t. series Übergangsreihe *f (im Periodensystem)*
t. state Übergangszustand *m*
t. temperature Umwandlungstemperatur *f*, Übergangstemperatur *f*; *(Plast)* Einfriertemperatur *f*; Sprungtemperatur *f*, Sprungpunkt *m (eines Supraleiters)*
t.-type system *(phys Ch)* Peritektikum *n*
transitional metal *s.* transition element
t. probability *s.* transition probability
transitory state Zwischenzustand *m*
translation *(Krist)* Translation *f*, Parallelverschiebung *f*
t. grating *s.* translational lattice
translational degree of freedom Translationsfreiheitsgrad *m*
t. energy Translationsenergie *f*
t. entropy Translationsentropie *f*
t. lattice *(Krist)* Translationsgitter *n*, Elementargitter *n*
t. motion Translation[sbewegung] *f*, fortschreitende Bewegung *f*
t. partition function Translationsverteilungsfunktion *f*
translocated herbicide translokales (translokal wirkendes, systemisches) Herbizid *n*
translocation Verlagerung *f*, Translokation *f*
t. of nutrients *(Bot)* Nährstoffverlagerung *f*
translucence, translucency Durchscheinbarkeit *f*, *(unvollständige)* Transparenz *f*
translucent durchscheinend, lichtdurchlässig, *(unvollkommen)* transparent
t. vitreous silica durchscheinendes (undurchsichtiges) Kieselglas *n*, Quarzgut *n*
translucid *s.* translucent
transmission 1. Übersendung *f*, Versand *m*, Beförderung *f*; 2. *(Tech)* Transmission *f (zur Kraftübertragung)*; *(Foto, Kolorimetrie oft unexakt für)* transmittance
t. belt Treibriemen *m*
t. belting Treibriemen *mpl*
t. grating Transmissionsgitter *n*
t. of detonation Detonationsübertragung *f*
t. of light Lichtdurchlässigkeit *f*
t. ratio *(Optik)* Durchlaßgrad *m*, Transmissionsgrad *m*
transmittance *(optische)* Durchlässigkeit *f*; *s.* transmission ratio
transmittancy *(Kolorimetrie)* Intensitätsverhältnis *n*; *s.* transmittance
transmitted light Durchlicht *n*, durchfallendes Licht *n*
t.-light technique Durchlichtmethode *f*
transmutation Umwandlung *f*, Stoffumwandlung *f*,

Elementumwandlung f, Verwandlung f, Transmutation f

transmute / to umwandeln, verwandeln

transparency Transparenz f, Durchsichtigkeit f, Lichtdurchlässigkeit f; Strahlendurchlässigkeit f; Dia[positiv] n

transparent durchsichtig, transparent, lichtdurchlässig

t. **chromium oxide** $Cr_2O_3 \cdot xH_2O$ Chromoxidgrün feurig n, Chromoxidhydratgrün n, Guignetgrün n, Mittlers Grün n, Viridian n, Smaragdgrün n

t. **glaze** (Ker) Transparentglasur f

t. **soap** transparente Seife f, Transparentseife f

t. **tissue paper** Glasseidenpapier n

t. **to infrared** infrarotdurchlässig

t. **vitreous silica** durchsichtiges (klares) Kieselglas n, Quarzglas n

transpiration Transpiration f, Wasserabgabe f, (Med auch) Ausdünstung f, Schweißabsonderung f

t. **coefficient** (Bio) Transpirationskoeffizient m

t. **method** Saturationsmethode f (zur Dampfdruckmessung)

transport air Trägerluft f

t. **number** (Ech) [Hittorfsche] Überführungszahl f

t. **reaction** Transportreaktion f

transuranic (transuranic, transuranium) element Transuran n

transversal extruder head Querspritzkopf m

transverse strength Biegefestigkeit f; Querbruchfestigkeit f

trap / to auffangen, abfangen, festhalten; einschließen (z.B. Gase in Kristallen)

trap Auffanggefäß n, Sammelgefäß n; Geruchsverschluß m, Wasserverschluß m, Siphon m; (Kch) Abscheider m, Falle f; (Kch) Trap m, Haftstelle f, Fangstelle f, Rekombinationszentrum n; Anzapfung f (in einem Rohrsystem); (Erdöl) Falle f (undurchlässige Gesteinsschicht)

trapped air Lufteinschluß m

trass (Geol) Traß m (trachytischer Tuff)

t. **cement** Traßzement m

Traube rule Traubesche Regel f

traumatic acid $HOOCCH_2(CH_2)_7CH = CHCOOH$ Traumatinsäure f, 2-Dodezendisäure f, 1-Dezen-1,10-dikarbonsäure f

Trauzl [lead block] test Bleiblockprobe f nach Trauzl, Trauzlsche Bleizylinderprobe f (zur Sprengstoffprüfung)

travel / to wandern

travel Lauf m, Wanderung f; s. t. rate

t. **rate** Laufgeschwindigkeit f, Wanderungsgeschwindigkeit f

travelling-belt extractor Extraktor (Extrakteur) m mit waagerechten Siebplattenförderern

t. **block** Flaschenzug[rollen]block m

t. **grate** Wanderrost m, Kettenrost m

t.**-pan filter** Bandzellenfilter n, Wandernutsche f, Kastenbandfilter n

t. **wire** Papiermaschinensieb n

traversellite (Min) Traversellit m

travertine Travertin m

tray Pfanne f; Trog m; Schale f, (bei Trocknern auch) Trockenschale f; Trockenblech n, Siebblech n, Trockenhorde f, Horde f (eines Trockners); Boden m (einer Destillationskolonne); (Pap) Siebwasserbehälter m, Siebwassersammelbecken n

t. **column** (Destillation) Bodenkolonne f, Bodensäule f

t. **dryer** Plattentrockner m

t. **efficiency** Bodenwirkungsgrad m (einer Destillationskolonne)

t. **thickener** Mehrkammereindicker m

t.**-to-tray calculation (procedure)** (Destillation) Berechnung f (der Mindesttrennstufenzahl) von Trennstufe zu Trennstufe

t. **tower** s. t. column

t. **truck** Hordenwagen m

t.**-truck dryer** Kanaltrockner m; Kammertrockner m (mit Hordenwagen)

t. **water** (Pap) Sieb[ab]wasser n, Abwasser n der Registerpartie

treacle (dicker schwarzbrauner) Sirup m, Melassesirup m, Melasse f; Sirup m

tread (Gum) Lauffläche f, Protektor m

t. **compound** s. t. stock

t. **extruder** (Gum) Laufstreifenspritzmaschine f, Laufflächenspritzmaschine f, Protektorspritzmaschine f

t. **head** (Gum) Laufflächenspritzkopf m, Protektorspritzkopf m

t. **mix** s. t. stock

t. **rubber** (Gum) Laufflächengummi m, Protektorgummi m

t. **stock** (Gum) Laufflächenmischung f

t. **wear** (Gum) Laufflächenabrieb m, Laufflächenabnutzung f

treat / to (chemisch) behandeln, versetzen; (mechanisch) bearbeiten

treater Behandlungsgefäß n

treating s. treatment

t. **process** Behandlungsverfahren n, (Erdöl auch) Nachbehandlungsverfahren n, Raffinationsverfahren n

t. **with copper** (Text) Kupfern n, Kupferung f

treatment Behandlung f, (Erdöl auch) Nachbehandlung f, Raffination f (i.e.S.); (mechanische) Bearbeitung f

to give a mild cracking t. mild (gelinde) kracken

t. **cosmetic (preparation)** kosmetisches Pflegemittel n

treble superphosphate (Landw) Doppelsuperphosphat n (Düngemittel aus Phosphorsäureaufschluß von Rohphosphaten)

tree / to (Krist) sich baumartig (als verästelte kristalline Masse) abscheiden, (i.e.S.) einen Bleibaum bilden

tree (Met) Bleibaum m

tremolite *(Min)* Tremolit *m*, Grammatit[strahlstein] *m (ein Inosilikat)*
tri. s. 1. trichloroethylene; 2. triclinic
triacetate Triazetat *n*
triacid dreisäurig *(Base)*; dreifachsauer *(saures Salz)*
triacid dreibasige (dreibasische, dreiwertige) Säure *f*
triacontane $C_{30}H_{62}$ Triakontan *n*
n-triacontanoic acid $C_{29}H_{59}COOH$ *n*-Triakontansäure *f*, Melissinsäure *f*
triad dreiwertig; *(Krist)* dreizählig, 3zählig
triad dreiwertiges Element *n*; dreiwertige Atomgruppe *f*; Triade *f (Dreiergruppe von Elementen mit ähnlichen Eigenschaften)*
t. axis *(Krist)* dreizählige Achse (Drehachse) *f*
t. grouping Darstellung *f* im Dreieckskoordinatensystem
t. prototropy *(auf reversibler Verschiebung eines Protons zum übernächsten Atom beruhende Prototropie)*
triads of Döbereiner Döbereiner-Triaden *fpl*
trial Versuch *m*, Probe *f*
through t. and error empirisch
t.-and-error procedure [empirisches] Näherungsverfahren *n*
t. dyeing Probefärbung *f*, Versuchsfärbung *f*
t. mix[ture] Probemischung *f*
triallyl cyanurate Triallylzyanurat *n*
triamidophosphoric acid $PO(NH_2)_3$ Triamidophosphorsäure *f*, Phosphorsäuretriamid *n*, Phosphoroxidtriamid *n*, Phosphoryltriamid *n*
triamine chelate Triaminchelat *n*
triammonium phosphate $(NH_4)_3PO_4$ Triammoniumphosphat *n*, Ammonium[ortho]phosphat *n*
triangular diagram Dreieckskoordinatensystem *n*
t. notch Thomson-Überfall *m*, dreieckiger Überfall *m (Mengenstrommessung)*
triarylmethane dye Triarylmethanfarbstoff *m*, Triphenylmethanfarbstoff *m*, Tritylfarbstoff *m*
triatomic dreiatomig
triazine Triazin *n (ein heterozyklischer Sechsring mit drei N-Atomen)*
triazole Triazol *n (ein heterozyklischer Fünfring mit drei N-Atomen)*
tribasic dreibasig, dreibasisch; *(bei Salzen mehrbasiger Säuren)* tertiär
t. acid *s.* triacid
t. calcium phosphate $Ca_3(PO_4)_2$ tertiäres Kalziumphosphat *n*, Trikalziumphosphat *n*, Kalzium[ortho]phosphat *n*
t. [ortho]phosphate M'_3PO_4 tertiäres (neutrales) Orthophosphat (Phosphat) *n*
t. potassium [ortho]phosphate K_3PO_4 tertiäres Kalium[ortho]phosphat *n*, Trikaliumphosphat *n*
t. sodium [ortho]phosphate Na_3PO_4 tertiäres Natrium[ortho]phosphat *n*, Trinatriumphosphat *n*
triboluminescence Tribolumineszenz *f*, Reibungslumineszenz *f*, Trennungsleuchten *n*

triboride Triborid *n*
tribromethyl alcohol *s.* tribromoethanol
tribromide Tribromid *n*
tribromoaniline $Br_3C_6H_2NH_2$ Tribromanilin *n*
tribromoethanol, tribromoethyl alcohol CBr_3CH_2OH Tribromäthanol *n*, Tribromäthylalkohol *m*
tribromomethane $CHBr_3$ Tribrommethan *n*, Bromoform *n*
tribromosilicane $SiHBr_3$ Tribromsilan *n*
tric. s. triclinic
tricalcic phosphate $Ca_3(PO_4)_2$ Trikalziumphosphat *n*, Kalzium[ortho]phosphat *n*
tricalcium aluminate $3CaO \cdot Al_2O_3$ Trikalziumaluminat *n*
t. orthoarsenate $Ca_3(AsO_4)_2$ Trikalzium[ortho]arsenat(V) *n*, Kalzium[ortho]arsenat(V) *n*
t. [ortho]phosphate *s.* tricalcic phosphate
tricarballylic acid $HOOC \cdot CH_2CH(COOH)CH_2COOH$ Trikarballylsäure *f*, 1,2,3-Propantrikarbonsäure *f*
tricarboxylic acid Trikarbonsäure *f*
t.-acid cycle Trikarbonsäurezyklus *m*, Zitronensäurezyklus *m*, Zitratzyklus *m*, Krebs-Zyklus *m*, Szent-Györgyi-Krebs-Zyklus *m*, Szent-Györgyi-Krebsscher Kreisprozeß *m*
trichite *(Krist)* Trichit *m*
trichloracetonitrile *s.* trichloroacetonitrile
trichloride Trichlorid *n*
trichloro derivative Trichlorderivat *n*
trichloroacetic acid Cl_3CCOOH Trichloressigsäure *f*, Trichloräthansäure *f*, TCA *n*
t. aldehyde $Cl_3C \cdot CH(OH)_2$ Trichlorazetaldehydhydrat *n*, Trichloräthylidenglykol *n*, 2,2,2-Trichlor-1,1-äthandiol *n*, Chloralhydrat *n*
trichloroacetone Cl_3CCOCH_3 Trichlorazeton *n*, 1,1,1-Trichlor-2-propanon *n*
trichloroacetonitrile $CCl_3CH(OH)CN$ Trichlorazetonitril *n*, Chlorzyanhydrin *n*
trichloroaldehyde CCl_3CHO Trichloraldehyd *m*, Trichlorazetaldehyd *m*, Trichloräthanal *n*, Chloral *n*
2,3,6-trichlorobenzoic acid $Cl_3C_6H_2COOH$ 2,3,6-Trichlorbenzoesäure *f*, *(als Herbizid auch)* 2,3,6-TBA *n*
trichloro-*tert*-butyl alcohol $(CH_3)_2C(OH)CCl_3$ Trichlorbutylalkohol *m*, Azetonchloroform *n*
trichlorocyanidine *s.* trichloro-sym-triazine
1,1,1-trichloroethane CH_3CCl_3 1,1,1-Trichloräthan *n*
1,1,2-trichloroethane $CH_2ClCHCl_2$ 1,1,2-Trichloräthan *n*
2,2,2-trichloro-1,1-ethanediol $CCl_3CH(OH)_2$ 2,2,2-Trichlor-1,1-äthandiol *n*, Trichloräthylidenglykol *n*, Chloralhydrat *n*, Trichlorazetaldehydhydrat *n*
trichloroethylene Trichloräthen *n*, Trichloräthylen *n*
trichloroethylidene glycol $Cl_3CCH(OH)_2$ Trichloräthylidenglykol *n*, 2,2,2-Trichlor-1,1-äthandiol *n*, Chloralhydrat *n*, Trichlorazetaldehydhydrat *n*

trichloromethane CHCl$_3$ Trichlormethan *n*, Chloroform *n*

2,4,5-trichlorophenoxyacetic acid Cl$_3$C$_6$H$_2$OCH$_2$COOH 2,4,5-Trichlorphenoxy-essigsäure *f*, *(als Herbizid auch)* 2,4,5-T *n*

2-(2,4,5-trichlorophenoxy)propionic acid 2-(2,4,5-Trichlorphenoxy)propionsäure *f*, *(als Herbizid auch)* Fenoprop *n*, 2,4,5-TP *n*

2,3,6-trichlorophenylacetic acid Cl$_3$C$_6$H$_2$CH$_2$COOH 2,3,6-Trichlorphenylessig-säure *f*, *(als Herbizid auch)* Fenac *n*

trichlorosil[ic]ane SiHCl$_3$ Trichlorsilan *n*

α-trichlorotoluene C$_6$H$_5$CCl$_3$ α-Trichlortoluol *n*, ω-Trichlormethylbenzol *n*, Phenylchloroform *n*, Benzotrichlorid *n*

trichloro-*sym*-triazine C$_3$Cl$_3$N$_3$ Trichlor-*sym*-triazin *n*, Zyanurchlorid *n*, Zyanursäurechlorid *n*

trichroism *(Krist)* Trichroismus *m*, Dreifarbigkeit *f*

trichromate M$^I_2$Cr$_3$O$_{10}$ Trichromat *n*

trickle / to 1. tröpfeln; rieseln, *(in dünnem Strom)* rinnen; sickern; 2. tröpfeln [lassen], träufeln; rieseln lassen; [ver]sickern lassen

t. down herabtröpfeln; herabrieseln, herabrinnen

trickle cooler Rieselkühler *m*

t. [low] process Trickle-Verfahren *n*, Trickle-Prozeß *m*, Trickle-Phase-Verfahren *n* *(zur hydrierenden Entschwefelung)*

trickling filter Tropfkörper *m* *(zur Abwasserreinigung)*

triclinic *(Krist)* triklin, trikl.

tricobalt tetroxide Co$_3$O$_4$ Kobalttetroxid *n*, Trikobalttetroxid *n*, Kobalt(II,III)-oxid *n*

Tricone mill Tricone-Mühle *f*, Dreikegelmühle *f*

tricopper monophosphide Cu$_3$P Trikupferphosphid *n*, Kupfer(I)-phosphid *n*

tricosane C$_{23}$H$_{48}$ Trikosan *n*

tricresyl phosphate (C$_6$H$_4$CH$_3$O)$_3$PO Phosphorsäuretrikresylester *m*, Trikresylphosphat *n*

tricyanogen chloride s. trichloro-*sym*-triazine

tridecane C$_{13}$H$_{28}$ Tridekan *n*

tridecanoic acid CH$_3$(CH$_2$)$_{11}$COOH Tridekansäure *f*, Dodekan-1-karbonsäure *f*

tridentate *(Komplexchemie)* dreizahnig, dreizähnig, dreizählig *(Koordinationswert des Liganden)*

tridymite Tridymit *m* *(eine Kristallform von Siliziumdioxid)*

Tridyne process *(Plast)* Tridyne-Verfahren *n*, zweistufiges Spritzpreßverfahren *n*

triene Trien *n* *(Kohlenwasserstoff mit dreifacher Kohlenstoffbindung)*

trientoxide M$_3$O *(Oxid mit Metall-Sauerstoff-Verhältnis 3:1)*

triethanolamine N(CH$_2$·CH$_2$OH)$_3$ Triäthanolamin *n*, Triäthylolamin *n*, β,β′,β′-Trioxytriäthylamin *n*, 2,2′,2″-Nitrolotriäthanol *n*

triethyl citrate C$_6$H$_5$O$_7$(C$_2$H$_5$)$_3$ Zitronensäure-triäthylester *m*, Triäthylzitrat *n*

t. phosphate (C$_2$H$_5$)$_3$PO$_4$ Phosphorsäuretri-äthylester *m*, Triäthylphosphat *n*

triethylaluminium Al(C$_2$H$_5$)$_3$ Triäthylaluminium *n*, Äthylaluminium *n*

triethylamine (C$_2$H$_5$)$_3$N Triäthylamin *n*

triethylborine B(CH$_2$CH$_3$)$_3$ Triäthylborin *n*, Bortriäthyl *n*, Boräthyl *n*

triethylene glycol (CH$_2$OCH$_2$CH$_2$OH)$_2$ Triäthylenglykol *n*, Triglykol *n*, 2,2′-Äthylendioxydiäthanol *n* .

trifluoride Trifluorid *n*

trifluoroacetic acid CF$_3$COOH Trifluoressigsäure *f*, Trifluoräthansäure *f*

trifluorochloroethylene CFCl=CF$_2$ Trifluor[mono]-chloräthylen *n*, Perfluorvinylchlorid *n*

trifluoroethanoic acid s. trifluoroacetic acid

trifluorosilicane SiHF$_3$ Trifluorsilan *n*

triformal, triformol s. 1,3,5-trioxane

trifunctional trifunktionell, dreifunktionell

t. unit trifunktionelle Einheit *f*, T-Einheit *f* *(Struktureinheit)*

trig. s. trigonal

trigermane Ge$_3$H$_8$ Trigerman *n*, Germaniumokta-hydrid *n*

triglyceride Triglyzerid *n*, Neutralglyzerid *n*

triglycol s. triethylene glycol

triglycollamic acid N(CH$_2$COOH)$_3$ Nitrilotries-sigsäure *f*, NTE

trigonal *(Krist)* trigonal, trig.

t. hybridization trigonale Hybridisierung *f* (Bastardisierung) *f*, sp^2-Hybridisierung *f*, sp^2-Ba-stardisierung *f*

trihalide Trihalogenid *n*

trihydrate Trihydrat *n*

trihydric alcohol dreiwertiger Alkohol *m*, Glyzerin *n* *(i.w.S.)*

t. phenol dreiwertiges Phenol *n*

trihydroxide Trihydroxid *n*

trihydroxyanthraquinone Trihydroxyanthrachinon *n*

1,2,3-trihydroxybenzene C$_6$H$_3$(OH)$_3$ 1,2,3-Tri-hydroxybenzol *n*, Pyrogallussäure *f*, Pyrogallol *n*

1,3,5-trihydroxybenzene C$_6$H$_3$(OH)$_3$ 1,3,5-Tri-hydroxybenzol *n*, Phlorogluzin *n*

2,3,4-trihydroxybenzoic acid (HO)$_3$C$_6$H$_2$COOH 2,3,4-Trihydroxybenzoesäure *f*, Pyrogallolkar-bonsäure *f*

2,4,5-t. acid (HO)$_3$C$_6$H$_2$COOH 2,4,5-Trihydroxy-benzoesäure *f*, 4-Hydroxygentisinsäure *f*

2,4,6-t. acid (HO)$_3$C$_6$H$_2$COOH 2,4,6-Trihydroxy-benzoesäure *f*, Phlorogluzinkarbonsäure *f*

3,4,5-t. acid (HO)$_3$C$_6$H$_2$COOH 3,4,5-Trihydroxy-benzoesäure *f*, Gallussäure *f*

1,2,3-trihydroxypropane CH$_2$OHCHOHCH$_2$OH 1,2,3-Trihydroxypropan *n*, 1,2,3-Propantriol *n*, Glyzerin *n*, Ölsüß *n*

triiodide Trijodid *n*

triiodomethane CHJ$_3$ Trijodmethan *n*, Jodoform *n*

triiodosilicane SiHJ$_3$ Trijodsilan *n*

trim / to beschneiden, besäumen; entgraten, putzen

trim *(beim Beschneiden von Rollen- und Bogenpapier abfallender)* Randstreifen *m*; Randbeschnitt *m*, Randabfall *m*

trimellitic acid $C_6H_3(COOH)_3$ Trimellit[h]säure f, 1,2,4-Benzoltrikarbonsäure f
trimer Trimer[es] n, trimere Verbindung f
trimeric trimer
trimeride s. trimer
trimerization Trimerisierung f, Trimerisation f
trimesic acid $C_6H_3(COOH)_3$ Trimesinsäure f, 1,3,5-Benzoltrikarbonsäure f
3,4,5-trimethoxybenzoic acid $(CH_3O)_3C_6H_2COOH$ 3,4,5-Trimethoxybenzoesäure f, Gallussäuretrimethyläther m
trimethylacetic acid $(CH_3)_3CCOOH$ Trimethylessigsäure f, Dimethylpropansäure f, Pivalinsäure f
trimethylamine $(CH_3)_3N$ Trimethylamin n
1,2,3-trimethylbenzene $C_6H_3(CH_3)_3$ 1,2,3-Trimethylbenzol n, Hemimellitol n, Hemellitol n
1,2,4-trimethylbenzene $C_6H_3(CH_3)_3$ 1,2,4-Trimethylbenzol n, Pseudokumol n
1,3,5-trimethylbenzene $C_6H_3(CH_3)_3$ 1,3,5-Trimethylbenzol n, Mesitylen n
trimethylborine, trimethylboron $B(CH_3)_3$ Trimethylbor n
trimethylbromomethane $(CH_3)_3CBr$ Trimethylbrommethan n, tert-Butylbromid n, 2-Brom-2-methylpropan n
trimethylcarbinol $(CH_3)_2COHCH_3$ Trimethylkarbinol n, tert-Butylalkohol m, tert-Butanol n, 2-Methylpropanol-(2) n
trimethylene Trimethylen n, Zyklopropan n
trimethylglycine, trimethylglycocoll Trimethylglykokoll n, Betain n, Oxyneurin n, Lyzin n (Alkaloid)
trimethylmethane $CH_3CH(CH_3)CH_3$ Trimethylmethan n, 2-Methylpropan n, Isobutan n, i-Butan n
trimethylquinoline Trimethylchinolin n
trimethylsilyl $(CH_3)_3Si-$ Trimethylsilyl n
trimmer (Plast) Beschneidemaschine f, Besäummaschine f; (Pap) Planschneider m, Guillotineschneider m, Buchbinderschneidemaschine f
trimming Beschneiden n, Besäumen n; Entgraten n, Putzen n
t. machine s. trimmer
trimolecular trimolekular
trimolybdate $M^I_2Mo_3O_{10}$ Trimolybdat n
trimyristin Trimyristin n, Myristinsäureglyzerylester m, Glyzerintrimyristat n
trinickelous orthophosphate $Ni_3(PO_4)_2$ Nikkel[ortho]phosphat n
trinitrate Trinitrat n
trinitration Trinitrierung f
trinitride Trinitrid n
2,4,6-trinitrophenol $C_6H_2(NO_2)_3OH$ 2,4,6-Trinitrophenol n, Pikrinsäure f
trinitrotoluene $C_6H_2(NO_2)_3CH_3$ Trinitrotoluol n, Tritol n, Trotyl n, TNT
triol Triol n (dreiwertiger Alkohol)
triose-phospho-dehydrogenase Triosephosphatdehydr[ogen]ase f
1,3,5-trioxane $(CH_2O)_3$ 1,3,5-Trioxan n, α-Trioxymethylen n

trioxide Trioxid n
trioxygen O_3 Trisauerstoff m, Ozon n
α-trioxymethylene s. 1,3,5-trioxane
tripalmitin $(C_{15}H_{31}COO)_3C_3H_5$ Tripalmitin n, Glyzerintripalmitat n, Glyzerintripalmitinsäureester m
tripeptide Tripeptid n (Peptid mit drei Aminosäureeinheiten)
tripestone (Min) Gekrösestein m (Kalziumsulfat)
triphenyl phosphate $(C_6H_5)_3PO_4$ Triphenylphosphat n, Phosphorsäuretriphenylester m
triphenylamine $(C_6H_5)_3N$ Triphenylamin n
t. dye Triphenylaminfarbstoff m
triphenylantimony $(C_6H_5)_3Sb$ Triphenylstibin n
triphenylborine, triphenylboron $(C_6H_5)_3B$ Triphenylbor n
triphenylcarbinol $(C_6H_5)_3COH$ Triphenylkarbinol n
triphenylene Triphenylen n, Isochrysen n, 9,10-Benzophenanthren n
triphenylmethane $(C_6H_5)_3CH$ Triphenylmethan n, Tritan n
t. dye Triphenylmethanfarbstoff m
triphenylmethyl s. trityl
triphenylmethylpenicillin Triphenylmethylpenizillin n
triphenylphosphine, triphenylphosphorus $(C_6H_5)_3P$ Triphenylphosphin n
triphenylrosaniline Triphenylrosanilin n
triphenylsilyl $(C_6H_5)_3Si-$ Triphenylsilyl n
triphenylstibine s. triphenylantimony
triphenyltin chloride $(C_6H_5)_3SnCl$ Triphenylzinnchlorid n
triphosphate $M^I_5P_3O_{10}$ Triphosphat n
triphosphopyridine nucleotide Triphosphopyridinnukleotid n, TPN$^+$, Nikotinamid-adenin-dinukleotidphosphat n, NADP$^+$, Kohydr[ogen]ase II f, Koferment II n
triphyline, triphylite, triphylline (Min) Triphylin m
triple dreifach
t. bond dreifache Bindung f, Dreifachbindung f
t.-bonded dreifach gebunden
t. chain (Bündel aus drei Polypeptidketten)
t. collision Dreierstoß m
t.-distilled dreifach destilliert
t. effect Tripeleffekt m
t.-effect evaporating unit s. t.-effect evaporator
t.-effect evaporator Tripeleffektverdampfer m, Dreikörperverdampfer m, Dreistufenverdampfer m
t.-linked s. t.-bonded
t. molecular collision s. t. collision
t. point Tripelpunkt m
t. salt Tripelsalz n
t. superphosphate (Landw) Doppelsuperphosphat n (Düngemittel aus Phosphorsäureaufschluß von Rohphosphaten)
triplet Triplett n, Triplettspektrallinie f
t. state Triplettzustand m
t. system Triplettsystem n
triplex board Triplexkarton m; Triplexpappe f

t. pump Drillingspumpe *f*
triplite *(Min)* Triplit *m*
triply bonded (bound, connected, linked) *s.*
triple-bonded
t. unsaturated dreifach ungesättigt
tripod Dreifuß *m*
t. retort stand Dreifußstativ *n*
tripolite *(Min)* Tripolit *m, (veraltet für)* Tripel *m*
tripotassium [ortho]phosphate K_3PO_4 Trikalium-
[ortho]phosphat *n*, Kalium[ortho]phosphat *n*
trippel *(Min)* Tripel *m*
tripper Abwurfwagen *m*, Bandabwurfwagen *m*,
Bandschleifenwagen *m*
trippkeite *(Min)* Trippkeit *m*
triptane Triptan *n*, 2,2,3-Trimethylbutan *n (sehr
klopffester Kraftstoff)*
trisaccharide $C_{18}H_{32}O_{16}$ Trisa[c]charid *n*
trisaziridinyl phosphine oxide Aziridinylphosphin-
oxid *n*, APO
trisazo dye Trisazofarbstoff *m*
trisilane $H_3Si \cdot SiH_2 \cdot SiH_3$ Trisilan *n*
trisilazane $H_3Si \cdot NH \cdot SiH_2 \cdot NH \cdot SiH_3$ Trisilazan *n*
trisilicane *s.* trisilane
trisilicate Trisilikat *n*
t. slag Trisilikatschlacke *f*
trisiloxane $H_3Si \cdot O \cdot SiH_2 \cdot O \cdot SiH_3$ Trisiloxan *n*
trisilthiane $H_3Si \cdot S \cdot SiH_2 \cdot S \cdot SiH_3$ Trisilthian *n*
trisodium [ortho]phosphate Na_3PO_4 Trinatrium-
[ortho]phosphat *n*, Natrium[ortho]phosphat *n*
t. salt Trinatriumsalz *n*
trispropylenediamine cobaltic chelate Tripro-
pylendiamin-kobalt(III)-chelat *n*
tristearin Tristearin *n*
trisubstituted trisubstituiert, dreifach substituiert
trisulphate $M^I_2S_3O_{10}$ Trisulfat *n*
trisulphide Trisulfid *n*
trisulphonic acid Trisulfo[n]säure *f*
tritactic polymer tritaktisches Polymer[es] *n*
triterpene $C_{30}H_{48}$ Triterpen *n*
trithionate $M^I_2S_3O_6$ Trithionat *n*
trithymotide Trithymotid *n*
tritium T, $^3_1$H Tritium *n*, schweres Deuterium *n*,
überschwerer Wasserstoff *n*
triton 1. *(Kch)* Triton *n*; 2. *s.* trinitrotoluene
tritopine Laudanidin *n, l-*Laudanin *n (Alkaloid)*
tritriacontane $C_{33}H_{68}$ Tritriakontan *n*
triturate / *to (feste Stoffe in kleinste Teilchen)*
pulverisieren, [zer]pulvern, zerreiben, verreiben,
zermahlen, zerstoßen
triturating roll Reiberwalze *f*
trituration Pulverisieren *n*, Pulvern *n*, Zerpulvern *n*,
Zerreiben *n*, Verreiben *n*, Zermahlen *n*, Zerstoßen
n (von festen Stoffen in kleinste Teilchen)
trityl $(C_6H_5)_3C—$ Trityl *n*, Triphenylmethyl *n*
t. chloride $(C_6H_5)_3CCl$ Tritylchlorid *n*
triuranium octoxide U_3O_8 Triuranoktoxid *n*, Uran-
(IV,VI)-oxid *n*
trivalency Dreiwertigkeit *f*, Trivalenz *f*
trivalent dreiwertig, trivalent
trivariant trivariant

trivial name Trivialname *m*
trögerite *(Min)* Trögerit *m (Uranylorthoarsenat)*
Tröger's base Trögersche Base *f*
trolley Förderwagen *m*
t. hearth kiln *(Ker)* Herdwagenofen *m*
trombone cooler Rieselkühler *m*
trommel [screen] Siebtrommel *f*, Trommelsieb *n*,
Rundsieb *n*
t. test Trommeltest *m*, Trommelprüfung *f*,
Trommelversuch *m*
trona, tronite *(Min)* Trona *m(f) (Trinatriumhydrogendikarbonat)*
troostite *(Min)* Troostit *m*
tropacocaine Tropakokain *n (Alkaloid)*
tropaeolin Tropäolin *n*
tropaeolin D Tropäolin D *n*, Methylorange *n*,
Orange III *n*, Helianthin *n (Natriumsalz der
4'-Dimethylaminoazobenzolsulfonsäure-(4))*
tropaeolin 0 Tropäolin 0 *n*, Tropäolin R *n*,
Resorzingelb *n*
tropaeolin 00 Tropäolin 00 *n*, Orange IV *n*,
Orange N *n*, Diphenylaminorange *n (Natriumsalz
der 4'-Anilinoazobenzolsulfonsäure-(4))*
tropaeolin 000 No. 1 Tropäolin 000 *n* Nr. 1,
Orange I *n*, α-Naphtholorange *n (Natriumsalz der
α-Naphtholazobenzolsulfonsäure-(4))*
tropaeolin R *s.* tropaeolin 0
tropaic acid *s.* tropic acid
tropane Tropan *n*, Hydrotropin *n*
t. alkaloid Tropa[n]alkaloid *n*, Tropinalkaloid *n*,
Tropein *n*
Tropenas [side-blown] converter Tropenas-Kon-
verter *m (seitlich blasender Kleinkonverter)*
tropeolin *s.* tropaeolin
tropic acid $C_6H_5CH(CH_2OH)COOH$ Tropasäure *f*,
2-Phenylhydrakrylsäure *f*, 2-Phenyl-3-hydroxy-
propionsäure *f*
tropical almond Katappenbaum *m*, Katappa-Terminalie *f*, Indischer Mandelbaum *m*, Terminalia
catappa L.
tropicalization Tropenfestmachen *n*
tropidine Tropidin *n*, Tropen *n*, Anhydrotropan *n*
tropine Tropin *n*, 3-Tropanol *n*, Tropan-(3α)-ol *n*,
8-Methyl-3-nortropanol *n*, *N*-Methyltropolin *n*
t. carboxylic acid Tropinkarbonsäure *f*, *l*-Ekgonin
n
tropinic acid Tropinsäure *f (Atropinabbauprodukt)*
tropinone $C_8H_{13}NO$ Tropinon *n*, 3-Tropanon *n*,
Tropan-(3)-on *n*
tropolone Tropolon *n (Zykloheptanderivat)*
trouble Störung *f*, Fehler *m*
t.-free störungsfrei
trough Mulde *f*, Trog *m*, Wanne *f*, Bottich *m*, Schiff
n; Rinne *f*, Gerinne *n*
t. classifier Stromklassierer *m*
t. conveyor dryer Muldentrockner *m*
t. mixer Trogmischer *m*, Muldenmischer *m*;
Trogkneter *m*
t.-shaped muldenförmig, trogförmig, wannen-
förmig

t. washer Rheorinne f, Stromrinne f, Schrägrinne f
troughed gemuldet, Mulden...
t. belt gemuldeter Gurt m
t.-belt conveyor gemuldeter Gurtförderer m
troughing Muldung f
t. idler Muldenrolle f
Trouton law (rule) Troutonsche Regel f
truck Wagen m, Hordenwagen m
t. chamber kiln s. t. kiln
t. dryer Kanaltrockner m; Kammertrockner m
t. kiln (Ker) Herdwagenofen m
true annealing vollständiges Ausglühen n, Vollständigglühen n (Umwandlungsglühung mit verzögerter Abkühlung)
t. density wahre (wirkliche) Dichte f, Feststoffdichte f
t. formula wahre Formel (Summenformel, Bruttoformel) f
t. free sulphur (Gum) freier Schwefel m, Freischwefel m
t. hemp Hanf m, Cannabis sativa L.
t. malachite (Min) Malachit m (Kupfer(II)-dihydroxidkarbonat)
t. perborate echtes (wahres) Perborat n, (veraltet für) echtes (wahres) Peroxoborat n
t. porosity wahre Porosität f
t. protein Reinprotein n
t. sapphire (Min) Sap[p]hir m (Aluminiumoxid)
t. solution echte (molekulare) Lösung f
t. solvent echtes (aktives) Lösungsmittel n
t. transference number wahre Überführungszahl f
t. vapour-phase [cracking] process True-Vapour-Phase-Verfahren n, True-Vapour-Phase-Krackprozeß m, TVP-Verfahren n
trunnion Zapfen m
t. discharge Hohlzapfenaustrag m
Truxillo coca Truxilla-Koka f, Peruanische Koka f (von Erythroxylum novogranatense (Morris) Hieron.)
trypan blue Trypanblau n
t. red Trypanrot n
trypanocidal trypanozid [wirkend]
tryparsamide Tryparsamid n, Na-N-Phenylglyzylamid-4-arsonat n
trypsin Trypsin n (zu den Proteinasen gehörendes Ferment)
t. method (Gum) Trypsin-Verfahren n (Wärmesensibilisierungsverfahren)
tryptamine Tryptamin n, 3-(2'-Aminoäthyl)indol n
tryptophan[e] Tryptophan n, Leuzyltryptophan n, Aminoindolpropionsäure f, β-Indolyl(3)-alanin n
Ts s. tosyl
T.S. s. 1. tub-sizing; 2. tub-sized; 3. total solids
tschermigite (Min) Tschermigit m, Ammon[iak]-alaun m (Ammoniumaluminiumsulfat-12-Wasser)
Tschitschibabin reaction Tschitschibabin-Reaktion f (zur Darstellung von 2-Aminopyridin)

Tschugaeff reaction Tschugajew-Reaktion f (zur Darstellung von Alkenen)
Tschugaeff-Zerewitinoff method Methode f von Tschugajew-Zerewitinow (zur Bestimmung aktiver Wasserstoffatome)
TSH s. thyrotrophic hormone
TSPP s. tetrasodium pyrophosphate
t.t. experiment s. test-tube experiment
T.T.T. diagram s. time-temperature-transformation diagram
tub Bottich m, Bütte f, Kübel m, Zuber m, Wanne f, Mulde f, Trog m, Kufe f, Faß n
t. colouring (Pap) Oberflächenfärbung f
t. size (Pap) Oberflächenleim m
t.-sized (Pap) oberflächengeleimt
t. sizing (Pap) Oberflächenleimung f
t.-sizing machine (Pap) Oberflächenleimmaschine f
β-tubaic acid Tubasäure f (Kumaronderivat)
tube 1. Rohr n; 2. Kerze f (eines Kerzenfilters); 3. Röhrenofen m, Röhrenerhitzer m (beim Tube-and-Tank-Verfahren)
t.-and-tank [cracking] process Tube-and-Tank-Verfahren n, Tube-and-Tank-Krackprozeß m
t. bundle Rohrbündel n
t. cleaner Rohrreiniger m
t. curare s. tubocurare
t.-drawing process (Glas) Röhrenziehverfahren n
t.-extruding press, t.-extrusion press (Gum) Schlauchspritzmaschine f
t. filter Kerzenfilter n
t. for sealing Bombenrohr n, Einschmelzrohr n, Schießrohr n
t. furnace 1. Röhrenofen m, Röhrenerhitzer m (für Destillations-, Krack- oder Kontaktprozesse); 2. (Ker, Met) Röhrenofen m, Rohrofen m; 3. (Labortechnik) Bombenofen m, Schießofen m
t. ice Röhreneis n
t. mill Rohrmühle f
t. rectifier Röhrengleichrichter m
t. roll (Pap) Registerwalze f
t. rotary dryer Röhrentrockner m
t. saturation (Zucker) Rohrsaturation f
t. sheet Rohrboden m
t.-side medium Rohrmedium n
t. still s. 1. t.-still plant; 2. t. furnace 1.
t.-still plant Röhrenofenanlage f, Pipestill-Anlage f
t. velocity Geschwindigkeit f im Rohr
t. wall Rohrwand[ung] f
t. welder, t.-welding machine (Plast) Rohrschweißgerät n
t. with bulbar enlargements Kugelrohr n, Kugelröhre f
tubeaxial fan Ringlüfter m
tubeless tyre (Gum) schlauchloser Reifen m
tuber s. tubing machine
tubing 1. Spritzen n von Schläuchen; 2. Schlauch m (ohne Gewebeeinlage), Schlauchstück n; 3. Rohre npl, Rohrleitung f; Rohrmaterial n

t. machine Spritzmaschine *f*, Schneckenpresse *f*

tubocurare Tubokurare *n*

tubular röhrenförmig, röhrenartig, Röhren..., Rohr...

t. boiler Röhrenkessel *m*

t. [bowl] centrifuge Röhrenzentrifuge *f*

t. condenser Röhrenkühler *m*

t. converter Röhrenkonverter *m*

t. cooler Röhrenkühler *m*

t. film Schlauchfolie *f*

t. filter *s.* tube filter

t. furnace *s.* tube furnace

t. heat exchanger Röhrenwärme[aus]tauscher *m*, Röhrenwärmeübertrager *m*, Rohrwärmeaustauscher *m*

t. heater 1. Rohrheizkörper *m*, Rohrheizelement *n*, Backerrohr *n*; 2. *s.* tube furnace 1.

t.-knit goods Rundgewirke *npl*, Rundgestricke *npl*, Schlauchware *f*

t. mill *s.* tube mill

t. parison schlauchförmiger Vorformling *m*, Külbel *n*

t. pasteurizer Röhrenerhitzer *m* *(für Milch)*

tuckstone *(Glas)* Nasenstein *m* *(im Schmelzofen)*

tufa [vulkanischer] Tuff *m*; Kalktuff *m*

tufaceous limestone Kalktuff *m*

tuffite *(Geol)* Tuffit *m*

tuille *s.* tweel

tulip electrode Fangtaschenelektrode *f*

Tully plant Tully-Anlage *f*, Tully-Generatorenanlage *f*

Tully process Tully-Verfahren *n* *(Schwelverfahren zur Wassergaserzeugung)*

tumble dryer Trommeltrockner *m*, Trockentrommel *f*, Zylindertrockner *m*, Trockenzylinder *m*, Walzentrockner *m*, Trockenwalze *f*

tumbler Drehtrommel *f*; Fallmischer *m*, Freifallmischer *m*; *(Pap)* Entrindungstrommel *f*; *(Pap)* Bleichtrommel *f*

t. mixer Fallmischer *m*, Freifallmischer *m*, *(i.e.S.)* Trommelmischer *m*, Mischtrommel *f*

t. test Trommeltest *m*, Trommelprüfung *f*, Trommelversuch *m*

tumbling Trommeln *n*, Rommeln *n*, *(Farb auch)* Trommellackieren *n*, Trommellackierung *f*, *(Plast auch)* Trommelpolieren *n*, Polieren *n* in Trommeln, *(Ker auch)* Scheuern *n*

t. barrel Lackiertrommel *f*, Poliertrommel *f*, Putztrommel *f*

t. mill Sturzmühle *f*

t. mixer *s.* tumbler mixer

tumorigenicity geschwulsterregende Wirkung *f*

tun Tonne *f*, Faß *n*; Maisch[e]bottich *m*, Maisch[e]-kessel *m*

tuna oil Thunfischlebertran *m*

tung oil Tungöl *n*, China-Holzöl *n*, [chinesisches] Holzöl *n* *(aus Samen von Aleurites fordii Hemsl.)*; *s.a.* Japanese tung oil

t.-oil green *(Komplexverbindung aus Kupferarsenit und verseiftem Tungöl; Insektizid)*

tungstate Wolframat *n*

t. phosphor Wolframatphosphor *m*

tungstein *(Min)* Tungstein *m*, *(veraltet für)* Scheelit *m* *(Kalziumwolframat)*

tungsten 1. W Wolfram *n*; 2. *s.* tungstein

t. arc lamp *s.* t. point lamp

t. bronze Wolframbronze *f*

t. carbide Wolframkarbid *n*

t. dibromide WBr_2 Wolframdibromid *n*, Wolfram(II)-bromid *n*

t. dichloride WCl_2 Wolframdichlorid *n*, Wolfram(II)-chlorid *n*

t. diiodide WJ_2 Wolframdijodid *n*, Wolfram(II)-jodid *n*

t. dioxide WO_2 Wolframdioxid *n*, Wolfram(IV)-oxid *n*

t. dioxydichloride WO_2Cl_2 Wolframdioxiddichlorid *n*

t. disulphide WS_2 Wolframdisulfid *n*, Wolfram(IV)-sulfid *n*

t. filament Wolframglühfaden *m*, Wolframleuchtfaden *m*

t. hexabromide WBr_6 Wolframhexabromid *n*, Wolfram(VI)-bromid *n*

t. hexachloride WCl_6 Wolframhexachlorid *n*, Wolfram(VI)-chlorid *n*

t. hexafluoride WF_6 Wolframhexafluorid *n*, Wolfram(VI)-fluorid *n*

t. metal powder Wolframmetallpulver *n*, metallisches Wolframpulver *n*

t. oxytetrachloride $WOCl_4$ Wolframoxidtetrachlorid *n*

t. pentabromide WBr_5 Wolframpentabromid *n*, Wolfram(V)-bromid *n*

t. pentachloride WCl_5 Wolframpentachlorid *n*, Wolfram(V)-chlorid *n*

t. point lamp Wolframpunktlichtlampe *f*, Wolframbogenlampe *f*

t. powder Wolframpulver *n*

t. steel Wolframstahl *m*

t. tetrachloride WCl_4 Wolframtetrachlorid *n*, Wolfram(IV)-chlorid *n*

t. tetraiodide WJ_4 Wolframtetrajodid *n*, Wolfram(IV)-jodid *n*

t. trioxide WO_3 Wolframtrioxid *n*, Wolfram(VI)-oxid *n*

t. trisulphide WS_3 Wolframtrisulfid *n*, Wolfram(VI)-sulfid *n*

tungstenite *(Min)* Tungstenit *m* *(Wolframdisulfid)*

tungstic Wolfram..., *(i.e.S.)* Wolfram(VI)-...

t. acid H_2WO_4 [gelbe] Wolframsäure *f*; 2. $H_2WO_4 \cdot H_2O$ [weiße] Wolframsäure *f*; 3. WO_3 Wolfram(VI)-oxid *n*

t.[-acid] anhydride *s.* t. oxide

t. ochre *s.* tungstite

t. oxide WO_3 Wolfram(VI)-oxid *n*, Wolframtrioxid *n*

tungsticyanide $M^I_3[W(CN)_8]$ Oktazyanowolframat(V) *n*

tungstite *(Min)* Tungstit *m*, Wolframocker *m*,

tungstoarsenate Wolframatoarsenat n
tungstoborate Wolframatoborat n
tungstoboric acid Wolframatoborsäure f
tungstocyanide $M^I_4[W(CN)_8]$ Oktazyanowolframat-(IV) n
tungstophosphate Wolframatophosphat n
tungstophosphoric acid Wolframatophosphorsäure f
tungstosilicic acid Wolframatokieselsäure f
tunnel cap Tunnelglocke f (einer Glockenboden-kolonne)
t. dryer (drying machine) Tunneltrockner m, Tunneltrockenanlage f, Kanaltrockner m
t. effect Tunneleffekt m
t. furnace (kiln) Tunnelofen m
tuno gum Carthagenakautschuk m (von Castilloa elastica Cero.)
tupelo gum Wasser-Tupelo m, Tupelobaum m, Nyssa aquatica L.
turanose Turanose f
turbid trübe; [dick]flüssig; schlammig
turbidimeter Trübungsmesser m, Turbidimeter n
turbidimetric turbidimetrisch
t. measurement s. turbidimetry
t. method turbidimetrische Methode f, turbidi-metrisches Verfahren n
t. titration turbidimetrische Titration f, Trübungs-titration f, (i.w.S.) Turbidimetrie f, Trübungs-messung f, Trübheitsmessung f
turbidimetry Turbidimetrie f, Trübungsmessung f, Trübheitsmessung f
turbidity Trübe f, Trübung f, Trübheit f
t. current Trübungsströmung f
t. measurement s. turbidimetry
t. point Trübungspunkt m
t. titration s. turbidimetric titration
turbidometer s. turbidimeter
turbine agitator s. t. mixer
t. blower s. turboblower
t. drilling Turbinen[dreh]bohren n, Turbinen-bohrung f, Turbobohren n
t. mixer Turbomischer m, Turborührer m, Turbinenrührer m
t. oil Turbinenöl n
t. pump Turbinenpumpe f, Pumpe f mit Leitrad; halbaxiale Pumpe f, Pumpe f in diagonaler Bauart
turbo [shelf] dryer Etagentrockner m (als Konvek-tionstrockner), Turbinentrockner m
turboblower Turbogebläse n, Kreiselgebläse n
turbocompressor Turbokompressor m
turbodissolver Turbolöser m
turbodrill[ing] Turbobohren n, Turbinen[dreh]-bohren n, Turbinenbohrung f
turboexpander Expansionsturbine f
turbogrid tray Gitter[rost]boden m, Turbogrid-boden m
turbojet fuel Düsenkraftstoff m, Düsentreibstoff m
turbostratic turbostratisch
turbulence Turbulenz f
turbulent turbulent, wirbelig

t. flow turbulente Strömung f
turgidity (Bio) Turgeszenz f
turgor [pressure] Turgor[druck] m, Wanddruck m
Turkey red Türkischrot n
T.-red dyeing Türkischrotfärberei f
T.-red oil Türkischrotöl n (sulfatiertes Rizinusöl)
T.-red oil soap Türkischrotölseife f
Turkish galls (Gerb) Türkische (Smyrnaer) Gallen fpl, Aleppogallen fpl (von Quercus infectoria Oliv.)
T. geranium oil (veraltet für) palmarosa oil
turmaline s. tourmaline
turmeric Kurkuma f, Gelbwurzel f, Curcuma longa L.
t. paper Kurkumapapier n (ein Reagenzpapier)
t. test Kurkumaprobe f
t. yellow Kurkumin n (Farbstoff der Kurkuma)
turn / to drehen (z.B. die Polarisationsebene); umschlagen nach (bei Indikatoren); [ab]drehen (z.B. Keramikerzeugnisse)
t. milky trüb werden, sich trüben
t. on (ein Gerät) anstellen, einschalten
t. up aufdrehen; (Brennerflamme) größer stellen
Turner's yellow Turners (Kasseler) Gelb n (basisches Bleichlorid wechselnder Zusammensetzung)
turning Drehen n (z.B. der Polarisationsebene); Umschlagen n (Umschlag m) nach (bei Indikatoren); Drehen n, Abdrehen n (z.B. von Keramikerzeugnissen)
t. point Umkehrpunkt m (von Schwingungen)
t. table s. turntable
turnings Drehspäne mpl
turnover Umklappen n (von Bindungen bei der Waldenschen Umkehrung)
t. number (Bioch) Wechselzahl f
t. plough Wendepflug m (Teil des Riffelkneters bei der Margarineherstellung)
t. time Verweilzeit f, Aufenthaltszeit f, Retentions-zeit f, Rückhaltezeit f, Haltezeit f, Stehzeit f, Standzeit f, Durchgangszeit f, Durchlaufzeit f
turntable Drehtisch m, (Ker auch) Drehscheibe f
turpentine Terpentin n(m) (Balsam von Nadelhöl-zern)
t. galls (Farb) Terpentingallen fpl, Pistaziengallen fpl, Judäaroben fpl (Blattgallen von Pistacia terebinthus L.)
t. oleoresin s. turpentine
t. substitute Terpentin[öl]ersatz m, Terpentinöl-surrogat n
turquoise (Min) Türkis m, Kal[l]ait m
tussah (tussur) silk Tussahseide f
tuyere Windform f, Winddüse f, Blasform f (des Hochofens)
t. cooler Windformkühlkasten m
t. level Windformebene f, Blasform[en]ebene f, Düsenebene f, Form[en]ebene f
t. opening Formöffnung f, Düsenöffnung f
t. stock Düsenstock m
t. zone Windformenzone f, Düsenzone f

TVP (T.V.P.) process s. true vapour-phase process
T.W. s. typewriting paper
Twaddell degree Twaddell-Grad m, Grad Twaddell m, °Tw, (fälschlich) Twaddle-Grad m
tweel Hubtür f
tweezers Pinzette f
twice-distilled doppelt (zweimal) destilliert
t.-rectified doppelt (zweimal) rektifiziert
twin [crystal] Zwilling[skristall] m
  t. curing unit (Gum) Doppel[einzel]heizer m
  t.-drum dryer Doppelwalzentrockner m
  t. grinding (Glas) Doppelbandschleifen n
  t. heater s. t. curing unit
  t. polishing (Glas) Doppelbandpolieren n
  t. press s. t. curing unit
  t.-rotor mixer Doppelwellenmuldenmischer m
  t. screw Doppelschnecke f
  t.-screw extruder Doppelschneckenextruder m, Zweischneckenextruder m
  t.-shell blender (mixer) V-Mischer m
  t.-tunnel dryer (Ker) Zwillingstunneltrockner m
  t.-tunnel kiln (Ker) Zwillingstunnelofen m
  t. tyre press (Gum) Doppel-Reifeneinzelheizer m
  t.-wire paper zweischichtiges (zweilagiges) Papier n mit zusammengegautschten Siebseiten
  t.-wire paper machine Papiermaschine f mit zwei Langsieben, Zweisiebpapiermaschine f
  t. worm (Am) Doppelschnecke f
twinkle / **to** flimmern, funkeln, aufblitzen, aufleuchten, szintillieren
twinkling Flimmern n, Funkeln n, Aufblitzen n, Aufleuchten n, Szintillieren n, Szintillation f
twinning (Krist) Zwillingsbildung f
twist [together] / **to** (Text) [ver]drehen, Drehung erteilen, zusammendrehen, zwirnen
twist 1. (Text) Drehung f, Verdrehung f, Drahtgebung f, Drahterteilung f, Zwirnen n, Zwirnung f; Draht m, Drall m, Torsion f; 2. Zwirn m
  t.-set (Text) drehungsfixiert, zwirnfixiert
twisting paper Dreheinschlagpapier n, Twistingpapier n
  t. vibration Torsionsschwingung f, Verdrehungsschwingung f, Drehschwingung f
twistless (Text) ohne Drehung, drehungsfrei, ungedreht, ungezwirnt
Twitchell process Twitchell-Verfahren n (Hydrolyse von Glyzeriden zu Fettsäuren und Glyzerin)
Twitchell reagent Twitchell-Reagens n, Twitchell-Reaktiv n (Gemisch von Sulfonsäuren, die man durch Sulfurierung eines Gemenges von Ölsäure oder Rizinusöl und Naphthalin oder Benzol erhält)
two-bath chrome tannage (tanning) Zweibadchromgerbung f
  t.-bath development (Foto) Zweibadentwicklung f, Zweischalenentwicklung f
  t.-bath fixation (Foto) Zweibadfixierung f
  t.-bath method Zweibadverfahren n
  t.-body collision Zweierstoß m
  t.-can Zweitopf m... (z.B. Washprimer), Zweikomponenten... (z.B. Anstrichfarbe)

t.-can paint Zweikomponentenfarbe f
t.-can system (Farb) Zweikomponentensystem n
t.-carbon acid C_2-Karbonsäure f, Fettsäure f mit zwei C-Atomen
t.-centre bond Zweizentrenbindung f
t.-centre molecular orbital Zweizentrenmolekülorbital n
t.-colour indicator zweifarbiger Indikator m
t.-component adhesive Zweikomponentenkleber m, Reaktionskleber m, Mischklebstoff m
t.-component system Zweikomponentensystem n, Zweistoffsystem n, binäres System n
t.-deck sifter Zweidecker-Siebmaschine f
t.-dimensional zweidimensional
t.-dimensional [paper] chromatography zweidimensionale Papierchromatografie f
t.-drum-type winder (Pap) Rollenschneidmaschine f mit Rollenaufwicklung auf zwei Tragwalzen
t.-electron bond Zweielektronenbindung f
t.-electron configuration Zweielektronenkonfiguration f
t.-electron reduction Zweielektronenreduktion f
t.-electron system Zweielektronensystem n
t.-emulsion process (system) Schlagen n auf Ansatz (Margarineherstellung)
t.-film theory Zweifilmtheorie f
t.-floor kilning Zweihordensystem n des Darrens
t.-fluid atomization Zweistoffzerstäubung f, Druckluftzerstäubung f, Zweistoffversprühung f, Druckluftversprühung f
t.-fluid nozzle Zweistoffdüse f, Druckluftdüse f, Druckluftzerstäuber m, Injektionszerstäuber m
t.-fluid nozzle atomization s. t.-fluid atomization
t.-hole rubber bung doppelt durchbohrter Gummistopfen m
t.-kernel zweikernig (Komplexverbindung)
t.-mash method Zweimaischverfahren n
t.-neck[ed] zweihalsig, Zweihals...
t.-neck[ed] flask Zweihalskolben m
t.-nozzle gun (spray gun) Doppel[spritz]pistole f (in der Farbspritztechnik)
t.-pack s. t.-can
t.-phase foam Zweiphasenschaum m, zweiphasiger Schaum m
t.-phase system Zweiphasensystem n, zweiphasiges System n
t.-piece die zweiteilige Matrize f
t.-ply board Duplexpappe f, zweischichtige Pappe f
t.-pocket grinder (Pap) Zweipressenschleifer m, 2-Pressen-Schleifer m
t.-position control Zweipunktregelung f
t.-product unit Zweiproduktenapparat m, Zweigutapparat m
t.-roll calender (Pap) Zweiwalzenkalander m
t.-roll mill Zweiwalzen-Walzwerk n; Zweiwalzenmühle f
t.-side coater (Pap) Streichmaschine f für doppelseitigen Strich, Streichmaschine f für zweiseitige Beschichtung

t.-sided *(Pap)* zweiseitig
t.-sided effect, t.-sidedness *(Pap)* Zweiseitigkeit *f*
t.-solvent process Zweilösungsmittelverfahren *n*, Duosolverfahren *n*, Duosolprozeß *m*
t.-stage zweistufig, Zweistufen...
t.-stage bleaching *(Pap)* Zweistufenbleiche *f*
t.-stage electrical precipitator Elektrofilter *n* (Elektroabscheider *m*) in Zweizonenanordnung
t.-stage process Zweistufenverfahren *n*, zweistufiges Verfahren *n*
t.-stage pump zweistufige Pumpe *f*
t.-stage resin Novolakharz *n*, Novola[c]k *m*
t.-step cure Vulkanisation *f* in zwei Stufen
t.-step process *s.* t.-stage process
t.-step reaction Zweistufenreaktion *f*
t.-step reaction mechanism Zweistufenmechanismus *m*
t.-table machine *(Glas)* Zweitischmaschine *f*
t.-throw pump Duplexpumpe *f*
t.-tier kiln *(Ker)* Zweietagenofen *m*, zweietagiger Ofen *m*
t.-tower [acid] system *(Pap)* Zweiturmsystem *n* *(Jensen-Türme)*
t.-way capillary stopcock kapillarer Zweiwegesperrhahn *m*, Zweiwegekapillarhahn *m*
t.-way chromatography *s.* t.-dimensional [paper] chromatography
t.-way cock *s.* t.-way stopcock
t.-way outlet Zweiwegemundstück *n*
t.-way paper chromatography zweidimensionale Papierchromatografie *f*
t.-way stopcock (tap) Zweiwege[sperr]hahn *m*
twofold axis of symmetry *(Krist)* zweizählige Achse (Drehachse) *f*
twyer *s.* tuyere
Tyler scale (series) *s.* Tyler standard screen (sieve) scale
Tyler standard screen (sieve) scale Tyler-Standardsiebskala *f*, Tyler-Normalsiebskala *f*, Tyler-Skala *f*, Tyler-Siebreihe *f*
tylophorine Tylophorin *n* *(Alkaloid)*
tympan paper Abschmutzmakulatur *f*, Abschmutzpapier *n*, Antimaculepapier *n*, Zwischenlagepapier *n*
Tyndall beam Tyndall-Licht *n*
Tyndall cone Tyndall-Kegel *m*
Tyndall effect Tyndall-Effekt *m*, Tyndall-Phänomen *n*, Faraday-Tyndall-Effekt *m*
Tyndall meter *s.* tyndallometer
Tyndall phenomenon *s.* Tyndall effect
tyndallometer Tyndall[o]meter *n*
type metal Letternmetall *n*, Schriftmetall *n*
t. of band Streifenart *f*, Mikrolithotype *f*
t. of bond Bindungstyp *m*, Bindungsart *f*
t. of coal Kohle[n]typ *m*, Kohle[n]art *f*
t. of emulsion Emulsionstyp *m*
t. of rubber Kautschuktyp *m*
t. paper *(veraltet für)* typewriting paper
n-t. semiconductor Halbleiter *m* vom n-Typ,

n[-Typ]-Halbleiter *m*, Elektronenüberschußhalbleiter *m*, Überschußhalbleiter *m*
typewriter paper *s.* typewriting paper
t.-ribbon ink Schreibmaschinenbänderfarbe *f*
typewriting paper Schreibmaschinenpapier *n*
typographic ink Hochdruckfarbe *f*, Druckfarbe *f* für Hochdruck
t. printing Hochdruck *m*
t. printing ink *s.* t. ink
t. process Hochdruckverfahren *n*
typomorphic mineral typomorphes Mineral *n*
tyramine Tyramin *n*, p-Hydroxyphenyläthylamin *n*
tyre *(Gum)* Reifen *m*
t. bead *(Gum)* Reifenwulst *m*
t.-building drum *(Gum)* Reifenwickeltrommel *f*, Wickeltrommel *f*; *s.* t.-building machine
t.-building machine *(Gum)* Reifenaufbaumaschine *f*, Reifenerzeugungsmaschine *f*, Reifenwickelmaschine *f*, Konfektioniermaschine *f*
t. compound *(Gum)* Reifenmischung *f*
t. cord *(Gum)* Reifenkord *m*
t. curing *(Gum)* Vulkanisieren *n* (Vulkanisation *f*) von Reifen, Reifenheizung *f*
t. factory *(Gum)* Reifenwerk *n*
t. industry *(Gum)* Reifenindustrie *f*
t. mould *(Gum)* Reifenform *f*
t. plant *s.* t. factory
t. shaping *(Gum)* Ausformen (Rundformen) *n* des Reifens
Tyrian purple Tyrischer (Antiker, Byzantinischer) Purpur *m*, Purpur *m* der Alten, 6,6'-Dibromindigo *m*
tyrocidine Tyrozidin *n* *(zyklisches Dekapeptid)*
tyrolite *(Min)* Tirolit *m*, Kupferschaum *m*
tyrosinase Tyrosinase *f*, Phenol[oxyd]ase *f*
tyrosine Tyrosin *n*, Tyr, p-Hydroxyphenylalanin *n*, p-Hydroxyphenyl-α-aminopropionsäure *f* α-Amino-β-(p-hydroxyphenyl)propionsäure *f*
tyrothricin Tyrothrizin *n* *(Antibiotikum)*
tysonite *(Min)* Tysonit *m*

U

U-link conveyor Stegkettenförderer *m* *(i.e.S.)* (Kette auf der Bahn gleitend)
U-seal Nutring *m*, Nutringdichtung *f*
U-shaped U-förmig
U-tube U-Rohr *n*
U-tube extractor Hildebrandt-Extraktor *m*, Hildebrandt-Extrakteur *m*
U-tube heat exchanger U-Rohr-Wärmeaustauscher *m*, U-Rohr-Wärmeübertrager *m*, Haarnadelwärmeaustauscher *m*, Haarnadelwärmeübertrager *m*
Ubbelohde drop[ping] point Tropfpunkt *m* nach Ubbelohde
Ubbelohde viscosimeter Ubbelohde-Viskosimeter *n*

UDP *s.* uridine diphosphate
UDPG *s.* uridinediphosphateglucose
Uhde cell Uhde-Zelle *f (eine Quecksilberzelle)*
ulexine Ulexin *n*, Zytisin *n*, Sophorin *n*, Baptitoxin *n (ein Lupinenalkaloid)*
ulexite *(Min)* Ulexit *m*
Ullmann condensation Ullmannsche Phenylanthranilsäuresynthese *f*
Ullmann reaction Ullmann-Reaktion *f*, Ullmannsche Reaktion *f (Diarylsynthese)*
ullmannite *(Min)* Ullmannit *m (Nickelantimonsulfid)*
ulmin Ulmin *n*
  u. brown Kohlebraun *n*, Kasseler Braun *n* (Erde *f*) *(bitumenhaltige erdige Braunkohle)*
  u. material Ulminstoff *m*
ultimate analysis Elementaranalyse *f*
  u. elongation Bruchdehnung *f*, Zerreißdehnung *f*, Reißdehnung *f*
  u. gum potentieller (möglicher) Gum *m*, potentielles (mögliches) Harz *n*, Harzneubildung *f*, Potential gum *m*
  u. line *(Spektralanalyse)* letzte (beständige) Linie *f*, Restlinie *f*, Grundlinie *f*, Nachweislinie *f*
  u. organic analysis organische Elementaranalyse *f*
  u. strength properties Festigkeitseigenschaften *fpl*
  u. tensile strength Zugfestigkeit *f*, Zerreißfestigkeit *f*, Reißfestigkeit *f*
  u. tensile [strength] test Zugversuch *m*, Zerreißprüfung *f*
ultra-accelerator Ultrabeschleuniger *m*
ultrabasic rock ultrabasisches Gestein *n*
ultrabasite *(Min)* Ultrabasit *m*, *(veraltet für)* Diaphorit *m*
ultracentrifuge Ultrazentrifuge *f*, Superzentrifuge *f*
ultrafast accelerator *s.* ultra-accelerator
ultrafilter / to durch Ultrafilter filtern (filtrieren)
ultrafilter Ultrafilter *n*
ultrafiltrate Ultrafiltrat *n*
ultrafiltration Ultrafiltration *f*, Ultrafiltrieren *n*
ultraforming Ultraformen *n*, Ultraforming *n (Abart des katalytischen Reformierens)*
ultralow volume method (spraying) ULV-Verfahren *n (wasserloses Versprühen von Pflanzenschutzmitteln mit nur 0,1 bis 0,6 l/ha Wirkstofflösung und Preßluft als Dispersionsmittel)*
ultramarine 1. *(Min)* Ultramarin *m*; 2. *s.* u. blue
  u. blue Ultramarin[blau] *n*
  u. yellow BaCrO$_4$ Ultramaringelb *n*, gelbes Ultramarin *n*, Barytgelb *n*, Steinbühler Gelb *n*, Bariumchromat *n*
ultramicroanalysis Ultramikroanalyse *f*, Submikrogrammethode *f*
ultramicrochemistry Ultramikrochemie *f*
ultramicrodetermination Ultramikrobestimmung *f*
ultramicroscope Ultramikroskop *n*
ultramicroscopic ultramikroskopisch

ultramine *s.* ultramarine yellow
ultrapure ultrarein, extrem rein
ultrapurification extreme Reinigung *f*, Hochreinigung *f*
ultrarapid accelerator *s.* ultra-accelerator
ultrared ultrarot, Ultrarot...-, UR-...-, infrarot, Infrarot..., IR-... *(Zusammensetzungen s. unter infrared)*
ultrashort wave Ultrakurzwelle *f*
ultrasonic Ultraschall...
  u. agglomerator Ultraschallsirene *f*, Hochfrequenzsirene *f (zur Staubabscheidung)*
  u. coagulation Ultraschallkoagulation *f*
  u. conching *(Lebm)* Konchieren *n* mit Ultraschallwellen
  u. field Ultraschallfeld *n*
  u. testing Ultraschallprüfung *f (Werkstoffprüfung)*
  u. treatment Ultraschallbehandlung *f*, Ultraschallbearbeitung *f*
  u. waves Ultraschallwellen *fpl*
  u. welding *(Plast)* Ultraschallschweißen *n*
ultraviolet ultraviolett, Ultraviolett...-, UV-...-
  u. absorber Ultraviolettabsorber *m*
  u. absorption Ultraviolettabsorption *f*
  u. analysis Ultraviolettanalyse *f*
  u. irradiation Ultraviolettbestrahlung *f*
  u. light ultraviolettes Licht *n*
  u. photography Ultraviolettfotografie *f*
  u. radiation Ultraviolettstrahlung *f*; Ultraviolettbestrahlung *f*
  u. resistance Ultraviolettbeständigkeit *f*
  u. spectrum Ultraviolettspektrum *n*
umbelliferone Umbelliferon *n*, Skimmetin *n*, Hydrangin *n*, 7-Hydroxykumarin *n*
umber Umbra *f*, Umber *m*, Umbraun *n*, Sepiabraun *n*, Erdbraun *n*, Kaledonischbraun *n*, Römischbraun *n (brauner Erdfarbstoff)*
UMP *s.* uridine monophosphate
Umpherston beater *(Pap)* Umpherston-Holländer *m*
U.M.R. *(Abk. für)* unsulphonated mineral residue
unaccelerated *(Gum)* nicht beschleunigt, beschleunigerfrei, ohne Beschleuniger
unaffected unangegriffen, nicht angegriffen; nicht beeinflußt
unaged ungealtert
unallowed gap Energielücke *f*, verbotenes Band *n*, verbotene Zone *f*, verbotener (nicht zugelassener) Energiebereich *m*
unalloyed unlegiert, rein, Rein-...
unaltered unverändert
unambiguous eindeutig
unannealed ungeglüht, nicht geglüht
unarrested entarretiert
unattacked unangegriffen, nicht angegriffen
unavailable *(Landw)* unzugänglich, nicht aufnehmbar, nicht verfügbar *(Nährstoffe)*
unbaffled nicht unterteilt, ohne Einbauten (Prallflächen, Leitbleche, Umlenkplatten)

unbalanced carrier Überschuß[ladungs]träger *m*
unbanded nicht streifig, ohne Streifen
unbarked *(Pap)* nicht entrindet, ungeschält
unbeaten *(Pap)* ungemahlen
unbleached ungebleicht
unbodied *(Farb)* unverdickt
unbranched unverzweigt, linear
unbreakable unzerbrechlich
unbuffered ungepuffert
unburned unverbrannt; ungebrannt
  u. combustible Unverbranntes *n*
  u. combustible (fuel) loss Verlust *m* durch Unverbranntes
unburnt *s.* unburned
uncased *(Erdöl)* unverrohrt
uncertainty principle Unschärferelation *f*, Unbestimmtheitsrelation *f*, Unbestimmtheitsbeziehung *f*, Ungenauigkeitsrelation *f*, Ungenauigkeitsbeziehung *f*, Unsicherheitsbeziehung *f*
unchanged unverändert; unangegriffen, nicht angegriffen; nicht umgesetzt (umgewandelt, umgeformt), unumgesetzt
uncharged ungeladen, ladungslos, [elektrisch] neutral
uncoagulated nicht koaguliert
uncoated mould *(Glas)* blanke Form *f*
uncoiling Entknäuelung *f (von Molekülen)*
uncombined frei, ungebunden
  u. heat freie (ungebundene) Wärme *f*
uncompressed ungepreßt
uncondensable nicht kondensierbar
unconjugated nichtkonjugiert
  u. double bond nichtkonjugierte Doppelbindung *f*
unconsolidated nichtverfestigt, unverfestigt, locker
uncontrolled crystallization ungeleitete Kristallisation *f*
unconverted unumgesetzt, nicht umgesetzt (umgewandelt, umgeformt)
uncooked ungekocht, roh; *(Zellstoff)* unaufgeschlossen
uncoordinated unkoordiniert
uncoupled nicht kompensiert
  u. spin nicht kompensierter Spin *m*
unctuous sich fettig anfühlend, fettig anzufühlen[d]
uncured *(Gum)* unvulkanisiert, unvernetzt, nicht vernetzt
undecanal $CH_3(CH_2)_9CHO$ Undekanal *n*, Hendekanal *n*, Undezylaldehyd *m*
undecane Undekan *n*, Hendekan *n*, *(i.e.S.)* $CH_3(CH_2)_9CH_3$ n-Undekan *n*, Undekan *n*, n-Hendekan *n*
undecanedioic acid $HOOC(CH_2)_9COOH$ Undekandisäure *f*, Hendekandisäure *f*, Nonan-1,9-dikarbonsäure *f*
undecanoic acid $CH_3(CH_2)_9COOH$ Undekansäure *f*, Hendekansäure *f*, *(bisher auch)* Undezylsäure *f*
undecanol Undekanol *n*, Hendekanol *n*

undecanone Undekanon *n*, Hendekanon *n*
undecene Undezen *n*, Hendezen *n*, Undezylen *n*
undecenoic acid Undezensäure *f*, Hendezensäure *f*, Undezylensäure *f*
undecenol Undezenol *n*, Hendezenol *n*
undecomposed unzersetzt
undecorticated ungeschält
undecyl Undezyl *n*, Hendezyl *n*, *(i.e.S.)* $CH_3(CH_2)_9CH_2—$ n-Undezyl *n*, Undezyl *n*, n-Hendezyl *n*
  u. alcohol $CH_3(CH_2)_9CH_2OH$ Undezylalkohol *m*, Alkohol C 11 *m*, 1-Undekanol *n*
  u. aldehyde *s.* undecanal
undecylamine Undezylamin *n*, Hendezylamin *n*, Aminoundekan *n*
undecylenate Undezylenat *n*
undecylene *s.* undecene
undecylenic acid *s.* undecenoic acid
undecylic acid *s.* undecanoic acid
  u. aldehyde *s.* undecanal
undecyne Undezin *n*, Hendezin *n*
undecynoic acid Undezinsäure *f*, Hendezinsäure *f*
under anhydrous conditions unter Wasserausschluß
  u. millstone Unterstein *m*, Bodenstein *m*
  u.-runner huller Unterläuferschälgang *m*
underbleach / to *(Pap)* unterbleichen, unvollständig bleichen
underbleaching *(Pap)* Unterbleichung *f*, Unterbleiche *f*, unvollständige Bleichung *f*
undercoat *(Farb)* Unteranstrich *m*; *(Farb)* Zwischenanstrich *m*
  u. material Anstrichfarbe *f* für Unteranstrich; Zwischenanstrichfarbe *f*
undercoater *s.* undercoat material
undercook / to *(Pap)* unterkochen, hartkochen, unvollständig kochen
undercooking *(Pap)* Unterkochung *f*, Hartkochung *f*, unvollständige Kochung *f*
undercure *(Plast)* Unterhärtung *f*, Untervulkanisation *f*; *(Gum)* Untervulkanisation *f*, Untervernetzung *f*
undercured *(Plast)* nicht ausgehärtet, ungenügend gehärtet; *(Gum)* untervulkanisiert, untervernetzt
undercut *(Plast)* Hinterschneidung *f*
underdevelop / to *(Foto)* unterentwickeln
underdevelopment *(Foto)* Unterentwicklung *f*
underdone *(Lebm)* halbgar
underdriven mit Antrieb von unten, von unten angetrieben
  u. centrifuge von unten angetriebene Zentrifuge *f*, stehende Zentrifuge *f*, Stehzentrifuge *f*
underdry / to *(Pap)* unvollständig (unzureichend) trocknen
underdrying *(Pap)* unvollständige (unzureichende) Trocknung *f*
underexpose / to *(Foto)* unterbelichten
underexposure *(Foto)* Unterbelichtung *f*
underfeed firing Unterschubfeuerung *f*
  u. stoker Unterschubrost *m*

u. stoking s. u. firing
underfiring s. underfeed firing
underflow Unterlauf m, Unterlaufprodukt n, (Flotation auch) Sinkgut n, Rückstand m, (Siebklassieren auch) Siebunterlauf m, Siebdurchlauf m, Siebdurchgang m
underglaze Unterglasur...
u. colour Unterglasurfarbe f
u. decoration Unterglasurdekor m
u. painting Unterglasurmalerei f
underground gasification Untertag[e]vergasung f, Lagerstättenvergasung f, Flözvergasung f, unterirdische Vergasung f, Vergasung f der Kohle im Flöz
u. leaching (Met) Laugung f in der Grube, Laugung f in situ, In-situ-Laugung f
u. line unterirdische (erdverlegte) Leitung f, Erdleitung f
u. mining Tiefbauförderung f, Tiefbau[betrieb] m, Untertagebau m, Förderung f im Tiefbau (Untertagebau)
u. reservoir s. u. storage tank
u. storage unterirdische Speicherung (Lagerung) f, Untertagespeicherung f
u. storage tank Untertagespeicher m, unterirdischer Speicher (Behälter) m
u. water Grundwasser n
u. work[ing] s. u. mining
underjet Unterbrenner m
u. cellar Düsenkeller m (eines Unterbrennerofens)
u. coke oven Unterbrenner[koks]ofen m, Koksofen m mit Unterbrennern
u. piping Unterbrennerleitungen fpl, Düsenleitungen fpl (eines Unterbrennerofens)
underneath the liquor unter den (bezw. dem) Flüssigkeitsspiegel, unter die (bzw. der) Flüssigkeitsoberfläche, unter das (bzw. dem) Niveau der Flüssigkeit
underretting (Text) unvollständige (zu kurze) Röste f
undersaturated untersättigt
undersaturation Untersättigung f
undersize [material] (Klassierung) Feinkorn n, Unterkorn n, (Siebklassierung auch) Siebfeines n, (Windsichten auch) Feingut n, Sichtfeines n
undervulcanization (Gum) Untervulkanisation f, Unternetzung f
undervulcanized (Gum) untervulkanisiert, unternetzt
undiluted unverdünnt
undissociated undissoziiert, nicht gespalten (zerfallen)
undistorted unverzerrt
undrained salts (Handelsname für sehr unreines Naphthalin)
undrawn (Text) ungereckt, unverstreckt (Chemiefaserstoffe)
undue unzulässig (z.B. Toxizität)
unesterified nicht verestert

unetched ungeätzt
unevaporated unverdampft
uneven distribution (Glas) ungleichmäßige Glasverteilung f (Fehler)
u. drying ungleich[mäßig]es Trocknen n
unexposed (Foto) unbelichtet
unfermentable unvergärbar, nicht vergärbar
unfilled ungefüllt, nicht gefüllt, füllstofffrei, ohne Füllstoff
unfilt[e]rable nicht filtrierbar (filtrierend)
unfinished unbearbeitet; (Gerb) nicht zugerichtet; (Pap) ungeglättet
unfired (Ker) roh
ungasified unvergast, nicht vergast
unglazed unglasiert, nicht glasiert; (Pap) ungeglättet
unground ungemahlen
unhairing (Gerb) Enthaarung f
u. knife (Gerb) Haareisen n, Enthaareisen n
u. machine (Gerb) Enthaar[ungs]maschine f
unhybridized unhybridisiert, nichthybridisiert
unhydrogenable nicht hydrierbar
uniaxial (Krist) [optisch] einachsig
unibivalent ein-zwei-wertig, 1-2-wertig
u. electrolyte ein-zwei-wertiger (1-2-wertiger) Elektrolyt m
unidentate (Komplexchemie) einzahnig, einzähnig, einzählig (Koordinationswert des Liganden)
unifining Unifining n (Variante der hydrierenden Raffination von Erdölprodukten)
u. process Unifining-Verfahren n
uniform gleichmäßig (z.B. Farbe)
u. corrosion ebenmäßige Korrosion f
unimolecular unimolekular, monomolekular, 1molekular
u. adsorbed layer monomolekulare Schicht (Adsorptionsschicht) f, monomolekularer Film (Oberflächenfilm) m, Monomolekularfilm m
u. film (layer) s. u. adsorbed layer
u. reaction unimolekulare (monomolekulare) Reaktion f
u. surface film s. u. adsorbed layer
uninegative einfachnegativ
uninflammability Flammwidrigkeit f, Nichtentflammbarkeit f
uninflammable flammwidrig, nicht entflammbar (brennbar)
union Vereinigung f, Verbindung f; Rohrverbindung f, Rohrverbindungsstück n, Schlauchverbindung f, Schlauchverbindungsstück n
u. dye Halbwollfarbstoff m, Mischfaserfarbstoff m
u. dyeing Färben n von Faserstoffmischungen, Unifärben n, Halbwollfärben n, Mischgewebefärben n
u. fabric Mischgewebe n
u. joint Verbindung f mit Überwurfmutter (Rohrverbindung)
unionized unionisiert, nichtionisiert
unipositive einfachpositiv

uniroll mill *(Farb)* Einwalzenmühle f, Einwalzenstuhl m, Einwalze f

Unisol process Unisolverfahren n *(zur Entschwefelung von Erdöldestillaten)*

unit Einheit f, Bezugseinheit f, Maßeinheit f; *(Chem, Krist)* Einheit f, Struktureinheit f, Baueinheit f; *(Tech)* Einheit f, Apparateeinheit f, Apparategruppe f

u. area Einheitsfläche f

u. cell Einheitszelle f, Elementarzelle f, Elementarkörper m

u. [electric] charge [elektrische] Elementarladung f, elektrisches Elementarquantum n, e

u. elongation Dehnung f je Längeneinheit, relative Dehnung f

u. of current Einheit f der Stromstärke, Stromstärkeeinheit f

u. of heat Wärmeeinheit f

u. operation [physikalische] Grundoperation f

u. plane *(Krist)* Hauptsymmetrieebene f

u. press s. u. vulcanizer

u. process [chemische] Grundreaktion f, Grundprozeß m

u. thickener Einkammereindicker m

u. tyre vulcanizer s. u. vulcanizer for tyres

u. value Einheitswert m

u. vulcanizer *(Gum)* Einzelheizer m

u. vulcanizer for tubes *(Gum)* Schlaucheinzelheizer m

u. vulcanizer for tyres *(Gum)* Reifeneinzelheizer m

unitarian, unitary unitär, unitarisch

u. bond unitarische (homöopolare, unpolare, einpolare, kovalente) Bindung f, Atombindung f, Elektronenpaarbindung f, Kovalenz f

Unitary thermal [polymerization] process Unitary-Thermal-Polymerization-Prozeß m *(kombiniertes Polymerisationsverfahren der M.W. Kellogg Co.)*

unite / to vereinigen, verbinden; sich vereinigen (verbinden)

uniunivalent ein-ein-wertig, 1-1-wertig

u. electrolyte ein-ein-wertiger (1-1-wertiger) Elektrolyt m

univalence, univalency Einwertigkeit f

univalent einwertig, monovalent

univariant univariant, monovariant, einfach frei

universal calender Vielzweckkalander m

u. developer *(Foto)* Universalentwickler m

u. emulsion *(Foto)* Universalemulsion f

u. gas constant universelle (allgemeine, molare) Gaskonstante f

u. indicator Universalindikator m

u. indicator paper Universalindikatorpapier n

unknown unbekannte Substanz f, unbekanntes Material n

unlabelled unmarkiert, nichtmarkiert

unleaded ungebleit, unverbleit *(Kraftstoff)*

unlike charge ungleichnamige Ladung f

unlime / to entkalken

unload / to entladen; entleeren; entlasten

unloaded s. unfilled

unloader knife Schälmesser n, Austragmesser n

unloading Entladen n, Entladung f; Entleeren n, Entleerung f; Entlastung f

u. valve Entlastungsventil n

unmalted ungemälzt

unmixed unvermischt

unmodified unmodifiziert, nichtmodifiziert, unverändert

unmordanted ungebeizt

unnatural unnatürlich, nicht in der Natur vorkommend

unofficial *(Pharm)* nicht offizinell

unordered ungeordnet, wirr

unorganized ferment ungeformtes Ferment n

unpacked unverpackt, nicht verpackt

unpaired ungepaart, unpaar[ig]

u. electron ungepaartes (unpaares) Elektron n

unpalatable unschmackhaft

unpigmented unpigmentiert, nichtpigmentiert; *(Gum)* s. unfilled

unplasticized unplasti[fi]ziert, weichmacherfrei, ohne Weichmacher

u. PVC unplastifiziertes (weichmacherfreies) PVC n; Hart-PVC n, PVC hart n, PVC-H

unpleasant unangenehm *(z.B. Geruch)*

unpressed ungepreßt

unprocessed unbehandelt, unbearbeitet, nicht verarbeitet, roh

unreacted unumgesetzt, nichtumgesetzt, nicht in Reaktion getreten, unreagiert, unverbraucht

unreactive reaktionslos, nichtreaktionsfähig

unreel / to abhaspeln, abwinden, abspulen, abwickeln

unrefined unraffiniert

unreinforced unverstärkt *(z.B. Kunststoff)*

unsalted ungesalzen, nicht gesalzen

unsaponifiable unverseifbar, nicht verseifbar

u. matter (residue) unverseifbarer Anteil m, Unverseifbares n, UV

unsaponified unverseift, nichtverseift

unsaturated un[ab]gesättigt; nicht völlig abgesättigt

u. acid ungesättigte Säure f

α,β-u. acid α,β-ungesättigte Säure f

u. aliphatic monocarboxylic acid ungesättigte aliphatische Monokarbonsäure f, Alkensäure f

u. compound ungesättigte Verbindung f

u.-surface drying Trocknen n bei stellenweise trockner Oberfläche

u. Weston cell ungesättigtes Weston-Normalelement (Weston-Element) n, Standard-Weston-Normalelement n, Weston-Standardelement n

unsaturateds Ungesättigte npl, ungesättigte Verbindungen fpl

unsaturation Ungesättigtheit f, Nichtsättigung f, Nichtgesättigtsein n

u. electron π-Elektron n

unseeded solution ungeimpfte Lösung f

unshared electron einsames (freies, nichtanteiliges, nichtbindendes) Elektron *n*, Nichtbindungselektron *n*
　u. electron pair einsames (freies) Elektronenpaar *n*
unshrinkable nichtschrumpfend, nichtkrumpfend, nichteinlaufend, schrumpffest, schrumpfbeständig, krumpffest
　u. finish Schrumpffestausrüstung *f*, Krumpffestausrüstung *f*, schrumpffreie Ausrüstung (Appretur) *f*
unsintered ungesintert
unsized *(Text)* ungeschlichtet; *(Pap)* ungeleimt
unslaked lime ungelöschter (gebrannter) Kalk *m*, Branntkalk *m*, Ätzkalk *m* *(Kalziumoxid)*
unsoured ungesäuert
unstabilized unstabilisiert
　u. gasoline unstabilisiertes (unstabiles, instabiles, wildes) Benzin *n*
unstable instabil, unstabil, unbeständig, labil
to make u. destabilisieren
　u. electron configuration unstabile Elektronenkonfiguration *f*
　u. gasoline s. unstabilized gasoline
　u. isotope instabiles (radioaktives) Isotop *n*, Radioisotop *n*, Radionuklid *n*
unstrained ungespannt
unsubstituted nichtsubstituiert
unsulphonated nichtsulfoniert
unsupported *(Plast)* trägerlos *(Film)*
　u. catalyst Katalysator *m* ohne Träger (Stützmaterial)
unsymmetrical unsymmetrisch, asymmetrisch
　u. compound unsymmetrische (asymmetrische) Verbindung *f*
　u. ether R·O·R′ gemischter (unsymmetrischer) Äther *m*
unsystematic name unsystematischer (nichtsystematischer, nichtrationeller, nichtsystematisch gebildeter, trivialer) Name *m*, unsystematische Benennung *f*, Trivialname *m*
untreated unbehandelt, unbearbeitet, nicht behandelt (verarbeitet), roh
　u. water Rohwasser *n*
untwisting machine *(Text)* Entwinder *m*
unusual sugar seltener Zucker *m*
unvulcanized *(Gum)* unvulkanisiert, unvernetzt, nichtvernetzt
unweatherable unverwitterbar
unwhizzed naphthalene *(Handelsname für sehr unreines Naphthalin)*
unwind / to *(Pap)* abrollen, abwickeln
unwind equipment *(Pap)* Abrollvorrichtung *f*
　u. stand *(Pap)* Abrollgestell *n*
unwinding *(Pap)* Abrollen *n*, Abwickeln *n*
　u. stand s. unwind stand
U.O.P. platforming process Platforming-Verfahren *n* der UOP (Universal Oil Products Co.)
up-oriented nach oben stehend
upas Upasbaum *m*, Antiaris toxicaria; Upas[gift] *n*

updraught firing aufsteigende Flammenführung *f*
　u. kiln Ofen *m* mit aufsteigender Flamme
updraw machine Aufwärtsziehmaschine *f* *(Glasrohrherstellung)*
　u. process Aufwärtsziehverfahren *n*, Schuller-Verfahren *n* *(Glasrohrherstellung)*
upend / to *(Kautschukmischung)* aufrollen
uperization Uperisation *f*, Ultrapasteurisation *f*
uperize / to uperisieren
upflow high-density tower *(Pap)* Dickstoff-Aufwärts[bleich]turm *m*, Aufwärts-Dickstoffturm *m*
upgrading Wertsteigerung *f*, Qualitätsverbesserung *f*
upkeep Erhaltung *f*, Unterhaltung *f*, Aufrechterhaltung *f*, Instandhaltung *f*
upper bainite oberer Bainit *m*, Oberbainit *m*, Gefüge *n* der oberen Zwischenstufe
　u. consolute (critical solution) temperature obere kritische Lösungstemperatur (Mischungstemperatur) *f*, oberer kritischer Lösungspunkt (Mischungspunkt) *m*
　u. explosion limit obere Explosionsgrenze *f*
　u. leather Oberleder *n*
　u. millstone Läufer[stein] *m*, Laufstein *m*
　u. part Oberteil *m*
　u. plunger Oberstempel *m* *(eines Preßwerkzeugs)*
　u. run of a deckle strap *(Pap)* oberer Deckelriemen *m*
　u. size obere Grenzkorngröße *f*
　u. stone s. u. millstone
upright *(Ker)* Brennstütze *f*, Stütze *f* *(Brennhilfsmittel)*
　u. digester *(Pap)* stehender Kocher *m*, Stehkocher *m*
　u. kiln Schachtofen *m*
uprun[ning] Aufwärtsgasen *n*, Aufwärtsgasung *f*, Gasen *n* (Gasung *f*) in aufsteigender Richtung, Gasen *n* von unten
upspout Steig[e]rohr *n*
upsteaming s. uprun[ning]
upstroke press Unterkolbenpresse *f*
uptake Aufnahme *f*, Aufnehmen *n*
　u. of liquid Flüssigkeitsaufnahme *f*
　u. of nutrients Nährstoffaufnahme *f*
　u. of water Wasseraufnahme *f*
upthrust Auftrieb *m*
upward current aufwärtsgerichtete Strömung *f*, Aufstrom *m*, aufsteigender (nach oben wirkender) Strom *m*
　u. velocity Geschwindigkeit *f* des Aufstroms
ur-acid Ursäure *f*, offenes Ureid *n*
uracil Urazil *n*, 2,4-Dioxotetrahydropyrimidin *n*
uraconite *(Min)* Urakonit *m*
uralite *(Min)* Uralit *m*
uralitization *(Min)* Uralitisierung *f*, Uralitisation *f*
uralitize / to *(Min)* uralitisieren
uramil Uramil *n*, Dialuramid *n*, 5-Aminobarbitursäure *f*

uramino NH_2CONH- Uramino..., *(veraltet für)* Ureido...

uranalysis *s.* urinalysis

uranate Uranat *n*

uranic Uran...; höherwertigem Uran entsprechend, *(meist)* Uran(VI)-...

  u. acid Uransäure *f*, *(i.e.S.)* H_2UO_4 *oder* $UO_2(OH)_2$ [normale] Uransäure *f*, Metauransäure *f*, Uranylhydroxid *n*

  u. ochre *(Min)* Uranocker *m*, Uranopilit *m*

  u. oxide UO_3 Uran(VI)-oxid *n*, Urantrioxid *n*

uranide *s.* uranoid

uraniferous uranhaltig

uranin[e] Uranin *n* *(Dinatriumsalz des Fluoreszeins)*

uraninite *(Min)* Uraninit *m* *(Uran(IV)-oxid)*

uranium U Uran *n*

  u. acetate $UO_2(CH_3COO)_2$ Uranazetat *n*, *(veraltet für)* Uranylazetat *n*

  u. ammonium carbonate *s.* uranyl ammonium carbonate

  u. fission Uranspaltung *f*

  u. hydride UH_3 Uranhydrid *n*

  u. mineral Uranmineral *n*

  u. nitrate $UO_2(NO_3)_2$ Urannitrat *n*, *(veraltet für)* Uranylnitrat *n*

  u. oxybromide *s.* uranyl bromide

  u. oxychloride *s.* uranyl chloride

  u. peroxide *s.* u. tetroxide

  u. series Uran-Radium-Zerfallsreihe *f*, Uran-Radium-Reihe *f*

  u. sesquisulphide U_2S_3 Diurantrisulfid *n*, Uran(III)-sulfid *n*

  u. tetroxide UO_4 *oder* $UO_2[O_2]$ Urantetroxid *n*, Uranperoxid *n*

  u. yellow $Na_2U_2O_7 \cdot 6H_2O$ Urangelb *n*, Natriumdiuranat *n*

uranniobite *(Min)* Uranniobit *m*, *(veraltet für)* Samarskit *m*

uranochre *s.* uranopilite

uranoid Uranoid *n*, *(bisher)* Uranid *n* *(Element einer Aktinoidenteilreihe)*

uranophane *(Min)* Uranophan *m* *(Kalziumuran-(VI)-hexahydroxiddiorthosilikat)*

uranopilite *(Min)* Uranopilit *m*, Uranocker *m*

uranosph[a]erite *(Min)* Uranosphärit *m*

uranotil *s.* uranophane

uranous Uran...; niederwertigem Uran entsprechend, *(meist)* Uran(IV)-...

  u. oxide UO_2 Uran(IV)-oxid *n*, Urandioxid *n*

  u. sulphate $U(SO_4)_2$ Uran(IV)-sulfat *n*

  u. uranate, u.-uranic oxide $UO_2 \cdot 2UO_3$ *oder* $U(UO_4)_2$ Uran(IV,VI)-oxid *n*, Triuranoktoxid *n*, Uran(IV)-uranat *n*

uranpyrochlore *(Min)* Uranpyrochlor *m*

uranvitriol *(Min)* Uranvitriol *m*, Johannit *m*

uranyl Uranyl *n* *(Atomgruppe oder Ion)*

  u. acetate $UO_2(CH_3COO)_2$ Uranylazetat *n*

  u. ammonium carbonate $2(NH_4)_2CO_3 \cdot UO_2CO_3$ Ammoniumuranylkarbonat *n*

  u. bromide UO_2Br_2 Uranylbromid *n*

  u. chloride UO_2Cl_2 Uranylchlorid *n*

  u. compound *s.* u. salt

  u. cyanoferrate(II), u. ferrocyanide $(UO_2)_2[Fe(CN)_6]$ Uranylhexazyanoferrat(II) *n*

  u. hydroxide $UO_2(OH)_2$ *oder* H_2UO_4 Uranylhydroxid *n*, [normale] Uransäure *f*, Metauransäure *f*

  u. nitrate $UO_2(NO_3)_2$ Uranylnitrat *n*

  u. oxide UO_3 Uranyloxid *n*, *(veraltet für)* Uran(VI)-oxid *n*, Urantrioxid *n*

  u. phosphate UO_2HPO_4 Uranylphosphat *n*, *(veraltet für)* Uranylhydrogenphosphat *n*

  u. salt Uranylsalz *n*, Uranylverbindung *f*, Dioxouran(VI)-salz *n*

urao *(Min)* Urao *n* *(veraltet; teils Trona, teils Thermonatrit)*

urate Urat *n* *(Salz der Harnsäure)*

p-urazine $C_2H_4O_2N_4$ 4-Urazin *n*, Diharnstoff *m*

urea NH_2CONH_2 Harnstoff *m*, Karbamid *n*, Kohlensäurediamid *n*; Harnstoff *m*, Harnstoffderivat *n*; *s.* u. resin

  u. addition compound Harnstoffadditionsverbindung *f*

  u. adduct Harnstoffaddukt *n*

  u.-bisulphite solubility *(Text)* Harnstoff-Bisulfit-Löslichkeit *f*, HBL

  u. calcium nitrate Harnstoffkalksalpeter *m*

  u. complex Harnstoffkomplex *m*, Harnstoffkomplexverbindung *f*

  u. cycle Harnstoffzyklus *m*, Ornithinzyklus *m*

  u. dewaxing Harnstoffentparaffinierung *f*, Harnstofftrennung *f*, Abscheidung *f* der *n*-Paraffinkohlenwasserstoffe durch Harnstoffextraktivkristallisation

  u. formaldehyde, u.-formaldehyde condensation product Harnstoff-Formaldehyd-Kondensat[ions]produkt] *n*, Harnstoffformaldehyd *n*

  u.-formaldehyde glue Harnstoff-Formaldehyd-Leim *m*

  u.-formaldehyde plastic Kunststoff *m* aus Harnstoff-Formaldehyd-Harz

  u.-formaldehyde resin Harnstoff-Formaldehyd-Harz *n*

  u. inclusion compound Harnstoffeinschlußverbindung *f*

  u. lattice Harnstoffgitter *n*

  u. molecular compound Harnstoffmolekülverbindung *f*

  u. nitrogen Harnstoffstickstoff *m*

  u. plastic Karbamidkunststoff *m*, Kunststoff *m* aus Harnstoffharz

  u. resin Harnstoffharz *n*, Karbamidharz *n*, Harnstoff-Aldehyd-Harz *n*; *s.* u.-formaldehyde resin

ureameter *s.* ureometer

urease Urease *f* *(harnstoffspaltendes Ferment)*

ureide Ureid *n*

ureido NH_2CONH- Ureido..., Karbamido...

  u. acid Ureidosäure *f*

ureidoacetic acid $NH_2CONHCH_2COOH$ Ureidoes-
sigsäure *f*, Hydantoinsäure *f*, *N*-Karbamoylglyzin
n
5-ureidohydantoin 5-Ureidohydantoin *n*, Allantoin
n
ureolytic harnstoffspaltend
ureometer Ureometer *n* *(Apparat zur Harn-
stoffbestimmung)*
urethan[e] Urethan *n*, Karbamidsäureester *m*,
(i.e.S.) $NH_2COOC_2H_5$ Urethan *n*, Äthylurethan *n*,
Karbamidsäureäthylester *m*; Polyurethan *n*, [po-
lymeres] Urethan *n*
 u. elastomer Polyurethanelastomer[es] *n*, Urethanelastomer[es] *n*
 u. foam Polyurethanschaum[stoff] *m*, Urethan-
schaumstoff *m*
 u. oil Urethanöl *n*
 u. rubber Polyurethankautschuk *m*, Urethankautschuk *m*
ureylene —NHC(=O)NH— Ureylen *n*
uric acid Harnsäure *f*, 2,6,8-Trihydroxypurin *n*,
Purin-2,6,8-triol *n*
 u.-acid calculus Harnsäurestein *m*
 u.-acid derivative Harnsäurederivat *n*
uricase Urikase *f*, Uratoxydase *f*, Urokooxydase *f*
uricolysis Urikolyse *f* *(Oxydation der Harnsäure
durch die Urikase zu Allantoin)*
uricolytic urikolytisch
 u. enzyme urikolytisches Ferment *n*
uridine Uridin *n*, Urazil-*D*-ribosid *n*
 u. diphosphate Uridindiphosphat *n*, UDP
 u. monophosphate Uridinmonophosphat *n*, Uridin[mono]phosphorsäure *f*, Uridylsäure *f* *(i.w.S.)*,
UMP
 u. 3'-monophosphate Uridin-3'-[mono]phosphat
n, Uridin-3'-[mono]phosphorsäure *f*, Uridylsäure
f, UMP-3'
 u. phosphate Uridinphosphat *n*, Uridinphos-
phorsäure *f*, *(i.e.S.)* s. u. monophosphate
 u. 3'-phosphate s. u. 3'-monophosphate
 u. triphosphate Uridintriphosphat *n*, Uridintri-
phosphorsäure *f*, UTP
**uridinediphosphateglucose,   uridinediphospho-
glucose** Uridindiphosphatglukose *f*, Uridindi-
phosphoglukose *f*, UDP-Glukose *f*, UDPG
uridinephosphoric acid s. uridine phosphate
uridylic acid s. 1. uridine monophosphate; 2.
uridine 3'-monophosphate
urinalysis Harnanalyse *f*
urinary calculus Harnstein *m*
 u. pigment Harnfarbstoff *m*
urine Harn *m*, Urin *m*
 u. analysis s. urinalysis
urinometer Urometer *n*, Harnwaage *f* *(Spin-
delaräometer zur Bestimmung der Wichte des
Harns)*
urobilin Urobilin *n*, Sterkobilin *n* *(Abbauprodukt
des Gallenfarbstoffs Bilirubin)*
urocan[in]ic acid Urokan[in]säure *f*, 4-Imidazol-
akrylsäure *f*

urochrome Urochrom *n* *(Urinfarbstoff)*
urochromogen Urochromogen *n* *(farblose Vorstufe
des Urochroms)*
uronic acid Uronsäure *f* *(eine Aldehydkarbonsäure
der Zuckerreihe)*
uroporphyrin Uroporphyrin *n* *(ein isomeres substituiertes Porphyrin)*
urunday extract *(Gerb)* Urundayextrakt *m(n)* *(von
Astronium balansae Engl.)*
use topically (in topical applications) / to *(Pharm)*
lokal anwenden (verwenden), zu Lokalbehand-
lungen verwenden
used catalyst Altkatalysator *m*
 u. detergent solution *(Text)* Schmutzflotte *f*,
schmutzbelastete (gebrauchte) Waschlauge *f*
 u. developer *(Foto)* gebrauchter Entwickler *m*
 u. salt *(Gerb)* *(bereits zur Häutekonservierung
verwendetes, wiedergewonnenes Salz)*
useful work Nutzarbeit *f*
usnetic acid Usnetinsäure *f*, Usnidinsäure *f*,
4,6-Dihydroxy-3,5-dimethyl-7-azetylkumaron-
essigsäure-(2) *f*; Usnetinsäure *f*, *(veraltet für)*
Lobarsäure *f*, Stereokaulsäure *f*
usnetinic acid Usnetininsäure *f*
usni[ni]c acid Usninsäure *f* *(ein Flechtenstoff)*
USP *(Abk. für)* 1. United States Pharmocopeia; 2.
United States Patent
USP grade *(den Forderungen des Arzneibuchs der
USA entsprechend)*
USS *(Abk. für)* United States Standard
usual dose *(Pharm)* Gebrauchsdosis *f*
uterotonic Uterustonikum *n*
utility boiler Kraftwerkkessel *m*, Kraftwerkdampf-
erzeuger *m*
utilization of sewage Abwasserverwertung *f*
UTP s. uridine triphosphate
uvarovite *(Min)* Uwarowit *m*, Kalkchromgranat *m*
(Kalziumchrom(III)-orthosilikat)
uvic acid $HOOC(CHOH)_2COOH$ Traubensäure *f*,
razemische Weinsäure (Säure) *f*, *dl*-Weinsäure *f*,
Paraweinsäure *f*; s. uvinic acid
uvinic acid $(CH_3)_2C_4HOCOOH$ Uvin[in]säure *f*,
Pyrotritarsäure *f*, 2,5-Dimethyl-3-furankarbon-
säure *f*
uwarowite s. uvarovite

V

V-belt Keilriemen *m*
V-clamp joint Verbindung *f* mit Schalen von
V-förmigem Querschnitt *(Rohrverbindung)*
V-ring, V-seal Kegel[dicht]ring *m*, Kegelring-
dichtung *f*
v-value Abbesche Zahl *f*, *v*-Wert *m*, reziproke
relative Dispersion *f*
Va-Purge process *(Pap)* Va-Purge-Verfahren *n*
Vacanceine red Vacanceine-Rot *n* *(erster β-Naph-
tholfarbstoff aus β-Naphtholnatrium und di-
azotiertem β-Naphthylamin)*

vacancy *(Krist)* Leerstelle f, Leerplatz m, Loch n, Lücke f, Gitterleerstelle f, Gitterloch n, Gitterlücke f, unbesetzter Gitterplatz m
v. pair Leerstellenpaar n
vacant frei, leer, unbesetzt, vakant
v. lattice position s. vacancy
v. [lattice] site s. vacancy
vaccenic acid Vakzensäure f, 11-Oktadezensäure f; Vakzensäure f, *trans*-11-Oktadezensäure f
vaccine Vakzine f, Vakzin n, Impfstoff m *(i.e.S.)*; Kuhpockenlymphe f
vacuometer s. vacuum gauge
vacuum Vakuum n, luftleerer (luftverdünnter) Raum m, Luftleere f
v.-and-blow process *(Glas)* Saugblaseverfahren n
v. bag *(Plast)* Vakuumsack m
v.-bag moulding *(Plast)* Vakuumgummisackverfahren n
v. blender s. v. kneader
v. box *(Pap)* Sauger m, Saug[er]kasten m
v. cement *(Plast)* Vakuumkitt m
v. chamber Vakuumarbeitsraum m
v. column Vakuumkolonne f
v. connection Vakuumanschluß m
v. cooler Vakuumkühler m
v. crystallization Vakuumkristallisation f
v. crystallizer Vakuumkristallisator m
v. deposition *(Plast)* Vakuumaufdampfung f
v. desiccator Vakuumexsikkator m
v. diecasting Vakuumdruckguß m, Unterdruckkokillenguß m
v. distillation Vakuumdestillation f, Destillation f im Vakuum, Destillation f unter vermindertem Druck
v.-distillation plant Vakuumdestillationsanlage f
v.-distilled vakuumdestilliert, im (unter) Vakuum destilliert
v. drum dryer Vakuumwalzentrockner m
v. drum filter Vakuumtrommelfilter n, Vakuumdrehfilter n
v.-dry / to unter (im) Vakuum trocknen
v. dryer Vakuumtrockenapparat m, Vakuumtrockner m; *(Pap)* Vakuumtrockenpartie f, Vakuumtrockner m
v. drying Vakuumtrocknen n, Vakuumtrocknung f
v.-drying oven Vakuumtrockenschrank m
v.-drying plant Vakuumtrockenanlage f, Vakuumtrocknungsanlage f
v. evaporation Vakuumverdampfung f
v. evaporator Vakuumverdampfer m, Vakuumverdampfapparat m
v. extrusion press *(Ker)* Vakuumstrangpresse f
v. filter Vakuumfilter n, Saugfilter n, Unterdruckfilter n
v. filtration Vakuumfiltrieren n, Vakuumfiltern n, Vakuumfiltration f
v. firing *(Ker)* Vakuumbrennen n
v. flask Vakuumkolben m

v. formability Vakuumverformbarkeit f
v. forming Vakuum[ver]formung f
v.-forming machine Vakuumformmaschine f
v. fumigation *(Vorratsschutz)* Vakuumbegasung f
v. gas oil Vakuumgasöl n
v. gauge Vakuummeßgerät n, Vakuummeter n
v. grease Vakuumfett n
v. headbox *(Pap)* Vakuumstoffauflauf m, vakuumgesteuerter Stoffauflauf m, Stoffauflauf m für Vakuumbetrieb
v. hopper Vakuumfülltrichter m
v. insulation Vakuumisolierung f
v. intake s. v. connection
v. jacket Vakuummantel m
v. kneader Vakuumkneter m
v. line Vakuumleitung f
v. melting Vakuumschmelzen n, Schmelzen n unter Vakuum
v. metallizing Vakuummetallisierung f
v. mixer Vakuummischer m
v. nutsche Vakuum[filter]nutsche f, Saugnutsche f
v. oven s. v.-drying oven
v. pan Vakuumpfanne f; Vakuumverdampfungsanlage f *(Vakuumsalzgewinnung)*
v. polarization Polarisation f des Vakuums, Vakuumpolarisation f
v. port s. v. connection
v. pump Vakuumpumpe f
v.-pump connection Vakuumpumpenanschluß m
v. residue Vakuumrückstand m, Rückstand m [bei] der Vakuumdestillation
v. shelf dryer Vakuumplattentrockner m, Vakuumheizplattentrockenschrank m
v. snap-back forming *(Plast)* Aufschrumpfen n unter Vakuum
v. spectrograph Vakuumspektrograf m
v. spectroscopy Vakuumspektroskopie f
v. still Vakuumdestillierapparat m; Vakuumkolonne f
v. tester Vakuumlecksuchgerät n
v.-tight vakuumdicht
v. tubing Vakuumschlauch m; Vakuumschläuche mpl
v. vessel Vakuumraum m
v. washer Vakuumwäscher m
val Val n, Grammäquivalent n
valence Valenz f, Wertigkeit f
v. angle Valenzwinkel m
v.-angle deviation Valenzwinkelabweichung f
v. band Valenzband n
v. bond Valenzbindung f
v.-bond approximation VB-Näherung f
v.-bond deviation s. v.-angle deviation
v.-bond method Valenzbindungsmethode f, Valenzstrukturmethode f, Methode f der Valenzstrukturen, VB-Methode f, Heitler-London-Slater-Pauling-Methode f, HLSP-Methode f, Spinmethode f, Elektronenpaarmethode f

v.-bond resonance Valenzbindungsresonanz *f*
v.-bond structure Valenzstruktur *f*
v.-bond theory Valenzbindungstheorie *f*, Valenz-strukturtheorie *f*, VB-Theorie *f*, Theorie (Bindungstheorie) *f* der Valenzstrukturen (Elektronenpaarbindungen), Heitler-London-Slater-Pauling-Theorie *f*, HLSP-Theorie *f*
v. change Valenzwechsel *m*, Wertigkeitswechsel *m*, Wertigkeitsänderung *f*
v.-chemical valenzchemisch
v. dash Valenzstrich *m*
v.-dash formula Valenzstrichformel *f*
v. deviation *s.* v.-angle deviation
v. electron Valenzelektron *n*, Leuchtelektron *n*, optisches Elektron *n*, Außenelektron *n*
v. electron pair Valenzelektronenpaar *n*
v. force Valenzkraft *f*
v. isomerization Valenzisomerisierung *f*
v. number Valenzzahl *f*
v. orbital Valenzorbital *n*
v. saturation Valenzabsättigung *f*
v. shell Valenzschale *f*
v. stage (state) Wertigkeitsstufe *f*, Valenzstufe *f*
v. structure *s.* v.-bond structure
v. tautomerism Valenztautomerie *f*
v. theory Valenztheorie *f*, Wertigkeitstheorie *f*, Valenzlehre *f*
valency *s.* valence
valentinite *(Min)* Valentinit *m*, Antimonblüte *f* *(Antimon(III)-oxid)*
valeraldehyde Valeraldehyd *m*, *(i.e.S.)* $CH_3(CH_2)_3CHO$ [*n*-]Valeraldehyd *m*, *n*-Amylaldehyd *m*, Pentanal *n*
valerate Valer[ian]at *n* *(Salz oder Ester einer Valeriansäure)*
valerian Baldrian *m*, Valeriana L.; Baldrianwurzel *f*
v. oil Baldrianöl *n* *(von Valeriana officinalis L.)*
valer[ian]ic acid Valeriansäure *f*, *(i.e.S.)* $CH_3(CH_2)_3COOH$ *n*-Valeriansäure *f*, [normale] Valeriansäure *f*, Baldriansäure *f*, Pentansäure *f*
v. aldehyde *s.* valeraldehyde
v. anhydride $[CH_3(CH_2)_3CO]_2O$ Valeriansäureanhydrid *n*
valeryl Valeryl *n*, *(i.e.S.)* $CH_3(CH_2)_3CO$— *n*-Valeryl *n*, Pentanoyl *n*
valerylene $CH_3C{\equiv}CCH_2CH_3$ Valerylen *n*, Äthylmethylazetylen *n*, 2-Pentin *n*
valine $(CH_3)_2CHCH(NH_2)COOH$ Valin *n*, α-Aminoisovaleriansäure *f*, α-Amino-β-methylbuttersäure *f*, 2-Amino-3-methylbutansäure *f*
Vallez filter Vallez-Filter *n*
valonea *(Gerb)* Valonea *f*, Wallonen *fpl (gerbstoffreiche Fruchtbecher mehrerer orientalischer Eichenarten)*
valoneaic acid Valoneasäure *f (Gerbstoffbaustein)*
valonia *s.* valonea
valonic acid *s.* valoneaic acid
value Wert *m*; Kennwert *m*, Wert *m*, Index *m*, Kennziffer *f*, Kennzahl *f*

v-**value** v-Wert *m*, Abbesche Zahl *f*, reziproke relative Dispersion *f*
Σ-**value** äquivalente Klärfläche *f (einer Zentrifuge)*
valve Ventil *n*, Absperrvorrichtung *f*
v. body Ventilgehäuse *n*, Ventilkörper *m*
v. bonnet Ventilhaube *f*
v. mixer Mischventil *n*
v. stem Ventilspindel *f*
v. tray Ventilboden *m*, Klappenboden *m*
v. trim Ventilauskleidung *f*
Van de Graaf accelerator Van-de-Graaf-Beschleuniger *m*
Van de Graaf generator Van-de-Graaf-Generator *m*, Van-de-Graaf-Bandgenerator *m*
van der Waals adsorption van-der-Waalssche (physikalische) Adsorption *f*
van der Waals attraction Van-der-Waals-Anziehung *f*, van-der-Waalssche Anziehung *f*, Van-der-Waals-Attraktion *f*
van der Waals attractive forces *s.* van der Waals forces [of attraction]
van der Waals bond Van-der-Waals-Bindung *f*, van-der-Waalssche (zwischenmolekulare) Bindung *f*
van der Waals constant van-der-Waalssche Konstante *f*
van der Waals crystal Van-der-Waals-Kristall *m*
van der Waals crystal aggregate van-der-Waalssches Kristallaggregat *n*
van der Waals equation [of state] van-der-Waalssche Zustandsgleichung (Gleichung) *f*
van der Waals forces [of attraction] Van-der-Waals-Kräfte *fpl*, van-der-Waalssche Kräfte (Anziehungskräfte) *fpl*
van der Waals isotherm van-der-Waalssche Isotherme *f*, Isotherme *f* nach der van-der-Waalsschen Gleichung (Formel)
van der Waals molecule Van-der-Waals-Molekül *n*, van-der-Waalssches Molekül *n*
van der Waals radius Van-der-Waals-Radius *m*, van-der-Waalscher Radius *m*
Van Slyke method Van-Slyke-Methode *f (zur gasvolumetrischen Bestimmung primärer aliphatischer Aminogruppen)*
vanadate Vanadat *n*
vanadic Vanadin...; höherwertigem Vanadin entsprechend, *(meist)* Vanadin(VI)-...
v. acid Vanadinsäure *f*
v. [acid] anhydride *s.* v. oxide
v. compound Verbindung *f* des höherwertigen Vanadins, *(meist)* Vanadin(V)-Verbindung *f*
v. fluoride VF_5 Vanadin(V)-fluorid *n*, Vanadinpentafluorid *n*
v. oxide V_2O_5 Vanadin(V)-oxid *n*, Divanadinpentoxid *n*, Vanadinpentoxid *n*
v. sulphide V_2S_5 Vanadin(V)-sulfid *n*, Vanadinpentasulfid *n*
vanadiferous vanadinhaltig
vanadinite *(Min)* Vanadinit *m*, Vanadinbleierz *n*

vanadite *(Min)* Vanadit *m*, Descloizit *m*; s. vanadinite
vanadium V Vanadin *n*, *(veraltet)* Vanadium *n*
  v. dichloride VCl_2 Vanadindichlorid *n*, Vanadin(II)-chlorid *n*
  v. diiodide VJ_2 Vanadindijodid *n*, Vanadin(II)-jodid *n*
  v. dioxide *s. v.* tetroxide
  v. disulphide V_2S_2 *oder* VS Vanadindisulfid *n*, Vanadinmonosulfid *n*, Vanadin(II)-sulfid *n*
  v. mica *(Min)* Vanadinglimmer *m*, Roscoelit *m*
  v. monosulphide *s. v.* disulphide
  v. oxybromide *s.* vanadyl monobromide
  v. oxydibromide *s.* vanadyl dibromide
  v. oxymonochloride *s.* vanadyl monochloride
  v. oxytribromide *s.* vanadyl tribromide
  v. pentafluoride VF_5 Vanadinpentafluorid *n*, Vanadin(V)-fluorid *n*
  v. pentasulphide V_2S_5 Vanadinpentasulfid *n*, Vanadin(V)-sulfid *n*
  v. pentoxide V_2O_5 Vanadinpentoxid *n*, Divanadinpentoxid *n*, Vanadin(V)-oxid *n*
  v. sesquioxide V_2O_3 Vanadintrioxid *n*, Divanadintrioxid *n*, Vanadin(III)-oxid *n*
  v. sesquisulphide V_2S_3 Vanadintrisulfid *n*, Divanadintrisulfid *n*, Vanadin(III)-sulfid *n*
  v. tetroxide V_2O_4 *oder* VO_2 Vanadintetroxid *n*, Vanadindioxid *n*, Vanadin(IV)-oxid *n*
  v. tribromide VBr_3 Vanadintribromid *n*, Vanadin(III)-bromid *n*
  v. trichloride VCl_3 Vanadintrichlorid *n*, Vanadin(III)-chlorid *n*
  v. trifluoride VF_3 Vanadintrifluorid *n*, Vanadin(III)-fluorid *n*
  v. trioxide V_2O_3 Vanadintrioxid *n*, Divanadintrioxid *n*, Vanadin(III)-oxid *n*
  v. trisulphide V_2S_3 Vanadintrisulfid *n*, Divanadin(III)-sulfid *n*
vanadous Vanadin...; niederwertigem Vanadin entsprechend, *(meist)* Vanadin(III)-...
  v. bromide VBr_3 Vanadin(III)-bromid *n*, Vanadintribromid *n*
  v. compound Verbindung *f* des niederwertigen Vanadins, *(meist)* Vanadin(III)-Verbindung *f*
  v. fluoride VF_3 Vanadin(III)-fluorid *n*, Vanadintrifluorid *n*
  v. oxide V_2O_3 Vanadin(III)-oxid *n*, Divanadintrioxid *n*, Vanadintrioxid *n*
  v. sulphide V_2S_3 Vanadin(III)-sulfid *n*, Divanadintrisulfid *n*, Vanadintrisulfid *n*
vanadyl Vanadyl *n*
  v. compound Vanadylverbindung *f*
  v. dibromide $VOBr_2$ Vanadyldibromid *n*, Vanadinoxiddibromid *n*, Vanadin(IV)-oxidbromid *n*
  v. dichloride $VOCl_2$ Vanadyldichlorid *n*, Vanadinoxiddichlorid *n*, Vanadin(IV)-oxidchlorid *n*
  v. monobromide VOBr Vanadyl[mono]bromid *n*, Vanadinoxid[mono]bromid *n*, Vanadin(III)-oxidbromid *n*
  v. monochloride VOCl Vanadyl[mono]chlorid *n*,

Vanadinoxid[mono]chlorid *n*, Vanadin(III)-oxidchlorid *n*
  v. salt Vanadylsalz *n*, Vanadinoxidsalz *n*
  v. sulphate $VOSO_4$ Vanadylsulfat *n*, Vanadin(IV)-oxidsulfat *n*
  v. tribromide $VOBr_3$ Vanadyltribromid *n*, Vanadinoxidtribromid *n*, Vanadin(V)-oxidbromid *n*
  v. trichloride $VOCl_3$ Vanadyltrichlorid *n*, Vanadinoxidtrichlorid *n*, Vanadin(V)-oxidchlorid *n*
vancomycin Vankomyzin *n* *(Antibiotikum)*
Vandyke brown Van-Dyck-Braun *n*, Van-Dyke-Braun *n (braun gefärbte Moorerde oder bitumenhaltige erdige Braunkohle)*
Vandyke red Van-Dyck-Rot *n*, Van-Dyke-Rot *n*, Florentiner Braun *n (Kupfer(II)-hexazyanoferrat(II))*
vane Schaufel *f*; Flügel *m*
  v. anemometer Flügelradanemometer *n*
  v. angle Schaufelwinkel *m*
  v. compressor Drehschieberverdichter *m*, Vielzellenverdichter *m*, Zellenverdichter *m*
  v. pump Flügel[zellen]pumpe *f*
vaneaxial fan Axiallüfter *m* mit Leiträdern
vaned-disk atomizer Zerstäuberscheibe (Sprühscheibe) *f* mit Schaufeln
vanilla bean Vanilleschote *f*
vanilal $C_6H_3(OH)(OC_2H_5)CHO$ Vanillal *n*, Bourbonal *n*, Protokatechualdehyd-3-äthyläther *m*
vanillaldehyde *s.* vanillin
vanillic acid $CH_3OC_6H_3(OH)COOH$ Vanillinsäure *f*, 4-Hydroxy-3-methoxybenzoesäure *f*
  v. aldehyde *s.* vanillin
vanillin $(CH_3O)(OH)C_6H_3CHO$ Vanillin *n*, Vanillaldehyd *m*, 4-Hydroxy-3-methoxybenzaldehyd *m*, Protokatechualdehyd-3-methyläther *m*
  v. sugar Vanillezucker *m*
vanishing cream *(Kosmet)* Vanishingcreme *f*, Tagescreme *f*
vanner Plan[en]herd *m*, Plachenherd *m*, Plachentisch *m*, Vanner *m*
van't Hoff equation Van't-Hoff-Gleichung *f*, van't-Hoffsche Gleichung *f*
van't Hoff equilibrium box van't-Hoffscher Gleichgewichtskasten *m*, Gleichgewichtskasten *m* nach van't Hoff
van't Hoff factor van't-Hoffscher Faktor *m*
van't Hoff isobar *s.* van't Hoff reaction isobar
van't Hoff isochore *s.* van't Hoff reaction isochore
van't Hoff isotherm *s.* van't Hoff reaction isotherm
van't Hoff law van't-Hoffsches Gesetz *n*, Gesetz *n* von van't Hoff
van't Hoff reaction box *s.* van't Hoff equilibrium box
van't Hoff reaction isobar van't-Hoffsche Reaktionsisobare *f*
van't Hoff reaction isochore van't-Hoffsche Reaktionsisochore *f*
van't Hoff reaction isotherm van't-Hoffsche Reaktionsisotherme *f*
vapor *s.* vapour
vaporability Verdampfbarkeit *f*, Verdampfungsfähigkeit *f*; Verdunstbarkeit *f*

vaporable, vaporizable verdampfbar; verdunstbar
vaporization Verdampfen *n*, Verdampfung *f*;
Verdunsten *n*, Verdunstung *f*
vaporize / to 1. verdampfen, in Dampf überführen;
verdunsten lassen; 2. verdampfen, *(unterhalb des
normalen Siedepunkts)* verdunsten
vaporizer Verdampfer *m*
vaporizing oil Motorenpetroleum *n*
vaporous dampfförmig
vapour Dampf *m*; Brüden[dampf] *m*
  v.-air mixture Dampf-Luft-Gemisch *n*, Luft-
  Wasserdampf-Gemisch *n*
  v. bubble Dampfblase *f*
  v. chamber Brüdenraum *m*, Dampfraum *m*
  v.-compression machine Kompressions[kälte]-
  maschine *f*
  v.-compression system Kompressions[dampf]-
  kälteanlage *f*, Kaltdampfverdichteranlage *f*
  v. condenser Kühler *m (für Dämpfe)*, Kondensator *m*
  v. curing *(Gum)* Kaltvulkanisation *f* nach dem
  „Dunstverfahren"
  v. degreasing Dampfentfettung *f*
  v. density Dampfdichte *f*, Gasdichte *f*
  v.-density determination Dampfdichtebestimmung *f*
  v.-density measurement Dampfdichtemessung *f*
  v.-density method Methode *f* (Verfahren *n*) der
  Dampfdichtebestimmung, Dampfdichtebestim-
  mungsmethode *f*
  v. dryer Dampftrockner *m*
  v. feed dampfförmiger Öleinsatz *m (beim
  Thermofor-Catalytic-Cracking)*
  v. hood Brüdenhaube *f*
  v.-liquid mixture Flüssigkeits-Dampf-Gemisch *n*,
  Dampf-Flüssigkeits-Gemisch *n*
  v. load Belastung *f (einer Rektifizierkolonne)*
  v. lock Förderstörung *f* im Kraftstoffleitungs-
  system durch Dampfblasenbildung
  v. permeability Wasserdampfdurchlässigkeit *f*
  v. phase Dampfphase *f*, dampfförmige Phase *f*,
  (bei der Hochdruckhydrierung meist) Gasphase
  f
  v.-phase catalyst Gasphasekatalysator *m*
  v.-phase chromatographic pattern Gaschroma-
  togramm *n*
  v.-phase chromatography Dampfphasenchro-
  matografie *f*; Gaschromatografie *f*
  v.-phase converter Gasphase[hydrier]ofen *m*
  v.-phase cracking Dampfphasekracken *n*, Gas-
  phasekracken *n*, Kracken *n* in der Dampfphase
  (Gasphase)
  v.-phase hydrogenation Gasphase[n]hydrierung
  f, Hydrierung *f* in [der] Gasphase
  v.-phase inhibitor Dampfphaseninhibitor *m*,
  Gasphaseninhibitor *m*, über die Dampfphase
  (Gasphase) wirkender Korrosionsschutzstoff *m*,
  VPI-Stoff *m*, VPI
  v.-phase inhibitor paper VPI-Korrosionsschutz-
  papier *n*, VPI-Papier *n*

v.-phase isomerization Dampfphasenisomerisie-
rung *f*, Gasphasenisomerisierung *f*, Isomerisierung *f* in der Dampfphase (Gasphase)
v.-phase nitration Dampfphasennitrierung *f*
v.-phase oxidation Dampfphasenoxydation *f*
v.-phase process Dampfphaseverfahren *n*, Gas-
phaseverfahren *n*
v.-phase transfer unit *(Destillation)* Übergangseinheit *f* in der Dampfphase
v. plating Vakuummetallisierung *f*, Hochvakuumbedampfung *f*, Herstellung *f* metallischer Überzüge nach Aufdampfverfahren (Vakuumaufdampfverfahren, Gasverfahren)
v. pressure Dampfdruck *m*
v.-pressure curve Dampfdruckkurve *f*
v.-pressure gradient Dampfdruckgefälle *n*
v.-pressure lowering Dampfdruckerniedrigung *f*
v.-pressure thermometer Dampfspannungsther-
mometer *n*, Tensionsthermometer *n*
v. rate Dampfgeschwindigkeit *f*
v. space s. v. chamber
v. state Dampfzustand *m*
v.-tight dampfdicht
v. tube Abzugsrohr *n*, Absaugvorrichtung *f (am
Aufschließgestell der Kjeldahl-Apparatur)*
v. velocity s. v. rate
var. s. variety
variable-displacement pump Verstellpumpe *f*
v.-speed drive stufenlos stellbarer Antrieb *m*
v. term variabler Term *m*, Laufterm *m*, Laufzahl
f
variance Varianz *f (Streuungsquadrat)*; *(phys Ch)*
Freiheit *f*, Freiheitsgrad *m*, *(i.e.S.)* Anzahl *f* der
Freiheiten (Freiheitsgrade)
variation Abweichung *f*, Veränderung *f*, Schwan-
kung *f*
v. in weight Masseveränderung *f*; Gewichts-
veränderung *f*
variety *(Min)* Varietät *f*, Var., V., Abart *f*;
Modifikation *f (z.B. des Schwefels)*
v. of coal Kohlevarietät *f*
variole *(Geol)* Variole *f (radialfasriges Kügelchen in
der Grundmasse des Variolits)*
variolite *(Geol)* Variolit *m*, Blatterstein *m*
variolitic *(Geol)* variolitisch
variscite *(Min)* Variscit *n (Aluminiumorthophosphat)*
varnish / to lackieren
varnish Lack *m*, *(besonders)* Klarlack *m*
  v.-makers' naphtha Lackbenzin *n*
  v. oil Lacköl *n*
  v. remover Lackentferner *m*, Abbeizmittel *n*
  v. resin Lackharz *n*
varnished fabric Lackgewebe *n*
  v. leather Lackleder *n*
  v. paper lackiertes Papier *n*, Lackpapier *n*, *(i.e.S.)*
  Firnispapier *n*
varnishing Lackieren *n*, Lackierung *f*; Lackbildung
f
  v. paper s. varnished paper

vasoconstrictive vasokonstriktorisch, gefäßkonstriktorisch, gefäßkontrahierend, gefäßverenge[r]nd

vasoconstrictor Vasokonstriktor *m*, gefäßverenge[r]ndes (gefäßkontrahierendes) Mittel *n*

vasodilator Vasodila[ta]tor *m*, Vasodilatans *n*, Gefäßerweiterungsmittel *n*, gefäßerweiterndes Mittel *n*

vasopressin Vasopressin *n* (*Hypophysenhinterlappenhormon*)

vat / to in der Küpe behandeln, [ver]küpen

vat Bottich *m*, Bütte *f*, großes Gefäß *n*; Trog *m*, Mulde *f*, Wanne *f*; (*Farb*) Küpe *f*; (*Gerb*) Geschirr *n*; (*Pap*) Bütte *f*, Trog *m*; (*Pap*) Siebtrog *m*, Stofftrog *m*, Rundsiebbütte *f* (*der Rundsiebpapiermaschine*); (*veraltet für*) pan of the beater

v. acid (*Farb*) Küpensäure *f*

v.-acid process Küpensäureverfahren *n*

v. colour (dye) Küpenfarbstoff *m*

v. dyeing Küpenfärberei *f*

v. liquor (*Farb*) Küpenflüssigkeit *f*

v. machine (*Pap*) Pappenmaschine *f*, Kartonmaschine *f*, (*i.e.S.*) Pappenrundsiebmaschine *f*, Rundsiebkartonmaschine *f*; (*Pap*) Rundsieb[papier]maschine *f*

v. mill Büttenpapierfabrik *f*; Papiermühle *f*

v. of pulp (*Pap*) Schöpfbütte *f*, Tauchbütte *f*

v. paper handgeschöpftes Papier *n*, Büttenpapier *n*, Handpapier *n*, Handbütten *n*, Echt-Bütten *n*, Schöpfpapier *n*

v. pasteurization (*Lebm*) Dauerpasteurisation *f*, Dauererhitzung *f* (*zur Pasteurisation*)

v. process (*Farb*) Verküpungsprozeß *m*, Verküpungsverfahren *n*

v.-sized (*Pap*) oberflächengeleimt

v. sizing (*Pap*) Oberflächenleimung *f*

vatlined board gedeckte (gegautschte) Pappe *f*

vatman (*Pap*) Schöpfer *m*

vattable (*Farb*) verküpbar

vatting (*Farb*) Verküpen *n*, Verküpung *f*

v. temperature Verküpungstemperatur *f*

VB approximation s. valence-bond approximation

VB method s. valence-bond method

VB theory s. valence-bond theory

vectorial vektoriell

vee-shaped V-förmig

v.-type mixer V-Mischer *m*

Vegard's law (rule) (*Krist*) Vegardsche Regel *f*

vegetable pflanzlich, vegetabilisch, Pflanzen...

v. adhesive s. v. glue

v. butter Pflanzenbutter *f*

v. drug pflanzliche Droge *f*

v. fat pflanzliches Fett *n*, Pflanzenfett *n*

v. fibre pflanzliche Faser *f*, Pflanzenfaser *f*; pflanzlicher Faserstoff *m*

v. glue pflanzlicher Leim *m*, Pflanzenleim *m*

v. gum Pflanzengummi *n*, Gummi *n*

v. ivory vegetabilisches Elfenbein *n* (*elfenbeinartiges Nährgewebe der Steinnuß*); Steinnuß *f*, Taguanuß *f*, Elfenbeinnuß *f* (*Frucht der Palme Phytelephas macrocarpa Ruiz et Pav.*)

v. juice Gemüsesaft *m*

v. lecithin Pflanzenlezithin *n*

v. margarine Pflanzenmargarine *f*, aus Pflanzenfetten hergestellte Margarine *f*

v. material (matter) Pflanzensubstanz *f*, Pflanzenstoff *m*, Pflanzenmaterial *n*, Pflanzenmasse *f*, pflanzliche (vegetabilische) Substanz *f*

v. oil pflanzliches Öl *n*, Pflanzenöl *n*

v. parchment vegetabilisches Pergament *n*, [echtes] Pergamentpapier *n*, Echtpergamentpapier *n*, Säurepergament *n*

v.-retanned mit pflanzlichen Gerbstoffen nachgegerbt (*Chromleder*)

v.-seed protection Gemüsesaatschutz *m*

v. spermaceti chinesisches Wachs *n*, Chinawachs *n*, Insektenwachs *n* (*durch Schildläuse abgeschiedenes Wachs von Fraxinus chinensis Roxb.*)

v. tallow Pflanzentalg *m*

v. tannage pflanzliche (vegetabilische) Gerbung *f*, Gerbung *f* mit pflanzlichen Gerbstoffen

v. wax vegetabilisches (Japanisches) Wachs *n* (*von Rhus succedanea L.*)

vehicle (*Farb*) Bindemittel *n*; (*Farb*) Bindemittellösung *f*

vein 1. Gang *m*, Ader *f*, Flöz *n*; 2. dünne Schliere *f* (*Glasfehler*)

v. chalk Faserkalk *m*

v. ore Gangerz *n*

veined gneiss Adergneis *m*

Vello process Vello-Verfahren *n* (*Glasröhrenziehverfahren*)

vellum (*Pap*) Schreibpergament *n*, Mönchspergament *n*

v. glaze (*Ker*) Velinglasur *f*

v. paper s. vellum

velocity Geschwindigkeit *f*

v. coefficient Geschwindigkeitskoeffizient *m*, Geschwindigkeitswert *m*, (*i.e.S.*) s. v. constant

v. component Geschwindigkeitskomponente *f*

v. constant Geschwindigkeitskonstante *f*, Reaktionsgeschwindigkeitskonstante *f*, spezifische Reaktionsgeschwindigkeit *f*

v. focusing Geschwindigkeitsfokussierung *f*

v. gradient Geschwindigkeitsgradient *m*, Geschwindigkeitsgefälle *n*

v. head Geschwindigkeits[druck]höhe *f*

v. meter Flügelradzähler *m*, Turbinenzähler *m* (*Mengenstrommessung*)

v. of dissociation Dissoziationsgeschwindigkeit *f*, Zerfallsgeschwindigkeit *f*

v. of elevation Steiggeschwindigkeit *f*

v. of esterification Veresterungsgeschwindigkeit *f*

v. of fall Fallgeschwindigkeit *f*, Sinkgeschwindigkeit *f*

v. of flame propagation Flammen[fortpflanzungs]geschwindigkeit *f*

v. of flow Fließgeschwindigkeit *f*, Strömungsgeschwindigkeit *f*, Durchflußgeschwindigkeit *f*, Durchlaufgeschwindigkeit *f*

v. of formation Bildungsgeschwindigkeit f
v. of hydrogenation Hydriergeschwindigkeit f
v. of ions Ionen[wanderungs]geschwindigkeit f
v. of light Lichtgeschwindigkeit f
v. of migration Wanderungsgeschwindigkeit f
v. of propagation Fortpflanzungsgeschwindigkeit f, Ausbreitungsgeschwindigkeit f
v. of racemization Razemisierungsgeschwindigkeit f
v. of reaction Reaktionsgeschwindigkeit f
v. of rearrangement Umlagerungsgeschwindigkeit f
v. of sedimentation Sedimentationsgeschwindigkeit f
v. of transformation Umwandlungsgeschwindigkeit f, Umsetzungsgeschwindigkeit f
v. selector Geschwindigkeitsselektor m
velour paper Velourpapier n, Plüschpapier n, Tuchpapier n, Samtpapier n
vena contracta Einschnürungsstelle f, Kontraktionsstelle f, Stelle f kleinsten Strömungsquerschnitts
veneer glue Furnierleim m
Venetian red Venetianischrot n
V. scarlet venezianischer Scharlach m
V. soap venezianische (Venezianer) Seife f
venite Venit m (Adergneis)
venous mercury electrode strömende Quecksilberelektrode f
vent 1. Entlüftung f; Entlüftungsöffnung f, Lüftungsöffnung f; 2. (Glas) Oberflächenriß m
v. channel Entlüftungskanal m, Luftkanal m
v. connection Entlüftungsanschluß m
v. extruder (Plast) Entgasungsextruder m, Entgasungsschneckenpresse f
v. gas Abgas n
v.-gas scrubber Abgaswäscher m, Abgaswascher m
v. line Entlüftungsleitung f
v. pipe Entlüftungsrohr n, Abzugsrohr n, Abzugslutte f
v. temperature Ablufttemperatur f, Abgastemperatur f
v. valve Entlüftungsventil n
vented extruder s. vent extruder
ventilation system Lüftungssystem n, Belüftungsanlage f
ventilator pipe Entlüftungsrohr n
venting (Plast) Lüften n, Entlüften n, Entlüftung f, Entgasen n, Entgasung f (des Werkzeugs, der Form)
Venturi meter Venturi-Messer m
Venturi scrubber Venturi-Wäscher m, Venturi-Wascher m, Venturi-Abscheider m
Venturi throat Verengung (engste Stelle) f am Venturi-Rohr
Venturi tube Venturi-Rohr n, Venturi-Düse f
Venturi washer s. Venturi scrubber
veratraldehyde $(CH_3O)_2C_6H_3CHO$ Veratrumaldehyd m, Veratraldehyd m, Protokatechualdehyddimethyläther m

veratric acid $(CH_3O)_2C_6H_3COOH$ Veratrumsäure f, 3,4-Dimethoxybenzoesäure f, Protokatechusäuredimethyläther m
veratrole $C_6H_4(OCH_3)_2$ Veratrol n, Pyrokatecholdimethyläther m, Brenzkatechindimethyläther m, 1,2-Dimethoxybenzol n
verbena oil (Kosmet) (ursprünglich) [echtes] Verbenaöl n (von Lippia triphylla (L'Hérit.) O. Kuntze); (jetzt) Lemongrasöl n, Zitronengrasöl n, Indisches Grasöl n, Verbenaöl n (von Cymbopogon flexuosus Stapf und C. citratus (DC.) Stapf)
Verdet constant Verdet-Konstante f
verdigris Grünspan m, spanisches Grün n (Gemisch aus basischen Kupfer(II)-azetaten; s. dagegen neutral verdigris)
verditer blue (Farb) Bergblau n, Bremer Blau n, Kupferblau n (basisches Kupferkarbonat)
vermicide Vermizid n, wurmtötendes Mittel n
vermiculite (Min) Vermikulit m (eine Gruppe von Phyllosilikaten)
vermifuge (Pharm) wurmabtreibendes Mittel n, Wurmmittel n, Helminthagogum n, Anthelminthikum n
vermil[l]ion (Farb) Vermillon[-Zinnober] m (gefälltes rotes Quecksilber(II)-sulfid)
vermontite (Min) Vermontit m, (veraltet für) Danait m (Abart von Arsenopyrit, Eisenarsensulfid)
Verona yellow Kasseler (Turners) Gelb n (basisches Bleichlorid wechselnder Zusammensetzung)
vertical basket extractor stehender Becherwerksextrakteur (Becherwerksextraktor) m
v. bucket elevator Senkrechtbecherwerk n
v. chamber Vertikalkammer f, Senkrechtkammer f (eines Kammerofens)
v. condenser senkrechter Kühler m
v. digester (Pap) stehender Kocher m, Stehkocher m
v. evaporator Vertikalrohrverdampfer m
v. extractor stehender Extrakteur (Extraktor) m
v.-flow electrical precipitator Vertikalelektrofilter n, Elektrofilterschlot m
v. flue Vertikalzug m, vertikaler (senkrechter) Zug m
v.-flue oven Horizontalkammerofen m mit senkrechten Heizzügen
v. kiln (Met) Schachtofen m
v. lehr (Glas) Kühlschacht m
v. oven Vertikalofen m, Senkrechtofen m, vertikaler (senkrechter) Ofen m (z.B. zur Gaserzeugung)
v. relationship Vertikalbeziehung f (im Periodensystem)
v. retort Vertikalretorte f, vertikale (senkrechte) Retorte f
v. screw mixer senkrechter Schneckenmischer m
v. sheet drawing process Vertikalziehverfahren (Senkrechtziehverfahren) n für Tafelglas
v. shell-and-tube condenser Steilrohr-Berieselungsverflüssiger m

v. tube evaporator Vertikalrohrverdampfer *m*
v. tube evaporator in sextuple effect Vertikalrohrverdampfer *m* in sechsstufiger Anlage
v. tube sweating stove *(Erdöl)* Zylinderzellenapparat *m (Schwitzapparat)*
vertically perforated brick Hochlochziegel *m*
Verven solution Vervens Reagens *n (aus CdJ₂ und KJ zum Nachweis von Alkaloiden)*
very dull *(Text)* tiefmatt
v. fine sand feinster Sand *m*, Staubsand *m*
v. inflammable sehr leicht brennbar
v. little soluble *s.* v. slightly soluble
v. low volume spray *s.* ultralow volume method
v. slightly soluble sehr wenig löslich
v. soluble sehr leicht löslich
vesicant blasenziehend
vesicant *(Med)* blasenziehendes Mittel *n*; *s.* vesicatory gas
vesicate / to *(Med)* zur Blasenbildung reizen, Blasen ziehen auf; Blasen bilden, sich mit Blasen bedecken
vesication *(Med)* Blasenbildung *f*; Blase *f (auf der Haut)*
vesicatory *s.* vesicant
v. gas blasenziehender gasförmiger Kampfstoff *m*
vesiculár blasig, *(Geol auch)* zellig
vessel Gefäß *n*, Behälter *m*
v. furnace *(Met)* Gefäßofen *m*
vesuvian[ite] *(Min)* Vesuvian *m (ein Sorosilikat)*
vesuvine brown Vesuvin *n*
vetiver oil Vetiveröl *n (von Vetivera zizanioides (L.)Nash)*
vetiverol *(Kosmet)* Vetiverol *n (Gemisch mehrerer Sesquiterpenalkohole aus Vetivera zizanioides (L.)Nash)*
vetivert *s.* vetiver oil
v. acetate *(Kosmet)* Vetiverylazetat *n (Estergemisch)*
vetivol *s.* vetiverol
VFA number *(Gum)* VFA-Zahl *f*
V.I. *s.* viscosity index
vibrate / to in Schwingungen versetzen, schwingen (vibrieren, oszillieren) lassen, schütteln, rütteln; schwingen, vibrieren, oszillieren, zittern
vibrated coarse concrete Rüttelgrobbeton *m*
v. concrete Rüttelbeton *m*, Vibrationsbeton *m*
vibrating ball mill schwingende Kugelmühle *f*, Schwingmühle *f*
v. conveyor Schwingförderer *m*, *(i.e.S.)* Schwingförderrinne *f*
v. conveyor dryer Vibrationsbandtrockner *m*, Schwingtrockner *m*
v. doctor *(Pap)* Vibrationsschaber *m*, traversierender Schaber *m*
v. mill *s.* v. ball mill
v. mudscreen *(Erdöl)* Vibrationsschüttelsieb *n*, Schüttelsieb *n (einer Rotary-Bohranlage)*
v. nozzle Vibrationsdüse *f*
v. screen Schwingsiebmaschine *f*, Schwingsieb

n, Vibrationssieb *n*, Vibratorsieb *n*, Schwingsortierer *m*, Vibrationssortierer *m*; *(Pap)* Vibrationsknotenfänger *m*
v. stress schwingende Beanspruchung *f*, Vibrationsbeanspruchung *f*
v. system Schwingungssystem *n*
v. tray dryer Vibrationsplattentrockner *m*, Hordenschwingtrockner *m*
vibration Schütteln *n*, Rütteln *n*; Schwingen *n*, Schwingung *f*, Vibrieren *n*, Vibration *f*, Oszillation *f*, Zittern *n*
v. amplitude Schwing[ungs]amplitude *f*; Schwing[ungs]weite *f*, Schwingweg *m*; Hublänge *f*, Hub *m*
v. ball mill *s.* vibrating ball mill
v. band Schwingungsbande *f*
v. filter Vibrationsfilter *n*
v. frequency Schwingungsfrequenz *f*
v. galvanometer Vibrationsgalvanometer *n*
v. mill *s.* vibrating ball mill
v.-proof erschütterungsfest
v.-rotation band Rotations-Schwingungs-Bande *f*
v.-rotation spectrum Rotations-Schwingungs-Spektrum *n*
v. spectrum Schwingungsspektrum *n*
vibrational Schwing[ungs]..., Vibrations..., oszillatorisch, Oszillations...
v. band *s.* vibration band
v. compaction Vibrationsverdichtung *f*
v. degree of freedom Schwingungsfreiheitsgrad *m*
v. energy Oszillationsenergie *f*, Schwingungsenergie *f*
v. ground state Schwingungsgrundzustand *m*
v. level Schwingungsniveau *n*
v. quantum number Schwingungsquantenzahl *f*, Vibrationsquantenzahl *f*
v.-rotational spectrum *s.* vibration-rotation spectrum
v. spectrum *s.* vibration spectrum
v. state Schwingungszustand *m*
v. transition Schwingungsübergang *m*
vibrationless erschütterungsfrei, schwingungsfrei
Vibratom Vibratom-Schwingmühle *f*
vibrator Rüttler *m*, Vibrator *m*, Schwingungs[ein]rüttler *m*
v. screen *s.* vibrating screen
vibratory ball mill *s.* vibrating ball mill
v. feeder Schwingförderer *m (als Zuteilvorrichtung)*
v. mill *s.* vibrating ball mill
vic *s.* vicinal
Vicat needle Vicat-Nadel *f*, Vicat-Apparat *m (Prüfgerät)*
Vicat needle point *s.* Vicat softening point
Vicat softening point (temperature) *(Plast)* Erweichungspunkt *m* nach Vicat, Formbeständigkeit *f* [in der Wärme] nach Vicat, Vicat-Zahl *f*
vicinal vizinal, vicinal, vic., benachbart, nach-

barständig, angrenzend, anliegend, anstoßend, Nachbar..., Neben..., Rand...
v. function Vizinalfunktion *f*
v. position Nachbarstellung *f*, 1,2,3-Stellung *f*
vicinity Umgebung *f*
Vickers [diamond pyramid] hardness Vickers-Härte *f*, HV
Vickers tester Vickers-Härteprüfer *m*
Vickery felt conditioner *(Pap)* Vickery-Filzinstandhalter *m*
victane Viktan *n*, Isobutylbenzol *n*
Victor Meyer method [of determining molecular weights] Bestimmung *f* der relativen Molekülmasse nach Victor Meyer
Victoria green Malachitgrün *n*, Neugrün *n*, Viktoriagrün *n*, Benzalgrün *n*, Benzoylgrün *n*, Bittermandelölgrün *n* *(ein Triphenylmethanfarbstoff); (Ker) (Glasurfarbe aus Kaliumdichromat mit wechselnden Zusätzen)*
Victoria red 1. Viktoriarot *n*, *(veraltet für)* Chromrot *n*, Chromzinnober *m*, Persischrot *n* *(basisches Bleichromat wechselnder Zusammensetzung)*; 2. *s.* vermilion
Vidal black *(Farb)* Vidal-Schwarz *n*
Vienna green Cu(CH₃COO)₂ · 3Cu(AsO₂)₂ Wiener (Schweinfurter) Grün *n*, Kupferarsenitazetat *n*
V. lime Wiener Kalk *m* *(pulverisierter Dolomit)*
view / to betrachten, beobachten
viewing aperture Schauöffnung *f*, Schauloch *n*
vigorous kräftig, stark, heftig, energisch, durchgreifend, lebhaft, stürmisch
vigoureux printing Vigoureuxdruck *m*, Kammzugdruck *m*, Melangedruck *m*
Villard effect Villard-Effekt *m*
vinaconic acid Vinakonsäure *f*, Äthylenmalonsäure *f*, 1,1-Zyklopropandikarbonsäure *f*
vinasse *(Gär)* Schlempe *f*
v. cinder *(Gär)* Schlempekohle *f*
vinegar Essig *m*
v. acid CH₃COOH Essigsäure *f*, Azetsäure *f*, Äthansäure *f*
v. factory Essigfabrik *f*
v. mother Essigmutter *f* *(gallertartig verquollene Essigbakterien)*
v. naphtha CH₃COOC₂H₅ Essigsäureäthylester *m*, Äthylazetat *n*, Essigester *m*
v. salt essigsaurer Kalk *m*, Graukalk *m*, rohes Kalziumazetat *n*
vinic acid C₂H₅OSO₃H Äthylschwefelsäure *f*
v. ether C₂H₅OC₂H₅ Äthyläther *m*, Diäthyläther *m*
vinol *s.* vinyl alcohol
vinyl Vinyl *n*
v. acetate CH₃COOCH=CH₂ Vinylazetat *n*
v. acetylene CH₂=CH·C≡CH Vinyläthin *n*, Vinylazetylen *n*
v. alcohol CH₂=CHOH Vinylalkohol *m*, Äthenol *n*
v. benzene C₆H₅CH=CH₂ Vinylbenzol *n*, Phenyläthen *n*, Styrol *n*

v. bromide CH₂=CHBr Vinylbromid *n*, Brom-äthen *n*
v. chloride CH₂=CHCl Vinylchlorid *n*, Chloräthen *n*
v. chloride dichlorethylene copolymer Vinylchlorid-Dichloräthen-Kopolymerisat *n*
v. compound Vinylverbindung *f*
v. cyanide CH₂=CHCN Vinylzyanid *n*, Akrylnitril *n*
v. ether CH₂=CHOCH=CH₂ Vinyläther *m*, Divinyläther *m*, Äthenyloxyäthen *n*
v. ethylene CH₂=CHCH=CH₂ Vinyläthen *n*, Vinyläthylen *n*, Butadien-(1,3) *n*, Divinyl *n*, Diäthen *n*
v. foam Vinylharzschaum *m*
v. formic acid CH₂=CHCOOH Vinylkarbonsäure *f*, Äthenkarbonsäure *f*, Propensäure *f*, Akrylsäure *f*
v. group —CH=CH₂ Vinylrest *m*, Vinylgruppe *f*, Äthenylgruppe *f*, Äthenylrest *m*
v. halide Vinylhalogenid *n*
v. lacquer Vinylharzlack *m*
v. plastic Vinyl[harz]kunststoff *m*, Vinoplast *m*
v. radical Vinylgruppe *f*, Vinylrest *m*, Vinylradikal *n*; [freies] Vinylradikal *n*
v. resin Vinylharz *n*
v. silicone Vinylsilikon *n*
v. sulphone dye Vinylsulfonfarbstoff *m*
v. sulphone group Vinylsulfongruppe *f*
v. trichloride Vinyltrichlorid *n*
vinylation Vinylierung *f*
vinylbenzene C₆H₅CH=CH₂ Vinylbenzol *n*, Phenyläthen *n*, Phenyläthylen *n*, Styrol *n*
vinylidene chloride CH₂=CCl₂ Vinylidenchlorid *n*, 1,1-Dichloräthen *n*
v. cyanide CH₂=C(CN)₂ Vinylidenzyanid *n*, 1,1-Dizyanäthen *n*
violanthrone Violanthron *n*, Dehydrodibenzanthron *n*
violaquercitrin Violaquerzitrin *n*, Querzetin-3-rutinosid *n*, Rutin *n*, Myrtikolorin *n*
violarite *(Min)* Violarit *m* *(ein Eisen-Nickel-Sulfid)*
violently explosive hochexplosiv
violeo salt *(Komplexchemie)* Violeosalz *n*
violet phosphorus violetter (Hittorfscher) Phosphor *m*
violuric acid Violursäure *f*, 5-Isonitrosobarbitursäure *f*, Isonitrosomalonylharnstoff *n*, Alloxanoxim-(5) *n*
viral protein Virusprotein *n*, Viruseiweiß *n*
virgin rein, unvermischt; roh, unbehandelt; *(Met)* gediegen
v. gasoline Rohbenzin *n*
v. material *(Plast)* frisch hergestelltes Material *n*
v. metal Primärmetall *n*
v. [olive] oil Jungfernöl *n* *(ohne Pressen gewonnenes Olivenöl)*
v. rubber Jungfernkautschuk *m* *(von Sapium thomsoni God.)*

v. sulphur gediegener (natürlicher) Schwefel *m*
v. zinc Rohzink *n*
virginium *(veraltet für)* francium
virial coefficient *(phys Ch)* Virialkoeffizient *m*
v. equation *(phys Ch)* Virialgleichung *f*, Virialsatz *m*
viricidal *s.* virucidal
v. agent *s.* virucide
viricide *s.* virucide
virucidal virentötend, virustötend, viruzid
virucide virentötendes (virustötendes, viruzides) Mittel (Präparat) *n*
visbreaking *s.* viscosity breaking
viscid klebrig; viskos, viskös, zäh[flüssig], dickflüssig
viscoelastic viskoelastisch
viscoelasticity Viskoelastizität *f*
viscometer *s.* viscosimeter
viscometry *s.* viscosimetry
viscose 1. Viskose...; 2. *s.* viscous
viscose Viskose *f*
v. dope Viskose[er]spinnlösung *f*
v. fibre Viskosefaser *f*; Viskosefaserstoff *m*, VI
v. film Viskosefolie *f*; Viskoseschicht *f*
v. modifier Viskosemodifier *m*, Viskosemodifizierungsmittel *n*, Viskosemodifikator *m*
v. process Viskoseverfahren *n*
v. rayon Viskoseseide *f*, Viskosereyon *n*
v. rayon process Viskoseverfahren *n*
v. ripening Viskosereifung *f*
v. [spinning] solution Viskose[er]spinnlösung *f*
v. staple fibre Viskosefaser *f*, Zellwolle *f*
viscosimeter Viskosimeter *n*
viscosimetry Viskosimetrie *f*, Viskositätsmessung *f*
viscosity Viskosität *f*, Zähigkeit *f*, Zähflüssigkeit *f*, innere Reibung *f*
v. apparatus Viskosimeter *n* mit Zubehör
v. breaking Viskositätsbrechen *n*, Viskositätsbrechung *f*, Visbreaking *n*
v. coefficient Viskositätskoeffizient *m*, Viskositätskonstante *f*, dynamische Viskosität *f*, Konstante *f* der inneren Reibung
v./density ratio Viskositäts-Dichte-Verhältnis *n*, kinematische Viskosität *f*
v. depressant Mittel *n* zur Herabsetzung der Viskosität
v. index Viskositätsindex *m*, V. I. *(Kennzahl für Schmieröle)*
v. index improver Viskositätsindexverbesserer *m*, Viskositätsindexerhöher *m*
v. modifier Viskositätsverbesserer *m*
v. number Viskositätszahl *f*
v. of non-Newton liquids Strukturviskosität *f*
v. pipette Kapillarviskosimeter *n* [nach Ostwald]
v. stability Viskositätsstabilität *f*, Viskositätsbeständigkeit *f*, Viskositätskonstanz *f*, Zähigkeitsstabilität *f*
v. stabilizer Viskositätsstabilisator *m*
v.-temperature coefficient Viskositäts-Tempera-

tur-Koeffizient *m*, Temperaturkoeffizient *m* der Viskosität (Zähigkeit), VTC
v.-temperature curve (slope) Viskositäts-Temperatur-Kurve *f*, VT-Kurve *f*
viscous viskos, viskös, zäh[flüssig], dickflüssig
v. elasticity Viskoelastizität *f*
v. filter mit viskoser Flüssigkeit benetztes *(meist* ölbenetztes) Filter *n*
v. flow viskoses Fließen *n*
v. force Zähigkeitskraft *f*, Reibungskraft *f* *(innere Reibung)*
viscousness *(veraltet für)* viscosity
visual colorimetry visuelle Kolorimetrie *f*
v. comparator *(Kolorimetrie)* visueller Komparator *m*
v. pigment Sehstoff *m*, *(i.e.S.) s.* v. purple
v. purple Sehpurpur *m*, Rhodopsin *n*
vital stain *(Bio)* Vitalfarbstoff *m*
v. staining *(Bio)* Vitalfärbung *f*
vitamin[e] Vitamin *n*
v. chemistry Vitaminchemie *f*
v. concentrate Vitaminkonzentrat *n*
v. content Vitamingehalt *m*
v. deficiency Vitaminmangel *m*; Vitaminmangelkrankheit *f*, Avitaminose *f*
v.-free vitaminfrei
v. potency Vitaminwirksamkeit *f*
vitamined vitamin[is]iert, mit Vitaminen angereichert; vitaminreich
vitaminization Vitamin[is]ieren *n*, Vitamin[is]ierung *f*
vitaminize / to vitamin[is]ieren, mit Vitaminen anreichern
vitellus Dotter *n*, Eidotter *n*, Eigelb *n*, Vitellus *m*
vitrain *(Kohlechemie)* Vitrit *m*, Vitrain *m*
v. lense Vitritlinse *f*
vitrainlike vitritähnlich
vitreous gläsern; glasartig, glasähnlich, glasig
v. china *(Ker)* Vitreous [China] *n*, Halbporzellan *n*, Porzellangut *n*
v. copper *(Min)* α-Chalkosin *m*, α-Kupferglanz *m* *(Kupfer(I)-sulfid)*
v. enamel Email *n*, Emaille *f*
v. fracture glasiger Bruch *m*
v. lustre Glasglanz *m*
v. malt Glasmalz *n*
v. phase glasige Phase *f*, Glasphase *f*
v. silica durchsichtiges (klares) Kieselglas *n*, Quarzglas *n*
v. silver Ag_2S Silbersulfid *n*; *(Min)* Silberglanz *m*, *(i.e.S.)* Argentit *m* *(kubischhexoktaedrisches Silbersulfid)*
v. slip *(Ker)* glasartiger Schlicker *m*
v. state glasiger (glasartiger) Zustand *m*, Glaszustand *m*, Glasphase *f*
vitric tuff Glastuff *m*
vitrifiable colour Emailfarbe *f*, Schmelzfarbe *f*
vitrification *(Ker)* Verglasung *f*; *(Geol)* Frittung *f*, Verglasung *f*, oberflächliche Anschmelzung *f*; *(Gum)* Glasumwandlung *f*, Umwandlung *f* zweiter Ordnung, Gamma-Umwandlung *f*, α-Anomalie *f*

v. range *(Ker)* Verglasungsbereich *m*
vitrified *(Ker)* verglast, glasartig
vitrify / to *(Ker)* verglasen, dicht werden
vitrinertite *(Kohlechemie)* Vitrinertit *m*
vitrinite *(Kohlechemie)* Vitrinit *m*
vitrinitic vitrinitisch, Vitrinit...
vitriol / to mit Schwefelsäure behandeln; *(Met)* mit Schwefelsäure beizen
vitriol Vitriol *n* *(kristallwasserhaltiges Sulfat eines zweiwertigen Metalls)*; s. v. oil
  v. oil Vitriolöl *n* *(i.w.S.)*, *(veraltet für)* Schwefelsäure *f*
vitriolated tartar K_2SO_4 Kaliumsulfat *n*
vitriolic acid *(veraltet für)* sulphuric acid, *(i.e.S.)* s. vitriol oil
  v. acid air *(historisch für)* sulphur dioxide
vitriolized bones *(mit Schwefelsäure behandeltes Knochenmehl; Düngemittel)*
vitroceramic, vitrokeram Vitrokeram *n*, Glaskeramik *f*, Keramik *f* aus Glas, glaskeramischer Stoff *m*
vitrophyric *(Geol)* vitro[por]phyrisch, glasig
vivianite *(Min)* Vivianit *m* *(Eisen(II)-orthophosphat)*
vivid lebhaft
v.m., VM s. volatile matter
V.M.P. naphtha s. varnish-makers' naphtha
vogesite *(Min)* Vogesit *m*, *(veraltet für)* Pyrop *m* *(Magnesiumaluminiumorthosilikat)*
void Hohlraum *m*, Leerstelle *f*; Pore *f*; *(Plast)* Hohlraum *m*, Blase *f (Fehler)*
  v. space Hohlraum *m*
  v. volume Volumen *n* der Zwischenräume, *(bei körnigem Gut auch)* Zwischenkornvolumen *n*
voidage relatives Porenvolumen *n*, relativer Porenraum *m*, Porosität *f*
Voith grinder *(Pap)* Stetigschleifer *m* Bauart Voith
volat. 1. s. volatile; 2. *(Abk. für)* volatilizes *(s. to volatilize)*
volatile flüchtig, leichtflüchtig, volatil, ätherisch *(z.B. Öl)*
  v. constituent flüchtige Komponente *f*, flüchtiger (leichtflüchtiger) Bestandteil *m*
  v. content s. v. matter content
  v. loss Verdampfungsverlust *m*
  v. matter flüchtige Substanz *f*, flüchtiger Stoff *m*, flüchtige Bestandteile *mpl*, Flüchtiges *n*
  v. matter content Gehalt *m* an flüchtigen Substanzen (Stoffen, Bestandteilen), Gehalt *m* an Flüchtigem
  v. oil ätherisches Öl *n*
volatileness s. volatility
volatiles s. volatile matter
volatility Flüchtigkeit *f*, Fugazität *f*
volatilizable leicht zu verflüchtigen[d], leicht verdampfbar
volatilization *(als Tätigkeit)* Verdampfung *f*, Verflüchtigung *f*, Vergasung *f* *(z.B. von Pflanzenschutzmitteln)*; *(als Vorgang)* Verflüchtigung *f*, Verdampfung *f*, *(unterhalb des normalen Siedepunkts)* Verdunstung *f*

volatilize / to verdampfen, verflüchtigen, *(z.B. Pflanzenschutzmittel)* vergasen; sich verflüchtigen, verdampfen, vergasen *(z.B. Pflanzenschutzmittel)*, *(unterhalb des normalen Siedepunkts)* verdunsten
volatize / to s. to volatilize
volborthite *(Min)* Volborthit *m* *(Kupfer(II)-vanadat)*
volcanic vulkanisch, Vulkan...
  v. ash vulkanische Asche *f*
  v. gases vulkanische Gase *npl*
  v. glass vulkanisches Glas *n*, Vulkanglas *n*
  v. rock vulkanisches Gestein *n*, Vulkanit *m*
  v. tuff vulkanischer Tuff *m*
Volhard method *(Volumetrie)* Endpunkt[s]bestimmung *f* nach Volhard
volt. *(Abk. für)* 1. volatile; 2. volatilizes
voltage *(Phys, ET)* Spannung *f*
  v. drop Spannungsabfall *m*
  v.-scan voltammetry spannungsgeregelte Voltametrie *f*
voltaic battery galvanische Batterie *f*
  v. cell (element) galvanisches (elektrochemisches) Element *n*, galvanische (elektrochemische) Zelle (Kette) *f*
voltaite Voltait *m* *(eisen(II)- und eisen(III)-haltiges Mineral)*
voltameter Voltameter *n*, Coulometer *n*
voltammeter Stromspannungsmesser *m*, [Universal-]Strom- und Spannungsmesser *m*
voltzine, voltzite Voltzin *m* *(sulfidisches arsenhaltiges Zinkmineral)*
volucrisporin Volucrisporin *n*, 2,5-Di-[m-hydroxyphenyl]1,4-benzochinon *n*
volume Volumen *n*, Rauminhalt *m*, Inhalt *m*, Raumteil *m*
  v. change Volumenänderung *f*
  v. concentration Volum[en]konzentration *f*
  v. contraction Volum[en]kontraktion *f*
  v. element Volum[en]element *n*, Raumelement *n*
  v. lifetime Volumenlebensdauer *f*, Lebensdauer *f* im Innern *(eines Halbleiterkristalls)*
  v. of filtrate Filtratvolumen *n*
  v. of pore space *(Bodenkunde)* Porenvolumen *n*, Hohlraumvolumen *n*
  v. of the molecules Molekülvolumen *n*, Eigenvolumen *n* der Moleküle (Molekeln)
  v. percentage Volum[en]prozent *n*, Vol. %, Raumhundertstel *n*
  v. resistance *(ET)* Durchgangswiderstand *m*
  v.-resistivity *(ET)* spezifischer Durchgangswiderstand *m*
  v. shrinkage *(Ker)* Volum[en]schwindung *f*, kubische (räumliche) Schwindung *f*
  v. stability Volumenbeständigkeit *f*, Raumbeständigkeit *f*
  v.-stable volumenbeständig, raumbeständig
  v. susceptibility *(Magnetochemie)* Volumensuszeptibilität *f*
volumenometer Volum[en]ometer *n*, Stereometer *n*

44*

volumetric volumetrisch, *(Analytik auch)* titrimetrisch, maßanalytisch, Titrations...
v. analysis *s.* volumetry
v. coulometer Volumencoulometer *n*
v. efficiency volumetrischer Wirkungsgrad *m* *(von Pumpen)*
v. factor *(Maßanalyse)* Faktor *m*
v. feeder (feeding device) *(Plast)* Volumendosiervorrichtung *f*
v. flask Meßkolben *m*, Maßkolben *m*, Meßflasche *f*
v. precipitation analysis Fällungs[maß]analyse *f*, Fällungstitration *f*
v. rate of flow Volumenstrom *m*
volumetry Volumetrie *f*, volumetrische Analyse *f*, Titrimetrie *f*, Maßanalyse *f*, Titrieranalyse *f*
volunteer *(Toxikologie)* freiwillige Versuchsperson *f*
volute casing Spiralgehäuse *n* *(bei Pumpen)*
v. pump Spiralgehäusepumpe *f*
Von Braun degradation Braunscher (von-Braunscher) Abbau *m* *(tertiärer Amine mit Bromzyanid)*
Vorce cell *(Dialyse)* Vorce-Zelle *f*
vortex Wirbel *m*, Strudel *m*, Spirale *f*, Vortex *m*
v. burner Wirbel[strom]brenner *m*
v. dryer Wirbeltrockner *m*
v. finder Wirbelsucher *m*
v. street Wirbelstraße *f*
vortexing Verwirbeln *n*, Verwirbelung *f*
vortrap Vortrap *m* *(ein Wirbelsichter)*
votator [chilling unit] Votator *m*, Votatoranlage *f* *(Margarineherstellung)*
v. margarine Votatormargarine *f*
v. [margarine] plant *s.* votator [chilling unit]
v. process Votatorverfahren *n* *(kontinuierliche Margarineherstellung)*
Votator crystallizer *(ein Ringkammerkühlungskristallisator)*
V.P.I. *s.* vapour-phase inhibitor
v.s. *s.* very soluble
V.S.P. *s.* Vicat softening point
v.s.s. *s.* very slightly soluble
vug[g], vugh *(Min)* Druse *f*
vulcanic *s.* volcanic
vulcanise / to *s.* to vulcanize
vulcanite Hartgummi *m*, Hartkautschuk *m*, Ebonit *m*
vulcanizable *(Gum)* vulkanisierbar, vernetzungsfähig, vernetzbar
vulcanizate *(Gum)* Vulkanisat *n*
v. properties Vulkanisateigenschaften *fpl*
vulcanization *(Gum)* Vulkanisation *f*, Vulkanisierung *f*, Vernetzung *f*, Heizung *f*
v. accelerator Vulkanisationsbeschleuniger *m*
v. by high-energy radiation Vulkanisation (Vernetzung) *f* durch energiereiche Strahlen (Strahlung), Strahlenvernetzung *f*
v. coefficient Vulkanisationskoeffizient *m*, Vulkanisationsgrad *m*, VK

v. curve Vulkanisationskurve *f*
v. in stages Stufenheizung *f*
v. process Vulkanisationsverfahren *n*; Vulkanisationsvorgang *m*
v. rate Vulkanisationsgeschwindigkeit *f*, Heizgeschwindigkeit *f*
v. reaction Vulkanisationsreaktion *f*
v. system Vulkanisationssystem *n*, Vernetzungssystem *n*, Vernetzersystem *n*, Vernetzerkombination *f*
v. temperature Vulkanisationstemperatur *f*, Heiztemperatur *f*
v. time Vulkanisationszeit *f*, Heizzeit *f*, Gesamtheizzeit *f*, *(i.e.S.)* Ausvulkanisationszeit *f*
vulcanize / to *(Gum)* vulkanisieren, vernetzen, heizen
v. at room temperature *(Gum)* kaltvulkanisieren, in der Kälte vulkanisieren, bei Raumtemperatur vulkanisieren
vulcanized fibre Vulkanfiber *f*, Fiber *f*
v. product vulkanisiertes Produkt *n*, Vulkanisat *n*
v. rubber vulkanisierter Kautschuk *m*, Kautschukvulkanisat *n*, Gummi *m*
v. rubber scrap (waste) Gummiabfälle *mpl*, Vulkanisatabfälle *mpl*, Abfallgummi *m*, Altgummi *m*
v. scrap (waste) rubber *s.* v. rubber scrap
vulcanizer *(Gum)* Vulkanisierapparat *m*, Vulkanisiergerät *n*, Vulkanisationsapparat *m*, Heizapparat *m*, Heizer *m*, Vulkanisator *m*
vulcanizing *s.* vulcanization
v. agent Vulkanisationsmittel *n*, Vulkanisationsagens *n*, Vulkanisiermittel *n*, Vernetzungsmittel *n*, Vernetzer *m*
v. apparatus *s.* vulcanizer
v. autoclave Vulkanisierkessel *m*, Vulkanisationskessel *m*; Autoklav *m*, Autoklav[en]presse *f*, Autoklavheizpresse *f*
v. boiler Vulkanisierkessel *m*, Vulkanisationskessel *m*
v. drum Vulkanisiertrommel *f*
v. ingredients *s.* vulcanization system; Vulkanisationsmittel *npl*, Vulkanisiermittel *npl*, Vulkanisationsagenzien *npl*
v. mould Vulkanisierform *f*
v. oven Vulkanisierofen *m*, Vulkanisationsofen *m*
v. pan *s.* v. boiler
v. press Vulkanisierpresse *f*
v. system *s.* vulcanization system
v. temperature *s.* vulcanization temperature
v. time *s.* vulcanization time
vulpinite *(Min)* Vulpinit *m* *(Kalziumsulfat)*
vultex *(Gum)* Vultex *m* *(vulkanisierter Latex)*

W

Wackenroder's solution Wackenrodersche Flüssigkeit *f* *(Gemisch von Polythionsäuren)*

wad 1. *(Min)* Wad *n(m)*, Manganschaum *m* *(Mangan(IV)-oxid)*; 2. Graphit *m*
w. of cotton Wattebausch *m*
wadding Zellstoffwatte *f*
Wagner-Meerwein rearrangement Wagner-Meerwein-Umlagerung *f*
wagnerite *(Min)* Wagnerit *m* *(fluoridhaltiges Magnesiumphosphat)*
wagon balance Gleiswaage *f*, Waggonwaage *f*
waist Einbuchtung *f*; Einschnürung *f*
Walden inversion Waldensche Umkehrung (Inversion) *f*
Walden law (rule) Waldensche Regel *f*
walk-in dryer begehbarer Trockner *m*
walkway Laufsteg *m*
wall Wand[ung] *f*, Mantel *m*
  w. board Wandpappe *f*
  w. catalysis Wandkatalyse *f*
  w. friction Wandreibung *f*
  w.-mounting draining board *(aufhängbares)* Abtropfbrett *n*, Ablaufbrett *n*, Trockenbrett *n*
  w. paper Tapetenpapier *n*
  w. reaction Wandreaktion *f*
  w. rock Nebengestein *n*, Seitengestein *n*, Nachbargestein *n*
  w. thickness Wanddicke *f*
  w. tile Wandfliese *f*
Wallace pocket meter *(Gum)* Wallace-Härtemesser *m*
Wallner lines *(Glas)* Wallner-Linien *fpl*
wallpaper *s.* wall paper
walnut oil Walnußöl *n* *(von Juglans regia L.)*
Wandel type strainer *(Pap)* Knotenfänger *m* von Wandel
wander / to wandern
wandoo [extract] *(Gerb)* *(Extrakt aus Holz und Rinde von Eucalyptus wandoo Blakely)*
war bronze *(Zinklegierung mit etwa 5 % Cu, 2 % Al und geringen Beimengungen von Pb und Sn)*
  w. gas gasförmiger Kampfstoff *m*, Kampfgas *n*
Warburg's enzyme Warburgsches gelbes Atmungsferment (Oxydationsferment) *n*
Ward bagasse furnace Ward-Feuerung *f*
warm / to [er]wärmen; vorwärmen; [be]heizen
  w. gently mäßig erwärmen
  w. in *(Glas)* vorwärmen, aufwärmen
  w. up *s.* to warm
warm ween
  w. bleach[ing] *(Pap)* Warmbleiche *f*, warme Bleiche *f*
  w.-dyeing colour *(Text)* Heißfärber *m*
  w. recovered make-up *(Erdöl)* warmes Umlauföl *n* *(beim Entwachsungsprozeß der Sharples Corp.)*
  w. spray Warmspritzen *n* *(in der Farbspritztechnik)*
  w.-tone developer *(Foto)* Warmtonentwickler *m*
  w.-up mill *s.* warming mill
warming Erwärmen *n*, Erwärmung *f*, Wärmen *n*; Vorwärmen *n*; Heizen *n*, Beheizen *n*, Beheizung *f*

w.-in *(Glas)* Anwärmen *n*, Aufwärmen *n*, Vorwärmen *n*
w. mill *(Gum)* Vorwärmwalzwerk *n*
w.-up *s.* warming
w.-up mill *s.* w. mill
w.-up period Aufheizperiode *f*, Aufheizabschnitt *m*
warmth Wärme *f*, Hitze *f* *(Empfindung)*
warning agent Warnstoff *m*
warp / to sich verziehen (krümmen, werfen), *(Ker auch)* aufwerfen
warp *(Text)* Kette *f*, Webkette *f*
  w. size *(Text)* Kettschlichte *f*
  w. sizing *(Text)* Kettschlichten *n*, Schlichten *n* der Kettfäden
  w. thread *(Text)* Kettfaden *m*
warpage, warping Verziehen *n*, Verzug *m*, Krümmen *n*, Werfen *n*, *(Ker auch)* Aufwerfen *n*
warringtonite *(Min)* Warringtonit *m*, *(veraltet für)* Brochantit *m* *(Kupfer(II)-hexahydroxidsulfat)*
wash / to [durch]waschen, auswaschen, spülen, reinigen; *(Foto)* wässern; *(Ker)* schlämmen; *(Aufbereitung)* läutern; *(Met)* plattieren; überziehen, überstreichen
  w. away wegwaschen
  w. countercurrently im Gegenstrom waschen
  w. down herabspülen
  w. out auswaschen
  w. thoroughly gründlich auswaschen, durchwaschen
  w. through durchspülen, durchwaschen
  w. together gemeinsam waschen
wash 1. Waschen *n*, Waschung *f*, Wäsche *f*, Durchwaschen *n*, Auswaschen *n*, Spülen *n*, Spülung *f*, Reinigung *f*; *(Foto)* Wässern *n*, Wässerung *f*; 2. *s. w.* liquid; *s. w.* water; *(Kosmet)* Wasser *n*, Lotion *f*; *(Landw)* Brühe *f*, Spritzbrühe *f* *(zur Schädlingsbekämpfung)*; Gärlösung *f*, Gärflüssigkeit *f*, Gärgut *n*, Maische *f*; 3. *(Met)* Plattierung *f*; *(flüssig aufgetragener)* Überzug *m*; 4. Goldsand *m*
  w. and wear product *(Text)* pflegeleichtes Erzeugnis *n*, Wasch-und-Trage-Erzeugnis *n*, Wash-and-wear-Erzeugnis *n*
  w. bottle *s.* washing bottle
  w. box *(Aufbereitungstechnik)* Setzkasten *m*, Setzmaschine *f*
  w. column Waschkolonne *f*
  w. drum Waschtrommel *f*
  w.-fast waschbar; *(Farb)* waschecht
  w. fastness Waschbarkeit *f*; *(Farb)* Waschechtheit *f*
  w. inlet Waschflüssigkeitszulauf *m*
  w. liquid (liquor) Waschflüssigkeit *f*, Waschlauge *f*, Reinigungsflüssigkeit *f*, *(Text auch)* Waschflotte *f*; *s. w.* water
  w. mill *s.* washing mill
  w. nozzle Waschflüssigkeitsdüse *f*
  w. oil Waschöl *n*
  w. primer *(Farb)* Haftgrundmittel *n*, Haft-

grundierung f, Wash-Primer m, Reaktions-Primer m, Reaktionsgrundierung f

w. solvent s. w. liquid

w. tank (Pap) Stoffgrube f, Stoffkasten m, Kochergrube f, Kocherbütte f, Blastank m, Ausblasbehälter m

w. test Waschprobe f

w. tub Waschbottich m, Waschkufe f, Waschbütte f, Waschwanne f

w.-up (Pap) Wäsche f

w. water Waschwasser n; Spülwasser n

washability Waschbarkeit f; Wasserlöslichkeit f

w. curve (Aufbereitungstechnik) Waschkurve f

washable waschbar, waschfest; wasserlöslich

washboard (Glas) Waschbrett n, Falten fpl (Oberflächenfehler)

washed kaolin geschlämmtes Kaolin n, Schlämmkaolin n

w. raw sugar affinierter Zucker m, Affinade f (im Affinierverfahren gewaschener Rohzucker)

washer 1. s. washing machine; (Pap) Waschholländer m; (Pap) Halbzeugholländer m; 2. Scheibe f, Ring m

w. beater (Pap) Waschholländer m

w.-cooler Waschkühler m, Kühl[er]wascher m

washery water Waschwasser n

washing Waschen n, Waschung f, Wäsche f, Durchwaschen n, Auswaschen n, Spülen n, Spülung f, Reinigung f; (Foto) Wässern n, Wässerung f; (Ker) Schlämmen n

w. agent Waschmittel n, (Text auch) Waschrohstoff m

w. aid Waschhilfsmittel n

w. bath (Text) Waschflotte f, Waschlauge f

w. blanket (Filtration) Waschband n

w. bottle 1. Waschflasche f, Gaswaschflasche f; 2. Spritzflasche f

w. cylinder Waschtrommel f

w. device Waschapparat m

w. drum s. w. cylinder

w. efficiency Waschkraft f, Waschwirkung f, Wascheffekt m, Reinigungswirkung f

w. engine (Pap) Waschholländer m

w. fastness s. wash fastness

w. liquid (liquor) s. wash liquid

w. machine Waschmaschine f, Waschapparat m, Wäscher m; (Gum) s. w. mill

w. mill (Gum) Waschwalzwerk n

w. of pulp Zellstoffwäsche f

w.-off Auswaschen n, Spülen n

w. oil Waschöl n

w.-out loss Auswaschverlust m

w. period Waschdauer f; (Foto) Wässerungszeit f

w. plate (Filtration) Waschplatte f

w. potcher (Pap) Waschholländer m

w. powder Waschpulver n

w. product s. w. agent

w. rate Strömungsgeschwindigkeit f der Waschflüssigkeit; (Foto) Schnelligkeit f der Auswässerung; (Foto) Auswässerungsgrad m

w. roll s. w. cylinder

w. soda $Na_2CO_3 \cdot 10H_2O$ Kristallsoda f, Waschsoda f, Soda f, Natriumkarbonat-10-Wasser n

w. solution Waschmittellösung f

w. spray Waschbrause f

w. tank Waschbottich m, Waschkasten m; (Foto) Wässerungstank m

w. thickener Wascheindicker m

w. time s. w. period

w. tower Waschturm m

w. tray thickener s. w. thickener

w. treatment Waschbehandlung f

w. water Waschwasser n; Spülwasser n

w. with alkali (Pap) Alkalibehandlung f, Alkaliextraktion f, alkalische Wäsche f, Alkaliwäsche f (Bleichstufe)

washings Waschwasser n; Waschflüssigkeit f

washout 1. Washout n (Auswaschung radioaktiver Schwebstoffe durch atmosphärische Niederschläge); 2. Washout m (Produkt von 1.)

washup (Pap) Wäsche f

wastage Verlust m; Verschleiß m, Abnutzung f; Schwinden n, Schwund m

waste Abfall m, Abfälle mpl, Abfallsubstanz f, Abgang m; Ausschuß m; Müll m; (Pap) Grobstoff m, Spuckstoff m, „Sauerkraut" n

w. acid Abfallsäure f

w.-acid plant (Farb) Abfallsäurebetrieb m

w.-acid-storage tank Lagertank (Vorratstank) m für Abfallsäure

w. crock (irdener) Abfallkübel m

w. disposal Abfallbeseitigung f, Müllbeseitigung f; Abwasserbeseitigung f

w. film (Foto) Filmabfall m

w. gas Abgas n, Rauchgas n

w.-gas flue Abgas[sammel]kanal m, Rauchgassammelkanal m

w. gas from furnace operation Verbrennungs[ab]gas n, Ofengas n

w.-gas side Abgasseite f

w. heat Ab[fall]wärme f, Abhitze f

w.-heat boiler Abhitze[dampf]kessel m

w.-heat oven Abhitze[koks]ofen m

w.-heat recovery Abwärmerückgewinnung f, Abhitzerückgewinnung f

w. jar Abfallkübel m

w. lime Abfallkalk m

w. line Abflußleitung f, Abzugsleitung f, Ablaßleitung f; Abwasserleitung f, Ausgußleitung f

w. liquor Ablauge f

w. liquor recovery (Pap) Ablaugengewinnung f, Ablaugenregeneration f, Laugenregeneration f

w. material s. waste; Altmaterial n, Altstoff m, Altwaren fpl; Versatzmaterial n, Versatzgut n, Versatz m

w. paper Altpapier n, Abfallpapier n, Papierabfälle mpl, Makulatur f, Ausschußpapier n, Ausschuß m

w. paper colouring Altpapierstoffärbung f

w. paper mill (vorwiegend Altpapier verarbeitende Papierfabrik, Schrenzpapierfabrik)

w. product Abfallprodukt *n*, Abfallgut *n*; *s.* waste
w. rock taubes Gestein *n*, Gangart *f*
w. rubber Abfallgummi *m*, Altgummi *m*, Gummiabfälle *mpl*, Vulkanisatabfälle *mpl*; *(Gum)* *(unvulkanisierte)* Fabrikationsabfälle *mpl*
w. soda lye *(Pap)* Schwarzlauge *f*
w. steam Abdampf *m*
w. stuff Bruchpapier *n*, Ausschußpapier *n*, Ausschuß *m*, Papierausschuß *m*, Kollerstoff *m*
w. tailing *(Aufbereitungstechnik)* Abgänge *mpl*, Berge *pl*
w. water Abwasser *n*
w. wood Abfallholz *n*
wasteful unvorteilhaft; ohne Nebenprodukt[en]gewinnung arbeitend, ohne (unter Verzicht auf) Nebenproduktgewinnung
watch / **to** beobachten, überwachen, beaufsichtigen
watch-case curing press *(Gum)* Einzelheizer *m*
w.-case vulcanizer *s.* w.-case curing press
w. glass Uhrglas *n*, Uhrglasschale *f*
w. oil Uhrenöl *n*, Uhrmacheröl *n*
water / **to** befeuchten, bewässern, [be]sprengen, [be]gießen; tränken; mit Wasser versorgen; *s.* to w. down
w. down *(z. B. Wein)* verwässern, mit Wasser verdünnen (strecken)
water Wasser *n*, *(i.e.S.)* H_2O [chemisch reines] Wasser *n*, Wasserstoffoxid *n*; *(Med)* Harn *m*, Urin *m*, Wasser *n*; Speichel *m*; *(Lebm)* Pflanzensaft *m*; Fruchtsaft *m*; *(Min)* Wasser *n* *(Vollkommenheit der Farblosigkeit bei Edelsteinen)*; *s.* waters
w.-absorbing wasseraufnehmend, wasseranziehend, wasserabsorbierend, hygroskopisch
w. absorption Wasserabsorption *f*, Wasseraufnahme *f*
w. addition Wasserzusatz *m*
w. analysis Wasseranalyse *f*, Wasseruntersuchung *f*
w. aspirator *s.* w.-jet pump
w. association Wasseranlagerung *f*
w.-attracting *s.* w.-absorbing
w. balance Wasserbilanz *f*
w.-base mud *(Erdöl)* Wasserspülung *f*
w.-base paint *s.* w. paint
w. bath Wasserbad *n*
w. bath rack Reagenzglasgestell *n* für Wasserbäder
w.-bearing wasserführend
w.-boiler reactor Wasser[kocher]reaktor *m*, Wasserkesselreaktor *m*
w. calorimeter Wasserkalorimeter *n*, Flüssigkeitskalorimeter *n*, Mischungskalorimeter *n*
w.-carriage system Entwässerungssystem *n*, Schmutzwasserkanalisation *f*
w. cell Diffusatzelle *f* *(bei der Dialyse gegen Wasser)*
w.-cement ratio Wasserzementfaktor *m*, Wasserzementwert *m*, WZ-Faktor *m*
w. channel Wasserkanal *m*

w. classification hydraulische Klassierung *f*
w. classifier hydraulischer Klassierer *m*
w. colour Aquarellfarbe *f*, Wasserfarbe *f*
w.-colour [drawing] paper Aquarellpapier *n*
w. conditioning Wasseraufbereitung *f*
w. consumption Wasserverbrauch *m*
w. contamination Wasserverunreinigung *f*, Wasserverschmutzung *f*
w. content Wassergehalt *m*
w.-cool / **to** mit Wasser kühlen
w.-cooled wassergekühlt
w.-cooled steel jacket wassergekühlter Stahlblechmantel *m*, Wassermantel *m* *(eines Schachtofens)*
w. cooling Wasserkühlung *f*, Kühlung *f* mit Wasser
w.-cross-linked *(Gum)* wasservernetzt
w. culture *(Bio)* Wasserkultur *f*
w. curtain Wasserschleier *m*, Wasservorhang *m*
w. demineralizer *s.* w. softener
w.-dilutable wasserverdünnbar, mit Wasser verdünnbar
w. distribution system Wasserverteilung[sanlage] *f*, Wasserverteilungssystem *n*
w. drive *(Erdöl)* Wassertrieb *m*, Wasserdruck *m*, Water-drive *m*
w.-drive field (reservoir) Wassertrieblagerstätte *f*, Lagerstätte *f* *(Erdölfeld n)* mit Wassertrieb
w. drop test Wassertropfenprobe *f*
w. droplet Wassertröpfchen *n*
w. ejector *s.* w.-operated ejector
w. equivalent Wasserwert *m* *(z.B. eines Kalorimeters)*
w. extract wäßriger Auszug *m*
w. flooding Wasserfluten *n*, Fluten *n*, Einpressen *n* von Wasser *(Sekundärverfahren zur Erdölförderung)*
w. flow Wasserstrom *m*
w. for industrial use Brauchwasser *n*, Betriebswasser *n*, Gebrauchswasser *n*, Fabrikationswasser *n*, Nutzwasser *n*
w. from the showers *(Pap)* Spritzwasser *n*
w. from the squirt trim *(Pap)* Siebrandspritzwasser *n*
w. from the tray *(Pap)* Sieb[ab]wasser *n*, Abwasser *n* der Registerpartie
w. gas Wassergas *n*
w.-gas coke Koks *m* zur Wassergaserzeugung
w.-gas equilibrium Wassergasgleichgewicht *n*
w.-gas generator Wassergasgenerator *m*, Wassergaserzeuger *m*
w.-gas machine Wassergasmaschine *f* *(Anlage zur diskontinuierlichen Wassergaserzeugung)*
w.-gas plant Wassergas[erzeugungs]anlage *f*
w.-gas process Wassergasprozeß *m*, Wassergasverfahren *n*
w.-gas reaction Wassergasreaktion *f* *(Reduktion von Wasser mittels Kohlenstoffs)*
w.-gas tar Wassergasteer *m*
w. glass Wasserglas *n*

w. glass cement Wasserglaskitt *m*
w. glass glue Wasserglasleim *m*
w. glass solution Wasserglaslösung *f*
w. hammer Druckstoß *m*, Wasserschlag *m*
w.-harden / to wasserhärten, in Wasser härten
w. hardening Wasserhärten *n*, Wasserhärtung *f*, Härten *n* in Wasser
w.-hardening steel wasserhärtender Stahl *m*, Wasserhärtestahl *m*, Wasserhärter *m*
w. hardness Wasserhärte *f*
w.-holding capacity *(Bodenkunde)* Wasserkapazität *f*, Wasserfassungsvermögen *n*, Wasserhaltevermögen *n*, Wasserhaltungsvermögen *n*
w. ice Wassereis *n*
w. imbibition Einlegen *n* in Wasser; Quellung *f* in Wasser; Wasseraufnahmewert *m*, Quellwert *m*
w. impurity Wasserverunreinigung *f*, Verunreinigung *f* (Schmutzstoff *m*) im Wasser
w.-in-fat emulsion Wasser-in-Fett-Emulsion *f*
w.-in-oil emulsion Wasser-in-Öl-Emulsion *f*
w.-in-oil type Wasser-in-Öl-Typ *m* *(von Emulsionen)*
w.-insoluble nichtwasserlöslich, wasserunlöslich, in Wasser unlöslich, nicht in Wasser löslich
w. jacket Wassermantel *m*
w.-jacketed wassergekühlt, mit Wassermantel, Wassermantel...
w. jet Wasserstrahl *m*
w.-jet aspirator (injector) *s.* w.-jet pump
w.-jet loom Wasserstrahl-Düsenwebmaschine *f*, hydraulische Düsenwebmaschine *f*
w.-jet [vacuum] pump Wasserstrahlpumpe *f*
w. leacheate wäßriger Auszug *m*
w. level Wasserspiegel *m*, Wasserstand *m*
w. level regulator Wasserstandsregler *m*
w. lime Wasserkalk *m*, hydraulischer Kalk *m* *(unter Wasser erhärtender Kalk)*; *s.* weakly hydraulic lime
w.-lines *(Pap)* Wasser[zeichen]linien *fpl*
w.-loving wasserliebend, wasserfreundlich, hydrophil
w. mangle *(Text)* Wasserkalander *m*
w. margarine Wassermargarine *f*
w. metabolism *(Bio)* Wasserstoffwechsel *m*, Wasserhaushalt *m*
w. meter Wasserzähler *m*, Wassermesser *m*, Wasseruhr *f*
w.-moderated *(Kch)* wassermoderiert, wassergebremst
w.-moderated reactor wassermoderierter Reaktor *m*, Wasserreaktor *m*
w. molecule Wassermolekül *n*
w. of capillarity Kapillarwasser *n*, Poren[saug]wasser *n*
w. of constitution Konstitutionswasser *n*, konstitutiv gebundenes Wasser *n*
w. of crystallization Kristallwasser *n*
w. of deliquescence von hygroskopischen Substanzen *(z.B. aus der Luft)* aufgenommenes Wasser *n*, *(i.e.S.)* hygroskopisches Wasser *n*

w. of hydration Hydratwasser *n*
w. of plasticity *(Ker)* Anmachwasser *n*, Plastizitätswasser *n*
w.-operated ejector Wasserstrahlpumpe *f*; Wasserstrahlgebläse *n*, Wasserstrahlverdichter *m*
w. paint Wasserfarbe *f*, wäßriges Anstrichmittel *n*
w. phase Wasserphase *f*, wäßrige Phase *f*
w. pollution *s.* w. contamination
w. proofing *s.* waterproofing
w. purification Wasserreinigung *f*
w.-purification process Wasserreinigungsverfahren *n*
w.-quench / to *(z.B. Koks)* durch (mit) Wasser ablöschen; *(Met)* in Wasser abschrecken (ablöschen)
w. quenching *(Met)* Wasserabschrecken *n*, Wasserabschreckung *f*, Wasserablöschung *f*, Abschrecken *n* in Wasser
w. removal Entwässerung *f*
w. repellency Wasserabweisungsvermögen *n*, wasserabweisende Kraft (Eigenschaft) *f*, wasserabweisendes Verhalten *n*, wasserabweisender Charakter *m*, Wasserabweisung *f*, Wasserabstoßung *f*, Abperleffekt *m*
w.-repellent wasserabweisend, wasserabstoßend, hydrophob
w. repellent Hydrophobier[ungs]mittel *n*, wasserabweisendes (wasserabstoßendes, hydrophobierendes) Mittel *n*
w.-repellent cement wasserabstoßender Zement *m*
w.-repellent finish wasserabweisende (wasserabstoßende) Imprägnierung (Appretur) *f*, hydrophobe Ausrüstung *f*
w. requirements Wasserbedarf *m*
w. resistance Wasserbeständigkeit *f*, Wasserfestigkeit *f*
w.-resistant wasserbeständig, wasserfest
w. ret[ting] *(Text)* Wasserröste *f*, Wasserrotte *f*
w.-saturated wassergesättigt
w. seal Wasserverschluß *m*, Wasserabschluß *m*, hydraulischer Verschluß *m*, Abschlußwasserschüssel *f*
w.-sealed mit Wasserverschluß
w.-sealed trough Wasserschüssel *f*
w.-seeking *s.* w.-absorbing
w.-slaked lime Löschkalk *m*, gelöschter Kalk *m*
w.-smoking *(Ker)* Schmauchen *n*
w.-smoking period *(Ker)* Schmauchperiode *f*
w. softener Wasserenthärter *m*, Wasserenthärtungsmittel *n*
w.-softener plant Wasserenthärtungsanlage *f*
w.-softening wasserenthärtend
w. softening Wasserenthärtung *f*
w.-softening plant *s.* w.-softener plant
w.-softening process Wasserenthärtungsverfahren *n*
w. solubility Wasserlöslichkeit *f*, Löslichkeit *f* in Wasser

w.-soluble wasserlöslich
w. solution wäßrige (wässerige) Lösung f
w. spraying Wasserbebrausung f, Wasserberegnung f, Wasserberieselung f, Besprühen (Übersprühen, Bebrausen, Abbrausen) n mit Wasser
w. stain (Farb) Wasserbeize f
w. storage (Pap) Lagerung f unter Wasser
w. supply Wasserversorgung f; Wasservorrat m
w.-supply line Wasser[versorgungs]leitung f
w.-supply plant Wasserversorgungsanlage f
w.-supply pump Wasserversorgungspumpe f, Speisepumpe f
w. surface Wasseroberfläche f
w. table Grundwasserspiegel m; Ölfeldwassergrenze f, Randwassergrenze f
w. tank Wassertank m, Wasserbehälter m
w.-thinned paint wasserverdünnbare (mit Wasser verdünnbare) Farbe f
w.-tight wasserdicht, wasserundurchlässig
w. tower Wasserturm m
w. transport (Bio) Wassertransport m, Wasserleitung f
w. treatment Wasserbehandlung f
w. tube Wasserrohr n
w.-tube boiler Wasserrohrkessel m, Wasserrohrdampferzeuger m
w. vapour Wasserdampf m
w. vapour absorption Wasserdampfaufnahme f, Feuchteaufnahme f
w. vapour permeability Wasserdampfdurchlässigkeit f
w. vapour pressure Wasserdampfdruck m
w. vapour resistance Wasserdampfundurchlässigkeit f
w. vapour transmission s. w. vapour permeability
w. varnish Wasserlack m
w. wash s. w. washing
w.-washed clay geschlämmtes Kaolin n, Schlämmkaolin n
w. washing Wasserwäsche f, Wäsche f (Waschen n, Auswaschen n) mit Wasser
w.-washing stage (Pap) Wäsche f mit Wasser
w.-white wasserhell
w. works Wasserwerk n
watercolour Aquarellfarbe f, Wasserfarbe f
waterholding capacity (Bodenkunde) Wasserkapazität f, Wasserfassungsvermögen n, Wasserhaltevermögen f, Wasserhaltungsvermögen n
watering Bewässern n, Bewässerung f; Verwässern n, Verwässerung f
waterleaf paper ungeleimtes Papier n
waterlogged (bis zum zulässigen Pegel) wassergefüllt, wassergesättigt, wasserdurchtränkt
watermark / to (Pap) mit Wasserzeichen versehen, Wasserzeichen einarbeiten
watermark (Pap) Wasserzeichen n
watermarked lines (Pap) Wasser[zeichen]linien fpl
w. paper Wasserzeichenpapier n
watermarking (Pap) Herstellung f (Einarbeiten n, Aufbringen n, Auftragen n) von Wasserzeichen, Wasserzeichenherstellung f

w. dandy [roll] (Pap) Wasserzeichenwalze f, Vordruckwalze f, Vorpreßwalze f, Egoutteur m
waterous barker (Pap) Naßentrindungsanlage f, (i.e.S.) Reibungsentrinder m System Thorne, Thorne-Entrindungsmaschine f, Thorne-Maschine f, Thorne-Anlage f
waterproof / to wasserdicht (wasserundurchlässig) machen; wasserdicht abschließen, kapseln
waterproof wasserdicht, wasserundurchlässig; wassergeschützt, wasserdicht abgeschlossen, gekapselt
w. finish wasserdichte Appretur f, Wasserdichtausrüstung f
waterproofing Wasserdichtmachen n, Wasserbeständigmachen n, Wasserfestmachen n, Trockenimprägnieren n
w. agent wasserdichtmachendes Mittel n, Wasserdichtmacher m
waters Wasser n, Wässer pl, (i.e.S.) Mineralwasser n, Mineralwässer pl; Toilettenwasser n
watertight wasserdicht, wasserundurchlässig
watery wäßrig, wässerig
w. pulp (Pap) Stoffsuspension f, Stoffwasser n
Watkin [heat] recorders (Ker) Watkin-Kennkörper mpl
Watkins [development] factor (Foto) Watkinsscher Entwicklungsfaktor m, Watkins-Faktor m, arithmetischer Entwicklungskoeffizient m
wattle bark Akazienrinde f, (Gerb meist unkorrekt) Mimosarinde f, Mimosenrinde f
wave Welle f
w. equation Wellengleichung f
w. function s. wavefunction
w. length s. wavelength
w. mechanics Wellenmechanik f
w. motion Wellenbewegung f
w. nature Wellennatur f
w. number Wellenzahl f
w. packet Wellenpaket n, Wellengruppe f
w. propagation Wellenausbreitung f, Wellenfortpflanzung f
w. property Welleneigenschaft f
w. theory Wellentheorie f
w. vector Wellen[zahl]vektor m
waved stress [sinusförmig] schwingende Beanspruchung f
wavefunction Wellenfunktion f
wavelength Wellenlänge f
w. range Wellenlängenbereich m
w. scale Wellenlängenskale f
wavellite (Min) Wavellit m (Aluminiumtrihydroxiddiorthophosphat-5-Wasser)
waving lotion Haarwellotion f
wavy sheet disintegration (Tech) Zerwellen n
wax / to wachsen, (i.e.S.) bohnern
wax Wachs n; Bienenwachs n; Paraffinwachs n, Festparaffin n, [festes] Paraffin n; Bohnerwachs n
w.-bearing wachshaltig, (i.w.S.) paraffinhaltig
w. cake Paraffinkuchen m

w. candle Wachskerze f
w. distillate Paraffindestillat n
w. emulsion *(Pap)* Paraffinemulsion f
w. finish Wachsappretur f
w.-free wachsfrei, *(i.w.S.)* paraffinfrei
w.-like wachsartig, wachsähnlich
w. oil Paraffindestillat n
w. paper Wachspapier n, gewachstes Papier n, *(i.w.S.)* paraffiniertes Papier n, Paraffinpapier n
w. pattern [verlorenes] Wachsmodell n, Wachsausschmelzmodell n
w. pencil Fettstift m
w. size *(Pap)* Wachsleim m, *(i.w.S.)* Paraffinleim m
waxed paper s. wax paper
waxing Wachsen n, *(i.e.S.)* Bohnern n
w. device Paraffineur m
waxlike wachsartig, wachsähnlich
waxy wachsartig, wachsähnlich; wächsern, aus Wachs
w. lustre *(Min)* Wachsglanz m
3-way cock Dreiweg[e]hahn m
weak acid schwache Säure f *(nach dem Dissoziationsgrad)*; *(Pap)* Halblösung f, Halbsäure f, Schwachsäure f
w.-acid tower s. w. tower
w. aqua arme Lösung f *(Kältetechnik)*
w. black liquor *(Pap)* Dünn[ab]lauge f
w. brine Dünnsole f
w. electrolyte schwacher Elektrolyt m
w. liquor *(Pap)* Schwachlauge f
w. tower *(Pap)* Nachturm m *(eines Zweiturmsystems)*
weakly acidic schwach sauer
w. caking coal schwachbackende (wenig bakkende) Kohle f, Kohle f mit schwachem (geringem) Backvermögen
w. coloured schwach [an]gefärbt, farbschwach, schwachfarbig
w. hydraulic lime Wasserkalk m *(Gehalt an Silikatbildnern 10 bis 15 %)*
w. sized paper Papier n mit schwacher Leimung
wear Verschleiß m, Abnutzung f
w. factor Verschleißfaktor m, Abnutzungsfaktor m
w. life Laufzeit f, Laufdauer f
w. resistance Abriebwiderstand m, Abriebbeständigkeit f, Abriebfestigkeit f, Widerstandsfähigkeit f gegen Abrieb, Abnutzungsbeständigkeit f, Verschleißwiderstand m, Verschleißfestigkeit f, *(Text meist)* Scheuerbeständigkeit f, Scheuerfestigkeit f
w.-resistant, w.-resisting abriebbeständig, abriebfest, abnutzungsbeständig, verschleißbeständig, verschleißfest, *(Text meist)* scheuerbeständig, scheuerfest
wearing quality *(Text)* Trageigenschaft f, Gebrauchseigenschaft f
w. surface *(Gum)* Lauffläche f, Protektor m
weather / to verwittern *(z.B. Minerale)*

weather-proof, w.-protected s. weatherproof
w. resistance Wetterbeständigkeit f, Wetterfestigkeit f, Witterungsbeständigkeit f
w.-resistant s. weatherproof
weatherability s. weather resistance
weatherable verwitterbar
weathered product Verwitterungsprodukt n
weathering Verwittern n, Verwitterung f; Bewettern n, Bewetterung f *(Lagern an der Luft)*; *(Ker)* Auswintern n, Wintern n; Aussommern n, Sommern n
w. fastness Wetterechtheit f, Bewetterungsechtheit f
w. resistance s. weather resistance
w. test Bewetterungsprüfung f, Bewetterungsversuch m, Bewitterungsversuch m
w. waste Verwitterungstrümmer pl, Verwitterungsschutt m
weatherproof wetterfest, wettergeschützt, wettersicher, wetterbeständig
web Gewebe n; *(Plast)* endlose Bahn f; *(Pap)* s. w. of paper
w.-calendered *(Pap)* rollengeglättet, rollensatiniert
w. calendering *(Pap)* Glätten (Satinieren) n von Rollenpapieren, Rollensatinage f
w. coating *(Plast)* Beschichten n endloser Bahnen
w. dryer *(Pap, Text)* Bahnentrockner m
w.-fed press *(Pap)* Rollendruckmaschine f
w. of fibre[s] s. w. of paper
w. of paper Papierbahn f, Stoffbahn f, Papiervlies n, Faserfilz m
w. of pulp Zellstoffbahn f
w. tension Papierzugspannung f, Papierbahnspannung f, Bahnspannung f
websterite *(Min)* Websterit m, *(veraltet für)* Aluminit m *(Aluminiumtetrahydroxidsulfit-7-Wasser)*
wedge Keil m; s. w. filter
w. cell Keilküvette f
w. filter *(Fotometrie)* Graukeil m
w. gate valve Keilschieber m
w. photometer Graukeilfotometer n
w.-shaped keilförmig
w. stilt *(Ker)* Dreifuß m *(Brennhilfsmittel)*
w. wire Profildraht m
w.-wire screen Keilspaltsieb n, Spaltsieb n
Wedge furnace Wedge-Ofen m, Mehretagenröstofen m nach Wedge
wedging Walken n, Homogenisieren n *(einer plastischen Masse)*
weed control (eradication) Unkrautbekämpfung f
weeding s. weed control
weedkiller Unkrautvertilgungsmittel n, Herbizid n
weeping Austreten n (Abscheidung f) von Flüssigkeit *(z.B. aus einem Gel)*, Flüssigkeitsabgabe f; Schwitzen n *(z.B. des Zements)*; *(Abfließen der Flüssigkeit von Siebböden, wenn die Dampfgeschwindigkeit die untere Belastungsgrenze unterschreitet)*

w. margarine zum Tränen neigende Margarine f
Weerman degradation Abbau m (Abbaumethode f) nach Weerman, Weermansche Reaktion f
weft thread (Text) Schußfaden m
weigh / to wiegen, eine Masse feststellen, wägen; wiegen, schwer sein, eine Masse haben; s. to w. out (up)
  w. out (up) abwiegen
weigh belt Bandwaage f
  w. scale Waagschale f
  w. tank Dosiertank m
weighable wiegbar, wägbar, mit feststellbarer Masse
weighed form Wägeform f, Wägungsform f (Gewichtsanalyse)
  w. object Wägegut n, [ab]gewogene Substanz f; Einwaage f
  w. to within ± 0.1 g auf ± 0,1 g genau gewogen
weighing Wiegen n, Feststellen n einer Masse, Wägen n, Wägung f, Wägevorgang m
  w. boat s. w. scoop
  w. bottle Wägegläschen n, Wägeglas n
  w. burette Wägebürette f
  w. error Wägefehler m, Wägungsfehler m
  w. room Wägezimmer n, Wägeraum m, Wägungszimmer n
  w. scoop Wägeschiffchen n
  w. tank Wiegebehälter m, Wägegefäß n, Tankwaage f
  w. tube Wägeröhrchen n
weight / to belasten, beschweren, (Pap auch) füllen, (Text auch) erschweren, chargieren
weight Masse f (Eigenschaft eines Körpers, in einem Schwerefeld ein Gewicht anzunehmen; Einheit: kg); Gewicht n (die von einer Masse ausgeübte Kraft; Einheit: kp); Massestück n, Wägestück n, (manchmal noch) Gewichtsstück n, Gewicht n (Körper von genau bestimmter Masse, der zum Wiegen eines anderen dient); s. atomic weight
  w. burette Wägebürette f
  w. change Masseänderung f, (bisher) Gewichtsänderung f
  w. feeder (feeding device) (Plast) Massedosiervorrichtung f
  w. increase Massenzunahme f
  w. loss Masseverlust m
  w. of a sheet of paper (Pap) Masse f je Bogen, (bisher) Bogengewicht n
  w. of paper Papiermasse f, (bisher) Papiergewicht n
  w. of pulp Faser[stoff]masse f
  w.-operated gewichtsbetätigt, lastbetätigt, Gewichts..., Last...
  w. per ream (Pap) Masse f je Ries, (bisher) Riesgewicht n
  w. percentage Masseprozent n, Masse%, (bisher) Gewichtsprozent n, Gew.%, Gewichtsanteile mpl
weighted mean (Statistik) gewogenes Mittel n
weighter s. weighting agent

weighting (Pap) Füllen n, Beschweren n; (Text) Erschweren n, Beschweren n, Chargieren n
  w. agent (Text) Erschwerungsmittel n
  w. function Gewichtsfunktion f (Regelungstechnik)
Weingärtner solution Weingärtners Tanninreaktiv n
weir Wehr n, Überlaufwehr n, Überlauf m, Stauwehr n, Ablaufwehr n; (Pap) Kropf m (beim Holländer)
  w. box (Pap) Überlaufkasten m
  w. level Wehrhöhe f (eines Glockenbodens)
Weissenberg method (technique) (Krist) Weißenberg-Verfahren n, Weißenberg-Böhm-Verfahren n
weissite (Min) Weissit m
weld / to schweißen; sich schweißen lassen, schweißbar sein
  w. together zusammenschweißen
weld Schweißnaht f, [geschweißte] Naht f; Schweißung f; Schweißstelle f
weldability Schweißbarkeit f
weldable schweißbar
welded pipe geschweißtes Rohr n
welding Schweißen n, Schweißung f
  w. electrode Schweißelektrode f
  w. flux Schweiß[fluß]mittel n
  w. glass Glas n für Schweißerschutzbrillen, [dunkles] Schweißerglas n
  w. gun Schweißpistole f
  w. heat Schweißwärme f; (manchmal unkorrekt für) welding temperature
  w.-neck flange Anschweißflansch m
  w. of plastics Plastschweißen n, Kunststoffschweißen n, Schweißen n von Plasten (Kunststoffen)
  w. powder Schweißpulver n
  w. rod Schweißstab m
  w. temperature Schweißtemperatur f
  w. torch Schweißbrenner m
  w. wire Schweißdraht m
Weldon [chlorine] process Weldon-Verfahren n
well 1. Brunnen m, Bohrloch n, Bohrung f; 2. Tauchrohr n
  w.-cooked pulp (Pap) weicher (weichgekochter, weit heruntergekochter) Zellstoff m
  w. head Bohrlochkopf m
  w.-milled (Pap) schmierig gemahlen
  w.-refined hochraffiniert
  w. water Brunnenwasser n
Wellman-Galusha producer Wellman-Galusha-Generator m
Wemco separator Wemco-Trommel[sink]scheider m
Werner complex (Komplexchemie) Werner-Komplex m, Wernerscher Komplex m
Werner-Pfleiderer mixer Werner-Pfleiderer-Kneter m, Werner-Pfleiderer-Innenmischer m
Werner theory Wernersche Theorie (Koordinationslehre) f
Werner-type coordination (Komplexchemie) Koordination f im Sinne von Werner

wernerite *(Min)* Wernerit *m*, Skapolith *m*
weslienite *(Min)* Weslienit *m*, *(veraltet für)* Roméit
m
West Indian elemi Chibouharz *n*, Gomartharz *n*
(Elemiharz von Bursera gummifera L.)
West type condenser Kühler *m* nach West,
West-Kühler *m*
Weston cell *s.* Weston normal cell
Weston figure *(Foto)* Weston-Zahl *f*
Weston normal cell Weston-Normalelement *n*,
Westonsches Normalelement *n*, Weston-Element
n
Weston standard cell Weston-Standardelement *n*,
Standard-Weston-Normalelement *n*, Standard-
Weston-Element *n*, ungesättigtes Weston-
Normalelement (Weston-Element) *n*
Westphal balance Westphalsche Waage *f*, *(i.w.S.)*
Mohr-Westphalsche Waage *f*
wet / to [be]netzen, befeuchten, anfeuchten, *(Ker*
auch) einsumpfen
w. out (up) *(Pap)* [be]feuchten
wet feucht, naß
w. adhesive Naßkleber *m*
w. analysis Naßanalyse *f*
w.-and-dry-bulb thermometer Psychrometer *n*,
Luftfeuchtigkeitsmesser *m*, Verdunstungsmesser *m*
w.-ashed feucht verascht
w. assay *(Met)* nasse Probe *f*, Naßprobe *f*
w. basis *s.* w.-weight basis
w. beating *(Pap)* schmierige Mahlung *f*,
Schmierigmahlung *f*
w. beet pulp *s.* w. pulp 1.
w.-blue leather *(vorgegerbtes oder vorbe-*
handeltes Leder aus indischen Ziegen- und
Zickelfellen)
w.-bottom furnace Schmelzfeuerung *f (Feuerung*
mit flüssigem Schlackenabzug)
w. break *(Pap)* Bahnriß *m* in der Naßpartie
w. broke *(Pap)* Naßausschuß *m*
w.-bulb temperature Feuchtkugeltemperatur *f*,
Temperatur *f* des feuchten Thermometers,
Kühlgrenztemperatur *f*
w.-bulb thermometer feuchtes Thermometer *n*
w. carbonizing *(Text)* nasses Karbonisieren *n*,
nasse Karbonisation *f*, Naßkarbonisation *f (von*
Wolle)
w. classification Naßklassieren *n*, Hydroklas-
sieren *n*, [nasses] Stromklassieren *n*, Trennen *n*
nach Gleichfälligkeit
w. classifier Naßklassierer *m*
w.-clean / to 1. naßreinigen, waschen; 2.
naßaufbereiten, auf nassem Wege aufbereiten
w. cleaner Naßreiniger *m*
w. cleaning 1. Naßreinigung *f*, Waschen *n*,
Wäsche *f*; 2. Naßaufbereitung *f*, Naßwäsche *f*,
nasse Aufbereitung *f*, Aufbereitung *f* auf nassem
Wege
w. collection Naßabscheiden *n*, Waschen *n*
w. collector Naßabscheider *m*, Skrubber *m*,
Wäscher *m*, Wascher *m*

w. collodion plate *(Foto)* Naß[kollodium]platte *f*
w. collodion process *(Foto)* nasses Kollodium-
verfahren *n*, Naßkollodiumverfahren *n*
w. compression nasse Kompression *f*
w. condenser nasser Mischkondensator *m*,
Einspritzkondensator *m*
w. cooling method Naßverfahren *n*, Duschverfahren *n (Margarineherstellung)*
w.-cyclone classifier Hydrozyklon *m*, Naßzyklon
m
w.-cylinder mill Trommelnaßmühle *f*
w. decatizing *(Text)* Naßdekatur *f*
w. discharge Naßaustrag *m*
w. end *s.* w. part
w.-end break *s.* w. break
w. end of the dryer section *(Pap)* Anfang *m* der
Trockenpartie
w.-end starch Rohstärke *f*, Grünstärke *f*
w. fastness *s.* w. strength
w. feed *s.* w. product
w. feeder Naßgutaufgabevorrichtung *f*, Naß-
gutzuführung *f*
w. felt *(Pap)* Naßfilz *m*
w.-felt roll *(Pap)* Naßfilzleitwalze *f*
w. galvanizing galvanisches Verzinken *n*
w. gas Naßgas *n*; *s.* w. natural gas
w. gas cleaning Naßgasreinigung *f*, Naßreinigung *(nasse Reinigung) f* eines Gases,
Gaswäsche *f*
w. gas meter nasser Gasmesser (Gaszähler) *m*
w. grinding *s.* w. milling 1.
w.-grinding mill Naßmühle *f*, Mühle *f* zur
Naß[ver]mahlung
w. jig Naßsetzmaschine *f*, hydraulische Setzmaschine *f*, hydraulischer Setzapparat *m*
w. lay-up *(Plast)* Handauflegeverfahren *n*
w. lecithin sludge Lezithinnaßschlamm *m*
w. liming *(Zucker)* nasse Scheidung *f*, Kalkmilchscheidung *f*, Naßscheidung *f*
w. machine *(Pap)* Holzschliffentwässerungsmaschine *f*; Zellstoffentwässerungsmaschine *f*
w. magnetic separator Naßmagnetscheider *m*
w. masterbatching *(Gum)* Herstellung *f* von
Masterbatches (Vormischungen) auf nassem
Wege
w. metallurgy Naßmetallurgie *f*, Hydrometallurgie *f*
w. milling [process] 1. Naß[ver]mahlen *n*,
Naß[ver]mahlung *f*; 2. *(Pap)* Mahlen *n (des Stoffs*
im Kollergang), Kollern *n*; Mahlen *n (des Stoffs*
im Holländer)
w. mixing method *s.* w. preparation
w. natural gas nasses (feuchtes) Erdgas (Naturgas) *n*, Naßgas *n*
w.-on-dry technique *(Text)* Naß-Trocken-
Verfahren *n*
w.-on-wet printing *(Text)* Naß-auf-Naß-Druck-
verfahren *n*, Naß-in-Naß-Druckverfahren *n (Film-*
druck)
w. pan Naßkoller[gang] *m*

w. part *(Pap)* Naßpartie *f*
w.-pit pump Tauchpumpe *f*, Unterwasserpumpe *f*
w.-plate process s. w. collodion process
w. precipitator Naßelektroabscheider *m*, Naßelektrofilter *n*
w. preparation *(Ker)* Naßaufbereitung *f*
w. press [machine] s. w. machine
w. pressing *(Ker)* Naßpressen *n*, plastisches Pressen *n*
w. process nasses Verfahren *n*, Naßverfahren *n*; Naßaufbereitung *f*, *(Ker auch)* Schlickerverfahren *n*
w.-process plant Naßaufbereitungsanlage *f*
w. product Feuchtgut *n*, Naßgut *n* *(bei der Trocknung)*
w. pulp 1. *(Zucker)* Naßschnitzel *npl*, Diffusionsschnitzel *npl*; 2. *(Pap)* schmieriger Stoff *m*
w. quenching Naßlöschen *n*, Naßlöschung *f (z.B. von Koks)*
w.-rendered fat Naßschmelzgrieben *fpl*
w.-rendered lard im Naßschmelzverfahren ausgeschmolzenes Fett *n*
w. rendering Naßschmelzen *n*, Naßschmelze *f* [auf Wasser]
w. rot Naßfäule *f*
w. salting *(Lebm)* Naßpökelung *f*, Naßpökeln *n*; *(Gerb)* *(Konservierung von Häuten durch Bestreuen mit Salz und anschließende Stapelung)*
w. screening Naßsieben *n*
w. sludge Naßschlamm *m*
w. spent grains Naßtreber *pl*
w. spinning Naßspinnen *n*, Naßspinnverfahren *n*
w. steam Naßdampf *m*
w. stock s. w. pulp 2.
w. strength Naßfestigkeit *f*, *(Text auch)* Naßechtheit *f*
w.-strength paper naßfestes (wasserfestes, wetterfestes) Papier *n*
w.-strength resin *(Pap)* Kunstharz *n* zur Naßfestleimung, Naßfestleim *m*
w. stuff s. w. pulp 2.
w.-surface temperature s. w.-bulb temperature
w. table Spülherd *m*
w. test naßchemische Untersuchung *f*
w. treatment Naßbehandlung *f*; *(Landw)* Naßbeize *f (von Saatgut)*
w. vapour s. w. steam
w. washing s. w. cleaning 2.
w. weight Masse *f* des feuchten Stoffs, Feuchtgutmasse *f*
w.-weight basis Bezugsbasis *f* Masse des feuchten Stoffs
Wetherill furnace *(Met)* Wetherill-Ofen *m*
wetness Feuchtigkeit *f*, Nässe *f*; *(Pap)* Schmierigkeit *f*
wettability Benetzbarkeit *f*, Netzbarkeit *f*
wettable [be]netzbar
w. powder *(Pflanzenschutz)* oberflächenaktives

Pulver *n*, Spritzpulver *n*, Suspensionsspritzmittel *n*
w. sulphur Netzschwefel *m*
wetted-wall absorber Dünnschichtabsorber *m*
w.-wall column leere Kolonne *f*, Rieselfilmkolonne *f*
wetter s. wetting agent
wetting Benetzen *n*, Benetzung *f*, Netzen *n*, Befeuchten *n*, Anfeuchten *n*, *(Ker auch)* Einsumpfen *n*
w. ability s. w. power
w. action netzende Wirkung *f*, Netzwirkung *f*
w. agent (aid) Benetzungsmittel *n*, Netzmittel *n*, Netzer *m*
w. characteristics s. w. properties
w. effect s. w. action
w. machine *(Pap)* Feuchteinrichtung *f*, Befeuchter *m*, Feucht[ungs]maschine *f*, Feuchtapparat *m*, Feuchter *m*
w.-out s. w.-up
w. power Benetzungsfähigkeit *f*, Netzfähigkeit *f*, Benetzungsvermögen *n*, Netzvermögen *n*, Netzkraft *f*
w. properties Netzeigenschaften *fpl*
w. tension Benetzungsspannung *f*
w.-up *(Pap)* Befeuchten *n*, Befeuchtung *f*, Feuchtung *f*
W.H. s. white heat
whale blubber Blubber *m*, Walspeck *m*
w.-liver oil Walleberöl *n*, Wallebertran *m*
w. oil Walöl *n*, Waltran *m*
Whatman paper Whatman-Papier *n* *(ein Filterpapier)*
WHC s. waterholding capacity
wheat Weizen *m*, Triticum L.
w. bran Weizenkleie *f*
w. flour Weizenmehl *n*
w.-flour starch s. w. starch
w. germ oil Weizenkeimöl *n*
w. malt Weizenmalz *m*
w. starch Weizenstärke *f*
Wheatstone bridge Wheatstone-Brücke *f*
Wheeler mill Wheeler-Mühle *f* *(eine Strahlmühle)*
whey Molke *f*, Molken *m*, Milchserum *n*, Käs[e]wasser *n*
w. butter Molkenbutter *f*
w. cheese Molkenkäse *m*
w. margarine Molkenmargarine *f*, mit Molke hergestellte Margarine *f*
w. powder Molkenpulver *n*, Trockenmolke *f*
w. protein Molkenprotein *n*, Milchserumprotein *n*
whipped cream geschlagener Rahm *m*, Schlagsahne *f*
w. margarine Vollrahmmargarine *f*
whipstock *(Erdöl)* Ablenkkeil *m*, Richtkeil *m*, Whipstock *m*
whirl sintering Wirbelsintern *n*
whisker Haarkristall *m*, Einkristallfaden *m*, Whisker *m*

whiskey *s.* whisky
whisking machine Schaum[schlag]maschine *f*
whisky Whisky *m*
white Weiß *n (Farbempfindung)*; weiß[färbend]er Farbstoff *m*, Weiß *n*
  w. amber amerikanischer Styrax (Storax) *m (Balsamharz von Liquidambar styraciflua L.)*
  w. antimony *(Min)* Valentinit *m*, Antimonblüte *f (Antimon(III)-oxid)*
  w. arsenic As_2O_3 Weißarsenik *n*, [weißes] Arsenik *n*, Arsentrioxid *n*, Arsen(III)-oxid *n*; *(Min)* Arsenblüte *f*, Arsenikblüte *f*, Arsenolith *m (Arsen(III)-oxid)*
  w. bole *(Min)* weißer (reiner) Ton *m*, Porzellanerde *f*, Kaolin *m*, China Clay *m(n)*
  w.-burning weißbrennend
  w. camphor oil weißes Kampferöl *n*, Kampferweißöl *n*
  w. carbon [black] Weißruß *m*
  w. casse *s.* w. haze
  w. cast iron weißes Gußeisen *n*, Weißguß *m*, Hartguß *m*
  w. cement weißer Zement *m*
  w. copperas *s.* w. vitriol
  w. crottle Pertusaria corallina (L.)Th.Fr. *(eine Farbflechte)*
  w. crystals Kristallzucker *m*, Sandzucker *m*, Streuzucker *m*
  w. dammar Pineyharz *n (von Vateria indica L.)*
  w. degree *s.* whiteness
  w.-discharge / to *(Text)* weißätzen
  w. discharge *(Text)* Weißätzung *f*, Weißätzen *n*; Weißätze *f*, Ätzweiß *n*
  w. factice *(Gum)* weißer Faktis *m*
  w. filler heller Füllstoff *m*
  w.-firing clay *(Ker)* weißbrennender Ton *m*
  w. flint Weißglas *n*
  w.-hard *(Ker)* knochentrocken
  w. haze weißer Bruch *m (Weinfehler)*
  w. heat Weißglut *f*
  w.-hot weißglühend
  w. iron pyrites *(Min)* Markasit *m (Eisendisulfid)*
  w. lead 1. Bleiweiß *n*, Karbonatbleiweiß *n (Bleihydroxidkarbonat)*; 2. *(Min)* Weißbleierz *n*, *(veraltet für)* Zerussit *m (Blei(II)-karbonat)*
  w. lead ore *s.* w. lead 2.
  w. leather Weißleder *n*, weißgares Leder *n*
  w. lime Weißkalk *m*
  w. liquor *(Pap)* Weißlauge *f*, Frischlauge *f*
  w.-liquor storage [tank] *(Pap)* Weißlaugenbehälter *m*, Weißlaugenvorratstank *m*
  w. malt helles Malz *n*
  w. maple Silberahorn *m*, Acer saccharinum L.
  w. massecuite Weißzuckerfüllmasse *f*, A-Füllmasse *f*, Erstproduktfüllmasse *f*
  w. metal Weißmetall *n*, Lagerweißmetall *n (Legierung zum Ausgießen von Lagern)*; Spurstein *m*, Weißmetall *n (praktisch eisenfreier Kupferstein mit 78 bis 80 % Kupfer)*
  w. mineral oil *s.* w. oil

  w. mustard Weißer Senf *m*, Sinapis alba L.
  w. nickel *(Min)* Weißnickelkies *m*, Chloanthit *m*
  w. oil Weißöl *n*
  w.-oil spray Petroleumemulsion *f (zur Schädlingsbekämpfung)*
  w. pan Weißzuckervakuumapparat *m*
  w. petroleum oil *s.* w. oil
  w. phosphorus weißer (gelber, farbloser) Phosphor *m*
  α-**w. phosphorus** kubisch kristallisierter weißer Phosphor *m*
  β-**w. phosphorus** hexagonal kristallisierter weißer Phosphor *m*
  w. pig iron weißes Roheisen *n*
  w. precipitate $Hg(NH_3)_2Cl_2$ schmelzbares weißes Präzipitat *n*; $Hg(NH_2)Cl$ unschmelzbares weißes Präzipitat *n*
  w. product *(Erdöl)* weißes (helles) Produkt *n*, Weißprodukt *n*
  w. reinforcing filler *(Gum)* heller Verstärkerfüllstoff (aktiver Füllstoff) *m*
  w. rot Weißfäule *f*
  w. rotting fungi Weißfäulepilze *mpl*
  w. rust weißer Rost *m*
  w. shellac weißer (weißgebleichter, gebleichter) Schellack *m*
  w. spirit White Sp[i]rit *m (ein Testbenzin)*
  w. squill *(Pharm)* weiße Meerzwiebel *f (von Urginea maritima (L.)Baker)*
  w. substitute *s.* w. factice
  w. sugar Weißzucker *m*, Verbrauchszucker *m*
  w. vitriol *(Min)* Zinkvitriol *m*, Goslarit *m (Zinksulfat-7-Wasser)*
  w. water *(Pap)* Abwasser *n*, Rück[lauf]wasser *n*, Kreislaufrückwasser *n*; Sieb[ab]wasser *n*
  w.-water chest *(Pap)* Abwasser[sammel]behälter *m*, Rückwasser[sammel]behälter *m*
  w.-water pump *(Pap)* Abwasserpumpe *f*, Rückwasserpumpe *f*
  w. wax weißes Wachs *n*, gebleichtes Bienenwachs *n*
  w. wine Weißwein *m*
whiten / to *(Pap)* bleichen, entfärben, den Weißgehalt erhöhen, auf eine höhere Weiße bleichen, einen hohen Weißzustand schaffen (herstellen); weiß werden
whiteness [degree] Weißgehalt *m*, Weißgrad *m*, Weiße *f*
whitening *(Pap)* Bleichen *n*, Bleichung *f*, Entfärben *n*, Entfärbung *f*, Schaffung (Herstellung) *f* eines hohen Weißgehalts, Erhöhung *f* des Weißgehalts; Weißwerden *n*
  w. agent *(Plast)* Weißtöner *m*
  w. stage *(Pap)* Endbleichstufe *f (in der der Stoff seinen höchsten Weißgehalt erreicht)*
whiteware Weißware *f (allgemeine Benennung für Ware aus weißem Scherben)*
  w. industry Porzellanindustrie *f*
whitewash 1. Sulfatblase *f (Glasfehler)*; 2. Tünche *f*, Weißtünche *f*

whiting *(fein)* gemahlene Kreide *f; (manchmal auch) (fein)* gemahlener Kalkstein *m*
w. dust (powder) Kreidepulver *n*
whizzed naphthalene *(durch Zentrifugieren)* gereinigtes Naphthalin *n*
whizzer Schleuder *f*, Zentrifuge *f*; Schleudertrockner *m*, Zentrifugaltrockenmaschine *f*, Trokkenzentrifuge *f*; Streuteller *m (eines Kreiselsichters)*
w. classifier Kreiselsichter *m*
whole casein Vollkasein *n*
w.-egg powder Volleipulver *n*
w. grain Vollkorn *n*
w. half-width of a spectral line s. width of a spectral line
w. meal Vollkornmehl *n*
w. milk Vollmilch *f*
w. stuff *(Pap)* Ganzstoff *m*, Ganzzeug *n*
w.-tyre reclaim *(Gum)* Ganzreifenregenerat, *n*
wick[-fed] lamp Dochtlampe *f*
wickability Saugfähigkeit *f*
wicket *(Ker)* gemauerte Ofentür *f*
widdrene Widdren *n (trizyklisches Sesquiterpen)*
widdrol Widdrol *n (Sesquiterpenalkohol)*
wide-angle X-ray pattern Weitwinkelaufnahme *f (kristallografische Röntgenaufnahme)*
w.-meshed weitmaschig, grobmaschig
w.-mouth bottle Weithalsflasche *f*
w.-mouth ware Weithalsartikel *mpl*, Weithalsgefäße *npl*
w.-mouthed flask Weithalskolben *m*
w.-neck bottle s. w.-mouth bottle
w.-neck flask s. w.-mouthed flask
w.-necked reagent bottle with conical shoulder Steilbrustflasche *f*
w.-stemmed funnel Pulvertrichter *m*
Widmannstätten figures (lines) Widmannstättensche Figuren *fpl*
Widmannstätten structure Widmannstättensches Gefüge *n*, Widmannstättensche Struktur *f*
width of a roll *(Pap)* Rollenbreite *f*
w. of a spectral line Linienbreite *f*
w. of the machine *(Pap)* Maschinenbreite *f*
w. of the wire *(Pap)* Siebbreite *f*
Wiedemann-Franz law Wiedemann-Franzsches Gesetz *n*
Wieland-Gumlich aldehyde Wieland-Gumlich-Aldehyd *m*, Karakurin(VII) *n*
Wien displacement law Wiensches Verschiebungsgesetz *n*
Wien effect Wien-Effekt *m*, Feldstärkeeffekt *m*
wiikite *(Min)* Wiikit *m*
Wijs [iodine monochloride] solution Wijs-Lösung *f*, Wijssche Lösung *f (Lösung von Jodmonochlorid in Eisessig zur Bestimmung der Jodzahl von Ölen und Fetten)*
wild:
to be w. kochen *(von unberuhigtem Stahl)*
w. fermentation Spontangärung *f*
w. gasoline wildes (unstabilisiertes, unstabiles, instabiles) Benzin *n*

w. hop Wildhopfen *m*
w. rubber Wildkautschuk *m*
w. silk Wildseide *f*, wilde Seide *f (z.B. Tussahseide)*
w. steel unberuhigter Stahl *m*
w. yeast wilde Hefe *f*, Wildhefe *f*
wildcat Bohrung *f* auf Neuland (unerforschtem Boden), Neubohrung *f*, Explorationsbohrloch *n*, Explorationsbohrung *f*, Untersuchungsbohrung *f*, Mutungsbohrung *f*, Wildcat-Bohrung *f*
Wilfley table Wilfley-Herd *m*
willemite *(Min)* Willemit *m (Zinkorthosilikat)*
Willgerodt reaction Willgerodt-Reaktion *f (Säureamidsynthese mittels Redoxamidierung von Alkyl-Aryl-Ketonen)*
Williams abrader *(Gum)* Williams-Abriebprüfer *m*, Williams-Prüfer *m*
Williams plastometer *(Gum)* Williams-Plastometer *n*, Plastometer *n* von Williams
Williams unit *(Text)* Williams-Einheit *f (eine Rollenkufe)*
Williamson ether synthesis Williamson-Synthese *f (von Äthern durch Alkoholatalkylierung)*
Williamson kiln Williamson-Ofen *m (Tunnelofen)*
Williamson synthesis s. Williamson ether synthesis
willow[ing machine] *(Pap)* Haderndrescher *m*
Willstätter nail Willstätter-Nagel *m (Filtrierhilfsmittel)*
Wilputte oven Wilputte-Ofen *m*, Wilputte-Koksofen *m*, Verbundkoksofen *m* der Wilputte Coke Oven Corp.
Wilson [cloud] chamber, Wilson cloud-track apparatus [Wilsonsche] Nebelkammer *f*, Wilson-Kammer *f*
wilting coefficient *(Landw)* Welkekoeffizient *m*
winch *(Text)* Haspel *f*, Winde *f*
w. back Haspelkufe *f*
w. dyeing machine Haspelfärbeapparat *m*
wind [up into a reel] / to *(Pap)* aufwickeln, aufrollen
wind belt Windmantel *m (eines Kupolofens)*
w.-blown sulphur Ventilatoschwefel *m*, Stäubeschwefel *m*
w. box Windkasten *m*; Saugkasten *m (eines Saugzugsinterapparats)*
w.-up arrangement (equipment) Aufwickelvorrichtung *f*, Aufrollvorrichtung *f*
winder *(Text)* Spulmaschine *f*; *(Pap)* Rollapparat *m*, Aufrollapparat *m*, Rollmaschine *f*, Roller *m*, *(i.e.S.)* Umroller *m*, Rollenschneidmaschine *f*, Rollenschneider *m*, Längsschneidemaschine *f*, Längsschneider *m*
w. drum *(Pap)* Tragwalze *f*, Tragtrommel *f*
w. shaft *(Pap)* Rollstange *f*, Aufrollstange *f*, Papierrollstange *f*
winding *(Text)* Aufwinden *n*, Wickeln *n*, Aufspulen *n*, Spulen *n*; *(Pap)* Aufrollen *n*, Aufwickeln *n*, Aufrollung *f*, Aufwicklung *f*, Papieraufrollung *f*, Papieraufwicklung *f*
w. and slitting machine *(Pap)* Umroller *m*, Rollenschneidmaschine *f*, Rollenschneider *m*, Längsschneidemaschine *f*, Längsschneider *m*

w. bobbin *(Text)* Aufnahmespule *f*, Aufwickelspule *f*

w. equipment *s.* wind-up arrangement

w. head *(Text)* Primärspule *f*, Trommelhülse *f*

w. machine *(Text)* Spulmaschine *f*

w. roll *(Pap)* aufzuwickelnde Papierrolle *f*

w. speed *(Text)* Spulgeschwindigkeit *f*, Aufspulgeschwindigkeit *f*, Aufwindegeschwindigkeit *f*

window glass Fensterglas *n*

w.-glass furnace Tafelglaswannenofen *m*

wine Wein *m*

w. body Körper (Extrakt, Stoff) *m* des Weines

w. ferment *s.* w. yeast

w.-fining agent Weinschönungsmittel *n*

w. gauge Weinwaage *f*, Önometer *n*

w. press Traubenpresse *f*, Traubenquetsche *f*, Kelter *f*

w. spirit C_2H_5OH Weingeist *m*, Weinspiritus *m*, Äthanol *n*

w. stone [roher] Weinstein *m*, Rohweinstein *m*, Kaliumhydrogentartrat *n*

w.-stones oil Weintraubenkernöl *n*, Traubenkernöl *n*

w. vinegar Weinessig *m*

w. yeast Weinhefe *f*

wing-shape policeman Spatenwischer *m* *(ein Gummiwischer)*

Winkler generator Winkler-Generator *m*, Winkler-Gaserzeuger *m*

Winkler-Koch process Winkler-Koch-Verfahren *n*, Winkler-Koch-Spaltverfahren *n*, Krackverfahren *n* nach Winkler und Koch

Winkler process Winkler-Verfahren *n*, Winkler-Vergasungsverfahren *n*, Winkler-Generatorverfahren *n*

winter butter Winterbutter *f*

w. cream Winterrahm *m*

w. wash *(Landw)* Winterspritzmittel *n*

wintering *(Ker)* Wintern *n*, Auswintern *n*

winterization Winterung *f*, Winterisation *f*, Demargarinisation *f*, Demargarinieren *n*, Entstearin[is]ierung *f*, Ausfrieren *n* *(von Öl)*

winterize / to *(Öl)* wintern, ausfrieren, entstearin[is]ieren, demargarinieren

wiped-film still Destillierapparat *m* mit Verteilerbürsten

wiper Abstreifer *m*, Schaber *m*, Ausräumer *m*; Verteilerbürste *f*, Fraktionierbürste *f* *(Destillation)*

wire Draht *m*; *(Pap)* Sieb *n*, Papiermaschinensieb *n*

w. bar *(Met)* Drahtbarren *m*

w. changing *(Pap)* Siebwechsel *m*

w. cloth Drahtgewebe *n*, Metallgewebe *n*; *(Pap)* Siebgewebe *n*

w.-cloth roll *(Pap)* Registerwalze *f*

w. coating Kabelmantel *m*, Kabelüberzug *m*

w. covering Umspritzen *n* von Drähten

w.-cut brick Strangpreßziegel *m*, geschnittener Ziegel *m*

w.-cut process plastisches Verfahren *n* *(Ziegelherstellung)*

w. edge *(Pap)* Siebrand *m*, Siebkante *f*

w. enamel Drahtemaillelack *m*, Drahtemaille *f*

w. end *(Pap)* Siebpartie *f*

w. frame *s.* w. table

w. gauze Drahtnetz *n*; *s.* w. cloth

w. glass Drahtglas *n*

w. guide *(Pap)* Sieblaufregler *m*

w.-guide roll *(Pap)* Sieblaufregulierwalze *f*; Siebleitwalze *f*, Leitwalze *f*

w.-leading roll *(Pap)* Siebleitwalze *f*, Leitwalze *f*

w. length *(Pap)* Sieblänge *f*

w. mark *(Pap)* Siebmarkierung *f*

w.-marked *(Pap)* siebmarkiert, mit Siebmarkierung

w. model Drahtrohrmodell *n*

w. part *(Pap)* Siebpartie *f*

w. pit *(Pap)* Siebwasserbehälter *m*, Siebwassersammelbecken *n*

w. roll *s.* w.-leading roll

w.-roll doctor *(Pap)* Schaber *m* an der Siebleitwalze

w. run *(Pap)* vorlaufender Siebtrum *m*

w. screen Drahtsieb *n*, Metallsieb *n*

w. side Siebseite *f*, Unterseite *f* *(des Papiers)*

w. sieve *s.* w. screen

w. spiral Drahtspirale *f*

w.-spiral-reinforced hose *(Gum)* Spiralschlauch *m*

w.-stretch roll *(Pap)* Siebspannwalze *f*

w. support Chromnickeldrahtdreieck *n* *(der Verkokungsapparatur nach Conradson)*

w. table *(Pap)* Siebtisch *m*

w. tension *(Pap)* Siebspannung *f*

w.-wound drahtumwickelt

wired glass *s.* wire glass

Wiswesser notation Wiswesser-Notation *f*

Wiswesser [notation] system Wiswesser-System *n*, Notationssystem *n* von Wiswesser

witch hazel Hamamelis *f*, Zaubernuß *f*, Hamamelis L.; Hamamelisblätter *npl*

withdraw / to *(Flüssigkeiten oder Gase)* abziehen, ableiten, abführen, absaugen, entnehmen, abnehmen

w. by pipette abpipettieren

w. electrons Elektronen aufnehmen

withdrawal Abziehen *n*, Ableiten *n*, Abführen *n*, Absaugen *n*, Entnehmen *n*, Abnehmen *n* *(von Flüssigkeiten oder Gasen)*

w. pump Abzugspumpe *f*, Absaugpumpe *f*

witherite *(Min)* Witherit *m* *(Bariumkarbonat)*

Wittig reaction Wittigsche Phosphin-Methylen-Reaktion *f* *(aufbauende Karbonylolefinierung)*

Wittig rearrangement Wittig-Umlagerung *f* *(von Äthern zu Karbinolaten)*

w/m., W/m. *s.* watermark

w/o emulsion, W/O emulsion *s.* water-in-oil emulsion

W/O type *s.* water-in-oil type

woad Waid *m*, Färberwaid *m*, Isatis tinctoria L.; Waid *m (Farbstoff aus den Blättern von Isatis tinctoria L.)*

w. vat Waidküpe *f*

wöhlerite *(Min)* Wöhlerit *m*

Wolff-Kishner reaction (reduction) Wolff-Kishner-Reaktion *f*, Wolff-Kishner-Reduktion *f*, Reduktion *f* nach Wolff-Kishner

Wolff rearrangement Wolffsche Umlagerung *f*

wolfram W Wolfram *n*

wolframate Wolframat *n*

wolframic Wolfram..., *(i.e.S.)* Wolfram(VI)-...

w. acid H_2WO_4 [gelbe] Wolframsäure *f*; $H_2WO_4 \cdot H_2O$ [weiße] Wolframsäure *f*; WO_3 Wolfram(VI)-oxid *n*

wolframine *(Min)* Wolframin *m*, *(veraltet für)* Tungstit *m (Kalziumwolframat)*

wolframite *(Min)* Wolframit *m*

wolfsbane Sturmhut *m*, Eisenhut *m*, Aconitum L.

wollastonite *(Min)* Wollastonit *m (Kalziummetasilikat)*

wood Holz *n*; Wald *m*

w. acid Holzsäure *f (bei thermischer Behandlung des Holzes entstehende organische Säure)*

w. adhesive *s.* w. glue

w. alcohol Holzgeist *m*, Holzspiritus *m*, Holzalkohol *m (roher Methylalkohol)*

w. ash Holzasche *f*

w. carbonization Holzverkohlung *f*, Holzentgasung *f*

w. charcoal Holzkohle *f*

w. chemistry Holzchemie *f*

w. chips *(Pap)* Hackspäne *mpl*, Hackschnitzel *npl*, Holzschnitzel *npl*, Kochschnitzel *npl*, Schnitzel *npl*; *(Plast)* Holzspäne *mpl*

w.-containing *(Pap)* holz[schliff]haltig

w.-containing paper holzhaltiges Papier *n*, Holzschliffpapier *n*

w. creosote Kreosot *n*, Holzteerkreosot *n*

w. distillation Holzdestillation *f*, Trockendestillation *f* des Holzes

w.-distillation gas *s.* w. gas

w. ether CH_3OCH_3 Holzäther *m*, *(veraltet für)* Dimethyläther *m*, Methyläther *m*, Methoxymethan *n*

w. fibre Holzfaser *f*

w.-fibre board Holzfaserplatte *f*

w.-fired kiln holzgefeuerter Ofen *m*

w. flour Holzmehl *n*

w.-free *(Pap)* holzfrei

w.-free paper holzfreies Papier *n*, Zellstoffpapier *n*

w. gas Holzgas *n*

w. glue Holzleim *m*

w. meal *s.* w. flour

w. naphtha *s.* w. alcohol

w. opal *(Min)* Holzopal *m*

w. peat Waldtorf *m*

w. preparation *(Pap)* Holzaufbereitung *f*, Holzvorbereitung *f*, Holzbearbeitung *f*

w. preservative Holzschutzmittel *n*

w. producer gas Holzgas *n (als Generatorgas)*

w. pulp *(Pap)* [mechanischer] Holzschliff *m*, [mechanischer] Holzstoff *m*, Holzmasse *f*, Schleifmasse *f*; Holzzellstoff *m*, Zellstoff *m*

w.-pulp board *(Pap)* Holzschliff *m* in Bogenform (Pappenform), Holzschliffpappe *f*, Holzschliffblätter *npl*; Zellstoff *m* in Bogenform (Pappenform), Zellstoffpappe *f*, Zellstoffblätter *npl*

w.-pulp cardboard Holzkarton *m (aus Holzschliff)*; Zellstoffkarton *m*

w.-pulp stock *(Pap)* [mechanischer] Holzschliff *m*, [mechanischer] Holzstoff *m*, Holzmasse *f*, Schleifmasse *f (in Breiform)*; Holzzellstoff *m (in Breiform)*, Zellstoffbrei *m*

w. room *(Pap)* Holzputzerei *f*, Holzaufbereitungsanlage *f*

w. rosin Wurzelharz *n*, Wurzelkolophonium *n*, Extraktionskolophonium *n*, Extraktionsharz *n*

w. speck Holzsplitter *m*, Splitter *m (Papierfehler)*

w. spirit *s.* w. alcohol

w. sugar Holzzucker *m (Produkt der Holzverzuckerung)*; Holzzucker *m*, Xylose *f*

w. tar Holzteer *m*

w.-tar creosote Kreosot *n*, Holzteerkreosot *n*

w. vinegar Holzessig *m*

w. weight *(Pap)* Holzmasse *f*, *(bisher)* Holzgewicht *n*

w. wool Holzwolle *f*

w. yard *(Pap)* Holz[lager]platz *m*

Woodall-Duckham continuous vertical retort Woodall-Duckham-Retorte *f*, Vertikalretorte *f* mit kontinuierlicher Beschickung nach Woodall-Duckham

Woodall-Duckham system Woodall-Duckham-Ofensystem *n*

wooden agitator Holzrührer *m*

w. cover Holzdeckel *m*

w. cribwork Holzauskleidung *f*

w. filter Holzfilter *n*

w. fluted roll Holzriffelwalze *f*

w. roll Holzwalze *f*

w. tub (vat) Holzbottich *m*, Holzbütte *f*

Wood's glass Wood-Glas *n*

Wood's metal Woodsches Metall *n*, Woods Legierung *f*

woodwardite *(Min)* Woodwardit *m (Aluminiumkupfer(II)-dodekahydroxidsulfat)*

woody holz[art]ig; *(Pap)* holz[schliff]haltig

w. brown coal, w. lignite holzartige (xylitische) Braunkohle *f*, Braunkohlenholz *n*, Xylit *m(n)*, Hylit *m(n)*, *(veraltet)* Lignit *m*, Braunkohlenlignit *m*, lignitische Braunkohle (Kohle) *f*

w. paper *s.* wood-containing paper

w. structure Holzstruktur *f*, Holzgefüge *n*

wool Wolle *f*

w. alcohol Wollfettalkohol *m*

w. dryer felt *(Pap)* Wolltrockenfilz *m*

w. dye Wollfarbstoff *m*

w. fabric Wollgewebe *n*, Wollstoff *m*

w. fat Wollfett n
w. felt (Pap) wollener Naßfilz m, Wollfilz m
w.-felt board Wollfilzpappe f
w.-felt paper Wollfilzpapier n
w. grease s. w. fat
w.-like wollartig, wollähnlich; vom Wolltyp
(W-Typ) (Chemiefaserstoff)
w. scouring Wollwäsche f
w.-scouring process Wollwaschverfahren n
w. type Wolltyp m, Wollart f, Wollsorte f
w.-washing machine Wollwaschmaschine f
w. wax Wollwachs n, Rohwollfett n
wool[l]en wollen, Woll...; Streichgarn...
w. dry felt s. wool dryer felt
w. fabric Wollgewebe n; Streichgarngewebe n
w. felt (Pap) wollener Naßfilz m, Wollfilz m
w. paper Kalanderwalzenpapier n
work / to arbeiten; (eine Apparatur) arbeiten lassen;
(Bgb) abbauen
w. for verarbeiten auf
w. up aufarbeiten
w. up into a reel (Pap) aufwickeln, aufrollen
work done geleistete (umgesetzte, gewonnene)
Arbeit f
w. done in shear Scherarbeit f
w. function [of electrons] Austrittsarbeit f,
Ablösearbeit f, Abtrennungsarbeit f
w.-harden / to kalthärten, kaltverfestigen, (durch
Kaltverformung) verfestigen
w. hardening Kalthärtung f, Kaltverfestigung f,
Verfestigung f (durch Kaltverformung)
w. of adhesion Adhäsionsarbeit f, Haftarbeit f
w. of cohesion Kohäsionsarbeit f
w. of compression Verdichtungsarbeit f, Kom-
pressionsarbeit f
w. of reaction (phys Ch) Reaktionsarbeit f,
[maximale] Nutzarbeit f, maximale Arbeit f [der
Reaktion]
w. of separation Abtrennungsarbeit f, Ablösearbeit f, Austrittsarbeit f
workability Bearbeitbarkeit f, Verarbeitbarkeit f,
Verarbeitungsfähigkeit f, (Ker auch) Verformbarkeit f
workable bearbeitbar, verarbeitbar, verarbeitungsfähig, (Ker auch) verformbar; (Bgb) abbaufähig,
abbauwürdig
working Arbeiten n; Kneten n (z.B. der Butter);
Gären n, Gärung f; (Bgb) Abbau m, Verhieb m
w. chamber (Glas) Arbeitswanne f, Entnahmeteil
m
w. electrode Arbeitselektrode f, Versuchselek-
trode f, Meßelektrode f
w. end s. w. chamber
w. level (Toxikologie) konventioneller Richtwert
m (für Toleranzdosen)
w. life Verwendbarkeitsdauer f (z.B. eines
Klebers)
w. pressure Arbeitsdruck m, Betriebsdruck m
w. range Verarbeitungsbereich m
w. solution Arbeitslösung f

w. substance Arbeitsstoff m
w. temperature Arbeitstemperatur f, Betriebstemperatur f
w.-up Aufarbeitung f
w. vat (Pap) Schöpfbütte f, Tauchbütte f
works chemist Betriebschemiker m
w. distillation Betriebsdestillation f
world consumption Weltverbrauch m
worm (Tech) Schnecke f
w. conveyor Förderschnecke f, Schneckenförderer m, Transportschnecke f
w. expander (Text) Spiralausbreiter m, Spi-
ralbreithalter m
w. feeder Schneckenspeiser m, Schneckenauf-
gabegerät n
w. gear Schneckengetriebe n, Schneckentrieb m
w. press Schneckenpresse f
w. shaft Schneckenwelle f
w.-type condenser Kondensator m mit Kühlschlange
wormwood oil Wermutöl n
wort Würze f, Bierwürze f
w. cooler Würzekühler m
w. cooling Würzekühlung f
w. copper Würze[koch]kessel m Würzepfanne f,
Braukessel m, Braupfanne f
w. pump Würzepumpe f
Woulfe bottle Woulfesche Flasche f
wound dressing (Pflanzenschutz) Wundverband m
wove paper Velinpapier n
woven asbestos Asbestgewebe n, Asbesttuch n
w.[-fabric] filter Tuchfilter n, Stoffilter n, Gewebefilter n; Gewebeabscheider m
w.-glass-fibre cloth Glasfasergewebe n
w. hose Gewebeschlauch m
w. roving Rovinggewebe n
WPC (Abk. für) World Petroleum Congress
wrap / to wickeln; einwickeln, umwickeln, einpacken, verpacken; bandagieren
wrapper Umhüllung f, Hülle f; Einwickelmaschine
f, Verpackungsmaschine f; (Pap) Rollenpackmaschine f; (Gum) Zwischenläufer m, Zwischenleinen n, Mitläufer m; s. wrapping paper
wrapping Wickeln n; Einwickeln n, Umwickeln n,
Einpacken n, Verpacken n; Bandagieren n;
Verpackungsmaterial n, Einwickelmaterial n
w. device (machine) Einwickelmaschine f,
Verpackungsmaschine f
w. mechanism Einwickelmechanismus m
w. paper Verpackungspapier n, Packpapier n,
Einpackpapier n; Einschlagpapier n, Einwik-
kelpapier n, Hüllpapier n (hochwertiges Verpackungspapier)
w. tissue Packseiden[papier] n, Einschlagseidenpapier n, Einwickelseidenpapier n
Wright slide-cell technique Wrightsche Objekt-
trägerzellentechnik f
wringer Scheidepresse f
wrinkle / to (Anstrichtechnik) runzeln, kräuseln,
Runzeln bilden; (Text) knittern

wrinkle *(Anstrichtechnik)* Runzel *f*; *(Text)* Knitter *m*, Falte *f*
w. finish *s.* w. varnish
w. resistance *(Text)* Knitterwiderstand *m*, Knitterfestigkeit *f*, Knitterechtheit *f*, Knitterfreiheit *f*
w. varnish Runzellack *m*, Kräusellack *m*
wrinkling *(Anstrichtechnik)* Runzeln *n*, Runzelung *f*, Kräuseln *n*, Kräuselung *f*, Runzelbildung *f*; *(Text)* Knittern *n*, Knitterung *f*
writing ink Schreibtinte *f*, Tinte *f*
w. paper Schreibpapier *n*
w. parchment tierisches (animalisches) Pergament *n*, Hautpergament *n*, Pergament *n*
wrought alloy Knetlegierung *f*
w. iron Schweißstahl *m*, *(veraltet)* Schweißeisen *n*
wulfenite *(Min)* Wulfenit *m*, Gelbbleierz *n* *(Bleimolybdat)*
Wulff-Bock crystallizer Wulff-Bock-Kristallisator *m*, [Wulff-Bocksche] Kristallisierwiege *f*
Wulff process Wulff-Verfahren *n* *(zur technischen Gewinnung des Azetylens)*
Würth producer Würth-Generator *m*, Würth-Abstichgaserzeuger *m*, Schlackenabstichgenerator *m* nach Würth
Wurtz-Fittig synthesis Wurtz-Fittig-Synthese *f*
Wurtz reaction (synthesis) Wurtzsche Reaktion (Synthese) *f*, Wurtz-Reaktion *f*, Wurtz-Synthese *f* *(Alkansynthese)*
wurtzite *(Min)* Wurtzit *m* *(Zinksulfid)*
WVT. *s.* water vapour transmission
w & w product *s.* wash and wear product
WWB *s.* wet-weight basis
wye Hosenrohr *n*

X

X-amorphous röntgenamorph
X irradiation Röntgenbestrahlung *f*
X radiation Röntgenstrahlung *f*
X-ray / to röntgen, [mit Röntgenstrahlen] durchleuchten, bestrahlen; ein Röntgenbild machen
X ray Röntgenstrahl *m*
X-ray absorption Röntgen[strahlen]absorption *f*, Absorption *f* der Röntgenstrahlen (Röntgenstrahlung)
X-ray absorption spectrum Röntgenabsorptionsspektrum *n*
X-ray analysis Röntgen[strahl]analyse *f*, *(i.e.S.)* Röntgenstrukturanalyse *f*, Strukturanalyse *f* mit Röntgenstrahlen
X-ray analysis of crystals (crystal structure) *s.* X-ray crystal-structure analysis
X-ray-analytical röntgenanalytisch
X-ray apparatus Röntgenapparat *m*, Röntgengerät *n*
X-ray beam Röntgen[strahlen]bündel *n*, Röntgenstrahlenbüschel *n*

X-ray camera Röntgenkamera *f*
X-ray chemical analysis röntgenchemische Analyse *f*, chemische Analyse *f* mit Röntgenstrahlen
X-ray crystal[-structure] analysis Röntgenkristallstrukturanalyse *f*, röntgenografische Kristallstrukturanalyse *f*, Kristallstrukturanalyse *f* mit Röntgenstrahlen, Röntgen[struktur]analyse *f* von Kristallen
X-ray crystallographic analysis *s.* X-ray crystal[-structure] analysis
X-ray crystallography Röntgenkristallografie *f*
X-ray data Röntgendaten *pl*, röntgenografische Daten *pl*
X-ray diagram *s.* X-ray pattern
X-ray diffraction Röntgendiffraktion *f*, Röntgen[strahl]beugung *f*, Beugung *f* der Röntgenstrahlen
X-ray diffraction analysis Röntgendiffraktionsanalyse *f*
X-ray diffraction diagram *s.* X-ray diffraction pattern
X-ray diffraction method Röntgendiffraktionsmethode *f*, Röntgenbeugungsmethode *f*
X-ray diffraction pattern (photograph, picture) Röntgendiffraktionsaufnahme *f*, Röntgendiffraktionsbild *n*, Röntgenbeugungsaufnahme *f*, Röntgenbeugungsbild *n*, Röntgenbeugungsdiagramm *n*
X-ray diffractometer Röntgendiffraktometer *n*
X-ray emission Röntgenemission *f*
X-ray emission spectrum *s.* X-ray spectrum
X-ray emulsion Röntgenemulsion *f*
X-ray equipment Röntgeneinrichtung *f*
X-ray examination *s.* X-ray investigation
X-ray film Röntgenfilm *m*
X-ray fluorescence Röntgenfluoreszenz *f*
X-ray fluorescence (fluorescent) analysis Röntgenfluoreszenzanalyse *f*
X-ray goniometer Röntgengoniometer *n*
X-ray image *s.* X-ray pattern
X-ray intensity Röntgen[strahl]intensität *f*, Intensität (Stärke) *f* der Röntgenstrahlen
X-ray interference Röntgen[strahl]interferenz *f*
X-ray investigation Röntgenuntersuchung *f*, röntgenografische Untersuchung *f*
X-ray irradiation *s.* X irradiation
X-ray level Röntgenterm *m*, Röntgenniveau *n*
X-ray measurement Röntgenmessung *f*, röntgenografische Messung *f*
X-ray metallography Röntgenmetallografie *f*
X-ray method Röntgen[strahlen]methode *f*, Röntgenverfahren *n*, röntgenografische Methode *f*, röntgenografisches Verfahren *n*
X-ray microanalysis Röntgen[strahl]mikroanalyse *f*
X-ray microscope Röntgen[strahl]mikroskop *n*
X-ray microscopy Röntgenmikroskopie *f*, Mikroskopie *f* mit Röntgenstrahlen
X-ray optics Röntgenoptik *f*

X-ray pattern (photograph) Röntgenaufnahme *f*, Röntgenbild *n*, Röntgen[fot]ogramm *n*, Röntgendiagramm *n*
X-ray-photographic röntgenografisch
X-ray photography Röntgen[fot]ografie *f*
X-ray picture *s.* X-ray pattern
X-ray plant Röntgenanlage *f*
X-ray powder camera Röntgenpulverkamera *f*, Pulverkamera *f*
X-ray powder method Pulververfahren *n*, [Debye-Scherrersche] Pulvermethode *f*, Debye-Scherrer-Verfahren *n*, Debye-Scherrer-Methode *f*
X-ray powder pattern (photograph) Pulveraufnahme *f*, Pulverdiagramm *n*, Debye-Scherrer-Röntgenaufnahme *f*, Debye-Scherrer-Aufnahme *f*, Debye-Scherrer-Röntgendiagramm *n*
X-ray powder spectroscopy Röntgenspektroskopie *f* nach dem Pulververfahren (Debye-Scherrer-Verfahren)
X-ray protection Röntgen[strahlen]schutz *m*
X-ray-protective glass Röntgenschutzglas *n*
X-ray radiation *s.* X radiation
X-ray range *s.* X-ray region
X-ray reflection (reflexion) Röntgenreflexion *f*, Reflexion *f* der Röntgenstrahlen
X-ray region Röntgengebiet *n*, Gebiet *n* der Röntgenstrahlen
X-ray result Röntgenbefund *m*, röntgenografischer Befund *m*
X-ray rotation photograph Röntgendrehaufnahme *f*, Drehkristallaufnahme *f*, Drehdiagramm *n*
X-ray scattering Röntgen[strahl]streuung *f*, Streuung *f* der Röntgenstrahlen
X-ray small-angle scattering Röntgenkleinwinkelstreuung *f*, RKS
X-ray spectrochemical analysis röntgenspektrochemische Analyse *f*
X-ray spectrogram Röntgenspektrogramm *n*
X-ray spectrograph Röntgenspektrograf *m*
X-ray spectrography Röntgenspektrografie *f*
X-ray spectrometer Röntgenspektrometer *n*
X-ray spectrometry Röntgenspektrometrie *f*
X-ray spectroscopic analysis röntgenspektroskopische Analyse *f*
X-ray spectroscopy Röntgenspektroskopie *f*
X-ray spectrum Röntgenspektrum *n*, Röntgenstrahlen[emissions]spektrum *n*
X-ray spectrum analysis Röntgenspektralanalyse *f*
X-ray structural analysis *s.* X-ray structure analysis
X-ray structure Röntgenstruktur *f*
X-ray structure analysis Röntgenstrukturanalyse *f*, Strukturanalyse *f* mit Röntgenstrahlen
X-ray study *s.* X-ray investigation
X-ray technique Röntgen[strahl]technik *f*
X-ray test[ing] Röntgenprüfung *f*, Prüfung *f* mit Röntgenstrahlen
X-ray tube Röntgenröhre *f*
X-ray unit Röntgeneinheit *f*, Einheit *f* der Röntgendosis

X-ray wavelength Röntgenwellenlänge *f*
X rays Röntgenstrahlen *mpl*, Röntgenlicht *n*
X unit X-Einheit *f*, XE
xanthan[e] *s.* xanthene
 x. hydride Xanthanwasserstoff *m*, Isopersulfozyansäure *f*, 5-Imino-1,2,4-dithiazolin-3-thion *n*
xanthanoic acid *s.* xanthene-9-carboxylic acid
xanthanol Xanthanol *n*, (veraltet für) Xanthen-9-ol *n*, Xanthenol *n*
xanthate / to xanth[ogen]ieren, sulfidieren
xanthate Xanth[ogen]at *n* (Salz oder Ester einer Xanthogensäure)
xanthatin Xanthatin *n* (Antibiotikum)
xanthating churn *s.* xanthator
xanthation Xanth[ogen]ieren *n*, Xanth[ogen]ierung *f*, Sulfidieren *n*, Sulfidierung *f*
xanthator Xanthatkneter *m*, Sulfidiertrommel *f*, Baratte *f*
xanthene Xanthen *n*, Xanthan *n*, Dibenzo-1,4-pyran *n*; Xanthen *n*, Xanthenderivat *n*
 x. dye[stuff] Xanthenfarbstoff *m*
 x. ring system Xanthenringsystem *n*
xanthene-9-carboxylic (xanthenecarboxylic) acid Xanthen-9-karbonsäure *f*, Xanthenkarbonsäure *f*, Xanthansäure *f*
xanthene-9-thione, xanthenethione Xanthen-9-thion *n*, Xan[then]thion *n*
xanthen-9-ol, xanthenol Xanthen-9-ol *n*, Xanthenol *n*, Xanthydrol *n*, 9-Hydroxyxanthen *n*
xanthen-9-one, xanthenone Xanthen-9-on *n*, Xanth[en]on *n*, 9-Oxoxanthen *n*
xanthenyl Xanth[en]yl *n* (Atomgruppe des Xanthens)
xanthic acid 1. *s.* xanthogenic acid; 2. HOCSSH Dithiokohlensäure *f*
 x. oxide Harnoxid *n*, (veraltet für) Xanthin *n*, 2,6-Dihydroxypurin *n*
xanthin Xanthin *n* (ein Karotinoidfarbstoff)
xanthine Xanthin *n*, 2,6-Dihydroxypurin *n*, 2,6-Purindion *n*; Xanthin *n*, Xanthinderivat *n*
 x. base Xanthinbase *f*
 x. oxidase Xanth[opter]inoxydase *f*, Xanthindehydrase *f*, Schardinger-Ferment *n*, Schardinger-Enzym *n*
xanthine-9-ribo[furano]side *s.* xanthosine
xanthione *s.* xanthene-9-thione
xanthogen Xanthogen *n*
xanthogenamide $NH_2CSOC_2H_5$ Xanthogenamid *n*
xanthogenate *s.* xanthate
xanthogenic acid Xanthogensäure *f*, (i.e.S.) C_2H_5OCSSH Xanthogensäure *f*, Äthylxanthogensäure *f*, Dithiokohlensäure-*O*-äthylester *m*
 x.-acid ester Xanthogensäureester *m*
xanthommatine Xanthommatin *n*, Xanthoommatin *n*
xanthomycin A Xanthomyzin A *n* (Antibiotikum aus Streptomyces-Stämmen)
xanthone *s.* xanthen-9-one
xanthonic acid Xanthonsäure *f*, (veraltet für) Xanthogensäure *f*

xanthophyll Xanthophyll *n (ein hydroxyliertes Karotinoid);* Xanthophyll *n,* Lutein *n,* 3,3'-Dihydroxy-α-karotin *n*
xanthophyllite *(Min)* Xanthophyllit *m*
xanthoproteic acid Xanthoproteinsäure *f*
　x. reaction Xanthoproteinreaktion *f*
　x. test Xanthoproteinprobe *f*
xanthoprotein Xanthoprotein *n*
　x. reaction *s.* xanthoproteic reaction
xanthopterin[e] Xanthopterin *n,* 2-Amino-4,6-pteridindiol *n,* 2-Amino-4,6-dihydroxypteridin *n*
xanthopurpurin Xanthopurpurin *n,* 1,3-Dihydroxy-9,10-anthrachinon *n*
xanthorhamnin Xanthorhamnin *n (Glykosid aus Gelbbeeren)*
xanthosine Xanthosin *n,* Xanthin-9-ribosid *n*
xanthotoxin Xanthotoxin *n,* 6-Hydroxy-7-methoxybenzofuran-5-akrylsäure-δ-lakton *n*
xanthurenic acid Xanthurensäure *f,* 4,8-Dihydroxychinaldinsäure *f,* 4,8-Dihydroxychinolin-2-karbonsäure *f*
xanthydrol *s.* xanthen-9-ol
xanthyl *s.* xanthenyl
xanthylium Xanthylium *n*
　x. salt Xanthyliumsalz *n*
xenate Xenat *n*
xenic acid H_6XeO_6 Xenonsäure *f*
xenoblast *(Geol)* Xenoblast *m*
xenoblastic *(Geol)* xenoblastisch
xenocryst *(Geol)* Fremdling *m,* Einschluß *m*
xenolith *(Geol)* Xenolith *m,* exogener (fremder, enallogener) Einschluß *m*
xenomorphic *(Geol)* xenomorph, allotriomorph, fremdgestaltig
xenon Xe Xenon *n*
xenotime *(Min)* Xenotim *m (Yttriumphosphat)*
xenyl $C_6H_5C_6H_4-$ Xenyl *n,* Biphenylyl *n*
xenylamine Xenylamin *n,* 4-Aminobiphenyl *n*
xerogel Xerogel *n*
xerographic xerografisch
　x. plate xerografische Platte *f*
xerography Xerografie *f*
xeroradiography Xeroradiografie *f*
ximenynic acid
　$CH_3(CH_2)_5CH=CHC≡C(CH_2)_7COOH$ Ximenynsäure *f,* Santalbsäure *f,* Oktadezen-(11)-in-(9)-säure *f*
xylan Xylan *n,* Holzgummi *n,* Hemizellulose A *f*
xylaric acid $HOOC(CHOH)_3COOH$ Xylarsäure *f,* Xylotrihydroxyglutarsäure *f*
xylene $C_6H_4(CH_3)_2$ Xylol *n,* Dimethylbenzol *n*
　x. light yellow Xylollichtgelb *n*
　x. musk Xylolmoschus *n*
xylenesulphonic acid Xylolsulfonsäure *f,* Dimethylbenzolsulfonsäure *f*
xylenol $(CH_3)_2C_6H_3OH$ Xylenol *n,* Dimethylphenol *n,* Dimethylhydroxybenzol *n,* Hydroxyxylol *n*
　x. resin Xylenolharz *n*
xylic acid *s.* xylylic acid
xylidine $(CH_3)_2C_6H_3NH_2$ Xylidin *n,* Dimethylanilin *n,* Aminodimethylbenzol *n,* Aminoxylol *n*

xylite Xylit *m(n),* Braunkohlenxylit *m(n),* Braunkohlenholz *n,* Hylit *m(n),* xylitische (holzartige) Braunkohle *f*
xylitol $CH_2OH(CHOH)_3CH_2OH$ Xylit *m (ein 1,2,3,4,5-Pentanpentol)*
xylobiose Xylobiose *f,* β-*D*-1,4-Xylopyranose *f*
xylodesose Xylodesose *f*
xylohexaose Xylohexaose *f*
xyloketose *s.* xylulose
xylol Xylol *n (als Handelsprodukt)*
xylonic acid Xylonsäure *f (eine 2,3,4,5-Tetrahydroxypentansäure)*
xylopentaose Xylopentaose *f*
D-xylopyranose *D*-Xylopyranose *f*
xyloquinone Xylochinon *n,* Dimethylchinon *n,* Dimethyl-1,4-zyklohexadiendion *n*
xylorcin[ol] Xylorzin *n,* Dihydroxyxylol *n,* Dimethylresorzin *n*
xylosazone *s.* xylose phenylosazone
xylose Xylose *f,* Holzzucker *m*
　x. phenylosazone Xylosephenylosazon *n,* Xylosazon *n*
xylotetraose Xylotetraose *f*
xylotrihydroxyglutaric acid Xylotrihydroxyglutarsäure *f*
xylotriose Xylotriose *f*
xyloyl $(CH_3)_2C_6H_3CO-$ Xyloyl *n (Atomgruppe der Xylylsäure)*
xylulose Xylulose *f,* Xyloketose *f*
xylyl $(CH_3)_2C_6H_3-$ Xylyl *n,* Dimethylphenyl *n*
　x. bromide $CH_3C_6H_4CH_2Br$ Xylylbromid *n,* α-Bromxylol *n*
　x. chloride $CH_3C_6H_4CH_2Cl$ Xylylchlorid *n,* α-Chlorxylol *n*
xylylene bromide $C_6H_4(CH_2Br)_2$ Xylylenbromid *n,* α,α'-Dibromxylol *n,* Bis(brommethyl)benzol *n*
　x. chloride $C_6H_4(CH_2Cl)_2$ Xylylenchlorid *n* α,α'-Dichlorxylol *n,* Bis(chlormethyl)benzol *n*
xylylic acid $(CH_3)_2C_6H_3COOH$ Xylylsäure *f,* Dimethylbenzoesäure *f,* Dimethylbenzolkarbonsäure *f*

Y

Y-shape connecting tube Y-Stück *n,* Y-förmiges Zwischenabzweigstück (Verbindungsstück) *n*
Y-valve Schrägsitzventil *n*
yagein Yagein *n,* Harmin *n (Alkaloid)*
yakka gum Akaroidharz *n,* Akaroidgummi *n,* Grasbaumharz *n (von australischen Xanthorrhoea-Arten)*
yangonin Yangonin *n (ein α-Pyronderivat)*
Yankee *s.* Yankee machine
Yankee dryer *(Pap)* Trockenzylinder *m* einer Selbstabnahmemaschine
Yankee machine Selbstabnahme[papier]maschine *f,* Yankee-Maschine *f*
yarn Garn *n,* Faden *m*

yarn

y.-dye / to im Garn färben
y. dyeing Garnfärben *n*, Färben *n* im (von) Garn
y.-dyeing machine Garnfärbeapparat *m*
yeast / to mit Hefe versetzen, Hefe zugeben; gären
yeast Hefe *f*, *(i.e.S.)* Hefepilz *m*, Verkaufshefe *f (z.B. Backhefe, Bierhefe)*
y. adenylic acid $C_{10}H_{14}O_7N_5P$ Hefeadenylsäure *f*, 3'(2')-Adenylsäure *f*, Adenosin-3'(2')-monophosphorsäure *f*
y. autolysate *s*. y. extract
y. cell *s*. y. plant
y. culture Hefekultur *f*
y.-cutting machine Hefeform- und -teilmaschine *f*
y. extract Hefe[n]extrakt *m*, Hefeautolysat *n*
y.-extruding machine Hefeformmaschine *f*
y. factory Hefefabrik *f*
y. fermentation Hefegärung *f*
y. growing Hefevermehrung *f*, Vermehrung *f* der Hefe[zellen]
y.-growing vat Hefeaufziehapparat *m*, Hefebirne *f*
y. head Schlaucherdecke *f (ein Gärstadium des Bieres)*
y. industry Hefeindustrie *f*
y.-leavened hefegetrieben
y. mash Hefenmaische *f*, Hefegut *n*
y. nucleic acid Hefenukleinsäure *f*
y. plant Hefepilz *m*, Hefezelle *f*
y. powder Trockenhefe *f*
y. preparation Hefepräparat *n*
y. propagator Hefekulturapparat *m*
y.-raised *s*. y.-leavened
y. separator Hefeseparator *m*, Hefeschleuder *f*
y. tub Hefebottich *m*, Hefebütte *f*, Hefewanne *f*
y. turbidity Hefetrübung *f*, Hefetrub *m*
y. water Hefewasser *n*
y. wine Hefewein *m*
y. works *s*. y. factory
yeastlike hef[eart]ig
yellow / to vergilben
yellow Gelb *n (Farbempfindung)*; gelb[färbend]er Farbstoff *m*, Gelb *n*
y. arsenic [sulphide] gelbes Schwefelarsen *n*, gelber Arsenik *m*, Auripigment *n*, Operment *n*, Rauschgelb *n*, Königsgelb *n*, Chinagelb *n*, Gelbglas *n*
y. berry Gelbbeere *f*
y. brass Gelbguß *m*, Messing *n*
y. Brazil wood gelbes Brasil[ien]holz *n*
y. cast gelbliche Tönung (Verfärbung) *f*, Gelbstich *m*, Gilbe *f*
y. copper [ore] *(Min)* Gelbkupfererz *n*, *(veraltet für)* Kupferkies *m*
y. copperas *(Min)* Copiapit *m (Magnesiumeisen(III)-dihydroxidhexasulfat)*
y. enzyme gelbes Ferment (Oxydationsferment) *n*, Flavinferment *n*, Flavinenzym *n*, Flavoproteid *n*
y. grass-tree gum *(Farb)* Gelbes Akaroidharz *n (von Xanthorrhoea hastilis R.Br.)*

y. heat Gelbglut *f*
y. lead ore *(Min)* Gelbbleierz *n*, Wulfenit *m (Bleimolybdat)*
y. lead oxide PbO Bleiglätte *f*, Blei(II)-oxid *n*
y. mechanical straw pulp [gelber] Strohstoff *m*, Gelbstrohstoff *m*
y. mercuric iodide HgJ_2 gelbes Quecksilber(II)-jodid *n*
y. mercuric oxide HgO gelbes (gefälltes) Quecksilber(II)-oxid *n*, *(Pharm auch)* gelbes Präzipitat *n*
y. ochre *(Min)* gelber Ocker *m*
y. phosphorus gelber (weißer, farbloser) Phosphor *m*
y. precipitate *s*. y. mercuric oxide
y. prussiate of potash $K_4[Fe(CN)_6]$ gelbes Blutlaugensalz *n*, Kaliumhexazyanoferrat(II) *n*
y. prussiate of soda (sodium) $Na_4[Fe(CN)_6]$ Natriumhexazyanoferrat(II) *n*
y. shade Gelbton *m*, gelber Farbton *m*, gelbe Farbnuance *f*
y. stain *(Glas)* Gelbbeize *f*, Gelbätze *f*
y. ultramarine gelbes Ultramarin *n (Handelsbezeichnung für einige chromhaltige Pigmente)*; gelbes Ultramarin *n (Gemisch von Zink- und Kalziumchromat)*; $BaCrO_4$ Ultramaringelb *n*, gelbes Ultramarin *n*, Barytgelb *n*, Steinbühler Gelb *n*, Bariumchromat *n*
y. ware *(Ker)* Gelbware *f*
y. wax gelbes Wachs *n*
yellowing Vergilben *n*, Vergilbung *f*, Gelbwerden *n*
yellowish tinge Gelbstich *m*, Gelbstichigkeit *f*
yenite *(Min)* Yenit *m*, *(veraltet für)* Ilvait *m*
yield / to 1. abgeben; liefern, [er]geben; 2. *(Werkstoffprüfung)* fließen
yield 1. Ausbeute *f*, Ertrag *m*, *(Met auch)* Ausbringen *n*; 2. *(Werkstoffprüfung)* Fließen *n*
y. phenomenon Fließerscheinung *f*
y. point Fließgrenze *f*, Streckgrenze *f*
y. point at elevated temperatures Warmstreckgrenze *f*
y. strain Fließdehnung *f*
y. strength praktische Fließgrenze *f*, Dehngrenze *f*
y. strength *(0,2% offset)* 0,2%-Dehngrenze *f*, 0,2%-Grenze *f*
y. stress Fließspannung *f*, Streckspannung *f*
y. value untere Fließgrenze (Streckgrenze) *f*; *(Farb)* Fließgrenze *f*, Fließwert *m*, Ausgiebigkeitswert *m*, Ausgiebigkeitsfaktor *m*
ylang-ylang oil Ylang-Ylang-Öl *n*
ylide Ylid *n*
y. reaction Ylidreaktion *f*
yohimbi alkaloid Yohimbealkaloid *n*
yohimbic acid Yohimbinsäure *f*, Yohimboasäure *f*, Alloyohimbensäure *f*
yohimbine Yohimbin *n*, Yohimboasäuremethylester *m*
y. hydrochloride Yohimbinhydro[gen]chlorid *n*
y. nitrate Yohimbin[hydrogen]nitrat *n*

y. oxindole Yohimbinoxindol *n*
yohimboaic acid *s.* yohimbic acid
yohimbone Yohimbon *n (Keton des Yohimbins)*
yolk 1. Eidotter *n,* Dotter *n,* Eigelb *n;* 2. *(Text)* Wollfett *n,* Wollschweiß *m*
y. powder Eigelbpulver *n*
Yorkshire grease *(rohes)* Wollfett *n*
young fustic ungarisches Gelbholz *n,* junger Fustik *m,* Jungfustik *m,* Fiset[te]holz *n,* Fustet *m*
y. wine Jungwein *m*
Young's modulus [of elasticity] Youngscher Modul *m,* Elastizitätsmodul *m,* E-Modul *m*
yperite (ClCH$_2$CH$_2$)$_2$S Yperit *n,* Schwefelyperit *n,* S-Yperit *n (Dichlordiäthylsulfid; Kampfstoff)*
ytterbia Yb$_2$O$_3$ Ytterbinerde *f,* Ytterbia *f, (veraltet für)* Ytterbium(III)-oxid *n*
ytterbic Ytterbium..., *(i.e.S.)* Ytterbium(III)...
y. chloride YbCl$_3$ Ytterbium(III)-chlorid *n,* Ytterbiumtrichlorid *n*
y. sulphate Yb$_2$(SO$_4$)$_3$ Ytterbium(III)-sulfat *n,* Ytterbiumsulfat *n*
ytterbite *(Min)* Ytterbit *m, (veraltet für)* Gadolinit *m (Yttrium-Eisen-Beryllosilikat)*
ytterbium Yb Ytterbium *n*
y. dichloride YbCl$_2$ Ytterbiumdichlorid *n,* Ytterbium(II)-chlorid *n*
y. earth Ytterbiumerde *f (als Sammelname)*
y. oxide Yb$_2$O$_3$ Ytterbium(III)-oxid *n*
y. trichloride YbCl$_3$ Ytterbiumtrichlorid *n,* Ytterbium(III)-chlorid *n*
ytterbous Ytterbium..., *(i.e.S.)* Ytterbium(II)-...
y. chloride YbCl$_2$ Ytterbium(II)-chlorid *n,* Ytterbiumdichlorid *n*
ytterite *(Min)* Ytterit *m*
yttria Y$_2$O$_3$ Yttererde *f,* Yttria *f, (veraltet für)* Yttriumoxid *n*
yttrialite *(Min)* Yttrialith *m*
yttric Yttrium...
yttrite *s.* ytterite
yttrium Y Yttrium *n*
y. acetate Y(CH$_3$COO)$_3$ Yttriumazetat *n*
y. bromide YBr$_3$ Yttriumbromid *n*
y. carbide YC$_2$ Yttriumkarbid *n*
y. carbonate Y$_2$(CO$_3$)$_3$ Yttriumkarbonat *n*
y. chloride YCl$_3$ Yttriumchlorid *n*
y. earth Yttererde *f (als Sammelname)*
y. fluoride YF$_3$ Yttriumfluorid *n*
y. hydroxide Y(OH)$_3$ Yttriumhydroxid *n*
y. iodide YJ$_3$ Yttriumjodid *n*
y. nitrate Y(NO$_3$)$_3$ Yttriumnitrat *n*
y. oxide Y$_2$O$_3$ Yttriumoxid *n*
y. sulphate Y$_2$(SO$_4$)$_3$ Yttriumsulfat *n*
yttrocerite *(Min)* Yttrozerit *m*
yttrotantalite *(Min)* Yttrotantalit *m*
yttrotitanite *(Min)* Yttrotitanit *m,* Keilhanit *m*
yucca fibre Yukkafaser *f*

Z

Z twist *(Gum)* Z-Drehung *f,* Rechtsdraht *m*
Z type of calender Z-Walzenkalander *m,* Z-Kalander *m,* Vierwalzenkalander *m* mit Z-Anordnung
Zachariasen's theory *(Glas)* Netzwerkhypothese *f* von W. H. Zachariasen
zaffar, zaffer, zaffir *s.* zaffre[e]
zaffre[e] Zaffer *m,* Saf[f]lor *m (geröstete Kobalterze)*
zamtite *(fälschlich für)* zaratite
Zanzibar aloe *(Pharm)* Sokotra-Aloe *f (von Aloe perryi Baker)*
Z. copal *(Farb)* Sansibarkopal *m (von Trachylobium verrucosum Oliv. und Tr. hornemannianum Heyne)*
zapota gum Chiclegummi *n,* Chicle *m (von Achras sapota L.)*
zaratite *(Min)* Zaratit *m (Nickeltetrahydroxidkarbonat)*
zavalite *(fälschlich für)* zaratite
zeaxanthin Zeaxanthin *n,* 3,3'-Dihydroxy-β-karotin
Zeeman effect Zeeman-Effekt *m*
zein Zein *n (Prolamin des Maises)*
z. plastic Zeinkunststoff *m*
z. staple Zeinfaser *f*
Zeisel determination [of methoxyl] *s.* Zeisel methoxyl determination
Zeisel method [for alkoxy groups] Zeisel-Methode *f,* Zeisel-Vieböck-Methode *f (zur Bestimmung der Alkoxylgruppen in Alkyl-Aryl-Äthern)*
Zeisel methoxyl determination Zeiselsche Methoxylbestimmung *f,* Zeisel-Bestimmung *f,* Bestimmung *f* der Methoxylgruppen nach Zeisel
Zeisel reaction Zeiselsche Reaktion *f*
zeolite *(Min)* Zeolith *m*
zeolitic zeolithisch
z. water zeolithisches (zeolithisch gebundenes) Wasser *n,* Zeolithwasser *n*
zeolitization Zeolithisierung *f,* Zeolithisation *f*
Zerewitinoff determination Zerewitinow-Bestimmung *f*
Zerewitinoff reagent Zerewitinows Reagens *n*
zero *s.* z. point
z. charge Nulladung *f*
z. concentration Konzentration Null *f*
z. group nullte Gruppe *f,* Gruppe 0 *f (im Periodensystem)*
z. line Nullinie *f (im Bandenspektrum)*
z. point Nullpunkt *m,* Ausgangspunkt *m (einer Skale)*
z.-point configuration Nullpunktskonfiguration *f*
z.-point configurational entropy Entropie *f* der Nullpunktskonfiguration
z.-point energy Nullpunktsenergie *f*
z.-point entropy Nullpunktsentropie *f*
z.-point vibration Nullpunktsschwingung *f*
z. porosity Nullporosität *f*
z. position Nullstellung *f*
z. tolerance Nulltoleranz *f*

z.-valent nullwertig

zeta potential Zeta-Potential *n*, ζ-Potential *n*, elektrokinetisches Potential *n*

Zettlitz kaolin Zettlitzer Kaolin *m*

zeunerite *(Min)* Zeunerit *m (Kupfer(II)-bisuranylorthoarsenat)*

zibet[h] Zibet *m (Riechstoffdroge)*

Ziegler catalyst Ziegler-Katalysator *m*, Ziegler-Kontakt *m*

Ziegler method Ziegler-Methode *f (zur Darstellung zyklischer Ketone)*

Ziegler process Ziegler-Verfahren *n*, Niederdruckverfahren *n* nach Ziegler *(ein Polymerisationsverfahren)*

Ziegler-type catalyst s. Ziegler catalyst

Ziervogel process Ziervogel-Prozeß *m (zur Silbergewinnung)*

zigzag arrangement Zickzackanordnung *f*

z. chain Zickzackkette *f*

z. kiln *(Ker)* Zickzackofen *m*

Zimmermann reaction Zimmermann-Reaktion *f*, Reaktion *f* nach Zimmermann

Zimmermann-Reinhardt solution Lösung *f* zur Eisenbestimmung nach Zimmermann-Reinhardt

zinc / to verzinken

zinc Zn Zink *n*

z. alkyl Zink[di]alkyl *n*

z. amalgam Zinkamalgam *n*

z. baryta white Barytzinkweiß *n*, Lithopone *f*, Lithopon *n*

z.-base auf Zinkbasis (Zinkgrundlage)

z. bichromate s. z. dichromate

z. blende *(Min)* Zinkblende *f*, Blende *f*, Sphalerit *m (Zinksulfid)*

z.-blende lattice Zinkblendegitter *n*

z. chloride $ZnCl_2$ Zinkchlorid *n*

z.-chloride solution Zinkchloridlösung *f*, *(Tech auch)* Chlorzinklauge *f*

z. chromate $ZnCrO_4$ Zink[mono]chromat *n*; s. z. yellow

z. chrome s. z. yellow

z. crown glass Zinkkronglas *n*

z. dichromate $ZnCr_2O_7$ Zinkdichromat(VI) *n*

z. diethyl $Zn(C_2H_5)_2$ Zink[di]äthyl *n*, Diäthylzink *n*, Zinkäthid *n*

z. dimethyl $Zn(CH_3)_2$ Zink[di]methyl *n*, Dimethylzink *n*, Zinkmethid *n*

z. dust Zinkstaub *m*

z.-dust distillation Zinkstaubdestillation *f*

z. electrode Zinkelektrode *f*

z. ethyl s. z. diethyl

z. ferrocyanide $Zn_2[Fe(CN)_6]$ Zinkhexazyanoferrat(II) *n*

z. fluorosilicate $Zn[SiF_6]$ Zinkhexafluorosilikat *n*, Zinksilikofluorid *n*

z. grey Zinkgrau *n*

z. hydride ZnH_2 Zinkwasserstoff *m*, Zinkhydrid *n*

z. hydrosulphite ZnS_2O_4 Zinkhydrosulfit *n*, *(veraltet für)* Zinkdithionit *n*

z.-hydrosulphite bleaching *(Pap)* Zinkhydrosulfitbleiche *f*

z.-lime vat *(Text)* Zink[staub]-Kalk-Küpe *f*

z. metasilicate $ZnSiO_3$ Zinktrioxosilikat *n*, Zinkmetasilikat *n*

z. methide (methyl) s. z. dimethyl

z. orthoarsenate $Zn_3(AsO_4)_2$ Zink[ortho]arsenat(V) *n*

z. orthophosphate $Zn_3(PO_4)_2$ Zink[ortho]phosphat *n*

z. orthosilicate Zn_2SiO_4 Zinkorthosilikat *n*, Zinktetroxosilikat *n*

z. oxide ZnO Zinkoxid *n*

z.-oxide catalyst Zinkoxidkatalysator *m*

z.-oxide smoke Zinkoxidrauch *m*

z. permanganate $Zn(MnO_4)_2$ Zinkpermanganat *n*, Zinkmanganat(VII) *n*

z. plating Verzinken *n*, Verzinkung *f*

z.-potassium chromate s. z. yellow

z.-potassium cyanide $K_2[Zn(CN)_4]$ Kaliumtetrazyanozinkat *n*

z. rhodanide s. z. thiocyanate

z. salt Zinksalz *n*

z. silicofluoride s. z. fluorosilicate

z. spar *(Min)* Zinkspat *m*, Smithsonit *m*

z. spinel *(Min)* Zinkspinell *m*, Gahnit *m (Zinkaluminat)*

z. sulphocyanate (sulphocyanide) s. z. thiocyanate

z. thiocyanate $Zn(SCN)_2$ Zinkthiozyanat *n*, Zinkrhodanid *n*

z. vapour Zinkdampf *m*

z. vitriol $ZnSO_4 \cdot 7H_2O$ Zinkvitriol *n*, Zinksulfat-7-Wasser *n*

z. white Zinkweiß *n (rohes Zinkoxid)*

z. yellow Zink[chrom]gelb *n*, Zink[kalium]chromat *n*, Zitronengelb *n (Pigment)*

zincate Zinkat *n*

zincic Zink...

zincing Verzinken *n*, Verzinkung *f*

zincite *(Min)* Zinkit *m*, Rotzinkerz *n (Zinkoxid)*

zinckenite *(Min)* Zinckenit *m (Blei(II)-antimon(III)-sulfid)*

zincking s. zincing

zincous s. zincic

zineb Zineb *n*, Zinkäthylenbisdithiokarbamat *n*

zingerone Zingeron *n*, 3-Methoxy-4-hydroxybenzylazeton *n*

zingiberene Zingiberen *n (ein monozyklischer Sesquiterpenkohlenwasserstoff)*

zingiberol Zingiberol *n (ein monozyklischer Sesquiterpenalkohol)*

zinkoferrite *(Min)* Zinkoferrit *m*, Ferrozinkit *m*, *(veraltet für)* Franklinit *m*

zinnwaldite *(Min)* Zinnwaldit *m*

zippeite *(Min)* Zippeit *m*, Uranblüte *f*

ziram Ziram *n*, Zinkdimethyldithiokarbamat *n*

zircon *(Min)* Zirkon *m (Zirkoniumorthosilikat)*

z. body *(Ker)* Zirkonmasse *f*

z. porcelain Zirkonporzellan *n*

z. sand Zirkonsand *m*

z. whiteware *(Ker)* Zirkonweißware *f*

zirconate Zirkonat n
zirconia ZrO_2 Zirkonerde f, Zirkonia f, (veraltet für) Zirkoniumdioxid n, Zirkonium(IV)-oxid n
zirconic Zirkonium...
  z. acid $Zr(OH)_4$ oder $ZrO_2 \cdot xH_2O$ Zirkonsäure f, (fälschlich für) Zirkonium(IV)-hydroxid n (oder) Zirkoniumoxidaquat n
  z. anhydride ZrO_2 Zirkonsäureanhydrid n, (veraltet für) Zirkonium(IV)-oxid n, Zirkonium[di]oxid n
zirconifluoride Fluorozirkonat n
zirconium Zr Zirkonium n
  z.-ammonium fluoride $(NH_4)_2[ZrF_6]$ Ammoniumhexafluorozirkonat n
  z. anhydride s. zirconic anhydride
  z. carbide ZrC Zirkonium[mono]karbid n
  z. dibromide $ZrBr_2$ Zirkoniumdibromid n, Zirkonium(II)-bromid n
  z. dichloride $ZrCl_2$ Zirkoniumdichlorid n, Zirkonium(II)-chlorid n
  z. dioxide ZrO_2 Zirkonium[di]oxid n, Zirkonium(IV)-oxid n
  z. fluoride ZrF_4 Zirkonium(IV)-fluorid n, Zirkoniumtetrafluorid n
  z. hydride Zirkoniumhydrid n
  z. hydroxide $Zr(OH)_4$ oder $ZrO_2 \cdot xH_2O$ Zirkonium(IV)-hydroxid n, Zirkoniumhydroxid n (oder) Zirkoniumoxidaquat n, Zirkonium[di]oxidhydrat n
  z. oxide ZrO_2 Zirkonium(IV)-oxid n, Zirkonium[di]oxid n; $ZrO_2 \cdot xH_2O$ Zirkoniumoxidaquat n, Zirkonium[di]oxidhydrat n
  z. oxybromide $ZrOBr_2$ Zirkoniumoxidbromid n
  z. oxychloride $ZrOCl_2$ Zirkoniumoxidchlorid n
  z. oxyiodide $ZrOJ_2$ Zirkoniumoxidjodid n
  z.-potassium fluoride $K_2[ZrF_6]$ Kaliumhexafluorozirkonat(IV) n
  z. silicate $ZrSiO_4$ Zirkoniumsilikat n
  z. sulphate $Zr(SO_4)_2$ Zirkonium(IV)-sulfat n
  z. sulphide ZrS_2 Zirkonium(IV)-sulfid n
  z. tannage Zirkongerbung f
  z. tetrachloride $ZrCl_4$ Zirkoniumtetrachlorid n, Zirkonium(IV)-chlorid n
  z. tetraiodide ZrJ_4 Zirkoniumtetrajodid n, Zirkonium(IV)-jodid n
  z. trichloride $ZrCl_3$ Zirkoniumtrichlorid n, Zirkonium(III)-chlorid n
zirconyl ZrO= (veraltet) Zirkonyl n
  z. bromide $ZrOBr_2$ Zirkonylbromid n, (veraltet für) Zirkoniumoxidbromid n
  z. chloride $ZrOCl_2$ Zirkonylchlorid n, (veraltet für) Zirkoniumoxidchlorid n
  z. hydroxide $ZrO(OH)_2$ Zirkonylhydroxid n, (veraltet für) Zirkoniumoxidhydroxid n
zoisite (Min) Zoisit m (Kalziumaluminiumhydroxidorthosilikat)
zonal distribution of minerals (Geol) zonale (zonare) Erzverteilung f (Verteilung f von Erzlagerstätten)
  z. structure (Geol) Zonarstruktur f, zonaler Aufbau m

zonary structure s. zonal structure
zone Zone f
  z. axis (Krist) Zonenachse f, Zonenkante f
  z. electrophoresis Zonenelektrophorese f
  z. length Zonenlänge f, Schmelzzonenlänge f, Länge f der Schmelzzone (beim Zonenschmelzen)
  z. levelling Zonenlegieren n, Zonennivellieren n, Zonenebnen n, Zonenplanieren n
  z.-melt / to zonenschmelzen
  z. melting Zonenschmelzen n, Zonenschmelzung f, Zonenschmelze f
  z.-melting apparatus Apparatur f zum Zonenschmelzen, Zonenschmelzanlage f, Zonenschmelzgerät n
  z.-melting furnace Zonenschmelzofen m
  z.-melting process Zonenschmelzverfahren n, Zonenschmelzmethode f; Zonenschmelzvorgang m
  z.-melting technique Zonenschmelzverfahren n, Zonenschmelzmethode f
  z.-melting unit s. z.-melting apparatus
  z. of cementation (Geol) Zementationszone f
  z. of combustion Verbrennungszone f
  z. of inhibition Hemmungszone f, Hemmungshof m (keimfreie Zone bei der Penizillinbestimmung)
  z. of oxidation Oxydationszone f
  z. of preheat Vorwärm[e]zone f, Vorheizzone f
  z. pass Zonen[schmelz]durchgang m
  z. precipitation Zonenfällung f
  z. purification s. z. refining
  z.-purify / to s. to z.-refine
  z.-refine / to zonenreinigen, durch Zonenschmelzen reinigen
  z. refiner s. z.-refining apparatus
  z. refining Zonenreinigen n, Zonenreinigung f
  z.-refining apparatus Apparatur f zur Zonenreinigung f, Zonenreinigungsanlage f
  z.-refining method Zonenreinigungsverfahren n, Zonenreinigungsmethode f
  z.-refining process Zonenreinigungsverfahren n, Zonenreinigungsmethode f; Zonenreinigungsvorgang m
  z.refining technique s. z.-refining method
  z.-refining unit s. z.-refining apparatus
  z. speed s. z.-travel rate
  z. travel Zonenwanderung f, Wandern (Hindurchwandern) n der Zonne (Schmelzzone) (beim Zonenschmelzen)
  z.-travel rate Zonengeschwindigkeit f, Geschwindigkeit f der Zonenwanderung, Wanderungsgeschwindigkeit f der Zone (Schmelzzone) (beim Zonenschmelzen)
  z. travelling s. z. travel
  z. width Zonenbreite f, Schmelzzonenbreite f (beim Zonenschmelzen)
zoning s. zonal structure
  z. speed s. zone-travel rate
zorgite (Min) Zorgit m
Zschocke disintegrator Zschocke-Desintegrator m, Desintegrator[gas]wäscher m nach Zschocke

zwitterion Zwitterion *n*, Ampho-Ion *n*
zygadenine Zygadenin *n (Alkaloid)*
zymase Zymase *f (aus Hefe isoliertes Ferment-gemisch)*
zymogen Zymogen *n*, Proferment *n*, Gärungserreger *m*

zymogenic, zymogenous zymogen, zymotisch, Gärung bewirkend (erregend)
zymology Zymologie *f (Lehre von der Gärung)*
zymotechnic[al] zymotechnisch, gärungstechnisch
zymotechnics Zymotechnik *f*, Gärungstechnik *f*

ANHANG

HINWEISE ZUR ANWENDUNG DER IUPAC-NOMENKLATUR
BEI DER BENENNUNG CHEMISCHER VERBINDUNGEN

Mit der Anwendung der IUPAC-Nomenklaturregeln bei der Benennung chemischer Verbindungen ergeben sich hinsichtlich der bisherigen deutschen Schreibweise chemischer Namen vor allem folgende Änderungen:

1. Die Schreibweise von Verbindungs- und Elementnamen mit k, z oder c wird der englischen Schreibweise weitestgehend angeglichen:

   Acetat, Cadmium, Caesium, Calcium, Carbonyl-,
   Cresol, Cellulose, Cobalt, Cyan-, Cyclo- usw.;

   dagegen wie bisher:

   Azo-, Keto-, Alkali-, Kalium, Zink.

2. Trivialnamen aromatischer Kohlenwasserstoffe auf -ol werden zu -en geändert:

   Benzen, Toluen, Styren, Cumen, Cymen
   Naphthalen   *statt*   Naphthalin.

3. Trivialnamen mehrwertiger Alkohole und einer Reihe von Phenolen enden — wie die der einwertigen Alkohole — konsequent auf -ol:

Glycerol	*statt*	Glyzerin
Sorbitol	*statt*	Sorbit
Mannitol	*statt*	Mannit
Resorcinol	*statt*	Resorzin.

4. Ethan   *statt*   Äthan
   (ebenso auch Ethen, Ethin, Ethyl- usw.).

5. Das Element Wismut wird in Angleichung an sein Symbol zu:

   Bismut.

   Name und Symbol des Jods werden gemäß der englischen Schreibweise (und dem griechischen Ursprungswort) zu:

   Iod, Symbol I.

   Die Elemente Ce, Cr, Nb, Nd, Pm, Ti, V und U erhalten wie im Englischen lateinische Endungen und werden zu:

   Cerium, Chromium, Niobium,
   Neodymium, Praseodymium,
   Titanium, Vanadium und
   Uranium.

6. Erscheinen im Namen einer Verbindung mehrere Präfixe für Substituenten, so werden sie *ohne* Berücksichtigung erforderlicher griechischer Zahlenpräfixe alphabetisch angeordnet.
   (Die bisherige Wahlmöglichkeit zwischen dieser Methode und der Ordnung von Substituenten nach steigender Größe oder Komplexität entfällt):

   Mono**c**hlortri**f**luorethen,
   2-**A**mino-4-**e**thyl-cyclohexandicarbonsäure
   2,3,6,7-Tetra**b**rom-1,4-di**m**ethyl-naphthalen.

7. Auch Stellungsangaben für funktionelle Gruppen und Lagebezeichnungen für Mehrfachbindungen werden den kennzeichnenden Suffixen unmittelbar vorangestellt:

4-Methyl-oct-2-en,

2,3,5-Trimethyl-hept-2-en.

Klein- und Großbuchstaben für Stellungsangaben wie o-, m-, p-, C-, S-, N- usw. werden nicht mehr durch eine andere Schriftart hervorgehoben.

Zur genaueren Information empfehlen wir:

1) LIEBSCHER, W., in Mitteilungsbl. Chem. Ges. DDR **25** (1978), S. 259—262.
2) HALLPAP, P., W. LIEBSCHER u. E. WIESNER, Nomenklatur organischer Verbindungen (Lehrprogrammbücher Hochschulstudium **8**), Leipzig 1978.
3) Handbuch zur Anwendung der Nomenklatur organisch-chemischer Verbindungen, hrsg. v. W. LIEBSCHER, Berlin 1979. (Dieses Werk unterrichtet auch über die noch gültigen Trivialnamen, deren Anzahl eingeschränkt wurde.)